Mathematical relations

$\pi = 3.141\,592\,653\,59\dots$
$e = 2.718\,281\,828\,46\dots$

Logarithms and exponentials

$\ln x + \ln y + \dots = \ln xy\dots$
$\ln x - \ln y = \ln(x/y)$
$a \ln x = \ln x^a$
$\ln x = (\ln 10) \log x = (2.302\,585 \dots) \log x$
$e^x e^y e^z \dots = e^{x+y+z+\dots}$
$e^x/e^y = e^{x-y}$
$(e^x)^a = e^{ax}$
$e^{\pm ix} = \cos x \pm i \sin x$

Taylor expansions

$$f(x) = \sum_{n=0}^{\infty} \frac{1}{n!} \left(\frac{d^n f}{dx^n} \right)_a (x-a)^n$$

$e^x = 1 + x + \frac{1}{2}x^2 + \dots$
$\ln x = (x-1) - \frac{1}{2}(x-1)^2 + \frac{1}{3}(x-1)^3 - \frac{1}{4}(x-1)^4 + \dots$
$\ln(1+x) = x - \frac{1}{2}x^2 + \frac{1}{3}x^3 \dots$
$\dfrac{1}{1+x} = 1 - x + x^2 \dots$

Derivatives

$d(f+g) = df + dg$
$d(fg) = f\,dg + g\,df$

$d\dfrac{f}{g} = \dfrac{1}{g}df - \dfrac{f}{g^2}dg$

$\dfrac{df}{dt} = \dfrac{df}{dg}\dfrac{dg}{dt}$

$\left(\dfrac{\partial y}{\partial x} \right)_z \left(\dfrac{\partial x}{\partial z} \right)_y \left(\dfrac{\partial z}{\partial y} \right)_x = -1$

$(\partial y/\partial x)_z = 1/(\partial x/\partial y)_z$

$\dfrac{dx^n}{dx} = nx^{n-1}$

$\dfrac{d}{dx}e^{ax} = ae^{ax}$

$\dfrac{d \ln x}{dx} = \dfrac{1}{x}$

Integrals

$$\int x^n\,dx = \frac{x^{n-1}}{n+1} + \text{constant}$$

$$\int \frac{1}{x}\,dx = \ln x + \text{constant}$$

$$\int_0^\infty x^n e^{-ax}\,dx = \frac{n!}{a^{n+1}}$$

$$\int \sin^2 ax\,dx = \tfrac{1}{2}x - (1/4a)\sin 2ax + \text{constant}$$

$$\int \sin ax \sin bx\,dx = \frac{\sin(a-b)x}{2(a-b)} - \frac{\sin(a+b)x}{2(a+b)} + \text{constant},$$

$\text{if } a^2 \neq b^2$

$$\operatorname{erf} z = \frac{2}{\pi^{1/2}} \int_z^0 e^{-y^2}\,dy$$

Useful relations

At 298.15 K

RT	2.4790 kJ mol^{-1}
RT/F	25.693 mV
$RT \ln 10/F$	59.160 mV
kT/hc	207.23 cm^{-1}
kT/e	25.693 meV
$V_m^{\ominus}$	2.4790×10^{-2} m^3 mol^{-1} = 24.790 L mol^{-1}

Conversion factors

1 eV	$1.602\,18 \times 10^{-19}$ J
	96.485 kJ mol^{-1}
	8065.5 cm^{-1}
1 cal	4.184* J
1 atm	101.325* kPa
	760* Torr
1 cm^{-1}	1.9864×10^{-23} J
1 D	$3.335\,64 \times 10^{-30}$ C m
1 Å	10^{-10} m*
1 T	10^4 G*

1 L atm = 101.325 J*
$\theta/^{\circ}\text{C} = T/\text{K} - 273.15^*$

* Exact value

Unit relations

Energy	1 J = 1 kg m^2 s^{-2}
	= 1 A V s
Force	1 N = 1 kg m s^{-2}
Pressure	1 Pa = 1 N m^{-2} = 1 kg m^{-1} s^{-2}
	= 1 J m^{-3}
Charge	1 C = 1 A s
Potential difference	1 V = 1 J C^{-1} = 1 kg m^2 s^{-3} A^{-1}

Prefixes

z	a	f	p	n	μ	m	c	d	Da	k	M	G	T
zepto	atto	femto	pico	nano	micro	milli	centi	deci	deka	kilo	mega	giga	tera
10^{-21}	10^{-18}	10^{-15}	10^{-12}	10^{-9}	10^{-6}	10^{-3}	10^{-2}	10^{-1}	10^{1}	10^{3}	10^{6}	10^{9}	10^{12}

General data and fundamental constants

Quantity	Symbol	Value	Power of ten	Units
Speed of light	c	2.997 924 58*	10^8	$m\ s^{-1}$
Elementary charge	e	1.602 176	10^{-19}	C
Faraday constant	$F = N_A e$	9.648 53	10^4	$C\ mol^{-1}$
Boltzmann constant	k	1.380 65	10^{-23}	$J\ K^{-1}$
Gas constant	$R = N_A k$	8.314 47		$J\ K^{-1}\ mol^{-1}$
		8.314 47	10^{-2}	$L\ bar\ K^{-1}\ mol^{-1}$
		8.205 74	10^{-2}	$L\ atm\ K^{-1}\ mol^{-1}$
		6.236 37	10	$L\ Torr\ K^{-1}\ mol^{-1}$
Planck constant	h	6.626 08	10^{-34}	$J\ s$
	$\hbar = h/2\pi$	1.054 57	10^{-34}	$J\ s$
Avogadro constant	N_A	6.022 14	10^{23}	mol^{-1}
Atomic mass unit	u	1.660 54	10^{-27}	kg
Mass				
electron	m_e	9.109 38	10^{-31}	kg
proton	m_p	1.672 62	10^{-27}	kg
neutron	m_n	1.674 93	10^{-27}	kg
Vacuum permittivity	$\varepsilon_0 = 1/c^2\mu_0$	8.854 19	10^{-12}	$J^{-1}\ C^2\ m^{-1}$
	$4\pi\varepsilon_0$	1.112 65	10^{-10}	$J^{-1}\ C^2\ m^{-1}$
Vacuum permeability	μ_0	4π*	10^{-7}	$J\ s^2\ C^{-2}\ m^{-1}\ (= T^2\ J^{-1}\ m^3)$
Magneton				
Bohr	$\mu_B = e\hbar/2m_e$	9.274 01	10^{-24}	$J\ T^{-1}$
nuclear	$\mu_N = e\hbar/2m_p$	5.050 78	10^{-27}	$J\ T^{-1}$
g value	g_e	2.002 32		
Bohr radius	$a_0 = 4\pi\varepsilon_0\hbar^2/m_e e^2$	5.291 77	10^{-11}	m
Fine-structure constant	$\alpha = \mu_0 e^2 c/2h$	7.297 35	10^{-3}	
	α^{-1}	1.370 36	10^2	
Second radiation constant	$c_2 = hc/k$	1.438 78	10^{-2}	m K
Stefan-Boltzmann constant	$\sigma = 2\pi^5 k^4/15h^3 c^2$	5.670 51	10^{-8}	$W\ m^{-2}\ K^{-4}$
Rydberg constant	$R = m_e e^4/8h^3 c\varepsilon_0^2$	1.097 37	10^5	cm^{-1}
Standard acceleration of free fall	g	9.806 65*		$m\ s^{-2}$
Gravitational constant	G	6.673	10^{-11}	$N\ m^2\ kg^{-2}$

* Exact value.

The Greek alphabet

A, α	alpha		N, ν	nu
B, β	beta		Ξ, ξ	xi
Γ, γ	gamma		O, o	omicron
Δ, δ	delta		Π, π	pi
E, ε	epsilon		P, ρ	rho
Z, ζ	zeta		Σ, σ	sigma
H, η	eta		T, τ	tau
Θ, θ	theta		Υ, υ	upsilon
I, ι	iota		Φ, ϕ	phi
K, κ	kappa		X, χ	chi
Λ, λ	lambda		Ψ, ψ	psi
M, μ	mu		Ω, ω	omega

ATKINS'
Physical Chemistry

www.oup.com/pchem7

ATKINS'
Physical Chemistry

Seventh Edition

Peter Atkins

Professor of Chemistry,
University of Oxford,
and Fellow of Lincoln College, Oxford

Julio de Paula

Professor of Chemistry,
Haverford College,
Haverford, Pennsylvania

OXFORD

UNIVERSITY PRESS

OXFORD
UNIVERSITY PRESS

Great Clarendon Street, Oxford OX2 6DP

Oxford University Press is a department of the University of Oxford.
It furthers the University's objective of excellence in research, scholarship,
and education by publishing worldwide in

Oxford New York

Athens Auckland Bangkok Bogotá Buenos Aires Calcutta
Cape Town Chennai Dar es Salaam Delhi Florence Hong Kong Istanbul
Karachi Kuala Lumpur Madrid Melbourne Mexico City Mumbai
Nairobi Paris São Paulo Shanghai Singapore Taipei Tokyo Toronto Warsaw

with associated companies in Berlin Ibadan

Oxford is a registered trade mark of Oxford University Press
in the UK and in certain other countries

Published in the United States
by Oxford University Press Inc., New York

First edition 1978
Second edition 1982
Third edition 1986
Fourth edition 1990
Fifth edition 1994
Sixth edition 1998
Seventh edition 2002

Library of Congress Cataloging in Publication Data

(Data available)

ISBN 0-19-879285-9

Typeset by Graphicraft Limited, Hong Kong
Printed in Italy by
LEGOPrint S.p.A., Trento, Italy

Preface for the student

Welcome to your study of physical chemistry, the subject that underlies all chemistry and increasingly biology, engineering, and medicine. This text will give you an introduction to the wide range of topics that constitute physical chemistry. It is unlikely that you will be expected to cover everything in the text, but in your later careers you may find it an invaluable reference to topics you did not cover as well as those you did.

We have to be honest: physical chemistry lies at the mathematical end of the spectrum that constitutes modern chemistry. We have done our best, though, to make the mathematical content as accessible and interesting as possible. First, the text is heavily illustrated, to help you grasp the qualitative aspects of a concept pictorially. Second, we have taken great care to describe why a particular mathematical step is being taken and, wherever appropriate, to interpret equations physically. You will find that many of the derivations are in the sections called *Justifications*. These *Justifications* enable you to read through a topic without getting bogged down in too much detail initially, and then return to the derivation once you have grasped the big picture. Because the *Justifications* are set off from the main text, they are in a convenient form for revision for examinations. Some of you might be advised to skip them altogether: take your instructor's advice. They are there to make the text authoritative and complete as a reference. The short marginal notes are designed to remind you of the mathematics being invoked. The background to these commentaries will be found in the *Further information* section at the back of the book.

We have provided a lot of practice material. There are numerous worked *Examples* throughout the text. Each one has a *Method* section which suggests how to approach the problem and collect the information you need. All end with a *Self-test* so you can test your understanding immediately after working through the example. You will also find many *Illustrations*. These are short worked examples, showing in a succinct way how a concept or equation is used. They too have *Self-tests*.

You may find the list of *Key ideas* at the end of each chapter particularly useful for review. We have collected here all the newly defined terms from the chapter, so the section acts as a glossary of important concepts. We have included the most important equations too. However, we do not recommend that you memorize these equations: it is far better to know how to derive an equation than simply to memorize it.

We draw your attention to the web site for this text at www.oup.com/pchem7. You will find a lot of additional information there, such as pointers to sources of data and further information about selected topics. One of the most important components of the site is the section we call *Living graphs*. These are dynamic versions of the graphs marked with the icon ◪ in the text, and will enable you to explore what happens as you change the parameters in an equation.

Physical chemistry is a demanding but richly rewarding subject. It underlies all chemistry, and is becoming increasingly important as the investigative and computational aspects of chemistry become more sophisticated. We wish you well with your studies, and hope that you find this text rewarding.

Oxford P.W.A.
Haverford J.de P.

Preface for the instructor

As has been the habit of the text, this edition has been substantially revised from its predecessor. We have aimed to make the text more flexible and to render it more accessible while retaining its authority and scope. At the same time we have sought to modernize the content and take into account the electronic aids that have become an essential part of teaching.

We have aimed for greater flexibility. The text is still divided into three parts, but material has been shifted around between chapters and the chapters have been reorganized. There is a continuing argument in the physical chemistry community about the choice of a 'quantum first' or a 'thermodynamics first' approach. One of the strategic aims of this revision is to make it possible to work through the text in a variety of orders. To help the instructor navigate, we have suggested two sequences at the end of this Preface. Less structured courses, such as some of those in Europe, will still be able to use the text for their short topic-centred courses.

We have increased the treatment of modern topics, and have noted the attention that biophysical chemistry is now receiving. That shift in attention is recognized by our inclusion in this edition of numerous *Boxes* which emphasize the application of physical chemistry to biological phenomena. However, physical chemistry is not just a subject underpinning biology: it is essential to chemical engineering, materials science, environmental science, and to the rest of chemistry. Therefore, we have also paid attention to applications in these areas. The use of *Boxes* to hold these applications away from the main exposition in the text itself is a part of our strategy to make the book flexible. Each box has two *Discussion questions*, so the material can be assigned and tested, and suggestions for *Further reading*.

We have aimed for a more accessible treatment of mathematics. The level of mathematical competence of students is of perennial but increasing concern. There is no denying that physical chemistry is a mathematical subject, but the manner in which the mathematics is handled is a crucial concern. We have adopted a number of strategies to help you and your students. First, the text has always been attractive to users for the way in which mathematics is motivated and interpreted physically. As in the sixth edition, we have continued the practice of putting key derivations into *Justifications*. The idea here is that a student can gain an overall picture of the subject before grappling with the derivations; the more recondite derivations can be ignored but are present for reference; the *Justifications* are readily available for review or assignment in lectures. In short, they are a component of the flexibility of the text, for their judicious use is a way of tailoring the level of the text to the needs of the course. Another mathematical feature is the use of marginal notes to remind the student of a key mathematical topic that is being invoked in a derivation. The background to these mathematical comments will be found in the revised and extended *Further information* section at the end of the text.

Users of earlier editions will see that there have been changes in the order of topics. The kinetic theory of gases is now treated in a single place, where molecular motion is introduced (Chapter 24). A new chapter on the Solid State (Chapter 23) collects a lot of material into one place which previously was dispersed and has added a discussion of mechanical and electrical properties, to act as a bridge to courses on material science. The old and rarely used chapter on electric and magnetic properties has been moved and repackaged as a chapter on Molecular Interactions (Chapter 21). The most notable additions to other parts of the text are the strengthening of quantum theory, the introduction of time-dependent perturbation theory as a way of unifying the presentation of spectroscopy, and the inclusion of a discussion of the mechanism of electron transfer reactions (Marcus theory). We have streamlined some of the more classical aspects of thermodynamics and its applications.

Each chapter now ends with a glossary of *Key ideas* and key equations. We hope this innovation will be regarded as a study aid rather than as an aid to rote memorization. Each chapter has a large number of structured worked *Examples*, each accompanied by a *Self-test*. There are also numerous *Illustrations*, which are short, unstructured examples of how a concept or equation is used. All these elements are numbered, so they can be assigned easily. As always, there are numerous end-of-chapter *Exercises* and more demanding *Problems*. The innovation in this edition is that the *Exercise* section begins with a selection of *Discussion questions*, designed to promote discussion and deeper understanding of the material. The *Problems* have also been extended by the inclusion of a new category: *Applications*. Solutions to the 'a' *Exercises* and the *Problems* with blue numbers are given in full in the *Student's Solutions Manual*; solutions to the 'b' *Exercises* and the *Problems* with black numbers are given in the *Instructor's Solutions Manual*. Both manuals have been prepared with great care by Charles Trapp with help from Carmen Giunta and Marshall Cady.

Electronic media are now an intrinsic part of pedagogy. This revision has a closely integrated *Web site* at www.oup.com/pchem7. On this site you will find:

- All the line-art drawings from the text, for use in lectures.
- *Living graphs*: dynamic versions of all the graphs in the text that are marked with a 'Living graph' icon ( ).
- Web addresses for sources of data and other links.

Marginal notes throughout the text alert the student to useful links to be found at the text's web site. In short, the web site is intended to be a closely integrated utility to enrich the text and to provide a dynamic resource.

Oxford   P.W.A.

Haverford   J.de P.

Acknowledgements

A book as extensive as this could not have been written without significant input from many individuals. We would like to reiterate our thanks to the hundreds of people who contributed to the first six editions. In addition, we have many new and old friends to thank for their contribution to this edition. First among them is Valerie Walters, Haverford College, who read the whole set of proofs through in great detail, and made many valuable suggestions. Our warm thanks go Charles Trapp, who has produced many editions of the *Solutions manuals* over the years, and has made numerous useful remarks. Carmen Giunta, one of his collaborators on the solutions manuals, also provided us with thoughtful input. Hideaki Chihara, Osaka, who has translated many editions into Japanese, has been particularly assiduous in sending us comments.

Many people gave their advice based on the sixth edition, and others reviewed the draft chapters as they emerged. We therefore wish to thank the following colleagues most warmly:

Warren Beck, Michigan State University
Dor Ben-Amotz, Purdue University
Isaac Bersuker, The University of Texas at Austin
Michael Blandamer, University of Leicester
Roland Böhmer, University of Mainz
Eric Borguet, University of Pittsburgh
Peter Budd, University of Manchester
David Cadenhead, University of Buffalo
Mark Dadmun, The University of Tennessee
Robert Dunbar, Case Western Reserve University
Thomas Engel, University of Washington
Paul Fletcher, University of Hull
Deborah Glover-Fischer, North Carolina Wesleyan College
James Goll, West Virginia University
Edward Grant, Purdue University
Susan Green, St. Olaf College

Arthur Halpern, Indiana State University
Judith Howard, University of Durham
Thomas J. Murphy, University of Maryland
Graham Jackson, University of Cape Town
Evguenii Kozliak, University of North Dakota
Matthew Kubasik, Fairfield University
Barbara Lawrence, Eastern Illinois University
Lawrence Lohr, University of Michigan
Peter Lykos, Illinois Institute of Technology
David Metcalf, University of Virginia
Gerald Morine, Bemidji State University
Gareth Morris, University of Manchester
Richard Moss, University of Southampton
Thomas Murphy, University of Maryland
Dwight Myers, East Central University
Breandán Ò Cochláin, National University of Ireland, Galway
John Parker, Heriot-Watt University
Dvora Perahia, Clemson University
Sanford Safron, Florida State University
Hans Sillescu, University of Mainz
Kathy Singfield, Saint Mary's University
Durwin Striplin, Davidson College
Göran Sundholm, Helsinki University of Technology
Peter Thistlethwaite, University of Melbourne
Alexander van Hook, University of Tennessee, Knoxville
Michael Wagner, George Washington University
Robert Wyatt, University of Texas at Austin
Nancy Xu, Old Dominion University
Lucy Ziurys, University of Arizona

We would also like to thank our two publishers, Oxford University Press and W.H. Freeman & Co., for their constant encouragement, advice, and assistance. Authors could not wish for a more congenial publishing environment.

Copyright acknowledgements

The permission of the respective copyright holders to reproduce tables of data and extracts from them is gratefully acknowledged as follows: Butterworths (28.1, 28.2 [© 1964]), Professor J.G. Calvert (17.1 [© 1966]), Dr J. Emsley (13.4 [© 1989]), Chapman and Hall (16.3 [© 1975], 2.5, 9.1 [© 1986]), CRC Press (1.4, 9.1, 10.2, 10.7, 14.3, 22.1 [© 1979]), Dover Publications Inc. (12.2 [© 1965]), Elsevier Scientific Publishing Company (22.1 [© 1978]), McGraw Hill Book Company (1.4, 1.5, 2.3, 3.1, 3.7, 4.1, 13.4, 16.2, 22.2, 24.3, 24.7 [© 1975], 5.2 [© 1961], 25.1, 25.2 [© 1965], 3.7 [© 1968]), National Bureau of Standards (2.5, 2.6 [© 1982]), Dr J. Nicholas (25.1, 25.2, 25.4 [© 1976]), Oxford University Press (25.1, 25.2, 25.4 [© 1975], 28.2, 28.3, 28.4 [© 1974]), Pergamon Press Ltd (22.3 [© 1961]), Prentice Hall Inc. (23.1, 23.3 [© 1971]), John Wiley and Sons Inc. (7.1 [© 1975], 21.6 [© 1954], 22.25 [© 1961]. The sources of the material are quoted at the foot of each table.

The permission of the following individuals is gratefully acknowledged: NMR spectroscopy, John Wiley and Sons Ltd, Professor H. Günther (18.7), Professor G. Ehrlich (28.29), Dr J. Evans (21.17), Dr A.J. Forty (28.4), Dr J. Foster (illustration in Box 12.1), Professor H. Ibach (28.9), Professor M. Karplus (27.23, 27.24), Professor D.A. King (28.26), Dr G. Morris (18.32, 18.46), Dr A.H. Narten (21.27), Oxford University Press (28.4, 28.5), Professor C.F. Quaite (illustration in Box 12.1), Professor M.W. Roberts (28.8, 28.11), Dr H.M. Rosenberg (28.5), Professor G.A. Somorjai (28.14, 28.17), Dr A. Stevens (23.17), Professor A.H. Zewail (illustration in Box 27.1).

Summary of contents

Conventions   xx

Part 1   Equilibrium   1

  1   The properties of gases   3
  2   The First Law: the concepts   30
  3   The First Law: the machinery   73
  4   The Second Law: the concepts   90
  5   The Second Law: the machinery   121
  6   Physical transformations of pure substances   135
  7   Simple mixtures   160
  8   Phase diagrams   193
  9   Chemical equilibrium   222
 10   Equilibrium electrochemistry   252

Part 2   Structure   291

 11   Quantum theory: introduction and principles   293
 12   Quantum theory: techniques and applications   325
 13   Atomic structure and atomic spectra   365
 14   Molecular structure   410
 15   Molecular symmetry   453
 16   Spectroscopy 1: rotational and vibrational spectra   481
 17   Spectroscopy 2: electronic transitions   538
 18   Spectroscopy 3: magnetic resonance   579
 19   Statistical thermodynamics: the concepts   628
 20   Statistical thermodynamics: the machinery   656
 21   Molecular interactions   686
 22   Macromolecules and aggregates   716
 23   The solid state   767

Part 3   Change   813

 24   Molecules in motion   815
 25   The rates of chemical reactions   862
 26   The kinetics of complex reactions   899
 27   Molecular reaction dynamics   944
 28   Processes at solid surfaces   977
 29   Dynamics of electron transfer   1013

Further information 1: Mathematical techniques   1048
Further information 2: Essential concepts of physics   1062
Data section   1071
Answers to exercises   1112
Answers to problems   1124
Index   1136

Contents

Conventions   xx

Part 1   Equilibrium   1

1   The properties of gases   3

The perfect gas   3

1.1   The states of gases   3
1.2   The gas laws   8

Real gases   16

1.3   Molecular interactions   16
1.4   The van der Waals equation   19
1.5   The principle of corresponding states   23
Checklist of key ideas   24
Further reading   25
Exercises   26
Problems   28

2   The First Law: the concepts   30

The basic concepts   30

2.1   Work, heat, and energy   31
2.2   The First Law   33

Work and heat   37

2.3   Expansion work   37
2.4   Heat transactions   41
2.5   Enthalpy   45
2.6   Adiabatic changes   52

Thermochemistry   55

2.7   Standard enthalpy changes   55
2.8   Standard enthalpies of formation   60
2.9   The temperature dependence of reaction enthalpies   62
Checklist of key ideas   64
Further reading   65
Exercises   66
Problems   69

3   The First Law: the machinery   73

State functions and exact differentials   73

3.1   State and path functions   73
3.2   Exact and inexact differentials   74

Thermodynamic consequences   75

3.3   Changes in internal energy   75
3.4   The temperature dependence of the enthalpy   79
3.5   The relation between C_V and C_p 85
Checklist of key ideas   86
Further reading   87
Exercises   87
Problems   88

4   The Second Law: the concepts   90

The direction of spontaneous change   90

4.1   The dispersal of energy   91
4.2   Entropy   92
4.3   Entropy changes accompanying specific processes   100
4.4   The Third Law of thermodynamics   106

Concentrating on the system   108

4.5   The Helmholtz and Gibbs energies   108
4.6   Standard molar Gibbs energies   114
Checklist of key ideas   115
Further reading   116
Exercises   116
Problems   118

5   The Second Law: the machinery   121

Combining the First and Second Laws   121

5.1   The fundamental equation   121
5.2   Properties of the internal energy   122

Properties of the Gibbs energy   124

5.3   General considerations   124
5.4   The variation of the Gibbs energy with temperature   126
5.5   The variation of the Gibbs energy with pressure   127
Checklist of key ideas   130
Further reading   131
Exercises   131
Problems   132

6   Physical transformations of pure substances   135

Phase diagrams   135

6.1   The stabilities of phases   135
6.2   Phase boundaries   136
6.3   Three typical phase diagrams   138

Phase stability and phase
transitions   140

6.4   The thermodynamic criterion of
      equilibrium   141
6.5   The dependence of stability on
      the conditions   141
6.6   The location of phase boundaries   144
6.7   The Ehrenfest classification of phase
      transitions   148

The physical liquid surface   150

6.8   Surface tension   150
6.9   Curved surfaces   151
6.10 Capillary action   153
      Checklist of key ideas   155
      Further reading   156
      Exercises   156
      Problems   157

7 Simple mixtures   160

The thermodynamic description
of mixtures   160

7.1   Partial molar quantities   161
7.2   The thermodynamics of mixing   166
7.3   The chemical potentials of liquids   168

The properties of solutions   172

7.4   Liquid mixtures   173
7.5   Colligative properties   175

Activities   182

7.6   The solvent activity   182
7.7   The solute activity   183
7.8   The activities of regular solutions   186
      Checklist of key ideas   187
      Further reading   188
      Exercises   189
      Problems   190

8 Phase diagrams   193

Phases, components, and degrees
of freedom   193

8.1   Definitions   193
8.2   The phase rule   195

Two-component systems   198

8.3   Vapour pressure diagrams   198
8.4   Temperature-composition diagrams   202
8.5   Liquid-liquid phase diagrams   204
8.6   Liquid-solid phase diagrams   208

      Checklist of key ideas   213
      Further reading   214
      Exercises   214
      Problems   218

9 Chemical equilibrium   222

Spontaneous chemical reactions   222

9.1   The Gibbs energy minimum   222
9.2   The description of equilibrium   226

The response of equilibria to
the conditions   233

9.3   How equilibria respond to pressure   234
9.4   The response of equilibria to
      temperature   235
9.5   The response of equilibria to pH   240
      Checklist of key ideas   246
      Further reading   246
      Exercises   247
      Problems   249

10 Equilibrium electrochemistry   252

The thermodynamic properties of ions
in solution   252

10.1 Thermodynamic functions of
      formation   253
10.2 Ion activities   256

Electrochemical cells   262

10.3 Half-reactions and electrodes   263
10.4 Varieties of cells   265
10.5 Standard potentials   270

Applications of standard potentials   274

10.6 The electrochemical series   274
10.7 The measurement of pH and pK_a 277
10.8 Thermodynamic functions   280
      Checklist of key ideas   282
      Further reading   283
      Exercises   284
      Problems   286

Part 2   Structure   291

11 Quantum theory: introduction
    and principles   293

The origins of quantum mechanics   293

11.1 The failures of classical physics   294
11.2 Wave-particle duality   299

The dynamics of microscopic systems   304

11.3 The Schrödinger equation   304
11.4 The Born interpretation of the
       wavefunction   306

Quantum mechanical principles   309

11.5 The information in a wavefunction   310
11.6 The uncertainty principle   317
       Checklist of key ideas   320
       Further reading   321
       Exercises   322
       Problems   323

12 Quantum theory: techniques and
    applications   325

Translational motion   325

12.1 A particle in a box   326
12.2 Motion in two and more dimensions   331
12.3 Tunnelling   334

Vibrational motion   338

12.4 The energy levels   338
12.5 The wavefunctions   339

Rotational motion   345

12.6 Rotation in two dimensions:
       the particle on a ring   345
12.7 Rotation in three dimensions:
       the particle on a sphere   349
12.8 Spin   354

Techniques of approximation   355

12.9 Time-independent perturbation theory   355
12.10 Time-dependent perturbation theory   358
       Checklist of key ideas   360
       Further reading   361
       Exercises   361
       Problems   363

13 Atomic structure and atomic spectra   365

The structure and spectra of
hydrogenic atoms   366

13.1 The structure of hydrogenic atoms   367
13.2 Atomic orbitals and their energies   372
13.3 Spectroscopic transitions and
       selection rules   381

The structures of many-electron atoms   383

13.4 The orbital approximation   384
13.5 Self-consistent field orbitals   392

The spectra of complex atoms   393

13.6 Quantum defects and ionization limits   395
13.7 Singlet and triplet states   395
13.8 Spin–orbit coupling   396
13.9 Term symbols and selection rules   399
       Checklist of key ideas   404
       Further reading   405
       Exercises   406
       Problems   408

14 Molecular structure   410

The Born–Oppenheimer approximation   410

Valence-bond theory   411

14.1 The hydrogen molecule   411
14.2 Homonuclear diatomic molecules   413
14.3 Polyatomic molecules   414

Molecular orbital theory   417

14.4 The hydrogen molecule-ion   418
14.5 The structures of diatomic molecules   422
14.6 Heteronuclear diatomic molecules   427

Molecular orbitals for polyatomic
systems   432

14.7 The Hückel approximation   433
14.8 Extended Hückel theory   438
14.9 Self-consistent field calculations   441
       Checklist of key ideas   446
       Further reading   447
       Exercises   448
       Problems   450

15 Molecular symmetry   453

The symmetry elements of objects   453

15.1 Operations and symmetry elements   454
15.2 The symmetry classification of
       molecules   456
15.3 Some immediate consequences of
       symmetry   459

Character tables   463

15.4 Character tables and symmetry labels   463
15.5 Vanishing integrals and orbital
       overlap   469
15.6 Vanishing integrals and selection rules   474
       Checklist of key ideas   476
       Further reading   477
       Exercises   477
       Problems   478

16 Spectroscopy 1: rotational and vibrational spectra 481

General features of spectroscopy 483

16.1 Experimental techniques 483
16.2 The intensities of spectral lines 491
16.3 Linewidths 495

Pure rotation spectra 497

16.4 Moments of inertia 497
16.5 The rotational energy levels 500
16.6 Rotational transitions 504
16.7 Rotational Raman spectra 507
16.8 Nuclear statistics and rotational states 510

The vibrations of diatomic molecules 511

16.9 Molecular vibrations 512
16.10 Selection rules 513
16.11 Anharmonicity 515
16.12 Vibration–rotation spectra 517
16.13 Vibrational Raman spectra of diatomic molecules 519

The vibrations of polyatomic molecules 520

16.14 Normal modes 520
16.15 Infrared absorption spectra of polyatomic molecules 523
16.16 Vibrational Raman spectra of polyatomic molecules 524
16.17 Symmetry aspects of molecular vibrations 526
Checklist of key ideas 529
Further reading 531
Exercises 532
Problems 534

17 Spectroscopy 2: electronic transitions 538

The characteristics of electronic transitions 539

17.1 The electronic spectra of diatomic molecules 539
17.2 The electronic spectra of polyatomic molecules 545

The fates of electronically excited states 550

17.3 Fluorescence and phosphorescence 550
17.4 Dissociation and predissociation 552

Lasers 553

17.5 General principles of laser action 554
17.6 Practical lasers 558
17.7 Applications of lasers in chemistry 562

Photoelectron spectroscopy 568

17.8 The technique 568
17.9 Ultraviolet photoelectron spectroscopy 569
17.10 X-ray photoelectron spectroscopy 570
Checklist of key ideas 571
Further reading 573
Exercises 574
Problems 575

18 Spectroscopy 3: magnetic resonance 579

The effect of magnetic fields on electrons and nuclei 579

18.1 The energies of electrons in magnetic fields 580
18.2 The energies of nuclei in magnetic fields 581
18.3 Magnetic resonance spectroscopy 583

Nuclear magnetic resonance 583

18.4 The NMR spectrometer 584
18.5 The chemical shift 585
18.6 The fine structure 590

Pulse techniques in NMR 599

18.7 The magnetization vector 599
18.8 Linewidths and rate processes 602
18.9 Spin decoupling 607
18.10 The nuclear Overhauser effect 608
18.11 Two-dimensional NMR 610
18.12 Solid-state NMR 614

Electron spin resonance 615

18.13 The ESR spectrometer 615
18.14 The g-value 617
18.15 Hyperfine structure 618
Checklist of key ideas 621
Further reading 623
Exercises 624
Problems 626

19 Statistical thermodynamics: the concepts 628

The distribution of molecular states 629

19.1 Configurations and weights 629
19.2 The molecular partition function 634

The internal energy and the entropy 639

19.3 The internal energy 640
19.4 The statistical entropy 642

The canonical partition function 647

19.5 The canonical ensemble 647
19.6 The thermodynamic information in the partition function 648

19.7 Independent molecules   649
      Checklist of key ideas   652
      Further reading   652
      Exercises   653
      Problems   654

20 Statistical thermodynamics: the machinery   656

      Fundamental relations   656

20.1 The thermodynamic functions   656
20.2 The molecular partition function   658

      Using statistical thermodynamics   667

20.3 Mean energies   667
20.4 Heat capacities   669
20.5 Equations of state   671
20.6 Residual entropies   672
20.7 Equilibrium constants   674
      Checklist of key ideas   681
      Further reading   682
      Exercises   682
      Problems   683

21 Molecular interactions   686

      Electric properties of molecules   686

21.1 Electric dipole moments   686
21.2 Polarizabilities   689
21.3 Relative permittivities   692
21.4 Refractive index   694

      Interactions between molecules   696

21.5 Interactions between dipoles   698
21.6 Repulsive and total interactions   705
21.7 Molecular interactions in gases   706
21.8 Molecular interactions in liquids   709
      Checklist of key ideas   713
      Further reading   715
      Exercises   715
      Problems   716

22 Macromolecules and aggregates   719

      Structure and dynamics   720

22.1 The different levels of structure   720
22.2 Random coils   721
22.3 The structure of proteins   725
22.4 The structure of nucleic acids   729
22.5 The stability of biological polymers   731

      Determination of size and shape   732

22.6 Mean molar masses   732
22.7 Mass spectrometry   735

22.8 Laser light scattering   736
22.9 Ultracentrifugation   742
22.10 Electrophoresis   745
22.11 Size-exclusion chromatography   747
22.12 Viscosity   748

      Self-assembly   750

22.13 Colloids   752
22.14 Micelles and biological membranes   754
22.15 Surface films   757
      Checklist of key ideas   760
      Further reading   762
      Exercises   763
      Problems   764

23 The solid state   767

      Crystal lattices   767

23.1 Lattices and unit cells   767
23.2 The identification of lattice planes   770
23.3 The investigation of structure   772
23.4 Neutron and electron diffraction   783

      Crystal structure   784

23.5 Metallic solids   784
23.6 Ionic solids   786
23.7 Molecular solids and covalent networks   789

      The properties of solids   791

23.8 Mechanical properties   791
23.9 Electrical properties   795
23.10 Magnetic properties   801
      Checklist of key ideas   805
      Further reading   807
      Exercises   808
      Problems   810

Part 3 Change   813

24 Molecules in motion   815

      Molecular motion in gases   815

24.1 The kinetic model of gases   816
24.2 Collisions with walls and surfaces   822
24.3 The rate of effusion   824
24.4 Transport properties of a perfect gas   826

      Molecular motion in liquids   832

24.5 Experimental results   832
24.6 The conductivities of electrolyte solutions 833
24.7 The mobilities of ions   835
24.8 Conductivities and ion–ion interactions   841

Diffusion 842

24.9 The thermodynamic view 842
24.10 The diffusion equation 846
24.11 Diffusion probabilities 852
24.12 The statistical view 853
Checklist of key ideas 854
Further reading 856
Exercises 856
Problems 859

25 The rates of chemical reactions 862

Empirical chemical kinetics 862

25.1 Experimental techniques 863
25.2 The rates of reactions 866
25.3 Integrated rate laws 871
25.4 Reactions approaching equilibrium 876
25.5 The temperature dependence of reaction rates 879

Accounting for the rate laws 882

25.6 Elementary reactions 882
25.7 Consecutive elementary reactions 883
25.8 Unimolecular reactions 890
Checklist of key ideas 892
Further reading 893
Exercises 894
Problems 896

26 The kinetics of complex reactions 899

Chain reactions 899

26.1 The rate laws of chain reactions 900
26.2 Explosions 902

Polymerization kinetics 904

26.3 Stepwise polymerization 904
26.4 Chain polymerization 906

Homogeneous catalysis 908

26.5 Features of homogeneous catalysis 909
26.6 Enzymes 909

Oscillating reactions 914

26.7 Autocatalysis 915
26.8 Autocatalytic mechanisms of oscillating reactions 916
26.9 Bistability 917
26.10 Chemical chaos 919

Photochemistry 920

26.11 Kinetics of photophysical and photochemical processes 927

26.12 Complex photochemical processes 934
Checklist of key ideas 936
Further reading 937
Exercises 938
Problems 940

27 Molecular reaction dynamics 944

Reactive encounters 944

27.1 Collision theory 945
27.2 Diffusion-controlled reactions 951
27.3 The material balance equation 954

Activated complex theory 956

27.4 The Eyring equation 956
27.5 Thermodynamic aspects 960

The dynamics of molecular collisions 963

27.6 Reactive collisions 963
27.7 Potential energy surfaces 966
27.8 Some results from experiments and calculations 967
Checklist of key ideas 971
Further reading 972
Exercises 973
Problems 974

28 Processes at solid surfaces 977

The growth and structure of solid surfaces 977

28.1 Surface growth 977
28.2 Surface composition 979

The extent of adsorption 987

28.3 Physisorption and chemisorption 988
28.4 Adsorption isotherms 989
28.5 The rates of surface processes 994

Catalytic activity at surfaces 999

28.6 Adsorption and catalysis 999
28.7 Examples of catalysis 1001
Checklist of key ideas 1006
Further reading 1007
Exercises 1008
Problems 1009

29 Dynamics of electron transfer 1013

Electron transfer in homogeneous systems 1014

29.1 Theory of electron transfer processes 1014
29.2 Experimental results 1018

Electron transfer in heterogeneous systems 1021

29.3	The electrode–solution interface	1021
29.4	The rate of charge transfer	1024
29.5	Voltammetry	1031
29.6	Electrolysis	1036
29.7	Working galvanic cells	1037
29.8	Corrosion	1039
	Checklist of key ideas	1041
	Further reading	1043
	Exercises	1043
	Problems	1045

Further information 1   Mathematical techniques 1048

Basic procedures 1048

1.1	Logarithms and exponentials	1048
1.2	Combinatorial functions	1048
1.3	Complex numbers and complex functions	1049
1.4	Vectors	1050

Calculus 1051

1.5	Differentiation and integration	1051
1.6	Power series and Taylor expansions	1052
1.7	Partial derivatives	1053
1.8	Undetermined multipliers	1054
1.9	Differential equations	1055

Matrix algebra 1057

1.10	Matrix addition and multiplication	1058
1.11	Simultaneous equations	1060

1.12	Eigenvalue equations	1059
	Further reading	1061

Further information 2   Essential concepts of physics 1062

Energy 1062

2.1	Kinetic and potential energy	1062
2.2	Energy units	1062

Classical mechanics 1063

2.3	The trajectory in terms of the energy	1063
2.4	Newton's second law	1064
2.5	Rotational motion	1064
2.6	The harmonic oscillator	1065

Waves 1066

2.7	The electromagnetic field	1066
2.8	Features of electromagnetic radiation	1067

Electrostatics 1068

2.9	The Coulomb interaction	1068
2.10	The Coulomb potential	1068
2.11	The strength of the electric field	1069
2.12	The dipole–dipole interaction	1069
2.13	Electric current and power	1070
	Further reading	1070

Data section	1071
Answers to exercises	1112
Answers to problems	1124
Index	1136

List of boxes

1.1	The gas laws and the weather	12
2.1	Differential scanning calorimetry	46
2.2	Food and energy reserves	59
4.1	The thermodynamics of refrigeration	98
4.2	Reaching very low temperatures	107
6.1	Supercritical fluids	137
7.1	Dialysis and protein binding	179
8.1	Liquid crystals	209
8.2	Ultrapurity and controlled impurity	211
9.1	Biological energy conversion: the synthesis of proteins and the oxidation of glucose	224
9.2	The extraction of metals from their oxides	239
10.1	Biological energy conversion: driving the synthesis of ATP with electron transfer reactions	275
10.2	Species-selective electrodes	279
11.1	Electron diffraction	301
12.1	Scanning microscopy	336
13.1	Spectroscopy of stars	394
16.1	Rotational and vibrational spectroscopy of interstellar space	482
16.2	Vibrational spectroscopy of biological cells	520
17.1	The photochemistry of vision	548
17.2	Spectroscopy of single molecules	563
18.1	Magnetic resonance imaging	612
18.2	Spin probes	616
19.1	Temperatures below zero	643
20.1	The helix–coil transition in polypeptides	674
21.1	Molecular recognition in biological systems	697
22.1	Molecular wires	750
23.1	Time-resolved X-ray diffraction techniques	773
23.2	Superconductors	796
24.1	The Sun as a ball of perfect gas	823
24.2	Transport across biological membranes	849
25.1	Kinetics of the helix–coil transition in polypeptides	865
26.1	The chemistry of stratospheric ozone	921
26.2	Photobiology and photomedicine	928
27.1	Femtochemistry	959
27.2	Controlling chemical reactions with lasers	964
28.1	Nanofabrication	986
28.2	Surface plasmon resonance	994

Conventions

Names of quantities

A **substance** is a distinct, pure form of matter.

The **amount of substance**, n (more colloquially 'number of moles' or 'chemical amount'), in a sample is reported in terms of the **mole** (mol): 1 mol is the amount of substance that contains as many objects (atoms, molecules, ions, or other specified entities) as there are atoms in exactly 12 g of carbon-12. This number is found experimentally to be approximately 6.02×10^{23} (see the endpapers for more precise values). If a sample contains N entities, the amount of substance it contains is $n = N/N_A$, where N_A is the Avogadro constant: $N_A = 6.02 \times 10^{23}$ mol^{-1}. Note that N_A is a quantity with units, not a pure number.

An **extensive property** is a property that depends on the amount of substance in the sample. Two examples are mass and volume. An **intensive property** is a property that is independent of the amount of substance in the sample. Examples are temperature, mass density (mass divided by volume), and pressure.

A **molar property**, X_m, is the value of an extensive property, X, of the sample divided by the amount of substance present in the sample: $X_m = X/n$. A molar property is intensive. An example is the **molar volume**, V_m, the volume of a sample divided by the amount of substance in the sample (the volume per mole). The one exception to the notation X_m is the **molar mass**, which is denoted M. The molar mass of an element is the mass per mole of its atoms. The molar mass of a molecular compound is the mass per mole of molecules, and the molar mass of an ionic compound is the mass per mole of formula units. A **formula unit** of an ionic compound is an assembly of ions corresponding to the chemical formula of the compound; so the formula unit NaCl consists of one Na$^+$ ion and one Cl$^-$ ion. The names *atomic weight* and *molecular weight* are still widely used in place of molar mass (often with the units omitted), but we shall not use them in this text.

The **molar concentration** ('molarity') of a solute in a solution is the amount of-substance of the solute divided by the volume of the solution. Molar concentration is usually expressed in moles per litre (mol L^{-1} or mol dm^{-3}; 1 L is identical to 1 dm^3). A solution in which the molar concentration of the solute is 1 mol L^{-1} is prepared by dissolving 1 mol of the solute in sufficient solvent to prepare 1 L of solution. Such a solution is widely called a '1 molar' solution and denoted 1 M. The term **molality** refers to the amount of substance of solute divided by the mass of solvent used to prepare the solution. Its units are typically moles of solute per kilogram of solvent (mol kg^{-1}).

Notational conventions

We use SI units and IUPAC conventions throughout, except in a small number of cases. These conventions are specified in the 'green book', *Quantities, units, and symbols in physical chemistry*, Blackwell Scientific, Oxford (1993).

The default numbering of equations is $(C.n)$, where C is the chapter; however, $[C.n]$ is used to denote a definition and $\{C.n\}$ is used to indicate that a variable x should be interpreted as $x/x^{\ominus}$, where $x^{\ominus}$ is a standard value. Cross-references within a chapter are denoted simply eqn n. A subscript r attached to an equation number indicates that the equation is valid only for a reversible change. A superscript $°$ indicates that the equation is valid for an ideal system, such as a perfect gas or an ideal solution.

The symbol $p^{\ominus}$ denotes 1 bar (10^5 Pa) exactly, and $b^{\ominus}$ denotes 1 mol kg^{-1} exactly.

When referring to temperature, T denotes a thermodynamic temperature (for instance, on the Kelvin scale) and θ a temperature on the Celsius scale.

For numerical calculations, unless otherwise specified, assume that zeros in data like 10, 100, 1000, etc. are significant (that is, interpret such data as 10., 100., 100., etc.).

Part 1
Equilibrium

Part 1 of the text develops the concepts that are needed for the discussion of equilibria in chemistry. Equilibria include physical change, such as fusion and vaporization, and chemical change, including electrochemistry. The discussion is in terms of thermodynamics, and particularly in terms of enthalpy and entropy. We see that we can obtain a unified view of equilibrium and the direction of spontaneous change in terms of the chemical potentials of substances. The chapters in Part 1 deal with the bulk properties of matter; those of Part 2 will show how these properties stem from the behaviour of individual atoms.

1   The properties of gases

2   The First Law: the concepts

3   The First Law: the machinery

4   The Second Law: the concepts

5   The Second Law: the machinery

6   Physical transformations of pure substances

7   Simple mixtures

8   Phase diagrams

9   Chemical equilibrium

10   Equilibrium electrochemistry

1

The properties of gases

The perfect gas

1.1 The states of gases

1.2 The gas laws

Real gases

1.3 Molecular interactions

1.4 The van der Waals equation

1.5 The principle of corresponding states

Checklist of key ideas

Further reading

Exercises

Problems

This chapter establishes the properties of gases that will be used throughout the text. It begins with an account of an idealized version of a gas, a perfect gas, and shows how its equation of state may be assembled experimentally. We then see how the properties of real gases differ from those of a perfect gas, and construct an equation of state that describes their properties.

The simplest state of matter is a **gas**, a form of matter that fills any container it occupies. Initially we consider only pure gases, but later in the chapter we see that the same ideas and equations apply to mixtures of gases too.

The perfect gas

We shall find it helpful to picture a gas as a collection of molecules (or atoms) in continuous random motion, with speeds that increase as the temperature is raised. A gas differs from a liquid in that, except during collisions, the molecules of a gas are widely separated from one another and move in paths that are largely unaffected by intermolecular forces.

1.1 The states of gases

The **physical state** of a sample of a substance, its physical condition, is defined by its physical properties. Two samples of a substance that have the same physical properties are in the same state. The state of a pure gas, for example, is specified by giving its volume, V, amount of substance (number of moles), n, pressure, p, and temperature, T. However, it has been established experimentally that it is sufficient to specify only three of these variables, for then the fourth variable is fixed. That is, it is an experimental fact that each substance is described by an **equation of state**, an equation that interrelates these four variables.

The general form of an equation of state is

$$p = f(T, V, n) \tag{1.1}$$

This equation tells us that, if we know the values of n, T, and V for a particular substance, then the pressure has a fixed value. Each substance is described by its own equation of state, but we know the explicit form of the equation in only a few special cases. One very important example is the equation of state of a 'perfect gas', which has the form

$$p = \frac{nRT}{V} \tag{1.2}$$

where R is a constant. Much of the rest of this chapter will examine the origin of this equation of state and its applications.

(a) Pressure

Pressure is defined as force divided by the area to which the force is applied. The greater the force acting on a given area, the greater the pressure. The origin of the force exerted by a gas is the incessant battering of the molecules on the walls of its container. The collisions are so numerous that they exert an effectively steady force, which is experienced as a steady pressure.

The SI unit of pressure, the *pascal* (Pa), is defined as 1 newton per metre-squared:

$$1 \text{ Pa} = 1 \text{ N m}^{-2} \tag{1.3a}$$

In terms of base units,

$$1 \text{ Pa} = 1 \text{ kg m}^{-1} \text{ s}^{-2} \tag{1.3b}$$

Several other units are still widely used (Table 1.1). A pressure of 10^5 Pa (1 bar) is the **standard pressure** for reporting data; we denote it $p^{\ominus}$.

Example 1.1 Calculating pressure

Suppose Isaac Newton weighed 65 kg. Calculate the pressure he exerted on the ground when wearing (a) boots with soles of total area 250 cm^2 in contact with the ground, (b) ice skates, of total area 2.0 cm^2.

Method Pressure is force divided by area ($p = F/A$), so the calculation depends on being able to calculate the force, F, that Newton exerts on the ground, and then to divide the force by the area, A, over which the force is exerted. To calculate the force, we need to know (from elementary physics) that the downward force an object of mass m exerts at the surface of the Earth is $F = mg$, where g is the acceleration of free fall, 9.81 m s^{-2}. Note that 1 N = 1 kg m s^{-2} and 1 Pa = 1 N m^{-2}. Convert the areas to square metres by using 1 cm^2 = 10^{-4} m^2. To reduce rounding errors, leave the evaluation of numerical expressions to the end of the calculation.

Table 1.1 Pressure units

Name	Symbol	Value
pascal	1 Pa	1 N m^{-2}, 1 kg m^{-1} s^{-2}
bar	1 bar	10^5 Pa
atmosphere	1 atm	101 325 Pa
torr	1 Torr	(101 325/760) Pa = 133.32 ... Pa
millimetre of mercury	1 mmHg	133.322 ... Pa
pound per square inch	1 psi	6.894 757 ... kPa

Answer The force exerted by Newton is

$$F = (65 \text{ kg}) \times (9.81 \text{ m s}^{-2}) = 65 \times 9.81 \text{ kg m s}^{-2} = 65 \times 9.81 \text{ N}$$

(This force works out to 6.4×10^2 N.) The force is the same whatever Newton's footwear. The areas 250 cm^2 and 2.0 cm^2 correspond to 2.50×10^{-2} m^2 and 2.0×10^{-4} m^2, respectively. The pressure he exerts in each case is:

(a) $\quad p = \dfrac{65 \times 9.81 \text{ N}}{2.50 \times 10^{-2} \text{ m}^2} = 2.6 \times 10^4 \text{ N m}^{-2} = 2.6 \times 10^4 \text{ Pa (26 kPa)}$

(b) $\quad p = \dfrac{65 \times 9.81 \text{ N}}{2.0 \times 10^{-4} \text{ m}^2} = 3.2 \times 10^6 \text{ Pa (3.2 MPa)}$

Comment One Newton exerts a force of much more than 1 newton. A pressure of 26 kPa corresponds to 0.26 atm and a pressure of 3.2 MPa corresponds to 32 atm.

Self-test 1.1 Calculate the pressure exerted by a mass of 1.0 kg pressing through the point of a pin of area 1.0×10^{-2} mm^2 at the surface of the Earth.

[0.98 GPa, 9.7×10^3 atm]

(a)

(b)

(c)

1.1 When a region of high pressure is separated from a region of low pressure by a movable wall, the wall will be pushed into one region or the other, as in (a) and (c). However, if the two pressures are identical, the wall will not move (b). The latter condition is one of mechanical equilibrium between the two regions.

If two gases are in separate containers that share a common movable wall (Fig. 1.1), the gas that has the higher pressure will tend to compress (reduce the volume of) the gas that has lower pressure. The pressure of the high-pressure gas will fall as it expands and that of the low-pressure gas will rise as it is compressed. There will come a stage when the two pressures are equal and the wall has no further tendency to move. This condition of equality of pressure on either side of a movable wall (a 'piston') is a state of **mechanical equilibrium** between the two gases. The pressure of a gas is therefore an indication of whether a container that contains the gas will be in mechanical equilibrium with another gas with which it shares a movable wall.

(b) The measurement of pressure

The pressure exerted by the atmosphere is measured with a **barometer**. The original version of a barometer (which was invented by Torricelli, a student of Galileo) was an inverted tube of mercury sealed at the upper end. When the column of mercury is in mechanical equilibrium with the atmosphere, the pressure at its base is equal to that exerted by the atmosphere. It follows that the height of the mercury column is proportional to the external pressure.

Example 1.2 Calculating the pressure exerted by a column of liquid

Derive an equation for the pressure at the base of a column of liquid of mass density ρ (rho) and height h at the surface of the Earth.

Method As in Example 1.1, $p = F/A$ and $F = mg$, so we need to know the mass of a column of liquid. The mass of a column of liquid is its mass density, ρ, multiplied by its volume, V. So, the first step is to calculate the volume of a cylindrical column of liquid.

1

(a) (b)

1.2 Two versions of a manometer used to measure the pressure of a sample of gas. (a) The height difference, h, of the two columns in the sealed-tube manometer is directly proportional to the pressure of the sample, and $p = \rho gh$, where ρ is the mass density of the liquid. (b) The difference in heights of the columns in the open-tube manometer is proportional to the difference in pressure between the sample and the atmosphere. In the example shown, the pressure of the sample is lower than that of the atmosphere.

Answer Let the column have cross-sectional area A; then its volume is Ah and its mass is $m = \rho Ah$. The force the column of this mass exerts at its base is

$$F = mg = \rho Ahg$$

The pressure at the base of the column is therefore

$$p = \frac{F}{A} = \frac{\rho Ahg}{A} = \rho gh$$

Comment Note that the pressure is independent of the shape and cross-sectional area of the column. The mass of the column increases as the area, but so does the area on which the force acts, so the two cancel.

Self-test 1.2 Derive an expression for the pressure at the base of a column of liquid of length l held at an angle θ (theta) to the vertical (**1**). $[p = \rho gl \cos \theta]$

The pressure of a sample of gas inside a container is measured with a **manometer** (Fig. 1.2). In its simplest form, a manometer is a U-tube filled with a liquid of low volatility (such as silicone oil). The pressure, p, of the gaseous sample balances the pressure exerted by the column of liquid, which is equal to ρgh if the column is of height h (see Example 1.2), plus the external pressure, p_{ex}, if one tube is open to the atmosphere:

$$p = p_{ex} + \rho gh \tag{1.4}$$

It follows that the pressure of the sample can be obtained by measuring the height of the column and noting the external pressure. More sophisticated techniques are used at lower pressures. Methods that avoid the complication of having to account for the vapour from the manometer fluid are also available. These superior methods include monitoring the deflection of a diaphragm, either mechanically or electrically, or monitoring the change in a pressure-sensitive electrical property. These techniques have displaced the use of manometers in laboratory experiments.

(c) Temperature

The concept of temperature springs from the observation that a change in physical state (for example, a change of volume) can occur when two objects are in contact with one another, as when a red-hot metal is plunged into water. Later (Section 2.1) we shall see that the change in state can be interpreted as arising from a flow of energy as heat from one object to another. The **temperature**, T, is the property that indicates the direction of the flow of energy through a thermally conducting, rigid wall. If energy flows from A to B when they are in contact, then we say that A has a higher temperature than B (Fig. 1.3).

It will prove useful to distinguish between two types of boundary that can separate the objects. A boundary is **diathermic** if a change of state is observed when two objects at different temperatures are brought into contact.[1] A metal container has diathermic walls. A boundary is **adiabatic** if no change occurs even though the two objects have different temperatures.

The temperature is a property that indicates whether two objects would be in 'thermal equilibrium' if they were in contact through a diathermic boundary. **Thermal equilibrium** is established if no change of state occurs when two objects A to B are in contact through a diathermic boundary. Suppose an object A (which we

---

1. The word dia is from the Greek for 'through'.

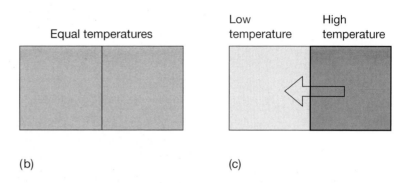

High temperature | Diathermic wall | Low temperature

Energy as heat

(a)

Equal temperatures

(b)

Low temperature | High temperature

(c)

1.3 Energy flows as heat from a region at a higher temperature to one at a lower temperature if the two are in contact through a diathermic wall, as in (a) and (c). However, if the two regions have identical temperatures, there is no net transfer of energy as heat even though the two regions are separated by a diathermic wall (b). The latter condition corresponds to the two regions being at thermal equilibrium.

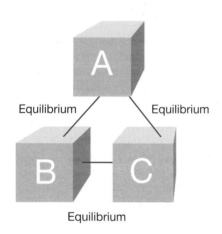

Equilibrium | Equilibrium

A

B | C

Equilibrium

1.4 The experience summarized by the Zeroth Law of thermodynamics is that an object A is in thermal equilibrium with B and B is in thermal equilibrium with C, then C is in thermal equilibrium with A.

can think of as a block of iron) is in thermal equilibrium with an object B (a block of copper), and that B is also in thermal equilibrium with another object C (a flask of water). Then it has been found experimentally that A and C will also be in thermal equilibrium when they are put in contact (Fig. 1.4). This observation is summarized by the **Zeroth Law of thermodynamics**:

> **If A is in thermal equilibrium with B, and B is in thermal equilibrium with C, then C is also in thermal equilibrium with A.**

The Zeroth Law justifies the concept of temperature and the use of a **thermometer**, a device for measuring the temperature. Thus, suppose that B is a glass capillary containing a liquid, such as mercury, that expands significantly as the temperature increases. Then, when A is in contact with B, the mercury column in the latter has a certain length. According to the Zeroth Law, if the mercury column in B has the same length when it is placed in contact with another object C, then we can predict that no change of state of A and C will occur when they are in contact. Moreover, we can use the length of the mercury column as a measure of the temperatures of A and C.

In the early days of thermometry (and still in laboratory practice today), temperatures were related to the length of a column of liquid, and the difference in lengths shown when the thermometer was first in contact with melting ice and then with boiling water was divided into 100 steps called 'degrees', the lower point being labelled 0. This procedure led to the **Celsius scale** of temperature. In this text, temperatures on the Celsius scale are denoted θ and expressed in *degrees Celsius* (°C). However, because different liquids expand to different extents, and do not always expand uniformly over a given range, thermometers constructed from different materials showed different numerical values of the temperature between their fixed points. The pressure of a gas, however, can be used to construct a **perfect-gas temperature scale** that is independent of the identity of the gas. The perfect-gas scale turns out to be identical to the **thermodynamic temperature scale** to be introduced in Section 4.2c, so we shall use the latter term from now on to avoid a proliferation of names. On the thermodynamic temperature scale, temperatures are denoted T and are normally reported in kelvins, K. Thermodynamic and Celsius temperatures are related by the exact expression

$$T/\text{K} = \theta/°\text{C} + 273.15 \tag{1.5}$$

This relation is the current definition of the Celsius scale in terms of the more fundamental Kelvin scale. It implies that a difference in temperature of 1°C is equivalent to a difference of 1 K.

Illustration 1.1

To express 25.00°C as a temperature in kelvins, we write

$$T/K = (25.00°C)/°C + 273.15 = 25.00 + 273.15 = 298.15$$

Note how the units are cancelled like numbers. This is the procedure called 'quantity calculus' in which a physical quantity (such as the temperature) is the product of a numerical value (25.00) and a unit (1°C). Multiplication of both sides by the unit K then gives $T = 298.15$ K.

1.2 The gas laws

The equation of state of a gas at low pressure was established by combining a series of empirical laws. We introduce these laws below, and then show how they can be combined into the single equation of state $pV = nRT$.

(a) The individual gas laws

Robert Boyle, acting on the suggestion of a correspondent John Townley, showed in 1661 that, to a good approximation, the pressure and volume of a fixed amount of gas at constant temperature are related by

$$pV = \text{constant} \tag{1.6}°$$

This relation is known as **Boyle's law**. That is, at constant temperature, the pressure of a sample of gas is inversely proportional to its volume, and the volume it occupies is inversely proportional to its pressure:

$$p \propto \frac{1}{V} \qquad V \propto \frac{1}{p} \tag{1.7}°$$

The variation of the pressure of a sample of gas as the volume is changed is depicted in Fig. 1.5. Each of the curves in the graph corresponds to a single temperature and hence is called an **isotherm**. According to Boyle's law, the isotherms of gases are hyperbolas. An alternative depiction, a plot of pressure against 1/volume, is shown in Fig. 1.6.

More recent experiments have shown that Boyle's law is valid only at low pressures, and that real gases obey the law only in the limit of the pressure approaching zero (which we write $p \to 0$). Boyle's law is therefore an example of a **limiting law**, a law that is strictly true only in a certain limit, in this case $p \to 0$. Equations valid in this limiting sense will be signalled by a ° on the equation number, as in eqn 6.

The molecular explanation of Boyle's law is that, if a sample of gas is compressed to half its volume, then twice as many molecules strike the walls in a given period of time than before it was compressed. As a result, the average force exerted on the walls is doubled. Hence, when the volume is halved the pressure of the gas is doubled, and $p \times V$ is a constant. The reason why Boyle's law applies to all gases regardless of their chemical identity (provided the pressure is low) is that at low pressures the average separation of molecules is so great that they exert no influence on one another and hence travel independently.

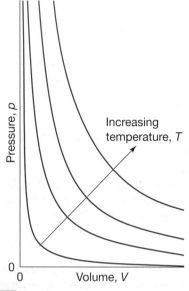

1.5 The pressure–volume dependence of a fixed amount of perfect gas at different temperatures. Each curve is a hyperbola (pV = constant) and is called an *isotherm*.

A hyperbola is a curve obtained by plotting y against x with xy = constant.

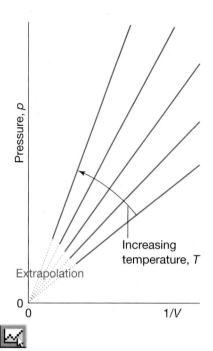

1.6 Straight lines are obtained when the pressure is plotted against $1/V$ at constant temperature.

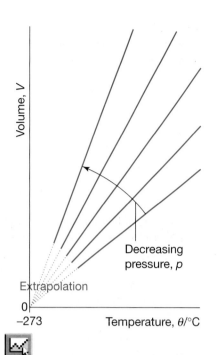

1.7 The variation of the volume of a fixed amount of gas with the temperature at constant pressure. Note that in each case the isobars extrapolate to zero volume at $-273°C$.

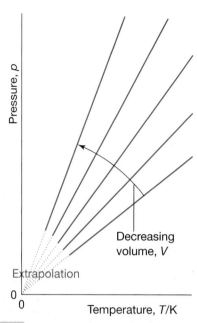

1.8 The pressure also varies linearly with the temperature at constant volume, and extrapolates to zero at $T = 0$ ($-273°C$).

The French scientist Guillaume Amontons established the next important property of gases, but his results were ignored for almost a century. Then Jacques Charles and Joseph Louis Gay-Lussac studied the effect of temperature on the volume of a sample of gas subjected to constant pressure. They found independently that the volume increased linearly with the temperature, whatever the identity of the gas, provided it was at low pressure. Specifically, they established that[2]

$$V = \text{constant} \times (\theta + 273.15°C) \qquad \text{(at constant pressure)} \qquad (1.8)°$$

The linear variation of volume with temperature summarized by this expression is illustrated in Fig. 1.7. The lines are examples of **isobars**, or lines showing the variation of properties at constant pressure.

Equation 1.8 suggests that the volume of any gas should extrapolate to zero at $\theta = -273.15°C$ and therefore that $-273.15°C$ is a natural zero of a fundamental temperature scale. As we have already indicated, a scale with 0 set at $-273.15°C$ is equivalent to the thermodynamic temperature scale devised by Kelvin, so **Charles's law** (which is also known as **Gay-Lussac's law**) can be written

$$V = \text{constant} \times T \qquad \text{(at constant pressure)} \qquad (1.9)°$$

An alternative version of Charles's law, in which the pressure of a sample of gas is monitored under conditions of constant volume, is

$$p = \text{constant} \times T \qquad \text{(at constant volume)} \qquad (1.10)°$$

This version of the law indicates that the pressure of a gas falls to zero as the temperature is reduced to zero (Fig. 1.8).

2. Recall that we use θ to denote temperatures on the Celsius scale.

The molecular explanation of Charles's law lies in the fact that raising the temperature of a gas increases the average speed of its molecules. The molecules collide with the walls more frequently and with greater impact. Therefore they exert a greater pressure on the walls of the container.

The final piece of experimental information we need is that, at a given pressure and temperature, the molar volume, $V_m = V/n$, the volume per mole of molecules, of a gas is approximately the same regardless of the identity of the gas. This observation implies that the volume of a sample of gas is proportional to the amount of molecules (in moles) present, and that the constant of proportionality is independent of the identity of the gas:

$$V = \text{constant} \times n \qquad \text{(at constant pressure and temperature)} \qquad (1.11)°$$

This conclusion is a modern form of **Avogadro's principle**, which states that equal volumes of gases at the same pressure and temperature contain the same numbers of molecules.[3]

(b) The perfect gas law

The empirical observations summarized by eqns 6–11 can be combined into a single expression:

$$pV = \text{constant} \times nT$$

This expression is consistent with Boyle's law ($pV = \text{constant}$) when n and T are constant, with both forms of Charles's law ($p \propto T$, $V \propto T$) when n and either V or p are held constant, and with Avogadro's principle ($V \propto n$) when p and T are constant. The constant of proportionality, which is found experimentally to be the same for all gases, is denoted R and called the **gas constant**. The resulting expression

$$pV = nRT \qquad (1.12)°$$

is the **perfect gas equation**. It is the approximate equation of state of any gas, and becomes increasingly exact as the pressure of the gas approaches zero. A gas that obeys eqn 12 exactly under all conditions is called a **perfect gas** (or *ideal gas*). A **real gas**, an actual gas, behaves more like a perfect gas the lower the pressure, and is described exactly by eqn 12 in the limit of $p \to 0$. The gas constant R can be determined by evaluating $R = pV/nT$ for a gas in the limit of zero pressure (to guarantee that it is behaving perfectly). However, a more accurate value can be obtained by measuring the speed of sound in a low-pressure gas and extrapolating its value to zero pressure. Table 1.2 lists the values of R in a variety of units.

The surface in Fig. 1.9 is a plot of the pressure of a fixed amount of perfect gas against its volume and thermodynamic temperature as given by eqn 12. The surface depicts the only possible states of a perfect gas: the gas cannot exist in states that do not correspond to points on the surface. The graphs in Figs. 1.5 and 1.7 correspond to sections through the surface (Fig. 1.10).

The perfect gas equation is of the greatest importance in physical chemistry because it is used to derive a wide range of relations that are used throughout thermodynamics. However, it is also of considerable practical utility for calculating the properties of a gas under a variety of conditions. For instance, the molar volume of a perfect gas under the conditions called **standard ambient temperature and pressure** (SATP), which means 298.15 K and 1 bar (that is, exactly 10^5 Pa), is easily

Table 1.2 The gas constant in various units

R
8.314 47 J K^{-1} mol^{-1}
8.205 74 × 10^{-2} L atm K^{-1} mol^{-1}
8.314 47 × 10^{-2} L bar K^{-1} mol^{-1}
8.314 47 Pa m^3 K^{-1} mol^{-1}
62.364 L Torr K^{-1} mol^{-1}
1.987 21 cal K^{-1} mol^{-1}

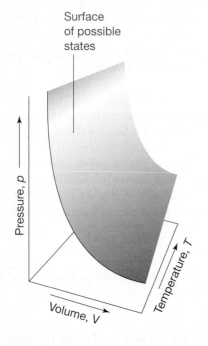

1.9 A region of the p, V, T surface of a fixed amount of perfect gas. The points forming the surface represent the only states of the gas that can exist.

3. Avogadro's principle is a principle rather than a law (a summary of experience) because it depends on the validity of a model, in this case the existence of molecules.

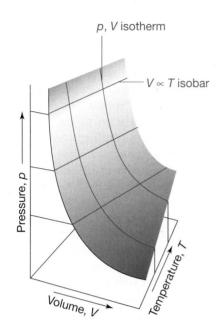

1.10 Sections through the surface shown in Fig. 1.9 at constant temperature give the isotherms shown in Fig. 1.5 and the isobars shown in Fig. 1.7.

	n	p	V	T
Initial	Same	100	Same	300
Final	Same	?	Same	500

2

calculated from $V_m = RT/p$ to be 24.789 L mol^{-1}. An earlier definition, **standard temperature and pressure** (STP), was 0°C and 1 atm; at STP, the molar volume of a perfect gas is 22.414 L mol^{-1}.

The perfect gas equation can be developed as follows to calculate the change in conditions when a constant amount of gas is subjected to different temperatures and pressures and allowed to occupy a different volume. First, write eqn 12 for the initial and final sets of conditions:

$$\frac{p_1 V_1}{T_1} = nR \qquad \frac{p_2 V_2}{T_2} = nR$$

Because n (in this problem) and R are both constant, it follows that

$$\frac{p_1 V_1}{T_1} = \frac{p_2 V_2}{T_2} \tag{1.13}°$$

This relation is sometimes called the **combined gas equation**. Among other applications, eqn 13 can be used to discuss processes in the atmosphere that give rise to the weather (Box 1.1).

Example 1.3 Using the combined gas equation

In an industrial process, nitrogen is heated to 500 K in a vessel of constant volume. If it enters the vessel at 100 atm and 300 K, what pressure would it exert at the working temperature if it behaved as a perfect gas?

Method The known and unknown data are summarized in (**2**). We expect the pressure to be greater on account of the increase in temperature. Moreover, because pressure is proportional to temperature, we can anticipate that the pressure will increase by a factor of T_2/T_1, where T_1 is the initial temperature and T_2 is the final temperature.

Answer Cancellation of the volumes (because $V_1 = V_2$) on each side of eqn 13 results in

$$\frac{p_1}{T_1} = \frac{p_2}{T_2}$$

which can be rearranged into

$$p_2 = \frac{T_2}{T_1} \times p_1$$

Substitution of the data then gives

$$p_2 = \frac{500\ \text{K}}{300\ \text{K}} \times (100\ \text{atm}) = 167\ \text{atm}$$

Comment Experiment shows that the pressure is actually 183 atm under these conditions, so the assumption that the gas is perfect leads to a 10 per cent error.

Self-test 1.3 What temperature would result in the same sample exerting a pressure of 300 atm? [900 K]

Box 1.1 The gas laws and the weather

The biggest sample of gas readily accessible to us is the atmosphere, a mixture of gases with the composition summarized in the table below. The composition is maintained moderately constant by diffusion and convection (winds, particularly the local turbulence called *eddies*) but the pressure and temperature vary with altitude and with the local conditions, particularly in the troposphere (the 'sphere of change'), the layer extending up to about 11 km.

The pressure and temperature vary with altitude. In the troposphere the average temperature is 15°C at sea level, falling to –57°C at the bottom of the tropopause at 11 km. This variation is much less pronounced when expressed on the Kelvin scale, ranging from 288 K to 216 K, an average of 268 K. If we suppose that the temperature has its average value all the way up to the tropopause, then the pressure varies with altitude, h, according to the *barometric formula*:

$$p = p_0 e^{-h/H}$$

where p_0 is the pressure at sea level and H is a constant approximately equal to 8 km. More specifically, $H = RT/Mg$, where M is the average molar mass of air and T is the temperature. The barometric formula fits the observed pressure distribution quite well even for regions well above the troposphere (see the first illustration). It implies that the pressure of the air and its density fall to half their sea-level value at $h = H \ln 2$, or 6 km.

Local variations of pressure, temperature, and composition in the troposphere are manifest as 'weather'. A small region of air is termed a *parcel*. First, we note that a parcel of warm air is less dense than the same parcel of cool air. As a parcel rises, it expands adiabatically (that is, without transfer of heat from its surroundings), so it cools. Cool air can absorb lower concentrations of water vapour than warm air, so the moisture forms clouds. Cloudy skies can therefore be associated with rising air and clear skies are often associated with descending air.

The motion of air in the upper altitudes may lead to an accumulation in some regions and a loss of molecules from other regions. The former result in the formation of regions of high pressure ('highs' or anticyclones) and the latter result in regions of low pressure ('lows', depressions, or cyclones). These regions are shown as H and L on the weather map opposite. The lines of constant pressure—differing by 4 mbar (400 Pa, about 3 Torr)—marked on it are called *isobars*. The elongated regions of high and low pressure are known, respectively, as *ridges* and *troughs*.

In meteorology, large-scale vertical movement is called *convection*. Horizontal pressure differentials result in the flow of air that we call *wind*. Because the Earth is rotating from west to east, winds are deflected towards the right in the Northern hemisphere and towards the left in the Southern hemisphere. Winds travel nearly parallel to the isobars, with low pressure to their left in the Northern hemisphere and to

The composition of dry air at sea level		
	Percentage	
Component	By volume	By mass
Nitrogen, N_2	78.08	75.53
Oxygen, O_2	20.95	23.14
Argon, Ar	0.93	1.28
Carbon dioxide, CO_2	0.031	0.047
Hydrogen, H_2	5.0×10^{-3}	2.0×10^{-4}
Neon, Ne	1.8×10^{-3}	1.3×10^{-3}
Helium, He	5.2×10^{-4}	7.2×10^{-5}
Methane, CH_4	2.0×10^{-4}	1.1×10^{-4}
Krypton, Kr	1.1×10^{-4}	3.2×10^{-4}
Nitric oxide, NO	5.0×10^{-5}	1.7×10^{-6}
Xenon, Xe	8.7×10^{-6}	1.2×10^{-5}
Ozone, O_3; summer	7.0×10^{-6}	1.2×10^{-5}
winter	2.0×10^{-6}	3.3×10^{-6}

The variation of atmospheric pressure with altitude, as predicted by the barometric formula and as suggested by the 'US Standard Atmosphere', which takes into account the variation of temperature with altitude.

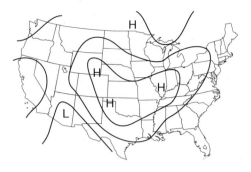

A typical weather map; in this case, for the United States on 1 January 2000.

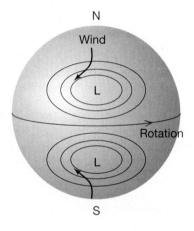

The flow of air ('wind') around regions of high and low pressure in the Northern and Southern hemispheres.

the right in the Southern hemisphere. At the surface, where wind speeds are lower, the winds tend to travel perpendicular to the isobars from high to low pressure. This differential motion results in a spiral outward flow of air clockwise in the Northern hemisphere around a high and an inward counterclockwise flow around a low (third illustration).

The air lost from regions of high pressure is restored as an influx of air converges into the region and descends. As we have seen, descending air is associated with clear skies. It also becomes warmer by compression as it descends, so regions of high pressure are associated with high surface temperatures. In winter, the cold surface air may prevent the complete fall of air, and result in a temperature *inversion*, with a layer of warm air over a layer of cold air. Geographical conditions may also trap cool air, as in Los Angeles, and the photochemical pollutants we know as *smog* may be trapped under the warm layer.

Discussion questions

1. The barometric formula relates the pressure of a gas of molar mass M at an altitude h to its pressure p_0 at sea level. Derive this relation by showing that the change in pressure dp for an infinitesimal change in altitude dh where the density is ρ is $dp = -\rho g dh$. Remember that ρ depends on the pressure. Evaluate the pressure difference between the top and bottom of (a) a laboratory vessel of height 15 cm, and (b) 1350 ft. Ignore temperature variations.

2. Ozone, O_3, is a trace atmospheric gas that plays an important role in screening the earth from harmful ultraviolet light, and the abundance of ozone is commonly reported in *Dobson units*. One Dobson unit is the thickness, in thousandths of a centimetre, of a column of gas if it were collected as a pure gas at 1.00 atm and 0°C. What amount of O_3 (in moles) is found in a column of atmosphere with a cross-sectional area of 1.00 dm^2 if the abundance is 250 Dobson units (a typical mid-latitude value)? In the seasonal Antarctic ozone hole, the column abundance drops below 100 Dobson units; how many moles of O_3 molecules are found in such a column of air above a 1.00 dm^2 area? Most atmospheric ozone is found between 10 and 50 km above the surface of the earth. If that ozone is spread uniformly through this portion of the atmosphere, what is the average molar concentration corresponding to (a) 250 Dobson units, (b) 100 Dobson units?

Further reading

S.K. Lower, Thermal physics (and some chemistry) of the atmosphere. *J. Chem. Educ.* **75**, 837 (1998).

R.P. Wayne, *Chemistry of atmospheres, an introduction to the chemistry of atmospheres of earth, the planets, and their satellites.* Oxford University Press (2000).

(c) Mixtures of gases

When dealing with gaseous mixtures, we often need to know the contribution that each component makes to the total pressure of the sample. In the nineteenth century John Dalton made observations that provide the answer and summarized them in **Dalton's law**:

> **The pressure exerted by a mixture of gases is the sum of the partial pressures of the gases.**

The **partial pressure** of a perfect gas is the pressure that it would exert if it occupied the container alone at the same temperature. That is, if a certain amount of H_2 exerts 25 kPa when present alone in a container, and an amount of N_2 exerts 80 kPa when present alone in the same container at the same temperature, then the total pressure when both are present is the sum of these two partial pressures, or 105 kPa (provided the individual gases and the mixture behave perfectly). More generally, if the partial pressure of a perfect gas A is p_A, that of a perfect gas B is p_B, and so on, then the total pressure when all the gases occupy the same container at the same temperature is

$$p = p_A + p_B + \cdots \tag{1.14}°$$

where, for each gas J,

$$p_J = \frac{n_J RT}{V} \tag{1.15}°$$

Example 1.4 Using Dalton's law

A container of volume 10.0 L holds 1.00 mol N_2 and 3.00 mol H_2 at 298 K. What is the total pressure in atmospheres if each component behaves as a perfect gas?

Method It follows from eqns 14 and 15 that the total pressure when two gases, A and B, occupy a container is

$$p = p_A + p_B = (n_A + n_B)\frac{RT}{V}$$

To obtain the answer expressed in atmospheres, use $R = 8.206 \times 10^{-2}$ L atm K^{-1} mol^{-1}.

Answer Under the stated conditions

$$p = (1.00 \text{ mol} + 3.00 \text{ mol}) \times \frac{(8.206 \times 10^{-2} \text{ L atm K}^{-1} \text{ mol}^{-1}) \times (298 \text{ K})}{10.0 \text{ L}}$$

$$= 9.78 \text{ atm}$$

Self-test 1.4 Calculate the total pressure when 1.00 mol N_2 and 2.00 mol O_2 are added to the same container (with the nitrogen and hydrogen still inside) at 298 K. [17.1 atm]

(d) Mole fractions and partial pressures

We can take a step closer to the discussion of mixtures of real gases by introducing the mole fraction of each component J. The **mole fraction**, x_J, is the amount of J expressed as a fraction of the total amount of molecules, n, in the sample:

$$x_J = \frac{n_J}{n} \qquad n = n_A + n_B + \cdots \tag{1.16}$$

When no J molecules are present, $x_J = 0$; when only J molecules are present, $x_J = 1$. A mixture of 1.0 mol N_2 and 3.0 mol H_2, and therefore of 4.0 mol molecules in all, consists of mole fractions 0.25 of N_2 and 0.75 of H_2. It follows from the definition of x_J that, whatever the composition of the mixture,

$$x_A + x_B + \cdots = 1 \tag{1.17}$$

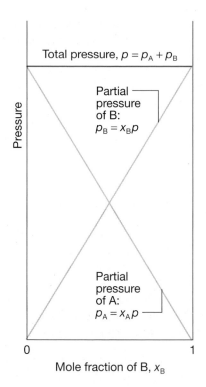

1.11 The partial pressures p_A and p_B of a binary mixture of (real or perfect) gases of total pressure p as the composition changes from pure A to pure B. The sum of the partial pressures is equal to the total pressure. If the gases are perfect, then the partial pressure is also the pressure that each gas would exert if it were present alone in the container.

Next, we define the **partial pressure**, p_J, of a gas J in a mixture (any gas, not just a perfect gas) as

$$p_J = x_J p \qquad [1.18]$$

where p is the total pressure of the mixture (Fig. 1.11). Only for a mixture of perfect gases can we identify p_J with the pressure that the gas J would exert if it were alone in the container and calculate its value from eqn 15. For real gases, the interactions between molecules result in the total pressure being slightly different from the sum of their partial pressures as defined in eqn 15.

It follows from eqns 17 and 18 that the sum of the partial pressures is equal to the total pressure:

$$p_A + p_B + \cdots = (x_A + x_B + \cdots)p = p \qquad (1.19)$$

This relation is true for both real and perfect gases.

Example 1.5 Calculating partial pressures

The mass percentage composition of dry air at sea level is approximately N_2: 75.5; O_2: 23.2; Ar: 1.3. What is the partial pressure of each component when the total pressure is 1.00 atm?

Method We expect species with a high mole fraction to have a proportionally high partial pressure. Partial pressures are defined by eqn 18. To use the equation, we need the mole fractions of the components. To calculate mole fractions, which are defined by eqn 16, we use the fact that the amount of molecules J of molar mass M_J in a sample of mass m_J is $n_J = m_J/M_J$. The mole fractions are independent of the total mass of the sample, so we can choose the latter to be 100 g (which makes the conversion from mass percentages very easy).

Answer The amounts of each type of molecule present in 100 g of air are

$$n(N_2) = 0.755 \times \frac{100\ \text{g}}{28.02\ \text{g mol}^{-1}} = \frac{0.755 \times 100}{28.02}\ \text{mol}$$

$$n(O_2) = 0.232 \times \frac{100\ \text{g}}{32.00\ \text{g mol}^{-1}} = \frac{0.232 \times 100}{32.00}\ \text{mol}$$

$$n(Ar) = 0.013 \times \frac{100\ \text{g}}{39.95\ \text{g mol}^{-1}} = \frac{0.013 \times 100}{39.95}\ \text{mol}$$

These three amounts work out as 2.69 mol, 0.725 mol, and 0.033 mol, respectively, for a total of 3.45 mol. The mole fractions and partial pressures (obtained by multiplying the mole fraction by the total pressure, 1.00 atm) are as follows:

	N_2	O_2	Ar
Mole fraction	0.780	0.210	0.0096
Partial pressure/atm	0.780	0.210	0.0096

Comment We have not had to assume that the gases are perfect: partial pressures are defined as $p_J = x_J p$ for any kind of gas.

Self-test 1.5 When carbon dioxide is taken into account, the mass percentages are 75.52 (N_2), 23.15 (O_2), 1.28 (Ar), and 0.046 (CO_2). What are the partial pressures when the total pressure is 0.900 atm?

[0.703, 0.189, 0.0084, 0.000 27 atm]

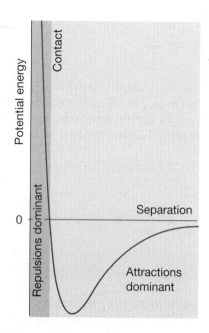

1.12 The variation of the potential energy of two molecules on their separation. High positive potential energy (at very small separations) indicates that the interactions between them are strongly repulsive at these distances. At intermediate separations, where the potential energy is negative, the attractive interactions dominate. At large separations (on the right) the potential energy is zero and there is no interaction between the molecules.

Real gases

Real gases do not obey the perfect gas law exactly. Deviations from the law are particularly important at high pressures and low temperatures, especially when a gas is on the point of condensing to liquid.

1.3 Molecular interactions

Real gases show deviations from the perfect gas law because molecules interact with each other. Repulsive forces between molecules assist expansion and attractive forces assist compression.

Repulsive forces are significant only when molecules are almost in contact: they are short-range interactions, even on a scale measured in molecular diameters (Fig. 1.12). Because they are short-range interactions, repulsions can be expected to be important only when the average separation of the molecules is small. This is the case at high pressure, when many molecules occupy a small volume. On the other hand, attractive intermolecular forces have a relatively long range and are effective over several molecular diameters. They are important when the molecules are fairly close together but not necessarily touching (at the intermediate separations in Fig. 1.12). Attractive forces are ineffective when the molecules are far apart (well to the right in Fig. 1.12). Intermolecular forces are also important when the temperature is so low that the molecules travel with such low mean speeds that they can be captured by one another.

At low pressure, when the sample occupies a large volume, the molecules are so far apart for most of the time that the intermolecular forces play no significant role, and the gas behaves virtually perfectly. At moderate pressure, when the average separation of the molecules is only a few molecular diameters, the attractive forces dominate the repulsive forces. In this case, the gas can be expected to be more compressible than a perfect gas because the forces help to draw the molecules together. At high pressure, when the average separation of the molecules is small, the repulsive forces dominate and the gas can be expected to be less compressible because now the forces help to drive the molecules apart.

(a) The compression factor

The **compression factor**, Z, of a gas is the ratio of its molar volume, V_m, to the molar volume of a perfect gas, V_m°, at the same pressure and temperature:

$$Z = \frac{V_m}{V_m^\circ}$$

[1.20a]

Because the molar volume of a perfect gas is equal to RT/p, an equivalent expression is

$$pV_m = RTZ$$

(1.20b)

We can explore the distance-dependence of intermolecular forces by plotting Z against pressure. Because for a perfect gas $Z = 1$ under all conditions, deviation of Z from 1 is a measure of departure from perfect behaviour.

Some experimental values of Z are plotted in Fig. 1.13. At very low pressures, all the gases shown have $Z \approx 1$ and behave nearly perfectly. At high pressures, all the gases have $Z > 1$, signifying that they have a larger molar volume than a perfect gas. Repulsive forces are now dominant. At intermediate pressures, most gases have

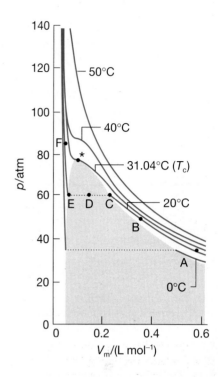

1.13 The variation of the compression factor, Z, with pressure for several gases at 0°C. A perfect gas has $Z = 1$ at all pressures. Notice that although the curves approach 1 as $p \to 0$, they do so with different slopes.

1.14 Experimental isotherms of carbon dioxide at several temperatures. The 'critical isotherm', the isotherm at the critical temperature, is at 31.04°C. The critical point is marked with a star.

$Z < 1$, indicating that the attractive forces are reducing the molar volume relative to that of a perfect gas.

(b) Virial coefficients

Figure 1.14 shows the experimental isotherms for carbon dioxide. At large molar volumes and high temperatures the real-gas isotherms do not differ greatly from perfect-gas isotherms. The small differences suggest that the perfect gas law is in fact the first term in an expression of the form

$$pV_m = RT(1 + B'p + C'p^2 + \cdots) \tag{1.21}$$

This expression is an example of a common procedure in physical chemistry, in which a simple law (in this case $pV = nRT$) is treated as the first term in a series in powers of a variable (in this case p). A more convenient expansion for many applications is

$$pV_m = RT\left(1 + \frac{B}{V_m} + \frac{C}{V_m^2} + \cdots\right) \tag{1.22}$$

These two expressions are two versions of the **virial equation of state**.[4] In each case, the term in parentheses can be identified with the compression factor, Z.

The coefficients B, C, ... , which depend on the temperature, are the second, third, ... **virial coefficients** (Table 1.3); the first virial coefficient is 1. The third virial coefficient, C, is usually less important than the second coefficient, B, in the sense that at typical molar volumes $C/V_m^2 \ll B/V_m$.

Table 1.3* Second virial coefficients, $B/(\text{cm}^3\ \text{mol}^{-1})$

	Temperature	
	273 K	600 K
Ar	−21.7	11.9
CO_2	−149.7	−12.4
N_2	−10.5	21.7
Xe	−153.7	−19.6

* More values are given in the *Data section*.

4. The name comes from the Latin word for force. The coefficients are sometimes denoted B_2, B_3, ...

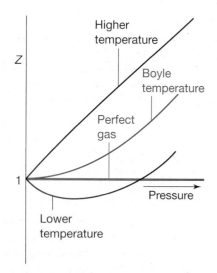

1.15 The compression factor, Z, approaches 1 at low pressures, but does so with different slopes. For a perfect gas, the slope is zero, but real gases may have either positive or negative slopes, and the slope may vary with temperature. At the Boyle temperature, the slope is zero and the gas behaves perfectly over a wider range of conditions than at other temperatures.

We can use the virial equation to demonstrate the important point that, although the equation of state of a real gas may coincide with the perfect gas law as $p \to 0$, not all of its properties necessarily coincide with those of a perfect gas in that limit. Consider, for example, the value of dZ/dp, the slope of the graph of compression factor against pressure. For a perfect gas $dZ/dp = 0$ (because $Z = 1$ at all pressures), but for a real gas

$$\frac{dZ}{dp} = B' + 2pC' + \cdots \to B' \text{ as } p \to 0 \tag{1.23}$$

However, B' is not necessarily zero, so the slope of Z with respect to p does not necessarily approach 0 (the perfect gas value). Because several properties depend on derivatives (as we shall see), the properties of real gases do not always coincide with the perfect gas values at low pressures. By a similar argument,

$$\frac{dZ}{d(1/V_m)} \to B \text{ as } V_m \to \infty, \text{ corresponding to } p \to 0 \tag{1.24}$$

Because the virial coefficients depend on the temperature, there may be a temperature at which $Z \to 1$ with zero slope at low pressure or high molar volume (Fig. 1.15). At this temperature, which is called the **Boyle temperature**, T_B, the properties of the real gas do coincide with those of a perfect gas as $p \to 0$. According to eqn 1.24, Z has zero slope as $p \to 0$ if $B = 0$, so we can conclude that $B = 0$ at the Boyle temperature. It then follows from eqn 1.22 that $pV_m \approx RT_B$ over a more extended range of pressures than at other temperatures because the first term after 1 (that is, B/V_m) in the virial equation is zero and C/V_m^2 and higher terms are negligibly small. For helium $T_B = 22.64$ K; for air $T_B = 346.8$ K; more values are given in Table 1.4.

(c) Condensation

Now consider what happens when we compress a sample of gas initially in the state marked A in Fig. 1.14 at constant temperature (by pushing in a piston). Near A, the pressure of the gas rises in approximate agreement with Boyle's law. Serious deviations from that law begin to appear when the volume has been reduced to B.

At C (which corresponds to about 60 atm for carbon dioxide) all similarity to perfect behaviour is lost, for suddenly the piston slides in without any further rise in pressure: this stage is represented by the horizontal line CDE. Examination of the contents of the vessel shows that just to the left of C a liquid appears, and there are two phases separated by a sharply defined surface. As the volume is decreased from C through D to E, the amount of liquid increases. There is no additional resistance to

Table 1.4* Critical constants of gases

	p_c/atm	V_c/(cm³ mol⁻¹)	T_c/K	Z_c	T_B/K
Ar	48.0	75.3	150.7	0.292	411.5
CO₂	72.9	94.0	304.2	0.274	714.8
He	2.26	57.8	5.2	0.305	22.64
O₂	50.14	78.0	154.8	0.308	405.9

*More values are given in the *Data section*.

the piston because the gas can respond by condensing. The pressure corresponding to the line CDE, when both liquid and vapour are present in equilibrium, is called the **vapour pressure** of the liquid at the temperature of the experiment.

At E, the sample is entirely liquid and the piston rests on its surface. Any further reduction of volume requires the exertion of considerable pressure, as is indicated by the sharply rising line to the left of E. Even a small reduction of volume from E to F requires a great increase in pressure.

(d) Critical constants

The isotherm at the temperature T_c (304.19 K, or 31.04°C for CO_2) plays a special role in the theory of the states of matter. An isotherm slightly below T_c behaves as we have already described: at a certain pressure, a liquid condenses from the gas and is distinguishable from it by the presence of a visible surface. If, however, the compression takes place at T_c itself, then a surface separating two phases does not appear and the volumes at each end of the horizontal part of the isotherm have merged to a single point, the **critical point** of the gas. The temperature, pressure, and molar volume at the critical point are called the **critical temperature**, T_c, **critical pressure**, p_c, and **critical molar volume**, V_c, of the substance. Collectively, p_c, V_c, and T_c are the **critical constants** of a substance (Table 1.4).

At and above T_c, the sample has a single phase that occupies the entire volume of the container. Such a phase is, by definition, a gas. Hence, the liquid phase of a substance does not form above the critical temperature. The critical temperature of oxygen, for instance, signifies that it is impossible to produce liquid oxygen by compression alone if its temperature is greater than 155 K: to liquefy oxygen—to obtain a fluid phase that does not occupy the entire volume—the temperature must first be lowered to below 155 K, and then the gas compressed isothermally. The single phase that fills the entire volume when $T > T_c$ may be much denser than we normally consider typical of gases, and the name **supercritical fluid** is preferred.

> The web site contains links to online databases of properties of gases.

1.4 The van der Waals equation

We can draw conclusions from the virial equations of state only by inserting specific values of the coefficients. It is often useful to have a broader, if less precise, view of all gases. Therefore, we introduce the approximate equation of state suggested by J.H. van der Waals in 1873. This equation is an excellent example of an expression that can be obtained by thinking scientifically about a mathematically complicated but physically simple problem; that is, it is a good example of 'model building'. Van der Waals himself proposed his equation on the basis of experimental evidence available to him in conjunction with rigorous thermodynamic arguments.

The **van der Waals equation** is

$$p = \frac{nRT}{V - nb} - a\left(\frac{n}{V}\right)^2 \tag{1.25a}$$

and a derivation is given in the *Justification* below. The equation is often written in terms of the molar volume $V_m = V/n$ as

$$p = \frac{RT}{V_m - b} - \frac{a}{V_m^2} \tag{1.25b}$$

The constants a and b are called the **van der Waals coefficients**. They are characteristic of each gas but independent of the temperature (Table 1.5).

Table 1.5[*] Van der Waals coefficients

	a/(atm L^2 mol^{-2})	b/(10^{-2} L mol^{-1})
Ar	1.337	3.20
CO_2	3.610	4.29
He	0.0341	2.38
Xe	4.137	5.16

[*] More values are given in the *Data section*.

Justification 1.1

The repulsive interactions between molecules are taken into account by supposing that they cause the molecules to behave as small but impenetrable spheres. The nonzero volume of the molecules implies that instead of moving in a volume V they are restricted to a smaller volume $V - nb$, where nb is approximately the total volume taken up by the molecules themselves. This argument suggests that the perfect gas law $p = nRT/V$ should be replaced by

$$p = \frac{nRT}{V - nb}$$

when repulsions are significant.

The pressure depends on both the frequency of collisions with the walls and the force of each collision. Both the frequency of the collisions and their force are reduced by the attractive forces, which act with a strength proportional to the molar concentration, n/V, of molecules in the sample. Therefore, because both the frequency and the force of the collisions are reduced by the attractive forces, the pressure is reduced in proportion to the square of this concentration. If the reduction of pressure is written as $-a(n/V)^2$, where a is a positive constant characteristic of each gas, the combined effect of the repulsive and attractive forces is the van der Waals equation of state as expressed in eqn 25.

In this *Justification* we have built the van der Waals equation using vague arguments about the volumes of molecules and the effects of forces. The equation can be derived in other ways, but the present method has the advantage that it shows how to derive the form of an equation out of general ideas. The derivation also has the advantage of keeping imprecise the significance of the coefficients a and b: they are much better regarded as empirical parameters than as precisely defined molecular properties.

Example 1.6 Using the van der Waals equation to estimate a molar volume

Estimate the molar volume of CO_2 at 500 K and 100 atm by treating it as a van der Waals gas.

Method To express eqn 1.25b as an equation for the molar volume, we need to rearrange it into

$$V_m^3 - \left(b + \frac{RT}{p}\right)V_m^2 + \left(\frac{a}{p}\right)V_m - \frac{ab}{p} = 0$$

Although closed expressions for the roots of a cubic equation can be given, they are very complicated. Unless analytical solutions are essential, it is usually more expedient to solve such equations with commercial software.

Answer According to Table 1.5, $a = 3.610$ L^2 atm mol^{-2} and $b = 4.29 \times 10^{-2}$ L mol^{-1}. Under the stated conditions, $RT/p = 0.410$ L mol^{-1}. The coefficients in the equation for V_m are therefore

$$b + RT/p = 0.453 \text{ L mol}^{-1}$$

$$a/p = 3.61 \times 10^{-2} \text{ (L mol}^{-1})^2$$

$$ab/p = 1.55 \times 10^{-3} \text{ (L mol}^{-1})^3$$

Therefore, on writing $x = V_m/(L\ mol^{-1})$, the equation to solve is

$$x^3 - 0.453x^2 + (3.61 \times 10^{-2})x - (1.55 \times 10^{-3}) = 0$$

The acceptable root is $x = 0.366$, which implies that $V_m = 0.366\ L\ mol^{-1}$.

Comment For a perfect gas under these conditions, the molar volume is $0.410\ L\ mol^{-1}$.

Self-test 1.6 Calculate the molar volume of argon at 100°C and 100 atm on the assumption that it is a van der Waals gas. [0.298 L mol⁻¹]

(a) The reliability of the equation

We now examine to what extent the van der Waals equation predicts the behaviour of real gases. It is too optimistic to expect a single, simple expression to be the true equation of state of all substances, and accurate work on gases must resort to the virial equation, use tabulated values of the coefficients at various temperatures, and analyse the systems numerically. The advantage of the van der Waals equation, however, is that it is analytical (that is, expressed symbolically) and allows us to draw some general conclusions about real gases. When the equation fails we must use one of the other equations of state that have been proposed (some are listed in Table 1.6), invent a new one, or go back to the virial equation.

That having been said, we can begin to judge the reliability of the equation by comparing the isotherms it predicts with the experimental isotherms for carbon dioxide in Fig. 1.14. Some calculated isotherms are shown in Figs 1.16 and 1.17. Apart from the oscillations below the critical temperature, they do resemble experimental isotherms quite well. The oscillations, the **van der Waals loops**, are unrealistic because they suggest that under some conditions an increase of pressure results in an increase of volume. Therefore, they are replaced by horizontal lines drawn so the loops define equal areas above and below the lines: this procedure is called the **Maxwell construction (3)**. The van der Waals coefficients, such as those in Table 1.5, are found by fitting the calculated curves to the experimental curves.

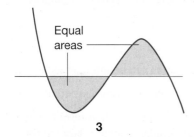

Equal areas

3

Table 1.6 Selected equations of state

	Equation	Reduced form*	Critical constants		
			p_c	V_c	T_c
Perfect gas	$p = \dfrac{RT}{V_m}$				
Van der Waals	$p = \dfrac{RT}{V_m - b} - \dfrac{a}{V_m^2}$	$p_r = \dfrac{8T_r}{3V_r - 1} - \dfrac{3}{V_r^2}$	$\dfrac{a}{27b^2}$	$3b$	$\dfrac{8a}{27bR}$
Berthelot	$p = \dfrac{RT}{V_m - b} - \dfrac{a}{TV_m^2}$	$p_r = \dfrac{8T_r}{3V_r - 1} - \dfrac{3}{T_r V_r^2}$	$\dfrac{1}{12}\left(\dfrac{2aR}{3b^3}\right)^{1/2}$	$3b$	$\dfrac{2}{3}\left(\dfrac{2a}{3bR}\right)^{1/2}$
Dieterici	$p = \dfrac{RTe^{-a/RTV_m}}{V_m - b}$	$p_r = \dfrac{e^2 T_r e^{-2/T_r V_r}}{2V_r - 1}$	$\dfrac{a}{4e^2 b^2}$	$2b$	$\dfrac{a}{4Rb}$
Virial (Kammerlingh Onnes)	$p = \dfrac{RT}{V_m}\left\{1 + \dfrac{B(T)}{V_m} + \dfrac{C(T)}{V_m^2} + \cdots\right\}$				

* Reduced variables are defined in Section 1.5.

1.16 The surface of possible states allowed by the van der Waals equation. Compare this surface with that shown in Fig. 1.9.

1.17 Van der Waals isotherms at several values of T/T_c. Compare these curves with those in Fig.1.14. The van der Waals loops are normally replaced by horizontal straight lines. The critical isotherm is the isotherm for $T/T_c = 1$.

(b) The features of the equation

The principal features of the van der Waals equation can be summarized as follows.

(1) Perfect gas isotherms are obtained at high temperatures and large molar volumes.

When the temperature is high, RT may be so large that the first term in eqn 1.25b greatly exceeds the second. Furthermore, if the molar volume is large in the sense $V_m \gg b$, then the denominator $V_m - b \approx V_m$. Under these conditions, the equation reduces to $p = RT/V_m$, the perfect gas equation.

(2) Liquids and gases coexist when cohesive and dispersing effects are in balance.

The van der Waals loops occur when both terms in eqn 1.25b have similar magnitudes. The first term arises from the kinetic energy of the molecules and their repulsive interactions; the second represents the effect of the attractive interactions.

(3) The critical constants are related to the van der Waals coefficients.

For $T < T_c$, the calculated isotherms oscillate, and each one passes through a minimum followed by a maximum. These extrema converge as $T \rightarrow T_c$ and coincide at $T = T_c$; at the critical point the curve has a flat inflexion (**4**). From the properties of curves, we know that an inflexion of this type occurs when both the first and second derivatives are zero. Hence, we can find the critical constants by calculating these derivatives and setting them equal to zero:

$$\frac{dp}{dV_m} = -\frac{RT}{(V_m - b)^2} + \frac{2a}{V_m^3} = 0$$

$$\frac{d^2p}{dV_m^2} = \frac{2RT}{(V_m - b)^3} - \frac{6a}{V_m^4} = 0$$

4

at the critical point. The solutions of these two equations are

$$V_c = 3b \qquad p_c = \frac{a}{27b^2} \qquad T_c = \frac{8a}{27Rb} \tag{1.26}$$

These relations can be tested by noting that the **critical compression factor**, Z_c, is predicted to be equal to

$$Z_c = \frac{p_c V_c}{RT_c} = \frac{3}{8} \tag{1.27}$$

for all gases. We see from Table 1.4 that although $Z_c < \frac{3}{8}$ (or 0.375), it is approximately constant (at 0.3) and the discrepancy is reasonably small.

1.5 The principle of corresponding states

An important general technique in science for comparing the properties of objects is to choose a related fundamental property of the same kind and to set up a relative scale on that basis. We have seen that the critical constants are characteristic properties of gases, so it may be that a scale can be set up by using them as yardsticks. We therefore introduce the **reduced variables** of a gas by dividing the actual variable by the corresponding critical constant:

$$p_r = \frac{p}{p_c} \qquad V_r = \frac{V_m}{V_c} \qquad T_r = \frac{T}{T_c} \tag{1.28}$$

If the reduced pressure of a gas is given, we can easily calculate its actual pressure by using $p = p_r p_c$, and likewise for the volume and temperature. Van der Waals, who first tried this procedure, hoped that gases confined to the same reduced volume, V_r, at the same reduced temperature, T_r, would exert the same reduced pressure, p_r. The hope was largely fulfilled (Fig. 1.18). The illustration shows the dependence of the compression factor on the reduced pressure for a variety of gases at various reduced temperatures. The success of the procedure is strikingly clear: compare this graph with Fig. 1.13, where similar data are plotted without using reduced variables. The observation that real gases at the same reduced volume and reduced temperature

1.18 The compression factors of four gases, including two from Fig. 1.13, plotted using reduced variables. The use of reduced variables organizes the data on to single curves.

exert the same reduced pressure is called the **principle of corresponding states**. The principle is only an approximation. It works best for gases composed of spherical molecules; it fails, sometimes badly, when the molecules are non-spherical or polar.

The van der Waals equation sheds some light on the principle. First, we express eqn 25b in terms of the reduced variables, which gives

$$p_r p_c = \frac{RT_r T_c}{V_r V_c - b} - \frac{a}{V_r^2 V_c^2}$$

Then we express the critical constants in terms of a and b by using eqn 26:

$$\frac{ap_r}{27b^2} = \frac{8aT_r}{27b(3bV_r - b)} - \frac{a}{9b^2 V_r^2}$$

which can be reorganized into

$$p_r = \frac{8T_r}{3V_r - 1} - \frac{3}{V_r^2} \tag{1.29}$$

This equation has the same form as the original, but the coefficients a and b, which differ from gas to gas, have disappeared. It follows that, if the isotherms are plotted in terms of the reduced variables (as we did in fact in Fig. 1.17 without drawing attention to the fact), then the same curves are obtained whatever the gas. This is precisely the content of the principle of corresponding states, so the van der Waals equation is compatible with it.

Looking for too much significance in this apparent triumph is mistaken, because other equations of state also accommodate the principle (Table 1.6). In fact, all we need are two parameters playing the roles of a and b, for then the equation can always be manipulated into reduced form. The observation that real gases obey the principle approximately amounts to saying that the effects of the attractive and repulsive interactions can each be approximated in terms of a single parameter. The importance of the principle is then not so much its theoretical interpretation but the way that it enables the properties of a range of gases to be coordinated on to a single diagram (for example, Fig. 1.18 instead of Fig. 1.13).

Checklist of key ideas

☐ gas, a form of matter that fills any container it occupies

The perfect gas

1.1 The states of gases

☐ physical state, the condition of a substance

☐ equation of state, an equation that interrelates pressure, volume, temperature, and amount of substance: $p = f(T, V, n)$

☐ pressure, force divided by the area to which the force is applied

☐ standard pressure, $p^{\ominus} = 1$ bar (10^5 Pa)

☐ mechanical equilibrium, the condition of equality of pressure on either side of a movable wall

☐ barometer, a device for measuring atmospheric pressure

☐ manometer, a device for measuring the pressure of gas in a system

☐ temperature, the property that indicates the direction of the flow of energy through a thermally conducting, rigid wall

☐ diathermic boundary, a boundary that permits the passage of energy as heat

☐ adiabatic boundary, a boundary that prevents the passage of energy as heat

☐ thermal equilibrium, a condition in which no change of state occurs when two objects A and B are in contact through a diathermic boundary

☐ Zeroth Law of thermodynamics: If A is in thermal equilibrium with B, and B is in thermal equilibrium with C, then C is also in thermal equilibrium with A

☐ thermometer, a device for measuring temperature

☐ Celsius scale, a temperature scale on which $\theta/^{\circ}\text{C} = T/\text{K} - 273.15$

☐ perfect-gas temperature scale, a temperature scale based on the properties of a perfect gas

□ thermodynamic temperature scale, a temperature scale defined in terms of thermodynamic quantities

1.2 The gas laws

□ Boyle's law: At constant temperature, the pressure of a sample of gas is inversely proportional to its volume, $p \propto 1/V$ at constant temperature

□ isotherm, curves in a graph corresponding to a single temperature

□ limiting law, a law that is strictly true only in a certain limit

□ isobar, curves in a graph corresponding to a single pressure

□ Charles's law: At constant pressure, the volume of a gas is proportional to the absolute temperature, $V \propto T$ at constant pressure

□ Avogadro's principle: Equal volumes of gases at the same pressure and temperature contain the same numbers of molecules

□ gas constant, R

□ perfect gas equation, $pV = nRT$

□ perfect gas, a gas that obeys $pV = nRT$ exactly under all conditions

□ real gas, an actual gas

□ standard ambient temperature and pressure (SATP), 298.15 K and 1 bar

□ standard temperature and pressure (STP), 0°C and 1 atm

□ combined gas equation, $p_1V_1/T_1 = p_2V_2/T_2$

□ Dalton's law: The pressure exerted by a mixture of gases is the sum of the partial pressures of the gases

□ mole fraction, the amount of J, n_J, expressed as a fraction of the total amount of molecules, n, in the sample, $x_J = n_J/n$

□ partial pressure of a perfect gas, the pressure a gas would exert if it occupied the container alone

□ partial pressure (in general), $p_J = x_J p$

Real gases

1.3 Molecular interactions

□ compression factor, $Z = V_m/V_m^\circ$

□ virial equation of state, $pV_m = RT(1 + B/V_m + C/V_m^2 + \cdots)$

□ virial coefficient, the coefficients $B, C, \ldots$

□ Boyle temperature, the temperature at which $dZ/dp = 0$ as $p \to 0$

□ vapour pressure, the pressure of a vapour in equilibrium with its condensed phase

□ critical point, the point at which the volumes at each end of the horizontal part of the isotherm have merged to a single point

□ critical temperature, the temperature at the critical point

□ critical pressure, the pressure at the critical point

□ critical molar volume, the molar volume at the critical point

□ critical constants, p_c, V_c, and T_c

□ supercritical fluid, a dense fluid phase above its critical temperature and pressure

1.4 The van der Waals equation

□ van der Waals equation, $p = nRT/(V - nb) - a(n/V)^2$

□ van der Waals coefficients, the parameters a and b

□ van der Waals loops, unrealistic oscillations in the van der Waals isotherms

□ Maxwell construction, the equal-area rule for the replacement of the van der Waals loops by straight lines

□ critical compression factor, the value of Z at the critical point

1.5 The principle of corresponding states

□ reduced variables, the actual variable divided by the corresponding critical constant

□ principle of corresponding states: Real gases at the same reduced volume and reduced temperature exert the same reduced pressure

Further reading

Articles of general interest

J.M. Alvaiño, J. Veguillas, and S. Valasco, Equations of state, collisional energy transfer, and chemical equilibrium in gases. *J. Chem. Educ.* **66**, 157 (1989).

J.G. Eberhart, The many faces of van der Waals's equation of state. *J. Chem. Educ.* **66**, 906 (1989).

J.G. Eberhart, A least-squares technique for determining the van der Waals parameters from the critical constants. *J. Chem. Educ.* **69**, 220 (1992).

E.F. Meyer and T.P. Meyer, Supercritical fluid: liquid, gas, both, or neither? A different approach. *J. Chem. Educ.* **63**, 463 (1986).

I.M. Mills, The choice of names and symbols for quantities in chemistry. *J. Chem. Educ.* **66**, 887 (1989).

J.L. Pauley and E.H. Davis, *P–V–T* isotherms of real gases: Experimental versus calculated values. *J. Chem. Educ.* **63**, 466 (1986).

G. Rhodes, Does a one-molecule gas obey Boyle's law? *J. Chem. Educ.* **69**, 16 (1992).

M. Ross, Equations of state. In *Encyclopedia of applied physics* (ed. G.L. Trigg), **6**, 291. VCH, New York (1993).

J.B. Ott, J.R. Coates, and H.T. Hall, Comparison of equations of state. *J. Chem. Educ.* **48**, 515 (1971).

Texts and sources of data and information

J.H. Dymond and E.B. Smith, *The virial coefficients of pure gases and mixtures.* Oxford University Press (1980).

A.D. McNaught and A. Wilkinson, *Compendium of chemical terminology.* Blackwell Scientific, Oxford (1997).

J.O. Hirschfelder, C.F. Curtiss, and R.B. Bird, *The molecular theory of gases and liquids.* Wiley, New York (1954).

I.M. Mills (ed.), *Quantities, units, and symbols in physical chemistry.* Blackwell Scientific, Oxford (1993).

B.W. Rossiter and R.C. Baetzgold (ed.) *Physical methods of chemistry*, VI. Wiley–Interscience, New York (1992).

A.J. Walton, *Three phases of matter.* Oxford University Press (1983).

Exercises

Discussion questions

1.1 (a) Explain how the experiments of Boyle, Charles (and others), and Avogadro led to the formulation of the perfect gas equation of state.

1.1 (b) Explain the term 'partial pressure' and explain why Dalton's law is a limiting law.

1.2 (a) Explain how the compression factor varies with pressure and temperature and describe how it reveals information about intermolecular interactions in real gases.

1.2 (b) What is the significance of the critical constants?

1.3 (a) Describe the formulation of the van der Waals equation and suggest a rationale for one other equation of state in Table 1.6.

1.3 (b) Explain how the van der Waals equation accounts for critical behaviour.

Numerical exercises

1.4 (a) A sample of air occupies 1.0 L at 25°C and 1.00 atm. What pressure is needed to compress it to 100 cm^3 at this temperature?

1.4 (b) A sample of carbon dioxide gas occupies 2.0 dm^3 at 20°C and 104 kPa. What pressure is needed to compress it to 250 cm^3 at this temperature?

1.5 (a) (a) Could 131 g of xenon gas in a vessel of volume 1.0 L exert a pressure of 20 atm at 25°C if it behaved as a perfect gas? If not, what pressure would it exert? (b) What pressure would it exert if it behaved as a van der Waals gas?

1.5 (b) (a) Could 25 g of argon gas in a vessel of volume 1.5 L exert a pressure of 2.0 bar at 30°C if it behaved as a perfect gas? If not, what pressure would it exert? (b) What pressure would it exert if it behaved as a van der Waals gas?

1.6 (a) A perfect gas undergoes isothermal compression, which reduces its volume by 2.20 L. The final pressure and volume of the gas are 3.78×10^3 Torr and 4.65 L, respectively. Calculate the original pressure of the gas in (a) Torr, (b) atm.

1.6 (b) A perfect gas undergoes isothermal compression, which reduces its volume by 1.80 dm^3. The final pressure and volume of the gas are 1.48×10^3 Torr and 2.14 dm^3, respectively. Calculate the original pressure of the gas in (a) Torr, (b) bar.

1.7 (a) To what temperature must a 1.0 L sample of a perfect gas be cooled from 25°C in order to reduce its volume to 100 cm^3?

1.7 (b) To what temperature must a sample of a perfect gas of volume 500 mL be cooled from 35°C in order to reduce its volume to 150 cm^3?

1.8 (a) A car tyre (that is, an automobile tire) was inflated to a pressure of 24 lb in^{-2} (1.00 atm = 14.7 lb in^{-2}) on a winter's day when the temperature was −5°C. What pressure will be found, assuming no leaks have occurred and that the volume is constant, on a subsequent summer's day when the temperature is 35°C? What complications should be taken into account in practice?

1.8 (b) A sample of hydrogen gas was found to have a pressure of 125 kPa when the temperature was 23°C. What can its pressure be expected to be when the temperature is 11°C?

1.9 (a) A sample of 255 mg of neon occupies 3.00 L at 122 K. Use the perfect gas law to calculate the pressure of the gas.

1.9 (b) A homeowner uses 4.00×10^3 m^3 of natural gas in a year to heat a home. Assume that natural gas is all methane, CH_4, and that methane is a perfect gas for the conditions of this problem, which are 1.00 atm and 20°C. What is the mass of gas used?

1.10 (a) In an attempt to determine an accurate value of the gas constant, R, a student heated a 20.000 L container filled with 0.251 32 g of helium gas to 500°C and measured the pressure as 206.402 cm of water in a manometer at 25°C. Calculate the value of R from these data. (The density of water at 25°C is 0.997 07 g cm^{-3}.)

1.10 (b) The following data have been obtained for oxygen gas at 273.15 K. Calculate the best value of the gas constant R from them and the best value of the molar mass of O_2.

p/atm	0.750 000	0.500 000	0.250 000
V_m/(L mol^{-1})	29.9649	44.8090	89.6384
ρ/(g L^{-1})	1.071 44	0.714 110	0.356 975

1.11 (a) At 500°C and 699 Torr, the mass density of sulfur vapour is 3.71 g L^{-1}. What is the molecular formula of sulfur under these conditions?

1.11 (b) At 100°C and 120 Torr, the mass density of phosphorus vapour is 0.6388 kg m^{-3}. What is the molecular formula of phosphorus under these conditions?

1.12 (a) Calculate the mass of water vapour present in a room of volume 400 m^3 that contains air at 27°C on a day when the relative humidity is 60 per cent.

1.12 (b) Calculate the mass of water vapour present in a room of volume 250 m^3 that contains air at 23°C on a day when the relative humidity is 53 per cent.

1.13 (a) Given that the density of air at 740 Torr and 27°C is 1.146 g L^{-1}, calculate the mole fraction and partial pressure of nitrogen and oxygen assuming that (a) air consists only of these two gases, (b) air also contains 1.0 mole per cent Ar.

1.13 (b) A gas mixture consists of 320 mg of methane, 175 mg of argon, and 225 mg of neon. The partial pressure of neon at 300 K is 66.5 Torr. Calculate (a) the volume and (b) the total pressure of the mixture.

1.14 (a) The density of a gaseous compound was found to be 1.23 g L^{-1} at 330 K and 150 Torr. What is the molar mass of the compound?

1.14 (b) In an experiment to measure the molar mass of a gas, 250 cm^3 of the gas was confined in a glass vessel. The pressure was 152 Torr at 298 K and, after correcting for buoyancy effects, the mass of the gas was 33.5 mg. What is the molar mass of the gas?

1.15 (a) The densities of air at −85°C, 0°C, and 100°C are 1.877 g L^{-1}, 1.294 g L^{-1}, and 0.946 g L^{-1}, respectively. From these data, and assuming that air obeys Charles's law, determine a value for the absolute zero of temperature in degrees Celsius.

1.15 (b) A certain sample of a gas has a volume of 20.00 L at 0°C and 1.000 atm. A plot of the experimental data of its volume against the Celsius temperature, θ, at constant p, gives a straight line of slope 0.0741 L (°C)$^{-1}$. From these data alone (without making use of the perfect gas law), determine the absolute zero of temperature in degrees Celsius.

1.16 (a) Calculate the pressure exerted by 1.0 mol C_2H_6 behaving as (a) a perfect gas, (b) a van der Waals gas when it is confined under the following conditions: (i) at 273.15 K in 22.414 L, (ii) at 1000 K in 100 cm^3. Use the data in Table 1.5.

1.16 (b) Calculate the pressure exerted by 1.0 mol H_2S behaving as (a) a perfect gas, (b) a van der Waals gas when it is confined under the following conditions: (i) at 273.15 K in 22.414 L, (ii) at 500 K in 150 cm^3. Use the data in Table 1.5.

1.17 (a) Estimate the critical constants of a gas with van der Waals parameters $a = 0.751$ atm L^2 mol^{-2} and $b = 0.0226$ L mol^{-1}.

1.17 (b) Estimate the critical constants of a gas with van der Waals parameters $a = 1.32$ atm L^2 mol^{-2} and $b = 0.0436$ L mol^{-1}.

1.18 (a) A gas at 250 K and 15 atm has a molar volume 12 per cent smaller than that calculated from the perfect gas law. Calculate (a) the compression factor under these conditions and (b) the molar volume of the gas. Which are dominating in the sample, the attractive or the repulsive forces?

1.18 (b) A gas at 350 K and 12 atm has a molar volume 12 per cent larger than that calculated from the perfect gas law. Calculate (a) the compression factor under these conditions and (b) the molar volume of the gas. Which are dominating in the sample, the attractive or the repulsive forces?

1.19 (a) In an industrial process, nitrogen is heated to 500 K at a constant volume of 1.000 m^3. The gas enters the container at 300 K and 100 atm. The mass of the gas is 92.4 kg. Use the van der Waals equation to determine the approximate pressure of the gas at its working temperature of 500 K. For nitrogen, $a = 1.352$ L^2 atm mol^{-2}, $b = 0.0387$ L mol^{-1}.

1.19 (b) Cylinders of compressed gas are typically filled to a pressure of 200 bar. For oxygen, what would be the molar volume at this pressure and 25°C based on (a) the perfect gas equation, (b) the van der Waals equation. For oxygen, $a = 1.364$ L^2 atm mol^{-2}, $b = 3.19 \times 10^{-2}$ L mol^{-1}.

1.20 (a) The density of water vapour at 327.6 atm and 776.4 K is 133.2 g dm^{-3}. (a) Determine the molar volume, V_m, of water and the compression factor, Z, from these data. (b) Calculate Z from the van der Waals equation with $a = 5.464$ L^2 atm mol^{-2} and $b = 0.0305$ L mol^{-1}.

1.20 (b) The density of water vapour at 1.00 bar and 383 K is 950.99 kg m^{-3}. (a) Determine the molar volume, V_m, of water and the compression factor, Z, from these data. (b) Calculate Z

from the van der Waals equation with $a = 5.464$ L^2 atm mol^{-2} and $b = 0.0305$ L mol^{-1}.

1.21 (a) Suppose that 10.0 mol C_2H_6(g) is confined to 4.860 L at 27°C. Predict the pressure exerted by the ethane from (a) the perfect gas and (b) the van der Waals equations of state. Calculate the compression factor based on these calculations. For ethane, $a = 5.507$ L^2 atm mol^{-2}, $b = 0.0651$ L mol^{-1}.

1.21 (b) At 300 K and 20 atm, the compression factor of a gas is 0.86. Calculate (a) the volume occupied by 8.2 mmol of the gas under these conditions and (b) an approximate value of the second virial coefficient B at 300 K.

1.22 (a) A vessel of volume 22.4 L contains 2.0 mol H_2 and 1.0 mol N_2 at 273.15 K. Calculate (a) the mole fractions of each component, (b) their partial pressures, and (c) their total pressure.

1.22 (b) A vessel of volume 22.4 L contains 1.5 mol H_2 and 2.5 mol N_2 at 273.15 K. Calculate (a) the mole fractions of each component, (b) their partial pressures, and (c) their total pressure.

1.23 (a) The critical constants of methane are $p_c = 45.6$ atm, $V_c = 98.7$ cm^3 mol^{-1}, and $T_c = 190.6$ K. Calculate the van der Waals parameters of the gas and estimate the radius of the molecules.

1.23 (b) The critical constants of ethane are $p_c = 48.20$ atm, $V_c = 148$ cm^3 mol^{-1}, and $T_c = 305.4$ K. Calculate the van der Waals parameters of the gas and estimate the radius of the molecules.

1.24 (a) Use the van der Waals parameters for chlorine to calculate approximate values of (a) the Boyle temperature of chlorine and (b) the radius of a Cl_2 molecule regarded as a sphere.

1.24 (b) Use the van der Waals parameters for hydrogen sulfide to calculate approximate values of (a) the Boyle temperature of the gas and (b) the radius of a H_2S molecule regarded as a sphere.

1.25 (a) Suggest the pressure and temperature at which 1.0 mol of (a) NH_3, (b) Xe, (c) He will be in states that correspond to 1.0 mol H_2 at 1.0 atm and 25°C.

1.25 (b) Suggest the pressure and temperature at which 1.0 mol of (a) H_2S, (b) CO_2, (c) Ar will be in states that correspond to 1.0 mol N_2 at 1.0 atm and 25°C.

1.26 (a) A certain gas obeys the van der Waals equation with $a = 0.50$ m^6 Pa mol^{-2}. Its volume is found to be 5.00×10^{-4} m^3 mol^{-1} at 273 K and 3.0 MPa. From this information calculate the van der Waals constant b. What is the compression factor for this gas at the prevailing temperature and pressure?

1.26 (b) A certain gas obeys the van der Waals equation with $a = 0.76$ m^6 Pa mol^{-2}. Its volume is found to be 4.00×10^{-4} m^3 mol^{-1} at 288 K and 4.0 MPa. From this information calculate the van der Waals constant b. What is the compression factor for this gas at the prevailing temperature and pressure?

Problems†

Numerical problems

1.1 A diving bell has an air space of 3.0 m³ when on the deck of a boat. What is the volume of the air space when the bell has been lowered to a depth of 50 m? Take the mean density of sea water to be 1.025 g cm⁻³ and assume that the temperature is the same as on the surface.

1.2 What pressure difference must be generated across the length of a 15 cm vertical drinking straw in order to drink a water-like liquid of density 1.0 g cm⁻³?

1.3 Recent communications with the inhabitants of Neptune have revealed that they have a Celsius-type temperature scale, but based on the melting point (0°N) and boiling point (100°N) of their most common substance, hydrogen. Further communications have revealed that the Neptunians know about perfect gas behaviour and they find that, in the limit of zero pressure, the value of pV is 28 L atm at 0°N and 40 L atm at 100°N. What is the value of the absolute zero of temperature on their temperature scale?

1.4 A meteorological balloon had a radius of 1.0 m when released at sea level at 20°C and expanded to a radius of 3.0 m when it had risen to its maximum altitude where the temperature was –20°C. What is the pressure inside the balloon at that altitude?

1.5 Deduce the relation between the pressure and mass density, ρ, of a perfect gas of molar mass M. Confirm graphically, using the following data on dimethyl ether at 25°C, that perfect behaviour is reached at low pressures and find the molar mass of the gas.

p/Torr	91.74	188.98	277.3	452.8	639.3	760.0
ρ/(g L⁻¹)	0.225	0.456	0.664	1.062	1.468	1.734

1.6 Charles's law is sometimes expressed in the form $V = V_0(1 + \alpha\theta)$, where θ is the Celsius temperature, α is a constant, and V_0 is the volume of the sample at 0°C. The following values for α have been reported for nitrogen at 0°C:

p/Torr	749.7	599.6	333.1	98.6
$10^3\alpha/(°C)^{-1}$	3.6717	3.6697	3.6665	3.6643

For these data calculate the best value for the absolute zero of temperature on the Celsius scale.

1.7 The molar mass of a newly synthesized fluorocarbon was measured in a gas microbalance. This device consists of a glass bulb forming one end of a beam, the whole surrounded by a closed container. The beam is pivoted, and the balance point is attained by raising the pressure of gas in the container, so increasing the buoyancy of the enclosed bulb. In one experiment, the balance point was reached when the fluorocarbon pressure was 327.10 Torr; for the same setting of the pivot, a balance was reached when CHF_3 ($M = 70.014$ g mol⁻¹) was introduced at 423.22 Torr. A repeat of the experiment with a different setting of the pivot required a pressure of 293.22 Torr of the fluorocarbon and 427.22 Torr of the CHF_3. What is the molar mass of the fluorocarbon? Suggest a molecular formula.

1.8 A constant-volume perfect gas thermometer indicates a pressure of 50.2 Torr at the triple point temperature of water (273.16 K). (a) What change of pressure indicates a change of 1.00 K at this temperature? (b) What pressure indicates a temperature of 100.00°C? (c) What change of pressure indicates a change of 1.00 K at the latter temperature?

1.9 A vessel of volume 22.4 L contains 2.0 mol H_2 and 1.0 mol N_2 at 273.15 K initially. All the H_2 reacted with sufficient N_2 to form NH_3. Calculate the partial pressures and the total pressure of the final mixture.

1.10 Calculate the molar volume of chlorine gas at 350 K and 2.30 atm using (a) the perfect gas law and (b) the van der Waals equation. Use the answer to (a) to calculate a first approximation to the correction term for attraction and then use successive approximations to obtain a numerical answer for part (b).

1.11 At 273 K measurements on argon gave $B = -21.7$ cm³ mol⁻¹ and $C = 1200$ cm⁶ mol⁻², where B and C are the second and third virial coefficients in the expansion of Z in powers of $1/V_m$. Assuming that the perfect gas law holds sufficiently well for the estimation of the second and third terms of the expansion, calculate the compression factor of argon at 100 atm and 273 K. From your result, estimate the molar volume of argon under these conditions.

1.12 Calculate the volume occupied by 1.00 mol N_2 using the van der Waals equation in the form of a virial expansion at (a) its critical temperature, (b) its Boyle temperature, and (c) its inversion temperature. Assume that the pressure is 10 atm throughout. At what temperature is the gas most perfect? Use the following data: $T_c = 126.3$ K, $a = 1.352$ L² atm mol⁻², $b = 0.0387$ L mol⁻¹.

1.13‡ The second virial coefficient of methane can be approximated by the empirical equation $B'(T) = a + be^{-c/T^2}$, where $a = -0.1993$ bar⁻¹, $b = 0.2002$ bar⁻¹, and $c = 1131$ K² with 300 K < T < 600 K. (a) What is the value of the Boyle temperature for methane? (b) Compare the molar volume of methane at 50 bar as predicted by both the virial equation and the perfect gas equation at (i) 298 K, (ii) 373 K.

1.14 The mass density of water vapour at 327.6 atm and 776.4 K is 1.332×10^2 g L⁻¹. Given that for water $T_c = 647.4$ K, $p_c = 218.3$ atm, $a = 5.464$ L² atm mol⁻², $b = 0.0305$ L mol⁻¹, and $M = 18.02$ g mol⁻¹, calculate (a) the molar volume. Then calculate the compression factor (b) from the data, (c) from the virial expansion of the van der Waals equation.

1.15 The critical volume and critical pressure of a certain gas are 160 cm³ mol⁻¹ and 40 atm, respectively. Estimate the critical temperature by assuming that the gas obeys the Berthelot equation of state. Estimate the radii of the gas molecules on the assumption that they are spheres.

1.16 Estimate the coefficients a and b in the Dieterici equation of state from the critical constants of xenon. Calculate the pressure exerted by 1.0 mol Xe when it is confined to 1.0 L at 25°C.

Theoretical problems

1.17 Show that the van der Waals equation leads to values of $Z < 1$ and $Z > 1$, and identify the conditions for which these values are obtained.

† Problems denoted with the symbol ‡ were supplied by Charles Trapp, Carmen Giunta, and Marshall Cady.

1.18 Express the van der Waals equation of state as a virial expansion in powers of $1/V_m$ and obtain expressions for B and C in terms of the parameters a and b. The expansion you will need is $(1 - x)^{-1} = 1 + x + x^2 + \cdots$. Measurements on argon gave $B = -21.7$ cm^3 mol^{-1} and $C = 1200$ cm^6 mol^{-2} for the virial coefficients at 273 K. What are the values of a and b in the corresponding van der Waals equation of state?

1.19‡ Derive the relation between the critical constants and the Dieterici equation parameters. Show that $Z_c = 2e^{-2}$ and derive the reduced form of the Dieterici equation of state. Compare the van der Waals and Dieterici predictions of the critical compression factor. Which is closer to typical experimental values?

1.20 A scientist proposed the following equation of state:

$$p = \frac{RT}{V_m} - \frac{B}{V_m^2} + \frac{C}{V_m^3}$$

Show that the equation leads to critical behaviour. Find the critical constants of the gas in terms of B and C and an expression for the critical compression factor.

1.21 Equations 1.21 and 1.22 are expansions in p and $1/V_m$, respectively. Find the relation between B, C and B', C''.

1.22 The second virial coefficient B' can be obtained from measurements of the density ρ of a gas at a series of pressures. Show that the graph of p/ρ against p should be a straight line with slope proportional to B'. Use the data on dimethyl ether in Problem 1.5 to find the values of B' and B at 25°C.

1.23 The equation of state of a certain gas is given by $p = RT/V_m + (a + bT)/V_m^2$, where a and b are constants. Find $(\partial V/\partial T)_p$.

1.24 The following equations of state are occasionally used for approximate calculations on gases: (gas A) $pV_m = RT(1 + b/V_m)$, (gas B) $p(V_m - b) = RT$. Assuming that there were gases that actually obeyed these equations of state, would it be possible to liquefy either gas A or B? Would they have a critical temperature? Explain your answer.

1.25 Derive an expression for the compression factor of a gas that obeys the equation of state $p(V - nb) = nRT$, where b and R are constants. If the pressure and temperature are such that $V_m = 10b$, what is the numerical value of the compression factor?

1.26‡ Amedeo Avogadro (*Journal de Physique* (1811)) noted that two volumes of hydrogen combine with one volume of oxygen to form two volumes of water vapour. He also gave the density of water vapour relative to that of air as 0.625 and hydrogen as 0.0732 (in units in which air had unit density). Use this information and Avogadro's principle to compute the molar masses of water vapour and oxygen relative to that of hydrogen.

1.27‡ The discovery of the element argon by Lord Rayleigh and Sir William Ramsay had its origins in Rayleigh's measurements of the density of nitrogen with an eye toward accurate determination of its molar mass. Rayleigh prepared some samples of nitrogen by chemical reaction of nitrogen-containing compounds; under his standard conditions, a glass globe filled with this 'chemical nitrogen' had a mass of 2.2990 g. He prepared other samples by removing oxygen, carbon dioxide, and water vapour from atmospheric air; under the same conditions, this 'atmospheric nitrogen' had a mass of 2.3102 g (Lord Rayleigh, *Royal Institution Proceedings* 14, 524 (1895)). With the hindsight of knowing accurate values for the molar masses of nitrogen and argon, compute the mole fraction of argon in the latter sample on the assumption that the former was pure nitrogen and the latter a mixture of nitrogen and argon.

1.28‡ A substance as elementary and well-known as argon still receives research attention. Stewart and Jacobsen have published a review of thermodynamic properties of argon (R.B. Stewart and R.T. Jacobsen, *J. Phys. Chem. Ref. Data* **18**, 639 (1989)) that included the following 300 K isotherm.

p/MPa	0.4000	0.5000	0.6000	0.8000	1.000
V_m/(L mol^{-1})	6.2208	4.9736	4.1423	3.1031	2.4795

p/MPa	1.500	2.000	2.500	3.000	4.000
V_m/(L mol^{-1})	1.6483	1.2328	0.983 57	0.817 46	0.609 98

(a) Compute the second virial coefficient, B, at this temperature. (b) Use nonlinear curve-fitting software to compute the third virial coefficient, C, at this temperature.

1.29‡ Show that the compression factor, Z, of a van der Waals gas can be expressed as $Z = V'_r/(V'_r - \frac{1}{8}) - 27/64T_r V'_r$, where $V'_r = p_c V_m/RT_c$ is the 'pseudoreduced volume', or alternatively as the solution of the cubic equation $Z^3 - \{(p_r/8T_r) + 1\}Z^2 + (27p_r/64T_r^2 - 27p_r^2/512T_r^3)Z - 27p_r^2/512T_r^3 = 0$. Solve this equation for Z for nitrogen, methane, propane, and ethene at $T_r = 1.2$ and $p_r = 3.0$ and compare to the value given in Fig. 1.18.

Applications: to atmospheric science

1.30 Balloons are still used to deploy sensors that monitor meteorological phenomena and the chemistry of the atmosphere. It is possible to investigate some of the technicalities of ballooning by using the perfect gas law. Suppose your balloon has a radius of 3.0 m and that it is spherical. (a) What amount of H_2 (in moles) is needed to inflate it to 1.0 atm in an ambient temperature of 25°C at sea level? (b) What mass can the balloon lift at sea level, where the density of air is 1.22 kg m^{-3}? (c) What would be the payload if He were used instead of H_2?

1.31‡ The preceding problem is most readily solved (see the *Solutions manual*) with the use of the Archimedes principle, which states that the lifting force is equal to the difference between the weight of the displaced air and the weight of the balloon. Prove the Archimedes principle for the atmosphere from the barometric formula. *Hint.* Assume a simple shape for the balloon, perhaps a right circular cylinder of cross-sectional area A and height h.

1.32‡ Chlorofluorocarbons such as CCl_3F and CCl_2F_2 have been linked to ozone depletion in Antarctica. As of 1994, these gases were found in quantities of 261 and 509 parts per trillion (10^{12}) by volume (World Resources Institute, *World Resources* 1996–97). Compute the molar concentration of these gases under conditions typical of (a) the mid-latitude troposphere (10°C and 1.0 atm) and (b) the Antarctic stratosphere (200 K and 0.050 atm).

1.33‡ The composition of the atmosphere is approximately 80 per cent nitrogen and 20 per cent oxygen by mass. At what height above the surface of the earth would the atmosphere become 90 per cent nitrogen and 10 per cent oxygen by mass? Assume that the temperature of the atmosphere is constant at 25°C. What is the pressure of the atmosphere at that height?

2

The First Law: the concepts

The basic concepts

2.1 Work, heat, and energy

2.2 The First Law

Work and heat

2.3 Expansion work

2.4 Heat transactions

2.5 Enthalpy

2.6 Adiabatic changes

Thermochemistry

2.7 Standard enthalpy changes

2.8 Standard enthalpies of formation

2.9 The temperature dependence of reaction enthalpies

Checklist of key ideas

Further reading

Exercises

Problems

This chapter introduces some of the basic concepts of thermodynamics. It concentrates on the conservation of energy—the experimental observation that energy can be neither created nor destroyed—and shows how the principle of the conservation of energy can be used to assess the energy changes that accompany physical and chemical processes. Much of this chapter examines the means by which a system can exchange energy with its surroundings in terms of the work it may do or the heat that it may produce. The target concept of the chapter is enthalpy, which is a very useful book-keeping property for keeping track of the heat output (or requirements) of physical processes and chemical reactions at constant pressure.

The release of energy can be used to provide heat when a fuel burns in a furnace, to produce mechanical work when a fuel burns in an engine, and to generate electrical work when a chemical reaction pumps electrons through a circuit. In chemistry, we encounter reactions that can be harnessed to provide heat and work, reactions that liberate energy, which is squandered (often to the detriment of the environment), but that give products we require, and reactions that constitute the processes of life. **Thermodynamics**, the study of the transformations of energy, enables us to discuss all these matters quantitatively and to make useful predictions.

The basic concepts

For the purposes of physical chemistry, the universe is divided into two parts, the system and its surroundings. The **system** is the part of the world in which we have a special interest. It may be a reaction vessel, an engine, an electrochemical cell, a biological cell, and so on. The **surroundings** comprise the region outside the system and are where we make our measurements. The type of system depends on the characteristics of the boundary that divides it from the surroundings (Fig. 2.1). If matter can be transferred through the boundary between the system and its surroundings the system is classified as **open**. If matter cannot pass through the boundary the system is classified as **closed**. Both open and closed systems can exchange energy with their surroundings. For example, a closed system can expand and thereby raise a weight in the surroundings; it may also transfer energy to them if they are at a lower temperature. An **isolated system** is a closed system that has neither mechanical nor thermal contact with its surroundings.

(a) Open

(b) Closed

(c) Isolated

2.1 (a) An open system can exchange matter and energy with its surroundings. (b) A closed system can exchange energy with its surroundings, but it cannot exchange matter. (c) An isolated system can exchange neither energy nor matter with its surroundings.

2.2 (a) A diathermic system is one that allows energy to escape as heat through its boundary if there is a difference in temperature between the system and its surroundings. (b) An adiabatic system is one that does not permit the passage of energy as heat through its boundary even if there is a temperature difference between the system and its surroundings.

2.1 Work, heat, and energy

The fundamental physical property in thermodynamics is work: **work** is done when an object is moved against an opposing force. Doing work is equivalent to raising a weight somewhere in the surroundings. An example of doing work is the expansion of a gas that pushes out a piston and raises a weight. A chemical reaction that drives an electric current through a resistance also does work, because the same current could be driven through a motor and used to raise a weight.

The **energy** of a system is its capacity to do work. When work is done on an otherwise isolated system (for instance, by compressing a gas or winding a spring), the capacity of the system to do work is increased, so the energy of the system is increased. When the system does work (when the piston moves out or the spring unwinds), the energy of the system is reduced and it can do less work than before.

Experiments have shown that the energy of a system may be changed by means other than work itself. When the energy of a system changes as a result of a temperature difference between the system and its surroundings we say that energy has been transferred as **heat**. When a heater is immersed in a beaker of water (the system), the capacity of the system to do work increases because hot water can be used to do more work than cold water. Not all boundaries permit the transfer of energy even though there is a temperature difference between the system and its surroundings. As we have seen (Section 1.1c), a boundary that does permit energy transfer as heat (such as steel and glass) is called **diathermic** and a boundary that does not permit energy transfer as heat is called **adiabatic** (Fig. 2.2).

An **exothermic process** is a process that releases energy as heat. All combustion reactions are exothermic. An **endothermic process** is a process that absorbs energy as heat. An example of an endothermic process is the vaporization of water. An endothermic process in a diathermic container results in energy flowing into the system as heat. An exothermic process in a similar diathermic container results in a release of energy as heat into the surroundings. When an endothermic process takes place in an adiabatic container, it results in a lowering of temperature of the system; an exothermic process results in a rise of temperature. These features are summarized in Fig. 2.3.

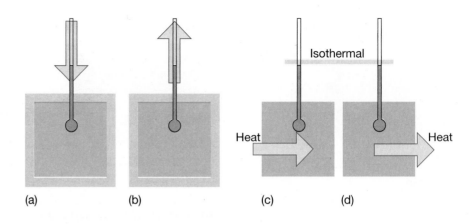

2.3 (a) When an endothermic process occurs in an adiabatic system, the temperature falls; (b) if the process is exothermic, then the temperature rises. (c) When an endothermic process occurs in a diathermic container, energy enters as heat from the surroundings, and the system remains at the same temperature. (d) If the process is exothermic, then energy leaves as heat, and the process is isothermal.

Molecular interpretation 2.1 Heat, work, and energy

In molecular terms, heat is the transfer of energy that makes use of *disorderly molecular motion*. The disorderly motion of molecules is called **thermal motion**. The thermal motion of the molecules in the hot surroundings stimulates the molecules in the cooler system to move more vigorously and, as a result, the energy of the system is increased. When a system heats its surroundings, molecules of the system stimulate the thermal motion of the molecules in the surroundings (Fig. 2.4).

In contrast, *work is the transfer of energy that makes use of organized motion* (Fig. 2.5). When a weight is raised or lowered, its atoms move in an organized way. The atoms

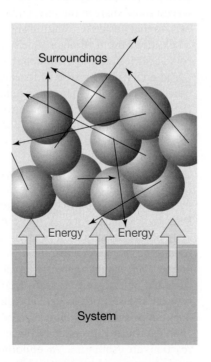

2.4 When energy is transferred to the surroundings as heat, the transfer stimulates disordered motion of the atoms in the surroundings. Transfer of energy from the surroundings to the system makes use of disordered motion (thermal motion) in the surroundings.

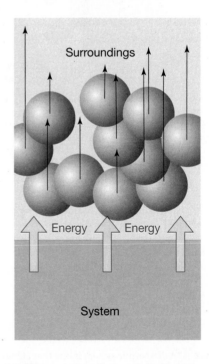

2.5 When a system does work, it stimulates orderly motion in the surroundings. For instance, the atoms shown here may be part of a weight that is being raised. The ordered motion of the atoms in a falling weight does work on the system.

in a spring move in an orderly way when it is wound; the electrons in an electric current move in an orderly direction when it flows. When a system does work it causes atoms or electrons in its surroundings to move in an organized way. Likewise, when work is done on a system, molecules in the surroundings are used to transfer energy to it in an organized way, as the atoms in a weight are lowered or a current of electrons is passed.

The distinction between work and heat is made in the surroundings. The fact that a falling weight may stimulate thermal motion in the system is irrelevant to the distinction between heat and work: work is identified as energy transfer making use of the organized motion of atoms in the surroundings, and heat is identified as energy transfer making use of thermal motion in the surroundings. In the compression of a gas, for instance, work is done as the particles of the compressing weight descend in an orderly way, but the effect of the incoming piston is to accelerate the gas molecules to higher average speeds. Because collisions between molecules quickly randomize their directions, the orderly motion of the atoms of the weight is in effect stimulating thermal motion in the gas. We observe the falling weight, the orderly descent of its atoms, and report that work is being done even though it is stimulating thermal motion.

The great revolution in physics that occurred in the opening years of the twentieth century was the realization that energy is *quantized*; that is, a particle can possess only certain energies, called its 'energy levels'. Broadly speaking, the heavier the particle and the greater its spatial freedom, the closer together are its energy levels. The number of particles, N_i, in a sample of N particles, that will be found in a state with an energy E_i when it is part of a system in thermal equilibrium at a temperature T is expressed by the *Boltzmann distribution*:

$$N_i = \frac{Ne^{-E_i/kT}}{q} \qquad q = \sum_i e^{-E_i/kT}$$

where k is *Boltzmann's constant*, a fundamental constant with the value 1.381×10^{-23} J K^{-1}. Boltzmann's constant is related to the gas constant by $R = N_A k$, where N_A is Avogadro's constant. One reason why the gas constant appears in so many expressions, including those apparently wholly unrelated to gases, is that it is really the fundamental constant k appearing in disguise. Broadly speaking, the Boltzmann distribution implies that the population of states decays exponentially with increasing energy, so most particles will be found in states of low energy and only a few will be found in states of high energy. Care must be taken with the exact interpretation though, because more than one state may correspond to the same energy: that is, an energy level may be degenerate and consist of several states. The function q is called the *partition function*. It plays a central role in the calculation of thermodynamic properties (Chapter 19).

2.2 The First Law

In thermodynamics, the total energy of a system is called its **internal energy**, U. The internal energy is the total kinetic and potential energy of the molecules in the system.[1] We denote by ΔU the change in internal energy when a system changes from an initial state i with internal energy U_i to a final state f of internal energy U_f:

$$\Delta U = U_f - U_i \qquad\qquad [2.1]$$

---

1. The internal energy does not include the kinetic energy arising from the motion of the system as a whole, such as its kinetic energy as it accompanies the Earth on its orbit round the Sun.

The internal energy is a **state function** in the sense that its value depends only on the current state of the system and is independent of how that state has been prepared. In other words, it is a function of the properties that determine the current state of the system. Changing any one of the state variables, such as the pressure, results in a change in internal energy. The internal energy is an extensive property.

Internal energy, heat, and work are all measured in the same units, the joule (J). The joule, which is named after the nineteenth-century scientist J.P. Joule, is defined as

$$1\,J = 1\,kg\,m^2\,s^{-2}$$

A joule is quite a small unit of energy: for instance, each beat of the human heart consumes about 1 J. Changes in molar internal energy, ΔU_m, are typically expressed in kilojoules per mole ($kJ\,mol^{-1}$). Certain other energy units are also used, but are more common in fields other than thermodynamics. Thus, 1 electronvolt (1 eV) is defined as the kinetic energy acquired when an electron is accelerated from rest through a potential difference of 1 V; the relation between electronvolts and joules is $1\,eV \approx 0.16\,aJ$ (where $1\,aJ = 10^{-18}\,J$). Many processes in chemistry have an energy of several electronvolts. Thus, the energy to remove an electron from a sodium atom is close to 5 eV. Calories (cal) and kilocalories (kcal) are still encountered:

$$1\,cal = 4.184\,J\,exactly$$

An energy of 1 cal is enough to raise the temperature of 1 g of water by 1°C.

Molecular interpretation 2.2 The internal energy of a gas

The key to relating the internal energy to the properties of individual molecules is the *equipartition theorem* of classical mechanics. There are two parts to this theorem, one qualitative, the other quantitative. First, we need to know that a 'quadratic contribution' to the energy means a contribution that can be expressed as the square of a variable, such as the position or the velocity. For example, the kinetic energy of an atom of mass m as it moves through space is

$$E_K = \tfrac{1}{2}mv_x^2 + \tfrac{1}{2}mv_y^2 + \tfrac{1}{2}mv_z^2$$

and there are three quadratic contributions to its energy. The equipartition theorem then states that, for a collection of particles at thermal equilibrium at a temperature T, *the average value of each quadratic contribution to the energy is the same and equal to $\tfrac{1}{2}kT$*, where k is Boltzmann's constant (see *Molecular interpretation 2.1*). The equipartition theorem is a conclusion from classical mechanics, and is applicable only when the effects of quantization can be ignored. In practice, it can be used for molecular translation and rotation but not vibration. At 25°C, $\tfrac{1}{2}kT = 2\,zJ$ (where $1\,zJ = 10^{-21}\,J$), or about 13 meV.

According to the equipartition theorem, the average energy of each term in the expression above is $\tfrac{1}{2}kT$. Therefore, the mean energy of the atoms is $\tfrac{3}{2}kT$ and the total energy of the gas (there being no potential energy contribution) is $\tfrac{3}{2}NkT$, or $\tfrac{3}{2}nRT$. We can therefore write

$$U_m = U_m(0) + \tfrac{3}{2}RT$$

where $U_m(0)$ is the molar internal energy at $T = 0$, when all translational motion has ceased and the sole contribution to the internal energy arises from the internal structure of the atoms. This equation shows that the internal energy of a perfect gas increases linearly with temperature. At 25°C, $\tfrac{3}{2}RT = 3.7\,kJ\,mol^{-1}$, so translational motion contributes about 4 $kJ\,mol^{-1}$ to the molar internal energy of a monatomic gas (the remaining contribution arises from the internal structure of the atoms).

Because the potential energy of interactions between the atoms (or molecules) of a perfect gas is zero, the internal energy is independent of how close they are together. Therefore, the internal energy is independent of the volume occupied by a perfect gas. A formal statement of this property is

$$\left(\frac{\partial U}{\partial V}\right)_T = 0$$

The internal energy of interacting molecules in condensed phases also has a contribution from the potential energy of their interaction. However, no simple expressions can be written down in general. Nevertheless, the crucial molecular point is that, as the temperature of a system is raised, the internal energy increases as the various modes of motion become more highly excited.

Self-test 2.1 Estimate the contribution of the rotational and translational motion of a nonlinear molecule to its molar internal energy. (see *Further information 2*) $[U_m = U_m(0) + 3RT; 3RT = 7.4\ \mathrm{kJ\ mol^{-1}}]$

(a) The conservation of energy

It has been found experimentally that the internal energy of a system may be changed either by doing work on the system or by heating it. Whereas we may know how the energy transfer has occurred (because we can see if a weight has been raised or lowered in the surroundings, indicating transfer of energy by doing work, or if ice has melted in the surroundings, indicating transfer of energy as heat), the system is blind to the mode employed. *Heat and work are equivalent ways of changing a system's internal energy.* A system is like a bank: it accepts deposits in either currency, but stores its reserves as internal energy. It is also found experimentally that, if a system is isolated from its surroundings, then no change in internal energy takes place. This summary of observations is now known as the **First Law of thermodynamics** and expressed as follows:

The internal energy of an isolated system is constant.

We cannot use a system to do work, leave it isolated for a month, and then come back expecting to find it restored to its original state and ready to do the same work again. The evidence for this property is that no 'perpetual motion machine' (a machine that does work without consuming fuel or some other source of energy) has ever been built.

These remarks may be summarized as follows. If we write w for the work done on a system, q for the energy transferred as heat to a system, and ΔU for the resulting change in internal energy, then it follows that

$$\Delta U = q + w \tag{2.2}$$

Equation 2 is the mathematical statement of the First Law, for it summarizes the equivalence of heat and work and the fact that the internal energy is constant in an isolated system (for which $q = 0$ and $w = 0$). The equation states that the change in internal energy of a closed system is equal to the energy that passes through its boundary as heat or work. It employs the 'acquisitive convention', in which $w > 0$ or $q > 0$ if energy is transferred to the system as work or heat and $w < 0$ or $q < 0$ if energy is lost from the system as work or heat. In other words, we view the flow of energy as work or heat from the system's perspective.

Illustration 2.1

If an electric motor produced 15 kJ of energy each second as mechanical work and lost 2 kJ as heat to the surroundings, then the change in the internal energy of the motor each second is

$$\Delta U = -2 \text{ kJ} - 15 \text{ kJ} = -17 \text{ kJ}$$

Suppose that, when a spring was wound, 100 J of work was done on it but 15 J escaped to the surroundings as heat. The change in internal energy of the spring is

$$\Delta U = +100 \text{ kJ} - 15 \text{ kJ} = +85 \text{ kJ}$$

(b) The formal statement of the First Law

The expression of the First Law that we have given is adequate for most purposes in thermodynamics. However, there are several unsatisfactory features about it, such as how we define and measure 'heat'. This section gives a more sophisticated version of the law, and shows how eqn 2 can be put on a firmer foundation. As the remainder of the text does not depend on this material, it is possible to omit it and go immediately to the following section ('Work and heat').

We begin by pretending that we do not know what we mean by 'energy'. We pretend that we know only what is meant by work, because we can observe a weight being raised or lowered in the surroundings. We also know how to measure work by noting the height through which the weight is raised. Throughout this section, work will be the fundamental, measurable quantity, and we define energy, heat, and the First Law in terms of work alone. We shall employ terms that have been established by the Zeroth Law of thermodynamics (Section 1.1), namely, state and temperature and the concepts of adiabatic and diathermic walls.

In an adiabatic system of a given composition, it is known experimentally that the same increase in temperature is brought about by the same quantity of any kind of work we do on the system. Thus, if 1 kJ of mechanical work is done on the system (by stirring it with rotating paddles, for instance), or 1 kJ of electrical work is done (by passing an electric current through a heater), and so on, then the same rise in temperature is produced. The following statement of the First Law of thermodynamics is a summary of a large number of observations of this kind:

> **The work needed to change an adiabatic system from one specified state to another specified state is the same however the work is done.**

This form of the law looks completely different from the form we gave before, but we shall now see how it implies eqn 2.

Suppose we do work w_{ad} on an adiabatic system to change it from an initial state i to a final state f. The work may be of any kind (mechanical or electrical) and may take the system through different intermediate states (different temperatures and pressures, for instance). We might (in ignorance of the First Law) think that we need to label w_{ad} with the path and to write w_{ad}(mechanical) or w_{ad}(electrical). However, the First Law tells us that w_{ad} is the same for all paths and depends only on the initial and final states. This conclusion is analogous to climbing a mountain: the height we must climb between any two points is independent of the path we take (Fig. 2.6). In mountain climbing we can attach a number, the altitude, A, to each point on the mountain and express the height, h, of the climb as a difference in altitudes:

$$h = A_f - A_i = \Delta A$$

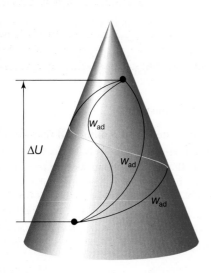

2.6 It is found that the same quantity of work must be done on an adiabatic system to achieve the same change of state even though different means of achieving that work may be used. This path independence implies the existence of a state function, the internal energy. The change in internal energy is like the change in altitude when climbing a mountain: its value is independent of path.

That is, in mountain climbing, the observation that h is independent of the path implies the existence of the state function A. The First Law has exactly the same implication. The fact that w_{ad} is independent of the path implies that to each state of the system we can attach a value of a quantity—we call it the 'internal energy', U—and express the work as a difference in internal energies:

$$\Delta w_{ad} = U_f - U_i = \Delta U \qquad [2.3]$$

This equation also shows that we can measure the change in the internal energy of a system by measuring the work needed to bring about the change in an adiabatic system.

(c) The mechanical definition of heat

Suppose we strip away the thermal insulation around the system and make it diathermic. The system is now in thermal contact with its surroundings as we drive it from the same initial state to the same final state. The change in internal energy is the same as before, because U is a state function, but we might find that the work we must do is not the same as before. Thus, whereas we might have needed to do 42 kJ of work when the system was in an adiabatic container, to achieve the same change of state we might now have to do 50 kJ of work. The difference between the work done in the two cases is *defined* as the heat absorbed by the system in the process:

$$q = w_{ad} - w \qquad [2.4]$$

In the present case, we would conclude that $q = 42$ kJ $- 50$ kJ $= -8$ kJ, and report that 8 kJ of energy had left the system as heat. We see that we now have a purely mechanical definition of heat in terms of work. We know how to measure work in terms of the height through which a weight falls, so we now also have a method for measuring heat in terms of work.

Finally, we can express eqn 4 in a more familiar way. Because we already know that ΔU is (by definition) equal to w_{ad}, the expression for the energy transferred to the system as heat is $q = \Delta U - w$. However, this expression is equivalent to eqn 2, the mathematical form of the First Law that we saw earlier.

Work and heat

The way can now be opened to powerful methods of calculation by switching attention to infinitesimal changes of state (such as infinitesimal change in temperature) and infinitesimal changes in the internal energy dU. Then, if the work done on a system is dw and the energy supplied to it as heat is dq, in place of eqn 2 we have

$$dU = dq + dw \qquad (2.5)$$

To use eqn 5 we must be able to relate dq and dw to events taking place in the surroundings.

2.3 Expansion work

We begin by discussing **expansion work**, the work arising from a change in volume. This type of work includes the work done by a gas as it expands and drives back the atmosphere. Many chemical reactions result in the generation or consumption of

External pressure, p_{ex}

External pressure, p_{ex}

Area, A

Distance moved, dz

Pressure, p

2.7 When a piston of area A moves out through a distance dz, it sweeps out a volume d$V = A$ dz. The external pressure p_{ex} is equivalent to a weight pressing on the piston, and the force opposing expansion is $F = p_{ex}A$.

gases (for instance, the thermal decomposition of calcium carbonate or the combustion of octane), and the thermodynamic characteristics of a reaction depend on the work it can do. The term 'expansion work' also includes work associated with negative changes of volume, that is, compression.

(a) The general expression for work

The calculation of expansion work starts from the definition used in physics, which states that the work required to move an object a distance dz against an opposing force of magnitude F is

$$\mathrm{d}w = -F\,\mathrm{d}z \qquad [2.6]$$

The negative sign tells us that, when the system moves an object against an opposing force, the internal energy of the system doing the work will decrease. Now consider the arrangement shown in Fig. 2.7, in which one wall of a system is a massless, frictionless, rigid, perfectly fitting piston of area A. If the external pressure is p_{ex}, the magnitude of the force acting on the outer face of the piston is $F = p_{ex}A$. When the system expands through a distance dz against an external pressure p_{ex}, it follows that the work done is d$w = -p_{ex}A$ dz. But A dz is the change in volume, dV, in the course of the expansion. Therefore, the work done when the system expands by dV against a pressure p_{ex} is

$$\mathrm{d}w = -p_{ex}\,\mathrm{d}V \qquad (2.7)$$

To obtain the total work done when the volume changes from V_i to V_f we integrate this expression between the initial and final volumes:

$$w = -\int_{V_i}^{V_f} p_{ex}\,\mathrm{d}V \qquad (2.8)$$

The force acting on the piston, $p_{ex}A$, is equivalent to a weight that is raised as the system expands.

If the system is compressed instead, then the same weight is lowered in the surroundings and eqn 8 can still be used, but now $V_f < V_i$. It is important to note that it is still the external pressure that determines the magnitude of the work. This somewhat perplexing conclusion seems to be inconsistent with the fact that the gas *inside* the container is opposing the compression. However, when a gas is compressed, the ability of the *surroundings* to do work is diminished by an amount determined by the weight that is lowered, and it is this energy that is transferred into the system.

Other types of work (for example, electrical work), which we shall call either **non-expansion work** or **additional work**, have analogous expressions, with each one the product of an intensive factor (the pressure, for instance) and an extensive factor (the change in volume). Some are collected in Table 2.1. For the present we continue with the work associated with changing the volume, the expansion work, and see what we can extract from eqn 7.

(b) Free expansion

By **free expansion** we mean expansion against zero opposing force. It occurs when $p_{ex} = 0$. According to eqn 7, d$w = 0$ for each stage of the expansion. Hence, overall:

$$w = 0 \qquad (2.9)$$

That is, no work is done when a system expands freely. Expansion of this kind occurs when a system expands into a vacuum.

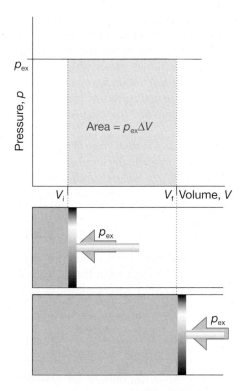

2.8 The work done by a gas when it expands against a constant external pressure, p_{ex}, is equal to the shaded area in this example of an indicator diagram.

The value of the integral $\int_a^b f(x)\,dx$ is equal to the area under the graph of $f(x)$ between $x = a$ and $x = b$.

Table 2.1 Varieties of work*

Type of work	dw	Comments	Units†
Expansion	$-p_{ex}\,dV$	p_{ex} is the external pressure dV is the change in volume	Pa m^3
Surface expansion	$\gamma\,d\sigma$	γ is the surface tension dσ is the change in area	$N\,m^{-1}$ m^2
Extension	$f\,dl$	f is the tension dl is the change of length	N m
Electrical	$\phi\,dq$	ϕ is the electric potential dq is the change in charge	V C

* In general, the work done on a system can be expressed in the form d$w = -F\,dz$, where F is a 'generalized force' and dz is a 'generalized displacement'.
† For work in joules (J). Note that 1 N m = 1 J and 1 V C = 1 J.

(c) Expansion against constant pressure

Now suppose that the external pressure is constant throughout the expansion. For example, the piston may be pressed on by the atmosphere, which exerts the same pressure throughout the expansion. A chemical example of this condition is the expansion of a gas formed in a chemical reaction. We can evaluate eqn 8 by taking the constant p_{ex} outside the integral:

$$w = -p_{ex}\int_{V_i}^{V_f} dV = -p_{ex}(V_f - V_i)$$

Therefore, if we write the change in volume as $\Delta V = V_f - V_i$,

$$w = -p_{ex}\Delta V \tag{2.10}$$

This result is illustrated graphically in Fig. 2.8, which makes use of the fact that an integral can be interpreted as an area. The magnitude of w, denoted $|w|$, is equal to the area beneath the horizontal line at $p = p_{ex}$ lying between the initial and final volumes. A p,V-graph used to compute expansion work is called an **indicator diagram**; James Watt first used one to indicate aspects of the operation of his steam engine.

(d) Reversible expansion

A **reversible change** in thermodynamics is a change that can be reversed by an infinitesimal modification of a variable. The key word 'infinitesimal' sharpens the everyday meaning of the word 'reversible' as something that can change direction. We say that a system is in **equilibrium** with its surroundings if an infinitesimal change in the conditions in opposite directions results in opposite changes in its state. One example of reversibility that we have encountered already is the thermal equilibrium of two systems with the same temperature. The transfer of energy as heat between the two is reversible because, if the temperature of either system is lowered infinitesimally, then energy flows into the system with the lower temperature. If the temperature of either system at thermal equilibrium is raised infinitesimally, then energy flows out of the hotter system.

Suppose a gas is confined by a piston and that the external pressure, p_{ex}, is set equal to the pressure, p, of the confined gas. Such a system is in mechanical equilibrium with its surroundings (as illustrated in Section 1.1) because an infinitesimal change in the external pressure in either direction causes changes in volume in opposite directions. If the external pressure is reduced infinitesimally, then the gas expands slightly. If the external pressure is increased infinitesimally, then the gas contracts slightly. In either case the change is reversible in the thermodynamic sense. If, on the other hand, the external pressure differs measurably from the internal pressure, then changing p_{ex} infinitesimally will not decrease it below the pressure of the gas and so will not change the direction of the process. Such a system is not in mechanical equilibrium with its surroundings and the expansion is thermodynamically irreversible.

To achieve reversible expansion we set p_{ex} equal to p at each stage of the expansion. In practice, this equalization could be achieved by gradually removing weights from the piston so that the downward force due to the weights always matches the changing upward force due to the pressure of the gas. When we set $p_{ex} = p$, eqn 7 becomes

$$dw = -p_{ex}\, dV = -p\, dV \qquad (2.11)_{rev}$$

(Equations valid only for reversible processes are labelled with a subscript 'rev'.) Although the pressure inside the system appears in this expression for the work, it does so only because p_{ex} has been set equal to p to ensure reversibility. The total work of reversible expansion is therefore

$$w = -\int_{V_i}^{V_f} p\, dV \qquad (2.12)_{rev}$$

We can evaluate the integral once we know how the pressure of the confined gas depends on its volume. Equation 12 is the link with the material covered in Chapter 1 for, if we know the equation of state of the gas, then we can express p in terms of V and evaluate the integral.

(e) Isothermal reversible expansion

Consider the isothermal, reversible expansion of a perfect gas. The expansion is made isothermal by keeping the system in thermal contact with its surroundings (which may be a constant-temperature bath). Because the equation of state is $pV = nRT$, we know that at each stage $p = nRT/V$, with V the volume at that stage of the expansion. The temperature T is constant in an isothermal expansion, so (together with n and R) it may be taken outside the integral. It follows that the work of reversible isothermal expansion of a perfect gas from V_i to V_f at a temperature T is

$$w = -nRT \int_{V_i}^{V_f} \frac{dV}{V} = -nRT \ln \frac{V_f}{V_i} \qquad (2.13)_{rev}^{\circ}$$

When the final volume is greater than the initial volume, as in an expansion, the logarithm in eqn 13 is positive and hence $w < 0$. In this case, the system has done work on the surroundings and the internal energy of the system has decreased as a result of the work it has done.[2] The equations also show that more work is done for a given change of volume when the temperature is increased. The greater pressure of the confined gas then needs a higher opposing pressure to ensure reversibility.

We can express the result of the calculation as an indicator diagram, for the magnitude of the work done is equal to the area under the isotherm $p = nRT/V$ (Fig. 2.9). Superimposed on the diagram is the rectangular area obtained for irreversible expansion against constant external pressure fixed at the same final value as that reached in the reversible expansion. More work is obtained when the expansion is reversible (the area is greater) because matching the external pressure to the inter-

2. We shall see later that there is a compensating influx of energy as heat, so overall the internal energy is constant for the isothermal expansion of a perfect gas.

2.9 The work done by a perfect gas when it expands reversibly and isothermally is equal to the area under the isotherm $p = nRT/V$. The work done during the irreversible expansion against the same final pressure is equal to the rectangular area shown slightly darker. Note that the reversible work is greater than the irreversible work.

nal pressure at each stage of the process ensures that none of the system's pushing power is wasted. We cannot obtain more work than for the reversible process because increasing the external pressure even infinitesimally at any stage results in compression. We may infer from this discussion that, because some pushing power is wasted when $p > p_{ex}$, the maximum work available from a system operating between specified initial and final states and passing along a specified path is obtained when the change takes place reversibly.

We have introduced the connection between reversibility and maximum work for the special case of a perfect gas undergoing expansion. Later (in Section 4.5c) we shall see that it applies to all substances and to all kinds of work.

Example 2.1 Calculating the work of gas production

Calculate the work done when 50 g of iron reacts with hydrochloric acid to produce hydrogen gas in (a) a closed vessel of fixed volume, (b) an open beaker at 25°C.

Method We need to judge the magnitude of the volume change and then to decide how the process occurs. If there is no change in volume, there is no expansion work however the process takes place. If the system expands against a constant external pressure, the work can be calculated from eqn 10. A general feature of processes in which a condensed phase changes into a gas is that the volume of the former may usually be neglected relative to that of the gas it forms.

Answer In (a) the volume cannot change, so no expansion work is done and $w = 0$. In (b) the gas drives back the atmosphere and therefore $w = -p_{ex}\Delta V$. We can neglect the initial volume because the final volume (after the production of gas) is so much larger and $\Delta V = V_f - V_i \approx V_f = nRT/p_{ex}$, where n is the amount of H_2 produced. Therefore,

$$w = -p_{ex}\Delta V \approx -p_{ex} \times \frac{nRT}{p_{ex}} = -nRT$$

Because the reaction is $Fe(s) + 2\,HCl(aq) \rightarrow FeCl_2(aq) + H_2(g)$, we know that 1 mol H_2 is generated when 1 mol Fe is consumed, and n can be taken as the amount of Fe atoms that react. Because the molar mass of Fe is 55.85 g mol⁻¹, it follows that

$$w \approx -\frac{50\ g}{55.85\ g\ mol^{-1}} \times (8.314\,51\ J\,K^{-1}\,mol^{-1}) \times (298.15\ K)$$

$$\approx -2.2\ kJ$$

The system (the reaction mixture) does 2.2 kJ of work driving back the atmosphere.

Comment Note that (for this perfect gas system) the external pressure does not affect the final result: the lower the pressure, the larger the volume occupied by the gas, so the effects cancel.

Self-test 2.2 Calculate the expansion work done when 50 g of water is electrolysed under constant pressure at 25°C. [−10 kJ]

2.4 Heat transactions

In general, the change in internal energy of a system is

$$dU = dq + dw_{exp} + dw_e \tag{2.14}$$

Resistance thermometer

Oxygen inlet

Firing leads

Water

Sample

Heater

2.10 A constant-volume bomb calorimeter. The 'bomb' is the central vessel, which is strong enough to withstand high pressures. The calorimeter (for which the heat capacity must be known) is the entire assembly shown here. To ensure adiabaticity, the calorimeter is immersed in a water bath with a temperature continuously readjusted to that of the calorimeter at each stage of the combustion.

where dw_e is work in addition (e for 'extra') to the expansion work, dw_{exp}. For instance, dw_e might be the electrical work of driving a current through a circuit. A system kept at constant volume can do no expansion work, so $dw_{exp} = 0$. If the system is also incapable of doing any other kind of work (if it is not, for instance, an electrochemical cell connected to an electric motor), then $dw_e = 0$ too. Under these circumstances:

$$dU = dq \qquad \text{(at constant volume, no additional work)} \tag{2.15}$$

We express this relation by writing $dU = dq_V$, where the subscript implies a change at constant volume. For a measurable change,

$$\Delta U = q_V \tag{2.16}$$

It follows that, by measuring the energy supplied to a constant-volume system as heat ($q > 0$) or obtained from it as heat ($q < 0$) when it undergoes a change of state, we are in fact measuring the change in its internal energy.

(a) Calorimetry

Calorimetry is the study of heat transfer during physical and chemical processes. A **calorimeter** is a device for measuring the heat transferred. The most common device for measuring ΔU is an **adiabatic bomb calorimeter** (Fig. 2.10). The process we wish to study—which may be a chemical reaction—is initiated inside a constant-volume container, the 'bomb'. The bomb is immersed in a stirred water bath, and the whole device is the calorimeter. The calorimeter is also immersed in an outer water bath. The water in the calorimeter and that of the outer bath are both monitored and adjusted to the same temperature. This arrangement ensures that there is no net loss of heat from the calorimeter to the surroundings (the bath) and hence that the calorimeter is adiabatic.

The change in temperature, ΔT, of the calorimeter is proportional to the heat that the reaction releases or absorbs. Therefore, by measuring ΔT we can determine q_V and hence find ΔU. The conversion of ΔT to q_V is best achieved by calibrating the calorimeter using a process of known energy output and determining the **calorimeter constant**, the constant C in the relation

$$q = C\Delta T \tag{2.17}$$

The calorimeter constant may be measured electrically by passing a current, I, from a source of known potential, $\mathcal{V}$, through a heater for a known period of time, t:

$$q = I\mathcal{V}t \tag{2.18}$$

Alternatively, C may be determined by burning a known mass of substance (benzoic acid is often used) that has a known heat output. With C known, it is simple to interpret an observed temperature rise as a release of heat.

Illustration 2.2

If we pass a current of 10.0 A from a 12 V supply for 300 s, then from eqn 18 the energy supplied as heat is

$$q = (10.0 \text{ A}) \times (12 \text{ V}) \times (300 \text{ s}) = 3.6 \times 10^4 \text{ A V s} = 36 \text{ kJ}$$

because 1 A V s = 1 J. If the observed rise in temperature is 5.5 K, then the calorimeter constant is $C = (36 \text{ kJ})/(5.5 \text{ K}) = 6.5 \text{ kJ K}^{-1}$.

(b) Heat capacity

The internal energy of a substance increases when its temperature is raised. The increase depends on the conditions under which the heating takes place and for the present we shall suppose that the sample is confined to a constant volume. For example, the sample may be a gas in a container of fixed volume. If the internal energy is plotted against temperature, then a curve like that in Fig. 2.11 may be obtained. The slope of the tangent to the curve at any temperature is called the **heat capacity** of the system at that temperature. The **heat capacity at constant volume** is denoted C_V and is defined formally as[3]

$$C_V = \left(\frac{\partial U}{\partial T}\right)_V$$

[2.19]

The notation is that of a 'partial derivative'. A partial derivative is a slope calculated with all variables except one held constant. In this case, the internal energy varies with the temperature and the volume of the sample, but we are interested only in its variation with the temperature, the volume being held constant (Fig. 2.12).

Partial derivatives are reviewed in *Further information 1*.

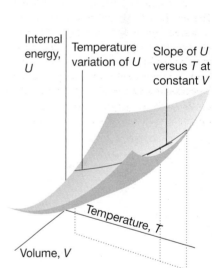

2.11 The internal energy of a system increases as the temperature is raised; this graph shows its variation as the system is heated at constant volume. The slope of the tangent to the curve at any temperature is the heat capacity at constant volume at that temperature. Note that, for the system illustrated, the heat capacity is greater at B than at A.

2.12 The internal energy of a system varies with volume and temperature, perhaps as shown here by the surface. The variation of the internal energy with temperature at one particular constant volume is illustrated by the curve drawn parallel to T. The slope of this curve at any point is the partial derivative $(\partial U/\partial T)_V$.

---

3. If the system can change its composition, it is necessary to distinguish between equilibrium and fixed-composition values of C_V. All applications in this chapter refer to a single substance, so this complication can be ignored.

Illustration 2.3

The heat capacity of a monatomic perfect gas can be calculated by inserting the expression for the internal energy derived in *Molecular interpretation 2.2*. There we saw that $U_m = U_m(0) + \frac{3}{2}RT$, so from eqn 19

$$C_{V,m} = \frac{\partial}{\partial T}(U_m(0) + \frac{3}{2}RT) = \frac{3}{2}R \tag{2.20}$$

The numerical value is 12.47 J K^{-1} mol^{-1}.

Heat capacities are extensive properties: 100 g of water, for instance, has 100 times the heat capacity of 1 g of water (and therefore requires 100 times the heat to bring about the same rise in temperature). The **molar heat capacity at constant volume**, $C_{V,m}$, is the heat capacity per mole of material, and is an intensive property (all molar quantities are intensive). Typical values of $C_{V,m}$ for polyatomic gases are close to 25 J K^{-1} mol^{-1}. For certain applications it is useful to know the **specific heat capacity** (more informally, the 'specific heat') of a substance, which is the heat capacity of the sample divided by the mass, usually in grams. The specific heat capacity of water at room temperature is close to 4 J K^{-1} g^{-1}. In general, heat capacities depend on the temperature and decrease at low temperatures. However, over small ranges of temperature at and above room temperature, the variation is quite small and for approximate calculations heat capacities can be treated as almost independent of temperature.

The heat capacity can be used to relate a change in internal energy to a change in temperature of a constant-volume system. It follows from eqn 19 that

$$dU = C_V\,dT \qquad \text{(at constant volume)} \tag{2.21a}$$

That is, an infinitesimal change in temperature brings about an infinitesimal change in internal energy, and the constant of proportionality is the heat capacity at constant volume. If the heat capacity is independent of temperature over the range of temperatures of interest, a measurable change of temperature, ΔT, brings about a measurable increase in internal energy, ΔU, where

$$\Delta U = C_V\Delta T \qquad \text{(at constant volume)} \tag{2.21b}$$

Because a change in internal energy can be identified with the heat supplied at constant volume (eqn 16), the last equation can be written

$$q_V = C_V\Delta T \tag{2.22}$$

This relation provides a simple way of measuring the heat capacity of a sample: a measured quantity of heat is supplied to the sample (electrically, for example), and the resulting increase in temperature is monitored. The ratio of the heat supplied to the temperature rise it causes is the heat capacity of the sample.

A large heat capacity implies that, for a given quantity of heat, there will be only a small increase in temperature (the sample has a large capacity for heat). An infinite heat capacity implies that there will be no increase in temperature however much heat is supplied. At a phase transition, such as at the boiling point of water, the temperature of a substance does not rise as heat is supplied: the energy is used to drive the endothermic phase transition, in this case to vaporize the water, rather than to increase its temperature. Therefore, at the temperature of a phase transition, the heat capacity of a sample is infinite. The properties of heat capacities close to phase transitions are treated more fully in Section 6.7.

2.13 When a system is subjected to constant pressure and is free to change its volume, some of the energy supplied as heat may escape back into the surroundings as work. In such a case, the change in internal energy is smaller than the energy supplied as heat.

This conclusion is the basis of the general 'product rule' for the differentiation of the product of two functions f and g: $d(fg) = f\,dg + g\,df$.

2.5 Enthalpy

The change in internal energy is not equal to the heat supplied when the system is free to change its volume. Under these circumstances some of the energy supplied as heat to the system is returned to the surroundings as expansion work (Fig. 2.13), so dU is less than dq. However, we shall now show that in this case the heat supplied at constant pressure is equal to the change in another thermodynamic property of the system, the enthalpy.

(a) The definition of enthalpy

The **enthalpy**, H, is defined as

$$H = U + pV \tag{2.23}$$

where p is the pressure of the system and V is its volume. Because U, p, and V are all state functions, the enthalpy is a state function too. As is true of any state function, the change in enthalpy, ΔH, between any pair of initial and final states is independent of the path between them.

Although the definition of enthalpy may appear arbitrary, it has important implications for thermochemisty. For instance, we show in the following *Justification* that eqn 23 implies that *the change in enthalpy is equal to the heat supplied at constant pressure* (so long as the system does no additional work):

$$dH = dq \quad \text{(at constant pressure, no additional work)} \tag{2.24a}$$

For a measurable change,

$$\Delta H = q_p \tag{2.24b}$$

Justification 2.1

For a general infinitesimal change in the state of the system, U changes to $U + dU$, p changes to $p + dp$, and V changes to $V + dV$, so from the definition in eqn 23, H changes from $U + pV$ to

$$H + dH = (U + dU) + (p + dp)(V + dV) = U + dU + pV + p\,dV + V\,dp + dp\,dV$$

The last term is the product of two infinitesimally small quantities and can therefore be neglected. As a result, after recognizing $U + pV = H$ on the right, we find that H changes to

$$H + dH = H + dU + p\,dV + V\,dp$$

and hence that

$$dH = dU + p\,dV + V\,dp$$

If we now substitute $dU = dq + dw$ into this expression, we get

$$dH = dq + dw + p\,dV + V\,dp$$

If the system is in mechanical equilibrium with its surroundings at a pressure p and does only expansion work, we can write $dw = -p\,dV$ and obtain

$$dH = dq + V\,dp$$

Now we impose the condition that the heating occurs at constant pressure by writing $dp = 0$. Then

$$dH = dq \quad \text{(at constant pressure, no additional work)}$$

as in eqn 24a.

Gas, vapour

Oxygen

Products

2.14 An adiabatic flame calorimeter consists of this component immersed in a stirred water bath. Combustion occurs as a known amount of reactant is passed through to fuel the flame, and the rise of temperature is monitored.

The result expressed in eqn 24 states that, when a system is subjected to a constant pressure, and only expansion work can occur, the change in enthalpy is equal to the energy supplied as heat. For example, if we supply 36 kJ of energy through an electric heater immersed in an open beaker of water, then the enthalpy of the water increases by 36 kJ and we write $\Delta H = +36$ kJ.

(b) The measurement of an enthalpy change

An enthalpy change can be measured calorimetrically by monitoring the temperature change that accompanies a physical or chemical change occurring at constant pressure. A calorimeter for studying processes at constant pressure is called an **isobaric calorimeter**. A simple example is a thermally insulated vessel open to the atmosphere: the heat released in the reaction is monitored by measuring the change in temperature of the contents. For a combustion reaction an **adiabatic flame calorimeter** may be used to measure ΔT when a given amount of substance burns in a supply of oxygen (Fig. 2.14). Another route to ΔH is to measure the internal energy change using a bomb calorimeter, and then to convert ΔU to ΔH. Because solids and liquids have small molar volumes, for them pV_m is so small that the molar enthalpy and molar internal energy are almost identical ($H_m = U_m + pV_m \approx U_m$). Consequently, if a process involves only solids or liquids, the values of ΔH and ΔU are almost identical. Physically, such processes are accompanied by a very small change in volume, the system does negligible work on the surroundings when the process occurs, so the energy supplied as heat stays entirely within the system. The most sophisticated way, however, is to use a **differential scanning calorimeter** (DSC), as explained in Box 2.1.

Box 2.1 **Differential scanning calorimetry**

A *differential scanning calorimeter* (DSC) measures the heat transferred to or from a sample at constant pressure during a physical or chemical change. The term 'differential' refers to the fact that the behaviour of the sample is compared to that of a reference material that does not undergo a physical or chemical change during the analysis. The term 'scanning' refers to the fact that the temperatures of the sample and reference material are increased, or scanned, systematically during the analysis.

A DSC consists of two small compartments that are heated electrically at a constant rate. The temperature, T, at time t during a linear scan is

$$T = T_0 + \alpha t$$

where T_0 is the initial temperature and α is the temperature scan rate (in kelvins per second, K s^{-1}). A computer controls the electrical power output in order to maintain the same temperature in the sample and reference compartments throughout the analysis (see the first illustration).

The temperature of the sample changes significantly relative to that of the reference material if a chemical or physical

Thermocouples

Sample   Reference

Heaters

A differential scanning calorimeter. The sample and a reference material are heated in separate but identical metal heat sinks. The output is the difference in power needed to maintain the heat sinks at equal temperatures as the temperature rises.

process involving heat transfer occurs in the sample during the scan. To maintain the same temperature in both compartments, excess heat is transferred to or from the sample during the process. For example, an endothermic process lowers the temperature of the sample relative to that of the reference and, as a result, the sample must be heated more strongly than the reference in order to maintain equal temperatures. It is customary to interpret the excess heat transferred to the sample, $q_{p,ex}$, in terms of an apparent change in the heat capacity at constant pressure of the sample, C_p, during the temperature scan. We shall see that this interpretation leads to a simple calculation of the enthalpy change associated with the process that occurs in the sample.

If no physical or chemical change occurs in the sample at temperature T, we use eqn 29 to write the heat transferred to the sample as $q_p = C_p \Delta T$, where $\Delta T = T - T_0$ and we have assumed that C_p is independent of temperature. The chemical or physical process requires the transfer of $q_p + q_{p,ex}$ to attain the same change in temperature of the sample. If we write the heat capacity of the sample as $C_p + C_{p,ex}$, it follows that

$$q_p + q_{p,ex} = (C_p + C_{p,ex})\Delta T$$

It follows that

$$C_{p,ex} = \frac{q_{p,ex}}{\Delta T} = \frac{q_{p,ex}}{\alpha t} = \frac{P_{ex}}{\alpha}$$

where $P_{ex} = q_{p,ex}/t$ is the excess electrical power necessary to equalize the temperature of the sample and reference compartments.

A DSC trace, also called a *thermogram*, consists of a plot of P_{ex} or $C_{p,ex}$ versus T (see the second illustration). Broad peaks in the thermogram indicate processes requiring heat transfer. From eqn 27, the enthalpy change associated with the process is

$$\Delta H = \int_{T_1}^{T_2} C_{p,ex}\, dT$$

where T_1 and T_2 are, respectively, the temperatures at which the process begins and ends. This relation shows that the enthalpy change is then the area under the curve of $C_{p,ex}$ versus T. With a DSC, enthalpy changes may be determined in samples of masses as low as 0.5 mg, which is a significant advantage over bomb or flame calorimeters, which require several grams of material.

Differential scanning calorimetry is used in the chemical industry to characterize polymers and in the biochemistry laboratory to assess the stability of proteins, nucleic acids, and membranes. Large molecules, such as synthetic or biological polymers, attain complex three-dimensional structures due to intra- and inter-molecular interactions, such as hydrogen bonding and hydrophobic interactions (Chapter 22). Disruption of these interactions is an endothermic process

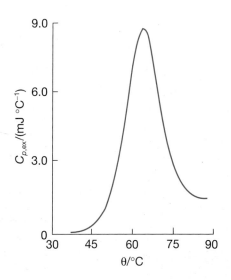

A thermogram for the protein ubiquitin at pH = 2.45. The protein retains its native structure up to about 45°C and then undergoes an endothermic conformational change. (Adapted from B. Chowdhry and S. LeHarne, *J. Chem. Educ.* **74**, 236 (1997).)

that can be studied with a DSC. For example, the thermogram shown in the illustration indicates that the protein ubiquitin retains its native structure up to about 45°C. At higher temperatures, the protein undergoes an endothermic conformational change that results in the loss of its three-dimensional structure. The same principles also apply to the study of structural integrity and stability of synthetic polymers, such as plastics.

Discussion questions

1. In many experimental thermograms, such as that shown in the illustration, the baseline below T_1 is at a different level from that above T_2. Explain this observation.

2. You have at your disposal a sample of pure polymer P and a sample of P that has just been synthesized in a large chemical reactor and that may contain impurities. Describe how you would use differential scanning calorimetry to determine the mole percentage composition of P in the allegedly impure sample.

Further reading

B. Chowdhry and S. LeHarne, Simulation and analysis of differential scanning calorimetry output: Protein unfolding studies. *J. Chem. Educ.* **74**, 236 (1997).

D.A. Skoog, F.J. Holler, and T.A. Nieman, *Principles of instrumental analysis.* Harcourt, Philadelphia (1998).

Example 2.2 Relating ΔH and ΔU

The internal energy change when 1.0 mol $CaCO_3$ in the form of calcite converts to aragonite is +0.21 kJ. Calculate the difference between the enthalpy change and the change in internal energy when the pressure is 1.0 bar given that the densities of the solids are 2.71 g cm^{-3} and 2.93 g cm^{-3}, respectively.

Method The starting point for the calculation is the relation between the enthalpy of a substance and its internal energy (eqn 23). The difference between the two quantities can be expressed in terms of the pressure and the difference of their molar volumes, and the latter can be calculated from their molar masses, M, and their mass densities, ρ, by using $\rho = M/V_m$.

Answer The change in enthalpy when the transition occurs is

$$\Delta H = H(\text{aragonite}) - H(\text{calcite})$$
$$= \{U(a) + pV(a)\} - \{U(c) + pV(c)\}$$
$$= \Delta U + p\{V(a) - V(c)\} = \Delta U + p\Delta V$$

The volume of 1.0 mol $CaCO_3$ (100 g) as aragonite is 34 cm^3, and that of 1.0 mol $CaCO_3$ as calcite is 37 cm^3. Therefore,

$$p\Delta V = (1.0 \times 10^5 \text{ Pa}) \times (34 - 37) \times 10^{-6} \text{ m}^3 = -0.3 \text{ J}$$

(because 1 Pa m^3 = 1 J). Hence,

$$\Delta H - \Delta U = -0.3 \text{ J}$$

which is only 0.1 per cent of the value of ΔU.

Comment It is usually justifiable to ignore the difference between the enthalpy and internal energy of condensed phases, except at very high pressures, when pV is no longer negligible.

Self-test 2.3 Calculate the difference between ΔH and ΔU when 1.0 mol of grey tin (density 5.75 g cm^{-3}) changes to white tin (density 7.31 g cm^{-3}) at 10.0 bar. At 298 K, $\Delta H = +2.1$ kJ.   [$\Delta H - \Delta U = -4.4$ J]

The enthalpy of a perfect gas is related to its internal energy by using $pV = nRT$ in the definition of H:

$$H = U + pV = U + nRT \tag{2.25}°$$

This relation implies that the change of enthalpy in a reaction that produces or consumes gas is

$$\Delta H = \Delta U + \Delta n_g RT \tag{2.26}°$$

where Δn_g is the change in the amount of gas molecules in the reaction.

Illustration 2.4

In the reaction $2 H_2(g) + O_2(g) \rightarrow 2 H_2O(l)$, 3 mol of gas-phase molecules is replaced by 2 mol of liquid-phase molecules so $\Delta n_g = -3$ mol. Therefore, at 298 K the enthalpy and internal energy changes taking place in the system are related by

$$\Delta H - \Delta U = (-3 \text{ mol}) \times RT \approx -7.4 \text{ kJ}$$

Note that the difference is expressed in kilojoules, not joules as in Example 2.2. The enthalpy change is smaller than the change in internal energy because, although heat escapes from the system when the reaction occurs, the system contracts when the liquid is formed, so energy is restored to it from the surroundings.

Example 2.3 Calculating a change in enthalpy

Water is heated to boiling under a pressure of 1.0 atm. When an electric current of 0.50 A from a 12 V supply is passed for 300 s through a resistance in thermal contact with boiling water, it is found that 0.798 g of water is vaporized. Calculate the molar internal energy and enthalpy changes at the boiling point (373.15 K).

Method Because the vaporization occurs at constant pressure, the enthalpy change is equal to the heat supplied by the heater. Therefore, the strategy is to calculate the heat supplied (from $q = I \mathcal{V} t$), express that as an enthalpy change, and then convert the result to a molar enthalpy change by division by the amount of H_2O molecules vaporized. To convert from enthalpy change to internal energy change, we assume that the vapour is a perfect gas and use eqn 26.

Answer The enthalpy change is

$$\Delta H = q_p = (0.50 \text{ A}) \times (12 \text{ V}) \times (300 \text{ s}) = +(0.50 \times 12 \times 300) \text{ J}$$

Here we have used 1 A V s = 1 J (see inside front cover). Because 0.798 g of water is $(0.798 \text{ g})/(18.02 \text{ g mol}^{-1}) = (0.798/18.02)$ mol H_2O, the enthalpy of vaporization per mole of H_2O is

$$\Delta H_m = +\frac{0.50 \times 12 \times 300 \text{ J}}{(0.798/18.02) \text{ mol}} = +41 \text{ kJ mol}^{-1}$$

In the process $H_2O(l) \rightarrow H_2O(g)$ the change in the amount of gas molecules is $\Delta n_g = +1$ mol, so

$$\Delta U_m = \Delta H_m - RT = +38 \text{ kJ mol}^{-1}$$

Comment The plus sign is added to positive quantities to emphasize that they represent an increase in internal energy or enthalpy. Notice that the internal energy change is smaller than the enthalpy change because energy has been used to drive back the surrounding atmosphere to make room for the vapour.

Self-test 2.4 The molar enthalpy of vaporization of benzene at its boiling point (353.25 K) is 30.8 kJ mol^{-1}. What is the molar internal energy change? For how long would the same 12 V source need to supply a 0.50 A current in order to vaporize a 10 g sample? [+27.9 kJ mol^{-1}, 660 s]

Enthalpy and internal energy changes may also be measured by noncalorimetric methods (see Chapters 9 and 10).

(c) The variation of enthalpy with temperature

The enthalpy of a substance increases as its temperature is raised. The relation between the increase in enthalpy and the increase in temperature depends on the

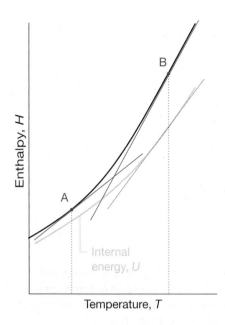

2.15 The slope of the tangent to a curve of the enthalpy of a system subjected to a constant pressure plotted against temperature is the constant-pressure heat capacity. The slope may change with temperature, in which case the heat capacity varies with temperature. Thus, the heat capacities at A and B are different. For gases, at a given temperature the slope of enthalpy versus temperature is steeper than that of internal energy versus temperature, and $C_{p,\mathrm{m}}$ is larger than $C_{V,\mathrm{m}}$.

conditions (for example, constant pressure or constant volume). The most important condition is constant pressure, and the slope of the tangent to a plot of enthalpy against temperature at constant pressure is called the **heat capacity at constant pressure**, C_p, at a given temperature (Fig. 2.15). More formally:

$$C_p = \left(\frac{\partial H}{\partial T}\right)_p \qquad [2.27]$$

The heat capacity at constant pressure is the analogue of the heat capacity at constant volume, and is an extensive property.[4] The **molar heat capacity at constant pressure**, $C_{p,\mathrm{m}}$, is the heat capacity per mole of material; it is an intensive property.

The heat capacity at constant pressure is used to relate the change in enthalpy to a change in temperature. For infinitesimal changes of temperature,

$$dH = C_p\, dT \qquad \text{(at constant pressure)} \qquad (2.28a)$$

If the heat capacity is constant over the range of temperatures of interest, then for a measurable increase in temperature

$$\Delta H = C_p \Delta T \qquad \text{(at constant pressure)} \qquad (2.28b)$$

Because an increase in enthalpy can be equated with the heat supplied at constant pressure, the practical form of the latter equation is

$$q_p = C_p \Delta T \qquad (2.29)$$

This expression shows us how to measure the heat capacity of a sample: a measured quantity of heat is supplied under conditions of constant pressure (as in a sample exposed to the atmosphere and free to expand), and the temperature rise is monitored.

The variation of heat capacity with temperature can sometimes be ignored if the temperature range is small; this approximation is highly accurate for a monatomic perfect gas (one of the noble gases). However, when it is necessary to take the variation into account, a convenient approximate empirical expression is

$$C_{p,\mathrm{m}} = a + bT + \frac{c}{T^2} \qquad (2.30)$$

The empirical parameters a, b, and c are independent of temperature (Table 2.2).

Table 2.2* Temperature variation of molar heat capacities, $C_{p,\mathrm{m}}/(\mathrm{J\ K^{-1}\ mol^{-1}}) = a + bT + c/T^2$

	a	$b/(10^{-3}\ \mathrm{K^{-1}})$	$c/(10^5\ \mathrm{K^2})$
C(s, graphite)	16.86	4.77	−8.54
CO_2(g)	44.22	8.79	−8.62
H_2O(l)	75.29	0	0
N_2(g)	28.58	3.77	−0.50

* More values are given in the *Data section* at the end.

---

4. As in the case of C_V, if the system can change its composition it is necessary to distinguish between equilibrium and fixed-composition values. All applications in this chapter refer to pure substances, so this complication can be ignored.

Example 2.4 Evaluating an increase in enthalpy with temperature

What is the change in molar enthalpy of N_2 when it is heated from 25°C to 100°C? Use the heat capacity information in Table 2.2.

Method The heat capacity of N_2 changes with temperature, so we cannot use eqn 28b (which assumes that the heat capacity of the substance is constant). Therefore, we must use eqn 28a, substitute eqn 30 for the temperature dependence of the heat capacity, and integrate the resulting expression from 25°C to 100°C.

Answer For convenience, we denote the two temperatures T_1 (298 K) and T_2 (373 K). The integrals we require are

$$\int_{H(T_1)}^{H(T_2)} dH = \int_{T_1}^{T_2} \left(a + bT + \frac{c}{T^2} \right) dT$$

Notice how the limits of integration correspond on each side of the equation: the integration over H on the left ranges from $H(T_1)$, the value of H at T_1, up to $H(T_2)$, the value of H at T_2, while on the right the integration over the temperature ranges from T_1 to T_2. Now we use the integrals

$$\int dx = x + \text{constant} \qquad \int x\, dx = \tfrac{1}{2}x^2 + \text{constant} \qquad \int \frac{dx}{x^2} = -\frac{1}{x} + \text{constant}$$

to obtain

$$H(T_2) - H(T_1) = a(T_2 - T_1) + \tfrac{1}{2}b(T_2^2 - T_1^2) - c\left(\frac{1}{T_2} - \frac{1}{T_1} \right)$$

Substitution of the numerical data results in

$$H(373\ \text{K}) = H(298\ \text{K}) + 2.20\ \text{kJ mol}^{-1}$$

If we had assumed a constant heat capacity of 29.14 J K^{-1} mol^{-1} (the value given by eqn 30 at 25°C), we would have found that the two enthalpies differed by 2.19 kJ mol^{-1}.

Self-test 2.5 At very low temperatures the heat capacity of a solid is proportional to T^3, and we can write $C_p = aT^3$. What is the change in enthalpy of such a substance when it is heated from 0 to a temperature T (with T close to 0)?

$[\Delta H = \tfrac{1}{4}aT^4]$

Integrals commonly encountered in physical chemistry are listed inside the front cover.

(d) The relation between heat capacities

Most systems expand when heated at constant pressure. Such systems do work on the surroundings and therefore some of the energy supplied to them as heat escapes back to the surroundings. As a result, the temperature of the system rises less than when the heating occurs at constant volume. A smaller increase in temperature implies a larger heat capacity, so we conclude that in most cases the heat capacity at constant pressure of a system is larger than its heat capacity at constant volume. There is a simple relation between the two heat capacities of a perfect gas:

$$C_p - C_V = nR \tag{2.31}°$$

(This relation is derived in Section 3.5.) It follows that the molar heat capacity of a perfect gas is about 8 J K^{-1} mol^{-1} larger at constant pressure than at constant

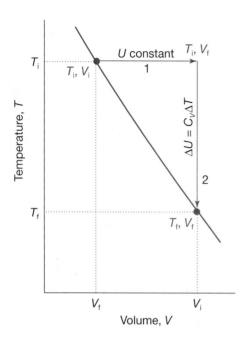

2.16 To achieve a change of state from one temperature and volume to another temperature and volume, we may consider the overall change as composed of two steps. In the first step, the system expands at constant temperature; there is no change in internal energy if the system consists of a perfect gas. In the second step, the temperature of the system is reduced at constant volume. The overall change in internal energy is the sum of the changes for the two steps.

volume. Because the heat capacity at constant volume of a monatomic gas is about $12 \, J \, K^{-1} \, mol^{-1}$, the difference is highly significant and must be taken into account.

2.6 Adiabatic changes

We are now equipped to deal with the changes that occur when a perfect gas expands adiabatically. A decrease in temperature should be expected: because work is done, the internal energy falls, and therefore the temperature of the working gas also falls. In molecular terms, the kinetic energy of the molecules falls as work is done, so their average speed decreases, and hence the temperature falls.

(a) The work of adiabatic change

The change in internal energy of a perfect gas when the temperature is changed from T_i to T_f and the volume is changed from V_i to V_f can be expressed as the sum of two steps (Fig. 2.16). In the first step, only the volume changes and the temperature is held constant at its initial value. However, because the internal energy of a perfect gas is independent of the volume the molecules occupy, the overall change in internal energy arises solely from the second step, the change in temperature at constant volume. Provided the heat capacity is independent of temperature, this change is

$$\Delta U = C_V(T_f - T_i) = C_V \Delta T \tag{2.32}$$

Because the expansion is adiabatic, we know that $q = 0$; because $\Delta U = q + w$, it then follows that $\Delta U = w_{ad}$. The subscript 'ad' denotes an adiabatic process. Therefore, by equating the two values we have obtained for ΔU, we obtain

$$w_{ad} = C_V \Delta T \tag{2.33}°$$

That is, the work done during an adiabatic expansion of a perfect gas is proportional to the temperature difference between the initial and final states. That is exactly what we expect on molecular grounds, because the mean kinetic energy is proportional to T, so a change in internal energy arising from temperature alone is also expected to be proportional to ΔT.

To use eqn 33, we need to relate the change in temperature to the change in volume (which we know). The most important type of adiabatic expansion (and the only kind that we need for later) is reversible adiabatic expansion, in which the external pressure is matched to the internal pressure throughout. For adiabatic, reversible expansion the initial and final temperatures are related by

$$V_f T_f^c = V_i T_i^c \qquad c = C_{V,m}/R \tag{2.34}°_{rev}$$

This expression is derived in the following *Justification* and the temperature dependence it implies is shown in Fig. 2.17. All we need do at this stage to find the work done by the system is to solve this relation for T_f,

$$T_f = T_i \left(\frac{V_i}{V_f} \right)^{1/c} \tag{2.35}°_{rev}$$

calculate ΔT, and substitute into eqn 33.

Justification 2.2

Consider a stage in the expansion when the pressure inside and out is p (the expansion is reversible, so the internal and external pressures are equal at all stages). The work done when the gas expands by dV is

2.17 The variation of temperature as a perfect gas is expanded reversibly and adiabatically. The curves are labelled with different values of $c = C_{V,m}/R$. Note that the temperature falls most steeply for gases with low molar heat capacity.

For the final stages of this derivation we use the relations $a \ln x = \ln x^a$ and $-\ln(x/y) = \ln(y/x)$.

$$dw = -p\, dV$$

However, for a perfect gas,

$$dU = C_V\, dT$$

(by the same argument as in the text for a macroscopic change). Therefore, because for an adiabatic change ($dq = 0$)

$$dU = dw + dq = dw$$

we can equate these two expressions for dU and write

$$C_V\, dT = -p\, dV$$

We are dealing with a perfect gas, so we can replace p by nRT/V and obtain

$$\frac{C_V\, dT}{T} = -\frac{nR\, dV}{V}$$

To integrate this expression we note that T is equal to T_i when V is equal to V_i, and is equal to T_f when V is equal to V_f at the end of the expansion. Therefore,

$$C_V \int_{T_i}^{T_f} \frac{dT}{T} = -nR \int_{V_i}^{V_f} \frac{dV}{V}$$

(We are taking C_V to be independent of temperature.) Then, because $\int dx/x = \ln x + \text{constant}$, we obtain

$$C_V \ln \frac{T_f}{T_i} = -nR \ln \frac{V_f}{V_i}$$

With $c = C_V/nR$ we obtain

$$\ln \left(\frac{T_f}{T_i} \right)^c = \ln \left(\frac{V_i}{V_f} \right)$$

which implies eqn 34.

Illustration 2.5

Consider the adiabatic, reversible expansion of 0.020 mol Ar, initially at 25°C, from 0.50 L to 1.00 L. The molar heat capacity of argon at constant volume is 12.48 J K^{-1} mol^{-1}, so $c = 1.501$. Therefore, from eqn 35,

$$T_f = (298\ \text{K}) \times \left(\frac{0.50\ \text{L}}{1.00\ \text{L}} \right)^{1/1.501} = 188\ \text{K}$$

It follows that $\Delta T = -110$ K, and therefore, from eqn 33, that

$$w_{ad} = C_V \Delta T = (0.020\ \text{mol}) \times (12.48\ \text{J K}^{-1}\ \text{mol}^{-1}) \times (-110\ \text{K}) = -27\ \text{J}$$

Note that temperature change is independent of the mass of gas but the work is not.

Self-test 2.6 Calculate the final temperature, the work done, and the change of internal energy when ammonia is used in a reversible adiabatic expansion from 0.50 L to 2.00 L, the other initial conditions being the same.

[195 K, −56 J, −56 J]

(b) Heat capacity ratio and adiabats

Now we consider the change in pressure that results from an adiabatic, reversible expansion of a perfect gas. We show in the *Justification* below that

$$pV^\gamma = \text{constant} \qquad (2.36)^\circ_{\text{rev}}$$

where the **heat capacity ratio**, γ, of a substance is defined as

$$\gamma = \frac{C_{p,m}}{C_{V,m}} \qquad [2.37]$$

Because the heat capacity at constant pressure is greater than the heat capacity at constant volume, $\gamma > 1$. For a perfect gas, it follows from eqn 31 that

$$\gamma = \frac{C_{V,m} + R}{C_{V,m}} = 1 + \frac{R}{C_{V,m}} \qquad (2.38)^\circ$$

For a monatomic perfect gas, $C_{V,m} = \frac{3}{2}R$ (see *Illustration 2.3*), so $\gamma = \frac{5}{3}$. For a gas of non-linear polyatomic molecules (which can rotate as well as translate), $C_{V,m} = 3R$, so $\gamma = \frac{4}{3}$.

Justification 2.3

The initial and final states of a perfect gas satisfy the perfect gas law regardless of how the change of state takes place, so we can use $pV = nRT$ to write

$$\frac{p_i V_i}{p_f V_f} = \frac{T_i}{T_f}$$

However, we know from (eqn 35) that for a reversible adiabatic change the temperature changes so as to satisfy

$$\frac{T_i}{T_f} = \left(\frac{V_f}{V_i}\right)^{1/c} = \left(\frac{V_f}{V_i}\right)^{\gamma-1}$$

where we use the definition of the heat capacity ratio (eqn 37). Then we combine the two expressions, to obtain

$$p_i V_i^\gamma = p_f V_f^\gamma$$

It follows that $pV^\gamma = \text{constant}$, as used in the text.

The curves of pressure versus volume for adiabatic change are known as **adiabats**, and one for a reversible path is illustrated in Fig. 2.18. Because $\gamma > 1$, an adiabat falls more steeply ($p \propto 1/V^\gamma$) than the corresponding isotherm ($p \propto 1/V$). The physical reason for the difference is that, in an isothermal expansion, energy flows into the system as heat and maintains the temperature; as a result, the pressure does not fall as much as in an adiabatic expansion.

Isotherm, $p \propto 1/V$
Adiabat, $p \propto 1/V^\gamma$

Pressure, p — Volume, V — Temperature, T

(a)

Isotherm
Adiabat
V_i V_f
(b) Volume, V

2.18 An adiabat depicts the variation of pressure with volume when a gas expands adiabatically. (a) An adiabat for a perfect gas undergoing reversible expansion. (b) Note that the pressure declines more steeply for an adiabat than it does for an isotherm because the temperature decreases in the former.

Illustration 2.6

When a sample of argon (for which $\gamma = \frac{5}{3}$) at 100 kPa expands reversibly and adiabatically to twice its initial volume the final pressure will be

$$p_f = p_i\left(\frac{V_i}{V_f}\right)^\gamma = (100 \text{ kPa}) \times (\tfrac{1}{2})^{5/3} = 31.5 \text{ kPa}$$

For an isothermal doubling of volume, the final pressure would be 50 kPa.

Thermochemistry

The study of the heat produced or required by chemical reactions is called **thermochemistry**. Thermochemistry is a branch of thermodynamics because a reaction vessel and its contents form a system, and chemical reactions result in the exchange of energy between the system and the surroundings. Thus we can use calorimetry to measure the heat produced or absorbed by a reaction, and can identify q with a change in internal energy (if the reaction occurs at constant volume) or a change in enthalpy (if the reaction occurs at constant pressure). Conversely, if we know the ΔU or ΔH for a reaction, we can predict the heat the reaction can produce.

We have already remarked that a process that releases heat is classified as exothermic and one that absorbs heat is classified as endothermic. Because the release of heat signifies a decrease in the enthalpy of a system (at constant pressure), we can now see that an exothermic process at constant pressure is one for which $\Delta H < 0$. Conversely, because the absorption of heat results in an increase in enthalpy, an endothermic process at constant pressure has $\Delta H > 0$.

2.7 Standard enthalpy changes

Changes in enthalpy are normally reported for processes taking place under a set of standard conditions. In most of our discussions we shall consider the **standard enthalpy change**, $\Delta H^{\ominus}$, the change in enthalpy for a process in which the initial and final substances are in their standard states:

> **The standard state of a substance at a specified temperature is its pure form at 1 bar.**[5]

For example, the standard state of liquid ethanol at 298 K is pure liquid ethanol at 298 K and 1 bar; the standard state of solid iron at 500 K is pure iron at 500 K and 1 bar. The standard enthalpy change for a reaction or a physical process is the difference between the products in their standard states and the reactants in their standard states, all at the same specified temperature.

As an example of a standard enthalpy change, the *standard enthalpy of vaporization*, $\Delta_{vap}H^{\ominus}$, is the enthalpy change per mole when a pure liquid at 1 bar vaporizes to a gas at 1 bar, as in

$$H_2O(l) \longrightarrow H_2O(g) \qquad \Delta_{vap}H^{\ominus}(373 \text{ K}) = +40.66 \text{ kJ mol}^{-1}$$

As implied by the examples, standard enthalpies may be reported for any temperature. However, the conventional temperature for reporting thermodynamic data is 298.15 K (corresponding to 25.00°C). Unless otherwise mentioned, all thermodynamic data in this text will refer to this conventional temperature.

(a) Enthalpies of physical change

The standard enthalpy change that accompanies a change of physical state is called the **standard enthalpy of transition** and is denoted $\Delta_{trs}H^{\ominus}$ (Table 2.3). The **standard enthalpy of vaporization**, $\Delta_{vap}H^{\ominus}$, is one example. Another is the **standard enthalpy of fusion**, $\Delta_{fus}H^{\ominus}$, the standard enthalpy change accompanying the conversion of a solid to a liquid, as in

$$H_2O(s) \longrightarrow H_2O(l) \qquad \Delta_{fus}H^{\ominus}(273 \text{ K}) = +6.01 \text{ kJ mol}^{-1}$$

5. The definition of standard state is more sophisticated for a real gas (Section 5.5) and for solutions (Section 10.2).

Table 2.3* Standard enthalpies of fusion and vaporization at the transition temperature, $\Delta_{trs}H^{\ominus}/(\text{kJ mol}^{-1})$

	T_f/K	Fusion	T_b/K	Vaporization
Ar	83.81	1.188	87.29	6.506
C_6H_6	278.61	10.59	353.2	30.8
H_2O	273.15	6.008	373.15	40.656
				44.016 at 298 K
He	3.5	0.021	4.22	0.084

* More values are given in the *Data section*.

As in this case, it is sometimes convenient to know the standard enthalpy change at the transition temperature as well as at the conventional temperature.

Because enthalpy is a state function, a change in enthalpy is independent of the path between the two states. This feature is of great importance in thermochemistry, for it implies that the same value of $\Delta H^{\ominus}$ will be obtained however the change is brought about (so long as the initial and final states are the same). For example, we can picture the conversion of a solid to a vapour either as occurring by sublimation (the direct conversion from solid to vapour),

$$H_2O(s) \longrightarrow H_2O(g) \qquad \Delta_{sub}H^{\ominus}$$

or as occurring in two steps, first fusion (melting) and then vaporization of the resulting liquid:

$$H_2O(s) \longrightarrow H_2O(l) \qquad \Delta_{fus}H^{\ominus}$$
$$H_2O(l) \longrightarrow H_2O(g) \qquad \Delta_{vap}H^{\ominus}$$
$$\text{Overall: } H_2O(s) \longrightarrow H_2O(g) \qquad \Delta_{fus}H^{\ominus} + \Delta_{vap}H^{\ominus}$$

Because the overall result of the indirect path is the same as that of the direct path, the overall enthalpy change is the same in each case (**1**), and we can conclude that (for processes occurring at the same temperature)

$$\Delta_{sub}H^{\ominus} = \Delta_{fus}H^{\ominus} + \Delta_{vap}H^{\ominus} \qquad (2.39)$$

An immediate conclusion is that, because all enthalpies of fusion are positive, the enthalpy of sublimation of a substance is greater than its enthalpy of vaporization (at a given temperature).

Another consequence of H being a state function is that the standard enthalpy changes of a forward process and its reverse must differ only in sign (**2**):

$$\Delta H^{\ominus}(A \to B) = -\Delta H^{\ominus}(B \to A) \qquad (2.40)$$

For instance, because the enthalpy of vaporization of water is +44 kJ mol⁻¹ at 298 K, its enthalpy of condensation at that temperature is −44 kJ mol⁻¹.

The different types of enthalpies encountered in thermochemistry are summarized in Table 2.4. We shall meet them again in various locations throughout the text.

(b) Enthalpies of chemical change

Now we consider enthalpy changes that accompany chemical reactions. There are two ways of reporting the change in enthalpy that accompanies a chemical reaction.

1

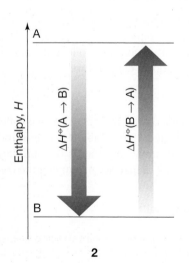

2

Table 2.4 Enthalpies of transition

Transition	Process	Symbol*
Transition	Phase $\alpha \longrightarrow$ phase β	$\Delta_{trs}H$
Fusion	$s \longrightarrow l$	$\Delta_{fus}H$
Vaporization	$l \longrightarrow g$	$\Delta_{vap}H$
Sublimation	$s \longrightarrow g$	$\Delta_{sub}H$
Mixing of fluids	Pure $\longrightarrow$ mixture	$\Delta_{mix}H$
Solution	Solute $\longrightarrow$ solution	$\Delta_{sol}H$
Hydration	$X^{\pm}(g) \longrightarrow X^{\pm}(aq)$	$\Delta_{hyd}H$
Atomization	Species(s, l, g) $\longrightarrow$ atoms(g)	$\Delta_{at}H$
Ionization	$X(g) \longrightarrow X^{+}(g) + e^{-}(g)$	$\Delta_{ion}H$
Electron gain	$X(g) + e^{-}(g) \longrightarrow X^{-}(g)$	$\Delta_{eg}H$
Reaction	Reactants $\longrightarrow$ products	$\Delta_{r}H$
Combustion	Compound(s, l, g) + O_2(g) $\longrightarrow$ CO_2(g), H_2O(l, g)	$\Delta_{c}H$
Formation	Elements $\longrightarrow$ compound	$\Delta_{f}H$
Activation	Reactants $\longrightarrow$ activated complex	$\Delta^{\ddagger}H$

* IUPAC recommendations. In common usage, the transition subscript is often attached to ΔH, as in ΔH_{trs}.

One is to write the **thermochemical equation**, a combination of a chemical equation and the corresponding change in standard enthalpy:

$$CH_4(g) + 2\,O_2(g) \longrightarrow CO_2(g) + 2\,H_2O(l) \qquad \Delta H^{\ominus} = -890 \text{ kJ}$$

$\Delta H^{\ominus}$ is the change in enthalpy when reactants in their standard states change to products in their standard states:

Pure, unmixed reactants in their standard states

$\longrightarrow$ pure, separated products in their standard states

Except in the case of ionic reactions in solution, the enthalpy changes accompanying mixing and separation are insignificant in comparison with the contribution from the reaction itself. For the combustion of methane, the standard value refers to the reaction in which 1 mol CH_4 in the form of pure methane gas at 1 bar reacts completely with 2 mol O_2 in the form of pure oxygen gas at 1 bar to produce 1 mol CO_2 as pure carbon dioxide at 1 bar and 2 mol H_2O as pure liquid water at 1 bar; the numerical value is for the reaction at 298 K.

Alternatively, we write the chemical equation and then report the **standard reaction enthalpy**, $\Delta_r H^{\ominus}$. Thus, for the combustion of methane, we write

$$CH_4(g) + 2\,O_2(g) \longrightarrow CO_2(g) + 2\,H_2O(l) \qquad \Delta_r H^{\ominus} = -890 \text{ kJ mol}^{-1}$$

For the reaction

$$2\,A + B \longrightarrow 3\,C + D$$

the standard reaction enthalpy is

$$\Delta_r H^{\ominus} = \{3H_m^{\ominus}(C) + H_m^{\ominus}(D)\} - \{2H_m^{\ominus}(A) + H_m^{\ominus}(B)\}$$

where $H_m^{\ominus}(J)$ is the standard molar enthalpy of species J at the temperature of interest. Note how the 'per mole' of $\Delta_r H^{\ominus}$ comes directly from the fact that molar enthalpies appear in this expression. We interpret the 'per mole' by noting the stoichiometric coefficients in the chemical equation. In this case 'per mole' in $\Delta_r H^\ominus$ means 'per 2 mole A', 'per mole B', 'per 3 mole C', or 'per mole D'. In general,

$$\Delta_r H^\ominus = \sum_{\text{Products}} v H_m^\ominus - \sum_{\text{Reactants}} v H_m^\ominus \qquad (2.41)$$

where in each case the molar enthalpies of the species are multiplied by their stoichiometric coefficients, v.[6]

Some standard reaction enthalpies have special names and a particular significance. For instance, the **standard enthalpy of combustion**, $\Delta_c H^\ominus$, is the standard reaction enthalpy for the complete oxidation of an organic compound to CO_2 gas and liquid H_2O if the compound contains C, H, and O, and to N_2 gas if N is also present. An example is the combustion of glucose:

$$C_6H_{12}O_6(s) + 6\,O_2(g) \longrightarrow 6\,CO_2(g) + 6\,H_2O(l) \qquad \Delta_c H^\ominus = -2808 \text{ kJ mol}^{-1}$$

The value quoted shows that 2808 kJ of heat is released when 1 mol $C_6H_{12}O_6$ burns under standard conditions (at 298 K). Some further values are listed in Table 2.5. We consider the thermochemistry of fuels and foods in Box 2.2.

(c) Hess's law

Standard enthalpies of individual reactions can be combined to obtain the enthalpy of another reaction. This application of the First Law is called **Hess's law**:

> **The standard enthalpy of an overall reaction is the sum of the standard enthalpies of the individual reactions into which a reaction may be divided.**

The individual steps need not be realizable in practice: they may be hypothetical reactions, the only requirement being that their chemical equations should balance. The thermodynamic basis of the law is the path-independence of the value of $\Delta_r H^\ominus$ and the implication that we may take the specified reactants, pass through any (possibly hypothetical) set of reactions to the specified products, and overall obtain the same change of enthalpy. The importance of Hess's law is that information about a reaction of interest, which may be difficult to determine directly, can be assembled from information on other reactions.

Table 2.5[*] Standard enthalpies of formation and combustion of organic compounds at 298 K

	$\Delta_f H^\ominus$/(kJ mol^{-1})	$\Delta_c H^\ominus$/(kJ mol^{-1})
Benzene, C_6H_6(l)	+49.0	−3268
Ethane, C_2H_6(g)	−84.7	−1560
Glucose, $C_6H_{12}O_6$(s)	−1274	−2808
Methane, CH_4(g)	−74.8	−890
Methanol, CH_3OH(l)	−238.7	−726

[*] More values are given in the *Data section*.

6. In this and similar expressions, all stoichiometric coefficients are positive. For a more sophisticated way of writing eqn 41, see Section 9.2.

Box 2.2 Food and energy reserves

The thermochemical properties of fuels and foods are commonly discussed in terms of their *specific enthalpy*, the magnitude of the enthalpy of combustion per gram of material or the *enthalpy density*, the magnitude of the enthalpy of combustion per litre of material. Thus, if the standard enthalpy of combustion is $\Delta_c H^{\ominus}$ and the molar mass of the compound is M, then the specific enthalpy is $|\Delta_c H^{\ominus}|/M$. The table lists the specific enthalpies of several fuels.

A typical 18–20 year old man requires a daily input of about 12 MJ; a woman of the same age needs about 9 MJ. If the entire consumption were in the form of glucose (which has a specific enthalpy of 16 kJ g^{-1}), that would require the consumption of 750 g of glucose for a man and 560 g for a woman. In fact, digestible carbohydrates have a slightly higher specific enthalpy (17 kJ g^{-1}) than glucose itself, so a carbohydrate diet is slightly less daunting than a pure glucose diet, as well as being more appropriate in the form of fibre, the indigestible cellulose that helps move digestion products through the intestine.

The specific enthalpy of fats, which are long-chain esters like tristearin (beef fat), is much greater than that of carbohydrates, at around 38 kJ g^{-1}, slightly less than the value for the hydrocarbon oils used as fuel (48 kJ g^{-1}). Fats are commonly used as an energy store, to be used only when the more readily accessible carbohydrates have fallen into short supply. In Arctic species, the stored fat also acts as a layer of insulation; in desert species (such as the camel), the fat is also a source of water, one of its oxidation products.

Proteins are also used as a source of energy, but their components, the amino acids, are often too valuable to squander in this way, and are used to construct other proteins instead. When proteins are oxidized (to urea, $CO(NH_2)_2$), the equivalent enthalpy density is comparable to that of carbohydrates.

The heat released by the oxidation of foods needs to be discarded in order to maintain body temperature within its typical range of 35.6–37.8°C. A variety of mechanisms contribute to this aspect of homeostasis. The general uniformity of temperature throughout the body is maintained largely by the flow of blood. When heat needs to be dissipated rapidly,

α-D-glucose

warm blood is allowed to flow through the capillaries of the skin, so producing flushing. Radiation is one means of discarding heat; another is evaporation and the energy demands of the enthalpy of vaporization of water. Evaporation removes about 2.4 kJ per gram of water perspired. When vigorous exercise promotes sweating (through the influence of heat selectors in the hypothalamus), 1–2 L of perspired water can be produced per hour, corresponding to a heat loss of 2.4–5.0 MJ h^{-1}.

Discussion questions

1. An average human produces about 10 MJ of heat each day through metabolic activity. If a human body were an isolated system of mass 65 kg with the heat capacity of water, what temperature rise would the body experience? Human bodies are actually open systems, and the main mechanism of heat loss is through the evaporation of water. What mass of water should be evaporated each day to maintain constant temperature?

2. The mean bond enthalpies of the C—C, C—H, C=O, and O—H bonds are 348, 412, 743, and 463 kJ mol^{-1}, respectively. The combustion of a fuel such as octane is exothermic because relatively weak bonds break to form relatively strong bonds. Use this information to justify why glucose (see the illustration) has a lower specific enthalpy than the lipid decanoic acid even though these compounds have similar molar masses.

Further reading

V. Smil, *Energies: An illustrated guide to the biosphere and civilization.* MIT Press, Cambridge, Mass. (1999).

O.R. Fennema (ed.), *Food chemistry.* Marcel Dekker, New York (1996).

Thermochemical properties of some fuels

Fuel	Combustion equation	$\Delta_c H^{\ominus}/(\text{kJ mol}^{-1})$	Specific enthalpy/(kJ g^{-1})	Enthalpy density/(kJ L^{-1})
Hydrogen	$H_2(g) + \frac{1}{2} O_2(g) \longrightarrow H_2O(l)$	−286	142	13
Methane	$CH_4(g) + 2\,O_2(g) \longrightarrow CO_2(g) + 2\,H_2O(l)$	−890	55	40
Octane	$C_8H_{18}(l) + \frac{25}{2} O_2(g) \longrightarrow 8\,CO_2(g) + 9\,H_2O(l)$	−5471	48	3.8×10^4
Methanol	$CH_3OH(l) + \frac{3}{2} O_2(g) \longrightarrow CO_2(g) + 2\,H_2O(l)$	−726	23	1.8×10^4

Example 2.5 Using Hess's law

The standard reaction enthalpy for the hydrogenation of propene,

$$CH_2{=}CHCH_3(g) + H_2(g) \longrightarrow CH_3CH_2CH_3(g)$$

is $-124\ kJ\ mol^{-1}$. The standard reaction enthalpy for the combustion of propane,

$$CH_3CH_2CH_3(g) + 5\ O_2(g) \longrightarrow 3\ CO_2(g) + 4\ H_2O(l)$$

is $-2220\ kJ\ mol^{-1}$. Calculate the standard enthalpy of combustion of propene.

Method Add or subtract the reactions given, together with any others needed, so as to reproduce the reaction required. Then add or subtract the reaction enthalpies in the same way. Additional data are in Table 2.5.

Answer The combustion reaction we require is

$$C_3H_6(g) + \tfrac{9}{2}O_2(g) \longrightarrow 3\ CO_2(g) + 3\ H_2O(l)$$

This reaction can be recreated from the following sum:

	$\Delta_r H^\ominus/(kJ\ mol^{-1})$
$C_3H_6(g) + H_2(g) \longrightarrow C_3H_8(g)$	-124
$C_3H_8(g) + 5\ O_2(g) \longrightarrow 3\ CO_2(g) + 4\ H_2O(l)$	-2220
$H_2O(l) \longrightarrow H_2(g) + \tfrac{1}{2}O_2(g)$	$+286$
$C_3H_6(g) + \tfrac{9}{2}O_2(g) \longrightarrow 3\ CO_2(g) + 3\ H_2O(l)$	-2058

Comment The skill to develop is the ability to assemble a given thermochemical equation from others.

Self-test 2.7 Calculate the enthalpy of hydrogenation of benzene from its enthalpy of combustion and the enthalpy of combustion of cyclohexane.

$$[-205\ kJ\ mol^{-1}]$$

Table 2.6* Standard enthalpies of formation of inorganic compounds, $\Delta_f H^\ominus/(kJ\ mol^{-1})$, at 298 K

$H_2O(l)$	-285.83	$H_2O_2(l)$	-187.78
$NH_3(g)$	-46.11	$N_2H_4(l)$	$+50.63$
$NO_2(g)$	$+33.18$	$N_2O_4(g)$	$+9.16$
$NaCl(s)$	-411.15	$KCl(s)$	-436.75

* More values are given in the *Data section*.

The web site contains links to online databases of thermochemical data.

2.8 Standard enthalpies of formation

The **standard enthalpy of formation**, $\Delta_f H^\ominus$, of a substance is the standard reaction enthalpy for the formation of the compound from its elements in their reference states. The **reference state** of an element is its most stable state at the specified temperature and 1 bar. For example, at 298 K the reference state of nitrogen is a gas of N_2 molecules, that of mercury is liquid mercury, that of carbon is graphite, and that of tin is the white (metallic) form. There is one exception to this general prescription of reference states: the reference state of phosphorus is taken to be white phosphorus despite this allotrope not being the most stable form but simply the more reproducible form of the element. Standard enthalpies of formation are expressed as enthalpies per mole of the compound. The standard enthalpy of formation of liquid benzene at 298 K, for example, refers to the reaction

$$6\ C(s, graphite) + 3\ H_2(g) \longrightarrow C_6H_6(l)$$

and is $+49.0\ kJ\ mol^{-1}$. The standard enthalpies of formation of elements in their reference states are zero at all temperatures because they are the enthalpies of such 'null' reactions as $N_2(g) \to N_2(g)$. Some enthalpies of formation are listed in Tables 2.5 and 2.6.

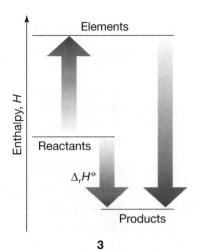

3

(a) The reaction enthalpy in terms of enthalpies of formation

Conceptually, we can regard a reaction as proceeding by decomposing the reactants into their elements and then forming those elements into the products (**3**). The value of $\Delta_r H^\ominus$ for the overall reaction is the sum of these 'unforming' and forming enthalpies. Because 'unforming' is the reverse of forming, the enthalpy of an unforming step is the negative of the enthalpy of formation. Hence, in the enthalpies of formation of substances, we have enough information to calculate the enthalpy of any reaction by using

$$\Delta_r H^\ominus = \sum_{\text{Products}} v \Delta_f H^\ominus - \sum_{\text{Reactants}} v \Delta_f H^\ominus \tag{2.42}$$

where in each case the enthalpies of formation of the species that occur are multiplied by their stoichiometric coefficients.

$C(H)_2(C)_2$

4

> **Illustration 2.7**
>
> The standard reaction enthalpy of $2\,HN_3(l) + 2\,NO(g) \rightarrow H_2O_2(l) + 4\,N_2(g)$ is calculated as follows:
>
> $$\Delta_r H^\ominus = \{\Delta_f H^\ominus(H_2O_2, l) + 4\Delta_f H^\ominus(N_2, g)\} - \{2\Delta_f H^\ominus(HN_3, l) + 2\Delta_f H^\ominus(NO, g)\}$$
> $$= \{-187.78 + 4(0)\}\,kJ\,mol^{-1} - \{2(264.0) + 2(90.25)\}\,kJ\,mol^{-1}$$
> $$= -896.3\,kJ\,mol^{-1}$$

(b) Group contributions

$C(C)_4$

5

We have seen how to construct standard reaction enthalpies by combining standard enthalpies of formation. The question that now arises is whether we can construct standard enthalpies of formation from a knowledge of the chemical constitution of the species. The short answer is that there is no thermodynamically exact way of expressing enthalpies of formation in terms of contributions from individual atoms and bonds. In the past, approximate procedures based on **mean bond enthalpies**, $\Delta H(A\!-\!B)$, the average enthalpy change associated with the breaking of a specific $A\!-\!B$ bond,

$$A\!-\!B(g) \longrightarrow A(g) + B(g) \qquad \Delta H(A\!-\!B)$$

have been used. However, this procedure is notoriously unreliable, in part because the $\Delta H(A\!-\!B)$ are average values for a series of related compounds. Nor does the approach distinguish between geometrical isomers, where the same atoms and bonds may be present but experimentally the enthalpies of formation might be significantly different.

Somewhat more reliable is an approach in which a molecule is regarded as being built up of **thermochemical groups**, an atom or physical group of atoms bound to at least two other atoms: two examples are shown as (**4**) and (**5**). The enthalpy of formation of the compound is then expressed (approximately, at least) as the sum of the contributions associated with all the thermochemical groups into which the molecule can be divided. A list of the appropriate values is given in Table 2.7 and the method is illustrated in the following example. Table 2.7 also contains information on heat capacity, which is approximately additive in a similar sense.

Table 2.7* Benson thermochemical groups

Group	$\Delta_f H^\ominus /$ (kJ mol^{-1})	$C_{p,m}^\ominus /$ (J K^{-1} mol^{-1})
$C(H)_3(C)$	-42.17	25.9
$C(H)_2(C)_2$	-20.7	22.8
$C(H)(C)_3$	-6.91	18.7
$C(C)_4$	$+8.16$	18.2

* More values are given in the *Data section*.

C(H)₃(C)

C(H)₂(C)₂

6

7

8

9

Example 2.6 Using the thermochemical group approach

Estimate the standard enthalpy of formation at 298 K of hexane in (a) the gas phase, (b) the liquid phase.

Method First, identify the thermochemical groups present in the molecule. Then add together the appropriate values in Table 2.7. To obtain the enthalpy of formation of the liquid, include the enthalpy of condensation of the vapour; that is, the negative of the enthalpy of vaporization of the liquid (Table 2.3).

Answer The decomposition of the molecule into groups is depicted in (**6**). There are two C(H)₃(C) groups and four C(H)₂(C)₂ groups; therefore

$$\Delta_f H^{\ominus}(C_6H_{14}, g) = 2(-42.17 \text{ kJ mol}^{-1}) + 4(-20.7 \text{ kJ mol}^{-1}) = -167.1 \text{ kJ mol}^{-1}$$

The enthalpy of vaporization of hexane is 33.1 kJ mol⁻¹; therefore (**7**),

$$\Delta_f H^{\ominus}(C_6H_{14}, l) = (-167.1 \text{ kJ mol}^{-1}) - (33.1 \text{ kJ mol}^{-1}) = -200.2 \text{ kJ mol}^{-1}$$

The experimental value is −198.7 kJ mol⁻¹.

Self-test 2.8 Estimate the standard enthalpy of formation of gaseous 2,2-dimethylpropane. [−162.6 kJ mol⁻¹]

(c) Enthalpies of formation and molecular modelling

It is difficult to estimate standard enthalpies of formation of conformational isomers. For example, it has been observed experimentally that the equatorial (**8**) and axial (**9**) conformers of methylcyclohexane have different standard enthalpies of formation even though they consist of the same thermochemical groups. This difference arises from the steric repulsions in the axial conformer, which raise its energy relative to that of the equatorial conformer.

Although it is possible to modify the thermochemical group approach to take steric interactions into account, computer-aided molecular modelling is becoming the technique of choice for estimating standard enthalpies of formation of molecules with complex three-dimensional structures and the 'group approach' is regarded as primitive and old-fashioned. Commercial software packages use the principles developed in Chapter 14 to calculate the standard enthalpy of formation of a conformer drawn on the computer screen. The difference between calculated standard enthalpies of formation of two conformers is then an estimate of the conformational energy difference. In the case of methylcyclohexane, the calculated conformational energy difference ranges from 5.9 to 7.9 kJ mol⁻¹, with the equatorial conformer having the lower standard enthalpy of formation. These estimates compare favourably with the experimental value of 7.5 kJ mol⁻¹. However, good agreement between calculated and experimental values is relatively rare. Computational methods almost always predict correctly which conformer is more stable but do not always predict the correct magnitude of the conformational energy difference.

2.9 The temperature dependence of reaction enthalpies

The standard enthalpies of many important reactions have been measured at different temperatures, and for serious work these accurate data must be used. However, in the absence of this information, standard reaction enthalpies at different

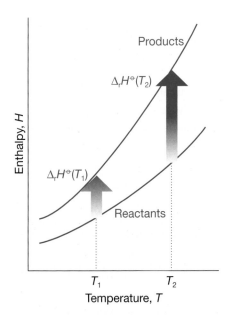

2.19 An illustration of the content of Kirchhoff's law. When the temperature is increased, the enthalpy of the products and the reactants both increase, but may do so to different extents. In each case, the change in enthalpy depends on the heat capacities of the substances. The change in reaction enthalpy reflects the difference in the changes of the enthalpies.

temperatures may be estimated from heat capacities and the reaction enthalpy at some other temperature (Fig. 2.19).

It follows from eqn 27 that, when a substance is heated from T_1 to T_2, its enthalpy changes from $H(T_1)$ to

$$H(T_2) = H(T_1) + \int_{T_1}^{T_2} C_p \, dT \tag{2.43}$$

(We have assumed that no phase transition takes place in the temperature range of interest.) Because this equation applies to each substance in the reaction, the standard reaction enthalpy changes from $\Delta_r H^{\ominus}(T_1)$ to

$$\Delta_r H^{\ominus}(T_2) = \Delta_r H^{\ominus}(T_1) + \int_{T_1}^{T_2} \Delta_r C_p^{\ominus} \, dT \tag{2.44}$$

where $\Delta_r C_p^{\ominus}$ is the difference of the molar heat capacities of products and reactants under standard conditions weighted by the stoichiometric numbers that appear in the chemical equation:

$$\Delta_r C_p^{\ominus} = \sum_{\text{Products}} v C_{p,m}^{\ominus} - \sum_{\text{Reactants}} v C_{p,m}^{\ominus} \tag{2.45}$$

Equation 44 is known as **Kirchhoff's law**. It is normally a good approximation to assume that $\Delta_r C_p$ is independent of the temperature, at least over reasonably limited ranges, as illustrated in the following example. Although the individual heat capacities may vary, their difference varies less significantly. In some cases the temperature dependence of heat capacities is taken into account by using eqn 30.

Example 2.7   Using Kirchhoff's law

The standard enthalpy of formation of gaseous H_2O at 298 K is −241.82 kJ mol^{-1}. Estimate its value at 100°C given the following values of the molar heat capacities at constant pressure: $H_2O(g)$: 33.58 J K^{-1} mol^{-1}; $H_2(g)$: 28.84 J K^{-1} mol^{-1}; $O_2(g)$: 29.37 J K^{-1} mol^{-1}. Assume that the heat capacities are independent of temperature.

Method   When $\Delta C_p^{\ominus}$ is independent of temperature in the range T_1 to T_2, the integral in eqn 44 evaluates to $(T_2 - T_1)\Delta_r C_p^{\ominus}$. Therefore,

$$\Delta_r H^{\ominus}(T_2) = \Delta_r H^{\ominus}(T_1) + (T_2 - T_1)\Delta_r C_p^{\ominus}$$

To proceed, write the chemical equation, identify the stoichiometric coefficients, and calculate $\Delta_r C_p^{\ominus}$ from the data.

Answer   The reaction is $H_2(g) + \frac{1}{2} O_2(g) \rightarrow H_2O(g)$, so

$$\Delta_r C_p^{\ominus} = C_{p,m}^{\ominus}(H_2O, g) - \{C_{p,m}^{\ominus}(H_2, g) + \tfrac{1}{2} C_{p,m}^{\ominus}(O_2, g)\}$$
$$= -9.94 \text{ J K}^{-1} \text{ mol}^{-1}$$

It then follows that

$$\Delta_f H^{\ominus}(373 \text{ K}) = -241.82 \text{ kJ mol}^{-1} + (75 \text{ K}) \times (-9.94 \text{ J K}^{-1} \text{ mol}^{-1})$$
$$= -242.6 \text{ kJ mol}^{-1}$$

Self-test 2.9   Estimate the standard enthalpy of formation of cyclohexene at 400 K from the data in Table 2.5.   [−163 kJ mol^{-1}]

Checklist of key ideas

☐ thermodynamics, the study of the transformations of energy

The basic concepts

☐ system, the part of the world in which we have a special interest

☐ surroundings, the region outside the system where we make our measurements

☐ open system, a system with a boundary through which matter can be transferred

☐ closed system, a system with a boundary through which matter cannot be transferred

☐ isolated system, a system with a boundary through which neither matter nor energy can be transferred

2.1 Work, heat, and energy

☐ work, the transfer of energy by motion against an opposing force

☐ energy, the capacity to do work

☐ heat, the transfer of energy as a result of a temperature difference

☐ diathermic boundary, a boundary that permits energy transfer as heat

☐ adiabatic boundary, a boundary that does not permit energy transfer as heat

☐ exothermic process, a process that releases energy as heat

☐ endothermic process, a process that absorbs energy as heat

☐ the molecular interpretation of heat and work, as random and ordered motion, respectively

☐ thermal motion, the random motion of molecules

2.2 The First Law

☐ internal energy, the total energy of a system

☐ state function, a property that depends only on the current state of the system and is independent of how that state has been prepared

☐ First Law of thermodynamics: The internal energy of an isolated system is constant, $\Delta U = q + w$

☐ alternative statement of First Law: The work needed to change an adiabatic system from one specified state to another specified state is the same however the work is done

☐ mechanical definition of heat, $q = w_{ad} - w$

Work and heat

2.3 Expansion work

☐ work, motion against an opposing force, $dw = -F\,dz$

☐ expansion work, work of expansion (or compression), $dw = -p_{ex}\,dV$

☐ non-expansion work or additional work, work other than expansion work

☐ free expansion, expansion into a vacuum, $w = 0$

☐ expansion against constant pressure, $w = -p_{ex}\Delta V$

☐ indicator diagram, a p,V-graph used to compute expansion work

☐ reversible change, a change that can be reversed by an infinitesimal modification of a variable

☐ equilibrium, a condition in which an infinitesimal change in a variable in opposite directions results in opposite changes in its state

☐ work of isothermal reversible expansion, $w = -nRT \ln(V_f/V_i)$

☐ maximum work, the work achieved in a reversible change

2.4 Heat transactions

☐ internal energy change and heat transfer at constant volume, $\Delta U = q_V$

☐ calorimetry, the study of heat transfer during physical and chemical processes

☐ calorimeter, a device for measuring the heat transferred

☐ adiabatic bomb calorimeter, a calorimeter in which there is no net loss of heat from the calorimeter to the surroundings

☐ calorimeter constant, the constant C in the relation $q = C\Delta T$

☐ heat capacity at constant volume, $C_V = (\partial U/\partial T)_V$

☐ molar heat capacity, the heat capacity divided by the amount of substance

☐ specific heat capacity, the heat capacity divided by the mass

☐ internal energy change and temperature rise, $dU = C_V\,dT$

2.5 Enthalpy

☐ enthalpy, $H = U + pV$

☐ enthalpy change and heat transfer at constant pressure, $\Delta H = q_p$

☐ relation between ΔU and ΔH, $\Delta H = \Delta U + \Delta n_g RT$

☐ heat capacity at constant pressure, $C_p = (\partial H/\partial T)_p$

☐ isobaric calorimeter, a calorimeter for studying processes at constant pressure

☐ adiabatic flame calorimeter, an isobaric calorimeter used to study the combustion of volatile compounds

☐ differential scanning calorimeter (DSC), an isobaric calorimeter in which the behaviour of a sample material is compared to the behaviour of a reference material

☐ relation between heat capacities of a perfect gas, $C_p - C_V = nR$

2.6 Adiabatic changes

☐ work of adiabatic change, $w_{ad} = C_V\Delta T$

☐ temperature change accompanying adiabatic change of a perfect gas, $T_f = T_i(V_i/V_f)^{1/c}$, $c = C_{V,m}/R$

☐ heat capacity ratio, $\gamma = C_{p,m}/C_{V,m}$

☐ pressure–volume relation for adiabatic change, $pV^\gamma = $ constant

☐ adiabat, curves of pressure versus volume for adiabatic, reversible changes

Thermochemistry

☐ thermochemistry, the study of the heat produced or required by chemical reactions

2.7 Standard enthalpy changes

☐ standard enthalpy change, the change in enthalpy for a process in which the initial and final substances are in their standard states

☐ standard state, the pure substance at 1 bar

☐ standard enthalpy of transition ($\Delta_{trs}H^\ominus$), the standard enthalpy change that accompanies a change of physical state

☐ standard enthalpy of vaporization, the standard enthalpy change accompanying the conversion of a solid or liquid to a vapour

- [ ] standard enthalpy of fusion, the standard enthalpy change accompanying the conversion of a solid to a liquid

- [ ] addition of enthalpy changes, as in $\Delta_{sub}H^{\ominus} = \Delta_{fus}H^{\ominus} + \Delta_{vap}H^{\ominus}$

- [ ] enthalpy change for reverse process, $\Delta H^{\ominus}(A \rightarrow B) = -\Delta H^{\ominus}(B \rightarrow A)$

- [ ] thermochemical equation, a combination of a chemical equation and the corresponding change in standard enthalpy

- [ ] standard reaction enthalpy, $\Delta_r H^{\ominus} = \sum_{Products} \nu H_m^{\ominus} - \sum_{Reactants} \nu H_m^{\ominus}$

- [ ] standard enthalpy of combustion, the standard reaction enthalpy for the complete oxidation of an organic compound to CO_2 gas and liquid H_2O if the compound contains C, H, and O, and to N_2 gas if N is also present

- [ ] Hess's law: The standard enthalpy of an overall reaction is the sum of the standard enthalpies of the individual reactions into which a reaction may be divided

2.8 Standard enthalpies of formation

- [ ] standard enthalpy of formation ($\Delta_f H^{\ominus}$), the standard reaction enthalpy for the formation of the compound from its elements in their reference states

- [ ] reference state, the most stable state of an element at the specified temperature and 1 bar

- [ ] combining enthalpies of formation, $\Delta_r H^{\ominus} = \sum_{Products} \nu \Delta_f H^{\ominus} - \sum_{Reactants} \nu \Delta_f H^{\ominus}$

- [ ] mean bond enthalpy, the average enthalpy change associated with the breaking of a specific A—B bond

- [ ] thermochemical group, an atom or physical group of atoms bound to at least two other atoms used in thermochemical calculations

2.9 The temperature dependence of reaction enthalpies

- [ ] Kirchhoff's law for the temperature dependence of reaction enthalpies, $$\Delta_r H^{\ominus}(T_2) = \Delta_r H^{\ominus}(T_1) + \int_{T_1}^{T_2} \Delta_r C_p^{\ominus} \, dT$$

Further reading

Articles of general interest

E.R. Boyko and J.F. Belliveau, Simplification of some thermochemical equations. *J. Chem. Educ.* **67**, 743 (1990).

F. Freeman, Z.M. Tsegai, M.L. Kasner, and W.J. Hehre, A comparison of the ab initio calculated and experimental conformational energies of alkylcyclohexanes. *J. Chem. Ed.* **77**, 661 (2000).

E.A. Gislason and N.C. Craig, General definitions of work and heat in thermodynamic processes. *J. Chem. Educ.* **64**, 670 (1987).

J. Güemez, S. Velasco, and M.A. Matías, Thermal coefficients and heat capacities in systems with chemical reaction: the Le Chatelier–Braun principle. *J. Chem. Educ.* **72**, 199 (1995).

I.H. Segel and L.D. Segel, Energetics of biological processes. In *Encyclopedia of applied physics* (ed. G.L. Trigg), **6**, 207. VCH, New York (1993).

T. Solomon, Standard enthalpies of formation of ions in solution. *J. Chem. Educ.* **68**, 41 (1991).

R.S. Treptow, Bond energies and enthalpies: an often neglected difference. *J. Chem. Educ.* **72**, 497 (1995).

R.S. Treptow, How thermodynamic data and equilibrium constants changed when the standard-state pressure became 1 bar. *J. Chem. Educ.* **76**, 212 (1999).

Texts and sources of data and information

S.W. Benson, *Thermochemical kinetics*. Wiley, New York (1976).

M.W. Chase, Jr. (ed.), *NIST-JANAF thermochemical tables*. Published as *J. Phys. Chem. Reference Data, Monograph no. 9*. American Institute of Physics, New York (1998).

M. Zabransky, V. Ruzicka Jr., V. Majer, and E.S. Domalski, *Heat capacity of liquids*. Published as *J. Phys. Chem. Reference Data, Monograph no. 6*. American Institute of Physics, New York (1996).

J.D. Cox and G. Pilcher, *Thermochemistry of organic and organometallic compounds*. Academic Press, New York (1970).

J.D. Cox, D.D. Wagman, and V.A. Medvedev, *CODATA key values for thermodynamics*. Hemisphere Publishing Corp., New York (1989).

W.E. Dasent, *Inorganic energetics*. Cambridge University Press (1982).

J.A. Dean (ed.), *Handbook of organic chemistry*. McGraw-Hill, New York (1987).

J. Emsley, *The elements*. Oxford University Press (1998).

E. Fermi, *Thermodynamics*. Dover, New York (1956).

E.F.G. Harrington, Thermodynamic data and their uses. *Chemical thermodynamics: Specialist Periodical Reports*. **1**, 31 (1973).

A.M. James and M.P. Lord (ed.), *Macmillan's chemical and physical data*. Macmillan, London (1992).

D.A. Johnson, *Some thermodynamic aspects of inorganic chemistry*. Cambridge University Press (1982).

I.M. Klotz and R.M. Rosenberg, *Chemical thermodynamics: basic theory and methods*. Wiley-Interscience, New York (2000).

G.N. Lewis and M. Randall, *Thermodynamics*. Revised by K.S. Pitzer and L. Brewer. McGraw-Hill, New York (1961).

M.L. McGlashan, *Chemical thermodynamics*. Academic Press, London (1979).

J.B. Pedley, R.D. Naylor, and S.P. Kirby, *Thermochemical data of organic compounds*. Chapman and Hall, London (1986).

B.W. Rossiter and R.C. Baetzgold (ed.) *Physical methods of chemistry*, **VI**. Wiley-Interscience (1992).

V. Smil, *Energies*. MIT Press, Cambridge (1999).

D.R. Stull and G.C. Sinke, *Thermodynamic properties of the elements*. American Chemical Society, Washington (1956).

D.R. Stull, E.F. Westrum, and G.C. Sinke, *The chemical thermodynamics of organic compounds*. Wiley-Interscience, New York (1969).

H.C. Van Ness, *Understanding thermodynamics*. Dover, New York (1983).

D.B. Wagman, W.H. Evans, V.B. Parker, R.H. Schumm, I. Halow, S.M. Bailey, K.L. Churney, and R.L. Nuttall, *The NBS tables of chemical thermodynamic properties*. Published as *J. Phys. Chem. Reference Data* **11**, Supplement 2 (1982).

D.R. Lide (ed.), *Handbook of chemistry and physics*, Vol. 81. CRC Press, Boca Raton (2000).

Exercises

Discussion questions

2.1 (a) Provide mechanical definitions of work and heat.

2.1 (b) Provide molecular interpretations for work and heat.

2.2 (a) Distinguish between expansion work against constant pressure and work of reversible expansion.

2.2 (b) Consider the reversible expansion of a perfect gas. Provide a physical interpretation for the fact that pV^{γ} = constant for an adiabatic change, whereas pV = constant for an isothermal change.

2.3 (a) Explain the difference between the change in internal energy and the change in enthalpy of a chemical or physical process.

2.3 (b) Describe at least two methods by which standard reaction enthalpies may be predicted. Discuss the advantages and disadvantages of each method.

Numerical exercises

Assume all gases are perfect unless stated otherwise. Note that 1 atm = 1.013 25 bar exactly. Unless otherwise stated, thermochemical data are for 298.15 K.

2.4 (a) Calculate the work done to raise a mass of 1.0 kg through 10 m on the surface of (a) the Earth (g = 9.81 m s^{-2}) and (b) the moon (g = 1.60 m s^{-2}).

2.4 (b) Calculate the work done to raise a mass of 5.0 kg through 100 m on the surface of (a) the Earth (g = 9.81 m s^{-2}) and (b) Mars (g = 3.73 m s^{-2}).

2.5 (a) Calculate the work needed for a 65 kg person to climb through 4.0 m on the surface of the Earth.

2.5 (b) Calculate the work needed for a bird of mass 120 g to fly to a height of 50 m from the surface of the Earth.

2.6 (a) A chemical reaction takes place in a container of cross-sectional area 100 cm^2. As a result of the reaction, a piston is pushed out through 10 cm against an external pressure of 1.0 atm. Calculate the work done by the system.

2.6 (b) A chemical reaction takes place in a container of cross-sectional area 50.0 cm^2. As a result of the reaction, a piston is pushed out through 15 cm against an external pressure of 121 kPa. Calculate the work done by the system.

2.7 (a) A sample consisting of 1.00 mol Ar is expanded isothermally at 0°C from 22.4 L to 44.8 L (a) reversibly, (b) against a constant external pressure equal to the final pressure of the gas, and (c) freely (against zero external pressure). For the three processes calculate q, w, ΔU, and ΔH.

2.7 (b) A sample consisting of 2.00 mol He is expanded isothermally at 22°C from 22.8 L to 31.7 L (a) reversibly, (b) against a constant external pressure equal to the final pressure of the gas, and (c) freely (against zero external pressure). For the three processes calculate q, w, ΔU, and ΔH.

2.8 (a) A sample consisting of 1.00 mol of monatomic perfect gas, for which $C_{V,m} = \frac{3}{2}R$, initially at p_1 = 1.00 atm and T_1 = 300 K, is heated reversibly to 400 K at constant volume. Calculate the final pressure, ΔU, q, and w.

2.8 (b) A sample consisting of 2.00 mol of perfect gas molecules, for which $C_{V,m} = \frac{5}{2}R$, initially at p_1 = 111 kPa and T_1 = 277 K, is heated reversibly to 356 K at constant volume. Calculate the final pressure, ΔU, q, and w.

2.9 (a) A sample of 4.50 g of methane occupies 12.7 L at 310 K. (a) Calculate the work done when the gas expands isothermally against a constant external pressure of 200 Torr until its volume has increased by 3.3 L. (b) Calculate the work that would be done if the same expansion occurred reversibly.

2.9 (b) A sample of argon of mass 6.56 g occupies 18.5 L at 305 K. (a) Calculate the work done when the gas expands isothermally against a constant external pressure of 7.7 kPa until its volume has increased by 2.5 L. (b) Calculate the work that would be done if the same expansion occurred reversibly.

2.10 (a) In the isothermal reversible compression of 52.0 mmol of a perfect gas at 260 K, the volume of the gas is reduced to one-third its initial value. Calculate w for this process.

2.10 (b) In the isothermal reversible compression of 1.77 mmol of a perfect gas at 273 K, the volume of the gas is reduced to 0.224 of its initial value. Calculate w for this process.

2.11 (a) A sample of 1.00 mol H_2O(g) is condensed isothermally and reversibly to liquid water at 100°C. The standard enthalpy of vaporization of water at 100°C is 40.656 kJ mol^{-1}. Find w, q, ΔU, and ΔH for this process.

2.11 (b) A sample of 2.00 mol CH_3OH(g) is condensed isothermally and reversibly to liquid at 64°C. The standard enthalpy of vaporization of methanol at 64°C is 35.3 kJ mol^{-1}. Find w, q, ΔU, and ΔH for this process.

2.12 (a) A strip of magnesium of mass 15 g is dropped into a beaker of dilute hydrochloric acid. Calculate the work done by the system as a result of the reaction. The atmospheric pressure is 1.0 atm and the temperature 25°C.

2.12 (b) A piece of zinc of mass 5.0 g is dropped into a beaker of dilute hydrochloric acid. Calculate the work done by the system as a result of the reaction. The atmospheric pressure is 1.1 atm and the temperature 23°C.

2.13 (a) Calculate the heat required to melt 750 kg of sodium metal at 371 K. The enthalpy of fusion of sodium is 2.601 kJ mol^{-1}.

2.13 (b) Calculate the heat required to melt 500 kg of potassium metal at 336 K. The enthalpy of fusion of potassium is 2.35 kJ mol^{-1}.

2.14 (a) A sample of a perfect gas was found to vary with temperature according to the expression $C_p/(\text{J K}^{-1})$ = 20.17 + 0.3665(T/K). Calculate q, w, ΔU, and ΔH when the temperature is raised from 25°C to 200°C (a) at constant pressure, (b) at constant volume.

2.14 (b) The constant-pressure heat capacity of a sample of a perfect gas was found to vary with temperature according to the expression $C_p/(\text{J K}^{-1})$ = 20.17 + 0.4001(T/K). Calculate q, w, ΔU, and ΔH when the temperature is raised from 0°C to 100°C (a) at constant pressure, (b) at constant volume.

2.15 (a) Calculate the final temperature of a sample of argon of mass 12.0 g that is expanded reversibly and adiabatically from 1.0 L at 273.15 K to 3.0 L.

2.15 (b) Calculate the final temperature of a sample of carbon dioxide of mass 16.0 g that is expanded reversibly and adiabatically from 500 mL at 298.15 K to 2.00 L.

2.16 (a) A sample of carbon dioxide of mass 2.45 g at 27.0°C is allowed to expand reversibly and adiabatically from 500 mL to 3.00 L. What is the work done by the gas?

2.16 (b) A sample of nitrogen of mass 3.12 g at 23.0°C is allowed to expand reversibly and adiabatically from 400 mL to 2.00 L. What is the work done by the gas?

2.17 (a) Calculate the final pressure of a sample of carbon dioxide that expands reversibly and adiabatically from 57.4 kPa and 1.0 L to a final volume of 2.0 L. Take $\gamma = 1.4$.

2.17 (b) Calculate the final pressure of a sample of water vapour that expands reversibly and adiabatically from 87.3 Torr and 500 mL to a final volume of 3.0 L. Take $\gamma = 1.3$.

2.18 (a) Calculate the final pressure of a sample of carbon dioxide of mass 2.4 g that expands reversibly and adiabatically from an initial temperature of 278 K and volume 1.0 L to a final volume of 2.0 L. Take $\gamma = 1.4$.

2.18 (b) Calculate the final pressure of a sample of water vapour of mass 1.4 g that expands reversibly and adiabatically from an initial temperature of 300 K and volume 1.0 L to a final volume of 3.0 L. Take $\gamma = 1.3$.

2.19 (a) Calculate the standard enthalpy of formation of butane at 25°C from its standard enthalpy of combustion.

2.19 (b) Calculate the standard enthalpy of formation of hexane at 25°C from its standard enthalpy of combustion.

2.20 (a) When 229 J of energy is supplied as heat to 3.0 mol Ar(g), the temperature of the sample increases by 2.55 K. Calculate the molar heat capacities at constant volume and constant pressure of the gas.

2.20 (b) When 178 J of energy is supplied as heat to 1.9 mol of gas molecules, the temperature of the sample increases by 1.78 K. Calculate the molar heat capacities at constant volume and constant pressure of the gas.

2.21 (a) A sample of a liquid of mass 25 g is cooled from 290 K to 275 K at constant pressure by the extraction of 1.2 kJ of energy as heat. Calculate q and ΔH and estimate the heat capacity of the sample.

2.21 (b) A sample of a liquid of mass 30.5 g is cooled from 288 K to 275 K at constant pressure by the extraction of 2.3 kJ of energy as heat. Calculate q and ΔH and estimate the heat capacity of the sample.

2.22 (a) When 3.0 mol O_2 is heated at a constant pressure of 3.25 atm, its temperature increases from 260 K to 285 K. Given that the molar heat capacity of O_2 at constant pressure is 29.4 J K^{-1} mol^{-1}, calculate q, ΔH, and ΔU.

2.22 (b) When 2.0 mol CO_2 is heated at a constant pressure of 1.25 atm, its temperature increases from 250 K to 277 K. Given that the molar heat capacity of CO_2 at constant pressure is 37.11 J K^{-1} mol^{-1}, calculate q, ΔH, and ΔU.

2.23 (a) A sample of 4.0 mol O_2 is originally confined in 20 L at 270 K and then undergoes adiabatic expansion against a constant pressure of 600 Torr until the volume has increased by a factor of 3.0. Calculate q, w, ΔT, ΔU, and ΔH. (The final pressure of the gas is not necessarily 600 Torr.)

2.23 (b) A sample of 5.0 mol CO_2 is originally confined in 15 L at 280 K and then undergoes adiabatic expansion against a constant pressure of 78.5 kPa until the volume has increased by a factor of 4.0. Calculate q, w, ΔT, ΔU, and ΔH. (The final pressure of the gas is not necessarily 78.5 kPa.)

2.24 (a) A sample consisting of 3.0 mol of perfect gas molecules at 200 K and 2.00 atm is compressed reversibly and adiabatically until the temperature reaches 250 K. Given that its molar constant-volume heat capacity is 27.5 J K^{-1} mol^{-1}, calculate q, w, ΔU, ΔH, and the final pressure and volume.

2.24 (b) A sample consisting of 2.5 mol of gas molecules at 220 K and 200 kPa is compressed reversibly and adiabatically until the temperature reaches 255 K. Given that its molar constant-volume heat capacity is 27.6 J K^{-1} mol^{-1}, calculate q, w, ΔU, ΔH, and the final pressure and volume.

2.25 (a) A sample consisting of 1.0 mol of perfect gas molecules with $C_V = 20.8$ J K^{-1} is initially at 3.25 atm and 310 K. It undergoes reversible adiabatic expansion until its pressure reaches 2.50 atm. Calculate the final volume and temperature and the work done.

2.25 (b) A sample consisting of 1.5 mol of perfect gas molecules with $C_{p,m} = 20.8$ J K^{-1} mol^{-1} is initially at 230 kPa and 315 K. It undergoes reversible adiabatic expansion until its pressure reaches 170 kPa. Calculate the final volume and temperature and the work done.

2.26 (a) Estimate the changes in volume that occur when a sample of mercury of volume 1.0 cm^3 is heated through 5.0 K at room temperature.

2.26 (b) Estimate the changes in volume that occur when a sample of iron of volume 5.0 cm^3 is heated through 10.0 K at room temperature.

2.27 (a) Consider a system consisting of 2.0 mol CO_2 (assumed to be a perfect gas) at 25°C confined to a cylinder of cross-section 10 cm^2 at 10 atm. The gas is allowed to expand adiabatically and irreversibly against a constant pressure of 1.0 atm. Calculate w, q, ΔU, ΔH, and ΔT when the piston has moved 20 cm.

2.27 (b) Consider a system consisting of 3.0 mol O_2 (assumed to be a perfect gas) at 25°C confined to a cylinder of cross-section 22 cm^2 at 820 kPa. The gas is allowed to expand adiabatically and irreversibly against a constant pressure of 110 kPa. Calculate w, q, ΔU, ΔH, and ΔT when the piston has moved 15 cm.

2.28 (a) A sample consisting of 65.0 g of xenon is confined in a container at 2.00 atm and 298 K and then allowed to expand adiabatically (a) reversibly to 1.00 atm, (b) against a constant pressure of 1.00 atm. Calculate the final temperature in each case.

2.28 (b) A sample consisting of 15.0 g of nitrogen is confined in a container at 220 kPa and 200 K and then allowed to expand adiabatically (a) reversibly to 110 kPa, (b) against a constant pressure of 110 kPa. Calculate the final temperature in each case.

2.29 (a) A certain liquid has $\Delta_{vap}H^{\ominus} = 26.0$ kJ mol^{-1}. Calculate q, w, ΔH, and ΔU when 0.50 mol is vaporized at 250 K and 750 Torr.

2.29 (b) A certain liquid has $\Delta_{vap}H^{\ominus} = 32.0$ kJ mol^{-1}. Calculate q, w, ΔH, and ΔU when 0.75 mol is vaporized at 260 K and 765 Torr.

2.30 (a) The standard enthalpy of formation of ethylbenzene is -12.5 kJ mol^{-1}. Calculate its standard enthalpy of combustion.

2.30 (b) The standard enthalpy of formation of phenol is -165.0 kJ mol^{-1}. Calculate its standard enthalpy of combustion.

2.31 (a) Calculate the standard enthalpy of hydrogenation of 1-hexene to hexane given that the standard enthalpy of combustion of 1-hexene is –4003 kJ mol^{-1}.

2.31 (b) Calculate the standard enthalpy of hydrogenation of 1-butene to butane given that the standard enthalpy of combustion of 1-butene is –2717 kJ mol^{-1}.

2.32 (a) The standard enthalpy of combustion of cyclopropane is –2091 kJ mol^{-1} at 25°C. From this information and enthalpy of formation data for $CO_2(g)$ and $H_2O(g)$, calculate the enthalpy of formation of cyclopropane. The enthalpy of formation of propene is +20.42 kJ mol^{-1}. Calculate the enthalpy of isomerization of cyclopropane to propene.

2.32 (b) From the following data, determine $\Delta_f H^\ominus$ for diborane, $B_2H_6(g)$, at 298 K:

(1) $B_2H_6(g) + 3\ O_2(g) \longrightarrow B_2O_3(s) + 3\ H_2O(g)$ $\quad \Delta_r H^\ominus = -1941$ kJ mol^{-1}
(2) $2\ B(s) + \frac{3}{2} O_2(g) \longrightarrow B_2O_3(s)$ $\quad \Delta_r H^\ominus = -2368$ kJ mol^{-1}
(3) $H_2(g) + \frac{1}{2} O_2(g) \longrightarrow H_2O(g)$ $\quad \Delta_r H^\ominus = -241.8$ kJ mol^{-1}

2.33 (a) Calculate the standard internal energy of formation of liquid methyl acetate from its standard enthalpy of formation, which is –442 kJ mol^{-1}.

2.33 (b) Calculate the standard internal energy of formation of urea from its standard enthalpy of formation, which is –333.51 kJ mol^{-1}.

2.34 (a) The temperature of a bomb calorimeter rose by 1.617 K when a current of 3.20 A was passed for 27.0 s from a 12.0 V source. Calculate the calorimeter constant.

2.34 (b) The temperature of a bomb calorimeter rose by 1.712 K when a current of 2.86 A was passed for 22.5 s from a 12.0 V source. Calculate the calorimeter constant.

2.35 (a) When 120 mg of naphthalene, $C_{10}H_8(s)$, was burned in a bomb calorimeter the temperature rose by 3.05 K. Calculate the calorimeter constant. By how much will the temperature rise when 10 mg of phenol, $C_6H_5OH(s)$, is burned in the calorimeter under the same conditions?

2.35 (b) When 2.25 mg of anthracene, $C_{14}H_{10}(s)$, was burned in a bomb calorimeter the temperature rose by 1.35 K. Calculate the calorimeter constant. By how much will the temperature rise when 135 mg of phenol, $C_6H_5OH(s)$, is burned in the calorimeter under the same conditions?

2.36 (a) Calculate the standard enthalpy of solution of AgCl(s) in water from the enthalpies of formation of the solid and the aqueous ions.

2.36 (b) Calculate the standard enthalpy of solution of AgBr(s) in water from the enthalpies of formation of the solid and the aqueous ions.

2.37 (a) The standard enthalpy of decomposition of the yellow complex H_3NSO_2 into NH_3 and SO_2 is +40 kJ mol^{-1}. Calculate the standard enthalpy of formation of H_3NSO_2.

2.37 (b) Given that the standard enthalpy of combustion of graphite is –393.51 kJ mol^{-1} and that of diamond is –395.41 kJ mol^{-1}, calculate the enthalpy of the graphite-to-diamond transition.

2.38 (a) The standard enthalpy of combustion of propane gas is –2220 kJ mol^{-1} and the standard enthalpy of vaporization of propane liquid is +15 kJ mol^{-1}. Calculate (a) the standard enthalpy and (b) the standard internal energy of combustion of liquid propane.

2.38 (b) The standard enthalpy of combustion of butane gas is –2878 kJ mol^{-1} and the standard enthalpy of vaporization of butane liquid is +21.0 kJ mol^{-1}. Calculate (a) the standard enthalpy and (b) the standard internal energy of combustion of liquid butane.

2.39 (a) Use standard enthalpies of formation to calculate the standard enthalpies of the following reactions:

(a) $2\ NO_2(g) \longrightarrow N_2O_4(g)$
(b) $NH_3(g) + HCl(g) \longrightarrow NH_4Cl(s)$

2.39 (b) Use standard enthalpies of formation to calculate the standard enthalpies of the following reactions:

(a) Cyclopropane(g) $\longrightarrow$ propene(g)
(b) $HCl(aq) + NaOH(aq) \longrightarrow NaCl(aq) + H_2O(l)$

2.40 (a) Given the reactions (1) and (2) below, determine (a) $\Delta_r H^\ominus$ and $\Delta_r U^\ominus$ for reaction (3), (b) $\Delta_f H^\ominus$ for both HCl(g) and $H_2O(g)$ all at 298 K. Assume all gases are perfect.

(1) $H_2(g) + Cl_2(g) \longrightarrow 2\ HCl(g)$ $\quad \Delta_r H^\ominus = -184.62$ kJ mol^{-1}
(2) $2\ H_2(g) + O_2(g) \longrightarrow 2\ H_2O(g)$ $\quad \Delta_r H^\ominus = -483.64$ kJ mol^{-1}
(3) $4\ HCl(g) + O_2(g) \longrightarrow Cl_2(g) + 2\ H_2O(g)$

2.40 (b) Given the reactions (1) and (2) below, determine (a) $\Delta_r H^\ominus$ and $\Delta_r U^\ominus$ for reaction (3), (b) $\Delta_f H^\ominus$ for both HI(g) and $H_2O(g)$ all at 298 K. Assume all gases are perfect.

(1) $H_2(g) + I_2(s) \longrightarrow 2\ HI(g)$ $\quad \Delta_r H^\ominus = +52.96$ kJ mol^{-1}
(2) $2\ H_2(g) + O_2(g) \longrightarrow 2\ H_2O(g)$ $\quad \Delta_r H^\ominus = -483.64$ kJ mol^{-1}
(3) $4\ HI(g) + O_2(g) \longrightarrow 2\ I_2(s) + 2\ H_2O(g)$

2.41 (a) For the reaction $C_2H_5OH(l) + 3\ O_2(g) \rightarrow 2\ CO_2(g) + 3\ H_2O(g)$, $\Delta_r U^\ominus = -1373$ kJ mol^{-1} at 298 K. Calculate $\Delta_r H^\ominus$.

2.41 (b) For the reaction $2\ C_6H_5COOH(s) + 13\ O_2(g) \rightarrow 12\ CO_2(g) + 6\ H_2O(g)$, $\Delta_r U^\ominus = -772.7$ kJ mol^{-1} at 298 K. Calculate $\Delta_r H^\ominus$.

2.42 (a) Calculate the standard enthalpies of formation of (a) $KClO_3(s)$ from the enthalpy of formation of KCl, (b) $NaHCO_3(s)$ from the enthalpies of formation of CO_2 and NaOH together with the following information:

$2\ KClO_3(s) \longrightarrow 2\ KCl(s) + 3\ O_2(g)$ $\quad \Delta_r H^\ominus = -89.4$ kJ mol^{-1}
$NaOH(s) + CO_2(g) \longrightarrow NaCO_3(s)$ $\quad \Delta_r H^\ominus = -127.5$ kJ mol^{-1}

2.42 (b) Calculate the standard enthalpy of formation of NOCl(g) from the enthalpy of formation of NO given in Table 2.5, together with the following information:

$2\ NOCl(g) \longrightarrow 2\ NO(g) + Cl_2(g)$ $\quad \Delta_r H^\ominus = +75.5$ kJ mol^{-1}

2.43 (a) Use the information in Table 2.5 to predict the standard reaction enthalpy of $2\ NO_2(g) \rightarrow N_2O_4(g)$ at 100°C from its value at 25°C.

2.43 (b) Use the information in Table 2.5 to predict the standard reaction enthalpy of $2\ H_2(g) + O_2(g) \rightarrow 2\ H_2O(l)$ at 100°C from its value at 25°C.

2.44 (a) From the data in Table 2.5, calculate $\Delta_r H^\ominus$ and $\Delta_r U^\ominus$ at (a) 298 K, (b) 378 K for the reaction $C(graphite) + H_2O(g) \rightarrow CO(g) + H_2(g)$. Assume all heat capacities to be constant over the temperature range of interest.

2.44 (b) Calculate $\Delta_r H^\ominus$ and $\Delta_r U^\ominus$ at 298 K and $\Delta_r H^\ominus$ at 348 K for the hydrogenation of ethyne (acetylene) to ethene (ethylene) from the enthalpy of combustion and heat capacity data in Tables 2.5 and 2.6. Assume the heat capacities to be constant over the temperature range involved.

2.45 (a) Set up a thermodynamic cycle for determining the enthalpy of hydration of Mg^{2+} ions using the following data: enthalpy of sublimation of Mg(s), +167.2 kJ mol^{-1}; first and second ionization enthalpies of Mg(g), 7.646 eV and 15.035 eV; dissociation enthalpy of Cl_2(g), +241.6 kJ mol^{-1}; electron-gain enthalpy of Cl(g), −3.78 eV; enthalpy of solution of $MgCl_2$(s), −150.5 kJ mol^{-1}; enthalpy of hydration of Cl$^-$(g), −383.7 kJ mol^{-1}.

2.45 (b) Set up a thermodynamic cycle for determining the enthalpy of hydration of Ca^{2+} ions using the following data: enthalpy of sublimation of Ca(s), +178.2 kJ mol^{-1}; first and second ionization enthalpies of Ca(g), 589.7 kJ mol^{-1} and 1145 kJ mol^{-1};

enthalpy of vaporization of bromine, +30.91 kJ mol^{-1}; dissociation enthalpy of Br_2(g), +192.9 kJ mol^{-1}; electron-gain enthalpy of Br(g), −331.0 kJ mol^{-1}; enthalpy of solution of $CaBr_2$(s), −103.1 kJ mol^{-1}; enthalpy of hydration of Br$^-$(g), −337 kJ mol^{-1}.

2.46 (a) Use the thermochemical groups in Table 2.7 to estimate the standard enthalpy of formation of (a) cyclohexane, (b) 1,4-dimethylhexane.

2.46 (b) Use the thermochemical groups in Table 2.7 to estimate the standard enthalpy of formation of (a) 2,2,4-trimethylpentane, (b) 2,2-dimethylpropane.

Problems†

Assume all gases are perfect unless stated otherwise. Note that 1 atm = 1.013 25 bar. Unless otherwise stated, thermochemical data are for 298.15 K.

Numerical problems

2.1 Calculate the heat needed to raise the temperature of the air in a house from 20°C to 25°C. Assume that the house contains 600 m^3 of air, which should be taken to be a perfect diatomic gas. The density of air is 1.21 kg m^{-3} at 20°C. Calculate ΔU and ΔH for the heating of the air.

2.2 Consider a perfect gas contained in a cylinder and separated by a frictionless adiabatic piston into two sections, A and B; section B is in contact with a water bath that maintains it at constant temperature. Initially, $T_A = T_B = 300$ K, $V_A = V_B = 2.00$ L, and $n_A = n_B = 2.00$ mol. Heat is supplied to Section A and the piston moves to the right reversibly until the final volume of Section B is 1.00 L. Calculate (a) the work done by the gas in Section A, (b) ΔU for the gas in Section B, (c) q for the gas in B, (d) ΔU for the gas in A, and (e) q for the gas in A. Assume $C_{V,m} = 20.0$ J K^{-1} mol^{-1}.

2.3 A sample consisting of 1 mol of a monatomic perfect gas (for which $C_{V,m} = \frac{3}{2}R$) is taken through the cycle shown in Fig. 2.20. (a) Determine the temperature at 1, 2, and 3. (b) Calculate q, w, ΔU, and ΔH for each step and for the overall cycle. If a numerical answer cannot be obtained from the information given, then write in +, −, 0, or ? as appropriate.

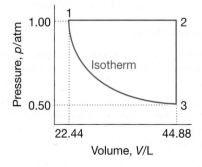

Fig. 2.20

2.4 A 5.0 g block of solid carbon dioxide is allowed to evaporate in a vessel of volume 100 cm^3 maintained at 25°C. Calculate the work done when the system expands (a) isothermally against a pressure of 1.0 atm, and (b) isothermally and reversibly to the same volume as in (a).

2.5 A sample consisting of 1.0 mol $CaCO_3$(s) was heated to 800°C, when it decomposed. The heating was carried out in a container fitted with a piston that was initially resting on the solid. Calculate the work done during complete decomposition at 1.0 atm. What work would be done if instead of having a piston the container was open to the atmosphere?

2.6 A sample consisting of 2.0 mol CO_2 occupies a fixed volume of 15.0 L at 300 K. When it is supplied with 2.35 kJ of energy as heat its temperature increases to 341 K. Assume that CO_2 is described by the van der Waals equation of state, and calculate w, ΔU, and ΔH.

2.7 A sample of 70 mmol Kr(g) expands reversibly and isothermally at 373 K from 5.25 cm^3 to 6.29 cm^3, and the internal energy of the sample is known to increase by 83.5 J. Use the virial equation of state up to the second coefficient $B = -28.7$ cm^3 mol^{-1} to calculate w, q, and ΔH for this change of state.

2.8 A piston exerting a pressure of 1.00 atm rests on the surface of water at 25°C. The pressure is reduced infinitesimally, and as a result 10.0 g of water evaporated and absorbed 22.2 kJ of heat. Calculate w, q, ΔU, and ΔH.

2.9 A new fluorocarbon of molar mass 102 g mol^{-1} was placed in an electrically heated vessel. When the pressure was 650 Torr, the liquid boiled at 78°C. After the boiling point had been reached, it was found that a current of 0.232 A from a 12.0 V supply passed for 650 s vaporized 1.871 g of the sample. Calculate the (molar) enthalpy and internal energy of vaporization.

2.10 A sample of 1.00 mol perfect gas molecules with $C_{p,m} = \frac{7}{2}R$ is put through the following cycle: (a) constant-volume heating to twice its initial temperature, (b) reversible, adiabatic expansion back to its initial temperature, (c) reversible isothermal compression back to 1.00 atm. Calculate q, w, ΔU, and ΔH for each step and overall.

2.11 An object is cooled by the evaporation of liquid methane at its normal boiling point (112 K). What volume of gas at 1.00 atm

† Problems denoted with the symbol ‡ were supplied by Charles Trapp, Carmen Giunta, and Marshall Cady.

pressure must be formed from the liquid in order to remove 32.5 kJ of energy as heat from the object?

2.12 The molar heat capacity of ethane is represented in the temperature range 298 K to 400 K by the empirical expression $C_{p,m}/(J\,K^{-1}\,mol^{-1}) = 14.73 + 0.1272(T/K)$. The corresponding expressions for C(s) and H_2(g) are given in Table 2.2. Calculate the standard enthalpy of formation of ethane at 350 K from its value at 298 K.

2.13 A sample of the sugar D-ribose ($C_5H_{10}O_5$) of mass 0.727 g was placed in a calorimeter and then ignited in the presence of excess oxygen. The temperature rose by 0.910 K. In a separate experiment in the same calorimeter, the combustion of 0.825 g of benzoic acid, for which the internal energy of combustion is -3251 kJ mol^{-1}, gave a temperature rise of 1.940 K. Calculate the internal energy of combustion of D-ribose and its enthalpy of formation.

2.14 The standard enthalpy of formation of the metallocene bis(benzene)chromium was measured in a calorimeter. It was found for the reaction $Cr(C_6H_6)_2$(s) $\rightarrow$ Cr(s) + 2 C_6H_6(g) that $\Delta_rU^{\ominus}(583\,K) = +8.0$ kJ mol^{-1}. Find the corresponding reaction enthalpy and estimate the standard enthalpy of formation of the compound at 583 K. The constant-pressure molar heat capacity of benzene is 140 J K^{-1} mol^{-1} in its liquid range and 28 J K^{-1} mol^{-1} as a gas.

2.15‡ When 1.3584 g of sodium acetate trihydrate was mixed into 100.0 mL of 0.2000 M HCl(aq) at 25°C in a solution calorimeter, its temperature fell by 0.397°C. This cooling occurred on account of the reaction:

$$H_3O^+(aq) + NaCH_3CO_2 \cdot 3H_2O(s)$$
$$\longrightarrow Na^+(aq) + CH_3COOH(aq) + 4\,H_2O(l)$$

The heat capacity of the calorimeter is 91.0 J K^{-1} and the heat capacity density of the acid solution is 4.144 J K^{-1} mL^{-1}. Determine the standard enthalpy of formation of the aqueous sodium cation. The standard enthalpy of formation of sodium acetate trihydrate is -1604 kJ mol^{-1}.

2.16‡ Since their discovery in 1985, fullerenes have received the attention of many chemical researchers. Kolesov *et al.* have recently reported the standard enthalpy of combustion and of formation of crystalline C_{60} based on calorimetric measurements (V.P. Kolesov, S.M. Pimenova, V.K. Pavlovich, N.B. Tamm, and A.A. Kurskaya, *J. Chem. Thermodynamics* **28**, 1121 (1996)). In one of their runs, they found the standard specific internal energy of combustion to be -36.0334 kJ g^{-1} at 298.15 K Compute $\Delta_cH^{\ominus}$ and $\Delta_fH^{\ominus}$ of C_{60}.

2.17‡ A thermodynamic study of DyCl$_3$ (E.H.P. Cordfunke, A.S. Booji, and M. Yu. Furkaliouk, *J. Chem. Thermodynamics* **28**, 1387 (1996)) determined its standard enthalpy of formation from the following information

(1) DyCl$_3$(s) $\longrightarrow$ DyCl$_3$(aq, in 4.0 M HCl(aq)) $\Delta_rH^{\ominus} = -180.06$ kJ mol^{-1}

(2) Dy(s) + 3 HCl(aq, 4.0 M) $\longrightarrow$ DyCl$_3$(aq, in 4.0 M HCl(aq)) + $\frac{3}{2}$H$_2$(g)
$\Delta_rH^{\ominus} = -699.43$ kJ mol^{-1}

(3) $\frac{1}{2}$ H$_2$(g) + $\frac{1}{2}$ Cl$_2$(g) $\longrightarrow$ HCl(aq, 4.0 M) $\Delta_rH^{\ominus} = -158.31$ kJ mol^{-1}

Determine $\Delta_fH^{\ominus}$(DyCl$_3$, s) from these data.

2.18‡ Silylene (SiH$_2$) is a key intermediate in the thermal decomposition of silicon hydrides such as silane (SiH$_4$) and disilane (Si$_2$H$_6$). Moffat *et al.* (H.K. Moffat, K.F. Jensen, and R.W. Carr, *J. Phys. Chem.* **95**, 145 (1991)) report $\Delta_fH^{\ominus}$(SiH$_2$) = +274 kJ mol^{-1}. If $\Delta_fH^{\ominus}$(SiH$_4$) = +34.3 kJ mol^{-1} and $\Delta_fH^{\ominus}$(Si$_2$H$_6$) = +80.3 kJ mol^{-1} (*CRC Handbook* (2000)), compute the standard enthalpies of the following reactions:

(a) SiH$_4$(g) $\longrightarrow$ SiH$_2$(g) + H$_2$(g)

(b) Si$_2$H$_6$(g) $\longrightarrow$ SiH$_2$(g) + SiH$_4$(g)

2.19‡ Silanone (SiH$_2$O) and silanol (SiH$_3$OH) are species believed to be important in the oxidation of silane (SiH$_4$). These species are much more elusive than their carbon counterparts. C.L. Darling and H.B. Schlegel (*J. Phys. Chem.* **97**, 8207 (1993)) report the following values (converted from calories) from a computational study: $\Delta_fH^{\ominus}$(SiH$_2$O) = -98.3 kJ mol^{-1} and $\Delta_fH^{\ominus}$(SiH$_3$OH) = -282 kJ mol^{-1}. Compute the standard enthalpies of the following reactions:

(a) SiH$_4$(g) + $\frac{1}{2}$ O$_2$(g) $\longrightarrow$ SiH$_3$OH(g)

(b) SiH$_4$(g) + O$_2$(g) $\longrightarrow$ SiH$_2$O(g) + H$_2$O(l)

(c) SiH$_3$OH(g) $\longrightarrow$ SiH$_2$O(g) + H$_2$(g)

Note that $\Delta_fH^{\ominus}$(SiH$_4$, g) = +34.3 kJ mol^{-1} (*CRC Handbook* (2000)).

2.20‡ *Polytropic processes* are defined as those that satisfy the condition $pV^n = C$, where C is a constant. In one experiment, 1.00 mol of 'air molecules' is compressed from 1.00 bar to 10.0 bar at 25°C by two different combinations of reversible polytropic processes: (1) heating at constant volume to the final pressure, followed by cooling at constant pressure, (2) adiabatic compression to the final volume, followed by cooling at constant volume. (a) Sketch these processes on a pV diagram and identify the value of n for each step in each process. (b) Calculate q, w, ΔU, and ΔH for each step in the processes and for the overall process. Note that the overall process can be accomplished in one isothermal reversible step. Assume air is a perfect diatomic gas with $C_{p,m} = \frac{7}{2}R$.

2.21‡ For reversible polytropic processes described by the general relation $pV^n = C$, derive the following expressions for work and heat

$$w = \frac{RT_1}{n-1}\left\{\left(\frac{p_2}{p_1}\right)^{(n-1)/n} - 1\right\}$$

$$q = \frac{(n-\gamma)RT_1}{(n-1)(\gamma-1)}\left\{\left(\frac{p_2}{p_1}\right)^{(n-1)/n} - 1\right\}$$

Show that these expressions reduce to already familiar expressions for $n = 0$, 1, γ, and ∞.

2.22‡ Ammonia is compressed in a piston-cylinder apparatus from an initial state of 30°C and 500 kPa to a final pressure of 1400 kPa. The following data were obtained during the process:

p/kPa	500	653	802	945	1100	1248	1400
V/L	1.25	1.08	0.96	0.84	0.72	0.60	0.50

(a) Is this a polytropic process? (See Problem 2.20 for the definition of a polytropic process.) If so, what is n? (b) Calculate the work done on the ammonia. (c) What is the final temperature?

Theoretical problems

2.23 In a machine of a particular design, the opposing force acting on a mass m varies as $F\sin(\pi x/a)$. Calculate the work needed to move the mass (a) from $x = 0$ to $x = a$ and (b) from $x = 0$ to $x = 2a$.

2.24 Calculate the work done during the isothermal reversible expansion of a gas that satisfies the virial equation of state, eqn 1.22. Evaluate (a) the work for 1.0 mol Ar at 273 K (for data, see Table 1.3) and (b) the same amount of a perfect gas. Let the expansion be from 500 cm^3 to 1000 cm^3 in each case.

2.25 With reference to Fig. 2.21, and assuming perfect gas behaviour, calculate: (a) the amount of gas molecules (in moles) in this system and its volume in states B and C, (b) the work done on the gas along the paths ACB and ADB, (c) The work done on the gas

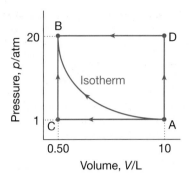

Fig. 2.21

along the isotherm AB, (d) q and ΔU for each of the three paths. Take $C_{V,m} = \frac{3}{2}R$ and $T = 313$ K.

2.26 Calculate the work done during the isothermal reversible expansion of a van der Waals gas. Account physically for the way in which the coefficients a and b appear in the final expression. Plot on the same graph the indicator diagrams for the isothermal reversible expansion of (a) a perfect gas, (b) a van der Waals gas in which $a = 0$ and $b = 5.11 \times 10^{-2}$ L mol^{-1}, and (c) $a = 4.2$ L^2 atm mol^{-2} and $b = 0$. (Use other values too if you have a computer available.) The values selected exaggerate the imperfections but give rise to significant effects on the indicator diagrams. Take $V_i = 1.0$ L, $n = 1.0$ mol, and $T = 298$ K.

2.27 When a system is taken from state A to state B along the path ACB in Fig. 2.22, 80 J of heat flows into the system and the system does 30 J of work. (a) How much heat flows into the system along path ADB if the work done is 10 J? (b) When the system is returned from state B to A along the curved path, the work done on the system is 20 J. Does the system absorb or liberate heat, and how much? (c) If $U_D - U_A = +40$ J, find the heat absorbed in the processes AD and DB.

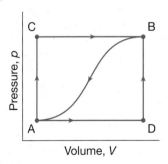

Fig. 2.22

2.28 Show that the value of ΔH for the adiabatic expansion of a perfect gas may be calculated by integration of $dH = V\,dp$, and evaluate the integral for reversible adiabatic expansion.

2.29 Express the work of isothermal reversible expansion of a van der Waals gas in reduced variables and find a definition of reduced work that makes the overall expression independent of the identity of the gas. Calculate the work of isothermal reversible expansion along the critical isotherm from V_c to xV_c.

Applications: to biology, geology, and chemical engineering

2.30 Glucose and fructose are simple sugars with the molecular formula $C_6H_{12}O_6$. Sucrose, or table sugar, is a complex sugar with molecular formula $C_{12}H_{22}O_{11}$ that consists of a glucose unit covalently bound to a fructose unit (a water molecule is released as a result of the reaction between glucose and fructose to form sucrose). (a) Calculate the energy released as heat when a typical table sugar cube of mass 1.5 g is burned in air. (b) To what height could you climb on the energy a table sugar cube provides assuming 25 per cent of the energy is available for work? (c) The mass of a typical glucose tablet is 2.5 g. Calculate the energy released as heat when a glucose tablet is burned in air. (d) To what height could you climb on the energy a cube provides assuming that 25 per cent of the energy is available for work?

2.31 In biological cells that have a plentiful supply of O_2, glucose is oxidized completely to CO_2 and H_2O by a process called *aerobic oxidation* (see Box 9.1). Muscle cells may be deprived of O_2 during vigorous exercise and, in that case, one molecule of glucose is converted to two molecules of lactic acid ($CH_3CH(OH)COOH$) by a process called *anaerobic glycolysis* (see Box 9.1). (a) When 0.3212 g of glucose was burned in a bomb calorimeter of calorimeter constant 641 J K^{-1} the temperature rose by 7.793 K. Calculate (i) the standard molar enthalpy of combustion, (ii) the standard internal energy of combustion, and (iii) the standard enthalpy of formation of glucose. (b) What is the biological advantage (in kilojoules per mole of energy released as heat) of complete aerobic oxidation compared with anaerobic glycolysis to lactic acid?

2.32 Geological conditions are sometimes so extreme that quantities neglected in normal laboratory experiments take on an overriding importance. For example, consider the formation of diamond under geophysically typical conditions. The density of graphite is 2.27 g cm^{-3} and that of diamond is 3.52 g cm^{-3} at a certain temperature and 500 kbar. By how much does $\Delta_{trs}U$ differ from $\Delta_{trs}H$ for the graphite-to-diamond transition?

2.33 Several characteristics of a hydrocarbon must be considered when it is being considered as a fuel. Among them are the specific enthalpy, the heat evolved divided by the mass of the sample, for the advantage of a high molar enthalpy of combustion may be eliminated if a large mass of fuel is to be transported. Use the data in Table 2.5 to calculate the following information on butane, pentane, and octane: (a) the heat output per mole, (b) the heat output per gram.

2.34‡ From the enthalpy of combustion data in Table 2.5 for the alkanes methane through octane, test the extent to which the relation $\Delta_c H^\ominus = k\{(M/(\text{g mol}^{-1})\}^n$ holds and find the numerical values for k and n. Predict $\Delta_c H^\ominus$ for decane and compare to the known value.

2.35 It is possible to investigate the thermochemical properties of hydrocarbons with molecular modelling methods. (a) Use electronic structure software to predict $\Delta_c H^\ominus$ values for the alkanes methane through pentane. To calculate $\Delta_c H^\ominus$ values, estimate the standard enthalpy of formation of $C_nH_{2(n+1)}$(g) by performing semi-empirical calculations (for example, AM1 or PM3 methods) and use experimental standard enthalpy of formation values for CO_2(g) and H_2O(l). (b) Compare your estimated values with the experimental values of $\Delta_c H^\ominus$ (Table 2.5) and comment on the reliability of the molecular modelling method. (c) Test the extent to which the relation $\Delta_c H^\ominus = k\{(M/(\text{g mol}^{-1})\}^n$ holds and find the numerical values for k and n.

2.36‡ Alkyl radicals are important intermediates in the combustion and atmospheric chemistry of hydrocarbons. Seakins *et al.* (P.W. Seakins, M.J. Pilling, J.T. Niiranen, D. Gutman, and L.N. Krasnoperov, *J. Phys. Chem.* **96**, 9847 (1992)) report $\Delta_f H^{\ominus}$ for a variety of alkyl radicals in the gas phase—information that is applicable to studies of pyrolysis and oxidation reactions of hydrocarbons. This information can be combined with thermodynamic data on alkenes to determine the reaction enthalpy for possible fragmentation of a large alkyl radical into smaller radicals and alkenes. Use the following set of data to compute the standard reaction enthalpies for three possible fates of the *tert*-butyl radical, namely,
(a) *tert*-$C_4H_9 \rightarrow sec$-C_4H_9, (b) *tert*-$C_4H_9 \rightarrow C_3H_6 + CH_3$,
(c) *tert*-$C_4H_9 \rightarrow C_2H_4 + C_2H_5$.

Species:	C_2H_5	sec-C_4H_9	$tert$-C_4H_9
$\Delta_f H^{\ominus}/(kJ\ mol^{-1})$	+121.0	+67.5	+51.3

2.37‡ N. Cohen (*J. Phys. Chem.* **96**, 9052 (1992)) has reported group additivity tables for the thermochemistry of alkyl radicals in the gas phase. He suggests computing enthalpies of formation based on the following bond dissociation energies ($\Delta U^{\ominus}$) for C—H bonds: primary (—(H)C(H)—H): 420.5 kJ mol^{-1}; secondary (—(C)C(H)—H): 410.5 kJ mol^{-1}; tertiary (—(C)C(C)—H): 398.3 kJ mol^{-1}. Estimate $\Delta_f H^{\ominus}$ of (a) C_2H_5, (b) sec-C_4H_9, and (c) *tert*-C_4H_9. ($\Delta_f H^{\ominus}$(2-methylpropane, g) = −134.2 kJ mol^{-1}.)

The First Law: the machinery

State functions and exact differentials

3.1 State and path functions

3.2 Exact and inexact differentials

Thermodynamic consequences

3.3 Changes in internal energy

3.4 The temperature dependence of the enthalpy

3.5 The relation between C_V and C_p

Checklist of key ideas

Further reading

Exercises

Problems

In this chapter we begin to unfold some of the power of thermodynamics by showing how to establish relations between different properties of a system. The procedure we use is based on the experimental fact that the internal energy and the enthalpy are state functions. We shall see that one very useful aspect of thermodynamics is that a property can be measured indirectly by measuring others and then combining their values. The relations we derive also enable us to discuss the liquefaction of gases and to establish the relation between the heat capacities of a substance at constant pressure and constant volume.

We saw in Section 2.2 that a 'state function' is a property that is independent of how a sample is prepared. In general, such properties are functions of variables that define the current state of the system, such as pressure and temperature. The internal energy and enthalpy are examples of state functions, for they depend on the current state of the system and are independent of its previous history. Properties that relate to the preparation of the state are called **path functions**. Examples of path functions are the work that is done in preparing a state and the energy transferred as heat. We do not speak of a system in a particular state as possessing work or heat. In each case, the energy transferred as work or heat relates to the path being taken, not the current state itself.

State functions and exact differentials

We can use the mathematical properties of state functions to draw far-reaching conclusions about the relations between physical properties and establish connections that may be completely unexpected. The practical importance of these results is that we can combine measurements of different properties to obtain the value of a property we require.

3.1 State and path functions

Consider a system undergoing the changes depicted in Fig. 3.1. The initial state of the system is i and in this state the internal energy is U_i. Work is done by the system as it expands adiabatically to a state f. In this state the system has an internal energy U_f and the work done on the system as it changes along Path 1 from i to f is w. Notice our use of language: U is a property of the state; w is a property of the path. Now

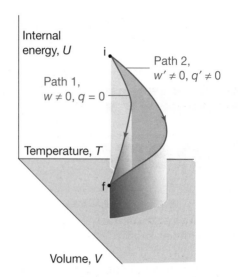

3.1 As the volume and temperature of a system are changed, the internal energy changes. An adiabatic and a nonadiabatic path are shown as Path 1 and Path 2, respectively: they correspond to different values of q and w but to the same value of ΔU.

consider another process, Path 2, in which the initial and final states are the same as those in Path 1 but in which the expansion is not adiabatic. The internal energy of both the initial and the final states are the same as before (because U is a state function). However, in the second path an energy q' enters the system as heat and the work w' is not the same as w. The work and the heat are path functions. In terms of the mountaineering analogy in Section 2.2, the change in altitude (a state function) is independent of the path, but the distance travelled (a path function) does depend on the path taken between the fixed endpoints.

3.2 Exact and inexact differentials

If a system is taken along a path (for example, by heating it), U changes from U_i to U_f, and the overall change is the sum (integral) of all the infinitesimal changes along the path:

$$\Delta U = \int_i^f dU \tag{3.1}$$

The value of ΔU depends on the initial and final states of the system but is independent of the path between them. This path independence of the integral is expressed by saying that dU is an 'exact differential'. In general, an **exact differential** is an infinitesimal quantity that, when integrated, gives a result that is independent of the path between the initial and final states.

When a system is heated, the total energy transferred as heat is the sum of all individual contributions at each point of the path:

$$q = \int_{i,\text{path}}^f dq \tag{3.2}$$

Notice the difference between this equation and eqn 1. First, we do not write Δq, because q is not a state function and the energy supplied as heat cannot be expressed as $q_f - q_i$. Secondly, we must specify the path of integration because q depends on the path selected (for example, an adiabatic path has $q = 0$, whereas a nonadiabatic path between the same two states would have $q \neq 0$). This path dependence is expressed by saying that dq is an 'inexact differential'. In general, an **inexact differential** is an infinitesimal quantity that, when integrated, gives a result that depends on the path between the initial and final states. Often dq is written $đq$ to emphasize that it is inexact.

The work done on a system to change it from one state to another depends on the path taken between the two specified states; for example, in general the work is different if the change takes place adiabatically and nonadiabatically. It follows that dw is an inexact differential. It is often written $đw$.

Example 3.1 Calculating work, heat, and internal energy

Consider a perfect gas inside a cylinder fitted with a piston. Let the initial state be T, V_i and the final state be T, V_f. The change of state can be brought about in many ways, of which the two simplest are the following: Path 1, in which there is free expansion against zero external pressure; Path 2, in which there is reversible, isothermal expansion. Calculate w, q, and ΔU for each process.

Method To find a starting point for a calculation in thermodynamics, it is often a good idea to go back to first principles, and to look for a way of expressing the quantity we are asked to calculate in terms of other quantities that are easier to calculate. Because the internal energy of a perfect gas arises only from the kinetic energy of its molecules, it is independent of the volume those molecules occupy; therefore, for any isothermal change, $\Delta U = 0$. We also know that in general $\Delta U = q + w$. The question depends on being able to combine the two expressions. We derived a number of expressions for the work done in a variety of processes in Chapter 2, and here we need to select the appropriate ones.

Answer Because $\Delta U = 0$ for both paths and $\Delta U = q + w$, in each case $q = -w$. The work of free expansion is zero (Section 2.3b); so in Path 1, $w = 0$ and $q = 0$. For Path 2, the work is given by eqn 2.13, so $w = -nRT \ln(V_f/V_i)$ and consequently $q = nRT \ln(V_f/V_i)$.

Self-test 3.1 Calculate the values of q, w, and ΔU for an irreversible isothermal expansion of a perfect gas against a constant nonzero external pressure.
$$[q = p_{ex}\Delta V, \ w = -p_{ex}\Delta V, \ \Delta U = 0]$$

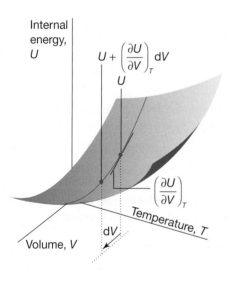

3.2 The partial derivative $(\partial U/\partial V)_T$ is the slope of U with respect to V with the temperature T held constant.

Partial derivatives are reviewed in *Further information 1*.

Thermodynamic consequences

The fact that a function is a state function has profound—and very useful—consequences. This chapter illustrates how to use the properties of state functions to establish unexpected relations between the physical properties of a system.

3.3 Changes in internal energy

We begin to unfold the consequences of dU being an exact differential by noting that, for a closed system of constant composition (the only type of system considered in this chapter), U is a function of volume and temperature.[1]

(a) General considerations

When V changes to $V + dV$ at constant temperature, U changes to

$$U' = U + \left(\frac{\partial U}{\partial V}\right)_T dV$$

The coefficient $(\partial U/\partial V)_T$, the slope of a plot of U against V at constant temperature, is the partial derivative of U with respect to V (Fig. 3.2). If, instead, T changes to $T + dT$ at constant volume (Fig. 3.3), the internal energy changes to

$$U' = U + \left(\frac{\partial U}{\partial T}\right)_V dT$$

---

1. U could be regarded as a function of V, T, and p but, because there is an equation of state, it is possible to express p in terms of V and T, so p is not an independent variable. We could choose p, T or p, V as independent variables, but V, T fit our purpose.

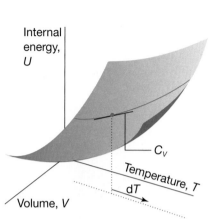

3.3 The partial derivative $(\partial U/\partial T)_V$ is the slope of U with respect to T with the volume V held constant.

3.4 An overall change in U, which is denoted dU, arises when both V and T are allowed to change. If second-order infinitesimals are ignored, the overall change is the sum of changes for each variable separately.

3.5 The constant-volume heat capacity, C_V, is the slope of U with respect to T with the volume V held constant.

Now suppose that V and T both change infinitesimally (Fig. 3.4) . The new internal energy, neglecting second-order infinitesimals (those proportional to $dV\,dT$), is

$$U' = U + \left(\frac{\partial U}{\partial V}\right)_T dV + \left(\frac{\partial U}{\partial T}\right)_V dT$$

As a result of the infinitesimal changes in conditions, the internal energy U' differs from U by the infinitesimal amount dU. Therefore, from the last equation we obtain the very important result that

$$dU = \left(\frac{\partial U}{\partial V}\right)_T dV + \left(\frac{\partial U}{\partial T}\right)_V dT \tag{3.3}$$

The interpretation of this equation is that, in a closed system of constant composition, any infinitesimal change in the internal energy is proportional to the infinitesimal changes of volume and temperature, the coefficients of proportionality being the partial derivatives.

A partial derivative is the slope of a graph of the property of interest against one of the variables on which it depends (recall Fig. 2.12), all the other variables being held constant. In many cases the slopes have a straightforward physical interpretation, and thermodynamics gets shapeless and difficult only when that meaning is not kept in sight. In the present case, we have already met $(\partial U/\partial T)_V$ in eqn 2.19, where we saw that it is the constant-volume heat capacity, C_V (Fig. 3.5). Therefore, we can write

$$dU = \left(\frac{\partial U}{\partial V}\right)_T dV + C_V\,dT \tag{3.4}$$

The other coefficient, $(\partial U/\partial V)_T$, plays a major role in thermodynamics because it is a measure of the variation of the internal energy of a substance as its volume is changed at constant temperature (Fig. 3.6). We shall denote it π_T (because it has the same dimensions as pressure), and call it the **internal pressure**:

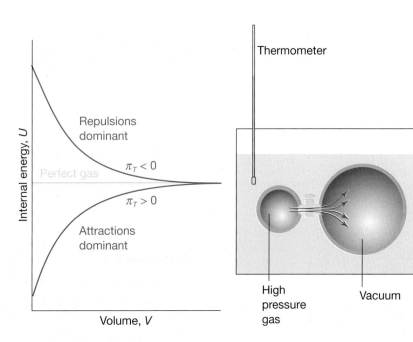

3.6 The internal pressure, π_T, is the slope of U with respect to V with the temperature T held constant.

3.7 For a perfect gas, the internal energy is independent of the volume (at constant temperature). If attractions are dominant in a real gas, the internal energy increases with volume because the molecules become further apart on average. If repulsions are dominant, the internal energy decreases as the gas expands.

3.8 A schematic diagram of the apparatus used by Joule in an attempt to measure the change in internal energy when a gas expands isothermally. The heat absorbed by the gas is proportional to the change in temperature of the bath.

$$\pi_T = \left(\frac{\partial U}{\partial V} \right)_T \tag{3.5}$$

We shall see that the internal pressure is a measure of the strength of the cohesive forces in the sample. Then

$$dU = \pi_T\, dV + C_V\, dT \tag{3.6}$$

If the internal energy increases ($dU > 0$) as the volume of the sample expands isotherm-ally ($dV > 0$), which is the case when there are attractive forces between the particles, a graph of internal energy against volume slopes upwards and $\pi_T > 0$ (Fig. 3.7). When there are no interactions between the molecules, the internal energy is independent of their separation and hence independent of the volume of the sample. We can therefore write

$$\text{For a perfect gas} \quad \left(\frac{\partial U}{\partial V} \right)_T = 0 \tag{3.7}°$$

or, more succinctly, $\pi_T = 0$. The statement $\pi_T = 0$ (that is, the internal energy is independent of the volume occupied by the sample) can be taken to be the definition of a perfect gas, for later we shall see that it implies the equation of state $pV = nRT$.

(b) The Joule experiment

James Joule thought that he could measure π_T by observing the change in temperature of a gas when it is allowed to expand into a vacuum. He used two metal vessels immersed in a water bath (Fig. 3.8). One was filled with air at about 22 atm and the other was evacuated. He then tried to measure the change in temperature of the water of the bath when a stopcock was opened and the air expanded into a vacuum. He observed no change in temperature.

The thermodynamic implications of the experiment are as follows. No work was done in the expansion into a vacuum, so $w = 0$. No heat entered or left the system (the gas) because the temperature of the bath did not change, so $q = 0$. Consequently, within the accuracy of the experiment, $\Delta U = 0$. It follows that U does not change much when a gas expands isothermally and therefore that $\pi_T = 0$.

Joule's experiment was crude. In particular, the heat capacity of the apparatus was so large that the temperature change that gases do in fact cause was too small to measure. His experiment was on a par with Boyle's: he extracted an essential limiting property of a gas, a property of a perfect gas, without detecting the small deviations characteristic of real gases.

Illustration 3.1

For ammonia, $\pi_{T,m} = 840$ Pa at 300 K and 1.0 bar where $\pi_{T,m} = (\partial U_m/\partial V)_T$, and $C_{V,m} = 27.32$ J K^{-1} mol^{-1}. The change in molar internal energy of ammonia when it is heated through 2.0 K and compressed through 100 cm^3 is approximately

$$\Delta U_m \approx (840 \text{ J m}^{-3} \text{ mol}^{-1}) \times (-100 \times 10^{-6} \text{ m}^3) + (27.32 \text{ J K}^{-1} \text{ mol}^{-1}) \times (2.0 \text{ K})$$
$$\approx -0.084 \text{ J mol}^{-1} + 55 \text{ J mol}^{-1} = 55 \text{ J mol}^{-1}$$

Note that the contribution of the heating term greatly dominates that of the compression term for this gas.

(c) Changes in internal energy at constant pressure

Partial derivatives have many useful properties and some that we shall draw on frequently are reviewed in *Further information 1*. Skilful use of them can often turn some unfamiliar quantity into a quantity that can be recognized, interpreted, or measured.

As an example, suppose we want to find out how the internal energy varies with temperature when the pressure of the system is kept constant. If we divide both sides of eqn 6 by dT and impose the condition of constant pressure on the resulting differentials, so dU/dT on the left becomes $(\partial U/\partial T)_p$, we obtain

$$\left(\frac{\partial U}{\partial T}\right)_p = \pi_T \left(\frac{\partial V}{\partial T}\right)_p + C_V \tag{3.8}$$

It is usually sensible in thermodynamics to inspect the output of a manipulation like this to see if it contains any recognizable physical quantity. The differential coefficient on the right in this expression is the slope of the plot of volume against temperature (at constant pressure). This property is normally tabulated as the **expansion coefficient**, α, of a substance,[2] which is defined as

$$\alpha = \frac{1}{V}\left(\frac{\partial V}{\partial T}\right)_p \tag{3.9}$$

A large value of α means that the volume of the sample responds strongly to changes in temperature. Table 3.1 lists some experimental values.

Table 3.1[*] Expansion coefficients (α) and isothermal compressibilities (κ_T).

Substance	$\alpha/(10^{-4}$ K$^{-1})$	$\kappa_T/(10^{-6}$ atm$^{-1})$
Benzene	12.4	92.1
Diamond	0.030	0.187
Lead	0.861	2.21
Water	2.1	49.6

[*] More values are given in the *Data section* at the end of this volume.

2. As for heat capacities, the expansion coefficient of a mixture depends on whether or not the composition is allowed to change. Throughout this chapter, we deal only with pure substances, so this complication can be disregarded.

Example 3.2 Using the expansion coefficient of a gas

Derive an expression for the expansion coefficient for a perfect gas.

Method The expansion coefficient is defined in eqn 9. To use this expression, substitute the expression for V in terms of T obtained from the equation of state for the gas. As implied by eqn 9, the pressure, p, is treated as a constant.

Answer Because $pV = nRT$, we can write

$$\alpha = \frac{1}{V}\left(\frac{\partial(nRT/p)}{\partial T}\right)_p = \frac{nR}{pV} = \frac{1}{T}$$

The higher the temperature, the less responsive is its volume to a change in temperature.

Self-test 3.2 Evaluate α for a gas for which the equation of state is $p = nRT/(V - nb)$. $\qquad [\alpha = (1 - b/V_\mathrm{m})/T]$

When we introduce the definition of α into the equation for $(\partial U/\partial T)_p$, we obtain

$$\left(\frac{\partial U}{\partial T}\right)_p = \alpha \pi_T V + C_V \tag{3.10}$$

This equation is entirely general (provided the system is closed and its composition is constant). It expresses the dependence of the internal energy on the temperature at constant pressure in terms of C_V, which can be measured in one experiment, in terms of α, which can be measured in another, and in terms of the quantity π_T. For a perfect gas, $\pi_T = 0$, so then

$$\left(\frac{\partial U}{\partial T}\right)_p = C_V \tag{3.11}°$$

That is, the constant-volume heat capacity of a perfect gas is equal to the slope of a plot of internal energy against temperature at constant pressure as well as (by definition) to the slope at constant volume.

3.4 The temperature dependence of the enthalpy

We can carry out a similar set of operations on the enthalpy, $H = U + pV$. The quantities U, p, and V are all state functions; therefore H is also a state function and dH is an exact differential.

(a) Changes in the enthalpy at constant volume

The variation of enthalpy with temperature at constant pressure is simply the constant-pressure heat capacity, C_p. The simplicity of this relation strongly suggests that H will prove to be a useful thermodynamic function when the pressure is under our control. We saw a sign of that in the relation $\Delta H = q_p$ (eqn 2.24). We shall therefore regard H as a function of p and T, and adapt the argument in Section 3.3 to find an expression for the variation of H with temperature at constant volume. This relation will prove useful for relating the heat capacities at constant pressure and volume and for a discussion of the liquefaction of gases.

By the same argument as for U (but with p in place of V) we find that for a closed system of constant composition,

$$dH = \left(\frac{\partial H}{\partial p}\right)_T dp + \left(\frac{\partial H}{\partial T}\right)_p dT \qquad (3.12)$$

We recognize the second coefficient as the definition of the constant-pressure heat capacity, C_p, so

$$dH = \left(\frac{\partial H}{\partial p}\right)_T dp + C_p\, dT \qquad (3.13)$$

The manipulation of this expression is slightly more involved than before, but we show in the following *Justification* that it implies that

$$\left(\frac{\partial H}{\partial T}\right)_V = \left(1 - \frac{\alpha\mu}{\kappa_T}\right)C_p \qquad (3.14)$$

where the **isothermal compressibility**, κ_T (kappa), is defined as

$$\kappa_T = -\frac{1}{V}\left(\frac{\partial V}{\partial p}\right)_T \qquad [3.15]$$

and the **Joule–Thomson coefficient**, μ (mu), is defined as

$$\mu = \left(\frac{\partial T}{\partial p}\right)_H \qquad [3.16]$$

Equation 14 applies to any substance. Because all the quantities that appear in it can be measured in suitable experiments, we now know how H varies with T when the volume of the sample is held constant.

Justification 3.1

First, divide eqn 13 through by dT and impose constant volume:

$$\left(\frac{\partial H}{\partial T}\right)_V = \left(\frac{\partial H}{\partial p}\right)_T\left(\frac{\partial p}{\partial T}\right)_V + C_p$$

> The Euler chain relation states that, for a differentiable function $z = z(x, y)$,
>
> $$\left(\frac{\partial y}{\partial x}\right)_z\left(\frac{\partial x}{\partial z}\right)_y\left(\frac{\partial z}{\partial y}\right)_x = -1$$

The third differential coefficient looks like something we ought to recognize, and is perhaps related to $(\partial V/\partial T)_p$, the expansion coefficient. It follows from Euler's chain relation that

$$\left(\frac{\partial p}{\partial T}\right)_V\left(\frac{\partial T}{\partial V}\right)_p\left(\frac{\partial V}{\partial p}\right)_T = -1$$

and therefore that

$$\left(\frac{\partial p}{\partial T}\right)_V = -\frac{1}{(\partial T/\partial V)_p(\partial V/\partial p)_T}$$

> The reciprocal identity states that
>
> $$\left(\frac{\partial y}{\partial x}\right)_z = \frac{1}{(\partial x/\partial y)_z}$$

Unfortunately, $(\partial T/\partial V)_p$ occurs instead of $(\partial V/\partial T)_p$. However, the reciprocal identity allows us to invert partial derivatives and to write

$$\left(\frac{\partial p}{\partial T}\right)_V = -\frac{(\partial V/\partial T)_p}{(\partial V/\partial p)_T} = \frac{\alpha}{\kappa_T}$$

Next, we must change $(\partial H/\partial p)_T$ into something recognizable. The chain relation lets us write this partial derivative as

$$\left(\frac{\partial H}{\partial p}\right)_T = -\frac{1}{(\partial p/\partial T)_H (\partial T/\partial H)_p}$$

Both derivatives may be brought up into the numerator:

$$\left(\frac{\partial H}{\partial p}\right)_T = -\left(\frac{\partial T}{\partial p}\right)_H \left(\frac{\partial H}{\partial T}\right)_p$$

and we can recognize both the constant-pressure heat capacity, C_p, and the Joule–Thomson coefficient μ as defined in eqn 16. Therefore,

$$\left(\frac{\partial H}{\partial p}\right)_T = -\mu C_p$$

When we use this expression and the expression for $(\partial p/\partial T)_V$ in the first equation in this *Justification*, we obtain eqn 14.

(b) The isothermal compressibility

The negative sign in the definition of κ_T, eqn 15, ensures that the compressibility is a positive quantity, because an increase of pressure, implying a positive dp, brings about a reduction of volume, a negative dV. The isothermal compressibility is obtained from the slope of the plot of volume against pressure at constant temperature (that is, it is proportional to the slope of an isotherm). Table 3.1 lists some values of κ_T. Its value for a perfect gas is obtained by substitution of the perfect gas equation of state into eqn 15, which gives

$$\kappa_T = -\frac{1}{V}\left(\frac{\partial(nRT/p)}{\partial p}\right)_T = -\frac{nRT}{V}\left(-\frac{1}{p^2}\right) = \frac{1}{p} \tag{3.17}°$$

This expression shows that, the higher the pressure of the gas, the lower its compressibility.

Example 3.3 Using the isothermal compressibility

The isothermal compressibility of water at 20°C and 1 atm is 4.96×10^{-5} atm^{-1}. What change of volume occurs when a sample of volume 50 cm^3 is subjected to an additional 1000 atm at constant temperature?

Method We know from the definition of κ_T that, for an infinitesimal change of pressure at constant temperature, the volume changes by

$$dV = \left(\frac{\partial V}{\partial p}\right)_T dp = -\kappa_T V\, dp$$

For a finite change in pressure, we need to integrate both sides. When confronted by an integration, it is often a useful first approximation (for substances other than gases) to suppose that the integrand is constant over the range of integration.

Thermocouples Gas at low pressure

Insulation

Porous barrier   Gas at high pressure

3.9 The apparatus used for measuring the Joule–Thomson effect. The gas expands through the porous barrier, which acts as a throttle, and the whole apparatus is thermally insulated. As explained in the text, this arrangement corresponds to an isenthalpic expansion (expansion at constant enthalpy). Whether the expansion results in a heating or a cooling of the gas depends on the conditions.

Answer The integrals we need to evaluate are

$$\int_{V_i}^{V_f} dV = -\int_{p_i}^{p_f} \kappa_T V \, dp$$

The integral on the left is ΔV. If we suppose that κ_T and V are approximately constant over the range of pressures of interest, we can write

$$\Delta V = -\kappa_T V \int_{p_i}^{p_f} dp = -\kappa_T V \, \Delta p$$

Substitution of the data into the last expression then gives

$$\Delta V = -(4.96 \times 10^{-5} \, \text{atm}^{-1}) \times (50 \, \text{cm}^3) \times (1000 \, \text{atm}) = -2.5 \, \text{cm}^3$$

Comment Because the compression results in a decrease in volume of only 5 per cent, the assumption of constant V and κ_T is probably acceptable as a first approximation. Note that very high pressures are needed to bring about significant changes of volume.

Self-test 3.3 A sample of copper of volume 50 cm³ is subjected to a additional pressure of 100 atm and a temperature increase of 5.0 K. Estimate the total change in volume.   [8.8 mm³]

(c) The Joule–Thomson effect

The analysis of the Joule–Thomson coefficient is central to the technological problems associated with the liquefaction of gases. We need to be able to interpret it physically and to measure it.

As shown in the *Justification* below, the cunning required to impose the constraint of constant enthalpy, so that the process is **isenthalpic**, was supplied by Joule and William Thomson (later Lord Kelvin). They let a gas expand through a porous barrier from one constant pressure to another, and monitored the difference of temperature that arose from the expansion (Fig. 3.9). The whole apparatus was insulated so that the process was adiabatic. They observed a lower temperature on the low pressure side, the difference in temperature being proportional to the pressure difference they maintained. This cooling by adiabatic expansion is now called the **Joule–Thomson effect**.

Justification 3.2

Here we show that the experimental arrangement results in expansion at constant enthalpy. Because all changes to the gas occur adiabatically,

$$q = 0, \quad \text{which implies} \quad \Delta U = w$$

Consider the work done as the gas passes through the barrier. We focus on the passage of a fixed amount of gas from the high pressure side, where the pressure is p_i, the temperature T_i, and the gas occupies a volume V_i (Fig. 3.10). The gas emerges on the low pressure side, where the same amount of gas has a pressure p_f, a temperature T_f, and occupies a volume V_f. The gas on the left is compressed isothermally by the upstream gas acting as a piston. The relevant pressure is p_i and the volume changes from V_i to 0; therefore, the work done on the gas is

$$w_1 = -p_i(0 - V_i) = p_i V_i$$

3.10 The thermodynamic basis of the Joule–Thomson expansion. The pistons represent the upstream and downstream gases, which maintain constant pressures either side of the throttle. The transition from the left-hand diagram to the right-hand diagram, which represents the passage of a given amount of gas through the throttle, occurs without change of enthalpy.

The gas expands isothermally on the right of the barrier (but possibly at a different constant temperature) against the pressure p_f provided by the downstream gas acting as a piston to be driven out. The volume changes from 0 to V_f, so the work done on the gas in this stage is

$$w_2 = -p_f(V_f - 0) = -p_f V_f$$

The total work done on the gas is the sum of these two quantities, or

$$w = w_1 + w_2 = p_i V_i - p_f V_f$$

It follows that the change of internal energy of the gas as it moves adiabatically from one side of the barrier to the other is

$$U_f - U_i = w = p_i V_i - p_f V_f$$

Reorganization of this expression gives

$$U_f + p_f V_f = U_i + p_i V_i, \qquad \text{or} \quad H_f = H_i$$

Therefore, the expansion occurs without change of enthalpy.

The property measured in the experiment is the ratio of the temperature change to the change of pressure, $\Delta T/\Delta p$. Adding the constraint of constant enthalpy and taking the limit of small Δp implies that the thermodynamic quantity measured is $(\partial T/\partial p)_H$, which is the Joule–Thomson coefficient, μ. In other words, the physical interpretation of μ is that it is the ratio of the change in temperature to the change in pressure when a gas expands under adiabatic conditions.

The modern method of measuring μ is indirect, and involves measuring the **isothermal Joule–Thomson coefficient**, the quantity

$$\mu_T = \left(\frac{\partial H}{\partial p}\right)_T \tag{3.18}$$

which is the slope of enthalpy against pressure at constant temperature (Fig. 3.11). As shown by the last equation in *Justification* 3.1, the two coefficients are related by:

$$\mu_T = -C_p \mu \tag{3.19}$$

To measure μ_T, the gas is pumped continuously at a steady pressure through a heat exchanger (which brings it to the required temperature), and then through a porous plug inside a thermally insulated container. The steep pressure drop is measured, and the cooling effect is exactly offset by an electric heater placed immediately after

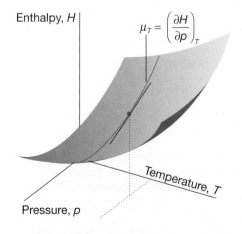

3.11 The isothermal Joule–Thomson coefficient is the slope of the enthalpy with respect to changing pressure, the temperature being held constant.

Table 3.2* Inversion temperatures (T_I), normal freezing (T_f) and boiling (T_b) points, and Joule–Thomson coefficients (μ) at 1 atm and 298 K

	T_I/K	T_f/K	T_b/K	μ/(K atm^{-1})
Ar	723	83.8	87.3	
CO_2	1500		194.7	1.11
He	40		4.2	−0.060
N_2	621	63.3	77.4	0.25

* More values are given in the *Data section*.

the plug (Fig. 3.12). The energy provided by the heater is monitored. Because the heat can be identified with the value of ΔH for the gas (because $\Delta H = q_p$), and the pressure change Δp is known, we can find μ_T from the limiting value of $\Delta H/\Delta p$ as $\Delta p \rightarrow 0$, and then convert it to μ. Table 3.2 lists some values obtained in this way.

Real gases have nonzero Joule–Thomson coefficients. Depending on the identity of the gas, the pressure, the relative magnitudes of the attractive and repulsive intermolecular forces, and the temperature, the sign of the coefficient may be either positive or negative (Fig. 3.13). A positive sign implies that dT is negative when dp is negative, in which case the gas cools on expansion. Gases that show a heating effect ($\mu < 0$) at one temperature show a cooling effect ($\mu > 0$) when the temperature is below their upper **inversion temperature**, T_I (Table 3.2, Fig. 3.14). As indicated in Fig. 3.14, a gas typically has two inversion temperatures, one at high temperature and the other at low.

The 'Linde refrigerator' makes use of Joule–Thompson expansion to liquefy gases (Fig. 3.15). The gas at high pressure is allowed to expand through a throttle; it cools and is circulated past the incoming gas. That gas is cooled, and its subsequent expansion cools it still further. There comes a stage when the circulating gas becomes so cold that it condenses to a liquid.

For a perfect gas, $\mu = 0$; hence, the temperature of a perfect gas is unchanged by Joule–Thomson expansion.[3] This characteristic points clearly to the involvement of intermolecular forces in determining the size of the effect. However, the Joule–

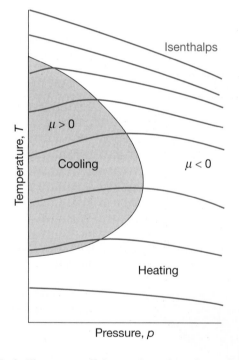

Isenthalps

$\mu > 0$

Cooling

$\mu < 0$

Temperature, T

Heating

Pressure, p

Gas flow

Heater

Thermometer   Porous plug   Thermometer

3.12 A schematic diagram of the apparatus used for measuring the isothermal Joule–Thomson coefficient. The electrical heating required to offset the cooling arising from expansion is interpreted as ΔH and used to calculate $(\partial H/\partial p)_T$ which is then converted to μ as explained in the text.

3.13 The sign of the Joule–Thomson coefficient, μ, depends on the conditions. Inside the boundary, the shaded area, it is positive and outside it is negative. The temperature corresponding to the boundary at a given pressure is the 'inversion temperature' of the gas at that pressure. For a given pressure, the temperature must be below a certain value if cooling is required, but if it becomes too low, the boundary is crossed again and heating occurs. Reduction of pressure under adiabatic conditions moves the system along one of the isenthalps, or curves of constant enthalpy. The inversion temperature curve runs through the points of the isenthalps where their slope changes from negative to positive.

3. Simple adiabatic expansion does cool a perfect gas, because the gas does work; recall Section 2.6.

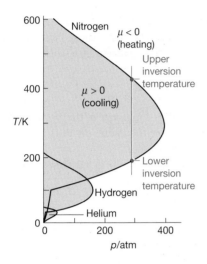

3.14 The inversion temperatures for three real gases, nitrogen, hydrogen, and helium.

Thomson coefficient of a real gas does not necessarily approach zero as the pressure is reduced even though the equation of state of the gas approaches that of a perfect gas. The coefficient behaves like the properties discussed in Section 1.3b in the sense that it depends on derivatives and not on p, V, and T themselves.

3.5 The relation between C_V and C_p

The constant-pressure heat capacity C_p differs from the constant-volume heat capacity C_V by the work needed to change the volume of the system. This work arises in two ways. One is the work of driving back the atmosphere; the other is the work of stretching the bonds in the material, including any weak intermolecular interactions. In the case of a perfect gas, the second makes no contribution. We shall now derive a general relation between the two heat capacities, and show that it reduces to the perfect gas result in the absence of intermolecular forces.

First, we carry through the calculation for a perfect gas. In this special case, we can use eqn 11 to express both heat capacities in terms of derivatives at constant pressure:

$$C_p - C_V = \left(\frac{\partial H}{\partial T}\right)_p - \left(\frac{\partial U}{\partial T}\right)_p \tag{3.20}$$

Then we introduce

$$H = U + pV = U + nRT$$

into the first term, which results in

$$C_p - C_V = \left(\frac{\partial U}{\partial T}\right)_p + nR - \left(\frac{\partial U}{\partial T}\right)_p = nR$$

This is the formal derivation of eqn 2.31.

In the *Justification* below we show that the general relation between the two heat capacities for any pure substance is

$$C_p - C_V = \frac{\alpha^2 TV}{\kappa_T} \tag{3.21}$$

This formula is a thermodynamic expression, which means that it applies to any substance (that is, that it is 'universally true'). It reduces to eqn 2.31 for a perfect gas when we set $\alpha = 1/T$ and $\kappa_T = 1/p$.

Justification 3.3

A useful rule when doing a problem in thermodynamics is to go back to first principles. In the present problem we do this twice, first by expressing C_p and C_V in terms of their definitions and then by inserting the definition $H = U + pV$:

$$C_p - C_V = \left(\frac{\partial H}{\partial T}\right)_p - \left(\frac{\partial U}{\partial T}\right)_V = \left(\frac{\partial U}{\partial T}\right)_p + \left(\frac{\partial (pV)}{\partial T}\right)_p - \left(\frac{\partial U}{\partial T}\right)_V$$

We have already calculated the difference of the first and third terms on the right, and eqn 9 lets us write this difference as $\alpha \pi_T V$. The factor αV gives the change in volume when the temperature is raised, and $\pi_T = (\partial U/\partial V)_T$ converts this change in volume into a change in internal energy. We can simplify the remaining term by noting that, because p is constant,

$$\left(\frac{\partial (pV)}{\partial T}\right)_p = p\left(\frac{\partial V}{\partial T}\right)_p = \alpha pV$$

3.15 The principle of the Linde refrigerator is shown in this diagram. The gas is recirculated, and so long as it is beneath its inversion temperature it cools on expansion through the throttle. The cooled gas cools the high-pressure gas, which cools still further as it expands. Eventually liquefied gas drips from the throttle.

The middle term of this expression identifies it as the contribution to the work of pushing back the atmosphere: $(\partial V/\partial T)_p$ is the change of volume caused by a change of temperature, and multiplication by p converts this expansion into work.

Collecting the two contributions gives

$$C_p - C_V = \alpha(p + \pi_T)V$$

As just remarked, the first term on the right (αpV) is a measure of the work needed to push back the atmosphere; the second term on the right, $\alpha\pi_T V$, is the work required to separate the molecules composing the system.

At this point we can go further by using the result we prove in *Justification 5.1* that

$$\pi_T = T\left(\frac{\partial p}{\partial T}\right)_V - p$$

When this expression is inserted in the last equation we obtain

$$C_p - C_V = \alpha TV\left(\frac{\partial p}{\partial T}\right)_V$$

The coefficient that appears here was encountered in *Justification 3.1*, where we saw that it is equal to α/κ_T. Therefore, this expression turns into eqn 21.

Because thermal expansivities α of liquids and solids are small, it is tempting to deduce from eqn 21 that for them $C_p \approx C_V$. But this is not always so, because the compressibility κ_T might also be small, so α^2/κ_T might be large. That is, although only a little work need be done to push back the atmosphere, a great deal of work may have to be done to pull atoms apart from one another as the solid expands. As an illustration, for water at 25°C, eqn 21 gives $C_{p,m} = 75.3 \text{ J K}^{-1} \text{ mol}^{-1}$ compared with $C_{V,m} = 74.8 \text{ J K}^{-1} \text{ mol}^{-1}$. In some cases, the two heat capacities differ by as much as 30 per cent.

Checklist of key ideas

☐ path function, a property that depends on the path between two states

State functions and exact differentials

3.1 State and path functions

3.2 Exact and inexact differentials

☐ exact differential, an infinitesimal quantity that, when integrated, gives a result that is independent of the path between the initial and final states

☐ inexact differential, an infinitesimal quantity that, when integrated, gives a result that depends on the path between the initial and final states

Thermodynamic consequences

3.3 Changes in internal energy

☐ general change in internal energy for a closed system, $dU = (\partial U/\partial V)_T \, dV + (\partial U/\partial T)_V \, dT$

☐ internal pressure, $\pi_T = (\partial U/\partial V)_T$

☐ internal energy and isothermal expansion of perfect gas, $(\partial U/\partial V)_T = 0$

☐ Joule experiment, the failure to observe a change in temperature when a gas expands freely into a vacuum

☐ expansion coefficient, $\alpha = (1/V)(\partial V/\partial T)_p$

☐ temperature dependence of internal energy at constant pressure, $(\partial U/\partial T)_p = \alpha\pi_T V + C_V$

3.4 The temperature dependence of the enthalpy

☐ isothermal compressibility, $\kappa_T = -(1/V)(\partial V/\partial p)_T$

☐ Joule–Thomson coefficient, $\mu = (\partial T/\partial p)_H$

☐ isenthalpic process, a process at constant enthalpy

☐ Joule–Thomson effect, the cooling of a gas by isenthalpic adiabatic expansion

☐ isothermal Joule–Thomson coefficient, $\mu_T = (\partial H/\partial p)_T = -C_p\mu$

☐ inversion temperature, the temperature at which the Joule–Thomson coefficient changes sign

3.5 The relation between C_V and C_p

☐ relation between C_V and C_p, in general $C_p - C_V = \alpha^2 TV/\kappa_T$

Further reading

Articles of general interest

R.A. Alberty, Legendre transforms in chemical thermodynamics. *Chem. Rev.* **94**, 1457 (1994).

E.W. Anacker, S.E. Anacker, and W.J. Swartz, Some comments on partial derivatives in thermodynamics. *J. Chem. Educ.* **64**, 674 (1987).

S.M. Blinder, Mathematical methods in elementary thermodynamics. *J. Chem. Educ.* **43**, 85 (1966).

G.A. Estèvez, K. Yang, and B.B. Dasgupta, Thermodynamic partial derivatives and experimentally measurable quantities. *J. Chem. Educ.* **66**, 890 (1989).

J. Wisniak, The Joule–Thomson coefficient for pure gases and their mixtures. *Chem. Educ.* **4**, 51 (1999).

Texts and sources of data and information

J.R. Barrante, *Applied mathematics for physical chemistry*. Prentice Hall, Upper Saddle River (1998).

I.M. Klotz and R.M. Rosenberg, *Chemical thermodynamics: basic theory and methods.* Wiley-Interscience, New York (2000).

M.L. McGlashan, *Chemical thermodynamics.* Academic Press, London (1979).

R.G. Mortimer, *Mathematics for physical chemistry.* Academic Press, San Diego (1999).

E. Steiner, *The chemistry maths book.* Oxford University Press (1996).

Exercises

Discussion questions

3.1 (a) Explain the significance of a physical observable being a state function.

3.1 (b) Compile a list of as many state functions as you can identify.

3.2 (a) Explain why the internal pressure of a perfect gas is zero.

3.2 (b) Suggest how the internal energy of a van der Waals gas should vary with volume at constant temperture.

3.3 (a) Explain the significance of the Joule experiment. What would Joule observe in a more sensitive apparatus?

3.3 (b) Explain the origin and implications of the Joule–Thomson effect.

Numerical exercises

Assume that all gases are perfect and that all data refer to 298.15 K unless stated otherwise.

3.4 (a) Show that the following functions have exact differentials: (a) $x^2y + 3y^2$, (b) $x \cos xy$.

3.4 (b) Show that the following functions have exact differentials: (a) x^3y^2, (b) $t(t + e^s) + s$.

3.5 (a) Let $z = ax^2y^3$. Find dz.

3.5 (b) Let $z = x/(1 + y)^2$. Find dz.

3.6 (a) (a) What is the total differential of $z = x^2 + 2y^2 - 2xy + 2x - 4y - 8$? (b) Show that $\partial^2z/\partial y\partial x = \partial^2z/\partial x\partial y$ for this function.

3.6 (b) (a) What is the total differential of $z = x^3 - 2xy^2 + 15$? (b) Show that $\partial^2z/\partial y\partial x = \partial^2z/\partial x\partial y$ for this function.

3.7 (a) Let $z = xy - y + \ln x + 2$. Find dz and show that it is exact.

3.7 (b) Let $z = x^2y + xy^2$. Find dz and show that it is exact.

3.8 (a) Express $(\partial C_V/\partial V)_T$ as a second-derivative of U and find its relation to $(\partial U/\partial V)_T$. From this relation show that $(\partial C_V/\partial V)_T = 0$ for a perfect gas.

3.8 (b) Express $(\partial C_p/\partial p)_T$ as a second-derivative of H and find its relation to $(\partial H/\partial p)_T$. From this relation show that $(\partial C_p/\partial p)_T = 0$ for a perfect gas.

3.9 (a) By direct differentiation of $H = U + pV$, obtain a relation between $(\partial H/\partial U)_p$ and $(\partial U/\partial V)_p$.

3.9 (b) Confirm that $(\partial H/\partial U)_p = 1 + p(\partial V/\partial U)_p$ by expressing $(\partial H/\partial U)_p$ as the ratio of two derivatives with respect to volume and then using the definition of enthalpy.

3.10 (a) Write an expression for dV given that V is a function of p and T. Deduce an expression for $d \ln V$ in terms of the expansion coefficient and the isothermal compressibility.

3.10 (b) Write an expression for dp given that p is a function of V and T. Deduce an expression for $d \ln p$ in terms of the expansion coefficient and the isothermal compressibility.

3.11 (a) The internal energy of a perfect monatomic gas relative to its value at $T = 0$ is $\frac{3}{2}nRT$. Calculate $(\partial U/\partial V)_T$ and $(\partial H/\partial V)_T$ for the gas.

3.11 (b) The internal energy of a perfect monatomic gas relative to its value at $T = 0$ is $\frac{3}{2}nRT$. Calculate $(\partial H/\partial p)_T$ and $(\partial U/\partial p)_T$ for the gas.

3.12 (a) The coefficient of thermal expansion, α, is defined in eqn 8 and the isothermal compressibility, κ_T, is defined in eqn 14. Starting from the expression for the total differential dV in terms of T and p, show that $(\partial p/\partial T)_V = \alpha/\kappa_T$.

3.12 (b) Evaluate α and κ_T for a perfect gas.

3.13 (a) When a certain freon used in refrigeration was expanded adiabatically from an initial pressure of 32 atm and 0°C to a final pressure of 1.00 atm, the temperature fell by 22 K. Calculate the Joule–Thomson coefficient, μ, at 0°C, assuming it remains constant over this temperature range.

3.13 (b) A vapour at 22 atm and 5°C was allowed to expand adiabatically to a final pressure of 1.00 atm; the temperature fell by 10 K. Calculate the Joule–Thomson coefficient, μ, at 5°C, assuming it remains constant over this temperature range.

3.14 (a) For a van der Waals gas, $\pi_T = a/V_m^2$. Calculate ΔU_m for the isothermal reversible expansion of nitrogen gas from an initial volume of 1.00 L to 24.8 L at 298 K. What are the values of q and w?

3.14 (b) For a van der Waals gas, $\pi_T = a/V_m^2$. Calculate ΔU_m for the isothermal reversible expansion of argon from an initial volume of 1.00 L to 22.1 L at 298 K. What are the values of q and w?

3.15 (a) The volume of a certain liquid varies with temperature as

$$V = V'\{0.75 + 3.9 \times 10^{-4}(T/K) + 1.48 \times 10^{-6}(T/K)^2\}$$

where V' is its volume at 300 K. Calculate its expansion coefficient, α, at 320 K.

3.15 (b) The volume of a certain liquid varies with temperature as

$$V = V'\{0.77 + 3.7 \times 10^{-4}(T/K) + 1.52 \times 10^{-6}(T/K)^2\}$$

where V' is its volume at 298 K. Calculate its expansion coefficient, α, at 310 K.

3.16 (a) The isothermal compressibility of copper at 293 K is 7.35×10^{-7} atm^{-1}. Calculate the pressure that must be applied in order to increase its density by 0.08 per cent.

3.16 (b) The isothermal compressibility of lead at 293 K is 2.21×10^{-6} atm^{-1}. Calculate the pressure that must be applied in order to increase its density by 0.08 per cent.

3.17 (a) Given that $\mu = 0.25$ K atm^{-1} for nitrogen, calculate the value of its isothermal Joule–Thomson coefficient. Calculate the energy that must be supplied as heat to maintain constant temperature when 15.0 mol N_2 flows through a throttle in an isothermal Joule–Thomson experiment and the pressure drop is 75 atm.

3.17 (b) Given that $\mu = 1.11$ K atm^{-1} for carbon dioxide, calculate the value of its isothermal Joule–Thomson coefficient. Calculate the energy that must be supplied as heat to maintain constant temperature when 12.0 mol CO_2 flows through a throttle in an isothermal Joule–Thomson experiment and the pressure drop is 55 atm.

3.18 (a) To design a particular kind of refrigerator we need to know the temperature reduction brought about by adiabatic expansion of the refrigerant gas. For one type of freon, $\mu = 1.2$ K atm^{-1}. What pressure difference is needed to produce a temperature drop of 5.0 K?

3.18 (b) For another type of freon (see previous exercise), $\mu = 13.3$ mK kPa^{-1}. What pressure difference is needed to produce a temperature drop of 4.5 K?

Problems†

Assume all gases are perfect unless stated otherwise. Unless otherwise stated, thermochemical data are for 298.15 K.

Numerical problems

3.1 The isothermal compressibility of lead is 2.21×10^{-6} atm^{-1}. Express this value in Pa^{-1}. A cube of lead of side 10 cm at 25°C was to be inserted in the keel of an underwater exploration TV camera, and its designers needed to know the stresses in the equipment. Calculate the change of volume of the cube at a depth of 1.000 km (disregarding the effects of temperature). Take the mean density of sea water as 1.03 g cm^{-3}. Given that the expansion coefficient of lead is 8.61×10^{-5} K^{-1} and that the temperature where the camera operates is –5°C, calculate the volume of the block taking the temperature into account too.

3.2 Calculate the change in (a) the molar internal energy and (b) the molar enthalpy of water when its temperature is raised by 10 K. Account for the difference between the two quantities.

3.3 The constant-volume heat capacity of a gas can be measured by observing the decrease in temperature when it expands adiabatically and reversibly. If the decrease in pressure is also measured, we can use it to infer the value of γ (the ratio of heat capacities, C_p/C_V) and hence, by combining the two values, deduce the constant-pressure heat capacity. A fluorocarbon gas was allowed to expand reversibly and adiabatically to twice its volume; as a

result, the temperature fell from 298.15 K to 248.44 K and its pressure fell from 1522.2 Torr to 613.85 Torr. Evaluate C_p.

3.4 A sample consisting of 1.00 mol of a van der Waals gas is compressed from 20.0 L to 10.0 L at 300 K. In the process, 20.2 kJ of work is done on the gas. Given that $\mu = \{(2a/RT) - b\}/C_{p,m}$, with $C_{p,m} = 38.4$ J K^{-1} mol^{-1}, $a = 3.60$ L^2 atm mol^{-2}, and $b = 0.44$ L mol^{-1}, calculate ΔH for the process.

3.5 Estimate γ (the ratio of heat capacities) for xenon at 100°C and 1.00 atm on the assumption that it is a van der Waals gas.

3.6‡ A cylindrical container of fixed total volume is divided into three sections, S_1, S_2, and S_3. The sections S_1 and S_2 are separated by an adiabatic piston whereas S_2 and S_3 are separated by a diathermic (heat-conducting) piston. The pistons can slide along the walls of the cylinder without friction. Each section of the cylinder contains 1.00 mol of a perfect diatomic gas. Initially, the gas pressure in all three sections is 1.00 bar and the temperature is 298 K. The gas in S_1 is heated slowly until the temperature of the gas in S_3 reaches 348 K. Find the final temperature, pressure, and volume, as well as the change in internal energy for each section. Determine the total energy supplied to the gas in S_1.

3.7‡ Solve Problem 3.6 for the case where both pistons are (a) adiabatic, (b) diathermic.

3.8‡ A gas obeying the equation of state $p(V - nb) = nRT$ is subjected to a Joule–Thomson expansion. Will the temperature increase, decrease, or remain the same?

† Problems denoted with the symbol ‡ were supplied by Charles Trapp and Carmen Giunta.

Theoretical problems

3.9 Determine whether or not $dz = xy\,dx + xy\,dy$ is exact by integrating it around the closed curve formed by the paths $y = x$ and $y = x^2$ between the points (0, 0) and (1, 1).

3.10 Decide whether $dq = (RT/p)\,dp - R\,dT$ is exact. Then determine whether multiplication of dq by $1/T$ is exact. Comment on the significance of your result.

3.11 Obtain the total differential of the function $w = xy + yz + xz$. Demonstrate that dw is exact by integration between the points (0, 0, 0) and (1, 1, 1) along the two different paths (1) $z = y = x$ and (2) $z = y = x^2$.

3.12 Derive the relation $C_V = -(\partial U/\partial V)_T(\partial V/\partial T)_U$ from the expression for the total differential of $U(T, V)$.

3.13 Starting from the expression for the total differential of $H(T, p)$, express $(\partial H/\partial p)_T$ in terms of C_p and the Joule–Thomson coefficient, μ.

3.14 Starting from the expression $C_p - C_V = T(\partial p/\partial T)_V(\partial V/\partial T)_p$, use the appropriate relations between partial derivatives to show that

$$C_p - C_V = \frac{T(\partial V/\partial T)_p^2}{(\partial V/\partial T)_T}$$

Evaluate $C_p - C_V$ for a perfect gas.

3.15 From an analysis of Joule's free expansion experiment, demonstrate that it is possible to calculate the change in internal energy of a perfect gas for any process by knowing only C_V and ΔT.

3.16 By the consideration of a suitable cycle involving a perfect gas, demonstrate that dq is an inexact differential and, therefore, that q is not a state function.

3.17 Use the fact that $(\partial U/\partial V)_T = a/V_m^2$ for a van der Waals gas to show that $\mu C_{p,m} \approx (2a/RT) - b$ by using the definition of μ and appropriate relations between partial derivatives. (*Hint*. Use the approximation $pV_m \approx RT$ when it is justifiable to do so.)

3.18 Obtain the expression for the total differential dp for a van der Waals gas in terms of dT and dV. Also obtain $(\partial V/\partial T)_p$. Demonstrate that dp is not an exact differential by integrating it from (T_1, V_1) to (T_2, V_2) along the two different paths, namely (1) $(T_1, V_1) \rightarrow (T_2, V_1) \rightarrow (T_2, V_2)$ and (2) $(T_1, V_1) \rightarrow (T_1, V_2) \rightarrow (T_2, V_2)$.

3.19 Take nitrogen to be a van der Waals gas with $a = 1.352\ L^2\ atm\ mol^{-2}$ and $b = 0.0387\ L\ mol^{-1}$, and calculate ΔH_m when the pressure on the gas is decreased from 500 atm to 1.00 atm at 300 K. For a van der Waals gas, $\mu = \{(2a/RT) - b\}/C_{p,m}$. Assume $C_{p,m} = \frac{7}{2}R$.

3.20 The pressure of a given amount of a van der Waals gas depends on T and V. Find an expression for dp in terms of dT and dV.

3.21 Rearrange the van der Waals equation of state to give an expression for T as a function of p and V (with n constant). Calculate $(\partial T/\partial p)_V$ and confirm that $(\partial T/\partial p)_V = 1/(\partial p/\partial T)_V$. Go on to confirm Euler's chain relation.

3.22 Calculate the isothermal compressibility and the expansion coefficient of a van der Waals gas. Show, using Euler's chain relation, that $\kappa_T R = \alpha(V_m - b)$.

3.23 Given that $\mu C_p = T(\partial V/\partial T)_p - V$, derive an expression for μ in terms of the van der Waals parameters a and b, and express it in terms of reduced variables. Evaluate μ at 25°C and 1.0 atm, when the molar volume of the gas is 24.6 L mol^{-1}. Use the expression obtained to derive a formula for the inversion temperature of a van der Waals gas in terms of reduced variables, and evaluate it for the xenon sample.

3.24 The thermodynamic equation of state $(\partial U/\partial V)_T = T(\partial p/\partial T)_V - p$ was quoted in the chapter. Derive its partner

$$\left(\frac{\partial H}{\partial p}\right)_T = -T\left(\frac{\partial V}{\partial T}\right)_p + V$$

from it and the general relations between partial differentials.

3.25 Show that for a van der Waals gas,

$$C_{p,m} - C_{V,m} = \lambda R \qquad \frac{1}{\lambda} = 1 - \frac{(3V_r - 1)^2}{4V_r^3 T_r}$$

and evaluate the difference for xenon at 25°C and 10.0 atm.

3.26 The speed of sound, c_s, in a gas of molar mass M is related to the ratio of heat capacities γ by $c_s = (\gamma RT/M)^{1/2}$. Show that $c_s = (\gamma p/\rho)^{1/2}$, where ρ is the mass density of the gas. Calculate the speed of sound in argon at 25°C.

3.27‡ A gas obeys the equation of state $V_m = RT/p + aT^2$ and its constant-pressure heat capacity is given by $C_{p,m} = A + BT + Cp$, where a, A, B, and C are constants independent of T and p. Obtain expressions for (a) the Joule–Thomson coefficient and (b) its constant-volume heat capacity.

Applications: to environmental science

3.28‡ In 1995, the Intergovernmental Panel on Climate Change considered a global average temperature rise of 1.0–3.5°C likely by the year 2100, with 2.0°C its best estimate (IPCC Second Assessment Synthesis of Scientific-Technical Information Relevant to Interpreting Article 2 of the UN Framework Convention on Climate Change (1995)). Predict the average rise in sea level due to thermal expansion of sea water based on temperature rises of 1.0°C, 2.0°C, and 3.5°C given that the volume of the Earth's oceans is $1.37 \times 10^9\ km^3$ and their surface area is $361 \times 10^6\ km^2$, and state the approximations that go into the estimates.

3.29‡ Concerns over the harmful effects of chlorofluorocarbons on stratospheric ozone have motivated a search for new refrigerants. One such alternative is 2,2-dichloro-1,1,1-trifluoroethane (refrigerant 123). B.A. Younglove and M. McLinden published a compendium of thermophysical properties of this substance (*J. Phys. Chem. Ref. Data* **23**, 7 (1994)), from which properties such as the Joule–Thomson coefficient μ can be computed. (a) Compute μ at 1.00 bar and 50°C given that $(\partial H/\partial p)_T = -3.29 \times 10^3\ J\ MPa^{-1}\ mol^{-1}$ and $C_{p,m} = 110.0\ J\ K^{-1}\ mol^{-1}$. (b) Compute the temperature change that would accompany adiabatic expansion of 2.0 mol of this refrigerant from 1.5 bar to 0.5 bar at 50°C.

3.30‡ Another alternative refrigerant (see preceding problem) is 1,1,1,2-tetrafluoroethane (refrigerant HFC-134a). R. Tillner-Roth and H.D. Baehr published a compendium of thermophysical properties of this substance (*J. Phys. Chem. Ref. Data* **23**, 657 (1994)), from which properties such as the Joule–Thomson coefficient μ can be computed. (a) Compute μ at 0.100 MPa and 300 K from the following data (all referring to 300 K):

p/MPa	0.080	0.100	0.12
specific enthalpy/(kJ kg^{-1})	426.48	426.12	425.76

(The specific constant-pressure heat capacity is 0.7649 kJ K^{-1} kg^{-1}.) (b) Compute μ at 1.00 MPa and 350 K from the following data (all referring to 350 K):

p/MPa	0.80	1.00	1.2
specific enthalpy/(kJ kg^{-1})	461.93	459.12	456.15

(The specific constant-pressure heat capacity is 1.0392 kJ K^{-1} kg^{-1}.)

4

The Second Law: the concepts

The direction of spontaneous change

4.1 The dispersal of energy

4.2 Entropy

4.3 Entropy changes accompanying specific processes

4.4 The Third Law of thermodynamics

Concentrating on the system

4.5 The Helmholtz and Gibbs energies

4.6 Standard molar Gibbs energies

Checklist of key ideas

Further reading

Exercises

Problems

The purpose of this chapter is to explain the origin of the spontaneity of physical and chemical change. We examine two simple processes and show how to define, measure, and use a property, the entropy, to discuss spontaneous changes quantitatively. The chapter also introduces a major subsidiary thermodynamic property, the Gibbs energy. The introduction of the Gibbs energy enables the spontaneity of a process to be expressed solely in terms of the properties of a system (instead of having to consider entropy changes in the system and its surroundings). The Gibbs energy also enables us to predict the maximum non-expansion work that a process can do.

Some things happen naturally; some things don't. A gas expands to fill the available volume, a hot body cools to the temperature of its surroundings, and a chemical re-action runs in one direction rather than another. Some aspect of the world deter-mines the **spontaneous** direction of change, the direction of change that does not require work to be done to bring the change about. We can confine a gas to a smaller volume, we can cool an object with a refrigerator, and we can force some reactions to go in reverse (as in the electrolysis of water). However, none of these processes happens spontaneously; each one must be brought about by doing work.

The recognition of two classes of process, spontaneous and non-spontaneous, is summarized by the **Second Law of thermodynamics**. This law may be expressed in a variety of equivalent ways. One statement was formulated by Kelvin:

> **No process is possible in which the sole result is the absorption of heat from a reservoir and its complete conversion into work.**

For example, it has proved impossible to construct an engine like that shown in Fig. 4.1, in which heat is drawn from a hot reservoir and completely converted into work. All real heat engines have both a hot source and a cold sink; some heat is always discarded into the cold sink and not converted into work. The Kelvin state-ment is a generalization of another everyday observation, that a ball at rest on a sur-face has never been observed to leap spontaneously upwards. An upward leap of the ball would be equivalent to the conversion of heat from the surface into work.

The direction of spontaneous change

What determines the direction of spontaneous change? It is not the total energy of the isolated system. The First Law of thermodynamics states that energy is conserved in any process, and we cannot disregard that law now and say that everything tends towards a state of lower energy: the total energy of an isolated system is constant.

Is it perhaps the energy of the *system* that tends towards a minimum? Two arguments show that this cannot be so. First, a perfect gas expands spontaneously into a vacuum, yet its internal energy remains constant as it does so. Secondly, if the energy of a system does happen to decrease during a spontaneous change, the energy of its surroundings must increase by the same amount (by the First Law). The increase in energy of the surroundings is just as spontaneous a process as the decrease in energy of the system.

When a change occurs, the total energy of an isolated system remains constant but it is parcelled out in different ways. Can it be, therefore, that the direction of change is related to the *distribution* of energy? We shall see that this idea is the key, and that spontaneous changes are always accompanied by a dispersal of energy into a more disordered form.

4.1 The dispersal of energy

We can begin to understand the role of the distribution of energy by thinking about a ball (the system) bouncing on a floor (the surroundings). The ball does not rise as high after each bounce because there are inelastic losses in the materials of the ball and floor and the kinetic energy of the ball's overall motion is converted into the energy of thermal motion. The direction of spontaneous change is towards a state in which the ball is at rest with all its energy degraded into the thermal motion of the atoms of the virtually infinite floor (Fig. 4.2).

A ball resting on a warm floor has never been observed to start bouncing. For bouncing to begin, something rather special would need to happen. In the first place, some of the thermal motion of the atoms in the floor would have to accumulate in a single, small object, the ball. This accumulation requires a spontaneous localization of energy from the myriad of vibrations of the atoms of the floor into the much smaller number of atoms that constitute the ball (Fig. 4.3). Furthermore, whereas the thermal motion is disorderly, for the ball to move upwards its atoms must all move in the same direction. The localization of random motion as orderly motion is so unlikely that we can dismiss it as virtually impossible.[1]

We appear to have found the signpost of spontaneous change: *we look for the direction of change that leads to more disorderly dispersal of the total energy of the isolated system.* This principle accounts for the direction of change of the bouncing ball, because its energy is dissipated as thermal motion of the atoms of the floor. The reverse process is not spontaneous because it is highly improbable that the disorderly distribution of energy will become organized into localized, uniform motion. A gas does not contract spontaneously, because to do so the disordered motion of its molecules would have to take them all into the same region of the container. The opposite change, spontaneous expansion, is a natural consequence of increasing disorder. An object does not spontaneously become warmer than its surroundings because it is highly improbable that the jostling of randomly vibrating atoms in the surroundings will lead to the accumulation of excess thermal motion in the object. The opposite change, the spreading of the object's energy into the surroundings as thermal motion, is natural.

It may seem very puzzling that collapse into disorder can account for the formation of such ordered substances as crystals or proteins. Nevertheless, in due course we shall see that organized structures can emerge as energy and matter disperse. We shall see, in fact, that collapse into disorder accounts for change in all its forms.

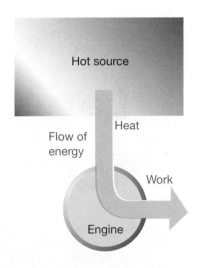

4.1 The Kelvin statement of the Second Law denies the possibility of the process illustrated here, in which heat is changed completely into work, there being no other change. The process is not in conflict with the First Law because energy is conserved.

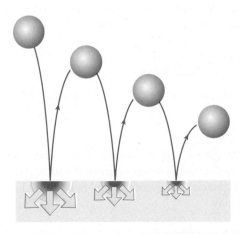

4.2 The direction of spontaneous change for a ball bouncing on a floor. On each bounce some of its energy is degraded into the thermal motion of the atoms of the floor, and that energy disperses. The reverse has never been observed to take place on a macroscopic scale.

---

1. Orderly motion, but on a much smaller scale, is observed as *Brownian motion*.

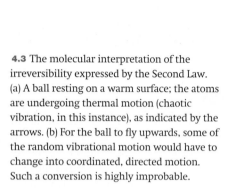

4.3 The molecular interpretation of the irreversibility expressed by the Second Law. (a) A ball resting on a warm surface; the atoms are undergoing thermal motion (chaotic vibration, in this instance), as indicated by the arrows. (b) For the ball to fly upwards, some of the random vibrational motion would have to change into coordinated, directed motion. Such a conversion is highly improbable.

(a)   (b)

4.2 Entropy

The First Law of thermodynamics led to the introduction of the internal energy, U. The internal energy is a state function that lets us assess whether a change is permissible: only those changes may occur for which the internal energy of an isolated system remains constant. The law that is used to identify the signpost of spontaneous change, the Second Law of thermodynamics, may also be expressed in terms of another state function, the **entropy**, S. We shall see that the entropy (which we shall define shortly, but is a measure of the molecular disorder of a system) lets us assess whether one state is accessible from another by a spontaneous change. The First Law uses the internal energy to identify *permissible* changes; the Second Law uses the entropy to identify the *spontaneous changes* among those permissible changes.

The Second Law of thermodynamics can be expressed in terms of the entropy:

The entropy of an isolated system increases in the course of a spontaneous change:

$$\Delta S_{tot} > 0 \tag{4.1}$$

where S_{tot} is the total entropy of the system and its surroundings. Thermodynamically irreversible processes (like cooling to the temperature of the surroundings and the free expansion of gases) are spontaneous processes, and hence must be accompanied by an increase in total entropy.

(a) The thermodynamic definition of entropy

The thermodynamic definition of entropy concentrates on the change in entropy, dS, that occurs as a result of a physical or chemical change (in general, as a result of a 'process'). The definition is motivated by the idea that a change in the extent to which energy is dispersed in a disorderly manner depends on how much energy is transferred as heat. As we have remarked, heat stimulates disorderly motion in the surroundings. Work, which stimulates uniform motion of atoms in the surroundings, does not change the degree of disorder and so does not change the entropy.

The thermodynamic definition of entropy is based on the expression

$$dS = \frac{dq_{rev}}{T} \qquad [4.2]$$

For a measurable change between two states i and f this expression integrates to

$$\Delta S = \int_{i}^{f} \frac{dq_{rev}}{T} \qquad (4.3)$$

That is, to calculate the difference in entropy between any two states of a system, we find a reversible path between them, and integrate the heat supplied at each stage of the path divided by the temperature at which the heat is supplied.

Molecular interpretation 4.1

The molecules in a system at high temperature are highly disorganized, either in terms of their locations or in terms of the occupation of their available translational, rotational, and vibrational energy states. A small additional transfer of energy will result in a relatively small additional disorder, much as sneezing in a busy street may be barely noticed. In contrast, the molecules in a system at low temperature have access to far fewer energy states (at $T = 0$, only the lowest state is accessible), and the same quantity of heat will have a pronounced effect on the degree of disorder, much as sneezing in a quiet library can be very disruptive. Hence, the change in entropy when a given quantity of heat is transferred will be greater when it is transferred to a cold body than when it is transferred to a hot body. This argument suggests that the change in entropy should be inversely proportional to the temperature at which the transfer takes place, as in eqn 2.

According to eqn 2, when the heat transferred is expressed in joules and the temperature is in kelvins, the units of entropy are joules per kelvin ($J\,K^{-1}$). Molar entropy, the entropy divided by the amount of substance, is expressed in joules per kelvin per mole ($J\,K^{-1}\,mol^{-1}$).[2]

Example 4.1 Calculating the entropy change for the isothermal expansion of a perfect gas

Calculate the entropy change of a sample of perfect gas when it expands isothermally from a volume V_i to a volume V_f.

Method The definition of entropy instructs us to find the heat absorbed for a reversible path between the stated initial and final states regardless of the actual manner in which the process takes place. A simplification is that the expansion is isothermal, so the temperature is a constant and may be taken outside the integral in eqn 3. The heat absorbed during a reversible isothermal expansion of a perfect gas can be calculated from $\Delta U = q + w$ and $\Delta U = 0$, which implies that $q = -w$ in general and therefore that $q_{rev} = -w_{rev}$ for a reversible change. The work of reversible isothermal expansion was calculated in Section 2.3 (eqn 2.13).

---

2. These are the same units as those of the gas constant, R, and molar heat capacities.

Answer Because the temperature is constant, eqn 3 becomes

$$\Delta S = \frac{1}{T} \int_{i}^{f} dq_{rev} = \frac{q_{rev}}{T}$$

From eqn 2.13, we know that

$$q_{rev} = -w_{rev} = nRT \ln \frac{V_f}{V_i}$$

Therefore, it follows that

$$\Delta S = nR \ln \frac{V_f}{V_i}$$

Comment As an illustration of this formula, when the volume of 1.00 mol of any perfect gas is doubled at any constant temperature

$$\Delta S = (1.00 \text{ mol}) \times (8.3145 \text{ J K}^{-1} \text{ mol}^{-1}) \times \ln 2 = +5.76 \text{ J K}^{-1}$$

Self-test 4.1 Calculate the change in entropy when the pressure of a perfect gas is changed isothermally from p_i to p_f. $[\Delta S = nR \ln(p_i/p_f)]$

We can use the definition in eqn 2 to formulate an expression for the change in entropy of the surroundings, ΔS_{sur}. Consider an infinitesimal transfer of heat dq_{sur} to the surroundings. The surroundings consist of a reservoir of constant volume, so the heat supplied to them can be identified with the change in their internal energy, dU_{sur}.[3] The internal energy is a state function, and dU_{sur} is an exact differential. As we have seen, these properties imply that dU_{sur} is independent of how the change is brought about and in particular is independent of whether the process is reversible or irreversible. The same remarks therefore apply to dq_{sur}, to which dU_{sur} is equal. Therefore, we can adapt the definition of entropy change in eqn 2 to write

$$dS_{sur} = \frac{dq_{sur,rev}}{T_{sur}} = \frac{dq_{sur}}{T_{sur}} \tag{4.4}$$

Furthermore, because the temperature of the surroundings is constant whatever the change, for a measurable change

$$\Delta S_{sur} = \frac{q_{sur}}{T_{sur}} \tag{4.5}$$

That is, regardless of how the change is brought about in the system, reversibly or irreversibly, we can calculate the change of entropy of the surroundings by dividing the heat transferred by the temperature at which the transfer takes place.

Equation 5 makes it very simple to calculate the changes in entropy of the surroundings that accompany any process. For instance, for any adiabatic change, $q_{sur} = 0$, so

$$\Delta S_{sur} = 0 \tag{4.6}$$

This expression is true however the change takes place, reversibly or irreversibly, provided no local hot spots are formed in the surroundings. That is, it is true so long as the surroundings remain in internal equilibrium. If hot spots do form, then the localized energy may subsequently disperse spontaneously and hence generate more entropy.

3. Alternatively, the surroundings can be regarded as being at constant pressure, in which case we could equate dq_{sur} to dH_{sur}.

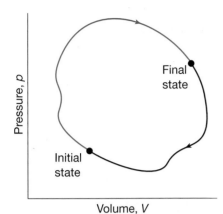

4.4 In a thermodynamic cycle, the overall change in a state function (from the initial state to the final state and then back to the initial state again) is zero.

Illustration 4.1

To calculate the entropy change in the surroundings when 1.00 mol $H_2O(l)$ is formed from its elements under standard conditions at 298 K, we use $\Delta H^{\ominus} = -286$ kJ from Table 2.6. The heat released is supplied to the surroundings, now regarded as being at constant pressure, so $q_{sur} = +286$ kJ. Therefore,

$$\Delta S_{sur} = \frac{2.86 \times 10^5 \text{ J}}{298 \text{ K}} = +960 \text{ J K}^{-1}$$

This strongly exothermic reaction results in an increase in the entropy of the surroundings as heat is released into them.

Self-test 4.2 Calculate the entropy change in the surroundings when 1.00 mol $N_2O_4(g)$ is formed from 2.00 mol $NO_2(g)$ under standard conditions at 298 K. $[+192 \text{ J K}^{-1}]$

(b) The entropy as a state function

Entropy is a state function. To prove this assertion, we need to show that the integral of dS is independent of path. To do so, it is sufficient to prove that the integral of eqn 2 around an arbitrary cycle is zero, for that guarantees that the entropy is the same at the initial and final states of the system regardless of the path taken between them (Fig. 4.4). That is, we need to show that

$$\oint \frac{dq_{rev}}{T} = 0 \tag{4.7}$$

where the symbol $\oint$ denotes integration around a closed path.

To prove eqn 7, we first consider the special cycle shown in Fig. 4.5. A **Carnot cycle**, which is named after the French engineer Sadi Carnot, consists of four reversible stages:

1 Reversible isothermal expansion from A to B at T_h; the entropy change is q_h/T_h, where q_h is the heat supplied to the system from the hot source.

2 Reversible adiabatic expansion from B to C. No heat leaves the system, so the change in entropy is zero. In the course of this expansion, the temperature falls from T_h to T_c, the temperature of the cold sink.

3 Reversible isothermal compression from C to D at T_c. Heat is released to the cold sink; the change in entropy of the system is q_c/T_c; in this expression q_c is negative.

4 Reversible adiabatic compression from D to A. No heat enters the system, so the change in entropy is zero. The temperature rises from T_c to T_h.

The total change in entropy around the cycle is

$$\oint dS = \frac{q_h}{T_h} + \frac{q_c}{T_c}$$

However, we show in the *Justification* below that

$$\frac{q_h}{q_c} = -\frac{T_h}{T_c} \tag{4.8}$$

Substitution of this relation into the preceding equation gives zero on the right, which is what we wanted to prove.

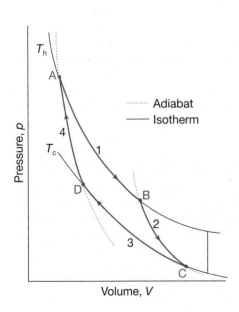

4.5 The basic structure of a Carnot cycle. In step 1, there is isothermal reversible expansion at the temperature T_h. Step 2 is a reversible adiabatic expansion in which the temperature falls from T_h to T_c. In Step 3 there is an isothermal reversible compression at T_c, and that isothermal step is followed by an adiabatic reversible compression, which restores the system to its initial state.

Justification 4.1

This *Justification* is based on the fact that the two temperatures in eqn 4.8 lie on the same adiabat in Fig. 4.5. As explained in Example 4.1, for a perfect gas:

$$q_h = nRT_h \ln \frac{V_B}{V_A} \qquad q_c = nRT_c \ln \frac{V_D}{V_C}$$

From the relations between temperature and volume for reversible adiabatic processes (eqn 2.34):

$$V_A T_h^c = V_D T_c^c \qquad V_C T_c^c = V_B T_h^c$$

Multiplication of the first of these expressions by the second gives

$$V_A V_C T_h^c T_c^c = V_D V_B T_h^c T_c^c$$

which simplifies to

$$\frac{V_A}{V_B} = \frac{V_D}{V_C}$$

Consequently,

$$q_c = nRT_c \ln \frac{V_D}{V_C} = nRT_c \ln \frac{V_A}{V_B} = -nRT_c \ln \frac{V_B}{V_A}$$

and therefore

$$\frac{q_h}{q_c} = \frac{nRT_h \ln(V_B/V_A)}{-nRT_c \ln(V_B/V_A)} = -\frac{T_h}{T_c}$$

as in eqn 8.

Hot source

T_h

q_h

+20

w

+5

−15

q_c

T_c

Cold sink

4.6 Suppose an energy q_h (e.g. 20 kJ) is supplied to the engine and q_c is lost from the engine (e.g. $q_c = -15$ kJ) and discarded into the cold reservoir. The work done by the engine is equal to $q_h + q_c$ (e.g. 20 kJ + (−15 kJ) = 5 kJ). The efficiency is the work done divided by the heat supplied from the hot source.

Now we need to show that the same conclusion applies to any material, not just a perfect gas. To do so, consider the **efficiency**, ε (epsilon), of a heat engine:

$$\varepsilon = \frac{\text{work performed}}{\text{heat absorbed}} = \frac{|w|}{q_h} \qquad [4.9]$$

The definition implies that, the greater the work output for a given supply of heat from the hot reservoir, the greater is the efficiency of the engine. We can express the definition in terms of the heat transactions alone, because (as shown in Fig. 4.6) the work supplied by the engine is the difference between the heat supplied by the hot reservoir and returned to the cold reservoir:

$$\varepsilon = \frac{q_h + q_c}{q_h} = 1 + \frac{q_c}{q_h} \qquad (4.10)$$

(Remember that $q_c < 0$.) It then follows that for a Carnot engine

$$\varepsilon_{rev} = 1 - \frac{T_c}{T_h} \qquad (4.11)$$

Now, the Second Law of thermodynamics implies that *all reversible engines have the same efficiency regardless of their construction*. To see the truth of this statement, suppose two reversible engines are coupled together and run between the same two reservoirs (Fig. 4.7). The working substances and details of construction of the two engines are entirely arbitrary. Initially, suppose that engine A is more efficient than engine B, and that we choose a setting of the controls that causes engine B to acquire

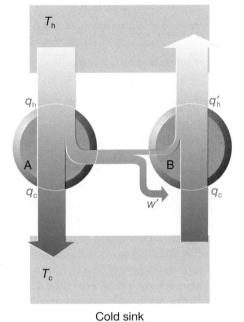

Hot source

T_h

q_h A q'_h B

q_c q_c

w'

T_c

Cold sink

(a)

Hot source

$q - q'$

w'

(b)

4.7 (a) The demonstration of the equivalence of the efficiencies of all reversible engines working between the same thermal reservoirs is based on the flow of energy represented in this diagram. (b) The net effect of the processes is the conversion of heat into work without there being a need for a cold sink: this is contrary to the Kelvin statement of the Second Law.

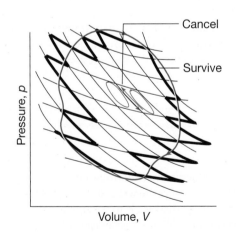

Cancel

Survive

Pressure, p

Volume, V

4.8 A general cycle can be divided into small Carnot cycles. The match is exact in the limit of infinitesimally small cycles. Paths cancel in the interior of the collection, and only the perimeter, an increasingly good approximation to the true cycle as the number of cycles increases, survives. Because the entropy change around every individual cycle is zero, the integral of the entropy around the perimeter is zero too.

the heat q_c from the cold reservoir and to release a certain quantity of heat into the hot reservoir. However, because engine A is more efficient than engine B, not all the work that A produces is needed for this process, and the difference can be used to do work. The net result is that the cold reservoir is unchanged, work has been produced, and the hot reservoir has lost a certain amount of energy. This outcome is contrary to the Kelvin statement of the Second Law, because some heat has been converted directly into work. In molecular terms, the disordered thermal motion of the hot reservoir has been converted into ordered motion characteristic of work. Because the conclusion is contrary to experience, the initial assumption that engines A and B can have different efficiencies must be false. It follows that the relation between the heat transfers and the temperatures (eqn 11) must also be independent of the working material, and therefore that eqn 8 is always true for any substance involved in a Carnot cycle.

To complete the argument, we note that any reversible cycle can be approximated as a collection of Carnot cycles and the cyclic integral around an arbitrary path is the sum of the integrals around each of the Carnot cycles (Fig. 4.8). This approximation becomes exact as the individual cycles are allowed to become infinitesimal. The entropy change around each individual cycle is zero (as demonstrated above), so the sum of entropy changes for all the cycles is zero. However, in the sum, the entropy change along any individual path is cancelled by the entropy change along the path it shares with the neighbouring cycle. Therefore, all the entropy changes cancel except for those along the perimeter of the overall cycle. That is,

$$\sum_{\text{all}} \frac{q_{\text{rev}}}{T} = \sum_{\text{perimeter}} \frac{q_{\text{rev}}}{T} = 0$$

In the limit of infinitesimal cycles, the non-cancelling edges of the Carnot cycles match the overall cycle exactly, and the sum becomes an integral. Equation 7 then follows immediately. This result implies that dS is an exact differential and therefore that S is a state function.

(c) The thermodynamic temperature

Suppose we have an engine that is working reversibly between a hot source at a temperature T_h and a cold sink at a temperature T; then we know from eqn 11 that

$$T = (1 - \varepsilon)T_h \tag{4.12}$$

This expression enabled Kelvin to define the **thermodynamic temperature scale** in terms of the efficiency of a heat engine. The zero of the scale occurs for a Carnot efficiency of 1. The size of the unit is entirely arbitrary, but on the Kelvin scale is defined by setting the temperature of the triple point of water as 273.16 K exactly. Then, if the heat engine has a hot source at the triple point of water, the temperature of the cold sink (the object we want to measure) is found by measuring the efficiency of the engine. This result is independent of the working substance (Box 4.1).

An additional point is that, as we saw in Section 2.2c, heat transferred can, in principle, be measured mechanically (in terms of the location of a weight). Therefore, it is possible, in principle at least, to use the distance moved by a weight to measure temperature. Kelvin's definition of the thermodynamic temperature scale puts the measurement of temperature on to a purely mechanical basis.

(d) The Clausius inequality

So far, we have verified that the entropy as defined in eqn 2 is a state function. We now need to verify that the entropy is a signpost of spontaneous change in the sense that $dS_{tot} \geq 0$ for any spontaneous change (with the equality applying for a reversible process).

Consider a system in thermal and mechanical contact with its surroundings at the same temperature, T. The system and the surroundings are not necessarily in mechanical equilibrium; for instance, a gas might have a greater pressure than its surroundings. Any change of state is accompanied by a change in entropy of the system, dS, and of the surroundings, dS_{sur}. Because the process might be irreversible, the total entropy will increase when a process occurs in the system, so we can write

$$dS + dS_{sur} \geq 0, \quad \text{or} \quad dS \geq -dS_{sur}$$

Box 4.1 **The thermodynamics of refrigeration**

The discussion of the text is the basis of the thermodynamic assessment of the power needed to cool objects in refrigerators. First, we consider the work required to cool an object, and refer to the illustration.

When an energy $|q_c|$ is removed from a cool source at a temperature T_c and then deposited in a warmer sink at a temperature T_h, as in a typical refrigerator, the change in entropy is

$$\Delta S = -\frac{|q_c|}{T_c} + \frac{|q_c|}{T_h} < 0$$

The process is not spontaneous because not enough entropy is generated in the warm sink to overcome the entropy loss from the hot source. To generate more entropy, energy must be added to the stream that enters the warm sink. Our task is to find the minimum energy that needs to be supplied. The outcome is expressed as the *coefficient of performance, c*:

$$c = \frac{\text{heat transferred}}{\text{work done}} = \frac{|q_c|}{|w|}$$

The less the work that is required to achieve a given transfer, the greater the coefficient of performance and the more efficient the refrigerator.

Because $|q_c|$ is removed from the cold source, and the work $|w|$ is added to the energy stream, the heat deposited in the hot sink is $|q_h| = |q_c| + |w|$. Therefore,

(a)

(b)

(a) The flow of energy as heat from a cold source to a hot sink is not spontaneous. As shown here, the entropy increase of the hot sink is smaller than the entropy increase of the cold source, so there is a net decrease in entropy. (b) The process becomes feasible if work is provided to add to the energy stream. Then the increase in entropy of the hot sink can be made to cancel the entropy decrease of the hot source.

$$\frac{1}{c} = \frac{|q_h| - |q_c|}{|q_c|} = \frac{|q_h|}{|q_c|} - 1$$

We can now use the result established in the text to express this result in terms of the temperatures alone, which is possible if the transfer is performed reversibly. This substitution leads to

$$c = \frac{T_c}{T_h - T_c}$$

for the thermodynamically optimum coefficient of performance. For a refrigerator withdrawing heat from ice-cold water ($T_c = 273$ K) in a typical environment ($T_h = 293$ K), $c = 14$, so, to remove 10 kJ (enough to freeze 30 g of water), requires an input of at least 0.71 kJ of work. Practical refrigerators, of course, have a lower coefficient of performance.

The work to *maintain* a low temperature is also relevant to the design of refrigerators. No thermal insulation is perfect, so there is always a flow of heat into the sample at a rate proportional to the temperature difference. If the rate at which heat leaks in is written $A(T_h - T_c)$, where A is a constant that depends on the size of the sample and the details of the insulation, then the minimum power, P, required to maintain the original temperature difference by pumping out that heat is

$$P = \frac{1}{c} \times A(T_h - T_c) = A \times \frac{(T_h - T_c)^2}{T_c}$$

We see that the power increases as the square of the temperature difference we are trying to maintain. For this reason, air-conditioners are much more expensive to run on hot days than on mild days.

Discussion questions

1. The calculation of the work that has to be done in order to lower the temperature of an object is a bit involved because the coefficient of performance changes with the temperature of the object. (a) Find an expression for the work needed to cool an object from T_i to T_f when the refrigerator is in a room at a temperature T_h. *Hint.* Write $dw = dq/c(T)$, relate dq to dT through the heat capacity C_p, and integrate the resulting expression. Assume that the heat capacity is independent of temperature in the range of interest. (b) Use the result in part (a) to calculate the work needed to freeze 250 g of water put into a refrigerator at 293 K. How long will it take when the refrigerator operates at 100 W?

2. The same expressions that apply to the treatment of refrigerators also describe the behaviour of heat pumps, where warmth is obtained from the back of a refrigerator while its front is being used to cool the outside world. Heat pumps are popular home heating devices because they are very efficient, as we can see by performing the following calculation. Compare the heat delivered to a room at 295 K by each of two methods: (a) direct conversion of 1.00 kJ of electrical energy in an electrical heater, and (b) use of 1.00 kJ of electrical energy to run a reversible heat pump with the outside at 260 K. Discuss the origin of the difference in the heat delivered to the interior of the house by the two methods.

Further reading

R.J. Dossat, *Principles of refrigeration*. Prentice-Hall, New Jersey (1997).

F.J. Hale, Heat engines and refrigerators. In *Encyclopedia of applied physics* (ed. G.L. Trigg), **7**, 303. VCH, New York (1993).

Because eqn 4 implies that $dS_{sur} = -dq/T$, where dq is the heat supplied to the system during the process (that is, $dq_{sur} = -dq$, because the heat that enters the system comes from the surroundings), it follows that for any change

$$dS \geq \frac{dq}{T} \tag{4.13}$$

This expression is the **Clausius inequality**.

Suppose the system is isolated from its surroundings. Then $dq = 0$, and the Clausius inequality implies that

$$dS \geq 0 \tag{4.14}$$

This is exactly the characteristic we need for the entropy to be the signpost of spontaneous change, for it tells us that in an *isolated* system the entropy cannot decrease when a spontaneous change takes place.

We can illustrate the content of the Clausius inequality in two simple cases. First, suppose a system undergoes an irreversible adiabatic change. Then $dq = 0$ and, by eqn 13, $dS \geq 0$. That is, for this type of spontaneous change, the entropy of the system has increased. Because no heat flows into the surroundings, their entropy remains constant and $dS_{sur} = 0$. Therefore, the total entropy of the system and its surroundings obeys $dS_{tot} \geq 0$.

Now consider irreversible isothermal expansion of a perfect gas. As we saw in Example 4.1, for such a change $dq = -dw$ (because $dU = 0$). If the gas expands freely into a vacuum, it does no work and $dw = 0$, which implies that $dq = 0$ too. Therefore, according to the Clausius inequality, $dS \geq 0$. Next, consider the surroundings. No heat is transferred into the surroundings, so $dS_{sur} = 0$. Therefore, in this case too $dS_{tot} \geq 0$.

Another type of irreversible process is spontaneous cooling. Consider a transfer of energy as heat from one system—the hot source—at a temperature T_h to another system—the cold sink—at a temperature T_c (Fig. 4.9). When $|dq|$ leaves the hot source, the entropy changes by $-|dq|/T_h$ (a decrease). When $|dq|$ enters the cold sink the entropy changes by $+|dq|/T_c$ (an increase). The overall change in entropy is therefore

$$dS = \frac{|dq|}{T_c} - \frac{|dq|}{T_h} = |dq|\left(\frac{1}{T_c} - \frac{1}{T_h}\right) \tag{4.15}$$

which is positive (because $T_h \geq T_c$). Hence, cooling (the transfer of heat from hot to cold) is spontaneous, as we know from experience. When the temperatures of the two systems are equal, $dS_{tot} = 0$: the two systems are then at thermal equilibrium.

4.3 Entropy changes accompanying specific processes

We now see how to calculate the entropy changes that accompany a variety of basic processes.

(a) The entropy of phase transition at the transition temperature

Because a change in the degree of molecular order occurs when a substance freezes or boils, we should expect the transition to be accompanied by a change in entropy. For example, when a substance vaporizes, a compact condensed phase changes into a widely dispersed gas, and we can expect the entropy of the substance to increase

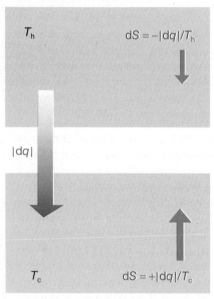

Hot source

T_h $dS = -|dq|/T_h$

$|dq|$

T_c $dS = +|dq|/T_c$

Cold sink

4.9 When energy leaves a hot reservoir as heat, the entropy of the reservoir decreases. When the same quantity of energy enters a cooler reservoir, the entropy increases by a larger amount. Hence, overall there is an increase in entropy and the process is spontaneous. Relative changes in entropy are indicated by the sizes of the arrows.

Table 4.1* Standard entropies (and temperatures) of phase transitions, $\Delta_{trs}S^{\ominus}/(J\,K^{-1}\,mol^{-1})$

	Fusion (at T_f)	Vaporization (at T_b)
Argon, Ar	14.17 (at 83.8 K)	74.53 (at 87.3 K)
Benzene, C_6H_6	38.00 (at 279 K)	87.19 (at 353 K)
Water, H_2O	22.00 (at 273.15 K)	109.0 (at 373.15 K)
Helium, He	4.8 (at 1.8 K and 30 bar)	19.9 (at 4.22 K)

* More values are given in the *Data section*.

considerably. The entropy of a solid substance increases when it melts to a liquid. The entropy also increases when the liquid phase turns into a gas.

Consider a system and its surroundings at the **normal transition temperature**, T_{trs}, the temperature at which two phases are in equilibrium at 1 atm. This temperature is 0°C (273 K) for ice in equilibrium with liquid water at 1 atm, and 100°C (373 K) for liquid water in equilibrium with its vapour at 1 atm. At the transition temperature, any transfer of heat between the system and its surroundings is reversible because the two phases in the system are in equilibrium. Because at constant pressure $q = \Delta_{trs}H$, the change in molar entropy of the system is[4]

$$\Delta_{trs}S = \frac{\Delta_{trs}H}{T_{trs}} \tag{4.16}$$

If the phase transition is exothermic ($\Delta_{trs}H < 0$, as in freezing or condensing), then the entropy change is negative. This decrease in entropy is consistent with the system becoming more ordered when a solid forms from a liquid. If the transition is endothermic ($\Delta_{trs}H > 0$, as in melting), then the entropy change is positive, which is consistent with the system becoming more disordered. Melting and vaporizing are endothermic processes, so both are accompanied by an increase in the system's entropy. This increase is consistent with liquids being more disordered than solids, and gases more disordered than liquids. Table 4.1 lists some experimental entropies of transition.

Table 4.2 lists in more detail the standard entropies of vaporization of several liquids at their boiling points. An interesting feature of the data is that a wide range of liquids give approximately the same standard entropy of vaporization (about 85 J K^{-1} mol^{-1}): this empirical observation is called **Trouton's rule**.

Molecular interpretation 4.2

The explanation of Trouton's rule is that a comparable amount of disorder is generated when any liquid evaporates and becomes a gas. Hence, all liquids can be expected to have similar standard entropies of vaporization.

Liquids that show significant deviations from Trouton's rule do so on account of the molecules in the liquid being arranged in a partially orderly manner. In such

---

4. Recall from Section 2.7 that $\Delta_{trs}H$ is an enthalpy change per mole of substance; so $\Delta_{trs}S$ is also a molar quantity.

Table 4.2* The standard entropies of vaporization of liquids

	$\Delta_{vap}H^{\ominus}/(kJ\ mol^{-1})$	$\theta_b/°C$	$\Delta_{vap}S^{\ominus}/(J\ K^{-1}\ mol^{-1})$
Benzene	+30.8	80.1	+87.2
Carbon tetrachloride	+30.00	76.7	+85.8
Cyclohexane	+30.1	80.7	+85.1
Hydrogen sulfide	+18.7	−60.4	+87.9
Methane	+8.18	−161.5	+73.2
Water	+40.7	100.0	+109.1

* More values are given in the *Data section*.

cases, a greater change of disorder occurs when the liquid evaporates than if the molecules were highly disordered in the liquid. An example is water, where the large entropy of vaporization reflects the presence of structure arising from hydrogen-bonding in the liquid. Hydrogen bonds tend to organize the molecules in the liquid so that they are less random than, for example, the molecules in liquid hydrogen sulfide (which is not hydrogen bonded).

Methane has an unusually low entropy of vaporization. A part of the reason is that the entropy of the gas itself is slightly low (186 J K^{-1} mol^{-1} at 298 K); the entropy of N$_2$ under the same conditions is 192 J K^{-1} mol^{-1}. As we shall see in Chapter 19, light molecules are difficult to excite into rotation; as a result, only a few rotational states are accessible at room temperature, and the disorder associated with the population of rotational states is low.

Example 4.2 Using Trouton's rule

Predict the standard molar enthalpy of vaporization of bromine given that it boils at 59.2°C.

Method We need to judge whether there is the likelihood of anomalous structural organization in the liquid phase or some anomaly in the gas phase. If there is not, it is permissible to use Trouton's rule in the form

$$\Delta_{vap}H^{\ominus} = T_b \times (85\ J\ K^{-1}\ mol^{-1})$$

Answer There is no hydrogen bonding in liquid bromine and Br$_2$ is a heavy molecule that is unlikely to display unusual behaviour in the gas phase, so it is probably safe to use Trouton's rule. Substitution of the data then gives

$$\Delta_{vap}H^{\ominus} = (332.4\ K) \times (85\ J\ K^{-1}\ mol^{-1}) = 2.8 \times 10^4\ J\ mol^{-1} = 28\ kJ\ mol^{-1}$$

The experimental value is 29.45 kJ mol^{-1}.

Self-test 4.3 Predict the enthalpy of vaporization of ethane from its boiling point, −88.6°C.

[16 kJ mol^{-1}]

4.10 The logarithmic increase in entropy of a perfect gas as it expands isothermally.

4.11 The logarithmic increase in entropy of a substance as it is heated at constant volume. Different curves correspond to different values of the constant-volume heat capacity (which is assumed constant over the temperature range) expressed as $C_{V,m}/R$.

(b) The expansion of a perfect gas

We established in Example 4.1 that the change in entropy of a perfect gas that expands isothermally from V_i to V_f is

$$\Delta S = nR \ln \frac{V_f}{V_i} \tag{4.17}°$$

Because S is a state function, the value of ΔS is independent of the path between the initial and final states, so this expression applies whether the change of state occurs reversibly or irreversibly. The logarithmic dependence of entropy on volume is illustrated in Fig. 4.10.

The *total* change in entropy, however, does depend on how the expansion takes place. If the change is reversible, the entropy change in the surroundings must be such as to give $\Delta S_{tot} = 0$. Therefore, in this case, ΔS_{sur} must be the negative of the expression in eqn 17. If the isothermal expansion occurs freely ($w = 0$) and irreversibly, then $q = 0$ (because $\Delta U = 0$). Consequently, $\Delta S_{sur} = 0$, and the total entropy change is given by eqn 17.

(c) The variation of entropy with temperature

We can use eqn 3 to calculate the entropy of a system at a temperature T_f from a knowledge of its entropy at a temperature T_i and the heat supplied to change its temperature from one value to the other:

$$S(T_f) = S(T_i) + \int_i^f \frac{dq_{rev}}{T} \tag{4.18}$$

We shall be particularly interested in the entropy change when the system is subjected to constant pressure (such as from the atmosphere) during the heating. Then, from the definition of constant-pressure heat capacity (eqn 2.27),

$$dq_{rev} = C_p \, dT$$

so long as the system is doing no non-expansion work. Consequently, at constant pressure:

$$S(T_f) = S(T_i) + \int_i^f \frac{C_p \, dT}{T} \tag{4.19}$$

The same expression applies at constant volume, but with C_p replaced by C_V. When C_p is independent of temperature in the temperature range of interest, we obtain

$$S(T_f) = S(T_i) + C_p \int_i^f \frac{dT}{T} = S(T_i) + C_p \ln \frac{T_f}{T_i} \tag{4.20}$$

with a similar expression for heating at constant volume. The logarithmic dependence of entropy on temperature is illustrated in Fig. 4.11.

Example 4.3 Calculating the entropy change

Calculate the entropy change when argon at 25°C and 1.00 atm in a container of volume 500 cm³ is allowed to expand to 1000 cm³ and is simultaneously heated to 100°C.

Method Because S is a state function, we are free to choose the most convenient path from the initial state. One such path is reversible isothermal expansion to the final volume, followed by reversible heating at constant volume to the final temperature. The entropy change in the first step is given by eqn 17 and that of the second step, provided C_V is independent of temperature, by eqn 20 (with C_V in place of C_p). In each case we need to know n, the amount of gas, and can calculate it from the perfect gas equation and the data for the initial state. The heat capacity at constant volume can be obtained from the value of $C_{p,m}$ in Table 2.6 and the relation $C_{p,m} - C_{V,m} = R$.

Answer The amount of Ar present (from $n = pV/RT$) is 0.0204 mol. The entropy change in the first step (expansion from 500 cm^3 to 1000 cm^3 at 298 K) is

$$\Delta S = nR \ln 2.00 = +0.118 \text{ J K}^{-1}$$

The entropy change in the second step, from 298 K to 373 K at constant volume, is

$$\Delta S = (0.0204 \text{ mol}) \times (12.48 \text{ J K}^{-1} \text{ mol}^{-1}) \times \ln\left(\frac{373 \text{ K}}{298 \text{ K}}\right)$$

$$= +0.057 \text{ J K}^{-1}$$

The overall entropy change, the sum of these two changes, is $\Delta S = +0.175 \text{ J K}^{-1}$.

Self-test 4.4 Calculate the entropy change when the same initial sample is compressed to 50.0 cm^3 and cooled to -25°C. $[-0.44 \text{ J K}^{-1}]$

(d) The measurement of entropy

The entropy of a system at a temperature T is related to its entropy at $T = 0$ by measuring its heat capacity C_p at different temperatures and evaluating the integral in eqn 20. The entropy of transition ($\Delta_{trs}H/T_{trs}$) must be added for each phase transition between $T = 0$ and the temperature of interest. For example, if a substance melts at T_f and boils at T_b, then its entropy above its boiling temperature is given by

$$S(T) = S(0) + \int_0^{T_f} \frac{C_p(s)dT}{T} + \frac{\Delta_{fus}H}{T_f} + \int_{T_f}^{T_b} \frac{C_p(l)dT}{T} + \frac{\Delta_{vap}H}{T_b} + \int_{T_b}^{T} \frac{C_p(g)dT}{T} \quad (4.21)$$

All the properties required, except $S(0)$, can be measured calorimetrically, and the integrals can be evaluated either graphically or, as is now more usual, by fitting a polynomial to the data and integrating the polynomial analytically. The procedure is illustrated in Fig. 4.12: the area under the curve of C_p/T against T is the integral required. Because $dT/T = d \ln T$, an alternative procedure is to evaluate the area under a plot of C_p against $\ln T$.

One problem with the measurement of entropy is the difficulty of measuring heat capacities near $T = 0$. There are good theoretical grounds for assuming that the heat capacity is proportional to T^3 when T is low (see Section 11.1c), and this dependence is the basis of the **Debye extrapolation**. In this method, C_p is measured down to as low a temperature as possible, and a curve of the form aT^3 is fitted to the data. That fit determines the value of a, and the expression $C_p = aT^3$ is assumed valid down to $T = 0$.

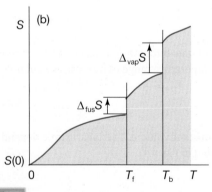

4.12 The calculation of an entropy from heat capacity data. (a) The variation of C_p/T with the temperature for a sample. (b) The entropy, which is equal to the area beneath the upper curve up to the corresponding temperature, plus the entropy of each phase transition passed.

Illustration 4.2

The standard molar entropy of nitrogen gas at 25°C has been calculated from the following data:

	$S_m^{\ominus}/(\text{J K}^{-1}\,\text{mol}^{-1})$
Debye extrapolation	1.92
Integration, from 10 K to 35.61 K	25.25
Phase transition at 35.61 K	6.43
Integration, from 35.61 K to 63.14 K	23.38
Fusion at 63.14 K	11.42
Integration, from 63.14 K to 77.32 K	11.41
Vaporization at 77.32 K	72.13
Integration, from 77.32 K to 298.15 K	39.20
Correction for gas imperfection	0.92
Total	192.06

Therefore,

$$S_m(298.15\text{ K}) = S_m(0) + 192.1\text{ J K}^{-1}\,\text{mol}^{-1}$$

Example 4.4 Calculating the entropy at low temperatures

The molar constant-pressure heat capacity of a certain solid at 10 K is 0.43 J K^{-1} mol^{-1}. What is its molar entropy at that temperature?

Method Because the temperature is so low, we can assume that the heat capacity varies with temperature as aT^3, in which case we can use eqn 19 to calculate the entropy at a temperature T in terms of the entropy at $T = 0$ and the constant a. When the integration is carried out, it turns out that the result can be expressed in terms of the heat capacity at the temperature T, so the data can be used directly to calculate the entropy.

Answer The integration required is

$$S(T) = S(0) + \int_0^T \frac{aT^3\,\mathrm{d}T}{T}$$

$$= S(0) + a\int_0^T T^2\,\mathrm{d}T = S(0) + \tfrac{1}{3}aT^3$$

However, because aT^3 is the heat capacity at the temperature T,

$$S(T) = S(0) + \tfrac{1}{3}C_p(T)$$

from which it follows that

$$S_m(10\text{ K}) = S_m(0) + 0.14\text{ J K}^{-1}\,\text{mol}^{-1}$$

Self-test 4.5 For metals, there is also a contribution to the heat capacity from the electrons that is linearly proportional to T when the temperature is low. Find its contribution to the entropy at low temperatures.

$$[S(T) = S(0) + C_p(T)]$$

4.4 The Third Law of thermodynamics

At $T = 0$, all energy of thermal motion has been quenched, and in a perfect crystal all the atoms or ions are in a regular, uniform array. The absence of both spatial disorder and thermal motion suggests that such materials also have zero entropy. This conclusion is consistent with the molecular interpretation of entropy, because $S = 0$ if there is only one way of arranging the molecules.

(a) The Nernst heat theorem

The experimental observation that turns out to be consistent with the view that the entropy of a regular array of molecules is zero at $T = 0$ is summarized by the **Nernst heat theorem**:

> **The entropy change accompanying any physical or chemical transformation approaches zero as the temperature approaches zero: $S \rightarrow 0$ as $T \rightarrow 0$ provided all the substances involved are perfectly ordered.**

As an example of the experimental evidence for this theorem, consider the entropy of the transition between orthorhombic sulfur, S(α), and monoclinic sulfur, S(β), which can be calculated from the transition enthalpy (-402 J mol^{-1}) at the transition temperature (369 K):

$$\Delta_{trs}S = S_m(\alpha) - S_m(\beta) = \frac{(-402 \text{ J mol}^{-1})}{369 \text{ K}} = -1.09 \text{ J K}^{-1} \text{ mol}^{-1}$$

The two individual entropies can also be determined by measuring the heat capacities from $T = 0$ up to $T = 369$ K. It is found that

$$S_m(\alpha) = S_m(\alpha, 0) + 37 \text{ J K}^{-1} \text{ mol}^{-1} \qquad S_m(\beta) = S_m(\beta, 0) + 38 \text{ J K}^{-1} \text{ mol}^{-1}$$

These two values imply that at the transition temperature

$$\Delta_{trs}S = S_m(\alpha, 0) - S_m(\beta, 0) - 1 \text{ J K}^{-1} \text{ mol}^{-1}$$

On comparing this value with the one above, we conclude that

$$S_m(\alpha, 0) - S_m(\beta, 0) \approx 0$$

in accord with the theorem.

It follows from the Nernst theorem that, if we arbitrarily ascribe the value zero to the entropies of elements in their perfect crystalline form at $T = 0$, then all perfect crystalline compounds also have zero entropy at $T = 0$ (because the change in entropy that accompanies the formation of the compounds, like the entropy of all transformations at that temperature, is zero). This conclusion is summarized by the **Third Law of thermodynamics**:

> **The entropy of all perfect crystalline substances is zero at $T = 0$.**

As far as thermodynamics is concerned, choosing this common value as zero is then a matter of convenience. The molecular interpretation of entropy, however, justifies the value $S = 0$ at $T = 0$. The Third Law has important implications for the attainment of very low temperatures (Box 4.2).

(b) Third-Law entropies

Entropies reported on the basis that $S(0) = 0$ are called **Third-Law entropies** (and often just 'entropies'). When the substance is in its standard state at the temperature T, the **standard (Third-Law) entropy** is denoted $S^{\ominus}(T)$. A list of values at 298 K is given in Table 4.3.

Table 4.3* Standard Third-Law entropies at 298 K

	$S_m^{\ominus}/(\text{J K}^{-1} \text{ mol}^{-1})$
Solids:	
Graphite, C(s)	5.7
Diamond, C(s)	2.4
Sucrose, $C_{12}H_{22}O_{11}$(s)	360.2
Iodine, I_2(s)	116.1
Liquids:	
Benzene, C_6H_6(l)	173.3
Water, H_2O(l)	69.9
Mercury, Hg(l)	76.0
Gases:	
Methane, CH_4(g)	186.3
Carbon dioxide, CO_2(g)	213.7
Hydrogen, H_2(g)	130.7
Helium, He(g)	126.2
Ammonia, NH_3(g)	192.4

* More values are given in the *Data section*.

Box 4.2 Reaching very low temperatures

The world record low temperature stands at about 280 pK. Gases may be cooled by Joule–Thomson expansion below their inversion temperature, and temperatures lower than 4 K (the boiling point of helium) can be reached by the evaporation of liquid helium by pumping rapidly through large diameter pipes. Temperatures as low as about 1 K can be reached in this way, but at lower temperatures helium is insufficiently volatile for this procedure to be effective; moreover, the superfluid phase begins to interfere with the cooling process by creeping round the apparatus.

Common methods used to reach very low temperatures include laser cooling (see *Discussion question 2*) and *adiabatic demagnetization*. In the absence of a magnetic field, the unpaired electrons of a paramagnetic material are orientated at random, but in the presence of a magnetic field there are more β spins ($m_s = -\frac{1}{2}$; these terms are explained in Section 12.8) than α spins ($m_s = +\frac{1}{2}$). In thermodynamic terms, the application of a magnetic field lowers the entropy of a sample (see the illustration), and at a given temperature, the entropy of a sample is lower when the field is on than when it is off.

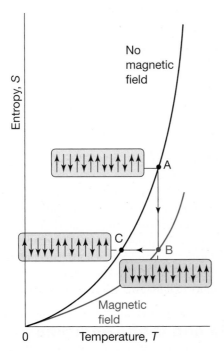

The technique of adiabatic demagnetization is used to attain very low temperatures. The upper curve shows the variation of the entropy of a paramagnetic system in the absence of an applied field. The lower curve shows the variation in entropy when a field is applied and has made the electron magnets more orderly. The isothermal magnetization step is from A to B; the adiabatic demagnetization step (at constant entropy) is from B to C.

A sample of paramagnetic material, such as a *d*- or *f*-metal complex, is cooled to about 1 K by using helium. Gadolinium(III) sulfate octahydrate, $Gd_2(SO_4)_3 \cdot 8H_2O$, has been used because each gadolinium ion carries several unpaired electrons but is separated from its neighbours by a coordination sphere of hydrating H_2O molecules. The sample is then exposed to a strong magnetic field while it is surrounded by helium, which provides thermal contact with the cold reservoir. This magnetization step is isothermal, and heat leaves the sample as the electron spins adopt the lower energy state (AB in the illustration). Thermal contact between the sample and the surroundings is now broken by pumping away the helium and the magnetic field is reduced to zero. This step is adiabatic and effectively reversible, so the state of the sample changes from B to C. At the end of this step the sample is the same as it was at A except that it now has a lower entropy. That lower entropy in the absence of a magnetic field corresponds to a lower temperature. That is, adiabatic demagnetization has cooled the sample.

Even lower temperatures can be reached if nuclear spins (which also behave like small magnets) are used instead of electron spins in the technique of *adiabatic nuclear demagnetization*. This technique was used to reach the current world record (in silver) of 280 pK.

Discussion questions

1. Suppose the two curves in the illustration could be expressed as aT^3 (field off) and bT^3 (field on). How many magnetization–demagnetization cycles would be needed to reduce the temperature of a sample of mass 0.10 g from (a) 1 K to 1 mK, (b) 1 K to 1 nK? Take $a = 6.2$ kJ K^{-4} g^{-1} and $b = 5.1$ kJ K^{-4} g^{-1}.

2. A group of 2000 rubidium atoms has been cooled to 20 nK. In the optical trapping technique, atoms in the gas phase are cooled by inelastic collisions with photons from intense laser beams, which act as walls of a very small container. (a) Assuming perfect gas behaviour, calculate the root mean square speed of and the pressure exerted by 1000 rubidium atoms trapped by laser beams in a volume of 1.0×10^{-15} m^3 at 0.10 mK (see Section 24.1). (b) What is the de Broglie wavelength of a rubidium atom travelling at the root mean square speed calculated in part (a) (see Section 11.2)?

Further reading

A.T. Serlemitsos, B.A. Warner, and S. Castles, Adiabatic demagnetization refrigerator for space use. *Advances in Cryogenics* **35B**, 1431 (1990).

C.N. Cohen-Tannoudji and W.D. Phillips, New mechanisms for laser cooling. *Physics Today*, 33 (October 1990).

The **standard reaction entropy**, $\Delta_r S^{\ominus}$, is defined, like the standard reaction enthalpy, as the difference between the molar entropies of the pure, separated products and the pure, separated reactants, all substances being in their standard states at the specified temperature:

$$\Delta_r S^{\ominus} = \sum_{\text{Products}} v S_m^{\ominus} - \sum_{\text{Reactants}} v S_m^{\ominus} \tag{4.22}$$

In this expression, each term is weighted by the appropriate stoichiometric coefficient. Standard reaction entropies are likely to be positive if there is a net formation of gas in a reaction, and are likely to be negative if there is a net consumption of gas.

Illustration 4.3

To calculate the standard reaction entropy of $H_2(g) + \frac{1}{2}O_2(g) \rightarrow H_2O(l)$ at 25°C, we use the data in Table 2.6 of the *Data section* to write

$$\begin{aligned}
\Delta_r S^{\ominus} &= S_m^{\ominus}(H_2O, l) - \{S_m^{\ominus}(H_2, g) + \tfrac{1}{2}S_m^{\ominus}(O_2, g)\} \\
&= 69.9 \, J\,K^{-1}\,mol^{-1} - \{130.7 + \tfrac{1}{2}(205.1)\} \, J\,K^{-1}\,mol^{-1} \\
&= -163.4 \, J\,K^{-1}\,mol^{-1}
\end{aligned}$$

The negative value is consistent with the conversion of two gases to a compact liquid.

Self-test 4.6 Calculate the standard reaction entropy for the combustion of methane to carbon dioxide and liquid water at 25°C. $\qquad$ [−243 J K⁻¹ mol⁻¹]

Concentrating on the system

Entropy is the basic concept for discussing the direction of natural change, but to use it we have to analyse changes in both the system and its surroundings. We have seen that it is always very simple to calculate the entropy change in the surroundings, and we shall now see that it is possible to devise a simple method for taking that contribution into account automatically. This approach focuses our attention on the system and simplifies discussions. Moreover, it is the foundation of all the applications of chemical thermodynamics that follow.

4.5 The Helmholtz and Gibbs energies

Consider a system in thermal equilibrium with its surroundings at a temperature T. When a change in the system occurs and there is a transfer of energy as heat between the system and the surroundings, the Clausius inequality, eqn 13, reads

$$dS - \frac{dq}{T} \geq 0 \tag{4.23}$$

We can develop this inequality in two ways according to the conditions (of constant volume or constant pressure) under which the process occurs.

(a) Criteria for spontaneity

First, consider heat transfer at constant volume. Then, in the absence of non-expansion work, we can write $dq_V = dU$; consequently

$$dS - \frac{dU}{T} \geq 0 \tag{4.24}$$

The importance of the inequality in this form is that it expresses the criterion for spontaneous change solely in terms of the state functions of the system. The inequality is easily rearranged to

$$T\, dS \geq dU \quad \text{(constant } V\text{, no additional work)}^5 \tag{4.25}$$

At either constant internal energy ($dU = 0$) or constant entropy ($dS = 0$), this expression becomes, respectively,

$$dS_{U,V} \geq 0 \qquad dU_{S,V} \leq 0 \tag{4.26}$$

where the subscripts indicate the constant conditions.

Equation 26 expresses the criteria for spontaneous change in terms of properties relating to the system. The first inequality states that, in a system at constant volume and constant internal energy (such as an isolated system), the entropy increases in a spontaneous change. That statement is essentially the content of the Second Law. The second inequality is less obvious, for it says that, if the entropy and volume of the system are constant, then the internal energy must decrease in a spontaneous change. Do not interpret this criterion as a tendency of the system to sink to lower energy. It is a disguised statement about entropy, and should be interpreted as implying that, if the entropy of the system is unchanged, then there must be an increase in entropy of the surroundings, which can be achieved only if the energy of the system decreases as energy flows out as heat.

When heat is transferred at constant pressure, and there is no work other than expansion work, we can write $dq_p = dH$ and obtain

$$T\, dS \geq dH \quad \text{(constant } p\text{, no additional work)} \tag{4.27}$$

At either constant enthalpy or constant entropy this inequality becomes, respectively,

$$dS_{H,p} \geq 0 \qquad dH_{S,p} \leq 0 \tag{4.28}$$

The interpretations of these inequalities are similar to those of eqn 26. The entropy of the system at constant pressure must increase if its enthalpy remains constant (for there can then be no change in entropy of the surroundings). Alternatively, the enthalpy must decrease if the entropy of the system is constant, for then it is essential to have an increase in entropy of the surroundings.

Because eqns 25 and 27 have the forms $dU - T\, dS \leq 0$ and $dH - T\, dS \leq 0$, respectively, they can be expressed more simply by introducing two more thermodynamic quantities. One is the **Helmholtz energy**, A, which is defined as

$$A = U - TS \tag{4.29}$$

The other is the **Gibbs energy**, G:

$$G = H - TS \tag{4.30}$$

All the symbols in these two definitions refer to the system.

---

5. Recall that 'additional work' is work other than expansion work.

When the state of the system changes at constant temperature, the two properties change as follows:

(a) $dA = dU - T\,dS$ (b) $dG = dH - T\,dS$ (4.31)

When we introduce eqns 25 and 27, respectively, we obtain the criteria of spontaneous change as

(a) $dA_{T,V} \leq 0$ (b) $dG_{T,p} \leq 0$ (4.32)

These inequalities are the most important conclusions from thermodynamics for chemistry. They are developed in subsequent sections and chapters.

(b) Some remarks on the Helmholtz energy

A change in a system at constant temperature and volume is spontaneous if $dA_{T,V} \leq 0$. That is, a change under these conditions is spontaneous if it corresponds to a decrease in the Helmholtz energy. Such systems move spontaneously towards states of lower A if a path is available. The criterion of equilibrium, when neither the forward nor reverse process has a tendency to occur, is

$$dA_{T,V} = 0 \qquad (4.33)$$

The expressions $dA = dU - T\,dS$ and $dA < 0$ are sometimes interpreted as follows. A negative value of dA is favoured by a negative value of dU and a positive value of $T\,dS$. This observation suggests that the tendency of a system to move to lower A is due to its tendency to move towards states of lower internal energy and higher entropy. However, this interpretation is false (even though it is a good rule of thumb for remembering the expression for dA) because the tendency to lower A is solely a tendency towards states of greater overall entropy. Systems change spontaneously if in doing so the total entropy of the system and its surroundings increases, not because they tend to lower internal energy. The form of dA may give the impression that systems favour lower energy, but that is misleading: dS is the entropy change of the system, $-dU/T$ is the entropy change of the surroundings (when the volume of the system is constant), and their total tends to a maximum.

(c) Maximum work

It turns out that A carries a greater significance than being simply a signpost of spontaneous change: *the change in the Helmholtz function is equal to the maximum work accompanying a process*:

$$dw_{max} = dA \qquad (4.34)$$

As a result, A is sometimes called the 'maximum work function', or the 'work function'.[6]

Justification 4.2

First we prove that a system does maximum work when it is working reversibly. (This conclusion was demonstrated in Section 2.3e for the expansion of a perfect gas; now we prove its universal validity.) We combine the Clausius inequality $dS \geq dq/T$ in the form $T\,dS \geq dq$ with the First Law, $dU = dq + dw$, and obtain

$$dU \leq T\,dS + dw$$

---

6. *Arbeit* is the German word for work; hence the symbol A.

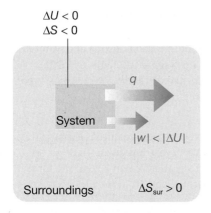

$\Delta U < 0$
$\Delta S < 0$

System

q

$|w| < |\Delta U|$

Surroundings $\Delta S_{sur} > 0$

4.13 In a system not isolated from its surroundings, the work done may be different from the change in internal energy. Moreover, the process is spontaneous if overall the entropy of the global, isolated system (the system plus its surroundings) increases. In the process depicted here, the entropy of the system decreases, so that of the surroundings must increase in order for the process to be spontaneous, which means that energy must pass from the system to the surroundings as heat. Therefore, less work than ΔU can be obtained.

(dU is smaller than the term of the right because we are replacing dq by T dS, which in general is larger.) This expression rearranges to

$$dw \geq dU - T \, dS$$

It follows that the most negative value of dw, and therefore the maximum energy that can be obtained from the system as work, is given by

$$dw_{max} = dU - T \, dS$$

and that this work is done only when the path is traversed reversibly (because then the equality applies). Because at constant temperature d$A = dU - T \, dS$, we conclude that d$w_{max} = dA$.

When a macroscopic isothermal change takes place in the system, eqn 34 becomes

$$w_{max} = \Delta A \tag{4.35}$$

with

$$\Delta A = \Delta U - T\Delta S \tag{4.36}$$

This expression shows that in some cases, depending on the sign of $T\Delta S$, not all the change in internal energy may be available for doing work. If the change occurs with a decrease in entropy (of the system), in which case $T\Delta S < 0$, then the right-hand side of this equation is not as negative as ΔU itself, and consequently the maximum work is less than ΔU. For the change to be spontaneous, some of the energy must escape as heat in order to generate enough entropy in the surroundings to overcome the reduction in entropy in the system (Fig. 4.13). In this case, Nature is demanding a tax on the internal energy as it is converted into work. This is the origin of the alternative name 'Helmholtz free energy' for A, because ΔA is that part of the change in internal energy that we are free to use to do work.

Molecular interpretation 4.3

Further insight into the relation between the work that a system can do and the Helmholtz energy is to recall that work is energy transferred to the surroundings as the uniform motion of atoms. We can interpret the expression $A = U - TS$ as showing that A is the total internal energy of the system, U, less a contribution that is stored in a disordered manner (the quantity TS). Because energy stored in a disordered manner cannot be used to achieve uniform motion in the surroundings, only the part of U that is not stored in a disordered way, the quantity $U - TS$, is available for conversion into work.

$\Delta U < 0$
$\Delta S > 0$

System

q

$|w| > |\Delta U|$

Surroundings $\Delta S_{sur} < 0$

4.14 In this process, the entropy of the system increases; hence we can afford to lose some entropy of the surroundings. That is, some of their energy may be lost as heat to the system. This energy can be returned to them as work. Hence the work done can exceed ΔU.

If the change occurs with an increase of entropy of the system (in which case $T\Delta S > 0$), the right-hand side of the equation is more negative than ΔU. In this case, the maximum work that can be obtained from the system is greater than ΔU. The explanation of this apparent paradox is that the system is not isolated and energy may flow in as heat as work is done. Because the entropy of the system increases, we can afford a reduction of the entropy of the surroundings yet still have, overall, a spontaneous process. Therefore, some heat (no more than the value of $T\Delta S$) may leave the surroundings and contribute to the work the change is generating (Fig. 4.14). Nature is now providing a tax refund.

Example 4.5 Calculating the maximum available work

When 1.000 mol $C_6H_{12}O_6$ (glucose) is oxidized to carbon dioxide and water at 25°C according to the equation $C_6H_{12}O_6(s) + 6\ O_2(g) \rightarrow 6\ CO_2(g) + 6\ H_2O(l)$, calorimetric measurements give $\Delta_r U^{\ominus} = -2808$ kJ mol^{-1} and $\Delta_r S = +182.4$ J K^{-1} mol^{-1} at 25°C. How much of this energy change can be extracted as (a) heat at constant pressure, (b) work?

Method We know that the heat released at constant pressure is equal to the value of ΔH, so we need to relate $\Delta_r H^{\ominus}$ to $\Delta_r U^{\ominus}$, which is given. To do so, we suppose that all the gases involved are perfect, and use eqn 2.26 in the form $\Delta_r H = \Delta_r U + \Delta \nu_g RT$. For the maximum work available from the process we use eqn 35.

Answer (a) Because $\Delta \nu_g = 0$, we know that $\Delta_r H^{\ominus} = \Delta_r U^{\ominus} = -2808$ kJ mol^{-1}. Therefore, at constant pressure, the energy available as heat is 2808 kJ mol^{-1}. (b) Because $T = 298$ K, the value of $\Delta_r A^{\ominus}$ is

$$\Delta_r A^{\ominus} = \Delta_r U^{\ominus} - T\Delta_r S^{\ominus} = -2862 \text{ kJ mol}^{-1}$$

Therefore, the combustion of 1.000 mol $C_6H_{12}O_6$ can be used to produce up to 2862 kJ of work.

Comment The maximum work available is greater than the change in internal energy on account of the positive entropy of reaction (which is partly due to the generation of a large number of small molecules from one big one). The system can therefore draw in energy from the surroundings (so reducing their entropy) and make it available for doing work.

Self-test 4.7 Repeat the calculation for the combustion of 1.000 mol $CH_4(g)$ under the same conditions, using data from Table 2.5 and Self-test 4.6.

$$[|q_p| = 890 \text{ kJ}, |w_{max}| = 813 \text{ kJ}]$$

(d) Some remarks on the Gibbs energy

The Gibbs energy (the 'free energy') is more common in chemistry than the Helmholtz energy because, at least in laboratory chemistry, we are usually more interested in changes occurring at constant pressure than at constant volume. The criterion $dG_{T,p} \leq 0$ carries over into chemistry as the observation that, *at constant temperature and pressure, chemical reactions are spontaneous in the direction of decreasing Gibbs energy*. Therefore, if we want to know whether a reaction is spontaneous, the pressure and temperature being constant, we assess the change in the Gibbs energy. If G decreases as the reaction proceeds, then the reaction has a spontaneous tendency to convert the reactants into products. If G increases, then the reverse reaction is spontaneous.

The existence of spontaneous endothermic reactions provides an illustration of the role of G. In such reactions, H increases, the system rises spontaneously to states of higher enthalpy, and $dH > 0$. Because the reaction is spontaneous we know that $dG < 0$ despite $dH > 0$; it follows that the entropy of the system increases so much that $T\,dS$ is strongly positive and outweighs dH in $dG = dH - T\,dS$. Endothermic reactions are therefore driven by the increase of entropy of the system, and this entropy change overcomes the reduction of entropy brought about in the surroundings by the inflow of heat into the system ($dS_{sur} = -dH/T$ at constant pressure).

(e) Maximum non-expansion work

The analogue of the maximum work interpretation of ΔA, and the origin of the name free energy, can be found for ΔG. In the *Justification* below, we show that, at constant temperature and pressure, the maximum additional (non-expansion) work, $w_{add,max}$, is given by the change in Gibbs energy:

$$\mathrm{d}w_{add,max} = \mathrm{d}G \qquad (4.37)$$

The corresponding expression for a measurable change is

$$w_{add,max} = \Delta G \qquad (4.38)$$

This expression is particularly useful for assessing the electrical work that may be produced by fuel cells and electrochemical cells, and we shall see many applications of it.

Justification 4.3

Because $H = U + pV$, for a general change in conditions, the change in enthalpy is

$$\mathrm{d}H = \mathrm{d}q + \mathrm{d}w + \mathrm{d}(pV)$$

The corresponding change in Gibbs energy ($G = H - TS$) is

$$\mathrm{d}G = \mathrm{d}H - T\,\mathrm{d}S - S\,\mathrm{d}T = \mathrm{d}q + \mathrm{d}w + \mathrm{d}(pV) - T\,\mathrm{d}S - S\,\mathrm{d}T$$

When the change is isothermal,

$$\mathrm{d}G = \mathrm{d}q + \mathrm{d}w + \mathrm{d}(pV) - T\,\mathrm{d}S$$

When the change is reversible, $\mathrm{d}w = \mathrm{d}w_{rev}$ and $\mathrm{d}q = \mathrm{d}q_{rev} = T\,\mathrm{d}S$, so for a reversible, isothermal process

$$\mathrm{d}G = T\,\mathrm{d}S + \mathrm{d}w_{rev} + \mathrm{d}(pV) - T\,\mathrm{d}S = \mathrm{d}w_{rev} + \mathrm{d}(pV)$$

The work consists of expansion work, which for a reversible change is given by $-p\,\mathrm{d}V$, and possibly some other kind of work (for instance, the electrical work of pushing electrons through a circuit or of raising a column of liquid); this additional work we denote $\mathrm{d}w_{add}$. Therefore, with $\mathrm{d}(pV) = p\,\mathrm{d}V + V\,\mathrm{d}p$,

$$\mathrm{d}G = (-p\,\mathrm{d}V + \mathrm{d}w_{add,rev}) + p\,\mathrm{d}V + V\,\mathrm{d}p = \mathrm{d}w_{add,rev} + V\,\mathrm{d}p$$

If the change occurs at constant pressure (as well as constant temperature), the last term disappears and $\mathrm{d}G = \mathrm{d}w_{add,rev}$. Therefore, at constant temperature and pressure, $\mathrm{d}w_{add,rev} = \mathrm{d}G$. However, because the process is reversible, the work done must now have its maximum value, so eqn 37 follows.

Example 4.6 Calculating the maximum non-expansion work of a reaction

How much energy is available for sustaining muscular and nervous activity from the combustion of 1.00 mol of glucose molecules under standard conditions at 37°C (blood temperature)? The standard entropy of reaction is $+182.4\ \mathrm{J\ K^{-1}\ mol^{-1}}$.

Method The non-expansion work available from the reaction is equal to the change in standard Gibbs energy for the reaction ($\Delta_r G^\ominus$, a quantity defined more fully below). To calculate this quantity, it is legitimate to ignore the temperature dependence of the reaction enthalpy, to obtain $\Delta_r H^\ominus$ from Table 2.5, and to substitute the data into $\Delta_r G^\ominus = \Delta_r H^\ominus - T\Delta_r S^\ominus$.

Answer Because the standard reaction enthalpy is -2808 kJ mol^{-1}, it follows that the standard reaction Gibbs energy is

$$\Delta_r G^\ominus = -2808 \text{ kJ mol}^{-1} - (310 \text{ K}) \times (182.4 \text{ J K}^{-1} \text{ mol}^{-1}) = -2865 \text{ kJ mol}^{-1}$$

Therefore, $w_{add,max} = -2865$ kJ for the combustion of 1 mol glucose molecules, and the reaction can be used to do up to 2865 kJ of non-expansion work.

Comment A person of mass 70 kg would need to do 2.1 kJ of work to climb vertically through 3.0 m; therefore, at least 0.13 g of glucose is needed to complete the task (and in practice significantly more).

Self-test 4.8 How much non-expansion work can be obtained from the combustion of 1.00 mol $CH_4(g)$ under standard conditions at 298 K? Use $\Delta_r S^\ominus = -243$ J K^{-1} mol^{-1}.

[818 kJ]

4.6 Standard molar Gibbs energies

Standard entropies and enthalpies of reaction can be combined to obtain the **standard Gibbs energy of reaction**, $\Delta_r G^\ominus$:

$$\Delta_r G^\ominus = \Delta_r H^\ominus - T\Delta_r S^\ominus \qquad [4.39]$$

The standard Gibbs energy of reaction is the difference in standard molar Gibbs energies of the products and reactants in their standard states at the temperature specified for the reaction as written. As in the case of standard reaction enthalpies, it is convenient to define the **standard Gibbs energy of formation**, $\Delta_f G^\ominus$ (or 'standard reaction Gibbs energy'), the standard reaction Gibbs energy for the formation of a compound from its elements in their reference states.[7] Standard free energies of formation of the elements in their reference states are zero, because their formation is a 'null' reaction. A selection of values for compounds is given in Table 4.4. From the values there, it is a simple matter to obtain the standard Gibbs energy of reaction by taking the appropriate combination:

$$\Delta_r G^\ominus = \sum_{\text{Products}} v \Delta_f G_m^\ominus - \sum_{\text{Reactants}} v \Delta_f G_m^\ominus \qquad (4.40)$$

with each term weighted by the appropriate stoichiometric coefficient.

Table 4.4* Standard Gibbs energies of formation at 298 K

	$\Delta_f G^\ominus/(\text{kJ mol}^{-1})$
Diamond, C(s)	+2.9
Benzene, C_6H_6(l)	+124.3
Methane, CH_4(g)	−50.7
Carbon dioxide, CO_2(g)	−394.4
Water, H_2O(l)	−237.1
Ammonia, NH_3(g)	−16.5
Sodium chloride, NaCl(s)	−384.1

* More values are given in the *Data section*.

Illustration 4.4

To calculate the standard Gibbs energy of the reaction $CO(g) + \frac{1}{2}O_2(g) \rightarrow CO_2(g)$ at 25°C, we write

$$\Delta_r G^\ominus = \Delta_f G^\ominus(CO_2, g) - \{\Delta_f G^\ominus(CO, g) + \tfrac{1}{2}\Delta_f G^\ominus(O_2, g)\}$$
$$= -394.4 \text{ kJ mol}^{-1} - \{(-137.2) + \tfrac{1}{2}(0)\} \text{ kJ mol}^{-1} = -257.2 \text{ kJ mol}^{-1}$$

Self-test 4.9 Calculate the standard reaction Gibbs energy for the combustion of $CH_4(g)$ at 298 K.

[−818 kJ mol^{-1}]

7. The reference state of an element was defined in Section 2.8.

The web site is a good source of links to online databases of thermochemical data.

Calorimetry (for ΔH directly, and for S via heat capacities) is only one of the ways of determining the values of Gibbs energies. They may also be obtained from equilibrium constants (Chapter 9) and electrochemical measurements (Chapter 10), and they may be calculated using data from spectroscopic observations (Chapter 20). The information in Tables 2.5 and 2.6 of the *Data section*, however, together with the machinery we shall now construct, is all we need in order to draw far-reaching conclusions about reactions and other processes of interest in chemistry.

Checklist of key ideas

- ☐ spontaneous direction of change, the direction of change that does not require work to be done to bring the change about

- ☐ Kelvin statement of the Second Law of thermodynamics: No process is possible in which the sole result is the absorption of heat from a reservoir and its complete conversion into work

The direction of spontaneous change

4.1 The dispersal of energy

- ☐ Spontaneity and the rise of disorder

- ☐ Collapse into disorder as the driving force of change

4.2 Entropy

- ☐ Second Law in terms of entropy: The entropy of an isolated system increases in the course of a spontaneous change: $\Delta S_{tot} > 0$

- ☐ thermodynamic definition of entropy, $dS = dq_{rev}/T$

- ☐ entropy change in the surroundings, $\Delta S_{sur} = q_{sur}/T_{sur}$

- ☐ entropy change for an adiabatic process, $\Delta S_{sur} = 0$

- ☐ Carnot cycle, a cycle composed of alternating isothermal and adiabatic reversible expansions and compressions

- ☐ efficiency of a heat engine, $\varepsilon = |w|/q_h$

- ☐ Carnot efficiency, $\varepsilon_{rev} = 1 - T_c/T_h$

- ☐ proof that entropy is a state function by showing that $\oint (1/T)dq_{rev} = 0$

- ☐ thermodynamic temperature scale, $T = (1 - \varepsilon)T_h$

- ☐ Kelvin scale, a thermodynamic temperature scale in which the triple point of water is set at 273.16 K

- ☐ Clausius inequality, $dS \geq dq/T$

4.3 Entropy changes accompanying specific processes

- ☐ normal transition temperature, T_{trs}, the temperature at which two phases are in equilibrium at 1 atm

- ☐ Entropy of transition at the transition temperature, $\Delta_{trs}S = \Delta_{trs}H/T_{trs}$

- ☐ Trouton's rule: many normal liquids have approximately the same standard entropy of vaporization (about 85 J K^{-1} mol^{-1})

- ☐ Entropy of isothermal expansion of a perfect gas, $\Delta S = nR \ln (V_f/V_i)$

- ☐ Variation of entropy with temperature,

$$S(T_f) = S(T_i) + \int_{T_i}^{T_f} (C_p/T)dT$$

- ☐ Measurement of entropy, from area under graph of C_p/T against T

- ☐ Debye extrapolation, $C_p = aT^3$ as $T \to 0$

4.4 The Third Law of thermodynamics

- ☐ Nernst heat theorem: The entropy change accompanying any physical or chemical transformation approaches zero as the temperature approaches zero: $\Delta S \to 0$ as $T \to 0$ provided all the substances involved are perfectly ordered

- ☐ Third Law of thermodynamics: The entropy of all perfect crystalline substances is zero at $T = 0$

- ☐ Third-Law entropy, an entropy reported on the basis that $S(0) = 0$

- ☐ standard reaction entropy, $\Delta_r S^\ominus = \sum_{Products} vS_m^\ominus - \sum_{Reactants} vS_m^\ominus$

Concentrating on the system

4.5 The Helmholtz and Gibbs energies

- ☐ criteria of spontaneity, $dS_{U,V} \geq 0$ and $dU_{S,V} \leq 0$

- ☐ Helmholtz energy, $A = U - TS$

- ☐ Gibbs energy, $G = H - TS$

- ☐ criteria of spontaneity, $dA_{T,V} \leq 0$ and $dG_{T,p} \leq 0$

- ☐ criterion of equilibrium at constant temperature and volume, $dA_{T,V} = 0$

- ☐ maximum work and the Helmholtz energy, $w_{max} = \Delta A$

- ☐ the criterion of equilibrium at constant temperature and pressure, $dG_{T,p} = 0$

- ☐ maximum additional (non-expansion) work and the Gibbs energy, $w_{add,max} = \Delta G$

4.6 Standard molar Gibbs energies

- ☐ standard Gibbs energy of reaction, $\Delta_r G^\ominus = \Delta_r H^\ominus - T\Delta_r S^\ominus$
$= \sum_{Products} vG_m^\ominus - \sum_{Reactants} vG_m^\ominus$

- ☐ standard Gibbs energy of formation $(\Delta_f G^\ominus)$, the standard reaction Gibbs energy for the formation of a compound from its elements in their reference states

- ☐ expressing $\Delta_r G^\ominus$ in terms of $\Delta_f G^\ominus$, $\Delta_r G^\ominus = \sum_{Products} v\Delta_f G^\ominus - \sum_{Reactants} v\Delta_f G^\ominus$

Further reading

Articles of general interest

M. Barón, With Clausius from energy to entropy. *J. Chem. Educ.* **66**, 1001 (1989).

N.C. Craig, Entropy analyses of four familiar processes. *J. Chem. Educ.* **65**, 760 (1988).

N.C. Craig, Entropy diagrams. *J. Chem. Educ.* **73**, 710 (1996).

W.H. Cropper, Walther Nernst and the last law. *J. Chem. Educ.* **64**, 3 (1987).

P. Djurdjevic and I. Gutman, A simple method for showing that entropy is a function of state. *J. Chem. Educ.* **65**, 399 (1988).

H. Frohlich, On the entropy change of surroundings of finite and infinite size. *J. Chem. Educ.* **73**, 716 (1996).

D.F.R. Gilson, Order end disorder and entropies of fusion. *J. Chem. Educ.* **69**, 23 (1992).

L. Glasser, Order, chaos, and all that! *J. Chem. Educ.* **66**, 997 (1989).

H.B. Hollinger and M.J. Zenzen, Thermodynamic irreversibility: 1. What is it? *J. Chem. Educ.* **68**, 31 (1991).

F.L. Lambert, Shuffled cards, messy desks, and disorderly dorm rooms—examples of entropy increase? Nonsense! *J. Chem. Educ.* **76**, 1385 (1999).

E.F. Meyer, The Carnot cycle revisited. *J. Chem. Educ.* **65**, 873 (1988).

P.G. Nelson, Derivation of the Second Law of thermodynamics from Boltzmann's distribution law. *J. Chem. Educ.* **65**, 390 (1988).

R.S. Ochs, Thermodynamics and spontaneity. *J. Chem. Educ.* **73**, 952 (1996).

K. Seidman and T.R. Michalik, The efficiency of reversible heat engines: The possible misinterpretation of a corollary to Carnot's theorem. *J. Chem. Educ.* **68**, 208 (1991).

J. Waser and V. Schomaker, A note on thermodynamic inequalities. *J. Chem. Educ.* **65**, 393 (1988).

Texts and sources of data and information

P.W. Atkins, *The second law*. Scientific American Books, New York (1984).

K.E. Bett, J.S. Rowlinson, and G. Saville, *Thermodynamics for chemical engineers*. MIT Press, Cambridge (1975).

J.B. Fenn, *Engines, energy, and entropy*. W.H. Freeman and Co., New York (1982).

J. Harte, *Consider a spherical cow: a course in environmental problem solving*. University Science Books, Sausalito (1988).

D.R. Gaskell, *Introduction to the thermodynamics of materials*. Taylor and Francis, London (1995).

D. Kondepudi and I. Prigogine, *Modern thermodynamics: from heat engines to dissipative structures*. Wiley, New York (1998).

H.-G. Lee, *Chemical thermodynamics for metals and materials*. Imperial College Press, London (1999).

J.M. Smith and H.C. Van Ness, *Introduction to chemical engineering thermodynamics*. McGraw–Hill, New York (1987).

Exercises

Discussion questions

4.1 (a) The evolution of life requires the organization of a very large number of molecules into biological cells. Does the formation of living organisms violate the Second Law of thermodynamics? State your conclusion clearly and present detailed arguments to support it.

4.1 (b) Justify Trouton's rule.

4.2 (a) Distinguish between a heat engine, a refrigerator, and a heat pump.

4.2 (b) You received an unsolicited proposal from a self-declared inventor who is seeking investors for the development of his latest idea: a device that uses heat extracted from the ground by a heat pump to boil water into steam that is used to heat a home and to power a steam engine that drives the heat pump. This procedure is potentially very lucrative because, after an initial extraction of energy from the ground, no fossil fuels would be required to keep the device running indefinitely. Would you invest in this idea? State your conclusion clearly and present detailed arguments to support it.

4.3 (a) The following expressions have been used to establish criteria for spontaneous change: $\Delta S_{tot} > 0$, $dS_{U,V} > 0$ and $dU_{S,V} < 0$, $dA_{T,V} < 0$, and $dG_{T,p} < 0$. Discuss the origin, significance, and applicability of each criterion.

4.3 (b) The following expressions have been used to establish criteria for equilibrium: $dA_{T,V} = 0$ and $dG_{T,p} = 0$. Discuss the origin, significance, and applicability of each criterion.

Numerical exercises

Assume that all gases are perfect and that data refer to 298.15 K unless otherwise stated.

4.4 (a) Calculate the change in entropy when 25 kJ of energy is transferred reversibly and isothermally as heat to a large block of iron at (a) 0°C, (b) 100°C.

4.4 (b) Calculate the change in entropy when 50 kJ of energy is transferred reversibly and isothermally as heat to a large block of copper at (a) 0°C, (b) 70°C.

4.5 (a) Calculate the molar entropy of a constant-volume sample of neon at 500 K given that it is 146.22 J K^{-1} mol^{-1} at 298 K.

4.5 (b) Calculate the molar entropy of a constant-volume sample of argon at 250 K given that it is 154.84 J K^{-1} mol^{-1} at 298 K.

4.6 (a) A sample consisting of 1.00 mol of a monatomic perfect gas with $C_{V,m} = \frac{3}{2}R$ is heated from 100°C to 300°C at constant pressure. Calculate ΔS (for the system).

4.6 (b) A sample consisting of 1.00 mol of a diatomic perfect gas with $C_{V,m} = \frac{5}{2}R$ is heated from 0°C to 100°C at constant pressure. Calculate ΔS (for the system).

4.7 (a) Calculate ΔS (for the system) when the state of 3.00 mol of a monatomic perfect gas, for which $C_{p,m} = \frac{5}{2}R$, is changed from 25°C and 1.00 atm to 125°C and 5.00 atm. How do you rationalize the sign of ΔS?

4.7 (b) Calculate ΔS (for the system) when the state of 2.00 mol of a diatomic perfect gas, for which $C_{p,m} = \frac{7}{2}R$, is changed from 25°C and 1.50 atm to 135°C and 7.00 atm. How do you rationalize the sign of ΔS?

4.8 (a) A sample consisting of 3.00 mol of a diatomic perfect gas at 200 K is compressed reversibly and adiabatically until its temperature reaches 250 K. Given that $C_{V,m} = 27.5$ J K^{-1} mol^{-1}, calculate q, w, ΔU, ΔH, and ΔS.

4.8 (b) A sample consisting of 2.00 mol of a diatomic perfect gas at 250 K is compressed reversibly and adiabatically until its temperature reaches 300 K. Given that $C_{V,m} = 27.5$ J K^{-1} mol^{-1}, calculate q, w, ΔU, ΔH, and ΔS.

4.9 (a) Calculate the increase in entropy when 1.00 mol of a monatomic perfect gas with $C_{p,m} = \frac{5}{2}R$ is heated from 300 K to 600 K and simultaneously expanded from 30.0 L to 50.0 L.

4.9 (b) Calculate the increase in entropy when 3.50 mol of a monatomic perfect gas with $C_{p,m} = \frac{5}{2}R$ is heated from 250 K to 700 K and simultaneously expanded from 20.0 L to 60.0 L.

4.10 (a) A system undergoes a process in which the entropy change is +2.41 J K^{-1}. During the process, 1.00 kJ of heat is added to the system at 500 K. Is the process thermodynamically reversible? Explain your reasoning.

4.10 (b) A system undergoes a process in which the entropy change is +5.51 J K^{-1}. During the process, 1.50 kJ of heat is added to the system at 350 K. Is the process thermodynamically reversible? Explain your reasoning.

4.11 (a) A sample of aluminium of mass 1.75 kg is cooled at constant pressure from 300 K to 265 K. Calculate (a) the energy that must be removed as heat and (b) the change in entropy of the sample.

4.11 (b) A sample of copper of mass 2.75 kg is cooled at constant pressure from 330 K to 275 K. Calculate (a) the energy that must be removed as heat and (b) the change in entropy of the sample.

4.12 (a) A sample of methane gas of mass 25 g at 250 K and 18.5 atm expands isothermally until its pressure is 2.5 atm. Calculate the change in entropy of the gas.

4.12 (b) A sample of nitrogen gas of mass 35 g at 230 K and 21.1 atm expands isothermally until its pressure is 4.3 atm. Calculate the change in entropy of the gas.

4.13 (a) A sample of perfect gas that initially occupies 15.0 L at 250 K and 1.00 atm is compressed isothermally. To what volume must the gas be compressed to reduce its entropy by 5.0 J K^{-1}?

4.13 (b) A sample of perfect gas that initially occupies 11.0 L at 270 K and 1.20 atm is compressed isothermally. To what volume must the gas be compressed to reduce its entropy by 3.0 J K^{-1}?

4.14 (a) Calculate the change in entropy when 50 g of water at 80°C is poured into 100 g of water at 10°C in an insulated vessel given that $C_{p,m} = 75.5$ J K^{-1} mol^{-1}.

4.14 (b) Calculate the change in entropy when 25 g of ethanol at 50°C is poured into 70 g of ethanol at 10°C in an insulated vessel, given that $C_{p,m} = 111.5$ J K^{-1} mol^{-1}.

4.15 (a) Calculate ΔH and ΔS_{tot} when two copper blocks, each of mass 10.0 kg, one at 100°C and the other at 0°C, are placed in contact in an isolated container. The specific heat capacity of copper is 0.385 J K^{-1} g^{-1} and may be assumed constant over the temperature range involved.

4.15 (b) Calculate ΔH and ΔS_{tot} when two iron blocks, each of mass 1.00 kg , one at 200°C and the other at 25°C, are placed in contact in an isolated container. The specific heat capacity of iron is 0.449 J K^{-1} g^{-1} and may be assumed constant over the temperature range involved.

4.16 (a) Consider a system consisting of 2.0 mol $CO_2(g)$, initially at 25°C and 10 atm and confined to a cylinder of cross-section 10.0 cm^2. It is allowed to expand adiabatically against an external pressure of 1.0 atm until the piston has moved outwards through 20 cm. Assume that carbon dioxide may be considered a perfect gas with $C_{V,m} = 28.8$ J K^{-1} mol^{-1} and calculate (a) q, (b) w, (c) ΔU, (d) ΔT, (e) ΔS.

4.16 (b) Consider a system consisting of 1.5 mol $CO_2(g)$, initially at 15°C and 9.0 atm and confined to a cylinder of cross-section 100.0 cm^2. The sample is allowed to expand adiabatically against an external pressure of 1.5 atm until the piston has moved outwards through 15 cm. Assume that carbon dioxide may be considered a perfect gas with $C_{V,m} = 28.8$ J K^{-1} mol^{-1}, and calculate (a) q, (b) w, (c) ΔU, (d) ΔT, (e) ΔS.

4.17 (a) The enthalpy of vaporization of chloroform ($CHCl_3$) is 29.4 kJ mol^{-1} at its normal boiling point of 334.88 K. Calculate (a) the entropy of vaporization of chloroform at this temperature and (b) the entropy change of the surroundings.

4.17 (b) The enthalpy of vaporization of methanol is 35.27 kJ mol^{-1} at its normal boiling point of 64.1°C. Calculate (a) the entropy of vaporization of methanol at this temperature and (b) the entropy change of the surroundings.

4.18 (a) Calculate the standard reaction entropy at 298 K of

  (a) $2 CH_3CHO(g) + O_2(g) \longrightarrow 2 CH_3COOH(l)$

  (b) $2 AgCl(s) + Br_2(l) \longrightarrow 2 AgBr(s) + Cl_2(g)$

  (c) $Hg(l) + Cl_2(g) \longrightarrow HgCl_2(s)$

4.18 (b) Calculate the standard reaction entropy at 298 K of

  (a) $Zn(s) + Cu^{2+}(aq) \longrightarrow Zn^{2+}(aq) + Cu(s)$

  (b) $C_{12}H_{22}O_{11}(s) + 12 O_2(g) \longrightarrow 12 CO_2(g) + 11 H_2O(l)$

4.19 (a) Combine the reaction entropies calculated in Exercise 4.18a with the reaction enthalpies, and calculate the standard reaction Gibbs energies at 298 K.

4.19 (b) Combine the reaction entropies calculated in Exercise 4.18b with the reaction enthalpies, and calculate the standard reaction Gibbs energies at 298 K.

4.20 (a) Use standard Gibbs energies of formation to calculate the standard reaction Gibbs energies at 298 K of the reactions in Exercise 4.18a.

4.20 (b) Use standard Gibbs energies of formation to calculate the standard reaction Gibbs energies at 298 K of the reactions in Exercise 4.18b.

4.21 (a) Calculate the standard Gibbs energy of the reaction $4\,HCl(g) + O_2(g) \rightarrow 2\,Cl_2(g) + 2\,H_2O(l)$ at 298 K, from the standard entropies and enthalpies of formation given in Table 2.6.

4.21 (b) Calculate the standard Gibbs energy of the reaction $CO(g) + CH_3OH(l) \rightarrow CH_3COOH(l)$ at 298 K, from the standard entropies and enthalpies of formation given in Tables 2.5 and 2.6.

4.22 (a) The standard enthalpy of combustion of solid phenol (C_6H_5OH) is -3054 kJ mol^{-1} at 298 K and its standard molar entropy is 144.0 J K^{-1} mol^{-1}. Calculate the standard Gibbs energy of formation of phenol at 298 K.

4.22 (b) The standard enthalpy of combustion of solid urea $(CO(NH_2)_2)$ is -632 kJ mol^{-1} at 298 K and its standard molar entropy is 104.60 J K^{-1} mol^{-1}. Calculate the standard Gibbs energy of formation of urea at 298 K.

4.23 (a) Calculate the change in the entropies of the system and the surroundings, and the total change in entropy, when a sample of nitrogen gas of mass 14 g at 298 K and 1.00 bar doubles its volume in (a) an isothermal reversible expansion, (b) an isothermal irreversible expansion against $p_{ex} = 0$, and (c) an adiabatic reversible expansion.

4.23 (b) Calculate the change in the entropies of the system and the surroundings, and the total change in entropy, when the volume of a sample of argon gas of mass 21 g at 298 K and 1.50 bar increases from 1.20 L to 4.60 L in (a) an isothermal reversible expansion, (b) an isothermal irreversible expansion against $p_{ex} = 0$, and (c) an adiabatic reversible expansion.

4.24 (a) Calculate the change in entropy when a monatomic perfect gas is compressed to half its volume and simultaneously heated to twice its initial temperature.

4.24 (b) Calculate the change in entropy when a diatomic perfect gas is compressed to one-third its volume and simultaneously heated to three times its initial temperature.

4.25 (a) Calculate the maximum non-expansion work per mole that may be obtained from a fuel cell in which the chemical reaction is the combustion of methane at 298 K.

4.25 (b) Calculate the maximum non-expansion work per mole that may be obtained from a fuel cell in which the chemical reaction is the combustion of propane at 298 K.

4.26 (a) Calculate the Carnot efficiency of a primitive steam engine operating on steam at 100°C and discharging at 60°C. (b) Repeat the calculation for a modern steam turbine that operates with steam at 300°C and discharges at 80°C.

4.26 (b) A certain heat engine operates between 1000 K and 500 K. (a) What is the maximum efficiency of the engine? (b) Calculate the maximum work that can be done for each 1.0 kJ of heat supplied by the hot source. (c) How much heat is discharged into the cold sink in a reversible process for each 1.0 kJ supplied by the hot source?

Problems[†]

Assume that all gases are perfect and that data refer to 298 K unless otherwise stated.

Numerical problems

4.1 Calculate the difference in molar entropy (a) between liquid water and ice at −5°C, (b) between liquid water and its vapour at 95°C and 1.00 atm. The differences in heat capacities on melting and on vaporization are 37.3 J K^{-1} mol^{-1} and −41.9 J K^{-1} mol^{-1}, respectively. Distinguish between the entropy changes of the sample, the surroundings, and the total system, and discuss the spontaneity of the transitions at the two temperatures.

4.2 The heat capacity of chloroform (trichloromethane, $CHCl_3$) in the range 240 K to 330 K is given by $C_{p,m}/(J K^{-1} mol^{-1}) = 91.47 + 7.5 \times 10^{-2} (T/K)$. In a particular experiment, 1.00 mol $CHCl_3$ is heated from 273 K to 300 K. Calculate the change in molar entropy of the sample.

4.3 A block of copper of mass 2.00 kg $(C_{p,m} = 0.385\,J K^{-1} mol^{-1})$ and temperature 0°C is introduced into an insulated container in which there is 1.00 mol $H_2O(g)$ at 100°C and 1.00 atm. (a) Assuming all the steam is condensed to water, what will be the final temperature of the system, the heat transferred from water to copper, and the entropy change of the water, copper, and the total system? (b) In fact, some water vapour is present at equilibrium. From the vapour pressure of water at the temperature calculated in (a), and assuming that the heat capacities of both gaseous and liquid water are constant and given by their values at that temperature, obtain an improved value of the final temperature, the heat transferred, and the various entropies. (*Hint.* You will need to make plausible approximations.)

4.4 Consider a perfect gas contained in a cylinder and separated by a frictionless adiabatic piston into two sections A and B. All changes in B are isothermal; that is, a thermostat surrounds B to keep its temperature constant. There is 2.00 mol of the gas in each section. Initially, $T_A = T_B = 300$ K, $V_A = V_B = 2.00$ L. Heat is added to Section A and the piston moves to the right reversibly until the final volume of Section B is 1.00 L. Calculate (a) ΔS_A and ΔS_B, (b) ΔA_A and ΔA_B, (c) ΔG_A and ΔG_B, (d) ΔS of the total system and its surroundings. If numerical values cannot be obtained, indicate whether the values should be positive, negative, or zero or are indeterminate from the information given. (Assume $C_{V,m} = 20$ J K^{-1} mol^{-1}.)

4.5 A Carnot cycle uses 1.00 mol of a monatomic perfect gas as the working substance from an initial state of 10.0 atm and 600 K. It expands isothermally to a pressure of 1.00 atm (step 1), and then adiabatically to a temperature of 300 K (step 2). This expansion is followed by an isothermal compression (step 3), and then an adiabatic compression (step 4) back to the initial state. Determine

† Problems denoted with the symbol ‡ were supplied by Charles Trapp, Carmen Giunta, and Marshall Cady.

the values of q, w, ΔU, ΔH, ΔS, ΔS_{tot}, and ΔG for each stage of the cycle and for the cycle as a whole. Express your answer as a table of values.

4.6 1.00 mol of a perfect gas at 27°C is expanded isothermally from an initial pressure of 3.00 atm to a final pressure of 1.00 atm in two ways: (a) reversibly, and (b) against a constant external pressure of 1.00 atm. Determine the values of q, w, ΔU, ΔH, ΔS, ΔS_{sur} and ΔS_{tot} for each path.

4.7 A sample of 1.00 mol of a perfect gas at 27°C and 1.00 atm is expanded adiabatically in two ways: (a) reversibly to 0.50 atm, and (b) against a constant external pressure of 0.50 atm. Determine the values of q, w, ΔU, ΔH, ΔS, ΔS_{sur} and ΔS_{tot} for each path where the data permit. Take $C_{V,m} = \frac{3}{2}R$.

4.8 A sample of 1.00 mol of a monatomic perfect gas with $C_{V,m} = \frac{3}{2}R$, initially at 298 K and 10 L, is expanded, with the surroundings maintained at 298 K, to a final volume of 20 L, in three ways: (a) isothermally and reversibly, (b) isothermally against a constant external pressure of 0.50 atm, (c) adiabatically against a constant external pressure of 0.50 atm. Calculate ΔS, ΔS_{sur}, ΔH, ΔT, ΔA, and ΔG for each path where the data permit.

4.9 The standard molar entropy of $NH_3(g)$ is 192.45 J K^{-1} mol^{-1} at 298 K, and its heat capacity is given by eqn 2.30 with the coefficients given in Table 2.2. Calculate the standard molar entropy at (a) 100°C and (b) 500°C.

4.10 A block of copper of mass 500 g and initially at 293 K is in thermal contact with an electric heater of resistance 1.00 kΩ and negligible mass. A current of 1.00 A is passed for 15.0 s. Calculate the change in entropy of the copper, taking $C_{p,m} = 24.4$ J K^{-1} mol^{-1}. The experiment is then repeated with the copper immersed in a stream of water that maintains its temperature at 293 K. Calculate the change in entropy of the copper and the water in this case.

4.11 Calculate the standard Helmholtz energy of formation, $\Delta_f A$, of $CH_3OH(l)$ at 298 K from the standard Gibbs energy of formation and the assumption that H_2 and O_2 are perfect gases.

4.12 Calculate the change in entropy when 200 g of (a) water at 0°C, (b) ice at 0°C is added to 200 g of water at 90°C in an insulated container.

4.13 Calculate (a) the maximum work and (b) the maximum non-expansion work that can be obtained from the freezing of supercooled water at –5°C and 1.0 atm. The densities of water and ice are 0.999 g cm^{-3} and 0.917 g cm^{-3}, respectively, at –5°C.

4.14 The molar heat capacity of lead varies with temperature as follows:

T/K	10	15	20	25	30	50
$C_{p,m}$/(J K^{-1} mol^{-1})	2.8	7.0	10.8	14.1	16.5	21.4

T/K	70	100	150	200	250	298
$C_{p,m}$/(J K^{-1} mol^{-1})	23.3	24.5	25.3	25.8	26.2	26.6

Calculate the standard Third-Law entropy of lead at (a) 0°C and (b) 25°C.

4.15 From standard enthalpies of formation, standard entropies, and standard heat capacities available from tables in the *Data section*, calculate: (a) the standard enthalpies and entropies at 298 K and 398 K for the reaction $CO_2(g) + H_2(g) \rightarrow CO(g) + H_2O(g)$. Assume that the heat capacities are constant over the temperature range involved.

4.16 The standard reaction Gibbs energy of

$$K_4[Fe(CN)_6] \cdot 3H_2O(s) \longrightarrow 4\,K^+(aq) + [Fe(CN)_6]^{4-}(aq) + 3\,H_2O(l)$$

is +26.120 kJ mol^{-1} (I.R. Malcolm, L.A.K. Staveley, and R.D. Worswick, *J. Chem. Soc. Faraday Trans. I*, 1532 (1973)). The enthalpy of solution of the trihydrate is +55.000 kJ mol^{-1}. Calculate (a) the standard molar entropy of the hexacyanoferrate(II) ion in water and (b) the standard reaction entropy given that the standard molar entropy of the solid trihydrate is 599.7 J K^{-1} mol^{-1} and that of the K$^+$ ion in water is 102.5 J K^{-1} mol^{-1}.

4.17 The heat capacity of anhydrous potassium hexacyanoferrate(II) varies with temperature as follows:

T/K	$C_{p,m}$/(J K^{-1} mol^{-1})	T/K	$C_{p,m}$/(J K^{-1} mol^{-1})
10	2.09	100	179.6
20	14.43	110	192.8
30	36.44	150	237.6
40	62.55	160	247.3
50	87.03	170	256.5
60	111.0	180	265.1
70	131.4	190	273.0
80	149.4	200	280.3
90	165.3		

Calculate the molar enthalpy relative to its value at $T = 0$ and the Third-Law entropy at each of these temperatures.

4.18 The compound 1,3,5-trichloro-2,4,6-trifluorobenzene is an intermediate in the conversion of hexachlorobenzene to hexafluorobenzene, and its thermodynamic properties have been examined by measuring its heat capacity over a wide temperature range (R.L. Andon and J.F. Martin, *J. Chem. Soc. Faraday Trans. I.* 871 (1973)). Some of the data are as follows:

T/K	14.14	16.33	20.03	31.15	44.08	64.81
$C_{p,m}$/(J K^{-1} mol^{-1})	9.492	12.70	18.18	32.54	46.86	66.36

T/K	100.90	140.86	183.59	225.10	262.99	298.06
$C_{p,m}$/(J K^{-1} mol^{-1})	95.05	121.3	144.4	163.7	180.2	196.4

Calculate the molar enthalpy relative to its value at $T = 0$ and the Third-Law entropy of the compound at these temperatures.

4.19‡ Given that $S_m^{\ominus} = 29.79$ J K^{-1} mol^{-1} for bismuth at 100 K and the following tabulated heat capacities data (D.G. Archer, *J. Chem. Eng. Data* **40**, 1015 (1995)), compute the standard molar entropy of bismuth at 200 K.

T/K	100	120	140	150	160	180	200
$C_{p,m}$/(J K^{-1} mol^{-1})	23.00	23.74	24.25	24.44	24.61	24.89	25.11

Compare the value to the value that would be obtained by taking the heat capacity to be constant at 24.44 J K^{-1} mol^{-1} over this range.

Theoretical problems

4.20 Show that the integral of dq_{rev}/T round a Carnot cycle is zero. Then show that the integral is negative if the isothermal reversible expansion stage is replaced by an isothermal irreversible expansion.

4.21 Prove that two reversible adiabatic paths can never cross. Assume that the energy of the system under consideration is a function of temperature only. (*Hint.* Suppose that two such paths can intersect, and complete a cycle with the two paths plus one isothermal path. Consider the changes accompanying each stage of the cycle and show that they conflict with the Kelvin statement of the Second Law.)

4.22 Represent the Carnot cycle on a temperature–entropy diagram and show that the area enclosed by the cycle is equal to the work done.

4.23 Find an expression for the change in entropy when two blocks of the same substance and of equal mass, one at the temperature T_h and the other at T_c, are brought into thermal contact and allowed to reach equilibrium. Evaluate the change for two blocks of copper, each of mass 500 g, with $C_{p,m} = 24.4$ J K^{-1} mol^{-1}, taking $T_h = 500$ K and $T_c = 250$ K.

4.24 A gaseous sample consisting of 1.00 mol molecules is described by the equation of state $pV_m = RT(1 + Bp)$. Initially at 373 K, it undergoes Joule–Thomson expansion from 100 atm to 1.00 atm. Given that $C_{p,m} = \frac{5}{2}R$, $\mu = 0.21$ K atm^{-1}, $B = -0.525$(K/T) atm^{-1}, and that these are constant over the temperature range involved, calculate ΔT and ΔS for the gas.

4.25 Prove that the perfect gas temperature scale and the thermodynamic temperature scale based on the Second Law of thermodynamics differ from each other by at most a constant numerical factor.

4.26 The definitions of the enthalpy, Gibbs energy, and Helmholtz energy have all been of the form $g = f + yz$. Show that the addition of the product yz is a general way of converting a function of x and y to a function of x and z in the sense that, if $df = a\,dx - z\,dy$, then $dg = a\,dx + y\,dz$.

4.27‡ It is found that the molar heat capacity of β-Ce$_2$Si$_2$O$_7$ at 1 bar in the range of temperatures 0–30 K is described by

$$C_{p,m}(T) = a(T/K)^3(1 - e^{-b/T^2})$$

where $a = 5.597 \times 10^{-4}$ J mol^{-1} K^{-4} and $b = 1701.8$ K^2 are empirical parameters. In the high-temperature range, 500–900 K, the molar heat capacity is described by

$$C_{p,m}(T)/(\text{J mol}^{-1}\,\text{K}^{-1}) = 222.3716 + 0.076\,790\,42(T/K)$$
$$- 3\,631\,886(T/K)^{-2}$$

No phase transition has been detected below 900 K. (a) Draw graphs of $S_m(T)$, $H_m(T) - H_m(0)$, $G_m(T) - G_m(0)$ in the range $0 \leq T \leq 30$ K. (b) According to the law of Dulong and Petit, the high-temperature heat capacity per mole of atoms of a crystalline material is 3R. How does the classical prediction compare with the experimental? (*Hint.* Use mathematical software or a spreadsheet to draw the graphs.)

4.28‡ Polytropic processes are those that satisfy the relation $pV^n = C$. Make schematic plots of polytropic processes on a pV and TS diagram for $n = 0, \pm 1, \gamma$(the heat capacity ratio), and $\pm\infty$. (*Hint.* Use mathematical software or a spreadsheet to draw the graphs.)

Applications: to chemical engineering and atmospheric science

4.29‡ Alkyl radicals are important intermediates in the combustion and atmospheric chemistry of hydrocarbons. N. Cohen (*J. Phys. Chem.* **96**, 9052 (1992)) reports group additivity tables for the thermochemistry of alkyl free radicals. A portion of the table follows. Use the table to estimate the standard molar entropies

of C$_2$H$_5$, sec-C$_4$H$_9$, and tert-C$_4$H$_9$. (For the structures of these species, see the problems in Chapter 3.) Note that $S_m^\ominus = S_{int}^\ominus - R \ln s$, where $S_{int}^\ominus$ is the so-called intrinsic molar entropy, computed by group additivity, and s is a symmetry number: $s = 6$ for C$_2$H$_5$, 3^2 for sec-C$_4$H$_9$, and 3^4 for tert-C$_4$H$_9$.

Group	$S_{int}^\ominus$/(J K^{-1} mol^{-1})
C—(C)(H)$_3$	126.8
·C—(C)(H)$_2$	135.9
·C—(C)$_2$(H)	59.3
·C—(C)$_3$	−29.2
C—(·C)(H)$_3$	126.8
C—(·C)(C)(H)$_2$	42.0

4.30‡ Use the following enthalpies of formation reported by Seakins *et al.* (P.W. Seakins, M.J. Pilling, J.T. Niiranen, D. Gutman, and L.N. Krasnoperov, *J. Phys. Chem.* **96**, 9847 (1992)) and entropies based on group additivity tables of Cohen (N. Cohen, *J. Phys. Chem.* **96**, 9052 (1992)) to compute $\Delta_r G^\ominus$ for three possible fates of the tert-butyl radical at 700 K, namely (a) tert-C$_4$H$_9$ → sec-C$_4$H$_9$, (b) tert-C$_4$H$_9$ → C$_3$H$_6$ + CH$_3$, (c) tert-C$_4$H$_9$ → C$_2$H$_4$ + C$_2$H$_5$.

Species	$\Delta_f H^\ominus$/(kJ mol^{-1})	$S_m^\ominus$/(J K^{-1} mol^{-1})
C$_2$H$_5$	+121.0	247.8
sec-C$_4$H$_9$	+67.5	336.6
tert-C$_4$H$_9$	+51.3	314.6

4.31 Suppose that an internal combustion engine runs on octane, for which the enthalpy of combustion is −5512 kJ mol^{-1} and take the mass of 1 gallon of fuel as 3 kg. What is the maximum height, neglecting all forms of friction, to which a car of mass 1000 kg can be driven on 1.00 gallon of fuel given that the engine cylinder temperature is 2000°C and the exit temperature is 800°C?

4.32 The cycle involved in the operation of an internal combustion engine is called the *Otto cycle*. Air can be considered to be the working substance and can be assumed to be a perfect gas. The cycle consists of the following steps: (1) reversible adiabatic compression from A to B, (2) reversible constant-volume pressure increase from B to C due to the combustion of a small amount of fuel, (3) reversible adiabatic expansion from C to D, and (4) reversible and constant-volume pressure decrease back to state A. Determine the change in entropy (of the system and of the surroundings) for each step of the cycle and determine an expression for the efficiency of the cycle, assuming that the heat is supplied in Step 2. Evaluate the efficiency for a compression ratio of 10:1. Assume that in state A, $V = 4.00$ L, $p = 1.00$ atm, and $T = 300$ K, that $V_A = 10V_B$, $p_C/p_B = 5$, and that $C_{p,m} = \frac{7}{2}R$.

4.33‡ Consider a Carnot engine operating in outer space between the temperatures T_h and T_c. The only way that the engine can discard heat at T_c is by radiation. The power radiated by the engine at T_c follows the Stefan–Boltzmann law (see Section 11.1), which for our purposes here is written $dq_c/dt = kAT_c^4$, where k is a constant related to the Stefan–Boltzmann constant. Find the ratio T_c/T_h that corresponds to a minimum area A of the radiator for a fixed power output and constant T_h.

5

The Second Law: the machinery

Combining the First and Second Laws

5.1 The fundamental equation

5.2 Properties of the internal energy

Properties of the Gibbs energy

5.3 General considerations

5.4 The variation of the Gibbs energy with temperature

5.5 The variation of the Gibbs energy with pressure

Checklist of key ideas

Further reading

Exercises

Problems

One of the principal applications of thermodynamics is to find relations between properties that might not be thought to be related. Several relations of this kind can be established by making use of the fact that the Gibbs energy is a state function. We also see how to derive expressions for the variation of the Gibbs energy with temperature and pressure and how to formulate expressions that are valid for real gases. These expressions will prove useful later when we need to discuss the effect of temperature and pressure on equilibrium constants.

The Gibbs energy, G, is of central importance to chemistry. In this chapter it begins to move to the centre of the stage. We have seen that, at constant temperature and pressure, a system tends to adjust so as to minimize the value of G. Therefore, the equilibrium state of a system corresponds to the state of minimum Gibbs energy. It follows that to know how the equilibrium state responds to changes in the conditions, such as how a melting temperature responds to changes in pressure or the equilibrium composition of a reaction mixture responds to changes in temperature, we need to know how G varies. To do so, we first consider how our new information about entropy can be combined with our knowledge about the internal energy that came from the First Law.

Combining the First and Second Laws

We have treated the First and Second Laws of thermodynamics as distinct statements. Both are relevant to the behaviour of matter, so we can bring the whole force of thermodynamics to bear on a problem by setting up a formulation that combines concepts from the two laws.

5.1 The fundamental equation

We have seen that the First Law of thermodynamics may be written

$$dU = dq + dw \tag{5.1}$$

For a reversible change in a closed system of constant composition, and in the absence of any additional (non-expansion) work,

$$dw_{rev} = -p \, dV \qquad dq_{rev} = T \, dS$$

The second of these two relations, of course, comes from the definition of entropy introduced in the discussion of the Second Law. Therefore, for a reversible change in a closed system,

$$dU = T\,dS - p\,dV \tag{5.2}$$

However, because dU is an exact differential, its value is independent of path. Therefore, the same value of dU is obtained whether the change is brought about irreversibly or reversibly. Consequently, *eqn 2 applies to any change—reversible or irreversible—of a closed system that does no additional (non-expansion) work*. We shall call this combination of the First and Second Laws the **fundamental equation**.

The fact that the fundamental equation applies to both reversible and irreversible changes may be puzzling at first sight. The reason is that only in the case of a reversible change may $T\,dS$ be identified with dq and $-p\,dV$ with dw. When the change is irreversible, $T\,dS > dq$ (the Clausius inequality) and $-p\,dV < dw$. The sum of dw and dq remains equal to the sum of $T\,dS$ and $-p\,dV$, provided the composition is constant.

5.2 Properties of the internal energy

Equation 2 shows that the internal energy of a closed system changes in a simple way when either S or V is changed ($dU \propto dS$ and $dU \propto dV$). These simple proportionalities suggest that U should be regarded as a function of S and V. We could regard U as a function of other variables, such as S and p or T and V, because they are all interrelated; but the simplicity of the fundamental equation suggests that $U(S, V)$ is the best choice.

The *mathematical* consequence of U being a function of S and V is that we can express an infinitesimal change dU in terms of changes dS and dV by

> Partial derivatives are reviewed in *Further information 1*.

$$dU = \left(\frac{\partial U}{\partial S}\right)_V dS + \left(\frac{\partial U}{\partial V}\right)_S dV \tag{5.3}$$

The two coefficients are the slopes of the plots of U against S and V, respectively. When this expression is compared to the *thermodynamic* relation, eqn 2, we see that, for systems of constant composition,

$$\left(\frac{\partial U}{\partial S}\right)_V = T \qquad \left(\frac{\partial U}{\partial V}\right)_S = -p \tag{5.4}$$

The first of these two equations is a purely thermodynamic definition of temperature as the ratio of the changes in the internal energy and entropy of a constant-volume, closed, constant-composition system. We are beginning to generate relations between the properties of a system and to discover the power of thermodynamics for establishing unexpected relations.

(a) The Maxwell relations

Suppose we can write an infinitesimal change in a function f as $df = g\,dx + h\,dy$ where g and h are functions of x and y. The mathematical criterion for df being an exact differential (in the sense that its integral is independent of path) is that

> The properties of differentials are reviewed in *Further information 1*.

$$\left(\frac{\partial g}{\partial y}\right)_x = \left(\frac{\partial h}{\partial x}\right)_y \tag{5.5}$$

Because the fundamental equation, eqn 2, is an expression for an exact differential, the functions multiplying dS and dV (namely T and $-p$) must pass this test. Therefore, it must be the case that

Table 5.1 The Maxwell relations

$$\left(\frac{\partial T}{\partial V}\right)_S = -\left(\frac{\partial p}{\partial S}\right)_V$$

$$\left(\frac{\partial T}{\partial p}\right)_S = \left(\frac{\partial V}{\partial S}\right)_p$$

$$\left(\frac{\partial p}{\partial T}\right)_V = \left(\frac{\partial S}{\partial V}\right)_T$$

$$\left(\frac{\partial V}{\partial T}\right)_p = -\left(\frac{\partial S}{\partial p}\right)_T$$

$$\left(\frac{\partial T}{\partial V}\right)_S = -\left(\frac{\partial p}{\partial S}\right)_V \tag{5.6}$$

We have generated a relation between quantities that, at first sight, would not seem to be related.

Equation 6 is an example of a **Maxwell relation**. However, apart from being unexpected, it does not look particularly interesting. Nevertheless, it does suggest that there may be other similar relations that are more useful. Indeed, we can use the fact that H, G, and A are all state functions to derive three more Maxwell relations. The argument to obtain them runs in the same way in each case: because H, G, and A are state functions, the expressions for dH, dG, and dA satisfy relations like eqn 6. All four relations are listed in Table 5.1 and we put them to work later in the chapter.

(b) The variation of internal energy with volume

The coefficient

$$\pi_T = \left(\frac{\partial U}{\partial V}\right)_T \tag{5.7}$$

which represents how the internal energy changes as the volume of a system is changed isothermally played a central role in the manipulation of the First Law, and in *Justification 3.3* we used the relation

$$\pi_T = T\left(\frac{\partial p}{\partial T}\right)_V - p \tag{5.8}$$

This relation is called a **thermodynamic equation of state** because it is an expression for pressure in terms of a variety of thermodynamic properties of the system. We are now ready to derive it by using a Maxwell relation.

Justification 5.1

We obtain an expression for the coefficient π_T by dividing both sides of eqn 3 by dV, imposing the constraint of constant temperature, which gives

$$\left(\frac{\partial U}{\partial V}\right)_T = \left(\frac{\partial U}{\partial S}\right)_V\left(\frac{\partial S}{\partial V}\right)_T + \left(\frac{\partial U}{\partial V}\right)_S$$

Next, we introduce the two relations in eqn 4, to obtain

$$\left(\frac{\partial U}{\partial V}\right)_T = T\left(\frac{\partial S}{\partial V}\right)_T - p$$

This equation is already beginning to look like the expression we want. The third Maxwell relation in Table 5.1 turns $(\partial S/\partial V)_T$ into $(\partial p/\partial T)_V$, which completes the proof of eqn 8

Example 5.1 Deriving a thermodynamic relation

Show thermodynamically that $\pi_T = 0$ for a perfect gas, and compute its value for a van der Waals gas.

Method Proving a result 'thermodynamically' means basing it entirely on general thermodynamic relations and equations of state, without drawing on molecular arguments (such as the existence of intermolecular forces). We know that for a perfect gas, $p = nRT/V$, so this relation should be used in eqn 8. Similarly, the van der Waals equation is given in Table 1.6, and for the second part of the question it should be used in eqn 8.

Answer Because $(\partial p/\partial T)_V = nR/V$ for a perfect gas (by differentiation of the equation of state), eqn 8 becomes

$$\pi_T = \frac{nRT}{V} - p = 0$$

The equation of state of a van der Waals gas is

$$p = \frac{nRT}{V - nb} - a\frac{n^2}{V^2}$$

Therefore, because a and b are independent of temperature, we can write

$$\left(\frac{\partial p}{\partial T}\right)_V = \frac{nR}{V - nb}$$

That is,

$$\pi_T = \frac{nRT}{V - nb} - \frac{nRT}{V - nb} + a\frac{n^2}{V^2} = a\frac{n^2}{V^2}$$

Comment This result for π_T implies that the internal energy of a van der Waals gas increases when it expands isothermally (that is, $(\partial U/\partial V)_T > 0$), and that the increase is related to the parameter a, which models the attractive interactions between the particles. A larger molar volume, corresponding to a greater average separation between molecules, implies weaker mean intermolecular attractions, so the total energy is greater.

Self-test 5.1 Calculate π_T for a gas that obeys the virial equation of state.

$$[\pi_T = RT^2(\partial B/\partial T)_V/V_m^2 + \cdots]$$

Properties of the Gibbs energy

The same arguments that we have used for U can be used for the Gibbs energy $G = H - TS$ and we can derive important expressions showing how G varies with pressure and temperature.

5.3 General considerations

When the system undergoes a change of state, G may change because H, T, and S change. For infinitesimal changes in each property,

$$dG = dH - T\,dS - S\,dT$$

Because $H = U + pV$, we know that

$$dH = dU + p\,dV + V\,dp$$

For a closed system doing no additional work, we can replace dU by the fundamental equation $dU = T\,dS - p\,dV$. The result of these steps is

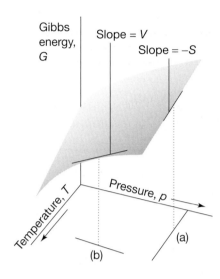

5.1 The variation of the Gibbs energy of a system with (a) temperature at constant pressure and (b) pressure at constant temperature. The slope of the former is equal to the negative of the entropy of the system and that of the latter is equal to the volume.

$$dG = (T \, dS - p \, dV) + p \, dV + V \, dp - T \, dS - S \, dT$$

That is, for a closed system in the absence of non-expansion work and at constant composition

$$dG = V \, dp - S \, dT \tag{5.9}$$

This expression, which shows that a change in G is proportional to a change in p or T, suggests that G may be best regarded as a function of p and T. It confirms that G is an important quantity in chemistry because the pressure and temperature are usually the variables under our control. In other words, G carries around the combined consequences of the First and Second Laws in a way that makes it particularly suitable for chemical applications.

The same argument that led to eqn 4, when applied to the exact differential dG, now gives

$$\left(\frac{\partial G}{\partial T}\right)_p = -S \qquad \left(\frac{\partial G}{\partial p}\right)_T = V \tag{5.10}$$

These relations show how the Gibbs energy varies with temperature and pressure (Fig. 5.1). Because S is positive, it follows that G decreases when the temperature is raised at constant pressure and composition. Moreover, the relation shows that G decreases most sharply when the entropy of the system is large. Therefore, the Gibbs energy of the gaseous phase of a substance, which has a high molar entropy, is more sensitive to temperature than its liquid and solid phases (Fig. 5.2). Because V is positive, G always increases when the pressure of the system is increased at constant temperature and composition. Because the molar volumes of gases are large, G is more sensitive to pressure for the gaseous phase of a substance than for its liquid and solid phases (Fig. 5.3).

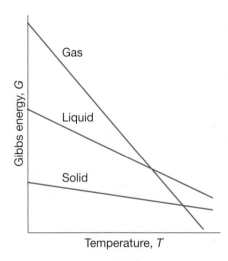

5.2 The variation of the Gibbs energy with the temperature is determined by the entropy. Because the entropy of the gaseous phase of a substance is greater than that of the liquid phase, and the entropy of the solid phase is smallest, the Gibbs energy changes most steeply for the gas phase, followed by the liquid phase, and then the solid phase of the substance.

> **Example 5.2 Calculating the effect of pressure on the Gibbs energy**
>
> Calculate the change in the molar Gibbs energy of (a) liquid water treated as an incompressible fluid and (b) water vapour treated as a perfect gas, when the pressure is increased isothermally from 1.0 bar to 2.0 bar at 298 K.
>
> **Method** In each case, we obtain the change in molar Gibbs energy by integration of eqn 9 with the temperature held constant (that is, setting d$T = 0$):
>
> $$G_m(p_f) - G_m(p_i) = \int_{p_i}^{p_f} V_m \, dp$$
>
> For an incompressible fluid, the molar volume is independent of the pressure, so V_m is a constant. For a perfect gas, the molar volume varies with pressure as $V_m = RT/p$, so this expression must be used in the integrand and the integration performed with RT a constant.
>
> **Answer** For the incompressible liquid, V_m is constant at 18.0 cm^3 mol^{-1}, so
>
> $$G_m(p_f) - G_m(p_i) = V_m \int_{p_i}^{p_f} dp = V_m \times (p_f - p_i)$$
> $$= (18.0 \times 10^{-6} \, m^3 \, mol^{-1}) \times (1.0 \times 10^5 \, Pa)$$
> $$= +1.8 \, J \, mol^{-1}$$

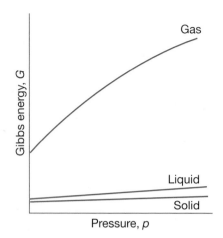

5.3 The variation of the Gibbs energy with the pressure is determined by the volume of the sample. Because the volume of the gaseous phase of a substance is greater than that of the same amount of liquid phase, and the entropy of the solid phase is smallest (for most substances), the Gibbs energy changes most steeply for the gas phase, followed by the liquid phase, and then the solid phase of the substance. Because the volumes of the solid and liquid phases of a substance are similar, they vary by similar amounts as the pressure is changed.

(because 1 Pa m^3 = 1 N m = 1 J). For a perfect gas:

$$G_m(p_f) - G_m(p_i) = \int_{p_i}^{p_f} \frac{RT}{p}\, dp = RT \ln \frac{p_f}{p_i}$$

$$= (2.48 \text{ kJ mol}^{-1}) \times \ln 2.0 = +1.7 \text{ kJ mol}^{-1}$$

Comment Note that G increases in both cases, and that the increase for a gas is 1000 times greater than for the liquid.

Self-test 5.2 Calculate the change in G_m for ice at –10°C, with density 0.917 g cm^{-3}, when the pressure is increased from 1.0 bar to 2.0 bar.

$$[+2.0 \text{ J mol}^{-1}]$$

5.4 The variation of the Gibbs energy with temperature

As we remarked in the introduction, because the equilibrium composition of a system depends on the Gibbs energy, to discuss the response of the composition to temperature we need to know how G varies with temperature.

The first expression in eqn 10 is our starting point for this discussion. Although it expresses the variation of G in terms of the entropy, we can express it in terms of the enthalpy by using the definition of G to write $S = (H - G)/T$. Then

$$\left(\frac{\partial G}{\partial T}\right)_p = \frac{G - H}{T} \tag{5.11}$$

We shall see later that the equilibrium constant of a reaction is related to G/T rather than to G itself,[1] and it turns out that the variation of this quantity with temperature is simpler than the temperature variation of G alone. In fact, it is easy to deduce from the last equation (see the *Justification* below) that

$$\left(\frac{\partial}{\partial T}\frac{G}{T}\right)_p = -\frac{H}{T^2} \tag{5.12a}$$

This expression is called the **Gibbs–Helmholtz equation**. It shows that, if we know the enthalpy of the system, then we know how G/T varies with temperature.

Justification 5.2

First, we write eqn 11 as

$$\left(\frac{\partial G}{\partial T}\right)_p - \frac{G}{T} = -\frac{H}{T}$$

Then we simplify the expression on the left by noting that

---

1. In Section 9.2b we derive the result that the equilibrium constant for a reaction is related to its standard reaction free energy by $\Delta_r G^{\ominus}/T = -R \ln K$.

$$\left(\frac{\partial}{\partial T}\frac{G}{T}\right)_p = \frac{1}{T}\left(\frac{\partial G}{\partial T}\right)_p + G\frac{\mathrm{d}}{\mathrm{d}T}\frac{1}{T} = \frac{1}{T}\left(\frac{\partial G}{\partial T}\right)_p - \frac{G}{T^2} = \frac{1}{T}\left\{\left(\frac{\partial G}{\partial T}\right)_p - \frac{G}{T}\right\}$$

When we substitute eqn 11 into this expression, we obtain eqn 12a.

Now note that we can use the chain rule of differentiation to write the left-hand side of eqn 12a as

$$\left(\frac{\partial(G/T)}{\partial T}\right)_p = \left(\frac{\partial(G/T)}{\partial(1/T)}\right)_p \frac{\mathrm{d}(1/T)}{\mathrm{d}T} = \left(\frac{\partial(G/T)}{\partial(1/T)}\right)_p \times \left(-\frac{1}{T^2}\right)$$

Substitution of this result into eqn 12a and multiplication of both sides by $-T^2$ gives an alternative form of the Gibbs–Helmholtz equation:

$$\left(\frac{\partial(G/T)}{\partial(1/T)}\right)_p = H \tag{5.12b}$$

The Gibbs–Helmholtz equation is most useful when it is applied to changes, including changes of physical state and chemical reactions at constant pressure. Then, because $\Delta G = G_f - G_i$ for the change of Gibbs energy between the final and initial states and because the equation applies to both G_f and G_i, we can write

$$\left(\frac{\partial}{\partial T}\frac{\Delta G}{T}\right)_p = -\frac{\Delta H}{T^2} \tag{5.13}$$

This equation shows that, if we know the change in enthalpy of a system undergoing a change, then we know how the change in Gibbs energy varies with temperature. As we shall see, this is a crucial piece of information in chemistry.

5.5 The variation of the Gibbs energy with pressure

To find the Gibbs energy at one pressure in terms of its value at another pressure, the temperature being constant, we set $\mathrm{d}T = 0$ in eqn 9 and integrate the remaining expression:

$$G(p_f) = G(p_i) + \int_{p_i}^{p_f} V\,\mathrm{d}p \tag{5.14}$$

This expression is applicable to any phase of matter.

(a) Liquids and solids

For a liquid or solid, the volume changes only slightly as the pressure changes (Fig. 5.4), so we can treat V as a constant and take it outside the integral. Then, for molar quantities,

$$G_m(p_f) = G_m(p_i) + V_m(p_f - p_i) = G_m(p_i) + V_m\Delta p \tag{5.15}$$

where $\Delta p = p_f - p_i$. Under normal laboratory conditions $V_m\Delta p$ is very small, as we saw in Example 5.2, and may be neglected. Hence, we may usually suppose that the Gibbs energies of solids and liquids are independent of pressure. However, if we are interested in geophysical problems, then, because pressures in the Earth's interior are huge, their effect on the Gibbs energy cannot be ignored. If the pressures are so great that there are substantial volume changes, we must use the complete expression, eqn 14.

The *chain rule* states that for a function $f = f(g)$, where $g = g(t)$,

$$\frac{\mathrm{d}f}{\mathrm{d}t} = \frac{\mathrm{d}f}{\mathrm{d}g}\frac{\mathrm{d}g}{\mathrm{d}t}$$

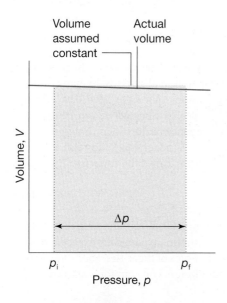

5.4 The difference in Gibbs energy of a solid or liquid at two pressures is equal to the rectangular area shown. We have assumed that the variation of volume with pressure is negligible.

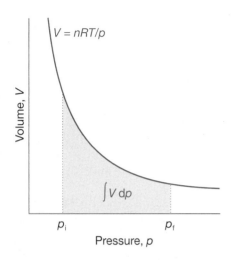

5.5 The difference in Gibbs energy for a perfect gas at two pressures is equal to the area shown below the perfect-gas isotherm.

Illustration 5.1

Suppose that for a certain phase transition of a solid $\Delta_{\text{trs}}V = +1.0\ \text{cm}^3\ \text{mol}^{-1}$. Then, for an increase in pressure to 3.0 Mbar, the Gibbs energy of the transition changes from $\Delta_{\text{trs}}G(1\ \text{bar})$ to

$$\Delta_{\text{trs}}G(3\ \text{Mbar}) = \Delta_{\text{trs}}G(1\ \text{bar}) + (1.0 \times 10^{-6}\ \text{m}^3\ \text{mol}^{-1}) \times (3.0 \times 10^{11}\ \text{Pa} - 1.0 \times 10^5\ \text{Pa})$$
$$= \Delta_{\text{trs}}G(1\ \text{bar}) + 3.0 \times 10^2\ \text{kJ mol}^{-1}$$

where we have used $1\ \text{Pa m}^3 = 1\ \text{J}$.

Self-test 5.3 Calculate the difference in molar Gibbs energy between the top and bottom of a column of mercury in a barometer of height 760 mm. The mass density of mercury is $13.6\ \text{g cm}^{-3}$. $\qquad$ $[+1.49\ \text{J mol}^{-1}]$

(b) Gases

The molar volumes of gases are large, so the Gibbs energy of a gas depends strongly on the pressure. Furthermore, because the volume also varies markedly with the pressure, we cannot treat it as a constant in the integral in eqn 14 (Fig. 5.5). For a perfect gas we substitute $V = nRT/p$ into the integral, and find

$$G(p_f) = G(p_i) + nRT \int_{p_i}^{p_f} \frac{\mathrm{d}p}{p} = G(p_i) + nRT \ln \frac{p_f}{p_i} \qquad (5.16)°$$

This expression shows that, when the pressure is increased tenfold at room temperature, the molar Gibbs energy increases by about 6 kJ mol⁻¹. It also follows from this equation that, if we set $p_i = p^{\ominus}$ (the standard pressure of 1 bar), then the Gibbs energy of a perfect gas at a pressure p is related to its standard value by

$$G(p) = G^{\ominus} + nRT \ln \frac{p}{p^{\ominus}} \qquad (5.17)°$$

The logarithmic dependence of the molar Gibbs energy on the pressure predicted by this expression is illustrated in Fig. 5.6.

At various stages in the development of physical chemistry it is necessary to switch from a consideration of idealized systems to real systems. In many cases it is desirable to preserve the form of the expressions that have been derived for an idealized system. Then deviations from the idealized behaviour can be expressed most simply. For instance, the pressure dependence of the molar Gibbs energy of a real gas might resemble that shown in Fig. 5.7. To adapt eqn 17 to this case, we replace the true pressure, p, by an effective pressure, called the **fugacity**,[2] f, and write

$$G_m = G_m^{\ominus} + RT \ln \frac{f}{p^{\ominus}} \qquad [5.18]$$

The fugacity, a function of the pressure and temperature, is defined so that this relation is exactly true. In later chapters we derive thermodynamically exact expressions in terms of the Gibbs energy, and therefore in terms of fugacities. For example, from elementary chemistry we know that the equilibrium constant for a reaction such as $H_2(g) + Br_2(g) \rightleftharpoons 2\ HBr(g)$ should be written

$$K = \frac{p_{HBr}^2}{p_{H_2} p_{Br_2}}$$

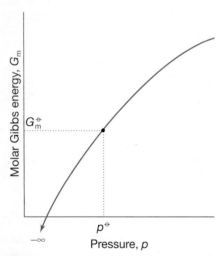

5.6 The molar Gibbs energy of a perfect gas is proportional to $\ln p$, and the standard state is reached at $p^{\ominus}$. Note that as $p \to 0$, the molar Gibbs energy becomes negatively infinite.

2. The name 'fugacity' comes from the Latin for 'fleetness' in the sense of 'escaping tendency'; fugacity has the same dimensions as pressure.

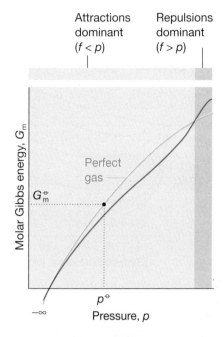

Attractions dominant ($f < p$)   Repulsions dominant ($f > p$)

$G_m^{\ominus}$

5.7 The molar Gibbs energy of a real gas. As $p \to 0$, the molar Gibbs energy coincides with the value for a perfect gas (shown by the pale line). When attractive forces are dominant (at intermediate pressures), the molar Gibbs energy is less than that of a perfect gas and the molecules have a lower 'escaping tendency'. At high pressures, when repulsive forces are dominant, the molar Gibbs energy of a real gas is greater than that of a perfect gas. Then the 'escaping tendency' is increased.

where p_J is the partial pressure of substance J; however, this expression is only an approximation. The thermodynamically exact expression is

$$K = \frac{f_{HBr}^2}{f_{H_2} f_{Br_2}}$$

where f_J is the fugacity of J.

Although expressions in terms of fugacities are exact, they are useful only if we know how to interpret fugacities in terms of actual pressures. We show in the *Justification* below that, if we write the fugacity as

$$f = \phi p \qquad \text{[5.19]}$$

where ϕ is the dimensionless **fugacity coefficient**, then ϕ is related to the compression factor, Z (Section 1.3a), of the gas between $p = 0$ and the pressure of interest by

$$\ln \phi = \int_0^p \frac{Z - 1}{p} \, dp \qquad (5.20)$$

Justification 5.3

Equation 14 is true for all gases whether real or perfect. Expressing it in terms of molar quantities and then using eqn 18 gives

$$\int_{p'}^{p} V_m \, dp = G_m - G_m' = RT \ln \frac{f}{f'}$$

In this expression, f is the fugacity when the pressure is p and f' is the fugacity when the pressure is p'. If the gas were perfect, we would write

$$\int_{p'}^{p} V_{\text{perfect,m}} \, dp = RT \ln \frac{p}{p'}$$

The difference between the two equations is

$$\int_{p'}^{p} (V_m - V_{\text{perfect,m}}) dp = RT \left(\ln \frac{f}{f'} - \ln \frac{p}{p'} \right)$$

which can be rearranged into

$$\ln \left(\frac{f}{p} \times \frac{p'}{f'} \right) = \frac{1}{RT} \int_{p'}^{p} (V_m - V_{\text{perfect,m}}) dp$$

When $p' \to 0$, the gas behaves perfectly and f' becomes equal to the pressure, p'. Therefore, $f'/p' \to 1$ as $p' \to 0$. If we take the limit, which means setting $f'/p' = 1$ on the left and $p' = 0$ on the right, the last equation becomes

$$\ln \frac{f}{p} = \frac{1}{RT} \int_0^p (V_m - V_{\text{perfect,m}}) dp$$

Then, with $\phi = f/p$,

$$\ln \phi = \frac{1}{RT} \int_0^p (V_m - V_{\text{perfect,m}}) dp$$

For a perfect gas, $V_{\text{perfect,m}} = RT/p$. For a real gas, $V_m = RTZ/p$. With these two substitutions, we obtain eqn 20.

5.8 The fugacity coefficient of a van der Waals gas plotted using the reduced variables of the gas. The curves are labelled with the reduced temperature $T_r = T/T_c$.

Table 5.2* The fugacity of nitrogen at 273 K

p/atm	f/atm
1	0.999 55
10	9.9560
100	97.03
1000	1839

* More values are given in the *Data section*.

We see from Fig. 1.13 that for most gases $Z < 1$ up to moderate pressures, but that $Z > 1$ at higher pressures. If $Z < 1$ throughout the range of integration, then the integrand in eqn 20 is negative and $\phi < 1$. This implies that $f < p$ (the molecules tend to stick together) and that the molar Gibbs energy of the gas is less than that of a perfect gas. At higher pressures, the range over which $Z > 1$ may dominate the range over which $Z < 1$. The integral is then positive, $\phi > 1$, and $f > p$ (the repulsive interactions are dominant and tend to drive the particles apart). Now the molar Gibbs energy of the gas is greater than that of the perfect gas at the same pressure. Figure 5.8, which has been calculated using the full van der Waals equation of state, shows how the fugacity coefficient depends on the pressure in terms of the reduced variables (Section 1.5). Because critical constants are available in Table 1.4, the graphs can be used for quick estimates of the fugacities of a wide range of gases. Table 5.2 gives some explicit values for nitrogen.

Checklist of key ideas

Combining the First and Second Laws

5.1 The fundamental equation

☐ fundamental equation,
$dU = T\,dS - p\,dV$

5.2 Properties of the internal energy

☐ Maxwell relations (Table 5.1)

☐ thermodynamic equation of state, an expression for pressure in terms of thermodynamic quantities,
$\pi_T = T(\partial p/\partial T)_V - p$

Properties of the Gibbs energy

5.3 General considerations

☐ general change in Gibbs energy,
$dG = V\,dp - S\,dT$

☐ the variation of Gibbs energy with pressure, $(\partial G/\partial p)_T = V$

☐ the variation of Gibbs energy with temperature, $(\partial G/\partial T)_p = -S$

5.4 The variation of the Gibbs energy with temperature

☐ Gibbs–Helmholtz equation,
$(\partial(G/T)/\partial T)_p = -H/T^2$

5.5 The variation of the Gibbs energy with pressure

☐ for a condensed phase,
$G_m(p_f) = G_m(p_i) + V_m\Delta p$

☐ for a perfect gas,
$G(p_f) = G(p_i) + nRT \ln (p_f/p_i)$

☐ fugacity, the effective pressure of a gas,
$G_m = G_m^\ominus + RT \ln(f/p^\ominus)$

☐ fugacity coefficient, $\phi = f/p$, with
$\ln \phi = \int_0^p \{(Z - 1)/p\}dp$

Further reading

Articles of general interest

L.L. Combs, An alternative view of fugacity. *J. Chem. Educ.* **69**, 218 (1992).

L. D'Alessio, On the fugacity of a van der Waals gas: An approximate expression that separates attractive and repulsive forces. *J. Chem. Educ.* **70**, 96 (1993).

R.M. Noyes, Thermodynamics of a process in a rigid container. *J. Chem. Educ.* **69**, 470 (1992).

R.M. Noyes, Application of the Gibbs function to chemical systems and subsystems. *J. Chem. Educ.* **73**, 404 (1996).

R.J. Tykodi, Spontaneity, accessibility, irreversibility, 'useful work': the availability

function, the Helmholtz function, and the Gibbs function. *J. Chem. Educ.* **72**, 103 (1995).

R.J. Tykodi, The Gibbs function, spontaneity, and walls. *J. Chem. Educ.* **73**, 398 (1996).

J.S. Winn, The fugacity of van der Waals gas. *J. Chem. Educ.* **65**, 772 (1988).

S.E. Wood and R. Battino, The Gibbs function controversy. *J. Chem. Educ.* **73**, 408 (1996).

Texts and sources of data and information

I.M. Klotz and R.M. Rosenberg, *Chemical thermodynamics: basic theory*

and methods. Wiley–Interscience, New York (2000).

G.N. Lewis and M. Randall, *Thermodynamics* (revised by K.S. Pitzer and L. Brewer). McGraw-Hill, New York (1961).

R.G. Mortimer, *Mathematics for physical chemistry.* Academic Press, San Diego (1999).

J.M. Smith and H.C. Van Ness, *Introduction to chemical engineering thermodynamics.* McGraw-Hill, New York (1987).

M.A. White, *Properties of materials.* Oxford University Press, Oxford (1999).

B.D. Wood, *Applications of thermodynamics.* Addison–Wesley, New York (1982).

Exercises

Discussion questions

5.1 (a) Discuss the physical interpretation of any one Maxwell relation.

5.1 (b) Account for the dependence of π_T of a van der Waals gas in terms of the significance of the parameters a and b.

5.2 (a) Suggest a physical interpretation of the dependence of the Gibbs energy on the pressure.

5.2 (b) Suggest a physical interpretation of the dependence of the Gibbs energy on the temperature.

5.3 (a) Explain the significance of the fugacity of a gas.

5.3 (b) Suggest a physical interpretation of the variation of the fugacity coefficient of a van der Waals gas as depicted in Fig. 5.8.

Numerical exercises

Assume all gases are perfect and that the temperature is 298.15 K unless stated otherwise.

5.4 (a) Express $(\partial S/\partial V)_T = (\partial p/\partial T)_V$ in terms of α and κ_T (see eqns 3.8 and 3.14 for definitions).

5.4 (b) Express $(\partial S/\partial p)_T = -(\partial V/\partial T)_p$ in terms of α (see eqn 3.8 for definition).

5.5 (a) Suppose that 3.0 mmol $N_2(g)$ occupies 36 cm^3 at 300 K and expands to 60 cm^3. Calculate ΔG for the process.

5.5 (b) Suppose that 2.5 mmol Ar(g) occupies 72 dm^3 at 298 K and expands to 100 dm^3. Calculate ΔG for the process.

5.6 (a) The change in the Gibbs energy of a certain constant-pressure process was found to fit the expression $\Delta G/J = -85.40 + 36.5(T/K)$. Calculate the value of ΔS for the process.

5.6 (b) The change in the Gibbs energy of a certain constant-pressure process was found to fit the expression $\Delta G/J = -73.1 + 42.8(T/K)$. Calculate the value of ΔS for the process.

5.7 (a) Calculate the change in Gibbs energy of 35 g of ethanol (mass density 0.789 g cm^{-3}) when the pressure is increased isothermally from 1 atm to 3000 atm.

5.7 (b) Calculate the change in Gibbs energy of 25 g of methanol (mass density 0.791 g cm^{-3}) when the pressure is increased isothermally from 100 kPa to 100 MPa.

5.8 (a) When 2.00 mol of a gas at 330 K and 3.50 atm is subjected to isothermal compression, its entropy decreases by 25.0 J K^{-1}. Calculate (a) the final pressure of the gas and (b) ΔG for the compression.

5.8 (b) When 3.00 mol of a gas at 230 K and 150 kPa is subjected to isothermal compression, its entropy decreases by 15.0 J K^{-1}. Calculate (a) the final pressure of the gas and (b) ΔG for the compression.

5.9 (a) Calculate the change in molar Gibbs energy of a perfect gas when its pressure is increased isothermally from 1.8 atm to 29.5 atm at 40°C.

5.9 (b) Calculate the change in molar Gibbs energy of a perfect gas when its pressure is increased isothermally from 92.0 kPa to 252.0 kPa at 50°C.

5.10 (a) The fugacity coefficient of a certain gas at 200 K and 50 bar is 0.72. Calculate the difference of its chemical potential from that of a perfect gas in the same state.

5.10 (b) The fugacity coefficient of a certain gas at 290 K and 2.1 MPa is 0.68. Calculate the difference of its chemical potential from that of a perfect gas in the same state.

5.11 (a) At 373 K, the second virial coefficient B of xenon is -81.7 cm^3 mol^{-1}. Calculate the value of B' and hence estimate the fugacity coefficient of xenon at 50 atm and 373 K.

5.11 (b) At 100 K, the second virial coefficient B of nitrogen is -160.0 cm^3 mol^{-1}. Calculate the value of B' and hence estimate the fugacity coefficient of nitrogen at 62 MPa and 100 K.

5.12 (a) Estimate the change in the Gibbs energy of 1.0 L of benzene when the pressure acting on it is increased from 1.0 atm to 100 atm.

5.12 (b) Estimate the change in the Gibbs energy of 1.0 L of water when the pressure acting on it is increased from 100 kPa to 300 kPa.

5.13 (a) Calculate the change in the molar Gibbs energy of hydrogen gas when its pressure is increased isothermally from 1.0 atm to 100.0 atm at 298 K.

5.13 (b) Calculate the change in the molar Gibbs energy of oxygen when its pressure is increased isothermally from 50.0 kPa to 100.0 kPa at 500 K.

5.14 (a) The molar Helmholtz energy of a certain gas is given by:

$$A_m = -\frac{a}{V_m} - RT \ln(V_m - b) + f(T)$$

where a and b are constants and $f(T)$ is a function of temperature only. Obtain the equation of state of the gas.

5.14 (b) The molar Gibbs energy of a certain gas is given by:

$$G_m = RT \ln p + A + Bp + \tfrac{1}{2}Cp^2 + \tfrac{1}{3}Dp^3$$

where A, B, C, and D are constants. Obtain the equation of state of the gas.

5.15 (a) Evaluate $(\partial S/\partial V)_T$ for a van der Waals gas. For an isothermal expansion, will ΔS be greater for a perfect gas or a van der Waals gas? Explain your conclusion.

5.15 (b) Evaluate $(\partial S/\partial V)_T$ for a Dieterici gas (Table 1.6). For an isothermal expansion, will ΔS be greater for a perfect gas or a Dieterici gas? Explain your conclusion.

Problems[†]

Numerical problems

5.1 Calculate $\Delta_r G^{\ominus}$(375 K) for the reaction 2 CO(g) + O$_2$(g) → 2 CO$_2$(g) from the value of $\Delta_r G^{\ominus}$ (298 K), $\Delta_r H^{\ominus}$ (298 K), and the Gibbs–Helmholtz equation.

5.2 Estimate the standard reaction Gibbs energy of N$_2$(g) + 3 H$_2$(g) → 2 NH$_3$(g) at (a) 500 K, (b) 1000 K from their values at 298 K.

5.3 At 200 K, the compression factor of oxygen varies with pressure as shown below. Evaluate the fugacity of oxygen at this temperature and 100 atm.

p/atm	1.0000	4.000 00	7.000 00	10.0000	40.00	70.00	100.0
Z	0.9971	0.987 96	0.978 80	0.969 56	0.8734	0.7764	0.6871

5.4‡ In an investigation of thermophysical properties of toluene (*J. Phys. Chem. Ref. Data* **18**, 1565 (1989)), R.D. Goodwin tabulated (among other quantities) the compression factor, Z, at several temperatures and pressures. From the following information, compute the fugacity coefficient of toluene at 600 K and (a) 30.0 bar and (b) 1000 bar.

p/bar	0.500	1.013	2.00	3.00	5.00	
Z	0.994 12	0.988 96	0.979 42	0.969 95	0.951 33	

p/bar	10.00	20.0	30.0	42.4	50.0	
Z	0.905 69	0.812 27	0.701 77	0.471 98	0.223 76	

p/bar	70.0	100.0	200	300	500	1000
Z	0.265 20	0.349 20	0.623 62	0.882 88	1.371 09	2.488 36

Theoretical problems

5.5 Show that, for a perfect gas, $(\partial U/\partial S)_V = T$ and $(\partial U/\partial V)_S = -p$.

5.6 Two of the four Maxwell relations were derived in the text, but two were not. Complete their derivation by showing that $(\partial S/\partial V)_T = (\partial p/\partial T)_V$ and $(\partial T/\partial p)_S = (\partial V/\partial S)_p$.

5.7 Use the Maxwell relations to express the derivatives $(\partial S/\partial V)_T$ and $(\partial V/\partial S)_p$ in terms of the expansion coefficient α and the isothermal compressibility κ_T.

5.8 Use the Maxwell relations and Euler's chain relation to express $(\partial p/\partial S)_V$ and $(\partial V/\partial S)_p$ in terms of the heat capacities, the expansion coefficient, and the isothermal compressibility.

5.9 Use the Maxwell relations to show that the entropy of a perfect gas depends on the volume as $S \propto R \ln V$.

5.10 Derive the thermodynamic equation of state

$$\left(\frac{\partial H}{\partial p}\right)_T = V - T\left(\frac{\partial V}{\partial T}\right)_p$$

Derive an expression for $(\partial H/\partial p)_T$ for (a) a perfect gas and (b) a van der Waals gas. In the latter case, estimate its value for 1.0 mol Ar(g) at 298 K and 10 atm. By how much does the enthalpy of the argon change when the pressure is increased isothermally to 11 atm?

5.11 Prove the following relation:

$$\left(\frac{\partial H}{\partial V}\right)_T = -V^2\left(\frac{\partial p}{\partial T}\right)_V\left(\frac{\partial(T/V)}{\partial V}\right)_p$$

---

† Problems denoted with the symbol ‡ were supplied by Charles Trapp, Carmen Giunta, and Marshall Cady.

5.12 Show that, if $B(T)$ is the second virial coefficient of a gas, and $\Delta B = B(T'') - B(T')$, $\Delta T = T'' - T'$, and T is the mean of T'' and T', then

$$\pi_T \approx \frac{RT^2 \Delta B}{V_m^2 \Delta T}$$

Estimate π_T for argon given that $B(250\ \mathrm{K}) = -28.0\ \mathrm{cm^3\ mol^{-1}}$ and $B(300\ \mathrm{K}) = -15.6\ \mathrm{cm^3\ mol^{-1}}$ at 275 K at (a) 1.0 atm, (b) 10.0 atm.

5.13 (a) Prove that the heat capacities C_V and C_p of a perfect gas are independent of both volume and pressure. May they depend on the temperature? (b) Deduce an expression for the dependence of C_V on volume of a gas that is described by the equation of state $pV_m/RT = 1 + B/V_m$.

5.14 The Joule coefficient, μ_J, is defined as $\mu_J = (\partial T/\partial V)_U$. Show that $\mu_J C_V = p - (\alpha T/\kappa_T)$.

5.15 Evaluate π_T for a Dieterici gas (Table 1.6). Justify physically the form of the expression obtained.

5.16‡ The second virial coefficient of methane can be approximated by the empirical equation $B'(T) = a + be^{-c/T^2}$, where $a = -0.1993\ \mathrm{bar^{-1}}$, $b = 0.2002\ \mathrm{bar^{-1}}$, and $c = 1131\ \mathrm{K^2}$ with $300\ \mathrm{K} < T < 600\ \mathrm{K}$. Derive an equation for the computation of each of the following properties for methane: (a) π_T at 25°C and 100 bar, (b) ΔS_m, ΔH_m, and ΔG_m, when the gas is compressed isothermally from 1 bar to 100 bar at 25°C. Assume that $B(T) = RTB'(T)$.

5.17 Instead of assuming that the volume of a condensed phase is constant when pressure is applied, assume only that the compressibility is constant. Show that when the pressure is changed isothermally by Δp, G changes to

$$G' = G + V_m \Delta p (1 - \tfrac{1}{2}\kappa_T \Delta p)$$

Assess the error in assuming that a solid is incompressible by applying this expression to the compression of copper when $\Delta p = 500\ \mathrm{atm}$. (For copper at 25°C, $\kappa_T = 0.8 \times 10^{-6}\ \mathrm{atm^{-1}}$ and $\rho = 8.93\ \mathrm{g\ cm^{-3}}$.)

5.18 Derive an expression for the standard reaction Gibbs energy $\Delta_r G^\ominus$ (the analogue of the standard reaction enthalpy) at a temperature T' in terms of its value $\Delta_r G^\ominus$ at T by using the Gibbs–Helmholtz equation and (a) assuming that $\Delta_r H^\ominus$ does not vary with temperature, (b) assuming instead that $\Delta_r C_p$ does not vary with temperature and using Kirchhoff's law.

5.19 The adiabatic compressibility, κ_S, is defined like κ_T (eqn 3.14) but at constant entropy. Show that for a perfect gas $p\gamma\kappa_S = 1$ (where γ is the ratio of heat capacities).

5.20 Show that, if S is regarded as a function of T and p, then

$$T\,dS = C_V\,dT + T\left(\frac{\partial p}{\partial T}\right)_V dV$$

Calculate the energy that must be transferred as heat to a van der Waals gas that expands reversibly and isothermally from V_i to V_f.

5.21 Suppose that S is regarded as a function of p and T. Show that

$$T\,dS = C_p\,dT - \alpha TV\,dp$$

Hence, show that the energy transferred as heat when the pressure on an incompressible liquid or solid is increased by Δp is equal to $-\alpha TV\Delta p$. Evaluate q when the pressure acting on 100 cm³ of mercury at 0°C is increased by 1.0 kbar. ($\alpha = 1.82 \times 10^{-4}\ \mathrm{K^{-1}}$.)

5.22‡ At 1 atm, liquid water has its maximum density at 4°C. What can be concluded about the variation of the entropy of liquid water with pressure at constant temperature at 3°C, 4°C, and 5°C?

5.23‡ You are told that the differential of pressure consistent with an equation of state is given by *one* of the following two expressions. Determine the equation of state.

$$dp = \frac{2nRT\,dV}{(V - nb)^2} + \frac{R(V - nb)\,dT}{nb^2}$$

$$dp = -\frac{nRT\,dV}{(V - nb)^2} + \frac{nR\,dT}{V - nb}$$

5.24 Suppose that the attractive interactions between gas particles can be neglected, and find an expression for the fugacity of a van der Waals gas in terms of the pressure. Estimate its value for ammonia at 10.00 atm and 298.15 K.

5.25 Find an expression for the fugacity coefficient when the attractive interaction is dominant in a van der Waals gas, and the pressure is low enough to make the approximation $4ap/(RT)^2 \ll 1$. Evaluate the fugacity for ammonia at 10.00 atm and 298.15 K.

5.26 Find an expression for the fugacity coefficient of a gas that obeys the equation of state

$$\frac{pV_m}{RT} = 1 + \frac{B}{V_m} + \frac{C}{V_m^2}$$

Use the resulting expression to estimate the fugacity of argon at 1.00 atm and 100 K using $B = -21.13\ \mathrm{cm^3\ mol^{-1}}$ and $C = 1054\ \mathrm{cm^6\ mol^{-2}}$.

5.27 Derive an expression for the fugacity coefficient of a gas that obeys the equation of state

$$\frac{pV_m}{RT} = 1 + \frac{qT}{V_m}$$

where q is a constant, and plot ϕ against $4pq/R$.

Applications: to biology, environmental science, and polymer science

5.28 At 298 K the standard enthalpy of combustion of sucrose is $-5797\ \mathrm{kJ\ mol^{-1}}$ and the standard Gibbs energy of the reaction is $-6333\ \mathrm{kJ\ mol^{-1}}$. Estimate the additional (non-expansion) work that may be obtained by raising the temperature to blood temperature, 37°C.

5.29‡ In 1995, the Intergovernmental Panel on Climate Change considered a global average temperature rise of 1.0–3.5°C likely by the year 2100, with 2.0°C its best estimate (IPCC Second Assessment Synthesis of Scientific–Technical Information Relevant to Interpreting Article 2 of the UN Framework Convention on Climate Change (1995)). Because water vapour is itself a greenhouse gas, the increase in water vapour content of the atmosphere is of some concern to climate change experts. Predict the relative increase in water vapour in the atmosphere based on a temperature rise of 2.0 K, assuming that the relative humidity remains constant. (The present global mean temperature is 290 K, and the equilibrium vapour pressure of water at that temperature is 0.0189 bar.)

5.30‡ Protocols have outlawed the use of certain CFC refrigerants. It has been suggested that sodium oxalate, $Na_2(CO_2)_2$, at elevated temperatures can be used to destroy existing CFC stockpiles. The mineralization of freon-12, for example, is

$$CF_2Cl_2(g) + 2\,Na_2(CO_2)_2(s) \longrightarrow 2\,NaF(s) + 2\,NaCl(s) + C(s) + 4\,CO_2(g)$$

(a) Without performing a calculation, what is your expectation concerning the spontaneity of the reaction? Why is the high temperature required? Why should reaction mixtures not be heated to temperatures much higher than 270°C? (b) Use library, internet, or database resources to find thermodynamic data that will make possible the calculation of $\Delta_r G^{\ominus}(T)$. Estimate the value of any thermodynamic property that cannot be found. Consult the web site of this text for a list of internet resources.

5.31‡ Nitric acid hydrates have received much attention as possible catalysts for heterogeneous reactions which bring about the Antarctic ozone hole. Worsnop *et al.* investigated the thermodynamic stability of these hydrates under conditions typical of the polar winter stratosphere (D.R. Worsnop, L.E. Fox, M.S. Zahniser, and S.C. Wofsy, *Science* **259**, 71 (1993)). They report thermodynamic data for the sublimation of mono-, di-, and trihydrates to nitric acid and water vapours, $HNO_3 \cdot nH_2O(s) \to HNO_3(g) + n\,H_2O(g)$, for $n = 1$, 2, and 3. Given $\Delta_r G^{\ominus}$ and $\Delta_r H^{\ominus}$ for these reactions at 220 K, use the Gibbs–Helmholtz equation to compute $\Delta_r G^{\ominus}$ at 190 K.

n	1	2	3
$\Delta_r G^{\ominus}$/(kJ mol^{-1})	46.2	69.4	93.2
$\Delta_r H^{\ominus}$/(kJ mol^{-1})	127	188	237

5.32‡ J. Gao and J.H. Weiner in their study of the origin of stress on the atomic level in dense polymer systems (*Science* **266**, 748 (1994)) observe that the tensile force required to maintain the length, l, of a long linear chain of N freely jointed links each of length, a, can be interpreted as arising from an entropic spring. For such a chain, $S(l) = -3kl^2/2Na^2 + C$, where k is the Boltzmann constant and C is a constant. Using thermodynamic relations of this and previous chapters, show that the tensile force obeys Hooke's law, $f = -k_f l$, if we assume that the energy U is independent of l.

5.33 The volume of a newly synthesized polymer was found to depend exponentially on the pressure as $V = V_0 e^{-p/p^*}$, where p is the excess pressure and p^* is a constant. Deduce an expression for the Gibbs energy of the polymer as a function of excess pressure. What is the natural direction of change of the compressed material when the pressure is relaxed?

6

Physical transformations of pure substances

Phase diagrams

6.1 The stabilities of phases

6.2 Phase boundaries

6.3 Three typical phase diagrams

Phase stability and phase transitions

6.4 The thermodynamic criterion of equilibrium

6.5 The dependence of stability on the conditions

6.6 The location of phase boundaries

6.7 The Ehrenfest classification of phase transitions

The physical liquid surface

6.8 Surface tension

6.9 Curved surfaces

6.10 Capillary action

Checklist of key ideas

Further reading

Exercises

Problems

The simplest applications of thermodynamics to chemically significant systems are to the discussion of the phase transitions of pure substances. We shall see that a phase diagram is a map of the pressures and temperatures at which each phase of a substance is the most stable. First, we describe the interpretation of empirically determined phase diagrams for a selection of materials. Then we turn to a consideration of the factors that determine the positions and shapes of the boundaries between the regions on a phase diagram. The practical importance of the expressions we derive is that they show how the vapour pressure of a substance varies with temperature and how the melting point varies with pressure. We shall see that the transitions between phases can be classified by noting how various thermodynamic functions change when the transition occurs. This chapter also introduces the chemical potential, a property that is at the centre of discussions of phase transitions and chemical reactions. Finally, we consider the thermodynamic discussion of liquid surfaces, such as surface tension and capillary action.

Vaporization, melting, and the conversion of graphite to diamond are all examples of changes of phase without change of chemical composition. In this chapter we describe such processes thermodynamically, using as the guiding principle the tendency of systems at constant temperature and pressure to minimize their Gibbs energy.

Phase diagrams

One of the most succinct ways of presenting the physical changes of state that a substance can undergo is in terms of its phase diagram. We present the concept in this section.

6.1 The stabilities of phases

A **phase** of a substance is a form of matter that is uniform throughout in chemical composition and physical state. Thus, we speak of solid, liquid, and gas phases of a substance, and of its various solid phases, such as the white and black allotropes of

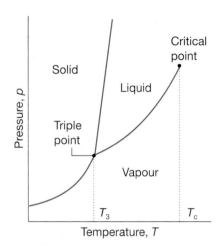

6.1 The general regions of pressure and temperature where solid, liquid, or gas is stable (that is, has minimum molar Gibbs energy) are shown on this phase diagram. For example, the solid phase is the most stable phase at low temperatures and high pressures. In the following paragraphs we locate the precise boundaries between the regions.

The web site is a good source of links to online databases of data on phase transitions.

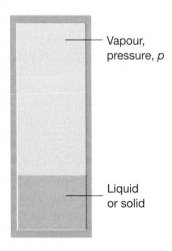

6.2 The vapour pressure of a liquid or solid is the pressure exerted by the vapour in equilibrium with the condensed phase.

phosphorus. A **phase transition**, the spontaneous conversion of one phase into another phase, occurs at a characteristic temperature for a given pressure. Thus, at 1 atm, ice is the stable phase of water below 0°C, but above 0°C liquid water is more stable. This difference indicates that below 0°C the Gibbs energy decreases as liquid water changes into ice and that above 0°C the Gibbs energy decreases as ice changes into liquid water. The **transition temperature**, T_{trs}, is the temperature at which the two phases are in equilibrium and the Gibbs energy is minimized at the prevailing pressure.

We must distinguish between the thermodynamic description of a phase transition and the rate at which the transition occurs. A transition that is predicted from thermodynamics to be spontaneous may occur too slowly to be significant in practice. For instance, at normal temperatures and pressures the molar Gibbs energy of graphite is lower than that of diamond, so there is a thermodynamic tendency for diamond to change into graphite. However, for this transition to take place, the C atoms must change their locations, which is an immeasurably slow process in a solid except at high temperatures. The rate of attainment of equilibrium is a kinetic problem and is outside the range of thermodynamics. In gases and liquids the mobilities of the molecules allow phase transitions to occur rapidly, but in solids thermodynamic instability may be frozen in. Thermodynamically unstable phases that persist because the transition is kinetically hindered are called **metastable phases**. Diamond is a metastable phase of carbon under normal conditions.

6.2 Phase boundaries

The **phase diagram** of a substance shows the regions of pressure and temperature at which its various phases are thermodynamically stable (Fig. 6.1). The lines separating the regions, which are called **phase boundaries**, show the values of p and T at which two phases coexist in equilibrium.

Consider a liquid sample of a pure substance in a closed vessel. The pressure of a vapour in equilibrium with the liquid is called the **vapour pressure** of the substance (Fig. 6.2). Therefore, the liquid–vapour phase boundary in a phase diagram shows how the vapour pressure of the liquid varies with temperature. Similarly, the solid–vapour phase boundary shows the temperature variation of the **sublimation vapour pressure**, the vapour pressure of the solid phase. The vapour pressure of a substance increases with temperature because at higher temperatures the Boltzmann distribution populates more heavily the states of higher energy corresponding to molecules that have escaped from their neighbours.

(a) Critical points and boiling points

When a liquid is heated in an open vessel, the liquid vaporizes from its surface. At the temperature at which its vapour pressure would be equal to the external pressure, vaporization can occur throughout the bulk of the liquid and the vapour can expand freely into the surroundings. The condition of free vaporization throughout the liquid is called **boiling**. The temperature at which the vapour pressure of a liquid is equal to the external pressure is called the **boiling temperature** at that pressure. For the special case of an external pressure of 1 atm, the boiling temperature is called the **normal boiling point**, T_b. With the replacement of 1 atm by 1 bar as standard pressure, there is some advantage in using the **standard boiling point** instead: this is the temperature at which the vapour pressure reaches 1 bar. Because 1 bar is slightly less than 1 atm (1.00 bar = 0.987 atm), the standard boiling point of a liquid is slightly lower than its normal boiling point. The normal boiling point of water is 100.0°C; its standard boiling point is 99.6°C.

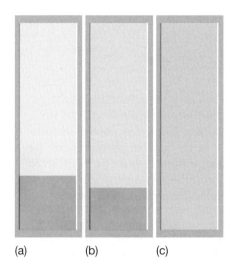

6.3 (a) A liquid in equilibrium with its vapour. (b) When a liquid is heated in a sealed container, the density of the vapour phase increases and that of the liquid decreases slightly. There comes a stage, (c), at which the two densities are equal and the interface between the fluids disappears. This disappearance occurs at the critical temperature. The container needs to be strong: the critical temperature of water is 374°C and the vapour pressure is then 218 atm.

Boiling does not occur when a liquid is heated in a closed vessel. Instead, the vapour pressure, and hence the density of the vapour, rise continuously as the temperature is raised (Fig. 6.3). At the same time, the density of the liquid decreases slightly as a result of its expansion. There comes a stage when the density of the vapour is equal to that of the remaining liquid and the surface between the two phases disappears. The temperature at which the surface disappears is the **critical temperature**, T_c, of the substance. We first encountered this property in Section 1.3d. The vapour pressure at the critical temperature is called the **critical pressure**, p_c. At and above the critical temperature, a single uniform phase called a **supercritical fluid** fills the container and an interface no longer exists. That is, above the critical temperature, the liquid phase of the substance does not exist (Box 6.1).

(b) Melting points and triple points

The temperature at which, under a specified pressure, the liquid and solid phases of a substance coexist in equilibrium is called the **melting temperature**. Because a substance melts at exactly the same temperature as that at which it freezes, the melting temperature of a substance is the same as its **freezing temperature**. The freezing temperature when the pressure is 1 atm is called the **normal freezing point**, T_f, and its freezing point when the pressure is 1 bar is called the **standard freezing point**. The normal and standard freezing points are negligibly different for most purposes. The normal freezing point is also called the **normal melting point**.

There is a set of conditions under which three different phases of a substance (typically solid, liquid, and vapour) all simultaneously coexist in equilibrium. These conditions are represented by the **triple point**, a point at which the three phase boundaries meet. The temperature at the triple point is denoted T_3. The triple point of a pure substance is outside our control: it occurs at a single definite pressure and

Box 6.1 Supercritical fluids

Supercritical carbon dioxide, scCO$_2$, is the centre of attention for an increasing number of solvent-based processes. The critical temperature 304.2 K (31.0°C) and pressure (72.9 atm) are readily accessible and carbon dioxide is cheap. The density of scCO$_2$ at its critical point is 0.45 g cm^{-3}. However, the transport properties of any supercritical fluid depend strongly on its density, which in turn is sensitive to the pressure and temperature. For instance, densities may be adjusted from a gas-like 0.1 g cm^{-3} to a liquid-like 1.2 g cm^{-3}. A useful rule of thumb is that the solubility of a solute is an exponential function of the density of the supercritical fluid, so small increases in pressure, particularly close to the critical point, can have very large effects on solubility.

A great advantage of scCO$_2$ is that there are no noxious residues once the solvent has been allowed to evaporate, so, coupled with its low critical temperature, scCO$_2$ is ideally suited to food processing and the production of pharmaceuticals. It is used, for instance, to remove caffeine from coffee. The supercritical fluid is also increasingly being used for dry cleaning, which avoids the use of carcinogenic and environmentally deleterious chlorinated hydrocarbons.

Supercritical CO$_2$ has been used since the 1960s as a mobile phase in *supercritical fluid chromatography* (SFC), but it fell out of favour when the more convenient technique of high-performance liquid chromatography (HPLC) was introduced. However, interest in SFC has returned, and there are separations possible in SFC that cannot easily be achieved by HPLC, such as the separation of lipids and of phospholipids. Samples as small as 1 pg can be analysed. The essential advantage of SFC is that diffusion coefficients in supercritical fluids are an order of magnitude greater than in liquids, so there is less resistance to the transfer of solutes through the column, with the result that separations may be effected rapidly or with high resolution.

The principal problem with scCO$_2$, though, is that the fluid is not a very good solvent and surfactants are needed to induce many potentially interesting solutes to dissolve. Indeed, scCO$_2$-based dry cleaning depends on the availability of cheap surfactants; so too does the use of scCO$_2$ as a solvent for homogeneous catalysts, such as metal complexes. There appear to be two principal approaches to solving the solubilizing problem. One solution is to use fluorinated and siloxane-based polymeric stabilizers, which allow polymerization reactions to proceed in scCO$_2$. The disadvantage of these stabilizers for commercial use is their great expense. An alternative and much cheaper approach is poly(ether-carbonate) copolymers. The copolymers can be made more soluble in scCO$_2$ by adjusting the ratio of ether and carbonate groups.

The critical temperature of water is 374°C and its pressure is 218 atm. The conditions for using scH$_2$O are therefore much more demanding than for scCO$_2$ and the properties of the fluid are highly sensitive to pressure. Thus, as the density of scH$_2$O decreases, the characteristics of a solution change from those of an aqueous solution through those of a non-aqueous solution and eventually to those of a gaseous solution. One consequence is that reaction mechanisms may change from ionic to radical.

Discussion questions

1.[1] The use of supercritical fluids as mobile phases in SFC depends on their properties as nonpolar solvents. The solubility parameter, δ, is defined as $(\Delta U_{\text{cohesive}}/V_{\text{m}})^{1/2}$, where $\Delta U_{\text{cohesive}}$ is the cohesive energy of the solvent, the energy per mole needed to increase the volume isothermally to an infinite value. Diethyl ether, carbon tetrachloride, and dioxane have solubility parameter ranges of 7–8, 8–9, and 10–11, respectively. (a) Derive a practical equation for the computa-

1. Provided by Marshal Cady and Charles Trapp.

tion of the isotherms for the reduced internal energy change, $\Delta U_{\text{r}}(T_{\text{r}}, V_{\text{r}})$ defined as

$$\Delta U_{\text{r}}(T_{\text{r}}, V_{\text{r}}) = \frac{U(T_{\text{r}}, V_{\text{r}}) - U(T_{\text{r}}, \infty)}{p_{\text{c}} V_{\text{c}}}$$

(b) Draw a graph of ΔU_{r} against p_{r} for the isotherms $T_{\text{r}} = 1$, 1.2, and 1.5 in the reduced pressure range for which $0.7 \leq V_{\text{r}} \leq 2$. (c) Draw a graph of δ against p_{r} for the carbon dioxide isotherms $T_{\text{r}} = 1$ and 1.5 in the reduced pressure range for which $1 \leq V_{\text{r}} \leq 3$. In what pressure range at $T_{\text{f}} = 1$ will carbon dioxide have solvent properties similar to those of liquid carbon tetrachloride? (*Hint.* Use mathematical software or a spreadsheet.)

2. The use of supercritical fluids for the extraction of a component from a complicated mixture is not confined to the decaffeination of coffee. Consult library and internet resources and prepare a discussion of the principles, advantages, disadvantages, and current uses of supercritical fluid extraction technology. (*Hint.* see *Further reading* below for ideas.)

Further reading

C.L. Phelps, N.G. Smart, and C.M. Wai, Past, present, and future applications of supercritical fluid extraction technology. *J. Chem. Educ.* **73**, 1163 (1996).

T.L. Chester, J.D. Pinkston, and D.E. Raynie, Supercritical fluid chromatography and extraction. *Analytical Chemistry* **70**, 301 (1998).

temperature characteristic of the substance. The triple point of water lies at 273.16 K and 611 Pa (6.11 mbar, 4.58 Torr), and the three phases of water (ice, liquid water, and water vapour) coexist in equilibrium at no other combination of pressure and temperature. This invariance of the triple point is the basis of its use in the definition of the thermodynamic temperature scale (Section 4.2c).

As we can see from Fig. 6.1, the triple point marks the lowest pressure at which a liquid phase of a substance can exist. If (as is common) the slope of the solid–liquid phase boundary is as shown in the diagram, then the triple point also marks the lowest temperature at which the liquid can exist; the critical temperature is the upper limit.

6.3 Three typical phase diagrams

We shall now see how these general features appear in the phase diagrams of pure substances.

(a) Water

Figure 6.4 is the phase diagram for water. The liquid–vapour boundary in the phase diagram summarizes how the vapour pressure of liquid water varies with temperature. It also summarizes how the boiling temperature varies with pressure: we simply read off the temperature at which the vapour pressure is equal to the prevailing atmospheric pressure. The solid–liquid boundary shows how the melting temperature

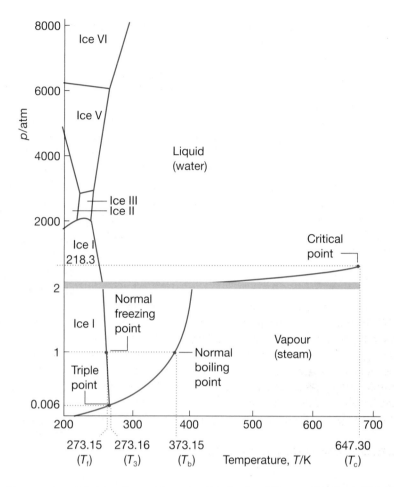

6.4 The experimental phase diagram for water showing the different solid phases. Note the change of vertical scale at 2 atm.

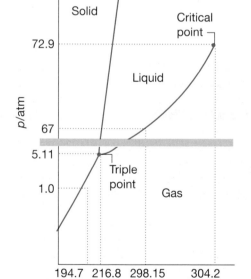

6.5 The experimental phase diagram for carbon dioxide. Note that, as the triple point lies at pressures well above atmospheric, liquid carbon dioxide does not exist under normal conditions (a pressure of at least 5.1 atm must be applied).

varies with the pressure. Its very steep slope indicates that enormous pressures are needed to bring about significant changes. Notice that the line has a negative slope up to 2 kbar, which means that the melting temperature falls as the pressure is raised. The reason for this almost unique behaviour can be traced to the decrease in volume that occurs on melting, and hence it being more favourable for the solid to transform into the liquid as the pressure is raised. The decrease in volume is a result of the very open molecular structure of ice: the H_2O molecules are held apart (as well as together) by the hydrogen bonds between them, but the structure partially collapses on melting, and the liquid is denser than the solid.

At high pressures, different structural forms of ice come into stability as the hydrogen bonds between molecules are modified by the stress. Some of these phases (which are called ice II, III, V, VI, and VII) melt at high temperatures.[2] Ice VII, for instance, melts at 100°C but exists only above 25 kbar. Note that five more triple points occur in the diagram other than the one where vapour, liquid, and ice I coexist. Each one occurs at a definite pressure and temperature that cannot be changed.

(b) Carbon dioxide

The phase diagram for carbon dioxide is shown in Fig. 6.5. The features to notice include the positive slope of the solid–liquid boundary (the direction of this line is

2. Ice IV was an illusion, or possibly a transient, not thermodynamically stable, phase.

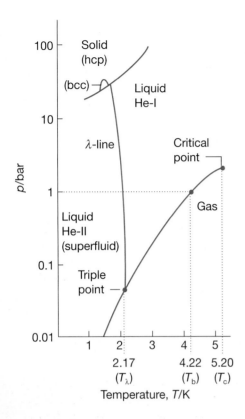

6.6 The phase diagram for helium (^{4}He). The λ-line marks the conditions under which the two liquid phases are in equilibrium. Helium-II is the superfluid phase. Note that a pressure of over 20 bar must be exerted before solid helium can be obtained. The labels hcp and bcc denote different solid phases in which the atoms pack together differently: hcp denotes hexagonal closed packing and bcc denotes body-centred cubic (see Section 21.1 for a description of these structures).

characteristic of most substances), which indicates that the melting temperature of solid carbon dioxide rises as the pressure is increased. Notice also that, as the triple point lies above 1 atm, the liquid cannot exist at normal atmospheric pressures whatever the temperature, and the solid sublimes when left in the open (hence the name 'dry ice'). To obtain the liquid, it is necessary to exert a pressure of at least 5.11 atm. Cylinders of carbon dioxide generally contain the liquid or compressed gas; at 25°C that implies a vapour pressure of 67 atm if both gas and liquid are present in equilibrium. When the gas squirts through the throttle it cools by the Joule–Thomson effect so, when it emerges into a region where the pressure is only 1 atm, it condenses into a finely divided snow-like solid.

(c) Helium

The phase diagram of helium is shown in Fig. 6.6. Helium behaves unusually at low temperatures. For instance, the solid and gas phases are never in equilibrium however low the temperature: the He atoms are so light that they vibrate with a large-amplitude motion even at very low temperatures, and the solid simply shakes itself apart. Solid helium can be obtained, but only by holding the atoms together by applying pressure.

When considering helium at low temperatures it is necessary to distinguish between the isotopes ^{3}He and ^{4}He, because quantum mechanical effects become important and these two isotopes differ not only in mass but also in the spin of their nuclei.[3] At this stage nuclear spin can be visualized as a spinning motion of the nucleus. A ^{4}He nucleus has zero spin; a ^{3}He nucleus has nonzero spin. As a result of quantum mechanical effects, pure helium-4 has a liquid–liquid phase transition at its **λ-line** (lambda line); the reason for this name is explained in Section 6.7. The liquid phase marked He-I behaves like a normal liquid. The other phase, He-II, is a **superfluid**. It is so called because it flows without viscosity. The phase diagram of helium-3 differs from the phase diagram of helium-4, but it also possesses a superfluid phase. Helium-3 is unusual in that the entropy of the liquid is lower than that of the solid, and melting is exothermic.

Phase stability and phase transitions

We shall now see how thermodynamic considerations can account for the features of the phase diagrams we have just described. All our considerations will be based on the Gibbs energy of a substance, and in particular on its molar Gibbs energy, G_m. In fact, this quantity will play such an important role in this chapter and the rest of the text, that we give it a special name and symbol, the **chemical potential**, μ (mu). For a one-component system, 'molar Gibbs energy' and 'chemical potential' are synonyms, so $\mu = G_m$, but in Chapter 7 we see that chemical potential has a broader significance and a more general definition. The name 'chemical potential' is also instructive: as we develop the concept, we shall see that μ is a measure of the potential that a substance has for bringing about change in a system. In this chapter, it reflects the potential of a substance to bring about physical change. In Chapter 9 we see that μ is the potential of a substance to bring about chemical change.

---

3. Nuclear spin is a quantum mechanical property that will be encountered in more detail in Part 2.

6.7 When two or more phases are in equilibrium, the molar Gibbs energy of a substance (and, in a mixture, a component) is the same in each phase and is the same at all points in each phase.

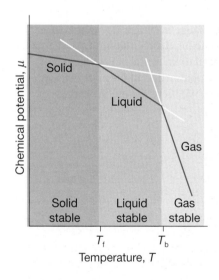

6.8 The schematic temperature dependence of the chemical potential of the solid, liquid, and gas phases of a substance (in practice, the lines are curved). The phase with the lowest molar Gibbs energy at a specified temperature is the most stable one at that temperature. The transition temperatures, the melting and boiling temperatures, are the temperatures at which the chemical potentials of the two phases are equal.

6.4 The thermodynamic criterion of equilibrium

We base our discussion on the following consequence of the Second Law: *at equilibrium, the chemical potential of a substance is the same throughout a sample, regardless of how many phases are present.* When the liquid and solid phases of a substance are in equilibrium, the chemical potential of the substance is the same throughout the system (Fig. 6.7).

To see the validity of this remark, consider a system in which the chemical potential of a substance is μ_1 at one location and μ_2 at another location. The locations may be in the same or in different phases. When an amount dn of the substance is transferred from one location to the other, the Gibbs energy of the system changes by $-\mu_1$dn when material is removed from location 1, and it changes by $+\mu_2$dn when that material is added to location 2. The overall change is therefore d$G = (\mu_2 - \mu_1)$ dn. If the chemical potential at location 1 is higher than that at location 2, the transfer is accompanied by a decrease in G, and so has a spontaneous tendency to occur. Only if $\mu_1 = \mu_2$ is there no change in G, and only then is the system at equilibrium. We conclude that the transition temperature, T_{trs}, is the temperature at which the chemical potentials of two phases are equal.

6.5 The dependence of stability on the conditions

At low temperatures, the solid phase of a substance has the lowest chemical potential and, provided the pressure is not too low, is usually the most stable at low temperatures. However, the chemical potentials of phases change with temperature in different ways, and as the temperature is raised the chemical potential of another phase (perhaps another solid phase, a liquid, or a gas) will fall below that of the solid. When that happens, a phase transition occurs if it is kinetically feasible to do so.

(a) The temperature dependence of phase stability

The temperature dependence of the Gibbs energy is expressed in terms of the entropy of the system by eqn 5.10 $((\partial G/\partial T)_p = -S)$. Because the chemical potential of a pure substance is equal to the molar Gibbs energy of that substance, it follows that

$$\left(\frac{\partial \mu}{\partial T}\right)_p = -S_m \tag{6.1}$$

This relation shows that, as the temperature is raised, the chemical potential of a pure substance decreases: $S_m > 0$ for all substances, so the slope of a plot of μ against T is negative.

Equation 1 implies that the slope of a plot of μ against temperature is steeper for gases than for liquids, because $S_m(g) > S_m(l)$. The slope is also steeper for a liquid than the corresponding solid, because $S_m(l) > S_m(s)$ almost always, on account of the greater disorder of a liquid. These features are illustrated in Fig. 6.8. The steep negative slope of $\mu(l)$ results in its falling below $\mu(s)$ when the temperature is high enough, and then the liquid becomes the stable phase: the solid melts. The chemical potential of the gas phase plunges steeply downwards as the temperature is raised (because the molar entropy of the vapour is so high), and there comes a temperature at which it lies lowest. Then the gas is the stable phase and the liquid vaporizes. When we bring about a phase transition, what we are actually doing is modifying the relative values of the chemical potentials of the phases. The easiest way of doing that is by changing the temperature of the sample.

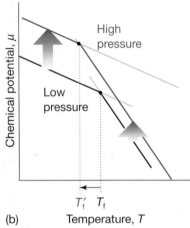

6.9 The pressure dependence of the chemical potential of a substance depends on the molar volume of the phase. The lines show schematically the effect of increasing pressure on the chemical potential of the solid and liquid phases (in practice, the lines are curved), and the corresponding effects on the freezing temperatures. (a) In this case the molar volume of the solid is less than that of the liquid and μ(s) increases less than μ(l). As a result, the freezing temperature rises. (b) Here the molar volume is greater for the solid than the liquid (as for water), μ(s) increases more strongly than μ(l), and the freezing temperature is lowered.

(b) The response of melting to applied pressure

Most substances melt at a higher temperature when subjected to pressure. It is as though the pressure is preventing the formation of the less dense liquid phase. Exceptions to this behaviour include water, for which the liquid is denser than the solid. Application of pressure to water encourages the formation of the liquid phase. That is, water freezes at a lower temperature when it is under pressure.

We can rationalize the response of melting temperatures to pressure as follows. The variation of the chemical potential with pressure is expressed (from eqn. 5.10) by

$$\left(\frac{\partial \mu}{\partial p}\right)_T = V_m \tag{6.2}$$

This equation shows that the slope of a plot of chemical potential against pressure is equal to the molar volume of the substance. An increase in pressure raises the chemical potential of any pure substance (because $V_m > 0$). In most cases, $V_m(l) > V_m(s)$ and the equation predicts that an increase in pressure increases the chemical potential of the liquid more than that of the solid. As shown in Fig. 6.9a, the effect of pressure in such a case is to raise the melting temperature slightly. For water, however, $V_m(l) < V_m(s)$, and an increase in pressure increases the chemical potential of the solid more than that of the liquid. In this case, the melting temperature is lowered slightly (Fig. 6.9b).

Example 6.1 Assessing the effect of pressure on the chemical potential

Calculate the effect on the chemical potentials of ice and water of increasing the pressure from 1.00 bar to 2.00 bar at 0°C. The density of ice is 0.917 g cm^{-3} and that of liquid water is 0.999 g cm^{-3} under these conditions.

Method From eqn 2, we know that the change in chemical potential of an incompressible substance when the pressure is changed by Δp is $\Delta \mu = V_m \Delta p$. Therefore, to answer the question, we need to know the molar volumes of the two phases of water. These values are obtained from the mass density, ρ, and the molar mass, M, by using $V_m = M/\rho$. We therefore use the expression

$$\Delta \mu = \frac{M \Delta p}{\rho}$$

Convert all data to SI units and use 1 Pa m^3 = 1 J.

Answer The molar mass of water is 18.02 g mol^{-1} (1.802 × 10^{-2} kg mol^{-1}); therefore,

$$\Delta \mu(\text{ice}) = \frac{(1.802 \times 10^{-2} \text{ kg mol}^{-1}) \times (1.00 \times 10^5 \text{ Pa})}{917 \text{ kg m}^{-3}} = +1.97 \text{ J mol}^{-1}$$

$$\Delta \mu(\text{water}) = \frac{(1.802 \times 10^{-2} \text{ kg mol}^{-1}) \times (1.00 \times 10^5 \text{ Pa})}{999 \text{ kg m}^{-3}} = +1.80 \text{ J mol}^{-1}$$

Comment The chemical potential of ice rises more sharply than that of water; so, if they are initially in equilibrium at 1 bar, there will be a tendency for the ice to melt at 2 bar.

Self-test 6.1 Calculate the effect of an increase in pressure of 1.00 bar on the liquid and solid phases of carbon dioxide (of molar mass 44.0 g mol^{-1}) in equilibrium with densities 2.35 g cm^{-3} and 2.50 g cm^{-3}, respectively.

[$\Delta \mu$(l) = +1.87 J mol^{-1}, $\Delta \mu$(s) = +1.76 J mol^{-1}; solid forms]

(c) The effect of applied pressure on vapour pressure

When pressure is applied to a condensed phase, its vapour pressure rises: in effect, molecules are squeezed out of the phase and escape as a gas. Pressure can be exerted on the condensed phases mechanically or by subjecting it to the applied pressure of an inert gas (Fig. 6.10); in the latter case, the vapour pressure is the partial pressure of the vapour in equilibrium with the condensed phase, and we speak of the **partial vapour pressure** of the substance. One complication (which we ignore here) is that if the condensed phase is a liquid, then the pressurizing gas might dissolve and change the properties of the liquid. Another complication is that the gas-phase molecules might attract molecules out of the liquid by the process of **gas solvation**, the attachment of molecules to gas-phase species.

The quantitative relation between the vapour pressure, p, when a pressure ΔP is applied and the vapour pressure, p^*, of the liquid in the absence of an additional pressure is

$$p = p^* e^{V_m \Delta P / RT} \tag{6.3a}$$

This equation shows how the vapour pressure increases when the pressure acting on the condensed phase is increased. If $V_m \Delta P / RT \ll 1$, where V_m is the molar volume of the liquid, the exponential function on the right of eqn 3 may be approximated by $1 + V_m \Delta P / RT$:

$$p \approx p^* \left(1 + \frac{V_m \Delta P}{RT} \right)$$

which rearranges to

$$\frac{p - p^*}{p^*} \approx \frac{V_m \Delta P}{RT} \tag{6.3b}$$

Calculations involving exponentials often become simpler if we note that $e^x = 1 + x + \frac{1}{2}x^2 + \cdots$. If $x \ll 1$, a good approximation is $e^x \approx 1 + x$.

Justification 6.1

We calculate the vapour pressure of a pressurized liquid by using the fact that at equilibrium the chemical potentials of the liquid and its vapour are equal: $\mu(l) = \mu(g)$. It follows that, for any change that preserves equilibrium, the resulting change in $\mu(l)$ must be equal to the change in $\mu(g)$; therefore, we can write $d\mu(g) = d\mu(l)$. When the pressure P on the liquid is increased by dP, the chemical potential of the liquid changes by $d\mu(l) = V_m(l) \, dP$. The chemical potential of the vapour changes by $d\mu(g) = V_m(g) \, dp$ where dp is the change in the vapour pressure we are trying to find. If we treat the vapour as a perfect gas, the molar volume can be replaced by $V_m(g) = RT/p$, and we obtain

$$d\mu(g) = \frac{RT \, dp}{p}$$

Next, we equate the changes in chemical potentials of the vapour and the liquid:

$$\frac{RT \, dp}{p} = V_m(l) \, dP$$

We can integrate this expression once we know the limits of integration.

When there is no additional pressure acting on the liquid, P (the pressure experienced by the liquid) is equal to the normal vapour pressure p^*; so, when $P = p^*$, $p = p^*$ too. When there is an additional pressure ΔP on the liquid, with the result that $P = p + \Delta P$, the vapour pressure is p (the value we want to find). The effect of pressure on the vapour pressure is so small that it is a good approximation to replace the p in

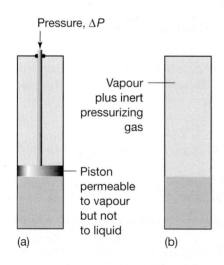

Pressure, ΔP

Vapour plus inert pressurizing gas

Piston permeable to vapour but not to liquid

(a) (b)

6.10 Pressure may be applied to a condensed phase either (a) by compressing the condensed phase or (b) by subjecting it to an inert pressurizing gas. When pressure is applied, the vapour pressure of the condensed phase increases.

$p + \Delta P$ by p^* itself, and to set the upper limit of the integral to $p^* + \Delta P$. The integrations required are therefore as follows:

$$RT \int_{p^*}^{p} \frac{dp}{p} = \int_{p^*}^{p^*+\Delta p} V_m(l) \, dP$$

We now assume that the molar volume of the liquid is the same throughout the small range of pressures involved. Then both integrations are straightforward, and lead to

$$RT \ln \frac{p}{p^*} = V_m(l) \Delta P$$

which rearranges to eqn 3.

Illustration 6.1

For water, which has density 0.997 g cm^{-3} at 25°C and therefore molar volume 18.1 cm^3 mol^{-1}, when subjected to an increase in pressure of 10 bar,

$$\frac{V_m \Delta P}{RT} = \frac{(1.81 \times 10^{-5} \text{ m}^3 \text{ mol}^{-1}) \times (1.0 \times 10^6 \text{ Pa})}{(8.3145 \text{ J K}^{-1} \text{ mol}^{-1}) \times (298 \text{ K})} = 7.3 \times 10^{-3}$$

where we have used 1 J = 1 Pa m^3. Because $V_m \Delta P / RT \ll 1$, eqn 3b can be used, and we obtain $(p - p^*)/p^* = 7.3 \times 10^{-3}$, an increase of 0.73 per cent.

Self-test 6.2 Calculate the effect of an increase in pressure of 100 bar on the vapour pressure of benzene at 25°C, which has density 0.879 g cm^{-3}.

[43 per cent; because the change is so large, use eqn 3a]

6.6 The location of phase boundaries

We can find the precise locations of the phase boundaries—the pressures and temperatures at which two phases can coexist—by making use of the fact that, when two phases are in equilibrium, their chemical potentials must be equal. Therefore, where the phases α and β are in equilibrium,

$$\mu_\alpha(p, T) = \mu_\beta(p, T) \tag{6.4}$$

By solving this equation for p in terms of T, we get an equation for the phase boundary.

(a) The slopes of the phase boundaries

It turns out to be simplest to discuss the phase boundaries in terms of their slopes, dp/dT.

Let p and T be changed infinitesimally, but in such a way that the two phases α and β remain in equilibrium. The chemical potentials of the phases are initially equal (the two phases are in equilibrium). They remain equal when the conditions are changed to another point on the phase boundary, where the two phases continue to be in equilibrium (Fig. 6.11). Therefore, the changes in the chemical potentials of the two phases must be equal and we can write $d\mu_\alpha = d\mu_\beta$. Because, from eqn 5.9, we know that

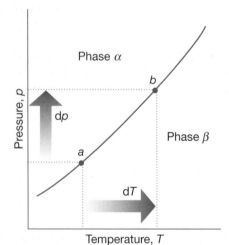

6.11 When pressure is applied to a system in which two phases are in equilibrium (at a), the equilibrium is disturbed. It can be restored by changing the temperature, so moving the state of the system to b. It follows that there is a relation between dp and dT that ensures that the system remains in equilibrium as either variable is changed.

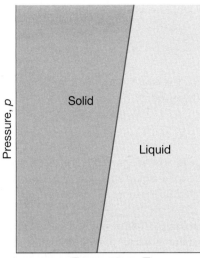

6.12 A typical solid–liquid phase boundary slopes steeply upwards. This slope implies that, as the pressure is raised, the melting temperature rises. Most substances behave in this way.

$$\mathrm{d}\mu = -S_\mathrm{m}\,\mathrm{d}T + V_\mathrm{m}\,\mathrm{d}p$$

for each phase, it follows that

$$-S_{\alpha,\mathrm{m}}\,\mathrm{d}T + V_{\alpha,\mathrm{m}}\,\mathrm{d}p = -S_{\beta,\mathrm{m}}\,\mathrm{d}T + V_{\beta,\mathrm{m}}\,\mathrm{d}p$$

where $S_{\alpha,\mathrm{m}}$ and $S_{\beta,\mathrm{m}}$ are the molar entropies of the phases and $V_{\alpha,\mathrm{m}}$ and $V_{\beta,\mathrm{m}}$ are their molar volumes. Hence

$$(V_{\beta,\mathrm{m}} - V_{\alpha,\mathrm{m}})\mathrm{d}p = (S_{\beta,\mathrm{m}} - S_{\alpha,\mathrm{m}})\mathrm{d}T \tag{6.5}$$

which rearranges into the **Clapeyron equation**:

$$\frac{\mathrm{d}p}{\mathrm{d}T} = \frac{\Delta_\mathrm{trs}S}{\Delta_\mathrm{trs}V} \tag{6.6}$$

In this expression $\Delta_\mathrm{trs}S = S_{\beta,\mathrm{m}} - S_{\alpha,\mathrm{m}}$ and $\Delta_\mathrm{trs}V = V_{\beta,\mathrm{m}} - V_{\alpha,\mathrm{m}}$ are the entropy and volume of transition. The Clapeyron equation is an exact expression for the slope of the phase boundary and applies to any phase equilibrium of any pure substance. Its importance is that we can use thermodynamic data to predict the appearance of phase diagrams and to understand their form. A more practical application is to the prediction of the response of freezing and boiling points to the application of pressure.

(b) The solid–liquid boundary

Melting (fusion) is accompanied by a molar enthalpy change $\Delta_\mathrm{fus}H$ and occurs at a temperature T. The molar entropy of melting at T is therefore $\Delta_\mathrm{fus}H/T$ (Section 4.3a), and the Clapeyron equation becomes

$$\frac{\mathrm{d}p}{\mathrm{d}T} = \frac{\Delta_\mathrm{fus}H}{T\Delta_\mathrm{fus}V} \tag{6.7}$$

where $\Delta_\mathrm{fus}V$ is the change in molar volume that occurs on melting. The enthalpy of melting is positive (the only exception is helium-3) and the volume change is usually positive and always small. Consequently, the slope $\mathrm{d}p/\mathrm{d}T$ is steep and usually positive (Fig. 6.12).

We can obtain the formula for the phase boundary by integrating $\mathrm{d}p/\mathrm{d}T$, assuming that $\Delta_\mathrm{fus}H$ and $\Delta_\mathrm{fus}V$ change so little with temperature and pressure that they can be treated as constant. If the melting temperature is T^* when the pressure is p^*, and T when the pressure is p, the integration required is

$$\int_{p^*}^{p} \mathrm{d}p = \frac{\Delta_\mathrm{fus}H}{\Delta_\mathrm{fus}V} \int_{T^*}^{T} \frac{\mathrm{d}T}{T}$$

Therefore, the approximate equation of the solid–liquid boundary is

$$p \approx p^* + \frac{\Delta_\mathrm{fus}H}{\Delta_\mathrm{fus}V} \ln\frac{T}{T^*} \tag{6.8}$$

This equation was originally obtained by yet another Thomson—James, the brother of William, Lord Kelvin. When T is close to T^*, the logarithm can be approximated by using

Calculations involving natural logarithms often become simpler if we note that $\ln(1 + x) = x - \frac{1}{2}x^2 + \frac{1}{3}x^3 \cdots$. If $x \ll 1$, a good approximation is $\ln(1 + x) \approx x$.

$$\ln\frac{T}{T^*} = \ln\left(1 + \frac{T - T^*}{T^*}\right) \approx \frac{T - T^*}{T^*}$$

Therefore,

$$p \approx p^* + \frac{\Delta_\mathrm{fus}H}{T^*\Delta_\mathrm{fus}V}(T - T^*) \tag{6.9}$$

This expression is the equation of a steep straight line when p is plotted against T (as in Fig. 6.12).

(c) The liquid–vapour boundary

The entropy of vaporization at a temperature T is equal to $\Delta_{vap}H/T$; the Clapeyron equation for the liquid–vapour boundary is therefore

$$\frac{dp}{dT} = \frac{\Delta_{vap}H}{T\Delta_{vap}V} \tag{6.10}$$

The enthalpy of vaporization is positive; $\Delta_{vap}V$ is large and positive. Therefore, dp/dT is positive, but it is much smaller than for the solid–liquid boundary. It follows that dT/dp is large, and hence that the boiling temperature is more responsive to pressure than the freezing temperature.

As an illustration of eqn 10, consider the cooking of food in water in a covered pot fitted with a valve that opens when the vapour pressure reaches a higher than ambient value. Upon heating, the vapour pressure of the water increases and boiling takes place only when the temperature reaches a value that opens the valve. This temperature is higher than for a liquid in a vessel at the ambient pressure. As a result, the food can be cooked at temperatures that are significantly higher than 100°C, and the process takes less time.[4]

Example 6.2 Estimating the effect of pressure on the boiling temperature

Estimate the typical size of the effect of increasing pressure on the boiling point of a liquid.

Method To use eqn 10 we need to estimate the right-hand side. At the boiling point, the term $\Delta_{vap}H/T$ is Trouton's constant (Section 4.3a). Because the molar volume of a gas is so much greater than the molar volume of a liquid, we can write

$$\Delta_{vap}V = V_m(g) - V_m(l) \approx V_m(g)$$

and take for $V_m(g)$ the molar volume of a perfect gas (at low pressures, at least).

Answer Trouton's constant has the value 85 J K^{-1} mol^{-1}. The molar volume of a perfect gas is about 25 L mol^{-1} at 1 atm and near but above room temperature. Therefore,

$$\frac{dp}{dT} \approx \frac{85 \text{ J K}^{-1} \text{ mol}^{-1}}{2.5 \times 10^{-2} \text{ m}^3 \text{ mol}^{-1}} = 3.4 \times 10^3 \text{ Pa K}^{-1}$$

We have used 1 J = 1 Pa m^3. This value corresponds to 0.034 atm K^{-1}, and hence to dT/dp = 29 K atm^{-1}. Therefore, a change of pressure of +0.1 atm can be expected to change a boiling temperature by about +3 K.

Self-test 6.3 Estimate dT/dp for water at its normal boiling point using the information in Table 4.2 and $V_m(g) = RT/p$.   [28 K atm^{-1}]

---

4. Like most chemical reactions, those associated with cooking of food occur at higher rates as the temperature increases. We discuss the temperature dependence of rates of chemical reactions in Chapter 25.

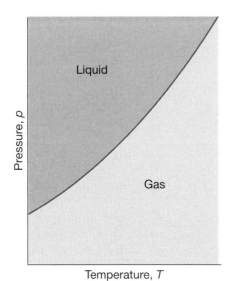

6.13 A typical liquid-vapour phase boundary. The boundary can be regarded as a plot of the vapour pressure against the temperature. Note that, in some depictions of phase diagrams in which a logarithmic pressure scale is used, the phase boundary has the opposite curvature (see Fig. 6.6). This phase boundary terminates at the critical point (not shown).

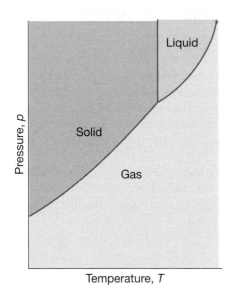

6.14 Near the point where they coincide (at the triple point), the solid-gas boundary has a steeper slope than the liquid-gas boundary because the enthalpy of sublimation is greater than the enthalpy of vaporization and the temperatures that occur in the Clausius-Clapeyron equation for the slope have similar values.

Because the molar volume of a gas is so much greater than the molar volume of a liquid, we can write $\Delta_{vap}V \approx V_m(g)$ (see the example above). Moreover, if the gas behaves perfectly, $V_m(g) = RT/p$. These two approximations turn the exact Clapeyron equation into

$$\frac{dp}{dT} = \frac{\Delta_{vap}H}{T(RT/p)}$$

which rearranges into the **Clausius–Clapeyron equation** for the variation of vapour pressure with temperature:

$$\frac{d \ln p}{dT} = \frac{\Delta_{vap}H}{RT^2} \tag{6.11}°$$

(We have used $dx/x = d \ln x$.) Like the Clapeyron equation, the Clausius–Clapeyron equation is important for understanding the appearance of phase diagrams, particularly the location and shape of the liquid–vapour and solid–vapour phase boundaries. Its practical significance is that it lets us predict how the vapour pressure varies with temperature and how the boiling temperature varies with pressure. For instance, if we also assume that the enthalpy of vaporization is independent of temperature, this equation integrates to

$$p = p^* e^{-\chi} \qquad \chi = \frac{\Delta_{vap}H}{R}\left(\frac{1}{T} - \frac{1}{T^*}\right) \tag{6.12}°$$

where p^* is the vapour pressure when the temperature is T^* and p the vapour pressure when the temperature is T. Equation 12 is the curve plotted as the liquid–vapour boundary in Fig. 6.13. The line does not extend beyond the critical temperature T_c, because above this temperature the liquid does not exist.

Illustration 6.2

Equation 12 can be used to estimate the vapour pressure of a liquid at any temperature from its normal boiling point, the temperature at which the vapour pressure is 1 atm (760 Torr). Thus, because the normal boiling point of benzene is 80°C (353 K) and (from Table 2.3), $\Delta_{vap}H^\ominus = 30.8$ kJ mol^{-1}, to calculate the vapour pressure at 20°C (293 K), we write

$$\chi = \frac{3.08 \times 10^4 \text{ J mol}^{-1}}{8.3145 \text{ J K}^{-1} \text{ mol}^{-1}}\left(\frac{1}{293 \text{ K}} - \frac{1}{353 \text{ K}}\right)$$

$$= \frac{3.08 \times 10^4}{8.3145}\left(\frac{1}{293} - \frac{1}{353}\right)$$

and substitute this value into eqn 12 with $p^* = 760$ Torr. The result is 89 Torr. The experimental value is 75 Torr.

(d) The solid–vapour boundary

The only difference between this case and the last is the replacement of the enthalpy of vaporization by the enthalpy of sublimation, $\Delta_{sub}H$. Because the enthalpy of sublimation is greater than the enthalpy of vaporization, the equation predicts a steeper slope for the sublimation curve than for the vaporization curve at similar temperatures, which is near where they meet (Fig. 6.14).

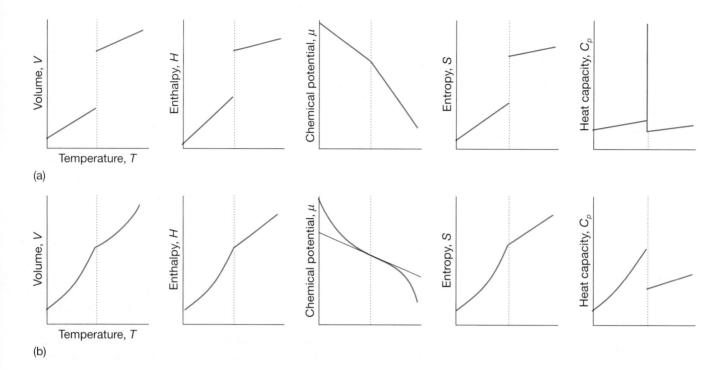

6.15 The changes in thermodynamic properties accompanying (a) first-order and (b) second-order phase transitions.

6.7 The Ehrenfest classification of phase transitions

There are many different types of phase transition, including the common examples of fusion and vaporization and the less common examples of solid–solid, conducting–superconducting, and fluid–superfluid transitions. We shall now see that it is possible to use thermodynamic properties of substances, and in particular the behaviour of the chemical potential, to classify phase transitions into different types. The classification scheme was originally proposed by Paul Ehrenfest, and is known as the **Ehrenfest classification**.

Many familiar phase transitions, like fusion and vaporization, are accompanied by changes of enthalpy and volume. These changes have implications for the slopes of the chemical potentials of the phases at either side of the phase transition. Thus, at the transition from a phase α to another phase β,

$$\left(\frac{\partial \mu_\beta}{\partial p}\right)_T - \left(\frac{\partial \mu_\alpha}{\partial p}\right)_T = V_{\beta,\mathrm{m}} - V_{\alpha,\mathrm{m}} = \Delta_{\mathrm{trs}}V$$

$$\left(\frac{\partial \mu_\beta}{\partial T}\right)_p - \left(\frac{\partial \mu_\alpha}{\partial T}\right)_p = -S_{\beta,\mathrm{m}} + S_{\alpha,\mathrm{m}} = -\Delta_{\mathrm{trs}}S = -\frac{\Delta_{\mathrm{trs}}H}{T_{\mathrm{trs}}}$$

(6.13)

Because $\Delta_{\mathrm{trs}}V$ and $\Delta_{\mathrm{trs}}H$ are nonzero for melting and vaporization, it follows that for such transitions the slopes of the chemical potential plotted against either pressure or temperature are different on either side of the transition (Fig. 6.15a). In other words, the first derivatives of the chemical potentials with respect to pressure and temperature are discontinuous at the transition.

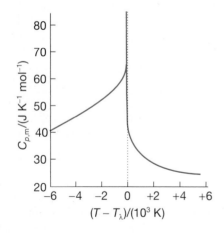

6.16 The λ-curve for helium, where the heat capacity rises to infinity. The shape of this curve is the origin of the name λ-transition.

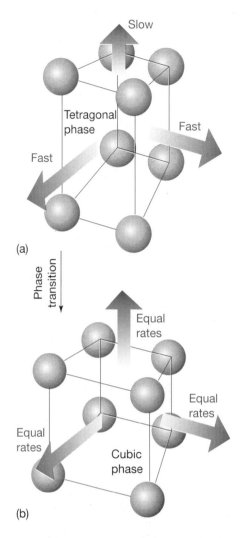

6.17 One version of a second-order phase transition in which (a) a tetragonal phase expands more rapidly in two directions than a third, and hence becomes a cubic phase, which (b) expands uniformly in three directions as the temperature is raised. There is no rearrangement of atoms at the transition temperature, and hence no enthalpy of transition.

A transition for which the first derivative of the chemical potential with respect to temperature is discontinuous is classified as a **first-order phase transition**. The constant-pressure heat capacity, C_p, of a substance is the slope of a plot of the enthalpy with respect to temperature. At a first-order phase transition, H changes by a finite amount for an infinitesimal change of temperature. Therefore, at the transition temperature the heat capacity is infinite. The physical reason is that heating drives the transition rather than raising the temperature. For example, boiling water stays at the same temperature even though heat is being supplied.

A **second-order phase transition** in the Ehrenfest sense is one in which the first derivative of μ with respect to temperature is continuous but its second derivative is discontinuous. A continuous slope of μ (a graph with the same slope on either side of the transition) implies that the volume and entropy (and hence the enthalpy) do not change at the transition (Fig. 6.15b). The heat capacity is discontinuous at the transition but does not become infinite there. An example of a second-order transition is the conducting–superconducting transition in metals at low temperatures.

The term **λ-transition** is applied to a phase transition that is not first-order yet the heat capacity becomes infinite at the transition temperature. Typically, the heat capacity of a system that shows such a transition begins to increase well before the transition (Fig. 6.16), and the shape of the heat capacity curve resembles the Greek letter lambda. This type of transition includes order–disorder transitions in alloys, the onset of ferromagnetism, and the fluid–superfluid transition of liquid helium.

Molecular interpretation 6.1

One type of second-order transition is associated with a change in symmetry of the crystal structure of a solid. Thus, suppose the arrangement of atoms in a solid is like that represented in Fig. 6.17a, with one dimension (technically, of the unit cell) longer than the other two, which are equal. Such a crystal structure is classified as tetragonal (see Section 23.1). Moreover, suppose the two shorter dimensions increase more than the long dimension when the temperature is raised. There may come a stage when the three dimensions become equal. At that point the crystal has cubic symmetry (Fig. 6.17b), and at higher temperatures it will expand equally in all three directions (because there is no longer any distinction between them). The tetragonal → cubic phase transition has occurred but, as it has not involved a discontinuity in the interaction energy between the atoms or the volume they occupy, the transition is not first-order.

The order–disorder transition in β-brass (CuZn) is an example of a λ-transition. The low-temperature phase is an orderly array of alternating Cu and Zn atoms. The high-temperature phase is a random array of the atoms (Fig. 6.18). At $T = 0$ the order is perfect, but islands of disorder appear as the temperature is raised. The islands form because the transition is cooperative in the sense that, once two atoms have exchanged locations, it is easier for their neighbours to exchange their locations. The islands grow in extent, and merge throughout the crystal at the transition temperature (742 K). The heat capacity increases as the transition temperature is approached because the cooperative nature of the transition means that it is increasingly easy for the heat supplied to drive the phase transition rather than to be stored as thermal motion.

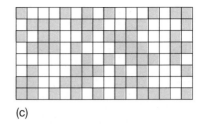

(a)   (b)   (c)

6.18 An order–disorder transition. (a) At $T = 0$, there is perfect order, with different kinds of atoms occupying alternate sites. (b) As the temperature is increased, atoms exchange locations and islands of each kind of atom form in regions of the solid. Some of the original order survives. (c) At and above the transition temperature, the islands occur at random throughout the sample.

The physical liquid surface

So far, we have concentrated on the properties of the boundaries in the phase diagram of a substance. However, the physical boundary between phases, such as the surface where solid is in contact with liquid or liquid is in contact with vapour, has interesting properties. In this section we concentrate on the liquid–vapour interface, which is interesting because it is so mobile. Chapter 28 deals with solid surfaces and their important role in catalysis.

6.8 Surface tension

Liquids tend to adopt shapes that minimize their surface area, for then the maximum number of molecules are in the bulk and hence surrounded by and interacting with neighbours. Droplets of liquids therefore tend to be spherical, because a sphere is the shape with the smallest surface-to-volume ratio. However, there may be other forces present that compete against the tendency to form this ideal shape and, in particular, gravity may flatten spheres into puddles or oceans.

Surface effects may be expressed in the language of Helmholtz and Gibbs energies. The link between these quantities and the surface area is the work needed to change the area by a given amount, and the fact that dA and dG are equal (under different conditions) to the work done in changing the energy of a system. The work needed to change the surface area, σ, of a sample by an infinitesimal amount $d\sigma$ is proportional to $d\sigma$, and we write

$$dw = \gamma \, d\sigma \qquad [6.14]$$

The constant of proportionality, γ, is called the **surface tension**; its dimensions are energy/area and its units are typically joules per metre squared ($J \, m^{-2}$). However, as in Table 6.1, values of γ are usually reported in newtons per metre ($N \, m^{-1}$, because $1 \, J = 1 \, N \, m$). The work of surface formation at constant volume and temperature can be identified with the change in the Helmholtz energy, and we can write

$$dA = \gamma \, d\sigma \qquad (6.15)$$

Because the Helmholtz energy decreases ($dA < 0$) if the surface area decreases ($d\sigma < 0$), surfaces have a natural tendency to contract. This is a more formal way of expressing what we have already described.

Table 6.1* Surface tensions of liquids at 293 K

	$\gamma/(mN \, m^{-1})$†
Benzene	28.88
Mercury	472
Methanol	22.6
Water	72.75

* More values are given in the *Data section*.
† Note that $1 \, N \, m^{-1} = 1 \, J \, m^{-2}$.

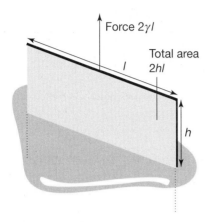

6.19 The model used for calculating the work of forming a liquid film when a wire of length l is raised and pulls the surface with it through a height h.

Example 6.3 Using the surface tension

Calculate the work needed to raise a wire of length l and to stretch the surface of a liquid through a height h in the arrangement shown in Fig. 6.19. Disregard gravitational potential energy.

Method According to eqn 14, the work required to create a surface area given that the surface tension does not vary as the surface is formed is $w = \gamma\sigma$. Therefore, all we need do is to calculate the surface area of the two-sided rectangle formed as the frame is withdrawn from the liquid.

Answer When the wire of length l is raised through a height h it increases the area of the liquid by twice the area of the rectangle (because there is a surface on each side). The total increase is therefore $2lh$ and the work done is $2\gamma lh$.

Comment The work can be expressed as a force × distance by writing it as $2\gamma l \times h$, and identifying γl as the opposing force on the wire of length l. This is why γ is called a tension and why its units are often chosen to be newtons per metre (N m^{-1}, so γl is a force in newtons).

Self-test 6.4 Calculate the work of creating a spherical cavity of radius r in a liquid of surface tension γ.   [$4\pi r^2 \gamma$]

6.9 Curved surfaces

The minimization of the surface area of a liquid may result in the formation of a curved surface, as in a bubble. We shall now see that there are two consequences of curvature, and hence of the surface tension, that are relevant to the properties of liquids. One is that the vapour pressure of a liquid depends on the curvature of its surface. The other is the capillary rise (or fall) of liquids in narrow tubes.

(a) Bubbles, cavities, and droplets

A **bubble** is a region in which vapour (and possibly air too) is trapped by a thin film; a **cavity** is a vapour-filled hole in a liquid. What are widely called 'bubbles' in liquids are therefore strictly cavities. True bubbles have two surfaces (one on each side of the film); cavities have only one. The treatments of both are similar, but a factor of 2 is required for bubbles to take into account the doubled surface area. A **droplet** is a small volume of liquid at equilibrium surrounded by its vapour (and possibly also air).

The pressure on the concave side of an interface, p_{in}, is always greater than the pressure on the convex side, p_{out}. This relation is expressed by the **Laplace equation**, which is derived in the following *Justification*:

$$p_{in} = p_{out} + \frac{2\gamma}{r} \tag{6.16}$$

Justification 6.2

The cavities in a liquid are at equilibrium when the tendency for their surface area to decrease is balanced by the rise of internal pressure which would then result. When the pressure inside a cavity is p_{in} and its radius is r, the outward force is

$$\text{pressure} \times \text{area} = 4\pi r^2 p_{in}$$

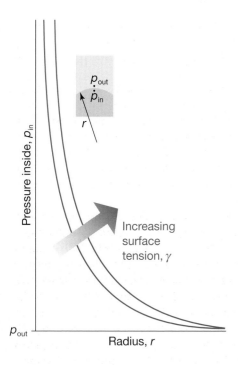

6.20 The dependence of the pressure inside a curved surface on the radius of the surface, for two different values of the surface tension.

The force inwards arises from the external pressure and the surface tension. The former has magnitude $4\pi r^2 p_{out}$. The latter is calculated as follows. The change in surface area when the radius of a sphere changes from r to $r + dr$ is

$$d\sigma = 4\pi(r + dr)^2 - 4\pi r^2 = 8\pi r \, dr$$

(The second-order infinitesimal, $(dr)^2$, is ignored.) The work done when the surface is stretched by this amount is therefore

$$dw = 8\pi\gamma r \, dr$$

As force × distance is work, the force opposing stretching through a distance dr when the radius is r is

$$F = 8\pi\gamma r$$

The total inward force is therefore $4\pi r^2 p_{out} + 8\pi\gamma r$. At equilibrium, the outward and inward forces are balanced, so we can write

$$4\pi r^2 p_{in} = 4\pi r^2 p_{out} + 8\pi\gamma r$$

which rearranges into eqn 16.

The Laplace equation shows that the difference in pressure decreases to zero as the radius of curvature becomes infinite (when the surface is flat, Fig. 6.20). Small cavities have small radii of curvature, so the pressure difference across their surface is quite large. For instance, a 'bubble' (actually, a cavity) of radius 0.10 mm in champagne implies a pressure difference of 1.5 kPa, which is enough to sustain a column of water of height 15 cm.

We saw in Section 6.5 that the vapour pressure of a liquid depends on the pressure applied to the liquid. Because curving a surface gives rise to a pressure differential of $2\gamma/r$, we can expect the vapour pressure above a curved surface to be different from that above a flat surface. By substituting this value of the pressure difference into eqn 3a we obtain the **Kelvin equation** for the vapour pressure of a liquid when it is dispersed as droplets of radius r:

$$p = p^* e^{2\gamma V_m / rRT} \tag{6.17}$$

The analogous expression for the vapour pressure inside a cavity can be written at once. The pressure of the liquid outside the cavity is less than the pressure inside, so the only change is in the sign of the exponent in the last expression.

(b) Nucleation

For droplets of water of radius 1 μm and 1 nm the ratios p/p^* at 25°C are about 1.001 and 3, respectively. The second figure, although quite large, is unreliable because at that radius the droplet is less than about 10 molecules in diameter and the basis of the calculation is suspect. The first figure shows that the effect is usually small; nevertheless it may have important consequences.

Consider, for example, the formation of a cloud. Warm, moist air rises into the cooler regions higher in the atmosphere. At some altitude the temperature is so low that the vapour becomes thermodynamically unstable with respect to the liquid and we expect it to condense into a cloud of liquid droplets. The initial step can be imagined as a swarm of water molecules congregating into a microscopic droplet. Because the initial droplet is so small it has an enhanced vapour pressure. Therefore, instead of growing it evaporates. This effect effectively stabilizes the vapour because an initial tendency to condense is overcome by a heightened tendency to evaporate.

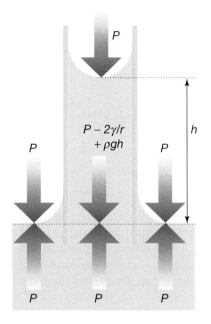

6.21 When a capillary tube is first stood in a liquid, the latter climbs up the walls, so curving the surface. The pressure just under the meniscus is less than that arising from the atmosphere by $2\gamma/r$. The pressure is equal at equal heights throughout the liquid provided the hydrostatic pressure (which is equal to ρgh) cancels the pressure difference arising from the curvature.

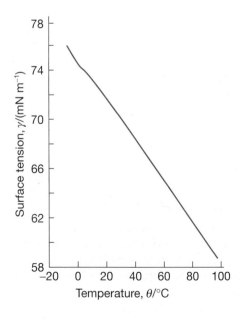

6.22 The variation of the surface tension of water with temperature.

The vapour phase is then said to be **supersaturated**. It is thermodynamically unstable with respect to the liquid but not unstable with respect to the small droplets that need to form before the bulk liquid phase can appear, so the formation of the latter by a simple, direct mechanism is hindered.

Clouds do form, so there must be a mechanism. Two processes are responsible. The first is that a sufficiently large number of molecules might congregate into a droplet so big that the enhanced evaporative effect is unimportant. The chance of one of these **spontaneous nucleation centres** forming is low, and in rain formation it is not a dominant mechanism. The more important process depends on the presence of minute dust particles or other kinds of foreign matter. These **nucleate** the condensation (that is, provide centres at which it can occur) by providing surfaces to which the water molecules can attach.

Liquids may be **superheated** above their boiling temperatures and **supercooled** below their freezing temperatures. In each case the thermodynamically stable phase is not achieved on account of the kinetic stabilization that occurs in the absence of nucleation centres. For example, superheating occurs because the vapour pressure inside a cavity is artificially low, so any cavity that does form tends to collapse. This instability is encountered when an unstirred beaker of water is heated, for its temperature may be raised above its boiling point. Violent bumping often ensues as spontaneous nucleation leads to bubbles big enough to survive. To ensure smooth boiling at the true boiling temperature, nucleation centres, such as small pieces of sharp-edged glass or bubbles (cavities) of air, should be introduced. A bubble chamber, a device for tracking elementary particles, depends on the nucleation of the evaporation of superheated liquid hydrogen by ionizing radiation.

6.10 Capillary action

The tendency of liquids to rise up capillary tubes (tubes of narrow bore), which is called **capillary action**, is a consequence of surface tension. Consider what happens when a glass capillary tube is first immersed in water or any liquid that has a tendency to adhere to the walls. The energy is lowest when a thin film covers as much of the glass as possible. As this film creeps up the inside wall it has the effect of curving the surface of the liquid inside the tube. This curvature implies that the pressure just beneath the curving meniscus is less than the atmospheric pressure by approximately $2\gamma/r$, where r is the radius of the tube and we assume a hemispherical surface. The pressure immediately under the flat surface outside the tube is p, the atmospheric pressure; but inside the tube under the curved surface it is only $p - 2\gamma/r$. The excess external pressure presses the liquid up the tube until hydrostatic equilibrium (equal pressures at equal depths) has been reached (Fig. 6.21).

(a) Capillary rise

The pressure exerted by a column of liquid of mass density ρ and height h is

$$p = \rho gh \tag{6.18}$$

This hydrostatic pressure matches the pressure difference $2\gamma/r$ at equilibrium. Therefore, the height of the column at equilibrium is obtained by equating $2\gamma/r$ and ρgh, which gives

$$h = \frac{2\gamma}{\rho gr} \tag{6.19}$$

This simple expression provides a reasonably accurate way of measuring the surface tension of liquids. Surface tension decreases with increasing temperature (Fig. 6.22).

6.23 The balance of forces that results in a contact angle, θ_c.

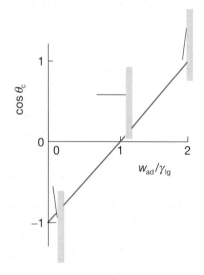

6.24 The variation of contact angle (shown by the semaphore-like object) as the ratio w_{ad}/γ_{lg} changes.

Illustration 6.3

If water at 25°C rises through 7.36 cm in a capillary of radius 0.20 mm, its surface tension at that temperature is

$$\gamma = \tfrac{1}{2}\rho ghr$$
$$= \tfrac{1}{2} \times (997.1 \text{ kg m}^{-3}) \times (9.81 \text{ m s}^{-2}) \times (7.36 \times 10^{-2} \text{ m}) \times (2.0 \times 10^{-4} \text{ m})$$
$$= 72 \text{ mN m}^{-1}$$

where we have used $1 \text{ kg m s}^{-2} = 1 \text{ N}$.

When the adhesive forces between the liquid and the material of the capillary wall are weaker than the cohesive forces within the liquid (as for mercury in glass), the liquid in the tube retracts from the walls. This retraction curves the surface with the concave, high-pressure side downwards. To equalize the pressure at the same depth throughout the liquid the surface must fall to compensate for the heightened pressure arising from its curvature. This compensation results in a capillary depression.

(b) The contact angle

In many cases there is a nonzero angle between the edge of the meniscus and the wall. If this contact angle is θ_c, then eqn 19 should be modified by multiplying the right hand side by $\cos \theta_c$.

The origin of the contact angle can be traced to the balance of forces at the line of contact between the liquid and the solid (Fig. 6.23). If the solid–gas, solid–liquid, and liquid–gas surface tensions (essentially the energy needed to create unit area of each of the interfaces) are denoted γ_{sg}, γ_{sl}, and γ_{lg}, respectively, then the vertical forces are in balance if

$$\gamma_{sg} = \gamma_{sl} + \gamma_{lg} \cos \theta_c \tag{6.20}$$

This expression solves to

$$\cos \theta_c = \frac{\gamma_{sg} - \gamma_{sl}}{\gamma_{lg}} \tag{6.21}$$

If we note that the work of adhesion of the liquid to the solid (per unit area of contact) is

$$w_{ad} = \gamma_{sg} + \gamma_{lg} - \gamma_{sl} \tag{6.22}$$

eqn 21 can be written

$$\cos \theta_c = \frac{w_{ad}}{\gamma_{lg}} - 1 \tag{6.23}$$

We now see that the liquid 'wets' (spreads over) the surface, corresponding to $0 < \theta_c < 90°$, when $1 < w_{ad}/\gamma_{lg} < 2$ (Fig. 6.24). The liquid does not wet the surface, corresponding to $90° < \theta_c < 180°$, when $0 < w_{ad}/\gamma_{lg} < 1$. For mercury in contact with glass, $\theta_c = 140°$, which corresponds to $w_{ad}/\gamma_{lg} = 0.23$, indicating a relatively low work of adhesion of the mercury to glass on account of the strong cohesive forces within mercury.

Checklist of key ideas

Phase diagrams

6.1 The stabilities of phases

☐ phase, a form of matter that is uniform throughout in chemical composition and physical state

☐ phase transition, the spontaneous conversion of one phase into another phase

☐ transition temperature, the temperature at which the two phases are in equilibrium

☐ metastable phase, a thermodynamically unstable phase that persists because the transition is kinetically hindered

6.2 Phase boundaries

☐ phase diagram, a diagram showing the regions of pressure and temperature at which its various phases are thermodynamically stable

☐ phase boundary, lines separating the regions in a phase diagram showing the values of p and T at which two phases coexist in equilibrium

☐ vapour pressure, the pressure of a vapour in equilibrium with the condensed (normally liquid) phase

☐ sublimation vapour pressure, the vapour pressure of a solid

☐ boiling, the condition of free vaporization throughout the liquid

☐ boiling temperature, the temperature at which the vapour pressure of a liquid is equal to the external pressure

☐ normal boiling point, the boiling point when the external pressure is 1 atm

☐ standard boiling point, the boiling point when the external pressure is 1 bar

☐ critical temperature, the temperature at which a liquid surface disappears and above which a liquid does not exist whatever the pressure

☐ critical pressure, the vapour pressure at the critical temperature

☐ supercritical fluid, a dense fluid phase above the critical temperature

☐ melting temperature or freezing temperature, the temperature at which, under a specified pressure, the liquid and solid phases of a substance coexist in equilibrium

☐ normal melting point or normal freezing point, the freezing temperature when the pressure is 1 atm

☐ standard freezing point, the freezing temperature when the pressure is 1 bar

☐ triple point, a point on a phase diagram at which the three phase boundaries meet and all three phases are in mutual equilibrium

6.3 Three typical phase diagrams

☐ the interpretation of the phase diagram for water

☐ the interpretation of the phase diagram for carbon dioxide

☐ the interpretation of the phase diagram for helium

☐ λ-line, the phase boundary between normal and superfluid liquid helium

☐ superfluid helium, a liquid phase that flows without viscosity

Phase stability and phase transitions

☐ chemical potential (of a pure substance), the molar Gibbs energy, $\mu = G_m$

6.4 The thermodynamic criterion of equilibrium

☐ the uniformity of chemical potential throughout a system at equilibrium

6.5 The dependence of stability on the conditions

☐ the variation of chemical potential with temperature, $(\partial\mu/\partial T)_p = -S_m$

☐ the variation of chemical potential with pressure, $(\partial\mu/\partial p)_T = V_m$

☐ partial vapour pressure, the partial pressure of the vapour of a component of a mixture in equilibrium with the condensed phase

☐ gas solvation, the attachment of molecules to gas phase species

☐ vapour pressure in the presence of applied pressure, $p = p^* e^{V_m \Delta P/RT}$

6.6 The location of phase boundaries

☐ Clapeyron equation, $dp/dT = \Delta_{trs}S/\Delta_{trs}V$

☐ the slope of the solid–liquid boundary, $dp/dT = \Delta_{fus}H/T\Delta_{fus}V$

☐ the slope of the liquid–vapour boundary, $dp/dT = \Delta_{vap}H/T\Delta_{vap}V$

☐ Clausius–Clapeyron equation, $d\ln p/dT = \Delta_{vap}H/RT^2$

6.7 The Ehrenfest classification of phase transitions

☐ Ehrenfest classification, a classification of phase transitions based on the behaviour of the chemical potential

☐ first-order phase transition, a transition for which the first derivative of the chemical potential with respect to temperature is discontinuous; at such a transition the heat capacity is infinite

☐ second-order phase transition, a transition for which the first derivative of the chemical potential with respect to temperature is continuous but its second derivative is discontinuous

☐ λ-transition, a phase transition that is not first-order but for which the heat capacity becomes infinite at the transition temperature

The physical liquid surface

6.8 Surface tension

☐ work of forming a surface, $dw = \gamma d\sigma$

☐ surface tension γ, the constant of proportionality between the work required to form a surface and the area of the surface

6.9 Curved surfaces

☐ bubble, a region in which vapour (and possibly air too) is trapped by a thin film

☐ cavity, a vapour-filled hole in a liquid

☐ droplet, a small volume of liquid at equilibrium surrounded by its vapour (and possibly also air)

☐ Laplace equation for the vapour pressure at a curved surface, $p_{in} = p_{out} + 2\gamma/r$

☐ Kelvin equation for the vapour pressure of droplets, $p = p^* e^{2\gamma V_m/rRT}$

☐ supersaturated phase, a phase that is thermodynamically unstable with respect to the liquid

☐ spontaneous nucleation centre, a location at which a sufficiently large number of molecules congregate into a droplet

☐ nucleate, provide surfaces to which molecules can attach and thereby induce condensation

☐ superheated, a liquid that has not boiled but is above its boiling temperature

☐ supercooled, a liquid that has not frozen but is below its freezing temperature

6.10 Capillary action

☐ capillary action, the tendency of liquids to rise up capillary tubes

☐ capillary rise and surface tension, $h = 2\gamma/\rho g r$

☐ contact angle and interfacial tension, $\cos\theta_c = (\gamma_{sg} - \gamma_{sl})/\gamma_{lg}$

☐ the criteria for surface wetting, $1 < w_{ad}/\gamma_{lg} < 2$, with $w_{ad} = \gamma_{sg} + \gamma_{lg} - \gamma_{sl}$

Further reading

Articles of general interest

W.B. Daniels, High-pressure techniques. In *Encyclopedia of applied physics* (ed. G.L. Trigg), **7**, 495. VCH, New York (1995).

B.B. Earl, The direct relation between altitude and boiling point. *J. Chem. Educ.* **67**, 45 (1990).

J.M. Honig, Mean field theory of phase transitions. *J. Chem. Educ.* **76**, 848 (1999).

E.K.H. Salje, Phase transitions, structural. In *Encyclopedia of applied physics* (ed. G.L. Trigg), **13**, 373. VCH, New York (1995).

J.M. Sanchez, Order–disorder transitions. In *Encyclopedia of applied physics* (ed. G.L. Trigg), **13**, 1. VCH, New York (1995).

K.M. Scholsky, Supercritical phase transitions at very high pressure. *J. Chem. Educ.* **66**, 989 (1989).

Texts and sources of data and information

J.R. Anderson, Determination of boiling and condensation temperatures. In *Techniques of Chemistry* (A. Weissberger and B.W. Rossiter, ed.) **5**, 199. Wiley, New York (1971).

T. Boublik, V. Fried, and E. Hála, *The vapor pressures of pure substances*. Elsevier, Amsterdam (1984).

W.D. Callister, Jr., *Materials science and engineering, an introduction*. Wiley, New York (2000).

J.J. Jasper, The surface tension of pure liquid compounds. *J. Phys. Chem. Ref. Data* **1**, 841 (1972).

A.N. Nesmeyanov. *Vapour pressure of the elements*. Pergamon Press, New York (1963).

J.S. Rowlinson and B. Widom, *Molecular theory of capillarity*. Clarendon Press, Oxford (1982).

S.I. Sandler, *Chemical and engineering thermodynamics*. Wiley, New York (1998).

E.L. Skau and J.C. Arthur, Determination of melting and freezing temperatures. In *Techniques of Chemistry* (A. Weissberger and B.W. Rossiter, ed.) **5**, 105. Wiley, New York (1971).

H.E. Stanley, Introduction to phase transitions and critical phenomena. Clarendon Press, Oxford (1971).

M.A. White, *Properties of materials*. Oxford University Press, Oxford (1999).

Exercises

Discussion questions

6.1 (a) Discuss the implications for phase stability of the variation of chemical potential with temperature and pressure.

6.1 (b) Suggest a physical interpretation of the phenomena of superheating and supercooling.

6.2 (a) Refer to Fig. 6.4 and describe the changes that would be observed when water vapour at 1.0 atm and 400 K is cooled at constant pressure to 260 K. Suggest the appearance of the cooling curve, a plot of temperature against time.

6.2 (b) Use the phase diagram in Fig. 6.5 to state what would be observed when a sample of carbon dioxide, initially at 1.0 atm and 298 K, is subjected to the following cycle: (a) isobaric (constant-pressure) heating to 320 K, (b) isothermal compression to 100 atm, (c) isobaric cooling to 210 K, (d) isothermal decompression to 1.0 atm, (e) isobaric heating to 298 K.

6.3 (a) Explain the significance of the Clapeyron equation and of the Clausius–Clapeyron equation.

6.3 (b) Distinguish between a first-order phase transition, a second-order phase transition, and a λ-transition.

Numerical exercises

6.4 (a) The vapour pressure of dichloromethane at 24.1°C is 400 Torr and its enthalpy of vaporization is 28.7 kJ mol⁻¹. Estimate the temperature at which its vapour pressure is 500 Torr.

6.4 (b) The vapour pressure of a substance at 20.0°C is 58.0 kPa and its enthalpy of vaporization is 32.7 kJ mol⁻¹. Estimate the temperature at which its vapour pressure is 66.0 kPa.

6.5 (a) The molar volume of a certain solid is 161.0 cm³ mol⁻¹ at 1.00 atm and 350.75 K, its melting temperature. The molar volume of the liquid at this temperature and pressure is 163.3 cm³ mol⁻¹. At 100 atm the melting temperature changes to 351.26 K. Calculate the enthalpy and entropy of fusion of the solid.

6.5 (b) The molar volume of a certain solid is 142.0 cm³ mol⁻¹ at 1.00 atm and 427.15 K, its melting temperature. The molar volume of the liquid at this temperature and pressure is 152.6 cm^3 mol^{-1}. At 1.2 MPa the melting temperature changes to 429.26 K. Calculate the enthalpy and entropy of fusion of the solid.

6.6 (a) The vapour pressure of a liquid in the temperature range 200 K to 260 K was found to fit the expression $\ln(p/\text{Torr}) = 16.255 - 2501.8/(T/\text{K})$. Calculate the enthalpy of vaporization of the liquid.

6.6 (b) The vapour pressure of a liquid in the temperature range 200 K to 260 K was found to fit the expression $\ln(p/\text{Torr}) = 18.361 - 3036.8/(T/\text{K})$. Calculate the enthalpy of vaporization of the liquid.

6.7 (a) The vapour pressure of benzene between 10°C and 30°C fits the expression $\log(p/\text{Torr}) = 7.960 - 1780/(T/\text{K})$. Calculate (a) the enthalpy of vaporization and (b) the normal boiling point of benzene.

6.7 (b) The vapour pressure of a liquid between 15°C and 35°C fits the expression $\log(p/\text{Torr}) = 8.750 - 1625/(T/\text{K})$. Calculate (a) the enthalpy of vaporization and (b) the normal boiling point of the liquid.

6.8 (a) When benzene freezes at 5.5°C its density changes from 0.879 g cm^{-3} to 0.891 g cm^{-3}. Its enthalpy of fusion is 10.59 kJ mol^{-1}. Estimate the freezing point of benzene at 1000 atm.

6.8 (b) When a certain liquid freezes at –3.65°C its density changes from 0.789 g cm^{-3} to 0.801 g cm^{-3}. Its enthalpy of fusion is 8.68 kJ mol^{-1}. Estimate the freezing point of the liquid at 100 MPa.

6.9 (a) In July in Los Angeles, the incident sunlight at ground level has a power density of 1.2 kW m^{-2} at noon. A swimming pool of area 50 m^2 is directly exposed to the sun. What is the maximum rate of loss of water assuming that all radiation is absorbed?

6.9 (b) Suppose the incident sunlight at ground level has a power density of 0.87 kW m^{-2} at noon. What is the maximum rate of loss of water from a lake of area 1.0 ha? (1 ha = 10^4 m^2.)

6.10 (a) An open vessel containing (a) water, (b) benzene, (c) mercury stands in a laboratory measuring 5.0 m × 5.0 m × 3.0 m at 25°C. What mass of each substance will be found in the air if there is no ventilation? (The vapour pressures are (a) 24 Torr, (b) 98 Torr, (c) 1.7 mTorr.)

6.10 (b) On a cold, dry morning after a frost, the temperature was –5°C and the partial pressure of water in the atmosphere fell to 2.0 Torr. Will the frost sublime? What partial pressure of water would ensure that the frost remained?

6.11 (a) Naphthalene, C$_{10}$H$_8$, melts at 80.2°C. If the vapour pressure of the liquid is 10 Torr at 85.8°C and 40 Torr at 119.3°C, use the Clausius–Clapeyron equation to calculate (a) the enthalpy of vaporization, (b) the normal boiling point, and (c) the entropy of vaporization at the boiling point.

6.11 (b) The boiling point of hexane is 69.0°C. Estimate (a) its enthalpy of vaporization and (b) its vapour pressure at 25°C and 60°C.

6.12 (a) Calculate the melting point of ice under a pressure of 50 bar. Assume that the density of ice under these conditions is approximately 0.92 g cm^{-3} and that of liquid water is 1.00 g cm^{-3}.

6.12 (b) Calculate the melting point of ice under a pressure of 10 MPa. Assume that the density of ice under these conditions is approximately 0.915 g cm^{-3} and that of liquid water is 0.998 g cm^{-3}.

6.13 (a) What fraction of the enthalpy of vaporization of water is spent on expanding the water vapour?

6.13 (b) What fraction of the enthalpy of vaporization of ethanol is spent on expanding its vapour?

6.14 (a) Calculate the vapour pressure of a spherical droplet of water of radius 10 nm at 20°C. The vapour pressure of bulk water at that temperature is 2.3 kPa and its density is 0.9982 g cm^{-3}.

6.14 (b) Calculate the vapour pressure of a spherical droplet of water of radius 20.0 nm at 35.0°C. The vapour pressure of bulk water at that temperature is 5.623 kPa and its density is 994.0 kg m^{-3}.

6.15 (a) The contact angle for water on clean glass is close to zero. Calculate the surface tension of water at 20°C given that at that temperature water climbs to a height of 4.96 cm in a clean glass capillary tube of internal radius 0.300 mm. The density of water at 20°C is 998.2 kg m^{-3}.

6.15 (b) The contact angle for water on clean glass is close to zero. Calculate the surface tension of water at 30°C given that at that temperature water climbs to a height of 9.11 cm in a clean glass capillary tube of internal diameter 0.320 mm. The density of water at 30°C is 0.9956 g cm^{-3}.

6.16 (a) Calculate the pressure differential of water across the surface of a spherical droplet of radius 200 nm at 20°C.

6.16 (b) Calculate the pressure differential of ethanol across the surface of a spherical droplet of radius 220 nm at 20°C. The surface tension of ethanol at that temperature is 22.39 mN m^{-1}.

Problems†

Numerical problems

6.1 The temperature dependence of the vapour pressure of solid sulfur dioxide can be approximately represented by the relation $\log(p/\text{Torr}) = 10.5916 - 1871.2/(T/\text{K})$ and that of liquid sulfur dioxide by $\log(p/\text{Torr}) = 8.3186 - 1425.7/(T/\text{K})$. Estimate the temperature and pressure of the triple point of sulfur dioxide.

6.2 Prior to the discovery that freon-12 (CF$_2$Cl$_2$) was harmful to the Earth's ozone layer, it was frequently used as the dispersing agent in spray cans for hair spray, etc. Its enthalpy of vaporization at its normal boiling point of –29.2°C is 20.25 kJ mol^{-1}. Estimate the pressure that a can of hair spray using freon-12 had to withstand at 40°C, the temperature of a can that has been standing in sunlight.

† Problems denoted by the symbol ‡ were supplied by Charles Trapp, Carmen Giunta, and Marshall Cady.

Assume that $\Delta_{vap}H$ is a constant over the temperature range involved and equal to its value at $-29.2°C$.

6.3 The enthalpy of vaporization of a certain liquid is found to be 14.4 kJ mol^{-1} at 180 K, its normal boiling point. The molar volumes of the liquid and the vapour at the boiling point are 115 cm^3 mol^{-1} and 14.5 dm^3 mol^{-1}, respectively. (a) Estimate dp/dT from the Clapeyron equation and (b) the percentage error in its value if the Clausius–Clapeyron equation is used instead.

6.4 Calculate the difference in slope of the chemical potential against temperature on either side of (a) the normal freezing point of water and (b) the normal boiling point of water. (c) By how much does the chemical potential of water supercooled to $-5.0°C$ exceed that of ice at that temperature?

6.5 Calculate the difference in slope of the chemical potential against pressure on either side of (a) the normal freezing point of water and (b) the normal boiling point of water. The densities of ice and water at 0°C are 0.917 g cm^{-3} and 1.000 g cm^{-3}, and those of water and water vapour at 100°C are 0.958 g cm^{-3} and 0.598 g L^{-1}, respectively. By how much does the chemical potential of water vapour exceed that of liquid water at 1.2 atm and 100°C?

6.6 The enthalpy of fusion of mercury is 2.292 kJ mol^{-1}, and its normal freezing point is 234.3 K with a change in molar volume of $+0.517$ cm^3 mol^{-1} on melting. At what temperature will the bottom of a column of mercury (density 13.6 g cm^{-3}) of height 10.0 m be expected to freeze?

6.7 50.0 L of dry air was slowly bubbled through a thermally insulated beaker containing 250 g of water initially at 25°C. Calculate the final temperature. (The vapour pressure of water is approximately constant at 23.8 Torr throughout, and its heat capacity is 75.5 J K^{-1} mol^{-1}. Assume that the air is not heated or cooled and that water vapour is a perfect gas.)

6.8 The vapour pressure, p, of nitric acid varies with temperature as follows:

$\theta/°C$	0	20	40	50	70	80	90	100
$p/$Torr	14.4	47.9	133	208	467	670	937	1282

What are (a) the normal boiling point and (b) the enthalpy of vaporization of nitric acid?

6.9 The vapour pressure of the ketone carvone ($M = 150.2$ g mol^{-1}), a component of oil of spearmint, is as follows:

$\theta/°C$	57.4	100.4	133.0	157.3	203.5	227.5
$p/$Torr	1.00	10.0	40.0	100	400	760

What are (a) the normal boiling point and (b) the enthalpy of vaporization of carvone?

6.10 Construct the phase diagram for benzene near its triple point at 36 Torr and 5.50°C using the following data: $\Delta_{fus}H = 10.6$ kJ mol^{-1}, $\Delta_{vap}H = 30.8$ kJ mol^{-1}, $\rho(s) = 0.891$ g cm^{-3}, $\rho(l) = 0.879$ g cm^{-3}.

6.11‡ In an investigation of the thermophysical properties of toluene (*J. Phys. Chem. Ref. Data* **18**, 1565 (1989)), R.D. Goodwin presented expressions for two coexistence curves. The solid–liquid coexistence curve is given by $p/$bar $= p_3/$bar $+ 1000 \times (5.60 + 11.727x)x$, where $x = T/T_3 - 1$ and the triple point pressure and temperature are $p_3 = 0.4362$ µbar and $T_3 = 178.15$ K, respectively. The liquid–vapour curve is given by:

$$\ln(p/\text{bar}) = -10.418/y + 21.157 - 15.996y + 14.015y^2 - 5.0120y^3 + 4.7224(1-y)^{1.70}$$

where $y = T/T_c = T/593.95$ K. (a) Plot the solid–liquid and liquid–vapour phase boundaries. (b) Estimate the standard melting point of toluene. (c) Estimate the standard boiling point of toluene. (d) Compute the standard enthalpy of vaporization of toluene, given that the molar volumes of the liquid and vapour at the normal boiling point are 0.12 L mol^{-1} and 30.3 L mol^{-1}, respectively.

6.12‡ In a study of the vapour pressure of chloromethane (*J. Chem. Eng. Data* **40**, 869 (1995)), A. Bah and N. Dupont-Pavlovsky presented data for the vapour pressure over solid chloromethane at low temperatures. Some of that data is shown below:

$T/$K	145.94	147.96	149.93	151.94	153.97	154.94
$p/$Pa	13.07	18.49	25.99	36.76	50.86	59.56

Estimate the standard enthalpy of sublimation of chloromethane at 150 K. (Take the molar volume of the vapour to be that of a perfect gas, and that of the solid to be negligible.)

6.13‡ The standard enthalpy of combustion of 7,8-benzoquinoline (benz) is -6556 ± 4 kJ mol^{-1} for the pure solid at 25°C. The combustion reaction is:

$$C_{13}H_9N(s) + \tfrac{61}{4}O_2(g) \longrightarrow 13\,CO_2(g) + \tfrac{9}{2}H_2O(l) + \tfrac{1}{2}N_2(g)$$

A restricted molecular flow apparatus has been used to determine the very small vapour pressures, p, of the solid by measuring solid mass loss, Δm, in measured time periods Δt, at a fixed temperature. Theoretical analysis of the apparatus yields the relation

$$p/\text{Torr} = 17.4\left(\frac{f(\Delta m/g)}{\alpha(A/\text{cm}^2)(\Delta t/s)}\right)\left(\frac{T/K}{M/(g\,mol^{-1})}\right)^{1/2}$$

where A is the surface area of the sample ($A = 3.64$ cm^2). The apparatus constants are $f = 8.99$ and $\alpha = 1$. The table below contains the experimental data. Use mathematical software or a spreadsheet to determine the enthalpy of sublimation and the entropy of sublimation for a vapour pressure of 1.00 Torr.

$T/$K	293.10	298.20	303.13	308.23	313.17	318.24	323.15
$\Delta t/$s	21 600	19 200	18 000	7140	4200	3600	1800
$\Delta m/$g	0.0481	0.0797	0.1275	0.0823	0.0801	0.1055	0.0877

Theoretical problems

6.14 Show that, for a transition between two incompressible solid phases, ΔG is independent of the pressure.

6.15 The change in enthalpy is given by $dH = C_p\,dT + V\,dp$. The Clapeyron equation relates dp and dT at equilibrium, and so in combination the two equations can be used to find how the enthalpy changes along a phase boundary as the temperature changes and the two phases remain in equilibrium. Show that $d(\Delta H/T) = \Delta C_p\,d\ln T$.

6.16 In the 'gas saturation method' for the measurement of vapour pressure, a volume V of gas (as measured at a temperature T and a pressure p) is bubbled slowly through the liquid, which is maintained at the temperature T, and a mass loss m is measured. Show that the vapour pressure, p, of the liquid is related to its molar mass, M, by $p = AmP/(1 + Am)$, where $A = RT/MPV$. The vapour pressure of geraniol ($M = 154.2$ g mol^{-1}), which is a component of oil of roses, was measured at 110°C. It was found that, when 5.00 L of nitrogen at 760 Torr was passed slowly through the heated liquid, the loss of mass was 0.32 g. Calculate the vapour pressure of geraniol.

6.17 Combine the barometric formula (stated in Box 1.1) for the dependence of the pressure on altitude with the Clausius–Clapeyron equation, and predict how the boiling temperature of a liquid depends on the altitude and the ambient temperature. Take the mean ambient temperature as 20°C and predict the boiling temperature of water at 3000 m.

6.18 Figure 6.8 gives a schematic representation of how the chemical potentials of the solid, liquid, and gaseous phases of a substance vary with temperature. All have a negative slope, but it is unlikely that they are truly straight lines as indicated in the illustration. Derive an expression for the curvatures (specifically, the second derivatives with respect to temperature) of these lines. Is there a restriction on the curvature of these lines? Which state of matter shows the greatest curvature?

6.19 The Clapeyron equation does not apply to second-order phase transitions, but there are two analogous equations, the *Ehrenfest equations*, that do. They are:

$$\frac{dp}{dT} = \frac{\alpha_2 - \alpha_1}{\kappa_{T,2} - \kappa_{T,1}}$$

$$\frac{dp}{dT} = \frac{C_{p,m2} - C_{p,m1}}{TV_m(\alpha_2 - \alpha_1)}$$

where α is the expansion coefficient, κ_T the isothermal compressibility, and the subscripts 1 and 2 refer to two different phases. Derive these two equations. Why does the Clapeyron equation not apply to second-order transitions?

6.20 For a first-order phase transition, to which the Clapeyron equation does apply, prove the relation

$$C_S = C_p - \frac{\alpha V \Delta_{trs}H}{\Delta_{trs}V}$$

where $C_S = (\partial q/\partial T)_S$ is the heat capacity along the coexistence curve of two phases.

Applications: to chemical engineering

6.21‡ A substance as well-known as methane still receives research attention because it is an important component of natural gas, a commonly used fossil fuel . Friend *et al.* have published a review of the thermophysical properties of methane (D.G. Friend, J.F. Ely, and H. Ingham, *J. Phys. Chem. Ref. Data* **18**, 583 (1989)), which included the following data describing the liquid–vapour phase boundary.

T/K	100	108	110	112	114	120	130	140
p/MPa	0.034	0.074	0.088	0.104	0.122	0.192	0.368	0.642

T/K	150	160	170	190
p/MPa	1.041	1.593	2.329	4.521

(a) Plot the liquid–vapour phase boundary. (b) Estimate the standard boiling point of methane. (c) Compute the standard enthalpy of vaporization of methane, given that the molar volumes of the liquid and vapour at the standard boiling point are 3.80×10^{-2} and 8.89 L mol^{-1}, respectively.

6.22‡ Diamond, an allotrope of carbon, is the hardest substance and the best conductor of heat yet characterized. For these reasons, diamond is used widely in industrial applications that require a strong abrasive. Unfortunately, it is difficult to synthesize diamond from the more readily available allotropes of carbon, such as graphite. To illustrate this point, calculated the pressure required to convert graphite into diamond at 25°C. The following data apply to 25°C and 100 kPa. Assume the specific volume, V_s, and κ_T are constant with respect to pressure changes.

	Graphite	Diamond
$\Delta_f G^{\ominus}/(kJ\ mol^{-1})$	0	+2.8678
$V_s/(cm^3\ g^{-1})$	0.444	0.284
κ_T/kPa	3.04×10^{-8}	0.187×10^{-8}

7

Simple mixtures

The thermodynamic description of mixtures

7.1 Partial molar quantities

7.2 The thermodynamics of mixing

7.3 The chemical potentials of liquids

The properties of solutions

7.4 Liquid mixtures

7.5 Colligative properties

Activities

7.6 The solvent activity

7.7 The solute activity

7.8 The activities of regular solutions

Checklist of key ideas

Further reading

Exercises

Problems

This chapter begins by developing the concept of chemical potential to show that it is a particular case of a class of properties called partial molar quantities. Then it explores how to use the chemical potential of a substance to describe the physical properties of a mixture. The underlying principle to keep in mind is that at equilibrium the chemical potential of a species is the same in every phase. We see, by making use of the experimental observations known as Raoult's and Henry's laws, how to express the chemical potential of a substance in terms of its mole fraction in a mixture. With this result established, we can calculate the effect of a solute on certain thermodynamic properties of a solution. These properties include the lowering of vapour pressure of the solvent, the elevation of its boiling point, the depression of its freezing point, and the origin of osmotic pressure. Finally, we see how to express the chemical potential of a substance in a real mixture in terms of a property known as the activity. We see how the activity may be measured, and conclude with a brief discussion of how the standard states of solutes and solvents are defined.

Chemistry deals with mixtures, including mixtures of substances that can react together. Therefore, we need to generalize the concepts introduced so far to deal with substances that are mingled together. As a first step towards dealing with chemical reactions (which are treated in Chapter 9), here we consider mixtures of substances that do not react together. At this stage we deal mainly with binary mixtures (mixtures of two components, A and B). We shall therefore often be able to simplify equations by making use of the relation $x_A + x_B = 1$. Another restriction of this chapter is that we consider mainly non-electrolyte solutions, in which the solute is not present as ions.

The thermodynamic description of mixtures

We have already seen that the partial pressure, which is the contribution of one component to the total pressure, is used to discuss the properties of mixtures of gases. For a more general description of the thermodynamics of mixtures we need to introduce other analogous 'partial' properties.

7.1 Partial molar quantities

The easiest partial molar property to visualize is the 'partial molar volume', the contribution that a component of a mixture makes to the total volume of a sample.

(a) Partial molar volume

Imagine a huge volume of pure water at 25°C. When a further 1 mol H_2O is added, the volume increases by 18 cm^3 and we can report that 18 cm^3 mol^{-1} is the molar volume of pure water. However, when we add 1 mol H_2O to a huge volume of pure ethanol, the volume increases by only 14 cm^3. The reason for the different increase in volume is that the volume occupied by a given number of water molecules depends on the identity of the molecules that surround them. In the latter case there is so much ethanol present that each H_2O molecule is surrounded by ethanol molecules, and the packing of the molecules results in the H_2O molecules increasing the volume by only 14 cm^3. The quantity 14 cm^3 mol^{-1} is the partial molar volume of water in pure ethanol. In general, the **partial molar volume** of a substance A in a mixture is the change in volume per mole of A added to a large volume of the mixture.

The partial molar volumes of the components of a mixture vary with composition because the environment of each type of molecule changes as the composition changes from pure A to pure B. It is this changing molecular environment, and the consequential modification of the forces acting between molecules, that results in the variation of the thermodynamic properties of a mixture as its composition is changed. The partial molar volumes of water and ethanol across the full composition range at 25°C are shown in Fig. 7.1.

The partial molar volume, V_J, of a substance J at some general composition is defined formally as follows:

$$V_J = \left(\frac{\partial V}{\partial n_J}\right)_{p,T,n'} \tag{7.1}$$

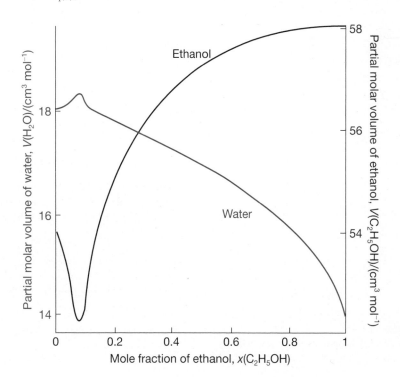

7.1 The partial molar volumes of water and ethanol at 25°C. Note the different scales (water on the left, ethanol on the right).

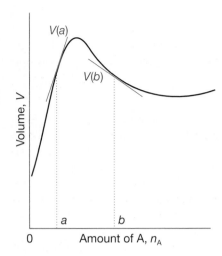

7.2 The partial molar volume of a substance is the slope of the variation of the total volume of the sample plotted against the composition. In general, partial molar quantities vary with the composition, as shown by the different slopes at the compositions a and b. Note that the partial molar volume at b is negative: the overall volume of the sample decreases as A is added.

where the subscript n' signifies that the amounts of all other substances present are constant.[1] The partial molar volume is the slope of the plot of the total volume as the amount of J is changed, the pressure, temperature, and amount of the other components being constant (Fig. 7.2). Its value depends on the composition, as we saw for water and ethanol. The definition in eqn 1 implies that, when the composition of the mixture is changed by the addition of dn_A of A and dn_B of B, then the total volume of the mixture changes by

$$dV = \left(\frac{\partial V}{\partial n_A}\right)_{p,T,n_B} dn_A + \left(\frac{\partial V}{\partial n_B}\right)_{p,T,n_A} dn_B = V_A\, dn_A + V_B\, dn_B \tag{7.2}$$

Once the partial molar volumes of the two components of a mixture at the composition (and temperature) of interest are known, we can state the total volume, V, of the mixture by using

$$V = n_A V_A + n_B V_B \qquad \textit{Euler's theorem.} \tag{7.3}$$

Justification 7.1

Consider a very large sample of the mixture of the specified composition. Then, when an amount n_A of A is added to the mixture, the composition remains virtually unchanged, the partial molar volume V_A is constant, and the volume of the sample changes by $n_A V_A$. When n_B of B is added, the volume changes by $n_B V_B$, for the same reason. The total change of volume is therefore $n_A V_A + n_B V_B$. The sample now occupies a larger volume, but the proportions of the components are still the same. At this stage, scoop out of the enlarged volume a sample containing n_A of A and n_B of B. Its volume is $n_A V_A + n_B V_B$. Because V is a state function, the same sample could have been prepared simply by mixing the appropriate amounts of A and B. This justifies eqn 3.

Partial molar volumes can be measured in several ways. One method is to measure the dependence of the volume on the composition and to fit the observed volume to a function of the amount of the substance. Once the function has been found, its slope can be determined at any composition of interest by differentiation.

7.3 The partial molar volume of ethanol as expressed by the polynomial in *Illustration 7.1*.

Illustration 7.1

The total volume of an ethanol solution at 25°C containing 1.000 kg of water is found to be given by the expression

$$V/\text{mL} = 1002.93 + 54.6664b - 0.363\,94b^2 + 0.028\,256b^3$$

where b is the numerical value of the molality. Because the amount of CH_3CH_2OH in moles is equal to the numerical value of the molality in moles per kilogram, we can write the partial molar volume of ethanol, V_E, as

$$V_E/(\text{mL mol}^{-1}) = \left(\frac{\partial(V/\text{mL})}{\partial b}\right)_{p,T,n_W} = 54.6664 - 2(0.363\,94)b + 3(0.028\,256)b^2$$

Figure 7.3 is a graph of this function.

1. The IUPAC recommendation is to denote a partial molar quantity by $\bar{X}$, but only when there is the possibility of confusion with the quantity X. For instance, the partial molar volume of NaCl in water could be written $\bar{V}(\text{NaCl, aq})$ to distinguish it from the volume of the solution, $V(\text{NaCl, aq})$.

Self-test 7.1 At 25°C, the density of a 50 per cent by mass ethanol/water solution is 0.914 g cm^{-3}. Given that the partial molar volume of water in the solution is 17.4 cm^3 mol^{-1}, what is the partial molar volume of the ethanol?

[56.4 cm^3 mol^{-1}]

Molar volumes are always positive, but partial molar quantities need not be. For example, the limiting partial molar volume of $MgSO_4$ in water (its partial molar volume in the limit of zero concentration) is -1.4 cm^3 mol^{-1}, which means that the addition of 1 mol $MgSO_4$ to a large volume of water results in a decrease in volume of 1.4 cm^3. The contraction occurs because the salt breaks up the open structure of water as the ions become hydrated, and it collapses slightly.

(b) Partial molar Gibbs energies

The concept of a partial molar quantity can be extended to any extensive state function. For a pure substance, the chemical potential is just another name for the molar Gibbs energy. For a substance in a mixture, the chemical potential is *defined* as being the partial molar Gibbs energy:

$$\mu_J = \left(\frac{\partial G}{\partial n_J} \right)_{p,T,n'} \tag{7.4}$$

That is, the chemical potential is the slope of a plot of Gibbs energy against the amount of the component J, with the pressure and temperature (and the amounts of the other substances) held constant (Fig. 7.4). By the same argument that led to eqn 3, it follows that the total Gibbs energy of a binary mixture is

$$G = n_A\mu_A + n_B\mu_B \tag{7.5}$$

where μ_A and μ_B are the chemical potentials at the composition of the mixture. That is, the chemical potential of a substance in a mixture is the contribution of that substance to the total Gibbs energy of the mixture.

The Gibbs energy depends on the composition, pressure, and temperature of the system. Thus, G may change when p, T, and the composition change and, for a system of components A, B, etc., the equation $dG = V\,dp - S\,dT$ becomes

$$dG = V\,dp - S\,dT + \mu_A\,dn_A + \mu_B\,dn_B + \cdots \tag{7.6}$$

This expression is the **fundamental equation of chemical thermodynamics**. Its implications and consequences are explored and developed in this and the next three chapters.

At constant pressure and temperature, eqn 6 simplifies to

$$dG = \mu_A\,dn_A + \mu_B\,dn_B + \cdots \tag{7.7}$$

We saw in Section 4.5e that under the same conditions $dG = dw_{add,max}$. Therefore, at constant temperature and pressure,

$$dw_{add,max} = \mu_A\,dn_A + \mu_B\,dn_B + \cdots \tag{7.8}$$

That is, additional (non-expansion) work can arise from the changing composition of a system. For instance, in an electrochemical cell, the chemical reaction is arranged to take place in two distinct sites (at the two electrodes). The electrical work the cell performs can be traced to its changing composition as products are formed from reactants.

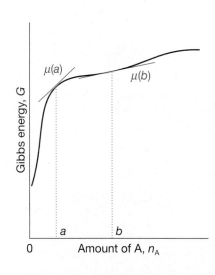

7.4 The chemical potential of a substance is the slope of the total Gibbs energy of a mixture with respect to the amount of substance of interest. In general, the chemical potential varies with composition, as shown for the two values at a and b. In this case, both chemical potentials are positive.

(c) The wider significance of the chemical potential

The chemical potential does more than show how G varies with composition. Because $G = U + pV - TS$, we can write a general infinitesimal change in U for a system of variable composition as

$$dU = -p\, dV - V\, dp + S\, dT + T\, dS + dG$$
$$= -p\, dV - V\, dp + S\, dT + T\, dS + (V\, dp - S\, dT + \mu_A\, dn_A + \mu_B\, dn_B + \cdots)$$
$$= -p\, dV + T\, dS + \mu_A\, dn_A + \mu_B\, dn_B + \cdots$$

This expression is the generalization of eqn 5.2 (that $dU = T\, dS - p\, dV$) to systems in which the composition may change. It follows that, at constant volume and entropy,

$$dU = \mu_A\, dn_A + \mu_B\, dn_B + \cdots$$

and hence that

$$\mu_J = \left(\frac{\partial U}{\partial n_J}\right)_{S,V,n'} \tag{7.9}$$

Therefore, not only does the chemical potential show how G changes when the composition changes, it also shows how the internal energy changes too (but under a different set of conditions). In the same way it is easy to deduce that

$$\text{(a)} \quad \mu_J = \left(\frac{\partial H}{\partial n_J}\right)_{S,p,n'} \qquad \text{(b)} \quad \mu_J = \left(\frac{\partial A}{\partial n_J}\right)_{V,T,n'} \tag{7.10}$$

Thus we see that the μ_J shows how all the extensive thermodynamic properties U, H, A, and G depend on the composition. This is why the chemical potential is so central to chemistry.

(d) The Gibbs–Duhem equation

Because the total Gibbs energy of a mixture is given by eqn 5 and the chemical potentials depend on the composition, when the compositions are changed infinitesimally we might expect G of a binary system to change by

$$dG = \mu_A\, dn_A + \mu_B\, dn_B + n_A\, d\mu_A + n_B\, d\mu_B$$

However, we have seen that at constant pressure and temperature a change in Gibbs energy is given by eqn 7. Because G is a state function, these two equations must be equal, which implies that at constant temperature and pressure

$$n_A\, d\mu_A + n_B\, d\mu_B = 0 \tag{7.11}$$

This equation is a special case of the **Gibbs–Duhem equation**:

$$\sum_J n_J\, d\mu_J = 0 \tag{7.12}$$

The significance of the Gibbs–Duhem equation is that the chemical potential of one component of a mixture cannot change independently of the chemical potentials of the other components. In a binary mixture, if one partial molar quantity increases, then the other must decrease:

$$d\mu_B = -\frac{n_A}{n_B}\, d\mu_A \tag{7.13}$$

The same line of reasoning applies to all partial molar quantities. We can see in Fig. 7.1, for example, that, where the partial molar volume of water increases, that of ethanol decreases. Moreover, as eqn 13 shows, and as we can see from Fig. 7.1, a small change in the partial molar volume of A corresponds to a large change in the partial molar volume of B if n_A/n_B is large, but the opposite is true when this ratio is small. In practice, the Gibbs–Duhem equation is used to determine the partial molar volume of one component of a binary mixture from measurements of the partial molar volume of the second component.

Example 7.1 Using the Gibbs–Duhem equation

The experimental values of the partial molar volume of $K_2SO_4(aq)$ at 298 K are given by the expression

$$V_{K_2SO_4}/(cm^3 \, mol^{-1}) = 32.280 + 18.216b^{1/2}$$

where b is the numerical value of the molality of K_2SO_4. Use the Gibbs–Duhem equation to derive an expression for the partial molar volume of water in the solution. The molar volume of pure water at 298 K is 18.079 cm³ mol⁻¹.

Method Let A denote K_2SO_4 and B denote H_2O, the solvent. The Gibbs–Duhem equation for the partial molar volumes of two components is

$$n_A \, dV_A + n_B \, dV_B = 0$$

This relation implies that

$$dV_B = -\frac{n_A}{n_B} \, dV_A$$

Therefore, V_B can be found by integration:

$$V_B = V_B^* - \int \frac{n_A}{n_B} \, dV_A$$

where V_B^* is the molar volume of pure B.

The first step is to change the variable V_A to the molality b, and then to integrate the right-hand side between $b = 0$ (pure B) and the molality of interest.

Answer It follows from the information in the question that, with A = K_2SO_4,

$$\frac{dV_A}{db} = 9.108b^{-1/2}$$

Therefore, the integration required is

$$V_B = V_B^* - 9.108 \int_0^b \frac{n_A}{n_B} b^{-1/2} \, db$$

However, we need to note that the molality is related to the amounts of A and B by

$$b = \frac{n_A}{n_B M_B}$$

where M_B is the molar mass of water (in kilograms per mole). Therefore,

$$V_B = V_B^* - 9.108 M_B \int_0^b b^{1/2} \, db$$

$$= V_B^* - \tfrac{2}{3}(9.108 M_B b^{3/2})$$

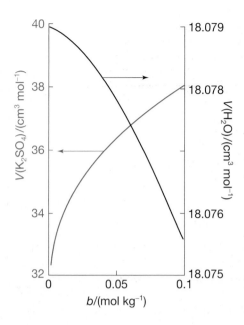

7.5 The partial molar volumes of the components of an aqueous solution of potassium sulfate.

It then follows, by substituting the data, that

$$V_B/(\text{cm}^3\,\text{mol}^{-1}) = 18.079 - 0.1094b^{3/2}$$

The partial molar volumes are plotted in Fig. 7.5.

Self-test 7.2 Repeat the calculation for a salt A for which $V_A/(\text{cm}^3\,\text{mol}^{-1}) = 6.218 + 5.146b - 7.147b^2$. $[V_B/(\text{cm}^3\,\text{mol}^{-1}) = 18.079 - 0.0464b^2 + 0.0859b^3]$

7.2 The thermodynamics of mixing

The dependence of the Gibbs energy of a mixture on its composition is given by eqn 5, and we know that at constant temperature and pressure systems tend towards lower Gibbs energy. This is the link we need in order to apply thermodynamics to the discussion of spontaneous changes of composition, as in the mixing of two substances. One simple example of a spontaneous mixing process is that of two gases introduced into the same container. The mixing is spontaneous, so it must correspond to a decrease in G. We shall now see how to express this idea quantitatively.

(a) The Gibbs energy of mixing

Let the amounts of two perfect gases in the two containers be n_A and n_B; both are at a temperature T and a pressure p. At this stage, the chemical potentials of the two gases have their 'pure' values, which are obtained by applying the definition $\mu = G_m$ to eqn 5.17:

$$\mu = \mu^\ominus + RT \ln \frac{p}{p^\ominus} \tag{7.14}°$$

where $\mu^\ominus$ is the **standard chemical potential**, the chemical potential of the pure gas at 1 bar.

The Gibbs energy of the total system is then given by eqn 5 as

$$G_i = n_A\mu_A + n_B\mu_B$$

$$= n_A\left(\mu_A^\ominus + RT\ln\frac{p}{p^\ominus}\right) + n_B\left(\mu_B^\ominus + RT\ln\frac{p}{p^\ominus}\right)$$

It will be much simpler notationally if we agree to let p denote the pressure relative to $p^\ominus$; that is, to replace $p/p^\ominus$ by p, for then we can write

$$G_i = n_A\{\mu_A^\ominus + RT\ln p\} + n_B\{\mu_B^\ominus + RT\ln p\} \tag{7.15}°$$

Equations for which this convention is used will be labelled {1}, {2}, ... ; to use the equations, we have to remember to replace p by $p/p^\ominus$ again. In practice, that simply means using the numerical value of p in bars. After mixing, the partial pressures of the gases are p_A and p_B, with $p_A + p_B = p$. The total Gibbs energy changes to

$$G_f = n_A\{\mu_A^\ominus + RT\ln p_A\} + n_B\{\mu_B^\ominus + RT\ln p_B\} \tag{7.16}°$$

The difference $G_f - G_i$, the Gibbs energy of mixing, $\Delta_{mix}G$, is therefore

$$\Delta_{mix}G = n_A RT\ln\frac{p_A}{p} + n_B RT\ln\frac{p_B}{p} \tag{7.17}°$$

At this point we may replace n_J by $x_J n$ and use Dalton's law (Section 1.2d) to write $p_J/p = x_J$ for each component, which gives

$$\Delta_{mix}G = nRT(x_A\ln x_A + x_B\ln x_B) \tag{7.18}°$$

7.6 The Gibbs energy of mixing of two perfect gases and (as discussed later) of two liquids that form an ideal solution. The Gibbs energy of mixing is negative for all compositions and temperatures, so perfect gases mix spontaneously in all proportions.

Because mole fractions are never greater than 1, the logarithms in this equation are negative, and $\Delta_{mix}G < 0$ (Fig. 7.6). The conclusion that $\Delta_{mix}G$ is negative for all compositions confirms that perfect gases mix spontaneously in all proportions. However, the equation extends common sense by allowing us to discuss the process quantitatively. We see, for instance, that $\Delta_{mix}G$ is directly proportional to the temperature but is independent of the total pressure.

Example 7.2 Calculating a Gibbs energy of mixing

A container is divided into two equal compartments (Fig. 7.7). One contains 3.0 mol H_2 at 25°C; the other contains 1.0 mol N_2 at 25°C. Calculate the Gibbs energy of mixing when the partition is removed. Assume perfect behaviour.

Method We proceed by calculating the initial Gibbs energy from the chemical potentials. To do so, we need the pressure of each gas. Write the pressure of nitrogen as p; then the pressure of hydrogen as a multiple of p can be found from the gas laws. Next, calculate the Gibbs energy for the system when the partition is removed. The volume of each gas doubles, so its partial pressure falls by a factor of 2.

Answer Given that the pressure of nitrogen is p, the pressure of hydrogen is $3p$; therefore, the initial Gibbs energy is

$$G_i = (3.0 \text{ mol})\{\mu^{\ominus}(H_2) + RT \ln 3p\} + (1.0 \text{ mol})\{\mu^{\ominus}(N_2) + RT \ln p\}$$

The partial pressure of nitrogen falls to $\frac{1}{2}p$ and that of hydrogen falls to $\frac{3}{2}p$. Therefore, the Gibbs energy changes to

$$G_f = (3.0 \text{ mol})\{\mu^{\ominus}(H_2) + RT \ln \tfrac{3}{2}p\} + (1.0 \text{ mol})\{\mu^{\ominus}(N_2) + RT \ln \tfrac{1}{2}p\}$$

The Gibbs energy of mixing is the difference of these two quantities:

$$\Delta_{mix}G = (3.0 \text{ mol})RT \ln \frac{\frac{3}{2}p}{3p} + (1.0 \text{ mol})RT \ln \frac{\frac{1}{2}p}{p}$$

$$= -(3.0 \text{ mol})RT \ln 2 - (1.0 \text{ mol})RT \ln 2$$

$$= -(4.0 \text{ mol})RT \ln 2 = -6.9 \text{ kJ}$$

Comment In this example, the value of $\Delta_{mix}G$ is the sum of two contributions: the mixing itself, and the changes in pressure of the two gases to their final pressure, $2p$. When 3.0 mol H_2 mixes with 1.0 mol N_2 at the same pressure, the change of Gibbs energy is -5.6 kJ and is independent of the initial common pressure.

Self-test 7.3 Suppose that 2.0 mol H_2 at 2.0 atm and 25°C and 4.0 mol N_2 at 3.0 atm and 25°C are mixed at constant volume. Calculate $\Delta_{mix}G$. What would be the value of $\Delta_{mix}G$ had the pressures been identical initially?

[−9.7 kJ, −9.5 kJ]

(b) Other thermodynamic mixing functions

Because $(\partial G/\partial T)_{p,n} = -S$, it follows immediately from eqn 18 that, for a mixture of perfect gases, the **entropy of mixing**, $\Delta_{mix}S$, is

$$\Delta_{mix}S = -\left(\frac{\partial \Delta_{mix}G}{\partial T}\right)_{p,n_A,n_B} = -nR(x_A \ln x_A + x_B \ln x_B) \tag{7.19}°$$

3.0 mol H_2	1.0 mol N_2
$3p$	p

3.0 mol H_2	1.0 mol N_2
$p(H_2) = 3p/2$	$p(N_2) = p/2$
	$2p$

7.7 The initial and final states considered in the calculation of the Gibbs energy of mixing of gases at different initial pressures.

7.8 The entropy of mixing of two perfect gases and (as discussed later) of two liquids that form an ideal solution. The entropy increases for all compositions and temperatures, so perfect gases mix spontaneously in all proportions. Because there is no transfer of heat to the surroundings when perfect gases mix, the entropy of the surroundings is unchanged. Hence, the graph also shows the total entropy of the system plus the surroundings when perfect gases mix.

Because $\ln x < 0$, it follows that $\Delta_{\text{mix}}S > 0$ for all compositions (Fig. 7.8). This increase in entropy is what we expect when one gas disperses into the other and the system becomes more disordered. The Gibbs energy of mixing in Example 7.2 was calculated as $-(4.0 \text{ mol})RT \ln 2$, so the corresponding entropy of mixing is $+(4.0 \text{ mol})R \ln 2$, or $+23 \text{ J K}^{-1}$.

We can calculate the isothermal, isobaric (constant pressure) **enthalpy of mixing**, $\Delta_{\text{mix}}H$, the enthalpy change accompanying mixing, of two perfect gases from $\Delta G = \Delta H - T\Delta S$. It follows from eqns 18 and 19 that

$$\Delta_{\text{mix}}H = 0 \tag{7.20}°$$

The enthalpy of mixing is zero, as we should expect for a system in which there are no interactions between the molecules forming the gaseous mixture. It follows that the whole of the driving force for mixing comes from the increase in entropy of the system, because the entropy of the surroundings is unchanged.

7.3 The chemical potentials of liquids

To discuss the equilibrium properties of liquid mixtures we need to know how the Gibbs energy of a liquid varies with composition. To calculate its value, we use the fact that, at equilibrium, the chemical potential of a substance present as a vapour must be equal to its chemical potential in the liquid.

(a) Ideal solutions

We shall denote quantities relating to pure substances by a superscript *, so the chemical potential of pure A is written μ_A^*, and as $\mu_A^*(l)$ when we need to emphasize that A is a liquid. Because the vapour pressure of the pure liquid is p_A^*, it follows from eqn 14 that the chemical potential of A in the vapour is $\mu_A^{\ominus} + RT \ln p_A^*$ (with p_A to be interpreted as the relative pressure $p_A/p^{\ominus}$). These two chemical potentials are equal at equilibrium (Fig. 7.9), so we can write

$$\mu_A^* = \mu_A^{\ominus} + RT \ln p_A^* \tag{7.21}$$

If another substance, a solute, is also present in the liquid, the chemical potential of A in the liquid is μ_A and its vapour pressure is p_A. In this case

$$\mu_A = \mu_A^{\ominus} + RT \ln p_A \tag{7.22}$$

Next, we combine these two equations to eliminate the standard chemical potential of the gas, and obtain

$$\mu_A = \mu_A^* + RT \ln \frac{p_A}{p_A^*} \tag{7.23}$$

In the final step we draw on additional experimental information about the relation between the ratio of vapour pressures and the composition of the liquid. In a series of experiments on mixtures of closely related liquids (such as benzene and methylbenzene), the French chemist François Raoult found that the ratio of the partial vapour pressure of each component to its vapour pressure as a pure liquid, p_A/p_A^*, is approximately equal to the mole fraction of A in the liquid mixture. That is, he established what we now call **Raoult's law**:

$$p_A = x_A p_A^* \tag{7.24}°$$

This law is illustrated in Fig. 7.10. Some mixtures obey Raoult's law very well, especially when the components are structurally similar (Fig. 7.11). Mixtures that obey the law throughout the composition range from pure A to pure B are called **ideal**

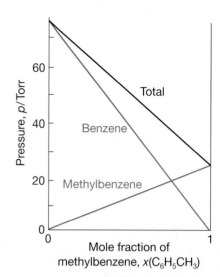

7.9 At equilibrium, the chemical potential of the gaseous form of a substance A is equal to the chemical potential of its condensed phase. The equality is preserved if a solute is also present. Because the chemical potential of A in the vapour depends on its partial vapour pressure, it follows that the chemical potential of liquid A can be related to its partial vapour pressure.

7.10 The total vapour pressure and the two partial vapour pressures of an ideal binary mixture are proportional to the mole fractions of the components.

7.11 Two similar liquids, in this case benzene and methylbenzene (toluene), behave almost ideally, and the variation of their vapour pressures with composition resembles that for an ideal solution.

solutions. When we write equations that are valid only for ideal solutions, we shall label them with a superscript °, as in eqn 24.

For an ideal solution, it follows from eqns 23 and 24 that

$$\mu_A = \mu_A^* + RT \ln x_A \qquad (7.25)°$$

This important equation can be used as the *definition* of an ideal solution (so that it implies Raoult's law rather than stemming from it). It is in fact a better definition than eqn 24 because it does not assume that the gas is perfect.

Molecular interpretation 7.1

The origin of Raoult's law can be understood in molecular terms by considering the rates at which molecules leave and return to the liquid. The law reflects the fact that the presence of a second component reduces the rate at which A molecules leave the surface of the liquid but does not inhibit the rate at which they return (Fig. 7.12).

The rate at which A molecules leave the surface is proportional to the number of them at the surface, which in turn is proportional to the mole fraction of A:

rate of vaporization = kx_A

where k is a constant of proportionality. The rate at which molecules condense is proportional to their concentration in the gas phase, which in turn is proportional to their partial pressure:

rate of condensation = $k'p_A$

At equilibrium, the rates of vaporization and condensation are equal, so

$$k'p_A = kx_A$$

It follows that

$$p_A = \frac{k}{k'}x_A$$

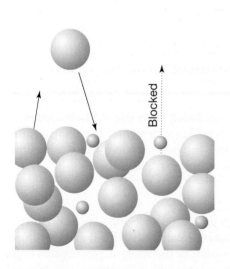

7.12 A pictorial representation of the molecular basis of Raoult's law. The large spheres represent solvent molecules at the surface of a solution (the uppermost line of spheres), and the small spheres are solute molecules. The latter hinder the escape of solvent molecules into the vapour, but do not hinder their return.

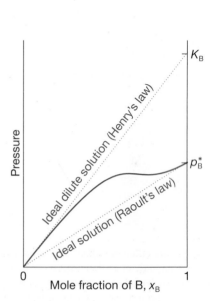

7.13 Strong deviations from ideality are shown by dissimilar liquids (in this case carbon disulfide and acetone (propanone)).

7.14 When a component (the solvent) is nearly pure, it has a vapour pressure that is proportional to the mole fraction with a slope p_B^* (Raoult's law). When it is the minor component (the solute) its vapour pressure is still proportional to the mole fraction, but the constant of proportionality is now K_B (Henry's law).

For the pure liquid, $x_A = 1$; so

$$p_A^* = \frac{k}{k'}$$

Equation 24 then follows by substitution of this relation into the previous one.

Some solutions depart significantly from Raoult's law (Fig. 7.13). Nevertheless, even in these cases the law is obeyed increasingly closely for the component in excess (the solvent) as it approaches purity. The law is therefore a good approximation for the properties of the solvent if the solution is dilute.

(b) Ideal-dilute solutions

In ideal solutions the solute, as well as the solvent, obeys Raoult's law. However, the English chemist William Henry found experimentally that, for real solutions at low concentrations, although the vapour pressure of the solute is proportional to its mole fraction, the constant of proportionality is not the vapour pressure of the pure substance (Fig. 7.14). **Henry's law** is:

$$p_B = x_B K_B \tag{7.26}°$$

In this expression x_B is the mole fraction of the solute and K_B is an empirical constant (with the dimensions of pressure) chosen so that the plot of the vapour pressure of B against its mole fraction is tangent to the experimental curve at $x_B = 0$.

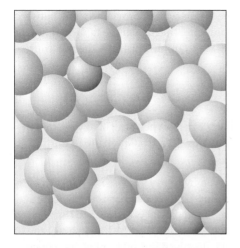

7.15 In a dilute solution, the solvent molecules (the blue spheres) are in an environment that differs only slightly from that of the pure solvent. The solute particles, however, are in an environment totally unlike that of the pure solute.

Mixtures for which the solute obeys Henry's law and the solvent obeys Raoult's law are called **ideal-dilute solutions**. We shall also label equations with a super-script ° when they have been derived from Henry's law.

Molecular interpretation 7.2

The difference in behaviour of the solute and solvent at low concentrations (as expressed by Henry's and Raoult's laws, respectively) arises from the fact that in a dilute solution the solvent molecules are in an environment very much like the one they have in the pure liquid (Fig. 7.15). In contrast, the solute molecules are surrounded by solvent molecules, which is entirely different from their environment when pure. Thus, the solvent behaves like a slightly modified pure liquid, but the solute behaves entirely differently from its pure state unless the solvent and solute molecules happen to be very similar. In the latter case, the solute also obeys Raoult's law.

Example 7.3 Investigating the validity of Raoult's and Henry's laws

The vapour pressures of each component in a mixture of propanone (acetone, A) and trichloromethane (chloroform, C) were measured at 35°C with the following results:

x_C	0	0.20	0.40	0.60	0.80	1
p_C/Torr	0	35	82	142	200	273
p_A/Torr	347	250	175	92	37	0

Confirm that the mixture conforms to Raoult's law for the component in large excess and to Henry's law for the minor component. Find the Henry's law constants.

Method Both Raoult's and Henry's laws are statements about the form of the graph of partial vapour pressure against mole fraction. Therefore, plot the partial vapour pressures against mole fraction. Raoult's law is tested by comparing the data with the straight line $p_J = x_J p_J^*$ for each component in the region in which it is in excess (and acting as the solvent). Henry's law is tested by finding a straight line $p_J = x_J K_J$ that is tangent to each partial vapour pressure at low x, where the component can be treated as the solute.

Answer The data are plotted in Fig. 7.16 together with the Raoult's law lines. Henry's law requires $K = 175$ Torr for propanone and $K = 165$ Torr for trichloromethane.

Comment Notice how the system deviates from both Raoult's and Henry's laws even for quite small departures from $x = 1$ and $x = 0$, respectively. We deal with these deviations in Section 7.5.

7.16 The experimental partial vapour pressures of a mixture of chloroform (trichloromethane) and acetone (propanone) based on the data in Example 7.3. The values of K are obtained by extrapolating the dilute solution vapour pressures as explained in the Example.

Self-test 7.4 The vapour pressure of chloromethane at various mole fractions in a mixture at 25°C was found to be as follows:

x	0.005	0.009	0.019	0.024
p/Torr	205	363	756	946

Estimate Henry's law constant.   [4×10^4 Torr]

The web site contains links to online databases of Henry's law constants.

Some Henry's law data are listed in Table 7.1. As well as providing a link between the mole fraction of solute and its partial pressure, the data in the table may also be used to calculate gas solubilities. The following example illustrates the procedure.

Table 7.1* Henry's law constants for gases in water at 298 K

	$K/(10\ \text{MPa})$	K/Torr
CO_2	0.167	1.25×10^6
H_2	7.12	5.34×10^7
N_2	8.68	6.51×10^7
O_2	4.40	3.30×10^7

* More values are given in the *Data section*.

Example 7.4 Using Henry's law

Estimate the molar solubility (the solubility in moles per litre) of oxygen in water at 25°C and a partial pressure of 160 Torr, its partial pressure in the atmosphere at sea level.

Method The mole fraction of solute is given by Henry's law as $x_J = p_J/K_J$, where p_J is the partial pressure of the gaseous solute J. All we need do is to calculate the mole fraction that corresponds to the stated partial pressure, and then interpret that mole fraction as a molar concentration. For the latter part of the calculation, we calculate the amount of O_2 dissolved in 1.00 kg of water (which corresponds to about 1.00 L water). The solution is dilute, so the expressions for the mole fraction can be simplified.

Answer Because the amount of O_2 dissolved is small, its mole fraction is

$$x_{O_2} = \frac{n_{O_2}}{n_{O_2} + n_{H_2O}} \approx \frac{n_{O_2}}{n_{H_2O}}$$

Hence,

$$n_{O_2} \approx x_{O_2} n_{H_2O} = \frac{p_{O_2} n_{H_2O}}{K} \approx \frac{(160\ \text{Torr}) \times (55.5\ \text{mol})}{3.30 \times 10^7\ \text{Torr}} = 2.69 \times 10^{-4}\ \text{mol}$$

The molality of the saturated solution is therefore $2.69 \times 10^{-4}\ \text{mol kg}^{-1}$, corresponding to a molar concentration of approximately $2.7 \times 10^{-4}\ \text{mol L}^{-1}$.

Self-test 7.5 Calculate the molar solubility of nitrogen in water exposed to air at 25°C; partial pressures were calculated in Example 1.5.

[0.51 mmol L^{-1}]

A knowledge of Henry's law constants for gases in blood and fats is important for the discussion of respiration, especially when the partial pressure of oxygen is abnormal, as in diving and mountaineering, and for the discussion of the action of gaseous anaesthetics. For example, the molar solubility of O_2 gas in blood decreases with decreasing atmospheric pressure and this impairs respiration at high altitudes. Conversely, the solubility of air in blood increases with increasing pressures. For this reason, a diver who has been underwater for a long period of time cannot be returned to the surface too rapidly: the sudden decrease in pressure lowers the solubility of gases and forms bubbles (emboli) in the bloodstream, with potentially fatal consequences.

The properties of solutions

In this section we consider the thermodynamics of mixing of liquids. First, we consider the simple case of mixtures of liquids that mix to form an ideal solution. In this way, we identify the thermodynamic consequences of molecules of one species mingling randomly with molecules of the second species. The calculation provides a background for discussing the deviations from ideal behaviour exhibited by real solutions.

7.4 Liquid mixtures

Thermodynamics can provide insight into the properties of liquid mixtures, and a few simple ideas can bring the whole field of study together.

(a) Ideal solutions

The Gibbs energy of mixing of two liquids to form an ideal solution is calculated in exactly the same way as for two gases (Section 7.2). The total Gibbs energy before liquids are mixed is

$$G_i = n_A \mu_A^* + n_B \mu_B^*$$

When they are mixed, the individual chemical potentials are given by eqn 25 and the total Gibbs energy is

$$G_f = n_A\{\mu_A^* + RT \ln x_A\} + n_B\{\mu_B^* + RT \ln x_B\}$$

Consequently, the Gibbs energy of mixing is

$$\Delta_{mix}G = nRT\{x_A \ln x_A + x_B \ln x_B\} \tag{7.27}°$$

where $n = n_A + n_B$. As for gases, it follows that the ideal entropy of mixing of two liquids is

$$\Delta_{mix}S = -nR\{x_A \ln x_A + x_B \ln x_B\} \tag{7.28}°$$

and the ideal enthalpy of mixing is zero.

Equation 27 is the same as that for two perfect gases and all the conclusions drawn there are valid here: the driving force for mixing is the increasing entropy of the system as the molecules mingle and the enthalpy of mixing is zero. It should be noted, however, that solution ideality means something different from gas perfection. In a perfect gas there are no interactions between molecules. In ideal solutions there are interactions, but the average A–B interactions in the mixture are the same as the average A–A and B–B interactions in the pure liquids. The variation of the Gibbs energy of mixing with composition is the same as that already depicted for gases in Fig. 7.6; the same is true of the entropy of mixing, Fig. 7.8.

Real solutions are composed of particles for which A–A, A–B, and B–B interactions are all different. Not only may there be an enthalpy change when liquids mix, but there may also be an additional contribution to the entropy arising from the way in which the molecules of one type might cluster together instead of mingling freely with the others. If the enthalpy change is large and positive or if the entropy change is adverse (because of a reorganization of the molecules that results in an orderly mixture), then the Gibbs energy might be positive for mixing. In that case, separation is spontaneous and the liquids may be immiscible. Alternatively, the liquids might be **partially miscible**, which means that they are miscible only over a certain range of compositions.

(b) Excess functions and regular solutions

The thermodynamic properties of real solutions are expressed in terms of the **excess functions**, X^E, the difference between the observed thermodynamic function of mixing and the function for an ideal solution. The **excess entropy**, S^E, for example, is defined as

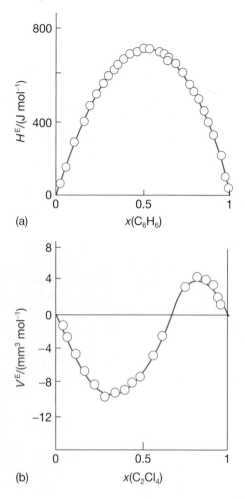

(a) $x(C_6H_6)$

(b) $x(C_2Cl_4)$

7.17 Experimental excess functions at 25°C. (a) H^E for benzene/cyclohexane; this graph shows that the mixing is endothermic (because $\Delta_{mix}H = 0$ for an ideal solution). (b) The excess volume, V^E, for tetrachloroethane/cyclopentane; this graph shows that there is a contraction at low tetrachloroethane mole fractions, but an expansion at high mole fractions (because $\Delta_{mix}V = 0$ for an ideal mixture).

$$S^E = \Delta_{mix}S - \Delta_{mix}S^{ideal} \qquad [7.29]$$

where $\Delta_{mix}S^{ideal}$ is given by eqn 28. The excess enthalpy and volume are both equal to the observed enthalpy and volume of mixing, because the ideal values are zero in each case.

Deviations of the excess energies from zero indicate the extent to which the solutions are nonideal. In this connection a useful model system is the **regular solution**, a solution for which $H^E \neq 0$ but $S^E = 0$. We can think of a regular solution as one in which the two kinds of molecules are distributed randomly (as in an ideal solution) but have different energies of interactions with each other. Figure 7.17 shows two examples of the composition dependence of excess functions.

We can make this discussion more quantitative by supposing that the excess enthalpy depends on composition as

$$H^E = n\beta RT x_A x_B \qquad (7.30)$$

where β is a parameter. We let $\beta = w/RT$, where w is an indication of the energy of AB interactions relative to AA and BB interactions. The function given by eqn 30 is plotted in Fig. 7.18, and we see it resembles the experimental curve in Fig. 7.17. If $\beta < 0$, mixing is exothermic and the solute–solvent interactions are more favourable than the solvent–solvent and solute–solute interactions. If $\beta > 0$, then the mixing is endothermic. Because the entropy of mixing has its ideal value for a regular solution, the excess Gibbs energy is equal to the excess enthalpy, and the Gibbs energy of mixing is

$$\Delta_{mix}G = nRT\{x_A \ln x_A + x_B \ln x_B + \beta x_A x_B\} \qquad (7.31)$$

Figure 7.19 shows how $\Delta_{mix}G$ varies with composition for different values of β. The important feature is that for $\beta > 2$ the graph shows two minima separated by a maximum. The implication of this observation is that, provided $\beta > 2$, then the system will separate spontaneously into two phases with compositions corresponding to the two minima, for that separation corresponds to a reduction in Gibbs energy. We take up this point in Sections 7.8 and 8.5.

x_A

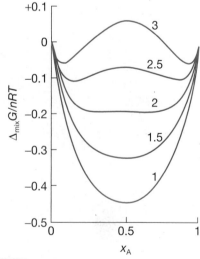

x_A

7.18 The excess enthalpy according to a model in which it is proportional to $\beta x_A x_B$, for different values of the parameter β.

7.19 The Gibbs energy of mixing for different values of the parameter β.

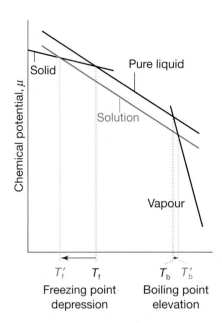

7.20 The chemical potential of a solvent in the presence of a solute. The lowering of the liquid's chemical potential has a greater effect on the freezing point than on the boiling point because of the angles at which the lines intersect (which are determined by entropies; recall Fig. 6.8).

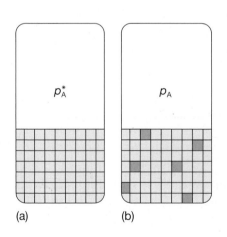

7.21 The vapour pressure of a pure liquid represents a balance between the increased disorder arising from vaporization and the decreased disorder of the surroundings. (a) Here the structure of the liquid is represented highly schematically by the grid of squares. (b) When solute (the dark squares) is present, the disorder of the condensed phase is relatively higher than that of the pure liquid, and there is a decreased tendency to acquire the disorder characteristic of the vapour.

7.5 Colligative properties

The properties we now consider are the elevation of boiling point, the depression of freezing point, and the osmotic pressure arising from the presence of a solute. In dilute solutions these properties depend only on the number of solute particles present, not their identity. For this reason, they are called **colligative properties** (denoting 'depending on the collection').

We assume throughout the following that the solute is not volatile, so it does not contribute to the vapour. We also assume that the solute does not dissolve in the solid solvent: that is, the pure solid solvent separates when the solution is frozen. The latter assumption is quite drastic, although it is true of many mixtures; it can be avoided at the expense of more algebra, but that introduces no new principles.

(a) The common features of colligative properties

All the colligative properties stem from the reduction of the chemical potential of the liquid solvent as a result of the presence of solute. The reduction is from μ_A^* for the pure solvent to $\mu_A^* + RT \ln x_A$ when a solute is present ($\ln x_A$ is negative because $x_A < 1$). There is no direct influence of the solute on the chemical potential of the solvent vapour and the solid solvent because the solute appears in neither the vapour nor the solid. As can be seen from Fig. 7.20, the reduction in chemical potential of the solvent implies that the liquid–vapour equilibrium occurs at a higher temperature (the boiling point is raised) and the solid–liquid equilibrium occurs at a lower temperature (the freezing point is lowered).

Molecular interpretation 7.3

The molecular origin of the lowering of the chemical potential is not the energy of interaction of the solute and solvent particles, because the lowering occurs even in an ideal solution (for which the enthalpy of mixing is zero). If it is not an enthalpy effect, it must be an entropy effect.

The pure liquid solvent has an entropy that reflects the disorder of its molecules. Its vapour pressure reflects the tendency of the solution towards greater entropy, which can be achieved if the liquid vaporizes to form a more disordered gas. When a solute is present, there is an additional contribution to the entropy of the liquid, even in an ideal solution. Because the entropy of the liquid is already higher than that of the pure liquid, there is a weaker tendency to form the gas (Fig. 7.21). The effect of the solute appears as a lowered vapour pressure, and hence a higher boiling point.

Similarly, the enhanced molecular randomness of the solution opposes the tendency to freeze. Consequently, a lower temperature must be reached before equilibrium between solid and solution is achieved. Hence, the freezing point is lowered.

The strategy for the quantitative discussion of the elevation of boiling point and the depression of freezing point is to look for the temperature at which, at 1 atm, one phase (the pure solvent vapour or the pure solid solvent) has the same chemical potential as the solvent in the solution. This is the new equilibrium temperature for the phase transition at 1 atm, and hence corresponds to the new boiling point or the new freezing point of the solvent.

7.22 The heterogeneous equilibrium involved in the calculation of the elevation of boiling point is between A in the pure vapour and A in the mixture, A being the solvent and B an involatile solute.

(b) The elevation of boiling point

The heterogeneous equilibrium of interest when considering boiling is between the solvent vapour and the solvent in solution at 1 atm (Fig. 7.22). We denote the solvent by A and the solute by B. The equilibrium is established at a temperature for which

$$\mu_A^*(g) = \mu_A^*(l) + RT \ln x_A \tag{7.32}$$

(The pressure of 1 atm is the same throughout, and will not be written explicitly.) We show in the *Justification* below that this equation implies that the presence of a solute at a mole fraction x_B causes an increase in normal boiling point from T^* to $T^* + \Delta T$, where

$$\Delta T = K x_B \qquad K = \frac{RT^{*2}}{\Delta_{vap}H} \tag{7.33}°$$

Justification 7.2

Equation 32 rearranges into

$$\ln(1 - x_B) = \frac{\mu_A^*(g) - \mu_A^*(l)}{RT} = \frac{\Delta_{vap}G}{RT}$$

where $\Delta_{vap}G$ is the Gibbs energy of vaporization of the pure solvent (A) and x_B is the mole fraction of the solute; we have used $x_A + x_B = 1$. We now write

$$\Delta_{vap}G = \Delta_{vap}H - T\Delta_{vap}S$$

and ignore the small temperature dependence of $\Delta_{vap}H$ and $\Delta_{vap}S$. Then,

$$\ln(1 - x_B) = \frac{\Delta_{vap}H}{RT} - \frac{\Delta_{vap}S}{R}$$

When $x_B = 0$, the boiling point is that of pure liquid A, T^*, and

$$\ln 1 = \frac{\Delta_{vap}H}{RT^*} - \frac{\Delta_{vap}S}{R}$$

Because $\ln 1 = 0$, the difference of the two equations is

$$\ln(1 - x_B) = \frac{\Delta_{vap}H}{R}\left(\frac{1}{T} - \frac{1}{T^*}\right)$$

We now suppose that the amount of solute present is so small that $x_B \ll 1$. We can then write $\ln(1 - x_B) \approx -x_B$ and hence obtain

$$x_B = \frac{\Delta_{vap}H}{R}\left(\frac{1}{T^*} - \frac{1}{T}\right)$$

Finally, because $T \approx T^*$, it also follows that

$$\frac{1}{T^*} - \frac{1}{T} = \frac{T - T^*}{TT^*} \approx \frac{\Delta T}{T^{*2}}$$

with $\Delta T = T - T^*$. The previous equation then rearranges into eqn 33.

Because eqn 33 makes no reference to the identity of the solute, only to its mole fraction, we conclude that the elevation of boiling point is a colligative property. The

Table 7.2* Cryoscopic and ebullioscopic constants

	$K_f/$ (K/(mol kg^{-1}))	$K_b/$ (K/(mol kg^{-1}))
Benzene	5.12	2.53
Camphor	40	
Phenol	7.27	3.04
Water	1.86	0.51

* More values are given in the *Data section*.

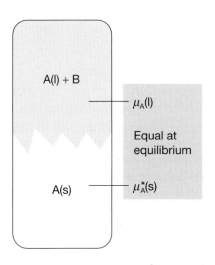

A(l) + B

$\mu_A(l)$

Equal at equilibrium

A(s)

$\mu_A^*(s)$

7.23 The heterogeneous equilibrium involved in the calculation of the lowering of freezing point is between A in the pure solid and A in the mixture, A being the solvent and B a solute that is insoluble in solid A.

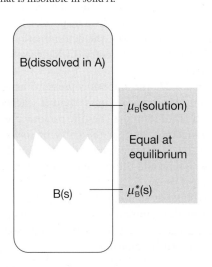

B(dissolved in A)

$\mu_B(solution)$

Equal at equilibrium

B(s)

$\mu_B^*(s)$

7.24 The heterogeneous equilibrium involved in the calculation of the solubility is between pure solid B and B in the mixture.

value of ΔT does depend on the properties of the solvent, and the biggest changes occur for solvents with high boiling points.[2] For practical applications of eqn 33, we note that the mole fraction of B is proportional to its molality, b, in the solution, and write

$$\Delta T = K_b b \qquad (7.34)$$

where K_b is the empirical **ebullioscopic constant** of the solvent (Table 7.2).

(c) The depression of freezing point

The heterogeneous equilibrium now of interest is between pure solid solvent A and the solution with solute present at a mole fraction x_B (Fig. 7.23). At the freezing point, the chemical potentials of A in the two phases are equal:

$$\mu_A^*(s) = \mu_A^*(l) + RT \ln x_A \qquad (7.35)$$

The only difference between this calculation and the last is the appearance of the solid's chemical potential in place of the vapour's. Therefore we can write the result directly from eqn 33:

$$\Delta T = K' x_B \qquad K' = \frac{RT^{*2}}{\Delta_{fus}H} \qquad (7.36)°$$

where ΔT is the freezing point depression, $T^* - T$, and $\Delta_{fus}H$ is the enthalpy of fusion of the solvent. Larger depressions are observed in solvents with low enthalpies of fusion and high melting points. When the solution is dilute, the mole fraction is proportional to the molality of the solute, b, and it is common to write the last equation as

$$\Delta T = K_f b \qquad (7.37)$$

where K_f is the empirical **cryoscopic constant** (Table 7.2). Once the cryoscopic constant of a solvent is known, the depression of freezing point may be used to measure the molar mass of a solute in the method known as cryoscopy; however, the technique is of little more than historical interest.

(d) Solubility

Although it is not strictly a colligative property, the solubility of a solute may be estimated by the same techniques as we have been using. When a solid solute is left in contact with a solvent, it dissolves until the solution is saturated. Saturation is a state of equilibrium, with the undissolved solute in equilibrium with the dissolved solute. Therefore, in a saturated solution the chemical potential of the pure solid solute, $\mu_B^*(s)$, and the chemical potential of B in solution, μ_B, are equal (Fig. 7.24). Because the latter is

$$\mu_B = \mu_B^*(l) + RT \ln x_B$$

we can write

$$\mu_B^*(s) = \mu_B^*(l) + RT \ln x_B \qquad (7.38)$$

This expression is the same as the starting equation of the last section, except that the quantities refer to the solute B, not the solvent A.

2. By Trouton's rule (Section 4.3a), $\Delta_{vap}H/T^*$ is a constant; therefore eqn 33 has the form $\Delta T \propto T^*$ and is independent of $\Delta_{vap}H$ itself.

7.25 The variation of solubility (the mole fraction of solute in a saturated solution) with temperature (T^* is the freezing temperature of the solute). Individual curves are labelled with the value of $\Delta_{fus}H/RT^*$.

The starting point is the same but the aim is different. In the present case, we want to find the mole fraction of B in solution at equilibrium when the temperature is T. Therefore, we start by rearranging the last equation into

$$\ln x_B = \frac{\mu_B^*(s) - \mu_B^*(l)}{RT} = -\frac{\Delta_{fus}G}{RT}$$

$$= -\frac{\Delta_{fus}H}{RT} + \frac{\Delta_{fus}S}{R}$$

At the melting point of the solute, T^*, we know that $\Delta_{fus}G = 0$, so $\Delta_{fus}G/RT^* = 0$ too; consequently, this term may be added to the right-hand side to give

$$\ln x_B = -\frac{\Delta_{fus}H}{RT} + \frac{\Delta_{fus}S}{R} + \frac{\Delta_{fus}H}{RT^*} - \frac{\Delta_{fus}S}{R}$$

$$= -\frac{\Delta_{fus}H}{RT} + \frac{\Delta_{fus}H}{RT^*}$$

It then follows that

$$\ln x_B = \frac{\Delta_{fus}H}{R}\left(\frac{1}{T^*} - \frac{1}{T}\right) \tag{7.39}°$$

Equation 39 is plotted in Fig. 7.25. It shows that the solubility of B decreases exponentially as the temperature is lowered from its melting point. The illustration also shows that solutes with high melting points and large enthalpies of melting have low solubilities at normal temperatures. However, the detailed content of eqn 39 should not be treated too seriously because it is based on highly questionable approximations, such as the ideality of the solution. One aspect of its approximate character is that it fails to predict that solutes will have different solubilities in different solvents, for no solvent properties appear in the expression.

(e) Osmosis

The phenomenon of **osmosis** (from the Greek word for 'push') is the spontaneous passage of a pure solvent into a solution separated from it by a **semipermeable membrane**, a membrane permeable to the solvent but not to the solute (Fig. 7.26). The **osmotic pressure**, Π, is the pressure that must be applied to the solution to stop the influx of solvent. Important examples of osmosis include transport of fluids through cell membranes, dialysis (Box 7.1), and **osmometry**, the determination of molar mass by the measurement of osmotic pressure. Osmometry is widely used to determine the molar masses of macromolecules.

In the simple arrangement shown in Fig. 7.27, the opposing pressure arises from the head of solution that the osmosis itself produces. Equilibrium is reached when the hydrostatic pressure of the column of solution matches the osmotic pressure. The complicating feature of this arrangement is that the entry of solvent into the solution results in its dilution, and so it is more difficult to treat than the arrangement in Fig. 7.26, in which there is no flow and the concentrations remain unchanged.

The thermodynamic treatment of osmosis depends on noting that, at equilibrium, the chemical potential of the solvent must be the same on each side of the membrane. As shown in the *Justification* below, this equality implies that for dilute solutions the osmotic pressure is given by the **van 't Hoff equation**:

$$\Pi = [B]RT \tag{7.40}°$$

where $[B] = n_B/V$ is the molar concentration of the solute.

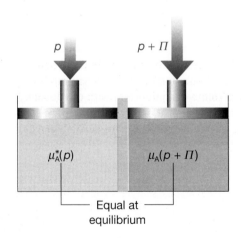

7.26 The equilibrium involved in the calculation of osmotic pressure, Π, is between pure solvent A at a pressure p on one side of the semipermeable membrane and A as a component of the mixture on the other side of the membrane, where the pressure is $p + \Pi$.

7.27 In a simple version of the osmotic pressure experiment, A is at equilibrium on each side of the membrane when enough has passed into the solution to cause a hydrostatic pressure difference.

Justification 7.3

On the pure solvent side the chemical potential of the solvent, which is at a pressure p, is $\mu_A^*(p)$. On the solution side, the chemical potential is lowered by the presence of the solute, which reduces the mole fraction of the solvent from 1 to x_A. However, the chemical potential of A is raised on account of the greater pressure, $p + \Pi$, that the solution experiences. At equilibrium the chemical potential of A is the same in both compartments, and we can write

$$\mu_A^*(p) = \mu_A(x_A, p + \Pi)$$

The presence of solute is taken into account in the normal way:

$$\mu_A(x_A, p + \Pi) = \mu_A^*(p + \Pi) + RT \ln x_A$$

We saw in Section 5.5 (eqn 5.14) how to take the effect of pressure into account:

$$\mu_A^*(p + \Pi) = \mu_A^*(p) + \int_p^{p+\Pi} V_m \, dp$$

where V_m is the molar volume of the pure solvent A. When these three equations are combined we get

$$-RT \ln x_A = \int_p^{p+\Pi} V_m \, dp$$

For dilute solutions, $\ln x_A$ may be replaced by $\ln(1 - x_B) \approx -x_B$. We may also assume that the pressure range in the integration is so small that the molar volume of the solvent is a constant. That being so, V_m may be taken outside the integral, giving

$$RTx_B = \Pi V_m$$

When the solution is dilute, $x_B \approx n_B/n_A$. Moreover, because $n_A V_m = V$, the total volume of the solvent, the equation simplifies to eqn 40.

Box 7.1 **Dialysis and protein binding**

In a dialysis experiment, a solution of macromolecules and smaller ions is placed in a bag made of a material that acts as a semipermeable membrane and the filled bag is immersed in a solvent. The membrane permits the passage of the small ions but not the macromolecules, so the former migrate through the membrane, leaving the macromolecules behind. Dialysis is used to study the binding of small molecules to macromolecules, such as an inhibitor to an enzyme, an antibiotic to DNA, and any other instance of cooperation or inhibition by small molecules attaching to large ones.

Suppose the molar concentration of the macromolecule M is [M] and the total concentration of small molecule A in the compartment containing the macromolecule is $[A]_{in}$. This total concentration is the sum of the concentrations of free A and bound A, which we write $[A]_{free}$ and $[A]_{bound}$, respectively. At equilibrium, the chemical potential of free A in the macromolecule solution is equal to the chemical potential of A in the solution on the other side of the membrane, where its concentration is $[A]_{out}$. The equality $\mu_{A,free} = \mu_{A,out}$ implies that $[A]_{free} = [A]_{out}$, provided the activity coefficient of A is the same in both solutions. Therefore, by measuring the concentration of A in the 'outside' solution, we can find the concentration of unbound A in the macromolecule solution and, from the difference $[A]_{in} - [A]_{free}$, which is equal to $[A]_{in} - [A]_{out}$, the concentration of bound A. The average number of A molecules bound to M molecules, ν, is then the ratio

$$\nu = \frac{[A]_{bound}}{[M]} = \frac{[A]_{in} - [A]_{out}}{[M]}$$

The bound and unbound A molecules are in equilibrium, $M + A \rightleftharpoons MA$, so their concentrations are related by an equilibrium constant K (see Chapter 9), where

$$K = \frac{[MA]}{[M]_{free}[A]_{free}} = \frac{[A]_{bound}}{([M] - [A]_{bound})[A]_{free}}$$

On division by [M], and replacement of $[A]_{free}$ by $[A]_{out}$, this expression becomes

$$K = \frac{v}{(1 - v)[A]_{out}}$$

If there are N *identical* and *independent* binding sites on each macromolecule, each macromolecule behaves like N separate smaller macromolecules, with the same value of K for each site. The average number of A molecules per site is v/N, so the last equation becomes

$$K = \frac{v/N}{\left(1 - \frac{v}{N}\right)[A]_{out}}$$

It then follows that

$$\frac{v}{[A]_{out}} = KN - Kv$$

This expression is the *Scatchard equation*. It implies that a plot of $v/[A]_{out}$ against v should be a straight line of slope $-K$ and intercept KN at $v = 0$ (see the illustration). From these two quantities, we can find the equilibrium constant for binding, the enthalpy of binding (from the temperature dependence of K and the van 't Hoff equation, Chapter 9), and the number of binding sites on each macromolecule. If a straight line is not obtained we can conclude that the binding sites are not equivalent or independent.

Discussion questions

1. The *Donnan equilibrium* determines the distribution of ions between two solutions in contact through a semipermeable membrane, in one of which there is a polyelectrolyte (a molecule bearing many charged groups, such as DNA). Consider the dialysis of a solution of a polyelectrolyte Na_vP, where the membrane is not permeable to the large anion P^{v-}. In such experiments, it is customary to add a high concentration of a salt such as NaCl to the solution inside and outside of the dialysis bag, so that the number of cations Na_vP provides is insignificant in comparison with the number supplied by the additional salt. Show that at equilibrium

$$[Na^+]_{in} - [Na^+]_{out} = \frac{v[P^{v-}][Na^+]_{in}}{2[Cl^-] + v[P^{v-}]}$$

$$[Cl^-]_{in} - [Cl^-]_{out} = -\frac{v[P^{v-}][Cl^-]_{in}}{2[Cl^-]}$$

where $[Cl^-] = \frac{1}{2}([Cl^-]_{in} + [Cl^-]_{out})$. Note that Na^+ ions accumulate inside the dialysis bag, which contains P^{v-}, and from a measurement of the ion concentrations it is possible to determine the net charge of the anion. *Hint.* Electrical neutrality must be preserved in the bulk on both sides of the membrane, so that $[Na^+]_{in} = [Cl^-]_{in} + v[P^{v-}]$ and $[Na^+]_{out} = [Cl^-]_{out}$.

2. For the experiment described in the previous question, show that the osmotic pressure is

$$\Pi = RT[P^{v-}](1 + B[P^{v-}]) \qquad B = \frac{v^2[Cl^-]_{out}}{4[Cl^-] + 2v[Cl^-]_{in}[P^{v-}]}$$

Finally, show that, when $[Cl^-]_{in}$ and $[Cl^-]_{out}$ are large relative to $[P^{v-}]$, $\Pi = RT[P^{v-}]$. Note that the osmotic pressure is independent of the value of v.

Further reading

K.E. van Holde, W.C. Johnson, and P.S. Ho, *Principles of physical biochemistry*. Prentice Hall, NJ (1998).

C.R. Cantor and P.R. Schimmel, *Biophysical chemistry, part III: The behavior of biological macromolecules*. W.H. Freeman, San Francisco (1980).

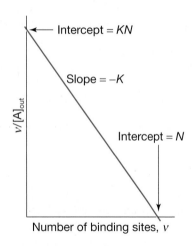

A Scatchard plot of $v/[A]_{out}$ against v. The slope is $-K$ and the intercept at $v = 0$ is KN.

Because the effect of osmotic pressure is so readily measurable and large, one of the most common applications of osmometry is to the measurement of molar masses of macromolecules, such as proteins and synthetic polymers. As these huge molecules dissolve to produce solutions that are far from ideal, it is assumed that the van 't Hoff equation is only the first term of a virial-like expansion:

$$\Pi = [B]RT\{1 + B[B] + \cdots\} \qquad (7.41)$$

7.28 The plot involved in the determination of molar mass by osmometry. The molar mass is calculated from the intercept at $c = 0$; in Chapter 23 we shall see that additional information comes from the slope.

The additional terms take the nonideality into account; the empirical constant B is called the **osmotic virial coefficient**.

Example 7.5 Using osmometry to determine the molar mass of a macromolecule

The osmotic pressures of solutions of poly(vinyl chloride), PVC, in cyclohexanone at 298 K are given below. The pressures are expressed in terms of the heights of solution (of mass density $\rho = 0.980$ g cm^{-3}) in balance with the osmotic pressure. Determine the molar mass of the polymer.

$c/(\text{g L}^{-1})$	1.00	2.00	4.00	7.00	9.00
h/cm	0.28	0.71	2.01	5.10	8.00

Method The osmotic pressure is measured at a series of mass concentrations, c, and a plot of Π/c against c is used to determine the molar mass of the polymer. We use eqn 41 with $[B] = c/M$ where c is the mass concentration of the polymer and M is its molar mass. The osmotic pressure is related to the hydrostatic pressure by $\Pi = \rho g h$ (Example 1.2) with $g = 9.81$ m s^{-2}. With these substitutions, eqn 41 becomes

$$\frac{h}{c} = \frac{RT}{\rho g M}\left(1 + \frac{Bc}{M} + \cdots\right)$$

$$= \frac{RT}{\rho g M} + \left(\frac{RTB}{\rho g M^2}\right)c + \cdots$$

Therefore, to find M, plot h/c against c, and expect a straight line with intercept $RT/\rho g M$ at $c = 0$.

Answer The data give the following values for the quantities to plot:

$c/(\text{g L}^{-1})$	1.00	2.00	4.00	7.00	9.00
$(h/c)/(\text{cm g}^{-1}\text{ L})$	0.28	0.36	0.503	0.729	0.889

The points are plotted in Fig. 7.28. The intercept is at 0.21. Therefore,

$$M = \frac{RT}{\rho g} \times \frac{1}{0.21\ \text{cm g}^{-1}\text{ L}}$$

$$= \frac{(8.3145\ \text{J K}^{-1}\text{ mol}^{-1}) \times (298\ \text{K})}{(980\ \text{kg m}^{-3}) \times (9.81\ \text{m s}^{-2})} \times \frac{1}{2.1 \times 10^{-3}\ \text{m}^4\ \text{kg}^{-1}}$$

$$= 1.2 \times 10^2\ \text{kg mol}^{-1}$$

where we have used 1 kg m^2 s^{-2} = 1 J.

Comment Molar masses of macromolecules are often reported in daltons (Da), with 1 Da = 1 g mol^{-1}. The macromolecule in this example has a molar mass of about 120 kDa. Modern osmometers give readings of osmotic pressure in pascals, so the analysis of the data is more straightforward and eqn 41 can be used directly.

Self-test 7.6 Estimate the depression of freezing point of the most concentrated of these solutions, taking K_f as about 10 K/(mol kg^{-1}).   [0.8 mK]

Syringe

Thermistor

Solvent

Insulation

7.29 A vapour-phase osmometer. The syringes introduce droplets of solvent and solution, respectively, on to the thermistors, and the difference in temperature (which arises from the different rates of vaporization) is noted.

(f) Vapour-phase osmometry

In the technique of **vapour-phase osmometry**, a droplet of solution is placed on one thermistor (a semiconductor temperature probe) and a droplet of pure solvent is placed on another thermistor (Fig. 7.29). The two droplets are surrounded by an atmosphere of solvent vapour. Because the vapour pressure of a solvent in solution is lower than when it is pure, the net rate of condensation on to a droplet of solution is greater than the rate of condensation on to a droplet of pure solvent. It follows that more heat is liberated in the solution droplet, and the rise in temperature is greater there than in the droplet of pure solvent. The difference in temperature is measured for a series of concentrations and extrapolated to zero concentration. After calibration by using samples of known molar mass, the molar mass of the sample can be inferred from the temperature difference between the two thermistors. The method is not particularly sensitive, and its use is confined to samples of low molar mass.

Activities

Now we see how to adjust the expressions developed earlier in the chapter to take into account deviations from ideal behaviour. In Section 5.5b we saw how the fugacity was introduced to take into account the effects of gas imperfections in a manner that resulted in the least upset of the form of equations. Here we see how the expressions encountered in the treatment of ideal solutions can also be preserved almost intact by introducing the concept of activity. It is important to be aware of the different definitions of standard states and activities, and they are summarized in Table 7.3. We shall put them to work in the next few chapters, when we shall see that using them is much easier than defining them.

7.6 The solvent activity

The general form of the chemical potential of a real or ideal solvent is given by a straightforward modification of eqn 23:

$$\mu_A = \mu_A^* + RT \ln \frac{p_A}{p_A^*} \qquad (7.42)°$$

Table 7.3 Standard states

Component	Basis	Standard state	Chemical potential	Limits
Solvent‡	Raoult	Pure solvent	$\mu = \mu^* + RT \ln a$ $a = p/p^*$ and $a = \gamma x$	$\gamma \to 1$ as $x \to 1$ (pure solvent)
Solute	Henry	(1) A hypothetical state of the pure solute	$\mu = \mu^\ominus + RT \ln a$ $a = p/K$ and $a = \gamma x$	$\gamma \to 1$ as $x \to 0$
		(2) A hypothetical state of the solute at molality $b^\ominus$	$\mu = \mu^\ominus + RT \ln a$ $a = \gamma b/b^\ominus$	$\gamma \to 1$ as $b \to 0$

‡ The implication of this entry is that the activity of a pure solid or liquid is l. The pressure in each case is 1 bar.

where p_A^* is the vapour pressure of pure A and p_A is the vapour pressure of A when it is a component of a solution. For an ideal solution, the solvent obeys Raoult's law at all concentrations and we write

$$\mu_A = \mu_A^* + RT \ln x_A \tag{7.43}°$$

The standard state of the solvent is the pure liquid (at 1 bar) and is obtained when $x_A = 1$. The form of the last equation can be preserved when the solution does not obey Raoult's law by writing

$$\mu_A = \mu_A^* + RT \ln a_A \tag{7.44}$$

The quantity a_A is the **activity** of A, a kind of 'effective' mole fraction, just as the fugacity is an effective pressure.

Because eqn 42 is true for both real and ideal solutions (the only approximation being the use of pressures rather than fugacities), we can conclude by comparing it with eqn 44 that

$$a_A = \frac{p_A}{p_A^*} \tag{7.45}$$

We see that there is nothing mysterious about the activity of a solvent: it can be determined experimentally simply by measuring the vapour pressure and then using eqn 45.

Illustration 7.2

The vapour pressure of 0.500 M KNO_3(aq) at 100°C is 749.7 Torr, so the activity of water in the solution at this temperature is

$$a_A = \frac{749.7 \text{ Torr}}{760.0 \text{ Torr}} = 0.9864$$

Because all solvents obey Raoult's law (that $p_A/p_A^* = x_A$) increasingly closely as the concentration of solute approaches zero, the activity of the solvent approaches the mole fraction as $x_A \to 1$:

$$a_A \to x_A \quad \text{as} \quad x_A \to 1 \tag{7.46}$$

As in the case of real gases, a convenient way of expressing this convergence is to introduce the **activity coefficient**, γ, by the definition

$$a_A = \gamma_A x_A \qquad \gamma_A \to 1 \quad \text{as} \quad x_A \to 1 \tag{7.47}$$

at all temperatures and pressures. The chemical potential of the solvent is then

$$\mu_A = \mu_A^* + RT \ln x_A + RT \ln \gamma_A \tag{7.48}$$

The standard state of the solvent, the pure liquid solvent at 1 bar, is established when $x_A = 1$.

7.7 The solute activity

The problem with defining activity coefficients and standard states for solutes is that they approach ideal-dilute (Henry's law) behaviour as $x_B \to 0$, not as $x_B \to 1$ (corresponding to pure solute). We shall show how to set up the definitions for a solute that obeys Henry's law exactly, and then show how to allow for deviations.

(a) Ideal-dilute solutions

A solute B that satisfies Henry's law has a vapour pressure given by $p_B = K_B x_B$, where K_B is an empirical constant. In this case, the chemical potential of B is

$$\mu_B = \mu_B^* + RT \ln \frac{p_B}{p_B^*}$$

$$= \mu_B^* + RT \ln \frac{K_B}{p_B^*} + RT \ln x_B$$

Both K_B and p_B^* are characteristics of the solute, so the second term may be combined with the first to give a new standard chemical potential:

$$\mu_B^{\ominus} = \mu_B^* + RT \ln \frac{K_B}{p_B^*} \tag{7.49}$$

It then follows that

$$\mu_B = \mu_B^{\ominus} + RT \ln x_B \tag{7.50}°$$

(b) Real solutes

We now permit deviations from ideal-dilute, Henry's law behaviour. For the solute, we introduce a_B in place of x_B in eqn 50, and obtain

$$\mu_B = \mu_B^{\ominus} + RT \ln a_B \tag{7.51}$$

The standard state remains unchanged in this last stage, and all the deviations from ideality are captured in the activity a_B. The value of the activity at any concentration can be obtained in the same way as for the solvent, but in place of eqn 45 we use

$$a_B = \frac{p_B}{K_B} \tag{7.52}$$

As for the solvent, it is sensible to introduce an activity coefficient through

$$a_B = \gamma_B x_B \tag{7.53}$$

Now all the deviations from ideality are captured in the activity coefficient γ_B. Because the solute obeys Henry's law as its concentration goes to zero, it follows that

$$a_B \to x_B \quad \text{and} \quad \gamma_B \to 1 \quad \text{as} \quad x_B \to 0 \tag{7.54}$$

at all temperatures and pressures. Deviations of the solute from ideality disappear as zero concentration is approached.

Example 7.6 Measuring activity

Use the information in Example 7.3 to calculate the activity and activity coefficient of chloroform in acetone at 25°C, treating it first as a solvent and then as a solute.

Method For the activity of chloroform as a solvent (the Raoult's law activity), form $a = p/p^*$ and $\gamma = a/x$. For its activity as a solute (the Henry's law activity), form $a = p/K$ and $\gamma = a/x$.

Answer Because $p^* = 273$ Torr and $K = 165$ Torr, we can construct the following table (with x the mole fraction of chloroform):

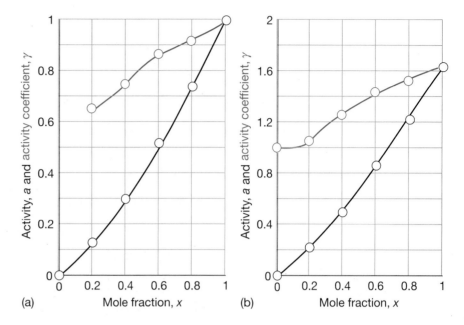

7.30 The variation of activity and activity coefficient of chloroform (trichloromethane) with composition according to (a) Raoult's law, (b) Henry's law.

(a) Mole fraction, x

(b) Mole fraction, x

x	0	0.20	0.40	0.60	0.80	1.00

From Raoult's law (chloroform regarded as the solvent):

a	0	0.13	0.30	0.52	0.73	1.00
γ		0.65	0.75	0.87	0.91	1.00

From Henry's law (chloroform regarded as the solute):

a	0	0.21	0.50	0.86	1.21	1.65
γ	1	1.05	1.25	1.43	1.51	1.65

These values are plotted in Fig. 7.30.

Comment Notice that $\gamma \to 1$ as $x \to 1$ in the Raoult's law case, but that $\gamma \to 1$ as $x \to 0$ in the Henry's law case.

Self-test 7.7 Calculate the activities and activity coefficients for acetone according to the two conventions.

[At $x_A = 0.60$, for instance, $a_R = 0.50$, $\gamma_R = 0.83$; $a_H = 1.00$, $\gamma_H = 1.67$]

(c) Activities in terms of molalities

The selection of a standard state is entirely arbitrary, so we are free to choose one that best suits our purpose and the description of the composition of the system. In chemistry, compositions are often expressed as molalities, b, in place of mole fractions. It therefore proves convenient to write

$$\mu_B = \mu_B^{\ominus} + RT \ln b_B \qquad \{7.55\}$$

where $\mu^{\ominus}$ has a different value from the standard values introduced earlier. According to this definition, the chemical potential of the solute has its standard value $\mu^{\ominus}$ when the molality of B is equal to $b^{\ominus}$ (that is, at 1 mol kg^{-1}). Note that as $b_B \to 0$, $\mu_B \to -\infty$; that is, as the solution becomes diluted, so the solute becomes increasingly stabilized. The practical consequence of this result is that it is very difficult to remove the last traces of a solute from a solution.

Now, as before, we incorporate deviations from ideality by introducing a dimensionless activity a_B, a dimensionless activity coefficient γ_B, and writing

$$a_B = \gamma_B \frac{b_B}{b^{\ominus}} \quad \text{where} \quad \gamma_B \to 1 \quad \text{as} \quad b_B \to 0 \qquad [7.56]$$

at all temperatures and pressures. The standard state remains unchanged in this last stage and, as before, all the deviations from ideality are captured in the activity coefficient γ_B. We then arrive at the following succinct expression for the chemical potential of a real solute at any molality:

$$\mu = \mu^{\ominus} + RT \ln a \qquad (7.57)$$

(d) The biological standard state

One important illustration of the ability to choose a standard state to suit the circumstances arises in biological applications. The conventional standard state of hydrogen ions (unit activity, corresponding to pH = 0)[3] is not appropriate to normal biological conditions. Therefore, in biochemistry it is common to adopt the **biological standard state**, in which pH = 7 (an activity of 10^{-7}, neutral solution) and to label the corresponding standard thermodynamic functions as $G^{\oplus}$, $H^{\oplus}$, $\mu^{\oplus}$, and $S^{\oplus}$ (some texts use $X^{\ominus\prime}$).

To find the relation between the thermodynamic and biological standard values of the chemical potential of hydrogen ions we need to note from eqn 57 that

$$\mu_{H^+} = \mu_{H^+}^{\ominus} + RT \ln a_{H^+} = \mu_{H^+}^{\ominus} - (RT \ln 10) \times \text{pH}$$

We have used $\ln x = (\ln 10)\log x$.

It follows that

$$\mu_{H^+}^{\oplus} = \mu_{H^+}^{\ominus} - 7RT \ln 10 \qquad (7.58)$$

At 298.15 K, $7RT \ln 10 = 39.96$ kJ mol^{-1}, so the two standard values differ by about 40 kJ mol^{-1}.

7.8 The activities of regular solutions

The material on regular solutions presented in Section 7.4 gives further insight into the origin of deviations from Raoult's law and its relation to activity coefficients. The starting point is the expression for the Gibbs energy of mixing for a regular solution (eqn 31). We show in the following *Justification* that eqn 31 implies that the activity coefficients are given by expressions of the form

$$\ln \gamma_A = \beta x_B^2 \qquad \ln \gamma_B = \beta x_A^2 \qquad (7.59)$$

These relations are called the **Margules equations**.

Justification 7.4

The Gibbs energy of mixing to form a nonideal solution is

$$\Delta_{\text{mix}}G = nRT\{x_A \ln a_A + x_B \ln a_B\}$$

This relation follows from the derivation of eqn 27 with activities in place of mole fractions. If each activity is replaced by γx, this expression becomes

$$\Delta_{\text{mix}}G = nRT\{x_A \ln x_A + x_B \ln x_B + x_A \ln \gamma_A + x_B \ln \gamma_B\}$$

3. pH = $-\log a_{H_3O^+}$ is introduced in Chapter 9; we presume that the concept is familiar from introductory chemistry courses.

7.31 The vapour pressure of a mixture based on a model in which the excess enthalpy is proportional to $\beta x_A x_B$. An ideal solution corresponds to $\beta = 0$ and gives a straight line, in accord with Raoult's law. Positive values of β give vapour pressures higher than ideal. Negative values of β give a lower vapour pressure.

Now we introduce the two expressions in eqn 59, and use $x_A + x_B = 1$, which gives

$$\Delta_{mix}G = nRT\{x_A \ln x_A + x_B \ln x_B + \beta x_A x_B^2 + \beta x_B x_A^2\}$$
$$= nRT\{x_A \ln x_A + x_B \ln x_B + \beta x_A x_B (x_A + x_B)\}$$
$$= nRT\{x_A \ln x_A + x_B \ln x_B + \beta x_A x_B\}$$

as required by eqn 31. Note, moreover, that the activity coefficients behave correctly for dilute solutions: $\gamma_A \to 1$ as $x_B \to 0$ and $\gamma_B \to 1$ as $x_A \to 0$.

At this point we can use the Margules equations to write the activity of A as

$$a_A = \gamma_A x_A = x_A e^{\beta x_B^2} = x_A e^{\beta(1-x_A)^2} \tag{7.60}$$

with a similar expression for a_B. The activity of A, though, is just the ratio of the vapour pressure of A in the solution to the vapour pressure of pure A, so we can write

$$p_A = \{x_A e^{\beta(1-x_A)^2}\}p_A^* \tag{7.61}$$

This function is plotted in Fig. 7.31. We see that $\beta = 0$, corresponding to an ideal solution, gives a straight line, in accord with Raoult's law (indeed, when $\beta = 0$, eqn 61 becomes $p_A = x_A p_A^*$, which is Raoult's law). Positive values of β (endothermic mixing, unfavourable solute–solvent interactions) give vapour pressures higher than ideal. Negative values of β (exothermic mixing, favourable solute–solvent interactions) give a lower vapour pressure. All the curves approach linearity and coincide with the Raoult's law line as $x_A \to 1$ and the exponential function in eqn 61 approaches 1. When $x_A \ll 1$, eqn 61 approaches

$$p_A = x_A e^{\beta} p_A^* \tag{7.62}$$

This expression has the form of Henry's law once we identify K with $e^{\beta} p_A^*$, which is different for each solute–solvent system. It follows that β can be determined from the ratio K/p_A^*.

Checklist of key ideas

The thermodynamic description of mixtures

7.1 Partial molar quantities

☐ partial molar volume, the change in volume per mole of A added to a large volume of the mixture; $V_J = (\partial V/\partial n_J)_{p,T,n'}$

☐ total volume of a mixture, $V = n_A V_A + n_B V_B$

☐ partial molar Gibbs energy and the definition of the chemical potential, $\mu_J = (\partial G/\partial n_J)_{p,T,n'}$

☐ total Gibbs energy of a mixture, $G = n_A \mu_A + n_B \mu_B$

☐ fundamental equation of chemical thermodynamics $dG = V\,dp - S\,dT + \mu_A\,dn_A + \mu_B\,dn_B + \cdots$

☐ chemical potential in terms of U, $\mu_J = (\partial U/\partial n_J)_{S,V,n'}$

☐ chemical potential in terms of H, $\mu_J = (\partial H/\partial n_J)_{S,p,n'}$

☐ chemical potential in terms of A, $\mu_J = (\partial A/\partial n_J)_{T,V,n'}$

☐ Gibbs–Duhem equation, $\sum_J n_J\,d\mu_J = 0$

7.2 The thermodynamics of mixing

☐ standard chemical potential of a perfect gas, the chemical potential of the pure gas at 1 bar

☐ chemical potential of a perfect gas, $\mu = \mu^{\ominus} + RT \ln(p/p^{\ominus})$

☐ Gibbs energy of mixing of two perfect gases, $\Delta_{mix}G = nRT(x_A \ln x_A + x_B \ln x_B)$

☐ entropy of mixing of two perfect gases, $\Delta_{mix}S = -nR(x_A \ln x_A + x_B \ln x_B)$

☐ enthalpy of mixing, the enthalpy change accompanying mixing: $\Delta_{mix}H = 0$ for perfect gases

7.3 The chemical potential of liquids

☐ Raoult's law, $p_A = x_A p_A^*$

☐ ideal solution, a solution in which all components obeys Raoult's law throughout the composition range

☐ chemical potential of a component of an ideal solution, $\mu_A = \mu_A^* + RT \ln x_A$

☐ Henry's law, $p_B = x_B K_B$

☐ ideal-dilute solution, mixtures for which the solute obeys Henry's law and the solvent obeys Raoult's law

The properties of solutions

7.4 Liquid mixtures

☐ Gibbs energy of mixing of two liquids,
$\Delta_{mix}G = nRT(x_A \ln x_A + x_B \ln x_B)$

☐ entropy of mixing,
$\Delta_{mix}S = -nR(x_A \ln x_A + x_B \ln x_B)$

☐ partially miscible liquids, liquids that are miscible only over a range of compositions

☐ excess function (X^E), the difference between the observed thermodynamic function of mixing and the function for an ideal solution

☐ excess entropy, $S^E = \Delta_{mix}S - \Delta_{mix}S^{ideal}$

☐ regular solution, a solution for which $H^E \neq 0$ but $S^E = 0$

7.5 Colligative properties

☐ colligative property, a property that depends only on the number of solute particles present, not their identity

☐ the elevation of boiling point, $\Delta T = K_b b$

☐ ebullioscopic constant, the constant K_b in $\Delta T = K_b b$

☐ the depression of freezing point, $\Delta T = K_f b$

☐ cryoscopic constant, the constant K_f in $\Delta T = K_f b$

☐ ideal solubility, $\ln x_B = (\Delta_{fus}H/R)(1/T^* - 1/T)$

☐ osmosis, the spontaneous passage of a pure solvent into a solution separated from it by a semipermeable membrane

☐ semipermeable membrane, a membrane permeable to the solvent but not to the solute

☐ osmotic pressure, the pressure that must be applied to the solution to stop the influx of solvent

☐ osmometry, the determination of molar mass by the measurement of osmotic pressure

☐ van 't Hoff equation for the osmotic pressure, $\Pi = [B]RT$

☐ osmotic virial coefficient, B in $\Pi = [B]RT(1 + B[B] + \cdots)$

☐ vapour-phase osmometry, a technique for measuring osmotic pressure

Activities

7.6 The solvent activity

☐ activity, $a_A = p_A/p_A^*$

☐ solvent activity and chemical potential, $\mu_A = \mu_A^* + RT \ln a_A$

☐ Raoult's law basis of activity and activity coefficient, $\gamma = a/x$

7.7 The solute activity

☐ chemical potential of a solute in an ideal-dilute solution, $\mu_B = \mu_B^\ominus + RT \ln a_B$

☐ solute activity and the Henry's law basis of activity coefficient, $\gamma = a/x$

☐ chemical potential of a solute in terms of its molality, $\mu_B = \mu_B^\ominus + RT \ln b_B$

☐ biological standard state, $\mu_{H^+}^\oplus = \mu_{H^+}^\ominus - 7RT \ln 10$

7.8 The activities of regular solutions

☐ Margules equations, $\ln \gamma_J = \beta x_J^2$

☐ activity of regular solution, $a_A = x_A e^{\beta(1-x_A)^2}$

Further reading

Articles of general interest

S.F. Ackley, Sea ice. In *Encyclopedia of applied physics* (ed. G.L. Trigg), **17**, 81. VCH, New York (1996).

M.E. Cardinali and C. Giomini. Boiling temperature vs. composition: An almost-exact explicit equation for a binary mixture following Raoult's law. *J. Chem. Educ.* **66**, 549 (1989).

J.J. Carroll, Henry's law: A historical view. *J. Chem. Educ.* **70**, 91 (1993).

M.J. Clugston, A mathematical verification of the second law of thermodynamics from the entropy of mixing. *J. Chem. Educ.* **67**, 203 (1990).

H.F. Franzen, The freezing point depression law in physical chemistry. *J. Chem. Educ.* **65**, 1077 (1988).

B. Freeman, Osmosis. In *Encyclopedia of applied physics* (ed. G.L. Trigg), **13**, 59. VCH, New York (1995).

E.F. Meyer, Thermodynamics of 'mixing' of ideal gases: a persistent pitfall. *J. Chem. Educ.* **64**, 676 (1987).

S. Sattar, Thermodynamics of mixing real gases. *J. Chem. Educ.* **77**, 1361 (2000).

R.L. Scott, Models for phase equilibria in fluid mixtures. *Acc. Chem. Res.* **20**, 97 (1987).

A. Vincent, Osmotic pressure and the effect of gravity on solutions. *J. Chem. Educ.* **73**, 998 (1996).

Texts and sources of data and information

C.E. Carraher, Jr., *Seymour/Carraher's polymer chemistry*. Marcel Dekker, New York (2000).

M.R.J. Dack, Solutions and solubilities. In *Techniques of chemistry* (A. Weissberger and B.W. Rossiter, ed.), **8**. Wiley, New York (1975).

I.M. Klotz and R.M. Rosenberg, *Chemical thermodynamics: basic theory and methods.* Wiley–Interscience, New York (2000).

W.J. Mader and L.T. Brady, Determination of solubility. In *Techniques of chemistry* (A. Weissberger and B.W. Rossiter, ed.), **5**, 257. Wiley, New York (1971).

J.N. Murrell and A.D. Jenkins, *Properties of liquids and solutions.* Wiley–Interscience, New York (1994).

J.R. Overton, Determination of osmotic pressure. In *Techniques of chemistry* (A. Weissberger and B.W. Rossiter, ed.), **5**, 309. Wiley, New York (1971).

J.S. Rowlinson and F.L. Swinton, *Liquids and liquid mixtures.* Butterworths, London (1982).

S.I. Sandler, *Chemical and engineering thermodynamics.* Wiley, New York (1998).

Exercises

Discussion questions

7.1 (a) State and justify the thermodynamic criterion for solution–vapour equilibrium.

7.1 (b) How is Raoult's law modified so as to describe the vapour pressure of real solutions?

7.2 (a) Explain what is meant by a regular solution.

7.2 (b) Explain the origin of colligative properties.

7.3 (a) Explain how colligative properties are used to determine molar mass.

7.3 (b) What is meant by the activity of a solute?

Numerical exercises

7.4 (a) The partial molar volumes of acetone (propanone) and chloroform (trichloromethane) in a mixture in which the mole fraction of $CHCl_3$ is 0.4693 are 74.166 $cm^3\ mol^{-1}$ and 80.235 $cm^3\ mol^{-1}$, respectively. What is the volume of a solution of mass 1.000 kg?

7.4 (b) The partial molar volumes of two liquids A and B in a mixture in which the mole fraction of A is 0.3713 are 188.2 $cm^3\ mol^{-1}$ and 176.14 $cm^3\ mol^{-1}$, respectively. The molar masses of the A and B are 241.1 $g\ mol^{-1}$ and 198.2 $g\ mol^{-1}$. What is the volume of a solution of mass 1.000 kg?

7.5 (a) At 25°C, the density of a 50 per cent by mass ethanol–water solution is 0.914 $g\ cm^{-3}$. Given that the partial molar volume of water in the solution is 17.4 $cm^3\ mol^{-1}$, calculate the partial molar volume of the ethanol.

7.5 (b) At 20°C, the density of a 20 per cent by mass ethanol–water solution is 968.7 $kg\ m^{-3}$. Given that the partial molar volume of ethanol in the solution is 52.2 $cm^3\ mol^{-1}$, calculate the partial molar volume of the water.

7.6 (a) At 300 K, the partial vapour pressures of HCl (that is, the partial pressure of the HCl vapour) in liquid $GeCl_4$ are as follows:

x_{HCl}	0.005	0.012	0.019
p_{HCl}/kPa	32.0	76.9	121.8

Show that the solution obeys Henry's law in this range of mole fractions, and calculate Henry's law constant at 300 K.

7.6 (b) At 310 K, the partial vapour pressures of a substance B dissolved in a liquid A are as follows:

x_B	0.010	0.015	0.020
p_B/kPa	82.0	122.0	166.1

Show that the solution obeys Henry's law in this range of mole fractions, and calculate Henry's law constant at 310 K.

7.7 (a) Predict the partial vapour pressure of HCl above its solution in liquid germanium tetrachloride of molality 0.10 $mol\ kg^{-1}$. For data, see Exercise 7.6a.

7.7 (b) Predict the partial vapour pressure of the component B above its solution in A in Exercise 7.6b when the molality of B is 0.25 $mol\ kg^{-1}$. The molar mass of A is 74.1 $g\ mol^{-1}$.

7.8 (a) Calculate the cryoscopic and ebullioscopic constants of tetrachloromethane.

7.8 (b) Calculate the cryoscopic and ebullioscopic constants of naphthalene.

7.9 (a) The vapour pressure of benzene is 400 Torr at 60.6°C, but it fell to 386 Torr when 19.0 g of an involatile organic compound was dissolved in 500 g of benzene. Calculate the molar mass of the compound.

7.9 (b) The vapour pressure of 2-propanol is 50.00 kPa at 338.8°C, but it fell to 49.62 kPa when 8.69 g of an involatile organic compound was dissolved in 250 g of 2-propanol. Calculate the molar mass of the compound.

7.10 (a) The addition of 100 g of a compound to 750 g of CCl_4 lowered the freezing point of the solvent by 10.5 K. Calculate the molar mass of the compound.

7.10 (b) The addition of 5.00 g of a compound to 250 g of naphthalene lowered the freezing point of the solvent by 0.780 K. Calculate the molar mass of the compound.

7.11 (a) The osmotic pressure of an aqueous solution at 300 K is 120 kPa. Calculate the freezing point of the solution.

7.11 (b) The osmotic pressure of an aqueous solution at 288 K is 99.0 kPa. Calculate the freezing point of the solution.

7.12 (a) Consider a container of volume 5.0 L that is divided into two compartments of equal size. In the left compartment there is nitrogen at 1.0 atm and 25°C; in the right compartment there is hydrogen at the same temperature and pressure. Calculate the entropy and Gibbs energy of mixing when the partition is removed. Assume that the gases are perfect.

7.12 (b) Consider a container of volume 250 mL that is divided into two compartments of equal size. In the left compartment there is argon at 100 kPa and 0°C; in the right compartment there is neon at the same temperature and pressure. Calculate the entropy and Gibbs energy of mixing when the partition is removed. Assume that the gases are perfect.

7.13 (a) Air is a mixture with a composition given in Self-test 1.5. Calculate the entropy of mixing when it is prepared from the pure (and perfect) gases.

7.13 (b) Calculate the Gibbs energy, entropy, and enthalpy of mixing when 1.00 mol C_6H_{14} (hexane) is mixed with 1.00 mol C_7H_{16} (heptane) at 298 K; treat the solution as ideal.

7.14 (a) What proportions of hexane and heptane should be mixed (a) by mole fraction, (b) by mass in order to achieve the greatest entropy of mixing?

7.14 (b) What proportions of benzene and ethylbenzene should be mixed (a) by mole fraction, (b) by mass in order to achieve the greatest entropy of mixing?

7.15 (a) Use Henry's law and the data in Table 7.1 to calculate the solubility (as a molality) of CO_2 in water at 25°C when its partial pressure is (a) 0.10 atm, (b) 1.00 atm.

7.15 (b) The mole fractions of N_2 and O_2 in air at sea level are approximately 0.78 and 0.21. Calculate the molalities of the solution formed in an open flask of water at 25°C.

7.16 (a) A water carbonating plant is available for use in the home and operates by providing carbon dioxide at 5.0 atm. Estimate the molar concentration of the soda water it produces.

7.16 (b) After some weeks of use, the pressure in the water carbonating plant mentioned in the previous exercise has fallen to 2.0 atm. Estimate the molar concentration of the soda water it produces at this stage.

7.17 (a) Calculate the freezing point of a glass of water of volume 250 cm³ sweetened with 7.5 g of sucrose.

7.17 (b) Calculate the freezing point of a glass of water of volume 200 cm³ in which 10 g of glucose has been dissolved.

7.18 (a) The enthalpy of fusion of anthracene is 28.8 kJ mol⁻¹ and its melting point is 217°C. Calculate its ideal solubility in benzene at 25°C.

7.18 (b) Predict the ideal solubility of lead in bismuth at 280°C given that its melting point is 327°C and its enthalpy of fusion is 5.2 kJ mol⁻¹.

7.19 (a) The osmotic pressures of solutions of polystyrene in toluene were measured at 25°C and the pressure was expressed in terms of the height of the solvent of density 1.004 g cm⁻³:

$c/(\text{g L}^{-1})$	2.042	6.613	9.521	12.602
h/cm	0.592	1.910	2.750	3.600

Calculate the molar mass of the polymer.

7.19 (b) The molar mass of an enzyme was determined by dissolving it in water, measuring the osmotic pressure at 20°C, and extrapolating the data to zero concentration. The following data were obtained:

$c/(\text{mg cm}^{-3})$	3.221	4.618	5.112	6.722
h/cm	5.746	8.238	9.119	11.990

Calculate the molar mass of the enzyme.

7.20 (a) Substances A and B are both volatile liquids with $p_A^* = 300$ Torr, $p_B^* = 250$ Torr, and $K_B = 200$ Torr (concentration

expressed in mole fraction). When $x_A = 0.9$, $b_B = 2.22$ mol kg⁻¹, $p_A = 250$ Torr, and $p_B = 25$ Torr. Calculate the activities and activity coefficients of A and B. Use the mole fraction, Raoult's law basis system, for A and the Henry's law basis system (both mole fractions and molalities) for B.

7.20 (b) Given that $p^*(\text{H}_2\text{O}) = 0.023\ 08$ atm and $p(\text{H}_2\text{O}) = 0.022\ 39$ atm in a solution in which 0.122 kg of a non-volatile solute ($M = 241$ g mol⁻¹) is dissolved in 0.920 kg water at 293 K, calculate the activity and activity coefficient of water in the solution.

7.21 (a) A dilute solution of bromine in carbon tetrachloride behaves as an ideal-dilute solution. The vapour pressure of pure CCl_4 is 33.85 Torr at 298 K. The Henry's law constant when the concentration of Br_2 is expressed as a mole fraction is 122.36 Torr. Calculate the vapour pressure of each component, the total pressure, and the composition of the vapour phase when the mole fraction of Br_2 is 0.050, on the assumption that the conditions of the ideal-dilute solution are satisfied at this concentration.

7.21 (b) Benzene and toluene form nearly ideal solutions. The boiling point of pure benzene is 80.1°C. Calculate the chemical potential of benzene relative to that of pure benzene when $x_{\text{benzene}} = 0.30$ at its boiling point. If the activity coefficient of benzene in this solution were actually 0.93 rather than 1.00, what would be its vapour pressure?

7.22 (a) By measuring the equilibrium between liquid and vapour phases of an acetone (A)–ethanol (E) solution at 57.2°C at 1.00 atm, it was found that $x_A = 0.400$ when $y_A = 0.516$. Calculate the activities and activity coefficients of both components in this solution on the Raoult's law basis. The vapour pressures of the pure components at this temperature are: $p_A^* = 786$ Torr and $p_E^* = 551$ Torr. (x_A is the mole fraction in the liquid and y_A the mole fraction in the vapour.)

7.22 (b) By measuring the equilibrium between liquid and vapour phases of a solution at 30°C at 1.00 atm, it was found that $x_A = 0.220$ when $y_A = 0.314$. Calculate the activities and activity coefficients of both components in this solution on the Raoult's law basis. The vapour pressures of the pure components at this temperature are: $p_A^* = 73.0$ kPa and $p_B^* = 92.1$ kPa. (x_A is the mole fraction in the liquid and y_A the mole fraction in the vapour.)

Problems[†]

Numerical problems

7.1 The following table gives the mole fraction of methylbenzene (A) in liquid and gaseous mixtures with butanone at equilibrium at 303.15 K and the total pressure p. Take the vapour to be perfect and calculate the partial pressures of the two components. Plot them against their respective mole fractions in the liquid mixture and find the Henry's law constants for the two components.

x_A	0		0.0898	0.2476	0.3577	0.5194	0.6036
y_A	0		0.0410	0.1154	0.1762	0.2772	0.3393
p/kPa	36.066		34.121	30.900	28.626	25.239	23.402

x_A	0.7188	0.8019	0.9105	1
y_A	0.4450	0.5435	0.7284	1
p/kPa	20.6984	18.592	15.496	12.295

7.2 The volume of an aqueous solution of NaCl at 25°C was measured at a series of molalities b, and it was found that the volume fitted the expression

$$V/\text{cm}^3 = 1003 + 16.62b + 1.77b^{3/2} + 0.12b^2$$

where V is the volume of a solution formed from 1.000 kg of water and b is to be understood as $b/b^\ominus$. Calculate the partial molar volume of the components in a solution of molality 0.100 mol kg⁻¹.

† Problems denoted with the symbol ‡ were supplied by Charles Trapp, Carmen Giunta, and Marshall Cady.

7.3 At 18°C the total volume of a solution formed from $MgSO_4$ and 1.000 kg of water fits the expression

$$V/cm^3 = 1001.21 + 34.69(b - 0.070)^2$$

where b is to be understood as $b/b^{\ominus}$. Calculate the partial molar volumes of the salt and the solvent when in a solution of molality 0.050 mol kg^{-1}.

7.4 The densities of aqueous solutions of copper(II) sulfate at 20°C were measured as set out below. Determine and plot the partial molar volume of $CuSO_4$ in the range of the measurements.

$m(CuSO_4)$/g	5	10	15	20
ρ/(g cm^{-3})	1.051	1.107	1.167	1.230

where $m(CuSO_4)$ is the mass of $CuSO_4$ dissolved in 100 g of solution.

7.5 What proportions of ethanol and water should be mixed in order to produce 100 cm^3 of a mixture containing 50 per cent by mass of ethanol? What change in volume is brought about by adding 1.00 cm^3 of ethanol to the mixture? (Use data from Fig. 7.1.)

7.6 Potassium fluoride is very soluble in glacial acetic acid and the solutions have a number of unusual properties. In an attempt to understand them, freezing point depression data were obtained by taking a solution of known molality and then diluting it several times (J. Emsley, *J. Chem. Soc. A*, 2702 (1971)). The following data were obtained:

b/(mol kg^{-1})	0.015	0.037	0.077	0.295	0.602
ΔT/K	0.115	0.295	0.470	1.381	2.67

Calculate the apparent molar mass of the solute and suggest an interpretation. Use $\Delta_{fus}H = 11.4$ kJ mol^{-1} and $T_f^* = 290$ K.

7.7 In a study of the properties of an aqueous solution of $Th(NO_3)_4$ (by A. Apelblat, D. Azoulay, and A. Sahar, *J. Chem. Soc. Faraday Trans.*, I, 1618, (1973)), a freezing point depression of 0.0703 K was observed for an aqueous solution of molality 9.6 mmol kg^{-1}. What is the apparent number of ions per formula unit?

7.8 The table below lists the vapour pressures of mixtures of iodoethane (I) and ethyl acetate (A) at 50°C. Find the activity coefficients of both components on (a) the Raoult's law basis, (b) the Henry's law basis with I as solute.

x_I	0	0.0579	0.1095	0.1918	0.2353	0.3718
p_I/Torr	0	20.0	52.7	87.7	105.4	155.4
p_A/Torr	280.4	266.1	252.3	231.4	220.8	187.9

x_I	0.5478	0.6349	0.8253	0.9093	1.0000
p_I/Torr	213.3	239.1	296.9	322.5	353.4
p_A/Torr	144.2	122.9	66.6	38.2	0

7.9 Plot the vapour pressure data for a mixture of benzene (B) and acetic acid (A) given below and plot the vapour pressure/composition curve for the mixture at 50°C. Then confirm that Raoult's and Henry's laws are obeyed in the appropriate regions. Deduce the activities and activity coefficients of the components on the Raoult's law basis and then, taking B as the solute, its activity and activity coefficient on a Henry's law basis. Finally, evaluate the excess Gibbs energy of the mixture over the composition range spanned by the data.

x_A	0.0160	0.0439	0.0835	0.1138	0.1714
p_A/Torr	3.63	7.25	11.51	14.2	18.4
p_B/Torr	262.9	257.2	249.6	244.8	231.8

x_A	0.2973	0.3696	0.5834	0.6604	0.8437	0.9931
p_A/Torr	24.8	28.7	36.3	40.2	50.7	54.7
p_B/Torr	211.2	195.6	153.2	135.1	75.3	3.5

7.10‡ Aminabhavi *et al.* examined mixtures of cyclohexane with various long-chain alkanes (T.M. Aminabhavi, V.B. Patil, M.I, Aralaguppi, J.D. Ortego, and K.C. Hansen, *J. Chem. Eng. Data* **41**, 526 (1996)). Among their data are the following measurements of the density of a mixture of cyclohexane and pentadecane as a function of the mole fraction of cyclohexane (x_c) at 298.15 K:

x_c	0.6965	0.7988	0.9004
ρ/(g cm^{-3})	0.7661	0.7674	0.7697

Compute the partial molar volume for each component in a mixture that has a mole fraction cyclohexane of 0.7988.

7.11‡ Comelli and Francesconi examined mixtures of propionic acid with various other organic liquids at 313.15 K (F. Comelli and R. Francesconi, *J. Chem. Eng. Data* **41**, 101 (1996)). They report the excess volume of mixing propionic acid with oxane as $V^E = x_1 x_2 \{a_0 + a_1(x_1 - x_2)\}$, where x_1 is the mole fraction of propionic acid, x_2 that of oxane, $a_0 = -2.4697$ cm^3 mol^{-1}, and $a_1 = 0.0608$ cm^3 mol^{-1}. The density of propionic acid at this temperature is 0.971 74 g cm^{-3}; that of oxane is 0.863 98 g cm^{-3}. (a) Derive an expression for the partial molar volume of each component at this temperature. (b) Compute the partial molar volume for each component in an equimolar mixture.

7.12‡ Francesconi, Lunelli, and Comelli studied the liquid–vapour equilibria of trichloromethane and 1,2-epoxybutane at several temperatures (R. Francesconi, B. Lunelli, and F. Comelli, *J. Chem. Eng. Data* **41**, 310 (1996)). Among their data are the following measurements of the mole fractions of trichloromethane in the liquid phase (x_T) and the vapour phase (y_T) at 298.15 K as a function of pressure.

p/kPa	23.40	21.75	20.25	18.75	18.15	20.25	22.50	26.30
x	0	0.129	0.228	0.353	0.511	0.700	0.810	1
y	0	0.065	0.145	0.285	0.535	0.805	0.915	1

Compute the activity coefficients of both components on the basis of Raoult's law.

7.13‡ Chen and Lee studied the liquid–vapour equilibria of cyclohexanol with several gases at elevated pressures (J.-T. Chen and M.-J. Lee, *J. Chem. Eng. Data* **41**, 339 (1996)). Among their data are the following measurements of the mole fractions of cyclohexanol in the vapour phase (y_{cyc}) and the liquid phase (x_{cyc}) at 393.15 K as a function of pressure.

p/bar	10.0	20.0	30.0	40.0	60.0	80.0
y_{cyc}	0.0267	0.0149	0.0112	0.009 47	0.008 35	0.009 21
x_{cyc}	0.9741	0.9464	0.9204	0.892	0.836	0.773

Determine the Henry's law constant of CO_2 in cyclohexanol, and compute the activity coefficient of CO_2.

7.14‡ Equation 39 indicates that solubility is an exponential function of temperature. The data in the table below give the solubility, S, of calcium acetate in water as a function of temperature.

θ/°C	0	20	40	60	80
S/(mol L^{-1})	36.4	34.9	33.7	32.7	31.7

Determine the extent to which the data fit the exponential $S = S_0 e^{\tau/T}$ and obtain values for S_0 and τ. Express these constants in terms of properties of the solute.

7.15‡ The excess enthalpy of mixing data for 1-propanol and tripropylamine at 25°C is found to fit the expression

$$H^E/(J\,mol^{-1}) = x_1(1-x_1)\sum_{r=0}^{3} A_r(1-2x_1)^r$$

where $A_r/(J\,mol^{-1}) = -1131, 1355, -855, 1127$ for $r = 0, 1, 2, 3$. (a) Draw a graph of H^E against x_1 and compare the graph shape to that expected for a solution in which $H^E = \beta RT x_1 x_2$. What do the graphs imply about intermolecular interaction energies? (b) At 25°C and infinite dilution the excess enthalpy of mixing 1-propanol in heptane is $H^E = +25.3$ kJ mol^{-1}, whereas for the mixing of tripropylamine in heptane $H^E = +121$ J mol^{-1}. Deduce the strength of the alcohol–amine hydrogen bond. Assume that intermolecular structural reorganization enthalpies cancel for the cycle. *Hint.* Use mathematical software.

7.16 The excess Gibbs energies of solutions of methylcyclohexane (MCH) and tetrahydrofuran (THF) at 303.15 K were found to fit the expression

$$G^E = RTx(1-x)\{0.4857 - 0.1077(2x-1) + 0.0191(2x-1)^2\}$$

where x is the mole fraction of the methylcyclohexane. Calculate the Gibbs energy of mixing when a mixture of 1.00 mol of MCH and 3.00 mol of THF is prepared.

Theoretical problems

7.17 The excess Gibbs energy of a certain binary mixture is equal to $gRTx(1-x)$ where g is a constant and x is the mole fraction of a solute A. Find an expression for the chemical potential of A in the mixture and sketch its dependence on the composition.

7.18 Use the Gibbs–Duhem equation to derive the Gibbs–Duhem–Margules equation

$$\left(\frac{\partial \ln f_A}{\partial \ln x_A}\right)_{p,T} = \left(\frac{\partial \ln f_B}{\partial \ln x_B}\right)_{p,T}$$

where f is the fugacity. Use the relation to show that, when the fugacities are replaced by pressures, if Raoult's law applies to one component in a mixture it must also apply to the other.

7.19 Use the Gibbs–Duhem equation to show that the partial molar volume (or any partial molar property) of a component B can be obtained if the partial molar volume (or other property) of A is known for all compositions up to the one of interest. Do this by proving that

$$V_B = V_B^* + \int_{V_A^*}^{V_A} \frac{x_A}{1-x_A}\,dV_A$$

Use the following data (which are for 298 K) to evaluate the integral graphically to find the partial molar volume of acetone at $x = 0.500$.

$x(CHCl_3)$	0	0.194	0.385	0.559	0.788	0.889	1.000
$V_m/(cm^3\,mol^{-1})$	73.99	75.29	76.50	77.55	79.08	79.82	80.67

7.20 Use the Gibbs–Helmholtz equation to find an expression for d ln x_A in terms of dT. Integrate d ln x_A from $x_A = 0$ to the value of interest, and integrate the right-hand side from the transition temperature for the pure liquid A to the value in the solution. Show that, if the enthalpy of transition is constant, then eqns 33 and 36 are obtained.

7.21 The 'osmotic coefficient', ϕ, is defined as $\phi = -(x_A/x_B)\ln a_A$. By writing $r = x_B/x_A$, and using the Gibbs–Duhem equation, show that

we can calculate the activity of B from the activities of A over a composition range by using the formula

$$\ln\left(\frac{a_B}{r}\right) = \phi - \phi(0) + \int_0^r \left(\frac{\phi-1}{r}\right)dr$$

7.22 Show that the osmotic pressure of a real solution is given by

$$\Pi V = -RT \ln a_A$$

Go on to show that, provided the concentration of the solution is low, this expression takes the form

$$\Pi V = \phi RT[B]$$

and hence that the osmotic coefficient, ϕ, (which is defined in Problem 7.21) may be determined from osmometry.

Applications: to biology and polymer science

7.23 Suppose that two equal volumes of 0.200 M NaCl(aq) are separated by a membrane and that an anionic protein P of mass 55 kg mol^{-1}, which cannot pass through the membrane, is added as its sodium salt Na$_6$P to a concentration of 50 g L^{-1} to the left-hand compartment of the apparatus. Calculate the molar concentrations of Na$^+$ and Cl$^-$ in each compartment.

7.24‡ Polymer scientists often report their data in rather strange units. For example, in the determination of molar masses of polymers in solution by osmometry, osmotic pressures are often reported in grams per square centimetre (g cm^{-2}) and concentrations in grams per cubic centimetre (g cm^{-3}). (a) With these choices of units, what would be the units of R in the van 't Hoff equation? (b) The data in the table below on the concentration dependence of the osmotic pressure of polyisobutene in chlorobenzene at 25°C have been adapted from J. Leonard and H. Daoust (*J. Polymer Sci.* **57**, 53 (1962)). From these data, determine the molar mass of polyisobutene by plotting Π/c against c. (c) Theta solvents are solvents for which the second osmotic coefficient is zero; for 'poor' solvents the plot is linear and for good solvents the plot is nonlinear. From your plot, how would you classify chlorobenzene as a solvent for polyisobutene? Rationalize the result in terms of the molecular structure of the polymer and solvent. (d) Determine the second and third osmotic virial coefficients by fitting the curve to the virial form of the osmotic pressure equation. (e) Experimentally, it is often found that the virial expansion can be represented as

$$\Pi/c = RT/M(1 + B'c + gB'^2c'^2 + \cdots)$$

and, in good solvents, the parameter g is often about 0.25. With terms beyond the second power ignored, obtain an equation for $(\Pi/c)^{1/2}$ and plot this quantity against c. Determine the second and third virial coefficients from the plot and compare to the values from the first plot. Does this plot confirm the assumed value of g?

$10^{-2}(\Pi/c)/(g\,cm^{-2}/g\,cm^{-3})$	2.6	2.9	3.6	4.3	6.0	12.0
$c/(g\,cm^{-3})$		0.0050	0.010	0.020	0.033	0.057	0.10

$10^{-2}(\Pi/c)/(g\,cm^{-2}/g\,cm^{-3})$	19.0	31.0	38.0	52	63
$c/(g\,cm^{-3})$		0.145	0.195	0.245	0.27	0.29

7.25‡ K. Sato, F.R. Eirich, and J.E. Mark (*J. Polymer Sci., Polym. Phys.* **14**, 619 (1976)) have reported the data in the table below for the osmotic pressures of polychloroprene ($\rho = 1.25$ g cm^{-3}) in toluene ($\rho = 0.858$ g cm^{-3}) at 30°C. Determine the molar mass of polychloroprene and its second osmotic virial coefficient.

$c/(mg\,cm^{-3})$	1.33	2.10	4.52	7.18	9.87
$\Pi/(N\,m^{-2})$	30	51	132	246	390

8

Phase diagrams

Phases, components, and degrees of freedom

8.1 Definitions

8.2 The phase rule

Two-component systems

8.3 Vapour pressure diagrams

8.4 Temperature-composition diagrams

8.5 Liquid–liquid phase diagrams

8.6 Liquid–solid phase diagrams

Checklist of key ideas

Further reading

Exercises

Problems

Phase diagrams for pure substances were introduced in Chapter 6. Now we develop their use systematically and show how they are rich summaries of empirical information about a wide range of systems. To set the stage, we introduce the famous phase rule of Gibbs, which shows the extent to which various parameters can be varied yet the equilibrium between phases preserved. With the rule established, we see how it can be used to discuss the phase diagrams that we met in the two preceding chapters. The chapter then introduces systems of gradually increasing complexity. In each case we shall see how the phase diagram for the system summarizes empirical observations on the conditions under which the various phases of the system are stable.

In this chapter we describe a systematic way of discussing the physical changes mixtures undergo when they are heated or cooled and when their compositions are changed. In particular, we see how to use phase diagrams to judge whether two substances are mutually miscible, whether an equilibrium can exist over a range of conditions, or whether the system must be brought to a definite pressure, temperature, and composition before equilibrium is established. Phase diagrams are of considerable commercial and industrial significance, particularly for semiconductors, ceramics, steels, and alloys. They are also the basis of separation procedures in the petroleum industry and of the formulation of foods and cosmetic preparations.

Phases, components, and degrees of freedom

All phase diagrams can be discussed in terms of a relationship, the phase rule, derived by J.W. Gibbs. We shall derive this rule first, and then apply it to a wide variety of systems. The phase rule requires a careful use of terms, so we begin by presenting a number of definitions.

8.1 Definitions

The term **phase** was introduced at the start of Chapter 6, where we saw that it signifies a state of matter that is uniform throughout, not only in chemical composition but also in physical state.[1] Thus we speak of the solid, liquid, and gas

---

1. The words are Gibbs's.

(a)

(b)

8.1 The difference between (a) a single-phase solution, in which the composition is uniform on a microscopic scale, and (b) a dispersion, in which regions of one component are embedded in a matrix of a second component.

phases of a substance, and of its various solid phases (as for black phosphorus and white phosphorus). The number of phases in a system is denoted P. A gas, or a gaseous mixture, is a single phase, a crystal is a single phase, and two totally miscible liquids form a single phase. A solution of sodium chloride in water is a single phase. Ice is a single phase ($P = 1$) even though it might be chipped into small fragments. A slurry of ice and water is a two-phase system ($P = 2$) even though it is difficult to map the boundaries between the phases. A system in which calcium carbonate undergoes thermal decomposition consists of two solid phases (one consisting of calcium carbonate and the other of calcium oxide) and one gaseous phase (consisting of carbon dioxide).

An alloy of two metals is a two-phase system ($P = 2$) if the metals are immiscible, but a single-phase system ($P = 1$) if they are miscible. This example shows that it is not always easy to decide whether a system consists of one phase or of two. A solution of solid A in solid B—a homogeneous mixture of the two substances—is uniform on a molecular scale. In a solution, atoms of A are surrounded by atoms of A and B, and any segment cut from the sample, however small, is representative of the composition of the whole.

A dispersion is uniform on a macroscopic scale but not on a microscopic scale, for it consists of grains or droplets of one substance in a matrix of the other. A small sample could come entirely from one of the minute grains of pure A and would not be representative of the whole (Fig. 8.1). Dispersions are important because, in many advanced materials (including steels), heat treatment cycles are used to achieve the precipitation of a fine dispersion of particles of one phase (such as a carbide phase) within a matrix formed by a saturated solid solution phase. The ability to control this microstructure resulting from phase equilibria makes it possible to tailor the mechanical properties of the materials to a particular application.

By a **constituent** of a system we mean a chemical species (an ion or a molecule) that is present. Thus, a mixture of ethanol and water has two constituents. A solution of sodium chloride has two constituents (sodium chloride and water) despite the fact that it contains Na^+ and Cl^- ions. The term constituent should be carefully distinguished from 'component', which has a more technical meaning. A **component** is a chemically independent constituent of a system. The number of components, C, in a system is the minimum number of independent species necessary to define the composition of all the phases present in the system.

When no reaction takes place, the number of components is equal to the number of constituents. Thus, pure water is a one-component system ($C = 1$), because we need only the species H_2O to specify its composition. Similarly, a mixture of ethanol and water is a two-component system ($C = 2$): we need the species H_2O and C_2H_5OH to specify its composition. When a reaction can occur under the conditions prevailing in the system, we need to decide the minimum number of species that, after allowing for reactions in which one species is synthesized from others, can be used to specify the composition of all the phases. Consider, for example, the equilibrium

$$CaCO_3(s) \rightleftharpoons CaO(s) + CO_2(g)$$
$$\text{Phase 1} \qquad \text{Phase 2} \qquad \text{Phase 3}$$

in which there are three phases. To specify the composition of the gas phase (Phase 3) we need the species CO_2, and to specify the composition of Phase 2 we need the species CaO. However, we not need an additional species to specify the composition of Phase 1 because its identity ($CaCO_3$) can be expressed in terms of the other two constituents by making use of the stoichiometry of the reaction. Hence, the system has three constituents but only two components ($C = 2$).

Example 8.1 Counting components

How many components are present in a system in which ammonium chloride undergoes thermal decomposition?

Method Begin by writing down the chemical equation for the reaction and identifying the constituents of the system (all the species present) and the phases. Then decide whether, under the conditions prevailing in the system, any of the constituents can be prepared from any of the other constituents. The removal of these constituents leaves the number of independent constituents. Finally, identify the minimum number of these independent constituents that are needed to specify the composition of all the phases.

Answer The chemical reaction is

$$NH_4Cl(s) \rightleftharpoons NH_3(g) + HCl(g)$$

There are three constituents and two phases (one solid, one gas). However, NH_3 and HCl are formed in fixed stoichiometric proportions by the reaction. Therefore, the compositions of both phases can be expressed in terms of the single species NH_4Cl. It follows that there is only one component in the system ($C = 1$).

Comment If additional HCl (or NH_3) were supplied to the system, the decomposition of NH_4Cl would not give the correct composition of the gas phase and HCl (or NH_3) would have to be invoked as a second component. A system that consists of hydrogen, oxygen, and water at room temperature has three components, despite it being possible to form H_2O from H_2 and O_2: under the conditions prevailing in the system, hydrogen and oxygen do not react to form water, so they are independent constituents.

Self-test 8.1 Give the number of components in the following systems: (a) water, allowing for its autoprotolysis, (b) aqueous acetic acid, (c) magnesium carbonate in equilibrium with its decomposition products.   [(a) 1, (b) 2, (c) 2]

The **variance**, F, of a system is the number of intensive variables that can be changed independently without disturbing the number of phases in equilibrium. In a single-component, single-phase system ($C = 1$, $P = 1$), the pressure and temperature may be changed independently without changing the number of phases, so $F = 2$. We say that such a system is **bivariant**, or that it has two **degrees of freedom**. On the other hand, if two phases are in equilibrium (a liquid and its vapour, for instance) in a single-component system ($C = 1$, $P = 2$), the temperature (or the pressure) can be changed at will, but the change in temperature (or pressure) demands an accompanying change in pressure (or temperature) to preserve the number of phases in equilibrium. That is, the variance of the system has fallen to 1.

8.2 The phase rule

In one of the most elegant calculations of the whole of chemical thermodynamics, J.W. Gibbs deduced the **phase rule**,[2] which is a general relation between the

---

2. Josiah Willard Gibbs spent most of his working life at Yale, and may justly be regarded as the originator of chemical thermodynamics. He reflected for years before publishing his conclusions, and then did so in precisely expressed papers in an obscure journal (*The Transactions of the Connecticut Academy of Arts and Sciences*). He needed interpreters before the power of his work was recognized and before it could be applied to industrial processes. He is regarded by many as the first great American theoretical scientist.

variance, F, the number of components, C, and the number of phases at equilibrium, P, for a system of any composition:

$$F = C - P + 2 \tag{8.1}$$

Justification 8.1

Consider first the special case of a one-component system. For two phases in equilibrium, we can write $\mu_J(\alpha) = \mu_J(\beta)$. Each chemical potential is a function of the pressure and temperature, so

$$\mu_J(\alpha, p, T) = \mu_J(\beta, p, T)$$

This is an equation relating p and T, so only one of these variables is independent (just as the equation $x + y = 2$ is a relation for y in terms of x: $y = 2 - x$). That conclusion is consistent with $F = 1$. For three phases in mutual equilibrium,

$$\mu_J(\alpha, p, T) = \mu_J(\beta, p, T) = \mu_J(\gamma, p, T)$$

This relation is actually two equations for two unknowns ($\mu_J(\alpha, p, T) = \mu_J(\beta, p, T)$ and $\mu_J(\beta, p, T) = \mu_J(\gamma, p, T)$), and therefore has a solution only for a single value of p and T (just as the pair of equations $x + y = 2$ and $3x - y = 4$ has the single solution $x = \frac{3}{2}$ and $y = \frac{1}{2}$). That conclusion is consistent with $F = 0$. Four phases cannot be in mutual equilibrium in a one-component system because the three equalities

$$\mu_J(\alpha, p, T) = \mu_J(\beta, p, T) \qquad \mu_J(\beta, p, T) = \mu_J(\gamma, p, T) \qquad \mu_J(\gamma, p, T) = \mu_J(\delta, p, T)$$

are three equations for two unknowns (p and T) and are not consistent (just as $x + y = 2$, $3x - y = 4$, and $x + 4y = 6$ has no solution).

Now consider the general case. We begin by counting the total number of intensive variables (properties that do not depend on the size of the system). The pressure, p, and temperature, T, count as 2. We can specify the composition of a phase by giving the mole fractions of $C - 1$ components. We need specify only $C - 1$ and not all C mole fractions because $x_1 + x_2 + \cdots + x_C = 1$, and all mole fractions are known if all except one are specified. Because there are P phases, the total number of composition variables is $P(C - 1)$. At this stage, the total number of intensive variables is $P(C - 1) + 2$.

At equilibrium, the chemical potential of a component J must be the same in every phase (Section 6.4):

$$\mu_J(\alpha) = \mu_J(\beta) = \cdots \quad \text{for } P \text{ phases}$$

That is, there are $P - 1$ equations of this kind to be satisfied for each component J. As there are C components, the total number of equations is $C(P - 1)$. Each equation reduces our freedom to vary one of the $P(C - 1) + 2$ intensive variables. It follows that the total number of degrees of freedom is

$$F = P(C - 1) + 2 - C(P - 1) = C - P + 2$$

which is eqn 1.

(a) One-component systems

For a one-component system, such as pure water, $F = 3 - P$. When only one phase is present, $F = 2$ and both p and T can be varied independently without changing the number of phases. In other words, a single phase is represented by an *area* on a phase

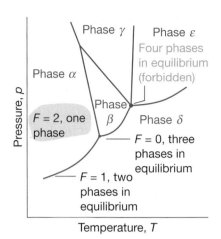

8.2 The typical regions of a one-component phase diagram. The lines represent conditions under which the two adjoining phases are in equilibrium. A point represents the unique set of conditions under which three phases coexist in equilibrium. Four phases cannot mutually coexist in equilibrium.

diagram. When two phases are in equilibrium $F = 1$, which implies that pressure is not freely variable if the temperature is set; indeed, at a given temperature, a liquid has a characteristic vapour pressure. It follows that the equilibrium of two phases is represented by a *line* in the phase diagram. Instead of selecting the temperature, we could select the pressure, but having done so the two phases would be in equilibrium at a single definite temperature. Therefore, freezing (or any other phase transition) occurs at a definite temperature at a given pressure.

When three phases are in equilibrium, $F = 0$ and the system is invariant. This special condition can be established only at a definite temperature and pressure that is characteristic of the substance and outside our control. The equilibrium of three phases is therefore represented by a *point*, the triple point, on a phase diagram. Four phases cannot be in equilibrium in a one-component system because F cannot be negative. These features are summarized in Fig. 8.2.

We can identify the features in Fig. 8.2 in the experimentally determined phase diagram for water (Fig. 8.3). This diagram summarizes the changes that take place as a sample, such as that at a, is cooled at constant pressure. The sample remains entirely gaseous until the temperature reaches b, when liquid appears. Two phases are now in equilibrium and $F = 1$. Because we have decided to specify the pressure, which uses up the single degree of freedom, the temperature at which this equilibrium occurs is not under our control. Lowering the temperature takes the system to c in the one-phase, liquid region. The temperature can now be varied around the point c at will, and only when ice appears at d does the variance become 1 again.

(b) Experimental procedures

Detecting a phase change is not always as simple as seeing water boil in a kettle, so special techniques have been developed. Two techniques are **thermal analysis**, which takes advantage of the effect of the enthalpy change during a first-order transition (Section 6.7), and differential scanning calorimetry (see Box 2.1). They are useful for solid–solid transitions, where simple visual inspection of the sample may be inadequate. In thermal analysis, a sample is allowed to cool and its temperature is monitored. At a first-order transition, heat is evolved and the cooling stops until the transition is complete. The cooling curve along the isobar *cde* in Fig. 8.3 therefore has the shape shown in Fig. 8.4. The transition temperature is obvious, and is used to mark point d on the phase diagram.

Modern work on phase transitions often deals with systems at very high pressures, and more sophisticated detection procedures must be adopted. Some of the highest pressures currently attainable are produced in a **diamond-anvil cell** like that illustrated in Fig. 8.5. The sample is placed in a minute cavity between two gemquality diamonds, and then pressure is exerted simply by turning the screw. The advance in design this represents is quite remarkable for, with a turn of the screw, pressures of up to about 1 Mbar can be reached which a few years ago could not be reached with equipment weighing tons.

The pressure is monitored spectroscopically by observing the shift of spectral lines in small pieces of ruby added to the sample, and the properties of the sample itself are observed optically through the diamond anvils. One application of the technique is to study the transition of covalent solids to metallic solids. Iodine, I_2, for instance, becomes metallic at around 200 kbar and makes a transition to a monatomic metallic solid at around 210 kbar. Studies such as these are relevant to the structure of material deep inside the Earth (at the centre of the Earth the pressure is around 5 Mbar) and in the interiors of the giant planets, where even hydrogen may be metallic.

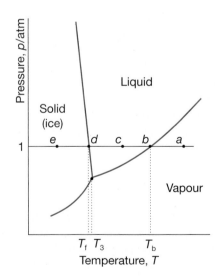

8.3 The phase diagram for water, a simplified version of Fig. 6.4. The label T_3 marks the temperature of the triple point, T_b the normal boiling point, and T_f the normal freezing point.

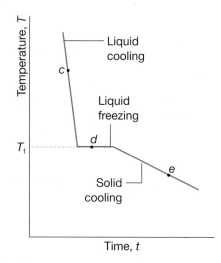

8.4 The cooling curve for the isobar *cde* in Fig. 8.3. The halt marked *d* corresponds to the pause in the fall of temperature while the first-order exothermic transition (freezing) occurs. This pause enables T_f to be located even if the transition cannot be observed visually.

Two-component systems

When two components are present in a system, $C = 2$ and $F = 4 - P$. If the temperature is constant, the remaining variance is $F' = 3 - P$, which has a maximum value of 2. (The prime on F indicates that one of the degrees of freedom has been discarded; in this case the temperature.) One of these two remaining degrees of freedom is the pressure and the other is the composition (as expressed by the mole fraction of one component). Hence, one form of the phase diagram is a map of pressures and compositions at which each phase is stable. Alternatively, the pressure could be held constant and the phase diagram depicted in terms of temperature and composition.

8.3 Vapour pressure diagrams

The partial vapour pressures of the components of an ideal solution of two volatile liquids are related to the composition of the liquid mixture by Raoult's law (Section 7.3a)

$$p_A = x_A p_A^* \qquad p_B = x_B p_B^* \tag{8.2}°$$

where p_A^* is the vapour pressure of pure A and p_B^* that of pure B. The total vapour pressure p of the mixture is therefore

$$p = p_A + p_B = x_A p_A^* + x_B p_B^* = p_B^* + (p_A^* - p_B^*)x_A \tag{8.3}°$$

This expression shows that the total vapour pressure (at some fixed temperature) changes linearly with the composition from p_B^* to p_A^* as x_A changes from 0 to 1 (Fig. 8.6).

(a) The composition of the vapour

The compositions of the liquid and vapour that are in mutual equilibrium are not necessarily the same. Common sense suggests that the vapour should be richer in the more volatile component. This expectation can be confirmed as follows. The partial pressures of the components are given by eqn 2. It follows from Dalton's law that the mole fractions in the gas, y_A and y_B, are

$$y_A = \frac{p_A}{p} \qquad y_B = \frac{p_B}{p} \tag{8.4}$$

Provided the mixture is ideal, the partial pressures and the total pressure may be expressed in terms of the mole fractions in the liquid by using eqn 2 for p_J and eqn 3 for the total vapour pressure p, which gives

$$y_A = \frac{x_A p_A^*}{p_B^* + (p_A^* - p_B^*)x_A} \qquad y_B = 1 - y_A \tag{8.5}°$$

Figure 8.7 shows the composition of the vapour plotted against the composition of the liquid for various values of $p_A^*/p_B^* > 1$. We see that in all cases $y_A > x_A$; that is, the vapour is richer than the liquid in the more volatile component. Note that, if B is non-volatile, so that $p_B^* = 0$ at the temperature of interest, then it makes no contribution to the vapour ($y_B = 0$).

Equation 3 shows how the total vapour pressure of the mixture varies with the composition of the liquid. Because we can relate the composition of the liquid to the composition of the vapour through eqn 5, we can now also relate the total vapour pressure to the composition of the vapour:

8.5 Ultrahigh pressures (up to about 2 Mbar) can be achieved using a diamond anvil. The sample, together with a ruby for pressure measurement and a drop of liquid for pressure transmission, are placed between two gem-quality diamonds. The principle of its action is like that of a nutcracker: the pressure is exerted by turning the screw by hand.

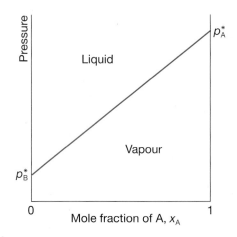

8.6 The variation of the total vapour pressure of a binary mixture with the mole fraction of A in the liquid when Raoult's law is obeyed.

8.7 The mole fraction of A in the vapour of a binary ideal solution expressed in terms of its mole fraction in the liquid, calculated using eqn 5 for various values of p_A^*/p_B^* (the label on each curve) with A more volatile than B. In all cases the vapour is richer than the liquid in A.

$$p = \frac{p_A^* p_B^*}{p_A^* + (p_B^* - p_A^*)y_A} \tag{8.6}°$$

This expression is plotted in Fig. 8.8.

(b) The interpretation of the diagrams

If we are interested in distillation, both the vapour and the liquid compositions are of equal interest. It is therefore sensible to combine the two preceding diagrams into one. The result is summarized in Fig. 8.9. The point a indicates the vapour pressure of a mixture of composition x_A, and the point b indicates the composition of the vapour that is in equilibrium with the liquid at that pressure. Note that when two phases are in equilibrium, $P = 2$ so $F' = 1$ (as usual, the prime indicating that one

8.8 The dependence of the vapour pressure of the same system as in Fig. 8.7, but expressed in terms of the mole fraction of A in the vapour by using eqn 6. Individual curves are labelled with the value of p_A^*/p_B^*.

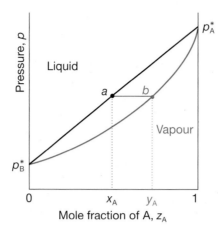

8.9 The dependence of the total vapour pressure of an ideal solution on the mole fraction of A in the entire system. A point between the two lines corresponds to both liquid and vapour being present; outside that region there is only one phase present. The mole fraction of A is denoted z_A, as explained below.

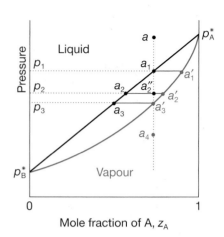

8.10 The points of the pressure–composition diagram discussed in the text. The vertical line through a is an *isopleth*, a line of constant composition of the entire system.

8.11 (a) A liquid in a container exists in equilibrium with its vapour. The superimposed fragment of the phase diagram shows the compositions of the two phases and their abundances (by the lever rule). (b) When the pressure is changed by drawing out a piston, the compositions of the phases adjust as shown by the tie line in the phase diagram. (c) When the piston is pulled so far out that all the liquid has vaporized and only the vapour is present, the pressure falls as the piston is withdrawn and the point on the phase diagram moves into the one-phase region.

degree of freedom, the temperature, has already been discarded). That is, if the composition is specified (so using up the only remaining degree of freedom), the pressure at which the two phases are in equilibrium is fixed.

A richer interpretation of the phase diagram is obtained if we interpret the horizontal axis as showing the *overall* composition, z_A, of the system. If the horizontal axis of the vapour pressure diagram is labelled with z_A, then all the points down to the solid diagonal line in the graph correspond to a system that is under such high pressure that it contains only a liquid phase (the applied pressure is higher than the vapour pressure), so $z_A = x_A$, the composition of the liquid. On the other hand, all points below the lower curve correspond to a system that is under such low pressure that it contains only a vapour phase (the applied pressure is lower than the vapour pressure), so $z_A = y_A$.

Points that lie between the two lines correspond to a system in which there are two phases present, one a liquid and the other a vapour. To see this interpretation, consider the effect of lowering the pressure on a liquid mixture of overall composition a in Fig. 8.10. The lowering of pressure can be achieved by drawing out a piston (Fig. 8.11). This degree of freedom is permitted by the phase rule because $F' = 2$ when $P = 1$, and even if the composition is selected one degree of freedom remains. The changes to the system do not affect the overall composition, so the state of the system moves down the vertical line that passes through a. This vertical line is called an **isopleth**, from the Greek words for 'equal abundance'. Until the point a_1 is reached (when the pressure has been reduced to p_1), the sample consists of a single liquid phase. At a_1 the liquid can exist in equilibrium with its vapour. As we have seen, the composition of the vapour phase is given by point a'_1. A line joining two points representing phases in equilibrium is called a **tie line**. The composition of the liquid is the same as initially (a_1 lies on the isopleth through a), so we have to conclude that at this pressure there is virtually no vapour present; however, the tiny amount of vapour that is present has the composition a'_1.

Now consider the effect of lowering the pressure to p_2, so taking the system to a pressure and overall composition represented by the point a''_2. This new pressure is below the vapour pressure of the original liquid, so it vaporizes until the vapour

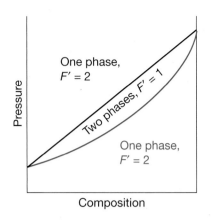

8.12 The general scheme of interpretation of a pressure–composition diagram (a vapour pressure diagram).

pressure of the remaining liquid falls to p_2. Now we know that the composition of such a liquid must be a_2. Moreover, the composition of the vapour in equilibrium with that liquid must be given by the point a_2' at the other end of the tie line. Note that two phases are now in equilibrium, so $F' = 1$ for all points between the two lines; hence, for a given pressure (such as at p_2) the variance is zero, and the vapour and liquid phases have fixed compositions (Fig. 8.12). If the pressure is reduced to p_3, a similar readjustment in composition takes place, and now the compositions of the liquid and vapour are represented by the points a_3 and a_3', respectively. The latter point corresponds to a system in which the composition of the vapour is the same as the overall composition, so we have to conclude that the amount of liquid present is now virtually zero, but the tiny amount of liquid present has the composition a_3. A further decrease in pressure takes the system to the point a_4; at this stage, only vapour is present and its composition is the same as the initial overall composition of the system (the composition of the original liquid).

(c) The lever rule

A point in the two-phase region of a phase diagram indicates not only qualitatively that both liquid and vapour are present, but represents quantitatively the relative amounts of each. To find the relative amounts of two phases α and β that are in equilibrium, we measure the distances l_α and l_β along the horizontal tie line, and then use the **lever rule** (Fig. 8.13):

$$n_\alpha l_\alpha = n_\beta l_\beta \tag{8.7}$$

Here n_α is the amount of phase α and n_β the amount of phase β. In the case illustrated in Fig. 8.13, because $l_\beta \approx 2l_\alpha$, the amount of phase α is about twice the amount of phase β.

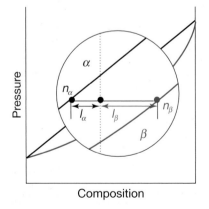

8.13 The lever rule. The distances l_α and l_β are used to find the proportions of the amounts of phases α (such as vapour) and β (for example, liquid) present at equilibrium. The lever rule is so called because a similar rule relates the masses at two ends of a lever to their distances from a pivot ($m_\alpha l_\alpha = m_\beta l_\beta$ for balance).

Justification 8.2

To prove the lever rule we write $n = n_\alpha + n_\beta$ and the overall amount of A as nz_A. The overall amount of A is also the sum of its amounts in the two phases:

$$nz_A = n_\alpha x_A + n_\beta y_A$$

Since also

$$nz_A = n_\alpha z_A + n_\beta z_A$$

by equating these two expressions it follows that

$$n_\alpha(x_A - z_A) = n_\beta(z_A - y_A)$$

which corresponds to eqn 7.

Illustration 8.1

At p_1 in Fig. 8.10, the ratio l_{vap}/l_{liq} is almost infinite for this tie line, so n_{liq}/n_{vap} is also almost infinite, and there is only a trace of vapour present. When the pressure is reduced to p_2, the value of l_{vap}/l_{liq} is about 0.3, so $n_{liq}/n_{vap} \approx 0.3$ and the amount of liquid is about 0.3 times the amount of vapour. When the pressure has been reduced to p_3, the sample is almost completely gaseous and because $l_{vap}/l_{liq} \approx 0$ we conclude that there is only a trace of liquid present.

8.4 Temperature–composition diagrams

To discuss distillation we need a **temperature–composition diagram**, a phase diagram in which the boundaries show the composition of the phases that are in equilibrium at various temperatures (and a given pressure, typically 1 atm). An example is shown in Fig. 8.14. Note that the liquid phase now lies in the lower part of the diagram.

(a) The distillation of mixtures

The region between the lines in Fig. 8.14 is a two-phase region where $F' = 1$. As usual, the prime indicates that one degree of freedom has been discarded; in this case, the pressure is being kept fixed, and hence at a given temperature the compositions of the phases in equilibrium are fixed. The regions outside the phase lines correspond to a single phase, so $F' = 2$, and the temperature and composition are both independently variable.

Consider what happens when a liquid of composition a_1 is heated. It boils when the temperature reaches T_2. Then the liquid has composition a_2 (the same as a_1) and the vapour (which is present only as a trace) has composition a'_2. The vapour is richer in the more volatile component A (the component with the lower boiling point). From the location of a_2, we can state the vapour's composition at the boiling point, and from the location of the tie line joining a_2 and a'_2 we can read off the boiling temperature (T_2) of the original liquid mixture.

In a **simple distillation**, the vapour is withdrawn and condensed. This technique is used to separate a volatile liquid from a non-volatile solute or solid. In **fractional distillation**, the boiling and condensation cycle is repeated successively. This technique is used to separate volatile liquids. We can follow the changes that occur by seeing what happens when the first condensate of composition a_3 is reheated. The phase diagram shows that this mixture boils at T_3 and yields a vapour of composition a'_3, which is even richer in the more volatile component. That vapour is drawn off, and the first drop condenses to a liquid of composition a_4. The cycle can then be repeated until in due course almost pure A is obtained.

The efficiency of a fractionating column is expressed in terms of the number of **theoretical plates**, the number of effective vaporization and condensation steps that are required to achieve a condensate of given composition from a given distillate. Thus, to achieve the degree of separation shown in Fig. 8.15a, the fractionating

The textbook's web site contains links to online databases of data on phase diagrams.

8.14 The temperature–composition diagram corresponding to an ideal mixture with the component A more volatile than component B. Successive boilings and condensations of a liquid originally of composition a_1 lead to a condensate that is pure A. The separation technique is called fractional distillation.

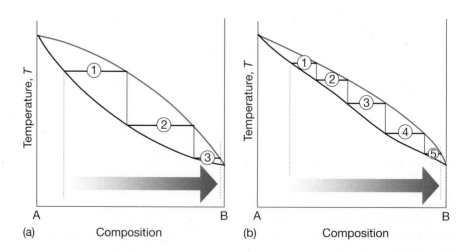

8.15 The number of theoretical plates is the number of steps needed to bring about a specified degree of separation of two components in a mixture. The two systems shown correspond to (a) 3, (b) 5 theoretical plates.

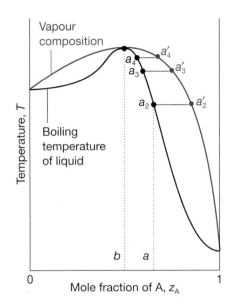

8.16 A high-boiling azeotrope. When the liquid of composition a is distilled, the composition of the remaining liquid changes towards b but no further.

column must correspond to three theoretical plates. To achieve the same separation for the system shown in Fig. 8.15b, in which the components have more similar partial pressures, the fractionating column must be designed to correspond to five theoretical plates.

(b) Azeotropes

Although many liquids have temperature–composition phase diagrams resembling the ideal version in Fig. 8.14, in a number of important cases there are marked deviations. A maximum in the phase diagram (Fig. 8.16) may occur when the favourable interactions between A and B molecules reduce the vapour pressure of the mixture below the ideal value: in effect, the A–B interactions stabilize the liquid. In such cases the excess Gibbs energy, G^E (Section 7.4), is negative (more favourable to mixing than ideal). Examples of this behaviour include trichloromethane/propanone and nitric acid/water mixtures. Phase diagrams showing a minimum (Fig. 8.17) indicate that the mixture is destabilized relative to the ideal solution, the A–B interactions then being unfavourable. For such mixtures G^E is positive (less favourable to mixing than ideal), and there may be contributions from both enthalpy and entropy effects. Examples include dioxane/water and ethanol/water mixtures.

Deviations from ideality are not always so strong as to lead to a maximum or minimum in the phase diagram, but when they do there are important consequences for distillation. Consider a liquid of composition a on the right of the maximum in Fig. 8.16. The vapour (at a_2') of the boiling mixture (at a_2) is richer in A. If that vapour is removed (and condensed elsewhere), then the remaining liquid will move to a composition that is richer in B, such as that represented by a_3, and the vapour in equilibrium with this mixture will have composition a_3'. As that vapour is removed, the composition of the boiling liquid shifts to a point such as a_4, and the composition of the vapour shifts to a_4'. Hence, as evaporation proceeds, the composition of the remaining liquid shifts towards B as A is drawn off. The boiling point of the liquid rises, and the vapour becomes richer in B. When so much A has been evaporated that the liquid has reached the composition b, the vapour has the same composition as the liquid. Evaporation then occurs without change of composition. The mixture is said to form an **azeotrope**.[3] When the azeotropic composition has been reached, distillation cannot separate the two liquids because the condensate has the same composition as the azeotropic liquid. One example of azeotrope formation is hydrochloric acid/water, which is azeotropic at 80 per cent by mass of water and boils unchanged at 108.6°C.

The system shown in Fig. 8.17 is also azeotropic, but shows its azeotropy in a different way. Suppose we start with a mixture of composition a_1, and follow the changes in the composition of the vapour that rises through a fractionating column (essentially a vertical glass tube packed with glass rings to give a large surface area). The mixture boils at a_2 to give a vapour of composition a_2'. This vapour condenses in the column to a liquid of the same composition (now marked a_3). That liquid reaches equilibrium with its vapour at a_3', which condenses higher up the tube to give a liquid of the same composition, which we now call a_4. The fractionation therefore shifts the vapour towards the azeotropic composition at b, but not beyond, and the azeotropic vapour emerges from the top of the column. An example is ethanol/water, which boils unchanged when the water content is 4 per cent by mass and the temperature is 78°C.

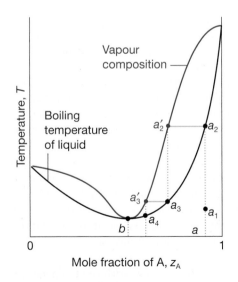

8.17 A low-boiling azeotrope. When the mixture at a is fractionally distilled, the vapour in equilibrium in the fractionating column moves towards b and then remains unchanged.

3. The name comes from the Greek words for 'boiling without changing'.

8.18 The distillation of (a) two immiscible liquids can be regarded as (b) the joint distillation of the separated components, and boiling occurs when the sum of the partial pressures equals the external pressure.

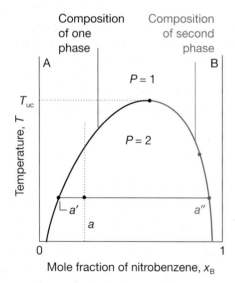

8.19 The temperature–composition diagram for hexane and nitrobenzene at 1 atm. The region below the curve corresponds to the compositions and temperatures at which the liquids are partially miscible. The upper critical temperature, T_{uc}, is the temperature above which the two liquids are miscible in all proportions.

(c) Immiscible liquids

Finally, we consider the distillation of two immiscible liquids, such as octane and water. At equilibrium, there is a tiny amount of A dissolved in B, and similarly a tiny amount of B dissolved in A: both liquids are saturated with the other component (Fig. 8.18a). As a result, the total vapour pressure of the mixture is close to $p = p_A^* + p_B^*$. If the temperature is raised to the value at which this total vapour pressure is equal to the atmospheric pressure, boiling commences and the dissolved substances are purged from their solution. However, this boiling results in a vigorous agitation of the mixture, so each component is kept saturated in the other component, and the purging continues as the very dilute solutions are replenished. This intimate contact is essential: two immiscible liquids heated in a container like that shown in Fig. 8.18b would not boil at the same temperature. The presence of the saturated solutions means that the 'mixture' boils at a lower temperature than either component would alone because boiling begins when the total vapour pressure reaches 1 atm, not when either vapour pressure reaches 1 atm. This distinction is the basis of **steam distillation**, which enables some heat-sensitive, water-insoluble organic compounds to be distilled at a lower temperature than their normal boiling point. The only snag is that the composition of the condensate is in proportion to the vapour pressures of the components, so oils of low volatility distil in low abundance.

8.5 Liquid–liquid phase diagrams

Now we consider temperature–composition diagrams for systems that consist of pairs of **partially miscible** liquids, which are liquids that do not mix in all proportions at all temperatures. An example is hexane and nitrobenzene. The same principles of interpretation apply as to liquid–vapour diagrams. When $P = 2$, $F' = 1$ (the prime denoting the adoption of constant pressure), and the selection of a temperature implies that the compositions of the immiscible liquid phases are fixed. When $P = 1$ (corresponding to a system in which the two liquids are fully mixed), both the temperature and the composition may be adjusted.

(a) Phase separation

Suppose a small amount of a liquid B is added to a sample of another liquid A at a temperature T'. It dissolves completely, and the binary system remains a single phase. As more B is added, a stage comes at which no more dissolves. The sample now consists of two phases in equilibrium with each other ($P = 2$), the most abundant one consisting of A saturated with B, the minor one a trace of B saturated with A. In the temperature–composition diagram drawn in Fig. 8.19, the composition of the former is represented by the point a' and that of the latter by the point a''. The relative abundances of the two phases are given by the lever rule.

When more B is added, A dissolves in it slightly. The compositions of the two phases in equilibrium remain a' and a'' because $P = 2$ implies that $F' = 0$, and hence that the compositions of the phases are invariant at a fixed temperature and pressure. However, the amount of one phase increases at the expense of the other. A stage is reached when so much B is present that it can dissolve all the A, and the system reverts to a single phase. The addition of more B now simply dilutes the solution, and from then on it remains a single phase.

The composition of the two phases at equilibrium varies with the temperature. For hexane and nitrobenzene, raising the temperature increases their miscibility. The two-phase system therefore becomes less extensive, because each phase in equilibrium is richer in its minor component: the A-rich phase is richer in B and the B-rich phase is richer in A. We can construct the entire phase diagram by repeating

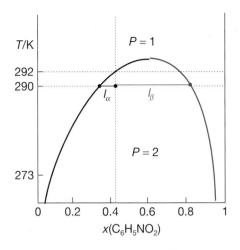

8.20 The temperature–composition diagram for hexane and nitrobenzene at 1 atm again, with the points and lengths discussed in the text.

the observations at different temperatures and drawing the envelope of the two-phase region.

Example 8.2 Interpreting a liquid–liquid phase diagram

A mixture of 50 g of hexane (0.59 mol C_6H_{14}) and 50 g nitrobenzene (0.41 mol $C_6H_5NO_2$) was prepared at 290 K. What are the compositions of the phases, and in what proportions do they occur? To what temperature must the sample be heated in order to obtain a single phase? Refer to Fig. 8.20.

Method The compositions of phases in equilibrium are given by the points where the tie line representing the temperature intersects the phase boundary. Their proportions are given by the lever rule (eqn 7). The temperature at which the components are completely miscible is found by following the isopleth upwards and noting the temperature at which it enters the one-phase region of the phase diagram.

Answer We denote hexane by H and nitrobenzene by N; refer to Fig. 8.20. The point $x_N = 0.41$, $T = 290$ K occurs in the two-phase region of the phase diagram. The horizontal tie line cuts the phase boundary at $x_N = 0.35$ and $x_N = 0.83$, so those are the compositions of the two phases. The ratio of amounts of each phase is equal to the ratio of the distances l_α and l_β:

$$\frac{n_\alpha}{n_\beta} = \frac{l_\beta}{l_\alpha} = \frac{0.83 - 0.41}{0.41 - 0.35} = \frac{0.42}{0.06} = 7$$

That is, there is about 7 times more hexane-rich phase than nitrobenzene-rich phase. Heating the sample to 292 K takes it into the single-phase region.

Comment Because the phase diagram has been constructed experimentally, these conclusions are not based on any assumptions about ideality. They would be modified if the system were subjected to a different pressure.

Self-test 8.2 Repeat the problem for 50 g of hexane and 100 g of nitrobenzene at 273 K.   [$x_N = 0.09$ and 0.95 in ratio 1:1.3; 294 K]

(b) Critical solution temperatures

The **upper critical solution temperature**, T_{uc}, is the highest temperature at which phase separation occurs.[4] Above the upper critical temperature the two components are fully miscible. This temperature exists because the greater thermal motion overcomes any potential energy advantage in molecules of one type being close together. One example is the nitrobenzene/hexane system shown in Fig. 8.20; an example of a solid solution is the palladium/hydrogen system, which shows two phases, one a solid solution of hydrogen in palladium and the other a palladium hydride, up to 300°C but forms a single phase at higher temperatures (Fig. 8.21).

The thermodynamic interpretation of the upper critical solution temperature focuses on the Gibbs energy of mixing and its variation with temperature. We saw in Section 7.4 that a simple model of a real solution results in a Gibbs energy of

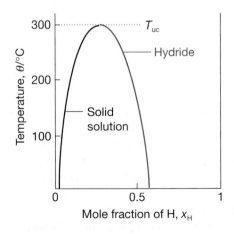

8.21 The phase diagram for palladium and palladium hydride, which has an upper critical temperature at 300°C.

4. The upper critical solution temperature is also called the 'upper consolute temperature'.

8.22 The temperature variation of the Gibbs energy of mixing of a system that is partially miscible at low temperatures. A system of composition in the region $P = 2$ forms two phases with compositions corresponding to the two local minima of the curve. This illustration is a duplicate of Fig. 7.19.

8.23 The location of the phase boundary as computed on the basis of the β-parameter model introduced in Section 7.4.

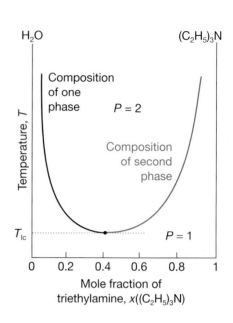

8.24 The temperature–composition diagram for water and triethylamine. This system shows a lower critical temperature at 292 K. The labels indicate the interpretation of the boundaries.

mixing that behaves as shown in Fig. 7.19. Provided the parameter $\beta = w/RT$ is greater than 2, the Gibbs energy of mixing has a double minimum (Fig. 8.22). As a result, for $\beta > 2$ we can expect phase separation to occur. The same model shows that the compositions corresponding to the minima are obtained by looking for the conditions at which $\partial \Delta_{mix}G/\partial x = 0$, and a simple manipulation of eqn 7.31 shows that we have to solve

> This expression is an example of a *transcendental equation*, an equation that does not have a solution that can be expressed in a closed form. The solutions can be found numerically by using mathematical software or by plotting the first term against the second and identifying the points of intersection as β is changed.

$$\ln \frac{x}{1 - x} + \beta(1 - 2x) = 0$$

The solutions are plotted in Fig. 8.23. We see that, as β decreases, which can be interpreted as an increase in temperature provided the intermolecular forces and w (the energy AB interactions relative to AA and BB interactions) remain constant, then the two minima move together and merge when $\beta = 2$. The upper critical solution temperature therefore corresponds to $w/2R$.

Some systems show a **lower critical solution temperature**, T_{lc}, below which they mix in all proportions and above which they form two phases.[5] An example is water and triethylamine (Fig. 8.24). In this case, at low temperatures the two components are more miscible because they form a weak complex; at higher temperatures the complexes break up and the two components are less miscible.

Some systems have both upper and lower critical solution temperatures. They occur because, after the weak complexes have been disrupted, leading to partial miscibility, the thermal motion at higher temperatures homogenizes the mixture again, just as in the case of ordinary partially miscible liquids. The most famous

---

5. The lower critical solution temperature is also called the 'lower consolute temperature'.

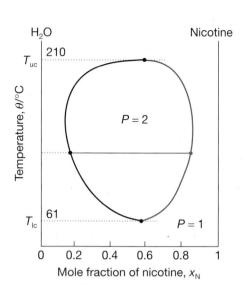

8.25 The temperature–composition diagram for water and nicotine, which has both upper and lower critical temperatures. Note the high temperatures for the liquid (especially the water): the diagram corresponds to a sample under pressure.

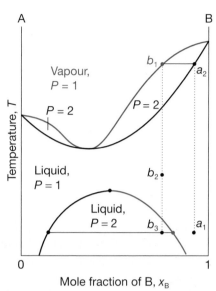

8.26 The temperature–composition diagram for a binary system in which the upper critical temperature is less than the boiling point at all compositions. The mixture forms a low-boiling azeotrope.

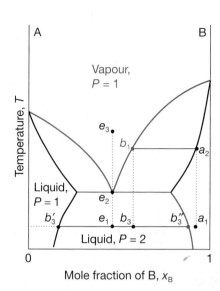

8.27 The temperature–composition diagram for a binary system in which boiling occurs before the two liquids are fully miscible.

example is nicotine and water, which are partially miscible between 6°C and 210°C (Fig. 8.25).

(c) The distillation of partially miscible liquids

Consider a pair of liquids that are partially miscible and form a low-boiling azeotrope. This combination is quite common because both properties reflect the tendency of the two kinds of molecule to avoid each other. There are two possibilities: one in which the liquids become fully miscible before they boil, the other in which boiling occurs before mixing is complete.

Figure 8.26 shows the phase diagram for two components that become fully miscible before they boil. Distillation of a mixture of composition a_1 leads to a vapour of composition b_1, which condenses to the completely miscible single-phase solution at b_2. Phase separation occurs only when this distillate is cooled to a point in the two-phase liquid region, such as b_3. This description applies only to the first drop of distillate. If distillation continues, the composition of the remaining liquid changes. In the end, when the whole sample has evaporated and condensed, the composition is back to a_1.

Figure 8.27 shows the second possibility, in which there is no upper critical solution temperature. The distillate obtained from a liquid initially of composition a_1 has composition b_3 and is a two-phase mixture. One phase has composition b_3' and the other has composition b_3''.

The behaviour of a system of composition represented by the isopleth e in Fig. 8.27 is interesting. A system at e_1 forms two phases, which persist (but with changing proportions) up to the boiling point at e_2. The vapour of this mixture has the same composition as the liquid (the liquid is an azeotrope). Similarly, condensing a vapour of composition e_3 gives a two-phase liquid of the same overall composition. At a fixed temperature, the mixture vaporizes and condenses like a single substance.

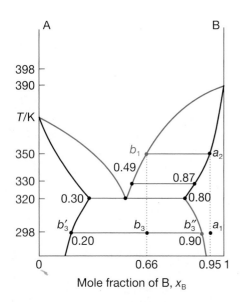

8.28 The points of the phase diagram in Fig. 8.27 that are discussed in Example 8.3.

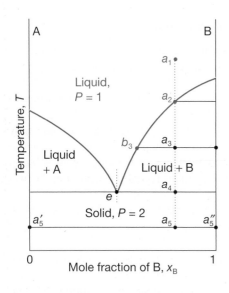

8.29 The temperature–composition phase diagram for two almost immiscible solids and their completely miscible liquids. Note the similarity to Fig. 8.27. The isopleth through e corresponds to the eutectic composition, the mixture with lowest melting point.

Example 8.3 Interpreting a phase diagram

State the changes that occur when a mixture of composition $x_B = 0.95$ (a_1) in Fig. 8.28 is boiled and the vapour condensed.

Method The area in which the point lies gives the number of phases; the compositions of the phases are given by the points at the intersections of the horizontal tie line with the phase boundaries; the relative abundances are given by the lever rule (eqn 7).

Answer The initial point is in the one-phase region. When heated it boils at 350 K (a_2) giving a vapour of composition $x_B = 0.66$ (b_1). The liquid gets richer in B, and the last drop (of pure B) evaporates at 390 K. The boiling range of the liquid is therefore 350 to 390 K. If the initial vapour is drawn off, it has a composition $x_B = 0.66$. This composition would be maintained if the sample were very large, but for a finite sample it shifts to higher values and ultimately to $x_B = 0.95$. Cooling the distillate corresponds to moving down the $x_B = 0.66$ isopleth. At 330 K, for instance, the liquid phase has composition $x_B = 0.87$, the vapour $x_B = 0.49$; their relative proportions are 1:3. At 320 K the sample consists of three phases: the vapour and two liquids. One liquid phase has composition $x_B = 0.30$; the other has composition $x_B = 0.80$ in the ratio 0.62:1. Further cooling moves the system into the two-phase region, and at 298 K the compositions are 0.20 and 0.90 in the ratio 0.82:1. As further distillate boils over, the overall composition of the distillate becomes richer in B. When the last drop has been condensed the phase composition is the same as at the beginning.

Self-test 8.3 Repeat the discussion, beginning at the point $x_B = 0.4$, $T = 298$ K.

8.6 Liquid–solid phase diagrams

Knowledge of the temperature–composition diagrams for solid mixtures guides the design of important industrial processes, such as the manufacture of semiconductors and liquid crystal displays (Boxes 8.1 and 8.2). In this section, we shall consider systems where solid and liquid phases may both be present at temperatures below the boiling point.

Consider the two-component liquid of composition a_1 in Fig. 8.29. The changes that occur may be expressed as follows.

1 $a_1 \rightarrow a_2$. The system enters the two-phase region labelled 'Liquid + B'. Pure solid B begins to come out of solution and the remaining liquid becomes richer in A.

2 $a_2 \rightarrow a_3$. More of the solid forms, and the relative amounts of the solid and liquid (which are in equilibrium) are given by the lever rule. At this stage there are roughly equal amounts of each. The liquid phase is richer in A than before (its composition is given by b_3) because some B has been deposited.

3 $a_3 \rightarrow a_4$. At the end of this step, there is less liquid than at a_3, and its composition is given by e. This liquid now freezes to give a two-phase system of pure B and pure A.

Box 8.1 Liquid crystals

A *mesophase* is a phase intermediate between solid and liquid. Mesophases are of great importance in biology, for they occur as lipid bilayers and in vesicular systems. A mesophase may arise when molecules have highly anisotropic shapes, such as being long and thin, as in (**1**), or disk-like, as in (**2**). When the solid melts, some aspects of the long-range order characteristic of the solid may be retained, and the new phase may be a *liquid crystal*, a substance having liquid-like imperfect long-range order in at least one direction in space but positional or orientational order in at least one other direction. A *thermotropic* liquid crystal displays a transition to the liquid crystalline phase as the temperature is changed. A *lyotropic* liquid crystal is a solution that displays a transition to the liquid crystalline phase as the composition is changed.

One type of retained long-range order gives rise to a *smectic phase* (from the Greek word for soapy), in which the molecules align themselves in layers (see the illustration). Other materials, and some smectic liquid crystals at higher

1

2

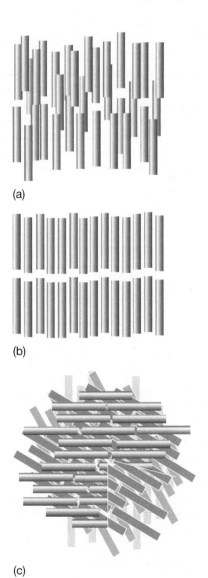

(a)

(b)

(c)

The arrangement of molecules in (a) the nematic phase, (b) the smectic phase, and (c) the cholesteric phase of liquid crystals. In the cholesteric phase, the stacking of layers continues to give a helical arrangement of molecules.

temperatures, lack the layered structure but retain a parallel alignment; this mesophase is called a *nematic phase* (from the Greek for thread, which refers to the observed defect structure of the phase). In the *cholesteric phase* (from the Greek for bile solid) the molecules lie in sheets at angles that change slightly between each sheet. That is, they form helical structures with a pitch that depends on the temperature. As a result, cholesteric liquid crystals diffract light and have colours that depend on the temperature. The strongly anisotropic optical properties of nematic liquid crystals, and their response to electric fields, form the basis of their use as data displays (LCDs).

In a 'twisted nematic' LCD, the liquid crystal is held between two flat plates about 10 μm apart. The inner surface of each plate is coated with a transparent conducting material, such as indium–tin oxide. The plates also have a surface that causes the liquid crystal to adopt a particular orientation at its interface and are typically set at 90° to each other but 270° in a 'supertwist' arrangement. The entire assembly is set between polarizing filters. The incident light passes through the outer polarizer; then its plane of polarization is rotated as it passes through the twisted nematic and, depending on the setting of the second polarizer, will pass through (if that

is how the second polarizer is arranged). When a potential difference is applied across the cell, the helical arrangement is lost and the plane of the light is no longer rotated and will be blocked by the second polarizer.

Although there are many liquid crystalline materials, some difficulty is often experienced in achieving a technologically useful temperature range for the existence of the mesophase. To overcome this difficulty, mixtures can be used. An example of the type of phase diagram that is then obtained is shown in the illustration. As can be seen, the mesophase exists over a wider range of temperatures than either liquid crystalline material alone.

Discussion questions

1. Disk-like molecules, such as (**2**), can form *discotic* liquid crystal mesophases that resemble a column, with the aromatic rings stacked one on top of the other and separated by very small distances (less than 0.5 nm). It has been proposed that such liquid crystals can act as 'molecular wires' for the conduction of electrons along the column. Explain the rationale for this proposal.

2. Some polymers can form liquid crystal mesophases with unusual physical properties. For example, liquid crystalline Kevlar (**3**) is strong enough to be the material of choice for bulletproof vests and is stable at temperatures up to 600 K. What molecular interactions contribute to the formation, thermal stability, and mechanical strength of liquid crystal mesophases in Kevlar?

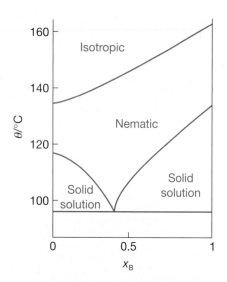

The phase diagram at 1 atm for a binary system of two liquid crystalline materials, 4,4.-dimethoxyazoxybenzene (A) and 4,4.-diethoxyazoxybenzene (B).

Further reading

P.J. Collings and M. Hird, *Introduction to liquid crystals: chemistry and physics*. Taylor & Francis, London (1997).

J.D. Lister and R. Shashidhar, Structure of liquid crystals. In *Encyclopedia of applied physics* (ed. G.L. Trigg), **8**, 515. VCH, New York (1994).

Box 8.2 Ultrapurity and controlled impurity

Advances in technology have called for materials of extreme purity. For example, semiconductor devices consist of almost perfectly pure silicon or germanium doped to a precisely controlled extent. For these materials to operate successfully, the impurity level must be kept down to less than 1 ppb (1 part in 10^9, which corresponds to a small grain of salt in 5 t of sugar).[6]

In the technique of *zone refining* the sample is in the form of a narrow cylinder. This cylinder is heated in a thin disk-like zone which is swept from one end of the sample to the other. The advancing liquid zone accumulates the impurities as it passes. In practice, a train of hot and cold zones are swept repeatedly from one end to the other as shown in the first illustration. The zone at the end of the sample is the impurity dump: when the heater has gone by, it cools to a dirty solid which can be discarded.

The technique makes use of the non-equilibrium properties of the system. It relies on the impurities being more soluble in the molten sample than in the solid, and sweeps them up by passing a molten zone repeatedly from one end to the other along a sample. The phase diagram in the second illustration gives some insight into the process. Consider a liquid (this represents the molten zone) on the isopleth through a_1, and let it cool without the entire sample coming to overall equilibrium. If the temperature falls to a_2 a solid of composition b_2 is deposited and the remaining liquid (the zone where the heater has moved on) is at a_2'. Cooling that liquid down an isopleth passing through a_2' deposits solid of composition b_3 and leaves liquid at a_3'. The process continues until the last drop of liquid to solidify is heavily contamin-

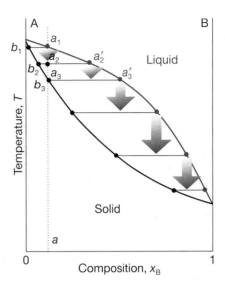

A binary temperature–composition diagram can be used to discuss zone refining, as explained in the text.

ated with B. There is plenty of everyday evidence that impure liquids freeze in this way. For example, an ice cube is clear near the surface but misty in the core: the water used to make ice normally contains dissolved air; freezing proceeds from the outside, and air is accumulated in the retreating liquid phase. It cannot escape from the interior of the cube, and so when that freezes it occludes the air in a mist of tiny bubbles.

A modification of zone refining is *zone levelling*. It is used to introduce controlled amounts of impurity (for example, of indium into germanium). A sample rich in the required dopant is put at the head of the main sample, and made molten. The zone is then dragged repeatedly in alternate directions through the sample, where it deposits a uniform distribution of the impurity.

Discussion questions

1. Use a phase diagram like that in the second illustration to indicate how zone levelling may be described.

2. The technique of *float zoning*, which is similar to zone refining, has produced very pure samples of silicon for use in the semiconductor industry. Consult a textbook of materials science or metallurgy and prepare a discussion of the principles, advantages, and disadvantages of float zoning.

Further reading

H.-G. Lee, *Chemical thermodynamics for metals and materials*. Imperial College Press, London (1999).

D.R. Gaskell, *Introduction to the thermodynamics of materials*. Taylor & Francis, London (1995).

The procedure for zone refining. (a) Initially, impurities are distributed uniformly along the sample. (b) After a molten zone is passed along the rod, the impurities are more concentrated at the right. In practice, a series of molten zones are passed along the rod from left to right.

6. 1 t = 10^3 kg.

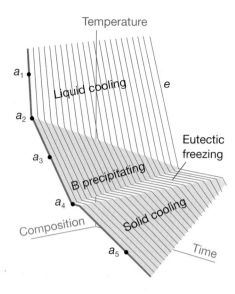

8.30 The cooling curves for the system shown in Fig. 8.29. For isopleth a, the rate of cooling slows at a_2 because solid B deposits from solution. There is a complete halt at a_4 while the eutectic solidifies. This halt is longest for the eutectic isopleth, e. The eutectic halt shortens again for compositions beyond e (richer in A). Cooling curves are used to construct the phase diagram.

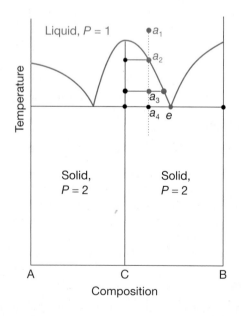

8.31 The phase diagram for a system in which A and B react to form a compound C = AB. This resembles two versions of Fig. 8.29 in each half of the diagram. The constituent C is a true compound, not just an equimolar mixture.

(a) Eutectics

The isopleth at e in Fig. 8.29 corresponds to the **eutectic** composition, the mixture with the lowest melting point.[7] A liquid with the eutectic composition freezes at a single temperature, without previously depositing solid A or B. A solid with the eutectic composition melts, without change of composition, at the lowest temperature of any mixture. Solutions of composition to the right of e deposit B as they cool, and solutions to the left deposit A: only the eutectic mixture (apart from pure A or pure B) solidifies at a single definite temperature ($F' = 0$ when $C = 2$ and $P = 3$) without gradually unloading one or other of the components from the liquid.

One technologically important eutectic is solder, which has mass composition of about 67 per cent tin and 33 per cent lead and melts at 183°C. The eutectic formed by 23 per cent NaCl and 77 per cent H_2O by mass melts at −21.1°C. When salt is added to ice under isothermal conditions (for example, when spread on an icy road) the mixture melts if the temperature is above −21.1°C (and the eutectic composition has been achieved). When salt is added to ice under adiabatic conditions (for example, when added to ice in a vacuum flask) the ice melts, but in doing so it absorbs heat from the rest of the mixture. The temperature of the system falls and, if enough salt is added, cooling continues down to the eutectic temperature. Eutectic formation occurs in the great majority of binary alloy systems, and is of great importance for the microstructure of solid materials. Although a eutectic solid is a two-phase system, it crystallizes out in a nearly homogeneous mixture of microcrystals. The two microcrystalline phases can be distinguished by microscopy and structural techniques such as X-ray diffraction.

Thermal analysis is a very useful practical way of detecting eutectics. We can see how it is used by considering the rate of cooling down the isopleth through a_1 in Fig. 8.29. The liquid cools steadily until it reaches a_2, when B begins to be deposited (Fig. 8.30). Cooling is now slower because the solidification of B is exothermic and retards the cooling. When the remaining liquid reaches the eutectic composition, the temperature remains constant ($F' = 0$) until the whole sample has solidified: this region of constant temperature is the eutectic halt. If the liquid has the eutectic composition e initially, the liquid cools steadily down to the freezing temperature of the eutectic, when there is a long **eutectic halt** as the entire sample solidifies (like the freezing of a pure liquid).

Monitoring the cooling curves at different overall compositions gives a clear indication of the structure of the phase diagram. The solid–liquid boundary is given by the points at which the rate of cooling changes. The longest eutectic halt gives the location of the eutectic composition and its melting temperature.

(b) Reacting systems

Many binary mixtures react to produce compounds, and technologically important examples of this behaviour include the III/V semiconductors, such as the gallium arsenide system, which forms the compound GaAs. Although three constituents are present, there are only two components because GaAs is formed from the reaction Ga + As $\rightleftharpoons$ GaAs. We shall illustrate some of the principles involved with a system that forms a compound C that also forms eutectic mixtures with the species A and B (Fig. 8.31).

A system prepared by mixing an excess of B with A consists of C and unreacted B. This is a binary B, C system, which we suppose forms a eutectic. The principal change from the eutectic phase diagram in Fig. 8.29 is that the whole of the phase diagram is squeezed into the range of compositions lying between equal amounts of A and B ($x_B = 0.5$, marked C in Fig. 8.31) and pure B. The interpretation of the

7. The name comes from the Greek words for 'easily melted'.

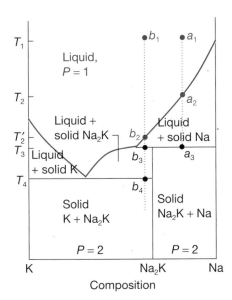

8.32 The phase diagram for an actual system (sodium and potassium) like that shown in Fig. 8.31, but with two differences. One is that the compound is Na_2K, corresponding to A_2B and not AB as in that illustration. The second is that the compound exists only as the solid, not as the liquid. The transformation of the compound at its melting point is an example of incongruent melting.

information in the diagram is obtained in the same way as for Fig. 8.29. The solid deposited on cooling along the isopleth a is the compound C. At temperatures below a_4 there are two solid phases, one consisting of C and the other of B.

(c) Incongruent melting

In some cases the compound C is not stable as a liquid. An example is the alloy Na_2K, which survives only as a solid (Fig. 8.32). Consider what happens as a liquid at a_1 is cooled:

1. $a_1 \rightarrow a_2$. Some solid Na is deposited, and the remaining liquid is richer in K.

2. $a_2 \rightarrow$ just below a_3. The sample is now entirely solid, and consists of solid Na and solid Na_2K.

Now consider the isopleth through b_1:

1. $b_1 \rightarrow b_2$. No obvious change occurs until the phase boundary is reached at b_2 when solid Na begins to deposit.

2. $b_2 \rightarrow b_3$. Solid Na deposits, but at b_3 a reaction occurs to form Na_2K: this compound is formed by the K atoms diffusing into the solid Na.

At this stage the liquid Na/K mixture is in equilibrium with a little solid Na_2K, but there is still no liquid compound.

3. $b_3 \rightarrow b_4$. As cooling continues, the amount of solid compound increases until at b_4 the liquid reaches its eutectic composition. It then solidifies to give a two-phase solid consisting of solid K and solid Na_2K.

If the solid is reheated, the sequence of events is reversed. No liquid Na_2K forms at any stage because it is too unstable to exist as a liquid. This behaviour is an example of **incongruent melting**, in which a compound melts into its components and does not itself form a liquid phase.

Checklist of key ideas

Phases, components, and degrees of freedom

8.1 Definitions

☐ phase, a state of matter that is uniform throughout, not only in chemical composition but also in physical state

☐ constituent, a chemical species (an ion or a molecule)

☐ component, a chemically independent constituent of a system

☐ variance, the number of intensive variables that can be changed independently without disturbing the number of phases in equilibrium

☐ degree of freedom, another name for variance

☐ bivariant system, a system with two degrees of freedom

8.2 The phase rule

☐ phase rule, $F = C - P + 2$

☐ application of the phase rule to one-component systems

☐ thermal analysis, a technique for detecting phase transitions that takes advantage of the effect of the enthalpy change during a first-order transition

☐ differential scanning calorimetry, a technique used for thermal analysis

☐ diamond anvil cell, a device used for high-pressure studies

Two-component systems

8.3 Vapour pressure diagrams

☐ vapour pressure of a mixture, $p = p_B^* + (p_A^* - p_B^*)x_A$

☐ composition of the vapour, $y_A = x_A p_A^* / \{p_B^* + (p_A^* - p_B^*)x_A\}$, $y_B = 1 - y_A$

☐ total vapour pressure of a mixture, $p = p_A^* p_B^* / \{p_A^* + (p_B^* - p_A^*)y_A\}$

☐ overall composition and the interpretation of phase diagrams

☐ isopleth, a line of constant composition

☐ tie line, a line joining two points representing phases in equilibrium

☐ lever rule, $n_\alpha l_\alpha = n_\beta l_\beta$

8.4 Temperature–composition diagrams

☐ temperature–composition diagram, a phase diagram in which the boundaries show the composition of the phases that are in equilibrium at various temperatures

☐ simple distillation, a process in which vapour is withdrawn and condensed

☐ fractional distillation, a process in which the boiling and condensation cycle is repeated successively

☐ theoretical plate, an effective vaporization and condensation step that is required to achieve a condensate of given composition from a given distillate

☐ azeotrope, a mixture that boils without change of composition

☐ the behaviour of immiscible liquids

8.5 Liquid–liquid phase diagrams

☐ partially miscible liquids, liquids that do not mix in all proportions at all temperatures

☐ phase separation and the interpretation of a liquid–liquid phase diagram

☐ upper critical solution temperature, the highest temperature at which phase separation occurs in a binary liquid mixture

☐ lower critical solution temperature, the temperature below which the components of a binary mixture mix in all proportions and above which they form two phases

☐ the interpretation of distillation of partially miscible liquids in terms of phase diagrams

8.6 Liquid–solid phase diagrams

☐ the interpretation of liquid–solid phase diagrams

☐ eutectic, the mixture with the lowest melting point; a liquid with the eutectic composition freezes at a single temperature

☐ eutectic halt, a delay in cooling while the eutectic freezes

☐ the effect of reactions on the appearance of phase diagrams

☐ incongruent melting, when a compound melts into its components and does not itself form a liquid phase

Further reading

Articles of general interest

J.S. Alper, The Gibbs phase rule revisited: interrelationships between components and phases. *J. Chem. Educ.* **76**, 1567 (1999).

R. Battino, The critical point and the number of degrees of freedom. *J. Chem. Educ.* **68**, 276 (1991).

W.B. Daniels, High-pressure techniques. In *Encyclopedia of applied physics* (ed. G.L. Trigg), **7**, 495. VCH, New York (1995).

A. Jayaraman, The diamond-anvil high-pressure cell. *Scientific American* **250** (4), 42 (1984).

N.K. Kildahl, Journey around a phase diagram. *J. Chem. Educ.* **71**, 1052 (1994).

J. Kochansky, Liquid systems with more than two immiscible phases. *J. Chem. Educ.* **68**, 655 (1991).

A.D. Pelton, Phase equilibria. In *Encyclopedia of applied physics* (ed. G.L. Trigg), **13**, 297. VCH, New York (1995).

R.M. Rosenberg, Description of regions in two-component phase diagrams. *J. Chem. Educ.* **76**, 2223 (1999).

R.J. Stead and K. Stead, Phase diagrams for ternary liquid systems. *J. Chem. Educ.* **67**, 385 (1990).

M. Zhao, Z. Wang, and L. Xiao, Determining the number of independent components by Brinkley's method. *J. Chem. Educ.* **69**, 539 (1992).

J. Zhou, W.S.W. Ho, and N.N. Li, Separation processes. In *Encyclopedia of applied physics* (ed. G.L. Trigg), **17**, 599. VCH, New York (1996).

Texts and sources of data and information

A. Alper, *Phase diagrams*, Vols. 1, 2, and 3. Academic Press, New York (1970).

W.D. Callister, Jr., *Materials science and engineering, an introduction*. Wiley, New York (2000).

M. Hillert, *Phase equilibria, phase diagrams and phase transformations: a thermodynamic basis*. Cambridge University Press (1998).

A. Reisman, *Phase equilibria*. Academic Press, New York (1970).

S.I. Sandler (ed.), *Models for thermodynamic and phase equilibria*. Marcel Dekker, New York (1994).

S.I. Sandler, *Chemical and engineering thermodynamics*. Wiley, New York (1998).

M.A. White, *Properties of materials*. Oxford University Press, Oxford (1999).

J. Wisniak, *Phase diagrams: A literature source book*. Elsevier, Amsterdam (1981–86).

Exercises

Discussion questions

8.1 (a) Define the following terms: phase, constituent, component, and degree of freedom.

8.1 (b) What factors determine the number of theoretical plates required to achieve a desired degree of separation in fractional distillation?

8.2 (a) Draw phase diagrams for the following types of systems. Label the regions and intersections of the diagrams, stating what

materials (possibly compounds or azeotropes) are present and whether they are solid, liquid, or gas. (a) One-component, pressure–temperature diagram; liquid density greater than that of solid. (b) Two-component, temperature–composition, solid–liquid diagram; one compound AB formed that melts congruently, negligible solid–solid solubility.

8.2 (b) Draw phase diagrams for the following types of systems. Label the regions and intersections of the diagrams, stating what materials (possibly compounds or azeotropes) are present and whether they are solid, liquid, or gas. (a) Two-component, temperature–composition, solid–liquid diagram; one compound of formula AB_2 that melts incongruently; negligible solid–solid solubility. (b) Two-component, constant temperature–composition, liquid–vapour diagram; formation of an azeotrope at $x_B = 0.333$, complete miscibility.

8.3 (a) Label the regions of the phase diagram in Fig. 8.33. State what substances (if compounds, give their formulas) exist in each region. Label each substance in each region as solid, liquid, or gas.

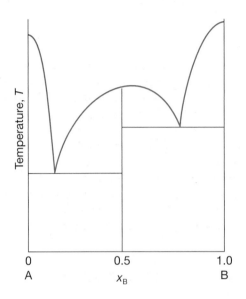

Fig. 8.33

8.3 (b) Label the regions of the phase diagram in Fig. 8.34. State what substances (if compounds, give their formulas) exist in each region. Label each substance in each region as solid, liquid, or gas.

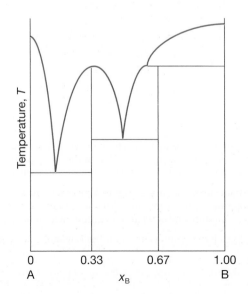

Fig. 8.34

Numerical exercises

8.4 (a) At 90°C, the vapour pressure of methylbenzene is 400 Torr and that of 1,2-dimethylbenzene is 150 Torr. What is the composition of a liquid mixture that boils at 90°C when the pressure is 0.50 atm? What is the composition of the vapour produced?

8.4 (b) At 90°C, the vapour pressure of 1,2-dimethylbenzene is 20 kPa and that of 1,3-dimethylbenzene is 18 kPa. What is the composition of a liquid mixture that boils at 90°C when the pressure is 19 kPa? What is the composition of the vapour produced?

8.5 (a) The vapour pressure of pure liquid A at 300 K is 575 Torr and that of pure liquid B is 390 Torr. These two compounds form ideal liquid and gaseous mixtures. Consider the equilibrium composition of a mixture in which the mole fraction of A in the vapour is 0.350. Calculate the total pressure of the vapour and the composition of the liquid mixture.

8.5 (b) The vapour pressure of pure liquid A at 293 K is 68.8 kPa and that of pure liquid B is 82.1 kPa. These two compounds form ideal liquid and gaseous mixtures. Consider the equilibrium composition of a mixture in which the mole fraction of A in the vapour is 0.612. Calculate the total pressure of the vapour and the composition of the liquid mixture.

8.6 (a) It is found that the boiling point of a binary solution of A and B with $x_A = 0.6589$ is 88°C. At this temperature the vapour pressures of pure A and B are 957.0 Torr and 379.5 Torr, respectively. (a) Is this solution ideal? (b) What is the initial composition of the vapour above the solution?

8.6 (b) It is found that the boiling point of a binary solution of A and B with $x_A = 0.4217$ is 96°C. At this temperature the vapour pressures of pure A and B are 110.1 kPa and 94.93 kPa, respectively. (a) Is this solution ideal? (b) What is the initial composition of the vapour above the solution?

8.7 (a) Dibromoethene (DE, $p_{DE}^* = 172$ Torr at 358 K) and dibromopropene (DP, $p_{DP}^* = 128$ Torr at 358 K) form a nearly ideal solution. If $z_{DE} = 0.60$, what is (a) p_{total} when the system is all liquid, (b) the composition of the vapour when the system is still almost all liquid?

8.7 (b) Benzene and toluene form nearly ideal solutions. At 20°C the vapour pressures of pure benzene and toluene are 74 Torr and 22 Torr, respectively. The solution is boiled by reducing the external pressure below the vapour pressure. Calculate (a) the pressure when boiling begins, (b) the composition of each component in the vapour, and (c) the vapour pressure when only a few drops of liquid remain. Assume that the rate of vaporization is low enough for the temperature to remain constant at 20°C.

8.8 (a) The following temperature/composition data were obtained for a mixture of octane (O) and methylbenzene (M) at 760 Torr, where x is the mole fraction in the liquid and y the mole fraction in the vapour at equilibrium.

$\theta/°C$	110.9	112.0	114.0	115.8	117.3	119.0	121.1	123.0
x_M	0.908	0.795	0.615	0.527	0.408	0.300	0.203	0.097
y_M	0.923	0.836	0.698	0.624	0.527	0.410	0.297	0.164

The boiling points are 110.6°C and 125.6°C for M and O, respectively. Plot the temperature/composition diagram for the mixture. What is the composition of the vapour in equilibrium with the liquid of composition (a) $x_M = 0.250$ and (b) $x_O = 0.250$?

8.8 (b) The following temperature/composition data were obtained for a mixture of two liquids A and B at 1.00 atm, where x is the mole fraction in the liquid and y the mole fraction in the vapour at equilibrium.

$\theta/°C$	125	130	135	140	145	150
x_A	0.91	0.65	0.45	0.30	0.18	0.098
y_A	0.99	0.91	0.77	0.61	0.45	0.25

The boiling points are 124°C for A and 155°C for B. Plot the temperature/composition diagram for the mixture. What is the composition of the vapour in equilibrium with the liquid of composition (a) $x_A = 0.50$ and (b) $x_B = 0.33$?

8.9 (a) State the number of components in the following systems. (a) NaH_2PO_4 in water at equilibrium with water vapour but disregarding the fact that the salt is ionized. (b) The same, but taking into account the ionization of the salt.

8.9 (b) State the number of components for a system in which $AlCl_3$ is dissolved in water, noting that hydrolysis and precipitation of $Al(OH)_3$ occur.

8.10 (a) Blue $CuSO_4 \cdot 5H_2O$ crystals release their water of hydration when heated. How many phases and components are present in an otherwise empty heated container?

8.10 (b) Ammonium chloride, NH_4Cl, decomposes when it is heated. (a) How many components and phases are present when the salt is heated in an otherwise empty container? (b) Now suppose that additional ammonia is also present. How many components and phases are present?

8.11 (a) A saturated solution of Na_2SO_4, with excess of the solid, is present at equilibrium with its vapour in a closed vessel. (a) How many phases and components are present? (b) What is the variance (the number of degrees of freedom) of the system? Identify the independent variables.

8.11 (b) Suppose that the solution referred to in Exercise 8.11a is not saturated. (a) How many phases and components are present? (b) What is the variance (the number of degrees of freedom) of the system? Identify the independent variables.

8.12 (a) Methylethyl ether (A) and diborane, B_2H_6 (B), form a compound that melts congruently at 133 K. The system exhibits two eutectics, one at 25 mole per cent B and 123 K and a second at 90 mole per cent B and 104 K. The melting points of pure A and B are 131 K and 110 K, respectively. Sketch the phase diagram for this system. Assume negligible solid–solid solubility.

8.12 (b) Sketch the phase diagram of the system NH_3/N_2H_4 given that the two substances do not form a compound with each other, that NH_3 freezes at –78°C and N_2H_4 freezes at +2°C, and that a eutectic is formed when the mole fraction of N_2H_4 is 0.07 and that the eutectic melts at –80°C.

8.13 (a) Figure 8.35 shows the phase diagram for two partially miscible liquids, which can be taken to be that for water (A) and 2-methyl-1-propanol (B). Describe what will be observed when a mixture of composition $x_B = 0.8$ is heated, at each stage giving the number, composition, and relative amounts of the phases present.

8.13 (b) Figure 8.36 is the phase diagram for silver and tin. Label the regions, and describe what will be observed when liquids of compositions a and b are cooled to 200 K.

Fig. 8.35

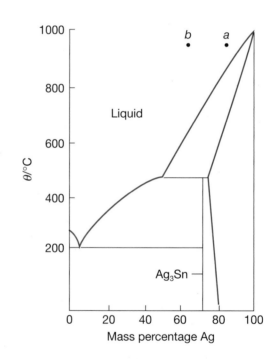

Fig. 8.36

8.14 (a) Indicate on the phase diagram in Fig. 8.37 the feature that denotes incongruent melting. What is the composition of the eutectic mixture and at what temperature does it melt?

8.14 (b) Indicate on the phase diagram in Fig. 8.38 the feature that denotes incongruent melting. What is the composition of the eutectic mixture and at what temperature does it melt?

8.15 (a) Sketch the cooling curves for the isopleths a and b in Fig. 8.37.

8.15 (b) Sketch the cooling curves for the isopleths a and b in Fig. 8.38.

Fig. 8.37

Fig. 8.38

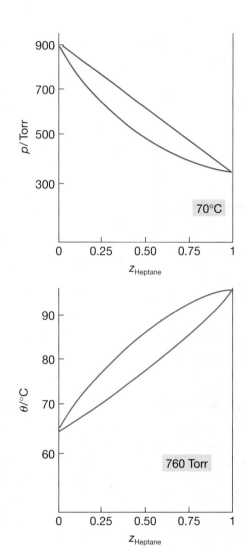

Fig. 8.39

8.16 (a) Use the phase diagram in Fig. 8.36 to state (a) the solubility of Ag in Sn at 800°C, (b) the solubility of Ag_3Sn in Ag at 460°C, (c) the solubility of Ag_3Sn in Ag at 300°C.

8.16 (b) Use the phase diagram in Fig. 8.37 to state (a) the solubility of B in A at 500°C, (b) the solubility of AB_2 in A at 390°C, (c) the solubility of AB_2 in B at 300°C.

8.17 (a) Figure 8.39 shows the experimentally determined phase diagrams for the nearly ideal solution of hexane and heptane. (a) Label the regions of the diagrams as to which phases are present. (b) For a solution containing 1 mol each of hexane and heptane, estimate the vapour pressure at 70°C when vaporization on reduction of the external pressure just begins. (c) What is the vapour pressure of the solution at 70°C when just one drop of liquid remains? (d) Estimate from the figures the mole fraction of hexane in the liquid and vapour phases for the conditions of part b. (e) What are the mole fractions for the conditions of part c? (f) At 85°C and 760 Torr, what are the amounts of substance in the liquid and vapour phases when $z_{heptane} = 0.40$?

8.17 (b) Uranium tetrafluoride and zirconium tetrafluoride melt at 1035°C and 912°C, respectively. They form a continuous series of solid solutions with a minimum melting temperature of 765°C and composition $x(ZrF_4) = 0.77$. At 900°C, the liquid solution of composition $x(ZrF_4) = 0.28$ is in equilibrium with a solid solution of composition $x(ZrF_4) = 0.14$. At 850°C the two compositions are 0.87 and 0.90, respectively. Sketch the phase diagram for this system and state what is observed when a liquid of composition $x(ZrF_4) = 0.40$ is cooled slowly from 900°C to 500°C.

8.18 (a) Methane (melting point 91 K) and tetrafluoromethane (melting point 89 K) do not form solid solutions with each other, and as liquids they are only partially miscible. The upper critical temperature of the liquid mixture is 94 K at $x(CF_4) = 0.43$ and the eutectic temperature is 84 K at $x(CF_4) = 0.88$. At 86 K, the phase in equilibrium with the tetrafluoromethane-rich solution changes from solid methane to a methane-rich liquid. At that temperature, the two liquid solutions that are in mutual equilibrium have the compositions $x(CF_4) = 0.10$ and $x(CF_4) = 0.80$. Sketch the phase diagram.

8.18 (b) Describe the phase changes that take place when a liquid mixture of 4.0 mol B_2H_6 (melting point 131 K) and 1.0 mol CH_3OCH_3 (melting point 135 K) is cooled from 140 K to 90 K. These substances form a compound $(CH_3)_2OB_2H_6$ that melts congruently at 133 K. The system exhibits one eutectic at $x(B_2H_6) = 0.25$ and 123 K and another at $x(B_2H_6) = 0.90$ and 104 K.

8.19 (a) Refer to the information in Exercise 8.18b and sketch the cooling curves for liquid mixtures in which $x(B_2H_6)$ is (a) 0.10, (b) 0.30, (c) 0.50, (d) 0.80, and (e) 0.95.

8.19 (b) Refer to the information in Exercise 8.18a and sketch the cooling curves for liquid mixtures in which $x(CF_4)$ is (a) 0.10, (b) 0.30, (c) 0.50, (d) 0.80, and (e) 0.95.

8.20 (a) Hexane and perfluorohexane show partial miscibility below 22.70°C. The critical concentration at the upper critical temperature is $x = 0.355$, where x is the mole fraction of C_6F_{14}. At 22.0°C the two solutions in equilibrium have $x = 0.24$ and $x = 0.48$, respectively, and at 21.5°C the mole fractions are 0.22 and 0.51. Sketch the phase diagram. Describe the phase changes that occur when perfluorohexane is added to a fixed amount of hexane at (a) 23°C, (b) 22°C.

8.20 (b) Two liquids, A and B, show partial miscibility below 52.4°C. The critical concentration at the upper critical temperature is $x = 0.459$, where x is the mole fraction of A. At 40.0°C the two solutions in equilibrium have $x = 0.22$ and $x = 0.60$, respectively, and at 42.5°C the mole fractions are 0.24 and 0.48. Sketch the phase diagram. Describe the phase changes that occur when B is added to a fixed amount of A at (a) 48°C, (b) 52.4°C.

Problems[†]

Numerical problems

8.1‡ 1-Butanol and chlorobenzene form a minimum-boiling azeotropic system. The mole fraction of 1-butanol in the liquid (x) and vapour (y) phases at 1.000 atm is given below for a variety of boiling temperatures (H. Artigas, C. Lafuente, P. Cea, F.M. Royo, and J.S. Urieta, *J. Chem. Eng. Data* **42**, 132 (1997)).

T/K	396.57	393.94	391.60	390.15	389.03	388.66	388.57
x	0.1065	0.1700	0.2646	0.3687	0.5017	0.6091	0.7171
y	0.2859	0.3691	0.4505	0.5138	0.5840	0.6409	0.7070

Pure chlorobenzene boils at 404.86 K. (a) Construct the chlorobenzene-rich portion of the phase diagram from the data. (b) Estimate the temperature at which a solution whose mole fraction of 1-butanol is 0.300 begins to boil. (c) State the compositions and relative proportions of the two phases present after a solution initially 0.300 1-butanol is heated to 393.94 K.

8.2‡ An *et al.* investigated the liquid–liquid coexistence curve of *N,N*-dimethylacetamide and heptane (X. An, H. Zhao, F. Fuguo, and W. Shen, *J. Chem. Thermodynamics* **28**, 1221 (1996)). Mole fractions of *N,N*-dimethylacetamide in the upper (x_1) and lower (x_2) phases of a two-phase region are given below as a function of temperature:

T/K	309.820	309.422	309.031	308.006	306.686
x_1	0.473	0.400	0.371	0.326	0.293
x_2	0.529	0.601	0.625	0.657	0.690

T/K	304.553	301.803	299.097	296.000	294.534
x_1	0.255	0.218	0.193	0.168	0.157
x_2	0.724	0.758	0.783	0.804	0.814

(a) Plot the phase diagram. (b) State the proportions and compositions of the two phases that form from mixing 0.750 mol

of *N,N*-dimethylacetamide with 0.250 mol of heptane at 296.0 K. To what temperature must the mixture be heated to form a single-phase mixture?

8.3‡ The following data have been obtained for the liquid–vapour equilibrium compositions of mixtures of nitrogen and oxygen at 100 kPa.

T/K	77.3	78	80	82	84	86	88	90.2
$x(O_2)$	0	10	34	54	70	82	92	100
$y(O_2)$	0	2	11	22	35	52	73	100
$p^*(O_2)/\text{Torr}$	154	171	225	294	377	479	601	760

Plot the data on a temperature–composition diagram and determine the extent to which it fits the predictions for an ideal solution by calculating the activity coefficients of O_2 at each composition.

8.4 Phosphorus and sulfur form a series of binary compounds. The best characterized are P_4S_3, P_4S_7, and P_4S_{10}, all of which melt congruently. Assuming that only these three binary compounds of the two elements exist, (a) draw schematically only the P/S phase diagram. Label each region of the diagram with the substance that exists in that region and indicate its phase. Label the horizontal axis as x_S and give the numerical values of x_S that correspond to the compounds. The melting point of pure phosphorus is 44°C and that of pure sulfur is 119°C. (b) Draw, schematically, the cooling curve for a mixture of composition $x_S = 0.28$. Assume that a eutectic occurs at $x_S = 0.2$ and negligible solid–solid solubility.

8.5 The table below gives the break and halt temperatures found in the cooling curves of two metals A and B. Construct a phase diagram consistent with the data of these curves. Label the regions of the diagram, stating what phases and substances are present. Give the probable formulas of any compounds that form.

---

† Problems denoted with the symbol ‡ were supplied by Charles Trapp, Carmen Giunta, and Marshall Cady.

$100x_B$	$\theta_{break}/°C$	$\theta_{halt,1}/°C$	$\theta_{halt,2}/°C$
0		1100	
10.0	1060	700	
20.0	1000	700	
30.0	940	700	400
40.0	850	700	400
50.0	750	700	400
60.0	670	400	
70.0	550	400	
80.0		400	
90.0	450	400	
100.0		500	

8.6 Consider the phase diagram in Fig. 8.40, which represents a solid–liquid equilibrium. Label all regions of the diagram according to the chemical species that exist in that region and their phases. Indicate the number of species and phases present at the points labelled b, d, e, f, g, and k. Sketch cooling curves for compositions $x_B = 0.16$, 0.23, 0.57, 0.67, and 0.84.

8.8 Sketch the phase diagram for the Mg/Cu system using the following information: $\theta_f(Mg) = 648°C$, $\theta_f(Cu) = 1085°C$; two intermetallic compounds are formed with $\theta_f(MgCu_2) = 800°C$ and $\theta_f(Mg_2Cu) = 580°C$; eutectics of mass percentage Mg composition and melting points 10 per cent (690°C), 33 per cent (560°C), and 65 per cent (380°C). A sample of Mg/Cu alloy containing 25 per cent Mg by mass was prepared in a crucible heated to 800°C in an inert atmosphere. Describe what will be observed if the melt is cooled slowly to room temperature. Specify the composition and relative abundances of the phases and sketch the cooling curve.

8.9‡ Figure 8.41 shows $\Delta_{mix}G(x_{Pb}, T)$ for a mixture of copper and lead. (a) What does the graph reveal about the miscibility of copper and lead and the spontaneity of solution formation? What is the variance (F) at (i) 1500 K, (ii) 1100 K? (b) Suppose that at 1500 K a mixture of composition (i) $x_{Pb} = 0.1$, (ii) $x_{Pb} = 0.7$, is slowly cooled to 1100 K. What is the equilibrium composition of the final mixture? Include an estimate of the relative amounts of each phase. (c) What is the solubility of (i) lead in copper, (ii) copper in lead at 1100 K?

8.10‡ The temperature–composition diagram for the Ca/Si binary system is shown in Fig. 8.42. (a) Identify eutectics, congruent

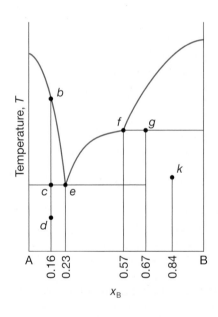

Temperature, T

x_B

Fig. 8.40

8.7 Solutions of 3-methylphenylamine (MP) were made up in glycerol and then warmed from room temperature. The mixture became turbid at θ_1 and then cleared at θ_2. Plot the phase diagram using the data below, and find the upper and lower critical temperatures.

Mass%	18	20	40	60	80	85
$\theta_1/°C$	48	18	8	10	19	25
$\theta_2/°C$	53	90	120	118	83	53

Mass% denotes the mass percentage composition of MP. State what happens as MP is added dropwise to glycerol at 60°C. State the number of phases present at each composition and their relative amounts.

Fig. 8.41

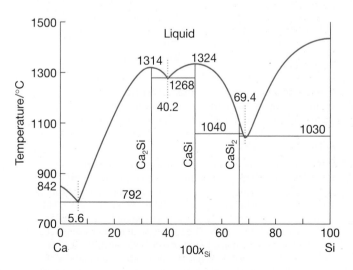

Fig. 8.42

melting compounds, and incongruent melting compounds.
(b) If a 20 per cent by atom composition melt of silicon at
1500°C is cooled to 1000°C, what phases (and phase composition)
would be at equilibrium? Estimate the relative amounts of each
phase. (c) Describe the equilibrium phases observed when an
80 per cent by atom composition Si melt is cooled to 1030°C.
What phases, and relative amounts, would be at equilibrium
at a temperature (i) slightly higher than 1030°C, (ii) slightly lower
than 1030°C? Draw a graph of the mole percentages of both Si(s)
and $CaSi_2(s)$ as a function of mole percentage of melt that is
freezing at 1030°C.

8.11 Iron(II) chloride (melting point 677°C) and potassium
chloride (melting point 776°C) form the compounds $KFeCl_3$
and K_2FeCl_4 at elevated temperatures. $KFeCl_3$ melts congruently
at 380°C and K_2FeCl_4 melts incongruently at 399°C. Eutectics
are formed with compositions $x = 0.38$ (melting point 351°C) and
$x = 0.54$ (melting point 393°C), where x is the mole fraction of $FeCl_2$.
The KCl solubility curve intersects the K_2FeCl_4 curve at $x = 0.34$.
Sketch the phase diagram. State the phases that are in equilibrium
when a mixture of composition $x = 0.36$ is cooled from 400°C to
300°C.

Theoretical problems

8.12 Show that two phases are in thermal equilibrium only if their
temperatures are the same.

8.13 Show that two phases are in mechanical equilibrium only if
their pressures are equal.

Applications: to chemical engineering

8.14 The compound *p*-azoxyanisole forms a liquid crystal. 5.0 g of
the solid was placed in a tube, which was then evacuated and sealed.
Use the phase rule to prove that the solid will melt at a definite
temperature and that the liquid crystal phase will make a transition
to a normal liquid phase at a definite temperature.

8.15 Magnesium oxide and nickel oxide withstand high
temperatures. However, they do melt when the temperature
is high enough and the behaviour of mixtures of the two is
of considerable interest to the ceramics industry. Draw the
temperature–composition diagram for the system using the data
below, where x is the mole fraction of MgO in the solid and y its
mole fraction in the liquid.

$\theta/°C$	1960	2200	2400	2600	2800
x	0	0.35	0.60	0.83	1.00
y	0	0.18	0.38	0.65	1.00

State (a) the melting point of a mixture with $x = 0.30$,
(b) the composition and proportion of the phases present
when a solid of composition $x = 0.30$ is heated to 2200°C,
(c) the temperature at which a liquid of composition $y = 0.70$
will begin to solidify.

8.16 The bismuth–cadmium phase diagram is of interest
in metallurgy, and its general form can be estimated from
expressions for the depression of the freezing point. Construct the
diagram using the following data: $T_f(Bi) = 544.5$ K, $T_f(Cd) = 594$ K,
$\Delta_{fus}H(Bi) = 10.88$ kJ mol^{-1}, $\Delta_{fus}H(Cd) = 6.07$ kJ mol^{-1}. The metals are
mutually insoluble as solids. Use the phase diagram to state what
would be observed when a liquid of composition $x(Bi) = 0.70$ is
cooled slowly from 550 K. What are the relative abundances of
the liquid and solid at (a) 460 K and (b) 350 K? Sketch the cooling
curve for the mixture.

8.17‡ Carbon dioxide at high pressure is used to separate various
compounds in citrus oil. The mole fraction of CO_2 in the liquid (x)
and vapour (y) at 323.2 K is given below for a variety of pressures
(Y. Iwai, T. Morotomi, K. Sakamoto, Y. Koga, and Y. Arai, *J. Chem. Eng.
Data* **41**, 951 (1996)).

p/MPa	3.94	6.02	7.97	8.94	9.27
x	0.2873	0.4541	0.6650	0.7744	0.8338
y	0.9982	0.9980	0.9973	0.9958	0.9922

(a) Plot the portion of the phase diagram represented by these data.
(b) State the compositions and relative proportions of the two
phases present after an equimolar gas mixture is compressed to
6.02 MPa at 323.2 K.

8.18‡ Large quantities of silicon are made by reducing SiO_2 with
carbon in arc furnaces at elevated temperatures. Many chemical
species may be present in these high-temperature SiO_2/graphite
mixtures, including C(s), SiO_2(s), SiO_2(l), Si(s), Si(l), SiC(s), SiO(g), and
CO(g). The table below contains information about the temperature
dependence of $G(T) - H_{SER} = a + bT$ for each of these species, and
several others, in the temperature range 1500–2500 K and at 1 bar,
where $G(T) - H_{SER}$ is the Gibbs energy at temperature T relative to
the enthalpies of the pure constituent elements in their reference
states at the conventional temperature. Use these data alone to
answer the following questions. Assume perfect gas behaviour and
ideal solutions when necessary.

	a/(kJ mol^{-1})	b/(kJ K^{-1} mol^{-1})
graphite	34.38	−0.040 54
Si(s)	44.281	−0.066 71
SiO_2(l)	−757.039	−0.184 56
Si(g)	484.667	−0.208 08
SiO(g)	−41.075	−0.281 23
O_2(g)	57.858	−0.268 56
SiO_2(s)	−818.448	−0.152 81
SiC(s)	2.521	−0.098 38
Si(l)	96.980	−0.097 43
SiO_2(g)	−209.039	−0.342 03
CO(g)	−54.581	−0.263 12
CO_2(g)	−302.873	−0.320 21

Calculate the normal melting point of silica and silicon.
Make graphs of the vapour pressures of each in the temperature
range 1500–2500 K and at a pressure of 1 bar. *Hint.* Use
mathematical software or a spreadsheet. For a continuation of
this story, see Problem 9.26.

8.19‡ The compound *tert*-amyl methyl ether (TAME) has potential
use as a gasoline antiknock agent. The Redlich–Kister empirical
equation for the excess Gibbs energy of mixing of methanol (1) and
TAME (2) is:

$$G^E/RT = x_1 x_2 \sum_{n=0}^{4} A_n (1 - 2x_1)^n$$

where $A_n = 1.4170, -0.052\,45, 0.1599, 0.1061, -0.091\,59$ for
$n = 0, 1, \ldots, 4$. The activity coefficients are given by

$$\ln \gamma_1 = x_2^2 \sum_{n=0}^{4} A_n f_n(x_1) \qquad \ln \gamma_2 = x_1^2 \sum_{n=0}^{4} (-1)^n A_n f_n(x_2)$$

where $f_n(x) = \{1 - 2(n + 1)x\}(1 - 2x)^{n-1}$. The vapour pressure and
molar volume of pure TAME are 6.09 kPa and 131.78 cm^3 mol^{-1},

respectively, at 288.15 K and those of methanol are 9.86 kPa and 40.254 cm^3 mol^{-1}. (a) Draw a graph of ln γ against x_1. Do the graphs reflect correct values at infinite dilutions of either TAME or methanol? (b) Draw a graph of $G^E(x_1)$. What does this graph reveal about the spontaneity of mixing? Does the curve have the shape expected for a regular solution? (c) Draw a graph of total vapour pressure, $p = p_1 + p_2$, and also show the curves for $p(x_1)$ and $p(y_1)$, where y is the mole fraction in the vapour. What is the vapour pressure and composition of the azeotrope? What is the equilibrium pressure when $x_1 = 0.2$? (d) Draw a graph of total vapour and partial vapour pressures. The curves would be $p(x_1)$, $p_1(x_1)$, and $p_2(x_1)$. What do these curves reveal about deviations from ideality? Calculate the Henry's law constant for both methanol and TAME. (e) Estimate the vapour pressure of pure TAME at 288.15 K when subjected to an applied pressure of 2.0 bar. *Hint.* Use mathematical software or a spreadsheet.

9 Chemical equilibrium

Spontaneous chemical reactions

9.1 The Gibbs energy minimum

9.2 The description of equilibrium

The response of equilibria to the conditions

9.3 How equilibria respond to pressure

9.4 The response of equilibria to temperature

9.5 The response of equilibria to pH

Checklist of key ideas

Further reading

Exercises

Problems

This chapter develops the concept of chemical potential and shows how it is used to account for the equilibrium composition of chemical reactions. The equilibrium composition corresponds to a minimum in the Gibbs energy plotted against the extent of reaction, and by locating this minimum we establish the relation between the equilibrium constant and the standard Gibbs energy of reaction. The thermodynamic formulation of equilibrium enables us to establish the quantitative effects of changes in pressure, temperature, and pH.

Chemical reactions move towards a dynamic equilibrium in which both reactants and products are present but have no further tendency to undergo net change. In some cases, the concentration of products in the equilibrium mixture is so much greater than the concentration of unchanged reactants that for all practical purposes the reaction is 'complete'. However, in many important cases the equilibrium mixture has significant concentrations of both reactants and products. In this chapter we see how to use thermodynamics to predict the equilibrium composition under any reaction conditions.

Spontaneous chemical reactions

We have seen that the direction of spontaneous change at constant temperature and pressure is towards lower values of the Gibbs energy, G. The idea is entirely general, and in this chapter we apply it to the discussion of reactions.

9.1 The Gibbs energy minimum

We locate the equilibrium composition of a reaction mixture by calculating the Gibbs energy of the reaction mixture and identifying the composition that corresponds to minimum G.

(a) The reaction Gibbs energy

Consider the equilibrium $A \rightleftharpoons B$. Even though this reaction looks trivial, there are many examples of it, such as the isomerization of pentane to 2-methylbutane and the conversion of L-alanine to D-alanine. Suppose an infinitesimal amount $d\xi$ of A turns into B, then the change in the amount of A present is $dn_A = -d\xi$ and the change in the amount of B present is $dn_B = +d\xi$. The quantity ξ is called the **extent of reaction**; it has the dimensions of amount of substance and is reported in moles.

When the extent of reaction changes by a finite amount $\Delta\xi$, the amount of A present changes from $n_{A,0}$ to $n_{A,0} - \Delta\xi$ and the amount of B changes from $n_{B,0}$ to $n_{B,0} + \Delta\xi$. So, if initially 2.0 mol A is present and we wait until $\Delta\xi = +1.5$ mol, the amount of A remaining will be 0.5 mol.

The **reaction Gibbs energy**, $\Delta_r G$, is defined as the slope of the graph of the Gibbs energy plotted against the extent of reaction:

$$\Delta_r G = \left(\frac{\partial G}{\partial \xi}\right)_{p,T} \tag{9.1}$$

Although Δ normally signifies a *difference* in values, here Δ_r signifies a *derivative*, the slope of G with respect to ξ. However, to see that there is a close relationship with the normal usage, suppose the reaction advances by $d\xi$. The corresponding change in Gibbs energy is

$$dG = \mu_A \, dn_A + \mu_B \, dn_B = -\mu_A \, d\xi + \mu_B \, d\xi = (\mu_B - \mu_A)d\xi$$

This equation can be reorganized into

$$\left(\frac{\partial G}{\partial \xi}\right)_{p,T} = \mu_B - \mu_A$$

That is,

$$\Delta_r G = \mu_B - \mu_A \tag{9.2}$$

We see that $\Delta_r G$ can also be interpreted as the difference between the chemical potentials of the reactants and products *at the composition of the reaction mixture*.

Because chemical potential varies with composition, the slope of the plot of Gibbs energy against extent of reaction changes as the reaction proceeds. Moreover, because the reaction runs in the direction of decreasing G (that is, down the slope of G versus ξ), we see from eqn 2 that the reaction A → B is spontaneous when $\mu_A > \mu_B$, whereas the reverse reaction is spontaneous when $\mu_B > \mu_A$. The slope is zero, and the reaction is spontaneous in neither direction, when

$$\Delta_r G = 0 \tag{9.3}$$

This condition occurs when $\mu_B = \mu_A$ (Fig. 9.1). It follows that, if we can find the composition of the reaction mixture that ensures $\mu_B = \mu_A$, then we can identify the composition of the reaction mixture at equilibrium.

(b) Exergonic and endergonic reactions

We can express the spontaneity of a reaction at constant temperature and pressure in terms of the reaction Gibbs energy:

If $\Delta_r G < 0$, the forward reaction is spontaneous.

If $\Delta_r G > 0$, the reverse reaction is spontaneous.

If $\Delta_r G = 0$, the reaction is at equilibrium.

A reaction for which $\Delta_r G < 0$ is called **exergonic** (from the Greek words for work-producing). The name signifies that, because the process is spontaneous, it can be used to drive another process, such as another reaction, or used to do non-expansion work. For example, the oxidation of carbohydrates can drive a number of biological processes, such as muscle contraction, synthesis of proteins, and brain activity (Box 9.2). A reaction for which $\Delta_r G > 0$ is called **endergonic** (signifying work-consuming). The reaction can be made to occur only by doing work on it, such as electrolysing water to reverse its spontaneous formation reaction. Reactions at equilibrium are spontaneous in neither direction: they are neither exergonic nor endergonic.

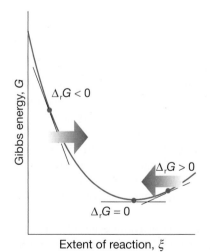

9.1 As the reaction advances (represented by motion from left to right along the horizontal axis) the slope of the Gibbs energy changes. Equilibrium corresponds to zero slope, at the foot of the valley.

Box 9.1 Biological energy conversion: the synthesis of proteins and the oxidation of glucose

In biological cells, the energy released by the oxidation of foods (Box 2.2) is stored in adenosine triphosphate (ATP, **1**). The essence of ATP's action is its ability to lose its terminal phosphate group by hydrolysis and to form adenosine diphosphate (ADP):

$$ATP(aq) + H_2O(l) \longrightarrow ADP(aq) + P_i(aq) + H_3O^+(aq)$$

P_i denotes an inorganic phosphate group, such as $H_2PO_4^-$. The biological standard values for ATP hydrolysis at 37°C (310 K, blood temperature) are $\Delta_r G^\ominus = -30$ kJ mol^{-1}, $\Delta_r H^\ominus = -20$ kJ mol^{-1}, and $\Delta_r S^\ominus = +34$ J K^{-1} mol^{-1}. The hydrolysis is therefore exergonic ($\Delta_r G^\ominus < 0$) under these conditions and 30 kJ mol^{-1} is available for driving other reactions. Moreover, because the reaction entropy is large, the reaction Gibbs energy is sensitive to temperature. In view of its exergonicity the ADP–phosphate bond has been called a 'high-energy phosphate bond'. The name is intended to signify a high tendency to undergo reaction, and should not be confused with 'strong' bond. In fact, even in the biological sense it is not of very 'high energy'. The action of ATP depends on it being intermediate in activity. Thus ATP acts as a phosphate donor to a number of acceptors (for instance, glucose), but is recharged by more powerful phosphate donors in a number of biochemical processes.

Here we use the oxidation of glucose to CO_2 and H_2O by O_2 as an example of how the breakdown of foods leads to the formation of ATP in the cell. The process begins with *glycolysis*, the partial oxidation of glucose by nicotinamide adenine dinucleotide (NAD$^+$, **2**) to pyruvate ion, $CH_3COCO_2^-$, continues with the *citric acid cycle*, which oxidizes pyruvate to CO_2, and ends with *oxidative phosphorylation*, which reduces O_2 to H_2O. Glycolysis is the main source of energy during *anaerobic metabolism*, a form of metabolism in which inhaled O_2 does not play a role. The citric acid cycle and oxidative phosphorylation are the main mechanisms for the extraction of energy from carbohydrates during *aerobic metabolism*, a form of metabolism in which inhaled O_2 does play a role.

Glycolysis occurs in the *cytosol*, the aqueous material encapsulated by the cell membrane, and consists of 10 enzyme-catalysed reactions. At blood temperature, $\Delta_r G^\ominus = -147$ kJ mol^{-1} for the oxidation of glucose by NAD$^+$ to pyruvate ions. The oxidation of one glucose molecule is coupled to the conversion of two ADP molecules to two ATP molecules, so the net reaction of glycolysis is:

1 ATP

NAD$^+$ R=H
NADP$^+$ R=PO$_3^{2-}$

2

$$C_6H_{12}O_6(aq) + 2\ NAD^+(aq) + 2\ ADP(aq) + 2\ P_i(aq) + 2\ H_2O(l)$$
$$\longrightarrow 2\ CH_3COCO_2^-(aq) + 2\ NADH(aq) + 2\ ATP(aq) + 2\ H_3O^+(aq)$$

The standard reaction Gibbs energy is $(-147) - 2(-30)$ kJ mol^{-1} $= -87$ kJ mol^{-1}. The reaction is exergonic, and therefore spontaneous: the oxidation of glucose is used to 'recharge' the ATP. In cells that are deprived of O_2 pyruvate ion is reduced to lactate ion, $CH_3C(OH)CO_2^-$, by NADH.[5] Very strenuous exercise, such as bicycle racing, can decrease sharply the concentration of O_2 in muscle cells and the condition known as muscle fatigue results from increased concentrations of lactate ion.

The standard Gibbs energy of combustion of glucose is -2880 kJ mol^{-1}, so terminating its oxidation at pyruvate is a poor use of resources. In the presence of O_2, pyruvate is

5. In yeast, the terminal products are ethanol and CO_2.

3

oxidized further during the citric acid cycle and oxidative phosphorylation, which occur in a special compartment of the cell called the *mitochondrion*. The citric acid cycle requires eight enzymes that drive the synthesis of ATP with the oxidation of pyruvate by NAD^+ and flavin adenine dinucleotide (FAD, **3**):

$$2\ CH_3COCO_2^-(aq) + 8\ NAD^+(aq) + 2\ FAD(aq) + 2\ ADP(aq)$$
$$+ 2\ P_i(aq) + 8\ H_2O(l) \longrightarrow 6\ CO_2(g) + 8\ NADH(aq) +$$
$$4\ H_3O^+(aq) + 2\ FADH_2(aq) + 2\ ATP(aq)$$

The NADH and $FADH_2$ go on to reduce O_2 during oxidative phosphorylation (Box 10.1), which also produces ATP. The citric acid cycle and oxidative phosphorylation generate as many as 38 ATP molecules for each glucose molecule consumed. Each mole of ATP extracts 30 kJ from the 2880 kJ supplied by 1 mol $C_6H_{12}O_6$ (180 g of glucose), so 1140 kJ is stored for later use. Therefore, aerobic oxidation of glucose is much more efficient than glycolysis.

In the cell, each ATP molecule can be used to drive an endergonic reaction for which $\Delta_r G^\ominus$ does not exceed +30 kJ mol^{-1}. For example, the biosynthesis of sucrose from glucose and fructose can be driven by plant enzymes because the reaction is endergonic to the extent $\Delta_r G^\ominus = +23$ kJ mol^{-1}. The biosynthesis of proteins is strongly endergonic, not only on account of the enthalpy change but also on account of the large decrease in entropy that occurs when many amino acids are assembled into a precisely determined sequence. For instance, the formation of a peptide link is endergonic, with $\Delta_r G^\ominus = +17$ kJ mol^{-1}, but the biosynthesis occurs indirectly and is equivalent to the consumption of three ATP molecules for each link. In a moderately small protein like myoglobin, with about 150 peptide links, the construction alone requires 450 ATP molecules, and therefore about 12 mol of glucose molecules for 1 mol of protein molecules.

Discussion questions

1. Here we investigate the molecular basis for the exergonicity of the ATP hydrolysis reaction. (a) The Gibbs energy of ATP hydrolysis becomes more negative with increasing pH (in the range 5 < pH < 9) and this observation has been used to support the hypothesis that electrostatic repulsion between adjacent phosphate groups is a factor that controls the exergonicity of ATP hydrolysis. Provide a rationale for the hypothesis and discuss how the pH dependence of $\Delta_r G$ for ATP hydrolysis supports it. Do these electrostatic effects contribute to the $\Delta_r H$ or $\Delta_r S$ terms that determine the exergonicity of the reaction? (b) The average P—O bond length in the phosphate ion is 154 pm, which is short relative to the value of 173 pm expected for a P—O single bond. This observation has been used to support the hypothesis that electron delocalization in the phosphate ion, one of the hydrolysis products, and the absence of electron delocalization in the triphosphate group of the reactant are factors that control the exergonicity of ATP hydrolysis. Provide a rationale for the hypothesis and discuss how the bond length data support it. Does electron delocalization contribute to the $\Delta_r H$ or $\Delta_r S$ terms that determine the exergonicity of the reaction? (c) The standard reaction entropy for the hydrolysis of polyphosphates is positive; this is thought to be one of the factors that controls the exergonicity of ATP hydrolysis. Why would an increase in entropy accompany the hydrolysis of a triphosphate group into a disphosphate and a phosphate group?

2. At blood temperature, $\Delta_r G^\ominus = -218$ kJ mol^{-1} and $\Delta_r H^\ominus = -120$ kJ mol^{-1} for the production of lactate ion during anaerobic glycolysis. Provide a molecular interpretation for the observation that the reaction is more exergonic than it is exothermic.

Further reading

C.K. Matthews, K.E. van Holde, and K.G. Ahern, *Biochemistry*. Addison Wesley Longman, San Francisco (2000).

D.A. Harris, *Bioenergetics at a glance*. Blackwell Science, Oxford (1995).

9.2 The description of equilibrium

With the background established, we are now ready to see how to apply thermodynamics to the description of chemical equilibrium.

(a) Perfect gas equilibria

When A and B are perfect gases we can use eqn 7.14 ($\mu = \mu^\ominus + RT \ln p$, with p interpreted as $p/p^\ominus$) to write

$$\Delta_r G = \mu_B - \mu_A = (\mu_B^\ominus + RT \ln p_B) - (\mu_A^\ominus + RT \ln p_A)$$

$$= \Delta_r G^\ominus + RT \ln \frac{p_B}{p_A} \qquad (9.4)°$$

If we denote the ratio of partial pressures by Q, we obtain

$$\Delta_r G = \Delta_r G^\ominus + RT \ln Q \qquad Q = \frac{p_B}{p_A} \qquad (9.5)°$$

The ratio Q is an example of a **reaction quotient**. It ranges from 0 (pure A) to infinity (pure B). The **standard reaction Gibbs energy**, $\Delta_r G^\ominus$, is defined (like the standard reaction enthalpy) as the difference in the standard molar Gibbs energies of the reactants and products. For our reaction

$$\Delta_r G^\ominus = G_{B,m}^\ominus - G_{A,m}^\ominus = \mu_B^\ominus - \mu_A^\ominus \qquad (9.6)$$

In Section 4.6 we saw that the difference in standard molar Gibbs energies of the products and reactants is equal to the difference in their standard Gibbs energies of formation, so in practice we calculate $\Delta_r G^\ominus$ from

$$\Delta_r G^\ominus = \Delta_f G^\ominus(B) - \Delta_f G^\ominus(A) \qquad (9.7)$$

At equilibrium $\Delta_r G = 0$. The ratio of partial pressures at equilibrium is denoted K, and eqn 5 becomes

$$0 = \Delta_r G^\ominus + RT \ln K$$

This expression rearranges to

$$RT \ln K = -\Delta_r G^\ominus \qquad K = \left(\frac{p_B}{p_A} \right)_{equilibrium} \qquad (9.8)°$$

This relation is a special case of one of the most important equations in chemical thermodynamics: it is the link between tables of thermodynamic data, such as those in the *Data section* at the end of this volume, and the chemically important equilibrium constant, K.

Molecular interpretation 9.1

In molecular terms, the minimum in the Gibbs energy, which corresponds to $\Delta_r G = 0$, stems from the Gibbs energy of mixing of the two gases. Hence, an important contribution to the position of chemical equilibrium is the mixing of the products with the reactants as the products are formed.

Consider a hypothetical reaction in which A molecules change into B molecules without mingling together. The Gibbs energy of the system changes from $G^\ominus(A)$ to $G^\ominus(B)$ in proportion to the amount of B that had been formed, and the slope of the plot of G against the extent of reaction is a constant and equal to $\Delta_r G^\ominus$ at all stages of the reaction (Fig. 9.2). There is no minimum in the graph (except at pure B).

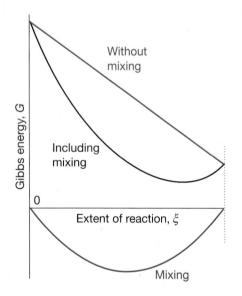

9.2 If the mixing of reactants and products is ignored, then the Gibbs energy changes linearly from its initial value (pure reactants) to its final value (pure products) and the slope of the line is $\Delta_r G^\ominus$. However, as products are produced, there is a further contribution to the Gibbs energy arising from their mixing (lowest curve). The sum of the two contributions has a minimum. That minimum corresponds to the equilibrium composition of the system.

However, in fact, the newly produced B molecules mix with the surviving A molecules. We have seen that the contribution of a mixing process to the change in Gibbs energy (eqn 7.18) is

$$\Delta_{mix}G = nRT(x_A \ln x_A + x_B \ln x_B)$$

This expression makes a U-shaped contribution to the total change in Gibbs energy. As can be seen from the illustration, there is now a minimum in the Gibbs energy, and its position corresponds to the equilibrium composition of the reaction mixture.

We see from eqn 8 that, when $\Delta_r G^\ominus > 0$, the equilibrium constant $K < 1$. Therefore, at equilibrium the partial pressure of A exceeds that of B, which means that the reactant A is favoured in the equilibrium. When $\Delta_r G^\ominus < 0$, the equilibrium constant $K > 1$, so at equilibrium the partial pressure of B exceeds that of A. Now the product B is favoured in the equilibrium.

(b) The general case of a reaction

We can easily extend the argument that led to eqn 8 to a general reaction. First, we need to generalize the concept of extent of reaction.

Consider the reaction $2 A + B \rightarrow 3 C + D$. A more sophisticated way of expressing the chemical equation is to write it in the symbolic form

$$0 = 3 C + D - 2 A - B$$

by subtracting the reactants from both sides (and replacing the arrow by an equals sign). This equation has the form

$$0 = \sum_J v_J J \tag{9.9}$$

where J denotes the substances and the v_J are the corresponding **stoichiometric numbers** in the chemical equation.[1] In our example, these numbers have the values

$$v_A = -2 \qquad v_B = -1 \qquad v_C = +3 \qquad v_D = +1$$

A stoichiometric number is positive for products and negative for reactants. Then we define ξ so that if it changes by $\Delta\xi$, then the change in the amount of any species J is $v_J \Delta\xi$.

Illustration 9.1

To express the equation

$$N_2(g) + 3 H_2(g) \longrightarrow 2 NH_3(g) \tag{9.10}$$

in the notation of eqn 9, we rearrange it to

$$0 = 2 NH_3(g) - N_2(g) - 3 H_2(g)$$

---

1. Stoichiometric *numbers* may be positive or negative; stoichiometric *coefficients* are always positive. Few, however, make the distinction between the two types of quantity.

and then identify the stoichiometric numbers as $v_{N_2} = -1$, $v_{H_2} = -3$, and $v_{NH_3} = +2$. Therefore, if initially there is 10 mol N_2 present, then when the extent of reaction changes from $\xi = 0$ to $\xi = 1$ mol, implying that $\Delta\xi = +1$ mol, the amount of N_2 changes from 10 mol to 9 mol. All the N_2 has been consumed when $\xi = 10$ mol. When $\Delta\xi = +1$ mol, the amount of H_2 changes by $-3 \times (1$ mol$)$ $= -3$ mol and the amount of NH_3 changes by $+2 \times (1$ mol$) = +2$ mol.

The reaction Gibbs energy, $\Delta_r G$, is defined in the same way as before, eqn 1. In the *Justification* below, we show that the Gibbs energy of reaction can always be written

$$\Delta_r G = \Delta_r G^{\ominus} + RT \ln Q \tag{9.11}$$

with the standard reaction Gibbs energy calculated from

$$\Delta_r G^{\ominus} = \sum_{\text{Products}} v\Delta_f G^{\ominus} - \sum_{\text{Reactants}} v\Delta_f G^{\ominus} \tag{9.12}$$

or, more formally,

$$\Delta_r G^{\ominus} = \sum_J v_J \Delta_f G^{\ominus}(J) \tag{9.13}$$

The reaction quotient, Q, has the form

$$Q = \frac{\text{activities of products}}{\text{activities of reactants}} \tag{9.14}$$

with each species raised to the power given by its stoichiometric coefficient. More formally, to write the general expression for Q we introduce the symbol Π to denote the product of what follows it (just as Σ denotes the sum), and define Q as

$$Q = \prod_J a_J^{v_J} \tag{9.15}$$

Because reactants have negative stoichiometric numbers, they automatically appear as the denominator when the product is written out explicitly. Recall from Table 7.3 that, for pure solids and liquids, the activity is 1, so such substances make no contribution to Q even though they may appear in the chemical equation.

Illustration 9.2

Consider the reaction

$$2\,A + 3\,B \rightarrow C + 2\,D$$

in which case $v_A = -2$, $v_B = -3$, $v_C = +1$, and $v_D = +2$. The reaction quotient is then

$$Q = \frac{a_C a_D^2}{a_A^2 a_B^3}$$

Justification 9.1

Consider the reaction with stoichiometric numbers v_J. When the reaction advances by $d\xi$, the amounts of reactants and products change by $dn_J = v_J\, d\xi$. The resulting infinitesimal change in the Gibbs energy at constant temperature and pressure is

$$dG = \sum_J \mu_J\, dn_J = \left(\sum_J v_J\mu_J\right) d\xi \tag{9.16}$$

It follows that

$$\Delta_r G = \left(\frac{\partial G}{\partial \xi}\right)_{p,T} = \sum_J v_J\mu_J \tag{9.17}$$

To make further progress, we note that the chemical potential of a species J is related to its activity by

$$\mu_J = \mu_J^\ominus + RT \ln a_J$$

When this expression is substituted into eqn 17 we obtain

$$\Delta_r G = \overbrace{\sum_J v_J\mu_J^\ominus}^{\Delta_r G^\ominus} + RT \sum_J v_J \ln a_J = \Delta_r G^\ominus + RT \sum_J \ln a_J^{v_J}$$

$$= \Delta_r G^\ominus + RT \ln \overbrace{\prod_J a_J^{v_J}}^{Q} = \Delta_r G^\ominus + RT \ln Q$$

with Q given by eqn 15.

We use $a \ln x = \ln x^a$ and $\ln x + \ln y + \cdots = \ln xy \cdots$.

Now we conclude the argument based on eqn 11. At equilibrium, the slope of G is zero: $\Delta_r G = 0$. The activities then have their equilibrium values and we can write

$$K = \left(\prod_J a_J^{v_J}\right)_{\text{equilibrium}} \tag{[9.18]}$$

This expression has the same form as Q, eqn 15, but is evaluated using equilibrium activities.[2] From now on, we shall not write the 'equilibrium' subscript explicitly, and will rely on the context to make it clear that for K we use equilibrium values and for Q we use the values at the specified stage of the reaction.

An equilibrium constant K expressed in terms of activities (or fugacities) is called a **thermodynamic equilibrium constant**. Note that, because activities are dimensionless numbers, the thermodynamic equilibrium constant is also dimensionless. In elementary applications, the activities that occur in eqn 18 are often replaced by the numerical values of molalities or molar concentrations, and fugacities are replaced by partial pressures. In either case, the resulting expressions are only approximations. The approximation is particularly severe for electrolyte solutions, for in them activity coefficients differ from 1 even in very dilute solutions.

At this point we set $\Delta_r G = 0$ in eqn 11 and replace Q by K. We immediately obtain

$$RT \ln K = -\Delta_r G^\ominus \tag{9.19}$$

2. For the special case treated in *Illustration 9.2*, $K = \left(\dfrac{a_C a_D^2}{a_A^2 a_B^3}\right)_{\text{equilibrium}}$.

This is an exact and highly important thermodynamic relation, for it enables us to predict the equilibrium constant of any reaction from tables of thermodynamic data, and hence to predict the equilibrium composition of the reaction mixture.[3]

Example 9.1 Calculating an equilibrium constant

Calculate the equilibrium constant for the ammonia synthesis reaction, eqn 10, at 298 K and show how K is related to the partial pressures of the species at equilibrium when the overall pressure is low enough for the gases to be treated as perfect.

Method Calculate the standard reaction Gibbs energy from eqn 12 and convert it to the value of the equilibrium constant by using eqn 19. The expression for the equilibrium constant is obtained from eqn 19 (or via eqn 14) and, because the gases are taken to be perfect, we replace each fugacity by a partial pressure.

Answer The standard Gibbs energy of the reaction is

$$\Delta_r G^\ominus = 2\Delta_f G^\ominus(NH_3, g) - \{\Delta_f G^\ominus(N_2, g) + 3\Delta_f G^\ominus(H_2, g)\}$$
$$= 2\Delta_f G^\ominus(NH_3, g) = 2 \times (-16.5 \text{ kJ mol}^{-1})$$

Then,

$$\ln K = -\frac{2 \times (-16.5 \times 10^3 \text{ J mol}^{-1})}{(8.3145 \text{ J K}^{-1} \text{ mol}^{-1}) \times (298 \text{ K})} = \frac{2 \times 16.5 \times 10^3}{8.3145 \times 298}$$

Hence, $K = 6.1 \times 10^5$. This result is thermodynamically exact. The thermodynamic equilibrium constant for the reaction is

$$K = \frac{a_{NH_3}^2}{a_{N_2} a_{H_2}^3} = \frac{(f_{NH_3}/p^\ominus)^2}{(f_{N_2}/p^\ominus)(f_{H_2}/p^\ominus)^3} = \frac{f_{NH_3}^2 p^{\ominus 2}}{f_{N_2} f_{H_2}^3}$$

and this ratio has exactly the value we have just calculated. At low overall pressures fugacities can be replaced by partial pressures and an approximate form of the equilibrium constant is

$$K = \frac{p_{NH_3}^2 p^{\ominus 2}}{p_{N_2} p_{H_2}^3}$$

Self-test 9.1 Evaluate the equilibrium constant for $N_2O_4(g) \rightleftharpoons 2 NO_2(g)$ at 298 K. [$K = 0.15$]

Example 9.2 Estimating the degree of dissociation at equilibrium

The standard Gibbs energy of reaction for the decomposition $H_2O(g) \rightarrow H_2(g) + \frac{1}{2}O_2(g)$ is $+118.08 \text{ kJ mol}^{-1}$ at 2300 K. What is the degree of dissociation of H_2O at 2300 K and 1.00 bar?

Method The equilibrium constant is obtained from the standard Gibbs energy of reaction by using eqn 19, so the task is to relate the degree of dissociation, α,

---

3. In Chapter 20 we shall see that the right-hand side may be expressed in terms of spectroscopic data; so this expression also provides a link between spectroscopy and equilibrium composition.

to K and then to find its numerical value. Proceed by expressing the equilibrium compositions in terms of α, and solve for α in terms of K. Because the standard Gibbs energy of reaction is large and positive, we can anticipate that K will be small, and hence that $\alpha \ll 1$, which opens the way to making approximations to obtain its numerical value.

Answer The equilibrium constant is obtained from eqn 19 in the form

$$\ln K = -\frac{\Delta_r G^\ominus}{RT} = -\frac{(+118.08 \times 10^3 \text{ J mol}^{-1})}{(8.3145 \text{ J K}^{-1} \text{ mol}^{-1}) \times (2300 \text{ K})} = -\frac{118.08 \times 10^3}{8.3145 \times 2300}$$

It follows that $K = 2.08 \times 10^{-3}$. The equilibrium composition can be expressed in terms of α by drawing up the following table:

	H_2O	H_2	O_2
Initial amount	n	0	0
Change to reach equilibrium	$-\alpha n$	$+\alpha n$	$+\frac{1}{2}\alpha n$
Amount at equilibrium	$(1-\alpha)n$	αn	$\frac{1}{2}\alpha n$
Mole fraction	$\dfrac{1-\alpha}{1+\frac{1}{2}\alpha}$	$\dfrac{\alpha}{1+\frac{1}{2}\alpha}$	$\dfrac{\frac{1}{2}\alpha}{1+\frac{1}{2}\alpha}$
Partial pressure	$\dfrac{(1-\alpha)p}{1+\frac{1}{2}\alpha}$	$\dfrac{\alpha p}{1+\frac{1}{2}\alpha}$	$\dfrac{\frac{1}{2}\alpha p}{1+\frac{1}{2}\alpha}$

The equilibrium constant is therefore

$$K = \frac{p_{H_2} p_{O_2}^{1/2}}{p_{H_2O}} = \frac{\alpha^{3/2} p^{1/2}}{(1-\alpha)(2+\alpha)^{1/2}}$$

In this expression, we have written p in place of $p/p^\ominus$, to keep the notation simple. Now make the approximation that $\alpha \ll 1$, and hence obtain

$$K \approx \frac{\alpha^{3/2} p^{1/2}}{2^{1/2}}$$

Under the stated condition, $p = 1.00$ bar (that is, $p/p^\ominus = 1.00$), so

$$\alpha \approx (2^{1/2}K)^{2/3} = 0.0205$$

That is, about 2 per cent of the water has decomposed.

Comment Always check that the approximation is consistent with the final answer. In this case $\alpha \ll 1$ in accord with the original assumption.

The web site contains links to online tools for the estimation of equilibrium constants of gas-phase reactions.

Self-test 9.2 Given that the standard Gibbs energy of reaction at 2000 K is +135.2 kJ mol^{-1} for the same reaction, suppose that steam at 200 kPa is passed through a furnace tube at that temperature. Calculate the mole fraction of O_2 present in the output gas stream.
[0.002 21]

(c) The relation between equilibrium constants

The only remaining problem is to express the thermodynamic equilibrium constant in terms of the mole fractions, x_J, or molalities, b_J, of the species. To do so, we need

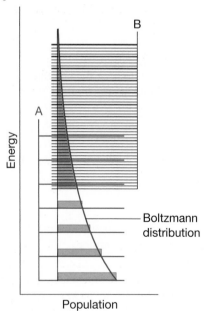

9.3 The Boltzmann distribution of populations over the energy levels of two species A and B with similar densities of energy levels; the reaction A → B is endothermic in this example. The bulk of the population is associated with the species A, so that species is dominant at equilibrium.

9.4 Even though the reaction A → B is endothermic, the density of energy levels in B is so much greater than that in A that the population associated with B is greater than that associated with A, so B is dominant at equilibrium.

to know the activity coefficients, and then to use $a_J = \gamma_J x_J$ or $a_J = \gamma_J b_J / b^\ominus$ (recalling that the activity coefficients depend on the choice). For example, in the latter case, for an equilibrium of the form A + B ⇌ C + D, where all four species are solutes, we write

$$K = \frac{a_C a_D}{a_A a_B} = \frac{\gamma_C \gamma_D}{\gamma_A \gamma_B} \times \frac{b_C b_D}{b_A b_B} = K_\gamma K_b \tag{9.20}$$

The activity coefficients must be evaluated at the equilibrium composition of the mixture, which may involve a complicated calculation, because the latter is known only if the equilibrium composition is already known. In elementary applications, and to begin the iterative calculation of the concentrations in a real example, the assumption is often made that the activity coefficients are all so close to unity that $K_\gamma = 1$. Then we obtain the result widely used in elementary chemistry that $K \approx K_b$, and equilibria are discussed in terms of molalities (or molar concentrations) themselves. In Chapter 10 we shall see a way of making better estimates of activity coefficients for equilibria involving ions.

Molecular interpretation 9.2

We can obtain a deeper insight into the origin and significance of the equilibrium constant by considering the Boltzmann distribution of molecules over the available states of a system composed of reactants and products (recall *Molecular interpretation 2.1*). When atoms can exchange partners, as in a reaction, the available states of the system include arrangements in which the atoms are present in the form of reactants and in the form of products: these arrangements have their characteristic sets of energy levels, but the Boltzmann distribution does not distinguish between their identities, only their energies. The atoms distribute themselves over both sets of energy levels in accord with the Boltzmann distribution (Fig. 9.3). At a given temperature, there will be a specific distribution of populations, and hence a specific composition of the reaction mixture.

It can be appreciated from the illustration that, if the reactants and products both have similar arrays of molecular energy levels, then the dominant species in a reaction mixture at equilibrium will be the species with the lower set of energy levels. However, the fact that the Gibbs energy occurs in the expression is a signal that entropy plays a role as well as energy. Its role can be appreciated by referring to Fig. 9.4. We see that, although the B energy levels lie higher than the A energy levels, in this instance they are much more closely spaced. As a result, their total population may be considerable and B could even dominate in the reaction mixture at equilibrium. Closely spaced energy levels correlate with a high entropy, so in this case we see that entropy effects dominate adverse energy effects. This competition is mirrored in eqn 19, as can be seen most clearly by using $\Delta_r G^\ominus = \Delta_r H^\ominus - T \Delta_r S^\ominus$ and writing it in the form

$$K = e^{-\Delta_r H^\ominus / RT} e^{\Delta_r S^\ominus / R} \tag{9.21}$$

Note that a positive reaction enthalpy results in a lowering of the equilibrium constant (that is, an endothermic reaction can be expected to have an equilibrium composition that favours the reactants). However, if there is positive reaction entropy, then the equilibrium composition may favour products, despite the endothermic character of the reaction.

(d) Equilibria in biological systems

We saw in Section 7.7d that for biological systems it is appropriate to adopt the biological standard state, in which pH = 7. It follows from eqn 7.58 that the relation between the thermodynamic and biological standard Gibbs energies of reaction for a reaction of the form

$$A + v\,H^+(aq) \rightarrow P \tag{9.22}$$

is

$$\Delta_r G^\oplus = \Delta_r G^\ominus + 7vRT \ln 10 \tag{9.23}$$

Note that there is no difference between the two standard values if hydrogen ions are not involved in the reaction ($v = 0$).

Illustration 9.3

Consider the reaction

$$NADH(aq) + H^+(aq) \longrightarrow NAD^+(aq) + H_2(g)$$

at 37°C, for which $\Delta_r G^\ominus = -21.8$ kJ mol^{-1}. NADH is the reduced form of nicotinamide adenine dinucleotide and NAD$^+$ is its oxidized form; the molecules play an important role in the later stages of the respiratory process (Box 10.2). It follows that, because $v = 1$ and $7 \ln 10 = 16.1$,

$$\Delta_r G^\oplus = -21.8 \text{ kJ mol}^{-1} + 16.1 \times (8.3145 \times 10^{-3} \text{ kJ K}^{-1} \text{ mol}^{-1}) \times (310 \text{ K})$$

$$= +19.7 \text{ kJ mol}^{-1}$$

Note that the biological standard value is opposite in sign (in this example) to the thermodynamic standard value.

Self-test 9.3 For a particular reaction of the form A → B + 2 H$^+$ in aqueous solution, it was found that $\Delta_r G^\ominus = +20$ kJ mol^{-1} at 28°C. Estimate the value of $\Delta_r G^\oplus$.   [−61 kJ mol^{-1}]

The response of equilibria to the conditions

Equilibria respond to changes in pressure, temperature, and concentrations of reactants and products. The equilibrium constant for a reaction is not affected by the presence of a catalyst or an enzyme (a biological catalyst). Catalysts increase the rate at which equilibrium is attained but do not affect its position (Sections 26.5 and 26.6). However, it is important to note that in industry reactions rarely reach equilibrium, partly on account of the rates at which reactants mix. As we shall see in Sections 26.7–10, catalysts can have some unexpected effects and may change the composition of the reaction mixture under non-equilibrium conditions.

9.3 How equilibria respond to pressure

The equilibrium constant depends on the value of $\Delta_r G^{\ominus}$, which is defined at a single, standard pressure. The value of $\Delta_r G^{\ominus}$, and hence of K, is therefore independent of the pressure at which the equilibrium is actually established. Formally, we may express this independence as

$$\left(\frac{\partial K}{\partial p}\right)_T = 0 \qquad (9.24)$$

The conclusion that K is independent of pressure does not necessarily mean that the equilibrium composition is independent of the pressure. However, before considering the consequences of pressure, we need to distinguish between the two ways in which pressure may be applied. The pressure within a reaction vessel can be increased by injecting an inert gas into it. However, so long as the gases are perfect, this addition of gas leaves all the partial pressures of the reacting gases unchanged: the partial pressure of a perfect gas is the pressure it would exert if it were alone in the container, so the presence of another gas has no effect. Put another way, the addition of an inert gas leaves the molar concentrations of the original gases unchanged, as they continue to occupy the same volume. It follows that pressurization by the addition of an inert gas has no effect on the equilibrium composition of the system (provided the gases are perfect). Alternatively, the pressure of the system may be increased by confining the gases to a smaller volume (that is, by compression). Now the partial pressures are changed. Put another way, their molar concentrations are modified because the volume the gases occupy is reduced.

We need to consider the role of compression and understand how changes in partial pressures can be consistent with the general result expressed in eqn 24 that the equilibrium constant itself is independent of the pressure. We shall see that compression can adjust the individual partial pressures of the reactants and products in such a way that, although each one changes, their ratio (as it appears in the equilibrium constant) remains the same. Consider, for instance, the perfect gas equilibrium $A \rightleftharpoons 2\,B$, for which the equilibrium constant is

$$K = \frac{p_B^2}{p_A p^{\ominus}}$$

The right-hand side of this expression remains constant only if an increase in p_A cancels an increase in the *square* of p_B. This relatively steep increase of p_A compared to p_B will occur if the equilibrium composition shifts in favour of A at the expense of B. Then the number of A molecules will increase as the volume of the container is decreased and its partial pressure will rise more rapidly than can be ascribed to a simple change in volume alone (Fig. 9.5).

The increase in the number of A molecules and the corresponding decrease in the number of B molecules brought about by compression is a special case of a principle proposed by the French chemist Henri Le Chatelier.[4] **Le Chatelier's principle** states that:

> **A system at equilibrium, when subjected to a disturbance, responds in a way that tends to minimize the effect of the disturbance**

The principle implies that, if a system at equilibrium is compressed, then the reaction will adjust so as to minimize the increase in pressure. This it can do by reducing the number of particles in the gas phase, which implies a shift $A \leftarrow 2\,B$.

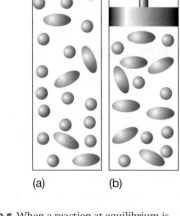

(a)   (b)

9.5 When a reaction at equilibrium is compressed (from *a* to *b*), the reaction responds by reducing the number of molecules in the gas phase (in this case by producing the dimers represented by the ellipses).

---

4. Le Chatelier also invented oxyacetylene welding.

9.6 The pressure dependence of the degree of dissociation, α, at equilibrium for an $A(g) \rightleftharpoons 2\,B(g)$ reaction for different values of the equilibrium constant K. The value $\alpha = 0$ corresponds to pure A; $\alpha = 1$ corresponds to pure B.

The quantitative treatment of the effect of compression set out in the *Justification* below leads to the conclusion that the extent of dissociation, α, of A into 2 B is

$$\alpha = \left(\frac{1}{1 + 4p/K} \right)^{1/2} \qquad (9.25)$$

where p is to be understood as $p/p^{\ominus}$. This formula shows that, even though K is independent of pressure, the amounts of A and B do depend on pressure (Fig. 9.6). It also shows that, as p is increased, α decreases, in accord with Le Chatelier's principle.

Justification 9.2

Suppose that there is an amount n of A present initially (and no B). At equilibrium the amount of A is $(1 - \alpha)n$ and the amount of B is $2\alpha n$. It follows that the mole fractions present at equilibrium are

$$x_A = \frac{(1 - \alpha)n}{(1 - \alpha)n + 2\alpha n} = \frac{1 - \alpha}{1 + \alpha} \qquad x_B = \frac{2\alpha}{1 + \alpha}$$

The equilibrium constant for the reaction (with p understood as $p/p^{\ominus}$) is

$$K = \frac{p_B^2}{p_A} = \frac{x_B^2 p^2}{x_A p} = \frac{4\alpha^2 p}{1 - \alpha^2}$$

This expression rearranges into eqn 25.

Illustration 9.4

To predict the effect of an increase in pressure on the composition of the ammonia synthesis at equilibrium, eqn 10, we note that the number of gas molecules decreases (from 4 to 2). So, Le Chatelier's principle predicts that an increase in pressure will favour the product. The equilibrium constant is

$$K = \frac{p_{NH_3}^2}{p_{N_2} p_{H_2}^3} = \frac{x_{NH_3}^2 p^2}{x_{N_2} x_{H_2}^3 p^4} = \frac{K_x}{p^2}$$

where K_x is the part of the equilibrium constant expression that contains the equilibrium mole fractions of reactants and products. Therefore, doubling the pressure must increase K_x by a factor of 4 to preserve the value of K.

Self-test 9.4 Predict the effect of a 10-fold pressure increase on the equilibrium composition of the reaction $3\,N_2(g) + H_2(g) \rightarrow 2\,N_3H(g)$.

[100-fold increase in K_x]

9.4 The response of equilibria to temperature

Le Chatelier's principle predicts that a system at equilibrium will tend to shift in the endothermic direction if the temperature is raised, for then energy is absorbed as heat and the rise in temperature is opposed. Conversely, an equilibrium can be expected to shift in the exothermic direction if the temperature is lowered, for then energy is released and the reduction in temperature is opposed. These conclusions can be summarized as follows:

Exothermic reactions: increased temperature favours the reactants.

Endothermic reactions: increased temperature favours the products.

We shall now justify these remarks and see how to express the changes quantitatively.

(a) The van 't Hoff equation

The van 't Hoff equation, which is derived in the *Justification* below, is an expression for the slope of a plot of the equilibrium constant (specifically, ln K) as a function of temperature. It may be expressed in either of two ways:

$$\text{(a)} \quad \frac{d \ln K}{dT} = \frac{\Delta_r H^\ominus}{RT^2} \qquad \text{(b)} \quad \frac{d \ln K}{d(1/T)} = -\frac{\Delta_r H^\ominus}{R} \tag{9.26}$$

Justification 9.3

From eqn 19, we know that

$$\ln K = -\frac{\Delta_r G^\ominus}{RT}$$

Differentiation of ln K with respect to temperature then gives

$$\frac{d \ln K}{dT} = -\frac{1}{R} \frac{d(\Delta_r G^\ominus / T)}{dT}$$

The differentials are complete because K and $\Delta_r G^\ominus$ depend only on temperature, not on pressure. To develop this equation we use the Gibbs–Helmholtz equation (eqn 5.13) in the form

$$\frac{d(\Delta_r G^\ominus / T)}{dT} = -\frac{\Delta_r H^\ominus}{T^2}$$

where $\Delta_r H^\ominus$ is the standard reaction enthalpy at the temperature T. Combining the two equations gives the van 't Hoff equation, eqn 26a. The second form of the equation is obtained by noting that

$$\frac{d(1/T)}{dT} = -\frac{1}{T^2}, \quad \text{so} \quad dT = -T^2 \, d(1/T)$$

Equation 26a shows that d ln $K/dT < 0$ (and therefore that $dK/dT < 0$) for a reaction that is exothermic under standard conditions ($\Delta_r H^\ominus < 0$). A negative slope means that ln K, and therefore K itself, decreases as the temperature rises. Therefore, as asserted above, in the case of an exothermic reaction the equilibrium shifts away from products. The opposite occurs in the case of endothermic reactions.

Some insight into the thermodynamic basis of this behaviour comes from the expression $\Delta_r G = \Delta_r H - T\Delta_r S$ written in the form $-\Delta_r G/T = -\Delta_r H/T + \Delta_r S$. When the reaction is exothermic, $-\Delta_r H/T$ corresponds to a positive change of entropy of the surroundings and favours the formation of products. When the temperature is raised, $-\Delta_r H/T$ decreases, and the increasing entropy of the surroundings has a less important role. As a result, the equilibrium lies less to the right. When the reaction is endothermic, the principal factor is the increasing entropy of the reaction system. The importance of the unfavourable change of entropy of the surroundings is reduced if the temperature is raised (because then $\Delta_r H/T$ is smaller), and the reaction is able to shift towards products.

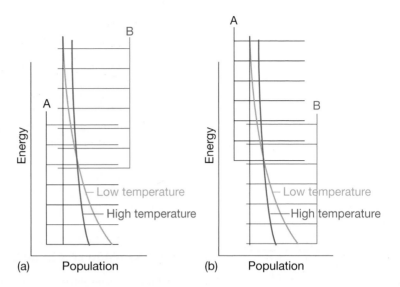

9.7 The effect of temperature on a chemical equilibrium can be interpreted in terms of the change in the Boltzmann distribution with temperature and the effect of that change in the population of the species. (a) In an endothermic reaction, the population of B increases at the expense of A as the temperature is raised. (b) In an exothermic reaction, the opposite happens.

Molecular interpretation 9.3

The typical arrangement of energy levels for an endothermic reaction is shown in Fig. 9.7a. When the temperature is increased, the Boltzmann distribution adjusts and the populations change as shown. The change corresponds to an increased population of the higher energy states at the expense of the population of the lower energy states. We see that the states that arise from the B molecules become more populated at the expense of the A molecules. Therefore, the total population of B states increases, and B becomes more abundant in the equilibrium mixture. Conversely, if the reaction is exothermic (Fig. 9.7b), then an increase in temperature increases the population of the A states (which start at higher energy) at the expense of the B states, so the reactants become more abundant.

Example 9.3 Measuring a reaction enthalpy

The data below show the temperature variation of the equilibrium constant of the reaction

$$Ag_2CO_3(s) \rightleftharpoons Ag_2O(s) + CO_2(g)$$

Calculate the standard reaction enthalpy of the decomposition.

T/K	350	400	450	500
K	3.98×10^{-4}	1.41×10^{-2}	1.86×10^{-1}	1.48

Method It follows from eqn 26b that, provided the reaction enthalpy can be assumed to be independent of temperature, a plot of $-\ln K$ against $1/T$ should be a straight line of slope $\Delta_r H^{\ominus}/R$.

Answer We draw up the following table:

T/K	350	400	450	500
$(10^3 \text{ K})/T$	2.86	2.50	2.22	2.00
$-\ln K$	7.83	4.26	1.68	−0.39

9.8 When $-\ln K$ is plotted against $1/T$, a straight line is expected with slope equal to $\Delta_r H^\ominus/R$. This is a noncalorimetric method for the measurement of reaction enthalpies.

These points are plotted in Fig. 9.8. The slope of the graph is $+9.6 \times 10^3$, so

$$\Delta_r H^\ominus = (+9.6 \times 10^3 \text{ K}) \times R = +80 \text{ kJ mol}^{-1}$$

Comment This is a noncalorimetric method of determining $\Delta_r H^\ominus$. A drawback is that the reaction enthalpy is actually temperature-dependent, so the plot is not expected to be perfectly linear. However, the temperature dependence is weak in many cases, so the plot is reasonably straight. In practice, the method is not very accurate, but it is often the only method available.

Self-test 9.5 The equilibrium constant of the reaction $2 SO_2(g) + O_2(g) \rightleftharpoons 2 SO_3(g)$ is 4.0×10^{24} at 300 K, 2.5×10^{10} at 500 K, and 3.0×10^4 at 700 K. Estimate the reaction enthalpy at 500 K.   [-200 kJ mol^{-1}]

(b)   The value of K at different temperatures

To find the value of the equilibrium constant at a temperature T_2 in terms of its value K_1 at another temperature T_1, we integrate eqn 26b between these two temperatures:

$$\ln K_2 - \ln K_1 = -\frac{1}{R} \int_{1/T_1}^{1/T_2} \Delta_r H^\ominus \, d(1/T) \tag{9.27}$$

If we suppose that $\Delta_r H^\ominus$ varies only slightly with temperature over the temperature range of interest, we may take it outside the integral. It follows that

$$\ln K_2 - \ln K_1 = -\frac{\Delta_r H^\ominus}{R}\left(\frac{1}{T_2} - \frac{1}{T_1}\right) \tag{9.28}$$

Illustration 9.5

To estimate the equilibrium constant for the synthesis of ammonia at 500 K from its value at 298 K (6.1×10^5 for the reaction as written in eqn 10) we use the standard reaction enthalpy, which can be obtained from Table 2.6 in the *Data section* by using $\Delta_r H^\ominus = 2\Delta_f H^\ominus(NH_3, g)$, and assume that its value is constant over the range of temperatures. Then, with $\Delta_r H^\ominus = -92.2$ kJ mol^{-1}, from eqn 28 we find

$$\ln K_2 = \ln(6.1 \times 10^5) - \frac{(-92.2 \times 10^3 \text{ J mol}^{-1})}{8.3145 \text{ J K}^{-1} \text{ mol}^{-1}}\left(\frac{1}{500 \text{ K}} - \frac{1}{298 \text{ K}}\right) = -1.71$$

It follows that $K_2 = 0.18$.

Self-test 9.6 The equilibrium constant for $N_2O_4(g) \rightleftharpoons 2 NO_2(g)$ was calculated in Self-test 9.1. Estimate its value at 100°C.   [15]

Knowledge of the temperature dependence of the equilibrium constant for a reaction can be useful in the design of laboratory and industrial processes. For example, synthetic chemists can improve the yield of a reaction by changing the temperature of the reaction mixture. Also, reduction of a metal oxide with carbon or carbon monoxide results in the extraction of the metal when the process is carried out at a temperature for which the equilibrium constant for the reaction is much greater than one (Box 9.2).

Box 9.2 The extraction of metals from their oxides

Metals can be obtained from their oxides by reduction with carbon or carbon monoxide if any of the equilibria

$$MO(s) + C(s) \rightleftharpoons M(s) + CO(g)$$

$$MO(s) + \tfrac{1}{2}\,C(s) \rightleftharpoons M(s) + \tfrac{1}{2}\,CO_2(g)$$

$$MO(s) + CO(g) \rightleftharpoons M(s) + CO_2(g)$$

lie to the right (i.e. have $K > 1$). As we shall see, these equilibria can be discussed in terms of the thermodynamic functions for the reactions

(i) $M(s) + \tfrac{1}{2}\,O_2(g) \longrightarrow MO(s)$

(ii) $\tfrac{1}{2}\,C(s) + \tfrac{1}{2}\,O_2(g) \longrightarrow \tfrac{1}{2}\,CO_2(g)$

(iii) $C(s) + \tfrac{1}{2}\,O_2(g) \longrightarrow CO(g)$

(iv) $CO(g) + \tfrac{1}{2}\,O_2(g) \longrightarrow CO_2(g)$

The temperature dependences of the standard Gibbs energies of reactions (i)–(iv) depend on the reaction entropy through $d\Delta_r G^{\ominus}/dT = -\Delta_r S^{\ominus}$. Because in reaction (iii) there is a net increase in the amount of gas, the standard reaction entropy is large and positive; therefore, its $\Delta_r G^{\ominus}$ decreases sharply with increasing temperature. In reaction (iv), there is a similar net decrease in the amount of gas, so $\Delta_r G^{\ominus}$ increases sharply with increasing temperature. In reaction (ii), the amount of gas is constant, so the entropy change is small and $\Delta_r G^{\ominus}$ changes only slightly with temperature. These remarks are summarized in the illustration, which is called an *Ellingham diagram*. Note that $\Delta_r G^{\ominus}$ decreases upwards!

At room temperature, $\Delta_r G^{\ominus}$ is dominated by the contribution of the reaction enthalpy ($T\Delta_r S^{\ominus}$ being relatively small), so the order of increasing $\Delta_r G^{\ominus}$ is the same as the order of increasing $\Delta_r H^{\ominus}$ (Al_2O_3 is most exothermic; Ag_2O is least). The standard reaction entropy is similar for all metals because in each case gaseous oxygen is eliminated and a compact, solid oxide is formed. As a result, the temperature dependence of the standard Gibbs energy of oxidation should be similar for all metals, as is shown by the similar slopes of the lines in the diagram. The kinks at high temperatures correspond to the evaporation of the metals; less pronounced kinks occur at the melting temperatures of the metals and the oxides.

Successful reduction of the oxide depends on the outcome of the competition of the carbon for the oxygen bound to the metal. The standard Gibbs energies for the reductions can be expressed in terms of the standard Gibbs energies for the reactions above:

$MO(s) + C(s) \longrightarrow M(s) + CO(g)$ $\Delta_r G^{\ominus} = \Delta_r G^{\ominus}(iii) - \Delta_r G^{\ominus}(i)$

$MO(s) + \tfrac{1}{2}\,C(s) \longrightarrow M(s) + \tfrac{1}{2}\,CO_2(g)$ $\Delta_r G^{\ominus} = \Delta_r G^{\ominus}(ii) - \Delta_r G^{\ominus}(i)$

$MO(s) + CO(g) \longrightarrow M(s) + CO_2(g)$ $\Delta_r G^{\ominus} = \Delta_r G^{\ominus}(iv) - \Delta_r G^{\ominus}(i)$

The equilibrium lies to the right if $\Delta_r G^{\ominus} < 0$. This is the case when the line for reaction (i) lies below (is more positive than) the line for one of the reactions (ii) to (iv).

The spontaneity of a reduction at any temperature can be predicted simply by looking at the diagram: a metal oxide is

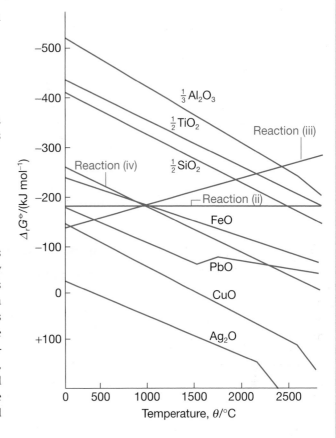

An Ellingham diagram for the discussion of metal ore reduction. Note that $\Delta_r G^{\ominus}$ is most negative at the top of the diagram.

reduced by any carbon reaction lying above it, because the overall reaction then has $\Delta_r G^{\ominus} < 0$. For example, CuO can be reduced to Cu at any temperature above room temperature. Even in the absence of carbon, Ag_2O decomposes when heated above 200°C because then the standard Gibbs energy for reaction (i) becomes positive (and the reverse reaction is then spontaneous). On the other hand, Al_2O_3 is not reduced by carbon until the temperature has been raised to above 2000°C.

Discussion questions

1. Use the Ellingham diagram in the illustration to identify the lowest temperature at which zinc oxide can be reduced to zinc metal by carbon.

2. How may an Ellingham diagram be used to decide whether one metal may be used to reduce the oxide of another metal?

Further reading

H.-G. Lee, *Chemical thermodynamics for metals and materials*. Imperial College Press, London (1999).

D.L. Johnson and G.B. Stracher (eds.), *Thermodynamic loop applications in materials systems*. Minerals, Metals, and Materials Society, Warrendale (1995).

9.5 The response of equilibria to pH

One of the most important examples of chemical equilibrium is the one that exists when acids and bases are present in solution. The proton transfer equilibria characteristic of solutions of Brønsted acids and bases[6] are established very rapidly (within a few nanoseconds) and are highly responsive to changes in the pH, where

$$pH = -\log a_{H_3O^+} \tag{9.29}$$

and H_3O^+ is the hydronium ion, a representation of the state of the proton in aqueous solution. At low concentrations, the activity of hydronium ions is approximately equal to the numerical value of their molality or molar concentration, so a determination of pH is an indication of hydronium ion concentration. However, many thermodynamic observables depend on pH itself, and there is no need to make this approximation and interpretation.

(a) Acid–base equilibria in water

We need three equilibrium constants to discuss the effect of pH on equilibria involving the transfer of protons between species. One is the acidity constant, K_a:

$$HA(aq) + H_2O(l) \rightleftharpoons H_3O^+(aq) + A^-(aq) \qquad K_a = \frac{a_{H_3O^+} a_{A^-}}{a_{HA}} \tag{9.30}$$

It is common to report values of K_a in terms of its negative common logarithm, pK_a:

$$pK_a = -\log K_a \tag{9.31}$$

The second equilibrium constant we require is the basicity constant, K_b, for a base B in water:

$$B(aq) + H_2O(l) \rightleftharpoons HB^+(aq) + OH^-(aq) \qquad K_b = \frac{a_{HB^+} a_{OH^-}}{a_B} \tag{9.32}$$

The third equilibrium to consider is the **autoprotolysis** of water, the equilibrium

$$2\,H_2O(l) \rightleftharpoons H_3O^+(aq) + OH^-(aq) \qquad K_w = a_{H_3O^+} a_{OH^-} \tag{9.33}$$

At 25°C, $K_w = 1.008 \times 10^{-14}$ ($pK_w = 14.00$), showing that only a few of the water molecules are ionized. If we introduce $pOH = -\log a_{OH^-}$ by analogy with pH, then it follows that

$$pK_w = pH + pOH \tag{9.34}$$

As may be verified by multiplying the expressions for K_a and K_b together, the acidity constant of the conjugate acid HB^+ is related to the basicity constant of the base B by

$$K_a K_b = K_w \tag{9.35}$$

The same relation holds between the acidity constant of an acid HA and the basicity constant of its conjugate base A^-. In each case, in elementary calculations it is common to replace activities by the numerical values of molar concentrations.

(b) The pH of acids and bases

The distinction between weak and strong acids and bases is an illustration of the different types of behaviour shown in Fig. 9.9: strong acids and bases are those for which the minimum Gibbs energy of the solution lies close to the products; weak

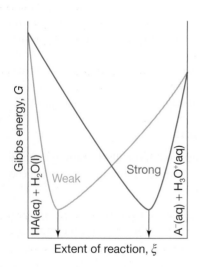

9.9 The Gibbs energy for a solution of a weak acid has a minimum that lies close to HA, and little deprotonation occurs at equilibrium. For a strong acid, the minimum lies close to products and deprotonation is almost complete.

6. It will be familiar from introductory chemistry that a Brønsted *acid* is a proton donor and a Brønsted *base* is a proton acceptor.

acids and bases are those for which the minimum lies close to the reactants. However, because the extent of proton transfer is so small (for typical solutions), an approximation is that the molar concentration of HA or B is unchanged from its nominal value. Moreover, because the molar concentrations of the species produced by the proton transfer are equal (to a good approximation),[7] the expression for K_a in eqn 30 simplifies to[8]

$$K_a \approx \frac{[H_3O^+]^2}{[HA]} \quad \text{or} \quad [H_3O^+] \approx (K_a[HA])^{1/2}$$

It then follows, by taking the negative logarithms of both sides, that

$$pH \approx \tfrac{1}{2}pK_a - \tfrac{1}{2}\log[HA] \tag{9.36}$$

A similar expression can be derived from eqn 32 for the pH of a solution of a weak base: eqn 32 is used to determine pOH and then eqn 34 is used to convert pOH to pH. The result is

$$pH = pK_w - \tfrac{1}{2}pK_b + \tfrac{1}{2}\log[B] \tag{9.37}$$

(c) Acid–base titrations

We shall suppose that we are titrating a volume V_A of a solution of a weak acid of nominal molar concentration A_0 (the analyte) with a solution of a strong base MOH of molar concentration B (the titrant). At the stoichiometric point of the titration,[9] enough titrant has been added for the amount of weak acid it supplies to match the amount of strong base in the analyte.

Our aim is to calculate the pH at any stage of the titration. The result can be calculated 'exactly' (see below) but it is more instructive to see how to make approximations at each stage. The 'exact' pH curve is shown in Fig. 9.10 together with the approximate formulas derived below.

When a volume V_B of the titrant has been added to the analyte the total volume of the analyte is $V = V_A + V_B$. The approximations we shall make are based on the fact that the acid is weak, and therefore that HA is more abundant than any A^- ions in the solution. Furthermore, when HA is present, it provides hydronium ions that greatly outnumber any that come from the autoprotolysis of water. Finally, when excess base is present, the OH^- ions it provides dominate any that come from the water autoprotolysis.

At the start of the titration, the analyte is a solution of a weak acid, so we can use eqn 36 to estimate its pH:

$$pH = \tfrac{1}{2}pK_a - \tfrac{1}{2}\log A_0 \tag{9.38}$$

After the addition of some base (but before the stoichiometric point is reached), the concentration of A^- ions stems almost entirely from the salt that is present as a result of the reaction

$$HA(aq) + OH^-(aq) \longrightarrow A^-(aq) + H_2O(l)$$

We may neglect the relatively small contribution of A^- ions from the weak acid that remains after the reaction. Therefore, $[A^-] = S$, where S is the concentration of the salt that is present after the reaction between HA and the titrant. The amount of HA molecules that remains is the original amount, A_0V_A, less the amount of HA

---

7. That is, in an acid solution $[H_3O^+] \approx [A^-]$ and in a basic solution $[OH^-] \approx [HB^+]$.
8. As before, the notation $\{\dots\}$ for the equation number reminds us that units (of [J]) have been discarded.
9. The stoichiometric point of a titration is still widely called the 'equivalence point'.

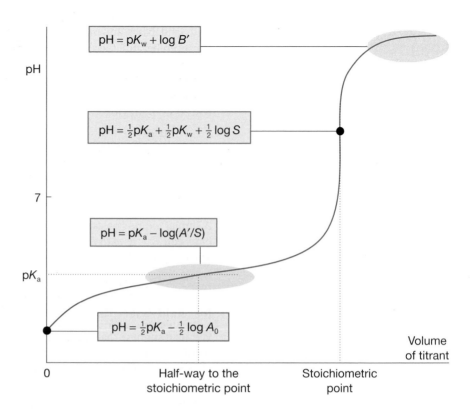

9.10 A summary of the regions of the pH curve of the titration of a weak acid with a strong base, and the equations used in different regions.

molecules that has been converted to salt, so the molar concentration of HA is $A' = (A_0 V_A / V) - S$. This calculation ignores the small additional loss of HA as a result of its deprotonation in solution. Hence[10]

$$K_a = \frac{a_{H_3O^+} a_{A^-}}{a_{HA}} \approx \frac{a_{H_3O^+} S}{A'} \tag{9.39}$$

It follows that

$$\text{pH} = \text{p}K_a - \log \frac{A'}{S} \tag{9.40}$$

This expression is called the **Henderson–Hasselbalch equation**. The general form of this equation, once we recognize that A' is the molar concentration of acid in the solution and S is the molar concentration of base, is

$$\text{pH} = \text{p}K_a - \log \frac{[\text{acid}]}{[\text{base}]} \tag{9.41}$$

When the molar concentrations of acid and salt (acting as the base) are equal,

$$\text{pH} = \text{p}K_a \tag{9.42}$$

Hence the pK_a of the acid can be measured directly from the pH of the mixture. In practice this is done by recording the pH during a titration and then examining the record for the pH half-way to the stoichiometric point (as indicated in Fig. 9.10).

At the stoichiometric point, the H_3O^+ ions in the solution stem from the influence of the OH^- ions on the autoprotolysis equilibrium, and the OH^- ions are produced by

---

10. We are making the doubtful approximation that the activity coefficient of the A⁻ ions is close to 1.

the proton transfer equilibrium from H_2O to A^-. Because only a small amount of HA is formed in this way, the concentration of A^- ions is almost exactly that of the salt, and we can write $[A^-] \approx S$. The number of OH^- ions that arise from the proton transfer equilibrium greatly outnumber those produced by the water autoprotolysis, so we can set $[HA] \approx [OH^-]$:

$$K_b = \frac{a_{HA}a_{OH^-}}{a_{A^-}} \approx \frac{[OH^-]^2}{S}$$

Then, because $K_a K_b = K_w$, we can conclude that

$$[OH^-] \approx \left(\frac{SK_w}{K_a}\right)^{1/2} \tag{9.43}$$

Therefore, after taking negative logarithms of both sides and using $pH = pK_w - pOH$,

$$pH = \tfrac{1}{2}pK_a + \tfrac{1}{2}pK_w + \tfrac{1}{2}\log S \tag{9.44}$$

The corresponding expression for the pH at the stoichiometric point of a titration of a weak base with a strong acid is

$$pH = \tfrac{1}{2}pK_a - \tfrac{1}{2}\log S \tag{9.45}$$

Illustration 9.6

The stoichiometric point of a titration of 25.00 mL of 0.100 M HClO(aq) with 0.100 M NaOH(aq) occurs when the molar concentration of NaClO is 0.050 mol L^{-1} (because the volume of the solution has increased from 25.00 mL to 50.00 mL). Therefore, the pH is

$$pH = \tfrac{1}{2} \times 7.43 + \tfrac{1}{2} \times 14.00 + \tfrac{1}{2}\log 0.050 = 10.1$$

Self-test 9.7 Estimate the pH at the stoichiometric point of a titration of 25.00 mL of 0.200 M NH_3(aq) with 0.300 M HCl(aq). [5.1]

When so much strong base has been added that the titration has been carried well past the stoichiometric point, the pH is determined by the excess base present. Then, $[H_3O^+] \approx K_w/[OH^-]$, so, if we write the molar concentration of excess base as B', this expression can be written

$$pH = pK_w + \log B' \tag{9.46}$$

In this expression, as in all the preceding ones, the molar concentrations must take into account the change of volume that occurs as the titrant is added to the analyte.

(d) The 'exact' form of the pH curve

All the foregoing has supposed that one particular species dominates the pH of the mixture and we have invoked approximations to estimate the pH of the solution. In fact, no such approximations are necessary, for it is quite easy to set up and solve the 'exact' equation for the pH of the solution. We refer to 'exact' rather than exact, because we continue to ignore activity coefficients and treat all activities as molar concentrations. As we show in the following *Justification*, at any stage of the titration

$$v = \frac{K_a A_0[H_3O^+] + (K_w - [H_3O^+]^2)([H_3O^+] + K_a)}{([H_3O^+] + K_a)([H_3O^+]^2 + B_0[H_3O^+] - K_w)} \qquad v = \frac{V_B}{V_A} \tag{9.47}$$

Justification 9.4

The five conditions we use are:

(a) Electrical neutrality: $[H_3O^+] + [M^+] = [A^-] + [OH^-]$, M^+ from added base

(b) Conservation of A groups: $[A^-] + [HA] = \dfrac{A_0 V_A}{V_A + V_B}$

where V_A is the (fixed) initial volume of acid and V_B is the volume of titrant (base) added.

(c) Concentration of M^+ groups: $[M^+] = \dfrac{B_0 V_B}{V_A + V_B}$

(d) Deprotonation equilibrium of HA: $[HA]K_a = [H_3O^+][A^-]$

(e) Autoprotolysis equilibrium: $K_w = [H_3O^+][OH^-]$

First, we express condition (b) in terms of $[H_3O^+]$ and $[A^-]$ by using condition (d) to eliminate $[HA]$:

$$[A^-] = \frac{A_0 K_a V_A}{(V_A + V_B)([H_3O^+] + K_a)}$$

Next, we use this relation and condition (c), and at the same time we use condition (e) and (a) to eliminate $[OH^-]$:

$$[H_3O^+] + \frac{B_0 V_B}{V_A + V_B} = \frac{K_a A_0 V_A}{(V_A + V_B)([H_3O^+] + K_a)} + \frac{K_w}{[H_3O^+]}$$

Now it is just a matter of algebra: we multiply through by $(V_A + V_B)[H_3O^+]$, and then divide through by V_A. The result, after a small rearrangement of terms and writing $v = V_B/V_A$ is eqn 47.

Equation 47 is a cubic equation in the hydronium ion concentration. However, modern mathematical software makes such expressions easy to plot, and a plot of pH against volume of titrant added is shown in Fig. 9.11 for a variety of values of K_a. We see that the curves have exactly the form expected in the regions treated by approximation.

(e) Using equilibria to stabilize pH: buffers

The slow variation of the pH in the vicinity of $S = A'$, when the molar concentrations of the salt and acid are equal, is the basis of **buffer action**, the ability of a solution to oppose changes in pH when small amounts of strong acids and bases are added to the solution. The physical basis of buffer action is that the existence of an abundant supply of A^- ions (because a salt is present) can remove any H_3O^+ ions brought by additional strong acid; moreover, the numerous HA molecules can supply H_3O^+ ions to react with any strong base that is added.

9.11 The 'exact' pH curves for the titration of a weak acid and strong base for different values of the pK_a of the acid.

Example 9.4 Estimating the pH of a buffer solution

Estimate the pH of an aqueous buffer solution that contains 0.200 mol L^{-1} KH_2PO_4 and 0.100 mol L^{-1} K_2HPO_4.

Table 9.1* Acidity constants in water at 298 K

	pK_{a1}	pK_{a2}	pK_{a3}
Acetic acid, CH_3COOH	4.76		
Ammonium ion, NH_4^+	9.25		
Carbonic acid, H_2CO_3	6.35	10.33	
Hydrocyanic acid	9.31		
Phosphoric acid, H_3PO_4	2.16	7.21	12.32

* More values are given in the *Data section*.

Method The pH of a solution of a weak acid and its salt can be estimated from the Henderson–Hasselbalch equation. To do so, we must identify the acid HA and its conjugate base A^-.

Answer In this example, the acid is the anion $H_2PO_4^-$ and its conjugate base is the anion HPO_4^{2-}:

$$H_2PO_4^-(aq) + H_2O(l) \rightleftharpoons H_3O^+(aq) + HPO_4^{2-}(aq)$$

The acidity constant we require is therefore pK_a for $H_2PO_4^-$ (the pK_{a2} of H_3PO_4) which from Table 9.1 is 7.21. Then, with $A' = 0.200$ mol L^{-1} and $S = 0.100$ mol L^{-1}, eqn 40 gives the pH of the solution as

$$pH = 7.21 - \log\frac{0.200}{0.100} = 6.91$$

Hence, the solution should buffer close to pH = 7.

Self-test 9.8 Calculate the pH of an aqueous buffer solution which contains 0.100 mol L^{-1} NH_3 and 0.200 mol L^{-1} NH_4Cl. [8.95; more realistically: 9]

(f) Using equilibria to detect changes in pH: indicators

The rapid change of pH near the stoichiometric point in a titration is the basis of indicator detection and of the more modern electronic detection techniques. An **acid–base indicator** is normally some large, water-soluble, weakly acidic organic molecule that can exist as acid (HIn) or conjugate base (In^-) forms that differ in colour. The two forms are in equilibrium in solution:

$$HIn(aq) + H_2O(l) \rightleftharpoons In^-(aq) + H_3O^+(aq)$$

If we make the usual assumption that the solution is so dilute that the activity of water is 1, then the equilibrium is described by the constant

$$K_{In} = \frac{a_{H_3O^+}a_{In^-}}{a_{HIn}} \tag{9.48}$$

We find the ratio of acid and base forms at a given pH by rearranging this expression to

$$\log\frac{[HIn]}{[In^-]} \approx pK_{In} - pH \tag{9.49}$$

Therefore, when $pH < pK_{In}$, the indicator is predominantly in its acidic form and has the corresponding colour; when the $pH > pK_{In}$, the indicator is mainly in its basic form. The **end point** is the pH of the solution when both forms are present in equal abundance, which occurs when $pH = pK_{In}$.

At the stoichiometric point of an acid–base titration, the pH changes sharply through several units, and if the pH passes through pK_{In} there is a pronounced colour change. With a well-chosen indicator, the end point coincides with the stoichiometric point of the titration.

Care must be taken to use an indicator that changes colour at the pH appropriate to the type of titration. Thus, in a weak acid–strong base titration, the stoichiometric point lies at the pH given by eqn 44, so an indicator that changes at that pH must be selected. Broadly speaking, an indicator with $pK_{In} > 7$ is required because the stoichiometric point lies at $pH > 7$. Similarly, in a strong acid–weak base titration, an indicator changing near the pH given by eqn 45 should be used. The stoichiometric points of such titrations lie at $pH < 7$, so an indicator with $pK_{In} < 7$ is required.

Checklist of key ideas

Spontaneous chemical reactions

9.1 The Gibbs energy minimum

☐ Extent of reaction (ξ): when the extent of reaction changes by a finite amount $\Delta\xi$, the amount of A present changes from $n_{A,0}$ to $n_{A,0} - \Delta\xi$

☐ reaction Gibbs energy, the slope of the graph of the Gibbs energy plotted against the extent of reaction: $\Delta_r G = (\partial G/\partial\xi)_{p,T}$

☐ relation of $\Delta_r G$ to the chemical potentials of the species, $\Delta_r G = \mu_B - \mu_A$

☐ condition of equilibrium, $\Delta_r G = 0$

☐ exergonic reaction, a reaction for which $\Delta_r G < 0$; such a reaction can be used to drive another process

☐ endergonic reaction, a reaction for which $\Delta_r G > 0$

9.2 The description of equilibrium

☐ general statement of a reaction, $0 = \sum_J \nu_J J$

☐ reaction quotient, $Q = \prod_J a_J^{\nu_J}$

☐ standard reaction Gibbs energy, $\Delta_r G^\ominus$

☐ equilibrium constant (K) in terms of $\Delta_r G^\ominus$, $\Delta_r G^\ominus = -RT \ln K$

☐ general expression for $\Delta_r G$ at an arbitrary stage of the reaction, $\Delta_r G = \Delta_r G^\ominus + RT \ln Q$

☐ $\Delta_r G^\ominus$ in terms of Gibbs energies of formation, $\Delta_r G^\ominus = \sum_{\text{Products}} \nu \Delta_f G^\ominus - \sum_{\text{Reactants}} \nu \Delta_f G^\ominus = \sum_J \nu_J \Delta_f G^\ominus(J)$

☐ thermodynamic equilibrium constant, an equilibrium constant K expressed in terms of activities (or fugacities): $K = (\prod_J a_J^{\nu_J})_{\text{equilibrium}}$

☐ relations between equilibrium constants, $K = K_\gamma K_b$

☐ equilibria in biological systems, $\Delta_r G^\oplus = \Delta_r G^\ominus + 7\nu RT \ln 10$

The response of equilibria to the conditions

☐ null effect of a catalyst

9.3 How equilibria respond to pressure

☐ response of K to changes in pressure, $(\partial K/\partial p)_T = 0$

☐ effect of pressure on composition

☐ Le Chatelier's principle: A system at equilibrium, when subjected to a disturbance, responds in a way that tends to minimize the effect of the disturbance

9.4 The response of equilibria to temperature

☐ Le Chatelier's principle for the effect of temperature: increased temperature favours the reactants in exothermic reactions and the products in endothermic reactions

☐ van 't Hoff equation for the effect of temperature on K, $\mathrm{d}\ln K/\mathrm{d}T = \Delta_r H^\ominus/RT^2$

☐ the equilibrium constant at one temperature in terms of its value at another temperature, $\ln K_2 - \ln K_1 = -(\Delta_r H^\ominus/R)(1/T_2 - 1/T_1)$

9.5 The response of equilibria to pH

☐ pH of a solution, $\text{pH} = -\log a_{H_3O^+}$

☐ acidity constant, $K_a = a_{H_3O^+} a_{A^-}/a_{HA}$

☐ the significance of $pK_a = -\log K_a$

☐ basicity constant, $K_b = a_{HB^+} a_{OH^-}/a_B$

☐ the relation between K_a and K_b, $K_a K_b = K_w$

☐ autoprotolysis constant, $K_w = a_{H_3O^+} a_{OH^-}$

☐ the relation between pH and pOH, $\text{pH} + \text{pOH} = pK_w$

☐ pH of a solution of a weak acid, $\text{pH} = \frac{1}{2} pK_a - \frac{1}{2}\log[\text{HA}]$

☐ changes in pH in the course of a titration, Fig. 9.10

☐ pH of a solution of a weak base, $\text{pH} = pK_w - \frac{1}{2} pK_b + \frac{1}{2}\log[\text{B}]$

☐ Henderson–Hasselbalch equation, $\text{pH} = pK_a - \log([\text{acid}]/[\text{base}])$

☐ the pH at the stoichiometric point of a titration of a weak acid with a strong base, $\text{pH} = \frac{1}{2} pK_a + \frac{1}{2} pK_w + \frac{1}{2}\log S$

☐ the pH after the stoichiometric point

☐ buffer action, the ability of a solution to oppose changes in pH when small amounts of strong acids and bases are added to the solution

☐ acid–base indicator, a water-soluble, weakly acidic organic molecule that can exist as acid (HIn) or conjugate base (In⁻) forms that differ in colour

☐ indicator constant, $K_{In} = a_{H_3O^+} a_{In^-}/a_{HIn}$

☐ selection of indicators, $pK_{In} \approx \text{pH}$ of stoichiometric point

Further reading

Articles of general interest

R.T. Allsop and N.H. George, Le Chatelier—a redundant principle? *Educ. in Chem.* **21**, 54 (1984).

K. Anderson, Practical calculation of the equilibrium constant and the enthalpy of reaction at different temperatures. *J. Chem. Educ.* **71**, 474 (1994).

S. Chaston, Thermodynamic diagrams. *J. Chem. Educ.* **76**, 216 (1999).

N.C. Craig, The chemists' delta. *J. Chem. Educ.* **64**, 668 (1987).

R. deLevie, Explicit expression of the general form of the titration curve in terms of concentration: Writing a single closed-form expression for the titration curve for a variety of titrations without using approximations or segmentation. *J. Chem. Educ.* **70**, 209 (1993).

M. Deumiè, B. Boulil, and O. Henri-Rousseau, On the minimum of the Gibbs free energy involved in chemical equilibrium: Some further comments on the stability of the system. *J. Chem. Educ.* **64**, 201 (1987).

A. Dumon, A. Lichanot, and E. Poquet, Describing chemical transformation: From the extent of reaction ξ to the reaction advancement ratio χ. *J. Chem. Educ.* **70**, 29 (1993).

J. Gold and V. Gold, Le Chatelier's principle and the laws of van 't Hoff. *Educ. in Chem.* **22**, 82 (1985).

H.R. Kemp, The effect of temperature and pressure on equilibria: A derivation of the van 't Hoff rules. *J. Chem. Educ.* **64**, 482 (1987).

S.R. Logan, Entropy of mixing and homogeneous equilibria. *Educ. in Chem.* **25**, 44 (1988).

J.J. MacDonald, Equilibrium, free energy, and entropy: Rates and differences. *J. Chem. Educ.* **67**, 380 (1990).

J.J. MacDonald, Equilibria and $\Delta G^{\ominus}$. *J. Chem. Educ.* **67**, 745 (1990).

R.M. Rosenberg, I.M. Klotz, Spontaneity and the equilibrium constant: advantages of the Planck function. *J. Chem. Educ.* **76**, 1448 (1999).

G. Schmitz, The uncertainty of pH. *J. Chem. Educ.* **71**, 117 (1994).

M.J. Schultz, Why equilibrium? Understanding the role of entropy of mixing. *J. Chem. Educ.* **76**, 1391 (1999).

R.S. Treptow, Free energy versus extent of reaction: understanding the difference between ΔG and $\partial G/\partial \xi$. *J. Chem. Educ.* **73**, 51 (1996).

H. Xijun and Y. Xiuping, Influences of temperature and pressure on chemical equilibrium in non-ideal systems. *J. Chem. Educ.* **68**, 295 (1991).

Texts and sources of data and information

M.J. Blandamer, *Chemical equilibria in solution.* Ellis Horwood/Prentice Hall, Hemel Hempstead (1992).

W.A. Cramer and D.A. Knaff, *Energy transduction in biological membranes, a textbook of bioenergetics.* Springer Verlag, New York (1990).

W.E. Dasent, *Inorganic energetics.* Cambridge University Press (1982).

K. Denbigh, *The principles of chemical equilibrium, with applications in chemistry and chemical engineering.* Cambridge University Press (1981).

R.N. Goldberg and Y.B. Tewari, Thermodynamics of enzyme-catalyzed reactions. *J. Phys. Chem. Ref. Data.* Part 1: **22**, 515 (1993). Part 2: **23**, 547 (1994). Part 3: **23**, 1035 (1994). Part 4: **24**, 1669 (1995). Part 5: **24**, 1765 (1995).

F.M. Harold, *The vital force: a study of bioenergetics.* W.H. Freeman and Co., New York (1986).

R. Holub and P. Vónka, *The chemical equilibrium of gaseous systems.* Reidel, Boston (1976).

D.A. Johnson, *Some thermodynamic aspects of inorganic chemistry.* Cambridge University Press (1982).

W.R. Smith and R.W. Missen, *Chemical reaction equilibrium analysis.* Wiley, New York (1982).

Exercises

Discussion questions

9.1 (a) Explain how the mixing of reactants and products affects the position of chemical equilibrium.

9.1 (b) Suggest how the thermodynamic equilibrium constant may respond differently to changes in pressure and temperature from the equilibrium constant expressed in terms of partial pressures.

9.2 (a) Explain Le Chatelier's principle in terms of thermodynamic quantities.

9.2 (b) Explain the physical basis of the van 't Hoff equation for the temperature dependence of K.

9.3 (a) Describe the changes in pH that take place during the titration of a weak acid with a strong base.

9.3 (b) Describe the changes in pH that take place during the titration of a weak base with a strong acid.

9.4 (a) What approximations are made in the derivation of the Henderson–Hasselbalch equation?

9.4 (b) Describe the basis of buffer action and indicator detection.

Numerical exercises

9.5 (a) The equilibrium constant for the isomerization of *cis*-2-butene to *trans*-2-butene is $K = 2.07$ at 400 K. Calculate the standard reaction Gibbs energy.

9.5 (b) The equilibrium constant for the dissociation of Br_2 at 1600 K is $K = 0.255$. Calculate the standard reaction Gibbs energy.

9.6 (a) The standard reaction Gibbs energy of the isomerization of *cis*-2-pentene to *trans*-2-pentene at 400 K is -3.67 kJ mol^{-1}. Calculate the equilibrium constant of the isomerization.

9.6 (b) The standard reaction Gibbs energy of the decomposition of $CaCO_3$ to CaO and CO_2 at 1173 K is $+0.178$ kJ mol^{-1}. Calculate the equilibrium constant of the isomerization.

9.7 (a) At 2257 K and 1.00 atm total pressure, water is 1.77 per cent decomposed at equilibrium by way of the reaction $2 H_2O(g) \rightleftharpoons 2 H_2(g) + O_2(g)$. Calculate (a) K, (b) $\Delta_r G^{\ominus}$, and (c) $\Delta_r G$ at this temperature.

9.7 (b) For the equilibrium, $N_2O_4(g) \rightleftharpoons 2 NO_2(g)$, the degree of dissociation, α_e, at 298 K is 0.201 at 1.00 bar total pressure. Calculate (a) $\Delta_r G$, (2) K, and (3) $\Delta_r G^{\ominus}$ at 298 K.

9.8 (a) Dinitrogen tetroxide is 18.46 per cent dissociated at 25°C and 1.00 bar in the equilibrium $N_2O_4(g) \rightleftharpoons 2 NO_2(g)$. Calculate (a) K at 25°C, (b) $\Delta_r G^{\ominus}$, (c) K at 100°C given that $\Delta_r H^{\ominus} = +57.2$ kJ mol^{-1} over the temperature range.

9.8 (b) Molecular bromine is 24 per cent dissociated at 1600 K and 1.00 bar in the equilibrium $Br_2(g) \rightleftharpoons 2 Br(g)$. Calculate (a) K at 25°C, (b) $\Delta_r G^{\ominus}$, (c) K at 2000°C given that $\Delta_r H^{\ominus} = +112$ kJ mol^{-1} over the temperature range.

9.9 (a) From information in the *Data section*, calculate the standard Gibbs energy and the equilibrium constant at (a) 298 K and (b) 400 K for the reaction $PbO(s) + CO(g) \rightleftharpoons Pb(s) + CO_2(g)$. Assume that the reaction enthalpy is independent of temperature.

9.9 (b) From information in the *Data section*, calculate the standard Gibbs energy and the equilibrium constant at (a) 25°C and (b) 50°C for the reaction $CH_4(g) + 3 Cl_2(g) \rightleftharpoons CHCl_3(l) + 3 HCl(g)$. Assume that the reaction enthalpy is independent of temperature.

9.10 (a) In the gas-phase reaction $2A + B \rightleftharpoons 3C + 2D$, it was found that, when 1.00 mol A, 2.00 mol B, and 1.00 mol D were mixed and allowed to come to equilibrium at 25°C, the resulting mixture contained 0.90 mol C at a total pressure of 1.00 bar. Calculate (a) the mole fractions of each species at equilibrium, (b) K_x, (c) K, and (d) $\Delta_r G^\ominus$.

9.10 (b) In the gas-phase reaction $A + B \rightleftharpoons C + 2D$, it was found that, when 2.00 mol A, 1.00 mol B, and 3.00 mol D were mixed and allowed to come to equilibrium at 25°C, the resulting mixture contained 0.79 mol C at a total pressure of 1.00 bar. Calculate (a) the mole fractions of each species at equilibrium, (b) K_x, (c) K, and (d) $\Delta_r G^\ominus$.

9.11 (a) The standard reaction enthalpy of $Zn(s) + H_2O(g) \rightarrow ZnO(s) + H_2(g)$ is approximately constant at $+224$ kJ mol^{-1} from 920 K up to 1280 K. The standard reaction Gibbs energy is $+33$ kJ mol^{-1} at 1280 K. Estimate the temperature at which the equilibrium constant becomes greater than 1.

9.11 (b) The standard enthalpy of a certain reaction is approximately constant at $+125$ kJ mol^{-1} from 800 K up to 1500 K. The standard reaction Gibbs energy is $+22$ kJ mol^{-1} at 1120 K. Estimate the temperature at which the equilibrium constant becomes greater than 1.

9.12 (a) The equilibrium constant of the reaction $2C_3H_6(g) \rightleftharpoons C_2H_4(g) + C_4H_8(g)$ is found to fit the expression

$$\ln K = -1.04 - \frac{1088}{T/K} + \frac{1.51 \times 10^5}{(T/K)^2}$$

between 300 K and 600 K. Calculate the standard reaction enthalpy and standard reaction entropy at 400 K.

9.12 (b) The equilibrium constant of a reaction is found to fit the expression

$$\ln K = -2.04 - \frac{1176}{T/K} + \frac{2.1 \times 10^7}{(T/K)^3}$$

between 400 K and 500 K. Calculate the standard reaction enthalpy and standard reaction entropy at 450 K.

9.13 (a) The standard reaction Gibbs energy of the isomerization of borneol ($C_{10}H_{17}OH$) to isoborneol in the gas phase at 503 K is $+9.4$ kJ mol^{-1}. Calculate the reaction Gibbs energy in a mixture consisting of 0.15 mol of borneol and 0.30 mol of isoborneol when the total pressure is 600 Torr.

9.13 (b) The equilibrium pressure of H_2 over solid uranium and uranium hydride, UH_3, at 500 K is 1.04 Torr. Calculate the standard Gibbs energy of formation of $UH_3(s)$ at 500 K.

9.14 (a) Calculate the percentage change in the equilibrium constant K_x of the reaction $H_2CO(g) \rightleftharpoons CO(g) + H_2(g)$ when the total pressure is increased from 1.0 bar to 2.0 bar at constant temperature.

9.14 (b) Calculate the percentage change in the equilibrium constant K_x of the reaction $CH_3OH(g) + NOCl(g) \rightleftharpoons HCl(g) + CH_3NO_2(g)$ when the total pressure is increased from 1.0 bar to 2.0 bar at constant temperature.

9.15 (a) The equilibrium constant for the gas-phase isomerization of borneol ($C_{10}H_{17}OH$) to isoborneol at 503 K is 0.106. A mixture consisting of 7.50 g of borneol and 14.0 g of isoborneol in a container of volume 5.0 L is heated to 503 K and allowed to come to equilibrium. Calculate the mole fractions of the two substances at equilibrium.

9.15 (b) The equilibrium constant for the reaction $N_2(g) + O_2(g) \rightleftharpoons 2 NO(g)$ is 1.69×10^{-3} at 2300 K. A mixture consisting of 5.0 g of nitrogen and 2.0 g of oxygen in a container of volume 1.0 L is heated to 2300 K and allowed to come to equilibrium. Calculate the mole fraction of NO at equilibrium.

9.16 (a) Use the data in Table 2.6 of the *Data section* to decide which of the following reactions have $K > 1$ at 298 K.

(a) $HCl(g) + NH_3(g) \rightleftharpoons NH_4Cl(s)$

(b) $2 Al_2O_3(s) + 3 Si(s) \rightleftharpoons 3 SiO_2(s) + 4 Al(s)$

(c) $Fe(s) + H_2S(g) \rightleftharpoons FeS(s) + H_2(g)$

9.16 (b) Use the data in Table 2.6 of the *Data section* to decide which of the following reactions have $K > 1$ at 298 K.

(a) $FeS_2(s) + 2 H_2(g) \rightleftharpoons Fe(s) + 2 H_2S(g)$

(b) $2 H_2O_2(l) + H_2S(g) \rightleftharpoons H_2SO_4(l) + 2 H_2(g)$

9.17 (a) Which of the equilibria in Exercise 9.16a are favoured (in the sense of K increasing) by a rise in temperature at constant pressure?

9.17 (b) Which of the equilibria in Exercise 9.16b are favoured (in the sense of K increasing) by a reduction in temperature at constant pressure?

9.18 (a) What is the standard enthalpy of a reaction for which the equilibrium constant is (a) doubled, (b) halved when the temperature is increased by 10 K at 298 K?

9.18 (b) What is the standard enthalpy of a reaction for which the equilibrium constant is (a) doubled, (b) halved when the temperature is increased by 15 K at 310 K?

9.19 (a) The standard Gibbs energy of formation of $NH_3(g)$ is -16.5 kJ mol^{-1} at 298 K. What is the reaction Gibbs energy when the partial pressure of the N_2, H_2, and NH_3 (treated as perfect gases) are 3.0 bar, 1.0 bar, and 4.0 bar, respectively? What is the spontaneous direction of the reaction in this case?

9.19 (b) The dissociation vapour pressure of NH_4Cl at 427°C is 608 kPa but at 459°C it has risen to 1115 kPa. Calculate (a) the equilibrium constant, (b) the standard reaction Gibbs energy, (c) the standard enthalpy, (d) the standard entropy of dissociation, all at 427°C. Assume that the vapour behaves as a perfect gas and that $\Delta H^\ominus$ and $\Delta S^\ominus$ are independent of temperature in the range given.

9.20 (a) Estimate the temperature at which $CaCO_3$(calcite) decomposes.

9.20 (b) Estimate the temperature at which $CuSO_4 \cdot 5H_2O$ undergoes dehydration.

9.21 (a) At the half-way point in the titration of a weak acid with a strong base the pH was measured as 5.40. (a) What is the acidity constant and the pK_a of the acid? (b) What is the pH of the solution that is 0.015 M in the acid?

9.21 (b) At the half-way point in the titration of a weak acid with a strong base the pH was measured as 4.82. (a) What is the acidity constant and the pK_a of the acid? (b) What is the pH of the solution that is 0.025 M in the acid?

9.22 (a) Calculate the pH of (a) 0.10 M NH_4Cl(aq), (b) 0.10 M $NaCH_3CO_2$(aq), (c) 0.100 M CH_3COOH(aq).

9.22 (b) Calculate the pH of (a) 0.10 M $NaHCO_2$(aq), (b) 0.20 M $NaC_6H_5CO_2$(aq), (c) 0.150 M HCN(aq).

9.23 (a) Calculate the pH at the stoichiometric point of the titration of 25.00 mL of 0.100 M lactic acid with 0.150 M NaOH(aq).

9.23 (b) Calculate the pH at the stoichiometric point of the titration of 25.00 mL of 0.100 M chlorous acid with 0.175 M NaOH(aq). The pK_a of chlorous acid is 1.96.

9.24 (a) Sketch the pH curve of a solution containing 0.10 M $NaCH_3CO_2$(aq) and a variable amount of acetic acid.

9.24 (b) Sketch the pH curve of a solution containing 0.15 M $NaC_6H_5CO_2$(aq) and a variable amount of benzoic acid.

9.25 (a) From the information in Table 9.1, select suitable buffers for (a) pH = 2.2 and (b) pH = 7.0.

9.25 (b) From the information in Table 9.1, select suitable buffers for (a) pH = 4.6 and (b) pH = 10.8.

Problems[†]

Numerical problems

9.1 The equilibrium constant for the reaction, $I_2(s) + Br_2(g) \rightleftharpoons$ 2 IBr(g) is 0.164 at 25°C. (a) Calculate $\Delta_r G^{\ominus}$ for this reaction. (b) Bromine gas is introduced into a container with excess solid iodine. The pressure and temperature are held at 0.164 atm and 25°C, respectively. Find the partial pressure of IBr(g) at equilibrium. Assume that all the bromine is in the liquid form and that the vapour pressure of iodine is negligible. (c) In fact, solid iodine has a measurable vapour pressure at 25°C. In this case, how would the calculation have to be modified?

9.2 Consider the decomposition of methane, CH_4(g), into the elements H_2(g) and C(s, graphite). (a) Given that $\Delta_f H^{\ominus}(CH_4, g) =$ −74.85 kJ mol^{-1} and that $\Delta_f S^{\ominus}(CH_4, g) = -80.67$ J K^{-1} mol^{-1} at 298 K, calculate the value of the equilibrium constant at 298 K. (b) Assuming that $\Delta_f H^{\ominus}$ is independent of temperature, calculate K at 50°C. (c) Calculate the degree of dissociation, α_e, of methane at 25°C and a total pressure of 0.010 bar. (d) Without doing any numerical calculations, explain how the degree of dissociation for this reaction will change as the pressure and temperature are varied.

9.3 The equilibrium pressure of H_2 over U(s) and UH_3(s) between 450 K and 715 K fits the expression

$$\ln(p/\text{Pa}) = 69.32 - \frac{1.464 \times 10^4}{T/K} - 5.65 \ln(T/K)$$

Find an expression for the standard enthalpy of formation of UH_3(s) and from it calculate $\Delta_r C_p^{\ominus}$.

9.4 The degree of dissociation, α_e, of CO_2(g) into CO(g) and O_2(g) at high temperatures was found to vary with temperature as follows:

T/K	1395	1443	1498
$\alpha_e/10^{-4}$	1.44	2.50	4.71

Assuming $\Delta_r H^{\ominus}$ to be constant over this temperature range, calculate K, $\Delta_r G^{\ominus}$, $\Delta_r H^{\ominus}$, and $\Delta_r S^{\ominus}$. Make any justifiable approximations.

9.5 The standard reaction enthalpy for the decomposition of $CaCl_2 \cdot NH_3$(s) into $CaCl_2$(s) and NH_3(g) is nearly constant at +78 kJ mol^{-1} between 350 K and 470 K. The equilibrium pressure of NH_3 in the presence of $CaCl_2 \cdot NH_3$ is 12.8 Torr at 400 K. Find an expression for the temperature dependence of $\Delta_r G^{\ominus}$ in the same range.

9.6 Calculate the equilibrium constant of the reaction CO(g) + H_2(g) $\rightleftharpoons H_2$CO(g) given that, for the production of liquid formaldehyde,

$\Delta_r G^{\ominus} = +28.95$ kJ mol^{-1} at 298 K and that the vapour pressure of formaldehyde is 1500 Torr at that temperature.

9.7 Acetic acid was evaporated in container of volume 21.45 cm^3 at 437 K and at an external pressure of 764.3 Torr, and the container was then sealed. The mass of acid present in the sealed container was 0.0519 g. The experiment was repeated with the same container but at 471 K, and it was found that 0.0380 g of acetic acid was present. Calculate the equilibrium constant for the dimerization of the acid in the vapour and the enthalpy of vaporization.

9.8 A sealed container was filled with 0.300 mol H_2(g), 0.400 mol I_2(g), and 0.200 mol HI(g) at 870 K and total pressure 1.00 bar. Calculate the amounts of the components in the mixture at equilibrium given that K = 870 for the reaction H_2(g) + I_2(g) $\rightleftharpoons$ 2 HI(g).

9.9 For the gas-phase reaction 2 A + B → 3 C + 2 D, it was found that, when 1.00 mol A, 2.00 mol B, and 1.00 mol D were mixed and came to equilibrium at 1.00 bar, the resulting mixture contained 0.90 mol C. Calculate the equilibrium constant.

9.10 The dissociation of I_2 can be monitored by measuring the total pressure, and three sets of results are as follows:

T/K	973	1073	1173
100p/atm	6.244	7.500	9.181
$10^4 n_I$	2.4709	2.4555	2.4366

where n_I is the amount of I atoms per mole of I_2 molecules in the mixture, which occupied 342.68 cm^3. Calculate the equilibrium constants of the dissociation and the standard enthalpy of dissociation at the mean temperature.

9.11‡ Alberty et al. report thermochemical properties of polycyclic aromatic hydrocarbons, many estimated by group additivity methods (R.A. Alberty, M.B. Chung, and A.K. Reif, J. Phys. Chem. Ref. Data **18**, 77 (1989)). Some of their data are shown below for three isomeric pyrenes of formula $C_{20}H_{12}$ at 1000 K.

Species	$\Delta_f H^{\ominus}$/(kJ mol^{-1})	$S_m^{\ominus}$/(J K^{-1} mol^{-1})
perylene	+253.2	987.9
benzo(e)pyrene	+253.2	993.7
benzo(a)pyrene	+262.4	999.4

What are the equilibrium mole fractions of these three isomers at 1000 K?

9.12‡ Thorn et al. recently carried out a study of Cl_2O(g) by photoelectron ionization (R.P. Thorn, L.J. Stief, S.-C. Kuo, and

---

† Problems denoted with the symbol ‡ were supplied by Charles Trapp, Carmen Giunta, and Marshall Cady.

R.B. Klemm, *J. Phys. Chem.* **100**, 14 178 (1996)). From their measurements, they report $\Delta_f H^{\ominus}(Cl_2O) = +77.2$ kJ mol^{-1}. They combined this measurement with literature data on the reaction $Cl_2O(g) + H_2O(g) \rightarrow 2 HOCl(g)$, for which $K = 8.2 \times 10^{-2}$ and $\Delta_r S^{\ominus} = +16.38$ J K^{-1} mol^{-1}, and with readily available thermodynamic data on water vapour to report a value for $\Delta_f H^{\ominus}(HOCl)$. Calculate that value. All quantities refer to 298 K.

9.13‡ The 1980s saw reports of $\Delta_f H^{\ominus}(SiH_2)$ ranging from 243 to 289 kJ mol^{-1}. For example, the lower value appeared was cited in the review article by R. Walsh (*Acc. Chem. Res.* **14**, 246 (1981)); Walsh now leans toward the upper end of the range (H.M. Frey, R. Walsh, and I.M. Watts, *J. Chem. Soc., Chem. Commun.* 1189 (1986)). The higher value was reported in S.-K. Shin and J.L. Beauchamp, *J. Phys. Chem.* **90**, 1507 (1986). If the standard enthalpy of formation is uncertain by this amount, by what factor is the equilibrium constant for the formation of SiH$_2$ from its elements uncertain at (a) 298 K, (b) 700 K?

Theoretical problems

9.14 Show that, if K_p increases with pressure, then K_{ϕ} must decrease, where $K = K_p K_{\phi}$, ϕ denoting the fugacity coefficient.

9.15 Express the equilibrium constant of a gas-phase reaction $A + 3 B \rightleftharpoons 2 C$ in terms of the equilibrium value of the extent of reaction, ξ, given that initially A and B were present in stoichiometric proportions. Find an expression for ξ as a function of the total pressure, p, of the reaction mixture and sketch a graph of the expression obtained.

9.16 When light passes through a cell of length l containing an absorbing gas at a pressure p, the absorption is proportional to pl. Consider the equilibrium $2 NO_2 \rightleftharpoons N_2O_4$, with NO_2 the absorbing species. Show that, when two cells of lengths l_1 and l_2 are used, and the pressures needed to obtain equal absorptions are p_1 and p_2, respectively, then the equilibrium constant is given by

$$K = \frac{(p_1 \rho^2 - p_2)^2}{\rho(\rho - 1)(p_2 - p_1 \rho)p^{\ominus}}$$

with $\rho = l_1/l_2$. The following data were obtained (R.J. Nordstrum and W.H. Chan, *J. Phys. Chem.* **80**, 847 (1976)):

Absorbance	p_1/Torr	p_2/Torr
0.05	1.00	5.47
0.10	2.10	12.00
0.15	3.15	18.65

with $l_1 = 395$ mm and $l_2 = 75$ mm. Determine the equilibrium constant of the reaction.

9.17 Find an expression for the standard reaction Gibbs energy at a temperature T' in terms of its value at another temperature T and the coefficients a, b, and c in the expression for the molar heat capacity listed in Table 2.2. Evaluate the standard Gibbs energy of formation of $H_2O(l)$ at 372 K from its value at 298 K.

9.18 Derive an 'exact' expression for $v = V_B/V_A$ for the titration of an aqueous solution of a weak base with an aqueous solution of a strong acid. *Hint.* See *Justification 9.4*.

Applications: to biology, environmental science, and chemical engineering

9.19 Hydrogen and carbon monoxide have been investigated for use in fuel cells, so their solubilities in molten salts are of interest. Their solubilities in a molten NaNO$_3$/KNO$_3$ mixture were examined

(E. Desimoni and P.G. Zambonin, *J. Chem. Soc. Faraday Trans.* 1, 2014 (1973)) with the following results:

$$\log s_{H_2} = -5.39 - \frac{768}{T/K} \qquad \log s_{CO} = -5.98 - \frac{980}{T/K}$$

where s is the solubility in mol cm^{-3} bar^{-1}. Calculate the standard molar enthalpies of solution of the two gases at 570 K.

9.20 Under biochemical standard conditions, aerobic respiration produces approximately 38 molecules of ATP per molecule of glucose that is completely oxidized. (a) What is the percentage efficiency of aerobic respiration under biochemical standard conditions? (b) The following conditions are more likely to be observed in a living cell: $p_{CO_2} = 5.3 \times 10^{-2}$ atm, $p_{O_2} = 0.132$ atm, [glucose] = 5.6×10^{-2} mol L^{-1}, [ATP] = [ADP] = [P$_i$] = 1.0×10^{-4} mol L^{-1}, pH = 7.4, $T = 310$ K. Assuming that activities can be replaced by the numerical values of molar concentrations, calculate the efficiency of aerobic respiration under these physiological conditions. (c) A typical diesel engine operates between $T_c = 873$ K and $T_h = 1923$ K with an efficiency that is approximately 75 per cent of the theoretical limit of $(1 - T_c/T_h)$ (see Section 4.2). Compare the efficiency of a typical diesel engine with that of aerobic respiration under typical physiological conditions (see part b). Why is biological energy conversion more or less efficient than energy conversion in a diesel engine?

9.21 It has been observed in living cells that the anion ATP^{4-} forms a 1:1 complex with the Mg^{2+} ion. The cation and anion form two coordinate covalent bonds: one that involves a negatively charged oxygen belonging to the terminal phosphate group of ATP^{4-} and another that involves a negatively charged oxygen belonging to the phosphate group adjacent to the terminal phosphate group of ATP^{4-}. Is hydrolysis of MgATP^{2-} more or less exergonic than hydrolysis of ATP^{4-} under identical intracellular conditions? Explain your answer. *Hint.* See Discussion question 1 of Box 9.1.

9.22‡ The dimerization of ClO in the Antarctic winter stratosphere is believed to play an important part in that region's severe seasonal depletion of ozone. The following equilibrium constants are based on measurements by R.A. Cox and G.D. Hayman (*Nature* **332**, 796 (1988)) on the reaction $2 ClO(g) \rightarrow (ClO)_2(g)$.

T/K	233	248	258	268	273
K	4.13×10^8	5.00×10^7	1.45×10^7	5.37×10^6	3.20×10^6

T/K	280	288	295	303
K	9.62×10^5	4.28×10^5	1.67×10^5	7.02×10^4

(a) Derive the values of $\Delta_r H^{\ominus}$ and $\Delta_r S^{\ominus}$ for this reaction. (b) Compute the standard enthalpy of formation and the standard molar entropy of $(ClO)_2$ given $\Delta_f H^{\ominus}(ClO) = +101.8$ kJ mol^{-1} and $S_m^{\ominus}(ClO) = 226.6$ J K^{-1} mol^{-1} (*CRC Handbook*, 2000).

9.23‡ Acid rain is an environmental concern in many parts of the world. In assessing the acidity of rainfall, it is important to have an idea of the acidity of natural rainwater. Assuming that natural rainwater (that is, uncontaminated with nitric or sulfuric acids) is in equilibrium with 3.6×10^{-4} atm CO_2 (Henry's law constant 1.25×10^6 Torr), what is the pH of natural rainwater? What would the pH of natural rainwater have been in pre-industrial times, when the partial pressure of CO_2 was about 2.8×10^{-4} atm? State your approximations explicitly.

9.24‡ Nitric acid hydrates have received much attention as possible catalysts for heterogeneous reactions that bring about the Antarctic ozone hole. Worsnop *et al.* investigated the thermodynamic stability of these hydrates under conditions typical of the polar winter stratosphere (D.R. Worsnop, L.E. Fox, M.S. Zahniser, and S.C. Wofsy, *Science* **259**, 71 (1993)). Standard reaction Gibbs energies can be computed for the following reactions at 190 K from their data:

(i) $H_2O(g) \longrightarrow H_2O(s)$ $\Delta_r G^\ominus = -23.6$ kJ mol^{-1}

(ii) $H_2O(g) + HNO_3(g) \longrightarrow HNO_3 \cdot H_2O(s)$ $\Delta_r G^\ominus = -57.2$ kJ mol^{-1}

(iii) $2 H_2O(g) + HNO_3(g) \longrightarrow HNO_3 \cdot 2H_2O(s)$ $\Delta_r G^\ominus = -85.6$ kJ mol^{-1}

(iv) $3 H_2O(g) + HNO_3(g) \longrightarrow HNO_3 \cdot 3H_2O(s)$ $\Delta_r G^\ominus = -112.8$ kJ mol^{-1}

Which solid is thermodynamically most stable at 190 K if $p_{H_2O} = 1.3 \times 10^{-7}$ bar and $p_{HNO_3} = 4.1 \times 10^{-10}$ bar? (*Hint.* Try computing $\Delta_r G^\ominus$ for each reaction under the prevailing conditions; if more than one solid forms spontaneously, examine $\Delta_r G^\ominus$ for the conversion of one solid to another.)

9.25‡ The ocean's ability to dissolve carbon is an important sink when modelling the carbon cycle. (a) Draw a graph of α_J versus pH for carbon dioxide solutions in the pH range 0–14 where the fraction α_J of species J is defined as

$$\alpha_J = \frac{[J]}{[\text{total carbon}]} = \frac{[J]}{[CO_2] + [H_2CO_3] + [H_2CO_3^-] + [CO_3^{2-}]}$$

Assume that the solutions are ideal. (b) What are the major chemical species at the physiological pH of 7.4 and equilibrium? (c) What is the maximum mass of ocean carbon dissolved as CO_2, H_2CO_3, HCO_3^-, CO_3^{2-} if the ocean has an average pH of 8, an average temperature of 298 K, and a volume of 1.37×10^{18} m^3? The partial pressure of atmospheric carbon dioxide is 3.3×10^{-4} atm. *Hint.* Use mathematical software.

9.26‡ Refer to Problem 8.18. (a) Draw an Ellingham diagram for the smelting reduction of silica with graphite in the temperature range 1500–2000 K and at 1 bar. What does the diagram reveal about the evolution of carbon dioxide during smelting? (b) Determine the equilibrium purity of silicon made by reducing silica with graphite at 2000 K and 1 bar. Assume that graphite is soluble in the silicon but that SiO_2 and SiC are not. Also determine the equilibrium partial pressures of SiO(g) and CO(g). Equilibria to consider include $SiO_2(l) + C(\text{melt}) \rightleftharpoons SiO(g) + CO(g)$, $SiO_2(l) + 2 C(\text{melt}) \rightleftharpoons Si(l) + 2 CO(g)$, and $SiO_2(l) + 3 C(\text{melt}) \rightleftharpoons SiC(s) + 2 CO(g)$. *Hint.* Use mathematical software.

9.27‡ Suppose that an iron catalyst at a particular manufacturing plant produces ammonia in the most cost-effective manner at 450°C when the pressure is such that $\Delta_r G$ for the reaction $\frac{1}{2}N_2(g) + \frac{2}{3}H_2(g) \rightarrow NH_3(g)$ is equal to –500 J mol^{-1}. (a) What pressure is needed? (b) Now suppose that a new catalyst is developed that is most cost-effective at 400°C when the pressure gives the same value of $\Delta_r G$. What pressure is needed when the new catalyst is used? What are the advantages of the new catalyst? Assume that (i) all gases are perfect gases or that (ii) all gases are van der Waals gases. Isotherms of $\Delta_r G(T, p)$ in the pressure range 100 atm $\leq p \leq$ 400 atm are needed to derive the answer. (c) Do the isotherms you plotted confirm Le Chatelier's principle concerning the response of equilibrium changes in temperature and pressure?

10

Equilibrium electrochemistry

The thermodynamic properties of ions in solution

10.1 Thermodynamic functions of formation

10.2 Ion activities

Electrochemical cells

10.3 Half-reactions and electrodes

10.4 Varieties of cells

10.5 Standard potentials

Applications of standard potentials

10.6 The electrochemical series

10.7 The measurement of pH and pK_a

10.8 Thermodynamic functions

Checklist of key ideas

Further reading

Exercises

Problems

The principles of thermodynamics established in the preceding chapters can be applied to solutions of electrolytes. One important extension of the previous material, however, is the need to take into account activity coefficients, for they differ significantly from 1 on account of the strong ionic interactions in electrolyte solutions. These coefficients are best treated as empirical quantities, but it is possible to estimate them in very dilute solutions. The bulk of this chapter is concerned with the description of the thermodynamic properties of reactions that take place in electrochemical cells, in which, as the reaction proceeds, it drives electrons through an external circuit. Thermodynamic arguments can be used to derive an expression for the electric potential of such cells and the potential can be related to their composition. There are two major topics developed in this connection. One is the definition and tabulation of standard potentials; the second is the use of these standard potentials to predict the equilibrium constants and other thermodynamic properties of chemical reactions.

Although we can discuss the thermodynamic properties of electrolyte solutions in terms of chemical potentials and activities in much the same way as solutions of non-electrolytes, electrolyte solutions have a number of distinctive features. One is the presence of strong interactions between ions in solution, which means that deviations from ideality are marked even in dilute systems. Therefore, we must equip ourselves with means of dealing with activity coefficients that differ significantly from 1. A second feature is that, because many reactions of ions involve the transfer of electrons, they can be studied (and utilized) by allowing them to take place in an electrochemical cell. Measurements like those described in this chapter provide data that are very useful for discussing the characteristics of electrolyte solutions and of ionic equilibria in solution.

The thermodynamic properties of ions in solution

Many of the concepts described in previous chapters carry over without change into the discussion of electrolyte solutions. In this section we focus on departures from ideality.

10.1 Thermodynamic functions of formation

The standard enthalpy and Gibbs energy of a reaction involving ions in solution are expressed in terms of standard enthalpies and Gibbs energies of formation listed in Table 2.6 in the *Data section*. These properties are used in exactly the same way as those for neutral compounds.

(a) Standard functions of formation of ions

The values of $\Delta_f H^{\ominus}$ and $\Delta_f G^{\ominus}$ refer to the formation of solutions of ions from the reference states of the parent elements. A major problem, however, is that solutions of cations cannot be prepared without their accompanying anions. Therefore, although the standard enthalpy of an overall reaction such as

$$Ag(s) + \tfrac{1}{2} Cl_2(g) \longrightarrow Ag^+(aq) + Cl^-(aq)$$

for which

$$\Delta_r H^{\ominus} = \Delta_f H^{\ominus}(Ag^+, aq) + \Delta_f H^{\ominus}(Cl^-, aq)$$

is meaningful and measurable (and found to be -61.58 kJ mol^{-1}), the enthalpies of the individual formation reactions are not measurable.

This problem is solved by defining one ion, conventionally the hydrogen ion, to have zero standard enthalpy and Gibbs energy of formation at all temperatures:

$$\Delta_f H^{\ominus}(H^+, aq) = 0 \qquad \Delta_f G^{\ominus}(H^+, aq) = 0 \qquad [10.1]$$

In essence, this definition adjusts the actual values of the enthalpies and Gibbs energies of formation of ions by a fixed amount, which is chosen so that the standard value for one of them, $H^+(aq)$, has the value zero. Then for the reaction

$$\tfrac{1}{2} H_2(g) + \tfrac{1}{2} Cl_2(g) \longrightarrow H^+(aq) + Cl^-(aq) \qquad \Delta_r G^{\ominus} = -131.23 \text{ kJ mol}^{-1}$$

we can write

$$\Delta_r G^{\ominus} = \Delta_f G^{\ominus}(H^+, aq) + \Delta_f G^{\ominus}(Cl^-, aq)$$
$$= \Delta_f G^{\ominus}(Cl^-, aq)$$

and hence identify $\Delta_f G^{\ominus}(Cl^-, aq)$ as -131.23 kJ mol^{-1}. All the Gibbs energies and enthalpies of formation of ions in Tables 10.1 and 2.6 were calculated in the same way.

Table 10.1* Standard thermodynamic functions of formation of ions in aqueous solution at 298 K

Ion	$\Delta_f H^{\ominus}$/(kJ mol^{-1})	$\Delta_f G^{\ominus}$/(kJ mol^{-1})
Cl$^-$	−167.2	−131.2
Cu^{2+}	+64.8	+65.5
H$^+$	0	0
K$^+$	−252.4	−283.3
Na$^+$	−240.1	−261.9
PO$_4^{3-}$	−1277.0	−1019.0

* More values are given in the *Data section*; see Table 2.6.

Illustration 10.1

With the value of $\Delta_f G^{\ominus}(Cl^-, aq)$ established, we can find the value of $\Delta_f G^{\ominus}(Ag^+, aq)$ from

$$Ag(s) + \tfrac{1}{2} Cl_2(g) \longrightarrow Ag^+(aq) + Cl^-(aq) \qquad \Delta_r G^{\ominus} = -54.12 \text{ kJ mol}^{-1}$$

which leads to $\Delta_f G^{\ominus}(Ag^+, aq) = +77.11$ kJ mol^{-1}.

Self-test 10.1 The standard enthalpy of formation of AgNO$_3$(aq) is -99.4 kJ mol^{-1} at 298 K. Calculate the standard enthalpy of formation of the nitrate ion in water by using the standard enthalpy of formation of Ag$^+$(aq) in Table 2.6.

$[-205.0$ kJ mol$^{-1}]$

(a)

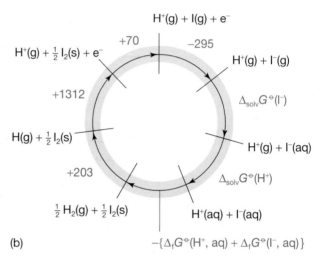

(b)

10.1 The thermodynamic cycles for the discussion of the Gibbs energies of solvation (hydration) and formation of (a) chloride ions, (b) iodide ions in aqueous solution. The sum of the changes in Gibbs energies around the cycle sum to zero because G is a state function.

(b)   Contributions to the Gibbs energy of formation

The factors responsible for the magnitude of the Gibbs energy of formation of an ion in solution can be identified by analysing it in terms of a thermodynamic cycle. As an illustration, we consider the standard Gibbs energy of formation of Cl^- in water, which is -131 kJ mol^{-1}. We do so by treating the formation reaction

$$\tfrac{1}{2}H_2(g) + \tfrac{1}{2}X_2(g) \longrightarrow H^+(aq) + X^-(aq)$$

as the outcome of the sequence of steps shown in Fig. 10.1 (with values taken from the *Data section*, particularly Tables 2.6, 13.4, and 13.5).[1] The sum of the Gibbs energies for all the steps around a closed cycle is zero, so

$$\Delta_f G^{\ominus}(Cl^-, aq) = 1272\ kJ\ mol^{-1} + \Delta_{solv}G^{\ominus}(H^+) + \Delta_{solv}G^{\ominus}(Cl^-)$$

---

1. The standard Gibbs energies of formation of the gas-phase ions are unknown. We have therefore used ionization energies and electron affinities (Section 13.4f), and have assumed that any differences from the Gibbs energies arising from conversion to enthalpy and the inclusion of entropies to obtain Gibbs energies in the formation of H^+ are cancelled by the corresponding terms in the electron gain of X. The conclusions from the cycles are therefore only approximate.

Table 10.2* Relative permittivities (dielectric constants) at 298 K

	ε_r
Ammonia	16.9
	22.4 (at −33°C)
Benzene	2.274
Ethanol	24.30
Water	78.54

* More values are given in the *Data section*.

An important point to note is that the value of $\Delta_f G^\ominus$ of an ion X is not determined by the properties of X alone but includes contributions from the dissociation, ionization, and hydration of hydrogen.

Gibbs energies of solvation of individual ions may be estimated from an equation derived by Max Born, who identified $\Delta_{solv}G^\ominus$ with the electrical work of transferring an ion from a vacuum into the solvent treated as a continuous dielectric of relative permittivity ε_r (Table 10.2). The resulting **Born equation**, which is derived in the *Justification* below, is

$$\Delta_{solv}G^\ominus = -\frac{z_i^2 e^2 N_A}{8\pi\varepsilon_0 r_i}\left(1 - \frac{1}{\varepsilon_r}\right) \tag{10.2}$$

where z_i is the charge number of the ion and r_i its radius (N_A is Avogadro's constant).[2] Note that $\Delta_{solv}G^\ominus < 0$, and that $\Delta_{solv}G^\ominus$ is strongly negative for small, highly charged ions in media of high relative permittivity. For water at 25°C,

$$\Delta_{solv}G^\ominus = -\frac{z_i^2}{(r_i/\text{pm})} \times (6.86 \times 10^4 \text{ kJ mol}^{-1}) \tag{10.3}$$

Justification 10.1

The electrical concepts required in this derivation are reviewed in *Further information 2*. The strategy of the calculation is to identify the Gibbs energy of solvation with the work of transferring an ion from a vacuum into the solvent. That work is calculated by taking the difference of the work of charging an ion when it is in the solution and the work of charging the same ion when it is in a vacuum. We model an ion as a sphere of radius r_i immersed in a medium of permittivity ε. When the charge of the sphere is q, the electric potential, ϕ, at its surface is

$$\phi = \frac{q}{4\pi\varepsilon r_i}$$

The work of bringing up a charge dq to the sphere is $\phi\, dq$. Therefore, the total work of charging the sphere from 0 to $z_i e$ is

$$w = \int_0^{z_i e} \phi\, dq = \frac{1}{4\pi\varepsilon r_i}\int_0^{z_i e} q\, dq = \frac{z_i^2 e^2}{8\pi\varepsilon r_i}$$

This electrical work of charging, when multiplied by Avogadro's constant, is the molar Gibbs energy for charging the ions.

The work of charging an ion in a vacuum is obtained by setting $\varepsilon = \varepsilon_0$, the vacuum permittivity. The corresponding value for charging the ion in a medium is obtained by setting $\varepsilon = \varepsilon_r \varepsilon_0$, where ε_r is the relative permittivity of the medium. It follows that the change in molar Gibbs energy that accompanies the transfer of ions from a vacuum to a solvent is the difference of these two quantities:

$$\Delta_{solv}G^\ominus = \frac{z_i^2 e^2 N_A}{8\pi\varepsilon r_i} - \frac{z_i^2 e^2 N_A}{8\pi\varepsilon_0 r_i}$$

which can easily be rearranged into eqn 2.

2. Ionic radii are given in Table 23.3.

Illustration 10.2

To see how closely the Born equation reproduces the experimental data, we calculate the difference in the values of $\Delta_f G^\ominus$ for Cl^- and I^- in water, for which $\varepsilon_r = 78.54$ at 25°C, given their radii as 181 pm and 220 pm (Table 23.3), respectively, is

$$\Delta_{solv} G^\ominus(Cl^-) - \Delta_{solv} G^\ominus(I^-) = -\left(\frac{1}{181} - \frac{1}{220}\right) \times (6.86 \times 10^4 \text{ kJ mol}^{-1})$$

$$= -67 \text{ kJ mol}^{-1}$$

This estimated difference is in good agreement with the experimental difference, which is −61 kJ mol⁻¹.

Self-test 10.2 Estimate the value of $\Delta_{solv} G^\ominus(Cl^-, aq) - \Delta_{solv} G^\ominus(Br^-, aq)$ from experimental data and from the Born equation.

[−27 kJ mol⁻¹ experimental; −29 kJ mol⁻¹ calculated]

(c) Standard entropies of ions in solution

Although the partial molar entropy of the solute in an electrolyte solution can be measured, there is no experimental way of ascribing a part of that entropy to the cations and a part to the anions. Therefore, yet again we are forced to define the partial molar entropy of one species and set up a table of values for other ions on that basis. The entropies of ions in solution are reported on a scale in which the standard entropy of the H^+ ions in water is taken as zero at all temperatures:

$$S^\ominus(H^+, aq) = 0 \tag{10.4}$$

The values based on this choice are listed in Table 10.3 and more fully in Table 2.6 in the *Data section*.

Because the entropies of ions in water are values relative to the hydrogen ion in water, they may be either positive or negative. A positive entropy means that an ion has a higher partial molar entropy than H^+ in water and a negative entropy means that the ion has a lower partial molar entropy than H^+ in water. For instance, the entropy of $Cl^-(aq)$ is +57 J K⁻¹ mol⁻¹ and that of $Mg^{2+}(aq)$ is −128 J K⁻¹ mol⁻¹. Partial molar ion entropies vary as expected on the basis that they are related to the degree to which the ions order the water molecules around them in the solution. Small, highly charged ions induce local structure in the surrounding water, and the disorder of the solution is decreased more than in the case of large, singly charged ions. The absolute, Third-law standard partial molar entropy of the proton in water can be estimated by proposing a model of the structure it induces, and there is some agreement on the value −21 J K⁻¹ mol⁻¹. The negative value indicates that the proton induces order in the solvent.

10.2 Ion activities

Interactions between ions are so strong that the approximation of replacing activities by molalities is valid only in very dilute solutions (less than 10⁻³ mol kg⁻¹ in total ion concentration) and in precise work activities themselves must be used.

(a) The definition of activity

We saw in Section 7.7c that the chemical potential of a solute in a real solution is related to its activity a by

Table 10.3* Standard entropies of ions in aqueous solution at 298 K

	$S_m^\ominus/(\text{J K}^{-1} \text{ mol}^{-1})$	
Cl^-	+56.5	
Cu^{2+}	−99.6	
H^+	0	by definition
K^+	−102.5	
Na^+	−59.0	
PO_4^{3-}	−221.8	

* More values are given in the *Data section*; see Table 2.6.

$$\mu = \mu^{\ominus} + RT \ln a \tag{10.5}$$

where the standard state is a hypothetical solution with molality $b^{\ominus} = 1 \text{ mol kg}^{-1}$ in which the ions are behaving ideally. The activity is related to the molality,[3] b, by

$$a = \gamma \frac{b}{b^{\ominus}} \tag{10.6}$$

where the activity coefficient, γ, depends on the composition, molality, and temperature of the solution. As the solution approaches ideality (in the sense of obeying Henry's law) at low molalities, the activity coefficient tends towards 1:

$$\gamma \to 1 \quad \text{and} \quad a \to b/b^{\ominus} \quad \text{as} \quad b \to 0 \tag{10.7}$$

Because all deviations from ideality are carried in the activity coefficient, the chemical potential can be written[4]

$$\mu = \mu^{\ominus} + RT \ln b + RT \ln \gamma = \mu^{\text{ideal}} + RT \ln \gamma \tag{10.8}$$

where μ^{ideal} is the chemical potential of the ideal-dilute solution of the same molality.

(b) Mean activity coefficients

If the chemical potential of a univalent cation M^+ is denoted μ_+ and that of a univalent anion X^- is denoted μ_-, the total molar Gibbs energy of the ions in the electrically neutral solution is the sum of these partial molar quantities. The molar Gibbs energy of an *ideal* solution is

$$G_m^{\text{ideal}} = \mu_+^{\text{ideal}} + \mu_-^{\text{ideal}} \tag{10.9}°$$

However, for a *real* solution of M^+ and X^- of the same molality,

$$G_m = \mu_+ + \mu_- = \mu_+^{\text{ideal}} + \mu_-^{\text{ideal}} + RT \ln \gamma_+ + RT \ln \gamma_- = G_m^{\text{ideal}} + RT \ln \gamma_+\gamma_- \tag{10.10}$$

All the deviations from ideality are contained in the last term.

There is no experimental way of separating the product $\gamma_+\gamma_-$ into contributions from the cations and the anions. The best we can do experimentally is to assign responsibility for the nonideality equally to both kinds of ion. Therefore, for a 1,1-electrolyte, we introduce the **mean activity coefficient** as the geometric mean of the individual coefficients:

The geometric mean of x and y is $(xy)^{1/2}$.

$$\gamma_{\pm} = (\gamma_+\gamma_-)^{1/2} \tag{10.11}$$

and express the individual chemical potentials of the ions as

$$\mu_+ = \mu_+^{\text{ideal}} + RT \ln \gamma_{\pm} \qquad \mu_- = \mu_-^{\text{ideal}} + RT \ln \gamma_{\pm} \tag{10.12}$$

The sum of these two chemical potentials is the same as before, eqn 10, but now the nonideality is shared equally.

We can generalize this approach to the case of a compound M_pX_q that dissolves to give a solution of p cations and q anions from each formula unit. The molar Gibbs energy of the ions is the sum of their partial molar Gibbs energies:

$$G_m = p\mu_+ + q\mu_- = G_m^{\text{ideal}} + pRT \ln \gamma_+ + qRT \ln \gamma_- \tag{10.13}$$

---

3. The IUPAC recommendation for the symbol is either m or b; whereas the former is more commonly used, b has the advantage that it is less likely to be confused with mass, m.

4. In accord with the convention adopted in earlier chapters, the notation has been simplified by interpreting b as $b/b^{\ominus}$ and labelling the equation { ... }.

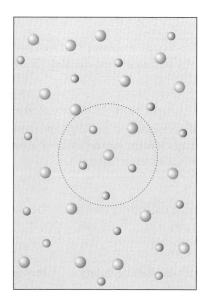

10.2 The picture underlying the Debye–Hückel theory is of a tendency for anions to be found around cations, and of cations to be found around anions (one such local clustering region is shown by the circle). The ions are in ceaseless motion, and the diagram represents a snapshot of their motion. The solutions to which the theory applies are far less concentrated than shown here.

If we introduce the mean activity coefficient

$$\gamma_\pm = (\gamma_+^p \, \gamma_-^q)^{1/s} \qquad s = p + q \qquad [10.14]$$

and write the chemical potential of each ion as

$$\mu_i = \mu_i^{\text{ideal}} + RT \ln \gamma_\pm \qquad (10.15)$$

we get the same expression as in eqn 13 for G_m when we write

$$G = p\mu_+ + q\mu_- \qquad (10.16)$$

However, both types of ion now share equal responsibility for the nonideality.

(c) The Debye–Hückel limiting law

The long range and strength of the Coulombic interaction between ions means that it is likely to be primarily responsible for the departures from ideality in ionic solutions and to dominate all the other contributions to nonideality. This domination is the basis of the **Debye–Hückel theory** of ionic solutions, which was devised by Peter Debye and Erich Hückel in 1923. We give here a qualitative account of the theory and its principal conclusions. The calculation itself, which is a profound example of how a seemingly intractable problem can be formulated and then resolved by drawing on physical insight, is outlined in the *Justification* below.

Oppositely charged ions attract one another. As a result, anions are more likely to be found near cations in solution, and vice versa (Fig. 10.2). Overall the solution is electrically neutral, but near any given ion there is an excess of counter ions (ions of opposite charge). Averaged over time, counter ions are more likely to be found near any given ion. This time-averaged, spherical haze around the central ion, in which counter ions outnumber ions of the same charge as the central ion, has a net charge equal in magnitude but opposite in sign to that on the central ion, and is called its **ionic atmosphere**. The energy, and therefore the chemical potential, of any given central ion is lowered as a result of its electrostatic interaction with its ionic atmosphere. This lowering of energy appears as the difference between the molar Gibbs energy G_m and the ideal value G_m^{ideal} of the solute, and hence can be identified with $RT \ln \gamma_\pm$. The stabilization of ions by their interaction with their ionic atmospheres is part of the explanation why chemists commonly use dilute solutions, in which the stabilization is less important, to achieve precipitation of ions from electrolyte solutions.

The model leads to the result that at very low concentrations the activity coefficient can be calculated from the **Debye–Hückel limiting law**

$$\log \gamma_\pm = -|z_+ z_-| A I^{1/2} \qquad (10.17)$$

where $A = 0.509$ for an aqueous solution at 25°C and I is the dimensionless **ionic strength** of the solution:

$$I = \tfrac{1}{2} \sum_i z_i^2 (b_i / b^\ominus) \qquad [10.18]$$

In this expression z_i is the charge number of an ion i (positive for cations and negative for anions) and b_i is its molality. The ionic strength occurs widely wherever ionic solutions are discussed, as we shall see. The sum extends over all the ions present in the solution. For solutions consisting of two types of ion at molalities b_+ and b_-,

$$I = \tfrac{1}{2}(b_+ z_+^2 + b_- z_-^2)/b^\ominus \qquad (10.19)$$

The ionic strength emphasizes the charges of the ions because the charge numbers occur as their squares. Table 10.4 summarizes the relation of ionic strength and molality in an easily usable form.

Table 10.4 Ionic strength and molality, $I = k \times b/b^\ominus$

k	X^-	X^{2-}	X^{3-}	X^{4-}
M^+	1	3	6	10
M^{2+}	3	4	15	12
M^{3+}	6	15	9	42
M^{4+}	10	12	42	16

For example, the ionic strength of an M_2X_3 solution of molality b, which is understood to give M^{3+} and X^{2-} ions in solution, is $15b/b^\ominus$.

Illustration 10.3

The mean activity coefficient of 5.0×10^{-3} mol kg^{-1} KCl(aq) at 25°C is calculated by writing

$$I = \tfrac{1}{2}(b_+ + b_-)/b^{\ominus} = b/b^{\ominus}$$

where b is the molality of the solution (and $b_+ = b_- = b$). Then, from eqn 17,

$$\log \gamma_{\pm} = -0.509 \times (5.0 \times 10^{-3})^{1/2} = -0.036$$

Hence, $\gamma_{\pm} = 0.92$. The experimental value is 0.927.

Self-test 10.3 Calculate the ionic strength and the mean activity coefficient of 1.00 mmol kg^{-1} CaCl$_2$(aq) at 25°C. [3.00 mmol kg^{-1}, 0.880]

Justification 10.2

Imagine a solution in which all the ions have their actual positions, but in which their Coulombic interactions have been turned off. The difference in molar Gibbs energy between the ideal and real solutions is equal to w_e, the electrical work of charging the system in this arrangement. Therefore, for a salt M$_p$X$_q$,

$$\ln \gamma_{\pm} = \frac{w_e}{sRT} \qquad s = p + q \tag{10.20}$$

It follows that we must first find the final distribution of the ions and then the work of charging them in that distribution.

The Coulomb potential at a distance r from an isolated ion of charge $z_i e$ in a medium of permittivity ε is

$$\phi_i = \frac{Z_i}{r} \qquad Z_i = \frac{z_i e}{4\pi\varepsilon} \tag{10.21}$$

The ionic atmosphere causes the potential to decay with distance more sharply than this expression implies. Such shielding is a familiar problem in electrostatics, and its effect is taken into account by replacing the Coulomb potential by the **shielded Coulomb potential**, an expression of the form

$$\phi_i = \frac{Z_i}{r} e^{-r/r_D} \tag{10.22}$$

where r_D is called the **Debye length**. When r_D is large, the shielded potential is virtually the same as the unshielded potential. When it is small, the shielded potential is much smaller than the unshielded potential, even for short distances (Fig. 10.3).

To calculate r_D, we need to know how the **charge density**, ρ_i, of the ionic atmosphere, the charge in a small region divided by the volume of the region, varies with distance from the ion. This step draws on another standard result of electrostatics, in which charge density and potential are related by **Poisson's equation**:

$$\nabla^2 \phi = -\frac{\rho}{\varepsilon} \tag{10.23}$$

Because we are considering only a spherical ionic atmosphere, we can use a simplified form of this equation in which the charge density varies only with distance from the central ion:

10.3 The variation of the shielded Coulomb potential with distance for different values of the Debye length, r_D/a. The smaller the Debye length, the more sharply the potential decays to zero. In each case, a is an arbitrary unit of length.

$$\frac{1}{r^2}\frac{d}{dr}\left(r^2\frac{d\phi_i}{dr}\right) = -\frac{\rho_i}{\varepsilon}$$

Substitution of the expression for the shielded potential, eqn 22, results in

$$r_D^2 = -\frac{\varepsilon\phi_i}{\rho_i} \tag{10.24}$$

To solve this equation we need to relate ρ_i and ϕ_i.

For the next step we draw on the fact that the energy of an ion depends on its closeness to the central ion, and then use the Boltzmann distribution (*Molecular interpretation* 2.1) to work out the probability that an ion will be found at each distance. The energy of an ion of charge $z_j e$ at a distance where it experiences the potential ϕ_i of the central ion i relative to its energy when it is far away in the bulk solution is its charge times the potential:

$$E = z_j e\phi_i$$

Therefore, according to the Boltzmann distribution, the ratio of the molar concentration, c_j, of ions at a distance r and the molar concentration in the bulk, c_j°, where the energy is zero, is:

$$\frac{c_j}{c_j^\circ} = e^{-E/kT}$$

The charge density, ρ_i, at a distance r from the ion i is the molar concentration of each type of ion multiplied by the charge per mole of ions, $z_i e N_A$. The quantity eN_A, the magnitude of the charge per mole of electrons, is the Faraday constant, F. It follows that

$$\rho_i = c_+ z_+ F + c_- z_- F = c_+^\circ z_+ F e^{-z_+ e\phi_i/kT} + c_-^\circ z_- F e^{-z_- e\phi_i/kT} \tag{10.25a}$$

At this stage we need to simplify the expression to avoid the awkward exponential terms. Because the average electrostatic interaction energy is small compared with kT we may write eqn 25a as

$$\rho_i = (c_+^\circ z_+ + c_-^\circ z_-)F - (c_+^\circ z_+^2 + c_-^\circ z_-^2)\frac{F^2\phi_i}{RT} + \cdots \tag{10.25b}$$

(To obtain this expression, we have replaced an e by F/N_A and recognized that $N_A k = R$.) The first term in the expansion is zero because it is the charge density in the bulk, uniform solution, and the solution is electrically neutral. The unwritten terms are assumed to be too small to be significant. The one remaining term can be expressed in terms of the ionic strength, eqn 18, by noting that in the dilute aqueous solutions we are considering there is little difference between molality and molar concentration, and $c \approx b\rho$, where ρ is the mass density of the solvent

$$c_+^\circ z_+^2 + c_-^\circ z_-^2 \approx (b_+^\circ z_+^2 + b_-^\circ z_-^2)\rho = 2Ib^\ominus\rho$$

With these approximations, eqn 25 becomes

$$\rho_i = -\frac{2\rho F^2 Ib^\ominus\phi_i}{RT}$$

We can now solve eqn 24 for r_D:

$$r_D = \left(\frac{\varepsilon RT}{2\rho F^2 Ib^\ominus}\right)^{1/2} \tag{10.26}$$

To calculate the activity coefficient we need to find the electrical work of charging the central ion when it is surrounded by its atmosphere. To do so, we need to know

Table 10.5* Mean activity coefficients in water at 298 K

$b/(\text{mol kg}^{-1})$	KCl	CaCl$_2$
0.001	0.966	0.888
0.01	0.902	0.732
0.1	0.770	0.524
1.0	0.607	0.725

* More values are given in the *Data section*.

the potential at the ion due to its atmosphere, ϕ_{atmos}. This potential is the difference between the total potential, given by eqn 22, and the potential due to the central ion itself:

$$\phi_{\text{atmos}} = \phi - \phi_{\text{central ion}} = Z_i\left(\frac{e^{-r/r_D}}{r} - \frac{1}{r}\right)$$

The potential at the central ion (at $r = 0$) is obtained by taking the limit of this expression as $r \to 0$, and is

$$\phi_{\text{atmos}}(0) = -\frac{Z_i}{r_D}$$

This expression shows us that the potential of the ionic atmosphere is equivalent to the potential arising from a single charge of equal magnitude but opposite sign to that of the central ion and located at a distance r_D from the ion. If the charge of the central ion were q and not $z_i e$, then the potential due to its atmosphere would be

$$\phi_{\text{atmos}}(0) = -\frac{q}{4\pi\varepsilon r_D}$$

The work of adding a charge dq to a region where the electrical potential is $\phi_{\text{atmos}}(0)$ is

$$dw_e = \phi_{\text{atmos}}(0)\,dq$$

Therefore, the total molar work of fully charging the ions is

$$w_e = N_A \int_0^{z_i e} \phi_{\text{atmos}}(0)\,dq = -\frac{N_A}{4\pi\varepsilon r_D}\int_0^{z_i e} q\,dq$$

$$= -\frac{N_A z_i^2 e^2}{8\pi\varepsilon r_D} = -\frac{z_i^2 F^2}{8\pi\varepsilon N_A r_D}$$

It follows from eqn 20 that the mean activity coefficient of the ions is

$$\ln \gamma_\pm = \frac{pw_{e,+} + qw_{e,-}}{sRT} = -\frac{(pz_+^2 + qz_-^2)F^2}{8\pi\varepsilon sN_A RTr_D}$$

However, for neutrality $pz_+ + qz_- = 0$;[5] therefore

$$\ln \gamma_\pm = -\frac{|z_+ z_-|F^2}{8\pi\varepsilon N_A RTr_D}$$

When we replace r_D by using eqn 26, and convert to common logarithms, this expression turns into eqn 17 with

$$A = \frac{F^3}{4\pi N_A \ln 10}\left(\frac{\rho b^{\ominus}}{2\varepsilon^3 R^3 T^3}\right)^{1/2} \tag{10.27}$$

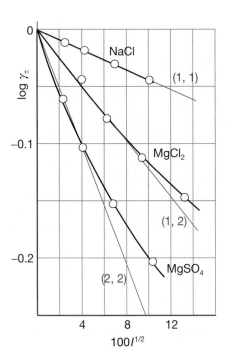

log $\gamma_\pm$

NaCl

(1, 1)

MgCl$_2$

(1, 2)

(2, 2)

MgSO$_4$

$100I^{1/2}$

10.4 An experimental test of the Debye–Hückel limiting law. Although there are marked deviations for moderate ionic strengths, the limiting slopes as $I \to 0$ are in good agreement with the theory, so the law can be used for extrapolating data to very low molalities.

The name 'limiting law' is applied to eqn 17 because ionic solutions of moderate molalities may have activity coefficients that differ from the values given by this expression, yet all solutions are expected to conform as $b \to 0$. Table 10.5 lists some experimental values of activity coefficients for salts of various valence types. Figure 10.4 shows some of these values plotted against $I^{1/2}$, and compares them with

5. For this step, multiply $pz_+ + qz_- = 0$ by p and also, separately, by q; add the two expressions and rearrange the result by using $p + q = s$ and $z_+ z_- = -|z_+ z_-|$.

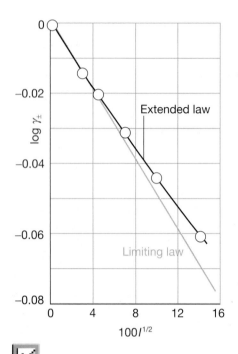

10.5 The extended Debye–Hückel law gives agreement with experiment over a wider range of molalities (as shown here for a 1,1-electrolyte), but it fails at higher molalities.

the theoretical straight lines calculated from eqn 17. The agreement at very low molalities (less than about 1 mmol kg^{-1}, depending on charge type) is impressive, and convincing evidence in support of the model. Nevertheless, the departures from the theoretical curves above these molalities are large, and show that the approximations are valid only at very low concentrations.

(d) The extended Debye–Hückel law

When the ionic strength of the solution is too high for the limiting law to be valid, the activity coefficient may be estimated from the **extended Debye–Hückel law**:

$$\ln \gamma_{\pm} = -\frac{A|z_+z_-|I^{1/2}}{1 + BI^{1/2}} \qquad (10.28)$$

where B is another dimensionless constant. Although B can be interpreted as a measure of the closest approach of the ions, it is best regarded as an adjustable empirical parameter. A curve drawn in this way is shown in Fig. 10.5. It is clear that eqn 28 accounts for some activity coefficients over a moderate range of dilute solutions (up to about 0.1 mol kg^{-1}); nevertheless it remains very poor near 1 mol kg^{-1}.

Current theories of activity coefficients for ionic solutes take an indirect route. They set up a theory for the dependence of the activity coefficient of the solvent on the concentration of the solute, and then use the Gibbs–Duhem equation (eqn 7.12) to estimate the activity coefficient of the solute. The results are reasonably reliable for solutions with molalities greater than about 0.1 mol kg^{-1} and are valuable for the discussion of mixed salt solutions, such as sea water.

Electrochemical cells

An **electrochemical cell** consists of two **electrodes**, or metallic conductors, in contact with an **electrolyte**, an ionic conductor (which may be a solution, a liquid, or a solid). An electrode and its electrolyte comprise an **electrode compartment**. The two electrodes may share the same compartment. The various kinds of electrode are summarized in Table 10.6 and illustrated in Fig. 10.6. Any 'inert metal' shown as part of the specification is present to act as a source or sink of electrons, but takes no other part in the reaction other than acting as a catalyst for it. If the electrolytes are different, the two compartments may be joined by a **salt bridge**, which is a tube containing a concentrated electrolyte solution in agar jelly that completes the electrical circuit and enables the cell to function.

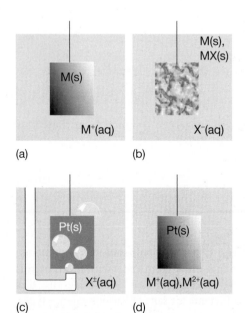

10.6 Typical electrode types: (a) metal–metal ion, (b) metal–insoluble salt, (c) gas, (d) redox electrodes.

Table 10.6 Varieties of electrode

Electrode type	Designation	Redox couple	Half-reaction
Metal/metal-ion	$M(s)|M^+(aq)$	M^+/M	$M^+(aq) + e^- \longrightarrow M(s)$
Gas electrode	$Pt(s)|X_2(g)|X^+(aq)$	X^+/X_2	$X^+(aq) + e^- \longrightarrow \frac{1}{2}X_2(g)$
	$Pt(s)|X_2(g)|X^-(aq)$	X_2/X^-	$\frac{1}{2}X_2(g) + e^- \longrightarrow X^-(aq)$
Metal/insoluble-salt	$M(s)|MX(s)|X^-(aq)$	$MX/M,X^-$	$MX(s) + e^- \longrightarrow M(s) + X^-(aq)$
Redox	$Pt(s)|M^+(aq),M^{2+}(aq)$	M^{2+}/M^+	$M^{2+}(aq) + e^- \longrightarrow M^+(aq)$

A **galvanic cell** is an electrochemical cell that produces electricity as a result of the spontaneous reaction occurring inside it. An **electrolytic cell** is an electrochemical cell in which a non-spontaneous reaction is driven by an external source of current.

10.3 Half-reactions and electrodes

It will be familiar from introductory chemistry courses that **oxidation** is the removal of electrons from a species and **reduction** is the addition of electrons to a species. A **redox reaction** is a reaction in which there is a transfer of electrons from one species to another. The electron transfer may be accompanied by other events, such as atom or ion transfer, but the net effect is electron transfer and hence a change in oxidation number of an element. The **reducing agent** (or 'reductant') is the electron donor; the **oxidizing agent** (or 'oxidant') is the electron acceptor.

(a) Half-reactions

Any redox reaction may be expressed as the difference of two reduction **half-reactions**, which are conceptual reactions showing the gain of electrons. For example, the reduction of Cu^{2+} ions by zinc can be expressed as the difference of the following two half-reactions:

$$Cu^{2+}(aq) + 2\,e^- \longrightarrow Cu(s) \qquad Zn^{2+}(aq) + 2\,e^- \longrightarrow Zn(s)$$

The difference of the two (copper − zinc) is

$$Cu^{2+}(aq) + Zn(s) \longrightarrow Cu(s) + Zn^{2+}(aq) \tag{10.29}$$

Even reactions that are not redox reactions may often be expressed as the difference of two reduction half-reactions.

Example 10.1 Expressing a reaction in terms of half-reactions

Express the dissolution of silver chloride in water as the difference of two reduction half-reactions.

Method First, write the overall chemical equation. Then select one of the reactants, and write a half-reaction in which the reactant is reduced to one of the products. Next, subtract that half-reaction from the overall reaction to identify the second half-reaction. In practice, it is easier to reverse the half-reaction and add it to the overall reaction. Write the resulting half-reaction as a reduction (by reversing it).

Answer The chemical equation of the overall reaction is

$$AgCl(s) \longrightarrow Ag^+(aq) + Cl^-(aq)$$

We select as one half-reaction the reduction of AgCl (more precisely, the reduction of the Ag(I) in AgCl to Ag(0)):

$$AgCl(s) + e^- \longrightarrow Ag(s) + Cl^-(aq)$$

Reversing this half-reaction gives

$$Ag(s) + Cl^-(aq) \longrightarrow AgCl(s) + e^-$$

and adding it to the overall reaction gives

$$AgCl(s) + Ag(s) + Cl^-(aq) \longrightarrow Ag^+(aq) + Cl^-(aq) + AgCl(s) + e^-$$

Now cancel species that appear on both sides of the arrow:

$$Ag(s) \longrightarrow Ag^+(aq) + e^-$$

Finally, reverse this half-reaction to obtain a reduction half-reaction:

$$Ag^+(aq) + e^- \longrightarrow Ag(s)$$

Comment There is no net change of oxidation number in dissolution of AgCl, so it is not a redox reaction.

Self-test 10.4 Express the formation of H_2O from H_2 and O_2 in acidic solution (a redox reaction) as the difference of two reduction half-reactions.

$$[4\,H^+(aq) + 4\,e^- \rightarrow 2\,H_2(g),\ O_2(g) + 4\,H^+(aq) + 4\,e^- \rightarrow 2\,H_2O(l)]$$

The reduced and oxidized species in a half-reaction form a **redox couple**. Thus, the redox couples mentioned so far are Cu^{2+}/Cu and Zn^{2+}/Zn. In general we write a couple as Ox/Red and the corresponding reduction half-reaction as

$$Ox + \nu\,e^- \longrightarrow Red \qquad (10.30)$$

We shall often find it useful to express the composition of an electrode compartment in terms of the reaction quotient, Q, for the half-reaction. This quotient is defined like the reaction quotient for the overall reaction, but the electrons are ignored. Thus, for the copper half-reaction $Cu^{2+}(aq) + 2\,e^- \rightarrow Cu(s)$, we write

$$Q = \frac{1}{a_{Cu^{2+}}}$$

We have used the fact that the pure metal (the standard state of the element) has unit activity (recall Table 7.3).

Illustration 10.4

The reaction quotient for the reduction of O_2 to H_2O in acid solution,

$$O_2(g) + 4\,H^+(aq) + 4\,e^- \longrightarrow 2\,H_2O(l)$$

is

$$Q = \frac{a_{H_2O}^2}{a_{H^+}^4 (f_{O_2}/p^{\ominus})} \approx \frac{p^{\ominus}}{a_{H^+}^4 p_{O_2}}$$

The approximations used in the second step are that the activity of water is 1 (because the solution is dilute) and the oxygen behaves as a perfect gas.

Self-test 10.5 Write the half-reaction and the reaction quotient for a chlorine gas electrode. $[Cl_2(g) + 2\,e^- \rightarrow 2\,Cl^-(aq),\ Q = a_{Cl^-}^2 p^{\ominus}/p_{Cl_2}]$

(b) Reactions at electrodes

The reduction and oxidation processes responsible for the overall reaction in a cell are separated in space: oxidation takes place in one electrode compartment and reduction takes place in the other compartment. As the reaction proceeds, the electrons released in the oxidation

10.7 When a spontaneous reaction takes place in a galvanic cell, electrons are deposited in one electrode (the site of oxidation, the anode) and collected from another (the site of reduction, the cathode), and so there is a net flow of current which can be used to do work. Note that the + sign of the cathode can be interpreted as indicating the electrode at which electrons enter the cell, and the – sign of the anode is where the electrons leave the cell.

$$Red_1 \longrightarrow Ox_1 + \nu\,e^-$$

at one electrode travel through the external circuit and re-enter the cell through the other electrode. There they bring about reduction:

$$Ox_2 + \nu\,e^- \longrightarrow Red_2$$

The electrode at which oxidation occurs is called the **anode**; the electrode at which reduction occurs is called the **cathode**.

In a galvanic cell, the cathode has a higher potential than the anode: the species undergoing reduction, Ox_2, withdraws electrons from its electrode (the cathode, Fig. 10.7), so leaving a relative positive charge on it (corresponding to a high potential). At the anode, oxidation results in the transfer of electrons to the electrode, so giving it a relative negative charge (corresponding to a low potential). In an electrolytic cell, the anode is still the location of oxidation (by definition), but now electrons must be withdrawn from the species in that compartment because oxidation does not occur spontaneously, and at the cathode there must be a supply of electrons to drive the reduction. Therefore, in an electrolytic cell the anode must be made relatively positive to the cathode (Fig. 10.8).

10.4 Varieties of cells

The simplest type of cell has a single electrolyte common to both electrodes (as in Fig. 10.7). In some cases it is necessary to immerse the electrodes in different electrolytes, as in the 'Daniell cell' in which the redox couple at one electrode is Cu^{2+}/Cu and at the other is Zn^{2+}/Zn (Fig. 10.9). In an **electrolyte concentration cell**, the electrode compartments are identical except for the concentrations of the electrolytes. In an **electrode concentration cell** the electrodes themselves have different

10.8 In an electrolytic cell, electrons are forced through the circuit by an external source. Although the cathode is still the site of reduction, it is now the negative electrode whereas the anode, the site of oxidation, is positive.

10.9 One version of the Daniell cell. The copper electrode is the cathode and the zinc electrode is the anode. Electrons leave the cell from the zinc and enter it again through the copper electrode.

10.10 The salt bridge, essentially an inverted U-tube full of concentrated salt solution in a jelly, has two opposing liquid junction potentials that almost cancel.

10.11 A typical cell for measuring a standard potential consists of a hydrogen electrode (on the left) and the electrode for the couple of interest (on the right).

concentrations, either because they are gas electrodes operating at different pressures or because they are amalgams (solutions in mercury) with different concentrations.

(a) Liquid junction potentials

In a cell with two different electrolyte solutions in contact, as in the Daniell cell, there is an additional source of potential difference across the interface of the two electrolytes. This potential is called the **liquid junction potential**, E_{lj}. Another example of a junction potential is that between different concentrations of hydrochloric acid. At the junction, the mobile H^+ ions diffuse into the more dilute solution. The bulkier Cl^- ions follow, but initially do so more slowly, which results in a potential difference at the junction. The potential then settles down to a value such that, after that brief initial period, the ions diffuse at the same rates. Electrolyte concentration cells always have a liquid junction; electrode concentration cells do not.

The contribution of the liquid junction to the potential can be reduced (to about 1 to 2 mV) by joining the electrolyte compartments through a salt bridge (Fig. 10.10). The reason for the success of the salt bridge is that the liquid junction potentials at either end are largely independent of the concentrations of the two dilute solutions, and so nearly cancel.

(b) Notation

In the notation for cells, phase boundaries are denoted by a vertical bar. For example, the cell in Fig. 10.11 is denoted by the 'cell diagram'

$$Pt(s) \mid H_2(g) \mid HCl(aq) \mid AgCl(s) \mid Ag(s)$$

A liquid junction is denoted by ⋮, so the cell in Fig. 10.9, is denoted

$$Zn(s) \mid ZnSO_4(aq) \vdots CuSO_4(aq) \mid Cu(s)$$

A double vertical line, ∥, denotes an interface for which it is assumed that the junction potential has been eliminated. Thus the cell in Fig. 10.10 is denoted

$$Zn(s) \mid ZnSO_4(aq) \parallel CuSO_4(aq) \mid Cu(s)$$

An electrolyte concentration cell in which the liquid junction potential is assumed to be eliminated is denoted

$$Pt(s) \mid H_2(g) \mid HCl(aq, b_1) \parallel HCl(aq, b_2) \mid H_2(g) \mid Pt(s)$$

(c) The cell reaction

The current produced by a galvanic cell arises from the spontaneous chemical reaction taking place inside it. The **cell reaction** is the reaction in the cell written on the assumption that the right-hand electrode is the cathode, and hence that the spontaneous reaction is one in which reduction is taking place in the right-hand compartment. Later we see how to predict if the right-hand electrode is in fact the cathode; if it is, then the cell reaction is spontaneous as written. If the left-hand electrode turns out to be the cathode, then the reverse of the corresponding cell reaction is spontaneous.

To write the cell reaction corresponding to a cell diagram, we first write the right-hand half-reaction as a reduction (because we have assumed that to be spontaneous). Then we subtract from it the left-hand reduction half-reaction (for, by implication, that electrode is the site of oxidation). Thus, in the cell

$$Zn(s) \mid ZnSO_4(aq) \parallel CuSO_4(aq) \mid Cu(s)$$

the two electrodes and their reduction half-reactions are

Right-hand electrode: $Cu^{2+}(aq) + 2\,e^- \longrightarrow Cu(s)$

Left-hand electrode: $Zn^{2+}(aq) + 2\,e^- \longrightarrow Zn(s)$

Hence, the overall cell reaction is the difference:

$$Cu^{2+}(aq) + Zn(s) \longrightarrow Cu(s) + Zn^{2+}(aq)$$

(d) The cell potential

A cell in which the overall cell reaction has not reached chemical equilibrium can do electrical work as the reaction drives electrons through an external circuit. The work that a given transfer of electrons can accomplish depends on the potential difference between the two electrodes. This potential difference is called the **cell potential** and is measured in volts, V. When the cell potential is large, a given number of electrons travelling between the electrodes can do a large amount of electrical work. When the cell potential is small, the same number of electrons can do only a small amount of work. A cell in which the overall reaction is at equilibrium can do no work, and then the cell potential is zero.

According to the discussion in Section 4.5e, we know that the maximum electrical work that a system (the cell) can do is given by the value of ΔG, and in particular that for a spontaneous process (in which both ΔG and w are negative) at constant temperature and pressure

$$w_{e,max} = \Delta G \tag{10.31}$$

Here we use w_e to denote electrical work, a specific kind of non-expansion work. It follows that, to draw thermodynamic conclusions from measurements of the work a cell can do, we must ensure that the cell is operating reversibly. Only then is it producing maximum work and only then can eqn 31 be used to relate that work to ΔG. Moreover, we saw in Section 9.1a that the reaction Gibbs energy is actually a property relating to a specified composition of the reaction mixture. Therefore, to measure $\Delta_r G$ we must ensure that the cell is operating reversibly at a specific, constant composition. Both these conditions are achieved by measuring the cell potential when it is balanced by an exactly opposing source of potential so that the cell reaction occurs reversibly, the composition is constant, and no current flows: in effect, the cell reaction is poised for change, but not actually changing. The resulting potential difference is called the **electromotive force** (emf), E, of the cell.

(e) The relation between E and $\Delta_r G$

The relation between the reaction Gibbs energy and the emf of the cell is

$$-\nu FE = \Delta_r G \tag{10.32}$$

where F is the **Faraday constant**, $F = eN_A$, the magnitude of the charge per mole of electrons. This equation, which is derived in the *Justification* below, is the key connection between electrical measurements on the one hand and thermodynamic properties on the other. It will be the basis of all that follows.

Justification 10.3

We consider the change in G when the cell reaction advances by an infinitesimal amount $d\xi$ at some composition. We saw in *Justification 9.1* that, at constant temperature and pressure, G changes by

$$dG = \sum_J \mu_J\, dn_J = \sum_J \nu_J \mu_J\, d\xi$$

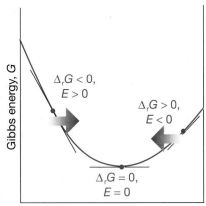

10.12 As explained in Chapter 9, a spontaneous reaction occurs in the direction of decreasing Gibbs energy. When expressed in terms of a cell potential, the spontaneous direction of change can be expressed in terms of the cell emf, E. The reaction is spontaneous as written (from left to right on the illustration) when $E > 0$. The reverse reaction is spontaneous when $E < 0$. When the cell reaction is at equilibrium, the cell potential is zero.

The reaction Gibbs energy, $\Delta_r G$, at the specified composition is

$$\Delta_r G = \left(\frac{\partial G}{\partial \xi}\right)_{p,T} = \sum_J \nu_J \mu_J$$

So we can write

$$dG = \Delta_r G \, d\xi$$

The maximum non-expansion (electrical) work that the reaction can do as it advances by $d\xi$ at constant temperature and pressure is therefore

$$dw_e = \Delta_r G \, d\xi$$

This work is infinitesimal, and the composition of the system is virtually constant when it occurs.

Suppose that the reaction advances by $d\xi$; then $\nu \, d\xi$ electrons must travel from the anode to the cathode. The total charge transported between the electrodes when this change occurs is $-\nu e N_A \, d\xi$ (because $\nu \, d\xi$ is the amount of electrons and the charge per mole of electrons is $-eN_A$). Hence, the total charge transported is $-\nu F \, d\xi$ because $eN_A = F$.

The work done when an infinitesimal charge $-\nu F \, d\xi$ travels from the anode to the cathode is equal to the product of the charge and the potential difference E (see Table 2.1 and *Further information 2*):

$$dw_e = -\nu F E \, d\xi$$

When we equate this relation to the one above, the advancement $d\xi$ cancels, and we obtain eqn 32.

It follows from eqn 32 that, by knowing the reaction Gibbs energy at a specified composition, we can state the cell emf at that composition. Note that a negative reaction Gibbs energy, corresponding to a spontaneous cell reaction, corresponds to a positive cell emf. Another way of looking at the content of eqn 32 is that it shows that the driving power of a cell (that is, its emf) is proportional to the slope of the Gibbs energy with respect to the extent of reaction. It is plausible that a reaction that is far from equilibrium (when the slope is steep) has a strong tendency to drive electrons through an external circuit (Fig. 10.12). When the slope is close to zero (when the cell reaction is close to equilibrium), the emf is small.

> **Illustration 10.5**
>
> To estimate the emf to expect for a typical cell we set $\Delta_r G \approx -100 \text{ kJ mol}^{-1}$ and $\nu = 1$; we find
>
> $$E \approx -\frac{(-100 \times 10^3 \text{ J mol}^{-1})}{1 \times (9.6 \times 10^4 \text{ C mol}^{-1})} \approx 1 \text{ V}$$
>
> We have used $1 \text{ J} = 1 \text{ C V}$.

(f) The Nernst equation

We can go on to relate the emf to the activities of the participants in the cell reaction. We know from eqn 9.11 that the reaction Gibbs energy is related to the composition of the reaction mixture by

10.13 The variation of cell emf with the value of the reaction quotient for the cell reaction for different values of v (the number of electrons transferred). At 298 K, $RT/F = 25.69$ mV, so the vertical scale refers to multiples of this value.

$$\Delta_r G = \Delta_r G^\ominus + RT \ln Q$$

where Q is the reaction quotient. It follows, on division of both sides by $-vF$, that

$$E = -\frac{\Delta_r G^\ominus}{vF} - \frac{RT}{vF} \ln Q$$

The first term on the right is written

$$E^\ominus = -\frac{\Delta_r G^\ominus}{vF} \qquad [10.33]$$

and called the **standard emf** of the cell. That is, the standard emf is the standard reaction Gibbs energy expressed as a potential (in volts). It follows that

$$E = E^\ominus - \frac{RT}{vF} \ln Q \qquad (10.34)$$

This equation for the emf in terms of the composition is called the **Nernst equation**; the dependence of cell potential on composition that it predicts is summarized in Fig. 10.13.

We see from eqn 34 that the standard emf (which will shortly move to centre stage of the exposition) can be interpreted as the emf when all the reactants and products are in their standard states, for then all activities are 1, so $Q = 1$ and $\ln Q = 0$. However, the fact that the standard emf is merely a disguised form of the standard reaction Gibbs energy should always be kept in mind and underlies all its applications.

Illustration 10.6

Because $RT/F = 25.7$ mV at 25°C, a practical form of the Nernst equation is

$$E = E^\ominus - \frac{25.7 \text{ mV}}{v} \ln Q$$

It then follows that, for a reaction in which $v = 1$, if Q is increased by a factor of 10, then the emf decreases by 59.2 mV.

(g) Concentration cells

We can use the Nernst equation to derive an expression for the emf of an electrolyte concentration cell. Consider the cell

$$M \,|\, M^+(aq, L) \,\|\, M^+(aq, R) \,|\, M$$

where the solutions L and R have different molalities. The cell reaction is

$$M^+(aq, R) \longrightarrow M^+(aq, L) \qquad Q = \frac{a_L}{a_R} \qquad v = 1$$

The standard emf of the cell is zero, because a cell cannot drive a current through a circuit when the two electrode compartments are identical (and, specifically, $\Delta_r G^\ominus = 0$ for the cell reaction). Therefore, the emf when the compartments have different concentrations is

$$E = -\frac{RT}{F} \ln \frac{a_L}{a_R} \approx -\frac{RT}{F} \ln \frac{b_L}{b_R} \qquad (10.35)$$

If R is the more concentrated solution, $E > 0$. Physically, the positive potential arises because positive ions tend to be reduced, so withdrawing electrons from the electrode. This process is dominant in the more concentrated right-hand electrode compartment.

One important example of a system that resembles this description is the biological cell membrane, which is more permeable to K^+ ions than to either Na^+ or Cl^- ions. In the enzyme $(Na^+–K^+)$-ATPase, the hydrolysis of ATP drives the pumping of ions across the membrane. For example, the concentration of K^+ inside an inactive nerve cell is about 20 times that on the outside, whereas the concentration of Na^+ outside the cell is about 10 times that on the inside. The difference in concentrations of ions results in a transmembrane potential difference of about -62 mV, with the negative sign denoting that the inside has a negative potential. This potential difference is also called the **resting potential** of the cell membrane.

The transmembrane potential difference plays a particularly interesting role in the transmission of nerve impulses. Upon receiving an impulse, which is called an **action potential**, a site in the nerve cell membrane becomes transiently permeable to Na^+ and the transmembrane potential difference changes. To propagate along a nerve cell, the action potential must change the transmembrane potential difference by at least 20 mV, to values that are less negative than -40 mV. Propagation occurs when an action potential in one site of the membrane triggers an action potential in an adjacent site, with sites behind the moving action potential returning to the resting potential.

(h) Cells at equilibrium

A special case of the Nernst equation has great importance in electrochemistry. Suppose the reaction has reached equilibrium; then $Q = K$, where K is the equilibrium constant of the cell reaction. However, a chemical reaction at equilibrium cannot do work, and hence it generates zero potential difference between the electrodes of a galvanic cell. Therefore, setting $E = 0$ and $Q = K$ in the Nernst equation gives

$$\ln K = \frac{vFE^{\ominus}}{RT} \tag{10.36}$$

This very important equation lets us predict equilibrium constants from measured standard cell potentials. However, before we use it extensively, we need to establish a further result.

Illustration 10.7

Because the standard potential of the Daniell cell is $+1.10$ V, the equilibrium constant for the cell reaction $Cu^{2+}(aq) + Zn(s) \rightarrow Cu(s) + Zn^{2+}(aq)$, for which $v = 2$, is $K = 1.5 \times 10^{37}$ at 298.15 K. We conclude that the displacement of copper by zinc goes virtually to completion.

10.5 Standard potentials

A galvanic cell is a combination of two electrodes, and each one can be considered as making a characteristic contribution to the overall cell potential. Although it is not possible to measure the contribution of a single electrode, we can define the

Table 10.7* Standard potentials at 298 K

Couple	$E^{\ominus}/V$
$Ce^{4+}(aq) + e^- \longrightarrow Ce^{3+}(aq)$	+1.61
$Cu^{2+}(aq) + 2e^- \longrightarrow Cu(s)$	+0.34
$AgCl(s) + e^- \longrightarrow Ag(s) + Cl^-(aq)$	+0.22
$2H^+(aq) + 2e^- \longrightarrow H_2(g)$	0
$Zn^{2+}(aq) + 2e^- \longrightarrow Zn(s)$	−0.76
$Na^+(aq) + e^- \longrightarrow Na(s)$	−2.71

* More values are given in the *Data section*.

potential of one of the electrodes as zero and then assign values to others on that basis. The specially selected electrode is the **standard hydrogen electrode** (SHE):

$$Pt(s)\,|\,H_2(g)\,|\,H^+(aq) \qquad E^{\ominus} = 0 \qquad\qquad [10.37]$$

at all temperatures. The **standard potential**, $E^{\ominus}$, of another couple is then assigned by constructing a cell in which it is the right-hand electrode and the standard hydrogen electrode is the left-hand electrode. For example, the standard potential of the Ag^+/Ag couple is the standard emf of the following cell:[6]

$$Pt(s)\,|\,H_2(g)\,|\,H^+(aq)\,\|\,Ag^+(aq)\,|\,Ag(s) \qquad E^{\ominus}(Ag^+/Ag) = E^{\ominus} = +0.80\ V$$

Likewise, the standard potential of the $AgCl/Ag,Cl^-$ couple is the standard emf of the following cell:

$$Pt(s)\,|\,H_2(g)\,|\,H^+(aq)\,\|\,Cl^-(aq)\,|\,AgCl(s)\,|\,Ag(s) \qquad E^{\ominus}(AgCl/Ag,Cl^-) = E^{\ominus} = +0.22\ V$$

Although a standard potential is written as though it refers to a half-reaction such as

$$AgCl(s) + e^- \longrightarrow Ag(s) + Cl^-(aq) \qquad E^{\ominus}(AgCl/Ag,Cl^-) = +0.22\ V$$

it should be understood that these equations are only shorthand for writing

$$AgCl(s) + \tfrac{1}{2} H_2(g) \longrightarrow Ag(s) + H^+(aq) + Cl^-(aq) \qquad E^{\ominus} = +0.22\ V$$

and that the standard potential is determined by properties of the hydrogen electrode as well as the couple to which the potential refers. Table 10.7 lists standard potentials at 298 K.

An important feature of standard emf of cells and standard potentials of electrodes is that they are unchanged if the chemical equation for the cell reaction or a half-reaction is multiplied by a numerical factor. A numerical factor increases the value of the standard Gibbs energy for the reaction. However, it also increases the number of electrons transferred by the same factor, and by eqn 33 the value of $E^{\ominus}$ remains unchanged.

(a) The cell emf in terms of individual standard potentials

The standard emf of a cell formed from any two electrodes is calculated by taking the difference of their standard potentials. This rule follows from the fact that a cell such as

$$Ag(s)\,|\,Ag^+(aq)\,\|\,Cl^-(aq)\,|\,AgCl(s)\,|\,Ag(s)$$

is equivalent to two cells joined back-to-back:

$$Ag(s)\,|\,Ag^+(aq)\,\|\,H^+(aq)\,|\,H_2(g)\,|\,Pt(s)—Pt(s)\,|\,H_2(g)\,|\,H^+(aq)\,\|\,Cl^-(aq)\,|\,AgCl(s)\,|\,Ag(s)$$

The overall potential of this composite cell, and therefore of the cell of interest, is

$$E^{\ominus} = E^{\ominus}(AgCl/Ag,Cl^-) - E^{\ominus}(Ag^+/Ag) = -0.58\ V$$

The standard potentials in Table 10.7 can all be used in the same way, and the standard emf of the cell is obtained by subtracting the standard potential of the left-hand electrode from the standard potential of the right-hand electrode:

$$E^{\ominus} = E^{\ominus}(\text{right}) - E^{\ominus}(\text{left}) \qquad\qquad (10.38)$$

Because $\Delta G^{\ominus} = -\nu F E^{\ominus}$, it then follows that, if the result gives $E^{\ominus} > 0$, then the corresponding cell reaction has $K > 1$.

---

6. Here and from now on, all values refer to 298 K.

Illustration 10.8

To calculate the equilibrium constant for the disproportionation
$2\,Cu^+(aq) \to Cu(s) + Cu^{2+}(aq)$ at 298.15 K we need

Right-hand electrode: $Cu(s)|Cu^+(aq)$ $Cu^+(aq) + e^- \longrightarrow Cu(aq)$ $E^\ominus = +0.52$ V
Left-hand electrode: $Pt(s)|Cu^{2+}(aq),Cu^+(aq)$ $Cu^{2+}(aq) + e^- \longrightarrow Cu^+(s)$
$$E^\ominus = +0.16 \text{ V}$$

The standard emf of the cell is therefore

$$E^\ominus = +0.52 \text{ V} - 0.16 \text{ V} = +0.36 \text{ V}$$

Then, because $v = 1$,

$$\ln K = \frac{0.36 \text{ V}}{0.025\,693 \text{ V}} = \frac{0.36}{0.025\,693}$$

Hence, $K = 1.2 \times 10^6$.

Example 10.2 Calculating an equilibrium constant

Evaluate the solubility constant of silver chloride (the equilibrium constant for the dissolution of $AgCl(s)$) and its solubility from cell potential data at 298.15 K.

Method We need to find an electrode combination that reproduces the solubility equilibrium, and then identify the solubility constant with the equilibrium constant of the cell reaction. Calculate the standard emf of the cell from the standard potentials and use the result in eqn 36. For the solubility, S, which is the concentration of solute at equilibrium, express the solubility product in terms of S and solve the resulting equation for S.

Answer We saw in Example 10.1 that the dissolution can be expressed as the difference of the following half-reactions:

$$AgCl(s) + e^- \longrightarrow Ag(s) + Cl^-(aq) \qquad E^\ominus = +0.22 \text{ V}$$
$$Ag^+(aq) + e^- \longrightarrow Ag(s) \qquad E^\ominus = +0.80 \text{ V}$$

The standard emf is therefore -0.58 V. Then, because $v = 1$,

$$\ln K = \frac{vE^\ominus}{RT/F} = \frac{1 \times (-0.58 \text{ V})}{2.5693 \times 10^{-2} \text{ V}} = -\frac{0.58}{2.5693 \times 10^{-2}}$$

Therefore, $K = 1.6 \times 10^{-10}$. Because $K = a_{Ag^+} a_{Cl^-} \approx S^2$, it follows that $S = 1.3 \times 10^{-5}$ mol kg^{-1}.

Self-test 10.6 Calculate the solubility constant and the solubility of mercury(I) chloride at 298.15 K. *Hint*. The mercury(I) ion is the diatomic species Hg_2^{2+}.

[2.6×10^{-18}, 8.7×10^{-7} mol kg^{-1}]

(b) The measurement of standard potentials

The procedure for measuring a standard potential can be illustrated by considering a specific case, the silver chloride electrode. The measurement is made on the 'Harned cell':

$$\text{Pt(s)}|\text{H}_2\text{(g)}|\text{HCl(aq)}|\text{AgCl(s)}|\text{Ag(s)} \qquad \tfrac{1}{2}\text{H}_2\text{(g)} + \text{AgCl(s)} \longrightarrow \text{HCl(aq)} + \text{Ag(s)}$$

for which the Nernst equation is

$$E = E^{\ominus}(\text{AgCl/Ag,Cl}^-) - \frac{RT}{F}\ln\frac{a_{\text{H}^+}a_{\text{Cl}^-}}{(f_{\text{H}_2}/p^{\ominus})^{1/2}}$$

We shall set $f = p^{\ominus}$ from now on, and for simplicity write the standard potential as $E^{\ominus}$. The activities can be expressed in terms of the molality b and the mean activity coefficient, $\gamma_{\pm}$, through eqn 11:

$$E = E^{\ominus} - \frac{RT}{F}\ln b^2 - \frac{RT}{F}\ln \gamma_{\pm}^2$$

This expression rearranges to

$$E + \frac{2RT}{F}\ln b = E^{\ominus} - \frac{2RT}{F}\ln \gamma_{\pm} \qquad\qquad (10.39)$$

From the Debye–Hückel limiting law for a 1,1-electrolyte, we know that $\ln \gamma_{\pm} \propto -b^{1/2}$ (the natural logarithm used here is proportional to the common logarithm that appears in eqn 17). Therefore, with the constant of proportionality in this relation written as $(F/2RT)C$:

$$E + \frac{2RT}{F}\ln b = E^{\ominus} + Cb^{1/2}$$

The expression on the left is evaluated at a range of molalities, plotted against $b^{1/2}$, and extrapolated to $b = 0$.[7] The intercept at $b^{1/2} = 0$ is the value of $E^{\ominus}$ for the silver/silver-chloride electrode.

> $\ln x = \ln 10 \log x \approx 2.303 \log x$.

Example 10.3 Determining the standard emf of a cell

The emf of the cell $\text{Pt(s)}|\text{H}_2\text{(g, }p^{\ominus}\text{)}|\text{HCl(aq, }b\text{)}|\text{AgCl(s)}|\text{Ag(s)}$ at 25°C has the following values:

$b/(10^{-3} b^{\ominus})$	3.215	5.619	9.138	25.63
E/V	0.520 53	0.492 57	0.468 60	0.418 24

Determine the standard emf of the cell.

Method Proceed as described above, but adjusted to the cell reaction. Start by writing the Nernst equation for the cell, and then express the activities that occur in Q in terms of the mean activity coefficient. The latter can be written as proportional to $b^{1/2}$ by using the Debye–Hückel limiting law. However, there is no need to write all the constants because the standard emf is obtained by extrapolation, as explained in the text.

Answer The ionic equation for the cell reaction is

$$\tfrac{1}{2}\text{H}_2\text{(g)} + \text{AgCl(s)} \longrightarrow \text{Ag(s)} + \text{H}^+\text{(aq)} + \text{Cl}^-\text{(aq)} \qquad Q = a_{\text{H}^+}a_{\text{Cl}^-} \qquad v = 1$$

because the solid is at unit activity and the gas is at 1 bar. The Nernst equation is therefore

---

7. In precise work, the $b^{1/2}$ term is brought to the left, and a higher order correction term from the extended Debye–Hückel law is used on the right.

10.14 The plot, and the extrapolation used for the experimental measurement of a standard cell emf. The intercept at $b^{1/2} = 0$ is $E^{\ominus}$.

$$E = E^{\ominus} - \frac{RT}{F} \ln a_{H^+} a_{Cl^-}$$

The activities are related to the molality, b, of HCl by

$$a_{H^+} a_{Cl^-} = \gamma_{\pm}^2 b_{H^+} b_{Cl^-} = \gamma_{\pm}^2 b^2$$

because $b_{H^+} = b_{Cl^-} = b$ for a fully deprotonated acid. Only a little work is then needed to convert the Nernst equation into

$$E + \frac{2RT}{F} \ln b = E^{\ominus} + C b^{1/2}$$

where C is a collection of constants that come from the limiting law. We now draw up the following table, using $2RT/F = 0.051\ 39$ V:

$b/(10^{-3}\ b^{\ominus})$	3.215	5.619	9.138	25.63
$(b/(10^{-3}\ b^{\ominus}))^{1/2}$	1.793	2.370	3.023	5.063
E/V	0.520 53	0.492 57	0.468 60	0.418 24
$E/V + 0.051\ 39 \ln b$	0.2256	0.2263	0.2273	0.2299

The data are plotted in Fig. 10.14; as can be seen, they extrapolate to $E^{\ominus} = 0.2232$ V.

Self-test 10.7 The data below are for the cell $Pt(s)|H_2(g, p^{\ominus})|HBr(aq, b)|AgBr(s)|Ag(s)$ at 25°C. Determine the standard emf of the cell.

$b/(10^{-4}\ b^{\ominus})$	4.042	8.444	37.19	
E/V	0.473 81	0.436 36	0.361 73	[0.071 V]

(c) The measurement of activity coefficients

Once the standard potential of an electrode in a cell is known, we can use it to determine the activities of the ions with respect to which it is reversible by measuring the cell emf with the ions at the concentration of interest. For example, the mean activity coefficient of the ions in hydrochloric acid of molality b is obtained from eqn 39 in the form

$$\ln \gamma_{\pm} = \frac{E^{\ominus} - E}{2RT/F} - \ln b \qquad \{10.40\}$$

once E has been measured.

Applications of standard potentials

Cell emfs are a convenient source of data on the Gibbs energies, enthalpies, and entropies of reactions. In practice the standard values of these quantities are the ones normally determined.

10.6 The electrochemical series

We have seen that for two redox couples Ox_1/Red_1 and Ox_2/Red_2, and the cell

$$Red_1, Ox_1 \| Red_2, Ox_2 \qquad E^{\ominus} = E_2^{\ominus} - E_1^{\ominus} \qquad (10.41a)$$

that the cell reaction

$$Red_1 + Ox_2 \longrightarrow Ox_1 + Red_2 \tag{10.41b}$$

is spontaneous as written if $E^{\ominus} > 0$, and therefore if $E_2^{\ominus} > E_1^{\ominus}$. Because in the cell reaction Red_1 reduces Ox_2, we can conclude that

Red_1 has a thermodynamic tendency to reduce Ox_2 if $E_1^{\ominus} < E_2^{\ominus}$

More briefly: low reduces high. The reactions of the electron transport chains of respiration and photosynthesis are good applications of this principle (Box 10.1).

Box 10.1 Biological energy conversion: driving the synthesis of ATP with electron transfer reactions

Many important biochemical processes involve electron transfer reactions. Here, we examine the electrochemical reactions associated with photosynthesis and the aerobic oxidation of glucose. These processes are related by the reactions

$$C_6H_{12}O_6(s) + 6\ O_2(g)$$

$$\xrightarrow[\text{photosynthesis}]{\text{glycolysis and the citric acid cycle}} 6\ CO_2(g) + 6\ H_2O(l)$$

The exergonic oxidation of one $C_6H_{12}O_6$ molecule requires the transfer of 24 electrons to six O_2 molecules. The half-reactions for the oxidation of glucose and the reduction of O_2 are

$$C_6H_{12}O_6(s) + 6\ H_2O(l) \longrightarrow 6\ CO_2(g) + 24\ H^+(aq) + 24\ e^-$$

$$6\ O_2(g) + 24\ H^+(aq) + 24\ e^- \longrightarrow 12\ H_2O(l)$$

The electrons do not flow directly from glucose to O_2. In the cell, glucose is oxidized to CO_2 by NAD^+ and FAD during glycolysis and the citric acid cycle (Box 9.1):

$$C_6H_{12}O_6(s) + 10\ NAD^+ + 2\ FAD + 4\ ADP + 4\ P_i^- + 2\ H_2O$$

$$\longrightarrow 6\ CO_2 + 10\ NADH + 2\ FADH_2 + 4\ ATP + 6\ H^+$$

In the mitochondrial electron transport chain, electrons from the powerful reducing agents NADH and $FADH_2$ pass through four membrane-bound protein complexes and two mobile electron carriers before reducing O_2 to H_2O. We shall see that the electron transfer reactions drive the synthesis of ATP at three of the membrane protein complexes.

In complex I (NADH:ubiquinone oxidoreductase), NADH is oxidized by ubiquinone (UQ, **1**) in a two-electron reaction:

$$H^+ + NADH + UQ \xrightarrow{\text{complex I}} NAD^+ + UQH_2$$

$$E^{\ominus} = +0.42\ V, \quad \Delta_r G^{\ominus} = -81\ kJ\ mol^{-1}$$

Additional UQ molecules are reduced by $FADH_2$ in complex II (succinate dehydrogenase):

$$FADH_2 + UQ \xrightarrow{\text{complex II}} FAD + UQH_2$$

$$E^{\ominus} = +0.015\ V, \quad \Delta_r G^{\ominus} = -2.9\ kJ\ mol^{-1}$$

Ubiquinone:　$R_1 = CH_3$, $R_2 = R_3 = OCH_3$
Plastoquinone:　$R_1 = H$, $R_2 = R_3 = CH_3$

1

Reduced ubiquinone migrates to complex III (ubiquinol: cytochrome c oxidoreductase), which catalyses the reduction of the protein cytochrome c (Cyt c). Cytochrome c contains the haem c group (**2**), the central iron ion of which can exist in oxidation states +3 and +2. The net reaction catalysed by complex III is

$$UQH_2 + 2\ Fe^{3+}(Cyt\ c) \xrightarrow{\text{complex III}} UQ + 2\ Fe^{2+}(Cyt\ c) + 2\ H^+$$

$$E^{\ominus} = +0.15\ V, \quad \Delta_r G^{\ominus} = -30\ kJ\ mol^{-1}$$

Reduced cytochrome c carries electrons from complex III to complex IV (ferrocytochrome c oxidase), where O_2 is reduced to H_2O:

$$2\ Fe^{2+}(Cyt\ c) + 2\ H^+ + \tfrac{1}{2}\ O_2 \xrightarrow{\text{complex IV}} 2\ Fe^{3+}(Cyt\ c) + H_2O$$

$$E^{\ominus} = +0.815\ V, \quad \Delta_r G^{\ominus} = -109\ kJ\ mol^{-1}$$

The reactions that occur in complexes I, III, and IV are sufficiently exergonic to drive the synthesis of ATP in the process called *oxidative phosphorylation*:

$$ADP + P_i + H^+ \longrightarrow ATP \quad \Delta_r G^{\ominus} = +30\ kJ\ mol^{-1}$$

2

3

Experiments show that 11 molecules of ATP are made for every three molecules of NADH and one molecule of $FADH_2$ that are oxidized by the mitochondrial electron transport chain. The ATP is then hydrolysed on demand to perform useful biochemical work throughout the cell.

In plant photosynthesis, solar energy drives the endergonic reduction of CO_2 to glucose, with concomitant oxidation of water to O_2 ($\Delta_r G^{\ominus} = +2880$ kJ mol^{-1}). The process takes place in the *chloroplast*, a special organelle of the plant cell. Electrons flow from reductant to oxidant via a series of electrochemical reactions that are coupled to the synthesis of ATP. First, the leaf absorbs solar energy and transfers it to membrane protein complexes known as photosystem I and photosystem II.[8] The absorption of energy from light decreases the reduction potential of special dimers of chlorophyll *a* molecules (**3**) known as P700 (in photosystem I) and P680 (in photosystem II). In their high-energy or excited states, P680 and P700 initiate electron transfer reactions that culminate in the oxidation of water to O_2 and the reduction of NADP$^+$ to NADPH:[9]

$$2\ NADP^+ + 2\ H_2O \xrightarrow{\ light\ } O_2 + 2\ NADPH + 2\ H^+$$

It is clear that energy from light is required to drive this reaction because, in the dark, $E^{\ominus} = -1.135$ V and $\Delta_r G^{\ominus} = +438.0$ kJ mol^{-1}.

Working together, photosystem I and the enzyme ferredoxin:NADP$^+$ oxidoreductase catalyse the light-induced

oxidation of NADP$^+$ to NADPH. The electrons required for this process come initially from P700 in its excited state. The resulting P700$^+$ is then reduced by the mobile carrier plastocyanin (Pc), a protein in which the bound copper ion can exist in oxidation states +2 and +1. The net reaction is

$$NADP^+ + 2\ Cu^+(Pc) + H^+$$

$$\xrightarrow{\ light,\ photosystem\ I\ } NADPH + 2\ Cu^{2+}(Pc)$$

Oxidized plastocyanin accepts electrons from reduced plastoquinone (PQ, **1**). The process is catalysed by the cytochrome b_6f complex, a membrane protein complex that resembles complex III of mitochondria:

$$PQH_2 + 2\ Cu^{2+}(Pc) \xrightarrow{\ cyt\ b_6f\ complex\ } PQ + 2\ H^+ + 2\ Cu^+(Pc)$$

$$E^{\ominus} = +0.370\ V, \quad \Delta_r G^{\ominus} = -71.4\ kJ\ mol^{-1}$$

This reaction is sufficiently exergonic to drive the synthesis of ATP in the process known as *photophosphorylation*.

Plastoquinone is reduced by water in a process catalysed by light and photosystem II. The electrons required for the reduction of plastoquinone come initially from P680 in its excited state. The resulting P680$^+$ is then reduced ultimately by water. The net reaction is

$$H_2O + PQ \xrightarrow{\ light,\ photosystem\ II\ } \tfrac{1}{2} O_2 + PQH_2$$

In this way, plant photosynthesis uses an abundant source of electrons (water) and of energy (the Sun) to drive the endergonic reduction of NADP$^+$, with concomitant synthesis of

---

8. See Section 26.11 and Box 26.2 for details of the energy transfer process.
9. NADP$^+$ is a form of NAD$^+$ with a phosphate group covalently attached to the carbohydrate portion of the molecule. See Box 9.1.

ATP. Experiments show that, for each molecule of NADPH formed in the chloroplast of green plants, one molecule of ATP is synthesized.

The ATP and NADPH molecules formed by the light-induced electron transfer reactions of plant photosynthesis participate directly in the reduction of CO_2 to glucose in the chloroplast:

$$6\ CO_2 + 12\ NADPH + 12\ ATP + 12\ H^+$$
$$\longrightarrow C_6H_{12}O_6 + 12\ NADP^+ + 12\ ADP + 12\ P_i^- + 6\ H_2O$$

In summary, electrochemical reactions mediated by membrane protein complexes harness energy in the form of ATP. Plant photosynthesis uses solar energy to transfer electrons from a poor reductant (water) to carbon dioxide. In the process, high-energy molecules (carbohydrates, such as glucose) are synthesized in the cell. Animals feed on the carbohydrates derived from photosynthesis. During aerobic metabolism, the O_2 released by photosynthesis as a waste product is used to oxidize carbohydrates to CO_2. This drives biological processes, such as biosynthesis, muscle contraction, cell division, and nerve conduction. Hence, the sustenance of life on Earth depends on a tightly regulated carbon–oxygen cycle that is driven by solar energy.

Discussion questions

1. In anaerobic bacteria, the source of carbon may be a molecule other than glucose and the final electron acceptor is some molecule other than O_2. Could a bacterium evolve to utilize the ethanol/nitrate pair instead of the glucose/O_2 pair as a source of metabolic energy?

2. The photosynthetic oxidation of water to O_2 occurs in an enzyme that contains four manganese ions, each of which can exist in oxidation states ranging from +2 to +4. The electrochemical production of one molecule of O_2 requires the oxidation of two molecules of water by a total of four electrons. However, the excited state of P680 can donate only one electron at a time to plastoquinone. Explain how electron transfer mediated by P680 can lead to the formation of a molecule of O_2 in photosystem II. *Hint.* See V.A. Szalai and G.W. Brudvig, How plants produce dioxygen. *American Scientist* **86**, 542 (1998).

Further reading

C.K. Matthews, K.E. van Holde, and K.G. Ahern, *Biochemistry*. Addison Wesley Longman, San Francisco (2000).

M. Saraste, Oxidative phosphorylation at the *fin de siècle*. *Science* **283**, 1488 (1999).

Table 10.8* The electrochemical series of the metals

Least strongly reducing
Gold
Platinum
Silver
Mercury
Copper
(Hydrogen)
Lead
Tin
Nickel
Iron
Zinc
Chromium
Aluminium
Magnesium
Sodium
Calcium
Potassium
Most strongly reducing

---

* The complete series can be inferred from Table 10.7.

Illustration 10.9

Because $E^{\ominus}(Zn^{2+}, Zn) = -0.76$ V $< E^{\ominus}(Cu^{2+}, Cu) = +0.34$ V, zinc has a thermodynamic tendency to reduce Cu^{2+} ions in aqueous solution.

Table 10.8 shows a part of the **electrochemical series**, the metallic elements (and hydrogen) arranged in the order of their reducing power as measured by their standard potentials in aqueous solution. A metal low in the series (with a lower standard potential) can reduce the ions of metals with higher standard potentials. This conclusion is qualitative. The quantitative value of K is obtained by doing the calculations we have described previously. For example, to determine whether zinc can displace magnesium from aqueous solutions at 298 K, we note that zinc lies above magnesium in the electrochemical series, so zinc cannot reduce magnesium ions in aqueous solution. Zinc can reduce hydrogen ions, because hydrogen lies higher in the series. However, even for reactions that are thermodynamically favourable, there may be kinetic factors that result in very slow rates of reaction.

10.7 The measurement of pH and pK_a

The half-reaction at a hydrogen electrode is

$$H^+(aq) + e^- \longrightarrow \tfrac{1}{2} H_2(g) \qquad Q = \frac{(f_{H_2}/p^{\ominus})^{1/2}}{a_{H^+}} \qquad \nu = 1$$

10.15 The glass electrode. It is commonly used in conjunction with a calomel electrode that makes contact with the test solution through a salt bridge.

If we let $f_{H_2} = p^{\ominus}$ and $E^{\ominus}(H^+/H_2) = 0$, the electrode potential is

$$E(H^+/H_2) = \frac{RT}{F} \ln a_{H^+} = -\frac{RT \ln 10}{F} \text{pH} \qquad (10.42a)$$

This expression makes sense physically. Increasing the pH above 0 (lowering the hydrogen ion activity below 1) decreases the tendency of the positive ions to become reduced at the electrode, so we should expect its potential to become negative. At 25°C, when $RT/F = 25.69$ mV, this relation becomes

$$E(H^+/H_2) = -59.16 \text{ mV} \times \text{pH} \qquad (10.42b)$$

Each unit increase in pH decreases the electrode potential by 59 mV.

The measurement of the pH of a solution is simple in principle, for it is based on the measurement of the potential of a hydrogen electrode immersed in the solution. The left-hand electrode of the cell is typically a saturated calomel ($Hg_2Cl_2(s)$) reference electrode with potential $E(\text{cal})$; the right-hand electrode is the hydrogen electrode with potential given by eqn 42. The pH of the cell is therefore

$$\text{pH} = \frac{E + E(\text{cal})}{(-59.16 \text{ }V)} \qquad (10.43)$$

The practical definition of the pH of a solution X is

$$\text{pH} = \text{pH(S)} - \frac{FE}{RT \ln 10} \qquad [10.44]$$

where E is the emf of the cell

$$Pt(s)\,|\,H_2(g)\,|\,S(aq)\,\|\,3.5\text{ м KCl}(aq)\,\|\,X(aq)\,|\,H_2(g)\,|\,Pt(s)$$

and S is a solution of standard pH. The currently recommended primary standards include a saturated aqueous solution of potassium hydrogen tartrate, which has pH = 3.557 at 25°C, and 0.0100 mol kg^{-1} disodium tetraborate, which has pH = 9.180 at that temperature.

In practice, indirect methods are much more convenient and the hydrogen electrode is replaced by the **glass electrode** (Fig. 10.15). This electrode is sensitive to hydrogen ion activity, and has a potential proportional to pH. It is filled with a phosphate buffer containing Cl$^-$ ions, and conveniently has $E = 0$ when the external medium is at pH = 7. The glass electrode is much more convenient to handle than the gas electrode itself and is calibrated using solutions of known pH.

The responsiveness of a glass electrode to the hydronium ion activity is a result of complex processes at the interface between the glass membrane and the solutions on either side of it. The membrane itself is permeable to Na$^+$ and Li$^+$ ions but not to H$^+$ ions. Therefore, the potential difference across the glass membrane must arise by a mechanism different from that responsible for biological transmembrane potentials. A clue to the mechanism comes from a detailed inspection of the glass membrane, for each face is coated with a thin layer of hydrated silica (Fig. 10.16). The hydrogen ions in the test solution modify this layer to an extent that depends on their activity in the solution, and the charge modification of the outside layer is transmitted to the inner layer by the Na$^+$ and Li$^+$ ions in the glass. The hydronium ion activity gives rise to a membrane potential by this indirect mechanism.

The electrochemical determination of pH opens up a route to the electrochemical determination of pK_a, for we saw in Section 9.5c that the pK_a of an acid is equal to the pH of a solution containing equal amounts of the acid and its conjugate base.

Electrodes can be constructed with potentials that are sensitive to the concentrations of ions other than the hydronium ion. Some examples of these **ion-selective electrodes** are discussed in Box 10.2.

10.16 A section through the wall of a glass electrode.

Box 10.2 Species-selective electrodes

A glass electrode is an example of an *ion-selective electrode*, an electrode that generates a potential in response to the presence of a solution of specific ions. Electrodes sensitive to hydrogen ions, and hence to pH, are typically glasses based on lithium silicate doped with heavy-metal oxides. The glass can be responsive to Na^+, K^+, and NH_4^+ ions by being doped with Al_2O_3 and B_2O_3.

A suitably adapted glass electrode can be used to detect the presence of certain gases. A simple form of a gas-sensing electrode consists of a glass electrode contained in an outer sleeve filled with an aqueous solution and separated from the test solution by a membrane that is permeable to gas. When a gas such as sulfur dioxide or ammonia diffuses into the aqueous solution, it modifies its pH, which in turn affects the potential of the glass electrode. The presence of an enzyme that converts a compound, such as urea or an amino acid, into ammonia, which then affects the pH, can be used to detect these organic compounds.

Somewhat more sophisticated devices are used as ion-selective electrodes that give potentials according to the presence of specific ions present in a test solution. In one arrangement, a porous lipophilic (hydrocarbon-attracting) membrane is attached to a small reservoir of a hydrophobic (water-repelling) liquid, such as dioctylphenylphosphonate, that saturates it (see the illustration). The liquid contains a chelating agent, such as $(RO)_2PO_2^-$ with R a C_8 to C_{18} chain, that acts as a kind of solubilizing agent for the ions with which it can form a complex. The chelated ions are able to migrate through the lipophilic membrane, and hence give rise to a transmembrane potential, which is detected by a silver/silver chloride electrode in the interior of the assembly. Electrodes of this construction can be designed to be sensit-ive to a variety of ionic species, including calcium, zinc, iron, lead, and copper ions.

In theory, the transmembrane potential should be determined entirely by differences in the activity of the species that the electrode was designed to detect. In practice, a small potential difference, called the *asymmetry potential*, is observed even when the activity of the test species is the same on both sides of the membrane.[10] Furthermore, all species-selective electrodes are sensitive to more than one species. For example, a Na^+ selective electrode also responds, albeit less effectively, to the activity of K^+ ions in the test solution. As a result of these effects, the potential of an electrode sensitive to species X^+ that is also susceptible to interference by species Y^+ is given by a modified form of the Nernst equation:

$$E = E_{ap} + \beta \frac{RT}{F} \ln(a_{X^+} + k_{X,Y} a_{Y^+})$$

where E_{ap} is the asymmetry potential, β is an experimental parameter that captures deviations from the Nernst equation, and $k_{X,Y}$ is the *selectivity coefficient* of the electrode and is related to the response of the electrode to the interfering species Y^+. A value of $\beta = 1$ indicates that the electrode responds to the activity of ions in solution in a way that is consistent with the Nernst equation and, in practice, most species-selective electrodes of high quality have $\beta \approx 1$. The selectivity coefficient, and hence interference effects, can be minimized when designing and manufacturing a species-selective electrode. For precise work, it is necessary to calibrate the response of the electrode by measuring E_{ap}, β, and $k_{X,Y}$ before performing experiments on solutions of unknown concentration of X^+.

Discussion questions

1. Suggest a calibration procedure for a Na^+-selective electrode, in which the values of E_{ap}, β, and $k_{X,Y}$ are measured experimentally. Assume that K^+ is the only interfering species.

2. Chemists and biochemists are often interested in measuring the concentration of an ion in solution, rather than its activity. The strategy for correlating directly the value of an electrode potential to a value of ion concentration consists of performing all measurements in solutions of constant and high ionic strength. Explain why this is a useful strategy.

Further reading

D.C. Harris, *Quantitative chemical analysis*. W.H. Freeman & Co., New York (1999).

J. Koryta, *Ions, electrodes, and membranes*. Wiley, New York (1992).

Silver/silver chloride electrode

Reservoir of hydrophobic liquid + chelating agent

Porous lipophilic membrane

The structure of an ion-selective electrode. Chelated ions are able to migrate through the lipophilic membrane.

---

10. The asymmetry potential is due to the fact that it is not possible to manufacture a membrane material that has the same structure and the same chemical properties throughout.

10.8 Thermodynamic functions

The standard emf of a cell is related to the standard reaction Gibbs energy through eqn 33 ($\Delta_r G^{\ominus} = -\nu F E^{\ominus}$). Therefore, by measuring $E^{\ominus}$ we can obtain this important thermodynamic quantity. Its value can then be used to calculate the Gibbs energy of formation of ions using the convention explained in Section 10.1a.

Illustration 10.10

The cell reaction taking place in

$$Pt(s)|H_2|H^+(aq)\|Ag^+(aq)|Ag(s) \qquad E^{\ominus} = +0.7996 \text{ V}$$

is

$$Ag^+(aq) + \tfrac{1}{2} H_2(g) \longrightarrow H^+(aq) + Ag(s) \qquad \Delta_r G^{\ominus} = -\Delta_f G^{\ominus}(Ag^+, aq)$$

Therefore, with $\nu = 1$, we find

$$\Delta_f G^{\ominus}(Ag^+, aq) = -(-FE^{\ominus}) = +77.15 \text{ kJ mol}^{-1}$$

which is in close agreement with the value in Table 2.6 of the *Data section*.

Example 10.4 Evaluating a standard potential from two others

Given that the standard potentials of the Cu^{2+}/Cu and Cu^+/Cu couples are +0.340 V and +0.522 V, respectively, evaluate $E^{\ominus}(Cu^{2+}, Cu^+)$.

Method First, we note that reaction Gibbs energies may be added (as in a Hess's law analysis of reaction enthalpies). Therefore, we should convert the $E^{\ominus}$ values to $\Delta G^{\ominus}$ values by using eqn 33, add them appropriately, and then convert the overall $\Delta G^{\ominus}$ to the required $E^{\ominus}$ by using eqn 33 again. This roundabout procedure is necessary because, as we shall see, although the factor F cancels, the factor ν in general does not.

Answer The electrode reactions are as follows:

(a) $Cu^{2+}(aq) + 2\,e^- \longrightarrow Cu(s) \qquad E^{\ominus} = +0.340 \text{ V}, \qquad$ so $\quad \Delta_r G^{\ominus} = -2(0.340 \text{ V})F$

(b) $Cu^+(aq) + e^- \longrightarrow Cu(s) \qquad E^{\ominus} = +0.522 \text{ V}, \qquad$ so $\quad \Delta_r G^{\ominus} = -(0.522 \text{ V})F$

The required reaction is

(c) $Cu^{2+}(aq) + e^- \longrightarrow Cu^+(aq) \qquad E^{\ominus} = -\Delta_r G^{\ominus}/F$

Because (c) = (a) − (b), the standard Gibbs energy of reaction (c) is

$$\Delta_r G^{\ominus} = \Delta_r G^{\ominus}(a) - \Delta_r G^{\ominus}(b) = (-0.158 \text{ V}) \times F$$

Therefore, $E^{\ominus} = +0.158$ V.

Comment Note that we cannot combine the $E^{\ominus}$ values directly and we must always work via $\Delta G^{\ominus}$. Alternatively, note that the generalization of the calculation illustrated above is

$$\nu_c E^{\ominus}(c) = \nu_a E^{\ominus}(a) - \nu_b E^{\ominus}(b)$$

Self-test 10.8 Calculate the standard potential of the Fe^{3+}/Fe couple from the values for the Fe^{3+}/Fe^{2+} and Fe^{2+}/Fe couples.   [−0.037 V]

The temperature coefficient of the standard cell emf, $dE^\ominus/dT$, gives the standard entropy of the cell reaction. This conclusion follows from the thermodynamic relation $(\partial G/\partial T)_p = -S$ and eqn 33, which combine to give[11]

$$\frac{dE^\ominus}{dT} = \frac{\Delta_r S^\ominus}{\nu F} \qquad (10.45)$$

Hence we have an electrochemical technique for obtaining standard reaction entropies and through them the entropies of ions in solution.

Finally, we can combine the results obtained so far and use them to obtain the standard reaction enthalpy:

$$\Delta_r H^\ominus = \Delta_r G^\ominus + T\Delta_r S^\ominus = -\nu F\left(E^\ominus - T\frac{dE^\ominus}{dT} \right) \qquad (10.46)$$

This expression provides a noncalorimetric method for measuring $\Delta_r H^\ominus$ and, through the convention $\Delta_f H^\ominus(H^+, aq) = 0$, the standard enthalpies of formation of ions in solution. Thus, electrical measurements can be used to calculate all the thermodynamic properties with which this chapter began.

Example 10.5 Using the temperature coefficient of the cell potential

The standard emf of the cell $Pt(s)|H_2(g)|HBr(aq)|AgBr(s)|Ag(s)$ was measured over a range of temperatures, and the data were fitted to the following polynomial:

$$E^\ominus/V = 0.071\,31 - 4.99 \times 10^{-4}(T/K - 298) - 3.45 \times 10^{-6}(T/K - 298)^2$$

Evaluate the standard reaction Gibbs energy, enthalpy, and entropy at 298 K.

Method The standard Gibbs energy of reaction is obtained by using eqn 33 after evaluating $E^\ominus$ at 298 K and by using $1\,V\,C = 1\,J$. The standard entropy of reaction is obtained by using eqn 45, which involves differentiating the polynomial with respect to T and then setting $T = 298$ K. The reaction enthalpy is obtained by combining the values of the standard Gibbs energy and entropy.

Answer At $T = 298$ K, $E^\ominus = +0.071\,31$ V, so

$$\Delta_r G^\ominus = -\nu F E^\ominus = -(1) \times (9.6485 \times 10^4\,C\,mol^{-1}) \times (+0.071\,31\,V)$$
$$= -6.880 \times 10^3\,V\,C\,mol^{-1} = -6.880\,kJ\,mol^{-1}$$

The temperature coefficient of the cell potential is

$$\frac{dE^\ominus}{dT} = -4.99 \times 10^{-4}\,V\,K^{-1} - 2(3.45 \times 10^{-6})(T/K - 298)\,V\,K^{-1}$$

At $T = 298$ K this expression evaluates to

$$\frac{dE}{dT} = -4.99 \times 10^{-4}\,V\,K^{-1}$$

So, from eqn 45, the reaction entropy is

$$\Delta_r S^\ominus = 1 \times (9.6485 \times 10^4\,C\,mol^{-1}) \times (-4.99 \times 10^{-4}\,V\,K^{-1})$$
$$= -48.2\,J\,K^{-1}\,mol^{-1}$$

---

11. The derivative is complete because $E^\ominus$, like $\Delta_r G^\ominus$, is independent of the pressure.

It then follows that

$$\Delta_r H^\ominus = \Delta_r G^\ominus + T\Delta_r S^\ominus = -6.880 \text{ kJ mol}^{-1} + (298 \text{ K}) \times (-0.0482 \text{ kJ K}^{-1} \text{ mol}^{-1})$$
$$= -21.2 \text{ kJ mol}^{-1}$$

Comment One difficulty with this procedure lies in the accurate measurement of small temperature coefficients of cell potential. Nevertheless, it is another example of the striking ability of thermodynamics to relate the apparently unrelated, in this case to relate electrical measurements to thermal properties.

Self-test 10.9 Predict the standard potential of the Harned cell at 303 K from tables of thermodynamic data. [+0.2222 V]

Checklist of key ideas

The thermodynamic properties of ions in solution

10.1 Thermodynamic functions of formation

☐ standard enthalpy of formation of ions in solution, the standard enthalpy change accompanying the formation of solutions of ions from the reference states of the parent elements

☐ standard enthalpy and Gibbs energy of formation of ions, based on $\Delta_f H^\ominus(\text{H}^+, \text{aq}) = 0$ and $\Delta_f G^\ominus(\text{H}^+, \text{aq}) = 0$

☐ contributions to $\Delta_f G^\ominus$, the use of a thermodynamic cycle

☐ Born equation,
$\Delta_{solv} G^\ominus = -(z_i^2 e^2 N_A / 8\pi\varepsilon_0 r_i)(1 - 1/\varepsilon_r)$

☐ standard (partial molar) entropies of ions in water, based on $S^\ominus(\text{H}^+, \text{aq}) = 0$

10.2 Ion activities

☐ the activity of an ion in solution,
$\mu = \mu^\ominus + RT \ln a$

☐ the activity coefficient, $a = \gamma b/b^\ominus$

☐ the chemical potential of an ion,
$\mu = \mu^{ideal} + RT \ln \gamma$

☐ mean activity coefficient, the geometric mean of the individual coefficients:
$\gamma_\pm = (\gamma_+^p \gamma_-^q)^{1/(p+q)}$

☐ total Gibbs energy of ions in solution,
$G = p\mu_+ + q\mu_-$

☐ Debye–Hückel theory, a theory of activity coefficients of electrolyte solutions based on the assumption that

coulombic interactions between ions are dominant

☐ ionic atmosphere, time-averaged, spherical haze of charge around the central ion, in which counter ions outnumber ions of the same charge as the central ion

☐ Debye–Hückel limiting law,
$\log \gamma_\pm = -|z_+ z_-| A I^{1/2}$

☐ shielded Coulomb potential,
$\phi_i = (z_i e / 4\pi\varepsilon_0 r)e^{-r/r_D}$

☐ Debye length, r_D, the distance in which the shielded Coulomb potential falls to $1/e$ of its unshielded value; a measure of the thickness of the ionic atmosphere

☐ charge density, ρ_i, the charge in a small region divided by the volume of the region

☐ Poisson's equation, an equation for calculating the potential arising from a given charge distribution, $\nabla^2\phi = -\rho/\varepsilon$

☐ ionic strength, $I = \frac{1}{2}\sum_i z_i^2 (b_i/b^\ominus)$

☐ extended Debye–Hückel law,
$\ln \gamma_\pm = -|z_+ z_-| A I^{1/2}/(1 + B I^{1/2})$

Electrochemical cells

☐ electrochemical cell, an arrangement in which a chemical reaction does electrical work or in which electrical work is used to bring about a chemical reaction

☐ electrode, a metallic conductor that makes contact with an electrolyte

☐ electrolyte, an ionic conductor

☐ electrode compartment, an electrode and its electrolyte

☐ salt bridge, a concentrated electrolyte solution in agar jelly that completes the electrical circuit between two electrode compartments

☐ galvanic cell, an electrochemical cell that produces electricity as a result of the spontaneous reaction occurring inside it.

☐ electrolytic cell, an electrochemical cell in which a non-spontaneous reaction is driven by an external source of current

10.3 Half-reactions and electrodes

☐ oxidation, the removal of electrons from a species

☐ reduction, the addition of electrons to a species

☐ redox reaction, a reaction in which there is a transfer of electrons from one species to another

☐ reducing agent (or 'reductant'), an electron donor in a redox reaction

☐ oxidizing agent (or 'oxidant'), an electron acceptor in a redox reaction

☐ half-reaction, a conceptual reaction showing the gain of electrons

☐ redox couple, the reduced and oxidized species in a half-reaction

☐ reaction quotient for a half-reaction, as for the reaction quotient for the overall reaction, but the electrons are ignored

- [ ] anode, the electrode at which oxidation occurs
- [ ] cathode, the electrode at which reduction occurs

10.4 Varieties of cells

- [ ] electrolyte concentration cell, a galvanic cell in which the electrode compartments are identical except for the concentrations of the electrolytes
- [ ] electrode concentration cell, a galvanic cell in which the electrodes themselves have different concentrations
- [ ] liquid junction potential, the potential difference across the interface of the two electrolytes
- [ ] cell notation: | phase boundary; : liquid junction; ‖ an interface at which the junction potential has been eliminated
- [ ] the convention for writing the cell reaction, the reaction in the cell written on the assumption that the right-hand electrode is the cathode
- [ ] cell potential, the potential difference between the two electrodes of a galvanic cell
- [ ] electromotive force (emf), the cell potential when it is balanced by an exactly opposing source of potential so that the cell reaction occurs reversibly, the composition is constant, and no current flows
- [ ] the cell potential in terms of the reaction Gibbs energy, $-\nu FE = \Delta_{\mathrm{r}}G$

- [ ] Faraday constant, $F = eN_{\mathrm{A}}$
- [ ] standard emf, the standard reaction Gibbs energy expressed as a potential: $E^{\ominus} = -\Delta_{\mathrm{r}}G^{\ominus}/\nu F$
- [ ] Nernst equation, the equation for the emf of a cell in terms of the composition: $E = E^{\ominus} - (RT/\nu F)\ln Q$
- [ ] transmembrane potential, $E = -(RT/F)\ln(b_{\mathrm{L}}/b_{\mathrm{R}})$
- [ ] resting potential, the potential difference across the membrane of a neuron at rest
- [ ] action potential, the potential difference across the membrane of a neuron when a signal is propagating
- [ ] equilibrium constant for the cell reaction, $\ln K = \nu FE^{\ominus}/RT$

10.5 Standard potentials

- [ ] standard hydrogen electrode (SHE), $\mathrm{Pt(s)|H_2(g)|H^+(aq)}$, $E^{\ominus} = 0$ at all temperatures
- [ ] standard potential of a couple ($E^{\ominus}$), the standard emf of a cell in which a couple forms the right-hand electrode and the standard hydrogen electrode is the left-hand electrode
- [ ] the standard emf in terms of the standard potentials of the electrodes, the difference of electrode potentials: $E^{\ominus} = E^{\ominus}(\text{right}) - E^{\ominus}(\text{left})$
- [ ] the graphical determination of the standard potential

- [ ] the electrochemical determination of activity coefficients

Applications of standard potentials

10.6 The electrochemical series

- [ ] the interpretation of the relative values of standard potentials, $\mathrm{Red_1}$ has a thermodynamic tendency to reduce $\mathrm{Ox_2}$ if $E_1^{\ominus} < E_2^{\ominus}$
- [ ] electrochemical series, the metallic elements (and hydrogen) arranged in the order of their reducing power as measured by their standard potentials in aqueous solution

10.7 The measurement of pH and pK_{a}

- [ ] the measurement of pH and pK_{a}, $\mathrm{pH} = \mathrm{pH(S)} - FE/(RT\ln 10)$
- [ ] glass electrode, an electrode sensitive to hydrogen ion activity
- [ ] ion-selective electrode, an electrode sensitive to the concentrations of ions other than the hydronium ion

10.8 Thermodynamic functions

- [ ] the temperature coefficient of cell potential, $\mathrm{d}E^{\ominus}/\mathrm{d}T = \Delta_{\mathrm{r}}S^{\ominus}/\nu F$
- [ ] the determination of standard reaction entropy and enthalpy, $\Delta_{\mathrm{r}}S^{\ominus} = \nu F\,\mathrm{d}E^{\ominus}/\mathrm{d}T$, $\Delta_{\mathrm{r}}H^{\ominus} = -\nu F(E^{\ominus} - T\,\mathrm{d}E^{\ominus}/\mathrm{d}T)$

Further reading

Articles of general interest

A.K. Covington, R.G. Bates, and D.A. Durst, Definition of pH scales, standard reference values, and related terminology. *Pure. Appl. Chem.* **57**, 531 (1985).

R.L. DeKock, Tendency of reaction, electrochemistry, and units. *J. Chem. Educ.* **73**, 955 (1996).

A.-S. Feiner and A.J. McEvoy, The Nernst equation. *J. Chem. Educ.* **71**, 493 (1994).

J.J. MacDonald, Cathodes, terminals, and signs. *Educ. in Chem.* **25**, 52 (1988).

Y. Marcus, Ionic radii in aqueous solutions. *Chem. Rev.* **88**, 1475 (1988).

P. Millet, Electric potential distribution in an electrochemical cell. *J. Chem. Educ.* **73**, 956 (1996).

P.J. Morgan and E. Gileadi, Alleviating the common confusion caused by polarity in electrochemistry. *J. Chem. Educ.* **66**, 912 (1989).

Texts and sources of data and information

R.R. Adžić and E.B. Yeager, Electrochemistry. In *Encyclopedia of applied physics* (ed. G.L. Trigg), **5**, 223. VCH, New York (1993).

M.S. Antelman, *The encyclopedia of chemical electrode potentials*. Plenum, New York (1982).

A.J. Bard (ed.), *Encyclopedia of electrochemistry of the elements.* Marcel Dekker (1973–1986).

A.J. Bard and L.R. Faulkner, *Electrochemical methods.* Wiley, New York (2000).

A.J. Bard, R. Parsons, and J. Jordan (ed.), *Standard potentials in aqueous solution.* Marcel Dekker, New York (1985).

J.O'M. Bockris, B.E. Conway, *et al.* (ed.), *Comprehensive treatise on electrochemistry.* 1–10. Plenum, New York (1980–85).

C.M.A. Brett and A.M.O. Brett, *Electroanalysis.* Oxford University Press (1998).

R.G. Compton and G.H.W. Sanders, *Electrode potentials.* Oxford University Press (1998).

W.A. Cramer and D.A. Knaff, *Energy transduction in biological membranes, a textbook of bioenergetics*. Springer-Verlag, New York (1990).

D.R. Crow, *Principles and applications of electrochemistry*. Blackie, London (1994).

J. Goodisman, *Electrochemistry: Theoretical foundations*. Wiley, New York (1987).

C.H. Hamann, A. Hamnett, and W. Vielstich, *Electrochemistry*. Wiley-VCH, Weinheim (1998).

D.A. Harris, *Bioenergetics at a glance*. Blackwell Science, Oxford (1995).

V.M.M. Lobo, *Handbook of electrolyte solutions, Parts A and B*. Elsevier, Amsterdam (1990).

D. Pletcher and F.C. Walsh, *Industrial electrochemistry*. Chapman and Hall, London (1993).

P. Rieger, *Electrochemistry*. Chapman and Hall, London (1993).

Exercises

Discussion questions

10.1 (a) Discuss the factors responsible for the magnitude of the Gibbs energy of formation of an ion in solution.

10.1 (b) Describe the general features of the Debye–Hückel theory of electrolyte solutions.

10.2 (a) Distinguish between galvanic and electrolytic cells.

10.2 (b) Distinguish between cell potential and electromotive force.

10.3 (a) Describe a method for the determination of the standard potential of a cell.

10.3 (b) Describe a method for the determination of the pH of an aqueous solution.

Numerical exercises

10.4 (a) Calculate $\Delta_r H^{\ominus}$ for the reaction $Zn(s) + CuSO_4(aq) \rightarrow ZnSO_4(aq) + Cu(s)$ from the information in Table 2.6 in the *Data section*.

10.4 (b) Calculate $\Delta_r H^{\ominus}$ for the reaction $NaCl(aq) + AgNO_3(aq) \rightarrow AgCl(s) + NaNO_3(aq)$ from the information in Table 2.6 in the *Data section*.

10.5 (a) Calculate the molar solubility of mercury(II) chloride at 25°C from standard Gibbs energies of formation.

10.5 (b) Calculate the molar solubility of lead(II) sulfide at 25°C where the standard Gibbs energy of formation is -98.7 kJ mol^{-1}.

10.6 (a) Estimate the standard Gibbs energy of formation of $F^-(aq)$ from the value for $Cl^-(aq)$, taking the radius of F^- as 131 pm.

10.6 (b) Estimate the standard Gibbs energy of formation of $NO_3^-(aq)$ from the value for $Cl^-(aq)$, taking the radius of the nitrate ion as 189 pm (its thermochemical radius).

10.7 (a) Relate the ionic strengths of (a) KCl, (b) $FeCl_3$, and (c) $CuSO_4$ solutions to their molalities, b.

10.7 (b) Relate the ionic strengths of (a) $MgCl_2$, (b) $Al_2(SO_4)_3$, and (c) $Fe_2(SO_4)_3$ solutions to their molalities, b.

10.8 (a) Calculate the ionic strength of a solution that is 0.10 mol kg^{-1} in KCl(aq) and 0.20 mol kg^{-1} in $CuSO_4$(aq).

10.8 (b) Calculate the ionic strength of a solution that is 0.040 mol kg^{-1} in $K_3[Fe(CN)_6]$(aq), 0.030 mol kg^{-1} in KCl(aq), and 0.050 mol kg^{-1} in NaBr(aq).

10.9 (a) Calculate the masses of (a) $Ca(NO_3)_2$ and, separately, (b) NaCl to add to a 0.150 mol kg^{-1} solution of KNO_3(aq) containing 500 g of solvent to raise its ionic strength to 0.250.

10.9 (b) Calculate the masses of (a) KNO_3 and, separately, (b) $Ba(NO_3)_2$ to add to a 0.110 mol kg^{-1} solution of KNO_3(aq) containing 500 g of solvent to raise its ionic strength to 1.00.

10.10 (a) What molality of $CuSO_4$ has the same ionic strength as 1.00 mol kg^{-1} KCl(aq)?

10.10 (b) What molality of $Al_2(SO_4)_3$ has the same ionic strength as 0.500 mol kg^{-1} $Ca(NO_3)_2$(aq)?

10.11 (a) Express the mean activity coefficient of the ions in a solution of $CaCl_2$ in terms of the activity coefficients of the individual ions.

10.11 (b) Express the mean activity coefficient of the ions in a solution of $Al_2(SO_4)_3$ in terms of the activity coefficients of the individual ions.

10.12 (a) Estimate the mean ionic activity coefficient and activity of a solution that is 0.010 mol kg^{-1} $CaCl_2$(aq) and 0.030 mol kg^{-1} NaF(aq).

10.12 (b) Estimate the mean ionic activity coefficient and activity of a solution that is 0.020 mol kg^{-1} NaCl(aq) and 0.035 mol kg^{-1} $Ca(NO_3)_2$(aq).

10.13 (a) The mean activity coefficient in a 0.500 mol kg^{-1} $LaCl_3$(aq) solution is 0.303 at 25°C. What is the percentage error in the value predicted by the Debye–Hückel limiting law?

10.13 (b) The mean activity coefficient in a 0.100 mol kg^{-1} $CaCl_2$(aq) solution is 0.524 at 25°C. What is the percentage error in the value predicted by the Debye–Hückel limiting law?

10.14 (a) The mean activity coefficients of HBr in three dilute aqueous solutions at 25°C are 0.930 (at 5.0 mmol kg^{-1}), 0.907 (at 10.0 mmol kg^{-1}), and 0.879 (at 20.0 mmol kg^{-1}). Estimate the value of B in the extended Debye–Hückel law.

10.14 (b) The mean activity coefficients of KCl in three dilute aqueous solutions at 25°C are 0.927 (at 5.0 mmol kg^{-1}), 0.902 (at 10.0 mmol kg^{-1}), and 0.816 (at 50.0 mmol kg^{-1}). Estimate the value of B in the extended Debye–Hückel law.

10.15 (a) For CaF_2, $K_s = 3.9 \times 10^{-11}$ at 25°C and the standard Gibbs energy of formation of CaF_2(s) is -1167 kJ mol^{-1}. Calculate the standard Gibbs energy of formation of CaF_2(aq).

10.15 (b) For PbI_2, $K_s = 1.4 \times 10^{-8}$ at 25°C and the standard Gibbs energy of formation of $PbI_2(s)$ is −173.64 kJ mol^{-1}. Calculate the standard Gibbs energy of formation of $PbI_2(aq)$.

10.16 (a) Consider a hydrogen electrode in aqueous HBr solution at 25°C operating at 1.15 atm. Calculate the change in the electrode potential when the molality of the acid is changed from 5.0 mmol kg^{-1} to 20.0 mol kg^{-1}. Activity coefficients are given in Exercise 10.14a.

10.16 (b) Consider a hydrogen electrode in aqueous HCl solution at 25°C operating at 105 kPa. Calculate the change in the electrode potential when the molality of the acid is changed from 5.0 mmol kg^{-1} to 50 mol kg^{-1}. Activity coefficients are given in Table 10.5.

10.17 (a) Devise a cell in which the cell reaction is $Mn(s) + Cl_2(g) \rightarrow MnCl_2(aq)$. Give the half-reactions for the electrodes and from the standard cell emf of 2.54 V deduce the standard potential of the Mn^{2+}/Mn couple.

10.17 (b) Devise a cell in which the cell reaction is $Cd(s) + Ni(OH)_3(s) \rightarrow Cd(OH)_2(s) + Ni(OH)_2(s)$. Give the half-reactions for the electrodes.

10.18 (a) Write the cell reactions and electrode half-reactions for the following cells:

(a) $Zn \,|\, ZnSO_4(aq) \,\|\, AgNO_3(aq) \,|\, Ag$

(b) $Cd \,|\, CdCl_2(aq) \,\|\, HNO_3(aq) \,|\, H_2(g) \,|\, Pt$

(c) $Pt \,|\, K_3[(CN)_6](aq), K_4[(CN)_6](aq) \,\|\, CrCl_3(aq) \,|\, Cr$

10.18 (b) Write the cell reactions and electrode half-reactions for the following cells:

(a) $Pt \,|\, Cl_2(g) \,|\, HCl(aq) \,\|\, K_2CrO_4(aq) \,|\, Ag_2CrO_4(s) \,|\, Ag$

(b) $Pt \,|\, Fe^{3+}(aq), Fe^{2+}(aq) \,\|\, Sn^{4+}(aq), Sn^{2+}(aq) \,|\, Pt$

(c) $Cu \,|\, Cu^{2+}(aq) \,\|\, Mn^{2+}(aq), H^+(aq) \,|\, MnO_2(s) \,|\, Pt$

10.19 (a) Devise cells in which the following are the reactions:

(a) $Zn(s) + CuSO_4(aq) \longrightarrow ZnSO_4(aq) + Cu(s)$

(b) $2\,AgCl(s) + H_2(g) \longrightarrow 2\,HCl(aq) + 2\,Ag(s)$

(c) $2\,H_2(g) + O_2(g) \longrightarrow 2\,H_2O(l)$

10.19 (b) Devise cells in which the following are the reactions:

(a) $2\,Na(s) + 2\,H_2O(l) \longrightarrow 2\,NaOH(aq) + H_2(g)$

(b) $H_2(g) + I_2(g) \longrightarrow 2\,HI(aq)$

(c) $H_3O^+(aq) + OH^-(aq) \longrightarrow H_2O(l)$

10.20 (a) Use standard potentials to calculate the standard emfs of the cells in Exercise 10.18a.

10.20 (b) Use standard potentials to calculate the standard emfs of the cells in Exercise 10.18b.

10.21 (a) Use standard potentials to calculate the standard emfs of the cells in Exercise 10.19a.

10.21 (b) Use standard potentials to calculate the standard emfs of the cells in Exercise 10.19b.

10.22 (a) Calculate the standard Gibbs energies at 25°C of the following reactions from the standard potential data in Table 10.7:

(a) $2\,Na(s) + 2\,H_2O(l) \longrightarrow 2\,NaOH(aq) + H_2(g)$

(b) $2\,K(s) + 2\,H_2O(l) \longrightarrow 2\,KOH(aq) + H_2(g)$

10.22 (b) Calculate the standard Gibbs energies at 25°C of the following reactions from the standard potential data in Table 10.7:

(a) $K_2S_2O_8(aq) + 2\,KI(aq) \longrightarrow I_2(s) + 2\,K_2SO_4(aq)$

(b) $Pb(s) + Zn(NO_3)_2(aq) \longrightarrow Pb(NO_3)_2(aq) + Zn(s)$

10.23 (a) The standard reaction Gibbs energy for $K_2CrO_4(aq) + 2\,Ag(s) + 2\,FeCl_3(aq) \rightarrow Ag_2CrO_4(s) + 2\,FeCl_2(aq) + 2\,KCl(aq)$ is −62.5 kJ mol^{-1} at 298 K. (a) Calculate the standard emf of the corresponding galvanic cell and (b) the standard potential of the $Ag_2CrO_4/Ag, CrO_4^{2-}$ couple.

10.23 (b) Two half-cell reactions may be combined in such a way as to form (a) a new half-cell reaction or (b) a complete cell reaction. Illustrate both (a) and (b) by using the half-cell reactions listed below and calculate $E^\ominus$ for both the new half-cell and complete cell reaction.

(i) $2\,H_2O(l) + 2\,e^- \longrightarrow H_2(g) + 2\,OH^-(aq)$ $\quad E_1^\ominus = -0.828$ V

(ii) $Ag^+(aq) + e^- \longrightarrow Ag(s)$ $\quad E_2^\ominus = +0.799$ V

10.24 (a) Calculate the standard potential of the couple $Ag_2S, H_2O/Ag, S^{2-}, O_2, H^+$ from the following data:

$Ag_2S(s) + 2\,e^- \longrightarrow 2\,Ag(s) + S^{2-}(aq)$ $\quad E^\ominus = -0.69$ V

$O_2(g) + 4\,H^+(aq) + 4\,e^- \longrightarrow 2\,H_2O(l)$ $\quad E^\ominus = +1.23$ V

10.24 (b) Consider the cell $Pt \,|\, H_2(g, p^\ominus) \,|\, HCl(aq) \,|\, AgCl(s) \,|\, Ag$, for which the cell reaction is $2\,AgCl(s) + H_2(g) \rightarrow 2\,Ag(s) + 2\,HCl(aq)$. At 25°C and a molality of HCl of 0.010 mol kg^{-1}, $E = +0.4658$ V. (a) Write the Nernst equation for the cell reaction, (b) Calculate $\Delta_r G$ for the cell reaction. (c) Assuming that the Debye–Hückel limiting law holds at this concentration, calculate $E^\ominus(AgCl, Ag)$.

10.25 (a) Use the Debye–Hückel limiting law and the Nernst equation to estimate the emf of the cell $Ag \,|\, AgBr(s) \,|\, KBr(aq, 0.050\ mol\ kg^{-1}) \,\|\, Cd(NO_3)_2(aq, 0.010\ mol\ kg^{-1}) \,|\, Cd$ at 25°C.

10.25 (b) Use the information in Tables 10.7 to calculate the standard emf of the cell $Ag \,|\, AgNO_3(aq) \,\|\, Fe(NO_3)_2(aq) \,|\, Fe$ and the standard Gibbs energy and enthalpy of the cell reaction at 25°C. Estimate the value of $\Delta_r G^\ominus$ at 35°C.

10.26 (a) Calculate the equilibrium constants of the following reactions at 25°C from standard potential data:

(a) $Sn(s) + Sn^{4+}(aq) \rightleftharpoons 2\,Sn^{2+}(aq)$

(b) $Sn(s) + 2\,AgCl(s) \rightleftharpoons SnCl_2(aq) + 2\,Ag(s)$

10.26 (b) Calculate the equilibrium constants of the following reactions at 25°C from standard potential data:

(a) $Sn(s) + CuSO_4(aq) \rightleftharpoons Cu(s) + SnSO_4(aq)$

(b) $Cu^{2+}(aq) + Cu(s) \rightleftharpoons 2\,Cu^+(aq)$

10.27 (a) Use the standard emfs of the couples $Au^+/Au (+1.69\ V)$, $Au^{3+}/Au (+1.40\ V)$, and $Fe^{3+}/Fe^{2+} (+0.77\ V)$ to calculate $E^\ominus$ and the equilibrium constant for the reaction $2\,Fe^{2+}(aq) + Au^{3+}(aq) \rightleftharpoons 2\,Fe^{3+}(aq) + Au^+(aq)$.

10.27 (b) Determine the standard emf of a cell in which the reaction is $Co^{3+}(aq) + 3\,Cl^-(aq) + 3\,Ag(s) \rightarrow 3\,AgCl(s) + Co(s)$ from the standard potentials of the couples $Ag/AgCl, Cl^- (+0.22\ V)$, $Co^{3+}/Co^{2+} (+1.81\ V)$, and $Co^{2+}/Co (-0.28\ V)$.

10.28 (a) The solubilities of AgCl and $BaSO_4$ in water are 1.34×10^{-5} mol kg^{-1} and 9.51×10^{-4} mol kg^{-1}, respectively, at 25°C. Calculate their solubility constants. Is there any significant difference when activity coefficients are ignored?

10.28 (b) The solubilities of AgI and Bi_2S_3 in water are 1.2×10^{-8} mol kg^{-1} and 1.6×10^{-20} mol kg^{-1}, respectively, at 25°C. Calculate their solubility constants. Is there any significant difference when activity coefficients are ignored?

10.29 (a) Derive an expression for the potential of an electrode for which the half-reaction is the reduction of $Cr_2O_7^{2-}$ ions to Cr^{3+} ions in acidic solution.

10.29 (b) Derive an expression for the potential of an electrode for which the half-reaction is the reduction of MnO_4^- ions to Mn^{2+} ions in acidic solution.

10.30 (a) The emf of the cell $Pt|H_2(g)|HCl(aq)|AgCl(s)|Ag$ is +0.322 V at 25°C. What is the pH of the electrolyte solution?

10.30 (b) The emf of the cell $Pt|H_2(g)|HI(aq)|AgI(s)|Ag$ is 1.00 V at 25°C. What is the pH of the electrolyte solution?

10.31 (a) The solubility of AgBr is 2.6 µmol kg^{-1} at 25°C. What is the emf of the cell $Ag|AgBr(aq)|AgBr(s)|Ag$ at that temperature?

10.31 (b) The solubility of AgI is 12 nmol kg^{-1} at 25°C. What is the emf of the cell $Ag|AgI(aq)|AgI(s)|Ag$ at that temperature?

10.32 (a) The emf of the cell $Ag|AgI(s)|AgI(aq)|Ag$ is +0.9509 V at 25°C. Calculate (a) the solubility of AgI and (b) its solubility constant.

10.32 (b) The emf of the cell $Bi|Bi_2S_3(s)|Bi_2S_3(aq)|Bi$ is +0.96 V at 25°C. Calculate (a) the solubility of Bi_2S_3 and (b) its solubility constant.

Problems[†]

Numerical problems

10.1 Devise a cell in which the overall reaction is $Pb(s) + Hg_2SO_4(s) \rightarrow PbSO_4(s) + 2\,Hg(l)$. What is its potential when the electrolyte is saturated with both salts at 25°C? The solubility constants of Hg_2SO_4 and $PbSO_4$ are 6.6×10^{-7} and 1.6×10^{-8}, respectively.

10.2 Given that $\Delta_r G^{\ominus} = -212.7$ kJ mol^{-1} for the reaction in the Daniell cell at 25°C, and $b(CuSO_4) = 1.0 \times 10^{-3}$ mol kg^{-1} and $b(ZnSO_4) = 3.0 \times 10^{-3}$ mol kg^{-1}, calculate (a) the ionic strengths of the solutions, (b) the mean ionic activity coefficients in the compartments, (c) the reaction quotient, (d) the standard cell emf, and (e) the cell emf. (Take $\gamma_+ = \gamma_- = \gamma_{\pm}$ in the respective compartments.)

10.3 Although the hydrogen electrode may be conceptually the simplest electrode and is the basis for our reference state of electrical potential in electrochemical systems, it is cumbersome to use. Therefore, several substitutes for it have been devised. One of these alternatives is the quinhydrone electrode (quinhydrone, Q·QH$_2$, is a complex of quinone, $C_6H_4O_2 = Q$, and hydroquinone, $C_6H_4O_2H_2 = QH_2$). The electrode half-reaction is $Q(aq) + 2\,H^+(aq) + 2\,e^- \rightarrow QH_2(aq)$, $E^{\ominus} = +0.6994$ V. If the cell $Hg|Hg_2Cl_2(s)|HCl(aq)|Q·QH_2|Au$ is prepared, and the measured cell potential is +0.190 V, what is the pH of the HCl solution? Assume that the Debye–Hückel limiting law is applicable.

10.4 Consider the cell, $Zn(s)|ZnCl_2(0.0050$ mol kg$^{-1})|Hg_2Cl_2(s)\,Hg(l)$, for which the cell reaction is $Hg_2Cl_2(s) + Zn(s) \rightarrow 2\,Hg(l) + 2\,Cl^-(aq) + Zn^{2+}(aq)$. Given that $E^{\ominus}(Zn^{2+}, Zn) = -0.7628$ V, $E^{\ominus}(Hg_2Cl_2, Hg) = +0.2676$ V, and that the emf is +1.2272 V, (a) write the Nernst equation for the cell. Determine (b) the standard emf, (c) $\Delta_r G$, $\Delta_r G^{\ominus}$, and K for the cell reaction, (d) the mean ionic activity and activity coefficient of $ZnCl_2$ from the measured emf, and (e) the mean ionic activity coefficient of $ZnCl_2$ from the Debye–Hückel limiting law. (f) Given that $(\partial E/\partial T)_p = -4.52 \times 10^{-4}$ V K^{-1}, calculate ΔS and ΔH.

10.5 The emf of the cell $Pt|H_2(g, p^{\ominus})|HCl(aq, b)|Hg_2Cl_2(s)|Hg(l)$ has been measured with high precision (G.J. Hills and D.J.G. Ives, *J. Chem. Soc.*, 311 (1951)) with the following results at 25°C:

$b/(mmol\ kg^{-1})$	1.6077	3.0769	5.0403	7.6938	10.9474
E/V	0.600 80	0.568 25	0.543 66	0.522 67	0.505 32

Determine the standard emf of the cell and the mean activity coefficient of HCl at these molalities. (Make a least-squares fit of the data to the best straight line.)

10.6 Careful measurements of the emf of the cell $Pt|H_2(g, p^{\ominus})|$ NaOH(aq, 0.0100 mol kg^{-1}), NaCl(aq, 0.011 25 mol kg^{-1})$|AgCl(s)|Ag$ have been reported (C.P. Bezboruah, M.F.G.F.C. Camoes, A.K. Covington, and J.V. Dobson, *J. Chem. Soc. Faraday Trans. I*, **69**, 949 (1973)). Among the data is the following information:

$\theta/°C$	20.0	25.0	30.0
E/V	1.047 74	1.048 64	1.049 42

Calculate pK_w at these temperatures and the standard enthalpy and entropy of the autoprotolysis of water at 25.0°C.

10.7 Measurements of the emf of cells of the type $Ag|AgX(s)MX(b_1)|M_xHg|MX(b_2)|AgX(s)|Ag$, where M_xHg denotes an amalgam and the electrolyte is an alkali metal halide dissolved in ethylene glycol, have been reported (U. Sen, *J. Chem. Soc. Faraday Trans. I* **69**, 2006 (1973)) and some values for LiCl are given below. Estimate the activity coefficient at the concentration marked * and then use this value to calculate activity coefficients from the measured cell potential at the other concentrations. Base your answer on the following version of the extended Debye–Hückel law:

$$\log \gamma = -\frac{AI^{1/2}}{1 - BI^{1/2}} + kI$$

with $A = 1.461$, $B = 1.70$, $k = 0.20$, and $I = b/b^{\ominus}$. For $b_2 = 0.091\ 41$ mol kg^{-1}:

$b_1/(mol\ kg^{-1})$	0.0555	0.091 41*	0.1652	0.2171	1.040	1.350
E/V	−0.0220	0.0000	0.0263	0.0379	0.1156	0.1336

10.8 Suppose the extended Debye–Hückel law for a 1,1-electrolyte is written in the simplified form $\log \gamma_{\pm} = -0.509I^{1/2} + kI$, where k is a constant and $I = b/b^{\ominus}$. Show that a plot of y against I, where

$$y = E + 0.1183 \log I - 0.0602I^{1/2}$$

should give a straight line with intercept $E^{\ominus}$ and slope −0.1183k. Apply the technique to the following data (at 25°C) on the cell $Pt|H_2(g, p^{\ominus})|HCl(aq, b)|AgCl(s)|Ag$:

$b/(mmol\ kg^{-1})$	123.8	25.63	9.138	5.619	3.215
E/mV	341.99	418.24	468.60	492.57	520.53

† Problems denoted with the symbol ‡ were supplied by Charles Trapp, Carmen Giunta, and Marshall Cady.

(a) Find the standard emf and the standard potential of the AgCl/Ag,Cl⁻ couple. (b) The emf was measured as 352.4 mV when $b = 100.0$ mol kg⁻¹. What is the pH and the mean ionic activity coefficient?

10.9 The mean activity coefficients for aqueous solutions of NaCl at 25°C are given below. Confirm that they support the Debye–Hückel limiting law and that an improved fit is obtained with the extended law.

$b/(\text{mmol kg}^{-1})$	1.0	2.0	5.0	10.0	20.0
$\gamma_\pm$	0.9649	0.9519	0.9275	0.9024	0.8712

10.10 The standard potential of the AgCl/Ag,Cl⁻ couple has been measured very carefully over a range of temperature (R.G. Bates and V.E. Bowers, *J. Res. Nat. Bur. Stand.* **53**, 283 (1954)) and the results were found to fit the expression

$$E^{\ominus}/\text{V} = 0.236\,59 - 4.8564 \times 10^{-4}(\theta/°C) - 3.4205 \times 10^{-6}(\theta/°C)^2$$
$$+ 5.869 \times 10^{-9}(\theta/°C)^3$$

Calculate the standard Gibbs energy and enthalpy of formation of Cl⁻(aq) and its entropy at 298 K.

10.11‡ The list of standard potentials compiled by S.G. Bratsch (*J. Phys. Chem. Ref. Data* **18**, 1 (1989)) shows that europium has the least negative reduction potential for the M³⁺/M couple in strong acid (pH = 0.000) of all the lanthanides, $E^{\ominus}(\text{Eu}^{3+}/\text{Eu}) = -1.991$ V, while the next easiest to reduce is ytterbium, with $E^{\ominus}(\text{Yb}^{3+}/\text{Yb}) = -2.19$ V. (a) Describe the criteria for a reducing agent which would deposit Eu metal from an acidic solution of M³⁺ lanthanide ions while leaving the other lanthanides in solution under standard conditions (unit activities and fugacities; pH = 0). Is there such a reagent in Table 10.7? (b) Now suppose the conditions are not standard. What is the maximum value of the ratio $a_{\text{Yb}^{3+}}/a_{\text{Eu}^{3+}}$ such that Eu still deposits spontaneously (just barely) but Yb does not?

10.12‡ (a) Derive a general relation for $(\partial E/\partial p)_{T,n}$ for electrochemical cells employing reactants in any state of matter. (b) E. Cohen and K. Piepenbroek (*Z. physik. Chem.* **167A**, 365 (1933)) calculated the change in volume for the reaction TlCl(s) + CNS⁻(aq) → TlCNS(s) + Cl⁻ at 30°C from density data and obtained $\Delta_r V = -2.666 \pm 0.080$ cm³ mol⁻¹. They also measured the emf of the cell Tl(Hg)|TlCNS(s)|KCNS:KCl|TlCl|Tl(Hg) at pressures up to 1500 atm. Their results are given in the following table:

p/atm	1.00	250	500	750	1000	1250	1500
E/mV	8.56	9.27	9.98	10.69	11.39	12.11	12.82

From this information, obtain $(\partial E/\partial p)_{T,n}$ at 30°C and compare to the value obtained from $\Delta_r V$. (c) Fit the data to a polynomial for E against p. How constant is $(\partial E/\partial p)_{T,n}$? (d) From the polynomial, estimate an effective isothermal compressibility for the cell as a whole.

10.13‡ Because the hydrogen electrode is the fundamental reference electrode, it is important to study the effect of pressure on cells containing it. W.R. Hainsworth, H.J. Rowley, and D.A. MacInnes (*J. Am. Chem. Soc.* **46**, 1437 (1924)) investigated the influence of pressure on the hydrogen/calomel cell, Pt|H₂(g, p)|HCl(0.1 м)|Hg₂Cl₂|Hg(l), up to pressures of 1000 atm at 25°C with the following results, where $\Delta E = E(p) - E(1\text{ atm})$.

p/atm	1.00	10	38	51	108	210	380	430
$\Delta E/\text{mV}$	0	29.5	47	51	61	69	79	82

p/atm	560	720	900	1020
$\Delta E/\text{mV}$	86	91	95	98

(a) Derive an expression for ΔE using the perfect gas law. Compare the expression with the experimental results. (b) Fit the values of ΔE and $(\partial E/\partial p)_{T,n}$ to a polynomial expression in p. (c) Obtain a theoretical equation for these quantities using both the virial form of the van der Waals equation and the empirical virial equation $Z = 1 + 0.000\,537(p/\text{atm}) + 3.5 \times 10^{-8}(p/\text{atm})^2$ used by Hainsworth *et al.* Plot the experimental and theoretical data for this cell on the same graph. How well do they fit? (d) Calculate the fugacity of H₂ from the cell potential data.

10.14 Use the data below to confirm that the Debye–Hückel limiting law correctly predicts the limiting values of the mean activity coefficient of acetic acid by demonstrating that pK'_a, where $K_a = K'_a K_\gamma$, plotted against $(\alpha b)^{1/2}$, where b is the molality of the acid and α its degree of deprotonation, should be a straight line.

$b/(\text{mmol kg}^{-1})$	0.0280	0.1114	0.2184	1.0283	2.414	5.9115
α	0.5393	0.3277	0.2477	0.1238	0.0829	0.0540

10.15‡ J.W. Larson (*J. Chem. Eng. Data* **40**, 1276 (1995)) studied the thermochemistry of vanadium(V) ions in aqueous solution. At neutral pH, the dominant species are $H_2VO_4^-$ and $V_4O_{12}^{4-}$. The equilibrium constant for the reaction $4\,H_2VO_4^-(\text{aq}) \rightarrow V_4O_{12}^{4-}(\text{aq}) + 4\,H_2O(\text{l})$ was determined to be 106.4 at 50°C. How is vanadium distributed between these two species at 50°C and unit ionic strength if the total vanadium concentration is 0.10 mmol kg⁻¹? Take the ionic strength to be 1.000 and use the following extended Debye–Hückel equation to estimate activity coefficients:

$$\log \gamma = -\frac{Az^2 I^{1/2}}{1 + I^{1/2}}$$

where $A = 0.5373$ at 50°C.

10.16‡ The table below summarizes the emf observed for the cell Pd|H₂(g, 1 bar)|BH(aq, b), B(aq, b)|AgCl(s)|Ag. Each measurement is made at equimolar concentrations of 2-pyridinium chloride (BH) and 2-aminopyridine (B). The data are for 25°C and it is found that $E^{\ominus} = 0.222\,51$ V. Use the data to determine pK_a for the acid at 25°C and the mean activity coefficient ($\gamma_\pm$) of BH as a function of molality (b) and ionic strength (I). Use the extended Debye–Hückel equation for the mean activity coefficient in the form:

$$-\log \gamma_\pm = \frac{AI^{1/2}}{1 + BI^{1/2}} - kb$$

where $A = 0.5091$ and B and k are parameters that depend upon the ions. Draw a graph of the mean activity coefficient with $b = 0.04$ mol kg⁻¹ and $0 \leq I \leq 0.1$.

$b/(\text{mol kg}^{-1})$	0.01	0.02	0.03	0.04	0.05
$E(25°C)/\text{V}$	0.744\,52	0.728\,53	0.719\,28	0.713\,14	0.708\,09

$b/(\text{mol kg}^{-1})$	0.06	0.07	0.08	0.09	0.10
$E(25°C)/\text{V}$	0.703\,80	0.700\,59	0.697\,90	0.695\,71	0.693\,38

Hint. Use mathematical software or a spreadsheet.

10.17‡ The table below summarizes p$K_a(T)$ for BH of Problem 10.16; add your value of p$K_a(25°C)$ calculated in Problem 10.16 to those of the table. Determine the coefficients a_0, a_1, and a_2 such that the regression equation p$K_a(T) = a_0 + a_1/(T/K) + a_2 \ln(T/K)$ gives the best least-squares fit of the data. Draw graphs of $\Delta_r G^{\ominus}(T)$, $\Delta_r H^{\ominus}(T)$, and $\Delta_r S^{\ominus}(T)$ for the proton transfer reaction.

$\theta/°C$	5	10	15	20	25	30	35	40
pK_a	7.177	7.060	6.949	6.841	?	6.640	6.543	6.449

10.18 The $Sb|Sb_2O_3(s)|OH^-(aq)$ electrode is reversible with respect to OH^- ions. Derive an expression for its potential in terms of (a) the pOH and (b) the pH of the solution. By how much does the potential change when the molality of NaOH(aq) in the electrode compartment is increased from 0.010 mol kg^{-1} to 0.050 mol kg^{-1} at 25°C? Use the Debye–Hückel limiting law to estimate any activity coefficients required.

10.19 Superheavy elements are now of considerable interest, particularly because signs of stability are starting to emerge with element 114, which has recently been made. Shortly before it was (falsely) believed that the first superheavy element had been discovered, an attempt was made to predict the chemical properties of ununpentium (Uup, element 115, O.L. Keller, C.W. Nestor, and B. Fricke, *J. Phys. Chem.* **78**, 1945 (1974)). In one part of the paper the standard enthalpy and entropy of the reaction $Uup^+(aq) + \frac{1}{2}H_2(g) \rightarrow Uup(s) + H^+(aq)$ were estimated from the following data: $\Delta_{sub}H^{\ominus}(Uup) = +1.5$ eV, $I(Uup) = 5.52$ eV, $\Delta_{hyd}H^{\ominus}(Uup^+, aq) = -3.22$ eV, $S^{\ominus}(Uup^+, aq) = +1.34$ meV K^{-1}, $S^{\ominus}(Uup, s) = 0.69$ meV K^{-1}. Estimate the expected standard potential of the Uup$^+$/Uup couple.

Theoretical problems

10.20 Show that the solubility, S, of a sparingly soluble 1:1 salt is related to its solubility constant by $S = K_s^{1/2} e^{1.172 S^{1/2}}$.

10.21 Suppose that a sparingly soluble salt MX has solubility constant K_s and solubility S. Show that in an ideal solution that is of concentration C of a freely soluble salt NX, the solubility of MX is changed to

$$S' = \frac{1}{2}(C^2 + 4K_s^2)^{1/2} - \frac{1}{2}C$$

and that $S' \approx K_s/C$ when K_s is small (in a sense to be specified).

10.22 Show that, if the ionic strength of a solution of the sparingly soluble salt MX and the freely soluble salt NX is dominated by the concentration C of the latter, and if it is valid to use the Debye–Hückel limiting law, the solubility S' in the mixed solution is given by

$$S' = \frac{K_s e^{4.606 A C^{1/2}}}{C}$$

when K_s is small (in a sense to be specified).

10.23 Show that the freezing-point depression of a real solution in which the solvent of molar mass M has activity a_A obeys

$$\frac{d \ln a_A}{d(\Delta T)} = -\frac{M}{K_f}$$

and use the Gibbs–Duhem equation to show that

$$\frac{d \ln a_B}{d(\Delta T)} = -\frac{1}{b_B K_f}$$

where a_B is the solute activity and b_B is its molality. Use the Debye–Hückel limiting law to show that the osmotic coefficient (ϕ, Problem 7.21) is given by $\phi = 1 - \frac{1}{3}A'I$ with $A' = 2.303A$ and $I = b/b^{\ominus}$.

Applications: to biology and chemical engineering

10.24 The addition of a small amount of a salt, such as $(NH_4)_2SO_4$, to a solution containing a charged protein increases the protein's solubility in water. This observation is called the *salting-in effect*. However, the addition of large amounts of salt can decrease the solubility of the protein to such an extent that the protein precipitates from solution. This observation is called the *salting-out effect* and is used widely by biochemists to isolate and purify proteins. Consider the equilibrium $PX_\nu(s) \rightleftharpoons P^{\nu+}(aq) + \nu X^-(aq)$, where $P^{\nu+}$ is a polycationic protein of charge $+\nu$ and X^- is its counter ion. Use Le Chatelier's principle and the physical principles behind the Debye–Hückel limiting law to provide a molecular interpretation for the salting-in and salting-out effects.

10.25 The standard potentials of proteins are not commonly measured by the methods described in this chapter because proteins often lose their native structure and function when they react on the surfaces of electrodes. In an alternative method, the oxidized protein is allowed to react with an appropriate electron donor in solution. The standard potential of the protein is then determined from the Nernst equation, the equilibrium concentrations of all species in solution, and the known standard potential of the electron donor. We shall illustrate this method with the protein cytochrome c. The one-electron reaction between cytochrome c, cyt, and 2,6-dichloroindophenol, D, can be followed spectrophotometrically because each of the four species in solution has a distinct colour, or absorption spectrum. We write the reaction as $cyt_{ox} + D_{red} \rightleftharpoons cyt_{red} + D_{ox}$, where the subscripts 'ox' and 'red' refer to oxidized and reduced states, respectively. (a) Consider $E_{cyt}^{\ominus}$ and $E_D^{\ominus}$ to be the standard potentials of cytochrome c and D, respectively. Show that, at equilibrium('eq'), a plot of $\ln([D_{ox}]_{eq}/[D_{red}]_{eq})$ versus $\ln([cyt_{ox}]_{eq}/[cyt_{red}]_{eq})$ is linear with slope of one and y-intercept $F(E_{cyt}^{\ominus} - E_D^{\ominus})/RT$, where equilibrium activities are replaced by the numerical values of equilibrium molar concentrations. (b) The following data were obtained for the reaction between oxidized cytochrome c and reduced D in a pH 6.5 buffer at 298 K. The ratios $[D_{ox}]_{eq}/[D_{red}]_{eq}$ and $[cyt_{ox}]_{eq}/[cyt_{red}]_{eq}$ were adjusted by titrating a solution containing oxidized cytochrome c and reduced D with a solution of sodium ascorbate, which is a strong reductant. From the data and the standard potential of D of 0.237 V, determine the standard potential cytochrome c at pH 6.5 and 298K.

$[D_{ox}]_{eq}/[D_{red}]_{eq}$	0.002 79	0.008 43	0.0257	0.0497
$[cyt_{ox}]_{eq}/[cyt_{red}]_{eq}$	0.0106	0.0230	0.0894	0.197

$[D_{ox}]_{eq}/[D_{red}]_{eq}$	0.0748	0.238	0.534
$[cyt_{ox}]_{eq}/[cyt_{red}]_{eq}$	0.335	0.809	1.39

10.26 Sodium fluoride is routinely added to public water supplies because it is known that fluoride ion can prevent tooth decay. In a fluoride-selective electrode used in the analysis of water samples a crystal of LaF$_3$ doped with Eu^{2+}, denoted as Eu^{2+}:LaF$_3$, provides a semipermeable barrier between the test solution and the solution inside the electrode (the filling solution), which contains 0.1 mol kg^{-1} NaF(aq) and 0.1 mol kg^{-1} NaCl(aq). A silver–silver chloride electrode immersed in the filling solution is connected to a potentiometer and the emf of the cell can be measured against an appropriate reference electrode. It follows that the half-cell for a fluoride-selective electrode is represented by

$$Ag(s)|AgCl(s)|NaCl(aq, b_1), NaF(aq, b_1)|Eu^{2+}:LaF_3(s)|F^-(aq, b_2)$$

where b_1 and b_2 are the molalities of fluoride ion in the filling and test solutions, respectively. (a) Derive an expression for the emf of this half-cell. (b) The fluoride-selective electrode just described is not sensitive to HF(aq). Hydroxide ion is the only interfering species, with $k_{F^-, OH^-} = 0.1$. Use this information and the fact that K_a of HF is 3.5×10^{-4} at 298 K to specify a range of pH values in which the electrode responds accurately to the activity of F$^-$ in the test solution at 298 K.

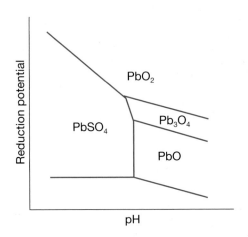

Reduction potential

PbO$_2$

PbSO$_4$

Pb$_3$O$_4$

PbO

pH

Fig. 10.17

10.27‡ A cell of the lead–acid battery may be represented as Pb(s)|PbSO$_4$(s)|H$_2$SO$_4$(aq)|PbO$_2$(s)|PbSO$_4$(s)|Pb(s) where the aqueous solution is about 35 per cent by mass sulfuric acid. The sulfuric

acid molality, b, is related to the solution density, d, at 25°C by the empirical equation:

$$b(H_2SO_4)/(\text{mol kg}^{-1}) = a(d - d_{25})/(\text{g cm}^{-3}) + c(d - d_{25})^2/(\text{g cm}^{-3})^2$$

where d_{25} is the density of pure water at 25°C, $a = 14.523$, and $c = 25.031$. (a) Draw graphs of molality, mass percentage, and molar concentration functions of density for densities in the range 1.0–1.4 g cm^{-3}. (b) Write balanced equations for the half-reactions and the cell reaction for the lead–acid battery. Determine $E^{\ominus}$, $\Delta_r G^{\ominus}$, $\Delta_r S^{\ominus}$, and $\Delta_r H^{\ominus}$ for the cell reaction at 25°C. Estimate $E^{\ominus}(15°C)$. Calculate the cell emf at 25°C when $Q = 6.0 \times 10^{-5}$. (c) Figure 10.17 shows a simplified 'Pourbaix diagram' for the redox chemistry of PbO$_2$/PbSO$_4$/Pb systems. The horizontal and slanted lines of the diagram are equilibrium half-cell potentials when all chemical species have unit activity. Vertical lines are non-redox equilibria. Prove that the redox potential lines at 25°C are given by $E = E^{\ominus} - (0.592 \text{ V}) (v_H/v)$ pH, where v_H is the number of hydrogen ions used in the half-reaction. Calculate $E(\text{PbO}_2/\text{PbSO}_4)$ at pH values of 5 and 8. What is $E(\text{PbSO}_4/\text{Pb})$ at any pH?

10.28 Fuel cells provide electrical power for spacecraft (as in the NASA space shuttles) and also show promise as power sources for automobiles. A fuel cell develops an electric potential from the chemical reaction between reagents supplied from an outside source. What is the emf of a cell fuelled by (a) hydrogen and oxygen, (b) the combustion of butane at 1.0 bar and 298 K?

Part 2

Structure

In Part 1 we examined the properties of bulk matter from the viewpoint of thermodynamics. In Part 2 we turn to the study of individual atoms and molecules from the viewpoint of quantum mechanics. The two viewpoints merge in Chapter 19.

11 **Quantum theory: introduction and principles**

12 **Quantum theory: techniques and applications**

13 **Atomic structure and atomic spectra**

14 **Molecular structure**

15 **Molecular symmetry**

16 **Spectroscopy 1: rotational and vibrational spectra**

17 **Spectroscopy 2: electronic transitions**

18 **Spectroscopy 3: magnetic resonance**

19 **Statistical thermodynamics: the concepts**

20 **Statistical thermodynamics: the machinery**

21 **Molecular interactions**

22 **Macromolecules and aggregates**

23 **The solid state**

11

Quantum theory: introduction and principles

The origins of quantum mechanics

11.1 The failures of classical physics

11.2 Wave–particle duality

The dynamics of microscopic systems

11.3 The Schrödinger equation

11.4 The Born interpretation of the wavefunction

Quantum mechanical principles

11.5 The information in a wavefunction

11.6 The uncertainty principle

Checklist of key ideas

Further reading

Exercises

Problems

This chapter introduces some of the basic principles of quantum mechanics. First, it reviews the experimental results that overthrew the concepts of classical physics. These experiments led to the conclusion that particles may not have an arbitrary energy and that the classical concepts of 'particle' and 'wave' blend together. The overthrow of classical mechanics inspired the formulation of a new set of concepts and led to the formulation of quantum mechanics. In quantum mechanics, all the properties of a system are expressed in terms of a wavefunction that is obtained by solving the Schrödinger equation. We see how to interpret wavefunctions. Finally, we introduce some of the techniques of quantum mechanics in terms of operators, and see that they lead to the uncertainty principle, one of the most profound departures from classical mechanics.

To understand the structures of individual atoms and molecules, we need to know how subatomic particles move in response to the forces they experience. It was once thought that the motion of atoms and subatomic particles could be expressed using **classical mechanics**, the laws of motion introduced in the seventeenth century by Isaac Newton, for these laws were very successful at explaining the motion of everyday objects and planets. However, towards the end of the nineteenth century, experimental evidence accumulated showing that classical mechanics failed when it was applied to very small particles, and it took until the 1920s to discover the appropriate concepts and equations for describing them. We describe the concepts of this new mechanics, which is called **quantum mechanics**, in this chapter, and apply them throughout the remainder of the text.

The origins of quantum mechanics

The basic principles of classical mechanics are reviewed in *Further information 2*. In brief, they show that classical physics: (1) predicts a precise trajectory for particles, with precisely specified locations and momenta at each instant, and (2) allows the translational, rotational, and vibrational modes of motion to be excited to any energy simply by controlling the forces that are applied. These conclusions agree with everyday experience. Everyday experience, however, does not extend to individual atoms, and careful experiments of the type described below have shown that

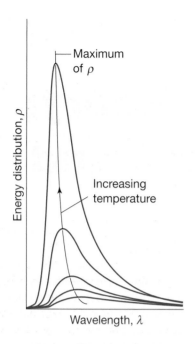

11.1 The energy distribution in a black-body cavity at several temperatures. Note how the energy density increases in the visible region as the temperature is raised, and how the peak shifts to shorter wavelengths. The total energy density (the area under the curve) increases as the temperature is increased (as T^4).

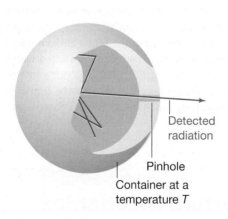

11.2 An experimental representation of a black body is a pinhole in an otherwise closed container. The radiation is reflected many times within the container and comes to thermal equilibrium with the walls at a temperature T. Radiation leaking out through the pinhole is characteristic of the radiation within the container.

classical mechanics fails when applied to the transfers of very small quantities of energy and to objects of very small mass.

11.1 The failures of classical physics

In this section we review some of the experimental evidence that showed that several concepts of classical mechanics are untenable. In particular, we shall see that observations of black-body radiation, heat capacities, and atomic and molecular spectra indicate that systems can take up energy only in discrete amounts.

(a) Black-body radiation

A hot object emits electromagnetic radiation. At high temperatures, an appreciable proportion of the radiation is in the visible region of the spectrum, and a higher proportion of short-wavelength blue light is generated as the temperature is raised. This behaviour is seen when a heated iron bar glowing red hot becomes white hot when heated further. The dependence is illustrated in Fig. 11.1, which shows how the energy output varies with wavelength at several temperatures. The curves are those of an ideal emitter called a **black body**, which is an object capable of emitting and absorbing all frequencies of radiation uniformly. A good approximation to a black body is a pinhole in an empty container maintained at a constant temperature, because any radiation leaking out of the hole has been absorbed and re-emitted inside so many times that it has come to thermal equilibrium with the walls (Fig. 11.2).

Figure 11.1 shows that the peak in the energy output shifts to shorter wavelengths as the temperature is raised. As a result, the short-wavelength tail of the energy distribution strengthens in the visible region and the perceived colour shifts towards the blue, as already mentioned. An analysis of the data led Wilhelm Wien (in 1893) to formulate the **Wien displacement law**:

$$T\lambda_{max} = \tfrac{1}{5}c_2 \qquad c_2 = 1.44 \text{ cm K} \tag{11.1}$$

where λ_{max} is the wavelength corresponding to the maximum of the distribution at a temperature T. The constant c_2 is called the **second radiation constant**. Using its value, we can predict that $\lambda_{max} \approx 2900$ nm at 1000 K.

A second feature of black-body radiation had been noticed in 1879 by Josef Stefan, who considered the total energy density, $\mathcal{E}$, the total electromagnetic energy in a region divided by the volume of the region ($\mathcal{E} = E/V$). The energy density of the electromagnetic field inside the container in Fig. 11.2 increases as the temperature is increased and, specifically, the **Stefan–Boltzmann law** states that

$$\mathcal{E} = aT^4 \tag{11.2a}$$

Ludwig Boltzmann's name is attached to this law because he explained it theoretically. An alternative form of the law is in terms of the **excitance**, M, the power emitted by a region of surface divided by the area of the surface: the excitance is a measure of the brightness of the emission.[1] Because the excitance is proportional to the energy density in the container, M is also proportional to T^4, and we can write

$$M = \sigma T^4 \qquad \sigma = 56.7 \text{ nW m}^{-2} \text{ K}^{-4} \tag{11.2b}$$

The constant σ is called the **Stefan–Boltzmann constant**. The Stefan–Boltzmann law implies that 1 cm^2 of the surface of a black body at 1000 K radiates about 6 W when all wavelengths of the emitted radiation are taken into account.

---

1. Power is the rate of supply of energy. Its SI units are watts, W: 1 W = 1 J s^{-1}.

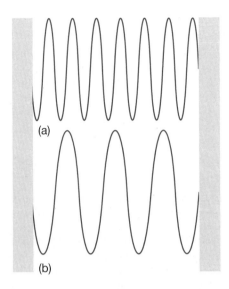

11.3 The electromagnetic vacuum can be regarded as able to support oscillations of the electromagnetic field. When a high-frequency, short-wavelength oscillator (a) is excited, that frequency of radiation is present. The presence of low-frequency, long-wavelength radiation (b) signifies that an oscillator of the corresponding frequency has been excited.

11.4 The Rayleigh–Jeans law (eqn 3) predicts an infinite energy density at short wavelengths. This approach to infinity is called the ultraviolet catastrophe.

The explanation of black-body radiation was a major challenge for nineteenth-century scientists, and in due course it was found to be beyond the capabilities of classical physics. The physicist Lord Rayleigh studied it theoretically from a classical viewpoint, and thought of the electromagnetic field as a collection of oscillators of all possible frequencies. He regarded the presence of radiation of frequency v (and therefore of wavelength $\lambda = c/v$, where c is the speed of light) as signifying that the electromagnetic oscillator of that frequency had been excited (Fig. 11.3). Rayleigh used the equipartition principle (Section 2.2) to calculate the average energy of each oscillator as kT. Then, with minor help from James Jeans, he arrived at the **Rayleigh–Jeans law**:[2]

$$\mathrm{d}\mathcal{E} = \rho \, \mathrm{d}\lambda \qquad \rho = \frac{8\pi kT}{\lambda^4} \tag{11.3}$$

where ρ is the proportionality constant between $\mathrm{d}\lambda$ and the energy density in that range of wavelengths; k is Boltzmann's constant ($k = 1.381 \times 10^{-23} \,\mathrm{J\,K^{-1}}$).

Unfortunately (for Rayleigh, Jeans, and classical physics), although the Rayleigh–Jeans law is quite successful at long wavelengths (low frequencies), it fails badly at short wavelengths (high frequencies). Thus, as λ decreases, ρ increases without going through a maximum (Fig. 11.4). The equation therefore predicts that oscillators of very short wavelength (corresponding to ultraviolet light, X-rays, and even γ-rays) are strongly excited even at room temperature. This absurd result, which implies that a large amount of energy is radiated in the high-frequency region of the electromagnetic spectrum, is called the **ultraviolet catastrophe**. According to classical physics, even cool objects should radiate in the visible and ultraviolet regions: according to classical physics, objects should glow in the dark; there should in fact be no darkness.

(b) The Planck distribution

The German physicist Max Planck studied black-body radiation from the viewpoint of thermodynamics. In 1900 he found that he could account for the experimental observations by proposing that the energy of each electromagnetic oscillator is limited to discrete values and cannot be varied arbitrarily. This proposal is quite contrary to the viewpoint of classical physics (on which the equipartition principle used by Rayleigh is based), in which all possible energies are allowed. The limitation of energies to discrete values is called the **quantization of energy**. In particular, Planck found that he could account for the observed distribution of energy if he supposed that the permitted energies of an electromagnetic oscillator of frequency v are integer multiples of hv:

$$E = nhv \qquad n = 0, 1, 2, \ldots \tag{11.4}$$

where h is a fundamental constant now known as **Planck's constant**.

On the basis of this assumption, Planck was able to derive the **Planck distribution**:[3]

$$\mathrm{d}\mathcal{E} = \rho \, \mathrm{d}\lambda \qquad \rho = \frac{8\pi hc}{\lambda^5 (e^{hc/\lambda kT} - 1)} \tag{11.5}$$

This expression fits the experimental curve very well at all wavelengths (Fig. 11.5), and the value of h, which is an undetermined parameter in the theory, may be obtained by varying its value until a best fit is obtained. The currently accepted value for h is $6.626\,08 \times 10^{-34}\,\mathrm{J\,s}$.

2. For references to the derivation of this expression, see *Further reading*.
3. For references to the derivation of this expression, see *Further reading*.

The Planck distribution resembles the Rayleigh–Jeans law (eqn 3) apart from the all-important exponential factor in the denominator. For short wavelengths, $hc/\lambda kT \gg 1$ and $e^{hc/\lambda kT} \to \infty$ faster than $\lambda^5 \to 0$; therefore $\rho \to 0$ as $\lambda \to 0$ or $\nu \to \infty$. Hence, the energy density approaches zero at high frequencies, in agreement with observation. For long wavelengths, $hc/\lambda kT \ll 1$, and the denominator in the Planck distribution can be replaced by

$$e^{hc/\lambda kT} - 1 = \left(1 + \frac{hc}{\lambda kT} + \cdots\right) - 1 \approx \frac{hc}{\lambda kT}$$

When this approximation is substituted into eqn 5, we find that the Planck distribution reduces to the Rayleigh–Jeans law.

The Planck distribution also accounts for the Stefan–Boltzmann and Wien laws. The former is obtained by integrating the energy density over all wavelengths from $\lambda = 0$ to $\lambda = \infty$, which gives

$$\mathcal{E} = \int_0^\infty \rho \, d\lambda = aT^4 \qquad a = \frac{4\sigma}{c}, \qquad \sigma = \frac{2\pi^5 k^4}{15c^2 h^3} \tag{11.6}$$

Substitution of the values of the fundamental constants gives $\sigma = 56.704$ nW m^{-2} K^{-4}, in accord with the experimental value. The Wien law is obtained by looking for the wavelength at which $d\rho/d\lambda = 0$, the condition for the maximum in the distribution. When we take the derivative, set it equal to zero, and make the approximation that the wavelength is so short that $\lambda \ll hc/kT$, we obtain

$$T\lambda_{max} = \frac{hc}{5k} \tag{11.7}$$

This result lets us identify the second radiation constant as $c_2 = hc/k = 1.439$ cm K, which is also in good agreement with experiment.

It is quite easy to see why Rayleigh's approach was unsuccessful and Planck's hypothesis was successful. The thermal motion of the atoms in the walls of the black body excites the oscillators of the electromagnetic field. According to classical mechanics, all the oscillators of the field share equally in the energy supplied by the walls, so even the highest frequencies are excited. The excitation of very high frequency oscillators results in the ultraviolet catastrophe. According to Planck's hypothesis, however, oscillators are excited only if they can acquire an energy of at least $h\nu$. This energy is too large for the walls to supply in the case of the very high frequency oscillators, so the latter remain unexcited. The effect of quantization is to reduce the contribution from the high-frequency oscillators, for they cannot be significantly excited with the energy available.

(c) Heat capacities

In the early nineteenth century, the French scientists Pierre-Louis Dulong and Alexis-Thérèse Petit determined the heat capacities of a number of monatomic solids.[4] On the basis of some somewhat slender experimental evidence, they proposed that the molar heat capacities of all monatomic solids are the same, and close to 25 J K^{-1} mol^{-1} (in modern units).

Dulong and Petit's law is easy to justify in terms of classical physics. If classical physics were valid, the equipartition principle could be used to calculate the heat capacity of a solid. According to this principle, the mean energy of an atom as it

---

4. As explained in Section 2.4b, the constant-volume heat capacity is defined as $C_V = (\partial U/\partial T)_V$. A small heat capacity indicates that a large rise in temperature results from a given transfer of energy.

Calculations involving exponentials often become simpler if we note that $e^x = 1 + x + \frac{1}{2}x^2 + \cdots$. If $x \ll 1$, a good approximation is $e^x \approx 1 + x$.

To derive this result, we make use of the following standard integral:

$$\int_0^\infty \frac{dx}{x^5(e^{1/x} - 1)} = \frac{\pi^4}{15}$$

11.5 The Planck distribution (eqn 5) accounts very well for the experimentally determined distribution of radiation. Planck's quantization hypothesis essentially quenches the contributions of high-frequency, short-wavelength oscillators. The distribution coincides with the Rayleigh–Jeans distribution at long wavelengths.

oscillates about its mean position in a solid is kT for each direction of displacement. As each atom can oscillate in three dimensions, the average energy of each atom is $3kT$; for N atoms the total energy is $3NkT$. The contribution of this motion to the molar internal energy is therefore

$$U_m = 3N_A kT = 3RT$$

because $N_A k = R$, the gas constant. The molar constant-volume heat capacity (eqn 2.19) is then predicted to be

$$C_{V,m} = \left(\frac{\partial U_m}{\partial T}\right)_V = 3R \tag{11.8}$$

This result, with $3R = 24.9 \text{ J K}^{-1} \text{ mol}^{-1}$, is in striking accord with Dulong and Petit's value.

Significant deviations from Dulong and Petit's law were observed when technological advances made it possible to measure heat capacities at low temperatures. It was found that the molar heat capacities of all metals are lower than $3R$ at low temperatures, and that the values approach zero as $T \to 0$. To account for these observations, Einstein (in 1905) assumed that each atom oscillated about its equilibrium position with a single frequency v. He then invoked Planck's hypothesis to assert that the energy of oscillation is confined to discrete values, and specifically to nhv, where n is an integer. Einstein first calculated the contribution of the oscillations of the atoms to the total molar energy of the metal (by a method described in Section 20.4) and obtained

$$U_m = \frac{3N_A hv}{e^{hv/kT} - 1}$$

in place of the classical expression $3RT$. Then he found the heat capacity by differentiating U_m with respect to T. The resulting expression is now known as the **Einstein formula**:

$$C_{V,m} = 3Rf^2 \qquad f = \frac{\theta_E}{T}\left(\frac{e^{\theta_E/2T}}{e^{\theta_E/T} - 1}\right) \tag{11.9}$$

The **Einstein temperature**, $\theta_E = hv/k$, is a way of expressing the frequency of oscillation of the atoms as a temperature: a high frequency corresponds to a high Einstein temperature.

At high temperatures (when $T \gg \theta_E$) the exponentials in f can be expanded as $1 + \theta_E/T + \cdots$ and higher terms ignored. The result is

$$f = \frac{\theta_E}{T}\left\{\frac{1 + \theta_E/2T + \cdots}{(1 + \theta_E/T + \cdots) - 1}\right\} \approx 1 \tag{11.10a}$$

Consequently, the classical result ($C_{V,m} = 3R$) is obtained at high temperatures. At low temperatures, when $T \ll \theta_E$,

$$f \approx \frac{\theta_E}{T}\left(\frac{e^{\theta_E/2T}}{e^{\theta_E/T}}\right) = \frac{\theta_E}{T}e^{-\theta_E/2T} \tag{11.10b}$$

The strongly decaying exponential function goes to zero more rapidly than $1/T$ goes to infinity; so $f \to 0$ as $T \to 0$, and the heat capacity therefore approaches zero too. We see that Einstein's formula accounts for the decrease of heat capacity at low temperatures. The physical reason for this success is that at low temperatures only a few oscillators possess enough energy to oscillate significantly. At higher temperatures, there is enough energy available for all the oscillators to become active: all $3N$ oscillators contribute, and the heat capacity approaches its classical value.

11.6 Experimental low-temperature molar heat capacities and the temperature dependence predicted on the basis of Einstein's theory. His equation (eqn 9) accounts for the dependence fairly well, but is everywhere too low.

11.7 Debye's modification of Einstein's calculation (eqn 11) gives very good agreement with experiment. For copper, $T/\theta_D = 2$ corresponds to about 170 K, so the detection of deviations from Dulong and Petit's law had to await advances in low-temperature physics.

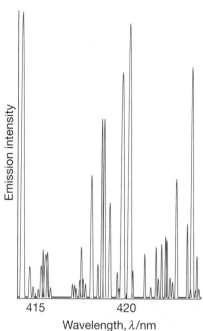

11.8 A region of the spectrum of radiation emitted by excited iron atoms consists of radiation at a series of discrete wavelengths (or frequencies).

11.9 When a molecule changes its state, it does so by absorbing radiation at definite frequencies. This spectrum is part of that due to the electronic, vibrational, and rotational excitation of sulfur dioxide (SO_2) molecules. This observation suggests that molecules can possess only discrete energies, not an arbitrary energy.

Figure 11.6 shows the temperature dependence of the heat capacity predicted by the Einstein formula. The general shape of the curve is satisfactory, but the numerical agreement is in fact quite poor. The poor fit arises from Einstein's assumption that all the atoms oscillate with the same frequency, whereas in fact they oscillate over a range of frequencies from zero up to a maximum value, ν_D. This complication is taken into account by averaging over all the frequencies present, the final result being the **Debye formula**:[5]

$$C_{V,m} = 3Rf \qquad f = 3\left(\frac{T}{\theta_D}\right)^3 \int_0^{\theta_D/T} \frac{x^4 e^x}{(e^x - 1)^2}\, dx \qquad (11.11)$$

where $\theta_D = h\nu_D/k$ is the **Debye temperature**. The integral in eqn 11 has to be evaluated numerically, but that is simple with mathematical software. The details of this modification, which, as Fig. 11.7 shows, gives improved agreement with experiment, need not distract us at this stage from the main conclusion, which is that quantization must be introduced in order to explain the thermal properties of solids.

(d) Atomic and molecular spectra

The most compelling evidence for the quantization of energy comes from the observation of the frequencies of radiation absorbed and emitted by atoms and molecules.

A typical atomic spectrum is shown in Fig. 11.8, and a typical molecular spectrum is shown in Fig. 11.9. The obvious feature of both is that radiation is emitted or absorbed at a series of discrete frequencies. This observation can be understood if the energy of the atoms or molecules is also confined to discrete values, for then

5. For the derivation of this expression, see the references in *Further reading*.

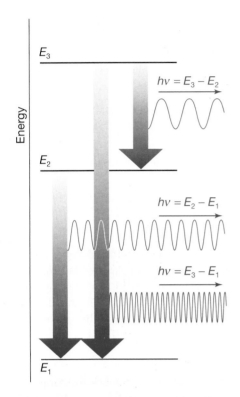

11.10 Spectral lines can be accounted for if we assume that a molecule emits a photon as it changes between discrete energy levels. Note that high-frequency radiation is emitted when the energy change is large.

energy can be discarded or absorbed only in discrete amounts (Fig. 11.10). Then, if the energy of an atom decreases by ΔE, the energy is carried away as radiation of frequency $v = \Delta E/h$, and a line appears in the spectrum. We develop the principles and applications of atomic spectroscopy in Chapter 13 and of molecular spectroscopy in Chapters 16–18.

11.2 Wave–particle duality

At this stage we have established that the energies of the electromagnetic field and of oscillating atoms are quantized. In this section we shall see the experimental evidence that led to the revision of two other basic concepts concerning natural phenomena. One experiment shows that electromagnetic radiation—which classical physics treats as wave-like—actually also displays the characteristics of particles. Another experiment shows that electrons—which classical physics treats as particles —also display the characteristics of waves.

(a) The particle character of electromagnetic radiation

The observation that electromagnetic radiation of frequency v can possess only the energies $0, hv, 2hv, \ldots$ suggests that it can be thought of as consisting of $0, 1, 2, \ldots$ particles, each particle having an energy hv. Then, if one of these particles is present, the energy is hv, if two are present the energy is $2hv$, and so on. These particles of electromagnetic radiation are now called **photons**. The observation of discrete spectra from atoms and molecules can be pictured as the atom or molecule generating a photon of energy hv when it discards an energy of magnitude ΔE, with $\Delta E = hv$.

Example 11.1 Calculating the number of photons

Calculate the number of photons emitted by a 100 W yellow lamp in 1.0 s. Take the wavelength of yellow light as 560 nm and assume 100 per cent efficiency.

Method Each photon has an energy hv, so the total number of photons needed to produce an energy E is E/hv. To use this equation, we need to know the frequency of the radiation (from $v = c/\lambda$) and the total energy emitted by the lamp. The latter is given by the product of the power (P, in watts) and the time interval for which the lamp is turned on ($E = P\Delta t$). In general, to avoid rounding and other numerical errors, it is best to carry out algebraic calculations first, and to substitute numerical values into a single, final formula.

Answer The number of photons is

$$N = \frac{E}{hv} = \frac{P\Delta t}{h(c/\lambda)} = \frac{\lambda P\Delta t}{hc}$$

Substitution of the data gives

$$N = \frac{(5.60 \times 10^{-7}\text{ m}) \times (100\text{ J s}^{-1}) \times (1.0\text{ s})}{(6.626 \times 10^{-34}\text{ J s}) \times (2.998 \times 10^{8}\text{ m s}^{-1})} = 2.8 \times 10^{20}$$

Comment Note that it would take nearly 40 min to produce 1 mol of these photons.

Self-test 11.1 How many photons does a monochromatic (single-frequency) infrared rangefinder of power 1 mW and wavelength 1000 nm emit in 0.1 s?

$[5 \times 10^{14}]$

11.11 In the photoelectric effect, it is found that no electrons are ejected when the incident radiation has a frequency below a value characteristic of the metal, and above that value, the kinetic energy of the photoelectrons varies linearly with the frequency of the incident radiation.

Further evidence for the particle-like character of radiation comes from the measurement of the energies of electrons produced in the **photoelectric effect**. This effect is the ejection of electrons from metals when they are exposed to ultraviolet radiation. The experimental characteristics of the photoelectric effect are as follows:

1 No electrons are ejected, regardless of the intensity of the radiation, unless its frequency exceeds a threshold value characteristic of the metal.

2 The kinetic energy of the ejected electrons increases linearly with the frequency of the incident radiation but is independent of the intensity of the radiation.

3 Even at low light intensities, electrons are ejected immediately if the frequency is above threshold.

Figure 11.11 illustrates the first and second characteristics.

These observations strongly suggest that the photoelectric effect depends on the ejection of an electron when it is involved in a collision with a particle-like projectile that carries enough energy to eject the electron from the metal. If we suppose that the projectile is a photon of energy $h\nu$, where ν is the frequency of the radiation, then the conservation of energy requires that the kinetic energy of the ejected electron should obey

$$\tfrac{1}{2}m_e\upsilon^2 = h\nu - \varPhi \tag{11.12}$$

In this expression $\varPhi$ is a characteristic of the metal called its **work function**, the energy required to remove an electron from the metal to infinity (Fig. 11.12). Photoejection cannot occur if $h\nu < \varPhi$ because the photon brings insufficient energy: this conclusion accounts for observation (1). Equation 12 predicts that the kinetic energy of an ejected electron should increase linearly with frequency, in agreement with observation (2). When a photon collides with an electron, it gives up all its energy, so we should expect electrons to appear as soon as the collisions begin, provided the photons have sufficient energy: this conclusion agrees with observation (3).

11.12 The photoelectric effect can be explained if it is supposed that the incident radiation is composed of photons that have energy proportional to the frequency of the radiation. (a) The energy of the photon is insufficient to drive an electron out of the metal. (b) The energy of the photon is more than enough to eject an electron, and the excess energy is carried away as the kinetic energy of the photoelectron (the ejected electron).

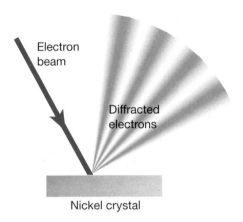

Electron beam

Diffracted electrons

Nickel crystal

11.13 The Davisson–Germer experiment. The scattering of an electron beam from a nickel crystal shows a variation of intensity characteristic of a diffraction experiment in which waves interfere constructively and destructively in different directions.

(b) The wave character of particles

Although contrary to the long-established wave theory of light, the view that light consists of particles had been held before, but discarded. No significant scientist, however, had taken the view that matter is wave-like. Nevertheless, experiments carried out in 1925 forced people to consider that possibility. The crucial experiment was performed by the American physicists Clinton Davisson and Lester Germer, who observed the diffraction of electrons by a crystal (Fig. 11.13). Diffraction is a characteristic property of waves because it occurs when there is interference between their peaks and troughs. Depending on whether the interference is constructive or destructive, the result is a region of enhanced or diminished intensity. Davisson and Germer's success was a lucky accident, because a chance rise of temperature caused their polycrystalline sample to anneal, and the ordered planes of atoms then acted as a diffraction grating. At almost the same time, G.P. Thomson, working in Scotland, showed that a beam of electrons was diffracted when passed through a thin gold foil. Electron diffraction is now the basis of a useful structural technique (Box 11.1).

The Davisson–Germer experiment, which has since been repeated with other particles (including molecular hydrogen), shows clearly that particles have wave-like properties. We have also seen that waves of electromagnetic radiation have particle-like properties. Thus we are brought to the heart of modern physics. When examined on an atomic scale, the classical concepts of particle and wave melt together, particles taking on the characteristics of waves, and waves the characteristics of particles.

Box 11.1 Electron diffraction

The wave-like character of electrons is the basis of the structural technique of *electron diffraction*, in which the diffraction pattern generated by interference when a beam of electrons passes through a sample is interpreted in terms of the locations of the atoms. *Electron microscopy* also makes use of the wave character of electrons, but is analogous to optical microscopy in that the image of an object is constructed from the pattern of scattering of a focused beam of electrons.

Electrons are scattered strongly by their interaction with the charges of electrons and nuclei, and so until recently could not be used to study the interiors of solid samples. However, they have been used for some time to study molecules in the gas phase, on surfaces, and in thin films. The application to surfaces, which is called *low energy electron diffraction* (LEED), is a major use of the technique and is discussed in Section 28.2e. Recent developments have extended electron diffraction techniques to solids, where they have certain advantages over X-ray diffraction (Chapter 23). For instance, they are applicable to very small samples, and so may be used when single-crystal X-ray diffraction is impractical or powder diffraction too complex to interpret. A sample size of about 10^4 unit cells can be used for electron diffraction studies on solids, which is several million times smaller than for X-ray crystallography.

In a typical gas-phase electron diffraction apparatus, electrons are emitted from a hot filament and are then accelerated through a potential gradient. They then pass through the stream of gas, and on to a fluorescent screen. The wavelength of electrons accelerated through a potential difference V is

$$\lambda = \frac{h}{(2m_e eV)^{1/2}}$$

(see Example 11.2). For an accelerating potential difference of 40 kV, the wavelength is 6.1 pm.

The gaseous sample presents all possible orientations of atom–atom separations to the electron beam. The resulting diffraction pattern consists of a series of concentric undulations on a background with an intensity that decreases steadily with increasing scattering angle (see the first illustration). The undulations are ascribed to *molecular scattering*, the sharply defined scattering from the nuclear positions. The background is due largely to *atomic scattering*. One way of levelling the total intensity and hence to emphasize the undulations is to insert a rotating heart-shaped disk in front of the screen (see the second illustration).

The scattering from a pair of nuclei separated by a distance R_{ij} and orientated at a definite angle to the incident beam can be calculated. The overall diffraction pattern is then calculated by allowing for all possible orientations of this pair of atoms. When the molecule consists of a number of atoms, we sum over the contribution from all pairs, and find that

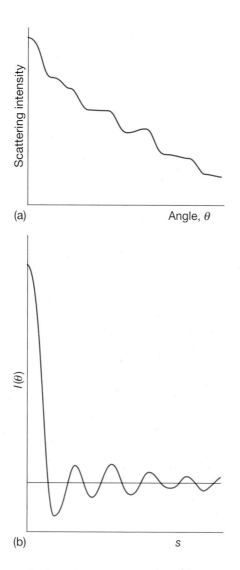

(a) Scattering intensity vs Angle, θ

(b) $I(\theta)$ vs s

The layout of an electron diffraction apparatus. The diffraction pattern is photographed from the fluorescent screen. A rotating heart-shaped sector emphasizes the scattering from the nuclear positions and suppresses the smoothly varying background due to scattering from the continuous electron distribution in the molecules.

(a) The scattering intensity consists of a smoothly varying background with undulation superimposed. (b) The undulations are emphasized if a sector is rotated in front of the screen and then the intensity plotted against $s = (4\pi/\lambda)\sin\frac{1}{2}\theta$.

the total intensity has an angular variation given by the *Wierl equation*:

$$I(\theta) = \sum_{i,j} f_i f_j \frac{\sin sR_{ij}}{sR_{ij}} \qquad s = \frac{4\pi}{\lambda}\sin\frac{1}{2}\theta$$

where λ is the wavelength of the electrons in the beam and θ is the scattering angle. The *electron scattering factor*, f, is a measure of the intensity of the electron scattering powers of the atoms. The Wierl equation is the equation for the appearance of the diffraction pattern caused by a molecule in the path of the electron beam. We encounter a very similar expression in the discussion of X-ray diffraction in Chapter 23 (eqn 23.6).

The electron diffraction pattern gives the distances between all possible pairs of atoms in the molecule (not just to those bonded together). When there are only a few atoms, the peaks can be analysed reasonably quickly, and the analysis proceeds by assuming a geometry and calculating the intensity pattern by using the Wierl equation. The best fit is then taken as the actual molecular geometry.

Discussion questions

1. Predict from the Wierl equation the positions of the first maximum and first minimum in the neutron and electron diffraction patterns of the Cl_2 molecules obtained with neutrons of wavelength 80 pm and electrons of wavelength 4.0 pm.

2. Use the Wierl equation to predict the appearance of the electron diffraction pattern of CCl_4 with an (as yet) undetermined C—Cl bond length but of known tetrahedral symmetry. Take $f_{Cl} = 17f$ and $f_C = 6f$ and note that $R(Cl, Cl) = (\frac{8}{3})^{1/2}R(C, Cl)$. Plot I/f^2 against $x = sR(C, Cl)$. In an actual experiment using 10.0 keV electrons the positions of the maxima occurred at $3°10'$, $5°22'$, and $7°54'$ and minima occurred at $1°46'$, $4°6'$, $6°40'$, and $9°10'$. What is the C—Cl bond length in CCl_4?

Further reading

R.M. Hanson and S.A. Bergman, Data-driven chemistry: Building models of molecular structure (literally) from electron diffraction data. *J. Chem. Educ.* **71**, 150 (1994).

L.C. Feldman and J.W. Mayer, *Fundamentals of surface and thin film analysis*. North-Holland, New York (1986).

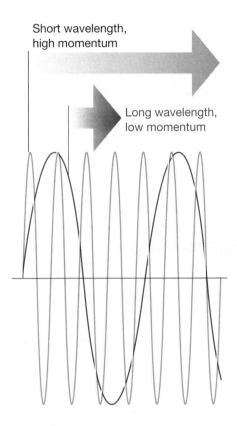

Short wavelength, high momentum

Long wavelength, low momentum

11.14 An illustration of the de Broglie relation between momentum and wavelength. A wave is associated with a particle (shortly this wave will be seen to be the wavefunction of the particle). A particle with high momentum has a wavefunction with a short wavelength, and vice versa.

Some progress towards coordinating these properties had already been made by the French physicist Louis de Broglie when, in 1924, he suggested that any particle, not only photons, travelling with a linear momentum p should have (in some sense) a wavelength given by the **de Broglie relation**:

$$\lambda = \frac{h}{p} \tag{11.13}$$

That is, a particle with a high linear momentum has a short wavelength (Fig. 11.14). Macroscopic bodies have such high momenta (even when they are moving slowly) that their wavelengths are undetectably small, and the wave-like properties cannot be observed.

Example 11.2 Estimating the de Broglie wavelength

Estimate the wavelength of electrons that have been accelerated from rest through a potential difference of 40 kV.

Method To use the de Broglie relation, we need to know the linear momentum, p, of the electrons. To calculate the linear momentum, we note that the energy acquired by an electron accelerated through a potential difference V is eV, where e is the magnitude of its charge. At the end of the period of acceleration, all the acquired energy is in the form of kinetic energy, $p^2/2m_e$, so we can determine p by setting $p^2/2m_e$ equal to eV. As before, carry through the calculation algebraically before substituting the data.

Answer The expression

$$\frac{p^2}{2m_e} = eV$$

solves to

$$p = (2m_e eV)^{1/2}$$

Then, from the de Broglie relation,

$$\lambda = \frac{h}{(2m_e eV)^{1/2}}$$

Substitution of the data and the fundamental constants (from inside the front cover) gives

$$\lambda = \frac{6.626 \times 10^{-34} \text{ J s}}{\{2 \times (9.109 \times 10^{-31} \text{ kg}) \times (1.609 \times 10^{-19} \text{ C}) \times (4.0 \times 10^4 \text{ V})\}^{1/2}}$$

$$= 6.1 \times 10^{-12} \text{ m}$$

where we have used 1 V C = 1 J.

Comment The wavelength of 6.1 pm is shorter than typical bond lengths in molecules (about 100 pm). Electrons accelerated in this way are used in the technique of electron diffraction for the determination of molecular structure (see Box 11.1).

Self-test 11.2 Calculate the wavelength of a neutron with a translational kinetic energy equal to kT at 300 K.

[178 pm]

We now have to conclude that not only has electromagnetic radiation the character classically ascribed to particles, but electrons (and all other particles) have the characteristics classically ascribed to waves. This joint particle and wave character of matter and radiation is called **wave–particle duality**. Duality strikes at the heart of classical physics, where particles and waves are treated as entirely distinct entities. We have also seen that the energies of electromagnetic radiation and of matter cannot be varied continuously, and that for small objects the discreteness of energy is highly significant. In classical mechanics, in contrast, energies could be varied continuously. Such total failure of classical physics for small objects implied that its basic concepts were false. A new mechanics had to be devised to take its place.

The dynamics of microscopic systems

Quantum mechanics acknowledges the wave–particle duality of matter by supposing that, rather than travelling along a definite path, a particle is distributed through space like a wave. This remark may seem mysterious at this stage: it will be interpreted more fully shortly. The mathematical representation of the wave that in quantum mechanics replaces the classical concept of trajectory is called a **wavefunction**, ψ (psi).

11.3 The Schrödinger equation

In 1926, the Austrian physicist Erwin Schrödinger proposed an equation for finding the wavefunction of any system. The **time-independent Schrödinger equation** for a particle of mass m moving in one dimension with energy E is

$$-\frac{\hbar^2}{2m}\frac{d^2\psi}{dx^2} + V(x)\psi = E\psi \tag{11.14}$$

The factor $V(x)$ is the potential energy of the particle at the point x; $\hbar$ (which is read h-cross or h-bar) is a convenient modification of Planck's constant:

$$\hbar = \frac{h}{2\pi} = 1.054\,57 \times 10^{-34}\ \text{J s} \tag{11.15}$$

For a partial justification of the form of this equation, see the *Justification* below. The discussions later in the chapter will help to overcome the apparent arbitrariness of this complicated expression. For the present, treat the equation as a quantum-mechanical postulate. Various ways of expressing the Schrödinger equation, of incorporating the time dependence of the wavefunction, and of extending it to more dimensions are collected in Table 11.1. In Chapter 12 we shall solve the equation for a number of important cases; in this chapter we are mainly concerned with its significance, the interpretation of its solutions, and seeing how it implies that energy is quantized.

Justification 11.1

Although the Schrödinger equation should be regarded as a postulate, like Newton's equations of motion, it can be seen to be plausible by noting that it implies the de Broglie relation for a freely moving particle in a region with constant potential energy V. After making the substitution $V(x) = V$, we can rearrange eqn 14 into

Table 11.1 The Schrödinger equation

For one-dimensional systems:

$$-\frac{\hbar^2}{2m}\frac{d^2\psi}{dx^2} + V(x)\psi = E\psi$$

where $V(x)$ is the potential energy of the particle and E is its total energy. For three-dimensional system

$$-\frac{\hbar^2}{2m}\nabla^2\psi + V\psi = E\psi$$

where V may depend on position and ∇^2 ('del squared') is

$$\nabla^2 = \frac{\partial^2}{\partial x^2} + \frac{\partial^2}{\partial y^2} + \frac{\partial^2}{\partial z^2}$$

In systems with spherical symmetry:

$$\nabla^2 = \frac{\partial^2}{\partial r^2} + \frac{2}{r}\frac{\partial}{\partial r} + \frac{1}{r^2}\Lambda^2$$

where

$$\Lambda^2 = \frac{1}{\sin^2\theta}\frac{\partial^2}{\partial\phi^2} + \frac{1}{\sin\theta}\frac{\partial}{\partial\theta}\sin\theta\frac{\partial}{\partial\theta}$$

In the general case the Schrödinger equation is written

$$H\psi = E\psi$$

Where H is the hamiltonian operator for the system:

$$H = -\frac{\hbar^2}{2m}\nabla^2 + V$$

For the evolution of a system with time, it is necessary to solve the time-dependent Schrödinger equation

$$H\Psi = i\hbar\frac{\partial\Psi}{\partial t}$$

Differential equations are reviewed in *Further information 1*.

$$\frac{d^2\psi}{dx^2} = -\frac{2m}{\hbar^2}(E - V)\psi$$

A solution of this equation is

$$\psi = e^{ikx} = \cos kx + i\sin kx \qquad k = \left\{\frac{2m(E - V)}{\hbar^2}\right\}^{1/2}$$

For this result, we have used the relation $e^{i\theta} = \cos\theta + i\sin\theta$, where $i = (-1)^{1/2}$.

Now we recognize that $\cos kx$ (or $\sin kx$) is a wave of wavelength $\lambda = 2\pi/k$, as can be seen by comparing $\cos kx$ with the standard form of a harmonic wave, $\cos(2\pi x/\lambda)$.[6] The quantity $E - V$ is equal to the kinetic energy of the particle, E_K, so $k = (2mE_K/\hbar^2)^{1/2}$, which implies that $E_K = k^2\hbar^2/2m$. Because $E_K = p^2/2m$, it follows that

$$p = k\hbar$$

Therefore, the linear momentum is related to the wavelength of the wavefunction by

$$p = \frac{2\pi}{\lambda} \times \frac{h}{2\pi} = \frac{h}{\lambda}$$

which is the de Broglie relation.

---

6. Waves and other essential concepts of physics are reviewed in *Further information 2*.

11.4 The Born interpretation of the wavefunction

It is a principal tenet of quantum mechanics that *the wavefunction contains all the dynamical information about the system it describes*. Here we concentrate on the information it carries about the location of the particle.

The interpretation of the wavefunction in terms of the location of the particle is based on a suggestion made by Max Born. He made use of an analogy with the wave theory of light, in which the square of the amplitude of an electromagnetic wave in a region is interpreted as its intensity and therefore (in quantum terms) as a measure of the probability of finding a photon present in the region. The **Born interpretation** of the wavefunction focuses on the square of the wavefunction (or the square modulus, $|\psi|^2 = \psi^*\psi$, if ψ is complex). It states that the value of $|\psi|^2$ at a point is proportional to the probability of finding the particle at that point. Specifically, for a one-dimensional system (Fig. 11.15):

> If the wavefunction of a particle has the value ψ at some point x, then the probability of finding the particle between x and $x + dx$ is proportional to $|\psi|^2 \, dx$.

Thus, $|\psi|^2$ is the **probability density**, and to obtain the probability it must be multiplied by the length of the infinitesimal region dx. The wavefunction ψ itself is called the **probability amplitude**. For a particle free to move in three dimensions (for example, an electron near a nucleus in an atom), the wavefunction depends on the point dr with coordinates x, y, and z, and the interpretation of $\psi(r)$ is as follows (Fig. 11.16):

> If the wavefunction of a particle has the value ψ at some point r, then the probability of finding the particle in an infinitesimal volume $d\tau = dx \, dy \, dz$ at that point is proportional to $|\psi|^2 \, d\tau$.

The Born interpretation does away with any worry about the significance of a negative (and, in general, complex) value of ψ because $|\psi|^2$ is real and never negative. There is no *direct* significance in the negative (or complex) value of a wavefunction: only the square modulus, a positive quantity, is directly physically significant, and both negative and positive regions of a wavefunction may correspond to a high probability of finding a particle in a region (Fig. 11.17). However, later we shall see that the presence of positive and negative regions of a wavefunction is of great *indirect* significance, because it gives rise to the possibility of constructive and destructive interference between different wavefunctions.

To form the complex conjugate, ψ^*, of a complex function, replace i wherever it occurs by $-i$. For instance, the complex conjugate of e^{ikx} is e^{-ikx}. If the wavefunction is real, $|\psi|^2 = \psi^2$.

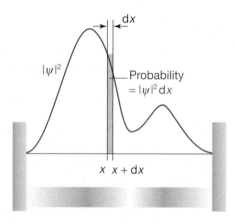

11.15 The wavefunction ψ is a probability amplitude in the sense that its square modulus ($\psi^*\psi$ or $|\psi|^2$) is a probability density. The probability of finding a particle in the region dx located at x is proportional to $|\psi|^2 \, dx$.

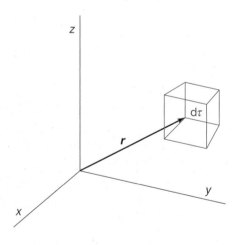

11.16 The Born interpretation of the wavefunction in three-dimensional space implies that the probability of finding the particle in the volume element $d\tau = dx \, dy \, dz$ at some location r is proportional to the product of $d\tau$ and the value of $|\psi|^2$ at that location.

Example 11.3 Interpreting a wavefunction

We shall see in Chapter 13 that the wavefunction of an electron in the lowest energy state of a hydrogen atom is proportional to e^{-r/a_0}, with a_0 a constant and r the distance from the nucleus. (Notice that this wavefunction depends only on this distance, not the angular position relative to the nucleus.) Calculate the relative probabilities of finding the electron inside a region of volume 1.0 pm³, which is small even on the scale of the atom, located at (a) the nucleus, (b) a distance a_0 from the nucleus.

Method The region of interest is so small on the scale of the atom that we can ignore the variation of ψ within it and write the probability, P, as proportional to the probability density (ψ^2; note that ψ is real) evaluated at the point of interest multiplied by the volume of interest, δV. That is, $P \propto \psi^2 \delta V$, with $\psi^2 \propto e^{-2r/a_0}$.

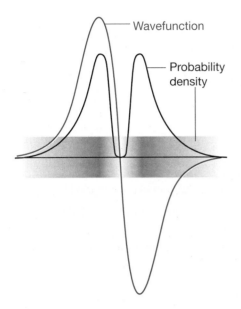

11.17 The sign of a wavefunction has no direct physical significance: the positive and negative regions of this wavefunction both correspond to the same probability distribution (as given by the square modulus of ψ and depicted by the density of shading).

Answer In each case $\delta V = 1.0 \text{ pm}^3$. (a) At the nucleus, $r = 0$, so

$$P \propto e^0 \times (1.0 \text{ pm}^3) = (1.0) \times (1.0 \text{ pm}^3)$$

(b) At a distance $r = a_0$ in an arbitrary direction,

$$P \propto e^{-2} \times (1.0 \text{ pm}^3) = (0.14) \times (1.0 \text{ pm}^3)$$

Therefore, the ratio of probabilities is $1.0/0.14 = 7.1$.

Comment Note that it is more probable (by a factor of 7) that the electron will be found at the nucleus than in the same volume element located at a distance a_0 from the nucleus. The negatively charged electron is attracted to the positively charged nucleus, and is likely to be found close to it.

Self-test 11.3 The wavefunction for the electron in its lowest energy state in the ion He$^+$ is proportional to e^{-2r/a_0}. Repeat the calculation for this ion. Any comment?   [55; more compact wavefunction]

(a) Normalization

A mathematical feature of the Schrödinger equation is that, if ψ is a solution, then so is $N\psi$, where N is any constant. This feature is confirmed by noting that ψ occurs in every term in eqn 14, so any constant factor can be cancelled. This freedom to vary the wavefunction by a constant factor means that it is always possible to find a **normalization constant**, N, such that the proportionality of the Born interpretation becomes an equality.

We find the normalization constant by noting that, for a normalized wavefunction $N\psi$, the probability that a particle is in the region dx is equal to $(N\psi^*)(N\psi)\,dx$ (we are taking N to be real). Furthermore, the sum over all space of these individual probabilities must be 1 (the probability of the particle being somewhere is 1). Expressed mathematically, the latter requirement is

$$N^2 \int \psi^*\psi \, dx = 1 \tag{11.16}$$

where the integral is over all the space accessible to the particle (for instance, from $-\infty$ to $+\infty$ if the particle can be anywhere in an infinite range). It follows that

$$N = \frac{1}{\left(\int \psi^*\psi \, dx \right)^{1/2}} \tag{11.17}$$

Therefore, by evaluating the integral, we can find the value of N and hence 'normalize' the wavefunction. From now on, unless we state otherwise, we always use wavefunctions that have been normalized to 1; that is, from now on we assume that ψ already includes a factor that ensures that (in one dimension)

$$\int \psi^*\psi \, dx = 1 \tag{11.18}$$

In three dimensions, the wavefunction is normalized if

$$\int \psi^*\psi \, dx \, dy \, dz = 1 \tag{11.19}$$

For systems with spherical symmetry, it is best to work in spherical polar coordinates r, θ, ϕ (Fig. 11.18): $x = r \sin\theta \cos\phi$, $y = r \sin\theta \sin\phi$, $z = r \cos\theta$. The volume element in spherical polar coordinates is $d\tau = r^2 \sin\theta \, dr \, d\theta \, d\phi$. To cover all space, the radius r ranges from 0 to ∞, the colatitude, θ, ranges from 0 to π, and the azimuth, ϕ, ranges from 0 to 2π (Fig. 11.19).

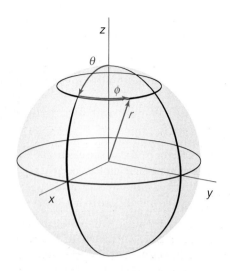

11.18 The spherical coordinates used for discussing systems with spherical symmetry.

Integrals commonly encountered in physical chemistry are listed inside the front cover.

or, more succinctly, if

$$\int \psi^* \psi \, d\tau = 1 \tag{11.20}$$

where $d\tau = dx \, dy \, dz$. In all such integrals, the integration is over all the space accessible to the particle.

Example 11.4 Normalizing a wavefunction

Normalize the wavefunction used for the hydrogen atom in Example 11.3.

Method We need to find the factor N that guarantees that the integral in eqn 20 is equal to 1. Because the wavefunction is spherically symmetrical, we work in spherical polar coordinates. A useful integral for calculations on atomic wavefunctions is

$$\int_0^\infty x^n e^{-ax} \, dx = \frac{n!}{a^{n+1}}$$

where $n!$ denotes a factorial: $n! = n(n-1)(n-2) \cdots 1$.

Answer The integration required is

$$\int \psi^* \psi \, d\tau = N^2 \int_0^\infty r^2 e^{-2r/a_0} \, dr \int_0^\pi \sin\theta \, d\theta \int_0^{2\pi} d\phi$$

$$= N^2 \times \tfrac{1}{4} a_0^3 \times 2 \times 2\pi = \pi a_0^3 N^2$$

Therefore, for this integral to equal 1,

$$N = \left(\frac{1}{\pi a_0^3} \right)^{1/2}$$

and the normalized wavefunction is

$$\psi = \left(\frac{1}{\pi a_0^3} \right)^{1/2} e^{-r/a_0}$$

Comment If Example 11.3 is now repeated, we can obtain the actual probabilities of finding the electron in the volume element at each location, not just their relative values. Given (from Section 13.1) that $a_0 = 52.9$ pm, the results are (a) 2.2×10^{-6}, corresponding to 1 chance in about 500 000 inspections of finding the electron in the test volume, and (b) 2.9×10^{-7}, corresponding to 1 chance in 3.4 million.

Self-test 11.4 Normalize the wavefunction given in Self-test 11.3.

$$[N = (8/\pi a_0^3)^{1/2}]$$

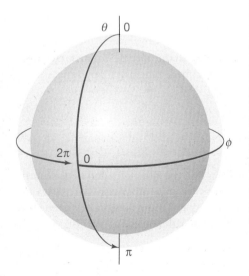

11.19 The surface of a sphere is covered by allowing θ to range from 0 to π, and then sweeping that arc around a complete circle by allowing ϕ to range from 0 to 2π.

The quantity $|\psi|^2 d\tau$ is a dimensionless probability and $d\tau$ has the dimensions of volume, (length)^d, where d is the number of spatial dimensions. Therefore, the dimensions of a normalized wavefunction are $1/(\text{length})^{d/2}$. Thus, in one spatial dimension, $d = 1$ and a normalized wavefunction has the dimensions of $1/(\text{length})^{1/2}$. For a three-dimensional system, the wavefunction has the dimensions of $1/(\text{length})^{3/2}$, as we saw in Example 11.4.

(b) Quantization

The Born interpretation puts severe restrictions on the acceptability of wavefunctions. The principal constraint is that ψ must not be infinite anywhere.[7] If it were, the integral in eqn 20 would be infinite and the normalization constant would be zero. The normalized function would then be zero everywhere, except where it is infinite, which would be unacceptable. The requirement that ψ is finite everywhere rules out many possible solutions of the Schrödinger equation, because many mathematically acceptable solutions rise to infinity and are therefore physically unacceptable. We shall meet several examples shortly.

The requirement that ψ is finite everywhere is not the only restriction implied by the Born interpretation. We could imagine (and in Section 12.6a will meet) a solution of the Schrödinger equation that gives rise to more than one value of $|\psi|^2$ at a single point. The Born interpretation implies that such solutions are unacceptable, because it would be absurd to have more than one probability that a particle is at the same point. This restriction is expressed by saying that the wavefunction must be *single-valued*; that is, have only one value at each point of space.

The Schrödinger equation itself also implies some mathematical restrictions on the type of functions that will occur. Because it is a second-order differential equation, the second derivative of ψ must be well-defined if the equation is to be applicable everywhere. We can take the second derivative of a function only if it is continuous (so there are no sharp steps in it, Fig. 11.20) and if its first derivative, its slope, is continuous (so there are no kinks).[8]

At this stage we see that ψ must be continuous, have a continuous slope, be single-valued, and be finite everywhere. An acceptable wavefunction cannot be zero everywhere, because the particle it describes must be somewhere. These are such severe restrictions that acceptable solutions of the Schrödinger equation do not in general exist for arbitrary values of the energy E. In other words, a particle may possess only certain energies, for otherwise its wavefunction would be physically unacceptable. That is, *the energy of a particle is quantized*. We can find the acceptable energies by solving the Schrödinger equation for motion of various kinds, and selecting the solutions that conform to the restrictions listed above. That is the task of the next chapter.

Quantum mechanical principles

We have claimed that a wavefunction contains all the information it is possible to obtain about the dynamical properties (for example, its location and momentum) of the particle. We have seen that the Born interpretation tells us as much as we can know about location, but how do we find any additional information?

11.20 The wavefunction must satisfy stringent conditions for it to be acceptable. (a) Unacceptable because it is not continuous, (b) unacceptable because its slope is discontinuous, (c) unacceptable because it is not single-valued, (d) unacceptable because it is infinite over a finite region.

7. Infinitely sharp spikes are acceptable provided they have zero width. A more complete formulation of the constraint is that the wavefunction must be infinite over any finite region and, specifically, that the integral of ψ^2 is finite. In elementary quantum mechanics the simpler restriction, to finite ψ, is sufficient.

8. There are cases, and we shall meet them, where acceptable wavefunctions have kinks. These cases arise when the potential energy has peculiar properties, such as rising abruptly to infinity. When the potential energy is smoothly well-behaved and finite, the slope of the wavefunction must be continuous; if the potential energy becomes infinite, then the slope of the wavefunction need not be continuous. There are only two cases of this behaviour in elementary quantum mechanics, and the peculiarity will be mentioned when we meet them.

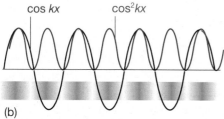

11.21 (a) The square modulus of a wavefunction corresponding to a definite state of linear momentum is a constant; so it corresponds to a uniform probability of finding the particle anywhere. (b) The probability distribution corresponding to the superposition of states of equal magnitude of linear momentum but opposite direction of travel.

We use the relation $\cos\theta = \frac{1}{2}(e^{i\theta} + e^{-i\theta})$.

11.5 The information in a wavefunction

The Schrödinger equation for a particle of mass m free to move parallel to the x-axis with zero potential energy is obtained from eqn 14 by setting $V = 0$, and is

$$-\frac{\hbar^2}{2m}\frac{d^2\psi}{dx^2} = E\psi \tag{11.21}$$

The solutions of this equation have the form

$$\psi = Ae^{ikx} + Be^{-ikx} \qquad E = \frac{k^2\hbar^2}{2m} \tag{11.22}$$

where A and B are constants. To verify that ψ is a solution of eqn 21, we simply substitute it into the left-hand side of the equation and confirm that we obtain $E\psi$.

(a) The probability density

Suppose that $B = 0$ in eqn 22,[9] then the wavefunction is simply

$$\psi = Ae^{ikx} \tag{11.23}$$

Where is the particle? We form the square modulus to find the probability density:

$$|\psi|^2 = (Ae^{ikx})^*(Ae^{ikx}) = (A^*e^{-ikx})(Ae^{ikx}) = |A|^2 \tag{11.24}$$

This probability is independent of x; so, wherever we look along the x-axis, there is an equal probability of finding the particle (Fig. 11.21a). In other words, if the wavefunction of the particle is given by eqn 23, then we cannot predict where we will find the particle. The same would be true if the wavefunction in eqn 22 had $A = 0$; then the probability density would be $|B|^2$, a constant.[10]

Now suppose that in the wavefunction $A = B$. Then eqn 22 becomes

$$\psi = A(e^{ikx} + e^{-ikx}) = 2A\cos kx \tag{11.25}$$

The probability density now has the form

$$|\psi|^2 = (2A\cos kx)^*(2A\cos kx) = 4|A|^2\cos^2 kx \tag{11.26}$$

This function is illustrated in Fig. 11.21b. As we see, the probability density periodically varies between 0 and $4|A|^2$. The locations where the probability density is zero correspond to *nodes* in the wavefunction: particles will never be found at the nodes. Specifically, a **node** is a point where a wavefunction passes *through* zero. The location where a wavefunction approaches zero without actually passing through zero is not a node. Nodes are defined in terms of the probability amplitude, the wavefunction itself. The probability density, of course, never passes through zero as it cannot be negative.

(b) Eigenvalues and eigenfunctions

Because the total energy of the particle is its kinetic energy, $p^2/2m$, it follows from eqn 22 that

$$p = k\hbar \tag{11.27}$$

This value is independent of the values of A and B.

---

9. We shall see later what determines the values of A and B; for the time being we can treat them as arbitrary constants.

10. It follows that if x is allowed to range from $-\infty$ to $+\infty$, the normalization constants, A or B, are 0. To avoid this embarrassing problem, x is allowed to range from $-L$ to $+L$, and L is allowed to go to infinity at the end of all calculations. We ignore this complication here.

To find a systematic way of extracting information from the wavefunction, we first note that any Schrödinger equation (such as those in eqns 14 and 21) may be written in the succinct form

$$H\psi = E\psi \tag{11.28}$$

with (in one dimension)

$$H = -\frac{\hbar^2}{2m}\frac{d^2}{dx^2} + V(x) \tag{11.29}$$

The quantity H is an **operator**, something that carries out a mathematical operation on the function ψ. In this case, the operation is to take the second derivative of ψ and (after multiplication by $-\hbar^2/2m$) to add the result to the outcome of multiplying ψ by V. The operator H plays a special role in quantum mechanics, and is called the **hamiltonian operator** after the nineteenth century mathematician William Hamilton.[11] The hamiltonian operator is the operator corresponding to the total energy of the system, the sum of the kinetic and potential energies. Consequently, we can infer that the first term in eqn 29 (the term proportional to the second derivative) must be the operator for the kinetic energy. When the Schrödinger equation is written as in eqn 28, it is seen to be an **eigenvalue equation**, an equation of the form

(Operator)(function) = (constant factor) × (same function)

If we denote a general operator by $\hat{\Omega}$ (where Ω is upper-case omega) and a constant factor by ω (lower-case omega), an eigenvalue equation has the form

$$\hat{\Omega}\psi = \omega\psi \tag{11.30}$$

The factor ω is called the **eigenvalue** of the operator $\hat{\Omega}$. The eigenvalue in eqn 28 is the energy. The function ψ is called an **eigenfunction** and is different for each eigenvalue. The eigenfunction in eqn 28 is the wavefunction corresponding to the energy E. It follows that another way of saying 'solve the Schrödinger equation' is 'find the eigenvalues and eigenfunctions of the hamiltonian operator for the system'. The wavefunctions are the eigenfunctions of the hamiltonian operator, and the corresponding eigenvalues are the allowed energies.

Example 11.5 Identifying an eigenfunction

Show that e^{ax} is an eigenfunction of the operator d/dx, and find the corresponding eigenvalue. Show that e^{ax^2} is not an eigenfunction of d/dx.

Method We need to operate on the function with the operator and check whether the result is a constant factor times the original function.

Answer For $\hat{\Omega} = d/dx$ and $\psi = e^{ax}$:

$$\hat{\Omega}\psi = \frac{d}{dx}e^{ax} = ae^{ax} = a\psi$$

---

11. Hamilton developed a form of classical mechanics that, it subsequently turned out, is well suited to the formulation of quantum mechanics and which shows very clearly the relation between the two theories.

Therefore e^{ax} is indeed an eigenfunction of d/dx, and its eigenvalue is a. For $\psi = e^{ax^2}$,

$$\hat{\Omega}\psi = \frac{d}{dx}e^{ax^2} = 2axe^{ax^2} = 2ax \times \psi$$

which is not an eigenvalue equation even though the same function ψ occurs on the right, because ψ is now multiplied by a variable factor ($2ax$), not a constant factor. Alternatively, if the right-hand side is written $2a(xe^{ax^2})$, we see that it is a constant times a *different* function.

Comment Much of quantum mechanics involves looking for functions that are eigenfunctions of a given operator, especially of the hamiltonian operator for the energy.

Self-test 11.5 Is the function $\cos ax$ an eigenfunction of (a) d/dx, (b) d^2/dx^2?

[(a) No, (b) yes]

The importance of eigenvalue equations is that the pattern

(Energy operator)ψ = (energy) $\times \psi$

exemplified by the Schrödinger equation is repeated for other **observables**, or measurable properties of a system, such as the momentum or the electric dipole moment. Thus, it is often the case that we can write

(Operator corresponding to an observable)ψ = (value of observable) $\times \psi$

The symbol $\hat{\Omega}$ in eqn 30 is then interpreted as an operator (for example, the hamiltonian, H) corresponding to an observable (for example, the energy), and the eigenvalue ω is the value of that observable (for example, the value of the energy, E). Therefore, if we know both the wavefunction ψ and the operator corresponding to the observable Ω of interest, and the wavefunction is an eigenfunction of the operator $\hat{\Omega}$, we can predict the outcome of an observation of the property Ω (for example, an atom's energy) by picking out the factor ω in the eigenvalue equation, eqn 30.

Quantum mechanical calculations are often simplified by using a property of wavefunctions called 'orthogonality'. To say that two functions are **orthogonal** means that

$$\int \psi_i^* \psi_j \, d\tau = 0 \tag{11.31}$$

A very general rule in quantum mechanics is that *eigenfunctions corresponding to different eigenvalues of the same operator are orthogonal*. For example, if ψ_1 corresponds to one energy, and ψ_2 corresponds to a different energy, then we know at once that the two functions are orthogonal and that the integral of their product is zero.

Strictly speaking, this rule applies only to *Hermitian operators*, which are operators for which $\int \psi_i^* \hat{\Omega} \psi_j \, d\tau = (\int \psi_j^* \hat{\Omega} \psi_i \, d\tau)^*$. We shall be dealing only with Hermitian operators.

Illustration 11.1

The wavefunctions $\sin x$ and $\sin 2x$ are eigenfunctions of the operator d^2/dx^2, with eigenvalues -1 and -4, respectively. To verify that the two wavefunctions are mutually orthogonal, we integrate the product $(\sin x)(\sin 2x)$ over all space, which we may take to span from $x = 0$ to $x = 2\pi$. A useful integral for this calculation is

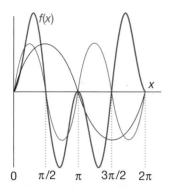

11.22 The integral of the function $f(x) = (\sin x)(\sin 2x)$ is equal to the area below the dark blue curve, and is zero, as can be inferred by symmetry.

$$\int \sin ax \sin bx \, dx = \frac{\sin(a-b)x}{2(a-b)} - \frac{\sin(a+b)x}{2(a+b)} + \text{constant}, \quad \text{if } a^2 \neq b^2$$

It follows that, for $a = 1$ and $b = 2$,

$$\int_0^{2\pi} (\sin x)(\sin 2x) \, dx = 0$$

Self-test 11.6 Plot the function $f(x) = (\sin x)(\sin 2x)$ and verify graphically that its integral from $x = 0$ to $x = 2\pi$ is equal to zero.

[The plot (Fig. 11.22) shows that

$$\int_0^{\pi/2} f(x) \, dx = \int_{3\pi/2}^{2\pi} f(x) \, dx = -\int_{\pi/2}^{\pi} f(x) \, dx = -\int_{\pi}^{3\pi/2} f(x) \, dx,$$

$$\text{so} \int_0^{2\pi} f(x) \, dx = 0.]$$

(c) Operators

To make these abstract procedures concrete, we need to set up and use the operator corresponding to a given observable. The procedure is summarized by the following rule:[12]

Observables, Ω, are represented by operators, $\hat{\Omega}$, built from the following position and momentum operators:

$$\hat{x} = x \times \qquad \hat{p}_x = \frac{\hbar}{i} \frac{d}{dx} \qquad\qquad [11.32]$$

That is, the operator for location along the x-axis is multiplication (of the wavefunction) by x and the operator for linear momentum parallel to the x-axis is proportional to taking the derivative (of the wavefunction) with respect to x.

Justification 11.2

To confirm that the operator for linear momentum in eqn 32 is plausible, we show here that it is consistent with the de Broglie relation. First, we set up the eigenvalue equation

$$\hat{p}_x \psi = p_x \psi \qquad\qquad (11.33)$$

in the form

$$\frac{\hbar}{i} \frac{d\psi}{dx} = p_x \psi$$

If the wavefunction is the one given in eqn 22 with $B = 0$,

$$\frac{\hbar}{i} \frac{d\psi}{dx} = \frac{\hbar}{i} A \frac{de^{ikx}}{dx} = \frac{\hbar}{i} A \times ike^{ikx} = k\hbar A e^{ikx} = k\hbar\psi$$

---

12. This rule applies to observables that depend on spatial variables; intrinsic properties, such as spin (see Section 12.8) are treated differently.

This is an eigenvalue equation, and by comparing it with eqn 30 we find that $p_x = +k\hbar$. Because $e^{ikx} = \cos kx + i \sin kx$, and we have seen that the harmonic functions $\cos kx$ and $\sin kx$ describe waves of wavelength $\lambda = 2\pi/k$, this linear momentum is equal to $p_x = +(2\pi/\lambda)(h/2\pi) = h/\lambda$, which is in accord with the de Broglie relation. However, the result is richer: the positive value implies that the linear momentum is directed towards positive x. Now suppose instead that the wavefunction is the one in eqn 22 with $A = 0$; then the same kind of calculation gives $p_x = -k\hbar$. It follows that a particle described by the second wavefunction has the same wavelength, magnitude of momentum, and the same kinetic energy as before, but the negative value of p_x implies that the motion is towards $-x$.

We use the definitions in eqn 32 to construct operators for other spatial observables. For example, suppose we wanted the operator for a potential energy of the form

$$V = \tfrac{1}{2}kx^2 \tag{11.34a}$$

with k a constant (later, we shall see that this potential describes the vibrations of atoms in molecules). Then it follows from eqn 32 that the operator corresponding to V is multiplication by x^2:

$$\hat{V} = \tfrac{1}{2}kx^2 \times \tag{11.34b}$$

In normal practice, the multiplication sign is omitted. To construct the operator for kinetic energy, we make use of the classical relation between kinetic energy and linear momentum, which in one dimension is

$$E_K = \frac{p_x^2}{2m} \tag{11.35a}$$

Then, using the operator for p_x in eqn 32 we find:

$$\hat{E}_K = \frac{1}{2m}\left(\frac{\hbar}{i}\frac{d}{dx}\right)\left(\frac{\hbar}{i}\frac{d}{dx}\right) = -\frac{\hbar^2}{2m}\frac{d^2}{dx^2} \tag{11.35b}$$

It follows that the operator for the total energy, the hamiltonian operator, is

$$\hat{H} = \hat{E}_K + \hat{V} = -\frac{\hbar^2}{2m}\frac{d^2}{dx^2} + \hat{V} \tag{11.36}$$

The expression for the kinetic energy operator, eqn 35b, gives another clue to the qualitative interpretation of a wavefunction and the structure of the Schrödinger equation. In mathematics, the second derivative of a function is a measure of its curvature: a large second derivative indicates a sharply curved function (Fig. 11.23). It follows that a sharply curved wavefunction is associated with a high kinetic energy, and one with a low curvature is associated with a low kinetic energy. This interpretation is consistent with the de Broglie relation, which predicts a short wavelength (a sharply curved wavefunction) when the linear momentum (and hence the kinetic energy) is high. However, it extends the interpretation to wavefunctions that do not spread through space and resemble those shown in Fig. 11.23. The curvature of a wavefunction in general varies from place to place. Wherever a wavefunction is sharply curved, its contribution to the total kinetic energy is large (Fig. 11.24). Wherever the wavefunction is not sharply curved, its contribution to the overall kinetic energy is low. As we shall shortly see, the observed kinetic energy of the particle is an integral of all the contributions of the kinetic energy from each region.

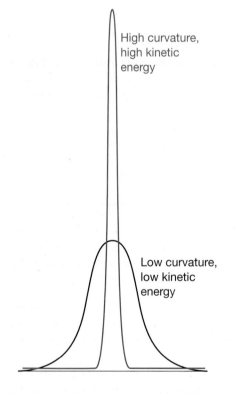

High curvature, high kinetic energy

Low curvature, low kinetic energy

11.23 Even if a wavefunction does not have the form of a periodic wave, it is still possible to infer from it the average kinetic energy of a particle by noting its average curvature. This illustration shows two wavefunctions: the sharply curved function corresponds to a higher kinetic energy than the less sharply curved function.

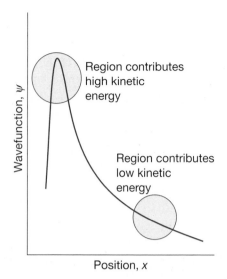

11.24 The observed kinetic energy of a particle is an average of contributions from the entire space covered by the wavefunction. Sharply curved regions contribute a high kinetic energy to the average; slightly curved regions contribute only a small kinetic energy.

In general, a linear combination of two functions f and g is $c_1 f + c_2 g$, where c_1 and c_2 are numerical coefficients, so a linear combination is a more general term than 'sum'. In a sum, $c_1 = c_2 = 1$.

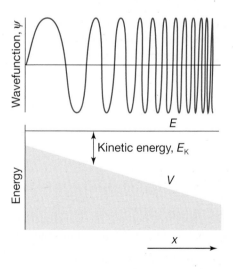

11.25 The wavefunction of a particle in a potential decreasing towards the right and hence subjected to a constant force to the right. Only the real part of the wavefunction is shown, the imaginary part is similar, but displaced to the right.

Hence, we can expect a particle to have a high kinetic energy if the average curvature of its wavefunction is high.

The association of high curvature with high kinetic energy will turn out to be a valuable guide to the interpretation of wavefunctions and the prediction of their shapes. For example, suppose we need to know the wavefunction of a particle with a given total energy and a potential energy that decreases with increasing x (Fig. 11.25). Because the difference $E - V = E_K$ increases from left to right, the wavefunction must become more sharply curved as x increases: its wavelength decreases as the local contributions to its kinetic energy increase. We can therefore guess that the wavefunction will look like the function sketched in the illustration, and more detailed calculation confirms this to be so.

(d) Superpositions and expectation values

Suppose now that the wavefunction is the one given in eqn 25 (with $A = B$). What is the linear momentum of the particle it describes? We quickly run into trouble if we use the operator technique. When we operate with $\hat{p}_x$, we find

$$\frac{\hbar}{i}\frac{d\psi}{dx} = \frac{2\hbar}{i}A\frac{d\cos kx}{dx} = -\frac{2k\hbar}{i}A\sin kx \tag{11.37}$$

This expression is not an eigenvalue equation, because the function on the right ($\sin kx$) is different from that on the left ($\cos kx$).

When the wavefunction of a particle is not an eigenfunction of an operator, the property to which the operator corresponds does not have a definite value. However, in the current example the momentum is not completely indefinite because the cosine wavefunction is a **linear combination**, or sum, of e^{ikx} and e^{-ikx}, and these two functions, as we have seen, individually correspond to definite momentum states. We say that the total wavefunction is a **superposition** of more than one wavefunction. Symbolically we can write the superposition as

$$\psi = \underset{\substack{\text{Particle}\\\text{with linear}\\\text{momentum}\\+k\hbar}}{\psi_{\rightarrow}} + \underset{\substack{\text{Particle}\\\text{with linear}\\\text{momentum}\\-k\hbar}}{\psi_{\leftarrow}}$$

The interpretation of this composite wavefunction is that, if the momentum of the particle is repeatedly measured in a long series of observations, then its magnitude will found to be $k\hbar$ in all the measurements (because that is the value for each component of the wavefunction). However, because the two component wavefunctions occur equally in the superposition, half the measurements will show that the particle is moving to the right ($p_x = +k\hbar$), and half the measurements will show that it is moving to the left ($p_x = -k\hbar$). According to quantum mechanics, we cannot predict in which direction the particle will in fact be found to be travelling; all we can say is that, in a long series of observations, there are equal probabilities of finding the particle travelling to the right and to the left.

The same interpretation applies to any wavefunction written as a linear combination of eigenfunctions of an operator. Thus, suppose the wavefunction is known to be a superposition of many different linear momentum eigenfunctions and written as the linear combination

$$\psi = c_1\psi_1 + c_2\psi_2 + \cdots = \sum_k c_k\psi_k \tag{11.38}$$

where the c_k are numerical coefficients and the ψ_k correspond to different momentum states. The functions ψ_k are said to form a **complete set** in the sense that any arbitrary function can be expressed as a linear combination of them. Then according to quantum mechanics:

1 When the momentum is measured, in a single observation one of the eigenvalues corresponding to the ψ_k that contribute to the superposition will be found.

2 The probability of measuring a particular eigenvalue in a series of observations is proportional to the square modulus ($|c_k|^2$) of the corresponding coefficient in the linear combination.

3 The average value of a large number of observations is given by the expectation value, $\langle\Omega\rangle$, of the operator $\hat{\Omega}$ corresponding to the observable of interest.

The **expectation value** of an operator $\hat{\Omega}$ is defined as

$$\langle\Omega\rangle = \int \psi^*\hat{\Omega}\psi \, d\tau$$

[11.39]

This formula is valid only for normalized wavefunctions. As we see in the *Justification* below, an expectation value is the weighted average of a large number of observations of a property.

Justification 11.3

If ψ is an eigenfunction of $\hat{\Omega}$ with eigenvalue ω, the expectation value of Ω is

$$\langle\Omega\rangle = \int \psi^*\hat{\Omega}\psi \, d\tau = \int \psi^*\omega\psi \, d\tau = \omega\int \psi^*\psi \, d\tau = \omega$$

because ω is a constant and may be taken outside the integral, and the resulting integral is equal to 1 for a normalized wavefunction. The interpretation of this expression is that, because every observation of the property Ω results in the value ω (because the wavefunction is an eigenfunction of $\hat{\Omega}$), the mean value of all the observations is also ω.

A wavefunction that is not an eigenfunction of the operator of interest can be written as a linear combination of eigenfunctions. For simplicity, suppose the wavefunction is the sum of two eigenfunctions (the general case, eqn 38, can easily be developed). Then

$$\langle\Omega\rangle = \int (c_1\psi_1 + c_2\psi_2)^*\hat{\Omega}(c_1\psi_1 + c_2\psi_2) \, d\tau$$

$$= \int (c_1\psi_1 + c_2\psi_2)^*(c_1\omega_1\psi_1 + c_2\omega_2\psi_2) \, d\tau$$

$$= c_1^*c_1\omega_1\int \psi_1^*\psi_1 \, d\tau + c_2^*c_2\omega_2\int \psi_2^*\psi_2 \, d\tau + c_1^*c_2\omega_2\int \psi_1^*\psi_2 \, d\tau + c_2^*c_1\omega_1\int \psi_2^*\psi_1 \, d\tau$$

The first two integrals on the right are both equal to 1 because the wavefunctions are normalized. Because ψ_1 and ψ_2 correspond to different eigenvalues, they are orthogonal, so we can conclude that

$$\langle\Omega\rangle = |c_1|^2\omega_1 + |c_2|^2\omega_2$$

This expression shows that the expectation value is the sum of the two eigenvalues weighted by the probabilities that each one will be found in a series of measurements. Hence, the expectation value is the weighted mean of a series of observations.

Example 11.6 Calculating an expectation value

Calculate the average value of the distance of an electron from the nucleus in the hydrogen atom in its state of lowest energy.

Method The average radius is the expectation value of the operator corresponding to the distance from the nucleus, which is multiplication by r. To evaluate $\langle r \rangle$, we need to know the normalized wavefunction (from Example 11.4) and then evaluate the integral in eqn 39.

Answer The average value is given by the expectation value

$$\langle r \rangle = \int \psi^* r \psi \, d\tau$$

which we evaluate by using spherical polar coordinates. Using the normalized function in Example 11.4 gives

$$\langle r \rangle = \frac{1}{\pi a_0^3} \int_0^\infty r^3 e^{-2r/a_0} \, dr \int_0^\pi \sin\theta \, d\theta \int_0^{2\pi} d\phi$$

$$= \frac{1}{\pi a_0^3} \times \frac{3! a_0^4}{2^4} \times 2 \times 2\pi = \tfrac{3}{2} a_0$$

Because $a_0 = 52.9$ pm (see Section 13.1), $\langle r \rangle = 79.4$ pm.

Comment The result means that, if a very large number of measurements of the distance of the electron from the nucleus are made, their mean value will be 79.4 pm. However, each different observation will give a different and unpredictable individual result because the wavefunction is not an eigenfunction of the operator corresponding to r.

Self-test 11.7 Evaluate the root mean square distance, $\langle r^2 \rangle^{1/2}$, of the electron from the nucleus in the hydrogen atom. $[3^{1/2} a_0 = 91.6$ pm$]$

The mean kinetic energy of a particle in one dimension is the expectation value of the operator given in eqn 35b. Therefore, we can write

$$\langle E_K \rangle = \int \psi^* \hat{E}_K \psi \, d\tau = -\frac{\hbar^2}{2m} \int \psi^* \frac{d^2\psi}{dx^2} \, d\tau \tag{11.40}$$

This conclusion confirms the previous assertion that the kinetic energy is a kind of average over the curvature of the wavefunction: we get a large contribution to the observed value from regions where the wavefunction is sharply curved (so $d^2\psi/dx^2$ is large) and the wavefunction itself is large (so that ψ^* is large too).

11.6 The uncertainty principle

We have seen that, if the wavefunction is Ae^{ikx}, then the particle it describes has a definite state of linear momentum, namely travelling to the right with momentum $p_x = +k\hbar$. However, we have also seen that the position of the particle described by this wavefunction is completely unpredictable. In other words, if the momentum is specified precisely, it is impossible to predict the location of the particle. This statement is one-half of a special case of the **Heisenberg uncertainty principle**, one of the most celebrated results of quantum mechanics:

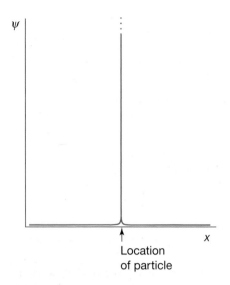

11.26 The wavefunction for a particle at a well-defined location is a sharply spiked function that has zero amplitude everywhere except at the particle's position.

It is impossible to specify simultaneously, with arbitrary precision, both the momentum and the position of a particle.

Before discussing the principle further, we must establish its other half: that, if we know the position of a particle exactly, then we can say nothing about its momentum. The argument draws on the idea of regarding a wavefunction as a superposition of eigenfunctions, and runs as follows.

If we know that the particle is at a definite location, its wavefunction must be large there and zero everywhere else (Fig. 11.26). Such a wavefunction can be created by superimposing a large number of harmonic (sine and cosine) functions, or, equivalently, a number of e^{ikx} functions. In other words, we can create a sharply localized wavefunction, called a **wave packet**, by forming a linear combination of wavefunctions that correspond to many different linear momenta. The superposition of a few harmonic functions gives a wavefunction that spreads over a range of locations (Fig. 11.27). However, as the number of wavefunctions in the superposition increases, the wave packet becomes sharper on account of the more complete interference between the positive and negative regions of the individual waves. When an infinite number of components is used, the wave packet is a sharp, infinitely narrow spike, which corresponds to perfect localization of the particle. Now the particle is perfectly localized. However, we have lost all information about its momentum because, as we saw above, a measurement of the momentum will give a result corresponding to any one of the infinite number of waves in the superposition, and which one it will give is unpredictable. Hence, if we know the location of the particle precisely (implying that its wavefunction is a superposition of an infinite number of momentum eigenfunctions), then its momentum is completely unpredictable.

A quantitative version of this result is

$$\Delta p \Delta q \geq \tfrac{1}{2}\hbar \tag{11.41}$$

In this expression Δp is the 'uncertainty' in the linear momentum parallel to the axis q, and Δq is the uncertainty in position along that axis. These 'uncertainties' are precisely defined, for they are the root mean square deviations of the properties from their mean values:

$$\Delta p = \{\langle p^2 \rangle - \langle p \rangle^2\}^{1/2} \qquad \Delta q = \{\langle q^2 \rangle - \langle q \rangle^2\}^{1/2} \tag{11.42}$$

If there is complete certainty about the position of the particle ($\Delta q = 0$), the only way that eqn 41 can be satisfied is for $\Delta p = \infty$, which implies complete uncertainty about the momentum. Conversely, if the momentum is known exactly ($\Delta p = 0$), then the position must be completely uncertain ($\Delta q = \infty$).

The p and q that appear in eqn 41 refer to the same direction in space. Therefore, whereas position on the x-axis and momentum parallel to the x-axis are restricted by the uncertainty relation, simultaneous location of position on x and motion parallel to y or z are not restricted.

Example 11.7 Using the uncertainty principle

The speed of a projectile of mass 1.0 g is known to within 1 μm s⁻¹. Calculate the minimum uncertainty in its position.

Method Estimate Δp from $m\Delta v$, where Δv is the uncertainty in the speed; then use eqn 41 to estimate the minimum uncertainty in position, Δq.

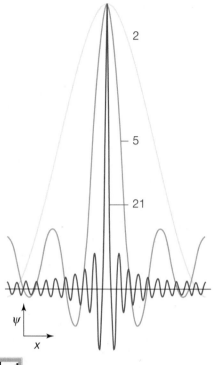

11.27 The wavefunction for a particle with an ill-defined location can be regarded as the superposition of several wavefunctions of definite wavelength that interfere constructively in one place but destructively elsewhere. As more waves are used in the superposition (as given by the numbers attached to the curves), the location becomes more precise at the expense of uncertainty in the particle's momentum. An infinite number of waves is needed to construct the wavefunction of a perfectly localized particle.

Answer The minimum uncertainty in position is

$$\Delta q = \frac{\hbar}{2m\Delta v}$$

$$= \frac{1.055 \times 10^{-34}\,\text{J s}}{2 \times (1.0 \times 10^{-3}\,\text{kg}) \times (1 \times 10^{-6}\,\text{m s}^{-1})} = 5 \times 10^{-26}\,\text{m}$$

where we have used $1\,\text{J} = 1\,\text{kg m}^2\,\text{s}^{-2}$.

Comment The uncertainty is completely negligible for all practical purposes concerning macroscopic objects. However, if the mass is that of an electron, the same uncertainty in speed implies an uncertainty in position far larger than the diameter of an atom; so the concept of a trajectory, the simultaneous possession of a precise position and momentum, is untenable.

Self-test 11.8 Estimate the minimum uncertainty in the speed of an electron in a one-dimensional region of length $2a_0$.   [500 km s^{-1}]

The Heisenberg uncertainty principle is more general than eqn 41 suggests. It applies to any pair of observables called **complementary observables**, which are defined in terms of the properties of their operators. Specifically, two observables Ω_1 and Ω_2 are complementary if

$$\hat{\Omega}_1(\hat{\Omega}_2\psi) \neq \hat{\Omega}_2(\hat{\Omega}_1\psi) \tag{11.43}$$

When the effect of two operators depends on their order (as this equation implies), we say that they do not **commute**.

Illustration 11.2

To show that the operators for position and momentum do not commute (and hence are complementary observables) we consider the effect of $\hat{x}\hat{p}_x$ (that is, the effect of $\hat{p}_x$ followed by the effect on the outcome of multiplication by x) on a wavefunction ψ:

$$\hat{x}\hat{p}_x\psi = x \times \frac{\hbar}{i}\frac{d\psi}{dx}$$

Next, we consider the effect of $\hat{p}_x\hat{x}$ on the same function (that is, the effect of multiplication by x followed by the effect of $\hat{p}_x$ on the outcome):

$$\hat{p}_x\hat{x}\psi = \frac{\hbar}{i}\frac{d}{dx}x\psi = \frac{\hbar}{i}\left(\psi + x\frac{d\psi}{dx}\right)$$

For this step we have used the standard rule about differentiating a product of functions (see *Further information 1*). The second expression is clearly different from the first, so the two operators do not commute.

The different outcomes of the effect of applying $\hat{\Omega}_1$ and $\hat{\Omega}_2$ in a different order are expressed by introducing the **commutator** of the two operators, which is defined as

$$[\hat{\Omega}_1, \hat{\Omega}_2] = \hat{\Omega}_1\hat{\Omega}_2 - \hat{\Omega}_2\hat{\Omega}_1 \tag{11.44}$$

We can conclude from *Illustration* 11.2 that the commutator of the operators for position and linear momentum is

$$[\hat{x}, \hat{p}_x] = i\hbar \tag{11.45}$$

This commutator is of such vital significance in quantum mechanics that it is taken as a fundamental distinction between classical mechanics and quantum mechanics. In fact, this commutator is taken as a postulate of quantum mechanics, and is used to justify the choice of the operators for position and linear momentum given in eqn 32.

With the concept of commutator established, the Heisenberg uncertainty principle can be given its most general form. For *any* two pairs of observables, Ω_1 and Ω_2, the uncertainties (to be precise, the root mean square deviations of their values from the mean) in simultaneous determinations are related by

$$\Delta\Omega_1 \Delta\Omega_2 \geq \tfrac{1}{2} |\langle [\hat{\Omega}_1, \hat{\Omega}_2] \rangle| \tag{11.46}$$

We obtain the special case of eqn 41 when we identify the observables with x and p_x and use eqn 45 for their commutator.

Complementary observables are observables with non-commuting operators. With the discovery that some pairs of observables are complementary (we meet more examples in the next chapter), we are at the heart of the difference between classical and quantum mechanics. Classical mechanics supposed, falsely as we now know, that the position and momentum of a particle could be specified simultaneously with arbitrary precision. However, quantum mechanics shows that position and momentum are complementary, and that we have to make a choice: we can specify position at the expense of momentum, or momentum at the expense of position.

The realization that some observables are complementary allows us to make considerable progress with the calculation of atomic and molecular properties; but it does away with some of the most cherished concepts of classical physics.

Checklist of key ideas

☐ classical mechanics, the laws of motion introduced in the seventeenth century by Isaac Newton

☐ quantum mechanics, the laws of motion introduced in the twentieth century by Heisenberg and Schrödinger

The origins of quantum mechanics

11.1 The failures of classical physics

☐ black body, an object capable of emitting and absorbing all frequencies of radiation uniformly

☐ Wien displacement law, $T\lambda_{max} = \tfrac{1}{5}c_2$

☐ second radiation constant, $c_2 = hc/k = 1.44$ cm K

☐ Stefan–Boltzmann law, $\mathcal{E} = aT^4$, $a = 4\sigma/c$

☐ excitance, the power emitted by a region of surface divided by the area of the surface

☐ Stefan–Boltzmann constant, $\sigma = 2\pi^5 k^4/15c^2 h^3 = 56.7$ nW m^{-2} K^{-4}

☐ Rayleigh–Jeans law, $d\mathcal{E} = \rho\, d\lambda$, $\rho = 8\pi kT/\lambda^4$

☐ quantization of energy, the limitation of energies to discrete values

☐ Planck's constant, $h = 6.626\ 08 \times 10^{-34}$ J s

☐ Planck distribution, $d\mathcal{E} = \rho\, d\lambda$, $\rho = (8\pi hc/\lambda^5)/(e^{hc/\lambda kT} - 1)$

☐ Dulong and Petit's law: the molar heat capacities of all monatomic solids are the same, and close to 25 J K^{-1} mol^{-1}

☐ Einstein formula, $C_{V,m} = 3Rf^2$, $f = (\theta_E/T)e^{\theta_E/2T}/(e^{\theta_E/T} - 1)$

☐ Einstein temperature, $\theta_E = h\nu/k$

☐ Debye formula, $C_{V,m} = 3Rf$, $$f = 3\left(\frac{T}{\theta_D}\right)^3 \int_0^{\theta_D/T} \frac{x^4 e^x}{(e^x - 1)^2}\, dx$$

☐ Debye temperature, $\theta_D = h\nu_D/k$

11.2 Wave–particle duality

☐ photoelectric effect, the ejection of electrons from metals when they are exposed to ultraviolet radiation: $\tfrac{1}{2}m_e v^2 = h\nu - \Phi$

☐ work function, the energy required to remove an electron from the metal to infinity

- Davisson–Germer experiment, the diffraction of electrons by a crystal
- electron diffraction, the diffraction of electrons by an object in their path
- de Broglie relation, $\lambda = h/p$
- wave–particle duality, the joint particle and wave character of matter and radiation

The dynamics of microscopic systems

- wavefunction, a mathematical function obtained by solving the Schrödinger equation and which contains all the dynamical information about a system

11.3 The Schrödinger equation

- time-independent Schrödinger equation, $-(\hbar^2/2m)(\mathrm{d}^2\psi/\mathrm{d}x^2) + V(x)\psi = E\psi$

11.4 The Born interpretation of the wavefunction

- Born interpretation, the value of $|\psi|^2$ at a point is proportional to the probability of finding the particle at that point
- probability density, the probability of finding a particle in a region divided by the volume of the region
- probability amplitude, the square-root of the probability density

- normalization constant, $N = 1/\{\int \psi^*\psi \, \mathrm{d}x\}^{1/2}$
- quantization, confinement of a dynamical observable to discrete values
- constraints on the wavefunction, the conditions a wavefunction must obey (be continuous, have a continuous first derivative, be single-valued, and be square-integrable)

Quantum mechanical principles

11.5 The information in a wavefunction

- node, a point where a wavefunction passes through zero
- operator, something that carries out a mathematical operation on a function
- hamiltonian operator, the operator for the total energy of a system, $H\psi = E\psi$
- eigenvalue, the constant ω in the eigenvalue equation $\hat{\Omega}\psi = \omega\psi$
- eigenfunction, the function ψ in the eigenvalue equation $\hat{\Omega}\psi = \omega\psi$
- eigenvalue equation, an equation of the form $\hat{\Omega}\psi = \omega\psi$
- observable, measurable properties of a system

- orthogonal functions, $\int \psi_i^*\psi_j \, \mathrm{d}\tau = 0$
- position operator, $\hat{x} = x \times$
- momentum operator, $\hat{p}_x = (\hbar/i)\,\mathrm{d}/\mathrm{d}x$
- linear combination of two functions, $c_1 f + c_2 g$
- superposition, a linear combination of wavefunctions
- complete set of functions, functions that can be used to express any arbitrary function as a linear combination
- expectation value, $\langle\Omega\rangle = \int \psi^*\hat{\Omega}\psi \, \mathrm{d}\tau$

11.6 The uncertainty principle

- Heisenberg uncertainty principle: It is impossible to specify simultaneously, with arbitrary precision, both the momentum and the position of a particle; $\Delta p \Delta q \geq \frac{1}{2}\hbar$
- wave packet, a sharply localized wavefunction
- complementary observables, observables corresponding to non-commuting operators
- commute, operators for which $[\hat{\Omega}_1, \hat{\Omega}_2] = 0$
- commutator, $[\hat{\Omega}_1, \hat{\Omega}_2] = \hat{\Omega}_1\hat{\Omega}_2 - \hat{\Omega}_2\hat{\Omega}_1$
- general form of the Heisenberg uncertainty principle: $\Delta\Omega_1\Delta\Omega_2 \geq \frac{1}{2}|\langle[\hat{\Omega}_1, \hat{\Omega}_2]\rangle|$

Further reading

Texts and sources of data and information

P.W. Atkins, *Quanta: A handbook of concepts.* Oxford University Press (1991).

P.W. Atkins and R.S. Friedman, *Molecular quantum mechanics.* Oxford University Press (1997).

D. Bohm, *Quantum theory.* Dover, New York (1989).

D.C. Cassidy, *Uncertainty: the life and science of Werner Heisenberg.* W.H. Freeman Co., New York (1992).

R.P. Feynman, R.B. Leighton, and M. Sands, *The Feynman lectures on physics.* Volume III. Addison–Wesley, Reading (1965).

C.S. Johnson, Jr. and L.G. Pedersen, *Problems and solutions in quantum chemistry and physics.* Dover, New York, 1986.

C. Kittel, *Introduction to solid state physics.* Wiley, New York (1996).

T.S. Kuhn, *Black-body theory and the quantum discontinuity 1894–1912.* Oxford University Press (1978).

M. Jammer, *The conceptual development of quantum mechanics.* McGraw-Hill, New York (1966).

W.J. Moore, *Schrödinger: life and thought.* Cambridge University Press (1989).

A. Pais, *Niels Bohr's times: in physics, philosophy, and polity.* Clarendon Press, Oxford (1991).

L. Pauling and E.B. Wilson, *Introduction to quantum mechanics with applications to chemistry.* Dover, New York (1985).

Exercises

Discussion questions

11.1 (a) Summarize the evidence that led to the introduction of quantum mechanics.

11.1 (b) Explain why Planck's introduction of quantization accounted for the properties of black-body radiation.

11.2 (a) Explain why Einstein's introduction of quantization accounted for the properties of heat capacities at low temperatures.

11.2 (b) Describe how a wavefunction determines the dynamical properties of a system and how those properties may be predicted.

11.3 (a) Account for the uncertainty relation between position and linear momentum in terms of the superposition of wavefunctions.

11.3 (b) Suggest how the general shape of a wavefunction can be predicted without solving the Schrödinger equation explicitly.

Numerical exercises

11.4 (a) Calculate the power radiated by a 2.0 m × 3.0 m section of the surface of a hot body at 1500 K.

11.4 (b) Calculate the power radiated by the surface of a cylindrical wire of length 5.0 cm and radius 0.12 mm that is heated to 3300 K by an electric current.

11.5 (a) Determine the wavelength of the radiation of the most intense electromagnetic radiation emitted from the surface of the star Sirius, which has a surface temperature of 11 000 K.

11.5 (b) Determine the wavelength of the radiation of the most intense electromagnetic radiation emitted from a furnace at 2500°C.

11.6 (a) Calculate the speed of an electron of wavelength 3.0 cm.

11.6 (b) Calculate the speed of a neutron of wavelength 3.0 cm.

11.7 (a) The fine-structure constant, α, plays a special role in the structure of matter; its approximate value is 1/137. What is the wavelength of an electron travelling at a speed αc, where c is the speed of light? (Note that the circumference of the first Bohr orbit in the hydrogen atom is 331 pm.)

11.7 (b) A certain diffraction experiment requires the use of electrons of wavelength 0.45 nm. Calculate the speed of the electrons.

11.8 (a) Calculate the linear momentum of photons of wavelength 750 nm. What speed does an electron need to travel to have the same linear momentum?

11.8 (b) Calculate the linear momentum of photons of wavelength 350 nm. What speed does a hydrogen molecule need to travel to have the same linear momentum?

11.9 (a) The energy required for the ionization of a certain atom is 3.44 aJ. The absorption of a photon of unknown wavelength ionizes the atom and ejects an electron with velocity 1.03 Mm s^{-1} (1 Mm = 10^6 m). Calculate the wavelength of the incident radiation.

11.9 (b) The energy required for the ionization of a certain atom is 5.12 aJ. The absorption of a photon of unknown wavelength ionizes

the atom and ejects an electron with velocity 345 km s^{-1}. Calculate the wavelength of the incident radiation.

11.10 (a) The speed of a certain proton is 0.45 Mm s^{-1}. If the uncertainty in its momentum is to be reduced to 0.0100 per cent, what uncertainty in its location must be tolerated?

11.10 (b) The speed of a certain electron is 995 km s^{-1}. If the uncertainty in its momentum is to be reduced to 0.0010 per cent, what uncertainty in its location must be tolerated?

11.11 (a) Calculate the energy per photon and the energy per mole of photons for radiation of wavelength (a) 600 nm (red), (b) 550 nm (yellow), (c) 400 nm (blue).

11.11 (b) Calculate the energy per photon and the energy per mole of photons for radiation of wavelength (a) 200 nm (ultraviolet), (b) 150 pm (X-ray), (c) 1.00 cm (microwave).

11.12 (a) Calculate the speed to which a stationary H atom would be accelerated if it absorbed each of the photons used in Exercise 11.11a.

11.12 (b) Calculate the speed to which a stationary ^{4}He atom (mass 4.0026 u) would be accelerated if it absorbed each of the photons used in Exercise 11.11b.

11.13 (a) A glow-worm of mass 5.0 g emits red light (650 nm) with a power of 0.10 W entirely in the backward direction. To what speed will it have accelerated after 10 y if released into free space and assumed to live?

11.13 (b) A photon-powered spacecraft of mass 10.0 kg emits radiation of wavelength 225 nm with a power of 1.50 kW entirely in the backward direction. To what speed will it have accelerated after 10.0 y if released into free space?

11.14 (a) A sodium lamp emits yellow light (550 nm). How many photons does it emit each second if its power is (a) 1.0 W, (b) 100 W?

11.14 (b) A laser used to read CDs emits red light of wavelength 700 nm. How many photons does it emit each second if its power is (a) 0.10 W, (b) 1.0 W?

11.15 (a) The peak of the Sun's emission occurs at about 480 nm; estimate the temperature of its surface.

11.15 (b) The peak of the emission from the hot iron in a steel furnace occurs at about 1600 nm; estimate the temperature of the steel.

11.16 (a) The work function for metallic cesium is 2.14 eV. Calculate the kinetic energy and the speed of the electrons ejected by light of wavelength (a) 700 nm, (b) 300 nm.

11.16 (b) The work function for metallic rubidium is 2.09 eV. Calculate the kinetic energy and the speed of the electrons ejected by light of wavelength (a) 650 nm, (b) 195 nm.

11.17 (a) Calculate the size of the quantum involved in the excitation of (a) an electronic oscillation of period 1.0 fs, (b) a molecular vibration of period 10 fs, (c) a pendulum of period 1.0 s. Express the results in joules and kilojoules per mole.

11.17 (b) Calculate the size of the quantum involved in the excitation of (a) an electronic oscillation of period 2.50 fs,

(b) a molecular vibration of period 2.21 fs, (c) a balance wheel of period 1.0 ms. Express the results in joules and kilojoules per mole.

11.18 (a) Calculate the de Broglie wavelength of (a) a mass of 1.0 g travelling at 1.0 cm s^{-1}, (b) the same, travelling at 100 km s^{-1}, (c) an He atom travelling at 1000 m s^{-1} (a typical speed at room temperature).

11.18 (b) Calculate the de Broglie wavelength of an electron accelerated from rest through a potential difference of (a) 100 V, (b) 1.0 kV, (c) 100 kV.

11.19 (a) Calculate the minimum uncertainty in the speed of a ball of mass 500 g that is known to be within 1.0 μm of a certain point on a bat. What is the minimum uncertainty in the position of a bullet of mass 5.0 g that is known to have a speed somewhere between 350.000 01 m s^{-1} and 350.000 00 m s^{-1}?

11.19 (b) An electron is confined to a linear region with a length of the same order as the diameter of an atom (about 100 pm). Calculate the minimum uncertainties in its position and speed.

11.20 (a) In an X-ray photoelectron experiment, a photon of wavelength 150 pm ejects an electron from the inner shell of an atom and it emerges with a speed of 21.4 Mm s^{-1}. Calculate the binding energy of the electron.

11.20 (b) In an X-ray photoelectron experiment, a photon of wavelength 121 pm ejects an electron from the inner shell of an atom and it emerges with a speed of 56.9 Mm s^{-1}. Calculate the binding energy of the electron.

Problems†

Numerical problems

11.1 The Planck distribution gives the energy in the wavelength range dλ at the wavelength λ. Calculate the energy density in the range 650 nm to 655 nm inside a cavity of volume 100 cm^3 when its temperature is (a) 25°C, (b) 3000°C.

11.2 The wavelength of the emission maximum from a small pinhole in an electrically heated container was determined at a series of temperatures, and the results are given below. Deduce a value for Planck's constant.

θ/°C	1000	1500	2000	2500	3000	3500
λ_{max}/nm	2181	1600	1240	1035	878	763

11.3 The Einstein frequency is often expressed in terms of an equivalent temperature θ_E, where $\theta_E = h\nu/k$. Confirm that θ_E has the dimensions of temperature, and express the criterion for the validity of the high-temperature form of the Einstein equation in terms of it. Evaluate θ_E for (a) diamond, for which $\nu = 46.5$ THz and (b) for copper, for which $\nu = 7.15$ THz. What fraction of the Dulong and Petit value of the heat capacity does each substance reach at 25°C?

11.4 The ground-state wavefunction for a particle confined to a one-dimensional box of length L is

$$\psi = \left(\frac{2}{L}\right)^{1/2} \sin\left(\frac{\pi x}{L}\right)$$

Suppose the box is 10.0 nm long. Calculate the probability that the particle is (a) between $x = 4.95$ nm and 5.05 nm, (b) between $x = 1.95$ nm and 2.05 nm, (c) between $x = 9.90$ nm and 10.00 nm, (d) in the right half of the box, (e) in the central third of the box.

11.5 The ground-state wavefunction of a hydrogen atom is

$$\psi = \left(\frac{1}{\pi a_0^3}\right)^{1/2} e^{-r/a_0}$$

where $a_0 = 53$ pm (the Bohr radius). (a) Calculate the probability that the electron will be found somewhere within a small sphere of radius 1.0 pm centred on the nucleus. (b) Now suppose that the same sphere is located at $r = a_0$. What is the probability that the electron is inside it?

11.6 The normalized wavefunctions for a particle confined to move on a circle are $\psi(\phi) = (1/2\pi)^{1/2} e^{-im\phi}$, where $m = 0, \pm1, \pm2, \pm3, \ldots$ and $0 \leq \phi \leq 2\pi$. Determine $\langle \phi \rangle$.

11.7 A particle is in a state described by the wavefunction $\psi(x) = (2a/\pi)^{1/4} e^{-\alpha x^2}$, where a is a constant and $-\infty \leq x \leq \infty$. Verify that the value of the product $\Delta p \Delta x$ is consistent with the predictions of the uncertainty principle.

11.8 A particle is in a state described by the wavefunction $\psi(x) = (2a)^{1/2} e^{-\alpha x}$, where a is a constant and $0 \leq x \leq \infty$. Determine the expectation value of the commutator of the position and momentum operators.

Theoretical problems

11.9 Derive Wien's law, that $\lambda_{max}T$ is a constant, from the Planck distribution, and deduce an expression for the constant.

11.10‡ Demonstrate explicitly that the Planck distribution reduces to the Rayleigh–Jeans law at long wavelengths.

11.11‡ Prior to Planck's derivation of the distribution law for black-body radiation, Wien found empirically a closely related distribution function that is very nearly but not exactly in agreement with the experimental results, namely, $\rho = (a/\lambda^5)e^{-b/\lambda kT}$. This formula shows small deviations from Planck's at long wavelengths. (a) By fitting Wien's empirical formula to Planck's at short wavelengths determine the constants a and b. (b) Demonstrate that Wien's formula is consistent with Wien's displacement law and with the Stefan–Boltzmann law.

11.12‡ Max Planck was the first to determine the Boltzmann constant, k, and his constant h from experimental data on

† Problems denoted with the symbol ‡ were supplied by Charles Trapp, Carmen Giunta, and Marshall Cady.

black-body radiation. Calculate values for k and h from the following data. The excitance M of a surface at 2000 K is 904.48 kW m^{-2}. At this temperature λ_{max} = 1.451 μm. *Hint.* Obtain λ_{max} from the Planck distribution by differentiation with respect to λ.

11.13 Normalize the following wavefunctions: (a) $\sin(n\pi x/L)$ in the range $0 \leq x \leq L$, (b) a constant in the range $-L \leq x \leq L$, (c) $e^{-r/a}$ in three-dimensional space, (d) $xe^{-r/2a}$ in three-dimensional space. *Hint.* The volume element in three dimensions is $d\tau = r^2\,dr\,\sin\theta\,d\theta\,d\phi$, with $0 \leq r < \infty$, $0 \leq \theta \leq \pi$, $0 \leq \phi \leq 2\pi$. A useful integral is

$$\int_0^\infty x^n e^{-ax}\,dx = \frac{n!}{a^{n+1}}$$

11.14 Two (unnormalized) excited state wavefunctions of the H atom are

(a) $\psi = \left(2 - \dfrac{r}{a_0}\right)e^{-r/2a_0}$

(b) $\psi = r\sin\theta\cos\phi\,e^{-r/2a_0}$

Normalize both functions to 1.

11.15 Identify which of the following functions are eigenfunctions of the operator d/dx: (a) e^{ikx}, (b) $\cos kx$, (c) k, (d) kx, (e) $e^{-\alpha x^2}$. Give the corresponding eigenvalue where appropriate.

11.16 Determine which of the following functions are eigenfunctions of the inversion operator i (which has the effect of making the replacement $x \rightarrow -x$): (a) $x^3 - kx$, (b) $\cos kx$, (c) $x^2 + 3x - 1$. State the eigenvalue of i when relevant.

11.17 Which of the functions in Problem 11.15 are (a) also eigenfunctions of d^2/dx^2 and (b) only eigenfunctions of d^2/dx^2? Give the eigenvalues where appropriate.

11.18 A particle is in a state described by the wavefunction

$\psi = (\cos\chi)e^{ikx} + (\sin\chi)e^{-ikx}$

where χ is a parameter. What is the probability that the particle will be found with a linear momentum (a) $+k\hbar$, (b) $-k\hbar$? What form would the wavefunction have if it were 90 per cent certain that the particle had linear momentum $+k\hbar$?

11.19 Evaluate the kinetic energy of the particle with wavefunction given in Problem 11.18.

11.20 Calculate the average linear momentum of a particle described by the following wavefunctions: (a) e^{ikx}, (b) $\cos kx$, (c) $e^{-\alpha x^2}$, where in each one x ranges from $-\infty$ to $+\infty$.

11.21 Evaluate the expectation values of r and r^2 for a hydrogen atom with wavefunctions given in Problem 11.14.

11.22 Calculate (a) the mean potential energy and (b) the mean kinetic energy of an electron in the ground state of a hydrogenic atom.

11.23 Use mathematical software to construct superpositions of cosine functions and explore how the wavefunction described becomes more localized as more components are included. Determine the probability that a given momentum will be observed. If you plot the superposition (which you should), set $x = 0$ at the centre of the screen and build the superposition there. Evaluate the root mean square location of the packet, $\langle x^2 \rangle^{1/2}$.

11.24 Determine the commutators of the operators (a) d/dx and x, (b) d/dx and x^2, (c) a and $a^\dagger$, where $a = (x + ip)/2^{1/2}$ and $a^\dagger = (x - ip)/2^{1/2}$.

Applications: to biology, environmental science, and astrophysics

11.25‡ The temperature of the Sun's surface is approximately 5800 K. On the assumption that the human eye evolved to be most sensitive at the wavelength of light corresponding to the maximum in the Sun's radiant energy distribution, determine the colour of light to which the eye is the most sensitive.

11.26‡ Solar energy strikes the top of the Earth's atmosphere at a rate of 343 W m^{-2}. About 30 per cent of this energy is reflected directly back into space by the Earth or the atmosphere. The Earth–atmosphere system absorbs the remaining energy and re-radiates it into space as black-body radiation. What is the average black-body temperature of the Earth? What is the wavelength of the most plentiful of the Earth's black-body radiation?

11.27‡ Problem 11.26 shows that the Earth's average temperature is maintained in a steady state through an energy balance between solar radiation absorbed by the Earth and black-body radiation emitted by the Earth and lost to space. Energy balances of this type are discussed in terms of energy flux, J, the energy passing through an area in an interval divided by the area and the duration of the interval and expressed in watts per square metre (W m^{-2}). (a) Prove that $J = \frac{1}{4}c\mathcal{E}$, where $\mathcal{E}$ is the isotropic black-body energy density (eq. 11.5). (*Hint.* Examine the radiation passing through area A in the time dt; use spherical coordinates centered on A and recognize that only volume elements within the hemisphere of radius $c\,dt$ contribute to the flux through A.) Also, determine $f(\tilde{v})$, where $dJ = f(\tilde{v})\,d\tilde{v}$ and prove that the Stefan–Boltzmann constant is given by eqn 11.6. Use $f(\tilde{v})$ to demonstrate graphically that the Earth's black-body emissions are in the infrared. (b) Consider an atmospheric model consisting of atmospheric nitrogen and oxygen only. The N_2 and O_2 molecules cannot absorb infrared radiation (Section 16.10). Determine the value of the Earth's surface temperature that is predicted by this model. It is found experimentally that the solar energy flux at the edge of the Earth's atmosphere is 0.1353 W cm^{-2} and that the fraction of the solar radiation scattered by gases and clouds of the atmosphere (the albedo) is 0.29. Consider that the magnitude of the solar radiation absorbed by the Earth equals the disk area of the Earth times the fraction of unscattered solar radiation times the solar radiation flux. The difference between the experimental value of the Earth's temperature (see Problem 11.26) and the temperature predicted by this model is the so-called greenhouse effect.

11.28‡ A star too small and cold to shine has been found by S. Kulkarni, K. Matthews, B.R. Oppenheimer, and T. Nakajima (*Science* **270**, 1478 (1995)). The spectrum of the object shows the presence of methane, which, according to the authors, would not exist at temperatures much above 1000 K. The mass of the star, as determined from its gravitational effect on a companion star, is roughly 20 times the mass of Jupiter. The star is considered to be a brown dwarf, the coolest ever found. (a) From available thermodynamic data, test the stability of methane at temperatures above 1000 K. (b) What is λ_{max} for this star? (c) What are the energy density and the excitance of the star relative to that of the Sun (6000 K)? (d) To determine whether the star will shine, estimate the fraction of the energy density of the star in the visible region of the spectrum.

12

Quantum theory: techniques and applications

Translational motion

12.1 A particle in a box

12.2 Motion in two and more dimensions

12.3 Tunnelling

Vibrational motion

12.4 The energy levels

12.5 The wavefunctions

Rotational motion

12.6 Rotation in two dimensions: the particle on a ring

12.7 Rotation in three dimensions: the particle on a sphere

12.8 Spin

Techniques of approximation

12.9 Time-independent perturbation theory

12.10 Time-dependent perturbation theory

To find the properties of systems according to quantum mechanics we need to solve the appropriate Schrödinger equation. This chapter presents the essentials of the solutions for three basic types of motion: translation, vibration, and rotation. We shall see that only certain wavefunctions and their corresponding energies are acceptable. Hence, quantization emerges as a natural consequence of the equation and the conditions imposed on it. The solutions bring to light a number of highly nonclassical, and therefore surprising, features of particles, especially their ability to tunnel into and through regions where classical physics would forbid them to be found. We also encounter a property of the electron, its spin, that has no classical counterpart. The chapter concludes with an introduction to the techniques used to obtain approximate solutions for systems for which exact solutions are unavailable.

The three basic modes of motion—translation (motion through space), vibration, and rotation—all play an important role in chemistry because they are ways in which molecules store energy. Gas-phase molecules, for instance, undergo translational motion, and their kinetic energy is a contribution to the total internal energy of a sample. Molecules can also store energy as rotational kinetic energy, and transitions between their rotational energy states can be observed spectroscopically. Energy is also stored as molecular vibration, and transitions between vibrational states are responsible for the appearance of infrared spectra.

Translational motion

Section 11.5 introduced the quantum mechanical description of free motion in one dimension. We saw there that the Schrödinger equation is

$$-\frac{\hbar^2}{2m}\frac{d^2\psi}{dx^2} = E\psi \tag{12.1a}$$

or more succinctly

$$H\psi = E\psi \qquad H = -\frac{\hbar^2}{2m}\frac{d^2}{dx^2} \tag{12.1b}$$

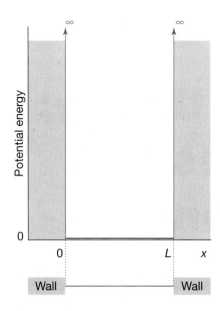

12.1 A particle in a one-dimensional region with impenetrable walls. Its potential energy is zero between $x = 0$ and $x = L$, and rises abruptly to infinity as soon as it touches the walls.

The general solutions are

$$\psi_k = Ae^{ikx} + Be^{-ikx} \qquad E_k = \frac{k^2\hbar^2}{2m} \tag{12.2}$$

Note that we are now labelling both the wavefunctions and the energies (that is, the eigenfunctions and eigenvalues of H) with the index k. We can verify that these functions are solutions by substituting ψ_k into the left-hand side of eqn 1a and showing that the result is equal to $E_k\psi_k$. In this case, all values of k, and therefore all values of the energy, are permitted. It follows that the translational energy of a free particle is not quantized.

We saw in Section 11.5c that a wavefunction of the form e^{ikx} describes a particle with linear momentum $p_x = +k\hbar$, corresponding to motion towards positive x (to the right), and that a wavefunction of the form e^{-ikx} describes a particle with the same magnitude of linear momentum but travelling towards negative x (to the left). That is, e^{ikx} is an eigenfunction of the operator $\hat{p}_x$ with eigenvalue $+k\hbar$, and e^{-ikx} is an eigenfunction with eigenvalue $-k\hbar$. In either state, $|\psi|^2$ is independent of x, which implies that the position of the particle is completely unpredictable. This conclusion is consistent with the uncertainty principle, because, if the momentum is certain, then the position cannot be specified (the operators $\hat{x}$ and $\hat{p}_x$ do not commute, Section 11.6).

12.1 A particle in a box

In this section, we consider a **particle in a box**, in which a particle of mass m is confined between two walls at $x = 0$ and $x = L$: the potential energy is zero inside the box but rises abruptly to infinity at the walls (Fig. 12.1). This model is an idealization of the potential energy of a gas-phase molecule that is free to move in a one-dimensional container. However, it is also the basis of the treatment of the electronic structure of metals (Chapter 23) and of a primitive treatment of conjugated molecules (Example 12.1). The particle in a box is also used in statistical thermodynamics in assessing the contribution of the translational motion of molecules to their thermodynamic properties (Chapter 19).

(a) The acceptable solutions

The Schrödinger equation for the region between the walls (where $V = 0$) is the same as for a free particle (eqn 1), so the general solutions given in eqn 2 are also the same. It is convenient to write them as

$$\psi_k(x) = C \sin kx + D \cos kx \qquad E_k = \frac{k^2\hbar^2}{2m} \tag{12.3}$$

We use $e^{\pm ix} = \cos x \pm i \sin x$ and absorb all numerical factors into the coefficients C and D.

For a free particle, any value of E_k corresponds to an acceptable solution. However, when the particle is confined within a region, the acceptable wavefunctions must satisfy certain **boundary conditions**, or constraints on the function at certain locations. It is physically impossible for the particle to be found with an infinite potential energy, so the wavefunction must be zero where V is infinite, at $x < 0$ and $x > L$. The continuity of the wavefunction then requires it to vanish just inside the well at $x = 0$ and $x = L$. That is, the boundary conditions are $\psi_k(0) = 0$ and $\psi_k(L) = 0$. These boundary conditions imply quantization, as we show informally first and then, following the *Justification*, more formally and generally.

Justification 12.1

For the informal demonstration of quantization, we consider each wavefunction to be a de Broglie wave that must fit within the container. The permitted wavelengths satisfy

$$L = n \times \tfrac{1}{2}\lambda \qquad n = 1, 2, \ldots$$

and therefore

$$\lambda = \frac{2L}{n} \qquad \text{with } n = 1, 2, \ldots$$

According to the de Broglie relation, these wavelengths correspond to the momenta

$$p = \frac{h}{\lambda} = \frac{nh}{2L}$$

The particle has only kinetic energy inside the box (where $V = 0$), so the permitted energies are

$$E = \frac{p^2}{2m} = \frac{n^2 h^2}{8mL^2} \qquad \text{with } n = 1, 2, \ldots \tag{12.4}$$

The more formal and widely applicable approach is as follows. Consider the wall at $x = 0$. According to eqn 3, $\psi(0) = D$ (because $\sin 0 = 0$ and $\cos 0 = 1$). But because $\psi(0) = 0$ we must have $D = 0$. It follows that the wavefunction must be of the form $\psi_k(x) = C \sin kx$. The value of ψ at the other wall (at $x = L$) is $\psi_k(L) = C \sin kL$, which must also be zero. Taking $C = 0$ would give $\psi_k(x) = 0$ for all x, which would conflict with the Born interpretation (the particle must be somewhere). Therefore, kL must be chosen so that $\sin kL = 0$, which is satisfied by

$$kL = n\pi \qquad n = 1, 2, \ldots \tag{12.5}$$

The value $n = 0$ is ruled out, because it implies $k = 0$ and $\psi_k(x) = 0$ everywhere (because $\sin 0 = 0$), which is unacceptable. Negative values of n merely change the sign of $\sin kL$ (because $\sin(-x) = -\sin x$). The wavefunctions are therefore

$$\psi_n(x) = C \sin(n\pi x/L) \qquad n = 1, 2, \ldots \tag{12.6}$$

(At this point we have started to label the solutions with the index n instead of k.) Because k and E_k are related by eqn 3, and k and n are related by eqn 5, it follows that the energy of the particle is limited to the values given in eqn 4.

Both approaches lead to the conclusion that the energy of the particle is quantized and that this quantization arises from the boundary conditions that ψ must satisfy if it is to be an acceptable wavefunction. This is a general conclusion: *the need to satisfy boundary conditions implies that only certain wavefunctions are acceptable, and hence restricts observables to discrete values.* So far, only energy has been quantized; shortly we shall see that other physical observables may also be quantized.

(b) Normalization

Before discussing the solution in more detail, we shall complete the derivation of the wavefunctions (which for a particle in a box are real; that is, do not contain i) by finding the normalization constant (here written C). To do so, we look for the value

To evaluate the integral, we use the standard form

$$\int \sin^2 ax \, dx = \tfrac{1}{2}x - (1/4a)\sin 2ax + \text{constant}.$$

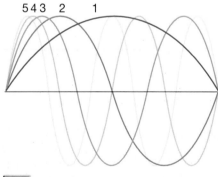

12.2 The allowed energy levels for a particle in a box. Note that the energy levels increase as n^2, and that their separation increases as the quantum number increases.

12.3 The first five normalized wavefunctions of a particle in a box. Each wavefunction is a standing wave, and successive functions possess one more half wave and a correspondingly shorter wavelength.

of C that ensures that the integral of ψ^2 over all the space available to the particle (i.e. from $x = 0$ to $x = L$) is equal to 1:

$$\int_0^L \psi^2 \, dx = C^2 \int_0^L \sin^2 \frac{n\pi x}{L} \, dx$$

$$= C^2 \times \frac{L}{2} = 1, \qquad \text{so} \quad C = \left(\frac{2}{L}\right)^{1/2}$$

for all n. Therefore, the complete solution to the problem is

$$E_n = \frac{n^2 h^2}{8mL^2} \qquad n = 1, 2, \ldots$$

$$\psi_n(x) = \left(\frac{2}{L}\right)^{1/2} \sin\left(\frac{n\pi x}{L}\right) \qquad \text{for} \quad 0 \leq x \leq L \tag{12.7}$$

The energies and wavefunctions are labelled with the 'quantum number' n. A **quantum number** is an integer (in some cases, as we shall see, a half-integer) that labels the state of the system. For a particle in a box there is an infinite number of acceptable solutions, and the quantum number n specifies the one of interest (Fig. 12.2). As well as acting as a label, a quantum number can often be used to calculate the energy corresponding to the state and to write down the wavefunction explicitly (in the present example, by using the information in eqn 7).

(c) The properties of the solutions

Figure 12.3 shows some of the wavefunctions of a particle in a box: they are all sine functions with the same amplitude but different wavelengths. Shortening the wavelength results in a sharper average curvature of the wavefunction and therefore an increase in the kinetic energy of the particle. Note that the number of nodes (points where the wavefunction passes *through* zero) also increases as n increases, and that the wavefunction ψ_n has $n-1$ nodes. Increasing the number of nodes between walls of a given separation increases the average curvature of the wavefunction and hence the kinetic energy of the particle.

The linear momentum of a particle in a box is not well defined because the wavefunction $\sin kx$ is a standing wave and, like the example of $\cos kx$ treated in Section 11.5d, not an eigenfunction of the linear momentum operator. However, each wavefunction is a superposition of momentum eigenfunctions:

$$\psi_n = \left(\frac{2}{L}\right)^{1/2} \sin \frac{n\pi x}{L} = \frac{1}{2i}\left(\frac{2}{L}\right)^{1/2} (e^{ikx} - e^{-ikx}) \qquad k = \frac{n\pi}{L} \tag{12.8}$$

It follows that measurement of the linear momentum will give the value $+k\hbar$ for half the measurements of momentum and $-k\hbar$ for the other half. This detection of opposite directions of travel with equal probability is the quantum mechanical version of the classical picture that a particle in a box rattles from wall to wall, and in any given period spends half its time travelling to the left and half travelling to the right.

Self-test 12.1 What is (a) the average value of the linear momentum of a particle in a box with quantum number n, (b) the average value of p^2?

[(a) $\langle p \rangle = 0$, (b) $\langle p^2 \rangle = n^2 h^2 / 4L^2$]

Because n cannot be zero, the lowest energy that the particle may possess is not zero (as would be allowed by classical mechanics, corresponding to a stationary particle) but

$$E_1 = \frac{h^2}{8mL^2} \tag{12.9}$$

This lowest, irremovable energy is called the **zero-point energy**. The physical origin of the zero-point energy can be explained in two ways. First, the uncertainty principle requires a particle to possess kinetic energy if it is confined to a finite region: the particle's location is not completely indefinite, so its momentum cannot be precisely zero. Hence it has nonzero kinetic energy. Second, if the wavefunction is to be zero at the walls, but smooth, continuous, and not zero everywhere, then it must be curved, and curvature in a wavefunction implies the possession of kinetic energy.

The separation between adjacent energy levels with quantum numbers n and $n + 1$ is

$$E_{n+1} - E_n = \frac{(n+1)^2 h^2}{8mL^2} - \frac{n^2 h^2}{8mL^2}$$

$$= (2n+1)\frac{h^2}{8mL^2} \tag{12.10}$$

This separation decreases as the length of the container increases, and is very small when the container has macroscopic dimensions. The separation of adjacent levels becomes zero when the walls are infinitely far apart. Atoms and molecules free to move in normal laboratory-sized vessels may therefore be treated as though their translational energy is not quantized. The translational energy of completely free particles (those not confined by walls) is not quantized.

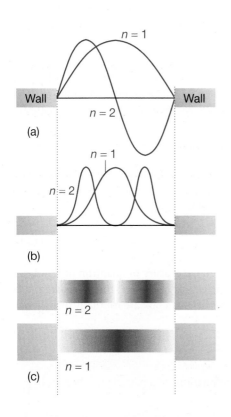

12.4 (a) The first two wavefunctions, (b) the corresponding probability distributions, and (c) a representation of the probability distribution in terms of the darkness of shading.

Illustration 12.1

For a container of length $L = 1.0$ nm, $h^2/8m_eL^2 = 6.0 \times 10^{-20}$ J (60 zJ). Therefore, the zero-point energy of an electron in such a box is $E_1 = 60$ zJ (corresponding to 0.37 eV). The minimum excitation energy is given by eqn 10 with $n = 1$, and is $3E_1$, or 1.8×10^2 zJ, which corresponds to 1.1 eV.

Self-test 12.2 Estimate a typical nuclear excitation energy by calculating the first excitation energy of a proton confined to a square well with a length roughly equal to the diameter of a nucleus (1 fm). [0.6 GeV]

The probability density for a particle in a box is

$$\psi^2(x) = \frac{2}{L}\sin^2\frac{n\pi x}{L} \tag{12.11}$$

and varies with position. The nonuniformity is pronounced when n is small (Fig. 12.4), but—provided we ignore the increasingly rapid oscillations—$\psi^2(x)$ becomes more uniform as n increases. The distribution at high quantum numbers reflects the classical result that a particle bouncing between the walls spends, on the average, equal times at all points. That the quantum result corresponds to the classical prediction at high quantum numbers is an illustration of the **correspondence principle**, which states that classical mechanics emerges from quantum mechanics as high quantum numbers are reached.

Example 12.1 Using the particle in a box solutions

The wavefunctions of an electron in a conjugated polyene (a molecule with alternating single and double bonds) can be approximated by particle-in-a-box wavefunctions. What is the probability, P, of locating the electron between $x = 0$ (the left-hand end of a molecule) and $x = 0.2$ nm in its lowest energy state in a conjugated molecule of length 1.0 nm?

Method The value of $\psi^2\,dx$ is the probability of finding the particle in the small region dx located at x; therefore, the total probability of finding the electron in the specified region is the integral of $\psi^2\,dx$ over that region. The wavefunction of the electron is given in eqn 7 with $n = 1$.

Answer The probability of finding the particle in a region between $x = 0$ and $x = l$ is

$$P = \int_0^l \psi_n^2\,dx = \frac{2}{L}\int_0^l \sin^2\frac{n\pi x}{L}\,dx$$

$$= \frac{l}{L} - \frac{1}{2n\pi}\sin\frac{2\pi nl}{L}$$

We then set $n = 1$ and $l = 0.2$ nm, which gives $P = 0.05$.

Comment The result corresponds to a chance of 1 in 20 of finding the electron in the region. As n becomes infinite, the sine term, which is multiplied by $1/n$, makes no contribution to P and the classical result, $P = l/L$, is obtained.

Self-test 12.3 Calculate the probability that an electron in the state with $n = 1$ will be found between $x = 0.25L$ and $x = 0.75L$ in a conjugated molecule of length L (with $x = 0$ at the left-hand end of the molecule). [0.82]

(d) Orthogonality and the bracket notation

We can now illustrate a property of wavefunctions first mentioned in Section 11.5. Two wavefunctions are **orthogonal** if the integral of their product vanishes. Specifically, the functions ψ_n and $\psi_{n'}$ are orthogonal if

$$\int \psi_n^*\psi_{n'}\,d\tau = 0 \tag{12.12}$$

where the integration is over all space. A general feature of quantum mechanics is that *wavefunctions corresponding to different energies are orthogonal*; therefore, we can be confident that all the wavefunctions of a particle in a box are mutually orthogonal.

Illustration 12.2

We can verify the orthogonality of wavefunctions of a particle in a box with $n = 1$ and $n = 3$ (Fig. 12.5):

$$\int_0^L \psi_1^*\psi_3\,dx = \frac{2}{L}\int_0^L \sin\frac{\pi x}{L}\sin\frac{3\pi x}{L}\,dx = 0$$

where we have used the standard integral given in *Illustration 11.1*.

12.5 Two functions are orthogonal if the integral of their product is zero. Here the calculation of the integral is illustrated graphically for two wavefunctions of a particle in a square well. The integral is equal to the total area beneath the graph of the product, and is zero.

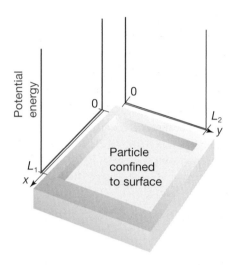

12.6 A two-dimensional square well. The particle is confined to the plane bounded by impenetrable walls. As soon as it touches the walls, its potential energy rises to infinity.

The integral in eqn 12 is often written

$$\langle n | n' \rangle = 0 \qquad (n' \neq n) \tag{12.13}$$

This **Dirac bracket notation** is much more succinct than writing out the integral in full. It also introduces the words 'bra' and 'ket' into the language of quantum mechanics. Thus, the **bra** $\langle n |$ corresponds to ψ_n^* and the **ket** $|n'\rangle$ corresponds to the wavefunction $\psi_{n'}$. When the bra and ket are put together as in eqn 13, the integration over all space is understood. Similarly, the normalization condition in eqn 11.20 becomes simply

$$\langle n | n \rangle = 1 \tag{12.14}$$

in bracket notation. These two expressions can be combined into one:

$$\langle n | n' \rangle = \delta_{nn'} \tag{12.15}$$

Here $\delta_{nn'}$, which is called the **Kronecker delta**, is 1 when $n' = n$ and 0 when $n' \neq n$. We shall see more of this notation later.

The property of orthogonality is of great importance in quantum mechanics because it enables us to eliminate a large number of integrals from calculations. Orthogonality plays a central role in the theory of chemical bonding (Chapter 14) and spectroscopy (Chapter 17). Sets of functions that are normalized and mutually orthogonal are called **orthonormal**. The wavefunctions in eqn 7 are orthonormal.

12.2 Motion in two and more dimensions

Next, we consider a two-dimensional version of the particle in a box. Now the particle is confined to a rectangular surface of length L_1 in the x-direction and L_2 in the y-direction; the potential energy is zero everywhere except at the walls, where it is infinite (Fig. 12.6). The wavefunction is now a function of both x and y and the Schrödinger equation is

$$-\frac{\hbar^2}{2m}\left(\frac{\partial^2 \psi}{\partial x^2} + \frac{\partial^2 \psi}{\partial y^2}\right) = E\psi \tag{12.16}$$

We need to see how to solve this *partial* differential equation, an equation in more than one variable.

(a) Separation of variables

Some partial differential equations can be simplified by the **separation of variables technique**, which divides the equation into two or more ordinary differential equations, one for each variable. An important application of this procedure, as we shall see, is the separation of the Schrödinger equation for the hydrogen atom into equations that describe the radial and angular variation of the wavefunction. The technique is particularly simple for a two-dimensional square well, as can be seen by testing whether a solution of eqn 16 can be found by writing the wavefunction as a product of functions, one depending only on x and the other only on y:

$$\psi(x, y) = X(x)Y(y) \tag{12.17}$$

With this substitution, we show in the *Justification* below that eqn 16 separates into two ordinary differential equations, one for each coordinate:

$$-\frac{\hbar^2}{2m}\frac{d^2X}{dx^2} = E_X X \qquad -\frac{\hbar^2}{2m}\frac{d^2Y}{dy^2} = E_Y Y \qquad E = E_X + E_Y \tag{12.18}$$

The quantity E_X is the energy associated with the motion of the particle parallel to the x-axis, and likewise for E_Y and motion parallel to the y-axis.

Justification 12.2

The first step in the justification of the separability of the wavefunction into the product of two functions X and Y is to note that, because X is independent of y and Y is independent of x, we can write

$$\frac{\partial^2 \psi}{\partial x^2} = \frac{\partial^2 XY}{\partial x^2} = Y \frac{d^2 X}{dx^2} \qquad \frac{\partial^2 \psi}{\partial y^2} = \frac{\partial^2 XY}{\partial y^2} = X \frac{d^2 Y}{dy^2}$$

Then eqn 16 becomes

$$-\frac{\hbar^2}{2m}\left(Y \frac{d^2 X}{dx^2} + X \frac{d^2 Y}{dy^2} \right) = EXY$$

When both sides are divided by XY, we can rearrange the resulting equation into

$$\frac{1}{X}\frac{d^2 X}{dx^2} + \frac{1}{Y}\frac{d^2 Y}{dy^2} = -\frac{2mE}{\hbar^2}$$

The first term on the left is independent of y, so if y is varied only the second term can change. But the sum of these two terms is a constant given by the right-hand side of the equation; therefore, even the second term cannot change when y is changed. In other words, the second term is a constant, which we write $-2mE_Y/\hbar^2$. By a similar argument, the first term is a constant when x changes, and we write it $-2mE_X/\hbar^2$, and $E = E_X + E_Y$. Therefore, we can write

$$\frac{1}{X}\frac{d^2 X}{dx^2} = -\frac{2mE_X}{\hbar^2} \qquad \frac{1}{Y}\frac{d^2 Y}{dy^2} = -\frac{2mE_Y}{\hbar^2}$$

which rearrange into the two ordinary (that is, single variable) differential equations in eqn 18.

Each of the two ordinary differential equations in eqn 18 is the same as the one-dimensional square-well Schrödinger equation. We can therefore adapt the results in eqn 7 without further calculation:

$$X_{n_1}(x) = \left(\frac{2}{L_1}\right)^{1/2} \sin \frac{n_1 \pi x}{L_1} \qquad Y_{n_2}(y) = \left(\frac{2}{L_2}\right)^{1/2} \sin \frac{n_2 \pi y}{L_2}$$

Then, because $\psi = XY$ and $E = E_X + E_Y$, we obtain

$$\psi_{n_1, n_2}(x, y) = \frac{2}{(L_1 L_2)^{1/2}} \sin \frac{n_1 \pi x}{L_1} \sin \frac{n_2 \pi y}{L_2} \qquad 0 \le x \le L_1, \quad 0 \le y \le L_2$$

$$E_{n_1, n_2} = \left(\frac{n_1^2}{L_1^2} + \frac{n_2^2}{L_2^2} \right) \frac{h^2}{8m} \tag{12.19}$$

with the quantum numbers taking the values $n_1 = 1, 2, \ldots$ and $n_2 = 1, 2, \ldots$ independently. Some of these functions are plotted in Fig. 12.7. They are the two-dimensional versions of the wavefunctions shown in Fig. 12.3. Note that two quantum numbers are needed in this two-dimensional problem, and in the Dirac bracket notation we denote the states by the ket $|n_1, n_2\rangle$.

We treat a particle in a three-dimensional box in the same way. The wavefunctions have another factor (for the z-dependence), and the energy has an additional

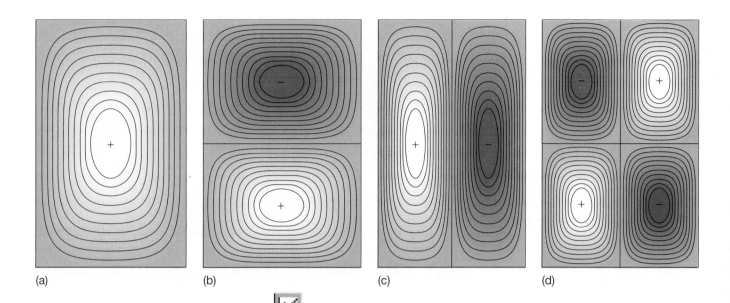

(a)   (b)   (c)   (d)

12.7 The wavefunctions for a particle confined to a rectangular surface depicted as contours of equal amplitude. (a) $n_1 = 1$, $n_2 = 1$, the state of lowest energy, (b) $n_1 = 1$, $n_2 = 2$, (c) $n_1 = 2$, $n_2 = 1$, and (d) $n_1 = 2$, $n_2 = 2$.

term in n_3^2/L_3^2. Solution of the Schrödinger equation by the separation of variables technique then gives

$$\psi_{n_1,n_2,n_3}(x,\,y,\,z) = \left(\frac{8}{L_1 L_2 L_3}\right)^{1/2} \sin\frac{n_1\pi x}{L_1}\, \sin\frac{n_2\pi y}{L_2}\, \sin\frac{n_3\pi z}{L_3}$$

$$0 \le x \le L_1, \quad 0 \le y \le L_2, \quad 0 \le z \le L_3$$

$$E_{n_1,n_2,n_3} = \left(\frac{n_1^2}{L_1^2} + \frac{n_2^2}{L_2^2} + \frac{n_3^2}{L_3^2}\right)\frac{h^2}{8m} \tag{12.20}$$

(b) Degeneracy

An interesting feature of the solutions for a particle in a two-dimensional box is obtained when the plane surface is square, with $L_1 = L_2 = L$. Then eqn 19 becomes

$$\psi_{n_1,n_2}(x,\,y) = \frac{2}{L}\sin\frac{n_1\pi x}{L}\,\sin\frac{n_2\pi y}{L} \qquad E_{n_1,n_2} = (n_1^2 + n_2^2)\frac{h^2}{8mL^2} \tag{12.21}$$

Consider the cases $n_1 = 1$, $n_2 = 2$ and $n_1 = 2$, $n_2 = 1$:

$$\psi_{1,2} = \frac{2}{L}\sin\frac{\pi x}{L}\,\sin\frac{2\pi y}{L} \qquad E_{1,2} = \frac{5h^2}{8mL^2}$$

$$\psi_{2,1} = \frac{2}{L}\sin\frac{2\pi x}{L}\,\sin\frac{\pi y}{L} \qquad E_{2,1} = \frac{5h^2}{8mL^2}$$

We see that different wavefunctions are **degenerate**, meaning that they correspond to the same energy. In this case, in which there are two degenerate wavefunctions, we say that the energy level $5(h^2/8mL^2)$ is 'doubly degenerate'. Alternatively, we say that the states $|1,2\rangle$ and $|2,1\rangle$ are degenerate. Degeneracy also occurs in three-dimensional boxes. We see from eqn 20, for instance, that, for a cubic box with sides of length L, the $|1,1,2\rangle$, $|1,2,1\rangle$, and $|2,1,1\rangle$ states all have the same energy, so the energy level $6(h^2/8mL^2)$ is triply degenerate.

12.8 The wavefunctions for a particle confined to a square surface. Note that one wavefunction can be converted into the other by a rotation of the box by 90°. The two functions correspond to the same energy. Degeneracy and symmetry are closely related.

The occurrence of degeneracy is related to the symmetry of the system. Figure 12.8 shows contour diagrams of the two degenerate functions $\psi_{1,2}$ and $\psi_{2,1}$. Because the box is square, we can convert one wavefunction into the other simply by rotating the plane by 90°. Interconversion by rotation through 90° is not possible when the plane is not square, and $\psi_{1,2}$ and $\psi_{2,1}$ are then not degenerate. Similar arguments account for the degeneracy of states in a cubic box. We shall see many other examples of degeneracy in the pages that follow (for example, in the hydrogen atom), and all of them can be traced to the symmetry properties of the system (see Section 15.4b).

12.3 Tunnelling

If the potential energy of a particle does not rise to infinity when it is in the walls of the container, and $E < V$, the wavefunction does not decay abruptly to zero. If the walls are thin (so that the potential energy falls to zero again after a finite distance), the wavefunction oscillates inside the box, varies smoothly inside the region representing the wall, and oscillates again on the other side of the wall outside the box (Fig. 12.9). Hence the particle might be found on the outside of a container even though according to classical mechanics it has insufficient energy to escape. Such leakage by penetration through classically forbidden zones is called **tunnelling**.

The Schrödinger equation can be used to calculate the probability of tunnelling of a particle of mass m incident on a finite barrier from the left. On the left of the barrier (for $x < 0$) the wavefunctions are those of a particle with $V = 0$, so from eqn 2 we can write

$$\psi = Ae^{ikx} + Be^{-ikx} \qquad k\hbar = (2mE)^{1/2} \tag{12.22}$$

The Schrödinger equation for the region representing the barrier (for $0 \leq x \leq L$), where the potential energy is the constant V, is

$$-\frac{\hbar^2}{2m}\frac{d^2\psi}{dx^2} + V\psi = E\psi \tag{12.23}$$

We shall consider particles that have $E < V$ (so, according to classical physics, the particle has insufficient energy to pass over the barrier), and therefore $V - E$ is positive. The general solutions of this equation are

$$\psi = Ce^{\kappa x} + De^{-\kappa x} \qquad \kappa\hbar = \{2m(V-E)\}^{1/2} \tag{12.24}$$

as we can readily verify by differentiating ψ twice with respect to x. The important feature to note is that the two exponentials are now real functions (as distinct from

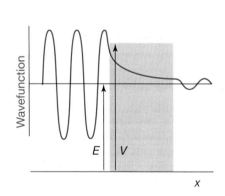

12.9 A particle incident on a barrier from the left has an oscillating wavefunction, but inside the barrier there are no oscillations (for $E < V$). If the barrier is not too thick, the wavefunction is nonzero at its opposite face, and so oscillations begin again there. (Only the real component of the wavefunction is shown.)

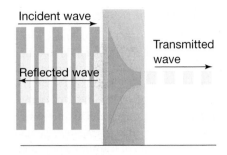

12.10 When a particle is incident on a barrier from the left, the wavefunction consists of a wave representing linear momentum to the right, a reflected component representing momentum to the left, a varying but not oscillating component inside the barrier, and a (weak) wave representing motion to the right on the far side of the barrier.

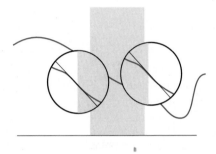

12.11 The wavefunction and its slope must be continuous at the edges of the barrier. The conditions for continuity enable us to connect the wavefunctions in the three zones and hence to obtain relations between the coefficients that appear in the solutions of the Schrödinger equation.

12.12 The transmission probabilities for passage through a barrier. The horizontal axis is the energy of the incident particle expressed as a multiple of the barrier height. The curves are labelled with the value of $L(2mV)^{1/2}/\hbar$. The graph on the left is for $E < V$ and that on the right for $E > V$. Note that $T > 0$ for $E < V$, whereas classically T would be zero. However, $T < 1$ for $E > V$, whereas classically T would be 1.

the complex, oscillating functions for the region where $V = 0$).[1] To the right of the barrier ($x > L$), where $V = 0$ again, the wavefunctions are

$$\psi = A'e^{ikx} + B'e^{-ikx} \qquad k\hbar = (2mE)^{1/2} \tag{12.25}$$

The complete wavefunction for a particle incident from the left consists of an incident wave, a wave reflected from the barrier, the exponentially changing amplitudes inside the barrier, and an oscillating wave representing the propagation of the particle to the right after tunnelling through the barrier successfully (Fig. 12.10). The acceptable wavefunctions have to obey the conditions set out in Section 11.4. In particular, they must be continuous at the edges of the barrier (at $x = 0$ and $x = L$, remembering that $e^0 = 1$):

$$A + B = C + D \qquad Ce^{\kappa L} + De^{-\kappa L} = A'e^{ikL} + B'e^{-ikL} \tag{12.26}$$

Their slopes (their first derivatives) must also be continuous there (Fig. 12.11):

$$ikA - ikB = \kappa C - \kappa D \qquad \kappa Ce^{\kappa L} - \kappa De^{-\kappa L} = ikA'e^{ikL} - ikB'e^{-ikL} \tag{12.27}$$

At this stage, we have four equations for the six unknown coefficients. If the particles are shot towards the barrier from the left, there can be no particles travelling to the left on the right of the barrier. Therefore, we can set $B' = 0$, which removes one more unknown. We cannot set $B = 0$ because some particles may be reflected back from the barrier toward negative x.

The probability that a particle is travelling towards positive x (to the right) on the left of the barrier is proportional to $|A|^2$, and the probability that it is travelling to the right on the right of the barrier is $|A'|^2$. The ratio of these two probabilities is called the **transmission probability**, T. After some algebra (see Problem 12.9) we find

$$T = \left\{ 1 + \frac{(e^{\kappa L} - e^{-\kappa L})^2}{16\varepsilon(1 - \varepsilon)} \right\}^{-1} \tag{12.28a}$$

where $\varepsilon = E/V$. This function is plotted in Fig. 12.12; the transmission coefficient for $E > V$ is shown there too. For high, wide barriers (in the sense that $\kappa L \gg 1$), eqn 28a simplifies to

1. Oscillating functions would be obtained if $E > V$.

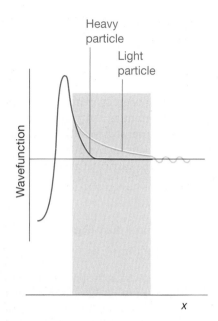

12.13 The wavefunction of a heavy particle decays more rapidly inside a barrier than that of a light particle. Consequently, a light particle has a greater probability of tunnelling through the barrier.

$$T \approx 16\varepsilon(1 - \varepsilon)e^{-2\kappa L} \tag{12.28b}$$

The transmission probability decreases exponentially with the thickness of the barrier and with $m^{1/2}$. It follows that particles of low mass are more able to tunnel through barriers than heavy ones (Fig. 12.13). Tunnelling is very important for electrons and muons, and moderately important for protons; for heavier particles it is less important. A number of effects in chemistry (for example, the isotope-dependence of some reaction rates) depend on the ability of the proton to tunnel more readily than the deuteron. The very rapid equilibration of proton transfer reactions (which were discussed in Chapter 9) is also a manifestation of the ability of protons to tunnel through barriers and transfer quickly from an acid to a base. Tunnelling of protons between acidic and basic groups is also an important feature of the mechanism of some enzyme-catalysed reactions. As we shall see in Chapter 29, electron tunnelling is one of the factors that determine the rates of electron transfer reactions at electrodes and in biological systems. The important technique of 'scanning tunnelling microscopy' (STM, Box 12.1) relies on the exponential dependence of electron tunnelling on the thickness of the region between a point and a surface.

Illustration 12.3

The ability of a proton to tunnel through a barrier contributes to the rapidity of proton transfer reactions in solution and therefore to the properties of acids and bases. To estimate the relative probabilities that a proton and a deuteron can tunnel through the same barrier of height 1.0 eV (1.6×10^{-19} J) and length 100 pm when their energy is 0.9 eV, we first evaluate

$$\kappa = \left\{ \frac{2(m/u) \times (1.67 \times 10^{-27} \text{ kg}) \times (1.6 \times 10^{-20} \text{ J})}{(1.055 \times 10^{-34} \text{ J s})^2} \right\}^{1/2} = \frac{(m/u)^{1/2}}{14 \text{ pm}}$$

The values of κ for a proton ($m = 1.0$ u) and a deuteron ($m = 2.0$ u) are 1/(14 pm) and 1/(9.9 pm), respectively, so $\kappa L \gg 1$ and eqn 28b can be used. The ratio of transmission probabilities is then

$$\frac{T_H}{T_D} = e^{-2(\kappa_H - \kappa_D)L} = 3.7 \times 10^2$$

(The ratio is very sensitive to rounding errors.) The result shows that the tunnelling probability of a proton (in the system specified) is nearly 80 million times greater than that of a deuteron. Therefore, we expect proton transfer reactions to be much faster than deuteron transfer reactions.

Self-test 12.4 Calculate the relative tunnelling probabilities when the barrier is twice as long, the other conditions being unchanged. $[1.4 \times 10^5]$

Box 12.1 Scanning microscopy

The central component in *scanning tunnelling microscopy* (STM) is a platinum–rhodium or tungsten needle, which is scanned across the surface of a conducting solid. When the tip of the needle is brought very close to the surface, electrons tunnel across the intervening space (see first illustration). In the constant-current mode of operation, the stylus moves up and down corresponding to the form of the surface, and the topography of the surface, including any adsorbates, can be mapped on an atomic scale. The vertical motion of the stylus is achieved by fixing it to a piezoelectric cylinder, which contracts or expands according to the potential difference it experiences. In the constant-z mode, the vertical position of

A scanning tunnelling microscope image of a liquid-crystal molecule (5-nonyl-2-nonoxylphenylpyrimidine) adsorbed on a graphite surface. (J.S. Foster, *et al.*, *Nature*, **338**, 137 (1988).

A representation of the tip of a scanning tunnelling microscope and the surface above which it lies. The tunnelling probability for electrons varies exponentially with the separation of the tip from the surface features, so the current (shown here by the blue band) is representative of atomic-scale variations in the topography of the surface.

The type of image that can be obtained with a scanning tunnelling microscope. The sample is of a silicon surface and the cliff is one atom high. (Sang-il Park and C.F. Quaite).

the stylus is held constant and the current is monitored. Because the tunnelling probability is very sensitive to the size of the gap, the microscope can detect tiny, atom-scale variations in the height of the surface.

An example of the kind of image obtained with a clean surface is shown in the second illustration, where the cliff is only one atom high. A spectacular demonstration of the power of the technique for displaying the shape of adsorbed

2. Problem supplied by Charles Trapp and Carmen Giunta.

species is shown in the third illustration, which shows a single molecule adsorbed on a surface.

In *atomic force microscopy* (AFM) a sharpened stylus attached to a beam is scanned across the surface. The force exerted by the surface and any adsorbate pushes or pulls on the stylus and deflects the beam. The deflection is monitored either by interferometry or by using a laser beam. Because no current is needed between the sample and the probe, the technique can be applied to nonconducting surfaces too.

Discussion questions

1. To get an idea of the distance dependence of the tunnelling current in STM, suppose that the wavefunction of the electron in the gap between sample and tip is given by $\psi = Be^{-\kappa x}$, where $\kappa = \{2m_e(V - E)/\hbar^2\}^{1/2}$; take $V - E = 2.0$ eV. By what factor would the current drop if the probe is moved from 0.50 nm to 0.60 nm from the surface?[2]

2. The potential energy of interaction between two charges q_1 and q_2 separated by a distance r is $V = q_1 q_2/4\pi\varepsilon_0 r$, where $\varepsilon_0 = 8.85 \times 10^{-12}$ C^2 J^{-1} m^{-1} is the vacuum permittivity (*Further information 2*). To get an idea of the magnitudes of forces measured by AFM, calculate the force acting between two electrons separated by 0.50 nm. By what factor does the force drop if the distance between the electrons increases to 0.60 nm? *Hint.* Note that $F = -dV/dr$.

Further reading

D. Bonnell (ed.), *Scanning probe microscopy and spectroscopy: theory, techniques, and applications.* Wiley, New York (2000).

T. Sakurai and Y. Watanabe (ed.), *Advances in scanning probe microscopy.* Springer, Berlin (2000).

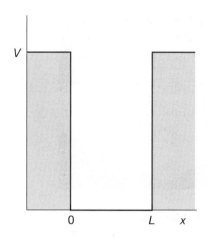

12.14 A potential well with a finite depth.

A problem related to the one just considered is that of a particle in a square well potential of finite depth (Fig. 12.14). In this kind of potential, the wavefunction penetrates into the walls, where it decays exponentially towards zero, and oscillates within the well. The wavefunctions are found by ensuring, as in the discussion of tunnelling, that they and their slopes are continuous at the edges of the potential. The two lowest energy solutions are shown in Fig. 12.15. A further difference from the solutions for an infinitely deep well is that there is only a finite number of bound states, states with $E < V$. Regardless of the depth and length of the well, there is always at least one bound state. Detailed consideration of the Schrödinger equation for the problem shows that in general the number of levels is equal to N, with

$$N - 1 < \frac{(8mVL)^{1/2}}{h} < N$$

where V is the depth of the well and L is its length.[3] We see that the deeper and wider the well, the greater the number of bound states. As the depth becomes infinite, so the number of bound states also becomes infinite, as we have already seen.

Vibrational motion

A particle undergoes **harmonic motion** if it experiences a restoring force proportional to its displacement:

$$F = -kx \qquad (12.29)$$

where k is the **force constant**: the stiffer the 'spring', the greater the value of k. Because force is related to potential energy by $F = -dV/dx$ (see *Further information 2*), the force in eqn 29 corresponds to a potential energy

$$V = \tfrac{1}{2}kx^2 \qquad (12.30)$$

This expression, which is the equation of a parabola (Fig. 12.16), is the origin of the term 'parabolic potential energy' for the potential energy characteristic of a harmonic oscillator. The Schrödinger equation for the particle is therefore

$$-\frac{\hbar^2}{2m}\frac{d^2\psi}{dx^2} + \tfrac{1}{2}kx^2\psi = E\psi \qquad (12.31)$$

12.4 The energy levels

Equation 31 is a standard equation in the theory of differential equations and its solutions are well known to mathematicians (see below).[4] Quantization of energy levels arises from the boundary conditions: the oscillator will not be found with infinitely large compressions or extensions, so the only allowed solutions are those for which $\psi = 0$ at $x = \pm\infty$. The permitted energy levels are

$$E_v = (v + \tfrac{1}{2})\hbar\omega \qquad \omega = \left(\frac{k}{m}\right)^{1/2} \qquad v = 0, 1, 2, \ldots \qquad (12.32)$$

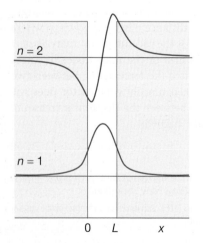

$n = 2$

$n = 1$

12.15 The lowest two bound-state wavefunctions for a particle in the well shown in Fig. 12.14.

3. For references to the derivation of this expression, see *Further reading*.
4. For the details of the solution, see *Further reading*.

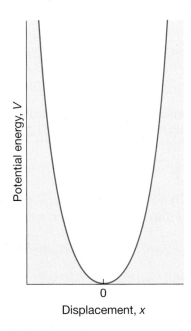

12.16 The parabolic potential energy $V = \frac{1}{2}kx^2$ of a harmonic oscillator, where x is the displacement from equilibrium. The narrowness of the curve depends on the force constant k: the larger the value of k, the narrower the well.

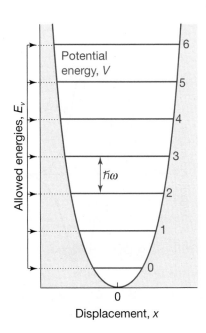

12.17 The energy levels of a harmonic oscillator are evenly spaced with separation $\hbar\omega$, with $\omega = (k/m)^{1/2}$. Even in its lowest state, an oscillator has an energy greater than zero.

Note that ω (omega) increases with increasing force constant and decreasing mass. It follows that the separation between adjacent levels is

$$E_{v+1} - E_v = \hbar\omega \tag{12.33}$$

which is the same for all v. Therefore, the energy levels form a uniform ladder of spacing $\hbar\omega$ (Fig. 12.17). The energy separation $\hbar\omega$ is negligibly small for macroscopic objects (with large mass), but is of great importance for objects with mass similar to that of atoms.

Illustration 12.4

The force constant of a typical X–H chemical bond is around 500 N m^{-1}. Because the mass of a proton is about 1.7×10^{-27} kg, $\omega \approx 5.4 \times 10^{14}$ s^{-1} (5.4×10^2 THz) and the separation of adjacent levels is $\hbar\omega \approx 5.7 \times 10^{-20}$ J (57 zJ, about 0.36 eV). This energy separation corresponds to 34 kJ mol^{-1}, which is chemically significant. The excitation of the vibration of the bond from one level to the level immediately above requires 57 zJ. Therefore, if it is caused by a photon, the excitation requires radiation of frequency $v = \Delta E/h = 86$ THz and wavelength $\lambda = c/v = 3.5$ μm. It follows that transitions between adjacent vibrational energy levels of molecules are stimulated by or emit infrared radiation (Chapter 16).

Because the smallest permitted value of v is 0, it follows from eqn 33 that a harmonic oscillator has a zero-point energy

$$E_0 = \frac{1}{2}\hbar\omega \tag{12.34}$$

For the typical molecular oscillator specified in *Illustration 12.4*, the zero-point energy is about 30 zJ, which corresponds to 0.2 eV, or 15 kJ mol^{-1}. The mathematical reason for the zero-point energy is that v cannot take negative values, for if it did the wavefunction would be ill-behaved. The physical reason is the same as for the particle in a square well: the particle is confined, its position is not completely uncertain, and therefore its momentum, and hence its kinetic energy, cannot be exactly zero. We can picture this zero-point state as one in which the particle fluctuates incessantly around its equilibrium position; classical mechanics would allow the particle to be perfectly still.

12.5 The wavefunctions

It is helpful at the outset to identify the similarities between the harmonic oscillator and the particle in a box, for then we shall be able to anticipate the form of the oscillator wavefunctions without detailed calculation. Like the particle in a box, a particle undergoing harmonic motion is trapped in a symmetrical well in which the potential energy rises to large values (and ultimately to infinity) for sufficiently large displacements (compare Figs 12.1 and 12.16). However, there are two important differences. First, because the potential energy climbs towards infinity only as x^2 and not abruptly, the wavefunction approaches zero more slowly at large displacements than for the particle in a box. Second, as the kinetic energy of the oscillator depends on the displacement in a more complex way (on account of the variation of the potential energy), the curvature of the wavefunction also varies in a more complex way.

The Hermite polynomials are solutions of the differential equation

$$H_v'' - 2yH_v' + 2vH_v = 0$$

where primes denote differentiation. They satisfy the recursion relation

$$H_{v+1} + 2yH_v - 2vH_{v-1} = 0$$

An important integral is

$$\int_{-\infty}^{\infty} H_{v'}H_v e^{-y^2}\, dy = \begin{cases} 0 & \text{if } v' \neq v \\ \pi^{1/2} 2^v v! & \text{if } v' = v \end{cases}$$

Table 12.1 The Hermite polynomials $H_v(y)$

v	H_v
0	1
1	$2y$
2	$4y^2 - 2$
3	$8y^3 - 12y$
4	$16y^4 - 48y^2 + 12$
5	$32y^5 - 160y^3 + 120y$
6	$64y^6 - 480y^4 + 720y^2 - 120$

(a) The form of the wavefunctions

The detailed solution of eqn 31 shows that the wavefunction for a harmonic oscillator has the form

$$\psi(x) = N \times (\text{polynomial in } x) \times (\text{bell-shaped Gaussian function})$$

where N is a normalization constant. A Gaussian function is a function of the form e^{-x^2} (Fig. 12.18). The precise form of the wavefunctions is

$$\psi_v(x) = N_v H_v(y)e^{-y^2/2} \qquad y = \frac{x}{\alpha} \qquad \alpha = \left(\frac{\hbar^2}{mk}\right)^{1/4} \tag{12.35}$$

The factor $H_v(y)$ is a **Hermite polynomial** (Table 12.1). For instance, because $H_0(y) = 1$, the wavefunction for the ground state (the lowest energy state, with $v = 0$) of the harmonic oscillator is

$$\psi_0(x) = N_0 e^{-y^2/2} = N_0 e^{-x^2/2\alpha^2} \tag{12.36}$$

It follows that the probability density is the bell-shaped Gaussian function

$$\psi_0^2(x) = N_0^2 e^{-x^2/\alpha^2} \tag{12.37}$$

The wavefunction and the probability distribution are shown in Fig. 12.19. Both curves have their largest values at zero displacement (at $x = 0$), so they capture the classical picture of the zero-point energy as arising from the ceaseless fluctuation of the particle about its equilibrium position.

The wavefunction for the first excited state of the oscillator, the state with $v = 1$, is obtained by noting that $H_1(y) = 2y$ (note that some of the Hermite polynomials are very simple functions!):

$$\psi_1(x) = N_1 \times 2y\, e^{-y^2/2} \tag{12.38}$$

This function has a node at zero displacement ($x = 0$), and the probability density has maxima at $x = \pm\alpha$, corresponding to $y = \pm 1$ (Fig. 12.20).

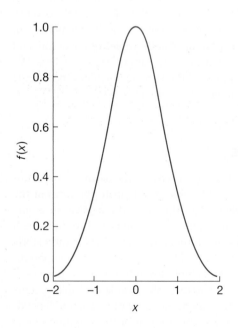

12.18 The graph of the Gaussian function, $f(x) = e^{-x^2}$.

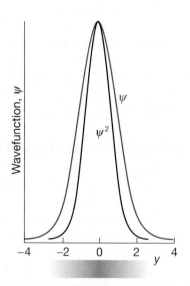

12.19 The normalized wavefunction and probability distribution (shown also by shading) for the lowest energy state of a harmonic oscillator.

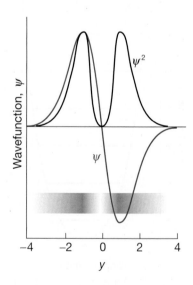

12.20 The normalized wavefunction and probability distribution (shown also by shading) for the first excited state of a harmonic oscillator.

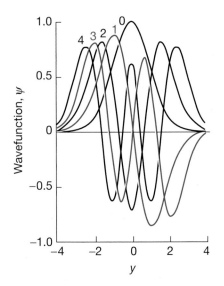

12.21 The normalized wavefunctions for the first five states of a harmonic oscillator. Even values of v are black; odd values are blue. Note that the number of nodes is equal to v and that alternate wavefunctions are symmetrical or antisymmetrical about $y = 0$ (zero displacement).

The shapes of several wavefunctions are shown in Fig. 12.21. The shading in Fig. 12.22, which represents the probability density, is based on the squares of these functions. At high quantum numbers, harmonic oscillator wavefunctions have their largest amplitudes near the turning points of the classical motion (the locations at which $V = E$, so the kinetic energy is zero). We see classical properties emerging in the correspondence limit of high quantum numbers, for a classical particle is most likely to be found at the turning points (where it travels most slowly) and is least likely to be found at zero displacement (where it travels most rapidly).

Example 12.2 Normalizing a harmonic oscillator wavefunction

Find the normalization constant for the harmonic oscillator wavefunctions.

Method Normalization is always carried out by evaluating the integral of $|\psi|^2$ over all space and then finding the normalization factor from eqn 11.17. The normalized wavefunction is then equal to $N\psi$. In this one-dimensional problem, the volume element is dx and the integration is from $-\infty$ to $+\infty$. The wavefunctions are expressed in terms of the dimensionless variable $y = x/\alpha$, so begin by expressing the integral in terms of y by using $dx = \alpha \, dy$. The integrals required are given above Table 12.1.

Answer The unnormalized wavefunction is

$$\psi_v(x) = H_v(y)e^{-y^2/2}$$

It follows from the integrals given above Table 12.1 that

$$\int_{-\infty}^{\infty} \psi_v^* \psi_v \, dx = \alpha \int_{-\infty}^{\infty} \psi_v^* \psi_v \, dy = \alpha \int_{-\infty}^{\infty} H_v^2(y)e^{-y^2} \, dy = \alpha \pi^{1/2} 2^v v!$$

where $v! = v(v-1)(v-2) \ldots 1$. Therefore,

$$N_v = \left(\frac{1}{\alpha \pi^{1/2} 2^v v!} \right)^{1/2}$$

Note that for a harmonic oscillator N_v is different for each value of v.

Comment The Hermite polynomials are members of a class of functions called *orthogonal polynomials*. These polynomials have a wide range of important properties that allow a number of quantum mechanical calculations to be done with relative ease. See *Further reading* for a reference to their properties.

12.22 The probability distributions for the first five states of a harmonic oscillator and the state with $v = 20$. Note how the regions of highest probability move towards the turning points of the classical motion as v increases.

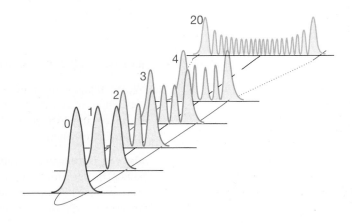

Self-test 12.5 Confirm, by explicit evaluation of the integral, that ψ_0 and ψ_1 are orthogonal.

[Evaluate the integral $\displaystyle\int_{-\infty}^{\infty} \psi_0^* \psi_1 \, dx$ by using the information above Table 12.1]

(b) The properties of oscillators

With the wavefunctions that are available, we can start calculating the properties of a harmonic oscillator. For instance, we can calculate the expectation values of an observable Ω by evaluating integrals of the type

$$\langle \Omega \rangle = \int_{-\infty}^{\infty} \psi_v^* \hat{\Omega} \psi_v \, dx \tag{12.39}$$

(Here and henceforth, the wavefunctions are all taken to be normalized to 1.) A tidier expression is obtained by using Dirac bracket notation, in which an integral is replaced by a bracket labelled with the quantum numbers of the states:

$$\langle v' | \hat{\Omega} | v \rangle = \int_{-\infty}^{\infty} \psi_{v'}^* \hat{\Omega} \psi_v \, dx \tag{12.40}$$

This bracket is also called a **matrix element** of the operator $\hat{\Omega}$ and written $\Omega_{v'v}$. Note how the operator stands between the bra and the ket (which may denote different states), in the place of the c in $\langle \text{bra}|c|\text{ket} \rangle$. An integration is implied whenever a complete bracket is written. In this notation, the expectation value is

$$\langle \Omega \rangle = \langle v | \hat{\Omega} | v \rangle \tag{12.41}$$

with the bra and the ket corresponding to the same state (with quantum number v and wavefunction ψ_v).

When the explicit wavefunctions are substituted, the integrals look fearsome, but the Hermite polynomials have many simplifying features. For instance, we show in the following example that the mean displacement, $\langle x \rangle$, and the mean square displacement, $\langle x^2 \rangle$, of the oscillator when it is in the state with quantum number v are

$$\langle x \rangle = 0 \qquad \langle x^2 \rangle = (v + \tfrac{1}{2}) \frac{\hbar}{(mk)^{1/2}} \tag{12.42}$$

The result for $\langle x \rangle$ shows that the oscillator is equally likely to be found on either side of $x = 0$ (like a classical oscillator). The result for $\langle x^2 \rangle$ shows that the mean square displacement increases with v. This increase is apparent from the probability densities in Fig. 12.22, and corresponds to the classical amplitude of swing increasing as the oscillator becomes more highly excited.

Example 12.3 Calculating properties of a harmonic oscillator

We can imagine the bending motion of a CO_2 molecule as a harmonic oscillation relative to the linear state of the molecule. We may be interested in the extent to which the molecule bends. Calculate the mean displacement of the oscillator when it is in a quantum state v.

Method Normalized wavefunctions must be used to calculate the expectation value. The operator for position along x is multiplication by the value of x (Section 11.5c). The resulting integral can be evaluated either by inspection (the integrand is the product of an odd and an even function), or by explicit evaluation using the formulas above Table 12.1. To give practice in this type of calculation, we illustrate the latter procedure. We shall need the relation $x = \alpha y$, which implies that $dx = \alpha\, dy$.

Answer The integral we require is

$$\langle x \rangle = \int_{-\infty}^{\infty} \psi_v^* x \psi_v \, dx = N_v^2 \int_{-\infty}^{\infty} (H_v e^{-y^2/2}) x (H_v e^{-y^2/2}) \, dx$$

$$= \alpha^2 N_v^2 \int_{-\infty}^{\infty} (H_v e^{-y^2/2}) y (H_v e^{-y^2/2}) \, dy$$

$$= \alpha^2 N_v^2 \int_{-\infty}^{\infty} H_v y H_v e^{-y^2} \, dy$$

Now use the recursion relation to form

$$yH_v = vH_{v-1} + \tfrac{1}{2}H_{v+1}$$

which turns the integral into

$$\int_{-\infty}^{\infty} H_v y H_v e^{-y^2} \, dy = v \int_{-\infty}^{\infty} H_{v-1} H_v e^{-y^2} \, dy + \tfrac{1}{2} \int_{-\infty}^{\infty} H_{v+1} H_v e^{-y^2} \, dy$$

Both integrals are zero, so $\langle x \rangle = 0$.

Comment As remarked in the text, the mean displacement is zero because the displacement occurs equally on either side of the equilibrium position. The following Self-test extends this calculation by examining the mean square displacement, which we can expect to be nonzero and to increase with increasing v.

Self-test 12.6 Calculate the mean square displacement $\langle x^2 \rangle$ of the particle from its equilibrium position. (Use the recursion relation twice.)   [eqn 42]

The mean potential energy of an oscillator, the expectation value of $V = \tfrac{1}{2}kx^2$, can now be calculated very easily:

$$\langle V \rangle = \langle \tfrac{1}{2}kx^2 \rangle = \tfrac{1}{2}(v + \tfrac{1}{2})\hbar \left(\frac{k}{m}\right)^{1/2} = \tfrac{1}{2}(v + \tfrac{1}{2})\hbar\omega \tag{12.43}$$

Because the total energy in the state with quantum number v is $(v + \tfrac{1}{2})\hbar\omega$, it follows that

$$\langle V \rangle = \tfrac{1}{2}E_v \tag{12.44}$$

The total energy is the sum of the potential and kinetic energies, so it follows at once that the mean kinetic energy of the oscillator is

$$\langle E_K \rangle = \tfrac{1}{2}E_v \tag{12.45}$$

The result that the mean potential and kinetic energies of a harmonic oscillator are equal (and therefore that both are equal to half the total energy) is a special case of the **virial theorem**:

If the potential energy of a particle has the form $V = ax^b$, then its mean potential and kinetic energies are related by

$$2\langle E_K \rangle = b\langle V \rangle \tag{12.46}$$

For a harmonic oscillator $b = 2$, so $\langle E_K \rangle = \langle V \rangle$, as we have found. The virial theorem is a short cut to the establishment of a number of useful results, and we shall use it again.

An oscillator may be found at extensions with $V > E$ that are forbidden by classical physics, for they correspond to negative kinetic energy. For example, it follows from the shape of the wavefunction (see the *Justification* below) that in its lowest energy state there is about an 8 per cent chance of finding an oscillator stretched beyond its classical limit and an 8 per cent chance of finding it with a classically forbidden compression. These tunnelling probabilities are independent of the force constant and mass of the oscillator. The probability of being found in classically forbidden regions decreases quickly with increasing v, and vanishes entirely as v approaches infinity, as we would expect from the correspondence principle. Macroscopic oscillators (such as pendulums) are in states with very high quantum numbers, so the probability that they will be found in a classically forbidden region is wholly negligible. Molecules, however, are normally in their vibrational ground states, and for them the probability is very significant.

Justification 12.3

According to classical mechanics, the turning point, x_{tp}, of an oscillator occurs when its kinetic energy is zero, which is when its potential energy $\frac{1}{2}kx^2$ is equal to its total energy E. This equality occurs when

$$x_{tp}^2 = \frac{2E}{k}, \quad \text{or} \quad x_{tp} = \pm \left(\frac{2E}{k} \right)^{1/2}$$

with E given by eqn 32. The probability of finding the oscillator stretched beyond a displacement x_{tp} is the sum of the probabilities $\psi^2 \, dx$ of finding it in any of the intervals dx lying between x_{tp} and infinity:

$$P = \int_{x_{tp}}^{\infty} \psi_v^2 \, dx$$

The variable of integration is best expressed in terms of $y = x/\alpha$, with $\alpha = (\hbar^2/mk)^{1/2}$ and then the turning point on the right lies at

$$y_{tp} = \frac{x_{tp}}{\alpha} = \left\{ \frac{2(v + \frac{1}{2})\hbar\omega}{\alpha^2 k} \right\}^{1/2} = (2v + 1)^{1/2}$$

For the state of lowest energy ($v = 0$), $y_{tp} = 1$ and the probability is

$$P = \int_{x_{tp}}^{\infty} \psi_0^2 \, dx = \alpha N_0^2 \int_1^{\infty} e^{-y^2} \, dy$$

The integral is a special case of the *error function*, erf z, which is defined as follows:

$$\text{erf } z = 1 - \frac{2}{\pi^{1/2}} \int_z^{\infty} e^{-y^2} \, dy$$

The values of this function are tabulated and available in mathematical software packages; a small selection of values is given in Table 12.2. In the present case

$$P = \tfrac{1}{2}(1 - \text{erf } 1) = \tfrac{1}{2}(1 - 0.843) = 0.079$$

It follows that, in 7.9 per cent of a large number of observations, any oscillator in the state $v = 0$ will be found stretched to a classically forbidden extent. There is the same

Table 12.2* The error function

z	erf z
0	0
0.01	0.0113
0.05	0.0564
0.10	0.1125
0.50	0.5205
1.00	0.8427
1.50	0.9661
2.0	0.9953

* More values are given in the *Data section*.

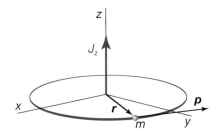

12.23 The angular momentum of a particle of mass m on a circular path of radius r in the xy-plane is represented by a vector J with the single nonzero component J_z of magnitude pr perpendicular to the plane.

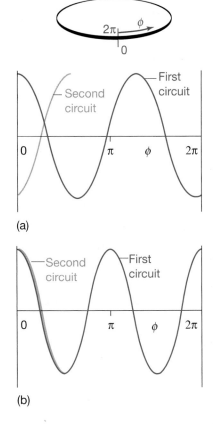

(a)

(b)

12.24 Two solutions of the Schrödinger equation for a particle on a ring. The circumference has been opened out into a straight line; the points at $\phi = 0$ and 2π are identical. The solution in (a) is unacceptable because it is not single-valued. Moreover, on successive circuits it interferes destructively with itself, and does not survive. The solution in (b) is acceptable: it is single-valued, and on successive circuits it reproduces itself.

probability of finding the oscillator with a classically forbidden compression. The total probability of finding the oscillator tunnelled into a classically forbidden region (stretched or compressed) is about 16 per cent. A similar calculation for the state with $v = 6$ shows that the probability of finding the oscillator outside the classical turning points has fallen to about 7 per cent.

Rotational motion

The treatment of rotational motion can be broken down into two parts. The first deals with motion in two dimensions and the second with rotation in three dimensions. It may be helpful to review the classical description of rotational motion given in *Further information 2*, particularly the concepts of moment of inertia and angular momentum.

12.6 Rotation in two dimensions: the particle on a ring

We consider a particle of mass m constrained to move in a circular path of radius r in the xy-plane (Fig. 12.23). The total energy is equal to the kinetic energy, because $V = 0$ everywhere. We can therefore write $E = p^2/2m$. According to classical mechanics, the **angular momentum**, J_z, around the z-axis (which lies perpendicular to the xy-plane) is $J_z = \pm pr$, so the energy can be expressed as $J_z^2/2mr^2$. Because mr^2 is the **moment of inertia**, I, of the mass on its path, it follows that

$$E = \frac{J_z^2}{2I} \qquad (12.47)$$

We shall now see that not all the values of the angular momentum are permitted in quantum mechanics, and therefore that both angular momentum and rotational energy are quantized.

(a) The qualitative origin of quantized rotation

Because $J_z = \pm pr$ and, from the de Broglie relation, $p = h/\lambda$, the angular momentum about the z-axis is

$$J_z = \pm \frac{hr}{\lambda}$$

Opposite signs correspond to opposite directions of travel. This equation shows that, the shorter the wavelength of the particle on a circular path of given radius, the greater the angular momentum of the particle. It follows that, if we can see why the wavelength is restricted to discrete values, then we shall understand why the angular momentum is quantized.

Suppose for the moment that λ can take an arbitrary value. In that case, the wavefunction depends on the azimuthal angle ϕ as shown in Fig. 12.24a. When ϕ increases beyond 2π, the wavefunction continues to change, but for an arbitrary wavelength it gives rise to a different value at each point, which is unacceptable (Section 11.4). An acceptable solution is obtained only if the wavefunction reproduces itself on successive circuits, as in Fig. 12.24b. Because only some wavefunctions have

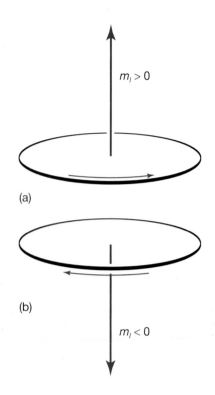

(a)

(b)

$m_l > 0$

$m_l < 0$

12.25 The angular momentum of a particle confined to a plane can be represented by a vector of length $|m_l|$ units along the z-axis and with an orientation that indicates the direction of motion of the particle. The direction is given by the right-hand screw rule.

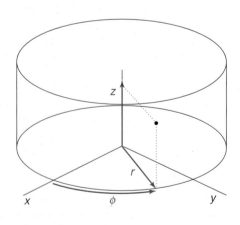

12.26 The cylindrical coordinates z, r, and ϕ for discussing systems with axial (cylindrical) symmetry. For a particle confined to the xy-plane, only r and ϕ can change.

this property, it follows that only some angular momenta are acceptable, and therefore that only certain rotational energies exist. Hence, the energy of the particle is quantized. Specifically, the only allowed wavelengths are

$$\lambda = \frac{2\pi r}{m_l}$$

with m_l, the conventional notation for this quantum number, taking integral values including 0.[5] The angular momentum is therefore limited to the values

$$J_z = \pm\frac{hr}{\lambda} = \frac{m_l hr}{2\pi r} = \frac{m_l h}{2\pi}$$

where we have allowed m_l to have positive or negative values. That is,

$$J_z = m_l\hbar \qquad m_l = 0, \pm 1, \pm 2, \ldots \tag{12.48}$$

Positive values of m_l correspond to rotation in a clockwise sense around the z-axis (as viewed in the direction of z, Fig. 12.25) and negative values of m_l correspond to counter-clockwise rotation around z. It then follows from eqn 47 that the energy is limited to the values

$$E = \frac{J_z^2}{2I} = \frac{m_l^2\hbar^2}{2I} \tag{12.49}$$

We shall see shortly that the corresponding normalized wavefunctions are

$$\psi_{m_l}(\phi) = \frac{e^{im_l\phi}}{(2\pi)^{1/2}} \tag{12.50}$$

The wavefunction with $m_l = 0$ is $\psi_0(\phi) = 1/(2\pi)^{1/2}$, and has the same value at all points on the circle.

We have arrived at a number of conclusions about rotational motion by cobbling together some classical notions and the de Broglie relation. Such a procedure can be very useful for establishing the general form (and, as in this case, the exact energies) for a quantum mechanical system. However, to be sure that the correct solutions have been obtained, and to obtain practice for more complex problems where this less formal approach is inadequate, we need to solve the Schrödinger equation explicitly. The formal solution is described in the *Justification* that follows.

Justification 12.4

The hamiltonian for a particle of mass m in a plane (with $V = 0$) is the same as that given in eqn 16:

$$H = -\frac{\hbar^2}{2m}\left(\frac{\partial^2}{\partial x^2} + \frac{\partial^2}{\partial y^2}\right)$$

and the Schrödinger equation is $H\psi = E\psi$, with the wavefunction a function of the angle ϕ. It is always a good idea to use coordinates that reflect the full symmetry of the system, so we introduce the coordinates r and ϕ (Fig. 12.26), where $x = r\cos\phi$ and $y = r\sin\phi$. By standard manipulations (see *Further reading*) we can write

$$\frac{\partial^2}{\partial x^2} + \frac{\partial^2}{\partial y^2} = \frac{\partial^2}{\partial r^2} + \frac{1}{r}\frac{\partial}{\partial r} + \frac{1}{r^2}\frac{\partial^2}{\partial \phi^2} \tag{12.51}$$

5. The value $m_l = 0$ corresponds to $\lambda = \infty$; a 'wave' of infinite wavelength has a constant height at all values of ϕ. See *Further information 1*.

However, because the radius of the path is fixed, the derivative with respect to r can be discarded. The hamiltonian then becomes

$$H = -\frac{\hbar^2}{2mr^2}\frac{d^2}{d\phi^2}$$

The moment of inertia $I = mr^2$ has appeared automatically, so H may be written

$$H = -\frac{\hbar^2}{2I}\frac{d^2}{d\phi^2} \tag{12.52}$$

and the Schrödinger equation is

$$\frac{d^2\psi}{d\phi^2} = -\frac{2IE}{\hbar^2}\psi \tag{12.53}$$

The normalized general solutions of the equation are

$$\psi_{m_l}(\phi) = \frac{e^{im_l\phi}}{(2\pi)^{1/2}} \qquad m_l = \pm\frac{(2IE)^{1/2}}{\hbar}$$

The quantity m_l is just a dimensionless number at this stage.

We now select the acceptable solutions from among these general solutions by imposing the condition that the wavefunction should be single-valued. That is, the wavefunction ψ must satisfy a **cyclic boundary condition**, and match at points separated by a complete revolution: $\psi(\phi + 2\pi) = \psi(\phi)$. On substituting the general wavefunction into this condition, we find

$$\psi_{m_l}(\phi + 2\pi) = \frac{e^{im_l(\phi+2\pi)}}{(2\pi)^{1/2}} = \frac{e^{im_l\phi}e^{2\pi im_l}}{(2\pi)^{1/2}} = \psi_{m_l}(\phi)e^{2\pi im_l}$$

As $e^{i\pi} = -1$, this relation is equivalent to

$$\psi_{m_l}(\phi + 2\pi) = (-1)^{2m_l}\psi(\phi) \tag{12.54}$$

Because we require $(-1)^{2m_l} = 1$, $2m_l$ must be a positive or a negative even integer (including 0), and therefore m_l must be an integer: $m_l = 0, \pm1, \pm2, \dots$.

$|m_l| = 2$

$|m_l| = 1$

$m_l = 0$

12.27 The real parts of the wavefunctions of a particle on a ring. As shorter wavelengths are achieved, the magnitude of the angular momentum around the z-axis grows in steps of $\hbar$.

(b) Quantization of rotation

We can summarize the conclusions so far as follows. The energy is quantized and restricted to the values given in eqn 49 ($E = m_l^2\hbar^2/2I$). The occurrence of m_l as its square means that the energy of rotation is independent of the sense of rotation (the sign of m_l), as we expect physically. In other words, states with a given value of $|m_l|$ are doubly degenerate, except for $m_l = 0$, which is non-degenerate. Although the result has been derived for the rotation of a single mass point, it also applies to any body of moment of inertia I constrained to rotate about one axis.

We have also seen that the angular momentum is quantized and confined to the values given in eqn 48 ($J_z = m_l\hbar$). The increasing angular momentum is associated with the increasing number of nodes in the real and imaginary parts of the wavefunction:[6] the wavelength decreases stepwise as $|m_l|$ increases, so the momentum with which the particle travels round the ring increases (Fig. 12.27). As shown in the following *Justification*, we can come to the same conclusion more formally by using the argument about the relation between eigenvalues and the values of observables established in Section 11.5.

6. The complex function $e^{im_l\phi}$ does not have nodes; however, it may be written as $\cos m_l\phi + i\sin m_l\phi$, and the real ($\cos m_l\phi$) and imaginary ($\sin m_l\phi$) components do have nodes.

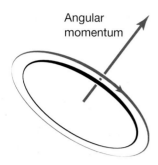

12.28 The basic ideas of the vector representation of angular momentum: the magnitude of the angular momentum is represented by the length of the vector, and the orientation of the motion in space by the orientation of the vector (using the right-hand screw rule).

Justification 12.5

In the discussion of translational motion in one dimension, we saw that the opposite signs in the wavefunctions e^{ikx} and e^{-ikx} correspond to opposite directions of travel, and that the linear momentum is given by the eigenvalue of the linear momentum operator. The same conclusions can be drawn here, but now we need the eigenvalues of the angular momentum operator. In classical mechanics the orbital angular momentum l_z about the z-axis is defined as[7]

$$l_z = xp_y - yp_x \qquad [12.55]$$

where p_x is the component of linear motion parallel to the x-axis and p_y is the component parallel to the y-axis. The operators for the two linear momentum components are given in eqn 11.32, so the operator for angular momentum about the z-axis, which we denote l_z, is

$$l_z = \frac{\hbar}{i}\left(x\frac{\partial}{\partial y} - y\frac{\partial}{\partial x}\right) \qquad (12.56)$$

When expressed in terms of the coordinates r and ϕ, by standard manipulations this equation becomes

$$l_z = \frac{\hbar}{i}\frac{\partial}{\partial \phi} \qquad (12.57)$$

With the angular momentum operator available, we can test the wavefunction in eqn 50. Disregarding the normalization constant, we find

$$l_z \psi_{m_l} = \frac{\hbar}{i}\frac{d\psi_{m_l}}{d\phi} = im_l\frac{\hbar}{i}e^{im_l\phi} = m_l\hbar\psi_{m_l} \qquad (12.58)$$

That is, ψ_{m_l} is an eigenfunction of l_z, and corresponds to an angular momentum $m_l\hbar$. When m_l is positive, the angular momentum is positive (clockwise when seen from below); when m_l is negative, the angular momentum is negative (counterclockwise when seen from below). These features are the origin of the vector representation of angular momentum, in which the magnitude is represented by the length of a vector and the direction of motion by its orientation (Fig. 12.28).

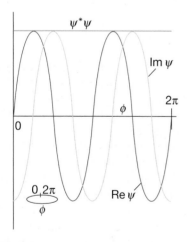

12.29 The probability density for a particle in a definite state of angular momentum is uniform, so there is an equal probability of finding the particle anywhere on the ring.

To locate the particle given its wavefunction in eqn 50, we form the probability density:

$$\psi_{m_l}^* \psi_{m_l} = \left(\frac{e^{im_l\phi}}{(2\pi)^{1/2}}\right)^* \left(\frac{e^{im_l\phi}}{(2\pi)^{1/2}}\right) = \left(\frac{e^{-im_l\phi}}{(2\pi)^{1/2}}\right)\left(\frac{e^{im_l\phi}}{(2\pi)^{1/2}}\right) = \frac{1}{2\pi}$$

Because this probability density is independent of ϕ, the probability of locating the particle somewhere on the ring is also independent of ϕ (Fig. 12.29). Hence the location of the particle is completely indefinite, and knowing the angular momentum precisely eliminates the possibility of specifying the particle's location. Angular momentum and angle are a pair of complementary observables (in the sense defined in Section 11.6), and the inability to specify them simultaneously with arbitrary precision is another example of the uncertainty principle.

---

7. The angular momentum in three dimensions is defined as $l = r \times p$; to obtain eqn 55, expand the vector product and identify the z-component. See *Further information 2*.

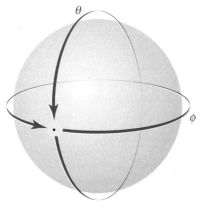

12.30 The wavefunction of a particle on the surface of a sphere must satisfy two cyclic boundary conditions; this requirement leads to two quantum numbers for its state of angular momentum.

12.7 Rotation in three dimensions: the particle on a sphere

We now consider a particle of mass m that is free to move anywhere on the surface of a sphere of radius r. We shall need the results of this calculation when we come to describe the states of electrons in atoms (Chapter 13) and of rotating molecules (Chapter 16). The requirement that the wavefunction should match as a path is traced over the poles as well as round the equator of the sphere surrounding the central point introduces a second cyclic boundary condition and therefore a second quantum number (Fig. 12.30).

(a) The Schrödinger equation

The hamiltonian for motion in three dimensions (Table 11.1) is

$$H = -\frac{\hbar^2}{2m}\nabla^2 + V \qquad \nabla^2 = \frac{\partial^2}{\partial x^2} + \frac{\partial^2}{\partial y^2} + \frac{\partial^2}{\partial z^2} \tag{12.59}$$

The symbol ∇^2 is a convenient abbreviation for the sum of the three second derivatives; it is called the **laplacian**, and read either 'del squared' or 'nabla squared'. For the particle confined to a spherical surface, $V = 0$ wherever it is free to travel, and the radius r is a constant. The wavefunction is therefore a function of the **colatitude**, θ, and the **azimuth**, ϕ (Fig. 12.31), and we write it $\psi(\theta, \phi)$. The Schrödinger equation is

$$-\frac{\hbar^2}{2m}\nabla^2\psi = E\psi \tag{12.60}$$

As shown in the following *Justification*, this partial differential equation can be simplified by the separation of variables procedure by expressing the wavefunction (for constant r) as the product

$$\psi(\theta, \phi) = \Theta(\theta)\Phi(\phi) \tag{12.61}$$

where Θ is a function only of θ and Φ is a function only of ϕ.

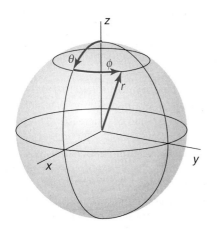

12.31 Spherical polar coordinates. For a particle confined to the surface of a sphere, only the colatitude, θ, and the azimuth, ϕ, can change.

Justification 12.6

The laplacian in spherical polar coordinates is (see *Further reading*)

$$\nabla^2 = \frac{\partial^2}{\partial r^2} + \frac{1}{r}\frac{\partial}{\partial r} + \frac{2}{r^2}\Lambda^2 \tag{12.62}$$

where the **legendrian**, Λ^2, is

$$\Lambda^2 = \frac{1}{\sin^2\theta}\frac{\partial^2}{\partial\phi^2} + \frac{1}{\sin\theta}\frac{\partial}{\partial\theta}\sin\theta\frac{\partial}{\partial\theta} \tag{12.63}$$

Because r is constant, we can discard the part of the laplacian that involves differentiation with respect to r, and so write the Schrödinger equation as

$$\frac{1}{r^2}\Lambda^2\psi = -\frac{2mE}{\hbar^2}\psi$$

or, because $I = mr^2$, as

$$\Lambda^2\psi = -\varepsilon\psi \qquad \varepsilon = \frac{2IE}{\hbar^2}$$

Table 12.3 The spherical harmonics $Y_{l,m_l}(\theta, \phi)$

l	m_l	Y_{l,m_l}
0	0	$\left(\dfrac{1}{4\pi}\right)^{1/2}$
1	0	$\left(\dfrac{3}{4\pi}\right)^{1/2}\cos\theta$
	± 1	$\mp\left(\dfrac{3}{8\pi}\right)^{1/2}\sin\theta\,e^{\pm i\phi}$
2	0	$\left(\dfrac{5}{16\pi}\right)^{1/2}(3\cos^2\theta - 1)$
	± 1	$\mp\left(\dfrac{15}{8\pi}\right)^{1/2}\cos\theta\sin\theta\,e^{\pm i\phi}$
	± 2	$\left(\dfrac{15}{32\pi}\right)^{1/2}\sin^2\theta\,e^{\pm 2i\phi}$
3	0	$\left(\dfrac{7}{16\pi}\right)^{1/2}(5\cos^3\theta - 3\cos\theta)$
	± 1	$\mp\left(\dfrac{21}{64\pi}\right)^{1/2}(5\cos^2\theta - 1)\sin\theta\,e^{\pm i\phi}$
	± 2	$\left(\dfrac{105}{32\pi}\right)^{1/2}\sin^2\theta\cos\theta\,e^{\pm 2i\phi}$
	± 3	$\mp\left(\dfrac{35}{64\pi}\right)^{1/2}\sin^3\theta\,e^{\pm 3i\phi}$

The spherical harmonics are orthogonal and normalized in the following sense:

$$\int_0^\pi \int_0^{2\pi} Y_{l',m_l'}(\theta, \phi)^* Y_{l,m_l}(\theta, \phi)\sin\theta\,d\theta\,d\phi = \delta_{l'l}\delta_{m_l'm_l}$$

An important 'triple integral' is

$$\int_0^\pi \int_0^{2\pi} Y_{l'',m_l''}(\theta, \phi)^* Y_{l',m_l'}(\theta, \phi) Y_{l,m_l}(\theta, \phi)\sin\theta\,d\theta\,d\phi$$

$= 0$ unless $m_l'' = m_l' + m_l$ and l'', l', and l can form a triangle.

To verify that this expression is separable, we substitute $\psi = \Theta\Phi$:

$$\frac{1}{\sin^2\theta}\frac{\partial^2(\Theta\Phi)}{\partial\phi^2} + \frac{1}{\sin\theta}\frac{\partial}{\partial\theta}\sin\theta\frac{\partial(\Phi\Theta)}{\partial\theta} = -\varepsilon\Theta\Phi$$

We now use the fact that Θ and Φ are each functions of one variable, so the partial derivatives become complete derivatives:

$$\frac{\Theta}{\sin^2\theta}\frac{d^2\Phi}{d\phi^2} + \frac{\Phi}{\sin\theta}\frac{d}{d\theta}\sin\theta\frac{d\Theta}{d\theta} = -\varepsilon\Theta\Phi$$

Division through by $\Theta\Phi$, multiplication by $\sin^2\theta$, and minor rearrangement give

$$\frac{1}{\Phi}\frac{d^2\Phi}{d\phi^2} + \frac{\sin\theta}{\Theta}\frac{d}{d\theta}\sin\theta\frac{d\Theta}{d\theta} + \varepsilon\sin^2\theta = 0$$

The first term on the left depends only on ϕ and the remaining two terms depend only on θ. We met a similar situation when discussing a particle on a rectangular surface (*Justification 12.2*) and, by the same argument, the complete equation can be separated. Thus, if we set the first term equal to the numerical constant $-m_l^2$ (a constant clearly chosen with an eye to the future), the separated equations are

$$\frac{1}{\Phi}\frac{d^2\Phi}{d\phi^2} = -m_l^2 \qquad \frac{\sin\theta}{\Theta}\frac{d}{d\theta}\sin\theta\frac{d\Theta}{d\theta} + \varepsilon\sin^2\theta = m_l^2$$

The first of these two equations is the same as in *Justification 12.4*, so it has the same solutions (eqn 50). The second is much more complicated to solve, but the solutions are tabulated as the *associated Legendre functions*. The cyclic boundary conditions on θ result in the introduction of a second quantum number, l, which identifies the acceptable solutions. The presence of the quantum number m_l in the second equation implies, as we see below, that the range of acceptable values of m_l is restricted by the value of l.

As indicated in *Justification 12.6*, solution of the Schrödinger equation shows that the acceptable wavefunctions are specified by two quantum numbers l and m_l which are restricted to the values

$$l = 0, 1, 2, \dots \qquad m_l = l, l-1, \dots, -l \qquad (12.64)$$

Note that the **orbital angular momentum quantum number** l is non-negative and that, for a given value of l, there are $2l + 1$ permitted values of the **magnetic quantum number**, m_l. The normalized wavefunctions are usually denoted $Y_{l,m_l}(\theta, \phi)$ and are called the **spherical harmonics** (Table 12.3).

Figure 12.32 is a representation of the spherical harmonics for $l = 0$ to 4 and $m_l = 0$ which emphasizes how the number of angular nodes (the positions at which the wavefunction passes through zero) increases as the value of l increases. There are no angular nodes around the z-axis for functions with $m_l = 0$, which corresponds to there being no component of orbital angular momentum about that axis. Figure 12.33 shows the distribution of the particle of a given angular momentum in more detail. In this representation, the value of $|Y_{l,m_l}|^2$ at each value of θ and ϕ is proportional to the distance of the surface from the origin. Note how, for a given value of l, the most probable location of the particle migrates towards the xy-plane as the value of $|m_l|$ increases.[8]

---

8. The real and imaginary components of the ϕ component of the wavefunctions, $e^{im_l\phi} = \cos m_l\phi + i\sin m_l\phi$, each have $|m_l|$ angular nodes, but these nodes are not seen when we plot the probability density, because $|e^{im_l\phi}|^2 = 1$.

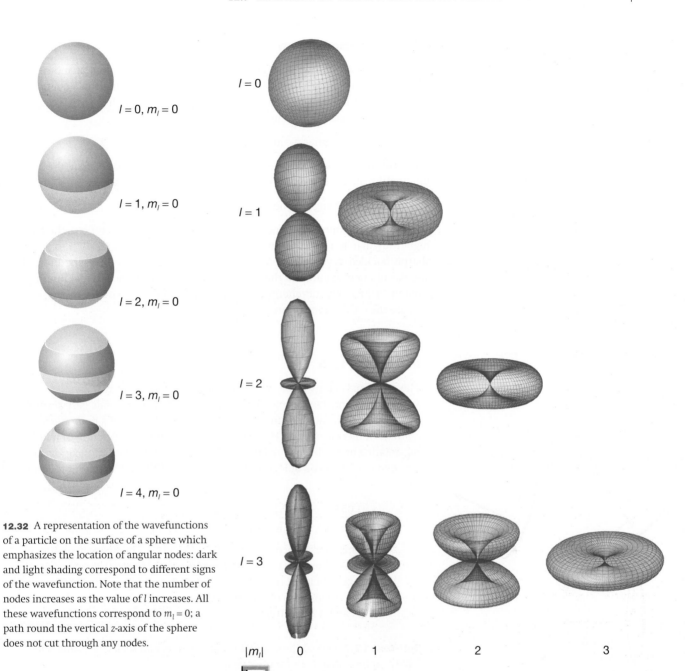

12.32 A representation of the wavefunctions of a particle on the surface of a sphere which emphasizes the location of angular nodes: dark and light shading correspond to different signs of the wavefunction. Note that the number of nodes increases as the value of l increases. All these wavefunctions correspond to $m_l = 0$; a path round the vertical z-axis of the sphere does not cut through any nodes.

12.33 A more complete representation of the wavefunctions for $l = 0, 1, 2,$ and 3. The distance of a point on the surface from the origin is proportional to the square modulus of the amplitude of the wavefunction at that point.

It also follows from the solution of the Schrödinger equation that the energy E of the particle is restricted to the values

$$E = l(l + 1)\frac{\hbar^2}{2I} \qquad l = 0, 1, 2, \dots \tag{12.65}$$

We see that the energy is quantized, and that it is independent of m_l. Because there are $2l + 1$ different wavefunctions (one for each value of m_l) that correspond to the same energy, it follows that a level with quantum number l is $(2l + 1)$-fold degenerate.

(b) Angular momentum

The energy of a rotating particle is related classically to its angular momentum J by $E = J^2/2I$ (see *Further information 2*). Therefore, by comparing this equation with eqn 65, we can deduce that, because the energy is quantized, then so to is the magnitude of the angular momentum, and confined to the values

$$\text{Magnitude of angular momentum} = \{l(l+1)\}^{1/2}\hbar \qquad l = 0, 1, 2, \dots \qquad (12.66a)$$

We have already seen (in the context of rotation in a plane) that the angular momentum about the z-axis is quantized, and that it has the values

$$z\text{-component of angular momentum} = m_l\hbar \qquad m_l = l, l-1, \dots, -l \qquad (12.66b)$$

The fact that the number of nodes in $\psi_{l,m_l}(\theta, \phi)$ increases with l reflects the fact that higher angular momentum implies higher kinetic energy, and therefore a more sharply buckled wavefunction. We can also see that the states corresponding to high angular momentum around the z-axis are those in which most nodal lines cut the equator: a high kinetic energy now arises from motion parallel to the equator because the curvature is greatest in that direction.

Illustration 12.5

The rotational energy levels of molecules are studied by microwave spectroscopy and Raman spectroscopy (Chapter 16), and provide very accurate bond lengths. We shall see in Chapter 16 that the rotation of a homonuclear diatomic molecule, such as H_2, can be represented by a single point of mass m_{H_2} rotating at a radius r equal to one-half of the bond length of the molecule. The moment of inertia of H_2 is then $I = m_{H_2}r^2 = 4.603 \times 10^{-48}$ kg m². It follows that

$$\frac{\hbar^2}{2I} = \frac{(1.054\,57 \times 10^{-34}\ \text{J s})^2}{2 \times (4.603 \times 10^{-48}\ \text{kg m}^2)} = 1.208 \times 10^{-21}\ \text{J}$$

or 1.208 zJ. This energy corresponds to 0.727 kJ mol⁻¹. The first few rotational energy levels are therefore 0 ($l = 0$), 2.416 zJ ($l = 1$), 7.248 zJ ($l = 2$), and 14.50 zJ ($l = 3$). The degeneracies of these levels are 1, 3, 5, and 7, respectively (from $2l + 1$), and the magnitudes of the angular momentum of the molecule are 0, $2^{1/2}\hbar$, $6^{1/2}\hbar$, and $(12)^{1/2}\hbar$ (from eqn 66a).

Self-test 12.7 Repeat the calculation for a deuterium molecule (same bond length, approximately twice the mass).

[Energies are smaller by a factor of two; same angular momenta and numbers of components]

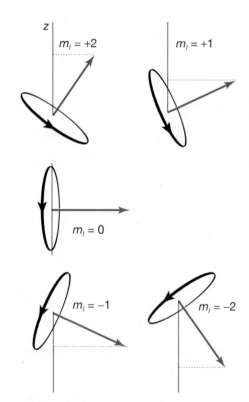

12.34 The permitted orientations of angular momentum when $l = 2$. We shall see soon that this representation is too specific because the azimuthal orientation of the vector (its angle around z) is indeterminate.

(c) Space quantization

The result that m_l is confined to the values $l, l-1, \dots, -l$ for a given value of l means that the component of angular momentum about the z-axis may take only $2l + 1$ values. If the angular momentum is represented by a vector of length proportional to its magnitude (that is, of length $\{l(l+1)\}^{1/2}$ units), then to represent correctly the value of the component of angular momentum, the vector must be oriented so that its projection on the z-axis is of length m_l units. In classical terms, this restriction means that the plane of rotation of the particle can take only a discrete range of orientations (Fig. 12.34). The remarkable implication is that *the orientation of a rotating body is quantized*.

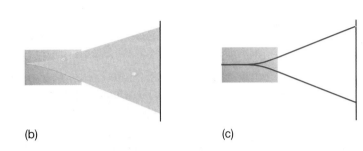

(a) (b) (c)

12.35 (a) The experimental arrangement for the Stern–Gerlach experiment: the magnet provides an inhomogeneous field. (b) The classically expected result. (c) The observed outcome using silver atoms.

The quantum mechanical result that a rotating body may not take up an arbitrary orientation with respect to some specified axis (for example, an axis defined by the direction of an externally applied electric or magnetic field) is called **space quantization**. It was confirmed by an experiment first performed by Otto Stern and Walther Gerlach in 1921, who shot a beam of silver atoms through an inhomogeneous magnetic field (Fig. 12.35). The idea behind the experiment was that a rotating, charged body behaves like a magnet and interacts with the applied field. According to classical mechanics, because the orientation of the angular momentum can take any value, the associated magnet can take any orientation. Because the direction in which the magnet is driven by the inhomogeneous field depends on the magnet's orientation, it follows that a broad band of atoms is expected to emerge from the region where the magnetic field acts. According to quantum mechanics, however, because the angular momentum is quantized, the associated magnet lies in a number of discrete orientations, so several sharp bands of atoms are expected.

In their first experiment, Stern and Gerlach appeared to confirm the classical prediction. However, the experiment is difficult because collisions between the atoms in the beam blur the bands. When the experiment was repeated with a beam of very low intensity (so that collisions were less frequent) they observed discrete bands, and so confirmed the quantum prediction.

(d) The vector model

Throughout the preceding discussion, we have referred to the z-component of angular momentum (the component about an arbitrary axis, which is conventionally denoted z), and have made no reference to the x- and y-components (the components about the two axes perpendicular to z). The reason for this omission is that because the operators for the three components do not commute with one another (Section 11.6); the uncertainty principle forbids the simultaneous, exact specification of more than one component (unless $l = 0$). In other words, l_x, l_y, and l_z are complementary observables. Therefore, if l_z is known, it is impossible to ascribe values to the other two components. It follows that the illustration in Fig. 12.34, which is summarized in Fig. 12.36a, gives a false impression of the state of the system, because it suggests definite values for the x- and y-components. A better picture must reflect the impossibility of specifying l_x and l_y if l_z is known.

The **vector model** of angular momentum uses pictures like that in Fig. 12.36b. The cones are drawn with side $\{l(l + 1)\}^{1/2}$ units, and represent the magnitude of the angular momentum. Each cone has a definite projection (of m_l units) on the z-axis, representing the system's precise value of l_z. The l_x and l_y projections, however, are indefinite. The vector representing the state of angular momentum can be thought of as lying with its tip on any point on the mouth of the cone. At this stage it should

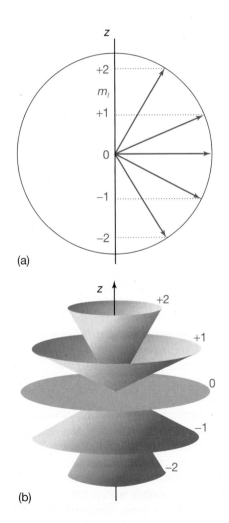

(a)

(b)

12.36 (a) A summary of Fig. 12.34. However, because the azimuthal angle of the vector around the z-axis is indeterminate, a better representation is as in (b), where each vector lies at an unspecified azimuthal angle on its cone.

not be thought of as sweeping round the cone; that aspect of the model will be added later when we allow the picture to convey more information.

12.8 Spin

Stern and Gerlach observed two bands of Ag atoms in their experiment. This observation seems to conflict with one of the predictions of quantum mechanics, because an angular momentum l gives rise to $2l + 1$ orientations, which is equal to 2 only if $l = \frac{1}{2}$, contrary to the conclusion that l must be an integer. The conflict was resolved by the suggestion that the angular momentum they were observing was not due to orbital angular momentum (the motion of an electron around the atomic nucleus) but arose instead from the motion of the electron about its own axis. This intrinsic angular momentum of the electron is called its **spin**. The explanation of the existence of spin emerged when Dirac combined quantum mechanics with special relativity and established the theory of relativistic quantum mechanics.

The spin of an electron about its own axis does not have to satisfy the same boundary conditions as those for a particle circulating around a central point, so the quantum number for spin angular momentum is subject to different restrictions. To distinguish this spin angular momentum from orbital angular momentum we use the **spin quantum number** s (in place of l; like l, s is a non-negative number) and m_s, the **spin magnetic quantum number**, for the projection on the z-axis. The magnitude of the spin angular momentum is $\{s(s + 1)\}^{1/2}\hbar$ and the component $m_s\hbar$ is restricted to the $2s + 1$ values

$$m_s = s, s - 1, \ldots, -s \tag{12.67}$$

The detailed analysis of the spin of a particle is sophisticated and shows that the property should not be taken to be an actual spinning motion. However, that picture can be very useful when used with care. For an electron it turns out that only one value of s is allowed, namely $s = \frac{1}{2}$, corresponding to an angular momentum of magnitude $(\frac{3}{4})^{1/2}\hbar = 0.866\hbar$. This spin angular momentum is an intrinsic property of the electron, like its rest mass and its charge, and every electron has exactly the same value: the magnitude of the spin angular momentum of an electron cannot be changed. The spin may lie in $2s + 1 = 2$ different orientations (Fig. 12.37). One orientation corresponds to $m_s = +\frac{1}{2}$ (this state is often denoted α or $\uparrow$); the other orientation corresponds to $m_s = -\frac{1}{2}$ (this state is denoted β or $\downarrow$).

The outcome of the Stern–Gerlach experiment can now be explained if we suppose that each Ag atom possesses an angular momentum due to the spin of a single electron, because the two bands of atoms then correspond to the two spin orientations. Why the atoms behave like this will be explained in Chapter 13.[9]

Like the electron, other elementary particles have characteristic, constant spin angular momenta. For example, protons and neutrons are spin-$\frac{1}{2}$ particles (that is, $s = \frac{1}{2}$) and invariably spin with angular momentum $(\frac{3}{4})^{1/2}\hbar = 0.866\hbar$. Because the masses of a proton and a neutron are so much greater than the mass of an electron, yet they all have the same spin angular momentum, the classical picture would be of these two particles spinning much more slowly than an electron. Some elementary particles have $s = 1$, and so have an intrinsic angular momentum of magnitude $2^{1/2}\hbar$. Some mesons are spin-1 particles (as are some atomic nuclei), but for our purposes the most important spin-1 particle is the photon. From the discussion in this chapter, we see that the photon has zero rest mass, zero charge, an energy $h\nu$, a linear momentum h/λ or $h\nu/c$, an intrinsic angular momentum of $2^{1/2}\hbar$, and travels at the speed c. We shall see the importance of photon spin in the next chapter.

12.37 An electron spin ($s = \frac{1}{2}$) can take only two orientations with respect to a specified axis. An α electron (top) is an electron with $m_s = +\frac{1}{2}$; a β electron (bottom) is an electron with $m_s = -\frac{1}{2}$. The vector representing the spin angular momentum lies at an angle of 55° to the z-axis (more precisely, the half-angle of the cones is $\arccos(\frac{1}{3})^{1/2}$).

9. It is already probably familiar from introductory chemistry, that the ground-state configuration of a silver atom is [Kr]$4d^{10}5s^1$, a single unpaired electron outside a closed shell.

Particles with half-integral spin are called **fermions** and those with integral spin (including 0) are called **bosons**. Thus, electrons and protons are fermions and photons are bosons. It is a very deep feature of nature that all the elementary particles that constitute matter are fermions whereas the fundamental particles that are responsible for the forces that bind fermions together are all bosons. Photons, for example, transmit the electromagnetic force that binds together electrically charged particles. Matter, therefore, is an assembly of fermions held together by forces conveyed by bosons.

To sum up, when we use the quantum numbers l and m_l we shall mean orbital angular momentum (circulation in space). When we use s and m_s we shall mean spin angular momentum (intrinsic angular momentum). When we use j and m_j we shall mean either (or, in some contexts to be described in Chapter 13, a combination of orbital and spin momenta).

Techniques of approximation

All the applications treated so far have had exact solutions. However, many problems —and almost all the problems of interest in chemistry—do not have exact solutions. To make progress with these problems we need to develop techniques of approximation. There are two major approaches, *variation theory* and *perturbation theory*. Variation theory is most commonly encountered in the context of molecular orbital theory, and we consider it there (Chapter 14). Here, we concentrate on perturbation theory.

12.9 Time-independent perturbation theory

In **perturbation theory**, we suppose that the hamiltonian for the problem we are trying to solve, H, can be expressed as the sum of a simple hamiltonian, $H^{(0)}$, which has known eigenvalues and eigenfunctions, and a contribution, $H^{(1)}$, which represents the extent to which the true hamiltonian differs from the 'model' hamiltonian:

$$H = H^{(0)} + H^{(1)} \tag{12.68}$$

In **time-independent perturbation theory**, the perturbation is always present and unvarying. For example, it might represent a dip in the potential energy of a particle in a box in some region along the length of the box.

In time-independent perturbation theory, we suppose that the true energy of the system differs from the energy of the simple system, and that we can write

$$E = E^{(0)} + E^{(1)} + E^{(2)} + \cdots \tag{12.69}$$

where $E^{(1)}$ is the 'first-order correction' to the energy, a contribution proportional to $H^{(1)}$, and $E^{(2)}$ is the 'second-order correction' to the energy, a contribution proportional to $H^{(1)2}$, and so on. The true wavefunction also differs from the 'simple' wavefunction, and we write

$$\psi = \psi^{(0)} + \psi^{(1)} + \psi^{(2)} + \cdots \tag{12.70}$$

In practice, we need to consider only the 'first-order correction' to the wavefunction, $\psi^{(1)}$. As we show in the *Justification* below, the first- and second-order corrections to the energy of the ground state (with the wavefunction ψ_0 and energy E_0), are

$$E_0^{(1)} = \int \psi_0^{(0)*} H^{(1)} \psi_0^{(0)} \, d\tau \tag{12.71}$$

and

$$E_0^{(2)} = \sum_{n \neq 0} \frac{\left| \int \psi_n^{(0)*} H^{(1)} \psi_0^{(0)} \, d\tau \right|^2}{E_0^{(0)} - E_n^{(0)}}$$

(12.72)

Justification 12.7

It helps to keep track of the order of the perturbation correction if we write

$$\psi = \psi^{(0)} + \lambda \psi^{(1)} + \lambda^2 \psi^{(2)} + \cdots$$

where the power of λ indicates the order of the correction. Likewise, we write

$$H = H^{(0)} + \lambda H^{(1)}$$

and

$$E = E^{(0)} + \lambda E^{(1)} + \lambda^2 E^{(2)} + \cdots$$

When these expressions are inserted into the Schrödinger equation, $H\psi = E\psi$, we obtain

$$(H^{(0)} + \lambda H^{(1)})(\psi^{(0)} + \lambda \psi^{(1)} + \lambda^2 \psi^{(2)} + \cdots)$$
$$= (E^{(0)} + \lambda E^{(1)} + \lambda^2 E^{(2)} + \cdots)(\psi^{(0)} + \lambda \psi^{(1)} + \lambda^2 \psi^{(2)} + \cdots)$$

which we can rewrite as

$$H^{(0)}\psi^{(0)} + \lambda(H^{(1)}\psi^{(0)} + H^{(0)}\psi^{(1)}) + \lambda^2(H^{(0)}\psi^{(2)} + H^{(1)}\psi^{(1)}) + \cdots$$
$$= E^{(0)}\psi^{(0)} + \lambda(E^{(0)}\psi^{(1)} + E^{(1)}\psi^{(0)}) + \lambda^2(E^{(2)}\psi^{(0)} + E^{(1)}\psi^{(1)} + E^{(0)}\psi^{(2)}) + \cdots$$

By comparing powers of λ, we find

Terms in λ^0: $H^{(0)}\psi^{(0)} = E^{(0)}\psi^{(0)}$

Terms in λ: $H^{(1)}\psi^{(0)} + H^{(0)}\psi^{(1)} = E^{(0)}\psi^{(1)} + E^{(1)}\psi^{(0)}$

Terms in λ^2: $H^{(0)}\psi^{(2)} + H^{(1)}\psi^{(1)} = E^{(2)}\psi^{(0)} + E^{(1)}\psi^{(1)} + E^{(0)}\psi^{(2)}$

and so on.

The equations we have derived are applicable to any state of the system. From now on we shall consider only the ground state ψ_0 with energy E_0. The first equation, which we now write

$$H^{(0)}\psi_0^{(0)} = E_0^{(0)}\psi_0^{(0)}$$

is the Schrödinger equation for the ground state of the unperturbed system, which we assume we can solve (for instance, it might be the equation for the ground state of the particle in a box, with the solutions given in eqn. 7). To solve the next equation, which is now written

$$H^{(1)}\psi_0^{(0)} + H^{(0)}\psi_0^{(1)} = E_0^{(0)}\psi_0^{(1)} + E_0^{(1)}\psi_0^{(0)}$$

we suppose that the first-order correction to the wavefunction can be expressed as a linear combination of the wavefunctions of the unperturbed system, and write

$$\psi_0^{(1)} = \sum_n c_n \psi_n^{(0)}$$

Substitution of this expression gives

$$H^{(1)}\psi_0^{(0)} + \sum_n c_n H^{(0)}\psi_n^{(0)} = \sum_n c_n E_0^{(0)}\psi_n^{(0)} + E_0^{(1)}\psi_0^{(0)}$$

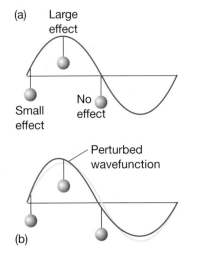

12.38 (a) The first-order energy is an average of the perturbation (represented by the hanging weights) over the unperturbed wavefunction. (b) The second-order energy is a similar average, but over the distortion induced by the perturbation.

We can isolate the term in $E_0^{(1)}$ by making use of the fact that the $\psi_n^{(0)}$ form a complete orthogonal and normalized set in the sense that

$$\int \psi_0^{(0)*}\psi_n^{(0)}\, d\tau = 0 \quad \text{if} \quad n \neq 0, \qquad \text{but} \quad 1 \quad \text{if} \quad n = 0$$

Therefore, when we multiply through by $\psi_0^{(0)*}$ and integrate over all space, we get

$$\underbrace{\int \psi_0^{(0)*}H^{(1)}\psi_0^{(0)}\, d\tau}_{} + \sum_n c_n \overbrace{\int \psi_0^{(0)*}H^{(0)}\psi_n^{(0)}\, d\tau}^{E_0^{(0)}\text{ if } n=0,\, 0\text{ otherwise}} = \sum_n c_n E_0^{(0)} \overbrace{\int \psi_0^{(0)*}\psi_n^{(0)}\, d\tau}^{1\text{ if } n=0,\, 0\text{ otherwise}} + E_0^{(1)} \overbrace{\int \psi_0^{(0)*}\psi_0^{(0)}\, d\tau}^{1}$$

That is,

$$\int \psi_0^{(0)*}H^{(1)}\psi_0^{(0)}\, d\tau = E_0^{(1)}$$

which is eqn 71.

To find the coefficients c_n, we multiply the same expression through by $\psi_k^{(0)*}$, where now $k \neq 0$, which gives

$$\int \psi_k^{(0)*}H^{(1)}\psi_0^{(0)}\, d\tau + \sum_n c_n \overbrace{\int \psi_k^{(0)*}H^{(0)}\psi_n^{(0)}\, d\tau}^{E_k^{(0)}\delta_{kn}} = \sum_n c_n E_0^{(0)} \overbrace{\int \psi_k^{(0)*}\psi_n^{(0)}\, d\tau}^{1\text{ if } n=k,\, 0\text{ otherwise}} + E_0^{(1)} \overbrace{\int \psi_k^{(0)*}\psi_0^{(0)}\, d\tau}^{0}$$

That is,

$$\int \psi_k^{(0)*}H^{(1)}\psi_0^{(0)}\, d\tau + c_k E_k^{(0)} = c_k E_0^{(0)}$$

which we can rearrange into

$$c_k = -\frac{\int \psi_k^{(0)*}H^{(1)}\psi_0^{(0)}\, d\tau}{E_k^{(0)} - E_0^{(0)}} \tag{12.73}$$

The second-order energy is obtained in a similar way using the results obtained here, but starting from the second-order expression (see *Further reading*).

We can understand the expression for $E^{(1)}$ as the average of the perturbation using the unperturbed wavefunction. An analogy is the shift in energy of vibration of a violin string when small weights are hung along its length. The weights hanging close to the nodes have little effect on its energy of vibration. Those hanging at the antinodes, however, have a pronounced effect (Fig. 12.38a). The second-order energy represents a similar average of the perturbation, but now the average is taken over the *perturbed* wavefunctions. In terms of the violin analogy, the average is now taken over the distorted waveform of the vibrating string, in which the nodes and antinodes are slightly shifted (Fig. 12.38b).

12.39 The potential energy for a particle in a box with a potential that varies as $-\varepsilon\sin(\pi x/L)$ across the floor of the box. We can expect the particle to accumulate more in the centre of the box (in the ground state at least) than in the unperturbed box.

Example 12.4 Using perturbation theory

Find the first-order correction to the ground-state energy for a particle in a well with a variation in the potential of the form $V = -\varepsilon\sin(\pi x/L)$ (Fig. 12.39).

Method Identify the first-order perturbation hamiltonian and evaluate $E_0^{(1)}$ from eqn 71. We can expect a small lowering of the energy because the average potential energy of the particle is lower in the distorted box.

Solution The perturbation hamiltonian is

$$H^{(1)} = -\varepsilon \sin(\pi x/L)$$

Therefore, the first-order correction to the energy is

$$E_0^{(1)} = \int_0^L \psi_1^{(0)} H^{(1)} \psi_1^{(0)} \, dx = -\frac{2\varepsilon}{L} \overbrace{\int_0^L \sin^3 \frac{\pi x}{L} \, dx}^{4L/3\pi} = -\frac{8\varepsilon}{3\pi}$$

As expected, the energy is lowered by the perturbation.

Self-test 12.8 Suppose that only $\psi_3^{(0)}$ contributes to the distortion of the wavefunction: calculate the coefficient c_3 and the second-order correction to the energy. $\qquad [c_3 = -8\varepsilon mL^2/15\pi\hbar^2, E_0^{(2)} = -64\varepsilon^2 mL^2/225\pi^2\hbar^2]$

12.10 Time-dependent perturbation theory

In **time-dependent perturbation theory**, the perturbation is either switched on and allowed to rise to its final value or is varying in time. Many of the perturbations encountered in chemistry are time-dependent. The most important is the effect of an oscillating electromagnetic field, which is responsible for **spectroscopic transitions**, changes in populations of quantized energy levels of an atom or molecule involving the absorption, emission, or scattering of electromagnetic radiation. When the atom or molecule changes its energy by ΔE, the difference is carried away as a photon of frequency ν, so

$$\Delta E = h\nu = \hbar\omega \tag{12.74}$$

The hamiltonian for the system has the form

$$H = H^{(0)} + H^{(1)}(t) \tag{12.75}$$

where $H^{(1)}(t)$ is the time-dependent perturbation. For example, the perturbation might correspond to the effect of the interaction between an oscillating electric field and an electric dipole moment, in which case we would write

$$H^{(1)}(t) = -\mu_z \mathcal{E} \cos \omega t \tag{12.76}$$

where ω is the frequency of the field and $\mathcal{E}$ is its amplitude. We suppose that the perturbation is absent until $t = 0$, and then it is turned on.

To cope with this type of problem, in which the perturbed wavefunction evolves with time, we need to solve the time-dependent Schrödinger equation,

$$H\Psi = i\hbar \frac{\partial \Psi}{\partial t} \tag{12.77}$$

We confirm in the *Justification* below that, if we write the first-order correction to the wavefunction as

$$\Psi_0^{(1)}(t) = \sum_n c_n(t)\Psi_n(t) = \sum_n c_n(t)\psi_n^{(0)} e^{-iE_n^{(0)}t/\hbar} \tag{12.78a}$$

then the coefficients in this expansion are given by

$$c_n(t) = \frac{1}{i\hbar} \int_0^t H_{n0}^{(1)}(t) e^{i\omega_{n0}t} \, dt \tag{12.78b}$$

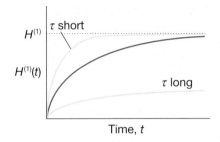

12.40 The time dependence of a slowly switched perturbation. A large value of τ corresponds to very slow switching.

Justification 12.8

The formal demonstration of eqn 78 is quite lengthy (see *Further reading*). Here we shall show that, given eqn 78b, then a perturbation that is switched on very slowly to a constant value gives the same expression for the coefficients as we obtained for time-independent perturbation theory. For such a perturbation, we write

$$H^{(1)}(t) = H^{(1)}(1 - e^{-t/\tau})$$

and take the time constant τ to be very long (Fig. 12.40). Substitution of this expression into eqn 78b gives

$$c_n(t) = \frac{1}{i\hbar} H^{(1)}_{n0} \int_0^t (1 - e^{-t/\tau}) e^{i\omega_{n0}t}\, dt = \frac{1}{i\hbar} H^{(1)}_{n0} \left\{ \frac{e^{i\omega_{n0}t} - 1}{i\omega_{n0}} - \frac{e^{(i\omega_{n0}-1/\tau)t} - 1}{i\omega_{n0} - 1/\tau} \right\}$$

At this point we suppose that the perturbation is switched slowly, in the sense that $\tau \gg 1/\omega_{n0}$. We also suppose that we are interested in the coefficients long after the perturbation has settled down into its final value, when $t \gg \tau$. Under these conditions the second term in braces becomes $-1/i\omega_{n0}$, so

$$c_n(t) = -\frac{H^{(0)}_{n0}}{\hbar\omega_{n0}} e^{i\omega_{n0}t}$$

Now we recognize that $\hbar\omega_{n0} = E_n^{(0)} - E_0^{(0)}$, which gives

$$c_n(t) = -\frac{H^{(0)}_{n0}}{E_n^{(0)} - E_0^{(0)}} e^{iE_n^{(0)}t/\hbar} e^{-iE_0^{(0)}t/\hbar}$$

When this expression is substituted into eqn 78a, we obtain the time-independent expression, eqn 73 (apart from an irrelevant overall phase factor).

In accord with the general rules for the interpretation of wavefunctions, the probability that the system will be found in the state n is proportional to the square modulus of the coefficient of the state, $|c_n(t)|^2$. Therefore, the rate of change of population of the state f due to transitions from state i is

$$w_{f\leftarrow i} = \frac{d|c_f|^2}{dt} = \frac{dc_f^* c_f}{dt} = c_f \frac{dc_f^*}{dt} + c_f^* \frac{dc_f}{dt}$$

Because the coefficients are proportional to the matrix elements of the perturbation, $w_{f\leftarrow i}$ is proportional to the square modulus of the matrix element of the perturbation between the two states:

$$w_{f\leftarrow i} \propto |H^{(1)}_{fi}|^2 \tag{12.79}$$

For instance, if the perturbation is that of the interaction of the electromagnetic field with a molecule (eqn 76), we would conclude that

$$w_{f\leftarrow i} \propto |\mu_{z,fi}|^2 \mathcal{E}^2 \tag{12.80}$$

Therefore, the rate of transition, and hence the intensity of absorption of the incident radiation, is proportional to the square of the **transition dipole moment**:

$$\mu_{z,fi} = \int \psi_f^* \mu_z \psi_i\, d\tau \tag{12.81}$$

The rate of transition is also proportional to $\mathcal{E}^2$, and therefore the intensity of the incident radiation (because the intensity is proportional to $\mathcal{E}^2$). This result will be the basis of most of the discussion of spectroscopy in Chapters 13 and 16–18 and of the kinetics of electron transfer in Chapter 29.

Checklist of key ideas

Translational motion

☐ wavefunction of free particle,
$\psi_k = Ae^{ikx} + Be^{-ikx}$, $E_k = k^2\hbar^2/2m$

12.1 Particle in a box

☐ particle in a box, a particle of mass m confined between two walls; the potential energy is zero inside the box but rises abruptly to infinity at the walls

☐ boundary conditions, constraints on the function at certain locations

☐ wavefunctions, $\psi_n(x) = (2/L)^{1/2}\sin(n\pi x/L)$

☐ energies, $E_n = n^2h^2/8mL^2$, $n = 1, 2, \ldots$

☐ quantum number, an integer (in some cases, a half-integer) that labels the state of the system

☐ zero-point energy, the lowest possible energy; $E_1 = h^2/8mL^2$

☐ correspondence principle: Classical mechanics emerges from quantum mechanics as high quantum numbers are reached

☐ orthogonality, the functions ψ_n and $\psi_{n'}$ are orthogonal if $\int \psi_n^*\psi_{n'}\,d\tau = 0$

☐ Dirac bracket notation,
$\langle n|\hat{\Omega}|n'\rangle = \int \psi_n^*\hat{\Omega}\psi_{n'}\,d\tau$

☐ bra, $\langle n|$

☐ ket, $|n\rangle$

☐ Kronecker delta, $\delta_{nn'} = 1$ when $n' = n$ and 0 when $n' \neq n$

☐ orthonormal functions, sets of functions that are normalized and mutually orthogonal

12.2 Motion in two or more dimensions

☐ separation of variables, $\psi(x, y) = X(x)Y(y)$

☐ wavefunctions, $\psi_{n_1,n_2}(x, y) = \{2/(L_1L_2)^{1/2}\}\sin(n_1\pi x/L_1)\sin(n_2\pi y/L_2)$

☐ energies, $E_{n_1,n_2} = (n_1^2/L_1^2 + n_2^2/L_2^2)h^2/8m$

☐ degenerate, different functions corresponding to the same energy

12.3 Tunnelling

☐ tunnelling, penetration into or through classically forbidden regions

☐ transmission probability,
$T = \{1 + (e^{\kappa L} - e^{-\kappa L})^2/16\varepsilon(1 - \varepsilon)\}^{-1}$
$\approx 16\varepsilon(1 - \varepsilon)e^{-2\kappa L}$, $\varepsilon = E/V$

Vibrational motion

☐ harmonic motion, motion in the presence of a restoring force proportional to the displacement, $F = -kx$

☐ force constant, the constant of proportionality in $F = -kx$

☐ parabolic potential energy, $V = \frac{1}{2}kx^2$

12.4 The energy levels

☐ energy levels, $E_v = (v + \frac{1}{2})\hbar\omega$, $\omega = (k/m)^{1/2}$, $v = 0, 1, 2, \ldots$

12.5 The wavefunctions

☐ wavefunctions, $\psi_v = N_vH_v(y)e^{-y^2/2}$, $y = x/\alpha$, $\alpha = (\hbar^2/mk)^{1/4}$

☐ Hermite polynomial, a polynomial solution to the differential equation $H_v'' - 2yH_v' + 2vH_v = 0$

☐ matrix element,
$\Omega_{v'v} = \langle v'|\hat{\Omega}|v\rangle = \int \psi_{v'}^*\hat{\Omega}\psi_v\,dx$

☐ mean displacement, $\langle x\rangle = 0$

☐ mean square displacement,
$\langle x^2\rangle = (v + \frac{1}{2})\hbar/(mk)^{1/2}$

☐ virial theorem: If the potential energy of a particle has the form $V = ax^b$, then its mean potential and kinetic energies are related by $2\langle E_K\rangle = b\langle V\rangle$

Rotational motion

12.6 Rotation in two dimensions: the particle on a ring

☐ angular momentum, the moment of linear momentum around a point

☐ moment of inertia of a particle of mass m on a path of radius r: $I = mr^2$

☐ cyclic boundary conditions, the requirement for a wavefunction to match at points separated by a complete revolution: $\psi(\phi + 2\pi) = \psi(\phi)$

☐ wavefunctions, $\psi_{m_l}(\phi) = (1/2\pi)^{1/2}e^{im_l\phi}$, $m_l = 0, \pm 1, \pm 2, \ldots$

☐ energies, $E = m_l^2\hbar^2/2I$

☐ orbital angular momentum operator, $l_z = xp_y - yp_x = (\hbar/i)\partial/\partial\phi$

☐ z-component of angular momentum, $l_z = m_l\hbar$

12.7 Rotation in three dimensions: the particle on a sphere

☐ laplacian, $\nabla^2 = \partial^2/\partial x^2 + \partial^2/\partial y^2 + \partial^2/\partial z^2$

☐ colatitude, θ, and azimuth, ϕ, the polar coordinates defined in Fig. 12.31

☐ Schrödinger equation,
$-(\hbar^2/2m)\nabla^2\psi = E\psi$

☐ legendrian, $\Lambda^2 = (1/\sin^2\theta)\partial^2/\partial\phi^2 + (1/\sin\theta)(\partial/\partial\theta)\sin\theta(\partial/\partial\theta)$

☐ orbital angular momentum quantum number, $l = 0, 1, 2, \ldots$

☐ spherical harmonics, the functions, $Y_{l,m_l}(\theta, \phi)$, Table 12.3

☐ energies, $E = l(l + 1)\hbar^2/2I$, $l = 0, 1, 2, \ldots$

☐ magnitude of angular momentum, $\{l(l + 1)\}^{1/2}\hbar$

☐ z-component of angular momentum, $m_l\hbar$, $m_l = l, l - 1, \ldots, -l$

☐ space quantization, the restriction of the component of angular momentum around an axis to discrete values

☐ vector model, a pictorial representation of angular momenta in terms of cones

12.8 Spin

☐ spin, an intrinsic angular momentum of a fundamental particle

☐ spin quantum number, s; for an electron $s = \frac{1}{2}$

☐ spin magnetic quantum number, $m_s = s, s - 1, \ldots, -s$; for an electron, $m_s = +\frac{1}{2}, -\frac{1}{2}$

☐ Stern–Gerlach experiment, an experiment to confirm the spin of an electron

☐ fermion, a particle with a half-integral spin quantum number

☐ boson, a particle with an integral spin quantum number

Techniques of approximation

12.9 Time-independent perturbation theory

☐ perturbation theory, a technique that supplies approximate solutions to the Schrödinger equation and in which the hamiltonian for the problem is expressed as a sum of simpler hamiltonians

☐ time-independent perturbation theory, the perturbation is always present and unvarying

☐ first-order correction to the energy, $E_0^{(1)} = \int \psi_0^{(0)*}H^{(1)}\psi_0^{(0)}\,d\tau$

□ second-order correction to the energy,

$$E_0^{(2)} = \sum_{n \neq 0} \frac{\left| \int \psi_n^{(0)*} H^{(1)} \psi_0^{(0)} \, d\tau \right|^2}{E_0^{(0)} - E_n^{(0)}}$$

□ first-order correction to the

wavefunction, $c_k = -\dfrac{\displaystyle\int \psi_k^{(0)*} H^{(1)} \psi_0^{(0)} \, d\tau}{E_k^{(0)} - E_0^{(0)}}$

12.10 Time-dependent perturbation theory

□ time-dependent perturbation theory, the perturbation is either switched on and allowed to rise to its final value or is varying in time

□ spectroscopic transitions, changes in populations of quantized energy levels of a system involving the absorption, emission, or scattering

of electromagnetic radiation, $\Delta E = h\nu$

□ first-order correction to the

wavefunction, $c_n(t) = \dfrac{1}{i\hbar} \displaystyle\int_0^t H_{n0}^{(1)}(t) e^{i\omega_{n0}t} \, dt$

□ transition rate: $w_{f\leftarrow i} \propto |H_{fi}^{(1)}|^2$

□ transition dipole moment:

$\mu_{z,fi} = \displaystyle\int \psi_f^* \mu_z \psi_i \, d\tau$

Further reading

Articles of general interest

H.F. Blanck, Introduction to a quantum mechanical harmonic oscillator using a modified particle-in-a-box problem. *J. Chem. Educ.* **69**, 98 (1992).

G.L. Breneman, The two-dimensional particle in a box. *J. Chem. Educ.* **67**, 866 (1990).

A. Cedillo, Quantum mechanical tunnelling through barriers: a spreadsheet approach. *J. Chem. Educ.* **77**, 528 (2000).

G.I. Gellene, Resonant states of a one-dimensional piecewise constant potential. *J. Chem. Educ.* **72**, 1015 (1995).

C.A. Hollingsworth, Accidental degeneracies of the particle in a box. *J. Chem. Educ.* **67**, 999 (1990).

W.-K. Li and S.M. Blinder, Particle in an equilateral triangle: exact solution of a nonseparable problem. *J. Chem. Educ.* **64**, 130 (1987).

Y.Q. Liang, H. Zhang, and Y.X. Dardenne, Momentum distributions for a particle in a box. *J. Chem. Educ.* **72**, 148 (1995).

T. McDermott and G. Henderson, Spherical harmonics in a cartesian frame. *J. Chem. Educ.* **67**, 915 (1990).

K.V. Mikkelsen and M.A. Ratner, Electron tunnelling in solid-state electron-transfer reactions. *Chem. Rev.* **87**, 113 (1987).

J.J.C. Mulder, Closed-form spherical harmonics: explicit polynomial expression for the associated legendre functions. *J. Chem. Educ.* **77**, 244 (2000).

C.C. Page, C.C. Moser, X. Chen, and P.L. Dutton, Natural engineering principles of electron tunnelling in biological oxidation–reduction. *Nature* **402**, 6757 (1999).

D. Ringe and G. Petsko, Quantum enzymology: tunnel vision. *Nature* **399**, 417 (1999).

K. Volkamer and M.W. Lerom, More about the particle-in-a-box system: the confinement of matter and the wave–particle dualism. *J. Chem. Educ.* **69**, 100 (1992).

Texts and sources of data and information

P.W. Atkins, *Quanta: A handbook of concepts.* Oxford University Press (1991).

P.W. Atkins and R.S. Friedman, *Molecular quantum mechanics.* Oxford University Press (1997).

D. Bohm, *Quantum theory.* Dover, New York (1989).

R.E. Christofferson, *Basic principles and techniques of molecular quantum mechanics.* Springer, New York (1989).

I.I. Goldman, V.D. Krivchenkov, and B.T. Geilikman. *Problems in quantum mechanics.* Dover, New York (1993).

C.S. Johnson, Jr. and L.G. Pedersen, *Problems and solutions in quantum chemistry and physics.* Dover, New York (1986).

I.N. Levine, *Quantum chemistry.* Prentice-Hall, Upper Saddle River (2000).

J.P. Lowe, *Quantum chemistry.* Academic Press, Boston (1993).

D.A. McQuarrie, *Quantum chemistry.* University Science Books, Mill Valley (1983).

L. Pauling and E.B. Wilson, *Introduction to quantum mechanics with applications to chemistry.* Dover, New York (1985).

P.J.E. Peebles, *Quantum mechanics.* Princeton University Press, Princeton (1992).

G.C. Schatz and M.A. Ratner, *Quantum mechanics in chemistry.* Ellis Horwood/Prentice Hall, Hemel Hempstead (1993).

Exercises

Discussion questions

12.1 (a) Discuss the physical origin of quantization energy for a particle confined to moving inside a one-dimensional box or on a ring.

12.1 (b) Discuss the correspondence principle and provide two examples.

12.2 (a) Define, justify, and provide examples of zero-point energy.

12.2 (b) Discuss the physical origins of quantum mechanical tunnelling. Why is tunnelling more likely to contribute to the mechanisms of electron transfer and proton transfer processes than to mechanisms of group transfer reactions, such as $AB + C \rightarrow A + BC$ (where A, B, and C are large molecular groups)?

12.3 (a) Distinguish between a fermion and a boson. Provide examples of each type of particle.

12.3 (b) Describe the essential steps in the derivation of the first-order and second-order corrections to the energy in time-independent perturbation theory.

Numerical exercises

12.4 (a) Calculate the energy separations in joules, kilojoules per mole, electronvolts, and reciprocal centimetres between the levels (a) $n = 2$ and $n = 1$, (b) $n = 6$ and $n = 5$ of an electron in a box of length 1.0 nm.

12.4 (b) Calculate the energy separations in joules, kilojoules per mole, electronvolts, and reciprocal centimetres between the levels (a) $n = 3$ and $n = 1$, (b) $n = 7$ and $n = 6$ of an electron in a box of length 1.50 nm.

12.5 (a) Calculate the probability that a particle will be found between $0.49L$ and $0.51L$ in a box of length L when it has (a) $n = 1$, (b) $n = 2$. Take the wavefunction to be a constant in this range.

12.5 (b) Calculate the probability that a particle will be found between $0.65L$ and $0.67L$ in a box of length L when it has (a) $n = 1$, (b) $n = 2$. Take the wavefunction to be a constant in this range.

12.6 (a) Calculate the expectation values of p and p^2 for a particle in the state $n = 1$ in a square-well potential.

12.6 (b) Calculate the expectation values of p and p^2 for a particle in the state $n = 2$ in a square-well potential.

12.7 (a) What are the most likely locations of a particle in a box of length L in the state $n = 3$?

12.7 (b) What are the most likely locations of a particle in a box of length L in the state $n = 5$?

12.8 (a) Consider a particle in a cubic box. What is the degeneracy of the level that has an energy three times that of the lowest level?

12.8 (b) Consider a particle in a cubic box. What is the degeneracy of the level that has an energy $\frac{14}{3}$ times that of the lowest level?

12.9 (a) Calculate the percentage change in a given energy level of a particle in a cubic box when the length of the edge of the cube is decreased by 10 per cent in each direction.

12.9 (b) A nitrogen molecule is confined in a cubic box of volume 1.00 m³. Assuming that the molecule has an energy equal to $\frac{3}{2}kT$ at $T = 300$ K, what is the value of $n = (n_x^2 + n_y^2 + n_z^2)^{1/2}$ for this molecule? What is the energy separation between the levels n and $n + 1$? What is its de Broglie wavelength? Would it be appropriate to describe this particle as behaving classically?

12.10 (a) Calculate the zero-point energy of a harmonic oscillator consisting of a particle of mass 2.33×10^{-26} kg and force constant 155 N m⁻¹.

12.10 (b) Calculate the zero-point energy of a harmonic oscillator consisting of a particle of mass 5.16×10^{-26} kg and force constant 285 N m⁻¹.

12.11 (a) For a harmonic oscillator consisting of a particle of mass 1.33×10^{-25} kg, the difference in adjacent energy levels is 4.82 zJ. Calculate the force constant of the oscillator.

12.11 (b) For a harmonic oscillator consisting of a particle of mass 2.88×10^{-25} kg, the difference in adjacent energy levels is 3.17 zJ. Calculate the force constant of the oscillator.

12.12 (a) Calculate the wavelength of a photon needed to excite a transition between neighbouring energy levels of a harmonic oscillator of mass equal to that of a proton (1.0078 u) and force constant 855 N m⁻¹.

12.12 (b) Calculate the wavelength of a photon needed to excite a transition between neighbouring energy levels of a harmonic oscillator of mass equal to that of an oxygen atom (15.9949 u) and force constant 544 N m⁻¹.

12.13 (a) Refer to Exercise 12.12a and calculate the wavelength that would result from doubling the mass of the particle.

12.13 (b) Refer to Exercise 12.12b and calculate the wavelength that would result from doubling the mass of the particle.

12.14 (a) Calculate the minimum excitation energies of (a) a pendulum of length 1.0 m on the surface of the Earth, (b) the balance-wheel of a clockwork watch ($v = 5$ Hz).

12.14 (b) Calculate the minimum excitation energies of (a) the 33 kHz quartz crystal of a watch, (b) the bond between two O atoms in O_2, for which $k = 1177$ N m⁻¹.

12.15 (a) Confirm that the wavefunction for the ground state of a one-dimensional linear harmonic oscillator given in Table 12.1 is a solution of the Schrödinger equation for the oscillator and that its energy is $\frac{1}{2}\hbar\omega$.

12.15 (b) Confirm that the wavefunction for the first excited state of a one-dimensional linear harmonic oscillator given in Table 12.1 is a solution of the Schrödinger equation for the oscillator and that its energy is $\frac{3}{2}\hbar\omega$.

12.16 (a) Assuming that the vibrations of a $^{35}Cl_2$ molecule are equivalent to those of a harmonic oscillator with a force constant $k = 329$ N m⁻¹, what is the zero-point energy of vibration of this molecule? The mass of a ^{35}Cl atom is 34.9688 u.

12.16 (b) Assuming that the vibrations of a $^{14}N_2$ molecule are equivalent to those of a harmonic oscillator with a force constant $k = 2293.8$ N m⁻¹, what is the zero-point energy of vibration of this molecule? The mass of a ^{14}N atom is 14.0031 u.

12.17 (a) The wavefunction, $\psi(\phi)$, for the motion of a particle in a ring is of the form $\psi = Ne^{im_l\phi}$. Determine the normalization constant, N.

12.17 (b) Confirm that wavefunctions for a particle in a ring with different values of the quantum number m_l are orthogonal.

12.18 (a) A point mass rotates in a circle with $l = 1$, Calculate the magnitude of its angular momentum and the possible projections of the angular momentum on an arbitrary axis.

12.18 (b) A point mass rotates in a circle with $l = 2$, Calculate the magnitude of its angular momentum and the possible projections of the angular momentum on an arbitrary axis.

12.19 (a) Draw scale vector diagrams to represent the states (a) $s = \frac{1}{2}$, $m_s = +\frac{1}{2}$, (b) $l = 1$, $m_l = +1$, (c) $l = 2$, $m_l = 0$.

12.19 (b) Draw the vector diagram for all the permitted states of a particle with $l = 6$.

Problems†

Numerical problems

12.1 Calculate the separation between the two lowest levels for an O_2 molecule in a one-dimensional container of length 5.0 cm. At what value of n does the energy of the molecule reach $\frac{1}{2}kT$ at 300 K, and what is the separation of this level from the one immediately below?

12.2 The mass to use in the expression for the vibrational frequency of a diatomic molecule is the effective mass $\mu = m_A m_B/(m_A + m_B)$, where m_A and m_B are the masses of the individual atoms. The following data on the infrared absorption wavenumbers (in cm^{-1}) of molecules is taken from *Spectra of diatomic molecules*, G. Herzberg, van Nostrand (1950):

$H^{35}Cl$	$H^{81}Br$	HI	CO	NO
2990	2650	2310	2170	1904

Calculate the force constants of the bonds and arrange them in order of increasing stiffness.

12.3 The rotation of a $^1H^{127}I$ molecule can be pictured as the orbital motion of an H atom at a distance 160 pm from a stationary I atom. (This is quite a good picture; to be precise, both atoms rotate around their common centre of mass, which is very close to the I nucleus.) Suppose that the molecule rotates only in a plane. Calculate the energy needed to excite the molecule into rotation. What, apart from 0, is the minimum angular momentum of the molecule?

12.4 Calculate the energies of the first four rotational levels of $^1H^{127}I$ free to rotate in three dimensions, using for its moment of inertia $I = \mu R^2$, with $\mu = m_H m_I/(m_H + m_I)$ and $R = 160$ pm.

12.5 A small step in the potential energy is introduced into the one-dimensional square-well problem as in Fig. 12.41. (a) Write a general expression for the first-order correction to the ground-state energy, $E_0^{(1)}$. (b) Evaluate the energy correction for $a = L/10$ (so the blip in the potential occupies the central 10 per cent of the well), with $n = 1$.

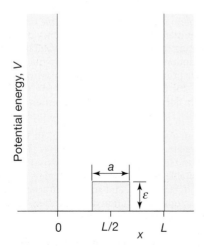

Fig. 12.41

12.6 We normally think of the one-dimensional well as being horizontal. Suppose it is vertical; then the potential energy of the particle depends on x because of the presence of the gravitational field. Calculate the first-order correction to the zero-point energy, and evaluate it for an electron in a box on the surface of the Earth. Account for the result. *Hint.* The energy of the particle depends on its height as mgh, where $g = 9.81$ m s^{-2}. Because g is so small, the energy correction is small; but it would be significant if the box were near a very massive star.

12.7 Calculate the second-order correction to the energy for the system described in Problem 12.6 and calculate the ground-state wavefunction. Account for the shape of the distortion caused by the perturbation. *Hint.* The following integrals are useful

$$\int x \sin ax \sin bx \, dx = -\frac{d}{da} \int \cos ax \sin bx \, dx$$

$$\int \cos ax \sin bx \, dx = \frac{\cos(a-b)x}{2(a-b)} - \frac{\cos(a+b)x}{2(a+b)} + \text{constant}$$

Theoretical problems

12.8 Set up the Schrödinger equation for a particle of mass m in a three-dimensional square well with sides L_1, L_2, and L_3. Show that the wavefunction is defined by three quantum numbers and that the Schrödinger equation is separable. Find the energy levels, and specialize the result to a cubic box of side L.

12.9 Derive eqn 12.28a, the expression for the transmission probability.

12.10‡ Consider the one-dimensional space in which a particle can experience one of three potentials depending upon its position. They are: $V = 0$ for $-\infty < x \leq 0$, 0, $V = V_2$ for $0 \leq x \leq L$, and $V = V_3$ for $L \leq x \leq \infty$. The particle wavefunction is to have both a component $e^{ik_1 x}$ that is incident upon the barrier V_2 and a reflected component $e^{-ik_1 x}$ in region 1 ($-\infty < x \leq 0$). In region 3 the wavefunction has only a forward component, $e^{ik_3 x}$, which represents a particle that has traversed the barrier. The energy of the particle, E, is somewhere in the range of $V_2 > E > V_3$. The transmission probability, T, is the ratio of the square modulus of the region 3 amplitude to the square modulus of the incident amplitude. (a) Base your calculation on the continuity of the amplitudes and the slope of the wavefunction at the locations of the zone boundaries and derive a general equation for T. (b) Show that the general equation for T reduces to eqn 12.28 in the high, wide barrier limit when $V_1 = V_3 = 0$. (c) Draw a graph of the probability of proton tunnelling when $V_3 = 0$, $L = 50$ pm, and $E = 10$ kJ mol^{-1} in the barrier range $E < V_2 < 2E$.

12.11 The wavefunction inside a long barrier of height V is $\psi = Ne^{-\kappa x}$. Calculate (a) the probability that the particle is inside the barrier and (b) the average penetration depth of the particle into the barrier.

12.12 Confirm that a function of the form e^{-gx^2} is a solution of the Schrödinger equation for the ground state of a harmonic oscillator and find an expression for g in terms of the mass and force constant of the oscillator.

---

† Problems denoted with the symbol ‡ were supplied by Charles Trapp, Carmen Giunta, and Marshall Cady.

12.13 Calculate the mean kinetic energy of a harmonic oscillator using the relations above Table 12.1.

12.14 Calculate the values of $\langle x^3 \rangle$ and $\langle x^4 \rangle$ for a harmonic oscillator using the relations above Table 12.1.

12.15 Determine the values of $\Delta x = (\langle x^2 \rangle - \langle x \rangle^2)^{1/2}$ and $\Delta p = (\langle p^2 \rangle - \langle p \rangle^2)^{1/2}$ for (a) a particle in a box of length L and (b) a harmonic oscillator. Discuss these quantities with reference to the uncertainty principle.

12.16 We shall see in Chapter 16 that the intensity of spectroscopic transitions between the vibrational states of a molecule are proportional to the square of the integral $\int \psi_{v'} x \psi_v \, dx$ over all space. Use the relations between Hermite polynomials given above Table 12.1 to show that the only permitted transitions are those for which $v' = v \pm 1$ and evaluate the integral in these cases.

12.17 Use the virial theorem to obtain an expression for the relation between the mean kinetic and potential energies of an electron in a hydrogen atom.

12.18 Evaluate the z-component of the angular momentum and the kinetic energy of a particle on a ring that is described by the (unnormalized) wavefunctions (a) $e^{i\phi}$, (b) $e^{-2i\phi}$, (c) $\cos \phi$, and (d) $(\cos \chi)e^{i\phi} + (\sin \chi)e^{-i\phi}$.

12.19 Confirm that the spherical harmonics (a) $Y_{0,0}$, (b) $Y_{2,-1}$, and (c) $Y_{3,+3}$ satisfy the Schrödinger equation for a particle free to rotate in three dimensions, and find its energy and angular momentum in each case.

12.20 Confirm that $Y_{3,+3}$ is normalized to 1. (The integration required is over the surface of a sphere.)

12.21 Derive an expression in terms of l and m_l for the half-angle of the apex of the cone used to represent an angular momentum according to the vector model. Evaluate the expression for an α spin. Show that the minimum possible angle approaches 0 as $l \to \infty$.

12.22 Show that the function $f = \cos ax \cos by \cos cz$ is an eigenfunction of ∇^2, and determine its eigenvalue.

12.23 Derive (in Cartesian coordinates) the quantum mechanical operators for the three components of angular momentum starting from the classical definition of angular momentum, $l = r \times p$. Show that any two of the components do not mutually commute, and find their commutator.

12.24 Starting from the operator $l_z = xp_y - yp_x$, prove that in spherical polar coordinates $l_z = -i\hbar \partial/\partial\phi$.

12.25‡ A particle is confined to move in a one-dimensional box of length L. (a) If the particle is classical, show that the average value of x is $\frac{1}{2}L$ and that the root-mean square value is $L/3^{1/2}$. (b) Show that, for large values of n, a quantum particle approaches the classical values. This result is an example of the *correspondence principle*, which states that, for very large values of the quantum numbers, the predictions of quantum mechanics approach those of classical mechanics.

Applications: to biology

12.26 To a first approximation, a π electron in a linear polyene may be considered to be a particle in a one-dimensional box. Some linear polyenes, of which β-carotene is an example, are important

biological co-factors that participate in processes as diverse as the absorption of solar energy in photosynthesis (Box 26.2) and protection against harmful biological oxidations. β-Carotene contains 22 conjugated C atoms and the average internuclear distance is 140 pm. In the ground state, each level up to $n = 11$ is occupied by two electrons. Calculate (a) the separation in energy between the ground state and the first excited state in which one electron occupies the state with $n = 12$, (b) the frequency of the radiation required to produce a transition between these two states, and (c) the total probability of finding an electron between C atoms 11 and 12 in the ground state of β-carotene.

12.27 When β-carotene is oxidized *in vivo*, it breaks in half and forms two molecules of retinal (vitamin A), which is a precursor to the pigment in the retina responsible for vision (Box 17.1). The conjugated system of retinal consists of 11 C atoms and one O atom. In the ground state of retinal, each level up to $n = 6$ is occupied by two electrons. Assuming an average internuclear distance of 140 pm, calculate (a) the separation in energy between the ground state and the first excited state in which one electron occupies the state with $n = 7$, and (b) the frequency of the radiation required to produce a transition between these two states. (c) Using your results from this problem and from Problem 12.26, choose among the words in parentheses to generate a rule for the prediction of frequency shifts in the absorption spectra of linear polyenes: *The absorption spectrum of a linear polyene shifts to (higher/lower) frequency as the number of conjugated atoms (increases/decreases).*

12.28 Many biological electron transfer reactions, such as those associated with biological energy conversion (Box 10.1), may be visualized as arising from electron tunnelling between protein-bound co-factors, such as cytochromes, quinones, flavins, and chlorophylls. This tunnelling occurs over distances that are often greater than 1.0 nm, with sections of protein separating electron donor from acceptor. For a specific combination of donor and acceptor, the rate of electron tunnelling is proportional to the transmission probability, with $\kappa \approx 7 \text{ nm}^{-1}$ (eqn 12.28). By what factor does the rate of electron tunnelling between two co-factors increase as the distance between them changes from 2.0 nm to 1.0 nm?

12.29 Carbon monoxide binds strongly to the Fe^{2+} ion of the haem group of the protein myoglobin (for an example of a haem group, see Box 10.1). Estimate the vibrational frequency of CO bound to myoglobin by using the data in Problem 12.2 and by making the following assumptions: the atom that binds to the haem group is immobilized, the protein is infinitely more massive than either the C or O atom, the C atom binds to the Fe^{2+} ion, and binding of CO to the protein does not alter the force constant of the $C\equiv O$ bond.

12.30 Of the four assumptions made in Problem 12.29, the last two are questionable. Suppose that the first two assumptions are still reasonable and that you have at your disposal a supply of myoglobin, a suitable buffer in which to suspend the protein, $^{12}C^{16}O$, $^{13}C^{16}O$, $^{12}C^{18}O$, $^{13}C^{18}O$, and an infrared spectrometer. Assuming that isotopic substitution does not affect the force constant of the $C\equiv O$ bond, describe a set of experiments that: (a) proves which atom, C or O, binds to the haem group of myoglobin, and (b) allows for the determination of the force constant of the $C\equiv O$ bond for myoglobin-bound carbon monoxide.

13

Atomic structure and atomic spectra

The structure and spectra of hydrogenic atoms

13.1 The structure of hydrogenic atoms

13.2 Atomic orbitals and their energies

13.3 Spectroscopic transitions and selection rules

The structures of many-electron atoms

13.4 The orbital approximation

13.5 Self-consistent field orbitals

The spectra of complex atoms

13.6 Quantum defects and ionization limits

13.7 Singlet and triplet states

13.8 Spin–orbit coupling

13.9 Term symbols and selection rules

Checklist of key ideas

Further reading

Exercises

Problems

We now use the principles of quantum mechanics introduced in the preceding two chapters to describe the internal structures of atoms. We see what experimental information is available from a study of the spectrum of atomic hydrogen. Then we set up the Schrödinger equation for an electron in an atom and separate it into angular and radial parts. The wavefunctions obtained are the 'atomic orbitals' of hydrogenic atoms. Next, we use these hydrogenic atomic orbitals to describe the structures of many-electron atoms. In conjunction with the Pauli exclusion principle, we account for the periodicity of atomic properties. The spectra of many-electron atoms are more complicated than those of hydrogen, but the same principles apply. We see in the closing sections of the chapter how such spectra are described by using term symbols, and the origin of the finer details of their appearance.

In this chapter we see how to use quantum mechanics to describe the **electronic structure** of an atom, the arrangement of electrons around a nucleus. The concepts we meet are of central importance for understanding the structures and reactions of atoms and molecules, and hence have extensive chemical applications. We need to distinguish between two types of atoms. A **hydrogenic atom** is a one-electron atom or ion of general atomic number Z; examples of hydrogenic atoms are H, He^+, Li^{2+}, and U^{91+}. A **many-electron atom** (or *polyelectronic atom*) is an atom or ion with more than one electron; examples include all neutral atoms other than H. So even He, with only two electrons, is a many-electron atom. Hydrogenic atoms are important because their Schrödinger equations can be solved exactly. They also provide a set of concepts that are used to describe the structures of many-electron atoms and, as we shall see in the next chapter, the structures of molecules too.

One of the principal experimental techniques for determining the electronic structures of atoms is **spectroscopy**, the detection and analysis of the electromagnetic radiation absorbed or emitted by a substance. The record of spectral intensity as a function of frequency (ν), wavelength (λ), or wavenumber ($\tilde{\nu} = \nu/c$)[1] of the radiation emitted or absorbed by an atom or a molecule is called its **spectrum** (from the Greek word for appearance). The spectrum of an atom consists of a series of 'lines', or sharply defined emission or absorption peaks.

---

1. The relation between these quantities is described in *Further information* 2.

13.1 The spectrum of atomic hydrogen. Both the observed spectrum and its resolution into overlapping series are shown. Note that the Balmer series lies in the visible region.

The structure and spectra of hydrogenic atoms

When an electric discharge is passed through gaseous hydrogen, the H_2 molecules are dissociated and the energetically excited H atoms that are produced emit light of discrete frequencies (Fig. 13.1). The first important contribution to the interpretation of this spectrum was made by the Swiss schoolteacher Johann Balmer, who pointed out in 1885 that (in modern terms) the wavenumbers of the lines in the visible region fit the expression

$$\tilde{\nu} \propto \frac{1}{2^2} - \frac{1}{n^2} \qquad n = 3, 4, \ldots$$

The lines this formula describes are now called the **Balmer series**. When further lines were discovered in the ultraviolet, giving the **Lyman series**, and in the infrared, the **Paschen series**, the Swedish spectroscopist Johannes Rydberg noted (in 1890) that all of them were described by the expression

$$\tilde{\nu} = R_H \left(\frac{1}{n_1^2} - \frac{1}{n_2^2} \right) \qquad R_H = 109\,677 \text{ cm}^{-1} \tag{13.1}$$

with $n_1 = 1$ (the Lyman series), 2 (the Balmer series), and 3 (the Paschen series), and that in each case $n_2 = n_1 + 1, n_1 + 2, \ldots$. The constant R_H is now called the **Rydberg constant** for the hydrogen atom.

Illustration 13.1

The line of longest wavelength (lowest wavenumber) in the Lyman series ($n_1 = 1$) is the one with $n_2 = 2$; its wavenumber is

$$\tilde{\nu} = R_H \left(\frac{1}{1^2} - \frac{1}{2^2} \right) = (109\,677 \text{ cm}^{-1}) \times \tfrac{3}{4} = 82\,258 \text{ cm}^{-1}$$

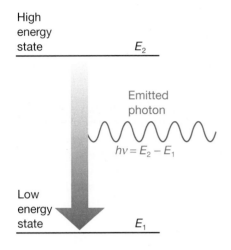

13.2 Energy is conserved when a photon is emitted, so the difference in energy of the atom before and after the emission event must be equal to the energy of the photon emitted.

Its wavelength is therefore

$$\lambda = \frac{1}{\tilde{v}} = \frac{1}{8.2258 \times 10^6 \text{ m}^{-1}} = 1.2157 \times 10^{-7} \text{ m}$$

or 121.57 nm, in the vacuum ultraviolet region of the spectrum.

Self-test 13.1 Calculate the shortest wavelength line in the Paschen series.

[821 nm]

The form of eqn 1 strongly suggests that the wavenumber of each spectral line can be written as the difference of two **terms**, each of the form

$$T_n = \frac{R_H}{n^2} \tag{13.2}$$

The **Ritz combination principle** states that *the wavenumber of any spectral line is the difference between two terms*. We say that two terms T_1 and T_2 'combine' to produce a spectral line of wavenumber

$$\tilde{v} = T_1 - T_2 \tag{13.3}$$

The Ritz combination principle applies to all types of atoms and molecules, but only for hydrogenic atoms do the terms have the simple form (constant)/n^2.

The Ritz combination principle is readily explained in terms of photons and the conservation of energy. Thus, a spectroscopic line arises from the transition of an atom from one energy level (a term) to another (another term) with the emission of the difference in energy as a photon (Fig. 13.2). This interpretation leads to the **Bohr frequency condition**, which states that, when an atom changes its energy by ΔE, the difference is carried away as a photon of frequency v, where

$$\Delta E = hv \tag{13.4}$$

Thus, if each spectroscopic term represents an energy hcT, the difference in energy when the atom undergoes a transition between two terms is $\Delta E = hcT_1 - hcT_2$, and the frequency of the light emitted is given by $v = cT_1 - cT_2$. This expression rearranges into the Ritz formula when expressed in terms of wavenumbers (on division by c).

Because spectroscopic observations show that electromagnetic radiation is absorbed and emitted by atoms only at certain wavenumbers, it follows that only certain energy states of atoms are permitted. Our tasks in the first part of this chapter are to determine the origin of this energy quantization, to find the permitted energy levels, and to account for the value of R_H.

13.1 The structure of hydrogenic atoms

The Coulomb potential energy of an electron in a hydrogenic atom of atomic number Z (and nuclear charge Ze) is

$$V = -\frac{Ze^2}{4\pi\varepsilon_0 r} \tag{13.5}$$

where r is the distance of the electron from the nucleus and ε_0 is the vacuum permittivity. The hamiltonian for the electron and a nucleus of mass m_N is therefore

$$H = \hat{E}_{K,electron} + \hat{E}_{K,nucleus} + \hat{V} = -\frac{\hbar^2}{2m_e}\nabla_e^2 - \frac{\hbar^2}{2m_N}\nabla_N^2 - \frac{Ze^2}{4\pi\varepsilon_0 r} \tag{13.6}$$

The subscripts on ∇^2 indicate differentiation with respect to the electron or nuclear coordinates.

(a) The separation of internal motion

Physical intuition suggests that the full Schrödinger equation ought to separate into two equations, one for the motion of the atom as a whole through space and the other for the motion of the electron relative to the nucleus. We have already solved the first of these equations, because it corresponds to the free translational motion of a particle of mass $m = m_e + m_N$ (Section 11.5). The initial strategy of the calculation is therefore to separate the relative motion of the electron and the nucleus from the motion of the atom as a whole. As we show in the *Justification* below, the resulting expression for the hamiltonian for the internal motion of the electron relative to the nucleus is

$$H = -\frac{\hbar^2}{2\mu}\nabla^2 - \frac{Ze^2}{4\pi\varepsilon_0 r} \qquad \frac{1}{\mu} = \frac{1}{m_e} + \frac{1}{m_N} \tag{13.7}$$

where differentiation is now with respect to the coordinates of the electron relative to the nucleus. The quantity μ is called the **reduced mass**. The reduced mass is very similar to the electron mass because m_N, the mass of the nucleus, is much larger than the mass of an electron, so $1/\mu \approx 1/m_e$. In all except the most precise work, the reduced mass can be replaced by m_e.

Justification 13.1

Consider a one-dimensional system in which the potential energy depends only on the separation of the two particles; the total energy is

$$E = \frac{p_1^2}{2m_1} + \frac{p_2^2}{2m_2} + V$$

where $p_1 = m_1\dot{x}_1$ and $p_2 = m_2\dot{x}_2$, the dot signifying differentiation with respect to time. The centre of mass (Fig. 13.3) is located at

$$X = \frac{m_1}{m}x_1 + \frac{m_2}{m}x_2 \qquad m = m_1 + m_2$$

and the separation of the particles is $x = x_1 - x_2$. It follows that

$$x_1 = X + \frac{m_2}{m}x \qquad x_2 = X - \frac{m_1}{m}x$$

The linear momenta of the particles can be expressed in terms of x and X:

$$p_1 = m_1\dot{x}_1 = m_1\dot{X} + \frac{m_1 m_2}{m}\dot{x} \qquad p_2 = m_2\dot{x}_2 = m_2\dot{X} - \frac{m_1 m_2}{m}\dot{x}$$

Then it follows that

$$\frac{p_1^2}{2m_1} + \frac{p_2^2}{2m_2} = \tfrac{1}{2}m\dot{X}^2 + \tfrac{1}{2}\mu\dot{x}^2$$

where μ is given in eqn 7. By writing $P = m\dot{X}$ for the linear momentum of the system as a whole and defining p as $\mu\dot{x}$, we find

$$E = \frac{P^2}{2m} + \frac{p^2}{2\mu} + V$$

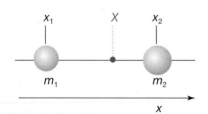

13.3 The coordinates used for discussing the separation of the relative motion of two particles from the motion of the centre of mass.

The corresponding hamiltonian (generalized to three dimensions) is therefore

$$H = -\frac{\hbar^2}{2m} \nabla_{\text{c.m.}}^2 - \frac{\hbar^2}{2\mu} \nabla^2 + V$$

where the first term differentiates with respect to the centre of mass coordinates and the second with respect to the relative coordinates.

Now we write the overall wavefunction as the product $\psi_{\text{total}} = \psi_{\text{c.m.}}\psi$, where the first factor is a function of only the centre of mass coordinates and the second is a function of only the relative coordinates. The overall Schrödinger equation, $H\psi_{\text{total}} = E_{\text{total}}\psi_{\text{total}}$, then separates by the argument that we have used in Sections 12.2a and 12.7:

$$-\frac{\hbar^2}{2m} \nabla_{\text{c.m.}}^2 \psi_{\text{c.m.}} = E_{\text{c.m.}}\psi_{\text{c.m.}} \qquad -\frac{\hbar^2}{2\mu} \nabla^2\psi + V\psi = E\psi$$

with $E_{\text{total}} = E_{\text{c.m.}} + E$.

From now on we consider only the internal, relative coordinates. The Schrödinger equation, $H\psi = E\psi$, is

$$-\frac{\hbar^2}{2\mu} \nabla^2\psi - \frac{Ze^2}{4\pi\varepsilon_0 r}\psi = E\psi \tag{13.8}$$

Because the potential energy is centrosymmetric (independent of angle), we can suspect that the equation is separable into radial and angular components. Therefore, we write

$$\psi(r, \theta, \phi) = R(r)Y(\theta, \phi) \tag{13.9}$$

and examine whether the Schrödinger equation can be separated into two equations, one for R and the other for Y. As shown in the *Justification* below, the equation does separate, and the equations we have to solve are

$$\Lambda^2 Y = -l(l + 1)Y \tag{13.10}$$

$$-\frac{\hbar^2}{2\mu}\left(\frac{\text{d}^2R}{\text{d}r^2} + \frac{2}{r}\frac{\text{d}R}{\text{d}r}\right) + V_{\text{eff}}R = ER \tag{13.11}$$

where

$$V_{\text{eff}} = -\frac{Ze^2}{4\pi\varepsilon_0 r} + \frac{l(l + 1)\hbar^2}{2\mu r^2} \tag{13.12}$$

Justification 13.2

The laplacian in three dimensions is given in eqn 12.62. It follows that the Schrödinger equation in eqn 8 is

$$-\frac{\hbar^2}{2\mu}\left(\frac{\partial^2}{\partial r^2} + \frac{2}{r}\frac{\partial}{\partial r} + \frac{1}{r^2}\Lambda^2\right)RY + VRY = ERY$$

Because R depends only on r and Y depends only on the angular coordinates, this equation becomes

$$-\frac{\hbar^2}{2\mu}\left(Y\frac{\text{d}^2R}{\text{d}r^2} + \frac{2Y}{r}\frac{\text{d}R}{\text{d}r} + \frac{R}{r^2}\Lambda^2 Y\right) + VRY = ERY$$

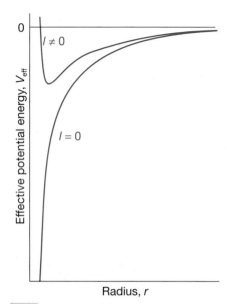

13.4 The effective potential energy of an electron in the hydrogen atom. When the electron has zero orbital angular momentum, the effective potential energy is the Coulombic potential energy. When the electron has nonzero orbital angular momentum, the centrifugal effect gives rise to a positive contribution, which is very large close to the nucleus. We can expect the $l = 0$ and $l \neq 0$ wavefunctions to be very different near the nucleus.

If we multiply through by r^2/RY, we obtain

$$-\frac{\hbar^2}{2\mu R}\left(r^2\frac{d^2R}{dr^2} + 2r\frac{dR}{dr}\right) + Vr^2 - \frac{\hbar^2}{2\mu Y}\Lambda^2 Y = Er^2$$

At this point we employ the usual argument. The term in Y is the only one that depends on the angular variables, so it must be a constant. When we write this constant as $\hbar^2 l(l+1)/2\mu$, eqn 11 follows immediately.

Equation 10 is the same as the Schrödinger equation for a particle free to move round a central point, and we considered it in Section 12.7. The solutions are the spherical harmonics (Table 12.3), and are specified by the quantum numbers l and m_l. We consider them in more detail shortly. Equation 11 is called the **radial wave equation**. The radial wave equation is the description of the motion of a particle of mass μ in a one-dimensional region where the potential energy is V_{eff}.

(b) The radial solutions

We can anticipate some features of the shapes of the radial wavefunctions by analysing the shape of V_{eff}. The first term in eqn 12 is the Coulomb potential energy of the electron in the field of the nucleus. The second term stems from the centrifugal force that arises from the angular momentum of the electron around the nucleus. When $l = 0$, the electron has no angular momentum, and the effective potential energy is purely Coulombic and attractive at all radii (Fig. 13.4). When $l \neq 0$, the centrifugal term gives a positive (repulsive) contribution to the effective potential energy. When the electron is close to the nucleus ($r \approx 0$), this repulsive term, which is proportional to $1/r^2$, dominates the attractive Coulombic component, which is proportional to $1/r$, and the net effect is an effective repulsion of the electron from the nucleus. The two effective potential energies, the one for $l = 0$ and the one for $l \neq 0$, are qualitatively very different close to the nucleus. However, they are similar at large distances because the centrifugal contribution tends to zero more rapidly than the Coulombic contribution. Therefore, we can expect the solutions with $l = 0$ and $l \neq 0$ to be quite different near the nucleus but similar far away from it.

We shall not go through the technical steps of solving the radial equation (see *Further reading*). It is sufficient to know that acceptable solutions can be found only for integral values of a quantum number n, and that the allowed energies are

$$E_n = -\frac{Z^2\mu e^4}{32\pi^2\varepsilon_0^2\hbar^2 n^2} \tag{13.13}$$

with $n = 1, 2, \ldots$.

The radial wave equation depends on l, and the radial wavefunctions, which depend on the values of both n and l (but not on m_l), all have the form

$$R(r) = (\text{polynomial in } r) \times (\text{decaying exponential in } r) \tag{13.14}$$

These functions are most simply written in terms of the dimensionless quantity ρ (rho), where

$$\rho = \frac{2Zr}{a_0} \qquad a_0 = \frac{4\pi\varepsilon_0\hbar^2}{m_e e^2} \tag{13.15}$$

The **Bohr radius**, a_0, has the value 52.9177 pm; it is so called because the same quantity appeared in Bohr's early model of the hydrogen atom as the radius of the electron orbit of lowest energy. Specifically, the radial wavefunctions for an electron with quantum numbers n and l are the (real) functions

Table 13.1 Hydrogenic radial wavefunctions

Orbital	n	l	$R_{n,l}$
$1s$	1	0	$2\left(\dfrac{Z}{a_0}\right)^{3/2} e^{-\rho/2}$
$2s$	2	0	$\dfrac{1}{2(2)^{1/2}}\left(\dfrac{Z}{a_0}\right)^{3/2}(2 - \tfrac{1}{2}\rho)e^{-\rho/4}$
$2p$	2	1	$\dfrac{1}{4(6)^{1/2}}\left(\dfrac{Z}{a_0}\right)^{3/2}\rho e^{-\rho/4}$
$3s$	3	0	$\dfrac{1}{9(3)^{1/2}}\left(\dfrac{Z}{a_0}\right)^{3/2}(6 - 2\rho + \tfrac{1}{9}\rho^2)e^{-\rho/6}$
$3p$	3	1	$\dfrac{1}{27(6)^{1/2}}\left(\dfrac{Z}{a_0}\right)^{3/2}(4 - \tfrac{1}{3}\rho)\rho e^{-\rho/6}$
$3d$	3	2	$\dfrac{1}{81(30)^{1/2}}\left(\dfrac{Z}{a_0}\right)^{3/2}\rho^2 e^{-\rho/6}$

The full wavefunction is $\psi = RY$, where Y is given in Table 12.3. In the table, $\rho = 2Zr/a_0$.

$$R_{n,l}(r) = N_{n,l}\left(\frac{\rho}{n}\right)^l L_{n,l}(\rho)e^{-\rho/2n} \tag{13.16}$$

where L is a polynomial in ρ called an *associated Laguerre polynomial*. Expressions for some radial wavefunctions are given in Table 13.1 and illustrated in Fig. 13.5. Note that, because R is proportional to ρ^l, all radial wavefunctions are zero at the nucleus unless $l = 0$.[2]

Illustration 13.2

To calculate the probability density for a $1s$-electron at the nucleus, we set $n = 1, l = 0, m_l = 0$ and evaluate ψ^2 at $r = 0$:

$$\psi_{1,0,0}(0, \theta, \phi) = R_{1,0}(0)Y_{0,0}(\theta, \phi) = 2\left(\frac{Z}{a_0}\right)^{3/2}\left(\frac{1}{4\pi}\right)^{1/2}$$

The probability density is therefore

$$\psi_{1,0,0}(0, \theta, \phi)^2 = \frac{Z^3}{\pi a_0^3}$$

which evaluates to $2.15 \times 10^{-6}\ \text{pm}^{-3}$ when $Z = 1$.

Self-test 13.2 Evaluate the probability density of the electron at the nucleus for a $2s$-electron.
$$[(Z/a_0)^3/8\pi]$$

---

2. The zero at $r = 0$ is not a radial node because the radial wavefunction does not pass through zero at that point (because r cannot be negative). Nodes at the nucleus are all angular nodes.

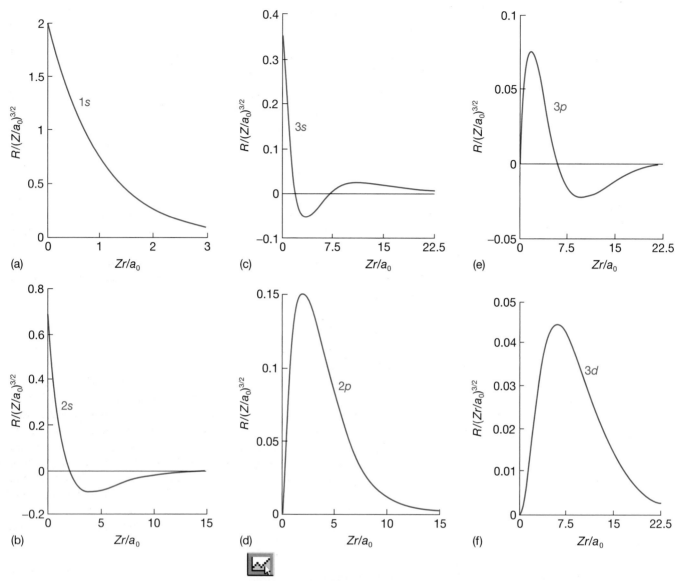

13.5 The radial wavefunctions of the first few states of hydrogenic atoms of atomic number Z. Note that the *s* orbitals have a nonzero and finite value at the nucleus. The horizontal scales are different in each case: orbitals with high principal quantum numbers are relatively distant from the nucleus.

13.2 Atomic orbitals and their energies

An **atomic orbital** is a one-electron wavefunction for an electron in an atom. Each hydrogenic atomic orbital is defined by three quantum numbers, designated n, l, and m_l. When an electron is described by one of these wavefunctions, we say that it 'occupies' that orbital. We could go on to say that the electron is in the state $|n,l,m_l\rangle$. For instance, an electron described by the wavefunction $\psi_{1,0,0}$ and in the state $|1,0,0\rangle$ is said to occupy the orbital with $n = 1$, $l = 0$, and $m_l = 0$.

The quantum number n is called the **principal quantum number**; it can take the values $n = 1, 2, 3, \ldots$ and determines the energy of the electron:

An electron in an orbital with quantum number n has an energy given by eqn 13.

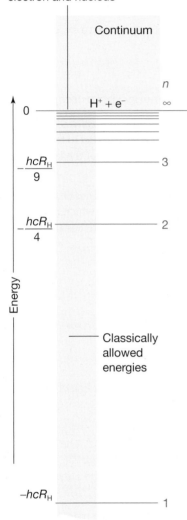

Energy of widely separated stationary electron and nucleus

Continuum

n

0 —— H$^+$ + e$^-$ —— ∞

$-\dfrac{hcR_H}{9}$ —————— 3

$-\dfrac{hcR_H}{4}$ —————— 2

Energy

—— Classically allowed energies

$-hcR_H$ —————— 1

13.6 The energy levels of a hydrogen atom. The values are relative to an infinitely separated, stationary electron and a proton.

The two other quantum numbers, l and m_l, come from the angular solutions, and specify the angular momentum of the electron around the nucleus:

> An electron in an orbital with quantum number l has an angular momentum of magnitude $\{l(l+1)\}^{1/2}\hbar$, with $l = 0, 1, 2, \ldots, n-1$.

> An electron in an orbital with quantum number m_l has a z-component of angular momentum $m_l\hbar$, with $m_l = 0, \pm1, \pm2, \ldots, \pm l$.

Note how the value of the principal quantum number, n, controls the maximum value of l, and how l in turn controls the range of values of m_l.

To define the state of an electron in a hydrogenic atom fully we need to specify not only the orbital it occupies but also its spin state. We saw in Section 12.8 that an electron possesses an intrinsic angular momentum that is described by the two quantum numbers s and m_s (the analogues of l and m_l). The value of s is fixed at $\frac{1}{2}$ for an electron, so we do not need to consider it further at this stage. However, m_s may be either $+\frac{1}{2}$ or $-\frac{1}{2}$, and to specify the electron's state in a hydrogenic atom we need to specify which of these values describes it. It follows that, to specify the state of an electron in a hydrogenic atom, we need to give the values of four quantum numbers, namely n, l, m_l, and m_s.

(a) The energy levels

The energy levels predicted by eqn 13 are depicted in Fig. 13.6. The energies, and also the separation of neighbouring levels, are proportional to Z^2, so the levels are four times as wide apart (and the ground state four times deeper in energy) in He$^+$ ($Z = 2$) than in H ($Z = 1$). All the energies given by eqn 13 are negative. They refer to the **bound states** of the atom, in which the energy of the atom is lower than that of the infinitely separated, stationary electron and nucleus (which corresponds to the zero of energy). There are also solutions of the Schrödinger equation with positive energies. These solutions correspond to **unbound states** of the electron, the states to which an electron is raised when it is ejected from the atom by a high-energy collision or photon (recall the unbound states of a particle in a finite well). The energies of the unbound electron are not quantized and form the continuum states of the atom.

Equation 13 is consistent with the spectroscopic result summarized by eqn 1, and we can identify the Rydberg constant for hydrogen ($Z = 1$) as

$$hcR_H = \frac{\mu_H e^4}{32\pi^2\varepsilon_0^2\hbar^2} \qquad [13.17]$$

where μ_H is the reduced mass for hydrogen. The **Rydberg constant** itself, R, is defined by the same expression except for the replacement of μ_H by the mass of an electron, m_e:

$$R_H = \frac{\mu_H}{m_e}R \qquad R = \frac{m_e e^4}{8\varepsilon_0^2 h^3 c} \qquad [13.18]$$

Insertion of the values of the fundamental constants into the expression for R_H gives almost exact agreement with the experimental value. The only discrepancies arise from the neglect of relativistic corrections, which the non-relativistic Schrödinger equation ignores.

(b) Ionization energies

The **ionization energy**, I, of an element is the minimum energy required to remove an electron from the ground state, the state of lowest energy, of one of its atoms. The ground state of hydrogen is the state with $n = 1$, which has energy

$$E_1 = -hcR_H$$

13.7 The plot of the data in Example 13.1 used to determine the ionization energy of an atom (in this case, of H).

The atom is ionized when the electron has been excited to the level corresponding to $n = \infty$ (see Fig. 13.6). Therefore, the energy that must be supplied is

$$I = hcR_H \qquad (13.19)$$

The value of I is 2.179 aJ (a, for atto, is the prefix that denotes 10^{-18}), which corresponds to 13.60 eV.

Example 13.1 Measuring an ionization energy spectroscopically

The emission spectrum of atomic hydrogen shows lines at 82 259, 97 492, 102 824, 105 292, 106 632, and 107 440 cm^{-1}, which correspond to transitions to the same lower state. Determine (a) the ionization energy of the lower state, (b) the value of the Rydberg constant.

Method The spectroscopic determination of ionization energies depends on the determination of the series limit, the wavenumber at which the series terminates and becomes a continuum. If the upper state lies at an energy $-hcR_H/n^2$, then, when the atom makes a transition to E_{lower}, a photon of wavenumber

$$\tilde{v} = -\frac{R_H}{n^2} - \frac{E_{lower}}{hc}$$

is emitted. However, because $I = -E_{lower}$, it follows that

$$\tilde{v} = \frac{I}{hc} - \frac{R_H}{n^2}$$

A plot of the wavenumbers against $1/n^2$ should give a straight line of slope $-R_H$ and intercept I/hc. Use a computer (or a calculator) to make a least-squares fit of the data to get a result that reflects the precision of the data.

Answer The wavenumbers are plotted against $1/n^2$ in Fig. 13.7. The (least-squares) intercept lies at 109 679 cm^{-1}, so the ionization energy is 2.1788 aJ (1312.1 kJ mol^{-1}). The slope is, in this instance, numerically the same, so $R_H = 109\ 679$ cm^{-1}.

Comment A similar extrapolation procedure can be used for many-electron atoms (see Section 13.6).

Self-test 13.3 The emission spectrum of atomic deuterium shows lines at 15 238, 20 571, 23 039, 24 380 cm^{-1}, which correspond to transitions to the same lower state. Determine (a) the ionization energy of the lower state, (b) the ionization energy of the ground state, (c) the mass of the deuteron (by expressing the Rydberg constant in terms of the reduced mass of the electron and the deuteron, and solving for the mass of the deuteron).

[(a) 328.1 kJ mol^{-1}, (b) 1312.4 kJ mol^{-1}, (c) 2.8×10^{-27} kg, a result very sensitive to R_D]

(c) Shells and subshells

All the orbitals of a given value of n are said to form a single **shell** of the atom. In a hydrogenic atom, all orbitals of given n, and therefore belonging to the same shell, have the same energy. It is common to refer to successive shells by letters:

$n =$ 1   2   3   4 ...

         K   L   M   N ...

Thus, all the orbitals of the shell with $n = 2$ form the L shell of the atom, and so on.

The orbitals with the same value of n but different values of l are said to form a **subshell** of a given shell. These subshells are generally referred to by letters:

$$l = \quad 0 \quad 1 \quad 2 \quad 3 \quad 4 \quad 5 \quad 6 \ldots$$
$$ s \quad p \quad d \quad f \quad g \quad h \quad i \ldots$$

The letters then run alphabetically (j is not used). Figure 13.8 is a version of Fig. 13.6 that shows the subshells explicitly. Because l can range from 0 to $n - 1$, giving n values in all, it follows that there are n subshells of a shell with principal quantum number n. Thus, when $n = 1$, there is only one subshell, the one with $l = 0$. When $n = 2$, there are two subshells, the 2s subshell (with $l = 0$) and the 2p subshell (with $l = 1$).

When $n = 1$ there is only one subshell, that with $l = 0$, and that subshell contains only one orbital, with $m_l = 0$ (the only value of m_l permitted). When $n = 2$, there are four orbitals, one in the s subshell with $l = 0$ and $m_l = 0$, and three in the $l = 1$ subshell with $m_l = +1, 0, -1$. When $n = 3$ there are nine orbitals (one with $l = 0$, three with $l = 1$, and five with $l = 2$). The organization of orbitals in the shells is summarized in Fig. 13.9. In general, the number of orbitals in a shell of principal quantum number n is n^2, so in a hydrogenic atom each energy level is n^2-fold degenerate.

13.8 The energy levels of the hydrogen atom showing the subshells and (in square brackets) the numbers of orbitals in each subshell. In hydrogenic atoms, all orbitals of a given shell have the same energy .

13.9 The organization of orbitals (white squares) into subshells (characterized by l) and shells (characterized by n).

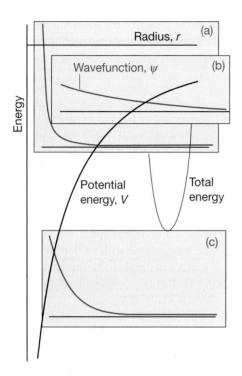

13.10 The balance of kinetic and potential energies that accounts for the structure of the ground state of hydrogen (and similar atoms). (a) The sharply curved but localized orbital has high mean kinetic energy, but low mean potential energy; (b) the mean kinetic energy is low, but the potential energy is not very favourable; (c) the compromise of moderate kinetic energy and moderately favourable potential energy. The parabola links the three panels and indicates schematically how the total energy varies.

(d) Atomic orbitals

The orbital occupied in the ground state is the one with $n = 1$ (and therefore with $l = 0$ and $m_l = 0$, the only possible values of these quantum numbers when $n = 1$). From Table 13.1 we can write (for $Z = 1$):

$$\psi = \frac{1}{(\pi a_0^3)^{1/2}} e^{-r/a_0} \tag{13.20}$$

This wavefunction is independent of angle and has the same value at all points of constant radius; that is, the 1s orbital is *spherically symmetrical*. The wavefunction decays exponentially from a maximum value of $1/(\pi a_0^3)^{1/2}$ at the nucleus (at $r = 0$). It follows that the most probable point at which the electron will be found is at the nucleus itself.

We can understand the general form of the ground-state wavefunction by considering the contributions of the potential and kinetic energies to the total energy of the atom. The closer the electron is to the nucleus on average, the lower its average potential energy. This dependence suggests that the lowest potential energy should be obtained with a sharply peaked wavefunction that has a large amplitude at the nucleus and is zero everywhere else (Fig. 13.10). However, this shape implies a high kinetic energy, because such a wavefunction has a very high average curvature. The electron would have very low kinetic energy if its wavefunction had only a very low average curvature. However, such a wavefunction spreads to great distances from the nucleus and the average potential energy of the electron will be correspondingly high. The actual ground-state wavefunction is a compromise between these two extremes: the wavefunction spreads away from the nucleus (so the expectation value of the potential energy is not as low as in the first example, but nor is it very high) and has a reasonably low average curvature (so the expectation of the kinetic energy is not very low, but nor is it as high as in the first example).

One way of depicting the probability density of the electron is to represent $|\psi|^2$ by density of shading (Fig. 13.11). A simpler procedure is to show only the **boundary surface**, the surface that captures a high proportion (typically about 90 per cent) of the electron probability. For the 1s orbital, the boundary surface is a sphere centred on the nucleus (Fig. 13.12).

All s orbitals are spherically symmetric, but differ in the number of radial nodes. For instance, the 2s orbital has a radial node where the polynomial factor (Table 13.1) is equal to zero:

$$2 - \frac{\rho}{2} = 0 \quad \text{at} \quad \rho = 4, \quad \text{which implies that} \quad r = \frac{2a_0}{Z}$$

13.11 Representations of the (a) 1s and (b) 2s hydrogenic atomic orbitals in terms of their electron densities (as represented by the density of shading).

(a) 1s

(b) 2s

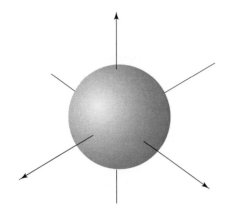

13.12 The boundary surface of an s orbital, within which there is a 90 per cent probability of finding the electron.

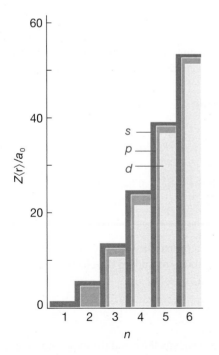

13.13 The variation of the mean radius of a hydrogenic atom with the principal and orbital angular momentum quantum numbers. Note that the mean radius lies in the order $d < p < s$ for a given value of n.

(remember that $\rho = 2Zr/a_0$). Hence, the $2s$ orbital of a hydrogenic atom with atomic number Z has a radial node at $2a_0/Z$ (see Fig. 13.5). Similarly, the $3s$ orbital has two nodes which are found by solving

$$6 - 2\rho + (\tfrac{1}{3}\rho)^2 = 0$$

One radial node is at $1.90a_0/Z$ and the other is at $7.10a_0/Z$ (see Fig. 13.5).

The energies of the s orbitals increase (the electron becomes less tightly bound) as n increases because the average distance of the electron from the nucleus increases. By the virial theorem with $b = -1$ (Section 12.5b, eqn 12.46), $\langle E_K \rangle = -\tfrac{1}{2}\langle V \rangle$ so, even though the average kinetic energy decreases as n increases, the total energy is equal to $\tfrac{1}{2}\langle V \rangle$, which becomes less negative as n increases.

Example 13.2 Calculating the mean radius of an orbital

Use hydrogenic orbitals to calculate the mean radius of a $1s$ orbital.

Method The mean radius is the expectation value

$$\langle r \rangle = \int \psi^* r \psi \, d\tau = \int r |\psi|^2 \, d\tau$$

We therefore need to evaluate the integral using the wavefunctions given in Tables 12.46 and 13.1 and $d\tau = r^2 \, dr \sin\theta \, d\theta \, d\phi$. The angular parts of the wavefunction are normalized in the sense that

$$\int_0^\pi \int_0^{2\pi} |Y_{l,m_l}|^2 \sin\theta \, d\theta \, d\phi = 1$$

The integral over r required is given in Example 11.6.

Answer With the wavefunction written in the form $\psi = RY$, the integration is

$$\langle r \rangle = \int_0^\infty \int_0^\pi \int_0^{2\pi} r R_{n,l}^2 |Y_{l,m_l}|^2 r^2 \, dr \sin\theta \, d\theta \, d\phi = \int_0^\infty r^3 R_{n,l}^2 \, dr$$

For a $1s$ orbital,

$$R_{1,0} = 2\left(\frac{Z^3}{a_0^3}\right)^{1/2} e^{-Zr/a_0}$$

Hence

$$\langle r \rangle = \frac{4Z^3}{a_0^3} \int_0^\infty r^3 e^{-2Zr/a_0} \, dr = \frac{3a_0}{2Z}$$

Comment The general expression for the mean radius of an orbital with quantum numbers l and n is

$$r_{n,l} = n^2 \left\{ 1 + \tfrac{1}{2}\left(1 - \frac{l(l+1)}{n^2}\right) \right\} \frac{a_0}{Z}$$

The variation with n and l is shown in Fig. 13.13. Note that, for a given principal quantum number, the mean radius decreases as l increases.

Self-test 13.4 Evaluate the mean radius (a) of a $3s$ orbital by integration and (b) of a $3p$ orbital by using the general formula. [(a) $27a_0/2Z$; (b) $25a_0/2Z$]

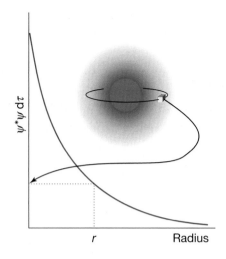

13.14 A constant-volume electron-sensitive detector (the small cube) gives its greatest reading at the nucleus, and a smaller reading elsewhere. The same reading is obtained anywhere on a circle of given radius: the *s* orbital is spherically symmetrical.

(e) Radial distribution functions

The wavefunction tells us, through the value of $|\psi|^2$, the probability of finding an electron in any region. We can imagine a probe with a volume $d\tau$ and sensitive to electrons, and which we can move around near the nucleus of a hydrogen atom. Because the probability density in the ground state of the atom is

$$|\psi|^2 \propto e^{-2Zr/a_0}$$

the reading from the detector decreases exponentially as the probe is moved out along any radius but is constant if the probe is moved on a circle of constant radius (Fig. 13.14).

Now consider the probability of finding the electron *anywhere* on a spherical shell of thickness dr at a radius r. The sensitive volume of the probe is now the volume of the shell (Fig. 13.15), which is $4\pi r^2\, dr$. The probability that the electron will be found between the inner and outer surfaces of this shell is the probability density at the radius r multiplied by the volume of the probe, or $|\psi|^2 \times 4\pi r^2\, dr$. This expression has the form $P(r)\, dr$, where

$$P(r) = 4\pi r^2 \psi^2 \tag{13.21}$$

For orbitals that are not spherically symmetrical, we have to use the more general expression

$$P(r) = r^2 R(r)^2 \tag{13.22}$$

where $R(r)$ is the radial wavefunction for the orbital in question.

Justification 13.3

The probability of finding an electron in a volume element $d\tau$ when its wavefunction is $\psi = RY$ is $|RY|^2\, d\tau$ with $d\tau = r^2\, dr \sin\theta\, d\theta\, d\phi$. The total probability of finding the electron at any angle at a constant radius is the integral of this probability over the surface of a sphere of radius r, and is written $P(r)\, dr$; so

$$P(r)\, dr = \int_0^\pi \int_0^{2\pi} R(r)^2 |Y(\theta, \phi)|^2 r^2\, dr \sin\theta\, d\theta\, d\phi$$

$$= r^2 R(r)^2\, dr \int_0^\pi \int_0^{2\pi} |Y(\theta, \phi)|^2 \sin\theta\, d\theta\, d\phi = r^2 R(r)^2\, dr$$

The last equality follows from the fact that the spherical harmonics are normalized (see Example 13.2). It follows that $P(r) = r^2 R(r)^2$, as stated in the text.

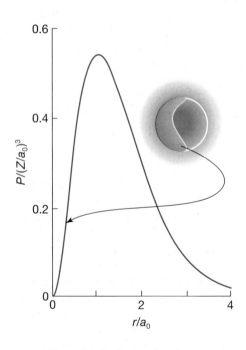

13.15 The radial distribution function P gives the probability that the electron will be found anywhere in a shell of radius r. For a 1*s* electron in hydrogen, P is a maximum when r is equal to the Bohr radius a_0. The value of P is equivalent to the reading that a detector shaped like a spherical shell would give as its radius is varied.

The **radial distribution function**, $P(r)$, is a probability density in the sense that, when it is multiplied by dr, it gives the probability of finding the electron anywhere in a shell of thickness dr at the radius r. For a 1*s* orbital,

$$P(r) = \frac{4Z^3}{a_0^3} r^2 e^{-2Zr/a_0} \tag{13.23}$$

Because r^2 increases with radius from zero at the nucleus, and the exponential term decreases towards zero at infinity, $P(0) = 0$ and $P(r) \rightarrow 0$ as $r \rightarrow \infty$ and passes through a maximum at an intermediate radius (see Fig. 13.15). The maximum of $P(r)$, which can be found by differentiation, marks the most probable radius at which the electron will be found, and for a 1*s* orbital in hydrogen occurs at $r = a_0$, the Bohr radius. When we carry through the same calculation for the radial distribution function of the 2s orbital in hydrogen, we find that the most probable radius is $5.2a_0 = 275$ pm. This larger value reflects the expansion of the atom as its energy increases.

Example 13.3 Calculating the most probable radius

Calculate the most probable radius, r^*, at which an electron will be found when it occupies a 1s orbital of a hydrogenic atom of atomic number Z, and tabulate the values for the one-electron species from H to Ne^{9+}.

Method We find the radius at which the radial distribution function of the hydrogenic 1s orbital has a maximum value by solving $dP/dr = 0$.

Answer The radial distribution function is given in eqn 23. It follows that

$$\frac{dP}{dr} = \frac{4Z^3}{a_0^3}\left(2r - \frac{2Zr^2}{a_0}\right)e^{-2Zr/a_0} = 0$$

at $r = r^*$. Therefore,

$$r^* = \frac{a_0}{Z}$$

Then, with $a_0 = 52.9$ pm,

	H	He^+	Li^{2+}	Be^{3+}	B^{4+}	C^{5+}	N^{6+}	O^{7+}	F^{8+}	Ne^{9+}
r^*/pm	52.9	26.5	17.6	13.2	10.6	8.82	7.56	6.61	5.88	5.29

Comment Notice how the 1s orbital is drawn towards the nucleus as the nuclear charge increases. At uranium the most probable radius is only 0.58 pm, almost 100 times closer than for hydrogen. (On a scale where $r^* = 10$ cm for H, $r^* = 1$ mm for U.) The electron then experiences strong accelerations and relativistic effects are important.

Self-test 13.5 Find the most probable distance of a 2s electron from the nucleus in a hydrogenic atom.

$$[(3 + 5^{1/2})a_0/Z]$$

(f) p orbitals

A p electron has nonzero orbital angular momentum. This momentum has a profound effect on the shape of the wavefunction close to the nucleus, for p orbitals have zero amplitude at $r = 0$. This difference from s orbitals can be understood classically in terms of the centrifugal effect of the orbital angular momentum, which tends to fling the electron away from the nucleus. It is also what we expect from the form of the effective potential energy shown in Fig. 13.4, which rises to infinity as $r \rightarrow 0$ and excludes the wavefunction from the nucleus. The same centrifugal effect appears in all orbitals with $l > 0$ (such as the d orbitals and the f orbitals). We see from eqn 16, in fact, that close to the nucleus a wavefunction is proportional to r^l, so p wavefunctions are proportional to r, d wavefunctions to r^2, and so on (Fig. 13.16). The increasingly strong dependence on r as l increases can be regarded classically as the outcome of increasing centrifugal effects arising from the increasing orbital angular momentum. As remarked previously, all orbitals with $l > 0$ have zero amplitude at the nucleus, and consequently zero probability density there.

The three $2p$ orbitals are distinguished by the three different values that m_l can take when $l = 1$. Because the quantum number m_l tells us the orbital angular momentum around an axis, these different values of m_l denote orbitals in which the

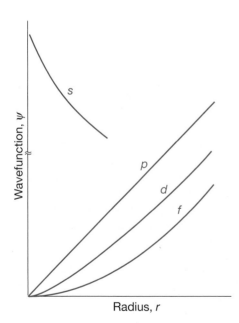

13.16 Close to the nucleus, p orbitals are proportional to r, d orbitals are proportional to r^2, and f orbitals are proportional to r^3. Electrons are progressively excluded from the neighbourhood of the nucleus as l increases. An s orbital has a finite, nonzero value at the nucleus.

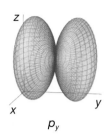

p_z p_x p_y

13.17 The boundary surfaces of p orbitals. A nodal plane passes through the nucleus and separates the two lobes of each orbital.

electron has different orbital angular momenta around an arbitrary z-axis but the same magnitude of that momentum (because l is the same for all three). The orbital with $m_l = 0$, for instance, has zero angular momentum around the z-axis. Its angular variation is proportional to $\cos\theta$, so the probability density, which is proportional to $\cos^2\theta$, has its maximum value on either side of the nucleus along the z-axis (at $\theta = 0$ and $180°$). The wavefunction of a $2p$ orbital with $m_l = 0$ is

$$\psi_{p_0} = R_{2,1}(r)Y_{1,0}(\theta, \phi) = \frac{1}{4(2\pi)^{1/2}}\left(\frac{Z}{a_0}\right)^{5/2} r\cos\theta\, e^{-Zr/2a_0} = r\cos\theta\, f(r)$$

where $f(r)$ is a function only of r. Because in spherical polar coordinates $z = r\cos\theta$, this wavefunction may also be written

$$\psi_{p_z} = zf(r) \tag{13.24}$$

All p orbitals with $m_l = 0$ have wavefunctions of this form regardless of the value of n. This way of writing the orbital is the origin of the name 'p_z orbital': its boundary surface is shown in Fig. 13.17. The wavefunction is zero everywhere in the plane passing through the nucleus and perpendicular to the z-axis, so this plane is a **nodal plane** of the orbital: the wavefunction changes sign on going from one side of the plane to the other.

The wavefunctions of $2p$ orbitals with $m_l = \pm 1$ have the following form:

$$\psi_{p_{\pm 1}} = R_{2,1}(r)Y_{1,\pm 1}(\theta, \phi) = \mp\frac{1}{8\pi^{1/2}}\left(\frac{Z}{a_0}\right)^{5/2} re^{-Zr/2a_0}\sin\theta\, e^{\pm i\phi} = \mp\frac{1}{2^{1/2}}r\sin\theta\, e^{\pm i\phi}f(r)$$

These functions do correspond to angular momentum about the z-axis: as we have seen (in Section 12.6b), wavefunctions with this ϕ dependence correspond to a particle with angular momentum either clockwise or counter-clockwise around the z-axis: $e^{+i\phi}$ corresponds to clockwise rotation when viewed from below, and $e^{-i\phi}$ corresponds to counter-clockwise rotation (from the same viewpoint). They have zero amplitude where $\theta = 0$ and $180°$ (along the z-axis) and maximum amplitude at $90°$, which is in the xy-plane. To draw the functions it is usual to take the real linear combinations

$$\psi_{p_x} = -\frac{1}{2^{1/2}}(p_{+1} - p_{-1}) = r\sin\theta\cos\phi\, f(r) = xf(r)$$

$$\tag{13.25}$$

$$\psi_{p_y} = \frac{i}{2^{1/2}}(p_{+1} + p_{-1}) = r\sin\theta\sin\phi\, f(r) = yf(r)$$

These linear combinations are standing waves with no net orbital angular momentum around the z-axis, as they are superpositions of states with equal and opposite values of m_l. The p_x orbital has the same shape as a p_z orbital, but it is directed along the x-axis (see Fig. 13.17); the p_y orbital is similarly directed along the y-axis. The wavefunction of any p orbital of a given shell can be written as a product of x, y, or z and the same radial function (which depends on the value of n).

Justification 13.4

In this remark, we justify the step of taking linear combinations of degenerate orbitals when we want to indicate a particular point. The freedom to do so rests on the fact that, whenever two or more wavefunctions correspond to the same energy, any linear combination of them is an equally valid solution of the Schrödinger equation.

Suppose ψ_1 and ψ_2 are both solutions of the Schrödinger equation with energy E; then we know that

$$H\psi_1 = E\psi_1 \qquad H\psi_2 = E\psi_2$$

Now consider the linear combination

$$\psi = c_1\psi_1 + c_2\psi_2$$

where c_1 and c_2 are arbitrary coefficients. Then it follows that

$$H\psi = H(c_1\psi_1 + c_2\psi_2) = c_1H\psi_1 + c_2H\psi_2 = c_1E\psi_1 + c_2E\psi_2 = E\psi$$

Hence, the linear combination is also a solution corresponding to the same energy E.

(g) *d* orbitals

When $n = 3$, l can be 0, 1, or 2. As a result, this shell consists of one $3s$ orbital, three $3p$ orbitals, and five $3d$ orbitals. The five d orbitals have $m_l = +2, +1, 0, -1, -2$ and correspond to five different angular momenta around the z-axis (but the same magnitude of angular momentum, because $l = 2$ in each case). As for the p orbitals, d orbitals with opposite values of m_l (and hence opposite senses of motion around the z-axis) may be combined in pairs to give real standing waves, and the boundary surfaces of the resulting shapes are shown in Fig. 13.18. The real combinations have the following forms:

$$\begin{aligned} d_{xy} &= xyf(r) \qquad d_{yz} = yzf(r) \qquad d_{zx} = zxf(r) \\ d_{x^2-y^2} &= \tfrac{1}{2}(x^2 - y^2)f(r) \qquad d_{z^2} = (1/2\sqrt{3})(3z^2 - r^2)f(r) \end{aligned} \tag{13.26}$$

13.3 Spectroscopic transitions and selection rules

The energies of the hydrogenic atoms are given by eqn 13. When the electron undergoes a **transition**, a change of state, from an orbital with quantum numbers n_1, l_1, m_{l1} to another (lower energy) orbital with quantum numbers n_2, l_2, m_{l2}, it undergoes a change of energy ΔE and discards the excess energy as a photon of electromagnetic radiation with a frequency v given by the Bohr frequency condition (eqn 4).

It is tempting to think that all possible transitions are permissible, and that a spectrum arises from the transition of an electron from any initial orbital to any other orbital. However, this is not so, because a photon has an intrinsic spin angular momentum corresponding to $s = 1$ (Section 12.8). The change in angular momentum of the electron must compensate for the angular momentum carried away by the photon. Thus, an electron in a d orbital ($l = 2$) cannot make a transition into an s orbital ($l = 0$) because the photon cannot carry away enough angular momentum. Similarly, an s electron cannot make a transition to another s orbital, because there would then be no change in the electron's angular momentum to make up for the

13.18 The boundary surfaces of d orbitals. Two nodal planes in each orbital intersect at the nucleus and separate the lobes of each orbital. The dark and light areas denote regions of opposite sign of the wavefunction.

angular momentum carried away by the photon. It follows that some spectroscopic transitions are **allowed**, meaning that they can occur, whereas others are **forbidden**, meaning that they cannot occur.

A **selection rule** is a statement about which transitions are allowed. They are derived (for atoms) by identifying the transitions that conserve angular momentum when a photon is emitted or absorbed. The selection rules for hydrogenic atoms are

$$\Delta l = \pm 1 \qquad \Delta m_l = 0, \pm 1 \tag{13.27}$$

The principal quantum number n can change by any amount consistent with the Δl for the transition, because it does not relate directly to the angular momentum.

Justification 13.5

We saw in Section 12.10 that the rate of transition between two states is proportional to the square of the transition dipole moment, μ_{fi}, between the initial and final states, where

$$\mu_{fi} = \langle f | \mu | i \rangle \tag{13.28}$$

and μ is the electric dipole moment operator. For a one-electron atom μ is multiplication by $-e\mathbf{r}$ with components $\mu_x = -ex$, $\mu_y = -ey$, and $\mu_z = -ez$. If the transition dipole moment is zero, the transition is forbidden; the transition is allowed if the transition moment is nonzero. Physically, the transition dipole moment is a measure of the dipolar 'kick' that the electron gives to or receives from the electromagnetic field.

To evaluate a transition dipole moment, we consider each component in turn. For example, for the z-component,

$$\mu_{z,fi} = -e\langle f | z | i \rangle = -e \int \psi_f^* z \psi_i \, d\tau \tag{13.29}$$

To evaluate the integral, we note from Table 12.3 that $z = (4\pi/3)^{1/2}rY_{1,0}$, so

$$\int \psi_f^* z \psi_i \, d\tau = \left(\frac{4\pi}{3}\right)^{1/2} \int_0^\infty R_{n_f,l_f} r R_{n_i,l_i} r^2 \, dr \int_0^\pi \int_0^{2\pi} Y_{l_f,m_{l,f}}^* Y_{1,0} Y_{l_i,m_{l,i}} \sin\theta \, d\theta \, d\phi$$

It follows from the properties of the spherical harmonics (Section 12.7) that the integral

$$\int_0^\pi \int_0^{2\pi} Y_{l_f,m_{l_f}}^* Y_{1,m} Y_{l_i,m_{l_i}} \sin\theta \, d\theta \, d\phi$$

is zero unless $l_f = l_i \pm 1$ and $m_{l,f} = m_{l,i} + m$. Because $m = 0$ in the present case, the angular integral, and hence the z-component of the transition dipole moment, is zero unless $\Delta l = \pm 1$ and $\Delta m_l = 0$, which is a part of the set of selection rules. The same procedure, but considering the x- and y-components, results in the complete set of rules.

13.19 A Grotrian diagram that summarizes the appearance and analysis of the spectrum of atomic hydrogen. The thicker the line, the more intense the transition.

Illustration 13.3

To identify the orbitals to which a 4d electron may make radiative transitions, we first identify the value of l and then apply the selection rule for this quantum number. Because $l = 2$, the final orbital must have $l = 1$ or 3. Thus, an electron may make a transition from a 4d orbital to any np orbital (subject to $\Delta m_l = 0, \pm 1$) and to any nf orbital (subject to the same rule). However, it cannot undergo a transition to any other orbital, so a transition to any ns orbital or to another nd orbital is forbidden.

Self-test 13.6 To what orbitals may a 4s-electron make radiative transitions?

[to np orbitals only]

The selection rules and the atomic energy levels jointly account for the structure of a **Grotrian diagram** (Fig. 13.19), which summarizes the energies of the states and the transitions between them. The thicknesses of the transition lines in the diagram denote their relative intensities in the spectrum.

The structures of many-electron atoms

The Schrödinger equation for a many-electron atom is highly complicated because all the electrons interact with one another. Even for a helium atom, with its two electrons, no analytical expression for the orbitals and energies can be given, and we are forced to make approximations. We shall adopt a simple approach based on what we already know about the structure of hydrogenic atoms. Later we shall see the kind of numerical computations that are currently used to obtain accurate wavefunctions and energies.

13.4 The orbital approximation

The wavefunction of a many-electron atom is a very complicated function of the coordinates of all the electrons, and we should write it $\Psi(\mathbf{r}_1, \mathbf{r}_2, \dots)$, where $\mathbf{r}_i$ is the vector from the nucleus to electron i. However, in the **orbital approximation** we suppose that a reasonable first approximation to this exact wavefunction is obtained by thinking of each electron as occupying its 'own' orbital, and writing

$$\Psi(\mathbf{r}_1, \mathbf{r}_2, \dots) = \psi(\mathbf{r}_1)\psi(\mathbf{r}_2)\dots \tag{13.30}$$

We can think of the individual orbitals as resembling the hydrogenic orbitals, but corresponding to nuclear charges modified by the presence of all the other electrons in the atom. This description is only approximate, but it is a useful model for discussing the chemical properties of atoms, and is the starting point for more sophisticated descriptions of atomic structure.

Justification 13.6

The orbital approximation would be exact if there were no interactions between electrons. To demonstrate the validity of this remark, we need to consider a system in which the hamiltonian for the energy is the sum of two contributions, one for electron 1 and the other for electron 2:

$$H = H_1 + H_2$$

In an actual atom (such as a helium atom), there is an additional term corresponding to the interaction of the two electrons, but we are ignoring that term. We shall now show that, if $\psi(\mathbf{r}_1)$ is an eigenfunction of H_1 with energy E_1, and $\psi(\mathbf{r}_2)$ is an eigenfunction of H_2 with energy E_2, then the product $\Psi(\mathbf{r}_1, \mathbf{r}_2) = \psi(\mathbf{r}_1)\psi(\mathbf{r}_2)$ is an eigenfunction of the combined hamiltonian H. To do so we write

$$
\begin{aligned}
H\Psi(\mathbf{r}_1, \mathbf{r}_2) &= (H_1 + H_2)\psi(\mathbf{r}_1)\psi(\mathbf{r}_2) = H_1\psi(\mathbf{r}_1)\psi(\mathbf{r}_2) + \psi(\mathbf{r}_1)H_2\psi(\mathbf{r}_2) \\
&= E_1\psi(\mathbf{r}_1)\psi(\mathbf{r}_2) + \psi(\mathbf{r}_1)E_2\psi(\mathbf{r}_2) = (E_1 + E_2)\psi(\mathbf{r}_1)\psi(\mathbf{r}_2) \\
&= E\Psi(\mathbf{r}_1, \mathbf{r}_2)
\end{aligned}
$$

where $E = E_1 + E_2$. This is the result we need to prove. However, if the electrons interact (as they do in fact), then the proof fails.

(a) The helium atom

The orbital approximation allows us to express the electronic structure of an atom by reporting its **configuration**, the list of occupied orbitals (usually, but not necessarily, in its ground state). Thus, as the ground state of a hydrogenic atom consists of the single electron in a $1s$ orbital, we report its configuration as $1s^1$.

The He atom has two electrons. We can imagine forming the atom by adding the electrons in succession to the orbitals of the bare nucleus (of charge $2e$). The first electron occupies a $1s$ hydrogenic orbital, but because $Z = 2$ that orbital is more compact than in H itself. The second electron joins the first in the $1s$ orbital, so the electron configuration of the ground state of He is $1s^2$.

(b) The Pauli principle

Lithium, with $Z = 3$, has three electrons. The first two occupy a $1s$ orbital drawn even more closely than in He around the more highly charged nucleus. The third electron, however, does not join the first two in the $1s$ orbital because that configuration is forbidden by the **Pauli exclusion principle**:

> **No more than two electrons may occupy any given orbital and, if two do occupy one orbital, then their spins must be paired.**

Electrons with paired spins, which we denote ↑↓, have zero net spin angular momentum because the spin of one electron is cancelled by the spin of the other. Specifically, one electron has $m_s = +\frac{1}{2}$, the other has $m_s = -\frac{1}{2}$ and they are orientated on their respective cones so that the resultant spin is zero (Fig. 13.20). The exclusion principle is the key to the structure of complex atoms, to chemical periodicity, and to molecular structure. It was proposed by Wolfgang Pauli in 1924 when he was trying to account for the absence of some lines in the spectrum of helium. Later he was able to derive a very general form of the principle from theoretical considerations.

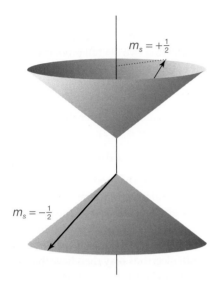

13.20 Electrons with paired spins have zero resultant spin angular momentum. They can be represented by two vectors that lie at an indeterminate position on the cones shown here but, wherever one lies on its cone, the other points in the opposite direction; their resultant is zero.

Justification 13.7

The Pauli exclusion principle in fact applies to any pair of identical fermions (particles with half integral spin). Thus it applies to protons, neutrons, and ^{13}C nuclei (all of which have spin $\frac{1}{2}$) and to ^{35}Cl nuclei (which have spin $\frac{3}{2}$). It does not apply to identical bosons (particles with integral spin), which include photons (spin 1), ^{12}C nuclei (spin 0). Any number of identical bosons may occupy the same orbital.

The Pauli exclusion principle is a special case of a general statement called the **Pauli principle**:

> **When the labels of any two identical fermions are exchanged, the total wavefunction changes sign. When the labels of any two identical bosons are exchanged, the total wavefunction retains the same sign.**

By 'total wavefunction' is meant the entire wavefunction, including the spin of the particles.

Consider the wavefunction for two electrons $\Psi(1, 2)$. The Pauli principle implies that it is a fact of nature (which has its roots in the theory of relativity) that the wavefunction must change sign if we interchange the labels 1 and 2 wherever they occur in the function:

$$\Psi(2, 1) = -\Psi(1, 2) \tag{13.31}$$

Suppose the two electrons in an atom occupy an orbital ψ, then in the orbital approximation the overall wavefunction is $\psi(1)\psi(2)$. To apply the Pauli principle, we must deal with the total wavefunction, the wavefunction including spin. There are several possibilities for two spins: both α, denoted $\alpha(1)\alpha(2)$, both β, denoted $\beta(1)\beta(2)$, and one α the other β, denoted either $\alpha(1)\beta(2)$ or $\alpha(2)\beta(1)$. Because we cannot tell which electron is α and which is β, in the last case it is appropriate to express the spin states as the (normalized) linear combinations

$$\sigma_+(1, 2) = (1/2^{1/2})\{\alpha(1)\beta(2) + \beta(1)\alpha(2)\}$$
$$\sigma_-(1, 2) = (1/2^{1/2})\{\alpha(1)\beta(2) - \beta(1)\alpha(2)\} \tag{13.32}$$

because these allow one spin to be α and the other β with equal probability. The total wavefunction of the system is therefore the product of the orbital part and one of the four spin states:

$$\psi(1)\psi(2)\alpha(1)\alpha(2) \qquad \psi(1)\psi(2)\beta(1)\beta(2) \qquad \psi(1)\psi(2)\sigma_+(1, 2) \qquad \psi(1)\psi(2)\sigma_-(1, 2)$$

The Pauli principle says that for a wavefunction to be acceptable (for electrons), it must change sign when the electrons are exchanged. In each case, exchanging the labels 1 and 2 converts the factor $\psi(1)\psi(2)$ into $\psi(2)\psi(1)$, which is the same, because the order of multiplying the functions does not change the value of the product. The same is true of $\alpha(1)\alpha(2)$ and $\beta(1)\beta(2)$. Therefore, the first two overall products are not allowed, because they do not change sign. The combination $\sigma_+(1, 2)$ changes to

$$\sigma_+(2, 1) = (1/2^{1/2})\{\alpha(2)\beta(1) + \beta(2)\alpha(1)\} = \sigma_+(1, 2)$$

because it is simply the original function written in a different order. The third overall product is therefore also disallowed. Finally, consider $\sigma_-(1, 2)$:

$$\sigma_-(2, 1) = (1/2^{1/2})\{\alpha(2)\beta(1) - \beta(2)\alpha(1)\} = -(1/2^{1/2})\{\alpha(1)\beta(2) - \beta(1)\alpha(2)\} = -\sigma_-(1, 2)$$

This combination does change sign (it is 'antisymmetric'). The product $\psi(1)\psi(2)\sigma_-(1, 2)$ also changes sign under particle exchange, and therefore it is acceptable.

Now we see that only one of the four possible states is allowed by the Pauli principle, and the one that survives has paired α and β spins.[3] This is the content of the Pauli exclusion principle. The exclusion principle is irrelevant when the orbitals occupied by the electrons are different, and both electrons may then have (but need not have) the same spin state. Nevertheless, even then the overall wavefunction must still be antisymmetric overall, and must still satisfy the Pauli principle itself.

In Li ($Z = 3$), the third electron cannot enter the $1s$ orbital because that orbital is already full: we say the K shell is **complete** and that the two electrons form a **closed shell**. Because a similar closed shell is characteristic of the He atom, we denote it [He]. The third electron is excluded from the K shell and must occupy the next available orbital, which is one with $n = 2$ and hence belonging to the L shell. However, we now have to decide whether the next available orbital is the $2s$ orbital or a $2p$ orbital, and therefore whether the lowest energy configuration of the atom is [He]$2s^1$ or [He]$2p^1$.

(c) Penetration and shielding

Unlike in hydrogenic atoms, the $2s$ and $2p$ orbitals (and, in general, all subshells of a given shell) are not degenerate in many-electron atoms. For reasons we shall now explain, s orbitals generally lie lower in energy than p orbitals of a given shell, and p orbitals lie lower than d orbitals.

An electron in a many-electron atom experiences a Coulombic repulsion from all the other electrons present. If it is at a distance r from the nucleus, it experiences a repulsion that can be represented by a point negative charge located at the nucleus and equal in magnitude to the total charge of the electrons within a sphere of radius r (Fig. 13.21). The effect of this point negative charge, when averaged over all the locations of the electron, is to reduce the full charge of the nucleus from Ze to $Z_{eff}e$, the **effective nuclear charge**. We say that the electron experiences a **shielded** nuclear charge, and the difference between Z and Z_{eff} is called the **shielding constant**, σ:

$$Z_{eff} = Z - \sigma \tag{13.33}$$

The electrons do not actually 'block' the full Coulombic attraction of the nucleus: the shielding constant is simply a way of expressing the net outcome of the nuclear attraction and the electronic repulsions in terms of a single equivalent charge at the centre of the atom.

The shielding constant is different for s and p electrons because they have different radial distributions (Fig. 13.22). An s electron has a greater **penetration** through

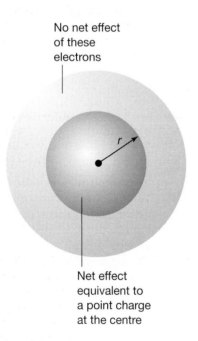

No net effect of these electrons

Net effect equivalent to a point charge at the centre

13.21 An electron at a distance r from the nucleus experiences a Coulombic repulsion from all the electrons within a sphere of radius r and which is equivalent to a point negative charge located on the nucleus. The negative charge reduces the effective nuclear charge of the nucleus from Ze to $Z_{eff}e$.

---

3. The distinction between σ_+ and σ_-, both of which have one α and one β spin, is explained in Section 13.7.

13.22 An electron in an *s* orbital (here a 3*s* orbital) is more likely to be found close to the nucleus than an electron in a *p* orbital of the same shell (note the closeness of the innermost peak of the 3*s* orbital to the nucleus at $r = 0$). Hence an *s* electron experiences less shielding and is more tightly bound than a *p* electron.

inner shells than a *p* electron, in the sense that it is more likely to be found close to the nucleus than a *p* electron of the same shell (the wavefunction of a *p* orbital, remember, is zero at the nucleus). Because only electrons inside the sphere defined by the location of the electron (in effect, the core electrons) contribute to shielding, an *s* electron experiences less shielding than a *p* electron. Consequently, by the combined effects of penetration and shielding, an *s* electron is more tightly bound than a *p* electron of the same shell. Similarly, a *d* electron penetrates less than a *p* electron of the same shell (recall that the wavefunction of a *d* orbital varies as r^2 close to the nucleus, whereas a *p* orbital varies as r), and therefore experiences more shielding.

Shielding constants for different types of electrons in atoms have been calculated from their wavefunctions obtained by numerical solution of the Schrödinger equation for the atom (Table 13.2). We see that, in general, valence-shell *s* electrons do experience higher effective nuclear charges than *p* electrons, although there are some discrepancies. We return to this point shortly.

The consequence of penetration and shielding is that the energies of subshells in a many-electron atom in general lie in the order

$$s < p < d < f$$

The individual orbitals of a given subshell remain degenerate because they all have the same radial characteristics and so experience the same effective nuclear charge.

We can now complete the Li story. Because the shell with $n = 2$ consists of two non-degenerate subshells, with the 2*s* orbital lower in energy than the three 2*p* orbitals, the third electron occupies the 2*s* orbital. This occupation results in the ground-state configuration $1s^2 2s^1$, with the central nucleus surrounded by a complete helium-like shell of two 1*s* electrons, and around that a more diffuse 2*s* electron. The electrons in the outermost shell of an atom in its ground state are called the **valence electrons** because they are largely responsible for the chemical bonds that the atom forms. Thus, the valence electron in Li is a 2*s* electron and its other two electrons belong to its core.

(d) The building-up principle

The extension of the procedure used for H, He, and Li to other atoms is called the **building-up principle**, or the *Aufbau principle*, from the German word for building up. The building-up principle proposes an order of occupation of the hydrogenic orbitals that accounts for the experimentally determined ground-state configurations of neutral atoms.[4]

We imagine the bare nucleus of atomic number Z, and then feed into the orbitals Z electrons in succession. The order of occupation is

1*s* 2*s* 2*p* 3*s* 3*p* 4*s* 3*d* 4*p* 5*s* 4*d* 5*p* 6*s*

and each orbital may accommodate up to two electrons. This order of occupation is approximately the order of energies of the individual orbitals, because, in general, the lower the energy of the orbital, the lower the total energy of the atom as a whole when that orbital is occupied. However, there are complicating effects arising from electron–electron repulsions that are important when the orbitals have very similar energies (such as the 4*s* and 3*d* orbitals near Ca and Sc), and we must take special care then.

We feed the Z electrons in succession into the orbitals subject to the demand of the exclusion principle that no more than two electrons can occupy any one orbital.

Table 13.2* Effective nuclear charge, $Z_{\text{eff}} = Z - \sigma$

Element	Z	Orbital	Z_{eff}
He	2	1*s*	1.6875
C	6	1*s*	5.6727
		2*s*	3.2166
		2*p*	3.1358

* More values are given in the *Data section* at the end of this volume.

4. Electron configurations are determined either spectroscopically or by measurements of magnetic properties.

Because an s subshell consists of only one orbital, up to two electrons may occupy it. A p subshell consists of three orbitals, so it can accommodate up to six electrons; a d subshell, consists of five orbitals and can accommodate up to ten electrons.

As an example, consider the carbon atom, for which $Z = 6$ and there are six electrons to accommodate. Two electrons enter and fill the $1s$ orbital, two enter and fill the $2s$ orbital, leaving two electrons to occupy the orbitals of the $2p$ subshell. Hence the ground-state configuration of C is $1s^2 2s^2 2p^2$ or, more succinctly, $[He]2s^2 2p^2$, with $[He]$ the helium-like $1s^2$ core. However, we can be more precise: we can expect the last two electrons to occupy different $2p$ orbitals because they will then be farther apart on average and repel each other less than if they were in the same orbital. Thus, one electron can be thought of as occupying the $2p_x$ orbital and the other the $2p_y$ orbital (the x, y, z designation is arbitrary, and it would be equally valid to use the complex forms of these orbitals), and the lowest energy configuration of the atom is $[He]2s^2 2p_x^1 2p_y^1$. The same rule applies whenever degenerate orbitals of a subshell are available for occupation. Thus, another rule of the building-up principle is:

> **Electrons occupy different orbitals of a given subshell before doubly occupying any one of them.**

Thus nitrogen ($Z = 7$) has the configuration $[He]2s^2 2p_x^1 2p_y^1 2p_z^1$, and only when we get to oxygen ($Z = 8$) is a $2p$ orbital doubly occupied, giving $[He]2s^2 2p_x^2 2p_y^1 2p_z^1$.

An additional point arises when electrons occupy orbitals singly, for there is then no requirement that their spins should be paired. We need to know whether the lowest energy is achieved when the electron spins are the same (both α, for instance, denoted ↑↑, if there are two electrons in question, as in C) or when they are paired (↑↓). This question is resolved by an empirical observation known as **Hund's maximum multiplicity rule**:

> **An atom in its ground state adopts a configuration with the greatest number of unpaired electrons.**

The explanation of Hund's rule is subtle, but it reflects the quantum mechanical property of **spin correlation**, that electrons with parallel spins behave as if they have a tendency to stay well apart, and hence repel each other less.[5] We can now conclude that, in the ground state of the carbon atom, the two $2p$ electrons have the same spin, that all three $2p$ electrons in the N atoms have the same spin, and that the two $2p$ electrons in different orbitals in the O atom have the same spin (the two in the $2p_x$ orbital are necessarily paired).

Justification 13.8

Suppose electron 1 is described by a wavefunction $\psi_a(\mathbf{r}_1)$ and electron 2 is described by a wavefunction $\psi_b(\mathbf{r}_2)$; then, in the orbital approximation the joint wavefunction of the electrons is the product $\Psi = \psi_a(\mathbf{r}_1)\psi_b(\mathbf{r}_2)$. However, this wavefunction is not acceptable, because it suggests that we know which electron is in which orbital, whereas we cannot keep track of electrons. According to quantum mechanics, the correct description is either of the two following wavefunctions:

$$\Psi_{\pm} = (1/2^{1/2})\{\psi_a(\mathbf{r}_1)\psi_b(\mathbf{r}_2) \pm \psi_b(\mathbf{r}_1)\psi_a(\mathbf{r}_2)\}$$

According to the Pauli principle (*Justification* 13.7), because Ψ_+ is symmetrical under particle interchange, it must be multiplied by an antisymmetric spin function (the one denoted σ_- in *Justification* 13.7). That combination corresponds to a spin-paired

---

5. The effect of spin correlation is to allow the atom to shrink slightly, so the electron–nucleus interaction is improved when the spins are parallel.

state. Conversely, Ψ_- is antisymmetric, so it must be multiplied by one of the three symmetric spin states. These three symmetric states correspond to electrons with parallel spins.[6]

Now consider the values of the two combinations when one electron approaches another, and $r_1 = r_2$. We see that Ψ_- vanishes, which means that there is zero probability of finding the two electrons at the same point in space when they have parallel spins. The other combination does not vanish when the two electrons are at the same point in space. Because the two electrons have different relative spatial distributions depending on whether their spins are parallel or not, it follows that their Coulombic interaction is different, and hence that the two states have different energies.

Neon, with $Z = 10$, has the configuration $[He]2s^2 2p^6$, which completes the L shell. This closed-shell configuration is denoted $[Ne]$, and acts as a core for subsequent elements. The next electron must enter the $3s$ orbital and begin a new shell, so a sodium atom, with $Z = 11$, has the configuration $[Ne]3s^1$. Like lithium with the configuration $[He]2s^1$, sodium has a single s electron outside a complete core.

This analysis has brought us to the origin of chemical periodicity. The L shell is completed by eight electrons, so the element with $Z = 3$ (Li) should have similar properties to the element with $Z = 11$ (Na). Likewise, Be ($Z = 4$) should be similar to $Z = 12$ (Mg), and so on, up to the noble gases He ($Z = 2$), Ne ($Z = 10$), and Ar ($Z = 18$).

Argon has complete $3s$ and $3p$ subshells and, as the $3d$ orbitals are high in energy, it counts as having a closed-shell configuration. Indeed, the $3d$ orbitals are so high in energy that the next electron (for K) occupies the $4s$ orbital, and the configuration of a K atom is analogous to that of an Na atom. The same is true of a Ca atom, which has the configuration $[Ar]4s^2$, the analogue of its congener Mg.

Ten electrons can be accommodated in the five $3d$ orbitals, which accounts for the electron configurations of scandium to zinc. Calculations of the type discussed in Section 13.5 show that for these atoms the energies of the $3d$ orbitals are always lower than the energy of the $4s$ orbital. However, spectroscopic results show that Sc has the configuration $[Ar]3d^1 4s^2$, instead of $[Ar]3d^3$ or $[Ar]3d^2 4s^1$. To understand this observation, we have to consider the nature of electron–electron repulsions in $3d$ and $4s$ orbitals. The most probable distance of a $3d$ electron from the nucleus is less than that for a $4s$ electron, so two $3d$ electrons repel each other more strongly than two $4s$ electrons. As a result, Sc has the configuration $[Ar]3d^1 4s^2$ rather than the two alternatives, for then the strong electron–electron repulsions in the $3d$ orbitals are minimized. The total energy of the atom is least despite the cost of allowing electrons to populate the high-energy $4s$ orbital (Fig. 13.23). The effect just described is generally true for scandium through zinc, so their electron configurations are of the form $[Ar]3d^n 4s^2$, where $n = 1$ for scandium and $n = 10$ for zinc.[7]

At gallium, the building-up principle is used in the same way as in preceding periods. Now the $4s$ and $4p$ subshells constitute the valence shell, and the period terminates with krypton. Because 18 electrons have intervened since argon, this period is the first 'long period' of the periodic table. The existence of the d-block elements (the 'transition metals') reflects the stepwise occupation of the $3d$ orbitals, and the subtle shades of energy differences and effects of electron–electron repulsion along this series give rise to the rich complexity of inorganic d-metal chemistry. A similar

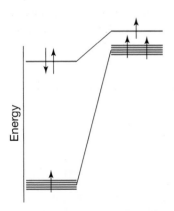

13.23 Strong electron–electron repulsions in the $3d$ orbitals are minimized in the ground state of Sc if the atom has the configuration $[Ar]3d^1 4s^2$ (shown on the left) instead of $[Ar]3d^2 4s^1$ (shown on the right). The total energy of the atom is lower when it has the $[Ar]3d^1 4s^2$ configuration despite the cost of populating the high-energy $4s$ orbital.

6. See Section 13.7 for an explanation of this point.

7. Two notable exceptions are Cr, with electron configuration $[Ar]3d^5 4s^1$, and Cu, with electron configuration $[Ar]3d^{10}4s^1$. See *Further reading* for a discussion of the theoretical basis for these exceptions.

intrusion of the f orbitals in Periods 6 and 7 accounts for the existence of the f block of the periodic table (the lanthanides and actinides).

(e) The configurations of ions

We derive the configurations of cations of elements in the s, p, and d blocks of the periodic table by removing electrons from the ground-state configuration of the neutral atom in a specific order. First, we remove valence p electrons, then valence s electrons, and then as many d electrons as are necessary to achieve the stated charge. For instance, because the configuration V is $[Ar]3d^34s^2$, the V^{2+} cation has the configuration $[Ar]3d^3$. It is reasonable that we remove the more energetic $4s$ electrons in order to form the cation, but it is not obvious why the $[Ar]3d^3$ configuration is preferred in V^{2+} over the $[Ar]3d^14s^2$ configuration, which is found in the isoelectronic Sc. Calculations show that the energy difference between $[Ar]3d^3$ and $[Ar]3d^14s^2$ depends on Z_{eff}. As Z_{eff} increases, transfer of a $4s$ electron to a $3d$ orbital becomes more favourable because the electron–electron repulsions are compensated by attractive interactions between the nucleus and the electrons in the spatially compact $3d$ orbital. Indeed, calculations reveal that, for a sufficiently large Z_{eff}, $[Ar]3d^3$ is lower in energy than $[Ar]3d^14s^2$. This conclusion explains why V^{2+} has a $[Ar]3d^3$ configuration and also accounts for the observed $[Ar]4s^03d^n$ configurations of the M^{2+} cations of Sc through Zn.

The configurations of anions of the p-block elements are derived by continuing the building-up procedure and adding electrons to the neutral atom until the configuration of the next noble gas has been reached. Thus, the configuration of the O^{2-} ion is achieved by adding two electrons to $[He]2s^22p^4$, giving $[He]2s^22p^6$, the same as the configuration of neon.

> The web site contains links to databases of atomic properties.

(f) Ionization energies and electron affinities

The minimum energy necessary to remove an electron from a many-electron atom in the gas phase is the **first ionization energy**, I_1, of the element.[8] The **second ionization energy**, I_2, is the minimum energy needed to remove a second electron (from the singly charged cation). The variation of the first ionization energy through the periodic table is shown in Fig. 13.24 and some numerical values are given in

13.24 The first ionization energies of the elements plotted against atomic number.

8. The symbol recommended by IUPAC for ionization energy is E_i; but this notation is open to confusion.

Table 13.3* First and second ionization energies, I_1/(kJ mol^{-1}) and I_2/(kJ mol^{-1})

H	1312	
He	2372	5251
Mg	738	1451
Na	496	4562

* More values are given in the *Data section*.

Table 13.3. In thermodynamic calculations we often need the **standard enthalpy of ionization**, $\Delta_{ion}H^{\ominus}$. As shown in the *Justification* below, the two are related by

$$\Delta_{ion}H^{\ominus}(T) = I + \tfrac{5}{2}RT \tag{13.34}$$

At 298 K, the difference between the ionization enthalpy and the corresponding ionization energy is 6.20 kJ mol^{-1}.

Justification 13.9

It follows from Kirchhoff's law (Section 2.9 and eqn 2.44) that the reaction enthalpy for

$$M(g) \rightarrow M^+(g) + e^-(g)$$

at a temperature T is related to the value at $T = 0$ by

$$\Delta_r H^{\ominus}(T) = \Delta_r H^{\ominus}(0) + \int_0^T \Delta_r C_p^{\ominus}\, dT$$

The molar constant-pressure heat capacity of each species in the reaction is $\tfrac{5}{2}R$, so $\Delta_r C_p^{\ominus} = +\tfrac{5}{2}R$. The integral in this expression therefore evaluates to $+\tfrac{5}{2}RT$. The reaction enthalpy at $T = 0$ is the same as the (molar) ionization energy, I. Equation 34 then follows. The same expression applies to each successive ionization step, so the overall ionization enthalpy for the formation of M^{2+} is

$$\Delta_r H^{\ominus}(T) = I_1 + I_2 + 5RT$$

The **electron affinity**, E_{ea}, is the energy released when an electron attaches to a gas-phase atom (Table 13.4). In a common, logical, but not universal convention (which we adopt), the electron affinity is positive if energy is released when the electron attaches to the atom (that is, $E_{ea} > 0$ implies that electron attachment is exothermic). It follows from a similar argument to that given in the *Justification* above that the **standard enthalpy of electron gain**, $\Delta_{eg}H^{\ominus}$, at a temperature T is related to the electron affinity by

$$\Delta_{eg}H^{\ominus}(T) = -E_{ea} - \tfrac{5}{2}RT \tag{13.35}$$

Note the change of sign. In typical thermodynamic cycles the $\tfrac{5}{2}RT$ that appears in eqn 35 cancels that in eqn 34, so ionization energies and electron affinities can be used directly. A final preliminary point is that the electron-gain enthalpy of a species X is the negative of the ionization enthalpy of its negative ion:

$$\Delta_{eg}H^{\ominus}(X) = -\Delta_{ion}H^{\ominus}(X^-) \tag{13.36}$$

As ionization energy is often easier to measure than electron affinity, this relation can be used to determine numerical values of the latter.

Ionization energies and electron affinities show periodicities, but the former is more regular and we concentrate on it. Lithium has a low first ionization energy: its outermost electron is well-shielded from the nucleus by the core ($Z_{eff} = 1.3$, compared with $Z = 3$) and it is easily removed. Beryllium has a higher nuclear charge than lithium, and its outermost electron (one of the two 2s electrons) is more difficult to remove: its ionization energy is higher. The ionization energy decreases between beryllium and boron because in the latter the outermost electron occupies a 2p orbital and is less strongly bound than if it had been a 2s electron. The ionization energy increases between boron and carbon because the latter's outermost electron

Table 13.4* Electron affinities, E_{ea}/(kJ mol^{-1})

Cl	349		
F	322		
H	73		
O	141	O$^-$	−844

* More values are given in the *Data section*.

is also 2*p* and the nuclear charge has increased. Nitrogen has a still higher ionization energy because of the further increase in nuclear charge.

There is now a kink in the curve that reduces the ionization energy of oxygen below what would be expected by simple extrapolation. The explanation is that at oxygen a 2*p* orbital must become doubly occupied, and the electron–electron repulsions are increased above what would be expected by simple extrapolation along the row. In addition, the loss of a 2*p* electron results in a configuration with a half-filled subshell (like that of N), which is an arrangement of low energy, so the energy of $O^+ + e^-$ is lower than might be expected, and the ionization energy is correspondingly low too. (The kink is less pronounced in the next row, between phosphorus and sulfur because their orbitals are more diffuse.) The values for oxygen, fluorine, and neon fall roughly on the same line, the increase of their ionization energies reflecting the increasing attraction of the more highly charged nuclei for the outermost electrons.

The outermost electron in sodium is 3*s*. It is far from the nucleus, and the latter's charge is shielded by the compact, complete neon-like core. As a result, the ionization energy of sodium is substantially lower than that of neon. The periodic cycle starts again along this row, and the variation of the ionization energy can be traced to similar reasons.

Electron affinities are greatest close to fluorine, for the incoming electron enters a vacancy in a compact valence shell and can interact strongly with the nucleus. The attachment of an electron to an anion (as in the formation of O^{2-} from O^-) is invariably endothermic, so E_{ea} is negative. The incoming electron is repelled by the charge already present. Electron affinities are also small, and may be negative, when an electron enters an orbital that is far from the nucleus (as in the heavier alkali metal atoms) or is forced by the Pauli principle to occupy a new shell (as in the noble gas atoms).

13.5 Self-consistent field orbitals

The central difficulty of the Schrödinger equation is the presence of the electron–electron interaction terms. The potential energy of the electrons is

$$V = -\sum_i \frac{Ze^2}{4\pi\varepsilon_0 r_i} + \tfrac{1}{2}\sum_{i,j}{}' \frac{e^2}{4\pi\varepsilon_0 r_{ij}} \tag{13.37}$$

The prime on the second sum indicates that $i \neq j$, and the factor of one-half prevents double-counting of electron pair repulsions (1 with 2 is the same as 2 with 1). The first term is the total attractive interaction between the electrons and the nucleus. The second term is the total repulsive interaction between the electrons; r_{ij} is the distance between electrons *i* and *j*. It is hopeless to expect to find analytical solutions of a Schrödinger equation with such a complicated potential energy term, but computational techniques are available that give very detailed and reliable numerical solutions for the wavefunctions and energies. The techniques were originally introduced by D.R. Hartree (before computers were available) and then modified by V. Fock to take into account the Pauli principle correctly. In broad outline, the **Hartree–Fock self-consistent field** (SCF) procedure is as follows.

Imagine that we have a rough idea of the structure of the atom. In the Ne atom, for instance, the orbital approximation suggests the configuration $1s^2 2s^2 2p^6$ with the orbitals approximated by hydrogenic atomic orbitals. Now consider one of the 2*p* electrons. A Schrödinger equation can be written for this electron by ascribing to it a potential energy due to the nuclear attraction and the repulsion from the other electrons. This equation has the form

13.25 The radial distribution functions for the orbitals of Na based on SCF calculations. Note the shell-like structure, with the $3s$ orbital outside the inner K and L shells.

$$-\frac{\hbar^2}{2m_e}\nabla_1^2\psi_{2p}(\mathbf{r}_1) - \frac{Ze^2}{4\pi\varepsilon_0 r_1}\psi_{2p}(\mathbf{r}_1)$$

$$+\frac{e^2}{2\pi\varepsilon_0}\left\{\sum_i\int\frac{\psi_i(\mathbf{r}_2)^*\psi_i(\mathbf{r}_2)}{r_{12}}\,d\tau_2\right\}\psi_{2p}(\mathbf{r}_1)$$

$$-\frac{e^2}{4\pi\varepsilon_0}\left\{\sum_i\int\frac{\psi_i(\mathbf{r}_2)^*\psi_{2p}(\mathbf{r}_2)}{r_{12}}\,d\tau_2\right\}\psi_i(\mathbf{r}_1)$$

$$=E_{2p}\psi_{2p}(\mathbf{r}_1)$$

(13.38)

The orbitals are labelled i, and the sums are over all the occupied orbitals. A similar equation can be written for the $1s$ and $2s$ orbitals in the atom.

Equation 38 is fearsome, but it can be interpreted by examining each term. The first term on the left is the usual kinetic energy contribution. The second term is the potential energy of attraction of the electron to the nucleus. The third term, which is called the **Coulomb operator**, is the potential energy of the electron of interest due to the charge density $-e|\psi_i(\mathbf{r}_2)|^2$ of the electrons in the other occupied orbitals. The fourth term, the **exchange operator**, takes into account the spin correlation effects discussed earlier. Note that, although the equation is for the $2p$ orbital in neon, it depends on the wavefunctions of all the other occupied orbitals in the atom.

There is no hope of solving eqn 38 analytically. However, it can be solved numerically if we guess an approximate form of the wavefunctions of all the orbitals except $2p$. The procedure is then repeated for the other orbitals in the atom, the $1s$ and $2s$ orbitals. This sequence of calculations gives the form of the $2p$, $2s$, and $1s$ orbitals, and in general they will differ from the set used initially to start the calculation.[9] These improved orbitals can be used in another cycle of calculation, and a second improved set of orbitals is obtained. The recycling continues until the orbitals and energies obtained are insignificantly different from those used at the start of the current cycle. The solutions are then self-consistent and accepted as solutions of the problem.

Figure 13.25 shows plots of some of the SCF Hartree–Fock radial distribution functions for sodium. They show the grouping of electron density into shells, as was anticipated by the early chemists, and the differences of penetration as discussed above. These SCF calculations therefore support the qualitative discussions that are used to explain chemical periodicity. They also considerably extend that discussion by providing detailed wavefunctions and precise energies.

The spectra of complex atoms

The spectra of atoms rapidly become very complicated as the number of electrons increases, but there are some important and moderately simple features (Box 13.1). The general idea is straightforward: lines in the spectrum (in either emission or absorption) occur when the atom undergoes a transition with a change of energy $|\Delta E|$, and emits or absorbs a photon of frequency $\nu = |\Delta E|/h$ and $\tilde{\nu} = |\Delta E|/hc$. Hence,

The web site contains links to databases of atomic spectra.

9. In practice, much more efficient procedures are used, and the equations for the wavefunctions are solved simultaneously.

we can expect the spectrum to give information about the energies of electrons in atoms. However, the actual energy levels are not given solely by the energies of the orbitals, because the electrons interact with one another in various ways, and there are contributions to the energy in addition to those we have already considered.

Box 13.1 Spectroscopy of stars

The bulk of stellar material consists of neutral and ionized forms of hydrogen and helium atoms, with helium being the product of 'hydrogen burning' by nuclear fusion. However, nuclear fusion also makes heavier elements. It is generally accepted that the outer layers of stars are composed of lighter elements, such as H, He, C, N, O, and Ne in both neutral and ionized forms. Heavier elements, including neutral and ionized forms of Si, Mg, Ca, S, and Ar, are found closer to the stellar core. The core itself contains the heaviest elements and ^{56}Fe is particularly abundant because it is a very stable nuclide. All of these elements are in the gas phase on account of the very high temperatures in stellar interiors. For example, the temperature is estimated to be 3.6 MK halfway to the centre of the Sun.

Astronomers use spectroscopic techniques to determine the chemical composition of stars because each element, and indeed each isotope of an element, has a characteristic spectral signature that is transmitted through space by the star's light. To understand the spectra of stars, we must first know why they shine. Nuclear reactions in the dense stellar interior generate radiation that travels to less dense outer layers. Absorption and re-emission of photons by the atoms and ions in the interior give rise to a quasi-continuum of radiation energy that is emitted into space by a thin layer of gas called the *photosphere*. To a good approximation, the distribution of energy emitted from a star's photosphere resembles the Planck distribution for a very hot black body (Section 11.1). For example, the energy distribution of our Sun's photosphere may be modelled by a Planck distribution with an effective temperature of 5800 K. Superimposed on the blackbody radiation continuum are sharp absorption and emission lines from neutral atoms and ions present in the photosphere. Analysis of stellar radiation with a spectrometer mounted on to a telescope, such as the Hubble Space Telescope, yields the chemical composition of the star's photosphere by comparison with known spectra of the elements. The data can also reveal the presence of small molecules, such as CN, C_2, TiO, and ZrO, in certain 'cold' stars, which are stars with relatively low effective temperatures.

The two outermost layers of a star are the *chromosphere*, a region just above the photosphere, and the *corona*, a region above the chromosphere that can be seen (with proper care) during eclipses. The photosphere, chromosphere, and corona comprise a star's 'atmosphere'. Our Sun's chromosphere is much less dense than its photosphere and its temperature is much higher, rising to about 10 000 K. The reasons for this increase in temperature are not fully understood. The temperature of the Sun's corona is very high, rising up to 1.5 MK, so black-body emission is strong from the X-ray to the radio frequency region of the spectrum. The spectrum of the Sun's corona is dominated by emission lines from electronically excited species, such as neutral atoms and a number of highly ionized species . The most intense emission lines in the visible range are from the Fe^{13+} ion at 530.3 nm, the Fe^{9+} ion at 637.4 nm, and the Ca^{4+} ion at 569.4 nm.

Because only light from the Sun's atmosphere reaches our telescopes, the overall chemical composition of a star must be inferred from theoretical work on its interior and from spectral analysis of its atmosphere. Data on the Sun indicate that it is 92 per cent hydrogen and 7.8 per cent helium. The remaining 0.2 per cent is due to heavier elements, among which C, N, O, Ne, and Fe are the most abundant. More advanced analysis of spectra also permit the determination of other properties of stars, such as their relative speeds (Problem 13.28) and their effective temperatures (Problem 16.29).

Discussion questions

1. Hydrogen is the most abundant element in all stars. However, only very weak absorption and emission lines due to neutral hydrogen are found in the spectra of stars with effective temperatures higher than 25 000 K. Account for this observation.

2. The distribution of isotopes of an element may yield clues about the nuclear reactions that occur in the interior of a star. Show that it is possible to use spectroscopy to confirm the presence of both $^4He^+$ and $^3He^+$ in a star by calculating the wavenumbers of the $n = 3 \rightarrow n = 2$ and of the $n = 2 \rightarrow n = 1$ transitions for each isotope.

Further reading

J. Audouze and G. Israël, *The Cambridge atlas of astronomy*. Cambridge University Press (1994).

C.R. Cowley, *An introduction to cosmochemistry*. Cambridge University Press (1995).

13.6 Quantum defects and ionization limits

One application of atomic spectroscopy is to the determination of ionization energies. However, we cannot use the procedure illustrated in Example 13.1 indiscriminately because the energy levels of a many-electron atom do not in general vary as $1/n^2$. If we confine attention to the outermost electrons, then we know that as a result of penetration and shielding, they experience a nuclear charge of slightly more than $1e$ because in a neutral atom the other $Z-1$ electrons cancel all but about one unit of nuclear charge. Typical values of Z_{eff} are a little more than 1, so we expect binding energies to be given by a term of the form $-hcR/n^2$, but lying slightly lower in energy than this formula predicts. We therefore introduce a **quantum defect**, δ, and write the energy as $-hcR/(n-\delta)^2$. The quantum defect is best regarded as a purely empirical quantity.

There are some excited states that are so diffuse that the $1/n^2$ variation is valid: these states are called **Rydberg states**. In such cases we can write

$$\tilde{v} = \frac{I}{hc} - \frac{R}{n^2} \tag{13.39}$$

and a plot of wavenumber against $1/n^2$ can be used to obtain I by extrapolation; in practice, one would use a linear regression fit using a computer. If the lower state is not the ground state (a possibility if we wish to generalize the concept of ionization energy), the ionization energy of the ground state can be determined by adding the appropriate energy difference to the ionization energy obtained as described here.

13.7 Singlet and triplet states

Suppose we were interested in the energy levels of a He atom, with its two electrons. We know that the ground-state configuration is $1s^2$, and can anticipate that an excited configuration will be one in which one of the electrons has been promoted into a $2s$ orbital, giving the configuration $1s^1 2s^1$. The two electrons need not be paired because they occupy different orbitals. According to Hund's maximum multiplicity rule, the state of the atom with the spins parallel lies lower in energy than the state in which they are paired. Both states are permissible, and can contribute to the spectrum of the atom.

Parallel and antiparallel (paired) spins differ in their overall spin angular momentum. In the paired case, the two spin momenta cancel each other, and there is zero net spin (as was depicted in Fig. 13.20). The paired-spin arrangement is called a **singlet**. Its spin state is the one we denoted σ_- in the discussion of the Pauli principle:

$$\sigma_-(1,2) = (1/2^{1/2})\{\alpha(1)\beta(2) - \beta(1)\alpha(2)\} \tag{13.40a}$$

The angular momenta of two parallel spins add together to give a nonzero total spin, and the resulting state is called a **triplet**. As illustrated in Fig. 13.26, there are three ways of achieving a nonzero total spin, but only one way to achieve zero spin. The three spin states are the symmetric combinations introduced earlier:

$$\alpha(1)\alpha(2) \qquad \sigma_+(1,2) = (1/2^{1/2})\{\alpha(1)\beta(2) + \beta(1)\alpha(2)\} \qquad \beta(1)\beta(2) \tag{13.40b}$$

The fact that the parallel arrangement of spins in the $1s^1 2s^1$ configuration of the He atom lies lower in energy than the antiparallel arrangement can now be expressed by saying that the triplet state of the $1s^1 2s^1$ configuration of He lies lower in energy than the singlet state. This is a general conclusion that applies to other atoms (and molecules), and *for states arising from the same configuration, the triplet state generally lies lower than the singlet state.* The origin of the energy difference lies in the effect of spin correlation on the Coulombic interactions between electrons, as we saw in

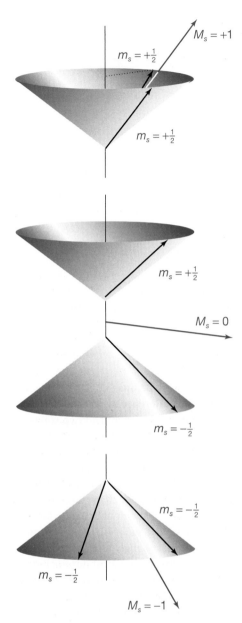

13.26 When two electrons have parallel spins, they have a nonzero total spin angular momentum. There are three ways of achieving this resultant, which are shown by these vector representations. Note that, although we cannot know the orientation of the spin vectors on the cones, the angle between the vectors is the same in all three cases, for all three arrangements have the same total spin angular momentum (that is, the resultant of the two vectors has the same length in each case, but points in different directions). Compare this diagram with Fig. 13.20, which shows the antiparallel case. Note that whereas two paired spins are precisely antiparallel, two 'parallel' spins are not strictly parallel.

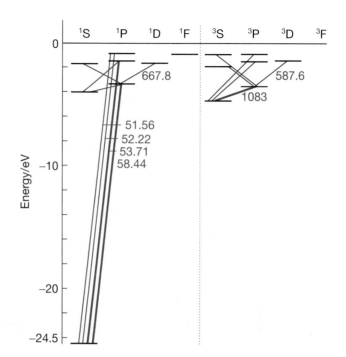

13.27 Part of the Grotrian diagram for a helium atom. Note that there are no transitions between the singlet and triplet levels.

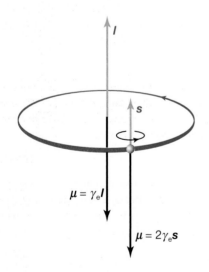

13.28 Angular momentum gives rise to a magnetic moment (μ). For an electron, the magnetic moment is antiparallel to the orbital angular momentum, but proportional to it. For spin angular momentum, there is a factor 2, which increases the magnetic moment to twice its expected value (see Section 18.1).

the case of Hund's rule for ground-state configurations. Because the Coulombic interaction between electrons in an atom is strong, the difference in energies between singlet and triplet states of the same configuration can be large. The two states of $1s^1 2s^1$ He, for instance, differ by 6421 cm^{-1} (corresponding to 0.80 eV).

The spectrum of atomic helium is more complicated than that of atomic hydrogen, but there are two simplifying features. One is that the only excited configurations it is necessary to consider are of the form $1s^1 nl^1$: that is, only one electron is excited. Excitation of two electrons requires an energy greater than the ionization energy of the atom, so the He$^+$ ion is formed instead of the doubly excited atom. Second, no radiative transitions take place between singlet and triplet states because the relative orientation of the two electron spins cannot change during a transition. Thus, there is a spectrum arising from transitions between singlet states (including the ground state) and between triplet states, but not between the two. Spectroscopically, helium behaves like two distinct species, and the early spectroscopists actually thought of helium as consisting of 'parahelium' and 'orthohelium'. The Grotrian diagram for helium in Fig. 13.27 shows the two sets of transitions.

13.8 Spin–orbit coupling

Electron spin has a further implication for the energies of atoms. Because an electron has spin angular momentum, and because moving charges generate magnetic fields, an electron has a magnetic moment that arises from its spin (Fig. 13.28). Similarly, an electron with orbital angular momentum (that is, an electron in an orbital with $l > 0$) is in effect a circulating current, and possesses a magnetic moment that arises from its orbital momentum. The interaction of the spin magnetic moment with the magnetic field arising from the orbital angular momentum is called **spin–orbit coupling**. The strength of the coupling, and its effect on the energy levels of the atom, depends on the relative orientations of the spin and orbital magnetic moments, and therefore on the relative orientations of the two angular momenta (Fig. 13.29).

(a) The total angular momentum

One way of expressing the dependence of the spin–orbit interaction on the relative orientation of the spin and orbital momenta is to say that it depends on the total angular momentum of the electron, the vector sum of its spin and orbital momenta. Thus, when the spin and orbital angular momenta are nearly parallel, the total angular momentum is high; when the two angular momenta are opposed, the total angular momentum is low.

The total angular momentum of an electron is described by the quantum numbers j and m_j, with $j = l + \frac{1}{2}$ (when the two angular momenta are in the same direction) or $j = l - \frac{1}{2}$ (when they are opposed, Fig. 13.30). The different values of j that can arise for a given value of l label **levels** of a term. For $l = 0$, the only permitted value is $j = \frac{1}{2}$ (the total angular momentum is the same as the spin angular momentum because there is no other source of angular momentum in the atom). When $l = 1$, j may be either $\frac{3}{2}$ (the spin and orbital angular momenta are in the same sense) or $\frac{1}{2}$ (the spin and angular momenta are in opposite senses).

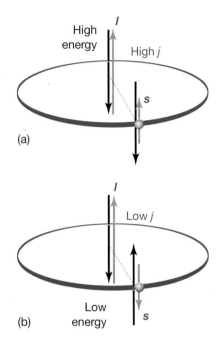

(a)

(b) Low energy

13.29 Spin–orbit coupling is a magnetic interaction between spin and orbital magnetic moments. When the angular momenta are parallel, as in (a), the magnetic moments are aligned unfavourably; when they are opposed, as in (b), the interaction is favourable. This magnetic coupling is the cause of the splitting of a configuration into levels.

Example 13.4 Identifying the levels of a configuration

Identify the levels that may arise from the configurations (a) d^1, (b) s^1.

Method In each case, identify the value of l and then the possible values of j. For these one-electron systems, the total angular momentum is the sum and difference of the orbital and spin momenta.

Answer (a) For a d electron, $l = 2$ and there are two levels in the configuration, one with $j = 2 + \frac{1}{2} = \frac{5}{2}$ and the other with $j = 2 - \frac{1}{2} = \frac{3}{2}$. (b) For an s electron $l = 0$, so only one level is possible, and $j = \frac{1}{2}$.

Self-test 13.7 Identify the levels of the configurations (a) p^1 and (b) f^1.

[(a) $\frac{3}{2}, \frac{1}{2}$; (b) $\frac{7}{2}, \frac{5}{2}$]

The dependence of the spin–orbit interaction on the value of j is expressed in terms of the **spin–orbit coupling constant**, A (which is typically expressed as a wavenumber). A quantum mechanical calculation leads to the result that the energies of the levels with quantum numbers s, l, and j are given by

$$E_{l,s,j} = \tfrac{1}{2}hcA\{j(j+1) - l(l+1) - s(s+1)\} \tag{13.41}$$

13.30 The coupling of the spin and orbital angular momenta of a d electron ($l = 2$) gives two possible values of j depending on the relative orientations of the spin and orbital angular momenta of the electron.

Justification 13.10

The energy of a magnetic moment $\boldsymbol{\mu}$ in a magnetic field $\boldsymbol{B}$ is equal to their scalar product $-\boldsymbol{\mu} \cdot \boldsymbol{B}$. If the magnetic field arises from the orbital angular momentum of the electron, it is proportional to $\boldsymbol{l}$; if the magnetic moment $\boldsymbol{\mu}$ is that of the electron spin, then it is proportional to $\boldsymbol{s}$. It then follows that the energy of interaction is proportional to the scalar product $\boldsymbol{s} \cdot \boldsymbol{l}$:

energy of interaction $= -\boldsymbol{\mu} \cdot \boldsymbol{B} \propto \boldsymbol{s} \cdot \boldsymbol{l}$

We take this expression to be the first-order perturbation contribution to the hamiltonian. Next, we note that the total angular momentum is the vector sum of the spin and orbital momenta: $\boldsymbol{j} = \boldsymbol{l} + \boldsymbol{s}$. The magnitude of the vector $\boldsymbol{j}$ is calculated by evaluating

$$\boldsymbol{j} \cdot \boldsymbol{j} = (\boldsymbol{l} + \boldsymbol{s}) \cdot (\boldsymbol{l} + \boldsymbol{s}) = \boldsymbol{l} \cdot \boldsymbol{l} + \boldsymbol{s} \cdot \boldsymbol{s} + 2\boldsymbol{s} \cdot \boldsymbol{l}$$

That is,

$$\boldsymbol{s} \cdot \boldsymbol{l} = \tfrac{1}{2}\{\boldsymbol{j}^2 - \boldsymbol{l}^2 - \boldsymbol{s}^2\}$$

This is a classical result. To make the transition to quantum mechanics, we treat all the quantities as operators, and write

$$\hat{\boldsymbol{s}} \cdot \hat{\boldsymbol{l}} = \tfrac{1}{2}\{\hat{\boldsymbol{j}}^2 - \hat{\boldsymbol{l}}^2 - \hat{\boldsymbol{s}}^2\} \tag{13.42}$$

At this point, we calculate the first-order correction to the energy by evaluating the expectation value (Section 12.9):

$$\begin{aligned} \langle j,l,s|\hat{\boldsymbol{s}} \cdot \hat{\boldsymbol{l}}|j,l,s\rangle &= \tfrac{1}{2}\langle j,l,s|\hat{\boldsymbol{j}}^2 - \hat{\boldsymbol{l}}^2 - \hat{\boldsymbol{s}}^2|j,l,s\rangle \\ &= \tfrac{1}{2}\{j(j+1) - l(l+1) - s(s+1)\}\hbar^2 \end{aligned} \tag{13.43}$$

Then, by inserting this expression into the formula for the energy, and writing the constant of proportionality as $hcA/\hbar^2$, we obtain eqn 41. The calculation of A is much more complicated: see *Further reading*.

Illustration 13.4

The unpaired electron in the ground state of an alkali metal atom has $l = 0$, so $j = \tfrac{1}{2}$. Because the orbital angular momentum is zero in this state, the spin–orbit coupling energy is zero (as is confirmed by setting $j = s$ and $l = 0$ in eqn 41). When the electron is excited to an orbital with $l = 1$, it has orbital angular momentum and can give rise to a magnetic field that interacts with its spin. In this configuration the electron can have $j = \tfrac{3}{2}$ or $j = \tfrac{1}{2}$, and the energies of these levels are

$$E_{3/2} = \tfrac{1}{2}hcA\{\tfrac{3}{2} \times \tfrac{5}{2} - 1 \times 2 - \tfrac{1}{2} \times \tfrac{3}{2}\} = \tfrac{1}{2}hcA$$
$$E_{1/2} = \tfrac{1}{2}hcA\{\tfrac{1}{2} \times \tfrac{3}{2} - 1 \times 2 - \tfrac{1}{2} \times \tfrac{3}{2}\} = -hcA$$

The corresponding energies are shown in Fig. 13.31. Note that the 'centre of gravity' of the levels is unchanged, because there are four states of energy $\tfrac{1}{2}hcA$ and two of energy $-hcA$.

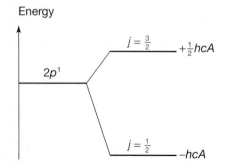

13.31 The levels of a ^{2}P term arising from spin–orbit coupling. Note that the low-j level lies below the high-j level in energy.

The strength of the spin–orbit coupling depends on the nuclear charge. To understand why this is so, imagine riding on the orbiting electron and seeing a charged nucleus apparently orbiting around us (like the Sun rising and setting). As a result, we find ourselves at the centre of a ring of current. The greater the nuclear charge, the greater this current, and therefore the stronger the magnetic field we detect. Because the spin magnetic moment of the electron interacts with this orbital magnetic field, it follows that, the greater the nuclear charge, the stronger the spin–orbit interaction. The coupling increases sharply with atomic number (as Z^4). Whereas it is only small in H (giving rise to shifts of energy levels of no more than about 0.4 cm^{-1}), in heavy atoms like Pb it is very large (giving shifts of the order of thousands of reciprocal centimetres).

$\tilde{v}/\mathrm{cm}^{-1}$

16 973 $^2\mathrm{P}_{3/2}$

17 cm^{-1}

16 956 $^2\mathrm{P}_{1/2}$

D$_1$ D$_2$

0    $^2\mathrm{S}_{1/2}$

$\tilde{v} \longrightarrow$

589.76 nm   589.16 nm

13.32 The energy-level diagram for the formation of the sodium D lines. The splitting of the spectral lines (by 17 cm^{-1}) reflects the splitting of the levels of the ^{2}P term.

(b) Fine structure

Two spectral lines are observed when the p electron of an electronically excited alkali metal atom undergoes a transition and falls into a lower s orbital. One line is due to a transition starting in a $j = \frac{3}{2}$ level and the other line is due to a transition starting in the $j = \frac{1}{2}$ level of the same configuration. The two lines are an example of the **fine structure** of a spectrum, the structure in a spectrum due to spin–orbit coupling. Fine structure can be clearly seen in the emission spectrum from sodium vapour excited by an electric discharge (for example, in one kind of street lighting). The yellow line at 589 nm (close to 17 000 cm^{-1}) is actually a doublet composed of one line at 589.76 nm (16 956.2 cm^{-1}) and another at 589.16 nm (16 973.4 cm^{-1}); the components of this doublet are the 'D lines' of the spectrum (Fig. 13.32). Therefore, in Na, the spin–orbit coupling affects the energies by about 17 cm^{-1}.

Example 13.5 Analysing a spectrum for the spin–orbit coupling constant

The origin of the D lines in the spectrum of atomic sodium is shown in Fig. 13.32. Calculate the spin–orbit coupling constant for the upper configuration of the Na atom.

Method We see from Fig. 13.32 that the splitting of the lines is equal to the energy separation of the $j = \frac{3}{2}$ and $\frac{1}{2}$ levels of the excited configuration. This separation can be expressed in terms of A by using eqn 41. Therefore, set the observed splitting equal to the energy separation calculated from eqn 41 and solve the equation for A.

Answer The two levels are split by

$$\Delta\tilde{v} = A\tfrac{1}{2}\{\tfrac{3}{2}(\tfrac{3}{2} + 1) - \tfrac{1}{2}(\tfrac{1}{2} + 1)\} = \tfrac{3}{2}A$$

The experimental value is 17.2 cm^{-1}; therefore

$$A = \tfrac{2}{3} \times (17.2 \text{ cm}^{-1}) = 11.5 \text{ cm}^{-1}$$

Comment The same calculation repeated for the other alkali metal atoms gives Li: 0.23 cm^{-1}, K: 38.5 cm^{-1}, Rb: 158 cm^{-1}, Cs: 370 cm^{-1}. Note the increase of A with atomic number (but more slowly than Z^4 for these many-electron atoms).

Self-test 13.8 The configuration … $4p^6 5d^1$ of rubidium has two levels at 25 700.56 cm^{-1} and 25 703.52 cm^{-1} above the ground state. What is the spin–orbit coupling constant in this excited state?   [1.18 cm^{-1}]

13.9 Term symbols and selection rules

We have used expressions such as 'the $j = \frac{3}{2}$ level of a configuration'. A **term symbol**, which is a symbol looking like $^2\mathrm{P}_{3/2}$ or $^3\mathrm{D}_2$, conveys this information much more succinctly. The convention of using lower-case letters to label orbitals and upper-case letters to label overall states applies throughout spectroscopy, not just to atoms.

A term symbol gives three pieces of information:

   1 The letter (for example, P or D in the examples) indicates the total orbital angular momentum quantum number, L.

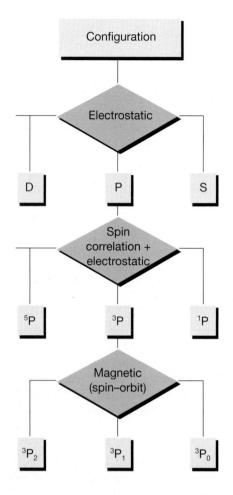

13.33 A summary of the types of interaction that are responsible for the various kinds of splitting of energy levels in atoms. For light atoms, magnetic interactions are small, but in heavy atoms they may dominate the electrostatic (charge–charge) interactions.

2 The left superscript in the term symbol (for example, the 2 in $^2P_{3/2}$) gives the multiplicity of the term.

3 The right subscript on the term symbol (for example, the $\frac{3}{2}$ in $^2P_{3/2}$) is the value of the total angular momentum quantum number, J.

We shall now say what each of these statements means; the contributions to the energies which we are about to discuss are summarized in Fig. 13.33.

(a) The total orbital angular momentum

When several electrons are present, it is necessary to judge how their individual orbital angular momenta add together or oppose each other. The **total orbital angular momentum quantum number**, L, tells us the magnitude of the angular momentum through $\{L(L+1)\}^{1/2}\hbar$. It has $2L+1$ orientations distinguished by the quantum number M_L, which can take the values L, $L-1$, ... , $-L$. Similar remarks apply to the **total spin quantum number**, S, and the quantum number M_S, and the **total angular momentum quantum number**, J, and the quantum number M_J.

The value of L (a non-negative integer) is obtained by coupling the individual orbital angular momenta by using the **Clebsch–Gordan series**:

$$L = l_1 + l_2, l_1 + l_2 - 1, \ldots, |l_1 - l_2| \tag{13.44}$$

The modulus signs are attached to $l_1 - l_2$ because L is non-negative. The maximum value, $L = l_1 + l_2$, is obtained when the two orbital angular momenta are in the same direction; the lowest value, $|l_1 - l_2|$, is obtained when they are in opposite directions. The intermediate values represent possible intermediate relative orientations of the two momenta (Fig. 13.34). For two p electrons (for which $l_1 = l_2 = 1$), $L = 2, 1, 0$. The code for converting the value of L into a letter is the same as for the s, p, d, f, ... designation of orbitals, but uses upper-case Roman letters:

L: 0 1 2 3 4 5 6 ...
     S P D F G H I ...

Thus, a p^2 configuration can give rise to D, P, and S terms. The terms differ in energy on account of the different spatial distribution of the electrons and the consequent differences in repulsion between them.

A closed shell has zero orbital angular momentum because all the individual orbital angular momenta sum to zero. Therefore, when working out term symbols, we need consider only the electrons of the unfilled shell. In the case of a single electron outside a closed shell, the value of L is the same as the value of l; so the configuration [Ne]$3s^1$ has only an S term.

13.34 The total angular orbital momenta of a p electron and a d electron correspond to $L=3$, 2, and 1 and reflect the different relative orientations of the two momenta.

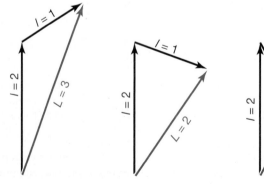

Example 13.6 Deriving the total orbital angular momentum of a configuration

Find the terms that can arise from the configurations (a) d^2, (b) p^3.

Method Use the Clebsch–Gordan series and begin by finding the minimum value of L (so that we know where the series terminates). When there are more than two electrons to couple together, use two series in succession: first couple two electrons, and then couple the third to each combined state, and so on.

Answer (a) Minimum value: $|l_1 - l_2| = |2 - 2| = 0$. Therefore,

$$L = 2 + 2, 2 + 2 - 1, \ldots, 0 = 4, 3, 2, 1, 0$$

corresponding to G, F, D, P, S terms, respectively. (b) First coupling: minimum value: $|1 - 1| = 0$. Therefore,

$$L' = 1 + 1, 1 + 1 - 1, \ldots, 0 = 2, 1, 0$$

Now couple l_3 with $L' = 2$, to give $L = 3, 2, 1$; with $L' = 1$, to give $L = 2, 1, 0$; and with $L' = 0$, to give $L = 1$. The overall result is

$$L = 3, 2, 2, 1, 1, 1, 0$$

giving one F, two D, three P, and one S terms.

Self-test 13.9 Repeat the question for the configurations (a) f^1d^1 and (b) d^3.

[(a) H, G, F, D, P; (b) I, 2H, 3G, 4F, 5D, 3P, S]

(b) The multiplicity

When there are several electrons to be taken into account, we must assess their total spin angular momentum quantum number, S (a non-negative integer or half integer).[10] Once again, we use the Clebsch–Gordan series in the form

$$S = s_1 + s_2, s_1 + s_2 - 1, \ldots, |s_1 - s_2| \tag{13.45}$$

to decide on the value of S, noting that each electron has $s = \frac{1}{2}$, which gives $S = 1, 0$ for two electrons (Fig. 13.35). If there are three electrons, the total spin angular momentum is obtained by coupling the third spin to each of the values of S for the first two spins, which results in $S = \frac{3}{2}$, and $S = \frac{1}{2}$.

The **multiplicity** of a term is the value of $2S + 1$. When $S = 0$ (as for a closed shell, like $1s^2$) the electrons are all paired and there is no net spin: this arrangement gives a singlet term, ^1S. A single electron has $S = s = \frac{1}{2}$, so a configuration such as $[Ne]3s^1$ can give rise to a doublet term, ^2S. The configuration $[Ne]3p^1$ likewise is a doublet, ^2P. When there are two unpaired electrons $S = 1$, so $2S + 1 = 3$, giving a triplet term, such as ^3D. We discussed the relative energies of singlets and triplets in Section 13.7 and saw that their energies differ on account of the different effects of spin correlation.

(c) The total angular momentum

As we have seen, the quantum number j tells us the relative orientation of the spin and orbital angular momenta of a single electron. The **total angular momentum quantum number**, J (a non-negative integer or half integer), does the same for several electrons. If there is a single electron outside a closed shell, $J = j$, with j either $l + \frac{1}{2}$ or $|l - \frac{1}{2}|$. The $[Ne]3s^1$ configuration has $j = \frac{1}{2}$ (because $l = 0$ and $s = \frac{1}{2}$), so the ^2S

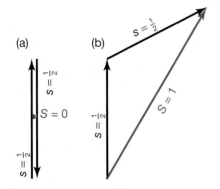

13.35 For two electrons (each of which has $s = \frac{1}{2}$), only two total spin states are permitted ($S = 0, 1$). The state with $S = 0$ can have only one value of M_S ($M_S = 0$) and is a singlet; the state with $S = 1$ can have any of three values of M_S (+1, 0, −1) and is a triplet. The vector representations of the singlet and triplet states are shown in Figs 13.20 and 13.26, respectively.

10. Distinguish italic S, the total spin quantum number, from Roman S, the term label.

term has a single level, which we denote $^2S_{1/2}$. The [Ne]$3p^1$ configuration has $l = 1$; therefore $j = \frac{3}{2}$ and $\frac{1}{2}$; the ^2P term therefore has two levels, $^2P_{3/2}$ and $^2P_{1/2}$. These levels lie at different energies on account of the magnetic spin–orbit interaction.

If there are several electrons outside a closed shell we have to consider the coupling of all the spins and all the orbital angular momenta. This complicated problem can be simplified when the spin–orbit coupling is weak (for atoms of low atomic number), for then we can use the **Russell–Saunders coupling** scheme. This scheme is based on the view that, if spin–orbit coupling is weak, then it is effective only when all the orbital momenta are operating cooperatively. We therefore imagine that all the orbital angular momenta of the electrons couple to give a total L, and that all the spins are similarly coupled to give a total S. Only at this stage do we imagine the two kinds of momenta coupling through the spin–orbit interaction to give a total J. The permitted values of J are given by the Clebsch–Gordan series

$$J = L + S, L + S - 1, \ldots, |L - S| \tag{13.46}$$

For example, in the case of the ^3D term of the configuration [Ne]$2p^13p^1$, the permitted values of J are 3, 2, 1 (because ^3D has $L = 2$ and $S = 1$), so the term has three levels, $^3D_3$, $^3D_2$, and $^3D_1$.

When $L > S$, the multiplicity is equal to the number of levels. For example, a ^2P term has the two levels $^2P_{3/2}$ and $^2P_{1/2}$, and ^3D has the three levels $^3D_3$, $^3D_2$, and $^3D_1$. However, this is not the case when $L < S$: the term ^2S, for example, has only the one level $^2S_{1/2}$.

Example 13.7 Deriving term symbols

Write the term symbols arising from the ground-state configurations of (a) Na and (b) F, and (c) the excited configuration $1s^22s^22p^13p^1$ of C.

Method Begin by writing the configurations, but ignore inner closed shells. Then couple the orbital momenta to find L and the spins to find S. Next, couple L and S to find J. Finally, express the term as $^{2S+1}\{L\}_J$, where $\{L\}$ is the appropriate letter. For F, for which the valence configuration is $2p^5$, treat the single gap in the closed-shell $2p^6$ configuration as a single particle.

Answer (a) For Na, the configuration is [Ne]$3s^1$, and we consider the single $3s$ electron. Because $L = l = 0$ and $S = s = \frac{1}{2}$, it is possible for $J = j = s = \frac{1}{2}$ only. Hence the term symbol is $^2S_{1/2}$. (b) For F, the configuration is [He]$2s^22p^5$, which we can treat as [Ne]$2p^{-1}$ (where the notation $2p^{-1}$ signifies the absence of a $2p$ electron). Hence $L = 1$, and $S = s = \frac{1}{2}$. Two values of $J = j$ are allowed: $J = \frac{3}{2}, \frac{1}{2}$. Hence, the term symbols for the two levels are $^2P_{3/2}$, $^2P_{1/2}$. (c) For C, the configuration is effectively $2p^13p^1$. This is a two-electron problem, and $l_1 = l_2 = 1$, $s_1 = s_2 = \frac{1}{2}$. It follows that $L = 2, 1, 0$ and $S = 1, 0$. The terms are therefore ^3D and ^1D, ^3P and ^1P, and ^3S and ^1S. For ^3D, $L = 2$ and $S = 1$; hence $J = 3, 2, 1$ and the levels are $^3D_3$, $^3D_2$, and $^3D_1$. For ^1D, $L = 2$ and $S = 0$, so the single level is $^1D_2$. The three levels of ^3P are $^3P_2$, $^3P_1$, and $^3P_0$, and the singlet is $^1P_1$. For the ^3S term there is only one level, $^3S_1$ (because $J = 1$ only), and the singlet term is $^1S_0$.

Comment The reason why we have treated an excited configuration of carbon is that in the ground configuration, $2p^2$, the Pauli principle forbids some terms, and deciding which survive (^1D, ^3P, ^1S, in fact) is quite complicated. That is, there is a distinction between 'equivalent electrons', which are electrons that occupy the same orbitals, and 'inequivalent electrons', which are electrons that occupy different orbitals. For information about how to deal with equivalent electrons, see *Further reading*.

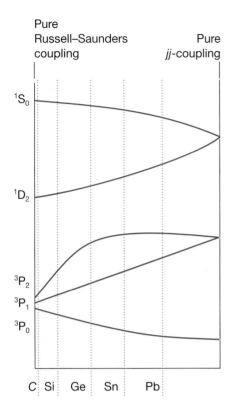

Pure
Russell–Saunders
coupling

Pure
jj-coupling

$^1S_0$

$^1D_2$

$^3P_2$
$^3P_1$
$^3P_0$

C Si Ge Sn Pb

13.36 The correlation diagram for some of the states of a two-electron system. All atoms lie between the two extremes but, the heavier the atom, the closer it lies to the pure *jj*-coupling case.

Self-test 13.10 Write down the terms arising from the configurations (a) $2s^1 2p^1$, (b) $2p^1 3d^1$.

[(a) $^3P_2$, $^3P_1$, $^3P_0$, $^1P_1$; (b) $^3F_4$, $^3F_3$, $^3F_2$, $^1F_3$, $^3D_3$, $^3D_2$, $^3D_1$, $^1D_2$, $^3P_2$, $^3P_1$, $^3P_0$, $^1P_1$]

Russell–Saunders coupling fails when the spin–orbit coupling is large (in heavy atoms). In that case, the individual spin and orbital momenta of the electrons are coupled into individual *j* values; then these momenta are combined into a grand total, *J*. This scheme is called ***jj*-coupling**. For example, in a p^2 configuration, the individual values of *j* are $\frac{3}{2}$ and $\frac{1}{2}$ for each electron. If the spin and the orbital angular momentum of each electron are coupled together strongly, it is best to consider each electron as a particle with angular momentum $j = \frac{3}{2}$ or $\frac{1}{2}$. These individual total momenta then couple as follows:

$$j_1 = \tfrac{3}{2} \quad \text{and} \quad j_2 = \tfrac{3}{2} \qquad J = 3, 2, 1, 0$$
$$j_1 = \tfrac{3}{2} \quad \text{and} \quad j_2 = \tfrac{1}{2} \qquad J = 2, 1$$
$$j_1 = \tfrac{1}{2} \quad \text{and} \quad j_2 = \tfrac{3}{2} \qquad J = 2, 1$$
$$j_1 = \tfrac{1}{2} \quad \text{and} \quad j_2 = \tfrac{1}{2} \qquad J = 1, 0$$

For heavy atoms, in which *jj*-coupling is appropriate, it is best to discuss their energies using these quantum numbers.

Although *jj*-coupling should be used for assessing the energies of heavy atoms, the term symbols derived from Russell–Saunders coupling can still be used as labels. To see why this procedure is valid, we need to examine how the energies of the atomic states change as the spin–orbit coupling increases in strength. Such a **correlation diagram** is shown in Fig. 13.36. It shows that there is a correspondence between the low spin–orbit coupling (Russell–Saunders coupling) and high spin–orbit coupling (*jj*-coupling) schemes, so the labels derived by using the Russell–Saunders scheme can be used to label the states of the *jj*-coupling scheme.

(d) Selection rules

Any state of the atom, and any spectral transition, can be specified by using term symbols. For example, the transitions giving rise to the yellow sodium doublet (which were shown in Fig. 13.32) are

$$3p^1\,^2P_{3/2} \to 3s^1\,^2S_{1/2} \qquad 3p^1\,^2P_{1/2} \to 3s^1\,^2S_{1/2}$$

By convention, the upper term precedes the lower. The corresponding absorptions are therefore denoted

$$^2P_{3/2} \leftarrow\, ^2S_{1/2} \qquad\quad ^2P_{1/2} \leftarrow\, ^2S_{1/2}$$

(The configurations have been omitted.)

We have seen that selection rules arise from the conservation of angular momentum during a transition and from the fact that a photon has a spin of 1. They can therefore be expressed in terms of the term symbols, because the latter carry information about angular momentum. A detailed analysis leads to the following rules:

$$\Delta S = 0 \quad \Delta L = 0, \pm 1 \quad \Delta l = \pm 1 \quad \Delta J = 0, \pm 1, \quad \text{but} \quad J = 0 \longleftrightarrow\!\!\!\!/\ J = 0 \tag{13.47}$$

The rule about ΔS (no change of overall spin) stems from the fact that the light does not affect the spin directly. The rules about ΔL and Δl express the fact that the orbital angular momentum of an individual electron must change (so $\Delta l = \pm 1$), but whether or not this results in an overall change of orbital momentum depends on the coupling.

The selection rules given above apply when Russell–Saunders coupling is valid (in light atoms). If we insist on labelling the terms of heavy atoms with symbols like ^{3}D, then we shall find that the selection rules progressively fail as the atomic number increases because the quantum numbers S and L become ill defined as jj-coupling becomes more appropriate. As explained above, Russell–Saunders term symbols are only a convenient way of labelling the terms of heavy atoms: they do not bear any direct relation to the actual angular momenta of the electrons in a heavy atom. For this reason, transitions between singlet and triplet states (for which $\Delta S = \pm 1$), while forbidden in light atoms, are allowed in heavy atoms.

Checklist of key ideas

☐ electronic structure of atoms, the arrangement of electrons around a nucleus

☐ hydrogenic atom, a one-electron atom or ion of general atomic number Z

☐ many-electron (polyelectronic) atom, an atom or ion with more than one electron

☐ spectroscopy, the detection and analysis of the electromagnetic radiation absorbed or emitted by a substance

☐ spectrum, the record of spectral intensity of the radiation emitted or absorbed by an atom or a molecule

The structure and spectra of hydrogenic atoms

☐ Balmer series, the lines arising from the transitions $n \to 2$ in the spectrum of atomic hydrogen

☐ Lyman series, the lines arising from the transitions $n \to 1$ in the spectrum of atomic hydrogen

☐ Paschen series, the lines arising from the transitions $n \to 3$ in the spectrum of atomic hydrogen

☐ Rydberg constant, the constant R_H in the expression $\tilde{v} = R_H(1/n_1^2 - 1/n_2^2)$

☐ term, a spectroscopic energy level

☐ Ritz combination principle, the wavenumber of any spectral line is the difference between two terms: $\tilde{v} = T_1 - T_2$

☐ Bohr frequency condition, $\Delta E = h\nu$

13.1 The structure of hydrogenic atoms

☐ reduced mass, $1/\mu = 1/m_e + 1/m_N$

☐ separation of variables, $\psi(r, \theta, \phi) = R(r)Y(\theta, \phi)$

☐ radial wave equation,
$$-\frac{\hbar^2}{2\mu}\left(\frac{d^2R}{dr^2} + \frac{2}{r}\frac{dR}{dr}\right) + V_{eff}R = ER$$

☐ energies, $E_n = -Z^2\mu e^4/32\pi^2\varepsilon_0^2\hbar^2 n^2$

☐ Bohr radius, $a_0 = 4\pi\varepsilon_0\hbar^2/m_e e^2$

☐ radial wavefunction, $R_{n,l}(r) = N_{n,l}(\rho/n)^l L_{n,l}(\rho)e^{-\rho/2n}$, $\rho = 2Zr/a_0$

13.2 Atomic orbitals and their energies

☐ atomic orbital, a one-electron wavefunction for an electron in an atom

☐ principal quantum number, the quantum number that determines the energy of an electron in a hydrogenic atom

☐ bound state, a state in which the energy of the atom is lower than that of the infinitely separated, stationary electron and nucleus

☐ unbound state, a state in which the energy of the atom is equal to or higher than that of the infinitely separated, stationary electron and nucleus

☐ Rydberg constant (theoretical value), $R_H = (\mu_H/m_e)R$, $R = m_e e^4/8\varepsilon_0^2 h^3 c$

☐ ionization energy, the minimum energy required to remove an electron from the ground state of an atom: $I = hcR_H$

☐ shell, all the orbitals of a given value of n

☐ subshell, orbitals with the same value of n but different values of l

☐ s-, p-, and d-orbitals, orbitals with $l = 0$, 1, and 2, respectively

☐ boundary surface, the surface that captures a high proportion of the electron probability

☐ radial distribution function, a probability density that, when it is multiplied by dr, gives the probability of finding the electron anywhere in a shell of thickness dr at the radius r; $P(r) = r^2R(r)^2$

☐ nodal plane, a plane dividing regions of opposite sign of the wavefunction

13.3 Spectroscopic transitions and selection rules

☐ transition, a change of state

☐ allowed and forbidden transitions, transitions allowed and disallowed by selection rules

☐ selection rule, a statement about which transitions are allowed

☐ transition dipole moment, $\mu_{fi} = \langle f|\mu|i\rangle$

☐ Grotrian diagram, a diagram summarizing the energies of the states and the transitions between them

The structures of many-electron atoms

13.4 The orbital approximation

☐ orbital approximation, the approximation of supposing that each electron occupies its 'own' orbital, $\Psi(r_1, r_2, \dots) = \psi(r_1)\psi(r_2)\dots$

☐ configuration, a list of occupied orbitals

☐ Pauli exclusion principle: No more than two electrons may occupy any given orbital, and if two do occupy one orbital, then their spins must be paired

☐ Pauli principle: When the labels of any two identical fermions are exchanged, the total wavefunction changes sign; when the labels of any two identical bosons are exchanged, the total wavefunction retains the same sign

☐ complete or closed shell (subshell), a shell (subshell) containing its maximum complement of electrons

- ☐ effective nuclear charge, Z_{eff}, the net charge experienced by an electron allowing for electron–electron repulsions
- ☐ shielding, the effective reduction in charge of a nucleus by surrounding electrons
- ☐ shielding constant, $\sigma = Z - Z_{eff}$
- ☐ penetration, the ability of an electron to be found inside inner shells and close to the nucleus
- ☐ valence electron, an electron in the outermost shell of an atom in its ground state
- ☐ building-up (*Aufbau*) principle, the procedure for filling atomic orbitals that leads to the ground-state configuration of an atom
- ☐ Hund's maximum multiplicity rule: An atom in its ground state adopts a configuration with the greatest number of unpaired electrons
- ☐ chemical periodicity, the periodic return of similar chemical properties with increasing atomic number
- ☐ configuration of ions, see Section 13.4(e)
- ☐ first ionization energy, the minimum energy necessary to remove an electron from a many-electron atom in the gas phase
- ☐ second ionization energy, the minimum energy necessary to remove an electron from an ionized many-electron atom in the gas phase
- ☐ standard enthalpy of ionization, $\Delta_{ion}H^{\ominus}(T) = I + \frac{5}{2}RT$
- ☐ electron affinity, the energy released when an electron attaches to a gas-phase atom

- ☐ standard enthalpy of electron gain, $\Delta_{eg}H^{\ominus}(T) = -E_{ea} - \frac{5}{2}RT$

13.5 Self-consistent field orbitals

- ☐ Hartree–Fock self-consistent field (SCF) procedure, see Section 13.5
- ☐ Coulomb operator, a contribution to the Schrödinger equation that takes into account the potential energy due to the charge density of the electrons in the other occupied orbitals
- ☐ exchange operator, a contribution to the Schrödinger equation that takes into account the effects of spin correlation

The spectra of complex atoms

13.6 Quantum defects and ionization limits

- ☐ quantum defect, an empirical correction to the energy, written as $-hcR/(n - \delta)^2$
- ☐ Rydberg states, highly diffuse excited states of atoms

13.7 Singlet and triplet states

- ☐ singlet, a state with $S = 0$; $\sigma_-(1, 2) = (1/2^{1/2})\{\alpha(1)\beta(2) - \beta(1)\alpha(2)\}$
- ☐ triplet, a state with $S = 1$; $\alpha(1)\alpha(2)$, $\sigma_+(1, 2) = (1/2^{1/2})\{\alpha(1)\beta(2) + \beta(1)\alpha(2)\}$, $\beta(1)\beta(2)$

13.8 Spin–orbit coupling

- ☐ spin–orbit coupling, the interaction of the spin magnetic moment with the magnetic field arising from the orbital angular momentum
- ☐ total angular momentum quantum number, j
- ☐ spin–orbit coupling constant, A in $E_{l,s,j} = \frac{1}{2}hcA\{j(j + 1) - l(l + 1) - s(s + 1)\}$

- ☐ fine structure, the structure in a spectrum due to spin–orbit coupling

13.9 Term symbols and selection rules

- ☐ term symbol, a symbolic specification of the state of an atom, $^{2S+1}\{L\}_J$
- ☐ total orbital angular momentum quantum number, L
- ☐ total spin quantum number, S
- ☐ total angular momentum quantum number, J
- ☐ Clebsch–Gordan series, $L = l_1 + l_2, l_1 + l_2 - 1, \ldots, |l_1 - l_2|$
- ☐ multiplicity, the number of levels of a term; the value of $2S + 1$
- ☐ level, a group of states of a term with a common value of J
- ☐ Russell–Saunders coupling, a coupling scheme based on the view that, if spin–orbit coupling is weak, then it is effective only when all the orbital momenta are operating cooperatively
- ☐ correlation diagram, a correspondence between the low spin–orbit coupling (Russell–Saunders coupling) and high spin–orbit coupling (jj-coupling) schemes
- ☐ jj-coupling, a coupling scheme based on the view that the individual spin and orbital momenta of the electrons are coupled into individual j values and these momenta are combined into a grand total, J
- ☐ selection rules, $\Delta S = 0$, $\Delta L = 0$, ± 1, $\Delta l = \pm 1$, $\Delta J = 0$, ± 1, but $J = 0 \longleftrightarrow J = 0$

Further reading

Articles of general interest

N. Agmon, Ionization potentials for isoelectronic series. *J. Chem. Educ.* **65**, 42 (1988).

R.D. Allendoerfer, Teaching the shapes of the hydrogen like and hybrid atomic orbitals. *J. Chem. Educ.* **67**, 37 (1990).

B. Bederson, Atoms. In *Encyclopedia of applied physics* (ed. G.L. Trigg), 2, 285. VCH, New York (1991).

J.L. Bills, Experimental 4s and 3d energies in atomic ground states. *J. Chem. Educ.* **75**, 589 (1998).

J.C.A. Boeyens, Understanding electron spin. *J. Chem. Educ.* **72**, 412 (1995).

K. Bonin and W. Happer, Atomic spectroscopy. In *Encyclopedia of applied physics* (ed. G.L. Trigg), **2**, 245. VCH, New York (1991).

G.L. Breneman, Order out of chaos: shapes of hydrogen orbitals. *J. Chem. Educ.* **65**, 31 (1988).

P. Cann, Ionization energies, parallel spins, and the stability of half-filled shells. *J. Chem. Educ.* **77**, 1056 (2000).

D.F. Fitts, Ladder operator treatment of the radial equation for the hydrogen like atom. *J. Chem. Educ.* **72**, 1066 (1995).

S.J. Gauerke and M.L. Campbell, A simple, systematic method for determining J levels for jj coupling. *J. Chem. Educ.* **71**, 457 (1994).

L. Guofan and M.L. Elizey Jr., Finding the terms of configurations of equivalent electrons by partitioning total spins. *J. Chem. Educ.* **64**, 771 (1987).

C.W. Haigh, The theory of atomic spectroscopy: *jj* coupling, intermediate coupling, and configuration interaction. *J. Chem. Educ.* **72**, 206 (1995).

M.J. Kaufman and C.G. Trowbridge, The ionization energy of helium. *J. Chem. Educ.* **76**, 88 (1999).

E.M.R. Kiremire, A numerical algorithm technique for deriving Russell–Saunders (R–S) terms. *J. Chem. Educ.* **64**, 951 (1987).

H. Kleindienst, A. Lüchow, and R. Benois, Pauli principle and permutation symmetry. *J. Chem. Educ.* **72**, 1019 (1995).

J. Mason. Periodic contractions among the elements: or, on being the right size. *J. Chem. Educ.* **65**, 17 (1988).

T.L. Meek, Electronegativities of the noble gases. *J. Chem. Educ.* **72**, 17 (1995).

M. Melrose and E.R. Scerri, Why the 4*s* orbital is occupied before the 3*d*. *J. Chem. Educ.* **74**, 498 (1996).

J.C. Morrison, A.W. Weiss, K. Kirby, and D. Cooper, Electronic structure of atoms and molecules. In *Encyclopedia of applied physics* (ed. G.L. Trigg), **6**, 45. VCH, New York (1993).

R.T. Myers, The periodicity of electron affinity. *J. Chem. Educ.* **67**, 307 (1990).

P.G. Nelson, Relative energies of 3*d* and 4*s* orbitals. *Educ. in Chem.* **29**, 84 (1992).

I. Novak, Electronic states and configurations: visualizing the difference. *J. Chem. Educ.* **76**, 135 (1999).

N.C. Pyper and M. Berry, Ionization energies revisited. *Educ. in Chem.* **27**, 135 (1990).

N.E. Robertson, S.O. Boggioni, and F.E. Henriquez, A simple mathematical approach to experimental ionization energies of atoms: a new method for estimating the extent of shielding. *J. Chem. Educ.* **71**, 101 (1994).

E.R. Scerri, Transition metal configurations and limitations of the orbital approximation. *J. Chem. Educ.* **66**, 481 (1989).

K.D. Sen, Electronegativity. In *Structure and bonding*, Vol. 6. Springer, New York (1987).

K.D. Sen, M. Slamet, and V. Sahni, Atomic shell structure in Hartree–Fock theory. *Chem. Phys. Letts.* **205**, 313 (1993).

N. Shenkuan, The physical basis of Hund's rule: orbital contraction effects. *J. Chem. Educ.* **69**, 800 (1992).

J.H. Summerfield, G.S. Beltrame, and J.G. Loeser, A simple model for understanding electron correlation methods. *J. Chem. Educ.* **76**, 1430 (1999).

L.G. Vanquickenborne, K. Pierloot, and D. Devoghel, Transition metals and the *Aufbau* principle. *J. Chem. Educ.* **71**, 469 (1994).

J.C. Wheeler, Electron affinities of the alkaline earth metals and the sign convention for electron affinity. *J. Chem. Educ.* **74**, 123 (1997).

Texts and sources of data and information

P.W. Atkins, *Quanta: A handbook of concepts.* Oxford University Press (1991).

P.W. Atkins and R.S. Friedman, *Molecular quantum mechanics.* Oxford University Press (1997).

S. Bashkin and J.O. Stonor, Jr., *Atomic energy levels and Grotrian diagrams.* North-Holland, Amsterdam (1975–1982).

P.F. Bernath, *Spectra of atoms and molecules.* Oxford University Press (1995).

E.U. Condon and H. Odabaşi, *Atomic structure.* Cambridge University Press (1980).

P.A. Cox, *Introduction to quantum theory and atomic structure.* Oxford Chemistry Primers, Oxford University Press (1996).

C.F. Fischer, *The Hartree–Fock method for atoms.* Wiley, New York (1977).

I.I. Gol'dman and V.D. Krivchenkov, *Problems in quantum mechanics.* Dover, New York (1993).

G. Herzberg, *Atomic spectra and atomic structure*, Dover, New York (1945).

C.S. Johnson, Jr. and L.G. Pedersen, *Problems and solutions in quantum chemistry and physics.* Dover, New York (1986).

D.R. Lide (ed.), *CRC Handbook of Chemistry and Physics*, Section 10. CRC Press, Boca Raton (2000).

P.R. Scott and W.G. Richards, *Energy levels in atoms and molecules.* Oxford Chemistry Primers, Oxford University Press (1994).

T.P. Softley, *Atomic spectra*, Oxford Chemistry Primers, Oxford University Press (1994).

Exercises

Discussion questions

13.1 (a) Describe the separation of variables procedure as it is applied to simplify the description of a hydrogenic atom free to move through space.

13.1 (b) List and describe the significance of the quantum numbers needed to specify the internal state of a hydrogenic atom.

13.2 (a) Specify and account for the selection rules for transitions in hydrogenic atoms.

13.2 (b) Explain the significance of (a) a boundary surface and (b) the radial distribution function for hydrogenic orbitals.

13.3 (a) Outline the electron configurations of many-electron atoms in terms of their location in the periodic table.

13.3 (b) Describe and account for the variation of first ionization energies along Period 2 of the periodic table.

13.4 (a) Describe the orbital approximation for the wavefunction of a many-electron atom. What are the limitations of the approximation?

13.4 (b) Explain the origin of spin–orbit coupling and how it affects the appearance of a spectrum.

Numerical exercises

13.5 (a) When ultraviolet radiation of wavelength 58.4 nm from a helium lamp is directed on to a sample of krypton, electrons are ejected with a speed of 1.59 Mm s^{-1}. Calculate the ionization energy of krypton.

13.5 (b) When ultraviolet radiation of wavelength 58.4 nm from a helium lamp is directed on to a sample of xenon, electrons are ejected with a speed of 1.79 Mm s^{-1}. Calculate the ionization energy of xenon.

13.6 (a) By differentiation of the 2s radial wavefunction, show that it has two extrema in its amplitude, and locate them.

13.6 (b) By differentiation of the 3s radial wavefunction, show that it has three extrema in its amplitude, and locate them.

13.7 (a) Locate the radial nodes in the 3s orbital of an H atom.

13.7 (b) Locate the radial nodes in the 3p orbital of an H atom.

13.8 (a) The wavefunction for the ground state of a hydrogen atom is Ne^{-r/a_0}. Determine the normalization constant N.

13.8 (b) The wavefunction for the 2s orbital of a hydrogen atom is $N(2 - r/a_0)e^{-r/2a_0}$. Determine the normalization constant N.

13.9 (a) Calculate the average kinetic and potential energies of an electron in the ground state of a hydrogen atom.

13.9 (b) Calculate the average kinetic and potential energies of a 2s electron in a hydrogenic atom of atomic number Z.

13.10 (a) Write down the expression for the radial distribution function of a 2s electron in a hydrogenic atom and determine the radius at which the electron is most likely to be found.

13.10 (b) Write down the expression for the radial distribution function of a 3s electron in a hydrogenic atom and determine the radius at which the electron is most likely to be found.

13.11 (a) What is the orbital angular momentum of an electron in the orbitals (a) 1s, (b) 3s, (c) 3d? Give the numbers of angular and radial nodes in each case.

13.11 (b) What is the orbital angular momentum of an electron in the orbitals (a) 4d, (b) 2p, (c) 3p? Give the numbers of angular and radial nodes in each case.

13.12 (a) Calculate the permitted values of j for (a) a d electron, (b) an f electron.

13.12 (b) Calculate the permitted values of j for (a) a p electron, (b) an h electron.

13.13 (a) An electron in two different states of an atom is known to have $j = \frac{3}{2}$ and $\frac{1}{2}$. What is its orbital angular momentum quantum number in each case?

13.13 (b) What are the allowed total angular momentum quantum numbers of a composite system in which $j_1 = 5$ and $j_2 = 3$?

13.14 (a) State the orbital degeneracy of the levels in a hydrogen atom that have energy (a) $-hcR_H$; (b) $-\frac{1}{9}hcR_H$; (c) $-\frac{1}{25}hcR_H$.

13.14 (b) State the orbital degeneracy of the levels in a hydrogenic atom (Z in parentheses) that have energy (a) $-4hcR_{atom}$, (2); (b) $-\frac{1}{4}hcR_{atom}$ (4), and (c) $-hcR_{atom}$ (5).

13.15 (a) What information does the term symbol $^1D_2$ provide about the angular momentum of an atom?

13.15 (b) What information does the term symbol $^3F_4$ provide about the angular momentum of an atom?

13.16 (a) At what radius does the probability of finding an electron at a point in the H atom fall to 50 per cent of its maximum value?

13.16 (b) At what radius in the H atom does the radial distribution function of the ground state have (a) 50 per cent, (b) 75 per cent of its maximum value.

13.17 (a) Which of the following transitions are allowed in the normal electronic emission spectrum of an atom: (a) $2s \rightarrow 1s$, (b) $2p \rightarrow 1s$, (c) $3d \rightarrow 2p$?

13.17 (b) Which of the following transitions are allowed in the normal electronic emission spectrum of an atom: (a) $5d \rightarrow 2s$, (b) $5p \rightarrow 3s$, (c) $5p \rightarrow 3f$?

13.18 (a) How many electrons can occupy the following subshells: (a) 1s, (b) 3p, (c) 3d, (d) 6g?

13.18 (b) How many electrons can occupy the following subshells: (a) 2s, (b) 4d, (c) 6f, (d) 6h?

13.19 (a) (a) Write the electronic configuration of the Ni^{2+} ion. (b) What are the possible values of the total spin quantum numbers S and M_S for this ion?

13.19 (b) (a) Write the electronic configuration of the V^{2+} ion. (b) What are the possible values of the total spin quantum numbers S and M_S for this ion?

13.20 (a) Suppose that an atom has (a) 2, (b) 3 electrons in different orbitals. What are the possible values of the total spin quantum number S? What is the multiplicity in each case?

13.20 (b) Suppose that an atom has (a) 4, (b) 5 electrons in different orbitals. What are the possible values of the total spin quantum number S? What is the multiplicity in each case?

13.21 (a) What atomic terms are possible for the electron configuration ns^1nd^1? Which term is likely to lie lowest in energy?

13.21 (b) What atomic terms are possible for the electron configuration np^1nd^1? Which term is likely to lie lowest in energy?

13.22 (a) What values of J may occur in the terms (a) ^1S, (b) ^2P, (c) ^3P? How many states (distinguished by the quantum number M_J) belong to each level?

13.22 (b) What values of J may occur in the terms (a) ^3D, (b) ^4D, (c) ^2G? How many states (distinguished by the quantum number M_J) belong to each level?

13.23 (a) Give the possible term symbols for (a) Li [He]2s^1, (b) Na [Ne]3p^1.

13.23 (b) Give the possible term symbols for (a) Sc [Ar]3$d^1$4s^2, (b) Br [Ar]3d^{10}4$s^2$4p^5.

Problems[†]

Numerical problems

13.1 The *Humphreys series* is another group of lines in the spectrum of atomic hydrogen. It begins at 12 368 nm and has been traced to 3281.4 nm. What are the transitions involved? What are the wavelengths of the intermediate transitions?

13.2 A series of lines in the spectrum of atomic hydrogen lies at 656.46 nm, 486.27 nm, 434.17 nm, and 410.29 nm. What is the wavelength of the next line in the series? What is the ionization energy of the atom when it is in the lower state of the transitions?

13.3 The Li^{2+} ion is hydrogenic and has a Lyman series at 740 747 cm^{-1}, 877 924 cm^{-1}, 925 933 cm^{-1}, and beyond. Show that the energy levels are of the form $-hcR/n^2$ and find the value of R for this ion. Go on to predict the wavenumbers of the two longest-wavelength transitions of the Balmer series of the ion and find the ionization energy of the ion.

13.4 A series of lines in the spectrum of neutral Li atoms rise from combinations of $1s^22p^1$ ^2P with $1s^2nd^1$ ^2D and occur at 610.36 nm, 460.29 nm, and 413.23 nm. The d orbitals are hydrogenic. It is known that the ^2P term lies at 670.78 nm above the ground state, which is $1s^22s^1$ ^2S. Calculate the ionization energy of the ground-state atom.

13.5‡ W.P. Wijesundera, S.H. Vosko, and F.A. Parpia (*Phys. Rev. A* **51**, 278 (1995)) attempted to determine the electron configuration of the ground state of lawrencium, element 103. The two contending configurations are $[Rn]5f^{14}7s^27p^1$ and $[Rn]5f^{14}6d7s^2$. Write down the term symbols for each of these configurations, and identify the lowest level within each configuration. Which level would be lowest according to a simple estimate of spin–orbit coupling?

13.6 The characteristic emission from K atoms when heated is purple and lies at 770 nm. On close inspection, the line is found to have two closely spaced components, one at 766.70 nm and the other at 770.11 nm. Account for this observation, and deduce what information you can.

13.7 Calculate the mass of the deuteron given that the first line in the Lyman series of H lies at 82 259.098 cm^{-1} whereas that of D lies at 82 281.476 cm^{-1}. Calculate the ratio of the ionization energies of H and D.

13.8 Positronium consists of an electron and a positron (same mass, opposite charge) orbiting round their common centre of mass. The broad features of the spectrum are therefore expected to be hydrogen-like, the differences arising largely from the mass differences. Predict the wavenumbers of the first three line of the Balmer series of positronium. What is the binding energy of the ground state of positronium?

13.9 The *Zeeman effect* is the modification of an atomic spectrum by the application of a strong magnetic field. It arises from the interaction between applied magnetic fields and the magnetic moments due to orbital and spin angular momenta (recall the evidence provided for electron spin by the Stern–Gerlach experiment, Section 12.8). To gain some appreciation for the so-called *normal Zeeman effect*, which is observed in transitions involving singlet states, consider a p electron, with $l = 1$ and $m_l = 0, \pm1$. In the absence of a magnetic field, these three states are degenerate. When a field of magnitude B is present, the degeneracy is removed and it is observed that the state with $m_l = +1$ moves up in energy by $\mu_B B$, the state with $m_l = 0$ is unchanged, and the state with $m_l = -1$ moves down in energy by $\mu_B B$, where $\mu_B = e\hbar/2m_e = 9.274 \times 10^{-24}$ J T^{-1} is the Bohr magneton (see Section 18.1). Therefore, a transition between a $^1S_0$ term and a $^1P_1$ term consists of three spectral lines in the presence of a magnetic field where, in the absence of the magnetic field, there is only one. (a) Calculate the splitting in reciprocal centimetres between the three spectral lines of a transition between a $^1S_0$ term and a $^1P_1$ term in the presence of a magnetic field of 2 T (where 1 T = 1 kg s^{-2} A^{-1}). (b) Compare the value you calculated in (a) with typical optical transition wavenumbers, such as those for the Balmer series of the H atom. Is the line splitting caused by the normal Zeeman effect relatively small or relatively large?

13.10 In 1976 it was mistakenly believed that the first of the 'superheavy' elements had been discovered in a sample of mica. Its atomic number was believed to be 126. What is the most probable distance of the innermost electrons from the nucleus of an atom of this element? (In such elements, relativistic effects are very important, but ignore them here.)

Theoretical problems

13.11 Is an electron further from the nucleus on average when it is in a 2s orbital or a 2p orbital?

13.12 What is the most probable point (not radius) at which a 2p electron will be found in the hydrogen atom?

13.13 Show by explicit integration that (a) hydrogenic 1s and 2s orbitals, (b) $2p_x$ and $2p_y$ orbitals are mutually orthogonal.

13.14‡ Use the radial wave equation for the hydrogen atom to demonstrate that the energies of the 2s and 2p orbitals are identical.

13.15‡ Explicit expressions for hydrogenic orbitals are given in Tables 13.1 and 12.3. (a) Verify both that the $3p_x$ orbital is normalized (to 1) and that $3p_x$ and $3d_{xy}$ are mutually orthogonal. (b) Determine the positions of both the radial nodes and nodal planes of the 3s, $3p_x$, and $3d_{xy}$ orbitals. (c) Determine the mean radius of the 3s orbital. (d) Draw a graph of the radial distribution function for the three orbitals (of part (b)) and discuss the significance of the graphs for interpreting the properties of many-electron atoms. (e) Create both xy-plane polar plots and boundary surface plots for these orbitals. Construct the boundary plots so that the distance from the origin to the surface is the absolute value of the angular part of the wavefunction. Compare the s, p, and d boundary surface plots with that of an f-orbital; e.g. $\psi_f \propto x(5z^2 - r^2) \propto \sin\theta(5\cos^2\theta - 1)\cos\phi$.

13.16 Determine whether the p_x and p_x orbitals are eigenfunctions of l_z. If not, does a linear combination exist that is an eigenfunction of l_z?

13.17 Show that l_z and l^2 both commute with the hamiltonian for a hydrogen atom. What is the significance of this result?

† Problems denoted with the symbol ‡ were supplied by Charles Trapp, Carmen Giunta, and Marshall Cady.

13.18 The 'size' of an atom is sometimes considered to be measured by the radius of a sphere that contains 90 per cent of the charge density of the electrons in the outermost occupied orbital. Calculate the 'size' of a hydrogen atom in its ground state according to this definition.

13.19‡ The diameters of atoms can be estimated from their densities in a condensed state. Calculate the diameters of hydrogen and uranium atoms in this manner from information in the *Data section*. One finds that all atoms are roughly the same size, with $r \approx 0.3 \pm 0.1$ nm. Why is that? In a plot of atomic radius against atomic number some periodicity is evident, but not to the extent seen in a plot of first ionization energies against atomic number. Explain this observation.

13.20 One of the most famous of the obsolete theories of the hydrogen atom was proposed by Bohr. It has been replaced by quantum mechanics but, by a remarkable coincidence (not the only one where the Coulomb potential is concerned), the energies it predicts agree exactly with those obtained from the Schrödinger equation. In the Bohr atom, an electron travels in a circle around the nucleus. The Coulombic force of attraction $(Ze^2/4\pi\varepsilon_0 r^2)$ is balanced by the centrifugal effect of the orbital motion. Bohr proposed that the angular momentum is limited to integral values of $\hbar$. When the two forces are balanced, the atom remains in a stationary state until it makes a spectral transition. Calculate the energies of a hydrogenic atom using the Bohr model.

13.21 The Bohr model of the atom is specified in Problem 13.20. What features of it are untenable according to quantum mechanics? How does the Bohr ground state differ from the actual ground state. Is there an experimental distinction between the Bohr and quantum mechanical models of the ground state?

13.22‡ In the Bohr model of the hydrogen atom, the electron orbits the nucleus at a distance of 52.9 pm. Calculate the speed of the electron in the first Bohr orbit. Considering the electron and proton to be classical charges, calculate the electrical field strength at the electron and the magnetic field strength at the proton. Compare to field strengths commonly available in the laboratory.

13.23 Atomic units of length and energy may be based on the properties of a particular atom. The usual choice is that of a hydrogen atom, with the unit of length being the Bohr radius, a_0, and the unit of energy being the energy of the 1s orbital. If the positronium atom (e^+, e^-) were used instead, with analogous definitions of units of length and energy, what would be the relation between these two sets of atomic units?

13.24‡ Dimensionless ratios that occur in the physical sciences are thought to be of fundamental significance. These ratios tend to be clustered around $(10^{20})^n$, where $n = 0, 1, 2, 3,$ and 4. One such ratio in the $n = 0$ group is the mass ratio of the two fundamental particles, the proton and the electron. Scientists are puzzled as to why this ratio should be close to 2000. The precise value of the ratio is determined by comparison of the atomic spectral lines in H and He$^+$. (a) Derive the following relations for the first line in any of the series (Lyman, Balmer, etc.) for H and He$^+$:

$$\gamma = \frac{\frac{1}{4}\tilde{\nu}_{He} - \tilde{\nu}_H}{\tilde{\nu}_H} = \frac{\mu_{He} - \mu_H}{\mu_H}$$

where μ is a reduced mass. (b) Calculate m_H/m_e from the following data:

$\lambda(n_2 \rightarrow n_1)$/nm	R_J/cm^{-1}
H	121.5664	109 677.7
He$^+$	30.3779	109 772.4

First do the calculation from the wavelength data, then derive a formula for the mass ratio in terms of the Rydberg constants R_J of the species J.

13.25 Some of the selection rules for hydrogenic atoms were derived in *Justification* 13.5. Complete the derivation by considering the x- and y-components of the electric dipole moment operator.

13.26‡ Stern–Gerlach splittings of atomic beams are small and require either large magnetic field gradients or long magnets for their observation. For a beam of atoms with zero orbital angular momentum, such as H or Ag, the deflection is given by $x = \pm(\mu_B L^2/4E_K)$ dB/dz, where μ_B is the Bohr magneton (Problem 13.9), L is the length of the magnet, E_K is the average kinetic energy of the atoms in the beam, and dB/dz is the magnetic field gradient across the beam. (a) Use the Maxwell–Boltzmann velocity distribution to show that the average translational kinetic energy of the atoms emerging as a beam from a pinhole in an oven at temperature T is $2kT$. (b) Calculate the magnetic field gradient required to produce a splitting of 1.00 mm in a beam of Ag atoms from an oven at 1000 K with a magnet of length 50 cm.

Applications: to astrophysics

13.27‡ Highly excited atoms have electrons with large principal quantum numbers. Such *Rydberg atoms* have unique properties and are of interest to astrophysicists. For hydrogen atoms with large n, derive a relation for the separation of energy levels. Calculate this separation for $n = 100$; also calculate the average radius, the geometric cross-section, and the ionization energy. Could a thermal collision with another hydrogen atom ionize this Rydberg atom? What minimum velocity of the second atom is required? Could a normal-sized neutral H atom simply pass through the Rydberg atom leaving it undisturbed? What might the radial wavefunction for a 100s orbital be like?

13.28 The spectrum of a star is used to measure its *radial velocity* with respect to the Sun, the component of the star's velocity vector that is parallel to a vector connecting the star's centre to the centre of the Sun. The measurement relies on the *Doppler effect*, in which radiation is shifted in frequency when the source is moving towards or away from the observer. When a star emitting electromagnetic radiation of frequency ν moves with a speed s relative to an observer, the observer detects radiation of frequency $\nu_{receding} = \nu f$ or $\nu_{approaching} = \nu/f$, where $f = \{(1 - s/c)/(1 + s/c)\}^{1/2}$ and c is the speed of light. It is easy to see that $\nu_{receding} < \nu$ and a receding star is characterized by a *red shift* of its spectrum with respect to the spectrum of an identical, but stationary source. Furthermore, $\nu_{approaching} > \nu$ and an approaching star is characterized by a *blue shift* of its spectrum with respect to the spectrum of an identical, but stationary source. In a typical experiment, ν is the frequency of a spectral line of an element measured in a stationary Earth-bound laboratory from a calibration source, such as an arc lamp (Section 16.1a). Measurement of the same spectral line in a star gives ν_{star} and the speed of recession or approach may be calculated from the value of ν and the equations above. (a) Three Fe I lines of the star HDE 271 182, which belongs to the Large Magellanic Cloud, occur at 438.882 nm, 441.000 nm, and 442.020 nm. The same lines occur at 438.392 nm, 440.510 nm, and 441.510 nm in the spectrum of an Earth-bound iron arc. Determine whether HDE 271 182 is receding from or approaching the Earth and estimate the star's radial speed with respect to the Earth. (b) What additional information would you need to calculate the radial velocity of HDE 271 182 with respect to the Sun?

14

Molecular structure

The Born–Oppenheimer approximation

Valence-bond theory

14.1 The hydrogen molecule

14.2 Homonuclear diatomic molecules

14.3 Polyatomic molecules

Molecular orbital theory

14.4 The hydrogen molecule-ion

14.5 The structures of diatomic molecules

14.6 Heteronuclear diatomic molecules

Molecular orbitals for polyatomic systems

14.7 The Hückel approximation

14.8 Extended Hückel theory

14.9 Self-consistent field calculations

Checklist of key ideas

Further reading

Exercises

Problems

The concepts developed in Chapter 13, particularly those of orbitals, can be extended to a description of the electronic structures of molecules. There are two principal quantum mechanical theories of molecular electronic structure. In valence-bond theory the starting point is the concept of the shared electron pair. We see how to write the wavefunction for such a pair, and how it may be extended to account for the structures of a wide variety of molecules. The theory introduces the concepts of σ and π bonds, promotion, and hybridization, which are used widely in chemistry. In molecular orbital theory (with which the bulk of the chapter is concerned), the concept of atomic orbital is extended to that of molecular orbital, which is a wavefunction that spreads over all the atoms in a molecule.

In this chapter we consider the origin of the strengths, numbers, and three-dimensional arrangement of chemical bonds between atoms. The quantum mechanical description of chemical bonding has become highly developed through the use of computers, and it is now possible to consider the structures of molecules of almost any complexity. We shall concentrate on the quantum mechanical description of the **covalent bond**, which was identified by G.N. Lewis (in 1916, before quantum mechanics was fully established) as an electron pair shared between two neighbouring atoms. We shall see, however, that the other principal type of bond, an **ionic bond**, in which the cohesion arises from the Coulombic attraction between ions of opposite charge, is also captured as a limiting case of a covalent bond between dissimilar atoms.

The Born–Oppenheimer approximation

All theories of molecular structure make the same simplification at the outset. Whereas the Schrödinger equation for a hydrogen atom can be solved exactly, an exact solution is not possible for any molecule because the simplest molecule consists of three particles (two nuclei and one electron). We therefore adopt the **Born–Oppenheimer approximation** in which it is supposed that the nuclei, being so much heavier than an electron, move relatively slowly and may be treated as stationary while the electrons move in their field. We can therefore think of the nuclei as being fixed at arbitrary locations, and then solve the Schrödinger equation for the wavefunction of the electrons alone.

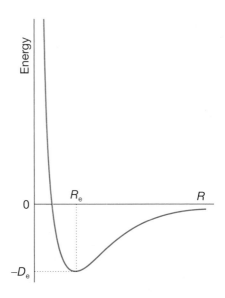

14.1 A molecular potential energy curve. The equilibrium bond length corresponds to the energy minimum.

The approximation is quite good for ground-state molecules, for calculations suggest that the nuclei in H_2 move through only about 1 pm while the electron speeds through 1000 pm, so the error of assuming that the nuclei are stationary is small. Exceptions to the approximation's validity include certain excited states of polyatomic molecules and the ground states of cations; both types of species are important when considering photoelectron spectroscopy (Section 17.8) and mass spectrometry.

The Born–Oppenheimer approximation allows us to select an internuclear separation in a diatomic molecule and then to solve the Schrödinger equation for the electrons at that nuclear separation. Then we choose a different separation and repeat the calculation, and so on. In this way we can explore how the energy of the molecule varies with bond length (and, in polyatomic molecules, with angles too), and obtain a **molecular potential energy curve** (Fig. 14.1).[1] It is called a *potential energy curve* because the kinetic energy of the stationary nuclei is zero. Once the curve has been calculated or determined experimentally (by using the spectroscopic techniques described in Chapters 16 and 17), we can identify the **equilibrium bond length**, R_e, the internuclear separation at the minimum of the curve, and the **bond dissociation energy**, D_0, which is closely related to the depth, D_e, of the minimum below the energy of the infinitely widely separated atoms.[2]

Valence-bond theory

The **valence-bond theory** (VB theory) was the first quantum mechanical theory of bonding to be developed. Although the theory has undergone much less computational development than molecular orbital theory, the language it introduced, which includes concepts such as spin pairing, σ and π bonds, and hybridization, is widely used throughout chemistry. That language is particularly widespread in the description of the properties and reactions of organic compounds.

14.1 The hydrogen molecule

The simplest molecule with an electron pair bond is H_2. We use this molecule to introduce the basic concepts of the theory.

(a) The spatial wavefunction

The wavefunction for an electron on each of two widely separated H atoms is

$$\psi = \psi_{H1s_A}(\mathbf{r}_1)\psi_{H1s_B}(\mathbf{r}_2)$$

if electron 1 is on atom A and electron 2 is on atom B. For simplicity, we shall write this wavefunction as $\psi = A(1)B(2)$. When the atoms are close, it is not possible to know whether it is electron 1 that is on A or electron 2. An equally valid description is therefore $\psi = A(2)B(1)$, in which electron 2 is on A and electron 1 is on B. When two

---

1. When more than one molecular parameter is changed in a polyatomic molecule, we obtain a potential energy surface.
2. The dissociation energy differs from the depth of the well by an energy equal to the zero-point vibrational energy of the bonded atoms: $D_0 = D_e - \frac{1}{2}\hbar\omega$, where ω is the vibrational frequency of the bond (Section 16.9).

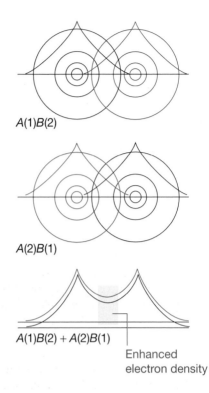

$A(1)B(2)$

$A(2)B(1)$

$A(1)B(2) + A(2)B(1)$

Enhanced electron density

14.2 It is very difficult to represent valence-bond wavefunctions because they refer to two electrons simultaneously. However, this illustration is an attempt. The atomic orbital for electron 1 is represented by the black contours, and that of electron 2 is represented by the blue contours. The top illustration represents $A(1)B(2)$, and the middle illustration represents the contribution $A(2)B(1)$. When the two contributions are superimposed, there is interference between the black contributions and between the blue contributions, resulting in an enhanced (two-electron) density in the internuclear region.

outcomes are equally probable, quantum mechanics instructs us to describe the true state of the system as a superposition of the wavefunctions for each possibility (Section 11.5d), so a better description of the molecule than either wavefunction alone is[3]

$$\psi = A(1)B(2) \pm A(2)B(1) \tag{14.1}$$

It turns out that the combination with lower energy is the one with a + sign, so the valence-bond wavefunction of the H_2 molecule is

$$\psi = A(1)B(2) + A(2)B(1) \tag{14.2}$$

The formation of the bond in H_2 can be pictured as due to the high probability that the two electrons will be found between the two nuclei and hence will bind them together. More formally, the wave pattern represented by the term $A(1)B(2)$ interferes constructively with the wave pattern represented by the contribution $A(2)B(1)$, and there is an enhancement in the value of the wavefunction in the internuclear region (Fig. 14.2).

The electron distribution described by the wavefunction in eqn 2 is called a **σ bond**. A σ bond has cylindrical symmetry around the internuclear axis, and is so called because, when viewed along the internuclear axis, it resembles a pair of electrons in an s orbital (and σ is the Greek equivalent of s). More precisely, the electrons in a σ bond have zero orbital angular momentum about the internuclear axis.[4]

The molecular potential energy curve for H_2 is calculated by changing the internuclear separation R and evaluating the expectation value of the energy at each selected separation. The resulting graph is shown in Fig. 14.3. The energy falls below that of two separated H atoms as the two atoms are brought within bonding distance and each electron is free to migrate to the other atom. However, the energy reduction is counteracted by an increase in energy from the Coulombic repulsion between the two positively charged nuclei. This positive contribution to the energy becomes large as R becomes small. Consequently, the total potential energy curve passes through a minimum and then climbs to a strongly positive value at small internuclear separations.

(b) The role of electron spin

So far, the electron spin has not played a role in the argument, yet a chemist's picture of a covalent bond is one in which the spins of two electrons pair as the atomic orbitals overlap. The origin of the role of spin is that the wavefunction given in eqn 2 can be formed only by a pair of electrons with opposed spins. Spin pairing is not an end in itself: it is a means of achieving a spatial wavefunction (and the probability distribution it implies) that corresponds to a low energy.

Justification 14.1

The Pauli principle requires the wavefunction of two electrons to change sign when the labels of the electrons are interchanged (see *Justification 13.7*). The total VB wavefunction for two electrons is

$$\psi(1, 2) = \{A(1)B(2) + A(2)B(1)\}\sigma(1, 2)$$

3. These linear combinations are not normalized.
4. Recall from Section 12.6 that the orbital angular momentum of an electron is related to the number of angular nodes in its wavefunction; but there are no angular nodes in the wavefunction of a σ bond, so it has zero orbital angular momentum.

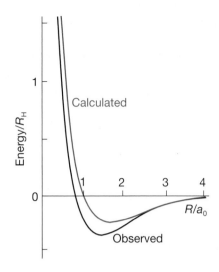

14.3 The molecular potential energy curve for the hydrogen molecule showing the variation of the energy of the molecule as the bond length is changed. The calculated curve refers to the valence-bond model.

where σ represents the spin component of the wavefunction. When the labels 1 and 2 are interchanged, this wavefunction becomes

$$\psi(2, 1) = \{A(2)B(1) + A(1)B(2)\}\sigma(2, 1) = \{A(1)B(2) + A(2)B(1)\}\sigma(2, 1)$$

The Pauli principle requires that $\psi(2, 1) = -\psi(1, 2)$, which is satisfied only if $\sigma(2, 1) = -\sigma(1, 2)$. The combination of two spins that has this property is

$$\sigma_-(1, 2) = (1/2^{1/2})\{\alpha(1)\beta(2) - \alpha(2)\beta(1)\}$$

which corresponds to paired electron spins (Section 13.7). Therefore, we conclude that the state of lower energy (and hence the formation of a chemical bond) is achieved if the electron spins are paired.

14.2 Homonuclear diatomic molecules

The essential features of valence-bond theory are the pairing of the electrons and the accumulation of electron density in the internuclear region that stems from that pairing. The same description can be applied to more complex molecules, such as homonuclear diatomic molecules. Nitrogen, N_2, is an example. To construct the valence-bond description of N_2, we consider the valence electron configuration of each atom, which is $2s^2 2p_x^1 2p_y^1 2p_z^1$. It is conventional to take the z-axis to be the internuclear axis, so we can imagine each atom as having a $2p_z$ orbital pointing towards a $2p_z$ orbital on the other atom (Fig. 14.4), with the $2p_x$ and $2p_y$ orbitals perpendicular to the axis. A σ bond is then formed by spin pairing between the two electrons in the two $2p_z$ orbitals. Its spatial wavefunction is given by eqn 2, but now A and B stand for the two $2p_z$ orbitals.

The remaining $2p$ orbitals cannot merge to give σ bonds as they do not have cylindrical symmetry around the internuclear axis. Instead, they merge to form two π bonds. A **π bond** arises from the spin pairing of electrons in two p orbitals that approach side-by-side (Fig. 14.5). It is so called because, viewed along the internuclear axis, a π bond resembles a pair of electrons in a p orbital (and π is the Greek equivalent of p). More precisely, an electron in a π bond has one unit of orbital angular momentum about the internuclear axis, for the wavefunction has one angular node.

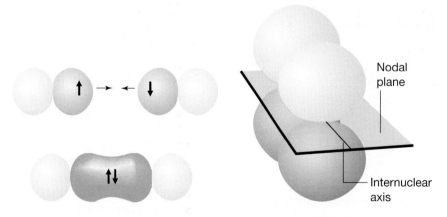

14.4 The orbital overlap and spin pairing between electrons in two collinear p orbitals that results in the formation of a σ bond.

14.5 A π bond results from orbital overlap and spin pairing between electrons in p orbitals with their axes perpendicular to the internuclear axis. The bond has two lobes of electron density separated by a nodal plane.

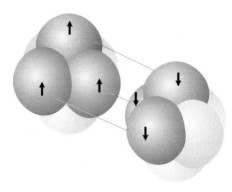

14.6 The structure of bonds in a nitrogen molecule: there is one σ bond and two π bonds. As explained later, the overall electron density has cylindrical symmetry around the internuclear axis.

There are two π bonds in N_2, one formed by spin pairing in two neighbouring $2p_x$ orbitals and the other by spin pairing in two neighbouring $2p_y$ orbitals. The overall bonding pattern in N_2 is therefore a σ bond plus two π bonds (Fig. 14.6), which is consistent with the Lewis structure :N≡N: for nitrogen.

14.3 Polyatomic molecules

Each σ bond in a polyatomic molecule is formed by the spin pairing of electrons in atomic orbitals with cylindrical symmetry about the relevant internuclear axis. Likewise, π bonds are formed by pairing electrons that occupy atomic orbitals of the appropriate symmetry.

The valence-bond description of H_2O will make this clear. The valence electron configuration of an O atom is $2s^2 2p_x^2 2p_y^1 2p_z^1$. The two unpaired electrons in the O2p orbitals can each pair with an electron in an H1s orbital, and each combination results in the formation of a σ bond (each bond has cylindrical symmetry about the respective O—H internuclear axis). Because the $2p_y$ and $2p_z$ orbitals lie at 90° to each other, the two σ bonds also lie at 90° to each other (Fig. 14.7). We can predict, therefore, that H_2O should be an angular molecule, which it is. However, the theory predicts a bond angle of 90°, whereas the actual bond angle is 104.5°.

Discrepancies between theory and experiment also appear when we consider larger molecules, such as NH_3. The valence electron configuration of an N atom is $2s^2 2p_x^1 2p_y^1 2p_z^1$, which suggests that three H atoms can form bonds by spin pairing with the electrons in the three half-filled 2p orbitals. The latter are perpendicular to each other, so we predict a trigonal pyramidal molecule with a bond angle of 90°. Experimental studies confirm the predicted shape, but the actual bond angle is 107°.

> **Self-test 14.1** Use valence-bond theory to suggest a shape for the hydrogen peroxide molecule, H_2O_2. [Each H—O—O bond 90°; experimental: 94.8°]

(a) Promotion

An apparent deficiency of valence-bond theory is its inability to account for carbon's tetravalence (its ability to form four bonds). The ground-state configuration of C is $2s^2 2p_x^1 2p_y^1$, which suggests that a carbon atom should be capable of forming only two bonds, not four. This deficiency is overcome by allowing for **promotion**, the excitation of an electron to an orbital of higher energy. Although electron promotion requires an investment of energy, that investment is worthwhile if the energy can be more than recovered from the greater strength or number of bonds that it allows to be formed. Promotion is not a 'real' process in which an atom somehow becomes excited and then forms bonds: it is a notional contribution to the overall energy change that occurs when bonds form.

In carbon, for example, the promotion of a 2s electron to a 2p orbital can be thought of as leading to the configuration $2s^1 2p_x^1 2p_y^1 2p_z^1$, with four unpaired electrons in separate orbitals. These electrons may pair with four electrons in orbitals provided by four other atoms (such as four H1s orbitals if the molecule is CH_4), and hence form four σ bonds. Although energy was required to promote the electron, it is more than recovered by the promoted atom's ability to form four bonds in place of the two bonds of the unpromoted atom. Promotion, and the formation of four bonds, is a characteristic feature of carbon because the promotion energy is quite small: the promoted electron leaves a doubly occupied 2s orbital and enters a vacant 2p orbital, hence significantly relieving the electron–electron repulsion it experiences in the former.

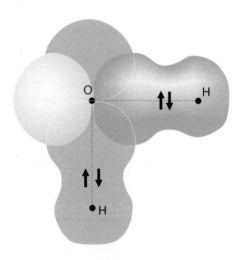

14.7 A first approximation to the valence-bond description of bonding in an H_2O molecule. Each σ bond arises from the overlap of an H1s orbital with one of the O2p orbitals. This model suggests that the bond angle should be 90°, which is significantly different from the experimental value.

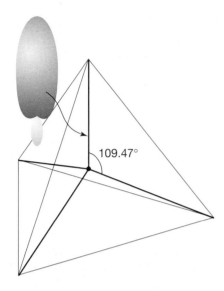

14.8 An sp^3 hybrid orbital formed from the superposition of s and p orbitals on the same atom. There are four such hybrids: each one points towards the corner of a regular tetrahedron. The overall electron density remains spherically symmetrical.

(b) Hybridization

The description of the bonding in CH_4 (and other alkanes) is still incomplete because it appears to imply the presence of three σ bonds of one type (formed from H1s and C2p orbitals) and a fourth σ bond of a distinctly different character (formed from H1s and C2s). This problem is overcome by realizing that the electron density distribution in the promoted atom is equivalent to the electron density in which each electron occupies a **hybrid orbital** formed by interference between the C2s and C2p orbitals. The origin of the hybridization can be appreciated by thinking of the four atomic orbitals, which are waves centred on a nucleus, as being like ripples spreading from a single point on the surface of a lake:[5] the waves interfere destructively and constructively in different regions, and give rise to four new shapes.

The specific linear combinations that give rise to four equivalent hybrid orbitals are

$$h_1 = s + p_x + p_y + p_z \qquad h_2 = s - p_x - p_y + p_z$$
$$h_3 = s - p_x + p_y - p_z \qquad h_4 = s + p_x - p_y - p_z \tag{14.3}$$

As a result of the interference between the component orbitals, each hybrid orbital consists of a large lobe pointing in the direction of one corner of a regular tetrahedron (Fig. 14.8). The angle between the axes of the hybrid orbitals is the tetrahedral angle, $\arccos\left(-\frac{1}{3}\right) = 109.47°$. Because each hybrid is built from one s orbital and three p orbitals, it is called an sp^3 **hybrid orbital**.

It is now easy to see how the valence-bond description of the CH_4 molecule leads to a tetrahedral molecule containing four equivalent C—H bonds. Each hybrid orbital of the promoted C atom contains a single unpaired electron; an H1s electron can pair with each one, giving rise to a σ bond pointing in a tetrahedral direction. For example, the (un-normalized) wavefunction for the bond formed by the hybrid orbital h_1 and the 1s_A orbital (with wavefunction that we shall denote A) is

$$\psi = h_1(1)A(2) + h_1(2)A(1)$$

Because each sp^3 hybrid orbital has the same composition, all four σ bonds are identical apart from their orientation in space (Fig. 14.9).

A further feature of hybridization is that a hybrid orbital has pronounced directional character, in the sense that it has an enhanced amplitude in the internuclear region. This directional character arises from the constructive interference between the s orbital and the positive lobes of the p orbitals (Fig. 14.10). As a result of the enhanced amplitude in the internuclear region, the bond strength is greater than for a bond formed from an s or p orbital alone. This increased bond strength is another factor that helps to repay the promotion energy.

Hybridization can also be used to describe the structure of an ethene molecule, $H_2C\text{=}CH_2$, and the torsional rigidity of double bonds. An ethene molecule is planar, with HCH and HCC bond angles close to 120°. To reproduce the σ bonding structure, we promote each C atom to a $2s^1 2p^3$ configuration. However, instead of using all four orbitals to form hybrids, we form sp^2 **hybrid orbitals** by the superposition of an s orbital and two p orbitals. As shown in Fig. 14.11, the three hybrid orbitals

$$h_1 = s + 2^{1/2}p_y$$
$$h_2 = s + \left(\tfrac{3}{2}\right)^{1/2}p_x - \left(\tfrac{1}{2}\right)^{1/2}p_y \tag{14.4}$$
$$h_3 = s - \left(\tfrac{3}{2}\right)^{1/2}p_x - \left(\tfrac{1}{2}\right)^{1/2}p_y$$

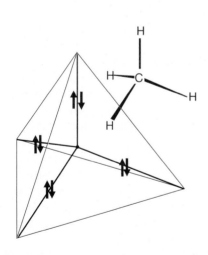

14.9 Each sp^3 hybrid orbital forms a σ bond by overlap with an H1s orbital located at the corner of the tetrahedron. This model accounts for the equivalence of the four bonds in CH_4.

5. It is admittedly difficult to imagine how a ripple resembling a p orbital could be contrived; but the general idea should be clear.

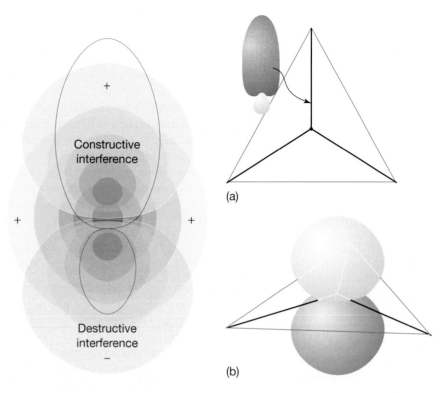

14.10 A more detailed representation of the formation of an sp^3 hybrid by interference between wavefunctions centred on the same atomic nucleus. (To simplify the representation, we have ignored the radial node of the $2s$ orbital.)

14.11 (a) An s orbital and two p orbitals can be hybridized to form three equivalent orbitals that point towards the corners of an equilateral triangle. (b) The remaining, unhybridized p orbital is perpendicular to the plane.

lie in a plane and point towards the corners of an equilateral triangle. The third $2p$ orbital ($2p_z$) is not included in the hybridization; its axis is perpendicular to the plane in which the hybrids lie. As always in superpositions, the proportion of each orbital in the mixture is given by the *square* of the corresponding coefficient. Thus, in the first of these hybrids the ratio of s to p contributions is 1:2. Similarly, the total p contribution in each of h_2 and h_3 is $\frac{3}{2} + \frac{1}{2} = 2$, so the ratio for these orbitals is also 1:2. The different signs of the coefficients ensure that constructive interference takes place in different regions of space, so giving the patterns in the illustration.

We can now describe the structure of $CH_2{=}CH_2$ as follows. The sp^2-hybridized C atoms each form three σ bonds by spin pairing with either the h_1 hybrid of the other C atom or with H1s orbitals. The σ framework therefore consists of C—H and C—C σ bonds at 120° to each other. When the two CH_2 groups lie in the same plane, the two electrons in the unhybridized p orbitals can pair and form a π bond (Fig. 14.12). The formation of this π bond locks the framework into the planar arrangement, for any rotation of one CH_2 group relative to the other leads to a weakening of the π bond (and consequently an increase in energy of the molecule).

A similar description applies to ethyne, HC≡CH, a linear molecule. Now the C atoms are **sp hybridized**, and the σ bonds are formed using hybrid atomic orbitals of the form

$$h_1 = s + p_z \qquad h_2 = s - p_z \tag{14.5}$$

14.12 A representation of the structure of a double bond in ethene; only the π bond is shown explicitly.

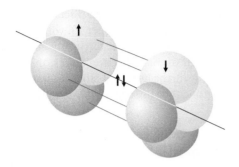

14.13 A representation of the structure of a triple bond in ethyne; only the π bonds are shown explicitly. The overall electron density has cylindrical symmetry around the axis of the molecule.

Table 14.1* Some hybridization schemes

Coordination number	Arrangement	Composition
2	Linear	sp, pd, sd
	Angular	sd
3	Trigonal planar	sp^2, p^2d
	Unsymmetrical planar	spd
	Trigonal pyramidal	pd^2
4	Tetrahedral	sp^3, sd^3
	Irregular tetrahedral	spd^2, p^3d, pd^3
	Square planar	p^2d^2, sp^2d
5	Trigonal bipyramidal	sp^3d, spd^3
	Tetragonal pyramidal	sp^2d^2, sd^4, pd^4, p^3d^2
	Pentagonal planar	p^2d^3
6	Octahedral	sp^3d^2
	Trigonal prismatic	spd^4, pd^5
	Trigonal antiprismatic	p^3d^3

* Source: H. Eyring, J. Walter, and G.E. Kimball, *Quantum chemistry*, Wiley (1944).

These two orbitals lie along the internuclear axis. The electrons in them pair either with an electron in the corresponding hybrid orbital on the other C atom or with an electron in one of the H1s orbitals. Electrons in the two remaining p orbitals on each atom, which are perpendicular to the molecular axis, pair to form two perpendicular π bonds (Fig. 14.13).

> **Self-test 14.2** Hybrid orbitals do not always form bonds. They may also contain lone pairs of electrons. Use valence-bond theory to suggest possible shapes for the hydrogen peroxide molecule, H_2O_2.
>
> [Each H—O—O bond angle is predicted to be approximately 109° (experimental: 94.8°); rotation around the O—O bond is possible, so the molecule interconverts between planar and nonplanar geometries at high temperatures.]

Other hybridization schemes, particularly those involving d orbitals, are often invoked in elementary work to be consistent with other molecular geometries (Table 14.1). The hybridization of N atomic orbitals always results in the formation of N hybrid orbitals, which may either form bonds or may contain lone pairs of electrons. For example, sp^3d^2 hybridization results in six equivalent hybrid orbitals pointing towards the corners of a regular octahedron and is sometimes invoked to account for the structure of octahedral molecules, such as SF_6.

Molecular orbital theory

In **molecular orbital theory** (MO theory), it is accepted that electrons should not be regarded as belonging to particular bonds but should be treated as spreading throughout the entire molecule. This theory has been more fully developed than VB theory and provides the language that is widely used in modern discussions of bonding. To introduce it, we follow the same strategy as in Chapter 13, where the one-electron H atom was taken as the fundamental species for discussing atomic structure and then developed into a description of many-electron atoms. In this chapter we use the simplest molecular species of all, the hydrogen molecule-ion, H_2^+, to introduce the essential features of bonding, and then use it as a guide to the structures of more complex systems.

14.4 The hydrogen molecule-ion

The hamiltonian for the single electron in H_2^+ is

$$H = -\frac{\hbar^2}{2m_e}\nabla_1^2 + V \qquad V = -\frac{e^2}{4\pi\varepsilon_0}\left(\frac{1}{r_{A1}} + \frac{1}{r_{B1}} - \frac{1}{R}\right) \tag{14.6}$$

where r_{A1} and r_{B1} are the distances of the electron from the two nuclei (**1**) and R is the distance between the two nuclei. In the expression for V, the first two terms in parentheses are the attractive contribution from the interaction between the electron and the nuclei; the remaining term is the repulsive interaction between the nuclei.

The one-electron wavefunctions obtained by solving the Schrödinger equation $H\psi = E\psi$ are called **molecular orbitals** (MO). A molecular orbital ψ gives, through the value of $|\psi|^2$, the distribution of the electron in the molecule. A molecular orbital is like an atomic orbital, but spreads throughout the molecule.

The Schrödinger equation can be solved for H_2^+ (within the Born–Oppenheimer approximation), but the wavefunctions are very complicated functions; moreover, the solution cannot be extended to polyatomic systems. Therefore, we adopt a simpler procedure that, while more approximate, can be extended readily to other molecules.

(a) Linear combinations of atomic orbitals

If an electron can be found in an atomic orbital belonging to atom A and also in an atomic orbital belonging to atom B, then the overall wavefunction is a superposition of the two atomic orbitals:

$$\psi_\pm = N(A \pm B) \tag{14.7}$$

where, for H_2^+, A denotes ψ_{H1s_A}, B denotes ψ_{H1s_B}, and N is a normalization factor. The technical term for the superposition in eqn 7 is a **linear combination of atomic orbitals** (LCAO). An approximate molecular orbital formed from a linear combination of atomic orbitals is called an **LCAO-MO**. A molecular orbital that has cylindrical symmetry around the internuclear axis, such as the one we are discussing, is called a **σ orbital** because it resembles an s orbital when viewed along the axis and, more precisely, because it has zero orbital angular momentum around the internuclear axis.

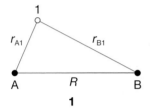

1

Example 14.1 Normalizing a molecular orbital

Normalize the molecular orbital ψ_+ in eqn 7.

Method We need to find the factor N such that

$$\int \psi^* \psi \, d\tau = 1$$

(a)

(b)

14.14 (a) The amplitude of the bonding molecular orbital in a hydrogen molecule-ion in a plane containing the two nuclei and (b) a contour representation of the amplitude.

2

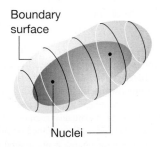

14.15 A general indication of the shape of the boundary surface of a σ orbital.

To proceed, substitute the LCAO into this integral, and make use of the fact that the atomic orbitals are individually normalized.

Answer When we substitute the wavefunction, we find

$$\int \psi^* \psi \, d\tau = N^2 \left\{ \int A^2 \, d\tau + \int B^2 \, d\tau + 2 \int AB \, d\tau \right\} = N^2(1 + 1 + 2S)$$

where $S = \int AB \, d\tau$. For the integral to be equal to 1, we require

$$N = \frac{1}{\{2(1 + S)\}^{1/2}}$$

Comment In H_2^+, $S \approx 0.59$, so $N = 0.56$.

Self-test 14.3 Normalize the orbital ψ_- in eqn 7.

$$[N = 1/\{2(1 - S)\}^{1/2}, \text{ so } N = 1.10]$$

Figure 14.14 shows the contours of constant amplitude for the two molecular orbitals in eqn 7, and Fig. 14.15 shows their boundary surfaces. Plots like these are readily obtained using commercially available software. The calculation is quite straightforward, because all we need do is feed in the mathematical forms of the two atomic orbitals, and let the program do the rest. In this case, we use

$$A = \frac{e^{-r_A/a_0}}{(\pi a_0^3)^{1/2}} \qquad B = \frac{e^{-r_B/a_0}}{(\pi a_0^3)^{1/2}} \tag{14.8}$$

and note that r_A and r_B are not independent (**2**):

$$r_B = \{r_A^2 + R^2 - 2r_A R \cos \theta\}^{1/2} \tag{14.9}$$

To make this plot, we have taken $N^2 = 0.31$ (Example 14.1).

(b) Bonding orbitals

According to the Born interpretation, the probability density of the electron in H_2^+ is proportional to the square modulus of its wavefunction. The probability density corresponding to the (real) wavefunction ψ_+ in eqn 7 is

$$\psi_+^2 = N^2(A^2 + B^2 + 2AB) \tag{14.10}$$

This probability density is plotted in Fig. 14.16.

An important feature of the probability density becomes apparent when we examine the internuclear region, where both atomic orbitals have similar amplitudes. According to eqn 10, the total probability density is proportional to the sum of

1 A^2, the probability density if the electron were confined to the atomic orbital A.

2 B^2, the probability density if the electron were confined to the atomic orbital B.

3 $2AB$, an extra contribution to the density.

This last contribution, the **overlap density**, is crucial, because it represents an enhancement of the probability of finding the electron in the internuclear region. The enhancement can be traced to the constructive interference of the two atomic orbitals: each has a positive amplitude in the internuclear region, so the total amplitude is greater there than if the electron were confined to a single atomic orbital.

We shall frequently use the result that *electrons accumulate in regions where atomic orbitals overlap and interfere constructively*. The accumulation of electron density between

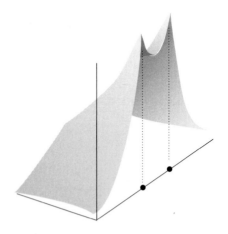

14.16 The electron density calculated by forming the square of the wavefunction used to construct Fig. 14.14. Note the accumulation of electron density in the internuclear region.

the nuclei puts the electron in a position where it interacts strongly with both nuclei. Hence, the energy of the molecule is lower than that of the separate atoms, where each electron can interact strongly with only one nucleus.[6]

The σ orbital we have described is an example of a **bonding orbital**, an orbital which, if occupied, helps to bind two atoms together. Specifically, we label it 1σ as it is the σ orbital of lowest energy. An electron that occupies a σ orbital is called a **σ electron**, and if that is the only electron present in the molecule (as in the ground state of H_2^+), then we report the configuration of the molecule as $1\sigma^1$.

The energy of the 1σ orbital decreases as R decreases from large values because electron density accumulates in the internuclear region as the constructive interference between the atomic orbitals increases (Fig. 14.17). However, at small separations there is too little space between the nuclei for significant accumulation of electron density there. In addition, the nucleus–nucleus repulsion (which is proportional to $1/R$) becomes large. As a result, the energy of the molecule rises at short distances, and there is a minimum in the potential energy curve. Calculations on H_2^+ give $R_e = 130$ pm and $D_e = 1.77$ eV (171 kJ mol^{-1}); the experimental values are 106 pm and 2.6 eV, so this simple LCAO-MO description of the molecule, while inaccurate, is not absurdly wrong.

Justification 14.2

To evaluate the energy of the bonding orbital, we calculate the expectation value of the hamiltonian. The expectation value of the hamiltonian in eqn 6 is

$$E_\pm = E_{H1s} + \frac{e^2}{4\pi\varepsilon_0 R} - \frac{j \pm k}{1 \pm S} \tag{14.11}$$

(We have included the energy for both linear combinations in eqn 7 in this expression; for the bonding orbital, use the upper sign throughout.) The integrals that appear in this expression are

$$S = \int AB\,d\tau = \left\{ 1 + \frac{R}{a_0} + \frac{1}{3}\left(\frac{R}{a_0}\right)^2 \right\} e^{-R/a_0}$$

$$j = \frac{e^2}{4\pi\varepsilon_0} \int \frac{A^2}{r_B}\,d\tau = \frac{e^2}{4\pi\varepsilon_0 R} \left\{ 1 - \left(1 + \frac{R}{a_0}\right) e^{-2R/a_0} \right\} \tag{14.12}$$

$$k = \frac{e^2}{4\pi\varepsilon_0} \int \frac{AB}{r_B}\,d\tau = \frac{e^2}{4\pi\varepsilon_0 a_0} \left(1 + \frac{R}{a_0}\right) e^{-R/a_0}$$

All three integrals are positive and decline exponentially towards zero at large internuclear separations. The integral j is a measure of the interaction between a

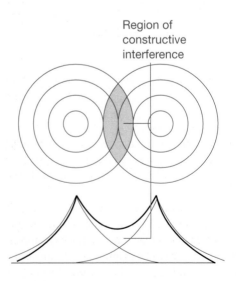

Region of constructive interference

14.17 A representation of the constructive interference that occurs when two H1s orbitals overlap and form a bonding σ orbital.

6. Unfortunately, this conventional explanation is probably incorrect in the case of H_2^+ (at least), because shifting an electron away from a nucleus into the internuclear region raises its potential energy. The modern explanation is more subtle, still controversial, and does not emerge from the simple LCAO treatment given here. It seems that, at the same time as the electron shifts into the internuclear region, the atomic orbitals shrink. This orbital shrinkage improves the electron–nucleus attraction more than it is decreased by the migration to the internuclear region, so there is a net lowering of potential energy. The kinetic energy of the electron is also modified because the curvature of the wavefunction is changed, but the change in kinetic energy is dominated by the change in potential energy. Throughout the following discussion we ascribe the strength of chemical bonds to the accumulation of electron density in the internuclear region. We leave open the question whether in molecules more complicated than H_2^+ the true source of energy lowering is that accumulation itself or some indirect but related effect.

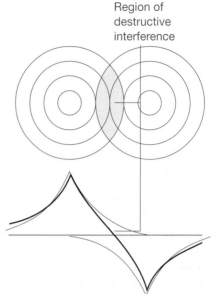

14.18 The calculated and experimental molecular potential energy curves for a hydrogen molecule-ion.

14.19 A representation of the destructive interference that occurs when two H1s orbitals overlap and form an antibonding σ^* orbital.

nucleus and electron density centred on the other nucleus; k is a measure of the interaction between a nucleus and the excess probability in the internuclear region arising from overlap. It is easy to use a mathematical software package to plot the energy as a function of R and hence to obtain Fig. 14.18.

(c) Antibonding orbitals

The linear combination ψ_- in eqn 7 corresponds to a higher energy than ψ_+. Because it is also a σ orbital we label it 2σ. This orbital has an internuclear nodal plane where A and B cancel exactly (Figs 14.19 and 14.20). The probability density is

$$\psi_-^2 = N^2(A^2 + B^2 - 2AB) \tag{14.13}$$

There is a reduction in probability density between the nuclei due to the $-2AB$ term (Fig. 14.21); in physical terms, there is destructive interference where the two atomic orbitals overlap.

The 2σ orbital is an example of an **antibonding orbital**, an orbital that, if occupied, contributes to a reduction in the cohesion between two atoms and helps to raise the energy of the molecule relative to the separated atoms. Antibonding orbitals are often labelled with an asterisk (*), so this particular orbital could also be denoted $2\sigma^*$ (and read '2 sigma star').

The destabilizing effect of an antibonding electron is partly due to the fact that it is excluded from the internuclear region, and hence is distributed largely outside the bonding region. In effect, whereas a bonding electron pulls two nuclei together, an antibonding electron pulls the nuclei apart (Fig. 14.22). Figure 14.18 also shows another feature that we draw on later: $|E_- - E_{H1s}| > |E_+ - E_{H1s}|$, which indicates that *the antibonding orbital is more antibonding than the bonding orbital is bonding*. This important conclusion stems in part from the presence of the nucleus–nucleus repulsion ($e^2/4\pi\varepsilon_0 R$): this contribution raises the energy of both molecular orbitals.

(a)

(b)

14.20 (a) The amplitude of the antibonding molecular orbital in a hydrogen molecule-ion in a plane containing the two nuclei and (b) a contour representation of the amplitude. Note the internuclear node.

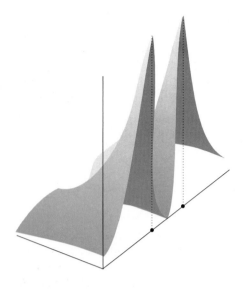

14.21 The electron density calculated by forming the square of the wavefunction used to construct Fig. 14.20. Note the elimination of electron density from the internuclear region.

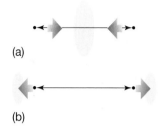

14.22 A partial explanation of the origin of bonding and antibonding effects. (a) In a bonding orbital, the nuclei are attracted to the accumulation of electron density in the internuclear region. (b) In an antibonding orbital, the nuclei are attracted to an accumulation of electron density outside the internuclear region.

14.23 A molecular orbital energy level diagram for orbitals constructed from the overlap of H1s orbitals; the separation of the levels corresponds to that found at the equilibrium bond length. The ground electronic configuration of H_2 is obtained by accommodating the two electrons in the lowest available orbital (the bonding orbital).

14.5 The structures of diatomic molecules

In Chapter 13 we used the hydrogenic atomic orbitals and the building-up principle to deduce the ground electronic configurations of many-electron atoms. We now do the same for many-electron diatomic molecules by using the H_2^+ molecular orbitals. The general procedure is to construct molecular orbitals by combining the available atomic orbitals. The electrons supplied by the atoms are then accommodated in the orbitals so as to achieve the lowest overall energy subject to the constraint of the Pauli exclusion principle, that no more than two electrons may occupy a single orbital (and then must be paired). As in the case of atoms, if several degenerate molecular orbitals are available, we add the electrons singly to each individual orbital before doubly occupying any one orbital (because that minimizes electron–electron repulsions). We also take note of Hund's maximum multiplicity rule (Section 13.4d) that, if electrons do occupy different degenerate orbitals, then a lower energy is obtained if they do so with parallel spins.

(a) The hydrogen and helium molecules

Consider H_2, the simplest many-electron diatomic molecule. Each H atom contributes a 1s orbital (as in H_2^+), so we can form the 1σ and $2\sigma^*$ orbitals from them, as we have seen already. At the experimental internuclear separation these orbitals will have the energies shown in Fig. 14.23, which is called a **molecular orbital energy level diagram**. Note that from two atomic orbitals we can build two molecular orbitals. In general, from N atomic orbitals we can build N molecular orbitals.

There are two electrons to accommodate, and both can enter 1σ by pairing their spins. The ground state configuration is therefore $1\sigma^2$ and the atoms are joined by a bond consisting of an electron pair in a bonding σ orbital. This approach shows that an electron pair, which was the focus of Lewis's account of chemical bonding, represents the maximum number of electrons that can enter a bonding molecular orbital.

The same argument shows why He does not form diatomic molecules.[7] Each He atom contributes a 1s orbital, so 1σ and $2\sigma^*$ molecular orbitals can be constructed.

---

7. Diatomic helium 'molecules' have been prepared: they consist of pairs of atoms held together by weak van der Waals forces of the type described in Chapter 21.

14.24 The ground electronic configuration of the hypothetical four-electron molecule He_2 has two bonding electrons and two antibonding electrons. It has a higher energy than the separated atoms, and so is unstable.

Table 14.2* Bond lengths

Bond	Order	R_e/pm
HH	1	74.14
NN	3	109.76
HCl	1	127.45
CH	1	*114*
CC	1	*154*
CC	2	*134*
CC	3	*120*

* More values will be found in the *Data section*. Numbers in italics are mean values for polyatomic molecules.

Table 14.3* Bond dissociation energies

Bond	Order	D_0/(kJ mol^{-1})
HH	1	432.1
NN	3	941.7
HCl	1	427.7
CH	1	*435*
CC	1	*368*
CC	2	*720*
CC	3	*962*

* More values will be found in the *Data section*. Numbers in italics are mean values for polyatomic molecules.

Although these orbitals differ in detail from those in H_2, the general shape is the same, and we can use the same qualitative energy level diagram in the discussion. There are four electrons to accommodate. Two can enter the 1σ orbital, but then it is full, and the next two must enter the $2\sigma^*$ orbital (Fig. 14.24). The ground electronic configuration of He_2 is therefore $1\sigma^2 2\sigma^{*2}$. We see that there is one bond and one antibond. Because an antibond is slightly more antibonding than a bond is bonding, an He_2 molecule has a higher energy than the separated atoms, so it is unstable relative to the individual atoms.

(b) Bond order

A measure of the net bonding in a diatomic molecule is its **bond order**, b:

$$b = \tfrac{1}{2}(n - n^*) \tag{14.14}$$

where n is the number of electrons in bonding orbitals and n^* is the number of electrons in antibonding orbitals. Thus each electron pair in a bonding orbital increases the bond order by 1 and each pair in an antibonding orbital decreases b by 1. For H_2, $b = 1$, corresponding to a single bond, H—H, between the two atoms. In He_2, $b = 0$, and there is no bond.

As we shall see, the bond order is a useful parameter for discussing the characteristics of bonds, because it correlates with bond length and bond strength. For bonds between atoms of a given pair of elements:

The greater the bond order, the shorter the bond.

The greater the bond order, the greater the bond strength.

Table 14.2 lists some typical bond lengths in diatomic and polyatomic molecules. The strength of a bond is measured by its bond dissociation energy, D_0, the energy required to separate the atoms to infinity.[8] Table 14.3 lists some experimental values of dissociation energies.

(c) Period 2 diatomic molecules

We now see how the concepts we have introduced apply to homonuclear diatomic molecules in general. In elementary treatments, only the orbitals of the valence shell are used to form molecular orbitals.

In Period 2, the valence orbitals are $2s$ and $2p$. A general principle of molecular orbital theory is that *all orbitals of the appropriate symmetry contribute to a molecular orbital*. Thus, to build σ orbitals, we form linear combinations of all atomic orbitals that have cylindrical symmetry about the internuclear axis. These orbitals include the $2s$ orbitals on each atom and the $2p_z$ orbitals on the two atoms (Fig. 14.25). The general form of the σ orbitals that may be formed is therefore

$$\psi = c_{A2s}\psi_{A2s} + c_{B2s}\psi_{B2s} + c_{A2p_z}\psi_{A2p_z} + c_{B2p_z}\psi_{B2p_z} \tag{14.15}$$

From these four atomic orbitals we can form four molecular orbitals of σ symmetry by an appropriate choice of the coefficients c.

---

8. Bond dissociation energies are commonly used in thermodynamic cycles, where bond enthalpies, $\Delta_{bond}H^{\ominus}$, should be used instead. It follows from the same kind of argument used in Section 13.4f concerning ionization enthalpies, that

$$X_2(g) \rightarrow 2\,X(g) \qquad \Delta_{bond}H^{\ominus}(T) = D_0 + \tfrac{3}{2}RT$$

To derive this relation, we have supposed that the molar constant-pressure heat capacity of X_2 is $\tfrac{7}{2}R$ (Chapter 19) for there is a contribution from two rotational modes as well as three translational modes.

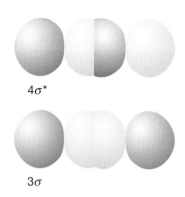

14.25 According to molecular orbital theory, σ orbitals are built from all orbitals that have the appropriate symmetry. In homonuclear diatomic molecules of Period 2, that means that two 2s and two $2p_z$ orbitals should be used. From these four orbitals, four molecular orbitals can be built.

14.26 A representation of the composition of bonding and antibonding σ orbitals built from the overlap of p orbitals. These illustrations are schematic.

The procedure for calculating the coefficients will be described in Section 14.7. At this stage we adopt a simpler route, and suppose that, because the 2s and $2p_z$ orbitals have distinctly different energies, they may be treated separately. That is, the four σ orbitals fall approximately into two sets, one consisting of two molecular orbitals of the form

$$\psi = c_{A2s}\psi_{A2s} + c_{B2s}\psi_{B2s} \qquad (14.16a)$$

and another consisting of two orbitals of the form

$$\psi = c_{A2p_z}\psi_{A2p_z} + c_{B2p_z}\psi_{B2p_z} \qquad (14.16b)$$

Because atoms A and B are identical, the energies of their 2s orbitals are the same, so the coefficients are equal (apart from a possible difference in sign); the same is true of the $2p_z$ orbitals. Therefore, the two sets of orbitals have the form $\psi_{A2s} \pm \psi_{B2s}$ and $\psi_{A2p_z} \pm \psi_{B2p_z}$.

The 2s orbitals on the two atoms overlap to give a bonding and an antibonding σ orbital (1σ and 2σ*, respectively) in exactly the same way as we have already seen for 1s orbitals. The two $2p_z$ orbitals directed along the internuclear axis overlap strongly. They may interfere either constructively or destructively, and give a bonding or antibonding σ orbital, respectively (Fig. 14.26). These two σ orbitals are labelled 3σ and 4σ*, respectively. In general, note how the numbering follows the order of increasing energy.[9]

(d) π orbitals

Now consider the $2p_x$ and $2p_y$ orbitals of each atom. These orbitals are perpendicular to the internuclear axis and may overlap broadside-on. This overlap may be constructive or destructive, and results in a bonding or an antibonding **π orbital** (Fig. 14.27). The notation π is the analogue of p in atoms for, when viewed along the axis of the molecule, a π orbital looks like a p orbital, and has one unit of orbital angular momentum around the internuclear axis. The two $2p_x$ orbitals overlap to give a bonding and antibonding π_x orbital, and the two $2p_y$ orbitals overlap to give

14.27 A schematic representation of the structure of π bonding and antibonding molecular orbitals.

9. In an alternative system of notation, 1σ and 2σ* are used to designate the molecular orbitals formed from the core 1s orbitals of the atoms; the orbitals we are considering would then be labelled from 3 to 6. We are ignoring orbitals formed from core orbitals.

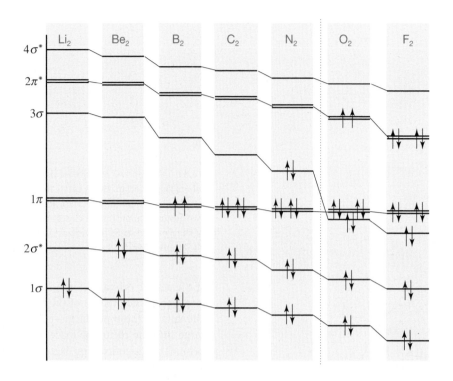

14.29 The variation of the orbital energies of Period 2 homonuclear diatomics.

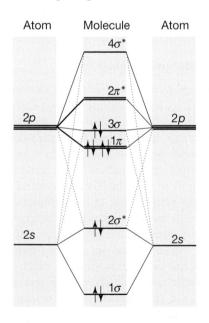

14.28 The molecular orbital energy level diagram for homonuclear diatomic molecules. As remarked in the text, this diagram should be used for O_2 and F_2.

14.30 An alternative molecular orbital energy level diagram for homonuclear diatomic molecules. As remarked in the text, this diagram should be used for diatomics up to and including N_2.

two π_y orbitals. The π_x and π_y bonding orbitals are degenerate; so too are their anti-bonding partners.[10]

In some cases, π orbitals are less strongly bonding than σ orbitals because their maximum overlap occurs off-axis. This relative weakness suggests that the molecular orbital energy level diagram ought to be as shown in Fig. 14.28. However, we must remember that we have constructed the diagram on the assumption that the $2s$ and $2p_z$ orbitals contribute to different sets of molecular orbitals whereas in fact all four atomic orbitals contribute jointly to the four σ orbitals. Hence, there is no guarantee that this order of energies should prevail, and it is found experimentally (by spectroscopy) and by detailed calculation that the order varies along Period 2 (Fig. 14.29). The order shown in Fig. 14.30 is appropriate as far as N_2, and Fig. 14.28 applies for O_2 and F_2. The relative order is controlled by the separation of the $2s$ and $2p$ orbitals in the atoms, which increases across the group. The consequent switch in order occurs at about N_2.

(e) The overlap integral

The extent to which two atomic orbitals on different atoms overlap is measured by the **overlap integral**, S:

$$S = \int \psi_A^* \psi_B \, d\tau \qquad [14.17]$$

10. Strictly, because we are dealing with molecules with cylindrical symmetry, we should consider the complex forms of the p orbitals, one corresponding to circulation about the internuclear axis clockwise and the other counter-clockwise. That is, we form $\pi_\pm \propto \pi_x \pm i\pi_y$, corresponding to angular momenta $\lambda\hbar$ with $\lambda = \pm 1$. Each complex orbital is like a cylindrical torus. Although it is conventional to draw the real forms, it should not be forgotten that each π orbital in a linear molecule corresponds to a cylindrical distribution of charge.

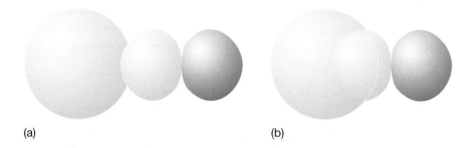

(a)   (b)

14.31 (a) When two orbitals are on atoms that are far apart, the wavefunctions are small where they overlap, so S is small. (b) When the atoms are closer, both orbitals have significant amplitudes where they overlap, and S may approach 1. Note that S will decrease again as the two atoms approach more closely than shown here, because the region of negative amplitude of the p orbital starts to overlap the positive amplitude of the s orbital. When the centres of the atoms coincide, $S = 0$.

14.32 The overlap integral, S, between two H1s orbitals as a function of their separation R.

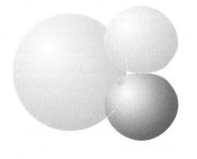

14.33 A p orbital in the orientation shown here has zero net overlap ($S = 0$) with the s orbital at all internuclear separations.

If the atomic orbital ψ_A on A is small wherever the orbital ψ_B on B is large, or vice versa, then the product of their amplitudes is everywhere small and the integral—the sum of these products—is small (Fig. 14.31). If ψ_A and ψ_B are simultaneously large in some region of space, then S may be large. If the two normalized atomic orbitals are identical (for example, 1s orbitals on the same nucleus), then $S = 1$. In some cases, simple formulas can be given for overlap integrals and the variation of S with bond length plotted (Fig. 14.32). It follows that $S = 0.59$ for two H1s orbitals at the equilibrium bond length in H_2^+, which is an unusually large value. Typical values for orbitals with $n = 2$ are in the range 0.2 to 0.3.

Now consider the arrangement in which an s orbital is superimposed on a p_x orbital of a different atom (Fig. 14.33). The integral over the region where the product of orbitals is positive exactly cancels the integral over the region where the product of orbitals is negative, so overall $S = 0$ exactly. Therefore, there is no net overlap between the s and p orbitals in this arrangement.

(f) The structures of homonuclear diatomic molecules

We show the general layout of the valence-shell atomic orbitals of Period 2 atoms on the left and right of the molecular orbital energy level diagrams in Figs 14.28 and 14.30. The lines in the middle are an indication of the energies of the molecular orbitals that can be formed by overlap of atomic orbitals: from the eight valence shell orbitals (four from each atom), we can form eight molecular orbitals. With the orbitals established, we can deduce the ground configurations of the molecules by adding the appropriate number of electrons to the orbitals and following the building-up rules. Anionic species (such as the peroxide ion, O_2^{2-}) need more electrons than the parent neutral molecules; cationic species (such as O_2^+) need fewer.

Consider N_2, which has 10 valence electrons. Two electrons pair, occupy, and fill the 1σ orbital; the next two occupy and fill the 2σ^* orbital. Six electrons remain. There are two 1π orbitals, so four electrons can be accommodated in them. The last two enter the 3σ orbital. The ground-state configuration of N_2 is therefore $1\sigma^2 2\sigma^{*2} 1\pi^4 3\sigma^2$ and the bond order is $\frac{1}{2}(8 - 2) = 3$. This bond order accords with the Lewis structure of the molecule (:N≡N:) and is consistent with its high dissociation energy (942 kJ mol^{-1}).

The ground-state electron configuration of O_2, with 12 valence electrons, is based on Fig. 14.28, and is $1\sigma^2 2\sigma^{*2} 3\sigma^2 1\pi^4 2\pi^{*2}$. Its bond order is 2. According to the building-up principle, however, the two 2π^* electrons occupy different orbitals: one will enter $2\pi_x^*$ and the other will enter $2\pi_y^*$. Because the electrons are in different orbitals, they will have parallel spins. Therefore, we can predict that an O_2 molecule will have a net spin angular momentum $S = 1$ and, in the language introduced in Section 13.7, be in a triplet state. Because electron spin is the source of a magnetic moment, we can go on to predict that oxygen should be paramagnetic.[11] This prediction, which valence-bond theory does not make, is confirmed by experiment.

An F_2 molecule has two more electrons than an O_2 molecule. Its configuration is therefore $1\sigma^2 2\sigma^{*2} 3\sigma^2 1\pi^4 2\pi^{*4}$ and $b = 1$. We conclude that F_2 is a singly bonded molecule, in agreement with its Lewis structure. The low bond order is consistent with its low dissociation energy (154 kJ mol^{-1}). The hypothetical molecule dineon, Ne_2, has two further electrons: its configuration is $1\sigma^2 2\sigma^{*2} 3\sigma^2 1\pi^4 2\pi^{*4} 4\sigma^{*2}$ and $b = 0$. The zero bond order is consistent with the monatomic nature of Ne.

Example 14.2 Judging the relative bond strengths of molecules and ions

Judge whether N_2^+ is likely to have a larger or smaller dissociation energy than N_2.

Method Because the molecule with the larger bond order is likely to have the larger dissociation energy, compare their electronic configurations and assess their bond orders.

Answer From Fig. 14.30, the electron configurations and bond orders are

$$N_2 \qquad 1\sigma^2 2\sigma^{*2} 1\pi^4 3\sigma^2 \qquad b = 3$$
$$N_2^+ \qquad 1\sigma^2 2\sigma^{*2} 1\pi^4 3\sigma^1 \qquad b = 2\tfrac{1}{2}$$

Because the cation has the smaller bond order, we expect it to have the smaller dissociation energy.

Comment The experimental dissociation energies are 942 kJ mol^{-1} for N_2 and 842 kJ mol^{-1} for N_2^+.

Self-test 14.4 Which can be expected to have the higher dissociation energy, F_2 or F_2^+?

$[F_2^+]$

14.6 Heteronuclear diatomic molecules

A **heteronuclear diatomic molecule** is a diatomic molecule formed from atoms of two different elements, such as CO and HCl. The electron distribution in the covalent bond between the atoms is not evenly shared because it is energetically favourable for the electron pair to be found closer to one atom than the other. This imbalance results in a **polar bond**, a covalent bond in which the electron pair is shared unequally by the two atoms. The bond in HF, for instance, is polar, with the electron pair closer to the F atom. The accumulation of the electron pair near the F atom results in that atom having a net negative charge, which is called a **partial negative charge** and denoted $\delta-$. There is a matching **partial positive charge**, $\delta+$, on the H atom.

---

11. A paramagnetic substance tends to move into a magnetic field; a diamagnetic substance tends to move out of one. Paramagnetism, the rarer property, arises when the molecules have unpaired electron spins. Both properties are discussed in more detail in Chapter 23.

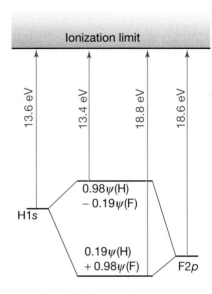

14.34 The atomic orbital energy levels of H and F atoms and the molecular orbitals they form.

(a) Polar bonds

A polar bond consists of two electrons in an orbital of the form

$$\psi = c_A A + c_B B \tag{14.18}$$

with unequal coefficients. The proportion of the atomic orbital A in the bond is $|c_A|^2$ and that of B is $|c_B|^2$. A nonpolar bond has $|c_A|^2 = |c_B|^2$ and a pure ionic bond has one coefficient zero (so the species A^+B^- would have $c_A = 0$ and $c_B = 1$). The atomic orbital with the lower energy makes the larger contribution to the bonding molecular orbital. The opposite is true of the antibonding orbital, for which the dominant component comes from the atomic orbital with higher energy.

These points can be illustrated by considering HF, and judging the energies of the atomic orbitals from the ionization energies of the atoms. The general form of the molecular orbitals is

$$\psi = c_H \psi_H + c_F \psi_F \tag{14.19}$$

where ψ_H is an H1s orbital and ψ_F is an F2p orbital. The H1s orbital lies 13.6 eV below the zero of energy (the separated proton and electron) and the F2p orbital lies at 18.6 eV (Fig. 14.34). Hence, the bonding σ orbital in HF is mainly F2p and the antibonding σ orbital is mainly H1s orbital in character. The two electrons in the bonding orbital are most likely to be found in the F2p orbital, so there is a partial negative charge on the F atom and a partial positive charge on the H atom.

(b) Electronegativity

The charge distribution in bonds is commonly discussed in terms of the **electronegativity**, χ (chi), of the elements involved. The electronegativity is a parameter introduced by Linus Pauling as a measure of the power of an atom to attract electrons to itself when it is part of a compound. Pauling used valence-bond arguments to suggest that an appropriate numerical scale of electronegativities could be defined in terms of bond dissociation energies, D, in electronvolts and proposed that the difference in electronegativities could be expressed as

$$|\chi_A - \chi_B| = 0.102\{D(A{-}B) - \tfrac{1}{2}[D(A{-}A) + D(B{-}B)]\}^{1/2} \tag{14.20}$$

Electronegativities based on this definition are called **Pauling electronegativities**. A list of Pauling electronegativities is given in Table 14.4. The most electronegative elements are those close to fluorine; the least are those close to caesium. It is found that, the greater the difference in electronegativities, the greater the polar character of the bond. The difference for HF, for instance, is 1.78; a C—H bond, which is commonly regarded as almost nonpolar, has an electronegativity difference of 0.51.

The American spectroscopist R.S. Mulliken proposed an alternative definition of electronegativity. He argued that an element is likely to be highly electronegative if it has a high ionization energy (so it will not release electrons readily) and a high electron affinity (so it is energetically favourable to acquire electrons). The **Mulliken electronegativity scale** is therefore based on the definition

$$\chi_M = \tfrac{1}{2}(I + E_{ea}) \tag{14.21}$$

where I is the ionization energy of the element and E_{ea} is its electron affinity (both in electronvolts, Section 13.4f).[12] The Mulliken and Pauling scales are approximately in line with each other. A reasonably reliable conversion between the two is $\chi_P = 1.35\chi_M^{1/2} - 1.37$.

Table 14.4 * Pauling electronegativities

Element	χ_P
H	2.2
C	2.6
N	3.0
O	3.4
F	4.0
Cl	3.2
Cs	0.79

* More values will be found in the *Data section*.

---

12. There are certain technical difficulties with this definition in connection with the electronic state chosen to represent the state of the atom in a compound.

(c) The variation principle

A more systematic way of discussing bond polarity and finding the coefficients in the linear combinations used to build molecular orbitals is provided by the **variation principle**:

> If an arbitrary wavefunction is used to calculate the energy, the value calculated is never less than the true energy.

This principle is the basis of all modern molecular structure calculations. The arbitrary wavefunction is called the **trial wavefunction**. The principle implies that, if we vary the coefficients in the trial wavefunction until the lowest energy is achieved (by evaluating the expectation value of the hamiltonian for each wavefunction), then those coefficients will be the best. We might get a lower energy if we use a more complicated wavefunction (for example, by taking a linear combination of several atomic orbitals on each atom), but we shall have the optimum (minimum energy) molecular orbital that can be built from the chosen **basis set**, the given set of atomic orbitals.

The method can be illustrated by the trial wavefunction in eqn 18. We show in the *Justification* below that the coefficients are given by the solutions of the two **secular equations**[13]

$$(\alpha_A - E)c_A + (\beta - ES)c_B = 0$$
$$(\beta - ES)c_A + (\alpha_B - E)c_B = 0 \tag{14.22}$$

The parameter α is called a **Coulomb integral**. It can be interpreted as the energy of the electron when it occupies A (for α_A) or B (for α_B), and is negative. In a homonuclear diatomic molecule, $\alpha_A = \alpha_B$. The parameter β is called a **resonance integral** (for classical reasons). It vanishes when the orbitals do not overlap, and at equilibrium bond lengths it is normally negative.

Justification 14.3

The trial wavefunction in eqn 18 is real but not normalized because at this stage the coefficients can take arbitrary values. Therefore, we can write $\psi^* = \psi$ but do not assume that $\int \psi^2 \, d\tau = 1$. The energy of the trial wavefunction is the expectation value of the energy operator (the hamiltonian, H, Section 11.5):

$$E = \frac{\displaystyle\int \psi^* H \psi \, d\tau}{\displaystyle\int \psi^* \psi \, d\tau} \tag{14.23}$$

We must search for values of the coefficients in the trial function that minimize the value of E. This is a standard problem in calculus, and is solved by finding the coefficients for which

$$\frac{\partial E}{\partial c_A} = 0 \qquad \frac{\partial E}{\partial c_B} = 0$$

The first step is to express the two integrals in terms of the coefficients. The denominator is

---

13. The name 'secular' is derived from the Latin word for age or generation. The term comes via astronomy, where the same equations appear in connection with slowly accumulating modifications of planetary orbits.

$$\int \psi^2 \, d\tau = \int (c_A A + c_B B)^2 \, d\tau$$

$$= c_A^2 \int A^2 \, d\tau + c_B^2 \int B^2 \, d\tau + 2c_A c_B \int AB \, d\tau = c_A^2 + c_B^2 + 2c_A c_B S$$

because the individual atomic orbitals are normalized and the third integral is the overlap integral S (eqn 17). The numerator is

$$\int \psi H \psi \, d\tau = \int (c_A A + c_B B) H (c_A A + c_B B) \, d\tau$$

$$= c_A^2 \int AHA \, d\tau + c_B^2 \int BHB \, d\tau + c_A c_B \int AHB \, d\tau + c_A c_B \int BHA \, d\tau$$

There are some complicated integrals in this expression, but we can combine them all into the parameters

$$\alpha_A = \int AHA \, d\tau \qquad \alpha_B = \int BHB \, d\tau \qquad \beta = \int AHB \, d\tau = \int BHA \, d\tau \qquad [14.24]$$

Then

$$\int \psi H \psi \, d\tau = c_A^2 \alpha_A + c_B^2 \alpha_B + 2c_A c_B \beta$$

The complete expression for E is

$$E = \frac{c_A^2 \alpha_A + c_B^2 \alpha_B + 2c_A c_B \beta}{c_A^2 + c_B^2 + 2c_A c_B S} \qquad (14.25)$$

Its minimum is found by differentiation with respect to the two coefficients and setting the results equal to 0. This involves elementary but slightly tedious work, and the end result is eqn 22.

To solve the secular equations for the coefficients we need to know the energy E of the orbital. As for any set of simultaneous equations, the secular equations have a solution if the **secular determinant**, the determinant of the coefficients, is zero (see *Further information 1*); that is, if

$$\begin{vmatrix} \alpha_A - E & \beta - ES \\ \beta - ES & \alpha_B - E \end{vmatrix} = 0 \qquad (14.26)$$

This determinant expands to a quadratic equation in E (see Illustration 14.1). Its two roots give the energies of the bonding and antibonding molecular orbitals formed from the atomic orbitals and, according to the variation principle, these roots are the best energies for the given basis set.

We need to know that a 2×2 determinant expands as follows:

$$\begin{vmatrix} a & b \\ c & d \end{vmatrix} = ad - bc$$

Illustration 14.1

To find the energies E of the bonding and antibonding orbitals of a homonuclear diatomic molecule set $\alpha_A = \alpha_B = \alpha$ in eqn 26 and get

$$\begin{vmatrix} \alpha - E & \beta - ES \\ \beta - ES & \alpha - E \end{vmatrix} = (\alpha - E)^2 - (\beta - ES)^2 = 0$$

The solutions of this equation are

$$E_\pm = \frac{\alpha \pm \beta}{1 \pm S}$$

Self-test 14.5 Find the coefficients corresponding to these two energies.

[See below; eqn 28]

The values of the coefficients in the linear combination are obtained by solving the secular equations using the two energies obtained from the secular determinant. The lower energy gives the coefficients for the bonding molecular orbital; the upper energy the coefficients for the antibonding molecular orbital. The secular equations give expressions for the ratio of the coefficients in each case, so we need a further equation in order to find their individual values. This equation is obtained by demanding that the best wavefunction should also be normalized. This condition means that, at this final stage, we must also ensure that

$$\int \psi^2 \, d\tau = c_A^2 + c_B^2 + 2c_A c_B S = 1 \tag{14.27}$$

(d) Two simple cases

The complete solutions of the secular equations are very cumbersome, even for 2×2 determinants, but there are two cases where the roots can be written down very simply.

We saw in Illustration 14.1 and its Self-test that, when the two atoms are the same, and we can write $\alpha_A = \alpha_B = \alpha$, the solutions are

$$E_+ = \frac{\alpha + \beta}{1 + S} \qquad c_A = \frac{1}{\{2(1 + S)\}^{1/2}} \qquad c_B = c_A$$

$$\tag{14.28}$$

$$E_- = \frac{\alpha - \beta}{1 - S} \qquad c_A = \frac{1}{\{2(1 - S)\}^{1/2}} \qquad c_B = -c_A$$

In this case, the bonding orbital has the form

$$\psi_+ = \frac{A + B}{\{2(1 + S)\}^{1/2}} \tag{14.29a}$$

and the corresponding antibonding orbital is

$$\psi_- = \frac{A - B}{\{2(1 - S)\}^{1/2}} \tag{14.29b}$$

in agreement with the discussion of homonuclear diatomics we have already given, but now with the normalization constant in place.

The second simple case is for a heteronuclear diatomic molecule but with $S = 0$ (a common approximation in elementary work). The secular determinant is then

$$\begin{vmatrix} \alpha_A - E & \beta \\ \beta & \alpha_B - E \end{vmatrix} = (\alpha_A - E)(\alpha_B - E) - \beta^2 = 0$$

The solutions can be expressed in terms of the parameter ζ (zeta), with

The function arctan x is the same as $\tan^{-1} x$.

$$\zeta = \tfrac{1}{2} \arctan \frac{2|\beta|}{\alpha_B - \alpha_A} \tag{14.30}$$

and are

$$E_- = \alpha_B - \beta \tan \zeta \qquad \psi_- = -A \sin \zeta + B \cos \zeta$$
$$E_+ = \alpha_A + \beta \tan \zeta \qquad \psi_+ = A \cos \zeta + B \sin \zeta$$

$$(14.31)$$

For $x \ll 1$, $\tan x \approx x$ and $\cot x \approx 1/x$. It follows that when $|\alpha_A - \alpha_B| \gg 2|\beta|$ we can write $\zeta \approx |\beta|/(\alpha_B - \alpha_A)$, which implies that $\tan \zeta \approx |\beta|/(\alpha_B - \alpha_A)$, and hence that $\cot \zeta \approx (\alpha_B - \alpha_A)/|\beta|$. Then (noting that $\beta/|\beta| = -1$) the energies of the two molecular orbitals are $E_- \approx \alpha_B$ and $E_+ \approx \alpha_A$. Because $\sin \zeta \approx \zeta$ and $\cos \zeta \approx 1$ when $\zeta \ll 1$, the orbitals are respectively almost pure B and almost pure A.

An important feature revealed by these solutions is that, as the difference $|\alpha_A - \alpha_B|$ increases, the value of ζ decreases.

When the energy difference is large the energies of the molecular orbitals differ only slightly from those of the atomic orbitals, which implies in turn that the bonding and antibonding effects are small. That is, the strongest bonding and antibonding effects are obtained when the two contributing orbitals have closely similar energies. The difference in energy between core and valence orbitals is the justification for neglecting the contribution of core orbitals to bonding. The core orbitals of one atom have a similar energy to the core orbitals of the other atom; but core–core interaction is largely negligible because the overlap between them (and hence the value of β) is so small.

Example 14.3 Calculating the molecular orbitals of HF

Calculate the wavefunctions and energies of the σ orbitals in the HF molecule, taking $\beta = -1.0$ eV and the following ionization energies: H1s: 13.6 eV; F2s: 40.2 eV; F2p: 18.6 eV.

Method Because the F2p and H1s orbitals are much closer in energy than the F2s and H1s orbitals, to a first approximation neglect the contribution of the F2s orbital. To use eqn 31, we need to know the values of the Coulomb integrals α_H and α_F. Because these integrals represent the energies of the H1s and F2p electrons, respectively, they are approximately equal to (the negative of) the ionization energies of the atoms. Calculate ζ from eqn 30 (with A identified as F and B as H), and then write the wavefunctions by using eqn 31.

Answer Setting $\alpha_H = -13.6$ eV and $\alpha_F = -18.6$ eV gives $\tan 2\zeta = 0.40$; so $\zeta = 10.9°$. Then

$$E_- = -13.4 \text{ eV} \qquad \psi_- = 0.98\psi_H - 0.19\psi_F$$
$$E_+ = -18.8 \text{ eV} \qquad \psi_+ = 0.19\psi_H + 0.98\psi_F$$

Comment Notice how the lower energy orbital (the one with energy -18.8 eV) has a composition that is more F2p orbital than H1s, and that the opposite is true of the higher energy, antibonding orbital.

Self-test 14.6 The ionization energy of Cl is 13.1 eV; find the form and energies of the σ orbitals in the HCl molecule using $\beta = -1.0$ eV.
$$[E_- = -12.3 \text{ eV}, \psi_- = -0.62\psi_H + 0.79\psi_{Cl}; E_+ = -14.4 \text{ eV}, \psi_+ = 0.79\psi_H + 0.62\psi_{Cl}]$$

Molecular orbitals for polyatomic systems

The molecular orbitals of polyatomic molecules are built in the same way as in diatomic molecules, the only difference being that we use more atomic orbitals to construct them. As for diatomic molecules, polyatomic molecular orbitals spread over the entire molecule. A molecular orbital has the general form

$$\psi = \sum_i c_i \psi_i \tag{14.32}$$

where ψ_i is an atomic orbital and the sum extends over all the valence orbitals of all the atoms in the molecule. To find the coefficients, we set up the secular equations and the secular determinant, just as for diatomic molecules, solve the latter for the energies, and then use these energies in the secular equations to find the coefficients of the atomic orbitals for each molecular orbital.

The principal difference between diatomic and polyatomic molecules lies in the greater range of shapes that are possible: a diatomic molecule is necessarily linear, but a triatomic molecule, for instance, may be either linear or angular with a characteristic bond angle. The shape of a polyatomic molecule—the specification of its bond lengths and its bond angles—can be predicted by calculating the total energy of the molecule for a variety of nuclear positions, and then identifying the conformation that corresponds to the lowest energy.

14.7 The Hückel approximation

Molecular orbital theory takes large molecules and extended aggregates of atoms, such as solid materials, into its stride. First we shall consider conjugated molecules, in which there is an alternation of single and double bonds along a chain of carbon atoms. Although the classification of an orbital as σ or π is strictly valid only in linear molecules, as will be familiar from introductory chemistry courses, it is also used to denote the local symmetry with respect to a given A—B bond axis.[14]

The π molecular orbital energy level diagrams of conjugated molecules can be constructed using a set of approximations suggested by Erich Hückel in 1931. In his approach, the π orbitals are treated separately from the σ orbitals, and the latter form a rigid framework that determines the general shape of the molecule. All the C atoms are treated identically, so all the Coulomb integrals α for the atomic orbitals that contribute to the π orbitals are set equal. For example, in ethene, we take the σ bonds as fixed, and concentrate on finding the energies of the single π bond and its companion antibond.

(a) Ethene and frontier orbitals

We express the π orbitals as LCAOs of the C2p orbitals that lie perpendicular to the molecular plane. In ethene, for instance, we would write

$$\psi = c_A A + c_B B \tag{14.33}$$

where the A is a C2p orbital on atom A, and so on. Next, the optimum coefficients and energies are found by the variation principle as explained in Section 14.6. That is, we have to solve the secular determinant, which in the case of ethene is eqn 26 with $\alpha_A = \alpha_B = \alpha$:

$$\begin{vmatrix} \alpha - E & \beta - ES \\ \beta - ES & \alpha - E \end{vmatrix} = 0 \tag{14.34}$$

The roots of this determinant can be found very easily (they are the same as those in Illustration 14.1). In a modern computation all the resonance integrals and overlap integrals would be included, but an indication of the molecular orbital energy level diagram can be obtained very readily if we make the following additional **Hückel approximations**:

---

14. In nonlinear molecules, there is no orbital angular momentum around the bond axis: the π orbital is a (real) standing wave with electron density on each side of the local molecular plane.

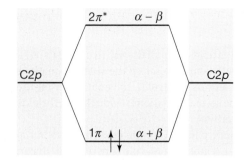

14.35 The Hückel molecular orbital energy levels of ethene. Two electrons occupy the lower π orbital.

1 All overlap integrals are set equal to zero.

2 All resonance integrals between non-neighbours are set equal to zero.

3 All remaining resonance integrals are set equal (to β).

These approximations are obviously very severe, but they let us calculate at least a general picture of the molecular orbital energy levels with very little work. The assumptions result in the following structure of the secular determinant:

1 All diagonal elements: $\alpha - E$.

2 Off-diagonal elements between neighbouring atoms: β.

3 All other elements: 0.

These approximations lead to

$$\begin{vmatrix} \alpha - E & \beta \\ \beta & \alpha - E \end{vmatrix} = (\alpha - E)^2 - \beta^2 = 0 \tag{14.35}$$

The roots of the equation are

$$E_{\pm} = \alpha \pm \beta \tag{14.36}$$

The + sign corresponds to the bonding combination (β is negative) and the – sign corresponds to the antibonding combination (Fig. 14.35).[15] The building-up principle then leads to the configuration $1\pi^2$, because each carbon atom supplies one electron to the π system. We can also estimate the $\pi^* \leftarrow \pi$ excitation energy ($2|\beta|$). The constant β is often left as an adjustable parameter; an approximate value for (C2p, C2p)-overlap π bonds is about -75 kJ mol^{-1}, corresponding to -0.8 eV.

The **highest occupied molecular orbital** in ethene, its HOMO, is the 1π orbital; the **lowest unfilled molecular orbital**, its LUMO, is the $2\pi^*$ orbital. These two orbitals jointly form the **frontier orbitals** of the molecule. The frontier orbitals are important because they are largely responsible for many of the chemical and spectroscopic properties of the molecule.

(b) The matrix formulation of the Hückel method

In preparation for making Hückel theory more sophisticated and readily applicable to bigger molecules, we need to reformulate it in terms of matrices and vectors (see *Further information 1*). We have seen that the secular equations that we have to solve for a two-atom system have the form

$$
\begin{aligned}
(H_{AA} - ES_{AA})c_A + (H_{AB} - ES_{AB})c_B = 0 \\
(H_{BA} - ES_{BA})c_A + (H_{BB} - ES_{BB})c_B = 0
\end{aligned}
\tag{14.37}
$$

(These expressions generalize eqn 22.) If we introduce the following matrices and column vectors

$$
\mathbf{H} = \begin{pmatrix} H_{AA} & H_{AB} \\ H_{BA} & H_{BB} \end{pmatrix} \qquad \mathbf{S} = \begin{pmatrix} S_{AA} & S_{AB} \\ S_{BA} & S_{BB} \end{pmatrix} \qquad \mathbf{c} = \begin{pmatrix} c_A \\ c_B \end{pmatrix}
\tag{14.38}
$$

then the two equations may be written more succinctly as

$$(\mathbf{H} - E\mathbf{S})\mathbf{c} = 0 \quad \text{or} \quad \mathbf{Hc} = \mathbf{ScE} \tag{14.39}$$

In the Hückel approximation, the **overlap matrix**, $\mathbf{S}$, is the unit matrix (with 1 on the diagonal and 0 elsewhere), but we leave the equation in its more general form for the moment. If there are N orbitals in the basis set (here there are two), then

15. To see the effect of neglecting overlap, compare the result obtained here with eqn 28.

there are N eigenvalues (E) and N corresponding column vectors. We label these eigenvalues and vectors with an index i, which means that we have to solve N equations of the form

$$Hc_i = Sc_iE_i \tag{14.40}$$

To proceed, we introduce the matrices

$$C = \begin{pmatrix} c_{1,A} & c_{2,A} \\ c_{1,B} & c_{2,B} \end{pmatrix} \qquad E = \begin{pmatrix} E_1 & 0 \\ 0 & E_2 \end{pmatrix} \tag{14.41}$$

for then the entire set of equations we have to solve can be expressed as

$$HC = SCE \tag{14.42}$$

In the Hückel approximation we neglect overlap and so set $S = 1$. Then

$$HC = CE$$

At this point, we multiply from the left by the inverse matrix C^{-1}, and find

$$C^{-1}HC = E \tag{14.43}$$

In other words, to find the eigenvalues E_i, we have to find a transformation of H that makes it diagonal. The diagonal elements then correspond to the eigenvalues E_i and the columns of the matrix C that brings about this diagonalization are the coefficients of the members of the basis set and hence give us the composition of the molecular orbitals.

3

Example 14.4 Finding the molecular orbitals by matrix diagonalization

Set up and solve the matrix equations within the Hückel approximation for the π-orbitals of butadiene (**3**).

Method The matrices will be four-dimensional for this four-atom system. Ignore overlap, and construct the matrix H by using the Hückel values α and β. Find the matrix C that diagonalizes H: for this step, use mathematical software. Full details are given in *Further information 1*.

Solution

$$H = \begin{pmatrix} H_{11} & H_{12} & H_{13} & H_{14} \\ H_{21} & H_{22} & H_{23} & H_{24} \\ H_{31} & H_{32} & H_{33} & H_{34} \\ H_{41} & H_{42} & H_{43} & H_{44} \end{pmatrix} = \begin{pmatrix} \alpha & \beta & 0 & 0 \\ \beta & \alpha & \beta & 0 \\ 0 & \beta & \alpha & \beta \\ 0 & 0 & \beta & \alpha \end{pmatrix}$$

Mathematical software then diagonalizes this matrix to

$$E = \begin{pmatrix} \alpha + 1.62\beta & 0 & 0 & 0 \\ 0 & \alpha + 0.62\beta & 0 & 0 \\ 0 & 0 & \alpha - 0.62\beta & 0 \\ 0 & 0 & 0 & \alpha - 1.62\beta \end{pmatrix}$$

and the matrix that achieves the diagonalization is

$$C = \begin{pmatrix} 0.372 & 0.602 & 0.602 & -0.372 \\ 0.602 & 0.372 & -0.372 & 0.602 \\ 0.602 & -0.372 & -0.372 & -0.602 \\ 0.372 & -0.602 & 0.602 & 0.372 \end{pmatrix}$$

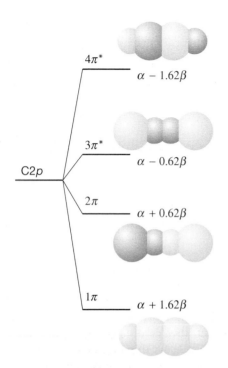

$4\pi^*$ $\alpha - 1.62\beta$

$3\pi^*$ $\alpha - 0.62\beta$

C2p

2π $\alpha + 0.62\beta$

1π $\alpha + 1.62\beta$

14.36 The Hückel molecular orbital energy levels of butadiene and the top view of the corresponding π orbitals. The four p electrons (one supplied by each C) occupy the two lower π orbitals. Note that the orbitals are delocalized.

We can conclude that the energies and molecular orbitals are

$E = \alpha + 1.62\beta$ $\quad \psi = 0.372\psi_1 + 0.602\psi_2 + 0.602\psi_3 + 0.372\psi_4$

$E = \alpha + 0.62\beta$ $\quad \psi = 0.602\psi_1 + 0.372\psi_2 - 0.372\psi_3 + 0.602\psi_4$

$E = \alpha - 0.62\beta$ $\quad \psi = 0.602\psi_1 - 0.372\psi_2 - 0.372\psi_3 + 0.602\psi_4$

$E = \alpha - 1.62\beta$ $\quad \psi = -0.372\psi_1 + 0.602\psi_2 - 0.602\psi_3 + 0.372\psi_4$

Comment Note that the orbitals are mutually orthogonal and, with overlap neglected, normalized.

Self-test 14.7 Repeat the exercise for the allyl radical, $CH_2\!=\!CH\!-\!\dot{C}H_2$.
$[E = \alpha + 2^{1/2}\beta, \alpha, \alpha - 2^{1/2}\beta; \psi = \frac{1}{2}\psi_1 + (\frac{1}{2})^{1/2}\psi_2 + \frac{1}{2}\psi_3,$
$\psi = (\frac{1}{2})^{1/2}\psi_1 - (\frac{1}{2})^{1/2}\psi_3, \psi = \frac{1}{2}\psi_1 - (\frac{1}{2})^{1/2}\psi_2 + \frac{1}{2}\psi_3]$

(c) Butadiene and π-electron binding energy

As we saw in the preceding example, the energies of the four LCAO-MOs for butadiene are

$$E = \alpha \pm 1.62\beta, \qquad \alpha \pm 0.62\beta \qquad (14.44)$$

These orbitals and their energies are drawn in Fig. 14.36. Note that, the greater the number of internuclear nodes, the higher the energy of the orbital. There are four electrons to accommodate, so the ground-state configuration is $1\pi^2 2\pi^2$. The frontier orbitals of butadiene are the 2π orbital (the HOMO, which is largely bonding) and the $3\pi^*$ orbital (the LUMO, which is largely antibonding). 'Largely' bonding means that an orbital has both bonding and antibonding interactions between various neighours, but the bonding effects dominate. 'Largely antibonding' indicates that the antibonding effects dominate.

An important point emerges when we calculate the total **π-electron binding energy**, E_π, the sum of the energies of each π electron, and compare it with what we find in ethene. In ethene the total energy is

$$E_\pi = 2(\alpha + \beta) = 2\alpha + 2\beta$$

In butadiene it is

$$E_\pi = 2(\alpha + 1.62\beta) + 2(\alpha + 0.62\beta) = 4\alpha + 4.48\beta$$

Therefore, the energy of the butadiene molecule lies lower by 0.48β (about -36 kJ mol^{-1}) than the sum of two individual π bonds. This extra stabilization of a conjugated system is called the **delocalization energy**. A closely related quantity is the **π-bond formation energy**, the energy released when a π bond is formed. Because the contribution of α is the same in the molecule as in the atoms, we can find the π-bond formation energy from the π-electron binding energy by writing

$$E_{bf} = E_\pi - N\alpha \qquad (14.45)$$

where N is the number of carbon atoms in the molecule. The π-bond formation energy in butadiene, for instance, is 4.48β.

Example 14.5 Estimating the delocalization energy

Use the Hückel approximation to find the energies of the π orbitals of cyclobutadiene, and estimate the delocalization energy.

Method Set up the secular determinant using the same basis as for butadiene, but note that atoms A and D are also now neighbours. Then solve for the roots of the secular equation and assess the total π-electron binding energy. For the delocalization energy, subtract from the total π-electron binding energy the energy of two π bonds.

Answer The hamiltonian matrix is

$$H = \begin{pmatrix} \alpha & \beta & 0 & \beta \\ \beta & \alpha & \beta & 0 \\ 0 & \beta & \alpha & \beta \\ \beta & 0 & \beta & \alpha \end{pmatrix}$$

Diagonalization gives the energies of the orbitals as

$$E = \alpha + 2\beta, \qquad \alpha, \qquad \alpha, \qquad \alpha - 2\beta$$

Four electrons must be accommodated. Two occupy the lowest orbital (of energy $\alpha + 2\beta$), and two occupy the doubly degenerate orbitals (of energy α). The total energy is therefore $4\alpha + 4\beta$. Two isolated π bonds would have an energy $4\alpha + 4\beta$; therefore, in this case, the delocalization energy is zero.

Self-test 14.8 Repeat the calculation for benzene. [See next subsection]

(d) Benzene and aromatic stability

The most notable example of delocalization conferring extra stability is benzene and the aromatic molecules based on its structure. Benzene is often expressed in a mixture of valence-bond and molecular orbital terms, with typically valence-bond language used for its σ framework and molecular orbital language used to describe its π electrons.

First, the valence-bond component. The six C atoms are regarded as sp^2 hybridized, with a single unhydridized perpendicular $2p$ orbital. One H atom is bonded by (Csp^2, H1s) overlap to each carbon atom, and the remaining hybrids overlap to give a regular hexagon of atoms (Fig. 14.37). The internal angle of a regular hexagon is $120°$, so sp^2 hybridization is ideally suited for forming σ bonds. We see that benzene's hexagonal shape permits strain-free σ bonding.

Now consider the molecular orbital component of the description. The six C2p orbitals overlap to give six π orbitals that spread all round the ring. Their energies are calculated within the Hückel approximation by diagonalizing the hamiltonian matrix

$$H = \begin{pmatrix} \alpha & \beta & 0 & 0 & 0 & \beta \\ \beta & \alpha & \beta & 0 & 0 & 0 \\ 0 & \beta & \alpha & \beta & 0 & 0 \\ 0 & 0 & \beta & \alpha & \beta & 0 \\ 0 & 0 & 0 & \beta & \alpha & \beta \\ \beta & 0 & 0 & 0 & \beta & \alpha \end{pmatrix}$$

The MO energies, the eigenvalues of this matrix, are simply

$$E = \alpha \pm 2\beta, \qquad \alpha \pm \beta, \qquad \alpha \pm \beta \tag{14.46}$$

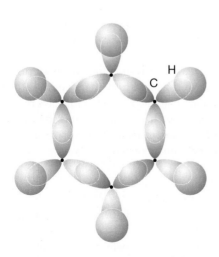

14.37 The σ framework of benzene is formed by the overlap of Csp^2 hybrids, which fit without strain into a hexagonal arrangement.

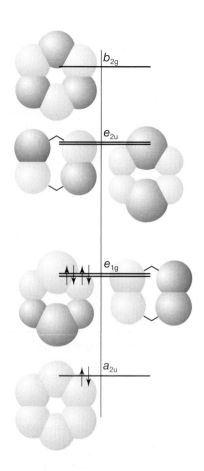

14.38 The Hückel orbitals of benzene and the corresponding energy levels. The symmetry labels are explained in Chapter 15. The bonding and antibonding character of the delocalized orbitals reflects the numbers of nodes between the atoms. In the ground state, only the bonding orbitals are occupied.

as shown in Fig. 14.38. The orbitals there have been given symmetry labels which we explain in Chapter 15. Note that the lowest energy orbital is bonding between all neighbouring atoms, the highest energy orbital is antibonding between each pair of neighbours, and the intermediate orbitals are a mixture of bonding, nonbonding, and antibonding character between adjacent atoms.

We now apply the building-up principle to the π system. There are six electrons to accommodate (one from each C atom), so the three lowest orbitals (a_{2u} and the doubly degenerate pair e_{1g}) are fully occupied, giving the ground-state configuration $a_{2u}^2 e_{1g}^4$. A significant point is that the only molecular orbitals occupied are those with net bonding character.

The π-electron energy of benzene is

$$E_\pi = 2(\alpha + 2\beta) + 4(\alpha + \beta) = 6\alpha + 8\beta$$

If we ignored delocalization and thought of the molecule as having three isolated π bonds, it would be ascribed a π-electron energy of only $3(2\alpha + 2\beta) = 6\alpha + 6\beta$. The delocalization energy is therefore $2\beta \approx -150$ kJ mol^{-1}, which is considerably more than for butadiene. The π-bond formation energy in benzene is 8β.

This discussion suggests that aromatic stability can be traced to two main contributions. First, the shape of the regular hexagon is ideal for the formation of strong σ bonds: the σ framework is relaxed and without strain. Second, the π orbitals are such as to be able to accommodate all the electrons in bonding orbitals, and the delocalization energy is large.

14.8 Extended Hückel theory

The first level of sophistication that we consider is **extended Hückel theory** (EHT), which is a significant improvement on the basic theory. The version of EHT formulated by Roald Hoffmann in 1963 predicts the shapes of molecular orbitals of hydrocarbons and boron hydrides. More sophisticated versions of the method are still in use today.

(a) The matrix formulation of the theory

First, EHT includes both σ and π orbitals, so it is not confined to planar conjugated hydrocarbons. Second, EHT does not ignore overlap but calculates all the overlap integrals explicitly and stores them as the matrix **S**. Formulas for overlap integrals between orbitals are available, so the software simply needs to evaluate them for the chosen geometry.

The diagonal elements of the hamiltonian matrix, corresponding to the α terms of Hückel theory, are set equal to the negative of the ionization energy of the atom, which for hydrogen is 13.60 eV. For other atoms, values are selected that give results in accord with experiment and that are close to the ionization energies of the orbitals. The off-diagonal elements of the hamiltonian matrix (corresponding to β and 0 in the Hückel theory) are assumed to be proportional to the overlap integral, and specifically

$$H_{ij} = \tfrac{1}{2}KS_{ij}(H_{ii} + H_{jj}) \tag{14.47}$$

where K is a constant commonly set equal to 1.75. Note that the calculation takes into account the presence of atoms of different elements.

From this point, the procedure is essentially the same as before. The complication that overlap is not negligible can be resolved in a number of ways. The most straightforward is to find the inverse of S, and to multiply eqn 42 from the left by it, which gives

$$S^{-1}HC = CE$$

and then to multiply this expression from the left by the inverse of C:

$$C^{-1}S^{-1}HC = E \tag{14.48}$$

Then, to find the eigenvalues, we need to find the matrix C that diagonalizes $S^{-1}H$ rather than H alone. There are more sophisticated ways of proceeding, but this is the essence of the calculation.

Example 14.6 Using the extended Hückel theory

Use EHT to calculate the energies of the three molecular orbitals in an H_3^+ molecule ion treated as an equilateral triangle of side $R = 3.25a_0$ (170 pm).

Method Set up the overlap matrix using the fact that the overlap between two H1s orbitals separated by R is given by the expression

$$S = \left\{ 1 + \frac{R}{a_0} + \frac{1}{3}\left(\frac{R}{a_0}\right)^2 \right\} e^{-R/a_0}$$

Then set up the hamiltonian matrix, using eqn 47 and the ionization energy of hydrogen, which is 13.60 eV (call this value $-\alpha$). Use mathematical software to find the energies and the coefficients of the orbitals.

Solution The overlap matrix is

$$S = \begin{pmatrix} 1 & S & S \\ S & 1 & S \\ S & S & 1 \end{pmatrix}$$

with S the value appropriate to the chosen bond length. The hamiltonian matrix is

$$H = \begin{pmatrix} \alpha & \alpha KS & \alpha KS \\ \alpha KS & \alpha & \alpha KS \\ \alpha KS & \alpha KS & \alpha \end{pmatrix} = \alpha \begin{pmatrix} 1 & KS & KS \\ KS & 1 & KS \\ KS & KS & 1 \end{pmatrix}$$

with $\alpha = -13.60$ eV. For $R = 3.25a_0$, $S = 0.30$. Then software gives the eigenvalues 0.679α, 0.679α, and 1.281α. We find one nondegenerate orbital and a pair of doubly degenerate orbitals. The matrix of eigenvectors is

$$C = \begin{pmatrix} 0.816 & 0.724 & 0.577 \\ -0.408 & -0.036 & 0.577 \\ -0.408 & 0.688 & 0.577 \end{pmatrix}$$

So, the orbital corresponding to the lowest energy (at 1.281α), the third eigenvalue, and hence the third column of eigenvectors, is

$$\psi = N\{0.577\psi_{1sA} + 0.577\psi_{1sB} + 0.577\psi_{1sC}\}$$

This orbital is not normalized to 1. To find the normalization constant, we write

$$(0.577N)^2 \left\{ \int \psi_{1sA}^2 \, d\tau + \int \psi_{1sB}^2 \, d\tau + \int \psi_{1sC}^2 \, d\tau \right.$$

$$\left. + 2\int \psi_{1sA}\psi_{1sB} \, d\tau + 2\int \psi_{1sB}\psi_{1sC} \, d\tau + 2\int \psi_{1sC}\psi_{1sA} \, d\tau \right\}$$

$$= 3(0.577N)^2(1 + 2S) = 1$$

It follows that $N = 0.791$ and the normalized orbital is

$$\psi = 0.456\psi_{1sA} + 0.456\psi_{1sB} + 0.456\psi_{1sC}$$

Self-test 14.9 Set up a simplified version of an EHT calculation of the electronic structure of H_2O using only $O2s$, $O2p_y$, $O2p_z$, and the two $H1s$ orbitals (where the molecular plane is the yz-plane). For the experimental bond angle and bond length, the overlap integrals required are $S(O2s, H1s) = 0.250$, $S(O2p_y, H1s) = 0.355$, and $S(O2p_z, H1s) = 0.275$, and the ionization energies are $I(O2s) = 12.0$ eV, $I(O2p) = 13.61$, and $I(H1s) = 13.60$ eV.

$[E/eV = (-2.23, -3.34, -17.01, -17.64, -16.78),$

$$C = \begin{pmatrix} -0.49 & -0.50 & 0.50 & 0.10 & 0.51 \\ -0.49 & 0.50 & -0.50 & 0.10 & 0.51 \\ 0.51 & 0 & 0 & 0.76 & 0.30 \\ 0 & 0.71 & 0.71 & 0 & 0 \\ 0.51 & 0 & 0 & -0.64 & 0.62 \end{pmatrix}]$$

(b) Population analysis

Where is the electron density concentrated in a molecule when electrons occupy molecular orbitals that are delocalized over all the atoms? To answer this type of question, we consider a simple two-atom molecular orbital of the form

$$\psi = c_A\psi_A + c_B\psi_B$$

To find the probability density for an electron that occupies this orbital, we form the square:

$$\psi^2 = c_A^2\psi_A^2 + c_B^2\psi_B^2 + 2c_Ac_B\psi_A\psi_B$$

The total probability of finding the electron in the molecule is 1, and is given by the integral of this probability density over all space:

$$1 = c_A^2 + c_B^2 + 2c_Ac_BS_{AB}$$

This expression strongly suggests that the total probability of finding the electron in the molecule is a sum of the probability of finding it in orbital A (the term c_A^2), the probability that it is in orbital B (the term c_B^2), and the probability that it is in the overlap region between the two atoms (the term $2c_Ac_BS_{AB}$). The last term is called the **overlap population** and this kind of apportioning of the population is called a **Mulliken population analysis** after R.S. Mulliken, who suggested it. In general, for an orbital expressed as a linear combination of atomic orbitals with coefficients c_i, the **atomic populations**, ρ_i, are

$$\rho_i = c_i^2 \tag{14.49a}$$

and the overlap populations are

$$\rho_{ij} = 2c_ic_jS_{ij} \tag{14.49b}$$

If two electrons occupy an orbital, then these contributions should be doubled. The overall populations are the sums of these quantities for all the occupied orbitals in the molecule multiplied by the number of electrons in each orbital. In one further step, we may apportion one-half of each overlap population to the neighbouring atoms and thereby obtain the **gross orbital population**.

Illustration 14.2

For H_3^+, with the geometry specified in Example 14.6, the total population on atom A (or any atom) is

$$\rho_A = 2 \times (0.456)^2 = 0.416$$

and the overlap population between atoms A and B (or any other pair of atoms)

$$\rho_{AB} = 2 \times 2 \times 0.456 \times 0.456 \times 0.300 = 0.250$$

The population matrix for this molecular ion is therefore

$$\rho = \begin{pmatrix} 0.416 & 0.250 & 0.250 \\ 0.250 & 0.416 & 0.250 \\ 0.250 & 0.250 & 0.416 \end{pmatrix}$$

The gross orbital population of atom A (and of any atom in this symmetrical molecule) is therefore

$$P_A = 0.416 + \tfrac{1}{2}(0.250 + 0.250) = 0.666$$

and the sum of these gross populations is 2, the number of electrons in the molecule.

Finally, we can assess the net charge on each atom by subtracting the gross orbital population (0.666) from the nuclear charge (+1), and so obtain +0.333 for each atom. Thus one-third of the overall charge is ascribed to each atom, as we expect by symmetry.

Current versions of EHT are easy to implement with a computer and give useful qualitative pictures of molecular orbitals in molecules with known structures. The most unfortunate drawback of the technique is its inability to predict the correct three-dimensional structures of even simple molecules. For example, EHT predicts that linear H_2O has a lower energy than angular H_2O.

The web site contains links to sites from which free software for EHT calculations can be downloaded.

14.9 Self-consistent field calculations

Many of the difficulties associated with EHT have been overcome by more sophisticated theories that not only calculate the shapes and energies of molecular orbitals but also predict with reasonable accuracy the structure and reactivity of molecules. The full treatment of molecular electronic structure is quite easy to formulate but difficult to implement. However, it has received an enormous amount of attention by computational chemists, and has become a keystone of modern chemical research. John Pople and Walter Kohn were awarded the Nobel Prize in Chemistry for 1998 for their contributions to the development of computational techniques for the elucidation of molecular structure.

(a) The Hartree–Fock equations

The starting point is to write down the many-electron wavefunction as a product of one-electron wavefunctions:

$$\psi = \psi_{a,\alpha}(1)\psi_{a,\beta}(2) \dots \psi_{z,\beta}(N)$$

This is the wavefunction for an N-electron closed-shell molecule in which electron 1 occupies molecular orbital ψ_a with spin α, electron 2 occupies molecular orbital ψ_a

with spin β, and so on. However, the wavefunction must satisfy the Pauli principle and change sign under the permutation of any pair of electrons. To achieve this behaviour, we write the wavefunction as a sum of all possible permutations with the appropriate sign:

$$\psi = \psi_{a,\alpha}(1)\psi_{a,\beta}(2) \dots \psi_{z,\beta}(N) - \psi_{a,\alpha}(2)\psi_{a,\beta}(1) \dots \psi_{z,\beta}(N) + \cdots$$

There are $N!$ terms in this sum, and the entire sum can be written as a determinant:

$$\psi = \frac{1}{\sqrt{N!}} \begin{vmatrix} \psi_{a\alpha}(1) & \psi_{a\beta}(1) & \dots & \dots & \psi_{z\beta}(1) \\ \psi_{a\alpha}(2) & \psi_{a\beta}(2) & \dots & \dots & \psi_{z\beta}(2) \\ \vdots & \vdots & & & \vdots \\ \vdots & \vdots & & & \vdots \\ \psi_{a\alpha}(N) & \psi_{a\beta}(N) & \dots & \dots & \psi_{z\beta}(N) \end{vmatrix} \tag{14.50a}$$

The initial factor ensures that the wavefunction is normalized if the component molecular orbitals are normalized. To save the tedium of writing out large determinants, the wavefunction is normally written simply as

$$\psi = (1/N!)^{1/2}\det|\psi_{a,\alpha}(1)\psi_{a,\beta}(2) \dots \psi_{z,\beta}(N)| \tag{14.50b}$$

When the determinantal wavefunction is combined with the variation principle (Section 14.6c), it turns out that the optimum wavefunctions, in the sense of corresponding to the lowest total energy, must satisfy the **Hartree–Fock equations**:

$$f_1\psi_{a,\sigma}(1) = \varepsilon\psi_{a,\sigma}(1) \tag{14.51}$$

where σ is either α or β. The **Fock operator** f_1 is

$$f_1 = h_1 + \sum_j \{2J_j(1) - K_j(1)\} \tag{14.52}$$

The three terms in this expression are the **core hamiltonian**

$$h_1 = -\frac{\hbar^2}{2m_e}\nabla_1^2 - \sum_n \frac{Z_n e^2}{4\pi\varepsilon_0 r_{n1}} \tag{14.53a}$$

the **Coulomb operator** J, where

$$J_j(1)\psi_a(1) = \left\{\int \psi_j^*(2)\psi_j(2)\left(\frac{e^2}{4\pi\varepsilon_0 r_{12}}\right)d\tau_2\right\}\psi_a(1) \tag{14.53b}$$

and the **exchange operator**, K, where

$$K_j(1)\psi_a(1) = \left\{\int \psi_j^*(2)\psi_a(2)\left(\frac{e^2}{4\pi\varepsilon_0 r_{12}}\right)d\tau_2\right\}\psi_j(1) \tag{14.53c}$$

Although the Hartree–Fock equations look deceptively simple, with the Fock operator looking like a hamiltonian, we see from these definitions that f actually depends on the wavefunctions of all the electrons. To proceed, we have to guess the initial form of the wavefunctions, use them in the definition of the Coulomb and exchange operators, and solve the Hartree–Fock equations. That process is then continued using the newly found wavefunctions until each cycle of calculation leaves the energies and wavefunctions unchanged to within a chosen criterion. This is the origin of the term **self-consistent field** (SCF) for this type of procedure.

The difficulty in this procedure is in the solution of the Hartree–Fock equations. To make progress, we have to express the wavefunctions as linear combinations of M atomic orbitals ϕ_i, and write

$$\psi_a = \sum_{i=1}^{N} c_{ia} \phi_i$$

As we show in the *Justification* below, the use of a linear combination like this leads to a set of equations that can be expressed in a matrix form known as the **Roothaan equations**:

$$FC = SC\varepsilon \tag{14.54}$$

where F is the matrix formed from the Fock operator:

$$F_{ij} = \int \phi_i^*(1) f_1 \phi_j(1) \, d\tau \tag{14.55a}$$

and S is the matrix of overlap integrals

$$S_{ij} = \int \phi_i^*(1) \phi_j(1) \, d\tau \tag{14.55b}$$

We see how closely the Roothaan equations resemble the matrix equations we used in the extended Hückel theory. The principal differences are that the Fock matrix F is much richer in structure, and in particular takes into account the requirements of the Pauli principle.

Justification 14.4

To construct the Roothaan equations we substitute the linear combination of atomic orbitals into eqn 14.51, which gives

$$f_1 \sum_{i=1}^{N} c_{i\alpha} \phi_i(1) = \varepsilon_\alpha \sum_{i=1}^{N} c_{i\alpha} \phi_i(1)$$

Now multiply from the left by $\phi_j^*(1)$ and integrate over the coordinates of electron 1:

$$\sum_{i=1}^{N} c_{i\alpha} \overbrace{\int \phi_j^*(1) f(1) \phi_i(1) \, d\mathbf{r}_1}^{F_{ji}} = \varepsilon_\alpha \sum_{i=1}^{N} c_{i\alpha} \overbrace{\int \phi_j^*(1) \phi_i(1) \, d\mathbf{r}_1}^{S_{ji}}$$

That is,

$$\sum_{i=1}^{N} c_{i\alpha} F_{ji} = \varepsilon_\alpha \sum_{i=1}^{N} c_{i\alpha} S_{ji}$$

This expression has the form of the matrix equation in eqn 54.

(b) Semi-empirical and *ab initio* methods

There are two main strategies for continuing the calculation from this point. In the **semi-empirical methods**, many of the integrals are estimated by appealing to spectroscopic data or physical properties such as ionization energies, and using a series of rules to set certain integrals equal to zero. In the ***ab initio* methods**, an attempt is made to calculate all the integrals that appear in the secular determinant. Both procedures employ a great deal of computational effort and, along with cryptanalysts and meteorologists, theoretical chemists are among the heaviest users of the fastest computers.

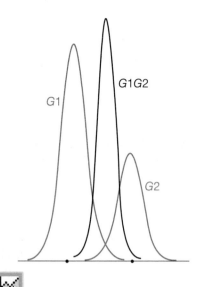

14.39 The product of two Gaussian functions (the blue curves) is itself a Gaussian function located between the two contributing Gaussians.

The Fock matrix has elements that consist of integrals of the form

$$(AB|CD) = \int A(1)B(1)\left(\frac{e^2}{4\pi\varepsilon_0 r_{12}}\right)C(2)D(2)\,d\tau_1\,d\tau_2 \tag{14.56}$$

where A, B, C, and D are atomic orbitals, which in general may be centred on different nuclei. It can be appreciated that, if there are several dozen atomic orbitals used to build the molecular orbitals, then there will be tens of thousands of integrals of this form to evaluate. One severe approximation is called **complete neglect of differential overlap** (CNDO), in which all integrals are set to zero unless A and B are the same orbitals centred on the same nucleus, and likewise for C and D. The surviving integrals are then adjusted until the energy levels are in good agreement with experiment. The more recent semi-empirical methods make less draconian decisions about which integrals are to be ignored, but they are all descendants of the early CNDO technique. These procedures are now readily available in commercial software packages and can be used with very little detailed knowledge of their mode of calculation. The packages also have sophisticated graphical output procedures, which enable one to analyse the shapes of orbitals and the distribution of electric charge in molecules. The latter is important when assessing, for instance, the likelihood that a given molecule will bind to an active site in an enzyme. Such studies can greatly reduce the time and cost of screening compounds for potential pharmacological activity.

Commercial packages are also available for *ab initio* calculations. Here the problem is to evaluate as efficiently as possible thousands of integrals. This task is greatly facilitated by expressing the atomic orbitals used in the LCAOs as linear combinations of Gaussian orbitals. A **Gaussian type orbital** (GTO) is a function of the form $e^{-\alpha r^2}$. The advantage of GTOs over the correct orbitals (which are proportional to $e^{-\alpha r}$) is that the product of two Gaussian functions is itself a Gaussian function that lies between the centres of the two contributing functions (Fig. 14.39). In this way, the four-centre integrals like that in eqn 56 become two-centre integrals of the form

$$(AB|CD) = \int X(1)\left(\frac{e^2}{4\pi\varepsilon_0 r_{12}}\right)Y(2)\,d\tau_1\,d\tau_2 \tag{14.57}$$

where X is the Gaussian corresponding to the product AB and Y is the corresponding Gaussian from CD. Integrals of this form are much easier and faster to evaluate numerically than the original four-centre integrals. Although more GTOs have to be used to simulate the atomic orbitals, there is an overall increase in speed of computation.

(c) Density functional theory

A technique that has gained considerable ground in recent years to become one of the most widely used techniques for the calculation of molecular structure is **density functional theory** (DFT). Its advantages include less demanding computational effort, less computer time, and—in some cases (particularly d-metal complexes)—better agreement with experimental values than is obtained from Hartree–Fock procedures.

The central focus of DFT is the electron density, ρ, rather than the wavefunction ψ. The 'functional' part of the name comes from the fact that the energy of the molecule is a function of the electron density, written $E[\rho]$, and the electron density is itself a function of position, $\rho(\mathbf{r})$, and in mathematics a function of a function is called a *functional*. The exact ground-state energy of an n-electron molecule is

$$E[\rho] = E_K + E_{P;e,N} + E_{P;e,e} + E_{XC}[\rho] \tag{14.58}$$

where E_K is the total electron kinetic energy, $E_{P;e,N}$ the electron–nucleus potential energy, $E_{P;e,e}$ the electron–electron potential energy, and $E_{XC}[\rho]$ the **exchange-correlation energy**, which takes into account all the effects due to spin. The orbitals used to construct the electron density from

$$\rho(\mathbf{r}) = \sum_{i=1}^{n} |\psi_i(\mathbf{r})|^2 \tag{14.59}$$

are calculated from the **Kohn–Sham equations**, which are found by applying the variation principle to the electron energy, and are like the Hartree–Fock equations except for a term V_{XC}, which is called the **exchange-correlation potential**:

$$\left\{ \overbrace{-\frac{\hbar^2}{2m_e}\nabla_1^2}^{\substack{\text{Kinetic}\\\text{energy}}} \overbrace{- \sum_{j=1}^{N} \frac{Z_j e^2}{4\pi\varepsilon_0 r_{j1}}}^{\substack{\text{Electron–nucleus}\\\text{attraction}}} + \overbrace{\int \frac{\rho(\mathbf{r}_2)e^2}{4\pi\varepsilon_0 r_{12}}\,\mathrm{d}\mathbf{r}_2}^{\substack{\text{Electron–electron}\\\text{repulsion}}} + \overbrace{V_{XC}(\mathbf{r}_1)}^{\substack{\text{Exchange–}\\\text{correlation}}} \right\} \psi_i(\mathbf{r}_1) = \varepsilon_i \psi_i(\mathbf{r}_1) \tag{14.60}$$

The exchange-correlation potential is the 'functional derivative' of the exchange-correlation energy:

$$V_{XC}[\rho] = \frac{\delta E_{XC}[\rho]}{\delta\rho} \tag{14.61}$$

The Kohn–Sham equations are solved iteratively and self-consistently. First, we guess the electron density. For this step it is common to use a superposition of atomic electron densities. Then the exchange-correlation potential is calculated by assuming an approximate form of the dependence of the exchange-correlation energy on the electron density and evaluating the functional derivative in eqn 61. For this step, the simplest approximation is the **local-density approximation** and to write

$$E_{XC}[\rho] = \int \rho(\mathbf{r})\varepsilon_{XC}[\rho(\mathbf{r})]\,\mathrm{d}\mathbf{r} \tag{14.62}$$

where ε_{XC} is the exchange-correlation energy per electron in a homogeneous gas of constant density. Next, the Kohn–Sham equations are solved to obtain an initial set of orbitals. This set of orbitals is used to obtain a better approximation to the electron density (from eqn 59) and the process is repeated until the density and the exchange-correlation energy are constant to within some tolerance.

(d) Graphical output

One of the most significant developments in computational chemistry has been the introduction of graphical representations of molecular orbitals and electron densities. The raw output of a molecular structure calculation is a list of the coefficients of the atomic orbitals in each molecular orbital and the energies of these orbitals. The graphical representation of a molecular orbital uses stylized shapes to represent the basis set, and then scales their size to indicate the coefficient in the linear combination. Different signs of the wavefunctions are represented by different colours: in the illustrations shown here, we use different shades of grey.

Once the coefficients are known, we can build up a representation of the electron density in the molecule by noting which orbitals are occupied and then forming the squares of those orbitals. The total electron density at any point is then the sum of the squares of the wavefunctions evaluated at that point. The outcome is commonly represented by an **isodensity surface**, a surface of constant total electron density (Fig. 14.40). As shown in the illustration, there are several styles of representing an isodensity

Consider the functional $G[f]$ where f is a function of x. When x changes to $x + \delta x$, the function changes to $f + \delta f$ and the functional changes to $G[f + \delta f]$. The functional derivative is then defined as

$$\frac{\delta G}{\delta f} = \lim_{\delta f \to 0} \frac{G[f + \delta f] - G[f]}{\delta f}$$

by analogy with the derivative of a function.

(a)

(b)

(c)

14.40 Various representations of an isodensity surface of ethanol: (a) as a solid form; (b) as a transparent form with a ball-and-stick representation of the molecule within; (c) as a mesh.

14.41 An elpot diagram of ethanol.

The web site contains links to sites from which free software for semi-empirical and *ab initio* calculations can be downloaded.

surface, as a solid form, as a transparent form with a ball-and-stick representation of the molecule within, or as a mesh. A related representation is a **solvent-accessible surface** in which the shape represents the shape of the molecule by imagining a sphere representing a solvent molecule rolling across the surface and plotting the locations of the centre of that sphere.

One of the most important aspects of a molecule other than its geometrical shape is the distribution of charge over its surface. The net charge at each point on an isodensity surface can be calculated by subtracting the charge due to the electron density at that point from the charge due to the nuclei: the result is an **electrostatic potential surface** (an 'elpot surface') in which net positive charge is shown in one colour and net negative charge is shown in another, with intermediate gradations of colour (Fig. 14.41).

Representations such as those we have illustrated are of critical importance in a number of fields. For instance, they are particularly important for studying the docking of a substrate molecule into an enzyme and for assessing the pharmacological activity of potential drugs.

Checklist of key ideas

☐ covalent bond, a bond formed by the sharing of a pair of electrons

☐ ionic bond, a bond in which the cohesion arises from the Coulombic attraction between ions of opposite charge

The Born–Oppenheimer approximation

☐ Born–Oppenheimer approximation, an approximation in which the nuclei are treated as stationary while the electrons move in their field

☐ molecular potential energy curve, a graph showing how the total energy of a molecule varies with bond length (or some other measure of nuclear arrangement)

☐ equilibrium bond length, the internuclear separation at the minimum of the molecular potential energy curve

☐ bond dissociation energy, the minimum energy to dissociate bonded atoms

Valence-bond theory

☐ valence-bond theory, a theory of bonding in which bonds are formed when electrons pair and orbitals merge

14.1 The hydrogen molecule

☐ valence-bond wavefunction, $\psi = A(1)B(2) + A(2)B(1)$

☐ σ bond, a bond with cylindrical symmetry around the internuclear axis

☐ electron pairing, electrons described by $\sigma_-(1, 2) = (1/2^{1/2})\{\alpha(1)\beta(2) - \alpha(2)\beta(1)\}$

14.2 Homonuclear diatomic molecules

☐ π bond, a bond that arises from the spin pairing of electrons in two p orbitals that approach side-by-side

14.3 Polyatomic molecules

☐ promotion, the excitation of an electron to an orbital of higher energy (in the context of valence-bond theory)

☐ hybrid orbital, a linear combination of atomic orbitals on the same atom

☐ sp-, sp^2-, and sp^3-hybrid orbitals, hybrids of s and p orbitals

Molecular orbital theory

14.4 The hydrogen molecule-ion

☐ molecular orbital, a one-electron wavefunction spreading across several atoms in a molecule

☐ linear combination of atomic orbitals (LCAO), $\psi_\pm = N(A \pm B)$

☐ LCAO-MO, a molecular orbital constructed as an LCAO

☐ σ orbital, an orbital with zero orbital angular momentum about the internuclear axis

☐ overlap density, the enhanced electron density arising from constructive interference in the internuclear region

☐ bonding orbital, an orbital which, when occupied, contributes to the stabilization of a molecule, $\psi = A + B$

☐ σ electron, an electron in a σ orbital

☐ antibonding orbital, an orbital which, when occupied, contributes to the destabilization of a molecule; $\psi = A - B$

14.5 The structures of diatomic molecules

☐ molecular orbital energy level diagram, a diagram showing the relative energies of the molecular orbitals of a molecule

☐ bond order, $b = \frac{1}{2}(n - n^*)$

☐ π orbital, a molecular orbital with one unit of angular momentum around the internuclear axis; in a nonlinear molecule, a molecular orbital with a nodal plane that includes the internuclear axis

☐ overlap integral, $S = \int \psi_A^* \psi_B \, d\tau$

☐ electron configurations of diatomic molecules, see Fig. 14.29

14.6 Heteronuclear diatomic molecules

☐ heteronuclear diatomic molecule, a diatomic molecule formed from atoms of two different elements

☐ polar bond, $\psi = c_A A + c_B B$ with $|c_A| \neq |c_B|$

☐ partial negative and positive charge, the charge resulting from an incomplete cancellation of the nuclear and electronic charges on an atom

- electronegativity, a measure of the power of an atom to attract electrons to itself when it is part of a compound

- Pauling electronegativity scale, $|\chi_A - \chi_B| = 0.102\{D(A-B) - \frac{1}{2}[D(A-A) + D(B-B)]\}^{1/2}$

- Mulliken electronegativity scale, $\chi_M = \frac{1}{2}(I + E_{ea})$

- variation principle: If an arbitrary wavefunction is used to calculate the energy, the value calculated is never less than the true energy

- trial wavefunction, a function with variable parameters used to model the true wavefunction in applications of the variation principle

- basis set, a set of functions used to model another function (particularly a molecular orbital)

- secular equations, a set of simultaneous equations which, when solved, give the coefficients and energies of LCAO-MOs

- Coulomb integral, $\alpha = H_{ii}$

- resonance integral, $\beta = H_{ij}$

- secular determinant, the determinant $|H - ES|$

Molecular orbitals for polyatomic systems

- molecular orbital, $\psi = \sum_i c_i \psi_i$

14.7 The Hückel approximation

- Hückel approximation: all overlap integrals equal to zero; all resonance integrals between non-neighbours equal to zero; all remaining resonance integrals are set equal (to β)

- highest occupied molecular orbital (HOMO)

- lowest unoccupied molecular orbital (LUMO)

- frontier orbitals, the HOMO and LUMO of a molecule

- hamiltonian matrix, the matrix H formed of all H_{ij}

- overlap matrix, the matrix S formed of all S_{ij}

- π-electron binding energy, the sum of the energies of each π electron

- π-bond formation energy, the energy released when a π bond is formed

- delocalization energy, the extra stabilization of a conjugated system

14.8 Extended Hückel theory

- extended Hückel theory, a procedure in which overlap integrals are calculated and $H_{ij} = \frac{1}{2}KS_{ij}(H_{ii} + H_{jj})$

- matrix formulation, $C^{-1}S^{-1}HC = E$; to find the eigenvalues, find the matrix C that diagonalizes $S^{-1}H$

- Mulliken population analysis, an analysis of the distribution of electrons in terms of the coefficients of LCAO-MOs

- atomic population, $\rho_i = c_i^2$

- overlap population, $\rho_{ij} = 2c_i c_j S_{ij}$

14.9 Self-consistent field calculations

- determinantal wavefunction, $\psi = (1/N!)^{1/2}\det|\psi_{a,\alpha}(1)\psi_{a,\beta}(2) \ldots \psi_{z,\beta}(N)|$

- Hartree–Fock equations, $f_1\psi_a(1) = \varepsilon_a\psi_a(1)$

- Fock operator, $f_1 = h_1 + \sum_j\{2J_j(1) - K_j(1)\}$

- core hamiltonian, $h_1 = -(\hbar^2/2m_e)\nabla_1^2 - \sum_n Z_n e^2/4\pi\varepsilon_0 r_{n1}$

- Coulomb operator, $J_j(1)\psi_a(1) = \left\{\int \psi_j^*(2)\psi_j(2)\left(\dfrac{e^2}{4\pi\varepsilon_0 r_{12}}\right)\mathrm{d}\tau_2\right\}\psi_a(1)$

- exchange operator, $K_j(1)\psi_a(1) = \left\{\int \psi_j^*(2)\psi_a(2)\left(\dfrac{e^2}{4\pi\varepsilon_0 r_{12}}\right)\mathrm{d}\tau_2\right\}\psi_j(1)$

- self-consistent field, a solution of the Hartee–Fock equations that is self-consistent

- Roothaan equations, $FC = SC\varepsilon$

- semi-empirical methods, procedures based on the estimation of molecular integrals from experimental information

- *ab initio* methods, procedures in which all integrals are calculated explicitly

- complete neglect of differential overlap (CNDO), an early semi-empirical procedure

- Gaussian type orbital (GTO), an atomic orbital with a Gaussian radial dependence, a function of the form $e^{-\alpha r^2}$

- density functional theory (DFT), an *ab initio* procedure based on the electron density

- exchange–correlation energy, E_{XC}, the contribution to the energy that takes into account all the effects due to spin in DFT

- Kohn–Sham equations, the basic equations of DFT, found by applying the variation principle to the electron energy, $\left\{-\dfrac{\hbar^2}{2m_e}\nabla_1^2 - \sum_{j=1}^N \dfrac{Z_j e^2}{4\pi\varepsilon_0 r_{j1}}\right.$ $\left. + \int \dfrac{\rho(r_2)e^2}{4\pi\varepsilon_0 r_{12}}\mathrm{d}r_2 + V_{XC}(r_1)\right\}\psi_i(r_1) = \varepsilon_i\psi_i(r_1)$

- exchange–correlation potential, V_{XC}, the functional derivative of the exchange–correlation energy, $V_{XC} = \delta E_{XC}/\delta\rho$

- local-density approximation, $E_{XC}[\rho] = \int \rho(r)\varepsilon_{XC}[\rho(r)]\,\mathrm{d}r$

- isodensity surface, a surface of constant total electron density

- solvent-accessible surface, the surface obtained by imagining a sphere representing a solvent molecule rolling across the surface and plotting the locations of the centre of that sphere

- electrostatic potential surface, a surface on which net charge is shown by gradations of colour

Further reading

Articles of general interest

G.L. Bendazzoli, The variational principle illustrated by simple examples. *J. Chem. Educ.* **70**, 913 (1992).

A.D. Buckingham and T.W. Rowlands, Can addition of a bonding electron weaken a bond? *J. Chem. Educ.* **68**, 282 (1991).

J.R. Dias, A facile Hückel molecular orbital solution of buckminsterfullerene using chemical graph theory. *J. Chem. Educ.* **66**, 1012 (1989).

P. Engelking, Molecules. In *Encyclopedia of applied physics* (ed. G.L. Trigg), **10**, 525. VCH, New York (1994).

J.B. Foresman, *Ab initio* techniques in chemistry: interpretation and visualization. In *Using computers in chemistry and chemical education* (ed. T.J. Zielinski and M.L. Swift), 243. American Chemical Society, Washington (1997).

G.A. Gallup, The Lewis electron-pair model, spectroscopy, and the role of the orbital picture in describing the electronic structure of molecules. *J. Chem. Educ.* **65**, 671 (1988).

A.F. Gaines and F.M. Page, The strengths of σ and π bonds in gaseous compounds. *J. Chem. Res.* (S) 200 (1980).

J.B. Holbrook, R. Salry-Grant, B.C. Smith, and T.V. Tandel, Lattice enthalpies of ionic halides, hydrides, oxides, and sulfides: second-electron affinities of atomic oxygen and sulfur. *J. Chem. Educ.* **67**, 304 (1990).

D.J. Klein and N. Trinajstić, Valence-bond theory and chemical structure. *J. Chem. Educ.* **67**, 633 (1990).

R.C. Mebane, S.A. Schanley, T.R. Rybolt, and C.D. Bruce, The correlation of physical properties of organic molecules with computed molecular surface areas. *J. Chem. Educ.* **76**, 688 (1999).

R.M. Metzger, Semiconductors to superconductors: organic lower-dimensional systems. In *Encyclopedia of applied physics* (ed. G.L. Trigg), **17**, 215 VCH, New York (1996).

J.C. Morrison, A.W. Weiss, K. Kirby, and D. Cooper, Electronic structure of atoms and molecules. In *Encyclopedia of applied physics* (ed. G.L. Trigg), **6**, 45. VCH, New York (1993).

E.I. von Nagy-Felsobuki, Hückel theory and photoelectron spectroscopy. *J. Chem. Educ.* **66**, 821 (1989).

S. Nordholm, Delocalization—the key concept of covalent bonding. *J. Chem. Educ.* **65**, 581 (1988).

R. Parson, Visualizing the variation principle: An intuitive approach to interpreting the theorem in geometric terms. *J. Chem. Educ.* **70**, 115 (1993).

A.B. Sannigrahi and T. Kar. Molecular orbital theory of bond order and valency. *J. Chem. Educ.* **65**, 674 (1988).

G.P. Shusterman and A.J. Shusterman, Teaching chemistry with electron density models. *J. Chem. Educ.* **74**, 771 (1997).

J. Simons, An experimental chemist's guide to ab initio quantum chemistry. *J. Phys. Chem.* **95**, 1017 (1991).

D.K. Smith, The antibonding effect. *J. Chem. Educ.* **77**, 780 (2000).

A.A. Woolf, Oxidation numbers and their limitations. *J. Chem. Educ.* **65**, 45 (1988).

Texts and sources of data and information

T.A. Albright and J.K. Burdett, *Problems in molecular orbital theory.* Oxford University Press (1992).

P.W. Atkins, *Quanta: A handbook of concepts.* Oxford University Press (1991).

P.W. Atkins and R.S. Friedman, *Molecular quantum mechanics.* Oxford University Press (1997).

J.K. Burdett, *Chemical bonding: a dialogue.* Wiley, New York (1996).

T. Clark and R. Koch, *The chemist's electronic book of orbitals.* Springer, Berlin (1999).

C.A. Coulson, *The shape and structure of molecules* (revised by R. McWeeny). Oxford University Press (1982).

R.L. deKock and H.B. Gray, *Chemical structure and bonding.* University Science Books, Mill Valley (1989).

G.H. Grant and W.G. Richards, *Computational chemistry.* Oxford Chemistry Primers, Oxford University Press (1995).

A. Hinchcliffe, *Modelling molecular structures.* Wiley, New York (2000).

Y. Jean, F. Volatron, and J.K. Burdett, *An introduction to molecular orbitals.* Oxford University Press, New York (1993).

C.S. Johnson, Jr. and L.G. Pedersen, *Problems and solutions in quantum chemistry and physics.* Dover, New York (1986).

M. Karplus and R.N. Porter, *Atoms and molecules: an introduction for students of physical chemistry.* Benjamin Cummings, Menlo Park (1970).

A.R. Leach, *Molecular modelling: principles and applications.* Longman, Harlow (2000).

I.N. Levine, *Quantum chemistry.* Prentice–Hall, Upper Saddle River (2000).

J.P. Lowe, *Quantum chemistry.* Academic Press, Boston (1993).

D.R. Lide (ed.), *CRC handbook of chemistry and physics*, Section 9, CRC Press, Boca Raton (2000).

Z.B. Maksić (ed.) *Theoretical models of chemical bonding.* Vols. 1–4, Springer-Verlag, Berlin (1990–1991).

J.N. Murrell, S.F.A. Kettle, and J.M. Tedder, *The chemical bond.* Wiley, New York (1985).

L. Pauling, *The nature of the chemical bond.* Cornell Univeristy Press, Ithaca (1960).

P.R. Scott and W.G. Richards, *Energy levels in atoms and molecules.* Oxford Chemistry Primers, Oxford University Press (1994).

D.F. Shriver, P.W. Atkins, and C.H. Langford, *Inorganic chemistry.* Oxford University Press, Oxford and W.H. Freeman Co, New York (1994).

A. Szabo and N.S. Ostlund, *Modern quantum chemistry: introduction to advanced electronic structure theory.* Dover, New York (1996).

B. Webster, *Chemical bonding theory.* Blackwell Scientific, Oxford (1990).

M.J. Winter, *Chemical bonding*, Oxford Chemistry Primers, Oxford University Press (1993).

A. Zewail (ed.), *The chemical bond: structure and dynamics.* Academic Press, San Diego (1992).

Exercises

Discussion questions

14.1 (a) Compare the approximations built into valence-bond theory and molecular orbital theory.

14.1 (b) Discuss the steps involved in the construction of sp^3, sp^2, and sp hybrid orbitals.

14.2 (a) Distinguish between the Pauling and Mulliken electronegativity scales.

14.2 (b) Discuss the steps involved in the calculation of the energy of a system by using the variation principle.

14.3 (a) Discuss the approximations built into the Hückel and extended Hückel methods.

14.3 (b) Distinguish between delocalization energy, π-electron binding energy, and π-bond formation energy.

14.4 (a) Distinguish between overlap population, atomic population, and gross orbital population.

14.4 (b) Distinguish between semi-empirical and *ab initio* methods of electronic structure determination. Then, discuss the similarities and distinctions between Hartree–Fock and DFT *ab initio* procedures.

Numerical exercises

14.5 (a) Give the ground-state electron configurations and bond orders of (a) Li_2, (b) Be_2, and (c) C_2.

14.5 (b) Give the ground-state electron configurations of (a) H_2^-, (b) N_2, and (c) O_2.

14.6 (a) Give the ground-state electron configurations of (a) CO, (b) NO, and (c) CN^-.

14.6 (b) Give the ground-state electron configurations of (a) ClF, (b) CS, and (c) O_2^-.

14.7 (a) From the ground-state electron configurations of B_2 and C_2, predict which molecule should have the greater bond dissociation energy.

14.7 (b) Which of the molecules N_2, NO, O_2, C_2, F_2, and CN would you expect to be stabilized by (a) the addition of an electron to form AB^-, (b) the removal of an electron to form AB^+?

14.8 (a) Sketch the molecular orbital energy level diagram for XeF and deduce its ground-state electron configurations. Is XeF likely to have a shorter bond length than XeF^+?

14.8 (b) Sketch the molecular orbital energy level diagrams for BrCl and deduce its ground-state electron configurations. Is BrCl likely to have a shorter bond length than $BrCl^-$?

14.9 (a) Use the electron configurations of NO and N_2 to predict which is likely to have the shorter bond length.

14.9 (b) Arrange the species O_2^+, O_2, O_2^-, O_2^{2-} in order of increasing bond length.

14.10 (a) Show that the sp^2 hybrid orbital $(s + 2^{1/2}p)/3^{1/2}$ is normalized to 1 if the s and p orbitals are normalized to 1.

14.10 (b) Normalize the molecular orbital $\psi_A + \lambda\psi_B$ in terms of the parameter λ and the overlap integral S.

14.11 (a) Confirm that the bonding and antibonding combinations $\psi_A \pm \psi_B$ are mutually orthogonal in the sense that their mutual overlap is zero.

14.11 (b) Suppose that a molecular orbital has the form $N(0.145A + 0.844B)$. Find a linear combination of the orbitals A and B that is orthogonal to this combination.

14.12 (a) Can the function $\psi = x(L - x)$ be used as a trial wavefunction for the $n = 1$ state of a particle with mass m in a one–dimensional box of length L? If the answer is yes, then express the energy of this trial wavefunction in terms of h, m, and L and compare it with the exact result (eqn 12.7). If the answer is no, explain why this is not a suitable trial wavefunction.

14.12 (b) Can the function $\psi = x^2(L - 2x)$ be used as a trial wavefunction for the $n = 1$ state of a particle with mass m in a

one–dimensional box of length L? If the answer is yes, then express the energy of this trial wavefunction in terms of h, m, and L and compare it with the exact result (eqn 12.7). If the answer is no, explain why this is not a suitable trial wavefunction.

14.13 (a) Suppose that the function $\psi = Ae^{-ar^2}$, with A being the normalization constant and a being an adjustable parameter, is used as a trial wavefunction for the 1s orbital of the hydrogen atom. Express the energy of this trial wavefunction as a function of the h, a, e, the electron charge, and μ, the effective mass of the H atom.

14.13 (b) Suppose that the function $\psi = Ae^{-ar^2}$, with A being the normalization constant and a being an adjustable parameter, is used as a trial wavefunction for the 1s orbital of the hydrogen atom. The energy of this trial wavefunction is

$$E = \frac{3a\hbar^2}{2\mu} - \frac{e^2}{\varepsilon_0}\left(\frac{a}{2\pi^3}\right)^{1/2}$$

where e is the electron charge, and μ is the effective mass of the H atom. What is the minimum energy associated with this trial wavefunction?

14.14 (a) Construct the molecular orbital energy level diagrams of ethene on the basis that the molecule is formed from the appropriately hybridized CH_2 or CH fragments.

14.14 (b) Construct the molecular orbital energy level diagrams of ethyne (acetylene) on the basis that the molecule is formed from the appropriately hybridized CH_2 or CH fragments.

14.15 (a) Write down the secular determinants for (a) linear H_3, (b) cyclic H_3 within the Hückel approximation.

14.15 (b) Predict the electronic configurations of (a) the benzene anion, (b) the benzene cation. Estimate the π-electron binding energy in each case.

14.16 (a) Write down the secular determinants for (a) anthracene (**4**), (b) phenanthrene (**5**) within the Hückel approximation and using the C2p orbitals as the basis set.

14.16 (b) Use mathematical software to estimate the π-electron binding energy of (a) anthracene (**4**), (b) phenanthrene (**5**) within the Hückel approximation.

4

5

Problems[†]

Numerical problems

14.1 Show that, if a wave $\cos kx$ centred on A (so that x is measured from A) interferes with a similar wave $\cos k'x$ centred on B (with x measured from B) a distance R away, then constructive interference occurs in the intermediate region when $k = k' = \pi/2R$ and destructive interference if $kR = \frac{1}{2}\pi$ and $k'R = \frac{3}{2}\pi$.

14.2 The overlap integral between two H1s orbitals on nuclei separated by a distance R is $S = \{1 + (R/a_0) + \frac{1}{3}(R/a_0)^2\}\,e^{-R/a_0}$. Plot this function for $0 \le R < \infty$.

14.3 Before doing the calculation below, sketch how the overlap between an s orbital and a σp orbital can be expected to depend on their separation. The overlap integral between an H1s orbital and a H2p orbital on nuclei separated by a distance R is $S = (R/a_0)\{1 + (R/a_0) + \frac{1}{3}(R/a_0)^2\}\,e^{-R/a_0}$. Plot this function, and find the separation for which the overlap is a maximum.

14.4 Calculate the total amplitude of the normalized bonding and antibonding LCAO-MOs that may be formed from two H1s orbitals at a separation of 106 pm. Plot the two amplitudes for positions along the molecular axis both inside and outside the internuclear region.

14.5 Repeat the calculation in Problem 14.4 but plot the probability densities of the two orbitals. Then form the difference density, the difference between ψ^2 and $\frac{1}{2}\{\psi_A^2 + \psi_B^2\}$.

14.6‡ Use the $2p_x$ and $2p_z$ hydrogenic atomic orbitals to construct simple LCAO descriptions of $2p\sigma$ and $2p\pi$ molecular orbitals. (a) Make a probability density plot, and both surface and contour plots of the xz-plane amplitudes of the $2p_z\sigma$ and $2p_z\sigma^*$ molecular orbitals. (b) Make surface and contour plots of the xz-plane amplitudes of the $2p_x\pi$ and $2p_x\pi^*$ molecular orbitals. Include plots for both of internuclear distances, R, of $10a_0$ and $3a_0$. Interpret the graphs, and describe why scientists are so interested in this graphical information.

14.7 Imagine a small electron-sensitive probe of volume 1.00 pm^3 inserted into an H_2^+ molecule-ion in its ground state. Calculate the probability that it will register the presence of an electron at the following positions: (a) at nucleus A, (b) at nucleus B, (c) half-way between A and B, (d) at a point 20 pm along the bond from A and 10 pm perpendicularly. Do the same for the molecule-ion the instant after the electron has been excited into the antibonding LCAO-MO.

14.8 The energy of H_2^+ with internuclear separation R is given by the expression

$$E = E_H - \frac{V_1 + V_2}{1 + S} + \frac{e^2}{4\pi\varepsilon_0 R}$$

where E_H is the energy of an isolated H atom; V_1 is the attractive potential energy between the electron centred on one nucleus and the charge of the other nucleus; V_2 is the attraction between the overlap density and one of the nuclei; S is the overlap integral. The values are given below. Plot the molecular potential energy curve and find the bond dissociation energy (in electronvolts) and the equilibrium bond length.

R/a_0	0	1	2	3	4
V_1/E_h	1.000	0.729	0.473	0.330	0.250
V_2/E_h	1.000	0.736	0.406	0.199	0.092
S	1.000	0.858	0.587	0.349	0.189

where $E_h = 27.3$ eV and $a_0 = 52.9$ pm and $E_H = -\frac{1}{2}E_h$.

14.9 The same data as in Problem 14.8 may be used to calculate the molecular potential energy curve for the antibonding orbital, which is given by

$$E = E_H - \frac{V_1 - V_2}{1 - S} + \frac{e^2}{4\pi\varepsilon_0 R}$$

Plot the curve.

14.10‡ The LCAO-MO constructed from normalized $1s$ hydrogenic wavefunctions centred on nuclei that are a distance R apart, and having g symmetry, does not describe the molecular hydrogen ion ground state accurately. It does provide insight into functional characteristics of wavefunctions, bonding, and numerical methods needed in quantum chemistry. The overlap, Coulomb, and resonance integrals of this LCAO-MO can be evaluated analytically to obtain the result given in eqns 14.11 and 14.12. In this problem we evaluate the integrals numerically for S, j, and k and compare the results with the values determined from the analytically integrated forms of eqn 14.12. (a) Use the LCAO-MO wavefunction and the H_2^+ hamiltonian to derive equations for the Coulomb and resonance integrals in terms of j and k; do not integrate j and k analytically. Using mathematical software or an electronic spreadsheet, evaluate the overlap, Coulomb, and resonance integrals numerically, and the total energy for the $1s\sigma$ MO in the range $a_0 < R < 4a_0$. Compare the results obtained through numerical integration with results obtained with the analytical equations. (b) Use the results of the numerical integrations to draw a graph of the total energy, $E(R)$, and determine the minimum of total energy, the equilibrium internuclear distance, and the spectroscopic dissociation energy (D_e).

14.11‡ J.G. Dojahn, E.C.M. Chen, and W.E. Wentworth (*J. Phys. Chem.* **100**, 9649 (1996)) characterized the potential energy curves of homonuclear diatomic halogen molecules and molecular anions. Among the properties they report are the equilibrium internuclear distance R_e, the vibrational wavenumber, $\tilde{v}$, and the dissociation energy, D_e:

Species	r_e/pm	$\tilde{v}$/cm^{-1}	D_e/eV
F_2	1.411	916.6	1.60
F_2^-	1.900	450.0	1.31

Rationalize these data in terms of molecular orbital configurations.

14.12‡ *Rydberg molecules* are molecules with an electron in an atomic orbital with principal quantum number n one higher than the valence shells of the constituent atoms. Speculate about the existence of 'hyper Rydberg' H_2 formed from two H atoms with 100s electrons. Make reasonable guesses about the binding energy, the equilibrium internuclear separation, the vibrational force constant,

---

† Problems denoted with the symbol ‡ were supplied by Charles Trapp, Carmen Giunta, and Marshall Cady.

and the rotational constant. Is such a molecule likely to exist under any circumstances?

14.13‡ Set up and solve the Hückel secular equations for the π electrons of NO_3^-. Express the energies in terms of the Coulomb integrals α_O and α_N and the resonance integral β. Determine the delocalization energy of the ion.

14.14 In the 'free electron molecular orbital' (FEMO) theory, the electrons in a conjugated molecule are treated as independent particles in a box of length L. Sketch the form of the two occupied orbitals in butadiene predicted by this model and predict the minimum excitation energy of the molecule. The tetraene CH_2=CHCH=CHCH=CHCH=CH_2 can be treated as a box of length $8R$, where $R \approx 140$ pm (as in this case, an extra half bond-length is often added at each end of the box). Calculate the minimum excitation energy of the molecule and sketch the HOMO and LUMO. Estimate the colour a sample of the compound is likely to appear in white light.

14.15‡ The FEMO theory (Problem 14.14) of conjugated molecules is rather crude and better results are obtained with simple Hückel theory. (a) Prove that for an open chain of N conjugated carbons the characteristic polynomial of the secular determinant, $P_N(x)$, where $x = (\alpha - E)/\beta$, obeys the recurrence relation $P_N = xP_{N-1} - P_{N-2}$, with $P_1 = x$ and $P_0 = 1$. (b) Determine a reasonable empirical estimate of the resonance integral for the homologous series consisting of ethene, butadiene, hexatriene, and octatetraene given that $\pi^* \leftarrow \pi$ ultraviolet absorptions from the HOMO to the LUMO occur at 61 500, 46 080, 39 750, and 32 900 cm^{-1}, respectively. (c) Calculate the π-electron delocalization energy, $E_{deloc} = E_\pi - n(\alpha + \beta)$, of octatetraene, where E_π is the total π-electron binding energy and n is the total number of π electrons.

14.16 Molecular orbital calculations based on semi-empirical and *ab initio* methods describe the spectroscopic properties of conjugated molecules a bit better than simple Hückel theory. (a) Using molecular modelling software[16] and the computational method of your choice (extended Hückel, semi-empirical, or *ab initio* methods), calculate the energy separation between the HOMO and LUMO of ethene, butadiene, hexatriene, and octatetraene. (b) Plot the HOMO–LUMO energy separations versus the experimental frequencies for $\pi^* \leftarrow \pi$ ultraviolet absorptions for these molecules (Problem 14.15). Use mathematical software to find the polynomial equation that best fits the data. (c) Discuss why the calibration procedure of part (b) is necessary. (d) Use your polynomial fit from part (b) to estimate the frequency of the $\pi^* \leftarrow \pi$ ultraviolet absorption of decapentaene from the calculated HOMO–LUMO energy separation.

14.17 Electronic excitation of a molecule may weaken or strengthen some bonds because bonding and antibonding characteristics differ between the HOMO and the LUMO. For example, a carbon–carbon bond in a linear polyene may have bonding character in the HOMO and antibonding character in the LUMO. Therefore, promotion of an electron from the HOMO to the LUMO weakens this carbon–carbon bond in the excited electronic state, relative to the ground electronic state. Display the HOMO and LUMO of each molecule in Problem 14.16 and discuss in detail any changes in bond order that accompany the $\pi^* \leftarrow \pi$ ultraviolet absorptions in these molecules.

14.18 As mentioned in Section 2.8c, molecular electronic structure methods may be used to estimate the standard enthalpy of formation of molecules in the gas phase. (a) Using molecular modelling software and a semi-empirical method of your choice, calculate the standard enthalpy of formation of ethene, butadiene, hexatriene, and octatetraene in the gas phase. (b) Consult a database of thermochemical data, such as the online sources listed in this textbook's web site, and, for each molecule in part (a), calculate the relative error between the calculated and experimental values of the standard enthalpy of formation. (c) A good thermochemical database will also report the uncertainty in the experimental value of the standard enthalpy of formation. Compare experimental uncertainties with the relative errors calculated in part (b) and discuss the reliability of your chosen semi-empirical method for the estimation of thermochemical properties of linear polyenes.

Theoretical problems

14.19 An sp^2 hybrid orbital that lies in the xy-plane and makes an angle of $120°$ to the x-axis has the form

$$\psi = \frac{1}{3^{1/2}}\left(s - \frac{1}{2^{1/2}}p_x + \frac{3^{1/2}}{2^{1/2}}p_y\right)$$

Use hydrogenic atomic orbitals to write the explicit form of the hybrid orbital. Show that it has its maximum amplitude in the direction specified.

14.20 Use the expressions in Problems 14.8 and 14.9 to show that the antibonding orbital is more antibonding than the bonding orbital is bonding at most internuclear separations.

14.21 Derive the expressions used in Problems 14.8 and 14.9 using the normalized LCAO-MOs for the H_2^+ molecule-ion. Proceed by evaluating the expectation value of the hamiltonian for the ion. Make use of the fact that ψ_A and ψ_B each individually satisfy the Schrödinger equation for an isolated H atom.

14.22 Take as a trial function for the ground state of the hydrogen atom (a) e^{-kr}, (b) e^{-kr^2} and use the variation principle to find the optimum value of k in each case. Identify the better wavefunction. The only part of the laplacian that need be considered is the part that involves radial derivatives (eqn 12.62).

14.23‡ A highly accurate description of the molecular hydrogen ion bond length, R_e, is provided by an MO constructed with the variation parameter η, within the LCAO-MO ground state consisting of two 1s hydrogenic orbitals centred upon nuclei A and B. With the nuclear separation R, the MO is

$$\psi = N(e^{-\eta r_A/a_0} + e^{-\eta r_B/a_0}) \qquad N = \left(\frac{\eta^3}{2\pi a_0^3(1 + S)}\right)^{1/2}$$

(a) Use this wavefunction and the variation principle to determine η, R_e, the electronic energy (E_{el}), the minimum total energy (E), and D_e. The electronic Hamiltonian does not contain the nuclear repulsion term. Draw graphs of $\eta(R)$, $E_{el}(R)$, and $E(R)$. With this MO it is found that the electronic energy is the sum of the expectation value for kinetic energy ($\eta^2 F_1$) and the expectation value for electron potential energy (ηF_2):

---

16. The web site contains links to sites from which free software for molecular orbital calculations can be downloaded.

$$E_{el} = \frac{e^2\{\eta^2 F_1(\omega) + \eta F_2(\omega)\}}{4\pi\varepsilon_0 a_0} \qquad \omega = \frac{\eta R}{a_0}$$

$$F_1(\omega) = \frac{1 + (1 + \omega - \frac{1}{3}\omega^2)e^{-\omega}}{2\{1 + S(\omega)\}}$$

$$F_2(\omega) = \frac{(1 + \omega)(e^{-2\omega} - 2\omega e^{-\omega} - 1)}{\omega\{1 + S(\omega)\}}$$

$$S(\omega) = (1 + \omega + \tfrac{1}{3}\omega^2)e^{-\omega}$$

(b) Check numerically to determine whether or not the virial theorem is satisfied by this solution. (c) Prove that the overlap integral, S, is correctly described by the above expression. The integration is facilitated with the ellipsoidal coordinates (μ, v, ϕ) defined by the following relations:

$$R\mu = r_A + r_B \qquad rv = r_A - r_B \qquad d\tau = \tfrac{1}{8}R^3(\mu^2 - v^2)\,d\mu\,dv\,d\phi$$

with $1 \leq \mu \leq \infty$, $-1 \leq v \leq 1$, and $0 \leq \phi \leq 2\pi$.

Applications: to astrophysics and biology

14.24‡ In Exercise 14.15a you were invited to set up the Hückel secular determinant for linear and cyclic H_3. The same secular determinant applies to the molecular ions H_3^+ and D_3^+. The molecular ion H_3^+ was discovered as long ago as 1912 by J.J. Thomson, but only more recently has the equivalent equilateral triangular structure been confirmed by M.J. Gaillard *et al.* (*Phys. Rev.* **A17**, 1797 (1978)). The molecular ion H_3^+ is the simplest polyatomic species with a confirmed existence and plays an important role in chemical reactions occurring in interstellar clouds that may lead to the formation of water, carbon monoxide, and ethyl alcohol. The H_3^+ ion has also been found in the atmospheres of Jupiter, Saturn, and Uranus. (a) Solve the Hückel secular equations for the energies of the H_3 system in terms of the parameters α and β, draw an energy level diagram for the orbitals, and determine the binding energies of H_3^{2+}, H_3^+, H_3, and H_3^-. (b) Accurate quantum mechanical calculations by G.D. Carney and R.N. Porter (*J. Chem. Phys.* **65**, 3547 (1976)) give the dissociation energy for the process $H_3^+(g) \rightarrow H(g) + H(g) + H^+(g)$ as 849 kJ mol^{-1}. From this information and data in Table 2.6, calculate the enthalpy of the reaction $H^+(g) + H_2(g) \rightarrow H_3^+(g)$. (c) From your equations and the information given, calculate a value for the resonance integral β in H_3^+. Then go on to calculate the bind energies of the other H_3 species in (a).

14.25‡ There is some indication that other hydrogen ring compounds and ions in addition to H_3 and D_3 species may play a role in interstellar chemistry. According to J.S. Wright and G.A. DiLabio (*J. Phys. Chem.* **96**, 10793 (1992)), H_5^-, H_6, and H_7^+ are particularly stable whereas H_4 and H_5^+ are not. Confirm these statements by performing Hückel calculations.

Performing

14.26 Molecular orbital calculations may be used to predict trends in the standard potentials of conjugated molecules, such as the

6

quinones and flavins, that are involved in biological electron-transfer reactions (Box 10.1). It is commonly assumed that decreasing the energy of the LUMO enhances the ability of a molecule to accept an electron into the LUMO, with an attendant increase in the value of the molecule's standard potential. Furthermore, a number of studies indicate that there is a linear correlation between the LUMO energy and the reduction potential of aromatic hydrocarbons (see, for example, J.P. Lowe, *Quantum chemistry*, Chapter 8, Academic Press (1993)). (a) The standard potentials at pH 7 for the one-electron reduction of methyl-substituted 1,4-benzoquinones (**6**) to their respective semiquinone radical anions are:

R_2	R_3	R_5	R_6	$E^{\ominus}$/V
H	H	H	H	0.078
CH$_3$	H	H	H	0.023
CH$_3$	H	CH$_3$	H	−0.067
CH$_3$	CH$_3$	CH$_3$	H	−0.165
CH$_3$	CH$_3$	CH$_3$	CH$_3$	−0.260

Using molecular modelling software and the computational method of your choice (extended Hückel, semi-empirical, or *ab initio* methods), calculate E_{LUMO}, the energy of the LUMO of each substituted 1,4-benzoquinone, and plot E_{LUMO} against $E^{\ominus}$. Do your calculations support a linear relation between E_{LUMO} and $E^{\ominus}$? (b) The 1,4-benzoquinone for which $R_2 = R_3 = CH_3$ and $R_5 = R_6 = OCH_3$ is a suitable model of ubiquinone, a component of the respiratory electron transport chain (Box 10.2). Determine E_{LUMO} of this quinone and then use your results from part (a) to estimate its standard potential. (c) The 1,4-benzoquinone for which $R_2 = R_3 = R_5 = CH_3$ and $R_6 = H$ is a suitable model of plastoquinone, a component of the photosynthetic electron transport chain (Box 10.1). Determine E_{LUMO} of this quinone and then use your results from part (a) to estimate its standard potential. Is plastoquinone expected to be a better or worse oxidizing agent than ubiquinone? (d) Based on your predictions and on basic concepts of biological electron transport (Box 10.1), suggest a reason why ubiquinone is used in respiration and plastoquinone is used in photosynthesis.

15 Molecular symmetry

The symmetry elements of objects

15.1 Operations and symmetry elements

15.2 The symmetry classification of molecules

15.3 Some immediate consequences of symmetry

Character tables

15.4 Character tables and symmetry labels

15.5 Vanishing integrals and orbital overlap

15.6 Vanishing integrals and selection rules

Checklist of key ideas

Further reading

Exercises

Problems

In this chapter we sharpen the concept of 'shape' into a precise definition of 'symmetry', and show that symmetry may be discussed systematically. We see how to classify any molecule according to its symmetry and how to use this classification to discuss molecular properties. After describing the symmetry properties of molecules themselves, we turn to a consideration of the effect of symmetry transformations on orbitals and see that their transformation properties can be used to set up a labelling scheme. These symmetry labels are used to identify integrals that necessarily vanish. One important integral is the overlap integral between two orbitals. By knowing which atomic orbitals may have nonzero overlap, we can decide which ones can contribute to molecular orbitals. We also see how to select linear combinations of atomic orbitals that match the symmetry of the nuclear framework. Finally, by considering the symmetry properties of integrals, we see that it is possible to derive the selection rules that govern spectroscopic transitions.

The systematic discussion of symmetry is called **group theory**. Much of group theory is a summary of common sense about the symmetries of objects. However, because group theory is systematic, its rules can be applied in a straightforward, mechanical way. In most cases the theory gives a simple, direct method for arriving at useful conclusions with the minimum of calculation, and this is the aspect we stress here. In some cases, though, they lead to unexpected results.

The symmetry elements of objects

Some objects are 'more symmetrical' than others. A sphere is more symmetrical than a cube because it looks the same after it has been rotated through any angle about any diameter. A cube looks the same only if it is rotated through certain angles about specific axes, such as 90°, 180°, or 270° about an axis passing through the centres of any of its opposite faces (Fig. 15.1), or by 120° or 240° about an axis passing through any of its opposite corners. Similarly, an NH_3 molecule is 'more symmetrical' than an H_2O molecule because NH_3 looks the same after rotations of 120° or 240° about the axis shown in Fig. 15.2, whereas H_2O looks the same only after a rotation of 180°.

An action that leaves an object looking the same after it has been carried out is called a **symmetry operation**. Typical symmetry operations include rotations, reflections, and inversions. There is a corresponding **symmetry element** for each

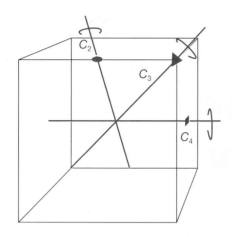

15.1 Some of the symmetry elements of a cube. The twofold, threefold, and fourfold axes are labelled with the conventional symbols.

15.2 (a) An NH_3 molecule has a threefold (C_3) axis and (b) an H_2O molecule has a twofold (C_2) axis. Both have other symmetry elements too.

1 CBrClFI

2 Benzene

symmetry operation, which is the point, line, or plane with respect to which the symmetry operation is performed. For instance, a rotation (a symmetry operation) is carried out around an axis (the corresponding symmetry element). We shall see that we can classify molecules by identifying all their symmetry elements, and grouping together molecules that possess the same set of symmetry elements. This procedure, for example, puts the trigonal pyramidal species NH_3 and SO_3^{2-} into one group and the angular species H_2O and SO_2 into another group.

15.1 Operations and symmetry elements

The classification of objects according to symmetry elements corresponding to operations that leave at least one common point unchanged gives rise to the **point groups**. There are five kinds of symmetry operation (and five kinds of symmetry element) of this kind. When we consider crystals (Chapter 23), we shall meet symmetries arising from translation through space. These more extensive groups are called **space groups**.

The **identity**, E, consists of doing nothing; the corresponding symmetry element is the entire object. Because every molecule is indistinguishable from itself if nothing is done to it, every object possesses at least the identity element. One reason for including the identity is that some molecules have only this symmetry element (**1**); another reason is technical and connected with the detailed formulation of group theory.

An **n-fold rotation** (the operation) about an **n-fold axis of symmetry**, C_n (the corresponding element), is a rotation through $360°/n$. The operation C_1 is a rotation through $360°$, and is equivalent to the identity operation E. An H_2O molecule has one twofold axis, C_2. An NH_3 molecule has one threefold axis, C_3, with which is associated two symmetry operations, one being $120°$ rotation in a clockwise sense and the other $120°$ rotation in a counter-clockwise sense.[1] A pentagon has a C_5 axis, with two (clockwise and counter-clockwise) rotations through $72°$ associated with it. It also has an axis denoted C_5^2, corresponding to two successive C_5 rotations; there are two such operations, one through $144°$ in a clockwise sense and the other through $144°$ in a counter-clockwise sense. A cube has three C_4 axes, four C_3 axes, and six C_2 axes. However, even this high symmetry is exceeded by a sphere, which possesses an infinite number of symmetry axes (along any diameter) of all possible integral values of n. If a molecule possesses several rotation axes, then the one (or more) with the greatest value of n is called the **principal axis**. The principal axis of a benzene molecule is the sixfold axis perpendicular to the hexagonal ring (**2**).

1. There is only one twofold rotation associated with a C_2 axis because clockwise and counter-clockwise $180°$ rotations are identical.

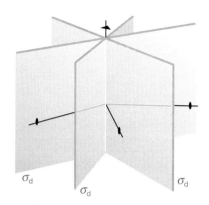

15.3 An H$_2$O molecule has two mirror planes. They are both vertical (i.e. contain the principal axis), so are denoted σ_v and σ_v'.

15.4 Dihedral mirror planes (σ_d) bisect the C_2 axes perpendicular to the principal axis.

A **reflection** (the operation) in a **mirror plane**, σ (the element), may contain the principal axis of a molecule or be perpendicular to it. If the plane is parallel to the principal axis, it is called 'vertical' and denoted σ_v. An H$_2$O molecule has two vertical planes of symmetry (Fig. 15.3) and an NH$_3$ molecule has three. A vertical mirror plane that bisects the angle between two C_2 axes is called a 'dihedral plane' and is denoted σ_d (Fig. 15.4). When the plane of symmetry is perpendicular to the principal axis it is called 'horizontal' and denoted σ_h. A C$_6$H$_6$ molecule has a C_6 principal axis and a horizontal mirror plane (as well as several other symmetry elements).

In an **inversion** (the operation) through a **centre of symmetry**, i (the element), we imagine taking each point in a molecule, moving it to the centre of the molecule, and then moving it out the same distance on the other side; that is, the point (x, y, z) is taken into the point $(-x, -y, -z)$. Neither an H$_2$O molecule nor an NH$_3$ molecule has a centre of inversion, but a sphere and a cube do have one. A C$_6$H$_6$ molecule does have a centre of inversion, as does a regular octahedron (Fig. 15.5); a regular tetrahedron and a CH$_4$ molecule do not.

An *n*-fold **improper rotation** (the operation) about an *n*-fold axis of improper rotation or an *n*-fold **improper rotation axis**, S_n, (the symmetry element) is composed of two successive transformations. The first component is a rotation through $360°/n$, and the second is a reflection through a plane perpendicular to the axis of that rotation; neither operation alone needs to be a symmetry operation. A CH$_4$ molecule has three S_4 axes (Fig. 15.6).

Centre of inversion, i

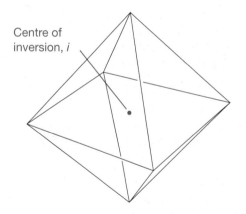

15.5 A regular octahedron has a centre of inversion (i).

15.6 (a) A CH$_4$ molecule has a fourfold improper rotation axis (S_4): the molecule is indistinguishable after a 90° rotation followed by a reflection across the horizontal plane, but neither operation alone is a symmetry operation. (b) The staggered form of ethane has an S_6 axis composed of a 60° rotation followed by a reflection.

(a)

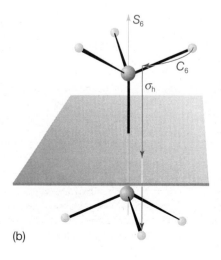

(b)

Table 15.1 The notation for point groups*

C_i	$\bar{1}$									
C_s	m									
C_1	1	C_2	2	C_3	3	C_4	4	C_6	6
		C_{2v}	$2mm$	C_{3v}	$3m$	C_{4v}	$4mm$	C_{6v}	$6mm$
		C_{2h}	$2/m$	C_{3h}	$\bar{6}$	C_{4h}	$4/m$	C_{6h}	$6/m$
		D_2	222	D_3	32	D_4	422	D_6	622
		D_{2h}	mmm	D_{3h}	$\bar{6}2m$	D_{4h}	$4/mmm$	D_{6h}	$6/mmm$
		D_{2d}	$\bar{4}2m$	D_{3d}	$\bar{3}m$	S_4	$\bar{4}/m$	S_6	$\bar{3}$
T	23	T_d	$\bar{4}3m$	T_h	$m3$				
O	432	O_h	$m3m$						

* In the International system (or Hermann–Mauguin system) for point groups, a number n denotes the presence of an n-fold axis and m denotes a mirror plane. A diagonal line / indicates that the mirror plane is perpendicular to the symmetry axis. It is important to distinguish symmetry elements of the same type but of different classes, as in $4/mmm$, in which there are three classes of mirror plane (σ_v, σ_h, and σ_d). A bar over a number indicates that the element is combined with an inversion. The only groups listed in this table are the so-called crystallographic point groups (Section 23.1).

COOH

HO

H

Centre of inversion

H

OH

HOOC

3 Meso-tartaric acid

4 Quinoline

C_2

O—————O

H

H

5 H_2O_2

σ_h

Cl

H

C

C_2

C

H

Cl

6 *trans*-CHCl=CHCl

15.2 The symmetry classification of molecules

To classify molecules according to their symmetries, we list their symmetry elements and collect together molecules with the same list of elements. This procedure puts CH_4 and CCl_4, which both possess the same symmetry elements as a regular tetrahedron, into the same group, and H_2O into another group.

The name of the group to which a molecule belongs is determined by the symmetry elements it possesses. There are two systems of notation (Table 15.1). The **Schoenflies system** (in which a name looks like C_{4v}) is more common for the discussion of individual molecules, and the **Hermann–Mauguin system**, or **International system** (in which a name looks like $4mm$), is used almost exclusively in the discussion of crystal symmetry.

(a) The groups C_1, C_i, and C_s

A molecule belongs to the group C_1 if it has no element other than the identity, as in (**1**). It belongs to C_i if it has the identity and the inversion alone (**3**), and to C_s if it has the identity and a mirror plane alone (**4**).

(b) The groups C_n, C_{nv}, and C_{nh}

A molecule belongs to the group C_n if it possesses an n-fold axis.[2] An H_2O_2 molecule has the elements E and C_2 (**5**), so it belongs to the group C_2.

If, in addition to the identity and a C_n axis, a molecule has n vertical mirror planes σ_v, then it belongs to the group C_{nv}. An H_2O molecule, for example, has the symmetry elements E, C_2, and $2\sigma_v$, so it belongs to the group C_{2v}. An NH_3 molecule has the elements E, C_3, and $3\sigma_v$, so it belongs to the group C_{3v}. A heteronuclear diatomic molecule such as HCl belongs to the group $C_{\infty v}$ because all rotations around the axis and reflections across the axis are symmetry operations. Other members of the group $C_{\infty v}$ include the linear OCS molecule and a cone.

Objects that, in addition to the identity and an n-fold principal axis, also have a horizontal mirror plane σ_h belong to the groups C_{nh}. An example is *trans*-CHCl=CHCl (**6**),

2. Note that symbol C_n is now playing a triple role: as the label of a symmetry element, a symmetry operation, and a group.

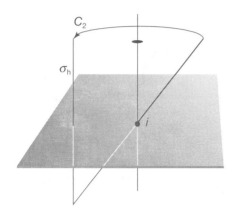

15.7 The presence of a twofold axis and a horizontal mirror plane jointly imply the presence of a centre of inversion in the molecule.

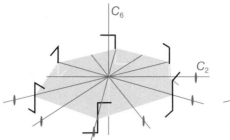

15.8 A molecule with n twofold rotation axes perpendicular to an n-fold rotation axis belongs to the group D_n.

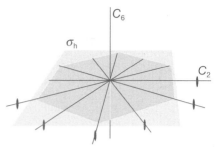

15.9 A molecule with a mirror plane perpendicular to a C_n axis, and with n twofold axes in the plane, belongs to the group D_{nh}.

7 B(OH)$_3$

8 Boron trifluoride, BF$_3$

which has the elements E, C_2, and σ_h, so belongs to the group C_{2h}; the molecule B(OH)$_3$ in the conformation shown in (**7**) belongs to the group C_{3h}. The presence of certain symmetry elements may be implied by the presence of others: thus, in C_{2h} the operations C_2 and σ_h jointly imply the presence of a centre of inversion (Fig. 15.7).

(c) The groups D_n, D_{nh}, and D_{nd}

A molecule that has an n-fold principal axis and n twofold axes perpendicular to C_n belongs to the group D_n (Fig. 15.8). A molecule belongs to D_{nh} if it also possesses a horizontal mirror plane (Fig. 15.9). The planar trigonal BF$_3$ molecule has the elements E, C_3, $3C_2$, and σ_h (with one C_2 axis along each B—F bond), so it belongs to D_{3h} (**8**). The C$_6$H$_6$ molecule has the elements E, C_6, $3C_2$, $3C_2'$, and σ_h together with some others that these elements imply,[3] so it belongs to D_{6h}. All homonuclear diatomic molecules, such as N$_2$, belong to the group $D_{\infty h}$ because all rotations around the axis are symmetry operations, as are end-to-end rotation and end-to-end reflection; $D_{\infty h}$ is also the group of the linear OCO and HCCH molecules and of a uniform cylinder. Other examples of D_{nh} molecules are shown in (**9**), (**10**), and (**11**).

A molecule belongs to the group D_{nd} if in addition to the elements of D_n it possesses n dihedral mirror planes σ_d. The twisted, 90° allene (**12**) belongs to D_{2d}, and the staggered conformation of ethane (**13**) belongs to D_{3d}.

9 Ethene, CH$_2$=CH$_2$ (D_{2h})

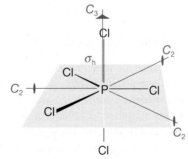

10 Phosphorus pentachloride, PCl$_5$ (D_{3h})

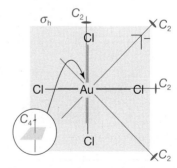

11 Tetrachloroaurate(III), [AuCl$_4$]$^-$ (D_{4h})

3. The prime on $3C_2'$ indicates that the three C_2 axes are different from the other three C_2 axes. In benzene, three of the C_2 axes bisect C—C bonds and the other three pass through vertices of the hexagon formed by the carbon framework of the molecule.

12 Allene, C_3H_2 (D_{2d})

13 Ethane, CH_3CH_3 (D_{3d})

14 Tetraphenylmethane, $C(C_6H_5)_4$ (S_4)

15 Buckminsterfullerene, C_{60} (*I*)

(a)

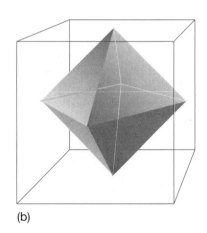

(b)

15.10 (a) Tetrahedral and (b) octahedral molecules are drawn in a way that shows their relation to a cube: they belong to the cubic groups T_d and O_h, respectively.

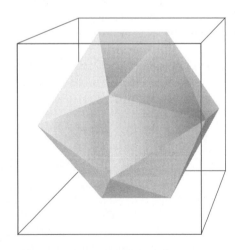

15.11 The relation of an icosahedron to a cube. The buckminsterfullerene molecule (**15**) is related to this object by cutting off each apex to form a regular pentagon.

(d) The groups S_n

Molecules that have not been classified into one of the groups mentioned so far, but which possess one S_n axis, belong to the group S_n. An example is tetraphenyl-methane, which belongs to the point group S_4 (**14**). Molecules belonging to S_n with $n > 4$ are rare. Note that the group S_2 is the same as C_i, so such a molecule will already have been classified as C_i.

(e) The cubic groups

A number of very important molecules (for example, CH_4 and SF_6) possess more than one principal axis. Most belong to the **cubic groups**, and in particular to the **tetrahedral groups** T, T_d, and T_h or to the **octahedral groups** O and O_h (Fig. 15.10). A few icosahedral (20-faced) molecules belonging to the **icosahedral group**, I (Fig. 15.11), are also known: they include some of the boranes and buckminster-fullerene, C_{60} (**15**). The groups T_d and O_h are the groups of the regular tetrahedron (for example, CH_4) and the regular octahedron (for example, SF_6), respectively. If the object possesses the rotational symmetry of the tetrahedron or the octahedron, but none of their planes of reflection, then it belongs to the simpler groups T or O (Fig. 15.12). The group T_h is based on T but also contains a centre of inversion (Fig. 15.13).

(a) *T*

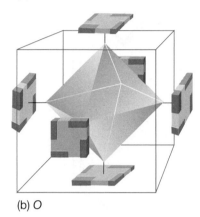

(b) *O*

15.12 Shapes corresponding to the point groups (a) *T* and (b) *O*. The presence of the decorated slabs reduces the symmetry of the object from T_d and O_h, respectively.

The web site contains links to interactive tutorials, where you use your web browser to assign point groups of molecules.

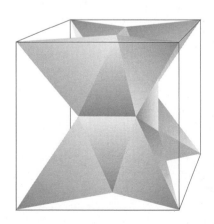

15.13 The shape of an object belonging to the group T_h.

16 Ruthenocene, $Ru(C_6H_5)_2$

17 Excited ferrocene, $Fe(C_6H_5)_2^*$

(f) The full rotation group

The **full rotation group**, R_3 (the 3 refers to rotation in three dimensions), consists of an infinite number of rotation axes with all possible values of n. A sphere and an atom belong to R_3, but no molecule does. Exploring the consequences of R_3 is a very important way of applying symmetry arguments to atoms, and is an alternative approach to the theory of orbital angular momentum.

Example 15.1 Identifying a point group of a molecule

Identify the point group to which a ruthenocene molecule (**16**) belongs.

Method The identification of a molecule's point group is simplified by referring to the flow diagram in Fig. 15.14 and the shapes shown in Fig. 15.15.

Answer The path to trace through the flow diagram in Fig. 15.14 is shown by a blue line; it ends at D_{nh}. Because the molecule has a fivefold axis, it belongs to the group D_{5h}.

Comment If the rings were staggered, as they are in an excited state of ferrocene that lies 4 kJ mol^{-1} above the ground state (**17**), the horizontal reflection plane would be absent, but dihedral planes would be present.

Self-test 15.1 Classify the pentagonal antiprismatic excited state of ferrocene (**17**). $[D_{5d}]$

15.3 Some immediate consequences of symmetry

Some statements about the properties of a molecule can be made as soon as its point group has been identified.

(a) Polarity

A **polar molecule** is one with a permanent electric dipole moment (HCl, O_3, and NH_3 are examples). If the molecule belongs to the group C_n with $n > 1$, it cannot possess a charge distribution with a dipole moment perpendicular to the symmetry

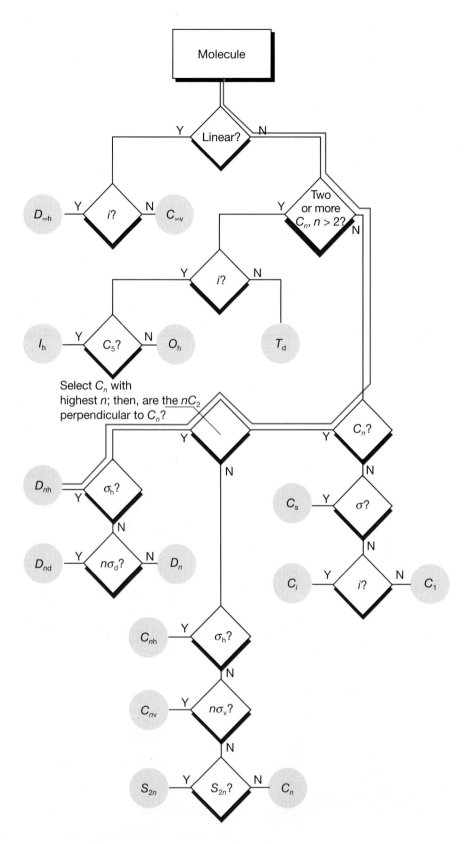

15.14 A flow diagram for determining the point group of a molecule. Start at the top and answer the question posed in each diamond (Y = yes, N = no).

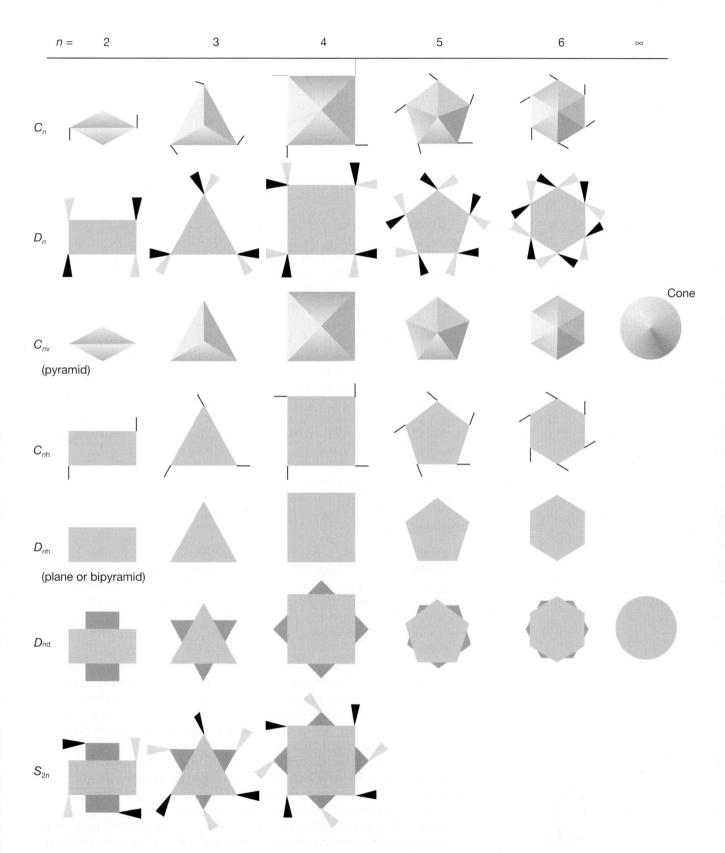

15.15 A summary of the shapes corresponding to different point groups. The group to which a molecule belongs can often be identified from this diagram without going through the formal procedure in Fig. 15.14.

18 L-Alanine, $NH_2CH(CH_3)COOH$ **19** Glycine, NH_2CH_2COOH

(a)   (b)

15.16 (a) A molecule with a C_n axis cannot have a dipole perpendicular to the axis, but (b) it may have one parallel to the axis. The arrows represent local contributions to the overall electric dipole, such as may arise from bonds between pairs of neighbouring atoms with different electronegativities.

axis because the symmetry of the molecule implies that any dipole that exists in one direction perpendicular to the axis is cancelled by an opposing dipole (Fig. 15.16a). For example, the perpendicular component of the dipole associated with one O—H bond in H_2O is cancelled by an equal but opposite component of the dipole of the second O—H bond, so any dipole that the molecule has must be parallel to the twofold symmetry axis. However, as the group makes no reference to operations relating the two ends of the molecule, a charge distribution may exist that results in a dipole along the axis (Fig. 15.16b), and H_2O has a dipole moment parallel to its twofold symmetry axis. The same remarks apply generally to the group C_{nv}, so molecules belonging to any of the C_{nv} groups may be polar. In all the other groups, such as C_{3h}, D, etc., there are symmetry operations that take one end of the molecule into the other. Therefore, as well as having no dipole perpendicular to the axis, such molecules can have none along the axis, for otherwise these additional operations would not be symmetry operations. We can conclude that *only molecules belonging to the groups C_n, C_{nv}, and C_s may have a permanent electric dipole moment.*

For C_n and C_{nv}, that dipole moment must lie along the symmetry axis. Thus ozone, O_3, which is angular and belongs to the group C_{2v}, may be polar (and is), but carbon dioxide, CO_2, which is linear and belongs to the group $D_{\infty h}$, is not.

(b) Chirality

A **chiral molecule** (from the Greek word for 'hand') is a molecule that cannot be superimposed on its mirror image. An **achiral molecule** is a molecule that can be superimposed on its mirror image. Chiral molecules are **optically active** in the sense that they rotate the plane of polarized light (a property discussed in more detail in Section 21.2). A chiral molecule and its mirror-image partner constitute an **enantiomeric pair** of isomers and rotate the plane of polarization in equal but opposite directions.

A molecule may be chiral, and therefore optically active, only if it does not possess an axis of improper rotation, S_n. However, we need to be aware that such an axis may be present under a different name, and be implied by other symmetry elements that are present. For example, molecules belonging to the groups C_{nh} possess an S_n axis implicitly because they possess both C_n and σ_h, which are the two components of an improper rotation axis. Any molecule containing a centre of inversion, i, also possesses an S_2 axis, because i is equivalent to C_2 in conjunction with σ_h, and that combination of elements is S_2 (Fig. 15.17). It follows that all molecules with centres of inversion are achiral and hence optically inactive. Similarly, because $S_1 = \sigma$, it follows that any molecule with a mirror plane is achiral.

A molecule may be chiral if it does not have a centre of inversion or a mirror plane, which is the case with the amino acid alanine (**18**), but not with glycine (**19**). However, a molecule may be achiral even though it does not have a centre of inversion. For example, the S_4 species (**20**) is achiral and optically inactive: though it lacks i (that is, S_2) it does have an S_4 axis.

20 $N(CH_2CH(CH_3)CH(CH_3)CH_2)_2^+$

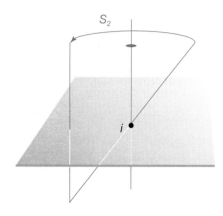

15.17 Some symmetry elements are implied by the other symmetry elements in a group. Any molecule containing an inversion also possesses at least an S_2 element because i and S_2 are equivalent.

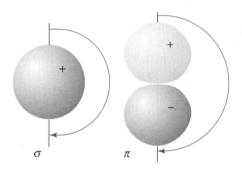

15.18 A rotation through 180° about the internuclear axis (perpendicular to the page) leaves the sign of a σ orbital unchanged but the sign of a π orbital is changed. In the language introduced in this chapter, the characters of the C_2 rotation are +1 and –1 for the σ and π orbitals, respectively.

	C_2	(i.e. rotation by 180°)
σ	+1	(i.e. no change of sign)
π	–1	(i.e. change of sign)

See *Further information 1* for a summary of the rules of matrix algebra.

Character tables

We shall now turn our attention away from the symmetries of molecules themselves and direct it towards the symmetry characteristics of orbitals that belong to the various atoms in a molecule. This material will enable us to discuss the formulation and labelling of molecular orbitals and selection rules in spectroscopy.

15.4 Character tables and symmetry labels

We saw in Chapter 14 that molecular orbitals of diatomic and linear polyatomic molecules are labelled σ, π, etc. These labels refer to the symmetries of the orbitals with respect to rotations around the principal symmetry axis of the molecule. Thus, a σ orbital does not change sign under a rotation through any angle, a π orbital changes sign when rotated by 180°, and so on (Fig. 15.18). The symmetry classifications σ and π can also be assigned to individual atomic orbitals in a linear molecule. For example, we can speak of an individual p_z orbital as having σ symmetry if the z-axis lies along the bond, because p_z is cylindrically symmetrical about the bond. This labelling of orbitals according to their behaviour under rotations can be generalized and extended to nonlinear polyatomic molecules, where there may be reflections and inversions to take into account as well as rotations.

(a) Representations and characters

Labels analogous to σ and π are used to denote the symmetries of orbitals in polyatomic molecules. These labels look like a, a_1, e, e_g, and we first encountered them in Fig. 14.38 in connection with the molecular orbitals of benzene. As we shall see, these labels indicate the behaviour of the orbitals under the symmetry operations of the relevant point group of the molecule.

A label is assigned to an orbital by referring to the **character table** of the group, a table that characterizes the different symmetry types possible in the point group. Thus, to assign the labels σ and π, we use the table shown in the margin. This table is a fragment of the full character table for a linear molecule. The entry +1 shows that the orbital remains the same and the entry –1 shows that the orbital changes sign under the operation C_2 at the head of the column (as illustrated in Fig. 15.18). So, to assign the label σ or π to a particular orbital, we compare the orbital's behaviour with the information in the character table.

The entries in a complete character table are derived by using the formal techniques of group theory and are called **characters**, χ (chi). These numbers characterize the essential features of each symmetry type in a way that we can illustrate by considering the C_{2v} molecule SO_2 and the valence p_x orbitals on each atom, which we shall denote p_S, p_A, and p_B (Fig. 15.19).

Under σ_v, the change $(p_S, p_B, p_A) \leftarrow (p_S, p_A, p_B)$ takes place. We can express this transformation by using matrix multiplication:

$$(p_S, p_B, p_A) = (p_S, p_A, p_B)\begin{pmatrix} 1 & 0 & 0 \\ 0 & 0 & 1 \\ 0 & 1 & 0 \end{pmatrix} = (p_S, p_A, p_B)D(\sigma_v) \tag{15.1}$$

The matrix $D(\sigma_v)$ is called a **representative** of the operation σ_v. Representatives take different forms according to the **basis**, the set of orbitals, that has been adopted.

We can use the same technique to find matrices that reproduce the other symmetry operations. For instance, C_2 has the effect $(-p_S, -p_B, -p_A) \leftarrow (p_S, p_A, p_B)$, and its representative is

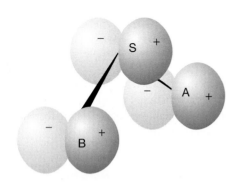

15.19 The three p_x orbitals that are used to illustrate the construction of a matrix representation in a C_{2v} molecule (SO_2).

$$D(C_2) = \begin{pmatrix} -1 & 0 & 0 \\ 0 & 0 & -1 \\ 0 & -1 & 0 \end{pmatrix} \tag{15.2}$$

The effect of σ_v' is $(-p_S, -p_A, -p_B) \leftarrow (p_S, p_A, p_B)$, and its representative is

$$D(\sigma_v') = \begin{pmatrix} -1 & 0 & 0 \\ 0 & -1 & 0 \\ 0 & 0 & -1 \end{pmatrix} \tag{15.3}$$

The identity operation has no effect on the basis, so its representative is the 3×3 unit matrix:

$$D(E) = \begin{pmatrix} 1 & 0 & 0 \\ 0 & 1 & 0 \\ 0 & 0 & 1 \end{pmatrix} \tag{15.4}$$

The set of matrices that represents *all* the operations of the group is called a **matrix representation**, Γ (uppercase gamma), of the group for the particular basis we have chosen. We denote this three-dimensional representation by $\Gamma^{(3)}$. The discovery of a matrix representation of the group means that we have found a link between symbolic manipulations of operations and algebraic manipulations of numbers.

The character of an operation in a particular matrix representation is the sum of the diagonal elements of the representative of that operation. Thus, in the basis we are illustrating, the characters of the representatives are

$D(E)$	$D(C_2)$	$D(\sigma_v)$	$D(\sigma_v')$
3	−1	1	−3

The character of an operation depends on the basis.

Inspection of the representatives shows that they are all of **block-diagonal form**:

$$D = \begin{pmatrix} \blacksquare & 0 & 0 \\ 0 & \blacksquare & \\ 0 & & \end{pmatrix}$$

The block-diagonal form of the representatives show us that the symmetry operations of C_{2v} never mix p_S with the other two functions. Consequently, the basis can be cut into two parts, one consisting of p_S alone and the other of (p_A, p_B). It is readily verified that the p_S orbital itself is a basis for the one-dimensional representation

$$D(E) = 1 \qquad D(C_2) = -1 \qquad D(\sigma_v) = 1 \qquad D(\sigma_v') = -1$$

which we shall call $\Gamma^{(1)}$. The remaining two basis functions are a basis for the two-dimensional representation $\Gamma^{(2)}$:

$$D(E) = \begin{pmatrix} 1 & 0 \\ 0 & 1 \end{pmatrix} \qquad D(C_2) = \begin{pmatrix} 0 & -1 \\ -1 & 0 \end{pmatrix} \qquad D(\sigma_v) = \begin{pmatrix} 0 & 1 \\ 1 & 0 \end{pmatrix} \qquad D(\sigma_v') = \begin{pmatrix} -1 & 0 \\ 0 & -1 \end{pmatrix}$$

These matrices are the same as those of the original three-dimensional representation, except for the loss of the first row and column. We say that the original three-dimensional representation has been **reduced** to the 'direct sum' of a one-dimensional representation 'spanned' by p_S, and a two-dimensional representation spanned by (p_A, p_B). This reduction is consistent with the common sense view that the central orbital plays a role different from the other two. We denote the reduction symbolically by writing

$$\Gamma^{(3)} = \Gamma^{(1)} + \Gamma^{(2)} \tag{15.5}$$

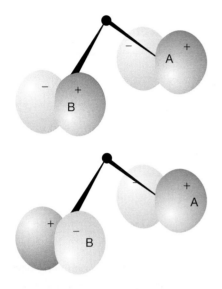

15.20 Two symmetry-adapted linear combinations of the basis orbitals shown in Fig. 15.19. The two combinations each span a one-dimensional irreducible representation, and their symmetry species are different.

The one-dimensional representation $\Gamma^{(1)}$ cannot be reduced any further, and is called an **irreducible representation** of the group (an 'irrep'). We can demonstrate that the two-dimensional representation $\Gamma^{(2)}$ is reducible (for this basis in this group) by switching attention to the linear combinations $p_1 = p_A + p_B$ and $p_2 = p_A - p_B$. These combinations are sketched in Fig. 15.20. The representatives in the new basis can be constructed from the old by noting, for example, that under σ_v, $(p_B, p_A) \leftarrow (p_A, p_B)$. In this way we find the following representation in the new basis:

$$D(E) = \begin{pmatrix} 1 & 0 \\ 0 & 1 \end{pmatrix} \quad D(C_2) = \begin{pmatrix} -1 & 0 \\ 0 & 1 \end{pmatrix} \quad D(\sigma_v) = \begin{pmatrix} 1 & 0 \\ 0 & -1 \end{pmatrix} \quad D(\sigma_v') = \begin{pmatrix} -1 & 0 \\ 0 & -1 \end{pmatrix}$$

The new representatives are all in block-diagonal form, and the two combinations are not mixed with each other by any operation of the group. We have therefore achieved the reduction of $\Gamma^{(2)}$ to the sum of two one-dimensional representations. Thus, p_1 spans

$$D(E) = 1 \quad D(C_2) = -1 \quad D(\sigma_v) = 1 \quad D(\sigma_v') = -1$$

which is the same one-dimensional representation as that spanned by p_S, and p_2 spans

$$D(E) = 1 \quad D(C_2) = 1 \quad D(\sigma_v) = -1 \quad D(\sigma_v') = -1$$

which is a different one-dimensional representation; we shall denote it $\Gamma^{(1)'}$.

At this point we have found two irreducible representations of the group C_{2v} (Table 15.2). The two irreducible representations are normally labelled B_1 and A_2, respectively. An A or a B is used to denote a one-dimensional representation; A is used if the character under the principal rotation is +1, and B is used if the character is –1. Subscripts are used to distinguish the irreducible representations if there is more than one of the same type: A_1 is reserved for the representation with character 1 for all operations. When higher dimensional irreducible representations are permitted, E denotes a two-dimensional irreducible representation and T a three-dimensional irreducible representation; all the irreducible representations of C_{2v} are one-dimensional.

There are, in fact, only two more species of irreducible representations of this group, for a surprising theorem of group theory states that

Number of symmetry species = number of classes (15.6)

Symmetry operations fall into the same **class** if they are of the same type (for example, rotations) and can be transformed into one another by a symmetry operation of the group. In C_{2v}, for instance, there are four classes (four columns in the character table), so there are only four species of irreducible representation. The character table in Table 15.2 therefore shows the characters of all the irreducible representations of this group.

Table 15.2* The C_{2v} character table

C_{2v}, 2mm	E	C_2	σ_v	σ_v'	$h = 4$	
A_1	1	1	1	1	z	z^2, y^2, x^2
A_2	1	1	–1	–1		xy
B_1	1	–1	1	–1	x	xz
B_2	1	–1	–1	1	y	yx

* More character tables are given at the end of the *Data section*.

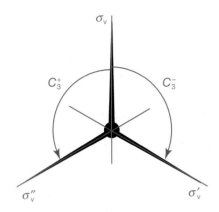

15.21 Symmetry operations in the same class are related to one another by the symmetry operations of the group. Thus, the three mirror frames shown here are related by threefold rotations, and the two rotations shown here are related by reflection in σ_v.

15.22 Typical symmetry-adapted linear combinations of orbitals in a C_{3v} molecule.

Table 15.3* The C_{3v} character table

C_{3v}, $3m$	E	$2C_3$	$3\sigma_v$	$h = 6$	
A_1	1	1	1	z	z^2, $x^2 + y^2$
A_2	1	1	−1		
E	2	−1	0	(x, y)	$(xy, x^2 - y^2)$, (xz, yz)

* More character tables are given at the end of the *Data section*.

(b) The structure of character tables

In general, the columns in a character table are labelled with the symmetry operations of the group. For instance, for the group C_{3v} the columns are headed E, C_3, and σ_v (Table 15.3). The numbers multiplying each operation are the numbers of members of each class. In the C_{3v} character table we see that the two threefold rotations (clockwise and counter-clockwise rotations by 120°) belong to the same class: they are related by a reflection (Fig. 15.21). The three reflections (one through each of the three vertical mirror planes) also lie in the same class: they are related by the threefold rotations. The two reflections of the group C_{2v} fall into different classes: although they are both reflections, one cannot be transformed into the other by any symmetry operation of the group.

The total number of operations in a group is called the **order**, h, of the group. The order of C_{3v}, for instance, is 6.

The rows under the labels for the operations summarize the symmetry properties of the orbitals. They are labelled with the **symmetry species** (the analogues of the labels σ and π). More formally, the symmetry species label the irreducible representations of the group, which are the basic types of behaviour that orbitals may show when subjected to the symmetry operations of the group, as we have illustrated for the group C_{2v}. By convention, irreducible representations are labelled with upper case Roman letters (such as A_1 and E) but the orbitals to which they apply are labelled with the lower case italic equivalents (so an orbital of symmetry species A_1 is called an a_1 orbital).[4] Examples of each type of orbital are shown in Fig. 15.22.

(c) Character tables and orbital degeneracy

The character of the identity operation E tells us the degeneracy of the orbitals. Thus, in a C_{3v} molecule, any orbital with a symmetry label a_1 or a_2 is non-degenerate. Any doubly degenerate pair of orbitals in C_{3v} must be labelled e because, in this group, only E symmetry species have characters greater than 1.

Because there are no characters greater than 2 in the column headed E in C_{3v}, we know that there can be no triply degenerate orbitals in a C_{3v} molecule. This last point is a powerful result of group theory, for it means that, with a glance at the character table of a molecule, we can state the maximum possible degeneracy of its orbitals.

---

4. Note that care must be taken to distinguish the identity element E (italic, a column heading) from the symmetry label E (roman, a row label).

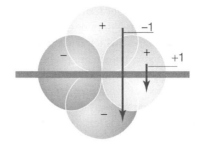

15.23 The two orbitals shown here have different properties under reflection through the mirror plane: one changes sign (character −1); the other does not (character +1).

Example 15.2 Using a character table to judge degeneracy

Can a trigonal planar molecule such as BF_3 have triply degenerate orbitals? What is the minimum number of atoms from which a molecule can be built that does display triple degeneracy?

Method First, identify the point group, and then refer to the corresponding character table in the *Data section*. The maximum number in the column headed by the identity E is the maximum orbital degeneracy possible in a molecule of that point group. For the second part, consider the shapes that can be built from two, three, etc. atoms, and decide which number can be used to form a molecule that can have orbitals of symmetry species T.

Answer Trigonal planar molecules belong to the point group D_{3h}. Reference to the character table for this group shows that the maximum degeneracy is 2, as no character exceeds 2 in the column headed E. Therefore, the orbitals cannot be triply degenerate. A tetrahedral molecule (symmetry group T) has an irreducible representation with a T symmetry species. The minimum number of atoms needed to build such a molecule is four (as in P_4, for instance).

Self-test 15.2 A buckminsterfullerene molecule, C_{60} (**15**), belongs to the icosahedral point group. What is the maximum possible degree of degeneracy of its orbitals? [5]

(d) Characters and operations

The characters in the rows labelled A and B and in the columns headed by symmetry operations other than the identity E indicate the behaviour of an orbital under the corresponding operations: a +1 indicates that an orbital is unchanged, and a −1 indicates that it changes sign. It follows that we can identify the symmetry label of the orbital by comparing the changes that occur to an orbital under each operation, and then comparing the resulting +1 or −1 with the entries in a row of the character table for the point group concerned.

For the rows labelled E or T (which refer to the behaviour of sets of doubly and triply degenerate orbitals, respectively), the characters in a row of the table are the sums of the characters summarizing the behaviour of the individual orbitals in the basis. Thus, if one member of a doubly degenerate pair remains unchanged under a symmetry operation but the other changes sign (Fig. 15.23), then the entry is reported as $\chi = 1 - 1 = 0$. Care must be exercised with these characters because the transformations of orbitals can be quite complicated; nevertheless, the sums of the individual characters are integers.

As an example, consider the $O2p_x$ orbital in H_2O. Because H_2O belongs to the point group C_{2v}, we know by referring to the C_{2v} character table (Table 15.2) that the labels available for the orbitals are a_1, a_2, b_1, and b_2. We can decide the appropriate label for $O2p_x$ by noting that under a 180° rotation (C_2) the orbital changes sign (Fig. 15.24), so it must be either B_1 or B_2, as only these two symmetry types have character −1 under C_2. The $O2p_x$ orbital also changes sign under the reflection σ'_v, which identifies it as B_1. As we shall see, any molecular orbital built from this atomic orbital will also be a b_1 orbital. Similarly, $O2p_y$ changes sign under C_2 but not under σ'_v; therefore, it can contribute to b_2 orbitals.

The behaviour of s, p, and d orbitals on a central atom under the symmetry operations of the molecule is so important that the symmetry species of these orbitals are

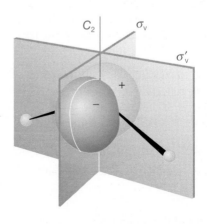

15.24 A p_x orbital on the central atom of a C_{2v} molecule and the symmetry elements of the group.

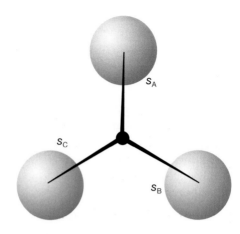

15.25 The three H1s orbitals used to construct symmetry-adapted linear combinations in a C_{3v} molecule such as NH_3.

generally indicated in a character table. To make these allocations, we look at the symmetry species of x, y, and z, which appear on the right-hand side of the character table. Thus, the position of z in Table 15.3 shows that p_z (which is proportional to $zf(r)$), has symmetry species A_1 in C_{3v}, whereas p_x and p_y (which are proportional to $xf(r)$ and $yf(r)$, respectively) are jointly of E symmetry. In technical terms, we say that p_x and p_y jointly **span** an irreducible representation of symmetry species E. An s orbital on the central atom always spans the fully symmetrical irreducible representation (typically labelled A_1 but sometimes A_1') of a group as it is unchanged under all symmetry operations.

The five d orbitals of a shell are represented by xy for d_{xy}, etc. and are also listed on the right of the character table. We can see at a glance that in C_{3v}, d_{xy} and $d_{x^2-y^2}$ on a central atom jointly belong to E and hence form a doubly degenerate pair.

(e) The classification of linear combinations of orbitals

So far, we have dealt with the symmetry classification of individual orbitals. The same technique may be applied to linear combinations of orbitals on atoms that are related by symmetry transformations of the molecule, such as the combination $\psi_1 = \psi_A + \psi_B + \psi_C$ of the three H1s orbitals in the C_{3v} molecule NH_3 (Fig. 15.25). This combination remains unchanged under a C_3 rotation and under any of the three vertical reflections of the group, so its characters are

$$\chi(E) = 1 \qquad \chi(C_3) = 1 \qquad \chi(\sigma_v) = 1$$

Comparison with the C_{3v} character table shows that ψ_1 is of symmetry species A_1, and therefore that it contributes to a_1 molecular orbitals in NH_3.

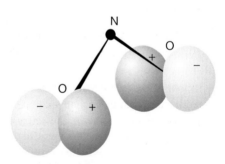

15.26 One symmetry-adapted linear combination of O2p_x orbitals in the C_{2v} NO_2^- ion.

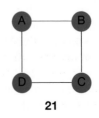

21

Example 15.3 Identifying the symmetry species of orbitals

Identify the symmetry species of the orbital $\psi = \psi_A - \psi_B$ in a C_{2v} NO_2 molecule, where ψ_A is an O2p_x orbital on one O atom and ψ_B that on the other O atom.

Method The negative sign in ψ indicates that the sign of ψ_B is opposite to that of ψ_A. We need to consider how the combination changes under each operation of the group, and then write the character as +1, −1, or 0 as specified above. Then we compare the resulting characters with each row in the character table for the point group, and hence identify the symmetry species.

Answer The combination is shown in Fig. 15.26. Under C_2, ψ changes into itself, implying a character of +1. Under the reflection σ_v, both orbitals change sign, so $\psi \to -\psi$, implying a character of −1. Under σ_v', $\psi \to -\psi$, so the character for this operation is also −1. The characters are therefore

$$\chi(E) = 1 \qquad \chi(C_2) = 1 \qquad \chi(\sigma_v) = -1 \qquad \chi(\sigma_v') = -1$$

These values match the characters of the A_2 symmetry species, so ψ can contribute to an a_2 orbital.

Self-test 15.3 Identify the symmetry type of the combination $\psi_A - \psi_B + \psi_C - \psi_D$ in a square planar array of H atoms of point group D_{4h} (**21**).

[B_{2g}]

(a)

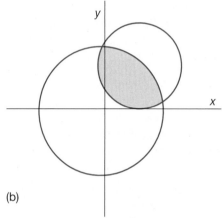

(b)

15.27 The value of an integral I (for example, an area) is independent of the coordinate system used to evaluate it, as in the two choices shown as (a) and (b). That is, I is a basis of a representation of symmetry species A_1 (or its equivalent).

15.28 A symmetry-adapted linear combination that belongs to the symmetry species E in a C_{3v} molecule such as NH_3. This combination can form a molecular orbital by overlapping with the p_x orbital on the central atom (the orbital with its axis parallel to the width of the page; see Fig. 15.31c).

15.5 Vanishing integrals and orbital overlap

Suppose we had to evaluate the integral

$$I = \int f_1 f_2 \, d\tau \tag{15.7}$$

where f_1 and f_2 are functions. For example, f_1 might be an atomic orbital A on one atom and f_2 an atomic orbital B on another atom, in which case I would be their overlap integral. If we knew that the integral is zero, we could say at once that a molecular orbital does not result from (A,B) overlap in that molecule. We shall now see that character tables provide a quick way of judging whether an integral is necessarily zero.

(a) The criteria for vanishing integrals

The key point in dealing with the integral I is that the value of any integral, and of an overlap integral in particular, is independent of the orientation of the molecule (Fig. 15.27). In group theory we express this point by saying that I *is invariant under any symmetry operation of the molecule*, and that each operation brings about the trivial transformation $I \rightarrow I$. Because the volume element $d\tau$ is invariant under any symmetry operation, it follows that the integral is nonzero only if the integrand itself, the product $f_1 f_2$, is unchanged by any symmetry operation of the molecular point group. If the integrand changed sign under a symmetry operation, the integral would be the sum of equal and opposite contributions, and hence would be zero. It follows that the only contribution to a nonzero integral comes from functions for which under any symmetry operation of the molecular point group $f_1 f_2 \rightarrow f_1 f_2$, and hence for which the characters of the operations are all equal to +1. Therefore, for I not to be zero, *the integrand $f_1 f_2$ must have symmetry species A_1* (or its equivalent in the specific molecular point group).

We use the following procedure to deduce the symmetry species spanned by the product $f_1 f_2$ and hence to see whether it does indeed span A_1.

1 Decide on the symmetry species of the individual functions f_1 and f_2 by reference to the character table, and write their characters in two rows in the same order as in the table.

2 Multiply the numbers in each column, writing the results in the same order.

3 Inspect the row so produced, and see if it can be expressed as a sum of characters from each column of the group. The integral must be zero if this sum does not contain A_1.

For example, if f_1 is the s_N orbital in NH_3 and f_2 is the linear combination $s_3 = s_B - s_C$ (Fig. 15.28), then, because s_N spans A_1 and s_3 is a member of the basis spanning E, we write

f_1:	1	1	1
f_2:	2	−1	0
$f_1 f_2$:	2	−1	0

The characters 2, −1, 0 are those of E alone, so the integrand does not span A_1. It follows that the integral must be zero. Inspection of the form of the functions (see Fig. 15.28) shows why this is so: s_3 has a node running through s_N. Had we taken $f_1 = s_N$ and $f_2 = s_1$ instead, where $s_1 = s_A + s_B + s_C$, then, because each spans A_1 with characters 1, 1, 1:

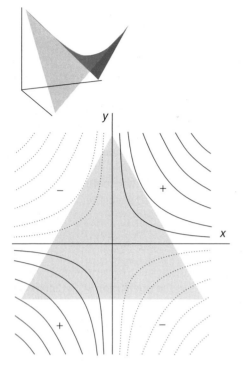

15.29 The integral of the function $f = xy$ over the tinted region is zero. In this case, the result is obvious by inspection, but group theory can be used to establish similar results in less obvious cases. The insert shows the shape of the function in three dimensions.

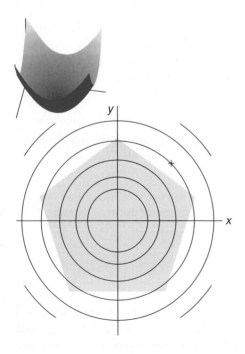

15.30 The integration of a function over a pentagonal region. The insert shows the shape of the function in three dimensions.

f_1:	1	1	1
f_2:	1	1	1
$f_1 f_2$:	1	1	1

The characters of the product are those of A_1 itself. Therefore, s_1 and s_N may have nonzero overlap. A short cut that works when f_1 and f_2 are bases for irreducible representations of a group is to note their symmetry species: if they are different, then the integral of their product must vanish; if they are the same, then the integral may be nonzero.

It is important to note that group theory is specific about when an integral must be zero, but integrals that it allows to be nonzero may be zero for reasons unrelated to symmetry. For example, the N—H distance in ammonia may be so great that the (s_1, s_N) overlap integral is zero simply because the orbitals are so far apart.

Example 15.4 Deciding if an integral must be zero (1)

May the integral of the function $f = xy$ be nonzero when evaluated over a region the shape of an equilateral triangle centred on the origin (Fig. 15.29)?

Method First, note that an integral over a single function f is included in the previous discussion if we take $f_1 = f$ and $f_2 = 1$ in eqn 7. Therefore, we need to judge whether f alone belongs to the symmetry species A_1 (or its equivalent) in the point group of the system. To decide that, we identify the point group and then examine the character table to see whether f belongs to A_1 (or its equivalent).

Answer An equilateral triangle has the point-group symmetry D_{3h}. If we refer to the character table of the group, we see that xy is a member of a basis that spans the irreducible representation E'. Therefore, its integral must be zero, because the integrand has no component that spans A_1'.

Self-test 15.4 Can the function $x^2 + y^2$ have a nonzero integral when integrated over a regular pentagon centred on the origin? [Yes, Fig. 15.30]

In many cases, the product of functions f_1 and f_2 spans a sum of irreducible representations. For instance, in C_{2v} we may find the characters 2, 0, 0, −2 when we multiply the characters of f_1 and f_2 together. In this case, we note that these characters are the sum of the characters for A_2 and B_1:

	E	C_{2v}	σ_v	σ_v'
A_2	1	1	−1	−1
B_1	1	−1	1	−1
$A_2 + B_1$	2	0	0	−2

To summarize this result we write the symbolic expression $A_2 \times B_1 = A_2 + B_1$, which is called the **decomposition of a direct product**.[5] Because the sum on the right

5. This expression is symbolic. The × and + signs in this expression are not ordinary multiplication and addition signs: formally, they denote technical procedures with matrices called a 'direct product' and a 'direct sum'.

does not include a component that is a basis for an irreducible representation of symmetry species A_1, we can conclude that the integral of $f_1 f_2$ over all space is zero in a C_{2v} molecule.

Whereas the decomposition of the characters 2, 0, 0, –2 can be done by inspection in this simple case, in other cases and more complex groups the decomposition is often far from obvious. For example, if we found the characters 8, –2, –6, 4, it would not be obvious that the sum contains A_1. Group theory, however, provides a systematic way of using the characters of the representation spanned by a product to find the symmetry species of the irreducible representations. The recipe is as follows:

1 Write down a table with columns headed by the symmetry operations of the group.

2 In the first row write down the characters of the symmetry species we want to analyse.

3 In the second row, write down the characters of the irreducible representation Γ we are interested in.

4 Multiply the two rows together, add the products together, and divide by the order of the group.

The resulting number is the number of times Γ occurs in the decomposition.

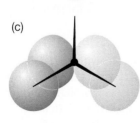

15.31 Orbitals of the same symmetry species may have non-vanishing overlap. This diagram illustrates the three bonding orbitals that may be constructed from (N2s, H1s) and (N2p, H1s) overlap in a C_{3v} molecule. (a) a_1; (b) and (c) the two components of the doubly degenerate e orbitals. (There are also three antibonding orbitals of the same species.)

Illustration 15.1

To find whether A_1 does indeed occur in the product with characters 8, –2, –6, 4 in C_{2v}, we draw up the following table:

	E	C_{2v}	σ_v	σ_v'	$h = 4$ (the order of the group)
$f_1 f_2$	8	–2	–6	4	(the characters of the product)
A_1	1	1	1	1	(the symmetry species we are interested in)
	8	–2	–6	4	(the product of the two sets of characters)

The sum of the numbers in the last line is 4; when that number is divided by the order of the group, we get 1, so A_1 occurs once in the decomposition. When the procedure is repeated for all four symmetry species, we find that $f_1 f_2$ spans $A_1 + 2A_2 + 5B_2$.

Self-test 15.5 Does A_2 occur among the symmetry species of the irreducible representations spanned by a product with characters 7, –3, –1, 5 in the group C_{2v}? [No]

(b) Orbitals with nonzero overlap

The rules just given let us decide which atomic orbitals may have nonzero overlap in a molecule. We have seen that s_N may have nonzero overlap with s_1 (the combination $s_A + s_B + s_C$), so bonding and antibonding molecular orbitals can form from (s_N, s_1) overlap (Fig. 15.31). The general rule is that *only orbitals of the same symmetry species may have nonzero overlap*, so only orbitals of the same symmetry species form bonding and antibonding combinations. It should be recalled from Chapter 14 that the selection of atomic orbitals that had mutual nonzero overlap is the central and initial step in the construction of molecular orbitals by the LCAO procedure. We are therefore at the point of contact between group theory and the material introduced in that chapter. The molecular orbitals formed from a particular set of atomic orbitals with nonzero overlap are labelled with the lower-case letter corresponding to the symmetry species. Thus, the (s_N, s_1)-overlap orbitals are called a_1 orbitals (or a_1^*, if we wish to emphasize that they are antibonding).

The s_2 and s_3 linear combinations have symmetry species E. Does the N atom have orbitals that have nonzero overlap with them (and give rise to e molecular orbitals)? Intuition (as supported by Fig. 15.31b and c) suggests that $N2p_x$ and $N2p_y$ should be suitable. We can confirm this conclusion by noting that the character table shows that in C_{3v}, the functions x and y jointly belong to the symmetry species E. Therefore, $N2p_x$ and $N2p_y$ also belong to E, so may have nonzero overlap with s_2 and s_3. This conclusion can be verified by multiplying the characters and finding that the product of characters can be expressed as the decomposition $E \times E = A_1 + A_2 + E$. The two e orbitals that result are shown in Fig. 15.31 (there are also two antibonding e orbitals).

We can see the power of the method by exploring whether any d orbitals on the central atom can take part in bonding. As explained earlier, reference to the C_{3v} character table shows that d_{z^2} has A_1 symmetry and that the pairs $(d_{x^2-y^2}, d_{xy})$ and (d_{yz}, d_{zx}) each transform as E. It follows that molecular orbitals may be formed by (s_1, d_{z^2}) overlap and by overlap of the s_2, s_3 combinations with the E d orbitals. Whether or not the d orbitals are in fact important is a question group theory cannot answer because the extent of their involvement depends on energy considerations, not symmetry.

Example 15.5 Determining which orbitals can contribute to bonding

The four H1s orbitals of methane span $A_1 + T_2$. With which of the C atom orbitals can they overlap? What bonding pattern would be possible if the C atom had d orbitals available?

Method Refer to the T_d character table (in the *Data section*) and look for s, p, and d orbitals spanning A_1 or T_2.

Answer An s orbital spans A_1, so it may have nonzero overlap with the A_1 combination of H1s orbitals. The C2p orbitals span T_2, so they may have nonzero overlap with the T_2 combination. The d_{xy}, d_{yz}, and d_{zx} orbitals span T_2, so they may overlap the same combination. Neither of the other two d orbitals span A_1 (they span E), so they remain nonbonding orbitals.

Comment It follows that in methane there are (C2s,H1s)-overlap a_1 orbitals and (C2p,H1s)-overlap t_2 orbitals. The C3d orbitals might contribute to the latter. The lowest energy configuration is probably $a_1^2 t_2^6$, with all bonding orbitals occupied.

Self-test 15.6 Consider the octahedral SF_6 molecule, with the bonding arising from overlap of S orbitals and a 2p orbital on each F directed towards the central S atom. The latter span $A_{1g} + E_g + T_{1u}$. What s orbitals have nonzero overlap? Suggest what the ground-state configuration is likely to be.

$$[3s(A_{1g}), 3p(T_{1u}), 3d(E_g); a_{1g}^2 t_{1u}^6 e_g^4]$$

(c) Symmetry-adapted linear combinations

So far, we have only asserted the forms of the linear combinations (such as s_1, etc.) that have a particular symmetry. Group theory also provides machinery that takes an arbitrary **basis**, or set of atomic orbitals (s_A, etc.), as input and generates combinations of the specified symmetry. Because these combinations are adapted to the symmetry of the molecule, they are called **symmetry-adapted linear combinations** (SALC). Symmetry-adapted linear combinations are the building blocks of LCAO molecular orbitals, for they include combinations such as those used to construct molecular orbitals in benzene. The construction of SALCs is the first step in any molecular orbital treatment of molecules.

The technique for building SALCs is derived by using the full power of group theory. We shall not show the derivation,[6] which is very lengthy, but present the main conclusions as a set of rules:

1 Construct a table showing the effect of each operation on each orbital of the original basis.

2 To generate the combination of a specified symmetry species, take each column in turn and:

   (i) Multiply each member of the column by the character of the corresponding operation.

   (ii) Add together all the orbitals in each column with the factors as determined in (i).

   (iii) Divide the sum by the order of the group.

For example, from the (s_N, s_A, s_B, s_C) basis in NH_3 we form the table shown in the margin. To generate the A_1 combination, we take the characters for A_1 (1, 1, 1, 1, 1, 1); then rules (i) and (ii) lead to

$$\psi \propto s_N + s_N + \cdots = 6s_N$$

The order of the group (the number of elements) is 6, so the combination of A_1 symmetry that can be generated from s_N is s_N itself. Applying the same technique to the column under s_A gives

$$\psi = \tfrac{1}{6}(s_A + s_B + s_C + s_A + s_B + s_C) = \tfrac{1}{3}(s_A + s_B + s_C)$$

The same combination is built from the other two columns, so they give no further information. The combination we have just formed is the s_1 combination we used before (apart from the numerical factor).

We now form the overall molecular orbital by forming a linear combination of all the SALCs of the specified symmetry species. In this case, therefore, the a_1 molecular orbital is

$$\psi = c_N s_N + c_1 s_1$$

This is as far as group theory can take us. The coefficients are found by solving the Schrödinger equation; they do not come directly from the symmetry of the system.

We run into a problem when we try to generate an SALC of symmetry species E, because, for representations of dimension 2 or more, the rules generate sums of SALCs. This problem can be illustrated as follows. In C_{3v}, the E characters are 2, −1, −1, 0, 0, 0, so the column under s_N gives

$$\psi = \tfrac{1}{6}(2s_N - s_N - s_N + 0 + 0 + 0) = 0$$

The other columns give

$$\tfrac{1}{6}(2s_A - s_B - s_C) \qquad \tfrac{1}{6}(2s_B - s_A - s_C) \qquad \tfrac{1}{6}(2s_C - s_B - s_A)$$

However, any one of these three expressions can be expressed as a sum of the other two (they are not 'linearly independent'). The difference of the second and third

	Original basis			
	s_N	s_A	s_B	s_C
Under E	s_N	s_A	s_B	s_C
C_3^+	s_N	s_B	s_C	s_A
C_3^-	s_N	s_C	s_A	s_B
σ_v	s_N	s_A	s_C	s_B
σ_v'	s_N	s_B	s_A	s_C
σ_v''	s_N	s_C	s_B	s_A

6. See *Further reading* for references to the derivation.

gives $\frac{1}{2}(s_B - s_C)$, and this combination and the first, $\frac{1}{6}(2s_A - s_B - s_C)$ are the two (now linearly independent) SALCs we have used in the discussion of e orbitals.

15.6 Vanishing integrals and selection rules

Integrals of the form

$$I = \int f_1 f_2 f_3 \, d\tau \tag{15.8}$$

are also common in quantum mechanics for they include matrix elements of operators (Section 11.5d), and it is important to know when they are necessarily zero. For the integral to be nonzero, *the product $f_1 f_2 f_3$ must span A_1 (or its equivalent) or contain a component that spans A_1*. To test whether this is so, the characters of all three functions are multiplied together in the same way as in the rules set out above.

Example 15.6 Deciding if an integral must be zero (2)

Does the integral $\int (3d_{z^2})x(3d_{xy}) \, d\tau$ vanish in a C_{2v} molecule?

Method We must refer to the C_{2v} character table (Table 15.2) and the characters of the irreducible representations spanned by $3z^2 - r^2$ (the form of the d_{z^2} orbital), x, and xy; then we can use the procedure set out above (with one more row of multiplication).

Answer We draw up the following table:

	E	C_2	σ_v	σ_v'	
$f_3 = d_{xy}$	1	1	−1	−1	A_2
$f_2 = x$	1	−1	1	−1	B_1
$f_1 = d_{z^2}$	1	1	1	1	A_1
$f_1 f_2 f_3$	1	−1	−1	1	

The characters are those of B_2. Therefore, the integral is necessarily zero.

Self-test 15.7 Does the integral $\int (2p_x)(2p_y)(2p_z) \, d\tau$ necessarily vanish in an octahedral environment?

[No]

We saw in Chapters 12 and 13, and will see in more detail in Chapters 16 and 17, that the intensity of a spectral line arising from a molecular transition between some initial state $|i\rangle$ with wavefunction ψ_i and a final state $|f\rangle$ with wavefunction ψ_f depends on the (electric) transition dipole moment, μ_{fi}. The z-component of this vector is defined through

$$\mu_{z,fi} = \langle f|\mu_z|i\rangle = -e \int \psi_f^* z \psi_i \, d\tau \tag{15.9}$$

where $-e$ is the charge of the electron. The transition moment has the form of the integral in eqn 8; so, once we know the symmetry species of the states, we can use group theory to formulate the selection rules for the transitions.

As an example, we investigate whether an electron in an a_1 orbital in H_2O (which belongs to the group C_{2v}) can make an electric dipole transition to a b_1 orbital (Fig. 15.32). We must examine all three components of the transition dipole

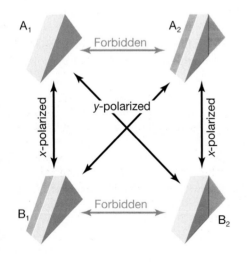

15.32 The polarizations of the allowed transitions in a C_{2v} molecule. The shading indicates the structure of the orbitals of the specified symmetry species. The perspective view of the molecule makes it look rather like a door-stop; however, from the side, each 'door-stop' is in fact an isosceles triangle.

moment, and take f_2 in eqn 8 as x, y, and z in turn. Reference to the C_{2v} character table shows that these components transform as B_1, B_2, and A_1, respectively. The three calculations run as follows:

	x-component				y-component				z-component			
	E	C_2	σ_v	σ_v'	E	C_2	σ_v	σ_v'	E	C_2	σ_v	σ_v'	
f_3	1	−1	1	−1	1	−1	1	−1	1	−1	1	−1	B_1
f_2	1	−1	1	−1	1	−1	−1	1	1	1	1	1	
f_1	1	1	1	1	1	1	1	1	1	1	1	1	A_1
$f_1 f_2 f_3$	1	1	1	1	1	1	−1	−1	1	−1	1	−1	

Only the first product (with $f_2 = x$) spans A_1, so only the x-component of the transition dipole moment may be nonzero. Therefore, we conclude that the electric dipole transitions between a_1 and b_1 are allowed. We can go on to state that the radiation emitted (or absorbed) is x-polarized and has its electric field vector in the x-direction, because that form of radiation couples with the x-component of a transition dipole.

Example 15.7 Deducing a selection rule

Is $p_x \rightarrow p_y$ an allowed transition in a tetrahedral environment?

Method We must decide whether the product $p_y q p_x$, with $q = x$, y, or z, spans A_1 by using the T_d character table.

Answer The procedure works out as follows:

	E	$8C_3$	$3C_2$	$6S_4$	$6\sigma_d$	
$f_3(p_y)$	3	0	−1	−1	1	T_2
$f_2(q)$	3	0	−1	−1	1	T_2
$f_1(p_x)$	3	0	−1	−1	1	T_2
$f_1 f_2 f_3$	27	0	−1	−1	1	

We can use the decomposition procedure described in Section 15.5a to deduce that A_1 occurs (once) in this set of characters, so $p_x \rightarrow p_y$ is allowed. Take care to include all the operators (as given by the numbers multiplying the symmetry element at the head of each column).

Comment A more detailed analysis (using the matrix representatives rather than the characters) shows that only $q = z$ gives an integral that may be nonzero, so the transition is z-polarized. That is, the electromagnetic radiation involved in the transition has its electric vector aligned in the z-direction.

Self-test 15.8 What are the allowed transitions, and their polarizations, of a b_1 electron in a C_{4v} molecule? $\qquad$ $[b_1 \rightarrow b_1(z); b_1 \rightarrow e(x, y)]$

The following chapters will show many more examples of the systematic use of symmetry. We shall see that the techniques of group theory greatly simplify the analysis of molecular structure and spectra.

Checklist of key ideas

☐ group theory, the systematic discussion of symmetry

The symmetry elements of objects

☐ symmetry operation, an action that leaves an object looking the same after it has been carried out

☐ symmetry element, the point, line, or plane with respect to which a symmetry operation is performed

15.1 Operations and symmetry elements

☐ point group, a group of symmetry operations that leaves at least one common point unchanged

☐ space group, a group of symmetry operations that includes translation through space

☐ identity, the operation of doing nothing

☐ n-fold rotation, a rotation through $360°/n$

☐ n-fold axis of symmetry, the symmetry element corresponding to an n-fold rotation

☐ principal axis, the symmetry axis with the largest value of n

☐ reflection, the operation of reflection in a plane

☐ mirror plane, the symmetry element associated with a reflection

☐ inversion, an operation in which each point in an object is moved to the centre of the object and then out the same distance on the other side

☐ centre of symmetry, the symmetry element associated with an inversion

☐ n-fold improper rotation, a rotation through $360°/n$ followed by a reflection through a plane perpendicular to the axis of rotation

☐ n-fold axis of improper rotation, the symmetry element associated with an n-fold improper rotation

15.2 The symmetry classification of molecules

☐ Schoenflies system, a notation for point groups commonly used for molecules (see Table 15.1)

☐ Hermann–Mauguin system, a notation for point groups commonly used for solids (see Table 15.1)

☐ International system, another name for the Hermann–Mauguin system

☐ cubic groups, collectively the tetrahedral and octahedral groups

☐ tetrahedral groups, the groups T, T_d, and T_h

☐ octahedral groups, the groups O and O_h

☐ icosahedral group, the group I

☐ full rotation group, the group consisting of an infinite number of rotation axes with all possible values of n

15.3 Some immediate consequences of symmetry

☐ polar molecule, a molecule with a permanent electric dipole moment; to be polar a molecule must belong to C_n, C_{nv}, or C_s (and have no higher symmetry)

☐ chiral molecule, a molecule that cannot be superimposed on its mirror image; a molecule may be chiral only if it does not possess an axis of improper rotation, S_n

☐ achiral molecule, a molecule that can be superimposed on its mirror image

☐ optically active, the property of being able to rotate the plane of polarization of light

☐ enantiomeric pair, a chiral molecule and its mirror-image partner

Character tables

15.4 Character tables and symmetry labels

☐ character table, a table that characterizes the different symmetry types possible in the point group; a table of symmetry species, symmetry elements, and their characters

☐ character, χ, numbers that characterize the essential features of each symmetry type; the sum of the diagonal elements of a matrix representative

☐ representative, the matrix $D(X)$ that brings about the transformation of the basis under the operation X

☐ basis, the set of functions on which the representative acts

☐ matrix representation, the set of representatives for all the operations of a group

☐ block-diagonal form, a matrix having its nonzero elements arranged in blocks along the principal diagonal

☐ reduced representation, a representation in which all the matrices have block-diagonal form

☐ irreducible representation, a representation that cannot be reduced further

☐ class of operation, symmetry operations that have the same type (e.g. rotations) and can be transformed into one another by a symmetry operation of the group

☐ order of group, the number of symmetry operations; number of symmetry species = number of classes

☐ symmetry species, labels for the irreducible representations of a group

☐ span, to act as a basis for an irreducible representation of given symmetry species

15.5 Vanishing integrals and orbital overlap

☐ criteria for $\int f_1 f_2 \, d\tau$ to be nonzero: the integrand $f_1 f_2$ must have symmetry species A_1 (or its equivalent in the specific molecular point group)

☐ decomposition of direct product, the reduction of a product of symmetry species to a sum of symmetry species, $\Gamma \times \Gamma' = \Gamma^{(1)} + \Gamma^{(2)} + \cdots$

☐ overlap and symmetry: only orbitals of the same symmetry species may have nonzero overlap

☐ symmetry-adapted linear combination (SALC), combinations of atomic orbitals adapted to the symmetry of the molecule and used as the building blocks for LCAO molecular orbitals

15.6 Vanishing integrals and selection rules

☐ criteria for $\int f_1 f_2 f_3 \, d\tau$ to be nonzero: the product $f_1 f_2 f_3$ must span A_1 (or its equivalent) or contain a component that spans A_1

☐ selection rules, the symmetry criteria for the non-vanishing of the transition moment between the initial and final states

Further reading

Articles of general interest

G.L. Breneman, Crystallographic symmetry point group notation flow chart. *J. Chem. Educ.* **64**, 216 (1987).

C. Contreras-Ortega, L. Vera, and E. Quiroz-Reyes, How great is the Great Orthogonality Theorem? *J. Chem. Educ.* **68**, 200 (1991).

C. Contreras-Ortega, L. Vera, and E. Quiroz-Reyes, More than one character table: a warning on the use of the rules of the irreducible representations and their characters. *J. Chem. Educ.* **72**, 821 (1995).

F. Chung and S. Sternberg, Mathematics and the buckyball. *American Scientist* **81**, 56 (1993).

M. Hamermesh, Group theory. In *Encyclopedia of applied physics* (ed. G.L. Trigg), **7**, 365. VCH, New York (1993).

F. Rioux, Quantum mechanics, group theory, and C$_{60}$. *J. Chem. Educ.* **71**, 464 (1994).

Texts and sources of data and information

T.A. Albright and J.K. Burdett, *Problems in molecular orbital theory*. Oxford University Press (1992).

P.W. Atkins, M.S. Child, and C.S.G. Phillips, *Tables for group theory*. Oxford University Press (1970).

P.W. Atkins and R.S. Friedman, *Molecular quantum mechanics*. Oxford University Press (1997).

D.M. Bishop, *Group theory and chemistry*. Dover, New York (1993).

R.L. Carter, *Molecular symmetry and group theory*. Wiley, New York (1998).

F.A. Cotton, *Chemical applications of group theory*. Wiley, New York (1990).

B.E. Douglas and C. Hollingsworth, *Symmetry in bonding and structure*. Academic Press, New York (1985).

R. Drago, *Physical methods for chemists*. Saunders, Philadelphia (1992).

D.C. Harris and M.D. Bertollucci, *Symmetry and spectroscopy: an introduction to vibrational and electronic spectroscopy*. Dover, New York (1989).

E. Heilbronner and J.D. Dunitz, *Reflections on symmetry: in chemistry, and elsewhere*. VCH, Weinheim (1993).

S.F.A. Kettle, Symmetry and structure: readable group theory for chemists. Wiley, New York (1995).

Exercises

Discussion questions

15.1 (a) Explain how a molecule is assigned to a point group.

15.1 (b) List the symmetry operations and the corresponding symmetry elements of the point groups.

15.2 (a) Explain the symmetry criteria that allow a molecule to be polar.

15.2 (b) Explain the symmetry criteria that allow a molecule to be optically active.

15.3 (a) Explain what is meant by (a) a representative and (b) a representation in the context of group theory.

15.3 (b) Explain the construction and content of a character table.

15.4 (a) Explain how spectroscopic selection rules arise and how they are formulated by using group theory.

15.4 (b) Outline how a direct product is expressed as a direct sum and how to decide whether the totally symmetric irreducible representation is present in the direct product.

Numerical exercises

15.5 (a) The CH_3Cl molecule belongs to the point group C_{3v}. List the symmetry elements of the group and locate them in the molecule.

15.5 (b) The CCl_4 molecule belongs to the point group T_d. List the symmetry elements of the group and locate them in the molecule.

15.6 (a) Which of the following molecules may be polar? (a) pyridine (C_{2v}), (b) nitroethane (C_s), (c) gas-phase $HgBr_2$ ($D_{\infty h}$), (d) $B_3N_3H_6$ (D_{3h}).

15.6 (b) Which of the following molecules may be polar? (a) CH_3Cl(C_{3v}), (b) $HW_2(CO)_{10}$(D_{4h}), (c) $SnCl_4$(T_d).

15.7 (a) Use symmetry properties to determine whether or not the integral $\int p_x z p_z \, d\tau$ is necessarily zero in a molecule with symmetry C_{4v}.

15.7 (b) Use symmetry properties to determine whether or not the integral $\int p_x z p_z \, d\tau$ is necessarily zero in a molecule with symmetry D_{6h}.

15.8 (a) Show that the transition $A_1 \rightarrow A_2$ is forbidden for electric dipole transitions in a C_{3v} molecule.

15.8 (b) Is the transition $A_{1g} \rightarrow E_{2u}$ forbidden for electric dipole transitions in a D_{6h} molecule?

15.9 (a) Show that the function xy has symmetry species B_2 in the group C_{4v}.

15.9 (b) Show that the function xyz has symmetry species A_1 in the group D_2.

15.10 (a) Molecules belonging to the point groups D_{2h} or C_{3h} cannot be chiral. Which elements of these groups rule out chirality?

15.10 (b) Molecules belonging to the point groups T_h or T_d cannot be chiral. Which elements of these groups rule out chirality?

15.11 (a) The group D_2 consists of the elements E, C_2, C_2', and C_2'', where the three twofold rotations are around mutually perpendicular axes. Construct the group multiplication table.

15.11 (b) The group C_{4v} consists of the elements E, $2C_4$, C_2, and $2\sigma_v$, $2\sigma_d$. Construct the group multiplication table.

15.12 (a) Identify the point groups to which the following objects belong: (a) a sphere, (b) an isosceles triangle, (c) an equilateral triangle, (d) an unsharpened cylindrical pencil.

15.12 (b) Identify the point groups to which the following objects belong: (a) a sharpened cylindrical pencil, (b) a three-bladed propellor, (c) a four-legged table, (d) yourself (approximately).

15.13 (a) List the symmetry elements of the following molecules and name the point groups to which they belong: (a) NO₂, (b) N₂O, (c) CHCl₃, (d) CH₂=CH₂, (e) *cis*-CHBr=CHBr, (f) *trans*-CHCl=CHCl.

15.13 (b) List the symmetry elements of the following molecules and name the point groups to which they belong: (a) naphthalene, (b) anthracene, (c) the three dichlorobenzenes.

15.14 (a) Assign (a) *cis*-dichloroethene and (b) *trans*-dichloroethene to point groups.

15.14 (b) Assign the following molecules to point groups: (a) HF, (b) IF₇ (pentagonal bipyramid), (c) XeO₂F₂ (see-saw), (d) Fe₂(CO)₉ (**22**), (e) cubane, C₈H₈, (f) tetrafluorocubane, C₈H₄F₄ (**23**).

15.15 (a) Which of the molecules in Exercises 15.13a and 15.14a can be (a) polar, (b) chiral?

15.15 (b) Which of the molecules in Exercises 15.13b and 15.14b can be (a) polar, (b) chiral?

15.16 (a) Consider the C_{2v} molecule NO₂. The combination $p_x(A) - p_x(B)$ of the two O atoms (with x perpendicular to the plane) spans A₂. Is there any orbital of the central N atom that can have a nonzero overlap with that combination of O orbitals? What would be the case in SO₂, where 3d orbitals might be available?

15.16 (b) Consider the C_{3v} ion NO₃⁻. Is there any orbital of the central N atom that can have a nonzero overlap with the combination $2p_z(A) - p_z(B) - p_z(C)$ of the three O atoms (with z perpendicular to the plane). What would be the case in SO₃, where 3d orbitals might be available?

15.17 (a) The ground state of NO₂ is A₁ in the group C_{2v}. To what excited states may it be excited by electric dipole transitions, and what polarization of light is it necessary to use?

15.17 (b) The ClO₂ molecule (which belongs to the group C_{2v}) was trapped in a solid. Its ground state is known to be B₁. Light polarized parallel to the y-axis (parallel to the OO separation) excited the molecule to an upper state. What is the symmetry of that state?

15.18 (a) What states of (a) benzene, (b) naphthalene may be reached by electric dipole transitions from their (totally symmetrical) ground states?

15.18 (b) What states of (a) anthracene, (b) coronene (**24**) may be reached by electric dipole transitions from their (totally symmetrical) ground states?

24 Coronene

15.19 (a) Write $f_1 = \sin \theta$ and $f_2 = \cos \theta$, and show by symmetry arguments using the group C_s that the integral of their product over a symmetrical range around $\theta = 0$ is zero.

15.19 (b) Determine whether the integral over f_1 and f_2 in Exercise 15.19a is zero over a symmetrical range about $\theta = 0$ in the group C_{3v}.

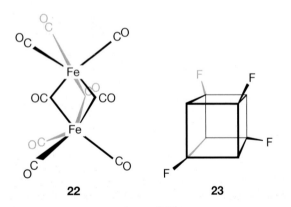

22   **23**

Problems[†]

15.1 List the symmetry elements of the following molecules and name the point groups to which they belong: (a) staggered CH₃CH₃, (b) chair and boat cyclohexane, (c) B₂H₆, (d) [Co(en)₃]³⁺, where en is ethylenediamine (ignore its detailed structure), (e) crown-shaped S₈. Which of these molecules can be (i) polar, (ii) chiral?

15.2 The group C_{2h} consists of the elements E, C_2, σ_h, i. Construct the group multiplication table and find an example of a molecule that belongs to the group.

15.3 The group D_{2h} has a C_2 axis perpendicular to the principal axis and a horizontal mirror plane. Show that the group must therefore have a centre of inversion.

15.4 Consider the H_2O molecule, which belongs to the group C_{2v}. Take as a basis the two H1s orbitals and the four valence orbital of the O atom and set up the 6×6 matrices that represent the group in this basis. Confirm by explicit matrix multiplication that the group multiplications (a) $C_2\sigma_v = \sigma'_v$ and (b) $\sigma_v\sigma'_v = C_2$. Confirm, by calculating the traces of the matrices: (a) that symmetry elements in the same class have the same character, (b) that the representation is reducible, and (c) that the basis spans $3A_1 + B_1 + 2B_2$.

15.5 Confirm that the z-component of orbital angular momentum is a basis for an irreducible representation of A_2 symmetry in C_{3v}.

15.6 The (one-dimensional) matrices $D(C_3) = 1$ and $D(C_2) = 1$, and $D(C_3) = 1$ and $D(C_2) = -1$ both represent the group multiplication $C_3C_2 = C_6$ in the group C_{6v} with $D(C_6) = +1$ and -1, respectively. Use the character tale to confirm these remarks. What are the representatives of σ_v and σ_d in each case?

15.7 Construct the multiplication table of the Pauli spin matrices, σ, and the 2×2 unit matrix:

$$\sigma_x = \begin{pmatrix} 0 & 1 \\ 1 & 0 \end{pmatrix} \quad \sigma_y = \begin{pmatrix} 0 & -i \\ i & 0 \end{pmatrix} \quad \sigma_z = \begin{pmatrix} 1 & 0 \\ 0 & -1 \end{pmatrix} \quad \sigma_0 = \begin{pmatrix} 1 & 0 \\ 0 & 1 \end{pmatrix}$$

Do the four matrices form a group under multiplication?

15.8 What irreducible representations do the four H1s orbitals of CH_4 span? Are there s and p orbitals of the central C atom that may form molecular orbitals with them? Could d orbitals, even if they were present on the C atom, play a role in orbital formation in CH_4?

15.9 Suppose that a methane molecule became distorted to (a) C_{3v} symmetry by the lengthening of one bond, (b) C_{2v} symmetry, by a kind of scissors action in which one bond angle opened and another closed slightly. Would more d orbitals become available for bonding?

15.10‡ B.A. Bovenzi and G.A. Pearse, Jr. (*J. Chem. Soc. Dalton Trans.*, 2763 (1997)) synthesized coordination compounds of the tridentate ligand pyridine-2, 6-diamidoxime ($C_7H_9N_5O_2$, **25**). Reaction with $NiSO_4$ produced a complex in which two of the essentially planar ligands are bonded at right angles to a single Ni atom. Name the point group and the symmetry operations of the resulting $[Ni(C_7H_9N_5O_2)_2]^{2+}$ complex cation.

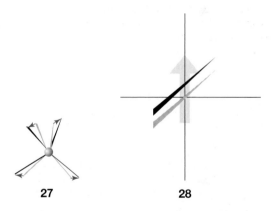

25

15.11‡ R. Eujen, B. Hoge, and D.J. Brauer (*Inorg. Chem.* **36**, 1464 (1997)) prepared and characterized several square-planar Ag(III) complex anions. In the complex anion [*trans*-Ag(CF_3)$_2$(CN)$_2$]$^-$, the Ag—CN groups are collinear. (a) Assuming free rotation of the CF_3 groups (that is, disregarding the AgCF and AgCH angles), name the point group of this complex anion. (b) Now suppose the CF_3 groups cannot rotate freely (because the ion was in a solid, for example). Structure (**26**) shows a plane that bisects the NC—Ag—CN axis and is perpendicular to it. Name the point group of the complex if each CF_3 group has a CF bond in that plane (so the CF_3 groups do not point to either CN group preferentially) and the CF_3 groups are (i) staggered, (ii) eclipsed.

26

15.12‡ A computational study by C.J. Marsden (*Chem. Phys. Lett.* **245**, 475 (1995)) of AM$_x$ compounds, where A is in Group 14 of the periodic table and M is an alkali metal, shows several deviations from the most symmetric structures for each formula. For example, most of the AM$_4$ structures were not tetrahedral but had two distinct values for MAM bond angles. They could be derived from a tetrahedron by a distortion shown in (**27**). (a) What is the point group of the distorted tetrahedron? (b) What is the symmetry species of the distortion considered as a vibration in the new, less symmetric group? Some AM$_6$ structures are not octahedral, but could be derived from an octahedron by translating a C—M—C axis as in (**28**). (c) What is the point group of the distorted octahedron? (d) What is the symmetry species of the distortion considered as a vibration in the new, less symmetric group?

27   **28**

15.13 The algebraic forms of the f orbitals are a radial function multiplied by one of the factors (a) $z(5z^2 - 3r^2)$, (b) $y(5y^2 - 3r^2)$, (c) $x(5x^2 - 3r^2)$, (d) $z(x^2 - y^2)$, (e) $y(x^2 - z^2)$, (f) $x(z^2 - y^2)$, (g) xyz. Identify the irreducible representations spanned by these orbitals in (a) C_{2v}, (b) C_{3v}, (c) T_d, (d) O_h. Consider a lanthanide ion at the centre of (a) a tetrahedral complex, (b) an octahedral complex. What sets of orbitals do the seven f orbitals split into?

15.14 Does the product xyz necessarily vanish when integrated over (a) a cube, (b) a tetrahedron, (c) a hexagonal prism, each centred on the origin?

15.15 The NO_2 molecule belongs to the group C_{2v}, with the C_2 axis bisecting the ONO angle. Taking as a basis the N2s, N2p, and O2p orbitals, identify the irreducible representations they span, and construct the symmetry-adapted linear combinations.

15.16 Construct the symmetry-adapted linear combinations of C2p_z orbitals for benzene, and use them to calculate the Hückel secular determinant. This procedure leads to equations that are much easier to solve than using the original orbitals. Show that the Hückel orbitals are those specified in Section 14.7d.

29 Phenanthrene

15.17 The phenanthrene molecule (**29**) belongs to the group C_{2v} with the C_2 axis perpendicular to the molecular plane. (a) Classify the irreducible representations spanned by the carbon $2p_z$ orbitals and find their symmetry-adapted linear combinations. (b) Use your results from part (a) to calculate the Hückel secular determinant. (c) What states of phenanthrene may be reached by electric dipole transitions from its (totally symmetrical) ground state?

15.18‡ In a spectroscopic study of C_{60}, F. Negri, G. Orlandi, and F. Zerbetto (*J. Phys. Chem.* **100**, 10849 (1996)) assigned peaks in the fluorescence spectrum. The molecule has icosahedral symmetry (I_h). The ground electronic state is A_{1g}, and the lowest-lying excited states are T_{1g} and G_g. (a) Are photon-induced transitions allowed from the ground state to either of these excited states? Explain your answer. (b) What if the transition is accompanied by a vibration that breaks the parity?

Applications: to astrophysics and biology

15.19‡ The H_3^+ molecular ion, which plays an important role in chemical reactions occurring in interstellar clouds, is known to be equilateral triangular. (a) Identify the symmetry elements and determine the point group of this molecule. (b) Take as a basis for a representation of this molecule the three H1s orbitals and set up the matrices that group in this basis. (c) Obtain the group multiplication table by explicit multiplication of the matrices. (d) Determine if the representation is reducible and, if so, give the irreducible representations obtained.

15.20‡ The H_3^+ molecular ion has recently been found in the interstellar medium and in the atmospheres of Jupiter, Saturn, and Uranus. The H_4 analogues have not yet been found, and the square planar structure is thought to be unstable with respect to vibration. Take as a basis for a representation of the point group of this

molecule the four H1s orbitals and determine if this representation is reducible.

15.21 Some linear polyenes, of which β-carotene is an example, are important biological co-factors that participate in processes as diverse as the absorption of solar energy in photosynthesis (Box 26.2) and protection against harmful biological oxidations. Use as a model of β-carotene a linear polyene containing 22 conjugated C atoms. (a) To what point group does this model of β-carotene belong? (b) Classify the irreducible representations spanned by the carbon $2p_z$ orbitals and find their symmetry-adapted linear combinations. (c) Use your results from part (b) to calculate the Hückel secular determinant. (d) What states of this model of β-carotene may be reached by electric dipole transitions from its (totally symmetrical) ground state?

15.22 The chlorophylls that participate in photosynthesis (Box 26.2) and the haem groups of cytochromes (Box 10.1) are derived from the porphine dianion group (**30**), which belongs to the D_{4h} point group. The ground electronic state is A_{1g} and the lowest-lying excited state is E_u. Is a photon-induced transition allowed from the ground state to the excited state? Explain your answer.

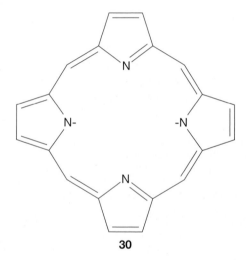

30

16

Spectroscopy 1: rotational and vibrational spectra

General features of spectroscopy

16.1 Experimental techniques

16.2 The intensities of spectral lines

16.3 Linewidths

Pure rotation spectra

16.4 Moments of inertia

16.5 The rotational energy levels

16.6 Rotational transitions

16.7 Rotational Raman spectra

16.8 Nuclear statistics and rotational states

The vibrations of diatomic molecules

16.9 Molecular vibrations

16.10 Selection rules

16.11 Anharmonicity

16.12 Vibration–rotation spectra

16.13 Vibrational Raman spectra of diatomic molecules

The vibrations of polyatomic molecules

16.14 Normal modes

16.15 Infrared absorption spectra of polyatomic molecules

16.16 Vibrational Raman spectra of polyatomic molecules

16.17 Symmetry aspects of molecular vibrations

Checklist of key ideas

Further reading

Exercises

Problems

The general strategy we adopt in the chapter is to set up expressions for the energy levels of molecules and then apply selection rules and considerations of populations to infer the form of spectra. Rotational energy levels are considered first, and we see how to derive expressions for their values and how to interpret rotational spectra in terms of molecular dimensions. Not all molecules can occupy all rotational states: we see the experimental evidence for this restriction and its explanation in terms of nuclear spin and the Pauli principle. Next, we consider the vibrational energy levels of diatomic molecules, and see that we can use the properties of harmonic oscillators developed in Chapter 12. Then we consider polyatomic molecules and find that their vibrations may be discussed as though they consisted of a set of independent harmonic oscillators, so the same approach as employed for diatomic molecules may be used. We also see that the symmetry properties of the vibrations of polyatomic molecules are helpful for deciding which modes of vibration can be studied spectroscopically.

The origin of spectral lines in molecular spectroscopy is the absorption, emission, or scattering of a photon when the energy of a molecule changes. The difference from atomic spectroscopy is that the energy of a molecule can change not only as a result of electronic transitions but also because it can undergo changes of rotational and vibrational state. Molecular spectra are therefore more complex than atomic spectra. However, they also contain information relating to more properties, and their analysis leads to values of bond strengths, lengths, and angles. They also provide a way of determining a variety of molecular properties, particularly molecular dimensions, shapes, and dipole moments. Molecular spectroscopy is also useful to astrophysicists and environmental scientists, for the chemical composition of interstellar space and of planetary atmospheres can be inferred from their rotational, vibrational, and electronic spectra (Box 16.1).

Pure rotational spectra, in which only the rotational state of a molecule changes, can be observed in the gas phase. Vibrational spectra of gaseous samples show features that arise from rotational transitions that accompany the excitation of vibration. Electronic spectra, which are described in Chapter 17, show features arising from simultaneous vibrational and rotational transitions. The simplest way of dealing with these complexities is to tackle each type of transition in turn, and then to see how simultaneous changes affect the appearance of spectra.

Box 16.1 Rotational and vibrational spectroscopy of interstellar space

Recent observations by the Cosmic Background Explorer (COBE) satellite support the long-held hypothesis that the distribution of energy in the current Universe can be modelled by a Planck distribution (eqn. 11.5) with $T = 2.726 \pm 0.001$ K, the bulk of the radiation spanning the microwave region of the spectrum. This *cosmic microwave background radiation* is thought to be the residue of energy released during the Big Bang, the event that brought the Universe into existence. Very small fluctuations in the background temperature are believed to account for the large-scale structure of the Universe.

The interstellar space in our galaxy is a little warmer than the cosmic background and consists largely of dust grains and gas clouds. The dust grains are carbon-based compounds and silicates of aluminium, magnesium, and iron, in which are embedded trace amounts of methane, water, and ammonia. Interstellar clouds are significant because it is from them that new stars, and consequently new planets, are formed. The hottest clouds are plasmas with temperatures of up to 10^6 K and densities of only about 3×10^3 particles m^{-3}. Colder clouds range from 0.1 to 1000 solar masses (1 solar mass $= 2 \times 10^{30}$ kg), have a density of about 5×10^5 particles m^{-3}, consist largely of hydrogen atoms, and have a temperature of about 80 K. There are also colder and denser clouds, some with masses greater than 500 000 solar masses, densities greater than 10^9 particles m^{-3}, and temperatures that can be lower than 10 K. They are also called *molecular clouds*, because they are composed primarily of H$_2$ and CO gas in

a proportion of about 10^5 to 1. There are also trace amounts of larger molecules.

It follows from the Boltzmann distribution (Section 2.1) and the low temperature of a molecular cloud that the vast majority of a cloud's molecules are in their vibrational and electronic ground states. However, rotational excited states are populated at 10–100 K and decay by spontaneous emission of radiation. As a result, the spectrum of the cloud in the radiofrequency and microwave regions consists of sharp lines corresponding to rotational transitions (see the illustration). The emitted light is collected by Earth-bound or space-borne *radiotelescopes*, telescopes with antennas and detectors optimized for the collection and analysis of radiation in the microwave–radiowave range of the spectrum. Earth-bound radiotelescopes are often located at the tops of high mountains, as atmospheric water vapour can reabsorb microwave radiation from space and hence interfere with the measurement.

Over 100 interstellar molecules have been identified via their rotational spectra, often by comparing radiotelescope data with spectra obtained in the laboratory or calculated by computational methods. The experiments have revealed the presence of trace amounts (with abundances of less than 10^{-8} relative to hydrogen) of neutral molecules, ions, and radicals. Examples of neutral molecules include hydrides, oxides (including water), sulfides, halogenated compounds, nitriles, hydrocarbons, aldehydes, alcohols, ethers, ketones, and amides (see *Further reading* for a complete list). The largest molecule detected by rotational spectroscopy is the nitrile HC$_{11}$N.

Rotational spectrum of the Orion nebula, showing spectral fingerprints of diatomic and polyatomic molecules present in the interstellar cloud. Adapted from G.A. Blake *et al.*, *Astrophys. J.* **315**, 621 (1987).

Interstellar space can also be investigated with vibrational spectroscopy by using a combination of telescopes and infrared detectors. The experiments are conducted primarily in space-borne telescopes because the Earth's atmosphere absorbs a great deal of infrared radiation (see Problem 16.30). In most cases, absorption by an interstellar species is detected against the background of infrared radiation emitted by a nearby star. The data can detect the presence of gas and solid water, CO, and CO_2 in molecular clouds. In certain cases, infrared emission can be detected, but these events are rare because interstellar space is too cold and does not provide enough energy to promote a significant number of molecules to vibrational excited states. However, infrared emissions can be observed if molecules are occasionally excited by high-energy photons emitted by hot stars in the vicinity of the cloud. For example, the polycyclic aromatic hydrocarbons hexabenzocoronene ($C_{48}H_{24}$) and circumcoronene ($C_{54}H_{18}$) have been identified from characteristic infrared emissions by excited normal modes composed primarily of C—H and C—C stretching motions.

Discussion questions

1. A. Dalgarno (in Chemistry in the interstellar medium, *Frontiers of Astrophysics* (ed. E.H. Avrett), Harvard University Press, Cambridge (1976)) notes that, although both CH and CN spectra show up strongly in the interstellar medium in the constellation Ophiuchus, the CN spectrum has become the standard for the determination of the temperature of the cosmic microwave background radiation. Demonstrate through a calculation why CH would not be as useful for this purpose as CN. The rotational constant B_0 for CH is 14.190 cm^{-1}.

2. The space immediately surrounding stars, also called the *circumstellar space*, is significantly warmer because stars are very intense black-body emitters with temperatures of several thousand kelvin (see Box 13.1). Discuss how such factors as cloud temperature, particle density, and particle velocity may affect the rotational spectrum of CO in an interstellar cloud. What new features in the spectrum of CO can be observed in gas ejected from and still near a star with temperatures of about 1000 K, relative to gas in a cloud with temperature of about 10 K? Explain how these features may be used to distinguish between circumstellar and interstellar material on the basis of the rotational spectrum of CO.

Further reading

C.R. Cowley, *An introduction to cosmochemistry*. Cambridge University Press (1995).

A.G.G.M. Tielens, Spectroscopy of interstellar molecules. In *Encyclopedia of spectroscopy and spectrometry* (ed. J.C. Lindon, G.E. Tranter, and J.L. Holmes), **2**, 943. Academic Press, San Diego (2000).

General features of spectroscopy

All types of spectra have some features in common, and we examine these first. We shall often need to use the relations between the frequency, v, wavelength, λ, and wavenumber, $\tilde{v}$, of electromagnetic radiation that are described in *Further information 2*:

$$\lambda = \frac{c}{v} \qquad \tilde{v} = \frac{v}{c} \tag{16.1}$$

The units of wavenumber are almost always chosen as reciprocal centimetres (cm^{-1}). Figure 16.1 summarizes the frequencies, wavelengths, and wavenumbers of the various regions of the electromagnetic spectrum and anticipates the type of transition that is characteristic of each region.

16.1 Experimental techniques

In **emission spectroscopy**, a molecule undergoes a transition from a state of high energy E_1 to a state of lower energy E_2 and emits the excess energy as a photon. In **absorption spectroscopy**, the net absorption[1] of nearly monochromatic

---

1. We say *net* absorption, because it will become clear that, when a sample is irradiated, both absorption and emission at a given frequency are stimulated, and the detector measures the difference, the net absorption.

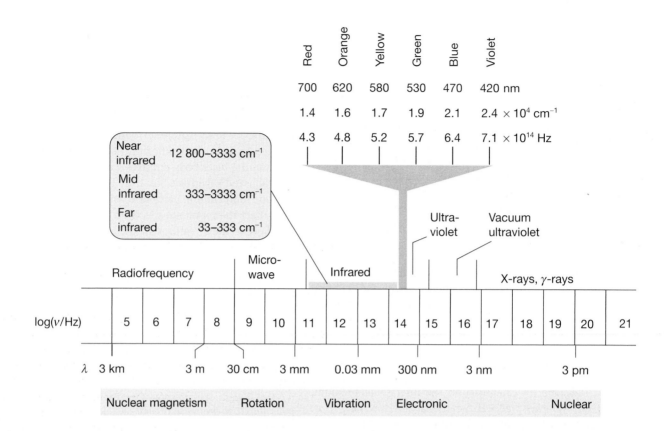

16.1 The electromagnetic spectrum and the classification of the spectral regions. The band at the bottom of the illustration indicates the types of transitions that absorb or emit in the various regions. ('Nuclear magnetism' refers to the types of transitions discussed in Chapter 18; 'nuclear' refers to transitions within the nucleus.)

(single-frequency) incident radiation is monitored as the radiation is swept over a range of frequencies. The energy, $h\nu$, of the photon emitted or absorbed, and therefore the frequency ν of the radiation emitted or absorbed, is given by the Bohr frequency condition

$$h\nu = E_1 - E_2 \tag{16.2}$$

Emission and absorption spectroscopy give the same information about energy level separations, but practical considerations generally determine which technique is employed. Emission spectroscopy, which we shall discuss in Chapter 17, is normally used for ultraviolet, visible, near-infrared, and X-ray spectroscopy. Here, we shall concentrate on absorption spectroscopy, which is widely employed in studies of electronic transitions, molecular rotations, and molecular vibrations.

Spectrometers are instruments that detect the frequency composition of electromagnetic radiation scattered, emitted, or absorbed by atoms and molecules. Figure 16.2 shows the general layout of an absorption spectrometer. Radiation from an appropriate source is directed toward a sample either by mirrors or by an optical fibre, which consists of one or more thin rods of material with a high refractive index. The optical fibre is flexible and can be used to deliver light to places that are difficult to reach, such as the interior of an organ of a living patient. In most spectrometers, light transmitted, emitted, or scattered by the sample is collected by mirrors or lenses and strikes a **dispersing element** that separates radiation into

16.2 The layout of a typical absorption spectrometer. The beams pass alternately through the sample and reference cells, and the detector is synchronized with them so that the relative absorption can be determined.

different frequencies. The intensity of light at each frequency is then analysed by a suitable detector.

(a) Sources of radiation

Spectrometers use sources of radiation that are either *polychromatic*, those spanning a wide range of frequencies, or *monochromatic*, those spanning a very narrow range of frequencies around a central value. Many commercial spectrometers use light sources that take advantage of black-body radiation from hot materials, which can provide radiation spanning a wide range of frequencies. For far-infrared radiation with $35 \text{ cm}^{-1} < \tilde{v} < 200 \text{ cm}^{-1}$, a typical source is a mercury arc inside a quartz envelope, most of the radiation being generated by the hot quartz. Either a *Nernst filament* or a *globar* is used as a source of mid-infrared radiation with $200 \text{ cm}^{-1} < \tilde{v} < 4000 \text{ cm}^{-1}$. The Nernst filament consists of a ceramic filament containing oxides of zirconium, yttrium, thallium, and cerium, which is typically heated to temperatures ranging from 1200 to 2000 K. The globar consists of a rod of silicon carbide, which is heated electrically to about 1500 K.

A *tungsten filament* gives out intense light in the range $320 \text{ nm} < \lambda < 2500 \text{ nm}$ when heated to about 3000 K. It is used in household light fixtures and in commercial spectrometers. A common problem associated with tungsten filaments is their tendency to decompose at operating temperatures higher than about 2500 K. In a *quartz–tungsten–halogen lamp*, iodine gas is used to extend the lifetime of the filament. Near the surface of the lamp's quartz envelope, iodine atoms and tungsten atoms ejected from the filament combine to make a variety of compounds, such as WI_2. As the tungsten–iodine compounds reach the hot filament, they decompose and replenish it with tungsten atoms, so retarding decomposition of the filament.

A *gas discharge lamp* is a common source of ultraviolet and visible radiation. In a *xenon discharge lamp*, an electrical discharge excites xenon atoms to excited states, which then emit ultraviolet radiation. When the pressure inside the lamp is less than about 1 kPa, the ouput of this lamp consists of sharp spectral lines from excited xenon. At higher pressures, the output of a xenon lamp consists of sharp lines superimposed on a broad, intense background due to emission from a *plasma*, a complex mixture of ions, formed from neutral atoms by the electrical discharge. High-pressure xenon lamps used in commercial spectrometers have emission profiles similar to that of a black body heated to 6000 K. In a *deuterium lamp*, excited D_2 molecules dissociate into electronically excited D atoms, which then emit intense radiation between 200–400 nm. Other discharge lamps make use of mercury, neon, and krypton, each with a distinct wavelength range of optimal emission.

The measurement of absorption spectra with a polychromatic source requires a dispersing element. However, a dispersing element is not necessary when the source

16.3 A synchrotron storage ring. The electrons injected into the ring from the linear accelerator and booster synchrotron are accelerated to high speed in the main ring. An electron in a curved path is subject to constant acceleration, and an accelerated charge radiates electromagnetic energy.

generates monochromatic radiation that can be tuned over a range of values. The *klystron* and the *Gunn diode* are electronic devices used to generate tunable microwave radiation. Lasers, which are discussed in more detail in Chapter 17, generate monochromatic electromagnetic radiation that can often be tuned over a range of frequencies; different types of laser are used to cover different regions of the electromagnetic spectrum.

For certain applications, synchrotron radiation from a *synchrotron storage ring* is appropriate. A synchrotron storage ring consists of an electron beam (actually a series of closely spaced packets of electrons) travelling in a circular path of several metres in diameter. As electrons travelling in a circle are constantly accelerated by the forces that constrain them to their path, they generate radiation (Fig. 16.3). Synchrotron radiation spans a wide range of frequencies, including the infrared and X-rays. Except in the microwave region, synchrotron radiation is much more intense than can be obtained by most conventional sources. The disadvantage of the source is that it is so large and costly that it is essentially a national facility, not a laboratory commonplace.

(b) The dispersing element

The dispersing element in most absorption spectrometers operating in the ultraviolet to near-infrared region of the spectrum is a **diffraction grating**, which consists of a glass or ceramic plate into which fine grooves have been cut and covered with a reflective aluminium coating. The grating causes interference between waves reflected from its surface, and constructive interference occurs when

$$n\lambda = d(\sin \theta - \sin \phi) \tag{16.3}$$

where $n = 1, 2, \ldots$ is the *diffraction order*, λ is the wavelength of the diffracted radiation, d is the distance between grooves, θ is the angle of incidence of the beam, and ϕ is the angle of emergence of the beam (Fig. 16.4). The *Illustration* below shows that, for a given n and θ, larger differences in ϕ are observed for different wavelengths when d is similar to the wavelength of radiation being analysed. Wide angular separation results in wide spatial separation between wavelengths some distance away from the grating, where a detector is placed.

16.4 One common dispersing element is a diffraction grating, which separates wavelengths spatially as a result of the scattering of light by fine grooves cut into a coated piece of glass. When a polychromatic light beam strikes the surface at an angle θ, several light beams of different wavelengths emerge at different angles ϕ (eqn 3).

Illustration 16.1

For $\lambda = 500$ nm, $n = 1$, and $\theta = 30.0°$, it follows from eqn 3 that $\phi = 0.0°$ and $26.7°$ for $d = 1000$ nm and 10 000 nm, respectively. If we repeat the calculation for $\lambda = 400$ nm, we obtain $\phi = 5.7°$ and $27.4°$ for $d = 1000$ nm and 10 000 nm, respectively. Therefore, a grating with $d = 1000$ nm disperses radiation in the 400–500 nm range over a wider range of angles than does a grating with $d = 10\,000$ nm.

In a **monochromator**, a narrow exit slit allows only a narrow range of wavelengths to reach the detector (Fig. 16.5). Turning the grating around an axis perpendicular to the incident and diffracted beams allows different wavelengths to be analysed; in this way, the absorption spectrum is built up one narrow wavelength range at a time. Typically, the grating is swept through an angle that investigates only the first order of diffraction ($n = 1$). In a **polychromator**, there is no slit and a broad range of wavelengths can be analysed simultaneously by *array detectors*, such as those discussed below.

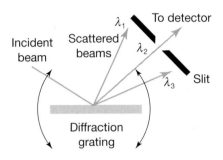

16.5 A polychromatic beam is dispersed by a diffraction grating into three component wavelengths λ_1, λ_2, and λ_3. In the configuration shown, only radiation with λ_2 passes through a narrow slit and reaches the detector. Rotating the diffraction grating in the direction shown by the double arrows allows λ_1 and λ_3 to reach the detector.

(c) Fourier transform techniques

Modern spectrometers, particularly those operating in the infrared and near-infrared, now almost always use **Fourier transform techniques** of spectral detection and analysis. The heart of a Fourier transform spectrometer is a *Michelson interferometer*, a device for analysing the frequencies present in a composite signal. The total signal from a sample is like a chord played on a piano, and the Fourier transform of the signal is equivalent to the separation of the chord into its individual notes, its spectrum.

A Michelson interferometer works by splitting the beam from the sample into two and introducing a varying path difference, p, into one of them (Fig. 16.6). When the two components recombine, there is a phase difference between them, and they interfere either constructively or destructively depending on the difference in path lengths. The detected signal oscillates as the two components alternately come into and out of phase as the path difference is changed (Fig. 16.7). If the radiation has wavenumber $\tilde{\nu}$, the intensity of the detected signal due to radiation in the range of wavenumbers $\tilde{\nu}$ to $\tilde{\nu} + d\tilde{\nu}$, which we denote $\Delta I(p, \tilde{\nu})\,d\tilde{\nu}$, varies with p as

$$I(p, \tilde{\nu})\,d\tilde{\nu} = I(\tilde{\nu})(1 + \cos 2\pi\tilde{\nu}p)\,d\tilde{\nu} \tag{16.4}$$

Hence, the interferometer converts the presence of a particular wavenumber component in the signal into a variation in intensity of the radiation reaching the detector. An actual signal consists of radiation spanning a large number of wavenumbers, and the total intensity at the detector, which we write $I(p)$, is the sum of contributions from all the wavenumbers present in the signal (Fig. 16.8):

$$I(p) = \int_0^\infty I(p, \tilde{\nu})\,d\tilde{\nu} = \int_0^\infty I(\tilde{\nu})(1 + \cos 2\pi\tilde{\nu}p)\,d\tilde{\nu} \tag{16.5}$$

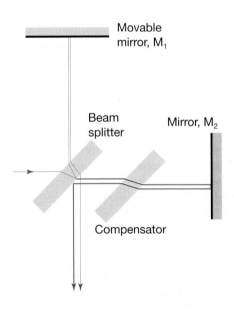

16.6 A Michelson interferometer. The beam-splitting element divides the incident beam into two beams with a path difference that depends on the location of the mirror M_1. The compensator ensures that both beams pass through the same thickness of material.

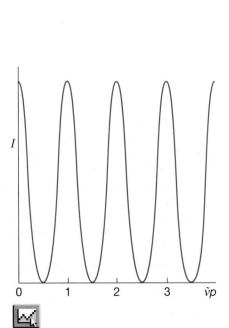

16.7 An interferogram produced as the path length p is changed in the interferometer shown in Fig. 16.6. Only a single-frequency component is present in the radiation.

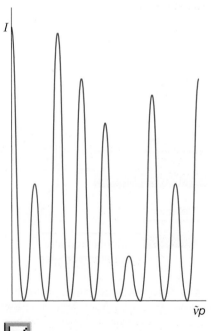

16.8 An interferogram obtained when several (in this case, three) frequencies are present in the radiation.

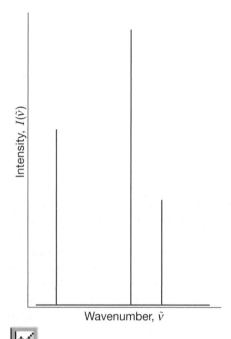

16.9 The three frequency components and their intensities that account for the appearance of the interferogram in Fig. 16.8. This spectrum is the Fourier transform of the interferogram, and is a depiction of the contributing frequencies.

The problem is to find $I(\tilde{\nu})$, the variation of intensity with wavenumber, which is the spectrum we require, from the record of values of $I(p)$. This step is a standard technique of mathematics, and is the 'Fourier transformation' step from which this form of spectroscopy takes its name. Specifically:

$$I(\tilde{\nu}) = 4 \int_0^\infty \{I(p) - \tfrac{1}{2}I(0)\} \cos 2\pi\tilde{\nu}p \, dp \tag{16.6}$$

where $I(0)$ is given by eqn 5 with $p = 0$. This integration is carried out in a computer connected to the spectrometer, and the output, $I(\tilde{\nu})$, is the absorption spectrum of the sample (Fig. 16.9).[2]

A major advantage of the Fourier transform procedure is that all the radiation emitted by the source is monitored continuously. This is in contrast to a spectrometer in which a monochromator discards most of the generated radiation. As a result, Fourier transform spectrometers have a higher sensitivity than conventional spectrometers.

(d) Raman spectroscopy

In **Raman spectroscopy**, molecular energy levels are explored by examining the frequencies present in the radiation scattered by molecules. In a typical experiment, a monochromatic incident laser beam is passed through the sample and the radiation scattered from the front face of the sample is monitored (Fig. 16.10). This detection geometry allows for the study of gases, pure liquids, solutions, suspensions, and solids. About 1 in 10^7 of the incident photons collide with the molecules, give up some of their energy, and emerge with a lower energy. These scattered photons constitute the lower-frequency **Stokes radiation** from the sample. Other incident photons may collect energy from the molecules (if they are already excited), and emerge as higher-frequency **anti-Stokes radiation**. The component of radiation scattered into the forward direction without change of frequency is called **Rayleigh radiation**.

Lasers are used as the radiation sources in Raman spectrometers for two reasons. First, the shifts in frequency of the scattered radiation from the incident radiation are quite small, so highly monochromatic radiation from a laser is required if the shifts are to be observed. Second, the intensity of scattered radiation is low, so intense incident beams, such as those from a laser, are needed. Raman spectra may be examined using visible and ultraviolet lasers, in which case a diffraction grating is used to distinguish between Rayleigh, Stokes, and anti-Stokes radiation. In most modern instruments Rayleigh radiation is absorbed by a *notch filter*, which absorbs a very narrow range of wavelengths and lets only Stokes and anti-Stokes radiation reach the detector. In cases where strong fluorescence emission from the sample or a contaminant interferes with weak Raman scattering from the sample, incident radiation in the near-infrared and Fourier transform techniques of detection are commonly used. In Fourier transform Raman spectrometers, radiation scattered by the sample passes through a notch filter and then through a Michelson interferometer. The total intensity I at the detector varies with the path difference p between the two beams in the interferometer and Fourier transformation of $I(p)$ results in the spectrum of the scattered radiation, showing intensity maxima at wavenumbers corresponding to Stokes and anti-Stokes lines. Raman spectroscopy is often complementary to infrared spectroscopy because, as we shall see, different selection rules are obeyed.

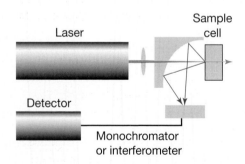

16.10 A common arrangement adopted in Raman spectroscopy. A laser beam first passes through a lens and then through a small hole in a mirror with a curved reflecting surface. The focused beam strikes the sample and scattered light is both deflected and focused by the mirror. The spectrum is analysed by a monochromator or an interferometer.

2. More precisely, it is the *transmission spectrum*, for the signal depends on the transmitted intensity. However, the absorption and transmission spectra carry the same information and the former term is normally employed.

(e) Detectors

The third component of a spectrometer is the **detector**, a device that converts radiation into an electric current or voltage for appropriate signal processing and display. Detectors may consist of a single radiation-sensing element or of several small elements arranged in one- or two-dimensional arrays.

The *photomultiplier tube* (PMT) is a sensing element in which the photoelectric effect (Section 11.2a) is used to generate an electrical signal proportional to the intensity of light that strikes the detector. In a PMT, photons first strike the *photocathode*, usually a metallic surface to which a large negative potential is applied. Each electron ejected from the photocathode is accelerated by a potential difference to another metallic surface, called the *dynode*, from which more electrons are ejected. The function of the dynode is to amplify the current produced at the photocathode. After travelling through a chain of dynodes, with each dynode at a more positive potential than the preceding dynode in the chain, all the ejected electrons are collected at a final metallic surface called the *anode*. Depending on how the detector is constructed, a PMT can produce up to 10^8 electrons per photon that strikes the photocathode. This sensitivity is important for the detection of light from weak sources, but can pose problems as well. At room temperature, a small number of electrons on the surfaces of the photocathode and dynodes have sufficient energy to be ejected even in the dark. When amplified through the dynode chain, these electrons give rise to a *dark current*, which interferes with measurements on the sample of interest. To minimize the dark current, it is common to lower the temperature of the PMT detector. The best operating temperature of a detector depends on how it is constructed and on how it is to be used. For applications that require the detection of very low levels of radiation, a PMT can be cooled to 77 K by immersion of the detector assembly into an insulated container filled with liquid N_2.

A common alternative to the PMT is the *photodiode*, a solid-state device that conducts electricity when struck by photons because light-induced electron transfer reactions in the detector material create mobile charge carriers (negatively charged electrons and positively charged 'holes'). Photodiodes are easy to manufacture, but not very sensitive. However, it is possible to improve their performance by accelerating the photo-generated electrons with a very large electrical potential difference. The high-energy electrons then collide with other atoms in the solid and ionize them, thus creating an avalanche of secondary charge carriers and increasing the sensitivity of the device toward photons. Photodiodes designed to operate in this manner are called *avalanche photodiodes*.

With appropriate choice of material, photodiodes can be used to detect light spanning a wide range of wavelengths. For example, silicon is sensitive in the visible region and germanium is used in most spectrometers operating in the near-infrared region of the spectrum, including Fourier transform Raman spectrometers.

The *charge-coupled device* (CCD) is a two-dimensional array of up to six million small photodiode detectors. With a CCD, a wide range of wavelengths that emerge from a polychromator are detected simultaneously, thus eliminating the need to measure light intensity one narrow wavelength range at a time. CCD detectors are used widely to measure absorption, emission, and Raman scattering. The CCD is also a critical component of some Raman microscopes.

The most common detectors found in commercial infrared spectrometers are sensitive in the mid-infrared region. The mercury–cadmium–telluride (MCT) detector is a *photovoltaic device* for which the potential difference changes upon exposure to infrared radiation. The deuterated triglycine sulfate (DTGS) detector is a *pyroelectric device*, in which the capacitance is sensitive to temperature and hence the presence of infrared radiation.

A microwave detector is typically a *crystal diode* consisting of a tungsten tip in contact with a semiconductor, such as germanium, silicon, or gallium arsenide. In microwave spectroscopy, as in other applications where the optical signal is very weak, the intensity of the radiation arriving at the detector is usually modulated, because alternating signals are easier to amplify than a steady signal. In some cases the beam is chopped by a rotating shutter. In other cases, the absorption characteristics of the sample itself are modulated. Ways of achieving the latter kind of modulation are described later in the chapter and in Chapter 18.

(f) Resolution

A number of factors determine a spectrometer's **resolution**, the smallest observable separation between two closely spaced spectral bands. For example, consider a spectral measurement conducted with a monochromator and a single-element detector, such as a PMT. We saw in *Illustration 16.1* that the ability of a diffraction grating to disperse light depends on the distance between the grating's grooves and on the wavelength of the incident radiation. Furthermore, the distance between the grating and the slit placed in front of the detector must be long enough and the slit's width must be narrow enough so full advantage can be taken of the grating's dispersing ability (see Fig. 16.5). It follows that a bad choice of grating, slit width, and detector placement may result in the failure to distinguish between closely spaced lines or to measure the actual linewidth of any one band in the spectrum.

The resolution of Fourier transform spectrometers is determined by the maximum path length difference, p_{max}, of the interferometer:

$$\Delta\tilde{\nu} = \frac{1}{2p_{max}}$$

(16.7)

To achieve a resolution of 0.1 cm^{-1} requires a maximum path length difference of 5 cm.

Assuming that all instrumental factors have been optimized, the highest resolution is obtained when the sample is gaseous and at such low pressure that collisions between the molecules are infrequent (see Section 16.3b). In liquids and solids, the actual linewidths can be so broad that the sample itself can limit the resolution.

(g) The sample

Gaseous samples are essential for rotational (microwave) spectroscopy, for only then can molecules rotate freely. To achieve sufficient absorption, path lengths through gaseous samples must be very long, of the order of metres; long path lengths are achieved by multiple passage of the beam between parallel mirrors at each end of the sample cavity.

The most common range for infrared spectroscopy is the mid-infrared (400–4000 cm^{-1}) and it is possible to measure either the radiation transmitted or reflected by the sample. Transmission measurements are useful for gaseous, liquid, and some powder samples. Ordinary glass and quartz absorb over most of this range, so some other material must be used as windows of the sample cell. Thus, the sample is typically held between windows of sodium chloride (which is transparent down to 625 cm^{-1}) or potassium bromide (which is transparent down to 400 cm^{-1}). Powders are prepared by grinding into a paste with 'Nujol', a hydrocarbon oil, or pressing into a solid disk (with powdered potassium bromide, for example). The infrared absorption spectra of solid samples that are opaque to infrared radiation are obtained by comparing the intensity of light reflected from the sample surface with the intensity of incident light.

In Raman spectroscopy, the spectrum is obtained with ultraviolet, visible, or near-infrared laser excitation, so sample cells made of a variety of materials may be used. For example, the vibrational Raman spectrum of a liquid may be obtained by holding it in a quartz or glass cell. Sometimes not even a sample cell is required because the technique can measure light scattered by the front face of an opaque sample that is held directly in front of the laser beam.

16.2 The intensities of spectral lines

The ratio of the transmitted intensity, I, to the incident intensity, I_0, at a given frequency is called the **transmittance**, T, of the sample at that frequency:

$$T = \frac{I}{I_0}$$

[16.8]

It is found empirically that the transmitted intensity varies with the length, l, of the sample and the molar concentration, $[J]$, of the absorbing species J in accord with the **Beer–Lambert law**:

$$I = I_0 10^{-\varepsilon[J]l}$$

(16.9)

The quantity ε is called the **molar absorption coefficient** (formerly, and still widely, the 'extinction coefficient'). The molar absorption coefficient depends on the frequency of the incident radiation and is greatest where the absorption is most intense. Its dimensions are 1/(concentration × length), and it is normally convenient to express it in litres per mole per centimetre (L mol^{-1} cm^{-1}). Alternative units are square centimetres per mole (cm^2 mol^{-1}). This change of units demonstrates that ε may be regarded as a molar cross-section for absorption and, the greater the cross-sectional area of the molecule for absorption, the greater its ability to block the passage of the incident radiation.

The form of eqn 9 suggests that it is sensible to introduce the **absorbance**, A, of the sample at a given wavenumber as

$$A = \log \frac{I_0}{I} \quad \text{or} \quad A = -\log T$$

[16.10]

Then the Beer–Lambert law becomes

$$A = \varepsilon[J]l$$

(16.11)

The product $\varepsilon[J]l$ was known formerly as the *optical density* of the sample.

Justification 16.1

The Beer–Lambert law is an empirical result. However, it is simple to account for its form. The reduction in intensity, dI, that occurs when light passes through a layer of thickness dl containing an absorbing species J at a molar concentration $[J]$ is proportional to the thickness of the layer, the concentration of J, and the intensity, I, incident on the layer (because the rate of absorption is proportional to the intensity, see below). We can therefore write

$$dI = -\kappa[J]I\, dl$$

where κ (kappa) is the proportionality coefficient, or equivalently

$$\frac{dI}{I} = -\kappa[J]\, dl$$

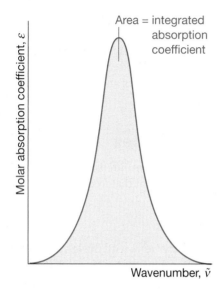

16.11 The integrated absorption coefficient of a transition is the area under a plot of the molar absorption coefficient against the wavenumber of the incident radition.

This expression applies to each successive layer into which the sample can be regarded as being divided. Therefore, to obtain the intensity that emerges from a sample of thickness l when the intensity incident on one face of the sample is I_0, we sum all the successive changes:

$$\int_{I_0}^{I} \frac{dI}{I} = -\kappa \int_{0}^{l} [J]\, dl$$

If the concentration is uniform, $[J]$ is independent of location, and the expression integrates to

$$\ln \frac{I}{I_0} = -\kappa [J] l$$

This expression gives the Beer–Lambert law when the logarithm is converted to base 10 by using $\ln x = (\ln 10)\log x$ and replacing κ by $\varepsilon \ln 10$.

Illustration 16.2

The Beer–Lambert law implies that the intensity of electromagnetic radiation transmitted through a sample at a given wavenumber decreases exponentially with the sample thickness and the molar concentration. If the transmittance is 0.1 for a path length of 1 cm (corresponding to a 90 per cent reduction in intensity), then it would be $(0.1)^2 = 0.01$ for a path of double the length (corresponding to a 99 per cent reduction in intensity overall).

The maximum value of the molar absorption coefficient, ε_{max}, is an indication of the intensity of a transition. However, as absorption bands generally spread over a range of wavenumbers, quoting the absorption coefficient at a single wavenumber might not give a true indication of the intensity of a transition. The **integrated absorption coefficient**, $\mathcal{A}$, is the sum of the absorption coefficients over the entire band (Fig. 16.11), and corresponds to the area under the plot of the molar absorption coefficient against wavenumber:

$$\mathcal{A} = \int_{\text{band}} \varepsilon(\tilde{v})\, d\tilde{v} \qquad [16.12]$$

For lines of similar widths, the integrated absorption coefficients are proportional to the heights of the lines.

(a) Absorption intensities

Einstein identified three contributions to the transitions between states. **Stimulated absorption** is the transition from a low-energy state to one of higher energy that is driven by the electromagnetic field oscillating at the transition frequency. We saw in Section 12.10 that the transition rate, w, is the rate of change of probability of the molecule being found in the upper state. We also saw that, the more intense the electromagnetic field (the more intense the incident radiation), the greater the rate at which transitions are induced and hence the stronger the absorption by the sample (Fig. 16.12). Einstein wrote the transition rate as

$$w = B\rho \qquad (16.13)$$

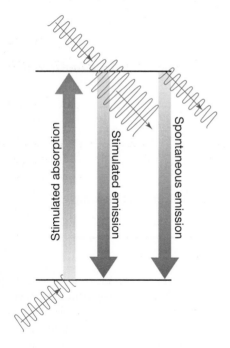

16.12 The processes that account for absorption and emission of radiation and the attainment of thermal equilibrium. The excited state can return to the lower state spontaneously as well as by a process stimulated by radiation already present at the transition frequency.

The constant B is the **Einstein coefficient of stimulated absorption** and $\rho\,\mathrm{d}v$ is the energy density of radiation in the frequency range v to $v + \mathrm{d}v$, where v is the frequency of the transition. When the molecule is exposed to black-body radiation from a source of temperature T, ρ is given by the Planck distribution (eqn 11.5):[3]

$$\rho = \frac{8\pi h v^3/c^3}{\mathrm{e}^{hv/kT} - 1} \tag{16.14}$$

For the time being, we can treat B as an empirical parameter that characterizes the transition: if B is large, then a given intensity of incident radiation will induce transitions strongly and the sample will be strongly absorbing. The **total rate of absorption**, W, the number of molecules excited during an interval divided by the duration of the interval, is the transition rate of a single molecule multiplied by the number of molecules N in the lower state: $W = Nw$.

Einstein considered that the radiation was also able to induce the molecule in the upper state to undergo a transition to the lower state, and hence to generate a photon of frequency v. Thus, he wrote the rate of this stimulated emission as

$$w' = B'\rho \tag{16.15}$$

where B' is the **Einstein coefficient of stimulated emission**. Note that only radiation of the same frequency as the transition can stimulate an excited state to fall to a lower state. However, Einstein realized that stimulated emission was not the only means by which the excited state could generate radiation and return to the lower state, and suggested that an excited state could undergo **spontaneous emission** at a rate that was independent of the intensity of the radiation (of any frequency) that is already present. He therefore wrote the total rate of transition from the upper to the lower state as

$$w' = A + B'\rho \tag{16.16}$$

The constant A is the **Einstein coefficient of spontaneous emission**. The overall rate of emission is

$$W' = N'(A + B'\rho) \tag{16.17}$$

where N' is the population of the upper state.

As demonstrated in the *Justification* below, Einstein was able to show that the two coefficients of stimulated absorption and emission are equal, and that the coefficient of spontaneous emission is related to them by

$$A = \left(\frac{8\pi h v^3}{c^3}\right)B \tag{16.18}$$

Justification 16.2

At thermal equilibrium, the total rates of emission and absorption are equal, so

$$NB\rho = N'(A + B'\rho)$$

This expression rearranges into

$$\rho = \frac{N'A}{NB - N'B'} = \frac{A/B}{N/N' - B'/B} = \frac{A/B}{\mathrm{e}^{hv/kT} - B'/B}$$

---

3. The slightly different form of the distribution stems from the fact that the ρ in eqn 11.5 is for the energy density written as $\rho\,\mathrm{d}\lambda$ whereas here it is written as $\rho\,\mathrm{d}v$, and $\mathrm{d}\lambda = (c/v^2)\,\mathrm{d}v$.

We have used the Boltzmann expression (Section 2.1) for the ratio of populations of states of energies E and E' in the last step:

$$\frac{N'}{N} = e^{-h\nu/kT} \qquad h\nu = E' - E$$

This result has the same form as the Planck distribution (eqn 14), which describes the radiation density at thermal equilibrium. Indeed, when we compare the two expressions for ρ, we can conclude that $B' = B$ and that A is related to B by eqn 18.

The growth of the importance of spontaneous emission with increasing frequency is a very important conclusion, as we shall see when we consider the operation of lasers (Section 17.5). The equality of the coefficients of stimulated emission and absorption implies that, if two states happen to have equal populations, then the rate of stimulated emission is equal to the rate of stimulated absorption, and there is then no net absorption.

Spontaneous emission can be largely ignored at the relatively low frequencies of rotational and vibrational transitions, and the intensities of these transitions can be discussed in terms of stimulated emission and absorption. Then the net rate of absorption is given by

$$W_{\text{net}} = NB\rho - N'B'\rho = (N - N')B\rho \tag{16.19}$$

and is proportional to the population difference of the two states involved in the transition.

(b) Selection rules and transition moments

We met the concept of a 'selection rule' in Sections 13.3 and 15.6 as a statement about whether a transition is forbidden or allowed. Selection rules also apply to molecular spectra, and the form they take depends on the type of transition. The underlying classical idea is that, for the molecule to be able to interact with the electromagnetic field and absorb or create a photon of frequency ν, it must possess, at least transiently, a dipole oscillating at that frequency. We saw in Section 12.10 that this transient dipole is expressed quantum mechanically in terms of the transition dipole moment, $\boldsymbol{\mu}_{\text{fi}}$, between states $|\text{i}\rangle$ and $|\text{f}\rangle$:

$$\boldsymbol{\mu}_{\text{fi}} = \langle \text{f}|\boldsymbol{\mu}|\text{i}\rangle = \int \psi_{\text{f}}^* \boldsymbol{\mu} \psi_{\text{i}} \, d\tau \tag{16.20}$$

where $\boldsymbol{\mu}$ is the electric dipole moment operator. The size of the transition dipole can be regarded as a measure of the charge redistribution that accompanies a transition: a transition will be active (and generate or absorb photons) only if the accompanying charge redistribution is dipolar (Fig. 16.13).

We know from time-dependent perturbation theory (Section 12.10) that the transition rate is proportional to $|\boldsymbol{\mu}_{\text{fi}}|^2$. It follows that the coefficient of stimulated absorption (and emission), and therefore the intensity of the transition, is also proportional to $|\boldsymbol{\mu}_{\text{fi}}|^2$. A detailed analysis gives

$$B = \frac{|\boldsymbol{\mu}_{\text{fi}}|^2}{6\varepsilon_0 \hbar^2} \tag{16.21}$$

Only if the transition moment is nonzero does the transition contribute to the spectrum. It follows that, to identify the selection rules, we must establish the conditions for which $\boldsymbol{\mu}_{\text{fi}} \neq 0$.

(a)

(b)

16.13 (a) When a $1s$ electron becomes a $2s$ electron, there is a spherical migration of charge; there is no dipole moment associated with this migration of charge; this transition is electric-dipole forbidden. (b) In contrast, when a $1s$ electron becomes a $2p$ electron, there is a dipole associated with the charge migration; this transition is allowed. (There are subtle effects arising from the sign of the wavefunction that give the charge migration a dipolar character, which this diagram does not attempt to convey.)

16.14 The Gaussian shape of a Doppler-broadened spectral line reflects the distribution of speeds in the sample at the temperature of the experiment. Notice that the line broadens as the temperature is increased.

A Gaussian function of the general form $y(x) = ae^{-(x-b)^2/2\sigma^2}$, where a, b, and σ are constants, has a maximum $y(b) = a$ and a width at half-height $\delta x = 2\sigma(2 \ln 2)^{1/2}$.

A **gross selection rule** specifies the general features a molecule must have if it is to have a spectrum of a given kind. For instance, we shall see that a molecule gives a rotational spectrum only if it has a permanent electric dipole moment. This rule, and others like it for other types of transition, will be explained in the relevant sections of the chapter. A detailed study of the transition moment leads to the **specific selection rules** that express the allowed transitions in terms of the changes in quantum numbers. We have already encountered examples of specific selection rules when discussing atomic spectra (Section 13.3), such as the rule $\Delta l = \pm 1$ for the angular momentum quantum number.

16.3 Linewidths

A number of effects contribute to the widths of spectroscopic lines. Some contributions to linewidths can be modified by changing the conditions, and to achieve high resolutions we need to know how to minimize these contributions. Other contributions cannot be changed, and represent an inherent limitation on resolution.

(a) Doppler broadening

One important broadening process in gaseous samples is the **Doppler effect**, in which radiation is shifted in frequency when the source is moving towards or away from the observer. When a source emitting electromagnetic radiation of frequency v moves with a speed s relative to an observer, the observer detects radiation of frequency[4]

$$v_{\text{receding}} = v\left(\frac{1 - s/c}{1 + s/c}\right)^{1/2} \qquad v_{\text{approaching}} = v\left(\frac{1 + s/c}{1 - s/c}\right)^{1/2} \tag{16.22}$$

where c is the speed of light. For non-relativistic speeds ($s \ll c$), these expressions simplify to

$$v_{\text{receding}} \approx \frac{v}{1 + s/c} \qquad v_{\text{approaching}} \approx \frac{v}{1 - s/c} \tag{16.23}$$

Molecules reach high speeds in all directions in a gas, and a stationary observer detects the corresponding Doppler-shifted range of frequencies. Some molecules approach the observer, some move away; some move quickly, others slowly. The detected spectral 'line' is the absorption or emission profile arising from all the resulting Doppler shifts. As shown in the following *Justification*, the profile reflects the distribution of molecular velocities parallel to the line of sight, which is a bell-shaped Gaussian curve. The Doppler line shape is therefore also a Gaussian (Fig. 16.14), and we show in the *Justification* that, when the temperature is T and the mass of the molecule is m, then the observed width of the line at half-height (in terms of frequency or wavelength) is

$$\delta v_{\text{obs}} = \frac{2v}{c}\left(\frac{2kT \ln 2}{m}\right)^{1/2} \qquad \delta\lambda_{\text{obs}} = \frac{2\lambda}{c}\left(\frac{2kT \ln 2}{m}\right)^{1/2} \tag{16.24}$$

For a molecule like N_2 at room temperature ($T \approx 300$ K), $dv/v \approx 2.3 \times 10^{-6}$. For a typical rotational transition wavenumber of 1 cm^{-1} (corresponding to a frequency of 30 GHz), the linewidth is about 70 kHz. Doppler broadening increases with temperature because the molecules acquire a wider range of speeds. Therefore, to obtain spectra of maximum sharpness, it is best to work with cold samples.

4. For a derivation of these expressions, see *Further reading*.

Justification 16.3

We know from the Boltzmann distribution (Section 2.1) that the probability that a gas molecule of mass m and speed s in a sample with temperature T has kinetic energy $E_K = \frac{1}{2}ms^2$ is proportional to $e^{-ms^2/2kT}$. The observed frequencies, v_{obs}, emitted or absorbed by the molecule are related to its speed by eqn 23:

$$v_{obs} = v\left(\frac{1}{1 \pm s/c}\right)$$

where v is the unshifted frequency. When $s \ll c$, the Doppler shift in the frequency is

$$v_{obs} - v \approx \pm vs/c$$

which implies a symmetrical distribution of observed frequencies with respect to molecular speeds. More specifically, the intensity I of a transition at v_{obs} is proportional to the probability of finding the molecule that emits or absorbs at v_{obs}, so it follows from the Boltzmann distribution and the expression for the Doppler shift that

$$I(v_{obs}) \propto e^{-mc^2(v_{obs}-v)^2/2v^2kT}$$

which has the form of a Gaussian function. The width at half-height can be calculated directly from the exponent (see the *Maths commentary*) to give eqn 24.

(b) Lifetime broadening

It is found that spectroscopic lines from gas-phase samples are not infinitely sharp even when Doppler broadening has been largely eliminated by working at low temperatures. The same is true of the spectra of samples in condensed phases and solution. This residual broadening is due to quantum mechanical effects. Specifically, when the Schrödinger equation is solved for a system that is changing with time, it is found that it is impossible to specify the energy levels exactly. If on average a system survives in a state for a time τ (tau), the **lifetime** of the state, then its energy levels are blurred to an extent of order δE, where

$$\delta E \approx \frac{\hbar}{\tau} \tag{16.25}$$

This expression is reminiscent of the Heisenberg uncertainty principle (eqn 11.46), and consequently this **lifetime broadening** is often called 'uncertainty broadening'. When the energy spread is expressed as a wavenumber through $\delta E = hc\delta\tilde{v}$, and the values of the fundamental constants introduced, this relation becomes

$$\delta\tilde{v} \approx \frac{5.3 \text{ cm}^{-1}}{\tau/\text{ps}} \tag{16.26}$$

No excited state has an infinite lifetime; therefore, all states are subject to some lifetime broadening and, the shorter the lifetimes of the states involved in a transition, the broader the corresponding spectral lines.

Two processes are responsible for the finite lifetimes of excited states. The dominant one for low-frequency transitions is **collisional deactivation**, which arises from collisions between molecules or with the walls of the container. If the **collisional lifetime**, the mean time between collisions, is τ_{col}, the resulting collisional

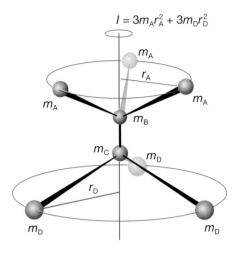

$$I = 3m_A r_A^2 + 3m_D r_D^2$$

16.15 The definition of moment of inertia. In this molecule there are three identical atoms attached to the B atom and three different but mutually identical atoms attached to the C atom. In this example, the centre of mass lies on an axis passing through the B and C atom, and the perpendicular distances are measured from this axis.

linewidth is $\delta E_{col} \approx \hbar/\tau_{col}$. Because $\tau_{col} = 1/z$, where z is the collision frequency, and from the kinetic model of gases (Section 24.1) we know that z is proportional to the pressure, we see that the collisional linewidth is proportional to the pressure. The collisional linewidth can therefore be minimized by working at low pressures.

The rate of spontaneous emission cannot be changed. Hence it is a natural limit to the lifetime of an excited state, and the resulting lifetime broadening is the **natural linewidth** of the transition. The natural linewidth is an intrinsic property of the transition, and cannot be changed by modifying the conditions. Natural linewidths depend strongly on the transition frequency (they increase with the coefficient of spontaneous emission A and therefore as v^3), so low-frequency transitions (such as the microwave transitions of rotational spectroscopy) have very small natural linewidths, and collisional and Doppler line-broadening processes are dominant. The natural lifetimes of electronic transitions are very much shorter than for vibrational and rotational transitions, so the natural linewidths of electronic transitions are much greater than those of vibrational and rotational transitions. For example, a typical electronic excited state natural lifetime is about 10^{-8} s (10 ns), corresponding to a natural width of about 5×10^{-4} cm^{-1} (15 MHz). A typical rotational state natural lifetime is about 10^3 s, corresponding to a natural linewidth of only 5×10^{-15} cm^{-1} (of the order of 10^{-4} Hz).

Pure rotation spectra

The general strategy we adopt for discussing molecular spectra and the information they contain is to find expressions for the energy levels of molecules and then to calculate the transition frequencies by applying the selection rules. We then predict the appearance of the spectrum by taking into account the transition moments and the populations of the states. In this section we illustrate the strategy by considering the rotational states of molecules.

16.4 Moments of inertia

The key molecular parameter we shall need is the **moment of inertia**, I, of the molecule (Section 12.6). The moment of inertia of a molecule is defined as the mass of each atom multiplied by the square of its distance from the rotational axis through the centre of mass of the molecule (Fig. 16.15):

$$I = \sum_i m_i r_i^2 \qquad [16.27]$$

where r_i is the perpendicular distance of the atom i from the axis of rotation. The moment of inertia depends on the masses of the atoms present and the molecular geometry, so we can suspect (and later shall see explicitly) that rotational spectroscopy will give information about bond lengths and bond angles.

In general, the rotational properties of any molecule can be expressed in terms of the moments of inertia about three perpendicular axes set in the molecule (Fig. 16.16). The convention is to label the moments of inertia I_a, I_b, and I_c, with the axes chosen so that $I_c \geq I_b \geq I_a$. For linear molecules, the moment of inertia around the internuclear axis is zero. The explicit expressions for the moments of inertia of some symmetrical molecules are given in Table 16.1.

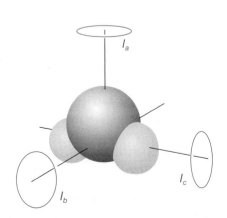

16.16 An asymmetric rotor has three different moments of inertia; all three rotation axes coincide at the centre of mass of the molecule.

Table 16.1 Moments of inertia†

1. Diatomics

$$I = \frac{m_A m_B}{m} R^2 = \mu R^2$$

2. Linear rotors

$$I = m_A R^2 + m_C R'^2 - \frac{(m_A R - m_C R')^2}{m}$$

$$I = 2m_A R^2$$

3. Symmetric rotors

$$I_\parallel = 2m_A R^2 (1 - \cos\theta)$$

$$I_\perp = m_A R^2 (1 - \cos\theta) + \frac{m_A}{m}(m_B + m_C)R^2(1 + 2\cos\theta)$$

$$\quad + \frac{m_C R'}{m}\left\{(3m_A + m_B)R' + 6m_A R[\tfrac{1}{3}(1 + 2\cos\theta)]^{1/2}\right\}$$

$$I_\parallel = 2m_A R^2 (1 - \cos\theta)$$

$$I_\perp = m_A R^2 (1 - \cos\theta) + \frac{m_A m_B}{m} R^2(1 + 2\cos\theta)$$

$$I_\parallel = 4m_A R^2$$

$$I_\perp = 2m_A R^2 + 2m_C R'^2$$

4. Spherical rotors

  $I = \tfrac{8}{3} m_A R^2$ $I = 4m_A R^2$

† In each case m is the total mass of the molecule.

1

Example 16.1 Calculating the moment of inertia of a molecule

Calculate the moment of inertia of an H_2O molecule around the axis defined by the bisector of the HOH angle (**1**). The HOH bond angle is 104.5° and the bond length is 95.7 pm.

Method According to eqn 27, the moment of inertia is the sum of the masses multiplied by the squares of their distances from the axis of rotation. The latter can be expressed by using trigonometry and the bond angle and bond length.

Answer From eqn 27,

$$I = \sum_i m_i r_i^2 = m_H r_H^2 + 0 + m_H r_H^2 = 2m_H r_H^2$$

If the bond angle of the molecule is denoted 2ϕ and the bond length is R, trigonometry gives $r_H = R \sin \phi$. It follows that

$$I = 2m_H R^2 \sin^2\phi$$

Substitution of the data gives

$$I = 2 \times (1.67 \times 10^{-27} \text{ kg}) \times (9.57 \times 10^{-11} \text{ m})^2 \times \sin^2 52.3° = 1.91 \times 10^{-47} \text{ kg m}^2$$

Comment The mass of the O atom makes no contribution to the moment of inertia for this mode of rotation as the atom is immobile while the H atoms circulate around it.

Self-test 16.1 Calculate the moment of inertia of a $CH^{35}Cl_3$ molecule around a rotational axis that contains the C—H bond. The C—Cl bond length is 177 pm and the HCCl angle is 107°; $m(^{35}Cl) = 34.97$ u. $[4.99 \times 10^{-45} \text{ kg m}^2]$

We shall suppose initially that molecules are **rigid rotors**, bodies that do not distort under the stress of rotation. Rigid rotors can be classified into four types (Fig. 16.17):

Spherical rotors have three equal moments of inertia (examples: CH_4, SiH_4, and SF_6).

Symmetric rotors have two equal moments of inertia (examples: NH_3, CH_3Cl, and CH_3CN).

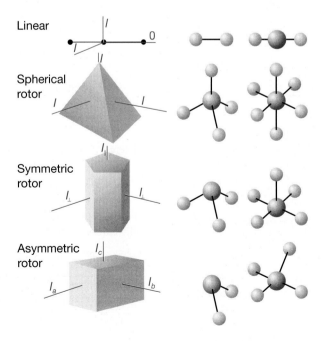

16.17 A schematic illustration of the classification of rigid rotors.

Linear rotors have one moment of inertia (the one about the molecular axis) equal to zero (examples: CO_2, HCl, OCS, and HC≡CH).

Asymmetric rotors have three different moments of inertia (examples: H_2O, H_2CO, and CH_3OH).

16.5 The rotational energy levels

The rotational energy levels of a rigid rotor may be obtained by solving the appropriate Schrödinger equation. Fortunately, however, there is a much less onerous short cut to the exact expressions that depends on noting the classical expression for the energy of a rotating body, expressing it in terms of the angular momentum, and then importing the quantum mechanical properties of angular momentum into the equations.

The classical expression for the energy of a body rotating about an axis a is

$$E_a = \tfrac{1}{2} I_a \omega_a^2 \tag{16.28}$$

where ω_a is the angular velocity (in radians per second, rad s^{-1}) about that axis and I_a is the corresponding moment of inertia. A body free to rotate about three axes has an energy

$$E = \tfrac{1}{2} I_a \omega_a^2 + \tfrac{1}{2} I_b \omega_b^2 + \tfrac{1}{2} I_c \omega_c^2$$

Because the classical angular momentum about the axis a is $J_a = I_a \omega_a$, with similar expressions for the other axes, it follows that

$$E = \frac{J_a^2}{2I_a} + \frac{J_b^2}{2I_b} + \frac{J_c^2}{2I_c} \tag{16.29}$$

This is the key equation. We described the quantum mechanical properties of angular momentum in Section 12.7b, and can now make use of them in conjunction with this equation to obtain the rotational energy levels.

(a) Spherical rotors

When all three moments of inertia are equal to some value I, as in CH_4 and SF_6, the classical expression for the energy is

$$E = \frac{J_a^2 + J_b^2 + J_c^2}{2I} = \frac{J^2}{2I}$$

where J is the magnitude of the angular momentum. We can immediately find the quantum expression by making the replacement

$$J^2 \rightarrow J(J+1)\hbar^2 \qquad J = 0, 1, 2, \dots$$

Therefore, the energy of a spherical rotor is confined to the values

$$E_J = J(J+1)\frac{\hbar^2}{2I} \qquad J = 0, 1, 2, \dots \tag{16.30}$$

The resulting ladder of energy levels is illustrated in Fig. 16.18. The energy is normally expressed in terms of the **rotational constant**, B, of the molecule, where

$$hcB = \frac{\hbar^2}{2I} \qquad \text{so} \qquad B = \frac{\hbar}{4\pi c I} \tag{16.31}$$

The expression for the energy is then

$$E_J = hcBJ(J+1) \qquad J = 0, 1, 2, \dots \tag{16.32}$$

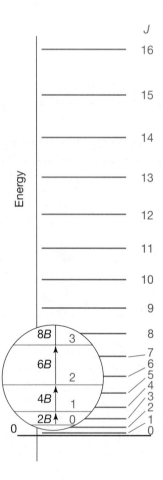

16.18 The rotational energy levels of a linear or spherical rotor. Note that the energy separation between neighbouring levels increases as J increases.

The rotational constant as defined by eqn 31 is a wavenumber.[5] The energy of a rotational state is normally reported as the **rotational term**, $F(J)$, a wavenumber, by division by hc:

$$F(J) = BJ(J + 1) \tag{16.33}$$

The separation of adjacent levels is

$$F(J) - F(J - 1) = 2BJ \tag{16.34}$$

Because the rotational constant decreases as I increases, we see that large molecules have closely spaced rotational energy levels. We can estimate the magnitude of the separation by considering CCl_4: from the bond lengths and masses of the atoms we find $I = 4.85 \times 10^{-45}$ kg m^2, and hence $B = 0.0577$ cm^{-1}.

(b) Symmetric rotors

In symmetric rotors, two moments of inertia are equal but different from the third (as in CH_3Cl, NH_3, and C_6H_6); the unique axis of the molecule is its **principal axis** (or *figure axis*). We shall write the unique moment of inertia (that about the principal axis) as $I_\parallel$ and the other two as $I_\perp$. If $I_\parallel > I_\perp$, the rotor is classified as **oblate** (like a pancake, and C_6H_6); if $I_\parallel < I_\perp$ it is classified as **prolate** (like a cigar, and CH_3Cl). The classical expression for the energy, eqn 29, becomes

$$E = \frac{J_b^2 + J_c^2}{2I_\perp} + \frac{J_a^2}{2I_\parallel}$$

This expression can be written in terms of $\mathcal{J}^2 = J_a^2 + J_b^2 + J_c^2$:

$$E = \frac{\mathcal{J}^2 - J_a^2}{2I_\perp} + \frac{J_a^2}{2I_\parallel} = \frac{\mathcal{J}^2}{2I_\perp} + \left(\frac{1}{2I_\parallel} - \frac{1}{2I_\perp}\right)J_a^2 \tag{16.35}$$

Now we generate the quantum expression by replacing $\mathcal{J}^2$ by $J(J + 1)\hbar^2$, where J is the angular momentum quantum number. We also know from the quantum theory of angular momentum (Section 12.7b) that the component of angular momentum about any axis is restricted to the values $K\hbar$, with $K = 0, \pm 1, \ldots, \pm J$. ($K$ is the quantum number used to signify a component on the principal axis; M_J is reserved for a component on an externally defined axis.) Therefore, we also replace J_a^2 by $K^2\hbar^2$. It follows that the rotational terms are

$$F(J, K) = BJ(J + 1) + (A - B)K^2 \qquad J = 0, 1, 2, \ldots \qquad K = 0, \pm 1, \ldots, \pm J \tag{16.36}$$

with

$$A = \frac{\hbar}{4\pi cI_\parallel} \qquad B = \frac{\hbar}{4\pi cI_\perp} \tag{16.37}$$

Equation 36 matches what we should expect for the dependence of the energy levels on the two distinct moments of inertia of the molecule. When $K = 0$, there is no component of angular momentum about the principal axis, and the energy levels depend only on $I_\perp$ (Fig. 16.19). When $K = \pm J$, almost all the angular momentum arises from rotation around the principal axis, and the energy levels are determined largely by $I_\parallel$. The sign of K does not affect the energy because opposite values of K correspond to opposite senses of rotation, and the energy does not depend on the sense of rotation.

$K \approx J$

(a)

$K = 0$

(b)

16.19 The significance of the quantum number K. (a) When $|K|$ is close to its maximum value, J, most of the molecular rotation is around the figure axis. (b) When $K = 0$ the molecule has no angular momentum about its principal axis: it is undergoing end-over-end rotation.

5. The definition of B as a wavenumber is convenient when we come to vibration–rotation spectra. However, for pure rotational spectroscopy it is more common to define B as a frequency. Then $B = \hbar/4\pi I$ and the energy is $E = hBJ(J + 1)$.

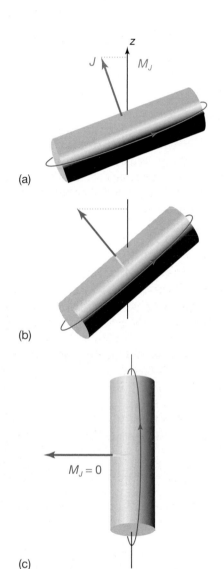

16.20 The significance of the quantum number M_J. (a) When M_J is close to its maximum value, J, most of the molecular rotation is around the laboratory z-axis. (b) an intermediate value of M_J. (c) When $M_J = 0$ the molecule has no angular momentum about the z-axis All three diagrams correspond to a state with $K = 0$; there are corresponding diagrams for different values of K, in which the angular momentum makes a different angle to the molecule's principal axis.

Example 16.2 Calculating the rotational energy levels of a molecule

A $^{14}NH_3$ molecule is a symmetric rotor with bond length 101.2 pm and HNH bond angle 106.7°. Calculate its rotational terms.

Method Begin by calculating the rotational constants A and B by using the expressions for moments of inertia given in Table 16.1. Then use eqn 36 to find the rotational terms.

Answer Substitution of $m_A = 1.0078$ u, $m_B = 14.0031$ u, $R = 101.2$ pm, and $\theta = 106.7°$ into the second of the symmetric rotor expressions in Table 16.1 gives $I_\parallel = 4.4128 \times 10^{-47}$ kg m² and $I_\perp = 2.8059 \times 10^{-47}$ kg m². Hence, $A = 6.344$ cm⁻¹ and $B = 9.977$ cm⁻¹. It follows from eqn 36 that

$$F(J, K)/\text{cm}^{-1} = 9.977J(J + 1) - 3.633K^2$$

Upon multiplication by c, $F(J, K)$ acquires units of frequency:

$$F(J, K)/\text{GHz} = 299.1J(J + 1) - 108.9K^2$$

Comment For $J = 1$, the energy needed for the molecule to rotate mainly about its figure axis ($K = \pm J$) is equivalent to 16.32 cm⁻¹ (489.3 GHz), but end-over-end rotation ($K = 0$) corresponds to 19.95 cm⁻¹ (598.1 GHz).

Self-test 16.2 A $CH_3{}^{35}Cl$ molecule has a C—Cl bond length of 178 pm, a C—H bond length of 111 pm, and an HCH angle of 110.5°. Calculate its rotational energy terms.

$$[F(J, K)/\text{cm}^{-1} = 0.444J(J + 1) + 4.58K^2;$$
$$\text{also } F(J, K)/\text{GHz} = 13.3J(J + 1) + 137K^2]$$

(c) Linear rotors

For a linear rotor (such as CO_2, HCl, and C_2H_2), in which the nuclei are regarded as mass points, the rotation occurs only about an axis perpendicular to the line of atoms and there is zero angular momentum around the line. Therefore, the component of angular momentum around the figure axis of a linear rotor is identically zero, and $K \equiv 0$ in eqn 36. The rotational terms of a linear molecule are therefore

$$F(J) = BJ(J + 1) \qquad J = 0, 1, 2, \ldots \tag{16.38}$$

This expression is the same as eqn 33 but we have arrived at it in a significantly different way: here $K \equiv 0$, but for a spherical rotor $A = B$.

(d) Degeneracies and the Stark effect

The energy of a symmetric rotor depends on J and K, and each level except those with $K = 0$ is doubly degenerate: the states with K and $-K$ have the same energy. However, we must not forget that the angular momentum of the molecule has a component on an external, laboratory-fixed axis. This component is quantized, and its permitted values are $M_J\hbar$, with $M_J = 0, \pm 1, \ldots, \pm J$, giving $2J + 1$ values in all (Fig. 16.20). The quantum number M_J does not appear in the expression for the energy, but it is necessary for a complete specification of the state of the rotor. Consequently, all $2J + 1$ orientations of the rotating molecule have the same energy. It follows that a symmetric rotor level is $2(2J + 1)$-fold degenerate for $K \neq 0$ and

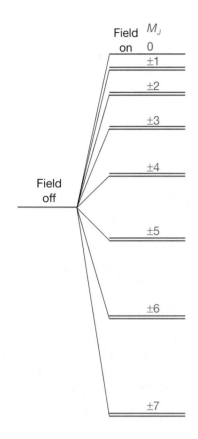

16.21 The effect of an electric field on the energy levels of a polar linear rotor. All levels are doubly degenerate except that with $M_J = 0$.

Centrifugal force

16.22 The effect of rotation on a molecule. The centrifugal force arising from rotation distorts the molecule, opening out bond angles and stretching bonds slightly. The effect is to increase the moment of inertia of the molecule and hence to decrease its rotational constant.

$(2J + 1)$-fold degenerate for $K = 0$. A linear rotor has K fixed at 0, but the angular momentum may still have $2J + 1$ components on the laboratory axis, so its degeneracy is $2J + 1$.

A spherical rotor can be regarded as a version of a symmetric rotor in which A is equal to B: The quantum number K may still take any one of $2J + 1$ values, but the energy is independent of which value it takes. Therefore, as well as having a $(2J + 1)$-fold degeneracy arising from its orientation in space, the rotor also has a $(2J + 1)$-fold degeneracy arising from its orientation with respect to an arbitrary axis in the molecule. The overall degeneracy of a symmetric rotor with quantum number J is therefore $(2J + 1)^2$. This degeneracy increases very rapidly: when $J = 10$, for instance, there are 441 states of the same energy.

The degeneracy associated with the quantum number M_J (the orientation of the rotation in space) is partly removed when an electric field is applied to a polar molecule (for example, HCl or NH_3), as illustrated in Fig. 16.21. The splitting of states by an electric field is called the **Stark effect**. For a linear rotor in an electric field $\mathcal{E}$, the energy of the state $|J, M_J\rangle$ is given by

$$E(J, M_J) = hcBJ(J + 1) + a(J, M_J)\mu^2\mathcal{E}^2 \qquad (16.39a)$$

where[6]

$$a(J, M_J) = \frac{J(J + 1) - 3M_J^2}{2hcBJ(J + 1)(2J - 1)(2J + 3)} \qquad (16.39b)$$

Note that the energy of a state with quantum number M_J depends on the square of the permanent electric dipole moment, μ. The observation of the Stark effect can therefore be used to measure this property, but the technique is limited to molecules that are sufficiently volatile to be studied by microwave spectroscopy. However, as spectra can be recorded for samples at pressures of only about 1 Pa, even some quite non-volatile substances may be studied. Sodium chloride, for example, can be studied as diatomic NaCl molecules at high temperatures.

(e) Centrifugal distortion

We have treated molecules as rigid rotors. However, the atoms of rotating molecules are subject to centrifugal forces that tend to distort the molecular geometry and change the moments of inertia (Fig. 16.22). The effect of centrifugal distortion on a diatomic molecule is to stretch the bond and hence to increase the moment of inertia. As a result, centrifugal distortion reduces the rotational constant and consequently the energy levels are slightly closer than the rigid-rotor expressions predict. The effect is usually taken into account largely empirically by subtracting a term from the energy and writing

$$F(J) = BJ(J + 1) - D_J J^2(J + 1)^2 \qquad (16.40)$$

The parameter D_J is the **centrifugal distortion constant**. It is large when the bond is easily stretched. The centrifugal distortion constant of a diatomic molecule is related to the vibrational wavenumber of the bond, $\tilde{\nu}$ (which, as we shall see later, is a measure of its stiffness), through the approximate relation (see Problem 16.23)

$$D_J \approx \frac{4B^3}{\tilde{\nu}^2} \qquad (16.41)$$

6. For a derivation of this expression, see *Further reading*.

16.23 To a stationary observer, a rotating polar molecule looks like an oscillating dipole that can stir the electromagnetic field into oscillation. This picture is the classical origin of the gross selection rule for rotational transitions.

Hence the observation of the convergence of the rotational levels as J increases can be interpreted in terms of the flexibility of the bond.

16.6 Rotational transitions

Typical values of B for small molecules are in the region of 0.1 to 10 cm^{-1} (for example, 0.356 cm^{-1} for NF_3 and 10.59 cm^{-1} for HCl), so rotational transitions lie in the microwave region of the spectrum. The transitions are detected by monitoring the net absorption of microwave radiation. Modulation of the transmitted intensity can be achieved by varying the energy levels with an oscillating electric field. In this **Stark modulation**, an electric field of about 10^5 V m^{-1} and a frequency of between 10 and 100 kHz is applied to the sample.

(a) Rotational selection rules

We have already remarked (Section 16.2) that the gross selection rule for the observation of a pure rotational spectrum is that a molecule must have a permanent electric dipole moment. That is, *for a molecule to give a pure rotational spectrum, it must be polar*. The classical basis of this rule is that a polar molecule appears to possess a fluctuating dipole when rotating, but a nonpolar molecule does not (Fig. 16.23). The permanent dipole can be regarded as a handle with which the molecule stirs the electromagnetic field into oscillation (and vice versa for absorption). Homonuclear diatomic molecules and symmetrical linear molecules such as CO_2 are rotationally inactive. Spherical rotors cannot have electric dipole moments unless they become distorted by rotation, so they are also inactive except in special cases. An example of a spherical rotor that does become sufficiently distorted for it to acquire a dipole moment is SiH_4, which has a dipole moment of about 8.3 μD by virtue of its rotation when $J \approx 10$ (for comparison, HCl has a permanent dipole moment of 1.1 D; molecular dipole moments and their units are discussed in Section 21.1). The pure rotational spectrum of SiH_4 has been detected by using long path lengths (10 m) through high-pressure (4 atm) samples.

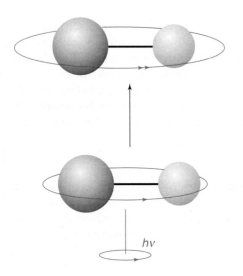

16.24 When a photon is absorbed by a molecule, the angular momentum of the combined system is conserved. If the molecule is rotating in the same sense as the spin of the incoming photon, then J increases by 1.

> **Illustration 16.3**
>
> Of the molecules N_2, CO_2, OCS, H_2O, $CH_2{=}CH_2$, C_6H_6, only OCS and H_2O are polar, so only these two molecules have microwave spectra.

> **Self-test 16.3** Which of the molecules H_2, NO, N_2O, CH_4 can have a pure rotational spectrum? [NO, N_2O]

The specific rotational selection rules are found by evaluating the transition dipole moment between rotational states. We show in the following *Justification* that, for a linear molecule, the transition moment vanishes unless the following conditions are fulfilled:

$$\Delta J = \pm 1 \qquad \Delta M_J = 0, \pm 1 \tag{16.42}$$

The transition $\Delta J = +1$ corresponds to absorption and the transition $\Delta J = -1$ corresponds to emission. The allowed change in J in each case arises from the conservation of angular momentum when a photon, a spin-1 particle, is emitted or absorbed (Fig. 16.24).

Justification 16.4

The total wavefunction for a molecule is

$$\psi_{total} = \psi_{c.m.}\psi$$

where $\psi_{c.m.}$ describes the motion of the centre of mass and ψ describes the internal motion of the molecule. If we neglect the effect of electron spin, the Born–Oppenheimer approximation allows us to write ψ as the product of an electronic part, $|\varepsilon\rangle$, a vibrational part, $|v\rangle$, and a rotational part, which for a diatomic molecule can be represented by the spherical harmonics $Y_{J,M_J}(\theta, \phi)$ (Section 12.7). The transition dipole moment for a spectroscopic transition is then

$$\mu_{fi} = \langle \varepsilon_f v_f Y_{J_f,M_{J,f}} | \mu | \varepsilon_i v_i Y_{J_i,M_{J,i}} \rangle$$

During a pure rotational transition the molecule does not change electronic or vibrational states, so that $\langle \varepsilon_f v_f | = \langle \varepsilon_i v_i | = \langle \varepsilon v |$ and we identify $\mu_{\varepsilon v} = \langle \varepsilon v | \mu | \varepsilon v \rangle$ with the *permanent* electric dipole moment of the molecule in the state εv. The transition dipole moment is then

$$\mu_{fi} = \langle Y_{J_f,M_{J,f}} | \mu_{\varepsilon v} | Y_{J_i,M_{J,i}} \rangle$$

The transition dipole moment has components $\mu_{\varepsilon v,x}$, $\mu_{\varepsilon v,y}$, and $\mu_{\varepsilon v,z}$, which, in spherical polar coordinates, are written in terms of μ_0, the magnitude of the vector, and the angles θ and ϕ as

$$\mu_{\varepsilon v,x} = \mu_0 \sin\theta\cos\phi \qquad \mu_{\varepsilon v,y} = \mu_0 \sin\theta\sin\phi \qquad \mu_{\varepsilon v,z} = \mu_0 \cos\theta$$

Here, we have taken the z-axis to be coincident with the figure axis. The transition dipole moment has three components, given by:

$$\mu_{fi,x} = \mu_0 \langle Y_{J_f,M_{J,f}} | \sin\theta\cos\phi | Y_{J_i,M_{J,i}} \rangle$$

$$\mu_{fi,y} = \mu_0 \langle Y_{J_f,M_{J,f}} | \sin\theta\sin\phi | Y_{J_i,M_{J,i}} \rangle$$

$$\mu_{fi,z} = \mu_0 \langle Y_{J_f,M_{J,f}} | \cos\theta | Y_{J_i,M_{J,i}} \rangle$$

We see immediately that the molecule must have a permanent dipole moment in order to have a rotational spectrum. This is the gross selection rule for rotational spectroscopy.

For the specific selection rules we need to examine the conditions for which the integrals do not vanish, and we must consider each component. For the z-component, we simplify the integral by using $\cos\theta \propto Y_{1,0}$ (Table 12.3). It follows that

> An important 'triple integral' involving the spherical harmonics is
>
> $$\int_0^\pi \int_0^{2\pi} Y_{l'',m_l''}(\theta, \phi)^* Y_{l',m_l'}(\theta, \phi) Y_{l,m_l}(\theta, \phi) \sin\theta\, d\theta\, d\phi = 0$$
>
> unless $m_l'' = m_l' + m_l$ and l'', l', and l can form a triangle.

$$\mu_{fi,z} \propto \langle Y_{J_f,M_{J,f}} | Y_{1,0} | Y_{J_i,M_{J,i}} \rangle$$

According to the properties of the spherical harmonics, this integral vanishes unless $J_f - J_i = \pm 1$ and $M_{J,f} - M_{J,i} = 0$. These are two of the selection rules stated in eqn 42.

For the x- and y-components, we use $\cos\phi = \frac{1}{2}(e^{i\phi} + e^{-i\phi})$ and $\sin\phi = -\frac{1}{2}i(e^{i\phi} - e^{-i\phi})$ to write $\sin\theta\cos\phi \propto Y_{1,1} + Y_{1,-1}$ and $\sin\theta\sin\phi \propto Y_{1,1} - Y_{1,-1}$. It follows that

$$\mu_{fi,x} \propto \langle Y_{J_f,M_{J,f}} | (Y_{1,1} + Y_{1,-1}) | Y_{J_i,M_{J,i}} \rangle$$

$$\mu_{fi,y} \propto \langle Y_{J_f,M_{J,f}} | (Y_{1,1} - Y_{1,-1}) | Y_{J_i,M_{J,i}} \rangle$$

According to the properties of the spherical harmonics, these integrals vanish unless $J_f - J_i = \pm 1$ and $M_{J,f} - M_{J,i} = \pm 1$. This completes the selection rules of eqn 42.

When the transition moment is evaluated for all possible relative orientations of the molecule to the line of flight of the photon, it is found that the total $J + 1 \leftrightarrow J$ transition intensity is proportional to

$$|\mu_{J+1,J}|^2 = \left(\frac{J+1}{2J+1}\right)\mu_0^2 \to \tfrac{1}{2}\mu_0^2 \qquad \text{for} \quad J \gg 1 \tag{16.43}$$

where μ_0 is the permanent electric dipole moment of the molecule. The intensity is proportional to the square of the permanent electric dipole moment, so strongly polar molecules give rise to much more intense rotational lines than less polar molecules.

For symmetric rotors, an additional selection rule states that $\Delta K = 0$. To understand this rule, consider the symmetric rotor NH_3, where the electric dipole moment lies parallel to the figure axis. Such a molecule cannot be accelerated into different states of rotation around the figure axis by the absorption of radiation, so $\Delta K = 0$.

(b) The appearance of rotational spectra

When these selection rules are applied to the expressions for the energy levels of a rigid symmetric or linear rotor, it follows that the wavenumbers of the allowed $J + 1 \leftarrow J$ absorptions are

$$\tilde{\nu}(J + 1 \leftarrow J) = 2B(J + 1) \qquad J = 0, 1, 2, \ldots \tag{16.44}$$

When centrifugal distortion is taken into account, the corresponding expression is

$$\tilde{\nu}(J + 1 \leftarrow J) = 2B(J + 1) - 4D_J(J+1)^3 \tag{16.45}$$

However, because the second term is typically very small compared with the first, the appearance of the spectrum closely resembles that predicted from eqn 44.

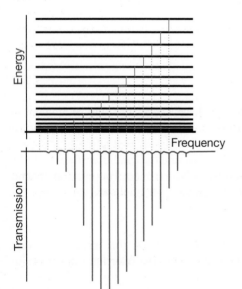

16.25 The rotational energy levels of a linear rotor, the transitions allowed by the selection rule $\Delta J = \pm 1$, and a typical pure rotational absorption spectrum (displayed here in terms of the radiation transmitted through the sample). The intensities reflect the populations of the initial level in each case and the strengths of the transition dipole moments.

Example 16.3 Predicting the appearance of a rotational spectrum

Predict the form of the rotational spectrum of NH_3.

Method We calculated the energy levels in Example 16.2. The NH_3 molecule is a polar symmetric rotor, so the selection rules $\Delta J = \pm 1$ and $\Delta K = 0$ apply. For absorption, $\Delta J = +1$ and we can use eqn 44. Because $B = 9.977$ cm^{-1}, we can draw up the following table for the $J + 1 \leftarrow J$ transitions.

J	0	1	2	3	...
$\tilde{\nu}/\text{cm}^{-1}$	19.95	39.91	59.86	79.82	...
$\tilde{\nu}/\text{GHz}$	598.1	1197	1795	2393	...

The line spacing is 19.95 cm^{-1} (598.1 GHz).

Self-test 16.4 Repeat the problem for $C^{35}ClH_3$ (see Self-test 16.2 for details).
[Lines of separation 0.888 cm^{-1} (26.6 GHz)]

The form of the spectrum predicted by eqn 44 is shown in Fig. 16.25. The most significant feature is that it consists of a series of lines with wavenumbers $2B$, $4B$, $6B$, ... and of separation $2B$. The measurement of the line spacing gives B, and hence the moment of inertia perpendicular to the principal axis of the molecule. Because the masses of the atoms are known, it is a simple matter to deduce the bond length of a diatomic molecule. However, in the case of a polyatomic molecule such as OCS or NH_3, the analysis gives only a single quantity, $I_\perp$, and it is not possible to infer both bond lengths (in OCS) or the bond length and bond angle (in NH_3). This difficulty can be overcome by using isotopically substituted molecules, such as ABC and A'BC; then, by assuming that $R(A—B) = R(A'—B)$, both A—B and B—C bond lengths can be extracted from the two moments of inertia. A famous example of this procedure is the study of OCS; the actual calculation is worked through in Problem 16.10. The assumption that bond lengths are unchanged by isotopic substitution is only an approximation, but it is a good approximation in most cases.

The intensities of spectral lines increase with increasing J and pass through a maximum before tailing off as J becomes large. The most important reason for the maximum in intensity is the existence of a maximum in the population of rotational levels. The Boltzmann distribution implies that the population of each state decays exponentially with increasing J, but the degeneracy of the levels increases. Specifically, the population of a rotational energy level J is given by the Boltzmann expression

$$N_J \propto N g_J e^{-E_J/kT}$$

where N is the total number of molecules and g_J is the degeneracy of the level J. The value of J corresponding to a maximum of this expression is found by treating J as a continuous variable, differentiating with respect to J, and then setting the result equal to zero. The result is(see Problem 16.25)

$$J_{max} \approx \left(\frac{kT}{2hcB} \right)^{1/2} - \frac{1}{2} \tag{16.46}$$

For a typical molecule (for example, OCS, with $B = 0.2$ cm^{-1}) at room temperature, $kT \approx 1000hcB$, so $J_{max} \approx 22$. However, it must be recalled that the intensity of each transition also depends on the value of J (eqn 43) and on the population difference between the two states involved in the transition (see Section 16.2). Hence the value of J corresponding to the most intense line is not necessarily quite the same as the value of J for the most highly populated level.

16.7 Rotational Raman spectra

The gross selection rule for rotational Raman transitions is *that the molecule must be anisotropically polarizable*. We begin by explaining what this means. A formal derivation of this rule is given in the following *Justification*.

The distortion of a molecule in an electric field is determined by its polarizability, α (Section 22.2). More precisely, if the strength of the field is $\mathcal{E}$, then the molecule acquires an induced dipole moment of magnitude

$$\mu = \alpha \mathcal{E} \tag{16.47}$$

in addition to any permanent dipole moment it may have. An atom is isotropically polarizable. That is, the same distortion is induced whatever the direction of the applied field. The polarizability of a spherical rotor is also isotropic. However, non-spherical rotors have polarizabilities that do depend on the direction of the field relative to the molecule, so these molecules are anisotropically polarizable (Fig. 16.26). The electron distribution in H_2, for example, is more distorted when the field is applied parallel to the bond than when it is applied perpendicular to it, and we write $\alpha_\parallel > \alpha_\perp$.

All linear molecules and diatomics (whether homonuclear or heteronuclear) have anisotropic polarizabilities, and so are rotationally Raman active. This activity is one reason for the importance of rotational Raman spectroscopy, for the technique can

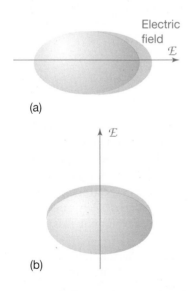

16.26 An electric field applied to a molecule results in its distortion, and the distorted molecule acquires a contribution to its dipole moment (even if it is nonpolar initially). The polarizability may be different when the field is applied (a) parallel or (b) perpendicular to the molecular axis (or, in general, in different directions relative to the molecule); if that is so, then the molecule has an anisotropic polarizability.

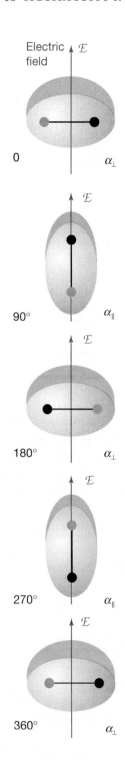

16.27 The distortion induced in a molecule by an applied electric field returns to its initial value after a rotation of only 180° (that is, twice a revolution). This is the origin of the $\Delta J = \pm 2$ selection rule in rotational Raman spectroscopy.

be used to study many of the molecules that are inaccessible to microwave spectroscopy. Spherical rotors such as CH_4 and SF_6, however, are rotationally Raman inactive as well as microwave inactive.[7]

We show in the *Justification* below that the specific rotational Raman selection rules are

Linear rotors: $\Delta J = 0, \pm 2$

Symmetric rotors: $\Delta J = 0, \pm 1, \pm 2;$ $\Delta K = 0$

(16.48)

The $\Delta J = 0$ transitions do not lead to a shift of the scattered photon's frequency in pure rotational Raman spectroscopy, and contribute to the unshifted Rayleigh radiation observed in the forward direction.[8]

Justification 16.5

We understand the origin of the gross and specific selection rules for rotational Raman spectroscopy by using a diatomic molecule as an example. The incident electric field of a wave of electromagnetic radiation of frequency ω_i, induces a molecular dipole moment that is given by

$$\mu_{ind} = \alpha \mathcal{E}(t) = \alpha \mathcal{E} \cos \omega_i t$$

If the molecule is rotating at a circular frequency ω_R, to an external observer its polarizability is also time-dependent (if it is anisotropic), and we can write

$$\alpha = \alpha_0 + \Delta\alpha \cos 2\omega_R t$$

where $\Delta\alpha = \alpha_\parallel - \alpha_\perp$ and α ranges from $\alpha_0 + \Delta\alpha$ to $\alpha_0 - \Delta\alpha$ as the molecule rotates. The 2 appears because the polarizability returns to its initial value twice each revolution (Fig. 16.27). Substituting this expression into the expression for the induced dipole moment gives

$$\mu_{ind} = (\alpha_0 + \Delta\alpha \cos 2\omega_R t) \times (\mathcal{E} \cos \omega_i t)$$
$$= \alpha_0 \mathcal{E} \cos \omega_i t + \mathcal{E}\Delta\alpha \cos 2\omega_R t \cos \omega_i t$$
$$= \alpha_0 \mathcal{E} \cos \omega_i t + \tfrac{1}{2}\mathcal{E}\Delta\alpha\{\cos(\omega_i + 2\omega_R)t + \cos(\omega_i - 2\omega_R)t\}$$

This calculation shows that the induced dipole has a component oscillating at the incident frequency (which generates Rayleigh radiation), and that it also has two components at $\omega_i \pm 2\omega_R$, which give rise to the shifted Raman lines. These lines appear only if $\Delta\alpha \neq 0$; hence the polarizability must be anisotropic for there to be Raman lines. This is the gross selection rule for rotational Raman spectroscopy. We also see that the distortion induced in the molecule by the incident electric field returns to its initial value after a rotation of 180° (that is, twice a revolution). This is the origin of the specific selection rule $\Delta J = \pm 2$.

We now use a quantum mechanical formalism to understand the selection rules. We use a strategy that is similar to that of *Justification 16.4*. First, we write the *x*-, *y*-, and *z*-components of the induced dipole moment as

$$\mu_{ind,x} = \mu_x \sin\theta \cos\phi$$
$$\mu_{ind,y} = \mu_y \sin\theta \sin\phi$$
$$\mu_{ind,z} = \mu_z \cos\theta$$

7. This inactivity does not mean that such molecules are never found in rotationally excited states. Molecular collisions do not have to obey such restrictive selection rules, and hence collisions between molecules can result in the population of any rotational state.
8. See Section 22.8 for the information present in the Rayleigh component of the scattered radiation.

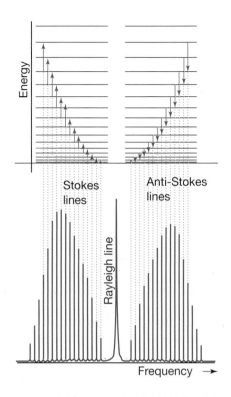

16.28 The rotational energy levels of a linear rotor and the transitions allowed by the $\Delta J = \pm 2$ Raman selection rules. The form of a typical rotational Raman spectrum is also shown. The Rayleigh line is much stronger than depicted in the figure; it is shown as a weaker line to improve visualization of the Raman lines.

where μ_x, μ_y, and μ_z are the components of the electric dipole moment of the molecule and the z-axis is coincident with the molecular figure axis. The incident electric field also has components along the x-, y-, and z-axes:

$$\mathcal{E}_x = \mathcal{E} \sin \theta \cos \phi \qquad \mathcal{E}_y = \mathcal{E} \sin \theta \sin \phi \qquad \mathcal{E}_z = \mathcal{E} \cos \theta$$

Using eqn 47, it follows that

$$\mu_{ind} = \alpha_\perp \mathcal{E}_x \sin \theta \cos \phi + \alpha_\perp \mathcal{E}_y \sin \theta \sin \phi + \alpha_\| \mathcal{E}_z \cos \theta = \alpha_\perp \mathcal{E} \sin^2\theta + \alpha_\| \mathcal{E} \cos^2\theta$$

By using the spherical harmonic $Y_{2,0}(\theta, \phi) = (5/16\pi)^{1/2}(3 \cos^2\theta - 1)$ and the relation $\sin^2\theta = 1 - \cos^2\theta$, it follows that:

$$\mu_{ind} = \left\{ \tfrac{1}{3}\alpha_\| + \tfrac{2}{3}\alpha_\perp + \tfrac{4}{3}\left(\frac{\pi}{5}\right)^{1/2} \Delta\alpha Y_{2,0}(\theta, \phi) \right\} \mathcal{E}$$

For a transition between two rotational states, we calculate the integral $\langle Y_{J_f, M_{J,f}} | \mu_{ind} | Y_{J_i, M_{J,i}} \rangle$, which has two components proportional to:

$$(\tfrac{1}{3}\alpha_\| + \tfrac{2}{3}\alpha_\perp)\langle Y_{J_f, M_{J,f}} | Y_{J_i, M_{J,i}} \rangle \qquad \text{and} \qquad \Delta\alpha\langle Y_{J_f, M_{J,f}} | Y_{2,0} Y_{J_i, M_{J,i}} \rangle$$

According to the properties of the spherical harmonics (Section 12.7), the first integral vanishes unless $J_f - J_i = 0$ and the second integral vanishes unless $J_f - J_i = \pm 2$ and $\Delta\alpha \neq 0$. These are the gross and specific selection rules for linear rotors.

We can predict the form of the Raman spectrum of a linear rotor by applying the selection rule $\Delta J = \pm 2$ to the rotational energy levels (Fig. 16.28). When the molecule makes a transition with $\Delta J = +2$, the scattered radiation leaves the molecule in a higher rotational state, so the wavenumber of the incident radiation, initially $\tilde{\nu}_i$, is decreased. These transitions account for the Stokes lines in the spectrum:

$$\tilde{\nu}(J+2 \leftarrow J) = \tilde{\nu}_i - \{F(J+2) - F(J)\} = \tilde{\nu}_i - 2B(2J+3) \tag{16.49a}$$

The Stokes lines appear to low frequency of the incident radiation and at displacements $6B$, $10B$, $14B$, … from $\tilde{\nu}_i$ for $J = 0, 1, 2, …$. When the molecule makes a transition with $\Delta J = -2$, the scattered photon emerges with increased energy. These transitions account for the anti-Stokes lines of the spectrum:

$$\tilde{\nu}(J-2 \leftarrow J) = \tilde{\nu}_i + \{F(J) - F(J-2)\} = \tilde{\nu}_i + 2B(2J-1) \tag{16.49b}$$

The anti-Stokes lines occur at displacements of $6B$, $10B$, $14B$, … (for $J = 2, 3, 4, … ;$ $J = 2$ is the lowest state that can contribute under the selection rule $\Delta J = -2$) to high frequency of the incident radiation. The separation of adjacent lines in both the Stokes and the anti-Stokes regions is $4B$, so from its measurement $I_\perp$ can be determined and then used to find the bond lengths exactly as in the case of microwave spectroscopy.

Example 16.4 Predicting the form of a Raman spectrum

Predict the form of the rotational Raman spectrum of $^{14}N_2$, for which $B = 1.99$ cm^{-1}, when it is exposed to monochromatic 336.732 nm laser radiation.

Method The molecule is rotationally Raman active because end-over-end rotation modulates its polarizability as viewed by a stationary observer. The Stokes and anti-Stokes lines are given by eqn 49.

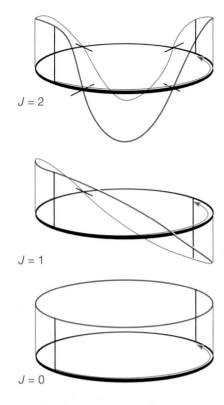

16.29 The symmetries of rotational wavefunctions (shown here, for simplicity as a two-dimensional rotor) under a rotation through 180°. Wavefunctions with J even do not change sign; those with J odd do change sign.

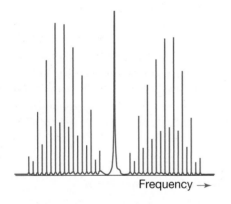

16.30 The rotational Raman spectrum of a diatomic molecule with two identical spin-$\frac{1}{2}$ nuclei shows an alternation in intensity as a result of nuclear statistics. The Rayleigh line is much stronger than depicted in the figure; it is shown as a weaker line to improve visualization of the Raman lines.

Answer Because $\lambda_i = 336.732$ nm corresponds to $\tilde{\nu}_i = 29\,697.2$ cm^{-1}, eqns 49a and 49b give the following line positions:

J	0	1	2	3
Stokes lines				
$\tilde{\nu}$/cm^{-1}	29 685.3	29 677.3	29 669.3	29 661.4
λ/nm	336.868	336.958	337.048	337.139
Anti-Stokes lines				
$\tilde{\nu}$/cm^{-1}			29 709.1	29 717.1
λ/nm			336.597	336.507

Comment There will be a strong central line at 336.732 nm accompanied on either side by lines of increasing and then decreasing intensity (as a result of transition moment and population effects). The spread of the entire spectrum is very small, so the incident light must be highly monochromatic.

Self-test 16.5 Repeat the calculation for the rotational Raman spectrum of NH$_3$ ($B = 9.977$ cm^{-1}).

[Stokes lines at 29 637.3, 29 597.4, 29 557.5, 29 517.6 cm^{-1}, anti-Stokes lines at 29 757.1, 29 797.0 cm^{-1}.]

16.8 Nuclear statistics and rotational states

If eqn 49 is used in conjunction with the rotational Raman spectrum of CO$_2$, the rotational constant is inconsistent with other measurements of C—O bond lengths. The results are consistent only if it is supposed that the molecule can exist in states with even values of J, so the Stokes lines are $2 \leftarrow 0$, $4 \leftarrow 2$, ... and not $5 \leftarrow 3$, $3 \leftarrow 1$, ...

The explanation of the missing lines is the Pauli principle and the fact that O nuclei are spin-0 bosons: just as the Pauli principle excludes certain electronic states, so too does it exclude certain molecular rotational states. The form of the Pauli principle given in *Justification 13.7* states that, when two identical bosons are exchanged, the overall wavefunction must remain unchanged in every respect, including sign. In particular, when a CO$_2$ molecule rotates through 180°, two identical O nuclei are interchanged, so the overall wavefunction of the molecule must remain unchanged. However, inspection of the form of the rotational wavefunctions (which have the same form as the s, p, etc. orbitals of atoms) shows that they change sign by $(-1)^J$ under such a rotation (Fig. 16.29). Therefore, only even values of J are permissible for CO$_2$, and hence the Raman spectrum shows only alternate lines.

The selective occupation of rotational states that stems from the Pauli principle is termed **nuclear statistics**. Nuclear statistics must be taken into account whenever a rotation interchanges equivalent nuclei. However, the consequences are not always as simple as for CO$_2$ because there are complicating features when the nuclei have nonzero spin: there may be several different relative nuclear spin orientations consistent with even values of J and a different number of spin orientations consistent with odd values of J. For molecular hydrogen and fluorine, for instance, with their two identical spin-$\frac{1}{2}$ nuclei, we show in the *Justification* below that there are three times as many ways of achieving a state with odd J than with even J, and there is a corresponding 3:1 alternation in intensity in their rotational Raman spectra (Fig. 16.30). In general, for a homonuclear diatomic molecule with nuclei of spin I, the numbers of ways of achieving states of odd and even J are in the ratio

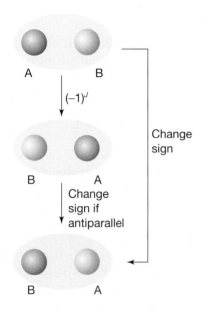

16.31 The interchange of two identical fermion nuclei results in the change in sign of the overall wavefunction. The relabelling can be thought of as occurring in two steps: the first is a rotation of the molecule; the second is the interchange of unlike spins (represented by the different colours of the nuclei). The wavefunction changes sign in the second step if the nuclei have antiparallel spins.

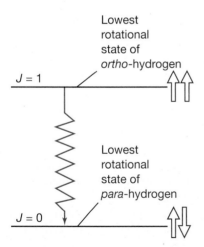

16.32 When hydrogen is cooled, the molecules with parallel nuclear spins accumulate in their lowest available rotational state, the one with $J = 1$. They can enter the lowest rotational state ($J = 0$) only if the spins change their relative orientation and become antiparallel. This is a slow process under normal circumstances, so energy is slowly released.

$$\frac{\text{Number of ways of achieving odd } J}{\text{Number of ways of achieving even } J} = \begin{cases} (I + 1)/I \text{ for half-integral spin nuclei} \\ I/(I + 1) \text{ for integral spin nuclei} \end{cases}$$
$$(16.50)$$

For hydrogen, $I = \frac{1}{2}$, and the ratio is 3:1. For N_2, with $I = 1$, the ratio is 1:2.

Justification 16.6

Hydrogen nuclei are fermions, so the Pauli principle requires the overall wavefunction to change sign under particle interchange. However, the rotation of an H_2 molecule through 180° has a more complicated effect than merely relabelling the nuclei, because it interchanges their spin states too if the nuclear spins are paired ($\uparrow\downarrow$) but not if they are parallel ($\uparrow\uparrow$).

For the overall wavefunction of the molecule to change sign when the spins are parallel, the rotational wavefunction must change sign. Hence, only odd values of J are allowed. In contrast, if the nuclear spins are paired, their wavefunction is $\alpha(A)\beta(B) - \alpha(B)\beta(A)$, which changes sign when α and β are exchanged in order to bring about a simple $A \leftrightarrow B$ interchange overall (Fig. 16.31). Therefore, for the overall wavefunction to change sign in this case requires the rotational wavefunction *not* to change sign. Hence, only even values of J are allowed if the nuclear spins are paired.

As there are three nuclear spin states with parallel spins (just like the triplet state of two parallel electrons, as in Fig. 13.26), but only one state with paired spins (the analogue of the singlet state of two electrons, see Fig. 13.20), it follows that the populations of the odd J and even J states should be in the ratio of 3:1, and hence the intensities of transitions originating in these levels will be in the same ratio.

Different relative nuclear spin orientations change into one another only very slowly, so an H_2 molecule with parallel nuclear spins remains distinct from one with paired nuclear spins for long periods. The two forms of hydrogen can be separated by physical techniques, and stored. The form with parallel nuclear spins is called **ortho-hydrogen** and the form with paired nuclear spins is called **para-hydrogen**. Because *ortho*-hydrogen cannot exist in a state with $J = 0$, it continues to rotate at very low temperatures and has an effective rotational zero-point energy (Fig. 16.32). This energy is of some concern to manufacturers of liquid hydrogen, for the slow conversion of *ortho*-hydrogen into *para*-hydrogen (which can exist with $J = 0$) as nuclear spins slowly realign releases rotational energy, which vaporizes the liquid. Techniques are used to accelerate the conversion of *ortho*-hydrogen to *para*-hydrogen to avoid this problem. One such technique is to pass hydrogen over a metal surface: the molecules adsorb on the surface as atoms, which then recombine in the lower energy *para*-hydrogen form.

The vibrations of diatomic molecules

In this section, we adopt the same strategy of finding expressions for the energy levels, establishing the selection rules, and then discussing the form of the spectrum. We shall also see how the simultaneous excitation of rotation modifies the appearance of a vibrational spectrum.

16.9 Molecular vibrations

We base our discussion on Fig. 16.33, which shows a typical potential energy curve (as in Fig. 14.1) of a diatomic molecule. In regions close to R_e (at the minimum of the curve) the potential energy can be approximated by a parabola, so we can write

$$V = \tfrac{1}{2}kx^2 \qquad x = R - R_e \tag{16.51}$$

where k is the **force constant** of the bond. The steeper the walls of the potential (the stiffer the bond), the greater the force constant.

To see the connection between the shape of the molecular potential energy curve and the value of k, note that we can expand the potential energy around its minimum by using a Taylor series:

$$V(x) = V(0) + \left(\frac{dV}{dx}\right)_0 x + \tfrac{1}{2}\left(\frac{d^2V}{dx^2}\right)_0 x^2 + \cdots \tag{16.52}$$

The term $V(0)$ can be set arbitrarily to zero. The first derivative of V is 0 at the minimum. Therefore, the first surviving term is proportional to the square of the displacement. For small displacements we can ignore all the higher terms, and so write

$$V(x) \approx \tfrac{1}{2}\left(\frac{d^2V}{dx^2}\right)_0 x^2 \tag{16.53}$$

Therefore, the first approximation to a molecular potential energy curve is a parabolic potential, and we can identify the force constant as

$$k = \left(\frac{d^2V}{dx^2}\right)_0 \tag{16.54}$$

We see that, if the potential energy curve is sharply curved close to its minimum, then k will be large. Conversely, if the potential energy curve is wide and shallow, then k will be small (Fig. 16.34).

The Schrödinger equation for the relative motion of two atoms of masses m_1 and m_2 with a parabolic potential energy is

$$-\frac{\hbar^2}{2m_{\text{eff}}}\frac{d^2\psi}{dx^2} + \tfrac{1}{2}kx^2\psi = E\psi \tag{16.55}$$

where m_{eff} is the **effective mass**:

$$m_{\text{eff}} = \frac{m_1 m_2}{m_1 + m_2} \tag{16.56}$$

These equations are derived in the same way as in *Justification 13.1*, but here the separation of variables procedure is used to separate the relative motion of the atoms from the motion of the molecule as a whole.[9]

The Schrödinger equation in eqn 55 is the same as eqn 12.31 for a particle of mass m undergoing harmonic motion. Therefore, we can use the results of Section 12.4 to write down the permitted vibrational energy levels:

$$E_v = (v + \tfrac{1}{2})\hbar\omega \qquad \omega = \left(\frac{k}{m_{\text{eff}}}\right)^{1/2} \qquad v = 0, 1, 2, \dots \tag{16.57}$$

The **vibrational terms** of a molecule, the energies of its vibrational states expressed in wavenumbers, are denoted $G(v)$, with $E_v = hcG(v)$, so

---

It is often useful to express a function $f(x)$ in the vicinity of $x = a$ as an infinite Taylor series of the form:

$$f(x) = f(a) + \left(\frac{df}{dx}\right)_a (x - a)$$

$$+ \frac{1}{2!}\left(\frac{d^2f}{dx^2}\right)_a (x - a)^2 + \cdots$$

$$= \sum_n \frac{1}{n!}\left(\frac{d^nf}{dx^n}\right)_a (x - a)^n$$

where $n = 0, 1, 2, \dots$.

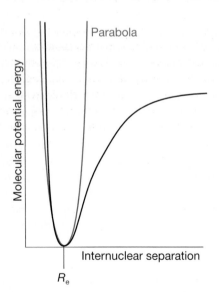

16.33 A molecular potential energy curve can be approximated by a parabola near the bottom of the well. The parabolic potential leads to harmonic oscillations. At high excitation energies the parabolic approximation is poor (the true potential is less confining), and is totally wrong near the dissociation limit.

Parabola

Molecular potential energy

Internuclear separation

R_e

---

9. In that context, the effective mass is called the 'reduced mass', and the name is widely used in this context too.

$$G(v) = (v + \tfrac{1}{2})\tilde{v} \qquad \tilde{v} = \frac{1}{2\pi c}\left(\frac{k}{m_{\text{eff}}}\right)^{1/2} \tag{16.58}$$

The vibrational wavefunctions are the same as those discussed in Section 12.5.

It is important to note that the vibrational terms depend on the effective mass of the molecule, not directly on its total mass. This dependence is physically reasonable for, if atom 1 were as heavy as a brick wall, then we would find $m_{\text{eff}} \approx m_2$, the mass of the lighter atom. The vibration would then be that of a light atom relative to that of a stationary wall (this is approximately the case in HI, for example, where the I atom barely moves and $m_{\text{eff}} \approx m_{\text{H}}$). For a homonuclear diatomic molecule $m_1 = m_2$, and the effective mass is half the total mass: $m_{\text{eff}} = \tfrac{1}{2}m$.

Illustration 16.4

An HCl molecule has a force constant of 516 N m^{-1}, a reasonably typical value. The effective mass of ^{1}H^{35}Cl is 1.63×10^{-27} kg (note that this mass is very close to the mass of the hydrogen atom, 1.67×10^{-27} kg, so the Cl atom is acting like a brick wall). These values imply $\omega = 5.63 \times 10^{14}$ s^{-1}, $v = 89.5$ THz (1 THz = 10^{12} Hz), $\tilde{v} = 2990$ cm^{-1}, $\lambda = 3.35$ mm. These characteristics correspond to electromagnetic radiation in the infrared region.

16.10 Selection rules

The gross selection rule for absorption of radiation by a molecular vibration is that *the electric dipole moment of the molecule must change when the atoms are displaced relative to one another.* Such vibrations are said to be **infrared active**. The classical basis of this rule is that the molecule can shake the electromagnetic field into oscillation if its dipole changes as it vibrates, and vice versa (Fig. 16.35); its formal basis is given in the following *Justification*. Note that the molecule need not have a permanent dipole: the rule requires only a change in dipole moment, possibly from zero. Some vibrations do not affect the molecule's dipole moment (for example, the stretching motion of a homonuclear diatomic molecule), so they neither absorb nor generate radiation: such vibrations are said to be **infrared inactive**. Homonuclear diatomic molecules are infrared inactive because their dipole moments remain zero however long the bond; heteronuclear diatomic molecules are infrared active.

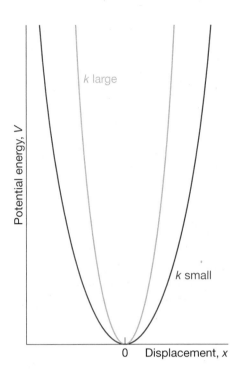

16.34 The force constant is a measure of the curvature of the potential energy close to the equilibrium extension of the bond. A strongly confining well (one with steep sides, a stiff bond) corresponds to high values of k.

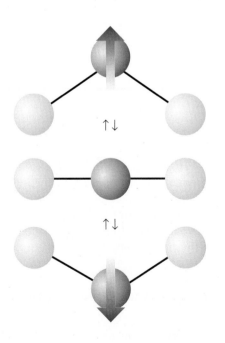

16.35 The oscillation of a molecule, even if it is nonpolar, may result in an oscillating dipole that can interact with the electromagnetic field.

Justification 16.7

The gross selection rule is based on an analysis of the transition dipole moment $\langle v_{\text{f}}|\boldsymbol{\mu}|v_{\text{i}}\rangle$. For simplicity, we shall consider a one-dimensional oscillator (like a diatomic molecule). The electric dipole moment operator depends on the location of all the electrons and all the nuclei in the molecule, so it varies as the internuclear separation changes (Fig. 16.36). If we think of the dipole moment as arising from two partial charges $\pm\delta q$ separated by a distance $R = R_{\text{e}} + x$, we can write its variation with displacement from the equilibrium separation, x, as

$$\mu = R\delta q = R_{\text{e}}\delta q + x\delta q = \mu_0 + x\delta q$$

where μ_0 is the electric dipole moment operator when the nuclei have their equilibrium separation. It then follows that, with f $\neq$ i,

$$\langle v_{\text{f}}|\boldsymbol{\mu}|v_{\text{i}}\rangle = \mu_0\langle v_{\text{f}}|v_{\text{i}}\rangle + \delta q\langle v_{\text{f}}|x|v_{\text{i}}\rangle$$

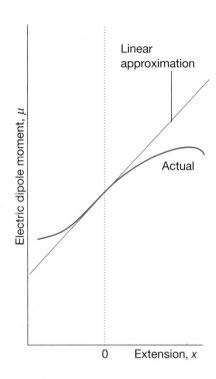

16.36 The electric dipole moment of a heteronuclear diatomic molecule varies as shown by the blue curve. For small displacements the change in dipole moment is proportional to the displacement.

An important integral involving Hermite polynomials is

$$\int_{-\infty}^{\infty} H_{v'} H_v e^{-y^2} dy = \begin{cases} 0 & \text{if } v' \neq v \\ \pi^{1/2} 2^v v! & \text{if } v' = v \end{cases}$$

The term proportional to μ_0 is zero because the states with different values of v are orthogonal. It follows that the transition dipole moment is

$$\langle v_f | \mu | v_i \rangle = \langle v_f | x | v_i \rangle \delta q$$

Because

$$\delta q = \frac{d\mu}{dx}$$

we can write the transition dipole moment more generally as

$$\langle v_f | \mu | v_i \rangle = \langle v_f | x | v_i \rangle \left(\frac{d\mu}{dx} \right)$$

and we see that the right-hand side is zero unless the dipole moment varies with displacement. This is the gross selection rule for vibrational spectroscopy.

The specific selection rule is determined by considering the value of $\langle v_f | x | v_i \rangle$. We need to write out the wavefunctions in terms of the Hermite polynomials given in Section 12.5 and then to use their properties (Example 12.3 should be reviewed, for it gives further details of the calculation). We note that $x = \alpha y$ with $\alpha = (\hbar^2/m_{\text{eff}} k)^{1/4}$ (eqn 12.35), and write

$$\langle v_f | x | v_i \rangle = N_{v_f} N_{v_i} \int_{-\infty}^{\infty} H_{v_f} x H_{v_i} e^{-y^2} dx = \alpha^2 N_{v_f} N_{v_i} \int_{-\infty}^{\infty} H_{v_f} y H_{v_i} e^{-y^2} dy$$

To evaluate the integral we use the recursion relation

$$yH_v = vH_{v-1} + \tfrac{1}{2}H_{v+1}$$

which turns the matrix element into

$$\langle v_f | x | v_i \rangle = \alpha^2 N_{v_f} N_{v_i} \left\{ v_i \int_{-\infty}^{\infty} H_{v_f} H_{v_i-1} e^{-y^2} dy + \tfrac{1}{2} \int_{-\infty}^{\infty} H_{v_f} H_{v_i+1} e^{-y^2} dy \right\}$$

The first integral is zero unless $v_f = v_i - 1$ and the second is zero unless $v_f = v_i + 1$. It follows that the transition dipole moment is zero unless $\Delta v = \pm 1$.

Illustration 16.5

Of the molecules N_2, CO_2, OCS, H_2O, $CH_2{=}CH_2$, and C_6H_6, all except N_2 possess at least one vibrational mode that results in a change of dipole moment, so all except N_2 can show a vibrational absorption spectrum. Not all the modes of complex molecules are vibrationally active. For example, the symmetric stretch of CO_2, in which the O—C—O bonds stretch and contract symmetrically is inactive because it leaves the dipole moment unchanged (at zero).

Self-test 16.6 Which of the molecules H_2, NO, N_2O, and CH_4 have infrared active vibrations?

[NO, N_2O, CH_4]

The specific vibrational selection rule, which is obtained from an analysis of the expression for the transition moment and the properties of integrals over harmonic oscillator wavefunctions (as shown in *Justification 16.7*), is

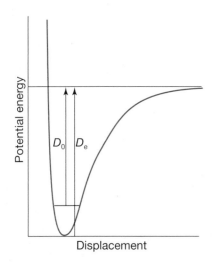

16.37 The dissociation energy of a molecule, D_e, differs from the depth of the potential well, D_0, on account of the zero-point energy of the vibrations of the bond.

$$\Delta v = \pm 1 \qquad (16.59)$$

Transitions for which $\Delta v = +1$ correspond to absorption and those with $\Delta v = -1$ correspond to emission.

It follows from the specific selection rules that the wavenumbers of allowed vibrational transitions, which are denoted $\Delta G_{v+\frac{1}{2}}$ for the transition $v + 1 \leftarrow v$, are

$$\Delta G_{v+\frac{1}{2}} = G(v + 1) - G(v) = \tilde{v} \qquad (16.60)$$

As we have seen, $\tilde{v}$ lies in the infrared region of the electromagnetic spectrum, so vibrational transitions absorb and generate infrared radiation.

At room temperature $kT/hc \approx 200\ cm^{-1}$, and most vibrational wavenumbers are significantly greater than $200\ cm^{-1}$. It follows from the Boltzmann distribution that almost all the molecules will be in their vibrational ground states initially. Hence, the dominant spectral transition will be the **fundamental transition**, $1 \leftarrow 0$. As a result, the spectrum is expected to consist of a single absorption line. If the molecules are formed in a vibrationally excited state, such as when vibrationally excited HF molecules are formed in the reaction $H_2 + F_2 \rightarrow 2\ HF^*$, the transitions $5 \rightarrow 4, 4 \rightarrow 3, \ldots$ may also appear (in emission). In the harmonic approximation, all these lines lie at the same frequency, and the spectrum is also a single line. However, as we shall now show, the breakdown of the harmonic approximation causes the transitions to lie at slightly different frequencies, so several lines are observed.

16.11 Anharmonicity

The vibrational terms in eqn 60 are only approximate because they are based on a parabolic approximation to the actual potential energy curve. A parabola cannot be correct at all extensions because it does not allow a bond to dissociate. At high vibrational excitations the swing of the atoms (more precisely, the spread of the vibrational wavefunction) allows the molecule to explore regions of the potential energy curve where the parabolic approximation is poor and additional terms in the Taylor expansion of V (eqn 52) must be retained. The motion then becomes **anharmonic**, in the sense that the restoring force is no longer proportional to the displacement. Because the actual curve is less confining than a parabola, we can anticipate that the energy levels become less widely spaced at high excitations.

(a) The convergence of energy levels

One approach to the calculation of the energy levels in the presence of anharmonicity is to use a function that resembles the true potential energy more closely. The **Morse potential energy** is

$$V = hcD_e\{1 - e^{-a(R-R_e)}\}^2 \qquad a = \left(\frac{m_{eff}\omega^2}{2hcD_e}\right)^{1/2} \qquad (16.61)$$

where D_e is the depth of the potential minimum (Fig. 16.37). Near the well minimum the variation of V with displacement resembles a parabola (as can be checked by expanding the exponential as far as the first term), but unlike a parabola, eqn 61 allows for dissociation at large displacements. The Schrödinger equation can be solved for the Morse potential and the permitted energy levels are

$$G(v) = (v + \tfrac{1}{2})\tilde{v} - (v + \tfrac{1}{2})^2 x_e\tilde{v} \qquad x_e = \frac{a^2\hbar}{2m_{eff}\omega} = \frac{\tilde{v}}{4D_e} \qquad (16.62)$$

The parameter x_e is called the **anharmonicity constant**. The number of vibrational levels of a Morse oscillator is finite, and $v = 0, 1, 2, \ldots, v_{max}$, as shown in Fig. 16.38

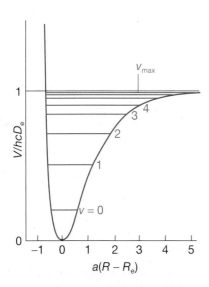

16.38 The Morse potential energy curve reproduces the general shape of a molecular potential energy curve. The corresponding Schrödinger equation can be solved, and the values of the energies obtained. The number of bound levels is finite.

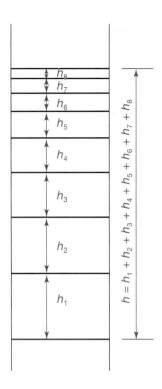

16.39 The dissociation energy is the sum of the separations of the vibrational energy levels up to the dissociation limit just as the length of a ladder is the sum of the separations of its rungs.

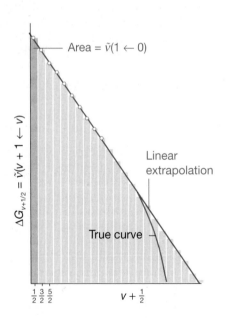

16.40 The area under a plot of transition wavenumber against vibrational quantum number is equal to the dissociation energy of the molecule. The assumption that the differences approach zero linearly is the basis of the Birge–Sponer extrapolation.

(see also Problem 16.28). The second term in the expression for G subtracts from the first with increasing effect as v increases, and hence gives rise to the convergence of the levels at high quantum numbers.

Although the Morse oscillator is quite useful theoretically, in practice the more general expression

$$G(v) = (v + \tfrac{1}{2})\tilde{v} - (v + \tfrac{1}{2})^2 x_e\tilde{v} + (v + \tfrac{1}{2})^3 y_e\tilde{v} + \cdots \qquad (16.63)$$

where x_e, y_e, … are empirical constants characteristic of the molecule, is used to fit the experimental data and to find the dissociation energy of the molecule. When anharmonicities are present, the wavenumbers of transitions with $\Delta v = +1$ are

$$\Delta G_{v+\frac{1}{2}} = \tilde{v} - 2(v + 1)x_e\tilde{v} + \cdots \qquad (16.64)$$

Equation 64 shows that, when $x_e \neq 0$, the transitions move to lower wavenumbers as v increases.

Anharmonicity also accounts for the appearance of additional weak absorption lines corresponding to the transitions $2 \leftarrow 0$, $3 \leftarrow 0$, … , even though these first, second, … **overtones** are forbidden by the selection rule $\Delta v = \pm 1$. The first overtone, for example, gives rise to an absorption at

$$G(v + 2) - G(v) = 2\tilde{v} - 2(2v + 3)x_e\tilde{v} + \cdots \qquad (16.65)$$

The reason for the appearance of overtones is that the selection rule is derived from the properties of harmonic oscillator wavefunctions, which are only approximately valid when anharmonicity is present. Therefore, the selection rule is also only an approximation. For an anharmonic oscillator, all values of Δv are allowed, but transitions with $\Delta v > 1$ are allowed only weakly if the anharmonicity is slight.

(b) The Birge–Sponer plot

When several vibrational transitions are detectable, a graphical technique called a **Birge–Sponer plot** may be used to determine the dissociation energy, D_0, of the bond. The basis of the Birge–Sponer plot is that the sum of successive intervals $\Delta G_{v+\frac{1}{2}}$ from the zero-point level to the dissociation limit is the dissociation energy:

$$D_0 = \Delta G_{1/2} + \Delta G_{3/2} + \cdots = \sum_v \Delta G_{v+\frac{1}{2}} \qquad (16.66)$$

just as the height of the ladder is the sum of the separations of its rungs (Fig. 16.39). The construction in Fig. 16.40 shows that the area under the plot of $\Delta G_{v+\frac{1}{2}}$ against $v + \frac{1}{2}$ is equal to the sum, and therefore to D_0. The successive terms decrease linearly when only the x_e anharmonicity constant is taken into account and the inaccessible part of the spectrum can be estimated by linear extrapolation. Most actual plots differ from the linear plot as shown in Fig. 16.39, so the value of D_0 obtained in this way is usually an overestimate of the true value.

Example 16.5 Using a Birge–Sponer plot

The observed vibrational intervals of H_2^+ lie at the following values for $1 \leftarrow 0$, $2 \leftarrow 1$, … respectively (in cm^{-1}): 2191, 2064, 1941, 1821, 1705, 1591, 1479, 1368, 1257, 1145, 1033, 918, 800, 677, 548, 411. Determine the dissociation energy of the molecule.

Method Plot the separations against $v + \frac{1}{2}$, extrapolate linearly to the point cutting the horizontal axis, and then measure the area under the curve.

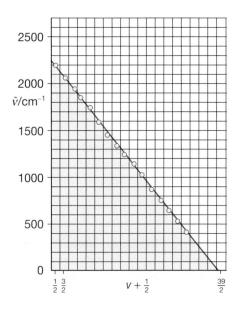

16.41 The Birge–Sponer plot used in Example 16.5. The area is obtained simply by counting the squares beneath the line or using the formula for the area of a right triangle.

Answer The points are plotted in Fig. 16.41, and a linear extrapolation is shown as a dotted line. The area under the curve (use the formula for the area of a triangle or count the squares) is 214. Each square corresponds to 100 cm^{-1} (refer to the scale of the vertical axis); hence the dissociation energy is 21 400 cm^{-1} (corresponding to 256 kJ mol^{-1}).

Self-test 16.7 The vibrational levels of HgH converge rapidly, and successive intervals are 1203.7, 965.6, 632.4, and 172 cm^{-1}. Estimate the dissociation energy.

[35.6 kJ mol^{-1}]

16.12 Vibration–rotation spectra

Each line of the high-resolution vibrational spectrum of a gas-phase heteronuclear diatomic molecule is found to consist of a large number of closely spaced components (Fig. 16.42). Hence, molecular spectra are often called **band spectra**. The separation between the components is of the order of 10 cm^{-1}, which suggests that the structure is due to rotational transitions accompanying the vibrational transition. A rotational change should be expected because classically we can think of the transition as leading to a sudden increase or decrease in the instantaneous bond length. Just as ice-skaters rotate more rapidly when they bring their arms in, and more slowly when they throw them out, so the molecular rotation is either accelerated or retarded by a vibrational transition.

(a) Spectral branches

A detailed analysis of the quantum mechanics of simultaneous vibrational and rotational changes shows that the rotational quantum number J changes by ± 1 during the vibrational transition of a diatomic molecule. If the molecule also possesses angular momentum about its axis, as in the case of the electronic orbital angular momentum of the paramagnetic molecule NO, then the selection rules also allow $\Delta J = 0$.

The appearance of the vibration–rotation spectrum of a diatomic molecule can be discussed in terms of the combined vibration–rotation terms, S:

$$S(v, J) = G(v) + F(J) \tag{16.67}$$

If we ignore anharmonicity and centrifugal distortion,

$$S(v, J) = (v + \tfrac{1}{2})\tilde{v} + BJ(J + 1) \tag{16.68}$$

16.42 A high-resolution vibration–rotation spectrum of HCl. The lines appear in pairs because H^{35}Cl and H^{37}Cl both contribute (their abundance ratio is 3:1). There is no Q branch, because $\Delta J = 0$ is forbidden for this molecule.

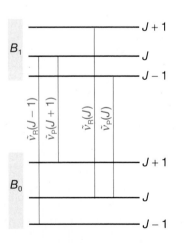

16.43 The formation of P, Q, and R branches in a vibration–rotation spectrum. The intensities reflect the populations of the initial rotational levels.

16.44 The method of combination differences makes use of the fact that some transitions share a common level.

In a more detailed treatment, B is allowed to depend on the vibrational state because as v increases the molecule swells slightly and the moment of inertia changes. We shall continue with the simple expression initially.

When the vibrational transition $v + 1 \leftarrow v$ occurs, J changes by ± 1 and in some cases by 0 (when $\Delta J = 0$ is allowed). The absorptions then fall into three groups called **branches** of the spectrum. The **P branch** consists of all transitions with $\Delta J = -1$:

$$\tilde{v}_P(J) = S(v + 1, J - 1) - S(v, J) = \tilde{v} - 2BJ \tag{16.69a}$$

This branch consists of lines at $\tilde{v} - 2B$, $\tilde{v} - 4B$, $\cdots$ with an intensity distribution reflecting both the populations of the rotational levels and the magnitude of the $J - 1 \leftarrow J$ transition moment (Fig. 16.43). The **Q branch** consists of all lines with $\Delta J = 0$, and its wavenumbers are all

$$\tilde{v}_Q(J) = S(v + 1, J) - S(v, J) = \tilde{v} \tag{16.69b}$$

for all values of J. This branch, when it is allowed (as in NO), appears at the vibrational transition wavenumber. In Fig. 16.43 there is a gap at the expected location of the Q branch because it is forbidden in HCl. The **R branch** consists of lines with $\Delta J = +1$:

$$\tilde{v}_R(J) = S(v + 1, J + 1) - S(v, J) = \tilde{v} + 2B(J + 1) \tag{16.69c}$$

This branch consists of lines displaced from $\tilde{v}$ to high wavenumber by $2B$, $4B$, $\dots$.

The separation between the lines in the P and R branches of a vibrational transition gives the value of B. Therefore, the bond length can be deduced without needing to take a pure rotational microwave spectrum. However, the latter is more precise.

(b) Combination differences

The rotational constant of the vibrationally excited state, B_1 (in general, B_v), is in fact slightly smaller than that of the ground vibrational state, B_0, because the anharmonicity of the vibration results in a slightly extended bond in the upper state. As a result, the Q branch (if it exists) consists of a series of closely spaced lines. The lines of the R branch converge slightly as J increases; and those of the P branch diverge:

$$\tilde{v}_P(J) = \tilde{v} - (B_1 + B_0)J + (B_1 - B_0)J^2$$
$$\tilde{v}_Q(J) = \tilde{v} + (B_1 - B_0)J(J + 1) \tag{16.70}$$
$$\tilde{v}_R(J) = \tilde{v} + (B_1 + B_0)(J + 1) + (B_1 - B_0)(J + 1)^2$$

To determine the two rotational constants individually, we use the method of **combination differences**. This procedure is used widely in spectroscopy to extract information about a particular state. It involves setting up expressions for the difference in the wavenumbers of transitions to a common state; the resulting expression then depends solely on properties of the other state.

As can be seen from Fig. 16.44, the transitions $\tilde{v}_R(J - 1)$ and $\tilde{v}_P(J + 1)$ have a common upper state, and hence can be anticipated to depend on B_0. Indeed, it is easy to show from eqn 70 that

$$\tilde{v}_R(J - 1) - \tilde{v}_P(J + 1) = 4B_0(J + \tfrac{1}{2}) \tag{16.71a}$$

Therefore, a plot of the combination difference against $J + \tfrac{1}{2}$ should be a straight line of slope $4B_0$, so the rotational constant of the molecule in the state $v = 0$ can be determined. (Any deviation from a straight line is a consequence of centrifugal distortion, so that effect can be investigated too.) Similarly, $\tilde{v}_R(J)$ and $\tilde{v}_P(J)$ have a common lower state, and hence their combination difference gives information about the upper state:

$$\tilde{v}_R(J) - \tilde{v}_P(J) = 4B_1(J + \tfrac{1}{2}) \tag{16.71b}$$

The two rotational constants of $^1H^{35}Cl$ found in this way are $B_0 = 10.440 \ cm^{-1}$ and $B_1 = 10.136 \ cm^{-1}$.

16.13 Vibrational Raman spectra of diatomic molecules

The gross selection rule for vibrational Raman transitions is that *the polarizability should change as the molecule vibrates*. As homonuclear and heteronuclear diatomic molecules swell and contract during a vibration, the control of the nuclei over the electrons varies, and hence the molecular polarizability changes. Both types of diatomic molecule are therefore vibrationally Raman active. The specific selection rule for vibrational Raman transitions in the harmonic approximation is $\Delta v = \pm 1$. The formal basis for the gross and specific selection rules is given in the following *Justification*.

Justification 16.8

The gross selection rule is based on an analysis of the transition dipole moment $\langle \varepsilon v_f | \boldsymbol{\mu} | \varepsilon v_i \rangle$, where, as in *Justification 16.4*, we use the Born–Oppenheimer approximation, and neglect the effect of rotation and electron spin. For simplicity, we consider a one-dimensional harmonic oscillator (like a diatomic molecule).

First, we use eqn 47 to write the transition dipole moment as

$$\boldsymbol{\mu}_{fi} = \langle \varepsilon v_f | \boldsymbol{\mu} | \varepsilon v_i \rangle = \langle \varepsilon v_f | \alpha | \varepsilon v_i \rangle \mathcal{E} = \langle v_f | \alpha(x) | v_i \rangle \mathcal{E}$$

where $\alpha(x) = \langle \varepsilon | \alpha | \varepsilon \rangle$ is the polarizability of the molecule, which we expect to be a function of small displacements x from the equilibrium bond length of the molecule. Next, we expand $\alpha(x)$ as a Taylor series, so the transition dipole moment becomes

$$\boldsymbol{\mu}_{fi} = \left\langle v_f \left| \alpha(0) + \left(\frac{d\alpha}{dx} \right)_0 x + \cdots \right| v_i \right\rangle \mathcal{E} = \langle v_f | v_i \rangle \alpha(0) \mathcal{E} + \left(\frac{d\alpha}{dx} \right)_0 \langle v_f | x | v_i \rangle \mathcal{E} + \cdots$$

The term containing $\langle v_f | v_i \rangle$ vanishes for $f \neq i$ because the harmonic oscillator wavefunctions are orthogonal. Therefore, the vibration is Raman active if $(d\alpha/dx)_0 \neq 0$ and $\langle v_f | x | v_i \rangle \neq 0$. Therefore, the polarizability of the molecule must change during the vibration; this is the gross selection rule of Raman spectroscopy. Also, we know from *Justification 16.7* that $\langle v_f | x | v_i \rangle \neq 0$ if $v_f - v_i = \pm 1$; this is the specific selection rule of Raman spectroscopy.

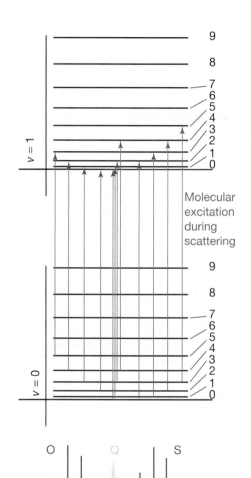

16.45 The formation of O, Q, and S branches in a vibration–rotation Raman spectrum of a linear rotor. Note that the frequency scale runs in the opposite direction to that in Fig. 16.43, because the higher energy transitions (on the right) extract more energy from the incident beam and leave it at lower frequency.

The lines to high frequency of the incident radiation, the anti-Stokes lines, are those for which $\Delta v = -1$. The lines to low frequency, the Stokes lines, correspond to $\Delta v = +1$. The intensities of the anti-Stokes and Stokes lines are governed largely by the Boltzmann populations of the vibrational states involved in the transition. It follows that anti-Stokes lines are usually weak because very few molecules are in an excited vibrational state initially.

In gas-phase spectra, the Stokes and anti-Stokes lines have a branch structure arising from the simultaneous rotational transitions that accompany the vibrational excitation (Fig. 16.45). The selection rules are $\Delta J = 0, \pm 2$ (as in pure rotational Raman

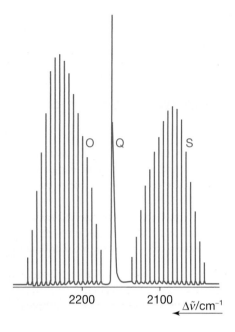

16.46 The structure of a vibrational line in the vibrational Raman spectrum of carbon monoxide, showing the O, Q, and S branches.

	$\tilde{v}/\mathrm{cm}^{-1}$	B/cm^{-1}	$k/(\mathrm{N\ m}^{-1})$

Table 16.2* Properties of diatomic molecules

	$\tilde{v}/\mathrm{cm}^{-1}$	B/cm^{-1}	$k/(\mathrm{N\ m}^{-1})$
$^{1}\mathrm{H}_2$	4400	60.86	575
$^{1}\mathrm{H}^{35}\mathrm{Cl}$	2991	10.59	516
$^{1}\mathrm{H}^{127}\mathrm{I}$	2309	6.61	313
$^{35}\mathrm{Cl}_2$	560	0.244	323

* More values are given in the *Data section*. See Tables 14.2 and 14.3 for related information (on D_e and R_e).

spectroscopy), and give rise to the **O branch** ($\Delta J = -2$), the **Q branch** ($\Delta J = 0$), and the **S branch** ($\Delta J = +2$):

$$\tilde{v}_O(J) = \tilde{v}_i - \tilde{v} - 2B + 4BJ$$
$$\tilde{v}_Q(J) = \tilde{v}_i - \tilde{v} \qquad\qquad (16.72)$$
$$\tilde{v}_S(J) = \tilde{v}_i - \tilde{v} - 6B - 4BJ$$

Note that, unlike in infrared spectroscopy, a Q branch is obtained for all linear molecules. The spectrum of CO, for instance, is shown in Fig. 16.46: the structure of the Q branch arises from the differences in rotational constants of the upper and lower vibrational states.

The information available from vibrational Raman spectra adds to that from infrared spectroscopy because homonuclear diatomics can also be studied. The spectra can be interpreted in terms of the force constants, dissociation energies, and bond lengths, and some of the information obtained is included in Table 16.2.

The vibrations of polyatomic molecules

There is only one mode of vibration for a diatomic molecule, the bond stretch. In polyatomic molecules there are several modes of vibration because all the bond lengths and angles may change and the vibrational spectra are very complex. Nonetheless, we shall see that infrared and Raman spectroscopy can be used to obtain information about the structure of systems as large as animal and plant tissues (Box 16.2).

16.14 Normal modes

We begin by calculating the total number of vibrational modes of a polyatomic molecule. We then see that we can choose combinations of these atomic displacements that give the simplest description of the vibrations.

(a) The number of vibrational modes

As shown in the *Justification* below, for a nonlinear molecule that consists of N atoms, there are $3N - 6$ independent modes of vibration. If the molecule is linear, there are $3N - 5$ independent vibrational modes.

Box 16.2 Vibrational spectroscopy of biological cells

Optical microscopes can now be combined with infrared and Raman spectrometers and the vibrational spectra of specimens as small as single biological cells obtained. The techniques of *vibrational microscopy* provide details of cellular events that cannot be observed with traditional light or electron microscopy.

Although infrared and Raman microscopes are complex devices, the principles behind their operation are simple: radiation illuminates a small area of the sample, and the transmitted, reflected, or scattered light is first collected by a microscope and then analysed by a spectrometer. The sample is then moved by very small increments along a plane perpendicular to the direction of illumination and the process is repeated until vibrational spectra for all sections of the sample are obtained. The size of a sample that can be studied by vibrational microscopy depends on a number of factors, such as the area of illumination and the excitance and wavelength of the incident radiation. Up to a point, the smaller the area that is illuminated, the smaller the area from which a spectrum can be obtained. High excitance is required to increase the rate of arrival of photons at the detector from small illuminated areas. For this reason, lasers and synchrotrons are the preferred radiation sources. Ultimately, the *diffraction limit* prevents the study of samples that are much smaller than the wavelength of light used as a probe (see *Discussion question 1*). In practice, two objects will appear as distinct images under a microscope if the distance between their centres is greater than the *Airy radius*, r_{Airy}:

$$r_{Airy} = 0.61 \frac{\lambda}{a}$$

where λ is the wavelength of the incident beam of radiation and a is the numerical aperture of the *objective lens*, the lens that collects light scattered by the object. The numerical aperture of the objective lens is defined as $a = n \sin \alpha$, where n is the refractive index of the lens material (the greater the refractive index, the greater the bending of a ray of light by the lens) and the angle α is the half-angle of the widest cone of scattered light that can be collected by the lens (so the lens collects light beams sweeping a cone with angle 2α). Use of the best equipment makes it possible to probe areas as small as 9 μm^2 by vibrational microscopy.

For Raman microscopy, the most common spectrometer system consists of a visible laser coupled to a polychromator and a CCD detector, although also in use are near-infrared Fourier transform spectrometers. The CCD detector can be used in a variation of Raman microscopy known as *Raman imaging*: a special optical filter allows only one Stokes line to reach the two-dimensional detector, which then contains a map of the distribution of the intensity of that line in the illuminated area.

For infrared microscopy, the radiation source is either a globar or a synchrotron, and a Fourier transform spectrometer is typically equipped with an MCT detector. Intense synchrotron radiation makes possible the study of very small samples. The illustration shows the infrared spectra of a single mouse cell, living and dying. Both spectra have features at 1545 cm^{-1} and 1650 cm^{-1} that are due to the peptide carbonyl groups of proteins and a feature at 1240 cm^{-1} that is due to the phosphodiester (PO$_2^-$) groups of lipids. The dying cell shows an additional absorption at 1730 cm^{-1}, which is due to the ester carbonyl group from an unidentified

Infrared absorption spectra of a single mouse cell: (dark blue) living cell, (light blue) dying cell. Adapted from N. Jamin *et al.*, *Proc. Natl. Acad. Sci. USA* **95**, 4837 (1998).

compound. From a plot of the intensities of individual absorption features as a function of position in the cell, it has been possible to map the distribution of proteins and lipids during cell division and cell death.

Vibrational microscopy has also been used in biomedical and pharmaceutical laboratories. Examples include the determination of the size and distribution of a drug in a tablet, the observation of conformational changes in proteins of cancerous cells upon administration of anti-tumour drugs, and the measurement of differences between diseased and normal tissue, such as diseased arteries and the white matter from brains of multiple sclerosis patients.

Discussion questions

1. Consult a physics textbook and prepare a discussion of the principles behind the diffraction limit. Can you think of ways in which the diffraction limit may be overcome? *Hint.* See Box 17.2.

2. Cellulose is a carbohydrate found in plants, such as alfalfa, which is used as food for farm animals. It is believed that the nutritional value of alfalfa decreases with increasing cellulose content. Assuming that you have at your disposal a sample of pure cellulose, describe a set of vibrational microscopy experiments that can help assess the content of cellulose in, and hence the nutritional value of, alfalfa in samples of animal food.

Further reading

P. Colarusso, L.H. Lidder, I.W. Levin, and E.N. Lewis, Raman and IR microspectroscopy. In *Encyclopedia of spectroscopy and spectrometry* (ed. J.C. Lindon, G.E. Tranter, and J.L. Holmes), **3**, 1945. Academic Press, San Diego (2000).

G. Turrell and J. Corset (ed.), *Raman microscopy: developments and applications*. Academic Press, London (1996).

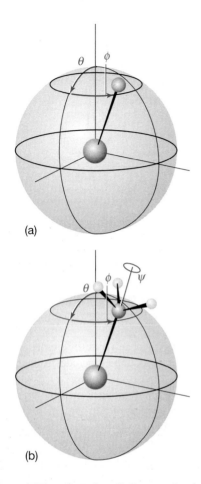

16.47 (a) The orientation of a linear molecule requires the specification of two angles. (b) The orientation of a nonlinear molecule requires the specification of three angles.

Justification 16.9

The total number of coordinates needed to specify the locations of N atoms is $3N$. Each atom may change its location by varying one of its three coordinates (x, y, and z), so the total number of displacements available is $3N$. These displacements can be grouped together in a physically sensible way. For example, three coordinates are needed to specify the location of the centre of mass of the molecule, so three of the $3N$ displacements correspond to the translational motion of the molecule as a whole. The remaining $3N - 3$ are non-translational 'internal' modes of the molecule.

Two angles are needed to specify the orientation of a linear molecule in space: in effect, we need to give only the latitude and longitude of the direction in which the molecular axis is pointing (Fig. 16.47a). However, three angles are needed for a non-linear molecule because we also need to specify the orientation of the molecule around the direction defined by the latitude and longitude (Fig. 16.47b). Therefore, two (linear) or three (nonlinear) of the $3N - 3$ internal displacements are rotational. This leaves $3N - 5$ (linear) or $3N - 6$ (nonlinear) displacements of the atoms relative to one another: these are the vibrational modes. It follows that the number of modes of vibration N_{vib} is $3N - 5$ for linear molecules and $3N - 6$ for nonlinear molecules.

Illustration 16.6

Water, H_2O, is a nonlinear triatomic molecule, and has three modes of vibration (and three modes of rotation); CO_2 is a linear triatomic molecule, and has four modes of vibration (and only two modes of rotation). Even a middle-sized molecule such as naphthalene ($C_{10}H_8$) has 48 distinct modes of vibration.

(b) Combinations of displacements

The next step is to find the best description of the modes. One choice for the four modes of CO_2, for example, might be the ones in Fig. 16.48a and c. This illustration shows the stretching of one bond (the mode v_L), the stretching of the other (v_R), and the two perpendicular bending modes (v_2). The description, while permissible, has a disadvantage: when one CO bond vibration is excited, the motion of the C atom sets the other CO bond in motion, so energy flows backwards and forwards between v_L and v_R. Moreover, the position of the centre of mass of the molecule varies in the course of either vibration.

The description of the vibrational motion is much simpler if linear combinations of v_L and v_R are taken. For example, one combination is v_1 in Fig. 16.48b: this mode is the **symmetric stretch**. In this mode, the C atom is buffeted simultaneously from each side and the motion continues indefinitely. Another mode is v_3, the **antisymmetric stretch**, in which the two O atoms always move in the same direction and opposite to that of the C atom. Both modes are independent in the sense that, if one is excited, then it does not excite the other. They are two of the 'normal modes' of the molecule, its independent, collective vibrational displacements. The two other normal modes are the bending modes v_2. In general, a **normal mode** is an independent, synchronous motion of atoms or groups of atoms that may be excited without leading to the excitation of any other normal mode.

The four normal modes of CO_2, and the N_{vib} normal modes of polyatomics in general, are the key to the description of molecular vibrations. Each normal

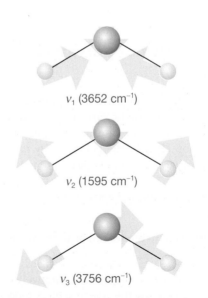

16.48 Alternative descriptions of the vibrations of CO_2. (a) The stretching modes are not independent, and if one C—O group is excited the other begins to vibrate. (b) The symmetric and antisymmetric stretches are independent, and one can be excited without affecting the other: they are normal modes. (c) The two perpendicular bending motions are also normal modes.

mode, q, behaves like an independent harmonic oscillator (if anharmonicities are neglected), so each has a series of terms

$$G_q(v) = (v + \tfrac{1}{2})\tilde{v}_q \qquad \tilde{v}_q = \frac{1}{2\pi c}\left(\frac{k_q}{m_q}\right)^{1/2} \tag{16.73}$$

where $\tilde{v}_q$ is the wavenumber of mode q and depends on the force constant k_q for the mode and on the effective mass m_q of the mode. The effective mass of the mode is a measure of the mass that is swung about by the vibration and in general is a complicated function of the masses of the atoms. For example, in the symmetric stretch of CO_2, the C atom is stationary, and the effective mass depends on the masses of only the O atoms. In the antisymmetric stretch and in the bends, all three atoms move, so all contribute to the effective mass. The three normal modes of H_2O are shown in Fig. 16.49: note that the predominantly bending mode (v_2) has a lower frequency than the others, which are predominantly stretching modes. It is generally the case that the frequencies of bending motions are lower than those of stretching modes. One point that must be appreciated is that only in special cases (such as the CO_2 molecule) are the normal modes purely stretches or purely bends. In general, a normal mode is a composite motion of simultaneous stretching and bending of bonds. Another point in this connection is that heavy atoms generally move less than light atoms in normal modes.

16.15 Infrared absorption spectra of polyatomic molecules

The gross selection rule for infrared activity is that *the motion corresponding to a normal mode should be accompanied by a change of dipole moment*. Deciding whether this is so can sometimes be done by inspection. For example, the symmetric stretch of CO_2 leaves the dipole moment unchanged (at zero, see Fig. 16.48), so this mode is infrared inactive. The antisymmetric stretch, however, changes the dipole moment because the molecule becomes unsymmetrical as it vibrates, so this mode is infrared active. Because the dipole moment change is parallel to the principal axis, the transitions arising from this mode are classified as **parallel bands** in the spectrum. Both bending modes are infrared active: they are accompanied by a changing dipole

16.49 The three normal modes of H_2O. The mode v_2 is predominantly bending, and occurs at lower wavenumber than the other two.

Table 16.3 Typical vibrational wavenumbers, $\tilde{v}/cm^{-1}$

C—H stretch	2850–2960
C—H bend	1340–1465
C—C stretch	700–1250
C=C stretch	1620–1680

* More values are given in the *Data section*.

The web site contains links to sites with free software for quantum mechanical calculations of frequencies and atomic displacements of normal modes of simple molecules.

The web site contains links to databases of infrared spectra.

perpendicular to the principal axis, so transitions involving them lead to a **perpendicular band** in the spectrum. The latter bands eliminate the linearity of the molecule, and as a result a Q branch is observed; a parallel band does not have a Q branch.

The active modes are subject to the specific selection rule $\Delta v_q = \pm 1$ in the harmonic approximation, so the wavenumber of the fundamental transition (the 'first harmonic') of each active mode is $\tilde{v}_q$. From the analysis of the spectrum, a picture may be constructed of the stiffness of various parts of the molecule: that is, we can establish its **force field**, the set of force constants corresponding to all the displacements of the atoms. The force field may also be estimated by using the semi-empirical and *ab initio* computational techniques described in Section 14.9. Superimposed on the simple force field scheme are the complications arising from anharmonicities and the effects of molecular rotation. Very often the sample is a liquid or a solid, and the molecules are unable to rotate freely. In a liquid, for example, a molecule may be able to rotate through only a few degrees before it is struck by another, so it changes its rotational state frequently. This random changing of orientation is called **tumbling**.

The lifetimes of rotational states in liquids are very short, so in most cases the rotational energies are ill-defined. Collisions occur at a rate of about $10^{13}\ s^{-1}$ and, even allowing for only a 10 per cent success rate in knocking the molecule into another rotational state, a lifetime broadening (eqn 26) of more than 1 cm^{-1} can easily result. The rotational structure of the vibrational spectrum is blurred by this effect, so the infrared spectra of molecules in condensed phases usually consist of broad lines spanning the entire range of the resolved gas-phase spectrum, and showing no branch structure.

One very important application of infrared spectroscopy to condensed phase samples, and for which the blurring of the rotational structure by random collisions is a welcome simplification, is chemical analysis. The vibrational spectra of different groups in a molecule give rise to absorptions at characteristic frequencies. Their intensities are also transferable between molecules. Consequently, the molecules in a sample can often be identified by examining its infrared spectrum and referring to a table of characteristic frequencies and intensities (Table 16.3).

16.16 Vibrational Raman spectra of polyatomic molecules

The normal modes of vibration of molecules are Raman active if they are accompanied by a changing polarizability. It is sometimes quite difficult to judge by inspection when this is so. The symmetric stretch of CO_2, for example, alternately swells and contracts the molecule: this motion changes the polarizability of the molecule, so the mode is Raman active. The other modes of CO_2 leave the polarizability unchanged, so they are Raman inactive.

A more exact treatment of infrared and Raman activity of normal modes leads to the **exclusion rule**:

> **If the molecule has a centre of symmetry, then no modes can be both infrared and Raman active.**

(A mode may be inactive in both.) Because it is often possible to judge intuitively if a mode changes the molecular dipole moment, we can use this rule to identify modes that are not Raman active. The rule applies to CO_2 but to neither H_2O nor CH_4 because they have no centre of symmetry. In general, it is necessary to use group theory to predict whether a mode is infrared or Raman active (Section 16.17).

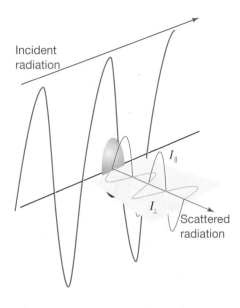

16.50 The definition of the planes used for the specification of the depolarization ratio, ρ, in Raman scattering.

(a) Depolarization

The assignment of Raman lines to particular vibrational modes is aided by noting the state of polarization of the scattered light. The **depolarization ratio**, ρ, of a line is the ratio of the intensities, I, of the scattered light with polarizations perpendicular and parallel to the plane of polarization of the incident radiation:

$$\rho = \frac{I_\perp}{I_\parallel} \qquad [16.74]$$

To measure ρ, the intensity of a Raman line is measured with a polarizing filter (a 'half-wave plate') first parallel and then perpendicular to the polarization of the incident beam. If the emergent light is not polarized, then both intensities are the same and ρ is close to 1; if the light retains its initial polarization, then $I_\perp = 0$, so $\rho = 0$ (Fig. 16.50). A line is classified as **depolarized** if it has ρ close to or greater than 0.75 and as **polarized** if $\rho < 0.75$. Only totally symmetrical vibrations give rise to polarized lines in which the incident polarization is largely preserved. Vibrations that are not totally symmetrical give rise to depolarized lines because the incident radiation can give rise to radiation in the perpendicular direction too.

(b) Resonance Raman spectra

A modification of the basic Raman effect involves using incident radiation that nearly coincides with the frequency of an electronic transition of the sample (Fig. 16.51). The technique is then called **resonance Raman spectroscopy**. It is characterized by a much greater intensity in the scattered radiation. Furthermore, because it is often the case that only a few vibrational modes contribute to the more intense scattering, the spectrum is greatly simplified.

Resonance Raman spectroscopy is used to study biological molecules that absorb strongly in the ultraviolet and visible regions of the spectrum. Examples include the pigments β-carotene and chlorophyll, which capture solar energy during plant photosynthesis (see Box 26.2). The resonance Raman spectra of Fig. 16.52 show vibrational transitions from only the few pigment molecules that are bound to very large proteins dissolved in an aqueous buffer solution. This selectivity arises from the fact that water (the solvent), amino acid residues, and the peptide group do not

16.51 (a) In conventional Raman spectroscopy, the incident radiation does not match an absorption frequency of the molecule, and there is only a 'virtual' transition to an excited state. (b) However, in the resonance Raman effect, the incident radiation has a frequency that coincides with a molecular transition.

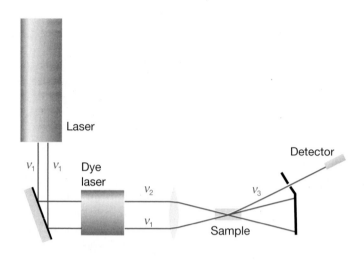

16.53 The experimental arrangement for the CARS experiment.

16.52 The resonance Raman spectra of a protein complex that is responsible for some of the initial electron transfer events in plant photosynthesis. (a) Laser excitation of the sample at 407 nm shows Raman bands due to both chlorophyll a and β-carotene bound to the protein because both pigments absorb light at this wavelength. (b) Laser excitation at 488 nm shows Raman bands from β-carotene only because chlorophyll a does not absorb light very strongly at this wavelength. (Adapted from D.F. Ghanotakis *et al.*, *Biochim. Biophys. Acta* **974**, 44 (1989).)

have electronic transitions at the laser wavelengths used in the experiment, so their *conventional* Raman spectra are weak compared to the enhanced spectra of the pigments. Comparison of the spectra in Figs 16.52a and 16.52b also shows that, with proper choice of excitation wavelength, it is possible to examine individual classes of pigments bound to the same protein: excitation at 488 nm, where β-carotene absorbs strongly, shows vibrational bands from β-carotene only, whereas excitation at 407 nm, where chlorophyll a and β-carotene absorb to similar extents, reveals features from both types of pigments.

(c) Coherent anti-Stokes Raman spectroscopy

The intensity of Raman transitions may be enhanced by **coherent anti-Stokes Raman spectroscopy** (CARS, Fig. 16.53). The technique relies on the fact that, if two laser beams of frequencies v_1 and v_2 pass through a sample, then they may mix together and give rise to coherent radiation of several different frequencies, one of which is

$$v_3 = 2v_1 - v_2 \tag{16.75}$$

Suppose that v_2 is varied until it matches any Stokes line from the sample, such as the one with frequency $v_1 - \Delta v$; then the coherent emission will have frequency

$$v_3 = 2v_1 - (v_1 - \Delta v) = v_1 + \Delta v \tag{16.76}$$

which is the frequency of the corresponding anti-Stokes line. This coherent radiation forms a narrow beam of high intensity.

An advantage of CARS is that it can be used to study Raman transitions in the presence of competing incoherent background radiation, and so can be used to observe the Raman spectra of species in flames. One example is the vibration–rotation CARS spectrum of N_2 gas in a methane–air flame shown in Fig. 16.54.

16.17 Symmetry aspects of molecular vibrations

One of the most powerful ways of dealing with normal modes, especially of complex molecules, is to classify them according to their symmetries. Each normal mode must belong to one of the symmetry species of the molecular point group, as discussed in Chapter 15.

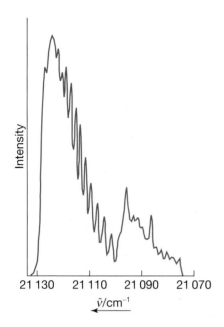

16.54 CARS spectrum of a methane-air flame at 2104 K. The peaks correspond to the Q branch of the vibration–rotation spectrum of N_2. (Adapted from J.F. Verdieck *et al.*, *J. Chem. Educ.* **59**, 495 (1982).)

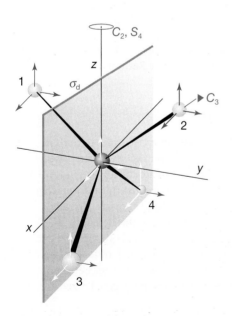

16.55 The atomic displacements of CH_4 and the symmetry elements used to calculate the characters.

Example 16.6 Identifying the symmetry species of a normal mode

Establish the symmetry species of the normal mode vibrations of CH_4, which belongs to the group T_d.

Method The first step in the procedure is to identify the symmetry species of the irreducible representations spanned by all the $3N$ displacements of the atoms, using the characters of the molecular point group. Find these characters by counting 1 if the displacement is unchanged under a symmetry operation, −1 if it changes sign, and 0 if it is changed into some other displacement. Next, subtract the symmetry species of the translations. Translational displacements span the same symmetry species as x, y, and z, so they can be obtained from the right-most column of the character table. Finally, subtract the symmetry species of the rotations, which are also given in the character table (and denoted there by R_x, R_y, or R_z).

Answer There are $3 \times 5 = 15$ degrees of freedom, of which $(3 \times 5) - 6 = 9$ are vibrations. Refer to Fig. 16.55. Under E, no displacement coordinates are changed, so the character is 15. Under C_3, no displacements are left unchanged, so the character is 0. Under the C_2 indicated, the z-displacement of the central atom is left unchanged, whereas its x- and y-components both change sign. Therefore $\chi(C_2) = 1 - 1 - 1 + 0 + 0 + \cdots = -1$. Under σ_d, the x- and z-displacements of C, H_3, and H_4 are left unchanged and the y-displacements are reversed; hence $\chi(\sigma_d) = 3 + 3 - 3 = 3$. Under the S_4 indicated, the z-displacement of the central atom is reversed, so $\chi(S_4) = -1$. The characters are therefore 15, 0, −1, −1, 3. By decomposing the direct product (Section 15.5a), we find that this representation spans $A_1 + E + T_1 + 3T_2$. The translations span T_2; the rotations span T_1. Hence, the nine vibrations span $A_1 + E + 2T_2$.

Comment The modes themselves are shown in Fig. 16.56. We shall see that symmetry analysis gives a quick way of deciding which modes are active.

Self-test 16.8 Establish the symmetry species of the normal modes of H_2O.

$$[2A_1 + B_2]$$

(a) Infrared activity of normal modes

It is best to use group theory to judge the activities of more complex modes of vibration. This is easily done by checking the character table of the molecular point group for the symmetry species of the irreducible representations spanned by x, y, and z, for their species are also the symmetry species of the components of the electric dipole moment. Then apply the following rule:

> If the symmetry species of a normal mode is the same as any of the symmetry species of x, y, or z, then the mode is infrared active.

Justification 16.10

The rule hinges on the form of the transition dipole moment between the ground state vibrational wavefunction, ψ_0, and that of the first excited state, ψ_1. The x-component is

$$\mu_{x,10} = \langle 1|\mu_x|0 \rangle = -e \int \psi_1^* x \psi_0 \, d\tau \tag{16.77}$$

16.56 Typical normal modes of vibration of a tetrahedral molecule. There are in fact two modes of symmetry species E and three modes of each T_2 symmetry species.

with similar expressions for the two other components of the transition moment. The ground state vibrational wavefunction is a Gaussian function of the form e^{-x^2}, so it is symmetrical in x. The wavefunction for the first excited state gives a non-vanishing integral only if it is proportional to x, for then the integrand is proportional to x^2 rather than to xy or xz. Consequently, the excited state wavefunction must have the same symmetry as the displacement x.

Example 16.7 Identifying infrared active modes

Which modes of CH_4 are infrared active?

Method Refer to the T_d character table to establish the symmetry species of x, y, and z for this molecule, and then use the rule given above.

Answer The functions x, y, and z span T_2. We found in Example 16.6 that the symmetry species of the normal modes are $A_1 + E + 2T_2$. Therefore, only the T_2 modes are infrared active.

Comment The distortions accompanying these modes lead to a changing dipole moment. The A_1 mode, which is inactive, is the symmetrical 'breathing' mode of the molecule.

Self-test 16.9 Which of the normal modes of H_2O are infrared active?

[All three]

(b) Raman activity of normal modes

Group theory provides an explicit recipe for judging the Raman activity of a normal mode. In this case, the symmetry species of the quadratic forms (x^2, xy, etc.) listed in the character table are noted (they transform in the same way as the polarizability), and then we use the following rule:

If the symmetry species of a normal mode is the same as the symmetry species of a quadratic form, then the mode may be Raman active.

Illustration 16.7

To decide which of the vibrations of CH_4 are Raman active, refer to the T_d character table. It was established in Example 16.6 that the symmetry species of the normal modes are $A_1 + E + 2T_2$. Because the quadratic forms span $A_1 + E + T_2$, all the normal modes are Raman active. By combining this information with that in Example 16.6, we see how the infrared and Raman spectra of CH_4 are assigned. The assignment of spectral features to the T_2 modes is straightforward because these are the only modes that are both infrared and Raman active. This leaves the A_1 and E modes to be assigned in the Raman spectrum. Measurement of the depolarization ratio distinguishes between these modes because the A_1 mode, being totally symmetric, is polarized and the E mode is depolarized.

Self-test 16.10 Which of the vibrational modes of H_2O are Raman active?

[All three]

Checklist of key ideas

General features of spectroscopy

16.1 Experimental techniques

- [ ] emission spectroscopy, spectroscopy based on the detection of a transition from a state of high energy to a state of lower energy

- [ ] absorption spectroscopy, spectroscopy based on the detection of the net absorption of nearly monochromatic incident radiation as the radiation is swept over a range of frequencies

- [ ] Bohr frequency condition, $h\nu = E_1 - E_2$

- [ ] spectrometers, instruments that detect the frequency composition of electromagnetic radiation scattered, emitted, or absorbed by atoms and molecules

- [ ] dispersing element, the component of a spectrometer that separates radiation into different frequencies

- [ ] diffraction grating, a glass or ceramic plate into which fine grooves have been cut and covered with a reflective aluminium coating

- [ ] polychromator, a device for analysing a broad range of wavelengths simultaneously

- [ ] Fourier transform techniques, a detection technique based on the mathematical procedure of taking a Fourier transform

- [ ] Raman spectroscopy, a technique in which molecular energy levels are explored by examining the frequencies present in scattered radiation

- [ ] Stokes radiation, scattered radiation at a lower frequency than the incident radiation

- [ ] anti-Stokes radiation, scattered radiation at a higher frequency than the incident radiation

- [ ] Rayleigh radiation, the component of radiation scattered into the forward direction without change of frequency

- [ ] detector, a device that converts radiation into an electric current or voltage for appropriate signal processing and display

- [ ] resolution, the smallest observable separation between two closely spaced spectral bands

16.2 The intensities of spectral lines

- [ ] transmittance, T, the ratio of the transmitted intensity, I, to the incident intensity, I_0, at a given frequency, $T = I/I_0$

- [ ] Beer–Lambert law: $I = I_0 10^{-\varepsilon[J]l}$

- [ ] molar absorption coefficient (extinction coefficient), the quantity ε in the Beer–Lambert law

- [ ] absorbance, A, of the sample at a given wavenumber, $A = -\log T$

- [ ] integrated absorption coefficient, $\mathcal{A}$, is the sum of the absorption coefficients over the entire band, $\mathcal{A} = \int_{\text{band}} \varepsilon(\tilde{\nu}) \, \mathrm{d}\tilde{\nu}$

- [ ] stimulated absorption (emission), the transition from a low (high) energy state to one of higher (lower) energy that is driven by the electromagnetic field oscillating at the transition frequency

- [ ] Einstein coefficient of stimulated absorption, the constant of proportionality in the expression for the rate of stimulated absorption, $w = B\rho$, $B = |\mu_{\text{fi}}|^2/6\varepsilon_0\hbar^2$

- [ ] total rate of absorption, W, the number of molecules excited during an interval divided by the duration of the interval, $W = Nw$

- [ ] Einstein coefficient of stimulated emission, the constant of proportionality in the expression for the rate of stimulated emission, $w' = B'\rho$

- [ ] spontaneous emission, radiative emission independent of the intensity of the radiation (of any frequency) that is already present

- [ ] Einstein coefficient of spontaneous emission, the function A in the expression $w' = A + B'\rho$

- [ ] relation between A, B, and B': $A = (8\pi h\nu^3/c^3)B$, $B' = B$

- [ ] gross selection rule, the specification of the general features a molecule must have if it is to have a spectrum of a given kind

- [ ] specific selection rule, a specification of the allowed transitions in terms of changes in quantum numbers

16.3 Linewidths

- [ ] Doppler effect, the shift in frequency of radiation when the source is moving towards or away from the observer, see eqn 16.22

- [ ] lifetime, τ, the time for which a state persists

- [ ] lifetime broadening, $\delta E \approx \hbar/\tau$

- [ ] collisional deactivation, deactivation that arises from collisions between molecules or with the walls of the container

- [ ] collisional lifetime, the mean time between collisions

- [ ] natural linewidth, the minimum linewidth; the linewidth due to spontaneous emission

Pure rotation spectra

16.4 Moments of inertia

- [ ] moment of inertia of a molecule, the mass of each atom multiplied by the square of its distance from the rotational axis through the centre of mass of the molecule, $I = \sum_i m_i r_i^2$

- [ ] rigid rotor, a body that does not distort under the stress of rotation

- [ ] spherical rotor, a rigid rotor with three equal moments of inertia

- [ ] symmetric rotor, a rigid rotor with two equal moments of inertia

- [ ] linear rotor, a rigid rotor with one moment of inertia equal to zero

- [ ] asymmetric rotor, a rigid rotor with three different moments of inertia

16.5 The rotational energy levels

- [ ] rotational constant, $B = \hbar/4\pi cI$

- [ ] rotational energy levels of a spherical rotor, $E_J = hcBJ(J+1)$, $J = 0, 1, 2, \ldots$

- [ ] rotational term, $F(J) = BJ(J+1)$

- [ ] principal axis (figure axis), the unique axis of a symmetric top

- [ ] oblate, a symmetric rotor with $I_\parallel > I_\perp$

- [ ] prolate, a symmetric rotor with $I_\parallel < I_\perp$

- [ ] rotational terms of a symmetric rotor, $F(J, K) = BJ(J+1) + (A-B)K^2$, $J = 0, 1, 2, \ldots$; $K = 0, \pm1, \ldots, \pm J$

- [ ] rotational terms of a linear rotor, $F(J) = BJ(J+1)$, $J = 0, 1, 2, \ldots$

- [ ] Stark effect, the splitting of states by an electric field

- [ ] centrifugal distortion constant, D_J, the empirical constant in the expression $F(J) = BJ(J+1) - D_J J^2(J+1)^2$ that takes into account centrifugal distortion, $D_J \approx 4B^3/\tilde{\nu}^2$

16.6 Rotational transitions

- [ ] Stark modulation, modulation of energy levels with an oscillating electric field

☐ rotational selection rule (gross): for a molecule to give a pure rotational spectrum, it must be polar

☐ rotational selection rule (specific): $\Delta J = \pm 1$, $\Delta M_J = 0, \pm 1$, $\Delta K = 0$

☐ rotational wavenumbers: $\tilde{\nu}(J + 1 \leftarrow J) = 2B(J + 1)$, $J = 0, 1, 2, \ldots$

☐ rotational wavenumbers including distortion, $\tilde{\nu}(J + 1 \leftarrow J) = 2B(J + 1) - 4D_J(J + 1)^3$, $J = 0, 1, 2, \ldots$

16.7 Rotational Raman spectra

☐ Raman selection rule (gross): the molecule must be anisotropically polarizable

☐ Raman selection rule (specific): Linear rotors: $\Delta J = 0, \pm 2$; Symmetric rotors: $\Delta J = 0, \pm 1, \pm 2$; $\Delta K = 0$

16.8 Nuclear statistics and rotational states

☐ nuclear statistics, the selective occupation of rotational states that stems from the Pauli principle

☐ relative intensities: see eqn 16.50

☐ *ortho*-hydrogen, dihydrogen with parallel nuclear spins

☐ *para*-hydrogen, dihydrogen with paired nuclear spins

The vibrations of diatomic molecules

16.9 Molecular vibrations

☐ force constant, the constant in Hooke's law of force; $k = (d^2V/dr^2)_0$

☐ effective mass, $m_{eff} = m_1 m_2 / (m_1 + m_2)$

☐ vibrational energy levels, $E_v = (v + \frac{1}{2})\hbar\omega$, $\omega = (k/m_{eff})^{1/2}$

☐ vibrational term, the energy of a vibrational state expressed in wavenumbers, $G(v) = (v + \frac{1}{2})\tilde{\nu}$, $\tilde{\nu} = (1/2\pi c)(k/m_{eff})^{1/2}$

16.10 Selection rules

☐ vibrational selection rule (gross): the electric dipole moment of the molecule must change when the atoms are displaced relative to one another

☐ infrared active, able to absorb infrared radiation

☐ vibrational selection rule (specific): $\Delta v = \pm 1$

☐ fundamental transition, $1 \leftarrow 0$

16.11 Anharmonicity

☐ anharmonic motion, oscillatory motion in which the restoring force is not proportional to the displacement

☐ Morse potential energy, $V = hcD_e\{1 - e^{-a(R-R_e)}\}^2$, $a = (m_{eff}\omega^2/2hcD_e)^{1/2}$

☐ vibrational terms of a Morse oscillator, $G(v) = (v + \frac{1}{2})\tilde{\nu} - (v + \frac{1}{2})^2 x_e\tilde{\nu}$

☐ anharmonicity constant, the constant $x_e = \tilde{\nu}/4D_e$

☐ vibrational terms (empirical form), $G(v) = (v + \frac{1}{2})\tilde{\nu} - (v + \frac{1}{2})^2 x_e\tilde{\nu} + (v + \frac{1}{2})^3 y_e\tilde{\nu} + \cdots$

☐ overtones, weak absorption lines corresponding to the transitions $2 \leftarrow 0, 3 \leftarrow 0, \ldots$

☐ Birge–Sponer plot, a graphical procedure for determining the dissociation energy of a bond

16.12 Vibration–rotation spectra

☐ band spectra, molecular spectra consisting of many closely spaced components

☐ P branch, vibration–rotation transitions with $\Delta J = -1$, $\tilde{\nu}_P(J) = S(v + 1, J - 1) - S(v, J) = \tilde{\nu} - 2BJ$

☐ Q branch, vibration–rotation transitions with $\Delta J = 0$, $\tilde{\nu}_Q(J) = S(v + 1, J) - S(v, J) = \tilde{\nu}$

☐ R branch, vibration–rotation transitions with $\Delta J = +1$, $\tilde{\nu}_R(J) = S(v + 1, J + 1) - S(v, J) = \tilde{\nu} + 2B(J + 1)$

☐ combination differences a procedure that involves setting up expressions for the difference in the wavenumbers of transitions to a common state

16.13 Vibrational Raman spectra of diatomic molecules

☐ vibrational Raman transitions (gross): the polarizability must change as the molecule vibrates

☐ vibrational Raman transitions (specific): $\Delta v = \pm 1$

☐ O branch, vibration–rotation Raman transitions with $\Delta J = -2$, $\tilde{\nu}_O(J) = \tilde{\nu}_i - \tilde{\nu} - 2B + 4BJ$

☐ Q branch, vibration–rotation Raman transitions with $\Delta J = 0$, $\tilde{\nu}_Q(J) = \tilde{\nu}_i - \tilde{\nu}$

☐ S branch, vibration–rotation Raman transitions with $\Delta J = +2$, $\tilde{\nu}_S(J) = \tilde{\nu}_i - \tilde{\nu} - 6B - 4BJ$

The vibrations of polyatomic molecules

16.14 Normal modes

☐ number of normal modes $3N - 6$ (nonlinear), $3N - 5$ (linear)

☐ symmetric stretch, a symmetrically stretching vibrational mode, Fig. 16.48

☐ antisymmetric stretch, a stretching mode, one-half of which is the mirror image of the other half, Fig. 16.48

☐ normal mode, an independent, synchronous motion of atoms or groups of atoms that may be excited without leading to the excitation of any other normal mode

16.15 Infrared absorption spectra of polyatomic molecules

☐ selection rule for infrared activity: the motion corresponding to a normal mode should be accompanied by a change of dipole moment

☐ parallel band, a vibrational transition in which the changing dipole moment is parallel to an axis

☐ perpendicular band, a vibrational transition in which the changing dipole moment is perpendicular to an axis

☐ force field, the set of force constants corresponding to all the displacements of the atoms

☐ tumbling, the random changing of orientation of a molecule

16.16 Vibrational Raman spectra of polyatomic molecules

☐ exclusion rule: If the molecule has a centre of symmetry, then no modes can be both infrared and Raman active

☐ depolarization ratio, ρ, the ratio of the intensities, I, of the scattered light with polarizations perpendicular and parallel to the plane of polarization of the incident radiation, $\rho = I_\perp / I_\parallel$

☐ depolarized line, a line with ρ close to or greater than 0.75

☐ polarized line, a line with $\rho < 0.75$

☐ resonance Raman spectroscopy, a Raman technique in which the frequency of the incident radiation nearly coincides with the frequency of an electronic transition of the sample

☐ coherent anti-Stokes Raman spectroscopy (CARS), a Raman technique that relies on the use of two incident beams of radiation

16.17 Symmetry aspects of molecular vibrations

☐ infrared activity: If the symmetry species of a normal mode is the same as any of the symmetry species of x, y, or z, then the mode is infrared active

☐ Raman activity: If the symmetry species of a normal mode is the same as the symmetry species of a quadratic form, then the mode may be Raman active

Further reading

Articles of general interest

M.-K. Ahn, A comparison of FTNMR and FTIR techniques. *J. Chem. Educ.* **66**, 802 (1989).

W.D. Bare, A more pedagogically sound treatment of Beer's law: a derivation based on a corpuscular-probability model. *J. Chem. Educ.* **77**, 929 (2000).

B.J. Bozlee, J.H. Luther, and M. Buraczewski, The infrared overtone intensity of a simple diatomic molecule: nitric oxide. *J. Chem. Educ.* **69**, 370 (1992).

C.W. David, IR vibration–rotation spectrum of the ammonia molecule. *J. Chem. Educ.* **73**, 46 (1996).

A.M. Ellis, Spectroscopic selection rules: the role of photon states. *J. Chem. Educ.* **76**, 1291 (1999).

M.J. Elrod, Greenhouse warming potentials from the infrared spectroscopy of atmospheric gases. *J. Chem. Educ.* **76**, 1702 (1999).

L. Glasser, Fourier transforms for chemists. Part I. Introduction to the Fourier transform. *J. Chem. Educ.* **64**, A228 (1987). Part II. Fourier transforms in chemistry and spectroscopy. *J. Chem. Educ.* **64**, A260 (1987).

P.L. Goodfriend, Diatomic vibrations revisited. *J. Chem. Educ.* **64**, 753 (1987).

D.K. Graff, Fourier and Hadamard: transforms in chemistry. *J. Chem. Educ.* **72**, 304 (1995).

E. Grunwald, J. Herzog, and C. Steel, Using Fourier transforms to understand spectral lineshapes. *J. Chem. Educ.* **72**, 210 (1995).

G. Henderson and B. Logsdon, Stark effects on rigid-rotor wavefunctions: a quantum description of dipolar rotors trapped in electric fields. *J. Chem. Educ.* **72**, 1021 (1995).

I.P. Herman, Raman scattering. In *Encyclopedia of applied physics* (ed. G.L. Trigg) **15**, 587. VCH, New York (1996).

N.K. Kildahl, Bond energy data summarized. *J. Chem. Educ.* **72**, 423 (1995).

A.R. Lacey, A student introduction to molecular vibrations. *J. Chem. Educ.* **64**, 756 (1987).

J.R. Lombardi, Radiation interaction with molecules. In *Encyclopedia of applied physics* (ed. G.L. Trigg) **15**, 509. VCH, New York (1996).

B.L. McLain, S.M. Clark, R.L. Gabriel, and D. Ben–Amotz, Educational applications of IR and Raman spectroscopy: a comparison of experiment and theory. *J. Chem. Educ.* **77**, 654 (2000).

K. Nakamoto and M.A. McKinney, Application of the correlation method to vibrational spectra of C_{60} and other fullerenes: predicting the number of IR- and Raman-active bands. *J. Chem. Educ.* **77**, 775 (2000).

W.D. Perkins, Fourier transform infrared spectroscopy: Part II. Advantages of FT-IR. *J. Chem. Educ.* **64**, A269 (1987).

D.A. Ramsay, Molecular spectroscopy. In *Encyclopedia of applied physics* (ed. G.L. Trigg) **10**, 491. VCH, New York (1994).

D.P. Strommen, Specific values of the depolarization ratio in Raman spectroscopy: Their origins and significance. *J. Chem. Educ.* **69**, 803 (1992).

G.J. Thomas, Jr., Raman spectroscopy of protein and nucleic acid assemblies. *Ann. Rev. Biophys. Biomol. Struct.* **28**, 1 (1999).

N.C. Thomas, The early history of spectroscopy. *J. Chem. Educ.* **68**, 631 (1991).

V.B.E. Thomsen, Why do spectral lines have a linewidth? *J. Chem. Educ.* **72**, 616 (1995).

R.S. Treptow, Bond energies and enthalpies: an often neglected difference. *J. Chem. Educ.* **72**, 497 (1995).

J.G. Verkade, A novel pictorial approach to teaching molecular motions in polyatomic molecules. *J. Chem. Educ.* **64**, 411 (1987).

R. Woods and G. Henderson, FTIR rotational spectroscopy. *J. Chem. Educ.* **64**, 921 (1987).

Texts and sources of data and information

P.W. Atkins and R.S. Friedman, *Molecular quantum mechanics*. Oxford University Press (1997).

P.F. Bernath, *Spectra of atoms and molecules*. Oxford University Press, New York (1995).

D.C. Harris and M.D. Bertollucci, *Symmetry and spectroscopy: an introduction to vibrational and electronic spectroscopy*. Dover, New York (1989).

D.B. Chase and J.F. Rabolt (ed.), *Fourier transform Raman spectroscopy: from concept to experiment*. Academic Press, San Diego (1994).

M. Diem, *Introduction to modern vibrational spectroscopy*. Wiley, New York (1993).

G. Duxbury, *Infrared vibration–rotation spectroscopy: from free radicals to the infrared sky*. Wiley, New York (2000).

R. Drago, *Physical methods for chemists*. Saunders, Philadelphia (1992).

W. Gordy and R.L. Cook, *Microwave molecular spectra*. Wiley, New York (1974).

H.-U. Gremlich and B. Yan, *Infrared and Raman spectroscopy of biological materials*. Marcel Dekker, New York (2001).

E. Hecht, *Optics*. Addison–Wesley, Menlo Park (1998).

G. Herzberg, *Molecular spectra and molecular structure I. Spectra of diatomic molecules*. Krieger, Malabar (1989).

G. Herzberg, *Molecular spectra and molecular structure II. Infrared and Raman spectra of polyatomic molecules*. Van Nostrand–Reinhold, New York (1945).

J.M. Hollas, *Modern spectroscopy*. Wiley, New York (1998).

J.M. Hollas, *High resolution spectroscopy*. Butterworth, London (1998).

K.P. Huber and G. Herzberg, *Molecular spectra and molecular structure IV. Constants of diatomic molecules*. Van Nostrand-Reinhold, New York (1979).

M.E. Jacox, *Vibrational and electronic energy levels of polyatomic transient molecules*. Journal of Physical and Chemical Reference Data, Monograph No. 3 (1994).

H.W. Kroto, *Molecular rotation spectra*. Dover, New York (1992).

D.R. Lide (ed.), *CRC Handbook of Chemistry and Physics*, Section 10, CRC Press, Boca Raton (2000).

J.C. Lindon, G.E. Tranter, and J.L. Holmes (ed.), *Encyclopedia of spectroscopy and spectrometry*. Academic Press, San Diego (2000).

J.L. McHale, *Molecular spectroscopy*. Prentice-Hall, Upper Saddle River (1999).

K. Nakamoto, *Infrared and Raman spectra of inorganic and coordination compounds. Part A. Theory and applications in inorganic chemistry; Part B. Applications in coordination, organometallic, and bioinorganic chemistry*. Wiley, New York (1997).

J.F. Ogilvie, *The vibrational and rotational spectrometery of diatomic molecules*. Academic Press, San Diego (1998).

J.W. Robinson (ed.), *Practical handbook of spectroscopy*. CRC Press, Boca Raton (1991).

B. Schrader, *Raman/IR atlas of organic compounds*. VCH, New York (1989).

G. Socrates, *Infrared and Raman characteristic group frequencies: tables and charts*. Wiley, New York (2000).

B. Stuart, *Biological applications of infrared spectroscopy*. Wiley, New York (1997).

C.H. Townes and A.L. Schawlow, *Microwave spectroscopy*. Dover, New York (1975).

R.P. Wayne, *Chemical instrumentation*, Oxford Chemistry Primers. Oxford University Press (1994).

S. Wilson (ed.), *Methods in computational chemistry*, Vol 4. Molecular vibrations. Plenum, New York (1992).

E.B. Wilson, J.C. Decius, and P.C. Cross, *Molecular vibrations*. Dover, New York (1980).

Exercises

Discussion questions

16.1 (a) Describe the principles behind the operation of a Fourier transform spectrometer.

16.1 (b) Describe the physical origins of linewidths in the absorption and emission spectra of gases, liquids, and solids.

16.2 (a) Discuss the physical origins of the gross selection rules for microwave and infrared spectroscopy.

16.2 (b) Discuss the physical origins of the gross selection rules for rotational and vibrational Raman spectroscopy.

16.3 (a) Consider a diatomic molecule that is highly susceptible to centrifugal distortion in its ground vibrational state. Do you expect excitation to high rotational energy levels to change the equilibrium bond length of this molecule? Justify your answer.

16.3 (b) Suppose that you wish to characterize the normal modes of benzene in the gas phase. Why is it important to obtain both infrared absorption and Raman spectra of your sample?

Numerical exercises

16.4 (a) Calculate the ratio of the Einstein coefficients of spontaneous and stimulated emission, A and B, for transitions with the following characteristics: (a) 70.8 pm X-rays, (b) 500 nm visible light, (c) 3000 cm^{-1} infrared radiation.

16.4 (b) Calculate the ratio of the Einstein coefficients of spontaneous and stimulated emission, A and B, for transitions with the following characteristics: (a) 500 MHz radiofrequency radiation, (b) 3.0 cm microwave radiation.

16.5 (a) What is the Doppler-shifted wavelength of a red (660 nm) traffic light approached at 80 km h^{-1}?

16.5 (b) At what speed of approach would a red (660 nm) traffic light appear green (520 nm)?

16.6 (a) Estimate the lifetime of a state that gives rise to a line of width (a) 0.10 cm^{-1}, (b) 1.0 cm^{-1}.

16.6 (b) Estimate the lifetime of a state that gives rise to a line of width (a) 100 MHZ, (b) 2.14 cm^{-1}.

16.7 (a) A molecule in a liquid undergoes about 1.0×10^{13} collisions in each second. Suppose (a) that every collision is effective in deactivating the molecule vibrationally and (b) that one collision in 100 is effective. Calculate the width (in cm^{-1}) of vibrational transitions in the molecule.

16.7 (b) A molecule in a gas undergoes about 1.0×10^{9} collisions in each second. Suppose (a) that every collision is effective in deactivating the molecule rotationally and (b) that one collision in 10 is effective. Calculate the width (in hertz) of rotational transitions in the molecule.

16.8 (a) Calculate the frequency of the $J = 4 \leftarrow 3$ transition in the pure rotational spectrum of $^{14}N^{16}O$. The equilibrium bond length is 115 pm.

16.8 (b) Calculate the frequency of the $J = 3 \leftarrow 2$ transition in the pure rotational spectrum of $^{12}C^{16}O$. The equilibrium bond length is 112.81 pm.

16.9 (a) If the wavenumber of the $J = 3 \leftarrow 2$ rotational transition of $^{1}H^{35}Cl$ considered as a rigid rotator is 63.56 cm^{-1}, what is (a) the moment of inertia of the molecule, (b) the bond length?

16.9 (b) If the wavenumber of the $J = 1 \leftarrow 0$ rotational transition of $^{1}H^{81}Br$ considered as a rigid rotator is 16.93 cm^{-1}, what is (a) the moment of inertia of the molecule, (b) the bond length?

16.10 (a) Given that the spacing of lines in the microwave spectrum of $^{27}Al^{1}H$ is constant at 12.604 cm^{-1}, calculate the moment of inertia and bond length of the molecule $(m(^{27}Al) = 26.9815$ u).

16.10 (b) Given that the spacing of lines in the microwave spectrum of $^{35}Cl^{19}F$ is constant at 1.033 cm^{-1}, calculate the moment of inertia and bond length of the molecule $(m(^{35}Cl) = 34.9688$ u, $m(^{19}F) = 18.9984$ u).

16.11 (a) The rotational constant of $^{127}I^{35}Cl$ is 0.1142 cm^{-1}. Calculate the ICl bond length $(m(^{35}Cl) = 34.9688$ u, $m(^{127}I) = 126.9045$ u).

16.11 (b) The rotational constant of $^{12}C^{16}O_2$ is 0.390 21 cm^{-1}. Calculate the bond length of the molecule $(m(^{12}C) = 12$ u exactly, $m(^{16}O) = 15.9949$ u).

16.12 (a) Determine the HC and CN bond lengths in HCN from the rotational constants $B(^{1}H^{12}C^{14}N) = 44.316$ GHz and $B(^{2}H^{12}C^{14}N) = 36.208$ GHz.

16.12 (b) Determine the CO and CS bond lengths in OCS from the rotational constants $B(^{16}O^{12}C^{32}S) = 6081.5$ MHz, $B(^{16}O^{12}C^{34}S) = 5932.8$ MHz.

16.13 (a) The wavenumber of the incident radiation in a Raman spectrometer is 20 487 cm^{-1}. What is the wavenumber of the scattered Stokes radiation for the $J = 2 \leftarrow 0$ transition of $^{14}N_2$?

16.13 (b) The wavenumber of the incident radiation in a Raman spectrometer is 20 623 cm^{-1}. What is the wavenumber of the scattered Stokes radiation for the $J = 4 \leftarrow 2$ transition of $^{16}O_2$?

16.14 (a) The rotational Raman spectrum of $^{35}Cl_2$ ($m(^{35}Cl) = 34.9688$ u) shows a series of Stokes lines separated by 0.9752 cm^{-1} and a similar series of anti-Stokes lines. Calculate the bond length of the molecule.

16.14 (b) The rotational Raman spectrum of $^{19}F_2$ ($m(^{19}F) = 18.9984$ u) shows a series of Stokes lines separated by 3.5312 cm^{-1} and a similar series of anti-Stokes lines. Calculate the bond length of the molecule.

16.15 (a) Which of the following molecules may show a pure rotational microwave absorption spectrum: (a) H_2, (b) HCl, (c) CH_4, (d) CH_3Cl, (e) CH_2Cl_2?

16.15 (b) Which of the following molecules may show a pure rotational microwave absorption spectrum: (a) H_2O, (b) H_2O_2, (c) NH_3, (d) N_2O?

16.16 (a) Which of the following molecules may show a pure rotational Raman spectrum: (a) H_2, (b) HCl, (c) CH_4, (d) CH_3Cl?

16.16 (b) Which of the following molecules may show a pure rotational Raman spectrum: (a) CH_2Cl_2, (b) CH_3CH_3, (c) SF_6, (d) N_2O?

16.17 (a) An object of mass 1.0 kg suspended from the end of a rubber band has a vibrational frequency of 2.0 Hz. Calculate the force constant of the rubber band.

16.17 (b) An object of mass 2.0 g suspended from the end of a spring has a vibrational frequency of 3.0 Hz. Calculate the force constant of the spring.

16.18 (a) Calculate the percentage difference in the fundamental vibration wavenumber of $^{23}Na^{35}Cl$ and $^{23}Na^{37}Cl$ on the assumption that their force constants are the same.

16.18 (b) Calculate the percentage difference in the fundamental vibration wavenumber of $^{1}H^{35}Cl$ and $^{2}H^{37}Cl$ on the assumption that their force constants are the same.

16.19 (a) The wavenumber of the fundamental vibrational transition of $^{35}Cl_2$ is 564.9 cm^{-1}. Calculate the force constant of the bond ($m(^{35}Cl) = 34.9688$ u).

16.19 (b) The wavenumber of the fundamental vibrational transition of $^{79}Br^{81}Br$ is 323.2 cm^{-1}. Calculate the force constant of the bond ($m(^{79}Br) = 78.9183$ u, $m(^{81}Br) = 80.9163$ u).

16.20 (a) Calculate the relative numbers of Cl_2 molecules ($\tilde{v} = 559.7$ cm^{-1}) in the ground and first excited vibrational states at (a) 298 K, (b) 500 K.

16.20 (b) Calculate the relative numbers of Br_2 molecules ($\tilde{v} = 321$ cm^{-1}) in the second and first excited vibrational states at (a) 298 K, (b) 800 K.

16.21 (a) The hydrogen halides have the following fundamental vibrational wavenumbers: 4141.3 cm^{-1} (HF); 2988.9 cm^{-1} (H^{35}Cl); 2649.7 cm^{-1} (H^{81}Br); 2309.5 cm^{-1} (H^{127}I). Calculate the force constants of the hydrogen–halogen bonds.

16.21 (b) From the data in Exercise 16.21a, predict the fundamental vibrational wavenumbers of the deuterium halides.

16.22 (a) For $^{16}O_2$, ΔG values for the transitions $v = 1 \leftarrow 0, 2 \leftarrow 0$, and $3 \leftarrow 0$ are, respectively, 1556.22, 3088.28, and 4596.21 cm^{-1}. Calculate $\tilde{v}$ and x_e. Assume y_e to be zero.

16.22 (b) For $^{14}N_2$, ΔG values for the transitions $v = 1 \leftarrow 0, 2 \leftarrow 0$, and $3 \leftarrow 0$ are, respectively, 2345.15, 4661.40, and 6983.73 cm^{-1}. Calculate $\tilde{v}$ and x_e. Assume y_e to be zero.

16.23 (a) The first five vibrational energy levels of HCl are at 1481.86, 4367.50, 7149.04, 9826.48, and 12 399.8 cm^{-1}. Calculate the dissociation energy of the molecule in reciprocal centimetres and electronvolts.

16.23 (b) The first five vibrational energy levels of HI are at 1144.83, 3374.90, 5525.51, 7596.66, and 9588.35 cm^{-1}. Calculate the dissociation energy of the molecule in reciprocal centimetres and electronvolts.

16.24 (a) Infrared absorption by $^{1}H^{81}Br$ gives rise to an R branch from $v = 0$. What is the wavenumber of the line originating from the rotational state with $J = 2$? Use the information in Table 16.2.

16.24 (b) Infrared absorption by $^{1}H^{127}I$ gives rise to an R branch from $v = 0$. What is the wavenumber of the line originating from the rotational state with $J = 2$? Use the information in Table 16.2.

16.25 (a) Which of the following molecules may show infrared absorption spectra: (a) H_2, (b) HCl, (c) CO_2, (d) H_2O?

16.25 (b) Which of the following molecules may show infrared absorption spectra: (a) CH_3CH_3, (b) CH_4, (c) CH_3Cl, (d) N_2?

16.26 (a) How many normal modes of vibration are there for the following molecules: (a) H_2O, (b) H_2O_2, (c) C_2H_4?

16.26 (b) How many normal modes of vibration are there for the following molecules: (a) C_6H_6, (b) $C_6H_6CH_3$, (c) HC≡C—C≡CH?

16.27 (a) Which of the three vibrations of an AB_2 molecule are infrared or Raman active when it is (a) angular, (b) linear?

16.27 (b) Which of the vibrations of an AB_3 molecule are infrared or Raman active when it is (a) trigonal planar, (b) trigonal pyramidal?

16.28 (a) Consider the vibrational mode that corresponds to the uniform expansion of the benzene ring. Is it (a) Raman, (b) infrared active?

16.28 (b) Consider the vibrational mode that corresponds to the boat-like bending of a benzene ring. Is it (a) Raman, (b) infrared active?

16.29 (a) The molecule CH_2Cl_2 belongs to the point group C_{2v}. The displacements of the atoms span $5A_1 + 2A_2 + 4B_1 + 4B_2$. What are the symmetries of the normal modes of vibration?

16.29 (b) A carbon disulfide molecule belongs to the point group $D_{\infty h}$. The nine displacements of the three atoms span $A_{1g} + A_{1u} + A_{2g} + 2E_{1u} + E_{1g}$. What are the symmetries of the normal modes of vibration?

Problems[†]

Numerical problems

16.1 Use mathematical software to evaluate the Planck distribution at any temperature and wavelength or frequency, and evaluate integrals for the energy density of the radiation between any two wavelengths. Calculate the total energy density in the visible region (700 nm to 400 nm) for a black body at (a) 1500 K, a typical operating temperature for globars, (b) 2500 K, a typical operating temperature for tungsten filament lamps, (c) 5800 K, the surface temperature of the Sun. What are the classical values at these temperatures?

16.2 Calculate the Doppler width (as a fraction of the transition wavelength) for any kind of transition in (a) HCl, (b) ICl at 25°C. What would be the widths of the rotational and vibrational transitions in these molecules (in MHz and cm^{-1}, respectively), given $B(ICl) = 0.1142\ cm^{-1}$ and $\tilde{v}(ICl) = 384\ cm^{-1}$ and additional information in Table 16.2.

16.3 The collision frequency of a molecule of mass m in a gas at a pressure p is $z = 4\sigma(kT/\pi m)^{1/2}p/kT$, where σ is the collision cross-section. Find an expression for the collision-limited lifetime of an excited state assuming that every collision is effective. Estimate the width of rotational transition in HCl ($\sigma = 0.30\ nm^2$) at 25°C and 1.0 atm. To what value must the pressure of the gas be reduced in order to ensure that collision broadening is less important than Doppler broadening?

16.4 The rotational constant of NH_3 is equivalent to 298 GHz. Compute the separation of the pure rotational spectrum lines in GHz, cm^{-1}, and mm, and show that the value of B is consistent with an N—H bond length of 101.4 pm and a bond angle of 106.78°.

16.5 The rotational constant for CO is 1.9314 cm^{-1} and 1.6116 cm^{-1} in the ground and first excited vibrational states, respectively. By how much does the internuclear distance change as a result of this transition?

16.6 Pure rotational Raman spectra of gaseous C_6H_6 and C_6D_6 yield the following rotational constants: $B(C_6H_6) = 0.189\ 60\ cm^{-1}$, $B(C_6D_6) = 0.156\ 81\ cm^{-1}$. The moments of inertia of the molecules about any axis perpendicular to the C_6 axis were calculated from these data as $I(C_6H_6) = 1.4759 \times 10^{-45}\ kg\ m^2$, $I(C_6D_6) = 1.7845 \times 10^{-45}\ kg\ m^2$. Calculate the CC, CH, and CD bond lengths.

16.7 Rotational absorption lines from $^1H^{35}Cl$ gas were found at the following wavenumbers (R.L. Hausler and R.A. Oetjen, *J. Chem. Phys.* **21**, 1340 (1953)): 83.32, 104.13, 124.73, 145.37, 165.89, 186.23, 206.60, 226.86 cm^{-1}. Calculate the moment of inertia and the bond length of the molecule. Predict the positions of the corresponding lines in $^2H^{35}Cl$.

16.8 Is the bond length in HCl the same as that in DCl? The wavenumbers of the $J = 1 \leftarrow 0$ rotational transitions for $H^{35}Cl$ and $^2H^{35}Cl$ are 20.8784 and 10.7840 cm^{-1}, respectively. Accurate atomic masses are 1.007 825 u and 2.0140 u for ^1H and ^2H, respectively. The mass of ^{35}Cl is 34.968 85 u. Based on this information alone, can you conclude that the bond lengths are the same or different in the two molecules?

16.9 Thermodynamic considerations suggest that the copper monohalides CuX should exist mainly as polymers in the gas phase, and indeed it proved difficult to obtain the monomers in sufficient abundance to detect spectroscopically. This problem was overcome by flowing the halogen gas over copper heated to 1100 K (E.L. Manson, F.C. de Lucia, and W. Gordy, *J. Chem. Phys.* **63**, 2724 (1975)). For CuBr the $J = 13-14$, $14-15$, and $15-16$ transitions occurred at 84 421.34, 90 449.25, and 96 476.72 MHz, respectively. Calculate the rotational constant and bond length of CuBr.

16.10 The microwave spectrum of $^{16}O^{12}CS$ (C.H. Townes, A.N. Holden, and F.R. Merritt, *Phys. Rev.* **74**, 1113 (1948)) gave absorption lines (in GHz) as follows:

J	1	2	3	4
^{32}S	24.325 92	36.488 82	48.651 64	60.814 08
^{34}S	23.732 33		47.462 40	

Use the expressions for moments of inertia in Table 16.1 and assume that the bond lengths are unchanged by substitution; calculate the CO and CS bond lengths in OCS.

16.11‡ In a study of the rotational spectrum of the linear FeCO radical, K. Tanaka, M. Shirasaka, and T. Tanaka (*J. Chem. Phys.* **106**, 6820 (1997)) report the following $J + 1 \leftarrow J$ transitions:

J	24	25	26
$\tilde{v}/m^{-1}$	214 777.7	223 379.0	231 981.2
J	27	28	29
$\tilde{v}/m^{-1}$	240 584.4	249 188.5	257 793.5

Evaluate the rotational constant of the molecule. Also, estimate the value of J for the most highly populated rotational energy level at 298 K and at 100 K.

16.12 The vibrational energy levels of NaI lie at the wavenumbers 142.81, 427.31, 710.31, and 991.81 cm^{-1}. Show that they fit the expression $(v + \frac{1}{2})\tilde{v} - (v + \frac{1}{2})^2 x_e \tilde{v}$, and deduce the force constant, zero-point energy, and dissociation energy of the molecule.

16.13 Predict the shape of the nitronium ion, NO_2^+, from its Lewis structure and the VSEPR model. It has one Raman active vibrational mode at 1400 cm^{-1}, two strong IR active modes at 2360 and 540 cm^{-1}, and one weak IR mode at 3735 cm^{-1}. Are these data consistent with the predicted shape of the molecule? Assign the vibrational wavenumbers to the modes from which they arise.

16.14 At low resolution, the strongest absorption band in the infrared absorption spectrum of $^{12}C^{16}O$ is centred at 2150 cm^{-1}. Upon closer examination at higher resolution, this band is observed to be split into two sets of closely spaced peaks, one on each side of the centre of the spectrum at 2143.26 cm^{-1}. The separation between the peaks immediately to the right and left of the centre is 7.655 cm^{-1}. Make the harmonic oscillator and rigid rotor approximations and calculate from these data: (a) the vibrational wavenumber of a CO molecule, (b) its molar zero-point vibrational energy, (c) the force constant of the CO bond, (d) the rotational constant B, and (e) the bond length of CO.

---

† Problems denoted with the symbol ‡ were supplied by Charles Trapp, Carmen Giunta, and Marshall Cady.

16.15 The HCl molecule is quite well described by the Morse potential with $D_e = 5.33$ eV, $\tilde{v} = 2989.7$ cm^{-1}, and $x_e\tilde{v} = 52.05$ cm^{-1}. Assuming that the potential is unchanged on deuteration, predict the dissociation energies (D_0) of (a) HCl, (b) DCl.

16.16 The Morse potential (eqn 16.61) is very useful as a simple representation of the actual molecular potential energy. When RbH was studied, it was found that $\tilde{v} = 936.8$ cm^{-1} and $x_e\tilde{v} = 14.15$ cm^{-1}. Plot the potential energy curve from 50 pm to 800 pm around $R_e = 236.7$ pm. Then go on to explore how the rotation of a molecule may weaken its bond by allowing for the kinetic energy of rotation of a molecule and plotting $V^* = V + hcBJ(J+1)$ with $B = \hbar/4\pi c\mu R^2$. Plot these curves on the same diagram for $J = 40$, 80, and 100, and observe how the dissociation energy is affected by the rotation. (Taking $B = 3.020$ cm^{-1} at the equilibrium bond length will greatly simplify the calculation.)

16.17‡ F. Luo, G.C. McBane, G. Kim, C.F. Giese, and W.R. Gentry (*J. Chem. Phys.* **98**, 3564 (1993)) reported experimental observation of the He$_2$ complex, a species that had escaped detection for a long time. The fact that the observation required temperatures in the neighbourhood of 1 mK is consistent with computational studies that suggest that hcD_e for He$_2$ is about 1.51×10^{-23} J, hcD_0 about 2×10^{-26} J, and R_e about 297 pm. (a) Estimate the fundamental vibrational wavenumber, force constant, moment of inertia, and rotational constant based on the harmonic oscillator and rigid-rotor approximations. (b) Such a weakly bound complex is hardly likely to be rigid. Estimate the vibrational wavenumber and anharmonicity constant based on the Morse potential.

16.18‡ High-resolution absorption lines of three infrared vibrational bands of carbon monoxide are summarized in the following table. (a) Make the J and m assignments for each line, using J to represent the initial rotational state and defining $m = -J$ for P branches and $m = J + 1$ for R branches. Derive an equation for the dependence of the spectral lines upon m including terms for anharmonicity, centrifugal distortion, and rotation–vibration coupling. The rotation–vibration perturbation of the energy of the state (v, J) is equal to $-a(v + \frac{1}{2})J(J+1)$, where a is the molecular rotation–vibration coupling constant. (b) Determine all molecular constants associated with the absorption line by performing an appropriate regression analysis of the data in the table. (c) Determine each of the following: the moment of inertia and bond length defined by the harmonic oscillator and rigid rotor models, I_e and R_e; the moments of inertia and bond lengths for the $v = 0, 1, 2, 3$ vibration states (define these by analogy to the harmonic oscillator and rigid rotor models), the depth of the Morse potential, D_e (the 'spectroscopic dissociation energy'), and D_0 (the bond dissociation energy).

Wavenumbers ($\tilde{v}$/cm^{-1}) of the infrared absorption lines for vibrational bands of carbon monoxide. From N. Mina-Camilde, C.I. Manzariares, and J.F. Caballero, *J. Chem. Educ.*, **73**, 804 (1996)

1 ← 0		2 ← 0		3 ← 0	
2059.6	2146.9	4169.8	4263.2	6253.1	6353.8
2063.9	2150.7	4174.6	4267.2	6259.3	6357.2
2068.8	2154.1	4180.0	4270.6	6264.6	6361.0
2073.1	2158.0	4184.8	4274.4	6270.4	6364.4
2077.4	2161.8	4190.1	4277.8	6276.2	6367.8
2081.8	2165.2	4194.9	4281.2	6281.5	6370.7
2086.1	2169.0	4199.7	4284.6	6286.8	6374.1
2090.5	2172.4	4204.1	4287.9	6291.6	6377.0
2094.3	2175.8	4208.9	4290.8	6296.9	6379.9
2098.7	2179.2	4213.7	4294.2	6303.7	6382.7
2103.0	2183.0	4218.0	4297.1	6307.1	6385.2
2106.8	2186.4	4222.4	4300.5	6311.4	6388.0
2111.2	2189.8	4226.7	4303.4	6316.2	6390.5
2115.5	2193.1	4231.1	4306.6	6321.0	6392.9
2119.4	2196.5	4235.4	4308.7	6325.4	6394.8
2123.2	2199.4	4239.7	4311.6	6329.7	6397.2
2127.1	2202.8	4244.1	4314.0	6334.0	6399.1
2131.4	2206.2	4247.9	4316.9	6337.9	6401.1
2135.3	2209.1	4251.8	4319.3	6342.2	6403.0
2139.2	2212.4	4255.6	4321.7	6346.1	6404.6

16.19 As mentioned in Section 16.15, the semi-empirical and *ab initio* methods discussed in Chapter 14 can be used to estimate the force field of a molecule. The molecule's vibrational spectrum can be simulated, and it is then possible to determine the correspondence between a vibrational frequency and the atomic displacements that give rise to a normal mode. (a) Using molecular modelling software[10] and the computational method of your choice (semi-empirical or *ab initio* methods), calculate the fundamental vibrational wavenumbers and visualize the vibrational normal modes of SO$_2$ in the gas phase. (b) The experimental values of the fundamental vibrational wavenumbers of SO$_2$ in the gas phase are 525 cm^{-1}, 1151 cm^{-1}, and 1336 cm^{-1}. Compare the calculated and experimental values. Even if agreement is poor, is it possible to establish a correlation between an experimental value of the vibrational wavenumber with a specific vibrational normal mode?

16.20 Consider the molecule CH$_3$Cl. (a) To what point group does the molecule belong? (b) How many normal modes of vibration does the molecule have? (c) What are the symmetries of the normal modes of vibration for this molecule? (d) Which of the vibrational modes of this molecule are infrared active? (e) Which of the vibrational modes of this molecule are Raman active?

16.21 Suppose that three conformations are proposed for the nonlinear molecule H$_2$O$_2$ (**2**, **3**, and **4**). The infrared absorption spectrum of gaseous H$_2$O$_2$ has bands at 870, 1370, 2869, and 3417 cm^{-1}. The Raman spectrum of the same sample has bands at 877, 1408, 1435, and 3407 cm^{-1}. All bands correspond to fundamental vibrational wavenumbers and you may assume that: (i) the 870 and 877 cm^{-1} bands arise from the same normal mode, and (ii) the 3417 and 3407 cm^{-1} bands arise from the same normal mode. (a) If H$_2$O$_2$ were linear, how many normal modes of vibration would it have? (b) Give the symmetry point group of each of the three proposed conformations of nonlinear H$_2$O$_2$. (c) Determine which of the

2   **3**   **4**

10. The web site contains links to molecular modelling freeware and to other sites where you may perform molecular orbital calculations directly from your web browser.

proposed conformations is inconsistent with the spectroscopic data. Explain your reasoning.

Theoretical problems

16.22 Show that the moment of inertia of a diatomic molecule composed of atoms of masses m_A and m_B and bond length R is equal to $m_{eff}R^2$, where $m_{eff} = m_A m_B/(m_A + m_B)$.

16.23 Derive eqn 16.41 for the centrifugal distortion constant D_J of a diatomic molecule of effective mass m_{eff}. Treat the bond as an elastic spring with force constant k and equilibrium length r_e that is subjected to a centrifugal distortion to a new length r_c. Begin the derivation by letting the particles experience a restoring force of magnitude $k(r_c - r_e)$ that is countered perfectly by a centrifugal force $m_{eff}\omega^2 r_c$, where ω is the angular velocity of the rotating molecule. Then introduce quantum mechanical effects by writing the angular momentum as $\{J(J+1)\}^{1/2}\hbar$. Finally, write an expression for the energy of the rotating molecule, compare it with eqn 16.40, and write an expression for D_J. For help with the classical aspects of this derivation, see *Further information 2*.

16.24 In the group theoretical language developed in Chapter 15, a spherical rotor is a molecule that belongs to a cubic or icosahedral point group, a symmetric rotor is a molecule with at least a threefold axis of symmetry, and an asymmetric rotor is a molecule without a threefold (or higher) axis. Linear molecules are linear rotors. Classify each of the following molecules as a spherical, symmetric, linear, or asymmetric rotor and justify your answers with group theoretical arguments: (a) CH_4, (b) CH_3CN, (c) CO_2, (d) CH_3OH, (e) benzene, (f) pyridine.

16.25 Derive an expression for the value of J corresponding to the most highly populated rotational energy level of a diatomic rotor at a temperature T remembering that the degeneracy of each level is $2J + 1$. Evaluate the expression for ICl (for which $B = 0.1142$ cm^{-1}) at 25°C. Repeat the problem for the most highly populated level of a spherical rotor, taking note of the fact that each level is $(2J + 1)^2$-fold degenerate. Evaluate the expression for CH_4 (for which $B = 5.24$ cm^{-1}) at 25°C.

16.26 The moments of inertia of the linear mercury(II) halides are very large, so the O and S branches of their vibrational Raman spectra show little rotational structure. Nevertheless, the peaks of both branches can be identified and have been used to measure the rotational constants of the molecules (R.J.H. Clark and D.M. Rippon, *J. Chem. Soc. Faraday Soc. II*, **69**, 1496 (1973)). Show, from a knowledge of the value of J corresponding to the intensity maximum that the separation of the peaks of the O and S branches is given by the Placzek–Teller relation $\delta\tilde{\nu} = (32BkT/hc)^{1/2}$. The following widths were obtained at the temperatures stated:

	HgCl$_2$	HgBr$_2$	HgI$_2$
θ/°C	282	292	292
$\delta\tilde{\nu}$/cm^{-1}	23.8	15.2	11.4

Calculate the bond lengths in the three molecules.

16.27‡ Use quantum mechanical concepts to explain the origin both of ammonia microwave absorptions at 0.8 cm^{-1} and 36 cm^{-1} and an infrared absorption at 1000 cm^{-1}. All are associated with the umbrella-like inversion of ammonia (see Fig. 16.57); the absorption at 1000 cm^{-1} is the lowest-energy IR transition associated with the inversion. (a) Prove that

$$C = 2B^{1/2} \qquad D = \frac{s_e}{\operatorname{arccosh} B^{1/4}}$$

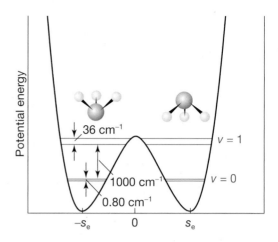

Fig. 16.57

for an assumed inversion potential of the form

$$V(s) = A\{1 + B\operatorname{sech}^4(s/D) - C\operatorname{sech}^2(s/D)\}$$

where s is the perpendicular distance of the nitrogen from the plane of the three hydrogens. At $s = s_e$ the potential has a minimum value equal to zero and the parameters A, B, C, and D are all positive. (b) Use the absorption lines and $s_e = 38.1$ pm to determine both the potential parameters and the value of the inversion potential barrier height, ε. Simplify the computations by assuming that the inversion mode vibrational wavefunctions are adequately described by harmonic oscillator wavefunctions. Also assume that the distances between hydrogen atoms remain constant during inversion. This approximation simplifies the effective mass to $m_{eff} = 3m_H m_N/(3m_H + m_N)$.

16.28 Confirm that a Morse oscillator has a finite number of bound states, the states with $V < hcD_e$. Determine the value of v_{max} for the highest bound state.

Applications: to biology, environmental science, and astrophysics

16.29 The protein haemerythrin is responsible for binding and carrying O_2 in some invertebrates. Each protein molecule has two Fe^{2+} ions that are in very close proximity and work together to bind one molecule of O_2. The Fe_2O_2 group of oxygenated haemerythrin is coloured and has an electronic absorption band at 500 nm. The resonance Raman spectrum of oxygenated haemerythrin obtained with laser excitation at 500 nm has a band at 844 cm^{-1} that has been attributed to the O—O stretching mode of bound $^{16}O_2$. (a) Why is resonance Raman spectroscopy and not infrared spectroscopy the method of choice for the study of the binding of O_2 to haemerythrin? (b) Proof that the 844 cm^{-1} band arises from a bound O_2 species may be obtained by conducting experiments on samples of haemerythrin that have been mixed with $^{18}O_2$, instead of $^{16}O_2$. Predict the fundamental vibrational wavenumber of the $^{18}O—^{18}O$ stretching mode in a sample of haemerythrin that has been treated with $^{18}O_2$. (c) The fundamental vibrational wavenumbers for the O—O stretching modes of O_2, O_2^- (superoxide anion), and O_2^{2-} (peroxide anion) are 1555, 1107, and 878 cm^{-1}, respectively. Explain this trend in terms of the electronic structures of O_2, O_2^-, and O_2^{2-}. (*Hint.* Review Section 14.5.) What are the bond orders of O_2, O_2^-, and O_2^{2-}? (d) Based on the data given above, which of the following species best describes the Fe_2O_2 group of haemerythrin: $Fe_2^{2+}O_2$,

5 **6**

7 **8**

$Fe^{2+}Fe^{3+}O_2^-$, or $Fe_2^{3+}O_2^{2-}$? Explain your reasoning. (d) The resonance Raman spectrum of haemerythrin mixed with $^{16}O^{18}O$ has two bands that can be attributed to the O—O stretching mode of bound oxygen. Discuss how this observation may be used to exclude one or more of the four proposed schemes (**5–8**) for binding of O_2 to the Fe_2 site of haemerythrin.

16.30‡ In Problem 11.27, we examined the origins of the greenhouse effect by considering an atmospheric model consisting of atmospheric nitrogen and oxygen only. Now, consider an atmospheric model consisting of nitrogen, oxygen, some water vapour, and some carbon dioxide. (a) Why is it that water and carbon dioxide are able to absorb some of the Earth's black-body radiation whereas nitrogen and oxygen cannot? Which vibrational modes are responsible for this absorption? (b) Water vapour shows strong absorption between 1300 cm^{-1} and 1900 cm^{-1} and also between 3550 cm^{-1} and 3900 cm^{-1}. Carbon dioxide shows strong absorption between 500 cm^{-1} and 725 cm^{-1} and also between 2250 cm^{-1} and 2400 cm^{-1}. Why are these bands so broad? (c) Assume that these gases absorb all radiation falling within these bands and calculate the average surface temperature predicted by this atmospheric model. What percentage of the greenhouse effect is explained by the presence of atmospheric water and carbon dioxide?

16.31‡ A mixture of carbon dioxide (2.1 per cent) and helium, at 1.00 bar and 298 K in a gas cell of length 10 cm, has an IR absorption band centred at 2349 cm^{-1} with absorbances, $A(\tilde{v})$, described by:

$$A(\tilde{v}) = \frac{a_1}{1 + a_2(\tilde{v} - a_3)^2} + \frac{a_4}{1 + a_5(\tilde{v} - a_6)^2}$$

where the coefficients are $a_1 = 0.932$, $a_2 = 0.005\,050$ cm^2, $a_3 = 2333$ cm^{-1}, $a_4 = 1.504$, $a_5 = 0.015\,21$ cm^2, $a_6 = 2362$ cm^{-1}. (a) Draw graphs of $A(\tilde{v})$ and $\varepsilon(\tilde{v})$. What is the origin of both the band and the band width? What are the allowed and forbidden transitions of this band? (b) Calculate the transition wavenumbers and absorbances of

the band with a simple harmonic oscillator–rigid rotor model and compare the result with the experimental spectra. The CO bond length is 116.2 pm. (c) Within what height, h, is basically all the IR emission from the Earth in this band absorbed by atmospheric carbon dioxide? The mole fraction of CO_2 in the atmosphere is 3.3×10^{-4} and $T/K = 288 - 0.0065(h/m)$ below 10 km. Draw a surface plot of the atmospheric transmittance of the band as a function of both height and wavenumber.

16.32 In Problem 13.28, we saw that Doppler shifts of atomic spectral lines are used to estimate the speed of recession or approach of a star. From the discussion in Section 16.3a, it is easy to see that Doppler broadening of an atomic spectral line depends on the temperature of the star that emits the radiation. A spectral line of $^{48}Ti^{8+}$ (of mass 47.95 u) in a distant star was found to be shifted from 654.2 nm to 706.5 nm and to be broadened to 61.8 pm. What is the speed of recession and the surface temperature of the star?

16.33‡ There is a gaseous interstellar cloud in the constellation Ophiuchus which is illuminated from behind by the star ζ-Ophiuci. Analysis of the Fraunhofer electronic-vibrational-rotational absorption lines obtained by H.S. Uhler and R.A. Patterson (*Astrophys. J.* **42**, 434 (1915)) shows the presence of CN molecules in the interstellar medium. A strong absorption line in the ultraviolet region at $\lambda = 387.5$ nm was observed corresponding to the transition $J = 0-1$. Unexpectedly, a second strong absorption line with 25 per cent of the intensity of the first was found at a slightly longer wavelength ($\Delta\lambda = 0.061$ nm) corresponding to the transition $J = 1-1$ (here allowed). Calculate the temperature of the CN molecules. Gerhard Herzberg, who was later to receive the Nobel Prize for his contributions to spectroscopy, calculated the temperature as 2.3 K. Although puzzled by this result, he did not realize its full significance. If he had, his prize might have been for the discovery of the cosmic microwave background radiation.

16.34‡ The H_3^+ ion has recently been found in the interstellar medium and in the atmospheres of Jupiter, Saturn, and Uranus. The rotational energy levels of H_3^+, an oblate symmetric rotor, are given by eqn 36, with C replacing A, when centrifugal distortion and other complications are ignored. Experimental values for vibrational-rotational constants are $\tilde{v}(E') = 2521.6$ cm^{-1}, $B = 43.55$ cm^{-1}, and $C = 20.71$ cm^{-1}. (a) Show that, for a nonlinear planar molecule (such as H_3^+), $I_C = 2I_B$. The rather large discrepancy with the experimental values is due to factors ignored in eqn 36. (b) Calculate an approximate value of the H–H bond length in H_3^+. (c) The value of R_e obtained from the best quantum mechanical calculations by J.B. Anderson (*J. Chem. Phys.* **96**, 3702 (1991)) is 87.32 pm. Use this result to calculate the values of the rotational constants B and C. (d) Assuming that the geometry and force constants are the same in D_3^+ and H_3^+, calculate the spectroscopic constants of D_3^+. The molecular ion D_3^+ was first produced by J.T. Shy, J.W. Farley, W.E. Lamb Jr, and W.H. Wing (*Phys. Rev. Lett* **45**, 535 (1980)) who observed the $v_2(E')$ band in the infrared.

17

Spectroscopy 2: electronic transitions

The characteristics of electronic transitions

17.1 The electronic spectra of diatomic molecules

17.2 The electronic spectra of polyatomic molecules

The fates of electronically excited states

17.3 Fluorescence and phosphorescence

17.4 Dissociation and predissociation

Lasers

17.5 General principles of laser action

17.6 Practical lasers

17.7 Applications of lasers in chemistry

Photoelectron spectroscopy

17.8 The technique

17.9 Ultraviolet photoelectron spectroscopy

17.10 X-ray photoelectron spectroscopy

Checklist of key ideas

Further reading

Exercises

Problems

Simple analytical expressions for the electronic energy levels of molecules cannot be given, so this chapter concentrates on the qualitative features of electronic transitions. A common theme throughout the chapter is that electronic transitions occur within a stationary nuclear framework. We pay particular attention to spontaneous radiative decay processes, which include fluorescence and phosphorescence. A specially important example of stimulated radiative decay is that responsible for the action of lasers, and we see how this stimulated emission may be achieved and employed. An extreme case of photon absorption is photoionization, in which an electron is expelled completely from a molecule. Photoionization spectroscopy is used to build up detailed pictures of orbital energies and the role of particular electrons in bonding. Hence it is an experimental technique for exploring the concepts encountered in Chapter 14.

The energies needed to change the electron distributions of molecules are of the order of several electronvolts (1 eV is equivalent to about 8000 cm^{-1} or 100 kJ mol^{-1}). Consequently, the photons emitted or absorbed when such changes occur lie in the visible and ultraviolet regions of the spectrum (Table 17.1). In some cases the relocation of electrons may be so extensive that it results in ionization or dissociation of the molecule. Photoionization techniques are not confined to single molecules: one of their major applications is to the study of the composition, structure, and properties of surfaces. We lay the groundwork for that kind of application in this chapter but the bulk of the discussion will be found in Chapter 28.

Table 17.1* Colour, frequency, and energy of light

Colour	λ/nm	ν/(10^{14} Hz)	E/(kJ mol^{-1})
Infrared	>1000	<3.0	<120
Red	700	4.3	170
Yellow	580	5.2	210
Blue	470	6.4	250
Ultraviolet	<300	>10	>400

* More values are given in the *Data section*.

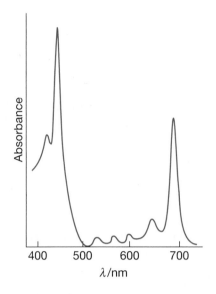

17.1 The absorption spectrum of chlorophyll in the visible region. Note that it absorbs in the red and blue regions, and that green light is not absorbed appreciably.

One of the revolutions that has occurred in physical chemistry in recent years is the application of lasers to spectroscopy and kinetics. Lasers have brought unprecedented precision to spectroscopy, made Raman spectroscopy a widely useful technique, and have made it possible to study chemical reactions on a femtosecond timescale. We shall see the principles of their action in this chapter.

The characteristics of electronic transitions

The nuclei in a molecule are subjected to different forces after an electronic transition has occurred, and the molecule may respond by starting to vibrate. The resulting vibrational structure of electronic transitions can be resolved for gaseous samples, but in a liquid or solid the lines usually merge together and result in a broad, almost featureless band (Fig. 17.1). Superimposed on the vibrational transitions that accompany the electronic transition of a molecule in the gas phase is an additional branch structure that arises from rotational transitions. The electronic spectra of gaseous samples are therefore very complicated, but rich in information.

17.1 The electronic spectra of diatomic molecules

We examine some general features of electronic transitions by using diatomic molecules as examples. We begin by labelling molecular orbitals according to their symmetry and assigning term symbols to ground and excited electronic states. Then we use the symmetry designations to formulate selection rules. Finally, we examine the origin of **vibronic transitions**, transitions that account for the vibrational structure in electronic spectra.

(a) Parity

The molecular orbitals of homonuclear diatomic molecules are labelled with a subscript g or u that specifies their **parity**, their behaviour under inversion. To decide on the parity, consider any point in a homonuclear diatomic molecule, and note the sign of the orbital at that point. Then imagine travelling on a straight line through the centre of the molecule to a point the same distance out on the other side; this process is called **inversion** and the central point is the **centre of inversion** (Fig. 17.2). If the orbital has the same sign, it has 'even parity' and is denoted g (from *gerade*, the German word for even). If the orbital has opposite sign, then it has 'odd parity' and is denoted u (from *ungerade*, uneven). The parity designation applies only to homonuclear diatomic molecules, because heteronuclear diatomic molecules (such as HCl) do not have a centre of inversion.

We see from Fig. 17.2 that a bonding σ orbital has even parity; so we write it σ_g; an antibonding σ orbital has odd parity and is written σ_u. A bonding π orbital has odd parity and is denoted π_u and an antibonding π orbital has even parity, denoted π_g.[1]

1. For simplicity of comparing homonuclear and heteronuclear molecules, we ignore the parity subscripts when numbering orbitals; however, a more formal convention is to number the g and u orbitals separately.

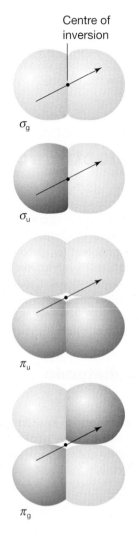

Centre of
inversion

σ_g

σ_u

π_u

π_g

17.2 The parity of an orbital is even (g) if its wavefunction is unchanged under inversion through the centre of symmetry of the molecule, but odd (u) if the wavefunction changes sign. Heteronuclear diatomic molecules do not have a centre of inversion, so for them the g, u classification is irrelevant.

(b) Term symbols

The term symbols of linear molecules (the analogues of the symbols ^2P, etc. for atoms) are constructed in a similar way to those for atoms, but now we must pay attention to the component of total orbital angular momentum about the internuclear axis, $\Lambda\hbar$. The value of $|\Lambda|$ is denoted by the symbols Σ, Π, Δ, ... for $|\Lambda| = 0, 1, 2, ...$, respectively. These labels are the analogues of S, P, D, ... for atoms. The value of Λ is the sum of the values of λ for the individual electrons in a molecule.[2] A single electron in a σ orbital has $\lambda = 0$: the orbital is cylindrically symmetrical and has no angular nodes when viewed along the internuclear axis. Therefore, if that is the only electron present, $\Lambda = 0$. The term symbol for H_2^+ is therefore Σ.

As in atoms, we use a superscript with the value of $2S + 1$ to denote the multiplicity of the term. The component of total spin angular momentum about the internuclear axis is denoted Σ, where $\Sigma = S, S - 1, S - 2, ... , -S$. For H_2^+, because there is only one electron, $S = s = \frac{1}{2}$ ($\Sigma = \pm\frac{1}{2}$) and the term symbol is $^2\Sigma$, a doublet term.[3] The overall parity of the term is added as a right subscript. For H_2^+, the parity of the only occupied orbital is g, so the term itself is also g, and in full dress is $^2\Sigma_g$. If there are several electrons, the overall parity is calculated by using

$$g \times g = g \qquad u \times u = g \qquad u \times g = u \qquad (17.1)$$

These rules are generated by interpreting g as +1 and u as −1. The term symbol for the ground state of any closed-shell homonuclear diatomic molecule is $^1\Sigma_g$ because the spin is zero (a singlet term in which all electrons paired), there is no orbital angular momentum from a closed shell, and the overall parity is g.

A π electron in a diatomic molecule has one unit of orbital angular momentum about the internuclear axis ($\lambda = \pm 1$) and, if it is the only electron outside a closed shell, gives rise to a Π term. If there are two π electrons (as in the ground state of O_2, with configuration $\pi_u^2\pi_u^2\pi_g^1\pi_g^1$) then the term symbol may be either Σ (if the electrons are travelling in opposite directions, which is the case if they occupy different π orbitals, one with $\lambda = +1$ and the other with $\lambda = -1$) or Δ (if they are travelling in the same direction, which is the case if they occupy the same π orbital, both $\lambda = +1$, for instance). For O_2, the two π electrons occupy different orbitals with parallel spins (a triplet term), so the ground term is $^3\Sigma$. The overall parity of the molecule is

$$(\text{closed shell}) \times g \times g = g$$

The term symbol is therefore $^3\Sigma_g$.

For Σ terms, a $\pm$ superscript denotes the behaviour of the molecular wavefunction under reflection in a plane containing the nuclei (Fig. 17.3). If, for convenience, we think of O_2 as having one electron in $2\pi_x$, which changes sign under reflection in the yz-plane, and the other electron in $2\pi_y$, which does not change sign under reflection in the same plane, the overall reflection symmetry is

$$(\text{closed shell}) \times (+) \times (-) = (-)$$

and the full term symbol of the ground electronic state of O_2 is $^3\Sigma_g^-$.

The term symbols of excited electronic states are constructed in a similar way. For example, the term symbol for the excited state of O_2 formed by placing two electrons in a $2\pi_x$ (or in a $2\pi_y$) orbital is $^1\Delta_g$ because $|\Lambda| = 2$ (two electrons in the same π orbital), the spin is zero (all electrons are paired), and the overall parity is (closed shell) $\times$ g $\times$ g = g. Table 17.2 and Fig. 17.4 summarize the configurations, term symbols, and energies of the ground and some excited states of O_2.

---

2. Recall from footnote 10 in Section 14.5d that $\lambda\hbar$ is the component of orbital angular momentum on the internuclear axis.

3. It is important to distinguish between the term symbol Σ and the quantum number Σ.

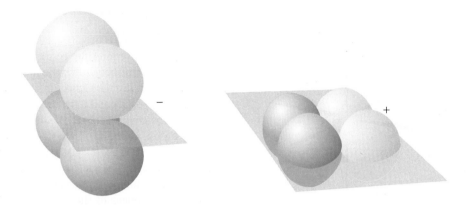

17.3 The + or − on a term symbol refers to the overall symmetry of a configuration under reflection in a plane containing the two nuclei.

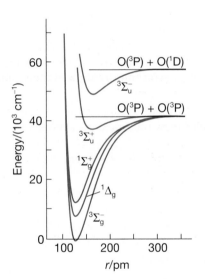

17.4 The electronic states of dioxygen.

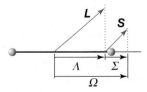

17.5 The coupling of spin and orbital angular momenta in a linear molecule: only the components along the internuclear axis are conserved.

Table 17.2 Properties of O_2 in its lower electronic states*

Configuration†	Term	Relative energy/cm^{-1}	$\tilde{\nu}$/cm^{-1}	R_e/pm
$\pi_u^2\pi_u^2\pi_g^1\pi_g^1$	$^3\Sigma_g^-$	0	1580	120.74
$\pi_u^2\pi_u^2\pi_g^2\pi_g^0$	$^1\Delta_g$	7 882.39	1509	121.55
$\pi_u^2\pi_u^2\pi_g^1\pi_g^1$	$^1\Sigma_g^+$	13 120.9	1433	122.68
$\pi_u^2\pi_u^1\pi_g^2\pi_g^1$	$^3\Sigma_u^+$	35 713	819	142
$\pi_u^2\pi_u^1\pi_g^2\pi_g^1$	$^3\Sigma_u^-$	49 363	700	160

* Adapted from G. Herzberg, *Spectra of diatomic molecules*, Van Nostrand, New York (1950) and D.C. Harris and M.D. Bertolucci, *Symmetry and spectroscopy: an introduction to vibrational and electronic spectroscopy*, Dover, New York (1989).

† The configuration $\pi_u^2\pi_u^1\pi_g^2\pi_g^1$ should also give rise to a $^3\Delta_u$ term, but electronic transitions to or from this state have not been observed.

(c) Selection rules

A number of selection rules govern which transitions between allowed states of a molecule will be observed in its electronic spectrum. The selection rule concerned with changes in angular momentum are

$$\Delta\Lambda = 0, \pm 1 \qquad \Delta S = 0 \qquad \Delta\Sigma = 0 \qquad \Delta\Omega = 0, \pm 1$$

where $\Omega = \Lambda + \Sigma$ is the quantum number for the total angular momentum (orbital and spin) along the internuclear axis (Fig. 17.5). As in atoms (Sections 13.3 and 13.9), the origins of these rules are conservation of angular momentum during a transition and the fact that a photon has a spin of 1.

There are two selection rules concerned with changes in symmetry. First, for Σ terms, only $\Sigma^+ \leftrightarrow \Sigma^+$ and $\Sigma^- \leftrightarrow \Sigma^-$ transitions are allowed. Second, the **Laporte selection rule** for centrosymmetric molecules (those with a centre of inversion) and atoms states that:

The only allowed transitions are transitions that are accompanied by a change of parity.

That is, u → g and g → u transitions are allowed, but g → g and u → u transitions are forbidden.

17.6 A *d–d* transition is parity-forbidden because it corresponds to a g–g transition. However, a vibration of the molecule can destroy the inversion symmetry of the molecule and the g, u classification no longer applies. The removal of the centre of symmetry gives rise to a vibronically allowed transition.

Justification 17.1

The last two selection rules result from the fact that the transition dipole moment in eqn 16.20 vanishes unless the integrand is invariant under all symmetry operations of the molecule. Hence, in a centrosymmetric molecule the integrand must have even (g) parity and in transitions involving Σ terms it must have + symmetry.

The three components of the dipole moment operator transform like x, y, and z, and are all u. Therefore, for a g → g transition, the overall parity of the transition dipole moment is g × u × g = u, so it must be zero. Likewise, for a u → u transition, the overall parity is u × u × u = u, so the transition dipole moment must also vanish. Hence, transitions without a change of parity are forbidden.

The z-component of the dipole moment operator has (+) symmetry. Therefore, for a (+) ↔ (−) transition, the overall symmetry of the transition dipole moment is (+) × (+) × (−) = (−), so it must be zero. Therefore, for Σ terms, $\Sigma^+ \leftrightarrow \Sigma^-$ transitions are not allowed.

A forbidden g → g transition can become allowed if the centre of symmetry is eliminated by an asymmetrical vibration, such as the one shown in Fig. 17.6. When the centre of symmetry is lost, g → g and u → u transitions are no longer parity-forbidden and become weakly allowed. A transition that derives its intensity from an asymmetrical vibration of a molecule is called a **vibronically allowed transition**.

Self-test 17.1 Which of the following electronic transitions are allowed in O_2: $^3\Sigma_g^- \leftrightarrow {}^1\Delta_g$, $^3\Sigma_g^- \leftrightarrow {}^1\Sigma_g^+$, $^3\Sigma_g^- \leftrightarrow {}^3\Delta_u$, $^3\Sigma_g^- \leftrightarrow {}^3\Sigma_u^+$, $^3\Sigma_g^- \leftrightarrow {}^3\Sigma_u^-$?

$$[^3\Sigma_g^- \leftrightarrow {}^3\Delta_u \text{ and } ^3\Sigma_g^- \leftrightarrow {}^3\Sigma_u^-]$$

(d) Vibrational structure

To examine the origin of vibrational structure in electronic spectra of molecules, we must consider the nature of nuclear motions during an electronic transition. This can be done by applying the **Franck–Condon principle**:

> **Because the nuclei are so much more massive than the electrons, an electronic transition takes place very much faster than the nuclei can respond.**

As a result of the transition, electron density is rapidly built up in new regions of the molecule and removed from others, and the initially stationary nuclei suddenly experience a new force field. They respond to the new force by beginning to vibrate, and (in classical terms) swing backwards and forwards from their original separation, which was maintained during the rapid electronic excitation. The stationary equilibrium separation of the nuclei in the initial electronic state therefore becomes a stationary turning point in the final electronic state (Fig. 17.7).

The quantum mechanical version of the Franck–Condon principle refines this picture. Before the absorption, the molecule is in the lowest vibrational state of its lowest electronic state (Fig. 17.8); the most probable location of the nuclei is at their equilibrium separation, R_e. The electronic transition is most likely to take place when the nuclei have this separation. When the transition occurs, the molecule is excited to the state represented by the upper curve. According to the Franck–Condon principle, the nuclear framework remains constant during this excitation, so we may imagine the transition as being up the vertical line in Fig. 17.7. The vertical line is the origin of the expression **vertical transition**, which is used to denote an electronic transition that occurs without change of nuclear geometry.

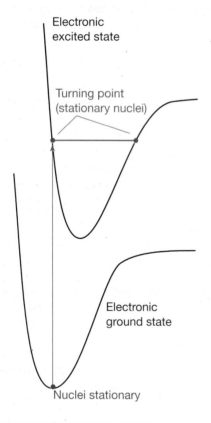

17.7 According to the Franck–Condon principle, the most intense vibronic transition is from the ground vibrational state to the vibrational state lying vertically above it. Transitions to other vibrational levels also occur, but with lower intensity.

The vertical transition cuts through several vibrational levels of the upper electronic state. The level marked * is the one in which the nuclei are most probably at the same initial separation R_e (because the vibrational wavefunction has maximum amplitude there), so this vibrational state is the most probable state for the termination of the transition. However, it is not the only accessible vibrational state because several nearby states have an appreciable probability of the nuclei being at the separation R_e. Therefore, transitions occur to all the vibrational states in this region, but most intensely to the state with a vibrational wavefunction that peaks most strongly near R_e.

The vibrational structure of the spectrum depends on the relative horizontal position of the two potential energy curves, and a long **vibrational progression**, a lot of vibrational structure, is stimulated if the upper potential energy curve is appreciably displaced horizontally from the lower. The upper curve is usually displaced to greater equilibrium bond lengths because electronically excited states usually have more antibonding character than electronic ground states.

The separation of the vibrational lines of an electronic absorption spectrum depends on the vibrational energies of the *upper* electronic state. Hence, electronic absorption spectra may be used to assess the force fields and dissociation energies of electronically excited molecules (for example, by using a Birge–Sponer plot, as in Problem 17.2).

(e) Franck–Condon factors

The quantitative form of the Franck–Condon principle is derived from the expression for the transition dipole moment, $\mu_{fi} = \langle f|\mu|i\rangle$. The dipole moment operator is a sum over all nuclei and electrons in the molecule:

$$\mu = -e \sum_i \mathbf{r}_i + e \sum_I Z_I \mathbf{R}_I \tag{17.2}$$

where the vectors are the distances from the centre of charge of the molecule. The intensity of the transition is proportional to the square modulus, $|\mu_{fi}|^2$, of the magnitude of the transition dipole moment (eqn 16.20), and we show in the *Justification* below that this intensity is proportional to the square modulus of the overlap integral, $S(v_f, v_i)$, between the vibrational states of the initial and final electronic states. This overlap integral is a measure of the match between the vibrational wavefunctions in the upper and lower electronic states: $S = 1$ for a perfect match and $S = 0$ when there is no similarity.

Justification 17.2

The overall state of the molecule consists of an electronic part, $|\varepsilon\rangle$, and a vibrational part, $|v\rangle$. Therefore, within the Born-Oppenheimer approximation, the transition dipole moment factorizes as follows:

$$\mu_{fi} = \left\langle \varepsilon_f v_f \left| \left\{ -e \sum_i \mathbf{r}_i + e \sum_I Z_I \mathbf{R}_I \right\} \right| \varepsilon_i v_i \right\rangle$$

$$= -e \sum_i \langle \varepsilon_f|\mathbf{r}_i|\varepsilon_i\rangle\langle v_f|v_i\rangle + e \sum_i Z_i \langle \varepsilon_f|\varepsilon_i\rangle\langle v_f|\mathbf{R}_I|v_i\rangle$$

The second term on the right of the last row is zero, because $\langle \varepsilon_f|\varepsilon_i\rangle = 0$ for two different electronic states (they are orthogonal). Therefore,

$$\mu_{fi} = -e \sum_i \langle \varepsilon_f|\mathbf{r}_i|\varepsilon_i\rangle\langle v_f|v_i\rangle = \mu_{\varepsilon_f,\varepsilon_i} S(v_f, v_i) \tag{17.3}$$

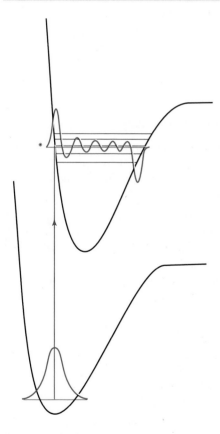

The web site contains links to on-line databases of electronic spectra of diatomic molecules.

17.8 In the quantum mechanical version of the Franck–Condon principle, the molecule undergoes a transition to the upper vibrational state that most closely resembles the vibrational wavefunction of the vibrational ground state of the lower electronic state. The two wavefunctions shown here have the greatest overlap integral of all the vibrational states of the upper electronic state and hence are most closely similar.

where

$$\boldsymbol{\mu}_{\varepsilon_f,\varepsilon_i} = -e \sum_i \langle \varepsilon_f | \boldsymbol{r}_i | \varepsilon_i \rangle \qquad S(v_f, v_i) = \langle v_f | v_i \rangle \tag{17.4}$$

The matrix element $\boldsymbol{\mu}_{\varepsilon_f,\varepsilon_i}$ is the transition dipole moment arising from the redistribution of electrons (and a measure of the 'kick' this redistribution gives to the electromagnetic field, and vice versa for absorption). The factor $S(v_f, v_i)$, is the overlap integral between the vibrational state $|v_i\rangle$ in the initial electronic state of the molecule, and the vibrational state $|v_f\rangle$ in the final electronic state of the molecule.

We saw in Section 12.10 that the transition intensity is proportional to the square of the magnitude of the transition dipole moment, so the intensity of an absorption is proportional to $|S(v_f, v_i)|^2$, which is known as the **Franck–Condon factor** for the transition. It follows that, the greater the overlap of the vibrational state wavefunction in the upper electronic state with the vibrational wavefunction in the lower electronic state, the greater the absorption intensity of that particular simultaneous electronic and vibrational transition. This conclusion is the basis of the illustration in Fig. 17.8, where we see that the vibrational wavefunction of the ground state has the greatest overlap with the vibrational states that have peaks at similar bond lengths in the upper electronic state.

Example 17.1 Calculating a Franck–Condon factor

Consider the transition from one electronic state to another, their bond lengths being R_e and R'_e and their force constants equal. Calculate the Franck–Condon factor for the 0–0 transition and show that the transition is most intense when the bond lengths are equal.

Method We need to calculate $S(0,0)$, the overlap integral of the two ground-state vibrational wavefunctions, and then take its square. The difference between harmonic and anharmonic vibrational wavefunctions is negligible for $v = 0$, so harmonic oscillator wavefunctions can be used (Table 12.1).

Answer We use the (real) wavefunctions

$$\psi_0 = \left(\frac{1}{\alpha\pi^{1/2}}\right)^{1/2} e^{-y^2/2} \qquad \psi'_0 = \left(\frac{1}{\alpha\pi^{1/2}}\right)^{1/2} e^{-y'^2/2}$$

where $y = (R - R_e)/\alpha$ and $y' = (R - R'_e)/\alpha$, with $\alpha = (\hbar^2/mk)^{1/4}$ (Section 12.5a). The overlap integral is

$$S(0,0) = \langle 0|0 \rangle = \int_{-\infty}^{\infty} \psi'_0 \psi_0 \, dR = \frac{1}{\pi^{1/2}} \int_{-\infty}^{\infty} e^{-(y^2+y'^2)/2} \, dy$$

We now write $\alpha z = R - \frac{1}{2}(R_e + R'_e)$, and manipulate this expression into

$$S(0,0) = \frac{1}{\pi^{1/2}} e^{-(R_e-R'_e)^2/4\alpha^2} \int_{-\infty}^{\infty} e^{-z^2} \, dz$$

The value of the integral is $\pi^{1/2}$. Therefore, the overlap integral is

$$S(0,0) = e^{-(R_e-R'_e)^2/4\alpha^2}$$

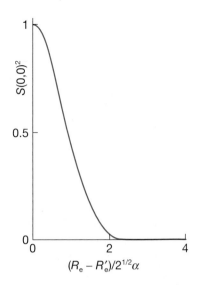

17.9 The Franck–Condon factor for the arrangement discussed in Example 17.1.

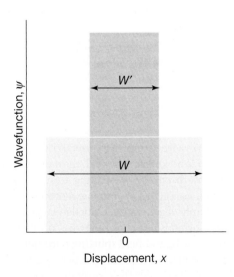

17.10 The model wavefunctions used in Self-test 17.2.

and the Franck–Condon factor is

$$S(0,0)^2 = e^{-(R_e - R'_e)^2/2\alpha^2}$$

This factor is equal to 1 when $R'_e = R_e$ and decreases as the equilibrium bond lengths diverge from each other (Fig. 17.9).

Comment For Br_2, $R_e = 228$ pm and there is an upper state with $R'_e = 266$ pm. Taking the vibrational wavenumber as 250 cm^{-1} gives $S(0,0)^2 = 5.1 \times 10^{-10}$, so the intensity of the 0-0 transition is only 5.1×10^{-10} what it would have been if the potential curves had been directly above each other.

Self-test 17.2 Suppose the vibrational wavefunctions can be approximated by rectangular functions of width W and W', centred on the equilibrium bond lengths (Fig. 17.10). Find the corresponding Franck–Condon factors when the centres are coincident and $W' < W$. $[S^2 = W'/W]$

(f) Rotational structure

Just as in vibrational spectroscopy, where a vibrational transition is accompanied by rotational excitation, so rotational transitions accompany the excitation of the vibrational excitation that accompanies electronic excitation. We therefore see P, Q, and R branches for each vibrational transition, and the electronic transition has a very rich structure. However, the principal difference is that electronic excitation can result in much larger changes in bond length than vibrational excitation causes alone, and the rotational branches have a more complex structure than in vibration-rotation spectra.

We suppose that the rotational constants of the electronic ground and excited states are B and B', respectively. The rotational energy levels of the initial and final states are

$$E(J) = hcBJ(J + 1) \qquad E(J') = hcB'J'(J' + 1)$$

and the rotational transitions occur at the following positions relative to the vibrational transition of wavenumber $\tilde{v}$ that they accompany:

P branch $(\Delta J = -1)$: $\tilde{v}_P(J) = \tilde{v} - (B' + B)J + (B' - B)J^2$

Q branch $(\Delta J = 0)$: $\tilde{v}_Q(J) = \tilde{v} + (B' - B)J(J + 1)$ (17.5)

R branch $(\Delta J = +1)$: $\tilde{v}_R(J) = \tilde{v} + (B' + B)(J + 1) + (B' - B)(J + 1)^2$

(These are the analogues of eqn 16.70.) First, suppose that the bond length in the electronically excited state is greater than that in the ground state; then $B' < B$ and $B' - B$ is negative. In this case the lines of the R branch converge with increasing J and, when J is such that $|B' - B|(J + 1) > B' + B$, the lines start to appear at decreasing wavenumber. That is, the R branch has a **band head** (Fig. 17.11a). When the bond is shorter in the excited state than in the ground state, $B' > B$ and $B' - B$ is positive. In this case, the lines of the P branch begin to converge and go through a head when J is such that $(B' - B)J > B' + B$ (Fig. 17.11b).

17.2 The electronic spectra of polyatomic molecules

The absorption of a photon can often be traced to the excitation of specific types of electrons or to electrons that belong to a small group of atoms in a polyatomic molecule. For example, when a carbonyl group (>C=O) is present, an absorption at

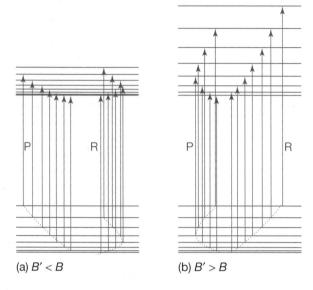

17.11 When the rotational constants of a diatomic molecule differ significantly in the initial and final states of an electronic transition, the P and R branches show a head. (a) The formation of a head in the R branch when $B' < B$; (b) the formation of a head in the P branch when $B' > B$.

(a) $B' < B$ (b) $B' > B$

Table 17.3* Absorption characteristics of some groups and molecules

Group	$\tilde{v}_{max}/cm^{-1}$	λ_{max}/nm	$\varepsilon_{max}/(L\ mol^{-1}\ cm^{-1})$
C=C ($\pi^* \leftarrow \pi$)	61 000	163	15 000
	57 300	174	5 500
C=O ($\pi^* \leftarrow n$)	37–35 000	270–290	10–20
H_2O ($\pi^* \leftarrow n$)	60 000	167	7 000

* More values are given in the *Data section*.

The web site contains links to databases of electronic spectra of polyatomic molecules.

about 290 nm is normally observed, although its precise location depends on the nature of the rest of the molecule. Groups with characteristic optical absorptions are called **chromophores** (from the Greek for 'colour bringer'), and their presence often accounts for the colours of substances (Table 17.3).

(a) *d–d* transitions

All five d orbitals of a given shell are degenerate in a free atom. In a d-metal complex, where the immediate environment of the atom is no longer spherical, the d orbitals are not all degenerate, and electrons can absorb energy by making transitions between them. In an octahedral complex, such as $[Ti(OH_2)_6]^{3+}$, the five d orbitals of the central atom are split into two sets (**1**), a triply degenerate set labelled t_{2g} and a doubly degenerate set labelled e_g. The three t_{2g} orbitals lie below the two e_g orbitals; the difference in energy is denoted Δ_O and called the **ligand-field splitting parameter** (the O denoting octahedral symmetry). The d orbitals also divide into two sets in a tetrahedral complex, but in this case the e orbitals lie below the t_2 orbitals and their separation is written Δ_T. Neither separation is large, so transitions between the two

1

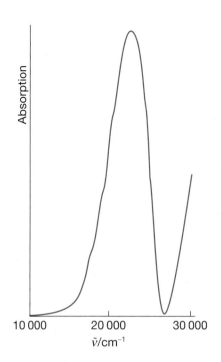

17.12 The electronic absorption spectrum of $[Ti(OH_2)_6]^{3+}$ in aqueous solution.

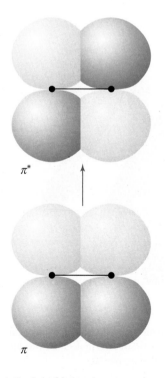

17.13 A C=C double bond acts as a chromophore. One of its important transitions is the $\pi^* \leftarrow \pi$ transition illustrated here, in which an electron is promoted from a π orbital to the corresponding antibonding orbital.

sets of orbitals typically occur in the visible region of the spectrum. The transitions are responsible for many of the colours that are so characteristic of d-metal complexes. As an example, the spectrum of $[Ti(OH_2)_6]^{3+}$ near 20 000 cm^{-1} (500 nm) is shown in Fig. 17.12, and can be ascribed to the promotion of its single d electron from a t_{2g} orbital to an e_g orbital. The wavenumber of the absorption maximum suggests that $\Delta_O \approx 20\ 000$ cm^{-1} for this complex, which corresponds to about 2.5 eV.

According to the Laporte rule (Section 17.1c), d–d transitions are parity-forbidden in octahedral complexes because they are g $\rightarrow$ g transitions (more specifically $e_g \leftarrow t_{2g}$ transitions). However, d–d transitions are vibronically allowed transitions because they are weakly allowed by asymmetrical vibrations such as that shown in Fig. 17.6.

(b) Charge-transfer transitions

A complex may absorb radiation as a result of the transfer of an electron from the ligands into the d orbitals of the central atom, or vice versa. In such **charge-transfer transitions** the electron moves through a considerable distance, which means that the transition dipole moment may be large and the absorption is correspondingly intense. This mode of chromophore activity is shown by the permanganate ion, MnO_4^-, and accounts for its intense violet colour (which arises from strong absorption within the range 420–700 nm). In this oxoanion, the electron migrates from an orbital that is largely confined to the O atom ligands to an orbital that is largely confined to the Mn atom. It is therefore an example of a **ligand-to-metal charge-transfer transition** (LMCT). The reverse migration, a **metal-to-ligand charge-transfer transition** (MLCT), can also occur. An example is the transfer of a d electron into the antibonding π orbitals of an aromatic ligand. The resulting excited state may have a very long lifetime if the electron is extensively delocalized over several aromatic rings, and such species can participate in photochemically induced redox reactions (see Section 26.11).

The intensities of charge-transfer transitions are proportional to the square of the transition dipole moment, in the usual way. We can think of the transition moment as a measure of the distance moved by the electron as it migrates from metal to ligand or vice versa, with a large distance of migration corresponding to a large transition dipole moment and therefore a high intensity of absorption. However, because the integrand in the transition dipole is proportional to the product of the initial and final wavefunctions, it is zero unless the two wavefunctions have nonzero values in the same region of space. Therefore, although large distances of migration favour high intensities, the diminished overlap of the initial and final wavefunctions for large separations of metal and ligands favour low intensities (see Problem 17.18). We encounter similar considerations when we examine electron transfer reactions (Section 29.1), which can be regarded as a special type of charge-transfer transition.

(c) $\pi^* \leftarrow \pi$ and $\pi^* \leftarrow n$ transitions

Absorption by a C=C double bond excites a π electron into an antibonding π^* orbital (Fig. 17.13). The chromophore activity is therefore due to a $\pi^* \leftarrow \pi$ transition (which is normally read 'π to π^* transition'). Its energy is about 7 eV for an unconjugated double bond, which corresponds to an absorption at 180 nm (in the ultraviolet). When the double bond is part of a conjugated chain, the energies of the molecular orbitals lie closer together and the $\pi^* \leftarrow \pi$ transition moves to longer wavelength; it may even lie in the visible region if the conjugated system is long enough. An important example of an $\pi^* \leftarrow \pi$ transition is provided by the photochemical mechanism of vision (Box 17.1).

Box 17.1 **The photochemistry of vision**

The eye is an exquisite photochemical organ that acts as a transducer, converting radiant energy into electrical signals that travel along neurons. Here we concentrate on the events taking place in the human eye, but similar processes occur in all animals. Indeed, a single type of protein, rhodopsin, is the primary receptor for light throughout the animal kingdom, which indicates that vision emerged very early in evolutionary history, no doubt because of its enormous value for survival.

Photons enter the eye through the cornea, pass through the ocular fluid that fills the eye, and fall on the retina. The ocular fluid is principally water, and passage of light through this medium is largely responsible for the *chromatic aberration* of the eye, the blurring of the image as a result of different frequencies being brought to slightly different focuses. Water, like any other transparent medium, disperses light because its refractive index varies with frequency (see Chapter 21). This effect can be traced to the connection between polarizability and refractive index: the greater the polarizability, the greater the refractive index and the greater the bending of a ray of light. Water is more polarizable in response to blue (high-frequency) light than red (low-frequency) light because each photon of blue light has more energy and can more readily interact with the water molecules. The chromatic aberration is reduced to some extent by the tinted region called the *macular pigment* that covers part of the retina. The pigments in this region are the carotene-like xanthophylls (**2**), which remove some of the blue light and hence help to sharpen the image. They also protect the photoreceptor molecules from too great a flux of potentially dangerous high-energy photons. The xanthophylls have delocalized electrons that spread along the chain of conjugated double bonds, and the π-to-π^* transition lies in the visible. Much the same type of protection is found in leaves, where carotene molecules help to protect the delicate chlorophyll molecules (see Problem 12.26).

About 57 per cent of the photons that enter the eye reach the retina; the rest are scattered or absorbed by the occular fluid. Here the primary act of vision takes place, in which the chromophore of a rhodopsin molecule absorbs a photon in another π-to-π^* transition. A rhodopsin molecule consists of an opsin protein molecule to which is attached an 11-*cis*-retinal molecule (**3**). The latter resembles half a carotene molecule, showing Nature's economy in its use of available materials. The attachment is by the formation of a Schiff's base, utilizing the —CHO group of the chromophore and the —NH$_2$ group from a lysine side-chain. The free 11-*cis*-retinal molecule absorbs in the ultraviolet, but attachment to the opsin protein molecule shifts the absorption into the visible region. The rhodopsin molecules are situated in the membranes of the rods and cones that cover the retina. The opsin molecule is anchored into the cell membrane by two hydrophobic groups and largely surrounds the chromophore (see illustration).

Immediately after the absorption of a photon, the 11-*cis*-retinal molecule undergoes photoisomerization into all-*trans*-retinal (**4**). Photoisomerization takes about 200 fs and about

The structure of the rhodopsin molecule, consisting of an opsin protein to which is attached an 11-*cis*-retinal molecule embedded in the space surrounded by the helical regions.

2 A xanthophyll

3 11-*cis*-retinal

4 All-*trans*-retinal

67 pigment molecules isomerize for every 100 photons that are absorbed. The process is able to occur because the π-to-π^* excitation of an electron loosens one of the π bonds (the one indicated by the arrow in the diagram), its torsional rigidity is lost, and one part of the molecule swings round into its new position. At that point, the molecule returns to its ground state, but is now trapped in its new conformation. The straightened tail of all-*trans*-retinal results in the molecule taking up more space than 11-*cis*-retinal did, so the molecule presses against the coils of the opsin molecule that surrounds it. Thus, in about 0.25–0.50 ms from the initial absorption event, the rhodopsin molecule is activated.

Now a sequence of biochemical events—the *biochemical cascade*—converts the energy accumulated in the activated form of the rhodopsin molecule into a pulse of electric potential that travels through the optical nerve into the optical cortex, where it is interpreted as a signal and incorporated into the web of events we call 'vision'. At the same time, the resting state of the rhodopsin molecule is restored by a series of nonradiative chemical events powered by ATP. The process involves the escape of all-*trans*-retinal as all-*trans*-retinol (in which —CHO has been reduced to —CH_2OH) from the opsin molecule by a process catalysed by the enzyme rhodopsin kinase and the attachment of another protein molecule, arrestin. The free all-*trans*-retinol molecule now undergoes enzyme-catalysed isomerization into 11-*cis*-retinol followed by dehydrogenation to form 11-*cis*-retinal, which is then delivered back into an opsin molecule. At this point, the cycle of excitation, photoisomerization, and regeneration is ready to begin again.

Discussion questions

1. Use molecule (**5**) as a model of the *trans* conformation of the chromophore found in rhodopsin. In this model, the methyl group bound to the nitrogen atom of the protonated Schiff's base replaces the protein. (a) Using molecular

5

modelling software and the computational method of your choice (extended Hückel, semi-empirical, or *ab initio* methods), calculate the energy separation between the HOMO and LUMO of (**5**). (b) Repeat the calculation for the 11-*cis* form of (**5**). (c) Based on your results from parts (a) and (b), do you expect the experimental frequency for the $\pi^* \leftarrow \pi$ visible absorption of the all-*trans* form of (**5**) to be higher or lower than that for the 11-*cis* form of (**5**)?

2. The flux of visible photons reaching Earth from the North Star is about 4×10^3 mm^{-2} s^{-1}. Of these photons, 30 per cent are absorbed or scattered by the atmosphere and 25 per cent of the surviving photons are scattered by the surface of the cornea of the eye. A further 9 per cent are absorbed inside the cornea. The area of the pupil at night is about 40 mm^2 and the response time of the eye is about 0.1 s. Of the photons passing through the pupil, about 43 per cent are absorbed in the ocular medium. How many photons from the North Star are focused on to the retina in 0.1 s? For a continuation of this story, see R.W. Rodieck, *The first steps in seeing*, Sinauer, Sunderland (1998).

Further reading

K. Palczewski, T. Kumasaka, T. Hori, C.A. Behnke, H. Motoshima, B.A. Fox, I.L. Trong, D.C. Teller, T. Okada, R.E. Stenkamp, M. Yamamoto, and M. Miyano, Crystal structure of rhodopsin: A G protein-coupled receptor. *Science* **289**, 739 (2000).

G.L. Zubay, *Biochemistry*, Chapter 29. McGraw–Hill. Boston (1998).

The transition responsible for absorption in carbonyl compounds can be traced to the lone pairs of electrons on the O atom. The Lewis concept of a 'lone pair' of electrons is represented in molecular orbital theory by a pair of electrons in an orbital confined largely to one atom and not appreciably involved in bond formation. One of these electrons may be excited into an empty π^* orbital of the carbonyl group (Fig. 17.14), which gives rise to a $\pi^* \leftarrow n$ transition (an '*n* to π^* transition'). Typical absorption energies are about 4 eV (290 nm). Because $\pi^* \leftarrow n$ transitions in carbonyls are symmetry-forbidden, the absorptions are weak.

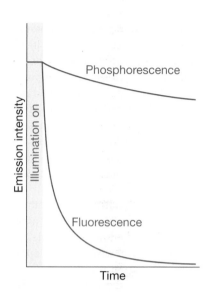

17.14 A carbonyl group (C=O) acts as a chromophore primarily on account of the excitation of a nonbonding O lone-pair electron to an antibonding CO π orbital.

17.15 The empirical (observation-based) distinction between fluorescence and phosphorescence is that the former is extinguished very quickly after the exciting source is removed, whereas the latter continues with relatively slowly diminishing intensity.

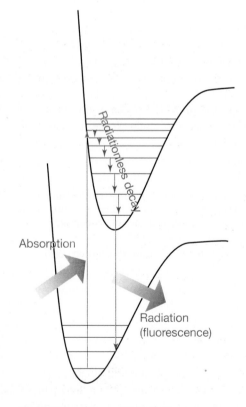

17.16 The sequence of steps leading to fluorescence. After the initial absorption, the upper vibrational states undergo radiationless decay by giving up energy to the surroundings. A radiative transition then occurs from the vibrational ground state of the upper electronic state.

The fates of electronically excited states

A **radiative decay process** is a process in which a molecule discards its excitation energy as a photon. A more common fate is **nonradiative decay**, in which the excess energy is transferred into the vibration, rotation, and translation of the surrounding molecules. This thermal degradation converts the excitation energy completely into thermal motion of the environment (that is, to 'heat'). An excited molecule may also take part in a chemical reaction, as we discuss in Chapter 26.

17.3 Fluorescence and phosphorescence

In **fluorescence**, spontaneous emission of radiation occurs within a few nanoseconds after the exciting radiation is extinguished (Fig. 17.15). In **phosphorescence**, the spontaneous emission may persist for long periods (even hours, but characteristically seconds or fractions of seconds). The difference suggests that fluorescence is a fast conversion of absorbed radiation into re-emitted energy, and that phosphorescence involves the storage of energy in a reservoir from which it slowly leaks.

(a) Fluorescence

Figure 17.16 shows the sequence of steps involved in fluorescence. The initial absorption takes the molecule to an excited electronic state, and if the absorption

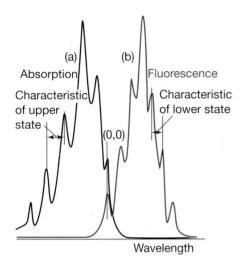

17.17 An absorption spectrum (a) shows a vibrational structure characteristic of the upper state. A fluorescence spectrum (b) shows a structure characteristic of the lower state; it is also displaced to lower frequencies (but the 0–0 transitions are coincident) and resembles a mirror image of the absorption.

spectrum were monitored it would look like the one shown in Fig. 17.17a. The excited molecule is subjected to collisions with the surrounding molecules, and as it gives up energy nonradiatively it steps down the ladder of vibrational levels to the lowest vibrational level of the electronically excited molecular state. The surrounding molecules, however, might now be unable to accept the larger energy difference needed to lower the molecule to the ground electronic state. It might therefore survive long enough to undergo spontaneous emission, and emit the remaining excess energy as radiation. The downward electronic transition is vertical (in accord with the Franck–Condon principle) and the fluorescence spectrum has a vibrational structure characteristic of the *lower* electronic state (Fig. 17.17b).

Provided they can be seen, the 0–0 absorption and fluorescence transitions can be expected to be coincident. The absorption spectrum arises from 1–0, 2–0, … transitions and occur at progressively higher wavenumber and with intensities governed by the Franck–Condon principle. The fluorescence spectrum arises from 0–0, 0–1, … *downward* transitions, and hence occur with decreasing wavenumbers. The 0–0 absorption and fluorescence peaks are not always exactly coincident because the solvent may interact differently with the solute in the ground and excited states (for instance, the hydrogen bonding pattern might differ). Because the solvent molecules do not have time to rearrange during the transition, the absorption occurs in an environment characteristic of the solvated ground state; however, the fluorescence occurs in an environment characteristic of the solvated excited state (Fig. 17.18).

Fluorescence occurs at a lower frequency than the incident radiation because the emissive transition occurs after some vibrational energy has been discarded into the surroundings. The vivid oranges and greens of fluorescent dyes are an everyday manifestation of this effect: they absorb in the ultraviolet and blue, and fluoresce in the visible. The mechanism also suggests that the intensity of the fluorescence ought to depend on the ability of the solvent molecules to accept the electronic and vibrational quanta. It is indeed found that a solvent composed of molecules with widely spaced vibrational levels (such as water) can in some cases accept the large quantum of electronic energy and so extinguish, or 'quench', the fluorescence. We examine the mechanisms of fluorescence quenching in Chapter 26.

17.18 The solvent can shift the fluorescence spectrum relative to the absorption spectrum. On the left we see that the absorption occurs with the solvent (the ellipses) in the arrangement characteristic of the ground electronic state of the molecule (the sphere). However, before fluorescence occurs, the solvent molecules relax into a new arrangement, and that arrangement is preserved during the subsequent radiative transition.

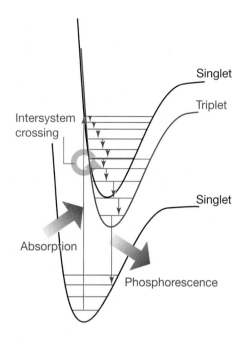

17.19 The sequence of steps leading to phosphorescence. The important step is the intersystem crossing, the switch from a singlet state to a triplet state brought about by spin–orbit coupling. The triplet state acts as a slowly radiating reservoir because the return to the ground state is spin-forbidden.

(b) Phosphorescence

Figure 17.19 shows the sequence of events leading to phosphorescence for a molecule with a singlet ground state. The first steps are the same as in fluorescence, but the presence of a triplet excited state plays a decisive role. The singlet and triplet excited states share a common geometry at the point where their potential energy curves intersect. Hence, if there is a mechanism for unpairing two electron spins (and achieving the conversion of ↑↓ to ↑↑), the molecule may undergo **intersystem crossing**, a nonradiative transition between states of different multiplicity, and become a triplet state. We saw in the discussion of atomic spectra (Section 13.9d) that singlet–triplet transitions may occur in the presence of spin–orbit coupling, and the same is true in molecules. We can expect intersystem crossing to be important when a molecule contains a moderately heavy atom (such as S), because then the spin–orbit coupling is large.

If an excited molecule crosses into a triplet state, it continues to deposit energy into the surroundings. However, it is now stepping down the triplet's vibrational ladder, and at the lowest energy level it is trapped because the triplet state is at a lower energy than the corresponding singlet (recall Hund's rule, Section 13.7). The solvent cannot absorb the final, large quantum of electronic excitation energy, and the molecule cannot radiate its energy because return to the ground state is spin-forbidden (Section 17.1). The radiative transition, however, is not totally forbidden because the spin–orbit coupling that was responsible for the intersystem crossing also breaks the selection rule. The molecules are therefore able to emit weakly, and the emission may continue long after the original excited state was formed.

The mechanism accounts for the observation that the excitation energy seems to get trapped in a slowly leaking reservoir. It also suggests (as is confirmed experimentally) that phosphorescence should be most intense from solid samples: energy transfer is then less efficient and intersystem crossing has time to occur as the singlet excited state steps slowly past the intersection point. The mechanism also suggests that the phosphorescence efficiency should depend on the presence of a moderately heavy atom (with strong spin–orbit coupling), which is in fact the case. The confirmation of the mechanism is the experimental observation (using the sensitive magnetic resonance techniques described in Chapter 18) that the sample is paramagnetic while the reservoir state, with its unpaired electron spins, is populated.

The various types of nonradiative and radiative transitions that can occur in molecules are often represented on a schematic **Jablonski diagram** of the type shown in Fig. 17.20.

17.4 Dissociation and predissociation

Another fate for an electronically excited molecule is **dissociation**, the breaking of bonds (Fig. 17.21). The onset of dissociation can be detected in an absorption spectrum by seeing that the vibrational structure of a band terminates at a certain energy. Absorption occurs in a continuous band above this **dissociation limit** because the final state is an unquantized translational motion of the fragments. Locating the dissociation limit is a valuable way of determining the bond dissociation energy.

In some cases, the vibrational structure disappears but resumes at higher photon energies. This **predissociation** can be interpreted in terms of the molecular potential energy curves shown in Fig. 17.22. When a molecule is excited to a vibrational level, its electrons may undergo a reorganization that results in it undergoing an **internal conversion**, a radiationless conversion to another state of the same multiplicity. An internal conversion occurs most readily at the point of intersection of the

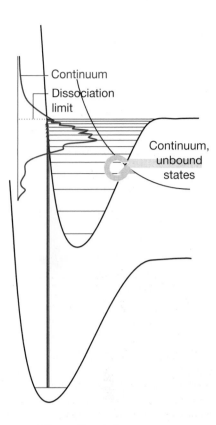

17.20 A Jablonski diagram (here, for naphthalene) is a simplified portrayal of the relative positions of the electronic energy levels of a molecule. Vibrational levels of a given electronic state lie above each other, but the relative horizontal locations of the columns bear no relation to the nuclear separations in the states. The ground vibrational states of each electronic state are correctly located vertically but the other vibrational states are shown only schematically. (IC: internal conversion; ISC: intersystem crossing.)

17.21 When absorption occurs to unbound states of the upper electronic state, the molecule dissociates and the absorption is a continuum. Below the dissociation limit the electronic spectrum shows a normal vibrational structure.

17.22 When a dissociative state crosses a bound state, as in the upper part of the illustration, molecules excited to levels near the crossing may dissociate. This process is called predissociation, and is detected in the spectrum as a loss of vibrational structure that resumes at higher frequencies.

two molecular potential energy curves, because there the nuclear geometries of the two states are the same. The state into which the molecule converts may be dissociative, so the states near the intersection have a finite lifetime, and hence their energies are imprecisely defined. As a result, the absorption spectrum is blurred in the vicinity of the intersection. When the incoming photon brings enough energy to excite the molecule to a vibrational level high above the intersection, the internal conversion does not occur (the nuclei are unlikely to have the same geometry). Consequently, the levels resume their well-defined, vibrational character with correspondingly well-defined energies, and the line structure resumes on the high-frequency side of the blurred region.

Lasers

Lasers have transformed chemistry as much as they have transformed the everyday world. In this section, we see some of the principles of their operation, and then explore their applications in chemistry. Lasers lie very much on the frontier of physics

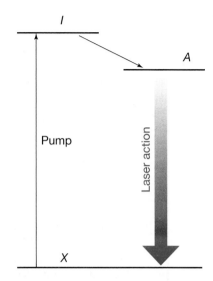

17.23 The transitions involved in one kind of three-level laser. The pumping pulse populates the intermediate state I, which in turn populates the laser state A. The laser transition is the stimulated emission $A \rightarrow X$.

and chemistry, for their operation depends on details of optics and, in some cases, of solid-state processes. We shall concentrate on the more chemical aspects of their operation, particularly the materials from which they are made and the events taking place within them.

17.5 General principles of laser action

The word laser is an acronym formed from **l**ight **a**mplification by **s**timulated **e**mission of **r**adiation. In stimulated emission, an excited state is stimulated to emit a photon by radiation of the same frequency; the more photons that are present, the greater the probability of the emission. The essential feature of laser action is positive-feedback: the more photons present of the appropriate frequency, the more photons of that frequency that will be stimulated to form.

(a) Population inversion

One requirement of laser action is the existence of a **metastable excited state**, an excited state with a long enough lifetime for it to participate in stimulated emission. Another requirement is the existence of a greater population in the metastable state than in the lower state where the transition terminates, for then there will be a net emission of radiation. Because at thermal equilibrium the opposite is true, it is necessary to achieve a **population inversion** in which there are more molecules in the upper state than in the lower.

One way of achieving population inversion is illustrated in Fig. 17.23. The molecule is excited to an intermediate state I, which then gives up some of its energy nonradiatively and changes into a lower state A; the laser transition is the return of A to the ground state X. Because three energy levels are involved overall, this arrangement leads to a **three-level laser**. In practice, I consists of many states, all of which can convert to the upper of the two laser states A. The $I \leftarrow X$ transition is stimulated with an intense flash of light in the process called **pumping**. The pumping is often achieved with an electric discharge through xenon or with the light of another laser. The conversion of I to A should be rapid, and the laser transitions from A to X should be relatively slow.

The disadvantage of this three-level arrangement is that it is difficult to achieve population inversion, because so many ground-state molecules must be converted to the excited state by the pumping action. The arrangement adopted in a **four-level laser** simplifies this task by having the laser transition terminate in a state A' other than the ground state (Fig. 17.24). Because A' is unpopulated initially, any population in A corresponds to a population inversion, and we can expect laser action if A is sufficiently metastable. Moreover, this population inversion can be maintained if the $A' \rightarrow X$ transitions are rapid, for these transitions will deplete any population in A' that stems from the laser transition, and keep the state A' relatively empty.

(b) Cavity and mode characteristics

The laser medium is confined to a cavity that ensures that only certain photons of a particular frequency, direction of travel, and state of polarization are generated abundantly. The cavity is essentially a region between two mirrors, which reflect the light back and forth. This arrangement can be regarded as a version of the particle in a box, with the particle now being a photon. As in the treatment of a particle in a box (Section 12.1), the only wavelengths that can be sustained satisfy

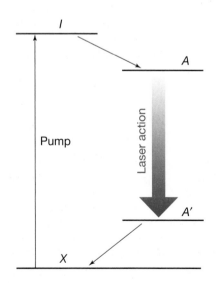

17.24 The transitions involved in a four-level laser. Because the laser transition terminates in an excited state (A'), the population inversion between A and A' is much easier to achieve.

$$n \times \tfrac{1}{2}\lambda = L \tag{17.6}$$

(a) Thermal equilibrium

Pump

(b) Population inversion

(c) Laser action

17.25 A schematic illustration of the steps leading to laser action. (a) The Boltzmann population of states, with more atoms in the ground state. (b) When the initial state absorbs, the populations are inverted (the atoms are pumped to the excited state). (c) A cascade of radiation then occurs, as one emitted photon stimulates another atom to emit, and so on. The radiation is coherent (phases in step).

where n is an integer and L is the length of the cavity. That is, only an integral number of half-wavelengths fit into the cavity; all other waves undergo destructive interference with themselves. In addition, not all wavelengths that can be sustained by the cavity are amplified by the laser medium (many fall outside the range of frequencies of the laser transitions), so only a few contribute to the laser radiation. These wavelengths are the **resonant modes** of the laser.

Photons with the correct wavelength for the resonant modes of the cavity and the correct frequency to stimulate the laser transition are highly amplified. One photon might be generated spontaneously, and travel through the medium. It stimulates the emission of another photon, which in turn stimulates more (Fig. 17.25). The cascade of energy builds up rapidly, and soon the cavity is an intense reservoir of radiation at all the resonant modes it can sustain. Some of this radiation can be withdrawn if one of the mirrors is partially transmitting.

The resonant modes of the cavity have various natural characteristics, and to some extent may be selected. Only photons that are travelling strictly parallel to the axis of the cavity undergo more than a couple of reflections, so only they are amplified, all others simply vanishing into the surroundings. Hence, laser light generally forms a beam with very low divergence. It may also be polarized, with its electric vector in a particular plane (or in some other state of polarization), by including a polarizing filter into the cavity or by making use of polarized transitions in a solid medium.

Laser radiation is **coherent** in the sense that the electromagnetic waves are all in step. In **spatial coherence** the waves are in step across the cross-section of the beam emerging from the cavity. In **temporal coherence** the waves remain in step along the beam. The latter is normally expressed in terms of a **coherence length**, l_C, the distance over which the waves remain coherent, and is related to the range of wavelengths, $\Delta\lambda$ present in the beam:

$$l_C = \frac{\lambda^2}{2\Delta\lambda} \tag{17.7}$$

If the beam were perfectly monochromatic, with strictly one wavelength present, then $\Delta\lambda$ would be zero, and the waves would remain in step for an infinite distance. When many wavelengths are present, the waves get out of step in a short distance and the coherence length is small. A typical light bulb gives out light with a coherence length of only about 400 nm; a He–Ne laser with $\Delta\lambda \approx 2$ pm has a coherence length of about 10 cm.

(c) Q-switching

A laser can generate radiation for as long as the population inversion is maintained. A laser can operate continuously when heat is easily dissipated, for then the population of the upper level can be replenished by pumping. Practical considerations govern whether or not continuous pumping is feasible, as we shall see when we consider some particular lasers. When overheating is a problem, the laser can be operated only in pulses, perhaps of microsecond or millisecond duration, so that the medium has a chance to cool or the lower state discard its population. However, it is sometimes desirable to have pulses of radiation rather than a continuous output, with a lot of power concentrated into a brief pulse. One way of achieving pulses is by **Q-switching**, the modification of the resonance characteristics of the laser cavity.[4]

4. The name comes from the 'Q-factor' used as a measure of the quality of a resonance cavity in microwave engineering.

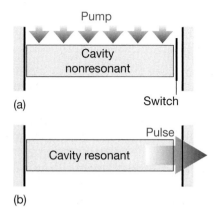

17.26 The principle of Q-switching. (a) The excited state is populated while the cavity is non-resonant. (b) Then the resonance characteristics are suddenly restored, and the stimulated emission emerges in a giant pulse.

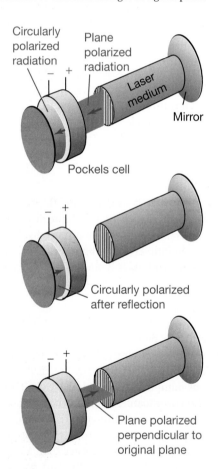

17.27 The principle of a Pockels cell. When light passes through a cell that is 'on', its plane of polarization is rotated and so the laser cavity is non-resonant (its Q-factor is reduced). When the cell is turned off, no change of polarization occurs, and the cavity becomes resonant.

Example 17.2 Relating the power and energy of a laser

A laser rated at 0.10 J can generate radiation in 3.0 ns pulses at a pulse repetition rate of 10 Hz. Assuming that the pulses are rectangular, calculate the peak power output and the average power output of this laser.

Method The power output is the energy released in an interval divided by the duration of the interval, and is expressed in watts (1 W = 1 J s^{-1}). To calculate the peak power output, P_{peak}, we divide the energy released during the pulse by the duration of the pulse. The average power output, $P_{average}$, is the total energy released by a large number of pulses divided by the duration of the time interval over which the total energy was measured. So, the average power is simply the energy released by one pulse multiplied by the pulse repetition rate.

Answer From the data,

$$P_{peak} = \frac{0.10\ J}{3.0 \times 10^{-9}\ s} = 3.3 \times 10^{7}\ J\ s^{-1}$$

That is, the peak power output is 33 MW. The pulse repetition rate is 10 Hz, so that ten pulses are emitted by the laser for every second of operation. It follows that the average power output is

$$P_{average} = 0.10\ J \times 10\ s^{-1} = 1.0\ J\ s^{-1} = 1.0\ W$$

Comment The peak power is much higher than the average power because this laser emits light for only 3.0×10^{-9} s during each second of operation.

Self-test 17.3 Calculate the peak power and average power output of a laser with a pulse energy of 2.0 mJ, a pulse duration of 30 ps (1 ps = 1×10^{-12} s), and a pulse repetition rate of 38 MHz.   [P_{peak} = 67 MW, $P_{average}$ = 76 kW]

The aim of Q-switching is to achieve a healthy population inversion in the absence of the resonant cavity, then to plunge the population-inverted medium into a cavity, and hence to obtain a sudden pulse of radiation. The switching may be achieved by impairing the resonance characteristics of the cavity in some way while the pumping pulse is active, and then suddenly to improve them (Fig. 17.26). One technique is to use a **Pockels cell**, which is an electro-optical device based on the ability of some crystals, such as those of potassium dihydrogenphosphate (KH_2PO_4), to convert plane-polarized light to circularly polarized light when an electrical potential difference is applied. If a Pockels cell is made part of a laser cavity, then its action and the change in polarization that occurs when light is reflected from a mirror convert light polarized in one plane into reflected light polarized in the perpendicular plane (Fig. 17.27). As a result, the reflected light does not stimulate more emission. However, if the cell is suddenly turned off, the polarization effect is extinguished and all the energy stored in the cavity can emerge as an intense pulse of stimulated emission. An alternative technique is to use a **saturable absorber**, typically a solution of a dye that loses its ability to absorb when many of its molecules have been excited by intense radiation. The dye then suddenly becomes transparent and the cavity becomes resonant. In practice, Q-switching can give pulses of about 5 ns duration.

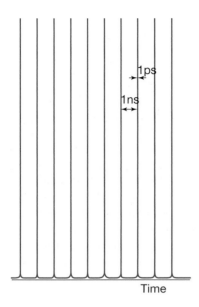

17.28 The output of a mode-locked laser consists of a stream of very narrow pulses separated by an interval equal to the time it takes for light to make a round trip inside the cavity.

17.29 The function derived in *Justification* 17.3 showing in more detail the structure of the pulses generated by a mode-locked laser.

(d) Mode locking

The technique of **mode locking** can produce pulses of picosecond duration and less. A laser radiates at a number of different frequencies, depending on the precise details of the resonance characteristics of the cavity and in particular on the number of half-wavelengths of radiation that can be trapped between the mirrors (the cavity modes). The resonant modes differ in frequency by multiples of $c/2L$ (as can be inferred from eqn 6 with $v = c/\lambda$). Normally, these modes have random phases relative to each other. However, it is possible to lock their phases together. Then interference occurs to give a series of sharp peaks, and the energy of the laser is obtained in picosecond bursts (Fig. 17.28). The sharpness of the peaks depends on the range of modes superimposed and, the wider the range, the narrower the pulses. In a laser with a cavity of length 30 cm, the peaks are separated by 2 ns. If 1000 modes contribute, the width of the pulses is 4 ps.

Justification 17.3

The general expression for a (complex) wave of amplitude $\mathcal{E}_0$ and frequency ω is $\mathcal{E}_0 e^{i\omega t}$. Therefore, each wave that can be supported by a cavity of length L has the form

$$\mathcal{E}_n(t) = \mathcal{E}_0 e^{2\pi(v + nc/2L)t}$$

where v is the lowest frequency. A wave formed by superimposing N modes with $n = 0, 1, \ldots, N - 1$ has the form

$$\mathcal{E}(t) = \sum_n \mathcal{E}_n(t) = \mathcal{E}_0 e^{2\pi ivt} \sum_{n=0}^{N-1} e^{i\pi nct/L}$$

The sum is a geometrical progression:

$$\sum_{n=0}^{N-1} e^{i\pi nct/L} = 1 + e^{i\pi ct/L} + e^{2i\pi ct/L} + \cdots$$

$$= \frac{\sin(N\pi ct/2L)}{\sin(\pi ct/2L)} \times e^{(N-1)i\pi ct/2L}$$

The intensity, I, of the radiation is proportional to the square modulus of the total amplitude, so

$$I \propto \mathcal{E}^*\mathcal{E} = \mathcal{E}_0^2 \frac{\sin^2(N\pi ct/2L)}{\sin^2(\pi ct/2L)}$$

This function is shown in Fig. 17.29. We see that it is a series of peaks with maxima separated by $t = 2L/c$, the round-trip transit time of the light in the cavity, and that the peaks become sharper as N is increased.

Mode locking is achieved by varying the Q-factor of the cavity periodically at the frequency $c/2L$. The modulation can be pictured as the opening of a shutter in synchrony with the round-trip travel time of the photons in the cavity, so only photons making the journey in that time are amplified. The modulation can be achieved by linking a prism in the cavity to a transducer driven by a radiofrequency source at a frequency $c/2L$. The transducer sets up standing-wave vibrations in the prism and modulates the loss it introduces into the cavity. Mode locking may also be accomplished passively by exploiting the **optical Kerr effect**, which arises from a change

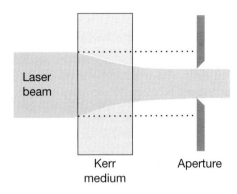

17.30 An illustration of the Kerr effect. An intense laser beam is focused inside a Kerr medium and passes through a small aperture in the laser cavity. The profile of a weak laser beam that is not focused by the Kerr medium is shown by the dashed lines. This effect may be used to mode-lock a laser, as explained in the text.

The web site contains links to databases on the optical properties of laser materials.

in refractive index of a well chosen medium, the **Kerr medium**, when it is exposed to intense electric fields such as those generated by intense laser pulses. The changes in refractive index result in the self-focusing of an intense laser pulse as it travels through the Kerr medium (Fig. 17.30): A Kerr medium is included in the cavity and next to it is a small aperture. The procedure makes use of the fact that the **gain**, the growth in intensity, of a frequency component of the radiation in the cavity is very sensitive to amplification and, once a particular frequency begins to grow, it can quickly dominate. When the power inside the cavity is low, a portion of the photons will be blocked by the aperture, creating a significant loss. A spontaneous fluctuation in intensity—a bunching of photons—may begin to turn on the optical Kerr effect and the changes in the refractive index of the Kerr medium will result in a **Kerr lens**, which is the self-focusing of the laser beam (as in Fig. 17.30). The bunch of photons can pass through and travel to the far end of the cavity, amplifying as it goes. The Kerr lens immediately disappears (if the medium is well chosen), but is re-created when the intense pulse returns from the mirror at the far end. In this way, that particular bunch of photons may grow to considerable intensity because it alone is stimulating emission in the cavity.

17.6 Practical lasers

Figure 17.31 summarizes the requirements for an efficient laser. In practice, the requirements can be satisfied by using a variety of different systems, and this section reviews some that are commonly available.[5] We also include some lasers that operate by using other than electronic transitions.

(a) Solid-state lasers

A **solid-state laser** is one in which the active medium is in the form of a single crystal or a glass. The first successful laser, the ruby laser built by Theodore Maiman in 1960, is an example (Fig. 17.32). Ruby is Al_2O_3 containing a small proportion of Cr^{3+} ions.[6] Ruby is a three-level system, and the ground state, which is also the lower level of the laser transition, is $^4A_2$ with three unpaired spins on each Cr^{3+} ion. The population inversion results from pumping a majority of the Cr^{3+} ions into an excited state by using an intense flash from another source, followed by a radiationless transition to another excited state. The pumping flash need not be monochromatic because the upper level actually consists of several states spanning a band of frequencies. The transition from the lower of the two excited states to the ground state ($^2E \rightarrow {}^4A_2$) is the laser transition, and gives rise to red 694 nm radiation. The population inversion is very difficult to sustain continuously, and in practice the ruby laser is pulsed. Typical pulses from a Q-switched ruby laser might consist of 2 J pulses persisting for 10 ns, corresponding to an average power of 0.2 GW per pulse.

The **neodymium laser** is an example of a four-level laser (Fig. 17.33). In one form it consists of Nd^{3+} ions at low concentration in yttrium aluminium garnet (YAG, specifically $Y_3Al_5O_{12}$), and is then known as a Nd–YAG laser. A neodymium laser operates at a number of wavelengths in the infrared, the band at 1064 nm being most common. The transition at 1064 nm is very efficient and the laser is capable of substantial power output. The power is great enough that focusing the beam on to a material may lead to the observation of **nonlinear optical phenomena**, which arise from changes in the optical properties of the substance in the presence of an intense

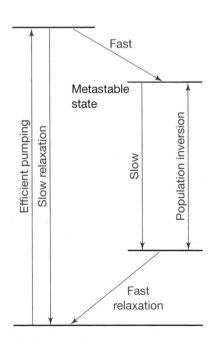

17.31 A summary of the features needed for efficient laser action.

5. We examine semiconductor lasers and light-emitting diodes in Chapter 22.
6. The normal green of Cr^{3+} is modified to red by the distortion of the local crystal field stemming from the replacement of an Al^{3+} ion by a slightly larger Cr^{3+} ion.

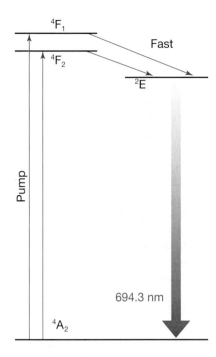

17.32 The transitions involved in a ruby laser. The laser medium, ruby, consists of Al_2O_3 doped with Cr^{3+} ions.

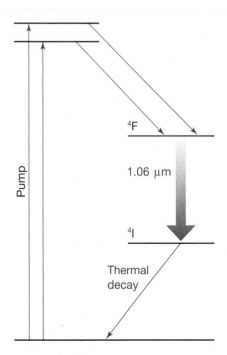

17.33 The transitions involved in a neodymium laser. The laser action takes place between two excited states, and the population inversion is easier to achieve than in the ruby laser.

electric field from electromagnetic radiation. The optical Kerr effect (Section 17.5) is a nonlinear optical phenomenon and so is **frequency doubling**, or **second harmonic generation**, in which an intense laser beam is converted to radiation with twice (and in general a multiple) of its initial frequency as it passes though a suitable material (see the *Justification*). Frequency doubling and tripling of a Nd–YAG laser produce green light at 532 nm and ultraviolet light at 355 nm, respectively.

Justification 17.4

We can account for frequency doubling by examining how a substance responds nonlinearly to incident radiation of frequency ω. Radiation of a particular frequency arises from oscillations of an electric dipole at that frequency and the incident electric field $\mathcal{E}$ induces an electric dipole of magnitude μ, in the substance. Allowing for nonlinear response, we can write

$$\mu = \alpha\mathcal{E} + \beta\mathcal{E}^2$$

The nonlinear terms can be expanded as follows if we suppose that the incident electric field is $\mathcal{E}_0 \cos \omega t$:

$$\beta\mathcal{E}^2 = \beta\mathcal{E}_0^2 \cos^2\omega t = \tfrac{1}{2}\beta\mathcal{E}_0^2(1 + \cos 2\omega t)$$

Hence, the nonlinear term contributes an induced electric dipole that oscillates at the frequency 2ω and which can act as a source of radiation of that frequency. Common materials that can be used for frequency doubling include crystals of potassium dihydrogenphosphate (KH_2PO_4), lithium niobate (Li_3NbO_4), and β-barium borate (β-BaB_2O_4).

The **titanium sapphire laser** consists of Ti^{3+} ions at low concentration in a crystal of sapphire (Al_2O_3). The electronic absorption spectrum of Ti^{3+} ion in sapphire is very similar to that shown in Fig. 17.12, with a broad absorption band centred at around 500 nm that arises from vibronically allowed d–d transitions of the Ti^{3+} ion in an octahedral environment provided by oxygen atoms of the host lattice. As a result, the emission spectrum of Ti^{3+} in sapphire is also broad and laser action occurs over a wide range of wavelengths (Fig. 17.34). Therefore, the titanium sapphire laser is an example of a **vibronic laser**, in which the laser transitions originate from vibronic transitions in the laser medium.

The titanium sapphire laser is usually pumped by another laser, such as a Nd–YAG laser or an argon-ion laser (Section 17.6b), and can be operated in both continuous or pulsed modes. Sapphire is a Kerr medium, and passively mode-locked titanium sapphire lasers produce energetic (20 mJ to 1 J) and very short (20–100 fs, 1 fs = 10^{-15} s) pulses. When considered together with broad wavelength tunability (700–1000 nm), these features of the titanium sapphire laser justify its wide use in modern spectroscopy and photochemistry.

Lasers that make use of the band structure of semiconducting solids are of great importance in modern technology and consumer products. They are described in Section 23.9.

(b) Gas lasers

Because gas lasers can be cooled by a rapid flow of the gas through the cavity, they can be used to generate high powers. The pumping is normally achieved using a gas that is different from the gas responsible for the laser emission itself.

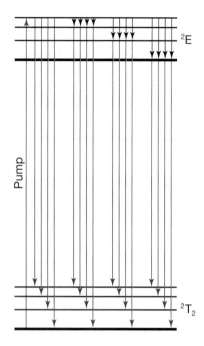

17.34 The transitions involved in a titanium sapphire laser. The laser medium consists of sapphire (Al_2O_3) doped with Ti^{3+} ions. Monochromatic light from a pump laser induces a $^2E \leftarrow {}^2T_2$ transition in a Ti^{3+} ion that resides in a site with octahedral symmetry. After radiationless vibrational excitation in the ^2E state, laser emission occurs from a very large number of closely spaced vibronic states of the medium. As a result, the titanium sapphire laser emits radiation over a broad spectrum that spans from about 700 nm to about 1000 nm.

In the **helium–neon laser** the active medium is a mixture of helium and neon in a mole ratio of about 5:1 (Fig. 17.35). The initial step is the excitation of an He atom to the metastable $1s^12s^1$ configuration by using an electric discharge (the collisions of electrons and ions cause transitions that are not restricted by electric-dipole selection rules). The excitation energy of this transition happens to match an excitation energy of neon, and during an He–Ne collision efficient transfer of energy may occur, leading to the production of highly excited, metastable Ne atoms with unpopulated intermediate states. Laser action generating 633 nm radiation (among about 100 other lines) then occurs.

The **argon-ion laser** (Fig. 17.36), one of a number of 'ion lasers', consists of argon at about 1 Torr, through which is passed an electric discharge. The discharge results in the formation of Ar^+ and Ar^{2+} ions in excited states, which undergo a laser transition to a lower state. These ions then revert to their ground states by emitting hard ultraviolet radiation (at 72 nm), and are then neutralized by a series of electrodes in the laser cavity. One of the design problems is to find materials that can withstand this damaging residual radiation. There are many lines in the laser transition because the excited ions may make transitions to many lower states, but two strong emissions from Ar^+ are at 488 nm (blue) and 514 nm (green); other transitions occur elsewhere in the visible region, in the infrared, and in the ultraviolet. The **krypton-ion laser** works similarly. It is less efficient, but gives a wider range of wavelengths, the most intense being at 647 nm (red), but it can also generate yellow, green, and violet lines. Both lasers are widely used in laser light shows (for this application

17.35 The transitions involved in a helium–neon laser. The pumping (of the neon) depends on a coincidental matching of the helium and neon energy separations, so excited He atoms can transfer their excess energy to Ne atoms during a collision.

17.36 The transitions involved in an argon-ion laser.

17.37 The transitions involved in a carbon dioxide laser. The pumping also depends on the coincidental matching of energy separations; in this case the vibrationally excited N_2 molecules have excess energies that correspond to a vibrational excitation of the antisymmetric stretch of CO_2. The laser transition is from $v_3 = 1$ to $v_1 = 1$.

argon and krypton are often used simultaneously in the same cavity) as well as laboratory sources of high-power radiation.

The **carbon dioxide laser** works on a slightly different principle (Fig. 17.37), for its radiation (between 9.2 μm and 10.8 μm, with the strongest emission at 10.6 μm, in the infrared) arises from vibrational transitions. Most of the working gas is nitrogen, which becomes vibrationally excited by electronic and ionic collisions in an electric discharge. The vibrational levels happen to coincide with the ladder of antisymmetric stretch (v_3, see Fig. 16.48) energy levels of CO_2, which pick up the energy during a collision. Laser action then occurs from the lowest excited level of v_3 to the lowest excited level of the symmetric stretch (v_1), which has remained unpopulated during the collisions. This transition is allowed by anharmonicities in the molecular potential energy. Some helium is included in the gas to help remove energy from this state and maintain the population inversion.

In a **nitrogen laser**, the efficiency of the stimulated transition (at 337 nm, in the ultraviolet, the transition $C^3\Pi_u \rightarrow B^3\Pi_g$) is so great that a single passage of a pulse of radiation is enough to generate laser radiation and mirrors are unnecessary: such lasers are said to be **superradiant**.

(c) Chemical and exciplex lasers

Chemical reactions may also be used to generate molecules with non-equilibrium, inverted populations. For example, the photolysis of Cl_2 leads to the formation of Cl atoms that attack H_2 molecules in the mixture and produce HCl and H. The latter then attacks Cl_2 to produce vibrationally excited ('hot') HCl molecules. Because the newly formed HCl molecules have non-equilibrium vibrational populations, laser action can result as they return to lower states. Such processes are remarkable examples of the direct conversion of chemical energy into coherent electromagnetic radiation.

The population inversion needed for laser action is achieved in a more underhand way in **exciplex lasers**,[7] for in these (as we shall see) the lower state does not effectively exist. This odd situation is achieved by forming an **exciplex**, a combination of two atoms or molecules that survives only in an excited state and which dissociates as soon as the excitation energy has been discarded. An example is a mixture of xenon, chlorine, and neon (which acts as a buffer gas). An electric discharge through the mixture produces excited Cl atoms, which attach to the Xe atoms to give the exciplex XeCl*. The exciplex survives for about 10 ns, which is time for it to participate in laser action at 308 nm (in the ultraviolet). As soon as XeCl* has discarded a photon, the atoms separate because the molecular potential energy curve of the ground state is dissociative, and the ground state of the exciplex cannot become populated (Fig. 17.38). The KrF* exciplex laser is another example: it produces radiation at 249 nm.

(d) Dye lasers

Gas lasers and most solid state lasers operate at discrete frequencies and, although the frequency required may be selected by suitable optics, the laser cannot be tuned continuously. The tuning problem is overcome by using a titanium sapphire laser (Section 17.6a) or a **dye laser**, which has broad spectral characteristics because the solvent broadens the vibrational structure of the transitions into bands. Hence, it is possible to scan the wavelength continuously (by rotating the diffraction grating in the cavity) and achieve laser action at any chosen wavelength. A commonly used dye is Rhodamine 6G in methanol (Fig. 17.39). As the gain is very high, only a short length of the optical path need be through the dye. The excited states of the active

---

7. The term 'excimer laser' is also widely encountered and used loosely when 'exciplex laser' is more appropriate. An exciplex has the form AB* whereas an excimer, an excited dimer, is AA*.

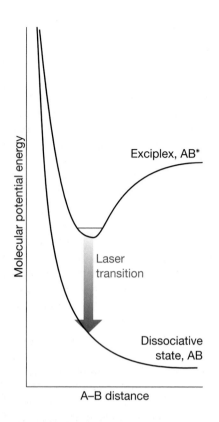

17.38 The molecular potential energy curves for an exciplex. The species can survive only as an excited state, because on discarding its energy it enters the lower, dissociative state. Because only the upper state can exist, there is never any population in the lower state.

17.39 The optical absorption spectrum of the dye Rhodamine 6G and the region used for laser action.

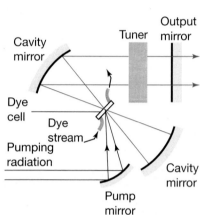

17.40 The configuration used for a dye laser. The dye is flowed through the cell inside the laser cavity. The flow helps to keep it cool and prevents degradation.

medium, the dye, are sustained by another laser or a flash lamp, and the dye solution is flowed through the laser cavity to avoid thermal degradation (Fig. 17.40).

17.7 Applications of lasers in chemistry

Laser radiation has five striking characteristics (Table 17.4). Each of them (sometimes in combination with the others) opens up interesting opportunities in spectroscopy, giving rise to 'laser spectroscopy' (see also Box 17.2), and in photochemistry, giving rise to 'laser photochemistry'.

Table 17.4 Characteristics of laser radiation and their chemical applications

Characteristic	Advantages	Applications
High power	Multiphoton process	Nonlinear spectroscopy Saturation spectroscopy
	Low detector noise	Improved sensitivity
	High scattering intensity	Raman spectroscopy
Monochromatic	High resolution State selection	Spectroscopy Isotope separation Photochemically precise State-to-state reaction dynamics
Collimated beam	Long path lengths Forward-scattering observable	Sensitivity Nonlinear Raman spectroscopy
Coherent	Interference between separate beams	CARS
Pulsed	Precise timing of excitation	Fast reactions Relaxation Energy transfer

Box 17.2 Spectroscopy of single molecules

There is great interest in the development of new experimental probes of very small specimens. On the one hand, our understanding of biochemical processes, such as enzymatic catalysis, protein folding, and the insertion of DNA into the cell's nucleus, will be enhanced if it is possible to visualize individual biopolymers at work. On the other hand, techniques that can probe the structure, dynamics, and reactivity of single molecules will be needed to advance research on nanometre-sized materials (Box 28.1). While scanning probe microscopy is a good probe of atoms and molecules on surfaces (Box 12.1), spectroscopic techniques may be used to study a wider variety of samples, from solids to flowing liquids.

We saw in Box 16.2 that it is possible to obtain the vibrational spectrum of samples with areas of more than 10 μm². The basic approach of illuminating a small area of the sample and collecting light with a microscope has also been used for many years to image biological cells by attaching a large number of small fluorescent molecules to proteins, nucleic acids, and membranes and then measuring the distribution of fluorescence intensity within the illuminated area. However, the diffraction limit prevents the visualization of samples that are smaller than one-half the wavelength of light used as a probe (Box 16.2). Most molecules—including biopolymers—have dimensions that are much smaller than visible wavelengths, so special techniques had to be developed to make single-molecule spectroscopy possible. We shall outline the most popular strategies; more comprehensive reviews of the field are listed in the *Further reading* section at the end of this box.

The bulk of the work done in the field of single-molecule spectroscopy is based on the collection by a microscope objective lens of laser-induced fluorescence emission from a sample. The laser is the radiation source of choice because it provides the high excitance required to increase the rate of arrival of photons onto the detector from small illuminated areas. Fluorescence emission occurs at a frequency that is lower than the frequency of the exciting laser radiation (Section 17.3). Therefore, with proper filtering to remove light due to Rayleigh scattering of the laser beam (Section 16.1d), it is possible to collect light from the sample that contains only the fluorescence spectrum of the molecule of interest. However, great care is required to eliminate fluorescent impurities from the sample.

Two techniques are commonly used to circumvent the diffraction limit. First, the concentration of the sample is kept so low that, on average, only one fluorescent molecule is in the illuminated area. Second, special strategies are used to illuminate very small volumes. In *near-field scanning optical microscopy* (NSOM), a very thin metal-coated optical fibre is used to deliver light to a small area. It is possible to construct

fibres with tip diameters in the range of 50 to 100 nm, which are indeed smaller than visible wavelengths. The fibre tip is placed very close to the sample, in a region known as the *near field*, where, according to classical physics, photons do not diffract. The illustration shows the image of a 4.5 μm × 4.5 μm sample of oxazine 720 dye molecules embedded in a polymer film and obtained with NSOM by measuring the fluorescence intensity as the tip is scanned over the film surface. Each peak corresponds to a single dye molecule.

In *far-field confocal microscopy*, laser light focused by an objective lens is used to illuminate about 1 μm³ of a very dilute sample placed beyond the near field. This illumination scheme is limited by diffraction and, as a result, data from far-field microscopy have less structural detail than data from NSOM. However, far-field microscopes are very easy to construct and the technique can be used to probe single molecules as long as there is one molecule, on average, in the illuminated area.

In the *wide-field epi illumination* method, light focused by an objective lens is used to illuminate a large area (for example, 100 μm × 100 μm) of a very dilute sample. A CCD detector (Section 16.1e) collects fluorescence emission from the molecules in the illuminated area. It is then possible to obtain a map of the distribution of fluorescent molecules in the illuminated area. This technique is similar to Raman imaging, which was discussed in Box 16.2.

Though still a relatively new technique, single-molecule spectroscopy has already been used to address important problems in chemistry and biology. Nearly all the techniques discussed in this text measure the average value of a property in a large ensemble of molecules. Single-molecule methods, which also include scanning probe microscopy, allow a chemist to study the nature of distributions of physical and chemical properties in an ensemble of molecules. For example, it is possible to measure the fluorescence lifetime of a molecule by moving the laser focus to a location on the sample that contains a molecule and then measuring the decay of fluorescence intensity after excitation with a pulsed laser. Such studies have shown that not every molecule in a sample has the same fluorescence lifetime, probably because each molecule interacts with its immediate environment in a slightly different way. These details are not apparent from conventional measurements of fluorescence lifetimes, in which many molecules are excited electronically and only an average lifetime for the ensemble can be measured.

Single-molecule spectroscopy is also having a positive impact on studies of biological processes. One example is the visualization of some of the steps involved in the synthesis of ATP by the enzyme ATPase, which we shall discuss in Box 24.1.

Image of a 4.5 μm × 4.5 μm sample of oxazine 720 dye molecules embedded in a polymer film and obtained with NSOM. Each peak corresponds to a single dye molecule. Reproduced with permission from X.S. Xie, *Acc. Chem. Res.* **29**, 598–606 (1996).

Discussion questions

1. Consider some of the precautions that must be taken when conducting single-molecule spectroscopy experiments. (a) What is the molar concentration of a solution in which there is, on average, one solute molecule in 1.0 μm^3 of solution? (b) It is important to use pure solvents in single-molecule spectroscopy because optical signals from fluorescent impurities in the solvent may mask optical signals from the solute. Suppose that water containing a fluorescent impurity with a molar mass of 100 g mol^{-1} is used as solvent and that the lot analysis indicates the presence of 1.0 × 10^{-7} kg of impurity per 1.00 kg of solvent. On average, how many impurity molecules will be present in 1.0 μm^3 of solution? You may take the density of water as 1.0 × 10^3 kg m^{-3}. Comment on the suitability of this solvent for single-molecule spectroscopy experiments.

2. Light-induced degradation of molecules, also called *photo-bleaching*, is a serious problem in single-molecule spectroscopy. A molecule of a fluorescent dye commonly used to label biopolymers can withstand about 10^6 excitations by photons before light-induced reactions destroy its π system and the molecule no longer fluoresces. For how long will a single dye molecule fluoresce while being excited by 1.0 mW of 488 nm radiation from a continuous-wave argon ion laser? You may assume that the dye has an absorption spectrum that peaks at 488 nm and that every photon delivered by the laser is absorbed by the molecule.

Further reading

S. Weiss, Fluorescence spectroscopy of single molecules. *Science* **283**, 1676 (1999).

X.S. Xie and J.K. Trautman, Optical studies of single molecules at room temperature. *Ann. Rev. Phys. Chem.* **49**, 441 (1998).

(a) Multiphoton spectroscopy

The large number of photons in an incident beam generated by a laser gives rise to a qualitatively different branch of spectroscopy, for the photon density is so high that more than one photon may be absorbed by a single molecule and gives rise to **multiphoton processes**. One application of multiphoton processes is that states inaccessible by conventional one-photon spectroscopy become observable because the overall transition occurs with no change of parity. For example, in one-photon spectroscopy, only g $\leftrightarrow$ u transitions are observable; in two-photon spectroscopy, however, the overall outcome of absorbing two photons is a g $\rightarrow$ g or a u $\rightarrow$ u transition.

(b) Raman spectroscopy

Raman spectroscopy was revitalized by the introduction of lasers. An intense excitation beam increases the intensity of scattered radiation, so the use of laser sources increases the sensitivity of Raman spectroscopy. A well defined beam also implies that the detector can be designed to collect only the radiation that has passed through a sample, and can be screened much more effectively against stray scattered light, which can obscure the Raman signal. The monochromaticity of laser radiation is also a great advantage, for it makes possible the observation of scattered light that differs by only fractions of reciprocal centimetres from the incident radiation. Such high resolution is particularly useful for observing the rotational structure of Raman lines because rotational transitions are of the order of a few reciprocal centimetres. Monochromaticity also allows observations to be made very close to absorption frequencies, giving rise to the techniques of Fourier-transform Raman spectroscopy (Section 16.1d) and resonance Raman spectroscopy (Section 16.16b).

The availability of nondivergent beams makes possible a qualitatively different kind of spectroscopy. The beam is so well-defined that it is possible to observe Raman transitions very close to the direction of propagation of the incident beam. This configuration is employed in the technique called **stimulated Raman spectroscopy**. In this form of spectroscopy, the Stokes and anti-Stokes radiation in the forward direction are powerful enough to undergo more scattering and hence give up or acquire more quanta of energy from the molecules in the sample. This multiple scattering results in lines of frequency $v_i \pm 2v_M$, $v_i \pm 3v_M$, and so on, where v_i is the frequency of the incident radiation and v_M the frequency of a molecular excitation.

(c) Precision-specified transitions

The monochromatic character of laser radiation is a very powerful characteristic because it allows us to excite specific states with very high precision. One consequence of state-specificity for photochemistry is that the illumination of a sample may be photochemically precise and hence efficient in stimulating a reaction, because its frequency can be tuned exactly to an absorption. The specific excitation of a particular excited state of a molecule may greatly enhance the rate of a reaction even at low temperatures. The rate of a reaction is generally increased by raising the temperature because the energies of the various modes of motion of the molecule are enhanced. However, this enhancement increases the energy of all the modes, even those that do not contribute appreciably to the reaction rate. With a laser we can excite the kinetically significant mode, so rate enhancement is achieved most efficiently. An example is the reaction

$$BCl_3 + C_6H_6 \longrightarrow C_6H_5-BCl_2 + HCl$$

which normally proceeds only above 600°C in the presence of a catalyst; exposure to 10.6 μm CO_2 laser radiation results in the formation of products at room temperature without a catalyst. The commercial potential of this procedure is considerable (provided laser photons can be produced sufficiently cheaply), because heat-sensitive compounds, such as pharmaceuticals, may perhaps be made at lower temperatures than in conventional reactions.

A related application is the study of **state-to-state reaction dynamics**, in which a specific state of a reactant molecule is excited and we monitor not only the rate at which it forms products but also the states in which they are produced. Studies such

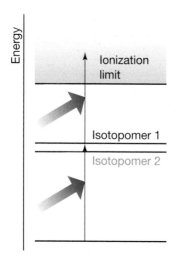

17.41 In one method of isotope separation, one photon excites an isotopomer to an excited state, and then a second photon achieves photoionization. The success of the first step depends on the nuclear mass.

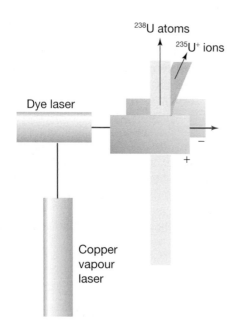

17.42 An experimental arrangement for isotope separation. The dye laser, which is pumped by a copper-vapour laser, photoionizes the U atoms selectively according to their mass, and the ions are deflected by the electric field applied between the plates.

as these give highly detailed information about the deployment of energy in chemical reactions (Chapter 27).

(d) Isotope separation

The precision state-selectivity of lasers is also of considerable potential for laser isotope separation. Isotope separation is possible because two **isotopomers**, or species that differ only in their isotopic composition, have slightly different energy levels and hence slightly different absorption frequencies.

One approach is to use **photoionization**, the ejection of an electron by the absorption of electromagnetic radiation. Direct photoionization by the absorption of a single photon does not distinguish between isotopomers because the upper level belongs to a continuum; to distinguish isotopomers it is necessary to deal with discrete states. At least two absorption processes are required. In the first step, a photon excites an atom to a higher state; in the second step, a photon achieves photoionization from that state (Fig. 17.41). The energy separation between the two states involved in the first step depends on the nuclear mass. Therefore, if the laser radiation is tuned to the appropriate frequency, only one of the isotopomers will undergo excitation and hence be available for photoionization in the second step. An example of this procedure is the photoionization of uranium vapour, in which the incident laser is tuned to excite ^{235}U but not ^{238}U. The ^{235}U atoms in the atomic beam are ionized in the two-step process; they are then attracted to a negatively charged electrode, and may be collected (Fig. 17.42). This procedure is being used in the latest generation of uranium separation plants.

Molecular isotopomers are used in techniques based on **photodissociation**, the fragmentation of a molecule following absorption of electromagnetic radiation. The key problem is to achieve both mass selectivity (which requires excitation to take place between discrete states) and dissociation (which requires excitation to continuum states). In one approach, two lasers are used: an infrared photon excites one isotopomer selectively to a higher vibrational level, and then an ultraviolet photon completes the process of photodissociation (Fig. 17.43). An alternative procedure is to make use of multiphoton absorption within the ground electronic state (Fig. 17.44); the efficiency of absorption of the first few photons depends on the match of their frequency to the energy level separations, so it is sensitive to nuclear mass. The absorbed photons open the door to a subsequent influx of enough photons to complete the dissociation process. The isotopomers $^{32}SF_6$ and $^{34}SF_6$ have been separated in this way.

In a third approach, a selectively vibrationally excited species may react with another species and give rise to products that can be separated chemically. This procedure has been employed successfully to separate isotopes of B, N, O, and, most efficiently, H. A variation on this procedure is to achieve selective **photoisomerization**, the conversion of a species to one of its isomers (particularly a geometrical isomer) on absorption of electromagnetic radiation. Once again, the initial absorption, which is isotope selective, opens the way to subsequent further absorption and the formation of a geometrical isomer that can be separated chemically. The approach has been used with the photoisomerization of CH_3NC to CH_3CN.

A different, more physical approach, that of **photodeflection**, is based on the recoil that occurs when a photon is absorbed by an atom and the linear momentum of the photon (which is equal to h/λ) is transferred to the atom. The atom is deflected from its original path only if the absorption actually occurs, and the incident radiation can be tuned to a particular isotope. The deflection is very small, so an atom must absorb dozens of photons before its path is changed sufficiently to allow collection. For instance, if a Ba atom absorbs about 50 photons of 550 nm light, it will be deflected by only about 1 mm after a flight of 1 m.

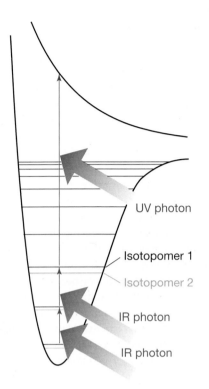

17.43 Isotopomers may be separated by making use of their selective absorption of infrared photons followed by photodissociation with an ultraviolet photon.

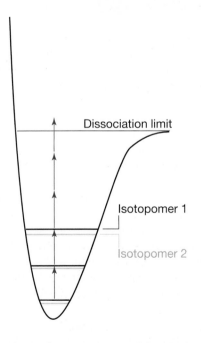

17.44 In an alternative scheme for separating isotopomers, multiphoton absorption of infrared photons is used to reach the dissociation limit of a ground electronic state.

(e) Time-resolved spectroscopy

The ability of lasers to produce pulses of very short duration is particularly useful in chemistry when we want to monitor processes in time. Q-switched lasers produce nanosecond pulses, which are generally fast enough to study reactions with rates controlled by the speed with which reactants can move through a fluid medium. However, when we want to study the rates at which energy is converted from one mode to another within a molecule, we need femtosecond and picosecond pulses. These timescales are available from mode-locked lasers. The shortest pulse currently reported is about 5 fs, corresponding to a packet of electromagnetic radiation of only a few wavelengths long.

In **time-resolved spectroscopy**, laser pulses are used to obtain the absorption, emission, or Raman spectrum of reactants, intermediates, products, and even transition states of reactions. It is also possible to study energy transfer, molecular rotations, vibrations, and conversion from one mode of motion to another. We shall see some of the information obtained from time–resolved spectroscopy in Chapters 25–29. Here, we will describe some of the experimental techniques that employ pulsed lasers.

The arrangement shown in Fig. 17.45 is often used to study ultrafast chemical reactions that can be initiated by light, such as the initial events of vision (Box 17.1). A strong and short laser pulse, the *pump*, promotes a molecule A to an excited electronic state A* that can either emit a photon (as fluorescence or phosphorescence) or react with another species B to yield a product C:

$$A + h\nu \rightarrow A^* \qquad \text{(absorption)}$$
$$A^* \rightarrow A \qquad \text{(emission)}$$
$$A^* + B \rightarrow [AB] \rightarrow C \qquad \text{(reaction)}$$

Here [AB] denotes either an intermediate or an activated complex. The rates of appearance and disappearance of the various species are determined by observing time-dependent changes in the absorption spectrum of the sample during the

17.45 A configuration used for time-resolved absorption spectroscopy, in which the same pulsed laser is used to generate a monochromatic pump pulse and, after continuum generation in a suitable liquid, a 'white' light probe pulse. The time delay between the pump and probe pulses may be varied by moving the motorized stage in the direction shown by the double arrow.

course of the reaction. This is done by passing a weak pulse of white light, the *probe*, through the sample at different times after the laser pulse. Pulsed 'white' light can be generated directly from the laser pulse by the nonlinear optical phenomenon of **continuum generation**, in which focusing an ultrafast laser pulse on a vessel containing a liquid such as water or carbon tetrachloride results in an outgoing beam with a wide distribution of frequencies. A time delay between the strong laser pulse and the 'white' light pulse can be introduced by allowing one of the beams to travel a longer distance before reaching the sample. For example, a difference in travel distance of $\Delta d = 3$ mm corresponds to a time delay $\Delta t = \Delta d/c \approx 10$ ps between two beams, where c is the speed of light. The relative distances travelled by the two beams in Fig. 17.45 are controlled by directing the 'white' light beam to a motorized stage carrying a pair of mirrors.

Variations of the arrangement in Fig. 17.45 allow for the observation of fluorescence decay kinetics of A* and time-resolved Raman spectra during the course of the reaction. The fluorescence lifetime of A* can be determined by exciting A as before and measuring the decay of the fluorescence intensity after the pulse with a fast photodetector system. In this case, continuum generation is not necessary. Time-resolved resonance Raman spectra of A, A*, B, [AB], or C can be obtained by initiating the reaction with a strong laser pulse of a certain wavelength and then, some time later, irradiating the sample with another laser pulse that can excite the resonance Raman spectrum of the desired species. Also in this case continuum generation is not necessary. Instead, the Raman excitation beam may be generated in a dye laser or by stimulated Raman scattering of the laser pulse in a medium such as $H_2(g)$, $D_2(g)$, or $CH_4(g)$.

17.46 An incoming photon carries an energy $h\nu$; an energy I_i is needed to remove an electron from an orbital i, and the difference appears as the kinetic energy of the electron.

Photoelectron spectroscopy

The technique of **photoelectron spectroscopy** (PES) measures the ionization energies of molecules when electrons are ejected from different orbitals, and uses the information to infer the orbital energies. The technique is also used to study solids, and in Chapter 28 we shall see the important information that it gives about species at or on surfaces.

17.8 The technique

Because energy is conserved when a photon ionizes a sample, the energy of the incident photon $h\nu$ must be equal to the sum of the ionization energy, I, of the sample and the kinetic energy of the **photoelectron**, the ejected electron (Fig. 17.46):

$$h\nu = \tfrac{1}{2}m_e v^2 + I \tag{17.8}$$

This equation (which is like the one used for the photoelectric effect, Section 11.2a) can be refined in two ways. First, photoelectrons may originate from one of a number of different orbitals, and each one has a different ionization energy. Hence, a series of different kinetic energies of the photoelectrons will be obtained, each one satisfying

$$h\nu = \tfrac{1}{2}m_e v^2 + I_i \tag{17.9}$$

where I_i is the ionization energy for ejection of an electron from an orbital i. Therefore, by measuring the kinetic energies of the photoelectrons, and knowing ν, these ionization energies can be determined. Photoelectron spectra are interpreted in terms of an approximation called **Koopmans' theorem**, which states that the ionization energy I_i is equal to the orbital energy of the ejected electron (formally: $I_i = -\varepsilon_i$). That is, we can identify the ionization energy with the energy of the orbital from which it is ejected. The theorem is only an approximation because it ignores the fact that the remaining electrons adjust their distributions when ionization occurs.

The ejection of an electron may leave an ion in a vibrationally excited state. Then not all the excess energy of the photon appears as kinetic energy of the photoelectron, and we should write

$$h\nu = \tfrac{1}{2}m_e v^2 + I_i + E^+_{\text{vib}} \tag{17.10}$$

where E^+_{vib} is the energy used to excite the ion into vibration. Each vibrational quantum that is excited leads to a different kinetic energy of the photoelectron, and gives rise to the vibrational structure in the photoelectron spectrum.

The ionization energies of molecules are several electronvolts even for valence electrons, so it is essential to work in at least the ultraviolet region of the spectrum and with wavelengths of less than about 200 nm. Much work has been done with radiation generated by a discharge through helium: the He(I) line $(1s^1 2p^1 \rightarrow 1s^2)$ lies at 58.43 nm, corresponding to a photon energy of 21.22 eV. Its use gives rise to the technique of **ultraviolet photoelectron spectroscopy** (UPS). When core electrons are being studied, photons of even higher energy are needed to expel them: X-rays are used, and the technique is denoted XPS. A modern version of PES makes use of synchrotron radiation (Section 16.1), which may be continuously tuned between UV and X-ray energies. The additional information that stems from the variation of the photoejection probability with wavelength is a valuable guide to the identity of the molecule and the orbital from which photoionization occurs.

Illustration 17.1

Photoelectrons ejected from N_2 with He(I) radiation had kinetic energies of 5.63 eV (1 eV = 8065.5 cm^{-1}). Helium(I) radiation of wavelength 58.43 nm has wavenumber 1.711×10^5 cm^{-1} and therefore corresponds to an energy of 21.22 eV. Then, from eqn 8, 21.22 eV = 5.63 eV + I_i, so I_i = 15.59 eV. This ionization energy is the energy needed to remove an electron from the HOMO of the N_2 molecule, the $3\sigma_g$ bonding orbital (see Fig. 14.30).

Self-test 17.4 Under the same circumstances, photoelectrons are also detected at 4.53 eV. To what ionization energy does that correspond? Suggest an origin.
[16.7 eV, $1\pi_u$]

The kinetic energies of the photoelectrons are measured using an electrostatic deflector that produces different deflections in the paths of the photoelectrons as they pass between charged plates (Fig. 17.47). As the field strength is increased, electrons of different speeds, and therefore kinetic energies, reach the detector. The electron flux can be recorded and plotted against kinetic energy to obtain the photoelectron spectrum.

17.9 Ultraviolet photoelectron spectroscopy

A typical photoelectron spectrum (of HBr) is shown in (Fig. 17.48). If we disregard the fine structure, we see that the HBr lines fall into two main groups. The least tightly bound electrons (with the lowest ionization energies and hence highest kinetic

17.47 A photoelectron spectrometer consists of a source of ionizing radiation (such as a helium discharge lamp for UPS and an X-ray source for XPS), an electrostatic analyser, and an electron detector. The deflection of the electron path caused by the analyser depends on their speed.

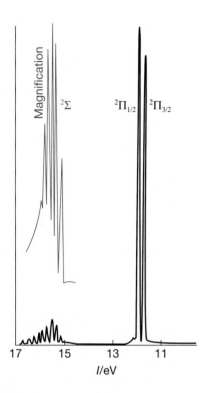

17.48 The photoelectron spectrum of HBr. The lowest ionization energy bands (Π) correspond to the ionization of a Br lone-pair electron. The higher ionization energy band (Σ) corresponds to the ionization of a bonding electron. The structure on the latter is due to the vibrational excitation of HBr$^+$ that results from the ionization.

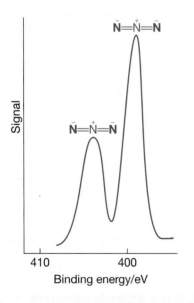

17.49 The photoelectron spectrum of solid NaN$_3$ excited by Al K-radiation, showing the region of N core ionization and the assignment. (K. Siegbahn *et al.*, *Science*, **176**, 245 (1972).)

energies when ejected) are those in the nonbonding lone pairs of Br (with $I = 11.8$ eV). The next ionization energy lies at 15.2 eV, and corresponds to the removal of an electron from the H—Br σ bond.

The HBr spectrum shows that ejection of a σ electron is accompanied by a long vibrational progression. The Franck–Condon principle would account for this progression if ejection were accompanied by an appreciable change of equilibrium bond length between HBr and HBr$^+$ because the ion is formed in a bond-compressed state, which is consistent with the important bonding effect of the σ electrons. The lack of much vibrational structure in the two bands labelled $^2\Pi$ is consistent with the nonbonding role of the Br$4p\pi$ lone pair of electrons, for the equilibrium bond length is little changed when one is removed.

Illustration 17.2

The highest kinetic-energy electrons in the spectrum of H$_2$O using 21.22 eV He radiation are about 9 eV and show a large vibrational spacing of 0.41 eV (or 3310 cm^{-1}), which is similar to the vibrational wavenumber of the symmetric stretching mode of the neutral H$_2$O molecule (3652 cm^{-1}). Therefore, we can suspect that the electron is ejected from an orbital that has little influence on the bonding in the molecule. That is, photoejection is from a largely nonbonding orbital.

17.10 X-ray photoelectron spectroscopy

In XPS, the energy of the incident photon is so great that electrons are ejected from inner cores of atoms. As a first approximation, core ionization energies are insensitive to the bonds between atoms because they are too tightly bound to be greatly affected by the changes that accompany bond formation, so core ionization energies are characteristic of the individual atom rather than the overall molecule. Consequently, XPS gives lines characteristic of the elements present in a compound or alloy. For instance, the K-shell ionization energies (eV) of the second-row elements are

Li	Be	B	C	N	O	F
50	110	190	280	400	530	690

Detection of one of these values (and values corresponding to ejection from other inner shells) indicates the presence of the corresponding element. This application is responsible for the alternative name **electron spectroscopy for chemical analysis** (ESCA). The technique is mainly limited to the study of surface layers (as we explore in Chapter 28) because, even though X-rays may penetrate into the bulk sample, the ejected electrons cannot escape except from within a few nanometres of the surface. Despite (or because of) this limitation, the technique is very useful for studying the surface state of heterogeneous catalysts, the differences between surface and bulk structures, and the processes that can cause damage to high-temperature superconductors and semiconductor wafers.

Whereas it is largely true that core ionization energies are unaffected by bond formation, it is not entirely true, and small but detectable shifts can be detected and interpreted in terms of the environments of the atoms. For example, the azide ion, N$_3^-$, gives the spectrum shown in Fig. 17.49. Although the spectrum lies in the region of 400 eV (and hence is typical of N1s electrons), it has a doublet structure with splitting 6 eV. This splitting can be understood by noting that the structure of the ion is

The web site contains links to a database of X-ray photoelectron spectra.

N=N=N, with more negative charge on the outer two N atoms than on the inner (the formal charges are (–1, +1, –1)). The presence of the negative charges on the terminal atoms lowers the core ionization energies, whereas the positive charge on the central atom raises it. This inequivalence of the atoms results in two lines in the spectrum with intensities in the ratio 2:1. Observations like this can be used to obtain valuable information about the presence of chemically inequivalent atoms of the same element.

For important applications of these photoionization techniques to surfaces and substances adsorbed on surfaces, see Chapter 28.

Checklist of key ideas

The characteristics of electronic transitions

17.1 The electronic spectra of diatomic molecules

☐ vibronic transition, a transition that accounts for the vibrational structure of electronic spectra

☐ parity, the behaviour of wavefunctions under inversion

☐ inversion, the process of projecting a point on a straight line through the centre of the molecule to a point the same distance out on the other side

☐ centre of inversion, the point in a molecule through which all points are inverted

☐ term symbol, $^{2S+1}\{|\Lambda|\}_{g,u}^{+/-}$

☐ selection rules, $\Delta\Lambda = 0, \pm 1, \Delta S = 0$, $\Delta\Sigma = 0, \Delta\Omega = 0, \pm 1$

☐ Laporte selection rule (for centrosymmetric molecules): The only allowed transitions are transitions that are accompanied by a change of parity

☐ vibronically allowed transition, a transition that derives its intensity from an asymmetrical vibration of a molecule

☐ Franck–Condon principle: Because the nuclei are so much more massive than the electrons, an electronic transition takes place very much faster than the nuclei can respond

☐ vertical transition, an electronic transition that occurs without change of nuclear geometry

☐ vibrational progression, vibrational structure of an electronic transition

☐ Franck–Condon factor, the quantity $|S(v_f, v_i)|^2$, $S(v_f, v_i) = \langle v_f|v_i\rangle$

☐ band head, the convergence of a P or R branch

17.2 The electronic spectra of polyatomic molecules

☐ chromophore, a group with a characteristic optical absorption

☐ ligand-field splitting parameter, Δ_O, the difference in energy between the t_{2g} and e_g orbitals in an octahedral d-metal complex

☐ charge-transfer transition, a transition in which an electron moves from metal to ligand (MLCT) or from ligand to metal (LMCT) in a complex

The fates of electronically excited states

☐ radiative decay process, a process in which a molecule discards its excitation energy as a photon

☐ nonradiative decay, a process in which the excess energy is transferred into the vibration, rotation, and translation of the surrounding molecules

17.3 Fluorescence and phosphorescence

☐ fluorescence, the spontaneous emission of radiation arising from a transition between states of the same multiplicity

☐ phosphorescence, the spontaneous emission of radiation arising from a transition between states of different multiplicity

☐ intersystem crossing, a nonradiative transition between states of different multiplicity

☐ Jablonski diagram, a schematic diagram of the various types of nonradiative and radiative transitions that can occur in molecules

17.4 Dissociation and predissociation

☐ dissociation, the breaking of bonds

☐ dissociation limit, the frequency at which vibrational structure is replaced by continuous absorption

☐ predissociation, dissociation that occurs before the dissociation limit is reached

☐ internal conversion, nonradiative transition between states of the same multiplicity

Lasers

17.5 General principles of laser action

☐ metastable excited state, an excited state with a long enough lifetime for it to participate in stimulated emission

☐ population inversion, an arrangement in which there are more molecules in the upper state than in the lower state

☐ three-level laser, a laser arrangement based on three participating states

☐ pumping, the transfer of population to an excited state

☐ four-level laser, a laser arrangement based on four participating states

☐ resonant modes, wavelengths that can be sustained by the cavity and contribute to the laser action

□ coherent, a term used to described radiation in which all electromagnetic waves are in phase

□ spatial coherence, the waves are in step across the cross-section of the beam

□ temporal coherence, the waves are in step along the beam

□ coherence length, the distance over which the waves remain coherent, $l_C = \lambda^2/2\Delta\lambda$

□ Q-switching, the modification of the resonance characteristics of the laser cavity and, consequently, of the laser output

□ Pockels cell, an electro-optical device based on the ability of some crystals to convert plane-polarized light to circularly polarized light when an electrical potential difference is applied

□ saturable absorber, a solution of a dye that loses its ability to absorb when many of its molecules have been excited by intense radiation

□ mode locking, a technique for producing pulses of picosecond duration and less

□ optical Kerr effect, the change in refractive index of a well chosen medium when it is exposed to intense electric fields

□ Kerr medium, the medium for displaying the Kerr effect

□ gain, the growth in intensity

□ Kerr lens, the self-focusing of the laser beam by using the Kerr effect

17.6 Practical lasers

□ solid-state laser, a laser in which the active medium is in the form of a single crystal or a glass

□ neodymium laser, a four-level laser that (for example) consists of Nd^{3+} ions at low concentration in yttrium aluminium garnet

□ nonlinear optical phenomena, optical phenomena that depend on the square and higher powers of the electric field strength of incident electromagneitc radiation

□ frequency doubling (or second harmonic generation), the process in which an intense laser beam is converted to radiation with twice its initial frequency as it passes though a suitable material

□ titanium sapphire laser, a laser that consists of Ti^{3+} ions at low concentration in a crystal of sapphire (Al_2O_3)

□ vibronic laser, a type of tunable laser in which the transitions originate from vibronic transitions in the laser medium

□ helium–neon laser, a laser in which the active medium is a mixture of helium and neon

□ argon-ion laser, a laser in which the active medium consists of argon ions

□ krypton-ion laser, a laser in which the active medium consists of krypton ions

□ carbon dioxide laser, a laser in which the active medium is carbon dioxide gas

□ nitrogen laser, a laser in which the active medium is nitrogen gas

□ superradiant, a laser in which the gain in intensity is so great that a single passage of a pulse of radiation is enough to generate laser radiation

□ exciplex laser, a laser that relies on the formation of an exciplex

□ exciplex, a combination of two atoms or molecules that survives only in an excited state and which dissociates as soon as the excitation energy has been discarded

□ dye laser, a type of tunable laser in which the active medium is a solution of dye molecules

17.7 Applications of lasers in chemistry

□ multiphoton processes, processes in which the photon density is so high that more than one photon may be absorbed by a single molecule

□ stimulated Raman spectroscopy, a form of Raman spectroscopy in which the Stokes and anti-Stokes radiation in the forward direction are powerful enough to undergo more scattering and hence give up or acquire more quanta of energy from the molecules in the sample

□ state-to-state reaction dynamics, a procedure in which a specific state of a reactant molecule is excited and the rate at which it forms products in a particular state is monitored

□ isotopomers, species that differ only in their isotopic composition

□ photoionization, the ejection of an electron by the absorption of electromagnetic radiation

□ photodissociation, the fragmentation of a molecule following absorption of electromagnetic radiation

□ photoisomerization, the conversion of a species to one of its isomers (particularly a geometrical isomer) on absorption of electromagnetic radiation

□ photodeflection, the recoil that occurs when a photon is absorbed by an atom

□ time-resolved spectroscopy, a technique in which laser pulses are used to obtain the absorption, emission, or Raman spectrum of reactants, intermediates, products, and transition states of reactions

□ continuum generation, a process in which focusing an ultrafast laser pulse on a vessel containing a liquid results in an outgoing beam with a wide distribution of frequencies

Photoelectron spectroscopy

□ photoelectron spectroscopy (PES), a technique based on the measurement of the ionization energies of molecules when electrons are ejected from different orbitals

17.8 The technique

□ photoelectron, an ejected electron with kinetic energy given by $h\nu = \frac{1}{2}m_e\upsilon^2 + I_i$

□ Koopmans' theorem: The ionization energy I_i is equal to the orbital energy of the ejected electron; $I_i = -\varepsilon_i$

□ ultraviolet photoelectron spectroscopy (UPS), PES in which ultraviolet radiation is used

17.9 Ultraviolet photoelectron spectroscopy

17.10 X-ray photoelectron spectroscopy

□ electron spectroscopy for chemical analysis (ESCA), PES in which X-rays are used to study the ejection of electrons from inner shells

Further reading

Articles of general interest

F. Ahmed, A good example of the Franck–Condon principle. *J. Chem. Educ.* **64**, 427 (1987).

M. Allan, Electron spectroscopic techniques in teaching. *J. Chem. Educ.* **64**, 418 (1987).

N. Bloembergen, Nonlinear optical instrumentation. *J. Chem. Educ.* **75**, 555 (1998).

P. Callis, Two-photon-induced fluorescence. *Ann. Rev. Phys. Chem.* **48**, 271 (1997).

E. Chen, R.A. Goldbeck, and D.S. Kliger. *Ann. Rev. Biophys. Biomol. Struct.* **26**, 327 (1997).

B.J. Duke and B. O'Leary, Non-Koopmans' molecules. *J. Chem. Educ.* **72**, 501 (1995).

P. Engelking, Laser photochemistry. In *Encyclopedia of applied physics* (ed. G.L. Trigg) **8**, 283. VCH, New York (1994).

F.J. Himpsell and I. Lindau, Photoemission and photoelectron spectra. In *Encyclopedia of applied physics* (ed. G.L. Trigg) **13**, 477. VCH, New York (1995).

R.M. Hochstrasser, Ultrafast spectroscopy of protein dynamics. *J. Chem. Educ.* **75**, 559 (1998).

P.L. Kelley and J.J. Zayhowski, Laser physics. In *Encyclopedia of applied physics* (ed. G.L. Trigg) **8**, 299. VCH, New York (1994).

L. Lessinger, Birge–Sponer extrapolation, and the electronic absorption spectrum of I_2. *J. Chem. Educ.* **71**, 388 (1994).

G. Mourou, Ultrahigh-intensity lasers: nonlinear optics in the relativistic regime for future applications in time-resolved chemistry. *J. Chem. Educ.* **75**, 565 (1998).

D.A. Ramsay, Molecular spectroscopy. In *Encyclopedia of applied physics* (ed. G.L. Trigg) **10**, 491. VCH. New York (1994).

G.M. Shalhoub, Visible spectra of conjugated dyes: integrating quantum chemical concepts with experimental data. *J. Chem. Educ.* **74**, 1317 (1997).

R.B. Snadden, The iodine spectrum revisited. *J. Chem. Educ.* **64**, 919 (1987).

O. Sorkhabi, W.M. Jackson, and I. Daizadeh, Cometary spectroscopy. *J. Chem. Educ.* **75**, 1472 (1998).

G. Steinmeyer, D.H. Sutter, L. Gallmann, N. Matuschek, and U. Keller, Frontiers in ultrashort pulse generation: pushing the limits in linear and nonlinear optics. *Science* **286**, 1507 (1999).

H. Takuma, Laser technology. In *Encyclopedia of applied physics* (ed. G.L. Trigg) **8**, 321. VCH, New York (1994).

J.C. Wright and T.J. Zielinski, Franck–Condon factors and their use in undergraduate quantum mechanics. *J. Chem. Educ.* **76**, 1367 (1999).

Texts and sources of data and information

D.L. Andrews, *Lasers in chemistry.* Springer-Verlag, Berlin (1997).

D.L. Andrews and A.A. Demidov (ed.), *An introduction to laser spectroscopy.* Plenum, New York (1995).

P.W. Atkins and R.S. Friedman, *Molecular quantum mechanics.* Oxford University Press (1997).

T.L. Barr, *Modern ESCA: the principles and practice of X-ray photoelectron spectroscopy.* CRC Press, Boca Raton (1994).

P.F. Bernath, *Spectra of atoms and molecules.* Oxford University Press, New York (1995).

C.R. Brundle and A.D. Baker (ed.) *Electron spectroscopy: theory, techniques, and applications*, Vols 1–4. Academic Press, London (1977–81).

W. Demtröder, *Laser spectroscopy: basic concepts and instrumentation.* Springer, Berlin (1996).

J.-C. Diels and W. Rudolph, *Ultrashort laser pulse phenomena: fundamentals, techniques, and applications on a femtosecond time scale.* Academic Press, San Diego (1996).

R. Drago, *Physical methods for chemists.* Saunders, Philadelphia (1992).

J.H.D. Eland, *Photoelectron spectra.* Butterworth, London (1984).

A.G. Gaydon, *Dissociation energies.* Chapman and Hall, London (1968).

D.C. Harris and M.D. Bertollucci, *Symmetry and spectroscopy: an introduction to vibrational and electronic spectroscopy.* Dover, New York (1989).

E. Hecht, *Optics.* Addison–Wesley, Menlo Park (1998).

B. Herman, *Fluorescence microscopy.* Springer, New York (1998).

G. Herzberg, *Molecular spectra and molecular structure I. Spectra of diatomic molecules.* Krieger, Malabar (1989).

G. Herzberg, *Molecular spectra and molecular structure III. Electronic spectra and electronic structure of polyatomic molecules.* Van Nostrand–Reinhold, New York (1966).

G. Herzberg, *The spectra and structures of simple free radicals: an introduction to molecular spectroscopy.* Dover, New York (1988).

J.M. Hollas, *Modern spectroscopy.* Wiley, New York (1998).

J.M. Hollas, *High resolution spectroscopy.* Butterworth, London (1998).

M.E. Jacox, *Vibrational and electronic energy levels of polyatomic transient molecules.* Journal of Physical and Chemical Reference Data, Monograph No. 3 (1994).

J.R. Lakowicz, *Principles of fluorescence spectroscopy.* Kluwer/Plenum, New York (1999).

D.R. Lide (ed.), *CRC Handbook of Chemistry and Physics*, Sections 9 and 10. CRC Press, Boca Raton (2000).

J.C. Lindon, G.E. Tranter, and J.L. Holmes (ed.), *Encyclopedia of spectroscopy and spectrometry.* Academic Press, San Diego (2000).

A.B. Myers and T.R. Rizzo (ed.), *Techniques in Chemistry XXIII: laser techniques in chemistry.* Wiley (1995).

J.L. McHale, *Molecular spectroscopy.* Prentice–Hall, Upper Saddle River (1999).

J.F. Moulder, W.F. Stickle, P.E. Sobol, and K. Bomben, *Handbook of X-ray photoelectron spectroscopy: a reference book of standard spectra for identification and interpretation of XPS data.* Physical Electronics, Eden Prairie (1995).

J.W. Robinson (ed.), *Practical handbook of spectroscopy.* CRC Press, Boca Raton (1991).

O. Svelto, *Principles of lasers.* Plenum, New York (1998).

V. Sunström (ed.), *Femtochemistry and femtobiology: ultrafast reaction dynamics at atomic-scale resolution.* Imperial College Press, London (1997).

G.R. Van Hecke and K.K. Karukstis, *A guide to lasers in chemistry.* Jones and Bartlett, Boston (1998).

C.E. Wayne and R.P. Wayne, *Photochemistry.* Oxford Chemistry Primers, Oxford University Press (1999).

Exercises

Discussion questions

17.1 (a) Explain the origin of the term symbol $^3\Sigma_g^+$ for the ground state of dioxygen.

17.1 (b) Explain the basis of the Franck–Condon principle and how it leads to the formation of a vibrational progression.

17.2 (a) How do the band heads in P and R branches arise? Could the Q branch show a head?

17.2 (b) Explain how colour can arise from molecules.

17.3 (a) Describe the mechanism of fluorescence. To what extent is a fluorescence spectrum not the exact mirror image of the corresponding absorption spectrum?

17.3 (b) What is the evidence for the correctness of the mechanism of fluorescence?

17.4 (a) Describe the principle of laser action, with actual examples.

17.4 (b) What features of laser radiation are applied in chemistry? Discuss two applications of lasers in chemistry.

Numerical exercises

17.5 (a) Where appropriate, give the parity of (a) π^* in F_2, (b) σ^* in NO, (c) δ in Tl_2, (d) δ^* in Fe_2.

17.5 (b) Give the parities of the six π molecular orbitals of benzene.

17.6 (a) The term symbol for the ground state of N_2^+ is $^2\Sigma_g^+$. What is the total spin and total orbital angular momentum of the molecule? Show that the term symbol agrees with the electron configuration that would be predicted using the building-up principle.

17.6 (b) One of the excited states of the C_2 molecule has the valence electron configuration $1\sigma_g^2 2\sigma_u^{*2} 1\pi_u^3 2\pi_g^1$. Give the multiplicity and parity of the term.

17.7 (a) The molar absorption coefficient of a substance dissolved in hexane is known to be 855 L mol^{-1} cm^{-1} at 270 nm. Calculate the percentage reduction in intensity when light of that wavelength passes through 2.5 mm of a solution of concentration 3.25 mmol L^{-1}.

17.7 (b) The molar absorption coefficient of a substance dissolved in hexane is known to be 327 L mol^{-1} cm^{-1} at 300 nm. Calculate the percentage reduction in intensity when light of that wavelength passes through 1.50 mm of a solution of concentration 2.22 mmol L^{-1}.

17.8 (a) A solution of an unknown component of a biological sample when placed in an absorption cell of path length 1.00 cm transmits 20.1 per cent of light of 340 nm incident upon it. If the concentration of the component is 0.111 mmol L^{-1}, what is the molar absorption coefficient?

17.8 (b) When light of wavelength 400 nm passes through 3.5 mm of a solution of an absorbing substance at a concentration 0.667 mmol L^{-1}, the transmission is 65.5 per cent. Calculate the molar absorption coefficient of the solute at this wavelength and express the answer in cm^2 mol^{-1}.

17.9 (a) The molar absorption coefficient of a solute at 540 nm is 286 L mol^{-1} cm^{-1}. When light of that wavelength passes through a 6.5 mm cell containing a solution of the solute, 46.5 per cent of the light was absorbed. What is the concentration of the solution?

17.9 (b) The molar absorption coefficient of a solute at 440 nm is 323 L mol^{-1} cm^{-1}. When light of that wavelength passes through a 7.50 mm cell containing a solution of the solute, 52.3 per cent of the light was absorbed. What is the concentration of the solution?

17.10 (a) The absorption associated with a particular transition begins at 230 nm, peaks sharply at 260 nm, and ends at 290 nm. The maximum value of the molar absorption coefficient is 1.21×10^4 L mol^{-1} cm^{-1}. Estimate the integrated absorption coefficient of the transition assuming a triangular lineshape (see eqn 16.12).

17.10 (b) The absorption associated with a certain transition begins at 199 nm, peaks sharply at 220 nm, and ends at 275 nm. The maximum value of the molar absorption coefficient is 2.25×10^4 L mol^{-1} cm^{-1}. Estimate the integrated absorption coefficient of the transition assuming an inverted parabolic lineshape (Fig. 17.50; use eqn 16.12).

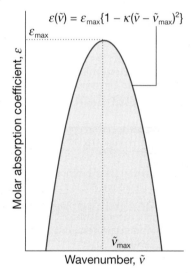

$$\varepsilon(\tilde{v}) = \varepsilon_{max}\{1 - \kappa(\tilde{v} - \tilde{v}_{max})^2\}$$

Molar absorption coefficient, ε / ε_{max} / $\tilde{v}_{max}$ / Wavenumber, $\tilde{v}$

Fig. 17.50

17.11 (a) The two compounds, 2,3-dimethyl-2-butene and 2,5-dimethyl-2,4-hexadiene, are to be distinguished by their ultraviolet absorption spectra. The maximum absorption in one compound occurs at 192 nm and in the other at 243 nm. Match the maxima to the compounds and justify the assignment.

17.11 (b) 1,3,5-hexatriene (a kind of 'linear' benzene) was converted into benzene itself. On the basis of a free-electron molecular orbital model (in which hexatriene is treated as a linear box and benzene as a ring), would you expect the lowest energy absorption to rise or fall in energy?

17.12 (a) The following data were obtained for the absorption by Br_2 in carbon tetrachloride using a 2.0 mm cell. Calculate the molar absorption coefficient of bromine at the wavelength employed:

$[Br_2]/(mol\ L^{-1})$	0.0010	0.0050	0.0100	0.0500
T/(per cent)	81.4	35.6	12.7	3.0×10^{-3}

17.12 (b) The following data were obtained for the absorption by a dye dissolved in methylbenzene using a 2.50 mm cell. Calculate the molar absorption coefficient of the dye at the wavelength employed:

[dye]/(mol L^{-1})	0.0010	0.0050	0.0100	0.0500
T/(per cent)	73	21	4.2	1.33×10^{-5}

17.13 (a) A 2.0-mm cell was filled with a solution of benzene in a non-absorbing solvent. The concentration of the benzene was 0.010 mol L^{-1} and the wavelength of the radiation was 256 nm (where there is a maximum in the absorption). Calculate the molar absorption coefficient of benzene at this wavelength given that the transmission was 48 per cent. What will the transmittance be in a 4.0-mm cell at the same wavelength?

17.13 (b) A 2.50-mm cell was filled with a solution of a dye. The concentration of the dye was 15.5 mmol L^{-1}. Calculate the molar absorption coefficient of benzene at this wavelength given that the transmission was 32 per cent. What will the transmittance be in a 4.50-mm cell at the same wavelength?

17.14 (a) A swimmer enters a gloomier world (in one sense) on diving to greater depths. Given that the mean molar absorption coefficient of sea water in the visible region is 6.2×10^{-5} L mol^{-1} cm^{-1}, calculate the depth at which a diver will experience (a) half the surface intensity of light, (b) one-tenth the surface intensity.

17.14 (b) Given that the maximum molar absorption coefficient of a molecule containing a carbonyl group is 30 L mol^{-1} cm^{-1} near 280 nm, calculate the thickness of a sample that will result in (a) half the initial intensity of radiation, (b) one-tenth the initial intensity.

17.15 (a) The electronic absorption bands of many molecules in solution have half-widths at half-height of about 5000 cm^{-1}. Estimate the integrated absorption coefficients of bands for which (a) $\varepsilon_{max} \approx 1 \times 10^4$ L mol^{-1} cm^{-1}, (b) $\varepsilon_{max} \approx 5 \times 10^2$ L mol^{-1} cm^{-1}.

17.15 (b) The electronic absorption band of a compound in solution had a Gaussian lineshape and a half-width at half-height of 4233 cm^{-1} and $\varepsilon_{max} = 1.54 \times 10^4$ L mol^{-1} cm^{-1}. Estimate the integrated absorption coefficient.

17.16 (a) The photoionization of H$_2$ by 21 eV photons produces H$_2^+$. Explain why the intensity of the $v = 2 \leftarrow 0$ transition is stronger than that of the $0 \leftarrow 0$ transition.

17.16 (b) The photoionization of F$_2$ by 21 eV photons produces F$_2^+$. Would you expect the $2 \leftarrow 0$ transition to be weaker or stronger than the $0 \leftarrow 0$ transition? Justify your answer.

Problems†

Numerical problems

17.1 The vibrational wavenumber of the oxygen molecule in its electronic ground state is 1580 cm^{-1}, whereas that in the first excited state (B $^3\Sigma_u^-$), to which there is an allowed electronic transition, is 700 cm^{-1}. Given that the separation in energy between the minima in their respective potential energy curves of these two electronic states is 6.175 eV, what is the wavenumber of the lowest energy transition in the band of transitions originating from the $v = 0$ vibrational state of the electronic ground state to this excited state? Ignore any rotational structure or anharmonicity.

17.2 A Birge–Sponer extrapolation yields 7760 cm^{-1} as the area under the curve for the B state of the oxygen molecule described in Problem 17.1. Given that the B state dissociates to ground-state atoms (at zero energy, ^{3}P) and 15 870 cm^{-1} (^{1}D) and the lowest vibrational state of the B state is 49 363 cm^{-1} above the lowest vibrational state of the ground electronic state, calculate the dissociation energy of the molecular ground state to the ground-state atoms.

17.3 The electronic spectrum of the IBr molecule shows two low-lying, well-defined convergence limits at 14 660 and 18 345 cm^{-1}. Energy levels for the iodine and bromine atoms occur at 0, 7598; and 0, 3685 cm^{-1}, respectively. Other atomic levels are at much higher energies. What possibilities exist for the numerical value of the dissociation energy of IBr? Decide which is the correct possibility by calculating this quantity from $\Delta_f H^{\ominus}$(IBr, g) = +40.79 kJ mol^{-1} and the dissociation energies of I$_2$(g) and Br$_2$(g) which are 146 and 190 kJ mol^{-1}, respectively.

17.4 In many cases it is possible to assume that an absorption band has a Gaussian lineshape (one proportional to e^{-x^2}) centred on the band maximum. Assume such a lineshape, and show that $A \approx 1.0645 c\varepsilon_{max}\Delta\tilde{\nu}_{1/2}$, where $\Delta\tilde{\nu}_{1/2}$ is the width at half-height. The absorption spectrum of azoethane (CH$_3$CH$_2$N$_2$) between 24 000 cm^{-1} and 34 000 cm^{-1} is shown in Fig. 17.51. First, estimate A for the band by assuming that it is Gaussian. Then integrate the absorption band graphically. The latter can be done either by ruling and counting squares, or by tracing the lineshape on to paper and weighing. A more sophisticated procedure would be to use mathematical software to fit a polynomial to the absorption band (or a Gaussian), and then to integrate the result analytically.

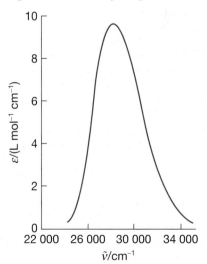

Fig. 17.51

† Problems denoted with the symbol ‡ were supplied by Charles Trapp and Carmen Giunta.

Fig. 17.52

Fig. 17.53

17.5 A lot of information about the energy levels and wavefunctions of small inorganic molecules can be obtained from their ultraviolet spectra. An example of a spectrum with considerable vibrational structure, that of gaseous SO_2 at 25°C, is shown in Fig. 17.52. Estimate the integrated absorption coefficient for the transition. What electronic states are accessible from the A_1 ground state of this C_{2v} molecule by electric dipole transitions?

17.6‡ J.G. Dojahn, E.C.M. Chen, and W.E. Wentworth (*J. Phys. Chem.* **100**, 9649 (1996)) characterized the potential energy curves of the ground and electronic states of homonuclear diatomic halogen anions. These anions have a $^2\Sigma_u^+$ ground state and $^2\Pi_g$, $^2\Pi_u$, and $^2\Sigma_g^+$ excited states. To which of the excited states are transitions by absorption of photons allowed? Explain.

17.7 A transition of particular importance in O_2 gives rise to the 'Schumann–Runge band' in the ultraviolet region. The wavenumbers (in cm^{-1}) of transitions from the ground state to the vibrational levels of the first excited state ($^3\Sigma_u^-$) are 50 062.6, 50 725.4, 51 369.0, 51 988.6, 52 579.0, 53 143.4, 53 679.6, 54 177.0, 54 641.8, 55 078.2, 55 460.0, 55 803.1, 56 107.3, 56 360.3, 56 570.6. What is the dissociation energy of the upper electronic state? (Use a Birge–Sponer plot.) The same excited state is known to dissociate into one ground-state O atom and one excited-state atom with an energy 190 kJ mol^{-1} above the ground state. (This excited atom is responsible for a great deal of photochemical mischief in the atmosphere.) Ground-state O_2 dissociates into two ground-state atoms. Use this information to calculate the dissociation energy of ground-state O_2 from the Schumann–Runge data.

17.8 Aromatic hydrocarbons and I_2 form complexes from which charge transfer electronic transitions are observed. The hydrocarbon acts an electron donor and I_2 as an electron acceptor. The energies

$h\nu_{max}$ of the charge-transfer transitions for a number of hydrocarbon–I_2 complexes are given below:

hydrocarbon	benzene	biphenyl	naphthalene
$h\nu_{max}$/eV	4.184	3.654	3.452

hydrocarbon	phenanthrene	pyrene	anthracene
$h\nu_{max}$/eV	3.288	2.989	2.890

Investigate the hypothesis that there is a correlation between the energy of the HOMO of the hydrocarbon (from which the electron comes in the charge-transfer transition) and $h\nu_{max}$. Use one of the molecular electronic structure methods discussed in Chapter 14 to determine the energy of the HOMO of each hydrocarbon in the data set.[8]

17.9 A certain molecule fluoresces at a wavelength of 400 nm with a half-life of 1.0 ns. It phosphoresces at 500 nm. If the ratio of the transition probabilities for stimulated emission for the S* → S to the T → S transitions is 1.0×10^5, what is the half-life of the phosphorescent state?

17.10‡ Refer to Fig. 17.32 and estimate the maximum laser power that can be delivered from a ruby crystal of length 5.0 cm and diameter 0.50 cm, in a pulse of duration 100 ns. 'Pink ruby' consists of about 0.050 per cent by mass Cr^{3+} and the mass density of Al_2O_3 is 3.97 g cm^{-3}. Assume that the pumping radiation is of sufficient intensity to pump all the chromium ions out of their ground state at a rate faster than they decay back to the ground state.

17.11 The photoelectron spectra of (a) N_2 and (b) CO are shown in Fig. 17.53. Ascribe the lines to the ionization processes involved

8. The web site contains links to molecular modelling freeware and to other sites where you may perform molecular orbital calculations directly from your web browser.

and classify the orbitals from which the electrons are ejected as bonding, nonbonding, or antibonding in the light of the extent of vibrational structure in the band. Analyse the bands near 4 eV in terms of the vibrational energy levels of the ions.

17.12 The photoelectron spectrum of NO can be described as follows (D.W. Turner, in *Physical methods in advanced inorganic chemistry* (ed. H.A.O. Hill and P. Day), Wiley, Chichester (1968)). Using He 584 pm (21.21 eV) radiation there is a single strong peak at kinetic energy 4.69 eV and a long series of 24 lines starting at 5.56 eV and ending at 2.2 eV. A shorter series of six lines begins at 12.0 eV and ends at 10.7 eV. Account for this spectrum.

17.13 The photoelectron spectrum of H_2O was obtained with 21.22 eV He radiation. A band near 7.0 eV shows a long vibrational series with spacing 0.125 eV. The bending mode of H_2O lies at 1596 cm^{-1}. What conclusions can you draw about the characteristics of the orbital occupied by the photoelectron?

Theoretical problems

17.14 Assume that the electronic states of the π electrons of a conjugated molecule can be approximated by the wavefunctions of a particle in a one-dimensional box, and that the dipole moment can be related to the displacement along this length by $\mu = -ex$. Show that the transition probability for the transition $2 \leftarrow 1$ is nonzero, whereas that for $3 \leftarrow 1$ is zero.

17.15 Use a group theoretical argument to decide which of the following transitions are electric-dipole allowed: (a) the $\pi^* \leftarrow \pi$ transition in ethene, (b) the $\pi^* \leftarrow n$ transition in a carbonyl group in a C_{2v} environment.

17.16 Suppose that you are a colour chemist and had been asked to intensify the colour of a dye without changing the type of compound, and that the dye in question was a polyene. Would you choose to lengthen or to shorten the chain? Would the modification to the length shift the apparent colour of the dye towards the red or the blue?

17.17 One measure of the intensity of a transition of frequency v is the *oscillator strength*, f, which is defined as

$$f = \frac{8\pi^2 m_e v |\boldsymbol{\mu}_{fi}|^2}{3he^2}$$

Consider an electron in an atom to be oscillating harmonically in one dimension (the three-dimensional version of this model was used in early attempts to describe atomic structure). The wavefunctions for such an electron are those in Table 12.1. Show that the oscillator strength for the transition of this electron from its ground state is exactly $\frac{1}{3}$.

17.18 Estimate the oscillator strength (see Problem 17.17) of a charge-transfer transition modelled as the migration of an electron from a H1s orbital on one atom to another H1s orbital on an atom a distance R away. Approximate the transition moment by $-eRS$ where S is the overlap integral of the two orbitals. Sketch the oscillator strength as a function of R using the curve for S given in Fig. 14.33. Why does the intensity fall to zero as R approaches 0 and infinity?

17.19 The line marked A in Fig. 17.54 is the fluorescence spectrum of benzophenone in solid solution in ethanol at low temperatures observed when the sample is illuminated with 360 nm light. What can be said about the vibrational energy levels of the carbonyl group in (a) its ground electronic state and (b) its excited electronic state? When naphthalene is illuminated with 360 nm light it does not absorb, but the line marked B in the illustration is the

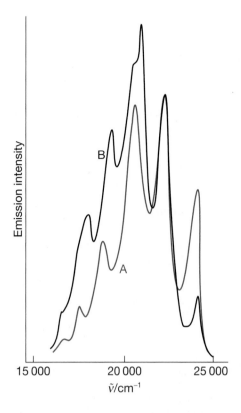

Fig. 17.54

phosphorescence spectrum of a solid solution of a mixture of naphthalene and benzophenone in ethanol. Now a component of fluorescence from naphthalene can be detected. Account for this observation.

17.20 The fluorescence spectrum of anthracene vapour shows a series of peaks of increasing intensity with individual maxima at 440 nm, 410 nm, 390 nm, and 370 nm followed by a sharp cut-off at shorter wavelengths. The absorption spectrum rises sharply form zero to a maximum at 360 nm with a trail of peaks of lessening intensity at 345 nm, 330 nm, and 305 nm. Account for these observations.

17.21 The Beer–Lambert law states that the absorbance of a sample at a wavenumber $\tilde{v}$ is proportional to the molar concentration [J] of the absorbing species J and to the length l of the sample (eqn 16.11). In this problem you will show that the intensity of fluorescence emission from a sample of J is also proportional to [J] and l. Consider a sample of J that is illuminated with a beam of intensity $I_0(\tilde{v})$ at the wavenumber $\tilde{v}$. Before fluorescence can occur, a fraction of $I_0(\tilde{v})$ must be absorbed and an intensity $I(\tilde{v})$ will be transmitted. However, not all of the absorbed intensity is emitted and the intensity of fluorescence depends on the fluorescence quantum yield, ϕ_f, the efficiency of photon emission. The fluorescence quantum yield ranges from 0 to 1 and is proportional to the ratio of the integral of the fluorescence spectrum over the integrated absorption coefficient. Because of a Stokes shift of magnitude $\Delta\tilde{v}_{Stokes}$, fluorescence occurs at a wavenumber $\tilde{v}_f$, with $\tilde{v}_f + \Delta\tilde{v}_{Stokes} = \tilde{v}$. It follows that the fluorescence intensity at $\tilde{v}_f$, $I_f(\tilde{v}_f)$, is proportional to ϕ_f and to the intensity of exciting radiation that is absorbed by J, $I_{abs}(\tilde{v}) = I_0(\tilde{v}) - I(\tilde{v})$. (a) Use the Beer–Lambert law to express $I_{abs}(\tilde{v})$ in terms of $I_0(\tilde{v})$, [J], l, and $\varepsilon(\tilde{v})$, the molar absorption coefficient of J at $\tilde{v}$. (b) Use your result from part (a) to show that $I_f(\tilde{v}_f) \propto I_0(\tilde{v})\varepsilon(\tilde{v})\phi_f[J]l$.

17.22 Spin angular momentum is conserved when a molecule dissociates into atoms. What atom multiplicities are permitted when (a) an O_2 molecule, (b) an N_2 molecule dissociates into atoms?

17.23 Show that, if a substance responds nonlinearly to two sources of radiation, one of frequency ω_1 and the other of frequency ω_2, then it may give rise to radiation of the sum and difference of the two frequencies. This nonlinear optical phenomenon is known as *frequency mixing* and is used to expand the wavelength range of lasers in laboratory applications, such as spectroscopy and photochemistry.

Applications: to environmental science and astrophysics

17.24‡ Ozone absorbs ultraviolet radiation in a part of the electromagnetic spectrum energetic enough to disrupt DNA in biological organisms and that is absorbed by no other abundant atmospheric constituent. This spectral range, denoted UV-B, spans the wavelengths of about 290 nm to 320 nm. The molar extinction coefficient of ozone over this range is given in the table below (W.B. DeMore, S.P. Sander, D.M. Golden, R.F. Hampson, M.J. Kurylo, C.J. Howard, A.R. Ravishankara, C.E. Kolb, and M.J. Molina, *Chemical kinetics and photochemical data for use in stratospheric modeling: Evaluation Number 11*, JPL Publication 94–26 (1994).

λ/nm	292.0	296.3	300.8	305.4	310.1	315.0	320.0
ε/(L mol^{-1} cm^{-1})	1512	865	477	257	135.9	69.5	34.5

Compute the integrated absorption coefficient of ozone over the wavelength range 290–320 nm. (*Hint.* $\varepsilon(\tilde{v})$ can be fitted to an exponential function quite well.)

17.25‡ The abundance of ozone is typically inferred from measurements of UV absorption and is often expressed in terms of *Dobson units* (DU): 1 DU is equivalent to a layer of pure ozone 10^{-3} cm thick at 1 atm and 0°C. Compute the absorbance of UV radiation at 300 nm expected for an ozone abundance of 300 DU (a typical value) and 100 DU (a value reached during seasonal Antarctic ozone depletions) given a molar absorption coefficient of 476 L mol^{-1} cm^{-1}.

17.26‡ G.C.G. Wachewsky, R. Horansky, and V. Vaida (*J. Phys. Chem.* **100**, 11 559 (1996)) examined the UV absorption spectrum of CH_3I, a species of interest in connection with stratospheric ozone chemistry. They found the integrated absorption coefficient to be dependent on temperature and pressure to an extent inconsistent with internal structural changes in isolated CH_3I molecules; they explained the changes as due to dimerization of a substantial fraction of the CH_3I, a process which would naturally be pressure- and temperature-dependent. (a) Compute the integrated absorption coefficient over a triangular lineshape in the range 31 250 to 34 483 cm^{-1} and a maximal molar absorption coefficient of 150 L mol^{-1} cm^{-1} at 31 250 cm^{-1}. (b) Suppose 1 per cent of the

CH_3I units in a sample at 2.4 Torr and 373 K exists as dimers. Compute the absorbance expected at 31 250 cm^{-1} in a sample cell of length 12.0 cm. (c) Suppose 18 per cent of the CH_3I units in a sample at 100 Torr and 373 K exists as dimers. Compute the absorbance expected at 31 250 cm^{-1} in a sample cell of length 12.0 cm; compute the molar absorption coefficient that would be inferred from this absorbance if dimerization was not considered.

17.27‡ The molecule Cl_2O_2 is believed to participate in the seasonal depletion of ozone over Antarctica. M. Schwell, H.-W. Jochims, B. Wassermann, U. Rockland, R. Flesch, and E. Rühl (*J. Phys. Chem.* **100**, 10 070 (1996)) measured the ionization energies of Cl_2O_2 by photoelectron spectroscopy in which the ionized fragments were detected using a mass spectrometer. From their data, we can infer that the ionization enthalpy of Cl_2O_2 is 11.05 eV and the enthalpy of the dissociative ionization $Cl_2O_2 \rightarrow Cl + OClO^+ + e^-$ is 10.95 eV. They used this information to make some inferences about the structure of Cl_2O_2. Computational studies had suggested that the lowest-energy isomer is ClOOCl, but that ClClO$_2$ (C_{2v}) and ClOClO are not very much higher in energy. The Cl_2O_2 in the photoionization step is the lowest-energy isomer, whatever its structure may be, and its enthalpy of formation had previously been reported as +133 kJ mol^{-1}. The Cl_2O_2 in the dissociative ionization step is unlikely to be ClOOCl, for the product can be derived from it only with substantial rearrangement. Given $\Delta_f H^{\ominus}(OClO^+) = +1096$ kJ mol^{-1} and $\Delta_f H^{\ominus}(e^-) = 0$, determine whether the Cl_2O_2 in the dissociative ionization is the same as that in the photoionization. If different, how much greater is its $\Delta_f H^{\ominus}$? Are these results consistent with or contradictory to the computational studies?

17.28‡ One of the principal methods for obtaining the electronic spectra of unstable radicals is to study the spectra of comets, which are almost entirely due to radicals. Many radical spectra have been found in comets, including that due to CN. These radicals are produced in comets by the absorption of far ultraviolet solar radiation by their parent compounds. Subsequently, their fluorescence is excited by sunlight of longer wavelength. The spectra of comet Hale–Bopp (C/1995 O1) have been the subject of many recent studies. One such study is that of the fluorescence spectrum of CN in the comet at large heliocentric distances by R.M. Wagner and D.G. Schleicher (*Science* **275**, 1918 (1997)), in which the authors determine the spatial distribution and rate of production of CN in the coma. The (0–0) vibrational band is centred on 387.6 nm and the weaker (1–1) band with relative intensity 0.1 is centred on 386.4 nm. The band heads for (0–0) and (1–1) are known to be 388.3 and 421.6 nm, respectively. From these data, calculate the energy of the excited S_1 state relative to the ground S_0 state, the vibrational wavenumbers and the difference in the vibrational wavenumbers of the two states, and the relative populations of the $v = 0$ and $v = 1$ vibrational levels of the S_1 state. Also estimate the effective temperature of the molecule in the excited S_1 state. Only eight rotational levels of the S_1 state are thought to be populated. Is that observation consistent with the effective temperature of the S_1 state?

18 Spectroscopy 3: magnetic resonance

The effect of magnetic fields on electrons and nuclei

18.1 The energies of electrons in magnetic fields

18.2 The energies of nuclei in magnetic fields

18.3 Magnetic resonance spectroscopy

Nuclear magnetic resonance

18.4 The NMR spectrometer

18.5 The chemical shift

18.6 The fine structure

Pulse techniques in NMR

18.7 The magnetization vector

18.8 Linewidths and rate processes

18.9 Spin decoupling

18.10 The nuclear Overhauser effect

18.11 Two-dimensional NMR

18.12 Solid-state NMR

Electron spin resonance

18.13 The ESR spectrometer

18.14 The g-value

18.15 Hyperfine structure

Checklist of key ideas

Further reading

Exercises

Problems

One of the most widely used spectroscopic procedures in chemistry makes use of the classical concept of resonance. The chapter begins with an account of conventional nuclear magnetic resonance, which shows how the resonance frequency of a magnetic nucleus is affected by its electronic environment and the presence of magnetic nuclei in its vicinity. Then we turn to the modern versions of NMR, which are based on the use of pulses of electromagnetic radiation and the processing of the resulting signal by Fourier transform techniques. The experimental techniques for electron spin resonance resemble those used in the early days of NMR. The information obtained, however, is very useful for the determination of the properties of species with unpaired electrons.

When two pendulums share a slightly flexible support and one is set in motion, the other is forced into oscillation by the motion of the common axle. As a result, energy flows between the two pendulums. The energy transfer occurs most efficiently when the frequencies of the two pendulums are identical. The condition of strong effective coupling when the frequencies of two oscillators are identical is called **resonance**.

Resonance is the basis of a number of everyday phenomena, including the response of radios to the weak oscillations of the electromagnetic field generated by a distant transmitter. In this chapter we explore some spectroscopic applications that, as originally developed (and in some cases still), depend on matching a set of energy levels to a source of monochromatic radiation and observing the strong absorption that occurs at resonance.

The effect of magnetic fields on electrons and nuclei

The Stern–Gerlach experiment (Section 12.8) provided evidence for electron spin. It turns out that many nuclei also possess spin angular momentum. Orbital and spin angular momenta give rise to magnetic moments, and to say that electrons and nuclei have magnetic moments means that, to some extent, they behave like small bar magnets. Therefore, we can expect that the application of a magnetic field should affect atoms and molecules. We first establish how the energies of an atom depend on the strength of an external field and then see how we can use magnetic field effects as the bases of spectroscopic techniques for the study of structure and dynamics of complex molecules.

18.1 The energies of electrons in magnetic fields

Electrons possess magnetic moments as a result of their orbital and spin angular momenta and these moments interact with an external magnetic field. Classically, the energy of a magnetic moment $\boldsymbol{\mu}$ in a magnetic field $\boldsymbol{B}$ is equal to the scalar product

$$E = -\boldsymbol{\mu} \cdot \boldsymbol{B} \tag{18.1}$$

Quantum mechanically, we write the hamiltonian as

$$H = -\hat{\boldsymbol{\mu}} \cdot \boldsymbol{B} \tag{18.2}$$

To write an expression for $\hat{\boldsymbol{\mu}}$, we recall from Section 13.8 that the magnetic moment is proportional to the angular momentum. Therefore, the magnetic moment operator is expected to be proportional to the angular momentum operator. For an electron possessing orbital angular momentum we write

$$\hat{\boldsymbol{\mu}} = \gamma_e \hat{\boldsymbol{l}} \quad \text{and} \quad H = -\gamma_e \boldsymbol{B} \cdot \hat{\boldsymbol{l}} \tag{18.3}$$

where $\hat{\boldsymbol{l}}$ is the orbital angular momentum operator and γ_e is a constant called the **magnetogyric ratio** of the electron. If the magnetic moment is treated as arising from the circulation of an electron of charge $-e$, standard electromagnetic theory gives

$$\gamma_e = -\frac{e}{2m_e} \tag{18.4}$$

The negative sign (arising from the sign of the electron's charge) shows that the orbital moment is antiparallel to its orbital angular momentum (as was depicted in Fig. 13.28).

Consider the interaction of the z-component of the orbital magnetic moment, μ_z, with a steady magnetic field B_0 along the z-direction.[1] It follows from eqn 3 that

$$\hat{\mu}_z = \gamma_e \hat{l}_z \quad \text{and} \quad H = -\hat{\mu}_z B_0 = -\gamma_e B_0 \hat{l}_z \tag{18.5}$$

where $\hat{\mu}_z$ is the component of the orbital magnetic moment operator along the z-direction and the operator $\hat{l}_z$ has eigenfunctions ψ_{m_l} with eigenvalues $m_l \hbar$. It follows that the z-component of the orbital magnetic moment and the energy of interaction are, respectively,

$$\mu_z = \gamma_e m_l \hbar \quad \text{and} \quad E = -\gamma_e m_l \hbar B_0 \tag{18.6}$$

From eqn 4, the possible values of μ_z are

$$\mu_z = -\frac{e}{2m_e} \times m_l \hbar = -\mu_B m_l \tag{18.7}$$

where the **Bohr magneton**, μ_B, is

$$\mu_B = \frac{e\hbar}{2m_e} \tag{18.8}$$

Its numerical value is 9.724×10^{-24} J T^{-1}. The Bohr magneton is often regarded as the fundamental quantum of magnetic moment. Therefore, in the presence of a

---

The scalar product (or 'dot product') of two vectors $\boldsymbol{a}$ and $\boldsymbol{b}$ is given by $\boldsymbol{a} \cdot \boldsymbol{b} = ab \cos \theta$, where a and b are the magnitudes of $\boldsymbol{a}$ and $\boldsymbol{b}$, respectively, and θ is the angle between $\boldsymbol{a}$ and $\boldsymbol{b}$.

---

1. B is actually the magnetic induction, and is measured in tesla, T; 1 T = 1 kg s^{-2} A^{-1}. The unit gauss, G, is also occasionally used: 1 T = 10^4 G.

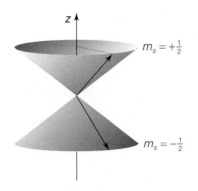

18.1 The interactions between the m_s states of an electron and an external magnetic field may be visualized as the precession of the vectors representing the angular momentum on the cones drawn here.

magnetic field, an electron in a state with quantum number m_l has an additional contribution to its energy given by

$$E_{m_l} = \mu_B m_l B_0 \qquad (18.9)$$

The spin magnetic moment of an electron, which has a spin quantum number $s = \frac{1}{2}$ (Section 12.8), is also proportional to its spin angular momentum. However, the spin magnetic moment and hamiltonian operators are, respectively,

$$\hat{\boldsymbol{\mu}} = g_e \gamma_e \hat{\boldsymbol{s}} \quad \text{and} \quad H = -g_e \gamma_e \boldsymbol{B} \cdot \hat{\boldsymbol{s}} \qquad g_e = 2.002\,319\ldots \qquad (18.10)$$

where $\hat{\boldsymbol{s}}$ is the spin angular momentum operator and the extra factor g_e is called the **g-value** of the electron. The g-value arises from relativistic effects and from interactions of the electron with the electromagnetic fluctuations in the vacuum that surround the electron.

Again we consider the interaction between the z-component of the spin magnetic moment with a magnetic field of magnitude B_0 in the z-direction. We write

$$\hat{\mu}_z = g_e \gamma_e \hat{s}_z \qquad \text{and} \qquad H = -g_e \gamma_e B_0 \hat{s}_z \qquad (18.11)$$

where the operator $\hat{s}_z$ has eigenfunctions α (also denoted as $\uparrow$) and β (also denoted as $\downarrow$) corresponding to the quantum numbers $m_s = +\frac{1}{2}$ and $m_s = -\frac{1}{2}$, respectively. It follows that, for an electron in a state m_s,

$$\mu_z = g_e \gamma_e m_s \hbar \quad \text{and} \quad E_{m_s} = -g_e \gamma_e m_s \hbar B_0 = g_e \mu_B m_s B_0 \qquad m_s = \pm \tfrac{1}{2} \qquad (18.12)$$

In the absence of a magnetic field, the states with different values of m_l and m_s are degenerate. When a field is present, the degeneracy is removed: the state with $m_s = +\frac{1}{2}$ moves up in energy by $\frac{1}{2}g_e \mu_B B_0$ and the state with $m_s = -\frac{1}{2}$ moves down by $\frac{1}{2}g_e \mu_B B_0$. The different energies arising from an interaction with an external field are sometimes represented on the vector model by picturing the vectors as **precessing**, or sweeping round their cones, with the rate of precession proportional to the energy of the state (Fig. 18.1).

18.2 The energies of nuclei in magnetic fields

The spin quantum number, I, of a nucleus is a fixed characteristic property of a nucleus and may be an integer or a half-integer but is never negative (Table 18.1). A nucleus with spin quantum number I has the following properties:

1 An angular momentum of magnitude $\{I(I + 1)\}^{1/2}\hbar$.

2 A component of angular momentum $m_I \hbar$ on an arbitrary axis, where $m_I = I$, $I - 1, \ldots, -I$.

3 If $I > 0$, a magnetic moment with a constant magnitude and an orientation that is determined by the value of m_I.

Table 18.1 Nuclear constitution and the nuclear spin quantum number

Number of protons	Number of neutrons	I
even	even	0
odd	odd	integer (1, 2, 3, …)
even	odd	half-integer ($\frac{1}{2}, \frac{3}{2}, \frac{5}{2}, \ldots$)
odd	even	half-integer ($\frac{1}{2}, \frac{3}{2}, \frac{5}{2}, \ldots$)

Table 18.2* Nuclear spin properties

Nuclide	Natural abundance/%	Spin I	g-factor, g_I	$\gamma/(10^7\ \mathrm{T}^{-1}\ \mathrm{s}^{-1})$
$^1\mathrm{n}$		$\frac{1}{2}$	−3.826	−18.32
$^1\mathrm{H}$	99.98	$\frac{1}{2}$	5.586	26.75
$^2\mathrm{H}$	0.02	1	0.857	4.10
$^{13}\mathrm{C}$	1.11	$\frac{1}{2}$	1.405	6.73
$^{14}\mathrm{N}$	99.64	1	0.404	1.93

* More values are given in the *Data section*. Note that $\gamma = g_I \mu_N / \hbar$.

According to the second property, the spin, and hence the magnetic moment, of the nucleus may lie in $2I + 1$ different orientations relative to an axis. A proton has $I = \frac{1}{2}$ and its spin may adopt either of two orientations; a $^{14}\mathrm{N}$ nucleus has $I = 1$ and its spin may adopt any of three orientations; both $^{12}\mathrm{C}$ and $^{16}\mathrm{O}$ have $I = 0$ and hence zero magnetic moment.

The energy of interaction between a nucleus with a magnetic moment $\boldsymbol{\mu}$ and an external magnetic field $\boldsymbol{B}$ may be calculated by using operators analogous to those of eqn 3:

$$\hat{\mu} = \gamma \hat{I} \quad \text{and} \quad H = -\gamma B \cdot \hat{I} \tag{18.13}$$

where γ is the characteristic value of the magnetogyric ratio for the nucleus under study (Table 18.2). The magnetic moment is sometimes expressed in terms of the **nuclear g-factor**, g_I, and the **nuclear magneton**, μ_N, by using

$$\gamma \hbar = g_I \mu_N \quad \mu_N = \frac{e\hbar}{2m_p} = 5.051 \times 10^{-27}\ \mathrm{J\ T}^{-1} \tag{18.14}$$

where m_p is the mass of the proton. Nuclear g-factors vary between −6 and +6 (see Table 18.2): positive values of g_I and γ denote a magnetic moment that is parallel to the spin; negative values indicate that the magnetic moment and spin are anti-parallel. The nuclear magneton is about 2000 times smaller than the Bohr magneton, so nuclear magnetic moments are about 2000 times weaker than the electron spin magnetic moment. The components of spin angular momentum relative to the z-axis are $m_I \hbar$, where $m_I = I, I − 1, \dots, −I$, and the corresponding components of the spin magnetic moment are

$$\mu_z = \gamma m_I \hbar \tag{18.15}$$

In a magnetic field B_0 in the z-direction, the $2I + 1$ orientations of the nucleus have different energies, which are given by

$$E_{m_I} = -\mu_z B_0 = -\gamma \hbar B_0 m_I \tag{18.16}$$

These energies are often expressed in terms of the **Larmor frequency**, ν_L:

$$E_{m_I} = -m_I h \nu_L \quad \nu_L = \frac{\gamma B_0}{2\pi} \tag{18.17}$$

For the remainder of this chapter we will assume that γ is positive, as is the case for the majority of nuclei. In such cases, the states with negative values of m_I lie above states with positive values of m_I. Equation 17 also shows that, the stronger the

Magnetic field off   Magnetic field on

18.2 Electron spin levels in a magnetic field. Note that the β state is lower in energy than the α state (because the magnetogyric ratio of an electron is negative). Resonance is achieved when the frequency of the incident radiation matches the frequency corresponding to the energy separation.

magnetic field, the higher the Larmor frequency. For protons, a field of 12 T corresponds to a Larmor frequency of about 500 MHz.

18.3 Magnetic resonance spectroscopy

The magnetic resonance experiment is the resonant absorption of radiation by nuclei or unpaired electrons exposed to a magnetic field. From eqn 12, the separation between the $m_s = -\frac{1}{2}$ and $m_s = +\frac{1}{2}$ levels of an electron spin in a magnetic field B_0 is

$$\Delta E = E_\alpha - E_\beta = g_e \mu_B B_0 \tag{18.18}$$

If the sample is exposed to radiation of frequency v, the energy separations come into resonance with the radiation when the frequency satisfies the **resonance condition** (Fig. 18.2):

$$h v = g_e \mu_B B_0 \tag{18.19}$$

At resonance there is strong coupling between the electron spins and the radiation, and strong absorption occurs as the spins make the transition $\beta \to \alpha$. **Electron spin resonance** (ESR), or **electron paramagnetic resonance** (EPR), is the study of molecules and ions containing unpaired electrons by observing the magnetic fields at which they come into resonance with monochromatic radiation. Magnetic fields of about 0.3 T (the value used in most commercial ESR spectrometers) correspond to resonance with an electromagnetic field of frequency 10 GHz (10^{10} Hz) and wavelength 3 cm. Because 3 cm radiation falls in the microwave region of the electromagnetic spectrum, ESR is a microwave technique.

The energy separation between the $m_I = +\frac{1}{2}$ ($\uparrow$ or α) and the $m_I = -\frac{1}{2}$ ($\downarrow$ or β) states of **spin-$\frac{1}{2}$ nuclei**, which are nuclei with $I = \frac{1}{2}$, is

$$\Delta E = E_\beta - E_\alpha = \frac{1}{2}\gamma\hbar B_0 - (-\frac{1}{2}\gamma\hbar B_0) = \gamma\hbar B_0 = h v_L \tag{18.20}$$

and resonant absorption occurs when the resonance condition (Fig. 18.3)

$$h v = \gamma\hbar B_0 = h v_L \tag{18.21}$$

is fulfilled. That is, an $\alpha \to \beta$ transition occurs at $v = v_L$. In its simplest form, **nuclear magnetic resonance** (NMR) is the study of the properties of molecules containing magnetic nuclei by applying a magnetic field and observing the frequency of the resonant electromagnetic field. Larmor frequencies of nuclei at the fields normally employed typically lie in the radiofrequency region of the electromagnetic spectrum, so NMR is a radiofrequency technique. For example, at 12 T, protons come into resonance at about 500 MHz (the Larmor frequency at that magnetic field).

For much of this chapter we consider spin-$\frac{1}{2}$ nuclei, but NMR is applicable to nuclei with any nonzero spin. As well as protons, which are the most common nuclei studied by NMR, spin-$\frac{1}{2}$ nuclei include ^{13}C, ^{19}F, and ^{31}P nuclei.

Field off   Field on

18.3 The nuclear spin energy levels of a spin-$\frac{1}{2}$ nucleus with positive magnetogyric ratio (for example, ^1H or ^{13}C) in a magnetic field. Resonance occurs when the energy separation of the levels matches the energy of the photons in the electromagnetic field.

Nuclear magnetic resonance

Although the NMR technique is simple in concept, NMR spectra can be highly complex. However, they have proved invaluable in chemistry, for they reveal so much structural information. A magnetic nucleus is a very sensitive, noninvasive probe of the surrounding electronic structure.

18.4 The layout of a typical NMR spectrometer. The link from the transmitter to the detector indicates that the high frequency of the transmitter is subtracted from the high frequency received signal to give a low frequency signal for processing.

18.4 The NMR spectrometer

An NMR spectrometer consists of the appropriate sources of radiofrequency electromagnetic radiation and a magnet that can produce a uniform, intense field. In simple instruments, the magnetic field is provided by a permanent magnet. For serious work, a **superconducting magnet** capable of producing fields of the order of 2 T and more is used (Fig. 18.4). The sample is placed in the cylindrically wound magnet. In some cases the sample is rotated rapidly to average out magnetic inhomogeneities. However, sample spinning can lead to irreproducible results, and is often avoided. Although a superconducting magnet operates at the temperature of liquid helium (4 K), the sample itself is normally at room temperature.

The use of high magnetic fields has several advantages, the most important being that they simplify the appearance of spectra and so allow them to be interpreted more readily. A further advantage is that the rate of energy uptake by the sample is greater in a high field. There are two contributions to this increase. One comes from the greater population difference between the upper and lower spin states at high fields, for the population difference is approximately proportional to B. The second contribution stems from the greater energy of each absorbed photon, which is also proportional to B. It follows that overall the signal is proportional to B^2.

Justification 18.1

According to the Boltzmann distribution, the ratio of populations is

$$\frac{N_\beta}{N_\alpha} = e^{-\Delta E/kT} \approx 1 - \frac{\Delta E}{kT}$$

where expansion of the exponential term is appropriate for $\Delta E \ll kT$, a condition usually met for electron and nuclear spins. It follows that

$$\frac{N_\alpha - N_\beta}{N_\alpha + N_\beta} \approx \frac{\Delta E}{2kT} = \frac{\gamma \hbar B}{2kT}$$

and the population difference is proportional to B. The energy of the photon absorbed when a nucleus makes a transition from its lower state to its higher state is $h\nu$; at resonance ν is equal to ν_L, and ν_L is proportional to B. Hence, at resonance, each photon has an energy that is proportional to B. The net rate of energy absorption is proportional to the population difference multiplied by the energy of each absorption event (the photon energy), so overall the net rate is proportional to B^2.

18.5 The chemical shift

Nuclear magnetic moments interact with the *local* magnetic field. The local field may differ from the applied field because the latter induces electronic orbital angular momentum (that is, the circulation of electronic currents) which gives rise to a small additional magnetic field δB at the nuclei. This additional field is proportional to the applied field, and it is conventional to write

$$\delta B = -\sigma B_0 \qquad [18.22]$$

where the dimensionless quantity σ is called the **shielding constant** of the nucleus (σ is usually positive but may be negative). The ability of the applied field to induce an electronic current in the molecule, and hence the strength of the resulting local magnetic field experienced by the nucleus, depends on the details of the electronic structure near the magnetic nucleus of interest, so nuclei in different chemical groups have different shielding constants. The calculation of reliable values of the shielding constant is very difficult, but trends in it are quite well understood and we concentrate on them.

(a) The δ-scale of chemical shifts

Because the total local field is

$$B_{\text{loc}} = B_0 + \delta B = (1 - \sigma)B_0 \qquad (18.23)$$

the Larmor frequency is

$$\nu_L = \frac{\gamma B_{\text{loc}}}{2\pi} = (1 - \sigma)\frac{\gamma B_0}{2\pi} \qquad (18.24)$$

This frequency is different for nuclei in different environments. Hence, different nuclei, even of the same element, come into resonance at different frequencies.

It is conventional to express the resonance frequencies in terms of an empirical quantity called the **chemical shift**, which is related to the difference between the resonance frequency, ν, of the nucleus in question and that of a reference standard, ν°:

$$\delta = \frac{\nu - \nu^\circ}{\nu^\circ} \times 10^6 \qquad [18.25]$$

The standard for protons is the proton resonance in tetramethylsilane ($Si(CH_3)_4$, commonly referred to as TMS), which bristles with protons and dissolves without reaction in many liquids. Other references are used for other nuclei. For ^{13}C, the reference frequency is the ^{13}C resonance in TMS; for ^{31}P it is the ^{31}P resonance in 85 per cent $H_3PO_4(aq)$. The advantage of the δ-scale is that shifts reported on it are independent of the applied field (because both numerator and denominator are proportional to the applied field).

Illustration 18.1

A nucleus with $\delta = 1.00$ (which is often, but unnecessarily, expressed as 1.00 ppm on account of the 10^6 in the definition of δ) in a spectrometer operating at 500 MHz will have a shift relative to the reference equal to

$$\nu - \nu^\circ = (500 \text{ MHz}) \times (1.00) \times 10^6 = 500 \text{ Hz}$$

In a spectrometer operating at 100 MHz, the shift relative to the reference would be only 100 Hz.

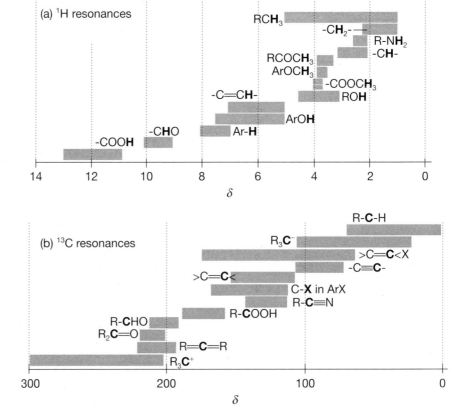

18.5 The range of typical chemical shifts for (a) ^1H resonances and (b) ^{13}C resonances.

The relation between δ and σ is obtained by substituting eqn 24 into eqn 25:

$$\delta = \frac{(1-\sigma)B_0 - (1-\sigma^\circ)B_0}{(1-\sigma^\circ)B_0} \times 10^6 = \frac{\sigma^\circ - \sigma}{1 - \sigma^\circ} \times 10^6 \approx (\sigma^\circ - \sigma) \times 10^6 \quad (18.26)$$

As the shielding, σ, gets smaller, δ *increases*. Therefore, we speak of nuclei with large chemical shifts as being strongly **deshielded**. Some typical chemical shifts are given in Fig. 18.5. As can be seen from the illustration, the nuclei of different elements have very different ranges of chemical shifts. The ranges exhibit the variety of electronic environments of the nuclei in molecules.

By convention, NMR spectra are plotted with δ increasing from right to left. Consequently, in a given applied magnetic field the Larmor frequency also increases from right to left. In a continuous wave (CW) spectrometer, in which the radiofrequency is held constant and the magnetic field is varied (a 'field sweep experiment'), the spectrum is displayed with the applied magnetic field increasing from left to right: a nucleus with a small chemical shift experiences a relatively low local magnetic field, so it needs a higher applied magnetic field to bring it into resonance with the radiofrequency field. Consequently, the right-hand (low chemical shift) end of the spectrum was previously known as the 'high field end' of the spectrum.

(b) Resonance of different groups of nuclei

The existence of a chemical shift explains the general features of the spectrum of ethanol shown in Fig. 18.6. The CH_3 protons form one group of nuclei with $\delta \approx 1$. The two CH_2 protons are in a different part of the molecule, experience a different local magnetic field, and resonate at $\delta \approx 3$. Finally, the OH proton is in another environment, and has a chemical shift of $\delta \approx 4$. The increasing value of δ (that is, the decrease

18.6 The ^{1}H-NMR spectrum of ethanol. The bold letters denote the protons giving rise to the resonance peak, and the step-like curve is the integrated signal.

in shielding) is consistent with the electron-withdrawing power of the O atom: it reduces the electron density of the OH proton most, and that proton is strongly deshielded. It reduces the electron density of the distant methyl protons least, and those nuclei are least deshielded.

The relative intensities of the signals (the areas under the absorption lines) can be used to help distinguish which group of lines corresponds to which chemical group. The determination of the area under an absorption line is referred to as the **integration** of the signal (just as any area under a curve may be determined by mathematical integration). Spectrometers can integrate the absorption automatically (as indicated in Fig. 18.6). In ethanol the group intensities are in the ratio 3:2:1 because there are three CH_3 protons, two CH_2 protons, and one OH proton in each molecule. Counting the number of magnetic nuclei as well as noting their chemical shifts helps to identify a compound present in a sample.

(c) The origin of shielding constants

The calculation of shielding constants is very difficult, even for small molecules, for it requires detailed information about the distribution of electron density in the ground and excited states and the excitation energies of the molecule. Some success has been achieved with the calculation for diatomic molecules and small molecules such as H_2O and CH_4. Large molecules, even proteins, are well within the scope of the calculation. Nevertheless, it is easier to understand the different contributions to chemical shifts by studying the large body of empirical information now available for large molecules.

The empirical approach supposes that the observed shielding constant is the sum of three contributions:

$$\sigma = \sigma(\text{local}) + \sigma(\text{neighbour}) + \sigma(\text{solvent}) \tag{18.27}$$

The **local contribution**, $\sigma(\text{local})$, is essentially the contribution of the electrons of the atom that contains the nucleus in question. The **neighbouring group contribution**, $\sigma(\text{neighbour})$, is the contribution from the groups of atoms that form the rest of the molecule. The **solvent contribution**, $\sigma(\text{solvent})$, is the contribution from the solvent molecules.

(d) The local contribution

It is convenient to regard the local contribution to the shielding constant as the sum of a **diamagnetic contribution**, σ_d, and a **paramagnetic contribution**, σ_p:

$$\sigma(\text{local}) = \sigma_d + \sigma_p \tag{18.28}$$

A diamagnetic contribution to $\sigma(\text{local})$ opposes the applied magnetic field and shields the nucleus in question, whereas a paramagnetic contribution to $\sigma(\text{local})$ reinforces the applied magnetic field and deshields the nucleus in question. Therefore, $\sigma_d > 0$ and $\sigma_p < 0$. The total local contribution is positive if the diamagnetic contribution dominates, and is negative if the paramagnetic contribution dominates.

The diamagnetic contribution arises from the ability of the applied field to generate a circulation of charge in the ground-state electron distribution of the atom. The circulation generates a magnetic field that opposes the applied field and hence shields the nucleus. The magnitude of σ_d depends on the electron density close to the nucleus and can be calculated from the **Lamb formula**:[2]

$$\sigma_d = \frac{e^2 \mu_0}{12 \pi m_e} \left\langle \frac{1}{r} \right\rangle \tag{18.29}$$

where μ_0 is the vacuum permeability (a fundamental constant, see inside the front cover) and r is the electron–nucleus distance.

18.7 The variation of chemical shielding with electronegativity. The shifts for the methylene protons agree with the trend expected with increasing electronegativity. However, to emphasize that chemical shifts are subtle phenomena, notice that the trend for the methyl protons is opposite to that expected. For these protons another contribution (the magnetic anisotropy of C—H and C—X bonds) is dominant.

> ### Illustration 18.2
>
> To calculate σ_d for the proton in a free H atom, we need to calculate the expectation value of $1/r$ for a hydrogen 1s orbital. Wavefunctions are given in Table 13.1, and a useful integral is given in Example 11.6. Because $d\tau = r^2\, dr \sin\theta\, d\theta\, d\phi$, we can write
>
> $$\left\langle \frac{1}{r} \right\rangle = \int \frac{\psi^*\psi}{r}\, d\tau = \frac{1}{\pi a_0^3} \int_0^{2\pi} d\phi \int_0^{\pi} \sin\theta\, d\theta \int_0^{\infty} r e^{-2r/a_0}\, dr = \frac{4}{a_0^3} \int_0^{\infty} r e^{-2r/a_0}\, dr = \frac{1}{a_0}$$
>
> Therefore,
>
> $$\sigma_d = \frac{e^2 \mu_0}{12 \pi m_e a_0}$$
>
> With the values of the fundamental constants inside the front cover, this expression evaluates to 1.78×10^{-5}.

The diamagnetic contribution is the only contribution in closed-shell free atoms. It is also the only contribution to the local shielding for distributions of charge that have spherical or cylindrical symmetry. Thus, it is the only contribution to the local shielding from inner cores of atoms, for cores remain spherical even though the atom may be a component of a molecule and its valence electron distribution highly distorted. The diamagnetic contribution is broadly proportional to the electron density of the atom containing the nucleus of interest. It follows that the shielding is decreased if the electron density on the atom is reduced by the influence of an electronegative atom nearby. That reduction in shielding translates into an increase in deshielding, and hence to an increase in the chemical shift δ as the electronegativity of a neighbouring atom increases (Fig. 18.7). That is, as the electronegativity increases, δ increases.

2. See *Further reading* for a derivation.

1

The local paramagnetic contribution, σ_p, arises from the ability of the applied field to force the electrons to circulate through the molecule by making use of orbitals that are unoccupied in the ground state. It is zero in free atoms and around the axes of linear molecules (such as ethyne, HC≡CH) where the electrons can circulate freely and a field applied along the internuclear axis is unable to force them into other orbitals. We can expect large paramagnetic contributions from small atoms in molecules with low-lying excited states. In fact, the paramagnetic contribution is the dominant local contribution for atoms other than hydrogen.

(e) Neighbouring group contributions

The neighbouring group contribution arises from the currents induced in nearby groups of atoms. Consider the influence of the neighbouring group X on the proton H in a molecule such as H—X. The applied field generates currents in the electron distribution of X and gives rise to an induced magnetic moment proportional to the applied field; the constant of proportionality is the magnetic susceptibility, χ (chi), of the group X.[3] The proton H is affected by this induced magnetic moment in two ways. First, the strength of the additional magnetic field the proton experiences is inversely proportional to the cube of the distance r between H and X. Second, the field at H depends on the anisotropy of the magnetic susceptibility of X, the variation of χ with the angle that X makes to the applied field. We assume that the magnetic susceptibility of X has two components, $\chi_\parallel$ and $\chi_\perp$, which are parallel and perpendicular to the axis of symmetry of X, respectively. The axis of symmetry of X makes an angle θ to the vector connecting X to H (**1**, where X is represented by the ellipse and H is represented by the circle).

To examine the effect of anisotropy of the magnetic susceptibility of X on the shielding constant, consider the case $\theta = 0$ for a molecule H—X that is free to tumble (**2** and **3**). Some of the time the H—X axis will be perpendicular to the applied field and then only $\chi_\perp$ will contribute to the induced magnetic moment that shields X from the applied field. The result is deshielding of the proton H, or $\sigma(\text{neighbour}) < 0$ (**2**). When the applied field is parallel to the H—X axis, only $\chi_\parallel$ contributes to the induced magnetic moment at X. The result is shielding of the proton H (**3**). We conclude that, as the molecule tumbles and the H—X axis takes all possible angles with respect to the applied field, the effects of anisotropic magnetic susceptibility do not average to zero because $\chi_\parallel \neq \chi_\perp$.

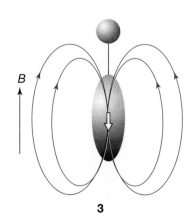

B

2

> **Self-test 18.1** For a tumbling H—X molecule, show that when $\theta = 90°$:
> (a) contributions from the $\chi_\perp$ component lead to shielding of H, or $\sigma(\text{neighbour}) > 0$, and (b) contributions from the $\chi_\parallel$ component lead to deshielding of H, or $\sigma(\text{neighbour}) < 0$. Comparison between the $\theta = 0°$ and $\theta = 90°$ cases shows that the patterns of shielding and deshielding by neighbouring groups depend not only on differences between $\chi_\parallel$ and $\chi_\perp$, but also the angle θ.
> [Draw diagrams similar to **2** and **3** where the $\chi_\perp$ component is parallel to the H—X axis and then analyse the problem as above.]

To a good approximation, the shielding constant $\sigma(\text{neighbour})$ depends on the distance r, the difference $\chi_\parallel - \chi_\perp$, as[4]

$$\sigma(\text{neighbour}) \propto (\chi_\parallel - \chi_\perp)\left(\frac{1 - 3\cos^2\theta}{r^3}\right) \tag{18.30}$$

3

3. Magnetic susceptibilities are discussed in Section 23.10.
4. See *Further reading* for a derivation of this expression.

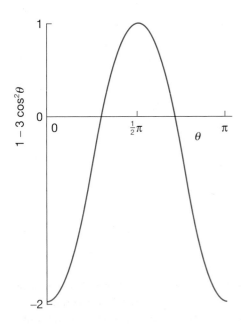

18.8 The variation of the function $1 - 3\cos^2\theta$ with the angle θ.

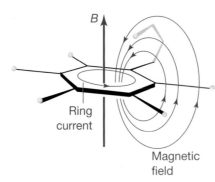

18.9 The shielding and deshielding effects of the ring current induced in the benzene ring by the applied field. Protons attached to the ring are deshielded but a proton attached to a substituent that projects above the ring is shielded.

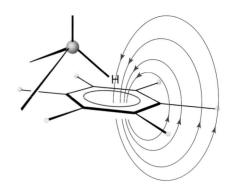

18.10 An aromatic solvent (benzene here) can give rise to local currents that shield or deshield a proton in a solvent molecule. In this relative orientation of the solvent and solute, the proton on the solute molecule is shielded.

where $\chi_{\parallel}$ and $\chi_{\perp}$ are both negative for a diamagnetic species such as the group X (Section 23.10). Equation 30 shows that the neighbouring group contribution may be positive or negative according to the relative magnitudes of the two magnetic susceptibilities and the relative orientation of the nucleus. The latter effect is easy to anticipate: if $54.7° < \theta < 125.3°$, then $1 - 3\cos^2\theta$ is positive, but it is negative otherwise (Fig. 18.8).

A special case of a neighbouring group effect is found in aromatic compounds. The strong anisotropy of the magnetic susceptibility of the benzene ring is ascribed to the ability of the field to induce a **ring current**, a circulation of electrons around the ring, when it is applied perpendicular to the molecular plane. Protons in the plane are deshielded (Fig. 18.9), but any that happen to lie above or below the plane (as members of substituents of the ring) are shielded.

(f) The solvent contribution

A solvent can influence the local magnetic field experienced by a nucleus in a variety of ways. Some of these effects arise from specific interactions between the solute and the solvent (such as hydrogen-bond formation and other forms of Lewis acid–base complex formation). The magnetic susceptibility of the solvent molecules, especially if they are aromatic, can also be the source of a local magnetic field. Moreover, if there are steric interactions that result in a loose but specific interaction between a solute molecule and a solvent molecule, then protons in the solute molecule may experience shielding or deshielding effects according to their location relative to the solvent molecule (Fig. 18.10). We shall see that the NMR spectra of species that contain protons with widely different chemical shifts are easier to interpret than those in which the shifts are similar, so the appropriate choice of solvent may help to simplify the appearance and interpretation of a spectrum.

18.6 The fine structure

The splitting of resonances into individual lines in Fig. 18.6 is called the **fine structure** of the spectrum. It arises because each magnetic nucleus may contribute to the local field experienced by the other nuclei and so modify their resonance frequencies. The strength of the interaction is expressed in terms of the **scalar spin coupling constant**, J, and reported in hertz (Hz). The scalar coupling constant is so called because the energy of interaction it describes is proportional to the scalar product of

18.11 The energy levels of an AX system. The four levels on the left are those of the two spins in the absence of spin–spin coupling. The four levels on the right show how a positive spin–spin coupling constant affects the energies. The transitions shown are for $\beta \leftarrow \alpha$ of A or X, the other nucleus (X or A, respectively) remaining unchanged.

the two interacting spins: $E \propto \mathbf{I}_1 \cdot \mathbf{I}_2$. The constant of proportionality in this expression is J. (More precisely, it is $hJ/\hbar^2$, because each angular momentum is proportional to $\hbar$.)

Spin coupling constants are independent of the strength of the applied field because they do not depend on the latter for their ability to generate local fields. If the resonance line of a particular nucleus is split by a certain amount by a second nucleus, then the resonance line of the second nucleus is split by the first to the same extent.

(a) The energy levels of coupled systems

It will be useful for later discussions to consider an NMR spectrum in terms of the energy levels of the nuclei and the transitions between them. In NMR, letters far apart in the alphabet (typically A and X) are used to indicate nuclei with very different chemical shifts; letters close together (such as A and B) are used for nuclei with similar chemical shifts. We shall consider first an AX system, a molecule that contains two spin-$\frac{1}{2}$ nuclei A and X with very different chemical shifts in the sense that the difference in chemical shift corresponds to a frequency that is large compared to their spin–spin coupling.

The energy level diagram for a single spin-$\frac{1}{2}$ nucleus and its single transition were shown in Fig. 18.3, and nothing more needs to be said. For a spin-$\frac{1}{2}$ AX system there are four spin states:

$$\alpha_A \alpha_X \quad \alpha_A \beta_X \quad \beta_A \alpha_X \quad \beta_A \beta_X$$

The energy depends on the orientation of the spins in the external magnetic field and, if spin–spin coupling is neglected,

$$E = -\gamma \hbar (1 - \sigma_A) B m_A - \gamma \hbar (1 - \sigma_X) B m_X = -h\nu_A m_A - h\nu_X m_X \tag{18.31}$$

where ν_A and ν_X are the Larmor frequencies of A and X and m_A and m_X are their quantum numbers. This expression gives the four lines on the left of Fig. 18.11. The spin–spin coupling depends on the relative orientation of the two nuclear spins, so it is proportional to the product $m_A m_X$; the constant of proportionality is hJ. Therefore, the energy including spin–spin coupling is

$$E = -h\nu_A m_A - h\nu_X m_X + hJ m_A m_X \tag{18.32}$$

If $J > 0$, a lower energy is obtained when $m_A m_X < 0$, which is the case if one spin is α and the other is β. A higher energy is obtained if both spins are α or both spins are β. The opposite is true if $J < 0$. The resulting energy level diagram (for $J > 0$) is shown on the right of Fig. 18.11. We see that the $\alpha\alpha$ and $\beta\beta$ states are both raised by $\frac{1}{4}hJ$ and that the $\alpha\beta$ and $\beta\alpha$ states are both lowered by $\frac{1}{4}hJ$.

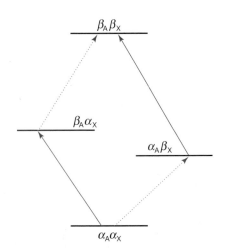

18.12 An alternative depiction of the energy levels and transitions shown in Fig. 18.11.

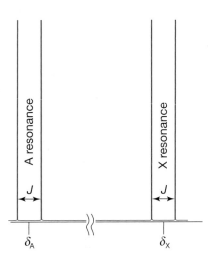

18.13 The effect of spin–spin coupling on an AX spectrum. Each resonance is split into two lines separated by J. The pairs of resonances are centred on the chemical shifts of the protons in the absence of spin–spin coupling.

18.14 The X resonance of an AX_2 species is also a doublet, because the two equivalent X nuclei behave like a single nucleus; however, the overall absorption is twice as intense as that of an AX species.

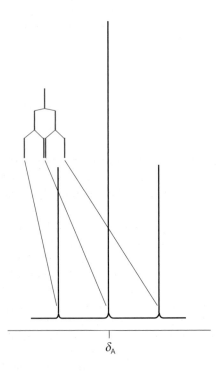

18.15 The origin of the 1:2:1 triplet in the A resonance of an AX_2 species. The resonance of A is split into two by coupling with one X nucleus (as shown in the inset), and then each of those two lines is split into two by coupling to the second X nucleus. Because each X nucleus causes the same splitting, the two central transitions are coincident and give rise to an absorption line of double the intensity of the outer lines.

When a transition of nucleus A occurs, nucleus X remains unchanged. Therefore, the A resonance is a transition for which $\Delta m_A = -1$ and $\Delta m_X = 0$. There are two such transitions, one in which $\beta_A \leftarrow \alpha_A$ occurs when the X nucleus is α_X, and the other in which $\beta_A \leftarrow \alpha_A$ occurs when the X nucleus is β_X. They are shown in Fig. 18.11 and in a slightly different form in Fig. 18.12. The energies of the transitions are

$$\Delta E = h\nu_A \pm \tfrac{1}{2}hJ \tag{18.33}$$

Therefore, the A resonance consists of a doublet of separation J centred on the chemical shift of A (Fig. 18.13).

Similar remarks apply to the X resonance, which consists of two transitions according to whether the A nucleus is α or β (as shown in Fig. 18.12). The transition energies are

$$\Delta E = h\nu_X \pm \tfrac{1}{2}hJ \tag{18.34}$$

It follows that the X resonance also consists of two lines of separation J, but they are centred on the chemical shift of X (as shown in Fig. 18.13).

(b) Patterns of coupling

We have seen that, in an AX system, spin–spin coupling will result in four lines in the NMR spectrum. Instead of a single line from A, we get a doublet of lines separated by J and centred on the chemical shift characteristic of A. The same splitting occurs in the X resonance: instead of a single line, the resonance is a doublet with splitting J (the same value as for the splitting of A) centred on the chemical shift characteristic of X. These features are summarized in Fig. 18.13.

A subtle point is that the X resonance in an AX_n species (such as an AX_2 or AX_3 species) is also a doublet with splitting J. As we shall explain below, *a group of equivalent nuclei resonates like a single nucleus*. The only difference for the X resonance of an AX_n species is that the intensity is n times as great as that of an AX species (Fig. 18.14). The A resonance in an AX_n species, though, is quite different from the A resonance in an AX species. For example, consider an AX_2 species with two equivalent X nuclei. The resonance of A is split into a doublet of separation J by one X, and each line of that doublet is split again by the same amount by the second X (Fig. 18.15). This

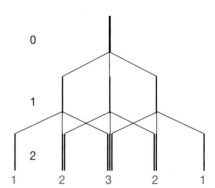

18.16 The origin of the 1:3:3:1 quartet in the A resonance of an AX_3 species. The third X nucleus splits each of the lines shown in Fig. 18.15 for an AX_2 species into a doublet, and the intensity distribution reflects the number of transitions that have the same energy.

18.17 The intensity distribution of the A resonance of an AX_n species can be constructed by considering the splitting caused by 1, 2, ... , n protons, as in Figs 18.15 and 18.16. The resulting intensity distribution has a binomial distribution and is given by the integers in the corresponding row of Pascal's triangle. Note that, although the lines have been drawn side-by-side for clarity, the members of each group are coincident. Four protons, in AX_4, split the A resonance into a 1:4:6:4:1 quintet.

18.18 The intensity distribution arising from spin–spin coupling with nuclei with $I = 1$ can be constructed similarly, but each successive nucleus splits the lines into three equal intensity components. Two equivalent spin-1 nuclei give rise to a 1:2:3:2:1 quintet.

4

splitting results in three lines in the intensity ratio 1:2:1 (because the central frequency can be obtained in two ways). The A resonance of an A_nX_2 species would also be a 1:2:1 triplet of splitting J, the only difference being that the intensity of the A resonance would be n times as great as that of AX_2.

Three equivalent X nuclei (an AX_3 species) split the resonance of A into four lines of intensity ratio 1:3:3:1 and separation J (Fig. 18.16). The X resonance, though, is still a doublet of separation J. In general, n equivalent spin-$\frac{1}{2}$ nuclei split the resonance of a nearby spin or group of equivalent spins into $n + 1$ lines with an intensity distribution given by Pascal's triangle (**4**). The easiest way of constructing the pattern of fine structure is to draw a diagram in which successive rows show the splitting of a subsequent proton. The procedure is illustrated in Fig. 18.17 and was used in Figs 18.15 and 18.16. It is easily extended to molecules containing nuclei with $I > \frac{1}{2}$ (Fig. 18.18).

Example 18.1 Accounting for the fine structure in a spectrum

Account for the fine structure in the NMR spectrum of the C—H protons of ethanol.

Method Consider how each group of equivalent protons (for example, three methyl protons) split the resonances of the other groups of protons. There is no splitting within groups of equivalent protons. Each splitting pattern can be decided by referring to Pascal's triangle.

Answer The three protons of the CH_3 group split the resonance of the CH_2 protons into a 1:3:3:1 quartet with a splitting J. Likewise, the two protons of the CH_2 group split the resonance of the CH_3 protons into a 1:2:1 triplet with the same splitting J. All the lines mentioned so far are split into doublets by the OH proton, but the splitting cannot be detected because the OH protons migrate rapidly from molecule to molecule and their effect averages to zero.

Self-test 18.2 What fine structure can be expected for the protons in $^{14}NH_4^+$? The spin quantum number of nitrogen is 1.   [1:1:1 triplet from N]

(c) The magnitudes of coupling constants

The scalar coupling constant of two nuclei separated by N bonds is denoted ^NJ, with subscripts for the types of nuclei involved. Thus, $^1J_{CH}$ is the coupling constant for a proton joined directly to a ^{13}C atom, and $^2J_{CH}$ is the coupling constant when the same two nuclei are separated by two bonds (as in ^{13}C—C—H). A typical value of $^1J_{CH}$ is in the range 120 to 250 Hz; $^2J_{CH}$ is between –10 and 20 Hz. Both ^3J and ^4J can give detectable effects in a spectrum, but couplings over larger numbers of bonds can generally be ignored. One of the longest range couplings that has been detected is $^9J_{HH} = 0.4$ Hz between the CH_3 and CH_2 protons in $CH_3C{\equiv}CC{\equiv}CC{\equiv}CCH_2OH$.

The sign of J_{XY} indicates whether the energy of two spins is lower when they are parallel ($J < 0$) or when they are antiparallel ($J > 0$). It is found that $^1J_{CH}$ is often positive, $^2J_{HH}$ is often negative, $^3J_{HH}$ is often positive, and so on. An additional point is that J varies with the angle between the bonds (Fig. 18.19). Thus, a $^3J_{HH}$ coupling constant is often found to depend on the dihedral angle ϕ (**5**) according to the **Karplus equation**:

$$J = A + B\cos\phi + C\cos 2\phi \tag{18.35}$$

with A, B, and C empirical constants with values close to +7 Hz, –1 Hz, and +5 Hz, respectively. It follows that the measurement of $^3J_{HH}$ in a series of related compounds can be used to determine their conformations. The coupling constant $^1J_{CH}$ also depends on the hybridization of the C atom, as the following values indicate:

	sp	sp^2	sp^3
$^1J_{CH}$/Hz	250	160	125

(d) The origin of spin–spin coupling

Spin–spin coupling is a very subtle phenomenon, and it is better to treat J as an empirical parameter than to use calculated values. However, we can get some insight into its origins, if not its precise magnitude—or always reliably its sign—by considering the magnetic interactions within molecules.

A nucleus with spin projection m_I gives rise to a magnetic field with z component B_{nuc} at a distance R, where, to a good approximation,[5]

$$B_{nuc} = -\frac{\gamma\hbar\mu_0}{4\pi R^3}(1 - 3\cos^2\theta)m_I \tag{18.36}$$

The angle θ is defined in (**6**). The magnitude of this field is about 0.1 mT when $R = 0.3$ nm, corresponding to a splitting of resonance signal of about 10^4 Hz, and is of the order of magnitude of the splitting observed in solid samples (see Section 18.12a).

In a liquid, the angle θ sweeps over all values as the molecule tumbles, and $1 - 3\cos^2\theta$ averages to zero. Hence the direct dipolar interaction between spins cannot

5

6

The *mean value theorem* states that average (or mean value) of a function $f(x)$ over the range $x = a$ to $x = b$ is

$$\frac{1}{b-a}\int_a^b f(x)\,dx$$

The volume element in polar coordinates is proportional to $\sin\theta\,d\theta$, and θ ranges from 0 to π. Therefore the average value of $(1 - 3\cos^2\theta)$ is

$$\frac{1}{\pi}\int_0^\pi (1 - 3\cos^2\theta)\sin\theta\,d\theta = 0.$$

5. See *Further reading*.

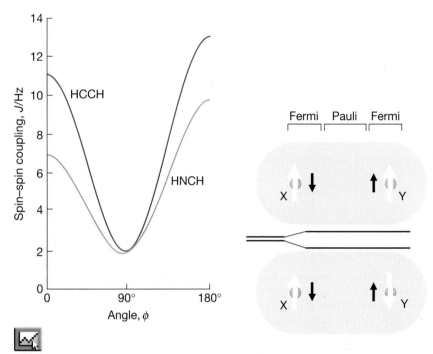

18.19 The variation of the spin–spin coupling constant $^3J_{HH}$ with angle (as defined in **5**) predicted by the Karplus equation for an HCCH group and an HNCH group.

18.20 The polarization mechanism for spin–spin coupling ($^1J_{HH}$). The two arrangements have slightly different energies. In this case, J is positive, corresponding to a lower energy when the nuclear spins are antiparallel.

18.21 The origin of the Fermi contact interaction. From far away, the magnetic field pattern arising from a ring of current (representing the rotating charge of the nucleus, the pale grey sphere) is that of a point dipole. However, if an electron can sample the field close to the region indicated by the sphere, the field distribution differs significantly from that of a point dipole. For example, if the electron can penetrate the sphere, then the spherical average of the field it experiences is not zero.

account for the fine structure of the spectra of rapidly tumbling molecules. The direct interaction does make an important contribution to the spectra of solid samples and is a very useful indirect source of structure information through its involvement in spin relaxation (Section 18.10).

Spin–spin coupling in molecules in solution can be explained in terms of the **polarization mechanism**, in which the interaction is transmitted through the bonds. The simplest case to consider is that of $^1J_{XY}$ where X and Y are spin-$\frac{1}{2}$ nuclei joined by an electron-pair bond (Fig. 18.20). The coupling mechanism depends on the fact that in some atoms it is favourable for the nucleus and a nearby electron spin to be parallel (both α or both β), but in others it is favourable for them to be antiparallel (one α and the other β). The electron–nucleus coupling is magnetic in origin, and may be either a dipolar interaction between the magnetic moments of the electron and nuclear spins or a **Fermi contact interaction**. A pictorial description of the Fermi contact interaction is as follows. First, we regard the magnetic moment of the nucleus as arising from the circulation of a current in a tiny loop with a radius similar to that of the nucleus (Fig. 18.21). Far from the nucleus the field generated by this loop is indistinguishable from the field generated by a point magnetic dipole. Close to the loop, however, the field differs from that of a point dipole. The magnetic interaction between this non-dipolar field and the electron's magnetic moment is the contact interaction. The lines of force depicted in Fig. 18.21 correspond to those for a proton with α spin. The lower energy state of an electron spin in such a field is the β state. In conclusion, the contact interaction depends on the very close approach of an electron to the nucleus and hence can occur only if the electron occupies an s orbital. We shall suppose that it is energetically favourable for an electron spin and a nuclear spin to be antiparallel (as is the case for a proton and an electron in a hydrogen atom).

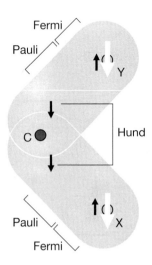

18.22 The polarization mechanism for $^2J_{HH}$ spin–spin coupling. The spin information is transmitted from one bond to the next by a version of the mechanism that accounts for the lower energy of electrons with parallel spins in different atomic orbitals (Hund's rule of maximum multiplicity). In this case, $J < 0$, corresponding to a lower energy when the nuclear spins are parallel.

If the X nucleus is α, a β electron of the bonding pair will tend to be found nearby (because that is energetically favourable for it). The second electron in the bond, which must have α spin if the other is β, will be found mainly at the far end of the bond (because electrons tend to stay apart to reduce their mutual repulsion). Because it is energetically favourable for the spin of Y to be antiparallel to an electron spin, a Y nucleus with β spin has a lower energy, and hence a lower Larmor frequency, than a Y nucleus with α spin. The opposite is true when X is β, for now the α spin of Y has the lower energy. In other words, the antiparallel arrangement of nuclear spins lies lower in energy than the parallel arrangement as a result of their magnetic coupling with the bond electrons. That is, $^1J_{HH}$ is positive.

To account for the value of $^2J_{XY}$, as in H—C—H, we need a mechanism that can transmit the spin alignments through the central C atom (which may be ^{12}C, with no nuclear spin of its own). In this case (Fig. 18.22), an X nucleus with β spin polarizes the electrons in its bond, and the β electron is likely to be found closer to the C nucleus. The more favourable arrangement of two electrons on the same atom is with their spins parallel (Hund's rule, Section 13.4d), so the more favourable arrangement is for the β electron of the neighbouring bond to be close to the C nucleus. Consequently, the α electron of that bond is more likely to be found close to the Y nucleus, and therefore that nucleus will have a lower energy if it is β. Hence, according to this mechanism, the lower Larmor frequency of Y will be obtained if its spin is parallel to that of X. That is, $^2J_{HH}$ is negative.

The coupling of nuclear spin to electron spin by the Fermi contact interaction is most important for proton spins, but it is not necessarily the most important mechanism for other nuclei. These nuclei may also interact by a dipolar mechanism with the electron magnetic moments and with their orbital motion, and there is no simple way of specifying whether J will be positive or negative.

(e) Equivalent nuclei

A group of nuclei are **chemically equivalent** if they are related by a symmetry operation of the molecule and have the same chemical shifts. Chemically equivalent nuclei are nuclei that would be regarded as 'equivalent' according to ordinary chemical criteria. Nuclei are **magnetically equivalent** if, as well as being chemically equivalent, they also have identical spin–spin interactions with any other magnetic nuclei in the molecule.

The difference between chemical and magnetic equivalence is illustrated by CH_2F_2 and $H_2C{=}CF_2$. In each of these molecules the protons are chemically equivalent: they are related by symmetry and undergo the same chemical reactions. However, although the protons in CH_2F_2 are magnetically equivalent, those in $CH_2{=}CF_2$ are not. One proton in the latter has spin-coupling interactions with a *cis* F nucleus, whereas the other proton has a *trans* interaction with it. In CH_2F_2 both protons are equally distant from the two F nuclei, so there is no distinction between them. Strictly speaking, the CH_3 protons in ethanol (and other compounds) are magnetically inequivalent on account of their different interactions with the CH_2 protons in the next group. However, they are in practice made magnetically equivalent by the rapid rotation of the CH_3 group, which averages out any differences. Magnetically inequivalent species can give very complicated spectra (for instance, the spectrum of $H_2C{=}CF_2$ consists of 24 lines), and we shall not consider them further.

An important feature of chemically equivalent magnetic nuclei is that, although they do couple together, the coupling has no effect on the appearance of the spectrum. The reason for the invisibility of the coupling is set out in the following *Justification*, but qualitatively it is that all allowed nuclear spin transitions are *collective* reorientations of groups of equivalent nuclear spins that do not change the relative orientations of the spins within the group (Fig. 18.23). Then, because the relative

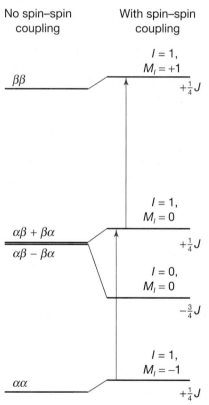

18.24 The energy levels of an A_2 system in the absence of spin–spin coupling are shown on the left. When spin–spin coupling is taken into account, the energy levels on the right are obtained. Note that the three states with total nuclear spin $I = 1$ correspond to parallel spins and give rise to the same increase in energy (J is positive); the one state with $I = 0$ (antiparallel nuclear spins) has a lower energy in the presence of spin–spin coupling. The only allowed transitions are those that preserve the angle between the spins, and so take place between the three states with $I = 1$. They occur at the same resonance frequency as they would have in the absence of spin–spin coupling.

18.23 (a) A group of two equivalent nuclei realigns as a group, without change of angle between the spins, when a resonant absorption occurs. Hence it behaves like a single nucleus and the spin–spin coupling between the individual spins of the group is undetectable. (b) Three equivalent nuclei also realign as a group without change of their relative orientations.

orientations of nuclear spins are not changed in any transition, the magnitude of the coupling between them is undetectable. Hence, an isolated CH_3 group gives a single, unsplit line because all the allowed transitions of the group of three protons occur without change of their relative orientations.

Justification 18.2

Consider an A_2 system of two chemically equivalent spin-$\frac{1}{2}$ nuclei. First, consider the energy levels in the absence of spin–spin coupling. There are four spin states which (just as for two electrons) can be classified according to their total spin I (the analogue of S for two electrons) and their total projection M_I on the z-axis. The states are analogous to those we developed for two electrons in singlet and triplet states:[6]

Spins parallel, $I = 1$: $M_I = +1$ $\alpha\alpha$

                  $M_I = 0$ $(1/2^{1/2})\{\alpha\beta + \beta\alpha\}$

                  $M_I = -1$ $\beta\beta$

Spins paired, $I = 0$: $M_I = 0$ $(1/2^{1/2})\{\alpha\beta - \beta\alpha\}$

The effect of a magnetic field on these four states is shown in Fig. 18.24: the energies of the two states with $M_I = 0$ are unchanged by the field because they are composed of equal proportions of α and β spins.

6. As in Section 13.7, the states we have selected are those with a definite resultant, and hence a well defined value of I. The + sign in $\alpha\beta + \beta\alpha$ signifies an in-phase alignment of spins and $I = 1$; the − sign in $\alpha\beta - \beta\alpha$ signifies an alignment out of phase by π, and hence $I = 0$. See Fig. 13.26.

The spin–spin coupling energy is proportional to the scalar product of the vectors representing the spins, and we write $E = (hJ/\hbar^2)\mathbf{I}_1 \cdot \mathbf{I}_2$. The scalar product can be expressed in terms of the total nuclear spin by noting that

$$I^2 = (\mathbf{I}_1 + \mathbf{I}_2)\cdot(\mathbf{I}_1 + \mathbf{I}_2) = I_1^2 + I_2^2 + 2\mathbf{I}_1 \cdot \mathbf{I}_2$$

and replacing the magnitudes by their quantum mechanical values:

$$\mathbf{I}_1 \cdot \mathbf{I}_2 = \tfrac{1}{2}\{I(I+1) - I_1(I_1+1) - I_2(I_2+1)\}\hbar^2$$

Then, because $I_1 = I_2 = \tfrac{1}{2}$, it follows that

$$E = \tfrac{1}{2}hJ\{I(I+1) - \tfrac{3}{2}\}$$

For parallel spins, $I = 1$ and $E = +\tfrac{1}{4}hJ$; for antiparallel spins $I = 0$ and $E = -\tfrac{3}{4}hJ$, as in Fig. 18.24. We see that three of the states move in energy in one direction and the fourth (the one with antiparallel spins) moves three times as much in the opposite direction. The resulting energy levels are shown on the right in Fig. 18.24.

The NMR spectrum of the A_2 species arises from transitions between the levels. However, the radiofrequency field affects the two equivalent protons equally, so it cannot change the orientation of one proton relative to the other; therefore, the transitions take place within the set of states that correspond to parallel spin (those labelled $I = 1$), and no spin-parallel state can change to a spin-antiparallel state (the state with $I = 0$). Put another way, the allowed transitions are subject to the selection rule $\Delta I = 0$. This selection rule is in addition to the rule $\Delta M_I = \pm 1$ that arises from the conservation of angular momentum and the unit spin of the photon. The allowed transitions are shown in Fig. 18.24: we see that there are only two transitions, and that they occur at the same resonance frequency that the nuclei would have in the absence of spin–spin coupling. Hence, the spin–spin coupling interaction does not affect the appearance of the spectrum.

(f) Strongly coupled nuclei

NMR spectra are usually much more complex than the foregoing simple analysis suggests. We have described the extreme case in which the differences in chemical shifts are much greater than the spin–spin coupling constants. In such cases it is simple to identify groups of magnetically equivalent nuclei and to think of the groups of nuclear spins as reorienting relative to each other. The spectra that result are called **first-order spectra**.

Transitions cannot be allocated to definite groups when the differences in their chemical shifts are comparable to their spin–spin coupling interactions. The complicated spectra that are then obtained are called **strongly coupled spectra** (or 'second-order spectra') and are much more difficult to analyse (Fig. 18.25). Because the difference in resonance frequencies increases with field, but spin–spin coupling constants are independent of it, a second-order spectrum may become simpler (and first-order) at high fields because individual groups of nuclei become identifiable again.

A clue to the type of analysis that is appropriate is given by the notation for the types of spins involved. Thus, an AX spin system (which consists of two nuclei with a large chemical shift difference) has a first-order spectrum. An AB system, on the other hand (with two nuclei of similar chemical shifts), gives a spectrum typical of a strongly coupled system. An AX system may have widely different Larmor frequencies because A and X are nuclei of different elements (such as ^{13}C and ^1H), in which case they form a **heteronuclear spin system**. AX may also denote a **homonuclear spin system** in which the nuclei are of the same element but in markedly different environments.

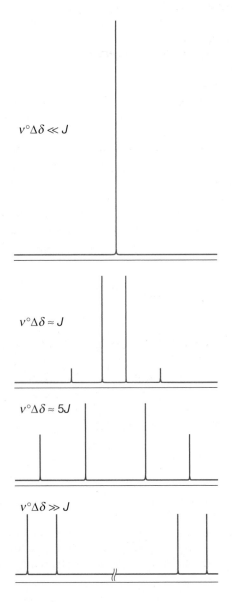

$\nu^\circ\Delta\delta \ll J$

$\nu^\circ\Delta\delta \approx J$

$\nu^\circ\Delta\delta \approx 5J$

$\nu^\circ\Delta\delta \gg J$

18.25 The NMR spectrum of an A_2 system (top) and an AX system (bottom) are simple, and give rise to 'first-order spectra'. At intermediate relative values of the chemical shift difference and the spin–spin coupling, complex 'strongly coupled' spectra are obtained. Note how the inner two lines of the bottom spectrum move together, grow in intensity, and form the single central line of the top spectrum. The two outer lines diminish in intensity and are absent in the top spectrum.

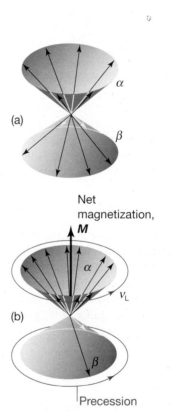

18.26 The vector model of angular momentum for a single spin-$\frac{1}{2}$ nucleus. The angle around the z-axis is indeterminate.

Pulse techniques in NMR

Modern methods of detecting the energy separation between nuclear spin states are more sophisticated than simply looking for the frequency at which resonance occurs. One of the best analogies that has been suggested to illustrate the difference between the old and new ways of observing an NMR spectrum is that of detecting the spectrum of vibrations of a bell. We could stimulate the bell with a gentle vibration at a gradually increasing frequency, and note the frequencies at which it resonated with the stimulation. A lot of time would be spent getting zero response when the stimulating frequency was between the bell's vibrational modes. However, if we were simply to hit the bell with a hammer, we would immediately obtain a clang composed of all the frequencies that the bell can produce. The equivalent in NMR is to monitor the radiation nuclear spins emit as they return to equilibrium after the appropriate stimulation. The resulting **Fourier-transform NMR** gives greatly increased sensitivity, so opening up the entire periodic table to the technique. Moreover, multiple-pulse FT-NMR gives chemists unparalleled control over the information content and display of spectra. We need to understand how the equivalent of the hammer blow is delivered and how the signal is monitored and interpreted. These features are generally expressed in terms of the vector model of angular momentum introduced in Section 12.7d.

18.7 The magnetization vector

Consider a sample composed of many identical spin-$\frac{1}{2}$ nuclei. As we saw in Section 12.7d, an angular momentum can be represented by a vector of length $\{I(I+1)\}^{1/2}$ units with a component of length m_I units along the z-axis. As the uncertainty principle does not allow us to specify the x- and y-components of the angular momentum, all we know is that the vector lies somewhere on a cone around the z-axis. For $I = \frac{1}{2}$, the length of the vector is $\frac{1}{2}\sqrt{3}$ and it makes an angle of 55° to the z-axis (Fig. 18.26).

In the absence of a magnetic field, the sample consists of equal numbers of α and β nuclear spins with their vectors lying at random angles on the cones. These angles are unpredictable, and at this stage we picture the spin vectors as stationary. The **magnetization**, M, of the sample, its net nuclear magnetic moment, is zero (Fig. 18.27a).

(a) The effect of the static field

Two changes occur in the magnetization when a magnetic field is present. First, the energies of the two orientations change, the α spins moving to low energy and the β spins to high energy (provided $\gamma > 0$). At 10 T, the Larmor frequency for protons is 427 MHz, and in the vector model the individual vectors are pictured as precessing at this rate. This motion is a pictorial representation of the difference in energy of the spin states (it is not an actual representation of reality). As the field is increased, the Larmor frequency increases and the precession becomes faster. Secondly, the populations of the two spin states (the numbers of α and β spins) change, and there will be more α spins than β spins. Because $h\nu_L/kT \approx 7 \times 10^{-5}$ for protons at 300 K and 10 T, there is only a tiny imbalance of populations, and it is even smaller for other nuclei with their smaller magnetogyric ratios. However, despite its smallness, the imbalance means that there is a net magnetization that we can represent by a vector M pointing in the z-direction and with a length proportional to the population difference (Fig. 18.27b).

18.27 The magnetization of a sample of spin-$\frac{1}{2}$ nuclei is the resultant of all their magnetic moments. (a) In the absence of an externally applied field, there are equal numbers of α and β spins at random angles around the z-axis (the field direction) and the magnetization is zero. (b) In the presence of a field, the spins precess around their cones (that is, there is an energy difference between the α and β states) and there are slightly more α spins than β spins. As a result, there is a net magnetization along the z-axis.

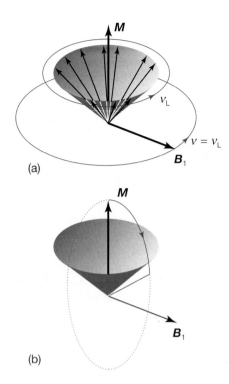

(a)

(b)

18.28 (a) In a resonance experiment, a circularly polarized radiofrequency magnetic field B_1 is applied in the xy-plane (the magnetization vector lies along the z-axis). (b) If we step into a frame rotating at the Larmor frequency, the radiofrequency field appears to be stationary if its frequency is the same as the Larmor frequency. When the two frequencies coincide, the magnetization vector of the sample begins to rotate around the direction of the B_1 field.

(b) The effect of the radiofrequency field

We now consider the effect of a circularly polarized radiofrequency field in the xy-plane. We have considered the electric component of this field in the other forms of spectroscopy that we have treated. However, in this chapter we consider only the magnetic component, for it is this component that interacts with the nuclear magnetic moment. The strength of the rotating magnetic field is B_1.

Suppose we choose the frequency of the oscillating field to be equal to the Larmor frequency of the spins. This choice is equivalent to selecting the resonance condition in the conventional experiment. The nuclei now experience a steady B_1 field because the rotating magnetic field is in step with the precessing spins (Fig. 18.28). Just as the spins precess about the strong static field B_0 at a frequency $\gamma B_0/2\pi$, so under the influence of the effectively stronger field B_1 they precess about B_1 at a frequency $\gamma B_1/2\pi$. If we apply the B_1 field in a pulse of duration $\pi/2\gamma B_1$, the magnetization precesses into the xy-plane, and we say that we have applied a **90° pulse** (or a '$\pi/2$ pulse'). The duration of the pulse depends on the strength of the B_1 field, but is typically of the order of microseconds. To a stationary external observer (a radiofrequency coil, Fig. 18.29), the magnetization vector is now rotating in the xy-plane at the Larmor frequency (about 430 MHz at 10 T). The rotating magnetization induces a 430 MHz signal in the coil, which can be amplified and processed. In practice, the processing takes place after subtraction of a constant high-frequency component, so that all the signal manipulation takes place at frequencies of a few kilohertz.

As time passes, the individual spins move out of step (partly because they are precessing at slightly different rates, as we shall explain later), so the magnetization vector shrinks exponentially with a time constant T_2 and induces an ever weaker signal in the detector coil. The form of the signal that we can expect is therefore the oscillating-decaying **free-induction decay** (FID) shown in Fig. 18.30, and the y-component of the magnetization varies as[7]

$$M_y(t) = M_0 \cos(2\pi\nu_L t)e^{-t/T_2} \tag{18.37}$$

We have considered the effect of a pulse applied at exactly the Larmor frequency. However, virtually the same effect is obtained off resonance, provided that the pulse is applied close to ν_L. If the difference in frequency is small compared to the inverse of the duration of the 90° pulse, the magnetization will end up in the xy-plane. Note that we do not need to know the Larmor frequency beforehand: the short pulse is the analogue of the hammer blow on the bell, exciting a range of frequencies. The detected signal shows that a particular resonant frequency is present.

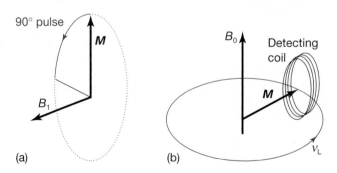

18.29 (a) If the radiofrequency field is applied for a certain time, the magnetization vector is rotated into the xy-plane. (b) To an external stationary observer (the coil), the magnetization vector is rotating at the Larmor frequency, and can induce a signal in the coil.

(a)

(b)

---

7. More precisely, in a static frame of reference such as the one used here the signal voltage is proportional to the derivative of M_y.

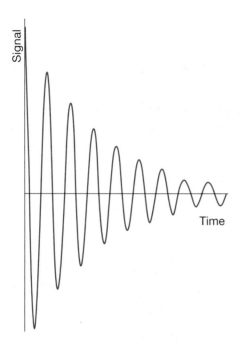

18.30 A simple free-induction decay of a sample of spins with a single resonance frequency.

Because $e^{\pm 2\pi i \nu t} = \cos(2\pi\nu t) \pm i\sin(2\pi\nu t)$, the expression $S(t) = \int_{-\infty}^{\infty} I(\nu)e^{-2\pi i\nu t}\,d\nu$ is a sum over harmonically oscillating functions, with each one weighted by the intensity $I(\nu)$.

18.31 (a) A free induction decay signal of a sample of AX species and (b) its analysis into its frequency components.

(c) Time- and frequency-domain signals

We can think of the magnetization vector of a homonuclear AX spin system with $J = 0$ as consisting of two parts, one formed by the A spins and the other by the X spins. When the 90° pulse is applied, both magnetization vectors are rotated into the xy-plane. However, because the A and X nuclei precess at different frequencies, they induce two signals in the detector coils, and the overall FID curve may resemble that in Fig. 18.31a. The composite FID curve is the analogue of the struck bell emitting a rich tone composed of all the frequencies at which it can vibrate.

The problem we must address is how to recover the resonance frequencies present in a free-induction decay. We encountered a similar problem when discussing Fourier-transform infrared spectra in Section 16.1c, where all the vibrational frequencies were detected at once. The same technique is used here. We know that the FID curve is a sum of oscillating functions, so the problem is to analyse it into its harmonic components.

The analysis of the FID curve is achieved by the standard mathematical technique of Fourier transformation. We start by noting that the signal $S(t)$ in the time domain, the total FID curve, is the sum (more precisely, the integral) over all the contributing frequencies

$$S(t) = \int_{-\infty}^{\infty} I(\nu)e^{-2\pi i\nu t}\,d\nu \tag{18.38}$$

We need $I(\nu)$, the spectrum in the frequency domain; it is obtained by evaluating the integral

$$I(\nu) = 2\mathrm{Re}\int_{0}^{\infty} S(t)e^{2\pi i\nu t}\,dt \tag{18.39}$$

where Re means take the real part of the following expression. This integral is very much like an overlap integral: it gives a nonzero value if $S(t)$ contains a component that matches the oscillating function $e^{2i\pi\nu t}$. The integration is carried out at a series of frequencies ν on a computer that is built into the spectrometer. When the signal in Fig. 18.31a is transformed in this way, we get the frequency-domain spectrum shown in Fig. 18.31b. One line represents the Larmor frequency of the A nuclei and the other that of the X nuclei.

The FID curve in Fig. 18.32 is obtained from a sample of ethanol. The frequency-domain spectrum obtained from it by Fourier transformation is the one that we have

(a)

(b)

Time

18.32 A free induction decay signal of a sample of ethanol. Its Fourier transform is the frequency-domain spectrum shown in Fig. 18.6. The total length of the image corresponds to about 1 s.

The web site contains links to databases of NMR spectra and to sites that allow for interactive simulation of NMR spectra.

already discussed (Fig. 18.6). We can now see why the FID curve in Fig. 18.32 is so complex: it arises from the precession of a magnetization vector that is composed of eight components, each with a characteristic frequency.

18.8 Linewidths and rate processes

The linewidths of NMR spectra, in common with other spectroscopic techniques, provide information about the rates of processes relating to the molecules in the sample. We have seen that the FID signal decreases with time, which implies that the component of the magnetization vector in the xy-plane must be shrinking. In this section we see some of the processes involved.

(a) Spin relaxation

There are two reasons why the component of the magnetization vector in the xy-plane shrinks. Both reflect the fact that the nuclear spins are not in thermal equilibrium with their surroundings (for then M lies parallel to z). The return to equilibrium is the process called **spin relaxation**.

At thermal equilibrium the spins have a Boltzmann distribution, with more α spins than β spins; however, a magnetization vector in the xy-plane immediately after a 90° pulse has equal numbers of α and β spins. This condition is known as **saturation**.

Now consider the effect of a **180° pulse**, which may be visualized as a flip of the net magnetization vector from one direction along the z-axis to the opposite direction. That is, the 180° pulse leads to population inversion of the spin system, which now has more β spins than α spins. After the pulse, the populations revert to their thermal equilibrium values exponentially. As they do so, the z-component of magnetization reverts to its equilibrium value M_0 with a time constant called the **longitudinal relaxation time**, T_1 (Fig. 18.33):

$$M_z(t) - M_0 \propto e^{-t/T_1} \tag{18.40}$$

Because this relaxation process involves giving up energy to the surroundings (the 'lattice') as β spins revert to α spins, the time constant T_1 is also called the **spin–lattice relaxation time**. Spin–lattice relaxation is caused by local magnetic fields that fluctuate at a frequency close to the resonance frequency of the $\alpha \rightarrow \beta$ transition. Such fields can arise from the tumbling motion of molecules in a fluid sample. If molecular tumbling is too slow or too fast compared to the resonance frequency, it will give rise to a fluctuating magnetic field with a frequency that is either too low

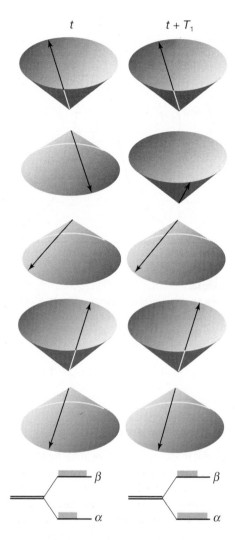

18.33 In longitudinal relaxation the spins relax back towards their thermal equilibrium populations. On the left we see the precessional cones representing spin-$\frac{1}{2}$ angular momenta, and they do not have their thermal equilibrium populations (there are more β spins than α spins). On the right, which represents the sample after a time T_1, the populations are those characteristic of a Boltzmann distribution. In actuality, T_1 is the time constant for relaxation to the arrangement on the right and $T_1 \ln 2$ is the half-life of the arrangement on the left.

or too high to stimulate a spin change from β to α, so T_1 will be long. Only if the molecule tumbles at about the resonance frequency will the fluctuating magnetic field be able to induce spin changes effectively, and only then will T_1 be short. The rate of molecular tumbling increases with temperature and with reducing viscosity of the solvent, so we can expect a dependence like that shown in Fig. 18.34.

A second aspect of spin relaxation is the fanning-out of the spins in the xy-plane if they precess at different rates (Fig. 18.35). The magnetization vector is large when all the spins are bunched together immediately after a 90° pulse. However, this orderly bunching of spins is not at equilibrium and, even if there were no spin-lattice relaxation, we would expect the individual spins to spread out until they were uniformly distributed with all possible angles around the z-axis. At that stage, the component of magnetization vector in the plane would be zero. The randomization of the spin directions occurs exponentially with a time constant called the **transverse relaxation time**, T_2:

$$M_y(t) \propto e^{-t/T_2} \qquad (18.41)$$

Because the relaxation involves the relative orientation of the spins, T_2 is also known as the **spin–spin relaxation time**. Any relaxation process that changes the balance between α and β spins will also contribute to this randomization, so the time constant T_2 is always less than or equal to T_1.

Local magnetic fields also affect spin–spin relaxation. When the fluctuations are slow, each molecule lingers in its local magnetic environment and the spin orientations randomize quickly around the applied field direction. If the molecules move rapidly from one magnetic environment to another, the effects of differences in local magnetic field average to zero: individual spins do not precess at very different rates, they can remain bunched for longer, and spin–spin relaxation does not take place as quickly. In other words, slow molecular motion corresponds to short T_2 and fast motion corresponds to long T_2 (as shown in Fig. 18.34). Calculations show that, when the motion is fast, $T_2 \approx T_1$.

If the y-component of magnetization decays with a time constant T_2, the spectral line is broadened (Fig. 18.36), and its width at half-height becomes

$$\Delta v_{1/2} = \frac{1}{\pi T_2} \qquad (18.42)$$

Typical values of T_2 in proton NMR are of the order of seconds, so linewidths of around 0.1 Hz can be anticipated, in broad agreement with observation.

So far, we have assumed that the equipment, and in particular the magnet, are perfect, and that the differences in Larmor frequencies arise solely from interactions within the sample. In practice, the magnet is not perfect, and the field is different at different locations in the sample. The inhomogeneity broadens the resonance, and in most cases this **inhomogeneous broadening** dominates the broadening we have discussed so far. It is common to express the extent of inhomogeneous broadening in terms of an **effective transverse relaxation time**, T_2^*, by using a relation like eqn 42, but writing

$$T_2^* = \frac{1}{\pi \Delta v_{1/2}} \qquad (18.43)$$

where $\Delta v_{1/2}$ is the observed width at half-height of a line with a Lorentzian shape of the form $I \propto 1/(1 + v^2)$. As an example, consider a line in a spectrum with a width of 10 Hz. It follows from eqn 43, that the effective transverse relaxation time is

$$T_2^* = \frac{1}{\pi \times (10 \text{ s}^{-1})} = 32 \text{ ms}$$

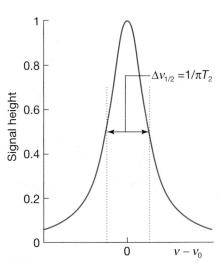

18.34 The variation of the two relaxation times with the rate at which the molecules move (either by tumbling or migrating through the solution). The horizontal axis can be interpreted as representing temperature or viscosity. Note that at rapid rates of motion, the two relaxation times coincide.

18.36 A Lorentzian absorption line. The width at half-height is inversely proportional to the parameter T_2 and, the longer the transverse relaxation time, the narrower the line.

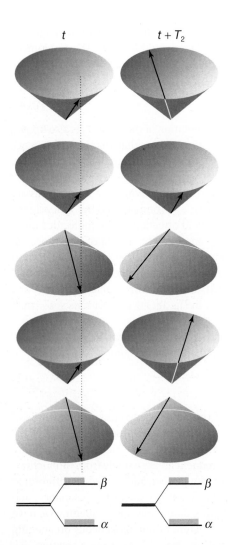

18.35 The transverse relaxation time, T_2, is the time it takes for the phases of the spins to become randomized (another condition for equilibrium) and to change from the orderly arrangement shown on the left to the disorderly arrangement on the right. Note that the populations of the states remain the same; only the relative phase of the spins relaxes. In actuality, T_2 is the time constant for relaxation to the arrangement on the right and $T_2 \ln 2$ is the half-life of the arrangement on the left.

(b) Conformational conversion and exchange processes

The appearance of an NMR spectrum is changed if magnetic nuclei can jump rapidly between different environments. Consider a molecule, such as N,N-dimethylformamide, that can jump between conformations; in its case, the methyl shifts depend on whether they are *cis* or *trans* to the carbonyl group (Fig. 18.37). When the jumping rate is low, the spectrum shows two sets of lines, one each from molecules in each conformation. When the interconversion is fast, the spectrum shows a single line at the mean of the two chemical shifts. At intermediate inversion rates, the line is very broad. This maximum broadening occurs when the lifetime, τ, of a conformation gives rise to a linewidth that is comparable to the difference of resonance frequencies, $\delta\nu$, and both broadened lines blend together into a very broad line. Coalescence of the two lines occurs when

$$\tau = \frac{\sqrt{2}}{\pi\delta\nu} \tag{18.44}$$

Example 18.2 Interpreting line broadening

The NO group in N,N-dimethylnitrosamine, $(CH_3)_2N\text{—}NO$, rotates about the N—N bond and, as a result, the magnetic environments of the two CH_3 groups are interchanged. In a 600-MHz spectrometer the two CH_3 resonances are separated by 390 Hz. At what rate of interconversion will the resonance collapse to a single line?

Method Use eqn 44 for the average lifetimes of the conformations. The rate of interconversion is the inverse of their lifetime.

18.37 When a molecule changes from one conformation to another, the positions of its protons are interchanged and jump between magnetically distinct environments.

Answer With $\delta \nu = 390$ Hz,

$$\tau = \frac{\sqrt{2}}{\pi \times (390 \ s^{-1})} = 1.2 \ ms$$

It follows that the signal will collapse to a single line when the interconversion rate exceeds about 830 s^{-1}.

Comment The dependence of the rate of collapse on the temperature is used to determine the energy barrier to interconversion.

Self-test 18.3 What would you deduce from the observation of a single line from the same molecule in a 300 MHz spectrometer?

[Conformation lifetime less than 2.3 ms]

A similar explanation accounts for the loss of fine structure in solvents able to exchange protons with the sample. For example, hydroxyl protons are able to exchange with water protons. When this **chemical exchange** occurs, a molecule ROH with an α-spin proton (we write this ROH$_\alpha$) rapidly converts to ROH$_\beta$ and then perhaps to ROH$_\alpha$ again because the protons provided by the solvent molecules in successive exchanges have random spin orientations. Therefore, instead of seeing a spectrum composed of contributions from both ROH$_\alpha$ and ROH$_\beta$ molecules (that is, a spectrum showing a doublet structure due to the OH proton) we see a spectrum that shows no splitting caused by coupling of the OH proton (as in Fig. 18.6). The effect is observed when the lifetime of a molecule due to this chemical exchange is so short that the lifetime broadening is greater than the doublet splitting. Because this splitting is often very small (a few hertz), a proton must remain attached to the same molecule for longer than about 0.1 s for the splitting to be observable. In water, the exchange rate is much faster than that, so alcohols show no splitting from the OH protons. In dry dimethylsulfoxide (DMSO), the exchange rate may be slow enough for the splitting to be detected.

(c) The measurement of T_1

The longitudinal relaxation time T_1 can be measured by the **inversion recovery technique**. The first step is to apply a 180° pulse to the sample. A 180° pulse is achieved by applying the B_1 field for twice as long as for a 90° pulse, so the magnetization vector precesses through 180° and points in the $-z$-direction (Fig. 18.38). No signal can be seen at this stage because there is no component of magnetization in the

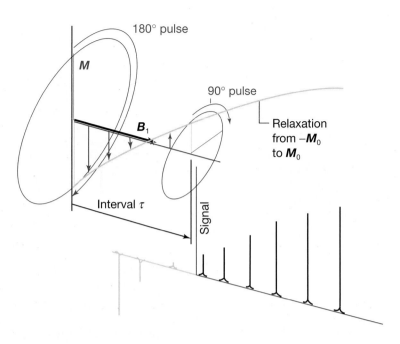

18.38 The result of applying a 180° pulse to the magnetization in the rotating frame and the effect of a subsequent 90° pulse. The amplitude of the frequency-domain spectrum varies with the interval between the two pulses because spin–lattice relaxation has time to occur.

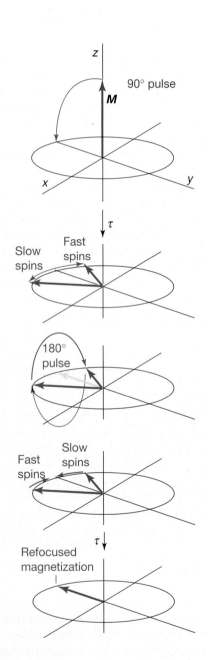

18.39 The sequence of pulses leading to the observation of a spin echo.

xy-plane (where the coil can detect it). The β spins begin to relax back into α spins, and the magnetization vector first shrinks and then grows exponentially back towards its thermal equilibrium value, M_z. After an interval τ, a 90° pulse is applied that rotates the magnetization into the *xy*-plane, where it generates an FID signal. The frequency-domain spectrum is then obtained by Fourier transformation.

The intensity of the spectrum obtained in this way depends on the length of the magnetization vector that is rotated into the *xy*-plane. The length of that vector changes exponentially as the interval between the two pulses is increased, so the intensity of the spectrum also changes exponentially with increasing τ. We can therefore measure T_1 by fitting an exponential curve to the series of spectra obtained after different values of τ.

(d) Spin echoes

The measurement of T_2 (as distinct from T_2^*) depends on the ability to eliminate the effects of inhomogeneous broadening. The cunning required is at the root of some of the most important advances that have been made in NMR since its introduction.

A **spin echo** is the magnetic analogue of an audible echo: transverse magnetization is created by a radiofrequency pulse, decays away, is reflected by a second pulse, and grows back to form an echo. The sequence of events is shown in Fig. 18.39. We can consider the overall magnetization as being made up of a number of different magnetizations, each of which arises from a **spin packet** of nuclei with very similar precession frequencies. The spread in these frequencies arises because the applied field B_0 is inhomogeneous, so different parts of the sample experience different fields. The precession frequencies also differ if there is more than one chemical shift present. As will be seen, the importance of a spin echo is that it can suppress the effects of both field inhomogeneities and chemical shifts.

First, a 90° pulse is applied to the sample. Let us follow events by using a frame of reference that is rotating at the same rate as the radiofrequency magnetic field of

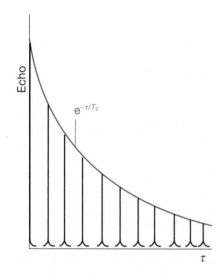

18.40 The exponential decay of spin echoes can be used to determine the transverse relaxation time.

the pulse, so B_1 is stationary along the x-axis and causes the magnetization to be rotated down into the xy-plane. The spin packets now begin to fan out because they have different Larmor frequencies, with some above the radiofrequency and some below. The detected signal depends on the resultant of the spin-packet magnetization vectors, and decays with a time constant T_2^* because of the combined effects of field inhomogeneity and spin–spin relaxation.

After an evolution period τ, a 180° pulse is applied to the sample; this time, about the y-axis of the rotating frame (the axis of the pulse is changed from x to y by a 90° phase shift of the radiofrequency radiation). The pulse rotates the magnetization vectors of the faster spin packets into the positions previously occupied by the slower spin packets, and vice versa. Thus, as the vectors continue to precess, the fast vectors are now behind the slow; the fan begins to close up again, and the resultant signal begins to grow back into an echo. At time 2τ, all the vectors will once more be aligned along the y-axis, and the fanning out caused by the field inhomogeneity is said to have been refocused; the spin echo has reached its maximum. Because the effects of field inhomogeneities have been suppressed by the refocusing, the echo signal will have been attenuated by the factor $e^{-2\tau/T_2}$ caused by spin–spin relaxation alone. After the time 2τ, the magnetization will continue to precess, fanning out once again, giving a resultant that decays with time constant T_2^*.

The important feature of the technique is that the size of the echo is independent of any local fields that remain constant during the two τ intervals. If a spin packet is 'fast' because it happens to be composed of spins in a region of the sample that experiences higher than average fields, then it remains fast throughout both intervals, and what it gains on the first interval it loses on the second interval. Hence, the size of the echo is independent of inhomogeneities in the magnetic field, for these remain constant. The true transverse relaxation arises from fields that fluctuate on a molecular timescale, and there is no guarantee that an individual 'fast' spin will remain 'fast' in the refocusing phase: the spins within the packets therefore spread with a time constant T_2. Hence, the effects of the true relaxation are not refocused, and the size of the echo decays with the time constant T_2 (Fig. 18.40).

18.9 Spin decoupling

Carbon-13 is a **dilute spin species** in the sense that it is unlikely that more than one ^{13}C nucleus will be found in any given small molecule (provided the sample has not been enriched with that isotope; the natural abundance of ^{13}C is only 1.1 per cent). Even in large molecules, although more than one ^{13}C nucleus may be present, it is unlikely that they will be close enough to give an observable splitting. Hence, it is not normally necessary to take into account ^{13}C—^{13}C spin–spin coupling within a molecule.

Protons are **abundant-spin species** in the sense that a molecule is likely to contain many of them. If we were observing a ^{13}C-NMR spectrum, we would obtain a very complex spectrum on account of the coupling of the one ^{13}C nucleus with many of the protons that are present. To avoid this difficulty, ^{13}C-NMR spectra are normally observed using the technique of **proton decoupling**. Thus, if the CH$_3$ protons of ethanol are irradiated with a second, strong, resonant radiofrequency pulse, they undergo rapid spin reorientations and the ^{13}C nucleus senses an average orientation. As a result, its resonance is a single line and not a 1:3:3:1 quartet. Proton decoupling has the additional advantage of enhancing sensitivity, because the intensity is concentrated into a single transition frequency instead of being spread over several transition frequencies (see Section 18.10). If care is taken to ensure that the other parameters on which the strength of the signal depends are kept constant, the intensities of proton-decoupled spectra are proportional to the

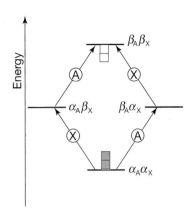

18.41 The energy levels of an AX system and an indication of their relative populations. Each blue square above the line represents an excess population and each white square below the line represents a population deficit. The transitions of A and X are marked.

number of ^{13}C nuclei present. The technique is widely used to characterize synthetic polymers.

18.10 The nuclear Overhauser effect

We have seen already that one advantage of protons in NMR is their high magnetogyric ratio, which results in relatively large Boltzmann population differences and hence appreciable resonance intensities. In the **nuclear Overhauser effect** (NOE), spin relaxation processes involving internuclear dipole–dipole interactions (eqn 36) are used to transfer this population advantage to another nucleus (such as ^{13}C or another proton), so that the latter's resonances are modified.[8]

To understand the effect, we consider the populations of the four levels of an AX system (these were shown in Fig. 18.12). At thermal equilibrium, the population of the $\alpha_A \alpha_X$ level is the greatest, and that of the $\beta_A \beta_X$ level is the least; the other two levels have almost the same energy and an intermediate population. The thermal equilibrium absorption intensities reflect these populations as shown in Fig. 18.13. Now consider the combined effect of saturating the X transition and spin relaxation. When we saturate the X transition, the populations of the X levels are equalized ($N_{\alpha_X} = N_{\beta_X}$) and all transitions involving $\alpha_X \leftrightarrow \beta_X$ spin flips are no longer observed. At this stage there is no change in the populations of the A levels. If that were all there were to happen, all we would see would be the loss of the X resonance and no effect on the A resonance.

Now consider the effect of spin relaxation. Relaxation can occur in a variety of ways if there is a dipolar interaction between the A and X spins. One possibility is for the magnetic field acting between the two spins to cause them both to flop from β to α, so the $\alpha_A \alpha_X$ and $\beta_A \beta_X$ states regain their thermal equilibrium populations. However, the populations of the $\alpha_A \beta_X$ and $\beta_A \alpha_X$ levels remain unchanged at the values characteristic of saturation. As we see from Fig. 18.42, the population difference between the states joined by transitions of A is now greater than at equilibrium, so the resonance absorption is enhanced. Another possibility is for the dipolar interaction between the two spins to cause α to flip to β and β to flop to α. This transition equilibrates the populations of $\alpha_A \beta_X$ and $\beta_A \alpha_X$ but leaves the $\alpha_A \alpha_X$ and $\beta_A \beta_X$ populations unchanged. Now we see from the illustration that the population differences in the states involved in the A transitions are decreased, so the resonance absorption is diminished.

Which effect wins? Does the NOE enhance the A absorption or does it diminish it? As in the discussion of relaxation times in Section 18.8, the efficiency of the intensity-enhancing $\beta_A \beta_X \leftrightarrow \alpha_A \alpha_X$ relaxation is high if the dipole field oscillates at a frequency close to the transition frequency, which in this case is about 2ν; likewise, the efficiency of the intensity-diminishing $\alpha_A \beta_X \leftrightarrow \beta_A \alpha_X$ relaxation is high if the dipole field is stationary (as there is no frequency difference between the initial and final states). A large molecule rotates so slowly that there is very little motion at 2ν, so we expect an intensity decrease (Fig. 18.43). A small molecule rotating rapidly can be expected to have substantial motion at 2ν, and a consequent enhancement of the signal. In practice, the enhancement lies somewhere between the two extremes and is reported in terms of the parameter η (eta), where

$$\eta = \frac{I_A - I_A^0}{I_A^0} \qquad [18.45]$$

8. In a dipole–dipole interaction between two nuclei, one nucleus influences the behaviour of another nucleus in much the same way that the orientation of a bar magnet is influenced by the presence of another bar magnet nearby. Dipole–dipole interactions are discussed in *Further information 2*.

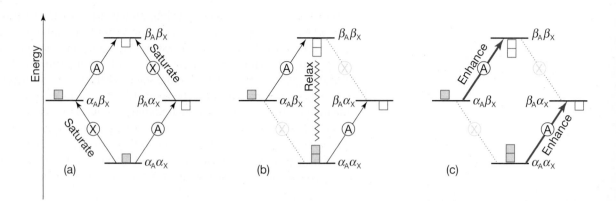

18.42 (a) When the X transition is saturated, the populations of its two states are equalized and the population excess and deficit become as shown (using the same symbols as in Fig. 18.41). (b) Dipole–dipole relaxation relaxes the populations of the highest and lowest states, and they regain their original populations. (c) The A transitions reflect the difference in populations resulting from the preceding changes, and are enhanced compared with those shown in Fig. 18.41.

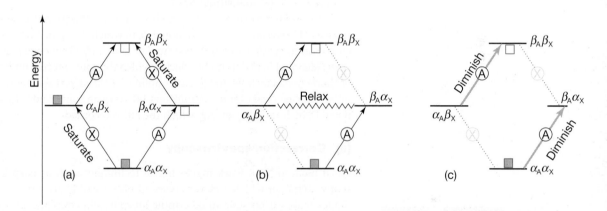

18.43 (a) When the X transition is saturated, just as in Fig. 18.42 the populations of its two states are equalized and the population excess and deficit become as shown. (b) Dipole–dipole relaxation relaxes the populations of the two intermediate states, and they regain their original populations. (c) The A transitions reflect the difference in populations resulting from the preceding changes, and are diminished compared with those shown in Fig. 18.41.

Here I_A° and I_A are the intensities of the NMR signals due to nucleus A before and after application of the radiofrequency pulse that saturates transitions due to the X nucleus. When A and X are nuclei of the same species, such as protons, η lies between −1 (diminution) and $+\frac{1}{2}$ (enhancement). However, η also depends on the values of the magnetogyric ratios of A and X. In the case of maximal enhancement it is possible to show that

$$\eta = \frac{\gamma_X}{2\gamma_A} \tag{18.46}$$

where γ_A and γ_X are the magnetogyric ratios of nuclei A and X, respectively. For ^{13}C coupled to a saturated proton, the ratio η evaluates to 1.99, which shows that an enhancement of about a factor of 2 can be achieved.

The NOE is also used to determine interproton distances. The Overhauser enhancement of a proton A generated by saturating a spin X depends on the fraction of A's spin–lattice relaxation that is caused by its dipolar interaction with X. Because the dipolar field is proportional to r^{-3}, where r is the internuclear distance, and the relaxation effect is proportional to the square of the field, and therefore to r^{-6}, the NOE may be used to determine the geometries of molecules in solution. The determination of the structure of a small protein in solution involves the use of several hundred NOE measurements, effectively casting a net over the protons present. The enormous importance of this procedure is that we can determine the conformation of polypeptides in an aqueous environment and do not need to try to make the single crystals that are essential for an X-ray diffraction investigation (Chapter 23).

18.11 Two-dimensional NMR

An NMR spectrum contains a great deal of information and, if many protons are present, is very complex. Even a first-order spectrum is complex, for the fine structure of different groups of lines can overlap. The complexity would be reduced if we could use two axes to display the data, with resonances belonging to different groups lying at different locations on the second axis. This separation is essentially what is achieved in **two-dimensional NMR**.

All two-dimensional NMR experiments use the **PEMD pulse sequence**, which consists of: a *preparation period* (denoted P), in which the spins first return to thermal equilibrium and then are excited by one or more radiofrequency pulses; an *evolution period* (denoted E) of duration t_1, during which the spins precess under the influence of their chemical shifts and scalar couplings; a *mixing period* (denoted M), in which pulses may be used to transfer information between spins; and a *detection period* (denoted D) of duration t_2, during which the FID is recorded.

(a) Correlation spectroscopy

Much modern NMR work makes use of techniques such as **correlation spectroscopy** (COSY) in which a clever choice of pulses and Fourier transformation techniques makes it possible to determine all spin–spin couplings in a molecule. The basic COSY experiment uses the simplest of all two-dimensional pulse sequences: a single 90° pulse to excite the spins at the end of the preparation period, and a mixing period containing just a second 90° pulse (Fig. 18.44).

The key to the COSY technique is the effect of the second 90° pulse, which can be illustrated by consideration of the four energy levels of an AX system (as shown in Fig. 18.12). At thermal equilibrium, the population of the $\alpha_A \alpha_X$ level is the greatest, and that of the $\beta_A \beta_X$ level is the smallest; the other two levels have almost the same energy and an intermediate population. After the first 90° pulse, the spins are no longer at thermal equilibrium. If a second 90° pulse is applied at a time t_1 that is short compared to the spin–lattice relaxation time T_1, the extra input of energy causes further changes in the populations of the four states. The changes in populations will depend on how far the individual magnetizations have precessed during the evolution period. For simplicity, let us consider a COSY experiment in which the second 90° pulse is split into two selective pulses, one applied to X and one to A. Depending on the evolution time t_1, the 90° pulse that excites X may leave the population differences across each of the two X transitions unchanged, inverted, or somewhere in between. Consider the extreme case in which one population difference is inverted and the other unchanged (Fig. 18.45). The 90° pulse that excites A will now generate an FID in which one of the two A transitions has increased in intensity, and the other has decreased. The overall effect is that precession of the X spins during

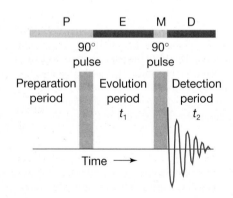

18.44 The pulse sequence used in correlation spectroscopy (COSY). The preparation period is much longer than either T_1 or T_2, so the spins have time to relax before the next cycle of pulses begins. A series of acquisitions of free-induction decays is taken during t_2 with a variable evolution time t_1. Fourier transformation on both variables t_1 and t_2 results in a two-dimensional spectrum, such as that shown in Fig. 18.46.

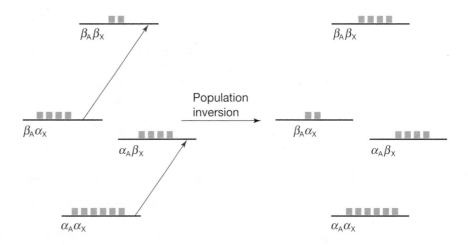

18.45 The change in the population of energy levels of an AX system that results from the second 90° pulse of a COSY experiment. Each blue square represents the same large number of spins. In this extreme example, a selective pulse inverts the populations of the $\beta_A\beta_X$ and $\beta_A\alpha_X$ levels and does not affect the populations of the $\alpha_A\alpha_X$ and $\alpha_A\beta_X$ levels. As a result, a 90° pulse that excites A generates an FID in which one of the two A transitions has increased in intensity, and the other has decreased.

the evolution period determines the amplitudes of the signals from the A spins obtained during the detection period. As the evolution time t_1 is increased, the intensities of the signals from A spins oscillate at rates determined by the frequencies of the two X transitions.

This transfer of information between spins is at the heart of two-dimensional NMR spectroscopy and leads to the correlation of different signals in a spectrum. In this case, information transfer tells us that there is a scalar coupling between A and X. If we conduct a series of experiments in which t_1 is incremented, Fourier transformation of the FIDs on t_2 yields a set of spectra $I(t_1, \nu_2)$ in which the A signal amplitudes oscillate as a function of t_1. A second Fourier transformation, this time on t_1, converts these oscillations into a two-dimensional spectrum $I(\nu_1, \nu_2)$. The signals are spread out in ν_1 according to their precession frequencies during the detection period. Thus, if we apply the COSY pulse sequence to our AX spin system (Fig. 18.44), the result is a two-dimensional spectrum that contains four groups of signals centred on the two chemical shifts in ν_1 and ν_2. Each group will show fine structure, consisting of a block of four signals separated by J_{AX}. The **diagonal peaks** are signals centred on (δ_A, δ_A) and (δ_X, δ_X) and lie along the diagonal $\nu_1 = \nu_2$ (from lower left to upper right). They arise from signals that did not change chemical shift between t_1 and t_2. The **cross peaks** (or *off-diagonal peaks*) are signals centred on (δ_A, δ_X) and (δ_X, δ_A) and owe their existence to the coupling between A and X. Consequently, cross peaks in COSY spectra allow us to map the couplings between spins and to trace out the bonding network in complex molecules. Figure 18.46 shows a simple example of a proton COSY spectrum of 1-nitropropane.

Our simplified description of the COSY experiment does not reveal some important details. For example, the second 90° pulse actually mixes the spin state transitions caused by the first 90° pulse (hence the term 'mixing period'). Each of the four transitions (two for A and two for X) generated by the first pulse can be converted into any of the other three, or into formally forbidden **multiple quantum transitions**, which have $|\Delta m| > 1$. The latter transitions cannot generate any signal in the receiver coil of the spectrometer, but their existence can be demonstrated by

18.46 Proton COSY spectrum of 1-nitropropane. The off-diagonal peaks show that the CH_3 protons are coupled to the central CH_2 protons, which in turn are coupled to the terminal CH_2NO_2 protons. (Spectrum provided by Prof. G. Morris.)

applying a third pulse to mix them back into the four observable single quantum transitions. Many modern NMR experiments exploit multiple quantum transition to filter out unwanted signals and to simplify spectra for interpretation.

Although information from two-dimensional NMR spectroscopy is trivial in an AX system, it can be of enormous help in the interpretation of more complex spectra. For example, a complex spectrum from a synthetic polymer or a protein would be impossible to interpret in one-dimensional NMR but can be interpreted reasonably rapidly by two-dimensional NMR. The techniques themselves are described in the books listed in *Further reading* at the end of the chapter.

(b) Nuclear Overhauser effect spectroscopy

Many different two-dimensional NMR experiments are based on the PEMD pulse sequence. We have seen that the nuclear Overhauser effect can provide information about internuclear distances through analysis of enhancement patterns in the NMR spectrum before and after saturation of selected resonances. In **nuclear Overhauser effect spectroscopy** (NOESY), the second 90° pulse of the COSY experiment is replaced by two 90° pulses separated by a time delay during which dipole–dipole

---

Box 18.1 **Magnetic resonance imaging**

One of the most striking applications of nuclear magnetic resonance is in medicine. *Magnetic resonance imaging* (MRI) is a portrayal of the concentrations of protons in a solid object. The technique relies on the application of specific pulse sequences to an object in an inhomogeneous magnetic field.

If an object containing hydrogen nuclei (a tube of water or a human body) is placed in an NMR spectrometer and exposed to a *homogeneous* magnetic field, then a single resonance signal will be detected. Now consider a flask of water in a magnetic field that varies linearly in the z-direction according to $B_0 + G_z z$, where G_z is the field gradient along the z-direction (see the first illustration). Then the water protons will be resonant at the frequencies

$$\nu_L(z) = \frac{\gamma}{2\pi}(B_0 + G_z z)$$

(Similar equations may be written for gradients along the x- and y-directions.) Application of a 90° radiofrequency pulse with $\nu = \nu_L(z)$ will result in a signal with an intensity that is proportional to the numbers of protons at the position z. This is an example of *slice selection*, the application of a selective 90° pulse that excites nuclei in a specific region, or slice, of the sample. It follows that the intensity of the NMR signal will be a projection of the numbers of protons on a line parallel to the field gradient. The image of a three-dimensional object such as a flask of water can be obtained if the slice selection technique is applied at different orientations (see the first illustration). In *projection reconstruction*, the projections can be analysed on a computer to reconstruct the three-dimensional distribution of protons in the object.

In a magnetic field that varies linearly over a sample, all the protons within a given slice (that is, at a given field value) come into resonance and give a signal of the corresponding intensity. The resulting intensity pattern is a map of the numbers in all the slices, and portrays the shape of the sample. Changing the orientation of the field shows the shape along the corresponding direction, and computer manipulation can be used to build up the three-dimensional shape of the sample.

In practice, the NMR signal is not obtained by direct analysis of the FID curve after application of a single 90° pulse. Instead, spin echoes are often detected with several variations of the 90°-τ-180° pulse sequence (Section 18.8d). In *phase encoding*, field gradients are applied during the evolution period and the detection period of a spin-echo pulse sequence. The first step consists of a 90° pulse that results in slice selection along the *z*-direction. The second step consists of application of a *phase gradient*, a field gradient along the *y*-direction, during the evolution period. At each position along the gradient, a spin packet will precess at a different Larmor frequency due to chemical shift effects and the field inhomogeneity, so each packet will dephase to a different extent by the end of the evolution period. We can control the extent of dephasing by changing the duration of the evolution period, so Fourier transformation on τ gives information about the location of a proton along the *y*-direction.[9] For each value of τ, the next steps are application of the 180° pulse and then of a *read gradient*, a field gradient along the *x*-direction, during detection of the echo. Protons at different positions along *x* experience different fields and will resonate at different frequencies. Therefore Fourier transformation of the FID gives different signals for protons at different positions along *x*.

A common problem with the techniques described above is image contrast, which must be optimized in order to show spatial variations in water content in the sample. One strategy for solving this problem takes advantage of the fact that the relaxation times of water protons are shorter for water in biological tissues than for the pure liquid. Furthermore, relaxation times from water protons are also different in healthy and diseased tissues. A T_1-*weighted image* is obtained by repeating the spin echo sequence before spin–lattice relaxation can return the spins in the sample to equilibrium. Under these conditions, differences in signal intensities are directly related to differences in T_1. A T_2-*weighted image* is obtained by using an evolution period τ that is relatively long. Each point on the image is an echo signal that behaves in the manner shown in Fig. 18.40, so signal intensities are strongly dependent on variations in T_2. However, allowing so much of the decay to occur leads to weak signals even for those protons with long spin–spin relaxation times. Another strategy involves the use of *contrast agents*, paramagnetic compounds that shorten the relaxation times of nearby protons. The technique is particularly useful in enhancing image contrast and in diagnosing disease if the contrast agent is distributed differently in healthy and diseased tissues.

The MRI technique is used widely to detect physiological abnormalities and to observe metabolic processes. With *functional MRI*, blood flow in different regions of the brain can be studied and related to the mental activities of the subject. The special advantage of MRI is that it can image *soft* tissues

The great advantage of MRI is that it can display soft tissue, such as in this cross-section through a patient's head. (Courtesy of the University of Manitoba.)

(see the second illustration), whereas X-rays are largely used for imaging hard, bony structures and abnormally dense regions, such as tumours. In fact, the invisibility of hard structures in MRI is an advantage, as it allows the imaging of structures encased by bone, such as the brain and the spinal cord. X-rays are known to be dangerous on account of the ionization they cause; the high magnetic fields used in MRI may also be dangerous but, apart from anecdotes about the extraction of loose fillings from teeth, there is no convincing evidence of their harmfulness, and the technique is considered safe.

Discussion questions

1. You are designing an MRI spectrometer. What field gradient (in microtesla per metre, μT m^{-1}) is required to produce a separation of 100 Hz between two protons separated by the long diameter of a human kidney (taken as 8 cm) given that they are in environments with $\delta = 3.4$? The radiofrequency field of the spectrometer is at 400 MHz and the applied field is 9.4 T.

2. Suppose a uniform disk-shaped organ is in a linear field gradient, and that the MRI signal is proportional to the number of protons in a slice of width δx at each horizontal distance *x* from the centre of the disk. Sketch the shape of the absorption intensity for the MRI image of the disk before any computer manipulation has been carried out.

Further reading

R.K. Hobbie, *Intermediate physics for medicine and biology*. Springer, New York (1997).

M.T. Vlaardingerbroek and J.A. de Boer, *Magnetic resonance imaging: theory and practice*. Springer, Berlin (1999).

---

9. For technical reasons, it is more common to vary the magnitude of the phase gradient. See *Further reading* for details.

interactions cause magnetization to be exchanged between neighbouring spins (Fig. 18.44). The result of double Fourier transformation is a spectrum in which the cross-peaks form a map of all the NOE interactions in a molecule. Because the nuclear Overhauser effect depends on the inverse sixth power of the separation between nuclei, NOESY data reveal internuclear distances up to about 50 nm.

18.12 Solid-state NMR

The principal difficulty with the application of NMR to solids is the low resolution that is characteristic of solid samples. Nevertheless, there are good reasons for seeking to overcome these difficulties. They include the possibility that a compound of interest is unstable in solution or that it is insoluble, so conventional solution NMR cannot be employed. Moreover, many species are intrinsically interesting as solids, and it is important to determine their structures and dynamics. Synthetic polymers are particularly interesting in this regard, and information can be obtained about the arrangement of molecules, their conformations, and the motion of different parts of the chain. This kind of information is crucial to an interpretation of the bulk properties of the polymer in terms of its molecular characteristics. Similarly, inorganic substances, such as the zeolites that are used as molecular sieves and shape-selective catalysts, can be studied using solid-state NMR, and structural problems can be resolved that cannot be tackled by X-ray diffraction.

Problems of resolution and linewidth are not the only features that plague NMR studies of solids. Because molecular rotation has almost ceased (except in special cases, including 'plastic crystals' in which the molecules continue to tumble), spin–lattice relaxation times are very long but spin–spin relaxation times are very short. Hence, in a pulse experiment, there is a need for lengthy delays—of several seconds—between successive pulses so that the spin system has time to revert to equilibrium. Even gathering the murky information may therefore be a lengthy process. Moreover, because lines are so broad, very high powers of radiofrequency radiation may be required to achieve saturation. Whereas solution pulse NMR uses transmitters of a few tens of watts, solid-state NMR may require transmitters rated at several hundreds of watts.

(a) The origins of linewidths in solids

There are two principal contributions to the linewidths of solids. One is the direct magnetic dipolar interaction between nuclear spins. As we saw in the discussion of spin–spin coupling, a nuclear magnetic moment will give rise to a local magnetic field, which points in different directions at different locations around the nucleus. If we are interested only in the component parallel to the direction of the applied magnetic field (because only this component has a significant effect), then we can use the classical expression (see *Further reading*) to write the magnitude of this magnetic field as

$$B_{nuc} = -\frac{\gamma \hbar \mu_0 m_I}{4\pi R^3}(1 - 3\cos^2\theta) \tag{18.47}$$

Unlike in solution, this field is not motionally averaged to zero. Many nuclei may contribute to the total local field experienced by a nucleus of interest, and different nuclei in a sample may experience a wide range of fields. Typical dipole fields are of the order of 10^{-3} T, which corresponds to splittings and linewidths of the order of 10^4 Hz.

A second source of linewidth is the anisotropy of the chemical shift. This anisotropy gives rise to the neighbour anisotropy effect described in Section 18.5e.

18.47 In magic angle spinning, the sample spins at 54.74° (that is, arccos $(\frac{1}{3})^{1/2}$) to the applied magnetic field. Rapid motion at this angle averages dipole–dipole interactions and chemical shift anisotropies to zero.

We have seen that chemical shifts arise from the ability of the applied field to generate electron currents in molecules. In general, this ability depends on the orientation of the molecule relative to the applied field. In solution, when the molecule is tumbling rapidly, only the average value of the chemical shift is relevant. However, the anisotropy is not averaged to zero for stationary molecules in a solid, and molecules in different orientations have resonances at different frequencies. The chemical shift anisotropy also varies with the angle between the applied field and the principal axis of the molecule as $1 - 3\cos^2\theta$.

(b) The reduction of linewidths

Fortunately, there are techniques available for reducing the linewidths of solid samples. One technique, **magic-angle spinning** (MAS), takes note of the $1 - 3\cos^2\theta$ dependence of both the dipole–dipole interaction and the chemical shift anisotropy. The 'magic angle' is the angle at which $1 - 3\cos^2\theta = 0$, and corresponds to 54.74°. In the technique, the sample is spun at high speed at the magic angle to the applied field (Fig. 18.47). All the dipolar interactions and the anisotropies average to the value they would have at the magic angle; but at that angle they are zero. The difficulty with MAS is that the spinning frequency must not be less than the width of the spectrum, which is of the order of kilohertz. However, gas-driven sample spinners that can be rotated at up to 25 kHz are now routinely available, and a considerable body of work has been done.

Pulsed techniques similar to those described in the previous section may also be used to reduce linewidths. The dipolar field of protons, for instance, may be reduced by a decoupling procedure. However, because the range of coupling strengths is so large, radiofrequency power of the order of 1 kW is required. Elaborate pulse sequences have also been devised that reduce linewidths by averaging procedures that make use of twisting the magnetization vector through an elaborate series of angles.

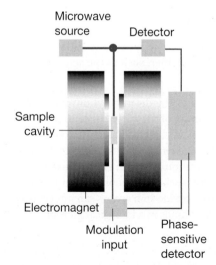

18.48 The layout of a continuous-wave ESR spectrometer. A typical magnetic field is 0.3 T, which requires 9 GHz (3 cm) microwaves for resonance.

Electron spin resonance

Electron spin resonance (ESR) is less widely applicable than NMR because it cannot be detected in normal, spin-paired molecules and the sample must possess unpaired electron spins. It is used to study radicals formed during chemical reactions or by radiation, radicals that act as probes of biological structure (Box 18.2), many d-metal complexes, and molecules in triplet states (such as those involved in phosphorescence, Section 17.3b). The sample may be a gas, a liquid, or a solid, but the free rotation of molecules in the gas phase gives rise to complications.

18.13 The ESR spectrometer

Both Fourier-transform (FT) and continuous-wave (CW) ESR spectrometers are available. The FT-ESR instrument is based on the concepts developed in Section 18.7, except that pulses of microwaves are used to excite electron spins in the sample. The layout of the more common CW-ESR spectrometer is shown in Fig. 18.48. It consists of a microwave source (a klystron or a Gunn oscillator), a cavity in which the sample is inserted in a glass or quartz container, a microwave detector, and an electromagnet with a field that can be varied in the region of 0.3 T. The ESR spectrum is obtained by monitoring the microwave absorption as the field is changed, and a

Box 18.2 Spin probes

We see in Sections 18.14 and 18.15 that anisotropy of the *g*-value and of the nuclear hyperfine interactions can be observed when a radical is immobilized, as in a solid. The illustration shows the variation of the lineshape of the ESR spectrum of the di-*tert*-butyl nitroxide radical (**7**) with temperature. At 292 K, the radical tumbles freely and isotropic hyperfine coupling to the ^{14}N nucleus gives rise to three sharp peaks. At 77 K, motion of the radical is restricted. Both isotropic and anisotropic hyperfine couplings determine the appearance of the spectrum, which now consists of three broad peaks.

A *spin probe* (or *spin label*) is a radical that interacts with a biopolymer and has an ESR spectrum that reports on the dynamical properties of the biopolymer. The ideal spin probe is one with an ESR spectrum that broadens significantly as its motion is restricted to a relatively small extent. Nitroxide spin probes have been used to show that the hydrophobic interiors of biological membranes are in fact very fluid and individual lipid molecules move laterally through the sheet-like structure of the membrane.

ESR spectra of the di-*tert*-butyl nitroxide radical at 292 K (top) and 77 K (bottom). Adapted from J.R. Bolton, in *Biological applications of electron spin resonance* (ed. H.M. Swartz, J.R. Bolton, and D.C. Borg), Wiley, New York (1972).

Just as chemical exchange can broaden proton NMR spectra (Section 18.8b), electron exchange between two radicals can broaden ESR spectra. Therefore, the distance between two spin probe molecules may be measured from the line-widths of their ESR spectra. The effect can be used in a number of biochemical studies. For example, the kinetics of association of two polypeptides labelled with the synthetic amino acid 2,2,6,6-tetramethylpiperidine-1-oxyl-4-amino-4-carboxylic acid (**8**) may be studied by measuring the line-width of the ESR spectrum of the label as a function of time. Alternatively, the thermodynamics of association may be studied by examining the temperature dependence of the ESR linewidth.

Discussion questions

1. Suggest how spin probes could be used to estimate the depth of a crevice in a biopolymer, such as the active site of an enzyme.

2. Sketch the ESR spectra of the di-*tert*-butyl nitroxide radical (**7**) at 292 K in the limits of very low concentration (at which electron exchange is negligible), moderate concentration (at which electron exchange effects begin to be observed), and high concentration (at which electron exchange effects predominate). Discuss how the observation of electron exchange between nitroxide spin probes can inform the study of lateral mobility of lipids in a biological membrane.

Further reading

E.J. Hustedt and A.H. Beth, Nitroxide spin–spin interactions. *Ann. Rev. Biophys. Biomol. Struct.* **28**, 129 (1999).

N. Kocherginsky, H.M. Swartz, and M. Sentjurc, *Nitroxide spin labels: reaction in biology and chemistry*. CRC Press, Boca Raton (1995).

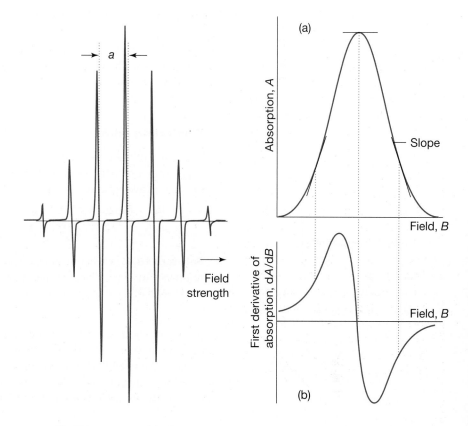

18.49 The ESR spectrum of the benzene radical anion, $C_6H_6^-$, in fluid solution. *a* is the hyperfine splitting of the spectrum; the centre of the spectrum is determined by the *g*-value of the radical.

18.50 When phase-sensitive detection is used, the signal is the first derivative of the absorption intensity. (a) The absorption, (b) the signal, the slope of the absorption signal at each point. Note that the peak of the absorption corresponds to the point where the derivative passes through zero.

typical spectrum (of the benzene radical anion, $C_6H_6^-$) is shown in Fig. 18.49. The peculiar appearance of the spectrum, which is in fact the first-derivative of the absorption, arises from the detection technique, which is sensitive to the slope of the absorption curve (Fig. 18.50).

18.14 The *g*-value

Equation 19 gives the resonance frequency for a transition between the $m_s = -\frac{1}{2}$ and the $m_s = +\frac{1}{2}$ levels of a 'free' electron in terms of the *g*-value $g_e \approx 2.0023$. The magnetic moment of an unpaired electron in a radical also interacts with an external field, but the *g*-value is different from that for a free electron because of local magnetic fields induced by the molecular framework of the radical. Consequently, the resonance condition is normally written as

$$h\nu = g\mu_B B_0 \tag{18.48}$$

where *g* is the **g-value** of the radical. Many organic radicals have *g*-values close to 2.0027; inorganic radicals have *g*-values typically in the range 1.9 to 2.1; paramagnetic *d*-metal complexes have *g*-values in a wider range (for example, 0 to 6).

The deviation of *g* from $g_e = 2.0023$ depends on the ability of the applied field to induce local electron currents in the radical, and therefore its value gives some information about electronic structure. In that sense, the g-value plays a similar role in ESR as the shielding constants in NMR. However, because g-values differ very little from g_e in many radicals (for example, 2.003 for H, 1.999 for NO_2, 2.01 for ClO_2), its main use in chemical applications is to aid the identification of the species present in a sample.

Illustration 18.3

The centre of the ESR spectrum of the methyl radical occurred at 329.40 mT in a spectrometer operating at 9.2330 GHz (radiation belonging to the X band of the microwave region). Its g-value is therefore

$$g = \frac{h\nu}{\mu_B B} = \frac{(6.626\,08 \times 10^{-34}\ \text{J s}) \times (9.2330 \times 10^9\ \text{s}^{-1})}{(9.2740 \times 10^{-24}\ \text{J T}^{-1}) \times (0.329\,40\ \text{T})} = 2.0027$$

Self-test 18.4 At what magnetic field would the methyl radical come into resonance in a spectrometer operating at 34.000 GHz (radiation belonging to the Q band of the microwave region)? [1.213 T]

Just as in the case of the chemical shift in NMR spectroscopy, the g-value is anisotropic: that is, its magnitude depends on the orientation of the radical with respect to the applied field. In solution, when the molecule is tumbling rapidly, only the average value of the g-value is observed. Therefore, anisotropy of the g-value is observed only for radicals trapped in solids.

18.15 Hyperfine structure

The most important features of ESR spectra are their **hyperfine structure**, the splitting of individual resonance lines into components. In general in spectroscopy, the term 'hyperfine structure' means the structure of a spectrum that can be traced to interactions of the electrons with nuclei other than as a result of the latter's point electric charge. The source of the hyperfine structure in ESR is the magnetic interaction between the electron spin and the magnetic dipole moments of the nuclei present in the radical.

(a) The effects of nuclear spin

Consider the effect on the ESR spectrum of a single H nucleus located somewhere in a radical. The proton spin is a source of magnetic field and, depending on the orientation of the nuclear spin, the field it generates adds to or subtracts from the applied field. The total local field is therefore

$$B_{\text{loc}} = B + am_I \qquad m_I = \pm\tfrac{1}{2} \tag{18.49}$$

where a is the **hyperfine coupling constant**. Half the radicals in a sample have $m_I = +\tfrac{1}{2}$, so half resonate when the applied field satisfies the condition

$$h\nu = g\mu_B(B + \tfrac{1}{2}a), \qquad \text{or} \qquad B = \frac{h\nu}{g\mu_B} - \tfrac{1}{2}a \tag{18.50}$$

The other half (which have $m_I = -\tfrac{1}{2}$) resonate when

$$h\nu = g\mu_B(B - \tfrac{1}{2}a), \qquad \text{or} \qquad B = \frac{h\nu}{g\mu_B} + \tfrac{1}{2}a \tag{18.51}$$

Therefore, instead of a single line, the spectrum shows two lines of half the original intensity separated by a and centred on the field determined by g (Fig. 18.51).

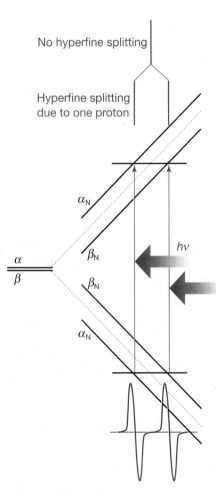

No hyperfine splitting

Hyperfine splitting due to one proton

α_N

β_N

$h\nu$

β_N

α
β

α_N

18.51 The hyperfine interaction between an electron and a spin-$\tfrac{1}{2}$ nucleus results in four energy levels in place of the original two. As a result, the spectrum consists of two lines (of equal intensity) instead of one. The intensity distribution can be summarized by a simple stick diagram. The diagonal lines show the energies of the states as the applied field is increased, and resonance occurs when the separation of states matches the fixed energy of the microwave photon.

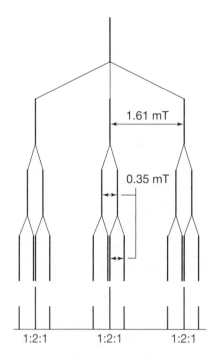

18.52 The analysis of the hyperfine structure of radicals containing one ^{14}N nucleus ($I = 1$) and two equivalent protons.

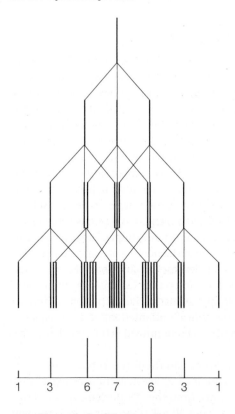

18.53 The analysis of the hyperfine structure of radicals containing three equivalent ^{14}N nuclei.

If the radical contains an ^{14}N atom ($I = 1$), its ESR spectrum consists of three lines of equal intensity, because the ^{14}N nucleus has three possible spin orientations, and each spin orientation is possessed by one-third of all the radicals in the sample. In general, a spin-I nucleus splits the spectrum into $2I + 1$ hyperfine lines of equal intensity.

When there are several magnetic nuclei present in the radical, each one contributes to the hyperfine structure. In the case of equivalent protons (for example, the two CH_2 protons in the radical CH_3CH_2) some of the hyperfine lines are coincident. It is not hard to show that, if the radical contains N equivalent protons, then there are $N + 1$ hyperfine lines with a binomial intensity distribution (i.e. the intensity distribution given by Pascal's triangle). The spectrum of the benzene radical anion in Fig. 18.49, which has seven lines with intensity ratio 1:6:15:20:15:6:1, is consistent with a radical containing six equivalent protons. More generally, if the radical contains N equivalent nuclei with spin quantum number I, then there are $2NI + 1$ hyperfine lines with a binomial intensity distribution (see *Further information 1*).

Example 18.3 Predicting the hyperfine structure of an ESR spectrum

A radical contains one ^{14}N nucleus ($I = 1$) with hyperfine constant 1.61 mT and two equivalent protons ($I = \frac{1}{2}$) with hyperfine constant 0.35 mT. Predict the form of the ESR spectrum.

Method We should consider the hyperfine structure that arises from each type of nucleus or group of equivalent nuclei in succession. So, split a line with one nucleus, then each of those lines is split by a second nucleus (or group of nuclei), and so on. It is best to start with the nucleus with the largest hyperfine splitting; however, any choice could be made, and the order in which nuclei are considered does not affect the conclusion.

Answer The ^{14}N nucleus gives three hyperfine lines of equal intensity separated by 1.61 mT. Each line is split into doublets of spacing 0.35 mT by the first proton, and each line of these doublets is split into doublets with the same 0.35 mT splitting (Fig. 18.52). The central lines of each split doublet coincide, so the proton splitting gives 1:2:1 triplets of internal splitting 0.35 mT. Therefore, the spectrum consists of three equivalent 1:2:1 triplets.

Comment Often it is quicker to realize that a group of equivalent protons gives a characteristic hyperfine pattern (two giving a 1:2:1 triplet, in this case), and to superimpose the patterns directly.

Self-test 18.5 Predict the form of the ESR spectrum of a radical containing three equivalent ^{14}N nuclei.   [Fig. 18.53]

The hyperfine structure of an ESR spectrum is a kind of fingerprint that helps to identify the radicals present in a sample. Moreover, because the magnitude of the splitting depends on the distribution of the unpaired electron near the magnetic nuclei present, the spectrum can be used to map the molecular orbital occupied by the unpaired electron. For example, because the hyperfine splitting in $C_6H_6^-$ is 0.375 mT, and one proton is close to a C atom with one-sixth the unpaired electron spin density (because the electron is spread uniformly around the ring), the hyperfine splitting caused by a proton in the electron spin entirely confined to a single adjacent C atom should be 6×0.375 mT = 2.25 mT. If in another aromatic radical we find a hyperfine splitting constant a, then the **spin density**, ρ, the probability that an unpaired electron is on the atom, can be calculated from the **McConnell equation**:

$$a = Q\rho \qquad\qquad (18.52)$$

with $Q = 2.25$ mT. In this equation, ρ is the spin density on a C atom and a is the hyperfine splitting observed for the H atom to which it is attached.

9

10

Illustration 18.4

The hyperfine structure of the ESR spectrum of the radical anion (naphthalene)$^{-}$ can be interpreted as arising from two groups of four equivalent protons. Those at the 1, 4, 5, and 8 positions in the ring have $a = 0.490$ mT and for those in the 2, 3, 6, and 7 positions have $a = 0.183$ mT. The densities obtained by using the McConnell equation are 0.22 and 0.08, respectively (**9**).

Self-test 18.6 The spin density in (anthracene)$^{-}$ is shown in (**10**). Predict the form of its ESR spectrum.

[A 1:2:1 triplet of splitting 0.43 mT split into a 1:4:6:4:1 quintet of splitting 0.22 mT, split into a 1:4:6:4:1 quintet of splitting 0.11 mT, $3 \times 5 \times 5 = 75$ lines in all]

(b) The origin of the hyperfine interaction

The hyperfine interaction is an interaction between the magnetic moments of the unpaired electron and the nuclei. There are two contributions to the interaction.

An electron in a p orbital does not approach the nucleus very closely, so it experiences a field that appears to arise from a point magnetic dipole. The resulting interaction is called the **dipole–dipole interaction**. The contribution of a magnetic nucleus to the local field experienced by the unpaired electron is given by an expression like that in eqn 36. A characteristic of this type of interaction is that it is anisotropic. Furthermore, just as in the case of NMR, the dipole–dipole interaction averages to zero when the radical is free to tumble. Therefore, hyperfine structure due to the dipole–dipole interaction is observed only for radicals trapped in solids.

An s electron is spherically distributed around a nucleus and so has zero average dipole–dipole interaction with the nucleus even in a solid sample. However, because an s electron has a nonzero probability of being at the nucleus, it is incorrect to treat the interaction as one between two point dipoles. An s electron has a Fermi contact interaction with the nucleus, which as we saw in Section 18.6d is a magnetic interaction that occurs when the point dipole approximation fails. The contact interaction is isotropic (that is, independent of the radical's orientation), and consequently is shown even by rapidly tumbling molecules in fluids (provided the spin density has some s-character).

The dipole–dipole interactions of p electrons and the Fermi contact interaction of s electrons can be quite large. For example, a $2p$ electron in a nitrogen atom experiences an average field of about 3.4 mT from the ^{14}N nucleus. A $1s$ electron in a hydrogen atom experiences a field of about 50 mT as a result of its Fermi contact interaction with the central proton. More values are listed in Table 18.3. The magnitudes of the contact interactions in radicals can be interpreted in terms of the

Table 18.3* Hyperfine coupling constants for atoms, a/mT

Nuclide	Isotropic coupling	Anisotropic coupling
^{1}H	50.8(1s)	
^{2}H	7.8(1s)	
^{14}N	55.2(2s)	3.4(2p)
^{19}F	1720(2s)	108.4(2p)

* More values are given in the *Data section*.

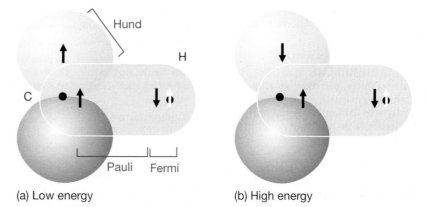

18.54 The polarization mechanism for the hyperfine interaction in π-electron radicals. The arrangement in (a) is lower in energy than that in (b), so there is an effective coupling between the unpaired electron and the proton.

(a) Low energy   (b) High energy

s orbital character of the molecular orbital occupied by the unpaired electron, and the dipole–dipole interaction can be interpreted in terms of the *p* character. The analysis of hyperfine structure therefore gives information about the composition of the orbital, and especially the hybridization of the atomic orbitals (see Problem 18.10).

We still have the source of the hyperfine structure of the $C_6H_6^-$ anion and other aromatic radical anions to explain. The sample is fluid, and as the radicals are tumbling the hyperfine structure cannot be due to the dipole–dipole interaction. Moreover, the protons lie in the nodal plane of the π orbital occupied by the unpaired electron, so the structure cannot be due to a Fermi contact interaction. The explanation lies in a **polarization mechanism** similar to the one responsible for spin–spin coupling in NMR. There is a magnetic interaction between a proton and the α electrons which results in one of the electrons tending to be found with a greater probability nearby (Fig. 18.54). The electron with opposite spin is therefore more likely to be close to the C atom at the other end of the bond. The unpaired electron on the C atom has a lower energy if it is parallel to that electron (Hund's rule favours parallel electrons on atoms), so the unpaired electron can detect the spin of the proton indirectly. Calculation using this model leads to a hyperfine interaction in agreement with the observed value of 2.25 mT.

Checklist of key ideas

- [ ] resonance, the condition of strong effective coupling when the frequencies of two oscillators are identical

The effect of magnetic fields on electrons and nuclei

18.1 The energies of electrons in magnetic fields

- [ ] magnetogyric ratio of the electron, $\gamma_e = -e/2m_e$
- [ ] Bohr magneton, $\mu_B = e\hbar/2m_e$ $= 9.724 \times 10^{-24}\,\text{J}\,\text{T}^{-1}$

- [ ] g-value of a free electron, the constant g_e in $\mu_z = g_e\gamma_e m_s\hbar$ that arises from relativistic effects and from interactions of the electron with the electromagnetic fluctuations in the vacuum that surrounds the electron

- [ ] precessing, sweeping of a vector representing a spin state of a particle around a cone

18.2 The energies of nuclei in magnetic fields

- [ ] magnetogyric ratios of nuclei, Table 18.2
- [ ] nuclear g-factor, Table 18.2

- [ ] nuclear magneton, $\mu_N = e\hbar/2m_p$ $= 5.051 \times 10^{-27}\,\text{J}\,\text{T}^{-1}$
- [ ] Larmor frequency, $\nu_L = \gamma B_0/2\pi$

18.3 Magnetic resonance spectroscopy

- [ ] resonance condition, $h\nu = g_e\mu_B B_0$ (for transitions between electron spin states), $h\nu = \gamma\hbar B_0 = h\nu_L$ (for transitions between nuclear spin states)

- [ ] electron spin resonance (ESR) or electron paramagnetic resonance (EPR), the study of molecules and ions containing unpaired electrons by observing the magnetic fields at which they come into resonance with monochromatic radiation

☐ spin-$\frac{1}{2}$ nuclei, nuclei with $I = \frac{1}{2}$

☐ nuclear magnetic resonance (NMR), the study of the properties of molecules containing magnetic nuclei by applying a magnetic field and observing the frequency of the resonant electromagnetic field

Nuclear magnetic resonance

18.4 The NMR spectrometer

☐ superconducting magnet, a magnet capable of producing magnetic fields of the order of 2 T when operated at very low temperatures

18.5 The chemical shift

☐ shielding constant, a dimensionless quantity related to local magnetic fields generated by electronic currents in molecular groups, $\sigma = \sigma(\text{local}) + \sigma(\text{neighbour}) + \sigma(\text{solvent})$

☐ chemical shift, a parameter related to the difference between the resonance frequency of a nucleus in question and that of a reference standard, $\delta = (v - v^\circ)/v^\circ \times 10^6$

☐ deshielded, a condition ascribed to nuclei with large chemical shift

☐ integration, the determination of the area under an absorption line

☐ local contribution to the shielding constant, the contribution of the electrons of the atom that contains the nucleus in question, $\sigma(\text{local}) = \sigma_d + \sigma_p$

☐ neighbouring group contribution to the shielding constant, the contribution from the groups of atoms that form the rest of the molecule, $\sigma(\text{neighbour}) \propto (\chi_\parallel - \chi_\perp)(1 - 3\cos^2\theta)/r^3$

☐ solvent contribution to the shielding constant, $\sigma(\text{solvent})$, the contribution from solvent molecules

☐ diamagnetic contribution to $\sigma(\text{local})$, σ_d the contribution that opposes the applied magnetic field and shields the nucleus in question

☐ paramagnetic contribution to $\sigma(\text{local})$, σ_p the contribution that reinforces the applied magnetic field and deshields the nucleus in question

☐ Lamb formula, $\sigma_d = e^2 \mu_0 \langle 1/r \rangle / 12\pi m_e$

☐ ring current, a circulation of electrons around an aromatic ring

18.6 The fine structure

☐ fine structure, the splitting of resonances into individual lines in a spectrum

☐ scalar coupling constant, J, a parameter that reports on the strength of the interaction between nuclear spins in a molecule

☐ Karplus equation, $J = A + B\cos\phi + C\cos 2\phi$

☐ polarization mechanism, a mechanism for spin–spin coupling in which the interaction is transmitted through bonds

☐ Fermi contact interaction, a mechanism for spin–spin coupling consisting of a magnetic interaction between a non-dipolar field from the nucleus and an electron's magnetic moment

☐ chemically equivalent, nuclei that are related by a symmetry operation of the molecule and the same chemical shifts

☐ magnetically equivalent, nuclei that are chemically equivalent and also have identical spin–spin interactions with other magnetic nuclei in the molecule

☐ first-order spectra, spectra in which the differences in chemical shifts of groups are much greater than the spin–spin coupling constants

☐ strongly coupled spectra (or second-order spectra), spectra in which the differences in chemical shifts of groups are comparable to their spin–spin coupling interactions

☐ heteronuclear spin system, a spin system consisting of nuclei of different elements

☐ homonuclear spin system, a spin system consisting of nuclei of the same element

Pulse techniques in NMR

☐ Fourier-transform NMR, a mode of NMR spectroscopy that monitors the response of the sample to excitation by one or more radiofrequency pulses

18.7 The magnetization vector

☐ magnetization, M, the net nuclear magnetic moment of a sample

☐ 90° pulse ($\pi/2$ pulse), a radiofrequency pulse that causes the magnetization originally in the z-axis to precess in the xy-plane

☐ free-induction decay (FID), the exponential decay of the signal in the detector coil after a 90° pulse

18.8 Linewidths and rate processes

☐ spin relaxation, the return of a spin system to thermal equilibrium

☐ saturation, equalization of the populations of the energy levels of a system

☐ 180° pulse, a radiofrequency pulse that leads to inversion of the populations of the energy levels of a system

☐ longitudinal relaxation time (or spin–lattice relaxation time), T_1, the time constant for the return of a spin system to equilibrium by giving up energy to the surroundings

☐ transverse relaxation time (or spin–spin relaxation time), T_2, the time constant for the return of a spin system to equilibrium via randomization of spin orientations

☐ inhomogeneous broadening, broadening of a spectral line caused by inhomogeneities in the applied magnetic field

☐ effective transverse relaxation time, T_2^*, the apparent value of the transverse relaxation time measured from spectral linewidths in the presence of inhomogeneous broadening, $T_2^* = 1/\pi\Delta v_{1/2}$

☐ chemical exchange, rapid exchange of nuclei between molecules in a sample

☐ inversion recovery technique, a technique for the measurement of T_1 that employs a 180°–τ–90° pulse sequence

☐ spin echo, a signal generated by a 90°–τ–180° pulse sequence

☐ spin packet, a group of nuclei with very similar precession frequencies

18.9 Spin decoupling

☐ dilute spin species, a spin species with low natural abundance

☐ abundant spin species, a spin species with high natural abundance

☐ proton decoupling, the removal of proton spin couplings from the NMR spectrum of another nucleus, such as ^{13}C

18.10 The nuclear Overhauser effect

☐ nuclear Overhauser effect, the transfer of population advantage from one nucleus to another by spin relaxation processes involving internuclear dipole–dipole interactions

18.11 Two-dimensional NMR

☐ two-dimensional NMR, a mode of NMR spectroscopy in which two axes are used to display spectral data, with resonances belonging to different groups lying at different locations on a two-dimensional contour map

☐ PEMD pulse sequence, a sequence of radiofrequency pulses consisting of a preparation period, an evolution period, a mixing period, and a detection period

☐ correlation spectroscopy (COSY), a technique in which all spin–spin couplings in a molecule are measured

☐ diagonal peaks, signals centred on the diagonal of a two-dimensional NMR spectrum

☐ cross peaks (or off-diagonal peaks), signals not centred on the diagonal of a two-dimensional NMR spectrum

☐ multiple quantum transitions, magnetic resonance transitions with $|\Delta m| > 1$

☐ nuclear Overhauser effect spectroscopy (NOESY), a technique in which internuclear distances are measured

18.12 Solid-state NMR

☐ magic-angle spinning (MAS), a solid-state NMR technique in which the sample is spun at an angle of 54.74° to the applied magnetic field

Electron spin resonance

18.13 The ESR spectrometer

18.14 The g-value

☐ g-value of a radical, the parameter g in the resonance condition $h\nu = g\mu_B B_0$

18.15 Hyperfine structure

☐ hyperfine structure, the splitting of individual lines in an ESR spectrum into components

☐ hyperfine coupling constant, a, a constant related to the magnetic interactions between nuclear spins and the unpaired electron spins in a radical

☐ spin density, ρ, the probability that an unpaired electron is on a specific atom in a molecule

☐ McConnell relation, $a = Q\rho$

☐ dipole–dipole interaction, the interaction between an electron spin and a nuclear spin relatively far away

☐ polarization mechanism, a mechanism for hyperfine interactions in which the unpaired electron can detect the spin of a proton indirectly via the bonding network of the radical

Further reading

Articles of general interest

M.-K. Ahn, A comparison of FTNMR and FTIR techniques. *J. Chem. Educ.* **66**, 802 (1989).

R.G. Barnes, Electron paramagnetic resonance. In *Encyclopedia of applied physics* (ed. G.L. Trigg) **5**, 475 VCH, New York (1993).

P.S. Beauchamp and R. Marquez, A general approach for calculating proton chemical shifts for methyl, methylene, and methine protons when there are one or more substituents within three carbons. *J. Chem. Educ.* **74**, 1483 (1997).

D.L. Bryce and R.E. Wasylishen, *Ab initio* calculations of NMR parameters for diatomic molecules. *J. Chem. Educ.* **78**, 124 (2001).

H. Dahn, Understanding chemical shifts in π systems: ^{13}C, ^{15}N, and ^{17}O NMR. *J. Chem. Educ.* **77**, 905 (2000).

J.H. Freed, New technologies in electron spin resonance. *Ann. Rev. Phys. Chem.* **51**, 655 (2000).

R. Freeman, Selective excitation in high-resolution NMR. *Chem. Rev.* **91**, 1397 (1991).

D.H. Grant, Paramagnetic susceptibility by NMR: the 'solvent correction' reexamined. *J. Chem. Educ.* **72**, 39 (1995).

Y. Habata and S. Akabori, Teaching ^1H NMR spectrometry using computer modeling, *J. Chem. Educ.* **78**, 121 (2001).

R.W. King and K.R. Williams. The Fourier transform in chemistry. Part 1. Nuclear magnetic resonance: Introduction. *J. Chem. Educ.* **66**, A213 (1989); Part 2. Nuclear magnetic resonance: The single pulse experiment. *Ibid.* A243; Part 3. Multiple-pulse experiments. *J. Chem. Educ.* **67**, A93 (1990); Part 4. NMR: Two-dimensional methods. *Ibid.* A125; A glossary of NMR terms. *Ibid.* A100.

T.D. Lash and S.S. Lash. The use of Pascal-like triangles in describing first-order coupling patterns. *J. Chem. Educ.* **64**, 315 (1987).

B.E. Mann, The analysis of first-order coupling patterns in NMR. *J. Chem. Educ.* **72**, 614 (1995).

R.H. Orcutt, A straightforward way to determine relative intensities of spin–spin splitting lines of equivalent nuclei in NMR spectra. *J. Chem. Educ.* **64**, 763 (1987).

J.C. Paniagua and A. Moyano, On the way of introducing some basic NMR aspects: from the classical and naive quantum models to the density-operator approach. *J. Chem. Educ.* **73**, 310 (1996).

L.J. Schwartz. A step-by-step picture of pulsed (time-domain) NMR. *J. Chem. Educ.* **65**, 959 (1988); idem. 752.

T.A. Shaler, Generalization of Pascal's triangle to nuclei of any spin. *J. Chem. Educ.* **68**, 853 (1991).

C. Suarez, Gas-phase NMR spectroscopy. *The Chemical Educator* **2** (1998).

B.H. Suits, Magnetic resonance spectrometers. In *Encyclopedia of applied physics* (ed. G.L. Trigg) **9**, 71. VCH, New York (1994).

D.J. Wink, Spin–lattice relaxation times in ^1H NMR spectroscopy. *J. Chem. Educ.* **66**, 810 (1989).

Texts and sources of data and information

A. Abragam and B. Bleaney, *Electron paramagnetic resonance of transition ions.* Dover, New York (1986).

R.J. Abraham, J. Fisher, and P. Lofthus, *Introduction to NMR spectroscopy.* Wiley, New York (1991).

N.M. Atherton, *Principles of electron spin resonance.* Ellis Horwood/Prentice Hall, Hemel Hempstead (1993).

P.W. Atkins and R.S. Friedman, *Molecular quantum mechanics.* Oxford University Press (1997).

E.D. Becker, *High resolution NMR: theory and chemical applications.* Academic Press, San Diego (2000).

D. Canet, *Nuclear magnetic resonance: concepts and methods.* Wiley, New York (1996).

H. Friebolin, *Basic one- and two-dimensional NMR spectroscopy*. Wiley-VCH, Weinheim (1998).

R.A. De Graaf, *In vivo NMR spectroscopy: principles and techniques*. Wiley, New York (1999).

R. Drago, *Physical methods for chemists*. Saunders, Philadelphia (1992).

J.N.S. Evans, *Biomolecular NMR spectroscopy*. Oxford University Press (1995).

R. Freeman, *A handbook of nuclear magnetic resonance spectroscopy*. Longman, London (1997).

R. Freeman, *Spin choreography: basic steps in high resolution NMR*. Oxford University Press (1998).

D.M. Grant and R.K. Harris (ed.), *Encyclopedia of nuclear magnetic resonance*. Vols. 1–8, Wiley, New York (1996).

H. Günther, *NMR spectroscopy: basic principles, concepts, and applications in chemistry*. Wiley, New York (1995).

P.J. Hore, *Nuclear magnetic resonance*. Oxford Chemistry Primers, Oxford University Press (1995).

D.R. Lide (ed.), *CRC Handbook of Chemistry and Physics*, Section 9. CRC Press, Boca Raton (2000).

J.C. Lindon, G.E. Tranter, and J.L. Holmes (eds.), *Encyclopedia of spectroscopy and spectrometry*. Academic Press, San Diego (2000).

C.P. Poole, Jr. and H.A. Farach (ed.), *Handbook of electron spin resonance: data sources, computer technology, relaxation, and ENDOR*. Vols. 1–2. Springer, Berlin (1997, 1999).

L. Que, Jr. (ed.), *Physical methods in bioinorganic chemistry: spectroscopy and magnetism*. University Science Books, Sausalito (2000).

J.K.M. Sanders and B.K. Hunter, *Modern NMR spectroscopy: a guide for chemists*. Oxford University Press (1993).

J.K.M. Sanders, E.C. Constable, and B.K. Hunter, *Modern NMR spectroscopy: a workbook of chemical problems*, Oxford University Press (1995).

Exercises

Discussion questions

18.1 (a) Discuss in detail the origins of the local, neighbouring group, and solvent contributions to the shielding constant.

18.1 (b) Discuss in detail the effects of a 90° pulse and of a 180° pulse on a system of spin-$\frac{1}{2}$ nuclei in a static magnetic field.

18.2 (a) Suggest a reason why the relaxation times of ^{13}C nuclei are typically much longer than those of ^1H nuclei.

18.2 (b) Discuss the origins of diagonal and cross peaks in the COSY spectrum of an AX system.

18.3 (a) Discuss how the Fermi contact interaction and the polarization mechanism contribute to spin–spin couplings in NMR and hyperfine interactions in ESR.

18.3 (b) Explain how the EPR spectrum of an organic radical can be used to pinpoint the molecular orbital occupied by the unpaired electron.

Numerical exercises

18.4 (a) What is the resonance frequency of a proton in a magnetic field of 14.1 T?

18.4 (b) What is the resonance frequency of a ^{19}F nucleus in a magnetic field of 16.2 T?

18.5 (a) ^{32}S has a nuclear spin of $\frac{3}{2}$ and a nuclear g-factor of 0.4289. Calculate the energies of the nuclear spin states in a magnetic field of 7.500 T.

18.5 (b) ^{14}N has a nuclear spin of 1 and a nuclear g-factor of 0.404. Calculate the energies of the nuclear spin states in a magnetic field of 11.50 T.

18.6 (a) Calculate the frequency separation of the nuclear spin levels of a ^{13}C nucleus in a magnetic field of 14.4 T given that the magnetogyric ratio is 6.73×10^7 T^{-1} s^{-1}.

18.6 (b) Calculate the frequency separation of the nuclear spin levels of a ^{14}N nucleus in a magnetic field of 15.4 T given that the magnetogyric ratio is 1.93×10^7 T^{-1} s^{-1}.

18.7 (a) In which of the following systems is the energy level separation the largest? (a) A proton in a 600 MHz NMR spectrometer, (b) a deuteron in the same spectrometer.

18.7 (b) In which of the following systems is the energy level separation the largest? (a) A ^{14}N nucleus in a 600 MHz NMR spectrometer, (b) an electron in a radical in a field of 0.300 T.

18.8 (a) Calculate the energy difference between the lowest and highest nuclear spin states of a ^{14}N nucleus in a 15.00 T magnetic field.

18.8 (b) Calculate the magnetic field needed to satisfy the resonance condition for unshielded protons in a 150.0 MHz radiofrequency field.

18.9 (a) Use Table 18.2 to predict the magnetic fields at which (a) ^1H, (b) ^2H, (c) ^{13}C come into resonance at (i) 250 MHz, (ii) 500 MHz.

18.9 (b) Use Table 18.2 to predict the magnetic fields at which (a) ^{14}N, (b) ^{19}F, and (c) ^{31}P come into resonance at (i) 300 MHz, (ii) 750 MHz.

18.10 (a) Calculate the relative population differences ($\delta N/N$) for protons in fields of (a) 0.30 T, (b) 1.5 T, and (c) 10 T at 25°C.

18.10 (b) Calculate the relative population differences ($\delta N/N$) for ^{13}C nuclei in fields of (a) 0.50 T, (b) 2.5 T, and (c) 15.5 T at 25°C.

18.11 (a) The first generally available NMR spectrometers operated at a frequency of 60 MHz; today it is not uncommon to use a spectrometer that operates at 800 MHz. What are the relative population differences of ^{13}C spin states in these two spectrometers at 25°C?

18.11 (b) What are the relative values of the chemical shifts observed for nuclei in the spectrometers mentioned in Exercise 18.11a in terms of (a) δ values, (b) frequencies?

18.12 (a) The chemical shift of the CH$_3$ protons in acetaldehyde (ethanal) is $\delta = 2.20$ and that of the CHO proton is 9.80. What is the difference in local magnetic field between the two regions of the molecule when the applied field is (a) 1.5 T, (b) 15 T?

18.12 (b) The chemical shift of the CH_3 protons in diethyl ether is $\delta = 1.16$ and that of the CH_2 protons is 3.36. What is the difference in local magnetic field between the two regions of the molecule when the applied field is (a) 1.9 T, (b) 16.5 T?

18.13 (a) Sketch the appearance of the ^1H-NMR spectrum of acetaldehyde (ethanal) using $J = 2.90$ Hz and the data in Exercise 18.12a in a spectrometer operating at (a) 250 MHz, (b) 500 MHz.

18.13 (b) Sketch the appearance of the ^1H-NMR spectrum of diethyl ether using $J = 6.97$ Hz and the data in Exercise 18.12b in a spectrometer operating at (a) 350 MHz, (b) 650 MHz.

18.14 (a) Two groups of protons are made equivalent by the isomerization of a fluxional molecule. At low temperatures, where the interconversion is slow, one group has $\delta = 4.0$ and the other has $\delta = 5.2$. At what rate of interconversion will the two signals merge in a spectrometer operating at 250 MHz?

18.14 (b) Two groups of protons are made equivalent by the isomerization of a fluxional molecule. At low temperatures, where the interconversion is slow, one group has $\delta = 5.5$ and the other has $\delta = 6.8$. At what rate of interconversion will the two signals merge in a spectrometer operating at 350 MHz?

18.15 (a) Sketch the form of the ^{19}F-NMR spectra of a natural sample of tetrafluoroborate ions, BF_4^-, allowing for the relative abundances of $^{10}BF_4^-$ and $^{11}BF_4^-$.

18.15 (b) From the data in Table 18.2, predict the frequency needed for ^{31}P-NMR in an NMR spectrometer designed to observe proton resonance at 500 MHz. Sketch the proton and ^{31}P resonances in the NMR spectrum of PH_4^+.

18.16 (a) Sketch the form of an $A_3M_2X_4$ spectrum, where A, M, and X are protons with distinctly different chemical shifts and $J_{AM} > J_{AX} > J_{MX}$.

18.16 (b) Sketch the form of an $A_2M_2X_5$ spectrum, where A, M, and X are protons with distinctly different chemical shifts and $J_{AM} > J_{AX} > J_{MX}$.

18.17 (a) Which of the following molecules have sets of nuclei that are chemically but not magnetically equivalent: (a) CH_3CH_3, (b) $CH_2{=}CH_2$?

18.17 (b) Which of the following molecules have sets of nuclei that are chemically but not magnetically equivalent: (a) $CH_2{=}C{=}CF_2$, (b) cis- and trans-$[Mo(CO)_4(PH_3)_2]$?

18.18 (a) The duration of a 90° or 180° pulse depends on the strength of the B_1 field. If a 90° pulse requires 10 μs, what is the strength of the B_1 field? How long would the corresponding 180° pulse require?

18.18 (b) The duration of a 90° or 180° pulse depends on the strength of the B_1 field. If a 180° pulse requires 12.5 μs, what is the strength of the B_1 field? How long would the corresponding 90° pulse require?

18.19 (a) What magnetic field would be required in order to use an ESR X-band spectrometer (9 GHz) to observe ^1H-NMR and a 300 MHz spectrometer to observe ESR?

18.19 (b) Some commercial ESR spectrometers use 8 mm microwave radiation (the Q band). What magnetic field is needed to satisfy the resonance condition?

18.20 (a) The centre of the ESR spectrum of atomic hydrogen lies at 329.12 mT in a spectrometer operating at 9.2231 GHz. What is the g-value of the atom?

18.20 (b) The centre of the ESR spectrum of atomic deuterium lies at 330.02 mT in a spectrometer operating at 9.2482 GHz. What is the g-value of the atom?

18.21 (a) A radical containing two equivalent protons shows a three-line spectrum with an intensity distribution 1:2:1. The lines occur at 330.2 mT, 332.5 mT, and 334.8 mT. What is the hyperfine coupling constant for each proton? What is the g-value of the radical given that the spectrometer is operating at 9.319 GHz?

18.21 (b) A radical containing three equivalent protons shows a four-line spectrum with an intensity distribution 1:3:3:1. The lines occur at 331.4 mT, 333.6 mT, 335.8 mT, and 338.0 mT. What is the hyperfine coupling constant for each proton? What is the g-value of the radical given that the spectrometer is operating at 9.332 GHz?

18.22 (a) A radical containing two inequivalent protons with hyperfine constants 2.0 mT and 2.6 mT gives a spectrum centred on 332.5 mT. At what fields do the hyperfine lines occur and what are their relative intensities?

18.22 (b) A radical containing three inequivalent protons with hyperfine constants 2.11 mT, 2.87 mT, and 2.89 mT gives a spectrum centred on 332.8 mT. At what fields do the hyperfine lines occur and what are their relative intensities?

18.23 (a) Predict the intensity distribution in the hyperfine lines of the ESR spectra of (a) $\cdot CH_3$, (b) $\cdot CD_3$.

18.23 (b) Predict the intensity distribution in the hyperfine lines of the ESR spectra of (a) $\cdot CH_2H_3$, (b) $\cdot CD_2CD_3$.

18.24 (a) The benzene radical anion has $g = 2.0025$. At what field should you search for resonance in a spectrometer operating at (a) 9.302 GHz, (b) 33.67 GHz?

18.24 (b) The naphthalene radical anion has $g = 2.0024$. At what field should you search for resonance in a spectrometer operating at (a) 9.312 GHz, (b) 33.88 GHz?

18.25 (a) The ESR spectrum of a radical with a single magnetic nucleus is split into four lines of equal intensity. What is the spin of the nucleus?

18.25 (b) The ESR spectrum of a radical with two equivalent nuclei of a particular kind is split into five lines of intensity ratio 1:2:3:2:1. What is the spin of the nuclei?

18.26 (a) Sketch the form of the hyperfine structures of radicals XH_2 and XD_2, where the nucleus X has $I = \frac{5}{2}$.

18.26 (b) Sketch the form of the hyperfine structures of radicals XH_3 and XD_3, where the nucleus X has $I = \frac{3}{2}$.

Problems[†]

Numerical problems

18.1 A scientist investigates the possibility of neutron spin resonance, and has available a commercial NMR spectrometer operating at 300 MHz. What field is required for resonance? What is the relative population difference at room temperature? Which is the lower energy spin state of the neutron?

18.2 Two groups of protons have $\delta = 4.0$ and $\delta = 5.2$ and are interconverted by a conformational change of a fluxional molecule. In a 60 MHz spectrometer the spectrum collapsed into a single line at 280 K but at 300 MHz the collapse did not occur until the temperature had been raised to 300 K. What is the activation energy of the interconversion?

18.3‡ Suppose that the FID in Fig. 18.30 was recorded in a 300 MHz spectrometer, and that the interval between maxima in the oscillations in the FID is 0.10 s. What is the Larmor frequency of the nuclei and the spin–spin relaxation time?

18.4‡ In a classic study of the application of NMR to the measurement of rotational barriers in molecules, P.M. Nair and J.D. Roberts (*J. Am. Chem. Soc.* **79**, 4565 (1957)) obtained the 40 MHz ^{19}F-NMR spectrum of $F_2BrCCBrCl_2$. Their spectra are reproduced in Fig. 18.55. At 193 K. the spectrum shows five resonance peaks. Peaks I and III are separated by 160 Hz, as are IV and V. The ratio of the integrated intensities of peak II to peaks I, III, IV, and V is approximately 10 to 1. At 273 K, the five peaks have collapsed into one. Explain the spectrum and its change with temperature. At what rate of interconversion will the spectrum collapse to a single line? Calculate the rotational energy barrier between the rotational isomers on the assumption that it is related to the rate of interconversion between the isomers.

18.5‡ Various versions of the Karplus equation (eqn 35) have been used to correlate data on vicinal proton coupling constants in systems of the type $R_1R_2CHCHR_3R_4$. The original version, (M. Karplus, *J. Am. Chem. Soc.* **85**, 2870 (1963)) is $^3J_{HH} = A\cos^2\phi_{HH} + B$.

When $R_3 = R_4 = H$, $^3J_{HH} = 7.3$ Hz; when $R_3 = CH_3$ and $R_4 = H$, $^3J_{HH} = 8.0$ Hz; when $R_3 = R_4 = CH_3$, $^3J_{HH} = 11.2$ Hz. Assume that only staggered conformations are important and determine which version of the Karplus equation fits the data better.

18.6‡ It might be unexpected that the Karplus equation, which was first derived for $^3J_{HH}$ coupling constants, should also apply to vicinal coupling between the nuclei of metals such as tin. T.N. Mitchell and B. Kowall (*Magn. Reson. Chem.* **33**, 325 (1995)) have studied the relation between $^3J_{HH}$ and $^3J_{SnSn}$ in compounds of the type $Me_3SnCH_2CHRSnMe_3$ and find that $^3J_{SnSn} = 78.86^3J_{HH} + 27.84$ Hz. (a) Does this result support a Karplus-type equation for tin? Explain your reasoning. (b) Obtain the Karplus equation for $^3J_{SnSn}$ and plot it as a function of the dihedral angle. (c) Draw the preferred conformation.

18.7 The angular NO_2 molecule has a single unpaired electron and can be trapped in a solid matrix or prepared inside a nitrite crystal by radiation damage of NO_2^- ions. When the applied field is parallel to the OO direction the centre of the spectrum lies at 333.64 mT in a spectrometer operating at 9.302 GHz. When the field lies along the bisector of the ONO angle, the resonance lies at 331.94 mT. What are the g-values in the two orientations?

18.8 The hyperfine coupling constant in $\cdot CH_3$ is 2.3 mT. Use the information in Table 18.3 to predict the splitting between the hyperfine lines of the spectrum of $\cdot CD_3$. What are the overall widths of the hyperfine spectra in each case?

18.9 The p-dinitrobenzene radical anion can be prepared by reduction of p-dinitrobenzene. The radical anion has two equivalent N nuclei ($I = 1$) and four equivalent protons. Predict the form of the ESR spectrum using $a(N) = 0.148$ mT and $a(H) = 0.112$ mT.

18.10 When an electron occupies a $2s$ orbital on an N atom it has a hyperfine interaction of 55.2 mT with the nucleus. The spectrum of NO_2 shows an isotropic hyperfine interaction of 5.7 mT. For what proportion of its time is the unpaired electron of NO_2 occupying a $2s$ orbital? The hyperfine coupling constant for an electron in a $2p$ orbital of an N atom is 3.4 mT. In NO_2 the anisotropic part of the hyperfine coupling is 1.3 mT. What proportion of its time does the unpaired electron spend in the $2p$ orbital of the N atom in NO_2? What is the total probability that the electron will be found on (a) the N atoms, (b) the O atoms? What is the hybridization ratio of the N atom? Does the hybridization support the view that NO_2 is angular?

18.11 The hyperfine coupling constants observed in the radical anions (**11**), (**12**), and (**13**) are shown (in millitesla, mT). Use the value for the benzene radical anion to map the probability of finding the unpaired electron in the π orbital on each C atom.

| 0°C |
| --30°C |
| --60°C |
| I II III IV V |
| --80°C   160 Hz |

Increasing magnetic field ⟶

18.55

NO₂

0.011

0.172 | 0.011

0.172

11

NO₂ (NO₂)

0.450 | 0.272

0.108

0.450 NO₂

12

† Problems denoted with the symbol ‡ were supplied by Charles Trapp and Carmen Giunta.

13

0.112, 0.112, 0.112, 0.112 (with NO$_2$ substituents)

Theoretical problems

18.12 Calculate σ_d for a hydrogenic atom with atomic number Z.

18.13 In this problem you will use the molecular electronic structure methods described in Chapter 14 to investigate the hypothesis that the magnitude of the ^{13}C chemical shift correlates with the net charge on a ^{13}C atom. (a) Using molecular modelling software[10] and the computational method of your choice (extended Hückel, semi-empirical, or *ab initio* methods), calculate the net charge at the C atom *para* to the substituents in this series of molecules: benzene, phenol, toluene, trifluorotoluene, benzonitrile, and nitrobenzene. (b) The ^{13}C chemical shifts of the *para* C atoms in each of the molecules that you examined in part (a) are given below:

Substituent	OH	CH$_3$	H	CF$_3$	CN	NO$_2$
δ	130.1	128.4	128.5	128.9	129.1	129.4

Is there a linear correlation between net charge and ^{13}C chemical shift of the *para* C atom in this series of molecules? (c) If you did find a correlation in part (b), use the concepts developed in this chapter to explain the physical origins of the correlation.

18.14 The z-component of the magnetic field at a distance R from a magnetic moment parallel to the z-axis is given by eqn 36. In a solid, a proton at a distance R from another can experience such a field and the measurement of the splitting it causes in the spectrum can be used to calculate R. In gypsum, for instance, the splitting in the H$_2$O resonance can be interpreted in terms of a magnetic field of 0.715 mT generated by one proton and experienced by the other. What is the separation of the protons in the H$_2$O molecule?

18.15 In a liquid, the dipolar magnetic field averages to zero: show this result by evaluating the average of the field given in eqn 36. *Hint.* The volume element is $\sin\theta\,d\theta\,d\phi$ in polar coordinates.

18.16 The shape of a spectral line, $I(\omega)$, is related to the free induction decay signal $G(t)$ by

$$I(\omega) = a\,\mathrm{Re}\int_0^\infty G(t)e^{i\omega t}\,dt$$

where a is a constant and 'Re' means take the real part of what follows. Calculate the lineshape corresponding to an oscillating, decaying function $G(t) = \cos\omega_0 t\,e^{-t/\tau}$.

18.17 In the language of Problem 18.16, show that, if $G(t) = (a\cos\omega_1 t + b\cos\omega_2 t)e^{-t/\tau}$, then the spectrum consists of two lines with intensities proportional to a and b and located at $\omega = \omega_1$ and ω_2, respectively.

18.18‡ The relative sensitivity of NMR lines for equal numbers of different nuclei at constant temperature for a given frequency is $R_\nu \propto (I+1)\mu\omega_0^2$, whereas for a given field it is $R_B \propto \{(I+1)/I^2\}\mu^3$. (a) From the data in Table 18.2, calculate these sensitivities for the deuteron, ^{13}C, ^{14}N, ^{19}F, and ^{31}P relative to the proton. (b) Derive the equation for R_B from the equation for R_ν.

Applications: to biology

18.19 When interacting with a large biopolymer or even larger organelle, a small molecule might not rotate freely in all directions and the dipolar interaction might not average to zero. Suppose a molecule is bound so that, although the vector separating two protons may rotate freely around the z-axis, the colatitude may vary only between 0 and θ'. Average the dipolar field over this restricted range of orientations and confirm that the average vanishes when $\theta' = \pi$ (corresponding to rotation over an entire sphere). What is the average value of the local dipolar field for the H$_2$O molecule in Problem 18.14 if it is bound to a biopolymer that enables it to rotate up to $\theta' = 30°$?

18.20 Suggest a reason why the spin–lattice relaxation time of benzene (a small molecule) in a mobile, deuterated hydrocarbon solvent increases with temperature whereas that of an oligonucleotide (a large molecule) decreases.

18.21 NMR spectroscopy may be used to determine the equilibrium constant for dissociation of a complex between a small molecule, such as an enzyme inhibitor I, and a protein, such as an enzyme E:

$$EI \rightleftharpoons E + I \qquad K_I = [E][I]/[EI]$$

In the limit of slow chemical exchange, the NMR spectrum of a proton in I would consist of two resonances: one at ν_I for free I and another at ν_{EI} for bound I. When chemical exchange is fast, the NMR spectrum of the same proton in I consists of a single peak with a resonance frequency ν given by:

$$\nu = f_I\nu_I + f_{EI}\nu_{EI}$$

where $f_I = [I]/([I] + [EI])$ and $f_{EI} = [EI]/([I] + [EI])$ are, respectively, the fractions of free I and bound I. For the purposes of analysing the data, it is also useful to define the frequency differences $\delta\nu = \nu - \nu_I$ and $\Delta\nu = \nu_{EI} - \nu_I$. (a) Show that, when the initial concentration of I, $[I]_0$, is much greater than the initial concentration of E, $[E]_0$, a plot of $[I]_0$ versus $\delta\nu^{-1}$ is a straight line with slope $[E]_0\Delta\nu$ and y-intercept $-K_I$.

18.22 The molecular electronic structure methods described in Chapter 14 may be used to predict the spin density distribution in a radical. Recent ESR studies have shown that the amino acid tyrosine participates in a number of biological electron transfer reactions, including the processes of water oxidation to O$_2$ in plant photosystem II and of O$_2$ reduction to water in cytochrome c oxidase (Box 10.1). During the course of these electron-transfer reactions, a tyrosine radical forms, with spin density delocalized over the side chain of the amino acid. (a) The phenoxy radical shown in (**14**) is a suitable model of the tyrosine radical. Using molecular modelling software and the computational method of your choice (extended Hückel, semi-empirical, or *ab initio* methods), calculate the spin densities at the O atom and at all of the C atoms in (**14**). (b) Predict the form of the ESR spectrum of (**14**).

14

---

10. The web site contains links to sites from which free molecular modelling software can be obtained.

19

Statistical thermodynamics: the concepts

The distribution of molecular states

19.1 Configurations and weights

19.2 The molecular partition function

The internal energy and the entropy

19.3 The internal energy

19.4 The statistical entropy

The canonical partition function

19.5 The canonical ensemble

19.6 The thermodynamic information in the partition function

19.7 Independent molecules

Checklist of key ideas

Further reading

Exercises

Problems

Statistical thermodynamics provides the link between the microscopic properties of matter and its bulk properties. Two key ideas are introduced in this chapter. The first is the Boltzmann distribution, which is used to predict the populations of states in systems at thermal equilibrium. In this chapter we see its derivation in terms of the distribution of particles over available states. The derivation leads naturally to the introduction of the partition function, which is the central mathematical concept of this and the next chapter. We see how to interpret the partition function and how to calculate it in a number of simple cases. We then see how to extract thermodynamic information from the partition function. In the final part of the chapter, we generalize the discussion to include systems that are composed of assemblies of interacting particles. Very similar equations are developed to those in the first part of the chapter, but they are much more widely applicable.

The preceding chapters of this part of the text have shown how the energy levels of molecules can be calculated, determined spectroscopically, and related to their structures. The next major step is to see how a knowledge of these energy levels can be used to account for the properties of matter in bulk. To do so, we now introduce the concepts of **statistical thermodynamics**, the link between molecular properties and bulk thermodynamic properties.

The crucial step in going from the quantum mechanics of individual molecules to the thermodynamics of bulk samples is to recognize that the latter deals with the *average* behaviour of large numbers of molecules. For example, the pressure of a gas depends on the average force exerted by its molecules, and there is no need to specify which molecules happen to be striking the wall at any instant. Nor is it necessary to consider the fluctuations in the pressure as different numbers of molecules collide with the wall at different moments. The fluctuations in pressure are very small compared with the steady pressure: it is highly improbable that there will be a sudden lull in the number of collisions, or a sudden surge. Fluctuations in other thermodynamic properties also occur, but for large numbers of particles they are negligible compared to the mean values.

This chapter introduces statistical thermodynamics in two stages. The first, the derivation of the Boltzmann distribution for individual particles, is of restricted applicability, but it has the advantage of taking us directly to a result of central importance in a straightforward and elementary way. We can *use* statistical thermodynamics once we have deduced the Boltzmann distribution. Then (in Section 19.5) we extend the arguments to systems composed of interacting particles.

The distribution of molecular states

We consider a closed system composed of N molecules. Although the total energy is constant at E, it is not possible to be definite about how that energy is shared between the molecules. Collisions result in the ceaseless redistribution of energy not only between the molecules but also among their different modes of motion. The closest we can come to a description of the distribution of energy is to report the **population** of a state, the average number of molecules that occupy it, and to say that on average there are n_i molecules in a state of energy ε_i. The populations of the states remain almost constant, but the precise identities of the molecules in each state may change at every collision.

The problem we address in this section is the calculation of the populations of states for any type of molecule in any mode of motion at any temperature. The only restriction is that the molecules should be independent, in the sense that the total energy of the system is a sum of their individual energies. We are discounting (at this stage) the possibility that in a real system a contribution to the total energy may arise from interactions between molecules. We also adopt the **principle of equal** *a priori* **probabilities**,[1] the assumption that all possibilities for the distribution of energy are equally probable. That is, we assume that vibrational states of a certain energy, for instance, are as likely to be populated as rotational states of the same energy.

One very important conclusion that will emerge from the following analysis is that the populations of states depend on a single parameter, the 'temperature'. That is, statistical thermodynamics provides a molecular justification for the concept of temperature and some insight into this crucially important quantity.

19.1 Configurations and weights

Any individual molecule may exist in states with energies $\varepsilon_0, \varepsilon_1, \dots$. We shall always take ε_0, the lowest state, as the zero of energy ($\varepsilon_0 = 0$), and measure all other energies relative to that state. To obtain the actual internal energy, U, we may have to add a constant to the calculated energy of the system. For example, if we are considering the vibrational contribution to the internal energy, then we must add the total zero-point energy of any oscillators in the sample.

(a) Instantaneous configurations

At any instant there will be n_0 molecules in the state with energy ε_0, n_1 with ε_1, and so on. The specification of the set of populations $n_0, n_1, \dots$ in the form $\{n_0, n_1, \dots\}$ is a statement of the instantaneous **configuration** of the system. The instantaneous configuration fluctuates with time because the populations change. We can picture a large number of different instantaneous configurations. One, for example, might be $\{N, 0, 0, \dots\}$, corresponding to every molecule being in its ground state. Another might be $\{N-2, 2, 0, 0, \dots\}$, in which two molecules are in the first excited state. The latter configuration is intrinsically more likely to be found than the former because it can be achieved in more ways: $\{N, 0, 0, \dots\}$ can be achieved in only one way, but $\{N-2, 2, 0, \dots\}$ can be achieved in $\frac{1}{2}N(N-1)$ different ways (Fig. 19.1).[2]

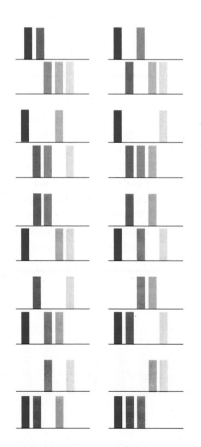

19.1 Whereas a configuration $\{5, 0, 0, \dots\}$ can be achieved in only one way, a configuration $\{3, 2, 0, \dots\}$ can be achieved in the ten different ways shown here, where the tinted blocks represent different molecules.

1. *A priori* means in this context loosely 'as far as one knows'. We have no reason to presume otherwise than that all states are equally likely to be occupied whatever their nature.
2. At this stage in the argument, we are ignoring the requirement that the total energy of the system should be constant (the second configuration has a higher energy than the first). The constraint of total energy is imposed later in this section.

Justification 19.1

One candidate for promotion to an upper state can be selected in N ways. There are $N - 1$ candidates for the second choice, so the total number of choices is $N(N - 1)$. However, we should not distinguish the choice (Jack, Jill) from the choice (Jill, Jack) because they lead to the same configurations. Therefore, only half the choices lead to distinguishable configurations, and the total number of distinguishable choices is $\frac{1}{2}N(N - 1)$.

If, as a result of collisions, the system were to fluctuate between the configurations $\{N, 0, 0, \dots\}$ and $\{N - 2, 2, 0, \dots\}$, it would almost always be found in the second, more likely state (especially if N were large). In other words, a system free to switch between the two configurations would show properties characteristic almost exclusively of the second configuration.

A general configuration $\{n_0, n_1, \dots\}$ can be achieved in W different ways, where W is called the **weight** of the configuration. The weight of the configuration $\{n_0, n_1, \dots\}$ is given by the expression

> In this expression, $x!$, x factorial, denotes $x(x - 1)(x - 2) \cdots 1$, and by definition $0! = 1$.

$$W = \frac{N!}{n_0!\, n_1!\, n_2! \cdots} \tag{19.1}$$

Equation 1 is a generalization of the formula $W = \frac{1}{2}N(N - 1)$, and reduces to it for the configuration $\{N - 2, 2, 0, \dots\}$.

Justification 19.2

Consider the number of ways of distributing N balls into bins. The first ball can be selected in N different ways, the next ball in $N - 1$ different ways for the balls remaining, and so on. Therefore, there are $N(N - 1) \dots 1 = N!$ ways of selecting the balls for distribution over the bins. However, if there are n_0 balls in the bin labelled ε_0, there would be $n_0!$ different ways in which the same balls could have been chosen (Fig. 19.2). Similarly, there are $n_1!$ ways in which the n_1 balls in the bin labelled ε_1 can be chosen, and so on. Therefore, the total number of distinguishable ways of distributing the balls so that there are n_0 in bin ε_0, n_1 in bin ε_1, etc. regardless of the order in which the balls were chosen is $N!/n_0!n_1! \dots$, which is the content of eqn 1.

$N = 18$

| 3! | 6! | 5! | 4! |

19.2 The 18 molecules shown here can be distributed into four receptacles (distinguished by the three vertical lines) in 18! different ways. However, 3! of the selections that put three molecules in the first receptacle are equivalent, 6! that put six molecules into the second receptacle are equivalent, and so on. Hence the number of distinguishable arrangements is 18!/3!6!5!4!.

Illustration 19.1

To calculate the number of ways of distributing 20 identical objects with the arrangement 1, 0, 3, 5, 10, 1, we note that the configuration is {1, 0, 3, 5, 10, 1} with $N = 20$; therefore the weight is

$$W = \frac{20!}{1!\,0!\,3!\,5!\,10!\,1!} = 9.31 \times 10^8$$

Self-test 19.1 Calculate the weight of the configuration in which 20 objects are distributed in the arrangement 0, 1, 5, 0, 8, 0, 3, 2, 0, 1. $[4.19 \times 10^{10}]$

It will turn out to be more convenient to deal with the natural logarithm of the weight, $\ln W$, rather than with the weight itself. We shall therefore need the expression

$$\ln W = \ln \frac{N!}{n_0!\,n_1!\,n_2!\,\cdots} = \ln N! - \ln(n_0!\,n_1!\,n_2!\,\cdots)$$

$$= \ln N! - (\ln n_0! + \ln n_1! + \ln n_2!\,\cdots)$$

$$= \ln N! - \sum_i \ln n_i!$$

where in the first line we have used $\ln(x/y) = \ln x - \ln y$ and in the second $\ln xy = \ln x + \ln y$. One reason for introducing $\ln W$ is that it is easier to make approximations. In particular, we can simplify the factorials by using **Stirling's approximation** in the form

$$\ln x! \approx x \ln x - x \tag{19.2}$$

Then the approximate expression for the weight is

$$\ln W = (N \ln N - N) - \sum_i (n_i \ln n_i - n_i) = N \ln N - \sum_i n_i \ln n_i \tag{19.3}$$

The final form of eqn 3 is derived by noting that the sum of n_i is equal to N, so the second and fourth terms in the second expression cancel.

The precise form of Stirling's approximation is

$$x! \approx (2\pi)^{1/2} x^{x+\frac{1}{2}} e^{-x}$$

and is in error by less than 1 per cent when x is greater than about 10. We deal with far larger values of x, and the simplified version in eqn 2 is adequate.

(b) The dominating configuration

We have seen that the configuration $\{N - 2, 2, 0, \dots\}$ dominates $\{N, 0, 0, \dots\}$, and it should be easy to believe that there may be other configurations that have a much greater weight than both. We shall see, in fact, that there is a configuration with so great a weight that it overwhelms all the rest in importance to such an extent that the system will almost always be found in it. The properties of the system will therefore be characteristic of that particular dominating configuration. This dominating configuration can be found by looking for the values of n_i that lead to a maximum value of W. Because W is a function of all the n_i, we can do this search by varying the n_i and looking for the values that correspond to $dW = 0$ (just as in the search for the maximum of any function), or equivalently a maximum value of $\ln W$. However, there are two difficulties with this procedure.

The first difficulty is that the only permitted configurations are those corresponding to the specified, constant, total energy of the system. This requirement rules out many configurations; $\{N, 0, 0, \dots\}$ and $\{N - 2, 2, 0, \dots\}$, for instance, have different energies, so both cannot occur in the same isolated system. It follows that, in looking for the configuration with the greatest weight, we must ensure that the configuration also satisfies the condition

$$\text{Constant total energy:} \qquad \sum_i n_i \varepsilon_i = E \qquad\qquad (19.4)$$

where E is the total energy of the system.

The second constraint is that, because the total number of molecules present is also fixed (at N), we cannot arbitrarily vary all the populations simultaneously. Thus, increasing the population of one state by 1 demands that the population of another state must be reduced by 1. Therefore, the search for the maximum value of W is also subject to the condition

$$\text{Constant total number of molecules:} \qquad \sum_i n_i = N \qquad\qquad (19.5)$$

(c) The Boltzmann distribution

We are looking for the set of numbers $n_0, n_1, \ldots$ for which W has its maximum value. We show in the following *Justification* that the populations in the configuration of greatest weight depend on the energy of the state according to the **Boltzmann distribution**:

$$\frac{n_i}{N} = \frac{e^{-\beta \varepsilon_i}}{\sum_i e^{-\beta \varepsilon_i}} \qquad\qquad (19.6a)$$

Equation 6a is the justification of the remark that a single parameter, here denoted β, determines the most probable populations of the states of the system. We shall see in Section 19.3b that

$$\beta = \frac{1}{kT} \qquad\qquad (19.6b)$$

where T is the thermodynamic temperature and k is the Boltzmann constant. In other words, the thermodynamic temperature is the unique parameter that governs the populations of states of a system at thermal equilibrium.

Justification 19.3

We have already remarked that it turns out to be simpler to find the condition for $\ln W$ being a maximum rather than dealing directly with W. Because $\ln W$ depends on all the n_i, when a configuration changes and the n_i change to $n_i + dn_i$, the function $\ln W$ changes to $\ln W + d \ln W$, where

$$d \ln W = \sum_i \left(\frac{\partial \ln W}{\partial n_i} \right) dn_i$$

At a maximum, $d \ln W = 0$. However, when the n_i change, they do so subject to the two constraints

$$\sum_i \varepsilon_i \, dn_i = 0 \qquad \sum_i dn_i = 0 \qquad\qquad (19.7)$$

The first constraint recognizes that the total energy must not change, and the second recognizes that the total number of molecules must not change. These two constraints prevent us from solving $d \ln W = 0$ simply by setting all $(\partial \ln W / \partial n_i) = 0$ because the dn_i are not all independent.

The way to take constraints into account was devised by the French mathematician Lagrange, and is called the **method of undetermined multipliers**. The technique is described in *Further information 1*. All we need here is the rule that a constraint should be multiplied by a constant and then added to the main variation equation. The variables are then treated as though they were all independent, and the constants are evaluated at the end of the calculation.

We employ the technique as follows. The two constraints are multiplied by the constants $-\beta$ and α, respectively (the minus sign in $-\beta$ has been included for future convenience), and then added to the expression for d ln W:

$$\text{d} \ln W = \sum_i \left(\frac{\partial \ln W}{\partial n_i} \right) \text{d}n_i + \alpha \sum_i \text{d}n_i - \beta \sum_i \varepsilon_i \, \text{d}n_i$$

$$= \sum_i \left\{ \left(\frac{\partial \ln W}{\partial n_i} \right) + \alpha - \beta \varepsilon_i \right\} \text{d}n_i$$

All the $\text{d}n_i$ are now treated as independent. Hence the only way of satisfying d ln $W = 0$ is to require that for each i,

$$\frac{\partial \ln W}{\partial n_i} + \alpha - \beta \varepsilon_i = 0 \tag{19.8}$$

when the n_i have their most probable values.

The expression for ln W is given in eqn 3. Differentiation of ln W with respect to n_i gives

$$\frac{\partial \ln W}{\partial n_i} = \frac{\partial (N \ln N)}{\partial n_i} - \sum_j \frac{\partial (n_j \ln n_j)}{\partial n_i}$$

The derivative of the first term is obtained as follows:

$$\frac{\partial (N \ln N)}{\partial n_i} = \left(\frac{\partial N}{\partial n_i} \right) \ln N + \frac{\partial N}{\partial n_i} = \ln N + 1$$

because $N = n_1 + n_2 + \cdots$ and its derivative with respect to any of the n is 1. The derivative of the second term is

> Here we use
>
> $$\frac{\partial \ln n_j}{\partial n_i} = \frac{1}{n_j} \left(\frac{\partial n_j}{\partial n_i} \right)$$
>
> Then, if $i \neq j$, n_j is independent of n_i, so $\partial n_j / \partial n_i = 0$. However, if $i = j$,
>
> $$\frac{\partial n_j}{\partial n_i} = \frac{\partial n_j}{\partial n_j} = 1$$

$$\sum_j \frac{\partial (n_j \ln n_j)}{\partial n_i} = \sum_j \left\{ \left(\frac{\partial n_j}{\partial n_i} \right) \ln n_j + n_j \left(\frac{\partial \ln n_j}{\partial n_i} \right) \right\}$$

$$= \sum_j \left(\frac{\partial n_j}{\partial n_i} \right) (\ln n_j + 1) = \ln n_i + 1$$

and therefore

$$\frac{\partial \ln W}{\partial n_i} = -(\ln n_i + 1) + (\ln N + 1) = -\ln \frac{n_i}{N}$$

It follows from eqn 8 that

$$-\ln \frac{n_i}{N} + \alpha - \beta \varepsilon_i = 0$$

and therefore that

$$\frac{n_i}{N} = e^{\alpha - \beta \varepsilon_i}$$

At this stage we note that

$$N = \sum_j n_j = Ne^\alpha \sum_j e^{-\beta\varepsilon_j}$$

(We are free to label the states with j instead of i.) Because the N cancels on each side of this equality, it follows that

$$e^\alpha = \frac{1}{\displaystyle\sum_j e^{-\beta\varepsilon_j}} \tag{19.9}$$

and eqn 6a follows immediately.

19.2 The molecular partition function

From now on we write the Boltzmann distribution as

$$p_i = \frac{e^{-\beta\varepsilon_i}}{q} \tag{19.10}$$

where p_i is the fraction of molecules in the state i, $p_i = n_i/N$, and q is the **molecular partition function**:

$$q = \sum_j e^{-\beta\varepsilon_j} \tag{[19.11]}$$

The sum in q is sometimes expressed slightly differently. It may happen that several states have the same energy, and so give the same contribution to the sum. If, for example, g_j states have the same energy ε_j (so the level is g_j-fold degenerate), we could write

$$q = \sum_{\text{levels } j} g_j e^{-\beta\varepsilon_j} \tag{19.12}$$

where the sum is now over energy levels (sets of states with the same energy), not individual states.

Example 19.1 Writing a partition function

Write an expression for the partition function of a linear molecule (such as HCl) treated as a rigid rotor.

Method To use eqn 12 we need to know (a) the energies of the levels, (b) the degeneracies, the number of states that belong to each level. Whenever calculating a partition function, the energies of the levels are expressed relative to 0 for the state of lowest energy. The energy levels of a rigid linear rotor were derived in Section 16.5c.

Answer From eqn 16.38, the energy levels of a linear rotor are $hcBJ(J+1)$, with $J = 0, 1, 2, \ldots$. The state of lowest energy has zero energy, so no adjustment need be made to the energies given by this expression. Each level consists of $2J+1$ degenerate states. Therefore,

$$q = \sum_{J=0}^{\infty} \overbrace{(2J+1)}^{g_J}\, e^{-\overbrace{\beta hcBJ(J+1)}^{\varepsilon_J}}$$

3ε

2ε

ε

0

19.3 The equally spaced infinite array of energy levels used in the calculation of the partition function. A harmonic oscillator has the same spectrum of levels.

Comment The sum can be evaluated numerically by supplying the value of B (from spectroscopy or calculation) and the temperature. For reasons explained in Section 20.2b, this expression applies only to unsymmetrical linear rotors (for example, HCl, not CO_2; in general, to $C_{\infty v}$ and not $D_{\infty h}$ species).

Self-test 19.2 Write the partition function for a two-level system, the lower state (at energy 0) being non-degenerate, and the upper state (at an energy ε) doubly degenerate.

$$[q = 1 + 2e^{-\beta\varepsilon}]$$

(a) An interpretation of the partition function

Some insight into the significance of a partition function can be obtained by considering how q depends on the temperature. When T is close to zero, the parameter $\beta = 1/kT$ is close to infinity. Then every term except one in the sum defining q is zero because each one has the form e^{-x} with $x \to \infty$. The exception is the term with $\varepsilon_0 \equiv 0$ (or the g_0 terms at zero energy if the ground state is g_0-fold degenerate), because then $\varepsilon_0/kT \equiv 0$ whatever the temperature, including zero. As there is only one surviving term when $T = 0$, and its value is g_0, it follows that

$$\lim_{T \to 0} q = g_0 \tag{19.13}$$

That is, at $T = 0$, the partition function is equal to the degeneracy of the ground state.

Now consider the case when T is so high that for each term in the sum $\varepsilon_j/kT \approx 0$. Because $e^{-x} = 1$ when $x = 0$, each term in the sum now contributes 1. It follows that the sum is equal to the number of molecular states, which in general is infinite:

$$\lim_{T \to \infty} q = \infty \tag{19.14}$$

In some idealized cases, the molecule may have only a finite number of states; then the upper limit of q is equal to the number of states. For example, if we were considering only the spin energy levels of a radical in a magnetic field, then there would be only two states ($m_s = \pm\frac{1}{2}$). The partition function for such a system can therefore be expected to rise towards 2 as T is increased towards infinity.

We see that *the molecular partition function gives an indication of the average number of states that are thermally accessible to a molecule at the temperature of the system*. At $T = 0$, only the ground level is accessible and $q = g_0$. At very high temperatures, virtually all states are accessible, and q is correspondingly large.

Example 19.2 Evaluating the partition function for a uniform ladder of energy levels

Evaluate the partition function for a molecule with an infinite number of equally spaced non-degenerate energy levels (Fig. 19.3). These levels can be thought of as the vibrational energy levels of a diatomic molecule in the harmonic approximation.

Method We expect the partition function to increase from 1 at $T = 0$ and approach infinity as $T \to \infty$. To evaluate eqn 11 explicitly, note that

$$1 + x + x^2 + \cdots = \frac{1}{1-x}$$

The sum of the infinite series $S = 1 + x + x^2 + \cdots$ is obtained by multiplying both sides by x, which gives $xS = x + x^2 + x^3 + \cdots = S - 1$. This relation reorganizes into

$$S = \frac{1}{1-x}$$

19.4 The partition function for the system shown in Fig. 19.3 (a harmonic oscillator) as a function of temperature.

19.5 The partition function for a two-level system as a function of temperature. The two graphs differ in the scale of the temperature axis to show the approach to 1 as $T \to 0$ and the slow approach to 2 as $T \to \infty$.

Answer If the separation of neighbouring levels is ε, the partition function is

$$q = 1 + e^{-\beta\varepsilon} + e^{-2\beta\varepsilon} + \cdots = 1 + e^{-\beta\varepsilon} + (e^{-\beta\varepsilon})^2 + \cdots = \frac{1}{1 - e^{-\beta\varepsilon}}$$

This expression is plotted in Fig. 19.4: notice that it rises from 1 to infinity as the temperature is raised, as anticipated.

Self-test 19.3 Find and plot an expression for the partition function of a system with one state at zero energy and another state at the energy ε.

$$[q = 1 + e^{-\beta\varepsilon}, \text{ Fig. 19.5}]$$

It follows from eqn 10 and the expression for q derived in Example 19.2 for a uniform ladder of states of spacing ε,

$$q = \frac{1}{1 - e^{-\beta\varepsilon}} \tag{19.15}$$

that the fraction of molecules in the state with energy ε_i is

$$p_i = (1 - e^{-\beta\varepsilon})e^{-\beta\varepsilon_i} \tag{19.16}$$

Figure 19.6 shows how p_i varies with temperature. At very low temperatures, where q is close to 1, only the lowest state is significantly populated. As the temperature is raised, the population breaks out of the lowest state, and the upper states become progressively more highly populated. At the same time, the partition function rises from 1 and its value gives an indication of the range of states populated. The name 'partition function' reflects the sense in which q measures how the total number of molecules is distributed—partitioned—over the available states.

The corresponding expressions for a two-level system derived in Self-test 19.3 are

$$p_0 = \frac{1}{1 + e^{-\beta\varepsilon}} \qquad p_1 = \frac{e^{-\beta\varepsilon}}{1 + e^{-\beta\varepsilon}} \tag{19.17}$$

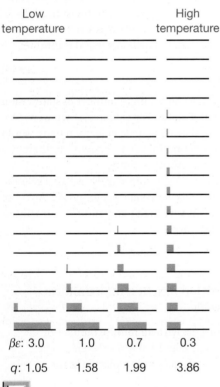

Low temperature			High temperature
$\beta\varepsilon$: 3.0	1.0	0.7	0.3
q: 1.05	1.58	1.99	3.86

19.6 The populations of the energy levels of the system shown in Fig. 19.3 at different temperatures, and the corresponding values of the partition function calculated in Example 19.2. Note that $\beta = 1/kT$.

19.7 The fraction of populations of the two states of a two-level system as a function of temperature (eqn 17). Note that, as the temperature approaches infinity, the populations of the two states become equal (and the fractions both approach 0.5).

These functions are plotted in Fig. 19.7. Notice how the populations tend towards equality ($p_0 = 0.5$, $p_1 = 0.5$) as $T \to \infty$. A common error is to suppose that all the molecules in the system will be found in the upper energy state when $T = \infty$; however, we see from eqn 17 that, as $T \to \infty$, the populations of states become equal. The same conclusion is true of multi-level systems too: as $T \to \infty$, all states become equally populated.

Example 19.3 Using the partition function to calculate a population

Calculate the proportion of I_2 molecules in their ground, first excited, and second excited vibrational states at 25°C. The vibrational wavenumber is 214.6 cm⁻¹.

Method Vibrational energy levels have a constant separation (in the harmonic approximation, Section 16.9), so the partition function is given by eqn 15 and the populations by eqn 16. To use the latter equation, we identify the index i with the quantum number v, and calculate p_v for $v = 0$, 1, and 2. At 298.15 K, $kT/hc = 207.226$ cm⁻¹.

Answer First, we note that

$$\beta\varepsilon = \frac{hc\tilde{v}}{kT} = \frac{214.6 \text{ cm}^{-1}}{207.226 \text{ cm}^{-1}} = 1.036$$

Then it follows from eqn 16 that the populations are

$$p_v = (1 - e^{-\beta\varepsilon})e^{-v\beta\varepsilon} = 0.645e^{-1.036v}$$

Therefore, $p_0 = 0.645$, $p_1 = 0.229$, $p_2 = 0.081$.

Comment The I—I bond is not stiff and the atoms are heavy: as a result, the vibrational energy separations are small and at room temperature several vibrational levels are significantly populated. The value of the partition function, $q = 1.55$, reflects this small but significant spread of populations.

Self-test 19.4 At what temperature would the $v = 1$ level of I_2 have (a) half the population of the ground state, (b) the same population as the ground state?　　　　　[(a) 445 K, (b) infinite]

(b) Approximations and factorizations

In general, exact analytical expressions for partition functions cannot be obtained. However, closed approximate expressions can often be found and prove to be very important in a number of chemical applications. For instance, the expression for the partition function for a particle of mass m free to move in a one-dimensional container of length X can be evaluated by making use of the fact that the separation of energy levels is very small and that large numbers of states are accessible at normal temperatures. As shown in the *Justification* below, in this case

$$q_X = \left(\frac{2\pi m}{h^2\beta} \right)^{1/2} X \tag{19.18}$$

This expression shows that the partition function for translational motion increases with the length of the box and the mass of the particle, for in each case the separation of the energy levels becomes smaller and more levels become thermally accessible.

For a given mass and length of the box, the partition function also increases with increasing temperature (decreasing β), because more states become accessible.

Justification 19.4

The energy levels of a molecule of mass m in a container of length X are given by eqn 12.7 with $L = X$:

$$E_n = \frac{n^2 h^2}{8mX^2} \qquad n = 1, 2, \dots$$

The lowest level ($n = 1$) has energy $h^2/8mX^2$, so the energies relative to that level are

$$\varepsilon_n = (n^2 - 1)\varepsilon \qquad \varepsilon = \frac{h^2}{8mX^2}$$

The sum to evaluate is therefore

$$q_X = \sum_{n=1}^{\infty} e^{-(n^2-1)\beta\varepsilon}$$

The translational energy levels are very close together in a container the size of a typical laboratory vessel; therefore, the sum can be approximated by an integral:

$$q_X = \int_1^{\infty} e^{-(n^2-1)\beta\varepsilon}\, dn$$

The extension of the lower limit to $n = 0$ and the replacement of $n^2 - 1$ by n^2 introduces negligible error but turns the integral into standard form. We make the substitution $x^2 = n^2\beta\varepsilon$, implying $dn = dx/(\beta\varepsilon)^{1/2}$, and therefore that

$$q_X = \left(\frac{1}{\beta\varepsilon}\right)^{1/2} \int_0^{\infty} e^{-x^2}\, dx = \left(\frac{1}{\beta\varepsilon}\right)^{1/2}\left(\frac{\pi^{1/2}}{2}\right) = \left(\frac{2\pi m}{h^2\beta}\right)^{1/2} X$$

Another useful feature of partition functions is used to derive expressions when the energy of a molecule arises from several different, independent sources: if the energy is a sum of contributions from independent modes of motion, then the partition function is a product of partition functions for each mode of motion. For instance, suppose the molecule we are considering is free to move in three dimensions. We take the length of the container in the y-direction to be Y and that in the z-direction to be Z. The total energy of a molecule ε is the sum of its translational energies in all three directions:

$$\varepsilon_{n_1 n_2 n_3} = \varepsilon_{n_1}^{(X)} + \varepsilon_{n_2}^{(Y)} + \varepsilon_{n_3}^{(Z)} \tag{19.19}$$

where n_1, n_2, and n_3 are the quantum numbers for motion in the x-, y-, and z-directions, respectively. Therefore, because $e^{a+b+c} = e^a e^b e^c$, the partition function factorizes as follows:

$$q = \sum_{\text{all } n} e^{-\beta\varepsilon_{n_1}^{(X)} - \beta\varepsilon_{n_2}^{(Y)} - \beta\varepsilon_{n_3}^{(Z)}} = \sum_{\text{all } n} e^{-\beta\varepsilon_{n_1}^{(X)}} e^{-\beta\varepsilon_{n_2}^{(Y)}} e^{-\beta\varepsilon_{n_3}^{(Z)}}$$

$$= \left(\sum_{n_1} e^{-\beta\varepsilon_{n_1}^{(X)}}\right)\left(\sum_{n_2} e^{-\beta\varepsilon_{n_2}^{(Y)}}\right)\left(\sum_{n_3} e^{-\beta\varepsilon_{n_3}^{(Z)}}\right) \tag{19.20}$$

$$= q_X q_Y q_Z$$

It is generally true that, if the energy of a molecule can be written as the sum of independent terms, the partition function is the corresponding product of individual contributions.

Equation 18 gives the partition function for translational motion in the *x*-direction. The only change for the other two directions is to replace the length *X* by the lengths *Y* or *Z*. Hence the partition function for motion in three dimensions is

$$q = \left(\frac{2\pi m}{h^2 \beta}\right)^{3/2} XYZ \tag{19.21}$$

The product of lengths *XYZ* is the volume, *V*, of the container, so we can write

$$q = \frac{V}{\Lambda^3} \qquad \Lambda = h\left(\frac{\beta}{2\pi m}\right)^{1/2} = \frac{h}{(2\pi m k T)^{1/2}} \tag{19.22}$$

The quantity Λ has the dimensions of length and is called the **thermal wavelength** of the molecule. The thermal wavelength decreases with increasing mass and temperature. As in the one-dimensional case, the partition function increases with the mass of the particle (as $m^{3/2}$) and the volume of the container (as *V*); for a given mass and volume, the partition function increases with temperature (as $T^{3/2}$).

Illustration 19.2

To calculate the translational partition function of an H_2 molecule confined to a 100 cm³ vessel at 25°C we use $m = 2.016$ u; then

$$\Lambda = \frac{6.626 \times 10^{-34} \text{ J s}}{\{2\pi \times (2.016 \times 1.6605 \times 10^{-27} \text{ kg}) \times (1.38 \times 10^{-23} \text{ J K}^{-1}) \times (298 \text{ K})\}^{1/2}}$$

$$= 7.12 \times 10^{-11} \text{ m}$$

where we have used 1 J = 1 kg m² s⁻². Therefore,

$$q = \frac{1.00 \times 10^{-4} \text{ m}^3}{(7.12 \times 10^{-11} \text{ m})^3} = 2.77 \times 10^{26}$$

About 10^{26} quantum states are thermally accessible, even at room temperature and for this light molecule. Many states are occupied if the thermal wavelength (which in this case is 71.2 pm) is small compared with the linear dimensions of the container.

Self-test 19.5   Calculate the translational partition function for a D_2 molecule under the same conditions.   [$q = 7.8 \times 10^{26}$, $2^{3/2}$ times larger]

The internal energy and the entropy

The importance of the molecular partition function is that it contains all the information needed to calculate the thermodynamic properties of a system of independent particles. In this respect, *q* plays a role in statistical thermodynamics very similar to that played by the wavefunction in quantum mechanics: *q* is a kind of thermal wavefunction.

19.3 The internal energy

We shall begin to unfold the importance of q by showing how to derive an expression for the internal energy of the system.

(a) The relation between U and q

The total energy of the system relative to the energy of the lowest state is

$$E = \sum_i n_i \varepsilon_i \tag{19.23}$$

Because the most probable configuration is so strongly dominating, we can use the Boltzmann distribution for the populations and write

$$E = \frac{N}{q} \sum_i \varepsilon_i e^{-\beta \varepsilon_i} \tag{19.24}$$

To manipulate this expression into a form involving only q we note that

$$\varepsilon_i e^{-\beta \varepsilon_i} = -\frac{\mathrm{d}}{\mathrm{d}\beta} e^{-\beta \varepsilon_i}$$

It follows that

$$E = -\frac{N}{q} \sum_i \frac{\mathrm{d}}{\mathrm{d}\beta} e^{-\beta \varepsilon_i} = -\frac{N}{q} \frac{\mathrm{d}}{\mathrm{d}\beta} \sum_i e^{-\beta \varepsilon_i} = -\frac{N}{q} \frac{\mathrm{d}q}{\mathrm{d}\beta} \tag{19.25}$$

19.8 The total energy of a two-level system (expressed as a multiple of $N\varepsilon$) as a function of temperature, on two temperature scales. The graph at the top shows the slow rise away from zero energy at low temperatures; the slope of the graph at $T = 0$ is 0 (that is, the heat capacity is zero at $T = 0$). The graph below shows the slow rise to 0.5 as $T \to \infty$ as both states become equally populated (see Fig. 19.7).

Illustration 19.3

From the two-level partition function $q = 1 + e^{-\beta \varepsilon}$, we can deduce that the total energy of N two-level systems is

$$E = -\left(\frac{N}{1 + e^{-\beta \varepsilon}} \right) \frac{\mathrm{d}}{\mathrm{d}\beta} \left(1 + e^{-\beta \varepsilon} \right) = \frac{N\varepsilon e^{-\beta \varepsilon}}{1 + e^{-\beta \varepsilon}} = \frac{N\varepsilon}{1 + e^{\beta \varepsilon}}$$

This function is plotted in Fig. 19.8. Notice how the energy is zero at $T = 0$, when only the lower state (at the zero of energy) is occupied, and rises to $\frac{1}{2}N\varepsilon$ as $T \to \infty$, when the two levels become equally populated.

There are several points in relation to eqn 25 that need to be made. Because $\varepsilon_0 = 0$ (remember that we measure all energies from the lowest available level), E should be interpreted as the value of the internal energy relative to its value at $T = 0$, $U(0)$. Therefore, to obtain the conventional internal energy U, we must add the internal energy at $T = 0$:

$$U = U(0) + E \tag{19.26}$$

Secondly, because the partition function may depend on variables other than the temperature (for example, the volume), the derivative with respect to β in eqn 25 is actually a *partial* derivative with these other variables held constant. The complete expression relating the molecular partition function to the thermodynamic internal energy of a system of independent molecules is therefore

$$U = U(0) - \frac{N}{q} \left(\frac{\partial q}{\partial \beta} \right)_V \tag{19.27a}$$

An equivalent form is obtained by noting that $dx/x = d \ln x$:

$$U = U(0) - N \left(\frac{\partial \ln q}{\partial \beta} \right)_V \tag{19.27b}$$

These two equations confirm that we need know only the partition function (as a function of temperature) to calculate the internal energy relative to its value at $T = 0$.

(b) The value of β

We now confirm that the parameter β, which we have anticipated is equal to $1/kT$, does indeed have that value. To do so, we compare the equipartition expression for the internal energy of a monatomic perfect gas, which from *Molecular interpretation* 2.2 we know to be

$$U = U(0) + \tfrac{3}{2}nRT \tag{19.28a}$$

with the value calculated from the translational partition function (see the following *Justification*), which is

$$U = U(0) + \frac{3N}{2\beta} \tag{19.28b}$$

It follows by comparing these two expressions that

$$\beta = \frac{N}{nRT} = \frac{nN_A}{nN_A kT} = \frac{1}{kT} \tag{19.29}$$

(We have used $N = nN_A$, where n is the amount of gas molecules, N_A is the Avogadro constant, and $R = N_A k$.) Although we have proved that $\beta = 1/kT$ by examining a very specific example, the translational motion of a perfect monatomic gas, the result is general.[3]

Justification 19.5

To use eqn 27, we introduce the translational partition function from eqn 22:

$$\left(\frac{\partial q}{\partial \beta} \right)_V = \left(\frac{\partial}{\partial \beta} \frac{V}{\Lambda^3} \right)_V = V \frac{d}{d\beta} \frac{1}{\Lambda^3} = -3 \frac{V}{\Lambda^4} \frac{d\Lambda}{d\beta}$$

Then we note from the formula for Λ in eqn 22 that

$$\frac{d\Lambda}{d\beta} = \frac{d}{d\beta} \left\{ \frac{h\beta^{1/2}}{(2\pi m)^{1/2}} \right\} = \frac{1}{2\beta^{1/2}} \times \frac{h}{(2\pi m)^{1/2}} = \frac{\Lambda}{2\beta}$$

and so obtain

$$\left(\frac{\partial q}{\partial \beta} \right)_V = -\frac{3V}{2\beta\Lambda^3}$$

Then, by eqn 27a,

$$U = U(0) - N \left(\frac{\Lambda^3}{V} \right) \left(-\frac{3V}{2\beta\Lambda^3} \right) = U(0) + \frac{3N}{2\beta}$$

---

3. See Example 20.1 for another demonstration that $\beta = 1/kT$ and *Further reading* for a general proof.

19.4 The statistical entropy

If it is true that the partition function contains all thermodynamic information, then it must be possible to use it to calculate the entropy as well as the internal energy. Because we know (from Section 4.2) that entropy is related to the dispersal of energy and that the partition function is a measure of the number of thermally accessible states, we can be confident that the two are indeed related.

We shall develop the relation between the entropy and the partition function in two stages. In the first stage, we justify one of the most celebrated equations in statistical thermodynamics, the **Boltzmann formula** for the entropy:

$$S = k \ln W \qquad [19.30]$$

In this expression, W is the weight of the most probable configuration of the system. In the second stage, we express W in terms of the partition function.

Justification 19.6

A change in the internal energy

$$U = U(0) + \sum_i n_i \varepsilon_i \qquad (19.31)$$

may arise from either a modification of the energy levels of a system (when ε_i changes to $\varepsilon_i + d\varepsilon_i$) or from a modification of the populations (when n_i changes to $n_i + dn_i$). The most general change is therefore

$$dU = dU(0) + \sum_i n_i\, d\varepsilon_i + \sum_i \varepsilon_i\, dn_i \qquad (19.32)$$

Because the energy levels do not change when a system is heated at constant volume (Fig. 19.9), in the absence of all changes other than heating

$$dU = \sum_i \varepsilon_i\, dn_i$$

We know from thermodynamics (and specifically from eqn 5.2) that under the same conditions

$$dU = dq_{rev} = T\, dS$$

Therefore,

$$dS = \frac{dU}{T} = k\beta \sum_i \varepsilon_i\, dn_i \qquad (19.33)$$

We also know that for changes in the most probable configuration (the only one we need consider)

$$\frac{\partial \ln W}{\partial n_i} + \alpha - \beta\varepsilon_i = 0$$

(this is eqn 8). After rearranging this expression to

$$\beta\varepsilon_i = \frac{\partial \ln W}{\partial n_i} + \alpha$$

we find that

$$dS = k \sum_i \left(\frac{\partial \ln W}{\partial n_i} \right) dn_i + k\alpha \sum_i dn_i$$

19.9 (a) When a system is heated, the energy levels are unchanged but their populations are changed. (b) When work is done on a system, the energy levels themselves are changed. The levels in this case are the one-dimensional particle-in-a-box energy levels of Chapter12: they depend on the size of the container and move apart as its length is decreased.

Heat

Work

But, because the number of molecules is constant, the sum over the dn_i is zero. Hence

$$dS = k \sum_i \left(\frac{\partial \ln W}{\partial n_i} \right) dn_i = k(d \ln W)$$

This relation strongly suggests the definition $S = k \ln W$, as in eqn 30.

The statistical entropy behaves in exactly the same way as the thermodynamic entropy. Thus, as the temperature is lowered, the value of W, and hence of S, decreases because fewer configurations are compatible with the total energy. In the limit $T \to 0$, $W = 1$, so $\ln W = 0$, because only one configuration (every molecule in the lowest level) is compatible with $E = 0$. It follows that $S \to 0$ as $T \to 0$, which is compatible with the Third Law of thermodynamics (see Box 19.1), that the entropies of all perfect crystals approach the same value as $T \to 0$ (Section 4.4a).

Box 19.1 Temperatures below zero

The Third Law of thermodynamics prohibits the achievement of absolute zero in a finite number of steps. However, it is possible to circumvent this restriction in systems that have a finite number of levels or in systems that are effectively finite because they have such weak coupling to their surroundings. In this box we consider a system of two-level systems isolated from the environment. The practical realization of such a system is a group of spin-$\frac{1}{2}$ nuclei that have very long relaxation times, such as the ^{19}F nuclei in cold solid LiF.

The Boltzmann distribution tells us that the ratio of populations in a two-level system at a temperature T is

$$\frac{N_+}{N_-} = e^{-\varepsilon/kT}$$

where ε is the separation of the two states. It follows that, if we can contrive the population of the upper state to exceed that of the lower state, then the temperature must have a negative value. Indeed, for a general population,

$$T = \frac{\varepsilon/k}{\ln(N_-/N_+)}$$

and the temperature is formally negative for all $N_+ > N_-$ (that is, whenever the population of the upper state exceeds that of the lower state).

All the statistical thermodynamic expressions we have derived apply to $T < 0$ as well as to $T > 0$, the difference being that states with $T < 0$ are not in thermal equilibrium and therefore have to be achieved by techniques that do not rely on the equalization of temperatures of the system and its surroundings. Pulse techniques in NMR can achieve non-equilibrium populations (Section 18.7) as can pumping procedures in laser technologies (Section 17.5). From now on, we shall suppose that these non-equilibrium distribu-

tions have been achieved, and will concentrate on the consequences.

The expressions for q, U, and S that we have derived in this chapter are applicable to $T < 0$ as well as to $T > 0$, and are shown in the first illustration. As will be apparent, the thermodynamic functions show sharp discontinuities on passing through zero, and $T = +0$ (corresponding to all population in the lower state) is quite distinct from $T = -0$, where all the population is in the upper state. The other thermodynamic functions (S and C_V) are continuous at $T = 0$. All these functions are continuous if we use $\beta = 1/kT$ as the dependent variable (see the second illustration), which shows that β is a more natural, if less human, variable than T. Note that $U \to 0$ as $\beta \to \infty$ (that is, as $T \to 0$, when only the lower state is occupied) and $U \to N\varepsilon$ as $\beta \to -\infty$ (that is, as $T \to -0$); we see that a state with $T = -0$ is 'hotter' than one with $T = +0$ (when only the upper state is occupied). The entropy of the system is zero on either side of $T = 0$, and rises to $Nk \ln 2$ as $T \to \pm\infty$. At the lowest positive temperature ($T = +0$, $\beta = \infty$) only one state is accessible (the lower state), and at the lowest negative temperature ($T = -0$, $\beta = -\infty$) only the upper state is accessible, so the entropy is zero in each case. The heat capacity is also symmetrical around $T = 0$. The deepest insight into this condition is obtained by using statistical thermodynamics to show that the heat capacity is proportional to the root mean square deviation of the energy from its average value:

$$C_V = \frac{(\langle E^2 \rangle - \langle E \rangle^2)^{1/2}}{kT^2}$$

(This is a special case of the *fluctuation–dissipation theorem*.) At $T = +0$ the r.m.s. deviation of the energy is zero as all the molecules are in the lower state, so $C_V = 0$. At $T = -0$, all the molecules are in the upper state, so the r.m.s. deviation is also zero.

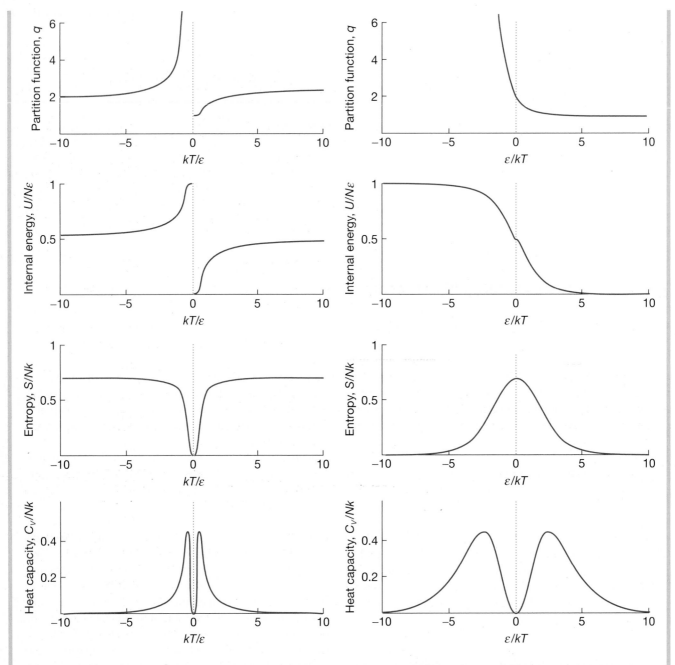

Various thermodynamic functions of a two-level system extended to negative temperatures.

Various thermodynamic functions of a two-level system extended to negative temperatures but plotted against $\beta = 1/kT$ (modified here to the dimensionless quantity ε/kT).

Further insight into the dependence of thermodynamic properties on temperature is obtained by noting the thermodynamic result (Section 5.2) that

$$\frac{1}{T} = \left(\frac{\partial S}{\partial U} \right)_V$$

When S is plotted against U for a two-level system (see the third illustration), we see that the entropy rises as energy is supplied to the system (as we would expect) provided that $T > 0$ (the thermal equilibrium regime). However, the entropy decreases as energy is supplied when $T < 0$. This conclusion is consistent with the thermodynamic definition of entropy,

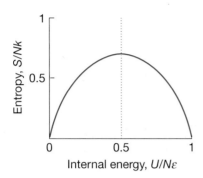

The variation of entropy with internal energy for a two-level system extended to negative temperatures.

$\mathrm{d}S = \mathrm{d}q_{rev}/T$. Physically, the increase in entropy for $T > 0$ corresponds to the increasing accessibility of the upper state, and the decrease for $T < 0$ corresponds to the shift towards population of the upper state alone as more energy is packed into the system.

The phenomenological laws of thermodynamics survive largely intact at negative temperatures. The First Law (in essence, the conservation of energy) is robust, and independent of how populations are distributed over states. The Second Law survives because the definition of entropy survives (as we have seen above). The efficiency of heat engines (Section 4.2), which is a direct consequence of the Second Law is still given by

$$\varepsilon = 1 - \frac{T_{cold}}{T_{hot}}$$

However, if the temperature of the cold reservoir is negative, the efficiency of the engine may be greater than 1. This condition corresponds to the amplification of signals achieved in lasers. Alternatively, $\varepsilon > 1$ implies that heat *can* be converted completely into work provided the heat is withdrawn from a reservoir at $T < 0$. If both reservoirs are at negative temperatures, then the efficiency is less than 1, as in the thermal equilibrium case treated in Chapter 3. The Third Law requires a slight amendment on account of the discontinuity of the populations across $T = 0$: it is impossible in a finite number of steps to cool any system down to +0 or to heat any system above −0.

Discussion questions

1. Discuss the operation of a refrigerator when either or both the reservoirs have a negative temperature.

2. Under what circumstances is it possible to speak of negative temperatures of an evenly spaced three-level system?

Further reading

R.S. Berry, S.A. Rice, and J. Ross, *Physical chemistry*. Oxford University Press (2000).

J. Wisniak, Negative absolute temperatures, a novelty. *J. Chem. Educ.* **77**, 518 (2000).

Now we relate the Boltzmann formula for the entropy to the partition function. To do so, we substitute the expression for ln W given in eqn 3 into eqn 30 and, as shown in the *Justification* below, obtain

$$S = \frac{U - U(0)}{T} + Nk \ln q \tag{19.34}$$

Justification 19.7

The first stage is to write

$$S = k \sum_i (n_i \ln N - n_i \ln n_i) = -k \sum_i n_i \ln \frac{n_i}{N} = -Nk \sum_i p_i \ln p_i$$

where $p_i = n_i/N$, the fraction of molecules in state i. It follows from eqn 10 that

$$\ln p_i = -\beta \varepsilon_i - \ln q$$

and therefore that

$$S = -Nk \left(-\beta \sum_i p_i \varepsilon_i - \sum_i p_i \ln q \right) = k\beta\{U - U(0)\} + Nk \ln q$$

19.10 The temperature variation of the entropy of the system shown in Fig. 19.3 (expressed here as a multiple of Nk). The entropy approaches zero as $T \rightarrow 0$, and increases without limit as $T \rightarrow \infty$.

We have used the fact that the sum over the p_i is equal to 1 and the sum over $Np_i\varepsilon_i$ is equal to $U - U(0)$ (see eqn 31). We have already established that $\beta = 1/kT$, so eqn 34 immediately follows.

Example 19.4 Calculating the entropy of a collection of oscillators

Calculate the entropy of a collection of N independent harmonic oscillators, and evaluate it using vibrational data for I_2 vapour at 25°C (Example 19.3).

Method To use eqn 34, we use the partition function for a molecule with evenly spaced vibrational energy levels, eqn 15. With the partition function available, the internal energy can be found by differentiation (as in eqn 27a), and the two expressions then combined to give S.

Answer The molecular partition function as given in eqn 15 is

$$q = \frac{1}{1 - e^{-\beta\varepsilon}}$$

The internal energy is obtained by using eqn 27a:

$$U - U(0) = -\frac{N}{q}\left(\frac{\partial q}{\partial \beta}\right)_V = \frac{N\varepsilon e^{-\beta\varepsilon}}{1 - e^{-\beta\varepsilon}} = \frac{N\varepsilon}{e^{\beta\varepsilon} - 1}$$

The entropy is therefore

$$S = Nk\left\{\frac{\beta\varepsilon}{e^{\beta\varepsilon} - 1} - \ln(1 - e^{-\beta\varepsilon})\right\}$$

This function is plotted in Fig. 19.10. For I_2 at 25°C, $\beta\varepsilon = 1.036$ (Example 19.3), so $S_m = 8.38$ J K^{-1} mol^{-1}.

Self-test 19.6 Evaluate the molar entropy of N two-level systems and plot the resulting expression. What is the entropy when the two states are equally thermally accessible?

$$[S/Nk = \beta\varepsilon/(1 + e^{\beta\varepsilon}) + \ln(1 + e^{-\beta\varepsilon}); \text{ see Fig. 19.11; } S = Nk \ln 2]$$

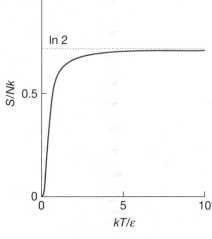

19.11 The temperature variation of the entropy of a two-level system (expressed as a multiple of Nk). As $T \rightarrow \infty$, the two states become equally populated and S approaches $Nk \ln 2$.

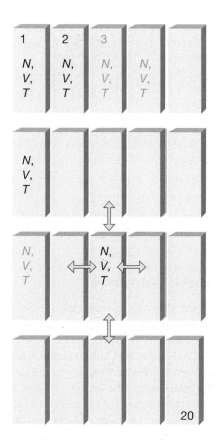

19.12 A representation of the canonical ensemble, in this case for $\tilde{N} = 20$. The individual replications of the actual system all have the same composition and volume. They are all in mutual thermal contact, and so all have the same temperature. Energy may be transferred between them as heat, and so they do not all have the same energy. The total energy $\tilde{E}$ of all 20 replications is a constant because the ensemble is isolated overall.

The canonical partition function

In this section we see how to generalize our conclusions to include systems composed of interacting molecules. We shall also see how to obtain the molecular partition function from the more general form of the partition function developed here.

19.5 The canonical ensemble

The crucial new concept we need when treating systems of interacting particles is the 'ensemble'. Like so many scientific terms, the term has basically its normal meaning of 'collection', but it has been sharpened and refined into a precise significance.

(a) The concept of ensemble

To set up an ensemble, we take a closed system of specified volume, composition, and temperature, and think of it as replicated $\tilde{N}$ times (Fig. 19.12). All the identical closed systems are regarded as being in thermal contact with one another, so they can exchange energy. The total energy of all the systems is $\tilde{E}$ and, because they are in thermal equilibrium with one another, they all have the same temperature, T. This imaginary collection of replications of the actual system with a common temperature is called the **canonical ensemble**.[4]

The important point about an ensemble is that it is a collection of imaginary replications of the system, so we are free to let the number of members be as large as we like; when appropriate, we can let $\tilde{N}$ become infinite.[5] The number of members of the ensemble in a state with energy E_i is denoted $\tilde{n}_i$, and we can speak of the configuration of the ensemble (by analogy with the configuration of the system used in Section 19.1) and its weight, $\tilde{W}$.

(b) Dominating configurations

Just as in Section 19.1, some of the configurations of the ensemble will be very much more probable than others. For instance, it is very unlikely that the whole of the total energy, $\tilde{E}$, will accumulate in one system. By analogy with the earlier discussion, we can anticipate that there will be a dominating configuration, and that we can evaluate the thermodynamic properties by taking the average over the ensemble using that single, most probable, configuration. In the **thermodynamic limit** of $\tilde{N} \to \infty$, this dominating configuration is overwhelmingly the most probable, and it dominates the properties of the system virtually completely.

---

4. The word 'canon' means 'according to a rule'. There are two other important ensembles. In the *microcanonical ensemble* the condition of constant temperature is replaced by the requirement that all the systems should have exactly the same energy: each system is individually isolated. In the *grand canonical ensemble* the volume and temperature of each system are the same, but they are open, which means that matter can be imagined as able to pass between the systems; the composition of each one may fluctuate, but now the chemical potential is the same in each system:

Microcanonical ensemble: N, V, E common

Canonical ensemble: N, V, T common

Grand canonical ensemble: μ, V, T common

5. Note that $\tilde{N}$ is unrelated to N, the number of molecules in the actual system; $\tilde{N}$ is the number of imaginary replications of that system.

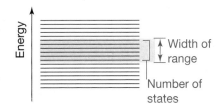

19.13 The energy density of states is the number of states in an energy range divided by the width of the range.

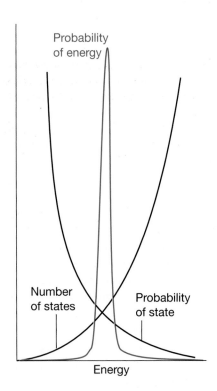

19.14 To construct the form of the distribution of members of the canonical ensemble in terms of their energies, we multiply the probability that any one is in a state of given energy, eqn 38, by the number of states corresponding to that energy (a steeply rising function). The product is a sharply peaked function at the mean energy, which shows that almost all the members of the ensemble have that energy.

The quantitative discussion follows the argument in Section 19.1 with the modification that N and n_i are replaced by $\tilde{N}$ and $\tilde{n}_i$. The weight of a configuration $\{\tilde{n}_0, \tilde{n}_1, \dots\}$ is

$$\tilde{W} = \frac{\tilde{N}!}{\tilde{n}_0!\,\tilde{n}_1!\,\dots} \tag{19.35}$$

The configuration of greatest weight, subject to the constraints that the total energy of the ensemble is constant at $\tilde{E}$ and that the total number of members is fixed at $\tilde{N}$, is given by the **canonical distribution**:

$$\frac{\tilde{n}_i}{\tilde{N}} = \frac{e^{-\beta E_i}}{Q} \qquad Q = \sum_i e^{-\beta E_i} \tag{19.36}$$

The quantity Q, which is a function of the temperature, is called the **canonical partition function**.

(c) Fluctuations from the most probable distribution

The shape of the canonical distribution in eqn 36 is only apparently an exponentially decreasing function of the energy of the system. We must appreciate that eqn 36 gives the probability of occurrence of members in a single state i of the entire system of energy E_i. There may in fact be numerous states with almost identical energies. For example, in a gas the identities of the molecules moving slowly or quickly can change without necessarily affecting the total energy. The density of states, the number of states in an energy range divided by the width of the range (Fig. 19.13), is a very sharply increasing function of energy. It follows that the probability of a member of an ensemble having a specified energy (as distinct from being in a specified state) is given by eqn 36, a sharply decreasing function, multiplied by a sharply increasing function (Fig. 19.14). Therefore, the overall distribution is a sharply peaked function. We conclude that most members of the ensemble have an energy very close to the mean value.

19.6 The thermodynamic information in the partition function

Like the molecular partition function, the canonical partition function carries all the thermodynamic information about a system. However, Q is more general than q because it does not assume that the molecules are independent. We can therefore use Q to discuss the properties of condensed phases and real gases where molecular interactions are important.

(a) The internal energy

If the total energy of the ensemble is $\tilde{E}$, and there are $\tilde{N}$ members, the average energy of a member is $E = \tilde{E}/\tilde{N}$. We use this quantity to calculate the internal energy of the system in the limit of $\tilde{N}$ (and $\tilde{E}$) approaching infinity:

$$U = U(0) + E = U(0) + \tilde{E}/\tilde{N} \qquad \text{as} \quad \tilde{N} \to \infty \tag{19.37}$$

The fraction, $\tilde{p}_i$, of members of the ensemble in a state i with energy E_i is given by the analogue of eqn 10 as

$$\tilde{p}_i = \frac{e^{-\beta E_i}}{Q} \tag{19.38}$$

It follows that the internal energy is given by

$$U = U(0) + \sum_i \tilde{p}_i E_i = U(0) + \frac{1}{Q} \sum_i E_i e^{-\beta E_i} \tag{19.39}$$

By the same argument that led to eqn 27,

$$U = U(0) - \frac{1}{Q}\left(\frac{\partial Q}{\partial \beta}\right)_V = U(0) - \left(\frac{\partial \ln Q}{\partial \beta}\right)_V \tag{19.40}$$

(b) The entropy

The total weight, $\tilde{W}$, of a configuration of the ensemble is the product of the average weight W of each member of the ensemble, $\tilde{W} = W^{\tilde{N}}$. Hence, we can calculate S from

$$S = k \ln W = k \ln \tilde{W}^{1/\tilde{N}} = \frac{k}{\tilde{N}} \ln \tilde{W} \tag{19.41}$$

It follows, by the same argument used in Section 19.4, that

$$S = \frac{U - U(0)}{T} + k \ln Q \tag{19.42}$$

19.7 Independent molecules

We shall now see how to recover the molecular partition function from the more general canonical partition function when the molecules are independent. When the molecules are independent and distinguishable (in the sense to be described), the relation between Q and q is

$$Q = q^N \tag{19.43}$$

Justification 19.8

The total energy of a collection of N independent molecules is the sum of the energies of the molecules. Therefore, we can write the total energy of a state i of the system as

$$E_i = \varepsilon_i(1) + \varepsilon_i(2) + \cdots + \varepsilon_i(N)$$

In this expression, $\varepsilon_i(1)$ is the energy of molecule 1 when the system is in the state i, $\varepsilon_i(2)$ the energy of molecule 2 when the system is in the same state i, and so on. The canonical partition function is then

$$Q = \sum_i e^{-\beta \varepsilon_i(1) - \beta \varepsilon_i(2) - \cdots - \beta \varepsilon_i(N)}$$

The sum over the states of the system can be reproduced by letting each molecule enter all its own individual states (although we meet an important proviso shortly). Therefore, instead of summing over the states i of the system, we can sum over all the individual states i of molecule 1, all the states i of molecule 2, and so on. This rewriting of the original expression leads to

$$Q = \left(\sum_i e^{-\beta \varepsilon_i}\right)\left(\sum_i e^{-\beta \varepsilon_i}\right) \cdots \left(\sum_i e^{-\beta \varepsilon_i}\right) = \left(\sum_i e^{-\beta \varepsilon_i}\right)^N = q^N$$

(a) Distinguishable and indistinguishable molecules

If all the molecules are identical and free to move through space, we cannot distinguish them and the relation $Q = q^N$ is not valid. Suppose that molecule 1 is in some state a, molecule 2 is in b, and molecule 3 is in c, then one member of the ensemble has an energy $E = \varepsilon_a + \varepsilon_b + \varepsilon_c$. This member, however, is indistinguishable from one formed by putting molecule 1 in state b, molecule 2 in state c, and molecule 3 in state a, or some other permutation. There are six such permutations in all, and $N!$ in general. In the case of indistinguishable molecules, it follows that we have counted too many states in going from the sum over system states to the sum over molecular states, so writing $Q = q^N$ overestimates the value of Q. The detailed argument is quite involved, but at all except very low temperatures it turns out that the correction factor is $1/N!$. Therefore:

(a) For distinguishable independent molecules: $Q = q^N$

(b) For indistinguishable independent molecules: $Q = q^N/N!$

$$(19.44)$$

For molecules to be indistinguishable, they must be of the same kind: an Ar atom is never indistinguishable from a Ne atom. Their identity, however, is not the only criterion. Each identical molecule in a crystal lattice, for instance, can be 'named' with a set of coordinates. Identical molecules in a lattice can therefore be treated as distinguishable because their sites are distinguishable, and we use eqn 44a. On the other hand, identical molecules in a gas are free to move to different locations, and there is no way of keeping track of the identity of a given molecule; we therefore use eqn 44b.

(b) The entropy of a monatomic gas

An important application of the previous material is the derivation (as shown in the *Justification* below) of the **Sackur–Tetrode equation** for the entropy of a monatomic gas:

$$S = nR \ln \left(\frac{e^{5/2} V}{n N_A \Lambda^3} \right) \qquad \Lambda = \frac{h}{(2\pi mkT)^{1/2}} \qquad (19.45a)$$

This equation implies that the molar entropy of a perfect gas of high molar mass is higher than that of one of low molar mass under the same conditions (because the former has more thermally accessible translational states). Because the gas is perfect, we can use the relation $V = nRT/p$ to express the entropy in terms of the pressure as

$$S = nR \ln \left(\frac{e^{5/2} kT}{p \Lambda^3} \right) \qquad (19.45b)$$

Justification 19.9

For a gas of independent molecules, Q may be replaced by $q^N/N!$, with the result that eqn 42 becomes

$$S = \frac{U - U(0)}{T} + Nk \ln q - k \ln N!$$

Because the number of molecules ($N = nN_A$) in a typical sample is large, we can use Stirling's approximation (eqn 2) to write

$$S = \frac{U - U(0)}{T} + nR \ln q - nR \ln N + nR$$

The only mode of motion for a gas of atoms is translation, and the partition function is $q = V/\Lambda^3$ (eqn 22), where Λ is the thermal wavelength. The internal energy is given by eqn 28, so the entropy is

$$S = \tfrac{3}{2}nR + nR\left(\ln\frac{V}{\Lambda^3} - \ln nN_A + 1 \right)$$

$$= nR\left(\ln e^{3/2} + \ln\frac{V}{\Lambda^3} - \ln nN_A + \ln e \right)$$

which rearranges into eqn 45.

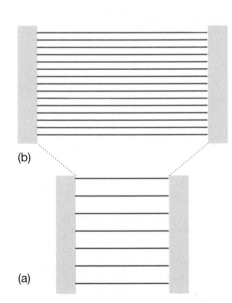

19.15 As the width of a container is increased (going from (a) to (b)), the energy levels become closer together (as $1/L^2$), and as a result more are thermally accessible at a given temperature. Consequently, the entropy of the system rises as the container expands.

Example 19.5 Using the Sackur–Tetrode equation

Calculate the standard molar entropy of gaseous argon at 25°C.

Method To calculate the molar entropy, S_m, from eqn 45b, divide both sides by n. To calculate the standard molar entropy, $S_m^\ominus$, set $p = p^\ominus$ in the expression for S_m:

$$S_m^\ominus = R\ln\left(\frac{e^{5/2}kT}{p^\ominus \Lambda^3} \right)$$

Answer The mass of an Ar atom is $m = 39.95$ u. At 25°C, its thermal wavelength is 16.0 pm (by the same kind of calculation as in Illustration 19.2). Therefore,

$$S_m^\ominus = R\ln\left\{ \frac{e^{5/2} \times (4.12 \times 10^{-21}\,\text{J})}{(10^5\,\text{N m}^{-2}) \times (1.60 \times 10^{-11}\,\text{m})^3} \right\}$$

$$= 18.6R = 155\,\text{J K}^{-1}\,\text{mol}^{-1}$$

Comment We can anticipate, on the basis of the number of accessible states for a lighter molecule, that the standard molar entropy of Ne is likely to be smaller than for Ar; its actual value is 17.60R at 298 K.

Self-test 19.7 Calculate the translational contribution to the standard molar entropy of H_2 at 25°C.

[14.2R]

The Sackur–Tetrode equation implies that, when a monatomic perfect gas expands isothermally from V_i to V_f, its entropy changes by

$$\Delta S = nR\ln(aV_f) - nR\ln(aV_i)$$

$$= nR\ln\frac{V_f}{V_i} \qquad\qquad (19.46)$$

where aV is the collection of quantities inside the logarithm of eqn 45a. This is exactly the expression we obtained by using classical thermodynamics (Example 4.1). Now, though, we see that that classical expression is in fact a consequence of the increase in the number of accessible translational states when the volume of the container is increased (Fig. 19.15).

Checklist of key ideas

☐ statistical thermodynamics, the link between molecular properties and bulk thermodynamic properties.

The distribution of molecular states

☐ population, the average number of molecules that occupy a state

☐ principle of equal *a priori* probabilities, the assumption that all possibilities for the distribution of energy are equally probable

19.1 Configurations and weights

☐ configuration, the specification of the set of populations $n_0, n_1, \ldots$ in the form $\{n_0, n_1, \ldots\}$

☐ weight, the number of ways in which a configuration can be achieved; $W = N!/n_0!n_1!n_2! \ldots$

☐ Stirling's approximation, $\ln x! \approx x \ln x - x$

☐ Boltzmann distribution, $n_i/N = e^{-\varepsilon_i/kT}/q$

☐ method of undetermined multipliers, a procedure for taking constraints into account when seeking the extremum of a function

19.2 The molecular partition function

☐ molecular partition function, $q = \sum_j e^{-\beta\varepsilon_j}$ or $q = \sum_{\text{levels } j} g_j e^{-\beta\varepsilon_j}$; $\beta = 1/kT$
  - for a two-level system, $q = 1 + e^{-\beta\varepsilon}$
  - for an evenly spaced ladder, $q = 1/(1 - e^{-\beta\varepsilon})$, $p_i = (1 - e^{-\beta\varepsilon})e^{-\beta\varepsilon_i}$
  - for a particle of mass m free to move in a region of length L: $q = L/\Lambda$
  - for a particle of mass m free to move in a region of volume V: $q = V/\Lambda^3$

☐ thermal wavelength, $\Lambda = h/(2\pi mkT)^{1/2}$

The internal energy and the entropy

19.3 The internal energy

☐ internal energy, $U(T) = U(0) + E$, $E = -(N/q)(\partial q/\partial\beta)_V = -N(\partial \ln q/\partial\beta)_V$
  - for a two-level system, $E = N\varepsilon/(1 + e^{\beta\varepsilon})$

19.4 The statistical entropy

☐ Boltzmann formula, $S = k \ln W$

☐ entropy in terms of the partition function, $S = \{U - U(0)\}/T + Nk \ln q$
  - for a two-level system, $S/Nk = \beta\varepsilon/(1 + e^{\beta\varepsilon}) + \ln(1 + e^{-\beta\varepsilon})$
  - for a uniform ladder: $S/Nk = \beta\varepsilon/(e^{\beta\varepsilon} - 1) - \ln(1 - e^{-\beta\varepsilon})$

The canonical partition function

19.5 The canonical ensemble

☐ canonical ensemble, an imaginary collection of replications of the actual system with a common temperature

☐ thermodynamic limit, the limit when $\tilde{N} \to \infty$

☐ canonical distribution, $\tilde{n}_i/\tilde{N} = e^{-\beta E_i}/\sum_j e^{-\beta E_j}$

☐ canonical partition function, $Q = \sum_i e^{-\beta E_i}$

19.6 The thermodynamic information in the partition function

☐ internal energy, $U = U(0) - (\partial \ln Q/\partial\beta)_V$

☐ entropy, $S = \{U - U(0)\}/T + k \ln Q$

19.7 Independent molecules

☐ distinguishable independent molecules: $Q = q^N$

☐ indistinguishable independent molecules: $Q = q^N/N!$

☐ Sackur–Tetrode equation, $S = nR \ln(e^{5/2}V/nN_A\Lambda^3) = nR \ln(e^{5/2}kT/p\Lambda^3)$

Further reading

Articles of general interest

C.W. David, On the Legendre transformation and the Sackur–Tetrode equation. *J. Chem. Educ.* **65**, 876 (1988).

A. Lotz, Simple statistical calculations of entropy changes. *Chem. Educator* **4**, 211 (1999).

S.A.C. McDowell, A simple derivation of the Boltzmann distribution. *J. Chem. Educ.* **76**, 1393 (1999).

P.G. Nelson, Statistical mechanical interpretation of entropy. *J. Chem. Educ.* **71**, 103 (1994).

Texts and sources of data and information

P.W. Atkins, *The second law.* Scientific American Books, New York (1984, revised 1994).

D. Chandler, *Introduction to modern statistical mechanics.* Oxford University Press (1987).

R.P.H. Gasser and W.G. Richards, *Introduction to statistical thermodynamics.* World Scientific, Singapore (1995).

C.E. Hecht, *Statistical mechanics and kinetic theory.* Dover, New York (1998).

T.L. Hill, *An introduction to statistical thermodynamics.* Dover, New York (1986).

Y.-K. Lim, *Problems and solutions on thermodynamics and statistical mechanics.* World Scientific, Singapore (1990).

A. Maczek, *Statistical thermodynamics.* Oxford Chemistry Primers, Oxford University Press (1998).

D.A. McQuarrie and J.D. Simon, *Molecular thermodynamics.* University Science Books, Sausalito (1999).

Exercises

Discussion questions

19.1 (a) Describe the essential steps in the derivation of the Boltzmann distribution.

19.1 (b) Describe the physical significance of the molecular partition function.

19.2 (a) Explain how the internal energy and entropy of a system composed of two levels vary with temperature.

19.2 (b) Explain the reasoning behind the derivation of the entropy in terms of the partition function.

19.3 (a) Enumerate the ways by which the parameter β may be identified with $1/kT$.

19.3 (b) Present arguments in favour of using β as a measure of temperature rather than T.

19.4 (a) Explain what is meant by an *ensemble* and why it is useful in statistical thermodynamics.

19.4 (b) Under what circumstances may identical particles be regarded as distinguishable?

Numerical exercises

19.5 (a) What are the relative populations of the states of a two-level system when the temperature is infinite?

19.5 (b) What is the temperature of a two-level system of energy separation equivalent to 300 cm^{-1} when the population of the upper state is one-half that of the lower state?

19.6 (a) Calculate the translational partition function at (a) 300 K and (b) 600 K of a molecule of molar mass 120 g mol^{-1} in a container of volume 2.00 cm^3.

19.6 (b) Calculate (a) the thermal wavelength, (b) the translational partition function of an Ar atom in a cubic box of side 1.00 cm at (i) 300 K and (ii) 3000 K.

19.7 (a) Calculate the ratio of the translational partition functions of D_2 and H_2 at the same temperature and volume.

19.7 (b) Calculate the ratio of the translational partition functions of xenon and helium at the same temperature and volume.

19.8 (a) A certain atom has a threefold degenerate ground level, a non-degenerate electronically excited level at 3500 cm^{-1}, and a threefold degenerate level at 4700 cm^{-1}. Calculate the partition function of these electronic states at 1900 K.

19.8 (b) A certain atom has a doubly degenerate ground level, a triply degenerate electronically excited level at 1250 cm^{-1}, and a doubly degenerate level at 1300 cm^{-1}. Calculate the partition function of these electronic states at 2000 K.

19.9 (a) Calculate the electronic contribution to the molar internal energy at 1900 K for a sample composed of the atoms specified in Exercise 19.8a.

19.9 (b) Calculate the electronic contribution to the molar internal energy at 2000 K for a sample composed of the atoms specified in Exercise 19.8b.

19.10 (a) A certain molecule has a non-degenerate excited state lying at 540 cm^{-1} above the non-degenerate ground state. At what temperature will 10 per cent of the molecules be in the upper state?

19.10 (b) A certain molecule has a doubly degenerate excited state lying at 360 cm^{-1} above the non-degenerate ground state. At what temperature will 15 per cent of the molecules be in the upper state?

19.11 (a) An electron spin can adopt either of two orientations in a magnetic field, and its energies are $\pm\mu_B B$, where μ_B is the Bohr magneton. Deduce an expression for the partition function and mean energy of the electron and sketch the variation of the functions with B. Calculate the relative populations of the spin states at (a) 4.0 K, (b) 298 K when $B = 1.0$ T.

19.11 (b) A nitrogen nucleus spin can adopt any of three orientations in a magnetic field, and its energies are $0, \pm\gamma_N\hbar B$, where γ_N is the magnetogyric ratio of the nucleus. Deduce an expression for the partition function and mean energy of the nucleus and sketch the variation of the functions with B. Calculate the relative populations of the spin states at (a) 1.0 K, (b) 298 K when $B = 20.0$ T.

19.12 (a) Consider a system of distinguishable particles having only two non-degenerate energy levels separated by an energy that is equal to the value of kT at 10 K. Calculate (a) the ratio of populations in the two states at (1) 1.0 K, (2) 10 K, and (3) 100 K, (b) the molecular partition function at 10 K, (c) the molar energy at 10 K, (d) the molar heat capacity at 10 K, (e) the molar entropy at 10 K.

19.12 (b) Consider a system of distinguishable particles having only three non-degenerate energy levels separated by an energy that is equal to the value of kT at 25.0 K. Calculate (a) the ratio of populations in the states at (1) 1.00 K, (2) 25.0 K, and (3) 100 K, (b) the molecular partition function at 25.0 K, (c) the molar energy at 25.0 K, (d) the molar heat capacity at 25.0 K, (e) the molar entropy at 25.0 K.

19.13 (a) At what temperature would the population of the first excited vibrational state of HCl be 1/e times its population of the ground state?

19.13 (b) At what temperature would the population of the first excited rotational level of HCl be 1/e times its population of the ground state?

19.14 (a) Calculate the standard molar entropy of neon gas at (a) 200 K, (b) 298.15 K.

19.14 (b) Calculate the standard molar entropy of xenon gas at (a) 100 K, (b) 298.15 K.

19.15 (a) Calculate the vibrational contribution to the entropy of Cl_2 at 500 K given that the wavenumber of the vibration is 560 cm^{-1}.

19.15 (b) Calculate the vibrational contribution to the entropy of Br_2 at 600 K given that the wavenumber of the vibration is 321 cm^{-1}.

19.16 (a) Identify the systems for which it is essential to include a factor of $1/N!$ on going from Q to q: (a) a sample of helium gas, (b) a sample of carbon monoxide gas, (c) a solid sample of carbon monoxide, (d) water vapour.

19.16 (b) Identify the systems for which it is essential to include a factor of $1/N!$ on going from Q to q: (a) a sample of carbon dioxide gas, (b) a sample of graphite, (c) a sample of diamond, (d) ice.

Problems†

Numerical problems

19.1‡ Consider a system A consisting of subsystems A_1 and A_2, for which $W_1 = 1 \times 10^{20}$ and $W_2 = 2 \times 10^{20}$. What is the number of configurations available to the combined system? Also, compute the entropies S, S_1, and S_2. What is the significance of this result?

19.2‡ Consider 1.00×10^{22} ^{4}He atoms in a box of dimensions 1.0 cm $\times 1.0$ cm $\times 1.0$ cm. Calculate the occupancy of the first excited level at 1.0 mK, 2.0 K, and 4.0 K. Do the same for ^{3}He. What conclusions might you draw from the results of your calculations?

19.3‡ By what factor does the number of available configurations increase when 100 J of energy is added to a system containing 1.00 mol of particles at constant volume at 298 K?

19.4‡ By what factor does the number of available configurations increase when 20 m^3 of air at 1.00 atm and 300 K is allowed to expand by 0.0010 per cent at constant temperature?

19.5 Explore the conditions under which the 'integral' approximation for the translational partition function is not valid by considering the translational partition function of an Ar atom in a cubic box of side 1.00 cm. Estimate the temperature at which, according to the integral approximation, $q = 10$ and evaluate the exact partition function at that temperature.

19.6 A certain atom has a doubly degenerate ground level pair and an upper level of four degenerate states at 450 cm^{-1} above the ground level. In an atomic beam study of the atoms it was observed that 30 per cent of the atoms were in the upper level, and the translational temperature of the beam was 300 K. Are the electronic states of the atoms in thermal equilibrium with the translational states?

19.7 (a) Calculate the electronic partition function of a tellurium atom at (i) 298 K, (ii) 5000 K by direct summation using the following data:

Term	Degeneracy	Wavenumber/cm^{-1}
Ground	5	0
1	1	4 707
2	3	4 751
3	5	10 559

(b) What proportion of the Te atoms are in the ground term and in the term labelled 2 at the two temperatures? (c) Calculate the electronic contribution to the standard molar entropy of gaseous Te atoms.

19.8 The four lowest electronic levels of a Ti atom are: $^3F_2$, $^3F_3$, $^3F_4$, and $^5F_1$, at 0, 170, 387, and 6557 cm^{-1}, respectively. There are many other electronic states at higher energies. The boiling point of titanium is 3287°C. What are the relative populations of these levels at the boiling point? *Hint.* The degeneracies of the levels are $2J + 1$.

19.9 The NO molecule has a doubly degenerate excited electronic level 121.1 cm^{-1} above the doubly degenerate electronic ground term. Calculate and plot the electronic partition function of NO from $T = 0$ to 1000 K. Evaluate (a) the term populations and

(b) the electronic contribution to the molar internal energy at 300 K. Calculate the electronic contribution to the molar entropy of the NO molecule at 300 K and 500 K.

19.10‡ J. Sugar and A. Musgrove (*J. Phys. Chem. Ref. Data* **22**, 1213 (1993)) have published tables of energy levels for germanium atoms and cations from Ge$^+$ to Ge^{+31}. The lowest-lying energy levels in neutral Ge are as follows:

	$^3P_0$	$^3P_1$	$^3P_2$	$^1D_2$	$^1S_0$
E/cm^{-1}	0	557.1	1410.0	7125.3	16 367.3

Calculate the electronic partition function at 298 K and 1000 K by direct summation. *Hint.* The degeneracy of a level is $2J + 1$.

19.11 Calculate, by explicit summation, the vibrational partition function and the vibrational contribution to the molar internal energy of I$_2$ molecules at (a) 100 K, (b) 298 K given that its vibrational energy levels lie at the following wavenumbers above the zero-point energy level: 0, 213.30, 425.39, 636.27, 845.93 cm^{-1}. What proportion of I$_2$ molecules are in the ground and first two excited levels at the two temperatures? Calculate the vibrational contribution to the molar entropy of I$_2$ at the two temperatures.

19.12‡ (a) The standard molar entropy of graphite at 298, 410, and 498 K is 5.69, 9.03, and 11.63 J K^{-1} mol^{-1}, respectively. If 1.00 mol C(graphite) at 298 K is surrounded by thermal insulation and placed next to 1.00 mol C(graphite) at 498 K, also insulated, how many configurations are there altogether for the combined but independent systems? (b) If the same two samples are now placed in thermal contact and brought to thermal equilibrium, the final temperature will be 410 K. (Why might the final temperature not be the average? It isn't.) How many configurations are there now in the combined system? Neglect any volume changes. (c) Demonstrate that this process is spontaneous.

Theoretical problems

19.13‡ Given that for gases the canonical partition function, Q, is related to the molecular partition function q by $Q = q^N/N!$, prove, using the expression for q and general thermodynamic relations, the perfect gas law $pV = nRT$.

19.14 A sample consisting of five molecules has a total energy 5ε. Each molecule is able to occupy states of energy $j\varepsilon$, with $j = 0, 1, 2, \ldots$. (a) Calculate the weight of the configuration in which the molecules are distributed evenly over the available states. (b) Draw up a table with columns headed by the energy of the states and write beneath them all configurations that are consistent with the total energy. Calculate the weights of each configuration and identify the most probable configurations.

19.15 A sample of nine molecules is numerically tractable but on the verge of being thermodynamically significant. Draw up a table of configurations for $N = 9$, total energy 9ε in a system with energy levels $j\varepsilon$ (as in Problem 19.14). Before evaluating the weights of the configurations, guess (by looking for the most 'exponential' distribution of populations) which of the configurations will turn out to be the most probable. Go on to calculate the weights and identify the most probable configuration.

† Problems denoted with the symbol ‡ were supplied by Charles Trapp, Carmen Giunta, and Marshall Cady.

19.16 The most probable configuration is characterized by a parameter we know as the 'temperature'. The temperatures of the system specified in Problems 19.14 and 19.15 must be such as to give a mean value of ε for the energy of each molecule and a total energy $N\varepsilon$ for the system. (a) Show that the temperature can be obtained by plotting p_j against j, where p_j is the (most probable) fraction of molecules in the state with energy $j\varepsilon$. Apply the procedure to the system in Problem 19.15. What is the temperature of the system when ε corresponds to 50 cm^{-1}? (b) Choose configurations other than the most probable, and show that the same procedure gives a worse straight line, indicating that a temperature is not well-defined for them.

19.17 A certain molecule can exist in either a non-degenerate singlet state or a triplet state (with degeneracy 3). The energy of the triplet exceeds that of the singlet by ε. Assuming that the molecules are distinguishable (localized) and independent, (a) obtain the expression for the molecular partition function. (b) Find expressions in terms of ε for the molar energy, molar heat capacity, and molar entropy of such molecules and calculate their values at $T = \varepsilon/k$.

19.18 Consider a system with energy levels $\varepsilon_j = j\varepsilon$ and N molecules. (a) Show that, if the mean energy per molecule is $a\varepsilon$, then the temperature is given by

$$\beta = \frac{1}{\varepsilon}\ln\left(1 + \frac{1}{a}\right)$$

Evaluate the temperature for a system in which the mean energy is ε, taking ε equivalent to 50 cm^{-1}. (b) Calculate the molecular partition function q for the system when its mean energy is $a\varepsilon$. (c) Show that the entropy of the system is

$$S/k = (1 + a)\ln(1 + a) - a\ln a$$

and evaluate this expression for a mean energy ε.

19.19 Consider Stirling's approximation for $\ln N!$ in the derivation of the Boltzmann distribution. What difference would it make if (a) a cruder approximation, $N! = N^N$, (b) the better approximation in the comment in Section 19.1a were used instead?

19.20‡ Consider the distribution of particles in a fluid (liquid) as a function of height in the fluid. Assume the particles have a slightly greater density than the fluid. (a) Show that the potential energy of such a particle is given by $V(h) = v(\rho - \rho_0)gh$, where ρ is the mass density of the particle, and ρ_0 is the mass density of the fluid. (b) Derive a formula for the number density of particles, N, in the fluid as a function of height. (c) Perrin (1906) found for gamboge gum grains in water with density 1.21×10^3 kg m^{-3} and volume 1.03×10^{-19} m^3 at 4°C that the number density decreased to half its value at $h = 1.23 \times 10^{-5}$ m. From this result determine the Boltzmann constant and the Avogadro constant.

Applications: to atmospheric science and astrophysics

19.21‡ Obtain the barometric formula (Problem 1.35) from the Boltzmann distribution. Recall that the potential energy of a particle at height h above the surface of the Earth is mgh. Convert the barometric formula from pressure to number density, $\mathcal{N}$. Compare the relative number densities, $\mathcal{N}(h)/\mathcal{N}(0)$, for O_2 and H_2O at $h = 8.0$ km, a typical cruising altitude for commercial aircraft.

19.22‡ Over time, planets lose their atmospheres unless they are replenished. A complete analysis of the overall process is very complicated and depends upon the radius of the planet, temperature, atmospheric composition, and other factors. Prove that the atmosphere of planets cannot be in an equilibrium state by demonstrating that the Boltzmann distribution leads to a uniform finite number density as $r \to \infty$. Hint. Recall that in a gravitational field the potential energy is $V(r) = -GMm/r$, where G is the gravitational constant, M is the mass of the planet, and m the mass of the particle.

19.23‡ Consider the electronic partition function of a perfect atomic hydrogen gas at a density of 1.99×10^{-4} kg m^{-3} and 5780 K. These are the mean conditions within the Sun's photosphere, the surface layer of the Sun, which is about 190 km thick. (a) Show that this partition function, which involves a sum over an infinite number of quantum states that are solutions to the isolated atomic hydrogen atom, is infinite. (b) Develop a theoretical argument for truncating the sum and estimate the maximum number of quantum states that contribute to the sum. (c) Calculate the equilibrium probability that an atomic hydrogen electron is in each quantum state. Are there any general implications concerning electronic states that will be observed for other atoms and molecules? Is it wise to apply these calculations in the study of the Sun's photosphere?

20 Statistical thermodynamics: the machinery

Fundamental relations

20.1 The thermodynamic functions

20.2 The molecular partition function

Using statistical thermodynamics

20.3 Mean energies

20.4 Heat capacities

20.5 Equations of state

20.6 Residual entropies

20.7 Equilibrium constants

Checklist of key ideas

Further reading

Exercises

Problems

In this chapter we apply the concepts of statistical thermodynamics to the calculation of chemically significant quantities. First, we establish the relations between thermodynamic functions and partition functions. Next, we show that the molecular partition function can be factorized into contributions from each mode of motion and establish the formulas for the partition functions for translational, rotational, and vibrational modes of motion and the contribution of electronic excitation. These contributions can be calculated from spectroscopic data. Finally, we turn to specific applications, which include the mean energies of modes of motion, the heat capacities of substances, and residual entropies. In the final section, we see how to calculate the equilibrium constant of a reaction and through that calculation understand some of the molecular features that determine the magnitudes of equilibrium constants and their variation with temperature.

A partition function is the bridge between thermodynamics, spectroscopy, and quantum mechanics. Once it is known, a partition function can be used to calculate thermodynamic functions, heat capacities, entropies, and equilibrium constants. It also sheds light on the significance of these properties.

Fundamental relations

In this section we see how to obtain any thermodynamic function once we know the partition function. Then we see how to calculate the molecular partition function, and through that the thermodynamic functions, from spectroscopic data.

20.1 The thermodynamic functions

We have already derived (in Chapter 19) the two expressions for calculating the internal energy and the entropy of a system from its canonical partition function, Q:

$$U - U(0) = -\left(\frac{\partial \ln Q}{\partial \beta}\right)_V \qquad S = \frac{U - U(0)}{T} + k \ln Q \qquad (20.1)$$

where $\beta = 1/kT$. If the molecules are independent, we can go on to make the substitutions $Q = q^N$ (for distinguishable molecules, as in a solid) or $Q = q^N/N!$ (for indistinguishable molecules, as in a gas). All the thermodynamic functions introduced in Part 1 are related to U and S, so we have a route to their calculation from Q. For later convenience, the expressions we derive here are collected at the end of the chapter.

(a) The Helmholtz energy

The Helmholtz energy, A, is defined as $A = U - TS$. This relation implies that $A(0) = U(0)$, so substitution for U and S by using eqn 1 leads to the very simple expression

$$A - A(0) = -kT \ln Q \tag{20.2}$$

(b) The pressure

It follows from classical thermodynamics, using an argument like that leading to eqn 5.10,[1] that the pressure, p, and the Helmholtz energy are related by

$$p = -\left(\frac{\partial A}{\partial V}\right)_T \tag{20.3}$$

Therefore,

$$p = kT\left(\frac{\partial \ln Q}{\partial V}\right)_T \tag{20.4}$$

This relation is entirely general, and may be used for any type of substance, including perfect gases, real gases, and liquids. Because Q is in general a function of the volume, temperature, and amount of substance, eqn 4 is an equation of state.

Example 20.1 Deriving an equation of state

Derive an expression for the pressure of a gas of independent particles.

Method We should suspect that the pressure is that given by the perfect gas law. To proceed systematically, substitute the explicit formula for Q for a gas of independent, indistinguishable molecules into eqn 4.

Answer For a gas of independent molecules, $Q = q^N/N!$ with $q = V/\Lambda^3$:

$$p = kT\left(\frac{\partial \ln Q}{\partial V}\right)_T = \frac{kT}{Q}\left(\frac{\partial Q}{\partial V}\right)_T = \frac{NkT}{q}\left(\frac{\partial q}{\partial V}\right)_T$$

$$= \frac{NkT\Lambda^3}{V} \times \frac{1}{\Lambda^3} = \frac{NkT}{V} = \frac{nRT}{V}$$

To derive this relation, we have used

$$\left(\frac{\partial q}{\partial V}\right)_T = \left(\frac{\partial(V/\Lambda^3)}{\partial V}\right)_T = \frac{1}{\Lambda^3}$$

and $NkT = nN_AkT = nRT$.

Comment The calculation shows that the equation of state of a gas of independent particles is indeed the perfect gas law. This calculation can be regarded as yet another way of deducing that $\beta = 1/kT$.

Self-test 20.1 Derive the equation of state of a sample for which $Q = q^N f/N!$, with $q = V/\Lambda^3$, where f depends on the volume. $[p = nRT/V + kT(\partial \ln f/\partial V)_T]$

---

1. Specifically, from $A = U - TS$, it follows that $dA = -p\,dV - S\,dT$; so $(\partial A/\partial V)_T = -p$.

(c) The enthalpy

At this stage we can use the expressions for U and p in the definition $H = U + pV$ to obtain an expression for the enthalpy, H, of any substance:

$$H - H(0) = -\left(\frac{\partial \ln Q}{\partial \beta}\right)_V + kTV\left(\frac{\partial \ln Q}{\partial V}\right)_T \tag{20.5}$$

We have already seen that $U - U(0) = \frac{3}{2}nRT$ for a gas of independent particles (eqn 19.28a), and have just shown that $pV = nRT$. Therefore, for such a gas,[2]

$$H - H(0) = \frac{5}{2}nRT \tag{20.6}°$$

(d) The Gibbs energy

One of the most important thermodynamic functions for chemistry is the Gibbs energy, $G = H - TS = A + pV$. We can now express this function in terms of the partition function by combining the expressions for A and p:

$$G - G(0) = -kT \ln Q + kTV\left(\frac{\partial \ln Q}{\partial V}\right)_T \tag{20.7}$$

This expression takes a simple form for a gas of independent molecules because pV in the expression $G = A + pV$ can be replaced by nRT:

$$G - G(0) = -kT \ln Q + nRT \tag{20.8}°$$

Furthermore, because $Q = q^N/N!$, and therefore $\ln Q = N \ln q - \ln N!$, it follows by using Stirling's approximation ($\ln N! \approx N \ln N - N$) that we can write

$$G - G(0) = -NkT \ln q + kT \ln N! + nRT = -nRT \ln q + kT(N \ln N - N) + nRT$$

$$= -nRT \ln \frac{q}{N} \tag{20.9}°$$

with $N = nN_A$. Now we see another interpretation of the Gibbs energy: it is proportional to the logarithm of the average number of thermally accessible states per molecule.

It will turn out to be convenient to define the **molar partition function**, $q_m = q/n$ (with units mol^{-1}), for then

$$G - G(0) = -nRT \ln \frac{q_m}{N_A} \tag{20.10}°$$

20.2 The molecular partition function

The energy of a molecule is the sum of contributions from its different modes of motion:[3]

$$\varepsilon_i = \varepsilon_i^T + \varepsilon_i^R + \varepsilon_i^V + \varepsilon_i^E \tag{20.11}$$

where T denotes translation, R rotation, V vibration, and E the electronic contribution. This separation is only approximate (except for translation) because the modes are not completely independent, but in most cases it is satisfactory. The separation of the electronic and vibrational motions, for example, is justified by the

---

2. Recall from Part 1 that we use a superscript ° on an equation number to denote a result valid only for a perfect gas.

3. The electronic contribution is not actually a 'mode of motion', but it is convenient to include it here.

Born–Oppenheimer approximation (Chapter 14), and the separation of the vibrational and rotational modes is valid to the extent that a molecule can be treated as a rigid rotor.

Given that the energy is a sum of independent contributions, the partition function factorizes into a product of contributions (recall Section 19.2b):

$$q = \sum_i e^{-\beta \varepsilon_i} = \sum_i e^{-\beta \varepsilon_i^T - \beta \varepsilon_i^R - \beta \varepsilon_i^V - \beta \varepsilon_i^E}$$

$$= \left(\sum_i e^{-\beta \varepsilon_i^T} \right) \left(\sum_i e^{-\beta \varepsilon_i^R} \right) \left(\sum_i e^{-\beta \varepsilon_i^V} \right) \left(\sum_i e^{-\beta \varepsilon_i^E} \right) = q^T q^R q^V q^E \qquad (20.12)$$

This factorization means that we can investigate each contribution separately.

(a) The translational contribution

The translational partition function of a molecule of mass m in a container of volume V was derived in Section 19.2:

$$q^T = \frac{V}{\Lambda^3} \qquad \Lambda = h \left(\frac{\beta}{2\pi m} \right)^{1/2} = \frac{h}{(2\pi m k T)^{1/2}} \qquad (20.13)$$

Notice that $q^T \to \infty$ as $T \to \infty$ because an infinite number of states becomes accessible as the temperature is raised. Even at room temperature $q^T \approx 2 \times 10^{28}$ for an O_2 molecule in a vessel of volume 100 cm^3.

The thermal wavelength, Λ, lets us judge whether the approximations that led to the expression for q^T are valid. The approximations are valid if many states are occupied, which requires V/Λ^3 to be large. That will be so if Λ is small compared with the linear dimensions of the container. For H_2 at 25°C, $\Lambda = 71$ pm, which is far smaller than any conventional container is likely to be (but comparable to pores in zeolites or cavities in clathrates). For O_2, a heavier molecule, $\Lambda = 18$ pm.

(b) The rotational contribution

As demonstrated in Example 19.1, the partition function of a nonsymmetrical (AB) linear rotor is

$$q^R = \sum_J (2J + 1) e^{-\beta h c B J (J+1)} \qquad (20.14)$$

The direct method of calculating q^R is to substitute the experimental values of the rotational energy levels into this expression and to sum the series numerically.

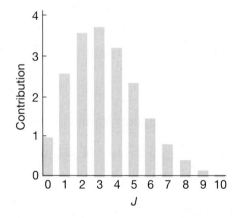

20.1 The contributions to the rotational partition function of an HCl molecule at 25°C. The vertical axis is the value of $(2J + 1)e^{-\beta h c B J(J+1)}$. Successive terms (which are proportional to the populations of the levels) pass through a maximum because the population of individual states decreases exponentially, but the degeneracy of the levels increases with J.

Example 20.2 Evaluating the rotational partition function explicitly

Evaluate the rotational partition function of $^1H^{35}Cl$ at 25°C, given that $B = 10.591$ cm^{-1}.

Method We use eqn 14 and evaluate it term by term. A useful relation is $kT/hc = 207.22$ cm^{-1} at 298.15 K. The sum is readily evaluated by using mathematical software.

Answer To show how successive terms contribute, we draw up the following table using $hcB/kT = 0.051\ 11$ (Fig. 20.1):

J	0	1	2	3	4	...	10
$(2J + 1)e^{-0.0511J(J+1)}$	1	2.71	3.68	3.79	3.24	...	0.08

The sum required by eqn 14 (the sum of the numbers in the second row of the table) is 19.9, hence $q^R = 19.9$ at this temperature. Taking J up to 50 gives $q^R = 19.902$.

Comment Notice that about ten J-levels are significantly populated but the number of populated states is larger on account of the $(2J + 1)$-fold degeneracy of each level. We shall shortly encounter the approximation that $q^R \approx kT/hcB$, which in the present case gives $q^R = 19.6$, in good agreement with the exact value and with much less work.

Self-test 20.2 Evaluate the rotational partition function for HCl at 0°C.

[18.26]

At room temperature $kT/hc \approx 200$ cm^{-1}. The rotational constants of many molecules are close to 1 cm^{-1} (Table 16.2) and often smaller.[4] It follows that many rotational levels are populated at normal temperatures. When this is the case, the partition function may be approximated by

Linear rotors: $\quad q^R = \dfrac{kT}{hcB}$ (20.15a)

Nonlinear rotors: $\quad q^R = \left(\dfrac{kT}{hc} \right)^{3/2} \left(\dfrac{\pi}{ABC} \right)^{1/2}$ (20.15b)

where A, B, and C are the rotational constants of the molecule. However, before using these expressions, read on (to eqns 17 and 18).

Justification 20.1

When many rotational states are occupied and kT is much larger than the separation between neighbouring states, the sum in the partition function can be approximated by an integral, much as we did for translational motion in *Justification 19.4*:

$$q^R = \int_0^\infty (2J + 1)e^{-\beta hcBJ(J+1)} \, dJ$$

Although this integral looks complicated, it can be evaluated without much effort by noticing that it can also be written as

$$q^R = -\frac{1}{\beta hcB} \int_0^\infty \left(\frac{d}{dJ} e^{-\beta hcBJ(J+1)} \right) dJ$$

Then, because the integral of a derivative of a function is the function itself,

$$q^R = -\frac{1}{\beta hcB} e^{-\beta hcBJ(J+1)} \Big|_0^\infty = \frac{1}{\beta hcB}$$

The calculation for a nonlinear molecule is along the same lines, but slightly trickier. First, we note that the energies of a symmetric rotor are

---

4. The very light H_2 molecule, for which $B = 60.9$ cm^{-1}, is one exception.

$$E_{J,K,M_J} = hcBJ(J+1) + hc(A-B)K^2$$

with $J = 0, 1, 2, \ldots$, $K = J, J-1, \ldots, -J$, and $M_J = J, J-1, \ldots, -J$. Instead of considering these ranges, we can cover the same values by allowing K to range from $-\infty$ to ∞, with J confined to $|K|, |K|+1, \ldots, \infty$ for each value of K. Because the energy is independent of M_J, and there are $2J+1$ values of M_J for each value of J, each value of J is $2J+1$-fold degenerate. It follows that the partition function

$$q = \sum_{J=0}^{\infty} \sum_{K=-J}^{J} \sum_{M_J=-J}^{J} e^{-E_{JKM_J}/kT}$$

can be written equivalently as

$$q = \sum_{K=-\infty}^{\infty} \sum_{J=|K|}^{\infty} (2J+1) e^{-E_{JKM_J}/kT}$$

$$= \sum_{K=-\infty}^{\infty} \sum_{J=|K|}^{\infty} (2J+1) e^{-hc\{BJ(J+1)+(A-B)K^2\}/kT}$$

$$= \sum_{K=-\infty}^{\infty} e^{-\{hc(A-B)/kT\}K^2} \sum_{J=|K|}^{\infty} (2J+1) e^{-hcBJ(J+1)/kT}$$

Now we assume that the temperature is so high that numerous states are occupied and that the sums may be approximated by integrals. Then

$$q = \int_{-\infty}^{\infty} e^{-\{hc(A-B)/kT\}K^2} \int_{|K|}^{\infty} (2J+1) e^{-hcBJ(J+1)/kT} \, dJ \, dK$$

As before, the integral over J can be recognized as the integral of the derivative of a function, which is the function itself, so

$$\int_{|K|}^{\infty} (2J+1) e^{-hcBJ(J+1)/kT} \, dJ = \int_{|K|}^{\infty} \left(-\frac{kT}{hcB} \right) \frac{d}{dJ} e^{-hcBJ(J+1)/kT} \, dJ$$

$$= \left(-\frac{kT}{hcB} \right) e^{-hcBJ(J+1)/kT} \Big|_{|K|}^{\infty} = \left(\frac{kT}{hcB} \right) e^{-hcB|K|(|K|+1)/kT}$$

$$\approx \left(\frac{kT}{hcB} \right) e^{-hcBK^2/kT}$$

In the last line we have supposed that $|K| \gg 1$ for most contributions. Now we can write

$$q = \frac{kT}{hcB} \int_{-\infty}^{\infty} e^{-\{hc(A-B)/kT\}K^2} e^{-hcBK^2/kT} \, dK$$

$$= \frac{kT}{hcB} \int_{-\infty}^{\infty} e^{-\{hcA/kT\}K^2} \, dK = \left(\frac{kT}{hcB} \right) \left(\frac{kT}{hcA} \right)^{1/2} \overbrace{\int_{-\infty}^{\infty} e^{-x^2} \, dx}^{\pi^{1/2}}$$

$$= \left(\frac{kT}{hc} \right)^{3/2} \left(\frac{\pi}{AB^2} \right)^{1/2}$$

For an asymmetric rotor, one of the As is replaced by C, to give eqn 15b.

Table 20.1* Rotational and vibrational temperatures

Molecule	Mode	θ_V/K	θ_R/K
H_2		6330	88
HCl		4300	9.4
I_2		309	0.053
CO_2	v_1	1997	0.561
	v_2	3380	
	v_3	960	

* For more values, see Table 16.2 in the *Data section*, and use $hc/k = 1.439$ K cm.

A useful way of expressing the temperature above which the rotational approximation is valid is to introduce the **rotational temperature**, $\theta_R = hcB/k$. Then 'high temperature' means $T \gg \theta_R$. Some typical values are shown in Table 20.1. The value for H_2 is abnormally high and we must be careful with the approximation for this molecule.

The general conclusion at this stage is that molecules with large moments of inertia (and hence small rotational constants and low rotational temperatures) have large rotational partition functions. The large value of q^R reflects the closeness in energy (compared with kT) of the rotational levels in large, heavy molecules, and the large number of them that are accessible at normal temperatures.

We must take care, however, not to include too many rotational states in the sum. For a homonuclear diatomic molecule or a symmetrical linear molecule (such as CO_2 or HC≡CH), a rotation through 180° results in an indistinguishable state of the molecule. Hence, the number of thermally accessible states is only half the number that can be occupied by a heteronuclear diatomic molecule, where rotation through 180° does result in a distinguishable state. Therefore, for a symmetrical linear molecule,

$$q^R = \frac{kT}{2hcB} \tag{20.16}$$

The equations for symmetrical and nonsymmetrical molecules can be combined into a single expression by introducing the **symmetry number**, σ, which is the number of indistinguishable orientations of the molecule. Then

$$q^R = \frac{kT}{\sigma hcB} \tag{20.17}$$

For a heteronuclear diatomic molecule $\sigma = 1$; for a homonuclear diatomic molecule or a symmetrical ($D_{\infty h}$) linear molecule, $\sigma = 2$.

Justification 20.2

The quantum mechanical origin of the symmetry number is the Pauli principle, which forbids the occupation of certain states. We saw in Section 16.8, for example, that H_2 may occupy rotational states with even J only if its nuclear spins are paired (*para*-hydrogen), and odd J states only if its nuclear spins are parallel (*ortho*-hydrogen). There are three states of *ortho*-H_2 to each value of J (because there are three parallel spin states of the two nuclei).

To set up the rotational partition function we note that 'ordinary' molecular hydrogen is a mixture of one part *para*-H_2 (with only its even-J rotational states occupied) and three parts *ortho*-H_2 (with only its odd-J rotational states occupied). Therefore, the average partition function per molecule is

$$q^R = \frac{1}{4}\left\{ \sum_{\text{even }J} (2J+1)e^{-\beta hcBJ(J+1)} + 3\sum_{\text{odd }J} (2J+1)e^{-\beta hcBJ(J+1)} \right\}$$

The odd-J states are more heavily weighted than the even-J states (Fig. 20.2). From the illustration we see that we would obtain approximately the same answer for the partition function (the sum of all the populations) if each J term contributed half its normal value to the sum. That is, the last equation can be approximated as

$$q^R = \frac{1}{2}\sum_J (2J+1)e^{-\beta hcBJ(J+1)}$$

and this approximation is very good when many terms contribute (at high temperatures).

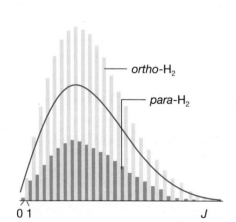

— ortho-H_2

— para-H_2

0 1 J

20.2 The values of the individual terms $(2J+1)e^{-\beta hcBJ(J+1)}$ contributing to the mean partition function of a 3:1 mixture of *ortho*- and *para*-H_2. The partition function is the sum of all these terms. At high temperatures, the sum is approximately equal to the sum of the terms over all values of J, each with a weight of $\frac{1}{2}$. This is the sum of the contributions indicated by the curve.

The same type of argument may be used for linear symmetrical molecules in which identical bosons are interchanged by rotation (such as CO_2). As pointed out in Section 16.8, if the nuclear spin of the bosons is 0, then only even-J states are admissible. Because only half the rotational states are occupied, the rotational partition function is only half the value of the sum obtained by allowing all values of J to contribute (Fig. 20.3).

The same care must be exercised for other types of symmetrical molecules, and for a nonlinear molecule we write

$$q^R = \frac{1}{\sigma}\left(\frac{kT}{hc}\right)^{3/2}\left(\frac{\pi}{ABC}\right)^{1/2} \tag{20.18}$$

Some typical values of the symmetry numbers required are given in Table 20.2. The value $\sigma(H_2O) = 2$ reflects the fact that a 180° rotation about its C_2 axis (the bisector of the H—O—H angle) interchanges two indistinguishable atoms. In NH_3, there are three indistinguishable orientations around its C_3 axis. For CH_4, any of three 120° rotations about any of its four C—H bonds leaves the molecule in an indistinguishable state, so the symmetry number is $3 \times 4 = 12$. For benzene, any of six orientations around its C_6 axis (the axis perpendicular to the plane of the molecule) leaves it apparently unchanged, as does a rotation of 180° around any of six C_2 axes in the plane of the molecule (three of which pass along each C—H bond and the remaining three pass through each C—C bond in the plane of the molecule).

A more formal way of arriving at the value of the symmetry number is to note that σ is the order (the number of elements) of the **rotational subgroup** of the molecule, the point group of the molecule with all but the identity and the rotations removed. The rotational subgroup of H_2O is $\{E, C_2\}$, so $\sigma = 2$. The rotational subgroup of NH_3 is $\{E, 2C_3\}$, so $\sigma = 3$. This recipe makes it easy to find the symmetry numbers for more complicated molecules. The rotational subgroup of CH_4 is obtained from the T character table as $\{E, 8C_3, 3C_2\}$, so $\sigma = 12$. For benzene, the rotational subgroup of D_{6h} is $\{E, 2C_6, 2C_3, C_2, 3C_2', 3C_2''\}$, so $\sigma = 12$.

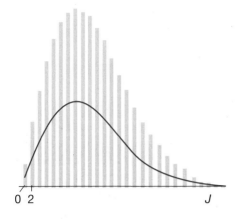

20.3 The relative populations of the rotational energy levels of CO_2. Only states with even J values are occupied. The full line shows the smoothed, averaged population of levels.

Table 20.2* Symmetry numbers

Molecule	σ
H_2O	2
NH_3	3
CH_4	12
C_6H_6	12

*For more values, see Table 16.2 in the *Data section*.

Example 20.3 Estimating a rotational partition function

Estimate the rotational partition function of ethene at 25°C given that $A = 4.828\ cm^{-1}$, $B = 1.0012\ cm^{-1}$, and $C = 0.8282\ cm^{-1}$.

Method Use eqn 18 with $kT/hc = 207.226\ cm^{-1}$. Next, identify the molecular point group (for example, by using the chart in Fig. 15.14 or the shapes in Fig. 15.15). The symmetry number is obtained by deciding on the rotational subgroup of the molecular point group, and counting the number of elements in the group.

Answer The point group of the molecule is D_{2h}. From that group's character table (see the *Data section*), the rotational subgroup of D_{2h} consists of the elements $\{E, C_{2x}, C_{2y}, C_{2z}\}$. The order of this subgroup is 4; therefore $\sigma = 4$. Then, because $ABC = 4.0033\ cm^{-3}$, it follows that $q^R = 661$.

Comment Ethene is quite a big molecule, the energy levels are close together (compared with kT at room temperature), and many are significantly populated at room temperature.

Self-test 20.3 Evaluate the rotational partition function of pyridine, C_5H_5N, at room temperature ($A = 0.2014$ cm^{-1}, $B = 0.1936$ cm^{-1}, $C = 0.0987$ cm^{-1}).

$[4.3 \times 10^4]$

(c) The vibrational contribution

The vibrational partition function of a molecule is calculated by substituting the measured vibrational energy levels into the exponentials appearing in the definition of q^V, and summing them numerically. In a polyatomic molecule each normal mode (Section 16.14) has its own partition function (provided the anharmonicities are so small that the modes are independent). The overall vibrational partition function is the product of the individual partition functions, and we can write $q^V = q^V(1)q^V(2) \ldots$, where $q^V(K)$ is the partition function for the Kth normal mode and is calculated by direct summation of the observed spectroscopic levels.

If the vibrational excitation is not too great, the harmonic approximation may be made, and the vibrational energy levels written as

$$E_v = (v + \tfrac{1}{2})hc\tilde{v} \qquad v = 0, 1, 2, \ldots \tag{20.19}$$

If, as usual, we measure energies from the zero-point level, then the permitted values are $\varepsilon_v = vhc\tilde{v}$ and the partition function is

$$q^V = \sum_v e^{-\beta vhc\tilde{v}} = \sum_v (e^{-\beta hc\tilde{v}})^v \tag{20.20}$$

(because $e^{ax} = (e^x)^a$). We met this sum in Example 19.2 (which is no accident: the ladder-like array of levels in Fig. 19.3 is exactly the same as that of a harmonic oscillator). The series can be summed in the same way, and gives

$$q^V = \frac{1}{1 - e^{-\beta hc\tilde{v}}} \tag{20.21}$$

This function is plotted in Fig. 20.4. In a polyatomic molecule, each normal mode gives rise to a partition function of this form.

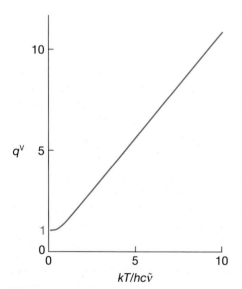

20.4 The vibrational partition function of a molecule in the harmonic approximation. Note that the partition function is linearly proportional to the temperature when the temperature is high ($T \gg \theta_V$).

Example 20.4 Calculating a vibrational partition function

The wavenumbers of the three normal modes of H_2O are 3656.7 cm^{-1}, 1594.8 cm^{-1}, and 3755.8 cm^{-1}. Evaluate the vibrational partition function at 1500 K.

Method Use eqn 21 for each mode, and then form the product of the three contributions. At 1500 K, $kT/hc = 1042.6$ cm^{-1}.

Answer We draw up the following table displaying the contributions of each mode:

Mode:	1	2	3
$\tilde{v}/$cm^{-1}	3656.7	1594.8	3755.8
$hc\tilde{v}/kT$	3.507	1.530	3.602
q^V	1.031	1.276	1.028

The overall vibrational partition function is therefore

$$q^V = 1.031 \times 1.276 \times 1.028 = 1.353$$

Comment The three normal modes of H_2O are at such high wavenumbers that even at 1500 K most of the molecules are in their vibrational ground state. However, there may be so many normal modes in a large molecule that their excitation may be significant even though each mode is not appreciably excited. For example, a nonlinear molecule containing 10 atoms has $3N - 6 = 24$ normal modes (Section 16.14a). If we assume a value of about 1.1 for the vibrational partition function of one normal mode, the overall vibrational partition function is about $q^V \approx (1.1)^{24} = 9.8$, which indicates significant vibrational excitation relative to a smaller molecule, such as H_2O.

Self-test 20.4 Repeat the calculation for CO_2, where the vibrational wavenumbers are 1388 cm^{-1}, 667.4 cm^{-1}, and 2349 cm^{-1}, the second being the doubly degenerate bending mode. [6.79]

In many molecules the vibrational wavenumbers are so great that $\beta hc\tilde{v} > 1$. For example, the lowest vibrational wavenumber of CH_4 is 1306 cm^{-1}, so $\beta hc\tilde{v} = 6.3$ at room temperature. C—H stretches normally lie in the range 2850 to 2960 cm^{-1}, so for them $\beta hc\tilde{v} \approx 14$. In these cases, $e^{-\beta hc\tilde{v}}$ in the denominator of q^V is very close to zero (for example, $e^{-6.3} = 0.002$), and the vibrational partition function for a single mode is very close to 1 ($q^V = 1.002$ when $\beta hc\tilde{v} = 6.3$), implying that only the zero-point level is significantly occupied.

Now consider the case of bonds so weak that $\beta hc\tilde{v} \ll kT$. When this condition is satisfied, the partition function may be approximated by expanding the exponential ($e^x = 1 + x + \cdots$):

$$q^V = \frac{1}{1 - (1 - \beta hc\tilde{v} + \cdots)} \tag{20.22}$$

That is, for weak bonds at high temperatures,

$$q^V = \frac{1}{\beta hc\tilde{v}} = \frac{kT}{hc\tilde{v}} \tag{20.23}$$

The temperatures for which eqn 23 is valid can be expressed in terms of the **vibrational temperature**, $\theta_V = hc\tilde{v}/k$ (Table 20.1). In terms of the vibrational temperature, 'high temperature' means $T \gg \theta_V$. The value for H_2 is abnormally high because the atoms are so light and the vibrational frequency is correspondingly high.

(d) The electronic contribution

Electronic energy separations from the ground state are usually very large, so for most cases $q^E = 1$. An important exception arises in the case of atoms and molecules having electronically degenerate ground states, in which case $q^E = g^E$, where g^E is the degeneracy of the electronic ground state. Alkali metal atoms, for example, have doubly degenerate ground states (corresponding to the two orientations of their electron spin), so $q^E = 2$.

Some atoms and molecules have low-lying electronically excited states. (At high enough temperatures, all atoms and molecules have thermally accessible excited states.) An example is NO, which has a configuration of the form $\ldots \pi^1$ (the molecule has one electron more than N_2). The orbital angular momentum may take two orientations with respect to the molecular axis (corresponding to circulation clockwise or counter-clockwise around the axis), and the spin angular momentum may also take two orientations, giving four states in all (Fig. 20.5). The energy of the two states in which the orbital and spin momenta are parallel (giving the $^2\Pi_{3/2}$ term) is slightly

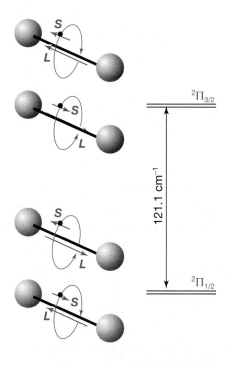

20.5 The doubly degenerate ground electronic level of NO (with the spin and orbital angular momentum around the axis in opposite directions) and the doubly degenerate first excited level (with the spin and orbital momenta parallel). The upper level is thermally accessible at room temperature.

$^2\Pi_{3/2}$

121.1 cm^{-1}

$^2\Pi_{1/2}$

20.6 The variation with temperature of the electronic partition function of an NO molecule. Note that the curve resembles that for a two-level system (Fig. 19.5), but rises from 2 (the degeneracy of the lower level) and approaches 4 (the total number of states) at high temperatures.

greater than that of the two other states in which they are antiparallel (giving the $^2\Pi_{1/2}$ term). The separation, which arises from spin–orbit coupling (Section 13.8), is only 121 cm^{-1}. Hence, at normal temperatures, all four states are thermally accessible. If we denote the energies of the two levels as $E_{1/2} = 0$ and $E_{3/2} = \varepsilon$, the partition function is

$$q^E = \sum_{\text{energy levels } j} g_j e^{-\beta\varepsilon_j} = 2 + 2\,e^{-\beta\varepsilon} \tag{20.24}$$

Figure 20.6 shows the variation of this function with temperature. At $T = 0$, $q^E = 2$, because only the doubly degenerate ground state is accessible. At high temperatures, q^E approaches 4 because all four states are accessible. At 25°C, $q^E = 3.1$.

(e) The overall partition function

The partition functions for each mode of motion of a molecule are collected in Table 20.3. The overall partition function is the product of each contribution. For a diatomic molecule with no low-lying electronically excited states and $T \gg \theta_R$,

$$q = g^E\left(\frac{V}{\Lambda^3}\right)\left(\frac{kT}{\sigma hcB}\right)\left(\frac{1}{1 - e^{-\beta hc\tilde{v}}}\right) \tag{20.25}$$

Table 20.3 Contributions to the molecular partition function*

Translation

$$q = \frac{V}{\Lambda^3} \qquad \Lambda = \left(\frac{h^2\beta}{2\pi m}\right)^{1/2}$$

$$\Lambda/\text{pm} = \frac{1749}{(T/K)^{1/2}(M/\text{g mol}^{-1})^{1/2}}$$

$$\frac{q_m^\ominus}{N_A} = \frac{kT}{p^\ominus\Lambda^3} = 2.561 \times 10^{-2}(T/K)^{5/2}(M/\text{g mol}^{-1})^{3/2}$$

Rotation
(a) Linear molecules

$$q = \frac{1}{\sigma hcB\beta} = \frac{0.6950}{\sigma} \times \frac{T/K}{(B/\text{cm}^{-1})}$$

(b) Nonlinear molecules

$$q = \frac{1}{\sigma}(1/hc\beta)^{2/3}\left(\frac{\pi}{ABC}\right)^{1/2} = \frac{1.0270}{\sigma} \times \frac{(T/K)^{3/2}}{(ABC/\text{cm}^{-3})^{1/2}}$$

Vibration

$$q = \frac{1}{1 - e^{-hc\tilde{v}\beta}} = \frac{1}{1 - e^{-a}} \qquad a = \frac{1.4388(\tilde{v}/\text{cm}^{-1})}{T/K}$$

Electronic

$$q = g_0$$

where g_0 is the degeneracy of the electronic ground state (when that is the only accessible level); at high temperatures, evaluate q explicitly.

---

* $\beta = 1/kT$. It is often useful to note that

$$\frac{hc}{k} = 1.438\,79 \text{ cm K}$$

See also inside front cover for further information.

Overall partition functions obtained in this way are approximate because they assume that the rotational levels are very close together and that the vibrational levels are harmonic. These approximations are avoided by using the energy levels identified spectroscopically and evaluating the sums explicitly.

Example 20.5 **Calculating a thermodynamic function from spectroscopic data**

Calculate the value of $G_m^\ominus - G_m^\ominus(0)$ for $H_2O(g)$ at 1500 K given that $A = 27.8778$ cm^{-1}, $B = 14.5092$ cm^{-1}, and $C = 9.2869$ cm^{-1} and the information in Example 20.4.

Method The starting point is eqn 10. For the standard value, we evaluate the translational partition function at $p^\ominus$ (that is, at 10^5 Pa exactly). The vibrational partition function was calculated in Example 20.4. Use the expressions in Table 20.3 for the other contributions.

Answer For this C_{2v} molecule, $\sigma = 2$. Because $m = 18.015$ u, $q_m^{T\ominus} = 1.706 \times 10^8$ mol^{-1}. For the vibrational contribution we have already found that $q^V = 1.352$. For the rotational contribution, $q^R = 486.7$. Therefore,

$$G_m^\ominus - G_m^\ominus(0) = -(8.3145 \text{ J K}^{-1} \text{ mol}^{-1}) \times (1500 \text{ K}) \times \ln\{(1.706 \times 10^8) \times 486.7 \times 1.352\}$$

$$= -317.3 \text{ kJ mol}^{-1}$$

Self-test 20.5 Repeat the calculation for CO_2. The vibrational data are given in Self-test 20.4; $B = 0.3902$ cm^{-1}. [-366.6 kJ mol^{-1}]

The web site contains links to online databases of atomic and molecular spectra.

Using statistical thermodynamics

We can now calculate any thermodynamic quantity from a knowledge of the energy levels of molecules: we have merged thermodynamics and spectroscopy. In this section, we indicate how to do the calculations for four important properties.

20.3 Mean energies

It is often useful to know the mean energy, $\langle \varepsilon \rangle$, of various modes of motion. When the molecular partition function can be factorized into contributions from each mode, the mean energy of each mode M is

$$\langle \varepsilon^M \rangle = -\frac{1}{q^M} \left(\frac{\partial q^M}{\partial \beta} \right)_V \qquad M = T, R, V, \text{ or } E \qquad (20.26)$$

(a) The mean translational energy

To see a pattern emerging, we consider first a one-dimensional system of length X, for which $q^T = X/\Lambda$, with $\Lambda = h(\beta/2\pi m)^{1/2}$. Then, if we note that Λ is a constant times $\beta^{1/2}$,

$$\langle \varepsilon^T \rangle = -\frac{\Lambda}{X}\left(\frac{\partial}{\partial \beta}\frac{X}{\Lambda}\right)_V = -\beta^{1/2}\frac{d}{d\beta}\left(\frac{1}{\beta^{1/2}}\right) = \frac{1}{2\beta} = \tfrac{1}{2}kT \tag{20.27}$$

For a molecule free to move in three dimensions, the analogous calculation leads to

$$\langle \varepsilon^T \rangle = \tfrac{3}{2}kT \tag{20.28}$$

Both conclusions are in agreement with the classical equipartition theorem (see Section 2.2) that the mean energy of each quadratic contribution to the energy is $\tfrac{1}{2}kT$. Furthermore, the fact that the mean energy is independent of the size of the container is consistent with the thermodynamic result that internal energy of a perfect gas is independent of its volume (Section 5.1b).

(b) The mean rotational energy

The mean rotational energy of a linear molecule is obtained from the partition function given in eqn 14. When the temperature is low ($T < \theta_R$), the series must be summed term by term, which gives

$$q^R = 1 + 3e^{-2\beta hcB} + 5e^{-6\beta hcB} + \cdots$$

Hence

$$\langle \varepsilon^R \rangle = \frac{hcB\,(6e^{-2\beta hcB} + 30e^{-6\beta hcB} + \cdots)}{1 + 3e^{-2\beta hcB} + 5e^{-6\beta hcB} + \cdots} \tag{20.29}$$

This function is plotted in Fig. 20.7. At high temperatures ($T \gg \theta_R$), q^R is given by eqn 17, and

$$\langle \varepsilon^R \rangle = -\frac{1}{q^R}\frac{dq^R}{d\beta} = -\sigma hc\beta B\frac{d}{d\beta}\frac{1}{\sigma hc\beta B} = \frac{1}{\beta} = kT \tag{20.30}$$

(q^R is independent of V, so the partial derivatives have been replaced by complete derivatives.) The high-temperature result is also in agreement with the equipartition theorem, for the classical expression for the energy of a linear rotor is $E_K = \tfrac{1}{2}I_\perp\omega_a^2 + \tfrac{1}{2}I_\perp\omega_b^2$. (There is no rotation around the line of atoms.) It follows from the equipartition theorem that the mean rotational energy is $2 \times \tfrac{1}{2}kT = kT$.

(c) The mean vibrational energy

The vibrational partition function in the harmonic approximation is given in eqn 21. Because q^V is independent of the volume, it follows that

$$\frac{dq^V}{d\beta} = \frac{d}{d\beta}\left(\frac{1}{1 - e^{-\beta hc\tilde{\nu}}}\right) = -\frac{hc\tilde{\nu}e^{-\beta hc\tilde{\nu}}}{(1 - e^{-\beta hc\tilde{\nu}})^2} \tag{20.31}$$

and hence that

$$\langle \varepsilon^V \rangle = \frac{hc\tilde{\nu}}{e^{\beta hc\tilde{\nu}} - 1} \tag{20.32}$$

The zero-point energy, $\tfrac{1}{2}hc\tilde{\nu}$, can be added to the right-hand side if the mean energy is to be measured from 0 rather than the lowest attainable level (the zero-point level). The variation of the mean energy with temperature is illustrated in Fig. 20.8. At high temperatures, when $T \gg \theta_V$, or $\beta hc\tilde{\nu} \ll 1$, the exponential functions can be expanded ($e^x = 1 + x + \cdots$) and all but the leading terms discarded. This approximation leads to

$$\langle \varepsilon^V \rangle = \frac{hc\tilde{\nu}}{(1 + \beta hc\tilde{\nu} + \cdots) - 1} \approx \frac{1}{\beta} = kT \tag{20.33}$$

20.7 The mean rotational energy of a nonsymmetrical linear rotor as a function of temperature. At high temperatures ($T \gg \theta_R$), the energy is linearly proportional to the temperature, in accord with the equipartition theorem.

20.8 The mean vibrational energy of a molecule in the harmonic approximation as a function of temperature. At high temperatures ($T \gg \theta_V$), the energy is linearly proportional to the temperature, in accord with the equipartition theorem.

This result is in agreement with the value predicted by the classical equipartition theorem, because the energy of a one-dimensional oscillator is $E = \frac{1}{2}mv_x^2 + \frac{1}{2}kx^2$ and the mean energy of each quadratic term is $\frac{1}{2}kT$.

20.4 Heat capacities

The constant-volume heat capacity is defined as $C_V = (\partial U/\partial T)_V$. The derivative with respect to T is converted into a derivative with respect to β by using

$$\frac{d}{dT} = \frac{d\beta}{dT}\frac{d}{d\beta} = -\frac{1}{kT^2}\frac{d}{d\beta} = -k\beta^2\frac{d}{d\beta} \tag{20.34}$$

It follows that

$$C_V = -k\beta^2\left(\frac{\partial U}{\partial \beta}\right)_V \tag{20.35}$$

Because the internal energy of a perfect gas is a sum of contributions, the heat capacity is also a sum of contributions from each mode. The contribution of mode M is

$$C_V^M = N\left(\frac{\partial \langle \varepsilon^M \rangle}{\partial T}\right)_V = -Nk\beta^2\left(\frac{\partial \langle \varepsilon^M \rangle}{\partial \beta}\right)_V \tag{20.36}$$

(a) The individual contributions

The temperature is always high enough (provided the gas is above its condensation temperature) for the mean translational energy to be $\frac{3}{2}kT$, the equipartition value. Therefore, the molar constant-volume heat capacity is

$$C_{V,m}^T = N_A\frac{d(\frac{3}{2}kT)}{dT} = \frac{3}{2}R \tag{20.37}$$

Translation is the only mode of motion for a monatomic gas, so for such a gas $C_{V,m} = \frac{3}{2}R = 12.47\ \mathrm{J\ K^{-1}\ mol^{-1}}$. This result is very reliable: helium, for example, has this value over a range of 2000 K. We saw in Section 3.3a that $C_{p,m} - C_{V,m} = R$, so for a monatomic perfect gas $C_{p,m} = \frac{5}{2}R$, and therefore

$$\gamma = \frac{C_p}{C_V} = \frac{5}{3} \tag{20.38}°$$

When the temperature is high enough for the rotations of the molecules to be highly excited (when $T \gg \theta_R$), we can use the equipartition value kT for the mean rotational energy (for a linear rotor) to obtain $C_{V,m} = R$. For nonlinear molecules, the mean rotational energy rises to $\frac{3}{2}kT$, so the molar rotational heat capacity rises to $\frac{3}{2}R$ when $T \gg \theta_R$. Only the lowest rotational state is occupied when the temperature is very low, and then rotation does not contribute to the heat capacity. We can calculate the rotational heat capacity at intermediate temperatures by differentiating the equation for the mean rotational energy (eqn 29). The resulting (untidy) expression, which is plotted in Fig. 20.9, shows that the contribution rises from zero (when $T = 0$) to the equipartition value (when $T \gg \theta_R$). Because the translational contribution is always present, we can expect the molar heat capacity of a gas of diatomic molecules ($C_{V,m}^T + C_{V,m}^R$) to rise from $\frac{3}{2}R$ to $\frac{5}{2}R$ as the temperature is increased above θ_R.

Molecular vibrations contribute to the heat capacity, but only when the temperature is high enough for them to be significantly excited. The equipartition mean energy is kT for each mode, so the maximum contribution to the molar heat capacity

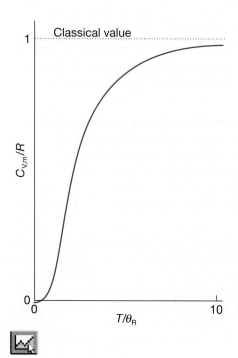

20.9 The temperature dependence of the rotational contribution to the heat capacity of a linear molecule.

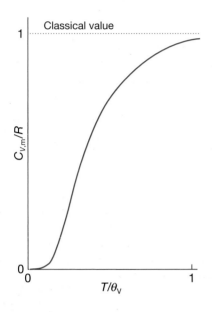

20.10 The temperature dependence of the vibrational heat capacity of a molecule in the harmonic approximation calculated by using eqn 39. Note that the heat capacity is within 10 per cent of its classical value for temperatures greater than θ_V.

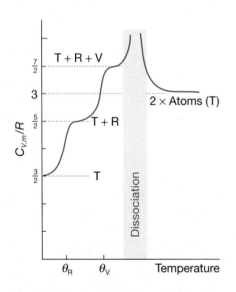

20.11 The general features of the temperature dependence of the heat capacity of diatomic molecules are as shown here. Each mode becomes active when its characteristic temperature is exceeded. The heat capacity becomes very large when the molecule dissociates because the energy is used to cause dissociation and not to raise the temperature. Then it falls back to the translation-only value of the atoms.

is R. However, it is very unusual for the vibrations to be so highly excited that equipartition is valid, and it is more appropriate to use the full expression for the vibrational heat capacity, which is obtained by differentiating eqn 32:

$$C_{V,m}^V = Rf^2 \qquad f = \frac{\theta_V}{T}\left(\frac{e^{-\theta_V/2T}}{1-e^{-\theta_V/T}}\right) \tag{20.39}$$

where $\theta_V = hc\tilde{v}/k$ is the vibrational temperature. The curve in Fig. 20.10 shows how the vibrational heat capacity depends on temperature. Note that, even when the temperature is only slightly above the vibrational temperature, the heat capacity is close to its equipartition value.[5]

(b) The overall heat capacity

The total heat capacity of a molecular substance is the sum of each contribution (Fig. 20.11). When equipartition is valid (when the temperature is well above the characteristic temperature of the mode, $T \gg \theta_M$) we can estimate the heat capacity by counting the numbers of modes that are active. In gases, all three translational modes are always active and contribute $\frac{3}{2}R$ to the molar heat capacity. If we denote the number of active rotational modes by v_R^* (so for most molecules at normal temperatures $v_R^* = 2$ for linear molecules, and 3 for nonlinear molecules), then the rotational contribution is $\frac{1}{2}v_R^*R$. If the temperature is high enough for v_V^* vibrational modes to be active, the vibrational contribution to the molar heat capacity is v_V^*R. In most cases $v_V^* \approx 0$. It follows that the total molar heat capacity is

$$C_{V,m} = \tfrac{1}{2}(3 + v_R^* + 2v_V^*)R \tag{20.40}$$

Example 20.6 **Estimating the molar heat capacity of a gas**

Estimate the molar constant-volume heat capacity of water vapour at 100°C. Vibrational wavenumbers are given in Example 20.4; the rotational constants of an H_2O molecule are 27.9, 14.5, and 9.3 cm^{-1}.

Method We need to assess whether the rotational and vibrational modes are active by computing their characteristic temperatures from the data (to do so, use $hc/k = 1.439$ cm K).

Answer The characteristic temperatures (in round numbers) of the vibrations are 5300 K, 2300 K, and 5400 K; the vibrations are therefore not excited at 373 K. The three rotational modes have characteristic temperatures 40 K, 21 K, and 13 K, so they are fully excited, like the three translational modes. The translational contribution is $\frac{3}{2}R = 12.5$ J K^{-1} mol^{-1}. Fully excited rotations contribute a further 12.5 J K^{-1} mol^{-1}. Therefore, a value close to 25 J K^{-1} mol^{-1} is predicted.

Comment The experimental value is 26.1 J K^{-1} mol^{-1}. The discrepancy is probably due to deviations from perfect gas behaviour.

Self-test 20.6 Estimate the molar constant-volume heat capacity of gaseous I_2 at 25°C ($B = 0.037$ cm^{-1}; see Table 16.2 for more data).　　　[29 J K^{-1} mol^{-1}]

---

5. Equation 39 is essentially the same as the Einstein formula for the heat capacity of a solid (eqn 11.9) with θ_V the Einstein temperature, θ_E. The only difference is that vibrations can take place in three dimensions in a solid.

20.5 Equations of state

The relation between p and Q in eqn 4 is a very important route to the equations of state of real gases in terms of intermolecular forces, for the latter can be built into Q. We have already seen (Example 20.1) that the partition function for a gas of independent particles leads to the perfect gas equation of state, $pV = nRT$. Real gases differ from perfect gases in their equations of state and we saw in Section 1.4b that their equations of state may be written

$$\frac{pV_m}{RT} = 1 + \frac{B}{V_m} + \frac{C}{V_m^2} + \cdots \tag{20.41}$$

where B is the second virial coefficient and C is the third virial coefficient.

The total kinetic energy of a gas is the sum of the kinetic energies of the individual molecules. Therefore, even in a real gas the canonical partition function factorizes into a part arising from the kinetic energy, which is the same as for the perfect gas, and a factor called the **configuration integral**, Z, which depends on the intermolecular potentials. We therefore write

$$Q = \frac{Z}{\Lambda^{3N}} \tag{20.42}$$

By comparing this equation with eqn 19.44 ($Q = q^N/N!$, with $q = V/\Lambda^3$), we see that for a perfect gas

$$Z = \frac{V^N}{N!} \tag{20.43}°$$

For a real gas, Z is related to the total potential energy E_p of interaction of all the particles by

$$Z = \frac{1}{N!} \int e^{-\beta E_p} \, \mathrm{d}\mathbf{r}_1 \, \mathrm{d}\mathbf{r}_2 \cdots \mathrm{d}\mathbf{r}_N \tag{20.44}$$

The physical origin of this term is that the probability of occurrence of each arrangement of molecules possible in the sample is given by a Boltzmann distribution in which the exponent is given by the potential energy corresponding to that arrangement.

Illustration 20.1

When the molecules do not interact with one another, $E_p = 0$ and hence $e^{-\beta E_p} = 1$. Then

$$Z = \frac{1}{N!} \int \mathrm{d}\mathbf{r}_1 \, \mathrm{d}\mathbf{r}_2 \cdots \mathrm{d}\mathbf{r}_N = \frac{V^N}{N!}$$

because $\int \mathrm{d}\mathbf{r} = V$, where V is the volume of the container. This result coincides with eqn 43.

When we consider only interactions between pairs of particles the configuration integral simplifies to

$$Z = \tfrac{1}{2} \int e^{-\beta E_p} \, \mathrm{d}\mathbf{r}_1 \, \mathrm{d}\mathbf{r}_2 \tag{20.45}$$

The second virial coefficient then turns out to be

$$B = -\frac{N_A}{2V} \int f \, d\mathbf{r}_1 \, d\mathbf{r}_2 \tag{20.46}$$

The quantity f is the **Mayer f-function**: it goes to zero when the two particles are so far apart that $E_p = 0$. When the intermolecular interaction depends only on the separation r of the particles and not on their relative orientation, as in the interaction of closed-shell atoms and tetrahedral and octahedral molecules, eqn 46 simplifies to

$$B = -2\pi N_A \int_0^\infty f r^2 \, dr \qquad f = e^{-\beta E_p} - 1 \tag{20.47}$$

The integral can be evaluated (usually numerically) by substituting an expression for the intermolecular potential energy.

Intermolecular potential energies are discussed in more detail in Chapter 21, where several expressions are developed for them. At this stage, we can illustrate how eqn 47 is used by considering the **hard-sphere potential**, which is infinite when the separation of the two molecules, r, is less than or equal to a certain value σ, and is zero for greater separations. Then

$$
\begin{aligned}
e^{-\beta E_p} = 0 \qquad f = -1 \qquad &\text{when} \quad r \le \sigma \quad (\text{and} \quad E_p = \infty) \\
e^{-\beta E_p} = 1 \qquad f = 0 \qquad &\text{when} \quad r > \sigma \quad (\text{and} \quad E_p = 0)
\end{aligned}
\tag{20.48}
$$

It follows from eqn 47 that the second virial coefficient is

$$B = 2\pi N_A \int_0^\sigma r^2 \, dr = \tfrac{2}{3}\pi N_A \sigma^3 \tag{20.49}$$

This calculation of B raises the question as to whether a potential can be found which, when the virial coefficients are evaluated, gives the van der Waals equation of state. Such a potential can be found: it consists of a hard-sphere repulsive core and a long-range, shallow attractive region. A further point is that, once a second virial coefficient has been calculated for a given intermolecular potential, it is possible to calculate other thermodynamic properties that depend on the form of the potential. For example, it is possible to calculate the isothermal Joule–Thomson coefficient, μ_T (Section 3.2c), from the thermodynamic relation

$$\lim_{p \to 0} \mu_T = B - T\frac{dB}{dT} \tag{20.50}$$

and from the result calculate the Joule–Thomson coefficient itself by using eqn 3.16.

20.6 Residual entropies

Entropies may be calculated from spectroscopic data; they may also be measured experimentally (Section 4.3d). In many cases there is good agreement, but in some the experimental entropy is less than the calculated value. One possibility is that the experimental determination failed to take a phase transition into account (and a contribution of the form $\Delta_{trs}H/T_{trs}$ incorrectly omitted from the sum). Another possibility is that some disorder is present in the solid even at $T = 0$. The entropy at $T = 0$ is then greater than zero and is called the **residual entropy**.

The origin and magnitude of the residual entropy can be explained by considering a crystal composed of AB molecules, where A and B are similar atoms (such as CO, with its very small electric dipole moment). There may be so little energy difference between ...AB AB AB AB... , ...AB BA BA AB... , and other arrangements that the

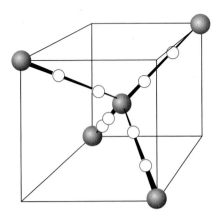

20.12 The possible locations of H atoms around a central O atom in an ice crystal are shown by the white spheres. Only one of the locations on each bond may be occupied by an atom, and two H atoms must be close to the O atom and two H atoms must be distant from it.

molecules adopt the orientations AB and BA at random in the solid. We can readily calculate the entropy arising from residual disorder by using the Boltzmann formula $S = k \ln W$. To do so, we suppose that two orientations are equally probable, and that the sample consists of N molecules. Because the same energy can be achieved in 2^N different ways (because each molecule can take either of two orientations), the total number of ways of achieving the same energy is $W = 2^N$. It follows that

$$S = k \ln 2^N = Nk \ln 2 = nR \ln 2 \tag{20.51}$$

We can therefore expect a residual molar entropy of $R \ln 2 = 5.8 \text{ J K}^{-1} \text{ mol}^{-1}$ for solids composed of molecules that can adopt either of two orientations at $T = 0$. If s orientations are possible, the residual molar entropy will be

$$S_m = R \ln s \tag{20.52}$$

An $FClO_3$ molecule, for example, can adopt four orientations with about the same energy, and the calculated residual molar entropy of $R \ln 4 = 11.5 \text{ J K}^{-1} \text{ mol}^{-1}$ is in good agreement with the experimental value ($10.1 \text{ J K}^{-1} \text{ mol}^{-1}$). For CO, the measured residual entropy is $5 \text{ J K}^{-1} \text{ mol}^{-1}$, which is close to $R \ln 2$, the value expected for a random structure of the form …CO CO OC CO OC OC… .

The residual entropy of ice is $3.4 \text{ J K}^{-1} \text{ mol}^{-1}$. This value can be explained in terms of the hydrogen-bonded structure of the solid. Each O atom is surrounded tetrahedrally by four H atoms, two of which are attached by short σ bonds, the other two being attached by long hydrogen bonds (Fig. 20.12). The randomness lies in which two of the four bonds are short, and an approximate analysis (see the *Justification* below) leads to $S_m(0) \approx R \ln \frac{3}{2} = 3.4 \text{ J K}^{-1} \text{ mol}^{-1}$, in good agreement with the experimental value.

Justification 20.3

Consider a sample of ice that consists of N H_2O molecules. Each of the $2N$ H atoms can be in one of two positions: either close to or far from an O atom (Fig. 20.13). There are therefore 2^{2N} possible arrangements. However, not all these arrangements are acceptable. Indeed, of the $2^4 = 16$ ways of arranging four H atoms around one O atom, only 6 have two short and two long OH distances and hence are acceptable. Therefore, the number of permitted arrangements is

$$W = 2^{2N} \left(\tfrac{6}{16} \right)^N = \left(\tfrac{3}{2} \right)^N$$

It then follows that the residual molar entropy is $R \ln\left(\frac{3}{2}\right)$, as stated in the text.

20.13 The six possible arrangements of H atoms in the locations identified in Fig. 20.12. Occupied locations are denoted by blue spheres and unoccupied locations by white spheres.

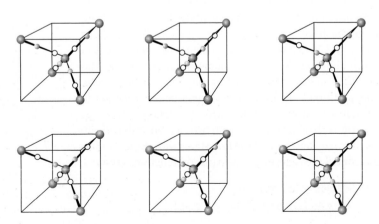

20.7 Equilibrium constants

The Gibbs energy of a gas of independent molecules is given by eqn 10 in terms of the molar partition function, $q_m = q/n$. The equilibrium constant K of a reaction is related to the standard Gibbs energy of reaction by $\Delta_r G^{\ominus} = -RT \ln K$. To calculate the equilibrium constant, we need to combine these two equations. We shall consider gas-phase reactions in which the equilibrium constant is expressed in terms of the partial pressures of the reactants and products. Approximate procedures are available for more complex systems in solution, such as the unwinding of a helical polypeptide (Box 20.1).

Box 20.1 The helix–coil transition in polypeptides

Proteins are polymers that attain well defined three-dimensional structures both in solution and in biological cells. They are *polypeptides* formed from different amino acids strung together by the *peptide link*, —CONH— (Section 22.3). Hydrogen bonds between amino acids of a polypeptide give rise to stable helical or sheet structures, which may collapse into a random coil when certain conditions are changed. For example, the synthetic polypeptide poly-γ-benzyl-glutamate is helical in a non-hydrogen-bonding solvent, but in a hydrogen-bonding solvent it forms a random coil.

The unwinding of a helix into a random coil is a *cooperative transition*, in which the polymer becomes increasingly more susceptible to structural changes once the process has begun. We examine here a model grounded in the principles of statistical thermodynamics that accounts for the cooperativity of the helix–coil transition in polypeptides.

To calculate the fraction of polypeptide molecules present as helix or coil we need to set up the partition function for the various states of the molecule. We could attempt a calculation of the electronic, vibrational, and rotational contributions to the molecular partition function of the entire polypeptide. However, the molecule is too large for this approach to be feasible. Instead, we express the partition function in terms of equilibrium constants for transitions between different states of the polypeptide chain. To illustrate the approach, consider a short polypeptide with four amino acids, each labelled with h if it contributes to a helical region or with c if it contributes to a random coil region. We let the states $hhhh$ and $cccc$ contribute terms q_0 and q_4, respectively, to the partition function q. Then we assume that each of the four states with one c amino acid contributes a term q_1. Similarly, each of the six states with two c amino acids contributes a term q_2, and each of the four states with three c amino acids contributes a term q_3. The partition function is then

$$q = q_0 + 4q_1 + 6q_2 + 4q_3 + q_4$$
$$= q_0(1 + 4q_1/q_0 + 6q_2/q_0 + 4q_3/q_0 + q_4/q_0)$$

From the Boltzmann distribution, we write $p_i = q_i/q$ and $q_i/q_0 = p_i/p_0$. Therefore, each q_i/q_0 term is a ratio of the fraction of molecules in a state with a number i of c amino acids over the fraction of molecules in a state with all h amino acids. That is, each q_i/q_0 term is a microscopic equilibrium constant, which we write as K_i. The partition function then becomes $q = q_0(1 + 4K_1 + 6K_2 + 4K_3 + K_4)$. If we let the $hhhh$ state be our reference state, so that $q_0 = 1$, it follows that

$$q = 1 + 4K_1 + 6K_2 + 4K_3 + K_4$$

Even for this relatively simple problem, the prospect of working with four different microscopic equilibrium constants is daunting. However, the problem simplifies if we make the reasonable assumption that the standard Gibbs energy associated with changing one h amino acid into one c amino acid has the same value $\Delta G^{\ominus}$ regardless of how many h or c amino acids are in the reactant or product state and regardless of where in the chain the conversion occurs. That is, $K_1 = e^{-\Delta G^{\ominus}/RT}$, $K_2 = e^{-2\Delta G^{\ominus}/RT} = K_1^2$, $K_3 = e^{-3\Delta G^{\ominus}/RT} = K_1^3$, and $K_4 = e^{-4\Delta G^{\ominus}/RT} = K_1^4$. The partition function then becomes, after using s for K_1:

$$q = 1 + 4s + 6s^2 + 4s^3 + s^4 = 1 + \sum_{i=1}^{4} N_i s^i$$

where s is often called the *stability parameter*, and N_i is the number of ways in which a state with a number i of c amino acids can be formed. This partition function has the form of the binomial expansion of $(1 + s)^4$ and N_i are the expansion coefficients. For a polypeptide with n amino acids, the partition function becomes

$$q = 1 + \sum_{i=1}^{n} N_i s^i$$

where $N_i = \dfrac{n!}{(n-i)! i!}$ are the coefficients of the binomial expansion of $(1 + s)^n$.

The binomial expansion of the expression $(1 + x)^n$ is:

$$(1 + x)^n = 1 + \sum_{i=1}^{n} N(n, i)x^i, \qquad N(n, i) = \frac{n!}{(n - i)!i!}$$

There are many ways in which a long polypeptide can change from a helix to a random coil. The simplest way to visualize the process is to use the *zipper model*, in which a conversion from h to c is allowed only if an amino acid adjacent to the one undergoing the conversion is already a c amino acid. For example, the zipper model allows a transition of the type $\dots hhhch\dots \rightarrow \dots hhhcc\dots$, but not a transition of the type $\dots hhhch\dots \rightarrow \dots hchch\dots$. The only exception to this rule is, of course, the very first conversion from h to c in a fully helical chain. It is easy to see that the zipper model does not allow non-contiguous coil regions to exist during the helix–coil transition. That is, states such as $\dots hhhccchhhhccc\dots$ are not allowed. This is a severe limitation of the model, as real polypeptide chains do not follow such strict rules during structural transitions. Nevertheless, we shall see that the zipper model does give some insight into the molecular details of the helix–coil transition.

Cooperativity is included in the zipper model by assuming that the first conversion from h to c, called the *nucleation step*, is less favourable then the remaining conversions and has microscopic equilibrium constant σs, where $\sigma \ll 1$. Each subsequent step is called a *propagation step* and has a microscopic equilibrium constant s. It is possible to show that the partition function is (*Discussion question 1*):

$$q = 1 + \sum_{i=1}^{n} N_i \sigma s^i$$

where N_i is the number of ways in which a state with a number i of c amino acids can be formed under the strictures of the zipper model. It turns out (*Discussion question 1*) that $N_i = n - i + 1$, so

$$q = 1 + \sigma(n + 1) \sum_{i=1}^{n} s^i - \sigma \sum_{i=1}^{n} is^i$$

After evaluating both geometric series (see the comment below), we write

$$q = 1 + \frac{\sigma s[s^{n+1} - (n + 1)s^n + 1]}{(s - 1)^2}$$

The geometric series evaluate to

$$\sum_{i=1}^{n} x^i = \frac{x^{n+1} - x}{x - 1}$$

$$\sum_{i=1}^{n} ix^i = \frac{x}{(x - 1)^2}[nx^{n+1} - (n + 1)x^n + 1]$$

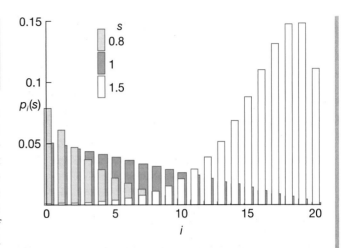

The distribution of p_i, the fraction of molecules that has a number i of c amino acids for $s = 0.8$ ($\langle i \rangle = 1.1$), 1.0 ($\langle i \rangle = 3.8$), and 1.5 ($\langle i \rangle = 15.9$), with $\sigma = 5.0 \times 10^{-3}$.

The fraction $p_i = q_i/q$ of molecules that has a number i of c amino acids is $p_i = [(n - i + 1)\sigma s^i]/q$ and the mean value of i is then $\langle i \rangle = \sum_i ip_i$. The first illustration shows the distribution of p_i for $s = 0.8$ ($\langle i \rangle = 1.1$), 1.0 ($\langle i \rangle = 3.8$), and 1.5 ($\langle i \rangle = 15.9$), with $\sigma = 5.0 \times 10^{-3}$. It is easy to see that most of the polypeptide chains remain largely helical when $s < 1$ and that most of the chains exist largely as random coils when $s > 1$. When $s = 1$, there is a more widespread distribution of length of random coil segments.

The *degree of conversion*, θ, of a polypeptide with n amino acids to a random coil is defined as

$$\theta = \frac{\langle i \rangle}{n} = \frac{\sum_{i=1}^{n} ip_i}{n} = \frac{\sum_{i=1}^{n} (n - i + 1)i\sigma s^i}{nq}$$

It is possible to show that the preceding expression may be written as (*Discussion question 1*):

$$\theta = \frac{1}{n} \frac{d}{d(\ln s)} \ln q$$

This is a general result that applies to any model of the helix–coil transition where the partition function q is expressed as a function of the stability parameter s. A more sophisticated model for the helix–coil transition must allow for helical segments to form in different regions of a long polypeptide chain, with the nascent helices being separated by shrinking coil segments. Calculations based on this more complete model, known as the *Zimm–Bragg model*, give the following expression for θ:

$$\theta = \frac{1}{2} \left(1 + \frac{(s - 1) + 2\sigma}{[(s - 1)^2 + 4s\sigma]^{1/2}} \right)$$

The second illustration shows plots of θ against s for several values of σ. The curves show the sigmoidal shape

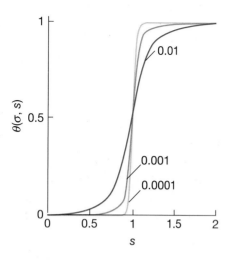

Plots of the degree of conversion θ, against s for several values of σ. The curves show the sigmoidal shape characteristic of cooperative behaviour.

the harder it is to get coil formation started, the sharper the transition from helix to coil.

Discussion questions

1. Complete some of the derivations in the preceding discussion. (a) Show that, within the tenets of the zipper model, $q = 1 + \sum_{i=1}^{n} N_i \sigma s^i$ and $N_i = n - i + 1$ is the number of ways in which an allowed state with a number i of c amino acids can be formed. (b) Using the zipper model, show that $\theta = (1/n)\mathrm{d}(\ln q)/\mathrm{d}(\ln s)$. *Hint.* As a first step, show that $\sum_i i(n - i + 1)\sigma s^i = s(\mathrm{d}q/\mathrm{d}s)$.

2. Investigate the effect of the parameter σ on the distribution of random coil segments in a polypeptide with $n = 20$ by plotting p_i against i for $s = 0.8$, 1.0, and 1.5, with $\sigma = 5.0 \times 10^{-2}$. Compare your results with those shown in the first illustration. What happens to the value of $\langle i \rangle$ as σ increases and s remains unchanged? Discuss the significance of any effect you discover.

Further reading

M. Daune, *Molecular biophysics: structures in motion*. Oxford University Press (1999).

K.E. van Holde, W.C. Johnson, and P.S. Ho, *Principles of physical biochemistry*. Prentice Hall, Upper Saddle River (1998).

characteristic of cooperative behaviour. There is a sudden surge of transition to a random coil as s passes through 1 and, the smaller the parameter σ, the greater the sharpness and hence the greater the cooperativity of the transition. That is,

(a) The relation between K and the partition function

To find an expression for the standard reaction Gibbs energy we need expressions for the standard molar Gibbs energies, $G^{\ominus}/n$, of each species. For these expressions, we need the value of the molar partition function when $p = p^{\ominus}$ (where $p^{\ominus} = 1$ bar): we denote this **standard molar partition function**, $q_m^{\ominus}$. Because only the translational component depends on the pressure, we can find $q_m^{\ominus}$ by evaluating the partition function with V replaced by $V_m^{\ominus}$, where $V_m^{\ominus} = RT/p^{\ominus}$. For a species J it follows that

$$G_m^{\ominus}(\mathrm{J}) = G_m^{\ominus}(\mathrm{J}, 0) - RT \ln \frac{q_{\mathrm{J,m}}^{\ominus}}{N_A} \tag{20.53}°$$

where $q_{\mathrm{J,m}}^{\ominus}$ is the standard molar partition function of J. By combining expressions like this one (as shown in the *Justification* below), it turns out that the equilibrium constant for the reaction

$$a\,\mathrm{A} + b\,\mathrm{B} \rightarrow c\,\mathrm{C} + d\,\mathrm{D}$$

is given by the expression

$D_0(\text{reactants})$

$D_0(\text{products})$

$\Delta_r E_0$

20.14 The definition of $\Delta_r E_0$ for the calculation of equilibrium constants.

$$K = \frac{(q_{\mathrm{C,m}}^{\ominus}/N_A)^c (q_{\mathrm{D,m}}^{\ominus}/N_A)^d}{(q_{\mathrm{A,m}}^{\ominus}/N_A)^a (q_{\mathrm{B,m}}^{\ominus}/N_A)^b} e^{-\Delta_r E_0/RT} \tag{20.54}$$

where $\Delta_r E_0$ is the difference in molar energies of the ground states of the products and reactants (this term is defined more precisely in the *Justification*), and is calculated from the bond dissociation energies of the species (Fig. 20.14).

Justification 20.4

The standard molar reaction Gibbs energy for the reaction is

$$\Delta_r G^\ominus = c G_m^\ominus(C) + d G_m^\ominus(D) - a G_m^\ominus(A) - b G_m^\ominus(B)$$

$$= c G_m^\ominus(C, 0) + d G_m^\ominus(D, 0) - a G_m^\ominus(A, 0) - b G_m^\ominus(B, 0)$$

$$- RT \left\{ c \ln \frac{q_{C,m}^\ominus}{N_A} + d \ln \frac{q_{D,m}^\ominus}{N_A} - a \ln \frac{q_{A,m}^\ominus}{N_A} - b \ln \frac{q_{B,m}^\ominus}{N_A} \right\}$$

Because $G(0) = U(0)$, the first term on the right is

$$\Delta_r E_0 = c U_m^\ominus(C, 0) + d U_m^\ominus(D, 0) - a U_m^\ominus(A, 0) - b U_m^\ominus(B, 0) \tag{20.55}$$

the reaction internal energy at $T = 0$ (a molar quantity).

Now we can write

$$\Delta_r G^\ominus = \Delta_r E_0 - RT \left\{ \ln \left(\frac{q_{C,m}^\ominus}{N_A} \right)^c + \ln \left(\frac{q_{D,m}^\ominus}{N_A} \right)^d - \ln \left(\frac{q_{A,m}^\ominus}{N_A} \right)^a - \ln \left(\frac{q_{B,m}^\ominus}{N_A} \right)^b \right\}$$

$$= \Delta_r E_0 - RT \ln \frac{(q_{C,m}^\ominus/N_A)^c (q_{D,m}^\ominus/N_A)^d}{(q_{A,m}^\ominus/N_A)^a (q_{B,m}^\ominus/N_A)^b}$$

$$= -RT \left\{ -\frac{\Delta_r E_0}{RT} + RT \ln \frac{(q_{C,m}^\ominus/N_A)^c (q_{D,m}^\ominus/N_A)^d}{(q_{A,m}^\ominus/N_A)^a (q_{B,m}^\ominus/N_A)^b} \right\}$$

At this stage we can pick out an expression for K by comparing this equation with $\Delta_r G^\ominus = -RT \ln K$, which gives

$$\ln K = -\frac{\Delta_r E^\ominus}{RT} + \ln \frac{(q_{C,m}^\ominus/N_A)^c (q_{D,m}^\ominus/N_A)^d}{(q_{A,m}^\ominus/N_A)^a (q_{B,m}^\ominus/N_A)^b}$$

This expression is easily rearranged into eqn 54 by taking antilogarithms of both sides. In terms of the stoichiometric numbers introduced in Section 9.1, we would write

$$K = \left\{ \prod_J \left(\frac{q_{J,m}^\ominus}{N_A} \right)^{\nu_J} \right\} e^{-\Delta_r E_0/RT}$$

(b) A dissociation equilibrium

We shall illustrate the application of eqn 54 to an equilibrium in which a diatomic molecule X_2 dissociates into its atoms:

$$X_2(g) \rightleftharpoons 2 X(g) \qquad K = \frac{p_X^2}{p_{X_2} p^\ominus} \tag{20.56}$$

According to eqn 54 (with $a = 1$, $b = 0$, $c = 2$, and $d = 0$):

$$K = \frac{(q_{X,m}^\ominus/N_A)^2}{q_{X_2,m}^\ominus/N_A} e^{-\Delta_r E_0/RT} = \frac{(q_{X,m}^\ominus)^2}{q_{X_2,m}^\ominus N_A} e^{-\Delta_r E_0/RT} \tag{20.57}$$

with

$$\Delta_r E_0 = 2U_m^\ominus(X, 0) - U_m^\ominus(X_2, 0) = D_0(X\!-\!X) \tag{20.58}$$

where $D_0(X\!-\!X)$ is the dissociation energy of the X—X bond. The standard molar partition functions of the atoms X are

$$q_{X,m}^\ominus = g_X\left(\frac{V_m^\ominus}{\Lambda_X^3}\right) = \frac{RT g_X}{p^\ominus \Lambda_X^3}$$

because $V_m^\ominus = RT/p^\ominus$; g_X is the degeneracy of the electronic ground state of X. The diatomic molecule X_2 also has rotational and vibrational degrees of freedom, so its standard molar partition function is

$$q_{X_2,m}^\ominus = g_{X_2}\left(\frac{V_m^\ominus}{\Lambda_{X_2}^3}\right)q_{X_2}^R q_{X_2}^V = \frac{RT g_{X_2} q_{X_2}^R q_{X_2}^V}{p^\ominus \Lambda_{X_2}^3}$$

where g_{X_2} is the degeneracy of the electronic ground state of X_2. It follows that the equilibrium constant is

$$K = \frac{kT g_X^2 \Lambda_{X_2}^3}{p^\ominus g_{X_2} q_{X_2}^R q_{X_2}^V \Lambda_X^6}\, e^{-D_0/RT} \tag{20.59}$$

where we have used $R/N_A = k$, the Boltzmann constant. All the quantities in this expression can be calculated from spectroscopic data. The Λs are defined in Table 20.3 and depend on the masses of the species and the temperature; the expressions for the rotational and vibrational partition functions are also available in Table 20.3 and depend on the rotational constant and vibrational wavenumber of the molecule.

Example 20.7 Evaluating an equilibrium constant

Evaluate the equilibrium constant for the dissociation $Na_2(g) \rightleftharpoons 2\,Na(g)$ at 1000 K from the following data: $B = 0.1547$ cm^{-1}, $\tilde{\nu} = 159.2$ cm^{-1}, $D_0 = 70.4$ kJ mol^{-1}. The Na atoms have doublet ground terms.

Method The partition functions required are specified in eqn 59. They are evaluated by using the expressions in Table 20.3. For a homonuclear diatomic molecule, $\sigma = 2$. In the evaluation of $kT/p^\ominus$ use $p^\ominus = 10^5$ Pa and 1 Pa m^3 = 1 J.

Answer The partition functions and other quantities required are as follows:

$\Lambda(Na_2) = 8.14$ pm $\Lambda(Na) = 11.5$ pm
$q^R(Na_2) = 2246$ $q^V(Na_2) = 4.885$
$g(Na) = 2$ $g(Na_2) = 1$

Then, from eqn 59

$$K = \frac{(1.38 \times 10^{-23}\ \text{J K}^{-1}) \times (1000\ \text{K}) \times 4 \times (8.14 \times 10^{-12}\ \text{m})^3}{(10^5\ \text{Pa}) \times 2246 \times 4.885 \times (1.15 \times 10^{-11}\ \text{m})^6} \times e^{-8.47}$$

$$= 2.42$$

where we have used 1 J = 1 kg m^2 s^{-2} and 1 Pa = 1 kg m^{-1} s^{-1}.

Comment For conversion to an equilibrium constant in terms of molar concentrations, K_c, use $[J] = p_J/RT$.

Self-test 20.7 Evaluate K at 1500 K.   [52]

(c) Contributions to the equilibrium constant

We are now in a position to appreciate the physical basis of equilibrium constants. To see what is involved, consider a simple $R \rightleftharpoons P$ gas-phase equilibrium (R for reactants, P for products).

Figure 20.15 shows two sets of energy levels; one set of states belongs to R, and the other belongs to P. The populations of the states are given by the Boltzmann distribution, and are independent of whether any given state happens to belong to R or to P. We can therefore imagine a single Boltzmann distribution spreading, without distinction, over the two sets of states. If the spacings of R and P are similar (as in Fig. 20.15), and P lies above R, the diagram indicates that R will dominate in the equilibrium mixture. However, if P has a high density of states (a large number of states in a given energy range, as in Fig. 20.16), then, even though its zero-point energy lies above that of R, the species P might still dominate at equilibrium.

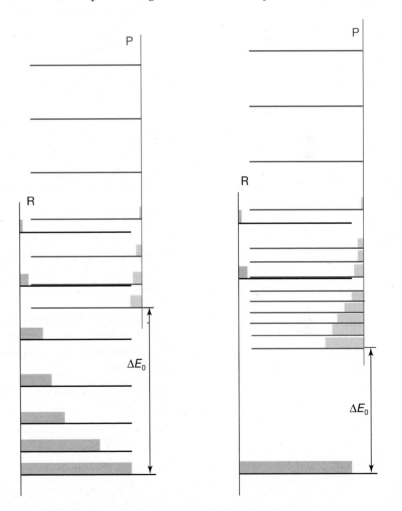

20.15 The array of R(eactants) and P(roducts) energy levels. At equilibrium all are accessible (to differing extents, depending on the temperature), and the equilibrium composition of the system reflects the overall Bolzmann distribution of populations. As ΔE_0 increases, R becomes dominant.

20.16 It is important to take into account the densities of states of the molecules. Even though P might lie well above R in energy (that is, ΔE_0 is large and positive), P might have so many states that its total population dominates in the mixture. In classical thermodynamic terms, we have to take entropies into account as well as enthalpies when considering equilibria.

It is quite easy to show (see the *Justification* below) that the ratio of numbers of R and P molecules at equilibrium is given by

$$\frac{N_P}{N_R} = \frac{q_P}{q_R} e^{-\Delta_r E_0/RT}$$

(20.60)

and therefore that the equilibrium constant for the reaction is

$$K = \frac{q_P}{q_R} e^{-\Delta_r E_0/RT}$$

(20.61)

just as would be obtained from eqn 54.[6]

Justification 20.5

The population in a state i of the composite (R, P) system is

$$n_i = \frac{N e^{-\beta \varepsilon_i}}{q}$$

where N is the total number of molecules. The total number of R molecules is the sum of these populations taken over the states belonging to R; these states we label r with energies ε_r. The total number of P molecules is the sum over the states belonging to P; these states we label P with energies ε_p' (the prime is explained in a moment):

$$N_R = \sum_r n_r = \frac{N}{q} \sum_r e^{-\beta \varepsilon_r}$$

$$N_P = \sum_p n_p = \frac{N}{q} \sum_p e^{-\beta \varepsilon_p'}$$

The sum over the states of R is its partition function, q_R, so

$$N_R = \frac{N q_R}{q}$$

The sum over the states of P is also a partition function, but the energies are measured from the ground state of the combined system, which is the ground state of R. However, because $\varepsilon_p' = \varepsilon_p + \Delta \varepsilon_0$, where $\Delta \varepsilon_0$ is the separation of zero-point energies (as in Fig. 20.16),

$$N_P = \frac{N}{q} \sum_p e^{-\beta(\varepsilon_p + \Delta \varepsilon_0)}$$

$$= \frac{N}{q} \left(\sum_p e^{-\beta \varepsilon_p} \right) e^{-\beta \Delta \varepsilon_0}$$

$$= \frac{N q_P}{q} e^{-\Delta_r E_0/RT}$$

The switch from $\Delta \varepsilon_0/k$ to $\Delta_r E_0/R$ in the last step is the conversion of molecular energies to molar energies.

The equilibrium constant of the $R \rightleftharpoons P$ reaction is proportional to the ratio of the numbers of the two types of molecule. Therefore,

---

6. For an $R \rightleftharpoons P$ equilibrium, the V factors in the partition functions cancel, so the appearance of q in place of $q^{\ominus}$ has no effect. In the case of a more general reaction, the conversion from q to $q^{\ominus}$ comes about at the stage of converting the pressures that occur in K to numbers of molecules.

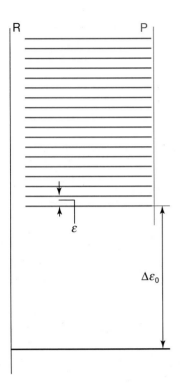

$$K = \frac{N_P}{N_R} = \frac{q_P}{q_R} e^{-\Delta_r E_0/RT}$$

as in eqn 61.

The content of eqn 61 can be seen most clearly by exaggerating the molecular features that contribute to it. We shall suppose that R has only a single accessible level, which implies that $q_R = 1$. We also suppose that P has a large number of evenly, closely spaced levels (Fig. 20.17). The partition function of P is then $q_P = kT/\varepsilon$. In this model system, the equilibrium constant is

$$K = \frac{kT}{\varepsilon} e^{-\Delta_r E_0/RT} \tag{20.62}$$

When $\Delta_r E_0$ is very large, the exponential term dominates and $K \ll 1$, which implies that very little P is present at equilibrium. When $\Delta_r E_0$ is small but still positive, K can exceed 1 because the factor kT/ε may be large enough to overcome the small size of the exponential term. The size of K then reflects the predominance of P at equilibrium on account of its high density of states. At low temperatures $K \ll 1$ and the system consists entirely of R. At high temperatures the exponential function approaches 1 and the pre-exponential factor is large. Hence P becomes dominant. We see that in this endothermic reaction (endothermic because P lies above R), a rise in temperature favours P, because its states become accessible. This behaviour is what we saw, from the outside, in Chapter 9.

The model also shows why the Gibbs energy, G, and not just the enthalpy, determines the position of equilibrium. It shows that the density of states (and hence the entropy) of each species as well as their relative energies controls the distribution of populations and hence the value of the equilibrium constant.

20.17 The model used in the text for a exploring the effects of energy separations and densities of states on equilibria. The products P can dominate provided ΔE_0 is not too large and P has an appreciable density of states.

Checklist of key ideas

Fundamental relations

20.1 The thermodynamic functions

□ internal energy, $U - U(0) = -(\partial \ln Q /\partial \beta)_V$

□ entropy, $S = \{U - U(0)\}/T + k \ln Q$

□ Helmholtz energy, $A - A(0) = -kT \ln Q$

□ pressure, $p = kT(\partial \ln Q /\partial V)_T$

□ enthalpy, $H - H(0) = -(\partial \ln Q /\partial \beta)_V + kTV(\partial \ln Q /\partial V)_T$

□ Gibbs energy, $G - G(0) = -kT \ln Q + kTV(\partial \ln Q /\partial V)_T$

□ Gibbs energy for a gas of independent particles, $G - G(0) = -nRT \ln(q_m/N_A)$

□ molar partition function, $q_m = q/n$

20.2 The molecular partition function

□ factorization, $q = q^T q^R q^V q^E$

□ translational contribution, $q^T = V/\Lambda^3$, $\Lambda = h/(2\pi mkT)^{1/2}$

□ rotational contribution: $q^R = kT/\sigma hcB$ (linear molecules), $q^R = (1/\sigma)(kT/hc)^{3/2}(\pi/ABC)^{1/2}$ (nonlinear molecules)

□ rotational temperature, $\theta_R = hcB/k$

□ symmetry number, σ, the number of indistinguishable orientations of a molecule; the order of the rotational subgroup of the molecule

□ rotational subgroup, the point group of a molecule with all but the identity and rotations removed

□ vibrational contribution, $q^V = 1/(1 - e^{-\beta hc\tilde{v}}) \approx kT/hc\tilde{v}$

□ vibrational temperature, $\theta_V = hc\tilde{v}/k$

□ electronic contribution, $q^E = g^E$ when only the g^E-fold degenerate ground state is accessible

Using statistical thermodynamics

20.3 Mean energies

□ mean energy, $\langle \varepsilon^M \rangle = -(1/q^M)(\partial q^M/\partial \beta)_V$

□ mean translational energy, $\langle \varepsilon^T \rangle = \frac{3}{2}kT$

□ mean rotational energy, $\langle \varepsilon^R \rangle = kT$ (linear molecule, $T \gg \theta_R$)

□ mean vibrational energy, $\langle \varepsilon^V \rangle = hc\tilde{v}/(e^{-hc\tilde{v}/kT} - 1) \approx kT$ ($T \gg \theta_V$)

20.4 Heat capacities

□ constant-volume heat capacity of mode M, $C_V^M = -Nk\beta^2(\partial \langle \varepsilon^M \rangle/\partial \beta)_V$

□ translational contribution, $C_{V,m}^T = \frac{3}{2}R$

□ rotational contribution, $C_{V,m}^R = R$ (linear molecules), $C_{V,m}^R = \frac{3}{2}R$ (nonlinear molecules) ($T \gg \theta_R$)

□ vibrational contribution, $C_{V,m}^V = Rf^2$, $f = (\theta_V/T)e^{-\theta_V/2T}/(1 - e^{-\theta_V/T})$

☐ overall heat capacity,
$C_{V,m} = \frac{1}{2}(3 + \nu_R^* + 2\nu_V^*)R$

20.5 Equations of state

☐ configuration integral, Z, in $Q = Z/\Lambda^{3N}$

☐ configuration integral for a perfect gas, $Z = V^N/N!$

☐ configuration integral for a real gas,
$Z = (1/N!)\int e^{-\beta E_P}\, d\mathbf{r}_1\, d\mathbf{r}_2 \ldots d\mathbf{r}_N$

☐ second virial coefficient,
$B = -(N_A/2V)\int f\, d\mathbf{r}_1\, d\mathbf{r}_2$

☐ Mayer f-function, $f = e^{-\beta E_P} - 1$

☐ second virial coefficient for isotropic potentials, $B = -2\pi N_A \int_0^\infty f r^2\, dr$

☐ hard-sphere potential, $f = -1$ when $r \le \sigma$, $f = 0$ when $r > \sigma$

☐ virial coefficient for hard-sphere potential, $B = \frac{2}{3}\pi N_A \sigma^3$

20.6 Residual entropies

☐ residual entropy, a nonzero entropy at $T = 0$ arising from molecular disorder, $S_m = R \ln s$

20.7 Equilibrium constants

☐ standard molar partition function, $q_m^\ominus$, the molar partition function when $p = p^\ominus$

☐ equilibrium constant,
$$K = \left\{ \prod_J (q_{J,m}^\ominus/N_A)^{\nu_J} \right\} e^{-\Delta_r E_0/RT}$$

☐ dissociation equilibrium,
$K = (kTg_X^2\Lambda_{X_2}^3/p^\ominus g_{X_2}q_{X_2}^R q_{X_2}^V \Lambda_X^6)e^{-D_0/RT}$

☐ equilibrium constant for model system,
$K = (kT/\varepsilon)\, e^{-\Delta_r E_0/RT}$

Further reading

Articles of general interest

C.W. David, A tractable model for studying solution thermodynamics. *J. Chem. Educ.* **64**, 484 (1987).

R.A. Alberty, The effect of a catalyst on the thermodynamic properties and partition functions of a group of isomers. *J. Chem. Educ.* **65**, 409 (1988).

M. Ross, Equations of state. In *Encyclopedia of applied physics* (ed. G.L. Trigg) **6**, 291 VCH. New York (1993).

Texts and sources of data and information

A. Ben-Naim, *Statistical thermodynamics for chemists and biologists.* Plenum, New York (1992).

D. Chandler, *Introduction to modern statistical mechanics.* Oxford University Press (1987).

N. Davidson, *Statistical thermodynamics.* McGraw-Hill, New York (1962).

C.E. Hecht, *Statistical mechanics and kinetic theory.* Dover, New York (1998).

T.L. Hill, *An introduction to statistical thermodynamics.* Dover, New York (1986).

Y.-K. Lim, *Problems and solutions on thermodynamics and statistical mechanics.* World Scientific, Singapore (1990).

K. Lucas, *Applied statistical thermodynamics.* Springer-Verlag, New York (1992).

D.A. McQuarrie and J.D. Simon, *Molecular thermodynamics.* University Science Books, Sausalito (1999).

Exercises

Discussion questions

20.1 (a) Discuss the limitations of the expressions $q^R = kT/hcB$, $q^V = kT/hc\tilde{\nu}$, and $q^E = g^E$.

20.1 (b) Explain the origin of the symmetry number.

20.2 (a) Explain the origin of residual entropy.

20.2 (b) Use concepts of statistical thermodynamics to describe the molecular features that determine the magnitudes of the constant-volume molar heat capacity of a molecular substance.

20.3 (a) Use concepts of statistical thermodynamics to describe the molecular features that lead to the equations of state of perfect and real gases.

20.3 (b) Use concepts of statistical thermodynamics to describe the molecular features that determine the magnitudes of equilibrium constants and their variation with temperature.

Numerical exercises

20.4 (a) Use the equipartition theorem to estimate the constant-volume molar heat capacity of (a) I_2, (b) CH_4, (c) C_6H_6 in the gas phase at 25°C.

20.4 (b) Use the equipartition theorem to estimate the constant-volume molar heat capacity of (a) O_3, (b) C_2H_6, (c) CO_2 in the gas phase at 25°C.

20.5 (a) Estimate the value of $\gamma = C_p/C_V$ for gaseous ammonia and methane. Do this calculation with and without the vibrational contribution to the energy. Which is closer to the expected experimental value at 25°C?

20.5 (b) Estimate the value of $\gamma = C_p/C_V$ for carbon dioxide. Do this calculation with and without the vibrational contribution to the energy. Which is closer to the expected experimental value at 25°C?

20.6 (a) Estimate the rotational partition function of HCl at (a) 25°C and (b) 250°C.

20.6 (b) Estimate the rotational partition function of O_3 at (a) 25°C and (b) 250°C.

20.7 (a) Give the symmetry number for each of the following molecules: (a) CO, (b) O_2, (c) H_2S, (d) SiH_4, and (e) $CHCl_3$.

20.7 (b) Give the symmetry number for each of the following molecules: (a) CO_2, (b) O_3, (c) SO_3, (d) SF_6, and (e) Al_2Cl_6.

20.8 (a) Calculate the rotational partition function of H_2O at 298 K from its rotational constants 27.878 cm^{-1}, 14.509 cm^{-1}, and 9.287 cm^{-1}. Above what temperature is the high-temperature approximation valid?

20.8 (b) Calculate the rotational partition function of SO_2 at 298 K from its rotational constants 2.027 36 cm^{-1}, 0.344 17 cm^{-1}, and 0.293 535 cm^{-1}. Above what temperature is the high-temperature approximation valid?

20.9 (a) From the results of Exercise 20.8a, calculate the rotational contribution to the molar entropy of gaseous water at 25°C.

20.9 (b) From the results of Exercise 20.8b, calculate the rotational contribution to the molar entropy of sulfur dioxide at 25°C.

20.10 (a) Calculate the rotational partition function of CH_4 (a) by direct summation of the energy levels at 298 K and 500 K, and (b) by the high-temperature approximation. Take B = 5.2412 cm^{-1}.

20.10 (b) Calculate the rotational partition function of CH_3CN (a) by direct summation of the energy levels at 298 K and 500 K, and (b) by the high-temperature approximation. Take A = 5.28 cm^{-1} and B = 0.307 cm^{-1}.

20.11 (a) The bond length of O_2 is 120.75 pm. Use the high-temperature approximation to calculate the rotational partition function of the molecule at 300 K.

20.11 (b) The NOF molecule is an asymmetric rotor with rotational constants 3.1752 cm^{-1}, 0.3951 cm^{-1}, and 0.3505 cm^{-1}. Calculate the rotational partition function of the molecule at (a) 25°C, (b) 100°C.

20.12 (a) Plot the molar heat capacity of a collection of harmonic oscillators as a function of T/θ_V, and predict the vibrational heat capacity of ethyne at (a) 298 K, (b) 500 K. The normal modes (and their degeneracies in parentheses) occur at wavenumbers 612(2), 729(2), 1974, 3287, and 3374 cm^{-1}.

20.12 (b) Plot the molar entropy of a collection of harmonic oscillators as a function of T/θ_V, and predict the standard molar

entropy of ethyne at (a) 298 K, (b) 500 K. For data, see the preceding exercise.

20.13 (a) A CO_2 molecule is linear, and its vibrational wavenumbers are 1388.2 cm^{-1}, 667.4 cm^{-1}, and 2349.2 cm^{-1}, the last being doubly degenerate and the others non-degenerate. The rotational constant of the molecule is 0.3902 cm^{-1}. Calculate the rotational and vibrational contributions to the molar Gibbs energy at 298 K.

20.13 (b) An O_3 molecule is angular, and its vibrational wavenumbers are 1110 cm^{-1}, 705 cm^{-1}, and 1042 cm^{-1}. The rotational constants of the molecule are 3.553 cm^{-1}, 0.4452 cm^{-1}, and 0.3948 cm^{-1}. Calculate the rotational and vibrational contributions to the molar Gibbs energy at 298 K.

20.14 (a) The ground level of Cl is $^2P_{3/2}$ and a $^2P_{1/2}$ level lies 881 cm^{-1} above it. Calculate the electronic contribution to the heat capacity of Cl atoms at (a) 500 K and (b) 900 K.

20.14 (b) The first electronically excited state of O_2 is $^1\Delta_g$ and lies 7918.1 cm^{-1} above the ground state, which is $^3\Sigma_g^-$. Calculate the electronic contribution to the molar Gibbs energy of O_2 at 400 K.

20.15 (a) The ground state of the Co^{2+} ion in $CoSO_4 \cdot 7H_2O$ may be regarded as $^4T_{9/2}$. The entropy of the solid at temperatures below 1 K is derived almost entirely from the electron spin. Estimate the molar entropy of the solid at these temperatures.

20.15 (b) Estimate the contribution of the spin to the molar entropy of a solid sample of a d-metal complex with $S = \frac{5}{2}$.

20.16 (a) Calculate the residual molar entropy of a solid in which the molecules can adopt (a) three, (b) five, (c) six orientations of equal energy at $T = 0$.

20.16 (b) Suppose that the hexagonal molecule $C_6H_nF_{6-n}$ has a residual entropy on account of the similarity of the H and F atoms. Calculate the residual for each value of n.

20.17 (a) Calculate the equilibrium constant of the reaction $I_2(g) \rightleftharpoons 2 I(g)$ at 1000 K from the following data for I_2: $\tilde{\nu}$ = 214.36 cm^{-1}, B = 0.0373 cm^{-1}, D_e = 1.5422 eV. The ground state of the I atoms is $^2P_{3/2}$, implying fourfold degeneracy.

20.17 (b) Calculate the value of K at 298 K for the gas-phase isotopic exchange reaction 2 $^{79}Br^{81}Br \rightleftharpoons {}^{79}Br^{79}Br + {}^{81}Br^{81}Br$. The Br_2 molecule has a non-degenerate ground state, with no other electronic states nearby. Base the calculation on the wavenumber of the vibration of $^{79}Br^{81}Br$, which is 323.33 cm^{-1}.

Problems[†]

Numerical problems

20.1 The NO molecule has a doubly degenerate electronic ground state and a doubly degenerate excited state at 121.1 cm^{-1}. Calculate the electronic contribution to the molar heat capacity of the molecule at (a) 50 K, (b) 298 K, and (c) 500 K.

20.2 Explore whether a magnetic field can influence the heat capacity of a paramagnetic molecule by calculating the electronic

contribution to the heat capacity of an NO_2 molecule in a magnetic field. Estimate the total constant-volume heat capacity using equipartition, and calculate the percentage change in heat capacity brought about by a 5.0 T magnetic field at (a) 50 K, (b) 298 K.

20.3 The energy levels of a CH_3 group attached to a larger fragment are given by the expression for a particle on a ring, provided the group is rotating freely. What is the high-temperature contribution

† Problems denoted with the symbol ‡ were supplied by Charles Trapp, Carmen Giunta, and Marshall Cady.

to the heat capacity and the entropy of such a freely rotating group at 25°C? The moment of inertia of CH_3 about its C_3 axis is 5.341×10^{-47} kg m^2.

20.4 Calculate the temperature dependence of the heat capacity of p-H_2 (in which only rotational states with even values of J are populated) at low temperatures on the basis that its rotational levels $J = 0$ and $J = 2$ constitute a system that resembles a two-level system except for the degeneracy of the upper level. Use $B = 60.864$ cm^{-1} and sketch the heat capacity curve. The experimental heat capacity of p-H_2 does in fact show a peak at low temperatures.

20.5 The pure rotational microwave spectrum of HCl has absorption lines at the following wavenumbers (in cm^{-1}): 21.19, 42.37, 63.56, 84.75, 105.93, 127.12, 148.31, 169.49, 190.68, 211.87, 233.06, 254.24, 275.43, 296.62, 317.80, 338.99, 360.18, 381.36, 402.55, 423.74, 444.92, 466.11, 487.30, 508.48. Calculate the rotational partition function at 25°C by direct summation.

20.6 Calculate the standard molar entropy of N_2(g) at 298 K from its rotational constant $B = 1.9987$ cm^{-1} and its vibrational wavenumber $\tilde{v} = 2358$ cm^{-1}. The thermochemical value is 192.1 J K^{-1} mol^{-1}. What does this suggest about the solid at $T = 0$?

20.7‡ J.G. Dojahn, E.C.M. Chen, and W.E. Wentworth (*J. Phys. Chem.* **100**, 9649 (1996)) characterized the potential energy curves of the ground and electronic states of homonuclear diatomic halogen anions. The ground state of F_2^- is $^2\Sigma_u^+$ with a fundamental vibrational wavenumber of 450.0 cm^{-1} and equilibrium internuclear distance of 190.0 pm. The first two excited states are at 1.609 and 1.702 eV above the ground state. Compute the standard molar entropy of F_2^- at 298 K.

20.8‡ In a spectroscopic study of buckminsterfullerene C_{60}, F. Negri, G. Orlandi, and F. Zerbetto (*J. Phys. Chem.* **100**, 10 849 (1996)) reviewed the wavenumbers of all the vibrational modes of the molecule. The wavenumber for the single A_u mode is 976 cm^{-1}; wavenumbers for the four threefold degenerate T_{1u} modes are 525, 578, 1180, and 1430 cm^{-1}; wavenumbers for the five threefold degenerate T_{2u} modes are 354, 715, 1037, 1190, and 1540 cm^{-1}; wavenumbers for six fourfold degenerate G_u modes are 345, 757, 776, 963, 1315, and 1410 cm^{-1}; and wavenumbers for the seven fivefold degenerate H_u modes are 403, 525, 667, 738, 1215, 1342, and 1566 cm^{-1}. How many modes have a vibrational temperature θ_V below 1000 K? Estimate the molar constant-volume heat capacity of C_{60} at 1000 K, counting as active all modes with θ_V below this temperature.

20.9‡ Treat carbon monoxide as a perfect gas and apply equilibrium statistical thermodynamics to the study of its properties, as specified below, in the temperature range 100–1000 K at 1 bar. $\tilde{v} = 2169.8$ cm^{-1}, $B = 1.931$ cm^{-1}, and $D_0 = 11.09$ eV; neglect anharmonicity and centrifugal distortion. (a) Examine the probability distribution of molecules over available rotational and vibrational states. (b) Explore numerically the differences, if any, between the rotational molecular partition function as calculated with the discrete energy distribution and that calculated with the classical, continuous energy distribution. (c) Calculate the individual contributions to $U_m(T) - U_m(100\text{ K})$, $C_{V,m}(T)$, and $S_m(T) - S_m(100\text{ K})$ made by the translational, rotational, and vibrational degrees of freedom.

20.10 Calculate and plot as a function of temperature, in the range 300 K to 1000 K, the equilibrium constant for the reaction CD_4(g) + HCl(g) $\rightleftharpoons$ CHD_3(g) + DCl(g) using the following data (numbers in parentheses are degeneracies): $\tilde{v}(CHD_3)$/cm^{-1} = 2993(1), 2142(1), 1003(3), 1291(2), 1036(2); $\tilde{v}(CD_4)$/cm^{-1} = 2109(1), 1092(2), 2259(3),

996(3); $\tilde{v}(HCl)$/cm^{-1} = 2991; $\tilde{v}(DCl)$/cm^{-1} = 2145; $B(HCl)$/cm^{-1} = 10.59; $B(DCl)$/cm^{-1} = 5.445; $B(CHD_3)$/cm^{-1} = 3.28; $A(CHD_3)$/cm^{-1} = 2.63; $B(CD_4)$/cm^{-1} = 2.63.

20.11 The exchange of deuterium between acid and water is an important type of equilibrium, and we can examine it using spectroscopic data on the molecules. Calculate the equilibrium constant at (a) 298 K and (b) 800 K for the gas-phase exchange reaction H_2O + DCl $\rightleftharpoons$ HDO + HCl from the following data: $\tilde{v}(H_2O)$/cm^{-1} = 3656.7, 1594.8, 3755.8; $\tilde{v}(HDO)$/cm^{-1} = 2726.7, 1402.2, 3707.5; $A(H_2O)$/cm^{-1} = 27.88; $B(H_2O)$/cm^{-1} = 14.51; $C(H_2O)$/cm^{-1} = 9.29; $A(HDO)$/cm^{-1} = 23.38; $B(HDO)$/cm^{-1} = 9.102; $C(HDO)$/cm^{-1} = 6.417; $B(HCl)$/cm^{-1} = 10.59; $B(DCl)$/cm^{-1} = 5.449; $\tilde{v}(HCl)$/cm^{-1} = 2991; $\tilde{v}(DCl)$/cm^{-1} = 2145.

Theoretical problems

20.12 Derive the Sackur–Tetrode equation for a monatomic gas confined to a two-dimensional surface, and hence derive an expression for the standard molar entropy of condensation to form a mobile surface film.

20.13‡ For H_2 at very low temperatures, only translational motion contributes to the heat capacity. At temperatures above $\theta_R = hcB/k$, the rotational contribution to the heat capacity becomes significant. At still higher temperatures, above $\theta_V = hv/k$, the vibrations contribute. At this latter temperature, dissociation of the molecule into the atoms must be considered. (a) Explain the origin of the expressions for θ_R and θ_V, and calculate their values for hydrogen. (b) Obtain an expression for the molar constant-pressure heat capacity of hydrogen at all temperatures taking into account the dissociation of hydrogen. (c) Make a plot of the molar constant-pressure heat capacity as a function of temperature in the high-temperature region where dissociation of the molecule is significant.

20.14 Derive expressions for the internal energy, heat capacity, entropy, Helmholtz energy, and Gibbs energy of a harmonic oscillator. Express the results in terms of the vibrational temperature, θ_V and plot graphs of each property against T/θ_V.

20.15‡ (a) Show that the number of molecules in any given rotational state of a linear molecule is given by $N_J = C(2J + 1)e^{-hcBJ(J+1)/kT}$, where C is a constant. (b) Use this result to derive eqn 16.46 for the value of J of the most highly populated rotational level. (c) Estimate the temperature at which the spectrum of HCl shown in Fig. 16.42 was obtained.

20.16 Although expressions like $\langle \varepsilon \rangle = -\text{d} \ln q/\text{d}\beta$ are useful for formal manipulations in statistical thermodynamics, and for expressing thermodynamic functions in neat formulas, they are sometimes more trouble than they are worth in practical applications. When presented with a table of energy levels, it is often much more convenient to evaluate the following sums directly:

$$q = \sum_j e^{-\beta\varepsilon_j} \qquad \dot{q} = \sum_j \beta\varepsilon_j e^{-\beta\varepsilon_j} \qquad \ddot{q} = \sum_j (\beta\varepsilon_j)^2 e^{-\beta\varepsilon_j}$$

(a) Derive expressions for the internal energy, heat capacity, and entropy in terms of these three functions. (b) Apply the technique to the calculation of the electronic contribution to the constant volume molar heat capacity of magnesium vapour at 5000 K using the following data:

Term	$^1S_0$	$^3P_0$	$^3P_1$	$^3P_2$	$^1P_1$	$^3S_0$
Degeneracy	1	1	3	5	3	3
$\tilde{v}$/cm^{-1}	0	21 850	21 870	21 911	35 051	41 197

20.17 Determine whether a magnetic field can influence the value of an equilibrium constant. Consider the equilibrium $I_2(g) \rightleftharpoons 2\,I(g)$ at 1000 K, and calculate the ratio of equilibrium constants $K(B)/K$, where $K(B)$ is the equilibrium constant when a magnetic field B is present and removes the degeneracy of the four states of the $^2P_{3/2}$ level. Data on the species are given in Exercise 20.17a. The electronic g-value of the atoms is $\frac{4}{3}$. Calculate the field required to change the equilibrium constant by 1 per cent.

20.18 The heat capacity ratio of a gas determines the speed of sound in it through the formula

$$c_s = \left(\frac{\gamma RT}{M} \right)^{1/2}$$

where $\gamma = C_p/C_V$ and M is the molar mass of the gas. Deduce an expression for the speed of sound in a perfect gas of (a) diatomic, (b) linear triatomic, (c) nonlinear triatomic molecules at high temperatures (with translation and rotation active). Estimate the speed of sound in air at 25°C.

Applications: to biology, materials science, environmental science, and astrophysics

20.19 An average human DNA molecule has 5×10^8 binucleotides (rungs on the DNA ladder) of four different kinds. If each rung were a random choice of one of these four possibilities, what would be the residual entropy associated with this typical DNA molecule?

20.20‡ R. Viswanathan, R.W. Schmude, Jr., and K.A. Gingerich (*J. Phys. Chem.* **100**, 10 784 (1996)) studied thermodynamic properties of several boron–silicon gas-phase species experimentally and theoretically. These species can occur in the high-temperature chemical vapour deposition of silicon-based semiconductors. Among the computations they reported was computation of the Gibbs energy of BSi(g) at several temperatures based on a $^4\Sigma^-$ ground state with equilibrium internuclear distance of 190.5 pm and fundamental vibrational wavenumber of 772 cm^{-1} and a $^2P_0$ first excited level 8000 cm^{-1} above the ground level. Compute the standard molar Gibbs energy $G_m^\ominus(2000\text{ K}) - G_m^\ominus(0)$.

20.21‡ The molecule Cl_2O_2, which is believed to participate in the seasonal depletion of ozone over Antarctica, has been studied by several means. M. Birk, R.R. Friedl, E.A. Cohen, H.M. Pickett, and S.P. Sander (*J. Chem. Phys.* **91**, 6588 (1989)) report its rotational constants (actually cB) as 13 109.4, 2409.8, and 2139.7 MHz. They also report that its rotational spectrum indicates a molecule with a symmetry number of 2.19. J. Jacobs, M. Kronberg, H.S.P. Möller, and H. Willner (*J. Amer. Chem. Soc.* 1106 (1994)) report its vibrational wavenumbers as 753, 542, 310, 127, 646, and 419 cm^{-1}. Compute $G_m^\ominus(200\text{ K}) - G_m^\ominus(0)$ of Cl_2O_2.

20.22‡ J. Hutter, H.P. Lüthi, and F. Diederich (*J. Amer. Chem. Soc.* **116**, 750 (1994)) examined the geometric and vibrational structure of several carbon molecules of formula C_n. Given that the ground state of C_3, a molecule found in interstellar space and in flames, is an angular singlet with moments of inertia 39.340, 39.032, and 0.3082 u Å^2 (where 1 Å = 10^{-10} m) and with vibrational wavenumbers of 63.4, 1224.5, and 2040 cm^{-1}, compute $G_m^\ominus(10.00\text{ K}) - G_m^\ominus(0)$ and $G_m^\ominus(1000\text{ K}) - G_m^\ominus(0)$ for C_3.

21

Molecular interactions

Electric properties of molecules

21.1 Electric dipole moments

21.2 Polarizabilities

21.3 Relative permittivities

21.4 Refractive index

Interactions between molecules

21.5 Interactions between dipoles

21.6 Repulsive and total interactions

21.7 Molecular interactions in gases

21.8 Molecular interactions in liquids

Checklist of key ideas

Further reading

Exercises

Problems

In this chapter we examine molecular interactions in gases and liquids and interpret them in terms of electric properties of molecules, such as electric dipole moments and polarizabilities, and some related properties, which include refractive index and optical activity. All these properties reflect the degree to which the nuclei of atoms exert control over the electrons in a molecule, either by causing electrons to accumulate in particular regions, or by permitting them to respond more or less strongly to the effects of external electric fields.

Molecular interactions are responsible for the unique properties of substances as simple as water and as complex as polymers. We begin our examination of molecular interactions by describing the electric properties of molecules, which may be interpreted in terms of concepts in electronic structure introduced in Chapter 14. We shall see that small imbalances of charge distributions in molecules allow them to interact with one another and with externally applied fields. One result of this interaction is the cohesion of molecules to form the bulk phases of matter. These interactions are also important for understanding the shapes adopted by biological and synthetic macromolecules, as we shall see in Chapter 22. The interaction between ions is treated in Chapter 10 (for solutions) and Chapter 23 (for solids).

Electric properties of molecules

Many of the electric properties of molecules can be traced to the competing influences of nuclei with different charges or the competition between the control exercised by a nucleus and the influence of an externally applied field. The former competition may result in an electric dipole moment. The latter may result in properties such as refractive index and optical activity.

21.1 Electric dipole moments

An **electric dipole** consists of two electric charges $+q$ and $-q$ separated by a distance R. This arrangement of charges is represented by a vector, the **electric dipole moment**, $\boldsymbol{\mu}$, that points from the negative charge to the positive charge (**1**).[1] The

---

1. In elementary chemistry, an electric dipole moment is represented by the arrow $+\rightarrow$ added to the Lewis structure for the molecule, with the + marking the positive end. Note that the direction of the arrow is opposite to that of $\boldsymbol{\mu}$.

1 Electric dipole

magnitude of $\boldsymbol{\mu}$ is $\mu = qR$. The magnitudes of dipole moments are still commonly reported in the non-SI unit debye, D, where[2]

$$1\,\text{D} = 3.335\,64 \times 10^{-30}\,\text{C m} \tag{21.1}$$

The dipole moment of a pair of charges $+e$ and $-e$ separated by 100 pm is 1.6×10^{-29} C m, corresponding to 4.8 D. Dipole moments of small molecules are typically about 1 D.

(a) Polar molecules

A **polar molecule** is a molecule with a permanent electric dipole moment. The permanent dipole moment stems from the partial charges on the atoms in the molecule that arise from differences in electronegativity or other features of bonding (Section 14.7). Nonpolar molecules acquire an induced dipole moment in an electric field on account of the distortion the field causes in their electronic distributions and nuclear positions; however, this induced moment is only temporary, and disappears as soon as the perturbing field is removed. Polar molecules also have their existing dipole moments temporarily modified by an applied field.

The Stark effect (Section 16.5) is used to measure the electric dipole moments of molecules for which a rotational spectrum can be observed. In many cases microwave spectroscopy cannot be used because the sample is not volatile, decomposes on vaporization, or consists of molecules that are so complex that their rotational spectra cannot be interpreted. In such cases the dipole moment may be obtained by measurements on a liquid or solid bulk sample using a method explained later.

All heteronuclear diatomic molecules are polar, and typical values of μ include 1.08 D for HCl and 0.42 D for HI (Table 21.1). A very approximate relation between the dipole moment and the difference in electronegativities of the two atoms, $\Delta\chi$, is

$$\mu/\text{D} \approx \Delta\chi \tag{21.2}$$

The more electronegative atom is normally the negative end of the dipole, but there are exceptions, particularly when antibonding orbitals are occupied. We remarked in Section 14.6 that the major contribution to an antibonding orbital is made by the atomic orbitals of the less electronegative atom. Therefore, if an antibonding orbital is occupied there may be so much electron density on the less electronegative atom that that atom has a partial negative charge. An example is CO ($\mu = 0.12$ D), in which the negative end of the dipole is on the carbon atom even though oxygen is more electronegative than carbon.

The interpretation and prediction of electric dipole moments is made even more complicated by the fact that a difference in atomic radii can result in an imbalance of electron density because the enhanced charge density associated with the overlap region lies closer to the nucleus of the smaller atom (Fig. 21.1). Such a **homopolar contribution** to the total dipole moment can arise even in the absence of a difference in electronegativity between the two atoms.

A polyatomic molecule is nonpolar if it fulfils certain symmetry criteria. We saw in Section 15.3a that a molecule is nonpolar if it belongs to a D point group or to one of the cubic or icosahedral point groups. We also saw that the dipole moment of a polar molecule with a symmetry axis cannot lie perpendicular to that axis (the dipole moment of NH_3, for instance, lies parallel to the molecular C_3 axis). The symmetry criterion is more important than the question of whether or not the atoms in

Table 21.1[*] Dipole moments (μ) and polarizability volumes (α')

	μ/D	$\alpha'/(10^{-30}\,\text{m}^3)$
CCl_4	0	10.5
H_2	0	0.819
H_2O	1.85	1.48
HCl	1.08	2.63
HI	0.42	5.45

[*] More values are given in the *Data section*.

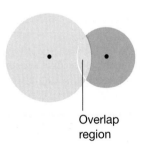

Overlap region

21.1 One of the contributions to the dipole moment of a molecule is the imbalance of charge arising from the overlap of orbitals of different radii. This diagram shows how the charge accumulation leads to a region of negative charge closer to the smaller atom.

2. The conversion factor stems from the original definition of the debye in terms of c.g.s. units: 1 D is the dipole moment of two equal and opposite charges of magnitude 1 e.s.u. separated by 1 Å. The debye is named after Peter Debye, a pioneer in the study of dipole moments of molecules.

2 Ozone, O_3

3 Addition of dipoles

C_{2v}

(a) $\mu_{obs} = 1.57$ D

D_{2h}

(b) $\mu_{obs} = 0$
$\mu_{calc} = 0$

C_{2v}

(c) $\mu_{obs} = 2.25$ D
$\mu_{calc} = 2.7$ D

C_{2v}

(d) $\mu_{obs} = 1.48$ D
$\mu_{calc} = 1.6$ D

21.2 The resultant dipole moments (grey) of the dichlorobenzene isomers (b to d) can be obtained approximately by vectorial addition of two chlorobenzene dipole moments (1.57 D).

the molecule are the same. Thus the homonuclear triatomic molecule O_3 (which is angular, with C_{2v} symmetry) is allowed to be polar by symmetry considerations, and in fact is polar because the electron density on the central O atom differs from that on the two outer O atoms. The dipole moment of the molecule lies parallel to the C_2 axis of the molecule (2). The heteronuclear triatomic molecule CO_2 (which is linear, with $D_{\infty h}$ symmetry) is strictly nonpolar by symmetry even though the C and O atoms have different electronegativities. The dipole moments associated with each CO bond point in opposite directions in CO_2, and cancel.

To a first approximation, the dipole moment of a polyatomic molecule can be resolved into contributions from various components (Fig. 21.2). Thus, 1,4-dichlorobenzene is nonpolar on account of the cancellation of the two equal but opposing moments associated with the presence of Cl atoms on opposite sides of the ring (the molecule has D_{2h} symmetry, so it is necessarily nonpolar). The isomer 1,2-dichlorobenzene (which has C_{2v} symmetry, with the C_2 axis lying along the bisector of the angle between the two C—Cl bonds) has a dipole moment that is approximately the resultant of two monochlorobenzene dipole moments arranged at 60°. The technique of vector addition can be applied with fair success to other series of related molecules, and the resultant of two dipole moments that make an angle θ to each other (3) is obtained from

$$\mu^2 = \mu_1^2 + \mu_2^2 + 2\mu_1\mu_2 \cos \theta \tag{21.3a}$$

When the two dipole moments have the same magnitude, this equation simplifies to

$$\mu = 2\mu_1 \cos \tfrac{1}{2}\theta \tag{21.3b}$$

(b) Polarization

The **polarization**, P, of a sample is the electric dipole moment density, the mean electric dipole moment of the molecules, $\langle \mu \rangle$, multiplied by the number density, $\mathcal{N}$:

$$P = \langle \mu \rangle \mathcal{N} \tag{21.4}$$

In the following pages we refer to the sample as a **dielectric**, by which is meant a polarizable, nonconducting medium.

The polarization of an isotropic fluid sample is zero in the absence of an applied field because the molecules adopt random orientations, so $\langle \mu \rangle = 0$. In the presence of a field, the dipoles become partially aligned because some orientations have lower energies than others. As a result, the electric dipole moment density is nonzero. We show in the *Justification* below that, at a temperature T

$$\langle \mu_z \rangle = \frac{\mu^2 \mathcal{E}}{3kT} \tag{21.5}$$

where z is the direction of the applied field $\mathcal{E}$. Moreover, as we shall see, there is an additional contribution from the dipole moment induced by the field.

Justification 21.1

The probability dp that a dipole has an orientation in the range θ to $\theta + d\theta$ is given by the Boltzmann distribution (Section 19.1c), which in this case is

$$dp = \frac{e^{-E(\theta)/kT} \sin \theta \, d\theta}{\displaystyle\int_0^\pi e^{-E(\theta)/kT} \sin \theta \, d\theta}$$

where $E(\theta)$ is the energy of the dipole in the field: $E(\theta) = -\mu\mathcal{E}\cos\theta$, with $0 \le \theta \le \pi$. The average value of the component of the dipole moment parallel to the applied electric field is therefore

$$\langle\mu_z\rangle = \int \mu\cos\theta\,dp = \mu\int\cos\theta\,dp = \frac{\mu\int_0^\pi e^{x\cos\theta}\cos\theta\sin\theta\,d\theta}{\int_0^\pi e^{x\cos\theta}\sin\theta\,d\theta}$$

with $x = \mu\mathcal{E}/kT$. The integral takes on a simpler appearance when we write $y = \cos\theta$ and note that $dy = -\sin\theta\,d\theta$.

$$\langle\mu_z\rangle = \frac{\mu\int_{-1}^1 ye^{xy}\,dy}{\int_{-1}^1 e^{xy}\,dy}$$

At this point we use

$$\int_{-1}^1 e^{xy}\,dy = \frac{e^x - e^{-x}}{x} \qquad \int_{-1}^1 ye^{xy}\,dy = \frac{e^x + e^{-x}}{x} - \frac{e^x - e^{-x}}{x^2}$$

It is now straightforward algebra to combine these two results and to obtain

$$\langle\mu_z\rangle = \mu L(x) \qquad L(x) = \frac{e^x + e^{-x}}{e^x - e^{-x}} - \frac{1}{x} \qquad x = \frac{\mu\mathcal{E}}{kT} \tag{21.6}$$

The function $L(x)$ is called the **Langevin function**.

Under most circumstances, x is very small (for example, if $\mu = 1$ D and $T = 300$ K, then x exceeds 0.01 only if the field strength exceeds 100 kV cm^{-1}, and most measurements are done at much lower strengths). When $x \ll 1$, the exponentials in the Langevin function can be expanded, and the largest term that survives is

$$L(x) = \tfrac{1}{3}x + \cdots \tag{21.7}$$

Therefore, the average molecular dipole moment is given by eqn 5.

Calculations involving exponentials often become simpler if we note that $e^x = 1 + x + \frac{1}{2}x^2 + \frac{1}{6}x^3 + \cdots$.

21.2 Polarizabilities

An applied electric field can distort a molecule as well as align its permanent electric dipole moment. The **induced dipole moment**, μ^*, is proportional to the field strength, $\mathcal{E}$, and we write[3]

$$\mu^* = \alpha\mathcal{E} \tag{21.8}$$

The constant of proportionality α is the **polarizability** of the molecule. The greater the polarizability, the larger is the induced dipole moment for a given applied field. When the applied field is very strong (as in laser beams), the induced moment is not strictly linear in the strength of the field, and we write

$$\mu^* = \alpha\mathcal{E} + \tfrac{1}{2}\beta\mathcal{E}^2 + \cdots \tag{21.9}$$

The coefficient β is the **hyperpolarizability** of the molecule.

3. We should use vector quantities and allow for the possibility that the induced dipole moment might not lie parallel to the applied field; for simplicity we discuss polarizabilities in terms of (scalar) magnitudes.

(a) Polarizability volumes

Polarizability has the units (coulomb-metre)^2 per joule, $C^2 \, m^2 \, J^{-1}$. That collection of units is awkward, so α is often expressed as a **polarizability volume**, α', by using the relation

$$\alpha' = \frac{\alpha}{4\pi\varepsilon_0} \tag{21.10}$$

where ε_0 is the vacuum permittivity. Because the units of $4\pi\varepsilon_0$ are coulomb-squared per joule per metre ($C^2 \, J^{-1} \, m^{-1}$), it follows that α' has the dimensions of volume (hence its name).[4] Polarizability volumes are similar in magnitude to actual molecular volumes (of the order of $10^{-30} \, m^3$, $1 \, Å^3$).

Some experimental polarizability volumes of molecules are given in Table 21.1. As shown in the *Justification* below, polarizability volumes correlate with the HOMO—LUMO separations in atoms and molecules. The electron distribution can be distorted readily if the LUMO lies close to the HOMO in energy, so the polarizability is then large. If the LUMO lies high above the HOMO, an applied field cannot perturb the electron distribution significantly, and the polarizability is low. Molecules with small HOMO—LUMO gaps are typically large, with numerous electrons.

Justification 21.2

When an electric field is increased by $d\mathcal{E}$, the energy of a molecule changes by $-\mu \, d\mathcal{E}$ and, if the molecule is polarizable, we interpret μ as μ^* (eqn 8). Therefore, the change in energy when the field is increased from 0 to $\mathcal{E}$ is

$$\Delta E = -\int_0^{\mathcal{E}} \mu^* \, d\mathcal{E} = -\int_0^{\mathcal{E}} \alpha \mathcal{E} \, d\mathcal{E} = -\tfrac{1}{2}\alpha\mathcal{E}^2$$

The contribution to the hamiltonian when a dipole moment is exposed to an electric field $\mathcal{E}$ in the z-direction is

$$H^{(1)} = -\mu_z \mathcal{E}$$

Comparison of these two expressions suggests that we should use second-order perturbation theory to calculate the energy of the system in the presence of the field, because then we shall obtain an expression proportional to $\mathcal{E}^2$. According to eqn 12.72, the second-order contribution to the energy is

$$E^{(2)} = \sum_n \frac{\left| \int \psi_n^* H^{(1)} \psi_0 \, d\tau \right|^2}{E_0^{(0)} - E_n^{(0)}} = \mathcal{E}^2 \sum_n \frac{\left| \int \psi_n^* \mu_z \psi_0 \, d\tau \right|^2}{E_0^{(0)} - E_n^{(0)}} = \mathcal{E}^2 \sum_n \frac{|\mu_{z,0n}|^2}{E_0^{(0)} - E_n^{(0)}}$$

where $\mu_{z,0n}$ is the *transition* dipole moment in the z-direction (eqn 12.81). By comparing the two expressions for the energy, we conclude that the polarizablity of the molecule in the z-direction is

$$\alpha = 2 \sum_n \frac{|\mu_{z,0n}|^2}{E_n^{(0)} - E_0^{(0)}} \tag{21.11}$$

---

4. When using older compilations of data, it is useful to note that polarizability volumes have the same numerical values as the 'polarizabilities' reported using c.g.s. electrical units, so the tabulated values previously called 'polarizabilities' can be used directly.

The content of eqn 11 can be appreciated by approximating the excitation energies by a mean value ΔE (an indication of the HOMO—LUMO separation), and supposing that the most important transition dipole moment is approximately equal to the charge of an electron multiplied by the radius, R, of the molecule. Then

$$\alpha \approx \frac{2e^2R^2}{\Delta E}$$

This expression shows that α increases with the size of the molecule and with the ease with which it can be excited (the smaller the value of ΔE).

If the excitation energy is approximated by the energy needed to remove an electron to infinity from a distance R from a single positive charge, we can write $\Delta E \approx e^2/4\pi\varepsilon_0 R$. When this expression is substituted into the equation above, both sides are divided by $4\pi\varepsilon_0$, and the factor of 2 ignored in this approximation, we obtain $\alpha' \approx R^3$, which is of the same order of magnitude as the molecular volume.

For all molecules other than those belonging to one of the cubic or icosahedral groups, the polarizability is anisotropic, by which is meant that its value depends on the orientation of the molecule relative to the field. The polarizability volume of benzene when the field is applied perpendicular to the ring is 12.3×10^{-30} m^3 and is 6.7×10^{-30} m^3 when the field is applied in the plane of the ring. The anisotropy of the polarizability determines whether a molecule is rotationally Raman active (Section 16.7).

(b) Polarization at high frequencies

When the applied field changes direction slowly, the permanent dipole moment has time to reorientate—the whole molecule rotates into a new direction—and follow the field. However, when the frequency of the field is high, a molecule cannot change direction fast enough to follow the change in direction of the applied field and the dipole moment then makes no contribution to the polarization of the sample. Because a molecule takes about 1 ps to turn through about 1 radian in a fluid, the loss of this contribution to the polarization occurs when measurements are made at frequencies greater than about 10^{11} Hz (in the microwave region). We say that the **orientation polarization**, the polarization arising from the permanent dipole moments, is lost at such high frequencies.

The next contribution to the polarization to be lost as the frequency is raised is the **distortion polarization**, the polarization that arises from the distortion of the positions of the nuclei by the applied field. The molecule is bent and stretched by the applied field, and the molecular dipole moment changes accordingly. The time taken for a molecule to bend is approximately the inverse of the molecular vibrational frequency, so the distortion polarization disappears when the frequency of the radiation is increased through the infrared. The disappearance of polarization occurs in stages: as shown in the *Justification* below, each successive stage occurs as the incident frequency rises above the frequency of a particular mode of vibration.

At even higher frequencies, in the visible region, only the electrons are mobile enough to respond to the rapidly changing direction of the applied field. The polarization that remains is now due entirely to the distortion of the electron distribution, and the surviving contribution to the molecular polarizability is called the **electronic polarizability**.

Justification 21.3

The quantum mechanical expression for the polarizability of a molecule in the presence of an electric field that is oscillating at a frequency ω in the z-direction is obtained by using time-dependent perturbation theory (Section 12.10 and *Further reading*) and is

$$\alpha(\omega) = \frac{2}{\hbar} \sum_n \frac{\omega_{n0} |\mu_{z,0n}|^2}{\omega_{n0}^2 - \omega^2}$$ (21.12)

The quantities in this expression (which is valid provided that ω is not close to ω_{n0}) are the same as those in the previous *Justification*, with $\hbar\omega_{n0} = E_n^{(0)} - E_0^{(0)}$. As $\omega \to 0$, the equation reduces to eqn 11 for the static polarizability. As ω becomes very high (and much higher than any excitation frequency of the molecule), the polarizability becomes

$$\alpha(\omega) = -\frac{2}{\hbar\omega^2} \sum_n \omega_{n0} |\mu_{0n}|^2 \to 0 \qquad \text{as} \quad \omega \to \infty$$

That is, when the incident frequency is much higher than any excitation frequency, the polarizability becomes zero. The argument applies to each type of excitation, vibrational as well as electronic, and accounts for the successive decreases in polarizability as the frequency is increased.

21.3 Relative permittivities

When two charges q_1 and q_2 are separated by a distance r in a vacuum, the potential energy of their interaction is[5]

$$V = \frac{q_1 q_2}{4\pi\varepsilon_0 r}$$ (21.13a)

When the same two charges are immersed in a medium (such as air or a liquid), their potential energy is reduced to

$$V = \frac{q_1 q_2}{4\pi\varepsilon r}$$ (21.13b)

where ε is the **permittivity** of the medium. The permittivity is normally expressed in terms of the dimensionless **relative permittivity**, ε_r, (which is also called the dielectric constant) of the medium:[6]

$$\varepsilon_r = \frac{\varepsilon}{\varepsilon_0}$$ [21.14]

The relative permittivity can have a very significant effect on the strength of the interactions between ions in solution. For instance, water has a relative permittivity of 78 at 25°C, so the interionic Coulombic interaction energy is reduced by nearly two orders of magnitude from its vacuum value. Some of the consequences of this reduction for electrolyte solutions were explored in Chapter 10.

---

5. The fundamentals of electrostatics are summarized in *Further information 2*.
6. The relative permittivity of a substance is measured by comparing the capacitance of a capacitor with and without the sample present, C and C_0, respectively) and using $\varepsilon_r = C/C_0$.

The relative permittivity of a substance is large if its molecules are polar or highly polarizable. The quantitative relation between the relative permittivity and the electric properties of the molecules is obtained by considering the polarization of a medium, and is expressed by the **Debye equation:**[7]

$$\frac{\varepsilon_r - 1}{\varepsilon_r + 2} = \frac{\rho P_m}{M} \tag{21.15}$$

where ρ is the mass density of the sample, M is the molar mass of the molecules, and P_m is the **molar polarization**, which is defined as

$$P_m = \frac{N_A}{3\varepsilon_0}\left(\alpha + \frac{\mu^2}{3kT}\right) \tag{21.16}$$

The term $\mu^2/3kT$ stems from the thermal averaging of the electric dipole moment in the presence of the applied field (eqn 5). The corresponding expression without the contribution from the permanent dipole moment is called the **Clausius–Mossotti** equation:

$$\frac{\varepsilon_r - 1}{\varepsilon_r + 2} = \frac{\rho N_A \alpha}{3M\varepsilon_0} \tag{21.17}$$

The Clausius–Mossotti equation is used when there is no contribution from permanent electric dipole moments to the polarization, either because the molecules are nonpolar or because the frequency of the applied field is so high that the molecules cannot orientate quickly enough to follow the change in direction of the field.

Equation 16 implies that the polarizability and permanent electric dipole moment of the molecules in a sample can be determined by measuring ε_r at a series of temperatures, calculating P_m, and plotting it against $1/T$. The slope of the graph is $N_A \mu^2/9\varepsilon_0 k$ and its intercept at $1/T = 0$ is $N_A \alpha/3\varepsilon_0$.

4 Camphor

Example 21.1 Determining dipole moment and polarizability

The relative permittivity of camphor (**4**) was measured at a series of temperatures with the results given below. Determine the dipole moment and the polarizability volume of the molecule.

$\theta/°C$	$\rho/(g\ cm^{-3})$	ε_r
0	0.99	12.5
20	0.99	11.4
40	0.99	10.8
60	0.99	10.0
80	0.99	9.50
100	0.99	8.90
120	0.97	8.10
140	0.96	7.60
160	0.95	7.11
200	0.91	6.21

Method According to eqn 15, we need to calculate $(\varepsilon_r - 1)/(\varepsilon_r + 2)$ at each temperature, and then multiply by M/ρ to form P_m. Next, from eqn 16, we

7. For the derivation of this and the following equations, see *Further reading*.

21.3 The plot of $P_m/(\text{cm}^3 \text{ mol}^{-1})$ against $(10^3 \text{ K})/T$ used in Example 21.1 for the determination of the polarizability and dipole moment of camphor.

should plot P_m against $1/T$ and expect a straight line. The intercept at $1/T = 0$ is $N_A \alpha/3\varepsilon_0 = (4\pi N_A/3)\alpha'$ and the slope is $N_A \mu^2/9\varepsilon_0 k$.

Answer For camphor, $M = 152.23$ g mol^{-1}. We can therefore use the data to draw up the following table:

$\theta/°C$	$(10^3 \text{ K})/T$	ε_r	$(\varepsilon_r - 1)/(\varepsilon_r + 2)$	$P_m/(\text{cm}^3 \text{ mol}^{-1})$
0	3.66	12.5	0.793	122
20	3.41	11.4	0.776	119
40	3.19	10.8	0.766	118
60	3.00	10.0	0.750	115
80	2.83	9.50	0.739	114
100	2.68	8.90	0.725	111
120	2.54	8.10	0.703	110
140	2.42	7.60	0.688	109
160	2.31	7.11	0.670	107
200	2.11	6.21	0.634	106

The points are plotted in Fig. 21.3. The intercept lies at 82.7, so $\alpha' = 3.3 \times 10^{-23}$ cm^3. The slope is 10.9, so $\mu = 4.46 \times 10^{-30}$ C m, corresponding to 1.34 D.

Comment Because the Debye equation describes molecules that are free to rotate, the data show that camphor, which does not melt until 175°C, is rotating even in the solid. It is an approximately spherical molecule.

Self-test 21.1 The relative permittivity of chlorobenzene is 5.71 at 20°C and 5.62 at 25°C. Assuming a constant density (1.11 g cm^{-3}), estimate its polarizability volume and dipole moment.   [1.4×10^{-23} cm^3, 1.2 D]

21.4 Refractive index

One of the optical properties of matter that we are almost in a position to explain is the ability of a prism to separate light into its component colours. This effect depends on the **refractive index**, n_r, of the medium, the ratio of the speed of light in a vacuum, c, to its speed c' in the medium:

$$n_r = \frac{c}{c'}$$ [21.18]

It follows from the Maxwell equations,[8] that the refractive index at a (visible or ultraviolet) specified frequency is related to the relative permittivity at that frequency by

$$n_r = \varepsilon_r^{1/2}$$ (21.19)

The molar polarization, P_m, and hence the molecular polarizability, α, can therefore be measured at frequencies typical of visible light (about 10^{15} to 10^{16} Hz) by measuring the refractive index of the sample (Table 21.2) and using the Clausius–Mossotti equation.

The refractive index is related to the molecular polarizability because the propagation of light through a medium can be imagined to occur by the incident light

Table 21.2* Refractive indices (at different wavelengths of light) relative to air at 20°C

	434 nm	589 nm	656 nm
C_6H_6(l)	1.524	1.501	1.497
CS_2(l)	1.675	1.628	1.618
H_2O(l)	1.340	1.333	1.331
KI(s)	1.704	1.666	1.658

* More values are given in the *Data section*.

8. The Maxwell equations describe the properties of electromagnetic radiation; they are not discussed in this text: see *Further reading*.

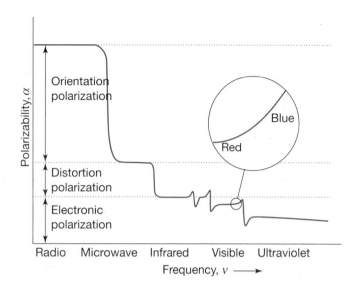

21.4 The general form of the variation of the polarizability with the frequency of the applied field. Note the considerable reduction in polarizability when the field is reversing direction so rapidly that the polar molecules cannot reorientate quickly enough to follow it. The inset shows the variation of the electronic polarizability in the visible region near an electronic excitation of the molecule.

inducing an oscillating dipole moment, which then radiates light of the same frequency. The newly generated radiation is delayed slightly by this process, so it propagates more slowly through the medium than through a vacuum. Because photons of high-frequency light carry more energy than those of low-frequency light, they can distort the electronic distributions of the molecules in their path more effectively. Therefore, after allowing for the loss of contributions from low-frequency modes of motion, we can expect the electronic polarizabilities of molecules, and hence the refractive index, to increase as the incident frequency rises towards an absorption frequency. This dependence on frequency is the origin of the dispersion of white light by a prism: the refractive index is greater for blue light than for red, and therefore the blue rays are bent more than the red. The term **dispersion** is a term carried over from this phenomenon to mean the variation of the refractive index, or of any property, with frequency. Figure 21.4 shows the typical dispersion of the polarizability of a sample.

The concept of refractive index is closely related to the property of optical activity. An **optically active** substance is a substance that rotates the plane of polarization of plane-polarized light. As shown in the *Justification* below, the rotation arises from the difference in the refractive indices for right and left circularly polarized light, n_R and n_L, respectively. By convention, in right-handed circularly polarized light the electric vector rotates clockwise as seen by an observer facing the oncoming beam (Fig. 21.5). A sample in which these two refractive indices are different is said to be **circularly birefringent**.

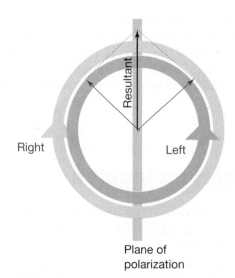

21.5 The superposition of left and right circularly polarized light as viewed from an observer facing the oncoming beam.

Justification 21.4

Before entering the medium, the beam is plane-polarized (that is, the electric field oscillates in a plane containing the propagation direction). This beam may be regarded as a superposition of two oppositely rotating circularly polarized components. On entering the medium, one component propagates faster than the other if their refractive indices are different (Fig. 21.6). If the sample is of length l, the difference in the times of passage is

$$\Delta t = \frac{l}{c_L} - \frac{l}{c_R}$$

L-component (slow)

$\Delta\theta$

R-component (fast)

21.6 Linearly polarized light entering the sample from the left can be regarded as the superposition of two counter-rotating circularly polarized components, with a definite phase relation. If one component propagates more rapidly than the other in the sample, when they emerge the phase relation is changed, and the resultant is plane-polarized light rotated through an angle $\Delta\theta$ to its original orientation.

where c_R and c_L are the speeds of the two components in the medium. In terms of the refractive indices, the difference is

$$\Delta t = (n_R - n_L)\frac{l}{c}$$

The phase difference between the two components when they emerge from the sample is therefore

$$\Delta\theta = 2\pi\nu\Delta t = \frac{2\pi c\Delta t}{\lambda} = (n_R - n_L)\frac{2\pi l}{\lambda}$$

where λ is the wavelength of the light. The two rotating electric vectors have a different phase when they leave the sample from the value they had initially, so their superposition gives rise to a plane-polarized beam rotated through an angle $\Delta\theta$ relative to the plane of the incoming beam. It follows that the angle of optical rotation is proportional to the difference in refractive index, $n_R - n_L$.

To explain why the refractive indices depend on the handedness of the light, we must examine why the polarizabilities depend on the handedness. One interpretation is that, if a molecule has a helical structure (including, if the molecule is small, a structure that can be regarded as being a fragment of a helix), its polarizability depends on whether or not the electric field of the incident radiation rotates in the same sense as the helix. Molecules having a helical structure are chiral, which is the criterion for optical activity discussed in Section 15.3b.

The angle of optical rotation varies with the frequency of the radiation. This variation is called **optical rotatory dispersion** (ORD). It arises from the individual dispersions of the polarizabilities (and refractive indices) for left- and right-circularly polarized radiation. The effect can be used to investigate the stereochemistry of molecules.

Associated with the differences in the two refractive indices (the circular birefringence of the medium) is a difference in absorption intensities I_R and I_L for right- and left-circularly polarized radiation. This difference is known as **circular dichroism** (CD). The CD spectrum of a sample is a plot of the variation of $I_L - I_R$ with frequency of the radiation. Circular dichroism is useful for determining the absolute configurations of d-metal complexes, because complexes with similar geometries have CD spectra with similar features. Analysis of the CD spectrum of a protein reveals its relative helical content because strong CD signals are expected from chiral structures such as helices.

Interactions between molecules

Van der Waals forces are the interactions between closed-shell molecules. They include the interactions between the partial charges of polar molecules and the repulsive interactions that prevent the complete collapse of matter to nuclear densities. The repulsive interactions arise from Coulombic repulsions and, indirectly, from the Pauli principle and the exclusion of electrons from regions of space where the orbitals of neighbouring species overlap. The balance of attractive and repulsive interactions between biopolymers or between a ligand and a biopolymer plays an important role in structural biology (Box 21.1).

Box 21.1 Molecular recognition in biological systems

Intermolecular interactions are responsible for the assembly of many biological structures. Hydrogen bonding and hydrophobic interactions are primarily responsible for the three-dimensional structures of biopolymers, such as proteins, nucleic acids, and cell membranes (Chapter 22). The binding of a ligand, or *guest*, to a biopolymer, or *host*, is also governed by intermolecular interactions. Examples of biological *host–guest complexes* include enzyme–substrate complexes, antigen–antibody complexes, and drug–receptor complexes. In all these cases, a site on the guest contains functional groups that can interact with complementary functional groups of the host. For example, a hydrogen bond donor group of the guest must be positioned near a hydrogen bond acceptor group of the host for tight binding to occur. It is generally true that many specific intermolecular contacts must be made in a biological host–guest complex and, as a result, a guest only binds hosts that are chemically similar. The strict rules governing molecular recognition of a guest by a host control every biological process, from metabolism to immunological response, and provide important clues for the design of effective drugs for the treatment of disease.

Interactions between nonpolar groups can be important in the binding of a guest to a host. For example, many enzyme active sites have hydrophobic pockets that bind nonpolar groups of a substrate. In addition to dispersion, repulsive, and hydrophobic interactions, π stacking interactions are also possible, in which the planar π systems of aromatic macrocycles lie one on top of the other, in a nearly parallel orientation. Such interactions are responsible for the stacking of hydrogen-bonded base pairs in DNA, as shown in the illustration. Some drugs with planar π systems, shown as a blue rectangle in the illustration, are effective because they intercalate between base pairs through π stacking interactions, causing the helix to unwind slightly and altering the function of DNA.

Coulombic interactions can be important in the interior of a biopolymer host, where the relative permittivity can be much lower than that of the aqueous exterior. For example, at physiological pH, amino acid side chains containing carboxylic acid or amine groups are negatively and positively charged, respectively, and can attract each other. Dipole-dipole interactions are also possible because many of the building blocks of biopolymers are polar, including the peptide link, —CONH— (see *Discussion question 2*). However, hydrogen bonding interactions are by far the most prevalent in a biological host–guest complexes. Many effective drugs on the market bind tightly and inhibit the action of enzymes that are associated with the progress of a disease. In many cases, a successful inhibitor will be able to form the same hydrogen bonds with the binding site that the normal substrate of the enzyme can form, except that the drug is chemically inert toward the enzyme. This strategy has been used in the design of protease inhibitor drugs that prevent the maturation of HIV in humans.

Discussion questions

1. Molecular orbital calculations may be used to predict structures of intermolecular complexes. Hydrogen bonds between purine and pyrimidine bases are responsible for the double helix structure of DNA (see Chapter 22). Consider methyl-adenine (**5**, with R = CH_3) and methyl-thymine (**6**, with R = CH_3) as models of two bases that can form hydrogen bonds in DNA. (a) Using molecular modelling software and the computational method of your choice (extended Hückel, semi-empirical, or *ab initio* methods), calculate the atomic charges of all atoms in methyl-adenine and methyl-thymine.

Some drugs with planar π systems, shown as a blue rectangle, intercalate between base pairs of DNA.

5

6

R denotes either the DNA backbone or a substitutent

(b) Based on your tabulation of atomic charges, identify the atoms in methyl-adenine and methyl-thymine that are likely to participate in hydrogen bonds. (c) Draw all possible adenine–thymine pairs that can be linked by hydrogen bonds, keeping in mind that linear arrangements of the A—H···B fragments are preferred in DNA. For this step, you may want to use your molecular modelling software to align the molecules properly. (d) Consult a biochemistry textbook and determine which of the pairs that you drew in part (c) occur naturally in DNA molecules. (e) Repeat parts (a)–(d) for cytosine and guanine, which also form base pairs in DNA (see Chapter 22 for the structures of these bases).

2. Molecular orbital calculations may be used to predict the dipole moments of molecules. (a) Using molecular modelling software and the computational method of your choice (extended Hückel, semi-empirical, or *ab initio* methods), calculate the dipole moment of the peptide link, modelled as a *trans-N*-methylacetamide (**7**). Plot the energy of interaction between these dipoles against the angle θ for $r = 3.0$ nm (see

7

eqn 24). (b) Compare the maximum value of the dipole-dipole interaction energy from part (a) to 20 kJ mol^{-1}, a typical value for the energy of a hydrogen-bonding interaction in biological systems.

Further reading

P.D. Beer, P.A. Gale, and D.K. Smith, *Supramolecular chemistry.* Oxford Chemistry Primers, Oxford University Press (1999).

U. Pindur, M. Haber, and K. Sattler, Antitumor active drugs as intercalators of deoxyribonucleic acid. *J. Chem. Educ.* **70**, 263 (1993).

21.7 The potential energy of interaction between a dipole and a point charge is the sum of the repulsion of like charges and the attraction of opposite charges. For a point dipole, $l \ll r$.

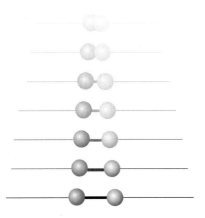

21.8 There are two contributions to the diminishing field of an electric dipole with distance (here seen from the side). The potentials of the charges decreases (shown here by a fading intensity) and the two charges appear to merge, so their combined effect approaches zero more rapidly than by the distance effect alone.

21.5 Interactions between dipoles

Most of the discussion in this section is based on the Coulombic potential energy of interaction between two charges (eqn 13a). We can easily adapt this expression to find the potential energy of a point charge and a dipole and to extend it to the interaction between two dipoles.

(a) The potential energy of interaction

We show in the *Justification* below that the potential energy of interaction between a point dipole $\mu_1 = q_1 l$ and the point charge q_2 in the arrangement shown in Fig. 21.7 is

$$V = -\frac{\mu_1 q_2}{4\pi\varepsilon_0 r^2}$$

(21.20)

With μ in coulomb-metre, q_2 in coulomb, and r in metres, V is obtained in joules. A **point dipole** is a dipole in which the separation between the charges is much smaller than the distance at which the dipole is being observed, $l \ll r$. The potential energy rises towards zero (the value at infinite separation of the charge and the dipole) more rapidly (as $1/r^2$) than that between two point charges (which varies as $1/r$) because, from the viewpoint of the point charge, the partial charges of the dipole seem to merge and cancel as the distance r increases (Fig. 21.8).

Justification 21.5

The sum of the potential energies of repulsion between like charges and attraction between opposite charges in the orientation shown in Fig. 21.7 is

$$V = \frac{1}{4\pi\varepsilon_0}\left(-\frac{q_1 q_2}{r - \frac{1}{2}l} + \frac{q_1 q_2}{r + \frac{1}{2}l}\right)$$

Because $l \ll r$ for a point dipole, this expression can be simplified by writing

$$V = \frac{q_1 q_2}{4\pi\varepsilon_0 r}\left(-\frac{1}{1-x} + \frac{1}{1+x}\right)$$

where $x = l/2r$. Expanding the terms in x and retaining only the leading term gives:

$$V = \frac{q_1 q_2}{4\pi\varepsilon_0 r}\{-(1 + x + \cdots) + (1 - x + \cdots)\} \approx -\frac{2xq_1 q_2}{4\pi\varepsilon_0 r} = -\frac{q_1 q_2 l}{4\pi\varepsilon_0 r^2}$$

With $\mu_1 = q_1 l$, this expression becomes eqn 20. This expression should be multiplied by $\cos\theta$ when the point charge lies at an angle θ to the axis of the dipole.

The following expansions are often useful: $\dfrac{1}{1+x} = 1 - x + x^2 - \cdots$

and $\dfrac{1}{1-x} = 1 + x + x^2 + \cdots$

21.9 The potential energy of interaction between two dipoles is the sum of the repulsions of like charges and the attractions of opposite charges. This illustration shows a collinear arrangement of dipoles.

Example 21.2 Calculating the interaction energy of two dipoles

Calculate the potential energy of interaction of two dipoles in the arrangement shown in Fig. 21.9 when their separation is r and $\theta = 0$.

Method We proceed in exactly the same way as in the *Justification*, but now the total interaction energy is the sum of four pairwise terms, two attractions between opposite charges, which contribute negative terms to the potential energy, and two repulsions between like charges, which contribute positive terms.

Answer The sum of the four contributions is

$$V = \frac{1}{4\pi\varepsilon_0}\left(-\frac{q_1 q_2}{r+l} + \frac{q_1 q_2}{r} + \frac{q_1 q_2}{r} - \frac{q_1 q_2}{r-l}\right) = -\frac{q_1 q_2}{4\pi\varepsilon_0 r}\left(\frac{1}{1+x} - 2 + \frac{1}{1-x}\right)$$

with $x = l/r$. As before, we expand the two terms in x and retain only the first surviving term, which is equal to $2x^2$. This step results in the expression

$$V = -\frac{2x^2 q_1 q_2}{4\pi\varepsilon_0 r}$$

Therefore, because $\mu_1 = q_1 l$ and $\mu_2 = q_2 l$, the potential energy of interaction in the alignment shown in the illustration is

$$V = -\frac{\mu_1 \mu_2}{2\pi\varepsilon_0 r^3}$$

Comment Notice that the interaction energy approaches zero more rapidly (as $1/r^3$) than for the previous case: now both interacting entities appear neutral to each other at large separations. See *Further information 2* for the general expression.

Self-test 21.2 Derive an expression for the potential energy when the dipoles are in the arrangement shown in Fig. 21.10. $[V = \mu_1 \mu_2 / 4\pi\varepsilon_0 r^3]$

21.10 A parallel arrangement of electric dipoles. (a) General case; (b) as used for Self-test 21.2.

Table 21.3 summarizes the various expressions for the interaction of charges and dipoles. It is quite easy to extend the formulas given there to obtain expressions for the energy of interaction of higher **multipoles**, or arrays of point charges (Fig. 21.11). Specifically, an *n*-**pole** is an array of point charges with an *n*-pole moment but no lower moment. Thus, a **monopole** is a point charge, and the monopole moment is what we normally call the overall charge. A dipole, as we have seen, is an array of charges that has no monopole moment (no net charge). A **quadrupole**

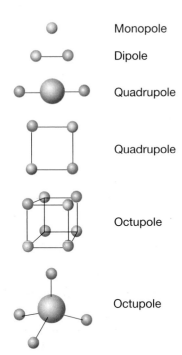

Monopole

Dipole

Quadrupole

Quadrupole

Octupole

Octupole

21.11 Typical charge arrays corresponding to electric multipoles. The field arising from an arbitrary finite charge distribution can be expressed as the superposition of the fields arising from a superposition of multipoles.

8 Carbon dioxide

9 Methane

Table 21.3 Multipole interaction potential energies

Interaction type	Distance dependence of potential energy	Typical energy/ (kJ mol^{-1})	Comment
Ion–ion	$1/r$	250	Only between ions*
Ion–dipole	$1/r^2$	15	
Dipole–dipole	$1/r^3$	2	Between stationary polar molecules
	$1/r^6$	0.6	Between rotating polar molecules
London (dispersion)	$1/r^6$	2	Between all types of molecules

The energy of a hydrogen bond A—H⋯B is typically 20 kJ mol^{-1} and occurs on contact for A, B = N, O, or F.
* Electrolyte solutions are treated in Chapter 10, ionic solids in Chapter 23.

consists of an array of point charges that has neither net charge nor dipole moment (as for CO_2 molecules, **8**). An **octupole** consists of an array of point charges that sum to zero and which has neither a dipole moment nor a quadrupole moment (as for CH_4 molecules, **9**). The feature to remember is that the interaction energy falls off more rapidly the higher the order of the multipole. For the interaction of an n-pole with an m-pole, the potential energy varies with distance as

$$V \propto \frac{1}{r^{n+m-1}} \tag{21.21}$$

The reason for the even steeper decrease with distance is the same as before: the array of charges appears to blend together into neutrality more rapidly with distance the higher the number of individual charges that contribute to the multipole. Note that a given molecule may have a charge distribution that corresponds to a superposition of several different multipoles.

(b) The electric field

The same kind of argument as that used to derive expressions for the potential energy can be used to establish the distance dependence of the strength of the electric field generated by a dipole. We shall need this expression when we calculate the dipole moment induced in one molecule by another.

The starting point for the calculation is the strength of the electric field[9] generated by a point electric charge:

$$\mathcal{E} = \frac{q}{4\pi\varepsilon_0 r^2} \tag{21.22}$$

9. The electric field is actually a vector, and we cannot simply add and subtract magnitudes without taking into account the directions of the fields. In the cases we consider, this will not be a complication because the two charges of the dipoles will be collinear and give rise to fields in the same direction. Be careful, though, with more general arrangements of charges.

21.12 The electric field of a dipole is the sum of the opposing fields from the positive and negative charges, each of which is proportional to $1/r^2$. The difference, the net field, is proportional to $1/r^3$.

The field generated by a dipole is the sum of the fields generated by each partial charge. For the point-dipole arrangement shown in Fig. 21.12, the same procedure that was used to derive the potential energy gives

$$\mathcal{E} = \frac{\mu}{2\pi\varepsilon_0 r^3} \tag{21.23}$$

The electric field of a multipole (in this case a dipole) decreases more rapidly with distance (as $1/r^3$ for a dipole) than a monopole (a point charge).

(c) Dipole–dipole interactions

The potential energy of interaction between two polar molecules is a complicated function of their relative orientation. When the two dipoles are parallel (as in Fig. 21.10a), the potential energy is simply (see *Further information 2*)

$$V = \frac{\mu_1\mu_2 f(\theta)}{4\pi\varepsilon_0 r^3} \qquad f(\theta) = 1 - 3\cos^2\theta \tag{21.24}$$

This expression applies to polar molecules in a fixed, parallel, orientation in a solid.

In a fluid of freely rotating molecules, the interaction between dipoles averages to zero because $f(\theta)$ changes sign as the orientation changes, and its average value is zero. Physically, the like partial charges of two freely rotating molecules are close together as much as the two opposite charges, and the repulsion of the former is cancelled by the attraction of the latter.

The interaction energy of two *freely* rotating dipoles is zero. However, because their mutual potential energy depends on their relative orientation, the molecules do not in fact rotate completely freely, even in a gas. In fact, the lower energy orientations are marginally favoured, so there is a nonzero average interaction between polar molecules. We show in the following *Justification* that the average potential energy of two rotating molecules that are separated by a distance r is

$$\langle V \rangle = -\frac{C}{r^6} \qquad C = \frac{2\mu_1^2\mu_2^2}{3(4\pi\varepsilon_0)^2 kT} \tag{21.25}$$

This expression describes the **Keesom interaction**.

> The mean value theorem states that average (or mean value) of a function $f(x)$ over the range $x = a$ to $x = b$ is
>
> $\dfrac{1}{b-a}\displaystyle\int_a^b f(x)\,dx$. The volume element
>
> in polar coordinates is proportional to $\sin\theta\,d\theta$, and θ ranges from 0 to π. Therefore the average value of $(1 - 3\cos^2\theta)$ is
>
> $\dfrac{1}{\pi}\displaystyle\int_0^\pi (1 - 3\cos^2\theta)\sin\theta\,d\theta = 0.$

Justification 21.6

The detailed calculation of the Keesom interaction energy is quite complicated, but the form of the final answer can be constructed quite simply. First, we note that the average interaction energy of two polar molecules rotating at a fixed separation r is given by

$$\langle V \rangle = \frac{\mu_1\mu_2\langle f\rangle}{4\pi\varepsilon_0 r^3}$$

where $\langle f \rangle$ now includes a weighting factor in the averaging that is equal to the probability that a particular orientation will be adopted. This probability is given by the Boltzmann distribution $p \propto e^{-E/kT}$, with E interpreted as the potential energy of interaction of the two dipoles in that orientation. That is,

$$p \propto e^{-V/kT} \qquad V = \frac{\mu_1\mu_2 f}{4\pi\varepsilon_0 r^3}$$

When the potential energy of interaction of the two dipoles is very small compared with the energy of thermal motion, we can use $V \ll kT$, expand the exponential function in p, and retain only the first two terms:

$$p \propto 1 - V/kT + \cdots$$

The weighted average of f is therefore

$$\langle f \rangle \propto \langle f \rangle_0 - \frac{\mu_1 \mu_2}{4\pi\varepsilon_0 kTr^3}\langle f^2 \rangle_0 + \cdots$$

where $\langle \cdots \rangle_0$ denotes an unweighted spherical average. The average value of f is zero, so the first term vanishes. However, the average value of f^2 is nonzero because f^2 is positive at all orientations, so we can write

$$\langle V \rangle = -\frac{\mu_1^2 \mu_2^2 \langle f^2 \rangle_0}{(4\pi\varepsilon_0)^2 kTr^6}$$

The average value $\langle f^2 \rangle_0$ turns out to be $\tfrac{2}{3}$ when the calculation is carried through in detail. The final result is that quoted in eqn 25.

The important features of eqn 25 are its negative sign (the average interaction is attractive), the dependence of the average interaction energy on the inverse sixth power of the separation, and its inverse dependence on the temperature. The last feature reflects the way that the greater thermal motion overcomes the mutual orientating effects of the dipoles at higher temperatures. The inverse sixth power arises from the inverse third power of the interaction potential energy that is weighted by the energy in the Boltzmann term, which is also proportional to the inverse third power of the separation.

At 25°C the average interaction energy for pairs of molecules with $\mu = 1$ D is about -0.07 kJ mol^{-1} when the separation is 0.5 nm. This energy should be compared with the average molar kinetic energy of $\tfrac{3}{2}RT = 3.7$ kJ mol^{-1} at the same temperature. The interaction energy is much smaller than the energies involved in the making and breaking of chemical bonds.

(d) Dipole–induced-dipole interactions

A polar molecule with dipole moment μ_1 can induce a dipole μ_2^* in a neighbouring polarizable molecule (Fig. 21.13). The induced dipole interacts with the permanent dipole of the first molecule, and the two are attracted together. The average interaction energy when the separation of the molecules is r is[10]

$$V = -\frac{C}{r^6} \qquad C = \frac{\mu_1^2 \alpha_2'}{4\pi\varepsilon_0} \tag{21.26}$$

where α_2' is the polarizability volume of molecule 2 and μ_1 is the permanent dipole moment of molecule 1.

The dipole–induced-dipole interaction energy is independent of the temperature because thermal motion has no effect on the averaging process. Moreover, like the dipole–dipole interaction, the potential energy depends on $1/r^6$: this distance dependence stems from the $1/r^3$ dependence of the field (and hence the magnitude of the induced dipole) and the $1/r^3$ dependence of the potential energy of interaction between

21.13 (a) A polar molecule (blue arrow) can induce a dipole (white arrow) in a nonpolar molecule, and (b) the latter's orientation follows the former's, so the interaction does not average to zero.

10. Note that the C in this expression is different from the C in eqn 25 and other expressions below: we are using the same symbol in C/r^6 to emphasize the similarity of form of each expression. For a derivation of this expression, see *Further reading*.

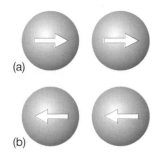

(a)

(b)

21.14 (a) In the dispersion interaction, an instantaneous dipole on one molecule induces a dipole on another molecule, and the two dipoles then interact to lower the energy. (b) The two instantaneous dipoles are correlated and, although they occur in different orientations at different instants, the interaction does not average to zero.

the permanent and induced dipoles. For a molecule with $\mu = 1$ D (such as HCl) near a molecule of polarizability volume $\alpha' = 10 \times 10^{-30}$ m^3 (such as benzene, Table 21.1), the average interaction energy is about -0.8 kJ mol^{-1} when the separation is 0.3 nm.

(e) Induced-dipole–induced-dipole interactions

Nonpolar molecules (including closed-shell atoms, such as Ar) attract one another even though neither has a permanent dipole moment. The abundant evidence for the existence of interactions between them is the formation of condensed phases of nonpolar substances, such as the condensation of hydrogen or argon to a liquid at low temperatures and the fact that benzene is a liquid at normal temperatures.

The interaction between nonpolar molecules arises from the transient dipoles that all molecules possess as a result of fluctuations in the instantaneous positions of electrons. To appreciate the origin of the interaction, suppose that the electrons in one molecule flicker into an arrangement that gives the molecule an instantaneous dipole moment μ_1^*. This dipole generates an electric field that polarizes the other molecule, and induces in that molecule an instantaneous dipole moment μ_2^*. The two dipoles attract each other and the potential energy of the pair is lowered. Although the first molecule will go on to change the size and direction of its instantaneous dipole, the electron distribution of the second molecule will follow; that is, the two dipoles are correlated in direction (Fig. 21.14). Because of this correlation, the attraction between the two instantaneous dipoles does not average to zero, and gives rise to an induced-dipole–induced-dipole interaction. This interaction is called either the **dispersion interaction** or the **London interaction** (for Fritz London, who first described it).

Polar molecules also interact by a dispersion interaction: such molecules also possess instantaneous dipoles, the only difference being that the time average of each fluctuating dipole does not vanish, but corresponds to the permanent dipole. Such molecules therefore interact both through their permanent dipoles and through the correlated, instantaneous fluctuations in these dipoles.

The strength of the dispersion interaction depends on the polarizability of the first molecule because the instantaneous dipole moment μ_1^* depends on the looseness of the control that the nuclear charge exercises over the outer electrons. The strength of the interaction also depends on the polarizability of the second molecule, for that polarizability determines how readily a dipole can be induced by another molecule. The actual calculation of the dispersion interaction is quite involved, but a reasonable approximation to the interaction energy is given by the **London formula**:

$$V = -\frac{C}{r^6} \qquad C = \tfrac{3}{2}\alpha_1'\alpha_2'\frac{I_1 I_2}{I_1 + I_2} \qquad (21.27)$$

where I_1 and I_2 are the ionization energies of the two molecules (Table 13.4). This interaction energy is also proportional to the inverse sixth power of the separation of the molecules. The dispersion interaction generally dominates all the interactions between molecules other than hydrogen bonds.

Illustration 21.1

For two CH$_4$ molecules, we can substitute $\alpha' = 2.6 \times 10^{-30}$ m^3 and $I \approx 700$ kJ mol^{-1} to obtain $V = -2$ kJ mol^{-1} for $r = 0.3$ nm. A very rough check on this figure is the enthalpy of vaporization of methane, which is 8.2 kJ mol^{-1}. However, this comparison is insecure, partly because the enthalpy of vaporization is a many-body quantity and partly because the long-distance assumption breaks down.

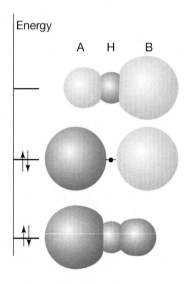

Energy

A   H   B

21.15 The molecular orbital interpretation of the formation of an A—H⋯B hydrogen bond. From the three A, H, and B orbitals, three molecular orbitals can be formed (their relative contributions are represented by the sizes of the spheres). Only the two lower energy orbitals are occupied, and there may therefore be a net lowering of energy compared with the separate AH and B species.

(f) Hydrogen bonding

The interactions described so far are universal in the sense that they are possessed by all molecules independent of their specific identity. However, there is a type of interaction possessed by molecules that have a particular constitution. A **hydrogen bond** is an attractive interaction between two species that arises from a link of the form A—H⋯B, where A and B are highly electronegative elements and B possesses a lone pair of electrons. Hydrogen bonding is conventionally regarded as being limited to N, O, and F but, if B is an anionic species (such as Cl⁻), it may also participate in hydrogen bonding. There is no strict cut-off for an ability to participate in hydrogen bonding, but N, O, and F participate most effectively.

The formation of a hydrogen bond can be regarded either as the approach between a partial positive charge of H and a partial negative charge of B or as a particular example of delocalized molecular orbital formation in which A, H, and B each supply one atomic orbital from which three molecular orbitals are constructed (Fig. 21.15). Recent experiments suggest that the hydrogen bonds in ice have significant covalent character and are more adequately described by a molecular orbital treatment. The A and H1s orbitals are those used to form the A—H bond in the AH molecule and the B orbital originally accommodates the lone pair on B. In the combined species, there are four electrons to accommodate (two from the A—H bond, two from the lone pair of B), and they occupy the two lowest molecular orbitals of the AHB fragment. Because the uppermost (most antibonding) orbital is vacant, it is feasible for the net effect to be a lowering of energy, and hence the formation of a hydrogen bond.

In practice, the strength of the bond is found to be about 20 kJ mol⁻¹. Because the bonding depends on orbital overlap, it is virtually a contact-like interaction that is turned on when AH touches B and is zero as soon as the contact is broken. If hydrogen bonding is present, it dominates the other intermolecular interactions. The properties of liquid and solid water, for example, are dominated by the hydrogen bonding between H_2O molecules. The structure of DNA and hence the transmission of genetic information is crucially dependent on the strength of hydrogen bonds between base pairs. The structural evidence for hydrogen bonding comes from noting that the internuclear distance between formally non-bonded atoms is less than their van der Waals contact distance, which suggests that a dominating attractive interaction is present. For example, the O—O distance in O—H⋯O is expected to be 280 pm on the basis of van der Waals radii, but is found to be 270 pm in typical compounds. Moreover, the H⋯O distance is expected to be 260 pm but is found to be only 170 pm.

Hydrogen bonds may be either symmetric or unsymmetric. In a symmetric hydrogen bond, the H atom lies midway between the two other atoms. This arrangement is rare, but occurs in F—H⋯F⁻, where both bond lengths are 120 pm. More common is the unsymmetrical arrangement, where the A—H bond is shorter than the H⋯B bond. Simple electrostatic arguments, treating A—H⋯B as an array of point charges (partial negative charges on A and B, partial positive on H), suggest that the lowest energy is achieved when the bond is linear, because then the two partial negative charges are furthest apart. The experimental evidence from structural studies support a linear or near-linear arrangement.

(g) The total attractive interaction

We shall consider molecules that are unable to participate in hydrogen bond formation. The total attractive interaction energy between rotating molecules is then the sum of the three van der Waals contributions discussed above. (Only the dispersion interaction contributes if both molecules are nonpolar.) In a fluid phase, all three contributions to the potential energy vary as the inverse sixth power of the separation of the molecules, so we may write

$$V = -\frac{C_6}{r^6} \tag{21.28}$$

where C_6 is a coefficient that depends on the identity of the molecules.

Although attractive interactions between molecules are often expressed as in eqn 28, we must remember that this equation has only limited validity. First, we have taken into account only dipolar interactions of various kinds, for they have the longest range and are dominant if the average separation of the molecules is large. However, in a complete treatment we should also consider quadrupolar and higher-order multipole interactions, particularly if the molecules do not have permanent electric dipole moments. Secondly, the expressions have been derived by assuming that the molecules can rotate reasonably freely. That is not the case in most solids, and in rigid media the dipole–dipole interaction is proportional to $1/r^3$ because the Boltzmann averaging procedure is irrelevant when the molecules are trapped into a fixed orientation.

A different kind of limitation is that eqn 28 relates to the interactions of pairs of molecules. There is no reason to suppose that the energy of interaction of three (or more) molecules is the sum of the pairwise interaction energies alone. The total dispersion energy of three closed-shell atoms, for instance, is given approximately by the **Axilrod–Teller formula**:

$$V = -\frac{C_6}{r_{AB}^6} - \frac{C_6}{r_{BC}^6} - \frac{C_6}{r_{CA}^6} + \frac{C'}{(r_{AB}r_{BC}r_{CA})^3} \tag{21.29a}$$

where

$$C' = a(3\cos\theta_A \cos\theta_B \cos\theta_C + 1) \tag{21.29b}$$

The parameter a is approximately equal to $\frac{3}{4}\alpha'C_6$; the angles θ are the internal angles of the triangle formed by the three atoms (**10**). The term in C' (which represents the non-additivity of the pairwise interactions) is negative for a linear arrangement of atoms (so that arrangement is stabilized) and positive for an equilateral triangular cluster. It is found that the three-body term contributes about 10 per cent of the total interaction energy in liquid argon.

21.6 Repulsive and total interactions

When molecules are squeezed together, the nuclear and electronic repulsions and the rising electronic kinetic energy begin to dominate the attractive forces. The repulsions increase steeply with decreasing separation in a way that can be deduced only by very extensive, complicated molecular structure calculations of the kind described in Chapter 14 (Fig. 21.16).

In many cases, however, progress can be made by using a greatly simplified representation of the potential energy, where the details are ignored and the general features expressed by a few adjustable parameters. One such approximation is the **hard-sphere potential**, in which it is assumed that the potential energy rises abruptly to infinity as soon as the particles come within a separation d:

$$V = \infty \quad \text{for} \quad r \leq d \qquad V = 0 \quad \text{for} \quad r > d \tag{21.30}$$

This very simple potential is surprisingly useful for assessing a number of properties. Another widely used approximation is the **Mie potential**:

$$V = \frac{C_n}{r^n} - \frac{C_m}{r^m} \tag{21.31}$$

10

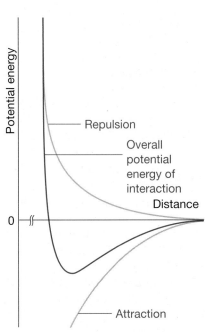

21.16 The general form of an intermolecular potential energy curve. At long range the interaction is attractive, but at close range the repulsions dominate.

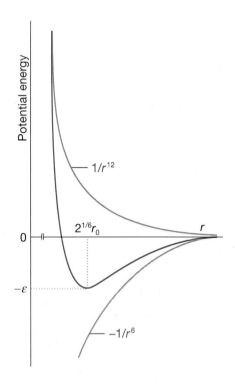

21.17 The Lennard-Jones potential, and the relation of the parameters to the features of the curve. The light lines are the two contributions.

with $n > m$. The first term represents repulsions and the second term attractions. The **Lennard-Jones potential** is a special case of the Mie potential with $n = 12$ and $m = 6$ (Fig. 21.17); it is often written in the form

$$V = 4\varepsilon \left\{ \left(\frac{r_0}{r} \right)^{12} - \left(\frac{r_0}{r} \right)^{6} \right\} \tag{21.32}$$

The two parameters are ε, the depth of the well,[11] and r_0, the separation at which $V = 0$ (Table 21.4). The well minimum occurs at $r_e = 2^{1/6} r_0$. Although the Lennard-Jones potential has been used in many calculations, there is plenty of evidence to show that $1/r^{12}$ is a very poor representation of the repulsive potential, and that an exponential form, e^{-r/r_0}, is greatly superior. An exponential function is more faithful to the exponential decay of atomic wavefunctions at large distances, and hence to the overlap that is responsible for repulsion. The potential with an exponential repulsive term and a $1/r^6$ attractive term is known as an **exp-6 potential**. These potentials can be used to calculate the virial coefficients of gases, as explained in Section 20.5, and through them various properties of real gases, such as the Joule–Thompson coefficient. The potentials are also used to model the structures of condensed fluids.

Until recently, the potential energy of interaction of molecules was of primary interest. However, with the advent of **atomic force microscopy** (AFM), in which the force between a molecular-sized probe and a surface is monitored (see Box 12.1), force is moving back into the centre of attention. As force, F, is the negative slope of potential, for a Lennard-Jones potential between individual molecules

$$F = -\frac{dV}{dr} = \frac{24\varepsilon}{r_0} \left\{ 2 \left(\frac{r_0}{r} \right)^{13} - \left(\frac{r_0}{r} \right)^{7} \right\} \tag{21.33}$$

The net attractive force is greatest at $r = \left(\frac{26}{7} \right)^{1/6} r_0$, or $1.244 r_0$, and at that distance is equal to $-144 \left(\frac{7}{26} \right)^{7/6} \varepsilon / 13 r_0$, or $-2.397 \varepsilon / r_0$. For typical parameters, the magnitude of this force is about 10 pN.

21.7 Molecular interactions in gases

Molecular interactions in the gas phase can be studied in **molecular beams**, which consist of a collimated, narrow stream of molecules travelling though an evacuated vessel. The beam is directed towards other molecules, and the scattering that occurs on impact is related to the intermolecular interactions.

(a) The basic principles of molecular beams

The basic arrangement for a molecular beam experiment is shown in Fig. 21.18. If the pressure of vapour in the source is increased so that the average distance molecules travel between collisions in the emerging beam is much shorter than the diameter of the pinhole, many collisions take place even outside the source. The net effect of these collisions, which give rise to **hydrodynamic flow**, is to transfer momentum into the direction of the beam. The molecules in the beam then travel with very similar speeds, so further downstream few collisions take place between them. This condition is called **molecular flow**. Because the spread in kinetic energy and hence in speeds is so small, the molecules are effectively in a state of very low translational temperature (Fig. 21.19). The translational temperature may reach as

Table 21.4* Lennard-Jones (12,6) parameters

	$(\varepsilon/k)/\text{K}$	r_0/pm
Ar	111.84	362.3
CCl_4	376.86	624.1
N_2	91.85	391.9
Xe	213.96	426.0

* More values are given in the *Data section*.
† ε is expressed as an effective temperature on division by Boltzmann's constant k.

11. Do not confuse the well depth parameter in the expression for the Lennard-Jones potential with the symbol for the permittivity of a medium (Section 21.3).

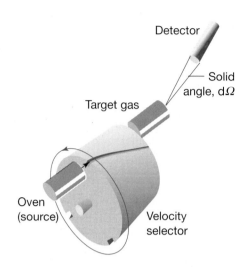

21.18 The basic arrangement of a molecular beam apparatus. The atoms or molecules emerge from a heated source, and pass through the velocity selector, a rotating slotted cylinder. The scattering occurs from the target gas (which might take the form of another beam), and the flux of particles entering the detector set at some angle is recorded.

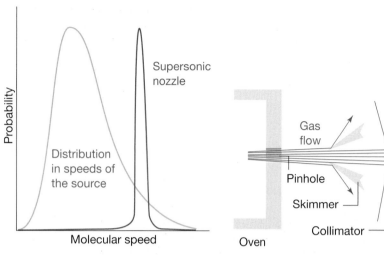

21.19 The shift in the mean speed and the width of the distribution brought about by use of a supersonic nozzle. See Section 24.1 for the precise form of the distribution of speeds in gases.

21.20 A supersonic nozzle skims off some of the molecules of the beam and leads to a beam with well defined velocity.

low as 1 K. Such jets are called **supersonic** because the average speed of the molecules in the jet is much greater than the speed of sound for the molecules that are not part of the jet.

A supersonic jet can be converted into a more parallel **supersonic beam** if it is 'skimmed' in the region of hydrodynamic flow and the excess gas pumped away. A skimmer consists of a conical nozzle shaped to avoid any supersonic shock waves spreading back into the gas and so increasing the translational temperature (Fig. 21.20). A jet or beam may also be formed by using helium or neon as the principal gas, and injecting molecules of interest into it in the hydrodynamic region of flow.

The low translational temperature of the molecules is reflected in the low rotational and vibrational temperatures of the molecules. In this context, a rotational or vibrational temperature means the temperature that should be used in the Boltzmann distribution to reproduce the observed populations of the states. However, as rotational modes equilibrate more slowly, and vibrational modes equilibrate even more slowly, the rotational and vibrational populations of the species correspond to somewhat higher temperatures, of the order of 10 K for rotation and 100 K for vibrations.

The target gas may be either a bulk sample or another molecular beam. The latter **crossed beam technique** gives a lot of information because the states of both the target and projectile molecules may be controlled. The intensity of the incident beam is measured by the **incident beam flux**, I, which is the number of particles passing through a given area in a given interval divided by the area and the duration of the interval.

The detectors may consist of a chamber fitted with a sensitive pressure gauge, a bolometer (a detector that responds to the incident energy by making use of the temperature dependence of resistance) or an ionization detector, in which the incoming molecule is first ionized and then detected electronically. The state of the scattered molecules may also be determined spectroscopically, and is of interest when the collisions change their vibrational or rotational states.

21.21 The definition of the solid angle, dΩ, for scattering.

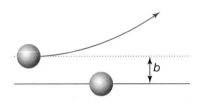

21.22 The definition of the impact parameter, b, as the perpendicular separation of the initial paths of the particles.

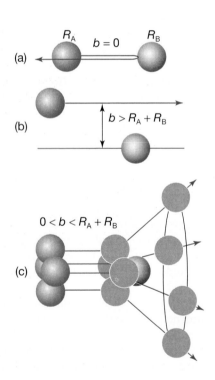

21.23 Three typical cases for the collisions of two hard spheres: (a) $b = 0$, giving backward scattering; (b) $b > R_A + R_B$, giving forward scattering; (c) $0 < b < R_A + R_B$, leading to scattering into one direction on a ring of possibilities. (The target molecule is taken to be so heavy that it remains virtually stationary.)

(b) The experimental observations

The primary experimental information from a molecular beam experiment is the fraction of the molecules in the incident beam that are scattered into a particular direction. The fraction is normally expressed in terms of dI, the rate at which molecules are scattered into a cone that represents the area covered by the 'eye' of the detector (Fig. 21.21). This rate is reported as the **differential scattering cross-section**, σ, the constant of proportionality between the value of dI and the intensity, I, of the incident beam, the number density of target molecules, $\mathcal{N}$, and the infinitesimal path length dx through the sample:

$$dI = \sigma I \mathcal{N} dx \qquad (21.34)$$

The value of σ (which has the dimensions of area) depends on the **impact parameter**, b, the initial perpendicular separation of the paths of the colliding molecules (Fig. 21.22), and the details of the intermolecular potential. The role of the impact parameter is most easily seen by considering the impact of two hard spheres (Fig. 21.23). If $b = 0$, the lighter projectile is on a trajectory that leads to a head-on collision, so the only scattering intensity is detected when the detector is at $\theta = \pi$. When the impact parameter is so great that the spheres do not make contact ($b > R_A + R_B$), there is no scattering and the scattering cross-section is zero at all angles except $\theta = 0$. Glancing blows, with $0 < b \leq R_A + R_B$, lead to scattering intensity in cones around the forward direction.

(c) Scattering effects

The scattering pattern of real molecules, which are not hard spheres, depends on the details of the intermolecular potential, including the anisotropy that is present when the molecules are non-spherical. The scattering also depends on the relative speed of approach of the two particles: a very fast particle might pass through the interaction region without much deflection, whereas a slower one on the same path might be temporarily captured and undergo considerable deflection (Fig. 21.24). The variation of the scattering cross-section with the relative speed of approach should therefore give information about the strength and range of the intermolecular potential.

A further point is that the outcome of collisions is determined by quantum, not classical, mechanics. The wave nature of the particles can be taken into account, at least to some extent, by drawing all classical trajectories that take the projectile particle from source to detector, and then considering the effects of interference between them.

Two quantum mechanical effects are of great importance. A particle with a certain impact parameter might approach the attractive region of the potential in such a way that the particle is deflected towards the repulsive core (Fig. 21.25), which then repels it out through the attractive region to continue its flight in the forward direction. Some molecules, however, also travel in the forward direction because they have impact parameters so large that they are undeflected. The wavefunctions of the particles that take the two types of path interfere, and the intensity in the forward direction is modified. The effect is called **quantum oscillation**. The same phenomenon accounts for the optical 'glory effect', in which a bright halo can sometimes be seen surrounding an illuminated object. (The coloured rings around the shadow of an aircraft cast on clouds by the sun, and often seen in flight, is an example of an optical glory.)

The second quantum effect we need consider is the observation of a strongly enhanced scattering in a non-forward direction. This effect is called **rainbow scattering** because the same mechanism accounts for the appearance of an optical

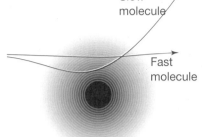

21.24 The extent of scattering may depend on the relative speed of approach as well as the impact parameter. The dark central zone represents the repulsive core; the fuzzy outer zone represents the long-range attractive potential.

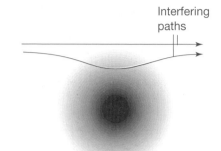

21.25 Two paths leading to the same destination will interfere quantum mechanically; in this case they give rise to quantum oscillations in the forward direction.

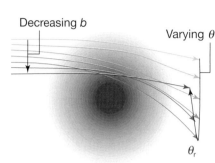

21.26 The interference of paths leading to rainbow scattering. The rainbow angle, θ_r, is the maximum scattering angle reached as b is decreased. Interference between the numerous paths at that angle modifies the scattering intensity markedly.

rainbow. The origin of the phenomenon is illustrated in Fig. 21.26. As the impact parameter decreases, there comes a stage at which the scattering angle passes through a maximum and the interference between the paths results in a strongly scattered beam. The **rainbow angle**, θ_r, is the angle for which $d\theta/db = 0$ and the scattering is strong.

Another phenomenon that can occur in certain beams is the capturing of one species by another. The vibrational temperature in supersonic beams is so low that **van der Waals molecules** may be formed, which are complexes of the form AB in which A and B are held together by van der Waals forces or hydrogen bonds. Large numbers of such molecules have been studied spectroscopically, including ArHCl, $(HCl)_2$, $ArCO_2$, and $(H_2O)_2$. More recently, van der Waals clusters of water molecules have been pursued as far as $(H_2O)_6$. The study of their spectroscopic properties gives detailed information about the intermolecular potentials involved.

21.8 Molecular interactions in liquids

The starting point for the discussion of solids is the well-ordered structures of perfect crystals, which will be discussed in Chapter 23. The starting point for the discussion of gases is the completely disordered distribution of the molecules of a perfect gas, as we saw in Chapter 1. Liquids lie between these two extremes. We shall see that the structural and thermodynamic properties of liquids depend on the nature of intermolecular interactions.

(a) The radial distribution function

The average relative locations of the particles of a liquid are expressed in terms of the **radial distribution function**, $g(r)$. This function is defined so that $g(r)r^2\, dr$ is the probability that a molecule will be found in the range dr at a distance r from another molecule. In a perfect crystal, $g(r)$ is a periodic array of sharp spikes, representing the certainty (in the absence of defects and thermal motion) that molecules (or ions) lie at definite locations. This regularity continues out to the edges of the crystal, so we say that crystals have **long-range order**. When the crystal melts, the long-range order is lost and, wherever we look at long distances from a given molecule, there is equal probability of finding a second molecule. Close to the first molecule, though,

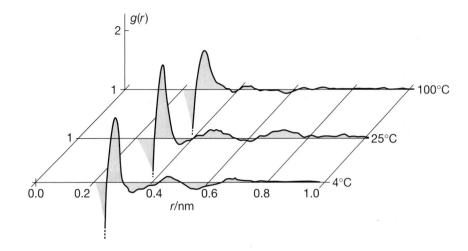

21.27 The radial distribution function of the oxygen atoms in liquid water at three temperatures. Note the expansion as the temperature is raised. (A.H. Narten, M.D. Danford, and H.A. Levy, *Discuss. Faraday. Soc.* **43**, 97 (1967).)

the nearest neighbours might still adopt approximately their original relative positions and, even if they are displaced by newcomers, the new particles might adopt their vacated positions. It is still possible to detect a sphere of nearest neighbours at a distance r_1, and perhaps beyond them a sphere of next-nearest neighbours at r_2. The existence of this **short-range order** means that the radial distribution function can be expected to oscillate at short distances, with a peak at r_1, a smaller peak at r_2, and perhaps some more structure beyond that.

The radial distribution function of the oxygen atoms in liquid water is shown in Fig. 21.27. Closer analysis shows that any given H_2O molecule is surrounded by other molecules at the corners of a tetrahedron. The form of $g(r)$ at 100°C shows that the intermolecular interactions (in this case, principally by hydrogen bonds) are strong enough to affect the local structure right up to the boiling point. Raman spectra indicate that in liquid water most molecules participate in either three or four hydrogen bonds. Infrared spectra show that about 90 per cent of hydrogen bonds are intact at the melting point of ice, falling to about 20 per cent at the boiling point.

(b) The calculation of *g(r)*

Because the radial distribution function can be calculated by making assumptions about the intermolecular interactions, it can be used to test theories of liquid structure. However, even a fluid of hard spheres without attractive interactions (a collection of ball-bearings in a container) gives a function that oscillates near the origin (Fig. 21.28), and one of the factors influencing, and sometimes dominating, the structure of a liquid is the geometrical problem of stacking together reasonably hard spheres. Indeed, the radial distribution function of a liquid of hard spheres shows more pronounced oscillations at a given temperature than any other type of liquid. The attractive part of the potential modifies this basic structure, but sometimes only quite weakly. One of the reasons behind the difficulty of describing liquids theoretically is the similar importance of both the attractive and repulsive (hard core) components of the potential.

The formal expression for the radial distribution function for molecules 1 and 2 in a fluid consisting of N particles is the somewhat fearsome equation

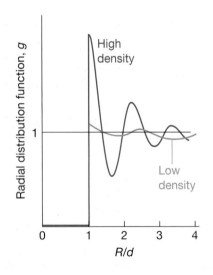

21.28 The radial distribution function for a simulation of a liquid using impenetrable hard spheres (ball-bearings).

$$g(r_{12}) = \frac{\int\int \cdots \int e^{-\beta V_N}\, d\mathbf{r}_3\, d\mathbf{r}_4 \cdots d\mathbf{r}_N}{N^2 \int\int \cdots \int e^{-\beta V_N}\, d\mathbf{r}_1\, d\mathbf{r}_2 \cdots d\mathbf{r}_N} \tag{21.35}$$

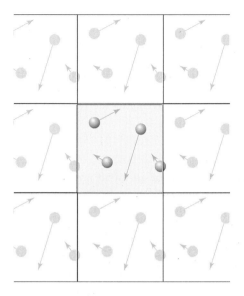

21.29 In a two-dimensional molecular dynamics simulation that uses periodic boundary conditions, when one particle leaves the cell its mirror image enters through the opposite face.

where $\beta = 1/kT$ and V_N is the N-particle potential energy. Although fearsome, this expression is nothing more than the Boltzmann distribution for the relative locations of two molecules in a field provided by all the other molecules in the system. There are several ways of building the intermolecular potential into the calculation of $g(r)$. Numerical methods take a box of about 10^3 particles (the number increases as computers grow more powerful), and the rest of the liquid is simulated by surrounding the box with replications of the original box (Fig. 21.29). Then, whenever a particle leaves the box through one of its faces, its image arrives through the opposite face. When calculating the interactions of a molecule in a box, it interacts with all the molecules in the box and all the periodic replications of those molecules and itself in the other boxes.

Once $g(r)$ is known it can be used to calculate the thermodynamic properties of liquids. For example, the contribution of the pairwise additive intermolecular potential, V_2, to the internal energy is given by the integral

$$U = \frac{2\pi N^2}{V} \int_0^\infty g(r)V_2 r^2 \, dr \tag{21.36}$$

That is, U is essentially the average two-particle potential energy weighted by $g(r)r^2 \, dr$, which is the probability that the pair of particles have a separation between r and $r + dr$. Likewise, the contribution that pairwise interactions make to the pressure is

$$\frac{pV}{nRT} = 1 - \frac{2\pi N}{kTV} \int_0^\infty g(r)v_2 r^2 \, dr \qquad v_2 = r\frac{dV_2}{dr} \tag{21.37}$$

The quantity v_2 is called the **virial** (hence the term 'virial equation of state'). To understand the physical content of this expression, we should rewrite it as

$$p = \frac{nRT}{V} - 2\pi \left(\frac{N}{V}\right)^2 \int_0^\infty g(r)v_2 r^2 \, dr$$

The first term on the right is the **kinetic pressure**, the contribution to the pressure from the impact of the molecules in free flight. The second term is essentially the internal pressure, $\pi_T = (\partial U/\partial V)_T$, introduced in Section 3.3a, representing the contribution to the pressure from the intermolecular forces. To see the connection, we should recognize $-dV_2/dr$ (in v_2) as the force required to move two molecules apart, and therefore $-r(dV_2/dr)$ as the work required to separate the molecules through a distance r. The second term is therefore the average of this work over the range of pairwise separations in the liquid as represented by the probability of finding two molecules at separations between r and $r + dr$, which is $g(r)r^2 \, dr$. In brief, the integral, when multiplied by the square of the number density, is the change in internal energy of the system as it expands and is therefore equal to the internal pressure.

(c) Monte Carlo methods

In the **Monte Carlo method**, the particles in the box are moved through small but otherwise random distances, and the change in total potential energy of the N particles in the box, ΔV_N, is calculated using one of the intermolecular potentials discussed in Sections 21.5 and 21.6. Whether or not this new configuration is accepted is then judged from the following rules:

1 If the potential energy is not greater than before the change, then the configuration is accepted.

2 If the potential energy is greater than before the change, the factor $e^{-\Delta V_N/kT}$ is compared with a random number between 0 and 1; if the factor is larger than the

random number, the configuration is accepted; if the factor is not larger, the configuration is rejected.

This procedure ensures that at equilibrium the probability of occurrence of any configuration is proportional to the Boltzmann factor $e^{-\Delta V_N/kT}$. The configurations generated in this way can then be used to construct $g(r)$ simply by counting the number of pairs of particles with a separation r and averaging the result over the whole collection of configurations.

(d) Molecular dynamics

In the **molecular dynamics** approach, the history of an initial arrangement is followed by calculating the trajectories of all the particles under the influence of the intermolecular potentials. Newton's laws are used to predict where each particle will be after a short time interval (about 1 fs, which is shorter than the average time between collisions), and then the calculation is repeated for tens of thousands of such steps. The time-consuming part of the calculation is the evaluation of the net force on the molecule arising from all the other molecules present in the system.

A molecular dynamics calculation gives a series of snapshots of the liquid, and $g(r)$ can be calculated as before. The temperature of the system is inferred by computing the mean kinetic energy of the particles and using the equipartition result that

$$\langle \tfrac{1}{2}mv_q^2 \rangle = \tfrac{1}{2}kT \tag{21.38}$$

for each coordinate q.

(e) The hydrophobic interaction

Nonpolar molecules do dissolve slightly in polar solvents, but strong interactions between solute and solvent are not possible and as a result it is found that each individual solute molecule is surrounded by a solvent cage (Fig. 21.30). To understand the consequences of this effect, consider the thermodynamics of transfer of a nonpolar hydrocarbon solute from a nonpolar solvent to water, a polar solvent. Experiments indicate that the process is endergonic ($\Delta_{\text{transfer}}G > 0$), as expected on the basis of the increase in polarity of the solvent, but exothermic ($\Delta_{\text{transfer}}H < 0$). Therefore, it is a large decrease in the entropy of the system ($\Delta_{\text{ransfer}}S < 0$) that accounts for the positive Gibbs energy of transfer. For example, the process

$$CH_4(\text{in } CCl_4) \longrightarrow CH_4(\text{aq})$$

has $\Delta_{\text{transfer}}G = +12 \text{ kJ mol}^{-1}$, $\Delta_{\text{transfer}}H = -10 \text{ kJ mol}^{-1}$, and $\Delta_{\text{transfer}}S = -75 \text{ J K}^{-1} \text{ mol}^{-1}$ at 298 K. Substances characterized by a positive Gibbs energy of transfer from a nonpolar to a polar solvent are called **hydrophobic**.

It is possible to quantify the hydrophobicity of a small molecular group R by defining the **hydrophobicity constant**, π, as

$$\pi = \log \frac{S}{S_0} \tag{21.39}$$

where S is the ratio of the molar solubility of the compound R—A in octanol, a nonpolar solvent, to that in water, and S_0 is the ratio of the molar solubility of the compound H—A in octanol to that in water. Therefore, positive values of π indicate hydrophobicity and negative values of π indicate hydrophilicity, the thermodynamic preference for water as a solvent. It is observed experimentally that the π values of most groups do not depend on the nature of A. However, measurements do suggest group additivity of π values. For example, the π values for R = CH_3, CH_2CH_3, $(CH_2)_2CH_3$, $(CH_2)_3CH_3$, and $(CH_2)_4CH_3$ are, respectively, 0.5, 1.0, 1.5, 2.0, and 2.5 and

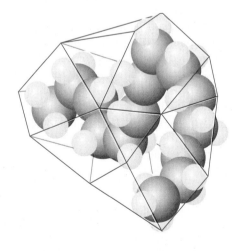

21.30 When a hydrocarbon molecule is surrounded by water, the H_2O molecules form a clathrate cage. As a result of this acquisition of structure, the entropy of the water decreases, so the dispersal of the hydrocarbon into the water is entropy-opposed; its coalescence is entropy-favoured.

we conclude that acyclic saturated hydrocarbons become more hydrophobic as the carbon chain length increases. This trend can be rationalized by $\Delta_{transfer}H$ becoming more positive and $\Delta_{transfer}S$ more negative as the number of carbon atoms in the chain increases.

At the molecular level, formation of a solvent cage around a hydrophobic molecule involves the formation of new hydrogen bonds among solvent molecules. This is an exothermic process and accounts for the negative values of $\Delta_{transfer}H$. On the other hand, the increase in order associated with formation of a very large number of small solvent cages decreases the entropy of the system and accounts for the negative values of $\Delta_{transfer}S$. However, when many solute molecules cluster together, fewer (albeit larger) cages are required and more solvent molecules are free to move. The net effect of formation of large clusters of hydrophobic molecules is then a decrease in the organization of the solvent and therefore a net *increase* in entropy of the system. This increase in entropy of the solvent is large enough to render spontaneous the association of hydrophobic molecules in a polar solvent.

The increase in entropy that results from fewer structural demands on the solvent placed by the clustering of nonpolar molecules is the origin of the **hydrophobic interaction**, which tends to stabilize groupings of hydrophobic groups in micelles and biopolymers (Chapter 22). The hydrophobic interaction is an example of an ordering process that is stabilized by a tendency toward greater disorder of the solvent.

Checklist of key ideas

Electric properties of molecules

21.1 Electric dipole moments

- [ ] electric dipole, two electric charges $+q$ and $-q$ separated by a distance R
- [ ] electric dipole moment, $\boldsymbol{\mu}$ the vector that points from $-q$ to $+q$ with magnitude $\mu = qR$
- [ ] polar molecule, a molecule with a permanent electric dipole moment
- [ ] dipole moment and electronegativity difference, $\mu/D \approx \Delta\chi$
- [ ] homopolar contribution, the contribution to the electric dipole moment that arises from a difference in atomic radii
- [ ] composition of electric dipole moments, $\mu^2 = \mu_1^2 + \mu_2^2 + 2\mu_1\mu_2 \cos\theta$
- [ ] polarization, the electric dipole moment density, $P = \langle\mu\rangle\mathcal{N}$
- [ ] dielectric, a polarizable, nonconducting medium
- [ ] Langevin function, $L(x) = (e^x + e^{-x})/(e^x - e^{-x}) - 1/x$, $x = \mu\mathcal{E}/kT$, with $\langle\mu_z\rangle = \mu L(x)$

21.2 Polarizabilities

- [ ] induced dipole moment, μ^*, the dipole moment induced by an applied electric field
- [ ] polarizability, the constant of proportionality α in $\mu^* = \alpha\mathcal{E}$
- [ ] hyperpolarizability, the coefficient β in $\mu^* = \alpha\mathcal{E} + \frac{1}{2}\beta\mathcal{E}^2 + \cdots$
- [ ] polarizability volume, $\alpha' = \alpha/4\pi\varepsilon_0$
- [ ] polarizability, perturbation expression: $\alpha = 2\sum_n |\mu_{z,0n}|^2/(E_n^{(0)} - E_0^{(0)})$
- [ ] orientation polarization, the polarization arising from the permanent dipole moments
- [ ] distortion polarization, the polarization arising from the distortion of the positions of the nuclei by the applied field
- [ ] electronic polarizability, the polarizability due to the distortion of the electron distribution

21.3 Relative permittivities

- [ ] permittivity, the quantity ε in the Coulomb potential energy, $V = q_1q_2/4\pi\varepsilon r$
- [ ] relative permittivity (dielectric constant), $\varepsilon_r = \varepsilon/\varepsilon_0$
- [ ] Debye equation, $(\varepsilon_r - 1)/(\varepsilon_r + 2) = \rho P_m/M$
- [ ] molar polarization, $P_m = (N_A/3\varepsilon_0)(\alpha + \mu^2/3kT)$
- [ ] Clausius–Mossotti equation, $(\varepsilon_r - 1)/(\varepsilon_r + 2) = \rho N_A\alpha/3M\varepsilon_0$

21.4 Refractive index

- [ ] refractive index, $n_r = c/c'$
- [ ] refractive index and relative permittivity, $n_r = \varepsilon_r^{1/2}$
- [ ] dispersion, the frequency dependence of a property
- [ ] optical activity, the ability to rotate the plane of polarization of light
- [ ] circular birefringent, a medium with different refractive indices for right- and left-circularly polarized light
- [ ] optical rotatory dispersion (ORD), the variation of the angle of optical rotation with frequency
- [ ] circular dichroism (CD), difference in absorption intensities for right- and left-circularly polarized light

Interactions between molecules

☐ van der Waals forces, interactions between closed-shell molecules

21.5 Interactions between dipoles

☐ interaction between a point dipole and a collinear point charge, $V = -\mu_1 q_2/4\pi\varepsilon_0 r^2$

☐ point dipole, a dipole in which the separation between the charges is much smaller than the distance at which the dipole is being observed

☐ interaction between two collinear point dipoles, $V = -\mu_1\mu_2/2\pi\varepsilon_0 r^3$

☐ multipole, an array of point charges

☐ n-pole, an array of point charges with an n-pole moment but no lower moment

☐ monopole, a point charge

☐ quadrupole, an array of point charges that has neither net charge nor dipole moment

☐ octupole, an array of point charges that sum to zero and which has neither a dipole moment nor a quadrupole moment

☐ multipole–multipole potential energy, $V \propto 1/r^{n+m-1}$

☐ electric field of point charge, $\mathcal{E} = q/4\pi\varepsilon_0 r^2$

☐ potential energy of two parallel point dipoles, $V = \mu_1\mu_2(1 - 3\cos^2\theta)/4\pi\varepsilon_0 r^3$

☐ Keesom interaction, $\langle V\rangle = -C/r^6$, $C = 2\mu_1^2\mu_2^2/3(4\pi\varepsilon_0)^2 kT$

☐ dipole–induced-dipole interaction, $\langle V\rangle = -C/r^6$, $C = \mu_1^2\alpha_2'/4\pi\varepsilon_0$

☐ dispersion interaction (London interaction), an induced-dipole–induced-dipole interaction

☐ London formula, $\langle V\rangle = -C/r^6$, $C = \frac{3}{2}\alpha_1'\alpha_2' I_1 I_2/(I_1 + I_2)$

☐ hydrogen bond, an attractive interaction between two species that arises from a link of the form A—H···B, where A and B are highly electronegative elements and B possesses a lone pair of electrons

☐ Axilrod–Teller formula,
$$V = -\frac{C_6}{r_{AB}^6} - \frac{C_6}{r_{BC}^6} - \frac{C_6}{r_{CA}^6} + \frac{C'}{(r_{AB}r_{BC}r_{CA})^3},$$
$$C' = a(3\cos\theta_A\cos\theta_B\cos\theta_C + 1)$$

21.6 Repulsive and total interactions

☐ hard-sphere potential, $V = \infty$ for $r \le d$; $V = 0$ for $r > d$

☐ Mie potential, $V = C_n/r^n - C_m/r^m$

☐ Lennard-Jones potential, $V = 4\varepsilon\{(r_0/r)^{12} - (r_0/r)^6\}$

☐ exp-6 potential, $V = 4\varepsilon\{e^{-r/r_0} - (r_0/r)^6\}$

☐ atomic force microscopy (AFM), a technique in which the force between a molecular-sized probe and a surface is monitored

21.7 Molecular interactions in gases

☐ molecular beam, a collimated, narrow stream of molecules travelling though an evacuated vessel

☐ hydrodynamic flow, net flow arising from intermolecular collisions

☐ molecular flow, collision-free flow

☐ supersonic jet, a stream of molecules in which the average speed of the molecules is much greater than the speed of sound for the molecules that are not part of the stream

☐ supersonic beam, a beam obtained when the region of hydrodynamic flow is skimmed from a supersonic jet and the excess gas pumped away

☐ crossed beam technique, a technique in which two molecular beams are incident at right angles

☐ incident beam flux, the number of particles passing through a given area in a given interval divided by the area and the duration of the interval

☐ differential scattering cross-section, σ, the constant of proportionality between the change in intensity and the intensity of the incident beam, the number density of target molecules, $\mathcal{N}$, and the infinitesimal path length dx through the sample: $dI = \sigma I\mathcal{N}\,dx$

☐ impact parameter, b, the initial perpendicular separation of the paths of the colliding molecules

☐ quantum oscillation, the modification of the scattering in the forward direction by interference between the wavefunctions of a particle along two different paths

☐ rainbow scattering, strongly enhanced scattering in a nonforward direction

☐ rainbow angle, the angle for which $d\theta/db = 0$ and the scattering is strong

☐ van der Waals molecules, complexes of the form AB in which A and B are held together by van der Waals forces or hydrogen bonds

21.8 Molecular interactions in liquids

☐ radial distribution function, $g(r)$, where $g(r)r^2\,dr$ is the probability that a molecule will be found in the range dr at a distance r from another molecule

☐ long-range order, regularity of arrangement that persists over large distances

☐ short-range order, regularity that ceases after a short distance

☐ internal energy of a fluid,
$$U = (2\pi N^2/V)\int_0^\infty g(r)V_2 r^2\,dr$$

☐ virial, $v_2 = r(dV_2/dr)$

☐ pressure of a fluid,
$$p = nRT/V - (2\pi N^2/V^2)\int_0^\infty g(r)v_2 r^2\,dr$$

☐ kinetic pressure, the contribution to the pressure from the impact of the molecules in free flight

☐ Monte Carlo method, a randomization technique constrained by the Boltzmann distribution

☐ molecular dynamics, a technique in which the history of an initial arrangement is followed by calculating the trajectories of all the particles under the influence of the intermolecular potentials

☐ hydrophobic, substances characterized by a positive Gibbs energy of transfer from a nonpolar to a polar solvent

☐ hydrophobicity constant, $\pi = \log(S/S_0)$

☐ hydrophobic interaction, an interaction that favours formation of clusters of hydrophobic groups in aqueous environments and that stems from changes in entropy of water molecules

Further reading

Articles of general interest

B.J. Alder and A.J. Ladd, Simulation by molecular dynamics. In *Encyclopedia of applied physics* (ed. G.L. Trigg), **18**, 281. VCH, New York (1997).

J.-L. Barrat and M.L. Klein, Molecular dynamics simulations of supercooled liquids near the glass transition. *Ann. Rev. Phys. Chem.* **42**, 23 (1991).

F. Caudros, I. Cachadiña, and W. Ahamuda, Determination of Lennard-Jones interaction parameters using a new procedure. *Molec. Engineering* **6**, 319 (1996).

F. Cuardos, A. Mulero, and P. Rubio, The perturbative theories of fluids as a modern version of van der Waals theory. *J. Chem. Educ.* **71**, 956 (1994).

N.E. Cusack, Structure of simple liquids. In *Encyclopedia of applied physics* (ed. G.L. Trigg), **8**, 533. VCH, New York (1994).

C.E. Dykstra, Electrical polarization in diatomic molecules. *J. Chem. Educ.* **65**, 198 (1988).

A. Gelessus, W. Thiel, and W. Weber, Multipoles and symmetry. *J. Chem. Educ.* **72**, 505 (1995).

M.S. Gordon and J.H. Jensen, Understanding the hydrogen bond using quantum chemistry. *Acc. Chem. Res.* **29**, 536–43 (1996).

H. Guerin, Influence of the well width on the third virial coefficient of the square-well intermolecular potential. *J. Chem. Educ.* **69**, 203 (1992).

J.M. Hunter and M.F. Jarrold, Molecular and atomic clusters. In *Encyclopedia of applied physics* (ed. G.L. Trigg), **10**, 411. VCH, New York (1994).

E.D. Isaacs, A. Shukla, P.M. Platzman, D.R. Hamann, B. Barbiellini, and C.A. Tulk, Covalency of the hydrogen bond in ice: a direct X-ray measurement. *Phys. Rev. Letts.* **82**, 600 (1999).

E. Saiz and M.P. Tarazona, Molecular dynamics and the water molecule. *J. Chem. Educ.* **74**, 1350 (1997).

C.P. Smyth, Determination of dipole moments. In *Techniques of chemistry* (ed. A. Weissberger and B.W. Rossiter), **4**, 351. Wiley-Interscience, New York (1972).

P.E.S. Wormer and A. van der Avoird, Intermolecular potentials, internal motions, and spectra of van der Waals and hydrogen-bonded complexes. *Chem. Rev.* **100**, 4109 (2000).

Texts and sources of data and information

M.P. Allen and D. Tildesley, *Computer simulation of liquids*. Oxford University Press (1989).

P.W. Atkins and R.S. Friedman, *Molecular quantum mechanics*. Oxford University Press (1997).

R.S. Berry, S.A. Rice, and J. Ross, *Physical chemistry*. Oxford University Press (2000).

G.W. Castellan, *Physical chemistry*. Benjamin Cummings, Menlo Park (1983).

H. Lodish, A. Berk, S.L. Zipursky, and P. Matsudaira, *Molecular cell biology*. W.H. Freeman & Co., New York (1999).

M.A.D. Fluendy and K.P. Lawley, *Chemical applications of molecular beam scattering*. Chapman and Hall, London (1973).

D. Frenkel and B. Smit, *Understanding molecular simulation: from algorithms to applications*. Academic Press, San Diego (1996).

A. Hinchliffe, *Chemical modeling: from atoms to liquids*. Wiley, New York (1999).

J.N. Israelachvili, *Intermolecular and surface forces*. Academic Press, New York (1998).

G.A. Jeffrey, *An introduction to hydrogen bonding*. Oxford University Press (1997).

A.R. Leach, *Molecular modelling: principles and applications*. Longman, Harlow (1996).

D.R. Lide (ed.), *CRC Handbook of Chemistry and Physics*, Sections 3, 4, 6, 9, 10, 12, and 13. CRC Press, Boca Raton (2000).

J.C. Lindon, G.E. Tranter, and J.L. Holmes (ed.), *Encyclopedia of spectroscopy and spectrometry*. Academic Press, San Diego (2000).

H.-J. Schneider and A. Yatsimirsky, *Principles and methods in supramolecular chemistry*. Wiley, Chichester (1999).

K.E. van Holde, W.C. Johnson, and P.S. Ho, *Principles of physical biochemistry*. Prentice Hall, Upper Saddle River (1998).

Exercises

Discussion questions

21.1 (a) Explain how the permanent dipole moment and the polarizability of a molecule arise.

21.1 (b) Explain why the polarizability of a molecule decreases at high frequencies.

21.2 (a) Describe the experimental procedures available for determining the electric dipole moment of a molecule.

21.2 (b) Account for the theoretical conclusion that many attractive interactions between molecules vary with their separation as $1/r^6$.

21.3 (a) Describe the formation of a hydrogen bond in terms of molecular orbitals.

21.3 (b) Describe how liquids are investigated using molecular dynamics, statistical thermodynamics, and Monte Carlo methods.

21.4 (a) Account for the hydrophobic interaction and discuss its manifestations.

21.4 (b) Describe how molecular beams are used to investigate intermolecular potentials.

Numerical exercises

21.5 (a) Which of the following molecules may be polar: ClF_3, O_3, H_2O_2?

21.5 (b) Which of the following molecules may be polar: SO_3, XeF_4, SF_4?

21.6 (a) The molar polarization of fluorobenzene vapour varies linearly with T^{-1}, and is 70.62 cm^3 mol^{-1} at 351.0 K and 62.47 cm^3 mol^{-1} at 423.2 K. Calculate the polarizability and dipole moment of the molecule.

21.6 (b) The molar polarization of the vapour of a compound was found to vary linearly with T^{-1}, and is 75.74 cm^3 mol^{-1} at 320.0 K and 71.43 cm^3 mol^{-1} at 421.7 K. Calculate the polarizability and dipole moment of the molecule.

21.7 (a) At 0°C, the molar polarization of liquid chlorine trifluoride is 27.18 cm^3 mol^{-1} and its density is 1.89 g cm^{-3}. Calculate the relative permittivity of the liquid.

21.7 (b) At 0°C, the molar polarization of a liquid is 32.16 cm^3 mol^{-1} and its density is 1.92 g cm^{-3}. Calculate the relative permittivity of the liquid. Take $M = 85.0$ g mol^{-1}.

21.8 (a) The refractive index of CH$_2$I$_2$ is 1.732 for 656 nm light. Its density at 20°C is 3.32 g cm^{-3}. Calculate the polarizability of the molecule at this wavelength.

21.8 (b) The refractive index of a compound is 1.622 for 643 nm light. Its density at 20°C is 2.99 g cm^{-3}. Calculate the polarizability of the molecule at this wavelength. Take $M = 65.5$ g mol^{-1}.

21.9 (a) The dipole moments of the bonds C—O and C=O are 1.2 D and 2.7 D, respectively. The bond lengths are 143 pm and 122 pm, respectively. Estimate the percentage ionic character of the bonds. How well do the results correlate with the electronegativity differences of the atoms in the bonds?

21.9 (b) The dipole moments of the bonds C—F and C—O are 1.4 D and 1.2 D, respectively. The bond lengths are 141 pm and 143 pm, respectively. Estimate the percentage ionic character of the bonds. How well do the results correlate with the electronegativity differences of the atoms in the bonds?

21.10 (a) The electric dipole moment of toluene (methylbenzene) is 0.4 D. Estimate the dipole moments of the three xylenes (dimethylbenzene). Which answer can you be sure about?

21.10 (b) Calculate the resultant of two dipole moments of magnitude 1.5 D and 0.80 D that make an angle of 109.5° to each other.

21.11 (a) Calculate the magnitude and direction of the dipole moment of the following arrangement of charges in the xy-plane: $3e$ at (0, 0), $-e$ at (0.32 nm, 0), and $-2e$ at an angle of 20° from the x-axis and a distance of 0.23 nm from the origin.

21.11 (b) Calculate the magnitude and direction of the dipole moment of the following arrangement of charges in the xy-plane: $4e$ at (0, 0), $-2e$ at (162 pm, 0), and $-2e$ at an angle of 30° from the x-axis and a distance of 143 pm from the origin.

21.12 (a) The polarizability volume of H$_2$O is 1.48×10^{-24} cm^3; calculate the dipole moment of the molecule (in addition to the permanent dipole moment) induced by an applied electric field of strength 1.0 kV cm^{-1}.

21.12 (b) The polarizability volume of NH$_3$ is 2.22×10^{-30} m^3; calculate the dipole moment of the molecule (in addition to the permanent dipole moment) induced by an applied electric field of strength 15.0 kV m^{-1}.

21.13 (a) The polarizability volume of H$_2$O at optical frequencies is 1.5×10^{-24} cm^3: estimate the refractive index of water. The experimental value is 1.33; what may be the origin of the discrepancy?

21.13 (b) The polarizability volume of a liquid of molar mass 72.3 g mol^{-1} and density 865 kg mol^{-1} at optical frequencies is 2.2×10^{-30} m^3: estimate the refractive index of the liquid.

21.14 (a) The dipole moment of chlorobenzene is 1.57 D and its polarizability volume is 1.23×10^{-23} cm^3. Estimate its relative permittivity at 25°C, when its density is 1.173 g cm^{-3}.

21.14 (b) The dipole moment of bromobenzene is 5.17×10^{-30} C m and its polarizability volume is approximately 1.5×10^{-29} m^3. Estimate its relative permittivity at 25°C, when its density is 1491 kg m^{-3}.

21.15 (a) A solution of an optically active substance shows an optical rotation of 250° in a cell of length 10 cm at 500 nm. What is the difference of the refractive indices of left and right circularly polarized light through this substance?

21.15 (b) A solution of an optically active substance shows an optical rotation of 192° in a cell of length 15 cm at 450 nm. What is the difference of the refractive indices of left and right circularly polarized light through this substance?

Problems[†]

Numerical problems

21.1 Suppose an H$_2$O molecule ($\mu = 1.85$ D) approaches an anion. What is the favourable orientation of the molecule? Calculate the electric field (in volts per metre) experienced by the anion when the water dipole is (a) 1.0 nm, (b) 0.3 nm, (c) 30 nm from the ion.

21.2 An H$_2$O molecule is aligned by an external electric field of strength 1.0 kV m^{-1} and an Ar atom ($\alpha' = 1.66 \times 10^{-24}$ cm^3) is brought up slowly from one side. At what separation is it energetically favourable for the H$_2$O molecule to flip over and point towards the approaching Ar atom?

21.3 The relative permittivity of chloroform was measured over a range of temperatures with the following results:

θ/°C	−80	−70	−60	−40	−20	0	20
ε_r	3.1	3.1	7.0	6.5	6.0	5.5	5.0
ρ/(g cm^{-3})	1.65	1.64	1.64	1.61	1.57	1.53	1.50

† Problems denoted with the symbol ‡ were supplied by Charles Trapp and Carmen Giunta.

The freezing point of chloroform is –64°C. Account for these results and calculate the dipole moment and polarizability volume of the molecule.

21.4 The relative permittivities of methanol (m.p. –95°C) corrected for density variation are given below. What molecular information can be deduced from these values? Take $\rho = 0.791$ g cm^{-3} at 20°C.

$\theta/°C$	–185	–170	–150	–140	–110	–80	–50	–20	0	20
ε_r	3.2	3.6	4.0	5.1	67	57	49	43	38	34

21.5 In his classic book *Polar molecules*, Debye reports some early measurements of the polarizability of ammonia. From the selection below, determine the dipole moment and the polarizability volume of the molecule.

T/K	292.2	309.0	333.0	387.0	413.0	446.0
$P_m/(cm^3\ mol^{-1})$	57.57	55.01	51.22	44.99	42.51	39.59

The refractive index of ammonia at 273 K and 100 kPa is 1.000 379 (for yellow sodium light). Calculate the molar polarizability of the gas at this temperature and at 292.2 K. Combine the value calculated with the static molar polarizability at 292.2 K and deduce from this information alone the molecular dipole moment.

21.6 Values of the molar polarization of gaseous water at 100 kPa as determined from capacitance measurements are given below as a function of temperature.

T/K	384.3	420.1	444.7	484.1	522.0
$P_m/(cm^3\ mol^{-1})$	57.4	53.5	50.1	46.8	43.1

Calculate the dipole moment of H_2O and its polarizability volume.

21.7‡ The notion of group additivity of thermodynamic properties was introduced in Chapter 2. Results of a computational study of the polarizabilities of the homologous series of linear saturated silanes $Si_{2N}H_{4N+2}$ raise the possibility of group additivity of polarizability. B. Champagne, E.A. Perpète, and J.-M. André (*J. Molec. Structure (Theochem)* **391**, 67 (1997)) report the following values for the electronic contribution to the polarizability, α, of $Si_{2N}H_{4N+2}$ based on *ab initio* quantum mechanical calculations:

N	1	2	3	4	5
$\alpha/(10^{-40}\ J^{-1}\ C\ m^2)$	3.495	7.766	12.40	17.18	22.04

N	6	7	8	9
$\alpha/(10^{-40}\ J^{-1}\ C\ m^2)$	26.92	31.82	36.74	41.63

Determine the average contribution to this quantity per additional Si_2H_4 group, and report the root mean square deviation between the reported values and the best group-additivity fit.

21.8‡ The noble gases and their complexes are favourite objects for the study of weak interatomic and intermolecular forces. J.-U. Grabow, A.S. Pine, G.T. Fraser, F.J. Lovas, R.D. Suenram, T. Emilsson, E. Arunan, and H.S. Gutowsky (*J. Chem. Phys.* **102**, 1181 (1995)) measured the pure rotational spectrum of ^{20}Ne^{40}Ar and reported its rotational constant (actually cB) as 2914.9 MHz. What is the internuclear separation of the complex? They also reported the centrifugal distortion constant (actually cD_J) as 231.01 kHz. Estimate the fundamental vibrational wavenumber and the force constant of the weak bond. ($m(^{40}Ar) = 39.963$ u; $m(^{20}Ne) = 19.992$ u.)

21.9‡ B.D. Shizgal (*J. Molec. Structure (Theochem)* **391**, 131 (1997)) reports the wavenumbers of vibrational transitions (based on quantum mechanical calculations) for NeAr to be 1909.3, 1060.3, and 386.6 m^{-1} for 1–0, 2–1, and 3–2, respectively. Determine the Morse potential parameters D_e and a for this complex.

21.10‡ F. Luo, G.C. MeBane, G. Kim, C.F. Giese, and W.R. Gentry (*J. Chem. Phys.* **98**, 3564 (1993)) reported experimental observation of the He$_2$ complex, a species that had escaped detection for a long time. The fact that the observation required temperatures in the neighbourhood of 1 mK is consistent with computational studies that suggest that hcD_e, for He$_2$ is about 1.51×10^{-23} J, hcD_0 about 2×10^{-26} J, and R about 297 pm. (a) Determine the Lennard-Jones parameters r_0, and ε and plot the Lennard-Jones potential for He–He interactions. (b) Plot the Morse potential given that $a = 5.79 \times 10^{10}$ m^{-1}.

21.11‡ D.D. Nelson, G.T. Fraser, and W. Klemperer (*Science* **238**, 1670 (1987)) examined several weakly bound gas-phase complexes of ammonia in search of examples in which the H atoms in NH$_3$ formed hydrogen bonds, but found none. For example, they found that the complex of NH$_3$ and CO$_2$ has the carbon atom nearest the nitrogen (299 pm away): the CO$_2$ molecule is at right angles to the C–N 'bond', and the H atoms of NH$_3$ are pointing away from the CO$_2$. The permanent dipole moment of this complex is reported as 1.77 D. If the N and C atoms are the centres of the negative and positive charge distributions, respectively, what is the magnitude of those partial charges (as multiples of e)?

21.12‡ From data in Table 21.1 calculate the molar polarization, relative permittivity, and refractive index of methanol at 20°C. Its density at that temperature is 0.7914 g cm^{-3}.

Theoretical problems

21.13 Calculate the potential energy of the interaction between two linear quadrupoles when they are (a) collinear, (b) parallel and separated by a distance r.

21.14 Show that, in a gas (for which the refractive index is close to 1), the refractive index depends on the pressure as $n_r = 1 + $ const $\times p$, and find the constant of proportionality. Go on to show how to deduce the polarizability volume of a molecule from measurements of the refractive index of a gaseous sample.

21.15 The refractive index of benzene is constant (at 1.51) from 0.4 GHz up to 0.55 GHz (in the microwave region of the spectrum), but then shows a series of oscillations between 1.47 and 1.54. Throughout the same frequency range, methylbenzene shows a higher refractive index (about 1.55), the same oscillations as in benzene, and additional oscillations between 1.52 and 1.56 near 0.4 GHz. Account for these observations.

21.16 Acetic acid vapour contains a proportion of planar, hydrogen-bonded dimers. The relative permittivity of pure liquid acetic acid is 7.14 at 290 K and increases with increasing temperature. Suggest an interpretation of the latter observation. What effect should isothermal dilution have on the relative permittivity of solutions of acetic acid in benzene?

21.17 Show that the mean interaction energy of N atoms of diameter d interacting with a potential energy of the form C_6/R^6 is given by $U = -2N^2 C_6/3Vd^3$, where V is the volume in which the molecules are confined and all effects of clustering are ignored. Hence, find a connection between the van der Waals parameter a and C_6, from $n^2 a/V^2 = (\partial U/\partial V)_T$.

21.18 Suppose the repulsive term in a Lennard-Jones (12,6)-potential is replaced by an exponential function of the form $e^{-r/d}$. Sketch the form of the potential energy and locate the distance at which it is a minimum.

21.19 The *cohesive energy density*, $\mathcal{U}$, is defined as U/V, where U is the mean potential energy of attraction within the sample and V its volume. Show that $\mathcal{U} = \frac{1}{2}\mathcal{N}\int V(R)\,d\tau$, where $\mathcal{N}$ is the number density of the molecules and $V(R)$ is their attractive potential energy and where the integration ranges from d to infinity and over all angles. Go on to show that the cohesive energy density of a uniform distribution of molecules that interact by a van der Waals attraction of the form $-C_6/R^6$ is equal to $(2\pi/3)(N_A^2/d^3M^2)\rho^2C_6$, where ρ is the mass density of the solid sample and M is the molar mass of the molecules.

21.20 Consider the collision between a hard-sphere molecule of radius R_1 and mass m, and an infinitely massive impenetrable sphere of radius R_2. Plot the scattering angle θ as a function of the impact parameter b. Carry out the calculation using simple geometrical considerations.

21.21 The dependence of the scattering characteristics of atoms on the energy of the collision can be modelled as follows. We suppose that the two colliding atoms behave as impenetrable spheres, as in Problem 21.20, but that the effective radius of the heavy atoms depends on the speed v of the light atom. Suppose its effective radius depends on v as R_2e^{-v/v^*}, where v^* is a constant. Take $R_1 = \frac{1}{2}R_2$ for simplicity and an impact parameter $b = \frac{1}{2}R_2$, and plot the scattering angle as a function of (a) speed, (b) kinetic energy of approach.

Applications: to biology

21.22 Phenylalanine (Phe, **11**) is a naturally occurring amino acid. What is the energy of interaction between its phenyl group and the electric dipole moment of a neighbouring peptide group? Take the distance between the groups as 4.0 nm and treat the phenyl group as a benzene molecule. The dipole moment of the peptide group is $\mu = 1.26$ D and the polarizability volume of benzene is $\alpha' = 1.04 \times 10^{-29}$ m^3.

21.23 Now consider the London interaction between the phenyl groups of two Phe residues (see Problem 21.22). (a) Estimate the potential energy of interaction between two such rings (treated as benzene molecules) separated by 1.0 nm. For the ionization energy, use $I = 5.0$ eV. (b) Given that force is the negative slope of the potential, calculate the distance dependence of the force acting between two nonbonded groups of atoms, such as the phenyl groups of Phe, in a polypeptide chain that can have a London dispersion interaction with each other. What is the separation at which the force between the phenyl groups (treated as benzene molecules) of two Phe residues is zero?

21.24 This problem gives an example of a *structure–activity relation* (SAR), in which it is possible to correlate the effect of a structural change in a compound with its biological function. The use of SARs can improve the design of drugs for the treatment of disease because it facilitates the prediction of the biological activity of a compound before it is synthesized. The binding of nonpolar groups of amino acids to hydrophobic sites in the interior of proteins is governed largely by hydrophobic interactions. (a) Consider a family of hydrocarbons R—H. The hydrophobicity constants, π, for R = CH$_3$, CH$_2$CH$_3$, (CH$_2$)$_2$CH$_3$, (CH$_2$)$_3$CH$_3$, and (CH$_2$)$_4$CH$_3$ are, respectively, 0.5, 1.0, 1.5, 2.0, and 2.5. Use these data to predict the π value for (CH$_2$)$_6$CH$_3$. (b) The equilibrium constants K_I for the dissociation of inhibitors (**12**) from the enzyme chymotrypsin were measured for different substituents R:

R	CH$_3$CO	CN	NO$_2$	CH$_3$	Cl
π	−0.20	−0.025	0.33	0.5	0.9
$\log K_I$	−1.73	−1.90	−2.43	−2.55	−3.40

Plot $\log K_I$ against π. Does the plot suggest a linear relationship? If so, what are the slope and intercept to the $\log K_I$ axis of the line that best fits the data? (c) Predict the value of K_I for the case R = H.

11

12

22

Macromolecules and aggregates

Structure and dynamics

22.1 The different levels of structure

22.2 Random coils

22.3 The structure of proteins

22.4 The structure of nucleic acids

22.5 The stability of biological polymers

Determination of size and shape

22.6 Mean molar masses

22.7 Mass spectrometry

22.8 Laser light scattering

22.9 Ultracentrifugation

22.10 Electrophoresis

22.11 Size-exclusion chromatography

22.12 Viscosity

Self-assembly

22.13 Colloids

22.14 Micelles and biological membranes

22.15 Surface films

Checklist of key ideas

Further reading

Exercises

Problems

Macromolecules exhibit a range of properties and problems that illustrate a wide variety of physical chemical principles. They need to be characterized in terms of their molar mass, their size, and their shape. However, the molecules are so large and the solutions they form depart so strongly from ideality, that techniques for accommodating these departures from ideality need to be developed. Another major problem concerns the influences that determine the shapes of the molecules. We consider a range of influences in this chapter, beginning with a structureless random coil and ending with the structurally precise forces that operate in polypeptides and nucleic acids. Atoms, small molecules, and macromolecules can form large assemblies that are held together by one or more of the molecular interactions described in Chapter 21. These assemblies, which include colloids and biological membranes, exhibit some of the typical properties of molecules but have their own characteristic features.

There are macromolecules everywhere, inside us and outside us. Some are natural: they include polysaccharides such as cellulose, polypeptides such as protein enzymes, and polynucleotides such as deoxyribonucleic acid (DNA). Others are synthetic: they include **polymers** such as nylon and polystyrene that are manufactured by stringing together and (in some cases) cross-linking smaller units known as **monomers**. Life in all its forms, from its intrinsic nature to its technological interaction with its environment, is the chemistry of macromolecules.

Macromolecules give rise to special problems that include the shapes and the lengths of polymer chains, the determination of their sizes, and the large deviations from ideality of their solutions. Natural macromolecules differ in certain respects from synthetic macromolecules, particularly in their composition and the resulting structure, but the two share a number of common properties. We concentrate on these common properties here. Another level of complexity arises when small molecules aggregate into large particles in a process that is called 'self-assembly' and give rise to aggregates. One example is the assembly of haemoglobin from four myoglobin-like polypeptides. A similar type of aggregation gives rise to a variety of disperse phases, which include colloids. The properties of these disperse phases resemble to a certain extent the properties of solutions of macromolecules, and we describe their common attributes in the final part of this chapter.

22.1 (a) A polymer adopts a highly organized helical conformation, an example of a secondary structure. The helix is represented as a cylinder. (b) Several helical segments connected by short random coils pack together, providing an example of tertiary structure.

Structure and dynamics

The concept of the 'structure' of a macromolecule takes on different meanings at the different levels at which we think about the arrangement of the chain or network of monomers. The term **configuration** refers to the structural features that can be changed only by breaking chemical bonds and forming new ones. Thus, the chains —A—B—C— and —A—C—B— have different configurations. The term **conformation** refers to the spatial arrangement of the different parts of a chain, and one conformation can be changed into another by rotating one part of a chain around a bond.

22.1 The different levels of structure

The **primary structure** of a macromolecule is the sequence of small molecular residues making up the polymer. The residues may form either a chain, as in polyethylene, or a more complex network, in which cross-links connect different chains, as in cross-linked polyacrylamide. In a synthetic polymer, virtually all the residues are identical and it is sufficient to name the monomer used in the synthesis. Thus, the repeating unit of polyethylene is —CH_2CH_2—, and the primary structure of the chain is specified by denoting it as —$(CH_2CH_2)_n$—.[1]

The concept of primary structure ceases to be trivial in the case of synthetic copolymers and biological macromolecules, for in general these substances are chains formed from different molecules. For example, proteins are **polypeptides** formed from different amino acids (about twenty occur naturally) strung together by the **peptide link**, —CONH—. The determination of the primary structure is then a highly complex problem of chemical analysis called **sequencing**. The **degradation** of a polymer is a disruption of its primary structure, when the chain breaks into shorter components.

The **secondary structure** of a macromolecule is the (often local) spatial arrangement of a chain. The secondary structure of an isolated molecule of polyethylene is a random coil, whereas that of a protein is a highly organized arrangement determined largely by hydrogen bonds, and taking the form of random coils, helices (Fig. 22.1a), or sheets in various segments of the molecule. The loss of secondary structure is called **denaturation**. When the hydrogen bonds in a protein are destroyed (for instance, by heating, as when cooking an egg) the structure denatures into a random coil.

The **tertiary structure** is the overall three-dimensional structure of a macromolecule. For instance, the hypothetical protein shown in Fig. 22.1b has helical regions connected by short random-coil sections. The helices interact to form a compact tertiary structure.[2]

The **quaternary structure** of a macromolecule is the manner in which large molecules are formed by the aggregation of others. Figure 22.2 shows how four molecular subunits, each with a specific tertiary structure, aggregate together. Quaternary structure can be very important in biology. For example, the oxygen-transport protein haemoglobin consists of four subunits that work together to take up and release O_2.

22.2 Several subunits with specific tertiary structures pack together, providing an example of quaternary structure.

1. More rigorously, the repeating unit is —CH_2— and the substance is polymethylene. However, the advantage of regarding the repeating unit as —CH_2CH_2— and naming it after its monomer is that derivatives, —$CHXCH_2$—, are seen to belong to the same family.
2. The interactions influencing the tertiary structure are discussed in Sections 22.3 and 22.4.

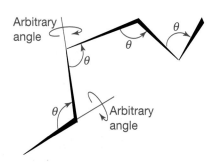

22.3 A freely jointed chain is like a three-dimensional random walk, each step being in an arbitrary direction but of the same length.

22.4 A better description is obtained by fixing the bond angle (for example, at the tetrahedral angle) and allowing free rotation about a bond direction.

22.2 Random coils

The most likely conformation of a chain of identical units not capable of forming hydrogen bonds or any other type of specific bond is a **random coil**. Polyethylene is a simple example. The random coil model is a helpful starting point for estimating the orders of magnitude of the hydrodynamic properties of polymers and denatured proteins in solution.

The simplest model of a random coil is a **freely jointed chain**, in which any bond is free to make any angle with respect to the preceding one (Fig. 22.3). We assume that the residues occupy zero volume, so different parts of the chain can occupy the same region of space. The model is obviously an oversimplification because a bond is actually constrained to a cone of angles around a direction defined by its neighbour (Fig. 22.4). In a hypothetical one-dimensional freely jointed chain all the residues lie in a straight line, and the angle between neighbours is either 0° or 180°. The residues in a three-dimensional freely jointed chain are not restricted to lie in a line or a plane.

(a) Measures of size

As shown in the following *Justification*, we can deduce the probability, P, that the ends of a one-dimensional freely jointed chain composed of N units of length l are a distance nl apart:

$$P = \left(\frac{2}{\pi N}\right)^{1/2} e^{-n^2/2N} \tag{22.1}$$

This function is plotted in Fig. 22.5 and can be used to calculate the probability that the ends of a three-dimensional freely jointed chain lie in the range R to $R + dR$. We write this probability as $f\,dR$, where

$$f = 4\pi\left(\frac{a}{\pi^{1/2}}\right)^3 R^2 e^{-a^2 R^2} \qquad a = \left(\frac{3}{2Nl^2}\right)^{1/2} \tag{22.2}$$

In some coils, the ends may be far apart whereas in others their separation is small.[3] An alternative interpretation of eqn 2 is to regard each coil in a sample as ceaselessly

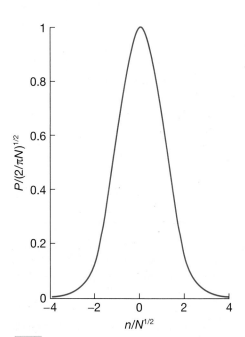

22.5 The probability distribution for the separation of the ends of a one-dimensional random coil. The separation of the ends is nl, where l is the bond length.

3. Here and elsewhere we are ignoring the fact that the chain cannot be longer than Nl. Although eqn 2 gives a nonzero probability for $R > Nl$, the values are so small that the errors in pretending that R can range up to infinity are negligible.

writhing from one conformation to another; then $f \, dR$ is the probability that at any instant the chain will be found with the separation of its ends between R and $R + dR$.

Justification 22.1

Consider a one-dimensional freely jointed polymer. We can specify the conformation of a molecule by stating the number of bonds pointing to the right (N_R) and the number pointing to the left (N_L). The distance between the ends of the chain is $(N_R - N_L)l$, where l is the length of an individual bond. We write $n = N_R - N_L$ and the total number of bonds as $N = N_R + N_L$.

The number of ways of forming a chain with a given end-to-end distance nl is the number of ways of having N_R right-pointing and N_L left-pointing bonds, and is given by the binomial coefficient

$$W = \frac{N!}{N_L! N_R!} = \frac{N!}{\{\frac{1}{2}(N + n)\}! \{\frac{1}{2}(N - n)\}!} \tag{22.3}$$

The probability that the separation is nl is

$$P = \frac{\text{number of polymers with } N_R \text{ bonds to the right}}{\text{total number of arrangements of bonds}}$$

$$= \frac{N!/N_R!(N - N_R)!}{2^N} = \frac{N!}{\{\frac{1}{2}(N + n)\}! \{\frac{1}{2}(N - n)\}! \, 2^N}$$

When the chain is compact in the sense that $n \ll N$, it is more convenient to evaluate $\ln P$: the factorials are then large and we can use Stirling's approximation (Section 19.1a) in the form

$$\ln x! \approx \ln(2\pi)^{1/2} + (x + \tfrac{1}{2}) \ln x - x$$

The result, after quite a lot of algebra, is

$$\ln P = \ln \left(\frac{2}{\pi N} \right)^{1/2} - \tfrac{1}{2}(N + n + 1) \ln (1 + v) - \tfrac{1}{2}(N - n + 1) \ln (1 - v)$$

where $v = n/N$. For a compact coil ($v \ll 1$) we use the approximation $\ln(1 \pm v) \approx \pm v - \tfrac{1}{2} v^2$ and so obtain

$$\ln P \approx \ln \left(\frac{2}{\pi N} \right)^{1/2} - \tfrac{1}{2} N v^2$$

which rearranges into eqn 1.

There are several measures of the geometrical size of a random coil. The **contour length**, R_c, is the length of the macromolecule measured along its backbone from atom to atom. For a polymer of N monomer units each of length l, the contour length is

$$R_c = Nl \tag{22.4}$$

The **root mean square separation**, R_{rms}, is a measure of the average separation of the ends of a random coil: it is the square root of the mean value of R^2, calculated by weighting each possible value of R^2 with the probability that R occurs:

$$R_{rms} = \left(\int_0^\infty R^2 f \, dR \right)^{1/2} = N^{1/2} l \tag{22.5}$$

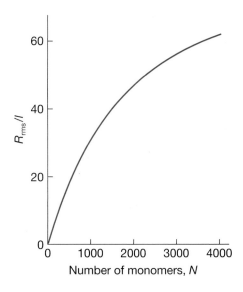

22.6 The variation of the root mean square separation of the ends of a three-dimensional random coil, R_{rms}, with the number of monomers.

We see that, as the number of monomer units increases, the root mean square separation of its end increases as $N^{1/2}$ (Fig. 22.6), and consequently its volume increases as $N^{3/2}$.

Another convenient measure of size is the **radius of gyration**, R_g, the radius of a thin hollow spherical shell of the same mass and moment of inertia as the molecule. It is calculated formally from the expression:[4]

$$R_g = \frac{1}{N}\left(\frac{1}{2}\sum_{ij} R_{ij}^2\right)^{1/2} \tag{22.6}$$

where R_{ij} is the separation of atoms i and j. The radius of gyration of the coil also increases as $N^{1/2}$:

$$R_g = \left(\frac{N}{6}\right)^{1/2} l \tag{22.7}$$

The radius of gyration may also be calculated for other geometries. For example, a solid uniform sphere of radius R has $R_g = (\frac{3}{5})^{1/2}R$, and a long thin uniform rod of length l has $R_g = (3)^{1/2}l/2$ for rotation about an axis perpendicular to the long axis. Table 22.1 lists some experimental values.

Example 22.1 Calculating the dimensions of a random coil

Calculate the mean separation of the ends of a freely jointed chain of N bonds of length l.

Method The general expression for the mean nth power of the end-to-end separation is

$$\langle R^n \rangle = \int_0^\infty R^n f \, dR$$

which should be used with $n = 1$ and f from eqn 2.

Answer The mean separation is

$$\langle R \rangle = 4\pi\left(\frac{a}{\pi^{1/2}}\right)^3 \int_0^\infty R^3 e^{-a^2 R^2} \, dR = \frac{2}{a\pi^{1/2}} = \left(\frac{8N}{3\pi}\right)^{1/2} l$$

The standard integral we have used is

$$\int_0^\infty x^3 e^{-a^2 x^2} \, dx = \frac{1}{2a^4}$$

Comment The result must be multiplied by a factor when the chain is not freely jointed (see below). The random coil model ignores the role of the solvent: a poor solvent will tend to cause the coil to tighten so that solute–solvent contacts are minimized; a good solvent does the opposite. Therefore, calculations based on this model are better regarded as lower bounds to the dimensions for a polymer in a good solvent and as an upper bound for a polymer in a poor solvent. The model is most reliable for a polymer in a bulk solid sample, where the coil is likely to have its natural dimensions.

Table 22.1* Radius of gyration

	$M/(\text{kg mol}^{-1})$	R_g/nm
Serum albumin	66	2.98
Polystyrene	3.2×10^3	50†
DNA	4×10^3	117

* More values are given in the *Data section*.
† In a poor solvent.

4. In Problem 22.17, this definition is shown to be equivalent to another and more easily visualized one in the case of a chain of identical atoms: the radius of gyration is the root mean square distance of the atoms from the centre of mass.

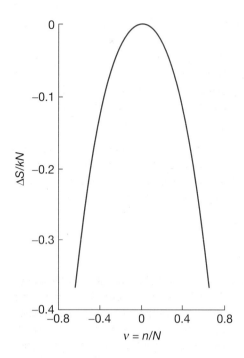

22.7 The change in molar entropy of a perfect elastomer as its extention changes; $v = 1$ corresponds to complete extension; $v = 0$, the conformation of highest entropy, corresponds to the random coil.

Self-test 22.1 Evaluate the root mean square separation of the ends of the chain. $[N^{1/2}l]$

(b) Conformational entropy

The random coil is the least structured conformation of a polymer chain and corresponds to the state of greatest entropy. Any stretching of the coil introduces order and reduces the entropy. Conversely, the formation of a random coil from a more extended form is a spontaneous process (provided enthalpy contributions do not interfere). As shown in the *Justification* below, we can use the same model to deduce that the change in **conformational entropy**, the statistical entropy arising from the arrangement of bonds, when a coil containing N bonds of length l is stretched or compressed by nl is

$$\Delta S = -\tfrac{1}{2}kN \ln\{(1 + v)^{1+v}(1 - v)^{1-v}\} \qquad v = n/N \tag{22.8}$$

This function is plotted in Fig. 22.7, and we see that minimum extension corresponds to maximum entropy.

Justification 22.2

The conformational entropy of the chain is $S = k \ln W$, where W is given by eqn 3. Therefore,

$$S/k = \ln N! - \ln\{\tfrac{1}{2}(N+n)\}! - \ln\{\tfrac{1}{2}(N-n)\}!$$

Because the factorials are large (except for large extensions), we can use Stirling's approximation to obtain

$$S/k = -\ln(2\pi)^{1/2} + (N+1)\ln 2 + (N + \tfrac{1}{2})\ln N - \tfrac{1}{2}\ln\{(N+n)^{N+n+1}(N-n)^{N-n+1}\}$$

The most probable conformation of the chain is the one with the ends close together ($n = 0$), as may be confirmed by differentiation. Therefore, the maximum entropy is

$$S/k = -\ln(2\pi)^{1/2} + (N+1)\ln 2 - \tfrac{1}{2}\ln N$$

The change in entropy when the chain is stretched or compressed by nl is therefore the difference of these two quantities, and the resulting expression is eqn 8.

(c) Constrained chains

The freely jointed chain model is improved by removing the freedom of bond angles to take any value. For long chains, we can simply take groups of neighbouring bonds and consider the direction of their resultant. Although each successive individual bond is constrained to a single cone of angle θ relative to its neighbour, the resultant of several bonds lies in a random direction. By concentrating on such groups rather than individuals, it turns out that for long chains the expressions for the root mean square separation and the radius of gyration given above should be multiplied by

$$F = \left(\frac{1 - \cos\theta}{1 + \cos\theta}\right)^{1/2} \tag{22.9}$$

For tetrahedral bonds, for which $\cos\theta = -\tfrac{1}{3}$ (that is, $\theta = 109.5°$), $F = 2^{1/2}$. Therefore:

$$R_{\text{rms}} = (2N)^{1/2}l \qquad R_{\text{g}} = \left(\frac{N}{3}\right)^{1/2} l \tag{22.10}$$

22.8 A random coil in three dimensions. This one contains about 200 units. The root mean square distance between the ends (R_{rms}) and the radius of gyration (R_g) are indicated.

Illustration 22.1

Consider a polyethylene chain with $M = 56$ kg mol^{-1}, corresponding to $N = 4000$. Because $l = 154$ pm for a C—C bond, we find $R_{rms} = 14$ nm and $R_g = 5.6$ nm (Fig. 22.8). This value of R_g means that, on average, the coils rotate like hollow spheres of radius 5.6 nm and mass equal to the molecular mass.

The model of a randomly coiled molecule is still an approximation, even after the bond angles have been restricted, because it does not take into account the impossibility of two or more atoms occupying the same place. Such self-avoidance tends to swell the coil, so (in the absence of solvent effects) it is better to regard R_{rms} and R_g as lower bounds to the actual values.

22.3 The structure of proteins

A protein is a polypeptide composed of linked α-amino acids, NH$_2$CHRCOOH, where R is one of about 20 groups. For a protein to function correctly, it needs to have a well defined conformation. For example, an enzyme has its greatest catalytic efficiency only when it is in a specific conformation. The amino acid sequence of a protein contains the necessary information to create the active conformation of the protein from a newly synthesized random coil. However, the prediction of the conformation from the primary structure, the so-called *protein folding problem*, is extraordinarily difficult and is still the focus of much research.

(a) The Corey–Pauling rules

The origin of the secondary structures of proteins is found in the rules formulated by Linus Pauling and Robert Corey in 1951. The essential feature is the stabilization of structures by hydrogen bonds involving the peptide link. The latter can act both as a donor of the H atom (the NH part of the link) and as an acceptor (the CO part). The **Corey–Pauling rules** are as follows (Fig. 22.9):

1 The four atoms of the peptide link lie in a relatively rigid plane.

The planarity of the link is due to delocalization of π electrons over the O, C, and N atoms and the maintenance of maximum overlap of their p orbitals.

2 The N, H, and O atoms of a hydrogen bond lie in a straight line (with displacements of H tolerated up to not more than 30° from the N—O vector).

3 All NH and CO groups are engaged in hydrogen bonding.

The rules are satisfied by two structures. One, in which hydrogen bonding between peptide links leads to a helical structure, is the **α-helix**. The other, in which hydrogen bonding between peptide links leads to a planar structure, is the **β-sheet**;[5] this form is the secondary structure of the protein fibroin, the constituent of silk.

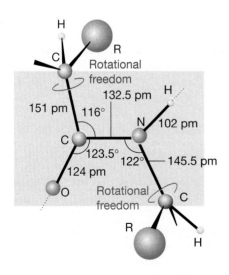

22.9 The dimensions that characterize the peptide link. The C—NH—CO—C atoms define a plane (the C—N bond has partial double-bond character), but there is rotational freedom around the C—CO and N—C bonds.

(b) The α-helix

The α-helix is illustrated in Fig. 22.10. Each turn of the helix contains 3.6 amino acid residues, so the period of the helix corresponds to 5 turns (18 residues). The pitch of a single turn (the distance between points separated by 360°) is 544 pm. The N—H···O bonds lie parallel to the axis and link every fourth group (so residue i is linked to residues $i - 4$ and $i + 4$). All the R groups point away from the major axis of the helix.

5. The β-sheet is often called the β-pleated sheet.

22.10 The polypeptide α-helix, with poly-L-glycine as an example. Carbon atoms are shown in grey, with nitrogen and oxygen atoms both in blue, and hydrogen atoms in white. See the web site for a full colour version. There are 3.6 residues per turn, and a translation along the helix of 150 pm per residue, giving a pitch of 540 pm. The diameter (ignoring side chains) is about 600 pm.

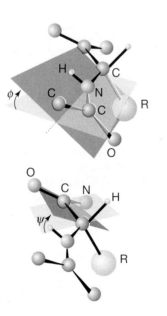

22.11 The definition of the torsional angles ψ and ϕ between two peptide units.

There is freedom for the helix to be arranged as either a right- or a left-handed screw, but the overwhelming majority of natural polypeptides are right-handed on account of the preponderance of the L-configuration of the naturally occurring amino acids, as we explain below. The reason for their preponderance is uncertain, but it may be related to the symmetries of fundamental particles and the nonconservation of parity (the fact that this universe behaves differently from its hypothetical mirror image).

(c) Conformational energy

A polypeptide chain adopts a conformation corresponding to a minimum Gibbs energy, which depends on the **conformational energy**, the energy of interaction between different parts of the chain, and the energy of interaction between the chain and surrounding solvent molecules. In the aqueous environment of biological cells, the outer surface of a protein molecule is covered by a mobile sheath of water molecules, and its interior contains pockets of water molecules. These water molecules play an important role in determining the conformation that the chain adopts through hydrophobic interactions and hydrogen bonding to amino acids in the chain.

The simplest calculations of the conformational energy of a polypeptide chain ignore entropy and solvent effects and concentrate on the total potential energy of all the interactions between nonbonded atoms. For example, these calculations predict that a right-handed α-helix of L-amino acids is marginally more stable than a left-handed helix of the same amino acids.

To calculate the energy of a conformation, we need to make use of many of the molecular interactions described in Chapter 21, and also of some additional interactions:

1 *Bond stretching.* Bonds are not rigid, and it may be advantageous for some bonds to stretch and others to be compressed slightly as parts of the chain press against one another. If we liken the bond to a spring, then the potential energy takes the form of Hooke's law

$$V_{\text{stretch}} = \tfrac{1}{2}k_{\text{stretch}}(R - R_{\text{e}})^2 \tag{22.11}$$

where R_{e} is the equilibrium bond length and k_{stretch} is the force constant, a measure of the stiffness of the bond in question.

2 *Bond bending.* An O—C—H bond angle (or some other angle) may open out or close in slightly to enable the molecule as a whole to fit together better. If the equilibrium bond angle is θ_{e}, we write

$$V_{\text{bend}} = \tfrac{1}{2}k_{\text{bend}}(\theta - \theta_{\text{e}})^2 \tag{22.12}$$

where k_{bend} is the force constant, a measure of how difficult it is to change the bond angle.

3 *Bond torsion.* There is a barrier to internal rotation of one bond relative to another (just like the barrier to internal rotation in ethane). Because the planar peptide link is relatively rigid, the geometry of a polypeptide chain can be specified by the two angles that two neighbouring planar peptide links make to each other. Figure 22.11 shows the two angles ϕ and ψ commonly used to specify this relative orientation. The sign convention is that a positive angle means that the front atom must be rotated clockwise to bring it into an eclipsed position relative to the rear atom. For an all-*trans* form of the chain, all ϕ and ψ are 180°. A helix is obtained when all the ϕ are equal and when all the ψ are equal. For a right-handed α-helix, all $\phi = -57°$ and all $\psi = -47°$. For a left-handed α-helix, both angles are positive. The torsional contribution to the total potential energy is

$$V_{\text{torsion}} = A(1 + \cos 3\phi) + B(1 + \cos 3\psi) \tag{22.13}$$

22.12 For large molecules, a plot of potential energy against the molecular geometry often show several local minima and a global minimum.

in which A and B are constants of the order of 1 kJ mol^{-1}. Because only two angles are needed to specify the conformation of a helix, and they range from $-180°$ to $+180°$, the torsional potential energy of the entire molecule can be represented on a **Ramachandran plot**, a contour diagram in which one axis represents ϕ and the other represents ψ.

4 *Interaction between partial charges.* If the partial charges q_i and q_j on the atoms i and j are known, a Coulombic contribution of the form $1/r$ can be included (Section 21.3):

$$V_{Coulomb} = \frac{q_i q_j}{4\pi\varepsilon r} \tag{22.14}$$

where ε is the permittivity of the medium in which the charges are embedded. Charges of $-0.28e$ and $+0.28e$ are assigned to N and H, respectively, and $-0.39e$ and $+0.39e$ to O and C, respectively. The interaction between partial charges does away with the need to take dipole–dipole interactions into account, for they are taken care of by dealing with each partial charge explicitly.

5 *Dispersive and repulsive interactions.* The interaction energy of two atoms separated by a distance r (which we know once ϕ and ψ are specified) can be given by the Lennard-Jones (12,6)-form (Section 21.6):

$$V_{LJ} = \frac{C}{r^{12}} - \frac{D}{r^6} \tag{22.15}$$

6 *Hydrogen bonding.* In some models of structure, the interaction between partial charges is judged to take into account the effect of hydrogen bonding. In other models, hydrogen bonding is added as another interaction of the form

$$V_{H\ bonding} = \frac{E}{r^{12}} - \frac{F}{r^{10}} \tag{22.16}$$

The total potential energy of a given conformation (ϕ, ψ) can be calculated by summing the contributions given by eqns 11–16 for all bond angles (including torsional angles) and pairs of atoms in the molecule. The procedure is known as a **molecular mechanics** simulation and is automated in commercially available molecular modelling software. For large molecules, plots of potential energy versus bond distance or bond angle often show several local minima and a global minimum (Fig. 22.12). The software packages include schemes for modifying the locations of the atoms and searching for these minima systematically.

Figure 22.13 shows the potential energy contours for the helical form of polypeptide chains formed from the nonchiral amino acid glycine (R = H) and the chiral amino acid L-alanine (R = CH$_3$). The contours were computed by summing all the

22.13 Contour plots of potential energy against the torsional angles ψ and ϕ, also known as Ramachandran diagrams, for (a) a glycyl residue of a polypeptide chain and (b) an alanyl residue. The darker the shading is, the lower the potential energy. The glycyl diagram is symmetrical, but regions I and II in the alanine diagram, which correspond to right- and left-handed helices, are unsymmetrical, and the minimum in region I lies lower than that in region II. (After D.A. Brant and P.J. Flory, *J. Mol. Biol.* **23**, 47 (1967).)

(a)

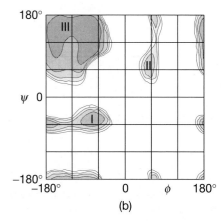

(b)

The web site contains links to sites where you may predict the secondary structure of a polypeptide by molecular mechanics simulations. There are also links to sites where you may visualize the structures of proteins and nucleic acids that have been obtained by experimental and theoretical methods.

contributions described above for each choice of angles, and then plotting contours of equal potential energy. The glycine map is symmetrical, with minima of equal depth at $\phi = -80°$, $\psi = +90°$ and at $\phi = +80°$, $\psi = -90°$. In contrast, the map for L-alanine is unsymmetrical, and there are three distinct low-energy conformations (marked I, II, III). The minima of regions I and II lie close to the angles typical of right- and left-handed α-helices, but the former has a lower minimum, which is consistent with the formation of right-handed helices from the naturally occurring L-amino acids.

The structure corresponding to the global minimum of a molecular mechanics simulation is a snapshot of the molecule at $T = 0$ because only the potential energy is included in the calculation; contributions to the total energy from kinetic energy are excluded. In a **molecular dynamics** simulation, the molecule is set in motion by heating it to a specified temperature.[6] The possible trajectories of all atoms under the influence of the intermolecular potentials are then calculated by integration of Newton's equations of motion (see *Further information 2*). These trajectories correspond to the conformations that the molecule can sample at the temperature of the simulation. At very low temperatures, the molecule cannot overcome some of the potential energy barriers given by eqns 11–16, atomic motion is restricted, and only a few conformations are possible. At high temperatures, more potential energy barriers can be overcome and more conformations are possible. Therefore, molecular dynamics calculations are useful tools for the visualization of the flexibility of polymers.

(d) The β-sheet

A β-sheet is formed by hydrogen bonding between two extended polypeptide chains (large absolute values of the torsion angles ϕ and ψ). Some of the R groups point above and some point below the sheet. Two types of structures can be distinguished from the pattern of hydrogen bonding between the constituent chains.

In an **antiparallel β-sheet** (Fig. 22.14a), $\phi = -139°$, $\psi = 113°$, and the N—H—O atoms of the hydrogen bonds form a straight line. This arrangement is a consequence of the antiparallel arrangement of the chains: every N—H bond on one chain is aligned with a C—O bond from another chain. Antiparallel β-sheets are very common in proteins. In a **parallel β-sheet** (Fig. 22.14b), $\phi = -119°$, $\psi = 113°$, and the N—H—O atoms of the hydrogen bonds are not perfectly aligned. This arrangement is a result of the parallel arrangement of the chains: each N—H bond on one chain is aligned with a N—H bond of another chain and, as a result, each C—O bond of one chain is aligned with a C—O bond of another chain. These structures are not common in proteins.

(e) Higher-order structures

As we have seen, covalent and non-covalent interactions may cause polypeptide chains with well defined secondary structures to fold into tertiary structures. Subunits with well defined tertiary structures may interact further to form quaternary structures.

Although we do not know all the rules that govern protein folding, a few general conclusions may be drawn from X-ray diffraction studies of water-soluble natural proteins and synthetic polypeptides. In an aqueous environment, the chains fold in such a way as to place nonpolar R groups in the interior (which is often not very accessible to solvent) and charged R groups on the surface (in direct contact with the polar solvent). A wide variety of structures can result from these broad rules. Among them, a **four-helix bundle** (Fig. 22.15), which is found in proteins such as cytochrome b_{562} (an electron transport protein), forms when each helix has a nonpolar region along its length. The four nonpolar regions pack together to form a nonpolar

(a)

(b)

22.14 The two types of β-sheets: (a) antiparallel ($\phi = -139°$, $\psi = 113°$), in which the N—H—O atoms of the hydrogen bonds form a straight line; (b) parallel ($\phi = -119°$, $\psi = 113°$), in which the N—H—O atoms of the hydrogen bonds are not perfectly aligned.

6. Molecular dynamics calculations are also discussed in Section 21.8d.

22.15 A four-helix bundle forms from the interactions between nonpolar amino acids on the surfaces of each helix, with the polar amino acids exposed to the aqueous environment of the solvent.

interior. Similarly, interconnected β-sheets may interact to form a **β-barrel** (Fig. 22.16), the interior of which is populated by nonpolar R groups and which has an exterior rich in charged residues. The retinol-binding protein of blood plasma, which is responsible for transporting vitamin A, is an example of a β-barrel structure.

Factors that promote the folding of proteins include covalent disulfide (—S—S—) links, Coulombic interactions between ions (which depend on the degree of protonation of groups and therefore on the pH), hydrogen bonding, van der Waals interactions, and hydrophobic interactions (Section 21.8e). The clustering of nonpolar, hydrophobic, amino acids into the interior of a protein is driven primarily by hydrophobic interactions.

22.4 The structure of nucleic acids

Nucleic acids are key components of the mechanism of storage and transfer of genetic information in biological cells. Deoxyribonucleic acid (DNA) contains the instructions for protein synthesis, which is carried out by different forms of ribonucleic acid (RNA). In this section, we discuss the main structural features of DNA and RNA.

Both DNA and RNA are *polynucleotides* (**1**), in which base–sugar–phosphate units are linked by phosphodiester bonds. In RNA the sugar is β-D-ribose and in DNA it is

R = OH (D-ribose)
H (2'-deoxy-D-ribose)

22.16 Eight antiparallel β-sheets, each represented by a blue arrow and linked by short random coils, fold together as a β-barrel. Nonpolar amino acids are in the interior of the barrel.

1

2 Adenine, A

3 Cytosine, C

4 Guanine, G

5 Thymine, T

6 Uracil, U

7 A—T base pair

8 C—G base pair

β-D-2-deoxyribose (as shown in **1**). The most common bases are adenine (A, **2**), cytosine (C, **3**), guanine (G, **4**), thymine (T, found in DNA only, **5**), and uracil (U, found in RNA only, **6**). At physiological pH, each phosphate group of the chain carries a negative charge and the bases are deprotonated and neutral. This charge distribution leads to two important properties. One is that the polynucleotide chain is a **polyelectrolyte**, a macromolecule with many different charged sites, with a large and negative overall surface charge. The second is that the bases can interact by hydrogen bonding, as shown for A—T (**7**) and C—G base pairs (**8**). The secondary and tertiary structures of DNA and RNA arise primarily from the pattern of this hydrogen bonding between bases of one or more chains.

In DNA, two polynucleotide chains wind around each other to form a double helix (Fig. 22.17). The chains are held together by links involving A—T and C—G base pairs that lie parallel to each other and perpendicular to the major axis of the helix. The structure is stabilized further by interactions between the planar π systems of the bases (Box 21.1). In B-DNA, the most common form of DNA found in biological cells, the helix is right-handed with a diameter of 2.0 nm and a pitch of 3.4 nm. Long stretches of DNA can fold further into a variety of tertiary structures. Two examples are shown in Fig. 22.18. Supercoiled DNA is found in the chromosome and can be visualized as the twisting of closed circular DNA (ccDNA), much like the twisting of a rubber band.

The extra —OH group in β-D-ribose imparts enough steric strain to a polynucleotide chain that stable double helices cannot form in RNA. Therefore, RNA exists primarily as single chains that can fold into complex structures by formation of A—U and G—C base pairs. One example of this is the structure of transfer RNA (tRNA), shown schematically in Fig. 22.19 in which base-paired regions are connected by loops and coils. Transfer RNAs help assemble polypeptide chains during protein synthesis in the cell.

22.5 The stability of biological polymers

The loss of their natural conformation by proteins and nucleic acids is called **denaturation**. It can be achieved by changing the temperature or by adding chemical agents. Cooking is an example of thermal denaturation. When eggs are cooked, the protein albumin is denatured irreversibly, collapsing into a structure that resembles a random coil. One example of chemical denaturation is the 'permanent waving' of hair, which is a result of the reorganization of the protein keratin in hair. Disulfide cross-links between the chains of keratin render the protein, and hence hair fibres, inflexible. Chemical reduction of the disulfide bonds unravels keratin, and allows hair to be shaped. Oxidation re-forms the disulfide bonds and sets the new shape. The 'permanence' is only temporary, however, because the structure of newly formed

22.17 DNA double helix, in which two polynucleotide chains are linked together by hydrogen bonds between adenine (A) and thymine (T) and between cytosine (C) and guanine (G).

Closed circular DNA

Supercoiled DNA

22.18 A long section of DNA may form closed circular DNA (ccDNA) by covalent linkage of the two ends of the chain. Twisting of ccDNA leads to the formation of supercoiled DNA.

22.19 Structure of a transfer RNA (tRNA).

9 Urea, $(NH_2)_2CO$

hair is genetically controlled. Other means of chemically denaturing a protein include the addition of compounds that form stronger hydrogen bonds than those within a helix or sheet. One example is urea (**9**), which competes for the NH and CO groups of a polypeptide. The action of acids or bases, which can protonate or deprotonate groups involved in hydrogen bonding or change the Coulombic interactions that determine the conformation of a protein, can also result in denaturation.

Closer examination of thermal denaturation reveals some of the chemical factors that determine protein and nucleic acid stability. Thermal denaturation is similar to the melting of synthetic polymers (Box 2.1 and Section 23.8c). Denaturation is a **cooperative process** in the sense that the biopolymer becomes increasingly more susceptible to denaturation once the process begins. This cooperativity is observed as a sharp step in a plot of fraction of unfolded polymer versus temperature (Box 20.1). The melting temperature, T_m, is the temperature at which the fraction of unfolded polymer is 0.5.

The thermal stability of DNA increases with the number of G—C base pairs in the sequence because each G—C base pair has three hydrogen bonds whereas each A—T base pair has only two. More energy is required to unravel a double helix that, on average, has more hydrogen bonding interactions per base pair.

Proteins are relatively unstable towards chemical and thermal denaturation. For example, $T_m = 320$ K for ribonuclease T_1 (an enzyme that cleaves RNA in the cell), which is low compared to the temperature at which the enzyme must operate (close to body temperature, 310 K). More surprisingly, the Gibbs energy for the unfolding of ribonuclease T_1 at pH 7.0 and 298 K is only 22.5 kJ mol^{-1}, which is comparable to the energy required to break a single hydrogen bond (about 20 kJ mol^{-1}). Therefore, unlike DNA, the stability of a protein does not increase in a simple way with the number of hydrogen-bonding interactions. While the reasons for the low stability of proteins are not known, the answer probably lies in a delicate balance of all intra- and intermolecular interactions that allow a protein to fold into its active conformation, as discussed in Section 22.3.

Determination of size and shape

X-ray diffraction (Chapter 23) can reveal the position of almost every heavy atom (that is, every atom other than hydrogen) even in very large molecules. However, there are several reasons why other techniques must also be used. In the first place, the sample might be a mixture of molecules with different chain lengths and extents of cross-linking, in which case sharp X-ray images are not obtained. Even if all the molecules in the sample are identical, it might prove impossible to obtain a single crystal. Furthermore, although work on proteins and DNA has shown how immensely interesting and motivating the data can be, the information is incomplete. For instance, what can be said about the shape of the molecule in its natural environment, a biological cell? What can be said about the response of its shape to changes in its environment?

22.6 Mean molar masses

A pure protein is **monodisperse**, meaning that it has a single, definite molar mass.[7] A synthetic polymer, however, is **polydisperse**, in the sense that a sample is a mixture of molecules with various chain lengths and molar masses. The various techniques that are used to measure molar masses result in different types of mean values of polydisperse systems.

The mean obtained from the determination of molar mass by osmometry (Section 7.5e) is the **number-average molar mass**, $\overline{M}_n$, which is the value obtained by weighting each molar mass by the number of molecules of that mass present in the sample:

$$\overline{M}_n = \frac{1}{N} \sum_i N_i M_i \tag{22.17}$$

where N_i is the number of molecules with molar mass M_i and there are N molecules in all. Viscosity measurements give the **viscosity-average molar mass**, $\overline{M}_v$, light-scattering experiments give the **weight-average molar mass**, $\overline{M}_w$ and sedimentation experiments give the **Z-average molar mass**, $\overline{M}_z$. Although such averages are often best left as empirical quantities, some may be interpreted in terms of the composition of the sample. Thus, the weight-average molar mass is the average calculated by weighting the molar masses of the molecules by the mass of each one present in the sample:

$$\overline{M}_w = \frac{1}{m} \sum_i m_i M_i \tag{22.18}$$

In this expression, m_i is the total mass of molecules of molar mass M_i and m is the total mass of the sample. Because $m_i = N_i M_i / N_A$, we can also express this average as

$$\overline{M}_w = \frac{\sum_i N_i M_i^2}{\sum_i N_i M_i} \tag{22.19}$$

This expression shows that the weight-average molar mass is proportional to the mean square molar mass. Similarly, the Z-average molar mass can be interpreted in terms of the mean cubic molar mass:

$$\overline{M}_z = \frac{\sum_i N_i M_i^3}{\sum_i N_i M_i^2} \tag{22.20}$$

Example 22.2 Calculating number and mass averages

Determine the number-average and the weight-average molar masses for a sample of poly(vinyl chloride) from the following data:

Molar mass interval/(kg mol^{-1})	Average molar mass within interval/(kg mol^{-1})	Mass of sample within interval/g
5–10	7.5	9.6
10–15	12.5	8.7
15–20	17.5	8.9
20–25	22.5	5.6
25–30	27.5	3.1
30–35	32.5	1.7

---

7. There may be small variations, such as one amino acid replacing another, depending on the source of the sample.

Method The relevant equations are eqns 17 and 18. Calculate the two averages by weighting the molar mass within each interval by the number and mass, respectively, of the molecule in each interval. Obtain the numbers in each interval by dividing the mass of the sample in each interval by the average molar mass for that interval. Because the number of molecules is proportional to amount of substance (the number of moles), the number-weighted average can be obtained directly from the amounts in each interval.

Answer The amounts in each interval are as follows:

Interval	5–10	10–15	15–20	20–25	25–30	30–35
Molar mass/(kg mol^{-1})	7.5	12.5	17.5	22.5	27.5	32.5
Amount/mmol	1.3	0.70	0.51	0.25	0.11	0.052
					Total:	2.92

The number-average molar mass is therefore

$$\overline{M}_n/(\text{kg mol}^{-1}) = \frac{1}{2.92}(1.3 \times 7.5 + 0.70 \times 12.5 + 0.51 \times 17.5 + 0.25 \times 22.5$$
$$+ 0.11 \times 27.5 + 0.052 \times 32.5) = 13$$

The weight-average molar mass is calculated directly from the data after noting that the total mass of the sample is 37.6 g:

$$\overline{M}_w/(\text{kg mol}^{-1}) = \frac{1}{37.6}(9.6 \times 7.5 + 8.7 \times 12.5 + 8.9 \times 17.5 + 5.6 \times 22.5$$
$$+ 3.1 \times 27.5 + 1.7 \times 32.5) = 16$$

Comment Note the significantly different values of the two averages. In this instance, $\overline{M}_w/\overline{M}_n = 1.2$.

Self-test 22.2 Evaluate the Z–average molar mass of the sample.

[19 kg mol^{-1}]

The ratio $\overline{M}_w/\overline{M}_n$ is called the **heterogeneity index** (or 'polydispersity index'). In the determination of protein molar masses we expect the various averages to be the same because the sample is monodisperse (unless there has been degradation). A synthetic polymer normally spans a range of molar masses and the different averages yield different values. Typical synthetic materials have $\overline{M}_w/\overline{M}_n \approx 4$. The term 'monodisperse' is conventionally applied to synthetic polymers in which this index is less than 1.1; commercial polyethylene samples might be much more heterogeneous, with a ratio close to 30. One consequence of a narrow molar mass distribution for synthetic polymers is often a higher degree of three-dimensional long-range order in the solid and therefore higher density and melting point (Section 22.5). The spread of values is controlled by the choice of catalyst and reaction conditions (Sections 26.1). In practice, it is found that long-range order is determined more by structural factors (branching, for instance) than by molar mass.

Average molar masses may be determined by osmotic pressure of polymer solutions (Section 7.5). The upper limit for the reliability of membrane osmometry is about 1000 kg mol^{-1}. A major problem for macromolecules of relatively low molar mass (less than about 10 kg mol^{-1}) is their ability to percolate through the membrane. One consequence of this partial permeability is that membrane osmometry tends to overestimate the average molar mass of a polydisperse mixture. Several

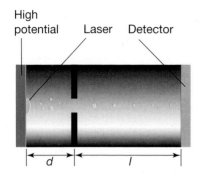

High
potential   Laser   Detector

22.20 Diagram of a matrix-assisted laser desorption/ionization time-of-flight (MALDI-TOF) mass spectrometer. A laser beam ejects macromolecules and ions from the solid matrix. The ionized macromolecules are accelerated by an electrical potential difference over a distance d and then travel through a drift region of length l. Ions with the smallest mass–to–charge number ratio (m/z) reach the detector first.

techniques for the determination of molar mass and polydispersity that are not so limited include mass spectrometry, laser light scattering, ultracentrifugation, electrophoresis, chromatography, and viscosity measurements.

22.7 Mass spectrometry

Mass spectrometry is among the most accurate techniques for the determination of molar masses. The procedure consists of ionizing the sample in the gas phase and then measuring the mass-to-charge number ratio (m/z) of all ions. Macromolecules present a challenge because it is difficult to produce gaseous ions of large species without fragmentation. However, two new techniques have emerged that circumvent this problem: **matrix-assisted laser desorption/ionization** (MALDI) and **electrospray ionization**. We shall discuss **MALDI-TOF mass spectrometry**, so called because the MALDI technique is coupled to a time-of-flight (TOF) ion detector.

Figure 22.20 shows a schematic view of a MALDI-TOF mass spectrometer. The macromolecule is first embedded in a solid matrix that often consists of an organic material such as *trans*-3-indoleacrylic acid and inorganic salts such as sodium chloride or silver trifluoroacetate. This sample is then irradiated with a pulsed laser, such as a nitrogen laser. The laser energy ejects electronically excited matrix ions, cations, and neutral macromolecules, thus creating a dense gas plume above the sample surface. The macromolecule is ionized by collisions and complexation with small cations, such as H^+, Na^+, and Ag^+.

In the TOF spectrometer, the ions are accelerated over a short distance d by an electrical field of strength $\mathcal{E}$ and then travel through a drift region of length l. The time, t, required for an ion of mass m and charge number z to reach the detector at the end of the drift region is (see the *Justification*):

$$t = l\left(\frac{m}{2ze\mathcal{E}d}\right)^{1/2} \tag{22.21}$$

where e is the fundamental charge. Because d, l, and $\mathcal{E}$ are fixed for a given experiment, the time of flight, t, of the ion is a direct measure of its m/z ratio, which is given by:

$$\frac{m}{z} = 2e\mathcal{E}d\left(\frac{t}{l}\right)^2 \tag{22.22}$$

Justification 22.3

Consider an ion of charge ze and mass m that is accelerated from rest by an electric field of strength $\mathcal{E}$ applied over a distance d. The kinetic energy, E_K, of the ion is

$$E_K = \tfrac{1}{2}mv^2 = ze\mathcal{E}d \tag{22.23}$$

where v is the speed of the ion. The drift region, l, and the time of flight, t, in the mass spectrometer are both sufficiently short that we can ignore acceleration and write $v = l/t$. Then substitution into eqn 23 gives

$$\tfrac{1}{2}m\left(\frac{l}{t}\right)^2 = ze\mathcal{E}d \tag{22.24}$$

Rearrangement of this equation gives eqn 21.

10 Poly(butylene adipate)

22.21 MALDI-TOF spectrum of a sample of poly(butylene adipate) with $\overline{M}_n = 4525$ g mol^{-1}. Adapted from Mudiman *et al.*, *J. Chem. Educ.*, **74**, 1288 (1997).

Figure 22.21 shows the MALDI-TOF mass spectrum of a polydisperse sample of poly(butylene adipate) (PBA, **10**). The MALDI technique produces mostly singly charged molecular ions that are not fragmented. Therefore, the multiple peaks in the spectrum arise from polymers of different lengths, with the intensity of each peak being proportional to the abundance of each polymer in the sample. Values of $\overline{M}_n$, $\overline{M}_w$, and the heterogeneity index can be calculated from the data. It is also possible to use the mass spectrum to verify the structure of a polymer, as shown in the following example.

Example 22.3 Interpreting the mass spectrum of a polymer

The mass spectrum in Fig. 22.21 consists of peaks spaced by 200 g mol^{-1}. The peak at 4113 g mol^{-1} corresponds to the polymer for which $n = 20$. From these data, verify that the sample consists of polymers with the general structure given by (**10**).

Method Because each peak corresponds to a different value of n, the molar mass difference, ΔM, between peaks corresponds to the molar mass, M, of the repeating unit (the group inside the brackets in **10**). Furthermore, the molar mass of the terminal groups (the groups outside the brackets in **10**) may be obtained from the molar mass of any peak, by using

M(terminal groups) = M(polymer with n repeating units) − $n\Delta M$ − M(cation)

where the last term corresponds to the molar mass of the cation that attaches to the macromolecule during ionization.

Answer The value of ΔM is consistent with the molar mass of the repeating unit shown in (**10**), which is 200 g mol^{-1}. The molar mass of the terminal group is calculated by recalling that Na$^+$ is the cation in the matrix:

M(terminal group) = 4113 g mol^{-1} − 20(200 g mol^{-1}) − 23 g mol^{-1} = 90 g mol^{-1}

The result is consistent with the molar mass of the —O(CH$_2$)$_4$OH terminal group (89 g mol^{-1}) plus the molar mass of the —H terminal group (1 g mol^{-1}).

Self-test 22.3 What would be the molar mass of the $n = 20$ polymer if silver trifluoroacetate were used instead of NaCl in the preparation of the matrix?

[4198 g mol^{-1}]

22.8 Laser light scattering

Large particles scatter light very efficiently. A familiar example is the light scattered by specks of dust in a sunbeam. Therefore, light scattering is a convenient method for the characterization of polymers, large aggregates (such as colloids), and biological

22.22 Rayleigh scattering from a sample of point-like particles. The intensity of scattered light depends on the angle θ between the incident and scattered beams.

systems from proteins to viruses. Unlike mass spectrometry, laser light scattering measurements may be performed in nearly intact samples; often the only preparation required is filtration of the sample.

(a) General principles of light scattering

When the oscillating electric field of electromagnetic radiation interacts with the electrons in a particle, an oscillating dipole moment develops with a magnitude proportional to the polarizability of the particle and the strength of the field (Section 21.1). Elastic light scattering is observed as the oscillating dipoles in the particle radiate at the same frequency as the frequency of the exciting electromagnetic radiation. The term *elastic* refers to the fact that the incident and scattered photons have the same frequency and hence the same energy. If the medium is perfectly homogeneous, as in a perfect crystal, the scattered waves interfere destructively in all directions except the direction of propagation of the exciting radiation. If the medium is inhomogeneous, as in an imperfect crystal or a solution of macromolecules, radiation is scattered into other directions as well.

Scattering of light by particles with diameters much smaller than the wavelength of the incident radiation is called **Rayleigh scattering** (Fig. 22.22). This type of scattering has several characteristic features.

1 The intensity of scattered light is proportional to λ^{-4}, so shorter wavelength radiation is scattered more intensely than longer wavelengths.[8]

2 The intensity of scattered light is proportional to the molar mass of the particle.

3 The intensity of scattered light depends on the scattering angle θ. In practice, data are collected at several angles to the incident laser beam (Example 22.4).

4 For very dilute solutions excited by plane-polarized light, the **Rayleigh ratio**, R_θ, a measure of the intensity of scattered light at a given scattering angle θ, is defined as

$$R_\theta = \frac{I}{I_0} \times \frac{r^2}{\sin^2\phi} \tag{22.25}$$

where I is the intensity of scattered light, I_0 is the intensity of incident light, r is the distance between the sample and the detector, and ϕ is the angle between the plane of polarization of the incident beam and the plane defined by the incident and scattered beams. For a beam plane-polarized perpendicular to the plane in Fig. 22.22, $\phi = 90°$.

8. The blue of the sky arises from the more intense scattering of the blue component of white sunlight by the molecules of the atmosphere.

As shown in the *Justification* below, for a solution of a polymer of mass concentration c_P,

$$R_\theta = K c_P \overline{M}_{\text{w}}, \qquad \text{with} \qquad K = \frac{4\pi^2 n_{\text{r,0}}^2 V (\text{d}n_\text{r}/\text{d}c_\text{P})^2}{\lambda^4 N_\text{A}} \qquad (22.26)$$

Here $n_{\text{r,0}}$ is the refractive index of the pure solvent, $(\text{d}n_\text{r}/\text{d}c_\text{P})$ is the change in refractive index of the solution with concentration of polymer, V is the volume of the sample, and N_A is the Avogadro constant.

Justification 22.4

The ratio of the intensity of the light scattered by a sample with polarizability α to the intensity of the incident light is (see *Further reading*)

$$\frac{I}{I_0} = \frac{\pi^2 \alpha^2}{\varepsilon_\text{r}^2 \lambda^4} \frac{\sin^2 \phi}{r^2}$$

where r is the distance between the sample and the detector. It follows from eqn 25 that

$$R_\theta = \frac{\pi^2 \alpha^2}{\varepsilon_\text{r}^2 \lambda^4}$$

The relation between the polarizability and the refractive index, n_r, of a solution is (see *Further reading*)

$$n_\text{r}^2 - n_{\text{r,0}}^2 = \frac{\mathcal{N} \alpha}{\varepsilon_\text{r}}$$

where $n_{\text{r,0}}$ is the refractive index of the solvent and $\mathcal{N}$ is the number density of polymer molecules. Because $\mathcal{N} = c_\text{P} N_\text{A}/M$ (where c_P is the mass concentration of the polymer and M is its molar mass), we have:

$$\alpha = \frac{\varepsilon_\text{r} M}{c_\text{P} N_\text{A}} (n_\text{r}^2 - n_{\text{r,0}}^2)$$

For a dilute solution, n_r differs little from $n_{\text{r,0}}$ and we can write:

$$n_\text{r} = n_{\text{r,0}} + \left(\frac{\text{d}n_\text{r}}{\text{d}c_\text{P}} \right) c_\text{P} + \cdots$$

It follows that

$$n_\text{r}^2 \approx n_{\text{r,0}}^2 + 2 n_{\text{r,0}} \left(\frac{\text{d}n_\text{r}}{\text{d}c_\text{P}} \right) c_\text{P}$$

and therefore that

$$\alpha = \frac{2\varepsilon_\text{r} n_{\text{r,0}} M}{N_\text{A}} \left(\frac{\text{d}n_\text{r}}{\text{d}c_\text{P}} \right)$$

The Rayleigh ratio for scattering by a single molecule now becomes

$$R_\theta = \frac{4\pi^2 n_{\text{r,0}}^2 M^2}{\lambda^4 N_\text{A}^2} \left(\frac{\text{d}n_\text{r}}{\text{d}c_\text{P}} \right)^2$$

For a sample of N molecules, we multiply the expression above by $N = c_\text{P} N_\text{A} V/M$ and obtain eqn 26 after substituting $\overline{M}_\text{w}$ for M.

(b) Scattering by large particles

When the wavelength of the incident radiation is comparable to the size of the scattering particles, scattering may occur from different sites of the same molecule and interference between scattered rays becomes important.[9] As a result, the scattering intensity is distorted from the form characteristic of small-particle, Rayleigh scattering of light given in eqn 26. A measure of the distortion is the **structure factor**, P_θ, measured at several angles θ.

If the molecule is regarded as composed of a number of atoms i at distances R_i from a convenient point, interference occurs between the radiation scattered by each pair. The scattering from all the particles is then calculated by allowing for contributions from all possible orientations of each pair of atoms in each molecule. This description is very much like the one used in the discussion of electron diffraction (Box 11.2), so we can expect the intensity pattern to be described by a kind of Wierl equation. Indeed, if there are N atoms in the macromolecule, and if all are assumed to have the same scattering power, then

$$P_\theta = \frac{1}{N^2} \sum_{i,j} \frac{\sin sR_{ij}}{sR_{ij}}, \qquad s = \frac{4\pi}{\lambda} \sin \tfrac{1}{2}\theta \qquad (22.27)$$

where R_{ij} is the separation of atoms i and j, and λ is the wavelength of the incident radiation. The observed Rayleigh ratio is then

$$R_\theta = P_\theta K c_p \overline{M}_w \qquad (22.28)$$

(c) Scattering by small particles

When the molecule is much smaller than the wavelength of the incident radiation in the sense that $sR_{ij} \ll 1$ (for example, if $R = 5$ nm, and $\lambda = 500$ nm, all the sR_{ij} are about 0.1), then we show in the *Justification* below that the deviation from Rayleigh scattering is proportional to the square of the radius of gyration, R_g, of the molecule:

$$P_\theta - 1 \propto R_g^2 \qquad (22.29)$$

Justification 22.5

When $sR_{ij} \ll 1$ we can use the expansion $\sin x = x - \tfrac{1}{6}x^3 + \cdots$ to write

$$\sin sR_{ij} = sR_{ij} - \tfrac{1}{6}(sR_{ij})^3 + \cdots$$

and then

$$P_\theta = \frac{1}{N^2} \sum_{i,j} \{1 - \tfrac{1}{6}(sR_{ij})^2 + \cdots \} = 1 - \frac{s^2}{6N^2} \sum_{i,j} R_{ij}^2 + \cdots$$

The sum over the squares of the separations gives the radius of gyration of the molecule (through eqn 6). Therefore

---

9. The effect accounts for the appearance of clouds, which, although we see them by scattered light, look white, not blue like the sky. The water molecules group together into droplets of a size comparable to the wavelength of light, and scatter cooperatively. Although blue light scatters more strongly, more molecules can contribute cooperatively when the wavelength is longer (as for red light), so the net result is uniform scattering for all wavelengths: white light scatters as white light. This paper looks white for the same reason. Cigarette smoke is blue before it is inhaled, but brownish after it is exhaled because the particles aggregate in the lungs.

$$P_\theta \approx 1 - \tfrac{1}{3}s^2 R_g^2 = 1 - \frac{16\pi^2 R_g^2 \sin^2 \tfrac{1}{2}\theta}{3\lambda^2} \qquad (22.30)$$

which shows that $P_\theta - 1$ is proportional to R_g^2.

(d) Scattering by nonideal solutions of polymers

The preceding discussion shows that structural properties, such as size and the weight-average molar mass of a macromolecule can be obtained from measurements of light scattering by a sample at several angles θ relative to the direction of propagation on an incident laser beam. However, eqn 28 applies only to ideal solutions. In practice, even relatively dilute polymer dispersions can deviate considerably from ideality. To take this into account, it is common to rewrite eqn 28 as

$$\frac{Kc_P}{R_\theta} = \frac{1}{P_\theta \overline{M}_W} + 2Bc_P \qquad (22.31)$$

where B is an empirical constant analogous to the osmotic virial coefficient (Section 7.5).

Macromolecules form strongly nonideal solutions: being so large, they displace a large quantity of solvent instead of replacing individual solvent molecules with negligible disturbance. In thermodynamic terms, the displacement and reorganization of solvent molecules implies that the entropy change is especially important when a macromolecule dissolves. Furthermore, its great bulk means that a macromolecule is unable to move freely through the solution because the molecule is excluded from the regions occupied by other solute molecules. There are also significant contributions to the Gibbs energy from the enthalpy of solution, largely because solvent–solvent interactions are more favourable than the macromolecule–solvent interactions that replace them.

The coefficient B arises from the effect of excluded volume, representing repulsive interactions, and attractive interactions. If we imagine a solution of a macromolecule being built by the successive addition of macromolecules to the solvent, each one being excluded by the ones that preceded it, then the value of B due to repulsions turns out to be:

$$B = \tfrac{1}{2}N_A v_p \qquad (22.32)$$

where v_p is the excluded volume due to a single molecule.

In broad terms, the excluded volume contributes to the excess entropy of solution (the entropy change in excess of the ideal value, Section 7.4), and the attractions and repulsions between macromolecules contribute to the excess enthalpy. For most solute–solvent systems there is a unique temperature (which is not always experimentally attainable) at which these effects cancel and the solution is virtually ideal. This temperature (the analogue of the Boyle temperature for real gases) is called the **Flory theta temperature**. At this temperature, B is zero. As an example, for polystyrene in cyclohexane the theta temperature is approximately 306 K, the exact value depending on the average molar mass of the polymer. A solution at its Flory theta temperature is called a **theta solution**. Because a theta solution behaves nearly ideally, its thermodynamic and structural properties are easier to describe even though the molar concentration is not low. In molecular terms, in a theta solution the molecules are in an unperturbed condition, whereas in other solutions expansion of the coiled molecule takes place as a result of interactions with the solvent.

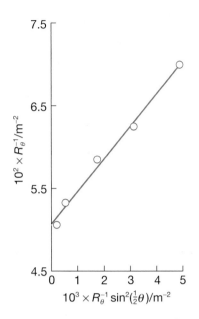

22.23 Plot of the data for Example 22.4.

Example 22.4 Determining the size of a polymer by light scattering

The following data for a sample of polystyrene in butanone were obtained at 20°C with plane-polarized light at $\lambda = 546$ nm.

$\theta/°$	26.0	36.9	66.4	90.0	113.6
R_θ/m^2	19.8	18.8	17.2	16.0	14.4

In a separate experiment, it was determined that $(dn_r/dc_p) = 2.14 \times 10^{-4}$ m^3 kg^{-1}. Other relevant experimental parameters are $n_{butanone} = 1.378$, $c_p = 312$ kg m^{-3}, and $\phi = 90°$. From this information, calculate R_g and $\overline{M}_w$ for the sample. Assume that B is negligibly small and that the polymer is small enough that eqn 30 holds.

Method Substituting the result of eqn 30 into eqn 28 we obtain, after some rearrangement:

$$\frac{1}{R_\theta} = \frac{1}{Kc_p\overline{M}_w} + \left(\frac{16\pi^2 R_g^2}{3\lambda^2}\right)\left(\frac{1}{R_\theta}\sin^2\tfrac{1}{2}\theta\right)$$

Hence, a plot of $1/R_\theta$ against $(1/R_\theta)\sin^2\tfrac{1}{2}\theta$ should be a straight line with slope $16\pi^2 R_g^2/3\lambda^2$ and y-intercept $1/Kc_p\overline{M}_w$.

Answer We construct a table of values of $1/R_\theta$ and $(1/R_\theta)\sin^2\tfrac{1}{2}\theta$ and plot the data (Fig. 22.23).

$10^2 \times R_\theta^{-1}/m^{-2}$	5.06	5.32	5.83	6.25	6.96
$10^3 \times R_\theta^{-1}\sin^2(\tfrac{1}{2}\theta)/m^{-2}$	2.56	5.33	17.5	31.3	48.7

The best straight line through the data has a slope of 0.391 and a y-intercept of 5.06×10^{-2}. From these values, we calculate $R_g = 4.71 \times 10^{-8}$ m = 47.1 nm and $\overline{M}_w = 987$ kg mol^{-1}.

Comment A more accurate method for more concentrated samples consists of conducting a series of experiments where R_θ versus θ data are obtained for several c_p values. From analysis of the entire data set, the R_g, $\overline{M}_w$, and B values are obtained.

Self-test 22.4 Estimate the length of the polymer, assuming that it is a rod. Comment on the validity of using eqn 30 to analyse the data.
[$l = 54.4$ nm. The length of the rod is 9.6% the wavelength of the exciting radiation. The use of eqn 30 appears to be justified.]

(e) Dynamic light scattering

A special laser scattering technique, **dynamic light scattering**, can be used to investigate the diffusion of polymers in solution. Consider two polymer molecules being irradiated by a laser beam. Suppose that at a time t the scattered waves from these particles interfere constructively at the detector, leading to a large signal. However, as the molecules move through the solution, the scattered waves may interfere destructively at another time t' and result in no signal. When this behaviour is extended to a very large number of molecules in solution, it results in fluctuations in light intensity that depend on the diffusion coefficient, D, which is a measure of the rate of molecular motion and is given by the **Stokes–Einstein relation:**[10]

10. For a full discussion of the diffusion coefficient and the derivation of the Stokes–Einstein relation, see Section 24.9e.

Table 22.2* Diffusion coefficients in water at 20°C

	$M/(\text{kg mol}^{-1})$	$D/(\text{m}^2\,\text{s}^{-1})$
Sucrose	0.342	4.59×10^{-10}
Lysozyme	14.1	1.04×10^{-10}
Haemoglobin	68	6.9×10^{-11}
Collagen	345	6.9×10^{-12}

* More values are given in the *Data section*.

$$D = \frac{kT}{f} \tag{22.33}$$

where f is the **frictional coefficient**, a measure of the forces that retard a molecule's motion. Table 22.2 lists some typical values of D. For a spherical particle of radius a in a solvent of viscosity η, the frictional coefficient is given by **Stokes's relation:**[11]

$$f = 6\pi a\eta \tag{22.34}$$

If the molecule is not spherical, we use appropriate values of f given in Table 22.3. Hence, analysis of intensity fluctuations in light scattered by a sample gives the diffusion coefficient and molecular size, in cases where the molecular shape is known. For dilute monodisperse systems of random coils, it has been found empirically that D is related to the molar mass M of the polymer by:

$$D = \beta_D M^{-0.6} \tag{22.35}$$

The coefficient β_D is obtained by determining D at fixed viscosity and temperature for a variety of standard samples with known molar masses. As we should expect, bulky random coils of high molar mass migrate more slowly (have a lower diffusion coefficient) through a solvent than random coils of low molar mass.

22.9 Ultracentrifugation

In a gravitational field, heavy particles settle towards the foot of a column of solution by the process called **sedimentation**. The rate of sedimentation depends on the strength of the field and on the masses and shapes of the particles. Spherical molecules (and compact molecules in general) sediment faster than rod-like or extended molecules. For example, DNA helices sediment much faster when they are denatured to a random coil, so sedimentation rates can be used to study denaturation. When the sample is at equilibrium, the particles are dispersed over a range of heights in accord with the Boltzmann distribution (because the gravitational field competes with the stirring effect of thermal motion). The spread of heights depends on the masses of the molecules, so the equilibrium distribution is another way to determine molar mass.

Sedimentation is normally very slow, but it can be accelerated by **ultracentrifugation**, a technique that replaces the gravitational field with a centrifugal field. The effect can be achieved in an ultracentrifuge, which is essentially a cylinder that can be rotated at high speed about its axis with a sample in a cell near its periphery (Fig. 22.24). Modern ultracentrifuges can produce accelerations equivalent to about 10^5 that of gravity ('$10^5\,g$'). Initially, the sample is uniform, but the 'top' (innermost) boundary of the solute moves outwards as sedimentation proceeds.

Table 22.3* Frictional coefficients and molecular geometry†

a/b	Prolate	Oblate
2	1.04	1.04
4	1.18	1.17
6	1.31	1.28
8	1.43	1.37
10	1.54	1.46

* More values and analytical expressions are given in the *Data section*.
† Entries are the ratio f/f_0 where $f_0 = 6\pi\eta c$ with $c = (ab^2)^{1/3}$ for prolate ellipsoids and $c = (a^2b)^{1/3}$ for oblate ellipsoids; $2a$ is the major axis and $2b$ is the minor axis.

(a) The rate of sedimentation

A solute particle of mass m has an effective mass $m_{\text{eff}} = bm$ on account of the buoyancy of the medium, with

$$b = 1 - \rho v_s \tag{22.36}$$

where ρ is the solution density, v_s is the partial specific volume of the solute ($v_s = (\partial V/\partial m_B)_T$, with m_B the total mass of solute), and ρv_s is the mass of solvent displaced per gram of solute. The solute particles at a distance r from the axis of a rotor

---

11. The formal definition of viscosity is given in Section 24.4; highly viscous liquids flow slowly and retard the motion of objects through them.

Blank
(balancing)

Sample

Rotor

(a)

ω

r

'Top'

Solution

'Bottom'

(b)

22.24 (a) An ultracentrifuge head. The sample on one side is balanced by a blank diametrically opposite. (b) Detail of the sample cavity: the 'top' surface is the inner surface, and the centrifugal force causes sedimentation towards the outer surface; a particle at a radius r experiences a force of magnitude $mr\omega^2$.

spinning at an angular velocity ω experience a centrifugal force of magnitude $m_{eff}r\omega^2$. The acceleration outwards is countered by a frictional force proportional to the speed, s, of the particles through the medium. This force is written fs, where f is the frictional coefficient (Section 22.8). The particles therefore adopt a **drift speed**, a constant speed through the medium, which is found by equating the two forces $m_{eff}r\omega^2$ and fs. The forces are equal when

$$s = \frac{m_{eff}r\omega^2}{f} = \frac{bmr\omega^2}{f} \tag{22.37}$$

The drift speed depends on the angular velocity and the radius, and it is convenient to define the **sedimentation constant**, S, as

$$S = \frac{s}{r\omega^2} \tag{22.38}$$

Then, because the average molecular mass is related to the average molar mass $\bar{M}_n$ through $m = \bar{M}_n/N_A$,

$$S = \frac{b\bar{M}_n}{fN_A} \tag{22.39}$$

On substituting the Stokes relation for spherical molecules (eqn 34), we obtain

$$S = \frac{b\bar{M}_n}{6\pi a\eta N_A} \tag{22.40}$$

and S may be used to determine either $\bar{M}_n$ or a. Again, if the molecules are not spherical, we use the appropriate value of f given in Table 22.3. As always when dealing with macromolecules, the measurements must be carried out at a series of concentrations and then extrapolated to zero concentration to avoid the complications that arise from the interference between bulky molecules.

Example 22.5 Determining a sedimentation constant

The sedimentation of the protein bovine serum albumin (BSA) was monitored at 25°C. The initial location of the solute surface was at 5.50 cm from the axis of rotation, and during centrifugation at 56 850 r.p.m. it receded as follows:

t/s	0	500	1000	2000	3000	4000	5000
r/cm	5.50	5.55	5.60	5.70	5.80	5.91	6.01

Calculate the sedimentation coefficient.

Method Equation 38 can be interpreted as a differential equation for $s = dr/dt$ in terms of r; so integrate it to obtain a formula for r in terms of t. The integrated expression, an expression for r as a function of t, will suggest how to plot the data and obtain from it the sedimentation constant.

Answer Equation 38 may be written

$$\frac{dr}{dt} = r\omega^2 S$$

This equation integrates to

$$\ln\frac{r}{r_0} = \omega^2 St$$

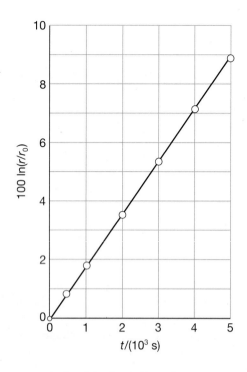

22.25 A plot of the data in Example 22.5.

It follows that a plot of $\ln(r/r_0)$ against t should be a straight line of slope $\omega^2 S$. Use $\omega = 2\pi v$, where v is in cycles per second, and draw up the following table:

t/s	0	500	1000	2000	3000	4000	5000
$10^2 \ln(r/r_0)$	0	0.905	1.80	3.57	5.31	7.19	8.87

The straight-line graph (Fig. 22.25) has slope 1.78×10^{-5}; so $\omega^2 S = 1.79 \times 10^{-5}\ \text{s}^{-1}$. Because $\omega = 2\pi \times (56\,850/60)\ \text{s}^{-1} = 5.95 \times 10^3\ \text{s}^{-1}$, it follows that $S = 5.02 \times 10^{-13}\ \text{s}$.

Comment We develop this result below. The unit 10^{-13} s is sometimes called a 'svedberg' and denoted Sv; in this case $S = 5.02$ Sv. Accurate results are obtained by extrapolating to zero concentration.

Self-test 22.5 Calculate the sedimentation constant given the following data (the other conditions being the same as above):

t/s	0	500	1000	2000	3000	4000	5000
r/cm	5.65	5.68	5.71	5.77	5.84	5.9	5.97

[3.11 Sv]

At this stage, it appears that we need to know the molecular radius a to obtain the molar mass from the value of S. Fortunately, this requirement can be avoided by drawing on the Stokes–Einstein relation (eqn 33) between f and the diffusion coefficient, D. The average molar mass is then:

$$\overline{M} = \frac{SRT}{bD} \tag{22.41}$$

This result is independent of the shape of the solute molecules.[12] It follows that we can find the molar mass by combining measurements of S and D by ultracentrifugation and dynamic light scattering, respectively.

(b) Sedimentation equilibria

The difficulty with using sedimentation rates to measure molar masses lies in the inaccuracies inherent in the determination of diffusion coefficients of polydisperse systems. This problem can be avoided by allowing the system to reach equilibrium, for the transport property D is then no longer relevant. As we show in the *Justification* below, the weight-average molar mass can be obtained from the ratio of concentrations of the macromolecules at two different radii in a centrifuge operating at angular frequency ω:

$$\overline{M}_w = \frac{2RT}{(r_2^2 - r_1^2)b\omega^2} \ln \frac{c_2}{c_1} \tag{22.42}$$

An alternative treatment of the data leads to the Z-average molar mass. The centrifuge is run more slowly in this technique than in the sedimentation rate method to avoid having all the solute pressed in a thin film against the bottom of the cell. At these slower speeds, several days may be needed for equilibrium to be reached.

12. We are not specifying which mean molar mass: which average is obtained depends on technical details of the experiment. See *Further reading*.

Justification 22.6

The distribution of particles is the outcome of the balance between the effect of the centrifugal force and the dispersing effect of diffusion down a concentration gradient. The kinetic energy of a particle of effective mass m at a radius r in a rotor spinning at a frequency ω is $\frac{1}{2}m\omega^2 r^2$, so the total chemical potential at a radius r is $\bar{\mu}(r) = \mu(r) - \frac{1}{2}m\omega^2 r^2$, where $\mu(r)$ is the contribution that depends on the concentration of solute. The condition for equilibrium is that the chemical potential is uniform, so

$$\left(\frac{\partial \bar{\mu}}{\partial r}\right)_T = \left(\frac{\partial \mu}{\partial r}\right)_T - M\omega^2 r = 0$$

with $M = mN_A$. To evaluate the partial derivative of μ, we write

$$\left(\frac{\partial \mu}{\partial r}\right)_T = \left(\frac{\partial \mu}{\partial p}\right)_{T,c}\left(\frac{\partial p}{\partial r}\right)_{T,c} + \left(\frac{\partial \mu}{\partial c}\right)_{T,p}\left(\frac{\partial c}{\partial r}\right)_{T,p} = Mv\omega^2 r\rho + RT\left(\frac{\partial \ln c}{\partial r}\right)_{T,p}$$

The first result follows from the fact that $(\partial \mu/\partial p)_T = V_m$, the partial molar volume, and $V_m = Mv$. It also makes use of the fact that the hydrostatic pressure at r is $p(r) = p(r_0) + \frac{1}{2}\rho\omega^2(r^2 - r_0^2)$, where r_0 is the radius of the surface of the liquid in the sample holder (that is, the location of its meniscus), with ρ the mass density of the solution. The concentration term stems from the expression $\mu = \mu^{\ominus} + RT \ln c$. The condition for equilibrium is therefore

$$Mr\omega^2(1 - v\rho) - RT\left(\frac{\partial \ln c}{\partial r}\right)_{T,p} = 0$$

and therefore, at constant temperature,

$$d \ln c = \frac{Mr\omega^2(1 - v\rho)\,dr}{RT}$$

This expression integrates to eqn 42 provided we identify M with $\overline{M}_W$.

22.10 Electrophoresis

Many macromolecules, such as DNA, are charged and move in response to an electric field. This motion is called **electrophoresis**. Electrophoretic mobility is a result of a constant drift speed, s, reached by an ion when the driving force $ze\mathcal{E}$ (where, as usual, ze is the net charge and $\mathcal{E}$ is the field strength) is matched by the frictional force fs. The drift speed (treated in detail in Section 24.7) is then:

$$s = \frac{ze\mathcal{E}}{f} \tag{22.43}$$

Therefore, the mobility of a macromolecule in an electric field depends on its net charge, size (and hence molar mass), and shape. The latter two factors are implied by the dependence of s on f.

Electrophoresis is a very valuable tool in the separation of biopolymers from complex mixtures, such as those resulting from fractionation of biological cells. We shall consider three techniques used by biochemists: gel electrophoresis, isoelectric focusing, and capillary electrophoresis. They differ in the manner in which the drift speeds of biomolecules are controlled in order to achieve separation of a mixture into its components.

(a) Gel electrophoresis

In **gel electrophoresis**, migration takes place through a gel slab. A common gel material for the study of proteins is cross-linked polyacrylamide. In most cases, the goal of the experiment is to separate a sample according to the molar masses of its components. However, eqn 43 states that shape and charge will also determine the drift speed. For example, proteins with the same size but different net charge travel along the slab at different speeds.

One way to avoid this problem and to effect a separation by molar mass is to denature the proteins in a controlled way. Sodium dodecyl sulfate is an anionic detergent that is very useful in this respect: it denatures proteins, whatever their initial shapes, into rods by forming a complex with them. Moreover, most proteins bind a constant amount of ion, so that the net charge per protein is well regulated. Under these conditions, different proteins in a mixture may be separated according to size only. The molar mass of each constituent protein is estimated by comparing its mobility in its rod-like complexes form with a standard sample of known molar mass. However, molar masses obtained by this method are not as accurate as those obtained by other techniques, such as MALDI-TOF and size-exclusion chromatography (Section 22.11).

(b) Isoelectric focusing

Naturally occurring macromolecules acquire a charge when dispersed in water. An important feature of proteins and other biopolymers is that their overall charge depends on the pH of the medium. For instance, in acidic environments protons attach to basic groups and the net charge is positive; in basic media the net charge is negative as a result of proton loss. At the **isoelectric point**, the pH is such that there is no net charge on the biopolymer. Consequently, the drift speed of a biopolymer depends on the pH of the medium, with $s = 0$ at the isoelectric point.

Isoelectric focusing is an electrophoresis method that exploits the change of drift speed with pH. Consider a mixture of distinct proteins dispersed in a medium with a pH gradient along the direction of an applied electric field. Each protein in the mixture will stop moving at a position in the gradient where the pH is equal to the isoelectric point. In this manner, the protein mixture can be separated into its components.

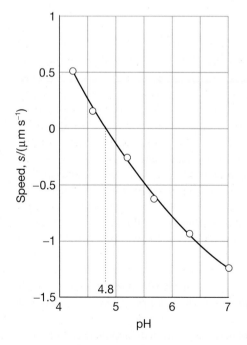

22.26 The plot of drift speed against pH by which the isoelectric point of a macromolecule can be determined: it corresponds to the pH at which the drift speed in the presence of an electric field is zero.

Example 22.6 Determining the isoelectric point of a protein

The drift speed of bovine serum albumin (BSA) under the influence of an electric field in aqueous solution was monitored at several values of pH, and the data are listed below (opposite signs indicate opposite directions of travel). What is the isoelectric point of the protein?

pH	4.20	4.56	5.20	5.65	6.30	7.00
Speed/($\mu m\ s^{-1}$)	+0.50	+0.18	−0.25	−0.65	−0.90	−1.25

Method The macromolecule has zero electrophoretic mobility when it is uncharged. Therefore, the isoelectric point is the pH at which it does not migrate in an electric field. We should therefore plot speed against pH and find by interpolation the pH of zero mobility.

Answer The data are plotted in Fig. 22.26. The drift speed is zero at pH = 4.8; hence pH = 4.8 is the isoelectric point.

Comment For some species, the isoelectric point must be obtained by extrapolation because the macromolecule might not be stable over the whole pH range.

Self-test 22.6 The following data were obtained for another protein:

pH	4.5	5.0	5.5	6.0
Speed/(μm s^{-1})	−0.10	−0.20	−0.30	−0.35

Estimate the pH of the isoelectric point. [4.3]

(c) Capillary electrophoresis

The drift speeds attained by polymers in traditional electrophoresis methods are rather low; as a result, several hours are often necessary to effect good separation of complex mixtures. According to eqn 43, one way to increase the drift speed is to increase the electric field strength. However, there are limits to this strategy because very large electric fields can heat the large surfaces of an electrophoresis apparatus unevenly, leading to a nonuniform distribution of electrophoretic mobilities and poor separation.

In **capillary electrophoresis**, the sample is dispersed in a medium (such as methylcellulose) and held in a thin glass or plastic tube with diameters ranging from 20 to 100 μm. The small size of the apparatus makes it easy to dissipate heat when large electric fields are applied. Excellent separations may be effected in minutes rather than hours. Each polymer fraction emerging from the capillary can be characterized further by other techniques, such as MALDI-TOF.

22.11 Size-exclusion chromatography

All the techniques discussed so far have certain drawbacks, including the time needed to obtain data, the high cost of necessary instrumentation, and the often awkward interpretation of the data. Many of these difficulties are circumvented by a technique that makes use of beads of porous polymeric material about 0.1 mm in diameter that capture molecules selectively, according to their size. In the technique of **size-exclusion chromatography** (SEC), or **gel permeation chromatography**, which is a fast and relatively inexpensive technique for molar mass determinations of polymers, a solution of the polymer sample is filtered through a column. The small molecules, which can permeate into the porous structure of the gel, require a long **elution time**, or time to pass through a particular length of column, whereas the larger ones, which are not captured, pass through rapidly. The average molar mass of a macromolecule may therefore be determined by observing its elution time in a column calibrated against standard samples.

The range of molar masses that can be determined by SEC can be altered by selecting columns made from polymers with different degrees of cross-linking (which influences the pore size of the gel) and different materials. Polystyrene gels are used for investigations of nonpolar polymers in nonpolar solvents, porous glass gels are used for more polar systems, and polyacrylamide gels can be used for proteins.

The most serious limitation of SEC comes from the fact that the elution time also depends on the shape of the polymer. Therefore, direct comparison between calibration standards and unknown samples can lead to errors in the determination of molar masses. This problem is common in SEC of proteins: two proteins with similar molar masses may have distinct shapes and hence different elution times on the

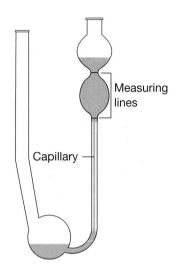

22.27 An Ostwald viscometer. The viscosity is measured by noting the time required for the liquid to drain between the two marks.

same column. However, SEC does give reasonably accurate molar mass values for a large number of proteins. Because elution is performed under pressure, molar mass determinations may be completed within a few minutes, in striking contrast to the time required for more classical techniques. Moreover, only a few milligrams of material are needed for reliable measurements.

22.12 Viscosity

The presence of a macromolecular solute increases the viscosity of a solution.[13] The effect is large even at low concentration, because big molecules affect the fluid flow over an extensive region surrounding them. At low concentrations the viscosity, η, of the solution is related to the viscosity of the pure solvent, η_0, by

$$\eta = \eta_0(1 + [\eta]c + \cdots) \tag{22.44}$$

The **intrinsic viscosity**, $[\eta]$, is the analogue of a virial coefficient (and has dimensions of 1/concentration). It follows from eqn 44 that

$$[\eta] = \lim_{c \to 0}\left(\frac{\eta - \eta_0}{c\eta_0}\right) = \lim_{c \to 0}\left(\frac{\eta/\eta_0 - 1}{c}\right) \tag{22.45}$$

Viscosities are measured in several ways. In the **Ostwald viscometer** shown in Fig. 22.27, the time taken for a solution to flow through the capillary is noted, and compared with a standard sample. The method is well suited to the determination of $[\eta]$ because the ratio of the viscosities of the solution and the pure solvent is proportional to the drainage time t and t_0 after correcting for different densities ρ and ρ_0:

$$\frac{\eta}{\eta_0} = \frac{t}{t_0} \times \frac{\rho}{\rho_0} \tag{22.46}$$

(In practice, the two densities are only rarely significantly different.) This ratio can be used directly in eqn 45. Viscometers in the form of rotating concentric cylinders are also used (Fig. 22.28), and the torque on the inner cylinder is monitored while the outer one is rotated. Such **rotating drum viscometers** have the advantage over the Ostwald type that the shear gradient between the cylinders is simpler than in the capillary and effects of the kind discussed shortly can be studied more easily.

There are many complications in the interpretation of viscosity measurements. Much of the work is based on empirical observations, and the determination of molar mass is usually based on comparisons with standard, nearly monodisperse sample. Some regularities are observed that help in the determination. For example, it is found that θ solutions of macromolecules often fit the **Mark–Kuhn–Houwink–Sakurada equation**:

$$[\eta] = K\overline{M}_{\mathrm{v}}^{a} \tag{22.47}$$

where K and a are constants that depend on the solvent and type of macromolecule (Table 22.4); the viscosity-average molar mass, $\overline{M}_{\mathrm{v}}$, appears in this expression. As an example, solutions of poly(γ-benzyl-L-glutamate) in its rod-like form have an intrinsic viscosity about four times greater than when it is denatured into random coils. Conversely, solutions of the enzyme ribonuclease are less viscous than solutions of the denatured form: this observation indicates that the folded protein is more compact than the denatured form.

22.28 A rotating drum viscometer. The torque on the inner drum is observed when the outer container is rotated.

13. As noted previously, the formal definition of viscosity is given in Section 24.4.

Table 22.4* Intrinsic viscosity

Macromolecule	Solvent	$\theta/°C$	$K/(cm^3\ g^{-1})$	a
Polystyrene	Benzene	25	9.5×10^{-3}	0.74
Polyisobutylene	Benzene	23	8.3×10^{-2}	0.50
Various proteins	Guanidine hydrochloride + $HSCH_2CH_2OH$		7.2×10^{-3}	0.66

* More values are given in the *Data section*.

Example 22.7 Using intrinsic viscosity to measure molar mass

The viscosities of a series of solutions of polystyrene in toluene were measured at 25°C with the following results:

$c/(g\ L^{-1})$	0	2.0	4.0	6.0	8.0	10.0
$\eta/(10^{-4}\ kg\ m^{-1}\ s^{-1})$	5.58	6.15	6.74	7.35	7.98	8.64

Calculate the intrinsic viscosity and estimate the molar mass of the polymer by using eqn 47 with $K = 3.80 \times 10^{-5}\ L\ g^{-1}$ and $a = 0.63$.

Method The intrinsic viscosity is defined in eqn 45; therefore, form this ratio at the series of data points and extrapolate to $c = 0$. Interpret $\overline{M}_v$ as $\overline{M}_v/(g\ mol^{-1})$ in eqn 47.

Answer We draw up the following table:

$c/(g\ L^{-1})$	0	2.0	4.0	6.0	8.0	10.0
η/η_0	1	1.102	1.208	1.317	1.430	1.549
$100[(\eta/\eta_0) - 1]/(c/g\ L^{-1})$		5.11	5.20	5.28	5.38	5.49

The points are plotted in Fig. 22.29. The extrapolated intercept at $c = 0$ is 0.0504, so $[\eta] = 0.0504\ L\ g^{-1}$. Therefore,

$$\overline{M}_v = \left(\frac{[\eta]}{K} \right)^{1/a} = 9.0 \times 10^4\ g\ mol^{-1}$$

Comment When $\eta \approx \eta_0$,

$$\ln \frac{\eta}{\eta_0} = \ln \left(1 + \frac{\eta - \eta_0}{\eta_0} \right) \approx \frac{\eta - \eta_0}{\eta_0} = \frac{\eta}{\eta_0} - 1$$

This relation is exact in the limit that η coincides with η_0, which is true when $c = 0$. Hence, $[\eta]$ can also be defined as the limit of $(1/c) \ln(\eta/\eta_0)$ as $c \to 0$. The intercept is identified more precisely by plotting both functions.

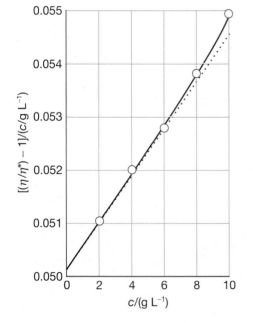

22.29 The plot used for the determination of intrinsic viscosity, which is taken from the intercept at $c = 0$; see Example 22.7.

Self-test 22.7 Evaluate the viscosity-average molar mass by using the second plotting technique. [90 kg mol^{-1}]

In some cases, the flow is non-Newtonian in the sense that the viscosity of the solution changes as the rate of flow increases. A decrease in viscosity with increasing rate of flow indicates the presence of long rod-like molecules that are orientated by the flow and hence slide past each other more freely. In some somewhat rare cases the stresses set up by the flow are so great that long molecules are broken up, with further consequences for the viscosity.

Self-assembly

Much of the material discussed in this chapter also applies to aggregates of particles that form by **self-assembly**, the spontaneous formation of complex structures of molecules or macromolecules held together by molecular interactions, such as Coulombic, dispersion, hydrogen bonding, or hydrophobic interactions. We have already encountered a few examples of self-assembly, such as the formation of liquid crystals (Box 8.1), of protein quaternary structures from two or more polypeptide chains, and of a DNA double helix from two polynucleotide chains. Now we concentrate on the specific properties of additional self-assembled systems, including small aggregates that are at the heart of detergent action, extended sheets like those forming biological cell membranes, and advanced materials for molecular electronics (Box 22.1).

Box 22.1 Molecular wires

Most of the macromolecules and self-assembled structures considered in this chapter are insulators, or very poor electrical conductors. However, a variety of newly developed macromolecular materials have electrical conductivities that rival those of silicon-based semiconductors and even metallic conductors. The hope is that these materials will form the basis for cheaper and smaller electronic devices, including micrometre- and even nanometre-scale transistors, sensors, and computer chips. We examine two examples: *carbon nanotubes*, in which carbon atoms assemble into thin cylinders that are both mechanically strong and highly conducting, and *conducting polymers*, in which extensively conjugated double bonds facilitate electron conduction along a polymer chain. The Nobel Prize in chemistry was awarded in 2000 to A.J. Heeger, A.J. McDiarmid, and H. Shirakawa for their pioneering work in the synthesis and characterization of conducting polymers.

Carbon nanotubes form under some of the same conditions that lead to the formation of fullerenes, such as C_{60} and C_{70}. In recent years, methods for more selective synthesis of nanotubes have been developed and they consist of different ways to condense a carbon plasma either in the presence or absence of a catalyst. The simplest structural motif is called a *single-walled nanotube* (SWNT) and is shown in the first illustration. In an SWNT, sp^2 hybridized carbon atoms form hexagonal rings reminiscent of the structure of the carbon sheets found in graphite. The tubes have diameters between 1 and 2 nm and lengths of several microns. The features shown in the first illustration have been confirmed by direct visualization with scanning tunnelling microscopy (Box 12.1). A *multi-walled nanotube* (MWNT) consists of several concentric SWNTs and its diameter varies between 2 and 25 nm.

The origin of electrical conductivity in carbon nanotubes is the delocalization of π bonds that form from unhydrized p orbitals. Recent studies have shown a correlation between structure and conductivity in SWNTs. The first illustration shows a SWNT that is a semiconductor, with conductivity intermediate between that of an insulator and of a metallic conductor.[14] If the hexagons are rotated by 60°, the resulting SWNT has the characteristics of a metallic conductor.

14. This is a simplified view of the properties of a semiconductor. See Section 23.9c for a more thorough discussion.

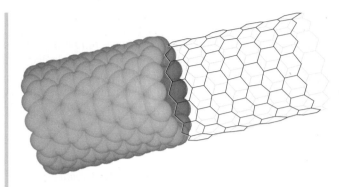

In a single-walled nanotube (SWNT), sp^2 hybridized carbon atoms form hexagonal rings that grow as tubes with diameters between 1 and 2 nm and lengths of several micrometres.

oxidation

The mechanism of migration of a partially localized cation radical, or polaron, in polyacetylene.

The mechanism of electrical conduction in conducting polymers is more complex. One example of a conducting polymer is polyacetylene, shown in the second illustration. Whereas the delocalized π bonds do suggest that electrons can move up and down the chain, the electrical conductivity of polyacetylene increases significantly when it is partially oxidized by I_2 and other strong oxidants. The product is a *polaron*, a partially localized cation radical that does not delocalize but rather travels through the chain, as shown in the second illustration. Further oxidation of the polymer forms either *bipolarons*, a dication that moves as a unit through the chain, or *solitons*, two separate cation radicals that move independently. Polarons and solitons contribute to the mechanism of charge conduction in polyacetylene.

Conducting polymers are slightly better electrical conductors than silicon semiconductors but are far worse than metallic conductors. They are currently used in a number of devices, such as electrodes in batteries, electrolytic capacitors, and sensors. Recent studies of photon emission by conducting polymers may lead to new technologies for light-emitting diodes and flat-panel displays. Like carbon nanotubes, conducting polymers show promise as molecular wires that can be incorporated into nanometre-sized electronic devices.

Discussion questions

1. Propose a design for a nanometre-sized transistor that uses a carbon nanotube as a component. Then compare your design with that tested by Tans *et al. Nature* **393**, 49 (1998).

2. The determination of the average molar masses of conducting polymers is an important part of their characterization. S. Holdcroft (*J. Polym. Sci., Polym. Phys.* **29**, 1585 (1991)) has determined the molar mases and Mark–Houwink constants for the electronically conducting polymer, poly(3-hexylthiophene) (P3HT) in tetrahydrofuran (THF) at 25°C by methods similar to those used for nonconducting polymers. The values for molar mass and intrinsic viscosity in the table below are adapted from their data. Determine the constants in the Mark–Kuhn–Houwink–Sakurada equation from these results and compare to the values obtained in your solution to Problem 22.9.[15]

$\overline{M}_v$/(kg mol^{-1})	3.8	11.1	15.3	58.8
$[\eta]$/(cm^3 g^{-1})	6.23	17.44	23.73	85.28

Further reading

J. Hu, T.W. Odom, and C.M. Lieber, Chemistry and physics in one dimension: synthesis and properties of nanowires and nanotubes. *Acc. Chem. Res.* **32**, 435 (1999).

S. Roth, *One-dimensional metals: physics and materials science*. VCH, New York (1995).

15. Problem supplied by Charles Trapp and Carmen Giunta.

22.13 Colloids

A **colloid**, or **disperse phase**, is a dispersion of small particles of one material in another. In this context, 'small' means something less than about 500 nm in diameter (about the wavelength of visible light). In general, colloidal particles are aggregates of numerous atoms or molecules, but are too small to be seen with an ordinary optical microscope. They pass through most filter papers, but can be detected by light-scattering and sedimentation.

(a) Classification and preparation

The name given to the colloid depends on the two phases involved. A **sol** is a dispersion of a solid in a liquid (such as clusters of gold atoms in water) or of a solid in a solid (such as ruby glass, which is a gold-in-glass sol, and achieves its colour by light scattering). An **aerosol** is a dispersion of a liquid in a gas (like fog and many sprays) or a solid in a gas (such as smoke): the particles are often large enough to be seen with a microscope. An **emulsion** is a dispersion of a liquid in a liquid (such as milk).

A further classification of colloids is as **lyophilic**, or solvent-attracting, and **lyophobic**, solvent-repelling. If the solvent is water, the terms **hydrophilic** and **hydrophobic**, respectively, are used instead. Lyophobic colloids include the metal sols. Lyophilic colloids generally have some chemical similarity to the solvent, such as —OH groups able to form hydrogen bonds. A **gel** is a semirigid mass of a lyophilic sol in which all the dispersion medium has penetrated into the sol particles.

The preparation of aerosols can be as simple as sneezing (which produces an imperfect aerosol). Laboratory and commercial methods make use of several techniques. Material (for example, quartz) may be ground in the presence of the dispersion medium. Passing a heavy electric current through a cell may lead to the sputtering (crumbling) of an electrode into colloidal particles. Arcing between electrodes immersed in the support medium also produces a colloid. Chemical precipitation sometimes results in a colloid. A precipitate (for example, silver iodide) already formed may be dispersed by the addition of a peptizing agent (for example, potassium iodide). Clays may be peptized by alkalis, the OH^- ion being the active agent.

Emulsions are normally prepared by shaking the two components together vigorously, although some kind of emulsifying agent usually has to be added to stabilize the product. This emulsifying agent may be a soap (the salt of a long-chain carboxylic acid) or other **surfactant** (surface active) species, or a lyophilic sol that forms a protective film around the dispersed phase. In milk, which is an emulsion of fats in water, the emulsifying agent is casein, a protein containing phosphate groups. It is clear from the formation of cream on the surface of milk that casein is not completely successful in stabilizing milk: the dispersed fats coalesce into oily droplets that float to the surface. This coagulation may be prevented by ensuring that the emulsion is dispersed very finely initially: intense agitation with ultrasonics brings this dispersion about, the product being 'homogenized' milk.

One way to form an aerosol is to tear apart a spray of liquid with a jet of gas. The dispersal is aided if a charge is applied to the liquid, for then electrostatic repulsions help to blast it apart into droplets. This procedure may also be used to produce emulsions, for the charged liquid phase may be directed into another liquid.

Colloids are often purified by dialysis (Box 7.1). The aim is to remove much (but not all, for reasons explained later) of the ionic material that may have accompanied their formation. As in the discussion of the Donnan effect in Box 7.1, a membrane (for example, cellulose) is selected that is permeable to solvent and ions, but not to the colloid particles. Dialysis is very slow, and is normally accelerated by applying an electric field and making use of the charges carried by many colloidal particles; the technique is then called **electrodialysis**.

(b) Structure and stability

Colloids are thermodynamically unstable with respect to the bulk. This instability can be expressed thermodynamically by noting that, because the change in Gibbs energy, dG, when the surface area of the sample changes by $d\sigma$ at constant temperature and pressure is $dG = \gamma d\sigma$, where γ is the interfacial surface tension (Section 6.10), it follows that $dG < 0$ if $d\sigma < 0$. The survival of colloids must therefore be a consequence of the kinetics of collapse: colloids are thermodynamically unstable but kinetically nonlabile.

At first sight, even the kinetic argument seems to fail: colloidal particles attract each other over large distances, so there is a long-range force that tends to condense them into a single blob. The reasoning behind this remark is as follows. The energy of attraction between two individual atoms i and j separated by a distance R_{ij}, one in each colloidal particle, varies with their separation as $1/R_{ij}^6$ (Section 21.5). The sum of all these pairwise interactions, however, decreases only as approximately $1/R^2$ (the precise variation depending on the shape of the particles and their closeness), where R is the separation of the centres of the particles. The sum has a much longer range than the $1/R^6$ dependence characteristic of individual particles and small molecules.

Several factors oppose the long-range dispersion attraction. For example, there may be a protective film at the surface of the colloid particles that stabilizes the interface and cannot be penetrated when two particles touch. Thus the surface atoms of a platinum sol in water react chemically and are turned into $—Pt(OH)_3H_3$, and this layer encases the particle like a shell. A fat can be emulsified by a soap because the long hydrocarbon tails penetrate the oil droplet but the carboxylate head groups (or other hydrophilic groups in synthetic detergents) surround the surface, form hydrogen bonds with water, and give rise to a shell of negative charge that repels a possible approach from another similarly charged particle.

(c) The electrical double layer

A major source of kinetic nonlability of colloids is the existence of an electric charge on the surfaces of the particles. On account of this charge, ions of opposite charge tend to cluster nearby, and an ionic atmosphere is formed, just as for ions (Section 10.2c).

We need to distinguish two regions of charge. First, there is a fairly immobile layer of ions that adhere tightly to the surface of the colloidal particle, and which may include water molecules (if that is the support medium). The radius of the sphere that captures this rigid layer is called the **radius of shear** and is the major factor determining the mobility of the particles. The electric potential at the radius of shear relative to its value in the distant, bulk medium is called the **zeta potential**, ζ, or the **electrokinetic potential**. Second, the charged unit attracts an oppositely charged atmosphere of mobile ions. The inner shell of charge and the outer ionic atmosphere is called the **electrical double layer**.

The theory of the stability of lyophobic dispersions was developed by B. Derjaguin and L. Landau and independently by E. Verwey and J.T.G. Overbeek, and is known as the **DLVO theory**. It assumes that there is a balance between the repulsive interaction between the charges of the electric double layers on neighbouring particles and the attractive interactions arising from van der Waals interactions between the molecules in the particles. The potential energy arising from the repulsion of double layers on particles of radius a has the form

$$V_{\text{repulsion}} = +\frac{Aa^2\zeta^2}{R}\,e^{-s/r_D} \tag{22.48}$$

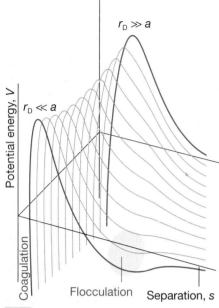

22.30 The potential energy of interaction as a function of the separation of the centres of the two particles and its variation with the ratio of the particle size to the thickness a of the electric double layer r_D. The regions labelled coagulation and flocculation show the dips in the potential energy curves where these processes occur.

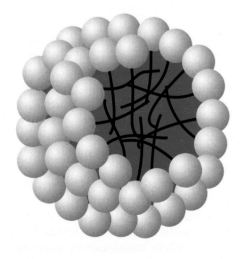

22.31 A schematic version of a spherical micelle. The hydrophilic groups are represented by spheres and the hydrophobic hydrocarbon chains are represented by the stalks; these stalks are mobile.

where A is a constant, ζ is the zeta potential,[16] R is the separation of centres, s is the separation of the surfaces of the two particles ($s = R - 2a$ for spherical particles of radius a), and r_D is the thickness of the double layer. This expression is valid for small particles with a thick double layer ($a \ll r_D$). When the double layer is thin ($r_D \ll a$), the expression is replaced by

$$V_{repulsion} = \tfrac{1}{2}Aa\zeta^2 \ln(1 + e^{-s/r_D}) \qquad (22.49)$$

In each case, the thickness of the double layer can be estimated from an expression like that derived for the thickness of the ionic atmosphere in the Debye–Hückel theory (eqn 10.27):

$$r_D = \left(\frac{\varepsilon RT}{2\rho F^2 I b^{\ominus}} \right)^{1/2} \qquad (22.50)$$

where I is the ionic strength of the solution, ρ its mass density, and $b^{\ominus} = 1 \text{ mol kg}^{-1}$. The potential energy arising from the attractive interaction has the form

$$V_{attraction} = -\frac{B}{s} \qquad (22.51)$$

where B is another constant. The variation of the total potential energy with separation is shown in Fig. 22.30.

At high ionic strengths, the ionic atmosphere is dense and the potential shows a secondary minimum at large separations. Aggregation of the particles arising from the stabilizing effect of this secondary minimum is called **flocculation**. The flocculated material can often be redispersed by agitation because the well is so shallow. **Coagulation**, the irreversible aggregation of distinct particles into large particles, occurs when the separation of the particles is so small that they enter the primary minimum of the potential energy curve and van der Waals forces are dominant.

The ionic strength is increased by the addition of ions, particularly those of high charge type, so such ions act as flocculating agents. This increase is the basis of the empirical **Schulze–Hardy rule**, that hydrophobic colloids are flocculated most efficiently by ions of opposite charge type and high charge number. The Al^{3+} ions in alum are very effective, and are used to induce the congealing of blood. When river water containing colloidal clay flows into the sea, the salt water induces flocculation and coagulation, and is a major cause of silting in estuaries. Metal oxide sols tend to be positively charged whereas sulfur and the noble metals tend to be negatively charged.

The primary role of the electric double layer is to confer kinetic non-lability. Colliding colloidal particles break through the double layer and coalesce only if the collision is sufficiently energetic to disrupt the layers of ions and solvating molecules, or if thermal motion has stirred away the surface accumulation of charge. This disruption may occur at high temperatures, which is one reason why sols precipitate when they are heated. The protective role of the double layer is the reason why it is important not to remove all the ions when a colloid is being purified by dialysis, and why proteins coagulate most readily at their isoelectric point.

22.14 Micelles and biological membranes

Surfactant molecules or ions can cluster together as **micelles**, which are colloid-sized clusters of molecules, for their hydrophobic tails tend to congregate, and their hydrophilic heads provide protection (Fig. 22.31).

16. The actual potential is that of the surface of the particles; there is some danger in identifying it with the zeta potential. See the references in *Further reading*.

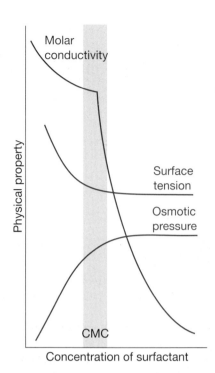

22.32 The typical variation of some physical properties of an aqueous solution of sodium dodecylsulfate close to the critical micelle concentration (CMC).

22.33 The cross-sectional structure of a spherical liposome.

The web site contains links to databases of thermodynamic properties of lipids.

(a) Micelle formation

Micelles form only above the **critical micelle concentration** (CMC) and above the **Krafft temperature**. The CMC is detected by noting a pronounced discontinuity in physical properties of the solution, particularly the molar conductivity (Fig. 22.32). The hydrocarbon interior of a micelle is like a droplet of oil. Nuclear magnetic resonance shows that the hydrocarbon tails are mobile, but slightly more restricted than in the bulk. Micelles are important in industry and biology on account of their solubilizing function: matter can be transported by water after it has been dissolved in their hydrocarbon interiors. For this reason, micellar systems are used as detergents, for organic synthesis, froth flotation, and petroleum recovery.

Nonionic surfactant molecules may cluster together in clumps of 1000 or more, but ionic species tend to be disrupted by the electrostatic repulsions between head groups and are normally limited to groups of less than about 100. The micelle population is often polydisperse, and the shapes of the individual micelles vary with concentration. Spherical micelles do occur, but micelles are more commonly flattened spheres close to the CMC.

Under certain experimental conditions, a **liposome** may form, with an inward pointing inner surface of molecules surrounded by an outward pointing outer layer (Fig. 22.33). Liposomes may be used to carry nonpolar drug molecules in blood. In concentrated solutions micelles formed from surfactant molecules may take the form of long cylinders and stack together in reasonably close-packed (hexagonal) arrays. These orderly arrangements of micelles are called **lyotropic mesomorphs**, and more colloquially 'liquid crystalline phases'.

The enthalpy of micelle formation reflects the contributions of interactions between micelle chains within the micelles and between the polar head groups and the surrounding medium. Consequently, enthalpies of micelle formation display no readily discernible pattern and may be positive (endothermic) or negative (exothermic). Many nonionic micelles form endothermically, with ΔH of the order of 10 kJ per mole of surfactant. That such micelles do form above the CMC indicates that the entropy change accompanying their formation must then be positive, and measurements suggest a value of about +140 J K^{-1} mol^{-1} at room temperature. The fact that the entropy change is positive even though the molecules are clustering together shows that hydrophobic interactions (Section 22.8e) are important in the formation of micelles.

(b) Membrane formation

Some micelles at concentrations well above the CMC form extended parallel sheets, called **lamellar micelles**, two molecules thick. The individual molecules lie perpendicular to the sheets, with hydrophilic groups on the outside in aqueous solution and on the inside in nonpolar media. Such lamellar micelles show a close resemblance to biological membranes, and are often a useful model on which to base investigations of biological structures.

Although lamellar micelles are convenient models of cell membranes, actual membranes are highly sophisticated structures. The basic structural element of a membrane is a phospholipid, such as phosphatidyl choline (**11**), which contains long hydrocarbon chains (typically in the range C_{14}—C_{24}) and a variety of polar groups, such as —CH$_2$CH$_2$N(CH$_3$)$_3^+$ in (**11**). The hydrophobic chains stack together to form an extensive bilayer about 5 nm across. The lipid molecules form layers instead of micelles because the hydrocarbon chains are too bulky to allow packing into nearly spherical clusters. Interspersed among the phospholipids are sterols, such as cholesterol (**12**), which is largely hydrophobic but does contain a hydrophilic —OH group. The presence of the sterols, which are present in different proportions in different

11 Phosphatidyl choline

12 Cholesterol

types of cell, prevents the hydrophobic chains of lipids from freezing into a solid array and, by disrupting the packing of the chains, spread the melting point of the membrane over a range of temperatures.

The bilayer is a highly mobile structure, as shown by ESR studies with spin-labelled phospholipids (Box 18.2). Not only are the hydrocarbon chains ceaselessly twisting and turning in the region between the polar groups, but the phospholipid and cholesterol molecules migrate over the surface. It is better to think of the membrane as a viscous fluid rather than a permanent structure, with a viscosity about 100 times that of water. In common with diffusional behaviour in general (see Section 24.10), the average distance a phospholipid molecule diffuses is proportional to the square-root of the time.[17] Typically, a phospholipid molecule migrates through about 1 μm (the diameter of a cell) in about 1 min.

Peripheral proteins are proteins attached to the bilayer. **Integral proteins** are proteins immersed in the mobile but viscous bilayer. These proteins may span the depth of the bilayer and consist of tightly packed α-helices or, in some cases, β-sheets containing hydrophobic residues that sit comfortably within the hydrocarbon region of the bilayer. There are two views of the motion of integral proteins in the bilayer. In the **fluid mosaic model** shown in Fig. 22.34 the proteins are mobile, but

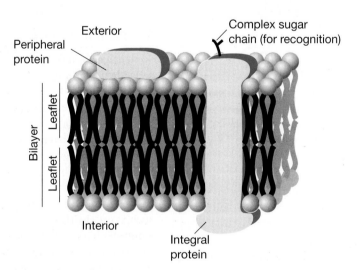

22.34 In the fluid mosaic model of a biological cell membrane, integral proteins diffuse through the lipid bilayer. In the alternative lipid raft model, a number of lipid and cholesterol molecules envelope and transport the protein around the membrane.

17. For a molecule confined to a two-dimensional plane, the average distance travelled in a time t is equal to $(4Dt)^{1/2}$.

22.35 A schematic diagram of the apparatus used to measure the surface pressure and other characteristics of a surface film. The surfactant is spread on the surface of the liquid in the trough, and then compressed horizontally by moving the compression barrier towards the mica float. The latter is connected to a torsion wire, so the difference in force on either side of the float can be monitored.

22.36 The variation of surface pressure with the area occupied by each surfactant molecule. The collapse pressures are indicated by the horizontal dotted lines.

their diffusion coefficients are much smaller than those of the lipids. In the **lipid raft model**, a number of lipid and cholesterol molecules form ordered structures, or 'rafts', that envelop proteins and help carry them to specific parts of the cell.

The mobility of the bilayer enables it to flow round a molecule close to the outer surface, to engulf it, and incorporate it into the cell by the process of *endocytosis*. Alternatively, material from the cell interior wrapped in cell membrane may coalesce with the cell membrane itself, which then withdraws and ejects the material in the process of *exocytosis*. The function of the proteins embedded in the bilayer, though, is to act as devices for transporting matter into and out of the cell in a more subtle manner. By providing hydrophilic channels through an otherwise alien hydrophobic environment, some proteins act as **ion channels** and **ion pumps** (Box 24.1).

22.15 Surface films

The compositions of surface layers have been investigated by the simple but technically elegant procedure of slicing thin layers off the surfaces of solutions and analysing their compositions. The physical properties of surface films have also been investigated. Surface films one molecule thick, such as that formed by a surfactant, are called **monolayers**. When a monolayer has been transferred to a solid support, it is called a **Langmuir–Blodgett film**, after Irving Langmuir and Katherine Blodgett, who developed experimental techniques for studying them.

(a) Surface pressure

The principal apparatus used for the study of surface monolayers is a **surface film balance** (Fig. 22.35). This device consists of a shallow trough and a barrier that can be moved along the surface of the liquid in the trough, and hence compress any monolayer on the surface. The **surface pressure**, π, the difference between the surface tension of the pure solvent and the solution ($\pi = \gamma^* - \gamma$) is measured by using a torsion wire attached to a strip of mica that rests on the surface and pressing against one edge of the monolayer. The parts of the apparatus that are in touch with liquids are coated in polytetrafluoroethylene to eliminate effects arising from the liquid–solid interface. In an actual experiment, a small amount (about 0.01 mg) of the surfactant under investigation is dissolved in a volatile solvent and then poured on to the surface of the water; the compression barrier is then moved across the surface and the surface pressure exerted on the mica bar is monitored.

Some typical results are shown in Fig. 22.36. One parameter obtained from the isotherms is the area occupied by the molecules when the monolayer is closely packed. This quantity is obtained from the extrapolation of the steepest part of the isotherm to the horizontal axis. As can be seen from the illustration, even though stearic acid (**13**) and isostearic acid (**14**) are chemically very similar (they differ only in the location of a methyl group at the end of a long hydrocarbon chain), they occupy significantly different areas in the monolayer. Neither, though, occupies as much area as the tri-*p*-cresyl phosphate molecule (**15**), which is like a wide bush rather than a lanky tree.

The second feature to note from Fig. 22.36 is that the tri-*p*-cresyl phosphate isotherm is much less steep than the stearic acid isotherms. This difference indicates that the tri-*p*-cresyl phosphate film is more compressible than the stearic acid films, which is consistent with their different molecular structures.

A third feature of the isotherms is the **collapse pressure**, the highest surface pressure. When the monolayer is compressed beyond the point represented by the collapse pressure, the monolayer buckles and collapses into a film several molecules thick. As can be seen from the isotherms in Fig. 22.36, stearic acid has a high collapse pressure, but that of tri-*p*-cresyl phosphate is significantly smaller, indicating a much weaker film.

(b) The thermodynamics of surface layers

A surfactant is active at the interface between two phases, such as at the interface between hydrophilic and hydrophobic phases. A surfactant accumulates at the interface, and modifies its surface tension and hence the surface pressure. To establish the relation between the concentration of surfactant at a surface and the change in surface tension it brings about, we consider two phases α and β in contact and suppose that the system consists of several components J, each one present in an overall amount n_J. If the components were distributed uniformly through the two phases right up to the interface, which is taken to be a plane of surface area σ, the total Gibbs energy, G, would be the sum of the Gibbs energies of both phases, $G = G(\alpha) + G(\beta)$. However, the components are not uniformly distributed because one may accumulate at the interface. As a result, the sum of the two Gibbs energies differs from G by an amount called the **surface Gibbs energy**, $G(\sigma)$:

$$G(\sigma) = G - \{G(\alpha) + G(\beta)\} \tag{22.52}$$

Similarly, if it is supposed that the concentration of a species J is uniform right up to the interface, then from its volume we would conclude that it contains an amount $n_J(\alpha)$ of J in phase α and an amount $n_J(\beta)$ in phase β. However, because a species may accumulate at the interface, the total amount of J differs from the sum of these two amounts by $n_J(\sigma) = n_J - \{n_J(\alpha) + n_J(\beta)\}$. This difference is expressed in terms of the **surface excess**, Γ_J:

$$\Gamma_J = \frac{n_J(\sigma)}{\sigma} \tag{22.53}$$

The surface excess may be either positive (an accumulation of J at the interface) or negative (a deficiency there).

The relation between the change in surface tension and the composition of a surface (as expressed by the surface excess) was derived by Gibbs. In the following *Justification* we derive the **Gibbs isotherm**, between the changes in the chemical potentials of the substances present in the interface and the change in surface tension:

$$d\gamma = -\sum_J \Gamma_J \, d\mu_J \tag{22.54}$$

Justification 22.7

A general change in G is brought about by changes in T, p, and the n_J:

$$dG = -S \, dT + V \, dp + \gamma \, d\sigma + \sum_J \mu_J \, dn_J$$

When this relation is applied to G, $G(\alpha)$, and $G(\beta)$ we find

$$dG(\sigma) = -S(\sigma) \, dT + \gamma \, d\sigma + \sum_J \mu_J \, dn_J(\sigma)$$

because at equilibrium the chemical potential of each component is the same in every phase, $\mu_J(\alpha) = \mu_J(\beta) = \mu_J(\sigma)$. Just as in the discussion of partial molar quantities (Section 7.1), the last equation integrates at constant temperature to

$$G(\sigma) = \gamma\sigma + \sum_J \mu_J n_J(\sigma)$$

13 Stearic acid, $C_{17}H_{35}COOH$

14 Isostearic acid, $C_{17}H_{35}COOH$

15 Tri-*p*-cresylphosphate

We are seeking a connection between the change of surface tension $d\gamma$ and the change of composition at the interface. Therefore, we use the argument which in Section 7.1 led to the Gibbs–Duhem equation (eqn 7.12), but this time we compare the expression

$$dG(\sigma) = \gamma \, d\sigma + \sum_J \mu_J \, dn_J(\sigma)$$

(which is valid at constant temperature) with the expression for the same quantity but derived from the preceding equation:

$$dG(\sigma) = \gamma \, d\sigma + \sigma \, d\gamma + \sum_J \mu_J \, dn_J(\sigma) + \sum_J n_J(\sigma) \, d\mu_J$$

The comparison implies that, at constant temperature,

$$\sigma \, d\gamma + \sum_J n_J(\sigma) \, d\mu_J = 0$$

Division by σ then gives eqn 54.

Now consider a simplified model of the interface in which the 'oil' and 'water' phases are separated by a geometrically flat surface. This approximation implies that only the surfactant, S, accumulates at the surface, and hence that Γ_{oil} and Γ_{water} are both zero. Then the Gibbs equation becomes

$$d\gamma = -\Gamma_S \, d\mu_S \tag{22.55}$$

For dilute solutions,

$$d\mu_S = RT d \ln c \tag{22.56}$$

where c is the molar concentration of the surfactant. It follows that

$$d\gamma = -RT\Gamma_s \frac{dc}{c}$$

at constant temperature, or

$$\left(\frac{\partial \gamma}{\partial c} \right)_T = -\frac{RT\Gamma_s}{c} \tag{22.57}$$

If the surfactant accumulates at the interface, its surface excess is positive and eqn 57 implies that $(\partial\gamma/\partial c)_T < 0$. That is, the surface tension decreases when a solute accumulates at a surface. Conversely, if the concentration dependence of γ is known, then the surface excess may be predicted and used to infer the area occupied by each surfactant molecule on the surface.

Checklist of key ideas

☐ polymer, a compound formed by linking together small molecules

☐ monomer, the small molecules linked together to form a polymer

Structure and dynamics

☐ configuration, the structural features that can be changed only by breaking chemical bonds and forming new ones

☐ conformation, the spatial arrangement of the different parts of a chain

22.1 The different levels of structure

☐ primary structure, the sequence of small molecular residues making up a polymer

☐ polypeptide, a macromolecule formed from amino acids

☐ peptide link, the group —CONH—

☐ sequencing, the determination of primary structure

☐ degradation, a disruption of primary structure

☐ secondary structure, the (often local) spatial arrangement of a chain

☐ denaturation, the loss of secondary structure

☐ tertiary structure, the overall three-dimensional structure of a macromolecule

☐ quaternary structure, the manner in which large molecules are formed by the aggregation of others

22.2 Random coils

☐ random coil, a conformation in which neighbouring groups adopt random angles to each other

☐ freely jointed chain, a conformation in which any bond is free to make any angle with respect to the preceding one

☐ contour length, the length of a macromolecule measured along its backbone: $R_c = Nl$

☐ root mean square separation, the square root of the mean value of R^2: $R_{rms} = N^{1/2}l$

☐ radius of gyration, the radius of a thin hollow spherical shell of the same mass and moment of inertia as the molecule: $R_g = (N/6)^{1/2}l$

☐ conformational entropy, the statistical entropy arising from the arrangement of bonds: $\Delta S = -\frac{1}{2}kN \ln\{(1+\nu)^{1+\nu}(1-\nu)^{1-\nu}\}$

22.3 The structure of proteins

☐ Corey–Pauling rules, rules explaining the origin of the secondary structures of proteins, Section 22.3(a)

☐ α-helix, a helical conformation formed by hydrogen bonding between peptide links

☐ β-sheet, a planar conformation formed by hydrogen bonding between peptide links

☐ conformational energy, the energy of interaction between different parts of the chain

☐ Ramachandran plot, a contour diagram of the conformational energy in which one axis represents ϕ and the other represents ψ

☐ molecular mechanics, a procedure in which the total potential energy of a given conformation is calculated by summing the contributions

☐ molecular dynamics, a procedure in which the possible trajectories of all atoms under the influence of the intermolecular potentials are calculated by integration of Newton's equations of motion

☐ antiparallel β-sheet, a β-sheet in which $\phi = -139°$, $\psi = 113°$, and the N—H—O atoms of the hydrogen bonds form a straight line

☐ parallel β-sheet, a β-sheet in which $\phi = -119°$, $\psi = 113°$, and the N—H—O atoms of the hydrogen bonds are not perfectly aligned

☐ four-helix bundle, the arrangement illustrated in Fig. 22.15

☐ β-barrel, the arrangement illustrated in Fig. 22.16

22.4 The structure of nucleic acids

☐ polyelectrolyte, a macromolecule with many different charged sites and an overall charge

22.5 The stability of biological polymers

☐ denaturation, the loss of their natural conformation by proteins and nucleic acids

☐ cooperative process, a process that becomes more probable the further it proceeds

Determination of size and shape

22.6 Mean molar masses

☐ monodisperse, a sample with a single, definite molar mass

☐ polydisperse, a sample with a wide range of molar masses

☐ number-average molar mass, $\overline{M}_n = (1/N)\sum_i N_i M_i$

☐ viscosity-average molar mass, the average molar mass obtained from viscosity measurements

☐ weight-average molar mass, the average molar mass obtained from light-scattering experiments: $\overline{M}_w = (1/m)\sum_i m_i M_i$

☐ Z-average molar mass, the average molar mass obtained from sedimentation measurements: $\overline{M}_Z = \sum_i N_i M_i^3 / \sum_i N_i M_i^2$

☐ heterogeneity index, $\overline{M}_w/\overline{M}_n$

22.7 Mass spectrometry

☐ matrix-assisted laser desorption/ionization (MALDI), a laser-based technique for ionization of macromolecules

☐ electrospray ionization, another technique for ionization of macromolecules

☐ MALDI-TOF mass spectrometry, a technique for the determination of molar masses of macromolecules that combines MALDI and time-of-flight mass spectrometry

22.8 Laser light scattering

☐ Rayleigh scattering, the scattering of light by particles with diameters much smaller than the wavelength of the incident radiation

☐ Rayleigh ratio, $R_\theta = (I/I_0)r^2/\sin^2\phi$

☐ structure factor,
$$P_\theta = \frac{1}{N^2}\sum_{i,j}\frac{\sin sR_{ij}}{sR_{ij}}, \qquad s = \frac{4\pi}{\lambda}\sin\frac{1}{2}\theta$$

☐ Flory theta temperature, the temperature at which attractive and repulsive interactions in a solution of a macromolecule cancel and the solution is virtually ideal

☐ theta solution, a solution at its theta temperature

☐ dynamic light scattering, a technique used to investigate the diffusion of polymers in solution

- ☐ Stokes–Einstein relation, $D = kT/f$
- ☐ frictional coefficient, f, a measure of the forces that retard a molecule's motion
- ☐ Stokes' relation, $f = 6\pi a \eta$

22.9 Ultracentrifugation

- ☐ sedimentation, the settling of particles to the foot of a column of solution
- ☐ ultracentrifugation, a sedimentation technique in which particles move under the influence of a centrifugal field
- ☐ drift speed, a constant speed through a medium when the driving and retarding forces are balanced
- ☐ sedimentation constant, $S = s/r\omega^2$
- ☐ sedimentation equilibria, the equilibrium distribution of particles in a gravitational or centrifugal field,
$$\overline{M}_w = \frac{2RT}{(r_2^2 - r_1^2)b\omega^2} \ln \frac{c_2}{c_1}$$

22.10 Electrophoresis

- ☐ electrophoresis, motion of macromolecules in an electric field
- ☐ gel electrophoresis, electrophoresis in which migration takes place through a gel slab
- ☐ isoelectric point, the pH at which there is no net charge on the biopolymer
- ☐ isoelectric focusing, an electrophoresis method that exploits the change of drift speed with pH
- ☐ capillary electrophoresis, an electrophoresis experiment in which the sample is dispersed in a medium held in a capillary tube

22.11 Size-exclusion chromatography

- ☐ size-exclusion chromatography (SEC), a technique that makes use of beads of porous polymeric material that capture molecules according to their size
- ☐ gel-permeation chromatography, another name for SEC
- ☐ elution time, the time required to pass through a particular length of column

22.12 Viscosity

- ☐ intrinsic viscosity, the coefficient $[\eta]$ in $\eta = \eta_0(1 + [\eta]c + \cdots)$
- ☐ Ostwald viscometer, a device for measuring viscosity from the time taken for a solution to flow through a capillary and compared with a standard sample
- ☐ rotating drum viscometer, a device for measuring viscosity that consists of rotating concentric cylinders

- ☐ Mark–Kuhn–Houwink–Sakurada equation, $[\eta] = K\overline{M}_v^a$

Self-assembly

- ☐ self-assembly, the spontaneous formation of complex structures of molecules or macromolecules held together by molecular interactions

22.13 Colloids

- ☐ colloids, a dispersion of small particles of one material in another
- ☐ disperse phase, a colloid
- ☐ sol, a dispersion of a solid in a liquid or of a solid in a solid
- ☐ aerosol, a dispersion of a liquid in a gas
- ☐ emulsion, a dispersion of a liquid in another liquid
- ☐ lyophilic, solvent-attracting
- ☐ lyophobic, solvent-repelling
- ☐ hydrophilic, water-attracting
- ☐ hydrophobic, water-repelling
- ☐ gel, a semirigid mass of a lyophilic sol in which all the dispersion medium has penetrated into the sol particles
- ☐ surfactant, a surface-active agent
- ☐ electrodialysis, dialysis in the presence of an electric field
- ☐ radius of shear, the radius of the sphere that captures a rigid layer of charge around a colloidal particle
- ☐ zeta potential, the electric potential at the radius of shear relative to its value in the distant, bulk medium
- ☐ electrokinetic potential, the zeta potential
- ☐ electrical double layer, the inner shell of charge and the outer ionic atmosphere around a colloidal particle
- ☐ DVLO theory, a theory of the stability of lyophobic dispersions
- ☐ flocculation, the reversible aggregation of colloidal particles
- ☐ coagulation, the irreversible aggregation of colloidal particles
- ☐ Schultze–Hardy rule: Hydrophobic colloids are flocculated most efficiently by ions of opposite charge type and high charge number

22.14 Micelles and biological membranes

- ☐ micelles, colloid-sized clusters of molecules

- ☐ critical micelle concentration (CMC), the concentration above which micelles form
- ☐ Krafft temperature, the temperature above which micelles form
- ☐ liposome, a vesicle with an inward pointing inner surface of molecules surrounded by an outward pointing outer layer
- ☐ lyotropic mesomorph, an orderly arrangement of micelles; a liquid crystalline phase
- ☐ lamellar micelle, an extended layer, two molecules thick
- ☐ integral protein, a protein embedded in a cell membrane
- ☐ peripheral protein, a protein attached to the surface of a cell membrane
- ☐ fluid mosaic model, a model of a cell membrane in which the proteins are mobile but have diffusion coefficients much smaller than those of the lipids
- ☐ lipid raft model, a model of a cell membrane in which a number of lipid and cholesterol molecules form ordered structures that envelop proteins and help carry them to specific parts of the cell
- ☐ ion channel, a hydrophilic channel through an otherwise hydrophobic environment
- ☐ ion pump, an ion channel through which ions are actively transported

22.15 Surface films

- ☐ monolayer, a single layer of molecules on a surface
- ☐ Langmuir–Blodgett film, a monolayer that has been transferred to a solid support
- ☐ surface film balance, a device for measuring surface pressure
- ☐ surface pressure, the difference between the surface tension of the pure solvent and the solution:
$\pi = \gamma^* - \gamma$
- ☐ collapse pressure, the highest surface pressure sustained by a surface film
- ☐ surface Gibbs energy, $G(\sigma) = G - \{G(\alpha) + G(\beta)\}$
- ☐ surface excess, $\Gamma_J = n_J(\sigma)/\sigma$
- ☐ Gibbs isotherm, $d\gamma = -\Sigma_J \Gamma_J d\mu_J$

Further reading

Articles of general interest

C.L. Cooper, Capillary electrophoresis. Part I. Theoretical and experimental background. *J. Chem. Educ.* **75**, 343 (1998).

C.L. Cooper and K.W. Whitaker, Capillary electrophoresis. Part II: applications. *J. Chem. Educ.* **75**, 347 (1998).

C.M. Guttman and B. Fanconi, Molecular properties of polymers. In *Encyclopedia of applied physics* (ed. G.L. Trigg), **16**, 549. VCH, New York (1996).

K. Simons and E. Ikonen, Functional rafts in cell membranes. *Nature* **387**, 567 (1997).

T.M. Laue and W.F. Stafford III, Modern applications of analytical ultracentrifugation. *Ann. Rev. Biophys. Biomol. Struct.* **28**, 75 (1999).

B.Y.H. Liu and D.Y.H. Pui, Aerosols. In *Encyclopedia of applied physics* (ed. G.L. Trigg), **1**, 415. VCH, New York (1991).

D.C. Muddiman, R. Bakhtiar, S.A. Hofstadler, and R.D. Smith, Matrix-assisted laser desorption/ionization mass spectrometry. *J. Chem. Educ.* **74**, 1288 (1997).

G. Paradossi, Size and shape of macromolecules. Calculation of the scattering function for simple geometries. *J. Chem. Educ.* **70**, 440 (1993).

N.C. Santos and M.A.R.B. Castanho, Teaching light scattering spectroscopy: the dimension and shape of the tobacco mosaic virus. *Biophys. J.* **71**, 1641 (1998).

M. Takeo, Disperse systems. In *Encyclopedia of applied physics* (ed. G.L. Trigg), **5**, 87. VCH, New York (1993).

Texts and sources of data and information

A.W. Adamson and A.P. Gast, *Physical chemistry of surfaces*. Wiley, New York (1997).

J. Barker, *Mass spectrometry*. Wiley, New York (1999).

B.J. Berne and R. Pecora, *Dynamic light scattering: with applications to chemistry, biology, and physics*. Dover, New York (2000).

C.E. Carraher, Jr., *Seymour/Carraher's polymer chemistry*, Marcel Dekker (2000).

M. Daune, *Molecular biophysics: structures in motion*. Oxford University Press (1999).

D.F. Evans and H. Wennerström, *The colloidal domain : where physics, chemistry, biology, and technology meet*. Wiley-VCH, New York (1999).

A. Fersht, *Structure and mechanism in protein science: a guide to enzyme catalysis and protein folding*. W.H. Freeman, New York (1999).

D. Frenkel and B. Smit, *Understanding molecular simulation*. Academic Press, San Diego (1996).

P. Flory, *Principles of polymer chemistry*. Cornell University Press, Ithaca (1953).

R.J. Hunter, *Foundations of colloid science*. Oxford University Press (2000).

C.S. Johnson and D.A. Gabriel, *Laser light scattering*. Dover, New York (1995).

A.R. Leach, *Molecular modelling: principles and applications*. Longman, Harlow (1997).

D.R. Lide (ed.), *CRC Handbook of Chemistry and Physics*, Sections 7 and 13. CRC Press, Boca Raton (2000).

J.C. Lindon, G.E. Tranter, and J.L. Holmes (ed.), *Encyclopedia of spectroscopy and spectrometry*. Academic Press, San Diego (2000).

S. Mori and H.G. Barth, *Size exclusion chromatography*. Springer, Berlin (1999).

Y. Moroi, *Micelles*. Plenum, New York (1992).

L.H. Sperling, *Introduction to physical polymer science*. Wiley-Interscience, New York (2001).

K.E. van Holde, W.C. Johnson, and P.S. Ho, *Principles of physical biochemistry*. Prentice Hall, Upper Saddle River (1998).

Exercises

Discussion questions

22.1 (a) Distinguish between contour length, root mean square separation, and radius of gyration of a random coil.

22.1 (b) Identify the terms in and limit the generality of the following expressions: (a) $\Delta S = -\frac{1}{2}kN \ln\{(1 + v)^{1+v}(1 - v)^{1-v}\}$, (b) $R_{rms} = (2N)^{1/2}l$, and (c) $R_g = (N/6)^{1/2}l$.

22.2 (a) Distinguish between molecular mechanics and molecular dynamics calculations. Why are these methods generally more popular in the field of polymer chemistry than the quantum mechanical procedures discussed in Chapter 14?

22.2 (b) It is possible to observe the cooperativity of biopolymer denaturation by determining the extent of denaturation as a function of some parameter that affects its stability, such as temperature or denaturant concentration. Consult library and reliable internet resources, such as those listed in the web site for this text, and write a brief report (similar in length and depth of coverage to one of the many *Boxes* in this text) summarizing the use of a spectroscopic technique of your choice in the study of protein denaturation. Some techniques for your consideration include: electronic absorption or emission spectroscopy (including circular dichroism spectroscopy, Section 21.4), vibrational spectroscopy (infrared absorption or Raman scattering), magnetic resonance spectroscopy (NMR or EPR). Your report should include: (a) a description of experimental methods; (b) a discussion of the information that can be obtained from the measurements; (c) an example from the chemical or biological literature of the use of the technique in protein stability work; (d) a brief discussion of the advantages and disadvantages of the technique of your choice over differential scanning calorimetry (Box 2.1), a very popular technique for the study of polymer stability. *Hint.* The theoretical treatment of the helix–coil transition given in Box 20.1 is useful.

22.3 (a) Distinguish between number-average, viscosity-average, weight-average, and Z-average molar masses. Discuss experimental techniques that can measure each of these properties.

22.3 (b) It is observed that the critical micelle concentration of sodium dodecyl sulfate in aqueous solution decreases as the concentration of added sodium chloride increases. Explain this effect.

22.4 (a) The fluidity of a lipid bilayer dispersed in aqueous solution depends on temperature and there are two important melting transitions. One transition is from a 'solid crystalline' state in which the hydrophobic chains are packed together tightly (hence move very little) to a 'liquid crystalline state', in which there is increased but still limited movement of the of the chains. The second transition, which occurs at a higher temperature than the first, is from the liquid crystalline state to a liquid state, in which the hydrophobic interactions holding the aggregate together are largely disrupted. (a) It is observed that the transition temperatures increase with the hydrophobic chain length and decrease with the number of C=C bonds in the chain. Explain these observations. (b) What effect is the inclusion of cholesterol likely to have on the transition temperatures of a lipid bilayer? Justify your answer.

22.4 (b) Explain the physical origins of surface activity by surfactant molecules.

Numerical exercises

22.5 (a) A polymer chain consists of 700 segments, each 0.90 nm long. If the chain were ideally flexible, what would be the r.m.s. separation of the ends of the chain?

22.5 (b) A polymer chain consists of 1200 segments, each 1.125 nm long. If the chain were ideally flexible, what would be the r.m.s. separation of the ends of the chain?

22.6 (a) Calculate the contour length (the length of the extended chain) and the root mean square separation (the end-to-end distance) for polyethylene with a molar mass of 280 kg mol^{-1}.

22.6 (b) Calculate the contour length (the length of the extended chain) and the root mean square separation (the end-to-end distance) for polypropylene of molar mass 174 kg mol^{-1}.

22.7 (a) Calculate the number-average molar mass and the mass-average molar mass of a mixture of equal amounts of two polymers, one having $M = 62$ kg mol^{-1} and the other $M = 78$ kg mol^{-1}.

22.7 (b) Calculate the number-average molar mass and the mass-average molar mass of a mixture of two polymers, one having $M = 62$ kg mol^{-1} and the other $M = 78$ kg mol^{-1}, with their amounts (numbers of moles) in the ratio 3 : 2.

22.8 (a) The radius of gyration of a long chain molecule is found to be 7.3 nm. The chain consists of C—C links. Assume the chain is randomly coiled and estimate the number of links in the chain.

22.8 (b) The radius of gyration of a long chain molecule is found to be 18.9 nm. The chain consists of links of length 450 pm. Assume the chain is randomly coiled and estimate the number of links in the chain.

22.9 (a) A solution consists of solvent, 30 per cent by mass of a dimer with $M = 30$ kg mol^{-1} and its monomer. What average molar mass would be obtained from measurement of (a) osmotic pressure, (b) light scattering?

22.9 (b) A solution consists of 25 per cent by mass of a trimer with $M = 22$ kg mol^{-1} and its monomer. What average molar mass would be obtained from measurement of: (a) osmotic pressure, (b) light scattering?

22.10 (a) Evaluate the rotational correlation time, $\tau_R = 4\pi a^3 \eta / 3kT$, for serum albumin in water at 25°C on the basis that it is a sphere of radius 3.0 nm. What is the value for a CCl_4 molecule in carbon tetrachloride at 25°C? (Viscosity data in Table 24.3 in the *Data section* at the end of this volume; take $a(CCl_4) = 250$ pm.)

22.10 (b) Evaluate the rotational correlation time, $\tau_R = 4\pi a^3 \eta / 3kT$, for a synthetic polymer in water at 20°C on the basis that it is a sphere of radius 4.5 nm.

22.11 (a) What is the relative rate of sedimentation for two spherical particles of the same density, but which differ in radius by a factor of 10?

22.11 (b) What is the relative rate of sedimentation for two spherical particles with densities 1.10 g cm^{-3} and 1.18 g cm^{-3} and which differ in radius by a factor of 8.4, the former being the larger?

22.12 (a) Human haemoglobin has a specific volume of 0.749×10^{-3} m^3 kg^{-1}, a sedimentation constant of 4.48 Sv, and a diffusion coefficient of 6.9×10^{-11} m^2 s^{-1}. Determine its molar mass from this information.

22.12 (b) A synthetic polymer has a specific volume of 8.01×10^{-4} m^3 kg^{-1}, a sedimentation constant of 7.46 Sv, and a diffusion coefficient of 7.72×10^{-11} m^2 s^{-1}. Determine its molar mass from this information.

22.13 (a) Find the drift speed of a particle of radius 20 μm and density 1750 kg m^{-3} which is settling from suspension in water (density 1000 kg m^{-3}) under the influence of gravity alone. The viscosity of water is 8.9×10^{-4} kg m^{-1} s^{-1}.

22.13 (b) Find the drift speed of a particle of radius 15.5 μm and density 1250 kg m^{-3} which is settling from suspension in water (density 1000 kg m^{-3}) under the influence of gravity alone. The viscosity of water is 8.9×10^{-4} kg m^{-1} s^{-1}.

22.14 (a) At 20°C the diffusion coefficient of a macromolecule is found to be 8.3×10^{-11} m^2 s^{-1}. Its sedimentation constant is 3.2 Sv in a solution of density 1.06 g cm^{-3}. The specific volume of the macromolecule is 0.656 cm^3 g^{-1}. Determine the molar mass of the macromolecule.

22.14 (b) At 20°C the diffusion coefficient of a macromolecule is found to be 7.9×10^{-11} m^2 s^{-1}. Its sedimentation constant is 5.1 Sv in a solution of density 997 kg m^{-3}. The specific volume of the macromolecule is 0.721 cm^3 g^{-1}. Determine the molar mass of the macromolecule.

22.15 (a) The data from a sedimentation equilibrium experiment performed at 300 K on a macromolecular solute in aqueous solution show that a graph of ln c against r^2 is a straight line with a slope of 729 cm^{-2}. The rotational rate of the centrifuge was 50 000 r.p.m. The specific volume of the solute is 0.61 cm^3 g^{-1}. Calculate the molar mass of the solute.

22.15 (b) The data from a sedimentation equilibrium experiment performed at 293 K on a macromolecular solute in aqueous solution show that a graph of ln c against $(r/cm)^2$ is a straight line with a slope of 821. The rotation rate of the centrifuge was 1080 Hz. The specific volume of the solute is 7.2×10^{-4} m^3 kg^{-1}. Calculate the molar mass of the solute.

22.16 (a) Calculate the radial acceleration (as so many g) in a cell placed at 6.0 cm from the centre of rotation in an ultracentrifuge operating at 80 000 r.p.m.

22.16 (b) Calculate the radial acceleration (as so many g) in a cell placed at 5.50 cm from the centre of rotation in an ultracentrifuge operating at 1.32 kHz.

Problems[†]

Numerical problems

22.1 In formamide as solvent, poly(γ-benzyl-L-glutamate) is found by light scattering experiments to have a radius of gyration proportional to M; in contrast, polystyrene in butanone has R_g proportional to $M^{1/2}$. Present arguments to show that the first polymer is a rigid rod, whereas the second is a random coil.

22.2 In a sedimentation experiment the position of the boundary as a function of time was found to be as follows:

t/min	15.5	29.1	36.4	58.2
r/cm	5.05	5.09	5.12	5.19

The rotation rate of the centrifuge was 45 000 r.p.m. Calculate the sedimentation constant of the solute.

22.3 Calculate the speed of operation (in r.p.m.) of an ultracentrifuge needed to obtain a readily measurable concentration gradient in a sedimentation equilibrium experiment. Take that gradient to be a concentration at the bottom of the cell about five times greater that at the top. Use $r_{top} = 5.0$ cm, $r_{bottom} = 7.0$ cm, $M \approx 10^5$ g mol^{-1}, $\rho v_s \approx 0.75$, $T = 298$ K.

22.4‡ A problem arises in the use of the Svedberg equation (eqn 41) for the determination of the molar masses of macromolecules due to the fact that values of S and D depend upon concentration. Consequently, accurate values for $\overline{M}_w$ must be obtained by extrapolation of the data to infinite dilution by using a virial expansion in the form

$$\frac{bD}{SRT} = \frac{1}{\overline{M}_w}(1 + 2B'c + 3gB'^2c^2 + \cdots)$$

where g is the parameter introduced in Problem 7.24. W.J. Closs, B.R. Jennings, and H.G. Gerrard (*Eur. Polymer J.* **4**, 639 (1968)) reported the data in the table below for polystyrene in cyclohexane at 35°C. The density of cyclohexane at this temperature, which can also be assumed to be that of the solution, is 0.765 g cm^{-3}, and the partial specific volume of polystyrene is 0.93 cm^3 g^{-1}. The dependence of the diffusion constant for these solutions has been determined empirically by T.A. King, A. Knox, W.I. Lee, and J.D.G. McAdam (*Polymer* **14**, 151 (1973)) to be given by the relation $D/(\text{cm}^2\,\text{s}^{-1}) = 1.3 \times 10^{-4}(\overline{M}_w/(\text{g mol}^{-1}))^{-0.497}$. Determine the molar mass of polystyrene in cyclohexane and the second virial coefficient, B'. Compare the molar mass obtained here to that calculated in Problem 22.26. Is there any reason for them to be the same?

c/(mg cm^{-3})	2.0	3.0	4.0	5.0	6.0	7.0
S/(10^{-13} s)	14.8	13.9	13.1	12.4	11.8	11.2

22.5 The concentration dependence of the viscosity of a polymer solution is found to be as follows:

c/(g L^{-1})	1.32	2.89	5.73	9.17
η/(g m^{-1} s^{-1})	1.08	1.20	1.42	1.73

The viscosity of the solvent is 0.985 g m^{-1} s^{-1}. What is the intrinsic viscosity of the polymer?

22.6 The times of flow of dilute solutions of polystyrene in benzene through a viscometer at 25°C are given in the table below. From these data, calculate the molar mass of the polystyrene samples. Since the solutions are dilute, assume that the densities of the solutions are the same as those of pure benzene. η(benzene) = 0.601×10^{-3} kg m^{-1} s^{-1} (0.601 cP) at 25°C.

c/(g L^{-1})	0	2.22	5.00	8.00	10.00
t/s	208.2	248.1	303.4	371.8	421.3

22.7 The viscosities of solutions of polyisobutylene in benzene were measured at 24°C (the θ temperature for the system) with the following results:

c/(g/10^2 cm^3)	0	0.2	0.4	0.6	0.8	1.0
η/(10^{-3} kg m^{-1} s^{-1})	0.647	0.690	0.733	0.777	0.821	0.865

Use the information in Table 22.4 to deduce the molar mass of the polymer.

22.8‡ Polystyrene in cyclohexane at 34.5°C forms a θ solution, with an intrinsic viscosity related to the molar mass by $[\eta] = KM^a$. The following data on polystyrene in cyclohexane are taken from L.J. Fetters, N. Hadjichristidis, J.S. Lindner, and J.W. Mays (*J. Phys. Chem. Ref. Data* **23**, 619 (1994)):

M/(kg mol^{-1})	10.0	19.8	106	249	359	860
$[\eta]$/(cm^3 g^{-1})	8.90	11.9	28.1	44.0	51.2	77.6

M/(kg mol^{-1})	1800	5470	9720	56 800
$[\eta]$/(cm^3 g^{-1})	113.9	195	275	667

Determine the parameters K and a. What is the molar mass of a polystyrene that forms a θ solution in cyclohexane with $[\eta] = 100$ cm^3 g^{-1}?

22.9‡ Standard polystyrene solutions of known average molar masses continue to be used for the calibration of many methods of characterizing polymer solutions. M. Kolinsky and J. Janca (*J. Polym. Sci., Polym. Chem.* **12**, 1181 (1974)) studied polystyrene in tetrahydrofuran (THF) for use in calibrating a gel permeation chromatograph. Their results for the intrinsic viscosity, $[\eta]$, as a function of average molar mass at 25°C are given in the table below. (a) Obtain the Mark–Houwink constants that fit these data. (b) Compare your values to those in Table 22.4 and Example 22.7. How might you explain the differences?

$\overline{M}_v$/(kg mol^{-1})	5.0	10.3	19.85	51	98.2	173	411	867
$[\eta]$/(cm^3 g^{-1})	5.2	8.8	14.0	27.6	43.6	67.0	125.0	206.7

22.10 The concentration dependence of the osmotic pressure of solutions of a macromolecule at 20°C was found to be as follows:

c/(g L^{-1})	1.21	2.72	5.08	6.60
Π/Pa	134	321	655	898

Determine the molar mass of the macromolecule and the osmotic virial coefficient.

22.11 The osmotic pressure of a fraction of poly(vinyl chloride) in a ketone solvent was measured at 25°C. The density of the solvent (which is virtually equal to the density of the solution) was 0.798 g cm^{-3}. Calculate the molar mass and the osmotic virial coefficient, B, of the fraction from the following data:

$c/(g/10^2 cm^3)$	0.200	0.400	0.600	0.088	1.000
h/cm	0.48	1.2	1.86	2.76	3.88

Theoretical problems

22.12 A polymerization process produced a Gaussian distribution of polymers in the sense that the proportion of molecules having a molar mass in the range M to $M + dM$ was proportional to $e^{-(M-\bar{M})^2/2\gamma}\,dM$. What is the number-average molar mass when the distribution is narrow?

22.13 Use eqn 2 to deduce expressions for (a) the root mean square separation of the ends of the chain, (b) the mean separation of the ends, and (c) their most probable separation. Evaluate these three quantities for a fully flexible chain with $N = 4000$ and $l = 154$ pm.

22.14 Construct a two-dimensional random walk by using a random number generating routine with mathematical software or electronic spreadsheet. Construct a walk of 50 and 100 steps. If there are many people working on the problem, investigate the mean and most probable separations in the plots by direct measurement. Do they vary as $N^{1/2}$?

22.15 Evaluate the radius of gyration, R_g, of (a) a solid sphere of radius a, (b) a long straight rod of radius a and length l. Show that in the case of a solid sphere of specific volume v_s,

$$R_g/nm \approx 0.056\,902 \times \{(v_s/cm^3\,g^{-1})(M/g\,mol^{-1})\}^{1/3}$$

Evaluate R_g for a species with $M = 100$ kg mol^{-1}, $v_s = 0.750$ cm^3 g^{-1}, and, in the case of the rod, of radius 0.50 nm.

22.16 The effective radius, a, of a random coil is related to its radius of gyration, R_g, by $a = \gamma R_g$, with $\gamma = 0.85$. Deduce an expression for the osmotic virial coefficient, B, in terms of the number of chain units for (a) a freely jointed chain, (b) a chain with tetrahedral bond angles. Evaluate B for $l = 154$ pm and $N = 4000$. Estimate B for a randomly coiled polyethylene chain of arbitrary molar mass, M, and evaluate it for $M = 56$ kg mol^{-1}. Use $B = \frac{1}{2}N_A v_P$, where v_P is the excluded volume due to a single molecule.

22.17 Radius of gyration is defined in eqn 6. Show that an equivalent definition is that R_g is the average root mean square distance of the atoms or groups (all assumed to be of the same mass); that is, that $R_g^2 = (1/N)\Sigma_j R_j^2$, where R_j is the distance of atom j from the centre of mass.

Applications: to biology and chemical engineering

22.18 In this problem you will use molecular mechanics software of your instructor's choice to gain some appreciation for the complexity of the calculations that lead to plots such as those in Fig. 22.13. Our model for the protein is the dipeptide (**16**) in which the terminal methyl groups replace the rest of the polypeptide chain. (a) Draw three initial conformers of (**16**) with R = H: one with $\phi = 75°$, $\psi = -65°$, a second with $\phi = \psi = 180°$, and a third with $\phi = 65°$, $\psi = 35°$. Use a molecular mechanics routine to optimize the geometry of each conformer and measure the total potential energy and the final ϕ and ψ angles in each case.[18] Did all of the initial conformers converge to the same final conformation? If not, what do these final conformers represent? Rationalize any observed

16

differences in total potential energy of the final conformers. (b) Use the approach in part (a) to investigate the case R = CH$_3$, with the same three initial conformers as starting points for the calculations. Rationalize any similarities and differences between the final conformers of the dipeptides with R = H and R = CH$_3$.

22.19 Calculate the excluded volume in terms of the molecular volume on the basis that the molecules are spheres of radius a. Evaluate the osmotic virial coefficient in the case of bushy stunt virus, $a = 14.0$ nm, and haemoglobin, $a = 3.2$ nm (see Problem 22.16). Evaluate the percentage deviation of the Rayleigh ratios of 1.00 g/(100 cm^3) solutions of bushy stunt virus ($M = 1.07 \times 10^4$ kg mol^{-1}) and haemoglobin ($M = 66.5$ kg mol^{-1}) from the ideal solution values. In eqn 22.28, let $P_\theta = 1$ and assume that both solutions have the same K value.

22.20 Use the information below and the expression for R_g of a solid sphere quoted in the Problem 22.15, to classify the species below as globular or rod-like.

	$M/(g\,mol^{-1})$	$v_s/(cm^3\,g^{-1})$	R_g/nm
Serum albumin	66×10^3	0.752	2.98
Bushy stunt virus	10.6×10^6	0.741	12.0
DNA	4×10^6	0.556	117.0

22.21 Suppose that a rod-like, 250-nm long DNA molecule undergoes a conformational change to a closed-circular (cc) form. (a) Use the information in Problem 22.20 and an incident wavelength $\lambda = 488$ nm to calculate the ratio of scattering intensities by each of these conformations, I_{rod}/I_{cc}, when $\theta = 20°$, $45°$, and $90°$. (b) Suppose that you wish to use light scattering as a technique for the study of conformational changes in DNA molecules. Based on your answer to part (a), at which angle would you conduct the experiments? Justify your choice.

22.22 In an ultracentrifugation experiment at 20°C on bovine serum albumin the following data were obtained: $\rho = 1.001$ g cm^{-3}, $v_s = 1.112$ cm^3 g^{-1}, $\omega/2\pi = 322$ Hz,

r/cm	5.0	5.1	5.2	5.3	5.4
$c/(mg\,cm^{-3})$	0.536	0.284	0.148	0.077	0.039

Evaluate the molar mass of the sample.

22.23 Sedimentation studies on haemoglobin in water gave a sedimentation constant $S = 4.5$ Sv at 20°C. The diffusion coefficient is 6.3×10^{-11} m^2 s^{-1} at the same temperature. Calculate the molar mass of haemoglobin using $v_s = 0.75$ cm^3 g^{-1} for its partial specific volume and $\rho = 0.998$ g cm^{-3} for the density of the solution. Estimate the effective radius of the haemoglobin molecule given that the viscosity of the solution is 1.00×10^{-3} kg m^{-1} s^{-1}.

18. Note to the instructor. Although any force field will work satisfactorily, the AMBER force field is strongly recommended, as it is optimized for calculations on biopolymers.

22.24 The rate of sedimentation of a recently isolated protein was monitored at 20°C and with a rotor speed of 50 000 r.p.m. The boundary receded as follows:

t/s	0	300	600	900	1200	1500	1800
r/cm	6.127	6.153	6.179	6.206	6.232	6.258	6.284

Calculate the sedimentation constant and the molar mass of the protein on the basis that its partial specific volume is $0.728 \text{ cm}^3 \text{ g}^{-1}$ and its diffusion coefficient is $7.62 \times 10^{-11} \text{ m}^2 \text{ s}^{-1}$ at 20°C, the density of the solution then being 0.9981 g cm^{-3}. Suggest a shape for the protein given that the viscosity of the solution is $1.00 \times 10^{-3} \text{ kg m}^{-1} \text{ s}^{-1}$ at 20°C.

22.25 Polystyrene is a synthetic polymer with the structure $-(CH_2CH(C_6H_5))_n-$. A batch of polydisperse polystyrene was prepared by initiating the polymerization with the *t*-butyl free radical (Section 26.4). As a result, the *t*-butyl group is expected to be covalently attached to the end of the final products. A sample from this batch was embedded in an organic matrix containing silver trifluoroacetate and the resulting MALDI-TOF spectrum consisted of a large number of peaks separated by 104 g mol^{-1}, with the most intense peak at $25\,578 \text{ g mol}^{-1}$. Comment on the purity of this sample and determine the number of $(CH_2CH(C_6H_5))$ units in the species that gives rise to the most intense peak in the spectrum.

22.26‡ A manufacturer of polystyrene beads claims that they have an average molar mass of 250 kg mol^{-1}. Solutions of these beads are studied by a physical chemistry student by dilute solution viscometry with an Ostwald viscometer in both the 'good' solvent toluene and the theta solvent cyclohexane. The drainage times, t_D,

as a function of concentration for the two solvents are given in the table below. (a) Fit the data to the virial equation for viscosity,

$$\eta = \eta^*(1 + [\eta]c + k'[\eta]^2c^2 + \cdots)$$

where k' is called the *Huggins constant* and is typically in the range 0.35–0.40. From the fit, determine the intrinsic viscosity and the Huggins constant. (b) Use the empirical Mark–Kuhn–Houwink–Sakurada equation (eqn 47) to determine the molar mass of polystyrene in the two solvents. For theta solvents, $a = 0.5$ and $K = 8.2 \times 10^{-5} \text{ L g}^{-1}$ for cyclohexane; for the good solvent toluene $a = 0.72$ and $K = 1.15 \times 10^{-5} \text{ L g}^{-1}$. (c) According to a general theory proposed by Kirkwood and Riseman, the root mean square end-to-end distance of a polymer chain in solution is related to $[\eta]$ by $[\eta] = \Phi\langle r^2\rangle^{3/2}/M$, where Φ is a universal constant with the value 2.84×10^{26} when $[\eta]$ is expressed in litres per gram and the distance is in metres. Calculate this quantity for each solvent. (d) From the molar masses calculate the average number of styrene $(C_6H_5CH=CH_2)$ monomer units, $\langle n\rangle$. (e) Calculate the length of a fully stretched, planar zigzag configuration, taking the C—C distance as 154 pm and the CCC bond angle to be 109°. (f) Use eqn 10 to calculate the radius of gyration, R_g. Also calculate $\langle r^2\rangle^{1/2} = n^{1/2}l$. Compare this result with that predicted by the Kirkwood–Riseman theory: which gives the better fit? (g) Compare your values for M to the results of Problem 7.24. Is there any reason why they should or should not agree? Is the manufacturer's claim valid?

c/(g L⁻¹ toluene)	0	1.0	3.0	5.0
t_D/s		8.37	9.11	10.72	12.52

c/(g L⁻¹ cyclohexane)	0	1.0	1.5	2.0
t_D/s		8.32	8.67	8.85	9.03

23

The solid state

Crystal lattices

23.1 Lattices and unit cells

23.2 The identification of lattice planes

23.3 The investigation of structure

23.4 Neutron and electron diffraction

Crystal structure

23.5 Metallic solids

23.6 Ionic solids

23.7 Molecular solids

The properties of solids

23.8 Mechanical properties

23.9 Electric properties

23.10 Magnetic properties

Checklist of key ideas

Further reading

Exercises

Problems

First, we see how to describe the regular arrangement of atoms in crystals and the symmetry of their arrangement. Then we consider the basic principles of X-ray diffraction, and see how the diffraction pattern can be interpreted in terms of the distribution of electron density in a unit cell. X-ray diffraction leads to information about the structures of metallic, ionic, and molecular solids, and we review some typical results and their rationalization in terms of atomic and ionic radii. With structures established, we move on to the properties of solids, and see how their mechanical, electrical, and magnetic properties stem from the constituent atoms.

The solid state includes most of the materials that make modern technology possible. It includes the wide varieties of steel that are used in architecture and engineering, the semiconductors and metallic conductors that are used in information technology and power distribution, the ceramics that increasingly are replacing metals, and the synthetic and natural polymers that are used in the textile industry and in the fabrication of many of the common objects of the modern world. The properties of solids stem, of course, from the arrangement and properties of the constituent atoms, and one of the challenges of this chapter is to see how a wide range of bulk properties, including rigidity, brittleness, and electrical conductivity, stem from the properties of atoms. One crucial aspect of this link is the pattern in which the atoms (and molecules) are stacked together, and we start this chapter with an examination of how the structures of solids are described and determined.

Crystal lattices

Early in the history of modern science it was suggested that the regular external form of crystals implied an internal regularity of their constituents. In this section we see how to describe the arrangement of atoms inside crystals.

23.1 Lattices and unit cells

A crystal is built up from regularly repeating 'structural motifs', which may be atoms, molecules, or groups of atoms, molecules, or ions. A **space lattice** is the pattern formed by points representing the locations of these motifs (Fig. 23.1). The space lattice is, in effect, an abstract scaffolding for the crystal structure. More formally, a space lattice is a three-dimensional, infinite array of points, each of which

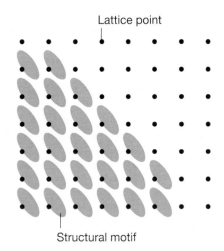

Lattice point

Structural motif

23.1 Each lattice point specifies the location of a structural motif (for example, a molecule or a group of molecules). The crystal lattice is the array of lattice points; the crystal structure is the collection of structural motifs arranged according to the lattice.

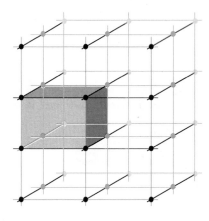

23.2 A unit cell is a parallel-sided (but not necessarily rectangular) figure from which the entire crystal structure can be constructed by using only translations (not reflections, rotations, or inversions).

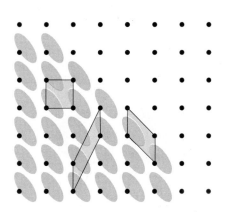

23.3 A unit cell can be chosen in a variety of ways, as shown here. It is conventional to choose the cell that represents the full symmetry of the lattice. In this rectangular lattice, the rectangular unit cell would normally be adopted.

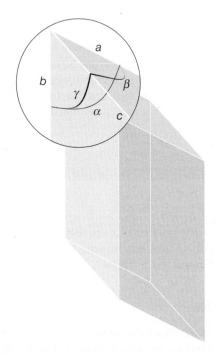

23.4 The notation for the sides and angles of a unit cell. Note that the angle α lies in the plane (a, b) and perpendicular to the axis c.

is surrounded in an identical way by its neighbours, and which defines the basic structure of the crystal. In some cases there may be a structural motif centred on each lattice point, but that is not necessary. The crystal structure itself is obtained by associating with each lattice point an identical structural motif.

The **unit cell** is an imaginary parellelepiped (parallel-sided figure) that contains one unit of the translationally repeating pattern (Fig. 23.2). A unit cell can be thought of as the fundamental region from which the entire crystal may be constructed by purely translational displacements (like bricks in a wall). A unit cell is commonly formed by joining neighbouring lattice points by straight lines (Fig. 23.3). Such unit cells are called **primitive**. It is sometimes more convenient to draw larger **non-primitive unit cells** that also have lattice points at their centres or on pairs of opposite faces. An infinite number of different unit cells can describe the same lattice, but we normally choose the one with sides that have the shortest lengths and that are most nearly perpendicular to one another. The lengths of the sides of a unit cell are denoted a, b, and c, and the angles between them are denoted α, β, and γ (Fig. 23.4).

Unit cells are classified into seven **crystal systems** by noting the rotational symmetry elements they possess. A *cubic unit cell*, for example, has four threefold axes in a tetrahedral array (Fig. 23.5). A *monoclinic unit cell* has one twofold axis; the unique axis is by convention the b-axis (Fig. 23.6). A *triclinic unit cell* has no rotational symmetry, and typically all three sides and angles are different (Fig. 23.7). Table 23.1 lists the **essential symmetries**, the elements that must be present for the unit cell to belong to a particular crystal system.

There are only 14 distinct space lattices in three dimensions. These **Bravais lattices** are illustrated in Fig. 23.8. It is conventional to portray these lattices by primitive unit cells in some cases and by non-primitive unit cells in others. A **primitive unit cell** (with lattice points only at the corners) is denoted P. A **body-centred unit cell** (I) also has a lattice point at its centre. A **face-centred unit cell** (F) has lattice points at its corners and also at the centres of its six faces. A **side-centred unit cell** (A, B, or C) has lattice points at its corners and at the centres of two opposite faces. For simple structures, it is often convenient to choose an atom belonging to the structural motif, or the centre of a molecule, as the location of a lattice point or the vertex of a unit cell, but that is not a necessary requirement.

23.5 A unit cell belonging to the cubic system has four threefold axes arranged tetrahedrally. The insert shows the threefold symmetry.

23.6 A unit belonging to the monoclinic system has a twofold axis (shown in more detail in the insert).

23.7 A triclinic unit cell has no axes of rotational symmetry.

Table 23.1 The seven crystal systems	
System	**Essential symmetries**
Triclinic	None
Monoclinic	One C_2 axis
Orthorhombic	Three perpendicular C_2 axes
Rhombohedral	One C_3 axis
Tetragonal	One C_4 axis
Hexagonal	One C_6 axis
Cubic	Four C_3 axes in a tetrahedral arrangement

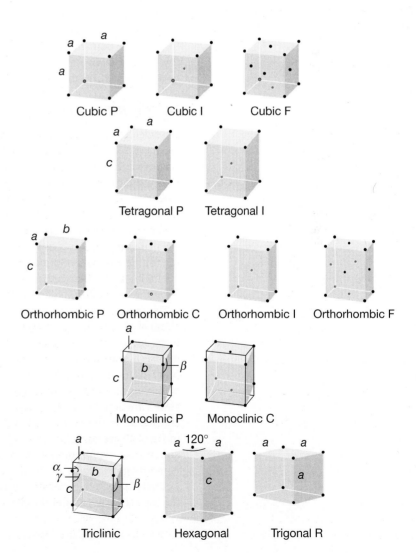

Cubic P   Cubic I   Cubic F

Tetragonal P   Tetragonal I

Orthorhombic P   Orthorhombic C   Orthorhombic I   Orthorhombic F

Monoclinic P   Monoclinic C

Triclinic   Hexagonal   Trigonal R

23.8 The fourteen Bravais lattices. The points are lattice points, and are not necessarily occupied by atoms. P denotes a primitive unit cell (R is used for a trigonal lattice), I a body-centred unit cell, F a face-centred unit cell, and C (or A or B) a cell with lattice points on two opposite faces.

23.9 Some of the planes that can be drawn through the points of a rectangular space lattice and their corresponding Miller indices (*hkl*): (a) (110), (b) (230), (c) ($\overline{1}$10), (d) (010).

23.2 The identification of lattice planes

The spacing of the lattice points in a crystal is an important quantitative aspect of its structure and its investigation. However, there are many different sets of planes (Fig. 23.9), and we need to be able to label them. Two-dimensional lattices are easier to visualize than three-dimensional lattices, so we shall introduce the concepts involved by referring to two dimensions initially, and then extend the conclusions by analogy to three dimensions.

(a) The Miller indices

Consider a two-dimensional rectangular lattice formed from a unit cell of sides a, b (as in Fig. 23.9). Each plane in the illustration (except the plane passing through the origin) can be distinguished by the distances at which it intersects the a and b axes. One way of labelling each set of parallel planes would therefore be to quote the smallest intersection distances. For example, we could denote the four sets in the illustration as $(1a, 1b)$, $(\frac{1}{2}a, \frac{1}{3}b)$, $(-1a, 1b)$, and $(\infty a, 1b)$. However, if we agree to quote distances along the axes as multiples of the lengths of the unit cell, then we can label the planes more simply as $(1, 1)$, $(\frac{1}{2}, \frac{1}{3})$, $(-1, 1)$, and $(\infty, 1)$. If the lattice in Fig. 23.9 is the top view of a three-dimensional orthorhombic lattice in which the unit cell has a length c in the z-direction, all four sets of planes intersect the z-axis at infinity. Therefore, the full labels are $(1, 1, \infty)$, $(\frac{1}{2}, \frac{1}{3}, \infty)$, $(-1, 1, \infty)$, and $(\infty, 1, \infty)$.

The presence of fractions and infinity in the labels is inconvenient. They can be eliminated by taking the reciprocals of the labels. As we shall see, taking reciprocals turns out to have further advantages. The **Miller indices**, (*hkl*), are the reciprocals of intersection distances (with fractions cleared by multiplying through by an appropriate factor, if taking the reciprocal results in a fraction). For example, the $(1, 1, \infty)$ planes in Fig. 23.9a are the (110) planes in the Miller notation. Similarly, the $(\frac{1}{2}, \frac{1}{3}, \infty)$ planes are denoted (230). Negative indices are written with a bar over the number, and Fig. 23.9c shows the ($\overline{1}$10) planes. The Miller indices for the four sets of planes in Fig. 23.9 are therefore (110), (230), ($\overline{1}$10), and (010). Figure 23.10 shows a three-dimensional representation of a selection of planes, including one in a lattice with non-orthogonal axes.

A helpful feature to remember is that, the smaller the absolute value of h in (*hkl*), the more nearly parallel the plane is to the a-axis.[1] The same is true of k and the

1. The {h00} planes are an exception.

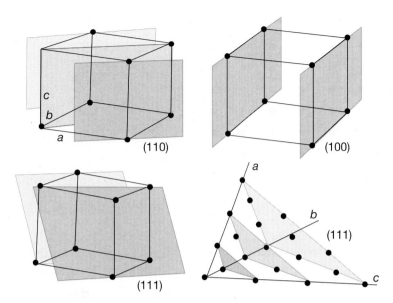

23.10 Some representative planes in three dimensions and their Miller indices. Note that a 0 indicates that a plane is parallel to the corresponding axis, and that the indexing may also be used for unit cells with non-orthogonal axes.

b-axis and l and the c-axis. When $h = 0$, the planes intersect the a-axis at infinity, so the $(0kl)$ planes are parallel to the a-axis. Similarly, the $(h0l)$ planes are parallel to b and the $(hk0)$ planes are parallel to c.

The notation (hkl) denotes an *individual* plane. To specify a *set* of parallel planes we use the notation $\{hkl\}$. Thus, we speak of the (110) plane in a lattice, and the set of all $\{110\}$ planes that lie parallel to the (110) plane.

(b) The separation of planes

The Miller indices are very useful for expressing the separation of planes. The separation of the $\{hk0\}$ planes in the square lattice shown in Fig. 23.11 is given by

$$\frac{1}{d_{hk0}^2} = \frac{h^2 + k^2}{a^2} \quad \text{or} \quad d_{hk0} = \frac{a}{(h^2 + k^2)^{1/2}} \tag{23.1}$$

By extension to three dimensions, the separation of the $\{hkl\}$ planes of a cubic lattice is given by

$$\frac{1}{d_{hkl}^2} = \frac{h^2 + k^2 + l^2}{a^2} \quad \text{or} \quad d_{hkl} = \frac{a}{(h^2 + k^2 + l^2)^{1/2}} \tag{23.2}$$

The corresponding expression for a general orthorhombic lattice is the generalization of this expression:

$$\frac{1}{d_{hkl}^2} = \frac{h^2}{a^2} + \frac{k^2}{b^2} + \frac{l^2}{c^2} \tag{23.3}$$

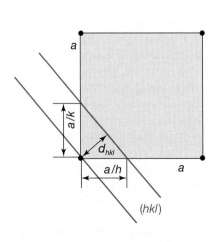

23.11 The dimensions of a unit cell and their relation to the plane passing through the lattice points.

Example 23.1 Using the Miller indices

Calculate the separation of (a) the $\{123\}$ planes and (b) the $\{246\}$ planes of an orthorhombic unit cell with $a = 0.82$ nm, $b = 0.94$ nm, and $c = 0.75$ nm.

Method For the first part, simply substitute the information into eqn 3. For the second part, instead of repeating the calculation, note that, if all three Miller indices are multiplied by n, then their separation is reduced by that factor (Fig. 23.12):

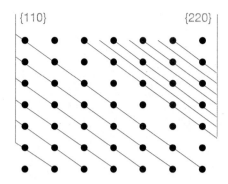

23.12 The separation of the {220} planes is half that of the {110} planes. In general, (the separation of the planes {nh, nk, nl} is n times smaller than the separation of the {hkl} planes.

$$\frac{1}{d_{nh,nk,nl}^2} = \frac{(nh)^2}{a^2} + \frac{(nk)^2}{b^2} + \frac{(nl)^2}{c^2} = n^2\left(\frac{h^2}{a^2} + \frac{k^2}{b^2} + \frac{l^2}{c^2}\right) = \frac{n^2}{d_{hkl}^2}$$

which implies that

$$d_{nh,nk,nl} = \frac{d_{hkl}}{n}$$

Answer Substituting the indices into eqn 3 gives

$$\frac{1}{d_{123}^2} = \frac{1^2}{(0.82\ \text{nm})^2} + \frac{2^2}{(0.94\ \text{nm})^2} + \frac{3^2}{(0.75\ \text{nm})^2} = 22\ \text{nm}^{-2}$$

Hence, $d_{123} = 0.21$ nm. It then follows immediately that d_{246} is one-half this value, or 0.11 nm.

Self-test 23.1 Calculate the separation of (a) the {133} planes and (b) the {399} planes in the same lattice. [0.19 nm, 0.063 nm]

23.3 The investigation of structure

A characteristic property of waves is that they interfere with one another, giving a greater displacement where peaks or troughs coincide and a smaller displacement where peaks coincide with troughs (Fig. 23.13). According to classical electromagnetic theory, the intensity of electromagnetic radiation is proportional to the square of the amplitude of the waves. Therefore, the regions of constructive or destructive interference show up as regions of enhanced or diminished intensities. The phenomenon of **diffraction** is the interference caused by an object in the path of waves, and the pattern of varying intensity that results is called the **diffraction pattern**. Diffraction occurs when the dimensions of the diffracting object are comparable to the wavelength of the radiation.

(a) X-ray diffraction

Wilhelm Röntgen discovered X-rays in 1895. Seventeen years later, Max von Laue suggested that they might be diffracted when passed through a crystal, for by then he had realized that their wavelengths are comparable to the separation of lattice planes. This suggestion was confirmed almost immediately by Walter Friedrich and Paul Knipping and has grown since then into a technique of extraordinary power. The bulk of this section will deal with the determination of structures using X-ray diffraction. The mathematical procedures necessary for the determination of structure from X-ray diffraction data are enormously complex, but such is the degree of integration of computers into the experimental apparatus that the technique is almost fully automated, even for large molecules and complex solids. The analysis is aided by molecular modelling techniques (Section 22.3c), which can guide the investigation towards a plausible structure.

X-rays are electromagnetic radiation with wavelengths of the order of 10^{-10} m. They are typically generated by bombarding a metal with high-energy electrons (Fig. 23.14). The electrons decelerate as they plunge into the metal and generate radiation with a continuous range of wavelengths called **Bremsstrahlung**.[2] Superimposed on the continuum are a few high-intensity, sharp peaks (Fig. 23.15). These peaks arise from collisions of the incoming electrons with the electrons in the inner shells of the atoms. A collision expels an electron from an inner shell, and an

(a)

(b)

23.13 When two waves are in the same region of space they interfere. Depending on their relative phase, they may interfere (a) constructively, to give an enhanced amplitude, or (b) destructively, to give a smaller amplitude. The component waves are shown in black and the resultant in blue.

2. *Bremse* is German for deceleration, *Strahlung* for ray.

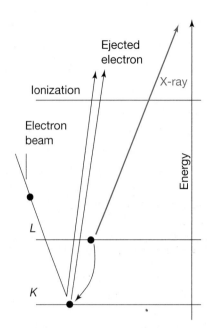

23.14 X-rays are generated by directing an electron beam on to a cooled metal target. Beryllium is transparent to X-rays (on account of the small number of electrons in each atom) and is used for the windows.

23.15 The X-ray emission from a metal consists of a broad, featureless Bremsstrahlung background, with sharp transitions superimposed on it. The label K indicates that the radiation comes from a transition in which an electron falls into a vacancy in the K shell of the atom.

23.16 The processes that contribute to the generation of X-rays. An incoming electron collides with an electron (in the K shell), and ejects it. Another electron (from the L shell in this illustration) falls into the vacancy and emits its excess energy as an X-ray photon.

electron of higher energy drops into the vacancy, emitting the excess energy as an X-ray photon (Fig. 23.16). If the electron falls into a K shell (a shell with $n = 1$), the X-rays are classified as **K-radiation**, and similarly for transitions into the L ($n = 2$) and M ($n = 3$) shells. Strong, distinct lines are labelled K_α, K_β, and so on. Increasingly, X-ray diffraction makes use of the radiation available from synchrotron sources (Section 16.1a), for its high intensity greatly enhances the sensitivity of the technique.

Von Laue's original method consisted of passing a broad-band beam of X-rays into a single crystal, and recording the diffraction pattern photographically. The idea behind the approach was that a crystal might not be suitably orientated to act as a diffraction grating for a single wavelength but, whatever its orientation, diffraction would be achieved for at least one of the wavelengths if a range of wavelengths was used. There is currently a resurgence of interest in this approach because synchrotron radiation spans a range of X-ray wavelengths (Box 23.1).

Box 23.1 Time-resolved X-ray diffraction techniques

The traditional X-ray diffraction techniques discussed in this chapter give only static pictures and are not useful in studies of dynamics and reactivity. This is because the Bragg rotation method requires stable crystals that do not change structure during long data acquisition times. For this reason, chemists have turned to the time-resolved spectroscopic techniques described in Section 17.7e in an attempt to follow chemical reactions and determine the structures of intermediates.

However, special time-resolved X-ray diffraction techniques have become available in recent years and it is now possible to make exquisitely detailed measurements of atomic motions during chemical and biochemical reactions.

Time-resolved X-ray diffraction techniques make use of synchrotron sources, which can emit intense polychromatic pulses of X-ray radiation with pulse widths varying from 100 ps to 200 ps (Section 16.1a). Instead of the Bragg method, the Laue method discussed in Section 23.3a is used because many reflections can be collected simultaneously, rotation of the sample is not required, and data acquisition times are short. However, good diffraction data cannot be obtained from a single X-ray pulse and reflections from several pulses must be averaged together. In practice, this averaging dictates the time resolution of the experiment, which is tens of microseconds or less.

The progress of a reaction may be studied either by real-time analysis of the evolving system or by trapping intermediates by chemical or physical means. Regardless of the strategy, all of the molecules in the crystal must be made to react at the same time, so special reaction initiation schemes are required. One way to initiate a reaction is to allow a solution containing one of the reactants to diffuse into a crystal containing the other reactant. This method is simple, but limited to relatively long reaction times, as diffusion of solutions into crystals large enough for crystallographic measurements is·of the order of seconds to minutes. Variations of the diffusion method can also be used to trap intermediates. One elegant example of the strategy is a study of the mechanism of action of the enzyme elastase, a digestive enzyme that cleaves peptide bonds selectively. The enzyme was rendered inactive by lowering the temperature of the crystal below the protein's glass transition temperature, T_g. Then, a solution containing the substrate was added to the crystal and the temperature raised above T_g. The reaction was allowed to proceed until an intermediate was formed, at which point the temperature was lowered very quickly below T_g, thus stopping the reaction and trapping the intermediate.

Another way to initiate a reaction is to use a pulsed laser as a trigger. It is common to use the pump-delay-probe technique already discussed in the context of time-resolved spectroscopy (Section 17.7e): a laser provides the pump pulse and after a delay that can be as short as a few nanoseconds, an X-ray pulse probes the sample. This method has the obvious advantage of being a tool for the study of ultrafast reactions, but is obviously limited to processes that can be induced by photon absorption.

A recent example of the power of time-resolved X-ray crystallography is the elucidation of structural changes that accompany the activation by light of the photoactive yellow protein of the bacterium *Ectothiorhodospira halophila*. Within 1 ns after absorption of a photon with $\lambda = 446$ nm, a protein-bound phenolate ion undergoes *trans–cis* isomerization to form the intermediate shown in the illustration. What follows is a series of rearrangements that include the ejection of the chromophore from its binding site deep in the protein, its return to the site, and reformation of the *cis* conformation. The physiological outcome of this cycle is a 'negative phototactic response', or movement of the organism away from light. Time-resolved X-ray diffraction studies in the

Photoisomerization of a protein-bound phenolate ion in the photoactive yellow protein of the bacterium *Ectothiorhodospira halophila*.

nanosecond to millisecond ranges identified a number of structural changes that follow electronic excitation of the chromophore with a laser pulse: isomerization, ejection, protonation of the exposed chromophore, and a number of amino acid motions.

Discussion questions

1. Intense X-ray radiation from synchrotron sources can damage samples by heating, generation of free radicals, and radiation-induced reduction. Account for the reduction of and the presence of radicals in a sample that has been exposed to X-rays.

2. The photodissociation of carbon monoxide from the haem protein myoglobin has been studied by time-resolved spectroscopy (for example, P. Anfinrud, R. de Vivie-Riedle, and V. Engel, *Proc. Natl. Acad. Sci. U.S.A.* **96**, 8328 (1999) and references cited therein) and time-resolved X-ray crystallography (for example, T.Y. Teng, V. Srajer, and K. Moffat, *Biochemistry* **36**, 12 087 (1997) and references cited therein). Consult the current literature and prepare a brief report comparing and contrasting the results obtained from spectroscopic and X-ray diffraction experiments.

Further reading

I. Schlichting, Crystallographic structure determination of unstable species. *Acc. Chem. Res.* **33**, 532 (2000).

B.L. Stoddard, New results using Laue diffraction and time-resolved crystallography. *Current Opinion in Structural Biology* **8**, 612 (1998).

23.17 X-ray powder photographs of (a) NaCl, (b) KCl and the indexed reflections. The smaller number of lines in (b) is a consequence of the similarity of the K⁺ and Cl⁻ scattering factors, as discussed later in the chapter.

The web site contains links to online databases of X-ray diffraction patterns.

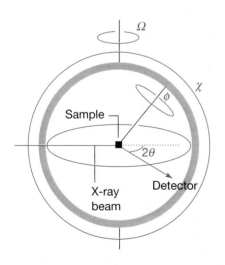

23.18 A four-circle diffractometer. The setting of the orientations (ϕ, χ, θ, and Ω) of the components is controlled by computer; each (hkl) reflection is monitored in turn, and their intensities are recorded.

An alternative technique to von Laue's was developed by Peter Debye and Paul Scherrer and independently by Albert Hull. They used monochromatic radiation and a powdered sample. When the sample is a powder, at least some of the crystallites will be orientated so as to give rise to diffraction. In modern powder diffractometers the intensities of the reflections are monitored electronically as the detector is rotated around the sample in a plane containing the incident ray (Fig. 23.17). Powder diffraction techniques are used to identify a sample of a solid substance by comparison of the positions of the diffraction lines and their intensities with diffraction patterns stored in a large data bank.[3] Powder diffraction data are also used to determine phase diagrams, for different solid phases result in different diffraction patterns, and to determine the relative amounts of each phase present in a mixture. The technique is also used for the initial determination of the dimensions and symmetries of unit cells.

The method developed by the Braggs (William and his son Lawrence, who later jointly won the Nobel Prize) is the foundation of almost all modern work in X-ray crystallography. They used a single crystal and a monochromatic beam of X-rays, and rotated the crystal until a reflection was detected. There are many different sets of planes in a crystal, so there are many angles at which a reflection occurs. The complete set of data consists of the list of angles at which reflections are observed and their intensities.

Single-crystal diffraction patterns are measured by using a **four-circle diffractometer** (Fig. 23.18). The computer linked to the diffractometer determines the unit cell dimensions and the angular settings of the diffractometer's four circles that are needed to observe any particular intensity peak in the diffraction pattern. The computer controls the settings, and moves the crystal and the detector for each

3. The powder diffraction file, which is maintained by the International Centre for Diffraction Data, ICDD, contains information on about 50 000 crystalline phases.

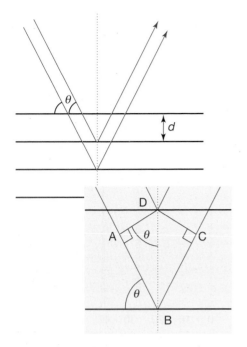

23.19 The conventional derivation of Bragg's law treats each lattice plane as a reflecting the incident radiation. The path lengths differ by AB + BC, which depends on the glancing angle, θ. Constructive interference (a 'reflection') occurs when AB + BC is equal to an integer number of wavelengths.

one in turn. At each setting, the diffraction intensity is measured, and background intensities are assessed by making measurements at slightly different settings. Computing techniques are now available that lead not only to automatic indexing but also to the automated determination of the shape, symmetry, and size of the unit cell. Moreover, several techniques are now available for sampling large amounts of data, including area detectors and image plates, which sample whole regions of diffraction patterns simultaneously.

(b) Bragg's law

An early approach to the analysis of diffraction patterns produced by crystals was to regard a lattice plane as a mirror, and to model a crystal as stacks of reflecting lattice planes of separation d (Fig. 23.19). The model makes it easy to calculate the angle the crystal must make to the incoming beam of X-rays for constructive interference to occur. It has also given rise to the name **reflection** to denote an intense beam arising from constructive interference.

Consider the reflection of two parallel rays of the same wavelength by two adjacent planes of a lattice, as shown in Fig. 23.19. One ray strikes point D on the upper plane but the other ray must travel an additional distance AB before striking the plane immediately below. Similarly, the reflected rays will differ in path length by a distance BC. The net path length difference of the two rays is then

$$AB + BC = 2d \sin \theta$$

where θ is the **glancing angle**. For many glancing angles the path-length difference is not an integer number of wavelengths, and the waves interfere largely destructively. However, when the path-length difference is an integer number of wavelengths (AB + BC = $n\lambda$), the reflected waves are in phase and interfere constructively. It follows that a reflection should be observed when the glancing angle satisfies **Bragg's law**:

$$n\lambda = 2d \sin \theta \qquad (23.4)$$

Reflections with $n = 2, 3, \ldots$ are called second-order, third-order, and so on; they correspond to path-length differences of 2, 3, $\ldots$ wavelengths. In modern work it is normal to absorb the n into d, to write the Bragg law as

$$\lambda = 2d \sin \theta \qquad (23.5)$$

and to regard the nth-order reflection as arising from the {nh, nk, nl} planes (see Example 23.1).

The primary use of Bragg's law is in the determination of the spacing between the layers in the lattice for, once the angle θ corresponding to a reflection has been determined, d may readily be calculated.

Example 23.2 Using Bragg's law

A first-order reflection from the {111} planes of a cubic crystal was observed at a glancing angle of 11.2° when Cu K_α X-rays of wavelength 154 pm were used. What is the length of the side of the unit cell?

Method The separation of the planes can be determined from Bragg's law. Because the crystal is cubic, the separation is related to the length of the side of the unit cell, a, by eqn 2, which may therefore be solved for a.

Answer According to eqn 5, the {111} planes responsible for the diffraction have separation

$$d_{111} = \frac{\lambda}{2 \sin \theta}$$

The separation of the {111} planes of a cubic lattice of side a is given by eqn 2 as

$$d_{111} = \frac{a}{3^{1/2}}$$

Therefore,

$$a = \frac{3^{1/2} \lambda}{2 \sin \theta} = \frac{3^{1/2} \times (154 \text{ pm})}{2 \sin 11.2°} = 687 \text{ pm}$$

Self-test 23.2 Calculate the angle at which the same crystal will give a reflection from the {123} planes. [24.8°]

Some types of unit cell give characteristic and easily recognizable patterns of lines. For example, in a cubic lattice of unit cell dimension a the spacing is given by eqn 2, so the angles at which the {hkl} planes give first-order reflections are given by

$$\sin \theta = \left(h^2 + k^2 + l^2 \right)^{1/2} \frac{\lambda}{2a}$$

The reflections are then predicted by substituting the values of h, k, and l:

{hkl}	{100}	{110}	{111}	{200}	{210}	{211}	{220}	{300}	{221}	{310} ...
$h^2 + k^2 + l^2$	1	2	3	4	5	6	8	9	9	10 ...

Notice that 7 (and 15, ...) is missing because the sum of the squares of three integers cannot equal 7 (or 15, ...). Therefore the pattern has absences that are characteristic of the cubic P lattice.

Self-test 23.3 Normally, experimental procedures measure 2θ rather than θ itself. A diffraction photograph of the element polonium gave lines at the following values of 2θ (in degrees) when 71.0 pm Mo X-rays were used: 12.1, 17.1, 21.0, 24.3, 27.2, 29.9, 34.7, 36.9, 38.9, 40.9, 42.8. Identify the unit cell and determine its dimensions. [cubic P; $a = 337$ pm]

(c) Scattering factors

To prepare the way to discussing modern methods of structural analysis we need to note that the scattering of X-rays is caused by the oscillations an incoming electromagnetic wave generates in the electrons of atoms, and heavy atoms give rise to stronger scattering than light atoms. This dependence on the number of electrons is expressed in terms of the **scattering factor**, f, of the element. If the scattering factor is large, then the atoms scatter X-rays strongly. The scattering factor of an atom is related to the electron density distribution in the atom, $\rho(r)$, by

$$f = 4\pi \int_0^\infty \rho(r) \frac{\sin kr}{kr} r^2 \, dr \qquad k = \frac{4\pi}{\lambda} \sin \theta \qquad (23.6)$$

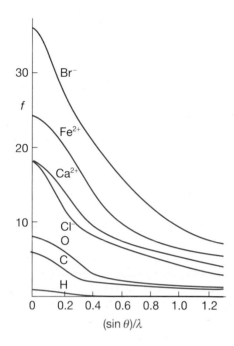

23.20 The variation of the scattering factor of atoms and ions with atomic number and angle. The scattering factor in the forward direction (at $\theta = 0$, and hence at $(\sin\theta)/\lambda = 0$) is equal to the number of electrons present in the species.

The value of f is greatest in the forward direction and smaller for directions away from the forward direction (Fig. 23.20). The detailed analysis of the intensities of reflections must take this dependence on direction into account (in single crystal studies as well as for powders). We show in the *Justification* below that, in the forward direction (for $\theta = 0$), f is equal to the total number of electrons in the atom.

Justification 23.1

As $\theta \to 0$, so $k \to 0$. Because $\sin x = x - \frac{1}{6}x^3 + \cdots$,

$$\lim_{x\to 0} \frac{\sin x}{x} = \lim_{x\to 0} \frac{x - \frac{1}{6}x^3 + \cdots}{x} = \lim_{x\to 0}(1 - \frac{1}{6}x^2 + \cdots) = 1$$

The factor $(\sin kr)/kr$ is therefore equal to 1 for forward scattering. It follows that in the forward direction

$$f = 4\pi \int_0^\infty \rho(r)\, r^2 \, \mathrm{d}r$$

The integral over the electron density ρ (the number of electrons in an infinitesimal region divided by the volume of the region) multiplied by the volume element $4\pi r^2 \, \mathrm{d}r$ is the total number of electrons, N_e, in the atom. Hence, in the forward direction, $f = N_e$. For example, the scattering factors of Na^+, K^+, and Cl^- are 10, 18, and 18, respectively.

The scattering factor is smaller in non-forward directions because $(\sin kr)/kr < 1$ for $\theta > 0$, so the integral is smaller than the value calculated above.

(d) The electron density

The problem we now address is how to interpret the data from a diffractometer in terms of the detailed structure of a crystal. To do so, we must go beyond Bragg's law.

If a unit cell contains several atoms with scattering factors f_j and coordinates ($x_j a$, $y_j b$, $z_j c$), then we show in the *Justification* below that the overall amplitude of a wave diffracted by the (hkl) planes is given by

$$F_{hkl} = \sum_j f_j \mathrm{e}^{\mathrm{i}\phi_{hkl}(j)} \qquad \text{where} \quad \phi_{hkl}(j) = 2\pi\left(hx_j + ky_j + lz_j\right) \tag{23.7}$$

The sum is over all the atoms in the unit cell. The quantity F_{hkl} is called the **structure factor**.

Justification 23.2

We begin by showing that, if in the unit cell there is an A atom at the origin and a B atom at the coordinates (xa, yb, zc), where x, y, and z lie in the range 0 to 1, then the phase difference, ϕ, between the hkl reflections of the A and B atoms is $\phi_{hkl} = 2\pi(hx + ky + lz)$.

Consider the crystal shown schematically in Fig. 23.21. The reflection corresponds to two waves from adjacent A planes, the phase difference of the waves being 2π. If there is a B atom at a fraction x of the distance between the two A planes, then it gives rise to a wave with a phase difference $2\pi x$ relative to an A reflection. To see this conclusion, note that, if $x = 0$, there is no phase difference; if $x = \frac{1}{2}$, the phase difference is π; if $x = 1$, the B atom lies where the lower A atom is and the phase

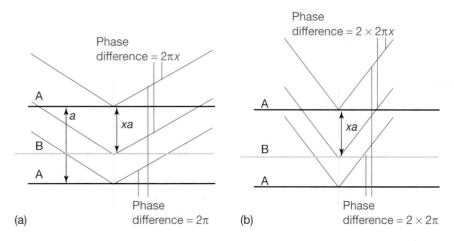

23.21 Diffraction from a crystal containing two kinds of atoms. (a) For a (100) reflection from the A planes, there is a phase difference of 2π between waves reflected by neighbouring planes. (b) For a (200) reflection, the phase difference is 4π. The reflection from a B plane at a fractional distance xa from an A plane has a phase that is x times these phase differences.

difference is 2π. Now consider a (200) reflection. There is now a $2 \times 2\pi$ difference between the waves from the two A layers, and if B were to lie at $x = 0.5$ it would give rise to a wave that differed in phase by 2π from the wave from the upper A layer. Thus, for a general fractional position x, the phase difference for a (200) reflection is $2 \times 2\pi x$. For a general ($h00$) reflection, the phase difference is therefore $h \times 2\pi x$. For three dimensions, this result generalizes to eqn 7.

The A and B reflections interfere destructively when the phase difference is π, and the total intensity is zero if the atoms have the same scattering power. For example, if the unit cells are cubic I with a B atom at $x = y = z = \frac{1}{2}$, then the A,B phase difference is $(h + k + l)\pi$. Therefore, all reflections for odd values of $h + k + l$ vanish because the waves are displaced in phase by π. Hence the diffraction pattern for a cubic I lattice can be constructed from that for the cubic P lattice (a cubic lattice without points at the centre of its unit cells) by striking out all reflections with odd values of $h + k + l$. Recognition of these **systematic absences** in a powder spectrum immediately indicates a cubic I lattice (Fig. 23.22).

If the amplitude of the waves scattered from A is f_A at the detector, that of the waves scattered from B is $f_B e^{i\phi_{hkl}}$, with ϕ_{hkl} the phase difference given in eqn 7. The total amplitude at the detector is therefore

$$F_{hkl} = f_A + f_B e^{i\phi_{hkl}}$$

Because the intensity is proportional to the square modulus of the amplitude of the wave, the intensity, I_{hkl}, at the detector is

$$I_{hkl} \propto F_{hkl}^* F_{hkl} = (f_A + f_B e^{-i\phi_{hkl}})(f_A + f_B e^{i\phi_{hkl}})$$

This expression expands to

$$I_{hkl} \propto f_A^2 + f_B^2 + f_A f_B(e^{i\phi_{hkl}} + e^{-i\phi_{hkl}}) = f_A^2 + f_B^2 + 2f_A f_B \cos \phi_{hkl}$$

The cosine term either adds to or subtracts from $f_A^2 + f_B^2$ depending on the value of ϕ_{hkl}, which in turn depends on h, k, and l and x, y, and z. Hence, there is a variation in the intensities of the lines with different hkl.

Face-centred cubic
h, k, l all even or all odd are present

Body-centred cubic
$h + k + l$ = odd are absent

Primitive cubic

(100)
(110)
(111)
(200)
(210)
(211)
(220)
(221), (300)
(310)
(311)
(222)
(320)
(321)
(400)
(410), (322)
(411), (330)
(331)
(420)
(421)
(332)
(422)
(502), (430)

23.22 The powder diffraction patterns and the systematic absences of three versions of a cubic cell. Comparison of the observed pattern with patterns like these enables the unit cell to be identified. The locations of the lines give the cell dimensions.

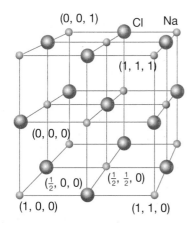

(0, 0, 1) Cl Na

(1, 1, 1)

(0, 0, 0)

$(\frac{1}{2}, 0, 0)$ $(\frac{1}{2}, \frac{1}{2}, 0)$

(1, 0, 0) (1, 1, 0)

23.23 The location of the atoms for the structure factor calculation in Example 23.3. The tinted circles are Na^+, the open circles are Cl^-.

Example 23.3 Calculating a structure factor

Calculate the structure factors for the unit cell in Fig. 23.23.

Method The structure factor is defined by eqn 7. To use this equation, consider the ions at the locations specified in Fig. 23.23. Write f^+ for the Na^+ scattering factor and f^- for the Cl^- scattering factor. Note that ions in the body of the cell contribute to the scattering with a strength f. However, ions on faces are shared between two cells (use $\frac{1}{2}f$), those on edges by four cells (use $\frac{1}{4}f$), and those at corners by eight cells (use $\frac{1}{8}f$). Two useful relations are

$$e^{i\pi} = -1 \qquad \cos\phi = \tfrac{1}{2}(e^{i\phi} + e^{-i\phi})$$

Answer From eqn 7, and summing over the coordinates of all 27 atoms in the illustration:

$$F_{hkl} = f^+\left(\tfrac{1}{8} + \tfrac{1}{8}e^{2\pi i l} + \cdots + \tfrac{1}{2}e^{2\pi i(\frac{1}{2}h + \frac{1}{2}k + l)}\right) + $$
$$f^-\left(e^{2\pi i(\frac{1}{2}h + \frac{1}{2}k + \frac{1}{2}l)} + \tfrac{1}{4}e^{2\pi i(\frac{1}{2}h)} + \cdots + \tfrac{1}{4}e^{2\pi i(\frac{1}{2}h + l)}\right)$$

To simplify this 27-term expression, we use

$$e^{2\pi i h} = e^{2\pi i k} = e^{2\pi i l} = 1$$

because h, k, and l are all integers:

$$F_{hkl} = f^+\{1 + \cos(h + k)\pi + \cos(h + l)\pi + \cos(k + l)\pi\} + $$
$$f^-\{(-1)^{h+k+l} + \cos k\pi + \cos l\pi + \cos h\pi\}$$

Then, because $\cos h\pi = (-1)^h$,

$$F_{hkl} = f^+\{1 + (-1)^{h+k} + (-1)^{h+l} + (-1)^{l+k}\} + f^-\{(-1)^{h+k+l} + (-1)^h + (-1)^k + (-1)^l\}$$

Now note that:

if h, k, and l are all even, $F_{hkl} = f^+\{1 + 1 + 1 + 1\} + f^-\{1 + 1 + 1 + 1\} = 4(f^+ + f^-)$

if h, k, and l are all odd, $F_{hkl} = 4(f^+ - f^-)$

if one index is odd and two are even, or vice versa, $F_{hkl} = 0$

Comment The hkl all-odd reflections are less intense than the hkl all-even. For $f^+ = f^-$, which is the case for identical atoms in a cubic P arrangement, the hkl all-odd have zero intensity, corresponding to the 'systematic absences' of cubic P unit cells.

Self-test 23.4 Which reflections cannot be observed for a cubic I lattice?

[for $h + k + l$ odd, $F_{hkl} = 0$]

The intensity of the (hkl) reflection is proportional to $|F_{hkl}|^2$, so in principle we can determine the structure factors experimentally by taking the square root of the corresponding intensities (but see below). Then, once we know all the structure factors F_{hkl}, we can calculate the electron density distribution, $\rho(r)$, in the unit cell by using the expression

$$\rho(\mathbf{r}) = \frac{1}{V}\sum_{hkl} F_{hkl}e^{-2\pi i(hx + ky + lz)} \qquad (23.8)$$

where V is the volume of the unit cell. Equation 8 is called a **Fourier synthesis** of the electron density.

23.24 The plot of the electron density calculated in Example 23.4.

23.25 The plot of the electron density calculated in Self-test 23.5. Note how a different choice of phases for the structure factors leads to a markedly different structure.

Example 23.4 Calculating an electron density by Fourier synthesis

Consider the {$h00$} planes of a crystal extending indefinitely in the x-direction. In an X-ray analysis the structure factors were found as follows:

h	0	1	2	3	4	5	6	7	8	9
F_h	16	−10	2	−1	7	−10	8	−3	2	−3

h	10	11	12	13	14	15
F_h	6	−5	3	−2	2	−3

(and $F_{-h} = F_h$). Construct a plot of the electron density projected on to the x-axis of the unit cell.

Method Because $F_{-h} = F_h$, it follows from eqn 8 that

$$V\rho(x) = \sum_{h=-\infty}^{\infty} F_h e^{-2\pi ihx} = F_0 + \sum_{h=1}^{\infty} \left(F_h e^{-2\pi ihx} + F_{-h} e^{2\pi ihx} \right)$$

$$= F_0 + \sum_{h=1}^{\infty} F_h \left(e^{-2\pi ihx} + e^{2\pi ihx} \right) = F_0 + 2 \sum_{h=1}^{\infty} F_h \cos 2\pi hx$$

and we evaluate the sum (truncated at $h = 15$) for points $0 \leq x \leq 1$ using mathematical software.

Answer The results are plotted in Fig. 23.24.

Comment The positions of three atoms can be discerned very readily. The more terms there are included, the more accurate the density plot. Terms corresponding to high values of h (short wavelength cosine terms in the sum) account for the finer details of the electron density; low values of h account for the broad features.

Self-test 23.5 Use mathematical software to experiment with different structure factors (including changing signs as well as amplitudes). For example, use the same values of F_h as above, but with positive signs for $h \geq 6$.

[Fig. 23.25]

(e) The phase problem

A problem with the procedure outlined above is that the observed intensity I_{hkl} is proportional to the square modulus $|F_{hkl}|^2$, so we cannot say whether we should use $+|F_{hkl}|$ or $-|F_{hkl}|$ in the sum in eqn 8. In fact, the difficulty is more severe for non-centrosymmetric unit cells because, if we write F_{hkl} as the complex number $|F_{hkl}|e^{i\alpha}$, where α is the phase of F_{hkl} and $|F_{hkl}|$ is its magnitude, then the intensity lets us determine $|F_{hkl}|$ but tells us nothing of its phase, which may lie anywhere from 0 to 2π. This ambiguity is called the **phase problem**; its consequences are illustrated by comparing Figs 23.24 and 23.25. Some way must be found to assign phases to the structure factors, for otherwise the sum for ρ could not be evaluated and the method would be useless.

The phase problem can be overcome to some extent by a variety of methods. One procedure that is widely used for inorganic materials with a reasonably small number of atoms in a unit cell and for organic molecules with a small number of heavy atoms, is the **Patterson synthesis**. Instead of the structure factors F_{hkl}, the values of $|F_{hkl}|^2$, which can be obtained without ambiguity from the intensities, are used in an expression that resembles eqn 8:

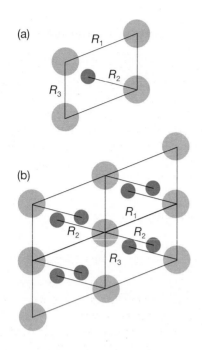

23.26 The Patterson synthesis corresponding to the pattern in (a) is the pattern in (b). The distance and orientation of each spot from the origin gives the orientation and separation of one atom–atom separation in (a). Some of the typical distances and their contribution to (b) are shown as R_1, etc.

$$P(\boldsymbol{r}) = \frac{1}{V} \sum_{hkl} |F_{hkl}|^2 e^{-2\pi i (hx+ky+lz)} \qquad (23.9)$$

The outcome of a Patterson synthesis is a map of the vector separations of the atoms (the distances and directions between atoms) in the unit cell. Thus, if atom A is at the coordinates (x_A, y_A, z_A) and atom B is at (x_B, y_B, z_B), then there will be a peak at $(x_A - x_B, y_A - y_B, z_A - z_B)$ in the Patterson map. There will also be a peak at the negative of these coordinates, because there is a vector from B to A as well as a vector from A to B. The height of the peak in the map is proportional to the product of the atomic numbers of the two atoms, $Z_A Z_B$. For example, if the unit cell has the structure shown in Fig. 23.26a, the Patterson synthesis would be the map shown in Fig. 23.26b, where the location of each spot relative to the origin gives the separation and relative orientation of each pair of atoms in the original structure.

Heavy atoms dominate the scattering because their scattering factors are large, of the order of their atomic numbers, and their locations may be deduced quite readily. The sign of F_{hkl} can now be calculated from the locations of the heavy atoms in the unit cell, and to a high probability the phase calculated for them will be the same as the phase for the entire unit cell. To see why this is so, we have to note that a structure factor of a centrosymmetric cell has the form

$$F = (\pm)f_{\text{heavy}} + (\pm)f_{\text{light}} + (\pm)f_{\text{light}} + \cdots \qquad (23.10)$$

where f_{heavy} is the scattering factor of the heavy atom and f_{light} the scattering factors of the light atoms. The f_{light} are all much smaller than f_{heavy}, and their phases are more or less random if the atoms are distributed throughout the unit cell. Therefore, the net effect of the f_{light} is to change F only slightly from f_{heavy}, and we can be reasonably confident that F will have the same sign as that calculated from the location of the heavy atom. This phase can then be combined with the observed $|F|$ (from the reflection intensity) to perform a Fourier synthesis of the full electron density in the unit cell, and hence to locate the light atoms as well as the heavy atoms.

Modern structural analyses make extensive use of **direct methods**. Direct methods are based on the possibility of treating the atoms in a unit cell as being virtually randomly distributed (from the radiation's point of view), and then to use statistical techniques to compute the probabilities that the phases have a particular value. It is possible to deduce relations between some structure factors and sums (and sums of squares) of others, which have the effect of constraining the phases to particular values (with high probability, so long as the structure factors are large). For example, the **Sayre probability relation** has the form

$$\text{sign of } \quad F_{h+h', k+k', l+l'} \quad \text{is probably equal to} \quad (\text{sign of } F_{hkl}) \times (\text{sign of } F_{h'k'l'}) \qquad (23.11)$$

For example, if F_{122} and F_{232} are both large and negative, then it is highly likely that F_{354}, provided it is large, will be positive.

(f) Structure refinement

In the final stages of the determination of a crystal structure, the parameters describing the structure (atom positions, for instance) are adjusted systematically to give the best fit between the observed intensities and those calculated from the model of the structure deduced from the diffraction pattern. This process is called **structure refinement**. Not only does the procedure give accurate positions for all the atoms in the unit cell, but it also gives an estimate of the errors in those positions and in the bond lengths and angles derived from them. The procedure also provides information on the vibrational amplitudes of the atoms.

23.4 Neutron and electron diffraction

According to the de Broglie relation, particles have wavelengths and may therefore undergo diffraction. Neutrons generated in a nuclear reactor and then slowed to thermal velocities have wavelengths similar to those of X-rays and may also be used for diffraction studies. For instance, a neutron generated in a reactor and slowed to thermal velocities by repeated collisions with a moderator (such as graphite) until it is travelling at about 4 km s^{-1} has a wavelength of about 100 pm. In practice, a range of wavelengths occurs in a neutron beam, but a monochromatic beam can be selected by diffraction from a crystal, such as a single crystal of germanium.

Example 23.5 Calculating the typical wavelength of thermal neutrons

Calculate the typical wavelength of neutrons that have reached thermal equilibrium with their surroundings at 373 K.

Method We need to relate the wavelength to the temperature. There are two linking steps. First, the de Broglie relation expresses the wavelength in terms of the linear momentum. Then the linear momentum can be expressed in terms of the kinetic energy, the mean value of which is given in terms of the temperature by the equipartition theorem (see Section 20.3).

Answer The de Broglie relation is $\lambda = h/p$. From the equipartition principle, we know that the mean translational kinetic energy of a neutron at a temperature T travelling in the x-direction is $E_K = \frac{1}{2}kT$. The kinetic energy is also equal to $p^2/2m$, where p is the momentum of the neutron and m is its mass. Hence, $p = (mkT)^{1/2}$. It follows that the neutron's wavelength is

$$\lambda = \frac{h}{(mkT)^{1/2}}$$

Therefore, at 373 K,

$$\lambda = \frac{6.626 \times 10^{-34} \text{ J s}}{\{(1.675 \times 10^{-27} \text{ kg}) \times (1.381 \times 10^{-23} \text{ J K}^{-1}) \times (373 \text{ K})\}^{1/2}} = 226 \text{ pm}$$

where we have used 1 J = 1 kg m^2 s^{-2}.

Self-test 23.6 Calculate the temperature needed for the average wavelength of the neutrons to be 100 pm.

$$[1.90 \times 10^3 \text{ K}]$$

Neutron diffraction differs from X-ray diffraction in two main respects. First, the scattering of neutrons is a nuclear phenomenon. Neutrons pass through the extra-nuclear electrons of atoms and interact with the nuclei through the 'strong force' that is responsible for binding nucleons together. As a result, the intensity with which neutrons are scattered is independent of the number of electrons and neighbouring elements in the periodic table may scatter neutrons with markedly different intensities. Neutron diffraction can be used to distinguish atoms of elements such as Ni and Co that are present in the same compound and to study order–disorder phase transitions in FeCo. A second difference is that neutrons possess a magnetic moment due to their spin. This magnetic moment can couple to the magnetic fields of atoms or ions in a crystal (if the ions have unpaired electrons) and modify the diffraction

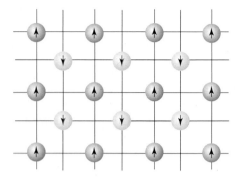

23.27 If the spins of atoms at lattice points are orderly, as in this antiferromagnetic material, where the spins of one set of atoms are aligned antiparallel to those of the other set, neutron diffraction detects two interpenetrating simple cubic lattices on account of the magnetic interaction of the neutron with the atoms, but X-ray diffraction would see only a single bcc lattice.

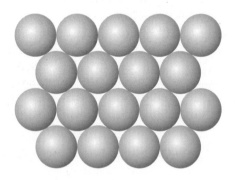

23.28 The first layer of close-packed spheres used to build a three-dimensional close-packed structure.

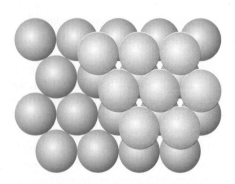

23.29 The second layer of close-packed spheres occupies the dips of the first layer. The two layers are the AB component of the close-packed structure.

pattern. One consequence is that neutron diffraction is well suited to the investigation of magnetically ordered lattices in which neighbouring atoms may be of the same element but have different orientations of their electronic spin (Fig. 23.27).

Electrons accelerated through a potential difference of 40 kV have wavelengths of about 6 pm, and so are also suitable for diffraction studies (see Box 11.1). However, their main application is to the study of surfaces, and we postpone their discussion until Chapter 28.

Crystal structure

The bonding within a solid may be of various kinds. Simplest of all (in principle) are elemental metals, where electrons are delocalized over arrays of identical cations and bind them together into a rigid but ductile and malleable whole.

23.5 Metallic solids

Most metallic elements crystallize in one of three simple forms, two of which can be explained in terms of hard spheres packing together in the closest possible arrangement.

(a) Close packing

Figure 23.28 shows a **close-packed** layer of identical spheres, one with maximum utilization of space. A close-packed three-dimensional structure is obtained by stacking such close-packed layers on top of one another. However, this stacking can be done in different ways, which result in close-packed **polytypes**, or structures that are identical in two dimensions (the close-packed layers) but differ in the third dimension.

In all polytypes, the spheres of second close-packed layer lie in the depressions of the first layer (Fig. 23.29). The third layer may be added in either of two ways. In one, the spheres are placed so that they reproduce the first layer (Fig. 23.30a), to give an ABA pattern of layers. Alternatively, the spheres may be placed over the gaps in the

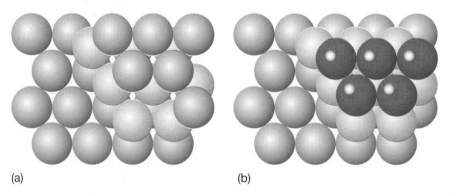

(a)   (b)

23.30 (a) The third layer of close-packed spheres might occupy the dips lying directly above the spheres in the first layer, resulting in an ABA structure, which corresponds to hexagonal close-packing. (b) Alternatively, the third layer might lie in the dips that are not above the spheres in the first layer, resulting in an ABC structure, which corresponds to cubic close-packing.

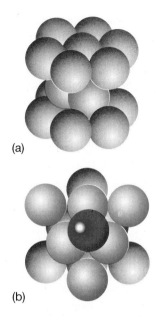

(a)

(b)

23.31 A fragment of the structure shown in Fig. 23.30 revealing the (a) hexagonal, (b) cubic symmetry. The tints on the spheres are the same as for the layers in Fig. 23.30.

Table 23.2 The crystal structures of some elements

Structure	Element
hcp*	Be, Cd, Co, He, Mg, Sc, Ti, Zn
fcc* (ccp, cubic F)	Ag, Al, Ar, Au, Ca, Cu, Kr, Ne, Ni, Pd, Pb, Pt, Rh, Rn, Sr, Xe
bcc (cubic I)	Ba, Cs, Cr, Fe, K, Li, Mn, Mo, Rb, Na, Ta, W, V
cubic P	Po

* Close-packed structures.

first layer (Fig. 23.30b), so giving an ABC pattern. Two polytypes are formed if the two stacking patterns are repeated in the vertical direction. If the ABA pattern is repeated, to give the sequence of layers ABABAB... , the spheres are **hexagonally close-packed** (hcp). Alternatively, if the ABC pattern is repeated, to give the sequence ABCABC... , the spheres are **cubic close-packed** (ccp). We can see the origins of these names by referring to Fig. 23.31. The ccp structure gives rise to a face-centred unit cell, so may also be denoted cubic F (or fcc, for face-centred cubic). It is also possible to have random sequences of layers; however, the hcp and ccp polytypes are the most important. Table 23.2 lists some elements possessing these structures.

The compactness of close-packed structures is indicated by their **coordination number**, the number of atoms immediately surrounding any selected atom, which is 12 in all cases. Another measure of their compactness is the **packing fraction**, the fraction of space occupied by the spheres, which is 0.740 (see the *Justification* below). That is, in a close-packed solid of identical hard spheres, only 26.0 per cent of the volume is empty space. The fact that many metals are close-packed accounts for their high densities.

Justification 23.3

To calculate a packing fraction of a ccp structure, we first calculate the volume of a unit cell, and then calculate the total volume of the spheres that fully or partially occupy it. The first part of the calculation is a straightforward exercise in geometry. The second part involves counting the fraction of spheres that occupy the cell.

Refer to Fig. 23.32. Because a diagonal of any face passes completely through one sphere and halfway through two other spheres, its length is $4R$. The length of a side is therefore $8^{1/2}R$ and the volume of the unit cell is $8^{3/2}R^3$. Because each cell contains the equivalent of $6 \times \frac{1}{2} + 8 \times \frac{1}{8} = 4$ spheres, and the volume of each sphere is $\frac{4}{3}\pi R^3$, the total occupied volume is $\frac{16}{3}\pi R^3$. The fraction of space occupied is therefore $\frac{16}{3}\pi R^3/8^{3/2}R^3 = 16\pi/8^{3/2}3$, or 0.740. Because an hcp structure has the same coordination number, its packing fraction is the same. The packing fractions of structures that are not close-packed are calculated similarly (see Exercises 23.19 and 23.22 and Problem 23.16).

23.32 The calculation of the packing fraction of an ccp unit cell.

(b) Less closely packed structures

As shown in Table 23.2, a number of common metals adopt structures that are less than close-packed. The departure from close packing suggests that factors such as specific covalent bonding between neighbouring atoms are beginning to influence

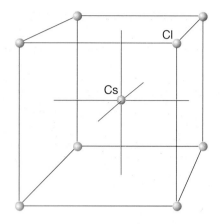

23.33 The caesium-chloride structure consists of two interpenetrating simple cubic arrays of ions, one of cations and the other of anions, so that each cube of ions of one kind has a counter ion at its centre.

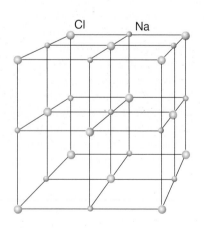

23.34 The rock-salt (NaCl) structure consists of two mutually interpenetrating slightly expanded face-centred cubic arrays of ions. The entire assembly shown here is the unit cell.

the structure and impose a specific geometrical arrangement. One such arrangement results in a cubic I (bcc, for body-centred cubic) structure, with one sphere at the centre of a cube formed by eight others. The coordination number of a bcc structure is only 8, but there are six more atoms not much further away than the eight nearest neighbours. The packing fraction of 0.68 is not much smaller than the value for a close-packed structure (0.74), and shows that about two-thirds of the available space is actually occupied.

23.6 Ionic solids

Two questions arise when we consider ionic solids: the structures adopted by the ions and the energetics of the crystal structure.

(a) Structure

When crystals of compounds of monatomic ions (such as NaCl and MgO) are modelled by stacks of hard spheres it is essential to allow for the different ionic radii (typically with the cations smaller than the anions) and different charges. The coordination number of an ion is the number of nearest neighbours of opposite charge; the structure itself is characterized as having (n_+, n_-) **coordination**, where n_+ is the coordination number of the cation and n_- that of the anion.

Even if, by chance, the ions have the same size, the problems of ensuring that the unit cells are electrically neutral makes it impossible to achieve 12-coordinate close-packed structures. As a result, ionic solids are generally less dense than metals. The best packing that can be achieved is the (8,8)-coordinate **caesium-chloride structure** in which each cation is surrounded by eight anions and each anion is surrounded by eight cations (Fig. 23.33). In this structure, an ion of one charge occupies the centre of a cubic unit cell with eight counter ions at its corners. The structure is adopted by CsCl itself and also by CaS, CsCN (with some distortion), and CuZn.

When the radii of the ions differ more than in CsCl, even eight-coordinate packing cannot be achieved. One common structure adopted is the (6,6)-coordinate **rock-salt structure** typified by NaCl (Fig. 23.34). In this structure, each cation is surrounded by six anions and each anion is surrounded by six cations. The rock-salt structure can be pictured as consisting of two interpenetrating slightly expanded cubic F (fcc) arrays, one of cations and the other of anions. This structure is adopted by NaCl itself and also by several other MX compounds, including KBr, AgCl, MgO, and ScN.

The switch from the caesium-chloride structure to the rock-salt structure occurs (sometimes) in accord with the **radius-ratio rule**, which is based on the value of the radius ratio, γ:

$$\gamma = \frac{r_{smaller}}{r_{larger}} \tag{23.12}$$

The two radii are those of the larger and smaller ions in the crystal. The radius-ratio rule is derived by considering the geometrical problem of packing the maximum number of hard spheres of one radius around a hard sphere of a different radius. The rule states that the caesium-chloride structure should be expected when

$$\gamma > 3^{1/2} - 1 = 0.732$$

and that the rock-salt structure should be expected when

$$2^{1/2} - 1 = 0.414 < \gamma < 0.732$$

For $\gamma < 0.414$, the most efficient packing leads to four-coordination of the type exhibited by the sphalerite (or zinc blende) form of ZnS (Fig. 23.35). The deviation of

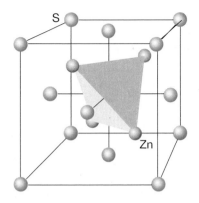

23.35 The structure of the sphalerite form of ZnS showing the location of the Zn atoms in the tetrahedral holes formed by the array of S atoms. (There is an S atom at the centre of the cube inside the tetrahedron of Zn atoms.)

Table 23.3* Ionic radii, R/pm

Na$^+$	102(6†), 116(8)
K$^+$	138(6), 151(8)
F$^-$	128(2), 131(4)
Cl$^-$	181 (close packing)

* More values are given in the *Data section*.
† Coordination number.

a structure from that expected on the basis of the radius-ratio rule is often taken to be an indication of a shift from ionic towards covalent bonding; however, a major source of unreliability is the arbitrariness of ionic radii and their variation with coordination number.

Ionic radii are derived from the distance between centres of adjacent ions in a crystal. However, we need to apportion the total distance between the two ions by defining the radius of one ion and then inferring the radius of the other ion. One scale that is widely used is based on the value 140 pm for the radius of the O^{2-} ion (Table 23.3). Other scales are also available (such as one based on F$^-$ for discussing halides), and it is essential not to mix values from different scales. Because ionic radii are so arbitrary, predictions based on them must be viewed cautiously.

(b) Energetics

The **lattice energy** of a solid is the difference in potential energy of the ions packed together in a solid and widely separated as a gas. The lattice energy is always positive; a high lattice energy indicates that the ions interact strongly with one another to give a tightly bonded solid. The **lattice enthalpy**, ΔH_L, is the change in molar enthalpy for the process

$$MX(s) \longrightarrow M^{z+}(g) + X^{z-}(g)$$

and its equivalent for other charge types and stoichiometries. The lattice enthalpy is equal to the lattice energy at $T = 0$; at normal temperatures they differ by only a few kilojoules per mole, and the difference is normally neglected.

Each ion in a solid experiences electrostatic attractions from all the other oppositely charged ions and repulsions from all the other like-charged ions. There are also repulsions arising from the overlap of the atomic orbitals of the ions and the role of the Pauli principle. The total Coulombic potential energy is the sum of all the electrostatic contributions. Each cation is surrounded by anions, and there is a large negative contribution from the attraction of the opposite charges. Beyond those nearest neighbours, there are cations that contribute a positive term to the total potential energy of the central cation. There is also a negative contribution from the anions beyond those cations, a positive contribution from the cations beyond them, and so on to the edge of the solid. These repulsions and attractions become progressively weaker as the distance from the central ion increases, but the net outcome of all these contributions is a lowering of energy.

First, consider a simple one-dimensional model of a solid consisting of a long line of uniformly spaced alternating cations and anions, with d the distance between their centres, the sum of the ionic radii (Fig. 23.36). If the charge numbers of the ions have the same absolute value (+1 and −1, or +2 and −2, for instance), then $z_1 = +z$, $z_2 = -z$, and $z_1 z_2 = -z^2$. The potential energy of the central ion is calculated by summing all the terms, with negative terms representing attractions to oppositely charged ions and positive terms representing repulsions from like-charged ions. For the interaction with ions extending in a line to the right of the central ion, the lattice energy is

$$E_P = \frac{1}{4\pi\varepsilon_0} \times \left(-\frac{z^2 e^2}{d} + \frac{z^2 e^2}{2d} - \frac{z^2 e^2}{3d} + \frac{z^2 e^2}{4d} - \cdots \right)$$

$$= -\frac{z^2 e^2}{4\pi\varepsilon_0 d} \left(1 - \frac{1}{2} + \frac{1}{3} - \frac{1}{4} + \cdots \right)$$

$$= -\frac{z^2 e^2}{4\pi\varepsilon_0 d} \times \ln 2$$

23.36 A line of alternating cations and ions used in the calculation of the Madelung constant in one dimension.

Calculations involving natural logarithms often become simpler if we note that $\ln x = (x-1) - \frac{1}{2}(x-1)^2 + \frac{1}{3}(x-1)^3 - \frac{1}{4}(x-1)^4 + \cdots$.

We have used the relation $1 - \frac{1}{2} + \frac{1}{3} - \frac{1}{4} + \cdots = \ln 2$. Finally, we multiply E_P by 2 to obtain the total energy arising from interactions on each side of the ion and then by the Avogadro constant, N_A, to obtain an expression for the lattice energy per mole of ions. The outcome is

$$E_P = -2\ln 2 \times \frac{z^2 N_A e^2}{4\pi\varepsilon_0 d}$$

with $d = r_{cation} + r_{anion}$. This energy is negative, corresponding to a net attraction. The calculation we have just performed can be extended to three-dimensional arrays of ions with different charges:

$$E_P = -A \times \frac{|z_1 z_2| N_A e^2}{4\pi\varepsilon_0 d} \tag{23.13}$$

The factor A is a positive numerical constant called the **Madelung constant**; its value depends on how the ions are arranged about one another. For ions arranged in the same way as in sodium chloride, $A = 1.748$. Table 23.4 lists Madelung constants for other common structures.

Quantum mechanical repulsion effects are taken into account by supposing that, because wavefunctions decay exponentially with distance at large distances from the nucleus, and repulsive interactions depend on the overlap of orbitals, the repulsive contribution to the potential energy has the form

$$E_P^* = N_A C' e^{-d/d^*} \tag{23.14}$$

with C' and d^* constants; the latter is commonly taken to be 34.5 pm. The total potential energy is the sum of E_P and E_P^*, and passes through a minimum when $d(E_P + E_P^*)/dd = 0$ (Fig. 23.37). A short calculation leads to the following expression for the minimum total potential energy (see Exercise 23.27a):

$$E_{P,min} = -\frac{N_A |z_A z_B| e^2}{4\pi\varepsilon_0 d}\left(1 - \frac{d^*}{d}\right) A \tag{23.15}$$

This expression is called the **Born–Mayer equation**. Provided we ignore zero-point contributions to the energy, we can identify the negative of this potential energy with the lattice energy. We see that large lattice energies are expected when the ions are highly charged (so $|z_A z_B|$ is large) and small (so d is small).

Experimental values of the lattice enthalpy (the enthalpy, rather than the energy) are obtained by using a **Born–Haber cycle**, a closed path of transformations starting and ending at the same point, one step of which is the formation of the solid compound from a gas of widely separated ions. A typical cycle, for potassium chloride, is shown in Fig. 23.38. It consists of the following steps (for convenience, starting at the elements):

Table 23.4 Madelung constants

Structural type*	A
Caesium chloride	1.763
Fluorite	2.519
Rock salt	1.748
Rutile	2.408
Sphalerite	1.638
Wurtzite	1.641

* For descriptions of the structural types not covered in this chapter, see references in *Further reading*.

	$\Delta H/(\text{kJ mol}^{-1})$	
1 Sublimation of K(s)	+89	[dissociation enthalpy of K(s)]
2 Dissociation of $\frac{1}{2}$ Cl$_2$(g)	+122	[$\frac{1}{2}$ × dissociation enthalpy of Cl$_2$(g)]
3 Ionization of K(g)	+418	[ionization enthalpy of K(g)]
4 Electron attachment to Cl(g)	−349	[electron gain enthalpy of Cl(g)]
5 Formation of solid from gas	$-\Delta H_L/(\text{kJ mol}^{-1})$	
6 Decomposition of compound	+437	[negative of enthalpy of formation of KCl(s)]

Because the sum of these enthalpy changes is equal to zero, we can infer from

$$89 + 122 + 418 - 349 - \Delta H_L/(\text{kJ mol}^{-1}) + 437 = 0$$

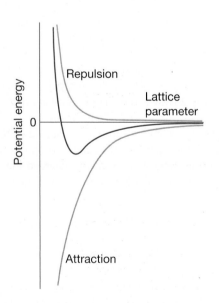

23.37 The contributions to the total potential energy of an ionic crystal.

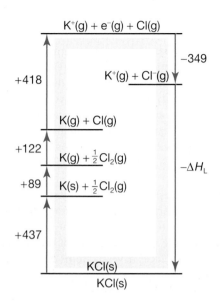

K$^+$(g) + e$^-$(g) + Cl(g)

+418

−349

K$^+$(g) + Cl$^-$(g)

K(g) + Cl(g)

+122

K(g) + $\frac{1}{2}$Cl$_2$(g)

−ΔH_L

+89

K(s) + $\frac{1}{2}$Cl$_2$(g)

+437

KCl(s)

KCl(s)

23.38 The Born–Haber cycle for KCl at 298 K. Enthalpies changes are in kilojoules per mole.

Table 23.5 Lattice enthalpies at 298 K, $\Delta H_L/(kJ\ mol^{-1})^*$

NaF	787
NaBr	751
MgO	3850
MgS	3406

* More values are given in the *Data section*.

that $\Delta H_L = +717$ kJ mol^{-1}. Some lattice enthalpies obtained in this way are listed in Table 23.5. As can be seen from the data, the trends in values are in general accord with the predictions of the Born–Mayer equation. Agreement is typically taken to imply that the ionic model of bonding is valid for the substance; disagreement implies that there is a covalent contribution to the bonding.

23.7 Molecular solids and covalent networks

X-ray diffraction studies of solids reveal a huge amount of information, including interatomic distances, bond angles, stereochemistry, and vibrational parameters. In this section we can do no more than hint at the diversity of types of solids found when molecules pack together or atoms link together in extended networks.

(a) Typical solids

In **covalent network solids**, covalent bonds in a definite spatial orientation link the atoms in a network extending through the crystal. The demands of directional bonding, which have only a small effect on the structures of many metals, now override the geometrical problem of packing spheres together, and elaborate and extensive structures may be formed. A famous example of a covalent solid is diamond (Fig. 23.39), in which each sp^3 hybridized C atom is bonded tetrahedrally to its four neighbours. Other examples are silicon, graphite, red phosphorus, and boron nitride.

Molecular solids, which are the subject of the overwhelming majority of modern structural determinations, are bonded together by van der Waals interactions (Chapter 21). The observed crystal structure is Nature's solution to the problem of condensing objects of various shapes into an aggregate of minimum energy (actually, for $T > 0$, of minimum Gibbs energy). The prediction of the structure is a very difficult task, and rarely possible. The problem is made more complicated by the role of hydrogen bonds, which in some cases dominate the crystal structure, as in ice (Fig. 23.40), but in others (for example, in phenol) distort a structure that is determined largely by the van der Waals interactions.

23.39 A fragment of the structure of diamond. Each C atom is tetrahedrally bonded to four neighbours. This framework-like structure results in a rigid crystal with a high thermal conductivity.

23.40 A fragment of the crystal structure of ice (ice-I). Each O atom is at the centre of a tetrahedron of four O atoms at a distance of 276 pm. The central O atom is attached by two short O—H bonds to two H atoms and by two long hydrogen bonds to the H atoms of two of the neighbouring molecules. Overall, the structure consists of planes of hexagonal puckered rings of H_2O molecules (like the chair form of cyclohexane).

1 D(+)-Tartaric acid

2 L(−)-Tartaric acid

(b) Absolute configurations

Although it has long been possible to separate enantiomers (mirror-image chiral isomers, Section 15.3b), it was not until X-ray diffraction techniques were developed that the absolute stereochemical configuration of an isomer could be determined. We now know, for example, that D-tartaric acid (**1**) is the isomer responsible for rotating light clockwise (that is, it is the (+)-isomer) and that L-tartaric acid (**2**) is the (−)-isomer. The X-ray method is not trivial, because enantiomers give almost identical diffraction patterns. The information about the absolute configuration is contained in small differences in diffraction intensities and is based on a technique developed by J.M. Bijvoet.

Consider first the diagrams in Fig. 23.41, which represent an idealized crystal and its mirror image. This model resembles the arrangement of Zn and S atoms in zinc blende, which was the first absolute configuration to be determined. The technique we are about to describe was used to show that the shiny (111) faces of the crystal have S atoms on the surface whereas the dull (111) faces have Zn atoms on the surface (Fig. 23.42). Each plane of atoms gives rise to a scattered wave, and their superpositions are shown in Fig. 23.41 (a and b). Note that the two superpositions have the same amplitude but differ in phase. The diffraction pattern therefore has the same intensity for each enantiomer and, at this stage, cannot be used to distinguish them.

The essence of the method is to use X-rays that are close to an absorption frequency of one species of atom in the sample. In the examination of ZnS, for instance, gold L_α radiation (127.6 pm) was used, which is close to the beginning of a zinc absorption band (which commences at 128.3 pm). In Bijvoet's development of this approach for a study of tartaric acid, a Rb atom is incorporated into the compound (he used sodium rubidium tartrate) with X-rays from a zirconium target. Atoms with absorptions close to the X-ray frequency introduce an extra phase shift in the scattered X-ray. A simple way to picture the additional phase shift is to imagine the X-rays as exciting the atom before being re-emitted, and being delayed in the process. The effect is called **anomalous scattering**.

23.41 The two versions (top and bottom) of the two layers of atoms represent enantiomers. The interference between their scattered waves results in composite waves that differ in phase (a and a′), but the absolute phase cannot be determined, and the intensities of the reflections are identical. If, however, the atoms represented by the larger spheres modify the phase of the waves they scatter, then the resultant superpositions differ in amplitude as well as phase (b and b′), the reflections have different intensities, and the absolute configuration can be established. The blue bands represent the waves scattered by the layers, with alternating positive and negative regions shown as light and dark. The width of a band indicates its intensity. The resultant in each instance is indicated by the grey band.

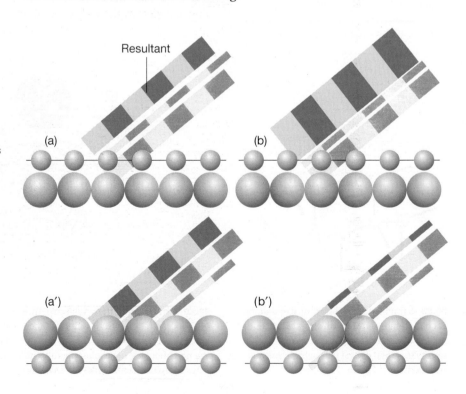

Resultant

(a)   (b)

(a′)   (b′)

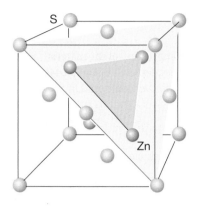

23.42 The (111) faces of the sphalerite crystal have either S atoms above Zn atoms or, as shown here, Zn atoms above S atoms.

The web site contains links to databases of properties of materials, such as metals and polymers.

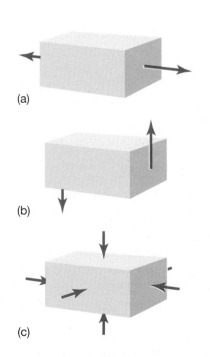

23.43 Types of stress applied to a body. (a) Uniaxial stress, (b) pure shear, (c) hydrostatic stress.

Suppose the layer represented by the large spheres in the crystal contains the anomalous scatterers; then the scattered waves are as shown in Fig. 23.41 (a′ and b′). The essential point is that the superpositions now differ slightly in amplitude, not only phase, so the diffracted intensities are slightly different in each case. Therefore, the enantiomers can in fact be distinguished because the scattering intensities differ.

Modern diffractometers are so sensitive that the incorporation of a heavy atom is no longer strictly necessary, except in the study of very large systems, such as proteins and protein complexes. It is now possible to detect the small intensity variations arising from the light atoms normally present. However, the procedure is much easier and more reliable if some moderately heavy atoms (such as S or Cl) are present. Anomalous scattering depends strongly on the wavelength of the X-radiation. Thus, atoms lighter than S and Cl give little effect for Mo K_α radiation, and until recently Cu K_α radiation had to be used.

The properties of solids

In this section we consider how the bulk properties of solids, particularly their mechanical, electrical, and magnetic properties, stem from the properties of their constituent atoms. The rational fabrication of modern materials depends crucially on an understanding of this link.

23.8 Mechanical properties

The fundamental concepts for the discussion of the mechanical properties of solids are stress and strain. The **stress** on an object is the applied force divided by the area to which it is applied. The **strain** is the resulting distortion of the sample. The general field of the relations between stress and strain is called **rheology**.

(a) Rheological properties

Stress may be applied in a number of different ways. Thus, **uniaxial stress** is a simple compression or extension in one direction (Fig. 23.43); **hydrostatic stress** is a stress applied simultaneously in all directions, as in a body immersed in a fluid. A **pure shear** is a stress that tends to push opposite faces of the sample in opposite directions.

A sample subjected to a small stress typically undergoes **elastic deformation** in the sense that it recovers its original shape when the stress is removed. For low stresses, the strain is linearly proportional to the stress. The response becomes non-linear at high stresses but may remain elastic. Above a certain threshold, the strain becomes **plastic** in the sense that recovery does not occur when the stress is re-moved. Plastic deformation occurs when bond breaking takes place and, in pure metals, typically takes place through the agency of dislocations. Brittle solids, such as ionic solids, exhibit sudden fracture as the stress focused by cracks causes them to spread catastrophically.

The response of a solid to an applied stress is commonly summarized by a number of coefficients of proportionality known as 'moduli':

$$\text{Young's modulus:} \quad E = \frac{\text{normal stress}}{\text{normal strain}} \qquad [23.16a]$$

$$\text{Bulk modulus:} \quad K = \frac{\text{pressure}}{\text{fractional change in volume}} \qquad [23.16b]$$

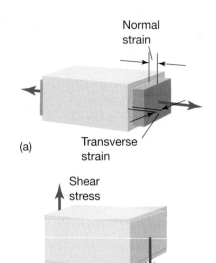

(a)

(b)

23.44 (a) Normal stress and the resulting strain, (b) shear stress. Poisson's ratio indicates the extent to which a body changes shape when subjected to a uniaxial stress.

The *chain rule* states that for a function $f = f(g)$, where $g = g(t)$,

$$\frac{df}{dt} = \frac{df}{dg}\frac{dg}{dt}$$

The *product rule* for the differentiation of the product of two functions f and g is

$$d(fg) = f\,dg + g\,df$$

Shear modulus: $G = \dfrac{\text{shear stress}}{\text{shear strain}}$ [23.16c]

where 'normal stress' refers to stretching and compression of the material, as shown in Fig. 23.44a and 'shear stress' refers to the stress depicted in Fig. 23.44b. The bulk modulus is the inverse of the compressibility, κ, first encountered in Section 3.4a:

$$\kappa = -\frac{1}{V}\left(\frac{\partial V}{\partial p}\right)_T \tag{23.17}$$

A third ratio indicates how the sample changes its shape (see Fig. 23.44):

Poisson's ratio: $v_{\text{p}} = \dfrac{\text{transverse strain}}{\text{normal strain}}$ [23.18]

These moduli are interrelated:

$$G = \frac{E}{2(1 + v_{\text{p}})} \qquad K = \frac{E}{3(1 - 2v_{\text{p}})} \tag{23.19}$$

We can use thermodynamic arguments to discover the relation of these moduli to the molecular properties of the solid. To do so, we combine eqn 17, rewritten for $K = 1/\kappa$, with the thermodynamic relation $p = -(\partial U/\partial V)_T$ (this is eqn 3.5), to obtain

$$K = V\left(\frac{\partial^2 U}{\partial V^2}\right)_T \tag{23.20}$$

This expression shows that the bulk modulus (and through eqn 19, the other two moduli) depends on the curvature of a plot of the internal energy against volume. To develop this conclusion, we note that the variation of internal energy with volume can be expressed in terms of its variation with a lattice parameter, R, such as the length of the side of a unit cell. We can write

$$\frac{\partial U}{\partial V} = \frac{\partial U}{\partial R}\frac{\partial R}{\partial V}$$

and so

$$\frac{\partial^2 U}{\partial V^2} = \frac{\partial U}{\partial R}\frac{\partial^2 R}{\partial V^2} + \frac{\partial^2 U}{\partial V \partial R}\frac{\partial R}{\partial V} = \frac{\partial U}{\partial R}\frac{\partial^2 R}{\partial V^2} + \frac{\partial^2 U}{\partial R^2}\left(\frac{\partial R}{\partial V}\right)^2$$

To calculate K at the equilibrium volume of the sample, we set $R = R_0$ and recognize that $\partial U/\partial R = 0$ at equilibrium, so

$$K_0 = V_0\left(\frac{\partial^2 U}{\partial R^2}\right)_{T,0}\left(\frac{\partial R}{\partial V}\right)_{T,0}^2$$

where the 0 denotes that the derivatives are evaluated at the equilibrium dimensions of the unit cell by setting $R = R_0$ after the derivative has been calculated. At this point we can write $V = aR^3$, where a is a constant that depends on the crystal structure, which implies that $\partial R/\partial V = 1/(3aR^2)$. Then, if the internal energy is given by a pairwise Lennard-Jones (12,6)-potential, eqn 21.32, we can write[4]

$$\left(\frac{\partial^2 U}{\partial R^2}\right)_{T,0} = \frac{72nN_A\varepsilon}{R_0^2}$$

4. To obtain this result, we have used the fact that, at equilibrium, $R = R_0$ and $\sigma^6/R_0^6 = \frac{1}{2}$, where σ is the scale parameter for the intermolecular potential (r_0 in eqn 21.32).

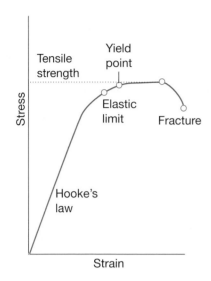

23.45 At small strains, a body obeys Hooke's law (stress proportional to strain) and is elastic (recovers its shape when the stress is removed). At high strains, the body is no longer elastic, may yield and become plastic. At even higher strains, the solid fails (at its limiting tensile strength) and finally fractures.

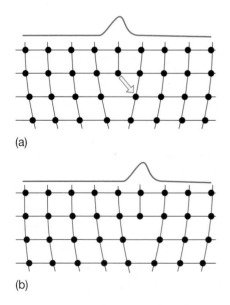

23.46 The motion of one region of a solid relative to another is facilitated by a defect, a cross-section through which is shown here. The effect of forming new bonds and breaking existing bonds in sequence as the defect migrates ((a) changing to (b)) is analogous to the motion of a ruck in a carpet facilitating its motion (as indicated by the blue line).

where n is the amount of substance in the sample of volume V_0. It then follows that

$$K = \frac{72nN_A\varepsilon}{9aR^3} = \frac{8nN_A\varepsilon}{V_0} = \frac{8N_A\varepsilon}{V_{0,m}} \tag{23.21}$$

where we have used $V_{0,m} = V_0/n$. We see that the bulk modulus is large (a stiff solid) if the potential well represented by the Lennard-Jones potential is deep and the solid is dense (so its molar volume is small).

(b) Plastic flow and elasticity

The typical behaviour of a solid under stress is illustrated in Fig. 23.45. For small strains, the stress–strain relation is a Hooke's law of force, with the strain directly proportional to the stress. For larger strains, though, dislocations begin to play a major role and the strain becomes plastic in the sense that the sample does not recover its original shape when the stress is removed.

The wholesale slip of one plane of atoms relative to its neighbour requires considerable force. An estimate of the stress required is $\frac{1}{6}G$, where G is the shear modulus. However, a dislocation permits stepwise migration of ranks of atoms, and much less force is needed in each step (Fig. 23.46). In this case, the role of dislocations in plastic deformation is like a ruck in a carpet: it is far easier to migrate a carpet by shifting a ruck through it than to slide the unrucked carpet across a surface.

The differing rheological characteristics of metals can be traced to the presence of **slip planes**, which are planes of atoms that under stress may slip or slide relative to one another. The slip planes of a ccp structure are the close-packed planes, and careful inspection of a unit cell shows that there are eight sets of slip planes in different directions. As a result, metals with cubic close-packed structures, like copper, are malleable: they can easily be bent, flattened, or pounded into shape. In contrast, a hexagonal close-packed structure has only one set of slip planes; and metals with hexagonal close packing, like zinc or cadmium, tend to be brittle.

Synthetic polymers are classified broadly as *elastomers*, *fibres*, and *plastics*, depending on their **crystallinity**, the degree of three-dimensional long-range order attained in the solid state. An **elastomer** is a flexible polymer that can expand or contract easily upon application of an external force. Elastomers are polymers with numerous cross-links that pull them back into their original shape when a stress is removed. They should have low Young's modulus when contracted, which means that the cohesive interactions between the chains should be weak. The weak directional constraints on silicon–oxygen bonds are responsible for the high elasticity of silicones. A **perfect elastomer**, a polymer in which the internal energy is independent of the extension of the random coil, can be modelled as a freely jointed chain. We saw in Section 22.2b that the contraction of an extended chain to a random coil is spontaneous in the sense that it corresponds to an increase in entropy; the entropy change of the surroundings is zero because no energy is released when the coil forms. In the following *Justification* we also see that the restoring force, F, of a one-dimensional perfect elastomer is

$$F = -\frac{kT}{2l}\ln\left(\frac{1+v}{1-v}\right) \qquad v = n/N \tag{23.22a}$$

where N is the total number of bonds of length l and the polymer is stretched or compressed by nl. This function is plotted in Fig. 23.47. At low extensions, when $v \ll 1$:

$$F \approx -\frac{vkT}{l} = -\frac{nkT}{Nl} \tag{23.22b}$$

23.47 The restoring force, F, of a one-dimensional perfect elastomer. For small strains, F is linearly proportional to the extension, corresponding to Hooke's law.

and the sample obeys Hooke's law: the restoring force is proportional to the displacement (which is proportional to n). For small displacements, therefore, the whole coil shakes with simple harmonic motion.

Justification 23.4

The work done on an elastomer when it is extended through a distance dx is $F\,dx$, where F is the restoring force. The change in internal energy is therefore

$$dU = T\,dS - p\,dV + F\,dx$$

It follows that

$$\left(\frac{\partial U}{\partial x}\right)_{T,V} = T\left(\frac{\partial S}{\partial x}\right)_{T,V} + F$$

In a perfect elastomer, as in a perfect gas, the internal energy is independent of the dimensions (at constant temperature), so $(\partial U/\partial x)_{T,V} = 0$. The restoring force is therefore

$$F = -T\left(\frac{\partial S}{\partial x}\right)_{T,V}$$

If now we substitute eqn 22.8 into this expression (we evade problems arising from the constraint of constant volume by supposing that the sample contracts laterally as it is stretched), we obtain

$$F = -\frac{T}{l}\left(\frac{\partial S}{\partial n}\right)_{T,V} = \frac{T}{Nl}\left(\frac{\partial S}{\partial v}\right)_{T,V} = -\frac{kT}{2l}\ln\left(\frac{1+v}{1-v}\right)$$

as in eqn 22a.

A **fibre** is a polymeric material that owes its strength to interactions between chains. One example is nylon-66 (Fig. 23.48). In contrast to elastomers, fibres need to have a high Young's modulus, which requires strong interactions between chains. Hydrogen bonding between chains, as in nylon, is one way to achieve a high modulus, and side chains are undesirable as they hinder the formation of ordered microcrystalline regions. Under certain conditions, nylon-66 can be prepared in a state of high crystallinity, in which hydrogen bonding between the peptide links of neighbouring chains results in an ordered array.

23.48 A fragment of two nylon-66 polymer chains showing the pattern of hydrogen bonds that are responsible for the cohesion between the chains.

23.49 The variation of specific volume with temperature of a synthetic polymer. The glass transition temperature, T_g, is at the point of intersection of extrapolations of the two linear parts of the curve.

A **plastic** is a polymer that can attain only a limited degree of crystallinity and as a result is neither as strong as a fibre nor as flexible as an elastomer. Certain materials, such as nylon-66, can be prepared either as a fibre or as a plastic. A sample of plastic nylon-66 may be visualized as consisting of crystalline hydrogen-bonded regions of varying size interspersed amongst amorphous, random coil regions. A single type of polymer may exhibit more than one characteristic, for to display fibrous character, the polymers need to be aligned; if the chains are not aligned, then the substance may be plastic. That is the case with nylon, poly(vinyl chloride), and the siloxanes.

(c) Melting and glass transitions in synthetic polymers

The crystallinity of synthetic polymers results from specific interactions, such as hydrogen bonding in nylon-66, that can be destroyed by thermal motion at sufficiently high temperatures. This change in crystallinity may be thought of as a kind of intramolecular melting from a crystalline solid to a more fluid random coil. Polymer melting also occurs at a specific **melting temperature**, T_m, which increases with the strength and number of intermolecular interactions in the material. Thus, polyethylene, which has chains that interact only weakly in the solid, has $T_m = 414$ K and nylon-66 fibres, in which there are strong hydrogen bonds between chains, has $T_m = 530$ K. High melting temperatures are desirable in most practical applications involving fibres and plastics.

All synthetic polymers undergo a transition from a state of high to low chain mobility at the **glass transition temperature**, T_g. To visualize the glass transition, we consider what happens to an elastomer as we lower its temperature. There is sufficient energy available at normal temperatures for limited bond rotation to occur and the flexible chains writhe. At lower temperatures, the amplitudes of the writhing motion decrease until a specific temperature, T_g, is reached at which motion is frozen completely and the sample forms a glass. Glass transition temperatures well below 300 K are desirable in elastomers that are to be used at normal temperatures. Both the glass transition temperature and the melting temperature of a polymer may be measured by differential scanning calorimetry (DSC, Box 2.1). Because the motion of the segments of a polymer chain increases at the glass transition temperature, T_g may also be determined from a plot of the specific volume of a polymer (the reciprocal of its mass density) against temperature (Fig. 23.49).

23.9 Electrical properties

We shall confine attention to electronic conductivity, but note that some ionic solids display ionic conductivity. Two types of solid are distinguished by the temperature dependence of their electrical conductivity:

A **metallic conductor** is a substance with a conductivity that decreases as the temperature is raised.

A **semiconductor** is a substance with a conductivity that increases as the temperature is raised.

A semiconductor generally has a lower conductivity than that typical of metals, but the magnitude of the conductivity is not the criterion of the distinction. It is conventional to classify semiconductors with very low electrical conductivities, such as most synthetic polymers, as **insulators**. We shall use this term, but it should be appreciated that it is one of convenience rather than one of fundamental significance. A **superconductor** is a solid that conducts electricity without resistance (Box 23.2).

Box 23.2 Superconductors

The resistance to flow of electrical current of a normal metallic conductor decreases smoothly with temperature but never vanishes. However, certain solids known as *superconductors* conduct electricity without resistance below a critical temperature, T_c. Following the discovery in 1911 that mercury is a superconductor below 4.2 K, the boiling point of liquid helium, physicists and chemists made slow but steady progress in the discovery of superconductors with higher values of T_c. Metals, such as tungsten, mercury, and lead, tend to have T_c values below about 10 K. Intermetallic compounds, such as Nb_3X (X = Sn, Al, or Ge), and alloys, such as Nb/Ti and Nb/Zr, have intermediate T_c values ranging between 10 K and 23 K. In 1986, *high-temperature superconductors* (HTSC) were discovered. Several ceramic materials containing oxocuprate motifs, Cu_mO_n, are now known with T_c values well above 77 K, the boiling point of the inexpensive refrigerant liquid nitrogen. For example, $HgBa_2Ca_2Cu_2O_8$ has T_c = 153 K.

Superconductors have unique magnetic properties as well. Some superconductors, classed as *Type I*, show abrupt loss of superconductivity when an applied magnetic field exceeds a critical value H_c characteristic of the material. It is observed that the value of H_c depends on temperature and T_c as

$$H_c(T) = H_c(0)\left(1 - \frac{T^2}{T_c^2} \right)$$

where $H_c(0)$ is the value of H_c as $T \to 0$. Type I superconductors are also completely diamagnetic below H_c, meaning that no magnetic field lines penetrate into the material. This complete exclusion of a magnetic field in a material is known as the *Meissner effect*, which can be visualized by the levitation of a superconductor above a magnet. *Type II* superconductors, which include the HTSCs, show a gradual loss of superconductivity and diamagnetism with increasing magnetic field.

There is a degree of periodicity in the elements that exhibit superconductivity. The metals iron, cobalt, nickel, copper, silver, and gold do not display superconductivity; nor do the alkali metals. It is observed that, for simple metals,

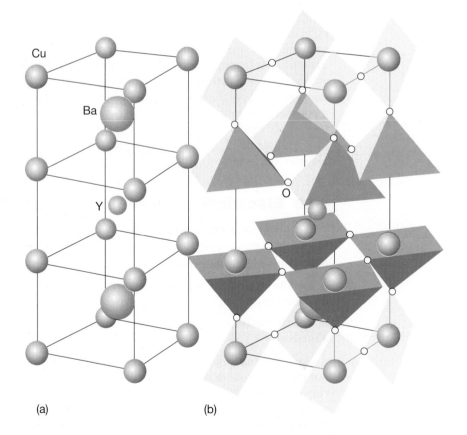

(a)   (b)

Structure of the $YBa_2Cu_3O_7$ superconductor. (a) Metal atom positions. (b) The polyhedra show the positions of oxygen atoms and indicate that the metal ions are in square planar and square-pyramidal coordination environments.

ferromagnetism and superconductivity never coexist, but in some of the oxocuprate superconductors ferromagnetism and superconductivity can coexist. One of the most widely studied oxocuprate superconductors $YBa_2Cu_3O_7$ (informally known as '123' on account of the proportions of the metal atoms in the compound) has the structure shown in the first illustration. The square-pyramidal CuO_5 units arranged as two-dimensional layers and the square planar CuO_4 units arranged in sheets are common structural features of oxocuprate HTSCs.

The mechanism of superconduction is well-understood for low-temperature materials but there is as yet no settled explanation of high-temperature superconductivity. The central concept of low-temperature superconduction is the existence of a *Cooper pair*, a pair of electrons that exists on account of the indirect electron–electron interactions fostered by the nuclei of the atoms in the lattice. Thus, if one electron is in a particular region of a solid, the nuclei there move toward it to give a distorted local structure (see the illustration below). Because that local distortion is rich in positive charge, it is favourable for a second electron to join the first. Hence, there is a virtual attraction between the two electrons, and they move together as a pair. The local distortion can be easily disrupted by thermal motion of the ions in the solid, so the virtual attraction occurs only at very low temperatures. A Cooper pair undergoes less scattering than an individual electron as it travels through the solid because the distortion caused by one electron can attract back the other electron should it be scattered out of its path in a collision.

The formation of a Cooper pair. One electron distorts the crystal lattice and the second electron has a lower energy if it goes to that region. These electron–lattice interactions effectively bind the two electrons into a pair.

Because the Cooper pair is stable against scattering, it can carry charge freely through the solid, and hence give rise to superconduction.

The Cooper pairs responsible for low-temperature superconductivity are likely to be important in HTSCs, but the mechanism for pairing is hotly debated. There is evidence implicating the arrangement of CuO_5 layers and CuO_4 sheets in the mechanism of high-temperature superconduction. It is believed that movement of electrons along the linked CuO_4 units accounts for superconductivity, whereas the linked CuO_5 units act as 'charge reservoirs' that maintain an appropriate number of electrons in the superconducting layers.

Superconductors can sustain large currents and, consequently, are excellent materials for the high-field magnets used in modern NMR spectroscopy (Chapter 18). However, the potential uses of superconducting materials are not limited to the field to chemical instrumentation. For example, HTSCs with T_c values near room temperature would be very efficient components of an electrical power transmission system, in which energy loss due to electrical resistance would be minimized. The appropriate technology is not yet available, but research in this area of materials science is active.

Discussion questions:

1. Lead has $T_c = 7.19$ K and $H_c = 63\ 901$ A m^{-1}. At what temperature does lead become superconducting in a magnetic field of 20 kA m^{-1}?

2.[5] P.G. Radaelli, M. Marezio, M. Perroux, S. de Brion, J.L. Tholence, Q. Huang, and A. Santoro (*Science* **265**, 380 (1994)) report the synthesis and structure of a material that becomes superconducting at temperatures below 45 K. The compound is based on a layered compound $Hg_2Ba_2YCu_2O_{8-\delta}$, which has a tetragonal unit cell with $a = 0.386\ 06$ nm and $c = 2.8915$ nm; each unit cell contains two formula units. The compound is made superconducting by partially replacing Y by Ca, accompanied by a change in unit cell volume by less than 1 per cent. Estimate the Ca content x in superconducting $Hg_2Ba_2Y_{1-x}Ca_xCu_2O_{7.55}$ given that the mass density of the compound is 7.651 g cm^{-3}.

Further reading

M.T. Weller, *Inorganic materials chemistry*. Oxford Chemistry Primers, Oxford University Press (1995).

M.A. White, *Properties of materials*. Oxford University Press (2000).

---

5. This problem was supplied by Charles Trapp and Carmen Giunta.

(a) The formation of bands

The central aspect of solids that determines their electrical properties is the distribution of their electrons. There are two models of this distribution. In one, the **nearly free-electron approximation**, the valence electrons are assumed to be trapped in a box with a periodic potential, with low energy corresponding to the locations of cations. In the **tight-binding approximation**, the valence electrons are assumed to occupy molecular orbitals delocalized throughout the solid. The latter model is more in accord with the discussion in the foregoing chapters, and we confine our attention to it.

We shall consider a one-dimensional solid, which consists of a single, infinitely long line of atoms, each one having one s orbital available for forming molecular orbitals. We can construct the LCAO-MOs of the solid by adding N atoms in succession to a line, and then infer the electronic structure using the building-up principle.

One atom contributes one s orbital at a certain energy (Fig. 23.50). When a second atom is brought up it overlaps the first and forms bonding and antibonding orbitals. The third atom overlaps its nearest neighbour (and only slightly the next-nearest), and from these three atomic orbitals, three molecular orbitals are formed: one is fully bonding, one fully antibonding, and the intermediate orbital is nonbonding between neighbours. The fourth atom leads to the formation of a fourth molecular orbital. At this stage, we can begin to see that the general effect of bringing up successive atoms is to spread the range of energies covered by the molecular orbitals, and also to fill in the range of energies with more and more orbitals (one more for each atom). When N atoms have been added to the line, there are N molecular orbitals covering a band of energies of finite width, and the Hückel secular determinant (Section 14.7) is

$$\begin{vmatrix} \alpha - E & \beta & 0 & 0 & 0 & \dots & 0 \\ \beta & \alpha - E & \beta & 0 & 0 & \dots & 0 \\ 0 & \beta & \alpha - E & \beta & 0 & \dots & 0 \\ 0 & 0 & \beta & \alpha - E & \beta & \dots & 0 \\ 0 & 0 & 0 & \beta & \alpha - E & \dots & 0 \\ \vdots & \vdots & \vdots & \vdots & \vdots & \dots & \vdots \\ 0 & 0 & 0 & 0 & 0 & \dots & \alpha - E \end{vmatrix} = 0$$

where β is now the (s,s) resonance integral. The theory of determinants applied to such a symmetrical example as this (technically a 'tridiagonal determinant') leads to the following expression for the roots:

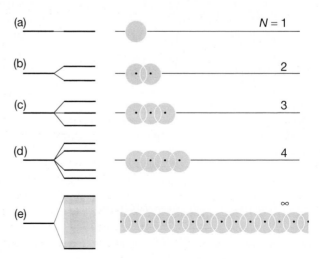

23.50 The formation of a band of N molecular orbitals by successive addition of N atoms to a line. Note that the band remains of finite width as N becomes infinite and, although it looks continuous, it consists of N different orbitals.

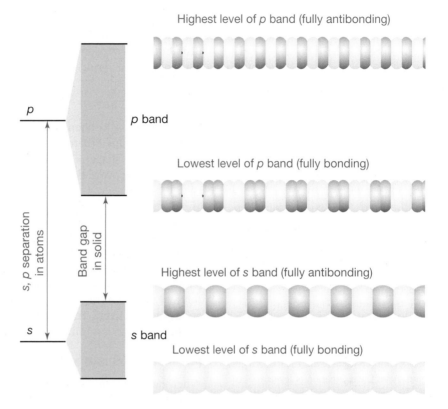

Highest level of p band (fully antibonding)

Lowest level of p band (fully bonding)

Highest level of s band (fully antibonding)

Lowest level of s band (fully bonding)

23.51 The overlap of s orbitals gives rise to an s band and the overlap of p orbitals gives rise to a p band. In this case, the s and p orbitals of the atoms are so widely spaced that there is a band gap. In many cases the separation is less and the bands overlap.

$$E_k = \alpha + 2\beta \cos \frac{k\pi}{N+1} \qquad k = 1, 2, \dots, N \qquad (23.23)$$

When N is infinitely large, the difference between neighbouring energy levels (the energies corresponding to k and k + 1) is infinitely small, but the band still has finite width overall:

$$E_N - E_1 \to 4\beta \qquad \text{as} \quad N \to \infty \qquad (23.24)$$

We can think of this band as consisting of N different molecular orbitals, the lowest-energy orbital (k = 1) being fully bonding, and the highest-energy orbital (k = N) being fully antibonding between adjacent atoms (Fig. 23.51). Similar bands form in three-dimensional solids.

The band formed from overlap of s orbitals is called the **s band**. If the atoms have p orbitals available, the same procedure leads to a **p band** (as shown in the upper half of Fig. 23.51). If the atomic p orbitals lie higher in energy than the s orbitals, then the p band lies higher than the s band, and there may be a **band gap**, a range of energies to which no orbital corresponds.

(b) The occupation of orbitals

Now consider the electronic structure of a solid formed from atoms each able to contribute one electron (for example, the alkali metals). There are N atomic orbitals and therefore N molecular orbitals packed into an apparently continuous band. There are N electrons to accommodate.

At T = 0, only the lowest $\frac{1}{2}N$ molecular orbitals are occupied (Fig. 23.52), and the HOMO is called the **Fermi level**. However, unlike in molecules, there are empty orbitals very close in energy to the Fermi level, so it requires hardly any energy to excite the uppermost electrons. Some of the electrons are therefore very mobile and give rise to electrical conductivity.

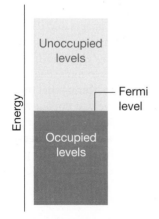

23.52 When N electrons occupy a band of N orbitals, it is only half full and the electrons near the Fermi level (the top of the filled levels) are mobile.

23.53 The Fermi–Dirac distribution, which gives the population of the levels at a temperature T. The high-energy tail decays exponentially towards zero. The curves are labelled with the value of μ/kT. The tinted grey region shows the occupation of levels at $T = 0$.

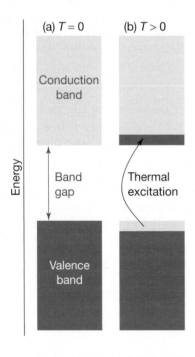

23.54 (a) When $2N$ electrons are present, the band is full and the material is an insulator at $T = 0$. (b) At temperatures above $T = 0$, electrons populate the levels of the upper *conduction band* and the solid is a semiconductor.

At temperatures above absolute zero, electrons can be excited by the thermal motion of the atoms. The population, P, of the orbitals is given by the **Fermi–Dirac distribution**, a version of the Boltzmann distribution that takes into account the effect of the Pauli principle:

$$P = \frac{1}{e^{(E-\mu)/kT} + 1} \qquad (23.25)$$

The quantity μ is the **chemical potential**,[6] which in this context is the energy of the level for which $P = \frac{1}{2}$ (note that the chemical potential decreases as the temperature increases). The shape of the Fermi–Dirac distribution is shown in Fig. 23.53. For energies well above μ, the 1 in the denominator can be neglected, and then

$$P \approx e^{-(E-\mu)/kT} \qquad (23.26)$$

The population now resembles a Boltzmann distribution, decaying exponentially with increasing energy. The higher the temperature, the longer the exponential tail.

The electrical conductivity of a metallic solid decreases with increasing temperature even though more electrons are excited into empty orbitals. This apparent paradox is resolved by noting that the increase in temperature causes more vigorous thermal motion of the atoms, so collisions between the moving electrons and an atom are more likely. That is, the electrons are scattered out of their paths through the solid, and are less efficient at transporting charge.

(c) Insulators and semiconductors

When each atom provides two electrons, the $2N$ electrons fill the N orbitals of the s band. The Fermi level now lies at the top of the band (at $T = 0$), and there is a gap before the next band begins (Fig. 23.54). As the temperature is increased, the tail of the Fermi–Dirac distribution extends across the gap, and electrons populate the empty orbitals of the upper band. They are now mobile, and the solid is an electrical conductor. In fact, it is a semiconductor, because the electrical conductivity depends on the number of electrons that are promoted across the gap, and that number increases as the temperature is raised. If the gap is large, though, very few electrons will be promoted at ordinary temperatures and the conductivity will remain close to zero, resulting in an insulator. Thus, the conventional distinction between an insulator and a semiconductor is related to the size of the band gap and is not an absolute distinction like that between a metal (incomplete bands at $T = 0$) and a semiconductor (full bands at $T = 0$).

Another method of increasing the number of charge carriers and enhancing the semiconductivity of a solid is to implant foreign atoms into an otherwise pure material. If these **dopants** can trap electrons, they withdraw electrons from the filled band, leaving holes which allow the remaining electrons to move (Fig. 23.55). This procedure gives rise to **p-type semiconductivity**, the p indicating that the holes are positive relative to the electrons in the band. Alternatively, a dopant might carry excess electrons (for example, phosphorus atoms introduced into germanium), and these additional electrons occupy otherwise empty bands, giving **n-type semiconductivity**, where n denotes the negative charge of the carriers. The preparation of doped but otherwise ultrapure materials was described in Section 8.7.

(d) Light-emitting diodes and semiconductor lasers

In this section we need to consider the properties of a **p–n junction**, the interface of the two types of semiconductor. The band structure at the junction is shown in Fig. 23.56. When a 'forward bias' is applied to the junction, in the sense that

---

6. Note that the chemical potential in eqn 25 has the dimensions of energy, not energy per mole.

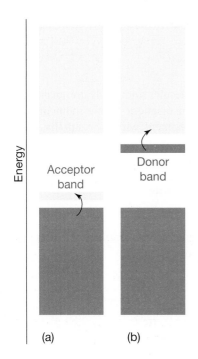

23.55 (a) A dopant with fewer electrons than its host can form a narrow band that accepts electrons from the valence band. The holes in the band are mobile and the substance is a p-type semiconductor. (b) A dopant with more electrons than its host forms a narrow band that can supply electrons to the conduction band. The electrons it supplies are mobile and the substance is an n-type semiconductor.

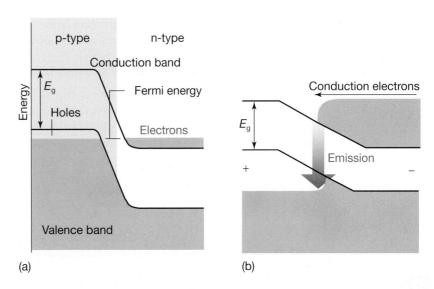

23.56 The structure of a diode junction (a) without bias, (b) with bias.

electrons are supplied through an external circuit to the n side of the junction, the electrons in the conduction band of the n-type semiconductor fall into the holes in the valence band of the p-type semiconductor. As they fall, they emit energy. In silicon semiconductors this energy is largely in the form of heat because the wavefunctions of the relevant states of the bands differ in linear momentum, so the transition can occur only if the electron transfers linear momentum to the lattice, and the device becomes warm. However, in some materials, most notably gallium arsenide, GaAs, the wavefunctions of the states involved correspond to the same linear momentum, so transition can occur without the lattice needing to be involved and the energy is emitted as light. Practical **light-emitting diodes** of this kind are widely used in electronic displays. Gallium arsenide itself emits infrared light, but the band gap is widened by incorporating phosphorus, and a material of composition approximately $GaAs_{0.6}P_{0.4}$ emits light in the red region of the spectrum.

A light-emitting diode is not a laser, because no resonance cavity and stimulated emission are involved. In **diode lasers**, light emission due to electron–hole recombination is employed as the basis of laser action. The population inversion can be sustained by sweeping away the electrons that fall into the holes of the p-type semiconductor, and a resonant cavity can be formed by using the high refractive index of the semiconducting material and cleaving single crystals so that the light is trapped by the abrupt variation of refractive index. One widely used material is $Ga_{1-x}Al_xAs$, which produces infrared laser radiation and is widely used in compact-disc (CD) players.

High-power diode lasers are also used to pump other lasers. One example is the pumping of Nd:YAG lasers by $Ga_{0.91}Al_{0.09}As/Ga_{0.7}Al_{0.3}As$ diode lasers. The Nd:YAG laser is often used to pump yet another laser, such as a Ti:sapphire laser. As a result, it is now possible to construct a laser system for steady-state or time-resolved spectroscopy entirely out of solid-state components.

23.10 Magnetic properties

The magnetic properties of metallic solids and semiconductors are best regarded as a part of physics as they depend strongly on the band structures of the material. Here we confine our attention largely to magnetic properties that stem from collections of individual molecules or ions. Much of the discussion applies to liquid- and gas-phase samples as well as to solids.

(a) Magnetic susceptibility

The magnetic and electric properties of molecules and solids are analogous. For instance, some molecules possess permanent magnetic dipole moments, and an applied magnetic field can induce a magnetic moment, with the result that the entire solid sample becomes magnetized. The analogue of the electric polarization, P, is the **magnetization**, $\mathcal{M}$, the average molecular magnetic dipole moment multiplied by the number density of molecules in the sample. The magnetization induced by a field of strength H is proportional to H, and we write

$$\mathcal{M} = \chi H \tag{23.27}$$

where χ is the dimensionless **volume magnetic susceptibility**. A closely related quantity is the **molar magnetic susceptibility**, χ_m:

$$\chi_m = \chi V_m \tag{23.28}$$

where V_m is the molar volume of the substance (we shall soon see why it is sensible to introduce this quantity). The **magnetic flux density**, B, is related to the applied field strength and the magnetization by

$$B = \mu_0(H + \mathcal{M}) = \mu_0(1 + \chi)H \tag{23.29}$$

where μ_0 is the vacuum permeability, $\mu_0 = 4\pi \times 10^{-7}\,\mathrm{J\,C^{-2}\,m^{-1}\,s^2}$. The magnetic flux density can be thought of as the density of magnetic lines of force permeating the medium. This density is increased if $\mathcal{M}$ adds to H (when $\chi > 0$), but the density is decreased if $\mathcal{M}$ opposes H (when $\chi < 0$). Materials for which χ is positive are called **paramagnetic**. Those for which χ is negative are called **diamagnetic**.

Just as polar molecules in fluid phases contribute a term proportional to $\mu^2/3kT$ to the electric polarization of a medium, so molecules with a permanent magnetic dipole moment of magnitude m contribute to the magnetization an amount proportional to $m^2/3kT$. However, unlike for polar molecules, this contribution to the magnetization is obtained even for paramagnetic species trapped in solids, because the direction of the spin of the electrons is typically not coupled to the orientation of the molecular framework. An applied field can also induce a magnetic moment by stirring up currents in the electron distribution like those responsible for the chemical shift in NMR (Section 18.5). The constant of proportionality between the induced moment and the applied field is called the **magnetizability**, ξ (xi), and the magnetic analogue of eqn 21.16 is

$$\chi = \mathcal{N}\mu_0\left(\xi + \frac{m^2}{3kT}\right) \tag{23.30}$$

We can now see why it is convenient to introduce χ_m, because the product of the number density $\mathcal{N}$ and the molar volume is the Avogadro constant, N_A:

$$\mathcal{N}V_m = \frac{NV_m}{V} = \frac{nN_A V_m}{nV_m} = N_A \tag{23.31}$$

Hence

$$\chi_m = N_A\mu_0\left(\xi + \frac{m^2}{3kT}\right) \tag{23.32}$$

and the density dependence of the susceptibility (which occurs in eqn 30 via $\mathcal{N} = N_A\rho/M$) has been eliminated. The expression for χ_m is in agreement with the empirical **Curie law**:

23.57 The arrangement of the Gouy balance for measuring magnetic susceptibilities. A paramagnetic sample appears to weigh more and a diamagnetic sample appears to weigh less when the magnetic field is on. In a modern alternative version (not shown), a sensitive balance is used to measure the force exerted by the sample on a suspended permanent magnet.

23.58 The arrangement used to determine magnetic susceptibility with a SQUID. The sample is moved upwards in small increments and the potential difference across the SQUID is measured.

$$\chi_m = A + \frac{C}{T} \tag{23.33}$$

with $A = N_A\mu_0\xi$ and $C = N_A\mu_0 m^2/3k$. As indicated above, and in contrast to electric moments, this expression applies to solids as well as fluid phases.

The magnetic susceptibility is traditionally measured with a **Gouy balance**. This instrument consists of a sensitive balance from which the sample hangs in the form of a narrow cylinder (Fig. 23.57) and lies between the poles of a magnet. If the sample is paramagnetic, it is drawn into the field, and its apparent weight is greater than when the field is off. A diamagnetic sample tends to be expelled from the field and appears to weigh less when the field is turned on. The balance is normally calibrated against a sample of known susceptibility. The modern version of the determination makes use of a **superconducting quantum interference device** (SQUID, Fig. 23.58). A SQUID takes advantage of the quantization of magnetic flux and the property of current loops in superconductors that, as part of the circuit, include a weakly conducting link through which electrons must tunnel. The current that flows in the loop in a magnetic field depends on the value of the magnetic flux, and a SQUID can be exploited as a very sensitive magnetometer.

Table 23.6 lists some experimental values. A typical paramagnetic volume susceptibility is about 10^{-3}, and a typical diamagnetic volume susceptibility is about $(-)10^{-5}$. The permanent magnetic moment can be extracted from susceptibility measurements by plotting χ against $1/T$.

(b) The permanent magnetic moment

The permanent magnetic moment of a molecule arises from any unpaired electron spins in the molecule. We saw in Section 13.10a that the magnitude of the magnetic moment of an electron is proportional to the magnitude of the spin angular momentum, $\{s(s + 1)\}^{1/2}\hbar$.

Table 23.6* Magnetic susceptibilities at 298 K

	$\chi/10^{-6}$	$\chi_m/(10^{-5}\ cm^3\ mol^{-1})$
$H_2O(l)$	−90	−160
NaCl(s)	−13.9	−38
Cu(s)	−9.6	−6.8
CuSO$_4$·5H$_2$O(s)	+176	+1930

* More values are given in the *Data section*.

$$m = g_e\{s(s + 1)\}^{1/2}\mu_B \qquad \mu_B = \frac{e\hbar}{2m_e} \tag{23.34}$$

where $g_e = 2.0023$ (see Section 18.1). If there are several electron spins in each molecule, they combine to a total spin S, and then $s(s + 1)$ should be replaced by $S(S + 1)$. It follows that the spin contribution to the molar magnetic susceptibility is

$$\chi_m = \frac{N_A g_e^2 \mu_0 \mu_B^2 S(S + 1)}{3kT} \tag{23.35}$$

This expression shows that the susceptibility is positive, so the spin magnetic moments contribute to the paramagnetic susceptibilities of materials. The contribution decreases with increasing temperature because the thermal motion randomizes the spin orientations. In practice, a contribution to the paramagnetism also arises from the orbital angular momenta of electrons: we have discussed the spin-only contribution.

Illustration 23.1

Consider a complex salt with three unpaired electrons per complex cation at 298 K, of mass density 3.24 g cm^{-3}, and molar mass 200 g mol^{-1}. First note that

$$\frac{N_A g_e^2 \mu_0 \mu_B^2}{3k} = 6.3001 \text{ cm}^3 \text{ K}^{-1} \text{ mol}^{-1}$$

Consequently,

$$\chi_m = 6.3001 \times \frac{S(S + 1)}{T/K} \text{ cm}^3 \text{ mol}^{-1}$$

Substitution of the data with $S = \frac{3}{2}$ gives $\chi_m = 7.9 \times 10^{-2}$ cm^3 mol^{-1}. Note that the density is not needed at this stage. To obtain the volume magnetic susceptibility, the molar susceptibility is divided by the molar volume $V_m = M/\rho$, where ρ is the mass density. In this illustration, $V_m = 61.7$ cm^3 mol^{-1}, so $\chi = 1.3 \times 10^{-3}$.

At low temperatures, some paramagnetic solids make a phase transition to a state in which large domains of spins align with parallel orientations. This cooperative alignment gives rise to a very strong magnetization and is called **ferromagnetism** (Fig. 23.59). In other cases, the cooperative effect leads to alternating spin orientations: the spins are locked into a low-magnetization arrangement to give an **antiferromagnetic phase**. The ferromagnetic phase has a nonzero magnetization in the absence of an applied field, but the antiferromagnetic phase has a zero magnetization because the spin magnetic moments cancel. The ferromagnetic transition occurs at the **Curie temperature**, and the antiferromagnetic transition occurs at the **Néel temperature**.

(c) Induced magnetic moments

An applied magnetic field induces the circulation of electronic currents. These currents give rise to a magnetic field that usually opposes the applied field, so the substance is diamagnetic. In a few cases the induced field augments the applied field, and the substance is then paramagnetic.

The great majority of molecules with no unpaired electron spins are diamagnetic. In these cases, the induced electron currents occur within the orbitals of the

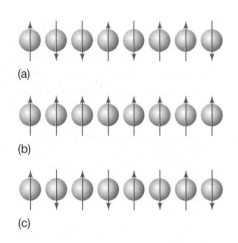

23.59 (a) In a paramagnetic material, the electron spins are aligned at random in the absence of an applied magnetic field. (b) In a ferromagnetic material, the electron spins are locked into a parallel alignment over large domains. (c) In an antiferromagnetic material, the electron spins are locked into an antiparallel arrangement. The latter two arrangements survive even in the absence of an applied field.

molecule that are occupied in its ground state. In the few cases in which molecules are paramagnetic despite having no unpaired electrons, the induced electron currents flow in the opposite direction because they can make use of unoccupied orbitals that lie close to the HOMO in energy. This orbital paramagnetism can be distinguished from spin paramagnetism by the fact that it is temperature-independent: this is why it is called **temperature-independent paramagnetism** (TIP).

We can summarize these remarks as follows. All molecules have a diamagnetic component to their susceptibility, but it is dominated by spin paramagnetism if the molecules have unpaired electrons. In a few cases (where there are low-lying excited states) TIP is strong enough to make the molecules paramagnetic even though their electrons are paired.

Checklist of key ideas

Crystal lattices

23.1 Lattices and unit cells

☐ space lattice, the pattern formed by points representing the locations of structural motifs (atoms, molecules, or groups of atoms, molecules, or ions)

☐ unit cell, an imaginary parellelepiped that contains one unit of a translationally repeating pattern

☐ primitive, a unit cell formed by joining neighbouring lattice points by straight lines

☐ crystal system, a classification based on the rotational symmetry elements of a unit cell, Table 23.1

☐ essential symmetry, the elements a unit cell must possess to belong to a particular crystal system

☐ Bravais lattice, the 14 distinct space lattices in three dimensions (Fig. 23.8)

☐ primitive unit cell, a unit cell with lattice points only at the corners

☐ body-centred unit cell, a unit cell with lattice points at the corners and at the centre

☐ face-centred unit cell, a unit cell with lattice points at the corners and on each face

☐ side-centred unit cell, a unit cell with lattice points at the corners and on two opposite faces

23.2 The identification of lattice planes

☐ Miller indices (hkl), indices that distinguish planes in a lattice

☐ separation of planes, $1/d_{hkl}^2 = h^2/a^2 + k^2/b^2 + l^2/c^2$

23.3 The investigation of structure

☐ diffraction, interference caused by an object in the path of waves

☐ diffraction pattern, the pattern of varying intensity that results from diffraction

☐ *Bremsstrahlung*, X-radiation generated by the deceleration of electrons

☐ *K*-radiation, X-radiation emitted when an electron falls into a K shell

☐ four-circle diffractometer, a device used in X-ray crystallography

☐ reflection, an intense beam arising from constructive interference

☐ glancing angle, θ, the angle of incidence of a beam of radiation

☐ Bragg's law, $\lambda = 2d \sin \theta$

☐ scattering factor, a measure of the ability of an atom to diffract radiation,

$$f = 4\pi \int_0^\infty \rho(r) \frac{\sin kr}{kr} r^2 \, dr \qquad k = \frac{4\pi}{\lambda} \sin \theta$$

☐ structure factor, $F_{hkl} = \sum_j f_j e^{i\phi_{hkl}(j)}$,

$\phi_{hkl}(j) = 2\pi(hx_j + ky_j + lz_j)$

☐ systematic absences, reflections with odd values of $h + k + l$ that are absent from the powder diffraction pattern

☐ Fourier synthesis, the construction of the electron density distribution from structure factors

$$\rho(r) = (1/V) \sum_{hkl} F_{hkl} \, e^{-2\pi i(hx+ky+lz)}$$

☐ phase problem, the ambiguity in phase of structure factors obtained from intensities

☐ Patterson synthesis, a map of interatomic vectors obtained by Fourier analysis of diffraction intensities

☐ direct methods, statistical techniques to compute the probabilities that phases have particular values

☐ Sayre probability relation, the sign of $F_{h+h',k+k',l+l'}$ is probably equal to (sign of F_{hkl}) × (sign of $F_{h'k'l'}$)

☐ structure refinement, the adjustment of structural parameters to give the best fit between the observed intensities and those calculated from the model of the structure deduced from the diffraction pattern

23.4 Neutron and electron diffraction

Crystal structure

23.5 Metallic solids

☐ close-packed, a layer of spheres with maximum utilization of space

☐ polytype, structures that are identical in two dimensions but differ in the third dimension

☐ hexagonally close-packed (hcp), the sequence of layers ABABAB...

☐ cubic close-packed (ccp), the sequence of layers ABCABC...

☐ coordination number, the number of nearest neighbours

☐ packing fraction, the fraction of space occupied by hard spheres

23.6 Ionic solids

☐ (n_+, n_-)-coordination, the number of nearest neighbours of opposite charge; n_+ is the coordination number of the cation and n_- that of the anion

☐ caesium-chloride structure, an ion of one charge occupies the centre of a cubic unit cell with eight counter ions at its corners, Fig. 23.33

☐ rock-salt structure, of two interpenetrating slightly expanded fcc arrays, one of cations and the other of anions, Fig. 23.34

☐ radius ratio, $\gamma = r_{smaller}/r_{larger}$

☐ radius-ratio rule, a rule suggesting which type of structure is likely based on the radius ratio

☐ lattice energy, the difference in potential energy of ions packed together in a solid and widely separated as a gas

☐ lattice enthalpy, ΔH_L, the change in molar enthalpy for the process $MX(s) \rightarrow M^{z+}(g) + X^{z-}(g)$

☐ Madelung constant, A, the constant characteristic of the symmetry of the lattice that occurs in the expression for the electrostatic contribution to the potential energy of ions

☐ Born–Mayer equation, for the total potential energy of an ionic crystal, $E_{P,min} = -(N_A|z_A z_B|e^2/4\pi\varepsilon_0 d)(1 - d/d^*)A$

☐ Born–Haber cycle, a closed path of transformations starting and ending at the same point, one step of which is the formation of the solid compound from a gas of widely separated ions

23.7 Molecular solids and covalent networks

☐ covalent network solid, a solid in which covalent bonds in a definite spatial orientation link the atoms in a network extending through the crystal

☐ molecular solid, a solid consisting of discrete molecules held together by van der Waals interactions

☐ anomalous scattering, an extra phase shift in X-rays scattered by atoms with absorptions close to the X-ray frequency

The properties of solids

23.8 Mechanical properties

☐ stress, the applied force divided by the area to which it is applied

☐ strain, the distortion of a sample resulting from an applied stress

☐ rheology, the study of the relation between stress and strain

☐ uniaxial stress, stress applied in one direction

☐ hydrostatic stress, a stress applied simultaneously in all directions

☐ pure shear, a stress that tends to push opposite faces of the sample in opposite directions

☐ elastic deformation, a deformation that disappears when the stress is removed

☐ plastic deformation, a strain from which recovery does not occur when the stress is removed

☐ Young's modulus, E = (normal stress)/(normal strain)

☐ bulk modulus, K = pressure/(fractional change in volume)

☐ shear modulus, G = (shear stress)/(shear strain)

☐ Poisson's ratio, ν_P = (transverse strain)/(normal strain)

☐ slip plane, planes of atoms that under stress may slip or slide relative to one another

☐ crystallinity, the degree of three-dimensional long-range order in the solid state

☐ elastomer, a flexible polymer that can expand or contract easily upon application of an external force

☐ perfect elastomer, a polymer in which the internal energy is independent of the extension of the random coil

☐ fibre, a polymer material that owes its strength to interactions between chains

☐ plastic, a polymer that can attain only a limited degree of crystallinity

☐ melting temperature, the temperature at which a polymer melts

☐ glass transition temperature, the temperature at which a transition occurs from a state of high to low chain mobility

23.9 Electrical properties

☐ metallic conductor, a substance with an electrical conductivity that decreases as the temperature is raised

☐ semiconductor, a substance with an electrical conductivity that increases as the temperature is raised

☐ insulator, a semiconductor with a very low electrical conductivity

☐ superconductor, a solid that conducts electricity without resistance

☐ nearly free electron approximation, a model of a metal in which the valence electrons are assumed to be trapped in a box with a periodic potential

☐ tight-binding approximation, a model of a metal in which the valence electrons are assumed to occupy molecular orbitals delocalized throughout the solid

☐ s and p bands, a band formed from overlap of s and p orbitals, respectively

☐ band gap, a range of energies to which no orbital corresponds

☐ Fermi level, the highest occupied molecular orbital in a solid at $T = 0$

☐ Fermi–Dirac distribution, $P = 1/(e^{(E-\mu)/kT} + 1)$

☐ chemical potential (of electron), the energy of the level for which $P = \frac{1}{2}$

☐ dopant, introduced impurity atoms

☐ p- and n-type semiconductivity, conduction by holes and particles, respectively

☐ p–n junction, a junction between p- and n-type semiconductors

☐ light-emitting diode, a semiconductor device that generates light

☐ diode laser, a semiconductor device in which light emission due to electron–hole recombination is employed as the basis of laser action

23.10 Magnetic properties

☐ magnetization, the magnetic dipole moment density, $\mathcal{M} = \chi H$

☐ volume magnetic susceptibility, the proportionality constant χ

☐ molar magnetic susceptibility, $\chi_m = \chi V_m$

☐ magnetic flux density, $B = \mu_0(H + \mathcal{M}) = \mu_0(1 + \chi)H$

☐ paramagnetic, a material for which χ is positive

☐ diamagnetic, a material for which χ is negative

☐ magnetizability, ξ, a measure of the extent to which a magnetic dipole moment may be induced in a molecule

☐ Curie law, $\chi_m = A + C/T$, $A = N_A \mu_0 \xi$ and $C = N_A \mu_0 m^2/3k$

☐ Gouy balance, a device for determining the magnetic susceptibility of a sample

☐ superconducting quantum interference device (SQUID), a superconducting device for determining the magnetic susceptibility of a sample

☐ ferromagnetism, strong, persistent magnetization arising from the cooperative alignment of spins

☐ antiferromagnetic phase, a phase in which spins are locked into a low-magnetization arrangement

☐ Curie temperature, the temperature of a ferromagnetic transition

☐ Néel temperature, the temperature of an antiferromagnetic transition

☐ temperature-independent paramagnetism (TIP), orbital paramagnetism

Further reading

Articles of general interest

M.G.D. Baumann, J.C. Wright, A.B. Ellis, T. Kuech, and G.C. Lisesnky, Diode lasers. *J. Chem. Educ.* **69**, 89 (1992).

J.P. Birk and P.R. Coffman, Finding the face-centred cube in the cubic close-packed structure. *J. Chem. Educ.* **69**, 953 (1992).

J.P. Chesick, Fourier transforms in chemistry. Part III. X-ray crystal structure analysis. *J. Chem. Educ.* **66**, 413 (1989).

F. Ebrahami and M.J. Kaufman, Structure of metals and alloys. In *Encyclopedia of applied physics* (ed. G.L. Trigg), **10**, 199. VCH, New York (1994).

S.R. Elliott, Structure of amorphous materials. In *Encyclopedia of applied physics* (ed. G.L. Trigg), **1**, 559. VCH, New York (1991).

J.H. Enemark, Introducing chemists to X-ray structure determination. *J. Chem. Educ.* **65**, 491 (1988).

L.M. Falicov, Diamagnetism. In *Encyclopedia of applied physics* (ed. G.L. Trigg), **4**, 557. VCH, New York (1992).

S. Foner, Measurement of magnetic properties and quantities. In *Encyclopedia of applied physics* (ed. G.L. Trigg), **9**, 463. VCH, New York (1994).

A.I. Goldman, Crystalline state. In *Encyclopedia of applied physics* (ed. G.L. Trigg), **4**, 365. VCH, New York (1992).

J.P. Glusker, Teaching crystallography to noncrystallographers. *J. Chem. Educ.* **65**, 474 (1988).

S.F.A. Kettle and L.J. Norrby, Really, your lattices are all primitive, Mr. Bravais! *J. Chem. Educ.* **70**, 959 (1993).

C.D. Graham, Jr., Magnetic materials. In *Encyclopedia of applied physics* (ed. G.L. Trigg), **9**, 1. VCH, New York (1994).

W.E. Hatfield, Magnetic measurements. In *Solid-state chemistry: techniques* (ed. A.K. Cheetham and P. Day). Oxford University Press (1990).

T. Li and J.H. Worrell, Construction of the seven basic crystallographic units. *J. Chem. Educ.* **66**, 73 (1989).

L.N. Mulay and I.L. Mulay, Static magnetic techniques and applications. In *Techniques of chemistry* (ed. B.W. Rossiter and J.E. Hamilton), **IIIB**, 133 (1989).

W.B. Pearson and C. Chieh, Crystallography. In *Encyclopedia of applied physics* (ed. G.L. Trigg), **4**, 385. VCH, New York (1992).

J. Pellicer, J.A. Manzanares, J. Zúñiga, P. Utrillas, and J. Fernández, Thermodynamics of rubber elasticity. *J. Chem. Educ.* **78**, 263 (2001).

A. Pisanty, The electronic structure of graphite: A chemist's introduction to band theory. *J. Chem. Educ.* **68**, 804 (1991).

C.G. Pope, X-ray diffraction and the Bragg equation. *J. Chem. Educ.* **74**, 129 (1997).

E. Schönherr, Crystal growth. In *Encyclopedia of applied physics* (ed. G.L. Trigg), **4**, 335. VCH, New York (1992).

S. Tiwari, Semiconductors, compound–electronic properties. In *Encyclopedia of applied physics* (ed. G.L. Trigg), **17**, 303. VCH, New York (1996).

J. Toofan, A simple expression between critical radius ratio and coordination number. *J. Chem. Educ.* **71**, 147 (1994).

T. Vogt, Neutron diffraction. In *Encyclopedia of applied physics* (ed. G.L. Trigg), **11**, 339. VCH, New York (1994).

Y. Yafet, Paramagnetism. In *Encyclopedia of applied physics* (ed. G.L. Trigg), **13**, 101. VCH, New York (1995).

Texts and sources of data and information

W.D. Callister, Jr., *Materials science and engineering: an introduction*. Wiley, New York (2000).

C.E. Carraher, Jr., *Seymour/Carraher's polymer chemistry*. Marcel Dekker (2000).

W. Clegg, *Crystal structure determination*. Oxford Chemistry Primers, Oxford University Press (1998).

P.A. Cox, *The electronic structure and chemistry of solids*. Oxford University Press (1987).

R. Drago, *Physical methods for chemists*. Saunders, Philadelphia (1992).

M.F.C. Ladd, *Chemical bonding in solids and fluids*. Ellis Horwood/Prentice Hall, Hemel Hempstead (1995).

W.A. Harrison, *Electronic structure and properties of solids: the physics of the chemical bond*. Dover, New York (1989).

C. Kittel, *Introduction to solid state physics*. Wiley, New York (1996).

D.R. Lide (ed.), *CRC Handbook of Chemistry and Physics*, Sections 3, 12, 13, and 15. CRC Press, Boca Raton (2000).

J.C. Lindon, G.E. Tranter, and J.L. Holmes (ed.), *Encyclopedia of spectroscopy and spectrometry*. Academic Press, San Diego (2000).

O. Svelto, *Principles of lasers*. Plenum, New York (1998).

G.R. Van Hecke and K.K. Karukstis, *A guide to lasers in chemistry*. Jones and Bartlett, Boston (1998).

I.M. Ward and D.W. Hedley, *An introduction to the mechanical properties of solid polymers*. Wiley, New York (1999).

M.T. Weller, *Inorganic materials chemistry*. Oxford Chemistry Primers, Oxford University Press (1995).

M.A. White, *Properties of materials*. Oxford University Press (2000).

Exercises

Discussion questions

23.1 (a) Explain how planes of lattice points are labelled.

23.1 (b) Describe the procedure for identifying the type and size of a cubic unit cell.

23.2 (a) What is meant by a systematic absence? How do they arise?

23.2 (b) Describe the phase problem and explain how it may be overcome.

23.3 (a) Describe the structures of elemental metallic solids in terms of the packing of hard spheres. To what extent is the hard-sphere model inaccurate?

23.3 (b) Describe the caesium-chloride and rock-salt structures in terms of the occupation of holes in expanded close-packed lattices.

23.4 (a) Explain how X-ray diffraction can be used to determine the absolute configuration of molecules.

23.4 (b) Explain how metallic conductors, semiconductors, and insulators are identified and explain their properties in terms of band theory.

23.5 (a) To what extent are the electric and magnetic properties of molecules analogous? How do they differ?

23.5 (b) Describe the characteristics of the Fermi–Dirac distribution. Why is it appropriate to call the parameter μ a chemical potential?

Numerical exercises

23.6 (a) Equivalent lattice points within the unit cell of a Bravais lattice have identical surroundings. What points within a face-centred cubic unit cell are equivalent to the point $(\frac{1}{2}, 0, 0)$?

23.6 (b) Equivalent lattice points within the unit cell of a Bravais lattice have identical surroundings. What points within a body-centred cubic unit cell are equivalent to the point $(\frac{1}{2}, 0, \frac{1}{2})$?

23.7 (a) Find the Miller indices of the planes that intersect the crystallographic axes at the distances $(2a, 3b, 2c)$ and $(2a, 2b, \infty c)$.

23.7 (b) Find the Miller indices of the planes that intersect the crystallographic axes at the distances $(1a, 3b, -c)$ and $(2a, 3b, 4c)$.

23.8 (a) Calculate the separations of the planes {111}, {211}, and {100} in a crystal in which the cubic unit cell has side 432 pm.

23.8 (b) Calculate the separations of the planes {121}, {221}, and {244} in a crystal in which the cubic unit cell has side 523 pm.

23.9 (a) The glancing angle of a Bragg reflection from a set of crystal planes separated by 99.3 pm is 20.85°. Calculate the wavelength of the X-rays.

23.9 (b) The glancing angle of a Bragg reflection from a set of crystal planes separated by 128.2 pm is 19.76°. Calculate the wavelength of the X-rays.

23.10 (a) What are the values of 2θ of the first three diffraction lines of bcc iron (atomic radius 126 pm) when the X-ray wavelength is 58 pm?

23.10 (b) What are the values of 2θ of the first three diffraction lines of fcc gold (atomic radius 144 pm) when the X-ray wavelength is 154 pm?

23.11 (a) Copper K_α radiation consists of two components of wavelengths 154.433 pm and 154.051 pm. Calculate the separation of the diffraction lines arising from the two components in a powder diffraction pattern recorded in a circular camera of radius 5.74 cm (with the sample at the centre) from planes of separation 77.8 pm.

23.11 (b) A synchrotron source produces X-radiation at a range of wavelengths. Consider two components of wavelengths 95.401 and 96.035 pm. Calculate the separation of the diffraction lines arising from the two components in a powder diffraction pattern recorded in a circular camera of radius 5.74 cm (with the sample at the centre) from planes of separation 82.3 pm.

23.12 (a) The compound Rb_3TlF_6 has a tetragonal unit cell with dimensions $a = 651$ pm and $c = 934$ pm. Calculate the volume of the unit cell.

23.12 (b) Calculate the volume of the hexagonal unit cell of sodium nitrate, for which the dimensions are $a = 1692.9$ pm and $c = 506.96$ pm.

23.13 (a) The orthorhombic unit cell of $NiSO_4$ has the dimensions $a = 634$ pm, $b = 784$ pm, and $c = 516$ pm, and the density of the solid is estimated as 3.9 g cm^{-3}. Determine the number of formula units per unit cell and calculate a more precise value of the density.

23.13 (b) An orthorhombic unit cell of a compound of molar mass 135.01 g mol^{-1} has the dimensions $a = 589$ pm, $b = 822$ pm, and $c = 798$ pm. The density of the solid is estimated as 2.9 g cm^{-3}. Determine the number of formula units per unit cell and calculate a more precise value of the density.

23.14 (a) The unit cells of $SbCl_3$ are orthorhombic with dimensions $a = 812$ pm, $b = 947$ pm, and $c = 637$ pm. Calculate the spacing, d, of the (411) planes.

23.14 (b) An orthorhombic unit cell has dimensions $a = 679$ pm, $b = 879$ pm, and $c = 860$ pm. Calculate the spacing, d, of the (322) planes.

23.15 (a) A substance known to have a cubic unit cell gives reflections with Cu K_α radiation (wavelength 154 pm) at glancing angles 19.4°, 22.5°, 32.6°, and 39.4°. The reflection at 32.6° is known to be due to the (220) planes. Index the other reflections.

23.15 (b) A substance known to have a cubic unit cell gives reflections with radiation of wavelength 137 pm at the glancing angles 10.7°, 13.6°, 17.7°, and 21.9°. The reflection at 17.7° is known to be due to the (111) planes. Index the other reflections.

23.16 (a) Potassium nitrate crystals have orthorhombic unit cells of dimensions $a = 542$ pm, $b = 917$ pm, and $c = 645$ pm. Calculate the glancing angles for the (100), (010), and (111) reflections using Cu K_α radiation (154 pm).

23.16 (b) Calcium carbonate crystals in the form of aragonite have orthorhombic unit cells of dimensions $a = 574.1$ pm, $b = 796.8$ pm, and $c = 495.9$ pm. Calculate the glancing angles for the (100), (010), and (111) reflections using radiation of wavelength 83.42 pm (from aluminium).

23.17 (a) Copper(I) chloride forms cubic crystals with four formula units per unit cell. The only reflections present in a powder photograph are those with either all even indices or all odd indices. What is the (Bravais) lattice type of the unit cell?

23.17 (b) A powder diffraction photograph from tungsten shows lines that index as (110), (200), (211), (220), (310), (222), (321), (400), … . Identify the (Bravais) lattice type of the unit cell.

23.18 (a) The coordinates, in units of a, of the atoms in a simple cubic lattice are (0, 0, 0), (0, 1, 0), (0, 0, 1), (0, 1, 1), (1, 0, 0), (1, 1, 0), (1, 0, 1), and (1, 1, 1). Calculate the structure factors F_{hkl} when all the atoms are identical.

23.18 (b) The coordinates, in units of a, of the atoms in a body-centred cubic lattice are (0, 0, 0), (0, 1, 0), (0, 0, 1), (0, 1, 1), (1, 0, 0), (1, 1, 0), (1, 0, 1), (1, 1, 1), and $(\frac{1}{2}, \frac{1}{2}, \frac{1}{2})$. Calculate the structure factors F_{hkl} when all the atoms are identical.

23.19 (a) Calculate the packing fraction for close-packed cylinders.

23.19 (b) Calculate the packing fraction for equilateral triangular rods stacked as shown in (**3**).

3

23.20 (a) Verify that the radius ratios for sixfold coordination is 0.414.

23.20 (b) Verify that the radius ratios for eightfold coordination is 0.732.

23.21 (a) From the data in Table 23.3 determine the radius of the smallest cation that can have (a) sixfold and (b) eightfold coordination with the O^{2-} ion.

23.21 (b) From the data in Table 23.3 determine the radius of the smallest cation that can have (a) sixfold and (b) eightfold coordination with the K^+ ion.

23.22 (a) Calculate the atomic packing factor for diamond.

23.22 (b) Calculate the atomic packing factor for a side-centred (C) cubic unit cell.

23.23 (a) Is there an expansion or a contraction as titanium transforms from hcp to body-centred cubic? The atomic radius of titanium is 145.8 pm in hcp but 142.5 pm in bcc.

23.23 (b) Is there an expansion or a contraction as iron transforms from hcp to bcc? The atomic radius of iron is 126 in hcp but 122 pm in bcc.

23.24 (a) In a Patterson synthesis, the spots correspond to the lengths and directions of the vectors joining the atoms in a unit cell. Sketch the pattern that would be obtained for a planar, triangular isolated BF_3 molecule.

23.24 (b) In a Patterson synthesis, the spots correspond to the lengths and directions of the vectors joining the atoms in a unit cell.

Sketch the pattern that would be obtained from the C atoms in an isolated benzene molecule.

23.25 (a) What velocity should neutrons have if they are to have wavelength 50 pm?

23.25 (b) Calculate the wavelength of neutrons that have reached thermal equilibrium by collision with a moderator at 300 K.

23.26 (a) What accelerating potential difference must be applied to electrons to generate a beam with wavelength 18 pm?

23.26 (b) Calculate the wavelengths of electrons that have been accelerated through (a) 1.0 kV, (b) 10 kV, (c) 40 kV.

23.27 (a) Derive the Born–Mayer equation (eqn 15) by calculating the energy at which $d(E_p + E_p^*)/dd = 0$, with E_p and E_p^* given by eqns 13 and 14, respectively.

23.27 (b) Calculate the lattice enthalpy of $MgBr_2$ from the following data:

	$\Delta H/(\text{kJ mol}^{-1})$
Sublimation of Mg(s)	+148
Ionization of Mg(g) to Mg^{2+}(g)	+2187
Vaporization of Br_2(l)	+31
Dissociation of Br_2(g)	+193
Electron attachment to Br(g)	−331
Formation of $MgBr_2$(s) from Mg(s) and Br_2(l)	−524

23.28 (a) Cotton consists of the polymer cellulose, which is a linear chain of glucose molecules. The chains are held together by hydrogen bonding. When a cotton shirt is ironed, it is first moistened, then heated under pressure. Explain this process.

23.28 (b) Sections of the solid fuel rocket boosters of the space shuttle *Challenger* were sealed together with O-ring rubber seals of circumference 11 m. These seals failed at 0°C, a temperature well above the crystallization temperature of the rubber. Speculate on why the failure occurred.

23.29 (a) Young's modulus for polyethylene at room temperature is 1.2 GPa. What strain will be produced when a mass of 1.0 kg is suspended from a polyethylene thread of diameter 1.0 mm?

23.29 (b) Young's modulus for iron at room temperature is 215 GPa. What strain will be produced when a mass of 10.0 kg is suspended from an iron wire of diameter 0.10 mm?

23.30 (a) Poisson's ratio for polyethylene is 0.45. What change in volume takes place when a cube of polyethylene of volume 1.0 cm³ is subjected to a uniaxial stress that produces a strain of 1.0 per cent?

23.30 (b) Poisson's ratio for lead is 0.41. What change in volume takes place when a cube of lead of volume 1.0 dm³ is subjected to a uniaxial stress that produces a strain of 2.0 per cent?

23.31 (a) The magnetic moment of $CrCl_3$ is $3.81\mu_B$. How many unpaired electrons does the Cr possess?

23.31 (b) The magnetic moment of Mn^{2+} in its complexes is typically $5.3\mu_B$. How many unpaired electrons does the ion possess?

23.32 (a) Calculate the molar susceptibility of benzene given that its volume susceptibility is -7.2×10^{-7} and its density 0.879 g cm⁻³ at 25°C.

23.32 (b) Calculate the molar susceptibility of cyclohexane given that its volume susceptibility is -7.9×10^{-7} and its density 811 kg m⁻³ at 25°C.

23.33 (a) According to Lewis theory, an O_2 molecule should be diamagnetic. However, experimentally it is found that $\chi_m/(m^3\,mol^{-1}) = (1.22 \times 10^{-5}\,K)/T$. Determine the number of unpaired spins in O_2. How is the problem of the Lewis structure resolved?

23.33 (b) Predict the molar susceptibility of nitrogen dioxide at 298 K. Why does the molar susceptibility of a sample of nitrogen dioxide gas decrease as it is compressed?

23.34 (a) Data on a single crystal of MnF_2 give $\chi_m = 0.1463\,cm^3\,mol^{-1}$ at 294.53 K. Determine the effective number of unpaired electrons in this compound and compare your result with the theoretical value.

23.34 (b) Data on a single crystal of $NiSO_4\cdot7H_2O$ give $\chi_m = 6.00 \times 10^{-8}\,m^3\,mol^{-1}$ at 298 K. Determine the effective number of unpaired

electrons in this compound and compare your result with the theoretical value.

23.35 (a) Estimate the spin-only molar susceptibility of $CuSO_4\cdot5H_2O$ at 25°C.

23.35 (b) Estimate the spin-only molar susceptibility of $MnSO_4\cdot4H_2O$ at 298 K.

23.36 (a) Approximately how large must the magnetic induction, B, be for the orientational energy of an $S = 1$ system to be comparable to kT at 298 K?

23.36 (b) Estimate the ratio of populations of the M_S states of a system with $S = 1$ in 15.0 T at 298 K.

Problems[†]

Numerical problems

23.1 In the early days of X-ray crystallography there was an urgent need to know the wavelengths of X-rays. One technique was to measure the diffraction angle from a mechanically ruled grating. Another method was to estimate the separation of lattice planes from the measured density of a crystal. The density of NaCl is 2.17 g cm^{-3} and the (100) reflection using Pd K_α radiation occurred at 6.0°. Calculate the wavelength of the X-rays.

23.2 The element polonium crystallizes in a cubic system. Bragg reflections, with X-rays of wavelength 154 pm, occur at sin $\theta = 0.225$, 0.316, and 0.388 from the (100), (110), and (111) sets of planes. The separation between the sixth and seventh lines observed in the powder diffraction pattern is larger than that between the fifth and sixth lines. Is the unit cell simple, body-centred, or face-centred? Calculate the unit cell dimension.

23.3 The unit cell dimensions of NaCl, KCl, NaBr, and KBr, all of which crystallize in face-centred cubic lattices, are 562.8 pm, 627.7 pm, 596.2 pm, and 658.6 pm, respectively. In each case, anion and cation are in contact along an edge of the unit cell. Does the data support the contention that ionic radii are constants independent of the counter ion?

23.4 The powder diffraction patterns of (a) tungsten, (b) copper obtained in a camera of radius 28.7 mm are shown in Fig. 23.60. Both were obtained with 154 pm X-rays and the scales are marked.

Identify the unit cell in each case, and calculate the lattice spacing. Estimate the metallic radii of W and Cu.

23.5 Elemental silver reflects X-rays of wavelength 154.18 pm at angles of 19.076°, 22.171°, and 32.256°. However, there are no other reflections at angles of less than 33°. Assuming a cubic unit cell, determine its type and dimension. Calculate the density of silver.

23.6 Genuine pearls consist of concentric layers of calcite crystals (CaCO$_3$) in which the trigonal axes are oriented along the radii. The nucleus of a cultured pearl is a piece of mother-of-pearl that has been worked into a sphere on a lathe. The oyster then deposits concentric layers of calcite on the central seed. Suggest an X-ray method for distinguishing between real and cultured pearls.

23.7 In their book *X-rays and crystal structures* (which begins 'It is now two years since Dr. Laue conceived the idea …') the Braggs give a number of simple examples of X-ray analysis. For instance, they report that the reflection from (100) planes in KCl occurs at 5°23′, but for NaCl it occurs at 6°0′ for X-rays of the same wavelength. If the side of the NaCl unit cell is 564 pm, what is the side of the KCl unit cell? The densities of KCl and NaCl are 1.99 g cm^{-3} and 2.17 g cm^{-3}, respectively. Do these values support the X-ray analysis?

23.8 The volume of a monoclinic unit cell is $abc \sin \beta$. Naphthalene has a monoclinic unit cell with two molecules per cell and sides in the ratio 1.377:1:1.436. The angle β is 122°49′ and the density of the solid is 1.152 g cm^{-3}. Calculate the dimensions of the cell.

23.9 Predict from the Wierl equation (Box 11.2) the positions of the first maximum and first minimum in the neutron and electron diffraction patterns of a Br$_2$ molecule obtained with neutrons of wavelength 78 pm wavelength and electrons of wavelength 4.0 pm.

23.10‡ B.A. Bovenzi and G.A. Pearse, Jr. (*J. Chem. Soc. Dalton Trans.* 2793–8 (1997)) synthesized coordination compounds of the tridentate ligand pyridine-2,6-diamidoxime (C$_7$H$_9$N$_5$O$_2$). The compound, which they isolated from the reaction of the ligand with CuSO$_4$ (aq), did not contain a [Cu(C$_7$H$_9$N$_5$O$_2$)$_2$]$^{2+}$ complex cation as expected. Instead, X-ray diffraction analysis revealed a linear polymer of formula [Cu(Cu(C$_7$H$_9$N$_5$O$_2$)(SO$_4$))·2H$_2$O]$_n$, which features bridging sulfate groups. The unit cell was primitive monoclinic with $a = 1.0427$ nm, $b = 0.8876$ nm, $c = 1.3777$ nm, and $\beta = 93.254°$.

(a)

|←— 1 cm —→|

(b)

Fig. 23.60

† Problems denoted with the symbol ‡ were supplied by Charles Trapp and Carmen Giunta.

4

The mass density of the crystals is 2.024 g cm^{-3}. How many monomer units are there per unit cell?

23.11‡ D. Sellmann, M.W. Wemple, W. Donaubauer, and F.W. Heinemann (*Inorg. Chem.* **36**, 1397 (1997)) describe the synthesis and reactivity of the ruthenium nitrido compound $[N(C_4H_9)_4][Ru(N)(S_2C_6H_4)_2]$. The ruthenium complex anion has the two 1,2-benzenedithiolate ligands (**4**) at the base of a rectangular pyramid and the nitrido ligand at the apex. Compute the mass density of the compound given that it crystallizes into an orthorhombic unit cell with $a = 3.6881$ nm, $b = 0.9402$ nm, and $c = 1.7652$ nm and eight formula units per cell. Replacing the ruthenium with an osmium results in a compound with the same crystal structure and a unit cell with a volume less than 1 per cent larger. Estimate the mass density of the osmium analogue.

23.12 Aided by the Born–Mayer equation for the lattice enthalpy and a Born–Haber cycle, show that formation of CaCl is an exothermic process (the sublimation enthalpy of Ca(s) is 176 kJ mol^{-1}). Show that an explanation for the nonexistence of CaCl can be found in the reaction enthalpy for the reaction $2\,CaCl(s) \rightarrow Ca(s) + CaCl_2$.

23.13‡ J.J. Dannenberg, D. Liotard, P. Halvick, and J.C. Rayez (*J. Phys. Chem.* **100**, 9631 (1996)) carried out theoretical studies of organic molecules consisting of chains of unsaturated four-membered rings. The calculations suggest that such compounds have large numbers of unpaired spins, and that they should therefore have unusual magnetic properties. For example, the lowest-energy state of the five-ring compound $C_{22}H_{14}$ (**5**) is computed to have $S = 3$, but the energies of $S = 2$ and $S = 4$ structures are each predicted to be 50 kJ mol^{-1} higher in energy. Compute the molar magnetic susceptibility of these three low-lying levels at 298 K. Estimate the molar susceptibility at 298 K if each level is present in proportion to its Boltzmann factor (effectively assuming that the degeneracy is the same for all three of these levels).

5

Theoretical problems

23.14 Show that the separation of the (*hkl*) planes in an orthorhombic crystal with sides a, b, and c is given by eqn 3.

23.15 Show that the volume of a triclinic unit cell of sides a, b, and c and angles α, β, and γ is

$$V = abc(1 - \cos^2\alpha - \cos^2\beta - \cos^2\gamma + 2\cos\alpha\cos\beta\cos\gamma)^{1/2}$$

Use this expression to derive expressions for monoclinic and orthorhombic unit cells. For the derivation, it may be helpful to use the result from vector analysis that $V = \boldsymbol{a} \cdot \boldsymbol{b} \times \boldsymbol{c}$ and to calculate V^2 initially.

23.16 Calculate the packing fractions of (a) a primitive cubic lattice, (b) a bcc unit cell, (c) an fcc unit cell.

23.17 The coordinates of the four I atoms in the unit cell of KIO$_4$ are $(0, 0, 0)$, $(0, \frac{1}{2}, \frac{1}{2})$, $(\frac{1}{2}, \frac{1}{2}, \frac{1}{2})$, $(\frac{1}{2}, 0, \frac{3}{4})$. By calculating the phase of the I reflection in the structure factor, show that the I atoms contribute no net intensity to the (114) reflection.

23.18 The coordinates, in units of a, of the A atoms, with scattering factor f_A, in a cubic lattice are $(0, 0, 0)$, $(0, 1, 0)$, $(0, 0, 1)$, $(0, 1, 1)$, $(1, 0, 0)$, $(1, 1, 0)$, $(1, 0, 1)$, and $(1, 1, 1)$. There is also a B atom, with scattering factor f_B, at $(\frac{1}{2}, \frac{1}{2}, \frac{1}{2})$. Calculate the structure factors F_{hkl} and predict the form of the powder diffraction pattern when (a) $f_A = f$, $f_B = 0$, (b) $f_B = \frac{1}{2}f_A$, and (c) $f_A = f_B = f$.

23.19 Consider the thermodynamic description of stretching rubber. The observables are the tension, t, and length, l (the analogues of p and V for gases). Because $dw = t\,dl$, the basic equation is $dU = T\,dS + t\,dl$. (The term $p\,dV$ is supposed negligible throughout.) If $G = U - TS - tl$, find expressions for dG and dA, and deduce the Maxwell relations

$$\left(\frac{\partial S}{\partial l}\right)_T = -\left(\frac{\partial t}{\partial T}\right)_l \qquad \left(\frac{\partial S}{\partial t}\right)_T = -\left(\frac{\partial l}{\partial T}\right)_t$$

Go on to deduce the equation of state for rubber

$$\left(\frac{\partial U}{\partial l}\right)_T = t - \left(\frac{\partial t}{\partial T}\right)_l$$

23.20 On the assumption that the tension required to keep a sample at a constant length is proportional to the temperature ($t = aT$, the analogue of $p \propto T$), show that the tension can be ascribed to the dependence of the entropy on the length of the sample. Account for this result in terms of the molecular nature of the sample.

23.21 For an isotropic substance, the elastic moduli and Poisson's ratio may be expressed in terms of two parameters λ and μ called the *Lamé constants*:

$$E = \frac{\mu(3\lambda + 2\mu)}{\lambda + \mu} \qquad K = \frac{3\lambda + 2\mu}{3} \qquad G = \mu \qquad \nu_P = \frac{\lambda}{2(\lambda + \mu)}$$

Use the Lamé constants to confirm the relations between G, K, and E given in eqn 19.

23.22 When energy levels in a band form a continuum, the density of energy levels $\rho(E)$, the number of levels in an energy range divided by the width of the range, may be written as $\rho(E) = dk/dE$, where dk is the change in the quantum number k and dE is the energy change. (a) Use eqn 23 to show that

$$\rho(E) = -\frac{(N + 1)/2\pi\beta}{\left\{1 - \left(\dfrac{E - \alpha}{2\beta}\right)^2\right\}^{1/2}}$$

where k, N, α, and β have the meanings described in Section 23.9a. (b) Use the expression above to show that $\rho(E)$ becomes infinite E approaches $\alpha \pm 2\beta$. That is, show that the density of energy levels increases towards the edges of the bands.

23.23 The magnetizability, ξ, and the volume and molar magnetic susceptibilities can be calculated from the wavefunctions of molecules. For instance, the magnetizability of a hydrogenic atom is given by the expression $\xi = -(e^2/6m_e)\langle r^2\rangle$, where $\langle r^2\rangle$ is the (expectation) mean value of r^2 in the atom. Calculate ξ and χ_m for the ground state of a hydrogenic atom.

23.24 Nitrogen dioxide, a paramagnetic compound, is in equilibrium with its dimer, dinitrogen tetroxide, a diamagnetic compound. Derive an expression in terms of the equilibrium constant, K, for the dimerization to show how the molar susceptibility varies with the pressure of the sample. Suggest how the susceptibility might be expected to vary as the temperature is changed at constant pressure.

23.25 An NO molecule has thermally accessible electronically excited states. It also has an unpaired electron, and so may be expected to be paramagnetic. However, its ground state is not paramagnetic because the magnetic moment of the orbital motion of the unpaired electron almost exactly cancels the spin magnetic moment. The first excited state (at 121 cm^{-1}) is paramagnetic because the orbital magnetic moment adds to, rather than cancels, the spin magnetic moment. The upper state has a magnetic moment of $2\mu_B$. Because the upper state is thermally accessible, the paramagnetic susceptibility of NO shows a pronounced temperature dependence even near room temperature. Calculate the molar paramagnetic susceptibility of NO and plot it as a function of temperature.

Applications: to biology and materials science

23.26 Although the crystallization of large biological molecules may not be as readily accomplished as that of small molecules, their crystal lattices are no different. Tobacco seed globulin forms face-centred cubic crystals with unit cell dimension of 12.3 nm and a density of 1.287 g cm^{-3}. Determine its molar mass.

23.27 The structures of crystalline macromolecules may be determined by X-ray diffraction techniques using methods similar to those for smaller molecules. Fully crystalline polyethylene has its chains aligned in an orthorhombic unit cell of dimensions 740 pm × 493 pm × 253 pm. There are two repeating CH$_2$CH$_2$ units per unit cell. Calculate the theoretical density of fully crystalline polyethylene. The actual density ranges from 0.92 to 0.95 g cm^{-3}.

23.28 The following table lists the glass transition temperatures, T_g, of several polymers. Discuss the reasons why the structure of the monomer unit has an effect on the value of T_g.

Polymer	Poly(oxymethylene)	Polyethylene
Structure	—(OCH$_2$)$_n$—	—(CH$_2$CH$_2$)$_n$—
T_g/K	198	253

Polymer	Poly(vinyl chloride)	Polystyrene
Structure	—(CH$_2$—CHCl)$_n$—	—(CH$_2$—CH(C$_6$H$_5$))$_n$—
T_g/K	354	381

23.29 Calculate the coefficient of thermal expansion of diamond given that the (111) reflection shifts from 22° 2′ 25″ to 21° 57′ 59″ on heating a crystal from 100 K to 300 K and 154.0562 pm X-rays are used.

23.30 The carbon–carbon bond length in diamond is 154.45 pm. If diamond were considered to be a close-packed structure of hard spheres with radii equal to half the bond length, what would be its expected density? The diamond lattice is face-centred cubic and its actual density is 3.516 g cm^{-3}. Can you explain the discrepancy?

Part 3
Change

Part 3 considers the processes by which change occurs. We prepare the ground for a discussion of the rates of reactions by considering the motion of molecules in gases and in liquids. Then we establish the precise meaning of reaction rate, and see how the overall rate, and the complex behaviour of some reactions, may be expressed in terms of elementary steps and the atomic events that take place when molecules meet. Characteristic physical and chemical events take place at surfaces, including catalysis, and we see how to describe them. A special type of surface is that of an electrode, and we shall see how to describe and understand the rate at which electrons are transferred between an electrode and species in solution.

24 **Molecules in motion**

25 **The rates of chemical reactions**

26 **The kinetics of complex reactions**

27 **Molecular reaction dynamics**

28 **Processes at solid surfaces**

29 **Dynamics of electron transfer**

24

Molecules in motion

Molecular motion in gases

24.1 The kinetic model of gases

24.2 Collisions with walls and surfaces

24.3 The rate of effusion

24.4 Transport properties of a perfect gas

Molecular motion in liquids

24.5 Experimental results

24.6 The conductivities of electrolyte solutions

24.7 The mobilities of ions

24.8 Conductivities and ion–ion interactions

Diffusion

24.9 The thermodynamic view

24.10 The diffusion equation

24.11 Diffusion probabilities

24.12 The statistical view

Checklist of key ideas

Further reading

Exercises

Problems

One of the simplest types of molecular motion to describe is the random motion of molecules of a perfect gas. We see that the kinetic theory can be used to account for the pressure of a gas and the rates at which molecules and energy migrate through gases. Molecular mobility is particularly important in liquids. Another simple kind of motion is the largely uniform motion of ions in solution in the presence of an electric field. Molecular and ionic motion have common features and, by considering them from a more general viewpoint, we derive expressions that govern the migration of properties through matter. One of the most useful consequences of this general approach is the formulation of the diffusion equation, which is an equation that shows how matter and energy spread through media of various kinds. Finally, we build a simple model for all types of molecular motion, in which the molecules migrate in a series of small steps, and see that it accounts for many of the properties of migrating molecules in both gases and condensed phases.

The general approach we describe in this chapter provides techniques for discussing the motion of all kinds of particles in all kinds of fluids. We set the scene by considering a simple type of motion, that of molecules in a perfect gas, and go on to see that molecular motion in liquids shows a number of similarities. We shall concentrate on the **transport properties** of a substance, its ability to transfer matter, energy, or some other property from one place to another. Four examples of transport properties are

Diffusion, the migration of matter down a concentration gradient.

Thermal conduction, the migration of energy down a temperature gradient.

Electric conduction, the migration of electric charge along an electrical potential gradient.

Viscosity, the migration of linear momentum down a velocity gradient.

It is convenient to include in the discussion **effusion**, the emergence of a gas from a container through a small hole.

Molecular motion in gases

Here we present the kinetic model of a perfect gas as a starting point for the discussion of its transport properties. In the **kinetic model** of gases we assume that the only contribution to the energy of the gas is from the kinetic energies of the

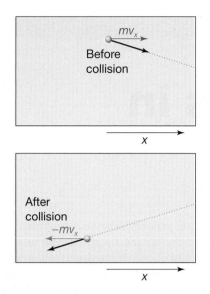

24.1 The pressure of a gas arises from the impact of its molecules on the walls. In an elastic collision of a molecule with a wall perpendicular to the x-axis, the x-component of velocity is reversed but the y- and z-components are unchanged.

molecules. The kinetic model is one of the most remarkable—and arguably most beautiful—models in physical chemistry for, from a set of very slender assumptions, powerful quantitative conclusions can be deduced.

24.1 The kinetic model of gases

The kinetic model is based on three assumptions:

   1 The gas consists of molecules of mass m in ceaseless random motion.

   2 The size of the molecules is negligible, in the sense that their diameters are much smaller than the average distance travelled between collisions.

   3 The molecules interact only through brief, infrequent, and elastic collisions.

An **elastic collision** is a collision in which the total translational kinetic energy of the molecules is conserved.

(a) Pressure and molecular speeds

From the very economical assumptions of the kinetic model, we show in the following *Justification* that the pressure and volume of the gas are related by

$$pV = \tfrac{1}{3}nMc^2 \tag{24.1}°$$

where $M = mN_A$, the molar mass of the molecules, and c is the **root mean square speed** of the molecules, the square root of the mean of the squares of the speeds, v, of the molecules:

$$c = \langle v^2 \rangle^{1/2} \tag{24.2}$$

Justification 24.1

Consider the arrangement in Fig. 24.1. When a particle of mass m that is travelling with a component of velocity v_x parallel to the x-axis collides with the wall on the right and is reflected, its linear momentum (the product of its mass and its velocity) changes from mv_x before the collision to $-mv_x$ after the collision (when it is travelling in the opposite direction). The x-component of momentum therefore changes by $2mv_x$ on each collision (the y- and z-components are unchanged). Many molecules collide with the wall in an interval Δt, and the total change of momentum is the product of the change in momentum of each molecule multiplied by the number of molecules that reach the wall during the interval.

Because a molecule with velocity component v_x can travel a distance $v_x\Delta t$ along the x-axis in an interval Δt, all the molecules within a distance $v_x\Delta t$ of the wall will strike it if they are travelling towards it (Fig. 24.2). It follows that, if the wall has area A, then all the particles in a volume $A \times v_x\Delta t$ will reach the wall (if they are travelling towards it). The number density of particles is nN_A/V, where n is the total amount of molecules in the container of volume V and N_A is Avogadro's constant, so the number of molecules in the volume $Av_x\Delta t$ is $(nN_A/V) \times Av_x\Delta t$.

At any instant, half the particles are moving to the right and half are moving to the left. Therefore, the average number of collisions with the wall during the interval Δt is $\tfrac{1}{2}nN_AAv_x\Delta t/V$. The total momentum change in that interval is the product of this number and the change $2mv_x$:

$$\text{Momentum change} = \frac{nN_A Av_x\Delta t}{2V} \times 2mv_x = \frac{nmAN_A v_x^2 \Delta t}{V} = \frac{nMAv_x^2\Delta t}{V}$$

where $M = mN_A$.

24.2 A molecule will reach the wall on the right within an interval Δt if it is within a distance $v_x\Delta t$ of the wall and travelling to the right.

Next, to find the force, we calculate the rate of change of momentum, which is this change of momentum divided by the interval Δt during which it occurs:

$$\text{Rate of change of momentum} = \frac{nMAv_x^2}{V}$$

This rate of change of momentum is equal to the force (by Newton's second law of motion). It follows that the pressure, the force divided by the area, is

$$\text{Pressure} = \frac{nMv_x^2}{V}$$

Not all the molecules travel with the same velocity, so the detected pressure, p, is the average (denoted $\langle \ldots \rangle$) of the quantity just calculated:

$$p = \frac{nM\langle v_x^2 \rangle}{V}$$

This expression already resembles the perfect gas equation of state.

Because the molecules are moving randomly and there is no net flow in a particular direction, the average speed along x is the same as that in the y- and z-directions. It follows that

$$c = (\langle v_x^2 \rangle + \langle v_y^2 \rangle + \langle v_z^2 \rangle)^{1/2} = (3\langle v_x^2 \rangle)^{1/2}$$

implying that $\langle v_x^2 \rangle = \frac{1}{3}c^2$. Equation 1 now follows immediately.

Equation 1 is one of the key results of the kinetic model. We see that, if the root mean square speed of the molecules depends only on the temperature, then at constant temperature

$$pV = \text{constant}$$

which is the content of Boyle's law. Moreover, for eqn 1 to be the equation of state of a perfect gas, its right-hand side must be equal to nRT. It follows that the root mean square speed of the molecules in a gas at a temperature T must be

$$c = \left(\frac{3RT}{M} \right)^{1/2} \tag{24.3}°$$

We can conclude that *the root mean square speed of the molecules of a gas is proportional to the square root of the temperature and inversely proportional to the square root of the molar mass.* That is, the higher the temperature, the higher the root mean square speed of the molecules, and at a given temperature, heavy molecules travel more slowly than light molecules. Sound waves are pressure waves, and for them to propagate the molecules of the gas must move to form regions of high and low pressure. Therefore, we should expect the root mean square speeds of molecules to be comparable to the speed of sound in air (340 m s^{-1}). The root mean square speed of N_2 molecules, for instance, is found from eqn 3 to be 515 m s^{-1}.

Equation 3 is an expression for the mean square speed of molecules. However, in an actual gas the speeds of individual molecules span a wide range, and the collisions in the gas continually redistribute the speeds among the molecules. Before a collision, a molecule may be travelling rapidly, but after a collision it may be accelerated to a very high speed, only to be slowed again by the next collision. The fraction of molecules that have speeds in the range v to $v + dv$ is proportional to the width of the range, and is written $f(v) \, dv$, where $f(v)$ is called the **distribution of speeds**.

24.3 The distribution of molecular speeds with temperature and molar mass. Note that the most probable speed (corresponding to the peak of the distribution) increases with temperature and with decreasing molar mass, and simultaneously the distribution becomes broader.

For any $a, x, y, z, \ldots$ we may write $a^{x+y+z+\cdots} = a^x a^y a^z \ldots$.

The precise form of f for molecules of a gas at a temperature T was derived by J.C. Maxwell, and is

$$f(v) = 4\pi \left(\frac{M}{2\pi RT} \right)^{3/2} v^2 e^{-Mv^2/2RT} \tag{24.4}$$

This expression is called the **Maxwell distribution of speeds**, and is derived in the following *Justification*. Figure 24.3 summarizes its main features. We see that the range of speeds broadens as the temperature increases. Lighter molecules also have a broader distribution of speeds than heavy molecules. To use eqn 4 to calculate the fraction of molecules in a given narrow range of speeds, Δv, we evaluate $f(v)$ at the speed of interest, then multiply it by the width of the range of speeds of interest; that is, we form $f(v)\Delta v$. To use the distribution to calculate the fraction in a range of speeds that is too wide to be treated as infinitesimal, we evaluate the integral:

$$\text{Fraction in the range } v_1 \text{ to } v_2 = \int_{v_1}^{v_2} f(v)\,dv \tag{24.5}$$

This integral is the area under the graph of f as a function of v and, except in special cases, has to be evaluated numerically by using mathematical software (Fig. 24.4).

Justification 24.2

The Boltzmann distribution is a key result of physical chemistry and is treated fully in Section 19.1. It implies that the fraction of molecules with velocity components v_x, v_y, v_z is proportional to an exponential function of their kinetic energy, which is

$$E = \tfrac{1}{2}mv_x^2 + \tfrac{1}{2}mv_y^2 + \tfrac{1}{2}mv_z^2$$

Therefore, we can write

$$f = Ke^{-E/kT} = Ke^{-(\frac{1}{2}mv_x^2 + \frac{1}{2}mv_y^2 + \frac{1}{2}mv_z^2)/kT} = Ke^{-mv_x^2/2kT}e^{-mv_y^2/2kT}e^{-mv_z^2/2kT}$$

where K is a constant of proportionality and $f\,dv_x\,dv_y\,dv_z$ is the fraction of molecules in the velocity range v_x to $v_x + dv_x$, v_y to $v_y + dv_y$, and v_z to $v_z + dv_z$. We see that the fraction factorizes into three factors, one for each axis, and we can write $f = f(v_x)f(v_y)f(v_z)$ with

$$f(v_x) = K^{1/3}e^{-mv_x^2/kT}$$

and likewise for the two other directions.

To determine the constant K, we note that a molecule must have a velocity somewhere in the range $-\infty < v_x < \infty$, so

$$\int_{-\infty}^{\infty} f(v_x)\,dv_x = 1$$

Substitution of the expression for $f(v_x)$ then gives

$$1 = K^{1/3} \int_{-\infty}^{\infty} e^{-mv_x^2/2kT}\,dv_x = K^{1/3}\left(\frac{2\pi kT}{m} \right)^{1/2}$$

Therefore, $K = (m/2\pi kT)^{3/2} = (M/2\pi RT)^{3/2}$, where M is the molar mass of the molecules. At this stage we know that

$$f(v_x) = \left(\frac{M}{2\pi RT} \right)^{1/2} e^{-Mv_x^2/2RT} \tag{24.6}$$

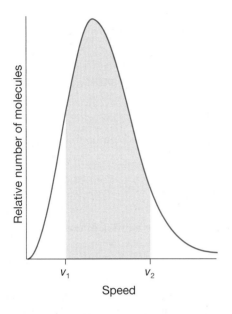

24.4 To calculate the probability that a molecule will have a speed in the range v_1 to v_2, we integrate the distribution between those two limits; the integral is equal to the area of the curve between the limits, as shown shaded here.

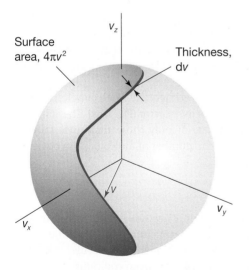

24.5 To evaluate the probability that a molecule has a speed in the range v to $v + dv$, we evaluate the total probability that the molecule will have a speed that is anywhere in a spherical shell of radius $v = (v_x^2 + v_y^2 + v_z^2)^{1/2}$ and thickness dv by summing the probabilities that it is in a volume element $dv_x\,dv_y\,dv_z$ at a distance v from the origin.

The probability that a molecule has a velocity in the range v_x to $v_x + dv_x$, v_y to $v_y + dv_y$, v_z to $v_z + dv_z$ is

$$f(v_x)f(v_y)f(v_z)\,dv_x\,dv_y\,dv_z = \left(\frac{M}{2\pi RT}\right)^{3/2} e^{-Mv^2/2RT}\,dv_x\,dv_y\,dv_z$$

where $v^2 = v_x^2 + v_y^2 + v_z^2$. The probability $f(v)\,dv$ that the molecules have a speed in the range v to $v + dv$ regardless of direction is the sum of the probabilities that the velocity lies in any of the volume elements $dv_x\,dv_y\,dv_z$ forming a spherical shell of radius v and thickness dv (Fig. 24.5). The sum of the volume elements on the right-hand side of the last equation is the volume of this shell, $4\pi v^2\,dv$. Therefore,

$$f(v) = 4\pi\left(\frac{M}{2\pi RT}\right)^{3/2} v^2 e^{-Mv^2/2RT}$$

as given in eqn 4.

Example 24.1 Calculating the mean speed of molecules in a gas

What is the mean speed, $\bar{c}$, of N_2 molecules in air at 25°C?

Method We are asked to calculate the mean speed, not the root mean square speed. A mean speed is calculated by multiplying each speed by the fraction of molecules that have that speed, and then adding all the products together. When the speed varies over a continuous range, the sum is replaced by an integral. To employ this approach here, we note that the fraction of molecules with a speed in the range v to $v + dv$ is $f(v)\,dv$, so the product of this fraction and the speed is $vf(v)\,dv$. The mean speed, $\bar{c}$, is obtained by evaluating the integral

$$\bar{c} = \int_0^\infty vf(v)\,dv$$

with f given in eqn 4.

Answer The integral required is

$$\bar{c} = 4\pi\left(\frac{M}{2\pi RT}\right)^{3/2}\int_0^\infty v^3 e^{-Mv^2/2RT}\,dv$$

$$= 4\pi\left(\frac{M}{2\pi RT}\right)^{3/2}\times\frac{1}{2}\left(\frac{2RT}{M}\right)^2 = \left(\frac{8RT}{\pi M}\right)^{1/2}$$

Substitution of the data then gives

$$\bar{c} = \left(\frac{8\times(8.3141\text{ J K}^{-1}\text{ mol}^{-1})\times(298\text{ K})}{\pi\times(28.02\times10^{-3}\text{ kg mol}^{-1})}\right)^{1/2} = 475\text{ m s}^{-1}$$

where we have used $1\text{ J} = 1\text{ kg m}^2\text{ s}^{-2}$. To evaluate the integral we have used the standard result from tables of integrals (or software) that

$$\int_0^\infty x^3 e^{-ax^2}\,dx = \frac{1}{2a^2}$$

Self-test 24.1 Evaluate the root mean square speed of the molecules by integration. You will need the integral

$$\int_0^\infty x^4 e^{-ax^2}\, dx = \frac{3}{8}\left(\frac{\pi}{a^5}\right)^{1/2}$$

$$[c = (3RT/M)^{1/2}, 515 \text{ m s}^{-1}]$$

As shown in Example 24.1, we can use the Maxwell distribution to evaluate the **mean speed**, $\bar{c}$, of the molecules in a gas:

$$\bar{c} = \left(\frac{8RT}{\pi M}\right)^{1/2} \tag{24.7}$$

We can identify the **most probable speed**, c^*, from the location of the peak of the distribution:

$$c^* = \left(\frac{2RT}{M}\right)^{1/2} \tag{24.8}$$

Figure 24.6 summarizes these results.

The **relative mean speed**, $\bar{c}_{rel}$, the mean speed with which one molecule approaches another, can also be calculated from the distribution:

$$\bar{c}_{rel} = 2^{1/2}\bar{c} \tag{24.9}$$

This result is much harder to derive, but the diagram in Fig. 24.7 should help to show that it is plausible. The last result can also be generalized to the relative mean speed of two dissimilar molecules of masses m_A and m_B:

$$\bar{c}_{rel} = \left(\frac{8kT}{\pi\mu}\right)^{1/2} \qquad \mu = \frac{m_A m_B}{m_A + m_B} \tag{24.10}$$

Note that the molecular masses (not the molar masses) and Boltzmann's constant, $k = R/N_A$, appear in this expression; the quantity μ is called the **reduced mass** of the molecules. Equation 10 turns into eqn 7 when the molecules are identical (that is, $m_A = m_B = m$, so $\mu = \frac{1}{2}m$).

The Maxwell distribution has been verified experimentally. For example, molecular speeds can be measured directly with a velocity selector (Fig. 24.8). The spinning cylinder has channels that permit the passage of only those molecules moving through them at the appropriate speed, and the number of molecules can be determined by collecting them at a detector.

(b) The collision frequency

A qualitative picture of the events taking place in a gas was first described in Section 1.1. The kinetic model enables us to make that picture more quantitative. In particular, it enables us to calculate the frequency with which molecular collisions occur and the distance a molecule travels on average between collisions.

We count a 'hit' whenever the centres of two molecules come within a distance d of each other, where d, the **collision diameter**, is of the order of the actual diameters of the molecules (for impenetrable hard spheres d is the diameter). As we show in the following *Justification*, we can use kinetic theory to deduce that the **collision frequency**, z, the number of collisions made by one molecule divided by the time interval during which the collisions are counted, when there are N molecules in a volume V is

$$z = \sigma\bar{c}_{rel}\mathcal{N} \tag{24.11}°$$

To find the location of this peak, differentiate f with respect to v and look for the value of v at which the derivative is zero (other than at $v = 0$ and $v = \infty$).

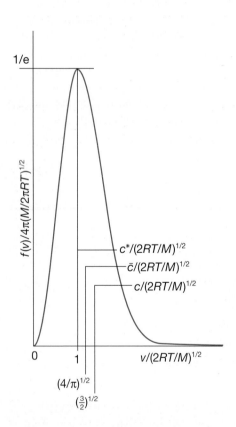

24.6 A summary of the conclusions that can be deduced from the Maxwell distribution for molecules of molar mass M at a temperature T: c^* is the most probable speed, $\bar{c}$ is the mean speed, and c is the root mean square speed.

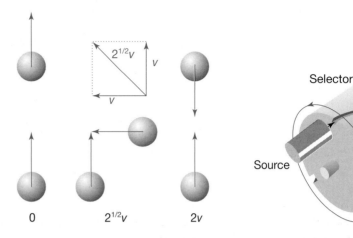

24.7 A simplified version of the argument to show that the mean relative speed of molecules in a gas is related to the mean speed. When the molecules are moving in the same direction, the mean relative speed is zero; it is $2v$ when the molecules are approaching each other. A typical mean direction of approach is from the side, and the mean speed of approach is then $2^{1/2}v$. The last direction of approach is the most characteristic, so the mean speed of approach can be expected to be about $2^{1/2}v$. This value is confirmed by more detailed calculation.

24.8 A velocity selector. The molecules are produced in the source (which may be an oven with a small hole in one wall), and travel in a beam towards the rotating channels. Only if the speed of a molecule is such as to carry it along the channel that rotates into its path will it reach the detector. Thus, the number of slow molecules can be counted by rotating the discs slowly, and the number of fast molecules counted by rotating the discs rapidly.

Table 24.1* Collision cross-sections

	σ/nm^2
Benzene, C_6H_6	0.88
Carbon dioxide, CO_2	0.52
Helium, He	0.21
Nitrogen, N_2	0.43

* More values are given in the *Data section*.

with $\mathcal{N} = N/V$ and $\bar{c}_{rel}$ given in eqn 10. The area $\sigma = \pi d^2$ is called the **collision cross-section** of the molecules. Some typical collision cross-sections are given in Table 24.1 (they are obtained by the techniques described in Section 21.7). In terms of the pressure,

$$z = \frac{\sigma \bar{c}_{rel} p}{kT} \qquad (24.12)°$$

Justification 24.3

We consider the positions of all the molecules except one to be frozen. Then we note what happens as one mobile molecule travels through the gas with a mean relative speed $\bar{c}_{rel}$ for a time Δt. In doing so it sweeps out a 'collision tube' of cross-sectional area $\sigma = \pi d^2$ and length $\bar{c}_{rel}\Delta t$, and therefore of volume $\sigma \bar{c}_{rel}\Delta t$ (Fig. 24.9). The number

24.9 In an interval Δt, a molecule of diameter d sweeps out a tube of radius d and length $\bar{c}_{rel}\Delta t$. As it does so it encounters other molecules with centres that lie within the tube, and each such encounter counts as one collision. In reality, the tube is not straight, but changes direction at each collision. Nevertheless, the volume swept out is the same, and this straightened version of the tube can be used as a basis of the calculation.

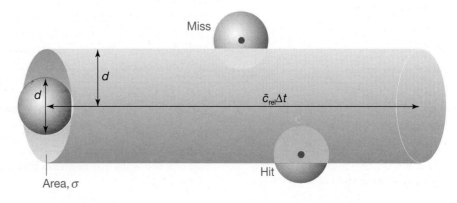

of stationary molecules with centres inside the collision tube is given by the volume of the tube multiplied by the number density $\mathcal{N} = N/V$, and is $\mathcal{N}\sigma\bar{c}_{rel}\Delta t$. The number of hits scored in the interval Δt is equal to this number, so the number of collisions divided by the time interval, is $\mathcal{N}\sigma\bar{c}_{rel}$. The expression in terms of the pressure of the gas is obtained by using the perfect gas equation to write

$$\frac{N}{V} = \frac{nN_A}{V} = \frac{pN_A}{RT} = \frac{p}{kT}$$

Equation 11 shows that, at constant volume, the collision frequency increases with increasing temperature. The reason for this increase is that the mean relative speed increases with temperature. Equation 12 shows that, at constant temperature, the collision frequency is proportional to the pressure. Such a proportionality is plausible, for the greater the pressure, the greater the number density of molecules in the sample, and the rate at which they encounter one another is greater even though their average speed remains the same. For an N_2 molecule in a sample at 1 atm and 25°C, $z \approx 5 \times 10^9$ s^{-1}, so a given molecule collides about 5×10^9 times each second. We are beginning to appreciate the timescale of events in gases.

(c) The mean free path

Once we have the collision frequency, we can calculate the **mean free path**, λ (lambda), the average distance a molecule travels between collisions. If a molecule collides with a frequency z, it spends a time $1/z$ in free flight between collisions, and therefore travels a distance $(1/z)\bar{c}$. Therefore, the mean free path is

$$\lambda = \frac{\bar{c}}{z} \tag{24.13}$$

Substitution of the expression for z in eqn 12 gives

$$\lambda = \frac{kT}{2^{1/2}\sigma p} \tag{24.14}$$

Doubling the pressure reduces the mean free path by half. A typical mean free path in nitrogen gas at 1 atm is 70 nm, or about 10^3 molecular diameters. Although the temperature appears in eqn 14, in a sample of constant volume, the pressure is proportional to T, so T/p remains constant when the temperature is increased. Therefore, the mean free path is independent of the temperature in a sample of gas in a container of fixed volume. The distance between collisions is determined by the number of molecules present in the given volume, not by the speed at which they travel.

In summary, a typical gas (N_2 or O_2) at 1 atm and 25°C can be thought of as a collection of molecules travelling with a mean speed of about 500 m s^{-1}. Each molecule makes a collision within about 1 ns, and between collisions it travels about 10^3 molecular diameters. The kinetic model of gases is valid (and the gas behaves nearly perfectly) if the diameter of the molecules is much smaller than the mean free path ($d \ll \lambda$), for then the molecules spend most of their time far from one another (Box 24.1).

24.2 Collisions with walls and surfaces

The key result for accounting for transport in the gas phase is the rate at which molecules strike an area (which may be an imaginary area embedded in the gas, or part of a real wall). The **collision flux**, Z_W, is the number of collisions with the area

Box 24.1 The Sun as a ball of perfect gas

The kinetic theory of gases is valid when the size of the particles is negligible compared with their mean free path. It may seem absurd, therefore, to expect the kinetic theory and, as a consequence, the perfect gas law, to be applicable to the dense matter of stellar interiors. In the Sun, for instance, the mean density is about 1.50 times that of liquid water at its centre and comparable to that of water about half way to its surface. However, we have to realize that the state of matter is that of a *plasma*, in which the electrons have been stripped from the atoms of hydrogen and helium that make up the bulk of the matter of stars. As a result, the particles making up the plasma have diameters comparable to those of nuclei, or about 10 fm. Therefore, a mean free path of only 0.1 pm satisfies the criterion for the validity of the kinetic theory and the perfect gas law. We can therefore use $pV = nRT$ as the equation of state for the stellar interior.

As for any perfect gas, the pressure in the interior of the Sun is related to the mass density, $\rho = m/V$, by $p = \rho RT/M$. Atoms are stripped of their electrons in the interior of stars so, if we suppose that the interior consists of ionized hydrogen atoms, the mean molar mass is one-half the molar mass of hydrogen, or 0.5 g mol^{-1} (the mean of the molar mass of H^+ and e^-, the latter being virtually 0). Half way to the centre of the Sun, the temperature is 3.6 MK and the mass density is 1.20 g cm^{-3} (slightly denser than water); so the pressure there works out as 7.2×10^{13} Pa, or about 720 million atmospheres.

We can combine this result with the expression for the pressure from kinetic theory ($p = \frac{1}{3}NMc^2/V$). Because the total kinetic energy of the particles is $E_K = \frac{1}{2}Nmc^2$, we can write $p = \frac{2}{3}E_K/V$. That is, the pressure of the plasma is related to the *kinetic energy density*, $\rho_K = E_K/V$, the kinetic energy of the molecules in a region divided by the volume of the region, by $p = \frac{2}{3}\rho_K$. It follows that the kinetic energy density half way to

the centre of the Sun is about 0.11 GJ cm^{-3}. In contrast, on a warm day (25°C) on Earth, the (translational) kinetic energy density of our atmosphere is only 0.15 J cm^{-3}.

Discussion questions

1. The principal components of the atmosphere of the Earth are diatomic molecules, which can rotate as well as translate. Given that the translational kinetic energy density of the atmosphere is 0.15 J cm^{-3}, what is the total kinetic energy density, including rotation?

2.[1] In the *standard model* of stellar structure (I. Nicholson, *The sun*; Rand McNally, New York (1982)), the interior of the Sun is thought to consist of 36 per cent H and 64 per cent He by mass, at a density of 158 g cm^{-3} very close to the centre. Both atoms are completely ionized. The approximate dimensions of the nuclei can be calculated from the formula $r_{nucleus} = 1.4 \times 10^{-15} A^{1/3}$ m, where A is the mass number. The size of the free electron, $r_e \approx 10^{-18}$ m, is negligible compared to the size of the nuclei. (a) Calculate the excluded volume in 1.0 cm^3 of the stellar interior and on that basis decide upon the applicability of the perfect gas law to this system. (b) The standard model suggests that the pressure in the stellar interior is 2.5×10^{11} atm. Calculate the temperature of the Sun's interior based on the perfect gas model. The generally accepted standard model value is 16 MK. (c) Would a van der Waals type of equation (with $a = 0$) give a better value for T?

Further reading

D.B. Clark, The ideal gas law at the center of the sum. *J. Chem. Educ.* **66**, 826 (1989).

C.R. Cowley, *An introduction to cosmochemistry*. Cambridge University Press (1995).

1. Problem supplied by Carmen Giunta and Charles Trapp.

---

in a given time interval divided by the area and the duration of the interval. The **collision frequency**, the number of hits per second, is obtained by multiplication of the collision flux by the area of interest. We show in the *Justification* below that the collision flux is

$$Z_W = \frac{p}{(2\pi mkT)^{1/2}} \tag{24.15}°$$

When $p = 100$ kPa (1.00 bar) and $T = 300$ K, $Z_W \approx 3 \times 10^{23}$ cm^{-2} s^{-1}.

Justification 24.4

Consider a wall of area A perpendicular to the x-axis (as in Fig. 24.2). If a molecule has $v_x > 0$ (that is, it is travelling in the direction of positive x), then it will strike the wall within an interval Δt if it lies within a distance $v_x \Delta t$ of the wall. Therefore, all molecules in the volume $Av_x \Delta t$, and with positive x-component of velocities, will strike the wall in the interval Δt. The total number of collisions in this interval is therefore the volume $Av_x \Delta t$ multiplied by the number density, $\mathcal{N}$, of molecules. However, to take account of the presence of a range of velocities in the sample, we must sum the result over all the positive values of v_x weighted by the probability distribution of velocities (eqn 6):

$$\text{Number of collisions} = \mathcal{N}A\Delta t \int_0^\infty v_x f(v_x)\, dx$$

The collision flux is the number of collisions divided by A and Δt, so

$$Z_W = \mathcal{N} \int_0^\infty v_x f(v_x)\, dx$$

Then, using the velocity distribution in eqn 6,

$$\int_0^\infty v_x f(v_x)\, dv_x = \left(\frac{m}{2\pi kT}\right)^{1/2} \int_0^\infty v_x e^{-mv_x^2/2kT}\, dv_x$$

$$= \left(\frac{kT}{2\pi m}\right)^{1/2}$$

Therefore,

$$Z_W = \mathcal{N}\left(\frac{kT}{2\pi m}\right)^{1/2} = \tfrac{1}{4}\bar{c}\mathcal{N} \tag{24.16}°$$

Substitution of $\mathcal{N} = nN_A/V = p/kT$ gives eqn 15.

24.3 The rate of effusion

The essential empirical observations on effusion are summarized by **Graham's law of effusion**, which states that the rate of effusion is inversely proportional to the square root of the molar mass. The basis of this result is that, as remarked above, the mean speed of molecules is inversely proportional to $M^{1/2}$, so the rate at which they strike the area of the hole is also inversely proportional to $M^{1/2}$. However, by using the expression for the rate of collisions, we can obtain a more detailed expression for the rate of effusion and hence use effusion data more effectively.

When a gas at a pressure p and temperature T is separated from a vacuum by a small hole, the rate of escape of its molecules is equal to the rate at which they strike the area of the hole (which is given by eqn 15). Therefore, for a hole of area A_0,

$$\text{Rate of effusion} = Z_W A_0 = \frac{pA_0}{(2\pi mkT)^{1/2}} = \frac{pA_0 N_A}{(2\pi MRT)^{1/2}} \tag{24.17}°$$

(In the last step we have used $R = N_A k$ and $M = mN_A$.) This rate is inversely proportional to $M^{1/2}$, in accord with Graham's law.

Example 24.2 Deducing the time dependence of the pressure inside an effusion oven

Derive an expression that shows how the pressure of a gas inside an effusion oven (a heated chamber with a small hole in one wall) varies with time if the oven is not replenished as the gas escapes.

Method The rate of effusion is proportional to the pressure of the gas in the container so, as gas effuses and the pressure falls, the rate of effusion will decrease. To find the explicit expression, set up a differential equation relating dp/dt to p, and then integrate it. The rate of effusion, as given by eqn 17, is the number of molecules that leave the container in a given interval divided by the duration of the interval. The first step is to relate the rate of change of pressure to the rate of change of number of molecules by using the perfect gas law in the form $pV = NkT$.

Differential equations are reviewed in *Further information 1*.

Answer The rate of change of pressure of a gas in a container at constant pressure and temperature is related to the rate of change of the number of molecules present by

$$\frac{dp}{dt} = \frac{kT}{V}\frac{dN}{dt}$$

The rate of change of the number of molecules is equal to the collision frequency with the hole, and that in turn is equal to the collision flux multiplied by the area of the hole:

$$\frac{dN}{dt} = -Z_W A_0 = -\frac{pA_0}{(2\pi mkT)^{1/2}}$$

Substitution of this expression into the one above gives

$$\frac{dp}{dt} = -\left(\frac{kT}{2\pi m}\right)^{1/2}\frac{pA_0}{V}$$

This expression integrates to

$$p = p_0 e^{-t/\tau} \qquad \tau = \left(\frac{2\pi m}{kT}\right)^{1/2}\frac{V}{A_0}$$

Comment The pressure falls exponentially towards zero; the decrease is faster the higher the temperature, the bigger the hole, and the lower the mass of the molecules.

Self-test 24.2 Show that $t_{1/2}$, the time required for the pressure to decrease to half its initial value, is independent of the initial pressure. $[t_{1/2} = \tau \ln 2]$

Equation 17 is the basis of the **Knudsen method** for the determination of the vapour pressures of liquids and solids, particularly of substances with very low vapour pressures. Thus, if the vapour pressure of a sample is p, and it is enclosed in a cavity with a small hole, then the rate of loss of mass from the container is proportional to p.

Example 24.3 Calculating the vapour pressure from a mass loss

Caesium (m.p. 29°C, b.p. 686°C) was introduced into a container and heated to 500°C. When a hole of diameter 0.50 mm was opened in the container for 100 s, a mass loss of 385 mg was measured. Calculate the vapour pressure of liquid caesium at 500 K.

Method The pressure of vapour is constant inside the container despite the effusion of atoms because the hot liquid metal replenishes the vapour. The rate of effusion is therefore constant, and given by eqn 17. To express the rate in terms of mass, multiply the number of atoms that escape by the mass of each atom.

Answer The mass loss Δm in an interval Δt is related to the collision flux by

$$\Delta m = Z_W A_0 m \Delta t$$

where A_0 is the area of the hole and m is the mass of one atom. It follows that

$$Z_W = \frac{\Delta m}{A_0 m \Delta t}$$

Because Z_W is related to the pressure by eqn 15, we can write

$$p = \left(\frac{2\pi RT}{M}\right)^{1/2} \frac{\Delta m}{A_0 \Delta t}$$

Because $M = 132.9 \text{ g mol}^{-1}$, substitution of the data gives $p = 11$ kPa (using $1 \text{ Pa} = 1 \text{ N m}^{-2} = 1 \text{ J m}^{-3}$), or 83 Torr.

Self-test 24.3 How long would it take 1.0 g of Cs atoms to effuse out of the oven under the same conditions?   [260 s]

24.4 Transport properties of a perfect gas

Transport properties are commonly expressed in terms of a number of 'phenomenological' equations, or equations that are empirical summaries of experimental observations. These phenomenological equations apply to all kinds of properties and media. In the following sections, we introduce the equations for the general case and then show how to calculate the parameters that appear in these phenomenological expressions.

(a) The phenomenological equations

The rate of migration of a property is measured by its **flux**, J, the quantity of that property passing through a given area in a given time interval divided by the area and the duration of the interval. If matter is flowing (as in diffusion), we speak of a **matter flux** of so many molecules per square metre per second; if the property is energy (as in thermal conduction), then we speak of the **energy flux** and express it in joules per square metre per second, and so on.

Experimental observations on transport properties show that the flux of a property is usually proportional to the first derivative of some other related property. For example, the flux of matter diffusing parallel to the z-axis of a container is found to be proportional to the first derivative of the concentration:

$$J(\text{matter}) \propto \frac{d\mathcal{N}}{dz} \tag{24.18}$$

where $\mathcal{N}$ is the number density of particles with units number per metre cubed (m^{-3}). The SI units of J are number per metre squared per second ($m^{-2}\ s^{-1}$). The proportionality of the flux of matter to the concentration gradient is sometimes called **Fick's first law of diffusion**: the law implies that, if the concentration varies steeply with position, then diffusion will be fast. There is no net flux if the concentration is uniform ($d\mathcal{N}/dz = 0$). Similarly, the rate of thermal conduction (the flux of the energy associated with thermal motion) is found to be proportional to the temperature gradient:

$$J(\text{energy}) \propto \frac{dT}{dz} \tag{24.19}$$

The SI units of this flux are joules per metre squared per second ($J\ m^{-2}\ s^{-1}$).

A positive value of J signifies a flux towards positive z; a negative value of J signifies a flux towards negative z. Because matter flows down a concentration gradient, from high concentration to low concentration, J is positive if $d\mathcal{N}/dz$ is negative (Fig. 24.10). Therefore, the coefficient of proportionality in eqn 19 must be negative, and we write it $-D$:

$$J(\text{matter}) = -D\frac{d\mathcal{N}}{dz} \tag{24.20}$$

The constant D is called the **diffusion coefficient**; its SI units are metre squared per second ($m^2\ s^{-1}$). Energy migrates down a temperature gradient, and the same reasoning leads to

$$J(\text{energy}) = -\kappa\frac{dT}{dz} \tag{24.21}$$

where κ is the **coefficient of thermal conductivity**. The SI units of κ are joules per kelvin per metre per second ($J\ K^{-1}\ m^{-1}\ s^{-1}$). Some experimental values are given in Table 24.2.

To see the connection between the flux of momentum and the viscosity, consider a fluid in a state of **Newtonian flow**, which can be imagined as occurring by a series of layers moving past one another (Fig. 24.11). The layer next to the wall of the vessel is stationary, and the velocity of successive layers varies linearly with distance, z, from the wall. Molecules ceaselessly move between the layers and bring with them the x-component of linear momentum they possessed in their original layer. A layer is retarded by molecules arriving from a more slowly moving layer because they have a low momentum in the x-direction. A layer is accelerated by molecules arriving from a more rapidly moving layer. We interpret the net retarding effect as the fluid's viscosity.

Because the retarding effect depends on the transfer of the x-component of linear momentum into the layer of interest, the viscosity depends on the flux of this x-component in the z-direction. The flux of the x-component of momentum is proportional to dv_x/dz because there is no net flux when all the layers move at the same velocity. We can therefore write

$$J(x\text{-component of momentum}) = -\eta\frac{dv_x}{dz} \tag{24.22}$$

The constant of proportionality, η, is the **coefficient of viscosity** (or simply 'the viscosity'). Its units are kilogram per metre per second ($kg\ m^{-1}\ s^{-1}$). Viscosities are often reported in poise (P), where $1\ P = 10^{-1}\ kg\ m^{-1}\ s^{-1}$. Some experimental values are given in Table 24.2.

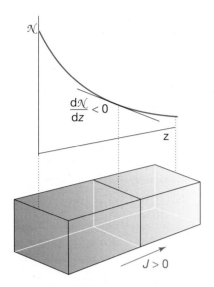

24.10 The flux of particles down a concentration gradient. Fick's first law states that the flux of matter (the number of particles passing through an imaginary window in a given interval divided by the area of the window and the length of the interval) is proportional to the density gradient at that point.

Table 24.2* Transport properties of gases at 1 atm

	$\kappa/(J\ K^{-1}\ m^{-1}\ s^{-1})$	$\eta/\mu P\dagger$	
	273 K	273 K	293 K
Ar	0.0163	210	223
CO_2	0.0145	136	147
He	0.1442	187	196
N_2	0.0240	166	176

* More values are given in the *Data section*.
† $\mu P = 10^{-7}\ kg\ m^{-1}\ s^{-1}$.

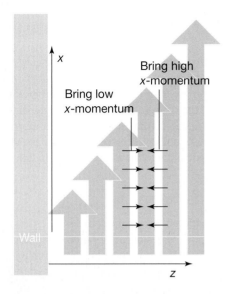

24.11 The viscosity of a fluid arises from the transport of linear momentum. In this illustration the fluid is undergoing laminar flow, and particles bring their initial momentum when they enter a new layer. If they arrive with high x-component of momentum they accelerate the layer; if they arrive with low x-component of momentum they retard the layer.

Table 24.3 Transport properties of perfect gases

Property	Transported quantity	Simple kinetic theory	Units
Diffusion	Matter	$D = \frac{1}{3}\lambda\bar{c}$	$m^2\,s^{-1}$
Thermal conductivity	Energy	$\kappa = \frac{1}{3}\lambda\bar{c}C_{V,m}[A]$ $= \dfrac{\bar{c}C_{V,m}}{(3\sqrt{2})\sigma N_A}$	$J\,K^{-1}\,m^{-1}\,s^{-1}$
Viscosity	Momentum	$\eta = \frac{1}{3}\lambda\bar{c}m\mathcal{N}$ $= \dfrac{m\bar{c}}{(3\sqrt{2})\sigma}$	$kg\,m^{-1}\,s^{-1}$

(b) Diffusion

We shall now see how the kinetic theory is used to justify Fick's law, and deduce the values of the transport coefficients of a perfect gas. As shown in the *Justification* below and summarized in Table 24.3, the kinetic theory leads to the result that, for a perfect gas,

$$D = \frac{1}{3}\lambda\bar{c} \tag{24.23}°$$

The mean free path, λ, decreases as the pressure is increased (Section 1.3b), so D decreases with increasing pressure and, as a result, the gas molecules diffuse more slowly. The mean speed, $\bar{c}$, increases with the temperature (Section 24.1a), so D also increases with temperature. As a result, molecules in a hot sample diffuse more quickly than those in a cool sample (for a given concentration gradient). Because the mean free path increases when the collision cross-section of the molecules decreases, the diffusion coefficient is greater for small molecules than for large molecules.

Justification 24.5

Consider the arrangement depicted in Fig. 24.12. On average, the molecules passing through the area A at $z = 0$ have travelled about one mean free path λ since their last collision. Therefore, the number density where they originated is $\mathcal{N}(z)$ evaluated at $z = -\lambda$. This number density is approximately

$$\mathcal{N}(-\lambda) = \mathcal{N}(0) - \lambda\left(\frac{d\mathcal{N}}{dz}\right)_0 \tag{24.24}$$

This relation, and others like it that follow, is based on the Taylor expansion of a function, $f(x) = f(0) + (df/dx)_0 x + \cdots$, truncated after the second term. The subscript 0 indicates that the slope df/dx should be evaluated at $x = 0$.

The average number of impacts on the imaginary window of area A_0 during an interval Δt is $Z_W A_0 \Delta t$, with $Z_W = \frac{1}{4}\mathcal{N}\bar{c}$ (eqn 16). Therefore, the flux from left to right, $J(L \to R)$, arising from the supply of molecules on the left, is

$$J(L \to R) = \frac{\frac{1}{4}A_0\mathcal{N}(-\lambda)\bar{c}\Delta t}{A_0\Delta t} = \frac{1}{4}\mathcal{N}(-\lambda)\bar{c} \tag{24.25a}$$

There is also a flux of molecules from right to left. On average, the molecules making the journey have originated from $z = +\lambda$ where the number density is $\mathcal{N}(\lambda)$. Therefore,

$$J(L \leftarrow R) = -\frac{1}{4}\mathcal{N}(\lambda)\bar{c} \tag{24.25b}$$

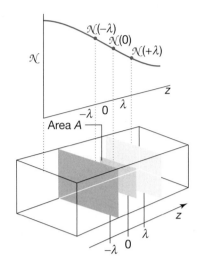

24.12 The calculation of the rate of diffusion of a gas considers the net flux of molecules through a plane of area A as a result of arrivals from on average a distance λ away in each direction, where λ is the mean free path.

24.13 One approximation ignored in the simple treatment is that some particles might make a long flight to the plane even though they are only a short perpendicular distance away, and therefore they have a higher chance of colliding during their journey.

The average number density at $z = +\lambda$ is approximately

$$\mathcal{N}(\lambda) = \mathcal{N}(0) + \lambda \left(\frac{d\mathcal{N}}{dz} \right)_0 \tag{24.26}$$

The net flux is

$$
\begin{aligned}
J_z &= J(L \rightarrow R) + J(L \leftarrow R) \\
&= \tfrac{1}{4}\bar{c} \left\{ \left[\mathcal{N}(0) - \lambda \left(\frac{d\mathcal{N}}{dz} \right)_0 \right] - \left[\mathcal{N}(0) + \lambda \left(\frac{d\mathcal{N}}{dz} \right)_0 \right] \right\} \\
&= -\tfrac{1}{2}\bar{c}\lambda \left(\frac{d\mathcal{N}}{dz} \right)_0
\end{aligned}
\tag{24.27}
$$

This equation shows that the flux is proportional to the first derivative of the concentration, in agreement with Fick's law.

At this stage it looks as though we can pick out a value of the diffusion coefficient by comparing eqns 20 and 27, so obtaining $D = \tfrac{1}{2}\lambda\bar{c}$. It must be remembered, however, that the calculation is quite crude, and is little more than an assessment of the order of magnitude of D. One aspect that has not been taken into account is illustrated in Fig. 24.13, which shows that, although a molecule may have begun its journey very close to the window, it could have a long flight before it gets there. Because the path is long, the molecule is likely to collide before reaching the window, so it ought to be added to the graveyard of other molecules that have collided. To take this effect into account involves a lot of work, but the end result is the appearance of a factor of $\tfrac{2}{3}$ representing the lower flux. The modification results in eqn 23.

(c) Thermal conduction

According to the kinetic theory of gases, and as shown in the *Justification* below, the thermal conductivity of a perfect gas A having molar concentration [A] is given by the expression

$$\kappa = \tfrac{1}{3}\lambda\bar{c}C_{V,\mathrm{m}}[\mathrm{A}] \tag{24.28}°$$

where $C_{V,\mathrm{m}}$ is the molar heat capacity at constant volume.

Justification 24.6

According to the equipartition theorem (Section 20.3), each molecule carries an average energy $\varepsilon = \nu kT$, where ν is a number of the order of 1. For monatomic particles, $\nu = \tfrac{3}{2}$. When one molecule passes through the imaginary window, it transports that energy on average. We suppose that the number density is uniform but that the temperature is not. On average, molecules arrive from the left after travelling a mean free path from their last collision in a hotter region, and therefore with a higher energy. Molecules also arrive from the right after travelling a mean free path from a cooler region. The two opposing energy fluxes are therefore

$$J(L \rightarrow R) = \tfrac{1}{4}\bar{c}\mathcal{N}\varepsilon(-\lambda) \qquad \varepsilon(-\lambda) = \nu k \left\{ T - \lambda \left(\frac{dT}{dz} \right)_0 \right\}$$

$$\tag{24.29}$$

$$J(L \leftarrow R) = -\tfrac{1}{4}\bar{c}\mathcal{N}\varepsilon(\lambda) \qquad \varepsilon(\lambda) = \nu k \left\{ T + \lambda \left(\frac{dT}{dz} \right)_0 \right\}$$

and the net flux is

$$J_z = J(L \rightarrow R) + J(L \leftarrow R) = -\tfrac{1}{2}vk\lambda\bar{c}\mathcal{N}\left(\frac{dT}{dz}\right)_0 \qquad (24.30)$$

As before, we multiply by $\tfrac{2}{3}$ to take long flight paths into account, and so arrive at

$$J_z = -\tfrac{1}{3}vk\lambda\bar{c}\mathcal{N}\left(\frac{dT}{dz}\right)_0 \qquad (24.31)$$

The energy flux is proportional to the temperature gradient, as we wanted to show. Comparison of this equation with eqn 21 shows that

$$\kappa = \tfrac{1}{3}vk\lambda\bar{c}\mathcal{N} \qquad (24.32)$$

Equation 28 then follows from $C_{V,m} = vkN_A$ for a perfect gas, where [A] is the molar concentration of A. For this step, we use $\mathcal{N} = N/V = nN_A/V = N_A[A]$.

Because λ is inversely proportional to the pressure, and hence inversely proportional to the molar concentration of the gas, it follows from eqn 28 that the thermal conductivity is independent of the pressure. The physical reason for this independence is that the thermal conductivity can be expected to be large when many molecules are available to transport the energy, but the presence of so many molecules limits their mean free path and they cannot carry the energy over a great distance. These two effects balance. The thermal conductivity is indeed found experimentally to be independent of the pressure, except when the pressure is very low, when $\kappa \propto p$. At low pressures λ exceeds the dimensions of the apparatus, and the distance over which the energy is transported is determined by the size of the container and not by the other molecules present. The flux is still proportional to the number of carriers, but the length of the journey no longer depends on λ, so $\kappa \propto [A]$, which implies that $\kappa \propto p$.

(d) The viscosity of a perfect gas

We have seen that viscosity is related to the flux of momentum. As shown in the *Justification* below, the expression obtained from the kinetic theory of gases is

$$\eta = \tfrac{1}{3}M\lambda\bar{c}[A] \qquad (24.33)°$$

where [A] is the molar concentration of the gas molecules and M is their molar mass.

Justification 24.7

Molecules travelling from the right in Fig. 24.14 (from a fast layer to a slower one) transport a momentum $mv_x(\lambda)$ to their new layer at $z = 0$; those travelling from the left transport $mv_x(-\lambda)$ to it. If it is assumed that the density is uniform, the collision flux is $\tfrac{1}{4}\mathcal{N}\bar{c}$. Those arriving from the right on average carry a momentum

$$mv_x(\lambda) = mv_x(0) + m\lambda\left(\frac{dv_x}{dz}\right)_0$$

Those arriving from the left bring a momentum

$$mv_x(-\lambda) = mv_x(0) - m\lambda\left(\frac{dv_x}{dz}\right)_0$$

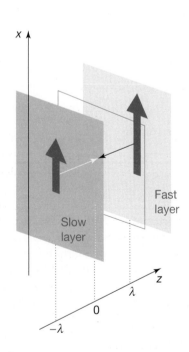

24.14 The calculation of the viscosity of a gas examines the net *x*-component of momentum brought to a plane from faster and slower layers on average a mean free path away in each direction.

The net flux of x-momentum in the z-direction is therefore

$$J = \tfrac{1}{4}\mathcal{N}\bar{c}\left\{\left[mv_x(0) - m\lambda\left(\frac{dv_x}{dz}\right)_0\right] - \left[mv_x(0) - m\lambda\left(\frac{dv_x}{dz}\right)_0\right]\right\}$$

$$= -\tfrac{1}{2}\mathcal{N}m\lambda\bar{c}\left(\frac{dv_x}{dz}\right)_0$$

The flux is proportional to the velocity gradient, as we wished to show. Comparison of this expression with eqn 22, and multiplication by $\tfrac{2}{3}$ in the normal way, leads to

$$\eta = \tfrac{1}{3}\mathcal{N}m\lambda\bar{c} \tag{24.34}$$

which can easily be converted into eqn 33.

(a) 300 K

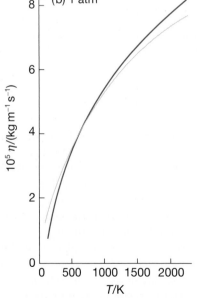

(b) 1 atm

24.15 The experimental results for (a) the pressure dependence of the viscosity of argon, and (b) its temperature dependence. The broken line in the latter is the calculated value. Fitting the observed and calculating curves is one way of determining the collision cross-section.

The viscosity is independent of the pressure: $\lambda \propto 1/p$ and $[A] \propto p$, implying that $\eta \propto \bar{c}$, independent of p. The physical reason is the same as for the thermal conductivity: more molecules are available to transport the momentum, but they carry it less far on account of the decrease in mean free path. An unexpected result is that, because $\bar{c} \propto T^{1/2}$, the viscosity coefficient is proportional to $T^{1/2}$. That is, the viscosity of a gas *increases* with temperature. This conclusion is explained when we remember that at high temperatures the molecules travel more quickly, so the flux of momentum is greater.[2]

There are two main techniques for measuring viscosities of gases. One technique depends on the rate of damping of the torsional oscillations of a disk hanging in the gas. The half-life of the decay of the oscillation depends on the viscosity and the design of the apparatus, and the apparatus needs to be calibrated. The other method is based on **Poiseuille's formula** for the rate of flow of a fluid through a tube of radius r:

$$\frac{dV}{dt} = \frac{(p_1^2 - p_2^2)\pi r^4}{16l\eta p_0} \tag{24.35}$$

where V is the volume flowing, p_1 and p_2 are the pressures at each end of the tube of length l, and p_0 is the pressure at which the volume is measured.

Such measurements confirm that the viscosities of gases are independent of pressure over a wide range. For instance, the results for argon from 10^{-3} atm to 10^2 atm are shown in Fig. 24.15, and we see that η is constant from about 0.01 atm to 20 atm. The measurements also confirm (to a lesser extent) the $T^{1/2}$ dependence. The dotted line in the illustration shows the calculated values using $\sigma = 22 \times 10^{-20}$ m^2, implying a collision diameter of 260 pm, in contrast to the van der Waals diameter of 335 pm obtained from the density of the solid. The agreement is not too bad, considering the simplicity of the model, especially the neglect of intermolecular forces.

Illustration 24.1

In a Poiseuille flow experiment to measure the viscosity of air at 298 K, the sample was allowed to flow through a tube of length 100 cm and internal diameter 1.00 mm. The high-pressure end was at 765 Torr and the low-pressure end was at 760 Torr. The volume was measured at the latter pressure.

2. As we shall see in Section 24.5, the viscosity of a liquid *decreases* with increase in temperature because intermolecular interactions must be overcome.

In 100 s, 90.2 cm^3 of air passed through the tube. The viscosity of air at 298 K is found by reorganizing the Poiseuille formula, eqn 35, into

$$\eta = \frac{(p_1^2 - p_2^2)\pi r^4}{16 l p_0 (dV/dt)}$$

and substituting the data (after converting the pressures to pascals by using 1 Torr = 133.3 Pa):

$$\eta = \frac{\{(765 \times 133.3 \text{ Pa})^2 - (760 \times 133.3 \text{ Pa})^2\} \times \pi \times (5.00 \times 10^{-4} \text{ m})^4}{16 \times (1.00 \times 10^{-1} \text{ m}) \times (760 \times 133.3 \text{ Pa}) \times \left(\dfrac{9.02 \times 10^{-5} \text{ m}^3}{100 \text{ s}} \right)}$$

$$= 1.82 \times 10^{-4} \text{ kg m}^{-1} \text{ s}^{-1}$$

where we have used 1 Pa = 1 kg m^{-1} s^{-2}. The kinetic theory expression gives $\eta = 1.4 \times 10^{-5}$ kg m^{-1} s^{-1}, so the agreement is quite good. Viscosities are commonly expressed in centipoise (cP) or (for gases) micropoise (μP), the conversion being 1 cP = 10^{-3} kg m^{-1} s^{-1}; the viscosity of air at 20°C is about 180 μP.

Self-test 24.4 What volume would be collected if the pressure gradient were doubled, other conditions remaining constant? [181 cm^3]

Molecular motion in liquids

We outlined what is currently known about the structure of simple liquids in Section 21.8. Here we consider a particularly simple type of motion through a liquid, that of an ion, and see that the information that motion provides can be used to infer the behaviour of uncharged species too.

24.5 Experimental results

The motion of molecules in liquids can be studied experimentally by a variety of methods. Relaxation time measurements in NMR and ESR (Section 18.5b) can be interpreted in terms of the mobilities of the molecules, and have been used to show that big molecules in viscous fluids typically rotate in a series of small (about 5°) steps, whereas small molecules in nonviscous fluids typically jump through about 1 radian (57°) in each step. Another important technique is **inelastic neutron scattering**, in which the energy neutrons collect or discard as they pass through a sample is interpreted in terms of the motion of its particles. The same technique is used to examine the internal dynamics of macromolecules.

More mundane than these experiments are viscosity measurements (Table 24.4). For a molecule to move in a liquid, it must acquire at least a minimum energy to escape from its neighbours. The probability that a molecule has at least an energy E_a is proportional to $e^{-E_a/RT}$, so the mobility of the molecules in the liquid should follow this type of temperature dependence. Because the coefficient of viscosity, η, is inversely proportional to the mobility of the particles, we should expect that

$$\eta \propto e^{E_a/RT} \tag{24.36}$$

(Note the positive sign of the exponent.) This expression implies that the viscosity should decrease sharply with increasing temperature. Such a variation is found

Table 24.4* Viscosities of liquids at 298 K, $\eta/(10^{-3}$ kg m^{-1} s^{-1})

Benzene	0.601
Mercury	1.55
Pentane	0.224
Water†	0.891

* More values are given in the *Data section*.
† The viscosity of water corresponds to 0.891 cP.

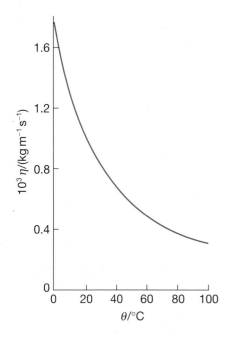

24.16 The experimental temperature dependence of the viscosity of water. As the temperature is increased, more molecules are able to escape from the potential wells provided by their neighbours, and so the liquid becomes more fluid. A plot of $\ln \eta$ against $1/T$ is a straight line (over a small range) with positive slope.

experimentally, at least over reasonably small temperature ranges (Fig. 24.16). The activation energy typical of viscosity is comparable to the mean potential energy of intermolecular interactions.

One problem with the interpretation of viscosity measurements is that the change in density of the liquid as it is heated makes a pronounced contribution to the temperature variation of the viscosity. Thus, the temperature dependence of viscosity at constant volume, when the density is constant, is much less than that at constant pressure. The intermolecular interactions between the molecules of the liquid govern the magnitude of E_a, but the problem of calculating it is immensely difficult and still largely unsolved. At low temperatures, the viscosity of water decreases as the pressure is increased. This behaviour is consistent with the rupture of hydrogen bonds.

24.6 The conductivities of electrolyte solutions

Further insight into the nature of molecular motion can be obtained by studying the motion of ions in solution, for ions can be dragged through the solvent by the application of a potential difference between two electrodes immersed in the sample. By studying the transport of charge through electrolyte solutions it is possible to build up a picture of the events that occur in them and, in some cases, to extrapolate the conclusions to species that have zero charge; that is, to neutral molecules.

(a) Conductance and conductivity

The fundamental measurement used to study the motion of ions is that of the electrical resistance, R, of the solution. The **conductance**, G, of a solution is the inverse of its resistance R: $G = 1/R$. As resistance is expressed in ohms, Ω, the conductance of a sample is expressed in Ω^{-1}. The reciprocal ohm used to be called the mho, but its official designation is now the siemens, S, and $1\ \text{S} = 1\ \Omega^{-1} = 1\ \text{C V}^{-1}\ \text{s}^{-1}$. The conductance of a sample decreases with its length l and increases with its cross-sectional area A. We therefore write

$$G = \frac{\kappa A}{l} \tag{24.37}$$

where κ is the **conductivity**. With the conductance in siemens and the dimensions in metres, it follows that the SI units of κ are siemens per metre (S m^{-1}).

The conductivity of a solution depends on the number of ions present, and it is normal to introduce the **molar conductivity**, Λ_m, which is defined as

$$\Lambda_m = \frac{\kappa}{c} \tag{24.38}$$

where c is the molar concentration of the added electrolyte. The SI unit of molar conductivity is siemens metre-squared per mole (S m^2 mol^{-1}), and typical values are about 10 mS m^2 mol^{-1} (where 1 mS = 10^{-3} S).

The molar conductivity is found to vary with the concentration. One reason for this variation is that the number of ions in the solution might not be proportional to the concentration of the electrolyte. For instance, the concentration of ions in a solution of a weak acid depends on the concentration of the acid in a complicated way, and doubling the concentration of the acid added does not double the number of ions. Secondly, because ions interact strongly with one another, the conductivity of a solution is not exactly proportional to the number of ions present.

The concentration dependence of molar conductivities indicates that there are two classes of electrolyte. The characteristic of a **strong electrolyte** is that its molar conductivity depends only slightly on the molar concentration (and in general decreases slightly as the concentration is increased, Fig. 24.17). The characteristic

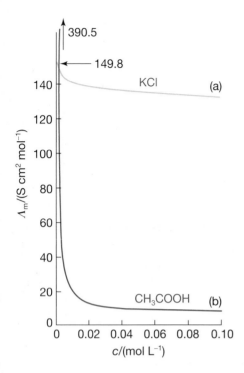

24.17 The concentration dependence of the molar conductivities of (a) a typical strong electrolyte (aqueous potassium chloride) and (b) a typical weak electrolyte (aqueous acetic acid).

Table 24.5* Limiting ionic conductivities in water at 298 K, $\lambda/(\text{mS m}^2 \text{ mol}^{-1})$

H^+	34.96	OH^-	19.91
Na^+	5.01	Cl^-	7.63
K^+	7.35	Br^-	7.81
Zn^{2+}	10.56	SO_4^{2-}	16.00

* More values are given in the *Data section*.

of a **weak electrolyte** is that its molar conductivity is normal at concentrations close to zero, but falls sharply to low values as the concentration increases. The classification depends on the solvent employed as well as the solute: lithium chloride, for example, is a strong electrolyte in water but a weak electrolyte in propanone.

(b) Strong electrolytes

Strong electrolytes are substances that are virtually fully ionized in solution, and include ionic solids and strong acids. As a result of their complete ionization, the concentration of ions in solution is proportional to the concentration of strong electrolyte added.

In an extensive series of measurements during the nineteenth century, Friedrich Kohlrausch showed that, at low concentrations, the molar conductivities of strong electrolytes vary linearly with the square root of the concentration:

$$\Lambda_m = \Lambda_m^\circ - \mathcal{K}c^{1/2} \tag{24.39}$$

This variation is called **Kohlrausch's law**. The constant Λ_m° is the **limiting molar conductivity**, the molar conductivity in the limit of zero concentration (when the ions are effectively infinitely far apart and do not interact with one another). The constant $\mathcal{K}$ is found to depend more on the stoichiometry of the electrolyte (that is, whether it is of the form MA, or M_2A, etc.) than on its specific identity. In due course we shall see that the $c^{1/2}$ dependence arises from interactions between ions: when charge is conducted ionically, ions of one charge are moving past the ions of interest and retard its progress.

Kohlrausch was also able to show that Λ_m° can be expressed as the sum of contributions from its individual ions. If the limiting molar conductivity of the cations is denoted λ_+ and that of the anions λ_-, then his **law of the independent migration of ions** states that

$$\Lambda_m^\circ = v_+\lambda_+ + v_-\lambda_- \tag{24.40}^\circ$$

where v_+ and v_- are the numbers of cations and anions per formula unit of electrolyte (for example, $v_+ = v_- = 1$ for HCl, NaCl, and $CuSO_4$, but $v_+ = 1$, $v_- = 2$ for $MgCl_2$). This simple result, which can be understood on the grounds that the ions migrate independently in the limit of zero concentration, lets us predict the limiting molar conductivity of any strong electrolyte from the data in Table 24.5.

Illustration 24.2

The limiting molar conductivity of $BaCl_2$ in water at 298 K is

$$\Lambda_m^\circ = (12.72 + 2 \times 7.63)\ \text{mS m}^2\ \text{mol}^{-1} = 27.98\ \text{mS m}^2\ \text{mol}^{-1}$$

(c) Weak electrolytes

Weak electrolytes are not fully ionized in solution. They include weak Brønsted acids and bases, such as CH_3COOH and NH_3. The marked concentration dependence of their molar conductivities arises from the displacement of the equilibrium

$$HA(aq) + H_2O(l) \rightleftharpoons H_3O^+(aq) + A^-(aq) \qquad K_a = \frac{a_{H_3O^+}a_{A^-}}{a_{HA}} \tag{24.41}$$

towards products at low molar concentrations.

The conductivity depends on the number of ions in the solution, and therefore on the **degree of ionization**, α, of the electrolyte; when referring to weak acids, we speak instead of the **degree of deprotonation**. It is defined so that, for the acid HA at a molar concentration c, at equilibrium

$$[H_3O^+] = \alpha c \qquad [A^-] = \alpha c \qquad [HA] = (1 - \alpha)c \tag{24.42}$$

If we ignore activity coefficients, the acidity constant, K_a, is approximately

$$K_a = \frac{\alpha^2 c}{1 - \alpha} \tag{24.43}°$$

from which it follows that

$$\alpha = \frac{K_a}{2c}\left\{\left(1 + \frac{4c}{K_a}\right)^{1/2} - 1\right\} \tag{24.44}°$$

The acid is fully deprotonated at infinite dilution, and its molar conductivity is then Λ_m°. Because only a fraction α is actually present as ions in the actual solution, the measured molar conductivity Λ_m is given by

$$\Lambda_m = \alpha \Lambda_m^\circ \tag{24.45}°$$

with α given by eqn 44.

Illustration 24.3

The molar conductivity of 0.0100 M $CH_3COOH(aq)$ at 298 K is $\Lambda_m = 1.65$ mS m^2 mol^{-1}. The degree of deprotonation, α, is calculated from eqn 45 with $\Lambda_m^\circ = 39.05$ mS m^2 mol^{-1} (Table 24.5). It follows that $\alpha = 0.0423$. The acidity constant, K_a, is calculated by substitution of α into eqn 43, which gives $K_a = 1.9 \times 10^{-5}$.

Self-test 24.5 The molar conductivity of 0.0250 M HCOOH(aq) is 4.61 mS m^2 mol^{-1}. Determine the pK_a of the acid. [3.44]

Once we know K_a, we can use eqns 44 and 45 to predict the concentration dependence of the molar conductivity. The result agrees quite well with the experimental curve in Fig. 24.17. More usefully, we can use the concentration dependence of Λ_m in measurements of the limiting molar conductance. First, we rearrange eqn 43 into

$$\frac{1}{\alpha} = 1 + \frac{\alpha c}{K_a} \tag{24.46}°$$

Then, by using eqn 45, we obtain **Ostwald's dilution law**:

$$\frac{1}{\Lambda_m} = \frac{1}{\Lambda_m^\circ} + \frac{\Lambda_m c}{K_a (\Lambda_m^\circ)^2} \tag{24.47}°$$

This equation implies that, if $1/\Lambda_m$ is plotted against $c\Lambda_m$, then the intercept at $c = 0$ will be $1/\Lambda_m^\circ$ (Fig. 24.18).

24.7 The mobilities of ions

To interpret conductivity measurements we need to know why ions move at different rates, why they have different molar conductivities, and why the molar conductivities of strong electrolytes decrease with the square root of the molar concentration. The central idea in this section is that, although the motion of an ion

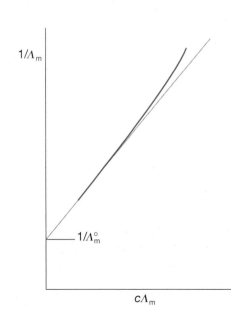

24.18 The graph used to determine the limiting value of the molar conductivity of a solution by extrapolation to zero concentration.

remains largely random, the presence of an electric field biases its motion, and the ion undergoes net migration through the solution.

(a) The drift speed

When the potential difference between two electrodes a distance l apart is $\Delta\phi$, the ions in the solution between them experience a uniform electric field of magnitude

$$\mathcal{E} = \frac{\Delta\phi}{l} \tag{24.48}$$

In such a field, an ion of charge[3] ze experiences a force of magnitude

$$\mathcal{F} = ze\mathcal{E} = \frac{ze\Delta\phi}{l} \tag{24.49}$$

A cation responds to the application of the field by accelerating towards the negative electrode and an anion responds by accelerating towards the positive electrode. However, this acceleration is short lived. As the ion moves through the solvent it experiences a frictional retarding force, $\mathcal{F}_{\text{fric}}$, proportional to its speed. If we assume that the Stokes formula (eqn 22.34) for a sphere of radius a and speed s applies even on a microscopic scale (and independent evidence from magnetic resonance suggests that it often gives at least the right order of magnitude), then we can write this retarding force as

$$\mathcal{F}_{\text{fric}} = fs \qquad f = 6\pi\eta a \tag{24.50}$$

The two forces act in opposite directions, and the ions quickly reach a terminal speed, the **drift speed**, when the accelerating force is balanced by the viscous drag. The net force is zero when

$$s = \frac{ze\mathcal{E}}{f} \tag{24.51}$$

It follows that the drift speed of an ion is proportional to the strength of the applied field. We write

$$s = u\mathcal{E} \tag{24.52}$$

where u is called the **mobility** of the ion (Table 24.6). Comparison of eqns 51 and 52 and use of eqn 50 shows that

$$u = \frac{ze}{f} = \frac{ze}{6\pi\eta a} \tag{24.53}$$

Table 24.6* Ionic mobilities in water at 298 K, $u/(10^{-8}\ \text{m}^2\ \text{s}^{-1}\ \text{V}^{-1})$

H^+	36.23	OH^-	20.64
Na^+	5.19	Cl^-	7.91
K^+	7.62	Br^-	8.09
Zn^{2+}	5.47	SO_4^{2-}	8.29

* More values are given in the *Data section*.

Illustration 24.4

For an order of magnitude estimate we can take $z = 1$ and a the radius of an ion such as Cs^+ (which might be typical of a smaller ion plus its hydration sphere), which is 170 pm. For the viscosity, we use $\eta = 1.0$ cP ($1.0 \times 10^{-3}\ \text{kg m}^{-1}\ \text{s}^{-1}$, Table 24.4). Then $u \approx 5 \times 10^{-8}\ \text{m}^2\ \text{V}^{-1}\ \text{s}^{-1}$. This value means that, when there is a potential difference of 1 V across a solution of length 1 cm (so $\mathcal{E} = 100\ \text{V m}^{-1}$), the drift speed is typically about 5 μm s^{-1}. That speed might seem slow, but not when expressed on a molecular scale, for it corresponds to an ion passing about 10^4 solvent molecules per second.

3. In this chapter we disregard the sign of the charge number and so avoid notational complications.

Because the drift speed governs the rate at which charge is transported, we might expect the conductivity to decrease with increasing solution viscosity and ion size. Experiments confirm these predictions for bulky ions (such as R_4N^+ and RCO_2^-) but not for small ions. For example, the molar conductivities of the alkali metal ions increase from Li^+ to Cs^+ (Table 24.6) even though the ionic radii increase. The paradox is resolved when we realize that the radius a in the Stokes formula is the **hydrodynamic radius** (or 'Stokes radius') of the ion, its effective radius in the solution taking into account all the H_2O molecules it carries in its hydration sphere. Small ions give rise to stronger electric fields than large ones,[4] so small ions are more extensively solvated than big ions. Thus, an ion of small ionic radius may have a large hydrodynamic radius because it drags many solvent molecules through the solution as it migrates. The hydrating H_2O molecules are often very labile, however, and NMR and isotope studies have shown that the exchange between the coordination sphere of the ion and the bulk solvent is very rapid.

The proton, although it is very small, has a very high molar conductivity (Table 24.6)! Proton and ^{17}O-NMR show that the times characteristic of protons hopping from one molecule to the next are about 1.5 ps, which is comparable to the time that inelastic neutron scattering shows it takes a water molecule to reorientate through about 1 rad (1 to 2 ps). According to the **Grotthuss mechanism**,[5] there is an effective motion of a proton that involves the rearrangement of bonds in a group of water molecules. However, the actual mechanism is still highly contentious. Attention now focuses on the $H_9O_4^+$ unit, in which the nearly trigonal planar H_3O^+ ion[6] is linked to three strongly solvating H_2O molecules. This cluster of atoms is itself hydrated, but the hydrogen bonds in the secondary sphere are weaker than in the primary sphere. It is envisaged that the rate-determining step is the cleavage of one of the weaker hydrogen bonds of this secondary sphere (Fig. 24.19a). After this bond cleavage has taken place, and the released molecule has rotated through a few degrees (a process that takes about 1 ps), there is a rapid adjustment of bond lengths and angles in the remaining cluster, to form an $H_5O_2^+$ cation of structure $H_2O\cdots H^+\cdots OH_2$ (Fig. 24.19b). Shortly after this reorganization has occurred, a new $H_9O_4^+$ cluster forms as other molecules rotate into a position where they can become members of a secondary hydration sphere, but now the positive charge is located one molecule to the right of its initial location (Fig. 24.19c). According to this model, there is no coordinated motion of a proton along a chain of molecules, simply a very rapid hopping between neighbouring sites, with a low activation energy. The model is consistent with the observation that the molar conductivity of protons increases as the pressure is raised, for increasing pressure ruptures the hydrogen bonds in water. The mobility of NH_4^+ in liquid ammonia is also anomalous and presumably occurs by an analogous mechanism.

(b) Mobility and conductivity

Ionic mobilities provide a link between measurable and theoretical quantities. As a first step we establish in the *Justification* below the following relation between an ion's mobility and its molar conductivity:

$$\lambda = zuF \tag{24.54}°$$

where F is the Faraday constant ($F = N_Ae$).

(a)

(b)

(c)

24.19 The mechanism of conduction by water as proposed by N. Agmon (*Chem. Phys. Letts.*, **244**, 456 (1995)). Proton transfer between neighbouring molecules occurs when one molecule rotates into such a position that an O—H⋯O hydrogen bond can flip into being an O⋯H—O hydrogen bond. See text for a description of the steps.

4. The electric field at the surface of a sphere of radius r is proportional to ze/r^2, so the smaller the radius the stronger the field.
5. The name of the mechanism is an allusion to an early view advanced by von Grotthuss, in which it was supposed that chains of dipoles were responsible for the transport of charge in water.
6. In the gas phase, H_3O^+ is trigonal pyramidal.

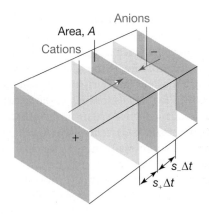

24.20 In the calculation of the current, all the cations within a distance $s_+\Delta t$ (that is, those in the volume $s_+A\Delta t$) will pass through the area A. The anions in the corresponding volume the other side of the window will also contribute to the current similarly.

Justification 24.8

To keep the calculation simple, we ignore signs in the following, and concentrate on the magnitudes of quantities: the direction of ion flux can always be decided by common sense.

Consider a solution of a fully dissociated strong electrolyte at a molar concentration c. Let each formula unit give rise to v_+ cations of charge z_+e and v_- anions of charge z_-e. The molar concentration of each type of ion is therefore vc (with $v = v_+$ or v_-), and the number density of each type is vcN_A. The number of ions of one kind that pass through an imaginary window of area A during an interval Δt is equal to the number within the distance $s\Delta t$ (Fig. 24.20), and therefore to the number in the volume $s\Delta tA$. (The same sort of argument was used in Section 24.1 in the discussion of the pressure of a gas.) The number of ions of that kind in this volume is equal to $s\Delta tAvcN_A$. The flux through the window (the number of this type of ion passing through the window divided by the area of the window and the duration of the interval) is therefore

$$J(\text{ions}) = \frac{s\Delta tAvcN_A}{A\Delta t} = svcN_A$$

Each ion carries a charge ze, so the flux of charge is

$$J(\text{charge}) = zsvceN_A = zsvcF$$

Because $s = u\mathcal{E}$, the flux is

$$J(\text{charge}) = zuvcF\mathcal{E}$$

The current, I, through the window due to the ions we are considering is the charge flux times the area:

$$I = JA = zuvcF\mathcal{E}A$$

Because the electric field is the potential gradient, $\Delta\phi/l$, we can write

$$I = \frac{zuvcFA\Delta\phi}{l} \tag{24.55}$$

Current and potential difference are related by Ohm's law, that

$$I = \frac{\Delta\phi}{R} = G\Delta\phi = \frac{\kappa A\Delta\phi}{l}$$

where we have used eqn 37 ($G = \kappa A/l$). Note that the proportionality of current to potential difference ($I \propto \Delta\phi$) is another example of a phenomenological flux equation like those introduced in Section 24.4a. Comparison of the last two expressions gives $\kappa = zuvcF$. Division by the molar concentration of ions, vc, then results in eqn 54.

Equation 54 applies to the cations and to the anions. Therefore, for the solution itself in the limit of zero concentration (when there are no interionic interactions),

$$\Lambda_m^\circ = (z_+u_+v_+ + z_-u_-v_-)F \tag{24.56}°$$

For a symmetrical z:z electrolyte (for example, $CuSO_4$ with $z = 2$), this equation simplifies to

$$\Lambda_m^\circ = z(u_+ + u_-)F \tag{24.57}°$$

Illustration 24.5

Earlier, we estimated the typical ionic mobility as 5×10^{-8} m^2 V^{-1} s^{-1}; so, with $z = 1$ for both the cation and anion, we can estimate that a typical limiting molar conductivity should be about 10 mS m^2 mol^{-1}, in accord with experiment. The experimental value for KCl, for instance, is 15 mS m^2 mol^{-1}.

(c) Transport numbers

The **transport number**, $t_\pm$, is defined as the fraction of total current carried by the ions of a specified type. For a solution of two kinds of ion, the transport numbers of the cations (t_+) and anions (t_-) are

$$t_\pm = \frac{I_\pm}{I} \qquad [24.58]$$

where $I_\pm$ is the current carried by the cation (I_+) or anion (I_-) and I is the total current through the solution. Because the total current is the sum of the cation and anion currents, it follows that

$$t_+ + t_- = 1 \qquad (24.59)$$

The **limiting transport number**, $t_\pm^\circ$, is defined in the same way but for the limit of zero concentration of the electrolyte solution. We shall consider only these limiting values from now on, for that avoids the problem of ionic interactions.

The current that can be ascribed to each type of ion is related to the mobility of the ion by eqn 55. Hence the relation between $t_\pm^\circ$ and $u_\pm$ is

$$t_\pm^\circ = \frac{z_+ v_+ u_\pm}{z_+ v_+ u_+ + z_- v_- u_-} \qquad (24.60)^\circ$$

However, because $z_+ v_+ = z_- v_-$ for any electrolyte, eqn 60 simplifies to

$$t_\pm^\circ = \frac{u_\pm}{u_+ + u_-} \qquad (24.61)^\circ$$

Moreover, because the ionic conductivities are related to the mobilities by eqn 54, it also follows from eqn 60 that

$$t_\pm^\circ = \frac{v_\pm \lambda_\pm}{v_+ \lambda_+ + v_- \lambda_-} = \frac{v_\pm \lambda_\pm}{\Lambda_m^\circ} \qquad (24.62)^\circ$$

and hence, for each type of ion,

$$v_\pm \lambda_\pm = t_\pm^\circ \Lambda_m^\circ \qquad (24.63)^\circ$$

Consequently, because there are independent ways of measuring transport numbers of ions, we can determine the individual ionic conductivities and (through eqn 54) the ionic mobilities.

(d) The measurement of transport numbers

One of the most accurate methods for measuring transport numbers is the **moving boundary method**, in which the motion of a boundary between two ionic solutions having a common ion is observed as a current flows.

Let MX be the salt of interest and NX a salt giving a denser solution. The solution of NX is called the **indicator solution**; it occupies the lower part of a vertical tube of cross-sectional area A (Fig. 24.21). The MX solution, which is called the **leading solution**, occupies the upper part of the tube. There is a sharp boundary between the two

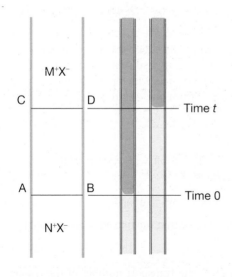

24.21 In the moving boundary method for the measurement of transport numbers the distance moved by the boundary is observed as a current is passed. All the M ions in the volume between AB and CD must have passed through CD if the boundary moves from AB to CD.

solutions. The indicator solution must be denser than the leading solution, and the mobility of the M ions must be greater than that of the N ions.[7] Thus, if any M ions diffuse into the lower solution, they will be pulled upwards more rapidly than the N ions around them, and the boundary will reform. The interpretation of the experiment makes use of the relation (see the *Justification* below) between the distance, *l*, moved by the boundary in the time Δt for which a current I is passed for a period Δt:

$$t_+ = \frac{z_+ clAF}{I\Delta t} \tag{24.64}$$

Hence, by measuring the distance moved, the transport number and hence the conductivity and mobility of the ions can be determined.

Justification 24.9

When a current I is passed for a time Δt, the boundary moves from AB to CD, so all the M ions in the volume between AB and CD must have passed through CD. That number is $clAN_A$, so the charge that the M ions transfer through the plane is $z_+ clAeN_A$. However, the *total* charge transferred when a current I flows for an interval Δt is $I\Delta t$. Therefore, the fraction due to the motion of the M ions, which is their transport number, is given by eqn 64.

In the **Hittorf method**, an electrolytic cell is divided into three compartments and a charge $I\Delta t$ is passed. An amount $I\Delta t/z_+ F$ of cations is discharged at the cathode but an amount $t_+ I\Delta t/z_+ F$ of cations migrates into the cathode compartment. The net change in the amount of cations in that compartment is therefore

$$\text{Net change} = (t_+ - 1)\frac{I\Delta t}{z_+ F} = -t_-\frac{I\Delta t}{z_+ F} \tag{24.65}$$

Hence, by measuring the change of composition in the cathode compartment, the anion transport number t_- can be deduced. Likewise, the change in composition of the anode compartment is $-t_+ I\Delta t/z_- F$, which gives the cation transport number, t_+.

Transport numbers may also be measured by using galvanic cells. In particular, the measurement is made on a **cell with transference**, which is a galvanic cell with a liquid boundary across which ions may pass from one electrode compartment to the other. An example is the cell

$$\text{Ag(s)}|\text{AgCl(s)}|\text{HCl}(m_1)|\text{HCl}(m_2)|\text{AgCl(s)}|\text{Ag(s)}$$

for which the emf is E_t and the electrodes are reversible with respect to anions (Cl^-). The corresponding cell without transference is

$$\text{Ag(s)}|\text{AgCl(s)}|\text{HCl}(m_1)|\text{H}_2\text{(g)}|\text{Pt(s)}|\text{H}_2\text{(g)}|\text{HCl}(m_2)|\text{AgCl(s)}|\text{Ag(s)}$$

and its emf is E. We show in the *Justification* below that the two emfs are related by

$$E_t = 2t_+ E \tag{24.66}$$

Therefore, comparison of the two emfs gives the transport number of the *counter ion* of the ion with respect to which the electrodes are reversible (in this case, the transport number of H^+).

---

7. One procedure is to add bromothymol blue indicator to a slightly alkaline solution of the ion of interest and to use a cadmium electrode at the lower end of the vertical tube. The electrode produces Cd^{2+} ions, which are slow moving and slightly acidic (the hydrated ion is a Brønsted acid), and the boundary is revealed by the colour change of the indicator.

Justification 24.10

The argument is similar to that used to analyse the Hittorf method. Consider the consequences of passing 1 mol of electrons through the cell with transport specified above. In the right-hand electrode compartment, 1 mol Cl^- is formed but t_- mol Cl^- migrates out of it across the junction. The net change is $(1 - t_-)$ mol $= t_+$ mol. At the same time, t_+ mol H^+ migrates into the compartment. In the left-hand electrode compartment 1 mol Cl^- is removed from solution (to form 1 mol AgCl), but t_- mol Cl^- flows in across the junction. The net change is therefore $(-1 + t_-)$ mol $= -t_+$ mol Cl^-. At the same time, t_+ mol H^+ flows out. The reaction Gibbs energy is therefore

$$\Delta_r G = t_+\{\mu(Cl^-, m_2) - \mu(Cl^-, m_1) + \mu(H^+, m_2) - \mu(H^+, m_1)\} = 2t_+RT \ln \frac{a_2}{a_1}$$

Because $\Delta_r G = -FE$, it follows that

$$E_t = -\frac{2t_+RT}{F} \ln \frac{a_2}{a_1}$$

For the same cell without transference, the Nernst equation gives

$$E_t = -\frac{RT}{F} \ln \frac{a_2}{a_1}$$

and the ratio of the two cell emfs is $2t_+$.

24.8 Conductivities and ion–ion interactions

The remaining problem is to account for the $c^{1/2}$ dependence of the Kohlrausch law (eqn 39). In Section 10.2c we saw something similar: the activity coefficients of ions at low concentrations also depend on $c^{1/2}$ and depend on their charge type rather than their specific identities. That $c^{1/2}$ dependence was explained in terms of the properties of the ionic atmosphere around each ion, and we can suspect that the same explanation applies here too.

To accommodate the effect of motion, we need to modify the picture of an ionic atmosphere as a spherical haze of charge. Because the ions forming the atmosphere do not adjust to the moving ion immediately, the atmosphere is incompletely formed in front of the moving ion and incompletely decayed behind the ion (Fig. 24.22). The overall effect is the displacement of the centre of charge of the atmosphere a short distance behind the moving ion. Because the two charges are opposite, the result is a retardation of the moving ion. This reduction of the ions' mobility is called the **relaxation effect**. A confirmation of the picture is obtained by observing the conductivities of ions at high frequencies, which are greater than at low frequencies: the atmosphere does not have time to follow the rapidly changing direction of motion of the ion, and the effect of the field averages to zero.

The ionic atmosphere has another effect on the motion of the ions. We have seen that the moving ion experiences a viscous drag. When the ionic atmosphere is present this drag is enhanced because the ionic atmosphere moves in an opposite direction to the central ion. The enhanced viscous drag, which is called the **electrophoretic effect**, reduces the mobility of the ions, and hence also reduces their conductivities.

The quantitative formulation of these effects is far from simple, but the **Debye–Hückel–Onsager theory** is an attempt to obtain quantitative expressions at about

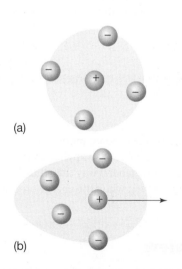

(a)

(b)

24.22 (a) In the absence of an applied field, the ionic atmosphere is spherically symmetric, but (b) when a field is present it is distorted and the centres of negative and positive charge no longer coincide. The attraction between the opposite charges retards the motion of the central ion.

Table 24.7 Debye–Hückel–Onsager coefficients for (1,1)-electrolytes at 298 K

Solvent	$A/(\text{mS m}^2 \text{ mol}^{-1}/(\text{mol L}^{-1})^{1/2})$	$B/(\text{mol L}^{-1})^{-1/2}$
Methanol	15.61	0.923
Propanone	32.8	1.63
Water	6.02	0.229

* More values are given in the *Data section*.

the same level of sophistication as the Debye–Hückel theory itself. The theory leads to a Kohlrausch-like expression in which

$$\mathcal{K} = A + B\Lambda_m^\circ \tag{24.67}$$

with

$$A = \frac{z^2 e F^2}{3\pi\eta}\left(\frac{2}{\varepsilon RT}\right)^{1/2} \qquad B = \frac{qz^3 e F}{24\pi\varepsilon RT}\left(\frac{2}{\varepsilon RT}\right)^{1/2} \tag{24.68}$$

where ε is the electric permittivity of the solvent (Section 22.1e) and $q = 0.586$ for a 1,1-electrolyte (Table 24.7). The slopes of the conductivity curves are predicted to depend on the charge type of the electrolyte, in accord with the Kohlrausch law, and some comparisons between theory and experiment are shown in Fig. 24.23. The agreement is quite good at very low molar concentrations (less than about 10^{-3} M, depending on the charge type).

Molecular dynamics calculations can also shed light on electric conduction. The key equation is a special case of a **Green–Kubo relation**, which expresses a transport property in terms of the fluctuations in microscopic properties of a system. The electrical conductivity is related to the fluctuations in the instantaneous electric current, j, in the sample that arises from variations in the velocities of the ions:

$$\kappa = \frac{1}{kTV}\int_0^\infty \langle j(0)j(t)\rangle\, dt \qquad j = \sum_i^N z_i e v_i \tag{24.69}$$

where v_i is the velocity of the ion i at a given instant and the angular brackets denote an average over the sample. If the ions are very mobile, there will be large fluctuations in the instantaneous currents in the sample, and the conductivity of the medium will be high. If the ions are locked into position, as in an ionic solid, there will be no instantaneous currents, and the ionic conductivity will be zero. The velocities of the ions are calculated explicitly in a molecular dynamics simulation, so the **correlation function**, the quantity $\langle j(0)j(t)\rangle$, can be evaluated reasonably simply by taking the average of a large number of such calculations with different initial conditions.

Diffusion

We are now in a position to extend the discussion of ionic motion to cover the migration of neutral molecules and of ions in the absence of an applied electric field. We shall do this by expressing ion motion in a more general way than hitherto, and will then discover that the same equations apply even when the charge on the particles is zero.

24.9 The thermodynamic view

We saw in Part 1 that, at constant temperature and pressure, the maximum non-expansion work that can be done per mole when a substance moves from a location where its chemical potential is μ to a location where its chemical potential is $\mu + d\mu$ is $dw = d\mu$. In a system in which the chemical potential depends on the position x,

$$dw = d\mu = \left(\frac{\partial\mu}{\partial x}\right)_{p,T} dx \tag{24.70}$$

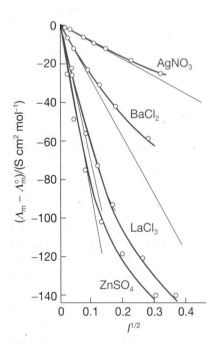

24.23 The dependence of molar conductivities on the square root of the ionic strength, and comparison (dotted lines) with the dependence predicted by the Debye–Hückel–Onsager theory.

We also saw in Chapter 2 (Table 2.1) that in general work can always be expressed in terms of an opposing force (which here we write $\mathcal{F}$), and that

$$\mathrm{d}w = -\mathcal{F}\,\mathrm{d}x \tag{24.71}$$

By comparing these two expressions, we see that the slope of the chemical potential can be interpreted as an effective force per mole of molecules. We write this **thermodynamic force** as

$$\mathcal{F} = -\left(\frac{\partial \mu}{\partial x}\right)_{p,T} \tag{24.72}$$

There is not necessarily a real force pushing the particles down the slope of the chemical potential. As we shall see, the force may represent the spontaneous tendency of the molecules to disperse as a consequence of the Second Law and the hunt for maximum entropy.

(a) The thermodynamic force of a concentration gradient

In a solution in which the activity of the solute is a, the chemical potential is

$$\mu = \mu^{\ominus} + RT \ln a$$

If the solution is not uniform the activity depends on the position and we can write

$$\mathcal{F} = -RT\left(\frac{\partial \ln a}{\partial x}\right)_{p,T} \tag{24.73}$$

A useful expression involving derivatives of logarithms is d ln y/dx = (1/y)(dy/dx).

If the solution is ideal, a may be replaced by the molar concentration c, and then

$$\mathcal{F} = -\frac{RT}{c}\left(\frac{\partial c}{\partial x}\right)_{p,T} \tag{24.74}°$$

Example 24.4 Calculating the thermodynamic force

Suppose the concentration of a solute decays exponentially along the length of a container. Calculate the thermodynamic force on the solute at 25°C given that the concentration falls to half its value in 10 cm.

Method According to eqn 74, the thermodynamic force is calculated by differentiating the concentration with respect to distance. Therefore, write an expression for the variation of the concentration with distance, and then differentiate it.

Answer The concentration varies with position as

$$c = c_0 e^{-x/\lambda}$$

where λ is the decay constant. Therefore,

$$\frac{\mathrm{d}c}{\mathrm{d}x} = -\frac{c}{\lambda}$$

Equation 74 then implies that

$$\mathcal{F} = \frac{RT}{\lambda}$$

We know that the concentration falls to $\frac{1}{2}c_0$ at $x = 10$ cm, so we can find λ from $\frac{1}{2} = e^{-(10\ cm)/\lambda}$. That is, $\lambda = (10\ cm/\ln 2)$. It follows that

$$\mathcal{F} = (8.314\ 51\ J\ K^{-1}\ mol^{-1}) \times (298\ K) \times \ln 2/(1.0 \times 10^{-1}\ m) = 17\ kN\ mol^{-1}$$

where we have used $1\ J = 1\ N\ m$.

Self-test 24.6 Calculate the thermodynamic force on the molecules of molar mass M in a vertical tube in a gravitational field on the surface of the Earth, and evaluate $\mathcal{F}$ for molecules of molar mass $100\ g\ mol^{-1}$. Comment on its magnitude relative to that just calculated.

$[\mathcal{F} = -Mg = -0.98\ N\ mol^{-1};$ the force arising from the concentration gradient greatly dominates that arising from the gravitational gradient]

(b) Fick's first law of diffusion

In Section 24.4a we saw that Fick's first law of diffusion (that the particle flux is proportional to the concentration gradient) could be deduced from the kinetic theory of gases. We shall now show that it can be deduced more generally and that it applies to the diffusion of species in condensed phases too.

We suppose that the flux of diffusing particles is motion in response to a thermodynamic force arising from a concentration gradient. The particles reach a steady drift speed, s, when the thermodynamic force, $\mathcal{F}$, is matched by the viscous drag. This drift speed is proportional to the thermodynamic force, and we write $s \propto \mathcal{F}$. However, the particle flux, J, is proportional to the drift speed, and the thermodynamic force is proportional to the concentration gradient, dc/dx. The chain of proportionalities ($J \propto s$, $s \propto \mathcal{F}$, and $\mathcal{F} \propto dc/dx$) implies that $J \propto dc/dx$, which is the content of Fick's law.

(c) The Einstein relation

Fick's law can be written[8]

$$J = -D\frac{dc}{dx} \tag{24.75}$$

In this expression, D is the diffusion coefficient and dc/dx is the slope of the molar concentration. The flux is related to the drift speed by

$$J = sc \tag{24.76}$$

This relation follows from the argument that we have used several times before. Thus, all particles within a distance $s\Delta t$, and therefore in a volume $s\Delta tA$, can pass through a window of area A in an interval Δt. Hence, the amount of substance that can pass through the window in that interval is $s\Delta tAc$. Therefore,

$$sc = -D\frac{dc}{dx}$$

If now we express dc/dx in terms of $\mathcal{F}$ by using eqn 74, we find

$$s = -\frac{D}{c}\frac{dc}{dx} = \frac{D\mathcal{F}}{RT} \tag{24.77}$$

---

8. This expression is derived from eqn 20 by dividing both sides by the Avogadro constant, which converts numbers into amounts (numbers of moles).

Therefore, once we know the effective force and the diffusion coefficient, D, we can calculate the drift speed of the particles (and vice versa) whatever the origin of the force.

There is one case where we already know the drift speed and the effective force acting on a particle: an ion in solution has a drift speed $s = u\mathcal{E}$ when it experiences a force $ez\mathcal{E}$ from an electric field of strength $\mathcal{E}$ (so $\mathcal{F} = N_A ez\mathcal{E} = zF\mathcal{E}$). Therefore, substituting these known values into eqn 77 gives

$$u\mathcal{E} = \frac{zF\mathcal{E}D}{RT}$$

and hence

$$u = \frac{zFD}{RT} \tag{24.78}$$

This equation rearranges into the very important result known as the **Einstein relation** between the diffusion coefficient and the ionic mobility:

$$D = \frac{uRT}{zF} \tag{24.79}°$$

On inserting the typical value $u = 5 \times 10^{-8}$ m² s⁻¹ V⁻¹, we find $D \approx 1 \times 10^{-9}$ m² s⁻¹ at 25°C as a typical value of the diffusion coefficient of an ion in water.

(d) The Nernst–Einstein equation

The Einstein relation provides a link between the molar conductivity of an electrolyte and the diffusion coefficients of its ions. First, by using eqns 54 ($\lambda = zuF$) and 78 we write

$$\lambda = zuF = \frac{z^2 DF^2}{RT} \tag{24.80}°$$

for each type of ion. Then, from $\Lambda_m^\circ = \nu_+ \lambda_+ + \nu_- \lambda_-$, the limiting molar conductivity is

$$\Lambda_m = (\nu_+ z_+^2 D_+ + \nu_- z_-^2 D_-)\frac{F^2}{RT} \tag{24.81}°$$

which is the **Nernst–Einstein equation**. An application of this equation is to the determination of ionic diffusion coefficients from conductivity measurements; another is to the prediction of conductivities using models of ionic diffusion (see below).

(e) The Stokes–Einstein equation

Equations 53 ($u = ez/f$) and 76 relate the mobility of an ion to the frictional force and to the diffusion coefficient, respectively. We can combine the two expressions using $eR/F = k$ into the **Stokes–Einstein equation**:

$$D = \frac{kT}{f} \tag{24.82}$$

If the frictional force is described by Stokes's law, then we also obtain a relation between the diffusion coefficient and the viscosity of the medium:

$$D = \frac{kT}{6\pi\eta a} \tag{24.83}$$

An important feature of eqn 82 (and of its special case, eqn 83) is that it makes no reference to the charge of the diffusing species. Therefore, the equation also applies

Table 24.8* Diffusion coefficients at 298 K, $D/(10^{-9}\,\mathrm{m^2\,s^{-1}})$

H^+ in water	9.31
I_2 in hexane	4.05
Na^+ in water	1.33
Sucrose in water	0.522

* More values are given in the *Data section*.

in the limit of vanishingly small charge; that is, it also applies to neutral molecules. Consequently, we may use viscosity measurements to estimate the diffusion coefficients for electrically neutral molecules in solution (Table 24.8). It must not be forgotten, however, that both equations depend on the assumption that the viscous drag is proportional to the speed.

Example 24.5 Interpreting the mobility of an ion

Use the experimental value of the mobility to evaluate the diffusion coefficient, the limiting molar conductivity, and the hydrodynamic radius of a sulfate ion in aqueous solution.

Method The starting point is the mobility of the ion, which is given in Table 24.6. The diffusion coefficient can then be determined from the Einstein relation, eqn 79. The ionic conductivity is related to the mobility by eqn 54. To estimate the hydrodynamic radius, a, of the ion, use the Stokes–Einstein relation to find f and the Stokes law to relate f to a.

Answer From Table 24.6, the mobility of SO_4^{2-} is $8.29 \times 10^{-8}\,\mathrm{m^2\,s^{-1}\,V^{-1}}$. It follows from eqn 79 that

$$D = \frac{uRT}{zF} = 1.1 \times 10^{-9}\,\mathrm{m^2\,s^{-1}}$$

From eqn 54 it follows that

$$\lambda = zuF = 16\,\mathrm{mS\,m^2\,mol^{-1}}$$

Finally, from $f = 6\pi\eta a$ using 0.891 cP (or $8.91 \times 10^{-4}\,\mathrm{kg\,m^{-1}\,s^{-1}}$) for the viscosity of water (Table 24.4):

$$a = \frac{kT}{6\pi\eta D} = 220\,\mathrm{pm}$$

Comment The bond length in SO_4^{2-} is 144 pm, so the radius calculated here is plausible and consistent with a small degree of solvation.

Self-test 24.7 Repeat the calculation for the NH_4^+ ion.

$$[1.96 \times 10^{-9}\,\mathrm{m^2\,s^{-1}},\ 7.4\,\mathrm{mS\,m^2\,mol^{-1}},\ 125\,\mathrm{pm}]$$

Experimental support for the relations derived above comes from conductivity measurements. In particular, **Walden's rule** is the empirical observation that the product $\eta\Lambda_m$ is very approximately constant for the same ions in different solvents (but there are numerous exceptions). Because $\Lambda_m \propto D$, and we have just seen that $D \propto 1/\eta$, we do indeed predict that $\Lambda_m \propto 1/\eta$, as Walden's rule implies. The usefulness of the rule, however, is muddied by the role of solvation: different solvents solvate the same ions to different extents, so both the hydrodynamic radius and the viscosity change with the solvent.

24.10 The diffusion equation

We now turn to the discussion of time-dependent diffusion processes, where we are interested in the spreading of inhomogeneities with time. One example is the temperature of a metal bar that has been heated at one end: if the source of heat is removed, then the bar gradually settles down into a state of uniform temperature.

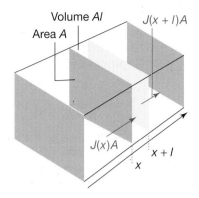

24.24 The net flux in a region is the difference between the flux entering from the region of high concentration (on the left) and the flux leaving to the region of low concentration (on the right).

When the source of heat is maintained and the bar can radiate, it settles down into a steady state of nonuniform temperature. Another example (and one more relevant to chemistry) is the concentration distribution in a solvent to which a solute is added. We shall focus on the description of the diffusion of particles, but similar arguments apply to the diffusion of physical properties, such as temperature. Our aim is to obtain an equation for the rate of change of the concentration of particles in an inhomogeneous region.

The central equation of this section is the **diffusion equation**,[9] which relates the rate of change of concentration at a point to the spatial variation of the concentration at that point:

$$\frac{\partial c}{\partial t} = D\frac{\partial^2 c}{\partial x^2}$$

(24.84)

We show in the following *Justification* that the diffusion equation follows from Fick's first law of diffusion.

Justification 24.11

Consider a thin slab of cross-sectional area A that extends from x to $x + l$ (Fig. 24.24). Let the concentration at x be c at the time t. The amount (number of moles) of particles that enter the slab in the infinitesimal interval dt is $JA\,dt$, so the rate of increase in molar concentration inside the slab (which has volume Al) on account of the flux from the left is

$$\frac{\partial c}{\partial t} = \frac{JA\,dt}{Al\,dt} = \frac{J}{l}$$

There is also an outflow through the right-hand window. The flux through that window is J', and the rate of change of concentration that results is

$$\frac{\partial c}{\partial t} = -\frac{J'A\,dt}{Al\,dt} = -\frac{J'}{l}$$

The net rate of change of concentration is therefore

$$\frac{\partial c}{\partial t} = \frac{J - J'}{l}$$

Each flux is proportional to the concentration gradient at the window. So, by using Fick's first law, we can write

$$J - J' = -D\frac{\partial c}{\partial x} + D\frac{\partial c'}{\partial x} = -D\frac{\partial c}{\partial x} + D\frac{\partial}{\partial x}\left\{c + \left(\frac{\partial c}{\partial x}\right)l\right\} = Dl\frac{\partial^2 c}{\partial x^2}$$

When this relation is substituted into the expression for the rate of change of concentration in the slab, we get eqn 84.

(a) The significance of the diffusion equation

The diffusion equation shows that the rate of change of concentration is proportional to the curvature (more precisely, to the second derivative) of the concentration with respect to distance. If the concentration changes sharply from point to point (if the distribution is highly wrinkled) then the concentration changes rapidly with time. If the curvature is zero, then the concentration is constant in time. If the

9. This equation is sometimes called 'Fick's second law of diffusion'.

concentration decreases linearly with distance, then the concentration at any point is constant because the inflow of particles is exactly balanced by the outflow.

The diffusion equation can be regarded as a mathematical formulation of the intuitive notion that there is a natural tendency for the wrinkles in a distribution to disappear. More succinctly: Nature abhors a wrinkle.

(b) Diffusion with convection

The transport of particles arising from the motion of a streaming fluid is called **convection**. If for the moment we ignore diffusion, then the flux of particles through an area A in an interval Δt when the fluid is flowing at a velocity v can be calculated in the way we have used several times before (by counting the particles within a distance $v\Delta t$), and is

$$J = \frac{cAv\Delta t}{A\Delta t} = cv \tag{24.85}$$

This J is called the **convective flux**. The rate of change of concentration in a slab of thickness l and area A is, by the same argument as before,

$$\frac{\partial c}{\partial t} = \frac{J - J'}{l} = \left\{ c - \left[c + \left(\frac{\partial c}{\partial x}\right)l \right] \right\} \frac{v}{l} = -v\frac{\partial c}{\partial x} \tag{24.86}$$

(We have assumed that the velocity does not depend on the position.)

When both diffusion and convection occur, the total change of concentration in a region is the sum of the two effects, and the **generalized diffusion equation** is

$$\frac{\partial c}{\partial t} = D\frac{\partial^2 c}{\partial x^2} - v\frac{\partial c}{\partial x} \tag{24.87}$$

A further refinement, which is important in chemistry, is the possibility that the concentrations of particles may change as a result of reaction. When reactions are included in eqn 87 (Section 27.2), we get a powerful differential equation for discussing the properties of reacting, diffusing, convecting systems and which is the basis of reactor design in chemical industry and of the utilization of resources in living cells.

(c) Solutions of the diffusion equation

The diffusion equation, eqn 84, is a second-order differential equation with respect to space and a first-order differential equation with respect to time. Therefore, we must specify two boundary conditions for the spatial dependence and a single initial condition for the time dependence.

As an illustration, consider a solvent in which the solute is initially coated on one surface of the container (for example, a layer of sugar on the bottom of a deep beaker of water). The single initial condition is that at $t = 0$ all N_0 particles are concentrated on the yz-plane (of area A) at $x = 0$. The two boundary conditions are derived from the requirements: (1) that the concentration must everywhere be finite and (2) that the total amount (number of moles) of particles present is n_0 (with $n_0 = N_0/N_A$) at all times. These requirements imply that the flux of particles is zero at the top and bottom surfaces of the system. Under these conditions it is found that

$$c(x, t) = \frac{n_0}{A(\pi Dt)^{1/2}} e^{-x^2/4Dt} \tag{24.88}$$

as may be verified by direct substitution. Figure 24.25 shows the shape of the concentration distribution at various times, and it is clear that the concentration spreads and tends to uniformity.

24.25 The concentration profiles above a plane from which a solute is diffusing. The curves are plots of eqn 88. The units of Dt and x are arbitrary, but are related so that Dt/x^2 is dimensionless. For example, if x is in metres, Dt would be in metres²; so, for $D = 10^{-9}$ m² s⁻¹, $Dt = 0.1$ m² corresponds to $t = 10^8$ s.

Another useful result is for a localized concentration of solute in a three-dimensional solvent (a sugar lump suspended in a large flask of water). The concentration of diffused solute is spherically symmetrical, and at a radius r is

$$c(r, t) = \frac{n_0}{8(\pi Dt)^{3/2}} e^{-r^2/4Dt} \tag{24.89}$$

Other chemically (and physically) interesting arrangements, such as transport of substances across biological membranes (Box 24.2), can be treated. In many cases the solutions are more cumbersome.

(d) The measurement of diffusion coefficients

The solutions of the diffusion equation are useful for experimental determinations of diffusion coefficients. In the **capillary technique**, a capillary tube, open at one end and containing a solution, is immersed in a well stirred larger quantity of solvent, and the change of concentration in the tube is monitored. The solute diffuses from the open end of the capillary at a rate that can be calculated by solving the diffusion equation with the appropriate boundary conditions, so D may be determined. In the **diaphragm technique**, the diffusion occurs through the capillary pores of a sintered glass diaphragm separating the well-stirred solution and solvent. The concentrations are monitored and then related to the solutions of the diffusion equation corresponding to this arrangement. Diffusion coefficients may also be measured by the dynamic light scattering technique described in Section 22.8 and by NMR.

Box 24.2 Transport across biological membranes

Controlled transport of molecules and ions across biological membranes is at the heart of a number of key cellular processes, such as the generation of action potentials in neurons (Section 10.4g), the transfer of glucose into red blood cells, and the synthesis of ATP by oxidative phosphorylation and photophosphorylation (Box 10.1). Here we examine in some detail the various ways in which non-electrolytes and electrolytes cross the lipid bilayer.

Suppose that a lipid bilayer provides a barrier that slows down the transfer of molecules or ions into or out of the cell. The thermodynamic tendency to transport a species A through the bilayer is partially determined by an activity gradient across the membrane, which results in a difference in molar Gibbs energy between the inside and the outside of the cell

$$\Delta G_m = G_{m,in} - G_{m,out} = RT \ln \frac{a_{in}}{a_{out}}$$

The equation implies that transport into the cell of either neutral or charged species is thermodynamically favourable if $a_{in} < a_{out}$ or, if we set the activity coefficients to unity, if $[A]_{in} < [A]_{out}$. If A is an ion, there is a second component of ΔG_m that arises from differences in Coulomb repulsions on each side of the bilayer and gives rise to a membrane

potential difference $\Delta\phi = \phi_{in} - \phi_{out}$. The final expression for ΔG is then

$$\Delta G_m = RT \ln \frac{[A]_{in}}{[A]_{out}} + zF\Delta\phi$$

where z is the ion charge number and F is Faraday's constant. This equation implies that there is a tendency, called *passive transport*, for a species to move down concentration and membrane potential gradients. It is also possible to move a species against these gradients, but the process must now be driven by the exergonic hydrolysis of ATP. This process is called *active transport*.

Consider the passive transport of an uncharged species A across a lipid bilayer of thickness l. To simplify the problem, we will assume that the concentration of A is always maintained at $[A] = [A]_0$ on one surface of the membrane and at $[A] = 0$ on the other surface, perhaps by a perfect balance between the rate of the process that produces A on one side and the rate of another process that consumes A completely on the other side. This is one example of a steady-state assumption, which will be discussed in more detail in Section 25.7. Then $\partial[A]/\partial t = 0$ and the diffusion equation simplifies to

$$D \frac{d^2[A]}{dx^2} = 0$$

where D is the diffusion coefficient and the steady-state assumption makes partial derivatives unnecessary. We use the boundary conditions $[A](0) = [A]_0$ and $[A](l) = 0$ to solve this differential equation and the result, which may be verified by differentiation, is

$$[A](x) = [A]_0 \left(1 - \frac{x}{l} \right)$$

which implies that the [A] decreases linearly inside the membrane. We now use Fick's first law to calculate the flux J of A through the membrane and the result is

$$J = D \frac{[A]_0}{l}$$

However, we need to modify this equation slightly to account for the fact that the concentration of A on the surface of a membrane is not always equal to the concentration of A measured in the bulk solution, which we assume to be aqueous. This difference arises from the significant difference in the solubility of A in an aqueous environment and in the solution-membrane interface. One way to deal with this problem is to define a *partition coefficient* κ as

$$\kappa = \frac{[A]_0}{[A]_s}$$

where $[A]_s$ is the concentration of A in the bulk aqueous solution. It follows that

$$J = \kappa D \frac{[A]_s}{l}$$

In spite of the assumptions that led to its final form, this equation describes adequately the passive transport of many non-electrolytes through membranes of blood cells.

In many cases the equation just derived underestimates the flux and the implication is that the membrane is more permeable than expected. However, because the permeability increases only for certain species and not others, this is evidence that transport can be mediated by carriers. One example is the transporter protein that carries glucose into cells.

A characteristic of a carrier C is that it binds to the transported species A and the dissociation of the AC complex is described by

$$AC \rightleftharpoons A + C \qquad K = \frac{[A][C]}{[AC]}$$

where we have used concentrations instead of activities. After writing $[C]_0 = [C] + [AC]$, where $[C]_0$ is the total concentration of carrier, it follows that

$$[AC] = \frac{[A][C]_0}{[A] + K}$$

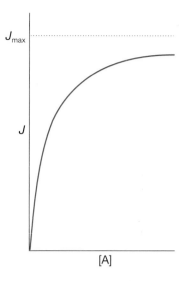

The flux of the species AC through a membrane varies with the concentration of the species A. The behaviour shown in the illustration and explained in the text is characteristic of mediated transport of A, with C as a carrier molecule.

We can now use the expression for J derived above to write an expression for the flux of the species AC through the membrane:

$$J = \frac{\kappa_{AC} D_{AC} [C]_0}{l} \frac{[A]}{[A] + K} = J_{max} \frac{[A]}{[A] + K}$$

where κ_{AC} and D_{AC} are the partition coefficient and diffusion coefficient of the species AC, respectively. We see from the first illustration that when $[A] \ll K$ the flux varies linearly with [A] and that the flux reaches a maximum value of $J_{max} = \kappa_{AC} D_{AC} [C]_0 / l$ when $[A] \gg K$. This behaviour is characteristic of mediated transport.

The transport of ions through bilayers also needs to be mediated because charged species do not partition well into the hydrophobic environment of the membrane. There are two mechanisms for ion transport: mediation by a carrier molecule or transport through a *channel former*, a protein that creates a hydrophilic pore through which the ion can pass. An example of a channel former is the polypeptide gramicidin A, which increases the membrane permeability to cations such as H^+, K^+, and Na^+.

Ion channels are proteins that effect the movement of specific ions down a membrane potential gradient. They are highly selective, so there is a channel protein for Ca^{2+}, another for Cl^-, and so on. The opening of the gate may be triggered by potential differences between the two sides of the membrane or by the binding of an *effector* molecule to a specific receptor site on the channel. It is now possible to measure currents and conductances through single ion channel proteins; they are of the order of 10^{-12} A and 10^{-11}–10^{-10} S, respectively.

Ions such as H^+, Na^+, K^+, and Ca^{2+} are often transported actively across membranes by integral proteins called *ion pumps*. Ion pumps are molecular machines that work by adopting conformations that are permeable to one ion but not others depending on the state of phosphorylation of the protein. Because protein phosphorylation requires dephosphorylation of ATP, the conformational change that opens or closes the pump is endergonic and requires the use of energy stored during metabolism. An example of an ion pump is the Na^+/K^+-ATPase already discussed in Section 10.4g in connection with neuronal action potentials.

Another important ion pump is the H^+-ATPase, which occurs in the inner membranes of the chloroplast and the mitochondrion (Box 10.1). The *chemiosmotic theory* proposed by Peter Mitchell explains how H^+-ATPases use the flux of protons through the membrane to power the phosphorylation of ADP to ATP. Some of the electron transfer reactions of respiration and photosynthesis also carry protons from one side of the membrane to the other, creating a proton gradient. We can estimate the Gibbs energy available for phosphorylation by writing

$$\Delta G_m = RT \ln \frac{[H^+]_{in}}{[H^+]_{out}} + F\Delta\phi$$

where we have used $z = +1$. After substituting $\Delta pH = pH_{in} - pH_{out} = -\log[H^+]_{in} + \log[H^+]_{out}$, it follows that

$$\Delta G_m = F\Delta\phi - 2.303RT\Delta pH$$

In the mitochondrion, $\Delta pH \approx -1.4$ and $\Delta\phi \approx 0.14$ V, so $\Delta G_m \approx +21.5$ kJ mol^{-1}. Because 31 kJ mol^{-1} is needed for phosphorylation, we conclude that at least 2 mol of protons (and probably much more) must flow through the membrane for the phosphorylation of 1 mol of ADP.

The coupling of the proton flow to the phosphorylation is itself of considerable interest and we present a model proposed by Paul Boyer.[10] Structural studies by John Walker show that the channel through which the protons flow is linked in tandem to a unit composed of six protein molecules arranged in pairs of α and β subunits to form three interlocked $\alpha\beta$ segments (see the second illustration). The conformations of the three pairs may be loose (L), tight (T), or open (O), and one of each type is present at each stage. A protein at the centre of the interlocked structure, the γ subunit shown as a white arrow, rotates and induces structural changes that cycle each of the three segments between L, T, and O conformations. At the start of a cycle, a T unit holds an ATP molecule. Then ADP and a P_i group migrate into the

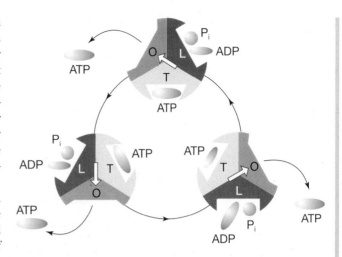

The mechanism of action of H^+-ATPase, a molecular motor that transports protons across the mitochondrial membrane and catalyses either the formation or hydrolysis of ATP.

L site and, as it closes into T, the earlier T site opens into O and releases its ATP. The ADP and P_i in the T site meanwhile condense into ATP, and the new L site is ready for the cycle to begin again. The proton flux drives the rotation of the γ subunit, and hence the conformational changes of the α/β segments, as well as providing the energy for the condensation reaction itself. Several key aspects of this mechanism have been confirmed experimentally. For example, the rotation of the γ subunit has been visualized directly by using single-molecule spectroscopy (Box 17.2).

Discussion questions

1. Provide a molecular interpretation for the observation that mediated transport leads to a maximum flux J_{max} when the concentration of the transported species becomes very large.

2. If the mitochondrial membrane potential were 70 mV, as is common for other membranes, how much ATP could be synthesized from the transport of 4 mol H^+, assuming the pH difference remains the same?

Further reading

C. Miller, Ion channels: doing hard chemistry with hard ions. *Curr. Op. Chem. Biol.* **4**, 148 (2000).

D. Stack, A.G.W. Leslie, and J.E. Walker, Molecular architecture of the rotary motor in ATP synthase. *Science* **286**, 1700 (1999).

10. Paul Boyer and John Walker received the Nobel Prize for chemistry in 1997 for their work on the mechanism of H^+-ATPase. They shared the award with Jens Skou, who was recognized for his work on the Na^+/K^+-ATPase. Peter Mitchell received the Nobel Prize for chemistry in 1978 for his chemiosmotic theory.

24.11 Diffusion probabilities

The solutions of the diffusion equation can be used to predict the concentration of particles (or the value of some other physical quantity, such as the temperature in a nonuniform system) at any location. We can also use them to calculate the net distance through which the particles diffuse in a given time.

> ### Example 24.6 Calculating the net distance of diffusion
>
> Calculate the net distance travelled on average by particles in a time t if they have a diffusion constant D.
>
> **Method** We need to calculate the probability that a particle will be found at a certain distance from the origin, and then calculate the average by weighting each distance by that probability.
>
> **Answer** The number of particles in a slab of thickness dx and area A at x, where the molar concentration is c, is $cAN_A\,dx$. The probability that any of the $N_0 = n_0 N_A$ particles is in the slab is therefore $cAN_A\,dx/N_0$. If the particle is in the slab, it has travelled a distance x from the origin. Therefore, the mean distance travelled by all the particles is the sum of each x weighted by the probability of its occurrence:
>
> $$\langle x \rangle = \int_0^\infty \frac{xcAN_A}{N_0}\,dx = \frac{1}{(\pi Dt)^{1/2}} \int_0^\infty x e^{-x^2/4Dt}\,dx = 2\left(\frac{Dt}{\pi}\right)^{1/2}$$
>
> **Comment** The average distance of diffusion varies as the square root of the lapsed time. If we use the Stokes–Einstein relation for the diffusion coefficient, the mean distance travelled by particles of radius a in a solvent of viscosity η is
>
> $$\langle x \rangle = \left(\frac{2kTt}{3\pi^2 \eta a}\right)^{1/2}$$
>
> **Self-test 24.8** Derive an expression for the root mean square distance travelled by diffusing particles in a time t. $[\langle x^2 \rangle^{1/2} = (2Dt)^{1/2}]$

As shown in Example 24.6, the average distance travelled by diffusing particle in a time t (in an arrangement like that illustrated in Fig. 24.25) is

$$\langle x \rangle = 2\left(\frac{Dt}{\pi}\right)^{1/2} \tag{24.90}$$

and the root mean square distance travelled in the same time is

$$\langle x^2 \rangle^{1/2} = (2Dt)^{1/2} \tag{24.91}$$

The latter is a valuable measure of the spread of particles when they can diffuse in both directions from the origin (for then $\langle x \rangle = 0$ at all times). The root mean square distance travelled by particles with a typical diffusion coefficient ($D = 5 \times 10^{-10}$ m^2 s^{-1}) is illustrated in Fig. 24.26, which shows how long it takes for diffusion to increase the net distance travelled on average to about 1 cm in an unstirred solution. The graph shows that diffusion is a very slow process (which is why solutions are stirred, to encourage mixing by convection).

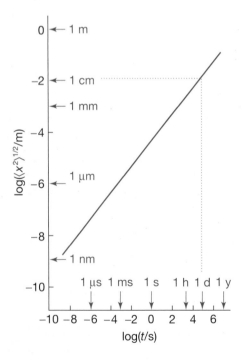

24.26 The root mean square distance covered by particles with $D = 5 \times 10^{-10}$ m^2 s^{-1}. Note the great slowness of diffusion.

24.12 The statistical view

An intuitive picture of diffusion is of the particles moving in a series of small steps and gradually migrating from their original positions. We shall explore this idea using a model in which the particles can jump through a distance λ in a time τ. The total distance travelled by a particle in a time t is therefore $t\lambda/\tau$. However, the particle will not necessarily be found at that distance from the origin. The direction of each step may be different, and the net distance travelled must take the changing directions into account.

If we simplify the discussion by allowing the particles to travel only along a straight line (the x-axis), and for each step (to the left or the right) to be through the same distance λ, then we obtain the **one-dimensional random walk**.[11] We show in the *Justification* below that the probability of a particle being at a distance x from the origin after a time t is

$$P = \left(\frac{2\tau}{\pi t}\right)^{1/2} e^{-x^2\tau/2t\lambda^2} \tag{24.92}$$

Justification 24.12

Consider a one-dimensional random walk in which each step is through a distance λ to the left or right. The net distance travelled after N steps is equal to the difference between the number of steps to the right (N_R) and to the left (N_L), and is $(N_R - N_L)\lambda$. We write $n = N_R - N_L$ and the total number of steps as $N = N_R + N_L$.

The number of ways of performing a walk with a given net distance of travel $n\lambda$ is the number of ways of making N_R steps to the right and N_L steps to the left, and is given by the binomial coefficient

$$W = \frac{N!}{N_L!N_R!} = \frac{N!}{\{\frac{1}{2}(N + n)\}!\{\frac{1}{2}(N - n)\}!}$$

The probability of the net distance walked being $n\lambda$ is

$$P = \frac{\text{number of paths with } N_R \text{ steps to the right}}{\text{total number of steps}}$$

$$= \frac{W}{2^N} = \frac{N!}{\{\frac{1}{2}(N + n)\}!\{\frac{1}{2}(N - n)\}!2^N}$$

The use of Stirling's approximation (Section 19.1a) in the form

$$\ln x! \approx \ln(2\pi)^{1/2} + (x + \tfrac{1}{2})\ln x - x$$

gives (after quite a lot of algebra)

$$\ln P = \ln\left(\frac{2}{\pi N}\right)^{1/2} - \tfrac{1}{2}(N + n + 1)\ln\left(1 + \frac{n}{N}\right) - \tfrac{1}{2}(N - n + 1)\ln\left(1 - \frac{n}{N}\right)$$

For small net distances ($n \ll N$) we can use the approximation $\ln(1 \pm x) \approx \pm x - \frac{1}{2}x^2$, and so obtain

$$\ln P \approx \ln\left(\frac{2}{\pi N}\right)^{1/2} - \frac{n^2}{2N}$$

---

11. The same model was used in the discussion of a one-dimensional random coil in Section 22.2a.

At this point, we note that the number of steps taken in a time t is $N = t/\tau$ and the net distance travelled from the origin is $x = n\lambda$. Substitution of these quantities into the equation just derived gives an expression that rearranges into eqn 92.

The differences of detail between eqns 88 and 92 arise from the fact that in the present calculation the particles can migrate in either direction from the origin. Moreover, they can be found only at discrete points separated by λ instead of being anywhere on a continuous line. The fact that the two expressions are so similar suggests that diffusion can indeed be interpreted as the outcome of a large number of steps in random directions.

We can now relate the coefficient D to the step length λ and the rate at which the jumps occur. Thus, by comparing the two exponents in eqn 88 and eqn 92 we can immediately write down the **Einstein–Smoluchowski equation**:

$$D = \frac{\lambda^2}{2\tau} \tag{24.93}$$

Illustration 24.6

Suppose that a SO_4^{2-} ion jumps through its own diameter each time it makes a move in an aqueous solution; then, because $D = 1.1 \times 10^{-9}\ m^2\ s^{-1}$ and $a = 210$ pm (as deduced from mobility measurements), it follows from $\lambda = 2a$ that $\tau = 80$ ps. Because τ is the time for one jump, the ion makes 1×10^{10} jumps per second.

The Einstein–Smoluchowski equation is the central connection between the microscopic details of particle motion and the macroscopic parameters relating to diffusion (for example, the diffusion coefficient and, through the Stokes–Einstein relation, the viscosity). It also brings us back full circle to the properties of the perfect gas. For if we interpret λ/τ as $\bar{c}$, the mean speed of the molecules, and interpret λ as a mean free path, then we can recognize in the Einstein–Smoluchowski equation exactly the same expression as we obtained from the kinetic theory of gases in *Justification 24.5*. That is, the diffusion of a perfect gas is a random walk with an average step size equal to the mean free path.

Checklist of key ideas

Molecular motion in gases

24.1 The kinetic model of gases

☐ transport property, the ability of a substance to transfer matter, energy, or some other property from one place to another

☐ diffusion, the migration of matter down a concentration gradient

☐ thermal conduction, the migration of energy down a temperature gradient

☐ electric conduction, the migration of electric charge along an electrical potential gradient

☐ viscosity, the migration of linear momentum down a velocity gradient

☐ effusion, the emergence of a gas from a container through a small hole

☐ kinetic model, a model of a gas in which the only contribution to the energy is from the kinetic energies of the molecules

☐ elastic collision, a collision in which the total translational kinetic energy of the molecules is conserved

☐ pressure of a gas, $pV = \frac{1}{3}nMc^2$

☐ root mean square speed, the square root of the mean of the squares of the speeds: $c = \langle v^2 \rangle^{1/2} = (3RT/M)^{1/2}$

☐ distribution of speeds, the function $f(v)$ which, through $f(v)\ dv$, gives the fraction of molecules that have speeds in the range v to $v + dv$

- [ ] Maxwell distribution of speeds, $f(v) = 4\pi(M/2\pi RT)^{3/2}v^2\,e^{-Mv^2/2RT}$

- [ ] mean speed, $\bar{c} = (8RT/\pi M)^{1/2}$

- [ ] most probable speed, $c^* = (2RT/M)^{1/2}$

- [ ] relative mean speed, $\bar{c}_{rel} = (8kT/\pi\mu)^{1/2}$

- [ ] reduced mass, $\mu = m_A m_B/(m_A + m_B)$

- [ ] collision diameter, the distance of approach corresponding to a collision

- [ ] collision frequency, the number of collisions made by a molecule in an interval divided by the length of the interval, $z = \sigma\bar{c}_{rel}\mathcal{N}$

- [ ] collision cross-section, $\sigma = \pi d^2$

- [ ] mean free path, the average distance a molecule travels between collisions: $\lambda = \bar{c}/z$

24.2 Collisions with walls and surfaces

- [ ] collision flux, Z_W, the number of collisions with an area in a given time interval divided by the area and the duration of the interval, $Z_W = p/(2\pi mkT)^{1/2}$

- [ ] collision frequency, the collision flux multiplied by the area of the region of interest

24.3 The rate of effusion

- [ ] Graham's law of effusion: the rate of effusion is inversely proportional to the square root of the molar mass

- [ ] rate of effusion: rate $= pA_0/(2\pi mkT)^{1/2}$

- [ ] Knudsen method, a method for the determination of the vapour pressures of liquids and solids

24.4 Transport properties of a perfect gas

- [ ] flux, the quantity of a property passing through a given area in a given time interval divided by the area and the duration of the interval

- [ ] matter flux, the flux of matter, $J(\text{matter}) \propto d\mathcal{N}/dz$

- [ ] energy flux, the flux of energy, $J(\text{energy}) \propto dT/dz$

- [ ] momentum flux, $J(\text{momentum}) \propto dv/dz$

- [ ] Fick's first law of diffusion: the flux of matter is proportional to the concentration gradient, $J(\text{matter}) = -D\,d\mathcal{N}/dz$

- [ ] diffusion coefficient, the coefficient in Fick's first law

- [ ] coefficient of thermal conductivity, the coefficient κ in $J(\text{energy}) = -\kappa\,dT/dz$

- [ ] Newtonian flow, flow that occurs by a series of layers moving past one another

- [ ] coefficient of viscosity, the coefficient η in $J(\text{momentum}) = -\eta\,dv_x/dz$

- [ ] diffusion coefficient for a perfect gas, $D = \frac{1}{3}\lambda\bar{c}$

- [ ] coefficient of thermal conductivity for a perfect gas, $\kappa = \frac{1}{3}\lambda\bar{c}C_{V,m}[A]$

- [ ] coefficient of viscosity of a perfect gas, $\eta = \frac{1}{3}M\lambda\bar{c}[A]$

- [ ] Poiseuille's formula for the rate of flow, $dV/dt = (p_1^2 - p_2^2)\pi r^4/16l\eta p_0$

Molecular motion in liquids

24.5 Experimental results

- [ ] inelastic neutron scattering, scattering in which neutrons collect or discard energy as they pass through a sample

24.6 The conductivities of electrolyte solutions

- [ ] conductance, G, the inverse of resistance

- [ ] conductivity, the constant κ in $G = \kappa A/l$

- [ ] molar conductivity, $\Lambda_m = \kappa/c$

- [ ] strong electrolyte, an electrolyte with a molar conductivity that varies only slightly with concentration

- [ ] weak electrolyte, an electrolyte with a molar conductivity that is normal at concentrations close to zero, but falls sharply to low values as the concentration increases

- [ ] Kohlrausch's law, for the concentration dependence of the molar conductivity of a strong electrolyte, $\Lambda_m = \Lambda_m^\circ - \mathcal{K}c^{1/2}$

- [ ] limiting molar conductivity, Λ_m°, the molar conductivity at zero concentration

- [ ] law of the independent migration of ions, $\Lambda_m^\circ = v_+\lambda_+ + v_-\lambda_-$

- [ ] degree of ionization (or deprotonation), the parameter α in $[H_3O^+] = \alpha c$, $[A^-] = \alpha c$, and $[HA] = (1 - \alpha)c$ for the deprotonation of a monoprotic acid

- [ ] Ostwald's dilution law, $1/\Lambda_m = 1/\Lambda_m^\circ + \Lambda_m c/K_a(\Lambda_m^\circ)^2$

24.7 The mobilities of ions

- [ ] drift speed, the terminal speed when an accelerating force is balanced by the viscous drag: $s = u\mathcal{E}$

- [ ] mobility of an ion, the coefficient u in the expression $s = u\mathcal{E}$; $u = ze/6\pi\eta a$

- [ ] hydrodynamic radius (Stokes radius), the effective radius of a particle in solution

- [ ] Grotthuss mechanism, a mechanism for the conduction of protons in solution

- [ ] ionic conductivity, the contribution of ions of one type to the molar conductivity: $\lambda = zuF$

- [ ] transport number, the fraction of total current carried by the ions of a specified type, $t_\pm = I_\pm/I$

- [ ] limiting transport number, the transport number at zero concentration, $t_\pm^\circ = u_\pm/(u_+ + u_-)$

- [ ] moving boundary method, a method for determining transport numbers based on the observation of the motion of a boundary between two ionic solutions having a common ion

- [ ] indicator solution, the solution that occupies the lower part of a vertical tube

- [ ] leading solution, the solution that occupies the upper part of a vertical tube

- [ ] Hittorf method, a method for determining transport numbers based on the determination of the changes in concentration when a current flows

- [ ] cell with transference, a cell with a liquid boundary

24.8 Conductivities and ion–ion interactions

- [ ] relaxation effect, the reduction of an ion's mobility due to distortion of the ionic atmosphere

- [ ] electrophoretic effect, the enhanced viscous drag due to the countercurrent of oppositely charged ions

- [ ] Debye–Hückel–Onsager theory, a theory of the concentration dependence of the molar conductivity of a strong electrolyte, $\mathcal{K} = A + B\Lambda_m^\circ$

- [ ] Green–Kubo relation, the relation between a transport property and fluctuations in microscopic properties of a system, $\kappa = \dfrac{1}{kTV}\displaystyle\int_0^\infty \langle j(0)j(t)\rangle\,dt$

- [ ] correlation function, the average of an observable compared at different times, $\langle j(0)j(t)\rangle$

Diffusion

24.9 The thermodynamic view

- [ ] thermodynamic force, $\mathcal{F} = -(\partial\mu/\partial x)_{p,T}$

- [ ] Einstein relation, $D = uRT/zF$

- [ ] Nernst–Einstein relation, $\Lambda_m = (v_+z_+^2 D_+ + v_-z_-^2 D_-)(F^2/RT)$

- [ ] Stokes–Einstein equation, $D = kT/f$

- [ ] Walden's rule: The product $\eta\Lambda_m$ is very approximately constant for the same ions in different solvents

24.10 The diffusion equation

- [ ] diffusion equation, the relation between the rate of change of concentration at a point and the spatial variation of the concentration at that point: $\partial c/\partial t = D\partial^2 c/\partial x^2$

☐ convection, the transport of particles arising from the motion of a streaming fluid

☐ convective flux, $J = cv$

☐ generalized diffusion equation, the diffusion equation including convection: $\partial c/\partial t = D\partial^2 c/\partial x^2 - v\partial c/\partial x$

☐ capillary technique, a technique for determining the diffusion coefficient

☐ diaphragm technique, a technique for determining the diffusion coefficient

24.11 Diffusion probabilities

☐ average distance travelled in diffusion, $\langle x \rangle = 2(Dt/\pi)^{1/2}$

☐ root mean square distance travelled in diffusion, $\langle x^2 \rangle^{1/2} = (2Dt)^{1/2}$

24.12 The statistical view

☐ one-dimensional random walk, a random walk along a straight line, $P = (2\tau/\pi t)^{1/2} e^{-x^2\tau/2t\lambda^2}$

☐ Einstein–Smoluchowski equation, $D = \lambda^2/2\tau$

Further reading

Articles of general interest

N. Agmon, The Grotthuss mechanism. *Chem. Phys. Letts.* **244**, 456 (1995).

P.J. Dunlop, B.J. Steel, and J.E. Lane, Experimental methods for studying diffusion in liquids, gases, and solids. In *Techniques of chemistry* (ed. A. Weissberger and B.W. Rossiter), **4**, 205. Wiley-Interscience, New York (1972).

B.L. Earl, Confusion in the expressions for transport coefficients. *J. Chem. Educ.* **66**, 147 (1989).

H.G. Herz, B.M. Braun, K.J. Müller, and R. Mauer, What is the physical significance of the pictures representing the Grotthuss H$^+$ conductance mechanism? *J. Chem. Educ.* **64**, 777 (1987).

T. Kenney. Graham's law: defining gas velocities. *J. Chem. Educ.* **67**, 871 (1990).

D.G. Leaist, Diffusion and ionic conduction in liquids. In *Encyclopedia of applied physics* (ed. G.L. Trigg), **5**, 661. VCH, New York (1993).

M.S.P. Sansom, I.H. Shrivastave, K.M. Ranatunga, and G.R. Smith, Simulations of ion channels: watching ions and water move. *Trends in Biol. Sci.* **25**, 368 (2000).

M. Spiro, Determination of transference numbers. In *Techniques of chemistry* (ed. A. Weissberger and B.W. Rossiter), **2A**, 205. Wiley-Interscience, New York (1971).

Texts and sources of data and information

A.J. Bard and L.R. Faulkner, *Electrochemical methods: fundamentals and applications.* Wiley, New York (2000).

R.B. Bird, W.E. Stewart, and E.N. Lightfoot, *Transport phenomena.* Wiley, New-York (1960).

J. Crank, *The mathematics of diffusion.* Clarendon Press, Oxford (1975).

C.E. Hecht, *Statistical mechanics and kinetic theory.* Dover, New York (1998).

J.O. Hirschfelder, *The molecular theory of gases and liquids.* Wiley, New York (1964).

W. Kauzmann, *Kinetic theory of gases.* Addison-Wesley, Reading (1966).

G.A. Krestov, N.P. Novosyolov, and I.S. Perelygin, *Ionic solvation.* Ellis Horwood/ Prentice Hall, Hemel Hempstead (1993).

D.R. Lide (ed.), *CRC Handbook of Chemistry and Physics,* Sections 5 and 6. CRC Press, Boca Raton (2000).

J.N. Murrell and A.D. Jenkins, *Properties of liquids and solutions.* Wiley-Interscience, New York (1994).

NIST thermodynamic and transport properties of pure fluids database, NIST standard reference database 12, National Institute of Standards and Testing, Gaithersburg (1995). This material is on CD-rom.

J.S. Rowlinson and F.L. Swinton, *Liquids and liquid mixtures.* Butterworths, London (1982).

D. Tabor, *Gases, liquids, and solids.* Cambridge University Press (1991).

K.E. van Holde, W.C. Johnson, and P.S. Ho, *Principles of physical biochemistry.* Prentice Hall, Upper Saddle River (1998).

A.J. Walton, *Three phases of matter.* Oxford University Press (1983).

Exercises

Discussion questions

24.1 (a) Use the kinetic theory to justify the following observations: (a) the rate of a reaction in the gas phase depends on the energy with which two molecules collide, which in turn depends on their speeds; (b) in the Earth's atmosphere, light gases, such as H_2 and He, are rare but heavier gases, such as O_2, CO_2, and N_2, are abundant.

24.1 (b) Provide a molecular interpretation for each of the following processes: diffusion, thermal conduction, electric conduction, and viscosity.

24.2 (a) Provide a molecular interpretation for the observation that the viscosity of a gas increases with temperature, whereas the viscosity of a liquid decreases with increasing temperature.

24.2 (b) Discuss the mechanism of proton conduction in liquid water.

24.3 (a) State the limits to the generality of the following expressions: (a) $J = -D(dc/dx)$, (b) $D = kT/f$, and (c) $D = kT/6\pi\eta a$.

24.3 (b) Discuss how nuclear magnetic resonance spectroscopy, inelastic neutron scattering, and dynamic light scattering may be used to measure the mobility of molecules in liquids.

Numerical exercises

24.4 (a) Determine the ratios of (a) the mean speeds, (b) the mean kinetic energies of H_2 molecules and Hg atoms at 20°C.

24.4 (b) Determine the ratios of (a) the mean speeds, (b) the mean kinetic energies of He atoms and Hg atoms at 25°C.

24.5 (a) A 1.0 L glass bulb contains 1.0×10^{23} H_2 molecules. If the pressure exerted by the gas is 100 kPa, what are (a) the temperature of the gas, (b) the root mean square speeds of the molecules. (c) Would the temperature be different if they were O_2 molecules?

24.5 (b) The best laboratory vacuum pump can generate a vacuum of about 1 nTorr. At 25°C and assuming that air consists of N_2 molecules with a collision diameter of 395 pm, calculate (a) the mean speed of the molecules, (b) the mean free path, (c) the collision frequency in the gas.

24.6 (a) At what pressure does the mean free path of argon at 25°C become comparable to the size of a 1 L vessel that contains it? Take $\sigma = 0.36$ nm^2.

24.6 (b) At what pressure does the mean free path of argon at 25°C become comparable to the diameters of the atoms themselves?

24.7 (a) At an altitude of 20 km the temperature is 217 K and the pressure 0.050 atm. What is the mean free path of N_2 molecules? ($\sigma = 0.43$ nm^2).

24.7 (b) At an altitude of 15 km the temperature is 217 K and the pressure 12.1 kPa. What is the mean free path of N_2 molecules? ($\sigma = 0.43$ nm^2).

24.8 (a) How many collisions does a single Ar atom make in 1.0 s when the temperature is 25°C and the pressure is (a) 10 atm, (b) 1.0 atm, (c) 1.0 µatm?

24.8 (b) How many collisions per second does an N_2 molecule make at an altitude of 15 km? (See Exercise 24.7b for data.)

24.9 (a) Calculate the mean free path of molecules in air using $\sigma = 0.43$ nm^2 at 25°C and (a) 10 atm, (b) 1.0 atm, (c) 1.0×10^{-6} atm.

24.9 (b) Calculate the mean free path of carbon dioxide molecules in air using $\sigma = 0.52$ nm^2 at 25°C and (a) 15 atm, (b) 1.0 bar, (c) 1.0 Torr.

24.10 (a) Use the Maxwell distribution of speeds to estimate the fraction of N_2 molecules at 500 K that have speeds in the range 290 to 300 m s^{-1}.

24.10 (b) Use the Maxwell distribution of speeds to estimate the fraction of CO_2 molecules at 300 K that have speeds in the range 200 to 250 m s^{-1}.

24.11 (a) A solid surface with dimensions 2.5 mm × 3.0 mm is exposed to argon gas at 90 Pa and 500 K. How many collisions do the Ar atoms make with this surface in 15 s?

24.11 (b) A solid surface with dimensions 3.5 mm × 4.0 cm is exposed to helium gas at 111 Pa and 1500 K. How many collisions do the He atoms make with this surface in 10 s?

24.12 (a) An effusion cell has a circular hole of diameter 2.50 mm. If the molar mass of the solid in the cell is 260 g mol^{-1} and its vapour pressure is 0.835 Pa at 400 K, by how much will the mass of the solid decrease in a period of 2.00 h?

24.12 (b) An effusion cell has a circular hole of diameter 3.00 mm. If the molar mass of the solid in the cell is 300 g mol^{-1} and its vapour pressure is 0.224 Pa at 450 K, by how much will the mass of the solid decrease in a period of 24.00 h?

24.13 (a) Calculate the flux of energy arising from a temperature gradient of 2.5 K m^{-1} in a sample of argon in which the mean temperature is 273 K.

24.13 (b) Calculate the flux of energy arising from a temperature gradient of 3.5 K m^{-1} in a sample of hydrogen in which the mean temperature is 260 K.

24.14 (a) Use the experimental value of the thermal conductivity of neon (Table 24.2) to estimate the collision cross-section of Ne atoms at 273 K.

24.14 (b) Use the experimental value of the thermal conductivity of nitrogen (Table 24.2) to estimate the collision cross-section of N_2 molecules at 298 K.

24.15 (a) In a double-glazed window, the panes of glass are separated by 5.0 cm. What is the rate of transfer of heat by conduction from the warm room (25°C) to the cold exterior (–10°C) through a window of area 1.0 m^2? What power of heater is required to make good the loss of heat?

24.15 (b) Two sheets of copper of area 1.50 m^2 are separated by 10.0 cm. What is the rate of transfer of heat by conduction from the warm sheet (50°C) to the cold sheet (–10°C). What is the rate of loss of heat?

24.16 (a) A manometer was connected to a bulb containing carbon dioxide under slight pressure. The gas was allowed to escape through a small pinhole, and the time for the manometer reading to drop from 75 cm to 50 cm was 52 s. When the experiment was repeated using nitrogen (for which $M = 28.02$ g mol^{-1}) the same fall took place in 42 s. Calculate the molar mass of carbon dioxide.

24.16 (b) A manometer was connected to a bulb containing nitrogen under slight pressure. The gas was allowed to escape through a small pinhole, and the time for the manometer reading to drop from 65.1 cm to 42.1 cm was 18.5 s. When the experiment was repeated using a fluorocarbon gas, the same fall took place in 82.3 s. Calculate the molar mass of the fluorocarbon.

24.17 (a) A space vehicle of internal volume 3.0 m^3 is struck by a meteor and a hole of radius 0.10 mm is formed. If the oxygen pressure within the vehicle is initially 80 kPa and its temperature 298 K, how long will the pressure take to fall to 70 kPa?

24.17 (b) A container of internal volume 22.0 m^3 was punctured, and a hole of radius 0.050 mm was formed. If the nitrogen pressure within the vehicle is initially 122 kPa and its temperature 293 K, how long will the pressure take to fall to 105 kPa?

24.18 (a) Use the experimental value of the coefficient of viscosity for neon (Table 24.2) to estimate the collision cross-section of Ne atoms at 273 K.

24.18 (b) Use the experimental value of the coefficient of viscosity for nitrogen (Table 24.2) to estimate the collision cross-section of the molecules at 273 K.

24.19 (a) Calculate the inlet pressure required to maintain a flow rate of 9.5×10^5 L h^{-1} of nitrogen at 293 K flowing through a pipe of length 8.50 m and diameter 1.00 cm. The pressure of gas as it leaves the tube is 1.00 bar. The volume of the gas is measured at that pressure.

24.19 (b) Calculate the inlet pressure required to maintain a flow rate of 8.70 cm^3 s^{-1} of nitrogen at 300 K flowing through a pipe of length 10.5 m and diameter 15 mm. The pressure of gas as it leaves the tube is 1.00 bar. The volume of the gas is measured at that pressure.

24.20 (a) Calculate the viscosity of air at (a) 273 K, (b) 298 K, (c) 1000 K. Take $\sigma \approx 0.40$ nm^2. (The experimental values are 173 μP at 273 K, 182 μP at 20°C, and 394 μP at 600°C.)

24.20 (b) Calculate the viscosity of benzene vapour at (a) 273 K, (b) 298 K, (c) 1000 K. Take $\sigma \approx 0.88$ nm^2.

24.21 (a) Calculate the thermal conductivities of (a) argon, (b) helium at 300 K and 1.0 mbar. Each gas is confined in a cubic vessel of side 10 cm, one wall being at 310 K and the one opposite at 295 K. What is the rate of flow of energy as heat from one wall to the other in each case?

24.21 (b) Calculate the thermal conductivities of (a) neon, (b) nitrogen at 300 K and 15 mbar. Each gas is confined in a cubic vessel of side 15 cm, one wall being at 305 K and the one opposite at 295 K. What is the rate of flow of energy as heat from one wall to the other in each case?

24.22 (a) The viscosity of carbon dioxide was measured by comparing its rate of flow through a long narrow tube (using Poiseuille's formula) with that of argon. For the same pressure differential, the same volume of carbon dioxide passed through the tube in 55 s as argon in 83 s. The viscosity of argon at 25°C is 208 μP; what is the viscosity of carbon dioxide? Estimate the molecular diameter of carbon dioxide.

24.22 (b) The viscosity of a chlorofluorocarbon (CFC) was measured by comparing its rate of flow through a long narrow tube (using Poiseuille's formula) with that of argon. For the same pressure differential, the same volume of the CFC passed through the tube in 72.0 s as argon in 18.0 s. The viscosity of argon at 25°C is 208 μP; what is the viscosity of the CFC? Estimate the molecular diameter of the CFC. Take $M = 200$ g mol^{-1}.

24.23 (a) Calculate the thermal conductivity of argon ($C_{V,m} = 12.5$ J K^{-1} mol^{-1}, $\sigma = 0.36$ nm^2) at room temperature (20°C).

24.23 (b) Calculate the thermal conductivity of nitrogen ($C_{V,m} = 20.8$ J K^{-1} mol^{-1}, $\sigma = 0.43$ nm^2) at room temperature (20°C).

24.24 (a) Calculate the diffusion constant of argon at 25°C and (a) 1.00 Pa, (b) 100 kPa, (c) 10.0 MPa. If a pressure gradient of 0.10 atm cm^{-1} is established in a pipe, what is the flow of gas due to diffusion?

24.24 (b) Calculate the diffusion constant of nitrogen at 25°C and (a) 10.0 Pa, (b) 100 kPa, (c) 15.0 MPa. If a pressure gradient of 0.20 bar m^{-1} is established in a pipe, what is the flow of gas due to diffusion?

24.25 (a) The mobility of a chloride ion in aqueous solution at 25°C is 7.91×10^{-8} m^2 s^{-1} V^{-1}. Calculate the molar ionic conductivity.

24.25 (b) The mobility of an acetate ion in aqueous solution at 25°C is 4.24×10^{-8} m^2 s^{-1} V^{-1}. Calculate the molar ionic conductivity.

24.26 (a) The mobility of a Rb$^+$ ion in aqueous solution is 7.92×10^{-8} m^2 s^{-1} V^{-1} at 25°C. The potential difference between two electrodes placed in the solution is 35.0 V. If the electrodes are 8.00 mm apart, what is the drift speed of the Rb$^+$ ion?

24.26 (b) The mobility of a Li$^+$ ion in aqueous solution is 4.01×10^{-8} m^2 s^{-1} V^{-1} at 25°C. The potential difference between two electrodes placed in the solution is 12.0 V. If the electrodes are 1.00 cm apart, what is the drift speed of the ion?

24.27 (a) What fraction of the total current is carried by Li$^+$ when current flows through an aqueous solution of LiBr at 25°C?

24.27 (b) What fraction of the total current is carried by Cl$^-$ when current flows through an aqueous solution of NaCl at 25°C?

24.28 (a) The limiting molar conductivities of KCl, KNO$_3$, and AgNO$_3$ are 14.99 mS m^2 mol^{-1}, 14.50 mS m^2 mol^{-1}, and 13.34 mS m^2 mol^{-1}, respectively (all at 25°C). What is the limiting molar conductivity of AgCl at this temperature?

24.28 (b) The limiting molar conductivities of NaI, NaCH$_3$CO$_2$, and Mg(CH$_3$CO$_2$)$_2$ are 12.69 mS m^2 mol^{-1}, 9.10 mS m^2 mol^{-1}, and 18.78 mS m^2 mol^{-1}, respectively (all at 25°C). What is the limiting molar conductivity of MgI$_2$ at this temperature?

24.29 (a) At 25°C the molar ionic conductivities of Li$^+$, Na$^+$, and K$^+$ are 3.87 mS m^2 mol^{-1}, 5.01 mS m^2 mol^{-1}, and 7.35 mS m^2 mol^{-1}, respectively. What are their mobilities?

24.29 (b) At 25°C the molar ionic conductivities of F$^-$, Cl$^-$, and Br$^-$ are 5.54 mS m^2 mol^{-1}, 7.635 mS m^2 mol^{-1}, and 7.81 mS m^2 mol^{-1}, respectively. What are their mobilities?

24.30 (a) The mobility of a NO$_3^-$ ion in aqueous solution at 25°C is 7.40×10^{-8} m^2 s^{-1} V^{-1}. Calculate its diffusion coefficient in water at 25°C.

24.30 (b) The mobility of a CH$_3$CO$_2^-$ ion in aqueous solution at 25°C is 4.24×10^{-8} m^2 s^{-1} V^{-1}. Calculate its diffusion coefficient in water at 25°C.

24.31 (a) The diffusion coefficient of CCl$_4$ in heptane at 25°C is 3.17×10^{-9} m^2 s^{-1}. Estimate the time required for a CCl$_4$ molecule to have a root mean square displacement of 5.0 mm.

24.31 (b) The diffusion coefficient of I$_2$ in hexane at 25°C is 4.05×10^{-9} m^2 s^{-1}. Estimate the time required for an iodine molecule to have a root mean square displacement of 1.0 cm.

24.32 (a) Estimate the effective radius of a sucrose molecule in water at 25°C given that its diffusion coefficient is 5.2×10^{-10} m^2 s^{-1} and that the viscosity of water is 1.00 cP.

24.32 (b) Estimate the effective radius of a glycine molecule in water at 25°C given that its diffusion coefficient is 1.055×10^{-9} m^2 s^{-1} and that the viscosity of water is 1.00 cP.

24.33 (a) The diffusion coefficient for molecular iodine in benzene is 2.13×10^{-9} m^2 s^{-1}. How long does a molecule take to jump through about one molecular diameter (approximately the fundamental jump length for translational motion)?

24.33 (b) The diffusion coefficient for CCl$_4$ in heptane is 3.17×10^{-9} m^2 s^{-1}. How long does a molecule take to jump through about one molecular diameter (approximately the fundamental jump length for translational motion)?

24.34 (a) What are the root mean square distances travelled by an iodine molecule in benzene and by a sucrose molecule in water at 25°C in 1.0 s?

24.34 (b) About how long, on average, does it take for the molecules in Exercise 24.34a to drift to a point (a) 1.0 mm, (b) 1.0 cm from their starting points?

Problems[†]

Numerical problems

24.1 Instead of the arrangement in Fig. 24.8, the speed of molecules can also be measured with a rotating slotted-disc apparatus, which consists of five coaxial 5.0 cm diameter disks separated by 1.0 cm, the slots in their rims being displaced by 2.0° between neighbours. The relative intensities, I, of the detected beam of Kr atoms for two different temperatures and at a series of rotation rates were as follows:

ν/Hz	20	40	80	100	120
I(40 K)	0.846	0.513	0.069	0.015	0.002
I(100 K)	0.592	0.485	0.217	0.119	0.057

Find the distributions of molecular velocities, $f(v_x)$, at these temperatures, and check that they conform to the theoretical prediction for a one-dimensional system.

24.2 Cars were timed by police radar as they passed in both directions below a bridge. Their velocities (kilometres per hour, numbers of cars in parentheses) to the east and west were as follows: 80 E (40), 85 E (62), 90 E (53), 95 E (12), 100 E (2); 80 W (38), 85 W (59), 90 W (60), 100 W (2). What are (a) the mean velocity, (b) the mean speed, (c) the root mean square speed?

24.3 A population consists of people of the following heights (in metres, numbers of individuals in brackets): 1.80 (1), 1.82 (2), 1.84 (4), 1.86 (7), 1.88 (10), 1.90 (15), 1.92 (9), 1.94 (4), 1.96 (0), 1.98 (1). What are (a) the mean height, (b) the root mean square height of the population?

24.4 Calculate the ratio of the thermal conductivities of gaseous hydrogen at 300 K to gaseous hydrogen at 10 K. Be circumspect, and think about the modes of motion that are thermally active at the two temperatures.

24.5 In the Knudsen method for the determination of vapour pressure, a weighed amount of a sample is heated inside a container, in the wall of which there is a small hole. The mass loss over a period of time can be related to the vapour pressure at the temperature of the experiment. If Δm is the mass lost in an interval Δt through a circular hole of radius R, find an expression relating the vapour pressure, p, to Δm and Δt. A Knudsen cell was used to determine the vapour pressure of germanium at 1000°C. During an interval of 7200 s the mass loss through a hole of radius 0.50 mm amounted to 43 μg. What is the vapour pressure of germanium at 1000°C? Assume the gas to be monatomic.

24.6 The nuclide ^{244}Bk (berkelium) decays by producing α particles, which capture electrons and form He atoms. Its half-life is 4.4 h. A sample of mass 1.0 mg was placed in a container of volume 1.0 cm^3 that was impermeable to α radiation, but there was also a hole of radius 2.0 μm in the wall. What is the pressure of helium at 298 K, inside the container after (a) 1.0 h, (b) 10 h?

24.7 An atomic beam is designed to function with (a) cadmium, (b) mercury. The source is an oven maintained at 380 K, there being a small slit of dimensions 1.0 cm × 1.0 cm × 10^{-3} cm. The vapour pressure of cadmium is 0.13 Pa and that of mercury is 152 kPa at this temperature. What is the atomic current (the number of atoms per unit time) in the beams?

24.8 Conductivities are often measured by comparing the resistance of a cell filled with the sample to its resistance when filled with some standard solution, such as aqueous potassium chloride. The conductivity of water is 76 mS m^{-1} at 25°C and the conductivity of 0.100 mol L^{-1} KCl(aq) is 1.1639 S m^{-1}. A cell had a resistance of 33.21 Ω when filled with 0.100 mol L^{-1} KCl(aq) and 300.0 Ω when filled with 0.100 mol L^{-1} CH$_3$COOH. What is the molar conductivity of acetic acid at that concentration and temperature?

24.9 The resistances of a series of aqueous NaCl solutions, formed by successive dilution of a sample, were measured in a cell with cell constant (the constant C in the relation $\kappa = C/R$) equal to 0.2063 cm^{-1}. The following values were found:

c/(mol L^{-1})	0.000 50	0.0010	0.0050	0.010	0.020	0.050
R/Ω	3314	1669	342.1	174.1	89.08	37.14

Verify that the molar conductivity follows the Kohlrausch law and find the limiting molar conductivity. Determine the coefficient $\mathcal{K}$. Use the value of $\mathcal{K}$ (which should depend only on the nature, not the identity of the ions) and the information that λ(Na$^+$) = 5.01 mS m^2 mol^{-1} and λ(I$^-$) = 7.68 mS m^2 mol^{-1} to predict (a) the molar conductivity, (b) the conductivity, (c) the resistance it would show in the cell, of 0.010 mol L^{-1} NaI(aq) at 25°C.

24.10 After correction for the water conductivity, the conductivity of a saturated aqueous solution of AgCl at 25°C was found to be 0.1887 mS m^{-1}. What is the solubility of silver chloride at this temperature?

24.11 What are the drift speeds of Li$^+$, Na$^+$, and K$^+$ in water when a potential difference of 10 V is applied across a 1.00-cm conductivity cell? How long would it take an ion to move from one electrode to the other? In conductivity measurements it is normal to use alternating current: what are the displacements of the ions in (a) centimetres, (b) solvent diameters, about 300 pm, during a half cycle of 1.0 kHz applied potential?

24.12 The mobilities of H$^+$ and Cl$^-$ at 25°C in water are 3.623×10^{-7} m^2 s^{-1} V^{-1} and 7.91×10^{-8} m^2 s^{-1} V^{-1}, respectively. What proportion of the current is carried by the protons in 1.0×10^{-3} M HCl(aq)? What fraction do they carry when the NaCl is added to the acid so that the solution is 1.0 mol L^{-1} in the salt? Note how concentration as well as mobility governs the transport of current.

24.13 In a moving boundary experiment on KCl the apparatus consisted of a tube of internal diameter 4.146 mm, and it contained aqueous KCl at a concentration of 0.021 mol L^{-1}. A steady current of 18.2 mA was passed, and the boundary advanced as follows:

Δt/s	200	400	600	800	1000
x/mm	64	128	192	254	318

Find the transport number of K$^+$, its mobility, and its ionic conductivity.

24.14 The proton possesses abnormal mobility in water, but does it behave normally in liquid ammonia? To investigate this question, a moving-boundary technique was used to determine the transport

---

† Problems denoted with the symbol ‡ were supplied by Charles Trapp, Carmen Giunta, and Marshall Cady.

number of NH_4^+ in liquid ammonia (the analogue of H_3O^+ in liquid water) at $-40°C$ (J. Baldwin, J. Evans, and J.B. Gill, *J. Chem. Soc. A*, 3389 (1971)). A steady current of 5.000 mA was passed for 2500 s, during which time the boundary formed between mercury(II) iodide and ammonium iodide solutions in ammonia moved 286.9 mm in a 0.013 65 mol kg^{-1} solution and 92.03 mm in a 0.042 55 mol kg^{-1} solution. Calculate the transport number of NH_4^+ at these concentrations, and comment on the mobility of the proton in liquid ammonia. The bore of the tube is 4.146 mm and the density of liquid ammonia is 0.682 g cm^{-3}.

24.15 A dilute solution of potassium permanganate in water at 25°C was prepared. The solution was in a horizontal tube of length 10 cm, and at first there was a linear gradation of intensity of the purple solution from the left (where the concentration was 0.100 mol L^{-1}) to the right (where the concentration was 0.050 mol L^{-1}). What are the magnitude and sign of the thermodynamic force acting on the solute (a) close to the left face of the container, (b) in the middle, (c) close to the right face? Give the force per mole and force per molecule in each case.

24.16 Estimate the diffusion coefficients and the effective hydrodynamic radii of the alkali metal cations in water from their mobilities at 25°C. Estimate the approximate number of water molecules that are dragged along by the cations. Ionic radii are given in Table 21.3.

24.17 Nuclear magnetic resonance can be used to determine the mobility of molecules in liquids. A set of measurements on methane in carbon tetrachloride showed that its diffusion coefficient is 2.05×10^{-9} m^2 s^{-1} at 0°C and 2.89×10^{-9} m^2 s^{-1} at 25°C. Deduce what information you can about the mobility of methane in carbon tetrachloride.

24.18 A concentrated sucrose solution is poured into a cylinder of diameter 5.0 cm. The solution consisted of 10 g of sugar in 5.0 cm^3 of water. A further 1.0 L of water is then poured very carefully on top of the layer, without disturbing the layer. Ignore gravitational effects, and pay attention only to diffusional processes. Find the concentration at 5.0 cm above the lower layer after a lapse of (a) 10 s, (b) 1.0 y.

24.19 In a series of observations on the displacement of rubber latex spheres of radius 0.212 μm, the mean square displacements after selected time intervals were on average as follows:

t/s	30	60	90	120
$10^{12}\langle x^2\rangle$/m^2	88.2	113.5	128	144

These results were originally used to find the value of the Avogadro constant, but there are now better ways of determining N_A, so the data can be used to find another quantity. Find the effective viscosity of water at the temperature of this experiment (25°C).

24.20‡ A.K. Srivastava, R.A. Samant, and S.D. Patankar (*J. Chem. Eng. Data* **41**, 431 (1996)) measured the conductance of several salts in a binary solvent mixture of water and a dipolar aprotic solvent 1,3-dioxolan-2-one (ethylene carbonate). They report the following conductances at 25°C in a solvent 80 per cent 1,3-dioxolan-2-one by mass:

NaI

c/(mmol L^{-1})	32.02	20.28	12.06	8.64	2.85	1.24	0.83
Λ_m/(S cm^2 mol^{-1})	50.26	51.99	54.01	55.75	57.99	58.44	58.67

KI

c/(mmol L^{-1})	17.68	10.88	87.19	2.67	1.28	0.83	0.19
Λ_m/(S cm^2 mol^{-1})	42.45	45.91	47.53	51.81	54.09	55.78	57.42

Calculate Λ_m° for NaI and KI in this solvent and $\lambda^\circ(Na) - \lambda^\circ(K)$. Compare your results to the analogous quantities in aqueous solution using Table 24.5 in the *Data section*.

24.21‡ A. Fenghour, W.A. Wakeham, V. Vesovic, J.T.R. Watson, J. Millat, and E. Vogel (*J. Phys. Chem. Ref. Data* **24**, 1649 (1995)) have compiled an extensive table of viscosity coefficients for ammonia in the liquid and vapour phases. Deduce the effective molecular diameter of NH_3 based on each of the following vapour-phase viscosity coefficients: (a) $\eta = 9.08 \times 10^{-6}$ kg m^{-1} s^{-1} at 270 K and 1.00 bar; (b) $\eta = 1.749 \times 10^{-5}$ kg m^{-1} s^{-1} at 490 K and 10.0 bar.

24.22‡ G. Bakale, K. Lacmann, and W.F. Schmidt (*J. Phys. Chem.* **100**, 12 477 (1996)) measured the mobility of singly charged C_{60}^- ions in a variety of nonpolar solvents. In cyclohexane at 22°C, the mobility is 1.1 cm^2 V^{-1} s^{-1}. Estimate the effective radius of the C_{60}^- ion. The viscosity of the solvent is 0.93×10^{-3} kg m^{-1} s^{-1}. *Comment.* The researchers interpreted the substantial difference between the radius so determined and the van der Waals radius of neutral C_{60} in terms of a solvation layer around the ion.

Theoretical problems

24.23 Start from the Maxwell–Boltzmann distribution and derive an expression for the most probable speed of a gas of molecules at a temperature T. Go on to demonstrate the validity of the equipartition conclusion that the average translational kinetic energy of molecules free to move in three dimensions is $\frac{3}{2}kT$.

24.24 Consider molecules that are confined to move in a plane (a two-dimensional gas). Calculate the distribution of speeds and determine the mean speed of the molecules at a temperature T.

24.25 A specially constructed velocity-selector accepts a beam of molecules from an oven at a temperature T but blocks the passage of molecules with a speed greater than the mean. What is the mean speed of the emerging beam, relative to the initial value, treated as a one-dimensional problem?

24.26 What is the proportion of gas molecules having (a) more than, (b) less than the root mean square speed? (c) What are the proportions having speeds greater and smaller than the mean speed?

24.27 Calculate the fractions of molecules in a gas that have a speed in a range Δv at the speed nc^* relative to those in the same range at c^* itself. This calculation can be used to estimate the fraction of very energetic molecules (which is important for reactions). Evaluate the ratio for $n = 3$ and $n = 4$.

24.28 Show how the ratio of two transport numbers t' and t'' for two cations in a mixture depends on their concentrations c' and c'', and their mobilities u' and u''.

24.29 Confirm that eqn 88 is a solution of the diffusion equation with the correct initial value.

24.30 The diffusion equation is valid when many elementary steps are taken in the time interval of interest; but the random walk calculation lets us discuss distributions for short times as well as for long. Use eqn 92 to calculate the probability of being six paces from the origin (that is, at $x = 6\lambda$) after (a) four, (b) six, (c) twelve steps.

24.31 Use mathematical software to calculate P in a one-dimensional random walk, and evaluate the probability of being at $x = n\lambda$ for $n = 6, 10, 14, \ldots, 60$. Compare the numerical value with the analytical value in the limit of a large number of steps. At what value of n is the discrepancy no more than 0.1 per cent?

24.32‡ A dilute solution of a weak (1,1)-electrolyte contains both neutral ion pairs and ions in equilibrium ($AB \rightleftharpoons A^+ + B^-$). Prove that molar conductivities are related to the degree of ionization by the equations:

$$\frac{1}{\Lambda_m} = \frac{1}{\Lambda_m(\alpha)} + \frac{(1-\alpha)\Lambda_m^\circ}{\alpha^2 \Lambda_m(\alpha)^2}$$

where Λ_m° is the molar conductivity at infinite dilution.

Applications: to astrophysics and biology

24.33 Calculate the escape velocity (the minimum initial velocity that will take an object to infinity) from the surface of a planet of radius R. What is the value for (a) the Earth, $R = 6.37 \times 10^6$ m, $g = 9.81$ m s^{-2}, (b) Mars, $R = 3.38 \times 10^6$ m, $m_{Mars}/m_{Earth} = 0.108$. At what temperatures do H_2, He, and O_2 molecules have mean speeds equal to their escape speeds? What proportion of the molecules have enough speed to escape when the temperature is (a) 240 K, (b) 1500 K? Calculations of this kind are very important in considering the composition of planetary atmospheres.

24.34‡ Interstellar space is quite different from the gaseous environments we commonly encounter on Earth. For instance, a typical density of the medium is about 1 atom cm^{-3} and that atom is typically H; the effective temperature due to stellar background radiation is about 10 000 K. Estimate the diffusion coefficient and thermal conductivity of H under these conditions. *Comment.* Energy is in fact transferred much more effectively by radiation.

24.35 Enrico Fermi, the great Italian scientist, was a master at making good approximate calculations based on little or no actual data. Hence, such calculations are often called 'Fermi calculations'. Do a Fermi calculation on how long it would take for a gaseous air-borne cold virus of molar mass 100 kg mol^{-1} to travel the distance between two conversing people 1.0 m apart by diffusion in still air.

24.36 The diffusion coefficient of a particular kind of tRNA molecule is $D = 1.0 \times 10^{-11}$ m^2 s^{-1} in the medium of a cell interior. How long does it take molecules produced in the cell nucleus to reach the walls of the cell at a distance 1.0 μm, corresponding to the radius of the cell?

24.37‡ In this problem, we examine a model for the transport of oxygen from air in the lungs to blood. First, show that, for the initial and boundary conditions $c(x, t) = c(x, 0) = c_0$, $(0 < x < \infty)$ and $c(0, t) = c_s$, $(0 \le t \le \infty)$ where c_0 and c_s are constants, the concentration, $c(x, t)$, of a species is given by

$$c(x, t) = c_0 + (c_s - c_0)\{1 - \mathrm{erf}(\xi)\} \qquad \xi(x, t) = \frac{x}{(4Dt)^{1/2}}$$

where $\mathrm{erf}(\xi)$ is the error function (*Justification 12.3*) and the concentration $c(x, t)$ evolves by diffusion from the yz-plane of constant concentration, such as might occur if a condensed phase is absorbing a species from a gas phase. Now draw graphs of concentration profiles at several different times of your choice for the diffusion of oxygen into water at 298 K (when $D = 2.10 \times 10^{-9}$ m^2 s^{-1}) on a spatial scale comparable to passage of oxygen from lungs through alveoli into the blood. Use $c_0 = 0$ and set c_s equal to the solubility of oxygen in water. *Hint.* Use mathematical software.

25

The rates of chemical reactions

Empirical chemical kinetics

25.1 Experimental techniques

25.2 The rates of reactions

25.3 Integrated rate laws

25.4 Reactions approaching equilibrium

25.5 The temperature dependence of reaction rates

Accounting for the rate laws

25.6 Elementary reactions

25.7 Consecutive elementary reactions

25.8 Unimolecular reactions

Checklist of key ideas

Further reading

Exercises

Problems

This chapter is the first of a sequence that explores the rates of chemical reactions. The chapter begins with a discussion of the definition of reaction rate and outlines the techniques for its measurement. The results of such measurements show that reaction rates depend on the concentration of reactants (and products) in characteristic ways that can be expressed in terms of differential equations known as rate laws. The solutions of these equations are used to predict the concentrations of species at any time after the start of the reaction. The form of the rate law also provides insight into the series of elementary steps by which a reaction takes place. The key task in this connection is the construction of a rate law from a proposed mechanism and its comparison with experiment. Simple elementary steps have simple rate laws, and these rate laws can be combined together by invoking one or more approximations. These approximations include the concept of the rate-determining stage of a reaction, the steady-state concentration of a reaction intermediate, and the existence of a pre-equilibrium.

This chapter introduces the principles of **chemical kinetics**, the study of reaction rates, by showing how the rates of reactions may be measured and interpreted. The remaining chapters of this part of the text then develop this material in more detail and apply it to more complicated or more specialized cases. The rate of a chemical reaction might depend on variables under our control, such as the pressure, the temperature, and the presence of a catalyst, and we may be able to optimize the rate by the appropriate choice of conditions. The study of reaction rates also leads to an understanding of the **mechanisms** of reactions, their analysis into a sequence of elementary steps.

Empirical chemical kinetics

The first steps in the kinetic analysis of reactions are to establish the stoichiometry of the reaction and identify any side reactions. The basic data of chemical kinetics are then the concentrations of the reactants and products at different times after a reaction has been initiated. The rates of most chemical reactions are sensitive to the temperature, so in conventional experiments the temperature of the reaction mixture must be held constant throughout the course of the reaction. This requirement puts severe demands on the design of an experiment. Gas-phase reactions, for instance, are often carried out in a vessel held in contact with a substantial block of metal. Liquid-phase reactions, including flow reactions, must be carried out in an efficient thermostat. Special efforts have to be made to study reactions at low temperatures, as in the study of the kinds of reactions that take place in interstellar clouds. Thus, supersonic expansion of the reaction gas can be used to attain temperatures as low as 10 K. For work in the liquid phase and the solid phase, very low temperatures are often reached by flowing cold liquid or cold gas around the reaction vessel. Alternatively, the entire reaction vessel is immersed in a thermally insulated container filled with a cryogenic liquid, such as liquid helium (for work at around 4 K) or liquid nitrogen (for work at around 77 K). Non-isothermal conditions are sometimes employed. For instance, the shelf-life of an expensive pharmaceutical may be explored by slowly raising the temperature of a single sample.

25.1 Experimental techniques

The method used to monitor concentrations depends on the species involved and the rapidity with which their concentrations change. Many reactions reach equilibrium over periods of minutes or hours, and several techniques may then be used to follow the changing concentrations.

(a) Monitoring the progress of a reaction

A reaction in which at least one component is a gas might result in an overall change in pressure in a system of constant volume, so its progress may be followed by recording the variation of pressure with time.

Example 25.1 Monitoring the variation in pressure

Predict how the total pressure varies during the gas-phase decomposition $2 N_2O_5(g) \rightarrow 4 NO_2(g) + O_2(g)$.

Method The total pressure (at constant volume and temperature and assuming perfect gas behaviour) is proportional to the number of gas-phase molecules. Therefore, because each mole of N_2O_5 gives rise to $\frac{5}{2}$ mol of gas molecules, we can expect the pressure to rise to $\frac{5}{2}$ times its initial value. To confirm this conclusion, express the progress of the reaction in terms of the fraction, α, of N_2O_5 molecules that have reacted.

Answer Let the initial pressure be p_0 and the initial amount of N_2O_5 molecules present be n. When a fraction α of the N_2O_5 molecules has decomposed, the amounts of the components in the reaction mixture are:

	N_2O_5	NO_2	O_2	Total
Amount:	$n(1-\alpha)$	$2\alpha n$	$\frac{1}{2}\alpha n$	$n(1+\frac{3}{2}\alpha)$

When $\alpha = 0$ the pressure is p_0, so at any stage the total pressure is

$$p = (1 + \tfrac{3}{2}\alpha)p_0$$

When the reaction is complete, the pressure will have risen to $\frac{5}{2}$ times its initial value.

Self-test 25.1 Repeat the calculation for $2 NOBr(g) \rightarrow 2 NO(g) + Br_2(g)$.

$$[p = (1 + \tfrac{1}{2}\alpha)p_0]$$

25.1 The arrangement used in the flow technique for studying reaction rates. The reactants are injected into the mixing chamber at a steady rate. The location of the spectrometer corresponds to different times after initiation.

Spectrophotometry, the measurement of absorption of radiation in a particular spectral region, is widely applicable, and is especially useful when one substance in the reaction mixture has a strong characteristic absorption in a conveniently accessible region of the electromagnetic spectrum. For example, the progress of the reaction

$$H_2(g) + Br_2(g) \longrightarrow 2\ HBr(g)$$

can be followed by measuring the absorption of visible light by bromine. A reaction that changes the number or type of ions present in a solution may be followed by monitoring the electrical conductivity of the solution. The replacement of neutral molecules by ionic products can result in dramatic changes in the conductivity, as in the reaction

$$(CH_3)_3CCl(aq) + H_2O(l) \longrightarrow (CH_3)_3COH(aq) + H^+(aq) + Cl^-(aq)$$

If hydrogen ions are produced or consumed, the reaction may be followed by monitoring the pH of the solution.

Other methods of determining composition include emission spectroscopy, mass spectrometry, gas chromatography, nuclear magnetic resonance, and electron spin resonance (for reactions involving radicals or paramagnetic d-metal ions).

(b) Application of the techniques

In a **real-time analysis** the composition of the system is analysed while the reaction is in progress. Either a small sample is withdrawn or the bulk solution is monitored. In the **flow method** the reactants are mixed as they flow together in a chamber (Fig. 25.1). The reaction continues as the thoroughly mixed solutions flow through the outlet tube, and observation of the composition at different positions along the tube is equivalent to the observation of the reaction mixture at different times after mixing. The disadvantage of conventional flow techniques is that a large volume of reactant solution is necessary. This makes the study of fast reactions particularly difficult because to spread the reaction over a length of tube the flow must be rapid. This disadvantage is avoided by the **stopped-flow technique**, in which the reagents are mixed very quickly in a small chamber fitted with a syringe instead of an outlet tube (Fig. 25.2). The flow ceases when the plunger of the syringe reaches a stop, and the reaction continues in the mixed solutions. Observations, usually using spectroscopic techniques such as ultraviolet–visible absorption, circular dichroism, and fluorescence emission, are made on the sample as a function of time. The technique allows for the study of reactions that occur on the millisecond to second timescale. The suitability of the stopped-flow method to the study of small samples means that it is appropriate for many biochemical reactions, and it has been widely used to study the kinetics of protein folding and enzyme action (Box 25.1).

Very fast reactions can be studied by **flash photolysis**, in which the sample is exposed to a brief flash of light that initiates the reaction, and then the contents of the reaction chamber are monitored. Most work is now done with lasers with photolysis pulse widths that range from femtoseconds to nanoseconds (Section 17.5). The apparatus used for flash photolysis studies is based on the experimental design for time-resolved spectroscopy (Section 17.7e). Reactions occurring on a picosecond or femtosecond timescale may be probed by using electronic absorption or emission, infrared absorption, or Raman scattering. The spectra are recorded at a series of times following laser excitation. The laser pulse can initiate the reaction by forming a reactive species, such as an excited electronic state of a molecule, a radical, or an ion. We discuss examples of excited state reactions in Chapter 26. An example of radical generation is the light-induced dissociation of $Cl_2(g)$ to yield Cl atoms that react with HBr to make HCl and Br according to the following sequence:

25.2 In the stopped-flow technique the reagents are driven quickly into the mixing chamber by the driving syringes and then the time dependence of the concentrations is monitored.

Box 25.1 Kinetics of the helix–coil transition in polypeptides

We saw in Box 20.1 that a simple statistical model accounts for the thermodynamic aspects of the helix–coil transition in polypeptides. The unfolding of a helix begins somewhere in the middle of the chain with a nucleation step, which is less favourable than the remaining helix to coil conversions, and continues in a cooperative fashion, in which the polymer becomes increasingly more susceptible to structural changes as more conversions occur. Here we examine the kinetics of the helix–coil transition, focusing primarily on experimental strategies and some recent results.

Earlier work on folding and unfolding of small polypeptides and large proteins relied primarily on rapid mixing and stopped-flow techniques (Section 25.1). In a typical stopped-flow experiment, a sample of the protein with a high concentration of a chemical denaturant, such as urea or guanidinium hydrochloride, is mixed with a solution containing a much lower concentration of the same denaturant. Upon entering the mixing chamber, the denaturant is diluted and the protein refolds. Unfolding is observed by mixing a sample of folded protein with a solution containing a high concentration of denaturant. These experiments are ideal for sorting out events in the millisecond timescale, such as the formation of contacts between helical segments in a large protein. However, the available data also indicate that, in a number of proteins, a significant portion of the folding process occurs in less than 1 ms, a time range not accessible by the stopped-flow technique. More recent temperature-jump and flash photolysis experiments have uncovered faster events. For example, at ambient temperature the formation of a loop between helical or sheet segments may be as fast as 1 μs and the formation of tightly packed cores with significant tertiary structure occurs in the 10–100 μs range. Among the fastest events are the formation and denaturation of helices and sheets from fully unfolded or folded peptide chains, respectively. Here, we examine how the laser-induced temperature-jump technique has been used in the study of the helix–coil transition.

The laser-induced temperature-jump technique takes advantage of the fact that proteins unfold, or melt, at high temperatures and each protein has a characteristic melting temperature (Section 22.5). Proteins also lose their native structures at very low temperatures, a process known as *cold denaturation*, and re-fold when the temperature in increased but kept significantly below the melting temperature. Hence, a temperature-jump experiment can be configured to monitor either folding or unfolding of a polypeptide, depending on the initial and final temperatures of the sample. The challenge of using melting or cold denaturation as the basis of kinetic measurements lies in increasing the temperature of the sample very quickly so fast relaxation proccess can be monitored. A number of clever strategies have been employed. In one example, a pulsed laser excites dissolved dye molecules that decay largely by internal conversion, or heat transfer to the solution. Another variation makes use of direct excitation of overtones of the O—H or O—D stretching modes of H_2O or D_2O, respectively, with a pulsed infrared laser. The latter strategy leads to temperature jumps in a small irradiated volume of about 20 K in less than 100 ps. Relaxation of the sample can then be probed by a variety of spectroscopic techniques, including absorption, emission, or Raman scattering. For example, the infrared absorption spectrum of a polypeptide is sensitive to polypeptide conformation, as the N—H stretching vibrations in the range 1630–1670 cm^{-1} are significantly different in the helix and coil forms.

Much of the kinetic work on the helix–coil transition has been conducted in small synthetic polypeptides rich in alanine, an amino acid that is known to stabilize helical structures. Both experimental results and statistical mechanical calculations suggest that the mechanism of unfolding consists of at least two steps: a very fast step in which amino acids at either end of a helical segment undergo transitions to coil regions and a slower rate-determining step that corresponds to the cooperative melting of the rest of the chain and loss of helical content. Using h and c to denote an amino acid residue belonging to a helical and coil region, respectively, the mechanism may be summarized as follows:

$hhhh \ldots \longrightarrow chhh \ldots$ very fast

$chhh \ldots \longrightarrow cccc \ldots$ rate-determining step

The rate-determining step is thought to account for the relaxation time of 160 ns measured with a laser-induced temperature jump between 282.5 K and 300.6 K in an alanine-rich polypeptide containing 21 amino acids. It is thought that the limitation on the rate of the helix–coil transition in this peptide arises from an activation energy barrier of 1.7 kJ mol^{-1} associated with nucleation events of the form $\ldots hhhh \ldots \to \ldots hhch \ldots$ in the middle of the chain. Therefore, nucleation is not only thermodynamically unfavourable but also kinetically slow. Models that use concepts of statistical thermodynamics also suggest that a $hhhh \ldots \to chhh \ldots$ transition at either end of a helical segment has a significantly lower activation energy on account of the converting amino acid not being flanked by h regions.

The time constant for the helix–coil transition has also been measured in proteins. In apomyoglobin (myoglobin lacking the haem co-factor), the unfolding of the helices appears to have a relaxation time of about 50 ns, even shorter than in synthetic peptides. It is difficult to interpret these results because we do not yet know how the amino acid sequence or interactions between helices in a folded protein affect the helix–coil relaxation time.

Discussion questions

1. Propose a set of experiments in which analysis of the line-shapes of NMR transitions (Section 18.8) is used to monitor fast events in protein folding or unfolding. What are the advantages and disadvantages of this NMR method over methods that use electronic or vibrational spectroscopy?

2. Consider a mechanism for the helix–coil transition in which nucleation occurs in the middle of the chain:

$$hhhh\ldots \rightleftharpoons hchh\ldots$$
$$hchh\ldots \rightleftharpoons cccc\ldots$$

We saw above that this type of nucleation is relatively slow, so neither step may be rate-determining. (a) Set up the rate equations for this alternate mechanism. (b) Apply the steady-state approximation and show that, under these circumstances, the mechanism is equivalent to $hhhh\ldots \rightleftharpoons cccc\ldots$ (c) Use your knowledge of experimental techniques and your results from parts (a) and (b) to support or refute the following statement: It is very difficult to obtain experimental evidence for intermediates in protein unfolding by performing simple rate measurements and one must resort to special time-resolved or trapping techniques to detect intermediates directly.

Further reading

R.H. Callender, R.B. Dyer, R. Gilmanshin, and W.H. Woodruff, Fast events in protein folding: the time evolution of primary processes. *Ann. Rev. Phys. Chem.* **49**, 173 (1998).

U. Mayor, C.M. Johnson, V. Daggett, and A.R. Fersht, Protein folding and unfolding in microseconds to nanoseconds by experiment and simulation. *Proc. Natl. Acad. Sci USA* **97**, 13518 (2000).

$$Cl_2 + h\nu \longrightarrow Cl + Cl$$
$$Cl + HBr \longrightarrow HCl^* + Br$$
$$HCl^* + M \longrightarrow HCl + M$$

Here HCl* denotes a thermally excited HCl molecule and M is a body (an unreactive molecule or the wall of the container) that removes the excess energy stored in HCl by the reaction between Cl and HBr.

In contrast to real-time analysis, **quenching methods** are based on stopping, or quenching, the reaction after it has been allowed to proceed for a certain time. In this way the composition is analysed at leisure and reaction intermediates may be trapped. These methods are suitable only for reactions that are slow enough for there to be little reaction during the time it takes to quench the mixture, so that reaction timescales on the order of hundreds of milliseconds can be studied with ease. In the **chemical quench flow method**, the reactants are mixed in much the same way as in the flow method but the reaction is quenched by another reagent, such as solution of acid or base, after the mixture has travelled along a fixed length of the outlet tube. Different reaction times can be selected by varying the flow rate along the outlet tube. An advantage of the chemical quench flow method over the stopped-flow method is that spectroscopic fingerprints are not needed in order to measure the concentration of reactants and products. Once the reaction has been quenched, the solution may be examined by rather 'slow' techniques, such as gel electrophoresis, mass spectrometry, and chromatography. In the **freeze quench method**, the reaction is quenched by cooling the mixture within milliseconds and the concentrations of reactants, intermediates, and products are measured spectroscopically.

25.2 The rates of reactions

Reaction rates depend on the composition and the temperature of the reaction mixture. The next few sections look at these observations in more detail.

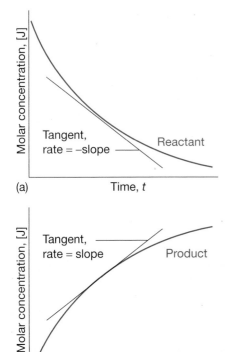

25.3 (a) The definition of (instantaneous) rate as the slope of the tangent drawn to the curve showing the variation of concentration with time. (b) For negative slopes, the sign is changed when reporting the rate, so all reaction rates are positive.

(a) The definition of rate

Consider a reaction of the form $A + 2B \rightarrow 3C + D$, in which at some instant the molar concentration of a participant J is [J]. The instantaneous **rate of consumption** of one of the reactants at a given time is $-d[R]/dt$, where R is A or B. This rate is a positive quantity (Fig. 25.3). The **rate of formation** of one of the products (C or D, which we denote P) is $d[P]/dt$ (note the difference in sign). This rate is also positive.

It follows from the stoichiometry for the reaction $A + 2B \rightarrow 3C + D$ that

$$\frac{d[D]}{dt} = \frac{1}{3}\frac{d[C]}{dt} = -\frac{d[A]}{dt} = -\frac{1}{2}\frac{d[B]}{dt}$$

so there are several rates connected with the reaction. The problem of having several possibly different rates to describe the same reaction is avoided by defining the unique **rate of reaction**, v, as the rate of change of the extent of reaction, ξ (the quantity introduced in Section 9.2b):

$$v = \frac{d\xi}{dt} \qquad [25.1a]$$

Because $v_J \xi = n_J - n_{J,0}$, the change in the extent of reaction is related to the change in the amount of each substance J by $v_J \, d\xi = dn_J$, so

$$v = \frac{1}{v_J}\frac{dn_J}{dt} \qquad (25.1b)$$

(Remember that v_J is negative for reactants and positive for products.) For a homogeneous reaction we divide by the (constant) volume of the system and write

$$v = \frac{1}{v_J}\frac{d[J]}{dt} \qquad (25.1c)$$

For a heterogeneous reaction, we divide instead by the surface area of the species and write

$$v = \frac{1}{v_J}\frac{d\sigma_J}{dt} \qquad (25.1d)$$

where σ_J is the surface density of J. In each case there is now a single rate for the entire reaction (for the chemical equation as written). With molar concentrations in moles per litre and time in seconds, reaction rates of homogeneous reactions are reported in moles per litre per second (mol L^{-1} s^{-1}) or related units. For heterogeneous reactions, rates are expressed in moles per square metre per second (mol m^{-2} s^{-1}) or related units.

Illustration 25.1

If the rate of formation of NO in the reaction $2\,NOBr(g) \rightarrow 2\,NO(g) + Br_2(g)$ is reported as 0.16 mmol L^{-1} s^{-1}, we use $v_{NO} = +2$ to report that $v = 0.080$ mmol L^{-1} s^{-1}. Because $v_{NOBr} = -2$ it follows that $d[NOBr]/dt = -0.16$ mmol L^{-1} s^{-1}. The rate of consumption of NOBr is therefore 0.16 mmol L^{-1} s^{-1}.

Self-test 25.2 The rate of change of molar concentration of CH$_3$ radicals in the reaction $2\,CH_3(g) \rightarrow CH_3CH_3(g)$ was reported as $d[CH_3]/dt = -1.2$ mol L^{-1} s^{-1} under particular conditions. What is (a) the rate of reaction and (b) the rate of formation of CH$_3$CH$_3$?   [(a) 0.60 mol L^{-1} s^{-1}, (b) 0.60 mol L^{-1} s^{-1}]

(b) Rate laws and rate constants

The rate of reaction is often found to be proportional to the concentrations of the reactants raised to a power. For example, the rate of a reaction may be found to be proportional to the molar concentrations of two reactants A and B, so we write

$$v = k[A][B] \tag{25.2}$$

with each concentration raised to the first power. The coefficient k is called the **rate constant** for the reaction. The rate constant is independent of the concentrations but depends on the temperature. An experimentally determined equation of this kind is called the **rate law** of the reaction. More formally, a rate law is an equation that expresses the rate of reaction as a function of the concentrations of all the species present in the overall chemical equation for the reaction at some time:

$$v = f([A],[B], \dots) \tag{25.3a}$$

For homogeneous gas-phase reactions, it is often more convenient to express the rate law in terms of partial pressures, which are related to molar concentrations by $p_J = RT[A]$. In this case, we write

$$v = f(p_A, p_B, \dots) \tag{25.3b}$$

The rate law of a reaction is determined experimentally, and cannot in general be inferred from the chemical equation for the reaction. The reaction of hydrogen and bromine, for example, has a very simple stoichiometry, $H_2(g) + Br_2(g) \rightarrow 2\,HBr(g)$, but its rate law is complicated:

$$v = \frac{k[H_2][Br_2]^{3/2}}{[Br_2] + k'[HBr]} \tag{25.4}$$

In certain cases the rate law does reflect the stoichiometry of the reaction; but that is either a coincidence or reflects a feature of the underlying reaction mechanism (see later).

A practical application of a rate law is that, once we know the law and the value of the rate constant, we can predict the rate of reaction from the composition of the mixture. Moreover, as we shall see later, by knowing the rate law, we can go on to predict the composition of the reaction mixture at a later stage of the reaction. Moreover, a rate law is a guide to the mechanism of the reaction, for any proposed mechanism must be consistent with the observed rate law.

(c) Reaction order

Many reactions are found to have rate laws of the form

$$v = k[A]^a[B]^b \dots \tag{25.5}$$

The power to which the concentration of a species (a product or a reactant) is raised in a rate law of this kind is the **order** of the reaction with respect to that species. A reaction with the rate law in eqn 2 is **first-order** in A and first-order in B. The **overall order** of a reaction with a rate law like that in eqn 5 is the sum of the individual orders, $a + b + \cdots$. The rate law in eqn 2 is therefore second-order overall.

A reaction need not have an integral order, and many gas-phase reactions do not. For example, a reaction having the rate law

$$v = k[A]^{1/2}[B] \tag{25.6}$$

is half-order in A, first-order in B, and three-halves-order overall. Some reactions obey a **zero-order rate law**, and therefore have a rate that is independent of the concentration of the reactant (so long as some is present). Thus, the catalytic decomposition of phosphine (PH_3) on hot tungsten at high pressures has the rate law

$$v = k \tag{25.7}$$

The PH_3 decomposes at a constant rate until it has almost entirely disappeared. Only heterogeneous reactions can have rate laws that are zero-order overall.

When a rate law is not of the form in eqn 5, the reaction does not have an overall order and may not even have definite orders with respect to each participant. Thus, although eqn 4 shows that the reaction of hydrogen and bromine is first-order in H_2, the reaction has an indefinite order with respect to both Br_2 and HBr and has no overall order.

These remarks point to three problems. First, we must see how to identify the rate law and obtain the rate constant from the experimental data. We concentrate on this aspect in this chapter. Second, we must see how to construct reaction mechanisms that are consistent with the rate law. We shall introduce the techniques of doing so in this chapter and develop them further in Chapter 26. Third, we must account for the values of the rate constants and explain their temperature dependence. We shall see a little of what is involved in this chapter, but leave the details until Chapter 27.

(d) The determination of the rate law

The determination of a rate law is simplified by the **isolation method** in which the concentrations of all the reactants except one are in large excess. If B is in large excess, for example, then to a good approximation its concentration is constant throughout the reaction. Although the true rate law might be $v = k[A][B]$, we can approximate [B] by $[B]_0$ and write

$$v = k'[A] \qquad k' = k[B]_0 \tag{25.8}$$

which has the form of a first-order rate law. Because the true rate law has been forced into first-order form by assuming that the concentration of B is constant, it is called a **pseudofirst-order rate law**. The dependence of the rate on the concentration of each of the reactants may be found by isolating them in turn (by having all the other substances present in large excess), and so constructing a picture of the overall rate law.

In the **method of initial rates**, which is often used in conjunction with the isolation method, the rate is measured at the beginning of the reaction for several different initial concentrations of reactants. We shall suppose that the rate law for a reaction with A isolated is $v = k[A]^a$; then its initial rate, v_0, is given by the initial values of the concentration of A, and we write $v_0 = k[A]_0^a$. Taking logarithms gives:

$$\log v_0 = \log k + a \log[A]_0 \tag{25.9}$$

For a series of initial concentrations, a plot of the logarithms of the initial rates against the logarithms of the initial concentrations of A should be a straight line with slope a.

Example 25.2 Using the method of initial rates

The recombination of iodine atoms in the gas phase in the presence of argon was investigated and the order of the reaction was determined by the method of initial rates. The initial rates of reaction of $2\,I(g) + Ar(g) \rightarrow I_2(g) + Ar(g)$ were as follows:

$[I]_0/(10^{-5}\ mol\ L^{-1})$		1.0	2.0	4.0	6.0
$v_0/(mol\ L^{-1}\ s^{-1})$	(a)	8.70×10^{-4}	3.48×10^{-3}	1.39×10^{-2}	3.13×10^{-2}
	(b)	4.35×10^{-3}	1.74×10^{-2}	6.96×10^{-2}	1.57×10^{-1}
	(c)	8.69×10^{-3}	3.47×10^{-2}	1.38×10^{-1}	3.13×10^{-1}

The Ar concentrations are (a) 1.0 mmol L^{-1}, (b) 5.0 mmol L^{-1}, and (c) 10 mmol L^{-1}. Determine the orders of reaction with respect to the I and Ar atom concentrations and the rate constant.

Method Plot the logarithm of the initial rate, $\log v_0$, against $\log[I]_0$ for a given concentration of Ar, and, separately, against $\log[Ar]_0$ for a given concentration of I. The slopes of the two lines are the orders of reaction with respect to I and Ar, respectively. The intercepts with the vertical axis at $\log[X] = 0$ give $\log k$.

Answer The plots are shown in Fig. 25.4. The slopes are 2 and 1 respectively, so the (initial) rate law is

$$v_0 = k[I]_0^2[Ar]_0$$

This rate law signifies that the reaction is second-order in [I], first-order in [Ar], and third-order overall. The intercept corresponds to $k = 9 \times 10^9\ mol^{-2}\ L^2\ s^{-1}$.

Comment The units of k come automatically from the calculation, and are always such as to convert the product of concentrations to concentration per unit time (for example, mol $L^{-1}\ s^{-1}$).

Self-test 25.3 The initial rate of a reaction depended on concentration of a substance J as follows:

$[J]_0/(mmol\ L^{-1})$	5.0	8.2	17	30
$v_0/(10^{-7}\ mol\ L^{-1}\ s^{-1})$	3.6	9.6	41	130

Determine the order of the reaction with respect to J and calculate the rate constant.

[2, $1.4 \times 10^{-2}\ L\ mol^{-1}\ s^{-1}$]

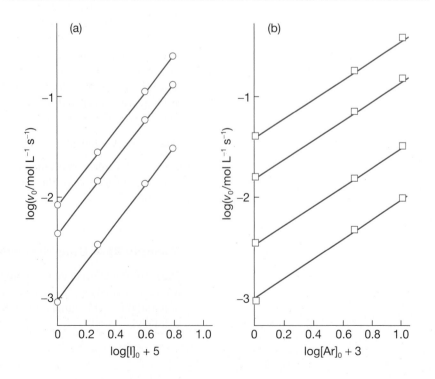

25.4 The plot of $\log v_0$ against (a) $\log[I]_0$ for a given $[Ar]_0$, and (b) $\log[Ar]_0$ for a given $[I]_0$.

25.5 The exponential decay of the reactant in a first-order reaction. The larger the rate constant, the more rapid the decay: here $k_{large} = 3k_{small}$.

The web site contains links to databases of rate constants of chemical reactions.

The method of initial rates might not reveal the full rate law, for the products may participate in the reaction and affect the rate. For example, products participate in the synthesis of HBr, because eqn 4 shows that the full rate law depends on the concentration of HBr. To avoid this difficulty, the rate law should be fitted to the data throughout the reaction. The fitting may be done, in simple cases at least, by using a proposed rate law to predict the concentration of any component at any time, and comparing it with the data. A law should also be tested by observing whether the addition of products or, for gas-phase reactions, a change in the surface-to-volume ratio in the reaction chamber affects the rate.

25.3 Integrated rate laws

Because rate laws are differential equations, we must integrate them if we want to find the concentrations as a function of time. Even the most complex rate laws may be integrated numerically. However, in a number of simple cases analytical solutions are easily obtained, and prove to be very useful. We shall examine a few of these simple cases here.

(a) First-order reactions

As shown in the *Justification* below, the first-order rate law for the consumption of a reactant A

$$\frac{d[A]}{dt} = -k[A] \tag{25.10a}$$

has the solution

$$\ln\left(\frac{[A]}{[A]_0}\right) = -kt \qquad [A] = [A]_0 e^{-kt} \tag{25.10b}$$

These two equations are versions of an **integrated rate law**, the integrated form of the rate law.

Justification 25.1

Equation 10a rearranges to

$$\frac{d[A]}{[A]} = -k\,dt$$

which can be integrated directly because k is a constant independent of t. Initially (at $t = 0$), the concentration of A is $[A]_0$, and at a later time t it is $[A]$, so we make these values the limits of the integrals and write

$$\int_{[A]_0}^{[A]} \frac{d[A]}{[A]} = -k \int_0^t dt$$

Because the integral of $1/x$ is $\ln x$, eqn 10b is obtained immediately.

Equation 10b shows that, if $\ln([A]/[A]_0)$ is plotted against t, then a first-order reaction will give a straight line of slope $-k$. Some rate constants determined in this way are given in Table 25.1. The second expression in eqn 10b shows that in a first-order reaction the reactant concentration decreases exponentially with time with a rate determined by k (Fig. 25.5).

Table 25.1* Kinetic data for first-order reactions

Reaction	Phase	$\theta/°C$	k/s^{-1}	$t_{1/2}$
$2\,N_2O_5 \longrightarrow 4\,NO_2 + O_2$	g	25	3.38×10^{-5}	5.70 h
$2\,N_2O_5 \longrightarrow 4\,NO_2 + O_2$	$Br_2(l)$	25	4.27×10^{-5}	4.51 h
$C_2H_6 \longrightarrow 2\,CH_3$	g	700	5.36×10^{-4}	21.6 min

* More values are given in the *Data section*.

Example 25.3 Analysing a first-order reaction

The variation in the partial pressure of azomethane with time was followed at 600 K, with the results given below. Confirm that the decomposition

$$CH_3N_2CH_3(g) \longrightarrow CH_3CH_3(g) + N_2(g)$$

is first-order in azomethane, and find the rate constant at 600 K.

t/s	0	1000	2000	3000	4000
$p/(10^{-2}\text{ Torr})$	8.20	5.72	3.99	2.78	1.94

Method As indicated in the text, to confirm that a reaction is first-order, plot $\ln([A]/[A]_0)$ against time and expect a straight line. Because the partial pressure of a gas is proportional to its concentration, an equivalent procedure is to plot $\ln(p/p_0)$ against t. If a straight line is obtained, its slope can be identified with $-k$.

Answer We draw up the following table:

t/s	0	1000	2000	3000	4000
$\ln(p/p_0)$	0	-0.360	-0.720	-1.082	-1.441

Figure 25.6 shows the plot of $\ln(p/p_0)$ against t. The plot is straight, confirming a first-order reaction, and its slope is $-3.6 \times 10^{-4}\text{ s}^{-1}$. Therefore, $k = 3.6 \times 10^{-4}\text{ s}^{-1}$.

Self-test 25.4 In a particular experiment, it was found that the concentration of N_2O_5 in liquid bromine varied with time as follows:

t/s	0	200	400	600	1000
$[N_2O_5]/(\text{mol L}^{-1})$	0.110	0.073	0.048	0.032	0.014

Confirm that the reaction is first-order in N_2O_5 and determine the rate constant. $[k = 2.1 \times 10^{-3}\text{ s}^{-1}]$

(b) Half-lives and time constants

A useful indication of the rate of a first-order chemical reaction is the **half-life**, $t_{1/2}$, of a substance, the time taken for the concentration of a reactant to fall to half its initial value. The time for [A] to decrease from $[A]_0$ to $\frac{1}{2}[A]_0$ in a first-order reaction is given by eqn 10b as

$$kt_{1/2} = -\ln\left(\frac{\frac{1}{2}[A]_0}{[A]_0}\right) = -\ln\frac{1}{2} = \ln 2$$

25.6 The determination of the rate constant of a first-order reaction: a straight line is obtained when ln [A] (or, as here, ln p) is plotted against t; the slope gives k.

Hence

$$t_{1/2} = \frac{\ln 2}{k} \tag{25.11}$$

($\ln 2 = 0.693$.) The main point to note about this result is that, for a first-order reaction, the half-life of a reactant is independent of its initial concentration. Hence, if the concentration of A at some *arbitrary* stage of the reaction is [A], then it will have fallen to $\frac{1}{2}$[A] after a further interval of $(\ln 2)/k$. Some half-lives are given in Table 25.1.

Another indication of the rate of a first-order reaction is the **time constant**, τ, the time required for the concentration of a reactant to fall to $1/e$ of its initial value. From eqn 10b it follows that

$$k\tau = -\ln\left(\frac{[A]_0/e}{[A]_0}\right) = -\ln\frac{1}{e} = 1$$

Hence, the time constant is the reciprocal of the rate constant:

$$\tau = \frac{1}{k} \tag{25.12}$$

(c) Second-order reactions

We show in the *Justification* below that the integrated form of the second-order rate law

$$\frac{d[A]}{dt} = -k[A]^2 \tag{25.13a}$$

is

$$\frac{1}{[A]} - \frac{1}{[A]_0} = kt \qquad [A] = \frac{[A]_0}{1 + kt[A]_0} \tag{25.13b}$$

Justification 25.2

To integrate eqn 13a we rearrange it into

$$\frac{d[A]}{[A]^2} = -k\,dt$$

The concentration of A is $[A]_0$ at $t = 0$ and [A] at a general time t later. Therefore,

$$-\int_{[A]_0}^{[A]} \frac{d[A]}{[A]^2} = k\int_0^t dt$$

Because the integral of $1/x^2$ is $-1/x$, we obtain eqn 13b by substitution of the limits.

The first expression in eqn 13b shows that to test for a second-order reaction we should plot $1/[A]$ against t and expect a straight line. The slope of the graph is k. Some rate constants determined in this way are given in Table 25.2. The second expression lets us predict the concentration of A at any time after the start of the reaction. It shows that the concentration of A approaches zero more slowly than in a first-order reaction with the same initial rate (Fig. 25.7).

25.7 The variation with time of the concentration of a reactant in a second-order reaction. The pale line is the corresponding decay in a first-order reaction with the same initial rate. For this illustration, $k_{large} = 3k_{small}$.

Table 25.2* Kinetic data for second-order reactions

Reaction	Phase	$\theta/°C$	$k/(L\ mol^{-1}\ s^{-1})$
$2\ NOBr \longrightarrow 2\ NO + Br_2$	g	10	0.80
$2\ I \longrightarrow I_2$	g	23	7×10^9
$CH_3Cl + CH_3O^-$	$CH_3OH(l)$	20	2.29×10^{-6}

* More values are given in the *Data section*.

It follows from eqn 13b by substituting $t = t_{1/2}$ and $[A] = \frac{1}{2}[A]_0$ that the half-life of a species A that is consumed in a second-order reaction is

$$t_{1/2} = \frac{1}{k[A]_0} \tag{25.14}$$

Therefore, unlike a first-order reaction, the half-life of a substance in a second-order reaction varies with the initial concentration. A practical consequence of this dependence is that species that decay by second-order reactions (which includes some environmentally harmful substances) may persist in low concentrations for long periods because their half-lives are long when their concentrations are low.

Another type of second-order reaction is one that is first-order in each of two reactants A and B:

$$\frac{d[A]}{dt} = -k[A][B] \tag{25.15}$$

Such a rate law cannot be integrated until we know how the concentration of B is related to that of A. For example, if the reaction is A + B → P, where P denotes products, and the initial concentrations are $[A]_0$ and $[B]_0$, then it is shown in the *Justification* below that, at a time t after the start of the reaction, the concentrations satisfy the relation

$$\ln\left(\frac{[B]/[B]_0}{[A]/[A]_0} \right) = ([B]_0 - [A]_0)kt \tag{25.16}$$

Therefore, a plot of the expression on the left against t should be a straight line from which k can be obtained. Note that, if $[A]_0 = [B]_0$, then the solutions are those already given in eqn 13b (but this solution cannot be found simply by setting $[A]_0 = [B]_0$ in eqn 16).

Justification 25.3

It follows from the reaction stoichiometry that, when the concentration of A has fallen to $[A]_0 - x$, the concentration of B will have fallen to $[B]_0 - x$ (because each A that disappears entails the disappearance of one B). It follows that

$$\frac{d[A]}{dt} = -k([A]_0 - x)([B]_0 - x)$$

Then, because $d[A]/dt = -dx/dt$, the rate law is

$$\frac{dx}{dt} = k([A]_0 - x)([B]_0 - x)$$

To solve an integral of the form $\int \dfrac{1}{(a-x)(b-x)}\,dx$, where a and b are constants, we use the method of partial fractions. First we write

$$\frac{1}{(a-x)(b-x)} = \frac{1}{b-a}\left(\frac{1}{a-x} - \frac{1}{b-x}\right)$$

and integrate the expression on the right. It follows that

$$\int \frac{dx}{(a-x)(b-x)}$$

$$= \frac{1}{b-a}\left[\int \frac{dx}{a-x} - \int \frac{dx}{b-x}\right]$$

$$= \frac{1}{b-a}\left(\ln\frac{1}{a-x} - \ln\frac{1}{b-x}\right) + \text{constant}$$

The initial condition is that $x = 0$ when $t = 0$; so the integration required is

$$\int_0^x \frac{dx}{([A]_0 - x)([B]_0 - x)} = k\int_0^t dt$$

The integral on the right is simply kt. The integral on the left is evaluated by using the method of partial fractions and by using $[A] = [A]_0$ and $[B] = [B]_0$ at $t = 0$ to give:

$$\int_0^x \frac{dx}{([A]_0 - x)([B]_0 - x)} = \frac{1}{[B]_0 - [A]_0}\left\{\ln\left(\frac{[A]_0}{[A]_0 - x}\right) - \ln\left(\frac{[B]_0}{[B]_0 - x}\right)\right\}$$

This expression can be simplified and rearranged into eqn 16 by combining the two logarithms and noting that $[A] = [A]_0 - x$ and $[B] = [B]_0 - x$.

Similar calculations may be carried out to find the integrated rate laws for other orders, and some are listed in Table 25.3.

Table 25.3 Integrated rate laws

Order	Reaction	Rate law*	$t_{1/2}$
0	$A \to P$	$v = k$	$\dfrac{[A]_0}{2k}$
		$kt = x$ for $0 \le x \le [A]_0$	
1	$A \to P$	$v = k[A]$	$\dfrac{\ln 2}{k}$
		$kt = \ln \dfrac{[A]_0}{[A]_0 - x}$	
2	$A \to P$	$v = k[A]^2$	$\dfrac{1}{k[A]_0}$
		$kt = \dfrac{x}{[A]_0([A]_0 - x)}$	
	$A + B \to P$	$v = k[A][B]$	
		$kt = \dfrac{1}{[B]_0 - [A]_0}\ln\dfrac{[A]_0([B]_0 - x)}{([A]_0 - x)[B]_0}$	
	$A + 2B \to P$	$v = k[A][B]$	
		$kt = \dfrac{1}{[B]_0 - 2[A]_0}\ln\dfrac{[A]_0([B]_0 - 2x)}{([A]_0 - x)[B]_0}$	
	$A \to P$ with autocatalysis	$v = k[A][P]$	
		$kt = \dfrac{1}{[A]_0 + [P]_0}\ln\dfrac{[A]_0([P]_0 + x)}{([A]_0 - x)[P]_0}$	
3	$A + 2B \to P$	$v = k[A][B]^2$	
		$kt = \dfrac{2x}{(2[A]_0 - [B]_0)([B]_0 - 2x)[B]_0}$	
		$+ \dfrac{1}{(2[A]_0 - [B]_0)^2}\ln\dfrac{[A]_0([B]_0 - 2x)}{([A]_0 - x)[B]_0}$	
$n \ge 2$	$A \to P$	$v = k[A]^n$	$\dfrac{2^{n-1} - 1}{(n-1)k[A]_0^{n-1}}$
		$kt = \dfrac{1}{n-1}\left\{\dfrac{1}{([A]_0 - x)^{n-1}} - \dfrac{1}{[A]_0^{n-1}}\right\}$	

* $x = [P]$, and $v = dx/dt$.

25.8 The approach of concentrations to their equilibrium values as predicted by eqn 20 for a reaction $A \rightleftharpoons B$ that is first-order in each direction, and for which $k = 2k'$.

25.4 Reactions approaching equilibrium

Because all the laws considered so far disregard the possibility that the reverse reaction is important, none of them describes the overall rate when the reaction is close to equilibrium. At that stage the products may be so abundant that the reverse reaction must be taken into account. In practice, however, most kinetic studies are made on reactions that are far from equilibrium, and the reverse reactions are unimportant.

(a) First-order reactions close to equilibrium

We can explore the variation of the composition with time close to chemical equilibrium by considering a reaction in which A forms B and both forward and reverse reactions are first-order (as in some isomerizations). The scheme we consider is

$$A \longrightarrow B \qquad v = k[A]$$
$$B \longrightarrow A \qquad v = k'[B] \tag{25.17}$$

The concentration of A is reduced by the forward reaction (at a rate $k[A]$) but it is increased by the reverse reaction (at a rate $k'[B]$). The net rate of change is therefore

$$\frac{d[A]}{dt} = -k[A] + k'[B] \tag{25.18}$$

If the initial concentration of A is $[A]_0$, and no B is present initially, then at all times $[A] + [B] = [A]_0$. Therefore,

$$\frac{d[A]}{dt} = -k[A] + k'([A]_0 - [A]) = -(k + k')[A] + k'[A]_0 \tag{25.19}$$

The solution of this first-order differential equation (as may be checked by differentiation) is

$$[A] = \frac{k' + ke^{-(k+k')t}}{k' + k}[A]_0 \tag{25.20}$$

Figure 25.8 shows the time dependence predicted by this equation.

As $t \to \infty$, the concentrations reach their equilibrium values, which are given by eqn 20 as:

$$[A]_{eq} = \frac{k'[A]_0}{k + k'} \qquad [B]_{eq} = [A]_0 - [A]_\infty = \frac{k[A]_0}{k + k'} \tag{25.21}$$

It follows that the equilibrium constant of the reaction is[1]

$$K = \frac{[B]_{eq}}{[A]_{eq}} = \frac{k}{k'} \tag{25.22}$$

Exactly the same conclusion can be reached—more simply, in fact—by noting that, at equilibrium, the forward and reverse rates must be the same, so

$$k[A]_{eq} = k'[B]_{eq} \tag{25.23}$$

This relation rearranges into eqn 22.

Equation 22 is very important, because it relates the thermodynamic quantity, the equilibrium constant, to quantities relating to rates. The practical importance of

1. This expression is only approximate because thermodynamic equilibrium constants are expressed in terms of activities, not concentrations.

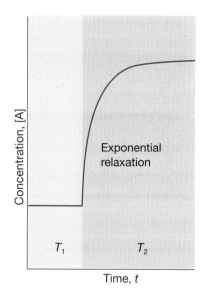

25.9 The relaxation to the new equilibrium composition when a reaction initially at equilibrium at a temperature T_1 is subjected to a sudden change of temperature, which takes it to T_2.

eqn 22 is that, if one of the rate constants can be measured, then the other may be obtained if the equilibrium constant is known.

For a more general reaction, the overall equilibrium constant can be expressed in terms of the rate constants for all the intermediate stages of the reaction mechanism:

$$K = \frac{k_a}{k_a'} \times \frac{k_b}{k_b'} \times \cdots$$

where the ks are the rate constants for the individual steps and the k's are for the corresponding reverse steps.

(b) Relaxation methods

The term **relaxation** denotes the return of a system to equilibrium. It is used in chemical kinetics to indicate that an externally applied influence has shifted the equilibrium position of a reaction, normally suddenly, and that the reaction is adjusting to the equilibrium composition characteristic of the new conditions (Fig. 25.9). We shall consider the response of reaction rates to a **temperature jump**, a sudden change in temperature. We know from Section 9.3a that the equilibrium composition of a reaction depends on the temperature (provided $\Delta_r H^\ominus$ is nonzero), so a shift in temperature acts as a perturbation on the system. One way of achieving a temperature jump is to discharge a capacitor through a sample made conducting by the addition of ions, but laser or microwave discharges can also be used. Temperature jumps of between 5 and 10 K can be achieved in about 1 µs with electrical discharges. The high energy output of pulsed lasers (Section 17.5) is sufficient to generate temperature jumps of between 10 and 30 K within nanoseconds in aqueous samples. Some equilibria are also sensitive to pressure, and **pressure-jump techniques** may then also be used.

When a sudden temperature increase is applied to a simple $A \rightleftharpoons B$ equilibrium that is first-order in each direction, we show in the *Justification* below that the composition relaxes exponentially to the new equilibrium composition:

$$x = x_0 e^{-t/\tau} \qquad \frac{1}{\tau} = k_a + k_b \tag{25.24}$$

where x is the departure from equilibrium at the new temperature and x_0 is the departure from equilibrium immediately after the temperature jump.

Justification 25.4

In the following analysis, we need to keep track of the fact that rate constants depend on temperature. At the initial temperature, when the rate constants are k_a' and k_b', the net rate of change of [A] is

$$\frac{d[A]}{dt} = -k_a'[A] + k_b'[B]$$

At equilibrium under these conditions, $d[A]/dt = 0$, and the concentrations are $[A]_{eq}'$ and $[B]_{eq}'$. Therefore,

$$k_a'[A]_{eq}' = k_b'[B]_{eq}'$$

When the temperature is increased suddenly, the rate constants change to k_a and k_b, but the concentrations of A and B remain for an instant at their old equilibrium values. As the system is no longer at equilibrium, it readjusts to the new equilibrium concentrations, which are now given by

$$k_a[A]_{eq} = k_b[B]_{eq}$$

and it does so at a rate that depends on the new rate constants.

We write the deviation of [A] from its new equilibrium value as x, so $[A] = x + [A]_{eq}$ and $[B] = [B]_{eq} - x$. The concentration of A then changes as follows:

$$\frac{d[A]}{dt} = -k_a(x + [A]_{eq}) + k_b(-x + [B]_{eq}) = -(k_a + k_b)x$$

because the two terms involving the equilibrium concentrations cancel. Because $d[A]/dt = dx/dt$, this equation is a first-order differential equation with the solution given in eqn 24.

Equation 24 shows that the concentrations of A and B relax into the new equilibrium at a rate determined by the sum of the two new rate constants. Because the equilibrium constant under the new conditions is $K \approx k_a/k_b$, its value may be combined with the relaxation time measurement to find the individual k_a and k_b.

Example 25.4 Analysing a temperature-jump experiment

The $H_2O(l) \rightarrow H^+(aq) + OH^-(aq)$ reaction relaxes to equilibrium with a relaxation time of 37 μs at 298 K and pH ≈ 7, and $pK_w = 14.01$. Given that the forward reaction is first-order and the reverse is second-order overall, calculate the rate constants for the forward and reverse reactions.

Method We need to derive an expression for the relaxation time, τ (the time constant for return to equilibrium), in terms of k_1 (forward, first-order reaction) and k_2 (reverse, second-order reaction). We can proceed as above, but it will be necessary to make the assumption that the deviation from equilibrium (x) is so small that terms in x^2 can be neglected. Relate k_1 and k_2 through the equilibrium constant, but be careful with units because K_w is dimensionless.

Answer The forward rate at the final temperature is $k_1[H_2O]$ and the reverse rate is $k_2[H^+][OH^-]$. The net rate of ionization of H_2O is

$$\frac{d[H_2O]}{dt} = -k_1[H_2O] + k_2[H^+][OH^-]$$

We write $[H_2O] = [H_2O]_{eq} + x$, $[H^+] = [H^+]_{eq} - x$, and $[OH^-] = [OH^-]_{eq} - x$, and obtain

$$\frac{dx}{dt} = -\{k_1 + k_2([H^+]_{eq} + [OH^-]_{eq})\}x - k_1[H_2O]_{eq} + k_2[H^+]_{eq}[OH^-]_{eq} + k_2x^2$$

$$\approx -\{k_1 + k_2([H^+]_{eq} + [OH^-]_{eq})\}x$$

where we have neglected the term in x^2 and used the equilibrium condition to eliminate the terms that are independent of x. It follows that

$$\frac{1}{\tau} = k_1 + k_2([H^+]_{eq} + [OH^-]_{eq})$$

The equilibrium condition is

$$k_1[H_2O]_{eq} = k_2[H^+]_{eq}[OH^-]_{eq}$$

From this expression it follows that

$$\frac{k_1}{k_2} = \frac{[H^+]_{eq}[OH^-]_{eq}}{[H_2O]_{eq}} = \frac{K_w \,(\text{mol L}^{-1})^2}{[H_2O]_{eq}} = \frac{K_w}{55.6}\,\text{mol L}^{-1}$$

because the molar concentration of pure water is 55.6 mol L^{-1}. If we write $K = K_w/55.6 = 1.8 \times 10^{-16}$, we obtain

$$\frac{1}{\tau} = k_2\{(K\,\text{mol L}^{-1}) + [H^+]_{eq} + [OH^-]_{eq}\} = k_2\{K + K_w^{1/2} + K_w^{1/2}\}\,\text{mol L}^{-1}$$

$$= (2.0 \times 10^{-7}) \times k_2\,\text{mol L}^{-1}$$

Hence,

$$k_2 = \frac{1}{(3.7 \times 10^{-5}\,\text{s}) \times (2.0 \times 10^{-7}\,\text{mol L}^{-1})} = 1.4 \times 10^{11}\,\text{L mol}^{-1}\,\text{s}^{-1}$$

It follows that

$$k_1 = k_2K\,\text{mol L}^{-1} = 2.4 \times 10^{-5}\,\text{s}^{-1}$$

Comment Notice how we keep track of units: K and K_w are dimensionless; k_2 is expressed in L mol^{-1} s^{-1} and k_1 is expressed in s^{-1}. The reaction is faster in ice, where $k_2 = 8.6 \times 10^{12}$ L mol^{-1} s^{-1}.

Self-test 25.5 Derive an expression for the relaxation time of a concentration when the reaction $A + B \rightleftharpoons C + D$ is second-order in both directions.
$$[1/\tau = k([A] + [B])_{eq} + k'([C] + [D])_{eq}]$$

25.5 The temperature dependence of reaction rates

The rate constants of most reactions increase as the temperature is raised. Many reactions in solution fall somewhere in the range spanned by the hydrolysis of methyl ethanoate (where the rate constant at 35°C is 1.82 times that at 25°C) and the hydrolysis of sucrose (where the factor is 4.13).

(a) The Arrhenius parameters

It is found experimentally for many reactions that a plot of ln k against $1/T$ gives a straight line. This behaviour is normally expressed mathematically by introducing two parameters, one representing the intercept and the other the slope of the straight line, and writing the **Arrhenius equation**

$$\ln k = \ln A - \frac{E_a}{RT} \tag{25.25}$$

The parameter A, which corresponds to the intercept of the line at $1/T = 0$ (at infinite temperature, Fig. 25.10), is called the **pre-exponential factor** or the **frequency factor**. The parameter E_a, which is obtained from the slope of the line ($-E_a/R$), is called the **activation energy**. Collectively, the two quantities are called the **Arrhenius parameters** (Table 25.4).

25.10 A plot of ln k against $1/T$ is a straight line when the reaction follows the behaviour described by the Arrhenius equation (eqn 25). The slope gives $-E_a/R$ and the intercept at $1/T = 0$ gives ln A.

Table 25.4* Arrhenius parameters

(1) First-order reactions	A/s^{-1}	$E_a/(kJ\ mol^{-1})$
$CH_3NC \longrightarrow CH_3CN$	3.98×10^{13}	160
$2\ N_2O_5 \longrightarrow 4\ NO_2 + O_2$	4.94×10^{13}	103.4
(2) Second-order reactions	$A/(L\ mol^{-1}\ s^{-1})$	$E_a/(kJ\ mol^{-1})$
$OH + H_2 \longrightarrow H_2O + H$	8.0×10^{10}	42
$NaC_2H_5O + CH_3I$ in ethanol	2.42×10^{11}	81.6

* More values are given in the *Data section*.

Example 25.5 Determining the Arrhenius parameters

The rate of the second-order decomposition of acetaldehyde (ethanal, CH_3CHO) was measured over the temperature range 700–1000 K, and the rate constants are reported below. Find E_a and A.

T/K	700	730	760	790	810	840	910	1000
$k/(L\ mol^{-1}\ s^{-1})$	0.011	0.035	0.105	0.343	0.789	2.17	20.0	145

Method According to eqn 25, the data can be analysed by plotting $\ln(k/L\ mol^{-1}\ s^{-1})$ against $1/(T/K)$ and getting a straight line. The slope of this line is $(-E_a/R)/K$ and the intercept at $1/T = 0$ is $\ln A$.

Answer We draw up the following table:

$(10^3\ K)/T$	1.43	1.37	1.32	1.27	1.23	1.19	1.10	1.00
$\ln(k/L\ mol^{-1}\ s^{-1})$	−4.51	−3.35	−2.25	−1.07	−0.24	0.77	3.00	4.98

Now plot $\ln k$ against $1/T$ (Fig. 25.11). The least-squares best fit of the line is with slope -2.27×10^4 and intercept 27.7. Therefore,

$$E_a = (2.27 \times 10^4\ K) \times (8.3145\ J\ K^{-1}\ mol^{-1}) = 189\ kJ\ mol^{-1}$$

$$A = e^{27.7}\ L\ mol^{-1}\ s^{-1} = 1.1 \times 10^{12}\ L\ mol^{-1}\ s^{-1}$$

Comment Note that A has the same units as k. The slopes and intercepts of graphs are always dimensionless, and care must be taken to relate the numerical value to the physical quantity by noting how the data have been plotted. In practice, A is obtained from one of the mid-range data values rather than using a lengthy extrapolation.

Self-test 25.6 Determine A and E_a from the following data:

T/K	300	350	400	450	500
$k/(L\ mol^{-1}\ s^{-1})$	7.9×10^6	3.0×10^7	7.9×10^7	1.7×10^8	3.2×10^8

$[8 \times 10^{10}\ L\ mol^{-1}\ s^{-1},\ 23\ kJ\ mol^{-1}]$

25.11 The Arrhenius plot using the data in Example 25.5.

The fact that E_a is given by the slope of the plot of $\ln k$ against $1/T$ means that, the higher the activation energy, the stronger the temperature dependence of the rate constant (that is, the steeper the slope). *A high activation energy signifies that the rate constant depends strongly on temperature.* If a reaction has zero activation energy, its rate is independent of temperature. In some cases the activation energy is negative, which indicates that the rate decreases as the temperature is raised. We shall see that such behaviour is a signal that the reaction has a complex mechanism.

The temperature dependence of some reactions is not Arrhenius-like, in the sense that a straight line is not obtained when ln k is plotted against $1/T$. However, it is still possible to define an activation energy as

$$E_a = RT^2 \left(\frac{\mathrm{d}\ln k}{\mathrm{d}T} \right) \qquad [25.26]$$

This definition reduces to the earlier one (as the slope of a straight line) for a temperature-independent activation energy. However, the definition in eqn 26 is more general than eqn 25, because it allows E_a to be obtained from the slope (at the temperature of interest) of a plot of ln k against $1/T$ even if the Arrhenius plot is not a straight line. Non-Arrhenius behaviour is sometimes a sign that quantum mechanical tunnelling is playing a significant role in the reaction (Section 25.7e).

(b) The interpretation of the parameters

For the present chapter we shall regard the Arrhenius parameters as purely empirical quantities that enable us to discuss the variation of rate constants with temperature. We begin by writing eqn 25 as

$$k = Ae^{-E_a/RT} \qquad (25.27)$$

To interpret E_a it is useful to consider how the potential energy changes in the course of a chemical reaction that begins with a collision between molecules of A and molecules of B (Fig. 25.12). As the reaction proceeds, A and B come into contact, distort, and begin to exchange or discard atoms. The **reaction coordinate** is the collection of motions, such as changes in interatomic distances and bond angles, that are directly involved in the formation of products from reactants. The potential energy rises to a maximum and the cluster of atoms that corresponds to the region close to the maximum is called the **activated complex**. After the maximum, the potential energy falls as the atoms rearrange in the cluster, and it reaches a value characteristic of the products. The climax of the reaction is at the peak of the potential energy, which corresponds to the activation energy E_a. Here two reactant molecules have come to such a degree of closeness and distortion that a small further distortion will send them in the direction of products. This crucial configuration is called the **transition state** of the reaction. Although some molecules entering the transition state might revert to reactants, if they pass through this configuration then it is inevitable that products will emerge from the encounter.[2]

We also conclude from the preceding discussion that *the activation energy is the minimum kinetic energy that reactants must have in order to form products*. For example, in a gas-phase reaction there are numerous collisions each second, but only a tiny proportion are sufficiently energetic to lead to reaction. The fraction of collisions with a kinetic energy in excess of an energy E_a is given by the Boltzmann distribution as $e^{-E_a/RT}$. Hence, the exponential factor in eqn 27 can be interpreted as the fraction of collisions that have enough kinetic energy to lead to reaction.

The pre-exponential factor is a measure of the rate at which collisions occur irrespective of their energy. Hence, the product of A and the exponential factor, $e^{-E_a/RT}$, gives the rate of *successful* collisions. We shall develop these remarks in Chapter 27 and see that they have their analogues for reactions that take place in liquids.

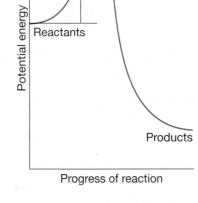

25.12 A potential energy profile for an exothermic reaction. The height of the barrier between the reactants and products is the activation energy of the reaction.

---

2. The terms activated complex and transition state or transition structure are often used as synonyms; however, we shall preserve a distinction.

Accounting for the rate laws

We now move on to the second stage of the analysis of kinetic data, their explanation in terms of a postulated reaction mechanism.

25.6 Elementary reactions

Most reactions occur in a sequence of steps called **elementary reactions**, each of which involves only a small number of molecules or ions. A typical elementary reaction is

$$H + Br_2 \longrightarrow HBr + Br$$

(We do not specify the phase of the species in the chemical equation for an elementary reaction.) This equation signifies that an H atom attacks a Br_2 molecule to produce an HBr molecule and a Br atom. The **molecularity** of an elementary reaction is the number of molecules coming together to react in an elementary reaction. In a **unimolecular reaction**, a single molecule shakes itself apart or its atoms into a new arrangement, as in the isomerization of cyclopropane (**1**) to propene (**2**). In a **bimolecular reaction**, a pair of molecules collide and exchange energy, atoms, or groups of atoms, or undergo some other kind of change. It is most important to distinguish molecularity from order: reaction order is an empirical quantity, and obtained from the experimental rate law; molecularity refers to an elementary reaction proposed as an individual step in a mechanism.

The rate law of a unimolecular elementary reaction is first-order in the reactant:

$$A \longrightarrow P \qquad \frac{d[A]}{dt} = -k[A] \tag{25.28}$$

where P denotes products (several different species may be formed). A unimolecular reaction is first-order because the number of A molecules that decay in a short interval is proportional to the number available to decay. (Ten times as many decay in the same interval when there are initially 1000 A molecules as when there are only 100 present.) Therefore, the rate of decomposition of A is proportional to its molar concentration.

An elementary bimolecular reaction has a second-order rate law:

$$A + B \longrightarrow P \qquad \frac{d[A]}{dt} = -k[A][B] \tag{25.29}$$

A bimolecular reaction is second-order because its rate is proportional to the rate at which the reactant species meet, which in turn is proportional to their concentrations. Therefore, if we have evidence that a reaction is a single-step, bimolecular process, we can write down the rate law (and then go on to test it). Bimolecular elementary reactions are believed to account for many homogeneous reactions, such as the dimerizations of alkenes and dienes and reactions such as

$$CH_3I(alc) + CH_3CH_2O^-(alc) \longrightarrow CH_3OCH_2CH_3(alc) + I^-(alc)$$

(where 'alc' signifies alcohol solution). There is evidence that the mechanism of this reaction is a single elementary step

$$CH_3I + CH_3CH_2O^- \longrightarrow CH_3OCH_2CH_3 + I^-$$

This mechanism is consistent with the observed rate law

$$v = k[CH_3I][CH_3CH_2O^-] \tag{25.30}$$

1 Cyclopropane

2 Propene

We shall see below how to combine a series of simple steps together into a mechanism and how to arrive at the corresponding rate law. For the present we emphasize that, *if the reaction is an elementary bimolecular process, then it has second-order kinetics, but if the kinetics are second-order, then the reaction might be complex.* The postulated mechanism can be explored only by detailed detective work on the system, and by investigating whether side products or intermediates appear during the course of the reaction. Detailed analysis of this kind was one of the ways, for example, in which the reaction $H_2(g) + I_2(g) \rightarrow 2\,HI(g)$ was shown to proceed by a complex reaction. For many years the reaction had been accepted on good, but insufficiently meticulous evidence as a fine example of a simple bimolecular reaction in which atoms exchanged partners during a collision.

25.7 Consecutive elementary reactions

Some reactions proceed through the formation of an intermediate (I), as in the consecutive unimolecular reactions

$$A \xrightarrow{\;k_a\;} I \xrightarrow{\;k_b\;} P$$

An example is the decay of a radioactive family, such as

$$^{239}U \xrightarrow{\;2.35\,min\;} {}^{239}Np \xrightarrow{\;2.35\,day\;} {}^{239}Pu$$

(The times are half-lives.) We can discover the characteristics of this type of reaction by setting up the rate laws for the net rate of change of the concentration of each substance.

(a) The variation of concentrations with time

The rate of unimolecular decomposition of A is

$$\frac{d[A]}{dt} = -k_a[A] \tag{25.31}$$

and A is not replenished. The intermediate I is formed from A (at a rate $k_a[A]$) but decays to P (at a rate $k_b[I]$). The net rate of formation of I is therefore

$$\frac{d[I]}{dt} = k_a[A] - k_b[I] \tag{25.32}$$

The product P is formed by the unimolecular decay of I:

$$\frac{d[P]}{dt} = k_b[I] \tag{25.33}$$

We suppose that initially only A is present, and that its concentration is $[A]_0$.

The first of the rate laws, eqn 31, is an ordinary first-order decay, so we can write

$$[A] = [A]_0 e^{-k_a t} \tag{25.34}$$

When this equation is substituted into eqn 32, we obtain after rearrangement

$$\frac{d[I]}{dt} + k_b[I] = k_a[A]_0 e^{-k_a t} \tag{25.35}$$

This differential equation has a standard form and, after setting $[I]_0 = 0$, the solution is

$$[I] = \frac{k_a}{k_b - k_a}(e^{-k_a t} - e^{-k_b t})[A]_0 \tag{25.36}$$

The solution of a first-order differential equation with the form

$$\frac{dy}{dx} + yf(x) = g(x)$$

is given by

$$e^{\int f(x)\,dx} y = \int e^{\int f(x)\,dx} g(x)\,dx + \text{constant}$$

Equation 35 is a special case of this standard form, with $f(x) = \text{constant}$.

25.13 The concentrations of A, I, and P in the consecutive reaction scheme $A \rightarrow I \rightarrow P$. The curves are plots of eqns 34, 36, and 37 with $k_a = 10k_b$. If the intermediate I is in fact the desired product, it is important to be able to predict when its concentration is greatest; see Example 25.6.

At all times $[A] + [I] + [P] = [A]_0$, so it follows that

$$[P] = \left\{ 1 + \frac{k_a e^{-k_b t} - k_b e^{-k_a t}}{k_b - k_a} \right\}[A]_0 \tag{25.37}$$

The concentration of the intermediate I rises to a maximum, and then falls to zero (Fig. 25.13). The concentration of the product P rises from zero towards $[A]_0$.

Example 25.6 Analysing consecutive reactions

Suppose that in an industrial batch process a substance A produces the desired compound I that goes on to decay to a worthless product C, each step of the reaction being first-order. At what time will I be present in greatest concentration?

Method The time dependence of the concentration of I is given by eqn 36. We can find the time at which [I] passes through a maximum, t_{max}, by calculating $d[I]/dt$ and setting the resulting rate equal to zero.

Answer It follows from eqn 36 that

$$\frac{d[I]}{dt} = -\frac{k_a [A]_0 (k_a e^{-k_a t} - k_b e^{-k_b t})}{k_b - k_a}$$

This rate is equal to zero when

$$k_a e^{-k_a t} = k_b e^{-k_b t}$$

Therefore,

$$t_{max} = \frac{1}{k_a - k_b} \ln \frac{k_a}{k_b}$$

Comment For a given value of k_a, as k_b increases both the time at which [I] is a maximum and the yield of I decrease.

Self-test 25.7 Calculate the maximum concentration of I and justify the last remark. $[[I]_{max}/[A]_0 = (k_a/k_b)^c, \, c = k_b/(k_b - k_a)]$

(b) The rate-determining step

Suppose now that $k_b \gg k_a$; then whenever an I molecule is formed it decays rapidly into P. Because

$$e^{-k_b t} \ll e^{-k_a t} \qquad k_b - k_a \approx k_b$$

eqn 37 reduces to

$$[P] \approx (1 - e^{-k_a t})[A]_0 \tag{25.38}$$

which shows that the formation of the final product P depends on only the *smaller* of the two rate constants. That is, the rate of formation of P depends on the rate at which I is formed, not on the rate at which I changes into P. For this reason, the step $A \rightarrow I$ is called the 'rate-determining step' of the reaction. Its existence has been likened to building a six-lane highway up to a single-lane bridge: the traffic flow is governed by the rate of crossing the bridge. Similar remarks apply to more complicated reaction mechanisms, and in general the **rate-determining step** is the

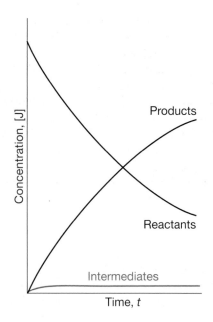

25.16 The basis of the steady-state approximation. It is supposed that the concentrations of intermediates remain small and hardly change during most of the course of the reaction.

Example 25.7 Using the steady-state approximation

Devise the rate law for the decomposition of N_2O_5,

$$2\,N_2O_5(g) \longrightarrow 4\,NO_2(g) + O_2(g)$$

on the basis of the following mechanism:

$$N_2O_5 \longrightarrow NO_2 + NO_3 \qquad k_a$$
$$NO_2 + NO_3 \longrightarrow N_2O_5 \qquad k_a'$$
$$NO_2 + NO_3 \longrightarrow NO_2 + O_2 + NO \qquad k_b$$
$$NO + N_2O_5 \longrightarrow NO_2 + NO_2 + NO_2 \qquad k_c$$

Method First identify the intermediates (species that occur in the reaction steps but do not appear in the overall reaction) and write expressions for their net rates of formation. Then, all net rates of change of the concentrations of intermediates are set equal to zero and the resulting equations are solved algebraically.

Answer The intermediates are NO and NO_3; the net rates of change of their concentrations are

$$\frac{d[NO]}{dt} = k_b[NO_2][NO_3] - k_c[NO][N_2O_5] \approx 0$$

$$\frac{d[NO_3]}{dt} = k_a[N_2O_5] - k_a'[NO_2][NO_3] - k_b[NO_2][NO_3] \approx 0$$

The net rate of change of concentration of N_2O_5 is

$$\frac{d[N_2O_5]}{dt} = -k_a[N_2O_5] + k_a'[NO_2][NO_3] - k_c[NO][N_2O_5]$$

and replacing the concentrations of the intermediates by using the equations above gives

$$\frac{d[N_2O_5]}{dt} = -\frac{2k_a k_b[N_2O_5]}{k_a' + k_b}$$

Comment The decomposition of N_2O_5 is problematic because its rate decreases more quickly than expected at low pressures. It is believed that this decrease is due to changes in the rate constants themselves (particularly k_a').

Self-test 25.8 Derive the rate law for the decomposition of ozone in the reaction $2\,O_3(g) \to 3\,O_2(g)$ on the basis of the (incomplete) mechanism

$$O_3 \longrightarrow O_2 + O \qquad k_a$$
$$O_2 + O \longrightarrow O_3 \qquad k_a'$$
$$O + O_3 \longrightarrow O_2 + O_2 \qquad k_b$$

$$[d[O_3]/dt = -2k_a k_b[O_3]^2/(k_a'[O_2] + k_b[O_3])]$$

(d) Pre-equilibria

From a simple sequence of consecutive reactions we now turn to a slightly more complicated mechanism in which an intermediate I reaches an equilibrium with the reactants A and B:

$$A + B \rightleftharpoons I \longrightarrow P \qquad\qquad (25.43)$$

The rate constants are k_a and k_a' for the forward and reverse reactions of the equilibrium and k_b for the final step. This scheme involves a **pre-equilibrium**, in which an intermediate is in equilibrium with the reactants. A pre-equilibrium arises when the rates of formation of the intermediate and its decay back into reactants are much faster than its rate of formation of products; thus, the condition is possible when $k_a' \gg k_b$ but not when $k_b \gg k_a'$. Because we assume that A, B, and I are in equilibrium, we can write

$$K = \frac{[I]}{[A][B]} \qquad K = \frac{k_a}{k_a'} \tag{25.44}$$

In writing these equations, we are presuming that the rate of reaction of I to form P is too slow to affect the maintenance of the pre-equilibrium (see the example below). The rate of formation of P may now be written:

$$\frac{d[P]}{dt} = k_b[I] = k_b K[A][B] \tag{25.45}$$

This rate law has the form of a second-order rate law with a composite rate constant:

$$\frac{d[P]}{dt} = k[A][B] \qquad k = k_b K = \frac{k_a k_b}{k_a'} \tag{25.46}$$

Example 25.8 Analysing a pre-equilibrium

Repeat the pre-equilibrium calculation but without ignoring the fact that I is slowly leaking away as it forms P.

Method Begin by writing the net rates of change of the concentrations of the substances and then invoke the steady-state approximation for the intermediate I. Use the resulting expression to obtain the rate of change of the concentration of P.

Answer The net rates of change of P and I are

$$\frac{d[P]}{dt} = k_b[I]$$

$$\frac{d[I]}{dt} = k_a[A][B] - k_a'[I] - k_b[I] \approx 0$$

The second equation solves to

$$[I] \approx \frac{k_a[A][B]}{k_a' + k_b}$$

When we substitute this into the expression for the rate of formation of P, we obtain

$$\frac{d[P]}{dt} \approx k[A][B] \qquad k = \frac{k_a k_b}{k_a' + k_b}$$

Comment This expression reduces to that in eqn 46 when the rate constant for the decay of I into products is much smaller than that for its decay into reactants, $k_b \ll k_a'$.

Self-test 25.9 Show that the pre-equilibrium mechanism in which $2\,A \rightleftharpoons I\,(K)$ followed by $I + B \rightarrow P\,(k_b)$ results in an overall third-order reaction.

$$[d[P]/dt = k_b K[A]^2[B]]$$

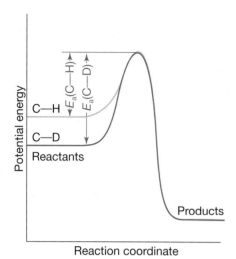

25.17 Changes in the reaction profile when a C—H bond undergoing cleavage is deuterated. In this illustration, the C—H and C—D bonds are modelled as simple harmonic oscillators. The only significant change is in the zero-point energy of the reactants, which is lower for C—D than for C—H. As a result, the activation energy is greater for C—D cleavage than for C—H cleavage.

(e) The kinetic isotope effect

The postulation of a plausible mechanism requires careful analysis of many experiments designed to determine the fate of atoms during the formation of products. Observation of the **kinetic isotope effect**, a decrease in the rate of a chemical reaction upon replacement of one atom in a reactant by a heavier isotope, facilitates the identification of bond-breaking events in the rate-determining step. A **primary kinetic isotope effect** is observed when the rate-determining step requires the scission of a bond involving the isotope. A **secondary kinetic isotope effect** is the reduction in reaction rate even though the bond involving the isotope is not broken to form product. In both cases, the effect arises from the increase in activation energy that accompanies the replacement of an atom by a heavier isotope on account of changes in the zero-point vibrational energies (Section 17.7d).

Here we consider the origin of the primary kinetic isotope effect in a reaction in which the rate-determining step is the scission of a C—H bond. The reaction coordinate corresponds to the stretching of the C—H bond and the potential energy profile is shown in Fig. 25.17. On deuteration, the dominant change is the reduction of the zero-point energy of the bond (because the deuterium atom is heavier). The whole reaction profile is not lowered, however, because the relevant vibration in the activated complex has a very low force constant, so there is little zero-point energy associated with the reaction coordinate in either isotopomeric form of the complex.

We assume that, to a good approximation, a change in the activation energy arises only from the change in zero-point energy of the stretching vibration, so

$$E_a(C—D) - E_a(C—H) = N_A\{\tfrac{1}{2}hc\tilde{v}(C—H) - \tfrac{1}{2}hc\tilde{v}(C—D)\} \tag{25.47}$$

where $\tilde{v}$ is the relevant vibrational wavenumber. From Section 16.9, we know that $\tilde{v}(C—D) = (\mu_{CH}/\mu_{CD})^{1/2}\tilde{v}(C—H)$, where μ is the relevant effective mass. It follows that

$$E_a(C—D) - E_a(C—H) = \tfrac{1}{2}N_A hc\tilde{v}(C—H)\left\{1 - \left(\frac{\mu_{CH}}{\mu_{CD}}\right)^{1/2}\right\} \tag{25.48}$$

If we assume further that the pre-exponential factor does not change upon deuteration, then the rate constants for the two species should be in the ratio

$$\frac{k(C—D)}{k(C—H)} = e^{-\lambda} \quad \text{with} \quad \lambda = \frac{hc\tilde{v}(C—H)}{2kT}\left\{1 - \left(\frac{\mu_{CH}}{\mu_{CD}}\right)^{1/2}\right\} \tag{25.49}$$

Note that $\lambda > 0$ because $\mu_{CD} > \mu_{CH}$ and that $k(C—D)/k(C—H)$ decreases with decreasing temperature.

Illustration 25.2

From infrared spectra, the fundamental vibrational wavenumber for stretching of a C—H bond is about 3000 cm^{-1}. From $\mu_{CH}/\mu_{CD} = 0.538$ and eqn 49, it follows that $k(C—D)/k(C—H) = 0.145$ at 298 K. We predict that at room temperature C—H cleavage should be about seven times faster than C—D cleavage, other conditions being equal. However, experimental values of $k(C—D)/k(C—H)$ can differ significantly from those predicted by eqn 49 on account of the severity of the assumptions in the model.

Secondary kinetic isotope effects arise from differences in the zero-point energies between reactants and an activated complex with significantly different structure. When a change in the zero-point energy of a single vibration involving motion of a C—H group can account for the change in activation energy upon deuteration, we write:

$$E_a(C—D) - E_a(C—H) =$$
$$N_A[\{\tfrac{1}{2}hc\tilde{v}^{\ddagger}(C—H) - \tfrac{1}{2}hc\tilde{v}(C—H)\} - \{\tfrac{1}{2}hc\tilde{v}^{\ddagger}(C—D) - \tfrac{1}{2}hc\tilde{v}(C—D)\}] \quad (25.50)$$

where $\tilde{v}^{\ddagger}$ and $\tilde{v}$ denote vibrational wavenumbers in the activated complex and reactant, respectively. With $\tilde{v}^{\ddagger}(C—D) = (\mu_{CH}/\mu_{CD})^{1/2}\tilde{v}^{\ddagger}(C—H)$ and $\tilde{v}(C—D) = (\mu_{CH}/\mu_{CD})^{1/2}\tilde{v}(C—H)$, it follows that

$$E_a(C—D) - E_a(C—H) = \tfrac{1}{2}N_A hc\{\tilde{v}^{\ddagger}(C—H) - \tilde{v}(C—H)\}\left\{1 - \left(\frac{\mu_{CH}}{\mu_{CD}}\right)^{1/2}\right\} \quad (25.51)$$

and

$$\frac{k(C—D)}{k(C—H)} = e^{-\lambda} \quad \text{with} \quad \lambda = \frac{hc\{\tilde{v}^{\ddagger}(C—H) - \tilde{v}(C—H)\}}{2kT}\left\{1 - \left(\frac{\mu_{CH}}{\mu_{CD}}\right)^{1/2}\right\} \quad (25.52)$$

Illustration 25.3

In the heterolytic dissociation

$$CHCl_3 \longrightarrow CHCl_2^+ + Cl^-$$

the activated complex resembles the product $CHCl_2^+$. From infrared spectra, the fundamental vibrational wavenumber for a bending motion involving the C—H group is about 800 cm^{-1} in $CHCl_3$ and about 1350 cm^{-1} in $CHCl_2^+$. Assuming that $\tilde{v}^{\ddagger}(C—H) = 1350$ cm^{-1} on account of the structural similarity between $CHCl_2^+$ and the activated complex, it follows from $\mu_{CH}/\mu_{CD} = 0.538$ and eqn 52 that $k(C—D)/k(C—H) = 0.709$ at 298 K. We predict that at room temperature the dissociation of $CHCl_3$ should be about 40 per cent faster than dissociation of $CDCl_3$. Comparison with the result from Illustration 25.2 shows that the secondary kinetic isotope effect leads to higher values of $k(C—D)/k(C—H)$ than does the primary kinetic isotope effect. This conclusion is supported by a number of experimental observations.

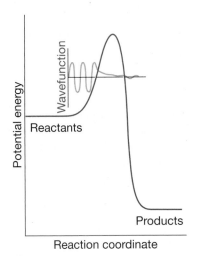

25.18 A proton can tunnel through the activation energy barrier that separates reactants from products, so the effective height of the barrier is reduced and the rate of the proton transfer reaction increases. The effect is represented by drawing the wavefunction of the proton near the barrier. Proton tunnelling is important only at low temperatures, when most of the reactants are trapped on the left of the barrier.

In some cases, substitution of deuterium for hydrogen results in values of $k(C—D)/k(C—H)$ that are too low to be accounted for by eqns 49 or 52, even when more complete models are used to predict ratios of rate constants. Such abnormal kinetic isotope effects are evidence for a path in which quantum mechanical tunnelling of hydrogen atoms takes place through the activation barrier (Fig. 25.18). We saw in Section 12.3 that the probability of tunnelling through a barrier decreases as the mass of the particle increases, so deuterium tunnels less efficiently through a barrier than hydrogen and its reactions are correspondingly slower. Quantum mechanical tunnelling can be the dominant process in reactions involving hydrogen atom or proton transfer when the temperature is so low that very few reactant molecules can overcome the activation energy barrier. We shall see in Chapter 29 that, because m_e is so small, tunnelling is also a very important contributor to the rates of electron transfer reactions.

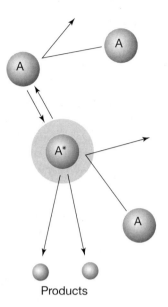

25.19 A representation of the Lindemann–Hinshelwood mechanism of unimolecular reactions. The species A is excited by collision with A, and the excited A molecule (A*) may either be deactivated by a collision with A or go on to decay by a unimolecular process to form products.

25.8 Unimolecular reactions

A number of gas-phase reactions follow first-order kinetics, as in the isomerization of cyclopropane mentioned earlier:

$$cyclo\text{-}C_3H_6 \longrightarrow CH_3CH{=}CH_2 \qquad v = k[cyclo\text{-}C_3H_6] \tag{25.53}$$

The problem with the interpretation of first-order rate laws is that presumably a molecule acquires enough energy to react as a result of its collisions with other molecules. However, collisions are simple bimolecular events, so how can they result in a first-order rate law? First-order gas-phase reactions are widely called 'unimolecular reactions' because they also involve an elementary unimolecular step in which the reactant molecule changes into the product. This term must be used with caution, though, because the overall mechanism has bimolecular as well as unimolecular steps.

(a) The Lindemann–Hinshelwood mechanism

The first successful explanation of unimolecular reactions was provided by Frederick Lindemann in 1921 and then elaborated by Cyril Hinshelwood. In the **Lindemann–Hinshelwood mechanism** it is supposed that a reactant molecule A becomes energetically excited by collision with another A molecule (Fig. 25.19):

$$A + A \longrightarrow A^* + A \qquad \frac{d[A^*]}{dt} = k_a[A]^2 \tag{25.54}$$

The energized molecule might lose its excess energy by collision with another molecule:

$$A + A^* \longrightarrow A + A \qquad \frac{d[A^*]}{dt} = -k_a'[A][A^*] \tag{25.55}$$

Alternatively, the excited molecule might shake itself apart and form products P. That is, it might undergo the unimolecular decay

$$A^* \longrightarrow P \qquad \frac{d[A^*]}{dt} = -k_b[A^*] \tag{25.56}$$

If the unimolecular step is slow enough to be the rate-determining step, the overall reaction will have first-order kinetics, as observed. This conclusion can be demonstrated explicitly by applying the steady-state approximation to the net rate of formation of A*:

$$\frac{d[A^*]}{dt} = k_a[A]^2 - k_a'[A][A^*] - k_b[A^*] \approx 0 \tag{25.57}$$

This equation solves to

$$[A^*] = \frac{k_a[A]^2}{k_b + k_a'[A]} \tag{25.58}$$

so the rate law for the formation of P is

$$\frac{d[P]}{dt} = k_b[A^*] = \frac{k_a k_b[A]^2}{k_b + k_a'[A]} \tag{25.59}$$

At this stage the rate law is not first-order. However, if the rate of deactivation by (A*,A) collisions is much greater than the rate of unimolecular decay, in the sense that

$$k_a'[A^*][A] \gg k_b[A^*] \qquad \text{or} \qquad k_a'[A] \gg k_b$$

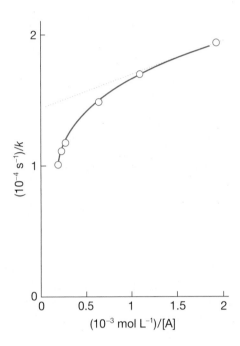

25.20 The pressure dependence of the unimolecular isomerization of *trans*-CHD=CHD showing a pronounced departure from the straight line predicted by eqn 63 based on the Lindemann–Hinshelwood mechanism.

then we can neglect k_b in the denominator of eqn 39 and obtain

$$\frac{d[P]}{dt} \approx k[A] \qquad k = k_a k_b / k'_a \tag{25.60}$$

Equation 60 is a first-order rate law, as we set out to show.

The Lindemann–Hinshelwood mechanism can be tested because it predicts that, as the concentration (and therefore the partial pressure) of A is reduced, the reaction should switch to overall second-order kinetics. Thus, when $k'_a[A] \ll k_b$, the rate law in eqn 59 is

$$\frac{d[P]}{dt} \approx k_a[A]^2 \tag{25.61}$$

The physical reason for the change of order is that at low pressures the rate-determining step is the bimolecular formation of A*. If we write the full rate law in eqn 59 as

$$\frac{d[P]}{dt} = k[A] \qquad k = \frac{k_a k_b [A]}{k_b + k'_a[A]} \tag{25.62}$$

then the expression for the effective rate-constant, k, can be rearranged to

$$\frac{1}{k} = \frac{k'_a}{k_a k_b} + \frac{1}{k_a[A]} \tag{25.63}$$

Hence, a test of the theory is to plot $1/k$ against $1/[A]$, and to expect a straight line.

Whereas the Lindemann–Hinshelwood mechanism agrees in general with the switch in order of unimolecular reactions, it does not agree in detail. Figure 25.20 shows a typical graph of $1/k$ against $1/[A]$. The graph has a pronounced curvature, corresponding to a larger value of k (a smaller value of $1/k$) at high pressures (low $1/[A]$) than would be expected by extrapolation of the reasonably linear low pressure (high $1/[A]$) data.

(b) The activation energy of a composite reaction

Although the rate of each step of a complex mechanism might increase with temperature and show Arrhenius behaviour, is that true of a composite reaction? To answer this question, we consider the high-pressure limit of the Lindemann–Hinshelwood mechanism as expressed in eqn 60. If each of the rate constants has an Arrhenius-like temperature dependence, we can use eqn 27 for each of them, and write

$$k = \frac{k_a k_b}{k'_a} = \frac{\left(A_a e^{-E_a(a)/RT}\right)\left(A_b e^{-E_a(b)/RT}\right)}{\left(A'_a e^{-E'_a(a)/RT}\right)}$$

$$= \frac{A_a A_b}{A'_a} e^{-\{E_a(a) + E_a(b) - E'_a(a)\}/RT} \tag{25.64}$$

That is, the composite rate constant k has an Arrhenius-like form with activation energy

$$E_a = E_a(a) + E_a(b) - E'_a(a) \tag{25.65}$$

Provided $E_a(a) + E_a(b) > E'_a(a)$, the activation energy is positive and the rate increases with temperature. However, it is conceivable that $E_a(a) + E_a(b) < E'_a(a)$ (Fig. 25.21), in which case the activation energy is negative and the rate will *decrease* as the temperature is raised. There is nothing remarkable about this behaviour: all it means is that the reverse reaction (corresponding to the deactivation of A*) is so sensitive to

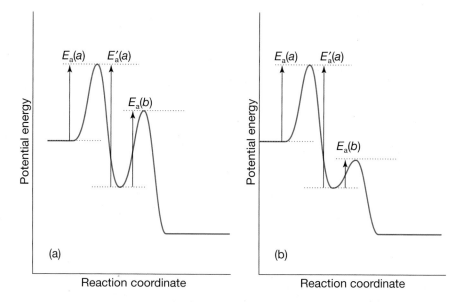

25.21 For a reaction with a pre-equilibrium, there are three activation energies to take into account, two referring to the reversible steps of the pre-equilibrium and one for the final step. The relative magnitudes of the activation energies determine whether the overall activation energy is (a) positive or (b) negative.

temperature that its rate increases sharply as the temperature is raised, and depletes the steady-state concentration of A^*. The Lindemann–Hinshelwood mechanism is an unlikely candidate for this type of behaviour because the deactivation of A^* has only a small activation energy, but there are reactions with analogous mechanisms in which a negative activation energy is observed.

When we examine the general rate law given in eqn 59, it is clear that the temperature dependence may be difficult to predict because each rate constant in the expression for k increases with temperature, and the outcome depends on whether the terms in the numerator dominate those in the denominator, or vice versa. The fact that so many reactions do show Arrhenius-like behaviour with positive activation energies suggests that their rate laws are in a 'simple' regime, like eqn 61 rather than eqn 59, and that the temperature dependence is dominated by the activation energy of the rate-determining stage.

Checklist of key ideas

☐ chemical kinetics, the study of reaction rates

☐ mechanism of reaction, the sequence of elementary steps involved in a reaction

Empirical chemical kinetics

25.1 Experimental techniques

☐ real-time analysis, a procedure in which the composition of a system is analysed while the reaction is in progress

☐ flow method, a procedure in which the composition of a system is analysed as the reactants flow into a mixing chamber

☐ stopped-flow technique, a procedure in which the reagents are mixed very quickly in a small chamber fitted with a syringe instead of an outlet tube

☐ flash photolysis, a procedure in which the reaction is initiated by a brief flash of light

☐ quenching methods, techniques based on stopping the reaction after it has been allowed to proceed for a certain time

☐ chemical quench flow method, a technique in which the reactants are mixed as in the flow method but the reaction is quenched by another reagent

☐ freeze quench method, a technique in which the reaction is quenched by cooling the mixture

25.2 The rates of reactions

☐ rate of consumption of a reactant R, $-d[R]/dt$

☐ rate of formation of a product P, $d[P]/dt$

☐ rate of reaction, $v = d\xi/dt$

☐ rate of homogeneous reaction, $v = (1/v_J)\, d[J]/dt$

☐ rate of heterogeneous reaction, $v = (1/v_J)\, d\sigma_J/dt$

- [ ] rate constant, the constant k in a rate law
- [ ] rate law, the rate as a function of concentration, $v = f([A], [B], \dots)$
- [ ] reaction order, the power to which the concentration of a species is raised in a rate law
- [ ] first-order reaction, a reaction with a rate law of the form $v = k[A]$
- [ ] second-order reaction, a reaction with a rate law of the form $v = k[A]^2$
- [ ] overall order, the sum of the orders in a rate law of the form $v = k[A]^a[B]^b \dots$
- [ ] zero-order rate law, a rate law of the form $v = k$
- [ ] isolation method, a procedure in which the concentrations of all the reactants except one are in large excess
- [ ] pseudofirst-order rate law, a rate law of the form $v = k'[A]$ with $k' = k[B]_0$ by maintaining B in large excess
- [ ] method of initial rates, a procedure in which the rate is measured at the beginning of the reaction for several different initial concentrations of reactants; $\log v_0 = \log k + a \log [A]_0$

25.3 Integrated rate laws

- [ ] integrated rate law, the integrated form of a rate law
- [ ] first-order integrated rate law, $\ln([A]/[A]_0) = -kt$, $[A] = [A]_0 e^{-kt}$
- [ ] half life, $t_{1/2} = (\ln 2)/k$ (first-order reaction), $t_{1/2} = 1/(k[A]_0)$ (second-order reaction)
- [ ] time constant, the time required for the concentration of a reactant to fall to $1/e$ of its initial value
- [ ] second-order integrated rate law, $1/[A] - 1/[A]_0 = kt$, $[A] = [A]_0/(1 + kt[A]_0)$

25.4 Reactions approaching equilibrium

- [ ] approach to equilibrium, $[A] = \{(k' + ke^{-(k+k')t})/(k + k')\}[A]_0$
- [ ] equilibrium constant, $K = k/k'$
- [ ] relaxation, the return to equilibrium

- [ ] temperature jump, a procedure in which a sudden temperature rise is imposed and the system returns to equilibrium
- [ ] pressure-jump techniques, as for temperature jump, but with a sudden change in pressure

25.5 The temperature dependence of reaction rates

- [ ] Arrhenius equation, $\ln k = \ln A - E_a/RT$
- [ ] pre-exponential factor (frequency factor), the parameter A in the Arrhenius equation
- [ ] activation energy, the parameter E_a in the Arrhenius equation; the minimum kinetic energy for reaction during a molecular encounter
- [ ] Arrhenius parameters, the parameters A and E_a
- [ ] generalized activation energy, $E_a = RT^2(\mathrm{d} \ln k/\mathrm{d}T)$
- [ ] reaction coordinate, the collection of motions that are directly involved in the formation of products
- [ ] activated complex, the cluster of atoms that corresponds to the region close to the maximum potential energy along the reaction coordinate
- [ ] transition state, a configuration of atoms in the activated complex which, if attained, leads to products

Accounting for the rate laws

25.6 Elementary reactions

- [ ] elementary reaction, a step in a reaction mechanism
- [ ] molecularity, the number of molecules coming together to react in an elementary reaction
- [ ] unimolecular reaction, an elementary reaction involving a single reactant molecule
- [ ] bimolecular reaction, an elementary reaction involving the encounter of two reactant molecules

25.7 Consecutive elementary reactions

- [ ] consecutive first-order reactions, a sequence of first-order reactions; see Fig. 25.13
- [ ] rate-determining step, the step in a mechanism that controls the overall rate of the reaction; commonly but not necessarily the slowest step
- [ ] steady-state approximation, the rates of change of concentrations of all reaction intermediates are negligibly small: $\mathrm{d}[I]/\mathrm{d}t \approx 0$
- [ ] induction period, the initial stage of a reaction during which reaction intermediates are formed
- [ ] pre-equilibrium, a state in which an intermediate is in equilibrium with the reactants and which arises when the rates of formation of the intermediate and its decay back into reactants are much faster than its rate of formation of products
- [ ] kinetic isotope effect, the decrease in the rate of a chemical reaction upon replacement of one atom in a reactant by a heavier isotope
- [ ] primary kinetic isotope effect, the kinetic isotope effect observed when the rate-determining step requires the scission of a bond involving the isotope: $k(C\!-\!D)/k(C\!-\!H) = e^{-\lambda}$, $\lambda = \{hc\tilde{v}(C\!-\!H)/2kT\}\{1 - (\mu_{CH}/\mu_{CD})^{1/2}\}$
- [ ] secondary kinetic isotope effect, the reduction in reaction rate even though the bond involving the isotope is not broken to form product: $k(C\!-\!D)/k(C\!-\!H) = e^{-\lambda}$, $\lambda = \{hc[\tilde{v}^{\ddagger}(C\!-\!H) - \tilde{v}(C\!-\!H)]/2kT\}\{1 - (\mu_{CH}/\mu_{CD})^{1/2}\}$

25.8 Unimolecular reactions

- [ ] Lindemann–Hinshelwood mechanism, a theory of 'unimolecular' reactions: $A + A \rightarrow A^* + A$, $A + A^* \rightarrow A + A$, $A^* \rightarrow P$; $\mathrm{d}[P]/\mathrm{d}t = k_a k_b[A]^2/(k_b + k_a'[A])$
- [ ] activation energies of composite reactions, $E_a = E_a(a) + E_a(b) - E_a'(a)$

Further reading

Articles of general interest

J. Andraos, How mathematics figures in chemistry: some examples. *J. Chem. Educ.* **76**, 258 (1999).

J. Andraos, A streamlined approach to solving simple and complex kinetic systems analytically. *J. Chem. Educ.* **76**, 1578 (1999).

M.N. Berberan-Santos and J.M.G. Martinho, Integration of kinetic rate equations by matrix methods. *J. Chem. Educ.* **67**, 375 (1990).

S. Bluestone and K.Y. Yan, A method to find the rate constants in chemical kinetics of a complex reaction. *J. Chem. Educ.* **72**, 884 (1995).

R.A.B. Bond, B.S. Martincigh, J.R. Mika, and R.H. Simoyi, The quasi-steady-state approximation: numerical validation. *J. Chem. Educ.* **75**, 1158 (1998).

R.W. Carr, Chemical kinetics. In *Encyclopedia of applied physics* (ed. G.L. Trigg), **3**, 345. VCH, New York (1992).

P.L. Corio, Chemical kinetics in constant-pressure systems. *J. Chem. Educ.* **76**, 1452 (1999).

T. Cvitaš, A new look at reaction rates. *J. Chem. Educ.* **76**, 1574 (1999).

J.G. Eberhardt and E. Levin, A simplified integration technique for reaction rate laws of integral order in several substances. *J. Chem. Educ.* **72**, 193 (1995).

G.I. Gellene, Application of kinetic approximations to the A + B ⇌ C reaction system. *J. Chem. Educ.* **72**, 196 (1995).

J.M. Goodman, How do approximations affect the solutions to kinetic equations? *J. Chem. Educ.* **76**, 275 (1999).

E. Levin and J.G. Eberhart, Simplified rate-law integration for reactions that are first-order in each of the two reactants. *J. Chem. Educ.* **66**, 705 (1989).

H. Maskill, The extent of reaction and chemical kinetics. *Educ. in Chem.* **21**, 122 (1984).

J.C. Reeve, Some provocative opinions on the terminology of chemical kinetics. *J. Chem. Educ.* **68**, 728 (1991).

S.R. Logan, The meaning and significance of 'the activation energy' of a chemical reaction. *Educ. in Chem.* **23**, 148 (1986).

Texts and sources of data and information

C.H. Bamford, C.F. Tipper, and R.G. Compton (ed.), *Comprehensive chemical kinetics*, Vols. 1–38. Elsevier, Amsterdam (1969–2001).

K.A. Connors, *Chemical kinetics*. VCH, Weinheim (1990).

K.J. Laidler, *Chemical kinetics*. Harper and Row, New York (1987).

J.C. Lindon, G.E. Tranter, and J.L. Holmes (ed.), *Encyclopedia of spectroscopy and spectrometry*. Academic Press, San Diego (2000).

S.R. Logan, *Fundamentals of chemical kinetics*. Longman, Harlow (1996).

NDRL/NIST solution kinetics database, NIST standard reference database 40, National Institute of Standards and Technology, Gaithersburg (1994). For the URL, see the web site for this book.

NIST chemical kinetics database, NIST standard reference database 17, National Institute of Standards and Technology, Gaithersburg (1998). For the URL, see the web site for this book.

M.J. Pilling and P.W. Seakins, *Reaction kinetics*. Oxford University Press (1996).

J.I. Steinfeld, J.S. Francisco, and W.L. Hase, *Chemical kinetics and dynamics*. Prentice Hall, Englewood Cliffs (1998).

Exercises

Discussion questions

25.1 (a) Write a brief report on a recent research article in which at least one of the following techniques was used to study the kinetics of a chemical reaction: stopped-flow techniques, flash photolysis, chemical quench-flow methods, freeze quench methods, temperature-jump methods, or pressure-jump methods. Your report should be similar in content and size to one of the *Boxes* found throughout this text.

25.1 (b) Describe the main features, including advantages and disadvantages, of the following experimental methods for determining the rate law of a reaction: the isolation method, the method of initial rates, and fitting data to integrated rate law expressions.

25.2 (a) Distinguish between reaction order and molecularity.

25.2 (b) Assess the validity of the following statement: the rate-determining step is the slowest step in a reaction mechanism.

25.3 (a) Distinguish between a pre-equilibrium approximation and a steady-state approximation.

25.3 (b) Define the terms in and limit the generality of the expression $\ln k = \ln A - E_a/RT$.

25.4 (a) Distinguish between a primary and a secondary kinetic isotope effect. Discuss how kinetic isotope effects in general can provide insight into the mechanism of a reaction.

25.4 (b) Discuss the limitations of the generality of the expression $k = k_a k_b[A]/(k_b + k_a'[A])$ for the effective rate constant of a unimolecular reaction A → P with the following mechanism: A + A ⇌ A* + A (k_a, k_a'), A* → P(k_b). Suggest an experimental procedure that may either support or refute the mechanism.

Numerical exercises

25.5 (a) The rate of the reaction A + 2 B → 3 C + D was reported as 1.0 mol L^{-1} s^{-1}. State the rates of formation and consumption of the participants.

25.5 (b) The rate of the reaction A + 3 B → C + 2 D was reported as 1.0 mol L^{-1} s^{-1}. State the rates of formation and consumption of the participants.

25.6 (a) The rate of formation of C in the reaction 2 A + B → 2 C + 3 D is 1.0 mol L^{-1} s^{-1}. State the reaction rate, and the rates of formation or consumption of A, B, and D.

25.6 (b) The rate of consumption of B in the reaction A + 3 B → C + 2 D is 1.0 mol L^{-1} s^{-1}. State the reaction rate, and the rates of formation or consumption of A, C, and D.

25.7 (a) The rate law for the reaction in Exercise 25.5a was found to be $v = k[A][B]$. What are the units of k? Express the rate law in terms of the rates of formation and consumption of (a) A, (b) C.

25.7 (b) The rate law for the reaction in Exercise 25.5b was found to be $v = k[A][B]^2$. What are the units of k? Express the rate law in terms of the rates of formation and consumption of (a) A, (b) C.

25.8 (a) The rate law for the reaction in Exercise 25.6a was reported as $d[C]/dt = k[A][B][C]$. Express the rate law in terms of the reaction rate; what are the units for k in each case?

25.8 (b) The rate law for the reaction in Exercise 25.6b was reported as $d[C]/dt = k[A][B][C]^{-1}$. Express the rate law in terms of the reaction rate; what are the units for k in each case?

25.9 (a) At 518°C, the rate of decomposition of a sample of gaseous acetaldehyde, initially at a pressure of 363 Torr, was 1.07 Torr s^{-1} when 5.0 per cent had reacted and 0.76 Torr s^{-1} when 20.0 per cent had reacted. Determine the order of the reaction.

25.9 (b) At 400 K, the rate of decomposition of a gaseous compound initially at a pressure of 12.6 kPa, was 9.71 Pa s^{-1} when 10.0 per cent had reacted and 7.67 Pa s^{-1} when 20.0 per cent had reacted. Determine the order of the reaction.

25.10 (a) At 518°C, the half-life for the decomposition of a sample of gaseous acetaldehyde (ethanal) initially at 363 Torr was 410 s. When the pressure was 169 Torr, the half-life was 880 s. Determine the order of the reaction.

25.10 (b) At 400 K, the half-life for the decomposition of a sample of a gaseous compound initially at 55.5 kPa was 340 s. When the pressure was 28.9 kPa, the half-life was 178 s. Determine the order of the reaction.

25.11 (a) The rate constant for the first-order decomposition of N_2O_5 in the reaction $2\,N_2O_5(g) \rightarrow 4\,NO_2(g) + O_2(g)$ is $k = 3.38 \times 10^{-5}$ s^{-1} at 25°C. What is the half-life of N_2O_5? What will be the pressure, initially 500 Torr, (a) 10 s, (b) 10 min after initiation of the reaction?

25.11 (b) The rate constant for the first-order decomposition of a compound A in the reaction $2\,A \rightarrow P$ is $k = 2.78 \times 10^{-7}$ s^{-1} at 25°C. What is the half-life of A? What will be the pressure, initially 32.1 kPa, (a) 10 s, (b) 10 min after initiation of the reaction?

25.12 (a) A second-order reaction of the type $A + B \rightarrow P$ was carried out in a solution that was initially 0.050 mol L^{-1} in A and 0.080 mol L^{-1} in B. After 1.0 h the concentration of A had fallen to 0.20 mol L^{-1}. (a) Calculate the rate constant. (b) What is the half-life of the reactants?

25.12 (b) A second-order reaction of the type $A + 2\,B \rightarrow P$ was carried out in a solution that was initially 0.075 mol L^{-1} in A and 0.030 mol L^{-1} in B. After 1.0 h the concentration of A had fallen to 0.045 mol L^{-1}. (a) Calculate the rate constant. (b) What is the half-life of the reactants?

25.13 (a) If the rate laws are expressed with (a) concentrations in moles per litre, (b) pressures in kilopascals, what are the units of the second-order and third-order rate constants?

25.13 (b) If the rate laws are expressed with (a) concentrations in molecules per metre cubed, (b) pressures in newtons per metre squared, what are the units of the second-order and third-order rate constants?

25.14 (a) The second-order rate constant for the reaction

$$CH_3COOC_2H_5(aq) + OH^-(aq) \longrightarrow CH_3CO_2^-(aq) + CH_3CH_2OH(aq)$$

is 0.11 L mol^{-1} s^{-1}. What is the concentration of ester after (a) 10 s, (b) 10 min when ethyl acetate is added to sodium hydroxide so that the initial concentrations are $[NaOH] = 0.050$ mol L^{-1} and $[CH_3COOC_2H_5] = 0.100$ mol L^{-1}?

25.14 (b) The second-order rate constant for the reaction $A + 2\,B \rightarrow C + D$ is 0.21 L mol^{-1} s^{-1}. What is the concentration of C after (a) 10 s, (b) 10 min when the reactants are mixed with initial concentrations of $[A] = 0.025$ mol L^{-1} and $[B] = 0.150$ mol L^{-1}?

25.15 (a) A reaction $2\,A \rightarrow P$ has a second-order rate law with $k = 3.50 \times 10^{-4}$ L mol^{-1} s^{-1}. Calculate the time required for the concentration of A to change from 0.260 mol L^{-1} to 0.011 mol L^{-1}.

25.15 (b) A reaction $2\,A \rightarrow P$ has a third-order rate law with $k = 3.50 \times 10^{-4}$ L^2 mol^{-2} s^{-1}. Calculate the time required for the concentration of A to change from 0.077 mol L^{-1} to 0.021 mol L^{-1}.

25.16 (a) The rate constant for the decomposition of a certain substance is 2.80×10^{-3} L mol^{-1} s^{-1} at 30°C and 1.38×10^{-2} L mol^{-1} s^{-1} at 50°C. Evaluate the Arrhenius parameters of the reaction.

25.16 (b) The rate constant for the decomposition of a certain substance is 1.70×10^{-2} L mol^{-1} s^{-1} at 24°C and 2.01×10^{-2} L mol^{-1} s^{-1} at 37°C. Evaluate the Arrhenius parameters of the reaction.

25.17 (a) The base-catalysed bromination of nitromethane-d$_3$ in water at room temperature (298 K) proceeds 4.3 times more slowly than the bromination of the undeuterated material. Account for this difference. Use $k_f(C—H) = 450$ N m^{-1}.

25.17 (b) Predict the order of magnitude of the isotope effect on the relative rates of displacement of (a) ^1H and ^3H, (b) ^{16}O and ^{18}O. Will raising the temperature enhance the difference? Take $k_f(C—H) = 450$ N m^{-1}, $k_f(C—O) = 1750$ N m^{-1}.

25.18 (a) Show that $t_{1/2} \propto 1/[A]^{n-1}$ for a reaction that is nth-order in A.

25.18 (b) Deduce an expression for the time it takes for the concentration of a substance to fall to one-third its initial value in an nth-order reaction.

25.19 (a) The effective rate constant for a gaseous reaction that has a Lindemann–Hinshelwood mechanism is 2.50×10^{-4} s^{-1} at 1.30 kPa and 2.10×10^{-5} s^{-1} at 12 Pa. Calculate the rate constant for the activation step in the mechanism.

25.19 (b) The effective rate constant for a gaseous reaction that has a Lindemann–Hinshelwood mechanism is 1.7×10^{-3} s^{-1} at 1.09 kPa and 2.2×10^{-4} s^{-1} at 25 Pa. Calculate the rate constant for the activation step in the mechanism.

25.20 (a) The pK_a of NH_4^+ is 9.25 at 25°C. The rate constant at 25°C for the reaction of NH_4^+ and OH^- to form aqueous NH_3 is 4.0×10^{10} L mol^{-1} s^{-1}. Calculate the rate constant for proton transfer to NH_3. What relaxation time would be observed if a temperature jump were applied to a solution of 0.15 mol L^{-1} $NH_3(aq)$ at 25°C?

25.20 (b) The equilibrium $A \rightleftharpoons B + C$ at 25°C is subjected to a temperature jump that slightly increases the concentrations of B and C. The measured relaxation time is 3.0 μs. The equilibrium constant for the system is 2.0×10^{-16} at 25°C, and the equilibrium concentrations of B and C at 25°C are both 2.0×10^{-4} mol L^{-1}. Calculate the rate constants for steps (1) and (2).

Problems[†]

Numerical problems

25.1 The data below applies to the formation of urea from ammonium cyanate, $NH_4CNO \rightarrow NH_2CONH_2$. Initially 22.9 g of ammonium cyanate was dissolved in in enough water to prepare 1.00 L of solution. Determine the order of the reaction, the rate constant, and the mass of ammonium cyanate left after 300 min.

t/min	0	20.0	50.0	65.0	150
m(urea)/g	0	7.0	12.1	13.8	17.7

25.2 The data below apply to the reaction, $(CH_3)_3CBr + H_2O \rightarrow (CH_3)_3COH + HBr$. Determine the order of the reaction, the rate constant, and the molar concentration of $(CH_3)_3CBr$ after 43.8 h.

t/h	0	3.15	6.20	10.00	18.30	30.80
$[(CH_3)_3CBr]/(10^{-2}$ mol L$^{-1})$	10.39	8.96	7.76	6.39	3.53	2.07

25.3 The thermal decomposition of an organic nitrile produced the following data:

$t/(10^3$ s)	0	2.00	4.00	6.00	8.00	10.00	12.00	∞
[nitrile]/(mol L^{-1})	1.10	0.86	0.67	0.52	0.41	0.32	0.25	0

Determine the order of the reaction and the rate constant. Note that this reaction does not go to completion.

25.4 The following data have been obtained for the decomposition of $N_2O_5(g)$ at 67°C according to the reaction $2 N_2O_5(g) \rightarrow 4 NO_2(g) + O_2(g)$. Determine the order of the reaction, the rate constant, and the half-life. It is not necessary to obtain the result graphically; you may do a calculation using estimates of the rates of change of concentration.

t/min	0	1	2	3	4	5
$[N_2O_5]/($mol L$^{-1})$	1.000	0.705	0.497	0.349	0.246	0.173

25.5 A first-order decomposition reaction is observed to have the following rate constants at the indicated temperatures. Estimate the activation energy.

$k/(10^{-3}$ s$^{-1})$	2.46	45.1	576
$\theta/°C$	0	20.0	40.0

25.6 The gas-phase decomposition of acetic acid at 1189 K proceeds by way of two parallel reactions:

(1) $CH_3COOH \longrightarrow CH_4 + CO_2$ $k_1 = 3.74$ s^{-1}

(2) $CH_3COOH \longrightarrow H_2C{=}C{=}O + H_2O$ $k_2 = 4.65$ s^{-1}

What is the maximum percentage yield of the ketene CH_2CO obtainable at this temperature?

25.7 The composition of a liquid-phase reaction $2 A \rightarrow B$ was followed by a spectrophotometric method with the following results:

t/min	0	10	20	30	40	∞
$[B]/($mol L$^{-1})$	0	0.089	0.153	0.200	0.230	0.312

Determine the order of the reaction and its rate constant.

25.8 The ClO radical decays rapidly by way of the reaction $2 ClO \rightarrow Cl_2 + O_2$. The following data have been obtained:

$t/(10^{-3}$ s)	0.12	0.62	0.96	1.60	3.20	4.00	5.75
$[ClO]/(10^{-6}$ mol L$^{-1})$	8.49	8.09	7.10	5.79	5.20	4.77	3.95

Determine the rate constant of the reaction and the average lifetime of a ClO radical.

25.9 Cyclopropane isomerizes into propene when heated to 500°C in the gas phase. The extent of conversion for various initial pressures has been followed by gas chromatography by allowing the reaction to proceed for a time with various initial pressures:

p_0/Torr	200	200	400	400	600	600
t/s	100	200	100	200	100	200
p/Torr	186	173	373	347	559	520

where p_0 is the initial pressure and p is the final pressure of cyclopropane. What are the order and rate constant for the reaction under these conditions?

25.10 The addition of hydrogen halides to alkenes has played a fundamental role in the investigation of organic reaction mechanisms. In one study (M.J. Haugh and D.R. Dalton, *J. Amer. Chem. Soc.* **97**, 5674 (1975)), high pressures of hydrogen chloride (up to 25 atm) and propene (up to 5 atm) were examined over a range of temperatures and the amount of 2-chloropropane formed was determined by NMR. Show that, if the reaction $A + B \rightarrow P$ proceeds for a short time δt, the concentration of product follows $[P]/[A] = k[A]^{m-1}[B]^n \delta t$ if the reaction is mth-order in A and nth-order in B. In a series of runs the ratio of [chloropropane] to [propene] was independent of [propene] but the ratio of [chloropropane] to [HCl] for constant amounts of propene depended on [HCl]. For $\delta t \approx 100$ h (which is short on the timescale of the reaction) the latter ratio rose from zero to 0.05, 0.03, 0.01 for p(HCl) = 10 atm, 7.5 atm, 5.0 atm, respectively. What are the orders of the reaction with respect to each reactant?

25.11 Use mathematical software or an electronic spreadsheet to examine the time dependence of [I] in the reaction mechanism $A \rightarrow I \rightarrow P$ (k_1, k_2). You may either integrate eqn 35 numerically (see *Further information 1*) or use eqn 36 directly. In all of the following calculations, use $[A]_0 = 1$ mol L^{-1} and a time range of 0 to 5 s. (a) Plot [I] against t for $k_1 = 10$ s^{-1} and $k_2 = 1$ s^{-1}. (b) Increase the ratio k_2/k_1 steadily by decreasing the value of k_1 and examine the plot of [I] against t at each turn. What approximation about d[I]/dt becomes increasingly valid?

25.12 Show that the following mechanism can account for the rate law of the reaction in Problem 25.10:

$$HCl + HCl \rightleftharpoons (HCl)_2 \qquad\qquad K_1$$
$$HCl + CH_3CH{=}CH_2 \rightleftharpoons \text{complex} \qquad K_2$$
$$(HCl)_2 + \text{complex} \longrightarrow CH_3CHClCH_3 + HCl + HCl \quad k \text{ (slow)}$$

What further tests could you apply to verify this mechanism?

25.13 In the experiments described in Problems 25.10 and 25.12 an inverse temperature dependence of the reaction rate was observed, the overall rate of reaction at 70°C being roughly one-third that at

---

[†] Problems denoted with the symbol ‡ were supplied by Charles Trapp, Carmen Giunta, and Marshall Cady.

19°C. Estimate the apparent activation energy and the activation energy of the rate-determining step given that the enthalpies of the two equilibria are both of the order of -14 kJ mol^{-1}.

25.14 The second-order rate constants for the reaction of oxygen atoms with aromatic hydrocarbons have been measured (R. Atkinson and J.N. Pitts, *J. Phys. Chem.* **79**, 295 (1975)). In the reaction with benzene the rate constants are 1.44×10^7 L mol^{-1} s^{-1} at 300.3 K, 3.03×10^7 L mol^{-1} s^{-1} at 341.2 K, and 6.9×10^7 L mol^{-1} s^{-1} at 392.2 K. Find the pre-exponential factor and activation energy of the reaction.

25.15 In Problem 25.9 the isomerization of cyclopropane over a limited pressure range was examined. If the Lindemann mechanism of first-order reactions is to be tested we also need data at low pressures. These have been obtained (H.O. Pritchard, R.G. Sowden, and A.F. Trotman-Dickenson, *Proc. R. Soc.* **A 217**, 563 (1953)):

p/Torr	84.1	11.0	2.89	0.569	0.120	0.067
$10^4\,k_{eff}$/s^{-1}	2.98	2.23	1.54	0.857	0.392	0.303

Test the Lindemann theory with these data.

25.16‡ P.W. Seakins, M.J. Pilling, L.T. Niiranen, D. Gutman, and L.N. Krasnoperov (*J. Phys. Chem.* **96**, 9847 (1992)) measured the forward and reverse rate constants for the gas-phase reaction $C_2H_5(g) + HBr(g) \rightarrow C_2H_6(g) + Br(g)$ and used their findings to compute thermodynamic parameters for C_2H_5. The reaction is bimolecular in both directions with Arrhenius parameters $A = 1.0 \times 10^9$ L mol^{-1} s^{-1}, $E_a = -4.2$ kJ mol^{-1} for the forward reaction and $k' = 1.4 \times 10^{11}$ L mol^{-1} s^{-1}, $E_a = 53.3$ kJ mol^{-1} for the reverse reaction. Compute $\Delta_f H^\ominus$, $S_m^\ominus$, and $\Delta_f G^\ominus$ of C_2H_5 at 298 K.

25.17‡ Here we consider the reaction $SiCl_3H(g) + Br_2(g) \rightarrow SiCl_3Br(g) + HBr(g)$. The temperature dependence of this reaction has been studied by measuring the degree of conversion, $\alpha(t)$, for initial mixtures of 4.0 volume percentage trichlorosilane and 10.0 volume percentage bromine in helium at 1 bar. The data at four different temperatures are as follows:

$\alpha(160°C, t) = 0.044$ at 6.07 min; 0.057 at 7.87 min; 0.076 at 10.00 min

$\alpha(180°C, t) = 0.137$ at 2.72 min; 0.188 at 3.91 min; 0.228 at 5.03 min; 0.273 at 6.67 min; 0.302 at 8.06 min

$\alpha(200°C, t) = 0.281$ at 1.39 min; 0.395 at 2.16 min; 0.516 at 3.36 min; 0.612 at 5.04 min

$\alpha(220°C, t) = 0.592$ at 0.67 min; 0.734 at 1.22 min; 0.866 at 2.16 min

For these reactants it is thought that the dissociation $Br_2(g) \rightarrow 2\,Br(g)$ is at equilibrium and that in the range 300–700 K, the equilibrium constant for the dissociation, can be approximated as $K = 10^{5.1}e^{-C/RT}$, where $C = 190.2$ kJ mol^{-1} and K is expressed in terms of concentrations in moles per centimetre cubed (mol cm^{-3}). No wall effects were observed. Reactions of HBr were discounted and no hexachlorodisilane was observed. (a) Assume a rate law of the form d$[SiCl_3Br]/$d$t = k[SiCl_3H]^a[Br_2]^b$ and determine the reaction orders a and b. (b) Determine the rate constant $k(T)$ for each of the four data sets. Evaluate the activation parameters. (c) Show that the proposed mechanism

$$SiCl_3H + \cdot Br \longrightarrow \cdot SiCl_3 + HBr$$
$$\cdot SiCl_3 + Br_2 \longrightarrow SiCl_3Br + \cdot Br$$

is consistent with the experimental rate law.

Theoretical problems

25.18 The reaction mechanism

$$A_2 \rightleftharpoons A + A \quad \text{(fast)}$$
$$A + B \longrightarrow P \quad \text{(slow)}$$

involves an intermediate A. Deduce the rate law for the reaction.

25.19 The equilibrium $A \rightleftharpoons B$ is first-order in both directions. Derive an expression for the concentration of A as a function of time when the initial molar concentrations of A and B are $[A]_0$ and $[B]_0$. What is the final composition of the system?

25.20 Derive an integrated expression for a second-order rate law $v = k[A][B]$ for a reaction of stoichiometry $2\,A + 3\,B \rightarrow P$.

25.21 Derive the integrated form of a third-order rate law $v = k[A]^2[B]$ in which the stoichiometry is $2\,A + B \rightarrow P$ and the reactants are initially present in (a) their stoichiometric proportions, (b) with B present initially in twice the amount.

25.22 Set up the rate equations for the reaction mechanism:

$$A \underset{k_a, k_a'}{\rightleftharpoons} B \underset{k_b, k_b'}{\rightleftharpoons} C$$

Show that the mechanism is equivalent to

$$A \underset{k_{eff}, k_{eff}'}{\rightleftharpoons} C$$

under specified circumstances.

25.23 Show that the ratio $t_{1/2}/t_{3/4}$, where $t_{1/2}$ is the half-life and $t_{3/4}$ is the time for the concentration of A to decrease to $\frac{3}{4}$ of its initial value (implying that $t_{3/4} < t_{1/2}$) can be written as a function of n alone, and can therefore be used as a rapid assessment of the order of a reaction.

25.24 Derive an equation for the steady-state rate of the sequence of reactions $A \rightleftharpoons B \rightleftharpoons C \rightleftharpoons D$, with [A] maintained at a fixed value and the product D removed as soon as it is formed.

25.25‡ For the second-order reaction $A + B \rightarrow$ Products, the rate of reaction, v, may be written

$$v = \frac{dx}{dt} = k([A]_0 - x)([B]_0 + x)$$

where x is the decrease in concentration of A or B as a result of reaction. Find an expression for the maximum rate and the conditions under which it applies. Draw a graph of v against x and, noting that v and x cannot be negative, identify the portion of the curve that corresponds to reality.

25.26‡ For the consecutive reaction $A \rightarrow B \rightarrow C$, Fig. 25.13 shows [B] plotted against time for $k_a = 10k_b$. For $[A]_0 = 1.0$ mol L^{-1} and $k_a = 1.0$ min^{-1}, plot [B] against t for $k_a/k_b = 5$, 1, and 0.5. For each case determine the time at which [B] reaches a maximum. For the only $A + B \rightarrow P$ reaction in Table 25.3, find an expression for x as a function of time.

Applications: to archaeology, biology, and environmental science

25.27 The half-life for the (first-order) radioactive decay of ^{14}C is 5730 y (it emits β rays with an energy of 0.16 MeV). An archaeological sample contained wood that had only 72 per cent of the ^{14}C found in living trees. What is its age?

25.28 One of the hazards of nuclear explosions is the generation of ^{90}Sr and its subsequent incorporation in place of calcium in bones. This nuclide emits β rays of energy 0.55 MeV, and has a half-life of 28.1 y. Suppose 1.00 μg was absorbed by a newly born child. How much will remain after (a) 18 y, (b) 70 y if none is lost metabolically?

25.29 Sucrose is readily hydrolysed to glucose and fructose in acidic solution. The hydrolysis is often monitored by measuring the angle of rotation of plane-polarized light passing through the solution. From the angle of rotation the concentration of sucrose can be determined. An experiment on the hydrolysis of sucrose in 0.50 M HCl(aq) produced the following data:

t/min	0	14	39	60	80
[sucrose]/(mol L^{-1})	0.316	0.300	0.274	0.256	0.238

t/min	110	140	170	210
[sucrose]/(mol L^{-1})	0.211	0.190	0.170	0.146

Determine the rate constant of the reaction and the average lifetime of a sucrose molecule.

25.30 Consider the following mechanism for renaturation of a double helix from its strands A and B:

$$A + B \rightleftharpoons \text{unstable helix} \quad \text{(fast)}$$
$$\text{Unstable helix} \longrightarrow \text{stable double helix} \quad \text{(slow)}$$

Derive the rate equation for the formation of the double helix and express the rate constant of the renaturation reaction in terms of the rate constants of the individual steps.

25.31‡ *Prebiotic reactions* are reactions that might have occurred under the conditions prevalent on the Earth before the first living creatures emerged and which can lead to analogues of molecules necessary for life as we now know it. To qualify, a reaction must proceed with favourable rates and equilibria. M.P. Robertson and S.I. Miller (*Science* **268**, 702(1995)) have studied the prebiotic synthesis of 5-substituted uracils, among them 5-hydroxymethyluracil (HMU). Amino acid analogues can be formed from HMU under prebiotic conditions by reaction with various nucleophiles, such as H_2S, HCN, indole, imidazole, etc. For the synthesis of HMU (the uracil analogue of serine) from uracil and formaldehyde (HCHO), the rate of addition is given by $\log\{k/(\text{L mol}^{-1}\text{ s}^{-1})\} = 11.75 - 5488/(T/K)$ (at pH = 7), and $\log K = -1.36 + 1794/(T/K)$. For this reaction, calculate the rates and equilibrium constants over a range of temperatures corresponding to possible prebiotic conditions, such as 0–50°C, and plot them against temperature. Also, calculate the activation energy and the standard reaction Gibbs energy and enthalpy at 25°C. Prebiotic conditions are not likely to be standard conditions. Speculate about how the actual values of the reaction Gibbs energy and enthalpy might differ from the standard values. Do you expect that the reaction would still be favourable?

25.32‡ Methane is a by-product of a number of natural processes (such as digestion of cellulose in ruminant animals and anaerobic decomposition of organic waste matter) and industrial processes (such as food production and fossil fuel use). Reaction with the hydroxyl radical OH is the main path by which CH_4 is removed from the lower atmosphere. T. Gierczak, R.K. Talukdar, S.C. Herndon, G.L. Vaghjiani, and A.R. Ravishankara (*J. Phys. Chem.* **A 101**, 3125 (1997)) measured the rate constants for the elementary bimolecular gas-phase reaction of methane with the hydroxyl radical over a range of temperatures of importance to atmospheric chemistry. Deduce the Arrhenius parameters A and E_a from the following measurements:

T/K	295	295	223	218
k/(10^6 L mol^{-1} s^{-1})	3.70	3.55	0.494	0.452

T/K	213	206	200	195
k/(10^6 L mol^{-1} s^{-1})	0.379	0.295	0.241	0.217

25.33‡ As we saw in Problem 25.32, reaction with the hydroxyl radical OH is the main path by which CH_4, a by-product of many natural and industrial processes, is removed from the lower atmosphere. T. Gierczak, R.K. Talukdar, S.C. Herndon, G.L. Vaghjiani, and A.R. Ravishankara (*J. Phys. Chem.* **A 101**, 3125 (1997)) measured the rate constants for the bimolecular gas-phase reaction $CH_4(g) + OH(g) \rightarrow CH_3(g) + H_2O(g)$ and found $A = 1.13 \times 10^9$ L mol^{-1} s^{-1} and $E_a = 14.1$ kJ mol^{-1} for the Arrhenius parameters. (a) Estimate the rate of consumption of CH_4. Take the average OH concentration to be 1.5×10^{-15} mol L^{-1}, that of CH_4 to be 4.0×10^{-8} mol L^{-1}, and the temperature to be –10°C. (b) Estimate the global annual mass of CH_4 consumed by this reaction (which is slightly less than the amount introduced to the atmosphere) given an effective volume for the Earth's lower atmosphere of 4×10^{21} L.

25.34‡ T. Gierczak, R.K. Talukdar, S.C. Herndon, G.L. Vaghjiani, and A.R. Ravishankara (*J. Phys. Chem.* **A 101**, 3125 (1997)) measured the rate constants for the bimolecular gas-phase reaction of methane with the hydroxyl radical in several isotopic variations. From their data, the following Arrhenius parameters can be obtained:

	A/(L mol^{-1} s^{-1})	E_a/(kJ mol^{-1})
$CH_4 + OH \longrightarrow CH_3 + H_2O$	1.13×10^9	14.1
$CD_4 + OH \longrightarrow CD_3 + DOH$	6.0×10^8	17.5
$CH_4 + OD \longrightarrow CH_3 + DOH$	1.01×10^9	13.6

Compute the rate constants at 298 K, and interpret the kinetic isotope effects.

25.35‡ The oxidation of HSO_3^- by O_2 in aqueous solution is a reaction of importance to the processes of acid rain formation and flue gas desulfurization. R.E. Connick, Y.-X. Zhang, S. Lee, R. Adamic, and P. Chieng (*Inorg. Chem.* **34**, 4543 (1995)) report that the reaction $2 HSO_3^- + O_2 \rightarrow 2 SO_4^{2-} + 2 H^+$ follows the rate law $v = k[HSO_3^-]^2[H^+]^2$. Given a pH of 5.6 and an oxygen molar concentration of 2.4×10^{-4} mol L^{-1} (both presumed constant), an initial HSO_3^- molar concentration of 5×10^{-5} mol L^{-1}, and a rate constant of 3.6×10^6 L^3 mol^{-3} s^{-1}, what is the initial rate of reaction? How long would it take for HSO_3^- to reach half its initial concentration?

Chain reactions

26.1 The rate laws of chain reactions

26.2 Explosions

Polymerization kinetics

26.3 Stepwise polymerization

26.4 Chain polymerization

Homogeneous catalysis

26.5 Features of homogeneous catalysis

26.6 Enzymes

Oscillating reactions

26.7 Autocatalysis

26.8 Autocatalytic mechanisms of oscillating reactions

26.9 Bistability

26.10 Chemical chaos

Photochemistry

26.11 Kinetics of photophysical and photochemical processes

26.12 Complex photochemical processes

Checklist of key ideas

Further reading

Exercises

Problems

26

The kinetics of complex reactions

This chapter extends the material introduced in Chapter 25 by showing how to deal with complex reaction mechanisms. First, we consider chain reactions and see that either complicated or simple rate laws can be obtained, depending on the conditions. Under certain circumstances, a chain reaction can become explosive, and we see some of the reasons for this behaviour. An important application of these more complicated techniques is to the kinetics of polymerization reactions. There are two major classes of polymerization process and the average molar mass of the product varies with time in distinctive ways. Second, we describe homogeneous catalysis and apply the associated concepts to enzyme-catalysed reactions. We also consider reactions in which the concentrations of the intermediates and products oscillate in time: although their rate laws may be well known, under certain circumstances the composition of a system may be unpredictable. Finally, we describe the principles of photochemistry.

Many reactions take place by mechanisms that involve several elementary steps. Some take place at a useful rate only after absorption of light or if a catalyst is present. Certain other reactions take place by mechanisms with positive or negative feedback, in which the products of the reaction influence the rate of formation of products. In this chapter we see how to develop the ideas introduced in Chapter 25 to deal with these special kinds of reactions.

Chain reactions

Many gas-phase reactions and liquid-phase polymerization reactions are **chain reactions**. In a chain reaction, a reaction intermediate produced in one step generates an intermediate in a subsequent step, then that intermediate generates another intermediate, and so on. The intermediates in a chain reaction are called **chain carriers**. In a **radical chain reaction** the chain carriers are radicals (species with unpaired electrons). Ions may also act as chain carriers. In nuclear fission the chain carriers are neutrons.

26.1 The rate laws of chain reactions

A chain reaction can have a simple rate law. As a first example, consider the **pyrolysis**, or thermal decomposition in the absence of air, of acetaldehyde (ethanal, CH_3CHO), which is found to be three-halves order in CH_3CHO:

$$CH_3CHO(g) \longrightarrow CH_4(g) + CO(g) \qquad v = k[CH_3CHO]^{3/2} \tag{26.1}$$

Some ethane is also detected. The **Rice–Herzfeld mechanism** for this reaction is as follows (the dot signifies an unpaired electron and marks a radical):

Initiation:	$CH_3CHO \longrightarrow \cdot CH_3 + \cdot CHO$	$v = k_i[CH_3CHO]$
Propagation:	$CH_3CHO + \cdot CH_3 \longrightarrow CH_4 + CH_3CO\cdot$	$v = k_p[CH_3CHO][\cdot CH_3]$
Propagation:	$CH_3CO\cdot \longrightarrow \cdot CH_3 + CO$	$v = k_p'[CH_3CO\cdot]$
Termination:	$\cdot CH_3 + \cdot CH_3 \longrightarrow CH_3CH_3$	$v = k_t[\cdot CH_3]^2$

The chain carriers $\cdot CH_3$ and $\cdot CHO$ are formed initially in the **initiation step**. To simplify the treatment, we shall ignore the subsequent reactions of $\cdot CHO$, except to note that it gives rise to the formation of CO and to the by-product H_2. The chain carrier $\cdot CH_3$ attacks other reactant molecules in the **propagation steps**, and each attack gives rise to a new carrier. Radicals combine and end the chain in the **termination step**.

To test the proposed mechanism we need to show that it leads to the observed rate law. According to the steady-state approximation (Section 25.7c), the net rate of change of the intermediates ($\cdot CH_3$ and $CH_3CO\cdot$) may be set equal to zero:

$$\frac{d[\cdot CH_3]}{dt} = k_i[CH_3CHO] - k_p[\cdot CH_3][CH_3CHO] + k_p'[CH_3CO\cdot] - 2k_t[\cdot CH_3]^2 = 0$$

$$\frac{d[CH_3CO\cdot]}{dt} = k_p[\cdot CH_3][CH_3CHO] - k_p'[CH_3CO\cdot] = 0$$

The sum of the two equations is

$$k_i[CH_3CHO] - 2k_t[\cdot CH_3]^2 = 0$$

which implies that the steady-state concentration of $\cdot CH_3$ radicals is

$$[\cdot CH_3] = \left(\frac{k_i}{2k_t}\right)^{1/2} [CH_3CHO]^{1/2} \tag{26.2}$$

It follows that the rate of formation of CH_4 is

$$\frac{d[CH_4]}{dt} = k_p[\cdot CH_3][CH_3CHO] = k_p\left(\frac{k_i}{2k_t}\right)^{1/2} [CH_3CHO]^{3/2} \tag{26.3}$$

which is in agreement with the three-halves order observed experimentally (eqn 1). However, this mechanism does not accommodate the formation of various known reaction by-products, such as propanone (CH_3COCH_3) and propanal (CH_3CH_2CHO).

In many cases, a chain reaction leads to a complicated rate law. An example is the **hydrogen–bromine reaction**:

$$H_2(g) + Br_2(g) \longrightarrow 2\,HBr(g) \qquad \frac{d[HBr]}{dt} = \frac{k[H_2][Br_2]^{3/2}}{[Br_2] + k'[HBr]} \tag{26.4}$$

The following mechanism has been proposed to account for this rate law (Fig. 26.1):

Initiation:	$Br_2 + M \longrightarrow Br\cdot + Br\cdot + M$	$v = k_i[Br_2][M]$

where M is either Br_2 or H_2. This step is an example of a **thermolysis**, a reaction initiated by heat or by vigorous intermolecular collisions.

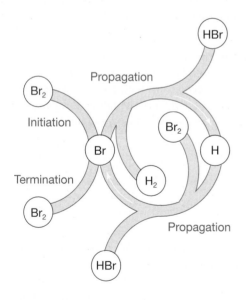

26.1 A schematic representation of the mechanism of the reaction between hydrogen and bromine. Note how the reactants and products are shown as arms to the circle, but the intermediates (H and Br) occur only within the circle. Similar diagrams are used to depict the action of catalysts.

Propagation: $Br\cdot + H_2 \longrightarrow HBr + H\cdot$ $v = k_p[Br\cdot][H_2]$

$H\cdot + Br_2 \longrightarrow HBr + Br\cdot$ $v = k_p'[H\cdot][Br_2]$

Retardation: $H\cdot + HBr \longrightarrow H_2 + Br\cdot$ $v = k_r[H\cdot][HBr]$

Termination: $Br\cdot + Br\cdot + M \longrightarrow Br_2 + M^*$ $v = k_t[Br\cdot]^2[M]$

A **retardation step** reduces the net rate of formation of product. In this case, the chain carrier H· attacks a molecule of HBr, the product. In the termination step, the third body M removes the energy of recombination. Other possible termination steps include the recombination of H atoms to form H_2 and combination of H and Br atoms. However, it turns out that only Br atom recombination is important. The net rate of formation of the product HBr is

$$\frac{d[HBr]}{dt} = k_p[Br\cdot][H_2] + k_p'[H\cdot][Br_2] - k_r[H\cdot][HBr]$$

We can now either analyse the rate equations numerically (see *Further information 1*) or look for approximate solutions and see if they agree with the empirical rate law. The following example illustrates the latter approach.

Example 26.1 Deriving the rate equation of a chain reaction

Derive the rate law for the formation of HBr according to the mechanism given above.

Method Make the steady-state approximation for the concentrations of any intermediates (H· and Br· in the present case) by setting the net rates of change of their concentrations equal to zero. Solve the resulting equations for the concentrations of the intermediates, and then use the resulting expressions in the equation for the net rate of formation of HBr.

Answer The net rates of formation of the two intermediates are

$$\frac{d[H\cdot]}{dt} = k_p[Br\cdot][H_2] - k_p'[H\cdot][Br_2] - k_r[H\cdot][HBr] = 0$$

$$\frac{d[Br\cdot]}{dt} = 2k_i[Br_2][M] - k_p[Br\cdot][H_2] + k_p'[H\cdot][Br_2] + k_r[H\cdot][HBr] - 2k_t[Br\cdot]^2[M] = 0$$

The steady-state concentrations of the intermediates are obtained by solving these two simultaneous equations and are

$$[Br\cdot] = \left(\frac{k_i}{k_t}\right)^{1/2}[Br_2]^{1/2} \qquad [H\cdot] = \frac{k_p(k_i/k_t)^{1/2}[H_2][Br_2]^{1/2}}{k_p'[Br_2] + k_r[HBr]}$$

Note that [M] has cancelled. When we substitute these concentrations into the expression for d[HBr]/dt, we obtain

$$\frac{d[HBr]}{dt} = \frac{2k_p(k_i/k_t)^{1/2}[H_2][Br_2]^{3/2}}{[Br_2] + (k_r/k_p')[HBr]}$$

This equation has the same form as the empirical rate law (eqn 4), so the two empirical rate constants can be identified as

$$k = 2k_p\left(\frac{k_i}{k_t}\right)^{1/2} \qquad k' = \frac{k_r}{k_p'}$$

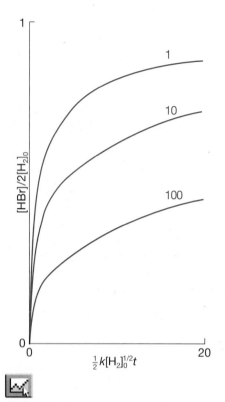

26.2 The numerical integration of the HBr rate law, Example 26.1, can be used to explore how the concentration of HBr changes with time. These runs began with stoichiometric proportions of H_2 and Br_2; the curves are labelled with the value of $2k' - 1$.

Comment The presence of HBr in the denominator of the rate law is a sign that it is acting as an inhibitor and reducing the rate of formation of product. Likewise, the presence of Br_2 stems from its role in the removal of the reactive ·H radicals from the chain. Numerical integration of the rate law with mathematical software shows the predicted time dependence of the concentration of HBr for this mechanism (Fig. 26.2).

Self-test 26.1 Deduce the rate law for the production of HBr when the initiation step is the photolysis, or light-induced decomposition, of Br_2 into two bromine atoms, Br·. Let the photolysis rate be $v = I_{abs}$, where I_{abs} is the intensity of absorbed radiation.　　　　　　　　　[See eqn 39, below]

26.2 Explosions

A **thermal explosion** is a very rapid reaction arising from a rapid increase of reaction rate with increasing temperature. The temperature of the system rises if the energy released by an exothermic reaction cannot escape, and the reaction goes faster. The acceleration of the rate results in an even faster rise of temperature, so the reaction becomes catastrophically fast. A **chain-branching explosion** occurs when the number of chain centres grows exponentially.

An example of both types of explosion is the reaction between hydrogen and oxygen:

$$2\,H_2(g) + O_2(g) \longrightarrow 2\,H_2O(g)$$

Although the net reaction is very simple, the mechanism is very complex and has not yet been fully elucidated. A chain reaction is involved, and the chain carriers include H·, ·O·, and ·OH. Some steps involving H· are:

Initiation:	$H_2 \longrightarrow H\cdot + H\cdot$	$v = \text{constant}\ (v_{\text{init}})$
Propagation:	$H_2 + \cdot OH \longrightarrow \cdot H + H_2O$	$v = k_p[H_2][\cdot OH]$
Branching:	$\cdot O_2\cdot + \cdot H \longrightarrow \cdot O\cdot + \cdot OH$	$v = k_b[\cdot O_2\cdot][H\cdot]$
	$\cdot O\cdot + H_2 \longrightarrow \cdot OH + \cdot H$	$v = k_b'[\cdot O\cdot][H_2]$
Termination:	$H\cdot + \text{wall} \longrightarrow \frac{1}{2}H_2$	$v = k_t[H\cdot]$
	$H\cdot + O_2 + M \longrightarrow HO_2\cdot + M^*$	$v = k_t'[H\cdot][O_2][M]$

A **branching step** is an elementary reaction that produces more than one chain carrier. Recall that an O atom has the ground-state configuration $[He]2s^2 2p^4$, with two unpaired electrons.

The occurrence of an explosion depends on the temperature and pressure of the system, and the **explosion regions** for the reaction, the conditions under which explosion occurs, are shown in Fig. 26.3. At very low pressures the system is outside the explosion region and the mixture reacts smoothly. At these pressures the chain carriers produced in the branching steps can reach the walls of the container where they combine. Increasing the pressure along a vertical line in the illustration takes the system through the **first explosion limit** (provided the temperature is greater than about 730 K). The chain carriers react before reaching the walls and the branching reactions are explosively efficient. The reaction is smooth when the pressure is above the **second explosion limit**. The concentration of molecules in the gas is then so great that the radicals produced in the branching reactions recombine via termination steps.

In the last elementary reaction of the mechanism, the third body (M) can remove excess energy. At low pressures three-particle collisions are unimportant and recombination is much slower. At higher pressures, when three-particle collisions are

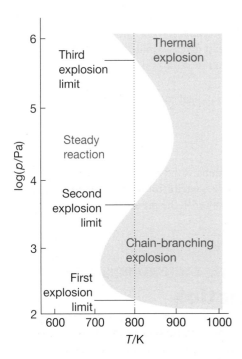

26.3 The explosion limits of the $H_2 + O_2$ reaction. In the explosive regions the reaction proceeds explosively when heated homogeneously.

26.4 The concentration of radicals in the fuel-rich regime of the hydrogen–oxygen reaction (a) under steady combustion conditions, (b) in the explosive region. For this graph, $\Delta k = k_{branch} - k_{term}$.

important, the explosive propagation of the chain slows down because the branching steps are diverted into simple propagation steps (see the following example). When the pressure is increased to above the **third explosion limit**, the reaction rate increases so much that a thermal explosion occurs.

Example 26.2 Examining the explosion behaviour of a chain reaction

For the reaction of hydrogen and oxygen described above, show that an explosion occurs when the rate of chain branching exceeds that of chain termination.

Method Identify the onset of explosion with the rapid increase in the concentration of radicals, and for simplicity identify that concentration with the concentration of H· atoms, which probably outnumber the highly reactive ·OH and ·O· radicals. Set up the corresponding rate laws for the reaction intermediates and then apply the steady-state approximation.

Answer The rate of formation of radicals, v_{rad}, is identified with $d[H\cdot]/dt$; therefore we write

$$v_{rad} = v_{init} + k_p[\cdot OH][H_2] - k_b[H\cdot][O_2] + k_b'[\cdot O\cdot][H_2] - k_t[H\cdot] - k_t'[H\cdot][O_2][M]$$

Applying the steady–state approximation to ·OH and ·O· gives

$$\frac{d[\cdot OH]}{dt} = -k_p[\cdot OH][H_2] + k_b[H\cdot][O_2] + k_b'[\cdot O\cdot][H_2] = 0$$

$$\frac{d[\cdot O\cdot]}{dt} = k_b[H\cdot][O_2] - k_b'[\cdot O\cdot][H_2] = 0$$

The solutions of these two algebraic equations are

$$[\cdot O\cdot] = \frac{k_b[H\cdot][O_2]}{k_b'[H_2]} \qquad [\cdot OH] = \frac{2k_b[H\cdot][O_2]}{k_p[H_2]}$$

The rate of formation of radicals is therefore

$$v_{rad} = v_{init} + (2k_b[O_2] - k_t - k_t'[O_2][M])[H\cdot]$$

We write $k_{branch} = 2k_b[O_2]$, a measure of the rate of the more important chain-branching step, and $k_{term} = k_t + k_t'[O_2][M]$, a measure of the rate of chain termination. Then,

$$\frac{d[H\cdot]}{dt} = v_{init} + (k_{branch} - k_{term})[H\cdot]$$

There are two solutions. At low O_2 concentrations, termination dominates branching, so $k_{term} > k_{branch}$. Then,

$$[H\cdot] = \frac{v_{init}}{k_{term} - k_{branch}}(1 - e^{-(k_{term} - k_{branch})t})$$

As can be seen from Fig. 26.4a, in this regime there is steady combustion of hydrogen. At high O_2 concentrations, branching dominates termination, or $k_{branch} > k_{term}$. Then,

$$[H\cdot] = \frac{v_{init}}{k_{branch} - k_{term}}(e^{(k_{branch} - k_{term})t} - 1)$$

There is now an explosive increase in the concentration of radicals (Fig. 26.4b).

> **Comment** Although the steady-state approximation does not hold under explosive conditions, the calculation at least gives an indication of the basis for the transition from smooth combustion to explosion.

> **Self-test 26.2** Calculate the variation in radical composition when the rates of branching and termination are equal.
>
> $$[[H\cdot] = v_{init}t]$$

Not all explosions are due to chain reactions. Solid-state explosions, such as the explosion of ammonium nitrate or TNT (2,4,6-trinitrotoluene), for instance, are simply decompositions that occur very rapidly with the production of large amounts of gas-phase molecules.

Polymerization kinetics

In **stepwise polymerization** any two monomers present in the reaction mixture can link together at any time and growth of the polymer is not confined to chains that are already forming (Fig. 26.5). As a result, monomers are removed early in the reaction and, as we shall see, the average molar mass of the product grows with time. In **chain polymerization** an activated monomer, M, attacks another monomer, links to it, then that unit attacks another monomer, and so on. The monomer is used up as it becomes linked to the growing chains (Fig. 26.6). High polymers are formed rapidly and only the yield, not the average molar mass, of the polymer is increased by allowing long reaction times.

26.3 Stepwise polymerization

Stepwise polymerization commonly proceeds by a condensation reaction, in which a small molecule (typically H_2O) is eliminated in each step. Stepwise polymerization is the mechanism of production of polyamides, as in the formation of nylon-66:

$$H_2N(CH_2)_6NH_2 + HOOC(CH_2)_4COOH \longrightarrow H_2N(CH_2)_6NHCO(CH_2)_4COOH + H_2O$$
$$\longrightarrow H-[NH(CH_2)_6NHCO(CH_2)_4CO]_n-OH$$

Polyesters and polyurethanes are formed similarly (the latter without elimination). A polyester, for example, can be regarded as the outcome of the stepwise condensation of a hydroxyacid HO—M—COOH. We shall consider the formation of a polyester from such a monomer, and measure its progress in terms of the concentration of the —COOH groups in the sample (which we denote A), for these groups gradually disappear as the condensation proceeds. Because the condensation reaction can occur between molecules containing any number of monomer units, chains of many different lengths can grow in the reaction mixture.

In the absence of a catalyst, we can expect the condensation to be overall second-order in the concentration of the —OH and —COOH (or A) groups, and write

$$\frac{d[A]}{dt} = -k[OH][A] \tag{26.5a}$$

However, because there is one —OH group for each —COOH group, this equation is the same as

26.5 In stepwise polymerization, growth can start at any pair of monomers, and so new chains begin to form throughout the reaction.

26.6 The process of chain polymerization. Chains grow as each chain acquires additional monomers.

$$\frac{d[A]}{dt} = -k[A]^2 \tag{26.5b}$$

If we assume that the rate constant for the condensation is independent of the chain length, then k remains constant throughout the reaction. The solution of this rate law is given by eqn 25.13, and is

$$[A] = \frac{[A]_0}{1 + kt[A]_0} \tag{26.6}$$

The fraction, p, of —COOH groups that have condensed at time t is, after application of eqn 6:

26.7 The average chain length of a polymer as a function of the fraction of reacted monomers, p. Note that p must be very close to 1 for the chains to be long.

$$p = \frac{[A]_0 - [A]}{[A]_0} = \frac{kt[A]_0}{1 + kt[A]_0} \tag{26.7}$$

Next, we calculate the **degree of polymerization**, which is defined as the average number of monomer residues per polymer molecule. This quantity is the ratio of the initial concentration of A, $[A]_0$, to the concentration of end groups, $[A]$, at the time of interest, because there is one –A group per polymer molecule. For example, if there were initially 1000 A groups and there are now only 10, each polymer must be 100 units long on average. Because we can express $[A]$ in terms of p (eqn 7), the average number of monomers per polymer molecule, $\langle n \rangle$, is

$$\langle n \rangle = \frac{[A]_0}{[A]} = \frac{1}{1 - p} \tag{26.8a}$$

This result is illustrated in Fig. 26.7. When we express p in terms of the rate constant k (eqn 7), we find

$$\langle n \rangle = 1 + kt[A]_0 \tag{26.8b}$$

The average length grows linearly with time. Therefore, the longer a stepwise polymerization proceeds, the higher the average molar mass of the product.

26.4 Chain polymerization

Chain polymerization occurs by addition of monomers to a growing polymer, often by a radical chain process. It results in the rapid growth of an individual polymer chain for each activated monomer. Examples include the addition polymerizations of ethene, methyl methacrylate, and styrene, as in

$$-CH_2CHX\cdot + CH_2{=}CHX \longrightarrow -CH_2CHXCH_2CHX\cdot$$

and subsequent reactions. The central feature of the kinetic analysis (which is summarized in the *Justification* below) is that the rate of polymerization is proportional to the square root of the initiator concentration:

$$v = k[I]^{1/2}[M] \tag{26.9}$$

Justification 26.1

There are three basic types of reaction step in a chain polymerization process:

(a) Initiation:

$$I \longrightarrow R\cdot + R\cdot \qquad v_i = k_i[I]$$
$$M + R\cdot \longrightarrow \cdot M_1 \qquad \text{(fast)}$$

where I is the initiator, R$\cdot$ the radical I forms, and $\cdot M_1$ is a monomer radical. We have shown a reaction in which a radical is produced, but in some polymerizations the initiation step leads to the formation of an ionic chain carrier. The rate-determining step is the formation of the radicals R$\cdot$ by homolysis of the initiator, so the rate of initiation is equal to the v_i given above.

(b) Propagation:

$$M + \cdot M_1 \longrightarrow \cdot M_2$$
$$M + \cdot M_2 \longrightarrow \cdot M_3$$
$$\vdots$$
$$M + \cdot M_{n-1} \longrightarrow \cdot M_n \qquad v_p = k_p[M][\cdot M]$$

If we assume that the rate of propagation is independent of chain size for sufficiently large chains, then we may use only the equation given above to describe the propagation process. Consequently, for sufficiently large chains, the rate of propagation is equal to the overall rate of polymerization.

Because this chain of reactions propagates quickly, the rate at which the total concentration of radicals grows is equal to the rate of the rate-determining initiation step. It follows that

$$\left(\frac{d[\cdot M]}{dt}\right)_{production} = 2fk_i[I] \tag{26.10}$$

where f is the fraction of radicals R$\cdot$ that successfully initiate a chain.

(c) Termination:

$$\cdot M_n + \cdot M_m \longrightarrow M_{n+m} \qquad \text{(mutual termination)}$$
$$\cdot M_n + \cdot M_m \longrightarrow M_n + M_m \qquad \text{(disproportionation)}$$
$$M + \cdot M_n \longrightarrow \cdot M + M_n \qquad \text{(chain transfer)}$$

In **mutual termination** two growing radical chains combine. In termination by **disproportionation** a hydrogen atom transfers from one chain to another, corresponding to the oxidation of the donor and the reduction of the acceptor. In **chain transfer**, a new chain initiates at the expense of the one currently growing.

Here we suppose that only mutual termination occurs. If we assume that the rate of termination is independent of the length of the chain, the rate law for termination is

$$v_t = k_t[\cdot M]^2$$

and the rate of change of radical concentration by this process is

$$\left(\frac{d[\cdot M]}{dt}\right)_{depletion} = -2k_t[\cdot M]^2$$

The steady-state approximation gives:

$$\frac{d[\cdot M]}{dt} = 2fk_i[I] - 2k_t[\cdot M]^2 = 0$$

The steady-state concentration of radical chains is therefore

$$[\cdot M] = \left(\frac{fk_i}{k_t}\right)^{1/2} [I]^{1/2} \tag{26.11}$$

Because the rate of propagation of the chains is the negative of the rate at which the monomer is consumed, we can write $v_p = -d[M]/dt$ and

$$v_p = k_p[\cdot M][M] = k_p\left(\frac{fk_i}{k_t}\right)^{1/2} [I]^{1/2}[M] \tag{26.12}$$

This rate is also the rate of polymerization, which has the form of eqn 9.

The **kinetic chain length**, v, is the ratio of the number of monomer units consumed per activated centre produced in the initiation step:

$$v = \frac{\text{number of monomer units consumed}}{\text{number of activated centres produced}} \tag{26.13}$$

The kinetic chain length can be expressed in terms of the rate expressions in *Justification 26.1*. To do so, we recognize that monomers are consumed at the rate that chains propagate. Then,

$$v = \frac{\text{rate of propagation of chains}}{\text{rate of production of radicals}}$$

By making the steady-state approximation, we set the rate of production of radicals equal to the termination rate. Therefore, we can write the expression for the kinetic chain length as

$$v = \frac{k_p[\cdot M][M]}{2k_t[M\cdot]^2} = \frac{k_p[M]}{2k_t[\cdot M]}$$

When we substitute the steady-state expression, eqn 11, for the radical concentration, we obtain

$$v = k[M][I]^{-1/2} \qquad k = \tfrac{1}{2}k_p(fk_ik_t)^{-1/2} \tag{26.14}$$

Consider a polymer produced by a chain mechanism with mutual termination. In this case, the average number of monomers in a polymer molecule, $\langle n \rangle$, produced by the reaction is the sum of the numbers in the two combining polymer chains. The average number of units in each chain is v. Therefore,

$$\langle n \rangle = 2v = 2k[M][I]^{-1/2} \tag{26.15}$$

with k given in eqn 14. We see that, the slower the initiation of the chain (the smaller the initiator concentration and the smaller the initiation rate constant), the greater the kinetic chain length, and therefore the higher the average molar mass of the polymer. Some of the consequences of molar mass for polymers were explored in Chapter 22: now we see how we can exercise kinetic control over them.

Homogeneous catalysis

A **catalyst** is a substance that accelerates a reaction but undergoes no net chemical change. The catalyst lowers the activation energy of the reaction by providing an alternative path that avoids the slow, rate-determining step of the uncatalysed reaction (Fig. 26.8). Catalysts can be very effective; for instance, the activation energy for the decomposition of hydrogen peroxide in solution is 76 kJ mol^{-1}, and the reaction is slow at room temperature. When a little iodide ion is added, the activation energy falls to 57 kJ mol^{-1} and the rate constant increases by a factor of 2000. **Enzymes**, which are biological catalysts, are very specific and can have a dramatic effect on the reactions they control. For example, the enzyme catalase reduces the activation energy for the decomposition of hydrogen peroxide to 8 kJ mol^{-1}, corresponding to an acceleration of the reaction by a factor of 10^{15} at 298 K.

A **homogeneous catalyst** is a catalyst in the same phase as the reaction mixture. For example, the decomposition of hydrogen peroxide in aqueous solution is catalysed by bromide ion or catalase (Sections 26.5 and 26.6). A **heterogeneous catalyst** is a catalyst in a different phase from the reaction mixture. For example, the hydrogenation of ethene to ethane, a gas-phase reaction, is accelerated in the presence of a solid catalyst such as palladium, platinum, or nickel. The metal provides a surface upon which the reactants bind; this binding facilitates encounters between reactants and increases the rate of the reaction. We examine heterogeneous catalysis in Chapter 28 and consider only homogeneous catalysis here.

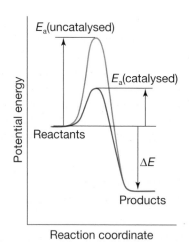

26.8 A catalyst provides a different path with a lower activation energy. The result is an increase in the rate of formation of products.

26.5 Features of homogeneous catalysis

We can obtain some idea of the mode of action of homogeneous catalysts by examining the kinetics of the bromide-catalysed decomposition of hydrogen peroxide:

$$2\,H_2O_2(aq) \longrightarrow 2\,H_2O(l) + O_2(g)$$

The reaction is believed to proceed through the following pre-equilibrium:[1]

$$H_3O^+ + H_2O_2 \rightleftharpoons H_3O_2^+ + H_2O \qquad K = \frac{[H_3O_2^+]}{[H_2O_2][H_3O^+]}$$

$$H_3O_2^+ + Br^- \longrightarrow HOBr + H_2O \qquad v = k[H_3O_2^+][Br^-]$$

$$HOBr + H_2O_2 \longrightarrow H_3O^+ + O_2 + Br^- \qquad \text{(fast)}$$

The second step is rate-determining. Therefore, we can obtain the rate law of the overall reaction by setting the overall rate equal to the rate of the second step and using the equilibrium constant to express the concentration of $H_3O_2^+$ in terms of the reactants. The result is

$$\frac{d[O_2]}{dt} = k_{eff}[H_2O_2][H_3O^+][Br^-]$$

with $k_{eff} = kK$, in agreement with the observed dependence of the rate on the Br^- concentration and the pH of the solution. The observed activation energy is that of the effective rate constant kK.

In **acid catalysis** the crucial step is the transfer of a proton to the substrate:

$$X + HA \longrightarrow HX^+ + A^- \qquad HX^+ \longrightarrow \text{products}$$

Acid catalysis is the primary process in the solvolysis of esters and keto–enol tautomerism. In **base catalysis**, a proton is transferred from the substrate to a base:

$$XH + B \longrightarrow X^- + BH^+ \qquad X^- \longrightarrow \text{products}$$

Base catalysis is the primary step in the isomerization and halogenation of organic compounds, and of the Claisen and aldol condensation reactions.

26.6 Enzymes

Enzymes are homogeneous biological catalysts. These ubiquitous compounds are special proteins or nucleic acids that contain an **active site**, which is responsible for binding the **substrate**, the reactant, and processing it into products. As is true of any catalyst, the active site returns to its original state after the products are released. The structure of the active site is specific to the reaction that it catalyses. For example, the active sites of the catalases have d-metal ions for binding of the substrate and for mediating the electron transfer reactions that lead to the decomposition of H_2O_2.

Enzyme-catalysed reactions are prone to inhibition by molecules that interfere with the formation of product. Many drugs for the treatment of disease function by inhibiting enzymes. For example, an important strategy in the treatment of acquired immune deficiency syndrome (AIDS) involves the steady administration of a specially designed protease inhibitor. The drug inhibits an enzyme that is key to the formation of the protein envelope surrounding the genetic material of the human immunodeficiency virus (HIV). Without a properly formed envelope, HIV cannot replicate in the host organism.

---

1. We have set the activity of H_2O in the equilibrium constant equal to 1 and assumed that the thermodynamic properties of the other substances are ideal.

26.9 The variation of the rate of an enzyme-catalysed reaction with substrate concentration. The approach to a maximum rate, v_{max}, for large [S] is explained by the Michaelis–Menten mechanism.

(a) The Michaelis–Menten mechanism of enzyme catalysis

Experimental studies of enzyme kinetics are typically conducted by monitoring the initial rate of product formation in a solution in which the enzyme is present at very low concentration. Indeed, enzymes are such efficient catalysts that significant accelerations may be observed even when their concentration is more than three orders of magnitude smaller than that of the substrate.

The principal features of many enzyme-catalysed reactions are as follows:

1 For a given initial concentration of substrate, $[S]_0$, the initial rate of product formation is proportional to the total concentration of enzyme, $[E]_0$.

2 For a given $[E]_0$ and low values of $[S]_0$, the rate of product formation is proportional to $[S]_0$.

3 For a given $[E]_0$ and high values of $[S]_0$, the rate of product formation becomes independent of $[S]_0$, reaching a maximum value known as the **maximum velocity**, v_{max}.

The **Michaelis–Menten mechanism** accounts for these features (Fig. 26.9). According to this mechanism, an enzyme–substrate complex is formed in the first step and either the substrate is released unchanged or after modification to form products:

$$E + S \rightleftharpoons ES \qquad k_a, k_a'$$
$$ES \longrightarrow P \qquad k_b \tag{26.16}$$

The rate of product formation according to this mechanism is

$$v = k_b[ES] \tag{26.17}$$

We can obtain the concentration of the enzyme–substrate complex by invoking the steady-state approximation and writing[2]

$$\frac{d[ES]}{dt} = k_a[E][S] - k_a'[ES] - k_b[ES] = 0$$

It follows that

$$[ES] = \left(\frac{k_a}{k_a' + k_b}\right)[E][S] \tag{26.18}$$

where [E] and [S] are the concentrations of *free* enzyme and substrate, respectively. To express the rate law in terms of the concentrations of enzyme and substrate added, we note that $[E]_0 = [E] + [ES]$. Moreover, because the substrate is typically in large excess relative to the enzyme, the free substrate concentration is approximately equal to the initial substrate concentration and we can write $[S] \approx [S]_0$. It then follows that:

$$[ES] = \frac{[E]_0}{1 + \left(\dfrac{k_a' + k_b}{k_a}\right)\dfrac{1}{[S]_0}}$$

By substituting this result into eqn 17, we obtain the **Michaelis–Menten equation** for the rate of an enzyme-catalysed reaction:

$$v = \frac{k_b[E]_0}{1 + \left(\dfrac{k_a' + k_b}{k_a}\right)\dfrac{1}{[S]_0}} \tag{26.19}$$

2. In their original formulation, Michaelis and Menten supposed that the initial step was a pre-equilibrium. The treatment as a steady-state problem was suggested later by G.E. Briggs and J.B.S. Haldane.

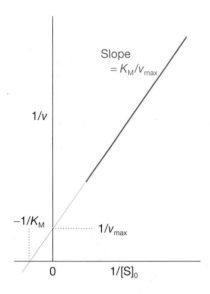

26.10 A Lineweaver–Burk plot for the analysis of an enzyme-catalysed reaction that proceeds by a Michaelis–Menten mechanism and the significance of the intercepts and the slope.

The constant $(k_a' + k_b)/k_a$ is characteristic of a given enzyme acting on a given substrate; it is called the **Michaelis constant** and written K_M. From eqn 18 it is also possible to express the Michaelis constant as $K_M = [E][S]/[ES]$. It follows that K_M has units of concentration.

Equation 19 shows that, in accord with experimental observations:

1 When $[S]_0 \ll K_M$, the rate is proportional to $[S]_0$:

$$v = \frac{k_a k_b}{k_a' + k_b}[S]_0[E]_0 \tag{26.20a}$$

2 When $[S]_0 \gg K_M$, the rate reaches its maximum value and is independent of $[S]_0$:

$$v = v_{max} = k_b[E]_0 \tag{26.20b}$$

Substitution of the definitions of K_M and v_{max} into eqn 19 gives:

$$v = \frac{v_{max}}{1 + K_M/[S]_0} \tag{26.21}$$

We can rearrange this expression into a form that is amenable to data analysis by linear regression:

$$\frac{1}{v} = \frac{1}{v_{max}} + \left(\frac{K_M}{v_{max}}\right)\frac{1}{[S]_0} \tag{26.22}$$

A **Lineweaver–Burk plot** is a plot of $1/v$ versus $1/[S]_0$, and according to eqn 22 it should yield a straight line with slope of K_M/v_{max}, a y-intercept at $1/v_{max}$, and an x-intercept at $-1/K_M$ (Fig. 26.10). The value of k_b is then calculated from the y-intercept and eqn 20b. However, the plot cannot give the individual rate constants k_a and k_a' that appear in the expression for K_M. The stopped-flow technique described in Section 25.1b can give the additional data needed, because we can find the rate of formation of the enzyme–substrate complex by monitoring the concentration after mixing the enzyme and substrate. This procedure gives a value for k_a, and k_a' is then found by combining this result with the values of k_b and K_M.

(b) The catalytic efficiency of enzymes

The **turnover number**, or **catalytic constant**, of an enzyme, k_{cat}, is the number of catalytic cycles (turnovers) performed by the active site in a given interval divided by the duration of the interval. This quantity has units of a first-order rate constant and, in terms of the Michaelis–Menten mechanism, is numerically equivalent to k_b, the rate constant for release of product from the enzyme–substrate complex. It follows from the identification of k_{cat} with k_b and from eqn 20b that

$$k_{cat} = k_b = \frac{v_{max}}{[E]_0} \tag{26.23}$$

The **catalytic efficiency**, ε (epsilon), of an enzyme is the ratio k_{cat}/K_M. The higher the value of ε, the more efficient is the enzyme. We can think of the catalytic activity as the effective rate constant of the enzymatic reaction. From $K_M = (k_a' + k_b)/k_a$ and eqn 23, it follows that

$$\varepsilon = \frac{k_{cat}}{K_M} = \frac{k_a k_b}{k_a' + k_b} \tag{26.24}$$

The efficiency reaches its maximum value of k_a when $k_b \gg k_a'$. Because k_a is the rate constant for the formation of a complex from two species that are diffusing freely in solution, the maximum efficiency is related to the maximum rate of diffusion of E

The web site contains links to databases of enzymes.

and S in solution. This limit (which is discussed further in Section 27.2) leads to rate constants of about 10^8–10^9 L mol^{-1} s^{-1} for molecules as large as enzymes at room temperature. The enzyme catalase has $\varepsilon = 4.0 \times 10^8$ L mol^{-1} s^{-1} and is said to have attained 'catalytic perfection'.

Example 26.3 Determining the catalytic efficiency of an enzyme

The enzyme carbonic anhydrase catalyses the hydration of CO_2 in red blood cells to give bicarbonate ion:

$$CO_2 + H_2O \longrightarrow HCO_3^- + H^+$$

The following data were obtained for the reaction at pH = 7.1, 273.5 K, and an enzyme concentration of 2.3 nmol L^{-1}:

$[CO_2]/(\text{mmol L}^{-1})$	1.25	2.5	5.0	20.0
rate/(mol L^{-1} s^{-1})	2.78×10^{-5}	5.00×10^{-5}	8.33×10^{-5}	1.67×10^{-4}

Determine the catalytic efficiency of carbonic anhydrase at 273.5 K.

Method Prepare a Lineweaver–Burk plot and determine the values of K_M and v_{max} graphically or from linear regression analysis. From eqn 23 and the enzyme concentration, calculate k_{cat} and the catalytic efficiency from eqn 24.

Answer Figure 26.11 shows the Lineweaver–Burk plot for the data. The slope is 40 s and the y-intercept is 4.0×10^3 L mol^{-1} s. Hence,

$$v_{max} = \frac{1}{4.0 \times 10^3 \text{ L mol}^{-1} \text{ s}} = 2.5 \times 10^{-4} \text{ mol L}^{-1} \text{ s}^{-1}$$

$$K_M = (2.5 \times 10^{-4} \text{ mol L}^{-1} \text{ s}^{-1}) \times (40 \text{ s}) = 1.0 \times 10^{-2} \text{ mol L}^{-1}$$

$$k_{cat} = \frac{2.5 \times 10^{-4} \text{ mol L}^{-1} \text{ s}^{-1}}{2.3 \times 10^{-9} \text{ mol L}^{-1}} = 1.1 \times 10^5 \text{ s}^{-1}$$

$$\varepsilon = \frac{k_{cat}}{K_M} = \frac{1.1 \times 10^5 \text{ s}^{-1}}{1.0 \times 10^{-2} \text{ mol L}^{-1}} = 1.1 \times 10^7 \text{ L mol}^{-1} \text{ s}^{-1}$$

Self-test 26.3 Calculate the substrate concentration required to decrease the rate to 50 per cent of v_{max}. Note that your answer is equal to K_M. This is a general result for the Michaelis–Menten mechanism. [1.0×10^{-2} mol L^{-1}]

26.11 The Lineweaver–Burk plot of the data for Example 26.3.

(c) Mechanisms of enzyme inhibition

An inhibitor, I, decreases the rate of product formation from the substrate by binding to the enzyme, to the ES complex, or to the enzyme and ES complex simultaneously. The most general kinetic scheme for enzyme inhibition is then:

$$E + S \rightleftharpoons ES \qquad k_a, k_a'$$

$$ES \longrightarrow E + P \qquad k_b$$

$$EI \rightleftharpoons E + I \qquad K_I = \frac{[E][I]}{[EI]} \tag{26.25a}$$

$$ESI \rightleftharpoons ES + I \qquad K_I' = \frac{[ES][I]}{[ESI]} \tag{26.25b}$$

The lower the values of K_I and K_I' the more efficient are the inhibitors. The rate of product formation is then given by eqn 17 above, since only ES leads to product. As shown in the following *Justification*, the rate of reaction in the presence of an inhibitor is

$$v = \frac{v_{max}}{\alpha' + \alpha K_M/[S]_0} \qquad (26.26)$$

where $\alpha = 1 + [I]/K_I$ and $\alpha' = 1 + [I]/K_I'$. This equation is very similar to the Michaelis–Menten equation for the uninhibited enzyme (eqn 21) and is also amenable to analysis by a Lineweaver–Burk plot:

$$\frac{1}{v} = \frac{\alpha'}{v_{max}} + \left(\frac{\alpha K_M}{v_{max}}\right)\frac{1}{[S]_0} \qquad (26.27)$$

Justification 26.2

By mass balance, the total concentration of enzyme is:

$$[E]_0 = [E] + [EI] + [ES] + [ESI]$$

By using eqns 25a and 25b and the definitions

$$\alpha = 1 + \frac{[I]}{K_I} \quad \text{and} \quad \alpha' = 1 + \frac{[I]}{K_I'}$$

it follows that

$$[E]_0 = [E]\alpha + [ES]\alpha'$$

By using $K_M = [E][S]/[ES]$ and assuming that $[S] \approx [S]_0$, we can write

$$[E]_0 = \frac{K_M[ES]}{[S]_0}\alpha + [ES]\alpha' = [ES]\left(\frac{\alpha K_M}{[S]_0} + \alpha'\right)$$

The expression for the rate of product formation is then:

$$v = k_b[ES] = \frac{k_b[E]_0}{\dfrac{\alpha K_M}{[S]_0} + \alpha'}$$

which, upon rearrangement, gives eqn 26.

(a) $\alpha > 1, \alpha' = 1$

$1/v$

$\alpha = \alpha' = 1$

0 $1/[S]$

(b) $\alpha = 1, \alpha' > 1$

$1/v$

$\alpha = \alpha' = 1$

0 $1/[S]$

(c) $\alpha = \alpha' > 1$

$1/v$

$\alpha = \alpha' = 1$

0 $1/[S]$

26.12 Lineweaver–Burk plots characteristic of the three major modes of enzyme inhibition: (a) competitive inhibition, (b) uncompetitive inhibition, and (c) non-competitive inhibition, showing the special case $\alpha = \alpha' > 1$.

There are three major modes of inhibition that give rise to distinctly different kinetic behaviour (Fig. 26.12). In **competitive inhibition** the inhibitor binds only to the active site of the enzyme and thereby inhibits the attachment of the substrate. This condition corresponds to $\alpha > 1$ and $\alpha' = 1$ (because ESI does not form). The slope of the Lineweaver–Burk plot increases by a factor of α relative to the slope for data on the uninhibited enzyme ($\alpha = \alpha' = 1$). The y-intercept does not change as a result of competitive inhibition. In **uncompetitive inhibition** the inhibitor binds to a site of the enzyme that is removed from the active site, but only if the substrate is already present. The inhibition occurs because ESI reduces the concentration of ES, the active type of complex. In this case $\alpha = 1$ (because EI does not form) and $\alpha' > 1$. The y-intercept of the Lineweaver–Burk plot increases by a factor of α' relative to the y-intercept for data on the uninhibited enzyme but the slope does not change. In **non-competitive inhibition**[3] the inhibitor binds to a site other than the active site,

3. Non-competitive inhibition is also known as *mixed inhibition*.

and its presence reduces the ability of the substrate to bind to the active site. Inhibition occurs at both the E and ES sites. This condition corresponds to $\alpha > 1$ and $\alpha' > 1$. Both the slope and y-intercept of the Lineweaver–Burk plot increase upon addition of the inhibitor. Figure 26.12c shows the special case of $K_I = K'_I$ and $\alpha = \alpha'$, which results in intersection of the lines at the x-axis.

In all cases, the efficiency of the inhibitor may be obtained by determining K_M and v_{max} from a control experiment with uninhibited enzyme and then repeating the experiment with a known concentration of inhibitor. From the slope and y-intercept of the Lineweaver–Burk plot for the inhibited enzyme (eqn 27), the mode of inhibition, the values of α or α', and the values of K_I or K'_I may be obtained.

Oscillating reactions

The concentrations of reactants, intermediates, and products of certain chemical reactions can vary periodically either in space or in time (Fig. 26.13). These oscillations occur in the presence of a feedback mechanism in which a product either increases or decreases the reaction rate. They can be sustained indefinitely only if the reaction is carried out under conditions that keep it far from equilibrium. This condition can be achieved by supplying reactants continuously and removing products continuously from a reaction vessel. Examples of such open systems are flow reactors (Sections 26.9 and 26.10) and biological cells. A biological cell, for instance, is an open system that can take in nutrients and excrete waste products of enzyme-catalysed reactions. Therefore, long-lasting chemical oscillations can occur only if a proper feedback mechanism is present. Such is the case in the glycolytic cycle, in which enzymes catalyse the breakdown of glucose with production of ATP and NADH. Feedback can occur if an enzyme of the glycolytic cycle, such as phosphofructokinase, is inhibited by high concentrations of ATP and activated by ADP. During the early stages of glycolysis, when ATP levels are low and ADP levels are high, the enzyme is active and the rate of production of NADH is high. However, as the concentration of ATP, a net product of glycolysis, rises in the cell, the enzyme becomes inhibited and the rate of production of NADH decreases. After some time, ATP levels drop again and a new cycle begins. Figure 26.14 shows how the concentration of NADH oscillates in yeast cells under anaerobic conditions.

26.13 Some reactions show oscillations in time; some show spatially periodic variations. This sequence of photographs shows the emergence of a spatial pattern.

26.14 Under anaerobic conditions, oscillations in the concentration of NADH can occur in yeast cells because of inhibition of the enzyme phosphofructokinase by high levels of ATP and its activation by high levels of ADP.

26.7 Autocatalysis

Autocatalysis is the catalysis of a reaction by the products. For example, in a reaction A → P it may be found that the rate law is

$$v = k[A][P] \tag{26.28}$$

and the reaction rate is proportional to the concentration of P.[4]

An example of autocatalysis is provided by the **Belousov–Zhabotinskii reaction** (BZ reaction), one of the first oscillating reactions to be reported and studied systematically. The BZ reaction occurs in a mixture of potassium bromate, malonic acid, and a cerium(IV) salt in acidic solution. Two steps of the mechanism, which will be discussed more fully in Section 26.8c, show the role of $HBrO_2$ as an autocatalyst:

$$BrO_3^- + HBrO_2 + H_3O^+ \longrightarrow 2\,BrO_2{}^{\cdot} + 2\,H_2O$$

$$2\,BrO_2{}^{\cdot} + 2\,Ce(III) + 2\,H_3O^+ \longrightarrow 2\,HBrO_2 + 2\,Ce(IV) + 2\,H_2O$$

The product $HBrO_2$ is a reactant in the first step and provides a feedback mechanism that enhances the rate of formation of $HBrO_2$.

Example 26.4 Calculating concentrations in an autocatalytic reaction

Integrate the rate equation for the A → P autocatalytic reaction in eqn 28.

Method The rate law is given in eqn 28, with $v = -d[A]/dt$. It is convenient to write $[A] = [A]_0 - x$, $[P] = [P]_0 + x$ and then to write down the expression for the rate of change of either species in terms of x.

Answer With the substitution indicated, the rate law becomes

$$\frac{dx}{dt} = k([A]_0 - x)([P]_0 + x)$$

Integration by using the relation

$$\frac{1}{([A]_0 - x)([P]_0 + x)} = \frac{1}{[A]_0 + [P]_0}\left(\frac{1}{[A]_0 - x} + \frac{1}{[P]_0 + x}\right)$$

(see also Section 25.3c) gives

$$\frac{1}{[A]_0 + [P]_0}\ln\left(\frac{([P]_0 + x)[A]_0}{[P]_0([A]_0 - x)}\right) = kt$$

We can rearrange this expression into

$$\frac{x}{[P]_0} = \frac{e^{at} - 1}{1 + be^{at}}$$

where $a = ([A]_0 + [P]_0)k$ and $b = [P]_0/[A]_0$.

Comment The solution is plotted in Fig. 26.15. The rate of reaction is slow initially (little P present), then fast (when P and A are both present), and finally slow again (when A has disappeared).

Self-test 26.4 At what time is the reaction rate a maximum?

$$[t_{max} = -(1/a)\ln b]$$

26.15 The concentration of product during the autocatalysed A → P reaction discussed in Example 26.4. Individual curves are labelled with the value of b.

4. The reaction gets started because there are usually other reaction routes for the formation of some P initially, which then takes part in the autocatalytic reaction proper.

26.16 The periodic variation of the concentrations of the intermediates X and Y in a Lotka–Volterra reaction. The system is in a steady state, but not at equilibrium.

26.17 An alternative representation of the periodic variation of [X] and [Y] is to plot one against the other. Then the system describes closed orbits; different orbits are obtained with different starting conditions.

26.8 **Autocatalytic mechanisms of oscillating reactions**

We now consider a few specific examples of chemical oscillations that result from autocatalysis.

(a) The Lotka–Volterra mechanism

We consider first an autocatalytic reaction of a particularly simple form that illustrates how chemical oscillations may occur. The actual chemical examples that have been discovered so far have a different mechanism, as we shall see. The **Lotka–Volterra mechanism** is as follows:

$$(a) \quad A + X \longrightarrow X + X \qquad \frac{d[A]}{dt} = -k_a[A][X]$$

$$(b) \quad X + Y \longrightarrow Y + Y \qquad \frac{d[X]}{dt} = -k_b[X][Y]$$

$$(c) \quad Y \longrightarrow B \qquad \frac{d[B]}{dt} = k_c[Y]$$

Steps (a) and (b) are autocatalytic. The concentration of A is held constant by supplying it to the reaction vessel as needed. The product B plays no part in the reaction once it has been produced, so it is unnecessary to remove it; in practice, though, it would normally be removed. These constraints leave [X] and [Y], the concentrations of the intermediates, as variables. Note that we are considering a **steady-state condition**, which is maintained by the flow of A into the reactor. This steady-state condition must not be confused with the steady-state approximation made earlier: in the present case we solve the rate equations exactly for the variable concentrations of X and Y, but hold [A] at an arbitrary but constant value.

The Lotka–Volterra equations can be solved numerically, and the results can be depicted in two ways. One way is to plot [X] and [Y] against time (Fig. 26.16). The same information is displayed more succinctly by plotting one concentration against the other (Fig. 26.17).

The periodic variation of the concentrations of the intermediates can be explained as follows. At some stage there may be only a little X present, but reaction (a) provides more, and the production of X autocatalyses the production of even more X. There is therefore a surge of X. However, as X is formed, reaction (b) can begin. It occurs slowly initially, because [Y] is small, but autocatalysis leads to a surge of Y. This surge, though, removes X, and so reaction (a) slows, and less X is produced. Because less X is now available, reaction (b) slows. As less Y becomes available to remove X, X has a chance to surge forward again, and so on.

(b) The brusselator

Another very interesting set of equations is called the **brusselator**:[5]

$$(a) \quad A \longrightarrow X \qquad \frac{d[X]}{dt} = k_a[A]$$

$$(b) \quad B + X \longrightarrow Y + C \qquad \frac{d[Y]}{dt} = k_b[B][X]$$

5. The mechanism is so-called because it is an oscillator that originated with Ilya Prigogine's group in Brussels.

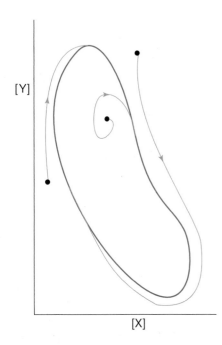

26.18 Some oscillating reactions approach a closed trajectory whatever their starting conditions. The closed trajectory (shown here in blue) is called a *limit cycle*.

(c) $X + X + Y \longrightarrow X + X + X$ $\qquad \dfrac{d[Y]}{dt} = -k_c[X]^2[Y]$

(d) $X \longrightarrow D$ $\qquad \dfrac{d[X]}{dt} = -k_d[X]$

Because the reactants (A and B) are maintained at constant concentration, the two variables are the concentrations of X and Y. These two concentrations may be calculated by solving the rate equations numerically, and the results are plotted in Fig. 26.18. The interesting feature is that, whatever the initial concentrations of X and Y, the system settles down into the same periodic variation of concentrations. The common trajectory to which the system migrates is called a **limit cycle**, and its period depends on the values of the rate constants. A limit cycle is an example of a structure called by mathematicians an **attractor** because it appears to attract to itself trajectories in its vicinity. In conventional chemistry, which deals with closed systems of reactants and products, the equilibrium state is an attractor determined by the minimum value of the Gibbs energy at a given temperature and pressure. In open systems maintained far from equilibrium, a limit cycle may act as an attractor; so, if the system is perturbed (by a surge of reactant, for instance), it relaxes back towards the time behaviour represented by the limit cycle.

(c) The oregonator

The mechanism of the BZ reaction was elucidated by Richard Noyes, and involves 18 elementary steps and 21 different chemical species. The main features of this awesomely complex mechanism can be reproduced by the following **oregonator**,[6] in which A stands for BrO_3^-, B for malonic acid, C for HOBr, P for additional oxidized and reduced products, X for $HBrO_2$, Y for Br^-, and Z for Ce^{4+}:

(a) $A + Y \longrightarrow X + C$

(b) $X + Y \longrightarrow C + C$

(c) $A + X \longrightarrow X + X + Z + Z$

(d) $X + X \longrightarrow A + C$

(e) $B + Z \longrightarrow Y + P$

The concentration of H_3O^+ is omitted because it is often held constant by buffering the reaction mixture. The concentrations of A, B, C, and D are also held constant by maintaining an influx of reactants and removal of products. The oscillations arise in a similar way to the brusselator, and can be traced to the autocatalysis in step (c) and the linkage between the reactions provided by the other steps.

26.9 Bistability

As shown in the *Justification* below, a number of chemical reactions carried out in flow reactors can have multiple steady states with very different chemical compositions. This possibility is depicted in Fig. 26.19 for a hypothetical reaction with two intermediates, X and Y. The compositions of the steady states are indicated by the lower and upper lines of a plot of [Y] versus [X]. The plot shows that, if the concentration of Y is at some high value in the reactor, and X is added, then the concentration of Y might decrease as shown by the upper line. If X is at some high value, then as Y is added the reaction might result in the slow increase of Y as shown by the lower line. However, a concentration may be reached at which the concentration of

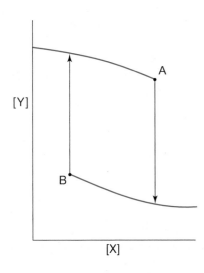

26.19 A system showing bistability. As the concentration of X is increased (by adding it to a reactor), the concentration of Y decreases along the upper curve but at A drops sharply to a low value. If X is then decreased, the concentration increases along the lower curve, but rises sharply to a high value at B.

6. It is so-called because Noyes and his group work in Oregon.

Y will jump from one curve to the other at points A and B (just as a supercooled liquid might suddenly solidify). This jump occurs because neither steady state is an equilibrium state in the thermodynamic sense: they occur in systems that are well removed from equilibrium, and the concentrations of X and Y represent the consequences of reactants continuously flowing into and of products flowing out of the reactor. The behaviour just described for a system with two stable steady states is known as **bistability**.

Justification 26.3

We demonstrate the possibility of multiple steady states by considering a hypothetical autocatalytic process in a flow reactor:

$$A + 2\,B \longrightarrow 3\,B \qquad \frac{d[A]}{dt} = -k[A][B]^2$$

The rate law for this reaction must take into account not only the chemical event (reaction between A and B) but also the inflow rate of A and the outflow rates of B and unreacted A. Let $[A]_0$ be the initial concentration of A at the inlet of the reactor and k_0 a rate constant for the flow of reactant (and equal to the reciprocal of the time spent by A in the reactor). Then,

$$\frac{d[A]}{dt} = -k[A][B]^2 + k_0[A]_0 - k_0[A] \tag{26.29}$$

where, on the right-hand side of the equation, the first term corresponds to the chemical reaction, the second term is the rate of flow into the reactor, and the third term is the rate of removal of unreacted A from the reactor.

Upon application of the steady-state approximation to eqn 29, substitution of the mass balance expression $[B] = [A]_0 - [A]$, and some rearrangement, we obtain:

$$([A]_0 - [A])\{k_0 - k[A]([A]_0 - [A])\} = 0$$

The roots of this equation are the steady-state concentrations of A, which we denote $[A]_{ss}$. Because the equation is cubic in $[A]$, it has multiple roots, which indicates a potential for coexistence of steady states in the system. However, we must ascertain that the values of $[A]_{ss}$ make physical sense.

The first root is $[A]_{ss} = [A]_0$, the unreacted state. We determine the remaining roots by setting $k_0 - k[A]([A]_0 - [A]) = 0$. It follows that:

$$k[A]^2 - k[A]_0[A] + k_0 = 0$$

The roots of this equation are:

$$[A]_{ss} = \tfrac{1}{2}[A]_0 \pm \frac{(k[A]_0^2 - 4k_0)^{1/2}}{2k^{1/2}}$$

The roots will be physically significant if they are real and positive, which is guaranteed if $k[A]_0^2 > 4k_0$. Therefore, the system will exhibit three steady states if the initial reaction rate far exceeds the flow rate of the reactor. Consequently, many autocatalytic reactions occurring in flow reactors can have multiple stable steady states.

Figure 26.19 does not describe an oscillation because it lacks a feedback mechanism that would cause, for example, the concentration of X to decrease to the value at point B upon a jump down from point A. Now consider the addition of a species Z that reacts with X and Y according to:

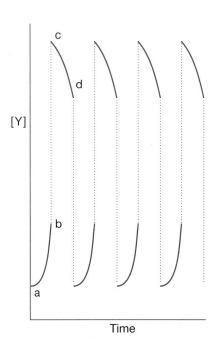

26.20 Chemical oscillation in a bistable system occurs as a result of the effect of a third substance Z which can react with X to produce Y and with Y to produce X. As a result, the system switches periodically between the upper and lower curves.

26.21 The leaping from one branch to another of a bistable system appears to the observer as a periodic surge and depletion of concentration.

In the absence of Z the flow of material might correspond to the stable state on the upper curve of Fig. 26.20. However, as Z reacts with Y to produce X the state of the system moves along the curve (to the right, as Y decreases and X increases) until the sudden transition occurs to the lower curve. Then Z reacts with X and produces Y, which means that the composition moves to the left along the lower curve. There comes a point, however, when the concentration of X has been reduced so much, and that of Y has risen so much, that there is a sudden transition to the upper curve, when the process begins again. The leaping from one stable state to another is manifest as the sudden surge or depletion of the concentration of a species (Fig. 26.21). Therefore, with the addition of chemical feedback, a bistable system can become oscillatory.

26.10 Chemical chaos

For some reactions, the solutions of rate laws cannot be used to predict the composition of the system from its initial composition. Instead of showing periodic oscillatory behaviour, the concentrations burst into chaotic oscillation, characterized by unpredictable amplitudes or unpredictable frequencies (Fig. 26.22). Such behaviour can be literally a matter of life and death for, should the heartbeat become chaotic (the phenomenon of fibrillation), then death may result. The type of chaos that stems from the structure of well defined (and often quite simple) nonlinear differential equations is called **deterministic chaos**, for the behaviour of the solutions is predictable but infinitely sensitive to the initial conditions.

26.22 The onset of chaos as a result of period-doubling. (a) Steady-state oscillation, (b) the oscillation after one period-doubling step, (c) the chaotic regime after many period-doubling steps.

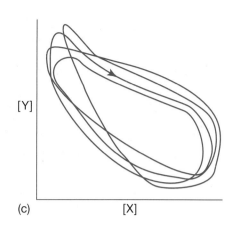

26.23 Successive trajectories in concentration space (a) before period-doubling, (b) after one period-doubling step, (c) after several period-doubling steps.

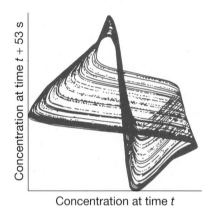

26.24 A representation of a strange attractor for a system showing chemical chaos.

There are several ways in which a reaction can be steered into a chaotic regime. For example, in certain systems the change in a parameter (such as a flow rate through the reaction vessel or the rate of stirring) may result in the successive doubling of the period of the limit cycle (Fig. 26.23). When this **period-doubling** occurs, the system must circulate twice round the cycle before an initial pair of concentrations is restored. Then, after a further modification of the parameter, the period doubles again, and the limit cycle circulates four times before repeating itself. As each period doubles, the periodicity of the motion becomes less apparent and finally appears like random fluctuations. The trajectory of the composition of the system is now highly complex, and may take a path that never retraces its path or crosses itself. Such a path is called a **strange attractor** (Fig. 26.24).

The term 'chaos' is in some respects misleading. First, the conditions under which certain systems of differential equations display period doubling can be specified exactly, and the progression through period-doubling cycles shows considerable regularity that is common to many classes of systems. In so far as the composition can be specified exactly, then a later composition can be predicted. Our inability to predict the composition of a system that is in a chaotic regime stems from our inability to know exactly the initial experimental conditions or measure the composition at an exact later instant. The unpredictability of a chaotic system lies not in the formulation or solution of the differential equations that describe the rates of processes, but in our ability to relate those solutions to the practical system of interest given the inherent imprecision of experimental observations.

Photochemistry

Many reactions can be initiated by the absorption of electromagnetic radiation by one of the mechanisms described in Section 17.2. The most important of all are the photochemical processes that capture the radiant energy of the Sun. Some of these reactions lead to the heating of the atmosphere during the daytime by absorption of ultraviolet radiation (Box 26.1). Others include the absorption of visible radiation during photosynthesis (Box 10.1). Without photochemical processes, the Earth would be simply a warm, sterile, rock. Table 26.1 summarizes common photochemical reactions.

Table 26.1 Examples of photochemical processes†

Process	General form		Example	
Ionization:	$A^* \longrightarrow A^+ + e^-$		$NO^* \longrightarrow NO^+ + e^-$	($\lambda = 134$ nm)
Electron transfer:	$A^* + B \longrightarrow A^+ + B^-$	(or $A^- + B^+$)	$[Ru(bpy)_3^{2+}]^* + Fe^{3+} \longrightarrow Ru(bpy)_3^{3+} + Fe^{2+}$	($\lambda = 452$ nm)
Dissociation:	$A^* \longrightarrow B + C$		$O_3^* \longrightarrow O_2 + O$	($\lambda < 1180$ nm)
	$A^* + B{-}C \longrightarrow A + B + C$		$Hg^* + CH_4 \longrightarrow Hg + CH_3 + H$	($\lambda = 254$ nm)
Addition:	$2\,A^* \longrightarrow B$			
	$A^* + B \longrightarrow AB$			($\lambda = 230$ nm)
Abstraction:	$A^* + B{-}C \longrightarrow A{-}B + C$		$Hg^* + H_2 \longrightarrow HgH + H$	($\lambda = 254$ nm)
Isomerization or rearrangement:	$A^* \longrightarrow A'$			($\lambda = 380$ nm; see Box 17.1)

† The excited state is denoted by an asterisk. In each example, the photochemical process is initiated by irradiation at the wavelength in parentheses.

Box 26.1 The chemistry of stratospheric ozone

The Earth's atmosphere contains primarily N_2 and O_2 gas, with low concentrations of a large number of other species of both natural and anthropogenic origins. Indeed, many of the natural trace constituents of our atmosphere participate in complex chemical reactions that have contributed to the proliferation of life on the planet. The development of industrial societies has added new components to the Earth's atmosphere and has led to significant changes in the concentrations of some natural trace species. The negative consequences of these changes for the environment are either already being felt or, more disturbingly, are yet to be felt in the next few decades. Careful kinetic analysis allows us to understand the origins of our complex atmosphere and point to ways in which environmental problems can be solved or avoided.

The Earth's atmosphere consists of layers, as shown in the first illustration. The pressure decreases as altitude increases (see Problems 1.35 and 19.21), but the variation of temperature with altitude is complex, owing to processes that capture radiant energy from the Sun. We focus on the *stratosphere*, a region spanning from 15 km to 50 km above the surface of the Earth, and on the chemistry of the trace component ozone, O_3.

In the *troposphere*, the region between the Earth's surface and the stratosphere, temperature decreases with increasing altitude. This behaviour may be understood in terms of a model in which the boundary between the troposphere and the stratosphere, also called the *tropopause*, is considered adiabatic. Then we know from Section 2.6 that, as atmospheric gases are allowed to expand from layers close to the surface to higher layers, the temperature varies with pressure, and hence height, as

$$\frac{T_{\text{low altitude}}}{T_{\text{high altitude}}} = \left(\frac{p_{\text{low altitude}}}{p_{\text{high altitude}}} \right)^c \qquad c = \frac{C_{p,m}}{C_{V,m}} - 1$$

The model predicts a decrease in temperature with increasing altitude because $C_{p,m}/C_{V,m} \approx \frac{7}{5}$ for air. In the stratosphere,

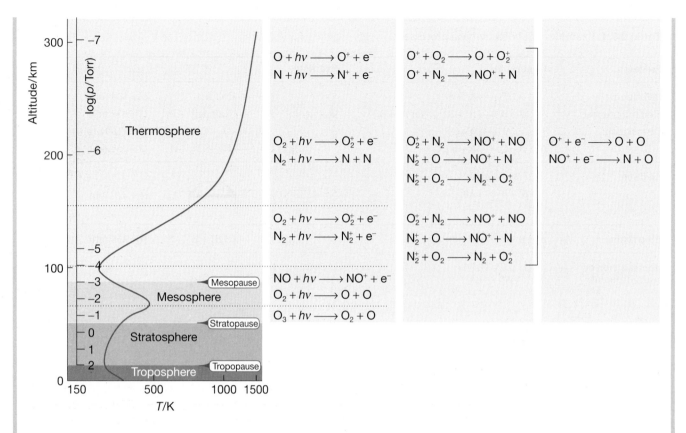

The temperature profile through the atmosphere and some of the reactions that occur in each region.

a *temperature inversion* is observed because of photochemical chain reactions that produce ozone from O_2. The *Chapman model* accounts for ozone formation and destruction in an atmosphere that contains only O_2:

Initiation: $O_2 + h\nu \longrightarrow O + O$

$185 \text{ nm} < \lambda < 220 \text{ nm}$ $\qquad v = k_1[O_2]$

Propagation: $O + O_2 + M \longrightarrow O_3 + M$

$\Delta H_r = -106.6 \text{ kJ mol}^{-1}$ $\qquad v = k_2[O][O_2][M]$

$O_3 + h\nu \longrightarrow O_2 + O$

$210 \text{ nm} < \lambda < 300 \text{ nm}$ $\qquad v = k_3[O_3]$

Termination: $O + O_3 \longrightarrow O_2 + O_2$

$\Delta H_r = -391.9 \text{ kJ mol}^{-1}$ $\qquad v = k_4[O][O_3]$

$O + O + M \longrightarrow O_2 + M$ $\qquad v = k_5[O]^2[M]$

where M is an arbitrary third body, such as O_2 in an 'oxygen-only' atmosphere, which helps to conserve angular momentum. The mechanism shows that absorption of radiation by O_2 and O_3 during the daytime leads to the production of reactive oxygen atoms, which, in turn, participate in exothermic reactions that are responsible for the heating of the stratosphere.

Using values of the rate constants that are applicable to stratospheric conditions, the Chapman model predicts a net formation of trace amounts of ozone, as seen in the second illustration (see also *Discussion question 1*). However, the model overestimates the concentration of ozone in the stratosphere because other trace species X contribute to catalytic enhancement of the termination step $O_3 + O \rightarrow O_2 + O_2$ according to

$X + O_3 \longrightarrow XO + O_2$

$XO + O \longrightarrow X + O_2$

The catalyst X can be H, OH, NO, or Cl. Chlorine atoms are produced by photolysis of CH_3Cl, which, in turn, is a by-product of reactions between Cl^- and decaying vegetation in oceans. Nitric oxide, NO, is produced in the stratosphere from reaction between excited oxygen atoms and N_2O, which is formed in the troposphere. The hydroxyl radical is a product, along with the methyl radical, of the reaction between excited oxygen atoms and methane gas, which is a by-product of a number of natural processes (such as digestion of cellulose in ruminant animals and anaerobic decomposition of organic waste matter) and industrial processes (such as food production and fossil fuel use). In spite of the

Net formation of ozone via the Chapman model in an stratospheric model containing only O_2, O, and O_3. The rate constants are consistent with reasonable stratospheric conditions. (a) Early reaction period after irradiation begins at $t = 0$. (b) Late reaction period, showing that the concentration of O atoms begins to level off after about four hours of continuous irradiation. (c) Late reaction period, showing that the ozone concentration also begins to level off similarly. For details of the calculation, see *Discussion question 1* and M.P. Cady and C.A. Trapp, *A Mathcad primer for physical chemistry*. Oxford University Press (1999).

presence of these catalysts, a natural stratosphere is still capable of maintaining a low concentration of ozone.

The chemistry outlined above shows that the photochemical reactions of the Chapman model account for absorption of a significant portion of solar ultraviolet radiation in the stratosphere. Hence, the surface of the Earth is bathed by

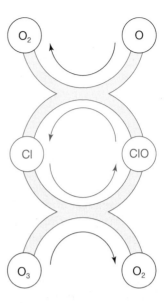

A catalytic cycle showing the propagation of ozone decomposition by chlorine atoms.

lower energy radiation, which does not damage biological tissue (see Box 26.2). However, some pollutants can lower the concentration of stratospheric ozone. For example, chlorofluorocarbons (CFCs) have been used as propellants and refrigerants over many years. As CFC molecules diffuse into the stratosphere, they are photolysed by ultraviolet radiation. For CF_2Cl_2, also known as CFC-12, the reaction is:

$$CF_2Cl_2 + h\nu \longrightarrow CF_2Cl + Cl$$

We already know that the resulting Cl atoms can participate in the decomposition of ozone according to the catalytic cycle shown in the third illustration. A number of experimental observations have linked this chemistry of CFCs to a dangerously rapid decline in the concentration of stratospheric ozone over the last three decades. Ozone depletion has increased the amount of ultraviolet radiation at the Earth's surface. The physiological consequences of prolonged exposure to ultraviolet radiation include DNA damage, genetic mutations, cell destruction, sunburning, and skin cancers. Consequently, there is concern that the depletion of stratospheric ozone may lead to an increase in mortality not only of animals but also the plants and lower organisms that form the base of the food chain.

Chlorofluorocarbons are being phased out according to international agreements and alternatives, such as the hydrofluorocarbon CH_2FCH_3, are already being used. However, the temperature inversion shown in the first illustration leads to trapping of gases in the troposphere, so CFCs are likely to continue to cause ozone depletion over many years as the molecules diffuse slowly into the stratosphere.

Discussion questions[7]

1. Use the Chapman model to explore the behaviour of a model atmosphere consisting of pure O_2 at 10 Torr and 298 K that is exposed to measurable frequencies and intensities of UV radiation. (a) Look up the values of k_2, k_4, and k_5 in a source such as the *CRC Handbook of chemistry and physics* or *Chemical kinetics and photochemical data for use in stratospheric modeling* (the URL is available in the web site of this book). The rate constants k_1 and k_3 depend upon the radiation conditions; assume values of 1.0×10^{-8} s^{-1} and 0.016 s^{-1}, respectively. If you cannot find a value for k_5, formulate chemically sound arguments for exclusion of the fifth step from the mechanism. (b) Write the rate expressions for the concentration of each chemical species. (c) Assume that the UV radiation is turned on at $t = 0$, and solve the rate expressions for the concentration of all species as a function of time over a period of 4 h. Examine relevant concentrations in the very early time period $t < 0.1$ s. State all assumptions. Is there any ozone present initially? Why must the pressure be low and the UV radiation intensities high for the production of ozone? Draw graphs of the time variations of both atomic oxygen and ozone on both the very short and the long timescales. What is the percentage of ozone after 4.0 h of irradiation? *Hint.* You will need a software package for solving a 'stiff' system of differential equations. Stiff differential equations have at least two rate constants with very different values and result in different behaviours on different timescales, so the solution usually requires that the total time period be broken into two or more periods; one may be very short and another very long. For help with using mathematical software to solve systems of differential equations, see M.P. Cady and C.A. Trapp, *A Mathcad primer for physical chemistry.* Oxford University Press (1999).

2.[7] Chlorine atoms react rapidly with ozone in the gas-phase bimolecular reaction $Cl + O_3 \rightarrow ClO + O_2$ with $k = (1.7 \times 10^{10}$ L mol^{-1} $s^{-1})e^{-260/(T/K)}$ (W.B. DeMore, S.P. Sander, D.M. Golden, R.F. Hampson, M.J. Kurylo, C.J. Howard, A.R. Ravishankara, C.E. Kolb, and M.J. Molina, Chemical kinetics and photochemical data for use in stratospheric modeling: Evaluation Number 11, JPL Publication 94–26 (1994)). Estimate the rate of this reaction at (a) 20 km, where $[Cl] = 5 \times 10^{-17}$ mol L^{-1}, $[O_3] = 8 \times 10^{-9}$ mol L^{-1}, and $T = 220$ K; (b) 45 km, where $[Cl] = 3 \times 10^{-15}$ mol L^{-1}, $[O_3] = 8 \times 10^{-11}$ mol L^{-1}, and $T = 270$ K.

Further reading

B.J. Finlayson-Pitts and J.N. Pitts, *Chemistry of the upper and lower atmosphere.* Academic Press, San Diego (2000).

R.P. Wayne, *Chemistry of atmospheres: an introduction to the chemistry of atmospheres of earth, the planets, and their satellites.* Oxford University Press (2000).

7. These problems were supplied by Charles Trapp, Carmen Giunta, and Marshall Cady.

26.11 Kinetics of photophysical and photochemical processes

Photochemical processes are initiated by the absorption of radiation by at least one component of a reaction mixture. In a **primary process**, products are formed directly from the excited state of a reactant. Examples include fluorescence (Section 17.3) and the *cis–trans* photoisomerization of retinal (Table 26.1, see also Box 17.1). Products of a **secondary process** originate from intermediates that are formed directly from the excited state of a reactant. Examples include photosynthesis and photochemical chain reactions (Section 26.12).

Competing with the formation of photochemical products is a host of primary photophysical processes that can deactivate the excited state (Table 26.2). Therefore, it is important to consider the timescales of excited state formation and decay before describing the mechanisms of photochemical reactions.

The web site contains links to databases on photochemical reactions.

(a) Timescales of photophysical processes

Electronic transitions caused by absorption of ultraviolet and visible radiation occur within 10^{-16}–10^{-15} s. We expect, then, that the upper limit for the rate constant of a first-order photochemical reaction is about 10^{16} s^{-1}. Fluorescence is slower than

Table 26.2 Common photophysical processes†

Primary absorption:	$S + h\nu \longrightarrow S^*$
Excited-state absorption:	$S^* + h\nu \longrightarrow S^{**}$
	$T^* + h\nu \longrightarrow T^{**}$
Fluorescence emission:	$S^* \longrightarrow S + h\nu$
Stimulated emission:	$S^* + h\nu \longrightarrow S + 2h\nu$
Intersystem crossing (ISC):	$S^* \longrightarrow T^*$
Phosphorescence emission:	$T^* \longrightarrow S + h\nu$
Internal conversion (IC):	$S^* \longrightarrow S$
Collision-induced emission:	$S^* + M \longrightarrow S + M + h\nu$
Collisional deactivation:	$S^* + M \longrightarrow S + M$
	$T^* + M \longrightarrow S + M$
Electronic energy transfer:	
singlet–singlet:	$S^* + S \longrightarrow S + S^*$
triplet–triplet:	$T^* + T \longrightarrow T + T^*$
Excimer formation:	$S^* + S \longrightarrow (SS)^*$
Energy pooling:	
singlet–singlet:	$S^* + S^* \longrightarrow S^{**} + S$
triplet–triplet:	$T^* + T^* \longrightarrow S^* + S$

† S denotes a singlet state and T a triplet state; M is a body that carries thermal energy. In photochemical studies, it is common to set 1 einstein = 1 mol photons.

absorption, with typical lifetimes of 10^{-12}–10^{-6} s. Therefore, the excited singlet state can initiate very fast photochemical reactions in the femtosecond (10^{-15} s) to picosecond (10^{-12} s) timescale. Examples of such ultrafast reactions are the initial events of vision (Box 17.1) and of photosynthesis. Typical intersystem crossing (ISC) and phosphorescence times for large organic molecules are 10^{-12}–10^{-4} s and 10^{-6}–10^{-1} s, respectively. As a consequence, excited triplet states are photochemically important. Indeed, because phosphorescence decay is several orders of magnitude slower than most typical reactions, species in excited triplet states can undergo a very large number of collisions with other reactants before deactivation. The interplay between reaction rates and excited state lifetimes is a very important factor in the determination of the kinetic feasibility of a photochemical process.

Illustration 26.1

To estimate whether the excited singlet or triplet state of the reactant is a suitable product precursor, we compare the emission lifetimes with the time constant, τ, of the reactant due to the chemical reaction. As an illustration, consider a unimolecular photochemical reaction with rate constant $k = 1.7 \times 10^4$ s^{-1} and time constant $\tau = 1/(1.7 \times 10^4 \text{ s}^{-1}) = 59$ μs that involves a reactant with an observed fluorescence lifetime of 1.0 ns and an observed phosphorescence lifetime of 1.0 ms. The excited singlet state is too short-lived and is not expected to be a major source of product in this reaction. On the other hand, the excited triplet state is a good candidate for a precursor.

(b) The primary quantum yield

We shall see that the rates of deactivation of the excited state by radiative, non-radiative, and chemical processes determine the yield of product in a photochemical reaction. The **primary quantum yield**, ϕ, is defined as the number of photophysical or photochemical events that lead to primary products divided by the number of photons absorbed by the molecule in the same interval. It follows that the primary quantum yield is also the rate of radiation-induced primary events divided by the rate of photon absorption. Because the rate of photon absorption is equal to the intensity of light absorbed by the molecule (Section 16.2), we write

$$\phi = \frac{\text{number of events}}{\text{number of photons absorbed}} = \frac{\text{rate of process}}{\text{intensity of light absorbed}} = \frac{v}{I_{abs}} \qquad [26.30]$$

A molecule in an excited state must either decay to the ground state or form a photochemical product. Therefore, the total number of molecules deactivated by radiative processes, nonradiative processes, and photochemical reactions must be equal to the number of excited species produced by absorption of light. We conclude that the sum of primary quantum yields ϕ_i for *all* photophysical and photochemical events i must be equal to 1, regardless of the number of reactions involving the excited state. It follows that

$$\sum_i \phi_i = \sum_i \frac{v_i}{I_{abs}} = 1 \qquad (26.31)$$

Illustration 26.2

For an excited singlet state that decays to the ground state only via the photophysical processes described in Section 26.11a, we write $\phi_f + \phi_{IC} + \phi_{ISC} + \phi_p = 1$, where ϕ_f, ϕ_{IC}, ϕ_{ISC}, and ϕ_p are the quantum yields of fluorescence, internal conversion, intersystem crossing, and phosphorescence, respectively. The quantum yield of photon emission by fluorescence and phosphorescence is $\phi_{emission} = \phi_f + \phi_p$, which is less than 1. If the excited singlet state also participates in a primary photochemical reaction with quantum yield ϕ_R, we write $\phi_f + \phi_{IC} + \phi_{ISC} + \phi_p + \phi_R = 1$.

We can now strengthen the link between reaction rates and primary quantum yield already established by eqns 30 and 31. By taking the constant I_{abs} out of the summation in eqn 31 and rearranging, we obtain $I_{abs} = \sum_i v_i$. Substituting this result into eqn 31 gives the general result

$$\phi_i = \frac{v_i}{\sum_i v_i} \qquad (26.32)$$

Therefore, the primary quantum yield may be determined directly from the experimental rates of *all* photophysical and photochemical processes that deactivate the excited state.

(c) Mechanism of decay of excited singlet states

Consider the formation and decay of an excited singlet state:

Absorption: $\qquad S + h v_i \longrightarrow S^* \qquad v_{abs} = I_{abs}$

Fluorescence: $\qquad S^* \longrightarrow S + h v_f \qquad v_f = k_f[S^*]$

Internal conversion: $\qquad$ $S^* \longrightarrow S$ $\qquad$ $v_{IC} = k_{IC}[S^*]$

Intersystem crossing: $\qquad$ $S^* \longrightarrow T^*$ $\qquad$ $v_{ISC} = k_{ISC}[S^*]$

in which S is an absorbing species, S^* an excited singlet state, T^* an excited triplet state, and $h\nu_i$ and $h\nu_f$ are the energies of the incident and fluorescent photons, respectively. If the absorbance of the sample is low and the incident light intensity is relatively high, we may assume that $[S^*]$ is small and constant. That is, we may invoke the steady-state approximation for $[S^*]$:

$$\frac{d[S^*]}{dt} = I_{abs} - k_f[S^*] - k_{IC}[S^*] - k_{ISC}[S^*] = I_{abs} - (k_f + k_{IC} + k_{ISC})[S^*] = 0$$

Consequently,

$$I_{abs} = (k_f + k_{IC} + k_{ISC})[S^*]$$

By using this expression and eqn 30, we obtain

$$\phi_f = \frac{v_f}{I_{abs}} = \frac{k_f[S^*]}{(k_f + k_{IC} + k_{ISC})[S^*]}$$

The final expression for the quantum yield of fluorescence is

$$\phi_f = \frac{k_f}{k_f + k_{IC} + k_{ISC}} \tag{26.33}$$

We can summarize the results of these calculations as follows. Under experimental conditions that justify the steady-state approximation for $[S^*]$, the sample emits fluorescence photons with a constant intensity, I, which is proportional to $k_f[S^*]$ and hence also proportional to ϕ_f. There are many commercial instruments designed to make such steady-state fluorescence measurements of I. When the light is turned off, the excited state decays exponentially:

$$[S^*]_t = [S^*]_0 \, e^{-t/\tau_0}$$

The observed fluorescence lifetime, τ_0, is given by:

$$\tau_0 = \frac{1}{k_f + k_{IC} + k_{ISC}} = \frac{\phi_f}{k_f}$$

The observed fluorescence lifetime can therefore be measured with a pulsed laser technique (Section 17.7). First, the sample is excited with a short light pulse from a laser at a wavelength where S absorbs strongly. Then, the decay of the fluorescence intensity after the pulse is monitored with a fast photodetector system.

(d) Quenching

The shortening of the lifetime of the excited state is called **quenching**. Quenching may be either a desired process, such as in energy or electron transfer (Box 26.2), or an undesired side reaction that can decrease the quantum yield of a desired photochemical process. Quenching effects may be studied by monitoring the emission from the excited state that is involved in the photochemical reaction.

The addition of a quencher, Q, opens an additional channel for deactivation of S^*:

Quenching: $\qquad$ $S^* + Q \longrightarrow S + Q$ $\qquad$ $v_Q = k_Q[Q][S^*]$

The steady-state approximation for $[S^*]$ now gives:

$$\frac{d[S^*]}{dt} = I_{abs} - (k_f + k_{IC} + k_{ISC} + k_Q[Q])[S^*] = 0$$

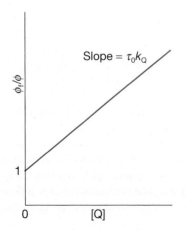

26.25 The format of a Stern–Volmer plot and the interpretation of the slope in terms of the rate constant for quenching and the observed fluorescence lifetime in the absence of quenching.

Slope = $\tau_0 k_Q$

ϕ_f/ϕ

1

0 $\qquad$ [Q]

Box 26.2 Photobiology and photomedicine

Up to 1 kJ s^{-1} m^{-2} of solar radiation reaches the Earth's surface, with the exact intensity depending on latitude, time of day, and weather. A significant amount of this energy is harnessed during photosynthesis (see Box 10.1). Other photochemical processes also occur in both photosynthetic and non-photosynthetic organisms. Among the beneficial processes in humans are vision, which is initiated by the photoisomerization of 11-*cis*-retinal in the protein rhodopsin (Section 17.1), and the biosynthesis of vitamin D$_3$ from 7-dehydrocholesterol in skin. Other processes, such as DNA damage caused by prolonged exposure to ultraviolet (UV) radiation, are deleterious to both higher and lower organisms. When controlled carefully, potentially harmful photochemical processes may be turned into beneficial forms of therapy. We shall explore plant photosynthesis, UV-induced DNA damage, and one variety of laser-based tumour therapy in some detail.

Plant photosynthesis

A large proportion of solar radiation with wavelengths below 400 nm and above 1000 nm is absorbed by atmospheric gases such as ozone and O$_2$, which absorb ultraviolet radiation, and CO$_2$ and H$_2$O, which absorb infrared radiation. As a result, plants, algae, and some species of bacteria evolved photosynthetic apparatus that capture visible and near-infrared radiation. Plants use radiation in the wavelength range of 400–700 nm to drive the transfer of electrons from water to carbon dioxide. We have already examined the thermodynamics of plant photosynthesis in Box 10.1; here we shall describe the kinetics of the capture and utilization of solar energy.

In the chloroplast, chlorophylls *a* and *b* (**1**) and carotenoids (of which β-carotene (**2**) is an example) bind to integral proteins called *light-harvesting complexes*, which absorb solar energy and transfer it to protein complexes known as *reaction centres*, where light-induced electron transfer reactions occur. The combination of a light-harvesting complex and a reaction centre complex is called a *photosystem*. Plants have two photosystems that drive the reduction of NADP$^+$ by water (Box 10.1):

$$H_2O + NADP^+ \xrightarrow{\text{light, photosystems I and II}} \tfrac{1}{2}O_2 + NADPH + H^+$$

Light-harvesting complexes bind large numbers of pigments in order to provide a sufficiently large area for capture of radiation. In photosystems I and II, absorption of a photon raises a chlorophyll or carotenoid molecule to an excited

1 Chlorophylls *a* (R = CH$_3$) and *b* (R = CHO)

2 β-Carotene

singlet state and within 0.1–5 ps the energy hops to a nearby pigment via the Förster mechanism (Section 26.11e). About 100–200 ps later, which corresponds to thousands of hops within the light-harvesting complex, more than 90 per cent of the absorbed energy reaches the reaction centre. There, a chlorophyll *a* dimer becomes electronically excited and initiates ultrafast electron transfer reactions. For example, the transfer of an electron from the excited singlet state of P680, the chlorophyll dimer of the photosystem II reaction centre, to its immediate electron acceptor, a phaeophytin *a* molecule,[8] occurs within 3 ps. Once the excited state of P680 has been quenched efficiently by this first reaction, subsequent steps that lead to the oxidation of water and reduction of plastoquinone occur more slowly, with reaction times varying from 200 ps to 1 ms. The electrochemical reactions within the photosystem I reaction centre also occur in this time interval.

In summary, the initial energy and electron transfer events of photosynthesis are under tight kinetic control.

---

8. Phaeophytin *a* is a chlorophyll *a* molecule where the central Mg^{2+} ion is replaced by two protons, which are bound to two of the pyrrole nitrogens in the ring.

Photosynthesis captures solar energy efficiently because the excited singlet state of chlorophyll is quenched rapidly by processes that occur with relaxation times that are much shorter than the fluorescence lifetime, which is about 5 ns in diethyl ether at room temperature.

DNA damage by UV radiation

We saw in Box 26.1 that the depletion of ozone from the Earth's stratosphere increases the level of UVB radiation at the surface, with deleterious consequences for the biosphere. One example of a biological effect induced by UVB radiation is DNA damage. The principal mechanism of DNA damage involves the photodimerization of adjacent thymine bases (3) to yield either a cyclobutane thymine dimer (left) or a so-called 6–4 photoproduct (right). The former has been linked directly to cell death and the latter may lead to DNA mutations and, consequently, to the formation of tumours. There are several natural mechanisms for protection from and repair of photochemical damage. For example, the enzyme DNA photolyase catalyses the destruction of cyclobutane thymine dimers. Also, ultraviolet radiation can induce the production of the pigment melanin (in a process more commonly known as tanning), which shields the skin from damage. However, repair and protective mechanisms become increasingly less effective with persistent and prolonged exposure to solar radiation.

Photodynamic therapy of tumour

In photodynamic therapy (PDT), laser radiation, which is usually delivered to the tumour via a fibre optic cable, is absorbed by a drug which, in its first excited triplet state ^3P, photosensitizes the formation of an excited singlet state of O_2, $^1O_2$. The $^1O_2$ molecules are very reactive and destroy cellular components and it is thought that cell membranes are the primary cellular targets. Hence, the following photochemical cycle leads to the shrinkage (and sometimes total destruction) of tumours:

Absorption:	$P + h\nu \longrightarrow P^*$
Intersystem crossing:	$P^* \longrightarrow {}^3P$
Photosensitization:	$^3P + {}^3O_2 \longrightarrow P + {}^1O_2$
Oxidation reactions:	$^1O_2 + \text{reactants} \longrightarrow \text{products}$

The photosensitizer is hence a 'photocatalyst' for the production of $^1O_2$. It is common practice to use a porphyrin photosensitizer, such as compounds derived from haematoporphyrin (4). However, much effort is being expended to develop better drugs with enhanced photochemical properties. A potential PDT drug must meet many criteria. From the point of view of pharmacological effectiveness, the drug must be soluble in tissue fluids, so it can be transported to the tumour through blood and secreted from the body through urine. The therapy should also result in very few side-effects. The drug must also have unique photophysical and photochemical properties. It must be activated photochemically at wavelengths that are not absorbed by blood and skin. In practice, this means that the drug should have a strong absorption band at $\lambda > 650$ nm. Drugs based on haematoporphyrin do not meet this criterion very well, so much effort is being expended in the design of novel porphyrin and related macrocycles with more desirable electronic properties. At the same time, the quantum yield of triplet formation and of $^1O_2$ formation must be high, so many drug molecules can be activated and many tumour oxidation reactions can occur during a short period of laser irradiation. Photodynamic therapy has been used successfully in the treatment of a number of cancers, including those of the lung, bladder, skin, oesophagus, and retina.

3

4

Discussion questions

1. The light-induced electron transfer reactions in photosynthesis occur because chlorophyll molecules (whether in monomeric or dimeric forms) are better reducing agents in their electronic excited states. Justify this observation with the help of molecular orbital theory.

2. The emission spectrum of a porphyrin dissolved in O_2-saturated water shows a strong band at 650 nm and a weak band at 1270 nm. In separate experiments, it was observed that the electronic absorption spectrum of the porphyrin sample showed bands at 420 nm and 550 nm, and the electronic absorption spectrum of O_2-saturated water showed no bands in the visible range of the spectrum (and therefore no emission spectrum when excited in the same range). Based on these data alone, make a preliminary assignment of the emission band at 1270 nm. Propose additional experiments that test your hypothesis. *Hint.* Review Section 17.3 and Problem 17.21.

Further reading

E. Kohen, R. Santus and J.G. Hirshberg, *Photobiology*. Academic Press, San Diego (1995).

J.B. Miller, Photodynamic therapy: the sensitization of cancer cells to light. *J. Chem. Educ.* **76**, 592 (1999).

and the fluorescence quantum yield in the presence of the quencher decreases to:

$$\phi = \frac{k_f}{k_f + k_{IC} + k_{ISC} + k_Q[Q]}$$

The ratio ϕ_f/ϕ is then given by:

$$\frac{\phi_f}{\phi} = 1 + \tau_0 k_Q[Q] \tag{26.34}$$

Therefore, a plot of the left-hand side of this expression against [Q] should be a straight line with slope $\tau_0 k_Q$. Such a plot is called a **Stern–Volmer plot** (Fig. 26.25). The method is quite general and may also be applied to the quenching of phosphorescence emission.

Because the fluorescence intensity and lifetime are both proportional to the fluorescence quantum yield, plots of I_0/I and τ_0/τ (where the subscript 0 indicates a measurement in the absence of quencher) versus [Q] should also be linear with the same slope and intercept as those shown for eqn 34.

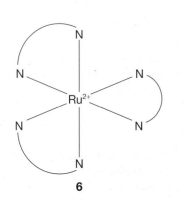

5

6

Example 26.5 Determining the quenching rate constant

The molecule 2,2′-bipyridine (**5**) forms a complex with the Ru^{2+} ion. Ruthenium(II) tris-(2,2′-bipyridyl), $Ru(bpy)_3^{2+}$ (**6**), has a strong metal-to-ligand charge transfer (MLCT) transition (Section 17.2) at 450 nm. The quenching of the $^*Ru(bpy)_3^{3+}$ excited state by $Fe(H_2O)_6^{3+}$ in acidic solution was monitored by measuring emission lifetimes at 600 nm. Determine the quenching rate constant for this reaction from the following data:

$[Fe(H_2O)_6^{3+}]/(10^{-4}\ mol\ L^{-1})$	0	1.6	4.7	7	9.4
$\tau/(10^{-7}\ s)$	6	4.05	3.37	2.96	2.17

Method Re-write the Stern–Volmer equation (eqn 34) for use with lifetime data; then fit the data to a straight-line, either graphically or by performing a linear-regression analysis with a computer.

Answer Upon substitution of τ_0/τ for ϕ_f/ϕ in eqn 34 and after rearrangement, we obtain:

$$\frac{1}{\tau} = \frac{1}{\tau_0} + k_Q[Q] \tag{26.35}$$

26.26 The Stern–Volmer plot of the data for Example 26.5.

Table 26.3 Values of R_0 for some donor–acceptor pairs*

Donor†	Acceptor	R_0/nm
Naphthalene	Dansyl	2.2
Dansyl	ODR	4.3
Pyrene	Coumarin	3.9
IAEDANS	FITC	4.9
Tryptophan	IAEDANS	2.2
Tryptophan	Haem	2.9

* Additional values may be found in: J.R. Lacowicz in *Principles of fluorescence spectroscopy*, Kluwer Academic/Plenum Publishers, New York (1999), p. 388.
† Abbreviations: Dansyl, 5-dimethylamino-1-naphthalenesulfonic acid; FITC, fluorescein 5-isothiocyanate; IAEDANS, 5-((((2-iodoacetyl)amino)ethyl)amino)naphthalene-1-sulfonic acid; ODR, octadecyl-rhodamine.

Figure 26.26 shows a plot of $1/\tau$ versus $[Fe^{3+}]$ and the results of a fit to eqn 35. The slope of the line is $k_Q = 2.8 \times 10^9$ L mol^{-1} s^{-1}.

Comment This example shows that measurements of emission lifetimes are preferred because they yield the value of k_Q directly. To determine the value of k_Q from intensity or quantum yield measurements, we need to make an independent measurement of τ_0.

Self-test 26.5 From the data above, predict the value of $[Fe^{3+}]$ required to decrease the intensity of $Ru(bpy)_3^{2+}$ emission to 50 per cent of the unquenched value. [6.0×10^{-4} mol L^{-1}]

Three common mechanisms for bimolecular quenching of an excited singlet (or triplet) state are:

Collisional deactivation:	$S^* + Q \longrightarrow S + Q$
Energy transfer:	$S^* + Q \longrightarrow S + Q^*$
Electron transfer:	$S^* + Q \longrightarrow S^+ + Q^-$ or $S^- + Q^+$

Collisional quenching is particularly efficient when Q is a heavy species, such as iodide ion, which receives energy from S* and then decays primarily by internal conversion to the ground state. This fact may be used to determine the accessibility of amino acid residues of a folded protein to solvent. For example, fluorescence from a tryptophan residue ($\lambda_{abs} \approx 290$ nm, $\lambda_{fluor} \approx 350$ nm) is quenched by iodide ion when the residue is on the surface of the protein and hence accessible to the solvent. Conversely, residues in the hydrophobic interior of the protein are not quenched effectively by I$^-$.

The quenching rate constant itself does not give much insight into the mechanism of quenching. For the system of Example 26.5, it is known that the quenching of the excited state of $Ru(bpy)_3^{2+}$ is a result of light-induced electron transfer to Fe^{3+}, but the quenching data do not allow us to prove the mechanism. However, there are some criteria that govern the relative efficiencies of energy and electron transfer.

(e) Energy transfer processes

According to the **Förster theory** of resonance energy transfer, which was proposed by T. Förster in 1959, energy transfer is efficient when:

1 The energy donor and acceptor are separated by a short distance, in the nanometre scale.

2 Photons emitted by the excited state of the donor can be absorbed directly by the acceptor.

We show in the following *Justification* that, for donor–acceptor systems that are held rigidly either by covalent bonds or by a protein 'scaffold', the efficiency of energy transfer, E_T, increases with decreasing distance, R, according to

$$E_T = \frac{R_0^6}{R_0^6 + R^6} \tag{26.36}$$

where R_0 is a parameter (with units of distance) that is characteristic of each donor–acceptor pair. Equation 36 has been verified experimentally and values of R_0 are available for a number of donor–acceptor pairs (Table 26.3).

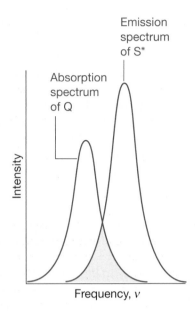

Emission spectrum of S*

Absorption spectrum of Q

Intensity

Frequency, v

26.27 According to the Förster theory, the rate of energy transfer from a molecule S* in an excited state to a quencher molecule Q is optimized at radiation frequencies in which the emission spectrum of S* overlaps with the absorption spectrum of Q, as shown in the shaded region.

The emission and absorption spectra of molecules span a range of wavelengths (Section 16.3), so the second requirement of the Förster theory is met when the emission spectrum of the donor molecule overlaps significantly with the absorption spectrum of the acceptor. In the overlap region, photons emitted by the donor have the proper energy to be absorbed by the acceptor (Fig. 26.27).

In many cases, it is possible to prove that energy transfer is the predominant mechanism of quenching if the excited state of the acceptor fluoresces or phosphoresces at a characteristic wavelength. In a pulsed laser experiment, the rise in fluorescence intensity from Q* with a characteristic time which is the same as that for the decay of the fluorescence of S* is often taken as indication of energy transfer from S to Q.

Justification 26.4

We visualize the energy transfer process $S^* + Q \rightarrow S + Q^*$ as follows. The oscillating electric field of electromagnetic radiation induces an oscillating electric dipole moment, μ_S, in S. Energy will be absorbed by S if the frequency of the incident radiation, v, is such that $\Delta E_S = hv$, where ΔE_S is the energy separation between the ground and excited electronic states of S and h is Planck's constant. This is the 'resonance condition' for absorption of radiation. The oscillating dipole on S now can affect electrons bound to a nearby Q molecule by inducing an oscillating dipole moment, μ_Q. If the frequency of oscillation of μ_S is such that $\Delta E_Q = hv$, then Q will absorb energy from S.

According to this picture, resonance energy transfer arises from the interaction between two oscillating dipoles with moments μ_S and μ_Q. From Section 21.5, the energy of the dipole–dipole interaction, $V_{\text{dipole–dipole}}$, is

$$V_{\text{dipole–dipole}} \propto \frac{\mu_S \mu_Q}{R^3}$$

where R is the distance between the dipoles. We saw in Section 12.10 that the rate of a transition from a state i to a state f at a radiation frequency v is proportional to the square modulus of the matrix element of the perturbation between the two states:

$$w_{f \leftarrow i} \propto |H_{fi}^{(1)}|^2$$

For energy transfer, the wavefunctions of the initial and final states may be denoted as $\psi_S^* \psi_Q$ and $\psi_S \psi_Q^*$, respectively, and $H^{(1)}$ may be written from $V_{\text{dipole–dipole}}$. It follows that the rate of energy transfer, w_T, at a fixed distance R is given by

$$w_T \propto \frac{1}{R^6}|\langle \psi_S \psi_{Q^*}|\mu_S \mu_Q|\psi_{S^*} \psi_Q \rangle|^2 = \frac{1}{R^6}|\langle \psi_S|\mu_S|\psi_{S^*}\rangle|^2 |\langle \psi_{Q^*}|\mu_Q|\psi_Q\rangle|^2$$

where we have used the fact that the terms related to S are functions of coordinates that are independent of those for the functions related to Q. In the last expression, the integrals are squares of transition dipole moments at the radiation frequency v, the first corresponding to emission of S* to S and the second to absorption of Q to Q*.

We interpret the expression for w_T as follows. The rate of energy transfer is proportional to R^{-6}, so it decreases sharply with increasing separation between the energy donor and acceptor. Furthermore, the energy transfer rate is optimized when both emission of radiation by S* and absorption of radiation by Q are efficient at the frequency v. Because the absorption and emission spectra of large molecules in condensed phases are broad, it follows that the energy transfer rate is optimal at radiation frequencies in which the emission spectrum of the donor and the absorption spectrum of the acceptor overlap significantly.

In practice, it is more convenient to measure the efficiency of energy transfer and not the rate itself. In much the same way that we defined the quantum yield as a ratio of rates, we can also define the efficiency of energy transfer, E_T, as the ratio

$$E_T = \frac{w_T}{w_T + w_0} \qquad w_0 = (k_f + k_{IC} + k_{ISC})[S^*] \qquad (26.37)$$

where w_0 is the rate of deactivation of S^* in the absence of the quencher. It is useful to express w_0 in terms of the parameter R_0, the characteristic distance at which $w_T = w_0$ for a specified pair of S and Q (Table 26.3). Using $w_T \propto R^{-6}$ and $w_0 \propto R_0^{-6}$, the expression for E_T rearranges into eqn 36.

The dependence of E_T on R forms the basis of **fluorescence resonance energy transfer** (FRET), a technique that can be used to measure distances in biological systems. In a typical FRET experiment, a site on a biopolymer or membrane is covalently labelled with an energy donor and another site is labelled covalently with an energy acceptor. In certain cases, the donor or acceptor may be natural constituents of the system, such as amino acid groups, co-factors, or enzyme substrates (see Illustration 26.3).

The efficiency, E_T, may be expressed in terms of the experimental fluorescence quantum yields ϕ_f and ϕ of the donor in the absence and presence of the acceptor, respectively. To show this, we use eqn 32 to write:

$$\phi_f = \frac{v_f}{w_0} \qquad \text{and} \qquad \phi = \frac{v_f}{w_0 + w_T}$$

where v_f is the rate of fluorescence. Using eqn 37 and after a little algebra, we obtain

$$E_T = 1 - \frac{\phi}{\phi_f} \qquad (26.38)$$

The distance between the labels is then calculated from the known value of R_0 and eqn 36. Several tests have shown that the FRET technique is useful for measuring distances ranging from 1 to 9 nm.

Illustration 26.3

The energy donor 1.5-I AEDANS (**7**) has a fluorescence quantum yield of 0.75 in aqueous solution. The visual pigment 11-*cis*-retinal (Box 17.1) is a quencher of fluorescence and $R_0 = 5.4$ nm for the 1.5-I AEDANS/11-*cis*-retinal pair. When an amino acid on the surface of the protein rhodopsin, which binds 11-*cis*-retinal in its interior, was labelled covalently with 1.5-AEDANS, the fluorescence quantum yield of the label decreased to 0.68. From eqn 38, we calculate $E_T = 1 - (0.68/0.75) = 0.093$ and from eqn 36 we calculate $R = 7.9$ nm, which is taken as the distance between the surface of the protein and 11-*cis*-retinal.

7

If donor and acceptor diffuse in solution or in the gas phase, Förster theory predicts that the efficiency of quenching by energy transfer increases as the average distance travelled between collisions of donor and acceptor decreases. That is, the quenching efficiency increases with concentration of quencher, as predicted by the Stern–Volmer equation.

(f) Electron transfer reactions

According to the **Marcus theory** of electron transfer, which was proposed by R.A. Marcus in 1965 and is discussed fully in Section 29.1, the rates of electron transfer (from ground or excited states) depend on:

1 The distance between the donor and acceptor, with electron transfer becoming more efficient as the distance between donor and acceptor decrease.

2 The reaction Gibbs energy, $\Delta_r G$, with electron transfer becoming more efficient as the reaction becomes more exergonic. For example, efficient photooxidation of S requires that the reduction potential of S* be lower than the reduction potential of Q.

3 The reorganization energy, the energy cost incurred by molecular rearrangements of donor, acceptor, and medium during electron transfer. The electron transfer rate is predicted to increase as this reorganization energy is matched closely by the reaction Gibbs energy.

Electron transfer can also be studied by time-resolved spectroscopy (Section 17.7e). The oxidized and reduced products often have electronic absorption spectra distinct from those of their neutral parent compounds. Therefore, the rapid appearance of such known features in the absorption spectrum after excitation by a laser pulse may be taken as indication of quenching by electron transfer.

26.12 Complex photochemical processes

Many photochemical processes have complex mechanisms that may be examined by considering the concepts developed above. In this section, we consider two examples: photochemical chain reactions and photosensitization.

(a) The overall quantum yield of a photochemical reaction

For complex reactions involving secondary processes, such as chain reactions initiated by photolysis, many reactant molecules might be consumed as a result of absorption of a single photon. The **overall quantum yield**, Φ, for such reactions, the number of reactant molecules consumed per photon absorbed, might exceed 1. In the photolysis of HI, for example, the processes are

$$HI + h\nu \longrightarrow H\cdot + I\cdot$$
$$H\cdot + HI \longrightarrow H_2 + I\cdot$$
$$I\cdot + I\cdot + M \longrightarrow I_2 + M^*$$

The overall quantum yield is 2 because the absorption of one photon leads to the destruction of two HI molecules. In a chain reaction, Φ may be very large, and values of about 10^4 are common. In such cases the chain acts as a chemical amplifier of the initial absorption step.

Example 26.6 Determining the quantum yield of a photochemical reaction

When a sample of 4-heptanone was irradiated for 100 s with 313 nm radiation with a power output of 50 W under conditions of total absorption, it was found that 2.8 mmol C_2H_4 was formed. What is the quantum yield for the formation of ethene?

Method First, calculate the amount of photons generated in an interval Δt: see Example 11.1. Then divide the amount of ethene molecules formed by the amount of photons absorbed.

Answer From Example 11.1, the amount (in moles) of photons absorbed is

$$n = \frac{P\Delta t}{(hc/\lambda)N_A}$$

If $n_{C_2H_4}$ is the amount of ethene formed, the quantum yield is

$$\Phi = \frac{n_{C_2H_4}}{n} = \frac{n_{C_2H_4}N_A hc}{\lambda P \Delta t}$$

$$= \frac{(2.8 \times 10^{-3}\ \text{mol}) \times (6.022 \times 10^{23}\ \text{mol}^{-1}) \times (6.626 \times 10^{-34}\ \text{J s}) \times (2.997 \times 10^8\ \text{m s}^{-1})}{(3.13 \times 10^{-7}\ \text{m}) \times (50\ \text{J s}^{-1}) \times (100\ \text{s})}$$

$$= 0.21$$

Self-test 26.6 The overall quantum yield for another reaction at 290 nm is 0.30. For what length of time must irradiation with a 100 W source continue in order to destroy 1.0 mol of molecules? [3.8 h]

(b) Rate laws of complex photochemical reactions

As an example of how to write a rate law for a complex photochemical process, consider the photochemical activation of the reaction

$$H_2(g) + Br_2(g) \longrightarrow 2\ HBr(g)$$

In place of the first step in the thermal reaction we have

$$Br_2 + h\nu \longrightarrow Br\cdot + Br\cdot \qquad v = I_{abs}$$

where I_{abs} is the number of photons of the appropriate frequency absorbed divided by the volume in which absorption occurs and the time interval. We are assuming a primary quantum yield of unity for the photodissociation of Br_2. It follows that I_{abs} should take the place of $k_i[Br_2][M]$ in the thermal reaction scheme, so from Example 26.1 we can write

$$\frac{d[HBr]}{dt} = \frac{2k_p(1/k_t[M])^{1/2}[H_2][Br_2]I_{abs}^{1/2}}{[Br_2] + (k_r/k_p')[HBr]} \tag{26.39}$$

We predict that the reaction rate should depend on the square root of the absorbed light intensity, which is confirmed experimentally.

(c) Photosensitization

The reactions of a molecule that does not absorb directly can be made to occur if another absorbing molecule is present, because the latter may be able to transfer its energy to the former during a collision. An example of this **photosensitization** is the reaction used to generate excited state O_2 in a type of tumour treatment known as *photodynamic therapy*. Another example is the reaction used to generate atomic hydrogen, the irradiation of hydrogen gas containing a trace of mercury vapour using radiation of wavelength 254 nm from a mercury discharge lamp. The Hg atoms are excited (to Hg*) by resonant absorption of the radiation, and then collide with H_2 molecules. Two reactions then take place:

$$Hg^* + H_2 \longrightarrow Hg + H\cdot + H\cdot$$
$$Hg^* + H_2 \longrightarrow HgH + H\cdot$$

The latter reaction is the initiation step for other mercury photosensitized reactions, such as the synthesis of formaldehyde from carbon monoxide and hydrogen:

$$H\cdot + CO \longrightarrow HCO\cdot$$

$$HCO\cdot + H_2 \longrightarrow HCHO + H\cdot$$

$$HCO\cdot + HCO\cdot \longrightarrow HCHO + CO$$

Note that the last step is termination by disproportionation rather than by combination.

Checklist of key ideas

Chain reactions

- ☐ chain reaction, a reaction in which an intermediate produced in one step generates an intermediate in a subsequent step, and so on
- ☐ chain carrier, the intermediates in a chain reaction
- ☐ radical chain reaction, a chain reaction in which the chain carriers are radicals

26.1 The rate laws of chain reactions

- ☐ pyrolysis, thermal decomposition in the absence of air
- ☐ Rice–Herzfeld mechanism, a chain reaction for the thermal decomposition of CH_3CHO
- ☐ initiation step, a step in which chain carriers are produced
- ☐ propagation step, a step in which a chain carrier reacts to produce another chain carrier
- ☐ termination step, a reaction in which radicals combine and end a chain or reactions
- ☐ hydrogen–bromine reaction, $d[HBr]/dt = k[H_2][Br_2]^{3/2}/([Br_2] + k'[HBr])$
- ☐ thermolysis, a reaction initiated by heat or by vigorous intermolecular collisions
- ☐ retardation step, a step in a chain reaction that reduces the net rate of formation of product

26.2 Explosions

- ☐ thermal explosion, a very rapid reaction arising from a rapid increase of reaction rate with increasing temperature
- ☐ chain-branching explosion, occurs when the number of chain centres grows exponentially
- ☐ branching step, an elementary reaction that produces more than one chain carrier

- ☐ explosion region, the conditions (of pressure and temperature) under which explosion occurs
- ☐ first explosion limit, the minimum pressure at a given temperature at which explosion occurs
- ☐ second explosion limit, the maximum pressure at a given temperature at which explosion occurs
- ☐ third explosion limit, the minimum pressure at a given temperature at which a thermal explosion occurs

Polymerization kinetics

- ☐ stepwise polymerization, a polymerization reaction in which any two monomers present in the reaction mixture can link together at any time and the growth of the polymer is not confined to chains that are already forming
- ☐ chain polymerization, a polymerization reaction in which an activated monomer attacks another monomer, links to it, then that unit attacks another monomer, and so on

26.3 Stepwise polymerization

- ☐ degree of polymerization, the average number of monomer residues per polymer molecule, $\langle n \rangle = 1/(1 - p)$, where p is the average number of monomers per polymer molecule; $\langle n \rangle = 1 + kt[A]_0$

26.4 Chain polymerization

- ☐ rate of chain polymerization, $v = k[I]^{1/2}[M]$
- ☐ mutual termination, an elementary reaction in which two growing radical chains combine
- ☐ disproportionation (in polymerization), an elementary reaction in which a hydrogen atom transfers from one chain to another

- ☐ chain transfer, an elementary reaction in which a new chain initiates at the expense of the one currently growing
- ☐ kinetic chain length, v, the ratio of the number of monomer units consumed per activated centre produced in the initiation step; $v = k[M][I]^{-1/2}$

Homogeneous catalysis

- ☐ catalyst, a substance that accelerates a reaction but undergoes no net chemical change
- ☐ enzyme, a biological catalyst
- ☐ homogeneous catalyst, a catalyst in the same phase as the reaction mixture
- ☐ heterogeneous catalyst, a catalyst in a different phase from the reaction mixture

26.5 Features of homogeneous catalysis

- ☐ acid catalysis, catalysis by the transfer of a proton from an acid to the substrate
- ☐ base catalysis, catalysis by the transfer of a proton from the substrate to a base

26.6 Enzymes

- ☐ active site, the region of an enzyme molecule at which reaction takes place
- ☐ substrate, the species that reacts in the presence of an enzyme
- ☐ maximum velocity, the greatest reaction rate for a given concentration of substrate; $v_{max} = k_b[E]_0$
- ☐ Michaelis–Menten mechanism, a mechanism for enzyme-catalysed reactions, $E + S \rightleftharpoons ES$, $ES \rightarrow P$
- ☐ Michaelis–Menten equation, $v = v_{max}/(1 + K_M/[S]_0)$
- ☐ Michaelis constant, $K_M = (k'_a + k_b)/k_a$
- ☐ Lineweaver–Burk plot, $1/v = 1/v_{max} + (K_M/v_{max})(1/[S]_0)$

☐ turnover number (catalytic constant), the number of catalytic cycles (turnovers) performed by the active site in a given interval divided by the duration of the interval; $k_{cat} = v_{max}/[E]_0$

☐ catalytic efficiency, $\varepsilon = k_{cat}/K_M$

☐ enzyme inhibition, $v = v_{max}/(\alpha' + \alpha K_M/[S]_0)$, $\alpha = 1 + [I]/K_I$, $\alpha' = 1 + [I]/K_I'$

☐ competitive inhibition, inhibition in which the inhibitor binds only to the active site of the enzyme; $\alpha > 1$ and $\alpha' = 1$

☐ uncompetitive inhibition, inhibition in which the inhibitor binds to a site of the enzyme that is removed from the active site, but only if the substrate is already present; $\alpha = 1$ and $\alpha' > 1$

☐ non-competitive inhibition, inhibition in which the inhibitor binds to a site other than the active site; $\alpha > 1$ and $\alpha' > 1$

Oscillating reactions

26.7 Autocatalysis

☐ autocatalysis, the catalysis of a reaction by the products

☐ Belousov–Zhabotinskii reaction, an oscillating reaction involving potassium bromate, malonic acid, and a cerium(IV) salt in acidic solution

26.8 Autocatalytic mechanisms of oscillating reactions

☐ Lotka–Volterra mechanism, a model of a reaction showing oscillating behaviour

☐ steady-state condition, a non-equilibrium state in which the time behaviour of a reaction mixture is maintained (with a composition that may be oscillating steadily)

☐ brusselator, a model of a reaction showing oscillating behaviour

☐ limit cycle (attractor), a common trajectory to which a system migrates

☐ oregonator, a model of a reaction showing oscillating behaviour; a mechanism for the BZ reaction

26.9 Bistability

☐ bistability, the ability to exist in two steady states

26.10 Chemical chaos

☐ deterministic chaos, temporal and spatial solutions that are predictable but infinitely sensitive to the initial conditions

☐ period-doubling, a process in which the system must circulate twice round the cycle before an initial pair of concentrations is restored

☐ strange attractor, a path that never retraces its path or crosses itself

Photochemistry

26.11 Kinetics of photophysical and photochemical processes

☐ primary process, a process in which products are formed directly from the excited state of a reactant

☐ secondary process, a process in which products originate from intermediates formed directly from the excited state of a reactant

☐ primary quantum yield, ϕ, the number of photophysical or photochemical events that lead to primary products divided by the number of photons absorbed by the molecule in the same interval, $\phi = v/I_{abs}$

☐ observed fluorescence lifetime, $\tau_0 = \phi_f/k_f$

☐ quenching, shortening of the lifetime of an excited state

☐ Stern–Volmer plot, $\phi_f/\phi = 1 + \tau_0 k_Q[Q]$

☐ Förster theory, a theory of resonance energy transfer, efficiency $E_T = R_0^6/(R_0^6 + R^6)$

☐ fluorescence resonance energy transfer (FRET), a technique used to measure distances in biological systems

☐ Marcus theory, a theory of electron transfer

26.12 Complex photochemical processes

☐ overall quantum yield, the number of reactant molecules consumed per photon absorbed

☐ photosensitization, a process in which a photon is absorbed by a second species and then the excitation energy is transferred to the molecule of interest

Further reading

Articles of general interest

J. Andraos, How mathematics figures in chemistry: some examples. *J. Chem. Educ.* **76**, 258 (1999).

S. Bluestone and K.Y. Yan, A method to find the rate constants in chemical kinetics of a complex reaction. *J. Chem. Educ.* **72**, 884 (1995).

J.G. Eberhardt and E. Levin, A simplified integration technique for reaction rate laws of integral order in several substances. *J. Chem. Educ.* **72**, 193 (1995).

I.R. Epstein, The role of flow systems in far-from-equilibrium dynamics. *J. Chem. Educ.* **66**, 191 (1989).

R.J. Field, The language of dynamics. *J. Chem. Educ.* **66**, 188 (1989).

R.J. Field and F.W. Schneider, Oscillating chemical reactions and nonlinear dynamics. *J. Chem. Educ.* **66**, 195 (1989).

G.I. Gellene, Application of kinetic approximations to the A + B ⇌ C reaction system. *J. Chem. Educ.* **72**, 196 (1995).

J.M. Goodman, How do approximations affect the solutions to kinetic equations? *J. Chem. Educ.* **76**, 275 (1999).

E. Hughes, Jr., Solving differential equations in kinetics by using power series. *J. Chem. Educ.* **66**, 46 (1989).

W. Jahnke and A.T. Winfree, Recipes for Belousov–Zhabotinskii reagents. *J. Chem. Educ.* **68**, 320 (1991).

K.J. Laidler, Rate-controlling step: a necessary or useful concept? *J. Chem. Educ.* **65**, 250 (1988).

S.R. Logan, Does a photochemical reaction have a reaction order? *J. Chem. Educ.* **74**, 1303 (1997).

F. Mata-Perez and J.F. Perez-Benito, The kinetic rate law for autocatalytic reactions. *J. Chem. Educ.* **64**, 925 (1987).

D.B. Northrop, On the meaning of K_m and V/K in enzyme kinetics. *J. Chem. Educ.* **75**, 1153 (1998).

R.M. Noyes, Some models of chemical oscillators. *J. Chem. Educ.* **66**, 190 (1989).

R.S. Ochs, Understanding enzyme inhibition. *J. Chem. Educ.* **77**, 1453 (2000).

E. Peacock–López, Introduction to chemical oscillation using a modified Lotka model. *Chem. Educator* **5**, 216 (2000).

R.T. Raines and D.E. Hansen, An intuitive approach to steady-state kinetics. *J. Chem. Educ.* **65**, 757 (1988).

H.G. Schuster, Chaotic phenomena. In *Encyclopedia of applied physics* (ed. G.L. Trigg), **3**, 189. VCH, New York (1992).

A.F. Shaaban, The integrated rate equation of the nitric oxide–oxygen reaction. *J. Chem. Educ.* **67**, 869 (1990).

L.J. Solzberg, Self-organization in chemistry. *J. Chem. Educ.* **66**, 187 (1989).

J.N. Spencer, Competitive and coupled reactions. *J. Chem. Educ.* **69**, 281 (1992).

G.F. Swiegers, Applying the principles of chemical kinetics to population growth problems. *J. Chem. Educ.* **70**, 364 (1993).

X. Tan, S. Lindenbaum, and N. Meltzer, A unified equation for chemical kinetics. *J. Chem. Educ.* **71**, 566 (1994).

S. Toby and F.S. Toby, The simulation of dynamic systems. *J. Chem. Educ.* **76**, 1584 (1999).

Texts and sources of data and information

M. Boudart, *Kinetics of chemical processes*. Butterworth, London (1991).

C.H. Bamford, C.F. Tipper, and R.G. Compton (ed.), *Comprehensive chemical kinetics*, Vols 1–38. Elsevier, Amsterdam (1969–2001).

M.P. Cady and C.A. Trapp, *A Mathcad primer for physical chemistry*. Oxford University Press (1999).

I.M. Campbell, *Introduction to synthetic polymers*. Oxford University Press (1997).

C.E. Carraher, Jr., *Seymour/Carraher's polymer chemistry*. Marcel Dekker (2000).

I.R. Epstein and J.A. Pojman, *An introduction to nonlinear chemical dynamics*. Oxford University Press (1998).

A. Fersht, *Structure and mechanism in protein science: a guide to enzyme catalysis and protein folding*. W.H. Freeman, New York (1999).

W.M. Horspool and P.-S. Song (ed.), *CRC handbook of organic photochemistry and photobiology*. CRC Press, Boca Raton (1995).

D.R. Lide (ed.), *CRC handbook of chemistry and physics*, Section 5. CRC Press, Boca Raton (2000).

J.C. Lindon, G.E. Tranter, and J.L. Holmes (ed.), *Encyclopedia of spectroscopy and spectrometry*. Academic Press, San Diego (2000).

S.R. Logan, *Chemical kinetics*. Longman, Harlow (1996).

S.L. Murov, I. Carmichael, and G.L. Hug, *Handbook of photochemistry*. Marcel Dekker, New York (1993).

NDRL/NIST solution kinetics database, NIST standard reference database 40, National Institute of Standards and Technology,

Gaithersburg (1994). The URL is available on the web site for this book.

NIST chemical kinetics database, NIST standard reference database 17, National Institute of Standards and Technology, Gaithersburg (1998). The URL is available on the web site for this book.

M.J. Pilling and P.W. Seakins, *Reaction kinetics*. Oxford University Press (1996).

S.K. Scott, *Oscillations, waves, and chaos in chemical kinetics*. Oxford Chemistry Primers, Oxford University Press (1994).

J.I. Steinfeld, J.S. Francisco, and W.L. Hase, *Chemical kinetics and dynamics*. Prentice Hall, Englewood Cliffs (1998).

N.J. Turro, *Modern molecular photochemistry*. University Science Books, Sausalito (1991).

G.R. Van Hecke and K.K. Karukstis, *A guide to lasers in chemistry*. Jones and Bartlett, Boston (1998).

K.E. van Holde, W.C. Johnson and P.H. Ho, *Principles of physical biochemistry*. Prentice Hall, Upper Saddle River (1998).

D. Voet and J.G. Voet, *Biochemistry*. Wiley, New York (1995).

R.P. Wayne, *Principles and applications of photochemistry*. Oxford University Press (1988).

C.E. Wayne and R.P. Wayne, *Photochemistry*. Oxford Chemistry Primers, Oxford University Press (1996).

Exercises

Discussion questions

26.1 (a) Identify any initiation, propagation, retardation, inhibition, and termination steps in the following chain mechanisms:

(a) (1) $AH \longrightarrow A\cdot + H\cdot$

    (2) $A\cdot \longrightarrow B\cdot + C$

    (3) $AH + B\cdot \longrightarrow A\cdot + D$

    (4) $A\cdot + B\cdot \longrightarrow P$

(b) (1) $A_2 \longrightarrow A\cdot + A\cdot$

    (2) $A\cdot \longrightarrow B\cdot + C$

    (3) $A\cdot + P \longrightarrow B\cdot$

    (4) $A\cdot + B\cdot \longrightarrow P$

26.1 (b) Bearing in mind distinctions between the mechanisms of stepwise and chain polymerization, describe ways in which it is possible to control the molar mass of a polymer by manipulating the kinetic parameters of polymerization.

26.2 (a) Discuss the features, advantages, and limitations of the Michaelis–Menten mechanism of enzyme action.

26.2 (b) Distinguish between competitive, non-competitive, and uncompetitive inhibition of enzymes. Discuss how these modes of inhibition may be detected experimentally.

26.3 (a) How can autocatalysis lead to chemical oscillations?

26.3 (b) Distinguish between a steady-state approximation and a steady-state condition. Discuss the role of bistability.

26.4 (a) Distinguish between the primary quantum yield and overall quantum yield of a chemical reaction. Describe an experimental procedure for the determination of the quantum yield.

26.4 (b) Discuss experimental procedures that make it possible to differentiate between quenching by energy transfer, collisions, or electron transfer.

Numerical exercises

In the following exercises and problems, it is recommended that rate constants are labelled with the number of the step in the proposed reaction mechanism, and any reverse steps are labelled similarly but with a prime.

26.5 (a) Derive the rate law for the decomposition of ozone in the reaction $2 O_3(g) \rightarrow 3 O_2(g)$ on the basis of the following proposed mechanism:

(1) $O_3 \rightleftharpoons O_2 + O$ k_1, k_1'
(2) $O + O_3 \longrightarrow O_2 + O_2$ k_2

26.5 (b) On the basis of the following proposed mechanism, account for the experimental fact that the rate law for the decomposition $2 N_2O_5(g) \rightarrow 4 NO_2(g) + O_2(g)$ is $v = k[N_2O_5]$.

(1) $N_2O_5 \rightleftharpoons NO_2 + NO_3$ k_1, k_1'
(2) $NO_2 + NO_3 \longrightarrow NO_2 + O_2 + NO$ k_2
(3) $NO + N_2O_5 \longrightarrow NO_2 + NO_2 + NO_2$ k_3

26.6 (a) A slightly different mechanism for the decomposition of N_2O_5 from that in Exercise 26.5b has also been proposed. It differs only in the last step, which is replaced by

(3) $NO + NO_3 \longrightarrow NO_2 + NO_2$ k_3

Show that this mechanism leads to the same overall rate law.

26.6 (b) Consider the following mechanism for the thermal decomposition of R_2:

(1) $R_2 \longrightarrow R + R$
(2) $R + R_2 \longrightarrow P_B + R'$
(3) $R' \longrightarrow P_A + R$
(4) $R + R \longrightarrow P_A + P_B$

where R_2, P_A, P_B are stable hydrocarbons and R and R' are radicals. Find the dependence of the rate of decomposition of R_2 on the concentration of R_2.

26.7 (a) Refer to Fig. 26.3 and determine the pressure range for a chain-branching explosion in the hydrogen–oxygen reaction at 800 K.

26.7 (b) Refer to Fig. 26.3 and determine the pressure range for a chain-branching explosion in the hydrogen–oxygen reaction at (a) 700 K, (b) 900 K.

26.8 (a) The condensation reaction of propanone, $(CH_3)_2CO$, in aqueous solution is catalysed by bases, B, which react reversibly with propanone to form the carbanion $C_3H_5O^-$. The carbanion then reacts with a molecule of propanone to give the product. A simplified version of the mechanism is

(1) $AH + B \longrightarrow BH^+ + A^-$
(2) $A^- + BH^+ \longrightarrow AH + B$
(3) $A^- + AH \longrightarrow$ product

where AH stands for propanone and A^- denotes its carbanion. Use the steady-state approximation to find the concentration of the

carbanion and derive the rate equation for the formation of the product.

26.8 (b) Consider the acid-catalysed reaction

(1) $HA + H^+ \rightleftharpoons HAH^+$ k_1, k_1', both fast
(2) $HAH^+ + B \longrightarrow BH^+ + AH$ k_2, slow

Deduce the rate law and show that it can be made independent of the specific term $[H^+]$.

26.9 (a) Consider the following chain mechanism:

(1) $AH \longrightarrow A\cdot + H\cdot$
(2) $A\cdot \longrightarrow B\cdot + C$
(3) $AH + B\cdot \longrightarrow A\cdot + D$
(4) $A\cdot + B\cdot \longrightarrow P$

Use the steady-state approximation to deduce that the decomposition of AH is first-order in AH.

26.9 (b) Consider the following chain mechanism:

(1) $A_2 \longrightarrow A\cdot + A\cdot$
(2) $A\cdot \longrightarrow B\cdot + C$
(3) $A\cdot + P \longrightarrow B\cdot$
(4) $A\cdot + B\cdot \longrightarrow P$

Use the steady-state approximation to deduce that the rate law for the consumption of A_2.

26.10 (a) The enzyme-catalysed conversion of a substrate at 25°C has a Michaelis constant of 0.035 mol L^{-1}. The rate of the reaction is 1.15×10^{-3} mol L^{-1} s^{-1} when the substrate concentration is 0.110 mol L^{-1}. What is the maximum velocity of this enzymolysis?

26.10 (b) The enzyme-catalysed conversion of a substrate at 25°C has a Michaelis constant of 0.042 mol L^{-1}. The rate of the reaction is 2.45×10^{-4} mol L^{-1} s^{-1} when the substrate concentration is 0.890 mol L^{-1}. What is the maximum velocity of this enzymolysis?

26.11 (a) In a photochemical reaction $A \rightarrow 2 B + C$, the quantum efficiency with 500 nm light is 2.1×10^2 mol einstein^{-1}. After exposure of 300 mmol of A to the light, 2.28 mmol of B is formed. How many photons were absorbed by A?

26.11 (b) In a photochemical reaction $A \rightarrow B + C$, the quantum efficiency with 500 nm light is 1.2×10^2 mol einstein^{-1}. After exposure of 200 mmol A to the light, 1.77 mmol B is formed. How many photons were absorbed by A?

26.12 (a) In an experiment to measure the quantum efficiency of a photochemical reaction, the absorbing substance was exposed to 490 nm light from a 100 W source for 45 min. The intensity of the transmitted light was 40 per cent of the intensity of the incident light. As a result of irradiation, 0.344 mol of the absorbing substance decomposed. Determine the quantum efficiency.

26.12 (b) In an experiment to measure the quantum efficiency of a photochemical reaction, the absorbing substance was exposed to 320 nm radiation from a 87.5 W source for 28.0 min. The intensity of the transmitted light was 0.257 that of the incident light. As a result of irradiation, 0.324 mol of the absorbing substance decomposed. Determine the quantum efficiency.

Problems[†]

Numerical problems

26.1 Studies of combustion reactions depend on knowing the concentrations of H atoms and HO radicals. Measurements on a flow system using ESR for the detection of radicals gave information on the reactions

$$(1) \ H + NO_2 \longrightarrow OH + NO \qquad k_1 = 2.9 \times 10^{10} \ \text{L mol}^{-1} \ \text{s}^{-1}$$
$$(2) \ OH + OH \longrightarrow H_2O + O \qquad k_2 = 1.55 \times 10^9 \ \text{L mol}^{-1} \ \text{s}^{-1}$$
$$(3) \ O + OH \longrightarrow O_2 + H \qquad k_3 = 1.1 \times 10^{10} \ \text{L mol}^{-1} \ \text{s}^{-1}$$

(J.N. Bradley, W. Hack, K. Hoyermann, and H.G. Wagner, *J. Chem. Soc. Faraday Trans.* I, 1889 (1973)). Using initial H atom and NO_2 concentrations of 4.5×10^{-10} mol cm^{-3} and 5.6×10^{-10} mol cm^{-3}, respectively, compute and plot curves showing the O, O_2, and OH concentrations as a function of time in the range 0–10 ns.

26.2 In a flow study of the reaction between O atoms and Cl_2 (J.N. Bradley, D.A. Whytock, and T.A. Zaleski, *J. Chem. Soc. Faraday Trans.* I, 1251 (1973)) at high chlorine pressures, plots of $\ln[O]_0/[O]$ against distances l along the flow tube, where $[O]_0$ is the oxygen concentration at zero chlorine pressure, gave straight lines. Given the flow velocity as 6.66 m s^{-1} and the data below, find the rate coefficient for the reaction $O + Cl_2 \rightarrow ClO + Cl$.

l/cm	0	2	4	6	8
$\ln([O]_0/[O])$	0.27	0.31	0.34	0.38	0.45
l/cm	10	12	14	16	18
$\ln([O]_0/[O])$	0.46	0.50	0.55	0.56	0.60

with $[O]_0 = 3.3 \times 10^{-8}$ mol L^{-1}, $[Cl_2] = 2.54 \times 10^{-7}$ mol L^{-1}, $p = 1.70$ Torr.

26.3‡ J.D. Chapple-Sokol, C.J. Giunta, and R.G. Gordon (*J. Electrochem. Soc.* **136**, 2993 (1989)) proposed the following radical chain mechanism for the initial stages of the gas-phase oxidation of silane by nitrous oxide:

$$(1) \ N_2O \longrightarrow N_2 + O$$
$$(2) \ O + SiH_4 \longrightarrow SiH_3 + OH$$
$$(3) \ OH + SiH_4 \longrightarrow SiH_3 + H_2O$$
$$(4) \ SiH_3 + N_2O \longrightarrow SiH_3O + N_2$$
$$(5) \ SiH_3O + SiH_4 \longrightarrow SiH_3OH + SiH_3$$
$$(6) \ SiH_3 + SiH_3O \longrightarrow (H_3Si)_2O$$

Label each step with its role in the chain. Use the steady-state approximation to show that this mechanism predicts the following rate law for SiH_4 consumption (provided k_1 and k_6 are in some sense small):

$$\frac{d[SiH_4]}{dt} = -k[N_2O][SiH_4]^{1/2}$$

26.4‡ Arylmercury compounds have played an ever-increasing role in organic synthesis in recent years because of their ability to accommodate almost all important organic functional groups and their chemical and thermal stability. Y. Wang (*Int. J. Chem. Kinet.* **25**, 91 (1993)) has studied the kinetics and mechanism of the dimerization of arylmercurials (A), with formulas XC_6H_4HgCl (X = H, p-CH$_3$, m-CH$_3$, o-CH$_3$, p-Cl, and others) catalysed by [ClRh(CO)$_2$]$_2$ (C) in hexamethylphosphoramide (HMPA). The observed dimerization in

each case is represented by the equation 2 ArHgCl $\rightarrow$ ArAr + HgCl$_2$ + Hg, and plots of $x/([A]_0 - x)$, where x is the molar concentration of the arylmercury compound reacted and $[A]_0$ the initial concentration, were found to be linear functions of time. In each case, $[A]_0 = 0.100$ mol L^{-1} and $[C] = 5.00 \times 10^{-4}$ mol L^{-1}. Concentration data as a function of time at 60°C for the compounds with X = H and Cl are

X = H $x/([A]_0 - x)$	0.14	0.48	0.63	0.84	1.12
t/min	25	85	115	150	205
X = p-Cl $x/([A]_0 - x)$	0.04	0.27	0.60	1.05	
t/min	30	123	236	405	

Similar results were obtained for the other compounds in the series. (a) Plot these data and confirm that the reactions are second-order. Would the first-order integrated rate law fit the data too? (b) The proposed mechanism for these reactions is

$$(1) \ [ClRh(CO)_2]_2 + ArHgCl \Longleftrightarrow [ArRh(CO)_2]_2 + HgCl_2 \quad k_1, k_1', \text{fast}$$
$$(2) \ [ArRh(CO)_2]_2 + ArHgCl \Longleftrightarrow Ar_2Rh(CO)_2HgCl \quad k_2, k_2', \text{slow}$$
$$(3) \ Ar_2Rh(CO)_2HgCl \longrightarrow [ClRh(CO)_2]_2 + ArAr + Hg \quad k_3, \text{fast}$$

Derive the rate law for the rate of disappearance of ArHgCl and determine the conditions under which this mechanism supports the experimental results.

26.5‡ The water formation reaction has been studied many times and continues to be of interest. Despite the many studies there is not uniform agreement on the mechanism. But, as explosions are known to occur at certain critical values of the pressure, any proposed mechanism that is to be considered plausible must be consistent with the existence of these critical explosion limits. One such plausible mechanism is that of Example 26.2. Another is the following

$$(1) \ H_2 \longrightarrow H + H$$
$$(2) \ H + O_2 \longrightarrow OH + O$$
$$(3) \ O + H_2 \longrightarrow OH + H$$
$$(4) \ H + O_2 \longrightarrow HO_2$$
$$(5) \ HO_2 + H_2 \longrightarrow H_2O + OH$$
$$(6) \ HO_2 + wall \longrightarrow destruction$$
$$(7) \ H + M \longrightarrow destruction$$

In a manner similar to that in Example 26.2, determine whether or not this mechanism can lead to explosions under appropriate conditions.

26.6‡ For many years the reaction $H_2(g) + I_2(g) \rightarrow 2 HI(g)$ and its reverse were assumed to be elementary bimolecular reactions. However, J.H. Sullivan (*J. Chem. Phys.* **46**, 73 (1967)) suggested that the following mechanism for the reaction, originally proposed by M. Bodenstein (*Z. Physik. Chem.* **29**, 56 (1898)), provides a better explanation of the experimental results:

$$(1) \ I_2 \longrightarrow I + I \qquad k_1, k_1'$$
$$(2) \ I + I + H_2 \longrightarrow HI + HI \qquad k_2$$

Obtain the expression for the rate of formation of HI based on this mechanism. Under what conditions does this rate law reduce to the one for the originally accepted mechanism?

---

† Problems denoted with the symbol ‡ were supplied by Charles Trapp, Carmen Giunta, and Marshall Cady.

26.7 The number of photons falling on a sample can be determined by a variety of methods, of which the classical one is chemical actinometry. The decomposition of oxalic acid $(COOH)_2$, in the presence of uranyl sulfate, $(UO_2)SO_4$, proceeds according to the sequence

(1) $\quad UO^{2+} + h\nu \longrightarrow (UO^{2+})^*$

(2) $\quad (UO^{2+})^* + (COOH)_2 \longrightarrow UO^{2+} + H_2O + CO_2 + CO$

with a quantum efficiency of 0.53 at the wavelength used. The amount of oxalic acid remaining after exposure can be determined by titration (with $KMnO_4$) and the extent of decomposition used to find the number of incident photons. In a particular experiment, the actinometry solution consisted of 5.232 g anhydrous oxalic acid, 25.0 cm³ water (together with the uranyl salt). After exposure for 300 s the remaining solution was titrated with 0.212 M $KMnO_4$(aq), and 17.0 cm³ was required for complete oxidation of the remaining oxalic acid. What is the rate of incidence of photons at the wavelength of the experiment? Express the answer in photons/second and einstein/second.

26.8 When benzophenone is illuminated with ultraviolet light it is excited into a singlet state. This singlet changes rapidly into a triplet, which phosphoresces. Triethylamine acts as a quencher for the triplet. In an experiment in methanol as solvent, the phosphorescence intensity varied with amine concentration as shown below. A time-resolved laser spectroscopy experiment had also shown that the half-life of the fluorescence in the absence of quencher is 29 μs. What is the value of k_q?

$[Q]/(\text{mol L}^{-1})$	0.0010	0.0050	0.0100
I_f/(arbitrary units)	0.41	0.25	0.16

26.9 An electronically excited state of Hg can be quenched by N_2 according to

$$Hg^*(g) + N_2(g, v = 0) \longrightarrow Hg(g) + N_2(g, v = 1)$$

in which energy transfer from Hg^* excites N_2 vibrationally. Fluorescence lifetime measurements of samples of Hg with and without N_2 present are summarized below ($T = 300$ K):

$p_{N_2} = 0.0$ atm

Relative fluorescence intensity	1.000	0.606	0.360	0.22	0.135
$\tau/\mu s$	0.0	5.0	10.0	15.0	20.0

$p_{N_2} = 9.74 \times 10^{-4}$ atm

Relative fluorescence intensity	1.000	0.585	0.342	0.200	0.117
$\tau/\mu s$	0.0	3.0	6.0	9.0	12.0

You may assume that all gases behave ideally. Determine the rate constant for the energy transfer process.

26.10 The Förster theory of resonance energy transfer and the basis for the FRET technique can be tested by performing fluorescence measurements on a series of compounds in which an energy donor and an energy acceptor are covalently linked by a rigid molecular linker of variable and known length. L. Stryer and R.P. Haugland (*Proc. Natl. Acad. Sci. USA* **58**, 719 (1967)) collected the following data on a family of compounds with the general composition dansyl-(L-prolyl)$_n$-naphthyl, in which the distance R between the naphthyl donor and the dansyl acceptor was varied from 1.2 nm to 4.6 nm by increasing the number of prolyl units in the linker:

R/nm	1.2	1.5	1.8	2.8	3.1	3.4	3.7	4.0	4.3	4.6
E_T	0.99	0.94	0.97	0.82	0.74	0.65	0.40	0.28	0.24	0.16

Are the data described adequately by eqn 36? If so, what is the value of R_0 for the naphthyl-dansyl pair?

Theoretical problems

26.11 The Rice–Herzfeld mechanism for the dehydrogenation of ethane is specified in Section 26.1, and it was noted there that it led to first-order kinetics. Confirm this remark, and find the approximations that lead to the rate law quoted there. How may the conditions be changed so that the reaction shows different orders?

26.12 The following mechanism has been proposed for the thermal decomposition of acetaldehyde (ethanal):

(1) $CH_3CHO \longrightarrow \cdot CH_3 + CHO$

(2) $\cdot CH_3 + CH_3CHO \longrightarrow CH_4 + \cdot CH_2CHO$

(3) $\cdot CH_2CHO \longrightarrow CO + \cdot CH_3$

(4) $\cdot CH_3 + \cdot CH_3 \longrightarrow CH_3CH_3$

Find an expression for the rate of formation of methane and the rate of disappearance of acetaldehyde.

26.13 Express the root mean square deviation $\{\langle M^2 \rangle - \langle \overline{M} \rangle^2\}^{1/2}$ of the molar mass of a condensation polymer in terms of p, and deduce its time dependence.

26.14 Calculate the ratio of the mean cube molar mass to the mean square molar mass in terms of (a) the fraction p, (b) the chain length.

26.15 Calculate the average polymer length in a polymer produced by a chain mechanism in which termination occurs by a disproportionation reaction of the form $M \cdot + \cdot M \rightarrow M + :M$.

26.16 Derive an expression for the time dependence of the degree of polymerization for a stepwise polymerization in which the reaction is acid-catalysed by the —COOH acid functional group. The rate law is $d[A]/dt = -k[A]^2[OH]$.

26.17 An autocatalytic reaction $A \rightarrow P$ is observed to have the rate law $d[P]/dt = k[A]^2[P]$. Solve the rate law for initial concentrations $[A]_0$ and $[P]_0$. Calculate the time at which the rate reaches a maximum.

26.18 Another reaction with the stoichiometry $A \rightarrow P$ has the rate law $d[P]/dt = k[A][P]^2$; integrate the rate law for initial concentrations $[A]_0$ and $[P]_0$. Calculate the time at which the rate reaches a maximum.

26.19 Use mathematical software to integrate the Lotka–Volterra equations and plot the concentration of Y against that of X. Explore the consequences of varying the starting concentrations for the integration. Identify the conditions (the concentrations of X and Y) corresponding to the steady state of the Lotka–Volterra equations (the point at the centre of the orbits).

26.20 Set up the overall rate equations for the concentrations of X and Y in terms of the oregonator and explore the periodic properties of the solutions.

26.21 Conventional equilibrium considerations do not apply when a reaction is being driven by light absorption. Thus the steady-state concentration of products and reactants might differ significantly from equilibrium values. For instance, suppose the reaction $A \rightarrow B$ is driven by light absorption, and that its rate is I_a, but that the reverse reaction $B \rightarrow A$ is bimolecular and second-order with a rate $k[B]^2$. What is the stationary state concentration of B? Why does this 'photostationary state' differ from the equilibrium state?

26.22 Derive an expression for the rate of disappearance of a species A in a photochemical reaction for which the mechanism is:

(1) initiation with light of intensity I: $\quad A \longrightarrow R \cdot + R \cdot$

(2) propagation: $\quad A + R \cdot \longrightarrow R \cdot + B$

(3) termination: $\quad R \cdot + R \cdot \longrightarrow R_2$

Hence, show that rate measurements will give only a combination of k_2 and k_3 if a steady state is reached, but that both may be obtained if a steady state is not reached.

26.23 The photochemical chlorination of chloroform in the gas has been found to follow the rate law $d[CCl_4]/dt = k[Cl_2]^{1/2} I_a^{1/2}$. Devise a mechanism that leads to this rate law when the chlorine pressure is high.

26.24 Photolysis of $Cr(CO)_6$ in the presence of certain molecules M, can give rise to the following reaction sequence:

(1) $Cr(CO)_6 + h\nu \longrightarrow Cr(CO)_5 + CO$

(2) $Cr(CO)_5 + CO \longrightarrow Cr(CO)_6$

(3) $Cr(CO)_5 + M \longrightarrow Cr(CO)_5M$

(4) $Cr(CO)_5M \longrightarrow Cr(CO)_5 + M$

Suppose that the absorbed light intensity is so weak that $I \ll k_4[Cr(CO)_5M]$. Find the factor f in the equation $d[Cr(CO)_5M]/dt = -f[Cr(CO)_5M]$. Show that a graph of $1/f$ against [M] should be a straight line.

Applications: to biology and environmental science

26.25 Models of population growth are analogous to chemical reaction rate equations. In the model due to Malthus (1798) the rate of change of the population N of the planet is assumed to be given by $dN/dt = $ births – deaths. The numbers of births and deaths are proportional to the population, with proportionality constants b and d. Obtain the integrated rate law. How well does it fit the (very approximate) data below on the population of the planet as a function of time?

Year	1750	1825	1922	1960	1974	1987	2000
$N/10^9$	0.5	1	2	3	4	5	6

26.26 Many enzyme-catalysed reactions are consistent with a modified version of the Michaelis–Menten mechanism in which the second step is also reversible. For this mechanism obtain an expression for the rate of formation of product and find its limiting behaviour for large and small concentrations of substrate.

26.27 The initial rate of O_2 production by the action of an enzyme on a substrate was measured for a range of substrate concentrations; the data are below. Evaluate the Michaelis constant for the reaction.

$[S]/(mol\ L^{-1})$	0.050	0.017	0.010	0.0050	0.0020
$v/(mm^3\ min^{-1})$	16.6	12.4	10.1	6.6	3.3

26.28 Enzyme-catalysed reactions are sometimes analysed by use of the *Eadie–Hofstee plot*, in which $v/[S]_0$ is plotted against v. Using the simple Michaelis–Menten mechanism, derive a relation between $v/[S]_0$ and v. Discuss how the values of K_M and v_{max} are obtained from analysis of the Eadie–Hofstee plot.

26.29 In general, the catalytic efficiency of an enzyme depends on the pH of the medium in which it operates. One way to account for this behaviour is to propose that the enzyme and the enzyme-substrate complex are active only in specific protonation states. This situation can be summarized by the following mechanism:

$$EH + S \rightleftharpoons ESH \qquad k_a, k_a'$$

$$ESH \longrightarrow EH + P \qquad k_b$$

$$EH \rightleftharpoons E^- + H^+ \qquad K_{E,a} = \frac{[E^-][H^+]}{[EH]}$$

$$EH_2^+ \rightleftharpoons EH + H^+ \qquad K_{E,b} = \frac{[EH][H^+]}{[EH_2^+]}$$

$$ESH \rightleftharpoons ES^- + H^+ \qquad K_{ES,a} = \frac{[ES^-][H^+]}{[ESH]}$$

$$ESH_2 \rightleftharpoons ESH + H^+ \qquad K_{ES,b} = \frac{[ESH][H^+]}{[ESH_2]}$$

in which only the EH and ESH forms are active. (a) For the mechanism above, show that

$$v = \frac{v'_{max}}{1 + K'_M/[S]_0}$$

with

$$v'_{max} = \frac{v_{max}}{1 + \frac{[H^+]}{K_{ES,b}} + \frac{K_{ES,a}}{[H^+]}} \qquad K'_M = K_M \frac{1 + \frac{[H^+]}{K_{E,b}} + \frac{K_{E,a}}{[H^+]}}{1 + \frac{[H^+]}{K_{ES,b}} + \frac{K_{ES,a}}{[H^+]}}$$

where v_{max} and K_M correspond to the form EH of the enzyme. (b) For pH values ranging from 0 to 14, plot v'_{max} against pH for a hypothetical reaction for which $v_{max} = 1.0 \times 10^{-6}$ mol L^{-1} s^{-1}, $K_{ES,b} = 1.0 \times 10^{-6}$ mol L^{-1}, and $K_{ES,a} = 1.0 \times 10^{-8}$ mol L^{-1}. Is there a pH at which v'_{max} reaches a maximum value? If so, determine the pH. (c) Redraw the plot in part (b) by using the same value of v_{max}, but $K_{ES,b} = 1.0 \times 10^{-4}$ mol L^{-1} and $K_{ES,a} = 1.0 \times 10^{-10}$ mol L^{-1}. Account for any differences between this plot and the plot from part (b).

26.30 The following results were obtained when the rate of an enzymolysis was monitored (a) without inhibitor, (b) with inhibitor at a concentration of 15 μmol L^{-1}:

$[S]_0/(10^{-4}\ mol\ L^{-1})$	1.0	3.0	7.0	12.0	18.0
(a) $v/(μmol\ L^{-1}\ s^{-1})$	0.49	0.95	1.3	1.5	1.6
(b) $v/(μmol\ L^{-1}\ s^{-1})$	0.27	0.52	0.71	0.81	0.86

Is the inhibition competitive, uncompetitive, or non-competitive?

26.31 Many biological and biochemical processes involve autocatalytic steps. In the SIR model of the spread and decline of infectious diseases the population is divided into three classes; the susceptibles, S, who can catch the disease, the infectives, I, who have the disease and can transmit it, and the removed class, R, who have either had the disease and recovered, are dead, are immune or isolated. The model mechanism for this process implies the following rate laws:

$$\frac{dS}{dt} = -rSI \qquad \frac{dI}{dt} = rSI - aI \qquad \frac{dR}{dt} = aI$$

What are the autocatalytic steps of this mechanism? Find the conditions on the ratio a/r that decide whether the disease will spread (an epidemic) or die out. Show that a constant population is built into this system, namely, that $S + I + R = N$, meaning that the timescales of births, deaths by other causes, and migration are assumed large compared to that of the spread of the disease.

26.32‡ Ultraviolet radiation photolyses O_3 to O_2 and O. Determine the rate at which ozone is consumed by 305 nm radiation in a layer of the stratosphere of thickness 1 km. The quantum efficiency is 0.94 at 220 K, the concentration about 8×10^{-9} mol L^{-1}, the molar absorption coefficient 260 L mol^{-1} cm^{-1}, and the flux of 305 nm radiation about 1×10^{14} photons cm^{-2} s^{-1}. Data from W.B. DeMore, S.P. Sander, D.M. Golden, R.F. Hampson, M.J. Kurylo, C.J. Howard, A.R. Ravishankara, C.E. Kolb, and M.J. Molina, *Chemical kinetics and* *photochemical data for use in stratospheric modeling: Evaluation Number 11,* JPL Publication 94–26 (1994).

26.33‡ Because of its importance in atmospheric chemistry, the thermal decomposition of nitric oxide, $2 NO(g) \rightarrow N_2(g) + O_2(g)$, has been amongst the most thoroughly studied of gas-phase reactions. The commonly accepted mechanism has been that of H. Wise and M.F. Freech (*J. Chem. Phys.* **22**, 1724 (1952)):

(1) $NO + NO \longrightarrow N_2O + O$ $\quad k_1$

(2) $O + NO \longrightarrow O_2 + N$ $\quad k_2$

(3) $N + NO \longrightarrow N_2 + O$ $\quad k_3$

(4) $O + O + M \longrightarrow O_2 + M$ $\quad k_4$

(5) $O_2 + M \longrightarrow O + O + M$ $\quad k_4'$

(a) Label the steps of this mechanism as initiation, propagation, etc. (b) Write down the full expression for the rate of disappearance of NO. What does this expression for the rate become on the basis of the assumptions that $v_2 = v_3$ when [N] reaches its steady-state concentration, that the rate of the propagation step is more rapid than the rate of the initiation step, and that oxygen atoms are in equilibrium with oxygen molecules? (c) Find an expression for the effective activation energy, $E_{a,eff}$, for the overall reaction in terms of the activation energies of the individual steps of the reaction. (d) Estimate $E_{a,eff}$ from the bond energies of the species involved. (e) It has been pointed out by R.J. Wu and C.T. Yeh (*Int. J. Chem. Kinet.* **28**, 89 (1996)) that the reported experimental values of $E_{a,eff}$ obtained by different authors have varied from 253 to 357 kJ mol^{-1}. They suggest that the assumption of oxygen atoms and oxygen molecules being in equilibrium is unwarranted and that the steady-state approximation needs to be applied to the entire mechanism. Obtain the overall rate law based on the steady-state approximation and find the forms that it assumes for low NO conversion (low O_2 concentration). (f) When the reaction conversion becomes significant, Wu and Yeh suggest that two additional elementary steps,

(6) $O_2 + M \longrightarrow O + O + M$ $\quad k_6$

(7) $NO + O_2 \longrightarrow O + NO_2$ $\quad k_7$

start to compete with step (1) as the initiation step. Obtain the rate laws based on these alternative mechanisms and again estimate the apparent activation energies. Is the range of these different theoretically estimated values for $E_{a,eff}$ consistent with the range of values obtained experimentally?

27

Molecular reaction dynamics

Reactive encounters

27.1 Collision theory

27.2 Diffusion-controlled reactions

27.3 The material-balance equation

Activated complex theory

27.4 The Eyring equation

27.5 Thermodynamic aspects

The dynamics of molecular collisions

27.6 Reactive collisions

27.7 Potential energy surfaces

27.8 Some results from experiments and calculations

Checklist of key ideas

Further reading

Exercises

Problems

The simplest quantitative account of reaction rates is in terms of collision theory, which can be used only for the discussion of reactions between simple species in the gas phase. Reactions in solution can be classified into two types: diffusion-controlled and activation-controlled. The former can be expressed quantitatively in terms of the diffusion equation. In activated complex theory, it is assumed that the reactant molecules form a complex that can be discussed in terms of the population of its energy levels. Activated complex theory inspires a thermodynamic approach to reaction rates, in which the rate constant is expressed in terms of thermodynamic parameters. This approach is useful for parametrizing the rates of reactions in solution. The highest level of sophistication is in terms of potential energy surfaces and the motion of molecules through these surfaces. As we shall see, such an approach gives a very intimate picture of the events that occur when reactions occur and is open to experimental study.

Now we are at the heart of chemistry. Here we examine the details of what happens to molecules at the climax of reactions. Extensive changes of structure are taking place and energies the size of dissociation energies are being redistributed among bonds: old bonds are being ripped apart and new bonds are being formed.

As may be imagined, the calculation of the rates of such processes from first principles is very difficult. Nevertheless, like so many intricate problems, the broad features can be established quite simply. Only when we enquire more deeply do the complications emerge. In this chapter we look at three levels of approach to the calculation of a rate constant for an elementary bimolecular reaction. Although a great deal of information can be obtained from gas-phase reactions, many reactions of interest take place in solution, and we shall also see to what extent their rates can be predicted.

Reactive encounters

In this section we consider two elementary approaches to the calculation of reaction rates, one relating to gas-phase reactions and the other to reactions in solution. Both approaches are based on the view that reactant molecules must meet, and that reaction takes place only if the molecules have a certain minimum energy. In the collision theory of bimolecular gas-phase reactions, which we mentioned briefly in Section 25.5b, products are formed only if the collision is sufficiently energetic; otherwise the colliding reactant molecules separate again. In solution, the reactant molecules may simply diffuse together and then acquire energy from their immediate surroundings while they are in contact.

27.1 Collision theory

We shall consider the bimolecular elementary reaction

$$A + B \longrightarrow P \qquad v = k_2[A][B] \tag{27.1}$$

where P denotes products, and aim to calculate the second-order rate constant k_2.

We can anticipate the general form of the expression for k_2 by considering the physical requirements for reaction. We can expect the rate v to be proportional to the rate of collisions, and therefore to the mean speed of the molecules, $\bar{c} \propto (T/M)^{1/2}$ where M is the molar mass of the molecules, their collision cross-section, σ, and the number densities of A and B:

$$v \propto \sigma(T/M)^{1/2}\mathcal{N}_A\mathcal{N}_B \propto \sigma(T/M)^{1/2}[A][B]$$

However, a collision will be successful only if the kinetic energy exceeds a minimum value, the activation energy, E_a, of the reaction. This requirement suggests that the rate constant should also be proportional to a Boltzmann factor of the form $e^{-E_a/RT}$. So we can anticipate that

$$k_2 \propto \sigma(T/M)^{1/2}e^{-E_a/RT}$$

However, not every collision will lead to reaction even if the energy requirement is satisfied, because the reactants may need to collide in a certain relative orientation. This 'steric requirement' suggests that a further factor, P, should be introduced, and that

$$k_2 \propto P\sigma(T/M)^{1/2}e^{-E_a/RT} \tag{27.2}$$

As we shall see in detail below, this expression has the form predicted by collision theory. It reflects three aspects of a successful collision:

$$k_2 \propto \text{steric requirement} \times \text{encounter rate} \times \text{minimum energy requirement}$$

(a) Collision rates in gases

We have anticipated that the reaction rate, and hence k_2, depends on the frequency with which molecules collide. The **collision density**, Z_{AB}, is the number of (A,B) collisions in a region of the sample in an interval of time divided by the volume of the region and the duration of the interval. The frequency of collisions of a single molecule in a gas was calculated in Section 24.1. As shown in the *Justification* below, that result can be adapted to deduce that

$$Z_{AB} = \sigma\left(\frac{8kT}{\pi\mu}\right)^{1/2} N_A^2[A][B] \tag{27.3}$$

where σ is the collision cross-section (Fig. 27.1)

$$\sigma = \pi d^2 \qquad d = \tfrac{1}{2}(d_A + d_B) \tag{27.4}$$

and μ is the reduced mass,

$$\mu = \frac{m_A m_B}{m_A + m_B} \tag{27.5}$$

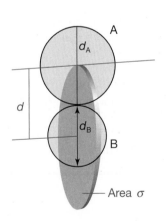

27.1 The collision cross-section for two molecules can be regarded to be the area within which the centre of the projectile molecule (A) must enter around the target molecule (B) in order for a collision to occur. If the diameters of the two molecules are d_A and d_B, the radius of the target area is $d = \tfrac{1}{2}(d_A + d_B)$ and the cross-section is πd^2.

Similarly, the collision density for like molecules at a molar concentration [A] is

$$Z_{AA} = \sigma \left(\frac{4kT}{\pi m_A} \right)^{1/2} N_A^2 [A]^2 \tag{27.6}$$

Collision densities may be very large. For example, in nitrogen at room temperature and pressure, with $d = 280$ pm, $Z = 5 \times 10^{34}$ m^{-3} s^{-1}.

Justification 27.1

It follows from eqn 24.11 that the collision frequency, z, for a single A molecule of mass m_A in a gas of other A molecules is

$$z = \sigma \bar{c}_{rel} \mathcal{N}_A \tag{27.7}$$

where $\mathcal{N}_A$ is the number density of A molecules and $\bar{c}_{rel}$ is their relative mean speed. As indicated in Section 24.1a

$$\bar{c}_{rel} = 2^{1/2} \bar{c} \qquad \bar{c} = \left(\frac{8kT}{\pi m} \right)^{1/2} \tag{27.8}$$

For future convenience, it is sensible to introduce $\mu = \frac{1}{2} m$ (for like molecules of mass m), and then to write

$$\bar{c}_{rel} = \left(\frac{8kT}{\pi \mu} \right)^{1/2} \tag{27.9}$$

This expression also applies to the mean relative speed of dissimilar molecules, provided that μ is interpreted as the reduced mass in eqn 5.

The total collision density is the collision frequency multiplied by the number density of A molecules:

$$Z_{AA} = \frac{1}{2} z \mathcal{N}_A = \frac{1}{2} \sigma \bar{c}_{rel} \mathcal{N}_A^2 \tag{27.10}$$

The factor of $\frac{1}{2}$ has been introduced to avoid double counting of the collisions (so one A molecule colliding with another A molecule is counted as one collision regardless of their actual identities). For collisions of A and B molecules present at number densities $\mathcal{N}_A$ and $\mathcal{N}_B$, the collision density is

$$Z_{AB} = \sigma \bar{c}_{rel} \mathcal{N}_A \mathcal{N}_B \tag{27.11}$$

Note that we have discarded the factor of $\frac{1}{2}$ because now we are considering an A molecule colliding with any of the B molecules as a collision.

The number density of a species J is $\mathcal{N}_J = N_A[J]$, where [J] is their molar concentration and N_A is the Avogadro constant. Equations 3 and 6 then follow.

(b) The energy requirement

According to collision theory, the rate of change in the molar concentration of A molecules is the product of the collision density and the probability that a collision occurs with sufficient energy. The latter condition can be incorporated by writing the collision cross-section as a function of the kinetic energy of approach of the two colliding species, and setting the cross-section, $\sigma(\varepsilon)$, equal to zero if the kinetic energy of approach is below a certain threshold value, ε_a. Later, we shall identify $N_A \varepsilon_a$ as E_a, the (molar) activation energy of the reaction. Then, for a collision with a specific relative speed of approach v_{rel} (not, at this stage, a mean value),

27.2 The variation of the reactive cross-section with energy as expressed by eqn 15. The data points are from experiments on the reaction H + D₂ → HD + D (K. Tsukiyama, B. Katz, and R. Bersohn, *J. Chem. Phys.*, **84**, 1934 (1986)).

$$\frac{d[A]}{dt} = -\sigma(\varepsilon)v_{rel}N_A[A][B] \tag{27.12}$$

The relative kinetic energy, ε, and the relative speed are related by $\varepsilon = \frac{1}{2}\mu v_{rel}^2$, so $v_{rel} = (2\varepsilon/\mu)^{1/2}$.[1] At this point we recognize that a wide range of approach energies is present in a sample, so we should average the expression just derived over a Boltzmann distribution of energies, write

$$\frac{d[A]}{dt} = -\left\{\int_0^\infty \sigma(\varepsilon)v_{rel}f(\varepsilon)\,d\varepsilon\right\}N_A[A][B] \tag{27.13}$$

and hence recognize the rate constant as

$$k_2 = N_A\int_0^\infty \sigma(\varepsilon)v_{rel}f(\varepsilon)\,d\varepsilon \tag{27.14}$$

Now suppose that the reactive collision cross-section is zero below ε_a. We show in the *Justification* below that, above ε_a, $\sigma(\varepsilon)$ varies as

$$\sigma(\varepsilon) = \left(1 - \frac{\varepsilon_a}{\varepsilon}\right)\sigma \tag{27.15}$$

This form of the energy dependence is broadly consistent with experimental determinations of the reaction between H and D₂ as determined by molecular beam measurements of the kind described later (Fig. 27.2). Then, in the *Justification* below, we also show that

$$k_2 = N_A\sigma\bar{c}_{rel}e^{-E_a/RT} \tag{27.16}$$

Justification 27.2

Consider two colliding molecules A and B with relative speed v_{rel} and relative kinetic energy $\varepsilon_{rel} = \frac{1}{2}\mu v_{rel}^2$ (Fig. 27.3). Intuitively we expect that a head-on collision between A and B will be most effective in bringing about a chemical reaction. Therefore, $v_{rel,A-B}$, the magnitude of the relative velocity component parallel to an axis that contains the vector connecting the centres of A and B, must be large. From trigonometry and the definitions of the distances b (the impact parameter) and d, and the angle θ given in Fig. 27.3, it follows that

$$v_{rel,A-B} = v_{rel}\cos\theta = v_{rel}\left(\frac{d^2 - b^2}{d^2}\right)^{1/2}$$

We assume that only the kinetic energy associated with the head-on component of the collision, ε, can lead to a chemical reaction. After squaring both sides of the equation above and multiplying by $\frac{1}{2}\mu$, it follows that

$$\varepsilon = \varepsilon_{rel}\frac{d^2 - b^2}{d^2}$$

The existence of an energy threshold, ε_a, for the formation of products implies that there is a maximum value of the impact parameter b, b_{max}, above which reactions do not occur. Setting $b = b_{max}$ and $\varepsilon = \varepsilon_a$ in the equation above gives

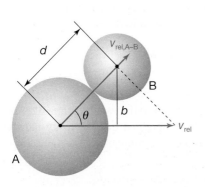

27.3 The parameters used in the calculation of the dependence of the collision cross-section on the relative kinetic energy of two molecules A and B.

---

1. See *Justification 13.1*. To arrive at this expression for the kinetic energy associated with the relative motion of two particles, it is necessary to separate the centre-of-mass coordinates from the internal coordinates of each particle.

$$b_{max}^2 = \left(1 - \frac{\varepsilon_a}{\varepsilon}\right)d^2$$

Substituting $\sigma(\varepsilon)$ for πb_{max}^2 and σ for πd^2 (Section 24.1b) in the equation above gives eqn 15. Note that the equation can be used only when $\varepsilon > \varepsilon_a$; for $\varepsilon < \varepsilon_a$, $\sigma = 0$.

We proceed with the calculation of the rate constant by considering the Maxwell–Boltzmann distribution of molecular speeds given in Section 24.1. It may be expressed in terms of the kinetic energy, ε, by writing $\varepsilon = \frac{1}{2}\mu v^2$, then $dv = d\varepsilon/(2\mu\varepsilon)^{1/2}$ and eqn 24.4 becomes

$$f(v)\,dv = 4\pi\left(\frac{\mu}{2\pi kT}\right)^{3/2}\left(\frac{2\varepsilon}{\mu}\right)e^{-\varepsilon/kT}\frac{d\varepsilon}{(2\mu\varepsilon)^{1/2}} \tag{27.17}$$

$$= 2\pi\left(\frac{1}{\pi kT}\right)^{3/2}\varepsilon^{1/2}e^{-\varepsilon/kT}\,d\varepsilon = f(\varepsilon)\,d\varepsilon$$

The integral we need to evaluate is therefore

$$\int_0^\infty \sigma(\varepsilon)v_{rel}f(\varepsilon)\,d\varepsilon = 2\pi\left(\frac{1}{\pi kT}\right)^{3/2}\int_0^\infty \sigma(\varepsilon)\left(\frac{2\varepsilon}{\mu}\right)^{1/2}\varepsilon^{1/2}e^{-\varepsilon/kT}\,d\varepsilon$$

$$= \left(\frac{8}{\pi\mu kT}\right)^{1/2}\left(\frac{1}{kT}\right)\int_0^\infty \varepsilon\sigma(\varepsilon)e^{-\varepsilon/kT}\,d\varepsilon$$

Now we introduce the approximation for $\sigma(\varepsilon)$ in eqn 15, and evaluate

$$\int_0^\infty \varepsilon\sigma(\varepsilon)e^{-\varepsilon/kT}\,d\varepsilon = \sigma\int_{\varepsilon_a}^\infty \varepsilon\left(1 - \frac{\varepsilon_a}{\varepsilon}\right)e^{-\varepsilon/kT}\,d\varepsilon = (kT)^2\sigma e^{-\varepsilon_a/kT}$$

We have made use of the fact that $\sigma = 0$ for $\varepsilon < \varepsilon_a$. It follows that

$$\int_0^\infty \sigma(\varepsilon)v_{rel}f(\varepsilon)\,d\varepsilon = \sigma\left(\frac{8kT}{\pi\mu}\right)^{1/2}e^{-\varepsilon_a/kT}$$

as in eqn 16 (with $\varepsilon_a/kT = E_a/RT$ and using eqn. 14).

Equation 16 has the Arrhenius form $k_2 = Ae^{-E_a/RT}$ provided the exponential temperature dependence dominates the weak square-root temperature dependence of the pre-exponential factor.[2] It follows that we can identify the activation energy, E_a, with the minimum kinetic energy along the line of approach that is needed for reaction, and that the pre-exponential factor is a measure of the rate at which collisions occur in the gas.

(c) The steric requirement

The simplest procedure for calculating k_2 is to use for σ the values obtained for non-reactive collisions (for example, typically those obtained from viscosity measurements) or from tables of molecular radii. Table 27.1 compares some values of the

---

2. From the general definition of activation energy in eqn 25.26 and eqn 16:

$$RT^2\frac{d\ln k}{dT} = E_a + \frac{1}{2}RT$$

so the activation energy is weakly temperature-dependent. Commonly, $E_a \gg RT$, and the $\frac{1}{2}RT$ can be neglected.

Table 27.1* Arrhenius parameters for gas-phase reactions

	$A/(\text{L mol}^{-1}\text{ s}^{-1})$		$E_a/(\text{kJ mol}^{-1})$	P
	Experiment	Theory		
$2\text{ NOCl} \longrightarrow 2\text{ NO} + 2\text{ Cl}$	9.4×10^9	5.9×10^{10}	102.0	0.16
$2\text{ ClO} \longrightarrow \text{Cl}_2 + \text{O}_2$	6.3×10^7	2.5×10^{10}	0.0	2.5×10^{-3}
$\text{H}_2 + \text{C}_2\text{H}_4 \longrightarrow \text{C}_2\text{H}_6$	1.24×10^6	7.4×10^{11}	180.0	1.7×10^{-6}
$\text{K} + \text{Br}_2 \longrightarrow \text{KBr} + \text{Br}$	1.0×10^{12}	2.1×10^{11}	0.0	4.8

* More values are given in the *Data section*.

pre-exponential factor calculated in this way with values obtained from Arrhenius plots (Section 25.5a). One of the reactions shows fair agreement between theory and experiment, but for others there are major discrepancies. In some cases the experimental values are orders of magnitude smaller than those calculated, which suggests that the collision energy is not the only criterion for reaction and that some other feature, such as the relative orientation of the colliding species, is important. Moreover, one reaction in the table has a pre-exponential factor larger than theory, which seems to indicate that the reaction occurs more quickly than the particles collide!

We can accommodate the disagreement between experiment and theory by introducing a **steric factor**, P, and expressing the **reactive cross-section**, σ^*, as a multiple of the collision cross-section, $\sigma^* = P\sigma$ (Fig. 27.4). Then the rate constant becomes

$$k_2 = P\sigma\left(\frac{8kT}{\pi\mu}\right)^{1/2} N_A e^{-E_a/RT} \tag{27.18}$$

This expression has the form we anticipated in eqn 2. The steric factor is normally found to be several orders of magnitude smaller than 1.

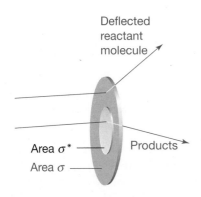

27.4 The collision cross-section is the target area that results in simple deflection of the projectile molecule; the reaction cross-section is the corresponding area for chemical change to occur on collision.

Example 27.1 Estimating a steric factor (1)

Estimate the steric factor for the reaction $\text{H}_2 + \text{C}_2\text{H}_4 \rightarrow \text{C}_2\text{H}_6$ at 628 K given that the pre-exponential factor is 1.24×10^6 L mol^{-1} s^{-1}.

Method To calculate P, we need to calculate the pre-exponential factor, A, by using eqn 18 and then compare the answer with experiment: the ratio is P. Table 24.1 lists collision cross-sections for non-reactive encounters. The best way to estimate the collision cross-section for dissimilar spherical species is to calculate the collision diameter for each one (from $\sigma = \pi d^2$), to calculate the mean of the two diameters, and then to calculate the cross-section for that mean diameter. However, as neither species is spherical, a simpler but more approximate procedure is just to take the average of the two collision cross-sections.

Answer The reduced mass of the colliding pair is

$$\mu = \frac{m_1 m_2}{m_1 + m_2} = 3.12 \times 10^{-27} \text{ kg}$$

because $m_1 = 2.016$ u for H_2 and $m_2 = 28.05$ u for C_2H_4 (the atomic mass unit, 1 u, is defined inside the front cover). Hence

$$\left(\frac{8kT}{\pi\mu}\right)^{1/2} = 2.66 \times 10^3 \text{ m s}^{-1}$$

From Table 24.1, $\sigma(H_2) = 0.27 \text{ nm}^2$ and $\sigma(C_2H_4) = 0.64 \text{ nm}^2$, giving a mean collision cross-section of $\sigma = 0.46 \text{ nm}^2$. Therefore,

$$A = \sigma \left(\frac{8kT}{\pi\mu}\right)^{1/2} N_A = 7.37 \times 10^{11} \text{ L mol}^{-1} \text{ s}^{-1}$$

Experimentally $A = 1.24 \times 10^6 \text{ L mol}^{-1} \text{ s}^{-1}$, so it follows that $P = 1.7 \times 10^{-6}$.

Comment The very small value of P is one reason why catalysts are needed to bring this reaction about at a reasonable rate. As a general guide, the more complex the molecules, the smaller the value of P.

Self-test 27.1 It is found for the reaction $NO + Cl_2 \rightarrow NOCl + Cl$ that $A = 4.0 \times 10^9 \text{ L mol}^{-1} \text{ s}^{-1}$ at 298 K. Use $\sigma(NO) = 0.42 \text{ nm}^2$ and $\sigma(Cl_2) = 0.93 \text{ nm}^2$ to estimate the P factor for the reaction. [0.018]

An example of a reaction for which it is possible to estimate the steric factor is $K + Br_2 \rightarrow KBr + Br$, with the experimental value $P = 4.8$. In this reaction, the distance of approach at which reaction occurs appears to be considerably larger than the distance needed for deflection of the path of the approaching molecules in a non-reactive collision. It has been proposed that the reaction proceeds by a **harpoon mechanism**. This brilliant name is based on a model of the reaction that pictures the K atom as approaching a Br_2 molecule, and when the two are close enough an electron (the harpoon) flips across from K to Br_2. In place of two neutral particles there are now two ions, so there is a Coulombic attraction between them: this attraction is the line on the harpoon. Under its influence the ions move together (the line is wound in), the reaction takes place, and $KBr + Br$ emerge. The harpoon extends the cross-section for the reactive encounter, and the reaction rate is greatly underestimated by taking for the collision cross-section the value for simple mechanical contact between $K + Br_2$.

Example 27.2 Estimating a steric factor (2)

Estimate the value of P for the harpoon mechanism by calculating the distance at which it is energetically favourable for the electron to leap from K to Br_2.

Method We should begin by identifying all the contributions to the energy of interaction between the colliding species. There are three contributions to the energy of the process $K + Br_2 \rightarrow K^+ + Br_2^-$. The first is the ionization energy, I, of K. The second is the electron affinity, E_{ea}, of Br_2. The third is the Coulombic interaction energy between the ions when they have been formed: when their separation is R, this energy is $-e^2/4\pi\varepsilon_0 R$. The electron flips across when the sum of these three contributions changes from positive to negative (that is, when the sum is zero).

Answer The net change in energy when the transfer occurs at a separation R is

$$E = I - E_{ea} - \frac{e^2}{4\pi\varepsilon_0 R}$$

The ionization energy I is larger than E_{ea}, so E becomes negative only when R has decreased to less than some critical value R^* given by

$$\frac{e^2}{4\pi\varepsilon_0 R^*} = I - E_{ea}$$

When the particles are at this separation, the harpoon shoots across from K to Br_2, so we can identify the reactive cross-section as $\sigma^* = \pi R^{*2}$. This value of σ^* implies that the steric factor is

$$P = \frac{\sigma^*}{\sigma} = \frac{R^{*2}}{d^2} = \left\{ \frac{e^2}{4\pi\varepsilon_0 d(I - E_{ea})} \right\}^2$$

where $d = R(K) + R(Br_2)$. With $I = 420$ kJ mol^{-1} (corresponding to 7.0×10^{-19} J), $E_{ea} \approx 250$ kJ mol^{-1} (corresponding to 4.2×10^{-19} J), and $d = 400$ pm, we find $P = 4.2$, in good agreement with the experimental value (4.8).

Self-test 27.2 Estimate the value of P for the harpoon reaction between Na and Cl_2 for which $d \approx 350$ pm; take $E_{ea} \approx 230$ kJ mol^{-1}.   [2.2]

Example 27.2 illustrates two points about steric factors. First, the concept of a steric factor is not wholly useless because in some cases its numerical value can be estimated. Second (and more pessimistically) most reactions are much more complex than $K + Br_2$, and we cannot expect to obtain P so easily. What we need is a more powerful theory that lets us calculate, and not merely guess, its value. We go some way to setting up that theory in Section 27.4.

27.2 Diffusion-controlled reactions

Encounters between reactants in solution occur in a very different manner from encounters in gases. Reactant molecules have to jostle their way through the solvent, so their encounter frequency is considerably less than in a gas. However, because a molecule also migrates only slowly away from a location, two reactant molecules that encounter each other stay near each other for much longer than in a gas. This lingering of one molecule near another on account of the hindering presence of solvent molecules is called the **cage effect**. Such an encounter pair may accumulate enough energy to react even though it does not have enough energy to do so when it first forms. The activation energy of a reaction is a much more complicated quantity in solution than in a gas because the encounter pair is surrounded by solvent and we need to consider the energy of the entire local assembly of reactant and solvent molecules.

(a) Classes of reaction

The complicated overall process can be divided into simpler parts by setting up a simple kinetic scheme. We suppose that the rate of formation of an encounter pair AB is first-order in each of the reactants A and B:

$$A + B \longrightarrow AB \qquad v = k_d[A][B] \tag{27.19}$$

As we shall see, k_d (where the d signifies diffusion) is determined by the diffusional characteristics of A and B. The encounter pair can break up without reaction or it can go on to form products P. If we suppose that both processes are pseudofirst-order reactions (with the solvent perhaps playing a role), then we can write

$$AB \longrightarrow A + B \qquad v = k_d'[AB] \tag{27.20}$$

and

$$AB \longrightarrow P \qquad v = k_a[AB] \tag{27.21}$$

The concentration of AB can now be found from the equation for the net rate of change of concentration of AB:

$$\frac{d[AB]}{dt} = k_d[A][B] - k_d'[AB] - k_a[AB] \approx 0 \tag{27.22}$$

This expression solves to

$$[AB] = \frac{k_d[A][B]}{k_a + k_d'} \tag{27.23}$$

The rate of formation of products is therefore

$$\frac{d[P]}{dt} \approx k_a[AB] = k_2[A][B] \qquad k_2 = \frac{k_a k_d}{k_a + k_d'} \tag{27.24}$$

Two limits can now be distinguished. If the rate of separation of the unreacted encounter pair is much slower than the rate at which it forms products, then $k_d' \ll k_a$ and the effective rate constant is

$$k_2 \approx \frac{k_a k_d}{k_a} = k_d \tag{27.25}$$

In this **diffusion-controlled limit**, the rate of reaction is governed by the rate at which the reactant molecules diffuse through the solvent. An indication that a reaction is diffusion-controlled is that its rate constant is of the order of 10^9 L mol^{-1} s^{-1} or greater. Because the combination of radicals involves very little activation energy, radical and atom recombination reactions are often diffusion-controlled.

An **activation-controlled reaction** arises when a substantial activation energy is involved in the reaction AB → P. Then $k_a \ll k_d'$ and

$$k_2 \approx \frac{k_a k_d}{k_d'} = k_a K \tag{27.26}$$

where K is the equilibrium constant for A + B $\rightleftharpoons$ AB. In this limit, the reaction proceeds at the rate at which energy accumulates in the encounter pair from the surrounding solvent. Some experimental data are given in Table 27.2.

(b) Diffusion and reaction

The rate of a diffusion-controlled reaction is calculated by considering the rate at which the reactants diffuse together. As shown in the *Justification* below, the rate constant for a reaction in which the two reactant molecules react if they come within a distance R^* of one another is

$$k_d = 4\pi R^* D N_A \tag{27.27}$$

where D is the sum of the diffusion coefficients the two reactant species in the solution.

Table 27.2* Arrhenius parameters for reactions in solution

	Solvent	$A/(\text{L mol}^{-1}\text{ s}^{-1})$	$E_a/(\text{kJ mol}^{-1})$
$(CH_3)_3CCl$ solvolysis	Water	7.1×10^{16}	100
	Ethanol	3.0×10^{13}	112
	Chloroform	1.4×10^4	45
$CH_3CH_2Br + OH^-$	Ethanol	4.3×10^{11}	89.5

* More values are given in the *Data section*.

Justification 27.3

According to the diffusion equation ($D_B\nabla^2[B] = \partial[B]/\partial t$, Section 24.10), the concentration of B when the system has reached a steady state ($\partial[B]/\partial t = 0$) satisfies $\nabla^2[B]_r = 0$, where the subscript r signifies a quantity that varies with the distance r. For a spherically symmetrical system, ∇^2 can be replaced by radial derivatives alone (see Table 11.1), so the equation satisfied by $[B]_r$ is

$$\frac{d^2[B]_r}{dr^2} + \frac{2}{r}\frac{d[B]_r}{dr} = 0 \tag{27.28}$$

The general solution of this equation is

$$[B]_r = a + \frac{b}{r} \tag{27.29}$$

We need two boundary conditions to pin down the values of the two constants. One condition is that $[B]_r$ has its bulk value $[B]$ as $r \to \infty$. The second condition is that the concentration of B is zero at $r = R^*$, the distance at which reaction occurs. It follows that $a = [B]$ and $b = -R^*[B]$, and hence that (for $r \geq R^*$)

$$[B]_r = \left(1 - \frac{R^*}{r}\right)[B] \tag{27.30}$$

Figure 27.5 illustrates the variation of concentration expressed by this equation.

The rate of reaction is the (molar) flux, J, of the reactant B towards A multiplied by the area of the spherical surface of radius R^*:

$$\text{Rate of reaction} = 4\pi R^{*2}J \tag{27.31}$$

From Fick's first law (eqn 24.18), the flux towards A is proportional to the concentration gradient, so at a radius R^*:

$$J = D_B\left(\frac{d[B]_r}{dr}\right)_{r=R^*} = \frac{D_B[B]}{R^*}$$

(A sign change has been introduced because we are interested in the flux towards decreasing values of r.) When this condition is substituted into the previous equation we obtain

$$\text{Rate of reaction} = 4\pi R^* D_B[B] \tag{27.32}$$

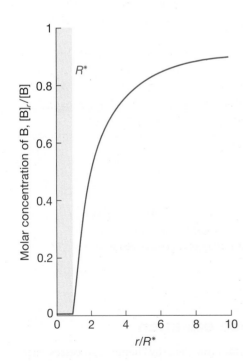

27.5 The concentration profile for reaction in solution when a molecule B diffuses towards another reactant molecule and reacts if it reaches R^*.

The rate of the diffusion-controlled reaction is equal to the average flow of B molecules to all the A molecules in the sample. If the bulk concentration of A is [A], the number of A molecules in the sample of volume V is $N_A[A]V$; the global flow of all B to all A is therefore $4\pi R^* D_B N_A[A][B]V$. Because it is unrealistic to suppose that all A are stationary; we replace D_B by the sum of the diffusion coefficients of the two species and write $D = D_A + D_B$. Then the rate of change of concentration of AB is

$$\frac{d[AB]}{dt} = 4\pi R^* D N_A[A][B] \tag{27.33}$$

Hence, the diffusion-controlled rate constant is as given in eqn 27.

We can take eqn 27 further by incorporating the Stokes–Einstein equation (eqn 24.83) relating the diffusion constant and the hydrodynamic radius R_A and R_B of each molecule in a medium of viscosity η:

$$D_A = \frac{kT}{6\pi\eta R_A} \qquad D_B = \frac{kT}{6\pi\eta R_B} \tag{27.34}$$

As these relations are approximate, little extra error is introduced if we write $R_A = R_B = \frac{1}{2}R^*$, which leads to

$$k_d = \frac{8RT}{3\eta} \tag{27.35}$$

(The R in this equation is the gas constant.) The radii have cancelled because, although the diffusion constants are smaller when the radii are large, the reactive collision radius is larger and the particles need travel a shorter distance to meet. In this approximation, the rate constant is independent of the identities of the reactants, and depends only on the temperature and the viscosity of the solvent.

Illustration 27.1

The rate constant for the recombination of I atoms in hexane at 298 K, when the viscosity of the solvent is 0.326 cP (with $1\,P = 10^{-1}\,kg\,m^{-1}\,s^{-1}$) is

$$k_d = \frac{8 \times (8.3145\,J\,K^{-1}\,mol^{-1}) \times (298\,K)}{3 \times (3.26 \times 10^{-4}\,kg\,m^{-1}\,s^{-1})}$$

$$= 2.0 \times 10^7\,m^3\,mol^{-1}\,s^{-1}$$

where we have used $1\,J = 1\,kg\,m^2\,s^{-2}$. Because $1\,m^3 = 10^3\,L$, this result corresponds to $2.0 \times 10^{10}\,L\,mol^{-1}\,s^{-1}$. The experimental value is $1.3 \times 10^{10}\,L\,mol^{-1}\,s^{-1}$, so the agreement is very good considering the approximations involved.

27.3 The material balance equation

The diffusion of reactants plays an important role in many chemical processes, such as the diffusion of O_2 molecules into red blood corpuscles and the diffusion of a gas towards a catalyst. We can have a glimpse of the kinds of calculations involved by considering the diffusion equation (Section 24.10) generalized to take into account the possibility that the diffusing, convecting molecules are also reacting.

(a) The formulation of the equation

Consider a small volume element in a chemical reactor (or a biological cell). The net rate at which J molecules enter the region by diffusion and convection is given by eqn 24.87:

$$\frac{\partial[J]}{\partial t} = D\frac{\partial^2[J]}{\partial x^2} - v\frac{\partial[J]}{\partial x}$$

(27.36)

The net rate of change of molar concentration due to chemical reaction is

$$\frac{\partial[J]}{\partial t} = -k[J]$$

(27.37)

if we suppose that J disappears by a pseudofirst-order reaction. Therefore, the overall rate of change of the concentration of J is

$$\frac{\partial[J]}{\partial t} = D\frac{\partial^2[J]}{\partial x^2} - v\frac{\partial[J]}{\partial x} - k[J]$$

(27.38)

Equation 38 is called the **material balance equation**. If the rate constant is large, then [J] will decline rapidly. However, if the diffusion constant is large, then the decline can be replenished as J diffuses rapidly into the region. The convection term, which may represent the effects of stirring, can sweep material either into or out of the region according to the sign of v.

(b) Solutions of the equation

The material balance equation is a second-order partial differential equation and is far from easy to solve in general. Some idea of how it is solved can be obtained by considering the special case in which there is no convective motion (as in an unstirred reaction vessel):

$$\frac{\partial[J]}{\partial t} = D\frac{\partial^2[J]}{\partial x^2} - k[J]$$

(27.39)

If the solution of this equation in the absence of reaction (that is, for $k = 0$) is [J], then the solution in the presence of reaction ($k > 0$) is

$$[J]^* = k\int_0^t [J]e^{-kt}\,dt + [J]e^{-kt}$$

(27.40)

We have already met one solution of the diffusion equation in the absence of reaction: eqn 24.88 is the solution for a system in which initially a layer of $n_0 N_A$ molecules is spread over a plane of area A:

$$[J] = \frac{n_0 e^{-x^2/4Dt}}{A(\pi Dt)^{1/2}}$$

(27.41)

When this expression is substituted into eqn 40 and the integral is evaluated, we obtain the concentration of J as it diffuses away from its initial surface layer and undergoes reaction in the solution above (Fig. 27.6).

Even this relatively simple example has led to an equation that is difficult to solve, and only in some special cases can the full material balance equation be solved analytically. Most modern work on reactor design and cell kinetics uses numerical methods to solve the equation, and detailed solutions for realistic environments can be obtained reasonably easily. Some of the interesting applications include the exploration of the spatial periodicity of autocatalytic reactions mentioned in Section 26.8.

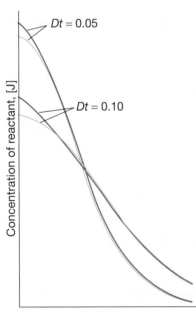

$Dt = 0.05$

$Dt = 0.10$

Concentration of reactant, [J]

Distance from plane, x

27.6 The concentration profiles for a diffusing, reacting system (for example, a column of solution) in which one reactant is initially in a layer at $x = 0$. In the absence of reaction (pale blue lines) the concentration profiles are the same as in Fig. 24.25.

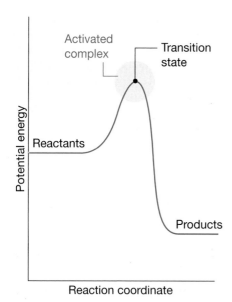

27.7 A reaction profile. The horizontal axis is the reaction coordinate, and the vertical axis is potential energy. The activated complex is the region near the potential maximum, and the transition state corresponds to the maximum itself.

Activated complex theory

We saw in Section 25.5b that an activated complex forms between reactants as they collide and begin to assume the nuclear and electronic configurations characteristic of products. We also saw that the change in potential energy associated with formation of the activated complex accounts for the activation energy of the reaction. We now consider a more detailed calculation of rate constants, which uses the concepts of statistical thermodynamics developed in Chapter 20. The approach we describe, which is called **activated complex theory** (ACT),[3] has the advantage that a quantity corresponding to the steric factor appears automatically, and P does not need to be grafted on to an equation as an afterthought. Activated complex theory is an attempt to identify the principal features governing the size of a rate constant in terms of a model of the events that take place during the reaction. There are several approaches to the calculation, all of which lead to the same final expression (see *Further reading*); here we present the simplest approach.

27.4 The Eyring equation

Activated complex theory pictures a reaction between A and B as proceeding through the formation of an activated complex, $C^{\ddagger}$, in a rapid pre-equilibrium (Fig. 27.7):[4]

$$A + B \rightleftharpoons C^{\ddagger} \qquad K^{\ddagger} = \frac{p_{C^{\ddagger}} p^{\ominus}}{p_A p_B}$$

When we express the partial pressures, p_J, in terms of the molar concentrations, [J], by using $p_J = RT[J]$, the concentration of activated complex is related to the (dimensionless) equilibrium constant by

$$[C^{\ddagger}] = \frac{RT}{p^{\ominus}} K^{\ddagger}[A][B] \tag{27.42}$$

The activated complex falls apart by unimolecular decay into products, P, with a rate constant $k^{\ddagger}$:

$$C^{\ddagger} \longrightarrow P \qquad v = k^{\ddagger}[C^{\ddagger}] \tag{27.43}$$

It follows that

$$v = k_2[A][B] \qquad k_2 = \frac{RT}{p^{\ominus}} k^{\ddagger} K^{\ddagger} \tag{27.44}$$

Our task is to calculate the unimolecular rate constant $k^{\ddagger}$ and the equilibrium constant $K^{\ddagger}$.

(a) The rate of decay of the activated complex

An activated complex can form products if it passes through the transition state (Section 25.5b). If its vibration-like motion along the reaction coordinate occurs with a frequency v, the frequency with which the cluster of atoms forming the complex approaches the transition state is also v. However, it is possible that not every oscillation along the reaction coordinate takes the complex through the transition state.

3. Activated complex theory is also widely referred to as *transition state theory*.
4. This chapter inevitably puts heavy demands on the letter K; the various meanings are summarized in Table 27.3 at the end of the chapter.

For instance, the centrifugal effect of rotations might also be an important contribution to the break-up of the complex, and in some cases the complex might be rotating too slowly, or rotating rapidly but about the wrong axis. Therefore, we suppose that the rate of passage of the complex through the transition state is proportional to the vibrational frequency along the reaction coordinate, and write

$$k^{\ddagger} = \kappa \nu \tag{27.45}$$

where κ is the **transmission coefficient**. In the absence of information to the contrary, κ is assumed to be about 1.

(b) The concentration of the activated complex

We saw in Section 20.7 how to calculate equilibrium constants from structural data. Equation 20.54 of that section can be used directly, which in this case gives

$$K^{\ddagger} = \frac{N_A q_{C^{\ddagger}}^{\ominus}}{q_A^{\ominus} q_B^{\ominus}} e^{-\Delta E_0/RT} \tag{27.46}$$

where $p^{\ominus} = 1$ bar and

$$\Delta E_0 = E_0(C^{\ddagger}) - E_0(A) - E_0(B) \tag{27.47}$$

The $q_J^{\ominus}$ are the standard molar partition functions, as defined in Section 20.1d. Note that the units of N_A and the q_J are mol^{-1}, so $K^{\ddagger}$ is dimensionless (as is appropriate for an equilibrium constant).

In the final step of this part of the calculation, we focus attention on the partition function of the activated complex. We have already assumed that a vibration of the activated complex $C^{\ddagger}$ tips it through the transition state. The partition function for this vibration is

$$q = \frac{1}{1 - e^{-h\nu/kT}} \tag{27.48}$$

where ν is its frequency (the same frequency that determines $k^{\ddagger}$). This frequency is much lower than for an ordinary molecular vibration because the oscillation corresponds to the complex falling apart (Fig. 27.8), so the force constant is very low.[5] Therefore, provided that $h\nu/kT \ll 1$, the exponential may be expanded and the partition function reduces to

$$q = \frac{1}{1 - \left(1 - \dfrac{h\nu}{kT} + \cdots\right)} \approx \frac{kT}{h\nu} \tag{27.49}$$

We can therefore write

$$q_{C^{\ddagger}} \approx \frac{kT}{h\nu} \bar{q}_{C^{\ddagger}} \tag{27.50}$$

where $\bar{q}$ denotes the partition function for all the other modes of the complex. The constant $K^{\ddagger}$ is therefore

$$K^{\ddagger} = \frac{kT}{h\nu} \bar{K}^{\ddagger} \qquad \bar{K}^{\ddagger} = \frac{N_A \bar{q}_{C^{\ddagger}}^{\ominus}}{q_A^{\ominus} q_B^{\ominus}} e^{-\Delta E_0/RT} \tag{27.51}$$

with $\bar{K}^{\ddagger}$ a kind of equilibrium constant, but with one vibrational mode of $C^{\ddagger}$ discarded.

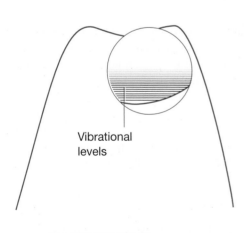

Vibrational
levels

Reaction coordinate

27.8 In an elementary depiction of the activated complex close to the transition state, there is a broad, shallow dip in the potential energy surface along the reaction coordinate. The complex vibrates harmonically and almost classically in this well. However, see footnote 5.

5. There is a real problem here. Figure 27.8 is probably an oversimplification, for in many cases there is no dip at the top of the barrier, and the curvature of the potential energy, and therefore the force constant, is negative. Formally, the vibrational frequency is then imaginary. We ignore this problem here, but see *Further reading*.

(c) The rate constant

We can now combine all the parts of the calculation into

$$k_2 = k^{\ddagger} \frac{RT}{p^{\ominus}} K^{\ddagger} = \kappa \nu \frac{kT}{h\nu} \frac{RT}{p^{\ominus}} \overline{K}^{\ddagger} \tag{27.52}$$

At this stage the unknown frequencies ν cancel, and after writing $\overline{K}_c^{\ddagger} = (RT/p^{\ominus})\overline{K}^{\ddagger}$, we obtain the **Eyring equation**:

$$k_2 = \kappa \frac{kT}{h} \overline{K}_c^{\ddagger} \tag{27.53}$$

The factor $\overline{K}_c^{\ddagger}$ is given by eqn 51 and the definition $\overline{K}_c^{\ddagger} = (RT/p^{\ominus})\overline{K}^{\ddagger}$ in terms of the partition functions of A, B, and C‡, so in principle we now have an explicit expression for calculating the second-order rate constant for a bimolecular reaction in terms of the molecular parameters for the reactants and the activated complex and the quantity κ.

The partition functions for the reactants can normally be calculated quite readily, using either spectroscopic information about their energy levels or the approximate expressions set out in Table 20.2. The difficulty with the Eyring equation, however, lies in the calculation of the partition function of the activated complex: C‡ is difficult to investigate spectroscopically (Box 27.1), and in general we need to make assumptions about its size, shape, and structure. We shall illustrate what is involved in one simple but significant case.

(d) The collision of structureless particles

Consider the case of two structureless particles A and B colliding to give an activated complex that resembles a diatomic molecule. Because the reactants J = A, B are structureless 'atoms', the only contributions to their partition functions are the translational terms:

$$q_J^{\ominus} = \frac{V_m^{\ominus}}{\Lambda_J^3} \qquad \Lambda_J = \frac{h}{(2\pi m_J kT)^{1/2}} \qquad V_m^{\ominus} = \frac{RT}{p^{\ominus}} \tag{27.54}$$

The activated complex is a diatomic cluster of mass $m_C = m_A + m_B$ and moment of inertia I. It has one vibrational mode, but that mode corresponds to motion along the reaction coordinate and therefore does not appear in $\bar{q}_{C^{\ddagger}}$. It follows that the standard molar partition function of the activated complex is

$$q_{C^{\ddagger}}^{\ominus} = \left(\frac{2IkT}{\hbar^2}\right) \frac{V_m^{\ominus}}{\Lambda_{C^{\ddagger}}^3} \tag{27.55}$$

The moment of inertia of a diatomic molecule of bond length r is μr^2, where $\mu = m_A m_B/(m_A + m_B)$ is the effective mass, so the expression for the rate constant is

$$k_2 = \kappa \frac{kT}{h} \frac{RT}{p^{\ominus}} \left(\frac{N_A \Lambda_A^3 \Lambda_B^3}{\Lambda_{C^{\ddagger}}^3 V_m^{\ominus}} \right) \left(\frac{2IkT}{\hbar^2} \right) e^{-\Delta E_0/RT}$$

$$= \kappa \frac{kT}{h} N_A \left(\frac{\Lambda_A \Lambda_B}{\Lambda_{C^{\ddagger}}} \right)^3 \left(\frac{2IkT}{\hbar^2} \right) e^{-\Delta E_0/RT} \tag{27.56}$$

$$= \kappa N_A \left(\frac{8kT}{\pi\mu} \right)^{1/2} \pi r^2 e^{-\Delta E_0/RT}$$

Finally, by identifying $\kappa \pi r^2$ as the reactive cross-section σ^*, we arrive at precisely the same expression as that obtained from simple collision theory (eqn 16).

Box 27.1 Femtochemistry

Until very recently there were no direct spectroscopic observations on activated complexes, for they have a very fleeting existence and often survive for only a few picoseconds. However, the development of femtosecond pulsed lasers (Section 17.5) and their application to chemistry in the form of *femtochemistry* has made it possible to make observations on species that have such short lifetimes that in a number of respects they resemble an activated complex. Ahmed Zewail received the Nobel Prize for Chemistry in 1999 for his pioneering work in femtochemistry.

In a typical experiment, a femtosecond pulse is used to excite a molecule to a dissociative state, and then a second femtosecond pulse is fired at an interval after the dissociating pulse. The frequency of the second pulse is set at an absorption of one of the free fragmentation products, so its absorption is a measure of the abundance of the dissociation product. For example, when ICN is dissociated by the first pulse, the emergence of CN from the photoactivated state can be monitored by watching the growth of the free CN absorption (or, more commonly, its laser-induced fluorescence). In this way it has been found that the CN signal remains zero until the fragments have separated by about 600 pm, which takes about 205 fs.

Some sense of the progress that has been made in the study of the intimate mechanism of chemical reactions can be obtained by considering the decay of the ion pair Na^+I^-. As shown in the first illustration, excitation of the ionic species with a femtosecond laser pulse forms an excited state that corresponds to a covalently bonded NaI molecule. The system can be described with two potential energy surfaces, one largely 'ionic' and another 'covalent', which cross at an internuclear separation of 693 pm. A short laser pulse is composed of a wide range of frequencies, which excite many vibrational states of NaI simultaneously. Consequently, the electronically excited complex exists as a superposition of states, or a localized wavepacket (Section 11.6), which oscillates between the 'covalent' and 'ionic' potential energy surfaces, as shown in the first illustration. The complex can also dissociate, shown as movement of the wavepacket toward very long internuclear separation along the dissociative surface. However, not every outward-going swing leads to dissociation because there is a chance that the I atom can be harpooned again, in which case it fails to make good its escape. The dynamics of the system is probed by a second laser pulse with a frequency that corresponds to the absorption frequency of the free Na product or to the frequency at which Na absorbs when it is a part of the complex. The latter frequency depends on the Na···I distance, so an absorption (in practice, a laser-induced fluorescence) is obtained each time the wavepacket returns to that separation.

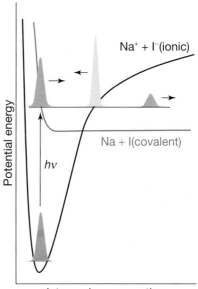

Excitation of the ion pair Na^+I^- forms an excited state with covalent character. Also shown is movement between a 'covalent' surface (in blue) and an 'ionic' surface (in black) of the wavepacket formed by laser excitation.

Femtosecond spectroscopic results for the reaction in which sodium iodide separates into Na and I. The full circles are the absorption of the electronically excited complex and the open circles the absorption of the free Na atoms (A.H. Zewail, *Science* **242**, 1645 (1988)).

A typical set of results is shown in the second illustration. The bound Na absorption intensity shows up as a series of pulses that recur in about 1 ps, showing that the wavepacket oscillates with about that period. The decline in intensity shows the rate at which the complex can dissociate as the two atoms swing away from each other. The free Na absorption also grows in an oscillating manner, showing the periodicity of wavepacket oscillation, each swing of which gives it a chance to dissociate. The precise period of the oscillation in NaI is 1.25 ps, corresponding to a vibrational wavenumber of 27 cm^{-1} (recall that the activated complex theory assumes that such a vibration has a very low frequency). The complex survives for about ten oscillations. In contrast, although the oscillation frequency of NaBr is similar, it barely survives one oscillation.

Femtosecond spectroscopy has also been used to examine analogues of the activated complex involved in bimolecular reactions. Thus, a molecular beam can be used to produce a van der Waals molecule (Section 21.7), such as IH···OCO. The HI bond can be dissociated by a femtosecond pulse, and the H atom is ejected towards the O atom of the neighbouring CO_2 molecule to form HOCO. Hence, the van der Waals molecule is a source of a species that resembles the activated complex of the reaction

$$H + CO_2 \longrightarrow [HOCO]^{\ddagger} \longrightarrow HO + CO$$

The probe pulse is tuned to the OH radical, which enables the evolution of [HOCO]‡ to be studied in real time. Femtosecond transition state spectroscopy has also been used to study more complex reactions, such as the Diels–Alder reaction, nucleophilic substitution reactions, and pericyclic addition and cleavage reactions. Biological process that are open to study by femtosecond spectroscopy include the energy-converting processes of photosynthesis (see Box 26.2) and the photostimulated processes of vision (see Box 17.1). In other experiments, the photoejection of carbon monoxide from myoglobin and the attachment of O_2 to the exposed site have been studied to obtain rate constants for the two processes (see *Discussion question 2* of Box 23.1).

Discussion questions

1. Consult literature sources and list the observed timescales during which the following processes occur: radiative

decay of excited electronic states, molecular rotational motion, molecular vibrational motion, proton transfer reactions, the initial event of vision, energy transfer in photosynthesis, the initial electron transfer events of photosynthesis, the helix-to-coil transition in polypeptides, harpoon reactions, and collisions in liquids.

2. This exercise gives some familiarity with the difficulties involved in predicting the structure of activated complexes. It also demonstrates the importance of femtosecond spectroscopy to our understanding of chemical dynamics because direct experimental observation of the activated complex removes much of the ambiguity of theoretical predictions. Consider the attack of H on D_2, which is one step in the $H_2 + D_2$ reaction. (a) Suppose that the H approaches D_2 from the side and forms a complex in the form of an isosceles triangle. Take the H—D distance as 30 per cent greater than in H_2 (74 pm) and the D—D distance as 20 per cent greater than in H_2. Let the critical coordinate be the antisymmetric stretching vibration in which one H—D bond stretches as the other shortens. Let all the vibrations be at about 1000 cm^{-1}. Estimate k_2 for this reaction at 400 K using the experimental activation energy of about 35 kJ mol^{-1}. (b) Now change the model of the activated complex in part (a) and make it linear. Use the same estimated molecular bond lengths and vibrational frequencies to calculate k_2 for this choice of model. (c) Clearly, there is much scope for modifying the parameters of the models of the activated complex. Use mathematical software or write and run a procedure that allows you to vary the structure of the complex and the parameters in a plausible way, and look for a model (or more than one model) that gives a value of k close to the experimental value, 4×10^5 L mol^{-1} s^{-1}.

Further reading

V. Sundström (ed.), *Femtochemistry and femtobiology: ultrafast reaction dynamics*. World Scientific, Singapore (1997).

A.H. Zewail, Femtochemistry: atomic-scale dynamics of the chemical bond. *J. Phys. Chem. A* **104**, 5660 (2000).

27.5 Thermodynamic aspects

The statistical thermodynamic version of activated complex theory rapidly runs into difficulties because only in some cases is anything known about the structure of the activated complex. However, the concepts that it introduces, principally that of an equilibrium between the reactants and the activated complex, have motivated a more general, empirical approach in which the activation process is expressed in terms of thermodynamic functions.

(a) Activation parameters

If we accept that $\bar{K}^{\ddagger}$ is an equilibrium constant (despite one mode of $C^{\ddagger}$ having been discarded), we can express it in terms of a **Gibbs energy of activation**, $\Delta^{\ddagger}G$, through the definition[6]

$$\Delta^{\ddagger}G = -RT \ln \bar{K}^{\ddagger} \qquad [27.57]$$

Then the rate constant becomes

$$k_2 = \kappa \frac{kT}{h} \frac{RT}{p^{\ominus}} e^{-\Delta^{\ddagger}G/RT} \qquad (27.58)$$

Because $G = H - TS$, the Gibbs energy of activation can be divided into an **entropy of activation**, $\Delta^{\ddagger}S$, and an **enthalpy of activation**, $\Delta^{\ddagger}H$, by writing

$$\Delta^{\ddagger}G = \Delta^{\ddagger}H - T\Delta^{\ddagger}S \qquad [27.59]$$

When eqn 59 is used in eqn 58 and κ is absorbed into the entropy term, we obtain

$$k_2 = B e^{\Delta^{\ddagger}S/R} e^{-\Delta^{\ddagger}H/RT} \qquad B = \frac{kT}{h} \frac{RT}{p^{\ominus}} \qquad (27.60)$$

The formal definition of activation energy, $E_a = RT^2(\partial \ln k/\partial T)$, then gives $E_a = \Delta^{\ddagger}H + 2RT$,[7] so

$$k_2 = e^2 B e^{\Delta^{\ddagger}S/R} e^{-E_a/RT} \qquad (27.61)$$

from which it follows that the Arrhenius factor A can be identified as

$$A = e^2 B e^{\Delta^{\ddagger}S/R} \qquad (27.62)$$

The entropy of activation is negative because two reactant species come together to form one species. However, if there is a reduction in entropy below what would be expected for the simple encounter of A and B, then A will be smaller than that expected on the basis of simple collision theory. Indeed, we can identify that additional reduction in entropy, $\Delta^{\ddagger}S_{steric}$, as the origin of the steric factor of collision theory, and write

$$P = e^{\Delta^{\ddagger}S_{steric}/R} \qquad (27.63)$$

Thus, the more complex the steric requirements of the encounter, the more negative the value of $\Delta^{\ddagger}S_{steric}$, and the smaller the value of P.

Gibbs energies, enthalpies, entropies, volumes, and heat capacities of activation are widely used to report experimental reaction rates, especially for organic reactions in solution. They are encountered when relationships between equilibrium constants and rates of reaction are explored using **correlation analysis**, in which $\ln K$ (which is equal to $-\Delta_r G^{\ominus}/RT$) is plotted against $\ln k$ (which is proportional to $-\Delta^{\ddagger}G/RT$). In many cases the correlation is linear, signifying that, as the reaction becomes thermodynamically more favourable, its rate constant increases (Fig. 27.9). This linear correlation is the origin of the alternative name **linear free energy relation** (LFER; see *Further reading*).

(b) Reactions between ions

The thermodynamic version of activated complex theory simplifies the discussion of reactions in solution. The statistical thermodynamic theory is very complicated to

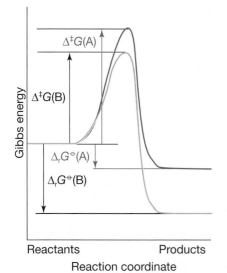

27.9 For a related series of reactions, as the magnitude of the standard reaction Gibbs energy increases, so the activation barrier decreases. The approximate linear correlation between $\Delta^{\ddagger}G$ and $\Delta_r G^{\ominus}$ is the origin of linear free energy relations.

6. All the $\Delta^{\ddagger}X$ in this section are *standard* thermodynamic quantities, $\Delta^{\ddagger}X^{\ominus}$, but we shall omit the standard state sign to avoid overburdening the notation.
7. For reactions in solution, $E_a = \Delta^{\ddagger}H + RT$.

apply because the solvent plays a role in the activated complex. In the thermodynamic approach we combine the rate law

$$\frac{d[P]}{dt} = k^{\ddagger}[C^{\ddagger}]$$

(27.64)

with the thermodynamic equilibrium constant

$$K = \frac{a_{C^{\ddagger}}}{a_A a_B} = K_\gamma \frac{[C^{\ddagger}]}{[A][B]} \qquad K_\gamma = \frac{\gamma_{C^{\ddagger}}}{\gamma_A \gamma_B}$$

(27.65)

Then

$$\frac{d[P]}{dt} = k_2[A][B] \qquad k_2 = \frac{k^{\ddagger}K}{K_\gamma}$$

(27.66)

If k_2° is the rate constant when the activity coefficients are 1 (that is, $k_2^\circ = k^{\ddagger}K$), we can write

$$k_2 = \frac{k_2^\circ}{K_\gamma}$$

(27.67)

At low concentrations the activity coefficients can be expressed in terms of the ionic strength, I, of the solution by using the Debye–Hückel limiting law (Section 10.2c, particularly eqn 10.17) in the form

$$\log \gamma_J = -A z_J^2 I^{1/2}$$

(27.68)

with $A = 0.509$ in aqueous solution at 298 K. Then

$$\log k_2 = \log k_2^\circ - A\{z_A^2 + z_B^2 - (z_A + z_B)^2\}I^{1/2} = \log k_2^\circ + 2A z_A z_B I^{1/2}$$

(27.69)

The charge numbers of A and B are z_A and z_B, so the charge number of the activated complex is $z_A + z_B$; the z_J are positive for cations and negative for anions.

Equation 69 expresses the **kinetic salt effect**, the variation of the rate constant of a reaction between ions with the ionic strength of the solution (Fig. 27.10). If the reactant ions have the same sign (as in a reaction between cations or between anions), then increasing the ionic strength by the addition of inert ions increases the rate constant. The formation of a single, highly charged ionic complex from two less highly charged ions is favoured by a high ionic strength because the new ion has a denser ionic atmosphere and interacts with that atmosphere more strongly. Conversely, ions of opposite charge react more slowly in solutions of high ionic strength. Now the charges cancel and the complex has a less favourable interaction with its atmosphere than the separated ions.

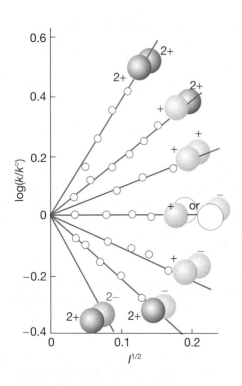

27.10 Experimental tests of the kinetic salt effect for reactions in water at 298 K. The ion types are shown as spheres, and the slopes of the lines are those given by the Debye–Hückel limiting law and eqn 69.

Example 27.3 Analysing the kinetic salt effect

The rate constant for the base hydrolysis of $[CoBr(NH_3)_5]^{2+}$ varies with ionic strength as tabulated below. What can be deduced about the charge of the activated complex in the rate-determining stage?

I	0.0050	0.0100	0.0150	0.0200	0.0250	0.0300
k/k°	0.718	0.631	0.562	0.515	0.475	0.447

Method According to eqn 69, plot $\log(k/k^\circ)$ against $I^{1/2}$, when the slope will give $1.02 z_A z_B$, from which we can infer the charges of the ions involved in the formation of the activated complex.

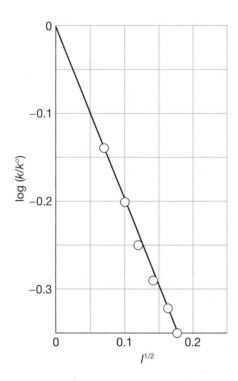

27.11 The experimental ionic strength dependence of the rate constant of a hydrolysis reaction: the slope gives information about the charge types involved in the activated complex of the rate-determining step. See Example 27.3.

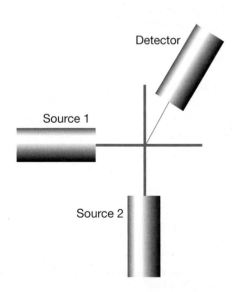

27.12 In a crossed-beam experiment, state-selected molecules are generated in two separate sources, and are directed perpendicular to one another. The detector responds to molecules (which may be product molecules if chemical reaction occurs) scattered into a chosen direction.

Answer Form the following table:

I	0.0050	0.0100	0.0150	0.0200	0.0250	0.0300
$I^{1/2}$	0.071	0.100	0.122	0.141	0.158	0.173
$\log(k/k^\circ)$	−0.14	−0.20	−0.25	−0.29	−0.32	−0.35

These points are plotted in Fig. 27.11. The slope of the (least squares) straight line is −2.04, indicating that $z_A z_B = -2$. Because $z_A = -1$ for the OH⁻ ion, if that ion is involved in the formation of the activated complex, then the charge number of the second ion is +2. This analysis suggests that the pentaamminebromocobalt(III) cation participates in the formation of the activated complex.

Comment The rate constant is also influenced by the relative permittivity of the medium.

Self-test 27.3 An ion of charge number +1 is known to be involved in the activated complex of a reaction. Deduce the charge number of the other ion from the following data:

I	0.0050	0.010	0.015	0.020	0.025	0.030
k/k°	0.930	0.902	0.884	0.867	0.853	0.841

[−1]

The dynamics of molecular collisions

We now come to the third and most detailed level of our examination of the factors that govern the rates of reactions.

27.6 Reactive collisions

Molecular beams allow us to study collisions between molecules in preselected energy states, and can be used to determine the states of the products of a reactive collision. Information of this kind is essential if a full picture of the reaction is to be built, because the rate constant is an average over events in which reactants in different initial states evolve into products in their final states.

(a) Experimental probes of reactive collisions

Detailed experimental information about the intimate processes that occur during reactive encounters comes from molecular beams, especially crossed molecular beams (Fig. 27.12). The detector for the products of the collision of two beams can be moved to different angles, so the angular distribution of the products can be determined. Because the molecules in the incoming beams can be prepared with different energies (for example, with different translational energies by using rotating sectors and supersonic nozzles, with different vibrational energies by using selective excitation with lasers, as shown in Box 27.2), and with different orientations (by using electric fields), it is possible to study the dependence of the success of collisions on these variables and to study how they affect the properties of the outcoming product molecules.

Box 27.2 Controlling chemical reactions with lasers

A long-standing goal of chemistry is to control the rate and distribution of products in chemical reactions, with an eye toward minimizing undesirable side reactions and improving the efficiency of industrial processes. We already have at our disposal a number of successful strategies for achieving this goal. For example, it is possible to synthesize a catalyst that accelerates a specific type of reaction but not others, so a desired product may be formed more quickly than an undesired product. However, this strategy is not very general, as a new catalyst needs to be developed for every reaction type of interest. A more ambitious and potentially more powerful strategy consists of using lasers to prepare specific states of reactant molecules that lead to a specific activated complex and hence a specific product; perhaps not even the major product isolated under ordinary laboratory conditions. Here we examine two ways in which the outcome of a chemical reaction can be affected by laser irradiation.

Some reactions may be controlled by exciting the reactants to different vibrational states. Consider the gas-phase reaction between H and HOD. It has been observed that H_2 and OD are the preferred products when thermally equilibrated H atoms react with vibrationally excited HOD molecules prepared by laser irradiation at a wavelength that excites the H—OD stretching mode from the $v = 0$ to the $v = 4$ energy level. On the other hand, when the same stretching mode is excited to the $v = 5$ energy level, HD and OH are the preferred products. This control strategy is commonly referred to as *mode-selective chemistry* and has been used to alter product distributions in a number of bimolecular reactions. However, the technique is limited to those cases in which energy can be deposited and remains localized in the desired vibrational mode of the reactant for a time that is much longer than the reaction time. This is difficult to achieve in large molecules, in which intramolecular vibrational relaxation redistributes energy among the many vibrational modes within a few picoseconds.

A strategy that seeks to avoid the problem of vibrational relaxation uses ultrafast lasers and is related closely to the techniques used for the spectroscopic detection of transition states (Box 27.1). Consider the reaction $I_2 + Xe \rightarrow XeI^* + I$, which occurs via a harpoon mechanism with a transition state denoted as $[Xe^+\cdots I^-\cdots I]$ (Section 27.1c). The reaction can be initiated by exciting I_2 to an electronic state at least 52 460 cm^{-1} above the ground state and then followed by measuring the time dependence of the chemiluminescence of XeI^*. To exert control over the yield of the product, a pair of femtosecond pulses can be used to induce the reaction. The first pulse excites the I_2 molecule to a low-energy and unreactive electronic state. We already know that excitation by a femtosecond pulse generates a wavepacket that can be treated as a particle traveling across the potential energy surface. In this case, the wavepacket does not have enough energy to react, but excitation by another laser pulse with the proper wavelength can provide the necessary additional energy. It follows that activated complexes with different geometries can be prepared by varying the time delay between the two pulses, as the partially localized wave packet will be at different locations along the potential energy surface as it evolves after being formed by the first pulse. Because the reaction occurs via the harpoon mechanism, the product yield is expected to be optimal if the second pulse is applied when the wave packet is at a point where the $Xe\cdots I_2$ distance is just right for electron transfer from Xe to I_2 to occur (see Example 27.2). This type of control of the $I_2 + Xe$ reaction has been demonstrated (see *Further reading*).

So far, the control techniques we have discussed have only been applied to reactions between relatively small molecules, with simple and well-understood potential energy surfaces. Extension of these techniques to the controlled synthesis of materials in routine laboratory work will require much more sophisticated knowledge of how laser pulses may be combined to stimulate a specific molecular response in a complex system.

Discussion questions

1. Another method for directing the outcome of a chemical reaction consists of using molecular beams to control the relative orientations of reactants during a collision. Consider the reaction $Rb + CH_3I \rightarrow RbI + CH_3$. How should CH_3I molecules and Rb atoms be oriented to maximize the production of RbI?

2. Consult literature sources and prepare a brief report on how lasers have been used to control one of the following reactions: $Cl + HOD$, $NH_3^+ + ND_3$, or $K + HF$.

Further reading

R.J. Gordon, L. Zhu, and I. Seideman, Coherent control of chemical reactions. *Acc. Chem. Res.* **32**, 1007 (1999).

R.N. Zare, Laser control of chemical reactions. *Science* **279**, 1875 (1998).

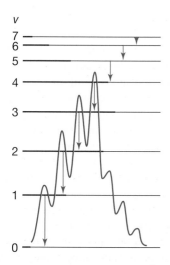

27.13 Infrared chemiluminescence from CO produced in the reaction O + CS → CO + S arises from the non-equilibrium populations of the vibrational states of CO and the radiative relaxation to equilibrium.

One method for examining the energy distribution in the products is **infrared chemiluminescence**, in which vibrationally excited molecules emit infrared radiation as they return to their ground states. By studying the intensities of the infrared emission spectrum, the populations of the vibrational states may be determined (Fig. 27.13). Another method makes use of **laser-induced fluorescence**. In this technique, a laser is used to excite a product molecule from a specific vibration–rotation level; the intensity of the fluorescence from the upper state is monitored and interpreted in terms of the population of the initial vibration–rotation state. When the molecules being studied do not fluoresce efficiently, coherent anti-Stokes Raman spectroscopy (CARS, Section 16.16) can be used to monitor the progress of reaction. **Multiphoton ionization** (MPI) techniques are also good alternatives for the study of weakly fluorescing molecules. In MPI, the absorption of several photons by a molecule results in ionization if the total photon energy is greater than the ionization energy of the molecule. One or more pulsed lasers are used to generate the molecular ions, which are commonly detected by time-of-flight mass spectrometry (TOF-MS, Section 22.7). The angular distribution of products can also be determined by **reaction product imaging**. In this technique, product ions are accelerated by an electric field towards a phosphorescent screen and light emitted from specific spots where the ions struck the screen is imaged by a charge-coupled device (CCD, Section 16.1). An important variant of MPI is **resonant multiphoton ionization** (REMPI), in which one or more photons promote a molecule to an electronically excited state and then additional photons are used to generate ions from the excited state. The power of REMPI lies in the fact that the experimenter can choose which reactant or product to study by tuning the laser frequency to the electronic absorption band of a specific molecule.

(b) State-to-state dynamics

The concept of collision cross-section was introduced in connection with collision theory in Section 27.1, where we saw that the second-order rate constant, k_2, can be expressed as a Boltzmann-weighted average of the reactive collision cross-section and the relative speed of approach. We shall write eqn 14 as

$$k_2 = \langle \sigma v_{rel} \rangle N_A \tag{27.70}$$

where the angle brackets denote a Boltzmann average. Molecular beam studies provide a more sophisticated version of this quantity, for they provide the **state-to-state cross-section**, $\sigma_{nn'}$, and hence the **state-to-state rate constant**, $k_{nn'}$:

$$k_{nn'} = \langle \sigma_{nn'} v_{rel} \rangle N_A \tag{27.71}$$

The rate constant k_2 is the sum of the state-to-state rate constant over all final states (because a reaction is successful whatever the final state of the products) and over a Boltzmann-weighted sum of initial states (because the reactants are initially present with a characteristic distribution of populations at a temperature T):

$$k_2 = \sum_{nn'} k_{nn'}(T) f_n(T) \tag{27.72}$$

where $f_n(T)$ is the Boltzmann factor at a temperature T. It follows that, if we can determine or calculate the state-to-state cross-sections for a wide range of approach speeds and initial and final states, then we have a route to the calculation of the rate constant for the reaction.

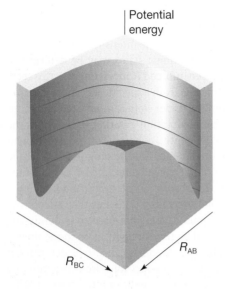

27.14 The potential energy surface for the $H + H_2 \rightarrow H_2 + H$ reaction when the atoms are constrained to be collinear.

27.7 Potential energy surfaces

One of the most important concepts for discussing beam results and calculating the state-to-state collision cross-section is the **potential energy surface** of a reaction, the potential energy as a function of the relative positions of all the atoms taking part in the reaction. Potential energy surfaces may be constructed from experimental data, with the techniques described in Section 27.6, and from results of *ab initio* molecular orbital calculations (Section 14.9b). The theoretical method requires the systematic calculation of the energies of the system in a large number of geometrical arrangements. Special computational techniques are used to take into account electron correlation, which arises from instantaneous interactions between electrons as they move closer to and farther from each other in a molecule or molecular cluster. Techniques that incorporate electron correlation are very time-consuming and, consequently, only reactions between relatively small particles, such as the reactions $H + H_2 \rightarrow H_2 + H$ and $H + H_2O \rightarrow OH + H_2$, are amenable to this type of theoretical treatment. An alternative is to use semi-empirical methods, in which results of calculations and experimental parameters are used to construct the potential energy surface.

To illustrate the features of a potential energy surface we consider the collision between an H atom and an H_2 molecule. Detailed calculations show that the approach of an atom along the H—H axis requires less energy for reaction than any other approach, so initially we confine our attention to a collinear approach. Two parameters are required to define the nuclear separations: one is the H_A—H_B separation R_{AB}, and the other is the H_B—H_C separation R_{BC}.

At the start of the encounter R_{AB} is infinite and R_{BC} is the H_2 equilibrium bond length. At the end of a successful reactive encounter R_{AB} is equal to the equilibrium bond length and R_{BC} is infinite. The total energy of the three-atom system depends on their relative separations, and can be found by doing a molecular orbital calculation. The plot of the total energy of the system against R_{AB} and R_{BC} gives the potential energy surface of this collinear reaction (Fig. 27.14). This surface is normally depicted as a contour diagram (Fig. 27.15).

When R_{AB} is very large, the variation in potential energy represented by the surface as R_{BC} changes are those of an isolated H_2 molecule as its bond length is altered. A section through the surface at $R_{AB} = \infty$, for example, is the same as the H_2 bonding potential energy curve drawn in Fig. 14.3. At the edge of the diagram where R_{BC} is very large, a section through the surface is the molecular potential energy curve of an isolated $H_A H_B$ molecule.

The actual path of the atoms in the course of the encounter depends on their total energy, the sum of their kinetic and potential energies. However, we can obtain an initial idea of the paths available to the system for paths that correspond to least potential energy. For example, consider the changes in potential energy as H_A approaches $H_B H_C$. If the H_B—H_C bond length is constant during the initial approach of H_A, then the potential energy of the H_3 cluster rises along the path marked A in Fig. 27.16. We see that the potential energy reaches a high value as H_A is pushed into the molecule and then decreases sharply as H_C breaks off and separates to a great distance. An alternative reaction path can be imagined (B) in which the H_B—H_C bond length increases while H_A is still far away. Both paths, although feasible if the molecules have sufficient initial kinetic energy, take the three atoms to regions of high potential energy in the course of the encounter.

The path of least potential energy is the one marked C, corresponding to R_{BC} lengthening as H_A approaches and begins to form a bond with H_B. The H_B—H_C bond relaxes at the demand of the incoming atom, and the potential energy climbs only as far as the saddle-shaped region of the surface, to the **saddle point** marked $C^\ddagger$. The encounter of least potential energy is one in which the atoms take route C up the

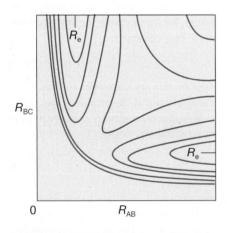

27.15 The contour diagram (with contours of equal potential energy) corresponding to the surface in Fig. 27.14. R_e marks the equilibrium bond length of an H_2 molecule (strictly, it relates to the arrangement when the third atom is at infinity).

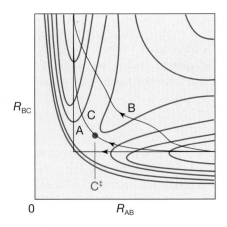

27.16 Various trajectories through the potential energy surface shown in Fig. 27.15. Path A corresponds to a route in which R_{BC} is held constant as H_A approaches; path B corresponds to a route in which R_{BC} lengthens at an early stage during the approach of H_A; path C is the route along the floor of the potential valley.

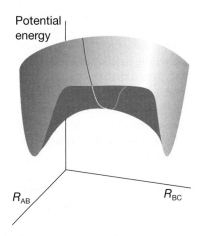

27.17 The transition state is a set of configurations (here, marked by the line across the saddle point) through which successful reactive trajectories must pass.

floor of the valley, through the saddle point, and down the floor of the other valley as H_C recedes and the new H_A—H_B bond achieves its equilibrium length. This path is the reaction coordinate we first met in Section 25.5.

We can now make contact with the activated complex theory of reaction rates. In terms of trajectories on potential surfaces, the transition state can be identified with a critical geometry such that every trajectory that goes through this geometry goes on to react (Fig. 27.17).

27.8 Some results from experiments and calculations

To travel successfully from reactants to products the incoming molecules must possess enough kinetic energy to be able to climb to the saddle point of the potential surface. Therefore, the shape of the surface can be explored experimentally by changing the relative speed of approach (by selecting the beam velocity) and the degree of vibrational excitation and observing whether reaction occurs and whether the products emerge in a vibrationally excited state (Fig. 27.18). For example, one

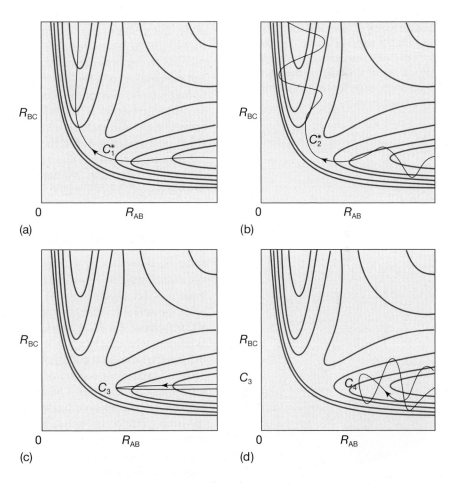

27.18 Some successful (*) and unsuccessful encounters. (a) C_1^* corresponds to the path along the foot of the valley; (b) C_2^* corresponds to an approach of A to a vibrating BC molecule, and the formation of a vibrating AB molecule as C departs. (c) C_3 corresponds to A approaching a non-vibrating BC molecule, but with insufficient translational kinetic energy; (d) C_4 corresponds to A approaching a vibrating BC molecule, but still the energy, and the phase of the vibration, is insufficient for reaction.

question that can be answered is whether it is better to smash the reactants together with a lot of translational kinetic energy or to ensure instead that they approach in highly excited vibrational states. Thus, is trajectory C_2^*, where the H_BH_C molecule is initially vibrationally excited, more efficient at leading to reaction than the trajectory C_1^*, in which the total energy is the same but has a high translational kinetic energy?

(a) The direction of attack and separation

Figure 27.19 shows the results of a calculation of the potential energy as an H atom approaches an H_2 molecule from different angles, the H_2 bond being allowed to relax to the optimum length in each case. The potential barrier is least for collinear attack, as we assumed earlier. (But we must be aware that other lines of attack are feasible and contribute to the overall rate.) In contrast, Fig. 27.20 shows the potential energy changes that occur as a Cl atom approaches an HI molecule. The lowest barrier occurs for approaches within a cone of half-angle 30° surrounding the H atom. The relevance of this result to the calculation of the steric factor of collision theory should be noted: not every collision is successful, because not every one lies within the reactive cone.

If the collision is sticky, so that when the reactants collide they orbit around each other, the products can be expected to emerge in random directions because all memory of the approach direction has been lost. A rotation takes about 1 ps, so if the collision is over in less than that time the complex will not have had time to rotate and the products will be thrown off in a specific direction. In the collision of K and I_2, for example, most of the products are thrown off in the forward direction.[8] This product distribution is consistent with the harpoon mechanism (Section 27.1c) because the transition takes place at long range. In contrast, the collision of K with CH_3I leads to reaction only if the molecules approach each other very closely. In this mechanism, K effectively bumps into a brick wall, and the KI product bounces out in the backward direction. The detection of this anisotropy in the angular distribution of products gives an indication of the distance and orientation of approach needed for reaction, as well as showing that the event is complete in less than 1 ps.

(b) Attractive and repulsive surfaces

Some reactions are very sensitive to whether the energy has been predigested into a vibrational mode or left as the relative translational kinetic energy of the colliding molecules. For example, if two HI molecules are hurled together with more than twice the activation energy of the reaction, then no reaction occurs if all the energy is translational. For $F + HCl \rightarrow Cl + HF$, for example, the reaction is about five times as efficient when the HCl is in its first vibrational excited state than when, although HCl has the same total energy, it is in its vibrational ground state.

The origin of these requirements can be found by examining the potential energy surface. Figure 27.21 shows an **attractive surface** in which the saddle point occurs early in the reaction coordinate. Figure 27.22 shows a **repulsive surface** in which the saddle point occurs late. A surface that is attractive in one direction is repulsive in the reverse direction.

Consider first the attractive surface. If the original molecule is vibrationally excited, then a collision with an incoming molecule takes the system along C. This

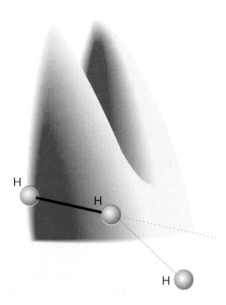

27.19 An indication of the anisotropy of the potential energy changes as H approaches H_2 with different angles of attack. The collinear attack has the lowest potential barrier to reaction. The surface indicates the potential energy profile along the reaction coordinate for each configuration.

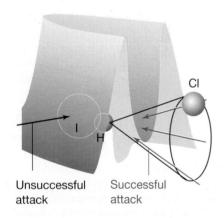

Unsuccessful attack

Successful attack

27.20 The potential energy barrier for the approach of Cl to HI. In this case, successful encounters occur only when Cl approaches within a cone surrounding the H atom.

8. There is a subtlety here. In molecular beam work the remarks normally refer to directions in a centre-of-mass coordinate system. The origin of the coordinates is the centre of mass of the colliding reactants, and the collision takes place when the molecules are at the origin. The way in which centre-of-mass coordinates are constructed and the events in them interpreted involves too much detail for our present purposes, but we should bear in mind that 'forward' and 'backward' have unconventional meanings. The details are explained in the books in *Further reading*.

27.21 An attractive potential energy surface. A successful encounter (C*) involves high translational kinetic energy and results in a vibrationally excited product.

27.22 A repulsive potential energy surface. A successful encounter (C*) involves initial vibrational excitation and the products have high translational kinetic energy. A reaction that is attractive in one direction is repulsive in the reverse direction.

path is bottled up in the region of the reactants, and does not take the system to the saddle point. If, however, the same amount of energy is present solely as translational kinetic energy, then the system moves along C* and travels smoothly over the saddle point into products. We can therefore conclude that reactions with attractive potential energy surfaces proceed more efficiently if the energy is in relative translational motion. Moreover, the potential surface shows that once past the saddle point the trajectory runs up the steep wall of the product valley, and then rolls from side to side as it falls to the foot of the valley as the products separate. In other words, the products emerge in a vibrationally excited state.

Now consider the repulsive surface (Fig. 27.22). On trajectory C the collisional energy is largely in translation. As the reactants approach, the potential energy rises. Their path takes them up the opposing face of the valley, and they are reflected back into the reactant region. This path corresponds to an unsuccessful encounter, even though the energy is sufficient for reaction. On C* some of the energy is in the vibration of the reactant molecule and the motion causes the trajectory to weave from side to side up the valley as it approaches the saddle point. This motion may be sufficient to tip the system round the corner to the saddle point and then on to products. In this case, the product molecule is expected to be in an unexcited vibrational state. Reactions with repulsive potential surfaces can therefore be expected to proceed more efficiently if the excess energy is present as vibrations. This is the case with the $H + Cl_2 \rightarrow HCl + Cl$ reaction, for instance.

(c) Classical trajectories

A clear picture of the reaction event can be obtained by using classical mechanics to calculate the trajectories of the atoms taking place in a reaction from a set of initial conditions, such as velocities, relative orientations, and internal energies of the reacting particles. The initial values used for the internal energy reflect the quantization of electronic, vibrational, and rotational energies in molecules but the features of quantum mechanics are not used explicitly in the calculation of the trajectory.

Figure 27.23 shows the result of such a calculation of the positions of the three atoms in the reaction $H + H_2 \rightarrow H_2 + H$, the horizontal coordinate now being time and the vertical coordinate the separations. This illustration shows clearly the vibration of the original molecule and the approach of the attacking atom. The

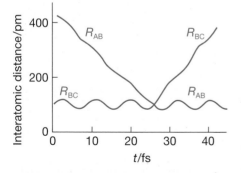

27.23 The calculated trajectories for a reactive encounter between A and a vibrating BC molecule leading to the formation of a vibrating AB molecule. This direct-mode reaction is between H and H_2. (M. Karplus, R.N. Porter, and R.D Sharma, *J. Chem. Phys.* **43**, 3258 (1965).)

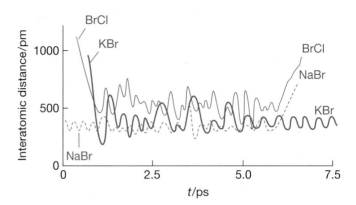

27.24 An example of the trajectories calculated for a complex-mode reaction, KCl + NaBr → KBr + NaCl, in which the collision cluster has a long lifetime. (P. Brumer and M. Karplus, *Faraday Disc. Chem. Soc.*, **55**, 80 (1973).)

reaction itself, the switch of partners, takes place very rapidly and is an example of a **direct mode process**. The newly formed molecule shakes, but quickly settles down to steady, harmonic vibration as the expelled atom departs. In contrast, Fig. 27.24 shows an example of a **complex mode process**, in which the activated complex survives for an extended period. The reaction in the illustration is the exchange reaction KCl + NaBr → KBr + NaCl. The tetratomic activated complex survives for about 5 ps, during which time the atoms make about 15 oscillations before dissociating into products.

(d) Quantum mechanical scattering theory

Classical trajectory calculations do not recognize the fact that the motion of atoms, electrons, and nuclei is governed by quantum mechanics. The concept of trajectory then fades and is replaced by the unfolding of a wavefunction that represents initially the reactants and finally products.

Complete quantum mechanical calculations of trajectories and rate constants are very onerous because it is necessary to take into account all the allowed electronic, vibrational, and rotational states populated by each atom and molecule in the system at a given temperature. It is common to define a 'channel' as a group of molecules in well-defined quantum mechanically allowed states. Then, at a given temperature, there are many channels that represent the reactants and many channels that represent possible products, with some transitions between channels being allowed but others not allowed. Furthermore, not every transition leads to a chemical reaction. For example, the process $H_2^* + OH \rightarrow H_2 + (OH)^*$, where the asterisk denotes an excited state, amounts to energy transfer between H_2 and OH, whereas the process $H_2^* + OH \rightarrow H_2O + H$ represents a chemical reaction. What complicates a quantum mechanical calculation of trajectories and rate constants even in this simple four-atom system is that many reacting channels present at a given temperature can lead to the desired products $H_2O + H$, which themselves may be formed as many distinct channels. The **cumulative reaction probability**, $N(E)$, at a fixed total energy E is then written as

$$N(E) = \sum_{i,j} P_{i,j}(E) \tag{27.73}$$

where $P_{i,j}(E)$ is the probability for a transition between a reacting channel i and a product channel j and the summation is over all possible transitions that lead to product. It is then possible to show that the rate constant is given by

$$k(T) = \frac{\int_0^\infty N(E)e^{-E/kT}\, dE}{hQ_r(T)} \tag{27.74}$$

where $Q_r(T)$ is the partition function density (the partition function divided by the volume) of the reactants at the temperature T. The significance of eqn 74 is that it provides a direct connection between an experimental quantity, the rate constant, and a theoretical quantity, $N(E)$.

Table 27.3 Summary of uses of k

Symbol	Significance
k	Boltzmann constant
k_2	Second-order rate constant
k_2°	Rate constant at zero ionic strength
$k_a, k_b, \ldots$	Rate constants for individual steps
$k_a', k_b', \ldots$	Rate constants for individual reverse steps
$k^\ddagger$	Rate constant for unimolecular decay of activated complex
K	Equilibrium constant (dimensionless)
K_γ	Ratio of activity coefficients
$K^\ddagger$	Proportionality constant between $[C^\ddagger]$ and $[A][B]$ (dimensions, 1/concentration)
κ	Transmission coefficient
$\bar{K}$	When multiplied by $p^\ominus/RT$, a type of equilibrium constant but with one vibrational mode discarded (dimensions, 1/concentration)
k_f	Force constant

Checklist of key ideas

Reactive encounters

27.1 Collision theory

☐ rate constant, $k_2 \propto$ steric requirement × encounter rate × minimum energy requirement

☐ collision density, the number of (A, B) collisions in a region of the sample in an interval of time divided by the volume of the region and the duration of the interval: $Z_{AB} = \sigma(8kT/\pi\mu)^{1/2}N_A^2[A][B]$

☐ general form of rate constant, $k_2 = N_A \int_0^\infty \sigma(\varepsilon)v_{rel}f(\varepsilon)\,d\varepsilon$

☐ collision cross-section, $\sigma(\varepsilon) = (1 - \varepsilon_a/\varepsilon)\sigma$

☐ steric factor, $P = \sigma^*/\sigma$

☐ reactive cross-section, σ^*, the area within which a molecule must approach another molecule for reaction to occur

☐ rate constant from collision theory, $k_2 = P\sigma(8kT/\pi\mu)^{1/2}N_A e^{-E_a/RT}$

☐ harpoon mechanism, a process in which electron transfer precedes atom extraction

27.2 Diffusion-controlled reactions

☐ cage effect, the lingering of one molecule near another on account of the hindering presence of solvent molecules

☐ diffusion-controlled limit, a reaction in which the rate is controlled by the rate at which reactant molecules encounter each other in solution; $k_d = 4\pi R^*DN_A \approx 8RT/3\eta$

☐ activation-controlled limit, a reaction in solution in which the rate is controlled by the rate of accumulating sufficient energy to react

27.3 The material balance equation

☐ material balance equation, $\partial[J]/\partial t = D\partial^2[J]/\partial x^2 - v\partial[J]/\partial x - k[J]$

Activated complex theory

☐ activated complex theory (ACT), a theory of rate constants for elementary bimolecular reactions

27.4 The Eyring equation

☐ transmission coefficient, κ, the constant of proportionality between the rate of passage of the complex through the transition state and the vibrational frequency along the reaction coordinate

☐ Eyring equation, $k_2 = \kappa(kT/h)\bar{K}_c^\ddagger$

27.5 Thermodynamic aspects

☐ Gibbs energy of activation, $\Delta^\ddagger G = -RT \ln \bar{K}^\ddagger$

☐ entropy and enthalpy of activation, $\Delta^\ddagger G = \Delta^\ddagger H - T\Delta^\ddagger S$

☐ correlation analysis, a procedure in which $\ln K$ is plotted against $\ln k$

☐ liner free energy relation (LFER), a linear relation obtained in correlation analysis

☐ kinetic salt effect, the effect of a change in ionic strength on the rate constant of a reaction; $\log k_2 = \log k_2^\circ + 2Az_A z_B I^{1/2}$

The dynamics of molecular collisions

27.6 Reactive collisions

- ☐ infrared chemiluminescence, a process in which vibrationally excited molecules emit infrared radiation as they return to their ground states

- ☐ laser-induced fluorescence, a technique in which a laser is used to excite a product molecule from a specific vibration–rotation level and then the intensity of fluorescence is monitored

- ☐ multiphoton ionization (MPI), a process in which the absorption of several photons by a molecule results in ionization

- ☐ reaction product imaging, a technique for the determination of the angular distribution of products

- ☐ resonant multiphoton ionization (REMPI), a technique in which one or more photons promote a molecule to an electronically excited state and then additional photons are used to generate ions from the excited state

- ☐ rate constant, $k_2 = \langle \sigma v_{rel} \rangle N_A$ $= \sum_{nn'} k_{nn'}(T) f_n(T)$

- ☐ state-to-state cross-section, $\sigma_{nn'}$, the cross-section for reaction in which a specified initial state changes into a specified final state

- ☐ state-to-state rate constant, the rate constant for a specified state-to-state reaction; $k_{nn'} = \langle \sigma_{nn'} v_{rel} \rangle N_A$

27.7 Potential energy surfaces

- ☐ potential energy surface, the potential energy as a function of the relative positions of all the atoms taking part in the reaction

- ☐ saddle point, the highest point on a potential energy surface encountered along the reaction coordinate

27.8 Some results from experiments and calculations

- ☐ attractive surface, a potential energy surface in which the saddle point occurs early on the reaction coordinate

- ☐ repulsive surface, a potential energy surface in which the saddle point occurs late on the reaction coordinate

- ☐ direct mode process, a bimolecular process in which the switch of partners takes place very rapidly

- ☐ complex mode process, a bimolecular process in which the activated complex survives for an extended period

- ☐ cumulative reaction probability, $N(E) = \sum_{i,j} P_{i,j}(E)$

Further reading

Articles of general interest

A.J. Alexander and R.N. Zare, Anatomy of elementary chemical reactions. *J. Chem. Educ.* **75**, 1105 (1998).

S.H. Bauer, General potential energy surfaces for catalytic processes. *J. Chem. Educ.* **76**, 440 (1999).

L.J. Butler, Chemical reaction dynamics beyond the Born–Oppenheimer approximation. *Ann. Rev. Phys. Chem.* **49**, 125 (1998).

R.W. Carr, Chemical kinetics, In *Encyclopedia of applied physics* (ed. G.L. Trigg), **3**, 345. VCH, New York (1992).

D.C. Clary, Quantum theory of chemical reaction dynamics. *Science* **279**, 1879 (1998).

F.F. Crim, Vibrational state control of bimolecular reactions: discovering and directing the chemistry. *Acc. Chem. Res.* **32**, 877 (1999).

R. de Levie, Stochastics, the basis of chemical dynamics. *J. Chem. Educ.* **77**, 771 (2000).

G.M. Fernández, J.A. Sordo, and T.L. Sordo, Analysis of potential energy surfaces. *J. Chem. Educ.* **65**, 665 (1988).

K.J. Laidler, Just what is a transition state? *J. Chem. Educ.* **65**, 540 (1988).

S.R. Logan, The meaning and significance of 'the activation energy' of a chemical reaction. *Educ. in Chem.* **23**, 148 (1986).

H. Maskill, The Arrhenius equation. *Educ. in Chem.* **27**, 111 (1990).

M.A. Smith, The nature of distribution functions for colliding systems: Calculation of averaged properties. *J. Chem. Educ.* **70**, 218 (1993).

A.H. Zewail, Femtochemistry: atomic-scale dynamics of the chemical bond using ultrafast lasers. *Angew. Chem. Int. Ed.* **39**, 2586 (2000).

Texts and sources of data and information

C.H. Bamford, C.F. Tipper, and R.G. Compton (ed.), *Comprehensive chemical kinetics*, Vols. 1–38. Elsevier, Amsterdam (1969–2001).

R.S. Berry, S.A. Rice, and J. Ross, *Physical chemistry*. Oxford University Press (2000).

R.B. Bernstein, *Chemical dynamics via molecular beam and laser techniques*. Clarendon Press, Oxford (1982).

G.D. Billing and K.V. Mikkelsen, *Molecular dynamics and chemical kinetics*. Wiley, New York (1996).

G.D. Billing and K.V. Mikkelsen, *Advanced molecular dynamics and chemical kinetics*. Wiley, New York (1997).

M.J. Blandamer, *Chemical equilibria in solution*. Ellis Horwood/Prentice Hall, Hemel Hempstead (1992).

M. Brouard, *Reaction dynamics*. Oxford Chemistry Primers, Oxford University Press (1998).

M.A. El-Sayed, I. Tanaka, and Y. Molin (ed.), *Ultrafast processes in chemistry and photobiology*. Blackwell Science, Oxford (1995).

R.G. Gilbert and S.C. Smith, *Theory of unimolecular and recombination reactions*. Blackwell Scientific, Oxford (1990).

D.M. Hirst, *Potential energy surfaces*. Taylor and Francis, London (1985).

R.D. Levine and R.B. Bernstein, *Molecular reaction dynamics and chemical reactivity*. Clarendon Press, Oxford (1987).

J.C. Lindon, G.E. Tranter, and J.L. Holmes (ed.), *Encyclopedia of spectroscopy and spectrometry*. Academic Press, San Diego (2000).

S.R. Logan, *Chemical kinetics*. Longman, Harlow (1996).

M.J. Pilling and P.W. Seakins, *Reaction kinetics*. Oxford University Press (1996).

J. Simons, *Energetic principles of chemical reactions*. Jones and Bartlett, Portola Valley (1983).

I.W.M. Smith, *Kinetics and dynamics of elementary gas reactions*. Butterworth, London (1980).

J.I. Steinfeld, J.S. Francisco, and W.L. Hase, *Chemical kinetics and dynamics*. Prentice Hall, Upper Saddle River (1999).

G.R. Van Hecke and K.K. Karukstis, *A guide to lasers in chemistry*. Jones and Bartlett, Boston (1998).

Exercises

Discussion questions

27.1 (a) Describe the essential features of the harpoon mechanism.

27.1 (b) Distinguish between a diffusion-controlled reaction and an activation-controlled reaction.

27.2 (a) Describe the formulation of the Eyring equation.

27.2 (b) Discuss the physical origin of the kinetic salt effect.

27.3 (a) Describe how the following techniques are used in the study of chemical dynamics: infrared chemiluminescence, laser-induced fluorescence, multiphoton ionization, resonant multiphoton ionization, reaction product imaging, and femtosecond spectroscopy.

27.3 (b) Justify the following statements. (a) Reactions with attractive potential energy surfaces proceed more efficiently if the energy is in relative translational motion. (b) Reactions with repulsive potential surfaces proceed more efficiently if the excess energy is present as vibrations.

Numerical exercises

27.4 (a) Calculate the collision frequency, z, and the collision density, Z, in ammonia, $R = 190$ pm, at 25°C and 100 kPa. What is the percentage increase when the temperature is raised by 10 K at constant volume?

27.4 (b) Calculate the collision frequency, z, and the collision density, Z, in carbon monoxide, $R = 180$ pm at 25°C and 100 kPa. What is the percentage increase when the temperature is raised by 10 K at constant volume?

27.5 (a) To use collision theory we must know the fraction of molecular collisions having at least the kinetic energy E_a along the line of flight. What is this fraction when (a) $E_a = 10$ kJ mol^{-1}, (b) $E_a = 100$ kJ mol^{-1} at (i) 300 K and (ii) 1000 K?

27.5 (b) To use collision theory we must know the fraction of molecular collisions having at least the kinetic energy E_a along the line of flight. What is this fraction when (a) $E_a = 15$ kJ mol^{-1}, (b) $E_a = 150$ kJ mol^{-1} at (i) 300 K and (ii) 800 K?

27.6 (a) Calculate the percentage increase in the fractions in Exercise 27.5a when the temperature is raised by 10 K.

27.6 (b) Calculate the percentage increase in the fractions in Exercise 27.5b when the temperature is raised by 10 K.

27.7 (a) Use the collision theory of gas-phase reactions to calculate the theoretical value of the second-order rate constant for the reaction $H_2(g) + I_2(g) \rightarrow 2 HI(g)$ at 650 K, assuming that it is elementary bimolecular. The collision cross-section is 0.36 nm^2, the reduced mass is 3.32×10^{-27} kg, and the activation energy is 171 kJ mol^{-1}.

27.7 (b) Use the collision theory of gas-phase reactions to calculate the theoretical value of the second-order rate constant for the reaction $D_2(g) + Br_2(g) \rightarrow 2 DBr(g)$ at 450 K, assuming that it is elementary bimolecular. Take the collision cross-section as 0.30 nm^2, the reduced mass as 3.930 u, and the activation energy as 200 kJ mol^{-1}.

27.8 (a) A typical diffusion coefficient for small molecules in aqueous solution at 25°C is 5×10^{-9} m^2 s^{-1}. If the critical reaction distance is 0.4 nm, what value is expected for the second-order rate constant for a diffusion-controlled reaction?

27.8 (b) Suppose that the typical diffusion coefficient for a reactant in aqueous solution at 25°C is 4.2×10^{-9} m^2 s^{-1}. If the critical reaction distance is 0.50 nm, what value is expected for the second-order rate constant for the diffusion-controlled reaction?

27.9 (a) Calculate the magnitude of the diffusion-controlled rate constant at 298 K for a species in (a) water, (b) pentane. The viscosities are 1.00×10^{-3} kg m^{-1} s^{-1}, and 2.2×10^{-4} kg m^{-1} s^{-1}, respectively.

27.9 (b) Calculate the magnitude of the diffusion-controlled rate constant at 298 K for a species in (a) decylbenzene, (b) concentrated sulfuric acid. The viscosities are 3.36 cP and 27 cP, respectively.

27.10 (a) Calculate the magnitude of the diffusion-controlled rate constant at 298 K for the recombination of two atoms in water, for which $\eta = 0.89$ cP. Assuming the concentration of the reacting species is 1.0 mmol L^{-1} initially, how long does it take for the concentration of the atoms to fall to half that value? Assume the reaction is elementary.

27.10 (b) Calculate the magnitude of the diffusion-controlled rate constant at 298 K for the recombination of two atoms in benzene, for which $\eta = 0.601$ cP. Assuming the concentration of the reacting species is 1.8 mmol L^{-1} initially, how long does it take for the concentration of the atoms to fall to half that value? Assume the reaction is elementary.

27.11 (a) For the gaseous reaction $A + B \rightarrow P$, the reactive cross-section obtained from the experimental value of the pre-exponential factor is 9.2×10^{-22} m^2. The collision cross-sections of A and B estimated from the transport properties are 0.95 and 0.65 nm^2, respectively. Calculate the P factor for the reaction.

27.11 (b) For the gaseous reaction $A + B \rightarrow P$, the reactive cross-section obtained from the experimental value of the pre-exponential factor is 8.7×10^{-22} m^2. The collision cross-sections of A and B estimated from the transport properties are 0.88 and 0.40 nm^2, respectively. Calculate the P factor for the reaction.

27.12 (a) Two neutral species, A and B, with diameters 588 pm and 1650 pm, respectively, undergo the diffusion-controlled reaction $A + B \rightarrow P$ in a solvent of viscosity 2.37×10^{-3} kg m^{-1} s^{-1} at 40°C. Calculate the initial rate d[P]/dt if the initial concentrations of A and B are 0.150 mol L^{-1} and 0.330 mol L^{-1}, respectively.

27.12 (b) Two neutral species, A and B, with diameters 442 pm and 885 pm, respectively, undergo the diffusion-controlled reaction $A + B \rightarrow P$ in a solvent of viscosity 1.27 cP at 20°C. Calculate the initial rate d[P]/dt if the initial concentrations of A and B are 0.200 mol L^{-1} and 0.150 mol L^{-1}, respectively.

27.13 (a) The reaction of propylxanthate ion in acetic acid buffer solutions has the mechanism $A^- + H^+ \rightarrow P$. Near 30°C the rate constant is given by the empirical expression $k_2 = (2.05 \times 10^{13})e^{-(8681\ K)/T}$ L mol^{-1} s^{-1}. Evaluate the energy and entropy of activation at 30°C.

27.13 (b) The reaction $A^- + H^+ \rightarrow P$ has a rate constant given by the empirical expression $k_2 = (8.72 \times 10^{12})e^{-(6134 \text{ K})/T}$ L mol^{-1} s^{-1}. Evaluate the energy and entropy of activation at 25°C.

27.14 (a) When the reaction in Exercise 27.13a occurs in a dioxane/water mixture which is 30 per cent dioxane by mass, the rate constant fits $k_2 = (7.78 \times 10^{14})e^{-(9134 \text{ K})/T}$ L mol^{-1} s^{-1} near 30°C. Calculate $\Delta^{\ddagger}G$ for the reaction at 30°C.

27.14 (b) A rate constant is found to fit the expression $k_2 = (6.45 \times 10^{13})e^{-(5375 \text{ K})/T}$ L mol^{-1} s^{-1} near 25°C. Calculate $\Delta^{\ddagger}G$ for the reaction at 25°C.

27.15 (a) The gas-phase association reaction between F_2 and IF_5 is first-order in each of the reactants. The energy of activation for the reaction is 58.6 kJ mol^{-1}. At 65°C the rate constant is 7.84×10^{-3} kPa^{-1} s^{-1}. Calculate the entropy of activation at 65°C.

27.15 (b) A gas-phase recombination reaction is first-order in each of the reactants. The energy of activation for the reaction is 49.6 kJ mol^{-1}. At 55°C the rate constant is 0.23 m^3 s^{-1}. Calculate the entropy of activation at 55°C.

27.16 (a) Calculate the entropy of activation for a collision between two structureless particles at 300 K, taking $M = 50$ g mol^{-1} and $\sigma = 0.40$ nm^2.

27.16 (b) Calculate the entropy of activation for a collision between two structureless particles at 500 K, taking $M = 78$ g mol^{-1} and $\sigma = 0.62$ nm^2.

27.17 (a) The pre-exponential factor for the gas-phase decomposition of ozone at low pressures is 4.6×10^{12} L mol^{-1} s^{-1} and its activation energy is 10.0 kJ mol^{-1}. What are (a) the entropy of activation, (b) the enthalpy of activation, (c) the Gibbs energy of activation at 298 K?

27.17 (b) The pre-exponential factor for a gas-phase decomposition of ozone at low pressures is 2.3×10^{13} L mol^{-1} s^{-1} and its activation energy is 30.0 kJ mol^{-1}. What are (a) the entropy of activation, (b) the enthalpy of activation, (c) the Gibbs energy of activation at 298 K?

27.18 (a) The rate constant of the reaction $H_2O_2(aq) + I^-(aq) + H^+(aq) \rightarrow H_2O(l) + HIO(aq)$ is sensitive to the ionic strength of the aqueous solution in which the reaction occurs. At 25°C, $k = 12.2$ L^2 mol^{-2} min^{-1} at an ionic strength of 0.0525. Use the Debye–Hückel limiting law to estimate the rate constant at zero ionic strength.

27.18 (b) At 25°C, $k = 1.55$ L^2 mol^{-3} min^{-1} at an ionic strength of 0.0241 for a reaction in which the rate-determining step involves the encounter of two singly charged cations. Use the Debye–Hückel limiting law to estimate the rate constant at zero ionic strength.

Problems†

Numerical problems

27.1 In the dimerization of methyl radicals at 25°C, the experimental pre-exponential factor is 2.4×10^{10} L mol^{-1} s^{-1}. What are (a) the reactive cross-section, (b) the P factor for the reaction if the C—H bond length is 154 pm?

27.2 Nitrogen dioxide reacts bimolecularly in the gas phase to give $2 \text{ NO} + O_2$. The temperature dependence of the second-order rate constant for the rate law $d[P]/dt = k[NO_2]^2$ is given below. What are the P factor and the reactive cross-section for the reaction?

T/K	600	700	800	1000
k/(cm^3 mol^{-1} s^{-1})	4.6×10^2	9.7×10^3	1.3×10^5	3.1×10^6

Take $\sigma = 0.60$ nm^2.

27.3 The diameter of the methyl radical is about 308 pm. What is the maximum rate constant in the expression $d[C_2H_6]/dt = k[CH_3]^2$ for second-order recombination of radicals at room temperature? 10 per cent of a 1.0 L sample of ethane at 298 K and 100 kPa is dissociated into methyl radicals. What is the minimum time for 90 per cent recombination?

27.4 The rates of thermolysis of a variety of *cis*- and *trans*-azoalkanes have been measured over a range of temperatures in order to settle a controversy concerning the mechanism of the reaction. In ethanol an unstable *cis*-azoalkane decomposed at a rate that was followed by observing the N_2 evolution, and this led to the rate constants listed below (P.S. Engel and D.J. Bishop, *J. Amer. Chem. Soc.* **97**, 6754 (1975)).

Calculate the enthalpy, entropy, energy, and Gibbs energy of activation at −20°C.

θ/°C	−24.82	−20.73	−17.02	−13.00	−8.95
$10^4 \times k$/s^{-1}	1.22	2.31	4.39	8.50	14.3

27.5 In an experimental study of a bimolecular reaction in aqueous solution, the second-order rate constant was measured at 25°C and at a variety of ionic strengths and the results are tabulated below. It is known that a singly charged ion is involved in the rate-determining step. What is the charge on the other ion involved?

I	0.0025	0.0037	0.0045	0.0065	0.0085
k/(L mol^{-1} s^{-1})	1.05	1.12	1.16	1.18	1.26

27.6 The rate constant of the reaction $I^-(aq) + H_2O_2(aq) \rightarrow H_2O(l) + IO^-(aq)$ varies slowly with ionic strength, even though the Debye–Hückel limiting law predicts no effect. Use the following data from 25°C to find the dependence of $\log k_r$ on the ionic strength:

I/(mol kg^{-1})	0.0207	0.0525	0.0925	0.1575
k_r/(L mol^{-1} min^{-1})	0.663	0.670	0.679	0.694

Evaluate the limiting value of k_r at zero ionic strength. What does the result suggest for the dependence of $\log \gamma$ on ionic strength for a neutral molecule in an electrolyte solution?

27.7 The total cross-sections for reactions between alkali metal atoms and halogen molecules are given in the table below (R.D. Levine and R.B. Bernstein, *Molecular reaction dynamics*, Clarendon Press, Oxford, 72 (1974)). Assess the data in terms of the harpoon mechanism.

$\dagger$ Problems denoted with the symbol $\ddagger$ were supplied by Charles Trapp, Carmen Giunta, and Marshall Cady.

σ^*/nm^2	Cl_2	Br_2	I_2
Na	1.24	1.16	0.97
K	1.54	1.51	1.27
Rb	1.90	1.97	1.67
Cs	1.96	2.04	1.95

Electron affinities are approximately 1.3 eV (Cl_2), 1.2 eV (Br_2), and 1.7 eV (I_2), and ionization energies are 5.1 eV (Na), 4.3 eV (K), 4.2 eV (Rb), and 3.9 eV (Cs).

27.8‡ M. Cyfert, B. Latko, and M. Wawrzeczyk (*Int. J. Chem. Kinet.* **28**, 103 (1996)) examined the oxidation of tris(1,10-phenanthroline)iron(II) by periodate in aqueous solution, a reaction that shows autocatalytic behavior. To assess the kinetic salt effect, they measured rate constants at a variety of concentrations of Na_2SO_4 far in excess of reactant concentrations and reported the following data:

$[Na_2SO_4]/(mol\ kg^{-1})$	0.2	0.15	0.1	0.05	0.25	0.0125	0.005
$k/(L^{1/2}\ mol^{-1/2}\ s^{-1})$	0.462	0.430	0.390	0.321	0.283	0.252	0.224

What can be inferred about the charge of the activated complex of the rate-determining step?

27.9‡ For the thermal decomposition of F_2O by the reaction $2\ F_2O(g) \rightarrow 2\ F_2(g) + O_2(g)$, J. Czarnowski and H.J. Schuhmacher (*Chem. Phys. Lett.* **17**, 235 (1972)) have suggested the following mechanism:

(1) $F_2O + F_2O \longrightarrow F + OF + F_2O$ k_1
(2) $F + F_2O \longrightarrow F_2 + OF$ k_2
(3) $OF + OF \longrightarrow O_2 + F + F$ k_3
(4) $F + F + F_2O \longrightarrow F_2 + F_2O$ k_4

(a) Using the steady-state approximation, show that this mechanism is consistent with the experimental rate law $-d[F_2O]/dt = k[F_2O]^2 + k'[F_2O]^{3/2}$. (b) The experimentally determined Arrhenius parameters in the range 501–583 K are $A = 7.8 \times 10^{13}$ L mol^{-1} s^{-1}, $E_a/R = 1.935 \times 10^4$ K for k and $A = 2.3 \times 10^{10}$ L mol^{-1} s^{-1}, $E_a/R = 1.691 \times 10^4$ K for k'. At 540 K, $\Delta_f H^\ominus(F_2O) = +24.41$ kJ mol^{-1}, $D(F\text{—}F) = 160.6$ kJ mol^{-1}, and $D(O\text{—}O) = 498.2$ kJ mol^{-1}. Estimate the bond dissociation energies of the first and second F—O bonds and the Arrhenius activation energy of reaction 2.

27.10‡ For the gas-phase reaction $A + A \rightarrow A_2$, the experimental rate constant, k_2, has been fitted to the Arrhenius equation with the pre-exponential factor $A = 4.07 \times 10^5$ L mol^{-1} s^{-1} at 300 K and an activation energy of 65.43 kJ mol^{-1}. Calculate $\Delta^\ddagger S$, $\Delta^\ddagger H$, $\Delta^\ddagger U$, and $\Delta^\ddagger G$ for the reaction.

27.11‡ One of the most historically most significant studies of chemical reaction rates was that by M. Bodenstein (*Z. physik. Chem.* **29**, 295 (1899)) of the gas-phase reaction $2\ HI(g) \rightarrow H_2(g) + I_2(g)$ and its reverse, with rate constants k and k', respectively. The measured rate constants as a function of temperature are

T/K	647	666	683	700	716	781
$k/(22.4\ L\ mol^{-1}\ min^{-1})$	0.230	0.588	1.37	3.10	6.70	105.9
$k'/(22.4\ L\ mol^{-1}\ min^{-1})$	0.0140	0.0379	0.0659	0.172	0.375	3.58

Demonstrate that these data are consistent with the collision theory of bimolecular gas-phase reactions.

27.12‡ The destructive oxidation of alkylarenes by aqueous permanganate may occur by attack upon the alkyl group or by attack upon the aromatic ring. In perchloric acid solution at pH ≥ 3 it is thought that permanganate anion attack upon the alkyl group predominates. However, at pH ≤ 0.3 permanganic acid attack upon

1

the ring prevails. In this problem, data that concerns the ring oxidation alone will be presented and analysed. A proposed mechanism for the ring destruction (**1**) envisions successive formation of a charge-transfer complex (CTC) and a σ-complex. It suggests that reaction activation energies for a homologous series of alkylarenes might be multilinear (that is, $f(x, y) = c_1 x + c_2 y + c_3$) in both basicity constants (K_b) and ionization energies (I). (a) Prove that, if the multilinear hypothesis is correct, the rate constant k is given by

$$\Delta pk(T) = \frac{aT_0 \Delta pK_b^\circ (cT - 1)}{T(cT_0 - 1)} - \frac{b\Delta I}{RT \ln 10}$$

where $\Delta pk = pk - pk_{ref}$, $\Delta pK_b = pK_b - pK_{b,ref}$, and $\Delta I = I - I_{ref}$; 'ref' values are those for a chosen reference compound within the series, and K_b° is the basicity constant at T_0. The parameters a, b, and c are temperature-independent properties, which may be treated as regression parameters and determined by fitting the equation to experimental rate constants. Assume that $\Delta(\Delta S_b) = c\Delta(\Delta H_b)$, where c is a temperature-independent constant, and $\Delta(\Delta X) = \Delta X - \Delta X_{ref}$. Specify any other assumption that you make. (b) At 20, 30, 40, 50, 60, and 70°C, $k/(10^{-2}\ L\ mol^{-1}\ s^{-1}) = 0.86, 2.5, 5.4, 13, 47,$ and 59, respectively, for 1,4-dimethylbenzene. Determine A and E_a. Derive equations for $\Delta^\ddagger H$ and $\Delta^\ddagger S$ that are appropriate for solution kinetics and calculate their values. (c) The table below contains kinetic properties for a homologous series (E.S. Rudakov, V.L. Lobachev, and E.V. Zaichuk, *Kinetics and Catalysis*, **37**, 500 (1996)). The Arrhenius parameters were determined with rate constants measured at the same temperatures as those of part (b). Determine the values of a, b, and c using methylbenzene as the reference compound. Develop a graphical method for evaluating the validity of the multilinear hypothesis.

Arene	$\log(A/(L\ mol^{-1}\ s^{-1}))$	$E_a/(kJ\ mol^{-1})$	pK_b at 273 K	I/eV
benzene	12.5	104	9.2	9.25
methylbenzene	11.4	84	6.3	8.82
1,2-dimethylbenzene	11.2	76	5.3	8.56
1,3-dimethylbenzene	10.5	68	3.2	8.56
1,4-dimethylbenzene	Part b	Part b	5.7	8.44
mesitylene	10.0	56	0.4	8.40
pseudocumene	10.5	62	2.9	8.27
durene	10.6	56	2.2	8.02

Theoretical problems

27.13 Confirm that eqn 40 is a solution of eqn 39, where $[J]_t$ is a solution of the same equation but with $k = 0$ and for the same initial conditions.

27.14 Evaluate [J]* numerically using mathematical software for integration in eqn 40, and explore the effect of increasing reaction rate constant on the spatial distribution of J.

27.15 Estimate the orders of magnitude of the partition functions involved in a rate expression. State the order of magnitude of q_m^T/N_A, q^R, q^V, q^E for typical molecules. Check that in the collision of two structureless molecules the order of magnitude of the pre-exponential factor is of the same order as that predicted by collision theory. Go on to estimate the P factor for a reaction in which $A + B \rightarrow P$, and A and B are nonlinear triatomic molecules.

27.16 Use the Debye–Hückel limiting law to show that changes in ionic strength can affect the rate of reaction catalysed by H^+ from the deprotonation of a weak acid. Consider the mechanism: $H^+(aq) + B(aq) \rightarrow P$, where H^+ comes from the deprotonation of the weak acid, HA. The weak acid has a fixed concentration. First show that $\log[H^+]$, derived from the ionization of HA, depends on the activity coefficients of ions and thus depends on the ionic strength. Then find the relationship between $\log(rate)$ and $\log[H^+]$ to show that the rate also depends on the ionic strength.

27.17 The Eyring equation can also be applied to physical processes. As an example, consider the rate of diffusion of an atom stuck to the surface of a solid. Suppose that in order to move from one site to another it has to reach the top of the barrier where it can vibrate classically in the vertical direction and in one horizontal direction, but vibration along the other horizontal direction takes it into the neighbouring site. Find an expression for the rate of diffusion, and evaluate it for W atoms on a tungsten surface ($E_a = 60$ kJ mol^{-1}). Suppose that the vibration frequencies at the transition state are (a) the same as, (b) one-half the value for the adsorbed atom. What is the value of the diffusion coefficient D at 500 K? (Take the site separation as 316 pm and $\nu = 1 \times 10^{11}$ Hz.)

27.18 Suppose now that the adsorbed, migrating species treated in Problem 27.17 is a spherical molecule, and that it can rotate classically as well as vibrate at the top of the barrier, but that at the adsorption site itself it can only vibrate. What effect does this have on the diffusion constant? Take the molecule to be methane, for which $B = 5.24$ cm^{-1}.

27.19 Show that the intensities of a molecular beam before and after passing through a chamber of length l containing inert

scattering atoms are related by $I = I_0 e^{-\mathcal{N}\sigma l}$, where σ is the collision cross-section and $\mathcal{N}$ the number density of scattering atoms.

27.20 In a molecular beam experiment to measure collision cross-sections it was found that the intensity of a CsCl beam was reduced to 60 per cent of its intensity on passage through CH$_2$F$_2$ at 10 μTorr, but that when the target was Ar at the same pressure the intensity was reduced only by 10 per cent. What are the relative cross-sections of the two types of collision? Why is one much larger than the other?

27.21‡ Show that bimolecular reactions between nonlinear molecules are much slower than those between atoms even when the activation energies of both reactions are equal. Use activated complex theory and make the following assumptions. (1) All vibrational partition functions are close to unity; (2) all rotational partition functions are approximately $1 \times 10^{1.5}$, which is a reasonable order of magnitude number; (3) the translational partition function for each species is 1×10^{26}.

Applications: to environmental science and biology

27.22‡ R. Atkinson (*J. Phys. Chem. Ref. Data* **26**, 215 (1997)) has reviewed a large set of rate constants relevant to the atmospheric chemistry of volatile organic compounds. The recommended rate constant for the bimolecular association of O$_2$ with an alkyl radical R at 298 K is 4.7×10^9 L mol^{-1} s^{-1} for R = C$_2$H$_5$ and 8.4×10^9 L mol^{-1} s^{-1} for R = cyclohexyl. Assuming no energy barrier, compute the steric factor, P, for each reaction. (*Hint.* Obtain collision diameters from collision cross-sections of similar molecules in the *Data section*.)

27.23‡ The compound α-tocopherol, a form of vitamin E, is a powerful antioxidant that may help to maintain the integrity of biological membranes. R.H. Bisby and A.W. Parker (*J. Amer. Chem. Soc.* **117**, 5664 (1995)) studied the reaction of photochemically excited duroquinone with the antioxidant in ethanol. Once the duroquinone was photochemically excited, a bimolecular reaction took place at a rate described as diffusion-limited. (a) Estimate the rate constant for a diffusion-limited reaction in ethanol. (b) The reported rate constant was 2.77×10^9 L mol^{-1} s^{-1}; estimate the critical reaction distance if the sum of diffusion constants is 1×10^{-9} m^2 s^{-1}.

28 Processes at solid surfaces

The growth and structure of solid surfaces

28.1 Surface growth

28.2 Surface composition

The extent of adsorption

28.3 Physisorption and chemisorption

28.4 Adsorption isotherms

28.5 The rates of surface processes

Catalytic activity at surfaces

28.6 Adsorption and catalysis

28.7 Examples of catalysis

Checklist of key ideas

Further reading

Exercises

Problems

In this chapter we see how solids grow at their surfaces and how the details of the structure and composition of solid surfaces can be determined experimentally. A major part of the material concerns the extent to which a surface is covered and the variation of the extent of coverage with the pressure and temperature. This material is used to discuss how surfaces affect the rate and course of chemical change by acting as the site of catalysis.

Processes at solid surfaces govern the viability of industry both constructively, as in catalysis, and destructively, as in corrosion. Chemical reactions at solid surfaces may differ sharply from reactions in the bulk, for reaction pathways of much lower activation energy may be provided, and hence result in catalysis. The concept of a solid surface has been extended in recent years with the availability of microporous materials as catalysts (Section 28.7).

We shall see that acronyms are widely used in surface studies; for convenience, a list of the acronyms used in this chapter is given at the end of the chapter.

The growth and structure of solid surfaces

In this section we see how surfaces are extended and crystals grow. The attachment of particles to a surface is called **adsorption**. The substance that adsorbs is the **adsorbate** and the underlying material that we are concerned with in this section is the **adsorbent** or **substrate**. The reverse of adsorption is **desorption**.

28.1 Surface growth

A simple picture of a perfect crystal surface is as a tray of oranges in a grocery store (Fig. 28.1). A gas molecule that collides with the surface can be imagined as a ping-pong ball bouncing erratically over the oranges. The molecule loses energy as it bounces, but it is likely to escape from the surface before it has lost enough kinetic energy to be trapped. The same is true, to some extent, of an ionic crystal in contact with a solution. There is little energy advantage for an ion in solution to discard some of its solvating molecules and stick at an exposed position on the surface.

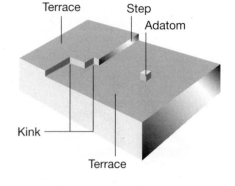

28.1 A schematic diagram of the flat surface of a solid. This primitive model is largely supported by surface-tunnelling microscope images (see Box 12.1).

28.2 Some of the kinds of defects that may occur on otherwise perfect terraces. Defects play an important role in surface growth and catalysis.

(a) The role of defects

The picture changes when the surface has defects, for then there are ridges of incomplete layers of atoms or ions. A typical type of surface defect is a **step** between two otherwise flat layers of atoms called **terraces** (Fig. 28.2). A step defect might itself have defects, for it might have kinks. When an atom settles on a terrace it bounces across it under the influence of the intermolecular potential, and might come to a step or a corner formed by a kink. Instead of interacting with a single terrace atom, the molecule now interacts with several, and the interaction may be strong enough to trap it. Likewise, when ions deposit from solution, the loss of the solvation interaction is offset by a strong Coulombic interaction between the arriving ions and several ions at the surface defect.

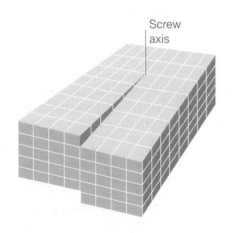

28.3 A screw dislocation occurs where one region of the crystal is pushed up through one or more unit cells relative to another region. The cut extends to the screw axis. As atoms lie along the step the dislocation rotates round the screw axis, and is not annihilated.

(b) Dislocations

Not all kinds of defect result in sustained surface growth. As the process of settling into ledges and kinks continues, there comes a stage when an entire lower terrace has been covered. At this stage the surface defects have been eliminated, and growth will cease. For continuing growth, a surface defect is needed that propagates as the crystal grows. We can see what form of defect this must be by considering the types of **dislocations**, or discontinuities in the regularity of the lattice, that exist in the bulk of a crystal. One reason for their formation may be that the crystal grows so quickly that its particles do not have time to settle into states of lowest potential energy before being trapped in position by the deposition of the next layer.

A special kind of dislocation is the **screw dislocation** shown in Fig. 28.3. Imagine a cut in the crystal, with the atoms to the left of the cut pushed up through a distance of one unit cell. The unit cells now form a continuous spiral around the end of the cut, which is called the **screw axis**. A path encircling the screw axis spirals up to the top of the crystal, and where the dislocation breaks through to the surface it takes the form of a spiral ramp.

The surface defect formed by a screw dislocation is a step, possibly with kinks, where growth can occur. The incoming particles lie in ranks on the ramp, and successive ranks reform the step at an angle to its initial position. As deposition continues the step rotates around the screw axis, and is not eliminated. Growth may

(a)

(b)

(c)

(d)

28.6 Counter-rotating screw dislocations on the same surface lead to the formation of terraces. Four stages of one cycle of growth are shown here. Subsequent deposition can complete each terrace.

28.4 The spiral growth arising from the propagation of a screw axis is clearly visible in this photograph of an alkane crystal. (B.R. Jennings and V.J. Morris, *Atoms in contact*, Clarendon Press, Oxford (1974); courtesy of Dr A.J. Forty.)

28.5 The spiral growth pattern is sometimes concealed because the terraces are subsequently completed by further deposition. This accounts for the appearance of this cadmium iodide crystal. (H.M. Rosenberg, *The solid state*, Clarendon Press, Oxford (1978).)

therefore continue indefinitely. Several layers of deposition may occur, and the edges of the spirals might be cliffs several atoms high (Fig. 28.4).

Propagating spiral edges can also give rise to flat terraces (Fig. 28.5). Terraces are formed if growth occurs simultaneously at neighbouring left- and right-handed screw dislocations (Fig. 28.6). Successive tables of atoms may form as counter-rotating defects collide on successive circuits, and the terraces formed may then fill up by further deposition at their edges to give flat crystal planes.

The rapidity of growth depends on the crystal plane concerned, and the slowest-growing faces dominate the appearance of the crystal. This feature is explained in Fig. 28.7, where we see that, although the horizontal face grows forward most rapidly, it grows itself out of existence, and the slower-growing faces survive.

28.2 Surface composition

Under normal conditions, a surface exposed to a gas is constantly bombarded with molecules and a freshly prepared surface is covered very quickly. Just how quickly can be estimated using the kinetic theory of gases and the expression (eqn 24.15) for the collision flux:

$$Z_W = \frac{p}{(2\pi mkT)^{1/2}} \tag{28.1a}$$

A practical form of this equation is

$$Z_W = \frac{Z_0(p/\text{Pa})}{\{(T/\text{K})(M/(\text{g mol}^{-1}))\}^{1/2}} \quad \text{with} \quad Z_0 = 2.63 \times 10^{24} \text{ m}^{-2} \text{ s}^{-1} \tag{28.1b}$$

where M is the molar mass of the gas. For air ($M \approx 29$ g mol^{-1}) at 1 atm and 25°C the collision flux is 3×10^{27} m^{-2} s^{-1}. Because 1 m^2 of metal surface consists of about 10^{19} atoms, each atom is struck about 10^8 times each second. Even if only a few collisions leave a molecule adsorbed to the surface, the time for which a freshly prepared surface remains clean is very short.

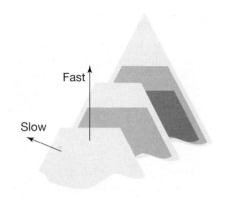

28.7 The slower-growing faces of a crystal dominate its final external appearance. Three successive stages of the growth are shown.

28.8 The X-ray photoelectron emission spectrum of a sample of gold contaminated with a surface layer of mercury. (M.W. Roberts and C.S. McKee, *Chemistry of the metal–gas interface*, Oxford (1978).)

(a) High-vacuum techniques

The obvious way to retain cleanliness is to reduce the pressure. When it is reduced to 10^{-4} Pa (as in a simple vacuum system) the collision flux falls to about 10^{18} m^{-2} s^{-1}, corresponding to one hit per surface atom in each 0.1 s. Even that is too brief in most experiments, and in **ultrahigh vacuum** (UHV) techniques pressures as low as 10^{-7} Pa (when $Z_W = 10^{15}$ m^{-2} s^{-1}) are reached on a routine basis and 10^{-9} Pa (when $Z_W = 10^{13}$ m^{-2} s^{-1}) are reached with special care. These collision fluxes correspond to each surface atom being hit once every 10^5 to 10^6 s, or about once a day.

The layout of a typical UHV apparatus is such that the whole of the evacuated part can be heated to 150–250°C for several hours to drive gas molecules from the walls. All the taps and seals are usually of metal so as to avoid contamination from greases. The sample is usually in the form of a thin foil, a filament, or a sharp point. Where there is interest in the role of specific crystal planes the sample is a single crystal with a freshly cleaved face. Initial surface cleaning is achieved either by heating it electrically or by bombarding it with accelerated gaseous ions. The latter procedure demands care because ion bombardment can shatter the surface structure and leave it an amorphous jumble of atoms. High-temperature annealing is then required to return the surface to an ordered state.

(b) Ionization techniques

Surface composition can be determined by a variety of ionization techniques. The same techniques can be used to detect any remaining contamination after cleaning and to detect layers of material adsorbed later in the experiment. Their common feature is that the **escape depth** of the electrons, the maximum depth from which ejected electrons come, is in the range 0.1–1.0 nm, which ensures that only surface species contribute.

One technique that may be used is photoelectron spectroscopy (Section 17.8), which in surface studies is normally called **photoemission spectroscopy**. X-rays or hard ultraviolet ionizing radiation of energy in the range 5–40 eV may be used, giving rise to the techniques denoted XPS and UPS, respectively. XPS, which examines inner-shell binding energies, is able to fingerprint the materials present (Fig. 28.8). UPS, which examines electrons ejected from valence shells, is more suited to establishing the bonding characteristics and the details of valence shell electronic structures of substances on the surface. Its usefulness is its ability to reveal which orbitals of the adsorbate are involved in the bond to the substrate. For instance, the principal difference between the photoemission results on free benzene and benzene adsorbed on palladium is in the energies of the π electrons. This difference is interpreted as meaning that the C_6H_6 molecules lie parallel to the surface and are attached to it by their π orbitals. In contrast, pyridine (C_6H_5N) stands more or less perpendicular to the surface, and is attached by a σ bond formed by the nitrogen lone pair.

In **secondary-ion mass spectrometry** (SIMS), the surface is ionized by bombardment with other ions and the secondary ions that emerge from the surface are detected by a mass spectrometer. Among the advantages of SIMS are the ability to detect adsorbed H and He atoms, which are not easily probed by XPS, and the high sensitivity of the mass spectrometer detector. A disadvantage is that SIMS analysis erodes the part of the sample that is bombarded. However, it is possible to control the degree of erosion to one or two monolayers by controlling the bombardment parameters.

(c) Vibrational spectroscopy

Several kinds of vibrational spectroscopy have been developed to study adsorbates and to show whether dissociation has occurred. Measurement of transmitted radiation is not practical in surfaces, which are typically too opaque to infrared or visible radiation. One technique that circumvents this problem is **reflection–absorption infrared spectroscopy** (RAIRS), in which the Fourier-transform infrared absorption spectrum of the adsorbate is obtained by comparing the intensity of the incident infrared beam with the intensity of infrared radiation reflected by the surface. Raman spectroscopy is better suited for studies of surfaces because it involves the detection of scattered radiation but spectral bands are typically very weak. However, **surface-enhanced Raman scattering** (SERS) is viable for surface studies: the strong enhancement of the Raman spectrum of the adsorbate can increase intensities by a factor as high as 10^6. The effect is due in part to local accumulations of electron density at the features of the roughened surface and at regions where bonding occurs. The SERS effect is also observed when molecules adsorb to colloidal particles of gold and silver, with the surface of the colloid fostering the enhancement of the Raman spectrum. Disadvantages of SERS include weak enhancement observed on flat single crystal surfaces and the fact that the technique works for only certain metals.

(d) Energy-loss spectroscopy

A hybrid version of photoemission spectroscopy and vibrational spectroscopy is **electron energy-loss spectroscopy** (EELS, or HREELS, where HR denotes high resolution) in which the energy loss suffered by a beam of electrons is monitored when they are reflected from a surface. As in Raman spectroscopy, the spectrum of energy loss can be interpreted in terms of the vibrational spectrum of the adsorbate. High resolution and sensitivity are attainable, and the technique is sensitive to light elements (to which X-ray techniques are insensitive). Very tiny amounts of adsorbate can be detected, and one report estimated that about 48 atoms of phosphorus were detected in one sample. As an example, Fig. 28.9 shows the EELS result for CO on the

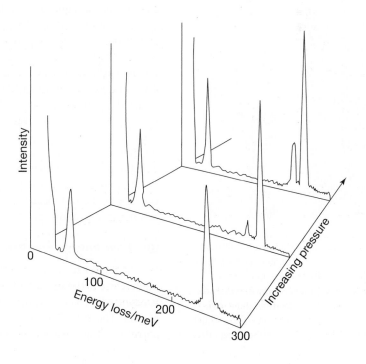

28.9 The electron energy-loss spectrum of CO adsorbed on Pt(111). The results for three different pressures are shown, and the growth of the additional peak at about 200 meV (1600 cm^{-1}) should be noted. (Spectra provided by Professor H. Ibach.)

1 Bridge site

(111) face of a platinum crystal as the extent of surface coverage increases. The main peak arises from CO attached perpendicular to the surface by a single Pt atom. As the coverage increases the neighbouring smaller peak increases in intensity. This peak is due to CO at a bridge site, attached to two Pt atoms, as in (**1**).

(e) Auger electron spectroscopy

A very important technique, which is widely used in the microelectronics industry, is **Auger electron spectroscopy** (AES). The **Auger effect** is the emission of a second electron after high-energy radiation has expelled another. The first electron to depart leaves a hole in a low-lying orbital, and an upper electron falls into it. The energy this releases may result either in the generation of radiation, which is called **X-ray fluorescence** (Fig. 28.10a), or in the ejection of another electron (Fig. 28.10b). The latter is the secondary electron of the Auger effect. The energies of the secondary electrons are characteristic of the material present, so the Auger effect effectively takes a fingerprint of the sample (Fig. 28.11). In practice, the Auger spectrum is normally obtained by irradiating the sample with an electron beam of energy in the range 1–5 keV rather than electromagnetic radiation. In **scanning Auger electron microscopy** (SAM), the finely focused electron beam is scanned over the surface and a map of composition is compiled; the resolution can reach below about 50 nm.

(f) Surface-extended X-ray absorption fine structure spectroscopy

The technique known as **surface-extended X-ray absorption fine structure spectroscopy** (SEXAFS) uses intense X-radiation from synchrotron sources (Section 16.1). Oscillations in X-ray absorbance are observed on the high-frequency side of the absorption edge (the start of an X-ray absorption band) of a substance. These oscillations arise from a quantum mechanical interference between the wavefunction of a photoejected electron and parts of that electron's wavefunction that are scattered by neighbouring atoms. If the waves interfere destructively, the photoelectron appears with lower probability and the X-ray absorption is correspondingly less. If the waves interfere constructively, then the photoelectron amplitude is higher, and the photoelectron has a higher probability of appearing; correspondingly, the X-ray absorption is greater. The oscillations therefore contain information about the number and distances of the neighbouring atoms. Such studies show that a solid's surface is much more plastic than had previously been thought, and that it undergoes **reconstruction**, or structural modification, in response to adsorbates that are present. An example is given in Fig. 28.12, which shows how a palladium surface is reconstructed in different ways and to different extents at different temperatures and with varying surface coverage of CO.

Extended X-ray absorption fine structure spectroscopy is also a useful technique for the study of solutions and suspensions, such as suspensions of proteins that bind *d*-metal ions. X-ray radiation of the proper wavelength is used to eject a photoelectron from the metal ion and the extended absorption fine structure contains information about the distances between the metal ion and nearby atoms, such as those belonging to ligands, or, if the protein is an enzyme, a bound substrate or inhibitor.

(g) Low-energy electron diffraction

One of the most informative techniques for determining the arrangement of the atoms close to the surface is **low-energy electron diffraction** (LEED). This technique is essentially electron diffraction (Box 11.2), but the sample is now the surface of a solid. The use of low-energy electrons (with energies in the range 10–200 eV, corresponding to wavelengths in the range 100–400 pm) ensures that the diffraction

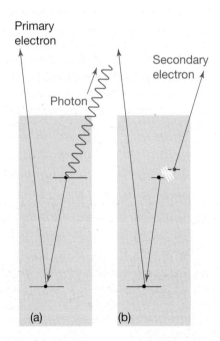

28.10 When an electron is expelled from a solid (a) an electron of higher energy may fall into the vacated orbital and emit an X-ray photon to produce X-ray fluorescence. Alternatively, (b) the electron falling into the orbital may give up its energy to another electron, which is ejected in the Auger effect.

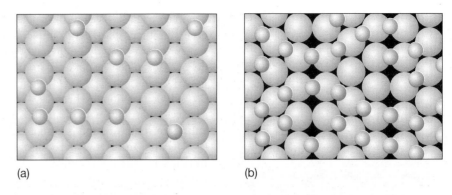

(a)   (b)

28.12 The restructuring of a surface that sometimes accompanies chemisorption. The diagrams illustrate the changes that occur when CO adsorbs on the (111) face of palladium. (a) Up to 0.3 monolayers, the surface structure is unchanged. (b) A metastable structure produced at low temperature when the surface is saturated.

28.11 An Auger spectrum of the same sample used for Fig. 28.8 taken before and after deposition of mercury. (M.W. Roberts and C.S. McKee, *Chemistry of the metal–gas interface*, Oxford (1978).)

is caused only by atoms on and close to the surface. The experimental arrangement is shown in Fig. 28.13, and typical LEED patterns, obtained by photographing the fluorescent screen through the viewing port, are shown in Fig. 28.14.

The LEED pattern portrays the two-dimensional structure of the surface. By studying how the diffraction intensities depend on the energy of the electron beam it is also possible to infer some details about the vertical location of the atoms and to measure the thickness of the surface layer, but the interpretation of LEED data is much more complicated than the interpretation of bulk X-ray data. The pattern is sharp if the surface is well-ordered for distances long compared with the wavelength of the incident electrons. In practice, sharp patterns are obtained for surfaces ordered to depths of about 20 nm and more. Diffuse patterns indicate either a poorly ordered surface or the presence of impurities. If the LEED pattern does not correspond to the pattern expected by extrapolation of the bulk surface to the surface, then either a reconstruction of the surface has occurred or there is order in the arrangement of an adsorbed layer.

LEED experiments show that the surface of a crystal rarely has exactly the same form as a slice through the bulk. As a general rule, it is found that metal surfaces are simply truncations of the bulk lattice, but the distance between the top layer of atoms and the one below is contracted by around 5 per cent. Semiconductors

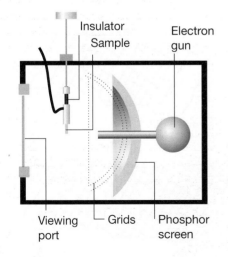

28.13 A schematic diagram of the apparatus used for a LEED experiment. The electrons diffracted by the surface layers are detected by the fluorescence they cause on the phosphor screen.

(a)   (b)

28.14 LEED photographs of (a) a clean platinum surface and (b) after its exposure to propyne, $CH_3C{\equiv}CH$. (Photographs provided by Professor G.A. Somorjai.)

148 pm

12 pm

130 pm

28.15 The structure of a surface close to the point of attachment of $CH_3C—$ to the (110) surface of rhodium at 300 K and the changes in positions of the metal atoms that accompany chemisorption.

generally have surfaces reconstructed to a depth of several layers. Reconstruction occurs in ionic solids. For example, in lithium fluoride the Li^+ and F^- ions close to the surface apparently lie on slightly different planes. An actual example of the detail that can now be obtained from refined LEED techniques is shown in Fig. 28.15 for $CH_3C—$ adsorbed on a (111) plane of rhodium.

A simple notation has been developed to describe the arrangement of atoms at a surface. An arrangement of atoms corresponding to the bulk unit cell is called the **substrate structure** and is designated (1×1). Thus, the substrate structure of the (111) face of a metal M would be designated M(111)-(1×1). An adsorbed species A that has the same structure would be denoted (1×1)-A. Therefore, a monolayer of oxygen atoms on the (111) face of silicon would be denoted Si(111)-(1×1)-O.

A surface structure with a unit cell side that is twice as large as the unit cell side of the substrate surface is denoted (2×2). In general, the relative size would be denoted $(n \times m)$, as illustrated for two common cases in Fig. 28.16. In a number of cases, the surface atoms form a lattice that is rotated with respect to the substrate. A surface structure rotated by 45°, for instance, would be denoted $R45°$, as in $(2^{1/2} \times 2^{1/2})R45°$.

The presence of terraces, steps, and kinks in a surface shows up in LEED patterns, and their surface density (the number of defects in a region divided by the area of the region) can be estimated. The importance of this type of measurement will emerge later. Three examples of how steps and kinks affect the pattern are shown in Fig. 28.17. The samples used were obtained by cleaving a crystal at different angles to a plane of atoms. Only terraces are produced when the cut is parallel to the plane,

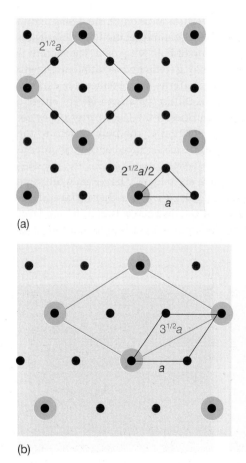

$2^{1/2}a$

$2^{1/2}a/2$

a

(a)

$3^{1/2}a$

a

(b)

28.16 The designation of two adsorbed surface layers: (a) $(2^{1/2} \times 2^{1/2})R45°$, (b) $(3^{1/2} \times 3^{1/2})R30°$.

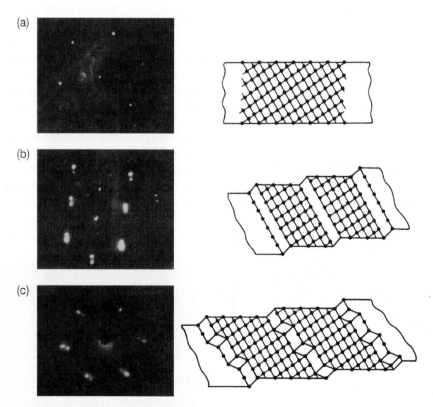

(a)

(b)

(c)

28.17 LEED patterns may be used to assess the defect density of a surface. The photographs correspond to a platinum surface with (a) low defect density, (b) regular steps separated by about six atoms, and (c) regular steps with kinks. (Photographs provided by Professor G.A. Samorjai.)

28.18 Visualization by STM of the reaction $SiH_3 \rightarrow SiH_2 + H$ on a 4.7 nm × 4.7 nm area of a Si(001) surface. (a) The Si(001) surface before exposure to $Si_2H_6(g)$. (b) Adsorbed Si_2H_6 dissociates into SiH_2(surface), on the left of the image, and SiH_3(surface), on the right. (c) After 8 min, SiH_3(surface) dissociates to SiH_2(surface) and H(surface). (Reproduced with permission from Y. Wang, M.J. Bronikowski, and R.J. Hamers, *Surface Science* **64**, 311 (1994).)

and the density of steps increases as the angle of the cut increases. The observation of additional structure in the LEED patterns, rather than blurring, shows that the steps are arrayed regularly.

(h) Scanning electron and scanning probe microscopy

Terraces, steps, kinks, and dislocations on a surface may be observed by **scanning electron microscopy** (SEM). In this technique, a focused beam of electrons with a well defined de Broglie wavelength replaces the lamp found in traditional light microscopes. Instead of glass or quartz lenses, magnetic fields are used to focus the beam. Electrons scattered from the small irradiated area are detected and the electrical signal is sent to a video screen. An image of the surface is then obtained by scanning the electron beam and the detector across the sample. As in traditional light microscopy, the wavelength of and the ability to focus the incident beam—in this case a beam of electrons—determine the microscope's resolution, the smallest distance between two objects that leads to two distinct images on the video screen.[1] Electron wavelengths in typical SEM instruments are smaller than 10 pm, but it is not possible to focus electrons well with magnetic lenses so that, in the end, the typical resolution of the SEM is about 50 nm—not enough for the resolution of individual atoms.

We saw in Box 12.1 that **scanning probe microscopy** (SPM) techniques, such as **scanning tunnelling microscopy** (STM) and **atomic force microscopy** (AFM), reveal the atomic details of structure of the surface and of adsorbates. It is possible to use STM to visualize chemical reactions as they happen on surfaces. Figure 28.18 shows the dissociation of SiH_3 adsorbed on to a Si(001) surface into adsorbed SiH_2 units and H atoms. The tip of the STM can also be used to manipulate adsorbed atoms on a surface, making possible the fabrication of complex and yet very tiny structures, such as nanometre-sized electronic devices (Box 28.1).

(i) Molecular beam techniques

Whereas many important studies have been carried out simply by exposing a surface to a gas, modern work is increasingly making use of **molecular beam scattering** (MBS). One advantage is that the activity of specific crystal planes can be

1. See Boxes 16.2 and 17.2 for more discussion of light microscopy techniques.

Box 28.1 **Nanofabrication**

Nanofabrication is the synthesis of *nanodevices*, nanometre-sized assemblies of atoms and molecules that can be used in technological applications. The field of nanofabrication is still new and nanodevices have not yet been incorporated into commercially available technologies. However, the future economic impact of nanotechnology could be very significant. For example, increased demand for very small digital electronic devices has driven the design of ever smaller and more powerful microprocessors. However, there is an upper limit on the density of electronic circuits that can be incorporated into silicon-based chips with current fabrication technologies. As the ability to process data increases with the number of circuits in a chip, it follows that soon chips and the devices that use them will have to become bigger if processing power is to increase indefinitely. One way to circumvent this problem is to fabricate devices from nanometre-sized components. Another advantage of making electronic nanodevices is the possibility of using quantum mechanical effects. For example, electron tunnelling between two conducting regions separated by a thin insulating region can increase the speed of electron conduction and, consequently, the processing speed in a digital nanoprocessor.

The study of nanodevices can also advance our basic understanding of chemical reactions. Nanometre-sized chemical reactors can serve as laboratories for the study of chemical reactions in constrained environments. Some of these reactions could become the foundation for the construction of nanometre-sized chemical sensors, with potential applications in medicine. For example, nanodevices with carefully designed biochemical properties could replace viruses and bacteria as the active species in vaccines.

A number of techniques have already been developed for the fabrication of nanometre-sized structures. The synthesis of *nanowires*, nanometre-sized atomic assemblies that conduct electricity, is a major step in the fabrication of nanodevices. In Box 22.1 we saw that conducting carbon nanotubes are formed by condensation of a carbon plasma. Silicon nanowires can be made by a similar technique, in which a pulsed laser beam is focused on to a solid target composed of silicon and iron. The laser ejects Fe and Si atoms from the surface of the target, forming a vapour phase that can condense into liquid $FeSi_n$ nanoclusters at sufficiently low temperatures. The phase diagram for this complex mixture shows that solid silicon and liquid $FeSi_n$ coexist at temperatures higher than 1473 K. Hence, it is possible to precipitate solid silicon from the mixture if the experimental conditions are controlled to maintain the $FeSi_n$ nanoclusters in a liquid state that is supersaturated with silicon. The silicon precipitate consists of nanowires with diameters of about 10 nm and lengths greater than 1 μm.

Germanium nanowires fabricated on to a silicon surface by molecular beam epitaxy. (Reproduced with permission from T. Ogino *et al. Acc. Chem. Res.* **32**, 447 (1999).)

Nanowires are also fabricated by *molecular beam epitaxy* (MBE), in which gaseous atoms or molecules are sprayed on to a crystalline surface in a UHV chamber. The result is formation of highly ordered structures of the adsorbed species. Through careful control of the chamber temperature and of the spraying process, it is possible to deposit thin films on to a surface or to create nanometre-sized assemblies with specific shapes. For example, the first illustration shows an AFM image of germanium nanowires on a silicon surface. The wires are about 2 nm high, 10–32 nm wide, and 10–600 nm long. It is also possible to deposit *quantum dots*, nanometre-sized boxes of atoms, on a surface. Semiconducting quantum dots could be important building blocks of nanometre-sized lasers.

Direct manipulation of atoms on a surface also leads to the formation of nanowires. The Coulomb attraction between an adatom and the tip of an STM can be exploited to move adatoms along a surface, arranging them into patterns, such as wires.

Molecular self-assembly has also been used as the basis for nanofabrication on surfaces. Of current interest are *self-assembled monolayers* (SAMs), ordered molecular aggregates that form a monolayer of material on a surface. To understand the formation of SAMs, consider exposing molecules such as alkyl thiols RSH, where R represents an alkyl chain, to an Au(0) surface. The thiols chemisorb on to the surface, forming $RS^-Au(I)$ adducts:

$$RSH + Au(0)_n \longrightarrow RS^-Au(I) \cdot Au(0)_{n-1} + \tfrac{1}{2}H_2$$

If R is a sufficiently long chain, van der Waals interactions between the adsorbed RS units lead to the formation of a highly ordered monolayer on the surface, as shown in the

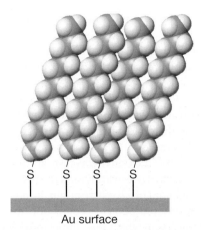

Self-assembled monolayers of alkylthiols formed on to a gold surface by chemisorption of the thiol groups and aggregation of the alkyl chains.

second illustration. It is observed that the Gibbs energy of formation of SAMs increases with the length of the alkyl chain, with each methylene group contributing 400–4000 J mol^{-1} to the overall Gibbs energy of formation.

The AFM can be used to manipulate SAMs into specific shapes on a surface by digging the microscope tip through the alkyl chains, bringing it into contact with the surface, and then moving SAMs around the surface. In one application of the technique, enzymes were bound to patterned SAMs. The experiment shows that it is possible to form

nanometre-sized reactors that take advantage of the catalytic properties of enzymes.

Discussion questions

1. The movement of atoms and ions on a surface depends on their ability to leave one position and stick to another, and therefore on the energy changes that occur. As an illustration, consider a two-dimensional square lattice of univalent positive and negative ions separated by 200 pm, and consider a cation on the upper terrace of this array. Calculate, by direct summation, its Coulombic interaction when it is in an empty lattice point directly above an anion. Now consider a high step in the same lattice, and let the cation move into the corner formed by the step and the terrace. Calculate the Coulombic energy for this position, and decide on the likely settling point for the cation.

2. Self-assembled monolayers (SAMs) are receiving more attention than Langmuir–Blodgett (LB) films as starting points for nanofabrication. How do SAMs differ from LB films and why are SAMs more useful than LB films in nanofabrication work?

Further reading

T. Ogino, H. Hibino, Y. Homma, Y. Kobayashi, K. Prabhakaran, K. Sumitomo, and H. Omi, Fabrication and integration of nano-structures on Si surfaces. *Acc. Chem. Res.* **32**, 443 (1999).

Y. Xia, J.A. Rodgers, K.E. Paul, and G.M. Whitesides, Unconventional methods for fabricating and patterning nanostructures. *Chem. Rev.* **99**, 1823 (1999).

investigated by directing the beam on to an orientated surface with known step and kink densities (as measured by LEED). Furthermore, if the adsorbate reacts at the surface the products (and their angular distributions) can be analysed as they are ejected from the surface and pass into a mass spectrometer. Another advantage is that the time of flight of a particle may be measured and interpreted in terms of its residence time on the surface. In this way a very detailed picture can be constructed of the events taking place during reactions at surfaces.

The extent of adsorption

The extent of surface coverage is normally expressed as the **fractional coverage**, θ:

$$\theta = \frac{\text{number of adsorption sites occupied}}{\text{number of adsorption sites available}} \qquad [28.2]$$

The fractional coverage is often expressed in terms of the volume of adsorbate adsorbed by $\theta = V/V_\infty$, where V_∞ is the volume of adsorbate corresponding to complete monolayer coverage. The **rate of adsorption**, $d\theta/dt$, is the rate of change of surface coverage, and can be determined by observing the change of fractional coverage with time.

Among the principal techniques for measuring $d\theta/dt$ are flow methods, in which the sample itself acts as a pump because adsorption removes particles from the gas. One commonly used technique is therefore to monitor the rates of flow of gas into and out of the system: the difference is the rate of gas uptake by the sample. Integration of this rate then gives the fractional coverage at any stage. In **flash desorption** the sample is suddenly heated (electrically) and the resulting rise of pressure is interpreted in terms of the amount of adsorbate originally on the sample. The interpretation may be confused by the desorption of a compound (for example, WO_3 from oxygen on tungsten). **Gravimetry**, in which the sample is weighed on a microbalance during the experiment, can also be used, as can radioactive tracers. In the latter case, the radioactivity of the sample is measured after exposure to an isotopically labelled gas.

28.3 Physisorption and chemisorption

Molecules and atoms can attach to surfaces in two ways. In **physisorption** (an abbreviation of 'physical adsorption'), there is a van der Waals interaction (for example, a dispersion or a dipolar interaction) between the adsorbate and the substrate. Van der Waals interactions have a long range but are weak, and the energy released when a particle is physisorbed is of the same order of magnitude as the enthalpy of condensation. Such small energies can be absorbed as vibrations of the lattice and dissipated as thermal motion, and a molecule bouncing across the surface will gradually lose its energy and finally adsorb to it in the process called **accommodation**. The enthalpy of physisorption can be measured by monitoring the rise in temperature of a sample of known heat capacity, and typical values are in the region of 20 kJ mol^{-1} (Table 28.1). This small enthalpy change is insufficient to lead to bond breaking, so a physisorbed molecule retains its identity, although it might be distorted by the presence of the surface.

In **chemisorption** (an abbreviation of 'chemical adsorption'), the molecules (or atoms) stick to the surface by forming a chemical (usually covalent) bond, and tend to find sites that maximize their coordination number with the substrate. The enthalpy of chemisorption is very much greater than that for physisorption, and typical values are in the region of 200 kJ mol^{-1} (Table 28.2). The distance between the surface and the closest adsorbate atom is also typically shorter for chemisorption than for physisorption. A chemisorbed molecule may be torn apart at the demand of the unsatisfied valencies of the surface atoms, and the existence of molecular fragments on the surface as a result of chemisorption is one reason why solid surfaces catalyse reactions.

Except in special cases, chemisorption must be exothermic. A spontaneous process requires $\Delta G < 0$. Because the translational freedom of the adsorbate is reduced when it is adsorbed, ΔS is negative. Therefore, in order for $\Delta G = \Delta H - T\Delta S$ to be negative, ΔH must be negative (that is, the process is exothermic). Exceptions may occur if the adsorbate dissociates and has high translational mobility on the surface. For example, H_2 adsorbs endothermically on glass because there is a large increase of translational entropy accompanying the dissociation of the molecules into atoms that move quite freely over the surface. In its case, the entropy change in the process $H_2(g) \rightarrow 2\,H(glass)$ is sufficiently positive to overcome the small positive enthalpy change.

Table 28.1* Maximum observed enthalpies of physisorption, $\Delta_{ad}H^\ominus/(kJ\ mol^{-1})$

CH_4	−21
H_2	−84
H_2O	−59
N_2	−21

* More values are given in the *Data section*.

Table 28.2* Enthalpies of chemisorption, $\Delta_{ad}H^\ominus/(kJ\ mol^{-1})$

Adsorbate	Adsorbent (substrate)		
	Cr	Fe	Ni
C_2H_4	−427	−285	−243
CO		−192	
H_2	−188	−134	
NH_3		−188	−155

* More values are given in the *Data section*.

The principal test for distinguishing chemisorption from physisorption used to be the enthalpy of adsorption. Values less negative than -25 kJ mol^{-1} were taken to signify physisorption, and values more negative than about -40 kJ mol^{-1} were taken to signify chemisorption. However, this criterion is by no means foolproof, and spectroscopic techniques that identify the adsorbed species are now available.

The enthalpy of adsorption depends on the extent of surface coverage, mainly because the adsorbate particles interact. If the particles repel each other (as for CO on palladium), the adsorption becomes less exothermic (the enthalpy of adsorption less negative) as coverage increases. Moreover, LEED studies show that such species settle on the surface in a disordered way until packing requirements demand order. If the adsorbate particles attract one another (as for O_2 on tungsten), then they tend to cluster together in islands, and growth occurs at the borders. These adsorbates also show order–disorder transitions when they are heated enough for thermal motion to overcome the particle–particle interactions, but not so much that they are desorbed.

28.4 Adsorption isotherms

The free gas and the adsorbed gas are in dynamic equilibrium, and the fractional coverage of the surface depends on the pressure of the overlying gas. The variation of θ with pressure at a chosen temperature is called the **adsorption isotherm**.

(a) The Langmuir isotherm

The simplest physically plausible isotherm is based on three assumptions:

  1 Adsorption cannot proceed beyond monolayer coverage.

  2 All sites are equivalent and the surface is uniform (that is, the surface is perfectly flat on a microscopic scale).

  3 The ability of a molecule to adsorb at a given site is independent of the occupation of neighbouring sites.

The third assumption supposes that there are no interactions between adsorbed molecules.

The dynamic equilibrium is

$$A(g) + M(surface) \rightleftharpoons AM(surface)$$

with rate constants k_a for adsorption and k_d for desorption. The rate of change of surface coverage due to adsorption is proportional to the partial pressure p of A and the number of vacant sites $N(1 - \theta)$, where N is the total number of sites:

$$\frac{d\theta}{dt} = k_a pN(1 - \theta) \qquad (28.3)$$

The rate of change of θ due to desorption is proportional to the number of adsorbed species, $N\theta$:

$$\frac{d\theta}{dt} = -k_d N\theta \qquad (28.4)$$

At equilibrium there is no net change (that is, the sum of these two rates is zero), and solving for θ gives the **Langmuir isotherm**:

$$\theta = \frac{Kp}{1 + Kp} \qquad K = \frac{k_a}{k_d} \qquad (28.5)$$

28.19 The plot of the data in Example 28.1. As illustrated here, the Langmuir isotherm predicts that a straight line should be obtained when p/V is plotted against p.

Example 28.1 Using the Langmuir isotherm

The data given below are for the adsorption of CO on charcoal at 273 K. Confirm that they fit the Langmuir isotherm, and find the constant K and the volume corresponding to complete coverage. In each case V has been corrected to 1.00 atm (101.325 kPa).

p/kPa	13.3	26.7	40.0	53.3	66.7	80.0	93.3
V/cm^3	10.2	18.6	25.5	31.5	36.9	41.6	46.1

Method From eqn 5,

$$Kp\theta + \theta = Kp$$

With $\theta = V/V_\infty$, where V_∞ is the volume corresponding to complete coverage, this expression can be rearranged into

$$\frac{p}{V} = \frac{p}{V_\infty} + \frac{1}{KV_\infty}$$

Hence, a plot of p/V against p should give a straight line of slope $1/V_\infty$ and intercept $1/KV_\infty$.

Answer The data for the plot are as follows:

p/kPa	13.3	26.7	40.0	53.3	66.7	80.0	93.3
$(p/kPa)/(V/cm^3)$	1.30	1.44	1.57	1.69	1.81	1.92	2.02

The points are plotted in Fig. 28.19. The (least squares) slope is 0.009 00, so $V_\infty = 111\ cm^3$. The intercept at $p = 0$ is 1.20, so

$$K = \frac{1}{(111\ cm^3) \times (1.20\ kPa\,cm^{-3})}$$

$$= 7.51 \times 10^{-3}\ kPa^{-1}$$

Comment At high surface coverage, the data would deviate from a straight line. The dimensions of K are 1/pressure.

Self-test 28.1 Repeat the calculation for the following data:

p/kPa	13.3	26.7	40.0	53.3	66.7	80.0	93.3
V/cm^3	10.3	19.3	27.3	34.1	40.0	45.5	48.0

$$[128\ cm^3,\ 6.69 \times 10^{-3}\ kPa^{-1}]$$

For adsorption with dissociation, the rate of adsorption is proportional to the pressure and to the probability that both atoms will find sites, which is proportional to the square of the number of vacant sites,

$$\frac{d\theta}{dt} = k_a p\{N(1 - \theta)\}^2 \tag{28.6}$$

The rate of desorption is proportional to the frequency of encounters of atoms on the surface, and is therefore second-order in the number of atoms present:

$$\frac{d\theta}{dt} = -k_d(N\theta)^2 \tag{28.7}$$

The condition for no net change leads to the isotherm

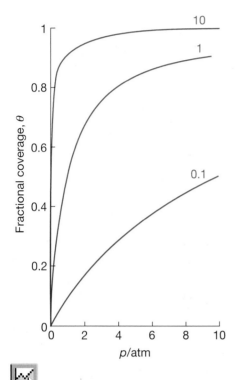

28.20 The Langmuir isotherm for dissociative adsorption, $X_2(g) \rightarrow 2\,X(\sigma)$, for different values of K.

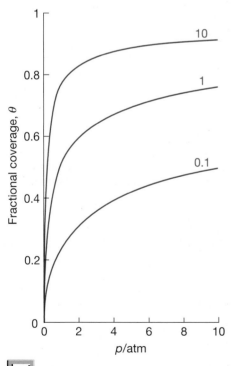

28.21 The Langmuir isotherm for non-dissociative adsorption for different values of K.

$$\theta = \frac{(Kp)^{1/2}}{1 + (Kp)^{1/2}} \tag{28.8}$$

The surface coverage now depends more weakly on pressure than for non-dissociative adsorption.

The shapes of the Langmuir isotherms with and without dissociation are shown in Figs 28.20 and 28.21. The fractional coverage increases with increasing pressure, and approaches 1 only at very high pressure, when the gas is forced on to every available site of the surface. Different curves (and therefore different values of K) are obtained at different temperatures, and the temperature dependence of K can be used to determine the **isosteric enthalpy of adsorption**, $\Delta_{ad}H^{\ominus}$, the standard enthalpy of adsorption at a fixed surface coverage. To determine this quantity we recognize that K is essentially an equilibrium constant, and then use the van 't Hoff equation (eqn 9.26) to write:

$$\left(\frac{\partial \ln K}{\partial T}\right)_{\theta} = \frac{\Delta_{ad}H^{\ominus}}{RT^2} \tag{28.9}$$

Example 28.2 Measuring the isosteric enthalpy of adsorption

The data below show the pressures of CO needed for the volume of adsorption (corrected to 1.00 atm and 273 K) to be 10.0 cm^3 using the same sample as in Example 28.1. Calculate the adsorption enthalpy at this surface coverage.

T/K	200	210	220	230	240	250
p/kPa	4.00	4.95	6.03	7.20	8.47	9.85

Method The Langmuir isotherm can be rearranged to

$$Kp = \frac{\theta}{1 - \theta}$$

Therefore, when θ is constant,

$$\ln K + \ln p = \text{constant}$$

It follows from eqn 9 that

$$\left(\frac{\partial \ln p}{\partial T}\right)_{\theta} = -\left(\frac{\partial \ln K}{\partial T}\right)_{\theta} = -\frac{\Delta_{ad}H^{\ominus}}{RT^2}$$

With $d(1/T)/dT = -1/T^2$, this expression rearranges to

$$\left(\frac{\partial \ln p}{\partial (1/T)}\right)_{\theta} = \frac{\Delta_{ad}H^{\ominus}}{R}$$

Therefore, a plot of $\ln p$ against $1/T$ should be a straight line of slope $\Delta_{ad}H^{\ominus}/R$.

Answer We draw up the following table:

T/K	200	210	220	230	240	250
$10^3/(T/\text{K})$	5.00	4.76	4.55	4.35	4.17	4.00
$\ln(p/\text{kPa})$	1.39	1.60	1.80	1.97	2.14	2.29

The points are plotted in Fig. 28.22. The slope (of the least squares fitted line) is -0.904, so

$$\Delta_{ad}H^{\ominus} = -(0.904 \times 10^3 \text{ K}) \times R = -7.52 \text{ kJ mol}^{-1}$$

28.22 The isosteric enthalpy of adsorption can be obtained from the slope of the plot of $\ln p$ against $1/T$, where p is the pressure needed to achieve the specified surface coverage. The data used are from Example 28.2.

28.23 Plots of the BET isotherm for different values of c. The value of V/V_{mon} rises indefinitely because the adsorbate may condense on the covered substrate surface.

Comment The value of K can be used to obtain a value of $\Delta_{ad}G^{\ominus}$, and then that value combined with $\Delta_{ad}H^{\ominus}$ to obtain the standard entropy of adsorption. The expression for $(\partial \ln p/\partial T)_\theta$ in this example is independent of the model for the isotherm.

Self-test 28.2 Repeat the calculation using the following data:

T/K	200	210	220	230	240	250
p/kPa	4.32	5.59	7.07	8.80	10.67	12.80

$$[-9.0 \text{ kJ mol}^{-1}]$$

(b) The BET isotherm

If the initial adsorbed layer can act as a substrate for further (for example, physical) adsorption, then, instead of the isotherm levelling off to some saturated value at high pressures, it can be expected to rise indefinitely. The most widely used isotherm dealing with multilayer adsorption was derived by Stephen Brunauer, Paul Emmett, and Edward Teller, and is called the **BET isotherm**:

$$\frac{V}{V_{mon}} = \frac{cz}{(1-z)\{1-(1-c)z\}} \quad \text{with} \quad z = \frac{p}{P^*} \tag{28.10}$$

In this expression, p^* is the vapour pressure above a layer of adsorbate that is more than one molecule thick and which resembles a pure bulk liquid, V_{mon} is the volume corresponding to monolayer coverage, and c is a constant, which is large when the enthalpy of desorption from a monolayer is large compared with the enthalpy of vaporization of the liquid adsorbate:

$$c = e^{(\Delta_{des}H^{\ominus} - \Delta_{vap}H^{\ominus})/RT} \tag{28.11}$$

Figure 28.23 illustrates the shapes of BET isotherms. They rise indefinitely as the pressure is increased because there is no limit to the amount of material that may condense when multilayer coverage may occur. A BET isotherm is not accurate at all pressures, but it is widely used in industry to determine the surface areas of solids.

Example 28.3 Using the BET isotherm

The data below relate to the adsorption of N_2 on rutile (TiO_2) at 75 K. Confirm that they fit a BET isotherm in the range of pressures reported, and determine V_{mon} and c.

p/kPa	0.160	1.87	6.11	11.67	17.02	21.92	27.29
V/mm^3	601	720	822	935	1046	1146	1254

At 75 K, $p^* = 76.0$ kPa. The volumes have been corrected to 1.00 atm and 273 K and refer to 1.00 g of substrate.

Method Equation 10 can be reorganized into

$$\frac{z}{(1-z)V} = \frac{1}{cV_{mon}} + \frac{(c-1)z}{cV_{mon}}$$

It follows that $(c-1)/cV_{mon}$ can be obtained from the slope of a plot of the expression on the left against z, and cV_{mon} can be found from the intercept at $z = 0$. The results can then be combined to give c and V_{mon}.

28.24 The BET isotherm can be tested, and the parameters determined, by plotting $z/(1 - z)V$ against $z = p/p^*$. The data are from Example 28.3.

Answer We draw up the following table:

p/kPa	0.160	1.87	6.11	11.67	17.02	21.92	27.29
$10^3 z$	2.11	24.6	80.4	154	224	288	359
$10^4 z/(1 - z)(V/\text{mm}^3)$	0.035	0.350	1.06	1.95	2.76	3.53	4.47

These points are plotted in Fig. 28.24. The least squares best line has an intercept at 0.0398, so

$$\frac{1}{cV_{\text{mon}}} = 3.98 \times 10^{-6} \text{ mm}^{-3}$$

The slope of the line is 1.23×10^{-2}, so

$$\frac{c - 1}{cV_{\text{mon}}} = (1.23 \times 10^{-2}) \times 10^3 \times 10^{-4} \text{ mm}^{-3} = 1.23 \times 10^{-3} \text{ mm}^{-3}$$

The solutions of these equations are $c = 310$ and $V_{\text{mon}} = 811 \text{ mm}^3$.

Comment At 1.00 atm and 273 K, 811 mm³ corresponds to 3.6×10^{-5} mol, or 2.2×10^{19} atoms. Because each atom occupies an area of about 0.16 nm², the surface area of the sample is about 3.5 m².

Self-test 28.3 Repeat the calculation for the following data:

p/kPa	0.160	1.87	6.11	11.67	17.02	21.92	27.29
V/cm³	235	559	649	719	790	860	950

[370, 615 cm³]

When $c \gg 1$, the BET isotherm takes the simpler form

$$\frac{V}{V_{\text{mon}}} = \frac{1}{1 - z} \tag{28.12}$$

This expression is applicable to unreactive gases on polar surfaces, for which $c \approx 10^2$ because $\Delta_{\text{des}}H^{\ominus}$ is then significantly greater than $\Delta_{\text{vap}}H^{\ominus}$ (eqn 11). The BET isotherm fits experimental observations moderately well over restricted pressure ranges, but it errs by underestimating the extent of adsorption at low pressures and by overestimating it at high pressures.

(c) Other isotherms

An assumption of the Langmuir isotherm is the independence and equivalence of the adsorption sites. Deviations from the isotherm can often be traced to the failure of these assumptions. For example, the enthalpy of adsorption often becomes less negative as θ increases, which suggests that the energetically most favourable sites are occupied first. Various attempts have been made to take these variations into account. The **Temkin isotherm**,

$$\theta = c_1 \ln(c_2 p) \tag{28.13}$$

where c_1 and c_2 are constants, corresponds to supposing that the adsorption enthalpy changes linearly with pressure. The **Freundlich isotherm**

$$\theta = c_1 p^{1/c_2} \tag{28.14}$$

corresponds to a logarithmic change. This isotherm attempts to incorporate the role of substrate–substrate interactions on the surface (see Problem 28.16).

Different isotherms agree with experiment more or less well over restricted ranges of pressure, but they remain largely empirical. Empirical, however, does not mean useless for, if the parameters of a reasonably reliable isotherm are known, reasonably reliable results can be obtained for the extent of surface coverage under various conditions. This kind of information is essential for any discussion of heterogeneous catalysis.

28.5 The rates of surface processes

The rates of surface processes may be studied by techniques described in Section 28.2 and Box 28.2. Another technique, **second harmonic generation** (SHG), is very important for the study of all types of surfaces, including thin films and liquid–gas interfaces. We saw in Section 17.6 that second harmonic generation is the conversion of an intense, pulsed laser beam to radiation with twice its initial frequency as it passes though a material. In addition to a number of crystals, surfaces are also suitable materials for SHG. Because pulsed lasers are the excitation sources, time-resolved measurements of the kinetics and dynamics of surface processes are possible over timescales as short as femtoseconds.

Box 28.2 Surface plasmon resonance

Surface plasmon resonance (SPR) is a technique in which the kinetics and thermodynamics of surface processes are measured by detecting changes in refractive index of a medium in contact with the surface. Plasmons are quantized and localized oscillations of electron density in metals. We saw in Section 23.9 that mobile electrons account for the electrical conductivity of metals. These mobile electrons form a dense 'gas' of charged particles, much like a plasma. Bombardment of the 'gas' by photons or an electron beam can cause transient changes in their distribution, with some regions becoming slightly more dense than others. Coulomb repulsion in the regions of high density causes electrons to move away from each other, lowering the density. The dynamics of this arrangement is similar to that found in any harmonic oscillator: the restoring force—in this case the Coulomb interaction —which tends to bring the system back to equilibrium gives rise to oscillations—in this case the plasmons—with well-defined frequencies. Plasmons can be excited in the bulk and on the surface of a metal. For example, the bulk and surface plasmon energies of aluminium are 15.3 eV and 10.3 eV, respectively.

It follows from this description that plasmons may also be visualized as electromagnetic waves that propagate through the solid. A surface plasmon also propagates away from the surface, but the amplitude of the wave, also called an *evanescent wave*, decreases exponentially with increasing distance from the surface.

The SPR technique is based on the absorption of energy from an incident beam of electromagnetic radiation by surface plasmons.[2] The resonance condition can be observed with appropriate choice of the wavelength and angle of incidence of the excitation beam. It is common practice to use a monochromatic beam and to vary the angle of incidence θ, as shown in the first illustration. The beam passes through a prism that strikes one side of a thin film of gold or silver. It is observed that the angle corresponding to resonance depends on the refractive index of the medium in direct contact with

The experimental arrangement for the observation of surface plasmon resonance, as explained in the text.

2. This energy loss mechanism occurs at much lower energies than those that give rise to the Auger effect (Section 28.2e).

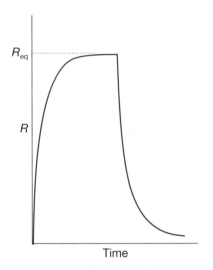

The time dependence of the SPR signal, R, showing the effect of binding of a ligand to a biopolymer adsorbed on to a surface. Binding leads to an increase in R until an equilibirum value, R_{eq}, is obtained. Passing a solution containing no ligand over the surface leads to dissociation and decrease in R.

the opposing side of the metallic film. This variation of the resonance angle with the state of the surface arises from the ability of the evanescent wave to interact with material a short distance away from the surface. For example, increasing the mass of material adsorbed on to the surface causes changes in the resonance angle. Hence, SPR can be used in the study of surface processes, such as the binding of molecules to a surface or binding of ligands to a biopolymer adsorbed on to the surface.

As an example of the SPR technique, we consider the study of the association of two polymers, A and B. In a typical experiment, a stream of solution containing a known concentration of A flows above the surface to which B is chemisorbed. The second illustration shows that the kinetics of binding of A to B may be followed by monitoring the time dependence of the SPR signal, denoted by R, which is typically the shift in resonance angle. The system is normally allowed to reach equilibrium, which is denoted by the plateau in the second illustration. Then, a solution containing no A is flowed above the surface and the AB complex dissociates. Again, analysis of the decay of the SPR signal reveals the kinetics of dissociation of the AB complex.

The equilibrium constant for formation of the AB complex can be measured directly from data of the type displayed in the second illustration. Consider the equilibrium

$$A + B \rightleftharpoons AB \qquad K = \frac{k_{on}}{k_{off}}$$

where k_{on} and k_{off} are the rate constants for formation and dissociation of the AB complex, and K is the equilibrium constant for formation of the AB complex. It follows that

$$\frac{d[AB]}{dt} = k_{on}[A][B] - k_{off}[AB]$$

In a typical SPR experiment, the flow rate of A is sufficiently high that $[A] = a_0$ is essentially constant. We can also write $[B] = b_0 - [AB]$ from mass-balance considerations, where b_0 is the total concentration of B. Finally, the SPR signal is often observed to be proportional to $[AB]$. The maximum value that R can have is $R_{max} \propto b_0$, which would be measured if all B molecules were ligated to A. We may then write

$$\frac{dR}{dt} = k_{on}a_0(R_{max} - R) - k_{off}R = k_{on}a_0R_{max} - (k_{on}a_0 + k_{off})R$$

At equilibrium $R = R_{eq}$ and $dR/dt = 0$. It follows that (after some algebra)

$$R_{eq} = R_{max}\left(\frac{a_0K}{a_0K + 1}\right)$$

Hence, the value of K can be obtained from measurements of R_{eq} for a series of a_0.

The SPR technique has been used in the study of thin films, metal–electrolyte surfaces, Langmuir–Blodgett films, and a number of biopolymer interactions, such as antibody–antigen and protein–DNA interactions. The most important advantage of SPR is its sensitivity; it is possible to measure the adsorption of nanograms of material on to a surface. For biological studies, the main disadvantage of SPR is its requirement for immobilization of at least one of the components of the system under study.

Discussion questions

1. Discuss how a plot of a_0/R_{eq} against a_0 may be used to evaluate R_{max} and K.

2. Show that, for the association part of the SPR experiment in the second illustration, $R(t) = R_{eq}(1 - e^{-k_{obs}t})$ and write an expression for k_{obs}. Then, derive an expression for $R(t)$ that applies to the dissociation part of the SPR experiment in the illustration.

Further reading

J.M. Brockman, B.P. Nelson, and R.M. Corn, Surface plasmon resonance imaging measurements of ultrathin organic films. *Ann. Rev. Phys. Chem.* **51**, 41 (2000).

P. Schuck, Use of surface plasmon resonance to probe the equilibrium and dynamic aspects of interactions between biological macromolecules. *Ann. Rev. Biophys. Biomol. Struct.* **26**, 541 (1997).

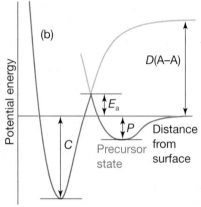

28.25 The potential energy profiles for the dissociative chemisorption of an A_2 molecule. In each case, P is the enthalpy of (non-dissociative) physisorption and C that for chemisorption (at $T = 0$). The relative locations of the curves determine whether the chemisorption is (a) not activated or (b) activated.

Figure 28.25 shows how the potential energy of a molecule varies with its distance from the substrate surface. As the molecule approaches the surface its energy falls as it becomes physisorbed into the **precursor state** for chemisorption. Dissociation into fragments often takes place as a molecule moves into its chemisorbed state (as was shown in Fig. 28.18), and after an initial increase of energy as the bonds stretch there is a sharp decrease as the adsorbate–substrate bonds reach their full strength. Even if the molecule does not fragment, there is likely to be an initial increase of potential energy as the bonds adjust as the molecule approaches the surface.

In most cases, therefore, we can expect there to be a potential energy barrier separating the precursor and chemisorbed states. This barrier, though, might be low, and might not rise above the energy of a distant, stationary particle (as in Fig. 28.25a). In this case, chemisorption is not an activated process and can be expected to be rapid. Many gas adsorptions on clean metals appear to be non-activated. In some cases the barrier rises above the zero axis (as in Fig. 28.25b); such chemisorptions are activated and slower than the non-activated kind. An example is H_2 on copper, which has an activation energy in the region of 20–40 kJ mol^{-1}.

One point that emerges from this discussion is that rates are not good criteria for distinguishing between physisorption and chemisorption. Chemisorption can be fast if the activation energy is small or zero, but it may be slow if the activation energy is large. Physisorption is usually fast, but it can appear to be slow if adsorption is taking place on a porous medium.

(a) The rate of adsorption

The rate at which a surface is covered by adsorbate depends on the ability of the substrate to dissipate the energy of the incoming particle as thermal motion as it crashes on to the surface. If the energy is not dissipated quickly, the particle migrates over the surface until a vibration expels it into the overlying gas or it reaches an edge. The proportion of collisions with the surface that successfully lead to adsorption is called the **sticking probability**, s:

$$s = \frac{\text{rate of adsorption of particles by the surface}}{\text{rate of collision of particles with the surface}} \qquad [28.15]$$

The denominator can be calculated from kinetic theory, and the numerator can be measured by observing the rate of change of pressure.

Values of s vary widely. For example, at room temperature CO has s in the range 0.1–1.0 for several d-metal surfaces, but for N_2 on rhenium $s < 10^{-2}$, indicating that more than a hundred collisions are needed before one molecule sticks successfully. Beam studies on specific crystal planes show a pronounced specificity: for N_2 on tungsten, s ranges from 0.74 on the (320) faces down to less than 0.01 on the (110) faces at room temperature. The sticking probability decreases as the surface coverage increases (Fig. 28.26). A simple assumption is that s is proportional to $1 - \theta$, the fraction uncovered, and it is common to write

$$s = (1 - \theta)s_0 \qquad (28.16)$$

where s_0 is the sticking probability on a perfectly clean surface. The results in the illustration do not fit this expression because they show that s remains close to s_0 until the coverage has risen to about 6×10^{13} molecules cm^{-2}, and then falls steeply. The explanation is probably that the colliding molecule does not enter the chemisorbed state at once, but moves over the surface until it encounters an empty site.

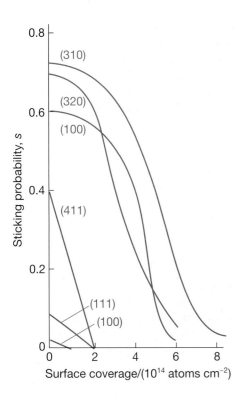

28.26 The sticking probability of N_2 on various faces of a tungsten crystal and its dependence on surface coverage. Note the very low sticking probability for the (110) and (111) faces. (Data provided by Professor D.A. King.)

(b) The rate of desorption

Desorption is always activated because the particles have to be lifted from the foot of a potential well. A physisorbed particle vibrates in its shallow potential well, and might shake itself off the surface after a short time. The temperature dependence of the first-order rate of departure can be expected to be Arrhenius-like, with an activation energy for desorption, E_d, comparable to the enthalpy of physisorption:

$$k_d = Ae^{-E_d/RT} \tag{28.17}$$

Therefore, the half-life for remaining on the surface has a temperature dependence

$$t_{1/2} = \frac{\ln 2}{k_d} = \tau_0 e^{E_d/RT} \qquad \tau_0 = \frac{\ln 2}{A} \tag{28.18}$$

(Note the positive sign in the exponent.) If we suppose that $1/\tau_0$ is approximately the same as the vibrational frequency of the weak particle–surface bond (about 10^{12} Hz) and $E_d \approx 25$ kJ mol^{-1}, then residence half-lives of around 10 ns are predicted at room temperature. Lifetimes close to 1 s are obtained only by lowering the temperature to about 100 K. For chemisorption, with $E_d = 100$ kJ mol^{-1} and guessing that $\tau_0 = 10^{-14}$ s (because the adsorbate–substrate bond is quite stiff), we expect a residence half-life of about 3×10^3 s (about an hour) at room temperature, decreasing to 1 s at about 350 K.

The desorption activation energy can be measured in several ways. However, we must be guarded in its interpretation because it often depends on the fractional coverage, and so may change as desorption proceeds. Moreover, the transfer of concepts such as 'reaction order' and 'rate constant' from bulk studies to surfaces is hazardous, and there are few examples of strictly first-order or second-order desorption kinetics (just as there are few integral-order reactions in the gas phase too).

If we disregard these complications, one way of measuring the desorption activation energy is to monitor the rate of increase in pressure when the sample is maintained at a series of temperatures, and to attempt to make an Arrhenius plot. A more sophisticated technique is **temperature-programmed desorption** (TPD) or **thermal desorption spectroscopy** (TDS). The basic observation is a surge in desorption rate (as monitored by a mass spectrometer) when the temperature is raised linearly to the temperature at which desorption occurs rapidly; but once the desorption has occurred there is no more adsorbate to escape from the surface, so the desorption flux falls again as the temperature continues to rise. The TPD spectrum, the plot of desorption flux against temperature, therefore shows a peak, the location of which depends on the desorption activation energy. There are three maxima in the example shown in Fig. 28.27, indicating the presence of three sites with different activation energies.

In many cases only a single activation energy (and a single peak in the TPD spectrum) is observed. When several peaks are observed they might correspond to adsorption on different crystal planes or to multilayer adsorption. For instance, Cd atoms on tungsten show two activation energies, one of 18 kJ mol^{-1} and the other of 90 kJ mol^{-1}. The explanation is that the more tightly bound Cd atoms are attached directly to the substrate, and the less strongly bound are in a layer (or layers) above the primary overlayer. Chemisorption cannot normally exceed monolayer coverage because a hydrocarbon gas, for instance, cannot chemisorb on to a surface already covered with its fragments; a metal adsorbate, on the other hand, as in this case, can provide a surface capable of further chemisorption.

Another example of a system showing two desorption activation energies is CO on tungsten, the values being 120 kJ mol^{-1} and 300 kJ mol^{-1}. The explanation is

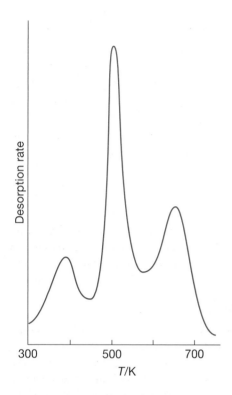

28.27 The flash desorption spectrum of H_2 on the (100) face of tungsten. The three peaks indicate the presence of three sites with different adsorption enthalpies and therefore different desorption activation energies. (P.W. Tamm and L.D. Schmidt, *J. Chem. Phys.*, **51**, 5352 (1969).)

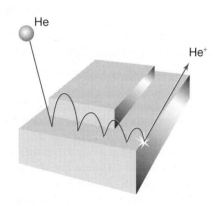

28.28 The events leading to an FIM image of a surface. The He atom migrates across the surface until it is ionized at an exposed atom, when it is pulled off by the externally applied potential. (The bouncing motion is due to the intermolecular potential, not gravity!)

believed to be the existence of two types of metal–adsorbate binding site, one involving a simple M—CO bond, the other adsorption with dissociation into individually adsorbed C and O atoms. In some cases two desorption peaks may occur even though there is only one type of site. This complication arises when there are significant interactions between the adsorbate particles with the result that at low surface coverages the enthalpy of adsorption is significantly different from its value at high coverages.

(c) Mobility on surfaces

A further aspect of the strength of the interactions between adsorbate and substrate is the mobility of the adsorbate. Mobility is often a vital feature of a catalyst's activity, because a catalyst might be impotent if the reactant molecules adsorb so strongly that they cannot migrate. The activation energy for diffusion over a surface need not be the same as for desorption because the particles may be able to move through valleys between potential peaks without leaving the surface completely. In general, the activation energy for migration is about 10–20 per cent of the energy of the surface–adsorbate bond, but the actual value depends on the extent of coverage. The defect structure of the sample (which depends on the temperature) may also play a dominant role because the adsorbed molecules might find it easier to skip across a terrace than to roll along the foot of a step, and these molecules might become trapped in vacancies in an otherwise flat terrace. Diffusion may also be easier across one crystal face than another, and so the surface mobility depends on which lattice planes are exposed.

Diffusion characteristics of an adsorbate can be examined by using STM to follow the change in surface characteristics or by **field-ionization microscopy** (FIM), which portrays the electrical characteristics of a surface by using the ionization of noble gas atoms to probe the surface (Fig. 28.28). An individual atom is imaged, the temperature is raised, and then lowered after a definite interval. A new image is then recorded, and the new position of the atom measured (Fig. 28.29). A sequence of images shows that the atom makes a random walk across the surface, and the diffusion coefficient, D, can be inferred from the mean distance, d, travelled in an interval τ by using the two-dimensional random walk expression $d = (D\tau)^{1/2}$. The value of D for different crystal planes at different temperatures can be determined directly in this way, and the activation energy for migration over each plane obtained from the Arrhenius-like expression

$$D = D_0 e^{-E_D/RT} \tag{28.19}$$

where E_D is the activation energy for diffusion. Typical values for W atoms on tungsten have E_D in the range 57–87 kJ mol^{-1} and $D_0 \approx 3.8 \times 10^{-11}$ m^2 s^{-1}. For CO on tungsten, the activation energy falls from 144 kJ mol^{-1} at low surface coverage to 88 kJ mol^{-1} when the coverage is high.

28.29 FIM micrographs showing the migration of Re atoms on rhenium during 3 s intervals at 375 K. (Photographs provided by Professor G. Ehrlich.)

Table 28.3* Activation energies of catalysed reactions

Reaction	Catalyst	$E_a/(\text{kJ mol}^{-1})$
2HI $\longrightarrow$	None	184
H$_2$ + I$_2$	Au	105
	Pt	59
2NH$_3$ $\longrightarrow$	None	350
N$_2$ + 3H$_2$	W	162

* More values are given in the *Data section*.

Catalytic activity at surfaces

A catalyst acts by providing an alternative reaction path with a lower activation energy (Table 28.3). It does not disturb the final equilibrium composition of the system, only the rate at which that equilibrium is approached. In this section we shall consider heterogeneous catalysis, in which (as mentioned in the introduction to Section 26.5) the catalyst and the reagents are in different phases. For simplicity, we shall consider only gas/solid systems.

Many catalysts depend on **coadsorption**, the adsorption of two or more species. One consequence of the presence of a second species may be the modification of the electronic structure at the surface of a metal. For instance, partial coverage of *d*-metal surfaces by alkali metals has a pronounced effect on the electron distribution and reduces the work function of the metal. Such modifiers can act as promoters (to enhance the action of catalysts) or as poisons (to inhibit catalytic action).

28.6 Adsorption and catalysis

Heterogeneous catalysis normally depends on at least one reactant being adsorbed (usually chemisorbed) and modified to a form in which it readily undergoes reaction. Often this modification takes the form of a fragmentation of the reactant molecules. In practice, the active phase is dispersed as very small particles of linear dimension less than 2 nm on a porous oxide support. **Shape-selective catalysts**, such as the zeolites, which have a pore size that can distinguish shapes and sizes at a molecular scale, have high internal specific surface areas, in the range of 100–500 m^2 g^{-1}.

(a) The Langmuir–Hinshelwood mechanism

In the **Langmuir–Hinshelwood mechanism** (LH mechanism) of surface-catalysed reactions, the reaction takes place by encounters between molecular fragments and atoms adsorbed on the surface. We therefore expect the rate law to be second-order in the extent of surface coverage:

$$A + B \longrightarrow P \qquad v = k\theta_A\theta_B \tag{28.20}$$

Insertion of the appropriate isotherms for A and B then gives the reaction rate in terms of the partial pressures of the reactants. For example, if A and B follow Langmuir isotherms, and adsorb without dissociation, so that

$$\theta_A = \frac{K_A p_A}{1 + K_A p_A + K_B p_B} \qquad \theta_B = \frac{K_B p_B}{1 + K_A p_A + K_B p_B} \tag{28.21}$$

then it follows that the rate law is

$$v = \frac{kK_A K_B p_A p_B}{(1 + K_A p_A + K_B p_B)^2} \tag{28.22}$$

The parameters K in the isotherms and the rate constant k are all temperature-dependent, so the overall temperature dependence of the rate may be strongly non-Arrhenius (in the sense that the reaction rate is unlikely to be proportional to $e^{-E_a/RT}$). The Langmuir–Hinshelwood mechanism is dominant for the catalytic oxidation of CO to CO$_2$.

(b) The Eley–Rideal mechanism

In the **Eley–Rideal mechanism** (ER mechanism) of a surface-catalysed reaction, a gas-phase molecule collides with another molecule already adsorbed on the surface. The rate of formation of product is expected to be proportional to the partial pressure, p_B, of the non-adsorbed gas B and the extent of surface coverage, θ_A, of the adsorbed gas A. It follows that the rate law should be

$$A + B \longrightarrow P \qquad v = kp_B\theta_A \tag{28.23}$$

The rate constant, k, might be much larger than for the uncatalysed gas-phase reaction because the reaction on the surface has a low activation energy and the adsorption itself is often not activated.

If we know the adsorption isotherm for A, we can express the rate law in terms of its partial pressure, p_A. For example, if the adsorption of A follows a Langmuir isotherm in the pressure range of interest, then the rate law would be

$$v = \frac{kKp_Ap_B}{1 + Kp_A} \tag{28.24}$$

If A were a diatomic molecule that adsorbed as atoms, we would substitute the isotherm given in eqn 8 instead.

According to eqn 24, when the partial pressure of A is high (in the sense $Kp_A \gg 1$), there is almost complete surface coverage, and the rate is equal to kp_B. Now the rate-determining step is the collision of B with the adsorbed fragments. When the pressure of A is low ($Kp_A \ll 1$), perhaps because of its reaction, the rate is equal to kKp_Ap_B; now the extent of surface coverage is important in the determination of the rate.

Almost all thermal surface-catalysed reactions are thought to take place by the LH mechanism, but a number of reactions with an ER mechanism have also been identified from molecular beam investigations. For example, the reaction between H(g) and D(ad) to form HD(g) is thought to be by an ER mechanism involving the direct collision and pick-up of the adsorbed D atom by the incident H atom. However, the two mechanisms should really be thought of as ideal limits, and all reactions lie somewhere between the two and show features of each one.

Example 28.4 Interpreting the kinetics of a catalysed reaction

The decomposition of phosphine (PH_3) on tungsten is first-order at low pressures and zeroth-order at high pressures. Account for these observations.

Method Write down a plausible rate law in terms of an adsorption isotherm and explore its form in the limits of high and low pressure.

Answer If the rate is supposed to be proportional to the surface coverage, we can write

$$v = k\theta = \frac{kKp}{1 + Kp}$$

where p is the pressure of phosphine. When the pressure is so low that $Kp \ll 1$,

$$v = kKp$$

and the decomposition is first-order. When $Kp \gg 1$,

$$v = k$$

and the decomposition is zeroth-order.

Comment Many heterogeneous reactions are first-order, which indicates that the rate-determining stage is the adsorption process.

Self-test 28.4 Suggest the form of the rate law for the deuteration of NH_3 in which D_2 adsorbs dissociatively and extensively (that is, $Kp \gg 1$, with p the partial pressure of D_2), and NH_3 (with partial pressure p') adsorbs at different sites. $\qquad [v = k(Kp)^{1/2}K'p'/(1 + K'p')]$

(c) Molecular beam studies

Molecular beam reactive scattering (MBRS) studies are able to provide detailed information about catalysed reactions. It has become possible to investigate how the catalytic activity of a surface depends on its structure as well as its composition. For instance, the cleavage of C—H and H—H bonds appears to depend on the presence of steps and kinks, and a terrace often has only minimal catalytic activity. The reaction $H_2 + D_2 \rightarrow 2\,HD$ has been studied in detail. For this reaction, terrace sites are inactive but one molecule in ten reacts when it strikes a step. Although the step itself might be the important feature, it may be that the presence of the step merely exposes a more reactive crystal face (the step face itself). Likewise, the dehydrogenation of hexane to hexene depends strongly on the kink density, and it appears that kinks are needed to cleave C—C bonds. These observations suggest a reason why even small amounts of impurities may poison a catalyst: they are likely to attach to step and kink sites, and so impair the activity of the catalyst entirely. A constructive outcome is that the extent of dehydrogenation may be controlled relative to other types of reactions by seeking impurities that adsorb at kinks and act as specific poisons.

Molecular beam studies can also be used to investigate the details of the reaction process, particularly by using **pulsed beams**, in which the beam is chopped into short slugs. The angular distribution of the products, for instance, can be used to assess the length of time that a species remains on the surface during the reaction, for a long residence time will result in a loss of memory of the incident beam direction.

28.7 Examples of catalysis

Almost the whole of modern chemical industry depends on the development, selection, and application of catalysts (Table 28.4). All we can hope to do in this section is to give a brief indication of some of the problems involved. Other than the ones we consider, these problems include the danger of the catalyst being poisoned by by-products or impurities, and economic considerations relating to cost and lifetime.

Table 28.4 Properties of catalysts

Catalyst	Function	Examples
Metals	Hydrogenation Dehydrogenation	Fe, Ni, Pt, Ag
Semiconducting oxides and sulfides	Oxidation Desulfurization	NiO, ZnO, MgO, Bi_2O_3/MoO_3, MoS_2
Insulating oxides	Dehydration	Al_2O_3, SiO_2, MgO
Acids	Polymerization Isomerization Cracking Alkylation	H_3PO_4, H_2SO_4, SiO_2/Al_2O_3, zeolites

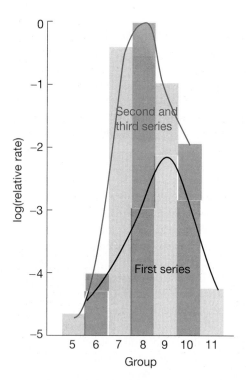

28.30 A volcano curve of catalytic activity arises because, although the reactants must adsorb reasonably strongly, they must not adsorb so strongly that they are immobilized. The lower curve refers to the first series of d-block metals, the upper curve to the second and third series d-block metals. The group numbers relate to the periodic table inside the back cover.

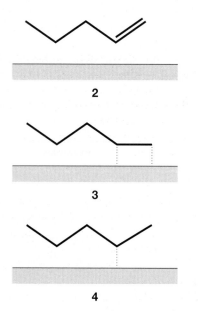

Table 28.5 Chemisorption abilities

	O_2	C_2H_2	C_2H_4	CO	H_2	CO_2	N_2
Ti, Cr, Mo, Fe	+	+	+	+	+	+	+
Ni, Co	+	+	+	+	+	+	–
Pd, Pt	+	+	+	+	+	–	–
Mn, Cu	+	+	+	+	±	–	–
Al, Au	+	+	+	+	–	–	–
Li, Na, K	+	+	–	–	–	–	–
Mg, Ag, Zn, Pb	+	–	–	–	–	–	–

+, Strong chemisorption; ±, chemisorption; –, no chemisorption. See G.C. Bond, *Heterogeneous catalysis*, Oxford University Press (1986) for further information.

(a) Catalytic activity

The activity of a catalyst depends on the strength of chemisorption as indicated by the 'volcano' curve in Fig. 28.30 (which is so-called on account of its general shape). To be active, the catalyst should be extensively covered by adsorbate, which is the case if chemisorption is strong. On the other hand, if the strength of the substrate–adsorbate bond becomes too great, the activity declines either because the other reactant molecules cannot react with the adsorbate or because the adsorbate molecules are immobilized on the surface. This pattern of behaviour suggests that the activity of a catalyst should initially increase with strength of adsorption (as measured, for instance, by the enthalpy of adsorption) and then decline, and that the most active catalysts should be those lying near the summit of the volcano. Most active metals are those that lie close to the middle of the d block.

Many metals are suitable for adsorbing gases, and the general order of adsorption strengths decreases along the series O_2, C_2H_2, C_2H_4, CO, H_2, CO_2, N_2. Some of these molecules adsorb dissociatively (for example, H_2). Elements from the d-block, such as iron, vanadium, and chromium, show a strong activity towards all these gases, but manganese and copper are unable to adsorb N_2 and CO_2. Metals towards the left of the periodic table (for example, magnesium and lithium) can adsorb (and, in fact, react with) only the most active gas (O_2). These trends are summarized in Table 28.5.

(b) Hydrogenation

An example of catalytic action is found in the hydrogenation of alkenes. The alkene (**2**) adsorbs by forming two bonds with the surface (**3**), and on the same surface there may be adsorbed H atoms. When an encounter occurs, one of the alkene–surface bonds is broken (forming (**4**) or (**5**)) and later an encounter with a second H atom releases the fully hydrogenated hydrocarbon, which is the thermodynamically more stable species.

The evidence for a two-stage reaction is the appearance of different isomeric alkenes in the mixture. The formation of isomers comes about because, while the hydrocarbon chain is waving about over the surface of the metal, an atom in the chain might chemisorb again to form (**6**) and then desorb to (**7**), an isomer of the original molecule. The new alkene would not be formed if the two hydrogen atoms attached simultaneously.

5

6

7

A major industrial application of catalytic hydrogenation is to the formation of edible fats from vegetable and animal oils. Raw oils obtained from sources such as the soya bean have the structure $CH_2(OOCR)CH(OOCR')CH_2(OOCR'')$, where R, R', and R'' are long-chain hydrocarbons with several double bonds. One disadvantage of the presence of many double bonds is that the oils are susceptible to atmospheric oxidation, and therefore are liable to become rancid. The geometrical configuration of the chains is responsible for the liquid nature of the oil, and in many applications a solid fat is at least much better and often necessary. Controlled partial hydrogenation of an oil with a catalyst carefully selected so that hydrogenation is incomplete and so that the chains do not isomerize (finely divided nickel, in fact), is used on a wide scale to produce edible fats. The process, and the industry, is not made any easier by the seasonal variation of the number of double bonds in the oils.

(c) Oxidation

Catalytic oxidation is also widely used in industry and in pollution control. Although in some cases it is desirable to achieve complete oxidation (as in the production of nitric acid from ammonia), in others partial oxidation is the aim. For example, the complete oxidation of propene to carbon dioxide and water is wasteful, but its partial oxidation to propenal (acrolein, $CH_2=CHCHO$) is the start of important industrial processes. Likewise, the controlled oxidations of ethene to ethanol, acetaldehyde, and (in the presence of acetic acid or chlorine) to vinyl acetate or vinyl chloride, are the initial stages of very important chemical industries.

Some of these reactions are catalysed by d-metal oxides of various kinds. The physical chemistry of oxide surfaces is very complex, as can be appreciated by considering what happens during the oxidation of propene to acrolein on bismuth molybdate. The first stage is the adsorption of the propene molecule with loss of a hydrogen to form the allyl radical, $CH_2=CHCH_2$. An O atom in the surface can now transfer to this radical, leading to the formation of acrolein (propenal, $CH_2=CHCHO$) and its desorption from the surface. The H atom also escapes with a surface O atom, and goes on to form H_2O, which leaves the surface. The surface is left with vacancies and metal ions in lower oxidation states. These vacancies are attacked by O_2 molecules in the overlying gas, which then chemisorb as O_2^- ions, so reforming the catalyst. This sequence of events, which is called the **Mars van Krevelen mechanism**, involves great upheavals of the surface, and some materials break up under the stress.

(d) Cracking and reforming

Many of the small organic molecules used in the preparation of all kinds of chemical products come from oil. These small building blocks of polymers, perfumes, and petrochemicals in general, are usually cut from the long-chain hydrocarbons drawn from the Earth as petroleum. The catalytically induced fragmentation of the long-chain hydrocarbons is called **cracking**, and is often brought about on silica–alumina catalysts. These catalysts act by forming unstable carbocations, which dissociate and rearrange to more highly branched isomers. These branched isomers burn more smoothly and efficiently in internal combustion engines, and are used to produce higher octane fuels.

Catalytic **reforming** uses a dual-function catalyst, such as a dispersion of platinum and acidic alumina. The platinum provides the metal function, and brings about dehydrogenation and hydrogenation. The alumina provides the acidic function, being able to form carbocations from alkenes. The sequence of events in catalytic reforming shows up very clearly the complications that must be unravelled if a reaction as important as this is to be understood and improved. The first step is the attachment of the long-chain hydrocarbon by chemisorption to the platinum. In this process

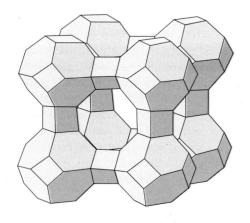

28.31 A framework representation of the general layout of the Si, Al, and O atoms in a zeolite material. Each vertex corresponds to a Si or Al atom and each edge corresponds to the approximate location of a O atom. Note the large central pore, which can hold cations, water molecules, or other small molecules.

first one and then a second H atom is lost, and an alkene is formed. The alkene migrates to a Brønsted acid site, where it accepts a proton and attaches to the surface as a carbocation. This carbocation can undergo several different reactions. It can break into two, isomerize into a more highly branched form, or undergo varieties of ring-closure. Then the adsorbed molecule loses a proton, escapes from the surface, and migrates (possibly through the gas) as an alkene to a metal part of the catalyst where it is hydrogenated. We end up with a rich selection of smaller molecules which can be withdrawn, fractionated, and then used as raw materials for other products.

(e) Zeolites

The concept of a solid surface has been extended in recent years with the availability of microporous materials, in which the surface effectively extends deep inside the solid. **Zeolites** are microporous aluminosilicates with the general formula $\{[M^{n+}]_{x/n} \cdot [H_2O]_m\}\{[AlO_2]_x[SiO_2]_y\}^{x-}$, where M^{n+} cations and H_2O molecules bind inside the cavities, or pores, of the Al—O—Si framework (Fig. 28.31). Small neutral molecules, such as CO_2, NH_3, and hydrocarbons (including aromatic compounds), can also adsorb to the internal surfaces and we shall see that this partially accounts for the utility of zeolites as catalysts.

Some zeolites for which $M = H^+$ are very strong acids and catalyse a variety of reactions that are of particular importance to the petrochemical industry. Examples include the dehydration of methanol to form hydrocarbons such as gasoline and other fuels:

$$x\text{CH}_3\text{OH} \xrightarrow{\text{zeolite}} (\text{CH}_2)_x + x\text{H}_2\text{O}$$

and the isomerization of *m*-xylene (**8**) to *p*-xylene (**9**). The catalytically important form of these acidic zeolites may be either a Brønsted acid (**10**) or a Lewis acid (**11**).

8 → (zeolite) → **9**

10

−H₂O ↓

11

Like enzymes, a zeolite catalyst with a specific compostion and structure is very selective toward certain reactants and products because only molecules of certain sizes can enter and exit the pores in which catalysis occurs. It is also possible that zeolites derive their selectivity from the ability to bind and to stabilize only transition states that fit properly in the pores. The analysis of the mechanism of zeolite catalysis is greatly facilitated by computer simulation of microporous systems, which shows how molecules fit in the pores, migrate through the connecting tunnels, and react at the appropriate active sites.

Table 28.6 Summary of acronyms

AES	Auger electron spectroscopy
AFM	Atomic force microscopy
BET isotherm	Brunauer, Emmett, Teller isotherm
EELS	Electron energy-loss spectroscopy
ER mechanism	Eley–Rideal mechanism
FIM	Field-ionization microscopy
HREELS	High-resolution electron energy-loss spectroscopy
LEED	Low-energy electron diffraction
LH mechanism	Langmuir–Hinshelwood mechanism
MBE	Molecular beam epitaxy
MBRS	Molecular beam reactive scattering
MBS	Molecular beam scattering
RAIRS	Reflection–absorption infrared spectroscopy
SAM	Scanning Auger electron microscopy
	Self-assembled monolayer
SEM	Scanning electron microscopy
SERS	Surface-enhanced Raman scattering
SEXAFS	Surface-extended X-ray absorption fine structure spectroscopy
SHG	Second harmonic generation
SIMS	Secondary ion mass spectrometry
SPM	Scanning probe microscopy
SPR	Surface plasmon resonance
STM	Scanning tunnelling microscopy
TDS	Thermal desorption spectroscopy
TPD	Temperature-programmed desorption
UHV	Ultrahigh vacuum
UPS	Ultraviolet photoemission spectroscopy
XPS	X-ray photoemission spectroscopy

Checklist of key ideas

The growth and structure of solid surfaces

☐ adsorption, the attachment of particles to a surface

☐ adsorbate, the substance adsorbed

☐ adsorbent (substrate), the substance on which another substance is adsorbed

☐ desorption, the detachment of an adsorbed substance

28.1 Surface growth

☐ step, a discontinuity between two otherwise flat layers

☐ terrace, a flat region of a surface

☐ dislocation, a discontinuity in the regularity of the lattice

☐ screw dislocation, a defect in which the unit cells now form a continuous spiral around the end of the cut

☐ screw axis, the axis of a screw dislocation

28.2 Surface composition

☐ ultrahigh vacuum (UHV), pressures lower than about 10^{-7} Pa

☐ escape depth, the maximum depth from which ejected electrons come

☐ photoemission spectroscopy, photoelectron spectroscopy applied to surfaces

☐ secondary-ion mass spectrometry (SIMS), a technique in which a surface is ionized by bombardment with other ions and the secondary ions that emerge from the surface are detected by a mass spectrometer

☐ reflection–absorption infrared spectroscopy (RAIRS), a technique for obtaining the infrared absorption spectrum of the adsorbate

☐ surface-enhanced Raman scattering (SERS), strong enhancement of the Raman spectrum of the adsorbate

☐ electron energy-loss spectroscopy (EELS), a technique in which the energy loss suffered by a beam of electrons is monitored when they are reflected from a surface

☐ Auger electron spectroscopy (AES), spectroscopy based on the Auger effect

☐ Auger effect, the emission of a second electron after high-energy radiation has expelled another

☐ X-ray fluorescence, the generation of fluorescence by the Auger effect

☐ scanning Auger electron microscopy (SAM), a technique for mapping the spatial variation over a surface

☐ surface-extended X-ray absorption fine structure spectroscopy (SEXAFS), spectroscopy that makes use of the oscillations in X-ray absorbance observed on the high-frequency side of an absorption edge

☐ reconstruction, modification of the substrate surface layers in response to adsorbates

☐ low-energy electron diffraction (LEED), electron diffraction by surfaces

☐ substrate structure, the arrangement of atoms corresponding to the bulk unit cell

☐ scanning electron microscopy (SEM), a microscopy technique that makes use of a focused beam of electrons with a well defined de Broglie wavelength

☐ scanning probe microscopy (SPM), the generic term for STM and AFM

☐ scanning tunnelling microscopy (STM), a technique for examining a surface with atomic resolution

☐ atomic force microscopy (AFM), a technique for examining and reorganizing a surface with atomic resolution

☐ molecular beam scattering (MBS), the scattering of a beam of adsorbate molecules by a surface

The extent of adsorption

☐ fractional coverage, θ the fraction of adsorption sites occupied

☐ rate of adsorption, the rate of change of fractional coverage; $d\theta/dt$

☐ flash desorption, a technique in which a sample is suddenly heated and the resulting rise of pressure is interpreted in terms of the amount of adsorbate originally on the sample

☐ gravimetry, the determination of fractional coverage by measurement of mass

28.3 Physisorption and chemisorption

☐ physisorption, adsorption by van der Waals interaction between the adsorbate and the substrate

☐ accommodation, the process of losing energy resulting in adsorption

☐ chemisorption, adsorption by the formation of a chemical bond

28.4 Adsorption isotherms

☐ adsorption isotherm, the relation between fractional coverage and partial pressure of a substrate

☐ Langmuir isotherm, $\theta = Kp/(1 + Kp)$, $K = k_a/k_d$

☐ isosteric enthalpy of adsorption, the standard enthalpy of adsorption at a fixed surface coverage; $\Delta_{ad}H^{\ominus} = RT^2(\partial \ln K/\partial T)_\theta$

☐ BET isotherm, $V/V_{mon} = cz/(1 - z)\{1 - (1 - c)z\}$, $z = p/p^*$

☐ Temkin isotherm, $\theta = c_1 \ln(c_2 p)$

☐ Freundlich isotherm, $\theta = c_1 p^{1/c_2}$

28.5 The rates of surface processes

☐ second harmonic generation (SHG), the process of generating radiation of twice the incident frequency (by a surface layer)

☐ precursor state, the initial state of an adsorbate molecule on a surface before it forms a chemical bond

☐ sticking probability, the proportion of collisions with a surface that lead to adsorption; $s = (1 - \theta)s_0$

☐ half-life for adsorption, $t_{1/2} = (\ln 2)/k_d$

☐ temperature-programmed desorption (TPD), the observation of a surge in desorption rate when the temperature is raised linearly

☐ thermal desorption spectroscopy (TDS), another name for temperature-programmed desorption

☐ field-ionization microscopy (FIM), a technique that portrays the electrical characteristics of a surface by using the ionization of noble gas atoms

Catalytic activity at surfaces

☐ coadsorption, the joint adsorption of two or more adsorbates

28.6 Adsorption and catalysis

☐ shape-selective catalyst, a solid with a pore size that can distinguish shapes and sizes at a molecular scale

☐ Langmuir–Hinshelwood mechanism, a reaction that takes place by encounters between molecular fragments and

atoms adsorbed on the surface;
$v = kK_AK_Bp_Ap_B/(1 + K_Ap_A + K_Bp_B)^2$

□ Eley–Rideal mechanism, a reaction in which a gas-phase molecule collides with another molecule already adsorbed on the surface; $v = kKp_Ap_B/(1 + Kp_A)$

□ molecular beam reactive scattering (MBRS), reactive scattering between a molecular beam and adsorbed molecules

□ pulsed beams, a technique in which a molecular beam is chopped into short slugs

28.7 Examples of catalysis

□ Mars van Krevelen mechanism, a mechanism for the oxidation of propene to acrolein on bismuth molybdate

□ cracking, the catalytically induced fragmentation of long-chain hydrocarbons

□ reforming, the catalytic isomerization and modification of hydrocarbons

□ zeolite, a microporous solid

Further reading

Articles of general interest

W. Beall Fowler, R. Phillips, and A.E. Carlsson, Point and extended defects in crystals. In *Encyclopedia of applied physics* (ed. G.L. Trigg), **14**, 317. VCH. New York (1996).

R.W. Carr, Chemical kinetics. In *Encyclopedia of applied physics* (ed. G.L. Trigg), **3**, 345. VCH, New York (1992).

R.R. Corderman, Auger spectroscopy. In *Encyclopedia of applied physics* (ed. G.L. Trigg), **2**, 297. VCH, New York (1991).

J.M. Cowley, Electron diffraction. In *Encyclopedia of applied physic* (ed. G.L. Trigg), **5**, 405. VCH, New York (1993).

A.W. Czanderna, Aspects of surface science for emphasis in the physical chemistry curriculum. In *Physical chemistry: developing a dynamic curriculum* (ed. R.W. Schwenz and R.J. Moore), 43. American Chemical Society, Washington (1993).

M.K. Weldon and C.M. Friend, Probing surface reaction mechanisms using chemical and vibrational methods: alkyl oxidation and reactivity of alcohols on transition metal surfaces. *Chem. Rev.* **96**, 1391 (1996).

A.J. Gellman, Transition states for surface-catalyzed chemistry. *Acc. Chem. Res.* **33**, 19 (2000).

L. Glasser, The BET isotherm in 3D. *Educ. in Chem.* **25**, 178 (1988).

F.J. Himpsell and I. Lindau, Photoemission and photoelectron spectra. In *Encyclopedia of applied physics* (ed. G.L. Trigg), **13**, 477. VCH, New York (1995).

B.B. Laird and A.D.J. Haymet, The crystal/liquid interface: structure and properties from computer simulation. *Chem. Rev.* **92**, 1819 (1992).

N. Mizuno and M. Misono, Heterogeneous catalysis. *Chem. Rev.* **98**, 199 (1998).

S.T. Oyama and G.A. Somorjai, Homogeneous, heterogeneous, and enzymatic catalysis. *J. Chem. Educ.* **65**, 765 (1988).

G.A. Somorjai, Modern surface science and surface technologies: an introduction. *Chem. Rev.* **96**, 1223 (1996).

E. Schönherr, Crystal growth. In *Encyclopedia of applied physics* (ed. G.L. Trigg), **4**, 335. VCH, New York (1992).

E. Shustorovich, Chemisorption theory: In search of the elephant. *Acc. Chem. Res.* **21**, 183 (1988).

J.C. Tully, Chemical dynamics at metal surfaces. *Ann. Rev. Phys. Chem.* **51**, 153 (2000).

N.-F. Zhou, The availability of a simple form of Gibbs adsorption equation for mixed surfactants. *J. Chem. Educ.* **66**, 137 (1989).

Texts and sources of data and information

A.W. Anderson and A. Gast, *Physical chemistry of surfaces*. Wiley, New York (1997).

G. Attard and C. Barnes, *Surfaces*. Oxford Chemistry Primers, Oxford University Press (1998).

C.H. Bamford, C.F. Tipper, and R.G. Compton (ed.), *Comprehensive chemical kinetics*, Vols. 1–38. Elsevier, Amsterdam (1969–2001).

G.C. Bond, *Heterogeneous catalysis: principles and applications*. Clarendon Press, Oxford (1986).

M. Boudart and G. Djéga-Mariadassou, *Kinetics of heterogeneous catalysis reactions*. Princeton University Press (1984).

M. Bowker, *The basis and applications of heterogeneous catalysis*. Oxford Chemistry Primers, Oxford University Press (1998).

G. Ertl, H. Knözinger, and J. Weitkamp, *Handbook of heterogeneous catalysis*. VCH, Weinheim (1997).

R.P.H. Gasser, *An introduction to chemisorption and catalysis by metals*. Clarendon Press, Oxford (1985).

B.C. Gates, *Catalytic chemistry*. Wiley, New York (1992).

J. Israelachvili, *Intermolecular and surface forces*. Academic Press, New York (1998).

J.C. Lindon, G.E. Tranter, and J.L. Holmes (ed.), *Encyclopedia of spectroscopy and spectrometry*. Academic Press, San Diego (2000).

S.R. Logan, *Chemical kinetics*. Longman, Harlow (1996).

K.L. Mittal (ed.), *Particles on surfaces*, Vol. 3. Plenum, New York (1991).

S.R. Morrison, *The chemical physics of surfaces*. Plenum, New York (1990).

M.J. Pilling and P.W. Seakins, *Reaction kinetics*. Oxford University Press (1996).

D.F. Shriver and P.W. Atkins, *Inorganic chemistry*. Oxford University Press, Oxford, and W.H. Freeman, New York (1999).

G.A. Somorjai, *Introduction to surface chemistry and catalysis*. Wiley, New York (1994).

J.M. Thomas and W.J. Thomas, *Principles and practice of heterogeneous catalysis*. VCH, Weinheim (1996).

J. Vickerman, *Surface analysis: techniques and applications*. Wiley, New York (1997).

M.T. Weller, *Inorganic materials chemistry*. Oxford Chemistry Primers, Oxford University Press (1996).

M.A. White, *Properties of materials*. Oxford University Press (1999).

A. Zangwill, *Physics at surfaces*. Cambridge University Press (1988).

V.P. Zhdanov, *Elementary physicochemical processes on solid surfaces*. Plenum, New York (1991).

Exercises

Discussion questions

28.1 (a) Distinguish between a step and a terrace.

28.1 (b) Describe how steps and terraces can be formed by dislocations.

28.2 (a) Describe the advantages and limitations of each of the spectroscopic techniques designated by the acronyms AES, EELS, HREELS, RAIRS, SERS, SEXAFS, SHG, UPS, and XPS.

28.2 (b) Describe the advantages and limitations of each of the microscopy, diffraction, and scattering techniques designated by the acronyms AFM, FIM, LEED, MBRS, MBS, SAM, SEM, and STM.

28.3 (a) Distinguish between the following adsorption isotherms: Langmuir, BET, Temkin, and Freundlich.

28.3 (b) Describe the essential features of the Langmuir–Hinshelwood, Eley–Rideal, and Mars van Krevelen mechanisms for surface-catalysed reactions.

28.4 (a) Account for the dependence of catalytic activity of a surface on the strength of chemisorption, as shown in Fig. 28.30.

28.4 (b) Discuss the unique physical and chemical properties of zeolites that make them useful heterogeneous catalysts.

Numerical exercises

28.5 (a) Calculate the frequency of molecular collisions per square centimetre of surface in a vessel containing (a) hydrogen, (b) propane at 25°C when the pressure is (i) 100 Pa, (ii) 0.10 μTorr.

28.5 (b) Calculate the frequency of molecular collisions per square centimetre of surface in a vessel containing (a) nitrogen, (b) methane at 25°C when the pressure is (i) 10.0 Pa, (ii) 0.150 μTorr.

28.6 (a) What pressure of argon gas is required to produce a collision rate of 4.5×10^{20} s^{-1} at 425 K on a circular surface of diameter 1.5 mm?

28.6 (b) What pressure of nitrogen gas is required to produce a collision rate of 5.00×10^{19} s^{-1} at 525 K on a circular surface of diameter 2.0 mm?

28.7 (a) Calculate the average rate at which He atoms strike a Cu atom in a surface formed by exposing a (100) plane in metallic copper to helium gas at 80 K and a pressure of 35 Pa. Crystals of copper are face-centred cubic with a cell edge of 361 pm.

28.7 (b) Calculate the average rate at which He atoms strike an iron atom in a surface formed by exposing a (100) plane in metallic iron to helium gas at 100 K and a pressure of 24 Pa. Crystals of iron are body-centred cubic with a cell edge of 145 pm.

28.8 (a) A monolayer of N_2 molecules (effective area 0.165 nm^2) is adsorbed on the surface of 1.00 g of an Fe/Al$_2$O$_3$ catalyst at 77 K, the boiling point of liquid nitrogen. Upon warming, the nitrogen occupies 2.86 cm^3 at 0°C and 760 Torr. What is the surface area of the catalyst?

28.8 (b) A monolayer of CO molecules (effective area 0.165 nm^2) is adsorbed on the surface of 1.00 g of an Fe/Al$_2$O$_3$ catalyst at 77 K, the boiling point of liquid nitrogen. Upon warming, the carbon monoxide occupies 4.25 cm^3 at 0°C and 1.00 bar. What is the surface area of the catalyst?

28.9 (a) The volume of oxygen gas at 0°C and 101 kPa adsorbed on the surface of 1.00 g of a sample of silica at 0°C was 0.284 cm^3 at 142.4 Torr and 1.430 cm^3 at 760 Torr. What is the value of V_{mon}?

28.9 (b) The volume of gas at 20°C and 1.00 bar adsorbed on the surface of 1.50 g of a sample of silica at 0°C was 1.60 cm^3 at 52.4 kPa and 2.73 cm^3 at 104 kPa. What is the value of V_{mon}?

28.10 (a) The enthalpy of adsorption of CO on a surface is found to be -120 kJ mol^{-1}. Is the adsorption physisorption or chemisorption? Estimate the mean lifetime of a CO molecule on the surface at 400 K.

28.10 (b) The enthalpy of adsorption of ammonia on a nickel surface is found to be -155 kJ mol^{-1}. Estimate the mean lifetime of an NH$_3$ molecule on the surface at 500 K.

28.11 (a) The average time for which an oxygen atom remains adsorbed to a tungsten surface is 0.36 s at 2548 K and 3.49 s at 2362 K. Find the activation energy for desorption. What is the pre-exponential factor for these tightly chemisorbed atoms?

28.11 (b) The chemisorption of hydrogen on manganese is activated, but only weakly so. Careful measurements have shown that it proceeds 35 per cent faster at 1000 K than at 600 K. What is the activation energy for chemisorption?

28.12 (a) The adsorption of a gas is described by the Langmuir isotherm with $K = 0.85$ kPa^{-1} at 25°C. Calculate the pressure at which the fractional surface coverage is (a) 0.15, (b) 0.95.

28.12 (b) The adsorption of a gas is described by the Langmuir isotherm with $K = 0.777$ kPa^{-1} at 25°C. Calculate the pressure at which the fractional surface coverage is (a) 0.20, (b) 0.75.

28.13 (a) A certain solid sample adsorbs 0.44 mg of CO when the pressure of the gas is 26.0 kPa and the temperature is 300 K. The mass of gas adsorbed when the pressure is 3.0 kPa and the temperature is 300 K is 0.19 mg. The Langmuir isotherm is known to describe the adsorption. Find the fractional coverage of the surface at the two pressures.

28.13 (b) A certain solid sample adsorbs 0.63 mg of CO when the pressure of the gas is 36.0 kPa and the temperature is 300 K. The mass of gas adsorbed when the pressure is 4.0 kPa and the temperature is 300 K is 0.21 mg. The Langmuir isotherm is known to describe the adsorption. Find the fractional coverage of the surface at the two pressures.

28.14 (a) For how long on average would an H atom remain on a surface at 298 K if its desorption activation energy were (a) 15 kJ mol^{-1}, (b) 150 kJ mol^{-1}? Take $\tau_0 = 0.10$ ps. For how long on average would the same atoms remain at 1000 K?

28.14 (b) For how long on average would an atom remain on a surface at 400 K if its desorption activation energy were (a) 20 kJ mol^{-1}, (b) 200 kJ mol^{-1}? Take $\tau_0 = 0.12$ ps. For how long on average would the same atoms remain at 800 K?

28.15 (a) A solid in contact with a gas at 12 kPa and 25°C adsorbs 2.5 mg of the gas and obeys the Langmuir isotherm. The enthalpy change when 1.00 mmol of the adsorbed gas is desorbed is +10.2 J.

What is the equilibrium pressure for the adsorption of 2.5 mg of gas at 40°C?

28.15 (b) A solid in contact with a gas at 8.86 kPa and 25°C adsorbs 4.67 mg of the gas and obeys the Langmuir isotherm. The enthalpy change when 1.00 mmol of the adsorbed gas is desorbed is +12.2 J. What is the equilibrium pressure for the adsorption of the same mass of gas at 45°C?

28.16 (a) Hydrogen iodide is very strongly adsorbed on gold but only slightly adsorbed on platinum. Assume the adsorption follows the Langmuir isotherm and predict the order of the HI decomposition reaction on each of the two metal surfaces.

28.16 (b) Suppose it is known that ozone adsorbs on a particular surface in accord with a Langmuir isotherm. How could you use the pressure dependence of the fractional coverage to distinguish between adsorption (a) without dissociation, (b) with dissociation into $O + O_2$, (c) with dissociation into $O + O + O$?

28.17 (a) Nitrogen gas adsorbed on charcoal to the extent of 0.921 cm^3 g^{-1} at 490 kPa and 190 K, but at 250 K the same amount

of adsorption was achieved only when the pressure was increased to 3.2 MPa. What is the enthalpy of adsorption of nitrogen on charcoal?

28.17 (b) Nitrogen gas adsorbed on a surface to the extent of 1.242 cm^3 g^{-1} at 350 kPa and 180 K, but at 240 K the same amount of adsorption was achieved only when the pressure was increased to 1.02 MPa. What is the enthalpy of adsorption of nitrogen on the surface?

28.18 (a) In an experiment on the adsorption of oxygen on tungsten it was found that the same volume of oxygen was desorbed in 27 min at 1856 K and 2.0 min at 1978 K. What is the activation energy of desorption? How long would it take for the same amount to desorb at (a) 298 K, (b) 3000 K?

28.18 (b) In an experiment on the adsorption of ethene on iron it was found that the same volume of the gas was desorbed in 1856 s at 873 K and 8.44 s at 1012 K. What is the activation energy of desorption? How long would it take for the same amount of ethene to desorb at (a) 298 K, (b) 1500 K?

Problems[†]

Numerical problems

28.1 In a study of the catalytic properties of a titanium surface it was necessary to maintain the surface free from contamination. Calculate the collision frequency per square centimetre of surface made by O_2 molecules at (a) 100 kPa, (b) 1.00 Pa and 300 K. Estimate the number of collisions made with a single surface atom in each second. The conclusions underline the importance of working at very low pressures (much lower than 1 Pa, in fact) in order to study the properties of uncontaminated surfaces. Take the nearest-neighbour distance as 291 pm.

28.2 Nickel is face-centred cubic with a unit cell of side 352 pm. What is the number of atoms per square centimetre exposed on a surface formed by (a) (100), (b) (110), (c) (111) planes? Calculate the frequency of molecular collisions per surface atom in a vessel containing (a) hydrogen, (b) propane at 25°C when the pressure is (i) 100 Pa, (ii) 0.10 μTorr.

28.3 The data below are for the chemisorption of hydrogen on copper powder at 25°C. Confirm that they fit the Langmuir isotherm at low coverages. Then find the value of K for the adsorption equilibrium and the adsorption volume corresponding to complete coverage.

p/Pa	25	129	253	540	1000	1593
V/cm^3	0.042	0.163	0.221	0.321	0.411	0.471

28.4 The data for the adsorption of ammonia on barium fluoride are reported below. Confirm that they fit a BET isotherm and find values of c and V_{mon}.

(a) $\theta = 0$°C, $p^* = 429.6$ kPa:

p/kPa	14.0	37.6	65.6	79.2	82.7	100.7	106.4
V/cm^3	11.1	13.5	14.9	16.0	15.5	17.3	16.5

(b) $\theta = 18.6$°C, $p^* = 819.7$ kPa:

p/kPa	5.3	8.4	14.4	29.2	62.1	74.0	80.1	102.0
V/cm^3	9.2	9.8	10.3	11.3	12.9	13.1	13.4	14.1

28.5 The following data have been obtained for the adsorption of H_2 on the surface of 1.00 g of copper at 0°C. The volume of H_2 below is the volume that the gas would occupy at STP (0°C and 1 atm).

p/atm	0.050	0.100	0.150	0.200	0.250
V/mL	1.22	1.33	1.31	1.36	1.40

Determine the volume of H_2 necessary to form a monolayer and estimate the surface area of the copper sample. The density of liquid hydrogen is 0.708 g cm^{-3}.

28.6 The adsorption of solutes on solids from liquids often follows a Freundlich isotherm. Check the applicability of this isotherm to the following data for the adsorption of acetic acid on charcoal at 25°C and find the values of the parameters c_1 and c_2.

[acid]/(mol L^{-1})	0.05	0.10	0.50	1.0	1.5
w_a/g	0.04	0.06	0.12	0.16	0.19

w_a is the mass adsorbed per unit mass of charcoal.

28.7 In some catalytic reactions the products may adsorb more strongly than the reacting gas. This is the case, for instance, in the catalytic decomposition of ammonia on platinum at 1000°C. As a

---

† Problems denoted with the symbol ‡ were supplied by Charles Trapp, Carmen Giunta, and Marshall Cady.

first step in examining the kinetics of this type of process, show that the rate of ammonia decomposition should follow

$$\frac{dp_{NH_3}}{dt} = -k_c \frac{p_{NH_3}}{p_{H_2}}$$

in the limit of very strong adsorption of hydrogen. Start by showing that, when a gas J adsorbs very strongly, and its pressure is p_J, that the fraction of uncovered sites is approximately $1/Kp_J$. Solve the rate equation for the catalytic decomposition of NH_3 on platinum and show that a plot of $F(t) = (1/t) \ln(p/p_0)$ against $G(t) = (p - p_0)/t$, where p is the pressure of ammonia, should give a straight line from which k_c can be determined. Check the rate law on the basis of the data below, and find k_c for the reaction.

t/s	0	30	60	100	160	200	250
p/kPa	13.3	11.7	11.2	10.7	10.3	9.9	9.6

28.8‡ N.E. Shafer and R.N. Zare in the article 'Through a beer glass darkly' (*Phys. Today* **44**, 48 (1991)) observe that the radii of bubbles ascending from the bottom of a glass of beer increase in size as they rise. The reason is that after the bottle of beer is opened the partial pressure of dissolved CO_2 in the beer is greater than the pressure of CO_2 in the bubble. Assuming that the pressure difference is approximately constant, the change in the number, N, of CO_2 molecules is proportional to the surface area of the bubble, that is, $dN/dt = 4\pi r^2 g$, where g is a constant and r is the time-dependent radius of the bubble. (a) Use the perfect gas law to show that the radius of the bubble increases linearly with time in accordance with $r = r_0 + vt$, where r_0 is the initial radius of the bubble and v is the rate of increase of its radius. Also find an express for v. (b) Shafer and Zare observe that the velocity and spacing between bubbles in a linear stream of bubbles proceeding from the bottom of the glass increases with height. Their data on radius and height as a function of time, with typical uncertainties are reproduced below.

Time/s	Radius, r/cm	Height, z/cm
0.00	0.017 ± 0.004	0.0 ± 0.2
0.54 ± 0.04	0.020	1.2
1.08	0.026	3.4
2.16	0.025	5.2
2.70	0.030	9.6
3.24	0.031	12.4
3.78	0.034	15.6

Test the goodness of fit of the radius data to the equation. From the fit, determine v and the constant g. Also, find an empirical relation between height and time and difference in height and time.

28.9‡ A. Akgerman and M. Zardkoohi (*J. Chem. Eng. Data* **41**, 185 (1996)) examined the adsorption of phenol from aqueous solution on to fly ash at 20°C. They fitted their observations to a Freundlich isotherm of the form $c_{ads} = Kc_{sol}^{1/n}$, where c_{ads} is the concentration of adsorbed phenol and c_{sol} is the concentration of aqueous phenol. Among the data reported are the following:

c_{sol}/(mg g^{-1})	8.26	15.65	25.43	31.74	40.00
c_{ads}/(mg g^{-1})	4.4	19.2	35.2	52.0	67.2

Determine the constants K and n. What further information would be necessary in order to express the data in terms of fractional coverage, θ?

28.10‡ C. Huang and W.P. Cheng (*J. Colloid Interface Sci.* **188**, 270 (1997)) examined the adsorption of the hexacyanoferrate(III) ion, $[Fe(CN)_6]^{3-}$, on γ-Al_2O_3 from aqueous solution. They modelled the

adsorption with a modified Langmuir isotherm, obtaining the following values of K at pH = 6.5:

T/K	283	298	308	318
$10^{-11} K$	2.642	2.078	1.286	1.085

Determine the isosteric enthalpy of adsorption, $\Delta_{ads}H^{\ominus}$, at this pH. The researchers also reported $\Delta_{ads}S^{\ominus} = +146$ J mol^{-1} K^{-1} under these conditions. Determine $\Delta_{ads}G^{\ominus}$.

28.11‡ M.-G. Olivier and R. Jadot (*J. Chem. Eng. Data* **42**, 230 (1997)) studied the adsorption of butane on silica gel. They report the following amounts of absorption (in moles per kilogram of silica gel) at 303 K:

p/kPa	31.00	38.22	53.03	76.38	101.97
n/(mol kg^{-1})	1.00	1.17	1.54	2.04	2.49
p/kPa	130.47	165.06	182.41	205.75	219.91
n/(mol kg^{-1})	2.90	3.22	3.30	3.35	3.36

Fit these data to a Langmuir isotherm, and determine the value of n that corresponds to complete coverage and the constant K.

28.12‡ The following data were obtained for the extent of adsorption, s, of acetone on charcoal from an aqueous solution of molar concentration, c, at 18°C.

c/(mmol L^{-1})	15.0	23.0	42.0	84.0	165	390	800
s/(mmol acetone/g charcoal)	0.60	0.75	1.05	1.50	2.15	3.50	5.10

Which isotherm fits this data best, Langmuir, Freundlich, or Temkin?

28.13‡ A series of experiments designed to measure the adsorption of methylene blue on activated carbon (J.H. Potgieter, *J. Chem. Educ.* **68**, 349 (1991)) uses 100.0 mL of 25.0 mg L^{-1} methylene blue solution. The respective amounts of activated carbon placed in solution were m/mg = 1, 5, 10, 12.5, 25, 30, and 100. After equilibration the corresponding methylene blue concentrations were found to be c/(mg L^{-1}) = 19.9, 10.8, 4.2, 3.5, 1.0, 0.70, and 0.20. Write the equations for the Langmuir and Freundlich adsorption isotherms in a form that is suitable for adsorption from solution. The equations should use concentration, c, and mass fraction adsorbed, x, where x is the mass of solute adsorbed divided by the mass of adsorbent. Perform suitable linear regression fits of the isotherms and critically evaluate the suitability of each fit; compare the statistics of the fits and provide evidence that one or the other is the better fit. Give a critique of the evidence. In addition to the linear relationship suggested in Section 28.4, write the Langmuir isotherm equation in a form for which x^{-1} versus c^{-1} is expected to be linear and perform the regression analysis. The regression statistics should include the correlation coefficient, the standard deviation or the coefficient of variation, the standard deviation of the intercept, and the standard deviation of the slope. Determine the maximum value of the mass fraction, x_{max}, for this activated carbon sample and use this property to estimate the specific surface area of the adsorbent. The model of methylene blue in Fig. 28.32 has the C, N, S frame drawn to scale and can be used to estimate the area occupied by the cation.

28.14 Although the attractive van der Waals interaction between individual molecules varies as R^{-6} the interaction of a molecule with a nearby solid (a homogeneous collection of molecules) varies as R^{-3}, where R is its vertical distance above the surface. Confirm this assertion. Calculate the interaction energy between an Ar atom and the surface of solid argon on the basis of a Lennard-Jones (6,12)-potential. Estimate the equilibrium distance of an atom above the surface.

100 pm

Fig. 28.32

Theoretical problems

28.15 Use the Gibbs adsorption isotherm (another name for eqn 22.54), to show that the volume adsorbed per unit area of solid, V_a/σ, is related to the pressure of the gas by $V_a = -(\sigma/RT)(\mathrm{d}\mu/\mathrm{d}\ln p)$, where μ is the chemical potential of the adsorbed gas.

28.16 If the dependence of the chemical potential of the gas on the extent of surface coverage is known, the Gibbs adsorption isotherm, eqn 22.54, can be integrated to give a relation between V_a and p, as in a normal adsorption isotherm. For instance, suppose that the change in the chemical potential of a gas when it adsorbs is of the form $\mathrm{d}\mu = -c_2(RT/\sigma)\,\mathrm{d}V_a$, where c_2 is a constant of proportionality: show that the Gibbs isotherm leads to the Freundlich isotherm in this case.

28.17 Finally we come full circle and return to the Langmuir isotherm. Find the form of $\mathrm{d}\mu$ that, when inserted in the Gibbs adsorption isotherm, leads to the Langmuir isotherm.

Applications to: chemical engineering and environmental science

28.18 The designers of a new industrial plant wanted to use a catalyst code-named CR-1 in a step involving the fluorination of butadiene. As a first step in the investigation they determined the form of the adsorption isotherm. The volume of butadiene adsorbed per gram of CR-1 at 15°C varied with pressure as given below. Is the Langmuir isotherm suitable at this pressure?

p/kPa	13.3	26.7	40.0	53.3	66.7	80.0
V/cm^3	17.9	33.0	47.0	60.8	75.3	91.3

Investigate whether the BET isotherm gives a better description of the adsorption of butadiene on CR-1. At 15°C, p^*(butadiene) = 200 kPa. Find V_{mon} and c.

28.19‡ In a study relevant to automobile catalytic converters, C.E. Wartnaby, A. Stuck, Y.Y. Yeo, and D.A. King (*J. Phys. Chem.* **100**, 12 483 (1996)) measured the enthalpy of adsorption of CO, NO, and O$_2$ on initially clean platinum 110 surfaces. They report $\Delta_{ads}H^\ominus$ for NO to be −160 kJ mol^{-1}. How much more strongly adsorbed is NO at 500°C than at 400°C?

28.20‡ The removal or recovery of volatile organic compounds (VOCs) from exhaust gas streams is an important process in environmental engineering. Activated carbon has long been used as an adsorbent in this process, but the presence of moisture in the stream reduces its effectiveness. M.-S. Chou and J.-H. Chiou (*J. Envir. Engrg. ASCE*, **123**, 437 (1997)) have studied the effect of moisture content on the adsorption capacities of granular activated carbon (GAC) for normal hexane and cyclohexane in air streams. From their data for dry streams containing cyclohexane, shown in the table below, they conclude that GAC obeys a Langmuir type model in which $q_{VOC,RH=0} = abc_{VOC}/(1 + bc_{VOC})$, where $q_{VOC} = m_{VOC}/m_{GAC}$, RH denotes relative humidity, a the maximum adsorption capacity, b is an affinity parameter, and c is the abundance in parts per million (ppm). The following table gives values of $q_{VOC,RH=0}$ for cyclohexane:

c/ppm	33.6°C	41.5°C	57.4°C	76.4°C	99°C
200	0.080	0.069	0.052	0.042	0.027
500	0.093	0.083	0.072	0.056	0.042
1000	0.101	0.088	0.076	0.063	0.045
2000	0.105	0.092	0.083	0.068	0.052
3000	0.112	0.102	0.087	0.072	0.058

(a) By linear regression of $1/q_{VOC,RH=0}$ against $1/c_{VOC}$, test the goodness of fit and determine values of a and b. (b) The parameters a and b can be related to $\Delta_{ads}H$, the enthalpy of adsorption, and $\Delta_b H$, the difference in activation energy for adsorption and desorption of the VOC molecules, through Arrhenius-type equations of the form $a = k_a \exp(-\Delta_{ads}H/RT)$ and $b = k_b \exp(-\Delta_b H/RT)$. Test the goodness of fit of the data to these equations and obtain values for k_a, k_b, $\Delta_{ads}H$, and $\Delta_b H$. (c) What interpretation might you give to k_a and k_b?

28.21‡ M.-S. Chou and J.-H. Chiou (*J. Envir. Engrg.*, *ASCE*, **123**, 437 (1997)) have studied the effect of moisture content on the adsorption capacities of granular activated carbon (GAC, Norit PK 1–3) for the volatile organic compounds (VOCs) normal hexane and cyclohexane in air streams. The following table shows the adsorption capacities ($q_{water} = m_{water}/m_{GAC}$) of GAC for pure water from moist air streams as a function of relative humidity (RH) in the absence of VOCs at 41.5°C.

RH	0.00	0.26	0.49	0.57	0.80	1.00
q_{water}	0.00	0.026	0.072	0.091	0.161	0.229

The authors conclude that the data at this and other temperatures obey a Freundlich-type isotherm, $q_{water} = k(RH)^{1/n}$. (a) Test this hypothesis for their data at 41.5°C and determine the constants k and n. (b) Why might VOCs obey the Langmuir model, but water the Freundlich model? (c) When both water vapour and cyclohexane were present in the stream the values given in the table below were determined for the ratio $r_{VOC} = q_{VOC}/q_{VOC,RH=0}$ at 41.5°C.

RH	0.00	0.10	0.25	0.40	0.53	0.76	0.81
r_{VOC}	1.00	0.98	0.91	0.84	0.79	0.67	0.61

The authors propose that these data fit the equation $r_{VOC} = 1 - q_{water}$. Test their proposal and determine values for k and n and compare to those obtained in part (b) for pure water. Suggest reasons for any differences.

28.22‡ The release of petroleum products by leaky underground storage tanks is a serious threat to clean ground water. BTEX compounds (benzene, toluene, ethylbenzene, and xylenes) are of primary concern due to their ability to cause health problems at low concentrations. D.S. Kershaw, B.C. Kulik, and S. Pamukcu (*J. Geotech. & Geoenvir. Engrg.* **123**, 324 (1997)) have studied the ability of ground tyre rubber to sorb (adsorb and absorb) benzene and *o*-xylene.

Though sorption involves more than surface interactions, sorption data is usually found to fit one of the adsorption isotherms. In this study, the authors have tested how well their data fit the linear ($q = Kc_{eq}$), Freundlich ($q = K_F c_{eq}^{1/n}$), and Langmuir ($q = K_L M c_{eq}/(1 + K_L c_{eq})$) type isotherms, where q is the mass of solvent sorbed per gram of ground rubber (in milligrams per gram), the Ks and M are empirical constants and c_{eq} the equilibrium concentration of contaminant in solution (in milligrams per litre). (a) Determine the units of the empirical constants. (b) Determine which of the isotherms best fits

the data in the table below for the sorption of benzene on ground rubber.

$c_{eq}/(mg\ L^{-1})$	97.10	36.10	10.40	6.51	6.21	2.48
$q/(mg\ g^{-1})$	7.13	4.60	1.80	1.10	0.55	0.31

(c) Compare the sorption efficiency of ground rubber to that of granulated activated charcoal, which for benzene has been shown to obey the Freundlich isotherm in the form $q = 1.0 c_{eq}^{1.6}$ with coefficient of determination $R^2 = 0.94$.

29

Dynamics of electron transfer

Electron transfer in homogeneous systems

29.1 Theory of electron transfer processes

29.2 Experimental results

Electron transfer in heterogeneous systems

29.3 The electrode–solution interface

29.4 The rate of charge transfer

29.5 Voltammetry

29.6 Electrolysis

29.7 Working galvanic cells

29.8 Corrosion

Checklist of key ideas

Further reading

Exercises

Problems

In this final chapter we examine one more example of chemical change, that resulting from the transfer of electrons. The approach is inspired by activated complex theory and draws on the languages of quantum mechanics and thermodynamics. First, we examine homogeneous systems and see that the rate of electron transfer depends on the distance between electron donor and acceptor, the standard Gibbs energy of reaction, and the energy associated with molecular rearrangements that accompany the transfer of charge. Second, we construct a model for heterogeneous systems, with special emphasis on description of processes at the solution–electrode interface. We use the model to derive a relation between the current density at the electrode and the overpotential, the difference between the electrode potential when a current is flowing and when it is not. The variation of current with overpotential can be used to infer details of the mechanism responsible for the redox process at the electrode. Finally, we analyse the kinetics of reactions that are responsible for power production in fuel cells and for corrosion.

The kinetic problem examined in this chapter is the rate at which oxidizable or reducible species—in short, **electroactive species**—can donate or accept electrons. We shall see that many factors control the rate of electron transfer processes even when the process is simple and does not involve the making or breaking of bonds.

Detailed knowledge of the factors that determine the rate of electron transfer leads to a better understanding of power production in biological cells and batteries, and of electron conduction in metals, semiconductors, and nanometre-sized electronic devices. Indeed, the economic consequences of electron transfer reactions are almost incalculable. Most of the modern methods of generating electricity are inefficient, and the development of fuel cells could revolutionize our production and deployment of energy. Today we produce energy inefficiently to produce goods that then decay by corrosion. Each step of this wasteful sequence could be improved by discovering more about the kinetics of electrochemical processes. Similarly, the techniques of organic and inorganic electrosynthesis, where an electrode is an active component of an industrial process, depend on intimate understanding of the kinetics of electron transfer processes.

In homogeneous systems, such as solutions, electron transfer occurs after the electron donor and acceptor form a complex. We shall see that the Gibbs energy of activation depends on the standard Gibbs energy of reaction and the energy associated with molecular rearrangements in the solvent and in the electroactive species that accompany the transfer of charge. The rate of electron transfer also depends on the distance between electron donor and acceptor. This distance dependence is easily observed in biological systems, where proteins fix the positions of electroactive cofactors. In heterogeneous systems, the rates of processes that occur at the interface between the phases, such as an electrode immersed in an ionic solution, are very important. A measure of this rate is the **current density**, j, the charge flux through a region (the electric current divided by the area of the region). We shall discuss the properties that control the current density and its consequences.

Electron transfer in homogeneous systems

We begin our examination with the factors that govern the rates of electron transfer in homogeneous systems. However, many of the ideas developed in this section also apply to the heterogeneous systems discussed later in the chapter.

Consider electron transfer from a donor species D to an acceptor species A in solution. First, D and A must diffuse through the solution and collide to form a complex DA, in which the donor and acceptor are separated by a distance comparable to r, the distance between the edges of each species. We assume that D, A, and DA take part in a pre-equilibrium:

$$D + A \rightleftharpoons DA \qquad K_1 = \frac{[DA]}{[D][A]} \tag{29.1a}$$

Next, electron transfer occurs in DA to yield D^+A^-:

$$DA \longrightarrow D^+A^- \qquad v = k_{et}[DA] \tag{29.1b}$$

The D^+A^- complex then breaks apart and the ions diffuse through the solution:

$$D^+A^- \rightleftharpoons D^+ + A^- \qquad K_2 = \frac{[D^+][A^-]}{[D^+A^-]} \tag{29.1c}$$

The focus of the following discussion is on the first-order electron transfer process $DA \rightarrow D^+A^-$. From activated complex theory (Section 27.4), it follows that

$$k_{et} = \kappa v e^{-\Delta^{\ddagger}G/RT} \tag{29.2}$$

where κ is the transmission coefficient, v is the vibrational frequency with which the activated complex approaches the transition state, and $\Delta^{\ddagger}G$ is the Gibbs energy of activation. Our first task is to write theoretical expressions for κv and $\Delta^{\ddagger}G$ by describing the motions of electrons and nuclei mathematically.

29.1 Theory of electron transfer processes

Our discussion concentrates on the following two key aspects of the theory:[1]

1 Electrons are transferred by tunnelling through a potential energy barrier, the height of which is partly determined by the ionization energies of the DA and D^+A^- complexes. Electron tunnelling influences the magnitude of κv.

1. The development of modern electron transfer theory began with independent work by R.A. Marcus, N.S. Hush, V.G. Levich, and R.R. Dogonadze between 1956 and 1959. Marcus received the Nobel Prize for chemistry in 1992 for his seminal contributions in this area.

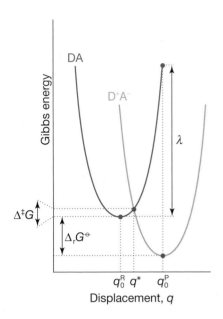

29.1 The Gibbs energy surfaces of the complexes DA and D⁺A⁻ involved in an electron transfer process are represented by parabolas characteristic of harmonic oscillators, with the displacement coordinate q corresponding to the changing geometries of the system. The plots also portray the Gibbs energy of activation, $\Delta^{\ddagger}G$, the standard reaction Gibbs energy, $\Delta_r G^{\ominus}$, and the reorganization energy, λ (discussed in Section 29.1b).

2 The complex DA and the solvent molecules surrounding it undergo structural rearrangements prior to electron transfer. The energy associated with these rearrangements and the standard reaction Gibbs energy determine $\Delta^{\ddagger}G$.

(a) Electron tunnelling

We saw in Section 17.1 that, according to the Franck–Condon principle, electronic transitions are so fast that they can be regarded as taking place in a stationary nuclear framework. This principle also applies to an electron transfer process in which an electron migrates from one energy surface, representing the dependence of the energy of DA on its geometry, to another representing the energy of D⁺A⁻. We can represent the potential energy (and the Gibbs energy) surfaces of the two complexes (the reactant complex, DA, and the product complex, D⁺A⁻) by the parabolas characteristic of harmonic oscillators, with the displacement coordinate corresponding to the changing geometries (Fig. 29.1). This coordinate represents a collective mode of the donor, acceptor, and solvent.

According to the Franck–Condon principle, the nuclei do not have time to move when the system passes from the reactant to the product surface as a result of the transfer of an electron. Therefore, electron transfer can occur only after thermal fluctuations bring the geometry of DA to q^* in Fig. 29.1, the value of the nuclear coordinate at which the two parabolas intersect.

The factor $\kappa\nu$ is a measure of the probability that the system will convert from reactants (DA) to products (D⁺A⁻) at q^* by electron transfer within the thermally excited DA complex. To understand the process, we must turn our attention to the effect that the rearrangement of nuclear coordinates has on electronic energy levels of DA and D⁺A⁻ for a given distance r between D and A (Fig. 29.2). Initially, the electron to be transferred occupies the HOMO of D, and the overall energy of DA is lower than that of D⁺A⁻ (Fig. 29.2a). As the nuclei rearrange to a configuration represented by q^* in Fig. 29.2, the highest occupied electronic level of DA and the lowest unoccupied electronic level of D⁺A⁻ become degenerate and electron transfer becomes energetically feasible (Fig. 29.2b). Over reasonably short distances r, the main mechanism of electron transfer is *tunnelling* through the potential energy barrier depicted in Fig. 29.2. The height of the barrier increases as the ionization energies of the DA and D⁺A⁻ complexes. After an electron moves from the HOMO of D to the LUMO of A, the system relaxes to the configuration represented by q_P° in Fig. 29.2 As shown in Fig. 29.2c, now the energy of D⁺A⁻ is lower than that of DA, reflecting the thermodynamic tendency for A to remain reduced and for D to remain oxidized.

The tunnelling event responsible for electron transfer is similar to that described in Section 12.3, except that in this case the electron tunnels from an electronic level of D, with wavefunction ψ_D, to an electronic level of A, with wavefunction ψ_A. We saw in Section 12.3 that the rate of an electronic transition from a level described by the wavefunction ψ_D to a level described by ψ_A is proportional to the square of the matrix element $\langle H_{DA}\rangle = \langle\psi_A|H_{DA}|\psi_D\rangle$, where H_{DA} is a hamiltonian that describes the coupling of the electronic wavefunctions. It turns out that in cases where the coupling is relatively weak we may write:

$$\langle H_{DA}\rangle^2 = \langle H_{DA}^{\circ}\rangle^2 e^{-\beta r} \tag{29.3}$$

where r is the edge-to-edge distance between D and A, β is a parameter that measures the sensitivity of the electronic coupling matrix element to distance, and $\langle H_{DA}^{\circ}\rangle$ is the value of the electronic coupling matrix element when D and A are in contact ($r = 0$). The exponential dependence on distance in eqn 3 is essentially the same as the exponential decrease in transmission probability through a potential energy barrier described in Section 12.3.

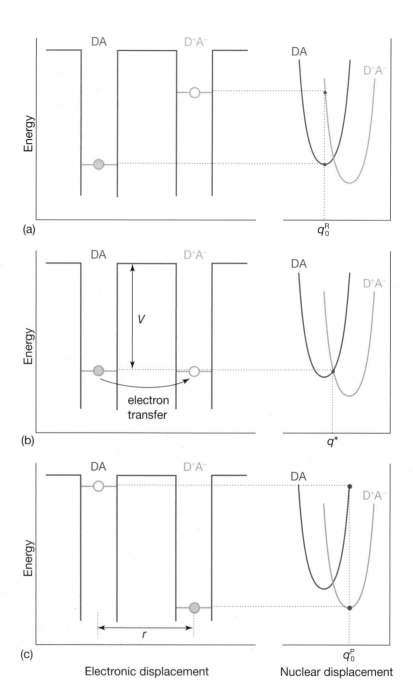

29.2 Correspondence between the electronic energy levels (shown on the left) and the nuclear energy levels (shown on the right) for the DA and D^+A^- complexes involved in an electron transfer process. (a) At the nuclear configuration denoted by q_0^R, the electron to be transferred in DA is in an occupied electronic energy level (denoted by a blue circle) and the lowest unoccupied energy level of D^+A^- (denoted by an unfilled circle) is of too high an energy to be a good electron acceptor. (b) As the nuclei rearrange to a configuration represented by q^*, DA and D^+A^- become degenerate and electron transfer occurs by tunnelling through the barrier of height V and width r, the edge-to-edge distance between donor and acceptor. (c) The system relaxes to the equilibrium nuclear configuration of D^+A^- denoted by q_0^P, in which the lowest unoccupied electronic level of DA is higher in energy than the highest occupied electronic level of D^+A^-. (Adapted from R.A. Marcus and N. Sutin, *Biochim. Biophys. Acta* **811**, 265 (1985).)

(b) The Gibbs energy of activation

We show in the *Justification* below that the Gibbs energy of activation is

$$\Delta^{\ddagger}G = \frac{(\Delta_r G^{\ominus} + \lambda)^2}{4\lambda} \tag{29.4}$$

where $\Delta_r G^{\ominus}$ is the standard reaction Gibbs energy for the electron transfer process DA $\rightarrow$ D^+A^-, and λ is the **reorganization energy**, the energy change associated with molecular rearrangements that must take place so that DA can take on the equilibrium geometry of D^+A^-. These molecular rearrangements include the relative reorientation of the D and A molecules in DA and the relative reorientation of the solvent molecules surrounding DA. Equation 4 shows that $\Delta^\ddagger G = 0$, with the implication that the reaction is not slowed down by an activation barrier, when $\Delta_r G^\ominus = -\lambda$, corresponding to the cancellation of the reorganization energy term by the standard reaction Gibbs energy.

Justification 29.1

The simplest way to derive an expression for the Gibbs energy of activation of electron transfer processes is to construct a model in which the surfaces for DA (the 'reactant complex', denoted R) and for D^+A^- (the 'product complex', denoted P) are described by classical harmonic oscillators (see *Further information 2*) with identical reduced masses μ and angular frequencies ω, but displaced minima, as shown in Fig. 29.1. The molar Gibbs energies $G_{m,R}(q)$ and $G_{m,P}(q)$ of the reactant and product complexes, respectively, may be written as

$$G_{m,R}(q) = \tfrac{1}{2}N_A\mu\omega^2(q - q_0^R)^2 + G_{m,R}(q_0^R) \tag{29.5a}$$

$$G_{m,P}(q) = \tfrac{1}{2}N_A\mu\omega^2(q - q_0^P)^2 + G_{m,P}(q_0^P) \tag{29.5b}$$

where q_0^R and q_0^P are the values of q at which the minima of the reactant and product parabolas occur, respectively. The standard reaction Gibbs energy for the electron transfer process $DA \rightarrow D^+A^-$ is $\Delta_r G^\ominus = G_{m,P}(q_0^P) - G_{m,R}(q_0^R)$, the difference in standard molar Gibbs energy between the minima of the parabolas. In Fig. 29.1, $\Delta_r G^\ominus < 0$.

We also note that q^*, the value of q corresponding to the transition state of the complex, may be written in terms of the parameter α, the fractional change in q:

$$q^* = q_0^R + \alpha(q_0^P - q_0^R) \tag{29.6}$$

We see from Fig. 29.1 that $\Delta^\ddagger G = G_{m,R}(q^*) - G_{m,R}(q_0^R)$. It then follows from eqns 5a, 5b, and 6 that

$$\Delta^\ddagger G = \tfrac{1}{2}N_A\mu\omega^2(q^* - q_0^R)^2 = \tfrac{1}{2}N_A\mu\omega^2\{\alpha(q_0^P - q_0^R)\}^2 \tag{29.7}$$

We now define the reorganization energy, λ, as

$$\lambda = \tfrac{1}{2}N_A\mu\omega^2(q_0^P - q_0^R)^2 \tag{29.8}$$

which can be interpreted as $G_{m,R}(q_0^P) - G_{m,R}(q_0^R)$ and, consequently, as the (Gibbs) energy required to deform the equilibrium configuration of DA to the equilibrium configuration of D^+A^- (as shown in Fig. 29.1). It follows from eqns 7 and 8 that

$$\Delta^\ddagger G = \alpha^2\lambda \tag{29.9}$$

Because $G_{m,R}(q^*) = G_{m,P}(q^*)$, it follows from eqns 5b, 8, and 9 that

$$\alpha^2\lambda = \tfrac{1}{2}N_A\mu\omega^2\{(\alpha - 1)(q_0^P - q_0^R)\}^2 + \Delta_r G^\ominus = (\alpha - 1)\lambda + \Delta_r G^\ominus \tag{29.10}$$

which implies that

$$\alpha = \tfrac{1}{2}\left(\frac{\Delta_r G^\ominus}{\lambda} + 1\right) \tag{29.11}$$

By combining eqns 9 and 11, we obtain eqn 4. We obtain an identical relation if we allow the harmonic oscillators to have different angular frequencies and hence different curvatures (see *Further reading*).

(c) The expression for the rate of electron transfer

The full expression for k_{et} turns out to be

$$k_{et} = \frac{2\langle H_{DA}\rangle^2}{h}\left(\frac{\pi^3}{4\lambda RT}\right)^{1/2} e^{-\Delta^{\ddagger}G/RT} \tag{29.12}$$

where $\langle H_{DA}\rangle^2$ can be approximated by eqn 3 and $\Delta^{\ddagger}G$ is given by eqn 4.[2]

Equation 12 has some limitations. First, it describes only those processes with weak electronic coupling between donor and acceptor. Weak coupling is observed when the electroactive species are sufficiently far apart that the wavefunctions ψ_A and ψ_D do not overlap extensively. An example of a weakly coupled system is the cytochrome c-cytochrome b_5 complex, in which the electroactive haem-bound iron ions shuttle between oxidation states +2 and +3 during electron transfer and are about 1.7 nm apart. Strong coupling is observed when the wavefunctions ψ_A and ψ_D overlap very extensively. Examples of strongly coupled systems are mixed-valence, binuclear d-metal complexes with the general structure $L_m M^{n+}$–B–$M^{p+}L_m$, in which the electroactive metal ions are separated by a bridging ligand B. In these systems, $r < 1.0$ nm. The weak coupling limit applies to a large number of electron transfer reactions, including those between proteins during metabolism (Box 10.1).[3] Second, the term $(\pi^3/4\lambda RT)^{1/2}\,e^{-\Delta^{\ddagger}G/RT}$ should be used only at high temperatures. At low temperatures, thermal fluctuations alone cannot bring the reactants to the transition state and activated complex theory, which is at the heart of the theory presented in this section, fails to account for any observed electron transfer. Electron transfer can still occur, but by *nuclear* tunnelling from the reactant to the product surfaces. We saw in Section 12.5 that the wavefunctions for the lower levels of the quantum mechanical harmonic oscillator extend significantly beyond classically allowed regions, so an oscillator can tunnel into a region of space in which another oscillator may be found. Full quantum mechanical treatments of electron transfer reactions replace the $(\pi^3/4\lambda RT)^{1/2}\,e^{-\Delta^{\ddagger}G/RT}$ term with Franck–Condon factors similar to those discussed in Section 17.1, which couple the *nuclear* wavefunctions and provide a measure of the contribution of nuclear tunnelling to the rate of electron transfer.

29.2 Experimental results

It is difficult to measure the distance dependence of k_{et} when the reactants are ions or molecules that are free to move in solution. In such cases, electron transfer occurs after a donor–acceptor complex forms and it is not possible to exert control over r, the edge-to-edge distance. The most meaningful experimental tests of the dependence of k_{et} on r are those in which the same donor and acceptor are positioned at a variety of distances, perhaps by covalent attachment to molecular linkers. Under these conditions, the term $e^{-\Delta^{\ddagger}G/RT}$ becomes a constant and, after taking the natural logarithm of eqn 12 and using eqn 3, we obtain

$$\ln k_{et} = -\beta r + \text{constant} \tag{29.13}$$

which implies that a plot of $\ln k_{et}$ against r should be a straight line with slope $-\beta$. The value of β depends on the medium through which the electron must travel from donor to acceptor. In a vacuum, $28\ \text{nm}^{-1} < \beta < 35\ \text{nm}^{-1}$, whereas $\beta \approx 9\ \text{nm}^{-1}$ when the intervening medium is a molecular link between donor and acceptor. Electron transfer between protein-bound cofactors can occur at distances of up to about 2.0 nm, a

---

2. See *Further reading* for references to the full derivation of eqn 12.
3. See *Further reading* for expressions that apply for the strong coupling limit.

1

$A =$ (a)

(b)

(c)

(d)

(e)

(f) $R_1 = R_2 = H$
(g) $R_1 = H$, $R_2 = Cl$
(h) $R_1 = R_2 = Cl$

relatively long distance on a molecular scale, with the protein providing an intervening medium between donor and acceptor.

There is, however, a great deal of controversy surrounding the interpretation of electron transfer data in proteins. Much of the available data may be interpreted with $\beta \approx 14 \text{ nm}^{-1}$, a value that appears to be insensitive to the primary and secondary structures of the protein but does depend slightly on the density of atoms in the section of protein that separates donor from acceptor. More detailed work on the specific effect of secondary structure suggests that $12.5 \text{ nm}^{-1} < \beta < 16.0 \text{ nm}^{-1}$ when the intervening medium consists primarily of α helices and $9.0 \text{ nm}^{-1} < \beta < 11.5 \text{ nm}^{-1}$ when the medium is primarily β sheet. Yet another view suggests that the electron takes specific paths through covalent bonds and hydrogen bonds that exist in the protein for the purpose of optimizing the rate of electron transfer.

The dependence of k_{et} on the standard reaction Gibbs energy has been investigated in systems where the edge-to-edge distance, the reorganization energy, and $\kappa \nu$ are constant for a series of reactions. Then we use eqn 4 to rewrite eqn 12 as

$$\ln k_{et} = -\frac{1}{4\lambda}\left(\frac{\Delta_r G^\ominus}{RT}\right)^2 - \frac{1}{2}\left(\frac{\Delta_r G^\ominus}{RT}\right) + \text{constant} \qquad (29.14)$$

and a plot of $\ln k_{et}$ (or $\log k_{et}$) against $\Delta_r G^\ominus$ (or $-\Delta_r G^\ominus$) is predicted to be shaped like a downward parabola. Equation 14 implies that the rate constant increases as $\Delta_r G^\ominus$ decreases but only up to $-\Delta_r G^\ominus = \lambda$. Beyond that, the reaction enters the **inverted region**, in which the rate constant decreases as the reaction becomes more exergonic ($\Delta_r G^\ominus$ becomes more negative). The inverted region has been observed in a series of special compounds in which the electron donor and acceptor are linked covalently to a molecular spacer of known and fixed size (Fig. 29.3).

The behaviour predicted by eqn 14 and observed experimentally in Fig. 29.3 can be explained by considering the dependence of the activation Gibbs energy on the standard Gibbs energy of electron transfer. We suppose that the energies of the reactant and product complexes can be characterized by parabolas with identical curvatures and fixed but distinct q_0^R and q_0^P. Now we let the minimum energy of the product complex change while keeping q_P° constant, which corresponds to changing the magnitude of $\Delta_r G^\ominus$. Figure 29.4 shows the effect of increasing the exergonicity of the process. In Fig. 29.4a we see that for a range of values of $\Delta_r G^\ominus$, $\Delta^\ddagger G > 0$ and the transition state is at $q_a^* > q_0^R$. As the process becomes more exergonic, the activation Gibbs energy decreases and the rate constant increases. (This behaviour is another example of a free-energy relation, first discussed in Section 27.5.) Figure 29.4b shows that when $\Delta^\ddagger G = 0$ and $q_b^* = q_0^R$ the rate constant for the process reaches a maximum as there is no activation barrier to overcome. According to eqn 4, this condition occurs when $-\Delta_r G^\ominus = \lambda$. Finally, Fig. 29.4c shows that as the process becomes even

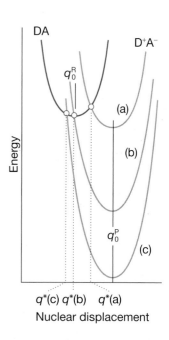

29.3 Variation of log k_{et} with $-\Delta_r G^\ominus$ for a series of compounds with the structures given in (**1**). Kinetic measurements were conducted in 2-methyltetrahydrofuran and at 296 K. The distance between donor (the reduced biphenyl group) and the acceptor is constant for all compounds in the series because the molecular linker remains the same. Each acceptor has a characteristic standard reduction potential, so it follows that the standard Gibbs energy for the electron transfer process is different for each compound in the series. The line is a fit to a version of eqn 14 and the maximum of the parabola occurs at $-\Delta_r G^\ominus = \lambda = 1.2$ eV = 1.2×10^2 kJ mol^{-1}. (Reproduced with permission from J.R. Miller, L.T. Calcaterra, and G.L. Closs, *J. Am. Chem. Soc.* **106**, 3047 (1984).)

29.4 (a) $\Delta^\ddagger G > 0$ and the transition state is at $q_a^* > q_R^\circ$. As the process becomes more exergonic, the activation Gibbs energy decreases and the rate constant increases. (b) When $\Delta^\ddagger G = 0$ and $q_b^* = q_0^R$ the rate constant for the process reaches a maximum as there is no activation barrier to overcome. (c) As the process becomes even more exergonic, $\Delta^\ddagger G$ becomes positive again but now the transition state is at $q_c^* < q_0^R$. The rate constant for the process decreases steadily as the activation barrier for the process increases with decreasing $\Delta_r G^\ominus$.

more exergonic, $\Delta^\ddagger G$ becomes positive again but now the transition state is at $q_c^* < q_0^R$. The rate constant for the process decreases steadily as the activation barrier for the process increases with decreasing $\Delta_r G^{\ominus}$. This is the explanation for the 'inverted region' observed in Fig. 29.3.

Some of the key features of electron transfer theory have been tested by experiments, showing in particular the predicted dependence of k_{et} on the standard reaction Gibbs energy and the edge-to-edge distance between electron donor and acceptor. The basic theory presented in Section 29.1 has been extended to transfer of other light particles, such as protons.

Electron transfer in heterogeneous systems

We now consider electron transfer in heterogeneous systems and focus on processes that occur at electrodes. As was the case in homogeneous systems, electron transfer on the surface of an electrode involves electron tunnelling. However, the electrode consists of a nearly infinite number of closely spaced electronic energy levels (Section 23.9). Furthermore, specific interactions with the electrode surface give the solute and solvent special properties that can be very different from those observed in the bulk of the solution. For this reason, we begin with a description of the electrode–solution interface. Then, we describe the kinetics of electrode processes by using a largely phenomenological (rather than strictly theoretical) approach that draws on the thermodynamic language inspired by activated complex theory.

29.3 The electrode–solution interface

The most primitive model of the boundary between the solid and liquid phases is as an **electrical double layer**, which consists of a sheet of positive charge at the surface of the electrode and a sheet of negative charge next to it in the solution (or vice versa).[4] We shall see that this arrangement creates an electrical potential difference, called the **Galvani potential difference**, between the bulk of the metal electrode and the bulk of the solution. More sophisticated models for the electrode–solution interface attempt to describe the gradual changes in the structure of the solution between two extremes: the charged electrode surface and the bulk of the solution.

(a) The structure of the interface

A more detailed picture of the interface can be constructed by speculating about the arrangement of ions and electric dipoles in the solution. In the **Helmholtz layer model** of the interface the solvated ions arrange themselves along the surface of the electrode but are held away from it by their hydration spheres (Fig. 29.5). The location of the sheet of ionic charge, which is called the **outer Helmholtz plane** (OHP), is identified as the plane running through the solvated ions. In this simple model, the electrical potential changes linearly within the layer bounded by the electrode surface on one side and the OHP on the other (see Exercise 29.7a). In a refinement of this model, ions that have discarded their solvating molecules and have become attached to the electrode surface by chemical bonds are regarded as forming the **inner Helmholtz plane** (IHP). The Helmholtz layer model ignores the disrupting effect of

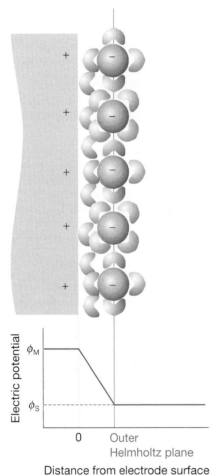

29.5 A simple model of the electrode–solution interface treats it as two rigid planes of charge. One plane, the outer Helmholtz plane (OHP), is due to the ions with their solvating molecules and the other plane is that of the electrode itself. The plot shows the dependence of the electric potential with distance from the electrode surface according to this model. Between the electrode surface and the OHP, the potential varies linearly from ϕ_M, the value in the metal, to ϕ_S, the value in the bulk of the solution.

4. We discussed the electrical double layer in connection with the properties of colloids (Section 22.13c).

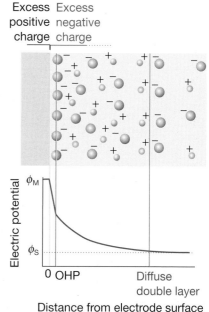

29.6 The Gouy–Chapman model of the electrical double layer treats the outer region as an atmosphere of counter-charge, similar to the Debye–Hückel theory of ion atmospheres. The plot of electrical potential against distance from the electrode surface shows the meaning of the diffuse double layer (see text for details).

29.7 A representation of the Stern model of the electrode–solution interface. The model incorporates the idea of an outer Helmholtz plane near the electrode surface and of a diffuse double layer further away from the surface.

thermal motion, which tends to break up and disperse the rigid outer plane of charge. In the **Gouy–Chapman model** of the **diffuse double layer**, the disordering effect of thermal motion is taken into account in much the same way as the Debye–Hückel model describes the ionic atmosphere of an ion (Section 10.2c) with the latter's single central ion replaced by an infinite, plane electrode.

Figure 29.6 shows how the local concentrations of cations and anions differ in the Gouy–Chapman model from their bulk concentrations. Ions of opposite charge cluster close to the electrode, and ions of the same charge are repelled from it. The modification of the local concentrations near an electrode implies that it might be misleading to use activity coefficients characteristic of the bulk to discuss the thermodynamic properties of ions near the interface. This is one of the reasons why measurements of the dynamics of electrode processes are almost always done using a large excess of supporting electrolyte (for example, a 1 mol L^{-1} solution of a salt, an acid, or a base). Under such conditions, the activity coefficients are almost constant because the inert ions dominate the effects of local changes caused by any reactions taking place.[5]

Neither the Helmholtz nor the Gouy–Chapman model is a very good representation of the structure of the double layer. The former overemphasizes the rigidity of the local solution; the latter underemphasizes its structure. The two are combined in the **Stern model**, in which the ions closest to the electrode are constrained into a rigid Helmholtz plane while outside that plane the ions are dispersed as in the Gouy–Chapman model (Fig. 29.7). Yet another level of sophistication is found in the **Grahame model**, which adds an inner Helmholtz plane to the Stern model.

5. The use of a concentrated solution also minimizes ion migration effects and ohmic losses.

(b) The electric potential at the interface

The potential at the interface can be analysed by imagining the separation of the electrode from the solution, but with the charges of the metal and the solution frozen in position. A positive test charge at great distances from the isolated electrode experiences a Coulomb potential that varies inversely with distance (Fig. 29.8). As the test charge approaches the electrode, which can be a metal or membrane electrode, it enters a region where the potential varies more slowly. This change in behaviour can be traced to the fact that the surface charge is not point-like but is spread over an area. At about 100 nm from the surface the potential varies only slightly with distance because the closer the point of observation is to the surface, although the potential from a given region of charge is stronger, a smaller area of surface is sampled (Fig. 29.9). The potential in this region is called the **outer potential**, ψ. As the test charge is taken through the skin of electrons on the surface of the electrode, the potential it experiences changes until the probe reaches the inner, bulk metal environment, where the potential is called the **inner potential**, ϕ. The difference between the inner and outer potentials is called the **surface potential**, χ.

A similar sequence of changes of potential is observed as a positive test charge is brought up to and through the solution surface. The potential changes to its outer value as the charge approaches the charged medium, then to its inner value as the probe is taken into the bulk.

Now consider bringing the electrode and solution back together again but without any change of charge distribution. The potential difference between points in the bulk metal and the bulk solution is the Galvani potential difference, $\Delta\phi$. Apart from a constant, this Galvani potential difference is the electrode potential that was at the centre of our discussion in Chapter 10. We shall ignore the constant, which cannot be measured anyway, and identify changes in $\Delta\phi$ with changes in electrode potential.

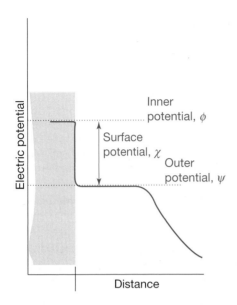

29.8 The variation of potential with distance from an electrode that has been separated from the electrolyte solution without there being an adjustment of charge. A similar diagram applies to the separated solution.

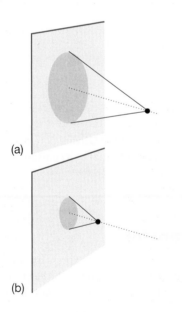

29.9 The origin of the distance-independence of the outer potential. (a) Far from the electrode, a point charge experiences a potential arising from a wide area but each contribution is weak. (b) Close to the electrode, the point charge experiences a potential arising from a small area but each contribution is strong. Provided the point charge is in a certain range of values (and specifically, where image charge effects can be ignored) the potential it experiences is largely independent of distance.

Justification 29.2

To demonstrate the relation between $\Delta\phi$ and E, consider the cell $Pt|H_2(g)|H^+(g)||M^+(aq)|M(s)$ and the half-reactions

$$M^+(aq) + e^- \longrightarrow M(s) \qquad H^+(aq) + e^- \longrightarrow \tfrac{1}{2} H_2(g)$$

The Gibbs energies of these two half-reactions can be expressed in terms of the chemical potentials, μ, of all the species. However, we must take into account the fact that the species are present in phases with different electric potentials. Thus, a cation in a region of positive potential has a higher chemical potential (is chemically more active in a thermodynamic sense) than in a region of zero potential.

The contribution of an electric potential to the chemical potential is calculated by noting that the electrical work of adding a charge ze to a region where the potential is ϕ is $ze\phi$, and therefore that the work per mole is $zF\phi$, where F is the Faraday constant. Because at constant temperature and pressure the maximum electrical work can be identified with the change in Gibbs energy (Section 10.2c), the difference in chemical potential of an ion with and without the electrical potential present is $zF\phi$. The chemical potential of an ion in the presence of an electric potential is called its **electrochemical potential**, $\bar{\mu}$. It follows that

$$\bar{\mu} = \mu + zF\phi \qquad [29.15]$$

where μ is the chemical potential of the species when the potential is zero. When $z = 0$ (a neutral species), the electrochemical potential is equal to the chemical potential.

To express the Gibbs energy for the half-reactions in terms of the electrochemical potentials of the species we note that the cations M^+ are in the solution where the inner potential is ϕ_S and the electrons are in the electrode where it is ϕ_M. It follows that

$$\Delta_r G_R = \bar{\mu}(M) - \{\bar{\mu}(M^+) + \bar{\mu}(e^-)\} = \mu(M) - \{\mu(M^+) + F\phi_S + \mu(e^-) - F\phi_M\}$$
$$= \mu(M) - \mu(M^+) - \mu(e^-) + F\Delta\phi_R$$

where $\Delta\phi_R = \phi_M - \phi_S$ is the Galvani potential difference at the right-hand electrode. Likewise, in the hydrogen half-reaction, the electrons are in the platinum electrode at a potential ϕ_{Pt} and the H^+ ions are in the solution where the potential is ϕ_S:

$$\Delta_r G_L = \tfrac{1}{2}\bar{\mu}(H_2) - \{\bar{\mu}(H^+) + \bar{\mu}(e^-)\} = \tfrac{1}{2}\mu(H_2) - \mu(H^+) - \mu(e^-) + F\Delta\phi_L$$

where $\Delta\phi_L = \phi_{Pt} - \phi_S$ is the Galvani potential difference at the left-hand electrode.

The overall reaction Gibbs energy is

$$\Delta_r G_R - \Delta_r G_L = \mu(M) + \mu(H^+) - \mu(M^+) - \tfrac{1}{2}\mu(H_2) + F(\Delta\phi_R - \Delta\phi_L)$$
$$= \Delta_r G + F(\Delta\phi_R - \Delta\phi_L)$$

where $\Delta_r G$ is the Gibbs energy of the cell reaction. When the cell is balanced against an external source of potential the entire system is at equilibrium. The overall reaction Gibbs energy is then zero,[6] and the last equation becomes

$$0 = \Delta_r G + F(\Delta\phi_R - \Delta\phi_L)$$

which rearranges to

$$\Delta_r G = -F(\Delta\phi_R - \Delta\phi_L) \tag{29.16}$$

If we compare this with the result established in Section 10.4e that

$$\Delta_r G = -FE \qquad E = E_R - E_L \tag{29.17}$$

we can conclude that[7]

$$E_R - E_L = \Delta\phi_R - \Delta\phi_L \tag{29.18}$$

This is the result we wanted to show, for it implies that the Galvani potential difference at each electrode can differ from the electrode potential by a constant at most; that constant cancels when the difference is taken.

29.4 The rate of charge transfer

Because an electrode reaction is heterogeneous, it is natural to express its rate as the flux of products, the amount of material produced over a region of the electrode surface in an interval of time divided by the area of the region and the duration of the interval.

(a) The rate laws

A first-order heterogeneous rate law has the form

$$\text{Product flux} = k[\text{species}] \tag{29.19}$$

6. The cell reaction itself is not necessarily at equilibrium: its tendency to change is balanced against the external source of potential and overall there is stalemate.
7. We are ignoring the effects of any metal–platinum and liquid junction potentials that may be present in an actual cell.

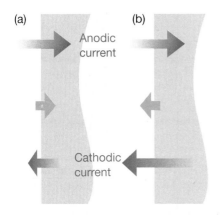

29.10 The net current density is defined as the difference between the cathodic and anodic contributions. (a) When $j_a > j_c$, the net current is anodic, and there is a net oxidation of the species in solution. (b) When $j_c > j_a$, the net current is cathodic, and the net process is reduction.

where [species] is the molar concentration of the relevant species in solution close to the electrode, just outside the double layer. The rate constant has dimensions of length/time (for example, cm s⁻¹). If the molar concentration of the oxidized and reduced materials outside the double layer are [Ox] and [Red] respectively, the rate of reduction of Ox, v_{Ox}, is

$$v_{Ox} = k_c[Ox] \tag{29.20a}$$

and the rate of oxidation of Red, v_{Red}, is

$$v_{Red} = k_a[Red] \tag{29.20b}$$

(The notation k_c and k_a is justified below.)

Now consider a reaction at the electrode in which an ion is reduced by the transfer of a single electron in the rate-determining step.[8] Then the net current density at the electrode is the difference between the current densities arising from the reduction of Ox and the oxidation of Red. Because the redox processes at the electrode involve the transfer of one electron per reaction event, the current densities, j, arising from the redox processes are the rates (as expressed above) multiplied by the charge transferred per mole of reaction, which is given by the Faraday constant. Therefore, there is a **cathodic current density** of magnitude

$$j_c = Fk_c[Ox] \quad \text{for} \quad Ox + e^- \longrightarrow Red \tag{29.21a}$$

arising from the reduction. There is also an opposing **anodic current density** of magnitude

$$j_a = Fk_a[Red] \quad \text{for} \quad Red \longrightarrow Ox + e^- \tag{29.21b}$$

arising from the oxidation.[9] The net current density at the electrode (Fig. 29.10) is the difference

$$j = j_a - j_c = Fk_a[Red] - Fk_c[Ox] \tag{29.22}$$

Note that, when $j_a > j_c$, so that $j > 0$, the current is anodic (Fig. 29.10a); when $j_c > j_a$, so that $j < 0$, the current is cathodic (Fig. 29.10b).

(b) The activation Gibbs energy

If a species is to participate in reduction or oxidation at an electrode, it must discard any solvating molecules, migrate through the electrode–solution interface, and adjust its hydration sphere as it receives or discards electrons. Likewise, a species already at the inner plane must be detached and migrate into the bulk. Because both processes are activated, we can expect to write their rate constants in the form suggested by activated complex theory (Section 27.4) as

$$k = Be^{-\Delta^{\ddagger}G/RT} \tag{29.23}$$

where $\Delta^{\ddagger}G$ is the activation Gibbs energy and B is a constant with the same dimensions as k.

When eqn 23 is inserted into eqn 22 we obtain

$$j = FB_a[Red]e^{-\Delta^{\ddagger}G_a/RT} - FB_c[Ox]e^{-\Delta^{\ddagger}G_c/RT} \tag{29.24}$$

8. The last phrase is important: in the deposition of cadmium, for instance; only one electron is transferred in the rate-determining step even though overall the deposition involves the transfer of two electrons.

9. Recall from Chapter 10 that the cathode is the site of reduction; hence the cathodic current is the current arising from reduction. The anode is the site of oxidation, and the anodic current is the current arising from oxidation.

29.11 The potential, ϕ, varies linearly between two plane parallel sheets of charge, and its effect on the Gibbs energy of the transition state depends on the extent to which the latter resembles the species at the inner or outer planes.

29.12 When the transition state resembles a species that has undergone reduction, the activation Gibbs energy for the anodic current is almost unchanged, but the full effect applies to the cathodic current. (a) Zero potential difference; (b) nonzero potential difference.

This expression allows the activation Gibbs energies to be different for the cathodic and anodic processes. That they are different is the central feature of the remaining discussion.

(c) The Butler–Volmer equation

Now we relate j to the Galvani potential difference, which varies across the electrode–solution interface as shown schematically in Fig. 29.11.

Consider the reduction reaction, $Ox + e^- \rightarrow Red$, and the corresponding reaction profile. If the transition state of the activated complex is product-like (as represented by the peak of the reaction profile being close to the electrode in Fig. 29.12), the activation Gibbs energy is changed from $\Delta^{\ddagger}G_c(0)$, the value it has in the absence of a potential difference across the double layer, to

$$\Delta^{\ddagger}G_c = \Delta^{\ddagger}G_c(0) + F\Delta\phi \tag{29.25}$$

Thus, if the electrode is more positive than the solution, $\Delta\phi > 0$, then more work has to be done to form an activated complex from Ox; in this case the activation Gibbs energy is increased. If the transition state is reactant-like (represented by the peak of the reaction profile being close to the outer plane of the double-layer in Fig. 29.13), then $\Delta^{\ddagger}G_c$ is independent of $\Delta\phi$. In a real system, the transition state has an intermediate resemblance to these extremes (Fig. 29.14) and the activation Gibbs energy for reduction may be written as

$$\Delta^{\ddagger}G_c = \Delta^{\ddagger}G_c(0) + \alpha F\Delta\phi \tag{29.26}$$

The parameter α is called the (cathodic) **transfer coefficient**, and lies in the range 0 to 1. Experimentally, α is often found to be about 0.5.

Now consider the oxidation reaction, $Red + e^- \rightarrow Ox$ and its reaction profile. Similar remarks apply. In this case, Red discards an electron to the electrode, so the

29.13 When the transition state resembles a species that has undergone oxidation, the activation Gibbs energy for the cathodic current is almost unchanged but the activation Gibbs energy for the anodic current is strongly affected. (a) Zero potential difference; (b) nonzero potential difference.

29.14 When the transition state is intermediate in its resemblance to reduced and oxidized species, as represented here by a peak located at an intermediate position as measured by α (with $0 < \alpha < 1$), both activation Gibbs energies are affected; here, $\alpha \approx 0.5$. (a) Zero potential difference; (b) nonzero potential difference.

extra work is zero if the transition state is reactant-like (represented by a peak close to the electrode). The extra work is the full $-F\Delta\phi$ if it resembles the product (the peak close to the outer plane). In general, the activation Gibbs energy for this anodic process is

$$\Delta^{\ddagger}G_a = \Delta^{\ddagger}G_a(0) - (1 - \alpha)F\Delta\phi \tag{29.27}$$

The two activation Gibbs energies can now be inserted in place of the values used in eqn 24 with the result that

$$j = FB_a[\text{Red}]e^{-\Delta^{\ddagger}G_a(0)/RT}e^{(1-\alpha)F\Delta\phi/RT} - FB_c[\text{Ox}]e^{-\Delta^{\ddagger}G_c(0)/RT}e^{-\alpha F\Delta\phi/RT} \tag{29.28}$$

This is an explicit, if complicated, expression for the net current density in terms of the potential difference.

The appearance of eqn 28 can be simplified. First, in a purely cosmetic step we write

$$f = \frac{F}{RT} \qquad [29.29]$$

Next, we identify the individual cathodic and anodic current densities:

$$\left. \begin{array}{l} j_a = FB_a[\text{Red}]e^{-\Delta^{\ddagger}G_a(0)/RT}e^{(1-\alpha)f\Delta\phi} \\ j_a = FB_c[\text{Ox}]e^{-\Delta^{\ddagger}G_c(0)/RT}e^{-\alpha f\Delta\phi} \end{array} \right\} \; j = j_a - j_c \qquad (29.30)$$

Example 29.1 Calculating the current density

Calculate the change in cathodic current density at an electrode when the potential difference changes from 1.0 V to 2.0 V at 25°C.

Method Equation 30 can be used to express the ratio of cathodic current densities j_c' and j_c when the potential differences are $\Delta\phi'$ and $\Delta\phi$. Then insert a typical value of α (namely, $\frac{1}{2}$) and the data.

Answer The ratio is

$$\frac{j_c'}{j_c} = e^{-\alpha f(\Delta\phi' - \Delta\phi)}$$

So, with $\alpha = \frac{1}{2}$ and $\Delta\phi' - \Delta\phi = 1.0$ V,

$$\alpha f \times (\Delta\phi' - \Delta\phi) = \frac{\frac{1}{2} \times (9.6485 \times 10^4 \text{ C mol}^{-1}) \times (1.0 \text{ V})}{(8.3145 \text{ J K}^{-1} \text{ mol}^{-1}) \times (298 \text{ K})} = \frac{9.6485 \times 10^4 \times 1.0}{2 \times 8.3145 \times 298}$$

Hence,

$$\frac{j_c'}{j_c} = e^{-\frac{9.6485 \times 10^4 \times 1.0}{2 \times 8.3145 \times 298}} = 4 \times 10^{-9}$$

where we have used 1 J = 1 V C.

Comment This huge change in current density occurs for a very mild and easily applied change of conditions. We can appreciate why the change is so great by realizing that a change of potential difference by 1 V changes the activation Gibbs energy by $(1 \text{ V}) \times F$, or about 50 kJ mol^{-1}, which has an enormous effect on the rates.

Self-test 29.1 Calculate the change in anodic current density under the same circumstances.

$$[j_a'/j_a = 3 \times 10^8]$$

If the cell is balanced against an external source, the Galvani potential difference, $\Delta\phi$, can be identified as the (zero-current) electrode potential, E, and we can write[10]

$$\begin{array}{l} j_a = FB_a[\text{Red}]e^{-\Delta^{\ddagger}G_a(0)/RT}e^{(1-\alpha)fE} \\ j_b = FB_c[\text{Ox}]e^{-\Delta^{\ddagger}G_c(0)/RT}e^{-\alpha fE} \end{array} \qquad (29.31)$$

---

10. We are assuming that we can identify the Galvani potential difference and the zero-current electrode potential. As explained earlier, they differ by a constant amount, which may be regarded as absorbed into the constant B.

When these equations apply, there is no net current at the electrode (as the cell is balanced), so the two current densities must be equal. From now on we denote them both as j_0, which is called the **exchange current density**.

When the cell is producing current (that is, when a load is connected between the electrode being studied and a second counter-electrode) the electrode potential changes from its zero-current value, E, to a new value, E', and the difference is the electrode's **overpotential**, η:

$$\eta = E' - E \tag{29.32}$$

Hence, $\Delta\phi$ changes to

$$\Delta\phi = E + \eta \tag{29.33}$$

It follows that the two current densities become

$$j_a = j_0 e^{(1-\alpha)f\eta} \qquad j_c = j_0 e^{-\alpha f\eta} \tag{29.34}$$

Then from eqn 28 we obtain the **Butler–Volmer equation**:

$$j = j_0 \{ e^{(1-\alpha)f\eta} - e^{-\alpha f\eta} \} \tag{29.35}$$

The Butler–Volmer equation is the basis of all that follows.

(d) The low overpotential limit

When the overpotential is so small that $f\eta \ll 1$ (in practice, η less than about 0.01 V) the exponentials in eqn 35 can be expanded by using $e^x = 1 + x + \cdots$ to give

$$j = j_0 \{ 1 + (1-\alpha)f\eta + \cdots - (1 - \alpha f\eta + \cdots) \} \approx j_0 f\eta \tag{29.36}$$

This equation shows that the current density is proportional to the overpotential, so at low overpotentials the interface behaves like a conductor that obeys Ohm's law. When there is a small positive overpotential the current is anodic ($j > 0$ when $\eta > 0$), and when the overpotential is small and negative the current is cathodic ($j < 0$ when $\eta < 0$). The relation can also be reversed to calculate the potential difference that must exist if a current density j has been established by some external circuit:

$$\eta = \frac{RTj}{Fj_0} \tag{29.37}$$

The importance of this interpretation will become clear below.

Illustration 29.1

The exchange current density of a $Pt(s)|H_2(g)|H^+(aq)$ electrode at 298 K is 0.79 mA cm^{-2}. Therefore, the current density when the overpotential is +5.0 mV is obtained by using eqn 36 and $f = F/RT = 1/(25.69 \text{ mV})$:

$$j = j_0 f\eta = \frac{(0.79 \text{ mA cm}^{-2}) \times (5.0 \text{ mV})}{25.69 \text{ mV}} = 0.15 \text{ mA cm}^{-2}$$

The current through an electrode of total area 5.0 cm^2 is therefore 0.75 mA.

Self-test 29.2   What would be the current at pH = 2.0, the other conditions being the same?   [−18 mA (cathodic)]

Current supply

V   A

Reference circuit

Counter electrode

Calomel reference electrode

Luggin capillary   Working electrode

29.15 The general arrangement for electrochemical rate measurements. The external source establishes a current between the working electrode and the counter electrode, and its effect on the potential difference of either of them relative to the reference electrode is observed. No current flows in the reference circuit.

(e) The high overpotential limit

When the overpotential is large and positive (in practice, $\eta \geq 0.12$ V), corresponding to the electrode being the anode in electrolysis, the second exponential in eqn 35 is much smaller than the first, and may be neglected. Then

$$j = j_0 e^{(1-\alpha)f\eta} \tag{29.38}$$

so

$$\ln j = \ln j_0 + (1-\alpha)f\eta \tag{29.39}$$

When the overpotential is large but negative (in practice, $\eta \leq -0.12$ V), corresponding to the cathode in electrolysis, the first exponential in eqn 35 may be neglected. Then

$$j = -j_0 e^{-\alpha f\eta} \tag{29.40}$$

so

$$\ln(-j) = \ln j_0 - \alpha f\eta \tag{29.41}$$

The plot of the logarithm of the current density against the overpotential is called a **Tafel plot**. The slope gives the value of α and the intercept at $\eta = 0$ gives the exchange current density.

The experimental arrangement used for a Tafel plot is shown in Fig. 29.15. A similar arrangement is typical of all kinds of electrochemical rate measurements. The current-carrying electrodes are the **working electrode**, the electrode of interest, and the **counter electrode**, which is necessary to complete the electrical circuit. The current flowing through them is controlled externally. If the area of the working electrode is A and the current is I, the current density across its surface is I/A. The potential difference across the interface cannot be measured directly, but the potential of the working electrode relative to a third electrode, the **reference electrode**, can be measured with a high impedance voltmeter, and no current flows in that half of the circuit. The reference electrode is in contact with the solution close to the working electrode through a 'Luggin capillary', which helps to eliminate any ohmic potential difference that might arise accidentally. Changing the current flowing through the working circuit causes a change of potential of the working electrode, and that change is measured with the voltmeter. The overpotential is then obtained by taking the difference between the potentials measured with and without a flow of current through the working circuit.

Example 29.2 Interpreting a Tafel plot

The data below refer to the anodic current through a platinum electrode of area 2.0 cm^2 in contact with an Fe^{3+}, Fe^{2+} aqueous solution at 298 K. Calculate the exchange current density and the transfer coefficient for the electrode process.

η/mV	50	100	150	200	250
I/mA	8.8	25.0	58.0	131	298

Method The anodic process is the oxidation Fe^{2+}(aq) $\rightarrow$ Fe^{3+}(aq) + e$^-$. To analyse the data, we make a Tafel plot (of $\ln j$ against η) using the anodic form (eqn 39). The intercept at $\eta = 0$ is $\ln j_0$ and the slope is $(1-\alpha)f$.

Answer Draw up the following table:

29.16 A Tafel plot is used to measure the exchange current density (given by the extrapolated intercept at $\eta = 0$) and the transfer coefficient (from the slope). The data are from Example 29.2.

η/mV	50	100	150	200	250
$j/mA\ cm^{-2}$	4.4	12.5	29.0	56.6	149
$\ln(j/mA\ cm^{-2})$	1.48	2.53	3.37	4.18	5.00

The points are plotted in Fig. 29.16. The high overpotential region gives a straight line of intercept 0.92 and slope 0.0163. From the former it follows that $\ln(j_0/mA\ cm^{-2}) = 0.92$, so $j_0 = 2.5\ mA\ cm^{-2}$. From the latter,

$$(1 - \alpha)\frac{F}{RT} = 0.0163\ mV^{-1}$$

so $\alpha = 0.58$.

Comment Note that the Tafel plot is nonlinear for $\eta < 100$ mV; in this region $\alpha f \eta = 2.3$ and the approximation that $\alpha f \eta \gg 1$ fails.

Self-test 29.3 Repeat the analysis using the following cathodic current data:

η/mV	−50	−100	−150	−200	−250	−300
I/mA	−0.3	−1.5	−6.4	−27.6	−118.6	−510

$[\alpha = 0.75, j_0 = 0.040\ mA\ cm^{-2}]$

Some experimental values for the Butler–Volmer parameters are given in Table 29.1. From them we can see that exchange current densities vary over a very wide range. For example, the N_2,N_3^- couple on platinum has $j_0 = 10^{-76}\ A\ cm^{-2}$, whereas the H^+, H_2 couple on platinum has $j_0 = 7.9 \times 10^{-4}\ A\ cm^{-2}$, a difference of 73 orders of magnitude. Exchange currents are generally large when the redox process involves no bond breaking (as in the $[Fe(CN)_6]^{3-},[Fe(CN)_6]^{4-}$ couple) or if only weak bonds are broken (as in Cl_2,Cl^-). They are generally small when more than one electron needs to be transferred, or when multiple or strong bonds are broken, as in the N_2,N_3^- couple and in redox reactions of organic compounds.

29.5 Voltammetry

The kinetics of electrode processes can be studied by **voltammetry**, in which the current is monitored as the potential of the electrode is changed, and by **chronopotentiometry**, in which the potential is monitored as the current flow is changed. Voltammetry may also be used to identify species present in solution and to determine their concentration.

Before we describe voltammetry in detail, we need to understand how electrode potentials vary with current. Electrodes with potentials that change only slightly when a current passes through them are classified as **non-polarizable**. Those with strongly current-dependent potentials are classified as **polarizable**. From the linearized equation (eqn 37) it is clear that the criterion for low polarizability is high exchange current density (so η may be small even though j is large). The calomel and H_2/Pt electrodes are both highly non-polarizable, which is one reason why they are so extensively used as reference electrodes in electrochemistry.

(a) Concentration polarization

One of the assumptions in the derivation of the Butler–Volmer equation is the negligible conversion of the electroactive species at low current densities, resulting in uniformity of concentration near the electrode. This assumption fails at high

Table 29.1* Exchange current densities and transfer coefficients at 298 K

Reaction	Electrode	$j_0/(A\ cm^{-2})$	α
$2\,H^+ + 2\,e^-$	Pt	7.9×10^{-4}	
$\longrightarrow H_2$	Ni	6.3×10^{-6}	0.58
	Pb	5.0×10^{-12}	
$Fe^{3+} + e^-$	Pt	2.5×10^{-3}	0.58
$\longrightarrow Fe^{2+}$			

* More values are given in the *Data section*.

current densities because the consumption of electroactive species close to the electrode results in a concentration gradient; diffusion of the species towards the electrode from the bulk is slow and may become rate-determining. A larger overpotential is then needed to produce a given current. This effect is called **concentration polarization** and its contribution to the total overpotential is called the **polarization overpotential**, η^c.

Consider a case for which the concentration polarization dominates all the rate processes and a redox couple of the type M^{z+}, M with the reduction $M^{z+} + ze^- \rightarrow M$. Under zero-current conditions, when the net current density is zero, the electrode potential is related to the activity, a, of the ions in the solution by the Nernst equation (eqn 10.35):

$$E = E^{\ominus} + \frac{RT}{zF} \ln a \tag{29.42}$$

As remarked earlier, electrode kinetics are normally studied using a large excess of support electrolyte so as to keep the mean activity coefficients approximately constant. Therefore, the constant activity coefficient in $a = \gamma c$ may be absorbed into E, and we write the **formal potential**, E°, of the electrode as

$$E^\circ = E^{\ominus} + \frac{RT}{zF} \ln \gamma \tag{29.43}$$

Then the electrode potential is

$$E = E^\circ + \frac{RT}{zF} \ln c \tag{29.44}$$

When the cell is producing current, the active ion concentration at the OHP changes to c' and the electrode potential changes to

$$E' = E^\circ + \frac{RT}{zF} \ln c' \tag{29.45}$$

The concentration overpotential is therefore

$$\eta^c = E' - E = \frac{RT}{zF} \ln\left(\frac{c'}{c}\right) \tag{29.46}$$

We now suppose that the solution has its bulk concentration, c, up to a distance δ from the outer Helmholtz plane, and then falls linearly to c. at the plane itself. This **Nernst diffusion layer** is illustrated in Fig. 29.17. The thickness of the Nernst layer (which is typically 0.1 mm, and strongly dependent on the condition of hydrodynamic flow due to any stirring or convective effects) is quite different from that of the electric double layer (which is typically less than 1 nm, and unaffected by stirring). The concentration gradient through the Nernst layer is

$$\frac{dc}{dx} = \frac{c' - c}{\delta} \tag{29.47}$$

29.17 In a simple model of the Nernst diffusion layer there is a linear variation in concentration between the bulk and the outer Helmholtz plane; the thickness of the layer depends strongly on the state of flow of the fluid. Note that the diffusion layer is much thicker relative to the OHP than shown here.

This gradient gives rise to a flux of ions towards the electrode, which replenishes the cations as they are reduced. The (molar) flux, J, is proportional to the concentration gradient, and according to Fick's first law (Section 24.4)

$$J = -D\left(\frac{\partial c}{\partial x}\right) \tag{29.48}$$

Therefore, the particle flux towards the electrode is

$$J = D\frac{c - c'}{\delta} \tag{29.49}$$

The cathodic current density towards the electrode is the product of the particle flux and the charge transferred per mole of ions, zF:

$$j = zFJ = zFD\frac{c - c'}{\delta} \qquad (29.50)$$

For instance, for the couple $[Fe(CN)_6]^{2-}/[Fe(CN)_6]^{3-}$, $z = 1$, but for Fe^{3+}/Fe, $z = 3$. The maximum rate of diffusion across the Nernst layer occurs when the gradient is steepest, which is when $c' = 0$. This concentration occurs when an electron from an ion that diffuses across the layer is snapped over the activation barrier and on to the electrode. No flow of current can exceed the **limiting current density**, j_{lim}, which is given by

$$j_{lim} = zFJ_{lim} = \frac{zFDc}{\delta} \qquad (29.51)$$

Example 29.3 Estimating the limiting current density

Estimate the limiting current density at 298 K for an electrode in a 0.10 M $Cu^{2+}(aq)$ unstirred solution in which the thickness of the diffusion layer is about 0.3 mm.

Method To use eqn 51, estimate D from the ionic conductivity $\lambda = 107$ S cm^2 mol^{-1} (Table 24.4) and the Nernst–Einstein equation (eqn 24.81).

Answer Because $\lambda = z^2F^2D/RT$, eqn 51 may be written

$$j_{lim} = \frac{cRT\lambda}{zF\delta}$$

Therefore, with $\delta = 0.3$ mm, $c = 0.10$ mol L^{-1}, $z = 2$, and $T = 298$ K, we obtain $j_{lim} = 5$ mA cm^{-2}. The result implies that the current towards an electrode of area 1 cm^2 electrode cannot exceed 5 mA in this (unstirred) solution.

Self-test 29.4 Evaluate the limiting current density for an Ag(s)|Ag$^+$(aq) electrode in 0.010 mol L^{-1} Ag$^+$(aq) at 298 K. Take $\delta = 0.03$ mm.

[5 mA cm^{-2}]

It follows from eqn 50 that the concentration c' is related to the current density at the double layer by

$$c' = c - \frac{j\delta}{zFD} \qquad (29.52)$$

Hence, as the current density is increased, the concentration falls below the bulk value. However, this decline in concentration is small when the diffusion constant is large, for then the ions are very mobile and can quickly replenish any ions that have been removed.

Finally, we substitute eqn 52 into eqn 46 and obtain the following expressions for the overpotential in terms of the current density, and vice versa:

$$\eta^c = \frac{RT}{zF}\ln\left(1 - \frac{j\delta}{zcFD}\right)$$

$$\qquad (29.53)$$

$$j = \frac{zcFD}{\delta}(1 - e^{zf\eta^c})$$

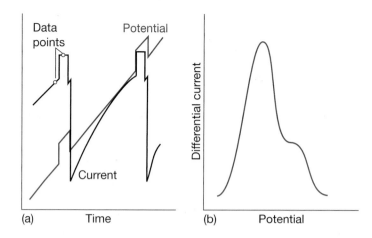

29.19 A differential pulse voltammetry experiment. (a) The potential is swept linearly as a mercury droplet grows on the end of a capillary dipping into the sample and then pulsed as shown by the blue line. The resulting current is shown as the black line and is sampled at the two points shown. (b) The data output is obtained as the difference of the currents at the two sampled points.

29.18 (a) The change of potential with time and (b) the resulting current/potential curve in a voltammetry experiment. The peak value of the current density is proportional to the concentration of electroactive species (for instance, [Ox]) in solution.

(b) Experimental techniques

The kind of output from **linear-sweep voltammetry** is illustrated in Fig. 29.18. Initially, the absolute value of the potential is low, and the cathodic current is due to the migration of ions in the solution. However, as the potential approaches the reduction potential of the reducible solute, the cathodic current grows. Soon after the potential exceeds the reduction potential the current rises and reaches a maximum value (as specified in eqn 51). This maximum current is proportional to the molar concentration of the species, so that concentration can be determined from the peak height after subtraction of an extrapolated baseline. In **differential pulse voltammetry** the current is monitored before and after a pulse of potential is applied, and the processed output is the slope of a curve like that obtained by linear-sweep voltammetry (Fig. 29.19). The area under the curve (in effect, the integral of the derivative displayed in the illustration) is proportional to the concentration of the species.

In **cyclic voltammetry** the potential is applied in a sawtooth manner to the working electrode and the current is monitored. A typical cyclic voltammogram is shown in Fig. 29.20. The shape of the curve is initially like that of a linear sweep experiment, but after reversal of the sweep there is a rapid change in current on account of the high concentration of oxidizable species close to the electrode that were generated on the reductive sweep. When the potential is close to the value required to oxidize the reduced species, there is a substantial anodic current until all the oxidation is complete, and the current returns to zero.

When the reduction reaction at the electrode can be reversed, as in the case of the $[Fe(CN)_6]^{3-}/[Fe(CN)_6]^{4-}$ couple, the cyclic voltammogram is broadly symmetric about the standard potential of the couple (as in Fig. 29.20b). The scan is initiated with $[Fe(CN)_6]^{3-}$ present in solution and, as the potential approaches $E^{\ominus}$ for the couple, the $[Fe(CN)_6]^{3-}$ near the electrode is reduced and current begins to flow. As the potential continues to change, the cathodic current begins to decline again because all the $[Fe(CN)_6]^{3-}$ near the electrode has been reduced and the current reaches its limiting value. The potential is now returned linearly to its initial value, and the reverse series of events occurs with the $[Fe(CN)_6]^{4-}$ produced during the forward scan now undergoing oxidation. The peak of current lies on the other side of $E^{\ominus}$, so the species

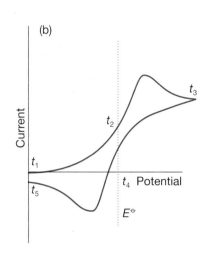

29.20 (a) The change of potential with time and (b) the resulting current/potential curve in a cyclic voltammetry experiment.

present and its standard potential can be identified, as indicated in the illustration, by noting the locations of the two peaks.

The overall shape of the curve gives details of the kinetics of the electrode process and the change in shape as the rate of change of potential is altered gives information on the rates of the processes involved. For example, the matching peak on the return phase of the sawtooth change of potential may be missing, which indicates that the oxidation (or reduction) is irreversible. The appearance of the curve may also depend on the timescale of the sweep for, if the sweep is too fast, some processes might not have time to occur. This style of analysis is illustrated in the following example.

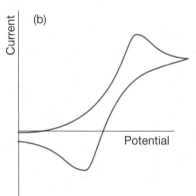

29.21 (a) When a non-reversible step in a reaction mechanism has time to occur, the cyclic voltammogram may not show the reverse oxidation or reduction peak. (b) However, if the rate of sweep is increased, the return step may be caused to occur before the irreversible step has had time to intervene, and a typical 'reversible' voltammogram is obtained.

Example 29.4 Analysing a cyclic voltammetry experiment

The electroreduction of *p*-bromonitrobenzene in liquid ammonia is believed to occur by the following mechanism:

$$BrC_6H_4NO_2 + e^- \longrightarrow BrC_6H_4NO_2^-$$
$$BrC_6H_4NO_2^- \longrightarrow \cdot C_6H_4NO_2 + Br^-$$
$$\cdot C_6H_4NO_2 + e^- \longrightarrow C_6H_4NO_2^-$$
$$C_6H_4NO_2^- + H^+ \longrightarrow C_6H_5NO_2$$

Suggest the likely form of the cyclic voltammogram expected on the basis of this mechanism.

Method Decide which steps are likely to be reversible on the timescale of the potential sweep: such processes will give symmetrical voltammograms. Irreversible processes will give unsymmetrical shapes as reduction (or oxidation) might not occur. However, at fast sweep rates, an intermediate might not have time to react, and a reversible shape will be observed.

Answer At slow sweep rates, the second reaction has time to occur, and a curve typical of a two-electron reduction will be observed, but there will be no oxidation peak on the second half of the cycle because the product, $C_6H_5NO_2$, cannot be oxidized (Fig. 29.21a). At fast sweep rates, the second reaction does not have time to take place before oxidation of the $BrC_6H_4NO_2^-$ intermediate starts to occur during the reverse scan, so the voltammogram will be typical of a reversible one-electron reduction (Fig. 29.21b).

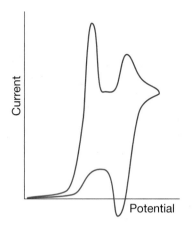

29.22 The cyclic voltammogram referred to in Self-test 29.5.

Self-test 29.5 Suggest an interpretation of the cyclic voltammogram shown in Fig. 29.22. The electroactive material is ClC_6H_4CN in acid solution; after reduction to $ClC_6H_4CN^-$, the radical anion may form C_6H_5CN irreversibly.

$[ClC_6H_4CN + e^- \rightleftharpoons ClC_6H_4CN^-, ClC_6H_4CN^- + H^+ + e^- \longrightarrow C_6H_5CN + Cl^-,$
$C_6H_5CN + e^- \rightleftharpoons C_6H_5CN^-]$

29.6 Electrolysis

To induce current to flow through an electrolytic cell and bring about a nonspontaneous cell reaction, the applied potential difference must exceed the zero-current potential by at least the **cell overpotential**. The cell overpotential is the sum of the overpotentials at the two electrodes and the ohmic drop (IR_s, where R_s is the internal resistance of the cell) due to the current through the electrolyte. The additional potential needed to achieve a detectable rate of reaction may need to be large when the exchange current density at the electrodes is small. For similar reasons, a working galvanic cell generates a smaller potential than under zero-current conditions. In this section we see how to cope with both aspects of the overpotential.

The rate of gas evolution or metal deposition during electrolysis can be estimated from the Butler–Volmer equation and tables of exchange current densities. The exchange current density depends strongly on the nature of the electrode surface, and changes in the course of the electrodeposition of one metal on another. A very crude criterion is that significant evolution or deposition occurs only if the overpotential exceeds about 0.6 V.

Example 29.5 Estimating the relative rates of electrolysis

Derive an expression for the relative rates of electrodeposition and hydrogen evolution in a solution in which both may occur.

Method We can calculate the ratio of the cathodic currents by using eqn 40. For simplicity, assume equal transfer coefficients.

Answer From eqn 40, where j' is the current density for electrodeposition and j is that for hydrogen evolution; j'_0 and j_0 are the corresponding exchange current densities.

$$\frac{j'}{j} = \frac{j'_0}{j_0} e^{(\eta - \eta')\alpha f}$$

Comment This equation shows that metal deposition is favoured by a large exchange current density and relatively high hydrogen evolution overvoltage (so $\eta - \eta'$ is positive and large). Note that $\eta < 0$ for a cathodic process, so $-\eta' > 0$.

Self-test 29.6 Deduce an expression for the ratio when the hydrogen evolution is limited by transport across a diffusion layer.

$[j'/j = (\delta j'_0/cFD)e^{-\alpha \eta' f}]$

A glance at Table 29.1 shows the wide range of exchange current densities for a metal/hydrogen electrode. The most sluggish exchange currents occur for lead and mercury, and the value of 1 pA cm^{-2} corresponds to a monolayer of atoms being replaced in about 5 y. For such systems, a high overpotential is needed to induce significant hydrogen evolution. In contrast, the value for platinum (1 mA cm^{-2})

corresponds to a monolayer being replaced in 0.1 s, so gas evolution occurs for a much lower overpotential.

The exchange current density also depends on the crystal face exposed. For the deposition of copper on copper, the (100) face has $j_0 = 1$ mA cm^{-2}, so for the same overpotential the (100) face grows at 2.5 times the rate of the (111) face, for which $j_0 = 0.4$ mA cm^{-2}.

29.7 Working galvanic cells

In working galvanic cells (those not balanced against an external potential), the overpotential leads to a smaller potential than under zero-current conditions. Furthermore, we expect the cell potential to decrease as current is generated because it is then no longer working reversibly and can therefore do less than maximum work.

(a) The cell potential

We shall consider the cell M|M$^+$(aq)||M$'^+$(aq)|M$'$ and ignore all the complications arising from liquid junctions. The potential of the cell is

$$E' = \Delta\phi_R - \Delta\phi_L \tag{29.54}$$

Because the cell potential differences differ from their zero-current values by overpotentials, we can write

$$\Delta\phi_X = E_X + \eta_X \tag{29.55}$$

where X is L or R for the left or right electrode, respectively. The cell potential is therefore

$$E' = E + \eta_R - \eta_L \tag{29.56a}$$

To avoid confusion about signs (η_R is negative, η_L is positive) and to emphasize that a working cell has a lower potential than a zero-current cell, we shall write this expression as

$$E' = E - |\eta_R| - |\eta_L| \tag{29.56b}$$

with E the cell emf. We should also subtract the ohmic potential difference IR_s, where R_s is the cell's internal resistance:

$$E' = E - |\eta_R| - |\eta_L| - IR_s \tag{29.57}$$

The ohmic term is a contribution to the cell's irreversibility—it is a thermal dissipation term—so the sign of IR_s is always such as to reduce the potential in the direction of zero.

The overpotentials in eqn 57 can be calculated from the Butler–Volmer equation for a given current, I, being drawn. We shall simplify the equations by supposing that the areas, A, of the electrodes are the same, that only one electron is transferred in the rate-determining steps at the electrodes, that the transfer coefficients are both $\frac{1}{2}$, and that the high-overpotential limit of the Butler–Volmer equation may be used. Then from eqns 40 and 57 we find

$$E' = E - IR_s - \frac{4RT}{F}\ln\left(\frac{I}{A\bar{j}}\right) \qquad \bar{j} = (j_{0L}j_{0R})^{1/2} \tag{29.58}$$

where j_{0L} and j_{0R} are the exchange current densities for the two electrodes.

The concentration overpotential also reduces the cell potential. If we use the Nernst diffusion layer model for each electrode, the total change of potential arising from concentration polarization is given by eqn 53 as

29.23 The dependence of the potential of a working galvanic cell on the current density being drawn (blue line) and the corresponding power output (black line) calculated by using eqns 60 and 61, respectively. Notice the sharp decline in power just after the maximum.

$$E' = E - \frac{RT}{zF} \ln \left\{ \left(1 - \frac{I}{Aj_{\text{lim,L}}} \right) \left(1 - \frac{I}{Aj_{\text{lim,R}}} \right) \right\} \qquad (29.59)$$

This contribution can be added to the one in eqn 58 to obtain a full (but still very approximate) expression for the cell potential when a current I is being drawn:

$$E' = E - IR_s - \frac{2RT}{zF} \ln g(I)$$

with

$$g(I) = \left(\frac{I}{A\bar{j}} \right)^{2z} \left\{ \left(1 - \frac{I}{Aj_{\text{lim,L}}} \right) \left(1 - \frac{I}{Aj_{\text{lim,R}}} \right) \right\}^{1/2} \qquad (29.60)$$

This equation depends on a lot of parameters, but an example of its general form is given in Fig. 29.23. Notice the very steep decline of working potential when the current is high and close to the limiting value for one of the electrodes.

Because the power, P, supplied by a galvanic cell is IE', from eqn 60 we can write

$$P = IE - I^2 R_s - \frac{2IRT}{zF} \ln g(I) \qquad (29.61)$$

The first term on the right is the power that would be produced if the cell retained its zero-current potential when delivering current. The second term is the power generated uselessly as heat as a result of the resistance of the electrolyte. The third term is the reduction of the potential at the electrodes as a result of drawing current.

The general dependence of power output on the current drawn is shown in Fig. 29.23 as the black line. Notice how maximum power is achieved just before the concentration polarization quenches the cell's performance. Information of this kind is essential if the optimum conditions for operating electrochemical devices are to be found and their performance improved.

(b) Fuel cells

A fuel cell operates like a conventional galvanic cell with the exception that the reactants are supplied from outside rather than forming an integral part of its construction. A fundamental and important example of a fuel cell is the hydrogen/oxygen cell, such as the ones used in the Apollo Moon missions (Fig. 29.24). One of the electrolytes used is concentrated aqueous potassium hydroxide maintained at 200°C and 20–40 atm; the electrodes may be porous nickel in the form of sheets of compressed powder. The cathode reaction is the reduction

$$O_2(g) + 2 H_2O(l) + 4 e^- \longrightarrow 4 OH^-(aq) \qquad E^\ominus = +0.40 \text{ V}$$

and the anode reaction is the oxidation

$$H_2(g) + 2 OH^-(aq) \longrightarrow 2 H_2O(l) + 2 e^-$$

For the corresponding reduction, $E^\ominus = -0.83$ V. Because the overall reaction

$$2 H_2(g) + O_2(g) \longrightarrow 2 H_2O(l)\} \qquad E^\ominus = +1.23 \text{ V}$$

is exothermic as well as spontaneous, it is less favourable thermodynamically at 200°C than at 25°C, so the cell potential is lower at the higher temperature. However, the increased pressure compensates for the increased temperature, and at 200°C and 40 atm $E \approx +1.2$ V.

One advantage of the hydrogen/oxygen system is the large exchange current density of the hydrogen reaction. Unfortunately, the oxygen reaction has an exchange current density of only about 0.1 nA cm^{-2}, which limits the current available from

29.24 A single cell of a hydrogen/oxygen fuel cell. In practice, a stack of many cells is used.

(a)

(b)

29.25 (a) A simple version of the corrosion process is that of a droplet of water, which is oxygen-rich near its boundary with air. The oxidation of the iron takes place in the region away from the oxygen because the electrons are transported through the metal. (b) The process may be modelled as a short-circuited electrochemical cell.

the cell. One way round the difficulty is to use a catalytic surface (to increase j_0) with a large surface area. One type of highly developed fuel cell has phosphoric acid as the electrolyte and operates with hydrogen and air at about 200°C; the hydrogen is obtained from a reforming reaction on natural gas. The power output of batteries of such cells has reached the order of 10 MW. Cells with molten carbonate electrolytes at about 600°C can make use of natural gas directly. Solid-state electrolytes are also used. They include one version in which the electrolyte is a solid polymeric ionic conductor at about 100°C, but in current versions it requires very pure hydrogen to operate successfully. Solid ionic conducting oxide cells operate at about 1000°C and can use hydrocarbons directly as fuel.

Electric storage cells operate as galvanic cells while they are producing electricity but as electrolytic cells while they are being charged by an external supply. The lead–acid battery is an old device, but one well suited to the job of starting cars (and the only one available). During charging the cathode reaction is the reduction of Pb^{2+} and its deposition as lead on the lead electrode. Deposition occurs instead of the reduction of the acid to hydrogen because the latter has a low exchange current density on lead. The anode reaction during charging is the oxidation of Pb(II) to Pb(IV), which is deposited as the oxide PbO_2. On discharge, the two reactions run in reverse. Because they have such high exchange current densities the discharge can occur rapidly, which is why the lead battery can produce large currents on demand.

29.8 Corrosion

A thermodynamic warning of the likelihood of corrosion is obtained by comparing the standard potentials of the metal reduction, such as

$$Fe^{2+}(aq) + 2\,e^- \longrightarrow Fe(s) \qquad E^{\ominus} = -0.44\ V$$

with the values for one of the following half-reactions:

In acidic solution:

   (a) $2\,H^+(aq) + 2\,e^- \longrightarrow H_2(g) \qquad E^{\ominus} = 0$

   (b) $4\,H^+(aq) + O_2(g) + 4\,e^- \longrightarrow 2\,H_2O(l) \qquad E^{\ominus} = +1.23\ V$

In basic solution:

   (c) $2\,H_2O(l) + O_2(g) + 4\,e^- \longrightarrow 4\,OH^-(aq) \qquad E^{\ominus} = +0.40\ V$

Because all three redox couples have standard potentials more positive than $E^{\ominus}(Fe^{2+}/Fe)$, all three can drive the oxidation of iron to iron(II). The electrode potentials we have quoted are standard values, and they change with the pH of the medium. For the first two:

$$E(a) = E^{\ominus}(a) + (RT/F)\ln a(H^+) = -(0.059\ V)pH$$
$$E(b) = E^{\ominus}(b) + (RT/F)\ln a(H^+) = 1.23\ V - (0.059\ V)pH$$

These expressions let us judge at what pH the iron will have a tendency to oxidize (see Chapter 10). A thermodynamic discussion of corrosion, however, only indicates whether a tendency to corrode exists. If there is a thermodynamic tendency, we must examine the kinetics of the processes involved to see whether the process occurs at a significant rate.

A model of a corrosion system is shown in Fig. 29.25a. It can be taken to be a drop of slightly acidic (or basic) water containing some dissolved oxygen in contact with the metal. The oxygen at the edges of the droplet, where the O_2 concentration is higher, is reduced by electrons donated by the iron over an area A. Those electrons are replaced by others released elsewhere as $Fe \rightarrow Fe^{2+} + 2\,e^-$. This oxidative release occurs over an area A' under the oxygen-deficient inner region of the droplet. The droplet acts as a short-circuited galvanic cell (Fig. 29.25b).

The rate of corrosion is measured by the current of metal ions leaving the metal surface in the anodic region. This flux of ions gives rise to the **corrosion current**, I_{corr}, which can be identified with the anodic current, I_a. We show in the justification below that the corrosion current is related to the cell potential of the corrosion couple by

$$I_{corr} = \bar{j}_0 \bar{A} e^{fE/4} \qquad \bar{j}_0 = (j_0 j_0')^{1/2} \qquad \bar{A} = (AA')^{1/2} \tag{29.62}$$

Justification 29.3

Because any current emerging from the anodic region must find its way to the cathodic region, the cathodic current, I_c, and the anodic current, I_a must both be equal to the corrosion current. In terms of the current densities at the oxidation and reduction sites, j and j', respectively, we can write

$$I_{corr} = jA = j'A' = (jj'AA')^{1/2} = \bar{j}\bar{A} \qquad \bar{j} = (jj')^{1/2} \qquad \bar{A} = (AA')^{1/2} \tag{29.63}$$

The Butler–Volmer equation is now used to express the current densities in terms of overpotentials. For simplicity we assume that the overpotentials are large enough for the high-overpotential limit (eqns 38 and 40) to apply, that polarization overpotential can be neglected, that the rate-determining step is the transfer of a single electron, and that the transfer coefficients are $\frac{1}{2}$. We also assume that, since the droplet is so small, there is negligible potential difference between the cathode and anode regions of the solution. Moreover, because it is short-circuited by the metal, the potential of the metal is the same in both regions, and so the potential difference between the metal and the solution is the same in both regions too; it is denoted $\Delta\phi_{corr}$. The overpotentials in the two regions are therefore

$$\eta = \Delta\phi_{corr} - \Delta\phi \qquad \eta' = \Delta\phi_{corr} - \Delta\phi'$$

and the magnitudes of the current densities are

$$j = j_0 e^{\eta f/2} = j_0 e^{f\Delta\phi_{corr}/2} e^{-f\Delta\phi/2}$$
$$j' = j_0' e^{-\eta' f/2} = j_0' e^{-f\Delta\phi_{corr}/2} e^{f\Delta\phi'/2}$$

These expressions can be substituted into the expression for I_{corr} and $\Delta\phi' - \Delta\phi$ replaced by the difference of electrode potentials E to give eqn 62.

The effect of the exchange current density on the corrosion rate can be seen by considering the specific case of iron in contact with acidified water. Thermodynamically, either hydrogen or oxygen reduction reaction (a) or (b) on p. 1039 is effective. However, the exchange current density of reaction (b) on iron is only about 10^{-14} A cm^{-2}, whereas for (a) it is 10^{-6} A cm^{-2}. The latter therefore dominates kinetically, and iron corrodes by hydrogen evolution in acidic solution.

For corrosion reactions with similar exchange current densities, eqn 62 predicts that the rate of corrosion is high when E is large. That is, rapid corrosion can be expected when the oxidizing and reducing couples have widely differing electrode potentials.

Several techniques for inhibiting corrosion are available. First, from eqn 62 we see that the rate of corrosion depends on the surfaces exposed: if either A or A' is zero, then the corrosion current is zero. This interpretation points to a trivial, yet often effective, method of slowing corrosion: cover the surface with some impermeable layer, such as paint, which prevents access of damp air. Paint also increases the effective solution resistance between the cathode and anode patches on the surface.

29.26 (a) In cathodic protection an anode of a more strongly reducing metal is sacrificed to maintain the integrity of the protected object (for example, a pipeline, bridge, or boat). (b) In impressed-current cathodic protection electrons are supplied from an external cell so that the object itself is not oxidized. The broken lines depict the completed circuit through the soil.

Unfortunately, this protection fails disastrously if the paint becomes porous. The oxygen then has access to the exposed metal and corrosion continues beneath the paintwork. Another form of surface coating is provided by **galvanizing**, the coating of an iron object with zinc. Because the latter's standard potential is –0.76 V, which is more negative than that of the iron couple, the corrosion of zinc is thermo-dynamically favoured and the iron survives (the zinc survives because it is protected by a hydrated oxide layer). In contrast, tin plating leads to a very rapid corrosion of the iron once its surface is scratched and the iron exposed because the tin couple ($E^{\ominus}$ = –0.14 V) oxidizes the iron couple ($E^{\ominus}$ = –0.44 V). Some oxides are inert kinetically in the sense that they adhere to the metal surface and form an impermeable layer over a fairly wide pH range. This **passivation**, or kinetic protection, can be seen as a way of decreasing the exchange currents by sealing the surface. Thus, aluminium is inert in air even though its standard potential is strongly negative (–1.66 V).

Another method of protection is to change the electric potential of the object by pumping in electrons that can be used to satisfy the demands of the oxygen reduction without involving the oxidation of the metal. In **cathodic protection**, the object is connected to a metal with a more negative standard potential (such as magnesium, –2.36 V). The magnesium acts as a **sacrificial anode**, supplying its own electrons to the iron and becoming oxidized to Mg^{2+} in the process (Fig. 29.26a). A block of magnesium replaced occasionally is much cheaper than the ship, building, or pipeline for which it is being sacrificed. In **impressed-current cathodic protection** (Fig. 29.26b) an external cell supplies the electrons and eliminates the need for iron to transfer its own.

Checklist of key ideas

Electron transfer in homogeneous systems

☐ electroactive species, an oxidizable or reducible species

☐ current density, j, the charge flux through a region; the electric current divided by the area of the region

29.1 Theory of electron transfer processes

☐ Gibbs energy of activation, $\Delta^{\ddagger}G = (\Delta_r G^{\ominus} + \lambda)^2/4\lambda$

☐ reorganization energy, the energy change associated with molecular rearrangements that must take place so that DA can take on the equilibrium geometry of D^+A^-.

☐ electron transfer rate constant, $k_{et} = (2\langle H_{DA}\rangle^2/h)(\pi^3/4\lambda RT)^{1/2}\,e^{-\Delta^{\ddagger}G/RT}$, $\langle H_{DA}\rangle^2 = \langle H_{DA}^{\circ}\rangle^2 e^{-\beta r}$

29.2 Experimental results

☐ experimental tests, $\ln k_{et} = -\beta r + \text{constant}$, $\ln k_{et} = -(1/4\lambda)(\Delta_r G^{\ominus})^2 - \tfrac{1}{2}\Delta_r G^{\ominus} + \text{constant}$

☐ inverted region, conditions under which the rate constant decreases as the reaction becomes more exergonic

Electron transfer in heterogeneous systems

29.3 The electrode–solution interface

☐ electrical double layer, a sheet of one charge at the surface of the electrode and a sheet of opposite charge next to it in the solution

☐ Galvani potential difference, the potential difference between the bulk of the metal electrode and the bulk of the solution

☐ Helmholtz layer model, a model of the double layer in which the solvated ions arrange themselves along the surface of the electrode but are held away from it by their hydration spheres

☐ outer Helmholtz plane (OHP), the outer sheet of charge

☐ inner Helmholtz plane (IHP), the inner sheet of charge

☐ Gouy–Chapman model, a model of the double layer in which the outer layer is diffuse

☐ diffuse double layer, an inner layer of charge and an outer ionic diffuse atmosphere

☐ Stern model, a model of the double layer in which the ions closest to the electrode are constrained into a rigid plane while outside that plane the ions are dispersed as in the Gouy–Chapman model

☐ Grahame model, a model that adds an inner Helmholtz plane to the Stern model

☐ outer potential, the potential far from the surface where the potential varies only slightly with distance

☐ inner potential, the potential in the interior of the metal electrode

☐ surface potential, the difference between the inner and outer potentials

☐ electrochemical potential, the chemical potential of an ion in the presence of an electric potential, $\bar{\mu} = \mu + zF\phi$

29.4 The rate of charge transfer

☐ cathodic current density, the current density due to reduction; $j_c = Fk_c[\text{Ox}]$

☐ anodic current density, the current density due to oxidation; $j_a = Fk_a[\text{Red}]$

☐ transfer coefficient, the parameter α in $\Delta^{\ddagger}G_c = \Delta^{\ddagger}G_c(0) + \alpha F\Delta\phi$

☐ exchange current density, j_0, the cathodic or anodic charge density at zero overpotential

☐ overpotential, the difference between the potential and the zero-current potential

☐ Butler–Volmer equation, $j = j_0\{e^{-(1-\alpha)f\eta} - e^{\alpha f\eta}\}, f = F/RT$

☐ low overpotential limit, $F\eta/RT \ll 1$, $\eta = RTj/j_0$

☐ high overpotential limit, $F\eta/RT \gg 1$, $\ln j = \ln j_0 + (1 - \alpha)f\eta$

☐ Tafel plot, a plot of the logarithm of the current density against the overpotential

☐ working electrode, the (current-carrying) electrode of interest

☐ counter electrode, a current-carrying electrode that completes a circuit with the working electrode

☐ reference electrode, the electrode relative to which the potential of the working electrode is measured

29.5 Voltammetry

☐ voltammetry, a technique in which the current is monitored as the potential of the electrode is changed

☐ chronopotentiometry, a technique in which the potential is monitored as the current flow is changed

☐ non-polarizable electrode, an electrode with a potential that changes only slightly when a current passes

☐ polarizable electrode, an electrode with a potential that changes strongly when a current passes

☐ concentration polarization, slow diffusion of a species towards the electrode from the bulk

☐ polarization overpotential, the overpotential arising from concentration polarization

☐ formal potential, $E^{\circ} = E^{\ominus} + (RT/zF) \ln \gamma$

☐ Nernst diffusion layer, the region near an electrode in which the concentration of a solution falls from its bulk

concentration to another value at the plane itself

☐ limiting current density, $j_{\lim} = zFDc/\delta$

☐ linear-sweep voltammetry, the observation of the variation in current as the potential of an electrode is changed linearly with time

☐ differential pulse voltammetry, a technique in which the current is monitored before and after the application of a pulse of potential

☐ cyclic voltammetry, the observation of the variation in current as the potential of an electrode is changed in a sawtooth manner

29.6 Electrolysis

☐ cell overpotential, the sum of the overpotentials at the two electrodes and the ohmic drop due to the current through the electrolyte

29.7 Working galvanic cells

☐ potential of working cell, $g(I) = (I/A\bar{j})^{2z}\{(1 - I/Aj_{\lim,L})(1 - I/Aj_{\lim,R})\}^{-1/2}$

☐ power supplied by working cell, $P = IE - I^2R_s - (2IRT/zF) \ln g(I)$

☐ fuel cell, a primary galvanic cell in which the reactants are supplied from outside

29.8 Corrosion

☐ corrosion current, a current proportional to the rate at which metal ions leave a metal surface in the anodic region during corrosion, $I_{\text{corr}} = \bar{j}_0\bar{A}e^{fE/4}$

☐ galvanizing, coating iron with a protective layer of zinc

☐ passivation, a protection procedure in which the metal surface is covered with a layer of metal oxides

☐ cathodic protection, a procedure in which the object to be proected is connected to a metal with a more negative standard potential

☐ sacrificial anode, a metal that supplies its own electrons and becomes oxidized in the process

☐ impressed-current cathodic protection, a protection procedure in which an external cell supplies electrons and eliminates the need for the protected metal to transfer its own

Further reading

Articles of general interest

R. Adžić and E.B. Yeager, Electrochemistry. In *Encyclopedia of applied physics* (ed. G.L. Trigg), **5**, 223. VCH, New York (1993).

J.L. Anderson, L.A. Coury, Jr., and J. Leddy, Dynamic electrochemistry: methodology and application. *Anal. Chem.* **72**, 4497 (2000).

J. Andraos, How mathematics figures in chemistry: some examples. *J. Chem. Educ.* **76**, 258 (1999).

V.L. Davidson, What controls the rate of interprotein electron transfer reactions. *Acc. Chem. Res.* **33**, 87 (2000).

A.T. Hubbard, Electrochemistry at well-characterized surfaces. *Chem. Rev.* **88**, 633 (1988).

C.C. Page, C.C. Moser, X. Chen, and P.L. Dutton, Natural engineering principles of electron tunnelling in biological oxidation–reduction. *Nature* **402**, 47 (1999).

R. Parsons, Electrical double layer: Recent experimental and theoretical developments. *Chem. Rev.* **90**, 813 (1990).

R.M. Wightman and D.O. Wipf, High-speed cyclic voltammetry. *Acc. Chem. Res.* **23**, 64 (1990).

J.R. Winkler, Electron tunneling pathways in proteins. *Curr. Op. Chem. Biol.* **4**, 192 (2000).

R. Ziessel, Photoinduced energy or electron transfer in supramolecular systems: applications to molecular wires and light-harvesting sensors. *J. Chem. Educ.* **74**, 673 (1997).

Texts and sources of data and information

C.H. Bamford, C.F. Tipper, and R.G. Compton (ed.), *Comprehensive chemical kinetics*, Vols. 1–38. Elsevier, Amsterdam (1969–2001).

A.J. Bard and L.R. Faulkner, *Electrochemical methods: fundamentals and applications*. Wiley, New York (2000).

R.S. Berry, S.A. Rice, and J. Ross, *Physical chemistry*. Oxford University Press (2000).

J.O'M. Bockris, R.E. White, and B.E. Conway (ed.), *Modern aspects of electrochemistry*. Vol. 33. Plenum, New York (1999).

J.O'M. Bockris and S.U.M. Khan, *Surface electrochemistry: a molecular level approach*. Plenum, New York (1993).

C.M.A. Brett and A.M.O. Brett, *Electrode potentials*. Oxford Chemistry Primers, Oxford University Press (1998).

D.R. Crow, *Principles and applications of electrochemistry*. Stanley Thornes, Cheltenham (1994).

A.C. Fisher, *Electrode dynamics*. Oxford Chemistry Primers, Oxford University Press (1998).

M.G. Fontanna and R.W. Staehle (ed.), *Advances in corrosion science and technology*. Plenum, New York (1980).

C.H. Hamann, W. Vielstich, and A. Hammett, *Electrochemistry*. Wiley–VCH, New York (1998).

J. Koryta, *Ions, electrodes, and membranes*. Wiley, New York (1991).

A.M. Kuznetsov and J. Ulstrup, *Electron transfer in chemistry and biology: an introduction to the theory*. Wiley, New York (1998).

D. Linden (ed.), *Handbook of batteries and cells*. McGraw-Hill, New York (1984).

S.R. Logan, *Chemical kinetics*. Longman, Harlow (1996).

C.D.S. Tuck, *Modern battery construction*. Ellis Horwood, New York (1991).

Exercises

Discussion questions

29.1 (a) Discuss how the following factors determine the rate of electron transfer in homogeneous systems: the distance between electron donor and acceptor, and the reorganization energy of redox active species and the surrounding medium.

29.1 (b) Consult the current literature on biological electron transfer and write a critical review of the experimental evidence for and against the existence of specific paths through covalent bonds and hydrogen bonds that optimize the rate of electron transfer in proteins.

29.2 (a) Discuss the main structural features of the electrical double layer.

29.2 (b) Define the terms in and limit the generality of the following expressions: (a) $j = j_0 f \eta$, (b) $j = j_0 e^{(1-\alpha) f \eta}$, and (c) $j = -j_0 e^{-\alpha f \eta}$.

29.3 (a) Distinguish between the electrical double layer and the Nernst diffusion layer.

29.3 (b) Discuss the technique of cyclic voltammetry and account for the characteristic shape of a cyclic voltammogram, such as those shown in Figs 29.20 and 29.21.

29.4 (a) Discuss the principles of operation of a fuel cell.

29.4 (b) Discuss the chemical origins of corrosion and useful strategies for preventing it.

Numerical exercises

29.5 (a) For a pair of electron donor and acceptor, $H_{AB} = 0.03$ cm^{-1}, $\Delta_r G^{\ominus} = -0.182$ eV, and $k_{et} = 30.5$ s^{-1} at 298 K. Estimate the value of the reorganization energy.

29.5 (b) For a pair of electron donor and acceptor, $k_{et} = 2.02 \times 10^5$ s^{-1} for $\Delta_r G^{\ominus} = -0.665$ eV. The standard reaction Gibbs energy changes to $\Delta_r G^{\ominus} = -0.975$ eV when a substituent is added to the electron acceptor and the rate constant for electron transfer changes to $k_{et} = 3.33 \times 10^6$ s^{-1}. The experiments were conducted at

298 K. Assuming that the distance between donor and acceptor is the same in both experiments, estimate the values of H_{AB} and λ.

29.6 (a) For a pair of electron donor and acceptor, $k_{et} = 2.02 \times 10^5$ s^{-1} when $r = 1.11$ nm and $k_{et} = 2.02 \times 10^5$ s^{-1} when $r = 1.23$ nm. Assuming that T, $\Delta_r G^{\ominus}$ and λ are the same in both experiments, estimate the value of β.

29.6 (b) Refer to the preceding exercise. Estimate the value of k_{et} when $r = 1.48$ nm.

29.7 (a) The Helmholtz model of the electric double layer is equivalent to a parallel plate capacitor. Hence the potential difference across the double layer is given by $\Delta\phi = \sigma d/\varepsilon$, where d is the distance between the plates and σ is the surface charge density. Assuming that this model holds for concentrated salt solutions calculate the magnitude of the electric field at the surface of silica in 5.0 M NaCl(aq) if the surface charge density is 0.10 C m^{-2}.

29.7 (b) Refer to the preceding exercise. Calculate the magnitude of the electric field at the surface of silica in 4.5 M NaCl(aq) if the surface charge density is 0.12 C m^{-2}.

29.8 (a) The transfer coefficient of a certain electrode in contact with M^{3+} and M^{4+} in aqueous solution at 25°C is 0.39. The current density is found to be 55.0 mA cm^{-2} when the overvoltage is 125 mV. What is the overvoltage required for a current density of 75 mA cm^{-2}?

29.8 (b) The transfer coefficient of a certain electrode in contact with M^{2+} and M^{3+} in aqueous solution at 25°C is 0.42. The current density is found to be 17.0 mA cm^{-2} when the overvoltage is 105 mV. What is the overvoltage required for a current density of 72 mA cm^{-2}?

29.9 (a) Determine the exchange current density from the information given in Exercise 29.8a.

29.9 (b) Determine the exchange current density from the information given in Exercise 29.8b.

29.10 (a) To a first approximation, significant evolution or deposition occurs in electrolysis only if the overpotential exceeds about 0.6 V. To illustrate this criterion determine the effect that increasing the overpotential from 0.40 V to 0.60 V has on the current density in the electrolysis of 1.0 M NaOH(aq), which is 1.0 mA cm^{-2} at 0.4 V and 25°C. Take $\alpha = 0.5$.

29.10 (b) Determine the effect that increasing the overpotential from 0.50 V to 0.60 V has on the current density in the electrolysis of 1.0 M NaOH(aq), which is 1.22 mA cm^{-2} at 0.50 V and 25°C. Take $\alpha = 0.50$.

29.11 (a) Use the data in Table 29.1 for the exchange current density and transfer coefficient for the reaction $2 H^+ + 2 e^- \rightarrow H_2$ on nickel at 25°C to determine what current density would be needed to obtain an overpotential of 0.20 V as calculated from (a) the Butler–Volmer equation, and (b) the Tafel equation. Is the validity of the Tafel approximation affected at higher overpotentials (of 0.4 V and more)?

29.11 (b) Use the data in Table 29.1 for the exchange current density and transfer coefficient for the reaction $Fe^{3+} + e^- \rightarrow Fe^{2+}$ on platinum at 25°C to determine what current density would be needed to obtain an overpotential of 0.30 V as calculated from (a) the Butler–Volmer equation, and (b) the Tafel equation. Is the validity of the Tafel approximation affected at higher overpotentials (of 0.4 V and more)?

29.12 (a) Estimate the limiting current density at an electrode in which the concentration of Ag$^+$ ions is 2.5 mmol L^{-1} at 25°C.

The thickness of the Nernst diffusion layer is 0.40 mm. The ionic conductivity of Ag$^+$ at infinite dilution and 25°C is 6.19 mS m^2 mol^{-1}.

29.12 (b) Estimate the limiting current density at an electrode in which the concentration of Mg^{2+} ions is 1.5 mmol L^{-1} at 25°C. The thickness of the Nernst diffusion layer is 0.32 mm. The ionic conductivity of Mg^{2+} at infinite dilution and 25°C is 10.60 mS m^2 mol^{-1}.

29.13 (a) A 0.10 M CdSO$_4$(aq) solution is electrolysed between a cadmium cathode and a platinum anode with a current density of 1.00 mA cm^{-2}. The hydrogen overpotential is 0.60 V. What will be the concentration of Cd^{2+} ions when evolution of H$_2$ just begins at the cathode? Assume all activity coefficients are unity.

29.13 (b) A 0.10 M FeSO$_4$(aq) solution is electrolysed between a magnesium cathode and a platinum anode with a current density of 1.50 mA cm^{-2}. The hydrogen overpotential is 0.60 V. What will be the concentration of Fe^{2+} ions when evolution of H$_2$ just begins at the cathode? Assume all activity coefficients are unity.

29.14 (a) A typical exchange current density, that for H$^+$ discharge at platinum, is 0.79 mA cm^{-2} at 25°C. What is the current density at an electrode when its overpotential is (a) 10 mV, (b) 100 mV, (c) −5.0 V? Take $\alpha = 0.5$.

29.14 (b) The exchange current density for a Pt|Fe^{3+},Fe^{2+} electrode is 2.5 mA cm^{-2}. The standard potential of the electrode is +0.77 V. Calculate the current flowing through an electrode of surface area 1.0 cm^2 as a function of the potential of the electrode. Take unit activity for both ions.

29.15 (a) Suppose that the electrode potential is set at 1.00 V. The exchange current density is 6.0×10^{-4} A cm^{-2} and $\alpha = 0.50$. Calculate the current density for the ratio of activities $a(Fe^{3+})/a(Fe^{2+})$ in the range 0.1 to 10.0 and at 25°C.

29.15 (b) Suppose that the electrode potential is set at 0.50 V. Calculate the current density for the ratio of activities $a(Cr^{3+})/a(Cr^{2+})$ in the range 0.1 to 10.0 and at 25°C.

29.16 (a) What overpotential is needed to sustain a current density of 20 mA cm^{-2} at a Pt|Fe^{3+},Fe^{2+} electrode in which both ions are at a mean activity $a = 0.10$?

29.16 (b) What overpotential is needed to sustain a current density of 15 mA cm^{-2} at a Pt|Ce^{4+},Ce^{3+} electrode in which both ions are at a mean activity $a = 0.010$? Take $j_0 = 6.0 \times 10^{-4}$ A cm^{-2}, $\alpha = 0.50$.

29.17 (a) How many electrons or protons are transported through the double layer in each second when the Pt,H$_2$|H$^+$,Pt|Fe^{3+},Fe^{2+}, and Pb,H$_2$|H$^+$ electrodes are at equilibrium at 25°C? Take the area as 1.0 cm^2 in each case. Estimate the number of times each second a single atom on the surface takes part in a electron transfer event, assuming an electrode atom occupies about (280 pm)^2 of the surface.

29.17 (b) How many electrons or protons are transported through the double layer in each second when the Cu,H$_2$|H$^+$ and Pt|Ce^{4+}, Ce^{3+} electrodes are at equilibrium at 25°C? Take the area as 1.0 cm^2 in each case. Estimate the number of times each second a single atom on the surface takes part in a electron transfer event, assuming an electrode atom occupies about (260 pm)^2 of the surface.

29.18 (a) What is the effective resistance at 25°C of an electrode interface when the overpotential is small? Evaluate it for 1.0 cm^2 (a) Pt,H$_2$|H$^+$, (b) Hg,H$_2$|H$^+$ electrodes.

29.18 (b) Evaluate the effective resistance at 25°C of an electrode interface for 1.0 cm^2 (a) Pb,H$_2$|H$^+$, (b) Pt|Fe^{2+},Fe^{3+} electrodes.

29.19 (a) State what happens when a platinum electrode in an aqueous solution containing both Cu^{2+} and Zn^{2+} ions at unit activity is made the cathode of an electrolysis cell.

29.19 (b) State what happens when a platinum electrode in an aqueous solution containing both Fe^{2+} and Ni^{2+} ions at unit activity is made the cathode of an electrolysis cell.

29.20 (a) What are the conditions that allow a metal to be deposited from aqueous acidic solution before hydrogen evolution occurs significantly at 293 K? Why may silver be deposited from aqueous silver nitrate?

29.20 (b) The overpotential for hydrogen evolution on cadmium is about 1 V at current densities of 1 mA cm^{-2}. Why may cadmium be deposited from aqueous cadmium sulfate?

29.21 (a) The exchange current density for H$^+$ discharge at zinc is about 50 pA cm^{-2}. Can zinc be deposited from a unit activity aqueous solution of a zinc salt?

29.21 (b) The standard potential of the $Zn^{2+}|Zn$ electrode is –0.76 V at 25°C. The exchange current density for H$^+$ discharge at platinum is 0.79 mA cm^{-2}. Can zinc be plated on to platinum at that temperature? (Take unit activities.)

29.22 (a) Can magnesium be deposited on a zinc electrode from a unit activity acid solution at 25°C?

29.22 (b) Can iron be deposited on a copper electrode from a unit activity acid solution at 25°C?

29.23 (a) Calculate the maximum (zero-current) potential difference of a nickel–cadmium cell, and the maximum possible power output when 100 mA is drawn at 25°C.

29.23 (b) Calculate the maximum (zero-current) potential difference of a lead–acid cell, and the maximum possible power output when 100 mA is drawn at 25°C.

29.24 (a) Calculate the thermodynamic limit to the zero-current potential of fuel cells operating on (a) hydrogen and oxygen, (b) methane and air. Use the Gibbs energy information in the *Data section*, and take the species to be in their standard states at 25°C.

29.24 (b) Calculate the thermodynamic limit to the zero-current potential of fuel cells operating on propane and air. Use the Gibbs energy information in the *Data section*, and take the species to be in their standard states at 25°C.

29.25 (a) The corrosion current density j_{corr} at an iron anode is 1.0 A m^{-2}. What is the corrosion rate in millimetres per year? Assume uniform corrosion.

29.25 (b) The corrosion current density j_{corr} at a zinc anode is 2.0 A m^{-2}. What is the corrosion rate in millimetres per year? Assume uniform corrosion.

Problems[†]

Numerical problems

29.1 Consider the reaction $D + A \rightleftharpoons D^+ + A^-$ with equilibrium constant $K = k_{12}/k_{21} = [D^+][A^-]/[D][A]$ in which k_{12} and k_{21} are, respectively, the *bimolecular* rate constants for formation of the free ions D^+ and A^- and for regeneration of the free neutral species D and A. The rate constant k_{12} is distinct from the *unimolecular* rate constant k_{et} of eqn 29.1b, which represents electron transfer within a DA complex to form a D^+A^- complex. The rate constant k_{12} may be determined experimentally or may be predicted by the *Marcus cross-relation*

$$k_{12} = (k_{11}k_{22}Kf)^{1/2}$$

where k_{11} and k_{22} are the experimental rate constants for the electron self-exchange processes $^*D + D^+ \rightarrow {}^*D^+ + D$ and $^*A + A^+ \rightarrow {}^*A^+ + A$, respectively, and f is a function of K, k_{11}, k_{22}, and the collision frequencies (see Problem 29.14 for a derivation of this equation). The asterisks are used to distinguish between two molecules of the same species. It is common to make the assumption that $f \approx 1$. Use the approximate form of the Marcus relation to estimate the rate constant for the reaction $Ru(bpy)_3^{3+} + Fe(H_2O)_6^{2+} \rightarrow Ru(bpy)_3^{2+} + Fe(H_2O)_6^{3+}$, where bpy stands for 4,4'-bipyridine. The following data are useful:

$$Ru(bpy)_3^{3+} + e^- \longrightarrow Ru(bpy)_3^{2+} \qquad E^{\ominus} = 1.26 \text{ V}$$
$$Fe(H_2O)_6^{3+} + e^- \longrightarrow Fe(H_2O)_6^{2+} \qquad E^{\ominus} = 0.77 \text{ V}$$
$$^*Ru(bpy)_3^{3+} + Ru(bpy)_3^{2+} \longrightarrow {}^*Ru(bpy)_3^{2+} + Ru(bpy)_3^{3+}$$
$$k_{Ru} = 4.0 \times 10^8 \text{ L mol}^{-1} \text{ s}^{-1}$$
$$^*Fe(H_2O)_6^{3+} + Fe(H_2O)_6^{2+} \longrightarrow {}^*Fe(H_2O)_6^{2+} + Fe(H_2O)_6^{3+}$$
$$k_{Fe} = 4.2 \text{ L mol}^{-1} \text{ s}^{-1}$$

29.2 In an experiment on the Pt|H$_2$|H$^+$ electrode in dilute H$_2$SO$_4$ the following current densities were observed at 25°C. Evaluate α and j_0 for the electrode.

η/mV	50	100	150	200	250
j/(mA cm^{-2})	2.66	8.91	29.9	100	335

How would the current density at this electrode depend on the overpotential of the same set of magnitudes but of opposite sign?

29.3 The standard potentials of lead and tin are –126 mV and –136 mV respectively at 25°C, and the overvoltages for their deposition are close to zero. What should their relative activities be in order to ensure simultaneous deposition from a mixture?

29.4 The limiting current density for the reaction $I_3^- + 2 e^- \rightarrow 3 I^-$ at a platinum electrode is 28.9 μA cm^{-2} when the concentration of KI is 6.6×10^{-4} mol L^{-1} and the temperature 25°C. The diffusion coefficient of I_3^- is 1.14×10^{-9} m^2 s^{-1}. What is the thickness of the diffusion layer?

29.5 Estimating the power output and potential of a cell under operating conditions is very difficult, but eqn 61 summarizes, in an approximate way, some of the parameters involved. As a first step in manipulating this expression, identify all the quantities that depend on the ionic concentrations. Express E in terms of the concentration and conductivities of the ions present in the cell. Estimate the parameters for Zn(s)|ZnSO$_4$(aq)||CuSO$_4$(aq)|Cu(s). Take electrodes of area 5 cm^2 separated by 5 cm. Ignore both potential differences and resistance of the liquid junction. Take the concentration as 1 mol L^{-1}, the temperature 25°C, and neglect activity coefficients.

† Problems denoted with the symbol ‡ were supplied by Charles Trapp, Carmen Giunta, and Marshall Cady.

Plot E as a function of the current drawn. On the same graph, plot the power output of the cell. What current corresponds to maximum power?

29.6 Consider a cell in which the current is activation-controlled. Show that the current for maximum power can be estimated by plotting $\log(I/I_0)$ and $c_1 - c_2 I$ against I (where $I_0 = A^2 j_0 j_0'$ and c_1 and c_2 are constants), and looking for the point of intersection of the curves. Carry through this analysis for the cell in Problem 29.5 ignoring all concentration overpotentials.

29.7‡ The rate of deposition of iron, v, on the surface of an iron electrode from an aqueous solution of Fe^{2+} has been studied as a function of potential, E, relative to the standard hydrogen electrode, by J. Konya (*J. Electroanal. Chem.* **84**, 83 (1977)). The values in the table below are based on the data obtained with an electrode of surface area 9.1 cm^2 in contact with a solution of concentration 1.70 μmol L^{-1} in Fe^{2+}. (a) Assuming unit activity coefficients, calculate the zero current potential of the Fe^{2+}/Fe cathode and the overpotential at each value of the working potential. (b) Calculate the cathodic current density, j_c, from the rate of deposition of Fe^{2+} for each value of E. (c) Examine the extent to which the data fit the Tafel equation and calculate the exchange current density.

$v/(\text{pmol s}^{-1})$	1.47	2.18	3.11	7.26
$-E/\text{mV}$	702	727	752	812

29.8‡ The thickness of the diffuse double layer according to the Gouy–Chapman model is given by eqn 22.50. Use this equation to calculate and plot the thickness as a function of concentration and electrolyte type at 25°C. For examples, choose aqueous solutions of NaCl and Na$_2$SO$_4$ ranging in concentration from 0.1 to 100 mmol L^{-1}.

29.9‡ V.V. Losev and A.P. Pchel'nikov (*Soviet Electrochem.* **6**, 34 (1970)) obtained the following current–voltage data for an indium anode relative to a standard hydrogen electrode at 293 K:

$-E/\text{V}$	0.388	0.365	0.350	0.335
$j/(\text{A m}^{-2})$	0	0.590	1.438	3.507

Use these data to calculate the transfer coefficient and the exchange current density. What is the cathodic current density when the potential is 0.365 V?

29.10‡ The redox reactions of quinones have been the subject of many studies over the years and they continue to be of interest to electrochemists. A relatively recent study is that of E. Kariv, J. Hermolin, and E. Gileadi (*Electrochim. Acta* **16**, 1437 (1971)) on methone (1,1-dimethyl-3,5-cyclohexanedione). The current–voltage data for the reduction of this compound (M) in anhydrous butanol on a mercury electrode are:

$-E/\text{V}$	1.50	1.58	1.63	1.72	1.87	1.98	≥ 2.10
$j/(\text{A m}^{-2})$	10	30	50	100	200	250	290

(a) How well do these data fit the empirical Tafel equation? (b) The authors postulate that the reduction product is the dimer HMMH formed by the following mechanism:

(1) M(sol) $\rightleftharpoons$ M(ads)

(2) M(ads) + H$^+$ + e$^-$ $\rightleftharpoons$ MH(ads)

(3) MH(ads) + MH(ads) $\rightleftharpoons$ HMMH

The affixes sol and ads refer to species in solution and on the surface of the electrode, respectively. Does this mechanism help to explain the current–voltage data?

29.11‡ Consider a one-electron transfer process for which the transfer is rate-determining. (a) Draw a graph that shows the

dependence of the relative current densities j_a/j_0, j_c/j_0, and j/j_0 on the overpotential. This is the polarization curve. How does variation of the transfer coefficient affects the polarization curve? (b) Draw a Tafel plot that shows both anodic and cathodic current densities for the platinum electrode process $Ce^{4+} + e^- \rightarrow Ce^{3+}$. Assume that the electron transfer is rate-determining. Explain the graphical methodology for using a Tafel plot of experimental data for evaluation of both the transfer coefficient and the exchange current density. Draw a graph that illustrates the methodology.

29.12‡ An early study of the hydrogen overpotential is that of H. Bowden and T. Rideal (*Proc. Roy. Soc.* **A120**, 59 (1928)), who measured the overpotential for H$_2$ evolution with a mercury electrode in dilute aqueous solutions of H$_2$SO$_4$ at 25°C. Determine the exchange current density and transfer coefficient, α, from their data:

$j/(\text{mA m}^{-2})$	2.9	6.3	28	100	250	630	1650	3300
η/V	0.60	0.65	0.73	0.79	0.84	0.89	0.93	0.96

Explain any deviations from the result expected from the Tafel equation.

29.13‡ The analytic description of reversible redox kinetics at a stationary electrode is well established (R.S. Nicholson and I. Shain, *Anal. Chem.*, **36**, 706 (1964)). For a linear-sweep voltammetry experiment with a constant potential scan rate, v, from E_{initial}, the equation for current density at a flat, non-reactive electrode with linear diffusion to and from the electrode is:

$$j(x) = zFc_{\text{Ox}}(\pi D_{\text{Ox}} z f v)^{1/2} \chi(x)$$

where $x = (E_{\text{initial}} - E)zf$, $f = RT/F$, and

$$\chi(x) = \frac{1}{\pi x^{1/2}(1 + \gamma\theta)} + \frac{1}{4\pi}\int_0^x \frac{1}{(x - y)^{1/2}\cosh^2(\frac{1}{2}\{\ln(\gamma\theta) - y\})}dy$$

$$\gamma = \left(\frac{D_{\text{Ox}}}{D_{\text{Red}}}\right)^{1/2} \qquad \theta = e^{f(E_{\text{initial}} - E^{\circ\prime})}$$

Far from the electrode, the bulk concentrations are c_{Ox} and $c_{\text{Red}} = 0$; at the electrode $j_{\text{Ox}} = -j_{\text{Red}}$. (a) Simulate the reversible electron transfer voltammogram for Ti^{3+}/Ti^{2+} reduction. Use $c_{\text{Ox}} = 1.0 \times 10^{-4}$ mol L^{-1} and $v = 10$ mV s^{-1}. Discuss the physical and chemical properties that cause the variations of the current density revealed by the simulation. How do the values of D_{Ox} and v affect the voltammogram? (b) Let E_{pc} be the voltammogram peak cathodic potential. Show that both $E_{\text{pc}} \approx E^{\circ\prime} - (28.5 \text{ mV})/z$ and $E_{\text{p/2}} - E_{\text{pc}} \approx 56.5$ mV. List all the experimental criteria for reversibility that are suggested by this simulation. $E_{\text{p/2}}$ is the potential at which $j = \frac{1}{2}j(E_{\text{pc}})$ and $E^{\circ\prime}$ is the standard formal potential at the ionic strength of the solution.

Theoretical problems

29.14 Derive the approximate form of the Marcus relation, $k_{12} = (k_{11}k_{22}K)^{1/2}$ (Problem 29.1), by following these steps. (a) Use eqn 4 to write expressions for $\Delta^{\ddagger}G_{12}$, $\Delta^{\ddagger}G_{11}$, and $\Delta^{\ddagger}G_{22}$, keeping in mind that $\Delta_r G^{\ominus} = 0$ for the electron self-exchange reactions. (b) Assume that the reorganization energy λ_{12} for the reaction D + A $\rightleftharpoons$ D$^+$ + A$^-$ is the average of the reorganization energies λ_{11} and λ_{22} of the electron self-exchange reactions. Then show that in the limit of small $\Delta_r G^{\ominus}$, or $\Delta_r G^{\ominus} \ll \lambda_{12}$,

$$\Delta^{\ddagger}G_{12} = \frac{\Delta^{\ddagger}G_{11} + \Delta^{\ddagger}G_{22} + \Delta_r G^{\ominus}_{12}}{2}$$

where $\Delta_r G_{12}^{\ominus}$ is the standard Gibbs energy for the reaction $D + A \rightleftharpoons D^+ + A^-$. (c) Use eqn 2 to write expressions for k_{11} and k_{22}. (d) Use eqn 2 and the result above to write an expression for k_{12}. (e) Complete the derivation by using the results from part (c), the relation $K = e^{-\Delta_r G_{12}^{\ominus}/RT}$, and assuming that all $\kappa\nu$ terms, which may be interpreted as collision frequencies, are identical.

29.15 If $\alpha = \frac{1}{2}$, an electrode interface is unable to rectify alternating current because the current density curve is symmetrical about $\eta = 0$. When $\alpha \neq \frac{1}{2}$, the magnitude of the current density depends on the sign of the overpotential, and so some degree of 'faradaic rectification' may be obtained. Suppose that the overpotential varies as $\eta = \eta_0 \cos \omega t$. Derive an expression for the mean flow of current (averaged over a cycle) for general α, and confirm that the mean current is zero when $\alpha = \frac{1}{2}$. In each case work in the limit of small η_0 but to second-order in $\eta_0 F/RT$. Calculate the mean direct current at 25°C for a 1.0 cm^2 hydrogen–platinum electrode with $\alpha = 0.38$ when the overpotential varies between ±10 mV at 50 Hz.

29.16 Now suppose that the overpotential is in the high overpotential region at all times even though it is oscillating. What waveform will the current across the interface show if it varies linearly and periodically (as a sawtooth waveform) between η_- and η_+ around η_0? Take $\alpha = \frac{1}{2}$.

29.17 Derive an expression for the current density at an electrode where the rate process is diffusion-controlled and η_c is known. Sketch the form of j/j_L as a function of η_c. What changes occur if anion currents are involved?

Applications to: biology and chemical engineering

29.18 A useful strategy for the study of electron transfer in proteins consists of attaching an electroactive species to the protein's surface and then measuring k_{et} between the attached species and an electroactive protein co-factor. J.W. Winkler and H.B. Gray (*Chem. Rev.* **92**, 369 (1992)) summarize data for cytochrome c (Box 10.1) modified by replacement of the haem iron by a zinc ion, resulting in a zinc-porphyrin (ZnP) moiety in the interior of the protein, and by attachment of a ruthenium ion complex to a surface histidine amino acid. The edge-to-edge distance between the electroactive species was thus fixed at 1.23 nm. A variety of ruthenium ion complexes with different standard reduction potentials were used. For each ruthenium-modified protein, either $Ru^{2+} \rightarrow ZnP^+$ or $ZnP^* \rightarrow Ru^{3+}$, in which the electron donor is an electronic excited state of the zinc-porphyrin formed by laser excitation, was monitored. This leads to different standard reaction Gibbs energies because the redox couples ZnP^+/ZnP and ZnP^+/ZnP^* have different standard potentials, with the electronically excited porphyrin being a more powerful reductant. Use the following data taken at 298 K to estimate the reorganization energy for this system:

$-\Delta_r G^{\ominus}$/eV	0.665	0.705	0.745	0.975	1.015	1.055
$k_{et}/(10^6 \text{ s}^{-1})$	0.657	1.52	1.12	8.99	5.76	10.1

29.19 The photosynthetic reaction centre of the purple photosynthetic bacterium *Rhodopseudomonas viridis* contains a number of bound cofactors that participate in electron transfer reactions. The following table shows data compiled by Moser *et al.*, *Nature* **355**, 796 (1992) on the rate constants for electron transfer between different cofactors and their edge-to-edge distances:

Reaction	$BChl^- \rightarrow BPh$	$BPh^- \rightarrow BChl_2^+$	$BPh^- \leftarrow Q_A$	cyt $c_{559} \rightarrow BChl_2^+$
r/nm	0.48	0.95	0.96	1.23
k_{et}/s^{-1}	1.58×10^{12}	3.98×10^9	1.00×10^9	1.58×10^8

Reaction	$Q_A^- \rightarrow Q_B$	$Q_A^- \rightarrow BChl_2^+$
r/nm	1.35	2.24
k_{et}/s^{-1}	3.98×10^7	63.1

(BChl, bacteriochlorophyll; BChl$_2$, bacteriochlorophyll dimer, functionally distinct from BChl; BPh, bacteriophaeophytin; Q_A and Q_B, quinone molecules bound to two distinct sites; cyt c_{559}, a cytochrome bound to the reaction centre complex). Are these data in agreement with the behaviour predicted by eqn 13? If so, evaluate the value of β.

29.20 The rate constant for electron transfer between a cytochrome c and the bacteriochlorophyll dimer of the reaction centre of the purple bacterium *Rhodobacter sphaeroides* (Problem 29.19) decreases with decreasing temperature in the range 300 K to 130 K. Below 130 K, the rate constant becomes independent of temperature. Account for these results.

29.21 For each group below, determine which metal has a thermodynamic tendency to corrode in moist air at pH = 7. Take as a criterion of corrosion a metal ion concentration of at least 10^{-6} mol L^{-1}.

(a) Fe, Cu, Pb, Al, Ag, Cr, Co

(b) Ni, Cd, Mg, Ti, Mn

29.22 Estimate the magnitude of the corrosion current for a patch of zinc of area 0.25 cm^2 in contact with a similar area of iron in an aqueous environment at 25°C. Take the exchange current densities as 1 μA cm^{-2} and the local ion concentrations as 1 μmol L^{-1}.

29.23 The corrosion potential of iron immersed in a de-aerated acidic solution of pH = 3 is −0.720 V as measured at 25°C relative to the standard calomel electrode with potential 0.2802 V. A Tafel plot of cathodic current density against overpotential yields a slope of 18 V^{-1} and the hydrogen ion exchange current density $j_0 = 0.10$ μA cm^{-2}. Calculate the corrosion rate in milligrams of iron per square centimetre per day (mg cm^{-2} d^{-1}).

Mathematical techniques

Basic procedures

1.1 Logarithms and exponentials

The **natural logarithm** of a number x is denoted $\ln x$, and is defined as the power to which $e = 2.718 \ldots$ must be raised for the result to be equal to x. It follows from the definition of logarithms that

$$\ln x + \ln y + \cdots = \ln xy \cdots \tag{FI1.1}$$

$$\ln x - \ln y = \ln(x/y) \tag{FI1.2}$$

$$a \ln x = \ln x^a \tag{FI1.3}$$

We also encounter the **common logarithm** of a number, $\log x$, the logarithm compiled with 10 in place of e. Common logarithms follow the same rules of addition and subtraction as natural logarithms. Common and natural logarithms are related by

$$\ln x = \ln 10 \log x \approx 2.303 \log x \tag{FI1.4}$$

The **exponential function**, e^x, plays a very special role in the mathematics of chemistry. The following properties are important:

$$e^x e^y e^z \cdots = e^{x+y+z+\cdots} \tag{FI1.5}$$

$$e^x/e^y = e^{x-y} \tag{FI1.6}$$

$$(e^x)^a = e^{ax} \tag{FI1.7}$$

1.2 Combinatorial functions

Combinatorial functions allow us to express the number of ways in which a system of particles may be configured; they are especially useful in statistical thermodynamics (Chapters 19 and 20). Consider a simple coin-toss problem. If n coins are tossed, the number $N(n, i)$ of outcomes that have i heads and $(n - i)$ tails, regardless of the order of the results, is given by the coefficients of the **binomial expansion** of $(1 + x)^n$:

$$(1 + x)^n = 1 + \sum_{i=1}^{n} N(n, i)x^i, \qquad N(n, i) = \frac{n!}{(n - i)!\, i!} \tag{FI1.8}$$

where $n!$ denotes a **factorial**, given by

$$n! = n(n-1)(n-2) \dots 1 \tag{FI1.9}$$

By definition $0! = 1$. The numbers $N(n, i)$, which are sometimes denoted $\begin{pmatrix} n \\ i \end{pmatrix}$, are also called **binomial coefficients**.

Suppose that, unlike the coin-toss problem, there are more than two possible results for each event. For example, there are six possible results for the roll of a die. For n rolls of the die, the number of ways, W, that correspond to n_1 occurrences of the number 1, n_2 occurrences of the number 2, and so on, is given by

$$W = \frac{n!}{n_1! \, n_2! \, n_3! \, n_4! \, n_5! \, n_6!}, \qquad n = \sum_{i=1}^{6} n_i$$

This is an example of a **multinomial coefficient**, which has the form

$$W = \frac{n!}{n_1! \, n_2! \dots n_m!}, \qquad n = \sum_{i=1}^{m} n_i \tag{FI1.10}$$

where W is the number of ways of achieving an outcome, n is the number of events, and m is the number of possible results. In Chapter 19 we use the multinomial coefficient to determine the number of ways to configure a system of identical particles given a specific distribution of particles into discrete energy levels.

In chemistry it is common to deal with a very large number of particles and outcomes and it is useful to express factorials in different ways. We can simplify factorials of large numbers by using **Stirling's approximation**:

$$n! \approx (2\pi)^{1/2} n^{n+\frac{1}{2}} e^{-n} \tag{FI1.11}$$

The approximation is in error by less than 1 per cent when n is greater than about 10. For very large values of n, it is possible to use another form of the approximation:

$$\ln n! \approx n \ln n - n \tag{FI1.12}$$

1.3 Complex numbers and complex functions

Complex numbers have the form

$$z = x + iy \tag{FI1.13}$$

where $i = (-1)^{1/2}$. The real numbers x and y are, respectively, the real and imaginary parts of z, denoted $\mathrm{Re}(z)$ and $\mathrm{Im}(z)$. We write the **complex conjugate** of z, z^*, by replacing i by $-i$:

$$z^* = x - iy \tag{FI1.14}$$

The **absolute value** or **modulus** of the complex number z is denoted $|z|$ and is given by:

$$|z| = (z^* z)^{1/2} = (x^2 + y^2)^{1/2} \tag{FI1.15}$$

The following rules apply for arithmetic operations involving complex numbers:

1 Addition. If $z = x + iy$ and $z' = x' + iy'$, then,

$$z \pm z' = (x \pm x') + i(y \pm y') \tag{FI1.16}$$

2 Multiplication. For z and z' defined above,

$$z \times z' = (x + iy)(x' + iy') = (xx' - yy') + i(xy' + yx') \tag{FI1.17}$$

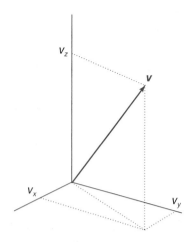

FI1.1 The vector v has components v_x, v_y, and v_z on the x-, y-, and z-axes with magnitudes v_x, v_y, and v_z, respectively.

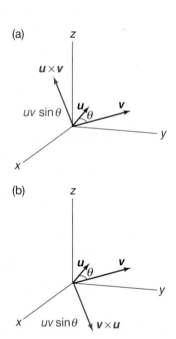

FI1.2 The direction of the cross-products of two vectors u and v with an angle θ between them: (a) $u \times v$ and (b) $v \times u$. Note that the cross-product, and the unit vector l of eqn 25, are perpendicular to both u and v but the direction depends on the order in which the product is taken.

3 Division. For z and z' defined above,

$$\frac{z}{z'} = \frac{z(z')^*}{|z'|^2} \tag{FI1.18}$$

Functions of complex arguments are useful in the discussion of wave equations (Chapter 11). We write the complex conjugate, f^*, of a complex function, f, by replacing i wherever it occurs by $-$i. For instance, the complex conjugate of e^{ix} is e^{-ix}.

Complex exponential functions may be written in terms of trigonometric functions. For example,

$$e^{\pm ix} = \cos x \pm i \sin x \tag{FI1.19}$$

which implies that

$$\cos x = \tfrac{1}{2}(e^{ix} + e^{-ix}) \tag{FI1.20}$$

$$\sin x = -i\tfrac{1}{2}(e^{ix} - e^{-ix}) \tag{FI1.21}$$

1.4 Vectors

A vector quantity has both magnitude and direction. The vector shown in Fig. FI1.1 has components on the x-, y-, and z-axes with magnitudes v_x, v_y, and v_z, respectively. The vector may be represented as

$$v = v_x i + v_y j + v_z k \tag{FI1.22}$$

where i, j, and k are **unit vectors**, vectors of magnitude 1, pointing along the positive directions on the x-, y-, and z-axes. The magnitude of the vector is denoted v or $|v|$ and is given by

$$v = (v_x^2 + v_y^2 + v_z^2)^{1/2} \tag{FI1.23}$$

Using this representation, we can define the following vector operations:

1 Addition and subtraction.[1] If $v = v_x i + v_y j + v_z k$ and $u = u_x i + u_y j + u_z k$, then

$$v \pm u = (v_x \pm u_x)i + (v_y \pm u_y)j + (v_z \pm u_z)k \tag{FI1.24}$$

2 Multiplication.[2] There are two ways to multiply vectors. In one procedure, the **cross-product** of two vectors u and v is a vector defined as

$$u \times v = (uv \sin \theta)l \tag{FI1.25a}$$

where θ is the angle between the two vectors and l is a unit vector perpendicular to both u and v, with a direction determined as in Fig. FI1.2. An equivalent definition is

$$u \times v = \begin{vmatrix} i & j & k \\ u_x & u_y & u_z \\ v_x & v_y & v_z \end{vmatrix} = (u_y v_z - u_z v_y)i - (u_x v_z - u_z v_x)j - (u_x v_y - u_y v_x)k \tag{FI1.25b}$$

where the structure in the middle is a determinant (see below). The second type of vector multiplication is the **scalar product** (or **dot product**) of two vectors u and v:

$$u \cdot v = uv \cos \theta \tag{FI1.26}$$

As its name suggests, the scalar product of two vectors is a scalar, a quantity with magnitude, not direction.

---

1. Graphical methods for vector addition and subtraction are demonstrated in Section 13.8.
2. Vector division is not defined.

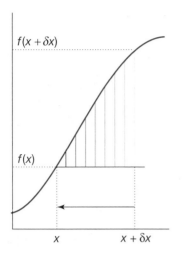

FI1.3 The slope of $f(x)$ at x, df/dx, is obtained by making a series of approximations to the value of $f(x + \delta x) - f(x)$ divided by the change in x, denoted δx, and allowing x to approach 0 (as indicated by the vertical lines getting closer to x).

Calculus

1.5 Differentiation and integration

Rates of change of functions—slopes of their graphs—are best discussed in terms of the infinitesimal calculus. The slope of a function, like the slope of a hill, is obtained by dividing the rise of the hill by the horizontal distance (Fig. FI1.3). However, because the slope may vary from point to point, we should make the horizontal distance between the points as small as possible. In fact, we let it become infinitesimally small—hence the name *infinitesimal* calculus. The values of a function f at two locations x and $x + \delta x$ are $f(x)$ and $f(x + \delta x)$, respectively. Therefore, the slope of the function f at x is the vertical distance, which we write δf, divided by the horizontal distance, which we write δx:

$$\text{Slope} = \frac{\text{rise in value}}{\text{horizontal distance}} = \frac{\delta f}{\delta x} = \frac{f(x + \delta x) - f(x)}{\delta x} \tag{FI1.27}$$

The slope at x itself is obtained by letting the horizontal distance become zero, which we write $\lim \delta x \to 0$. In this limit, the δ is replaced by a d, and we write

$$\text{Slope at } x = \frac{df}{dx} = \lim_{\delta x \to 0} \frac{f(x + \delta x) - f(x)}{\delta x} \tag{FI1.28}$$

To work out the slope of any function, we work out the expression on the right: this process is called **differentiation** and the expression for df/dx is the **derivative** of the function f with respect to the variable x. Some important derivatives are given inside the front cover of the text. Most of the functions encountered in chemistry can be differentiated by using the following rules:[3]

Rule 1 For two functions f and g:

$$d(f + g) = df + dg \tag{FI1.29}$$

Rule 2 (the product rule) For two functions f and g:

$$d(fg) = f\,dg + g\,df \tag{FI1.30}$$

Rule 3 (the quotient rule) For two functions f and g:

$$d\frac{f}{g} = \frac{1}{g}\,df - \frac{f}{g^2}\,dg \tag{FI1.31}$$

Rule no. 4 (the chain rule) For a function $f = f(g)$, where $g = g(t)$,

$$\frac{df}{dt} = \frac{df}{dg}\frac{dg}{dt} \tag{FI1.32}$$

The area under a graph of any function f is found by the techniques of **integration**. For instance, the area under the graph of the function f drawn in Fig. FI1.4 can be written as the value of f evaluated at a point multiplied by the width of the region, δx, and then all those products $f(x)\,\delta x$ summed over all the regions:

$$\text{Area between } a \text{ and } b = \sum f(x)\,\delta x$$

When we allow δx to become infinitesimally small, written dx, and sum an infinite number of strips, we write:

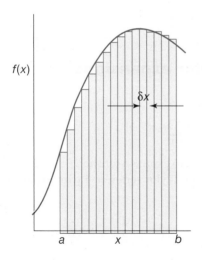

FI1.4 The shaded area is equal to the definite integral of $f(x)$ between the limits a and b.

3. In these expressions, derivatives df/dx are written as df.

$$\text{Area between } a \text{ and } b = \int_a^b f(x)\,\mathrm{d}x \tag{FI1.33}$$

The elongated S symbol on the right is called the **integral** of the function f. When written as $\int$ alone, it is the **indefinite integral** of the function. When written with limits (as in eqn 33), it is the **definite integral** of the function. The definite integral is the indefinite integral evaluated at the upper limit (b) minus the indefinite integral evaluated at the lower limit (a). The **mean value theorem** states that the average value of a function $f(x)$ over the range $x = a$ to $x = b$ is

$$\text{Average value of } f(x) \text{ from } a \text{ to } b = \frac{1}{b-a}\int_a^b f(x)\,\mathrm{d}x \tag{FI1.34}$$

Integration is the inverse of differentiation. That is, if we integrate a function and then differentiate the result, we get back the original function. Some important integrals are given on the front back cover of the text. Many other standard forms are found in tables (see *Further reading*) and it is also possible to calculate definite and indefinite integrals with mathematical software. Two integration techniques are useful:

Technique 1 (integration by parts)   For two functions f and g:

$$\int f\frac{\mathrm{d}g}{\mathrm{d}x}\,\mathrm{d}x = fg - \int g\frac{\mathrm{d}f}{\mathrm{d}x}\,\mathrm{d}x \tag{FI1.35}$$

Technique 2 (method of partial fractions)   To solve an integral of the form $\int \frac{1}{(a-x)(b-x)}\,\mathrm{d}x$, where a and b are constants, we write

$$\frac{1}{(a-x)(b-x)} = \frac{1}{b-a}\left(\frac{1}{a-x} - \frac{1}{b-x}\right)$$

and integrate the expression on the right. It follows that

$$\int \frac{\mathrm{d}x}{(a-x)(b-x)} = \frac{1}{b-a}\left[\int \frac{\mathrm{d}x}{a-x} - \int \frac{\mathrm{d}x}{b-x}\right] \tag{FI1.36}$$

$$= \frac{1}{b-a}\left(\ln\frac{1}{a-x} - \ln\frac{1}{b-x}\right) + \text{constant}$$

1.6 Power series and Taylor expansions

A **power series** has the form

$$c_0 + c_1(x-a) + c_2(x-a)^2 + \cdots + c_n(x-a)^n + \cdots = \sum_{n=0}^{\infty} c_n(x-a)^n \tag{FI1.37}$$

where c_n and a are constants. It is often useful to express a function $f(x)$ in the vicinity of $x = a$ as a special power series called the **Taylor series**, or **Taylor expansion**, which has the form:

$$f(x) = f(a) + \left(\frac{\mathrm{d}f}{\mathrm{d}x}\right)_a (x-a) + \frac{1}{2!}\left(\frac{\mathrm{d}^2 f}{\mathrm{d}x^2}\right)_a (x-a)^2 + \cdots + \frac{1}{n!}\left(\frac{\mathrm{d}^n f}{\mathrm{d}x^n}\right)_a (x-a)^n \tag{FI1.38}$$

$$= \sum_{n=0}^{\infty} \frac{1}{n!}\left(\frac{\mathrm{d}^n f}{\mathrm{d}x^n}\right)_a (x-a)^n$$

The following Taylor expansions are often useful:

$$\frac{1}{1+x} = 1 - x + x^2 \cdots$$

$$e^x = 1 + x + \tfrac{1}{2}x^2 + \cdots$$

$$\ln x = (x-1) - \tfrac{1}{2}(x-1)^2 + \tfrac{1}{3}(x-1)^3 - \tfrac{1}{4}(x-1)^4 + \cdots$$

$$\ln(1+x) = x - \tfrac{1}{2}x^2 + \tfrac{1}{3}x^3 \cdots$$

If $x \ll 1$, then $(1+x)^{-1} \approx 1 - x$, $e^x \approx 1 + x$, and $\ln(1+x) \approx x$.

1.7 Partial derivatives

A **partial derivative** of a function of more than one variable, such as $f(x, y)$, is the slope of the function with respect to one of the variables, all the other variables being held constant (see Fig. 3.2). Although a partial derivative shows how a function changes when one variable changes, it may be used to determine how the function changes when more than one variable changes by an infinitesimal amount. Thus, if f is a function of x and y, then when x and y change by dx and dy, respectively, f changes by

$$df = \left(\frac{\partial f}{\partial x}\right)_y dx + \left(\frac{\partial f}{\partial y}\right)_x dy \tag{FI1.39}$$

where the symbol ∂ is used (instead of d) to denote a partial derivative. The quantity df is also called the **differential** of f. For example, if $f = ax^3y + by^2$, then

$$\left(\frac{\partial f}{\partial x}\right)_y = 3ax^2y \qquad \left(\frac{\partial f}{\partial y}\right)_x = ax^3 + 2by$$

Then, when x and y undergo infinitesimal changes, f changes by

$$df = 3ax^2y\, dx + (ax^3 + 2by)\, dy$$

Partial derivatives may be taken in any order:

$$\frac{\partial^2 f}{\partial x \partial y} = \frac{\partial^2 f}{\partial y \partial x} \tag{FI1.40}$$

For the function f given above, it is easy to verify that

$$\left(\frac{\partial}{\partial y}\left(\frac{\partial f}{\partial x}\right)_y\right)_x = 3ax^2 \qquad \left(\frac{\partial}{\partial x}\left(\frac{\partial f}{\partial y}\right)_x\right)_y = 3ax^2$$

In the following, z is a variable on which x and y depend (for example, x, y, and z might correspond to p, V, and T).

Relation 1   When x is changed at constant z:

$$\left(\frac{\partial f}{\partial x}\right)_z = \left(\frac{\partial f}{\partial x}\right)_y + \left(\frac{\partial f}{\partial y}\right)_x \left(\frac{\partial y}{\partial x}\right)_z \tag{FI1.41}$$

Relation 2

$$\left(\frac{\partial y}{\partial x}\right)_z = \frac{1}{(\partial x/\partial y)_z} \tag{FI1.42}$$

Relation 3

$$\left(\frac{\partial x}{\partial y}\right)_z = -\left(\frac{\partial x}{\partial z}\right)_y \left(\frac{\partial z}{\partial y}\right)_x \tag{FI1.43}$$

By combining this relation and Relation 2 we obtain the **Euler chain relation**:

$$\left(\frac{\partial y}{\partial x}\right)_z \left(\frac{\partial x}{\partial z}\right)_y \left(\frac{\partial z}{\partial y}\right)_x = -1 \tag{FI1.44}$$

Relation 4 This relation establishes whether or not df is an **exact differential**.

$$\mathrm{d}f = g(x, y)\, \mathrm{d}x + h(x, y)\, \mathrm{d}y \quad \text{is exact if} \quad \left(\frac{\partial g}{\partial y}\right)_x = \left(\frac{\partial h}{\partial x}\right)_y \tag{FI1.45}$$

If df is exact, its integral between specified limits is independent of the path.

1.8 Undetermined multipliers

Suppose we need to find the maximum (or minimum) value of some function f that depends on several variables x_1, x_2, … , x_n. When the variables undergo a small change from x_i to $x_i + \delta x_i$ the function changes from f to $f + \delta f$, where

$$\delta f = \sum_i^n \left(\frac{\partial f}{\partial x_i}\right)\delta x_i \tag{FI1.46}$$

At a minimum or maximum, $\delta f = 0$, so then

$$\sum_i^n \left(\frac{\partial f}{\partial x_i}\right)\delta x_i = 0 \tag{FI1.47}$$

If the x_i were all independent, all the δx_i would be arbitrary, and this equation could be solved by setting each $(\partial f/\partial x_i) = 0$ individually. When the x_i are not all independent, the δx_i are not all independent, and the simple solution is no longer valid. We proceed as follows.

Let the constraint connecting the variables be an equation of the form $g = 0$. For example, in Chapter 19, one constraint was $n_0 + n_1 + \cdots = N$, which can be written

$$g = 0, \quad \text{with} \quad g = (n_0 + n_1 + \cdots) - N$$

The constraint $g = 0$ is always valid, so g remains unchanged when the x_i are varied:

$$\delta g = \sum_i \left(\frac{\partial g}{\partial x_i}\right)\delta x_i = 0 \tag{FI1.48}$$

Because δg is zero, we can multiply it by a parameter, λ, and add it to eqn 47:

$$\sum_i^n \left\{\left(\frac{\partial f}{\partial x_i}\right) + \lambda\left(\frac{\partial g}{\partial x_i}\right)\right\}\delta x_i = 0 \tag{FI1.49}$$

This equation can be solved for one of the δx, δx_n for instance, in terms of all the other δx_i. All those other δx_i ($i = 1, 2, … , n - 1$) are independent, because there is only one constraint on the system. But here is the trick: λ is arbitrary; therefore we can choose it so that the coefficient of δx_n in eqn 49 is zero. That is, we choose λ so that

$$\left(\frac{\partial f}{\partial x_n}\right) + \lambda\left(\frac{\partial g}{\partial x_n}\right) = 0 \tag{FI1.50}$$

Then eqn 49 becomes

$$\sum_i^{n-1} \left\{\left(\frac{\partial f}{\partial x_i}\right) + \lambda\left(\frac{\partial g}{\partial x_i}\right)\right\}\delta x_i = 0 \tag{FI1.51}$$

Now the $n-1$ variations δx_i *are* independent, so the solution of this equation is

$$\left(\frac{\partial f}{\partial x_i}\right) + \lambda\left(\frac{\partial g}{\partial x_i}\right) = 0 \qquad i = 1, 2, \ldots, n-1 \tag{FI1.52}$$

However, eqn 50 has exactly the same form as this equation, so the maximum or minimum of f can be found by solving

$$\left(\frac{\partial f}{\partial x_i}\right) + \lambda\left(\frac{\partial g}{\partial x_i}\right) = 0 \qquad i = 1, 2, \ldots, n \tag{FI1.53}$$

The use of this approach was illustrated in the text for two constraints and therefore two undetermined multipliers λ_1 and λ_2 (α and $-\beta$).

The multipliers λ cannot always remain undetermined. One approach is to solve eqn 50 instead of incorporating it into the minimization scheme. In Chapter 19 we used the alternative procedure of keeping λ undetermined until a property was calculated for which the value was already known. Thus, we found that $\beta = 1/kT$ by calculating the internal energy of a perfect gas.

1.9 Differential equations

(a) Ordinary differential equations

An **ordinary differential equation** is a relation between derivatives of a function of one variable and the function itself, as in

$$a\frac{d^2y}{dx^2} + b\frac{dy}{dx} + cy = 0 \tag{FI1.54}$$

The coefficients a, b, etc. may be functions of x. The **order** of the equation is the order of the highest derivative that occurs in it, so eqn 54 is a second-order equation. Only rarely in science is a differential equation of order higher than 2 encountered. A solution of a differential equation is an expression for y as a function of x. The process of solving a differential equation is commonly termed 'integration', and in simple cases simple integration can be employed to find $y(x)$. A **general solution** of a differential equation is the most general solution of the equation and is expressed in terms of a number of constants. When the constants are chosen to accord with certain specified **initial conditions** (if one variable is the time) or certain **boundary conditions** (to fulfil certain spatial restrictions on the solutions), we obtain the **particular solution** of the equation. A first-order differential equation requires the specification of *one* boundary (or initial) condition; a second-order differential equation requires the specification of *two* such conditions, and so on.

First-order differential equations may often be solved by direct integration. For example, the equation

$$\frac{dy}{dx} = axy$$

with a constant may be rearranged into

$$\frac{dy}{y} = ax\,dx$$

and then integrated to

$$\ln y = \tfrac{1}{2}ax^2 + A$$

where A is a constant. If we know that $y = y_0$ when $x = 0$ (for instance), then it follows that $A = \ln y_0$, and hence the particular solution of the equation is

$$\ln y = \tfrac{1}{2}ax^2 + \ln y_0$$

This expression rearranges to

$$y = y_0 e^{ax^2/2}$$

First-order equations of a more complex form can often be solved by an appropriate substitution. For example, it is sensible to try the substitution $y = sx$, and to change the variables from x and y to x and s. An alternative useful transformation is to write $x = u + a$ and $y = v + b$, and then to select a and b to simplify the form of the resulting expression.

Solutions to complex differential equations may also be found by referring to tables (see *Further reading*). For example, first-order equations of the form

$$\frac{dy}{dx} + yf(x) = g(x) \tag{FI1.55}$$

appear in the study of chemical kinetics. The solution is given by

$$ye^{\int f(x)\,dx} = \int e^{\int f(x)\,dx} g(x)\,dx + \text{constant} \tag{FI1.56}$$

Mathematical software is now capable of finding analytical solutions of a wide variety of differential equations.

Second-order differential equations are in general much more difficult to solve than first-order equations. One powerful approach commonly used to lay siege to second-order differential equations is to express the solution as a power series:

$$y = \sum_{n=0}^{\infty} c_n x^n \tag{FI1.57}$$

and then to use the differential equation to find a relation between the coefficients. This approach results, for instance, in the Hermite polynomials that form part of the solution of the Schrödinger equation for the harmonic oscillator (Section 12.5). All the second-order differential equations that occur in this text can be found tabulated in compilations of solutions or can be solved with mathematical software, and the specialized techniques that are needed to establish the form of the solutions may be found in mathematical texts.

(b) Numerical integration of differential equations

Many of the differential equations that describe physical phenomena are so complicated that their solutions cannot be cast as functions. In such cases, we resort to numerical methods, in which approximations are made in order to integrate the differential equation. Software packages are now readily available that can be used to solve almost any equation numerically. The general form of such programs to solve $df/dx = g(x)$, for instance, replaces the infinitesimal quantity $df = g(x)\,dx$ by the small quantity $\Delta f = g(x)\Delta x$, so that

$$f(x + \Delta x) \approx f(x) + g(x)\Delta x$$

and then proceeds numerically to step along the x-axis, generating $f(x)$ as it goes. The actual algorithms adopted are much more sophisticated than this primitive scheme, but stem from it. Among the simple numerical methods, the fourth-order **Runge–Kutta method** is one of the most accurate.

The *Further reading* section lists monographs that discuss the derivation of the fourth-order Runge–Kutta method. Here we illustrate the procedure with a first-order differential equation of the form:

$$\frac{dy}{dx} = f(x, y) \tag{FI1.58}$$

One example of this differential equation in chemical kinetics is eqn 25.32, which describes the time dependence of the concentration of an intermediate I in the reaction sequence $A \rightarrow I \rightarrow P$.

To obtain an approximate value of the integral of eqn 58, we proceed by rewriting it in terms of finite differences instead of differentials:

$$\frac{\Delta y}{\Delta x} = f(x, y)$$

where Δy may also be written as $y(x + \Delta x) - y(x)$. The fourth-order Runge–Kutta method is based on the following approximation:

$$y(x + \Delta x) = y(x) + \tfrac{1}{6}(k_1 + 2k_2 + 2k_3 + k_4) \tag{FI1.59}$$

where

$$k_1 = f(x, y)\Delta x \tag{FI1.60a}$$

$$k_2 = f(x + \tfrac{1}{2}\Delta x, y + \tfrac{1}{2}k_1)\Delta x \tag{FI1.60b}$$

$$k_3 = f(x + \tfrac{1}{2}\Delta x, y + \tfrac{1}{2}k_2)\Delta x \tag{FI1.60c}$$

$$k_4 = f(x + \Delta x, y + k_3)\Delta x \tag{FI1.60d}$$

Therefore, if we know the functional form of $f(x, y)$ and $y(0)$, we can use eqns 60(a–d) to calculate values of y for a range of x values. The process can be automated easily with an electronic spreadsheet or with mathematical software. The accuracy of the calculation increases with decreasing values of the increment Δx.

(c) Partial differential equations

A **partial differential equation** is a differential in more than one variable. An example is

$$\frac{\partial^2 y}{\partial t^2} = a\frac{\partial^2 y}{\partial x^2} \tag{FI1.61}$$

with y a function of the two variables x and t. In certain cases, partial differential equations may be separated into ordinary differential equations. Thus, the Schrödinger equation for a particle in a two-dimensional square well (Section 12.2) may be separated by writing the wavefunction, $\psi(x, y)$, as the product $X(x)Y(y)$, which results in the separation of the second-order partial differential equation into two second-order differential equations in the variables x and y. A good guide to the likely success of such a **separation of variables** procedure is the symmetry of the system.

Matrix algebra

A **matrix** is an array of numbers. Matrices may be combined together by addition or multiplication according to generalizations of the rules for ordinary numbers. Most numerical matrix manipulations are now carried out with mathematical software.

Consider a square matrix M of n^2 numbers arranged in n columns and n rows. These n^2 numbers are the **elements** of the matrix, and may be specified by stating the row, r, and column, c, at which they occur. Each element is therefore denoted M_{rc}. For example, in the matrix

$$M = \begin{pmatrix} 1 & 2 \\ 3 & 4 \end{pmatrix}$$

the elements are $M_{11} = 1$, $M_{12} = 2$, $M_{21} = 3$, and $M_{22} = 4$. This is an example of a 2×2 matrix. The **determinant**, $|M|$, of this matrix is

$$|M| = \begin{vmatrix} 1 & 2 \\ 3 & 4 \end{vmatrix} = 1 \times 4 - 2 \times 3 = -2$$

A **diagonal matrix** is a matrix in which the only nonzero elements lie on the major diagonal (the diagonal from M_{11} to M_{nn}). Thus, the matrix

$$D = \begin{pmatrix} 1 & 0 & 0 \\ 0 & 2 & 0 \\ 0 & 0 & 1 \end{pmatrix}$$

is diagonal. The condition may be written

$$M_{rc} = m_r \delta_{rc} \tag{FI1.62}$$

where δ_{rc} is the **Kronecker delta**, which is equal to 1 for $r = c$ and to 0 for $r \neq c$. In the above example, $m_1 = 1$, $m_2 = 2$, and $m_3 = 1$. The **unit matrix**, **1** (and occasionally **I**), is a special case of a diagonal matrix in which all nonzero elements are 1.

The **transpose** of a matrix M is denoted M^T and is defined by

$$M^T_{mn} = M_{nm} \tag{FI1.63}$$

Thus, for the matrix M we have been using,

$$M^T = \begin{pmatrix} 1 & 3 \\ 2 & 4 \end{pmatrix}$$

Matrices are very useful in chemistry. They simplify some mathematical tasks, such as solving systems of simultaneous equations, the treatment of molecular symmetry (Chapter 15), and quantum mechanical calculations (Chapter 14).

1.10 Matrix addition and multiplication

Two matrices M and N may be added to give the sum $S = M + N$, according to the rule

$$S_{rc} = M_{rc} + N_{rc} \tag{FI1.64}$$

(that is, corresponding elements are added). Thus, with M given above and

$$N = \begin{pmatrix} 5 & 6 \\ 7 & 8 \end{pmatrix}$$

the sum is

$$S = \begin{pmatrix} 1 & 2 \\ 3 & 4 \end{pmatrix} + \begin{pmatrix} 5 & 6 \\ 7 & 8 \end{pmatrix} = \begin{pmatrix} 6 & 8 \\ 10 & 12 \end{pmatrix}$$

Two matrices may also be multiplied to give the product $P = MN$ according to the rule

$$P_{rc} = \sum_n M_{rn} N_{nc} \tag{FI1.65}$$

For example, with the matrices given above,

$$P = \begin{pmatrix} 1 & 2 \\ 3 & 4 \end{pmatrix} \begin{pmatrix} 5 & 6 \\ 7 & 8 \end{pmatrix}$$

$$= \begin{pmatrix} 1 \times 5 + 2 \times 7 & 1 \times 6 + 2 \times 8 \\ 3 \times 5 + 4 \times 7 & 3 \times 6 + 4 \times 8 \end{pmatrix}$$

$$= \begin{pmatrix} 19 & 22 \\ 43 & 50 \end{pmatrix}$$

It should be noticed that in general $MN \neq NM$, and matrix multiplication is in general non-commutative.

The **inverse** of a matrix M is denoted M^{-1}, and is defined so that

$$MM^{-1} = M^{-1}M = 1 \tag{FI1.66}$$

The inverse of a matrix can be constructed by using mathematical software, but in simple cases the following procedure can be carried through without much effort:

1 Form the determinant of the matrix. For example, for our matrix M, $|M| = -2$.

2 Form the transpose of the matrix. For example, $M^{\mathrm{T}} = \begin{pmatrix} 1 & 3 \\ 2 & 4 \end{pmatrix}$.

3 Form $\tilde{M}'$, where $\tilde{M}'_{rc}$ is the **cofactor** of the element M_{rc}; that is, it is the determinant formed from M with the row r and column c deleted. For example, $\tilde{M}' = \begin{pmatrix} 4 & -2 \\ -3 & 1 \end{pmatrix}$.

4 Construct the inverse as $M^{-1} = \tilde{M}'/|M|$. For example,

$$M^{-1} = \frac{1}{-2} \begin{pmatrix} 4 & -2 \\ -3 & 1 \end{pmatrix} = \begin{pmatrix} -2 & 1 \\ \frac{3}{2} & -\frac{1}{2} \end{pmatrix}$$

1.11 Simultaneous equations

A set of n simultaneous equations

$$a_{11}x_1 + a_{12}x_2 + \cdots + a_{1n}x_n = b_1$$
$$a_{21}x_1 + a_{22}x_2 + \cdots + a_{2n}x_n = b_2$$
$$\vdots \tag{FI1.67}$$
$$a_{n1}x_1 + a_{n2}x_2 + \cdots + a_{nn}x_n = b_n$$

can be written in matrix notation if we introduce the **column vectors** x and b:

$$x = \begin{pmatrix} x_1 \\ x_2 \\ \vdots \\ x_n \end{pmatrix} \qquad b = \begin{pmatrix} b_1 \\ b_2 \\ \vdots \\ b_n \end{pmatrix}$$

Then, with the a matrix of coefficients a_{rc}, the n equations may be written as

$$ax = b \tag{FI1.68}$$

The formal solution is obtained by multiplying both sides of this matrix equation by a^{-1}, for then

$$x = a^{-1}b \tag{FI1.69}$$

1.12 Eigenvalue equations

An **eigenvalue equation** is a special case of eqn 68 in which

$$ax = \lambda x \tag{FI1.70}$$

where λ is a constant, the **eigenvalue**, and x is the **eigenvector**. In general, there are n eigenvalues $\lambda^{(i)}$, and they satisfy the n simultaneous equations

$$(a - \lambda 1)x = 0 \tag{FI1.71}$$

There are n corresponding eigenvectors $x^{(i)}$. Equation 71 has a solution only if the determinant of the coefficients is zero. However, this determinant is just $|a - \lambda 1|$, so the n eigenvalues may be found from the solution of the **secular equation**:

$$|a - \lambda 1| = 0 \tag{FI1.72}$$

The n eigenvalues the secular equation yields may be used to find the n eigenvectors. These eigenvectors (which are $n \times 1$ matrices), may be used to form an $n \times n$ matrix X. Thus, because

$$x^{(1)} = \begin{pmatrix} x_1^{(1)} \\ x_2^{(1)} \\ \vdots \\ x_n^{(1)} \end{pmatrix} \qquad x^{(2)} = \begin{pmatrix} x_1^{(2)} \\ x_2^{(2)} \\ \vdots \\ x_n^{(2)} \end{pmatrix} \qquad \text{etc.}$$

we may form the matrix

$$X = (x^{(1)}, x^{(2)}, \ldots, x^{(n)}) = \begin{pmatrix} x_1^{(1)} & x_1^{(2)} & \cdots & x_1^{(n)} \\ x_2^{(1)} & x_2^{(2)} & \cdots & x_1^{(n)} \\ \vdots & \vdots & & \vdots \\ x_n^{(1)} & x_n^{(2)} & \cdots & x_n^{(n)} \end{pmatrix}$$

so that $X_{rc} = x_r^{(c)}$. If further we write $\Lambda_{rc} = \lambda_r \delta_{rc}$, so that Λ is a diagonal matrix with the elements $\lambda_1, \lambda_2, \ldots, \lambda_n$ along the diagonal, then all the eigenvalue equations $ax^{(i)} = \lambda_i x^{(i)}$ may be confined into the single equation

$$aX = X\Lambda \tag{FI1.73}$$

because this expression is equal to

$$\sum_n a_{rn} X_{nc} = \sum_n X_{rn} \Lambda_{nc}$$

or

$$\sum_n a_{rn} x_n^{(c)} = \sum_n x_r^{(n)} \lambda_n \delta_{nc} = \lambda_c x_r^{(c)}$$

as required. Therefore, if we form X^{-1} from X, we construct a **similarity transformation**

$$\Lambda = X^{-1}aX \tag{FI1.74}$$

that makes a diagonal (because Λ is diagonal). It follows that if the matrix X that causes $X^{-1}aX$ to be diagonal is known, then the problem is solved: the diagonal matrix so produced has the eigenvalues as its only nonzero elements, and the matrix X used to bring about the transformation has the corresponding eigenvectors as its columns. The solutions of eigenvalue equations are best found by using mathematical software.

Further reading

J.R. Barrante, *Applied mathematics for physical chemistry*. Prentice Hall, Upper Saddle River (1998).

M.P. Cady and C.A. Trapp, *A Mathcad primer for physical chemistry*. W.H. Freeman and Company, New York (2000).

W.H. Cropper, *Mathematical computer programs for physical chemistry*. Springer, New York (1998).

D.R. Lide (ed.), *CRC handbook of chemistry and physics*, Appendix A. CRC Press, Boca Raton (2000).

R.G. Mortimer, *Mathematics for physical chemistry*. Academic Press, San Diego (1999).

D.S. Sivia and S.G. Rawlings, *Foundations of science mathematics*. Oxford Chemistry Primers, Oxford University Press (1999).

D. Zwillinger (ed.), *CRC standard mathematical tables and formulae*. CRC Press, Boca Raton (1996).

Essential concepts of physics

Energy

The central concept of all explanations in physical chemistry, as in so many other branches of physical science, is that of **energy**, the capacity to do work. We make use of the apparently universal law of nature that *energy is conserved*; that is, energy can neither be created nor destroyed. Although energy can be transferred from one location to another, the total energy is constant.

2.1 Kinetic and potential energy

The **kinetic energy**, E_K, of a body is the energy the body possesses as a result of its motion. For a body of mass m travelling at a speed v,

$$E_K = \tfrac{1}{2}mv^2 \tag{FI2.1}$$

The **potential energy**, E_P or V, of a body is the energy it possesses as a result of its position. The zero of potential energy is arbitrary. For example, the gravitational potential energy of a body is often set to zero at the surface of the Earth; the electrical potential energy of two charged particles is set to zero when their separation is infinite. No universal expression for the potential energy can be given because it depends on the type of interaction the body experiences. One example that gives rise to a simple expression is the potential energy of a body of mass m in the gravitational field close to the surface of the Earth (a gravitational field acts on the mass of a body). If the body is at a height h above the surface of the Earth, then its potential energy is mgh, where g is a constant called the **acceleration of free fall**, $g = 9.81$ m s^{-2}, and $V = 0$ at $h = 0$ (the arbitrary zero mentioned previously).

The total energy is the sum of the kinetic and potential energies of a particle:

$$E = E_K + E_P \tag{FI2.2}$$

2.2 Energy units

The SI unit of energy is the *joule* (J), which is defined as

$$1\text{ J} = 1\text{ kg m}^2\text{ s}^{-2} \tag{FI2.3}$$

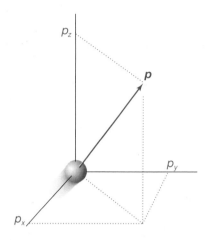

Calories (cal) and kilocalories (kcal) are still encountered in the chemical literature: by definition, 1 cal = 4.184 J. An energy of 1 cal is enough to raise the temperature of 1 g of water by 1°C.

The rate of change of energy is called the **power**, P, expressed as joules per second, or *watt*, W:

$$1\ W = 1\ J\ s^{-1} \tag{FI2.4}$$

Classical mechanics

Classical mechanics describes the behaviour of objects in terms of two equations. One expresses the fact that the total energy is constant in the absence of external forces; the other expresses the response of particles to the forces acting on them.

2.3 The trajectory in terms of the energy

The **velocity**, v, of a particle is the rate of change of its position:

$$v = \frac{dr}{dt} \tag{FI2.5}$$

The velocity is a vector, with both direction and magnitude. The magnitude of the velocity is the **speed**, v. The **linear momentum**, p, of a particle of mass m is related to its velocity, v, by

$$p = mv \tag{FI2.6}$$

Like the velocity vector, the linear momentum vector points in the direction of travel of the particle (Fig. FI2.1). In terms of the linear momentum, the total energy of a particle is

$$E = \frac{p^2}{2m} + V(x) \tag{FI2.7}$$

This equation can be used to show that a particle will have a definite **trajectory**, or definite position and momentum at each instant. For example, consider a particle free to move in one direction (along the x-axis) in a region where $V = 0$ (so the energy is independent of position). Because $v = dx/dt$, it follows from eqns 6 and 7 that

$$\frac{dx}{dt} = \left(\frac{2E_K}{m}\right)^{1/2} \tag{FI2.8}$$

A solution of this equation is

$$x(t) = x(0) + \left(\frac{2E_K}{m}\right)^{1/2} t \tag{FI2.9}$$

The linear momentum is a constant:

$$p(t) = mv(t) = m\frac{dx}{dt} = (2mE_K)^{1/2} \tag{FI2.10}$$

Hence, if we know the initial position and momentum, we can predict all later positions and momenta exactly.

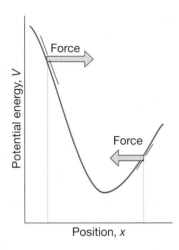

FI2.2 The force acting on a particle is determined by the slope of the potential energy at each point. The force points in the direction of lower potential energy.

2.4 Newton's second law

The **force**, F, experienced by a particle free to move in one dimension is related to its potential energy, V, by

$$F = -\frac{dV}{dx}$$
(FI2.11a)

This relation implies that the direction of the force is towards decreasing potential energy (Fig. FI2.2). In three dimensions,

$$\boldsymbol{F} = -\nabla V \qquad \nabla = \boldsymbol{i}\frac{\partial}{\partial x} + \boldsymbol{j}\frac{\partial}{\partial y} + \boldsymbol{k}\frac{\partial}{\partial z}$$
(FI2.11b)

Newton's second law of motion states that the rate of change of momentum is equal to the force acting on the particle. In one dimension:

$$\frac{dp}{dt} = F$$
(FI2.12a)

Because $p = m(dx/dt)$ in one dimension, it is sometimes more convenient to write this equation as

$$m\frac{d^2x}{dt^2} = F$$
(FI2.12b)

The second derivative, d^2x/dt^2, is the **acceleration** of the particle, its rate of change of velocity (in this instance, along the x-axis). It follows that, if we know the force acting everywhere and at all times, then solving eqn 12 will also give the trajectory. This calculation is equivalent to the one based on E, but is more suitable in some applications. For example, it can be used to show that if a particle of mass m is initially stationary and is subjected to a constant force F for a time τ, then its kinetic energy increases from zero to

$$E_K = \frac{F^2\tau^2}{2m}$$
(FI2.13)

and then remains at that energy after the forces ceases to act. Because the applied force, F, and the time, τ, for which it acts may be varied at will, the solution implies that the energy of the particle may be increased to any value.

2.5 Rotational motion

The rotational motion of a particle about a central point is described by its **angular momentum**, J. The angular momentum is a vector: its magnitude gives the rate at which a particle circulates and its direction indicates the axis of rotation (Fig. FI2.3). The magnitude of the angular momentum, J, is given by the expression

$$J = I\omega$$
(FI2.14)

where ω is the **angular velocity** of the body, its rate of change of angular position (in radians per second), and I is the **moment of inertia**. The analogous roles of m and I, of v and ω, and of p and J in the translational and rotational cases, respectively, should be remembered, because they provide a ready way of constructing and recalling equations. For a point particle of mass m moving in a circle of radius r, the moment of inertia about the axis of rotation is given by the expression

$$I = mr^2$$
(FI2.15)

FI2.3 The angular momentum of a particle is represented by a vector along the axis of rotation and perpendicular to the plane of rotation. The length of the vector denotes the magnitude of the angular momentum. The direction of motion is clockwise to an observer looking in the direction of the vector.

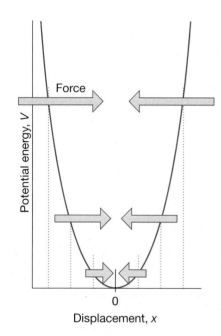

FI2.4 The force acting on a particle that undergoes harmonic motion. The force is directed towards zero displacement and is proportional to the displacement. The corresponding potential energy is parabolic (proportional to x^2).

To accelerate a rotation it is necessary to apply a **torque**, T, a twisting force. Newton's equation is then

$$\frac{dJ}{dt} = T \tag{FI2.16}$$

If a constant torque is applied for a time τ, the rotational energy of an initially stationary body is increased to

$$E_K = \frac{T^2\tau^2}{2I} \tag{FI2.17}$$

The implication of this equation is that an appropriate torque and period for which it is applied can excite the rotation to an arbitrary energy.

2.6 The harmonic oscillator

A **harmonic oscillator** consists of a particle that experiences a restoring force proportional to its displacement from its equilibrium position:

$$F = -kx \tag{FI2.18}$$

An example is a particle joined to a rigid support by a spring. The constant of proportionality k is called the **force constant**, and the stiffer the spring the greater the force constant. The negative sign in F signifies that the direction of the force is opposite to that of the displacement (Fig. FI2.4).

The motion of a particle that undergoes harmonic motion is found by substituting the expression for the force, eqn 18, into Newton's equation, eqn 12b. The resulting equation is

$$m\frac{d^2x}{dt^2} = -kx$$

A solution is

$$x(t) = A \sin \omega t \qquad p(t) = m\omega A \cos \omega t \qquad \omega = (k/m)^{1/2} \tag{FI2.19}$$

These solutions show that the position of the particle varies **harmonically** (that is, as $\sin \omega t$) with a frequency $v = \omega/2\pi$. They also show that the particle is stationary ($p = 0$) when the displacement, x, has its maximum value, A, which is called the **amplitude** of the motion.

The total energy of a classical harmonic oscillator is proportional to the square of the amplitude of its motion. To confirm this remark we note that the kinetic energy is

$$E_K = \frac{p^2}{2m} = \frac{(m\omega A \cos \omega t)^2}{2m} = \tfrac{1}{2}m\omega^2 A^2 \cos^2 \omega t \tag{FI2.20}$$

Then, because $\omega = (k/m)^{1/2}$, this expression may be written

$$E_K = \tfrac{1}{2}kA^2 \cos^2 \omega t \tag{FI2.21}$$

The force on the oscillator is $F = -kx$, so it follows from the relation $F = -dV/dx$ that the potential energy of a harmonic oscillator is

$$V = \tfrac{1}{2}kx^2 = \tfrac{1}{2}kA^2 \sin^2 \omega t \tag{FI2.22}$$

The total energy is therefore

$$E = \tfrac{1}{2}kA^2 \cos^2 \omega t + \tfrac{1}{2}kA^2 \sin^2 \omega t = kA^2 \tag{FI2.23}$$

(We have used $\cos^2\omega t + \sin^2\omega t = 1$.) That is, the energy of the oscillator is constant and, for a given force constant, is determined by its maximum displacement. It follows that the energy of an oscillating particle can be raised to any value by stretching the spring to any desired amplitude A. Note that the frequency of the motion depends only on the inherent properties of the oscillator (as represented by k and m) and is independent of the energy; the amplitude governs the energy, through $E = \frac{1}{2}kA^2$, and is independent of the frequency. In other words, the particle will oscillate at the same frequency regardless of the amplitude of its motion.

Waves

Waves are disturbances that travel through space with a finite velocity. Examples of disturbances include the collective motion of water molecules in ocean waves and of gas particles in sound waves. Waves can be characterized by a **wave equation**, a differential equation that describes the motion of the wave in space and time. **Harmonic waves** are waves with displacements that can be expressed as sine or cosine functions. These concepts are used in classical physics to describe the wave character of electromagnetic radiation, which is the focus of the following discussion.

2.7 The electromagnetic field

In classical physics, electromagnetic radiation is understood in terms of the **electromagnetic field**, an oscillating electric and magnetic disturbance that spreads as a harmonic wave through empty space, the vacuum. The wave travels at a constant speed called the *speed of light*, c, which is about 3×10^8 m s^{-1}. As its name suggests, an electromagnetic field has two components, an **electric field** that acts on charged particles (whether stationary of moving) and a **magnetic field** that acts only on moving charged particles. The electromagnetic field is characterized by a **wavelength**, λ (lambda), the distance between the neighbouring peaks of the wave, and its **frequency**, ν (nu), the number of times per second at which its displacement at a fixed point returns to its original value (Fig. FI2.5). The frequency is measured in *hertz*,

FI2.5 (a) The wavelength, λ, of a wave is the peak-to-peak distance. (b) The wave is shown travelling to the right at a speed c. At a given location, the instantaneous amplitude of the wave changes through a complete cycle (the four dots show half a cycle) as it passes a given point. The frequency, ν, is the number of cycles per second that occur at a given point. Wavelength and frequency are related by $\lambda\nu = c$.

(a)

(b)

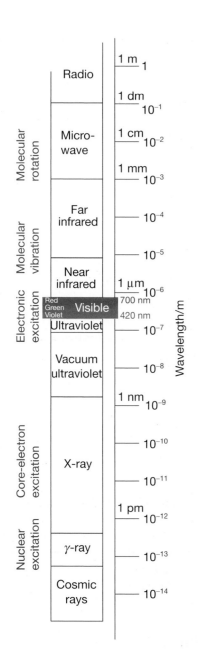

FI2.6 The regions of the electromagnetic spectrum and the types of excitation that give rise to each region.

where 1 Hz = 1 s^{-1}. The wavelength and frequency of an electromagnetic wave are related by

$$\lambda v = c \tag{FI2.24}$$

Therefore, the shorter the wavelength, the higher the frequency. The characteristics of a wave are also reported by giving the **wavenumber**, $\tilde{v}$ (nu tilde), of the radiation, where

$$\tilde{v} = \frac{v}{c} = \frac{1}{\lambda} \tag{FI2.25}$$

A wavenumber can be interpreted as the number of complete wavelengths in a given length. Wavenumbers are normally reported in reciprocal centimetres (cm^{-1}), so a wavenumber of 5 cm^{-1} indicates that there are 5 complete wavelengths in 1 cm. The classification of the electromagnetic field according to its frequency and wavelength is summarized in Fig. FI2.6.

2.8 Features of electromagnetic radiation

Consider an electromagnetic disturbance travelling along the x-direction with wavelength λ and frequency v. The functions that describe the oscillating electric field, $\mathcal{E}(x, t)$, and magnetic field, $B(x, t)$, may be written as

$$\mathcal{E}(x, t) = \mathcal{E}_0 \cos\{2\pi vt - (2\pi/\lambda)x + \phi\} \tag{FI2.26a}$$

$$B(x, t) = B_0 \cos\{2\pi vt - (2\pi/\lambda)x + \phi\} \tag{FI2.26b}$$

where $\mathcal{E}_0$ and B_0 are the amplitudes of the electric and magnetic fields, respectively, and the parameter ϕ is the **phase** of the wave, which varies from $-\pi$ to π and gives the relative location of the peaks of two waves. If two waves, in the same region of space, with the same wavelength are shifted by $\phi = \pi$ or $-\pi$ (so the peaks of one wave coincide with the troughs of the other), then the resultant wave will have diminished amplitudes. The waves are said to interfere destructively. A value of $\phi = 0$ (coincident peaks) corresponds to constructive interference, or the enhancement of the amplitudes.

Equations 26a and 26b represent electromagnetic radiation that is **plane polarized**; it is so called because the electric and magnetic fields each oscillate in a single plane (in this case the xy- and xz-planes, respectively, Fig. FI2.7). The plane of polarization may be orientated in any direction around the direction of propagation (the x-direction in Fig. FI2.7), with the electric and magnetic fields perpendicular to that direction (and perpendicular to each other). An alternative mode of polarization is **circular polarization**, in which the electric and magnetic fields rotate around the direction of propagation in either a clockwise or a counter-clockwise sense but remain perpendicular to it and each other.

It is easy to show by differentiation that eqns 26a and 26b satisfy the following equations:

$$\frac{\partial^2}{\partial x^2}\psi(x, t) = -\frac{4\pi^2}{\lambda^2}\psi(x, t) \qquad \frac{\partial^2}{\partial t^2}\psi(x, t) = -4\pi^2 v^2 \psi(x, t) \tag{FI2.27}$$

where $\psi(x, t)$ is either $\mathcal{E}(x, t)$ or $B(x, t)$.

According to classical electromagnetic theory, the intensity of electromagnetic radiation is proportional to the square of the amplitude of the wave. For example, the light detectors discussed in Chapter 16 are based on the interaction between the electric field of the incident radiation and the detecting element, so light intensities are proportional to $\mathcal{E}_0^2$.

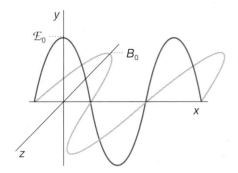

FI2.7 Electromagnetic radiation consists of a wave of electric and magnetic fields perpendicular to the direction of propagation (in this case the x-direction), and mutually perpendicular to each other. This illustration shows a plane-polarized wave, with the electric and magnetic fields oscillating in the xy- and xz-planes, respectively.

Electrostatics

Electrostatics is the study of the interactions of stationary electric charges. The elementary charge, the magnitude of charge carried by a single electron or proton, is $e \approx 1.60 \times 10^{-19}$ C. The magnitude of the charge per mole is the Faraday constant: $F = N_A e = 9.65 \times 10^4$ C mol^{-1}.

2.9 The Coulomb interaction

If a point charge q_1 is at a distance r in a vacuum from another point charge q_2, then their potential energy is

$$V = \frac{q_1 q_2}{4\pi\varepsilon_0 r} \tag{FI2.29}$$

The constant ε_0 is the **vacuum permittivity**, a fundamental constant with the value 8.85×10^{-12} C^2 J^{-1} m^{-1}. (Note that with r in metres, m, and the charges in coulombs, C, the potential energy is in joules, J.) This very important relation is called the **Coulomb potential energy** and the interaction it describes is called the **Coulomb interaction** of two charges. The Coulomb potential energy is equal to the work that must be done to bring up a charge q_1 from infinity to a distance r from a charge q_2.

It follows from eqns 11 and 29 that the electrical force, F, exerted by a charge q_1 on a second charge q_2 has magnitude

$$F = \frac{q_1 q_2}{4\pi\varepsilon_0 r^2} \tag{FI2.30}$$

The force itself is a vector directed along the line joining the two charges. With charge in coulombs and distance in metres, the force is obtained in newtons.

In a medium other than a vacuum, the potential energy of interaction between two charges is reduced, and the vacuum permittivity is replaced by the **permittivity**, ε, of the medium. It is common to express the permittivity as a multiple of the vacuum permittivity, and to write $\varepsilon = \varepsilon_r \varepsilon_0$, where ε_r is the **relative permittivity** (or dielectric constant) of the medium. For water at 25°C, $\varepsilon_r = 78.54$.

2.10 The Coulomb potential

The potential energy of a charge q_1 in the presence of another charge q_2 can be expressed in terms of the **Coulomb potential**, ϕ:

$$V = q_1 \phi \qquad \phi = \frac{q_2}{4\pi\varepsilon_0 r} \tag{FI2.31}$$

The units of potential are joules per coulomb, J C^{-1} so, when ϕ is multiplied by a charge in coulombs, the result is in joules. The combination joules per coulomb occurs widely in electrostatics, and is called a *volt*, V:

$$1\,\text{V} = 1\,\text{J C}^{-1} \tag{FI2.32}$$

If there are several charges q_2, q_3, ... present in the system, the total potential experienced by the charge q_1 is the sum of the potential generated by each charge:

$$\phi = \phi_2 + \phi_3 + \cdots \tag{FI2.33}$$

When the charge distribution is more complex than a single point-like object, the Coulomb potential is described in terms of a charge density, ρ. With charge in coulombs and length in metres, the charge density is expressed in coulomb per metre-cubed ($C\ m^{-3}$). The electric potential arising from a charge distribution with density ρ is the solution to **Poisson's equation**:

$$\nabla^2\phi = -\rho/\varepsilon_0 \tag{FI2.34}$$

where $\nabla^2 = (\partial^2/\partial x^2 + \partial^2/\partial y^2 + \partial^2/\partial z^2)$. If the distribution is spherically symmetrical, then so too is ϕ and eqn 34 reduces to the form used in *Justification 10.2*.

2.11 The strength of the electric field

Just as the potential energy of a charge q_1 can be written $V = q_1\phi$, so the magnitude of the force on q_1 can be written $F = q_1\mathcal{E}$, where $\mathcal{E}$ is the magnitude of the electric field strength arising from q_2 or from some more general charge distribution. The electric field strength (which, like the force, is actually a vector quantity) is the negative gradient of the electric potential:

$$\mathcal{E} = -\nabla\phi \tag{FI2.35}$$

2.12 The dipole–dipole interaction

An important problem in physical chemistry is the calculation of the potential energy of interaction between two point dipoles with moments $\boldsymbol{\mu}_1$ and $\boldsymbol{\mu}_2$, separated by a vector $\boldsymbol{r}$ (Sections 21.5 and 26.11). From classical electromagnetic theory, the potential energy of $\boldsymbol{\mu}_2$ in the electric field $\mathcal{E}_1$ generated by $\boldsymbol{\mu}_1$ is given by the dot product

$$V = -\mathcal{E}_1 \cdot \boldsymbol{\mu}_2 \tag{FI2.36}$$

To calculate $\mathcal{E}_1$, we consider a distribution of point charges q_i located at x_i, y_i, and z_i from the origin (that is, at r_i). It follows from eqns 31 and 33 that the Coulomb potential due to this distribution at a point with coordinates x, y, and z is:

$$\phi = \sum_i \frac{q_i}{4\pi\varepsilon_0} \frac{1}{\{(x - x_i)^2 + (y - y_i)^2 + (z - z_i)^2\}^{1/2}} \tag{FI2.37}$$

where $\boldsymbol{r}$ is the location of the point of interest and the r_i are the locations of the charges q_i. If we suppose that all the charges are close to the origin (in the sense that $r_i \ll r$), we can use a Taylor expansion to write

$$\phi(\boldsymbol{r}) = \sum_i \frac{q_i}{4\pi\varepsilon_0}\left\{ \frac{1}{r} + \left(\frac{\partial\{(x - x_i)^2 + (y - y_i)^2 + (z - z_i)^2\}^{1/2}}{\partial x_i} \right)_{x_i=0} x_i + \cdots \right\}$$

$$= \sum_i \frac{q_i}{4\pi\varepsilon_0}\left\{ \frac{1}{r} + \frac{xx_i}{r^3} + \cdots \right\} \tag{FI2.38}$$

the ellipses ($\ldots$) include the terms arising from derivatives with respect to y_i and z_i and higher derivatives. If the charge distribution is electrically neutral, the first term disappears because $\sum_i q_i = 0$. Next we note that $\sum_i q_i x_i = \mu_x$, and likewise for the y- and z-components. That is,

$$\phi = \frac{1}{4\pi\varepsilon_0 r^3}(\mu_x x + \mu_y y + \mu_z z)$$

$$= \frac{1}{4\pi\varepsilon_0 r^3}\boldsymbol{\mu}_1 \cdot \boldsymbol{r} \tag{FI2.39}$$

From eqn 35, the electric field strength is

$$\mathcal{E}_1 = \frac{1}{4\pi\varepsilon_0}\nabla\frac{\boldsymbol{\mu}_1\cdot\boldsymbol{r}}{r^3} = -\frac{\boldsymbol{\mu}_1}{4\pi\varepsilon_0 r^3} - \frac{\boldsymbol{\mu}_1\cdot\boldsymbol{r}}{4\pi\varepsilon_0}\nabla\frac{1}{r^3} \tag{FI2.40}$$

It follows from eqns 36 and 40 that

$$V = \frac{\boldsymbol{\mu}_1\cdot\boldsymbol{\mu}_2}{4\pi\varepsilon_0 r^3} - 3\frac{(\boldsymbol{\mu}_1\cdot\boldsymbol{r})(\boldsymbol{\mu}_2\cdot\boldsymbol{r})}{4\pi\varepsilon_0 r^5} \tag{FI2.41}$$

For the arrangement shown in Fig. 21.10, in which $\boldsymbol{\mu}_1\cdot\boldsymbol{r} = \mu_1 r\cos\theta$ and $\boldsymbol{\mu}_2\cdot\boldsymbol{r} = \mu_2 r\cos\theta$, eqn 41 becomes:

$$V = \frac{\mu_1\mu_2 f(\theta)}{4\pi\varepsilon_0 r^3} \qquad f(\theta) = 1 - 3\cos^2\theta \tag{FI2.42}$$

which is eqn 21.24.

2.13 Electric current and power

The motion of charge gives rise to an electric **current**, I. Electric current is measured in *ampere*, A, where

$$1\,\text{A} = 1\,\text{C s}^{-1} \tag{FI2.43}$$

If the current flows from a region of potential ϕ_i to ϕ_f, through a **potential difference** $\Delta\phi = \phi_f - \phi_i$, the rate of doing work is the current (the rate of transfer of charge) multiplied by the potential difference, $I\Delta\phi$. The rate of doing electrical work is the **electrical power**, P, so

$$P = I\Delta\phi \tag{FI2.44}$$

With current in amperes and the potential difference in volts, the power works out in watts. The total energy, E, supplied in an interval Δt is the power (the rate of energy supply) multiplied by the duration of the interval:

$$E = P\Delta t = I\Delta\phi\Delta t \tag{FI2.45}$$

The energy is obtained in joules with the current in amperes, the potential difference in volts, and the time in seconds.

Further reading

R.P. Feynman, R.B. Leighton, and M. Sands, *The Feynman lectures on physics*, Vols 1–3. Addison–Wesley, Reading (1966).

G.A.D. Ritchie and D.S. Sivia, *Foundations of physics for chemists*. Oxford Chemistry Primers, Oxford University Press (2000).

W.S. Warren, *The physical basis of chemistry*. Academic Press, San Diego (2000).

R. Wolfson and J.M. Pasachoff, *Physics for scientists and engineers with modern physics*. Addison–Wesley, Reading (1999).

Data section

The following tables reproduce and expand the data given in the short tables in the text, and follow their numbering. Standard states refer to a pressure of $p^{\ominus} = 1$ bar. The general references are as follows:

AIP: D.E. Gray (ed.), *American Institute of Physics handbook*. McGraw-Hill, New York (1972).

AS: M. Abramowitz and I.A. Stegun (ed.), *Handbook of mathematical functions*. Dover, New York (1963).

E: J. Emsley, *The elements*. Oxford University Press (1991).

HCP: D.R. Lide (ed.), *Handbook of chemistry and physics*. CRC Press, Boca Raton (2000).

JL: A.M. James and M.P. Lord, *Macmillan's chemical and physical data*. Macmillan, London (1992).

KL: G.W.C. Kaye and T.H. Laby (ed.), *Tables of physical and chemical constants*. Longman, London (1973).

LR: G.N. Lewis and M. Randall, revised by K.S. Pitzer and L. Brewer, *Thermodynamics*. McGraw-Hill, New York (1961).

NBS: *NBS tables of chemical thermodynamic properties*, published as *J. Phys. and Chem. Reference Data*, **11**, Supplement 2 (1982).

RS: R.A. Robinson and R.H. Stokes, *Electrolyte solutions*. Butterworth, London (1959).

TDOC: J.B. Pedley, J.D. Naylor, and S.P. Kirby, *Thermochemical data of organic compounds*. Chapman & Hall, London (1986).

Contents

The following is a directory of all tables in the text; those included in this *Data section* are marked with an asterisk. The remainder will be found on the pages indicted in parentheses.

Physical properties of selected materials*
Masses and natural abundances of selected nuclides*
1.1   Pressure units (4)
1.2   The gas constant in various units (10)
        The composition of dry air at sea level (12)
1.3   Second virial coefficients*
1.4   Critical constants of gases*
1.5   Van der Waals coefficients*
1.6   Selected equations of state (21)

2.1   Varieties of work (39)
2.2   Temperature variation of molar heat capacities*
2.3   Standard enthalpies of fusion and vaporization at the transition temperature*

2.4   Enthalpies of transition [notation] (57)
2.5   Thermodynamic data for organic compounds*
        Thermochemical properties of some fuels (59)
2.6   Thermodynamic data for elements and inorganic compounds*
2.6a   Standard enthalpies of hydration at infinite dilution*
2.6b   Standard ion hydration enthalpies*
2.7   Benson thermochemical groups*

3.1   Expansion coefficients and isothermal compressibilities*
3.2   Inversion temperatures, normal freezing and boiling points, and Joule–Thomson coefficients*

4.1   Standard entropies (and temperatures) of phase transitions*
4.2   Standard entropies of vaporization of liquids*
4.3   Standard Third-Law entropies [see Tables 2.5 and 2.6]*

4.4 Standard Gibbs energies of formation [see Tables 2.5 and 2.6]*

5.1 The Maxwell relations (123)
5.2 The fugacity coefficient of nitrogen*

6.1 Surface tensions of liquids*

7.1 Henry's law constants for gases*
7.2 Cryoscopic and ebullioscopic constants*
7.3 Standard states [summary of definitions] (182)

9.1 Acidity constants for aqueous solutions*

10.1 Standard thermodynamic functions of ions in aqueous solution [see Table 2.6]*
10.2 Relative permittivities (dielectric constants)*
10.3 Standard molar entropies of ions in aqueous solution [see Table 2.6]
10.4 Ionic strength and molality (258)
10.5 Mean activity coefficients in water*
10.6 Varieties of electrodes (262)
10.7 Standard potentials*
10.8 The electrochemical series of metals [see also Table 10.7] (277)

11.1 The Schrödinger equation (305)

12.1 The Hermite polynomials (340)
12.2 The error function*
12.3 The spherical harmonics (350)

13.1 Hydrogenic radial wavefunctions (371)
13.2 Screening constants for atoms*
13.3 Ionization energies*
13.4 Electron affinities*

14.1 Some hybridization schemes (417)
14.2 Bond lengths*
14.3a Bond dissociation enthalpies*
14.3b Mean bond enthalpies*
14.4 Pauling and Mulliken electronegativities*

15.1 The notation for point groups (456)
15.2 The C_{2v} character table (465)
15.3 The C_{3v} character table (466)

16.1 Moments of inertia [formulas] (498)
16.2 Properties of diatomic molecules*
16.3 Typical vibrational wavenumbers*

17.1 Colour, frequency, and energy of light*
17.2 Ground and excited states of O_2 (541)
17.3 Absorption characteristics of some groups and molecules*
17.4 Characteristics of laser radiation and its chemical application (562)

18.1 Nuclear constitution and the nuclear spin quantum number (581)
18.2 Nuclear spin properties*
18.3 Hyperfine coupling constants for atoms*

20.1 Rotational and vibrational temperatures [see also Table 16.2] (662)
20.2 Symmetry numbers [see also Table 16.2] (663)
20.3 Contributions to the molecular partition functions [summary of formulas] (666)

21.1 Dipole moments, polarizabilities, and polarizability volumes*
21.2 Refractive indices relative to air*
21.3 Multipole interaction potential energies (700)
21.4 Lennard-Jones (12,6)-potential parameters*

22.1 Radius of gyration of some macromolecules*
22.2 Diffusion coefficients of macromolecules in water*
22.3 Frictional coefficients and molecular geometry*
22.4 Intrinsic viscosity*

23.1 The seven crystal systems (769)
23.2 The crystal structures of some elements (785)
23.3 Ionic radii*
23.4 Madelung constants (788)
23.5 Lattice enthalpies*
23.6 Magnetic susceptibilities*

24.1 Collision cross-sections*
24.2 Transport properties of gases*
24.3 Transport properties of perfect gases [summary of formulas] (828)
24.4 Viscosities of liquids*
24.5 Limiting ionic conductivities in water*
24.6 Ionic mobilities in water*
24.7 Debye–Hückel–Onsager coefficients for (1,1)-electrolytes*
24.8 Diffusion coefficients*

25.1 Kinetic data for first-order reactions*
25.2 Kinetic data for second-order reactions*
25.3 Integrated rate laws (875)
25.4 Arrhenius parameters*

26.1 Examples of photochemical processes (921)
26.2 Common photophysical processes (925)
26.3 Values of R_0 for donor–acceptor pairs (931)

27.1 Arrhenius parameters for gas-phase reactions*
27.2 Arrhenius parameters for reactions in solution [see Table 25.4]*
27.3 Summary of uses of k [notation] (971)

28.1 Maximum observed enthalpies of physisorption*
28.2 Enthalpies of chemisorption*
28.3 Activation energies of catalysed reactions*
28.4 Properties of catalysts (1001)
28.5 Chemisorption abilities (1002)
28.6 Summary of acronyms (1005)

29.1 Exchange current densities and transfer coefficients*

Character tables

Physical properties of selected materials

Elements	$\rho/(g\,cm^{-3})$ at 293 K†	T_f/K	T_b/K	Inorganic compounds	$\rho/(g\,cm^{-3})$ at 293 K†	T_f/K	T_b/K
Aluminium(s)	2.698	933.5	2740	$CaCO_3$(s, calcite)	2.71	1612	1171d
Argon(g)	1.381	83.8	87.3	$CuSO_4 \cdot 5H_2O$(s)	2.284	383($-H_2O$)	423($-5H_2O$)
Boron(s)	2.340	2573	3931	HBr(g)	2.77	184.3	206.4
Bromine(l)	3.123	265.9	331.9	HCl(g)	1.187	159.0	191.1
Carbon(s, gr)	2.260	3700s		HI(g)	2.85	222.4	237.8
Carbon(s, d)	3.513			H_2O(l)	0.997	273.2	373.2
Chlorine(g)	1.507	172.2	239.2	D_2O(l)	1.104	277.0	374.6
Copper(s)	8.960	1357	2840	NH_3(g)	0.817	195.4	238.8
Fluorine(g)	1.108	53.5	85.0	KBr(s)	2.750	1003	1708
Gold(s)	19.320	1338	3080	KCl(s)	1.984	1049	1773s
Helium(g)	0.125		4.22	NaCl(s)	2.165	1074	1686
Hydrogen(g)	0.071	14.0	20.3	H_2SO_4(l)	1.841	283.5	611.2
Iodine(s)	4.930	386.7	457.5	**Organic compounds**			
Iron(s)	7.874	1808	3023	Acetaldehyde, CH_3CHO(l, g)	0.788	152	293
Krypton(g)	2.413	116.6	120.8	Acetic acid, CH_3COOH(l)	1.049	289.8	391
Lead(s)	11.350	600.6	2013	Acetone, $(CH_3)_2CO$(l)	0.787	178	329
Lithium(s)	0.534	453.7	1620	Aniline, $C_6H_5NH_2$(l)	1.026	267	457
Magnesium(s)	1.738	922.0	1363	Anthracene, $C_{14}H_{10}$(s)	1.243	490	615
Mercury(l)	13.546	234.3	629.7	Benzene, C_6H_6(l)	0.879	278.6	353.2
Neon(g)	1.207	24.5	27.1	Carbon tetrachloride, CCl_4(l)	1.63	250	349.9
Nitrogen(g)	0.880	63.3	77.4	Chloroform, $CHCl_3$(l)	1.499	209.6	334
Oxygen(g)	1.140	54.8	90.2	Ethanol, C_2H_5OH(l)	0.789	156	351.4
Phosphorus(s, wh)	1.820	317.3	553	Formaldehyde, HCHO(g)		181	254.0
Potassium(s)	0.862	336.8	1047	Glucose, $C_6H_{12}O_6$(s)	1.544	415	
Silver(s)	10.500	1235	2485	Methane, CH_4(g)		90.6	111.6
Sodium(s)	0.971	371.0	1156	Methanol, CH_3OH(l)	0.791	179.2	337.6
Sulfur(s, α)	2.070	386.0	717.8	Naphthalene, $C_{10}H_8$(s)	1.145	353.4	491
Uranium(s)	18.950	1406	4018	Octane, C_8H_{18}(l)	0.703	216.4	398.8
Xenon(g)	2.939	161.3	166.1	Phenol, C_6H_5OH(s)	1.073	314.1	455.0
Zinc(s)	7.133	692.7	1180	Sucrose, $C_{12}H_{22}O_{11}$(s)	1.588	457d	

d: decomposes;   s: sublimes;   Data: AIP, E, HCP, KL;   † For gases, at their boiling points.

Masses and natural abundances of selected nuclides

Nuclide		m/u	Abundance/%
H	^{1}H	1.0078	99.985
	^{2}H	2.0140	0.015
He	^{3}He	3.0160	0.000 13
	^{4}He	4.0026	100
Li	^{6}Li	6.0151	7.42
	^{7}Li	7.0160	92.58
B	^{10}B	10.0129	19.78
	^{11}B	11.0093	80.22
C	^{12}C	12*	98.89
	^{13}C	13.0034	1.11
N	^{14}N	14.0031	99.63
	^{15}N	15.0001	0.37
O	^{16}O	15.9949	99.76
	^{17}O	16.9991	0.037
	^{18}O	17.9992	0.204
F	^{19}F	18.9984	100
P	^{31}P	30.9738	100
S	^{32}S	31.9721	95.0
	^{33}S	32.9715	0.76
	^{34}S	33.9679	4.22
Cl	^{35}Cl	34.9688	75.53
	^{37}Cl	36.9651	24.4
Br	^{79}Br	78.9183	50.54
	^{81}Br	80.9163	49.46
I	^{127}I	126.9045	100

* Exact value.

Table 1.3 Second virial coefficients, $B/(cm^3\ mol^{-1})$

	100 K	273 K	373 K	600 K
Air	−167.3	−13.5	3.4	19.0
Ar	−187.0	−21.7	−4.2	11.9
CH_4		−53.6	−21.2	8.1
CO_2		−142	−72.2	−12.4
H_2	−2.0	13.7	15.6	
He	11.4	12.0	11.3	10.4
Kr		−62.9	−28.7	1.7
N_2	−160.0	−10.5	6.2	21.7
Ne	−6.0	10.4	12.3	13.8
O_2	−197.5	−22.0	−3.7	12.9
Xe		−153.7	−81.7	−19.6

Data: AIP, JL. The values relate to the expansion in eqn 1.22 of Section 1.3b; convert to eqn 1.21 using $B' = B/RT$.
For Ar at 273 K, $C = 1200\ cm^6\ mol^{-1}$

Table 1.4 Critical constants of gases

	p_c/atm	$V_c/(cm^3\ mol^{-1})$	T_c/K	Z_c	T_B/K
Ar	48.00	75.25	150.72	0.292	411.5
Br_2	102	135	584	0.287	
C_2H_4	50.50	124	283.1	0.270	
C_2H_6	48.20	148	305.4	0.285	
C_6H_6	48.6	260	562.7	0.274	
CH_4	45.6	98.7	190.6	0.288	510.0
Cl_2	76.1	124	417.2	0.276	
CO_2	72.85	94.0	304.2	0.274	714.8
F_2	55	144			
H_2	12.8	64.99	33.23	0.305	110.0
H_2O	218.3	55.3	647.4	0.227	
HBr	84.0	363.0			
HCl	81.5	81.0	324.7	0.248	
He	2.26	57.76	5.21	0.305	22.64
HI	80.8	423.2			
Kr	54.27	92.24	209.39	0.291	575.0
N_2	33.54	90.10	126.3	0.292	327.2
Ne	26.86	41.74	44.44	0.307	122.1
NH_3	111.3	72.5	405.5	0.242	
O_2	50.14	78.0	154.8	0.308	405.9
Xe	58.0	118.8	289.75	0.290	768.0

Data: AIP, KL

Table 1.5 Van der Waals coefficients

	a/(atm L^2 mol^{-2})	b/(10^{-2} L mol^{-1})		a/(atm L^2 mol^{-2})	b/(10^{-2} L mol^{-1})
Ar	1.337	3.20	H_2S	4.484	4.34
C_2H_4	4.552	5.82	He	0.0341	2.38
C_2H_6	5.507	6.51	Kr	5.125	1.06
C_6H_6	18.57	11.93	N_2	1.352	3.87
CH_4	2.273	4.31	Ne	0.205	1.67
Cl_2	6.260	5.42	NH_3	4.169	3.71
CO	1.453	3.95	O_2	1.364	3.19
CO_2	3.610	4.29	SO_2	6.775	5.68
H_2	0.2420	2.65	Xe	4.137	5.16
H_2O	5.464	3.05			

Data: HCP

Table 2.2 Temperature variation of molar heat capacities†

	a	b/(10^{-3} K^{-1})	c/(10^5 K^2)
Monatomic gases			
	20.78	0	0
Other gases			
Br_2	37.32	0.50	−1.26
Cl_2	37.03	0.67	−2.85
CO_2	44.22	8.79	−8.62
F_2	34.56	2.51	−3.51
H_2	27.28	3.26	0.50
I_2	37.40	0.59	−0.71
N_2	28.58	3.77	−0.50
NH_3	29.75	25.1	−1.55
O_2	29.96	4.18	−1.67
Liquids (from melting to boiling)			
$C_{10}H_8$, naphthalene	79.5	0.4075	0
I_2	80.33	0	0
H_2O	75.29	0	0
Solids			
Al	20.67	12.38	0
C (graphite)	16.86	4.77	−8.54
$C_{10}H_8$, naphthalene	−115.9	3.920×10^3	0
Cu	22.64	6.28	0
I_2	40.12	49.79	0
NcCl	45.94	16.32	0
Pb	22.13	11.72	0.96

† For $C_{p,m}$/(J K^{-1} mol^{-1}) $= a + bT + c/T^2$
Source: LR.

Table 2.3 Standard enthalpies of fusion and vaporization at the transition temperature, $\Delta_{trs}H^{\ominus}/(\text{kJ mol}^{-1})$

	T_f/K	Fusion	T_b/K	Vaporization		T_f/K	Fusion	T_b/K	Vaporization
Elements					CO_2	217.0	8.33	194.6	25.23 s
Ag	1234	11.30	2436	250.6	CS_2	161.2	4.39	319.4	26.74
Ar	83.81	1.188	87.29	6.506	H_2O	273.15	6.008	373.15	40.656
Br_2	265.9	10.57	332.4	29.45					44.016 at 298 K
Cl_2	172.1	6.41	239.1	20.41	H_2S	187.6	2.377	212.8	18.67
F_2	53.6	0.26	85.0	3.16	H_2SO_4	283.5	2.56		
H_2	13.96	0.117	20.38	0.916	NH_3	195.4	5.652	239.7	23.35
He	3.5	0.021	4.22	0.084	**Organic compounds**				
Hg	234.3	2.292	629.7	59.30	CH_4	90.68	0.941	111.7	8.18
I_2	386.8	15.52	458.4	41.80	CCl_4	250.3	2.5	350	30.0
N_2	63.15	0.719	77.35	5.586	C_2H_6	89.85	2.86	184.6	14.7
Na	371.0	2.601	1156	98.01	C_6H_6	278.61	10.59	353.2	30.8
O_2	54.36	0.444	90.18	6.820	C_6H_{14}	178	13.08	342.1	28.85
Xe	161	2.30	165	12.6	$C_{10}H_8$	354	18.80	490.9	51.51
K	336.4	2.35	1031	80.23	CH_3OH	175.2	3.16	337.2	35.27
Inorganic compounds									37.99 at 298 K
CCl_4	250.3	2.47	349.9	30.00	C_2H_5OH	158.7	4.60	352	43.5

Data: AIP;   s denotes sublimation

Table 2.5 Thermodynamic data for organic compounds (all values are for 298 K)

	M/(g mol^{-1})	$\Delta_f H^{\ominus}$/(kJ mol^{-1})	$\Delta_f G^{\ominus}$/(kJ mol^{-1})	$S_m^{\ominus}$/(J K^{-1} mol^{-1})†	$C_{p,m}^{\ominus}$/(J K^{-1} mol^{-1})	$\Delta_c H^{\ominus}$/(kJ mol^{-1})
C(s) (graphite)	12.011	0	0	5.740	8.527	−393.51
C(s) (diamond)	12.011	+1.895	+2.900	2.377	6.113	−395.40
CO_2(g)	44.040	−393.51	−394.36	213.74	37.11	
Hydrocarbons						
CH_4(g), methane	16.04	−74.81	−50.72	186.26	35.31	−890
CH_3(g), methyl	15.04	+145.69	+147.92	194.2	38.70	
C_2H_2(g), ethyne	26.04	+226.73	+209.20	200.94	43.93	−1300
C_2H_4(g), ethene	28.05	+52.26	+68.15	219.56	43.56	−1411
C_2H_6(g), ethane	30.07	−84.68	−32.82	229.60	52.63	−1560
C_3H_6(g), propene	42.08	+20.42	+62.78	267.05	63.89	−2058
C_3H_6(g), cyclopropane	42.08	+53.30	+104.45	237.55	55.94	−2091
C_3H_8(g), propane	44.10	−103.85	−23.49	269.91	73.5	−2220
C_4H_8(g), 1-butene	56.11	−0.13	+71.39	305.71	85.65	−2717
C_4H_8(g), cis-2-butene	56.11	−6.99	+65.95	300.94	78.91	−2710
C_4H_8(g), trans-2-butene	56.11	−11.17	+63.06	296.59	87.82	−2707
C_4H_{10}(g), butane	58.13	−126.15	−17.03	310.23	97.45	−2878
C_5H_{12}(g), pentane	72.15	−146.44	−8.20	348.40	120.2	−3537
C_5H_{12}(l)	72.15	−173.1				
C_6H_6(l), benzene	78.12	+49.0	+124.3	173.3	136.1	−3268

Table 2.5 (Continued)

	$M/(\text{g mol}^{-1})$	$\Delta_f H^\ominus/(\text{kJ mol}^{-1})$	$\Delta_f G^\ominus/(\text{kJ mol}^{-1})$	$S_m^\ominus/(\text{J K}^{-1}\text{ mol}^{-1})$†	$C_{p,m}^\ominus/(\text{J K}^{-1}\text{ mol}^{-1})$	$\Delta_c H^\ominus/(\text{kJ mol}^{-1})$
Hydrocarbons (Continued)						
C_6H_6(g)	78.12	+82.93	+129.72	269.31	81.67	−3302
C_6H_{12}(l), cyclohexane	84.16	−156	+26.8	204.4	156.5	−3920
C_6H_{14}(l), hexane	86.18	−198.7		204.3		−4163
$C_6H_5CH_3$(g), methylbenzene (toluene)	92.14	+50.0	+122.0	320.7	103.6	−3953
C_7H_{16}(l), heptane	100.21	−224.4	+1.0	328.6	224.3	
C_8H_{18}(l), octane	114.23	−249.9	+6.4	361.1		−5471
C_8H_{18}(l), iso-octane	114.23	−255.1				−5461
$C_{10}H_8$(s), naphthalene	128.18	+78.53				−5157
Alcohols and phenols						
CH_3OH(l), methanol	32.04	−238.66	−166.27	126.8	81.6	−726
CH_3OH(g)	32.04	−200.66	−161.96	239.81	43.89	−764
C_2H_5OH(l), ethanol	46.07	−277.69	−174.78	160.7	111.46	−1368
C_2H_5OH(g)	46.07	−235.10	−168.49	282.70	65.44	−1409
C_6H_5OH(s), phenol	94.12	−165.0	−50.9	146.0		−3054
Carboxylic acids, hydroxy acids, and esters						
HCOOH(l), formic	46.03	−424.72	−361.35	128.95	99.04	−255
CH_3COOH(l), acetic	60.05	−484.5	−389.9	159.8	124.3	−875
CH_3COOH(aq)	60.05	−485.76	−396.46	178.7		
$CH_3CO_2^-$(aq)	59.05	−486.01	−369.31	+86.6	−6.3	
$(COOH)_2$(s), oxalic	90.04	−827.2			117	−254
C_6H_5COOH(s), benzoic	122.13	−385.1	−245.3	167.6	146.8	−3227
$CH_3CH(OH)COOH$(s), lactic	90.08	−694.0				−1344
$CH_3COOC_2H_5$(l), ethyl acetate	88.11	−479.0	−332.7	259.4	170.1	−2231
Alkanals and alkanones						
HCHO(g), methanal	30.03	−108.57	−102.53	218.77	35.40	−571
CH_3CHO(l), ethanal	44.05	−192.30	−128.12	160.2		−1166
CH_3CHO(g)	44.05	−166.19	−128.86	250.3	57.3	−1192
CH_3COCH_3(l), propanone	58.08	−248.1	−155.4	200.4	124.7	−1790
Sugars						
$C_6H_{12}O_6$(s), α-D-glucose	180.16	−1274				−2808
$C_6H_{12}O_6$(s), β-D-glucose	180.16	−1268	−910	212		
$C_6H_{12}O_6$(s), β-D-fructose	180.16	−1266				−2810
$C_{12}H_{22}O_{11}$(s), sucrose	342.30	−2222	−1543	360.2		−5645
Nitrogen compounds						
$CO(NH_2)_2$(s), urea	60.06	−333.51	−197.33	104.60	93.14	−632
CH_3NH_2(g), methylamine	31.06	−22.97	+32.16	243.41	53.1	−1085
$C_6H_5NH_2$(l), aniline	93.13	+31.1				−3393
$CH_2(NH_2)COOH$(s), glycine	75.07	−532.9	−373.4	103.5	99.2	−969

Data: NBS, TDOC; † Standard entropies of ions may be either positive or negative because the values are relative to the entropy of the hydrogen ion.

Table 2.6 Thermodynamic data for elements and inorganic compounds (all values relate to 298 K)

	$M/(\text{g mol}^{-1})$	$\Delta_f H^{\ominus}/(\text{kJ mol}^{-1})$	$\Delta_f G^{\ominus}/(\text{kJ mol}^{-1})$	$S_m^{\ominus}/(\text{J K}^{-1} \text{mol}^{-1})\dagger$	$C_{p,m}^{\ominus}/(\text{J K}^{-1} \text{mol}^{-1})$
Aluminium (aluminum)					
Al(s)	26.98	0	0	28.33	24.35
Al(l)	26.98	+10.56	+7.20	39.55	24.21
Al(g)	26.98	+326.4	+285.7	164.54	21.38
Al^{3+}(g)	26.98	+5483.17			
Al^{3+}(aq)	26.98	−531	−485	−321.7	
Al_2O_3(s, α)	101.96	−1675.7	−1582.3	50.92	79.04
$AlCl_3$(s)	133.24	−704.2	−628.8	110.67	91.84
Argon					
Ar(g)	39.95	0	0	154.84	20.786
Antimony					
Sb(s)	121.75	0	0	45.69	25.23
SbH_3(g)	124.77	+145.11	+147.75	232.78	41.05
Arsenic					
As(s, α)	74.92	0	0	35.1	24.64
As(g)	74.92	+302.5	+261.0	174.21	20.79
As_4(g)	299.69	+143.9	+92.4	314	
AsH_3(g)	77.95	+66.44	+68.93	222.78	38.07
Barium					
Ba(s)	137.34	0	0	62.8	28.07
Ba(g)	137.34	+180	+146	170.24	20.79
Ba^{2+}(aq)	137.34	−537.64	−560.77	+9.6	
BaO(s)	153.34	−553.5	−525.1	70.43	47.78
$BaCl_2$(s)	208.25	−858.6	−810.4	123.68	75.14
Beryllium					
Be(s)	9.01	0	0	9.50	16.44
Be(g)	9.01	+324.3	+286.6	136.27	20.79
Bismuth					
Bi(s)	208.98	0	0	56.74	25.52
Bi(g)	208.98	+207.1	+168.2	187.00	20.79
Bromine					
Br_2(l)	159.82	0	0	152.23	75.689
Br_2(g)	159.82	+30.907	+3.110	245.46	36.02
Br(g)	79.91	+111.88	+82.396	175.02	20.786
Br^-(g)	79.91	−219.07			
Br^-(aq)	79.91	−121.55	−103.96	+82.4	−141.8
HBr(g)	90.92	−36.40	−53.45	198.70	29.142
Cadmium					
Cd(s, γ)	112.40	0	0	51.76	25.98
Cd(g)	112.40	+112.01	+77.41	167.75	20.79
Cd^{2+}(aq)	112.40	−75.90	−77.612	−73.2	
CdO(s)	128.40	−258.2	−228.4	54.8	43.43
$CdCO_3$(s)	172.41	−750.6	−669.4	92.5	

Table 2.6 (Continued)

	$M/(\text{g mol}^{-1})$	$\Delta_f H^\ominus/(\text{kJ mol}^{-1})$	$\Delta_f G^\ominus/(\text{kJ mol}^{-1})$	$S_m^\ominus/(\text{J K}^{-1} \text{mol}^{-1})†$	$C_{p,m}^\ominus/(\text{J K}^{-1} \text{mol}^{-1})$
Caesium (cesium)					
Cs(s)	132.91	0	0	85.23	32.17
Cs(g)	132.91	+76.06	+49.12	175.60	20.79
Cs^+(aq)	132.91	−258.28	−292.02	+133.05	−10.5
Calcium					
Ca(s)	40.08	0	0	41.42	25.31
Ca(g)	40.08	+178.2	+144.3	154.88	20.786
Ca^{2+}(aq)	40.08	−542.83	−553.58	−53.1	
CaO(s)	56.08	−635.09	−604.03	39.75	42.80
$CaCO_3$(s) (calcite)	100.09	−1206.9	−1128.8	92.9	81.88
$CaCO_3$(s) (aragonite)	100.09	−1207.1	−1127.8	88.7	81.25
CaF_2(s)	78.08	−1219.6	−1167.3	68.87	67.03
$CaCl_2$(s)	110.99	−795.8	−748.1	104.6	72.59
$CaBr_2$(s)	199.90	−682.8	−663.6	130	
Carbon (for 'organic' compounds of carbon, see Table 2.5)					
C(s) (graphite)	12.011	0	0	5.740	8.527
C(s) (diamond)	12.011	+1.895	+2.900	2.377	6.113
C(g)	12.011	+716.68	+671.26	158.10	20.838
C_2(g)	24.022	+831.90	+775.89	199.42	43.21
CO(g)	28.011	−110.53	−137.17	197.67	29.14
CO_2(g)	44.010	−393.51	−394.36	213.74	37.11
CO_2(aq)	44.010	−413.80	−385.98	117.6	
H_2CO_3(aq)	62.03	−699.65	−623.08	187.4	
HCO_3^-(aq)	61.02	−691.99	−586.77	+91.2	
CO_3^{2-}(aq)	60.01	−677.14	−527.81	−56.9	
CCl_4(l)	153.82	−135.44	−65.21	216.40	131.75
CS_2(l)	76.14	+89.70	+65.27	151.34	75.7
HCN(g)	27.03	+135.1	+124.7	201.78	35.86
HCN(l)	27.03	+108.87	+124.97	112.84	70.63
CN^-(aq)	26.02	+150.6	+172.4	+94.1	
Chlorine					
Cl_2(g)	70.91	0	0	223.07	33.91
Cl(g)	35.45	+121.68	+105.68	165.20	21.840
Cl^-(g)	34.45	−233.13			
Cl^-(aq)	35.45	−167.16	−131.23	+56.5	−136.4
HCl(g)	36.46	−92.31	−95.30	186.91	29.12
HCl(aq)	36.46	−167.16	−131.23	56.5	−136.4
Chromium					
Cr(s)	52.00	0	0	23.77	23.35
Cr(g)	52.00	+396.6	+351.8	174.50	20.79
CrO_4^{2-}(aq)	115.99	−881.15	−727.75	+50.21	
$Cr_2O_7^{2-}$(aq)	215.99	−1490.3	−1301.1	+261.9	

Table 2.6 (Continued)

	$M/(\text{g mol}^{-1})$	$\Delta_f H^{\ominus}/(\text{kJ mol}^{-1})$	$\Delta_f G^{\ominus}/(\text{kJ mol}^{-1})$	$S_m^{\ominus}/(\text{J K}^{-1} \text{ mol}^{-1})\dagger$	$C_{p,m}^{\ominus}/(\text{J K}^{-1} \text{ mol}^{-1})$
Copper					
Cu(s)	63.54	0	0	33.150	24.44
Cu(g)	63.54	+338.32	+298.58	166.38	20.79
Cu$^+$(aq)	63.54	+71.67	+49.98	+40.6	
Cu^{2+}(aq)	63.54	+64.77	+65.49	−99.6	
Cu$_2$O(s)	143.08	−168.6	−146.0	93.14	63.64
CuO(s)	79.54	−157.3	−129.7	42.63	42.30
CuSO$_4$(s)	159.60	−771.36	−661.8	109	100.0
CuSO$_4$·H$_2$O(s)	177.62	−1085.8	−918.11	146.0	134
CuSO$_4$·5H$_2$O(s)	249.68	−2279.7	−1879.7	300.4	280
Deuterium					
D$_2$(g)	4.028	0	0	144.96	29.20
HD(g)	3.022	+0.318	−1.464	143.80	29.196
D$_2$O(g)	20.028	−249.20	−234.54	198.34	34.27
D$_2$O(l)	20.028	−294.60	−243.44	75.94	84.35
HDO(g)	19.022	−245.30	−233.11	199.51	33.81
HDO(l)	19.022	−289.89	−241.86	79.29	
Fluorine					
F$_2$(g)	38.00	0	0	202.78	31.30
F(g)	19.00	+78.99	+61.91	158.75	22.74
F$^-$(aq)	19.00	−332.63	−278.79	−13.8	−106.7
HF(g)	20.01	−271.1	−273.2	173.78	29.13
Gold					
Au(s)	196.97	0	0	47.40	25.42
Au(g)	196.97	+366.1	+326.3	180.50	20.79
Helium					
He(g)	4.003	0	0	126.15	20.786
Hydrogen (see also deuterium)					
H$_2$(g)	2.016	0	0	130.684	28.824
H(g)	1.008	+217.97	+203.25	114.71	20.784
H$^+$(aq)	1.008	0	0	0	0
H$^+$(g)	1.008	+1536.20			
H$_2$O(s)	18.015			37.99	
H$_2$O(l)	18.015	−285.83	−237.13	69.91	75.291
H$_2$O(g)	18.015	−241.82	−228.57	188.83	33.58
H$_2$O$_2$(l)	34.015	−187.78	−120.35	109.6	89.1
Iodine					
I$_2$(s)	253.81	0	0	116.135	54.44
I$_2$(g)	253.81	+62.44	+19.33	260.69	36.90
I(g)	126.90	+106.84	+70.25	180.79	20.786
I$^-$(aq)	126.90	−55.19	−51.57	+111.3	−142.3
HI(g)	127.91	+26.48	+1.70	206.59	29.158

Table 2.6 (Continued)

	$M/(\text{g mol}^{-1})$	$\Delta_f H^{\ominus}/(\text{kJ mol}^{-1})$	$\Delta_f G^{\ominus}/(\text{kJ mol}^{-1})$	$S_m^{\ominus}/(\text{J K}^{-1} \text{mol}^{-1})\dagger$	$C_{p,m}^{\ominus}/(\text{J K}^{-1} \text{mol}^{-1})$
Iron					
Fe(s)	55.85	0	0	27.28	25.10
Fe(g)	55.85	+416.3	+370.7	180.49	25.68
Fe^{2+}(aq)	55.85	−89.1	−78.90	−137.7	
Fe^{3+}(aq)	55.85	−48.5	−4.7	−315.9	
Fe_3O_4(s) (magnetite)	231.54	−1118.4	−1015.4	146.4	143.43
Fe_2O_3(s) (haematite)	159.69	−824.2	−742.2	87.40	103.85
FeS(s, α)	87.91	−100.0	−100.4	60.29	50.54
FeS_2(s)	119.98	−178.2	−166.9	52.93	62.17
Krypton					
Kr(g)	83.80	0	0	164.08	20.786
Lead					
Pb(s)	207.19	0	0	64.81	26.44
Pb(g)	207.19	+195.0	+161.9	175.37	20.79
Pb^{2+}(aq)	207.19	−1.7	−24.43	+10.5	
PbO(s, yellow)	223.19	−217.32	−187.89	68.70	45.77
PbO(s, red)	223.19	−218.99	−188.93	66.5	45.81
PbO_2(s)	239.19	−277.4	−217.33	68.6	64.64
Lithium					
Li(s)	6.94	0	0	29.12	24.77
Li(g)	6.94	+159.37	+126.66	138.77	20.79
Li^+(aq)	6.94	−278.49	−293.31	+13.4	68.6
Magnesium					
Mg(s)	24.31	0	0	32.68	24.89
Mg(g)	24.31	+147.70	+113.10	148.65	20.786
Mg^{2+}(aq)	24.31	−466.85	−454.8	−138.1	
MgO(s)	40.31	−601.70	−569.43	26.94	37.15
$MgCO_3$(s)	84.32	−1095.8	−1012.1	65.7	75.52
$MgCl_2$(s)	95.22	−641.32	−591.79	89.62	71.38
Mercury					
Hg(l)	200.59	0	0	76.02	27.983
Hg(g)	200.59	+61.32	+31.82	174.96	20.786
Hg^{2+}(aq)	200.59	+171.1	+164.40	−32.2	
Hg_2^{2+}(aq)	401.18	+172.4	+153.52	+84.5	
HgO(s)	216.59	−90.83	−58.54	70.29	44.06
Hg_2Cl_2(s)	472.09	−265.22	−210.75	192.5	102
$HgCl_2$(s)	271.50	−224.3	−178.6	146.0	
HgS(s, black)	232.65	−53.6	−47.7	88.3	
Neon					
Ne(g)	20.18	0	0	146.33	20.786

Table 2.6 (Continued)

	$M/(\text{g mol}^{-1})$	$\Delta_f H^{\ominus}/(\text{kJ mol}^{-1})$	$\Delta_f G^{\ominus}/(\text{kJ mol}^{-1})$	$S_m^{\ominus}/(\text{J K}^{-1}\text{ mol}^{-1})\dagger$	$C_{p,m}^{\ominus}/(\text{J K}^{-1}\text{ mol}^{-1})$
Nitrogen					
$N_2(g)$	28.013	0	0	191.61	29.125
$N(g)$	14.007	+472.70	+455.56	153.30	20.786
$NO(g)$	30.01	+90.25	+86.55	210.76	29.844
$N_2O(g)$	44.01	+82.05	+104.20	219.85	38.45
$NO_2(g)$	46.01	+33.18	+51.31	240.06	37.20
$N_2O_4(g)$	92.01	+9.16	+97.89	304.29	77.28
$N_2O_5(s)$	108.01	−43.1	+113.9	178.2	143.1
$N_2O_5(g)$	108.01	+11.3	+115.1	355.7	84.5
$HNO_3(l)$	63.01	−174.10	−80.71	155.60	109.87
$HNO_3(aq)$	63.01	−207.36	−111.25	146.4	−86.6
$NO_3^-(aq)$	62.01	−205.0	−108.74	+146.4	−86.6
$NH_3(g)$	17.03	−46.11	−16.45	192.45	35.06
$NH_3(aq)$	17.03	−80.29	−26.50	111.3	
$NH_4^+(aq)$	18.04	−132.51	−79.31	+113.4	79.9
$NH_2OH(s)$	33.03	−114.2			
$HN_3(l)$	43.03	+264.0	+327.3	140.6	43.68
$HN_3(g)$	43.03	+294.1	+328.1	238.97	98.87
$N_2H_4(l)$	32.05	+50.63	+149.43	121.21	139.3
$NH_4NO_3(s)$	80.04	−365.56	−183.87	151.08	84.1
$NH_4Cl(s)$	53.49	−314.43	−202.87	94.6	
Oxygen					
$O_2(g)$	31.999	0	0	205.138	29.355
$O(g)$	15.999	+249.17	+231.73	161.06	21.912
$O_3(g)$	47.998	+142.7	+163.2	238.93	39.20
$OH^-(aq)$	17.007	−229.99	−157.24	−10.75	−148.5
Phosphorus					
$P(s, wh)$	30.97	0	0	41.09	23.840
$P(g)$	30.97	+314.64	+278.25	163.19	20.786
$P_2(g)$	61.95	+144.3	+103.7	218.13	32.05
$P_4(g)$	123.90	+58.91	+24.44	279.98	67.15
$PH_3(g)$	34.00	+5.4	+13.4	210.23	37.11
$PCl_3(g)$	137.33	−287.0	−267.8	311.78	71.84
$PCl_3(l)$	137.33	−319.7	−272.3	217.1	
$PCl_5(g)$	208.24	−374.9	−305.0	364.6	112.8
$PCl_5(s)$	208.24	−443.5			
$H_3PO_3(s)$	82.00	−964.4			
$H_3PO_3(aq)$	82.00	−964.8			
$H_3PO_4(s)$	94.97	−1279.0	−1119.1	110.50	106.06
$H_3PO_4(l)$	94.97	−1266.9			
$H_3PO_4(aq)$	94.97	−1277.4	−1018.7	−222	
$PO_4^{3-}(aq)$	94.97	−1277.4	−1018.7	−221.8	
$P_4O_{10}(s)$	283.89	−2984.0	−2697.0	228.86	211.71
$P_4O_6(s)$	219.89	−1640.1			

Table 2.6 (Continued)

	$M/(\text{g mol}^{-1})$	$\Delta_f H^{\ominus}/(\text{kJ mol}^{-1})$	$\Delta_f G^{\ominus}/(\text{kJ mol}^{-1})$	$S_m^{\ominus}/(\text{J K}^{-1}\text{ mol}^{-1})$†	$C_{p,m}^{\ominus}/(\text{J K}^{-1}\text{ mol}^{-1})$
Potassium					
K(s)	39.10	0	0	64.18	29.58
K(g)	39.10	+89.24	+60.59	160.336	20.786
K$^+$(g)	39.10	+514.26			
K$^+$(aq)	39.10	−252.38	−283.27	+102.5	21.8
KOH(s)	56.11	−424.76	−379.08	78.9	64.9
KF(s)	58.10	−576.27	−537.75	66.57	49.04
KCl(s)	74.56	−436.75	−409.14	82.59	51.30
KBr(s)	119.01	−393.80	−380.66	95.90	52.30
KI(s)	166.01	−327.90	−324.89	106.32	52.93
Silicon					
Si(s)	28.09	0	0	18.83	20.00
Si(g)	28.09	+455.6	+411.3	167.97	22.25
SiO$_2$(s, α)	60.09	−910.94	−856.64	41.84	44.43
Silver					
Ag(s)	107.87	0	0	42.55	25.351
Ag(g)	107.87	+284.55	+245.65	173.00	20.79
Ag$^+$(aq)	107.87	+105.58	+77.11	+72.68	21.8
AgBr(s)	187.78	−100.37	−96.90	107.1	52.38
AgCl(s)	143.32	−127.07	−109.79	96.2	50.79
Ag$_2$O(s)	231.74	−31.05	−11.20	121.3	65.86
AgNO$_3$(s)	169.88	−129.39	−33.41	140.92	93.05
Sodium					
Na(s)	22.99	0	0	51.21	28.24
Na(g)	22.99	+107.32	+76.76	153.71	20.79
Na$^+$(aq)	22.99	−240.12	−261.91	59.0	46.4
NaOH(s)	40.00	−425.61	−379.49	64.46	59.54
NaCl(s)	58.44	−411.15	−384.14	72.13	50.50
NaBr(s)	102.90	−361.06	−348.98	86.82	51.38
NaI(s)	149.89	−287.78	−286.06	98.53	52.09
Sulfur					
S(s, α) (rhombic)	32.06	0	0	31.80	22.64
S(s, β) (monoclinic)	32.06	+0.33	+0.1	32.6	23.6
S(g)	32.06	+278.81	+238.25	167.82	23.673
S$_2$(g)	64.13	+128.37	+79.30	228.18	32.47
S^{2-}(aq)	32.06	+33.1	+85.8	−14.6	
SO$_2$(g)	64.06	−296.83	−300.19	248.22	39.87
SO$_3$(g)	80.06	−395.72	−371.06	256.76	50.67
H$_2$SO$_4$(l)	98.08	−813.99	−690.00	156.90	138.9
H$_2$SO$_4$(aq)	98.08	−909.27	−744.53	20.1	−293
SO$_4^{2-}$(aq)	96.06	−909.27	−744.53	+20.1	−293
HSO$_4^-$(aq)	97.07	−887.34	−755.91	+131.8	−84

Table 2.6 (Continued)

	$M/(\text{g mol}^{-1})$	$\Delta_f H^{\ominus}/(\text{kJ mol}^{-1})$	$\Delta_f G^{\ominus}/(\text{kJ mol}^{-1})$	$S_m^{\ominus}/(\text{J K}^{-1} \text{mol}^{-1})\dagger$	$C_{p,m}^{\ominus}/(\text{J K}^{-1} \text{mol}^{-1})$
Sulfur (Continued)					
$H_2S(g)$	34.08	−20.63	−33.56	205.79	34.23
$H_2S(aq)$	34.08	−39.7	−27.83	121	
$HS^-(aq)$	33.072	−17.6	+12.08	+62.08	
$SF_6(g)$	146.05	−1209	−1105.3	291.82	97.28
Tin					
$Sn(s, \beta)$	118.69	0	0	51.55	26.99
$Sn(g)$	118.69	+302.1	+267.3	168.49	20.26
$Sn^{2+}(aq)$	118.69	−8.8	−27.2	−17	
$SnO(s)$	134.69	−285.8	−256.9	56.5	44.31
$SnO_2(s)$	150.69	−580.7	−519.6	52.3	52.59
Xenon					
$Xe(g)$	131.30	0	0	169.68	20.786
Zinc					
$Zn(s)$	65.37	0	0	41.63	25.40
$Zn(g)$	65.37	+130.73	+95.14	160.98	20.79
$Zn^{2+}(aq)$	65.37	−153.89	−147.06	−112.1	46
$ZnO(s)$	81.37	−348.28	−318.30	43.64	40.25

Source: NBS; † Standard entropies of ions may be either positive or negative because the values are relative to the entropy of the hydrogen ion.

Table 2.6a Standard enthalpies of hydration at infinite dilution, $\Delta_{hyd}H^{\ominus}/(\text{kJ mol}^{-1})$

	Li^+	Na^+	K^+	Rb^+	Cs^+
F^-	−1026	−911	−828	−806	−782
Cl^-	−884	−783	−685	−664	−640
Br^-	−856	−742	−658	−637	−613
I^-	−815	−701	−617	−596	−572

Entries refer to $X^+(g) + Y^-(g) \rightarrow X^+(aq) + Y^-(aq)$.
Data: Principally J.O'M. Bockris and A.K.N. Reddy, *Modern electrochemistry*, Vol. 1. Plenum Press, New York (1970).

Table 2.6b Standard ion hydration enthalpies, $\Delta_{hyd}H^{\ominus}/(\text{kJ mol}^{-1})$ at 298 K

Cations

H^+	(−1090)	Ag^+	−464	Mg^{2+}	−1920
Li^+	−520	NH_4^+	−301	Ca^{2+}	−1650
Na^+	−405			Sr^{2+}	−1480
K^+	−321			Ba^{2+}	−1360
Rb^+	−300			Fe^{2+}	−1950
Cs^+	−277			Cu^{2+}	−2100
				Zn^{2+}	−2050
				Al^{3+}	−4690
				Fe^{3+}	−4430

Anions

OH^-	−460						
F^-	−506	Cl^-	−364	Br^-	−337	I^-	−296

Entries refer to $X^{\pm}(g) \rightarrow X^{\pm}(aq)$ based on $H^+(g) \rightarrow H^+(aq)$;
$\Delta H^{\ominus} = -1090 \text{ kJ mol}^{-1}$.
Data: Principally J.O'M. Bockris and A.K.N. Reddy, *Modern electrochemistry*, Vol. 1. Plenum Press, New York (1970).

Table 2.7 Benson thermochemical groups

Group	$\Delta_f H^{\ominus}/(\text{kJ mol}^{-1})$	$S_m^{\ominus}/(\text{J K}^{-1} \text{mol}^{-1})$	$C_{p,m}^{\ominus}/(\text{J K}^{-1} \text{mol}^{-1})$
C(H)$_3$(C)	−42.17	127.2	25.9
C(H)$_2$(C)$_2$	−20.7	39.4	22.8
C(H)(C)$_3$	−6.19	−50.50	18.7
C(C)$_4$	+8.16	−146.9	18.2
C(Cl)(H)$_2$(C)	−65.7	158	37
C(Br)(H)$_2$(C)	−22	169	38
C(I)(H)$_2$(C)	+37		
C(Cl)(H)(C)$_2$	−60.2	74.1	36
C(Br)(H)(C)$_2$	−9.6	84.9	
C(Cl)(H)(C)$_3$	−53.1	−32	36
C(Br)(H)(C)$_3$		−18	39
C(Cl)$_3$(C)		210	66.1
[O—(C)(H)] + [C—(O)(H)$_3$]	−48.1	59.5	10.5
[O—(C)(C)] + 2[C—(O)(H)$_3$]	−45.3	69.4	15.7
[C—(O)(C)(H)$_2$] + 2[C—(O)(H)$_3$]	+2.0	−20.4	−1.1

Data: S.W. Benson, *Thermochemical kinetics*. McGraw-Hill, New York (1976).

Table 3.1 Expansion coefficients, α, and isothermal compressibilities, κ_T

	$\alpha/(10^{-4} \text{K}^{-1})$	$\kappa_T/(10^{-6} \text{atm}^{-1})$
Liquids		
Benzene	12.4	92.1
Carbon tetrachloride	12.4	90.5
Ethanol	11.2	76.8
Mercury	1.82	38.7
Water	2.1	49.6
Solids		
Copper	0.501	0.735
Diamond	0.030	0.187
Iron	0.354	0.589
Lead	0.861	2.21

The values refer to 20°C
Data: AIP(α), KL(κ_T).

Table 3.2 Inversion temperatures, normal freezing and boiling points, and Joule–Thomson coefficients at 1 atm and 298 K

	T_I/K	T_f/K	T_b/K	$\mu_{JT}/(\text{K atm}^{-1})$
Air	603			0.189 at 50°C
Argon	723	83.8	87.3	
Carbon dioxide	1500	194.7s		1.11 at 300 K
Helium	40		4.22	−0.062
Hydrogen	202	14.0	20.3	−0.03
Krypton	1090	116.6	120.8	
Methane	968	90.6	111.6	
Neon	231	24.5	27.1	
Nitrogen	621	63.3	77.4	0.27
Oxygen	764	54.8	90.2	0.31

s: sublimes
Data: AIP, JL, and M.W. Zemansky, *Heat and thermodynamics*. McGraw–Hill, New York (1957).

Table 4.1 Standard entropies (and temperatures) of phase transitions, $\Delta_{trs}S^{\ominus}/(J\ K^{-1}\ mol^{-1})$

	Fusion (at T_f)	Vaporization (at T_b)
Ar	14.17 (at 83.8 K)	74.53 (at 87.3 K)
Br_2	39.76 (at 265.9 K)	88.61 (at 332.4 K)
C_6H_6	38.00 (at 278.6 K)	87.19 (at 353.2 K)
CH_3COOH	40.4 (at 289.8 K)	61.9 (at 391.4 K)
CH_3OH	18.03 (at 175.2 K)	104.6 (at 337.2 K)
Cl_2	37.22 (at 172.1 K)	85.38 (at 239.0 K)
H_2	8.38 (at 14.0 K)	44.96 (at 20.38 K)
H_2O	22.00 (at 273.2 K)	109.0 (at 373.2 K)
H_2S	12.67 (at 187.6 K)	87.75 (at 212.0 K)
He	4.8 (at 1.8 K and 30 bar)	19.9 (at 4.22 K)
N_2	11.39 (at 63.2 K)	75.22 (at 77.4 K)
NH_3	28.93 (at 195.4 K)	97.41 (at 239.73 K)
O_2	8.17 (at 54.4 K)	75.63 (at 90.2 K)

Data: AIP

Table 4.2 Standard entropies of vaporization of liquids at their normal boiling point

	$\Delta_{vap}H^{\ominus}/(kJ\ mol^{-1})$	$\theta_b/°C$	$\Delta_{vap}S^{\ominus}/(J\ K^{-1}\ mol^{-1})$
Benzene	30.8	80.1	+87.2
Carbon disulfide	26.74	46.25	+83.7
Carbon tetrachloride	30.00	76.7	+85.8
Cyclohexane	30.1	80.7	+85.1
Decane	38.75	174	+86.7
Dimethyl ether	21.51	−23	+86
Ethanol	38.6	78.3	+110.0
Hydrogen sulfide	18.7	−60.4	+87.9
Mercury	59.3	356.6	+94.2
Methane	8.18	−161.5	+73.2
Methanol	35.21	65.0	+104.1
Water	40.7	100.0	+109.1

Data: JL

Table 4.3 Standard Third-Law entropies at 298 K: see Tables 2.5 and 2.6

Table 4.4 Standard Gibbs energies of formation at 298 K: see Tables 2.5 and 2.6

Table 5.2 The fugacity coefficient of nitrogen at 273 K

p/atm	ϕ	p/atm	ϕ
1	0.999 55	300	1.0055
10	0.9956	400	1.062
50	0.9812	600	1.239
100	0.9703	800	1.495
150	0.9672	1000	1.839
200	0.9721		

Data: LR

Table 6.1 Surface tensions of liquids at 293 K

	$\gamma/(\text{mN m}^{-1})$
Benzene	28.88
Carbon tetrachloride	27.0
Ethanol	22.8
Hexane	18.4
Mercury	472
Methanol	22.6
Water	72.75
	72.0 at 25°C
	58.0 at 100°C

Data: KL

Table 7.1 Henry's law constants for gases at 298 K, $K/(10 \text{ MPa})$

	Water	Benzene
CH_4	4.19	0.0569
CO_2	0.167	0.0114
H_2	7.12	0.367
N_2	8.68	0.239
O_2	4.40	

Data: R.J. Silbey and R.A. Alberty, *Physical chemistry*, Wiley, New York (2001).

Table 7.2 Cryoscopic and ebullioscopic constants

	$K_f/(\text{K kg mol}^{-1})$	$K_b/(\text{K kg mol}^{-1})$
Acetic acid	3.90	3.07
Benzene	5.12	2.53
Camphor	40	
Carbon disulfide	3.8	2.37
Carbon tetrachloride	30	4.95
Naphthalene	6.94	5.8
Phenol	7.27	3.04
Water	1.86	0.51

Data: KL

Table 9.1 Acidity constants for aqueous solutions at 298 K. (a) In order of acid strength

Acid	HA	A$^-$	K_a	pK_a
Hydriodic	HI	I$^-$	10^{11}	-11
Hydrobromic	HBr	Br$^-$	10^9	-9
Hydrochloric	HCl	Cl$^-$	10^7	-7
Sulfuric	H$_2$SO$_4$	HSO$_4^-$	10^2	-2
Perchloric*	HClO$_4$	ClO$_4^-$	4.0×10^1	-1.6
Hydronium ion	H$_3$O$^+$	H$_2$O	1	0.0
Oxalic	(COOH)$_2$	HOOCCO$_2^-$	5.6×10^{-2}	1.25
Sulfurous	H$_2$SO$_3$	HSO$_3^-$	1.4×10^{-2}	1.85
Hydrogensulfate ion	HSO$_4^-$	SO$_4^{2-}$	1.0×10^{-2}	1.99
Phosphoric	H$_3$PO$_4$	H$_2$PO$_4^-$	6.9×10^{-3}	2.16
Glycinium ion	$^+$NH$_3$CH$_2$COOH	NH$_2$CH$_2$COOH	4.5×10^{-3}	2.35
Hydrofluoric	HF	F$^-$	6.3×10^{-4}	3.20
Formic	HCOOH	HCO$_2^-$	1.8×10^{-4}	3.75
Hydrogenoxalate ion	HOOCCO$_2^-$	C$_2$O$_4^{2-}$	1.5×10^{-5}	3.81
Lactic	CH$_3$CH(OH)COOH	CH$_3$CH(OH)CO$_2^-$	1.4×10^{-4}	3.86
Acetic (ethanoic)	CH$_3$COOH	CH$_3$CO$_2^-$	1.4×10^{-5}	4.76
Butanoic	CH$_3$CH$_2$CH$_2$COOH	CH$_3$CH$_2$CH$_2$CO$_2^-$	1.5×10^{-5}	4.83
Propanoic	CH$_3$CH$_2$COOH	CH$_3$CH$_2$CO$_2^-$	1.4×10^{-5}	4.87
Anilinium ion	C$_6$H$_5$NH$_3^+$	C$_6$H$_5$NH$_2$	1.3×10^{-5}	4.87
Pyridinium ion	C$_5$H$_5$NH$^+$	C$_6$H$_5$N	5.9×10^{-6}	5.23
Carbonic	H$_2$CO$_3$	HCO$_3^-$	4.5×10^{-7}	6.35
Hydrosulfuric	H$_2$S	HS$^-$	8.9×10^{-8}	7.05
Dihydrogenphosphate ion	H$_2$PO$_4^-$	HPO$_4^{2-}$	6.2×10^{-8}	7.21
Hypochlorous	HClO	ClO$^-$	4.0×10^{-8}	7.40
Hydrazinium ion	NH$_2$NH$_3^+$	NH$_2$NH$_2$	8×10^{-9}	8.1
Hypobromous	HBrO	BrO$^-$	2.8×10^{-9}	8.55
Hydrocyanic	HCN	CN$^-$	6.2×10^{-10}	9.21
Ammonium ion	NH$_4^+$	NH$_3$	5.6×10^{-10}	9.25
Boric*	B(OH)$_3$	B(OH)$_4^-$	5.4×10^{-10}	9.27
Trimethylammonium ion	(CH$_3$)$_3$NH$^+$	(CH$_3$)$_3$N	1.6×10^{-10}	9.80
Phenol	C$_6$H$_5$OH	C$_6$H$_5$O$^-$	1.0×10^{-10}	9.99
Hydrogencarbonate ion	HCO$_3^-$	CO$_3^{2-}$	4.8×10^{-11}	10.33
Hypoiodous	HIO	IO$^-$	3×10^{-11}	10.5
Ethylammonium ion	CH$_3$CH$_2$NH$_3^+$	CH$_3$CH$_2$NH$_2$	2.2×10^{-11}	10.65
Methylammonium ion	CH$_3$NH$_3^+$	CH$_3$NH$_2$	2.2×10^{-11}	10.66
Dimethylammonium ion	(CH$_3$)$_2$NH$_2^+$	(CH$_3$)$_2$NH	1.9×10^{-11}	10.73
Triethylammonium ion	(CH$_3$CH$_2$)$_3$NH$^+$	(CH$_3$CH$_2$)$_3$N	1.8×10^{-11}	10.75
Diethylammonium ion	(CH$_3$CH$_2$)$_2$NH$_2^+$	(CH$_3$CH$_2$)$_2$NH	1.4×10^{-11}	10.84
Hydrogenarsenate ion	HAsO$_4^{2-}$	AsO$_4^{3-}$	5.1×10^{-12}	11.29
Hydrogenphosphate ion	HPO$_4^{2-}$	PO$_4^{3-}$	4.8×10^{-13}	12.32
Hydrogensulfide ion	HS$^-$	S^{2-}	1.0×10^{-19}	19.00

* At 293 K.

Table 9.1 Acidity constants for aqueous solutions at 298 K. (b) In alphabetical order

Acid	HA	A$^-$	K_a	pK_a
Acetic (ethanoic)	CH_3COOH	$CH_3CO_2^-$	1.4×10^{-5}	4.76
Ammonium ion	NH_4^+	NH_3	5.6×10^{-10}	9.25
Anilinium ion	$C_6H_5NH_3^+$	$C_6H_5NH_2$	1.3×10^{-5}	4.87
Boric*	$B(OH)_3$	$B(OH)_4^-$	5.4×10^{-10}	9.27
Butanoic	$CH_3CH_2CH_2COOH$	$CH_3CH_2CH_2CO_2^-$	1.5×10^{-5}	4.83
Carbonic	H_2CO_3	HCO_3^-	4.5×10^{-7}	6.35
Diethylammonium ion	$(CH_3CH_2)_2NH_2^+$	$(CH_3CH_2)_2NH$	1.4×10^{-11}	10.84
Dihydrogenphosphate ion	$H_2PO_4^-$	HPO_4^{2-}	6.2×10^{-8}	7.21
Dimethylammonium ion	$(CH_3)_2NH_2^+$	$(CH_3)_2NH$	1.9×10^{-11}	10.73
Ethylammonium ion	$CH_3CH_2NH_3^+$	$CH_3CH_2NH_2$	2.2×10^{-11}	10.65
Formic	$HCOOH$	HCO_2^-	1.8×10^{-4}	3.75
Glycinium ion	$^+NH_3CH_2COOH$	NH_2CH_2COOH	4.5×10^{-3}	2.35
Hydrazinium ion	$NH_2NH_3^+$	NH_2NH_2	8×10^{-9}	8.1
Hydriodic	HI	I^-	10^{11}	−11
Hydrobromic	HBr	Br^-	10^9	−9
Hydrochloric	HCl	Cl^-	10^7	−7
Hydrocyanic	HCN	CN^-	6.2×10^{-10}	9.21
Hydrofluoric	HF	F^-	6.3×10^{-4}	3.20
Hydrogenarsenate ion	$HAsO_4^{2-}$	AsO_4^{3-}	5.1×10^{-12}	11.29
Hydrogencarbonate ion	HCO_3^-	CO_3^{2-}	4.8×10^{-11}	10.33
Hydrogenoxalate ion	$HOOCCO_2^-$	$C_2O_4^{2-}$	1.5×10^{-5}	3.81
Hydrogenphosphate ion	HPO_4^{2-}	PO_4^{3-}	4.8×10^{-13}	12.32
Hydrogensulfate ion	HSO_4^-	SO_4^{2-}	1.0×10^{-2}	1.99
Hydrogensulfide ion	HS^-	S^{2-}	1.0×10^{-19}	19.00
Hydronium ion	H_3O^+	H_2O	1	0.0
Hydrosulfuric	H_2S	HS^-	8.9×10^{-8}	7.05
Hypobromous	$HBrO$	BrO^-	2.8×10^{-9}	8.55
Hypochlorous	$HClO$	ClO^-	4.0×10^{-8}	7.40
Hypoiodous	HIO	IO^-	3×10^{-11}	10.5
Lactic	$CH_3CH(OH)COOH$	$CH_3CH(OH)CO_2^-$	1.4×10^{-4}	3.86
Methylammonium ion	$CH_3NH_3^+$	CH_3NH_2	2.2×10^{-11}	10.66
Oxalic	$(COOH)_2$	$HOOCCO_2^-$	5.6×10^{-2}	1.25
Perchloric*	$HClO_4$	ClO_4^-	4.0×10^1	−1.6
Phenol	C_6H_5OH	$C_6H_5O^-$	1.0×10^{-10}	9.99
Phosphoric	H_3PO_4	$H_2PO_4^-$	6.9×10^{-3}	2.16
Propanoic	CH_3CH_2COOH	$CH_3CH_2CO_2^-$	1.4×10^{-5}	4.87
Pyridinium ion	$C_5H_5NH^+$	C_6H_5N	5.9×10^{-6}	5.23
Sulfuric	H_2SO_4	HSO_4^-	10^2	−2
Sulfurous	H_2SO_3	HSO_3^-	1.4×10^{-2}	1.85
Triethylammonium ion	$(CH_3CH_2)_3NH^+$	$(CH_3CH_2)_3N$	1.8×10^{-11}	10.75
Trimethylammonium ion	$(CH_3)_3NH^+$	$(CH_3)_3N$	1.6×10^{-10}	9.80

* At 293 K.

Table 10.1 Standard thermodynamic functions of ions in aqueous solution at 298 K. See Table 2.6

Table 10.2 Relative permittivities (dielectric constants) at 293 K

Nonpolar molecules		Polar molecules	
Methane (at −173°C)	1.655	Water	78.54 (at 298 K)
			80.10
Carbon tetrachloride	2.238	Ammonia	16.9 (at 298 K)
			22.4 at −33°C
Cyclohexane	2.024	Hydrogen sulfide	9.26 at −85°C
			5.93 (at 283 K)
Benzene	2.283	Methanol	33.0
		Ethanol	25.3
		Nitrobenzene	35.6

Data: HCP

Table 10.3 Standard molar entropies of ions in aqueous solution at 298 K. See Table 2.6

Table 10.5 Mean activity coefficients in water at 298 K

$b/b^{\ominus}$	HCl	KCl	$CaCl_2$	H_2SO_4	$LaCl_3$	$In_2(SO_4)_3$
0.001	0.966	0.966	0.888	0.830	0.790	
0.005	0.929	0.927	0.789	0.639	0.636	0.16
0.01	0.905	0.902	0.732	0.544	0.560	0.11
0.05	0.830	0.816	0.584	0.340	0.388	0.035
0.10	0.798	0.770	0.524	0.266	0.356	0.025
0.50	0.769	0.652	0.510	0.155	0.303	0.014
1.00	0.811	0.607	0.725	0.131	0.387	
2.00	1.011	0.577	1.554	0.125	0.954	

Data: RS, HCP, and S. Glasstone, *Introduction to electrochemistry*. Van Nostrand (1942).

Table 10.7 Standard potentials at 298 K. (a) In electrochemical order

Reduction half-reaction	$E^{\ominus}/V$	Reduction half-reaction	$E^{\ominus}/V$
Strongly oxidizing		$Cu^{2+} + e^- \longrightarrow Cu^+$	+0.16
$H_4XeO_6 + 2H^+ + 2e^- \longrightarrow XeO_3 + 3H_2O$	+3.0	$Sn^{4+} + 2e^- \longrightarrow Sn^{2+}$	+0.15
$F_2 + 2e^- \longrightarrow 2F^-$	+2.87	$AgBr + e^- \longrightarrow Ag + Br^-$	+0.07
$O_3 + 2H^+ + 2e^- \longrightarrow O_2 + H_2O$	+2.07	$Ti^{4+} + e^- \longrightarrow Ti^{3+}$	0.00
$S_2O_8^{2-} + 2e^- \longrightarrow 2SO_4^{2-}$	+2.05	$2H^+ + 2e^- \longrightarrow H_2$	0, by definition
$Ag^{2+} + e^- \longrightarrow Ag^+$	+1.98	$Fe^{3+} + 3e^- \longrightarrow Fe$	−0.04
$Co^{3+} + e^- \longrightarrow Co^{2+}$	+1.81	$O_2 + H_2O + 2e^- \longrightarrow HO_2^- + OH^-$	−0.08
$H_2O_2 + 2H^+ + 2e^- \longrightarrow 2H_2O$	+1.78	$Pb^{2+} + 2e^- \longrightarrow Pb$	−0.13
$Au^+ + e^- \longrightarrow Au$	+1.69	$In^+ + e^- \longrightarrow In$	−0.14
$Pb^{4+} + 2e^- \longrightarrow Pb^{2+}$	+1.67	$Sn^{2+} + 2e^- \longrightarrow Sn$	−0.14
$2HClO + 2H^+ + 2e^- \longrightarrow Cl_2 + 2H_2O$	+1.63	$AgI + e^- \longrightarrow Ag + I^-$	−0.15
$Ce^{4+} + e^- \longrightarrow Ce^{3+}$	+1.61	$Ni^{2+} + 2e^- \longrightarrow Ni$	−0.23
$2HBrO + 2H^+ + 2e^- \longrightarrow Br_2 + 2H_2O$	+1.60	$Co^{2+} + 2e^- \longrightarrow Co$	−0.28
$MnO_4^- + 8H^+ + 5e^- \longrightarrow Mn^{2+} + 4H_2O$	+1.51	$In^{3+} + 3e^- \longrightarrow In$	−0.34
$Mn^{3+} + e^- \longrightarrow Mn^{2+}$	+1.51	$Tl^+ + e^- \longrightarrow Tl$	−0.34
$Au^{3+} + 3e^- \longrightarrow Au$	+1.40	$PbSO_4 + 2e^- \longrightarrow Pb + SO_4^{2-}$	−0.36
$Cl_2 + 2e^- \longrightarrow 2Cl^-$	+1.36	$Ti^{3+} + e^- \longrightarrow Ti^{2+}$	−0.37
$Cr_2O_7^{2-} + 14H^+ + 6e^- \longrightarrow 2Cr^{3+} + 7H_2O$	+1.33	$Cd^{2+} + 2e^- \longrightarrow Cd$	−0.40
$O_3 + H_2O + 2e^- \longrightarrow O_2 + 2OH^-$	+1.24	$In^{2+} + e^- \longrightarrow In^+$	−0.40
$O_2 + 4H^+ + 4e^- \longrightarrow 2H_2O$	+1.23	$Cr^{3+} + e^- \longrightarrow Cr^{2+}$	−0.41
$ClO_4^- + 2H^+ + 2e^- \longrightarrow ClO_3^- + H_2O$	+1.23	$Fe^{2+} + 2e^- \longrightarrow Fe$	−0.44
$MnO_2 + 4H^+ + 2e^- \longrightarrow Mn^{2+} + 2H_2O$	+1.23	$In^{3+} + 2e^- \longrightarrow In^+$	−0.44
$Br_2 + 2e^- \longrightarrow 2Br^-$	+1.09	$S + 2e^- \longrightarrow S^{2-}$	−0.48
$Pu^{4+} + e^- \longrightarrow Pu^{3+}$	+0.97	$In^{3+} + e^- \longrightarrow In^{2+}$	−0.49
$NO_3^- + 4H^+ + 3e^- \longrightarrow NO + 2H_2O$	+0.96	$U^{4+} + e^- \longrightarrow U^{3+}$	−0.61
$2Hg^{2+} + 2e^- \longrightarrow Hg_2^{2+}$	+0.92	$Cr^{3+} + 3e^- \longrightarrow Cr$	−0.74
$ClO^- + H_2O + 2e^- \longrightarrow Cl^- + 2OH^-$	+0.89	$Zn^{2+} + 2e^- \longrightarrow Zn$	−0.76
$Hg^{2+} + 2e^- \longrightarrow Hg$	+0.86	$Cd(OH)_2 + 2e^- \longrightarrow Cd + 2OH^-$	−0.81
$NO_3^- + 2H^+ + e^- \longrightarrow NO_2 + H_2O$	+0.80	$2H_2O + 2e^- \longrightarrow H_2 + 2OH^-$	−0.83
$Ag^+ + e^- \longrightarrow Ag$	+0.80	$Cr^{2+} + 2e^- \longrightarrow Cr$	−0.91
$Hg_2^{2+} + 2e^- \longrightarrow 2Hg$	+0.79	$Mn^{2+} + 2e^- \longrightarrow Mn$	−1.18
$Fe^{3+} + e^- \longrightarrow Fe^{2+}$	+0.77	$V^{2+} + 2e^- \longrightarrow V$	−1.19
$BrO^- + H_2O + 2e^- \longrightarrow Br^- + 2OH^-$	+0.76	$Ti^{2+} + 2e^- \longrightarrow Ti$	−1.63
$Hg_2SO_4 + 2e^- \longrightarrow 2Hg + SO_4^{2-}$	+0.62	$Al^{3+} + 3e^- \longrightarrow Al$	−1.66
$MnO_4^{2-} + 2H_2O + 2e^- \longrightarrow MnO_2 + 4OH^-$	+0.60	$U^{3+} + 3e^- \longrightarrow U$	−1.79
$MnO_4^- + e^- \longrightarrow MnO_4^{2-}$	+0.56	$Sc^{3+} + 3e^- \longrightarrow Sc$	−2.09
$I_2 + 2e^- \longrightarrow 2I^-$	+0.54	$Mg^{2+} + 2e^- \longrightarrow Mg$	−2.36
$Cu^+ + e^- \longrightarrow Cu$	+0.52	$Ce^{3+} + 3e^- \longrightarrow Ce$	−2.48
$I_3^- + 2e^- \longrightarrow 3I^-$	+0.53	$La^{3+} + 3e^- \longrightarrow La$	−2.52
$NiOOH + H_2O + e^- \longrightarrow Ni(OH)_2 + OH^-$	+0.49	$Na^+ + e^- \longrightarrow Na$	−2.71
$Ag_2CrO_4 + 2e^- \longrightarrow 2Ag + CrO_4^{2-}$	+0.45	$Ca^{2+} + 2e^- \longrightarrow Ca$	−2.87
$O_2 + 2H_2O + 4e^- \longrightarrow 4OH^-$	+0.40	$Sr^{2+} + 2e^- \longrightarrow Sr$	−2.89
$ClO_4^- + H_2O + 2e^- \longrightarrow ClO_3^- + 2OH^-$	+0.36	$Ba^{2+} + 2e^- \longrightarrow Ba$	−2.91
$[Fe(CN)_6]^{3-} + e^- \longrightarrow [Fe(CN)_6]^{4-}$	+0.36	$Ra^{2+} + 2e^- \longrightarrow Ra$	−2.92
$Cu^{2+} + 2e^- \longrightarrow Cu$	+0.34	$Cs^+ + e^- \longrightarrow Cs$	−2.92
$Hg_2Cl_2 + 2e^- \longrightarrow 2Hg + 2Cl^-$	+0.27	$Rb^+ + e^- \longrightarrow Rb$	−2.93
$AgCl + e^- \longrightarrow Ag + Cl^-$	+0.22	$K^+ + e^- \longrightarrow K$	−2.93
$Bi^{3+} + 3e^- \longrightarrow Bi$	+0.20	$Li^+ + e^- \longrightarrow Li$	−3.05

Table 10.7 Standard electrode potentials at 298 K. (b) In alphabetical order

Reduction half-reaction	$E^{\ominus}/V$	Reduction half-reaction	$E^{\ominus}/V$
$Ag^+ + e^- \longrightarrow Ag$	+0.80	$I_2 + 2e^- \longrightarrow 2I^-$	+0.54
$Ag^{2+} + e^- \longrightarrow Ag^+$	+1.98	$I_3^- + 2e^- \longrightarrow 3I^-$	+0.53
$AgBr + e^- \longrightarrow Ag + Br^-$	+0.0713	$In^+ + e^- \longrightarrow In$	−0.14
$AgCl + e^- \longrightarrow Ag + Cl^-$	+0.22	$In^{2+} + e^- \longrightarrow In^+$	−0.40
$Ag_2CrO_4 + 2e^- \longrightarrow 2Ag + CrO_4^{2-}$	+0.45	$In^{3+} + 2e^- \longrightarrow In^+$	−0.44
$AgF + e^- \longrightarrow Ag + F^-$	+0.78	$In^{3+} + 3e^- \longrightarrow In$	−0.34
$AgI + e^- \longrightarrow Ag + I^-$	−0.15	$In^{3+} + e^- \longrightarrow In^{2+}$	−0.49
$Al^{3+} + 3e^- \longrightarrow Al$	−1.66	$K^+ + e^- \longrightarrow K$	−2.93
$Au^+ + e^- \longrightarrow Au$	+1.69	$La^{3+} + 3e^- \longrightarrow La$	−2.52
$Au^{3+} + 3e^- \longrightarrow Au$	+1.40	$Li^+ + e^- \longrightarrow Li$	−3.05
$Ba^{2+} + 2e^- \longrightarrow Ba$	+2.91	$Mg^{2+} + 2e^- \longrightarrow Mg$	−2.36
$Be^{2+} + 2e^- \longrightarrow Be$	−1.85	$Mn^{2+} + 2e^- \longrightarrow Mn$	−1.18
$Bi^{3+} + 3e^- \longrightarrow Bi$	+0.20	$Mn^{3+} + e^- \longrightarrow Mn^{2+}$	+1.51
$Br_2 + 2e^- \longrightarrow 2Br^-$	+1.09	$MnO_2 + 4H^+ + 2e^- \longrightarrow Mn^{2+} + 2H_2O$	+1.23
$BrO^- + H_2O + 2e^- \longrightarrow Br^- + 2OH^-$	+0.76	$MnO_4^- + 8H^+ + 5e^- \longrightarrow Mn^{2+} + 4H_2O$	+1.51
$Ca^{2+} + 2e^- \longrightarrow Ca$	−2.87	$MnO_4^- + e^- \longrightarrow MnO_4^{2-}$	+0.56
$Cd(OH)_2 + 2e^- \longrightarrow Cd + 2OH^-$	−0.81	$MnO_4^{2-} + 2H_2O + 2e^- \longrightarrow MnO_2 + 4OH^-$	+0.60
$Cd^{2+} + 2e^- \longrightarrow Cd$	−0.40	$Na^+ + e^- \longrightarrow Na$	−2.71
$Ce^{3+} + 3e^- \longrightarrow Ce$	−2.48	$Ni^{2+} + 2e^- \longrightarrow Ni$	−0.23
$Ce^{4+} + e^- \longrightarrow Ce^{3+}$	+1.61	$NiOOH + H_2O + e^- \longrightarrow Ni(OH)_2 + OH^-$	+0.49
$Cl_2 + 2e^- \longrightarrow 2Cl^-$	+1.36	$NO_3^- + 2H^+ + e^- \longrightarrow NO_2 + H_2O$	−0.80
$ClO^- + H_2O + 2e^- \longrightarrow Cl^- + 2OH^-$	+0.89	$NO_3^- + 4H^+ + 3e^- \longrightarrow NO + 2H_2O$	+0.96
$ClO_4^- + 2H^+ + 2e^- \longrightarrow ClO_3^- + H_2O$	+1.23	$NO_3^- + H_2O + 2e^- \longrightarrow NO_2^- + 2OH^-$	+0.10
$ClO_4^- + H_2O + 2e^- \longrightarrow ClO_3^- + 2OH^-$	+0.36	$O_2 + 2H_2O + 4e^- \longrightarrow 4OH^-$	+0.40
$Co^{2+} + 2e^- \longrightarrow Co$	−0.28	$O_2 + 4H^+ + 4e^- \longrightarrow 2H_2O$	+1.23
$Co^{3+} + e^- \longrightarrow Co^{2+}$	+1.81	$O_2 + e^- \longrightarrow O_2^-$	−0.56
$Cr^{2+} + 2e^- \longrightarrow Cr$	−0.91	$O_2 + H_2O + 2e^- \longrightarrow HO_2^- + OH^-$	−0.08
$Cr_2O_7^{2-} + 14H^+ + 6e^- \longrightarrow 2Cr^{3+} + 7H_2O$	+1.33	$O_3 + 2H^+ + 2e^- \longrightarrow O_2 + H_2O$	+2.07
$Cr^{3+} + 3e^- \longrightarrow Cr$	−0.74	$O_3 + H_2O + 2e^- \longrightarrow O_2 + 2OH^-$	+1.24
$Cr^{3+} + e^- \longrightarrow Cr^{2+}$	−0.41	$Pb^{2+} + 2e^- \longrightarrow Pb$	−0.13
$Cs^+ + e^- \longrightarrow Cs$	−2.92	$Pb^{4+} + 2e^- \longrightarrow Pb^{2+}$	+1.67
$Cu^+ + e^- \longrightarrow Cu$	+0.52	$PbSO_4 + 2e^- \longrightarrow Pb + SO_4^{2-}$	−0.36
$Cu^{2+} + 2e^- \longrightarrow Cu$	+0.34	$Pt^{2+} + 2e^- \longrightarrow Pt$	+1.20
$Cu^{2+} + e^- \longrightarrow Cu^+$	+0.16	$Pu^{4+} + e^- \longrightarrow Pu^{3+}$	+0.97
$F_2 + 2e^- \longrightarrow 2F^-$	+2.87	$Ra^{2+} + 2e^- \longrightarrow Ra$	−2.92
$Fe^{2+} + 2e^- \longrightarrow Fe$	−0.44	$Rb^+ + e^- \longrightarrow Rb$	−2.93
$Fe^{3+} + 3e^- \longrightarrow Fe$	−0.04	$S + 2e^- \longrightarrow S^{2-}$	−0.48
$Fe^{3+} + e^- \longrightarrow Fe^{2+}$	+0.77	$S_2O_8^{2-} + 2e^- \longrightarrow 2SO_4^{2-}$	+2.05
$[Fe(CN)_6]^{3-} + e^- \longrightarrow [Fe(CN)_6]^{4-}$	+0.36	$Sc^{3+} + 3e^- \longrightarrow Sc$	−2.09
$2H^+ + 2e^- \longrightarrow H_2$	0, by definition	$Sn^{2+} + 2e^- \longrightarrow Sn$	−0.14
$2H_2O + 2e^- \longrightarrow H_2 + 2OH^-$	−0.83	$Sn^{4+} + 2e^- \longrightarrow Sn^{2+}$	+0.15
$2HBrO + 2H^+ + 2e^- \longrightarrow Br_2 + 2H_2O$	+1.60	$Sr^{2+} + 2e^- \longrightarrow Sr$	−2.89
$2HClO + 2H^+ + 2e^- \longrightarrow Cl_2 + 2H_2O$	+1.63	$Ti^{2+} + 2e^- \longrightarrow Ti$	−1.63
$H_2O_2 + 2H^+ + 2e^- \longrightarrow 2H_2O$	+1.78	$Ti^{3+} + e^- \longrightarrow Ti^{2+}$	−0.37
$H_4XeO_6 + 2H^+ + 2e^- \longrightarrow XeO_3 + 3H_2O$	+3.0	$Ti^{4+} + e^- \longrightarrow Ti^{3+}$	0.00
$Hg_2^{2+} + 2e^- \longrightarrow 2Hg$	+0.79	$Tl^+ + e^- \longrightarrow Tl$	−0.34
$Hg_2Cl_2 + 2e^- \longrightarrow 2Hg + 2Cl^-$	+0.27	$U^{3+} + 3e^- \longrightarrow U$	−1.79
$Hg^{2+} + 2e^- \longrightarrow Hg$	+0.86	$U^{4+} + e^- \longrightarrow U^{3+}$	−0.61
$2Hg^{2+} + 2e^- \longrightarrow Hg_2^{2+}$	+0.92	$V^{2+} + 2e^- \longrightarrow V$	−1.19
$Hg_2SO_4 + 2e^- \longrightarrow 2Hg + SO_4^{2-}$	+0.62	$V^{3+} + e^- \longrightarrow V^{2+}$	−0.26
		$Zn^{2+} + 2e^- \longrightarrow Zn$	−0.76

Table 12.2 The error function

z	erf z	z	erf z
0	0	0.45	0.475 48
0.01	0.011 28	0.50	0.520 50
0.02	0.022 56	0.55	0.563 32
0.03	0.033 84	0.60	0.603 86
0.04	0.045 11	0.65	0.642 03
0.05	0.056 37	0.70	0.677 80
0.06	0.067 62	0.75	0.711 16
0.07	0.078 86	0.80	0.742 10
0.08	0.090 08	0.85	0.770 67
0.09	0.101 28	0.90	0.796 91
0.10	0.112 46	0.95	0.820 89
0.15	0.168 00	1.00	0.842 70
0.20	0.222 70	1.20	0.910 31
0.25	0.276 32	1.40	0.952 28
0.30	0.328 63	1.60	0.976 35
0.35	0.379 38	1.80	0.989 09
0.40	0.428 39	2.00	0.995 32

Data: AS

Table 13.2 Screening constants for atoms; values of $Z_{eff} = Z - \sigma$ for neutral ground-state atoms

	H							He
$1s$	1							1.6875

	Li	Be	B	C	N	O	F	Ne
$1s$	2.6906	3.6848	4.6795	5.6727	6.6651	7.6579	8.6501	9.6421
$2s$	1.2792	1.9120	2.5762	3.2166	3.8474	4.4916	5.1276	5.7584
$2p$			2.4214	3.1358	3.8340	4.4532	5.1000	5.7584

	Na	Mg	Al	Si	P	S	Cl	Ar
$1s$	10.6259	11.6089	12.5910	13.5745	14.5578	15.5409	16.5239	17.5075
$2s$	6.5714	7.3920	8.3736	9.0200	9.8250	10.6288	11.4304	12.2304
$2p$	6.8018	7.8258	8.9634	9.9450	10.9612	11.9770	12.9932	14.0082
$3s$	2.5074	3.3075	4.1172	4.9032	5.6418	6.3669	7.0683	7.7568
$3p$			4.0656	4.2852	4.8864	5.4819	6.1161	6.7641

Data: E. Clementi and D.L. Raimondi, *Atomic screening constants from SCF functions*.
IBM Res. Note NJ-27 (1963). *J. chem. Phys* **38**, 2686 (1963).

Table 13.3 Ionization energies, $I/(\text{kJ mol}^{-1})$

H							He
1312.0							2372.3
							5250.4
Li	Be	B	C	N	O	F	Ne
513.3	899.4	800.6	1086.2	1402.3	1313.9	1681	2080.6
7298.0	1757.1	2427	2352	2856.1	3388.2	3374	3952.2
Na	Mg	Al	Si	P	S	Cl	Ar
495.8	737.7	577.4	786.5	1011.7	999.6	1251.1	1520.4
4562.4	1450.7	1816.6	1577.1	1903.2	2251	2297	2665.2
		2744.6		2912			
K	Ca	Ga	Ge	As	Se	Br	Kr
418.8	589.7	578.8	762.1	947.0	940.9	1139.9	1350.7
3051.4	1145	1979	1537	1798	2044	2104	2350
		2963	2735				
Rb	Sr	In	Sn	Sb	Te	I	Xe
403.0	549.5	558.3	708.6	833.7	869.2	1008.4	1170.4
2632	1064.2	1820.6	1411.8	1794	1795	1845.9	2046
		2704	2943.0	2443			
Cs	Ba	Tl	Pb	Bi	Po	At	Rn
375.5	502.8	589.3	715.5	703.2	812	930	1037
2420	965.1	1971.0	1450.4	1610			
		2878	3081.5	2466			

Data: E

Table 13.4 Electron affinities, $E_{\text{ea}}/(\text{kJ mol}^{-1})$

H							He
72.8							−21
Li	Be	B	C	N	O	F	Ne
59.8	≤0	23	122.5	−7	141	322	−29
					−844		
Na	Mg	Al	Si	P	S	Cl	Ar
52.9	≤0	44	133.6	71.7	200.4	348.7	−35
					−532		
K	Ca	Ga	Ge	As	Se	Br	Kr
48.3	2.37	36	116	77	195.0	324.5	−39
Rb	Sr	In	Sn	Sb	Te	I	Xe
46.9	5.03	34	121	101	190.2	295.3	−41
Cs	Ba	Tl	Pb	Bi	Po	At	Rn
45.5	13.95	30	35.2	101	186	270	−41

Data: E

Table 14.2 Bond lengths, R_e/pm

(a) Bond lengths in specific molecules

Br_2	228.3
Cl_2	198.75
CO	112.81
F_2	141.78
H_2^+	106
H_2	74.138
HBr	141.44
HCl	127.45
HF	91.680
HI	160.92
N_2	109.76
O_2	120.75

(b) Mean bond lengths from covalent radii*

H	37						
C	77(1) 67(2) 60(3)	N	74(1) 65(2)	O	66(1) 57(2)	F	64
Si	118	P	110	S	104(1) 95(2)	Cl	99
Ge	122	As	121	Se	104	Br	114
		Sb	141	Te	137	I	133

* Values are for single bonds except where indicated otherwise (values in parentheses). The length of an A—B covalent bond (of given order) is the sum of the corresponding covalent radii.

Table 14.3a Bond dissociation enthalpies, $\Delta H^{\ominus}$(A—B)/(kJ mol^{-1}) at 298 K

Diatomic molecules

H—H	436	F—F	155	Cl—Cl	242	Br—Br	193	I—I	151
O=O	497	C=O	1076	N≡N	945				
H—O	428	H—F	565	H—Cl	431	H—Br	366	H—I	299

Polyatomic molecules

H—CH_3	435	H—NH_2	460	H—OH	492	H—C_6H_5	469
H_3C—CH_3	368	H_2C=CH_2	720	HC≡CH	962		
HO—CH_3	377	Cl—CH_3	352	Br—CH_3	293	I—CH_3	237
O=CO	531	HO—OH	213	O_2N—NO_2	54		

Data: HCP, KL

Table 14.3b Mean bond enthalpies, $\Delta H^{\ominus}(A\text{—}B)/(\text{kJ mol}^{-1})$

	H	C	N	O	F	Cl	Br	I	S	P	Si
H	436										
C	412	348(i)									
		612(ii)									
		838(iii)									
		518(a)									
N	388	305(i)	163(i)								
		613(ii)	409(ii)								
		890(iii)	946(iii)								
O	463	360(i)	157	146(i)							
		743(ii)		497(ii)							
F	565	484	270	185	155						
Cl	431	338	200	203	254	242					
Br	366	276				219	193				
I	299	238				210	178	151			
S	338	259			496	250	212		264		
P	322									201	
Si	318		374	466							226

(i) Single bond, (ii) double bond, (iii) triple bond, (a) aromatic.
Data: HCP and L. Pauling, *The nature of the chemical bond*. Cornell University Press (1960).

Table 14.4 Pauling (*italics*) and Mulliken electronegativities

H							He
2.20							
3.06							
Li	**Be**	**B**	**C**	**N**	**O**	**F**	**Ne**
0.98	*1.57*	*2.04*	*2.55*	*3.04*	*3.44*	*3.98*	
1.28	1.99	1.83	2.67	3.08	3.22	4.43	4.60
Na	**Mg**	**Al**	**Si**	**P**	**S**	**Cl**	**Ar**
0.93	*1.31*	*1.61*	*1.90*	*2.19*	*2.58*	*3.16*	
1.21	1.63	1.37	2.03	2.39	2.65	3.54	3.36
K	**Ca**	**Ga**	**Ge**	**As**	**Se**	**Br**	**Kr**
0.82	*1.00*	*1.81*	*2.01*	*2.18*	*2.55*	*2.96*	*3.0*
1.03	1.30	1.34	1.95	2.26	2.51	3.24	2.98
Rb	**Sr**	**In**	**Sn**	**Sb**	**Te**	**I**	**Xe**
0.82	*0.95*	*1.78*	*1.96*	*2.05*	*2.10*	*2.66*	*2.6*
0.99	1.21	1.30	1.83	2.06	2.34	2.88	2.59
Cs	**Ba**	**Tl**	**Pb**	**Bi**			
0.79	*0.89*	*2.04*	*2.33*	*2.02*			

Data: Pauling values: A.L. Allred, *J. Inorg. Nucl. Chem.* **17**, 215 (1961); L.C. Allen and J.E. Huheey, *ibid.*, **42**, 1523 (1980). Mulliken values: L.C. Allen, *J. Am. Chem. Soc.* **111**, 9003 (1989). The Mulliken values have been scaled to the range of the Pauling values.

Table 16.2 Properties of diatomic molecules

	$\tilde{v}_0/cm^{-1}$	θ_V/K	B/cm^{-1}	θ_R/K	r/pm	$k/(N\ m^{-1})$	$D/(kJ\ mol^{-1})$	σ
$^1H_2^+$	2321.8	3341	29.8	42.9	106	160	255.8	2
$^1H_2$	4400.39	6332	60.864	87.6	74.138	574.9	432.1	2
$^2H_2$	3118.46	4487	30.442	43.8	74.154	577.0	439.6	2
$^1H^{19}F$	4138.32	5955	20.956	30.2	91.680	965.7	564.4	1
$^1H^{35}Cl$	2990.95	4304	10.593	15.2	127.45	516.3	427.7	1
$^1H^{81}Br$	2648.98	3812	8.465	12.2	141.44	411.5	362.7	1
$^1H^{127}I$	2308.09	3321	6.511	9.37	160.92	313.8	294.9	1
$^{14}N_2$	2358.07	3393	1.9987	2.88	109.76	2293.8	941.7	2
$^{16}O_2$	1580.36	2274	1.4457	2.08	120.75	1176.8	493.5	2
$^{19}F_2$	891.8	1283	0.8828	1.27	141.78	445.1	154.4	2
$^{35}Cl_2$	559.71	805	0.2441	0.351	198.75	322.7	239.3	2
$^{12}C^{16}O$	2170.21	3122	1.9313	2.78	112.81	1903.17	1071.8	1
$^{79}Br^{81}Br$	323.2	465	0.0809	10.116	283.3	245.9	190.2	1

Data: AIP

Table 16.3 Typical vibrational wavenumbers, $\tilde{v}/cm^{-1}$

C—H stretch	2850–2960
C—H bend	1340–1465
C—C stretch, bend	700–1250
C=C stretch	1620–1680
C≡C stretch	2100–2260
O—H stretch	3590–3650
H-bonds	3200–3570
C=O stretch	1640–1780
C≡N stretch	2215–2275
N—H stretch	3200–3500
C—F stretch	1000–1400
C—Cl stretch	600–800
C—Br stretch	500–600
C—I stretch	500
CO_3^{2-}	1410–1450
NO_3^-	1350–1420
NO_2^-	1230–1250
SO_4^{2-}	1080–1130
Silicates	900–1100

Data: L.J. Bellamy, *The infrared spectra of complex molecules* and *Advances in infrared group frequencies*. Chapman and Hall.

Table 17.1 Colour, frequency, and energy of light

Colour	λ/nm	$v/(10^{14}\ Hz)$	$\tilde{v}/(10^4\ cm^{-1})$	E/eV	$E/(kJ\ mol^{-1})$
Infrared	>1000	<3.00	<1.00	<1.24	<120
Red	700	4.28	1.43	1.77	171
Orange	620	4.84	1.61	2.00	193
Yellow	580	5.17	1.72	2.14	206
Green	530	5.66	1.89	2.34	226
Blue	470	6.38	2.13	2.64	254
Violet	420	7.14	2.38	2.95	285
Near ultraviolet	300	10.0	3.33	4.15	400
Far ultraviolet	<200	>15.0	>5.00	>6.20	>598

Data: J.G. Calvert and J.N. Pitts, *Photochemistry*. Wiley, New York (1966).

Table 17.3 Absorption characteristics of some groups and molecules

Group	$\tilde{\nu}_{max}/(10^4\ cm^{-1})$	λ_{max}/nm	$\varepsilon_{max}/(L\ mol^{-1}\ cm^{-1})$
$C{=}C\ (\pi^* \leftarrow \pi)$	6.10	163	1.5×10^4
	5.73	174	5.5×10^3
$C{=}O\ (\pi^* \leftarrow n)$	3.7–3.5	270–290	10–20
$-N{=}N-$	2.9	350	15
	>3.9	<260	Strong
$-NO_2$	3.6	280	10
	4.8	210	1.0×10^4
C_6H_5-	3.9	255	200
	5.0	200	6.3×10^3
	5.5	180	1.0×10^5
$[Cu(OH_2)_6]^{2+}(aq)$	1.2	810	10
$[Cu(NH_3)_4]^{2+}(aq)$	1.7	600	50
$H_2O\ (\pi^* \leftarrow n)$	6.0	167	7.0×10^3

Table 18.2 Nuclear spin properties

Nuclide	Natural abundance %	Spin I	Magnetic moment μ/μ_N	g-value	$\gamma/(10^7\ T^{-1}\ s^{-1})$	NMR frequency at 1 T, ν/MHz
$^1n^*$		$\frac{1}{2}$	−1.9130	−3.8260	−18.324	29.164
^1H	99.9844	$\frac{1}{2}$	2.792 85	5.5857	26.752	42.576
^2H	0.0156	1	0.857 44	0.857 45	4.1067	6.536
$^3H^*$		$\frac{1}{2}$	2.978 96	−4.2553	−20.380	45.414
^{10}B	19.6	3	1.8006	0.6002	2.875	4.575
^{11}B	80.4	$\frac{3}{2}$	2.6886	1.7923	8.5841	13.663
^{13}C	1.108	$\frac{1}{2}$	0.7024	1.4046	6.7272	10.708
^{14}N	99.635	1	0.403 76	0.403 56	1.9328	3.078
^{17}O	0.037	$\frac{5}{2}$	−1.893 79	−0.7572	−3.627	5.774
^{19}F	100	$\frac{1}{2}$	2.628 87	5.2567	25.177	40.077
^{31}P	100	$\frac{1}{2}$	1.1316	2.2634	10.840	17.251
^{33}S	0.74	$\frac{3}{2}$	0.6438	0.4289	2.054	3.272
^{35}Cl	75.4	$\frac{3}{2}$	0.8219	0.5479	2.624	4.176
^{37}Cl	24.6	$\frac{3}{2}$	0.6841	0.4561	2.184	3.476

* Radioactive.
μ is the magnetic moment of the spin state with the largest value of m_I: $\mu = g_I \mu_N I$ and μ_N is the nuclear magneton (see inside front cover).
Data: KL and HCP

Table 18.3 Hyperfine coupling constants for atoms, a/mT

Nuclide	Spin	Isotropic coupling	Anisotropic coupling
^{1}H	$\frac{1}{2}$	50.8($1s$)	
^{2}H	1	7.8($1s$)	
^{13}C	$\frac{1}{2}$	113.0($2s$)	6.6($2p$)
^{14}N	1	55.2($2s$)	4.8($2p$)
^{19}F	$\frac{1}{2}$	1720($2s$)	108.4($2p$)
^{31}P	$\frac{1}{2}$	364($3s$)	20.6($3p$)
^{35}Cl	$\frac{3}{2}$	168($3s$)	10.0($3p$)
^{37}Cl	$\frac{3}{2}$	140($3s$)	8.4($3p$)

Data: P.W. Atkins and M.C.R. Symons, *The structure of inorganic radicals*. Elsevier, Amsterdam (1967).

Table 21.1 Dipole moments, polarizabilities, and polarizability volumes

	$\mu/(10^{-30}$ C m)	μ/D	$\alpha/(10^{-30}$ m^3)	$\alpha'/(10^{-40}$ J^{-1} C^2 m^2)
Ar	0	0	1.66	1.85
C_2H_5OH	5.64	1.69		
$C_6H_5CH_3$	1.20	0.36		
C_6H_6	0	0	10.4	11.6
CCl_4	0	0	10.5	11.7
CH_2Cl_2	5.24	1.57	6.80	7.57
CH_3Cl	6.24	1.87	4.53	5.04
CH_3OH	5.70	1.71	3.23	3.59
CH_4	0	0	2.60	2.89
$CHCl_3$	3.37	1.01	8.50	9.46
CO	0.390	0.117	1.98	2.20
CO_2	0	0	2.63	2.93
H_2	0	0	0.819	0.911
H_2O	6.17	1.85	1.48	1.65
HBr	2.67	0.80	3.61	4.01
HCl	3.60	1.08	2.63	2.93
He	0	0	0.20	0.22
HF	6.37	1.91	0.51	0.57
HI	1.40	0.42	5.45	6.06
N_2	0	0	1.77	1.97
NH_3	4.90	1.47	2.22	2.47
1,2-$C_6H_4(CH_3)_2$	2.07	0.62		

Data: HCP and C.J.F. Böttcher and P. Bordewijk, *Theory of electric polarization*. Elsevier, Amsterdam (1978).

Table 21.2 Refractive indices relative to air at 20°C

	434 nm	589 nm	656 nm
Benzene	1.5236	1.5012	1.4965
Carbon tetrachloride	1.4729	1.4676	1.4579
Carbon disulfide	1.6748	1.6276	1.6182
Ethanol	1.3700	1.3618	1.3605
KCl(s)	1.5050	1.4904	1.4973
KI(s)	1.7035	1.6664	1.6581
Methanol	1.3362	1.3290	1.3277
Methylbenzene	1.5170	1.4955	1.4911
Water	1.3404	1.3330	1.3312

Data: AIP

Table 22.1 Radius of gyration of some macromolecules

	$M/(\text{kg mol}^{-1})$	R_g/nm
Serum albumin	66	2.98
Myosin	493	46.8
Polystyrene	3.2×10^3	50 (in poor solvent)
DNA	4×10^3	117.0
Tobacco mosaic virus	3.9×10^4	92.4

Data: C. Tanford, *Physical chemistry of macromolecules*. Wiley, New York (1961).

Table 22.2 Diffusion coefficients of macromolecules in water at 20°C

	$M/(\text{kg mol}^{-1})$	$D/(10^{-10} \text{ m}^2 \text{ s}^{-1})$
Sucrose	0.342	4.586
Ribonuclease	13.7	1.19
Lysozyme	14.1	1.04
Serum albumin	65	0.594
Haemoglobin	68	0.69
Urease	480	0.346
Collagen	345	0.069
Myosin	493	0.116

Data: C. Tanford, *Physical chemistry of macromolecules*. Wiley, New York (1961).

Table 21.4 Lennard-Jones (12,6)-potential parameters

	$(\varepsilon/k)/\text{K}$	r_0/pm
Ar	111.84	362.3
C_2H_2	209.11	463.5
C_2H_4	200.78	458.9
C_2H_6	216.12	478.2
C_6H_6	377.46	617.4
CCl_4	378.86	624.1
Cl_2	296.27	448.5
CO_2	201.71	444.4
F_2	104.29	357.1
Kr	154.87	389.5
N_2	91.85	391.9
O_2	113.27	365.4
Xe	213.96	426.0

Source: F. Cuadros, I. Cachadiña, and W. Ahamuda, *Molec. Engineering*, **6**, 319 (1996).

Table 22.3 Frictional coefficients and molecular geometry

Major axis/Minor axis	Prolate	Oblate
2	1.04	1.04
3	1.11	1.10
4	1.18	1.17
5	1.25	1.22
6	1.31	1.28
7	1.38	1.33
8	1.43	1.37
9	1.49	1.42
10	1.54	1.46
50	2.95	2.38
100	4.07	2.97

Data: K.E. Van Holde, *Physical biochemistry*. Prentice-Hall, Englewood Cliffs (1971).

Sphere; radius a, $c = a f_0$

Prolate ellipsoid; major axis $2a$, minor axis $2b$, $c = (ab^2)^{1/3}$

$$f = \left\{ \frac{(1 - b^2/a^2)^{1/2}}{(b/a)^{2/3} \ln\{[1 + (1 - b^2/a^2)^{1/2}]/(b/a)\}} \right\} f_0$$

Oblate ellipsoid; major axis $2a$, minor axis $2b$, $c = (a^2 b)^{1/3}$

$$f = \left\{ \frac{(a^2/b^2 - 1)^{1/2}}{(a/b)^{2/3} \arctan[(a^2/b^2 - 1)^{1/2}]} \right\} f_0$$

Long rod; length l, radius a, $c = (3a^2/4)^{1/3}$

$$f = \left\{ \frac{(1/2a)^{2/3}}{(3/2)^{1/3}\{2 \ln(l/a) - 0.11\}} \right\} f_0$$

In each case $f_0 = 6\pi\eta c$ with the appropriate value of c.

Table 22.4 Intrinsic viscosity

Macromolecule	Solvent	$\theta/°C$	$K/(10^{-3}\ cm^3\ g^{-1})$	a
Polystyrene	Benzene	25	9.5	0.74
	Cyclohexane	34†	81	0.50
Polyisobutylene	Benzene	23†	83	0.50
	Cyclohexane	30	26	0.70
Amylose	0.33 M KCl(aq)	25†	113	0.50
Various proteins‡	Guanidine hydrochloride + $HSCH_2CH_2OH$		7.16	0.66

† The θ temperature.
‡ Use $[\eta] = KN^a$; N is the number of amino acid residues.
Data: K.E. Van Holde, *Physical biochemistry*. Prentice-Hall, Englewood Cliffs (1971).

Table 23.3 Ionic radii (r/pm)†

$Li^+(4)$	$Be^{2+}(4)$	$B^{3+}(4)$	N^{3-}	$O^{2-}(6)$	$F^-(6)$		
59	27	12	171	140	133		
$Na^+(6)$	$Mg^{2+}(6)$	$Al^{3+}(6)$	P^{3-}	$S^{2-}(6)$	$Cl^-(6)$		
102	72	53	212	184	181		
$K^+(6)$	$Ca^{2+}(6)$	$Ga^{3+}(6)$	$As^{3-}(6)$	$Se^{2-}(6)$	$Br^-(6)$		
138	100	62	222	198	196		
$Rb^+(6)$	$Sr^{2+}(6)$	$In^{3+}(6)$		$Te^{2-}(6)$	$I^-(6)$		
149	116	79		221	220		
$Cs^+(6)$	$Ba^{2+}(6)$	$Tl^{3+}(6)$					
167	136	88					

***d*-block elements (high-spin ions)**

$Sc^{3+}(6)$	$Ti^{4+}(6)$	$Cr^{3+}(6)$	$Mn^{3+}(6)$	$Fe^{2+}(6)$	$Co^{3+}(6)$	$Cu^{2+}(6)$	$Zn^{2+}(6)$
73	60	61	65	63	61	73	75

† Numbers in parentheses are the coordination numbers of the ions. Values for ions without a coordination number stated are estimates.
Data: R.D. Shannon and C.T. Prewitt, *Acta Cryst.* **B25**, 925 (1969).

Table 23.5 Lattice enthalpies, $\Delta H_L^{\ominus}/(\text{kJ mol}^{-1})$

	F	Cl	Br	I
Halides				
Li	1037	852	815	761
Na	926	787	752	705
K	821	717	689	649
Rb	789	695	668	632
Cs	750	676	654	620
Ag	969	912	900	886
Be		3017		
Mg		2524		
Ca		2255		
Sr		2153		

Oxides

MgO	3850	CaO	3461	SrO	3283	BaO	3114

Sulfides

MgS	3406	CaS	3119	SrS	2974	BaS	2832

Entries refer to $MX(s) \rightarrow M^+(g) + X^-(g)$.
Data: Principally D. Cubicciotti, *J. Chem. Phys.* **31**, 1646 (1959).

Table 23.6 Magnetic susceptibilities at 298 K

	$\chi/10^{-6}$	$\chi_m/(10^{-4}\ \text{cm}^3\ \text{mol}^{-1})$
Water	−90	−16.0
Benzene	−7.2	−6.4
Cyclohexane	−7.9	−8.5
Carbon tetrachloride	−8.9	−8.4
NaCl(s)	−13.9	−3.75
Cu(s)	−96	−6.8
S(s)	−12.9	−2.0
Hg(l)	−28.5	−4.2
$CuSO_4 \cdot 5H_2O(s)$	+176	+192
$MnSO_4 \cdot 4H_2O(s)$	+2640	$+2.79 \times 10^3$
$NiSO_4 \cdot 7H_2O(s)$	+416	+600
$FeSO_4(NH_4)_2SO_4 \cdot 6H_2O(s)$	+755	$+1.51 \times 10^3$
Al(s)	+22	+2.2
Pt(s)	+262	+22.8
Na(s)	+7.3	+1.7
K(s)	+5.6	+2.5

Data: KL and $\chi_m = \chi M/\rho$.

Table 24.1 Collision cross-sections, σ/nm^2

Ar	0.36
C_2H_4	0.64
C_6H_6	0.88
CH_4	0.46
Cl_2	0.93
CO_2	0.52
H_2	0.27
He	0.21
N_2	0.43
Ne	0.24
O_2	0.40
SO_2	0.58

Data: KL

Table 24.2 Transport properties of gases at 1 atm

	$\kappa/(\text{J K}^{-1}\ \text{m}^{-1}\ \text{s}^{-1})$	$\eta/\mu\text{P}$	
	273 K	273 K	293 K
Air	0.0241	173	182
Ar	0.0163	210	223
C_2H_4	0.0164	97	103
CH_4	0.0302	103	110
Cl_2	0.079	123	132
CO_2	0.0145	136	147
H_2	0.1682	84	88
He	0.1442	187	196
Kr	0.0087	234	250
N_2	0.0240	166	176
Ne	0.0465	298	313
O_2	0.0245	195	204
Xe	0.0052	212	228

Data: KL

Table 24.4 Viscosities of liquids at 298 K, $\eta/(10^{-3} \text{ kg m}^{-1} \text{ s}^{-1})$

Benzene	0.601
Carbon tetrachloride	0.880
Ethanol	1.06
Mercury	1.55
Methanol	0.553
Pentane	0.224
Sulfuric acid	27
Water†	0.891

† The viscosity of water over its entire liquid range is represented with less than 1 per cent error by the expression

$$\log(\eta_{20}/\eta) = A/B,$$
$$A = 1.370\,23(t-20) + 8.36 \times 10^{-4}(t-20)^2$$
$$B = 109 + t \qquad t = \theta/°C$$

Convert $\text{kg m}^{-1} \text{ s}^{-1}$ to centipoise (cP) by multiplying by 10^3 (so $\eta \approx 1$ cP for water).
Data: AIP, KL

Table 24.5 Limiting ionic conductivities in water at 298 K, $\lambda/(\text{mS m}^2 \text{ mol}^{-1})$

Cations		Anions	
Ba^{2+}	12.72	Br^-	7.81
Ca^{2+}	11.90	$CH_3CO_2^-$	4.09
Cs^+	7.72	Cl^-	7.635
Cu^{2+}	10.72	ClO_4^-	6.73
H^+	34.96	CO_3^{2-}	13.86
K^+	7.350	$(CO_2)_2^{2-}$	14.82
Li^+	3.87	F^-	5.54
Mg^{2+}	10.60	$[Fe(CN)_6]^{3-}$	30.27
Na^+	5.010	$[Fe(CN)_6]^{4-}$	44.20
$[N(C_2H_5)_4]^+$	3.26	HCO_2^-	5.46
$[N(CH_3)_4]^+$	4.49	I^-	7.68
NH_4^+	7.35	NO_3^-	7.146
Rb^+	7.78	OH^-	19.91
Sr^{2+}	11.89	SO_4^{2-}	16.00
Zn^{2+}	10.56		

Data: KL, RS

Table 24.6 Ionic mobilities in water at 298 K, $u/(10^{-8} \text{ m}^2 \text{ s}^{-1} \text{ V}^{-1})$

Cations		Anions	
Ag^+	6.24	Br^-	8.09
Ca^{2+}	6.17	$CH_3CO_2^-$	4.24
Cu^{2+}	5.56	Cl^-	7.91
H^+	36.23	CO_3^{2-}	7.46
K^+	7.62	F^-	5.70
Li^+	4.01	$[Fe(CN)_6]^{3-}$	10.5
Na^+	5.19	$[Fe(CN)_6]^{4-}$	11.4
NH_4^+	7.63	I^-	7.96
$[N(CH_3)_4]^+$	4.65	NO_3^-	7.40
Rb^+	7.92	OH^-	20.64
Zn^{2+}	5.47	SO_4^{2-}	8.29

Data: Principally Table 24.4 and $u = \lambda/zF$.

Table 24.7 Debye–Hückel–Onsager coefficients for (1,1)-electrolytes at 25°C

Solvent	$A/(\text{mS m}^2 \text{ mol}^{-1}/ (\text{mol L}^{-1})^{1/2})$	$B/(\text{mol L}^{-1})^{-1/2}$
Acetone (propanone)	3.28	1.63
Acetonitrile	2.29	0.716
Ethanol	8.97	1.83
Methanol	15.61	0.923
Nitrobenzene	4.42	0.776
Nitromethane	111	0.708
Water	6.020	0.229

Data: J.O'M. Bockris and A.K.N. Reddy, *Modern electrochemistry*. Plenum, New York (1970).

Table 24.8 Diffusion coefficients at 25°C, $D/(10^{-9}\ m^2\ s^{-1})$

Molecules in liquids					Ions in water			
I_2 in hexane	4.05	H_2 in $CCl_4(l)$	9.75		K^+	1.96	Br^-	2.08
in benzene	2.13	N_2 in $CCl_4(l)$	3.42		H^+	9.31	Cl^-	2.03
CCl_4 in heptane	3.17	O_2 in $CCl_4(l)$	3.82		Li^+	1.03	F^-	1.46
Glycine in water	1.055	Ar in $CCl_4(l)$	3.63		Na^+	1.33	I^-	2.05
Dextrose in water	0.673	CH_4 in $CCl_4(l)$	2.89				OH^-	5.03
Sucrose in water	0.5216	H_2O in water	2.26					
		CH_3OH in water	1.58					
		C_2H_5OH in water	1.24					

Data: AIP and (for the ions) $\lambda = zuF$ in conjunction with Table 24.5.

Table 25.1 Kinetic data for first-order reactions

	Phase	$\theta/°C$	k/s^{-1}	$t_{1/2}$
$2\ N_2O_5 \longrightarrow 4\ NO_2 + O_2$	g	25	3.38×10^{-5}	5.70 h
	$HNO_3(l)$	25	1.47×10^{-6}	131 h
	$Br_2(l)$	25	4.27×10^{-5}	4.51 h
$C_2H_6 \longrightarrow 2\ CH_3$	g	700	5.36×10^{-4}	21.6 min
Cyclopropane $\longrightarrow$ propene	g	500	6.71×10^{-4}	17.2 min
$CH_3N_2CH_3 \longrightarrow C_2H_6 + N_2$	g	327	3.4×10^{-4}	34 min
Sucrose $\longrightarrow$ glucose + fructose	aq(H^+)	25	6.0×10^{-5}	3.2 h

g: High pressure gas-phase limit.
Data: Principally K.J. Laidler, *Chemical kinetics*. Harper & Row, New York (1987); M.J. Pilling and P.W. Seakins, *Reaction kinetics*. Oxford University Press (1995); J. Nicholas, *Chemical kinetics*. Harper & Row, New York (1976). See also JL.

Table 25.2 Kinetic data for second-order reactions

	Phase	$\theta/°C$	$k/(L\ mol^{-1}\ s^{-1})$
$2\ NOBr \longrightarrow 2\ NO + Br_2$	g	10	0.80
$2\ NO_2 \longrightarrow 2\ NO + O_2$	g	300	0.54
$H_2 + I_2 \longrightarrow 2\ HI$	g	400	2.42×10^{-2}
$D_2 + HCl \longrightarrow DH + DCl$	g	600	0.141
$2\ I \longrightarrow I_2$	g	23	7×10^9
	hexane	50	1.8×10^{10}
$CH_3Cl + CH_3O^-$	methanol	20	2.29×10^{-6}
$CH_3Br + CH_3O^-$	methanol	20	9.23×10^{-6}
$H^+ + OH^- \longrightarrow H_2O$	water	25	1.35×10^{11}
	ice	-10	8.6×10^{12}

Data: Principally K.J. Laidler, *Chemical kinetics*. Harper & Row, New York (1987); M.J. Pilling and P.W. Seakins, *Reaction kinetics*. Oxford University Press (1995); J. Nicholas, *Chemical kinetics*. Harper & Row, New York (1976).

Table 25.4 Arrhenius parameters

First-order reactions	A/s^{-1}	$E_a/(\text{kJ mol}^{-1})$
Cyclopropane $\longrightarrow$ propene	1.58×10^{15}	272
$CH_3NC \longrightarrow CH_3CN$	3.98×10^{13}	160
cis-CHD=CHD $\longrightarrow$ $trans$-CHD=CHD	3.16×10^{12}	256
Cyclobutane $\longrightarrow 2\,C_2H_4$	3.98×10^{13}	261
$C_2H_5I \longrightarrow C_2H_4 + HI$	2.51×10^{17}	209
$C_2H_6 \longrightarrow 2\,CH_3$	2.51×10^{7}	384
$2\,N_2O_5 \longrightarrow 4\,NO_2 + O_2$	4.94×10^{13}	103
$N_2O \longrightarrow N_2 + O$	7.94×10^{11}	250
$C_2H_5 \longrightarrow C_2H_4 + H$	1.0×10^{13}	167

Second-order, gas-phase	$A/(\text{L mol}^{-1}\,\text{s}^{-1})$	$E_a/(\text{kJ mol}^{-1})$
$O + N_2 \longrightarrow NO + N$	1×10^{11}	315
$OH + H_2 \longrightarrow H_2O + H$	8×10^{10}	42
$Cl + H_2 \longrightarrow HCl + H$	8×10^{10}	23
$2\,CH_3 \longrightarrow C_2H_6$	2×10^{10}	ca. 0
$NO + Cl_2 \longrightarrow NOCl + Cl$	4.0×10^{9}	85
$SO + O_2 \longrightarrow SO_2 + O$	3×10^{8}	27
$CH_3 + C_2H_6 \longrightarrow CH_4 + C_2H_5$	2×10^{8}	44
$C_6H_5 + H_2 \longrightarrow C_6H_6 + H$	1×10^{8}	ca. 25

Second-order, solution	$A/(\text{L mol}^{-1}\,\text{s}^{-1})$	$E_a/(\text{kJ mol}^{-1})$
$C_2H_5ONa + CH_3I$ in ethanol	2.42×10^{11}	81.6
$C_2H_5Br + OH^-$ in water	4.30×10^{11}	89.5
$C_2H_5I + C_2H_5O^-$ in ethanol	1.49×10^{11}	86.6
$CH_3I + C_2H_5O^-$ in ethanol	2.42×10^{11}	81.6
$C_2H_5Br + OH^-$ in ethanol	4.30×10^{11}	89.5
$CO_2 + OH^-$ in water	1.5×10^{10}	38
$CH_3I + S_2O_3^{2-}$ in water	2.19×10^{12}	78.7
Sucrose + H_2O in acidic water	1.50×10^{15}	107.9
$(CH_3)_3CCl$ solvolysis		
in water	7.1×10^{16}	100
in methanol	2.3×10^{13}	107
in ethanol	3.0×10^{13}	112
in acetic acid	4.3×10^{13}	111
in chloroform	1.4×10^{4}	45
$C_6H_5NH_2 + C_6H_5COCH_2Br$		
in benzene	91	34

Data: Principally J. Nicholas, *Chemical kinetics*. Harper & Row, New York (1976) and A.A. Frost and R.G. Pearson, *Kinetics and mechanism*. Wiley, New York (1961).

Table 27.1 Arrhenius parameters for gas-phase reactions

	$A/(L\ mol^{-1}\ s^{-1})$		$E_a/(kJ\ mol^{-1})$	P
	Experiment	Theory		
$2\ NOCl \longrightarrow 2\ NO + Cl_2$	9.4×10^9	5.9×10^{10}	102.0	0.16
$2\ NO_2 \longrightarrow 2\ NO + O_2$	2.0×10^9	4.0×10^{10}	111.0	5.0×10^{-2}
$2\ ClO \longrightarrow Cl_2 + O_2$	6.3×10^7	2.5×10^{10}	0.0	2.5×10^{-3}
$H_2 + C_2H_4 \longrightarrow C_2H_6$	1.24×10^6	7.4×10^{11}	180	1.7×10^{-6}
$K + Br_2 \longrightarrow KBr + Br$	1.0×10^{12}	2.1×10^{11}	0.0	4.8

Data: Principally M.J. Pilling and P.W. Seakins, *Reaction kinetics*. Oxford University Press (1995).

Table 27.2 Arrhenius parameters for reactions in solution. See Table 25.4

Table 28.1 Maximum observed enthalpies of physisorption, $\Delta_{ad}H^{\ominus}/(kJ\ mol^{-1})$

C_2H_2	−38	H_2	−84
C_2H_4	−34	H_2O	−59
CH_4	−21	N_2	−21
Cl_2	−36	NH_3	−38
CO	−25	O_2	−21
CO_2	−25		

Data: D.O. Haywood and B.M.W. Trapnell, *Chemisorption*. Butterworth (1964).

Table 28.2 Enthalpies of chemisorption, $\Delta_{ad}H^{\ominus}/(kJ\ mol^{-1})$

Adsorbate	Adsorbent (substrate)											
	Ti	Ta	Nb	W	Cr	Mo	Mn	Fe	Co	Ni	Rh	Pt
H_2		−188			−188	−167	−71	−134			−117	
N_2		−586						−293				
O_2						−720					−494	−293
CO	−640							−192	−176			
CO_2	−682	−703	−552	−456	−339	−372	−222	−225	−146	−184		
NH_3				−301				−188		−155		
C_2H_4		−577		−427	−427			−285		−243	−209	

Data: D.O. Haywood and B.M.W. Trapnell, *Chemisorption*. Butterworth (1964).

Table 28.3 Activation energies of catalysed reactions

	Catalyst	$E_a/(kJ\ mol^{-1})$
$2\ HI \longrightarrow H_2 + I_2$	None	184
	Au(s)	105
	Pt(s)	59
$2\ NH_3 \longrightarrow N_2 + 3\ H_2$	None	350
	W(s)	162
$2\ N_2O \longrightarrow 2\ N_2 + O_2$	None	245
	Au(s)	121
	Pt(s)	134
$(C_2H_5)_2O$ pyrolysis	None	224
	$I_2(g)$	144

Data: G.C. Bond, *Heterogeneous catalysis.* Clarendon Press, Oxford (1986).

Table 29.1 Exchang current densities and transfer coefficients at 298 K

Reaction	Electrode	$j_0/(A\ cm^{-2})$	α
$2\ H^+ + 2\ e^- \longrightarrow H_2$	Pt	7.9×10^{-4}	
	Cu	1×10^{-6}	
	Ni	6.3×10^{-6}	0.58
	Hg	7.9×10^{-13}	0.50
	Pb	5.0×10^{-12}	
$Fe^{3+} + e^- \longrightarrow Fe^{2+}$	Pt	2.5×10^{-3}	0.58
$Ce^{4+} + e^- \longrightarrow Ce^{3+}$	Pt	4.0×10^{-5}	0.75

Data: Principally J.O'M. Bockris and A.K.N. Reddy, *Modern electrochemistry.* Plenum, New York (1970).

Character tables

The groups C_1, C_s, C_i

C_1 (1)	E	$h = 1$
A	1	

$C_s = C_h$ (m)	E	σ_h	$h = 2$	
A′	1	1	x, y, R_z	x^2, y^2, z^2, xy
A″	1	−1	z, R_x, R_y	yz, xz

$C_i = S_2$ (1̄)	E	i	$h = 2$	
A_g	1	1	R_x, R_y, R_z	$x^2, y^2, z^2, xy, xz, yz$
A_u	1	−1	x, y, z	

The groups C_{nv}

C_{2v}, 2mm	E	C_2	σ_v	σ_v'	$h = 4$	
A_1	1	1	1	1	z, z^2, x^2, y^2	
A_2	1	1	−1	−1	xy	R_z
B_1	1	−1	1	−1	x, xz	R_y
B_2	1	−1	−1	1	y, yz	R_x

$C_{3v}, 3m$	E	$2C_3$	$3\sigma_v$	$h = 6$	
A_1	1	1	1	$z, z^2, x^2 + y^2$	
A_2	1	1	−1		R_z
E	2	−1	0	$(x, y), (xy, x^2 - y^2)\ (xz, yz)$	(R_x, R_y)

$C_{4v}, 4mm$	E	C_2	$2C_4$	$2\sigma_v$	$2\sigma_d$	$h = 8$	
A_1	1	1	1	1	1	$z, z^2, x^2 + y^2$	
A_2	1	1	1	−1	1		R_z
B_1	1	1	−1	1	−1	$x^2 - y^2$	
B_2	1	1	−1	−1	1	xy	
E	2	−2	0	0	0	$(x, y), (xz, yz)$	(R_x, R_y)

C_{5v}	E	$2C_5$	$2C_5^2$	$5\sigma_v$	$h = 10, \alpha = 72°$	
A_1	1	1	1	1	$z, z^2, x^2 + y^2$	
A_2	1	1	1	−1		R_z
E_1	2	$2\cos\alpha$	$2\cos 2\alpha$	0	$(x, y), (xz, yz)$	(R_x, R_y)
E_2	2	$2\cos 2\alpha$	$2\cos\alpha$	0	$(xy, x^2 - y^2)$	

$C_{6v}, 6mm$	E	C_2	$2C_3$	$2C_6$	$3\sigma_d$	$3\sigma_v$	$h = 12$	
A_1	1	1	1	1	1	1	$z, z^2, x^2 + y^2$	
A_2	1	1	1	1	−1	1		R_z
B_1	1	−1	1	−1	−1	1		
B_2	1	−1	1	−1	1	−1		
E_1	2	−2	−1	1	0	0	$(x, y), (xz, yz)$	(R_x, R_y)
E_2	2	2	−1	−1	0	0	$(xy, x^2 - y^2)$	

$C_{\infty v}$	E	$2C_\phi$†	$\infty\sigma_v$	$h = \infty$	
$A_1 (\Sigma^+)$	1	1	1	$z, z^2, x^2 + y^2$	
$A_2 (\Sigma^-)$	1	1	−1		R_z
$E_1 (\Pi)$	2	$2\cos\phi$	0	$(x, y), (xz, yz)$	(R_x, R_y)
$E_2 (\Delta)$	2	$2\cos 2\phi$	0	$(xy, x^2 - y^2)$	

† There is only one member of this class if $\phi = \pi$.

The groups D_n

D_2, 222	E	C_2^z	C_2^y	C_2^x	$h = 4$	
A_1	1	1	1	1	x^2, y^2, z^2	
B_1	1	1	−1	−1	z, xy	R_z
B_2	1	−1	1	−1	y, xz	R_y
B_3	1	−1	−1	1	x, yz	R_x

D_3, 32	E	$2C_3$	$3C_2'$	$h = 6$	
A_1	1	1	1	$z^2, x^2 + y^2$	
A_2	1	1	−1	z	R_z
E	2	−1	0	$(x, y), (xz, yz), (xy, x^2 - y^2)$	(R_x, R_y)

D_4, 422	E	C_2	$2C_4$	$2C_2'$	$2C_2''$	$h = 8$	
A_1	1	1	1	1	1	$z^2, x^2 + y^2$	
A_2	1	1	1	−1	−1	z	R_z
B_1	1	1	−1	1	−1	$x^2 - y^2$	
B_2	1	1	−1	−1	1	xy	
E	2	−2	0	0	0	$(x, y), (xz, yz)$	(R_x, R_y)

The groups D_{nh}

D_{3h}, $\bar{6}2m$	E	σ_h	$2C_3$	$2S_3$	$3C_2'$	$3\sigma_v$	$h = 12$	
A_1'	1	1	1	1	1	1	$z^2, x^2 + y^2$	
A_2'	1	1	1	1	−1	−1		R_z
A_1''	1	−1	1	−1	1	−1		
A_2''	1	−1	1	−1	−1	1	z	
E'	2	2	−1	−1	0	0	$(x, y), (xy, x^2 - y^2)$	
E''	2	−2	−1	1	0	0	(xz, yz)	(R_x, R_y)

D_{4h}, $4/mmm$	E	$2C_4$	C_2	$2C_2'$	$2C_2''$	i	$2S_4$	σ_h	$2\sigma_v$	$2\sigma_d$	$h = 16$	
A_{1g}	1	1	1	1	1	1	1	1	1	1	$x^2 + y^2, z^2$	
A_{2g}	1	1	1	−1	−1	1	1	1	−1	−1		R_z
B_{1g}	1	−1	1	1	−1	1	−1	1	1	−1	$x^2 - y^2$	
B_{2g}	1	−1	1	−1	1	1	−1	1	−1	1	xy	
E_g	2	0	−2	0	0	2	0	−2	0	0	(xz, yz)	(R_x, R_y)
A_{1u}	1	1	1	1	1	−1	−1	−1	−1	−1		
A_{2u}	1	1	1	−1	−1	−1	−1	−1	1	1	z	
B_{1u}	1	−1	1	1	−1	−1	1	−1	−1	1		
B_{2u}	1	−1	1	−1	1	−1	1	−1	1	−1		
E_u	2	0	−2	0	0	−2	0	2	0	0	(x, y)	

D_{5h}	E	$2C_5$	$2C_5^2$	$5C_2$	σ_h	$2S_5$	$2S_5^3$	$5\sigma_v$	$h = 20$	$\alpha = 72°$
A_1'	1	1	1	1	1	1	1	1	$x^2 + y^2, z^2$	
A_2'	1	1	1	−1	1	1	1	−1		R_z
E_1'	2	$2\cos\alpha$	$2\cos 2\alpha$	0	2	$2\cos\alpha$	$2\cos 2\alpha$	0	(x, y)	
E_2'	2	$2\cos 2\alpha$	$2\cos\alpha$	0	2	$2\cos 2\alpha$	$2\cos\alpha$	0	(x^2-y^2, xy)	
A_1''	1	1	1	1	−1	−1	−1	−1		
A_2''	1	1	1	−1	−1	−1	−1	1	z	
E_1''	2	$2\cos\alpha$	$2\cos 2\alpha$	0	−2	$-2\cos\alpha$	$-2\cos 2\alpha$	0	(xz, yz)	(R_x, R_y)
E_2''	2	$2\cos 2\alpha$	$2\cos\alpha$	0	−2	$-2\cos 2\alpha$	$-2\cos\alpha$	0		

$D_{\infty h}$	E	$2C_\phi$	$\infty C_2'$	i	$2iC_\infty$	iC_2'	$h = \infty$	
$A_{1g}\ (\Sigma_g^+)$	1	1	1	1	1	1	$z^2, x^2 + y^2$	
$A_{1u}\ (\Sigma_u^+)$	1	1	1	−1	−1	−1	z	
$A_{2g}\ (\Sigma_g^-)$	1	1	−1	1	1	−1		R_z
$A_{2u}\ (\Sigma_u^-)$	1	1	−1	−1	1	1		
$E_{1g}\ (\Pi_g)$	2	$2\cos\phi$	0	2	$-2\cos\phi$	0	(xz, yz)	(R_x, R_y)
$E_{1u}\ (\Pi_u)$	2	$2\cos\phi$	0	−2	$2\cos\phi$	0	(x, y)	
$E_{2g}\ (\Delta_g)$	2	$2\cos 2\phi$	0	2	$2\cos 2\phi$	0	$(xy, x^2 - y^2)$	
$E_{2u}\ (\Delta_u)$	2	$2\cos 2\phi$	0	−2	$-2\cos 2\phi$	0		
$\vdots$								

The cubic groups

$T_d, \overline{4}3m$	E	$8C_3$	$3C_2$	$6\sigma_d$	$6S_4$	$h = 24$	
A_1	1	1	1	1	1	$x^2 + y^2 + z^2$	
A_2	1	1	1	−1	−1		
E	2	−1	2	0	0	$(3z^2 - r^2, x^2 - y^2)$	
T_1	3	0	−1	−1	1		(R_x, R_y, R_z)
T_2	3	0	−1	1	−1	$(x, y, z), (xy, xz, yz)$	

O_h (m3m)	E	$8C_3$	$6C_2$	$6C_2$	$3C_2$ $(= C_4^2)$	i	$6S_4$	$8S_6$	$3\sigma_h$	$6\sigma_d$	$h = 48$	
A_{1g}	1	1	1	1	1	1	1	1	1	1		$x^2 + y^2 + z^2$
A_{2g}	1	1	−1	−1	1	1	−1	1	1	−1		
E_g	2	−1	0	0	2	2	0	−1	2	0		$(2z^2 - x^2 - y^2, x^2 - y^2)$
T_{1g}	3	0	−1	1	−1	3	1	0	−1	−1	(R_x, R_y, R_z)	
T_{2g}	3	0	1	−1	−1	3	−1	0	−1	1		(xy, yz, xy)
A_{1u}	1	1	1	1	1	−1	−1	−1	−1	−1		
A_{2u}	1	1	−1	−1	1	−1	1	−1	−1	1		
E_u	2	−1	0	0	2	−2	0	1	−2	0		
T_{1u}	3	0	−1	1	−1	−3	−1	0	1	1	(x, y, z)	
T_{2u}	3	0	1	−1	−1	−3	1	0	1	−1		

The icosahedral group

I	E	$12C_5$	$12C_5^2$	$20C_3$	$15C_2$	$h=60$	
A	1	1	1	1	1		$z^2+y^2+z^2$
T_1	3	$\frac{1}{2}(1+\sqrt{5})$	$\frac{1}{2}(1-\sqrt{5})$	0	−1	(x, y, z) (R_x, R_y, R_z)	
T_2	3	$\frac{1}{2}(1-\sqrt{5})$	$\frac{1}{2}(1+\sqrt{5})$	0	−1		
G	4	−1	−1	1	0		
G	5	0	0	−1	1		$(2z^2-x^2-y^2,\ x^2-y^2,\ xy,\ yz,\ zx)$

Further information: P.W. Atkins, M.S. Child, and C.S.G. Phillips, *Tables for group theory*. Oxford University Press (1970).

Answers to exercises

Detailed solutions to the (a) exercises can be found in the *Student's Solutions Manual to accompany Atkins' Physical Chemistry*, seventh edition, by P.W. Atkins, C.A. Trapp., M. Cady, and C. Giunta

Answers to 'a' exercises

1.4   10 atm.
1.5   (a) 24 atm; (b) 22 atm.
1.6   (a) 2.57 kTorr; (b) 3.38 atm.
1.7   30 K.
1.8   30 lb in^{-2}.
1.9 4.20×10^{-2} atm.
1.10 8.3147 J K^{-1} mol^{-1}.
1.11 S_8.
1.12   6.2 kg.
1.13   (a) 0.758, 0.242, 561 Torr, 179 Torr;
       (b) 0.780, 0.210, 0.010, 577 Torr, 155 Torr, 7 Torr.
1.14   169 g mol^{-1}.
1.15 -272°C.
1.16   (a) 1.0 atm, 8.2×10^2 atm; (b) 1.0 atm, 1.7×10^3 atm.
1.17   67.8 mL mol^{-1}, 54.5 atm, 120 K.
1.18   (a) 0.88; (b) 1.2 L.
1.19   140 atm.
1.20   (a) 0.1353 L mol^{-1}, 0.6957; (b) 0.658.
1.21   (a) 50.7 atm; (b) 35.2 atm, 0.695.
1.22   (a) 0.67, 0.33; (b) 2.0 atm, 1.0 atm; (c) 3.0 atm.
1.23   32.9 cm^3 mol^{-1}, 1.33 L^2 atm mol^{-2}, 0.24 nm.
1.24   (a) 1.41×10^3 K; (b) 0.59 nm.
1.25   (a) 3.64 kK, 8.7 atm; (b) 2.60 kK, 4.5 atm; (c) 46.7 K, 0.18 atm.
1.26 4.6×10^{-5} m^3 mol^{-1}, 0.66.

2.4   (a) 98 J; (b) 16 J.
2.5   2.6 kJ.
2.6 -1.0×10^2 J.
2.7   (a) $\Delta U = 0$, $\Delta H = 0$, $q = +1.57$ kJ, $w = -1.57$ kJ;
      (b) $\Delta U = 0$, $\Delta H = 0$, $q = +1.13$ kJ, $w = -1.13$ kJ; (c) all 0.
2.8   1.33 atm, $\Delta U = +1.25$ kJ, $w = 0$, $q = +1.25$ kJ.
2.9   (a) -88 J; (b) -167 J.
2.10   +123 J.
2.11 $\Delta U = -37.55$ kJ, $\Delta H = -40.656$ kJ, $q = -40.656$ kJ, $w = +3.10$ kJ.
2.12 -1.5 kJ.

2.13   84.8 MJ.

2.14   (a) $\Delta U = +26.8$ kJ, $\Delta H = +28.3$ kJ, $q = +28.3$ kJ, $w = -1.45$ kJ; (b) $\Delta U = +26.8$ kJ, $\Delta H = +28.3$ kJ, $q = +26.8$ kJ, $w = 0$.

2.15   131 K.

2.16   194 J.

2.17   22 kPa.

2.18   0.45 atm.

2.19 -125 kJ mol^{-1}.

2.20 $C_{p,m} = 30$ J K^{-1} mol^{-1}, $C_{V,m} = 22$ J K^{-1} mol^{-1}.

2.21 $q_p = -1.2$ kJ, $\Delta H = -1.2$ kJ, $C_p = 80$ J K^{-1}.

2.22 $\Delta U = +1.6$ kJ, $\Delta H = +2.2$ kJ, $q = +2.2$ kJ.

2.23 $w = -3.2$ kJ, $q = 0$, $\Delta T = -38$ K, $\Delta U = -3.2$ kJ, $\Delta H = -4.5$ kJ.

2.24 $w = +4.1$ kJ, $q = 0$, $\Delta U = +4.1$ kJ, $\Delta H = +5.4$ kJ, $p_f = 5.2$ atm, $V_f = 11.8$ L.

2.25   9.4 L, 288 K, -0.46 kJ.

2.26 $+0.9$ mm^3.

2.27 $q = 0$, $w = \Delta U = -20$ J, $\Delta T = -0.35$ K, $\Delta H = -26$ J.

2.28   (a) 226 K; (b) 238 K.

2.29 $\Delta U = +12$ kJ, $\Delta H = +13$ kJ, $q = +13$ kJ, $w = -1.0$ kJ.

2.30 -4564.7 kJ mol^{-1}.

2.31 -126 kJ mol^{-1}.

2.32 $+53$ kJ mol^{-1}, -33 kJ mol^{-1}.

2.33 -432 kJ mol^{-1}.

2.34   641 J K^{-1}.

2.35   1.58 kJ K^{-1}, $+2.05$ K.

2.36 $+65.49$ kJ mol^{-1}.

2.37 -383 kJ mol^{-1}.

2.38   (a) -2205 kJ mol^{-1}; (b) -2200 kJ mol^{-1}.

2.39   (a) -57.20 kJ mol^{-1}; (b) -176.01 kJ mol^{-1}.

2.40   (a) -114.40 kJ mol^{-1}, -109.44 kJ mol^{-1}; (b) -92.31 kJ mol^{-1}, -241.82 kJ mol^{-1}.

2.41 -1368 kJ mol^{-1}.

2.42   (a) -392.1 kJ mol^{-1}; (b) -946.6 kJ mol^{-1}; (c) $+52.5$ kJ mol^{-1}.

2.43 -56.98 kJ mol^{-1}.

2.44   (a) $+131.29$ kJ mol^{-1}, 128.81 kJ mol^{-1}; (b) $+132.56$ kJ mol^{-1}, 129.42 kJ mol^{-1}.

2.45 -1892.2 kJ mol^{-1}.

2.46   (a) -124.2 kJ mol^{-1}; (b) -222.46 kJ mol^{-1}.

3.4   (a) $\partial^2 f/\partial y \partial x = \partial/\partial y(2xy) = 2x$, $\partial^2 f/\partial x \partial y = \partial/\partial x(x^2 + 6y) = 2x$;
    (b) $\partial^2 f/\partial y \partial x = \partial/\partial y(\cos xy - xy \sin xy)$
$$= -x \sin xy - x \sin xy - x^2 y \cos xy$$
$$= -2x \sin xy - x^2 y \cos xy,$$
$$\partial^2 f/\partial x \partial y = \partial/\partial x(-x^2 \sin xy) = -2x \sin xy - x^2 y \cos xy.$$

3.5 $\mathrm{d}z = 2axy^3\,\mathrm{d}x + 3ax^2 y^2\,\mathrm{d}y$.

3.6   (a) $\mathrm{d}z = (2x - 2y + 2)\,\mathrm{d}x + (4y - 2x - 4)\,\mathrm{d}y$.

3.7 $\mathrm{d}z = (y + 1/x)\,\mathrm{d}x + (x - 1)\,\mathrm{d}y$.

3.8 $(\partial C_V/\partial V)_T = (\partial(\partial U/\partial V)_T/\partial T)_V$.

3.9 $(\partial H/\partial U)_p = 1 + p(\partial V/\partial U)_p$.

3.10 $\mathrm{d}V = (\partial V/\partial p)_T\,\mathrm{d}p + (\partial V/\partial T)_p\,\mathrm{d}T$; $\mathrm{d}\ln V = -\kappa_T\,\mathrm{d}p + \alpha\,\mathrm{d}T$.

3.11   0, 0.

3.13   0.71 K atm^{-1}.

3.14 $\Delta U_m = +131$ J mol^{-1}, $q = +8.05 \times 10^3$ J mol^{-1}, $w = -7.92 \times 10^3$ J mol^{-1}.

3.15 1.31×10^{-3} K^{-1}.

3.16 1×10^3 atm.

3.17 -7.2 J atm^{-1} mol^{-1}, $+8.1$ kJ.

3.18 -4.2 atm.

4.4   (a) $+92$ J K^{-1}; (b) $+67$ J K^{-1}.

4.5   152.67 J K^{-1} mol^{-1}.

4.6 $+8.92$ J K^{-1}.

4.7 -22.1 J K^{-1}.

4.8 $w = +4.1$ kJ, $q = 0$, $\Delta U = +4.1$ kJ, $\Delta H = +5.4$ kJ, $\Delta S = 0$.

4.9 $+12.9$ J K^{-1}.

4.10   Not reversible.

4.11   (a) -54.9 kJ; (b) -195 J K^{-1}.

4.12 $+26$ J K^{-1}.

4.13   6.6 L.

4.14 $+2.8$ J K^{-1}.

4.15 ΔH(overall) $= 0$, ΔH(individual) $= \pm 1.9 \times 10^2$ kJ, ΔS (overall) $= +93.4$ J K^{-1}.

4.16   (a) $q = 0$; (b) $w = -20$ J; (c) $\Delta U = -20$ J; (d) $\Delta T = -0.35$ K; (e) $\Delta S = +0.60$ J K^{-1}.

4.17   (a) $+87.8$ J K^{-1} mol^{-1}; (b) -87.8 J K^{-1} mol^{-1}.

4.18   (a) -386.1 J K^{-1} mol^{-1}; (b) $+92.6$ J K^{-1} mol^{-1}; (c) -153.1 J K^{-1} mol^{-1}.

4.19   (a) -521.5 kJ mol^{-1}; (b) 25.8 kJ mol^{-1}; (c) -178.7 kJ mol^{-1}.

4.20   (a) -522.1 kJ mol^{-1}; (b) $+25.78$ kJ mol^{-1}; (c) -178.6 kJ mol^{-1}.

4.21 -93.05 kJ mol^{-1}.

4.22 -50 kJ mol^{-1}.

4.23   (a) $+2.9$ J K^{-1}; -2.9 J K^{-1}, 0; (b) $+2.9$ J K^{-1}, 0, $+2.9$ J K^{-1}; (c) 0, 0, 0.

4.24 $\Delta S = n(C_{V,m} - R)\ln 2$.

4.25   817.90 kJ mol^{-1}.

4.26   0.11, 0.38.

5.4 $(\partial S/\partial V)_T = \alpha/\kappa_T$.

5.5 -3.8 J.

5.6 -36.5 J K^{-1}.

5.7 $+10$ kJ.

5.8   (a) 15.7 atm; (b) $+8.25$ kJ.

5.9 $+7.3$ kJ mol^{-1}.

5.10 -0.55 kJ mol^{-1}.

5.11 -2.63×10^{-8} Pa^{-1}, 0.88.

5.12 $+10$ kJ.

5.13 $+11$ kJ mol^{-1}.

5.14 $p = RT/(V_m - b) - a/V_m^2$.

5.15 $(\partial S/\partial V)_T = nR/(V - nb)$, ΔS greater for van der Waals gas.

6.2   At 373 K, water vapour condenses to liquid. At 273 K, liquid water freezes. Ice remains at 260 K. There is a pause in the rate of cooling at 373 K and at 273 K.

6.4   303 K (30°C).

6.5 $+16$ kJ mol^{-1}, $+45.2$ J K^{-1} mol^{-1}.

6.6 $+20.80$ K J mol^{-1}.

6.7   (a) $+34.08$ kJ mol^{-1}; (b) 350.5 K.

6.8   281.8 K (8.7°C).

6.9   25 g s^{-1}.

6.10 (a) 1.7 kg; (b) 31 kg; (c) 1.4 g.

6.11 (a) +49 kJ mol^{-1}; (b) 216°C; (c) +99 J K^{-1} mol^{-1}.

6.12 272.80 K.

6.13 0.076 30 (7.6 per cent).

6.14 2.6 kPa.

6.15 72.8 mN m^{-1}.

6.16 728 kPa.

7.4   886.8 cm^3.

7.5   56.3 cm^3 mol^{-1}.

7.6   6.4 MPa.

7.7   0.13 MPa.

7.8 K_f = 32 K kg mol^{-1}, K_b = 5.22 K kg mol^{-1}.

7.9   82 g mol^{-1}.

7.10 381 g mol^{-1}.

7.11 −0.09°C.

7.12 +1.2 J K^{-1}, −0.35 kJ.

7.13 +4.7 J K^{-1} mol^{-1}.

7.14 (a) 1:1; (b) 0.8600.

7.15 (a) 3.4 mmol kg^{-1}; (b) 34 mmol kg^{-1}.

7.16 0.17 mol L^{-1}.

7.17 −0.16°C.

7.18 24 g kg^{-1}.

7.19 87 kg mol^{-1}.

7.20 Raoult's law basis: a_A = 0.833, γ_A = 0.93; Henry's law basis (concentration in mole fractions), a_B = 0.125, γ_B = 1.25; (concentration as molality) a_B = 2.8, γ_B = 1.25.

7.21 $p(CCl_4)$ = 32.2 Torr, $p(Br_2)$ = 6.1 Torr, p(Total) = 38.3 Torr, $y(CCl_4)$ = 0.841, $y(Br_2)$ = 0.16.

7.22 a_A = 0.499, a_M = 0.668, γ_A = 1.25, γ_M = 1.11.

8.4 x_A = 0.920, y_A = 0.968.

8.5   440 Torr, x_A = 0.268.

8.6   (a) yes; (b) y_A = 0.830.

8.7   (a) 154 Torr; (b) y_{DE} = 0.67.

8.8   (a) y_M = 0.36; (b) y_M = 0.82.

8.9   (a) 2; (b) 2.

8.10 2, 2.

8.11 (a) 3, 2; (b) 1.

8.13 At b_3, C = 2, P = 2, F = 2. The compositions of the phases are x_A = 0.18 and x_A = 0.70. At b_2, C = 2, P = 1, F = 3. Between the liquid line and b_1, C = 2, P = 2, F = 2. Above b_1, C = 1, P = 1, F = 3.

8.14 Incongruent melting occurs at 460°C. The composition of the eutectic is 4 per cent by mass of silver; it melts at θ_e = 215°C.

8.16 (a) ≈ 80 per cent silver by mass; (b) compound decomposes; (c) ≈ 80 per cent silver by mass.

8.17 (b) 620 Torr; (c) 490 Torr; (d) x_{Hexane} = 0.50, y_{Hexane} = 0.72; (e) y_{Hexane} = 0.50, x_{Hexane} = 0.30; (f) n_{vap} ≈ 1.7 mol, n_{liq} ≈ 0.3 mol.

8.20 (a) The mixture has a single liquid phase at all compositions.

(b) At $x(C_6F_{14})$ = 0.24, two liquid phases separate with compositions x = 0.24 and x = 0.48. At x > 0.48, a single phase forms.

9.5   −2.42 kJ mol^{-1}.

9.6   3.01.

9.7   (a) 2.85×10^{-6}; (b) +240 kJ mol^{-1}; (c) 0.

9.8   (a) 0.1411; (b) +4.855 kJ mol^{-1}; (c) 14.556.

9.9   (a) −68.26 kJ mol^{-1}, 9.2×10^{11}; (b) −69.7 kJ mol^{-1}, 1.3×10^9.

9.10 (a) 0.087 (A), 0.370 (B), 0.196 (C), 0.438 (D); (b) 0.33; (c) 0.33; (d) +2.8 kJ mol^{-1}.

9.11 1.5 kK.

9.12 +2.77 kJ mol^{-1}, −16.5 J K^{-1} mol^{-1}.

9.13 +12.3 kJ mol^{-1}.

9.14 50 per cent.

9.15 0.904, 0.096.

9.16 (a, c).

9.17 (b).

9.18 (a) +53 kJ mol^{-1}; (b) −53 kJ mol^{-1}.

9.19 −14.38 kJ mol^{-1}, product formation.

9.20 1110 K.

9.21 (a) 5.40; (b) 3.61.

9.22 (a) 5.13; (b) 8.88; (c) 2.88.

9.23 8.3.

9.25 (a) Na_2HPO_4/H_3PO_4; (b) $H_2PO_4^-/HPO_4^{2-}$.

10.4 −218.66 kJ mol^{-1}.

10.5 1.25×10^{-5} mol L^{-1}.

10.6 −291 kJ mol^{-1}.

10.7 (a) $I(KCl) = b/b^{\ominus}$; (b) $I(FeCl_3) = 6b/b^{\ominus}$; (c) $I(CuSO_4) = 4b/b^{\ominus}$.

10.8 0.90.

10.9 (a) 2.73 g; (b) 2.92 g.

10.10 0.25 mol kg^{-1}.

10.11 $\gamma_{\pm} = (\gamma_+ \gamma_-^2)^{1/3}$.

10.12 0.56.

10.13 1×10^4 per cent.

10.14 2.01.

10.15 −1108 kJ mol^{-1}.

10.16 +34.2 mV.

10.17 −1.18 V.

10.18 (a) $Ag^+(aq) + e^- \rightarrow Ag(s)$, $Zn^{2+}(aq) + 2e^- \rightarrow Zn(s)$, $2Ag^+(aq) + Zn(s) \rightarrow 2Ag(s) + Zn^{2+}(aq)$, +1.56 V;
(b) $H^+(aq) + e^- \rightarrow \frac{1}{2}H_2(g)$, $Cd^{2+}(aq) + 2e^- \rightarrow Cd(s)$, $Cd(s) + 2H^+(aq) \rightarrow Cd^{2+}(aq) + H_2(g)$, +0.40 V;
(c) $Cr^{3+}(aq) + 3e^- \rightarrow Cr(s)$, $[Fe(CN)_6]^{3-}(aq) + e^- \rightarrow [Fe(CN)_6]^{4-}(aq)$, $Cr^{3+}(aq) + 3[Fe(CN)_6]^{4-}(aq) \rightarrow Cr(s) + 3[Fe(CN)_6]^{3-}(aq)$, −1.10 V.

10.19 (a) Zn(s)|ZnSO$_4$(aq)||CuSO$_4$(aq)|Cu(s), +1.10 V;
(b) Pt|H$_2$(g)|HCl(aq)|AgCl(s)|Ag(s), +0.22 V;
(c) Pt|H$_2$(g)|H$^+$(aq), H$_2$O(l)|O$_2$(g)|Pt, +1.23 V.

10.20 (a) +1.56 V; (b) +0.40 V; (c) −1.10 V.

10.21 (a) +1.10 V; (b) +0.22 V; (c) +1.23 V.

10.22 (a) −363 kJ mol^{-1}; (b) −405 kJ mol^{-1}.

10.23 (a) +0.324 V; (b) +0.45 V.

10.24 +1.92 V.

10.25 −0.62 V.

10.26 (a) 6.5×10^9; (b) 1.5×10^{12}.

10.27 +0.49 V; 4×10^{16}.

10.28 $1.80 \times 10^{-10} \to 1.78 \times 10^{-10}$, $9.04 \times 10^{-7} \to 5.1 \times 10^{-7}$.

10.29 $E = E^{\ominus} - (RT/6F)\ln[(a(\mathrm{Cr}^{3+})^2)/(a(\mathrm{Cr}_2\mathrm{O}_7^{2-})a(\mathrm{H}^+)^{14})]$.

10.30 0.86.

10.31 0.

10.32 (a) 1×10^{-8} mol kg^{-1}; (b) 1×10^{-16}.

11.4 1.7 MW.

11.5 262 nm.

11.6 2.42 cm s^{-1}.

11.7 332 pm.

11.8 8.83×10^{-28} kg m s^{-1}, 0.969 km s^{-1}.

11.9 50.6 nm.

11.10 0.70 nm.

11.11 $E/(10^{-19}$ J), $E/($kg mol$^{-1})$: (a) 3.31, 199; (b) 3.61, 218; (c) 4.97, 299.

11.12 (a) 0.66 m s^{-1}; (b) 0.72 m s^{-1}; (c) 0.99 m s^{-1}.

11.13 21 m s^{-1}.

11.14 (a) 2.8×10^{18}; (b) 2.8×10^{20}.

11.15 6 kK.

11.16 (a) no ejection; (b) 3.19×10^{-19} J, 837 km s^{-1}.

11.17 (a) 7×10^{-19} J, 400 kJ mol^{-1}; (b) 7×10^{-20} J, 40 kJ mol^{-1}; (c) 7×10^{-34} J, 4×10^{-13} kJ mol^{-1}.

11.18 (a) 6.6×10^{-29} m; (b) 6.6×10^{-36} m; (c) 99.7 pm.

11.19 1.1×10^{-28} m s^{-1}, 1×10^{-27} m.

11.20 1.12×10^{-15} J.

12.4 (a) 1.81×10^{-19} J, 110 kJ mol^{-1}, 1.1 eV, 9.1×10^3 cm^{-1}; (b) 6.6×10^{-19} J, 400 kJ mol^{-1}, 4.1 eV, 3.3×10^4 cm^{-1}.

12.5 (a) 0.04; (b) 0.

12.6 0, $h^2/4L^2$.

12.7 $L/6$, $L/2$, $5L/6$.

12.8 3.

12.9 23 per cent.

12.10 4.30×10^{-21} J.

12.11 278 N m^{-1}.

12.12 2.63 μm.

12.13 3.72 μm.

12.14 (a) 3.3×10^{-34} J; (b) 3.3×10^{-33} J.

12.16 5.61×10^{-21} J.

12.17 $N = 1/(2\pi)^{1/2}$.

12.18 1.49×10^{-34} J s; 0, $\pm 1.05 \times 10^{-34}$ J s.

13.5 14.0 eV.

13.6 $r = 4a_0$, 0.

13.7 101 pm, 376 pm.

13.8 $N = 2/a_0^{3/2}$.

13.9 $\langle V \rangle = 2E(1s)$, $\langle T \rangle = -E(1s)$.

13.10 $r^* = 5.24a_0/Z$.

13.11 (Angular momentum/$\hbar$, angular nodes, radial nodes) = (a) 0, 0, 0; (b) 0, 0, 2; (c) $6^{1/2}$, 2, 0.

13.12 (a) $\frac{5}{2}$, $\frac{3}{2}$; (b) $\frac{7}{2}$, $\frac{5}{2}$.

13.13 1, 1

13.14 (a) 1; (b) 9; (c) 25.

13.15 $L = 2$, $S = 0$, $J = 2$.

13.16 $r = 0.35a_0$.

13.17 (b, c).

13.18 (a) 2; (b) 6; (c) 10; (d) 18.

13.19 (a) [Ar]$3d^8$; (b) $S = 1$, $M_S = 0$, ± 1, $S = 0$, $M_S = 0$.

13.20 (a) 1 (3), 0 (1); (b) $\frac{3}{2}$ (4), $\frac{1}{2}$ (2), $\frac{1}{2}$ (2).

13.21 $^3\mathrm{D}_3$, $^3\mathrm{D}_2$, $^3\mathrm{D}_1$, $^1\mathrm{D}_2$ with $^3\mathrm{D} < {}^1\mathrm{D}$.

13.22 (a) 0 (1); (b) $\frac{3}{2}$ (4), $\frac{1}{2}$ (2); (c) 2 (5), 1 (3), 2 (1).

13.23 (a) $^2\mathrm{S}_{1/2}$; (b) $^2\mathrm{P}_{3/2}$, $^2\mathrm{P}_{1/2}$.

14.5 (a) $1\sigma^2$ (1); (b) $1\sigma^2 2\sigma^{*2}$ (0); (c) $1\sigma^2 2\sigma^{*2} 1\pi^4$ (2).

14.6 (a) $1\sigma^2 2\sigma^{*2} 1\pi^4 3\sigma^2$; (b) $1\sigma^2 2\sigma^{*2} 1\pi^4 3\sigma^2 2\pi^{*1}$; (c) $1\sigma^2 2\sigma^{*2} 1\pi^4 3\sigma^2$.

14.7 C_2.

14.8 XeF$^+$ is shorter.

14.9 N$_2$ is shorter.

14.12 $E_{\mathrm{trial}} = 5\hbar^2/mL^2 = 1.013\,E_{\mathrm{true}}$.

14.13 $E_{\mathrm{trial}} = 3a\hbar^2/2\mu - (e^2/a_0)(a/2\pi^3)^{1/2}$.

15.5 E, C_3, $3\sigma_v$.

15.6 (a, b).

15.7 Yes.

15.10 i, σ_h.

15.12 (a) R_3; (b) C_{2v}; (c) D_{3h}; (d) $D_{\infty h}$.

15.13 (a) C_{2v}; (b) $C_{\infty v}$; (c) C_{3v}; (d) D_{2h}.

15.14 (a) C_{2v}; (b) C_{2h}.

15.15 Polar: NO$_2$, N$_2$O, CHCl$_3$, and cis-CHBr=CHBr. Chiral: none.

15.16 No, d_{xy}.

15.17 B$_1(x)$, B$_2(y)$, A$_1(z)$.

15.18 (a) E$_{1u}$, A$_{2u}$; (b) B$_{3u}$, B$_{2u}$, B$_{1u}$.

16.4 (a) 0.0469 J m^{-3} s; (b) 1.33×10^{-13} J m^{-3} s; (c) 4.50×10^{-16} J m^{-3} s.

16.5 $0.999\,999\,925 \times 660$ nm.

16.6 (a) 5×10 ps; (b) 5 ps.

16.7 (a) 53 cm^{-1}; (b) 0.53 cm^{-1}.

16.8 0.409 THz.

16.9 (a) 2.642×10^{-47} kg m^2; (b) 127.4 pm.

16.10 4.442×10^{-47} kg m^2, 165.9 pm.

16.11 232.1 pm.

16.12 106.5 pm, 115.6 pm.

16.13 20 475 cm^{-1}.

16.14 198.9 pm.

16.15 b, d, and cis-CH$_2$Cl$_2$.

16.16 a, b and d.

16.17 0.16 kN m^{-1}.

16.18 1.089 per cent.

16.19 328.7 N m^{-1}.

16.20 (a) 0.067; (b) 0.20.

16.21 HF (967.0 N m^{-1}), HCl (515.6 N m^{-1}), HBr (411.8 N m^{-1}), HI (314.2 N m^{-1}).

16.22 $1580.38\ cm^{-1}$, 7.644×10^{-3}.

16.23 5.15 eV.

16.24 $2699.77\ cm^{-1}$.

16.25 b, c, and d.

16.26 (a) 3; (b) 6; (c) 12.

16.27 (a) All; (b) symmetric stretch: Raman, antisymmetric stretch and bends: IR.

16.28 Raman active.

16.29 $4A_1 + A_2 + 2B_1 + 2B_2$.

17.5 (a) g; (c) g; (d) u.

17.6 $\frac{1}{2}$, 0.

17.7 80 per cent.

17.8 $6.28 \times 10^3\ L\ mol^{-1}\ cm^{-1}$.

17.9 $1.5\ mmol\ L^{-1}$.

17.10 $5.44 \times 10^7\ L\ mol^{-1}\ cm^{-2}$.

17.11 Diene: 243 nm; butene: 192 nm.

17.12 $450\ L\ mol^{-1}\ cm^{-1}$.

17.13 $159\ L\ mol^{-1}\ cm^{-1}$, 23 per cent.

17.14 (a) 0.9 m; (b) 3 m.

17.15 (a) $5 \times 10^7\ L\ mol^{-1}\ cm^{-2}$; (b) $2.5 \times 10^6\ L\ mol^{-1}\ cm^{-2}$.

18.4 600 MHz.

18.5 $(-1.625 \times 10^{-26}\ J) \times m_I$.

18.6 154 MHz.

18.7 (a) proton.

18.8 $6.116 \times 10^{-26}\ J$.

18.9 (i) (a) 5.87 T; (b) 38.3 T; (c) 23.4 T.
(ii) (a) 11.7 T; (b) 76.6 T; (c) 46.8 T.

18.10 (a) 1×10^{-6}; (b) 5.1×10^{-6}; (c) 3.4×10^{-5}.

18.11 13.

18.12 (a) 11 μT; (b) 110 μT.

18.14 $6.7 \times 10^2\ s^{-1}$.

18.17 (b).

18.18 0.59 mT, 20 μs.

18.19 0.2 kT, 10 mT.

18.20 2.0022.

18.21 2.3 mT, 2.003.

18.22 330.2 mT, 332.2 mT, 332.8 mT, 334.8 mT, 1:1:1:1.

18.23 (a) 1:3:3:1; (b) 1:3:6:7:6:3:1.

18.24 (a) 331.9 mT; (b) 1.201 T.

18.25 $\frac{3}{2}$.

19.5 1.

19.6 (a) 2.57×10^{27}; (b) 7.26×10^{27}.

19.7 2.83.

19.8 3.156.

19.9 $2.45\ kJ\ mol^{-1}$.

19.10 354 K.

19.11 (a) 0.71; (b) 0.996.

19.12 (a) 5×10^{-5}, 0.4, 0.905; (b) 1.4; (c) $22\ J\ mol^{-1}$; (d) $1.6\ J\ K^{-1}\ mol^{-1}$; (e) $4.8\ J\ K^{-1}\ mol^{-1}$.

19.13 4303 K.

19.14 (a) $138\ J\ K^{-1}\ mol^{-1}$; (b) $146\ J\ K^{-1}\ mol^{-1}$.

19.15 $5.18\ J\ K^{-1}\ mol^{-1}$.

19.16 a, b, and d.

20.4 (a) $5R/2$; (b) $3R$; (c) $3R$.

20.5 NH_3: 1.33 and 1.11 (1.31); CH_4: 1.33 and 1.08 (1.31).

20.6 (a) 19.6; (b) 34.3.

20.7 (a) 1; (b) 2; (c) 2; (d) 12; (e) 3.

20.8 43.1, 23.36 K.

20.9 $43.76\ J\ K^{-1}\ mol^{-1}$.

20.10 (a) 36.95, 80.08; (b) 36.7, 79.7.

20.11 72.5.

20.12 (a) $14.93\ J\ K^{-1}\ mol^{-1}$; (b) $25.65\ J\ K^{-1}\ mol^{-1}$.

20.13 $-13.8\ kJ\ mol^{-1}$, $-0.20\ kJ\ mol^{-1}$.

20.14 (a) $0.236R$; (b) $0.193R$.

20.15 $11.5\ J\ K^{-1}\ mol^{-1}$.

20.16 (a) $9\ J\ K^{-1}\ mol^{-1}$; (b) $13\ J\ K^{-1}\ mol^{-1}$; (c) $15\ J\ K^{-1}\ mol^{-1}$.

20.17 3.70×10^{-3}.

21.5 O_3 and H_2O_2, but rotation about the O—O bond in H_2O_2 averages out the polarity.

21.6 $1.01 \times 10^{-39}\ J^{-1}\ C^2\ m^2$ ($9.1 \times 10^{-24}\ cm^3$), 1.7 D.

21.7 4.8.

21.8 $1.42 \times 10^{-39}\ J^{-1}\ C^2\ m^2$ ($1.28 \times 10^{-23}\ cm^3$).

21.9 17 per cent, 23 per cent; no correlation.

21.10 (a) 0 (by symmetry); (b) 0.7 D; (c) 0.4 D.

21.11 37 D at 11.7° to x-axis.

21.12 4.9 μD.

21.13 1.34.

21.14 18.

21.15 6.9×10^{-6}.

22.5 24 nm.

22.6 3.08 μm, 3.08 nm.

22.7 $70\ kg\ mol^{-1}$, $71\ kg\ mol^{-1}$.

22.8 1.4×10^4.

22.9 (a) $18\ kg\ mol^{-1}$; (b) $20\ kg\ mol^{-1}$.

22.10 24 ns, 14 pm.

22.11 100.

22.12 $63\ kg\ mol^{-1}$.

22.13 $0.73\ mm\ s^{-1}$.

22.14 $31\ kg\ mol^{-1}$.

22.15 $3.4\ Mg\ mol^{-1}$.

22.16 4.3×10^5 g.

23.6 $(1, \frac{1}{2}, 0)$, $(1, 0, \frac{1}{2})$, $(\frac{1}{2}, \frac{1}{2}, \frac{1}{2})$.

23.7 (323), (110).

23.8 249 pm, 176 pm, 432 pm.

23.9 70.7 pm.

23.10 16°, 23°, 28°.

23.11 0.215 cm.

23.12 $3.96 \times 10^{-28}\ m^3$.

23.13 4, 4.01 g cm^{-3}.

23.14 190 pm.

23.15 (111), (200), (311).

23.16 8.17°, 4.82°, 11.75°.

23.17 fcc.

23.18 $F_{hkl} = f$.

23.19 0.9069.

23.20 0.414.

23.21 (a) 58.0 pm; (b) 102 pm.

23.22 0.340.

23.23 +1.6 per cent.

23.25 7.9 km s^{-1}.

23.26 4.6 kV.

23.29 0.010.

23.30 9.3×10^{-4} cm^3.

23.31 3.

23.32 -6.4×10^{-5} cm^3 mol^{-1}.

23.33 2.

23.34 4.326.

23.35 +0.016 cm^3 mol^{-1}.

23.36 222 T.

24.4 (a) 9.975; (b) 1.

24.5 (a) 72 K; (b) 0.95 km s^{-1}; (c) 72 K.

24.6 81 mPa.

24.7 9.7×10^{-7} m.

24.8 (a) 5×10^{10} s^{-1}; (b) 5×10^9 s^{-1}; (c) 5×10^3 s^{-1}.

24.9 (a) 6.7 nm; (b) 67 nm; (c) 6.7 cm.

24.10 9.06×10^{-3}.

24.11 1.9×10^{20}.

24.12 104 mg.

24.13 4.1×10^{-2} J m^{-2} s^{-1}.

24.14 0.056 nm^2.

24.15 17 W, 17 W.

24.16 43 g mol^{-1}.

24.17 30 h.

24.18 0.142 nm^2.

24.19 205 kPa.

24.20 (a) 130 μP; (b) 130 μP; (c) 240 μP.

24.21 (a) 5.4 mJ K^{-1} m^{-1} s^{-1}, 8.1 mW;
       (b) 29 mJ K^{-1} m^{-1} s^{-1}, 44 mW.

24.22 138 μP, 390 pm.

24.23 5.4×10^{-3} J K^{-1} m^{-1} s^{-1}.

24.24 (a) 1.1 m^2 s^{-1}, 4.4×10^2 mol m^{-2} s^{-1};
       (b) 1.1×10^{-5} m^2 s^{-1}, 4.4×10^{-3} mol m^{-2} s^{-1};
       (c) 1.1×10^{-7} m^2 s^{-1}, 4.4×10^{-5} mol m^{-2} s^{-1}.

24.25 7.63×10^{-3} S m^2 mol^{-1}.

24.26 347 μm s^{-1}.

24.27 0.331.

24.28 13.83 mS m^2 mol^{-1}.

24.29 4.01×10^{-8} m^2 V^{-1} s^{-1}, 5.19×10^{-8} m^2 V^{-1} s^{-1},
       7.62×10^{-8} m^2 V^{-1} s^{-1}.

24.30 1.90×10^{-9} m^2 s^{-1}.

24.31 1.3 ks.

24.32 420 pm.

24.33 27 ps.

24.34 113 μm; 5.594×10^{-5} m.

25.5 C: 3.0 mol L^{-1} s^{-1}, D: 1.0 mol L^{-1} s^{-1}, A: 1.0 mol L^{-1} s^{-1},
      B: 2.0 mol L^{-1} s^{-1}.

25.6 v: 0.50 mol L^{-1} s^{-1}, D: 1.5 mol L^{-1} s^{-1}, A: 1.0 mol L^{-1} s^{-1},
      B: 0.50 mol L^{-1} s^{-1}.

25.7 L mol^{-1} s^{-1}; (a) k[A][B]; (b) $3k$[A][B].

25.8 $v = \frac{1}{2}k$[A][B][C], L^2 mol^{-2} s^{-1}.

25.9 2.

25.10 2.

25.11 10.3 ks; (a) 499.7 Torr; (b) 480 Torr.

25.12 (a) 4.1×10^{-3} L mol^{-1} s^{-1}; (b) A: 2.6 ks, B: 7.4 ks.

25.13 (a) L mol^{-1} s^{-1}, L^2 mol^{-2} s^{-1}; (b) kPa^{-1} s^{-1}, kPa^{-2} s^{-1}.

25.14 (a) 45, 95 mmol L^{-1}; (b) 1, 51 mmol L^{-1}.

25.15 124 ks.

25.16 64.9 kJ mol^{-1}, 4.32×10^8 mol L^{-1} s^{-1}.

25.17 $k_D/k_H \approx 0.15$.

25.19 1.9 MPa^{-1} s^{-1}.

25.20 7.1×10^5 s^{-1}, 7.61 ns.

26.1 (b) initiation, (3) retardation, (4) termination

26.5 $v = k_1 k_2 [O_3]^2/(k_1'[O_2] + k_2[O_3])$.

26.7 0.16 to 4.0 kPa.

26.8 d[P]/dt = $(k_1/k_3$[AH]^2[B])/$(k_2$[BH]$^+$ + k_3[AH]).

26.10 1.52 mmol L^{-1} s^{-1}.

26.11 3.3×10^{18}.

26.12 0.518.

27.4 9.5×10^9 s^{-1}, 1.2×10^{35} m^{-3} s^{-1}, 1.7%.

27.5 (a) 0.018, 0.30; (b) 3.9×10^{-18}, 6.0×10^{-6}.

27.6 (a) 13 per cent, 1.2 per cent; (b) 130 per cent, 12 per cent.

27.7 1.7×10^{-2} L mol^{-1} s^{-1}.

27.8 3×10^{10} L mol^{-1} s^{-1}.

27.9 (a) 6.61×10^6 m^3 mol^{-1} s^{-1}; (b) 3.0×10^7 m^3 mol^{-1} s^{-1}.

27.10 7.4×10^9 L mol^{-1} s^{-1}; 0.14 μs.

27.11 1.2×10^{-3}.

27.12 1.9×10^8 mol L^{-1} s^{-1}.

27.13 69.7 kJ mol^{-1}, -25 J K^{-1} mol^{-1}.

27.14 +71.9 kJ mol^{-1}.

27.15 -96.6 J K^{-1} mol^{-1}.

27.16 -76 J K^{-1} mol^{-1}.

27.17 (a) -45.8 J K^{-1} mol^{-1}; (b) +5.0 kJ mol^{-1}; (c) +18.7 kJ mol^{-1}.

27.18 20.9 L^2 mol^{-2} min^{-1}.

28.5 (a) 1.07×10^{25} m^{-2} s^{-1}, 1.4×10^{18} m^{-2} s^{-1};
      (b) 2.35×10^{24} m^{-2} s^{-1}, 3.1×10^{17} m^{-2} s^{-1}.

28.6 1.3×10^4 Pa.

28.7 3.4×10^5 s^{-1}.

28.8 12.7 m^2.

28.9 20.5 cm^3.

28.10 Chemisorption, 50 s.

28.11 $E_d = 610$ kJ mol^{-1}, $\tau_0 = 1.13 \times 10^{-13}$ s, $A = 6.15 \times 10^{12}$ s^{-1}.

28.12 (a) 0.21 kPa; (b) 22 kPa.

28.13 0.83, 0.36.

28.14 (a) 40 ps, 0.6 ps; (b) 20 Ts, 7 μs.

28.15 15 kPa.

28.16 0 on gold, 1 on platinum.

28.17 -13 kJ mol^{-1}.

28.18 650 kJ mol^{-1}; (a) 1.1×10^{97} min; (b) 2.9×10^{-6} min.

29.5 1.2 eV.

29.6 12 nm^{-1}.

29.7 0.24 GV m^{-1}.

29.8 138 mV.

29.9 2.8 mA cm^{-2}.

29.10 Increase $\times 50$.

29.11 (a) 0.17 mA cm^{-2}; (b) 0.17 mA cm^{-2}.

29.12 0.99 A m^{-2}.

29.13 0.2 μmol L^{-1}.

29.14 (a) 0.31 mA cm^{-2}; (b) 5.41 mA cm^{-2}; (c) -2.19 mA cm^{-2}.

29.15 $a(Fe^{3+})/a(Fe^{2+}) = 0.1$: 684 mA cm^{-2};
$a(Fe^{3+})/a(Fe^{2+}) = 1$: 215 mA cm^{-2};
$a(Fe^{3+})/a(Fe^{2+}) = 10$: 68 mA cm^{-2}.

29.16 108 mV.

29.17 (a) 4.9×10^{15} cm^{-2} s^{-1}, 3.8 s^{-1}; (b) 1.6×10^{16} cm^{-2} s^{-1}, 12 s^{-1}; (c) 3.1×10^7 cm^{-2} s^{-1}, 2.4×10^{-8} s^{-1}.

29.18 (a) 33 Ω; (b) 33 GΩ.

29.21 Yes.

29.22 No.

29.23 $+1.30$ V, 0.13 W.

29.24 (a) $+1.23$ V; (b) $+1.06$ V.

29.25 1.2 mm year^{-1}.

Answers to 'b' exercises

1.4 832 kPa.

1.5 (a) 10 bar; (b) 10.4 bar.

1.6 (a) 8.04×10^2 Torr; (b) 1.07 bar.

1.7 92.4 K.

1.8 120 kPa.

1.9 2.67×10^3 kg.

1.10 8.206 15 $\times 10^{-2}$ L atm mol^{-1} K^{-1}, 31.9987 g mol^{-1}.

1.11 P$_4$.

1.12 2.6 kg.

1.13 (a) 3.14 L; (b) 212 Torr.

1.14 16.4 g mol^{-1}.

1.15 -270°C.

1.16 (a) 1.0 atm, 270 atm; (b) 0.99 atm, 190 atm.

1.17 0.131 L mol^{-1}, 25.7 atm, 109 K.

1.18 (a) 1.12, repulsive; (b) 2.7 L mol^{-1}.

1.19 (a) 0.124 L mol^{-1}; (b) 0.112 L mol^{-1}.

1.20 (a) $V_m = 31.73$ L mol^{-1}, $z = 0.996$; (b) 0.995.

1.21 (a) 8.7 mL; (b) -0.15 L mol^{-1}.

1.22 (a) 0.63, 0.37; (b) $p(N_2) = 2.5$ atm, $p(H_2) = 1.5$ atm; (c) 4.0 atm.

1.23 $a = 3.16$ L^2 atm mol^{-2}, $b = 0.493$ L mol^{-1}, $r = 1.94 \times 10^{-10}$ m.

1.24 (a) 1259 K; (b) 0.129 nm.

1.25 (a) 2.6 atm, 881 K; (b) 2.2 atm, 718 K; (c) 1.4 atm, 356 K.

1.26 0.13 L mol^{-1}, 0.67.

2.4 (a) 4.9×10^3 J; (b) 1.9×10^3 J.

2.5 59 J.

2.6 -91 J.

2.7 $\Delta U = 0$, $\Delta H = 0$; (a) $w = -1.62$ kJ, $q = +1.62$ kJ; (b) $w = -1.38$ kJ, $q = +1.38$ kJ; (c) $w = q = 0$.

2.8 $p_2 = 143$ kPa, $w = 0$, $q = \Delta U = +3.28$ kJ.

2.9 (a) -19 J; (b) -52.8 J.

2.10 $+6.01$ J.

2.11 $q = \Delta H = -70.6$ kJ, $w = +5.60 \times 10^3$ J, $\Delta U = -65.0$ kJ.

2.12 -188 J.

2.13 $+30.1$ MJ.

2.14 (a) $q = \Delta H = +14.9$ kJ, $w = -831$ J, $\Delta U = +14.1$ kJ; (b) $q = \Delta U = +14.1$ kJ, $w = 0$, $\Delta H = +14.9$ kJ.

2.15 200 K.

2.16 -325 J.

2.17 8.5 Torr.

2.18 $p_i = 1.9$ atm, $p_f = 0.46$ atm.

2.19 -199 kJ mol^{-1}.

2.20 $C_{p,m} = 53$ J K^{-1} mol^{-1}, $C_{V,m} = 45$ J K^{-1} mol^{-1}.

2.21 $q_p = \Delta H = -2.3$ kJ, $C = 0.18$ kJ K^{-1}.

2.22 $\Delta H = q = +2.0$ kJ mol^{-1}, $\Delta U = +1.6$ kJ mol^{-1}.

2.23 $q = 0$, $w = \Delta U = -3.5$ kJ, $\Delta T = -24$ K, $\Delta H = -4.5$ kJ.

2.24 $q = 0$, $w = \Delta U = +2.4$ kJ, $\Delta H = +3.1$ kJ, $V_f = 14$ L, $p_f = 3.8 \times 10^5$ Pa.

2.25 20 L, 275 K, -0.75 kJ.

2.26 1.8×10^{-3} cm^3.

2.27 $q = 0$, $w = \Delta U = -36$ J, $\Delta T = -0.57$ K, $\Delta H = -50$ J.

2.28 (a) 164 K; (b) 171 K.

2.29 $q = \Delta H = +24$ kJ, $w = -1.6$ kJ, $\Delta H = +22.4$ kJ.

2.30 -3053.6 kJ mol^{-1}.

2.31 -126 kJ mol^{-1}.

2.32 -1152 kJ mol^{-1}.

2.33 -324.83 kJ mol^{-1}.

2.34 451 J K^{-1}.

2.35 69.3 J K^{-1}, 63.1 K.

2.36 $+84.40$ kJ mol^{-1}.

2.37 $+1.90$ kJ mol^{-1}.

2.38 (a) -2857 kJ mol^{-1}; (b) -2851 kJ mol^{-1}.

2.39 (a) -32.88 kJ mol^{-1}; (b) -55.84 kJ mol^{-1}.

2.40 (a) -589.56 kJ mol^{-1}, -582.13 kJ mol^{-1}; (b) $+26.48$ kJ mol^{-1}, -241.82 kJ mol^{-1}.

2.41 -760.3 kJ mol^{-1}.

2.42 $+52.5$ kJ mol^{-1}.

2.43 -566.93 kJ mol^{-1}.

2.44 -175 kJ mol^{-1}, -173 kJ mol^{-1}, -176 kJ mol^{-1}.

2.45 -1587 kJ mol^{-1}.

2.46 (a) -229.6 kJ mol^{-1}; (b) -160.5 kJ mol^{-1}.

3.5 $dz = dx/[(1 + y)^2 - 2x\,dy/(1 + y)^3]$.

3.6   (a) $(3x^2 - 2y^2)\,dx - 4xy\,dy$.

3.7 $dz = (2xy + y^2)\,dx + (x^2 + 2xy)\,dy$.

3.8 $(\partial C_p/\partial p)_T = [\partial(\partial H/\partial p)_T/\partial T]_p$.

3.10 $dp = (\partial p/\partial V)_T\,dV + (\partial V/\partial T)_V\,dT$,
      $d\ln p = (1/p\kappa_T)(\alpha\,dT - dV/V)$.

3.11   0, 0.

3.12 $\kappa_T = 1/p$, $\alpha = 1/T$.

3.13   0.48 K atm^{-1}.

3.14 $\Delta V_m = 129$ J mol^{-1}, $q = +7.747$ kJ mol^{-1},
      $w = -7.62$ kJ mol^{-1}.

3.15 1.27×10^{-3} K^{-1}.

3.16 3.6×10^2 atm.

3.17 -41.2 J atm^{-1} mol^{-1}, 27.2 kJ.

3.18 -340 kPa.

4.4   (a) 1.8×10^2 J K^{-1}; (b) 1.5×10^2 J K^{-1}.

4.5   152.64 J K^{-1} mol^{-1}.

4.6   9.08 J K^{-1}.

4.7 -7.3 J K^{-1}.

4.8 $q = 0$, $\Delta S = 0$, $\Delta U = w = +2.75$ kJ, $\Delta H = +3.58$ kJ.

4.9   76.9 J K^{-1}.

4.10   Not reversible.

4.11   (a) -58.2 kJ; (b) -193 J K^{-1}.

4.12   17 J K^{-1}.

4.13   6.00 L.

4.14   0.2 J K^{-1}.

4.15 $\Delta H_{total} = 0$, $\Delta S_{total} = +24$ J K^{-1}.

4.16   (a) 0; (b) -230 J; (c) -230 J; (d) -5.3 K; (e) $+3.21$ J K^{-1}.

4.17   (a) $+104.6$ J K^{-1}; (b) -104.6 J K^{-1}.

4.18   (a) -21.0 J K^{-1} mol^{-1}; (b) $+512$ J K^{-1} mol^{-1}.

4.19   (a) -212.40 kJ mol^{-1}; (b) -5798 kJ mol^{-1}.

4.20   (a) -212.55 kJ mol^{-1}; (b) -5798 kJ mol^{-1}.

4.21 -86.2 kJ mol^{-1}.

4.22 -197 kJ mol^{-1}.

4.23   (a) $+3.0$ J K^{-1}, -3.0 J K^{-1}, 0;
      (b) $+3.0$ J K^{-1}, 0, $+3.0$ J K^{-1}; (c) 0, 0, 0.

4.24 $\Delta S = \frac{3}{2}nR \ln 3$.

4.25   2108.11 kJ mol^{-1}.

4.26   (a) 0.500; (b) 0.50 kJ; (c) 0.5 kJ.

5.4   Both (Maxwell relation) $= -\alpha V$.

5.5 -2.0 J.

5.6 -42.8 J K^{-1}.

5.7   3.2 kJ.

5.8   (a) 274 kPa; (b) 3.45 kJ.

5.9   2.71 kJ mol^{-1}.

5.10 -0.93 kJ mol^{-1}.

5.11 -1.92×10^{-7} Pa^{-1}, 10^{-6}.

5.12   200 J.

5.13 $+2.88$ kJ mol^{-1}.

5.14 $V = (RT/p)(1 + B'p/RT + C'p^2/RT + D'p^3/RT)$.

5.15 $\left(\dfrac{\partial S}{\partial V}\right)_T = \dfrac{R\left(1 + \dfrac{a}{RV_mT}\right)e^{-a/RV_mT}}{V_m - b}$

ΔS for a Dieterici gas may be greater or less than for a perfect gas, depending on temperature: at high T, greater; at low T, less.

6.2   (a) gas expands; (b) gas contracts; (c) gas freezes; (d) solid sublimes; (e) gas expands.

6.4   23°C.

6.5   2.4 kJ mol^{-1}, 5.5 J K^{-1} mol^{-1}.

6.6   25.25 kJ mol^{-1}.

6.7   (a) 31.11 kJ mol^{-1}; (b) 276.9 K.

6.8   272 K.

6.9   3.6 kg s^{-1}.

6.10   Yes, $p \geq 3$ Torr.

6.11   (a) 29 kJ mol^{-1}; (b) 0.22 atm, 0.76 atm.

6.12   272.40 K.

6.13 6.73×10^{-2}.

6.14   5.92 kPa.

6.15 7.12×10^{-2} N m^{-1}.

6.16 2.04×10^5 Pa.

7.4   843.5 cm^3.

7.5   18 cm^3.

7.6 8.2×10^3 kPa.

7.7 1.5×10^2 kPa.

7.8   7.1 K kg mol^{-1}, 4.99 K kg mol^{-1}.

7.9   270 g mol^{-1}.

7.10   178 g mol^{-1}.

7.11 -0.077°C.

7.12 6.34×10^{-2} J K^{-1}, -17.3 J.

7.13 -3.43 kJ, $+11.5$ J K^{-1}, 0.

7.14   (a) $n_B/n_E = 1$; (b) $m_B/m_E = 0.7358$.

7.15 N_2: 0.51 mmol kg^{-1}; O_2: 0.27 mmol kg^{-1}.

7.16   0.067 mol L^{-1}.

7.17 -0.52°C.

7.18   11 kg Pb/1 kg Bi.

7.19   14.0 kg mol^{-1}.

7.20 $\alpha = 0.9701$, $\gamma = 0.980$.

7.21 -3.54 kJ mol^{-1}, 212 Torr.

7.22 $a_A = 0.436$, $a_B = 0.755$, $\gamma_A = 1.98$, $\gamma_B = 0.968$.

8.4 $x_A = 0.5$, $y_A = 0.5$.

8.5   73.4 kPa, $x_A = 0.653$.

8.6   (a) yes; (b) $y_A = 0.458$, $y_B = 0.542$.

8.7   (a) 48 Torr; (b) $y_B = 0.77$, $y_T = 0.23$; (c) 34 Torr.

8.8   (a) $y_A = 0.81$; (b) $x_A = 0.67$, $y_A = 0.925$.

8.9   3.

8.10   (a) $C = 1$, $P = 2$; (b) $C = 2$, $P = 2$.

8.11   (a) $C = 2$, $P = 2$; (b) $F = 2$.

8.14 $x_B \approx 0.53$ at T_2 and $x_B = 0.82$ at T_3.

8.16   (a) $x_B \approx 0.75$; (b) $x_{AB_2} \approx 0.8$; (c) $x_{AB_2} \approx 0.6$.

8.17 A solid solution with $x(ZrF_4) = 0.24$ appears at 855°C. The solid solution continues to form, and its ZrF_4 content increases until it reaches $x(ZrF_4) = 0.40$ at 820°C. At that temperature, the entire sample is solid.

9.5 $+18.18$ kJ mol^{-1}.

9.6 0.98.

9.7 (a) 0; (b) 0.168; (c) 4.41 kJ mol^{-1}.

9.8 (a) 0.24; (b) $+19$ kJ mol^{-1}; (c) 2.96.

9.9 (a) -308.84 kJ mol^{-1}, 1.3×10^{54};
(b) -306.52 kJ mol^{-1}, 3.5×10^{49}.

9.10 (a) 0.178 (A), 0.31 (B), 0.116 (C), 0.674 (D);
(b) 9.6; (c) 9.6; (d) -5.6 kJ mol^{-1}.

9.11 1.4×10^3 K.

9.12 $+7.2$ kJ mol^{-1}, -21 J K^{-1} mol^{-1}.

9.13 -41.0 kJ mol^{-1}.

9.14 No change.

9.15 1.6×10^{-2}.

9.16 (b).

9.17 (b).

9.18 (a) $+39$ kJ mol^{-1}; (b) -39 kJ mol^{-1}.

9.19 (a) 9.24; (b) -12.9 kJ mol^{-1}; (c) $+161$ kJ mol^{-1};
(d) $+248$ J K^{-1} mol^{-1}.

9.20 397 K.

9.21 (a) 1.5×10^{-5}, 4.82; (b) 3.21.

9.22 (a) 8.37; (b) 8.74; (c) 5.07.

9.23 7.38.

9.25 (a) aniline and anilinium ion;
(b) ethylammonium ion and ethylamine.

10.4 -65.49 kJ mol^{-1}.

10.5 9.3×10^{-15} mol kg^{-1}.

10.6 -363 kJ mol^{-1}.

10.7 (a) $3b/b^{\ominus}$; (b) $15b/b^{\ominus}$; (c) $15b/b^{\ominus}$.

10.8 0.320.

10.9 (a) 45 g; (b) 38.8 g.

10.10 0.100 mol kg^{-1}.

10.11 $\gamma_{\pm} = (\gamma_{+}^2 \gamma_{-}^3)^{1/5}$.

10.12 0.661.

10.13 47.1 per cent.

10.14 1.3.

10.15 -128.8 kJ mol^{-1}.

10.16 $+56.3$ mV.

10.17 Cd(s)|CD(OH)$_2$(s)|OH$^-$(aq)|Ni(OH)$_2$(s)|Ni(OH)$_3$(s)|Pt,
L: Cd(s) + 2OH$^-$(aq) → Cd(OH)$_2$(s) + 2e$^-$,
R: Ni(OH)$_3$(s) + e$^-$ → Ni(OH)$_2$(s) + OH$^-$.

10.18 (a) Ag$_2$CrO$_4$(s) + 2e$^-$ → 2Ag(s) + CrO$_4^{2-}$(aq),
Cl$_2$(g) + 2e$^-$ → 2Cl$^-$(aq), Ag$_2$CrO$_4$(s) +
2Cl$^-$(aq) → 2Ag(s) + CrO$_4^{2-}$(aq) + Cl$_2$(g), -0.91 V;
(b) Sn^{4+}(aq) + 2e$^-$ → Sn^{2+}(aq),
Fe^{3+}(aq) + e$^-$ → Fe^{2+}(aq), Sn^{4+}(aq) +
2Fe^{2+}(aq) → Sn^{2+}(aq) + 2Fe^{3+}(aq), -0.62 V;
(c) MnO$_2$(s) + 4H$^+$(aq) + 2e$^-$ → Mn^{2+}(aq) + 2H$_2$O(l),
Cu^{2+}(aq) + 2e$^-$ → Cu(s), Cu(s) + MnO$_2$(s) + 4H$^+$(aq)
→ Cu^{2+}(aq) + Mn^{2+}(aq) + 2H$_2$O(l), $+0.89$ V.

10.19 (a) Na(s)|NaOH(aq)|H$_2$(g)|Pt, $+1.88$ V;
(b) Pt|H$_2$(g)|HI(aq)|I$_2$(s)|Pt, $+0.54$ V;
(c) Pt|H$_2$(g)|H$_2$O(l)|H$_2$(g)|Pt 0.83 V.

10.20 (a) -0.91 V; (b) -0.62 V; (c) $+0.89$ V.

10.21 (a) $+1.88$ V; (b) $+0.54$ V; (c) $+0.83$ V.

10.22 (a) -291 kJ mol^{-1}; (b) $+122$ kJ mol^{-1}.

10.23 (a) -2.455 V; (b) $+1.627$ V.

10.24 (a) $E = E^{\ominus} - (2RT/F) \ln(\gamma_{\pm}b)$; (b) -89.89 kJ mol^{-1}; (c) $+0.223$ V.

10.25 -1.24 V, $+239$ kJ mol^{-1}, $+300.3$ kJ mol^{-1},
$+237$ kJ mol^{-1} at 35°C.

10.26 1.7×10^{16}; (b) 8.2×10^{-7}.

10.27 $+0.20$ V.

10.28 AgI: 1.4×10^{-16}; Bi$_2$S$_3$: 1.1×10^{-97}; no significant difference.

10.29 $E = E^{\ominus} - (RT/5F) \ln\{a(Mn^{2+})/[a(MnO_4^-)a(H^+)^8]\}$.

10.30 9.72.

10.31 0.

10.32 (a) 10^{-20} mol L^{-1}; (b) 10^{-98}.

11.4 2.5×10^2 W.

11.5 1.15 μm.

11.6 1.3×10^{-5} m s^{-1}.

11.7 1.6×10^6 m s^{-1}.

11.8 1.89×10^{-27} kg m s^{-1}, 0.565 m s^{-1}.

11.9 38.4 nm.

11.10 5.8×10^{-6} m.

11.11 (a) 9.93×10^{-19} J, 5.98×10^5 J mol^{-1}; (b) 1.32×10^{-15} J,
7.98×10^8 J mol^{-1}; (c) 1.99×10^{-23} J, 12.0 J mol^{-1}.

11.12 (a) 0.499 m s^{-1}; (b) 665 m s^{-1}; (c) 9.98×10^{-6} m s^{-1}.

11.13 158 m s^{-1}.

11.14 (a) 3.52×10^{17} s^{-1}; (b) 3.52×10^{18} s^{-1}.

11.15 1800 K.

11.16 (a) 0; (b) 6.84×10^{-19} J, 1.23×10^6 m s^{-1}.

11.17 (a) 2.65×10^{-19} J, 160 kJ mol^{-1}; (b) 3.00×10^{-19} J,
181 kJ mol^{-1}; (c) 6.62×10^{-31} J, 4.0×10^{-10} kJ mol^{-1}.

11.18 (a) 1.23×10^{-10} m; (b) 3.9×10^{-11} m; (c) 3.88×10^{-12} m.

11.19 $\Delta x = 100$ pm, $\Delta v = 5.8 \times 10^5$ m s^{-1}.

11.20 1.67×10^{-16} J.

12.4 (a) 2.14×10^{-19} J, 1.29×10^2 kJ mol^{-1}, 1.34 eV,
1.08×10^4 cm^{-1}; (b) 3.48×10^{-19} J, 2.10×10^2 kJ mol^{-1},
2.17 eV, 1.75×10^4 cm^{-1}.

12.5 (a) 0.03; (b) 0.03.

12.6 $\langle p \rangle = 0$, $\langle p^2 \rangle = h^2/L^2$.

12.7 $L/10, 3L/10, L/2, 7L/10, 9L/10$.

12.8 6.

12.9 $n = 7.26 \times 10^{10}$, $\Delta E = 1.71 \times 10^{-31}$ J, $\lambda = 27.5$ pm; may be treated classically.

12.10 3.92×10^{-21} J.

12.11 260 N m^{-1}.

12.12 13.2 μm.

12.13 18.7 μm.

12.14 (a) 2.2×10^{-29} J; (b) 3.14×10^{-20} J.

12.16 2.3421×10^{-20} J.

12.18   Magnitude: 2.58×10^{-34} J s^{-1}; projections:
0, $\pm 1.0546 \times 10^{-34}$ J s, and $2.110\,94 \times 10^{-34}$ J s.

13.5 1.94×10^{-18} J.

13.6 $r = 11.5a_0/Z$, $3.53a_0/Z$, 0.

13.7 $r = 0$, $6a_0$.

13.8 $N = \frac{1}{4}(2\pi a_0^3)^{-1/2}$.

13.9 $\langle E_K \rangle = \hbar^2 Z^2/8ma_0^2$, $\langle V \rangle = -Z^2 e^2/16\pi\varepsilon_0 a_0$.

13.10 $P_{3s} = 4\pi r^2 (1/4\pi)(1/243)(Z/a_0)^3 (6 - 6\rho + \rho^2)^2 e^{-\rho}$, $p = 2Zn/3a_0$
$r = 0.74a_0/Z$, $4.19a_0/Z$, and $13.08\, a_0/Z$.

13.11   (a) 2.45×10^{-34} J s, 2, 1; (b) 1.49×10^{-34} J s, 1, 0;
(c) 1.49×10^{-34} J s, 1, 1.

13.12   (a) $\frac{1}{2}$, $\frac{3}{2}$; (b) $\frac{9}{2}$, $\frac{11}{2}$.

13.13   8, 7, 6, 5, 4, 3, 2.

13.14   (a) 1; (b) 64; (c) 25.

13.15 $S = 1$, $L = 3$, $J = 4$, $(2S + 1) = 3$ is the multiplicity.

13.16   (a) 110 pm, 20.1 pm; (b) 86 pm, 29.4 pm.

13.17   (b).

13.18   (a) 2; (b) 10; (c) 14; (d) 22.

13.19 $1s^2 2s^2 2p^6 3s^2 3p^6 3d^3$, S: $\frac{1}{2}$ or $\frac{3}{2}$, M_S: $\pm\frac{1}{2}$, $\pm\frac{3}{2}$.

13.20   (a) 2 (5), 1 (3), 0 (1); (b) $\frac{5}{2}$ (6), $\frac{3}{2}$ (4), $\frac{1}{2}$ (2).

13.21 $^1F_3$; $^3F_4$, $^3F_3$, $^3F_2$; $^1D_2$; $^3D_3$, $^3D_2$, $^3D_1$; $^1P_1$; $^3P_2$, $^3P_1$, $^3P_0$; $^3F_2$ is lowest.

13.22   (a) $J = 3, 2, 1$ (7, 5, 3 states);
(b) $J = \frac{7}{2}, \frac{5}{2}, \frac{3}{2}, \frac{1}{2}$ (8, 6, 4, 2 states); (c) $J = \frac{9}{2}, \frac{7}{2}$ (10, 8).

13.23   (a) $^2D_{5/2}$, $^2D_{3/2}$; (b) $^2P_{3/2}$, $^2P_{1/2}$.

14.5   (a) $1\sigma^2 2\sigma^{*1}$; (b) $1\sigma^2 2\sigma^{*2} 1\pi^4 3\sigma^2$; (c) $1\sigma^2 2\sigma^{*2} 3\sigma^2 1\pi^4 2\pi^{*2}$.

14.6   (a) $1\sigma^2 2\sigma^{*2} 3\sigma^2 1\pi^4 2\pi^{*4}$; (b) $1\sigma^2 2\sigma^{*2} 1\pi^4 2\pi^{*2}$;
(c) $1\sigma^2 2\sigma^{*2} 3\sigma^2 1\pi^4 2\pi^{*3}$.

14.7   (a) C_2, CN; (b) NO, O_2, F_2.

14.8 $1\sigma^2 1\pi^4 2\pi^{*2}$, BrCl is shorter.

14.9 O_2^+, O_2, O_2^-, O_2^{2-}.

14.10 $N = 1/(1 + 2\lambda S + \lambda^2)^{1/2}$.

14.11 $N(0.844A - 0.145B)$.

14.12   Not appropriate because of boundary conditions.

14.13 $-\mu e^4/12\pi^3\varepsilon_0^2\hbar^2$.

14.15   (a) $C_6H_6^-$; $a_{2u}^2 e_{1g}^4 e_{2u}^1$; $E = 7\alpha + 7\beta$
(b) $C_6H_6^+$; $a_{2u}^2 e_{1g}^3$; $E = 5\alpha + 7\beta$.

14.16   (a) $19.313\,68\beta$; (b) $19.448\,24\beta$.

15.5   4 C_3 axes (each C—Cl axis), 4 C_2 axes (bisecting
Cl—C—Cl angles), 3 S_4 axes (same as C_2 axes), and
6 dihedral mirror planes (each Cl—C—Cl plane).

15.6   (a).

15.7   necessarily 0.

15.8   forbidden.

15.10 T_d: (S_4 and S_1); T_h: (S_2).

15.12   (a) $C_{\infty v}$; (b) D_3; (c) C_{4v}; (d) C_s.

15.13   (a) D_{2h}; (b) D_{2h}; (c) $C_{2v}(1, 2)$, $C_{2v}(1, 3)$, $D_{2h}(1, 4)$.

15.14   (a) $C_{\infty v}$; (b) D_{5h}; (c) C_{2v}; (d) D_{3h}; (e) O_h; (f) T_d.

15.15   (a) o-dichlorobenzene, m-dichlorobenzene HF, XeO_2F_2;
(b) none.

15.16 NO_3^-: p_x and p_y; SO_3; all d-orbitals except d_{z^2}.

15.17 A_2.

15.18   (a) B_{3u}, B_{2u}, B_{1u}, (b) A_{2u}, E_{1u}.

15.19   yes.

16.4   (a) 7.73×10^{-32} J m^{-3} s; (b) 6.2×10^{-28} J m^{-3} s.

16.5 6.36×10^7 m s^{-1}.

16.6   (a) 1.59 ns; (b) 2.48 ps.

16.7   (a) 1.6×10^2 MHz; (b) 16 MHz.

16.8 3.4754×10^{-11} s^{-1}.

16.9   (a) 3.307×10^{-47} kg m^2; (b) 141.4 pm.

16.10 5.420×10^{-46} kg m^2, 162.8 pm.

16.11   116.21 pm.

16.12   116.1 pm, 155.9 pm.

16.13   20 603 cm^{-1}.

16.14   141.78 pm.

16.15   all.

16.16   all but (c) SF_6.

16.17   0.71 N m^{-1}.

16.18   28.4 per cent.

16.19   245.9 N m^{-1}.

16.20   (a) 0.212; (b) 0.561.

16.21   3002.3 cm^{-1} (DF), 2143.7 cm^{-1} (DCl), 1885.8 cm^{-1} (DBr),
1640.1 cm^{-1} (DI).

16.22   2374.05 cm^{-1}, 6.087×10^{-3}.

16.23 3.235×10^4 cm^{-1}, 4.01 eV.

16.24   2347.16 cm^{-1}.

16.25   all but (d) N_2.

16.26   (a) 30; (b) 42; (c) 13.

16.27   (a) IR: A_2'', E', Raman: A_1', E'; (b) IR: A_1, E, Raman: A_1, E.

16.28   (a) inactive; (b) active.

16.29 $A_{1g} + A_{2g} + E_{1u}$.

17.5 a_{2u}, e_{1g}, e_{2u}, b_{2g}; hence parities are u, g, u, g.

17.6   3, u.

17.7   22.2 per cent.

17.8 7.9×10^5 cm^2 mol^{-1}.

17.9 1.33×10^{-3} mol L^{-1}.

17.10 1.56×10^8 L mol^{-1} cm^{-2} (1.56×10^9 mmol^{-1}).

17.11   rise.

17.12   552 L mol^{-1} cm^{-1}.

17.13   128 L mol^{-1} cm^{-1}, 0.13.

17.14   (a) 0.020 cm; (b) 0.033 cm.

17.15 1.39×10^8 L mol^{-1} cm^{-2} (1.39×10^9 mmol^{-1}).

17.16   stronger.

18.4   649 MHz.

18.5 $E_{\pm 1} = \mp 2.35 \times 10^{-26}$ J, 0.

18.6   47.3 MHz.

18.7   (b).

18.8   3.523 T.

18.9   (a) 97.5 T, 244 T; (b) 7.49 T, 18.7 T; (c) 17.4 T, 43.5 T.

18.10   (a) 4.3×10^{-7}; (b) 2.2×10^{-6}; (c) 1.34×10^{-5}.

18.11   (a) independent of B and v; (b) 13.

18.12   (a) 4.2×10^{-6} T; (b) 3.63×10^{-5} T.

18.13   spectrum appears narrower at 650 MHz.

18.14 2.9×10^3 s^{-1}.

18.15   203 MHz.

18.17   neither.

18.18 9.40×10^{-4} T, 6.25 μs.

18.19   1.3 T.

18.20   2.0022.

18.21   2.2 mT, 1.992.

18.22   eight lines at $(332.8 \pm 1.055 \pm 1.435 \pm 1.445)$ mT, all of equal intensity (in a high resolution spectrometer).

18.23   (a) a triplet (1:2:1) of quartets (1:3:3:1); (b) a quintet (1:2:3:2:1) of septets (1:3:6:7:6:3:1).

18.24   (a) 332.3 mT; (b) 1209 mT.

18.25   1.

19.5   623 K.

19.6   (a) 15.9 pm, 5.04 pm; (b) 2.47×10^{26}, 7.82×10^{27}.

19.7   187.9.

19.8   4.006.

19.9   7.605 kJ mol^{-1}.

19.10   213 K.

19.11   (a) 0.997, 0.994; (b) 0.999 99, 0.999 98.

19.12   (a) 1.00 K: $n_2/n_1 = 1.39 \times 10^{11}$, $n_3/n_1 = 1.93 \times 10^{-22}$; 25.0 K: $n_2/n_1 = 0.368$, $n_3/n_1 = 0.135$; 100 K: $n_2/n_1 = 0.779$, $n_3/n_1 = 0.607$; (b) 1.503; (c) 88.3 J mol^{-1}; (d) 3.53 J K^{-1} mol^{-1}; (e) 6.92 J K^{-1} mol^{-1}.

19.13   7.62 K.

19.14   (a) 147 J K^{-1} mol^{-1}; (b) 169.6 J K^{-1} mol^{-1}.

19.15   10.7 J K^{-1} mol^{-1}.

19.16   (a).

20.4   (a) 3R, 6R; (b) 3R, 21R; (c) 2.5R, 6.5R. (In each case the first value assumes no vibrational contribution; the second a full vibrational contribution.)

20.5   with: 1.15; without: 1.40; experimental: 1.29.

20.6   (a) 143; (b) 251.

20.7   (a) 2; (b) 2; (c) 6; (d) 24; (e) 4.

20.8   5840, 0.8479 K.

20.9   84.57 J K^{-1} mol^{-1}.

20.10   (a) 2.50×10^3, 5.43×10^3; (b) the same.

20.11   (a) 8.03×10^3; (b) 1.13×10^4.

20.12   (a) 5.70 J K^{-1} mol^{-1}; (b) 14.83 J K^{-1} mol^{-1}.

20.13 -20.1 kJ mol^{-1}, -110 J mol^{-1}.

20.14 -3.65 kJ mol^{-1}.

20.15   14.90 J K^{-1} mol^{-1}.

20.17 ≈ 0.25.

21.5 SF_4.

21.6 $\alpha = 2.55 \times 10^{-39}$ C^2 m^2 J^{-1}, $\mu = 3.23 \times 10^{-30}$ C m (3.23 D).

21.7   8.97.

21.8 3.40×10^{-40} C^2 m^2 J^{-1}.

21.9 μ(C—F) $> \mu$(C—O).

21.10   1.4 D.

21.11 9.45×10^{-29} C m, 194.0°.

21.12 3.71×10^{-36} C m.

21.13   1.10.

21.14   16.

21.15 3.2×10^{-6}.

22.5   38.97 nm.

22.6 1.26×10^{-6} m, 1.97×10^{-8} m.

22.7 $\overline{M}_n = 68$ kg mol^{-1}, $\overline{M}_w = 69$ kg mol^{-1}.

22.8 1.06×10^4.

22.9   (a) 8.8 kg mol^{-1}; (b) 11 kg mol^{-1}.

22.10 9.4×10^{-8} s.

22.11   71.

22.12   120 kg mol^{-1}.

22.13 1.47×10^{-4} m s^{-1}.

22.14   56 kg mol^{-1}.

22.15 3.1×10^3 kg mol^{-1}.

22.16 3.9×10^5 g.

23.6 $(\frac{1}{2}, \frac{1}{2}, 0)$, $(0, \frac{1}{2}, \frac{1}{2})$.

23.7 $(3, 1, \overline{3})$, $(6, 4, 3)$.

23.8   214 pm, 174 pm, 87.2 pm.

23.9   86.7 pm.

23.10   38.2°, 44.4°, 64.6°.

23.11   0.054 cm.

23.12   1.2582 nm^3.

23.13   5, 2.90 g cm^{-3}.

23.14   182 pm.

23.15   (110), (110) (1, 1, 1), (200).

23.16   4.166°, 3.000°, 7.057°.

23.17   body-centred cubic.

23.18   2f for $h + k + l$ even; 0 for $h + k + l$ odd.

23.19 $\frac{2}{3}$.

23.21   (a) 57 pm; (b) 111 pm.

23.22   0.370.

23.23   contraction.

23.25   252 pm.

23.26   (a) 39 pm; (b) 12 pm; (c) 6.1 pm.

23.27   2421 kJ mol^{-1}.

23.29 5.8×10^{-2}.

23.30   0.003 dm^3.

23.31   5.

23.32 -8.2×10^{-4} cm^3 mol^{-1}.

23.33 1.58×10^{-8} m^3 mol^{-1}, dimerization occurs.

23.34   2.52.

23.35 1.85×10^{-7} m^3 mol^{-1}.

23.36   0.935.

24.4   (a) 7.079; (b) 1.

24.5   (a) 475 m s^{-1}; (b) 40 km; (c) 0.01 s^{-1}.

24.6 2.4×10^7 Pa.

24.7 4.1×10^{-7} m.

24.8 9.9×10^8 s^{-1}.

24.9   (a) 3.7×10^{-9} m; (b) 5.5×10^{-8} m; (c) 4.1×10^{-5} m.

24.10 9.6×10^{-2}.

24.11 1.1×10^{21}.

24.12 4.89×10^{-4} kg.

24.13 0.17 J m^{-2} s^{-1}.

24.14 1.61×10^{-19} m^2.

24.15 22 J s^{-1}.

24.16 554 g mol^{-1}.

24.17 1.5×10^4 s.

24.18 3.00×10^{-19} m^2.

24.19 1.00×10^5 Pa.

24.20 (a) 0.95×10^{-5} kg m^{-1} s^{-1}; (b) 0.99×10^{-5} kg m^{-1} s^{-1}; (c) 1.81×10^{-5} kg m^{-1} s^{-1}.

24.21 (a) 0.0114 J m^{-1} s^{-1} K^{-1}, 0.017 J s^{-1}; (b) 9.0×10^{-3} J m^{-1} s^{-1} K^{-1}, 0.014 J s^{-1}.

24.22 52.0×10^{-7} kg m^{-1} s^{-1}, 923 pm.

24.23 9.0×10^{-3} J m^{-1} s^{-1} K^{-1}.

24.24 (a) 0.107 m^2 s^{-1}, 0.87 mol m^{-2} s^{-1}; (b) 1.07×10^{-5} m^2 s^{-1}, 8.7×10^{-5} mol m^{-2} s^{-1}; (c) 7.13×10^{-8} m^2 s^{-1}, 5.8×10^{-7} mol m^{-2} s^{-1}.

24.25 4.09 mS m^2 mol^{-1}.

24.26 4.81×10^{-5} m^2 V^{-1} s^{-1}.

24.27 0.604.

24.28 25.96 mS m^2 mol^{-1}.

24.29 5.74×10^{-8} m^2 V^{-1} s^{-1}, 7.913×10^{-8} m^2 V^{-1} s^{-1}, 8.09×10^{-8} m^2 V^{-1} s^{-1}.

24.30 1.09×10^{-9} m^2 s^{-1}.

24.31 4.1×10^3 s.

24.32 207 pm.

24.33 20 ps.

24.34 (a) 78 s; (b) 17 ms.

25.5 A: 1.0 mol L^{-1} s^{-1}, B: 3.0 mol L^{-1} s^{-1}, C: 1.0 mol L^{-1} s^{-1}, D: 2.0 mol L^{-1} s^{-1}.

25.6 rate of: reaction, 0.33 mol L^{-1}s^{-1}, formation of C, 0.33 mol L^{-1}s^{-1}; formation of D, 0.66 mol L^{-1}s^{-1}; consumption of A, 0.33 mol L^{-1} s^{-1}.

25.7 k: L^2 mol^{-2} s^{-1}; (a) d[A]/d$t = -k$[A][B]^2; (b) d[C]/d$t = k$[A][B]^2.

25.8 k: s^{-1}, k[A][B][C]$^{-1}$.

25.9 2.00.

25.10 0.

25.11 1.80×10^6 s; (a) 31.5 kPa; (b) 29.0 kPa.

25.12 (a) 3.5×10^{-3} L mol^{-1} s^{-1}; (b) A: 2.4 h; B: 0.44 h.

25.13 m^3 molecule^{-1} s^{-1}, m^6 molecule^{-2} s^{-1}; Pa^{-1} s^{-1}, Pa^{-2} s^{-1}.

25.14 (a) 6.5×10^{-3} mol L^{-1}; (b) 0.025 mol L^{-1}.

25.15 1.5×10^6 s.

25.16 $E_a = 9.9$ kJ mol^{-1}; $A = 0.94$ L mol^{-1} s^{-1}.

25.17 (a) $k_T/k_H \approx 0.06$; (b) $k_{18}/k_{16} \approx 0.89$.

25.18 $\{(3^{n-1} - 1)/k(n - 1)\}$[A]$_0^{(1-n)}$.

25.19 9.9×10^{-6} s^{-1} Pa^{-1}.

25.20 $k_r = 1.7 \times 10^{-7}$ s^{-1}, $k_f = 8.3 \times 10^8$ L mol^{-1} s^{-1}.

26.5 k[N$_2$O$_5$] $= k_1 k_2$[N$_2$O$_5$]$/(k_1' + k_2)$.

26.6 $-k_1$[R$_2$] $- k_2(k_1/k_4)^{1/2}$[R$_2$]$^{3/2}$.

26.7 (a) does not occur; (b) 1.3×10^2 Pa to 3×10^4 Pa.

26.8 $(k_1 k_2 K_a^{1/2}/k_1')$[HA]$^{3/2}$[B].

26.10 2.57×10^{-4} mol L^{-1} s^{-1}.

27.4 6.64×10^9 s^{-1}, 8.07×10^{34} m^3 s^{-1}, 1.6 per cent.

27.5 (a) 2.4×10^{-3}, 0.10; (b) 7.7×10^{-27}, 1.6×10^{-10}.

27.6 (a) 1.2, 1.03; (b) 7.4, 1.3.

27.7 1.7×10^{-12} L mol^{-1} s^{-1}.

27.8 3.2×10^7 m^3 mol^{-1} s^{-1}.

27.9 (a) 1.97×10^6 m^3 mol^{-1} s^{-1}; (b) 2.4×10^5 m^3 mol^{-1} s^{-1}.

27.10 1.10×10^7 m^3 mol^{-1} s^{-1}, 5.05×10^{-8} s.

27.11 2.22×10^{-3}.

27.12 1.54×10^8 mol L^{-1} s^{-1}.

27.13 48.52 kJ mol^{-1}, -32.2 J K^{-1} mol^{-1}.

27.14 46.8 kJ mol^{-1}.

27.15 -93 J K^{-1} mol^{-1}.

27.16 -80.0 J K^{-1} mol^{-1}.

27.17 (a) -24.1 J K^{-1} mol^{-1}; (b) 27.5 kJ mol^{-1}; (c) 34.7 kJ mol^{-1}.

27.18 1.08 L^2 mol^{-2} min^{-1}.

28.5 (a) 2.88×10^{23} m^{-2} s^{-1}, 5.75×10^{17} m^{-2} s^{-1}; (b) 3.81×10^{24} m^{-2} s^{-1}, 7.60×10^{17} m^{-2} s^{-1}.

28.6 7.3×10^2 Pa.

28.7 6.6×10^4 s^{-1}.

28.8 18.8 m^2.

28.9 9.7 cm^3.

28.10 200 s.

28.11 3.7 kJ mol^{-1}.

28.12 (a) 0.32 kPa; (b) 3.9 kPa.

28.13 0.75, 0.25.

28.14 (a) 4.9×10^{-11} s, 2.4×10^{-12} s; (b) 1.6×10^{13} s, 1.4 s.

28.15 6.50 kPa.

28.17 -6.40 kJ mol^{-1}.

28.18 $E_d = 2.85 \times 10^5$ J mol^{-1}; (a) 1.48×10^{36} s; (b) 1.38×10^{-4} s.

29.5 $H_{AB} = 9.39 \times 10^{-24}$ J; $\lambda = 1.53$ eV.

29.6 1.4×10^3 s^{-1}.

29.7 2.8×10^8 V m^{-1}.

29.8 0.37 V.

29.9 1.6 mA cm^{-2}.

29.10 8.5 mA cm^{-2}.

29.11 (a) 0.34 A cm^{-2}; (b) 0.34 A cm^{-2}.

29.12 1.3 A m^{-2}.

29.13 4×10^{-6} mol L^{-1}.

29.14 $(2.5$ mA cm$^{-2}) [e^{0.42E'/f}(3.41 \times 10^{-6}) - e^{-0.58E'/f}(3.55 \times 10^7)]$.

29.16 0.61 V.

29.17 Cu, H$_2$|H$^+$: 6.2×10^{-12} s^{-1} cm^{-2}, 4.2×10^{-3} s^{-1}; Pt|Ce^{4+}, Ce^{3+}: 2.54×10^{14} s^{-1} cm^{-2}, 0.17 s^{-1}.

29.18 (a) 5.1 GΩ; (b) 10 Ω.

29.21 no.

29.22 no.

29.23 1.80 V, 0.180 W.

29.24 $0.976\,75$ V.

29.25 1.5 mm y^{-1}.

Answers to problems

Detailed solutions for selected problems (indicated with an asterisk) can be found in the *Student's Solutions Manual to accompany Atkins' Physical Chemistry*, seventh edition, by P.W. Atkins, C.A. Trapp, M. Cady, and C. Giunta.

1.1* 0.50 m^3.

1.2 1.5 kPa.

1.3* $-233°\text{N}$.

1.4 3.2×10^{-2} atm.

1.5* $p = \rho RT/M$, 45.9 g mol^{-1}.

1.6 $-272.95°\text{C}$.

1.7 102 g mol^{-1}, CH_2FCF_3 or CHF_2CHF_2.

1.8* (a) 0.184 Torr; (b) 68.6 Torr; (c) 0.184 Torr.

1.9 0.33 atm (N$_2$), 0 (H$_2$), 1.33 atm (NH$_3$), 1.66 atm.

1.10 (a) 12.5 L mol^{-1}; (b) 12.3 L mol^{-1}.

1.11* 0.927, 0.208 L.

1.12* (a) 0.944 L mol^{-1}, $Z = 0.911$; (b) 2.67 L mol^{-1}, $Z = 0.992$;
    (c) 5.11 L mol^{-1}, $Z = 1.002$. N$_2$ is closest to perfection at T$_I$.

1.13 (a) 501 K;
    (b) virial: 0.456 L mol^{-1}, 0.599 L mol^{-1}; perfect: 0.496 L mol^{-1}, 0.621 L mol^{-1}.

1.14* (a) 0.1353 L mol^{-1}; (b) 0.6957;
    (c) 0.59 (from expansion in $1/V_m$), 0.72 (from expansion in p).

1.15 210 K, 0.28 nm.

1.16 5.649 L^2 atm mol^{-1}, 59.4 cm^3 mol^{-1}, 21 atm.

1.17* $Z < 1$ when $\dfrac{a}{RT} > b$ and attractive forces predominate; $Z > 1$ when $\dfrac{a}{RT} < b$
    and repulsive forces predominate

1.18 $B = b - a/RT$, $C = b^2$, 1.26 L^2 atm mol^{-2}, 34.6 cm^3 mol^{-1}.

1.19* $V_c = 2b$, $T_c = a/4Rb$, $P_c = (a/4b^2)e^{-2}$, $Z_c = 0.271$, Z_c (vdw) $= 0.375$.

1.20* $V_c = 3C/B$, $T_c = B^2/3RC$, $p_c = B^3/27C^2$, $\frac{1}{3}$.

1.21* $B' = B/RT$, $C' = (C - B^2)/R^2T^2$.

1.22 -0.18 atm^{-1}, -4.4 L mol^{-1}.

1.23 $(dV_m/dT)_p = (RV_m + b)/(2pV_m + RT)$.

1.24* no.

1.25 1.11.

1.26* 8.54, 15.1.

1.27 0.011.

1.28* (a) $B = -1.32 \times 10^{-2}$ L mol^{-1};
    (b) $B = -1.51 \times 10^{-2}$ L mol^{-1}, $C = 1.07 \times 10^{-3}$ L^2 mol^{-2}.

1.29 $Z = 0.611$ for all gases; the experimental value from Fig. 1.27 is close to 0.55.

1.30*   (a) 4.6×10^3 mol; (b) 1.3×10^2 kg; (c) 1.2×10^2 kg.

1.32*   (a) 1.1×10^{-11} mol L^{-1}, 2.2×10^{-11} mol L^{-1};
      (b) 8.0×10^{-13} mol L^{-1}, 1.6×10^{-12} mol L^{-1}.

1.33*   51.5 km, 3.0×10^{-3} bar.

2.1* $q = \Delta H = \Delta U = 2.6$ MJ.

2.2*   (a) -3.46 kJ; (b) 0; (c) -3.46 kJ; (d) $+24.0$ kJ; (e) $+27.5$ kJ.

2.3* $T_1 = 273$ K, $T_2 = 546$ K, $T_3 = 273$ K; Step 1→2:
      $w = -2.27$ kJ, $q = +5.67$ kJ, $\Delta U = +3.40$ kJ, $\Delta H = +5.67$ kJ;
      Step 2→3: $w = 0$, $q = -3.40$ kJ, $\Delta U = -3.40$ kJ, $\Delta H = -5.67$ kJ;
      Step 3→1: $w = +1.57$ kJ, $q = -1.57$ kJ, $\Delta U = 0$, $\Delta H = 0$;
      Cycle: $w = -0.70$ kJ, $q = +0.70$ kJ, $\Delta U = 0$, $\Delta H = 0$.

2.4   (a) -0.27 kJ; (b) -0.94 kJ.

2.5 -8.9 kJ, -8.9 kJ.

2.6* $w = 0$, $\Delta U = +2.35$ kJ, $\Delta H = +3.03$ kJ.

2.7 $w = -25$ J, $q = +109$ J, $\Delta H = +97$ J.

2.8 $w = -1.7$ kJ, $\Delta U = +20.5$ kJ, $\Delta H = +22.2$ kJ,
      $\Delta_{vap}H = +40$ kJ mol^{-1}.

2.9* $+98.7$ kJ mol^{-1}, $+95.8$ kJ mol^{-1}.

2.10*   (a) $w = 0$, $q = +6.19$ kJ, $\Delta U = +6.19$ kJ, $\Delta H = +8.67$ kJ;
      (b) $w = -6.19$ kJ, $q = 0$, $\Delta U = -6.19$ kJ, $\Delta H = -8.67$ kJ;
      (c) $w = +4.29$ kJ, $q = -4.29$ kJ, $\Delta U = 0$, $\Delta H = 0$;
         cycle: $w = -1.90$ kJ, $q = +1.90$ kJ, $\Delta U = 0$, $\Delta H = 0$.

2.11   36.5 L.

2.12* -87.33 kJ mol^{-1}.

2.13* -2.13 MJ mol^{-1}, -1.267 MJ mol^{-1}.

2.14 $+17.7$ kJ mol^{-1}, $+116.0$ kJ mol^{-1}.

2.15* $\Delta_f H^{\ominus}(Na^+, aq) = -241$ kJ mol^{-1}.

2.16* $-25\,968$ kJ mol^{-1}, $+2357$ kJ mol^{-1}.

2.17 -994.30 kJ mol^{-1}.

2.18*   (a) $+240$ kJ mol^{-1}; (b) $+228$ kJ mol^{-1}.

2.19   (a) -101.8 kJ mol^{-1}; (b) -344.2 kJ mol^{-1};
      (c) $+44.0$ kJ mol^{-1}.

2.20*   (a) (1) isochoric step: $n = \infty$; isobaric step: $n = 0$;
         (2) adiabatic step: $n = \gamma$; isochoric step: $n = \infty$;
      (b) (1) isochoric step: $w = 0$, $\Delta U = q = +55.8$ kJ,
         $\Delta H = +78.1$ kJ; isobaric step: $w = +22.3$ kJ,
         $\Delta U = -55.8$ kJ, $\Delta H = q = -78.1$ kJ; overall:
         $w = -q = 22.3$ kJ, $\Delta U = \Delta H = 0$;
         (2) adiabatic step: $q = 0$, $\Delta U = w = +9.37$ kJ,
         $\Delta H = +13.1$ kJ; isochoric step: $w = 0$,
         $\Delta U = q = -9.37$ kJ, $\Delta H = -13.1$ kJ; overall:
         $w = -q = 9.37$ kJ, $\Delta U = \Delta H = 0$.

2.22*   (a) no, as a plot of $\ln p$ against $\ln V$ does not yield a
      straight line; (b) numerical integration yields
      $w = +685$ J; (c) fitting the data to the van der Waals
      equation yields $T = 350$ K.

2.23   (a) $-2Fa/\pi$; (b) 0.

2.24* $w = -nRT \ln(V_2/V_1) + n^2RTB(1/V_2 - 1/V_1) + \frac{1}{2}n^3RTC(1/V_2^2 - 1/V_1^2)$; (a) -1.5 kJ; (b) -1.6 kJ.

2.25   (a) 0.39 mol, 0.50 L, 0.50 L; (b) $+19$ kJ; (c) $+3.0$ kJ;
      (d) $\Delta U = 0$, for all 3 paths; path ACB: $q = -9.5 \times 10^2$ J;
      path ADB: $q = -1.9 \times 10^4$ J; path AB: $q = -3.0 \times 10^3$ J.

2.26* $w = -nRT \ln\{(V_2 - nb)/(V_1 - nb)\} - n^2a(1/V_2 - 1/V_1)$.
      (a) -1.7 kJ; (b) -1.8 kJ; (c) -1.5 kJ.

2.27   (a) $+60$ kJ; (b) -70 J; (c) $+10$ J, $+50$ J.

2.28* $\Delta H = nC_{p,m}(T_f - T_i)$.

2.29 $w_r = 3bw/a$; $w_r = -\frac{8}{9}nT_r \ln\{(V_{r,2} - \frac{1}{3})/(V_{r,1} - \frac{1}{3})\}$
      $- n(1/V_{r,2} - 1/V_{r,1})$; $w_r/n = -\frac{8}{9}\ln\{\frac{1}{2}(3x - 1)\} - 1/x + 1$.

2.30   (a) -25 kJ; (b) 9.7 m; (c) 39 kJ; (d) 15 m.

2.31*   (a) -2.80 MJ mol^{-1};
      (b) more exothermic by 5376 kJ mol^{-1}.

2.32* $\Delta_{trs}H - \Delta_{trs}U = -93.9$ kJ mol^{-1}.

2.33*   (a) 2878, 3537, 5471 kJ mol^{-1}; (b) 49.51, 49.02, 47.89 kJ g^{-1}.

2.34* $k = -67.44$ kJ mol^{-1}, $n = 0.9253$, $R = 1.000$ (an excellent
      fit). Predicted $\Delta_c H^{\ominus}$(decane) $= -6625.5$ kJ mol^{-1};
      experimental value, -6772.5 kJ mol^{-1}; error,
      2.17 per cent.

2.35   (a) Calculated valued at PM3 level (kJ mol^{-1}):
         -910, -1569, -2225, -2882, -3540;
      (b) agreement is good;
      (c) $n = 0.903$, $k = -73.7$ kJ mol^{-1}.

2.36*   (a) 16.2 kJ mol^{-1}; (b) 114.6 kJ mol^{-1}; (c) 122.0 kJ mol^{-1}.

2.37   (a) 120.3 kJ mol^{-1}; (b) 68.9 kJ mol^{-1}; (c) 48.1 kJ mol^{-1}.

3.1 2.18×10^{-11} Pa^{-1}, -0.220 cm^3, 997.2 cm^3.

3.2*   (a) $+0.75$ kJ mol^{-1}; (b) $+0.75$ kJ mol^{-1}.

3.3*   41.40 J K^{-1} mol^{-1}.

3.4* -30.5 J mol^{-1}.

3.5   1.67.

3.6* $T_1 = 842$ K, $T_2 = T_3 = 348$ K; $p_1 = p_2 = p_3 = 1.72$ bar;
      $V_1 = 40.7$ L, $V_2 = V_3 = 16.8$ L; $\Delta U_1 = 11.3$ kJ,
      $\Delta U_2 = \Delta U_3 = 1.04$ kJ; ΔU (total) $= 13.4$ kJ.

3.7   (a) Same answers as for problem 3.6. It does not matter
      whether the piston between chambers 2 and 3 is
      diathermic or adiabatic as long as that between
      chambers 1 and 2 is adiabatic.
      (b) $T_1 = T_2 = T_3 = 348$ K, $p_1 = p_2 = p_3 = 1.17$ bar,
      $V_1 = V_2 = V_3 = 24.8$ L, $\Delta V_1 = \Delta V_2 = \Delta V_3 = 1.04$ kJ,
      $\Delta V_{total} = 3.12$ kJ.

3.8*   increase.

3.9*   not exact.

3.10*   not exact; dq/T is exact.

3.11 $dw = (y + z)\, dx + (x + z)\, dy + (x + y)\, dz$.

3.13 $(\partial H/\partial p)_T = -\mu C_p$.

3.14* $C_{p,m} - C_{V,m} = R$.

3.18 $dp = \{R/(V_m - b)\}\, dT + \{2a/V_m^3 - RT/(V_m - b)^2\}\, dV_m$,
      $(\partial V/\partial T_p = RV^3(V - b)/(RTV^3 - 2a(V - b)^2)$.

3.19* $+3.60$ kJ.

3.20 $dp = \{a(V_m - 2b)/V_m^3 - p\}\, dV_m/(V_m - b) + (p + a/V_m^2)\, dT/T$.

3.21 $T = p(V - nb)/nR + na(V - nb)/RV^2$, $(\partial T/\partial p)_V = (V_m - b)R$.

3.22* $\alpha = RV^2(V - nb)/\{RTV^3 - 2na(V - nb)^2\}$,
      $\kappa_T = V^2(V - nb)^2/\{RTV^3 - 2na(V - nb)^2\}$.

3.23 $\mu C_p = (1 - b\zeta/V_m)V/(\zeta - 1)$, $\zeta = RTV_m^3/2a(V_m - b)^2$,
      1.41 K atm^{-1},
      $T_1 = (\frac{27}{4})T_c(1 - b/V_m)^2$, 2021 K.

3.24* $-T\left(\dfrac{\partial V}{\partial T}\right)_p + V.$

3.25 $9.2\ \text{J K}^{-1}\ \text{mol}^{-1}.$

3.26* $322\ \text{m s}^{-1}.$

3.27   (a) $\mu = \alpha T^2/C_p$; (b) $C_V = C_p - R(1 + 2apT/R)^2.$

3.28* $0.80\ \text{m}, 1.6\ \text{m}, 2.8\ \text{m}.$

3.29   (a) $29.9\ \text{K MPa}^{-1}$; (b) $-2.99\ \text{K}.$

3.30*   (a) $23.5\ \text{K MPa}^{-1}$; (b) $14.0\ \text{K MPa}^{-1}.$

4.1   (a) $-21.3\ \text{J K}^{-1}\ \text{mol}^{-1}, +21.7\ \text{J K}^{-1}\ \text{mol}^{-1}$;
       (b) $-111.2\ \text{J K}^{-1}\ \text{mol}^{-1}, -1.5\ \text{J K}^{-1}\ \text{mol}^{-1}.$

4.2 $+11\ \text{J K}^{-1}.$

4.3*   (a) $57.0°\text{C}, -43.9\ \text{kJ}, -118\ \text{J K}^{-1}, +146\ \text{J K}^{-1}, +28\ \text{J K}^{-1}$;
       (b) $49.9°\text{C}.$

4.4*   (a) $+50.74\ \text{J K}^{-1}, -11.5\ \text{J K}^{-1}$;
       (b) ΔA_A is indeterminate, $\Delta A_\text{B} = +3.46\ \text{kJ}$;
       (c) ΔG_A is indeterminate, $\Delta G_\text{B} = +3.46\ \text{kJ}$;
       (d) $+39.2\ \text{J K}^{-1}, -39.2\ \text{J K}^{-1}.$

4.5*   Step 1: $w = -11.5\ \text{kJ}, q = +11.5\ \text{kJ}, \Delta U = 0, \Delta H = 0,$
          $\Delta S = +19.1\ \text{J K}^{-1}, \Delta S_\text{tot} = 0$;
       Step 2: $w = -3.74\ \text{kJ}, q = 0, \Delta U = -3.74\ \text{kJ},$
          $\Delta H = -6.23\ \text{kJ}, \Delta S = 0, \Delta S_\text{sur} = 0,$
       Step 3: $w = +5.74\ \text{kJ}, q = -5.74\ \text{kJ}, \Delta U = 0, \Delta H = 0,$
          $\Delta S = -19.1\ \text{J K}^{-1}, \Delta S_\text{tot} = 0$;
       Step 4: $w = +3.74\ \text{kJ}, q = 0, \Delta U = +3.74\ \text{kJ},$
          $\Delta H = +6.23\ \text{kJ}, \Delta S = 0, \Delta S_\text{tot} = 0$;
       Cycle: $w = -5.8\ \text{kJ},$
          $q = +5.8\ \text{kJ}, \Delta U = 0, \Delta H = 0, \Delta S = 0, \Delta S_\text{tot} = 0.$

4.6*   Path (a): $w = -2.74\ \text{kJ}, q = +2.74\ \text{kJ}, \Delta U = 0, \Delta H = 0,$
          $\Delta S = +9.13\ \text{J K}^{-1}, \Delta S_\text{sur} = -9.13\ \text{J K}^{-1}, \Delta S_\text{tot} = 0$;
       Path (b): $w = -1.66\ \text{kJ}, q = +1.66\ \text{kJ}, \Delta U = 0, \Delta H = 0,$
          $\Delta S = +9.13\ \text{J K}^{-1}, \Delta S_\text{sur} = -5.53\ \text{J K}^{-1},$
          $\Delta S_\text{tot} = +3.60\ \text{J K}^{-1}.$

4.7*   Path (a): $q = 0, w = -9.1 \times 10^2\ \text{J}, \Delta H = -1.5\ \text{kJ}, \Delta S = 0,$
          $\Delta S_\text{sur} = 0, \Delta S_\text{tot} = 0$;
       Path (b): $q = 0, w = -7.5\ \text{kJ}, \Delta H = -1.2\ \text{kJ},$
          $\Delta S = +1.12\ \text{J K}^{-1}, \Delta S_\text{sur} = 0, \Delta S_\text{tot} = +1.12\ \text{J K}^{-1}.$

4.8   Process (a): $\Delta S = +5.8\ \text{J K}^{-1}, \Delta S_\text{sur} = -5.8\ \text{J K}^{-1}, \Delta H = 0,$
          $\Delta T = 0, \Delta A = -1.7\ \text{kJ}, \Delta G = -1.7\ \text{kJ}$;
       Process (b): $\Delta S = +5.8\ \text{J K}^{-1}, \Delta S_\text{sur} = -1.7\ \text{J K}^{-1}, \Delta H = 0,$
          $\Delta T = 0, \Delta A = -1.7\ \text{J K}^{-1}, \Delta G = -1.7\ \text{kJ}$;
       Process (c): $\Delta S = +3.9\ \text{J K}^{-1}, \Delta S_\text{sur} = 0, \Delta H = -0.84\ \text{kJ},$
          $\Delta T = -41\ \text{K}, \Delta A$ and ΔG indeterminate.

4.9   (a) $200.7\ \text{J K}^{-1}\ \text{mol}^{-1}$; (b) $232.0\ \text{J K}^{-1}\ \text{mol}^{-1}.$

4.10 $+45.4\ \text{J K}^{-1}, +51.2\ \text{J K}^{-1}.$

4.11* $-160.07\ \text{kJ mol}^{-1}.$

4.12   (a) $+17.0\ \text{J K}^{-1}$; (b) $+36\ \text{J K}^{-1}.$

4.13*   (a) $0.11\ \text{kJ mol}^{-1}$; (b) $0.11\ \text{kJ mol}^{-1}.$

4.14*   (a) $63.88\ \text{J K}^{-1}\ \text{mol}^{-1}$; (b) $66.08\ \text{J K}^{-1}\ \text{mol}^{-1}.$

4.15   at 298 K: $+41.16\ \text{kJ mol}^{-1}, +42.08\ \text{J K}^{-1}\ \text{mol}^{-1}$;
       at 398 K: $+40.84\ \text{kJ mol}^{-1}, +41.08\ \text{J K}^{-1}\ \text{mol}^{-1}.$

4.16*   (a) $+76.9\ \text{J K}^{-1}\ \text{mol}^{-1}$; (b) $+96.864\ \text{J K}^{-1}\ \text{mol}^{-1}.$

4.17   at 200 K: $32.00\ \text{kJ mol}^{-1}, 293.5\ \text{J K}^{-1}\ \text{mol}^{-1}.$

4.18* $+34.4\ \text{kJ mol}^{-1}, 243\ \text{J K}^{-1}\ \text{mol}^{-1}$ at 298 K.

4.19* $46.60\ \text{J K}^{-1}\ \text{mol}^{-1}, 46.73\ \text{J K}^{-1}\ \text{mol}^{-1}.$

4.23 $\Delta S = nC_{p,\text{m}} \ln(T_\text{f}/T_\text{h}) + nC_{p,\text{m}} \ln(T_\text{f}/T_\text{c}), +22.6\ \text{J K}^{-1}.$

4.24* $-21\ \text{K}, +35.9\ \text{J K}^{-1}\ \text{mol}^{-1}.$

4.27 $C_{p,\text{m}} = 274\ \text{J K}^{-1}\ \text{mol}^{-1}$ (Dulong and Petit);
          $287\ \text{J K}^{-1}\ \text{mol}^{-1}$ (Experimental).

4.29* $+247.8\ \text{J K}^{-1}\ \text{mol}^{-1}, +336.6\ \text{J K}^{-1}\ \text{mol}^{-1}, +314.7\ \text{J K}^{-1}\ \text{mol}^{-1}.$

4.30   (a) $+0.8\ \text{kJ mol}^{-1}$; (b) $+11.9\ \text{kJ mol}^{-1}$;
       (c) $+15.0\ \text{kJ mol}^{-1}.$

4.31 $7.8\ \text{km}.$

4.32*   Step 1: $\Delta S = 0, \Delta S_\text{sur} = 0$;
       Step 2: $\Delta S = +33\ \text{J K}^{-1}, \Delta S_\text{sur} = -33\ \text{J K}^{-1}$;
       Step 3: $\Delta S = 0, \Delta S_\text{sur} = 0$;
       Step 4: $\Delta S = -33\ \text{J K}^{-1}, \Delta S_\text{sur} = +33\ \text{J K}^{-1}.$

4.33 $\dfrac{4}{3}.$

5.1* $-501\ \text{kJ mol}^{-1}.$

5.2   (a) $+7\ \text{kJ mol}^{-1}$; (b) $+107\ \text{kJ mol}^{-1}.$

5.3* $73\ \text{atm}.$

5.4*   (a) 0.750; (b) $0.372.$

5.7* $(\partial S/\partial V)_T = \alpha/\kappa_T$ and $(\partial V/\partial S)_p = \alpha T V/C_p.$

5.8 $(\partial p/\partial S)_V = \alpha T/\kappa_T C_V.$

5.10   (a) $(\partial H/\partial P)_T = 0$;
       (b) $(\partial H/\partial p)_T = \{nb - (2na/RT)\lambda^2\}/\{1 - (2na/RTV)\lambda^2\}$
          $= -8.3\ \text{J atm}^{-1}; \Delta H = -8\ \text{J}; \lambda = 1 - b/V_\text{m}.$

5.12   (a) $3.0 \times 10^{-3}\ \text{atm}$; (b) $0.30\ \text{atm}.$

5.13 $(\partial C_V/\partial V)_T = (RT/V_\text{m}^2)(\partial^2(BT)/\partial T^2)_V.$

5.15 $\pi_T = ap/RTV_\text{m}.$

5.16* $\pi_T = 4.85\ \text{J L}^{-1}, \Delta S_\text{m} = -41.1\ \text{J K}^{-1}\ \text{mol}^{-1},$
       $\Delta H_\text{m} = -1.23\ \text{kJ mol}^{-1}, \Delta G_\text{m} = 11.0\ \text{kJ mol}^{-1}.$

5.17 0.02 per cent.

5.18*   (a) $\Delta_\text{r}G' = \tau\Delta_\text{r}G + (1 - \tau)\Delta_\text{r}H$;
       (b) $\Delta_\text{r}G' = \tau\Delta_\text{r}G + (1 - \tau)\Delta_\text{r}H - T\Delta_\text{r}C_p) - T'\Delta_\text{r}C_p \ln \tau, \tau = T'/T.$

5.20* $q_\text{rev} = nRT \ln\{(V_\text{f} - nb)/(V_\text{i} - nb)\}.$

5.21 $-0.50\ \text{kJ}.$

5.22*   As pressure increases, entropy increases, remains
       constant, and decreases, respectively.

5.23*   The second option leads to $p = nRT/(V - nb) + \text{constant}.$

5.24* $f = 10.2\ \text{atm}.$

5.25 $\phi = \text{e}^{-ap/(RT)^2}; f = 9.237\ \text{atm}.$

5.26* $\ln \phi = Bp/RT + (C - B^2)p^2/2R^2T^2 + \cdots, 0.999\ \text{atm}.$

5.27 $\phi = 2\text{e}^{\lambda^{1/2}-1}/(1 + \lambda^{1/2}), \lambda = 1 + 4pq/R.$

5.28 $-21\ \text{kJ mol}^{-1}.$

5.29*   13 per cent.

5.30   (a) spontaneous; (b) $\Delta_\text{r}G^{\ominus} = -566\ \text{kJ mol}^{-1}.$

5.31 $+57.2\ \text{kJ mol}^{-1}, +85.6\ \text{kJ mol}^{-1}, +112.8\ \text{kJ mol}^{-1}.$

5.33* $G' = G + p^*V_0(1 - \text{e}^{-p/p^*}),$ expansion.

6.1* $196.0\ \text{K}, 11.1\ \text{Torr}.$

6.2* $9\ \text{atm}.$

6.3   (a) $+5.56\ \text{kPa K}^{-1}$; (b) 2.5 per cent.

6.4*   (a) $-22.0\ \text{J K}^{-1}\ \text{mol}^{-1}$;
       (b) $-109.0\ \text{J K}^{-1}\ \text{mol}^{-1}$;
       (c) $+110\ \text{J mol}^{-1}.$

6.5   (a) -1.63 cm^3 mol^{-1}; (b) $+30.1$ L mol^{-1}, 0.6 kJ mol^{-1}.

6.6*   234.4 K.

6.7   22°C.

6.8*   (a) 357 K (84°C); (b) $+38.0$ kJ mol^{-1}.

6.9   (a) 227.5°C; (b) $+53$ kJ mol^{-1}.

6.11   (b) 178.18 K; (c) 383.6 K; (d) $+33.0$ kJ mol^{-1}.

6.12 $+31.6$ kJ mol^{-1}.

6.13* $\Delta_{sub}H(1\ \text{Torr}) = +80.84$ kJ mol^{-1},
$\Delta_{sub}S(1\ \text{Torr}) = +201.1$ J K^{-1} mol^{-1}.

6.16   9.8 Torr.

6.17 $1/T_h = 1/T_b + Mgh/T\Delta_{vap}H$, 363 K (90°C).

6.18* $(\partial^2\mu/\partial T^2)_p = -C_{p,m}/T$.

6.19*   Both ΔV and ΔS are zero.

6.21*   (b) 112 K; (c) $+8.07$ kJ mol^{-1}.

6.22 $+1.60 \times 10^4$ bar.

7.1* $K_A = 15.58$ kPa, $K_B = 47.03$ kPa.

7.2*   17.5 cm^3 mol^{-1} (NaCl), 18.07 cm^3 mol^{-1} (H$_2$O).

7.3 -1.4 cm^3 mol^{-1} (MgSO$_4$), 18.04 cm^3 mol^{-1} (H$_2$O).

7.4*   12.0 cm^3 mol^{-1}.

7.5   57.9 mL ethanol, 45.8 mL water, 0.96 cm^3.

7.6*   36 to 62 g mol^{-1}.

7.7   4.

7.8*   (b) $K_A = 450$ Torr, $K_I = 465$ Torr.

7.10* $V_{cyclo} = 109.0$ cm^3 mol^{-1}, $V_{polymer} = 279.3$ cm^3 mol^{-1}.

7.11   (a) $V_1 = V_{m,1} + a_0x_2^2 + a_1(3x_1 - x_2)x_2^2$,
$V_2 = V_{m,2} + a_0x_1^2 + a_1(x_1 - 3x_2)x_1^2$;
(b) $V_1 = 75.63$ cm^3 mol^{-1}, $V_2 = 99.06$ cm^3 mol^{-1}.

7.12*   For $x_T = 0.228, 0.511, 0.810$, $\gamma_T = 0.490, 0.723, 0.966$, and $\gamma_E = 1.031, 0.920, 0.497$.

7.13 $K_H = 371$ bar; at $p = 60.0$ bar, $\gamma_{CO_2} = 0.98$.

7.14* $S_0 = 19.89$ mol L^{-1}, $\tau = 165$ K; fits well, $R = 0.999\ 78$.

7.15   (b) -24.9 kJ mol^{-1}.

7.16 -4.6 kJ.

7.17* $\mu_A = \mu_A^* - RT \ln x_A + gRTx_B^2$.

7.19* $V_B(x_A, x_B) = V_B(0, 1) - \int_{V_A(0)}^{V_A(x_A)}\{x_A\ dV_A/(1 - x_A)\}$,
80.4 cm^3 mol^{-1}.

7.23* $[\text{Cl}^-]_R = [\text{Na}^+]_R = 0.201$ mol L^{-1};
$[\text{Cl}^-]_L = [\text{Na}^+]_L - 6[\text{P}] = 0.199$ mol L^{-1}.

7.24   (a) g cm mol^{-1} K^{-1}; (b) 1.1×10^5 g mol^{-1}; (c) 'good';
(d) $B' = 21.4$ cm^3 g^{-1}, $C' = 211$ cm^6 g^{-2};
(e) $(\pi/c)^{1/2} = (RT/\bar{M}_n)^{1/2} \times (1 + \frac{1}{2}B'c)$; $B' = 28.0$ cm^3 g^{-1}, $C' = 196$ cm^6 g^{-2}; yes.

7.25* 1.26×10^5 g mol^{-1}, 1.23×10^4 L mol^{-1}.

8.1*   (b) 391.0 K; (c) $n_{liq}/n_{vap} = 0.532$.

8.2   (b) $n_{left}/n_{right} = 0.093$; 302.5 K at $x = 0.750$.

8.3*   Temperature (γ): 78 K (0.9); 80 K (1.08); 82 K (1.04); 84 K (1.00); 86 K (0.99); 88 K (0.99); 90.2 K (0.99).
To within experimental uncertainties the solution appears ideal.

8.4*   (a) $\chi_S = 0.43(\text{P}_4\text{S}_3), 0.64(\text{P}_4\text{S}_7), 0.71(\text{P}_4\text{S}_{10})$.

8.5*   A compound with probable formula A$_3$B exists. It melts incongruently at 700°C. The proportions of A and B in the product are dependent upon the overall composition and the temperature. A eutectic exists at 400°C and $x_B \approx 0.83$.

8.6   The number of distinct chemical species (as opposed to components) and phases present at the indicated points are, respectively $b(3, 2)$, $d(2, 2)$, $e(4, 3)$, $f(4, 3)$, $g(4, 3)$, $k(2, 2)$.

8.7* $T_{uc} = 122$°C, $T_{lc} = 8$°C.

8.8   MgCu$_2$: 16 per cent Mg by mass; Mg$_2$Cu: 43 per cent Mg by mass.

8.9*   (a) $F(1500\ \text{K}) = 3$, $F(1100\ \text{K}) = 2$;
(c) 0.67 g Pb per g Cu; 0.050 g Cu per g Pb.

8.10   (a) eutectics: 40.2 at % Si at 1268°C; 69.4 at % Si at 1030°C congruent melting compounds: Ca$_2$Si, 1314°C; CaSi, 1324°C incongruent melting compound: CaSi$_2$ at 1040°C;
(b) $n_{Ca}/n_{liq} = 2.86$;
(c) $n_{Si}/n_{liq} = 0.53$ at slightly above 1030°C;
$n_{Si}/n_{CaSi_2} = 0.665$ at slightly below 1030°C.

8.11*   No phases are in equilibrium as the system cools. The phases through which the system passes upon cooling are: liquid, liquid + solid K$_2$FeCl$_4$, solid K$_2$FeCl$_4$ + solid KFeCl$_3$.

8.15*   (a) 2150°C; (b) x(MgO) = 0.35, y(MgO) = 0.18, ratio 0.4;
(c) 2640°C.

8.16   (a) $n(l)/n(s) = 5$; (b) no liquid.

8.17   (b) $n_{liq}/n_{vap} = 10.85$.

8.18*   (a) mp(SiO$_2$) = 1934.1 K, mp(Si) = 1715.5 K.

8.19   (a) Yes, both; (b) unfavorable, approximately regular;
(c) $\chi_{methanol} = \xi_{methanol} = 0.682$, $p = 11.59$ kPa;
at $\chi_{methanol} = 0.2$, p = 10.00 kPa;
(d) $K_{methanol} = 45.1$ kPa, $K_{TAME} = 25.3$ kPa; (e) 6.16 kPa.

9.1*   (a) $+4.48$ kJ mol^{-1}; (b) 0.101 atm.

9.2   (a) 1.24×10^{-9}; (b) 1.29×10^{-8}; (c) 1.8×10^{-4};
(d) as pressure increases, α decreases; as temperature increases, α increases.

9.3 $\Delta_f H^{\ominus} = -(2.196 \times 10^4\ \text{K} - 8.84T)R$, $8.48R$.

9.4* $\Delta_r H^{\ominus} = +3.00 \times 10^5$ J mol^{-1} and $\Delta_r S^{\ominus} = +102$ J K^{-1} mol^{-1} throughout this temperature range;
at 1395 K: $K = 1.22 \times 10^{-6}$, $\Delta_r G^{\ominus} = +158$ kJ mol^{-1};
at 1443 K: $K = 2.80 \times 10^{-6}$, $\Delta_r G^{\ominus} = +153$ kJ mol^{-1};
at 1498 K: $K = 7.23 \times 10^{-6}$, $\Delta_r G^{\ominus} = +147$ kJ mol^{-1}.

9.5 $\Delta_r G^{\ominus}(T)/(\text{kJ mol}^{-1}) = 78 - 0.161(T/\text{K})$.

9.6* 1.69×10^{-5}.

9.7   5.71, -103 kJ mol^{-1}.

9.8*   7 mmol H$_2$, 107 mmol I$_2$, 786 mmol HI.

9.9   0.33.

9.10 1.800×10^{-3} (at 973 K), 1.109×10^{-2} (at 1073 K), 4.848×10^{-2} (at 1173 K), $+158$ kJ mol^{-1}.

9.11*   0.23, 0.46, 0.30.

9.12*   +76.8 kJ mol^{-1}.

9.13   (a) 1.2×10^8; (b) 2.7×10^3.

9.15 $\xi = 1 - 1/(1 + ap/p^{\ominus})^{1/2}$.

9.16   0.140.

9.17 $\Delta_r G' = \Delta_r G + (T - T')\Delta_r S + \alpha\Delta a + \beta\Delta b + \gamma\Delta c$,
     $\alpha = T' - T - T' \ln(T'/T)$, $\beta = \frac{1}{2}(T'^2 - T^2) - T'(T' - T)$,
     $\gamma = 1/T - 1/T' + \frac{1}{2}T'(1/T'^2 - 1/T^2)$; -225.31 kJ mol^{-1}.

9.18 $v = \dfrac{B_0 K_b[\text{OH}^-] + (K_w - [\text{OH}^-]^2)([\text{OH}^-] + K_b)}{([\text{OH}^-] + K_b)([\text{OH}^-]^2 + A_0[\text{OH}^-] - K_w)}$.

9.19*   +14.7 kJ mol^{-1}, +18.8 kJ mol^{-1}.

9.20   (a) 40%; (b) 74%;
     (c) 41% for diesel compared to 74% for aerobic respiration.

9.21   less.

9.22   (a) -72.4 kJ mol^{-1}, -144 J K^{-1} mol^{-1};
     (b) +131.2 kJ mol^{-1}, +309.2 J K^{-1} mol^{-1}.

9.23*   now: 5.64; then: 5.70.

9.24   the trihydrate.

9.25*   (b) maximum carbon mass in ocean $\approx 2.67 \times 10^{12}$ tonne.

9.26   5.3 per cent carbon impurity; $p_{\text{CO}} = 0.207$ bar; $p_{\text{SiO}} = 0.559$ bar.

9.27*   (a) $p(450°\text{C}) = 156.5$ bar, $p(400°\text{C}) = 81.8$ bar;
     (b) $p(450°\text{C}) = 132.5$ bar, $p(400°\text{C}) = 210$ bar.

10.1   Pb(s)|PbSO$_4$(s)|PbSO$_4$(aq)|Hg$_2$SO$_4$(aq)|Hg$_2$SO$_4$(s)|Hg(l), +1.03 V.

10.2*   (a) 4.0×10^{-3}, 1.2×10^{-2}; (b) 0.74, 0.60; (c) 5.9;
     (d) +1.102 V; (e) +1.079 V.

10.3*   2.0.

10.4*   (a) $E = E^{\ominus} - (38.54 \text{ mV}) \times \{\ln(4^{1/3}b) + \ln \gamma_\pm\}$;
     (b) 1.0304 V; (c) 6.84×10^{34}; (d) 0.763; (e) 0.75;
     (f) -87.2 J K^{-1} mol^{-1}, -262.4 kJ mol^{-1}.

10.5*   +0.268 38 V.

10.6   14.23 at 20.0°C, +74.9 kJ mol^{-1}, +80.0 kJ mol^{-1}, -17.1 J K^{-1} mol^{-1}.

10.7   0.533.

10.8*   (a) +0.2223 V, +0.2223 V; (b) 1.10, 0.796.

10.10 -131.25 kJ mol^{-1}, +56.7 J K^{-1} mol^{-1}, -167.10 kJ mol^{-1}.

10.11   (a) -1.991 V $> E^{\ominus} > -2.19$ V, scandium;
     (b) 1.4×10^{10}.

10.12   (a) $(\partial E/\partial p)_{T,n} = -(\Delta_r V/\nu F)$;
     (b) 2.84×10^{-6} V atm^{-1} from fit, 2.80×10^{-6} V atm^{-1} from $\Delta_r V$;
     (c) $E_{30°\text{C}}/\text{V} = 0.008\,56 + 2.84 \times 10^{-6}p/(\text{atm})$. A second-order polynomial fit is slightly better;
     (d) $\kappa_T = +3.2 \times 10^{-7}$ atm^{-1}.

10.13*   (a) $\Delta E = -(RT/\nu F) \ln p/(\text{atm})$; fits well below 100 atm, but deviates above that pressure;
     (b) $\Delta E = (RT/\nu F)\{\ln p/(\text{atm}) + Cp\}$, $(\partial E/\partial p)_T = (RT/\nu F)\{1/p + C\}$, $C = 6.665 \times 10^{-4}$ atm^{-1}, $R = 0.999\,40$;
     (c) From empirical virial equation:
     $\Delta E = (RT/\nu F)\ln p/(\text{atm}) + 0.000\,537(p - 1)/(\text{atm})$
     $+ 1.75 \times 10^{-8}(p^2 - 1)/(\text{atm}^2)$;

(d) p, $\phi/\phi_{1\text{ atm}}$: 1.00 atm, 1.00; 10 atm, 0.994; 38 atm, 1.02; 51 atm, 1.04; 108 atm, 1.07; 210 atm, 1.03; 380 atm, 1.23; 430 atm, 1.38; 560 atm, 1.44; 720 atm, 1.64; 900 atm, 1.81; 1020 atm, 2.02.

10.15 $b(\text{H}_2\text{VO}_4^-) = 0.0048$ mol kg^{-1}, $b(\text{V}_4\text{O}_{12}^{4-}) = 0.0013$ mol kg^{-1}.

10.16*   pK_a(25°C) = 6.736.

10.17* $a_0 = -14.102$, $a_1 = 2461$ K, $a_2 = 2.209$.

10.18   (a) $E = E^{\ominus} + (2.303RT/F)\text{pOH}$;
     (b) $E = E^{\ominus} + (2.303RT/F) \times (\text{p}K_w - \text{pH})$;
     (c) -37.6 mV.

10.19 -1.2 V.

10.25   (b) +0.206 V.

10.26* $5 < \text{pH} < 8$.

10.27   (b) $E^{\ominus} = 2.05$ V, $\Delta_r G^{\ominus} = -395.6$ kJ mol^{-1},
     $\Delta_r H^{\ominus} = -359.5$ kJ mol^{-1}, $\Delta_r S^{\ominus} = 121$ J K^{-1} mol^{-1};
     $E^{\ominus}(15°\text{C}) = 2.051$ V; $E(25°\text{C}, Q) = 2.175$ V;
     (c) $E(\text{pH} = 5) = 1.100$ V; $E(\text{pH} = 8) = 0.745$ V;
     $E(\text{PbSO}_4/\text{Pb}) = E^{\ominus} - 0.359$ V.

10.28*   (a) +1.23 V; (b) +1.09 V.

11.1*   (a) 1.6×10^{-33} J m^{-3}; (b) 2.5×10^{-4} J m^{-3}.

11.2* 6.29×10^{-34} J s.

11.3   (a) 2231 K, 0.031R; (b) 343 K, 0.897R.

11.4*   (a) 0.020; (b) 0.007; (c) 7×10^{-6}; (d) 0.5; (e) 0.61.

11.5   (a) 9.0×10^{-6}; (b) 1.2×10^{-6}.

11.6* π.

11.7 $\hbar/2$.

11.8*   i$\hbar$.

11.9 $\lambda_{\max}T = hc/5k$.

11.11*   (a) $a = 8\pi hc$, $b = hc$; (b) $c_2 = hc/k$.

11.12 $k = 1.382 \times 10^{-23}$ J K^{-1}, $h = 6.69 \times 10^{-34}$ J s.

11.13*   (a) $N = (2/L)^{1/2}$; (b) $1/c(2L)^{1/2}$; (c) $1/(\pi a^3)^{1/2}$; (d) $1/(32\pi a^5)^{1/2}$.

11.14   (a) $N = 1/(32\pi a_0^3)^{1/2}$; (b) $N = 1/(32\pi a_0^5)^{1/2}$.

11.15   (a) ik; (c) 0.

11.16   (a) -1; (b) +1.

11.17*   (a) $-k^2$; (b) $-k^2$; (c) 0; (d) 0.

11.18*   (a) $\cos^2\chi$; (b) $\sin^2\chi$; (c) $0.95\text{e}^{\text{i}kx} \pm 0.32\text{e}^{-\text{i}kx}$.

11.19 $\hbar^2 k^2/2m$.

11.20   (a) $k\hbar$; (b) 0; (c) 0.

11.21*   (a) $6a_0$, $42a_0^2$; (b) $5a_0$, $30a_0^2$.

11.22*   (a) $-e^2/4\pi\varepsilon_0 a_0$; (b) $\hbar^2/2m_e a_0^2$.

11.24*   (a) 1; (b) $2x$; (c) $\hbar$.

11.25*   500 nm, blue green.

11.26   255 K, 11.3 μm.

11.27   (a) $f(\tilde{v}) = \dfrac{2\pi hc^2\tilde{v}^3}{\text{e}^{hc\tilde{v}/kT} - 1}$; (b) 255 K.

11.28*   (a) Assuming $\Delta_r C_p \approx$ constant, methane becomes unstable above ≈ 825 K; the authors' statement is confirmed;
     (b) $\lambda_{\max}$ (1000 K) = 2880 nm;
     (c) ρ(brown dwarf)/ρ(Sum) = 8.8×10^{-3} at $\lambda_{\max}$ (brown dwarf), M(brown dwarf)/M(Sun) = 7.7×10^{-4};
     (d) 2.31×10^{-7}, it hardly shines.

12.1* 1.24×10^{-39} J, 2.2×10^9, 1.8×10^{-30} J.

12.2* CO (1900 N m^{-1}) > NO (1600 N m^{-1}) > HCl (516 N m^{-1}) > HBr (412 N m^{-1}) > HI (314 N m^{-1}).

12.3* 1.30×10^{-22} J, $\pm\hbar$.

12.4 $E/(10^{-22}$ J$) = 0$, 2.62, 7.86, 15.72.

12.5* (a) $\dfrac{\varepsilon a}{L} + \dfrac{\varepsilon}{\pi}\sin\left(\dfrac{\pi a}{L}\right)$; (b) 0.1984ε.

12.6 $\frac{1}{2}mgL$.

12.7* With $a = mgL/(h^2/8mL^2)$, we have $E^{(2)} = -0.010\,83\,amgL$, $\Psi^{(1)} = a\{0.0600\,\Psi_2 + 0.000\,96\,\Psi_4 + 0.000\,13\,\Psi_6 + \cdots\}$.

12.8 $E = (h^2/8m) \times (n_1^2/L_1^2 + n_2^2/L_2^2 + n_3^2/L_3^2)$.

12.10 (a) $T = \dfrac{4k_1^2 k_2^2}{(a^2 + b^2)\sinh(k_2 b) + b^2}$
where $(a^2 + b^2) = (k_1^2 + k_2^2)(k_2^2 + k_3^2)$, $b^2 = k_2^2(k_2 + k_3)^2$.

12.11* (a) $N^2/2\kappa$; (b) $N^2/4\kappa^2$.

12.12 $g = \frac{1}{2}(mk/\hbar^2)^{1/2}$.

12.13* $\langle T \rangle = \frac{1}{2}(v + \frac{1}{2})\hbar\omega$.

12.14 0, $\frac{3}{4}(2v^2 + 2v + 1)\alpha^4$.

12.15* (a) $L((1/12) - (1/2\pi^2 n^2))^{1/2}$, $nh/2L$;
(b) $\{(v\frac{1}{2})\hbar/\omega m\}^{1/2}$, $\{(v + \frac{1}{2})\hbar\omega m\}^{1/2}$.

12.16* $\mu_{v+1,v} = \alpha\{(v + 1)/2\}^{1/2}$, $\mu_{v-1,v} = \alpha(v/2)^{1/2}$.

12.17 $\langle T \rangle = -\frac{1}{2}\langle V \rangle$.

12.18 (a) $+\hbar$, $\hbar^2/2I$; (b) $-2\hbar$, $2\hbar^2/I$; (c) 0, $\hbar^2/2I$; (d) $\hbar\cos 2\chi$, $\hbar^2/2I$.

12.19* (a) 0, 0; (b) $3\hbar^2/I$, $6^{1/2}\hbar$; (c) $6\hbar^2/I$, $2(3^{1/2})\hbar$.

12.21* $\cos\theta = m_l/\{l(l + 1)\}^{1/2}$, $54°44'$.

12.22 $-(a^2 + b^2 + c^2)$.

12.23* $l_x = (\hbar/i)(y\partial/\partial z - z\partial/\partial y)$ and cyclic permutations, $[l_x, l_y] = i\hbar l_z$.

12.25 (b) $\langle x^2 \rangle_n^{1/2} = [L^2/3 - \frac{1}{4}(n\pi/L)^2]^{1/2}$; as $n \to \infty$, $\langle x^2 \rangle_n^{1/2} \to L/\sqrt{3}$.

12.26* (a) 1.60×10^{-19} J; (b) 2.42×10^{14} Hz;
(c) total number of electrons = 1.123.

12.27 (a) 3.30×10^{-19} J; (b) 4.98×10^{14} s^{-1}; (c) lower, increases.

12.28* 1.2×10^6.

12.29 2.68×10^{14} s^{-1}.

13.1* $n_2 \to 6$, 7503, 5908, 5129, ... 3908 nm.

13.2 397.13 nm, 3.40 eV.

13.3* $987\,663$ cm^{-1}, $137\,175$ cm^{-1}, $185\,187$ cm^{-1}, 122.5 eV.

13.4 5.39 eV.

13.5 (1) $^2P_{1/2}$(lower) and $^2P_{3/2}$, (2) $^2D_{3/2}$(lower) and $^2D_{5/2}$; simple estimate suggests $^2D_{3/2}$ is the ground state.

13.6* $A = 38.50$ cm^{-1}.

13.7 3.3429×10^{-27} kg, $I_D/I_H = 1.000\,272$.

13.8* 7621 cm^{-1}, $10\,228$ cm^{-1}, $11\,552$ cm^{-1}, 6.80 eV.

13.9* (a) 0.9 cm^{-1}; (b) small.

13.10 0.420 pm.

13.11* $2s$.

13.12 106 pm.

13.14 $E_{2s} = E_{2p} = -(1/4)(Z^2\hbar^2/2\mu a_0^2)$.

13.15 (b) $3s$ radial nodes: $\rho = 3 \pm 3^{1/2}$, no nodal plane;
$3p_x$ radial nodes: $\rho = 0$ and 4, yz nodal plane ($\phi = 90°$);
$3d_{xy}$ radial node: $\rho = 0$ only, xz nodal plane ($\phi = 0$) and yz nodal plane ($\phi = 90°$); (c) $13.5a_0/Z$.

13.16* $p_x \pm ip_y$.

13.17* Exact knowledge of energy, total orbital angular momentum, and its projection along an arbitrary axis are possible.

13.18* $2.66a_0$.

13.19* $d_H \approx 0.314$ nm, $d_U \approx 0.301$ nm.

13.20 $E = -(Z^2 e^4 m_e/32\pi^2\varepsilon_0^2\hbar^2)(1/n^2)$.

13.22* $v = 2.19 \times 10^6$ m s^{-1}, $\varepsilon = 5.14 \times 10^{11}$ V m^{-1}, $\mathcal{H} = 9.98 \times 10^6$ A m^{-1}.

13.23* $a_{Ps} = 2a_0$, $E_{1,Ps} = \frac{1}{2}E_{1,H}$.

13.24* (b) 1663 (from wavelength data), 1840 (from Rydberg constants).

13.25 $\Delta l = \pm 1$, $\Delta m_l = 0$ or ± 1.

13.26 (b) 23.8 T m^{-1}.

13.27* $\Delta E = 4.29 \times 10^{-24}$ J $= 0.216$ cm^{-1}, $\langle r \rangle_{100} = 529$ nm, $I_{100} = 10.9677$ cm^{-1}; yes, thermal energy is 207 cm^{-1} and is sufficient, $v_{min} = 511$ m s^{-1}.

13.28 (a) receding, $s = 3.381 \times 10^5$ m s^{-1}.

14.3* $R_{max} = 2.1a_0$.

14.7 (a) 8.6×10^{-7}, 2.0×10^{-6}; (b) 8.6×10^{-7}, 2.0×10^{-6};
(c) 3.7×10^{-7}, 0; (d) 4.9×10^{-7}, 5.5×10^{-7}.

14.8* 1.9 eV, 130 pm.

14.10 (a) $\alpha = -\frac{1}{2} - j + \frac{1}{R}$, $\beta = (\frac{1}{R} - \frac{1}{2})S - k$;
(b) $E_{min} = -15.367$ eV, $R_e = 131.864$ pm, $D_e = 1.7612$ eV.

14.12 Thermal motion would cause the molecule to break apart; it is not likely to exist for more than one vibrational period.

14.13 $E = \alpha_0$(twice), $\frac{1}{2}\{(\alpha_0 + \alpha_N) \pm [(\alpha_0 - \alpha_N)^2 + 12\beta^2]^{1/2}\}$;
E(delocalization) $= [(\alpha_0 - \alpha_N)^2 + 12\beta^2]^{1/2} - [(\alpha_0 - \alpha N)^3 + 4\beta^2]^{1/2}$.

14.14* $\Delta E = 2.7$ eV, $\lambda = 460$ nm, orange.

14.15* (b) $\beta \approx -21\,135$ cm^{-1}; (c) $E_{deloc} = 1.5176\beta$.

14.16* (b) $\Delta E/eV \approx 3.3534 + 1.3791 \times 10^{-4}\,\tilde{v}/$cm^{-1};
(d) $30\,937$ cm^{-1}.

14.18* (b) at the PM$_3$ level: C_2H_4, 32.6%; C_4H_6, 19.3%; C_8H_8, 12.2%; C_8H_{10}, 16.6%.

14.19 $\dfrac{1}{4}\left(\dfrac{1}{6\pi}\right)^{1/2}\left(\dfrac{Z}{a_0}\right)^{3/2}\left\{2 - \dfrac{l}{2}\left(1 + \dfrac{\cos\phi - \sqrt{3}\sin\phi}{\sqrt{2}}\sin\theta\right)\right\} \times e^{-f/4}$.

14.21 $E = E_t = E_H - \dfrac{v_1 + v_2}{1 + S} + \dfrac{e^2}{4\pi\varepsilon_0}\left(\dfrac{1}{R}\right)$.

14.22* (a) $E = -hc\mathcal{R}_H$; (b) $-(8/3\pi)hc\mathcal{R}_H$.

14.23 (a) $E_{min} = -15.96$ eV, $R_e = 106.009$ pm, $D_e = 2.354$ eV, $\eta = 1.238$; (b) virial theorem is satisfied.

14.24* (a) $E = (\alpha - \beta)$ (twice), $\alpha + 2\beta$; binding energies: $\alpha + 2\beta$, $2\alpha + 4\beta$, $3\alpha + 3\beta$, $4\alpha + 2\beta$, respectively;
(b) $\Delta_r H = -413$ kJ mol^{-1}, slightly less than binding energy of H_2 (435.94 kJ mol^{-1});
(c) $\beta = (-849$ kJ mol$^{-1} - 2\alpha)/4$; H_3^{2+}: -425 kJ mol^{-1}; H_3^+: -849 kJ mol^{-1}; H_3: $3(\alpha/2 - 212$ kJ mol$^{-1})$; H_3^-: $3\alpha - 425$ kJ mol^{-1}.

14.26* (a) linear relation exists; (b) -0.122 V; (c) -0.174, better.

15.1*   (a) D_{3d}; (b) D_{3d}, C_{2v}; (c) D_{2h}; (d) D_3; (e) D_{4d}.

15.2*   *trans*-CHCl=CHCl.

15.3 $C_2\sigma_h = i$.

15.6   representation 1: $\mathbf{D}(\sigma_v) = \mathbf{D}(\sigma_d) = +1$ or -1;
representation 2: $\mathbf{D}(\sigma_v) = -\mathbf{D}(\sigma_d) = +1$ or -1.

15.7*   the matrices do not form a group.

15.8 $A_1 + T_2$, s and p, (d_{xy}, d_{yz}, d_{zx}) span T_2.

15.9   (a) all five d orbitals; (b) all except $d_{xy}(A_2)$.

15.10   group: S_4; operations: S_4, C_2.

15.11   (a) D_{2h}; (b) (i) C_{2h}, (ii) C_{2v}.

15.12   (a) D_{2d}; (b) A_1; (c) C_{4v}; (d) A_1.

15.13*   (a) $2A_1 + A_2 + 2B_1 + 2B_2$; (b) $A_1 + 3E$; (c) $A_1 + T_1 + T_2$;
(d) $A_{2u} + T_{1u} + T_{2u}$.

15.14   (a) yes; (b) no; (c) yes.

15.15*   irreducible representations: $4A_1 + 2B_1 + 3B_2 + A_2$.

15.17   (a) $7A_2 + 7B_1$;
(c) $A_1 \leftarrow A_1$(z-polarized), $B_2 \leftarrow B_2$(y-polarized).

15.18*   neither transition allowed; with vibration transition to T_{1g} becomes allowed, to G_g remains foridden.

15.19*   (a) symmetry elements; E, $2C_3$, $3C_2$, σ_h, $2S_3$, $3\sigma_v$; point group: D_{3h}; (d) reduces to: $A_1' + E'$.

15.20*   The basis spans $A_{1g} + B_{1g} + E_u$.

15.21   (a) C_{2h}; (b) $11A_u + 11B_g$ spanned; the $11A_u$ SALCs are $\frac{1}{2}(a + a')$, $\frac{1}{2}(b + b')$, ... $\frac{1}{2}(k + k')$; the $11B_g$ SALCs are the same combinations; no Bu combinations;
(d) $B_u \leftarrow A_g$ transitions are allowed with x or y polarization.

15.22 z-polarized transition not allowed, x- and y-polarized transitions are allowed.

16.1   (a) 2.13×10^{-6} J m^{-3} (2.206 J m^{-3}, classical);
(b) 9.88×10^{-4} J m^{-3} (3.676 J m^{-3}, classical);
(c) 3.15×10^{-1} J m^{-3} (8.528 J m^{-3}, classical).

16.2   Doppler width as fraction: (a) 2.1×10^{-6}, (b) 9.7×10^{-7}
widths: (a) 1.3 MHz, 0.0063 cm^{-1};
        (b) 6.6 kHz, 0.0004 cm^{-1}.

16.3   700 MHz, 1 Torr.

16.4*   596 GHz, 19.9 cm^{-1}, 0.503 mm, $B = 9.941$ cm^{-1}.

16.5   From 112.83 pm to 123.52 pm.

16.6* R(CC) $= 139.6$ pm, R(CH) $= R$(CD) $= 108.5$ pm.

16.7* 2.728×10^{-47} kg m^2, 129.5 pm; DCl lines at 10.56, 21.11, 31.67, ... cm^{-1}.

16.8   HCl: 128.393 pm, DCl: 128.13 pm.

16.9* R(CuBr) $= 218$ pm.

16.10 R(CO) $= 116.28$ pm, R(CS) $= 155.97$ pm.

16.11*   14.35 m^{-1}, 26, 15.

16.12 $k = 93.8$ N m^{-1}, 142.81 cm^{-1}, 3.36 eV.

16.13*   linear, $\tilde{v}_1 = 1400$ cm^{-1}, $\tilde{v}_2$ (bend): 540 cm^{-1},
$\tilde{v}_3$: 2360 cm^{-1} combination band, $\tilde{v}_1 + \tilde{v}_3$: 3735 cm^{-1}.

16.14   (a) 2143.26 cm^{-1}; (b) 12.8195 kJ mol^{-1};
(c) 1.855 63 kN m^{-1}; (d) 1.91 cm^{-1}; (e) 113 pm.

16.15   (a) 5.15 eV; (b) 5.20 eV.

16.17*   (a) 152 m^{-1}, 2.72×10^{-4} kg s^{-2}, 2.93×10^{-46} kg m^2, 95.5 m^{-1}; (b) 293 m^{-1}, 0.96.

16.18*   (b) $\tilde{v} = 2169.3$ cm^{-1}, $\chi_e\tilde{v} = 13.14$ cm^{-1}, $B = 1.930$ cm^{-1}, $a = 0.017\,47$ cm^{-1}, $D_J = 4.178 \times 10^{-6}$ cm^{-1}.
(c) $I_e = 1.450 \times 10^{-46}$ kg m^2;
$R_e = 112.8$ pm;
$D_e = 89\,533$ cm^{-1}, $D_0 = 88\,452$ cm^{-1};
$I_{v=0} = 1.457 \times 10^{-46}$ kg m^2, $R_{v=0} = 113.1$ pm;
$I_{v=2} = 1.484 \times 10^{-46}$ kg m^2, $R_{v=2} = 114.1$ pm;
$I_{v=1} = 1.470 \times 10^{-46}$ kg m^2, $R_{v=1} = 113.6$ pm;
$I_{v=3} = 1.498 \times 10^{-46}$ kg m^2, $R_{v=3} = 114.7$ pm.

16.19   (a) Density functional calculations with PC Spartan Pro™ yield the following values A_1(525 cm^{-1}); A_1(1152 cm^{-1}); B_2(1336 cm^{-1}).
(b) Comparison of DFT results with experimental is good.

16.20*   (a) C_{3v}; (b) 9; (c) $2A_1 + A_2 + 3E$; (d) all but A_2; (e) all but A_2.

16.21   (a) 7; (b) 2(C_{2h}), 3(C_{2v}), 4(C_2); (c) structure 2 is inconsistent with data.

16.24*   (a) spherical rotor, (b) symmetric rotor, (c) linear rotor, (d) asymmetric rotor, (e) symmetric rotor, (f) asymmetric rotor.

16.25* $J_{max} = (kT/2hcB)^{1/2} - \frac{1}{2}$, 30, $J_{max} = (kT/hcB)^{1/2} - \frac{1}{2}$, 6.

16.26   230 pm, 240 pm, 250 pm.

16.27*   (b) $A/hc = 46\,119$ cm^{-1}, $B = 1.4275$, $C = 2.3896$, $D = 88.99$ pm, $\varepsilon/hc = 1750$ cm^{-1}.

16.28   the greatest integer less than $v = 2De/\tilde{v} - \frac{1}{2}$.

16.29   (a) O—O stretching mode is IR inactive or at best weak;
(b) 796 cm^{-1}; (c) 2, 1.5, and 1 respectively;
(d) $Fe_2^{3+}O_2^{2-}$; (e) 5 and 6 can be excluded.

16.30*   (c) 281.9 K, 81.1 per cent.

16.31   (c) $h \leq 30$ m.

16.32* 2.30×10^7 m s^{-1}, 8.34×10^5 K.

16.33*   2.35 K.

16.34   (b) 87.64 pm (from B), 89.86 pm (from C), 88.7 pm (average);
(c) $B = 43.87$ cm^{-1}, $C = 21.91$ cm^{-1};
(d) $\tilde{v}_2(D_3^+) = 1783.0$ cm^{-1}, $B(D_3^+) = 21.80$ cm^{-1}, $C(D_3^+) = 10.37$ cm^{-1}.

17.1*   49 364 cm^{-1}.

17.2*   5.1147 eV.

17.3   14 660 cm^{-1}.

17.4* 4.8×10^4 L mol^{-1} cm^{-2}.

17.5 1.1×10^6 L mol^{-1} cm^{-2}.

17.6* $^2\Sigma_g^+$, $^2\Pi_g$.

17.7*   5.06 eV.

17.8   There is a good correlation.

17.9   0.2 ms.

17.10*   65 MW.

17.13   Ejected electron bonded between the two hydrogens of the water molecule.

17.15   (a) allowed; (b) forbidden.

17.16   lengthen; to blue.

17.18* $f = (RS/a_0)^2f_0$.

17.21   (a) $I_{abs} = I_0(1 - 10^{-\varepsilon[J]l})$.

17.22 (a) 3 + 1, 3 + 3; (b) 4 + 4, 2 + 2.

17.24 1.24×10^5 L mol^{-1} cm^{-2}.

17.25 6.37, 2.12.

17.26* (a) 2.42×10^5 L mol^{-1} cm^{-2}; (b) 0.185; (c) 6.97, 135 L mol^{-1} cm^{-1}.

17.27 28 kJ mol^{-1} greater, consistent.

17.28* $V_1 - V_0 = 3.1938$ eV, $\tilde{v}_1 - \tilde{v}_0 = 79.538$ cm^{-1}, $\tilde{v}_1 = 2113.8$ cm^{-1}, $\tilde{v}_0 = 2034.3$ cm^{-1}, relative populations = 10, $T_{eff} = 1.3 \times 10^3$ K.

18.1* 10.3 T, 2.42×10^{-5}, β.

18.2 57 kJ mol^{-1}.

18.3* 300×10^6 Hz $\pm$ 10 Hz, 0.29 s.

18.4* 4×10^2 s^{-1}, 3.7 kJ mol^{-1}, 16 kJ mol^{-1}.

18.5 both fit data equally well.

18.6* (a) yes; (b) $^3J_{SnSn}$/Hz = 580 − 79 cos ϕ + 395 cos 2ϕ; (c) a staggered conformation with the two SnMe$_3$ groups at 180° to each other relative to the C—C bond.

18.7* 1.992, 2.002.

18.8 6.9 mT, 2.1 mT.

18.8* −0.89 mT.

18.9* 1:2:3:2:1 quintet of 1:4:6:4:1 quintets.

18.10 0.10, 0.38; (a) 0.48; (b) 0.52; 3.8, 131°.

18.11 (7) $\rho(3) = \rho(6) = 0.005$, $\rho(4) = \rho(5) = 0.076$; (8) $\rho(2) = \rho(4) = 0.200$, $\rho(3) = 0.048$, $\rho(6) = 0.121$; (9) all $\rho = 0.050$.

18.12* $\sigma_d = Ze^2\mu_0/12\pi m_e a_0$.

18.13* (b) There is a roughly linear correlation if phenol is omitted from the data set.

18.14* 158 pm.

18.16* $I = \frac{1}{2}A\tau/\{1 + (\omega_0 - \omega)^2\tau^2\}$.

18.18* (a) R_v(nuclide)/R_v(proton): 0.409, 0.251, 0.193, 0.941, 0.405; $R_{\mathcal{B}}$(nuclide)/$R_{\mathcal{B}}$(proton) = 0.009 65, 0.015 90, 0.001 01, 0.833 50, 0.066 54.

18.19* average dipolar field = 0.58 mT.

18.21 $[I]_0 = \dfrac{[E]_0\Delta\nu}{\delta\nu} - K$.

18.22* (a) Using density functional methods $P_2 = P_6 = 0.27$; $P_3 = P_5 = 0.10$. (b) $a_2 = a_6 = 0.61$ mT; $a_3 = a_5 = 0.23$ mT; $a_4 = 0.46$ mT.

19.1* $W = 2 \times 10^{40}$, $S = 1.282 \times 10^{-21}$ J K^{-1}, $S_1 = 0.637 \times 10^{-21}$ J K^{-1}, $S_2 = 0.645 \times 10^{-21}$ J K^{-1}. Entropy is an extensive property.

19.2* ^{4}He: 6.30×10^5, 7, 2; ^{3}He: 9.63×10^5, 11, 4.

19.3* $\Delta W/W \approx 2.4 \times 10^{25}$.

19.4 $\Delta W/W \approx 4.8 \times 10^{21}$.

19.5* 3.5 fK, 7.41.

19.6 no.

19.7* (a) 5.00, 6.26; (b) 1.00 at 298 K, 0.80 at 5000 K; 6.5×10^{-11} at 298 K and 0.12 at 5000 K; (c) 13.38 J K^{-1} mol^{-1}, 18.07 J K^{-1} mol^{-1}.

19.8 0.257, 0.336, 0.396, 0.011.

19.9* (a) 0.64, 0.36; (b) 0.52 kJ mol^{-1}.

19.10 1.209, 3.004.

19.11 (a) 1.049, 123 J mol^{-1}, 1.63 J K^{-1} mol^{-1}; (b) 1.55, 1.348 kJ mol^{-1}, 8.17 J K^{-1} mol^{-1}; proportions are: p_0 = (a) 0.953, (b) 0.645; p_1 = (a) 0.044, (b) 0.230; p_2 = (a) 0.002, (b) 0.083.

19.12* (a) $10^{5.44\times10^{23}}$; (b) $10^{5.69\times10^{23}}$, the heat capacity is temperature-dependent; (c) this is a spontaneous process, $\Delta S_{U,V} > 0$.

19.14* (a) 1; (b) {2,2,0,1,0,0} and {2,1,2,0,0,0}.

19.15 {4,2,2,1,0,0,0,0,0,0}.

19.16 (a) 160 K.

19.17* (a) $q = 1 + 3e^{-\varepsilon/kT}$, 2.104; (b) 0.5245$RT$, 2.074 J K^{-1} mol^{-1}, 10.55 J K^{-1} mol^{-1}.

19.18* (a) 104 K; (b) $q = 1 + a$; (c) $2Nk \ln 2$.

19.19 (a) no difference; (b) no measurable difference.

19.20* (b) ln $N(0)/N(h) = \nu(\rho - \rho_0) gh/kT$; (c) $k = 1.36 \times 10^{-23}$ J K^{-1}; $N_A = 6.11 \times 10^{23}$ mol^{-1}.

19.21 O$_2$: 0.36, H$_2$O: 0.57.

19.23 (b) $n_{max} \approx 28$.

20.1* (a) 0.351R; (b) 0.079R; (c) 0.029R.

20.2 (a) 0.1 per cent; (b) 4×10^{-3} per cent.

20.3* 4.2, 15 J K^{-1} mol^{-1}.

20.5* 19.89.

20.6* 191.4 J K^{-1} mol^{-1}, residual entropy is negligible.

20.7* 199.4 J K^{-1} mol^{-1}.

20.8* 28, 258 J K^{-1} mol^{-1}.

20.11* (a) 3.89; (b) 2.41.

20.12* $S_m = R \ln(2\pi e^2 m\sigma_m/h^2 N_A\beta)$, $\sigma_m = \sigma/n$; $\Delta S_m = R \ln\{(\sigma_m/V_m)(h^2\beta/2\pi me)^{1/2}\}$.

20.13 (a) $\theta_R = 87.55$ K, $\theta_V = 6330$ K; (b) $C_{V,m}$(equil.mixt.)] $= 2\alpha C_{V,m}$(H) + $(1 - \alpha)$ $C_{V,m}$(H$_2$), $\alpha = [K/(K + 4)]^{1/2}$, $C_{V,m}$(H) = 3R/2, $C_{V,m}$(H$_2$) = $\frac{5}{2}R$ + $\{(\theta_V/T)$ $[e^{-(\theta_v/2T)}/(1 - e^{-(\theta_v/T)})]\}^2R$, $K = \{kT\Lambda^3(H_2)/pq^R(H_2)q^V(H_2)\Lambda^6(H)\}e^{-(D_0/RT)}$.

20.14 $U - U(0) = H - H(0) = (N\hbar\omega)/(e^x - 1)$, $C_V = kN\{x^2e^x/(e^x - 1)^2\}$, $S = Nk\{x/(e^x - 1) - \ln(1 - e^{-x})\}$, $A - A(0) = G - G(0) = NkT \ln(1 - e^{-x})$.

20.15 (c) $T \approx 374$ K.

20.16* (b) 5.41 J K^{-1} mol^{-1}.

20.17 100 T.

20.18* 350 m s^{-1}.

20.19* 9.57×10^{-15} J K^{-1}.

20.20 513.5 kJ mol^{-1}.

20.21* 45.76 kJ mol^{-1}.

20.22 0.6608 kJ mol^{-1}, 241.5 kJ mol^{-1}.

21.1* (a) 0.11 GV m^{-1}; (b) 4 GV m^{-1}; (c) 4 kV m^{-1}.

21.2 2.4 nm.

21.3* 1.2×10^{-23} cm^3, 0.86 D.

21.4 1.38×10^{-23} cm^3, 0.34 D.

21.5* 2.24×10^{-24} cm^3, 1.58 D.

21.6   1.85 D, 1.36×10^{-24} cm^3.

21.7 4.80×10^{-40} J^{-1} C m^2, 0.26×10^{-40} J^{-1} C m^2.

21.8* $R = 360.71$ pm, $\tilde{v} = 21.84$ cm^{-1}, $k = 0.3746$ Nm^{-1}.

21.9   46.07 cm^{-1}, 1.769×10^8 cm^{-1}.

21.10*   (a) 1.51×10^{-23} J, 265 pm.

21.11   0.123.

21.12   8.14 cm^3 mol^{-1}, 1.76, 1.33.

21.13*   (a) $6Q_1Q_2/\pi\varepsilon_0 r^5$; (b) $9Q_1Q_2/4\pi\varepsilon_0 r^5$.

21.14* $n_r = 1 + p(2\pi\alpha'/kT)$.

21.16*   the relative permittivity should decrease.

21.17 $a = 2\pi N_A^2 C_6/3d^3$.

21.18   when $A = \sigma = 1$, a minimum occurs at $r = 1.63$.

21.22*   1.0 J mol^{-1}.

21.23*   (a) -39 J mol^{-1}; (b) ∞.

21.24   (a) 3.5;

      (b) Linear with slope $= -1.49$ and intercept $= -1.95$;

      (c) 1.12×10^{-2}.

22.1   PBLG is rod-like; polystyrene is a random coil.

22.2*   5.0 Sv.

22.3   3500 r.p.m.

22.4 $M_w = 1.23 \times 10^6$ g mol^{-1}, $B' = 3.72 \times 10^{-2}$ cm^3 mg^{-1}. The average molar masses are different and there is no reason for them to be the same; different preparations of the same polymer can lead to much different average molar masses.

22.5*   0.0716 L g^{-1}.

22.6*   0.21 Mg mol^{-1}.

22.7*   158 kg mol^{-1}.

22.8* $K = 2.73$ cm^3 g^{-1} kg$^{-1/2}$ mol$^{1/2}$, $a = 0.500$, $M = 1.34 \times 10^3$ kg mol^{-1}.

22.9*   (a) $K = 0.0117$ cm^3 g^{-1}, $a = 0.717$,

      (b) THF is polar. The constants depend upon both solute and solvent and their interactions.

22.10*   23.1 kg mol^{-1}, 1.02 m^3 mol^{-1}.

22.11   155 kg mol^{-1}, 13.7 m^3 mol^{-1}.

22.12 $\overline{M}_n = \overline{M} + (2\gamma/\pi)^{1/2}$.

22.13*   (a) $lN^{1/2}$, 9.74 nm;

      (b) $(8N/3\pi)^{1/2}l$, 8.97 nm;

      (c) $(2N/3)^{1/2}l$, 7.95 nm.

22.15   (a) $(\tfrac{3}{5})^{1/2}a$; (b) $l/2(3)^{1/2}$, 2.40 nm, 46 nm.

22.16*   (a) $B = 4.22 \times 10^{23}$ mol$^{-1} \times (l\sqrt{N})^3$; 0.39 m^3 mol^{-1};

      (b) $B = 1.19 \times 10^{24}$ mol$^{-1} \times (l\sqrt{N})^3$; 1.1 m^3 mol^{-1}.

22.18*   (a) Two of three converged; different conformations represent local energy minima.

      (b) None converged with the MMFF force field.

22.19* $v_p = 8\,v_{mol}$;

      BSV: 28 m^3 mol^{-1}, 2.6%.

      Hb: 0.33 m^3 mol^{-1}, 5%.

22.20*   serum albumin and bushy stunt virus resemble spheres; DNA does not.

22.22   65.6 kg mol^{-1}.

22.23   69 kg mol^{-1}, 3.4 nm.

22.24   5.14 Sv, 60.1 kg mol^{-1}.

22.26   (a) toluene: 0.086 L g^{-1}, 0.37; cyclohexane: 0.042 L g^{-1}, 0.35; (b) toluene: 2.4×10^5 g mol^{-1}; cyclohexane: 2.6×10^5 g mol^{-1}; (c) toluene: 42 nm; cyclohexane: 34 nm; (d) toluene: 2.3×10^3; cyclohexane: 2.5×10^3; (e) toluene: 5.8×10^2 nm; cyclohexane: 6.2×10^2 nm; (f) toluene: 2.1×10^2 nm, 7.4 nm; cyclohexane: 2.2×10^2 nm, 7.7 nm; (g) no reason for them to agree; the manufacturer's claim is valid.

23.1   118 pm.

23.2*   P, 342 pm.

23.3*   yes.

23.4   (a) bcc, 315 pm, 136 pm; (b) fcc, 364 pm, 129 pm.

23.5*   10.51 g cm^{-3}.

23.7*   628 pm, yes.

23.8   834 pm, 606 pm, 870 pm.

23.9*   neutron: 0°, 14.0°; electron: 0°, 0.72°.

23.10   4.

23.11*   1.385 g cm^{-3}, 1.578 g cm^{-3}.

23.12 $\Delta_f H^{\ominus} = -373$ kJ mol^{-1}.

23.13* $s = 2$, 0.127 cm^3 mol^{-1};

      $s = 3$, 0.254 cm^3 mol^{-1};

      $s = 4$, 0.423 cm^3 mol^{-1};

      Overall molar susceptibility corresponds to $s = 3$ state value.

23.15 $V(\text{monoclinic}) = abc \sin \beta$;

      $V(\text{orthorhombic}) = abc$;

23.16   (a) 0.5236; (b) 0.6802; (c) 0.7405.

23.17   0.

23.18   (a) no absences; (b) alternation ($h + k + l$ odd or even); (c) $h + k + l$ odd missing.

23.19* $dG = -SdT - ldt$; $dA = -SdT - tdl$; $\left(\dfrac{\partial U}{\partial l}\right)_t = -T\left(\dfrac{\partial t}{\partial T}\right)_l + t$

23.20* $t = -T\left(\dfrac{\partial S}{\partial l}\right)_T$

23.22 $P(E) = -\dfrac{\dfrac{N+1}{2\pi\beta}}{\left[1 - \left(\dfrac{E - \alpha}{2\beta}\right)^2\right]^{1/2}}.$

23.23 $\xi = -e^2 a_0^2/2m_e$, $\chi_m = -N_A \mu_0 e^2 a_0^2/2m_e$.

23.24*   the susceptibility varies as $\alpha = 1 - d$, where $d =$ degree of dimerization; $d = 1 - \left(\dfrac{1}{4(p/k) + 1}\right)^{1/2}$;

      susceptibility decreases as pressure increases; susceptibility increases as temperature increases.

23.25 $\chi_m = (25.2 \text{ cm}^3 \text{ mol}^{-1})/\{(T/K) \times [1 + e^{174(T/K)}]\}$.

23.26* 3.61×10^5 g mol^{-1}.

23.27*   1.01 g cm^{-3}.

23.29 $\alpha_{volume} = 4.8 \times 10^{-5}$ K^{-1}, $\alpha_{linear} = 1.6 \times 10^{-5}$ K^{-1}.

23.30*   7.654 g cm^{-3}.

24.2* (a) 2.8 km h^{-1}; (b) 86 km h^{-1}; (c) 86 km h^{-1}.

24.3 (a) 1.89 m; (b) 1.89 m.

24.4 9.1.

24.5* 7.3 mPa.

24.6* (a) 100 Pa; (b) 24 Pa.

24.7 (a) 2×10^{14} s^{-1}; (b) 1×10^{20} s^{-1}.

24.8* 5.3 S cm^2 mol^{-1}.

24.9* $\Lambda_m^\circ = 12.6$ mS m^2 mol^{-1}; $\kappa = 7.30$ mS m^2 mol^{-1} M$^{1/2}$; (a) 11.96 mS m^2 mol^{-1}; (b) 119.6 mS m^{-1}; (c) 172.5 Ω.

24.10 1.36×10^{-5} mol L^{-1}.

24.11* 40 μm s^{-1}, 52 μm s^{-1}, 76 μm s^{-1}; 250 s, 190 s, 130 s; (a) 13 nm, 17 nm, 24 nm. (b) 43, 55, 81.

24.12 0.82, 0.0028.

24.13* 0.48, 7.5×10^{-8} m^2 s^{-1} V^{-1}, 72 S cm^2 mol^{-1}.

24.14 0.278, 0.278.

24.15* (a) 12 kN mol^{-1}, 2.1×10^{-20} N molecule^{-1}; (b) 17 kN mol^{-1}, 2.8×10^{-20} N molecule^{-1}; (c) 25 kN mol^{-1}, 4.1×10^{-20} N molecule^{-1}.

24.16* Li$^+$: 4 water molecules; Na$^+$: 1 to 2 water molecules.

24.17 $E_a = 9.3$ kJ mol^{-1}.

24.18* (a) $\approx$ 0, (b) 0.063 м.

24.19 1.2×10^{-3} kg m^{-1} s^{-1}.

24.20* Λ_m°(NaI) = 60.7 S cm^2 mol^{-1}, Λ_m°(KI) = 58.9 S cm^2 mol^{-1}, λ°(Na$^+$) $- \lambda^\circ$(K$^+$) = 1.8 S cm^2 mol^{-1}; the analogous quantities in water are, respectively, 126.9, 150.3, and -23.4 S cm^2 mol^{-1}.

24.21 (a) 368 pm, (b) 307 pm.

24.22 830 pm.

24.23* $c^* = (2kT/m)^{1/2}$.

24.24* $c_{mean} = (\pi kT/2m)^{1/2}$.

24.25 $v = 0.47 v_{initial}$.

24.26* (a) 0.39; (b) 0.61; (c) 0.47, 0.53.

24.27 3.02×10^{-3}, 4.9×10^{-6}.

24.28 $t'/t'' = c'u'z'/c''u''z''$.

24.30* (a) 0; (b) 0.016; (c) 0.054.

24.31 $n > 60$.

24.33 Escape velocities: (a) 11.2 km s^{-1}, (b) 5.0 km s^{-1}. Temperatures/10^3 K for H$_2$, He, and O$_2$ respectively: (a) 11.9, 23.7, 190; (b) 2.4, 4.8, 38. Proportions (Earth) for H$_2$, He, O$_2$ respectively: at 240 K, 3.7×10^{-27}, 4×10^{-54}, 0; at 1500 K, 1.5×10^{-4}, 1.0×10^{-8}, 0. (Mars): at 240 K, 1.1×10^{-5}, 4.9×10^{-11}, 0; at 1500 K, 0.25, 0.044, 4.5×10^{-14}.

24.34* 1.6×10^{16} m^2 s^{-1}, 0.34 J K^{-1} m^{-1} s^{-1}.

24.35* 100 Ms.

24.36* 1.7×10^{-2} s.

25.1* 2, 59 mL mol^{-1} min^{-1}, 2.94 g.

25.2 1, 1.51×10^{-5} s^{-1}, 9.82 mmol L^{-1}.

25.3* 1, 1.2×10^{-4} s^{-1}.

25.4 1, 5.84×10^{-3} s^{-1}, 1.98 min.

25.5* 97.0 kJ mol^{-1}.

25.6* 55.4 per cent.

25.7 1, 2.8×10^{-4} s^{-1}.

25.8 2.37×10^7 L mol^{-1} s^{-1}.

25.9* 1, 7.2×10^{-4} s^{-1}.

25.10* propene: 1; HCl: 3.

25.12* rate = kK_2K_1[HCl]^3[CH$_3$CH=CH$_2$].

25.13 -18 kJ mol^{-1}, $+10$ kJ mol^{-1}.

25.14* 1.14×10^{10} L mol^{-1} s^{-1}, 16.7 kJ mol^{-1}.

25.15 deviates from theory at low pressures.

25.16* 121.2 kJ mol^{-1}, 247.0 J K^{-1} mol^{-1}, 148.3 kJ mol^{-1}.

25.17* (a) $a = 1$, $b = \frac{1}{2}$;
(b) in units of cm$^{3/2}$ mol$^{-1/2}$ s^{-1}
k(433.15 K) = 0.077
k(453.15 K) = 0.504
k(473.15 K) = 2.26
k(493.15 K) = 12.0.
$E_a = 148$ kJ mol^{-1}.
$A = 5.00 \times 10^{16}$ cm$^{3/2}$ mol$^{-3/2}$ s^{-1}.

25.18 $v = k_2 K^{1/2}[A_2]^{1/2}[B]$.

25.19* $[B]_\infty/[A]_\infty = k/k'$.

25.20 $kt = [1/(3A_0 - 2B_0)] \times \ln\{(2x - A_0)B_0/[A_0(3x - B_0)]\}$.

25.21* (a) $kt = \dfrac{2x(A_0 - x)}{A_0^2(A_0 - 2x)^2}$;

(b) $kt = \dfrac{2x}{A_0^2(A_0 - 2x)} + \dfrac{1}{A_0^2}\ln\dfrac{A_0 - 2x}{A_0 - x}$.

25.22* when B is a reactive intermediate.

25.23 $t_{1/2}/t_{3/4} = (2^{n-1} - 1)/\{(\frac{4}{3})^{n-1} - 1\}$.

25.24 $R_1 = R_2 = R_3 = [A]_0 k_1 \times \left(1 - \left(\dfrac{k_1'}{k_2 + k_1' - \left(\dfrac{k_2' k_2}{k_2' + k_3} \right)} \right) \right)$.

25.25 $v_{max} = k(([A]_0 = [B]_0)/2)^2$ for $[B]_0 \leq [A]_0$, $(x/[A]_0) \leq 1$ corresponds to reality.

25.26 2.01 min, 1 min; 0.693 min; $x = \dfrac{[B]_0 - [A]_0 f(t)}{1 - f(t)}$, $f(t) = e^{-\{[A]_0 - [B]_0 kt\}}$.

25.27* 2720 y.

25.28 (a) 0.642 μg; (b) 0.177 μg.

25.29* 3.65×10^{-3} min^{-1}, 274 min.

25.30 $v = k[A][B]$, $k = k_1 k_2/k_2'$.

25.31* $E_a = 105$ kJ mol^{-1}, $\Delta G = -26.6$ kJ mol^{-1}, $\Delta H = -34.3$ kJ mol^{-1}; still favourable under prebiotic conditions.

25.32* 1.03×10^9 L mol^{-1} s^{-1}, 13.9 kJ mol^{-1}.

25.33 (a) 1.1×10^{-16} mol L^{-1} s^{-1}; (b) 2.2×10^{11} kg or 220 Tg.

25.34* $k_1 = 3.82 \times 10^6$ L mol^{-1} s^{-1}; $k_2 = 5.1 \times 10^5$ L mol^{-1} s^{-1}; $k_3 = 4.17 \times 10^6$ L mol^{-1} s^{-1}.

25.35* 9×10^{-10} mol L^{-1} s^{-1}, 3×10^5 s.

26.2 5.0×10^7 L mol^{-1} s^{-1}.

26.4* (a) For both X = H and X = Cl, the fit of the data to the second-order integrated rate expression is superior to the fit to the first-order rate expressions as evidenced by comparing correlation coefficients and standard deviations of the slopes and intercepts.
(b) rate = $-k_2 k_1$[(ClRh(CO)$_2$)$_2$][ArHgCl]^2/[HgCl$_2$].

26.5   Yes, when $k_1[O_2]/\{(k_1 + k_3)[O_2] + k_6[M]\} > 1$.

26.6*   When step (b) is rate-determining and step (a) is a rapid equilibrium so that [l] is in a steady state.

26.7* 1.9×10^{20} s^{-1}, 3.1×10^{-4} einstein s^{-1}.

26.8 5.1×10^6 L mol^{-1} s^{-1}.

26.9* $k_q = 1.98 \times 10^9$ L mol^{-1} s^{-1}.

26.10*   3.52 nm.

26.12 $\dfrac{d[CH_4]}{dt} = k_b\left(\dfrac{k_a}{2k_d}\right)^{1/2}[CH_3CHO]^{3/2}$.

      $\dfrac{d[CH_3CHO]}{dt} = -k_b\left(\dfrac{k_a}{2k_d}\right)^{1/2}[CH_3CHO]^{3/2}$.

26.13* $\delta M = M\{kt[A]_0(1 + kt[A]_0)\}^{1/2}$.

26.14   (a) $M(1 + 4p + p^2)/(1 - p^2)$;
      (b) $(6\langle n \rangle^2 - 6\langle n \rangle + 1)\langle n \rangle$.

26.15* $\langle n \rangle = v = k[\cdot M][I]^{1/2}$

26.16 $\langle n \rangle = (1 + 2kt[A]_0)^{1/2}$

26.17* $([A]_0 + [P]_0)^2 kt_{max} = \tfrac{1}{2} - p - \ln 2p$.

26.18 $([A]_0 + [P]_0)^2 kt_{max} = (2 - p)/2p + \ln(2/p)$.

26.19* $[X] = k_c/k_b$, $[Y] = k_a[A]/k_b$.

26.21 $[B] = (I_a/k)^{1/2} \propto [A]^{1/2}$.

26.22* $\dfrac{d[A]}{dt} = -I - k_2\left(\dfrac{I}{k_3}\right)^{1/2}[A]$.

26.24* $f = k_2 k_4[CO]/\{k_2[CO] + k_3[M]\}$.

26.25 $N(t) = N_0 e^{(b-d)t}$; fits data with $R^2 = 0.983$.

26.26 $d[P]/dt = \{k_b[E]_0[S] + (k_a'k_b'/k_a)[E]_0[P]\}/\{K_M + [S] + (k_b'k_a)[P]\}$.
      For [S] large: $d[P]/dt = k_b[E]_0$; For [S] small: $d[P]/dt = k_a'[E]_0[P]/(k_P + [P])$, where $k_P = (k_a' + k_b)/k_b'$.

26.27*   0.010 mol L^{-1}.

26.28 $\dfrac{v}{[s]_0} = \dfrac{v_{max}}{k_M} - \dfrac{v}{k_M}$.

26.29*   (b) pH = 7.0.

26.30*   Non-competitive.

26.31*   Step 1 is autocatalytic. $a/r < S_0$: infection spreads; $a/r > S_0$: infection dies out.

26.32 5.9×10^{-13} mol L^{-1} s^{-1}.

26.33*   (a) Initiation, propagation, propagation, termination, initiation; (b) $d[NO]/dt = -2k_b(k_{-d}/k_d)^{1/2}[O_2]^{1/2}[NO]$;
      (c) $E_{a,eff} = E_b + \tfrac{1}{2}E_{-d} - \tfrac{1}{2}E_d$;
      (d) $E_{a,eff} = 381$ kJ mol^{-1}, consistent with high end of range;
      (e) $d[NO]/dt = -2k_b(k_a/2k_d[M])^{1/2}[NO]^2$;
      (f) $d[NO]/dt = -2k_b(k_e/2k_d[M])^{1/2}[O]^{1/2}[NO]^{3/2}$, where k_e is the rate constant for NO + O$_2$ → O + NO$_2$; $E_{a,eff} = 253$ kJ mol^{-1}, consistent with low end of range.

27.1   (a) 0.044 nm^2; (b) 0.15.

27.2*   0.007, 0.0040 nm^2.

27.3 1.7×10^{11} L mol^{-1} s^{-1}, 3.6 ns.

27.4*   +83.8 kJ mol^{-1}, +19.1 J K^{-1} mol^{-1}, 85.9 kJ mol^{-1}, +79.0 kJ mol^{-1}.

27.5*   2–.

27.6   0.658 L mol^{-1} min^{-1}.

27.8*   A complex of two univalent ions of the same sign.

27.9* $k = k_1$, $k' = k_2\left(\dfrac{k_1}{k_2}\right)^{1/2}$;
      (b) $D(FO\text{---}F) \approx E_{a1} = 160.9$ kJ mol^{-1}, $D(O\text{---}F) = 224.4$ kJ mol^{-1}, $E_{a2} = 60$ kJ mol^{-1}.

27.10   –148 J K^{-1} mol^{-1}, 60.44 kJ mol^{-1}, 62.9 kJ mol^{-1}, 104.8 kJ mol^{-1}, respectively.

27.11*   In k fits the Arrhenius equation with $A = 3.12 \times 10^{14}$ L mol^{-1} s^{-1}, $E_a = 193$ kJ mol^{-1}, $R = 0.999\,76$; ln k' fits with $A' = 7.29 \times 10^{11}$ L mol^{-1} s^{-1}, $E_a' = 175$ kJ mol^{-1}, $R = 0.998\,48$.

27.12   (b) $A = 1.223 \times 10^{11}$ L mol^{-1} s^{-1}; $E_a = 73.8$ kJ mol^{-1};
      (c) a = 0.413, b = –0.192, c = 1.39×10^{-3} K^{-1}.

27.15* $P = 5.2 \times 10^{-6}$.

27.16 $\log v \propto I^{1/2}$.

27.17*   (a) 2.7×10^{-15} m^2 s^{-1}; (b) 1.1×10^{-14} m^2 s^{-1}.

27.18 2×10^{-13} m^2 s^{-1} if $v^{\ddagger} = v$; 9×10^{-13} m^2 s^{-1} if $v^{\ddagger} = \tfrac{1}{2}v$.

27.20   5.

27.21* k_2 (atoms)/k_2(molecules) $\approx 3 \times 10^7$.

27.22 1.6×10^{-3}, 1.8×10^{-3}.

27.23* 6.23×10^9 L mol^{-1} s^{-1}, 0.37 nm.

25.34 $k_1 = 3.82 \times 10^6$ L mol^{-1} s^{-1}, $k_2 = 5.1 \times 10^5$ L mol^{-1} s^{-1}, $k_3 = 4.17 \times 10^6$ L mol^{-1} s^{-1}, $k_2/k_1 = 0.13$.

28.1* $Z_w = 2.69 \times 10^{23}$ cm^{-2} s^{-1} at 100 kPa, $Z_w = 2.69 \times 10^{18}$ cm^{-2} s^{-1} at 1.000 Pa, $Z_{atom} = 2.0 \times 10^8$ s^{-1} at 100 kPa, $Z_{atom} = 2.0 \times 10^3$ s^{-1} at 1.000 Pa.

28.2   (a) 1.61×10^{15} cm^{-2};
      (b) 1.14×10^{15} cm^{-2};
      (c) 1.86×10^{15} cm^{-2}.

28.3*   0.37 Torr^{-1}.

28.4   (a) 164, 13.1 cm^3; (b) 264, 12.5 cm^3.

28.5*   1.4 mL, 5.9 m^2.

28.6   2.4, 0.16.

28.7   0.02 Torr s^{-1}.

28.8*   (a) $v = gkT/p$; (b) $R = 0.959$ for the linear equation which yields $g = 1.0 \times 10^{21}$ m^{-3} s^{-1}. $z(t)$ is seen to be nonlinear; $z(t) = a(e^{bt} - 1)$, with $a = 5.71$ cm, $b = 0.35$ s^{-1}, $R = 0.994$.

28.9 $K = 0.138$ mg g^{-1}, $n = 0.58$, amount corresponding to monolayer coverage must be known.

28.10*   –20.1 kJ mol^{-1}, –63.6 kJ mol^{-1}.

28.11 $n = 5.78$ mol kg^{-1}, $K = 7.02$ Pa^{-1}.

28.12   Regression analysis provides the following coefficients of correlation: R(Langmuir) = 0.973; R(Freundlich) = 0.999\,94; R(Temkin) = 0.9590. The fit to the Freundlich isotherm is best.

28.13* $X_{max} = 0.380$, specific surface area of carbon = 922 m^2 mol^{-1}.

28.14* $U \propto C_6/R^3$, 294 pm.

28.17 $d\mu' = (RTV_\infty/\sigma)\,d\ln(1-\theta)$.

28.18 BET better; 75.4 cm^3, 3.98.

28.19* 40.4.

28.20* (a) R values range from 0.975 to 0.998; the fit is good at all temperatures.
(b) $k_a = 3.68 \times 10^{-3}$, $k_b = 2.62 \times 10^{-5}$ ppm^{-1}, $\Delta_{ad}H = -8.67$ kJ mol^{-1}, $\Delta_b H = -15.7$ kJ mol^{-1}.

28.21* (a) $k = 0.2289$ and $n = 0.6180$; $R = 0.9995$.
(c) $k = 0.5227$, $n = 0.7273$; $R = 0.996$.

28.22 (a) $K(g_R L^{-1})^{-1}$, $K_F[(mg)^{(1-1/n)}g_R^{-1} L^{-1/n}]$, $K_L[mg\,L^{-1})^{-1}]$, $M(mg\,g_R^{-1})$;
(b) R(linear) = 0.9612, R(Freundlich) = 0.9682, R(Langmuir) = 0.9690; on that basis alone the fits are equally satisfactory, but not good. The Langmuir isotherm can be eliminated, as it gives a negative value for K_L; the fit to the Freundlich isotherm has a large standard deviation. Hence, the linear isotherm seems best, but the Freundlich isotherm is preferred for this kind of system.
(c) $q_{rubber}/q_{charcoal} = 0.164 c_{eq}^{-0.46}$; hence much worse.

29.1* 5.68×10^8 L mol^{-1} s^{-1}.

29.2* 0.38, 0.78 mA cm^{-2}.

29.3 $a(Sn^{2+}) \approx 2.2 a(Pb^{2+})$.

29.4* 0.25 mm.

29.5* 87 mA.

29.7* (a) $E_0 = -0.618$ V, $\eta/mV = -84, -109, -134, -194$;
(b) $j_c/(\mu A\,cm^{-2}) = 0.0324, 0.0469, 0.0663, 0.154$;
(c) $j_0 = 0.009\,97\,\mu A\,cm^{-2}$, $\alpha = 0.363$; excellent fit, $R = 0.999\,94$.

29.9 0.50, 0.150 A m^{-2}, -0.038 A m.

29.10* (a) No linear region exists; the Tafel equation cannot be used to calculate j_0 and α.

29.12 $j_0 = 2.00 \times 10^{-5}$ mA m^{-2}, $\alpha = 0.498$; $R = 0.9990$. There are no significant deviations.

29.14 (c) $k_{11} = \kappa_{11}\nu_{11}\exp\left(\dfrac{-\Delta^\ddagger G_{11}}{RT}\right)$, $k_{22} = \kappa_{22}\nu_{22}\exp\left(\dfrac{-\Delta^\ddagger G_{22}}{RT}\right)$;

(d) $k_{12} = \kappa_{12}\nu_{12}\exp\left(\dfrac{-\Delta^\ddagger G}{RT}\right)$

$\approx \kappa_{12}\nu_{12}\exp\left(\dfrac{-\Delta^\ddagger G_{11} - \Delta^\ddagger G_{22} - \Delta_r G_{12}^\ominus}{2RT}\right)$;

(e) $k_{12} = (k_{11}k_{22}K)^{1/2}$.

29.15* 7.2 μA.

29.17 $j/j_L = 1 - \exp(F\eta_c/RT)$.

29.18* 2.3 eV.

29.19 Yes, 13.4 nm^{-1}.

29.21 (a) all, (b) all.

29.22 6 μA.

29.23* 0.28 mg cm^{-2} d^{-1}.

Answers to Box questions

1.1.1 (a) 0.00; (b) -0.05.

1.1.2 At mid latitudes: $n = 1.2 \times 10^{-3}$ mol, $n/V = 2.8 \times 10^{-9}$ mol L^{-1}. In the ozone layer: $n = 4.46 \times 10^{-4}$ mol, $n/V = 1.1 \times 10^{-9}$ mol L^{-1}.

2.2.1 +37 K, 4.09 kg.

4.1.1 (a) $nC_{p,m}|T_f - T_i - T_i \ln(T_f/T_i)|$; (b) 6.86 KJ; 68.6 s, but this time is meaningless.

4.1.2 8.43 times as efficient as direct conversion.

4.2.1 (a) 106; (b) 318.

4.2.2 (a) $c = 0.171$ m s^{-1}, $p = 1.40$ nPa;
(b) $\lambda = 27.3$ nm.

6.1.1 (a) $U_r(T_r, p_r) = -\displaystyle\int_{V_r}^\infty \frac{2p_r(T_r, V_r)}{T_r V_r}\,dV_r$;
(c) $0.85 < p_r < 0.40$ when $T_r = 1$.

9.1.1 1300°C.

10.1.1 Yes.

11.1.1 For neutrons: 5.8^B, 1.7^B. For electrons: 0.3^B, 0.9^B.

11.1.2 (a) $R_{CCl} = 177$ pm; (b) fcc, $a = 364$ pm, = 129 pm.

12.1.1 0.235.

12.1.2 $F = 9.2 \times 10^{-10}$ N, factor = 0.694.

13.1.1 $K = 60$, therefore there are few H atoms present and the hydrogen spectrum will be weak.

16.1.1 B_0(CN, calculated) = 1.892 cm^{-1}; thus it has transitions that span the background radiation.

17.1.1 Values depend on method, but all show lower frequency for **5** than for the 11-cis form.

17.2.1 (a) 1.7 nmol^{-1}; (b) 6.0 H 10^2 molecules.

18.1.1 29 μT m^{-1}.

18.2.1 The deeper the crevice, the broader the ESR line.

23.2.1 6.0 K.

23.2.2 $X = 0.41$.

24.1.1 0.25 J cm^{-3}.

26.1.1 (c) 0.0123% O_3.

27.2.1 Rb atoms must hit the I side of CH_3I.

28.1.1 (a) -0.11; (b) -0.51.

Index

90° pulse 600
180° pulse 602

ab initio method 443
absence, systematic 779
absolute
  configuration, determination 790
  value 1049
absorbance 491
absorption 926
  characteristics, table 546
  spectroscopy 483
abstraction 921
abundant spin species 607
acceleration 1064
  of free fall 1062
accommodation 988
acetaldehyde decomposition 900
acetone, activity 185
achiral molecule 462
acidity constant
  introduced 240
  table 245
acid–base
  equilibria 240
  indicator 245
  titration 241
acrolein 1003
acronym summary table 1005
ACT 956
action potential 270
activated complex 881, 956
  theory 956
activation energy 879
  catalysis 999
  composite reaction 891
  general definition 881
  interpretation 881
  negative 891
activation parameter 961
activation-controlled limit 952
active site 909
active transport 849
activity
  defined 183
  ions 257
  regular solution 186
activity coefficient
  defined 183
  mean 257
  measurement 274
adatom 978
addition 921
additional work 38

adenosine 730
adiabat 54
adiabatic 6
  bomb calorimeter 42
  boundary 31
  change 52
  demagnetization 107
  flame calorimeter 46
  nuclear demagnetization 107
ADP phosphorylation 851
adsorbate 977
adsorbent 977
adsorption 977
  dissociative 990
    isotherm 989
AEDANS 933
aerobic metabolism 224
aerosol 752
AES 982
AFM 337, 706, 985
AIDS 909
air composition 12
Airy radius 521
alanine 462
alkylation 1001
all-*trans*-retinal 548
allene symmetry 458
allowed transition 382
  and symmetry 475
α-helix 725
ammonia inversion 536
Amontons, Guillaume 9
amount of substance *xx*
amplitude 1065
anaerobic metabolism 224
angular momentum 1064
  eigenvalues 348
  introduced 345
  operator 348
  quantization 352
angular velocity 1064
anharmonicity 515
anharmonicity constant 515
anode 265
anode (in PMT) 489
anodic current density 1025
anomalous scattering 790
anthracene radical anion 620
anti-Stokes radiation 488
antibonding orbital 421
anticyclone 12
antiferromagnetic material 784
antiferromagnetic phase 804
antiparallel β-sheet 728

antisymmetric stretch 522
apomyoglobin 865
approximation
  Born–Oppenheimer 410
  Hückel 433
  local-density 445
  parabolic 512
argon-ion laser 560
Arrhenius
  equation 879
  parameters 879, 949, 953
associated Laguerre polynomial 371
asymmetric rotor 500
asymmetry potential 279
AT base pair 730
atmosphere
  chemical reactions 922
  properties 12
  structure 921
atmosphere (unit) 4
atomic force microscopy 337, 706,
  985
atomic orbital 372, 376
  mean radius 377
  organization into shells 375
atomic population 440
atomic spectra 298
atomic weight *xx*
ATP 224
  hydrolysis 851
ATPase 851
attractive surface 968
attractor 917
Aufbau principle, see building-up
  principle
Auger
  effect 982
  electron spectroscopy 982
autocatalysis 915
autoprotolysis 240
avalanche photodiode 489
average
  chain length 906
  value 316
Avogadro's principle 10
Axilrod–Teller formula 705
azeotrope 203
azide ion 570
azimuth 349

Balmer series 366
band
  formation 798
  gap 799

head 545
  spectra 517
bar 4
barometer 5
barometric formula 12, 655
barrier penetration 334
base pair 730
basicity constant 240
basis
  of representation 472
  set 429
beer 1010
Beer–Lambert law 491
Belousov–Zhabotinskii reaction 915
Benson group 61
benzene
  excess enthalpy 174
  orbitals and energies 437
  π-electron energy 438
  radical anion, ESR spectrum 617
  structure 437
benzophenone, fluorescence
    spectrum 577
Berthelot equation 21
BET isotherm 992
β parameter 641
β-barrel 729
β-carotene 928
β-pleated sheet, see β sheet
β-sheet 725, 728
bimolecular reaction 882
binomial
  coefficient 1049
  expansion 1048
biochemical cascade 549
biological
  energy conversion 224, 275
  membrane transport 849
  standard state 186
  system, equilibrium 233
bipolaron 751
bipyridine 930
birefringent 695
Birge–Sponer plot 516
bistability 918
bivariant 195
black body radiation 294
block-diagonal matrix 464
Blodgett, K. 757
blue sky 737
body-centred unit cell 768
Bohr
  frequency condition 367
  magneton 580

model 409
radius 370
boiling 136
boiling point elevation 176
boiling temperature 136
effect of pressure 146
Boltzmann distribution 632
and chemical equilibrium 232
introduced 33
ionic atmosphere 260
Boltzmann formula (entropy) 642
Boltzmann's constant 33
bomb calorimeter 42
bond 410
bending in polymers 726
cleavage, kinetic isotope effect 888
dissociation energy 411, 423
dissociation enthalpy 423
enthalpy, mean 61
length
correlation with bond order 423
determination 507
equilibrium 411
table 423
order 423
strength, correlation with bond order 423
stretching in polymers 726
torsion 726
bonding 420
role of overlap 420
orbital 420
Born
equation 255
interpretation 306
Born–Haber cycle 788
Born–Mayer equation 788
Born–Oppenheimer approximation 410
boron trifluoride symmetry 457
boson 355
Pauli principle 385
bouncing ball 91
boundary
condition 325, 1055
cyclic 347
and quantization 309
surface 376
Boyer, P. 851
Boyle temperature 18
Boyle, Robert 8
Boyle's law 8
bra 331
bracket notation 331
Brackett series 366
Bragg, L. 775
Bragg, W. 775
Bragg's law 776
brain MRI 613
branch of spectrum 518
branching step 902
Bravais lattice 768, 769
Bremsstrahlung 772
bridge site 982
bromide-ion catalysis 909
Brønsted
acid 240
base 240
Brunauer, S. 992

brusselator 916
bubble 151
buckminsterfullerene symmetry 458
buffer action 244
building-up principle 387
bulk modulus 791
buoyancy effect 742
butadiene
Hückel method 435
orbitals and energies 436
Butler–Volmer equation 1029
BZ reaction 915

caesium-chloride structure 786
cage effect 951
calcium/silicon phase diagram 219
calculus 1051
calomel electrode 278
calorie 34, 1063
calorimeter 42
adiabatic flame 46
constant 42
differential scanning 46
calorimetry 42
camphor 693
canonical
distribution 648
ensemble 647
partition function
entropy 649
internal energy 648
relation to molecular partition function 650
capillary
action 153
electrophoresis 747
rise 153
technique 849
carbon
nanotube 750
reductant 239
term symbol 402
carbon dioxide
isotherm 17
laser 561
normal modes 522
phase diagram 139
supercritical 137
carbon monoxide
adsorbed 981
photoelectron spectrum 576
Raman spectrum 520
reductant 239
vibrational bands 535
carbonic anhydrase 912
carbonyl group, chromophore 549
carbon–carbon double bond chromophore 547
Carnot
cycle 95
efficiency 96
carotene 480, 928
carotenoid 928
CARS 526
catalysis 908, 999
catalyst 908, 999
microporous 1004
properties 1001

catalytic
constant 911
cycle, chlorine 923
efficiency 911
cathode 265
cathodic
current density 1025
protection 1041
cavity 151
cavity mode characteristics 554
CCD 489
cell
biological 520
emf, and standard potentials 271
membrane 755
notation 266
overpotential 1036
potential 267
reaction 266
with transference 840
working 1038
Celsius scale 7
centre
of inversion 539
of symmetry 455
centrifugal distortion constant 503
cesium, see caesium
CFC 923
CG base pair 730
chain
carrier 899
mechanism 906
polymerization 904
reaction 899
rule 792, 1051
transfer 907
chain–branching explosion 902
channel (reaction) 970
channel former 850
Chapman model 922
character 463
table 463
charge density 1069
ionic solution 259
charge transfer rate 1024
charge-coupled device 489
charge-transfer transition 547
Charles, Jacques 9
Charles's law 9
chemical
bond 410
chaos 919
equilibrium, criterion 223
exchange 605
kinetics 862
potential 163, 800
and enthalpy 164
and Helmholtz energy 164
and internal energy 164
introduced 140
of liquids 168
uniformity 141
variation with pressure 142
variation with temperature 141
quench flow method 866
reaction, formal notation 227
shift 585
typical values 586
chemically equivalent nuclei 596
chemiosmotic theory 851

chemisorption 988
ability 1002
chiral molecule, symmetry 462
chlorine
atmospheric effects 922
catalytic cycle 923
reaction with hydrogen iodide 968
chlorofluorocarbon 923
chloroform, activity 185
chloroform, see trichloromethane
chlorophyll 928
absorption spectrum 539
chloroplast 276, 928
cholesteric phase 210
cholesterol 756
chromatography 747
chromophore 546
chromosphere 394
chronopotentiometry 1031
circular dichroism 696
circularly
birefringent 695
polarized 1067
circumcoronene 483
circumstellar space 483
cis-retinal 548
citric acid cycle 224
Clapeyron equation 145
class (of operation) 465
classical mechanics
failures 294
introduced 293
clathrate cage 712
Clausius inequality 98
Clausius–Clapeyron equation 147
Clebsch–Gordan series 400
close-packed structure 784
closed
shell 386
system 30
cloud 12
colour 739
CMC 755
CNDO 444
coadsorption 999
coagulation 754
COBE 482
coefficient
of performance 98
of thermal conduction 827
perfect gas 829
of viscosity 827
perfect gas 830
cofactor 1059
coherence length 555
coherent
anti-Stokes Raman spectroscopy 526
radiation 555
colatitude 349
cold denaturation 865
collapse pressure 757, 758
colligative properties
definition 175
origin 175
collision
cross-section 821, 945
energy dependence 947
density 945
diameter 820

flux 822, 979
  frequency 820, 823, 946
  structureless particles 958
  theory 945
  tube 821
collision-induced emission 925
collisional
  deactivation 496, 925, 931
  lifetime 496
colloid 752
  stability 753
colour
  and energy 538
  and frequency 484, 538
column vector 1059
combination difference 518
combinatorial function 1048
combine 367
combined gas equation 11
common logarithm 1048
commutator 319
commute 319
competitive inhibition 913
complementary observables 319
complete
  neglect of differential overlap 444
  set 315
  shell 386
complex
  conjugate 1049
  mode process 970
  number 1049
component 194
compression factor 16
  critical 23
  and fugacity 129
  real gases 17
concentration
  cell 265, 269
  polarization 1032
  profile 848, 955
condensation reaction 904
conductance 833
conductivity 833
  ion size 837
  proton 837
configuration 720
  of atom 384
  dominating 631
  of system 629
configuration integral 671
conformation 720
conformational
  conversion 604
  energy 726
  entropy 724
consecutive reactions 883
conservation of energy 35
constant
  anharmonicity 515
  Boltzmann 33
  catalytic 911
  centrifugal distortion 503
  Faraday 267
  force 512
  hydrophobicity 712
  Madelung 788
  Michaelis 911
  normalization 307
  Planck 295

  rate 868
  rotational 500
  Rydberg 366, 373
  second radiation 294
  spin–orbit coupling 397
  Stefan–Boltzmann 294
constituent 194
constrained chain 724
contact angle 154
continuum generation 568
contour length 722
convection 12, 848
convective flux 848
cooling
  curve 212
  spontaneous 100
Cooper pair 797
cooperative
  process 732
  transition 674
coordinates
  cylindrical 346
  spherical polar 308, 349
coordination number 785
core hamiltonian 442
Corey, R. 725
Corey–Pauling rules 725
corona 394
coronene 478
correlation
  analysis 961
  diagram 403
  function 842
  spectroscopy 610
correspondence principle 329
corresponding states 24
corrosion 1039
Cosmic Background Explorer 482
cosmic microwave background
  radiation 482
COSY 610
Coulomb
  integral 429
  interaction 1068
  operator 393, 442
  potential 1068
    hydrogenic atom 367
  potential energy 1068
counter electrode 1030
covalent
  bond 410
  network solid 789
cracking 1001, 1003
critical
  compression factor 23
  constants 19, 18
  micelle
    concentration 755
    formation 755
  molar volume 19
  point 19
  pressure 19, 137
  solution temperature 205
  temperature 19, 137
cross peak 611
cross-product 1050
cross-section, state-to-state 965
crossed beam
  experiment 963
  technique 707

cryoscopic constant 177
crystal
  diode 490
  structure elements 785
  system 768, 769
crystallinity 793
cubic
  close packing 785
  group 458
  unit cell 768, 769
cultured pearl 810
cumulative reaction probability
  970
Curie
  law 802
  temperature 804
current 1070
  density 1014
curvature and kinetic energy 315
cycle
  Born–Haber 788
  Carnot 95
  Otto 120
cyclic
  boundary conditions 347
  voltammetry 1034
cyclobutadiene, π-bonding energy
  437
cyclohexane, excess enthalpy 174
cyclone 12
cyclopropane isomerization 882
cylindrical coordinates 346
cystosine 730
cytosol 224

d orbital 381
  hybridization 417
d–d transition 542, 546
D-lines 399
Dalton, John 13
Dalton's law 13
Daniell cell 265
dark current 489
Davisson, Clinton 301
Davisson–Germer experiment 301
de Broglie
  relation 303
  wavelength 303
Debye
  equation 693
  extrapolation 104
  formula 298
  length 259
  T^3 law 104
  temperature 298
debye unit 687
Debye, P. 687, 775
Debye–Hückel
  limiting law 258, 962
  theory 258
Debye–Hückel-Onsager theory 841
Debye–Scherrer method 775
decomposition of direct product 470
defect 793, 978
defect density (surface) 984
definite integral 1052
degeneracy
  and symmetry 334, 466
  rotational 502
degenerate 333

degradation 720
degree
  of conversion 675
  of dissociation 230
    and pressure 235
  of freedom (phase rule) 195
  of ionization 834
  of polymerization 906
  of protonation 835
dehydration 1001
dehydrogenation 1001
delocalization energy 436
δ scale 585
denaturation 720, 731, 865
density functional theory 444
density of states, band theory 811
depolarization ratio 525
depolarized line 525
depression 12
  of freezing point 177
Derjaguin, B. 753
deshielded 586
desorption 977
  flux 997
  temperature-programmed 997
desulfurization 1001
detection period 610
detector 489
determinant 430, 1058
deterministic chaos 919
deuterium lamp 485
deviations from perfect behaviour
  16
DFT 444
DHO theory 841
diagonal
  matrix 1058
  peak 611
diagonalization 1060
dialysis 179, 752
diamagnetic
  contribution 588
  substance 802
diamond structure 789
diamond-anvil cell 197
diaphragm technique 849
diathermic 6
  boundary 31
diatomic molecule structure (VB)
  413
diberyllium 424
diboron 424
dichloroethene 456
dielectric 688
Dieterici equation 21
differential 1053
  equation 1055
  overlap 444
  pulse voltammetry 1034
  scanning calorimeter 46
  scattering cross-section 708
differentiation 1051
diffraction 772
  grating 486
  limit 521, 563
  order 486
  pattern 772
    powder 779
diffractometer 775
diffuse double layer 1022

diffusion 815
coefficient 827
and coefficient of viscosity 845
equation 847, 953
generalized 848
Fick's first law 827, 844
Fick's second law 847
mean distance 852
probability 852
reaction 952, 955
surface 998
thermodynamic view 842
diffusion-controlled limit 952
difluorine, see fluorine
dihedral groups 457
dilithium 424
dilute spin species 607
dinitrobenzene radical anion 626
dinitrogen pentoxide decomposition 863, 886
dinitrogen, see nitrogen
diode laser 801
dioxygen, see oxygen
dipolar magnetic field 594, 614
dipole
moment
determination 503
induced 689
resultant 688
potential energy 698
dipole–dipole interaction 620, 701, 1069
dipole-induced-dipole interaction 702
Dirac bracket notation 331
direct
method 782
mode process 970
sum 464
discotic liquid crystal 210
dislocation 978
disperse phase 719, 752
dispersing element 484
dispersion 695
interaction 703
displacement
average 342
law 294
disproportionation (radical chain) 907
dissociation 552, 921
energy, determination 516
equilibrium 677
limit 552
dissociative adsorption 990
distillation 202
partially miscible liquids 207
distinguishable molecules 650
distortion polarization 691
distribution of speeds 817
disulfide link 729
DLVO theory 753
DNA
radius of gyration 723
structure 729
supercoiled 731
thermal stability 732
DNA damage 929
Dobson unit 13, 578
dominating configuration 631

Donnan
effect 752
equilibrium 180
dopant 800
Doppler
broadening 495
effect 409, 495
dot product 1050
double helix 731
drift speed 743, 836
droplet 151
DSC 46
DU 578
duality 304
Dulong, Pierre-Louis 296
Dulong–Petit law 296
dye laser 561
dynamic light scattering 741
dynode 489

Eadie–Hofstee plot 942
ebullioscopic constant 177
eclipse 394
Ectothiorhodospira halophila 774
edible fat 1003
EELS 981
effect
Donnan 752
Doppler 409
electrophoretic 841
Joule–Thomson 82
kinetic isotope 888
nuclear Overhauser 608
optical Kerr 557
photoelectric 300
relaxation 841
Zeeman 408
effective
mass 512
nuclear charge 386
potential energy 370
rate constant 891
transverse relaxation time 603
effector molecule (ion transport) 850
efficiency 96
effusion 815
effusion rate 824
egg, cooked 731
Ehrenfest
classification 148
equations 159
EHT 438
eigenfunction 311
eigenvalue 311, 1060
equation 311, 1060
eigenvector 1060
einstein (unit) 925
Einstein
coefficient
spontaneous emission 493
stimulated absorption 493
stimulated emission 493
relation 845
temperature 297
Einstein–Smoluchowski equation 854
elastase 774
elastic
collision 816
deformation 791

elastomer 793
electric
conduction 815
dipole 686
dipole moment 686
and electronegativity 687
field 1066
point charge 700
point dipole 701
field strength 1069
electrical
conductivity, temperature dependence 800
double layer 753, 1021
heating 42
power 1070
properties 795
electroactive species 1013
electrochemical
cell 262
with transference 840
potential 1023
series 277
electrode
compartment 262
concentration cell 265
electron transfer 1021
varieties 262
electrodialysis 752
electrokinetic potential 753
electrolysis 1036
relative rate 1036
electrolyte concentration cell 265
electrolytic cell 263
electromagnetic
field 1066
radiation 1067
spectrum 484, 1067
electromotive force 267
electron
affinity 391
density 444
Fourier synthesis 780
diffraction 301, 783
energy-loss spectroscopy 981
microscopy 301
paramagnetic resonance 583
scattering factor 302
spectroscopy for chemical analysis 570
spin, bond formation 412
spin resonance 583, 615
transfer 931, 1013
biological 275
electrode 1021
photochemical 921
rate equation 1018
tunnelling 1015
wavelength 301
electronegativity 428
electric dipole moment 687
electronic
polarizability 691
structure 365
electrophoresis 745
electrophoretic effect 841
electrospray ionization 735
electrostatic potential surface 446
electrostatics 1068

element of matrix 1058
elementary reaction 882
consecutive 883
elements, crystal structure 785
Eley–Rideal mechanism 1000
Ellingham diagram 239
elpot diagram 446
elution time 747
emboli 172
emf 267
temperature coefficient 281
emission spectroscopy 483
Emmett, P. 992
emulsion 752
enantiomeric pair 462
encounter pair 951
end point 245
end-to-end separation 721
endergonic reaction 223
endocytosis 757
endothermic
process 31
reactions
spontaneous 112
energy 1062
conformational 726
conservation of 35
dispersal of 91
electron in magnetic field 580
flux 826
introduced 31
level
A_2 system 597
AX system 591, 608
hydrogen atom 373
linear rotor 502
particle in a box 328
particle in a two-dimensional square well 332
particle on a sphere 351
symmetric rotor 501
vibration 512
separation, and system size 329
nucleus in magnetic field 583
pooling 925
quantization 295
transfer 931
ensemble 647
enthalpy
of activation 961
change
combining values 56
heat transfer at constant pressure 45
measurement 46
of reverse process 56
relation to internal energy change 48
of chemisorption 988
defined 45
density 59
of electron gain 391
of formation and molecular modelling 62
of ionization 391
of mixing, gases 168
from partition function 658
of physisorption 988
of transition, notation 57

variation with pressure and
    temperature 80
variation with temperature 50
entropy
    of activation 961
    Boltzmann formula 642
    canonical partition function 649
    change of surroundings 94
    conformational 724
    dependence on volume 651
    expansion of perfect gas 103
    hydrophobic effect 713
    increasing 92
    introduced 92
    at low temperatures 105
    measurement 104
    of mixing
        gases 167
        liquids 173
    nitrogen 105
    partition function 645, 656
    perfect elastomer 724
    solvent cage 713
    as a state function 95
    statistical 642
    thermodynamic definition 93
    two-level system 646
    uniform ladder 646
    variation with temperature 103
enzyme 908, 909
    inhibition 912
epitaxy 986
EPR 583
equation
    Berthelot 21
    Born 255
    Born–Mayer 788
    Butler–Volmer 1029
    Clapeyron 145
    Clausius–Mossotti 693
    Debye 693
    Dieterici 21
    diffusion 847
    eigenvalue 311
    Einstein 845
    Einstein–Smoluchowski 854
    Gibbs–Duhem 164
    Gibbs–Helmholtz 126
    Henderson–Hasselbalch 242
    Karplus 594
    Laplace 151
    Margules 186
    Mark–Kuhn–Houwink–Sakurada
        748
    material balance 955
    McConnell 620
    Michaelis–Menten 910
    Nernst 269
    Nernst–Einstein 845
    of state 3
        partition function 657, 671
        table 21
        thermodynamic 123
        virial 17
        perfect gas 10
    Poisson's 259, 1069
    Sackur–Tetrode 650
    Schrödinger 304
    secular 429
    Stokes–Einstein 741, 845
    thermochemical 57

time-dependent Schrödinger
    305
van der Waals 19
van 't Hoff 178, 991
    Wierl 302
equilibrium 39
    approach to 876
    biological system 233
    bond length 411
    constant
        contributions 679
        density of states 679
        from standard potentials 271
        molecular interpretation 232
        partition function 676
        rate constants 876
        relation between 232
        relation to standard Gibbs
            energy of reaction 229
        and standard cell emf 270
        temperature dependence 238
        thermodynamic 229
    Donnan 180
    effect of temperature 235
    mechanical 5
    response to pressure 234
    role of mixing 226
    thermal 6
    thermodynamic criterion 141
equipartition theorem 34
equivalence of heat and work
    35
equivalent nuclei 593
ER mechanism 1000
error function 344
ESCA 570
escape
    depth 980
    velocity 861
ESR 583, 615
essential symmetry 768
estuary silting 754
ethane, symmetry elements 455, 458
ethanol
    azeotrope 203
    elpot diagram 446
    FID 601
    isodensity surface 445
    NMR spectrum 587
    partial molar volume 161
ethene
    structure 433
    symmetry 457
    VB structure 416
ethyne, VB structure 416
Euler chain relation 80, 1054
eutectic 212
    halt 212
exact differential 74, 1054
excess
    enthalpy
        benzene 174
        cyclohexane 174
    entropy 173
    function 173
exchange
    current density 1029
    operator 393, 442
exchange-correlation
    energy 445
    potential 445

excimer 561
    formation 925
exciplex 561
    laser 561
excitance 294, 521
excited-state absorption 925
exclusion rule 524
exergonic reaction 223
exocytosis 757
exothermic process 31
exp-6 potential 706
expansion
    coefficient
        defined 78
        perfect gas 79
    work 37
expectation value 316
explosion 902
    model 903
    region 902
exponential
    decay 871
    function 1048
extended
    Debye–Hückel law 262
    Hückel theory 438
extensive property xx
extent
    of adsorption 987
    of reaction 222
extinction coefficient, see molar
        absorption coefficient
Eyring equation 958

f orbital, angular form 479
face-centred unit cell 768
factorial 1049
far-field confocal microscopy 563
Faraday constant 267
FEMO 451
femtochemistry 959
femtosecond spectroscopy 959
Fermi
    calculation 861
    contact interaction 595, 621
    level 799
fermion 355
    Pauli principle 385
Fermi–Dirac distribution 800
ferrocene symmetry 459
ferromagnetism 804
fibre 793, 794
fibrillation 919
Fick's
    first law of diffusion 827, 844
    second law of diffusion 847
FID 600
field sweep experiment 586
field-ionization microscopy 998
FIM 998
fine structure 399
    NMR 590
finite well 338
first
    explosion limit 902
    ionization energy 390
    Law 35
        formal statement 36
        mathematical statement 35
first-order
    energy 355

phase transition 149
    reaction 868
    spectra 598
    wavefunction 357
flame CARS 527
flash
    desorption 988, 997
    photolysis 864
float zoning 211
flocculation 754
Florey theta temperature 740
flow method 864
fluctuation 648
fluid mosaic model 756
fluorescence 550, 925, 926
    effect of solvent 551
    resonance energy 933
    spectrum 551
fluorine, configuration 424
flux 826
Fock operator 442
Fock, V. 392
food, thermochemical properties 59
forbidden transition 382
    and symmetry 475
force 1064
    between molecules 706
    constant 338, 512
        negative 957
    thermodynamic 843
    field 524
formal potential 1032
formula unit xx
Förster theory 931
Förster, T. 931
four-circle diffractometer 775
four-helix bundle 728
four-level laser 554
Fourier
    synthesis of electron density 780
    transform 601
        NMR 599
        technique 487
fractional
    coverage 987
    distillation 202
framework representation 1004
Franck–Condon
    principle 542
        electron transfer 1015
    factor 543
free
    energy, see Gibbs energy
    expansion 38
    particle wavefunction 310, 325
free-electron molecular orbital
        method 451
free-induction decay 600
freely jointed chain 721
freeze quench method 866
freezing
    point depression 177
    temperature 137
frequency 1066
    doubling 559
    factor 879
FRET 933
Freundlich isotherm 993
frictional coefficient 742
Friedrich, W. 772
frontier orbital 434

fuel
  cell 1038
  thermochemical properties 59
fuel-rich regime 903
fugacity 128
  coefficient 129
    van der Waals gas 130
full rotation group 459
functional 444
  derivative 445
  MRI 613
fundamental
  equation 122
  equation of chemical
    thermodynamics 163
  transition 515

g symmetry 539
g-value
  electron 581, 617
  nucleus 582
Galvani potential difference 1021,
  1023
galvanic cell 263
  working 1037
galvanizing 1041
γ-ray 484
gas
  constant, table 10
  defined 3
  discharge lamp 485
  kinetic model 815
  laws 8
  mixture 13
  saturation method 158
  solubility 172
  solvation 143
  spontaneous expansion 100
gauss 580
Gaussian
  function 495
  type orbital 444
Gay-Lussac, Joseph Louis 9
Gay-Lussac's law 9
gel 752
  electrophoresis 746
general solution 1055
generalized diffusion equation 848
gerade 539
Gerlach, Walther 353
germanium nanowire 986
Germer, Lester 301
Gibbs energy
  of activation 961
  of activation (electron transfer)
    1026
  of activation (Marcus theory) 1016
  biological processes 225
  and composition 163
  defined 109
  electron transfer 1015
  partial molar 163
  partition function 658
  perfect gas 128
  properties 124
  surface 758
  variation with pressure 125, 127
  variation with temperature 126
  energy of mixing 166
    liquids 173
    regular solution 174

Gibbs isotherm 758
Gibbs, J.W. 195
Gibbs–Duhem equation 164
Gibbs–Helmholtz equation 126
glancing angle 776
glass
  electrode 278
  transition temperature 795
globar 485
glucose structure 59
glycine 462
glycolysis 224, 914
glycolytic cycle oscillation 914
Gouy balance 803
Gouy–Chapman model 1022
Grahame model 1022
Graham's law of effusion 824
grand canonical ensemble 647
graphical output 445
gravimetry 988
greenhouse effect 537
Green–Kubo relation 842
gross
  orbital population 440
  selection rule 495
Grotrian diagram
  atomic hydrogen 383
  helium 396
Grotthuss mechanism 837
group
  classification
    flow diagram 460
    typical shapes 461
  theory 453
GTO 444
guanine 730
Gunn
  diode 486
  oscillator 615
gyromagnetic ratio, see
    magnetogyric ratio

haematoporphyrin 929
haemerythrin 536
hair 731
half-life
  first-order reaction 872
  second-order reaction 874
half-life (adsorption) 997
half-reaction 263
hamiltonian 305
  matrix 437
  operator 311, 314
  rotational motion 346
hard-sphere potential 705
  radial distribution function 710
harmonic 516
  motion 338, 1065
  oscillator
    classical treatment 1065
    energy levels 338
    wavefunctions 338
    zero-point energy 339
  wave 1066
harmonics 515
Harned cell 272
harpoon mechanism 950
Hartree, D.R. 392
Hartree–Fock
  equation 442
  self-consistent field 392

heart attack 919
heat
  introduced 31
  mechanical definition 37
  molecular interpretation 32
  and work, equivalence 35
heat capacity 43
  constant pressure 50
  constant volume 43
  Debye formula 298
  dissociation 670
  Einstein 297
  isobaric 50
  linear molecule 669
  molar 44, 50
  overall 670
  partition function 669
  ratio 53
  relation between C_p and C_V 51, 85
  solids 296
  specific 44
  variation with temperature 50
  vibrational 670
Heisenberg uncertainty principle 317
helium
  atom 384
  Grotrian diagram 396
  monatomic character 422
  phase diagram 140
  superfluid 140
helium–neon laser 560
helix–coil transition 674, 865
Helmholtz
  energy
    and surface formation 150
    defined 109
    maximum work 110
    molecular interpretation 110
    partition function 657
  layer 1021
Henderson–Hasselbalch equation
  242
Henry's law 170
  constant, table 172
Hermann–Mauguin system 456
Hermite polynomial 340
Hermitian operator 312
Hess's law 58
heterogeneity index 734
heterogeneous catalyst 908
heterolytic dissociation, isotope
    effect 889
heteronuclear diatomic molecule
    427
heteronuclear spin system 598
hexabenzocoronene 483
hexagonal
  close packing 785
  unit cell 769
hexane, partially miscible 205
high
  field end 586
  overpotential limit 1030
high-boiling azeotrope 203
high-performance liquid
    chromatography 137
high-temperature superconductor 796
highest occupied molecular orbital
    434
Hittorf method 840
HIV 909

HOMO 434
homogeneous catalyst 908
homogenized milk 752
homonuclear
  diatomic molecule, structure 425
  spin system 598
homopolar contribution 687
Hooke's law 793
host–guest complex 697
HREELS 981
Hückel
  approximation 433
  method, matrix formulation 434
  theory, solids 798
human immunodeficiency virus
    909
Humphreys series 408
Hund's rule 388, 596
hybrid orbital 415
hybridization 415
  d orbital 417
  table 417
hydrochloric acid, azeotrope 203
hydrodynamic
  flow 706
  radius 837
hydrogen
  atom, energy levels 373
  bond 704, 727
  endothermic adsorption 988
  Grotrian diagram 383
  molecular orbital 422
  molecule, structure 411
  molecule-ion 418
    analysis of bonding 420
    energies 420
  ortho and para 662
  spectrum 366
  triatomic cation 439, 480, 537
    population analysis 441
hydrogen bromide, photoelectron
    spectrum 570
hydrogen chloride, rotational
    spectrum 517
hydrogen fluoride, structure 428
hydrogen peroxide 456
  decomposition 909
hydrogenation 1001
  mechanism 1002
hydrogenic
  atom 365
    Coulomb potential 367
    radial wavefunction 371
    Schrödinger equation 367
    separation of internal variables
      369
  wavefunction, radial 372
hydrogen–bromine reaction 868
  rate law 900
hydrogen–deuterium reaction 947
hydrogen–oxygen reaction
    mechanism 902
hydronium ion 240
hydrophilic 752
hydrophobic 712, 752
  interaction 712, 713
hydrophobicity constant 712
hydrostatic
  pressure 6
  stress 791
hyperbola 8

hyperfine
  coupling constant table 620
  splitting ESR 617
hyperpolarizability 689

ICDD 775
ice
  high pressure polymorphs 139
  residual entropy 673
  structure 789
icosahedron 458
ideal
  gas, see perfect gas
  solubility 178
  solution 168
ideal–dilute solution 171
identity 454
IHP 1021
immiscible liquids 204
impact parameter 708
impressed-current cathodic
    protection 1041
incident beam flux 707
incongruent melting 213
indefinite integral 1052
indicator 245
  diagram 39
  solution 839
indistinguishable molecules 650
induced
  dipole moment 689
  magnetic moment 804
induced-dipole–induced-dipole
    interaction 703
induction period 885
inelastic neutron scattering 832
inexact differential 74
infrared
  active 513
    normal modes 527
  chemiluminescence 965
  inactive 513
  radiation 484
inhomogeneous broadening 603
initial
  condition 1055
  rate method 869
initiation step 900
inner
  Helmholtz plane 1021
  potential 1023
insulator 795
integral 1052
  criteria for zero 469
  protein 756
integrated
  absorption coefficient 492
  rate law
    first order 871
    second-order reaction 873
    table 875
integration 1051
  in NMR 587
  by parts 1052
intensity maximum (rotational
    spectra) 507
intensive property xx
interaction energy, summary table
    700
interferogram 487

Intergovernmental Panel on Climate
    Change 133
internal
  conversion 552, 553, 927
  energy
    and volume 35
    canonical partition function 648
    effect of temperature 44
    equipartition theorem 34
    heat transfer at constant volume
      42
    introduced 33
    partition function 640, 656
    properties 122
    two-level system 640
    variation with temperature and
      volume 76
    variation with volume 123
  motion separation 368
  pressure 123
    defined 76
    perfect gas 77, 124
    van der Waals gas 124
International Centre for Diffraction
    Data 775
International system (point groups)
    456
interstellar
  clouds 482
  molecules 482
intersystem crossing 552, 925, 927
intrinsic viscosity 748
inverse matrix 1059
inversion 455, 539
  ammonia 536
inversion recovery technique
    605
inversion temperature 84
inverted region 1019
ion
  bombardment 980
  channel 757, 850
  configuration 390
  pump 757
ion-selective electrode 278
ionic
  atmosphere 258
    colloidal particle 754
  bond 410
  conductivity 837
  mobility 836
  ionic conductivity 837
  radius 787
  solid, structure 786
  strength 258
    effect on rate 962
ionization
  degree of 834
  energy 390
    atom 373
    determination 374
    periodicity 390
  photochemical 921
IPCC 133
irreducible representation 465
  notation 465
irreversibility 92
ISC 925
iscosahedral group 458
isenthalpic process 82

isobar 9, 12
isobaric
  calorimeter 46
  heat capacity 50
isodensity surface 445
isoelectric
  focussing 746
  point 746
isolated system 30
isolation method 869
isomerization 921, 1001
isopleth 200
isosteric enthalpy of adsorption 991
isotherm 8
  adsorption 989
  carbon dioxide 17
  Freundlich 993
  Gibbs 758
  Langmuir 989
  Temkin 993
isothermal
  compressibility
    defined 80
    perfect gas 81
  Joule–Thomson coefficient 83
  partition function 672
isotherms, van der Waals 22
isotope separation 566
isotopic substitution (rotational
    spectra) 507
isotopomer 566

Jablonski diagram 552
jj-coupling 403
joule 34, 1062
Joule experiment 77
Joule, J.P. 34
Joule–Thomson
  coefficient 80
    partition function 672
    table 84
  effect 82

K-radiation 773
Karplus equation 594
Keesom interaction 701
Kelvin equation 152
keratin 731
Kerr
  lens 558
  medium 558
ket 331
Kevlar 210
kinetic
  chain length 907
  energy 1062
    operator 314
    density 823
  isotope effect 888
  model of gases 815
  pressure 711
  salt effect 962
kink 978
Kirchhoff's law 63
Kirkwood–Riseman theory 759
klystron 486, 615
KMT, see kinetic model of gases 815
Knipping, P. 772
Knudsen method 825
Kohlrausch's law 834

Kohn, W. 441
Kohn–Sham equations 445
Koopmans' theorem 569
Krafft temperature 755
Kronecker delta 331, 1058
krypton-ion laser 560

Lagrange method 633
Laguerre polynomial 371
Lamb formula 588
Λ quantum number 424
λ line 140
λ transition 149
Lamé constants 811
lamellar micelle 755
Landau, L. 753
Langevin function 689
Langmuir isotherm 989
Langmuir, I. 757
Langmuir–Blodgett film 757
Langmuir–Hinshelwood mechanism
    999
Laplace equation 151
laplacian 305, 349
  operator 369
Laporte selection rule 541
Larmor frequency 582
laser 554
  characteristics 562
  control of reactions 964
  power 556
  Raman spectroscopy 565
  spectroscopy 562
laser-induced fluorescence 965
lattice
  energy 787
  enthalpy 787
  plane
    identification 770
    separation 771
law
  Beer–Lambert 491
  Boyle's 8
  Charles's 9
  Dalton's 13
  Dulong–Petit's 296
  Fick's first 827, 844
  Fick's second 847
  gas 8
  Gay-Lussac's 9
  Graham's 824
  Henry's 170
  Hess's 58
  independent migration of ions 834
  Kirchhoff's 63
  Kohlrausch's 834
  limiting 8
  Ostwald's 835
  Raoult's 168
  rate 868
  Rayleigh–Jeans 295
  Stefan–Boltzmann 294
  T^3 104
  Wien displacement 294
LCAO 418
LCAO-MO 418
LCD 210
Le Chatelier's principle 234
  effect of compression 234
  effect of temperature 235

leading solution 839
LEED 982
legendrian 305, 349
Lennard-Jones potential 706, 727
level of term 397
lever rule 201
Lewis, G.N. 410
LH mechanism 999
lifetime 496
  broadening 496
ligand-field splitting parameter 546
ligand-to-metal charge-transfer
    transition 547
light scattering 736, 739
  polymer solution 740
light-emitting diode 801
light-harvesting complex 928
limit cycle 917
limiting
  current density 1033
  law 8, 258
    Debye–Hückel 261
  molar conductivity 834
  transport number 839
Linde refrigerator 85
Lindemann–Hinshelwood
    mechanism 890
linear
  combination 315
    of atomic orbitals 418
    degenerate functions 381
    symmetry classification 468
  free energy relation 961
  molecule, heat capacity 669
  momentum 1063
    operator 313
  rotor 500
    energy levels 502
    partition function 660
linear-sweep voltammetry 1034
Lineweaver–Burk plot 911
linewidth 495
lipid raft model 757
liposome 755
liquid
  crystal 209, 755
    phase diagram 210
  junction potential 266
  viscosity temperature dependence
    832
liquid–liquid phase diagram 204
liquid–solid phase diagram 208
liquid–vapour boundary,
    location 146
LMCT 547
local contribution 587
local-density approximation 445
logarithm 1048
London
  formula 703
  interaction 703
long-range order 709
longitudinal relaxation time 602
Lorentzian absorption line 604
Lotka–Volterra mechanism 916
low
  energy electron diffraction 301
  overpotential limit 1029
low-boiling azeotrope 203
low-energy electron diffraction 982

lower
  consolute temperature 206
  critical solution temperature 206
lowest unoccupied molecular orbital
    434
Luggin capillary 1030
LUMO 434
Lyman series 366
lyophilic 752
lyophobic 752
lyotropic mesomorph 755

macromolecule 719
  size and shape 732
  solution nonideality 740
Madelung constant 788
magic angle 615
magic-angle spinning 615
magnetic
  field 1066
  flux density 802
  moment
    orbital 396
    spin 396
    induced 804
  properties 801
  quantum number 350, 373
  resonance imaging 612
  scattering 784
  susceptibility 589, 802
magnetically equivalent nuclei 596
magnetizability 802
magnetization 802
  NMR 599
magnetogyric ratio, electron 580
MALDI-TOF mass spectrometry 735
manometer 6
many-electron atom 365
  electronic structure 383
  Schrödinger equation 393
  selection rule 403
Marcus
  inverted region 1019
  theory 934, 1014
Marcus, R.A. 934
Margules equation 186
Mark-Kuhn-Houwink-Sakurada
    equation 748
Mars van Krevelen mechanism 1003
MAS 615
mass spectrometry 735
material balance equation 955
matrix 1057
  addition 1058
  diagonalization 435
  element 342
  multiplication 1058
  representation 464
matrix-assisted desorption/ionization
    735
matter flux 826
maximum
  additional work 113
  non-expansion work 113
  velocity (enzymolysis) 910
  work, Helmholtz energy 110
Maxwell
  construction 21
  distribution of speeds 818
  relations 123

Mayer f-function 672
MBE 986
MBRS 1001
MBS 985
McConnell equation 620
mean
  activity coefficient 257
    table 261
  bond enthalpy 61
  energy 667
  free path 822, 854
  rotational energy 668
  separation 723
  speed 820
    molecules 819
  translational energy 667
  value theorem 594, 701,
    1052
  vibrational energy 668
mechanical
  equilibrium 5
  properties 791
Meissner effect 796
melting
  effect of pressure 142
  temperature 137, 795
  thermodynamic aspects 141
membrane osmometry 734
meso-tartaric acid 456
mesopause 922
mesophase 209
mesosphere 922
metal, reduction of ores 239
metal-to-ligand charge-transfer
    transition 547
metallic conductor 795
metallic solid 784
metastable
  excited state 554
  phase 136
methane
  anomalous entropy of vaporization
    102
  normal mode analysis 527
  promotion 414
  symmetry elements 455
  VB theory 414
method
  of initial rates 869
  of undetermined multipliers
    633
micelle 754
Michaelis constant 911
Michaelis-menten
  equation 910
  mechanism 910
Michelson interferometer 487
microcanonical ensemble 647
microporous catalyst 1004
microwave radiation 484
Mie potential 705
milk 752
Miller indices 770
mirror plane 455
Mitchell, P. 851
mitochondrial electron transport
    275
mitochondrion 225
  ion transport 851
mixed inhibition 913

mixing
  driving force 173
  Gibbs energy 166, 173
mixing period 610
MLCT 547
MO
  see molecular orbital
  theory 417
mobility 836
  surface 998
mode locking 557
mode-selective chemistry 964
modulus 791, 1049
molality xx
molar (unit) xx
molar
  absorption coefficient 491
  concentration xx
  conductivity 833
    limiting 834
  heat capacity 44
  magnetic susceptibility 802
  mass xx
    from osmometry 181
  partition function 676
  polarization 693
  property xx
  volume xx
molarity, see molar concentration
mole xx
  fraction 14
molecular
  aggregate 719
  beam 706
  beam epitaxy 986
  beam reactive scattering 1001
  beam scattering (surfaces) 985
  dynamics 711, 712, 728
  flow 706
  interactions 16
  mechanics 727
  modelling 62
  motion, liquids 832
  orbital 418
    energy level diagram 422
    homonuclear diatomics 424
    polyatomic molecules 432
    solids 798
    theory 417
  partition function 634
  potential energy curve 411
  recognition 697
  solid 789
  spectra 298
  vibration 512
  weight xx
  wire 750
molecularity 882
moment of inertia 497, 1064
  introduced 345
  table 498
momentum transport 827
monochromatic 485
monochromator 486
monoclinic unit cell 768, 769
monodisperse 732
monolayer 757
monomer 719
monopole 699
Monte Carlo method 711

Morse potential energy 515
most probable
   radius 379
   speed 820
mouse cell 521
moving boundary method 839
MPI 965
MRI 612
Mulliken
   population analysis 440
   scale 428
Mulliken, R.S. 440
multi-walled nanotube 750
multinomial coefficient 1049
multiphoton
   ionization 965
   processes 564
multiple
   quantum transition 611
   steady states 918
multiplicity 401, 540
multipole 699
mutual termination 907
MWNT 750

n-fold
   axis of symmetry 454
   improper rotation 455
   improper rotation axis 455
   rotation 454
n-pole 699
(n_+,n_-) coordination 786
n-to-π^* transition 547
n-type semiconductivity 800
NAD 224
NADP 224
nanodevice 986
nanofabrication 986
nanotechnology 986
nanotube 750
   electric conductivity 750
nanowire 986
naphthalene radical anion 620
natural
   linewidth 497
   logarithm 1048
near field 563
near-field scanning optical
    microscopy 563
nearly free-electron approximation
   798
Néel temperature 804
negative temperatures 643
neighbouring group contribution
   587, 589
nematic phase 210
neodymium laser 558
Nernst
   diffusion layer 1032
   equation 269
   filament 485
   heat theorem 106
Nernst–Einstein equation 845
nerve propagation 270
net rate of absorption 494
network solid 789
neuron 270
neutron
   diffraction 783
   scattering 832

Newtonian flow 827
Newton's second law of motion
   1064
nitric oxide, partition function 665
nitrobenzene, partially miscible 205
nitrogen
   configuration 424
   entropy 105
   fugacity 130
   laser 561
   molecule, structure 413
   photoelectron spectrum 576
   sticking probability 997
nitrogen oxide, see nitric oxide
nitropropane, COSY spectrum 611
nitroxide spin label 616
NMR 583
   spectrometer 584
nodal plane 380
node 310
NOE 608
NOESY 612
non-competitive inhibition 913
non-expansion work 38
non-polarizable electrode 1031
non-primitive unit cell 768
nonlinear
   optical phenomena 558
   rotor, partition function 660
nonradiative decay 550
normal
   boiling point 136
   freezing point 137
   melting point 137
   mode 520, 522
    symmetry species 527
    tetrahedal molecule 528
   transition temperature 101
   Zeeman effect 408
normalization constant 307
North Star, photon flux 549
Noyes, R. 917
NSOM 563
nuclear
   g-factor 582
   magnetic resonance 583
   magneton 582
   Overhauser effect 608
    spectroscopy 612
   population 584
   spin
    nuclear constitution 581
    properties, table 582
   statistics 510
nucleate 153
nucleation 152, 675
nucleic acid 729
Nujol 490
number-average molar mass 733
numerical integration 1056
nylon-66 794
   formation 904

O branch 520
objective lens 521
oblate 501
observable 312
octahedral
   complex, d orbital splitting 546
   group 458

octupole 700
OHP 1021
one-dimensional random walk 853
open system 30
operator
   angular momentum 348
   commutator 319
   construction 313
   Coulomb 393, 442
   exchange 393, 442
   Fock 442
   hamiltonian 311, 314
   Hermitian 312
   introduced 311
   kinetic energy 314
   laplacian 369
   linear momentum 313
   position 313
   potential energy 314
opsin 548
optical
   density, see absorbance
   Kerr effect 557
   rotatory dispersion 696
   trapping 107
optically active 462, 695
orbital
   angular momentum
    around internuclear axis 424
    quantum number 350, 373
   approximation 384
   atomic 372, 376
   energy, diatomic molecules 424
   magnetic moment 396
   molecular 418
   overlap 413
   symmetry species 468
ORD 696
order of group 466
order–disorder phase transition 150,
   783
ordinary differential equation 1055
oregonator 917
orientation
   polarization 691
   specification 522
Orion nebula 482
ortho-hydrogen 511, 662
orthogonal 312, 330
orthorhombic unit cell 769
oscillating reaction 914
oscillator strength 577
osmometry 178, 734
   vapour-phase 182
osmosis 178
osmotic virial coefficient 181
Ostwald viscometer 748
Ostwald's dilution law 835
Otto cycle 120
outer
   Helmholtz plane 1021
   potential 1023
overall
   partition function 666
   quantum yield 934
   reaction 868
Overbeek, J.T.G. 753
overlap 413
   density 419
   integral 424

matrix 434
   population 440
   and symmetry 469
overpotential 1029
overtone 516
oxidation 263, 1001
   catalytic 1003
oxidative phosphorylation 224, 275
oxygen
   configuration 424
   electronic energy levels table 541
   paramagnetism 427
   term symbol 540
ozone 921
   decomposition 886
   depletion 923
   hole 578
   layer 13
   polarity 688

P branch 518
p band 799
p orbital 379
   boundary surfaces 380
   real and complex forms 380
p-type semiconductivity 800
packing fraction 785
paired electrons 385
palladium hydride, phase diagram
   205
para-hydrogen 511, 662
parabolic approximation 512
parallel
   band 523
   β-sheet 728
paramagnetic 802
   contribution 588
   substance 427
parcel (of air) 12
parity 539
partial
   charge 427
    interaction 727
   derivative 1053
    introduced 43
   differential equation 1057
   fraction 875, 1052
   molar Gibbs energy 163
   molar volume 161
    polynomial expression 162
   pressure
    perfect gas 14
    real gas 15
   specific volume 742
   vapour pressure 143
partially miscible liquid 173
   distillation 207
particle
   in a box 325
   in a finite well 338
   in a two-dimensional well 331
   on a sphere 351
particle-hole formalism 402
particular solution 1055
partition
   coefficient 850
   function
    contributions, table 666
    density 971
    electronic 665

enthalpy 658
entropy 645, 656
equation of state 657, 671
factorization 659
Gibbs energy 658
heat capacity 669
Helmholtz energy 657
internal energy 640, 656
interpretation 635
introduced 33
molecular 634, 658
overall 666
pressure 657
rotational 659
rotor 660
translational 637, 659
two-level system 636
uniform ladder 636
vibrational 664
pascal 4
Pascal's triangle 593
Paschen series 366
passivation 1041
passive transport 849
path function 73
Patterson synthesis 781
Pauli
    exclusion principle 385
    principle 385
        bond formation 412
    spin matrices 479
Pauling scale 428
Pauling, L. 725
PDT 929
PEMD pusle sequence 610
penetration 386
    probability 344
peptide link 674, 720
peptizing agent 752
perfect
    elastomer 793
    elastomer entropy 724
    gas 10
        adiabat 54
        coefficient of thermal
            conduction 829
        coefficient of viscosity 830
        diffusion coefficient 828
        entropy 103
        equation 10
        expansion coefficient 79
        Gibbs energy 128
        isothermal compressibility 81
        temperature scale 7
        transport properties 826
Period 2 diatomic molecules 423
period-doubling 920
periodicity of ionization energy
    390
peripheral protein 756
permanent waving 731
permeation chromatography 747
permittivity 692, 1068
    relative 692
perpendicular band 524
perturbation theory 355
    time-dependent 358
    time-independent 355
PES 568
Petit, Alexis-Thérèse 296

pH 240
    buffer solution 244
    curve 242
        exact form 243
    measurement 278
    of weak acid 241
phaeophytin 928
phase 135, 193
    boundary 136
        equations for 144
    diagram
        calcium/silicon 219
        carbon dioxide 139
        helium 140
        introduced 136
        liquid crystal 210
        liquid–liquid 204
        liquid–solid 208
        palladium hydride 205
        reacting system 212
        regular solution 206
        silver/tin 216
        sodium–potassium 213
        triple point 197
        water 138
    problem 781
    rule 195
        one-component system 196
        two-component system 198
    separation 204
    transition 136
        Ehrenfest classification 148
        entropy 101
        high pressure 197
        order–disorder 783
phase-sensitive detection 617
phenanthrene 480
phenomenological transport
    equations 826
phenoxy radical 627
phenylalanine 718
phosphatidyl choline 756
phospholipid 756
phosphorescence 550, 552, 925
phosphorus pentachloride,
    symmetry 457
phosphorylation 851
photobiology 928
photobleaching 564
photocatalyst 929
photocathode 489
photochemical
    process 924, 925
    table 925
    reaction, rate law 935
photochemistry 920
photodeflection 566
photodiode 489
photodissociation 566
photodynamic
    theory 935
    therapy 929
photoejection 300
photoelectric effect 300
photoelectron 568
    spectroscopy 568
photoemission spectroscopy 980
photoionization 566
photoisomerization 566
photomedicine 928

photomultiplier tube 489
photon 299
photophosphorylation 276
photoproduct 929
photosensitization 935
photosensitizer 929
photosphere 394
photosynthesis 275, 928
photosystem 928
photosystems I and II 276
phototactic response 774
photovoltaic device 489
physical state 3
physisorption 988
π bond 413
    formation energy 436
π orbital 424
π-electron binding energy 436
π-to-π^* transition 547
$\pi/2$ pulse 600
pK_a
    defined 240
    determination 242
    measurement 278
Planck distribution 295
    starlight 394
Planck, Max 295
Planck's constant 295
plane polarized 1067
plasma 485
plasmon 994
plastic 793, 795
plastocyanin 276
plastoquinone 276
PMT 489
p–n junction 800
Pockels cell 556
pOH 240
point
    dipole 698
    group 454
    group notation 456
Poiseuille's formula 831
Poisson's
    equation 259, 1069
    ratio 792
polar
    bond 427
    molecule 459, 687
        symmetry 462
polarizability 508, 689
    electronic 691
    frequency dependence 692
    molecular structure 691
    volume 690
polarizable electrode 1031
polarization 688
    mechanism
        ESR 621
        NMR 595
    overpotential 1032
polarized line 525
polaron 751
poly(butylene adipate) 736
polyamide 904
polyatomic molecule, VB theory 414
polychromatic 485
polychromator 486
polydisperse 732
polydispersity index 734

polyelectrolyte 730
polyelectronic atom see many-
    electron atom 365
polyester 904
polyethylene 720
polymer 719
    size light scattering 741
polymerization 1001
    kinetics 904
polypeptide 674, 720
    helix–coil transition 865
polystyrene, radius of gyration 723
polytropic process 70
polytype 784
Pople, J. 441
population 629
    analysis 440
    inversion 554
    two-level system 637
porphine group 480
porphyrin 929
position operator 313
position–momentum uncertainty
    318
potassium sulfate, partial molar
    volume 166
potential
    difference 1070
    energy 1062
        adsorption 996
        dipole 698
        effective 370
        intermolecular 16
        multipole interaction 700
        operator 314
        profile 881
        surface 966
        trajectory 969
        transition state 967
    of working cell 1038
Pourbaix diagram 289
powder diffraction pattern 775, 779
power 1063
    for refrigeration 99
    series 1052
    working cell 1038
pre-equilibrium 887
pre-exponential factor 879
prebiotic reactions 898
precess 581
precision-specified transitions 565
precursor state (adsorption) 996
predissociation 552
preparation period 610
pressure
    change during reaction 863
    curvature of surface 151
    defined 4
    effect on boiling temperature
        146
    hydrostatic 6
    kinetic energy density 823
    kinetic theory 816
    measurement 5
    partition function 657
    time dependence with effusion
        825
    units, table 4
pressure-jump technique 877
Prigogine, I. 916

primary
  kinetic isotope effect 888
  process 924
  quantum yield 926
  structure 720
primitive unit cell 768
principal
  axis 454, 501
  quantum number 372
principle
  Avogadro's 10
  building-up 387
  correspondence 329
  of corresponding states 24
  equal *a priori* probabilities 629
  Franck-Condon 542
  Le Chatelier's 234
  of corresponding states 24
  Pauli 385
  Pauli exclusion 385
  Ritz combination 367
  uncertainty 317
probability
  amplitude 306
  density 306
product rule 45, 1051
projection reconstruction 612
prolate 501
promotion 414
propagation step 900
propanone, see acetone
propenal 1003
protein
  binding 179
  electron transfer 1019
  structure 725
proton
  conductivity 837
  decoupling 607
  pump 851
  tunnelling 889
  in water 253
pseudofirst-order rate law 869
pulsed
  beam 1001
  laser 555
pumping 554
pure shear 791
pyroelectric device 489
pyrolysis 900

Q branch 518, 520
Q-switching 555
quadrupole 699
quantization
  angular momentum 352
  energy 295
  from boundary conditions 309
  rotation 347
  space 353
quantum
  defect 395
  mechanics 293
  number
    introduced 328
    magnetic 350, 373
    orbital angular momentum 350, 373
    particle in a box 328
    principal 372

rotational 502
spin 354
total angular momentum 400, 401
total orbital angular momentum 400
total spin 400, 401
oscillation 708
yield 934
quaternary structure 720
quenching 927
  methods 866
quinoline 456
quotient rule 1051

R branch 518
radial
  distribution function 378
  distribution function (fluids) 709
  node 376
  wave equation 370
radiative decay 550
radical chain reaction 899
radiofrequency
  pulse 600
  radiation 484
radius
  of gyration 723, 739
  hydrodynamic 837
  of shear 753
  Stokes 837
radius-ratio rule 786
rainbow
  angle 709
  scattering 708
RAIRS 981
Ramachandran plot 727
Raman
  activity, normal mode 528
  imaging 521
  scattering 981
  spectroscopy 488
random
  coil 721
  walk 853
Raoult's law 168
rate
  of adsorption 988, 996
  consumption 867
  constant 868
    activated complex theory 958
    collision theory 946
    state-to-state 965
  of desorption 997
  effusion 824
  equation, electron transfer 1018
  formation 867
  law 868
    hydrogen-bromine reaction 900
    photochemical reaction 935
  reaction 867
rate-determining step 884
Rayleigh
  radiation 488
  ratio 737
  scattering 737
Rayleigh-Jeans law 295
RDS 885

reaction
  control by laser 964
  coordinate 881
  enthalpy, from equilibrium constant 237
  formal notation 228
  Gibbs energy, and cell emf 267
    defined 223
  mechanism 862
  order 868
  product imaging 965
  profile, RDS 885
  quotient
    and cell emf 269
    defined 226
  rate 867
    response to temperature 880
reactive
  collisions 963
  cross-section 949
real gas 10, 16
real-time analysis 864
rearrangement 921
reciprocal
  centimetre 483
  identity 80
reconstruction 982
redox
  couple 264
  reaction 263
reduced
  mass 368, 820
    see also effective mass 512
  representation 464
  variables 23
reducing agent 263
reduction 263
reference
  electrode 1030
  state 60
reflection 455
  symmetry 540
  X-ray diffraction 776
reflection–absorption infrared spectroscopy 981
reforming 1003
refractive index 694
  dispersion 695
  frequency dependence 695
refrigeration 107
  thermodynamics of 98
regular solution 174
  activity 186
  Gibbs energy of mixing 174
  phase diagram 206
  vapour pressure 187
relative
  mean speed 820
  permittivity 692, 1068
    refractive index 694
    table 255
relaxation
  effect 841
  time, variation with rate 604
REMPI 965
reorganization energy 1016
representative matrix 463
repulsive surface 968
residual entropy 672
resistance 833

resolution 490
resonance 579
  condition 583
  integral 429
  Raman spectra 526
  Raman spectroscopy 525
resonant
  mode (laser) 555
  multiphoton ionization 965
resting potential 270
retardation step 901
retina 548
retinal 548
retinol 549
reversible change 39
rheology 791
rhodamine 6G 561
rhodopsin 548
Rice-Hertzfeld mechanism 900
ridge 12
rigid rotor 499
ring current 590
Ritz combination principle 367
RNA
  structure 729
rock-salt structure 786
rods and cones 548
Röntgen, W. 772
root mean square
  separation 722
  speed 816
Roothaan equations 443
rotating drum viscometer 748
rotation
  on a sphere 349
  quantization 347
rotational
  constant 500
  degeneracy 502
  energy level 500
  motion 345
    hamiltonian 346
  quantum number 502
  Raman spectra 507
    selection rule 508
  selection rule 504
  spectra, appearance 506
  state lifetime 524
  structure, vibronic transition 545
  subgroup 663
  temperature 662
  term 501
  transition 504
rubber, thermodynamics 811
ruby
  glass 752
  laser 558
rule
  chain 792
  Hund's 596
  lever 201
  phase 195
  radius-ratio 786
  Schulze-Hardy 754
  Trouton's 101
  Walden's 846
Runge-Kutta method 1056
Russell-Saunders coupling 402
ruthenocene symmetry 459

Rydberg
  atom 409
  constant 366, 373
  state 395

s band 799
S branch 520
s orbital 376
Sackur–Tetrode equation 650
sacrificial anode 1041
saddle point 966
SALC 473
salt bridge 262, 266
SAM 982, 986
sample preparation (spectroscopy)
  490
SATP 10
saturable absorber 556
saturation 602
Sayre probability relation 782
scalar
  coupling constant 590
  product 580, 1050
scanning
  Auger electron spectroscopy 982
  electron microscopy 985
  probe microscopy 985
  tunnelling microscopy 336, 985
Scatchard plot 180
scattering 737
  factor 777
  quantum effects 708
  small particles 739
SCF 442
  procedure 392
Scherrer, P. 775
Schrödinger equation
  hydrogenic atom 367
  many-electron atom 393
  molecular vibration 512
  rotational motion 349
  time-dependent 305, 358
  time-independent 304
Schulze–Hardy rule 754
screw
  axis 978, 979
  dislocation 978
SEC 747
second
  explosion limit 902
  harmonic generation 559, 994
  ionization energy 390
  Law of thermodynamics 90
  order spectra 598
  radiation constant 294
second-order
  energy 356
  phase transition 149
secondary
  kinetic isotope effect 888
  process 924
  structure 720
secondary-ion mass spectroscopy
  980
secular
  determinant 430
  equation 429, 1060
  solution 431
sedimentation 742
  constant 743

equilibrium 744
selection rule 382, 494
  diatomic molecules 541
  Laporte 541
  many-electron atoms 403
  molecular vibration 513
  rotational 504
  symmetry basis 474
  vibrational Raman 519
  vibrations of polyatomic molecules
    523
selectivity coefficient (electrode) 279
self-assembled monolayer 986
self-assembly 750
self-consistent field 442
SEM 985
semi-empirical method 443
semiconductor 795
semipermeable membrane 178
separation
  internal motion 368
  of variables 331, 1057
sequencing 720
SERS 981
serum albumin, radius of gyration 723
SEXAFS 982
SFC 137
shape-selective catalysis 999
shear modulus 792
shell 374
SHG 994
shielded
  Coulomb potential 259
  nuclear charge 386
shielding constant 386, 585
  constant origin 587
  variation with electronegativity
    588
Shoenflies system 456
short-range order 710
side-centred unit cell 768
σ bond 412
σ electron 420
σ orbital 418
silane adsorbed (STM image) 985
silver/tin phase diagram 216
similarity transformation 1060
simple distillation 202
SIMS 980
simultaneous equations 1059
single-molecule spectroscopy 563
single-walled nanotube 750
singlet 395
  vector diagram 385
singlet–singlet energy transfer 925
singlet–triplet transition 552
size-exclusion chromatography 747
Skou, J. 851
sky colour 737
slice selection 612
slip plane 793
smectic phase 209
smog 13
smoke 752
  colour 739
sodium azide 570
sodium D-lines 399
sodium–potassium phase diagram
  213
sol 752

solder 212
solid-state
  explosion 904
  laser 558
  NMR 614
solids, heat capacity 296
solid–liquid boundary
  location 145
solid–vapour boundary
  location 147
soliton 751
solubility
  gas 172
  ideal 178
  parameter (SCF) 138
solution, regular 174
solvent cage 713
solvent contribution 587
  NMR 590
source of radiation 485
sp hybrid orbital 416
sp^2 hybrid orbital 415
sp^3 hybrid orbital 415
space
  group 454
  lattice 767
  quantization 353
span (a representation) 464, 468
spatial coherence 555
species-selective electrode 279
specific
  enthalpy 59
  heat capacity 44
  selection rule 495
spectral
  line 299
  line intensity 491
spectrometer 484
spectrophotometry 864
spectroscopic transition,
    perturbation theory 358
spectroscopy 365
  interstellar molecules 482
spectrum 365
  atomic hydrogen 366
  complex atoms 393
speed 1063
  most probable 820
  root mean square 816
sphalerite 787
spherical
  harmonic 350
    triple integral 505
  polar coordinates 308, 349
  rotor 499
spherically symmetrical 376
spin 354
  correlation 389
  decoupling 607
  density 620
  echo 606
  label 616
  magnetic moment 396
  magnetic quantum number 354
  matrices 479
  packet 606
  probe 616
  quantum number 354
  relaxation 602
  state, two electrons 385

spin-$\frac{1}{2}$ nucleus 583
spin–lattice relaxation time 602
spin–orbit coupling 396, 552
  constant 397
  nuclear charge 398
spin–spin coupling
  origin 594
  relaxation time 603
spiral growth 979
SPM 985
spontaneous
  change 90
    criteria 109
  cooling 100
  emission 493
  nucleation centre 153
SPR 994
square-integrable 309
SQUID 803
stability parameter 674
standard
  ambient temperature and pressure
    10
  boiling point 136
  chemical potential
    defined 166
  emf 269
    determination 273
  enthalpy
    change 55
    of combustion 58
    of electron gain 391
    of formation 60
    of formation of ions 253
    of fusion 55
    of ionization 391
    of sublimation 56
    of vaporization 55
  entropy
    of ions 256
    of transition 101
    of vaporization 102
  freezing point 137
  Gibbs energy
    of formation 114, 254
    of formation of ions 253
    of reaction 114
    of solvation 255
  molar partition function 676
  potential 271
    determination 273
    from two others 280
  pressure 4
  reaction enthalpy 57
    and cell reaction 281
    from standard enthalpy of
      formation 61
    variation with temperature 63
  reaction entropy 108
    and cell reaction 281
  reaction Gibbs energy 226
  state
    biological 186
    defined 55
    summary table 182
  temperature and pressure 11
star
  composition 394
  radial velocity 409
  spectra 394

Stark
  effect 503
  modulation 504
state function 34
state-to-state
  cross-section 965
  rate constant 965
  reaction dynamics 565
statistical thermodynamics 628
steady-state
  approximation 885
  condition 916
steam distillation 204
Stefan–Boltzmann
  constant 294
  law 294
stellar
  composition 394
  structure 823
step (surface) 978
stepwise polymerization 904
steric factor 945, 949
Stern model 1022
Stern, Otto 353
Stern–Gerlach experiment 353
Stern–Volmer plot 930
sticking probability 996
stimulated
  absorption 492
  emission 925
  Raman spectroscopy 565
Stirling's approximation 631, 1049
STM 336, 985
stoichiometric
  number 227
  point 242
Stokes
  radiation 488
  radius 837
  relation 742
Stokes–Einstein equation 741, 845
stopped-flow method 864
STP 11
strain 791
strange attractor 920
stratopause 922
stratosphere 921
stress 791
strong electrolyte 833
strongly coupled
  nuclei 598
  spectra 598
structure
  factor 778
  refinement 782
sublimation vapour pressure 136
subshell 375
  relative energies 387
substance xx
substrate 909, 977
  structure 984
sulfur, entropy 106
sulfur dioxide, absorption spectrum
  576
Sun
  corona 394
  perfect gas model 823
supercoiled DNA 731
superconducting
  magnet 584
  quantum interference device 803

superconductor 795, 796
supercooled 153
supercritical fluid 19, 137
  chromatography 137
  water 138
superfluid 140
superheated 153
superposition 315
superradiant 561
supersaturated 153
supersonic
  beam 707
  jet 707
  nozzle 707
surface
  composition 979
  excess 758
  film
    thermodynamics 758
    balance 757
  formation 150
  Gibbs energy 758
  growth 977
  layer designation 984
  migration 998
  mobility 998
  plasmon resonance 994
  potential 1023
  pressure 757
  reconstruction 982
  tension 150
    measurement 154
surface-enhanced Raman scattering
  981
surface-extended absorption fine
    structure spectroscopy
    982
surfactant 752
surroundings 30
  entropy change 94
svedberg 744
SWNT 750
symmetric
  rotor 499, 501
  stretch 522
symmetry 453
  and degeneracy 334
  element 453
  number 662
  operation 453
  species 465, 466
symmetry-adapted linear
    combination 473
synchrotron 521
  storage ring 486
system 30
systematic absence 779

T_1-weighted image 613
T_2-weighted image 613
T^3 law 104
Tafel plot 1030
tartaric acid 790
Taylor
  expansion 828, 1052
  series 512, 1052
TDS 997
Teller, E. 992
Temkin isotherm 993
temperature 6
  coefficient of emf 281

introduced 6
inversion 13, 922
scale 7
  thermodynamic 98
temperature-independent
    paramagnetism 805
temperature-jump method 865, 877
temperature-programmed
    desorption 997
temperature–composition diagram
  202
temporal coherence 555
term 367
  symbol 399
    diatomic molecule 540
termination step 900
terrace 978
tertiary structure 720
tesla 580
tetrachloroaurate(III) ion symmetry
  457
tetragonal unit cell 769
tetrahedral
  bond 415
  group 458
tetraphenylmethane symmetry 458
theorem
  equipartition 34
  mean value 594
  virial 343
theoretical plates 202
thermal
  analysis 197
  conduction 815
  decomposition 900
  desorption spectroscopy 997
  equilibrium 6
  explosion 902
  motion 32
  neutrons 783
thermochemical
  equation 57
  group 61
thermochemistry 55
thermodynamic
  cycle, ions in solution 254
  equation of state 123
  equilibrium constant 229
  force 843
  functions, electrochemical
    determination 280
  temperature scale 7, 98
thermodynamics
  First Law 35
  fundamental equation 163
  introduced 30
  Second Law 90
  Third Law 106
  Zeroth Law 7
thermogram 47
thermolysis 900
thermometer 7
thermosphere 922
thermotropic liquid crystal 209
theta solution 740
third
  explosion limit 903
  Law of thermodynamics 106
Third-Law entropy 106
Thomson, G.P. 301
three-level laser 554

tie line 200
tight-binding approximation 798
time constant (of reaction) 873
time-dependent
  perturbation theory 358
  Schrödinger equation 305, 358
time-independent
  perturbation theory 355
  Schrödinger equation 304
time-of-flight detector 735
time-resolved spectroscopy 567
TIP 805
titanium sapphire laser 559
titanium(III) spectrum 547
TMS 585
TNT 904
TOF spectrometry 735
torque 1065
torr 4
Torricelli 5
torsional angle definition 726
total
  angular momentum 397
    quantum number 400, 401
  orbital angular momentum
    quantum number linear
      molecule 540
    quantum number 400
  rate of absorption 493
  spin quantum number 400, 401
Townley, John 8
TPD 997
trajectory 1063
  potential energy surface 969
trans-retinal 548
transfer coefficient 1026
transfer-RNA 731
transition 381
  dipole moment 474, 494
    from perturbation theory 359
  molecular vibration 513
  moment, atomic 382
  rate 359
  state 881
    potential energy surface 967
    theory, see activated complex
      theory
  temperature 136
translational
  motion 325
  partition function 637, 639
transmembrane potential 279
transmission
  coefficient 957
  probability 335
transmittance 491
transport
  biological membrane 849
  number 839
  property 815
    perfect gas 826
transpose 1058
transverse relaxation time 603
tri-p-cresylphosphate 758
trial wavefunction 429
triclinic unit cell 768, 769
  volume 811
tridiagonal determinant 798
triethylamine miscibility 206
trigonal unit cell 769
trinitrotoluene 904

triple point 137
  phase diagram 197
triplet 395
  vector diagram 395
triplet–triple energy transfer 925
tRNA 731
tropopause 921
troposphere 12, 921
trough 12
Trouton's rule 101
tumbling 524
tumour 929
tungsten filament 485
tungsten–halogen lamp 485
tunnelling 334, 1015
  probability 344
turnover number 911
twisted nematic liquid crystal 210
two-dimensional
  NMR 610
  well 331
two-level system
  entropy 646
  internal energy 640
  negative temperature 643
  partition function 636
  population 637
  thermodynamic properties 643
Type-I superconductor 796
Type-II superconductor 796
tyrosine 730

u symmetry 539
ubiquinone 275
ubiquitin thermogram 47
UHV 980
ultracentrifugation 742
ultrahigh vacuum technique 980
ultraviolet
  catastrophe 295
  photoelectron spectroscopy 569
  radiation 484
  radiation damage 929
unbound state 373
uncertainty
  broadening, see lifetime
    broadening
  position–momentum 318
  principle 317
uncompetitive inhibition 913
undetermined multiplier 633, 1054
ungerade 539
uniaxial stress 791
unimolecular reaction 882, 890
unit
  cell 768
  matrix 1058
  vector 1050
upper
  consolute temperature 205
  critical solution temperature
    205
UPS 569, 980

uracil 730
urea 732
vacuum
  permittivity 1068
  ultraviolet radiation 484
valence-bond theory 411
van der Waals
  coefficients, table 19
  equation 19
  forces 696
  isotherms 22
  loops 21
  molecules 709
van 't Hoff equation
  adsorption 991
  chemical equilibrium 236
  osmosis 178
vanishing integral 469
vaporization, thermodynamic
  aspects 141
vapour
  composition 198
  pressure
    and activity 183
    diagram 198
    effect of applied pressure 143
    and mass loss 826
    and phase diagram 136
    introduced 19
    of mixture 168
    Raoult's law 168
    regular solution 187
    variation with temperature 147
vapour-phase osmometry 182
variance 195
variation principle 429
VB
  theory 411
  wavefunction 412
vector 1050
  diagram
    singlet state 385
    triplet state 395
  model 353
  spin-$\frac{1}{2}$ nucleus 599
velocity 1063
  selector 707, 821
vertical transition 542
Verway, E. 753
vibrational
  microscopy 520
  mode, number 520
  motion 338
  progression 543
  Raman spectra 519
  structure 542
  temperature 665
  term 512
  wavenumber table 524
vibration–rotation spectra 517
vibronic
  laser 559
  transition 539

vibronically allowed transition 542
virial 711
  coefficients 17
    osmotic 181
  equation of state 17, 671
  coefficient
  theorem 343
viscometer 748
viscosity 815, 827
  polymer solution 748
  pressure dependence (gas) 831
  temperature dependence (gas)
    831
  temperature dependence (liquid)
    832
viscosity-average molar mass 733
vision 548
volcano curve 1002
volt 1068
voltammetry 1031
volume
  element 306
  from partial molar volumes 162
  magnetic susceptibility 802
von Laue, M. 772

Walden's rule 846
Walker, J. 851
water
  anomalous entropy of vaporization
    102
  ice structure 789
  normal modes 523
  partial molar volume 161
  phase diagram 138
  radial distribution function
    710
  surface tension and temperature
    153
  symmetry elements 455
  VB theory 414
  viscosity 833
watt 1063
wave 1066
  equation 1066
  packet 318
wavefunction 325
  conditions on 309
  dimensions 308
  first-order 357
  free particle 310, 325
  interpretation 306
  introduced 304
  near nucleus 379
  normalized 307
  particle in a box 328
  particle in a finite well 338
  particle on a sphere 351
  particle in a two-dimensional
    square well 332
wavelength 483, 1066
  de Broglie 303
  electron 301

wavenumber 365, 483, 1067
wave–particle duality 304
weak electrolyte 834
weather 12
  map 13
weight (of configuration) 630
weight-average molar mass
  733
wetting 154
wide-field epi illumination method
  564
Wien displacement law 294
Wierl equation 302
wind 12
work
  additional 38
  adiabatic change 52
  against constant pressure 39
  electrical 39
  expansion 37, 38
  extension 39
  free expansion 38
  function 300
  gas production 41
  introduced 31
  maximum 110
  molecular interpretation 32
  non-expansion 38
  reversible expansion 40
  surface expansion 39
  table 39
working
  cell 1038
  electrode 1030
wrinkle, Nature's abhorrence of
  848

X-ray 484, 772
  diffraction 772
    time-resolved 773
  fluorescence 982
xanthophyll 548
xenon discharge lamp 485
XPS 980
xylene isomerization 1004

YAG laser 801
Young's modulus 791

Z-average molar mass 733
Zeeman effect 408
zeolite 1004
zero-order reaction 868
zero-point energy
  harmonic oscillator 339
  particle in a box 329
Zeroth Law of thermodynamics 7
zeta potential 753
Zimm–Bragg model 675
zipper model 675
zone
  levelling 211
  refining 211

The Periodic Table

Period	1	2	3	4	5	6	7	8	9	10	11	12	13/III	14/IV	15/V	16/VI	17/VII	18/VIII
1	1 H 1.0079																	2 He 4.003
2	3 Li 6.941	4 Be 9.012											5 B 10.81	6 C 12.01	7 N 14.01	8 O 16.00	9 F 19.00	10 Ne 20.18
3	11 Na 22.99	12 Mg 24.31											13 Al 26.98	14 Si 28.09	15 P 30.97	16 S 32.07	17 Cl 35.45	18 Ar 39.95
4	19 K 39.10	20 Ca 40.08	21 Sc 44.96	22 Ti 47.87	23 V 50.94	24 Cr 52.00	25 Mn 54.94	26 Fe 55.85	27 Co 58.93	28 Ni 58.68	29 Cu 63.55	30 Zn 65.39	31 Ga 69.72	32 Ge 72.61	33 As 74.92	34 Se 78.96	35 Br 79.90	36 Kr 83.80
5	37 Rb 85.47	38 Sr 87.62	39 Y 88.91	40 Zr 91.22	41 Nb 92.91	42 Mo 95.94	43 Tc (98)	44 Ru 101.07	45 Rh 102.91	46 Pd 106.42	47 Ag 107.87	48 Cd 112.41	49 In 114.82	50 Sn 118.71	51 Sb 121.76	52 Te 127.60	53 I 126.90	54 Xe 131.29
6	55 Cs 132.91	56 Ba 137.33	La- Lu	72 Hf 178.49	73 Ta 180.95	74 W 183.84	75 Re 186.21	76 Os 190.23	77 Ir 192.22	78 Pt 195.08	79 Au 196.97	80 Hg 200.59	81 Tl 204.38	82 Pb 207.2	83 Bi 208.98	84 Po (209)	85 At (210)	86 Rn (222)
7	87 Fr (223)	88 Ra (226)	Ac- Lr	104 Rf (261)	105 Db (262)	106 Sg (263)	107 Bh (262)	108 Hs (265)	109 Mt (266)	110 Uun	111 Uuu	112 Uub	113 Uut					

s block   *d block*   *p block*

Lanthanides

57 La 138.91	58 Ce 140.12	59 Pr 140.91	60 Nd 144.24	61 Pm (145)	62 Sm 150.36	63 Eu 151.96	64 Gd 157.25	65 Tb 158.93	66 Dy 162.50	67 Ho 164.93	68 Er 167.26	69 Tm 168.93	70 Yb 173.04	71 Lu 174.97

Actinides

89 Ac (227)	90 Th 232.04	91 Pa 231.04	92 U 238.03	93 Np (237)	94 Pu (244)	95 Am (243)	96 Cm (247)	97 Bk (247)	98 Cf (251)	99 Es (252)	100 Fm (257)	101 Md (258)	102 No (259)	103 Lr (262)

f block